D1212101

Encyclopedia of Fluid Mechanics

VOLUME 2

Dynamics of Single-Fluid Flows and Mixing

Gulf Publishing Company
Book Division
Houston, London, Paris, Tokyo

Encyclopedia of Fluid Mechanics

VOLUME 2

Dynamics of Single-Fluid Flows and Mixing

N. P. Cheremisinoff, Editor

in collaboration with—

K. Ando
K. Arai
J. Berlamont
P. N. Cheremisinoff
N. L. Coleman
A. T. Conlisk
R. Conti
A. O. Demuren
K. Endoh
L. J. Forney
T. Fukuda
V. K. Garg
A. Gianetto
M. M. Gibson
M. A. Gill
R. Grimshaw
F. G. Hammitt
M. I. Haque
J. Heestand
L. Hering
T. Hirose
Y. Jaluria
K. Kataoka
R. Kern
Y. Kitamura
S. Komori
M. Kondo
J. Kubie
M. Kuriyama
B. Lakshminarayana
T. Maruyama
J. H. Masliyah
J. A. McCorquodale
R. A. McIntire
J. W. Meader
A. N. Menendez
Z. Meyer
H. Miyashiro
Y. Murakami
K. Nandakumar
J. C. Nicholas
R. Norscini
N. S. Nosseir
J. Y. Oldshue
S. Ono
J. Osterwalder
Peerless Pump Co.
N. Rajaratnam
W. Resnick
P. Rice
S. Saito
B. P. Sangal
J. A. Schetz
M. Y. Su
K. Takahashi
T. Takahashi
G. Vetter
J. Villermaux
B. J. West

Encyclopedia of Fluid Mechanics

VOLUME 2

Dynamics of Single-Fluid Flows and Mixing

Library of Congress Cataloging in Publication Data

Main entry under title:

Encyclopedia of fluid mechanics.

Includes indexes.
Contents: v. 1. Flow phenomena and measurement—v. 2. Dynamics of single-fluid flows and mixing.
1. Fluid mechanics—Dictionaries. I. Cheremisinoff, Nicholas P.
TA357.E53 1985 620.1′06 85-9742

ISBN 0-87201-513-0 (v. 1)
ISBN 0-87201-514-9 (v. 2)

ISBN 0-87201-514-9

CONTENTS

CONTRIBUTORS TO THIS VOLUME ix
(For a note about the editor please see page xii)

PREFACE xiii

SECTION I: CHANNEL AND FREE SURFACE FLOWS

1. Theory of Solitary Waves in Shallow Fluids 3
R. Grimshaw

2. Statistics of Deep Water Surface Waves 26
B. J. West

3. Turbulence Structure and Scalar Diffusion in Thermally Stratified Open-Channel Flow 47
S. Komori

4. Form Resistance in Open-Channel Flow in the Presence of Ripples and Dunes 71
M. I. Haque

5. Unstable Turbulent Channel Flow 98
J. Berlamont

6. Hydraulic Jumps and Internal Flows 122
J. A. McCorquodale

7. Wave Attenuation in Open-Channel Flow 174
A. N. Menendez and R. Norscini

8. Straight Sediment Stable Channels 203
M. A. Gill

9. Two-Dimensional Channel Flows Over Rough Surfaces 220
N. L. Coleman

10. Three-Dimensional Deep-Water Waves 236
M. Y. Su

11. Estimating Peak Flows 258
B. P. Sangal

SECTION II: MIXING PHENOMENA AND PRACTICES

12. Hydrodynamics of Laminar Buoyant Jets 317
Y. Jaluria

13. Impinging Jets 349
N. S. Nosseir

14. Hydrodynamics of Confined Coaxial Jets 367
M. M. Gibson

15. Tubulent Mixing and Diffusion of Jets 391
N. Rajaratnam

16. Hydrodynamics of Jets in Cross Flow 406
J. A. Schetz

17. Modeling Turbulent Jets in Cross Flow 430
A. O. Demuren

18. Batchwise Jet Mixing in Tanks 466
P. Rice

19. Stability of Jets in Liquid-Liquid Systems 474
Y. Kitamura and T. Takahashi

20. Modeling Turbulent Jets with Variable Density 511
K. Kataoka

21. Jet Mixing of Fluids in Vessels 544
T. Maruyama

22. Mixing in Loop Reactors 557
Y. Murakami, T. Hirose, and S. Ono

23. Vertical Circulation in Density-Stratified Reservoirs 572
Z. Meyer

24. Rollover in Stratified Holding Tanks 637
J. W. Meader and J. Heestand

25. Jet Injection for Optimum Pipeline Mixing 660
L. J. Forney

26. Swirl Flow Generated by Twisted Pipes and Tape Inserts 691
K. Nandakumar and J. H. Masliyah

27. Micromixing Phenomena in Stirred Reactors 707
J. Villermaux

28. Backmixing in Stirred Vessels 772
K. Ando, T. Fukuda and K. Endoh

29. Industrial Mixing Equipment 803
J. Y. Oldshue

30. Simulation of Mechanically Agitated, Liquid-Liquid Reaction Vessels 851
W. Resnick

31. Solid Agitation and Mixing in Aqueous Systems 886
R. Conti and A. Gianetto

32. Mixing and Agitation of Viscous Fluids 901
S. Saito, K. Arai, K. Takahashi and M. Kuriyama

33. Use of Helical Ribbon Blenders for Non-Newtonian Materials 927
S. Saito, K. Arai, K. Takahashi and M. Kuriyama

SECTION III: FLUID TRANSPORT EQUIPMENT

34. Pump Classifications and Design Features 951
N. P. Cheremisinoff

35. System Analysis for Pumping Equipment Selection 1001
Technical Staff of Peerless Pump, subsidiary of Sterling Co., Montebello, California, USA.

36. Design and Operation of Multistage Centrifugal Pumps 1038
H. Miyashiro and M. Kondo

37. Oscillating Displacement Pumps 1057
G. Vetter and L. Hering

38. Cavitation and Erosion: Monitoring and Correlating Methods 1119
F. G. Hammitt

39. Sizing of Centrifugal Pumps and Piping 1138
R. Kern

40. Fluid Dynamics of Inducers 1152
B. Lakshminarayana

41. Hydrodynamics of Outflow from Vessels 1187
J. Kubie

42. Design Features of Fans, Blowers, and Compressors 1208
N. P. Cheremisinoff and P. N. Cheremisinoff

43. Analysis of Axial Flow Turbines 1335
V. K. Garg

44. Guidelines for Efficiency Scaling Process of Hydraulic Turbomachines with Different Technical Roughnesses of Flow Passages 1355
J. Osterwalder and L. Hippe

45. Stabilizing Turbomachinery with Pressure Dam Bearings ... 1375
J. C. Nicholas

46. Fluid Dynamics and Design of Gas Centrifuges 1393
A. T. Conlisk

47. Rupture Disc Sizing and Selection 1434
R. A. McIntire

INDEX 1487

CONTRIBUTORS TO THIS VOLUME

K. Ando, Department of Chemical Engineering, Muroran Institute of Technology, Muroran, Japan.

K. Arai, Department of Chemical Engineering, Yamagata University, Yonezawa, Japan.

J. Berlamont, Laboratorium Voor Hydraulica, Katholieke Universiteit Te Leuven, Heverlee, Belgium.

N. P. Cheremisinoff, Exxon Chemical Co., Linden, New Jersey, USA.

P. N. Cheremisinoff, Department of Civil and Environmental Engineering, New Jersey Institute of Technology, Newark, New Jersey, USA.

N. L. Coleman, USDA Sedimentation Laboratory, Oxford, Mississippi, USA.

A. T. Conlisk, Department of Mechanical Engineering, Ohio State University, Columbus, Ohio, USA.

R. Conti, Dipartimento de Scienza dei Materiali e Ingeneria Chimica, Politecnico di Torino, Italy.

A. O. Demuren, University of Karlsruhe, Karlsruhe, Federal Republic of Germany.

K. Endoh, Department of Chemical Process Engineering, Hokkaido University, Sapporo, Japan.

L. J. Forney, School of Chemical Engineering, Georgia Institute of Technology, Atlanta, Georgia, USA.

T. Fukuda, Government Industrial Development Laboratory, Hokkaido, Toyohira-ku, Sapporo, Japan.

V. K. Garg, Department of Mechanical Engineering, Indian Institute of Technology, Kanpur, India.

A. Gianetto, Dipartimento di Scienza dei Materiali e Ingeneria Chimica, Politecnico di Torino, Italy.

M. M. Gibson, Department of Mechanical Engineering, Imperial College of Science & Technology, London, England.

M. A. Gill, Hydraulic Engineering, Water Resources & Environmental Department, Ahmadu Bello University, Zaria, Nigeria.

R. Grimshaw, Department of Mathematics, University of Melbourne, Victoria, Australia.

F. G. Hammitt, Department of Mechanical Engineering and Applied Mechanics, University of Michigan, Ann Arbor, Michigan, USA.

M. I. Haque, Department of Civil, Mechanical and Environmental Engineering, The George Washington University, Washington, DC, USA.

J. Heestand, Cabot Corporation, Billerica Technical Center, Billerica, Massachusetts, USA.

L. Hering, American Lewa Inc., Natick, Massachusetts, USA.

L. Hippe, Department of Hydraulic Machines and Plants, Technical University, Darmstadt, Federal Republic of Germany.

T. Hirose, Department of Industrial Chemistry, Kumamoto University, Japan.

Y. Jaluria, Mechanical and Aerospace Engineering Department, Rutgers University, New Brunswick, New Jersey, USA.

K. Kataoka, Department of Chemical Engineering, Kobe University, Rokkodai, Kobe, Japan.

R. Kern, Hoffman-La Roche, Inc., Nutley, New Jersey, USA.

Y. Kitamura, Department of Industrial Chemistry, Okayama University, Okayama, Japan.

S. Komori, The National Institute for Environmental Studies, Ibaraki, Japan.

M. Kondo, Mechanical Engineering Research Laboratory, Tsuchiura Works, Hitachi Ltd., Tokyo, Japan.

J. Kubie, Central Electricity Generating Board, Health and Safety Department, London, England.

M. Kuriyama, Department of Chemical Engineering, Yamagata University, Yonezawa, Japan.

B. Lakshminarayana, Department of Aerospace Engineering, Pennsylvania State University, University Park, Pennsylvania, USA.

T. Maruyama, Department of Chemical Engineering, Kyoto University, Kyoto, Japan.

J. H. Masliyah, Department of Chemical Engineering, University of Alberta, Edmonton, Alberta, Canada.

J. A. McCorquodale, Department of Civil Engineering, University of Windsor, Ontario, Canada.

R. A. McIntire, Fike Metal Products, Corp., Blue Springs, Missouri, USA.

J. W. Meader, Department of Chemical Engineering, Worcester Polytechnic Institute, Worcester, Massachusetts, USA.

A. N. Menendez, Iowa Institute of Hydraulic Research, University of Iowa, Iowa City, Iowa, USA.

Z. Meyer, Water Engineering Institute, Technical University of Szczecin, Szczecin, Poland.

H. Miyashiro, Mechanical Engineering Research Laboratory, Tsuchiura Works, Hitachi Ltd., Tokyo, Japan.

Y. Murakami, Department of Chemical Engineering, Kyushu University, Kyushu, Japan.

K. Nandakumar, Department of Chemical Engineering, University of Alberta, Edmonton, Alberta, Canada.

J. C. Nicholas, Turbodyne Division, McGraw-Edison Co., Wellsville, New York, USA.

R. Norscini, Laboratorio de Hidraulica Aplicada, Instituto Nacional de Ciencia y Tecnica Hidricas, Ezeiza, Argentina.

N. S. Nosseir, Department of Aerospace Engineering and Engineering Mechanics, San Diego State University, San Diego, California, USA.

J. Y. Oldshue, Mixing Equipment Co., Inc., Rochester, New York, USA.

S. Ono, Section of Production Technology, Shoei Chemical Industry Co., Ltd., Japan.

J. Osterwalder, Department of Hydraulic Machines and Plants, Technical University, Darmstadt, Federal Republic of Germany.

Peerless Pump Co., A Sterling Co., Montebello, California, USA.

N. Rajaratnam, Department of Civil Engineering, University of Alberta, Edmonton, Alberta, Canada.

W. Resnick, Department of Chemical Engineering, Technion-Israel Institute of Technology, Haifa, Israel.

P. Rice, Department of Chemical Engineering, University of Technology, Loughborough Leicestershire, England.

S. Saito, Department of Chemical Engineering, Yamagata University, Yonezawa, Japan.

B. P. Sangal, National Hydrology Research Institute, Ottawa, Canada.

J. A. Schetz, Department of Aerospace and Ocean Engineering, Virginia Polytechnic Institute & State University, Blacksburg, Virginia, USA.

M. Y. Su, National Ocean Research and Development Activity, Department of the Navy, NSTL Station, Mississippi, USA.

K. Takahashi, Department of Chemical Engineering, Yamagata University, Yonezawa, Japan.

T. Takahashi, Department of Industrial Chemistry, Okayama University, Okayama, Japan.

G. Vetter, Erlangen University, Erlangen, Federal Republic of Germany.

J. Villermaux, Laboratoire des Sciences du Genie Chimique, CNRS-ENSIC, Nancy, France.

B. J. West, Center for Studies of Nonlinear Dynamics, La Jolla Institute, La Jolla, California, USA.

ABOUT THE EDITOR

Nicholas P. Cheremisinoff heads the product development group in the Elastomers Technology Division of Exxon Chemical Company. Previously, he led the Reactor and Fluid Dynamics Modeling Group at Exxon Research and Engineering Company. He received his B.S., M.S., and Ph.D. degrees in chemical engineering from Clarkson College of Technology, and he is also a member of a number of professional societies including AIChE, Tau Beta Pi, and Sigma Xi.

PREFACE

Volume 2, *Dynamics of Single Fluid Flows and Mixing,* is divided into three sections. Section I, "Channel and Free Surface Flows," contains 11 chapters covering the dynamics of open-channel flow systems. The distinguishing feature of this class of flow system, as contrasted with pipe flow, is that the cross-sectional area is free to change in accordance with dynamic conditions instead of being fixed. The study of these flows is of importance to both industrial applications and in the control of natural and manmade waterways. The chapters presented provide both a qualitative and rigorous treatment of free-surface behavior and bulk flow instabilities.

Section II, "Mixing Phenomena and Practices," contains 22 chapters. The theories of convective mixing turbulence and diffusion are reviewed with discussions heavily oriented towards engineering applications. Topics covered include convective diffusion, hydrodynamics of jet mixing and instability, density and thermal stratification and rollover, turbulent mixing in irregular-geometry pipe flows, pipeline mixing, mixing in industrial reactors, and mechanical agitation of Newtonian and non-Newtonian fluids. The section provides detailed discussions of both industrial hardware and scale-up principles. Theory and phenomenological descriptions are given for a well-balanced presentation.

Section III, "Fluid Transport Equipment," comprises 14 chapters. This section is also heavily design oriented, covering the hydraulics of pipe flow, pump system scale-up, transport of compressible fluids, and turbomachinery design. This section provides extensive hydraulic data and design practices/guidelines for process-oriented engineers.

This second volume presents the efforts of more than 60 specialists and organizations. Additionally, the experience and opinions of scores of engineers and researchers who reviewed and refereed the material presented are also incorporated. Each contributor is to be regarded as responsible for the statements and recommendations in his chapter. These individuals are to be congratulated for devoting their time and efforts to producing this volume, for without their efforts this work could not have become a reality. Special thanks is also expressed to Gulf Publishing Company for the production of this series.

Nicholas P. Cheremisinoff

ENCYCLOPEDIA OF FLUID MECHANICS

VOLUME 1: FLOW PHENOMENA AND MEASUREMENT

Transport Properties and Flow Instability
Flow Dynamics and Frictional Behavior
Flow and Turbulence Measurement

VOLUME 2: DYNAMICS OF SINGLE-FLUID FLOWS AND MIXING

Channel and Free Surface Flows
Mixing Phenomena and Practices
Fluid Transport Equipment

VOLUME 3: GAS-LIQUID FLOWS

Properties of Dispersed and Atomized Flows
Flow Regimes, Hold-Up, and Pressure Drop
Reactors and Industrial Applications

VOLUME 4: SOLIDS AND GAS-SOLIDS FLOWS

Properties of Particulates and Powders
Particle-Gas Flows
Fluidization and Industrial Applications
Particulate Capture and Classification

VOLUME 5: SLURRY FLOW TECHNOLOGY

Slurry and Suspension Flow Properties
Unit Operations of Slurry Flows

VOLUME 6: COMPLEX FLOW PHENOMENA AND MODELING

Special Topics in Complex and Multiphase Flows
Transport Phenomena in the Environment
Flow Simulation and Modeling

SECTION I

CHANNEL AND FREE SURFACE FLOWS

CONTENTS

CHAPTER 1. THEORY OF SOLITARY WAVES IN SHALLOW FLUIDS 3

CHAPTER 2. STATISTICS OF DEEP WATER SURFACE WAVES 26

CHAPTER 3. TURBULENCE STRUCTURE AND SCALAR DIFFUSION IN THERMALLY-STRATIFIED OPEN-CHANNEL FLOW 47

CHAPTER 4. FORM RESISTANCE IN OPEN-CHANNEL FLOW IN THE PRESENCE OF RIPPLES AND DUNES 71

CHAPTER 5. UNSTABLE TURBULENT CHANNEL FLOW 98

CHAPTER 6. HYDRAULIC JUMPS AND INTERNAL FLOWS 122

CHAPTER 7. WAVE ATTENUATION IN OPEN CHANNEL FLOW 174

CHAPTER 8. STRAIGHT SEDIMENT STABLE CHANNELS 203

CHAPTER 9. TWO-DIMENSIONAL CHANNEL FLOWS OVER ROUGH SURFACES 220

CHAPTER 10. THREE-DIMENSIONAL DEEP-WATER WAVES 236

CHAPTER 11. ESTIMATING PEAK FLOWS 258

CHAPTER 1

THEORY OF SOLITARY WAVES IN SHALLOW FLUIDS

R. Grimshaw

Department of Mathematics
University of Melbourne
Parkville, Victoria, Australia

CONTENTS

INTRODUCTION, 3

KORTEWEG-DE VRIES THEORY, 6

LARGE-AMPLITUDE SOLITARY WAVES, 10

SOLITARY WAVE INTERACTIONS, 11

SLOWLY VARYING SOLITARY WAVES, 14

SOLITARY WAVES IN DENSITY STRATIFIED FLUIDS, 18

NOTATION, 21

REFERENCES, 22

INTRODUCTION

Solitary waves are nonlinear waves of permanent form and long wavelength. They occur because of a balance between nonlinear wave-steepening effects and linear dispersive effects. Solitary waves possess a number of distinctive properties which contrast strongly with the properties of linear waves. For many years the solitary wave was regarded as a curiosity and largely ignored. However, its central role in the theory of long, nonlinear waves is now generally accepted. The main reason for this has been the discovery that certain nonlinear wave equations are exactly integrable, and that solutions describing a train of solitary waves can be obtained using the inverse scattering transform. Indeed the solitary wave can be regarded as the archetypal solution of these equations. Further, these nonlinear wave equations are generic as they are model equations for many different physical systems. Concomitant with these theoretical developments has come the recognition that solitary waves are ubiquitous in nature, and readily produced in the laboratory for a variety of physical systems. Within the hydrodynamic context the key model equation for the study of long nonlinear waves is the Korteweg-de Vries equation, whose derivation and properties will be described later in this chapter. The reader who wishes to pursue the exciting theoretical developments associated with the inverse scattering transform should consult the review articles by Scott et al. [1], and Miura [2], and the textbooks by Whitham [3] and Ablowitz and Segur [4].

In this survey of solitary waves in shallow fluids we shall give a brief historical introduction followed by an account of the Korteweg-de Vries theory. We shall then give a discussion of large amplitude solitary waves, a brief account of solitary wave interactions, and a description of how solitary waves are affected by friction and such geometrical effects as variable fluid depth. Finally we shall give an account of solitary interfacial waves, and solitary waves in density stratified fluids. There is some overlap between the contents of this chapter and the recent survey article by Miles

[5], and the reader is recommended to consult this article which also contains an extensive list of references.

The first documented observation of a solitary wave was made by John Scott Russell [6] in 1834:

> I believe I shall best introduce this phaenomenon by describing the circumstances of my own first acquaintance with it. I was observing the motion of a boat which was rapidly drawn along a narrow channel by a pair of horses, when the boat suddenly stopped—not so the mass of water in the channel which it had put in motion; it accumulated round the prow of the vessel in a state of violent agitation, then suddenly leaving it behind, rolled forward with great velocity, assuming the form of a large solitary elevation, a rounded, smooth and well-defined heap of water, which continued its course along the channel apparently without change of form or diminution of speed. I followed it on horseback, and overtook it still rolling on at a rate of some eight or nine miles an hour, preserving its original figure some thirty feet long and a foot to a foot and a half in height. Its height gradually diminished, and after a chase of one or two miles I lost it in the windings of the channel. Such, in the month of August 1834, was my first chance interview with that singular and beautiful phaenomenon ....
>
> *John Scott Russell (1845)*

Russell subsequently undertook some laboratory experiments in which he produced solitary waves in a wave tank by either releasing an impounded elevation of water at one end of the tank or by dropping a weight at one end of the tank. Russell observed that all solitary waves were waves of elevation and that the wave speed increases with the wave amplitude. Russell's observations caused some controversy at the time of their publication as they apparently conflicted with Airy's nonlinear shallow-water theory which predicts that nonlinear waves of elevation will steepen and eventually break. The conflict arises because the nonlinear shallow-water theory neglects dispersion which generally tends to prevent wave-steepening. The controversy over Russell's observations was resolved in the 1870s by Boussinesq [7], and independently by Rayleigh [8], who showed that allowance for the effects of the vertical acceleration of fluid particles (which is responsible for the leading order effects of dispersion) as well as the effects of nonlinearity, leads to a wave of permanent form. They obtained the well-known solitary wave solution

$$\eta = a\,\mathrm{sech}^2[k(x - ct)] \qquad (1)$$

where η = the free-surface displacement above the undisturbed level h
x = the horizontal coordinate in the direction of wave propagation
t = the time (see Figure 1)

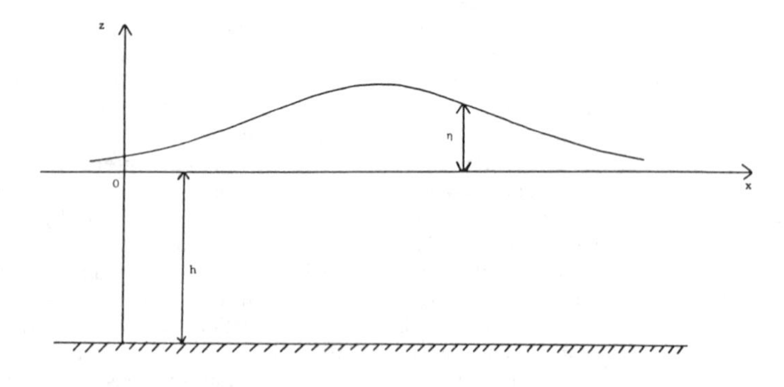

Figure 1. The coordinate system.

The wave speed c is given by

$$F^2 = c^2/c_0^2 = 1 + a/h, \qquad c_0^2 = gh \tag{2}$$

Here c_0 is the linear long wave speed and F is the Froude number. Note that solitary waves are supercritical. Equation 2 agrees with Russell's empirical formula for the wave speed as a function of wave amplitude deduced from his observations. The parameter k^{-1} is a characteristic length for the solitary wave and is given by

$$a/h = \tfrac{4}{3}h^2k^2 \tag{3}$$

Equations 1, 2, and 3 describe the three essential features of the solitary wave. These are, respectively, the sech^2-profile, the increase of wave speed with wave amplitude, and the decrease in length with wave amplitude. Equation 3 also shows that the solitary wave is necessarily a wave of elevation. Further, from Equation 3 we can form the dimensionless parameter a/h^3k^2 which is often called the Ursell number, following the recognition by Ursell [9] that it is a direct measure of the significance of nonlinearity vis-a-vis that of dispersion.

Some two decades after the work of Boussinesq and Rayleight, Korteweg and de Vries [10] derived the following model equation to describe the unidirectional propagation of long waves in water whose undisturbed level is h.

$$\eta_t + c_0\eta_x + \frac{3}{2}\frac{c_0}{h}\eta\eta_x + \frac{1}{6}c_0h^2\eta_{xxx} = 0 \tag{4}$$

Here the subscripts denote partial derivatives and the notation is described in Figure 1. This equation is now known as the Korteweg-de Vries equation (or KdV equation for short) and has become much celebrated since the discovery by Gardner et al. [11, 12] in 1967 that it is exactly integrable using the inverse scattering transform technique. It is readily verified that the solitary wave (Equation 1) satisfies the KdV equation (Equation 4), provided that Equation 3 holds, and the wave speed c is given by

$$c/c_0 = 1 + \tfrac{1}{2}a/h \tag{5}$$

Note that this expression for the wave speed agrees with Equation 2 to the lowest order in wave amplitude a/h. The KdV equation also has a family of periodic traveling wave solutions, of the form $\eta = \eta(x - ct)$. They are called cnoidal waves and in the limit of infinite wavelength the solitary wave (Equation 1) is recovered. For further details see the text by Whitham [3]. Korteweg and de Vries [10] in fact derived a more general form of Equation 4 which incorporates the effects of surface tension T, such that the coefficient h^2 in the coefficient of η_{xxx} is replaced by $(h^2 - 3T/\rho g)$. When the corresponding change is made in Equation 2 we see that the solitary wave is a wave of elevation for $T < \frac{1}{3}\rho gh^2$ and is supercritical, but for $T > \frac{1}{3}\rho gh^2$ it is a wave of depression and is subcritical. When $T \approx \frac{1}{3}\rho h^2$ the KdV equation must be amended by the inclusion of a fifth-order derivative η_{xxxxx} [13]. However, from numerical integration of the fully nonlinear equations, Hunter and Vanden-Broeck [13] have recently conjectured that the KdV equation does not provide an accurate description of periodic gravity-capillary waves for $0 < T < \frac{1}{3}\rho gh^2$, and that also the solitary wave cannot be obtained as the continuous limit of periodic waves as the wavelength tends to infinity. The reason for this seems to be that the presence of surface tension T introduces a large number of small scale "dimples" on the wave profile thus invalidating the long wave hypothesis used in the derivation of KdV equation.

In spite of this early derivation of the KdV equation it was not until the 1960s that much interest was shown in it, or in the solitary wave. The current interest began with the discovery in 1965 by Zabusky and Kruskal [14] that the solitary wave solutions of the KdV equation interact elastically. By obtaining numerical solutions of Equation 4 they discovered that when a solitary wave of a given amplitude and speed overtakes a solitary wave of smaller amplitude, and hence smaller speed, there is a nonlinear interaction after which two solitary waves emerge from the interaction with identical amplitudes and speeds, but with the larger wave now in the front position (see Figure 3). Because of the analogy with particles, Zabusky and Kruskal called these solutions *solitons*, and suggested that the KdV equation has N-soliton solutions, in which the asymptotic state as $t \to \infty$ consists of a train of N amplitude-ordered solitary waves. These initial discoveries by Zabusky and Kruskal were followed by the discovery that the KdV equation possesses a number of other remarkable properties; including its exact integrability by the inverse scattering transform technique [11, 12]. Further, it was discovered that the KdV equation was not alone in possessing these remarkable properties, and many other nonlinear wave equations are now known to possess similar properties (see the review article by Scott et al. [1] and the texts by Whitham [3], or Ablowitz and Segur [4]). Simultaneously came the recognition that the KdV equation is a model equation for the propagation of long nonlinear equations, with applications to many different physical systems.

Since Russell's initial experiments, there have been a number of experiments performed with the purpose of verifying the Boussinesq profile (Equation 1) and the wave speed formula (Equation 2). A particularly comprehensive set of experiments are those reported by Daily and Stephan [15, 16]. They established that Equations 1 and 2 fit the observations quite well for values of a/h up to 0.4–0.5. Subsequent experiments have been largely aimed at determining the validity of the KdV equation (Equation 4) as a model for the evolution of long nonlinear waves in shallow water. For instance Zabusky and Galvin [17] and Hammack and Segur [18] have shown that, with due allowance for the effects of friction, the KdV equation provides an accurate model for describing the evolution of long nonlinear waves for a wide range of initial conditions.

KORTEWEG-DE VRIES THEORY

Consider an inviscid incompressible fluid of constant density which is bounded below by a horizontal, impermeable bed of infinite lateral extent, whose undisturbed depth is h (see Figure 1). We shall assume that the flow is two-dimensional and irrotational and hence the velocity components in the horizontal and vertical directions are given by $u = \phi_x$ and $v = \phi_y$ respectively where $\phi(x, y, t)$ is the velocity potential. At the bottom, $y = -h$, v must vanish. At the free surface $y = \eta(x, t)$ the kinematic and dynamic boundary conditions are respectively

$$\eta_t + u\eta_x = v \text{ on } y = \eta \tag{6}$$

$$\phi_t + \tfrac{1}{2}(\phi_x^2 + \phi_y^2) + g\eta = 0 \text{ on } y = \eta \tag{7}$$

In deriving the dynamic boundary condition (Equation 7), which expresses the constancy of the pressure on the free surface, the Bernoulli relation is used.

To derive the KdV equation from these fully nonlinear equations of motion the following procedure is adopted. First, we identify two small parameters. One, α, measures nonlinearity and is the ratio of wave amplitude to the undisturbed depth; the other, ϵ, measures dispersion and is the ratio of the undisturbed depth to a typical wavelength. The KdV equation is derived under the hypothesis that $\alpha = \epsilon^2$, which represents the appropriate balance between nonlinearity and dispersion. This balance can be anticipated from the fact that linear sinusoidal waves of speed c and wavenumber κ possess a dispersion relation $c = c(\kappa)$, which in the limit $\kappa \to 0$ becomes $c = c_0(1 - \frac{1}{6}\kappa^2 h^2 + \cdots)$ where $c_0 = \sqrt{gh}$. Identifying κh with ϵ we see that the leading order dispersive term is $0(\epsilon^2)$, and in the nonlinear theory, dispersive terms must balance terms which are relatively $0(\alpha)$. Further it can be shown that, relative to a frame of reference moving with the linear long wave speed c_0, the waves evolve on a time-scale proportional to ϵ^{-3}. Many authors have used these hypotheses to derive the KdV equation. The approach used here is based on a multi-scale expansion used by Benney [19] in a more general context. We put

$$\theta = \epsilon(x - c_0 t), \qquad s = \epsilon^3 t \tag{8}$$

and seek an expansion of the form

$$\eta = \alpha A(\theta, s) + \alpha^2 A_1(\theta, s) + 0(\alpha^3) \tag{9}$$

with a similar expansion for ϕ. It is important to emphasize that the procedure seeks out those waves moving to the right and hence the KdV equation is a uni-directional wave equation. Since the velocity potential $\phi(y, \theta, s)$ satisfies Laplace's equation it follows that, using the bottom boundary condition,

$$\phi = \epsilon\psi(\theta, s) - \tfrac{1}{2}\epsilon^3(y + h)^2\psi_{\theta\theta}(\theta, s) + \tfrac{1}{24}\epsilon^5(y + h)^4\psi_{\theta\theta\theta\theta}(\theta, s) + 0(\epsilon^7)$$

where $\psi = B(\theta, s) + \varepsilon^2 B_1(\theta, s) + 0(\epsilon^4)$ (10)

Substituting the expansions Equations 9 and 10 into Equations 6 and 7 we obtain, at leading order

$$c_0 A = hB_\theta \tag{11}$$

At this stage, we have recovered the linear long-wave theory for waves moving to the right. At the next order, we find that

$$-c_0 A_{1\theta} + hB_{1\theta\theta} + A_s + (A_\theta B_\theta + AB_{\theta\theta}) - \tfrac{1}{6}h^3 B_{\theta\theta\theta\theta} = 0 \tag{12a}$$

$$gA_1 - c_0 B_{1\theta} + B_s + \tfrac{1}{2}B_\theta^2 + \tfrac{1}{2}c_0 h^2 B_{\theta\theta\theta} = 0 \tag{12b}$$

Eliminating A_1, B_1 between these two equations, and using Equation 11, we obtain the KdV equation:

$$A_s + \frac{3}{2}\frac{c_0}{h} AA_\theta + \frac{1}{6}c_0 h^2 A_{\theta\theta\theta} = 0 \tag{13}$$

Taking into account the change of notation embodied in Equations 8 and 9, this agrees with Equation 4. Note that the coefficient of the dispersive term $A_{\theta\theta\theta}$ is precisely that required by the linear dispersion relation.

As a preliminary to discussing the theory of the KdV Equation 13 we make a further change of variables [18]:

$$r = \frac{\theta}{h} \tag{14a}$$

$$\tau = \frac{1}{6}(g/h)^{1/2} s \tag{14b}$$

$$f(r, \tau) = \frac{3}{2}\frac{A(\theta, s)}{h} \tag{14c}$$

In terms of these nondimensional variables Equation 13 becomes the canonical KdV equation:

$$f_\tau + 6ff_r + f_{rrr} = 0 \tag{15}$$

The solitary wave solution Equation 1 reduces to

$$f = 2\kappa^2 \operatorname{sech}^2[\kappa(r - 4\kappa^2\tau)] \tag{16}$$

showing that a solitary wave of characteristic length κ^{-1} has amplitude $2\kappa^2$ and speed $4\kappa^2$ (relative to the linear long wave speed).

Let us now consider the initial-value problem for the KdV equation (Equation 15) for initial data $f(r, 0)$ that vanishes sufficiently rapidly as $r \to \pm\infty$. Gardner et al. [11, 12] have discovered the remarkable property that this initial value problem can be solved by using f(r, 0) as the input potential for a linear scattering problem for the Schrödinger equation, and then using the scattering data to form a linear integral equation, whose solution finally yields f(r, τ). This is the celebrated inverse scattering transform technique which provides a linear algorithm for solving the nonlinear KdV equation (Equation 15). For comprehensive accounts of this theory, see the texts by Whitham [3] or Ablowitz and Segur [4]. No less remarkable are the consequences of this theory for the asymptotic behavior of f(r, τ) as $\tau \to \infty$. We shall give a brief summary of some of the main results based largely on the sequence of papers by Segur [20] and Hammack and Segur [18, 21].

- An arbitrary initial disturbance f(r, 0) which vanishes sufficiently rapidly as $r \to \pm\infty$ evolves as $\tau \to \infty$ into a finite number (N) of solitons and a dispersive oscillatory tail. The oscillatory tail ultimately separates from the solitons and decays as $\tau \to \infty$.
- When the net volume V per unit width is finite and positive,

$$V = \int_{-\infty}^{\infty} f(r, 0)\, dr > 0 \tag{17}$$

  then $N \geq 1$. In general N increases with V. If $f(r, 0) \leq 0$ then $N = 0$ and the asymptotic solution is just the oscillatory tail.
- Each soliton is ultimately described by Equation 16 where κ takes one of the values $\kappa_1, \kappa_2, \ldots, \kappa_N (\kappa_1 > \kappa_2 > \cdots \kappa_N)$. Here $\kappa_1, \kappa_2, \ldots, \kappa_N$ are determined by solving the linear eigenvalue problem

$$\psi_{rr} + [\lambda + f(r, 0)]\psi = 0, \qquad \psi \to 0 \text{ as } r \to \pm\infty \tag{18}$$

  for the negative real eigenvalues, $\lambda = -\kappa^2$. Equation 18 is the linear Schrödinger equation whose scattering data determine f(r, τ). For the solitons the scattering data are $\kappa_1, \ldots, \kappa_N$ together with the normalization constants $d_1, \ldots, d_N$ associated with the corresponding eigenfunctions; these are defined so that $\psi_n \sim d_n \exp(-\kappa_n r)$ when $r \to \infty$, where ψ_n is normalized so that the integral of ψ_n^2 is exactly one. The continuous spectrum corresponds to $\lambda > 0$ and determines the oscillatory tail.
- As $\tau \to \infty$ the asymptotic state is, apart from the oscillatory tail,

$$f \sim \sum_{n=1}^{N} 2\kappa_n^2 \operatorname{sech}^2[\kappa_n(r + r_n - 4\kappa_n^2\tau)] \tag{19}$$

  The phase shifts r_n can also be found from the scattering data of Equation 18. Note that the solitons described by Equation 19 are amplitude ordered. A typical solution of the initial value problem for the KdV equation (Equation 15) is shown in Figure 2.
- For $V > 0$ the asymptotic state (Equation 19) is achieved for dimensional times $t \gg t_s$ where $t_s = (g/h)^{1/2} U_0^2 V^{-3}$ is a 'sorting time' defined by Hammack and Segur [18], subject to the small amplitude, long wave hypotheses that $\eta_0 h^{-1} \ll 1$ and $h^2 \ell^{-2} \ll 1$. Here η_0 is a measure of the height of the initial disturbance, ℓ is a characteristic horizontal length of the initial disturbance and $U_0 = \eta_0 \ell^2 h^{-3}$ is an Ursell number for the initial disturbance.

It is apparent from these results that the solitary wave is the archetypal solution of the KdV equations and dominates the long-time behavior of the solution. It is also clear from these results that the solitary wave is extremely robust, at least with respect to perturbations along the wave direction. Kadomtsev and Petviashvili [22] have shown that the solitary wave is neutrally stable with respect to small, transverse disturbances. Observations support the view that solitary waves are stable entities able to survive moderate perturbations. Experimental confirmation of the theoretical results described above has been reported by Zabusky and Galvin [17] and Hammack and Segur [18]. In particular, the KdV equation accurately predicts the number N of evolving solitons, and with due allowance for the effects of friction, can also predict the amplitudes of the larger solitons. However, experimental confirmation of the oscillatory tail has been frustrated by frictional effects.

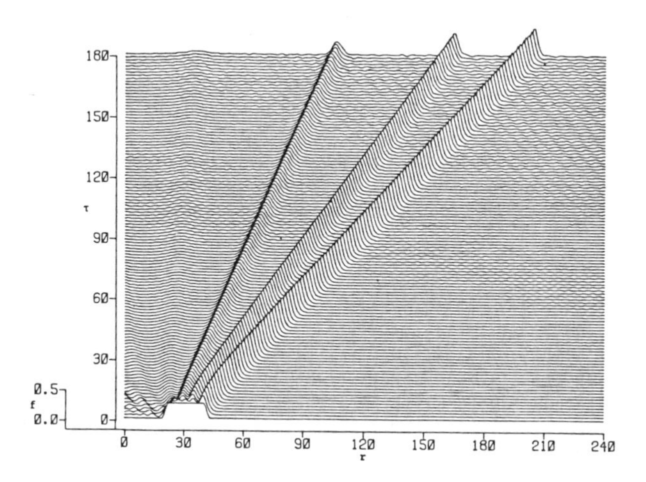

Figure 2. A plot of the evolution of three solitons from the initial condition shown. The plot is obtained by numerical integration of the KdV equation (Equation 15). The small-scale oscillations seen entering the domain from the right-hand boundary are due to numerical noise. However the oscillations in the bottom left-hand corner are part of the oscillatory tail.

We conclude this section with a simple illustration of the KdV theory. Let

$$f(r, 0) = f_0 \operatorname{sech}^2(r/r_0) \tag{20}$$

In this case the eigenvalues are given by [3, 5]

$$\kappa_n = (2r_0)^{-1}[(1 + 4f_0r_0^2)^{1/2} - (2n - 1)], \qquad 1 < n \leq N \tag{21}$$

while the number N of solitons is the largest integer less than $\frac{1}{2}[(1 + 4f_0r_0^2)^{1/2} + 1]$. There is just one soliton for $0 < f_0r_0^2 < 2$, and an additional soliton appears as $f_0r_0^2$ passes through the critical values 2, 6, 12, . . . , $N(N - 1)$. $\kappa_N = 0$ at each critical value, and is therefore trivial in Equation 19.

An interesting application of this example is to the fissioning of a solitary wave which meets a sudden change of depth [23, 24, 5]. Consider a solitary wave (Equation 1) traveling in water whose undisturbed depth is h_0. This then satisfies the KdV equation (Equation 4) in which h is replaced by h_0 and consequently Equation 3 is replaced $3a = 4k^2h_0^3$. Then suppose that the undisturbed depth changes continuously from h_0 at $x = -L$ to h at $x = 0$, and then remains at h for $x > 0$. If it is assumed that the change from h_0 to h is sufficiently rapid compared to the KdV length scale for the solitary wave to be unchanged in shape ($kL \ll 1$), but sufficiently gradual for the linear Green's law to be applicable ($kL \gg a/h$), then near $x = 0$ the solitary wave has become

$$a(h_0/h)^{1/4} \operatorname{sech}^2[(3a/4h_0^3)^{1/2}(h_0/h)^{1/2}x] \tag{22}$$

Here we recall that according to the linear Green's law the amplitude change is proportional to $h^{-1/4}$ and the wavelength change is proportional to $h^{-1/2}$. However, subsequent behavior is now governed by the KdV equation (Equation 13) for which Equation 22 is an initial condition. The example just described shows that N solitons are formed whose amplitudes are given by

$$\frac{1}{4} a\left(\frac{h}{h_0}\right)^2 [1 + 8(h_0/h)^{9/4})^{1/2} - (2n - 1)]^2 \tag{23}$$

while the number N of solitons is the largest integer for which the bracketed quantity in Equation 23 is positive. For instance there is just one soliton for $h_0/h < 1$, two solitons for $1 < h_0/h < 3^{4/9}$, three solitons for $3^{4/9} < h_0/h < 6^{4/9}$, and so on. Madsen and Mei [25] observed fissioning of a solitary wave propagating onto a shelf in a series of numerical and laboratory experiments, for which Tappert and Zabusky [23] and Johnson [24] report reasonable agreement with the theory just outlined. Byrne [26] and Gallagher [27] have reported field evidence of secondary crest formation when a shoreward-propagating swell encounters a submerged topographic feature such as an offshore sand bar or submerged reef. It seems likely that these field observations are examples of fissioning.

LARGE-AMPLITUDE SOLITARY WAVES

The Boussinesq solitary wave (Equation 1) is an approximate solution of the fully nonlinear equations of motion for an inviscid incompressible fluid of constant density. It is the first term in an asymptotic expansion in powers of $\alpha = a/h$ where a is the amplitude at the wave crest. There have been numerous attempts to extend this expansion to higher order. The third-order expansion was obtained by Grimshaw [28], and is given by

$$\eta/h = \alpha \operatorname{sech}^2 \theta - \tfrac{3}{4}\alpha \operatorname{sech}^2 \theta \tanh^2 \theta + \alpha^3(\tfrac{5}{8} \operatorname{sech}^2 \theta \tanh^2 \theta - \tfrac{101}{80} \operatorname{sech}^4 \theta \tanh^2 \theta) + \cdots \quad (24)$$

where $\theta = k(x - ct)$.

The wave speed c is given by

$$F^2 = c^2/c_0^2 = 1 + \alpha - \tfrac{1}{20}\alpha^2 - \tfrac{3}{70}\alpha^3 + \cdots \quad (25)$$

while the characteristic length k^{-1} is given by

$$kh = (\tfrac{3}{4}\alpha)^{1/2}(1 - \tfrac{5}{8}\alpha + \tfrac{71}{128}\alpha^2 + \cdots) \quad (26)$$

We note here the useful exact relation due to Stokes [29],

$$F^2 = \frac{\tan 2kh}{2kh} \quad (27)$$

which is derived from the fully nonlinear equations by noting that when $|\theta| \to \infty$, η is proportional to $\exp(-2k|\theta|)$. Fenton [30] extended this expansion to the ninth order using computer-aided calculations, and Longuet-Higgins and Fenton [31] continued the expansion numerically to the fourteenth order. However, attempts to extrapolate these series expansions to value of α_m have not been successful; here α_m is the amplitude ratio for the highest wave. The main reason for this is the fact, discovered by Longuet-Higgins and Fenton [31], that the speed c (and hence F^2) attains its maximum for a value of α less than α_m; the maximum value of F is 1.294 and is attained at $\alpha = 0.80$, whereas $\alpha_m = 0.833$ when $F = 1.291$. Also the mass, momentum, and energy exhibit similar maxima as functions of α. Longuet-Higgins and Fenton [31] attempted to circumvent these difficulties by introducing the expansion parameter

$$\omega = 1 - \frac{q^2}{c_0^2} \quad (28)$$

where q is the velocity at the wave crest in a frame of reference moving with the wave; ω varies monotonically from 0 to 1 as α varies from 0 to α_m and satisfies the useful identity

$$\omega = 2\alpha - (F^2 - 1) \quad (29)$$

which is a consequence of the Bernoulli relation on the free surface (see Equation 7). Note that when $\alpha = \alpha_m$, $q = 0$, $\omega = 1$, and $F^2 = 2\alpha_m$. They also improve convergence by introducing Padé approximates. Nevertheless, on the basis of numerical integrations of the fully nonlinear equations which are described later, it now seems that the series expansions and their extrapolations are not accurate beyond the maximum of F^2 (i.e., for $\omega > 0.92$ and $\alpha > 0.80$). For values of ω and α less than these critical values, Longuet-Higgins and Fenton's [31] results are sufficiently accurate for most practical purposes. Comparison with the laboratory measurements of Daily and Stephan [15, 16] shows good agreement for F^2 as a function of α over the range of α used in the laboratory ($0 < \alpha < 0.62$). Indeed for $\alpha < 0.5$ the theoretical predictions and the laboratory measurements differ little from the Boussinesq expression (Equation 2). Further, the Boussinesq sech^2-profile (Equation 1) fits the observations and the more accurate theoretical predictions rather well for $\alpha < 0.5$. Thus in practice small or moderate amplitude solitary waves are well described by the Boussinesq theory.

An alternative to the series-expansion is to seek numerical solutions to the fully nonlinear equations. The most common approach has been to use potential theory to convert the nonlinear free surface conditions of Equations 6 and 7 into a nonlinear integral equation, which is then solved numerically. For reviews of this approach, see Miles [5] or Schwartz and Fenton [32]. With a continuing improvement in computing capacity these numerical solutions are now capable of giving accurate results for the whole range of α up to α_m. For instance, Byatt-Smith and Longuet-Higgins [33] and Hunter and Vanden-Broeck [34] have confirmed that F^2 (as a function of α or ω) has a maximum of 1.294 at $\omega = 0.92$ and $\alpha = 0.80$. For the highest wave ($\alpha = \alpha_m$ when $q = 0$ and $\omega = 1$) it was shown by Stokes [35] that the wave crest develops a sharp corner whose enclosed angle is 120°. The recent numerical calculations of Williams [36] and Hunter and Vanden-Broeck [34] show that $\alpha_m = 0.8332$ and the corresponding value of F is 1.291. This result for α_m differs from many earlier calculations which predicted $\alpha_m \approx 0.827$ [5]. However, these earlier calculations were not sufficiently accurate because they did not include sufficient resolution near the wave crest. For α just below α_m the wave profile near the crest is smooth but contains slopes greater than 30°, which is the maximum slope for the highest wave. All these results are primarily of theoretical interest as the laboratory observations of Daily and Stephan [15, 16] suggest that wave breaking occurs when α exceeds the range 0.6–0.7.

Existence theorems for the existence of the solitary wave as a solution of the fully nonlinear equations were first obtained by Lavrent'ev [37, 38] and Friedrichs and Hyers [39] for small amplitude waves. These results confirmed that the Boussinesq approximation (Equation 1) is indeed the first term in the asymptotic series in the amplitude ratio α. Existence theorems for solitary waves of all amplitudes up to and including that of the highest wave have been elusive. Recently, however, a series of papers by Amick and Toland [40, 41] and Amick, Fraenkel, and Toland [42] have established the existence of the solitary wave for all α up to, and including α_m. They also prove that the highest wave has the crest angle of 120° predicted by Stokes [35] and that the solitary wave is the limit of a family of periodic waves. However, we note that this last result may not be valid when the effects of capillarity are included [13].

SOLITARY WAVE INTERACTIONS

Following Miles [43, 44] solitary wave interactions can be divided into two classes. *Weak* interactions are those for which the difference in speeds of the two interacting waves, calculated along the trajectory of either wave, is $O(1)$ with respect to c_0. *Strong* interactions are those for which the difference in speeds is $O(\alpha)$ with respect to c_0, where α is a measure of the amplitude ratio of either wave. Specifically if ψ is the angle between the wave directions, then the interaction is weak when $|\psi| \gg \alpha$ and strong when ψ is $O(\alpha)$. The distinction between the two classes is that for weak interactions, the interaction time is relatively short, both solitary waves emerged unchanged to $O(\alpha)$ and the interaction is $O(\alpha^2)$, while for strong interactions, the interaction time is relatively long and the interaction is an $O(\alpha)$ term.

The simplest example of a weak interaction is the collision of two solitary waves moving in opposite directions. The theory for this case has been discussed by many authors [43], most recently and comprehensively by Su and Mirie [45]. To second-order in amplitude, their solution is

$$\eta/h = \alpha_1 \operatorname{sech}^2 \theta_1 + \alpha_2 \operatorname{sech}^2 \theta_2 - \tfrac{3}{4}\alpha_1^2 \operatorname{sech}^2 \theta_1 \tanh^2 \theta_1 - \tfrac{3}{4}\alpha_2^2 \operatorname{sech}^2 \theta_2 \tanh^2 \theta_2 + \tfrac{1}{2}\alpha_1\alpha_2 \operatorname{sech}^2 \theta_1 \operatorname{sech}^2 \theta_2 \quad (30)$$

where $\theta_1 = k_1[x - c_1 t + \tfrac{1}{2}h(\tfrac{1}{3}\alpha_2)^{1/2}(1 + \tanh \theta_2)]$

$\theta_2 = k_2[x + c_2 t + \tfrac{1}{2}h(\tfrac{1}{3}\alpha_1)^{1/2}(1 + \tanh \theta_1)]$

Here the wave speeds $c_{1,2}$ and characteristic lengths $k_{1,2}$ are given by Equations 25 and 26 with α replaced by $\alpha_{1,2}$ respectively. Note that when either of $\alpha_{1,2}$ are zero, Equation 30 reduces to Equation 25 as far as the second-order terms. When $\alpha_1 = \alpha_2$ the solution of Equation 30 provides a model of the run-up of a solitary wave against a vertical wall; the maximum run-up for an incident wave of amplitude ratio α is $(2\alpha + \frac{1}{2}\alpha^2)$ and hence is greater than the value 2α predicted by a linear theory. The most significant features of the interaction described by Equation 30 are the interaction terms of $0(\alpha_1, \alpha_2)$, and the phase shifts in the expressions for $\theta_{1,2}$. For instance, considering the wave whose amplitude is α_1 the phase shift is $k_1 h(\frac{1}{3}\alpha_2)^{1/2}$ and is found by fixing $x - c_1 t$ and taking the limits $\theta_2 \to \pm\infty$. Su and Mirie [45] have extended Equation 30 to the third order in amplitude, and discovered that after interaction each solitary wave is accompanied by a trailing dispersive wave train. This has the consequence that the phase shifts after interaction are spatially dependent.

Numerical solutions by Mirie and Su [46] and Fenton and Rienecker [47] for the head-on collision of two identical solitary waves (i.e. $\alpha_1 = \alpha_2$) have confirmed that the post-interaction solitary waves are unsteady; with due allowance for the effect of this unsteadiness on the phase shift calculation they find reasonable agreement with the theoretically predicted phase shifts, and very good agreement with the theoretically predicted run-up. However, laboratory experiments by Maxworthy [48] on the head-on collision of two identical solitary waves, while confirming the theoretically predicted run-up, found that the phase shift was amplitude-independent in contradiction to the theoretical predictions. The most likely explanation of this disagreement is that the post-interaction phase shift is spatially dependent, making unambiguous measurements difficult to obtain. To conclude this discussion of weak interactions we refer the reader to Miles [43] who has obtained formulas analogous to Equation 30 for oblique interactions when ψ, the angle between the wave directions, is such that $|\psi| \gg \alpha_{1,2}$.

Strong interactions between solitary waves traveling in the same direction are succinctly described by the *exact* N-soliton of the canonical KdV equation (Equation 15). This is obtained from the inverse scattering transform when the continuous spectrum of the linear scattering problem (Equation 18) is empty, and the scattering data consists only of the negative eigenvalues $-\kappa_1^2, \ldots, -\kappa_N^2$ together with the normalization constants $d_1, \ldots, d_N$ of the corresponding eigenfunctions. The N-soliton solution can also be obtained directly from the KdV equation by Hirota's [49] method. It is given by [1–4, 20]

$$f = 2\frac{\partial^2}{\partial r^2} \log(\det(I + P)) \quad (31)$$

where I is the N × N identity matrix, and P is an N × N matrix given by

$$P = \left[\frac{d_n d_m}{\kappa_n + \kappa_m} \exp[-(\kappa_m + \kappa_n)r + 4(\kappa_n^3 + \kappa_m^3)\tau]\right] \quad (32)$$

When $|\tau| \to \infty$, the asymptotic behavior of f is given by Equation 19 where the phase shifts are now designated as $r_n^{\pm}$ according as $\tau \to \pm\infty$. Thus the N-soliton solution describes a sequence of solitary wave interactions where, after each interaction, each solitary wave emerges unchanged in shape or speed and the only remnant of the interaction is a phase shift. The total phase shift for each soliton is given by

$$2\kappa_n(r_n^+ - r_n^-) = \sum_{m=n+1}^{N} \log\left(\frac{\kappa_n + \kappa_m}{\kappa_n - \kappa_m}\right)^2 - \sum_{m=1}^{n-1} \log\left(\frac{\kappa_m + \kappa_n}{\kappa_m - \kappa_n}\right)^2 \quad (33)$$

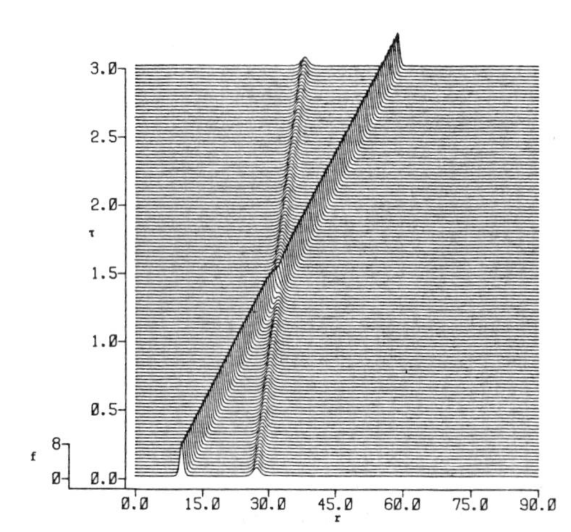

Figure 3. A plot of the 2-soliton interaction for the case $\kappa_1 = 1$ and $\kappa_2 = 2$ (see Equation 34).

and is simply the sum of the phase shifts of each pair-wise interaction. The two-soliton interaction (N = 2) is typical, and is given by

$$f = 2\frac{\partial^2}{\partial r^2}\log[1 + \exp\phi_1 + \exp\phi_2 + \exp(\phi_1 + \phi_2 + r_{12})] \tag{34}$$

where $\phi_n = -2\kappa_n(r - 4\kappa_n^2\tau) + \log(d_n^2/2\kappa_n)$

$$r_{12} = -\log\left(\frac{\kappa_1 + \kappa_2}{\kappa_1 - \kappa_2}\right)^2$$

Note that $\mp r_{12}/2\kappa_{1,2}$ are the phase shifts of each soliton. Equation 34 is plotted in Figure 3 for typical values of the parameters. Weidman and Maxworthy [50] have reported experimental confirmation of the phase shifts and interaction mechanism predicted by the two-soliton solution, although Fenton and Rienecker [47] report some minor deviations in a numerical integration of the fully nonlinear equations.

Strong oblique interactions have been studied by Miles [43, 44] and are best described by the exact N-soliton solution of the two-dimensional KdV equation, or Kadomtsev-Petviashvili equation [22]. This is derived by noting that for strong oblique interactions the angle between the wave directions is $0(\alpha)$. Hence if the x-axis is along the axis of the wave directions, the appropriate scaling for the transverse z-direction is $Z = \epsilon^2 z$. When this is added to the scaled variables (Equation 8) it may be shown that with η given by Equation 9, $A(\theta, s, Z)$ satisfies the Kadomtsev-Petviashvili equation.

$$\left(A_s + \frac{3}{2}\frac{c_0}{h}AA_\theta + \frac{1}{6}c_0h^2A_{\theta\theta\theta}\right)_\theta + \frac{1}{2}c_0A_{zz} = 0 \tag{35}$$

With the further change of variables (Equation 14) supplemented by $\zeta = Z/h$, this becomes the canonical equation

$$(f_\tau + 6ff_r + f_{rrr})_r + 3f_{\zeta\zeta} = 0 \tag{36}$$

The fundamental solitary wave solution of Equation 36 is

$$f = 2\kappa^2\operatorname{sech}^2\kappa(r + p\zeta - (4\kappa^2 + 3p^2)\tau) \tag{37}$$

Note that p measures the slope of the wave direction to the r-axis, and when $p = 0$ this reduces to the solitary wave solution (Equation 16) of the KdV equation (Equation 15). The N-soliton solution of Equation 36 is given by similar expressions to Equations 31 and 32 [4]. In particular, the two-soliton solution is again given by Equation 34 where now $\phi_{1,2}$ and r_{12} are given by

$$\phi_n = -2\kappa_n[r - p_n\zeta - (4\kappa_n^2 + 3p_n^2)\tau] + d_n \tag{38a}$$

$$r_{12} = -\log\left[\frac{4(\kappa_1 + \kappa_2)^2 - (p_1 - p_2)^2}{4(\kappa_1 - \kappa_2)^2 - (p_1 - p_2)^2}\right] \tag{38b}$$

Here d_n is a constant and r_{12} again measures the phase shift of each soliton.

The most striking feature of this solution is that it is singular whenever

$$4(\kappa_1 - \kappa_2)^2 < (p_1 - p_2)^2 < 4(\kappa_1 + \kappa_2)^2 \tag{39}$$

as then r_{12} is a complex number, and the corresponding solution has a singularity. This implies that the interaction of two solitary waves cannot be observed whenever Equation 39 is satisfied, since the singular solutions are probably unstable. Further, the boundaries of this singular régime,

$$(p_1 - p_2)^2 = 4(\kappa_1 \pm \kappa_2)^2 \tag{40}$$

is precisely the condition for the resonant interaction of three solitary waves. Thus, for the solitary wave (Equation 37) we define the wavenumber (k, m) and frequency ω by $k = \kappa, m = \kappa p$ and $\omega = -\kappa(4\kappa^2 + 3p^2)$. Then it may be verified that Equation 40 is the resonance condition.

$$k_3 = k_1 \pm k_2, \qquad m_3 = m_1 \pm m_2, \qquad \omega_3 = \omega_1 \pm \omega_2 \tag{41}$$

When Equation 40 is satisfied, $r_{12} \to \mp\infty$ and the phase shift is infinite. Miles [44] has shown that the corresponding solutions describe the resonant interaction of two solitary waves to produce a third.

Consider next the oblique reflection of a solitary wave from a vertical wall. If ψ_i is the angle between the incident wave direction and the wall, then when ψ_i is O(1), the reflected wave is an identical wave which makes an angle $(-\psi_i)$ with the wall. The theory is that for weak oblique interactions, and the solution has been obtained by Miles [43]. However, when ψ_i is $O(\alpha^{1/2})$ observations show that "Mach reflection" occurs, in which the apex of the incident and reflected waves moves away from the wall, and is joined to the wall by a third solitary wave, the "Mach stem" [44, 51]. Miles [44] has proposed that this "Mach reflection" can be described by the resonant interaction of the solitary waves. If the wall is aligned with r-axis, let the incident wave be described by $\kappa_1 = 1$ and $p_1 = -2\beta$ where $\psi_i \approx \alpha^{1/2}p_1$. Then the reflected wave has $\kappa_2 = \beta$ and $p_2 = 2$, while the "Mach stem" is given by $\kappa_3 = 1 + \beta$ and $p_3 = 0$. This solution is valid for $p_1^2 < 3$, after which regular reflection holds. However, experiments by Melville [51] have failed to give unambiguous confirmation of this model of "Mach reflection."

SLOWLY VARYING SOLITARY WAVES

The solitary wave Equation 1, which is a solution of the KdV equation (Equation 4), is applicable only for unidirectional propagation in a uniform friction-free environment. In particular, Equation 4 is applicable for propagation in a rectangular channel of uniform depth h and width b. Solitary waves in channels of nonrectangular cross-section have been discussed by Peregrine [52] who showed that there is again a solitary wave of the form Equation 1 satisfying a KdV equation similar to Equation 4, but with different coefficients for the terms η_x, $\eta\eta_x$ and η_{xxx}. In particular c_0 is now given by $c_0^2 = gS/b$ in place of Equation 2 where S is the cross-sectional area and b is the width of the free surface in the undisturbed fluid.

However, when the channel varies in the direction of propagation the solitary wave will deform in response to the changing environment. The KdV equation, Equation 4 or 13, must now be generalized to incorporate the effects of the varying environment. First, let us consider the case when the depth h varies slowly in the direction of wave propagation. The generalized KdV equation for this case has been derived and discussed by Ostrovsky and Pelinovsky [53], Kakutani [54], and Johnson [24, 55, 56]. It is now supposed that $h = h(X)$ where $X = \epsilon^3 x$ and we recall that ϵ^2 is a measure of the wave amplitude, and ϵ a measure of dispersive effects (see Equations 8 and 9). The choice of scale for the depth variation is motivated by the requirement that environmental changes should be in balance with the effects of nonlinearity and dispersion. The phase variable θ and evolution time variable s used previously (see Equations 8 and 9) are now replaced by

$$s = \int_0^X \frac{dX'}{c_0(X')}, \qquad \theta = \frac{1}{\epsilon^2}(s - \epsilon^3 t) \tag{42}$$

where we note that since $h = h(X)$, $c_0 = (gh)^{1/2}$ is a function of X. The generalized KdV equation which now replaces Equation 13 is

$$A_s + \frac{1}{2}\frac{\gamma_s}{\gamma} A + \frac{3}{2h} A A_\theta + \frac{1}{6}\frac{h^2}{c_0^2} A_{\theta\theta\theta} = 0 \tag{43}$$

where $\quad \gamma = c_0$

Note that h and c_0 are functions of s. The significance of the coefficient γ which measures the effect of the varying environment is that γA^2 is proportional to the wave energy flux in the X-direction. In a rectangular channel of depth h and width b where b also varies, $\gamma = c_0 b$ [57–59].

Unlike the constant-coefficient KdV equation (Equation 13) the variable coefficient equation (Equation 43) does not in general possess exact solutions analogous to the N-soliton solution of Equation 13. However, two contrasting approximate solutions can be obtained. When the environmental variation is rapid compared to the KdV length scale ($\gamma_s/\gamma \gg 1$ in Equation 43) then γA^2 is approximately conserved over short distances. This is just the linear Green's law and governs the change in wave amplitude over the region where γ varies. A solitary wave encountering such a region fissions. This process has been described previously in this chapter and the formula (23) gives the amplitudes of the solitons formed when there is a change of depth from h_0 to h. Numerical solutions of Equation 43 by Johnson [24] and the experiments by Madsen and Mei [25] confirm the phenomenon of fissioning. In complete contrast let us consider the case when the environmental change is gradual compared to the KdV length scale ($\gamma_s/\gamma \ll 1$ in Equation 43). An approximate solution in this case is the slowly varying solitary wave, which can be constructed either by a multiscale perturbation technique applied directly to Equation 43 [55, 56, 60] or by an adaption of the inverse scattering transform technique [61, 62]. Both methods produce the same result, with the slowly varying solitary wave described by:

$$A = a\,\text{sech}^2 k\psi \tag{44}$$

where $\quad \psi = \theta - \int_0^s V\,ds$

$$V = \tfrac{1}{2}a/h$$

$$a/h = \tfrac{4}{3}h^2k^2/c_0^2$$

Note that here Vc_0 is the nonlinear phase speed correction and c_0k^{-1} is a characteristic length scale. The definitions differ from Equations 2 and 3 since here θ (Equation 43) differs from the phase used previously (see Equation 8) by the factor c_0. The slow variation in amplitude is most readily found by observing that Equation 43 possesses the conservation law

$$\frac{\partial}{\partial s}\int_{-\infty}^{\infty} \gamma A^2\,d\theta = 0 \tag{45}$$

which expresses conservation of wave energy flux. Substituting Equation 44 into Equation 45 we obtain the law

$$\gamma a^2/k = \text{constant} \tag{46}$$

For the case when only the depth varies, $\gamma = c_0$, and we find that $a \propto h^{-1}$. This result was apparently first obtained by Boussinesq [5] and has subsequently been rederived by many authors (for instance [53–56, 63]). Comparison with experiments by Camfield and Street [64] suggests that the prediction $a \propto h^{-1}$ is reasonably good for sufficiently small bottom slopes. When the width b also varies, $\gamma = c_0 b$ and we find from Equation 46 that $a \propto h^{-1} b^{-2/3}$ [57–59]. Experiments by Chang, Melville, and Miles [65] in a channel of constant depth but varying width have confirmed this prediction for a diverging channel; for a converging channel they found some departures from the theoretical prediction which was attributed to nonlinear distortion of the variables s, θ (i.e. for increasing wave amplitudes the true wave speed c departs from c_0 and should be used in Equation 42 in place of c_0).

An interesting feature of the slowly varying solitary wave is that Equation 43 also possesses the conservation law

$$\frac{\partial}{\partial s}\int_{-\infty}^{\infty} \gamma^{1/2} A d\theta = 0 \tag{47}$$

It is readily seen that with the amplitude a given by Equation 46, Equation 47 cannot be satisfied by the solitary wave alone. The resolution of this is that the slowly varying solitary wave loses mass to a shelf, or tail, which is a linear long wave trailing behind the solitary wave. Its amplitude at the rear of the solitary wave is given by [56, 60–62].

$$A_1(s) = 2hk_s/k^2 \tag{48}$$

Here it follows from Equations 44 and 46 that $k \propto h^{-3/2} b^{-1/3}$. Note that when k increases with s, the shelf has the same sign as the solitary wave. The shelf occupies the region $\psi < 0$ and is given by

$$\left[\frac{\gamma(s_0)}{\gamma(s)}\right]^{1/2} A_1(s_0) \tag{49}$$

where $\quad \theta = \int_0^{s_0} V\, ds$

Throughout this discussion it has been implicitly assumed that h and b are constants for $s \geq 0$ and vary smoothly for $s > 0$. It then follows from Equation 49 that the shelf extends from the solitary wave located at $\psi = 0$ (see Equation 44) to the location $\theta = 0$, at which point it has zero amplitude. It can now be verified that the slowly varying solitary wave (Equation 44), together with the shelf (Equation 49) will satisfy the conservation law (Equation 47). Numerical solutions of Equation 43 by Chang, Melville, and Miles [65] and Knickerbocker and Newell [66] have confirmed the existence of the shelf.

Miles [67] has drawn attention to the curious fact that the conservation law (Equation 47) is not the conservation law for mass flux, which is

$$\frac{\partial}{\partial s}\int_{-\infty}^{\infty} c_0 b A\, d\theta = 0 \tag{50}$$

It must be emphasized that this is not a conservation law for the generalized KdV equation (Equation 43), and hence cannot be satisfied by solutions of that equation. It has been shown by Miles [67] that to resolve this dilemma and satisfy Equation 50 a reflected linear long wave must be added to the solution; the amplitude of this reflected wave is very small, and is $O(\alpha A_1)$. The reflected wave propagates with speed c_0 in the negative x-direction, and is not encompassed in the theory leading to the generalized KdV equation (Equation 43).

Next, let us consider cylindrical solitary waves. These arise when the hypothesis of unidirectional propagation is replaced by the assumption that the waves are axisymmetric and that propagation occurs in the radial direction. If r is the radial coordinate, the phase variable θ and time evolution variable s used previously (see Equations 8 and 9) are now replaced by

$$s = \frac{\epsilon^3 r}{c_0}, \qquad \theta = \epsilon\left(\frac{r}{c_0} - t\right) \tag{51}$$

where we are now assuming for simplicity that h is constant. Nevertheless note the similarity with the definitions, Equation 42. The cylindrical KdV equation is then Equation 43 with $\gamma = s$ [68, 69]. For the relationship between this equation, the Kadomstev-Petviashvili equation (Equation 35) and a non-axisymmetric KdV equation, see Johnson [69]. The slowly varying solitary wave solution of the cylindrical KdV equation is now given by Equations 44 and 46 with $\gamma = s$; it follows that $a \propto s^{-2/3}$ [70].

The dissipation of solitary waves has been discussed by Miles [5] who identifies a number of possible mechanisms. The most significant of these for laboratory experiments is viscous dissipation in the bottom and side-wall boundary layers. In the field scattering due to bottom roughness may also contribute to solitary wave damping. Here we shall consider only the damping due to a laminar bottom boundary layer, for which the predicted decay rate was obtained by Keulegan [71]. It has been shown by Miles [72] and Grimshaw [73] that the KdV equation (Equation 13) is replaced by

$$A_s + \frac{3}{2}\frac{c_0}{h} A A_\theta + \frac{1}{6} c_0 h^2 A_{\theta\theta\theta} + \frac{1}{2}\left(\frac{\nu c_0}{\epsilon^5 h^2}\right)^{1/2} \upsilon(A) = 0 \tag{52}$$

where the operator υ is defined by

$$\upsilon(A) = \frac{1}{2\pi}\int_{-\infty}^{\infty} (-i\kappa)^{1/2} \exp(i\kappa\theta)\mathscr{F}(A)\, d\kappa \tag{53a}$$

$$\mathscr{F}(A) = \int_{-\infty}^{\infty} \exp(i\kappa\theta) A\, d\theta \tag{53b}$$

Here ν is the kinematic viscosity and is assumed to be $0(\epsilon^5)$; also we recall that s and θ are defined by Equation 8. Again we adopt the slowly varying wave hypothesis, with the result that Equation 46 is replaced by [74].

$$\frac{\partial}{\partial s}\left(\frac{a^2}{k}\right) + \frac{3}{4} f\left(\frac{\nu c_0 k}{\epsilon^5 h^2}\right)^{1/2}\left(\frac{a^2}{k}\right) = 0 \tag{54}$$

where $\quad f = \int_{-\infty}^{\infty} \operatorname{sech}^2 \psi\, \upsilon(\operatorname{sech}^2 \psi)\, d\psi$

The constant f can be evaluated using Parseval's theorem for Fourier transforms, and we find that $f \approx 0.72$. Using Equation 3 the differential equation for a in Equation 54 can be integrated with the result that the predicted amplitude decay is given by

$$a = a_0\left[1 + 0.084\left(\frac{a_0}{h}\right)^{1/4}\left(\frac{\nu c_0}{\epsilon^5 h^3}\right)^{1/2} s\right]^{-4} \tag{55}$$

Here a_0 is the initial amplitude. This is the expression obtained by Keulegan [71]. Hammack and Segur [18] and Weidman and Maxworthy [50] have tested Equation 55 against laboratory measurements and report that it provides a reasonably accurate prediction of solitary wave decay, provided the effect of the side-walls is included by multiplying the coefficient of s in Equation 55 by $(1 + 2h/b)$ where b is the width of the channel.

SOLITARY WAVES IN DENSITY STRATIFIED FLUIDS

The distinguishing feature of the solitary wave is the balance between nonlinearity and dispersion. Indeed the KdV equation is a model equation for a variety of physical systems. Hence it should be no surprise that solitary waves can propagate in the waveguide formed by density stratification. These internal solitary waves are a common occurrence on the pycnocline of lakes, fjords or in coastal waters [74, 75]; they also occur on inversion layers in the atmosphere [76]. For recent surveys of the theory of these waves, which are more comprehensive than the discussion that follows see Grimshaw [73] or Redekopp [77].

The simplest model of internal solitary waves is provided by a two-layer fluid in which the upper (lower) layer has density $\rho_1(\rho_2)$ with $\rho_2 > \rho_1$, and the upper (lower) fluid is bounded above (below) by a rigid plane at $y = h_1(-h_2)$. The interface, whose undisturbed position is $y = 0$, is given by $y = \eta(x, t)$. For this system the linear long wave speed is c_0 where

$$c_0^2 = g(\rho_2 - \rho_1)h_1h_2/\rho_1h_2 + \rho_2h_1 \tag{56}$$

To obtain the KdV equation for this system we must first identify the small parameters α and ϵ. Here α, which measures nonlinearity, is the ratio of wave amplitude to the undisturbed depth $h = h_1 + h_2$, while ϵ, which measures dispersion is the ratio of h to a typical wavelength. With the KdV hypothesis $\alpha = \epsilon^2$, we again introduce the scaled variables (Equation 8) and the expansion (Equation 9) and obtain the KdV equation [78, 79].

$$A_s + \mu AA_\theta + \delta A_{\theta\theta\theta} = 0 \tag{57}$$

Here the coefficients μ and δ are given by

$$\mu = \tfrac{3}{2}c_0(\rho_2h_1^2 - \rho_1h_2^2)/h_1h_2(\rho_1h_2 + \rho_2h_1) \tag{58a}$$

$$\delta = \tfrac{1}{6}c_0h_1h_2(\rho_2h_2 + \rho_1h_1)/(\rho_1h_2 + \rho_2h_1) \tag{58b}$$

When the upper boundary is free the interfacial displacement again satisfies the KdV equation Equation 57, but the expressions for c_0 (Equation 56), μ and δ (Equation 58) are given by different expressions. However, in the Boussinesq approximation when $\sigma = 2(\rho_2 - \rho_1)/(\rho_2 + \rho_1) \rightarrow 0$ but $g\sigma$ (and hence c_0) remains finite, the expressions for c_0, μ and δ remain valid to within terms of $O(\sigma)$. The solitary wave solution of the KdV equation Equation 57 is

$$A = a \operatorname{sech}^2 k(\theta - Vs) \tag{59}$$

where $\quad V = \tfrac{1}{3}\mu a = 4\delta k^2$

Here V is the nonlinear phase speed correction, and k^{-1} is a characteristic length scale. Note that from Equation 58, $\mu >$ or < 0, based on $\rho_2h_1^2 >$ or $< \rho_1h_2^2$ and the solitary wave is accordingly a wave of elevation or depression. When $\rho_2h_1^2 \approx \rho_1h_2^2$ the coefficient $\mu \approx 0$, and it is necessary to include a cubic nonlinear term in Equation 57. The balance between nonlinearity and dispersion is then $\alpha \approx \epsilon$ where μ is $O(\epsilon)$, and the solitary wave (Equation 59) is replaced by an expression of the form

$$a \operatorname{sech}^2 \zeta/(1 - b \tanh^2 \zeta)$$

where $\quad \zeta = k(\theta - Vs)$

$\quad V = \tfrac{1}{3}\mu a/(1 + b) = 4\delta k^2$

Here b is determined by the coefficient of the cubic nonlinear term. For further details see Miles [80], Kakutani and Yamasaki [81], or Gear and Grimshaw [82].

It was pointed out by Benjamin [83] and Davis and Acrivos [84] that when either of the depths h_1 or h_2 become large the theory just described fails and must be replaced by a different theory based on a different balance between nonlinearity and dispersion. To be specific, let us suppose that h_2 is large, and $0(\epsilon^{-1})$; we put $H = \epsilon h_2$. Here ϵ is the ratio of h_1 to a typical wavelength, and measures dispersion. The measure of nonlinearity is α which is now the ratio of wave amplitude to h_1. The balance between nonlinearity and dispersion is now $\alpha = \epsilon$, and can be anticipated from the fact that the linear dispersion relation is now $c = c_0 - \gamma\kappa \coth\kappa H + \cdots$, whereas in the KdV theory it is $c = c_0 - \delta\kappa^2 + \cdots$. We again introduce the scaled variables (Equation 8) and the expansion (Equation 9) and obtain the following equation

$$A_s + \mu A A_\theta + \gamma \mathscr{L}(A_\theta) = 0 \tag{60}$$

where $$\mathscr{L}(A) = -\frac{1}{2\pi}\int_{-\infty}^{\infty} \kappa \coth(\kappa H) \exp(i\kappa\theta)\mathscr{F}(A)\, d\kappa$$

Here we recall that $\mathscr{F}(A)$ is the Fourier transform of A defined in Equation 53. Here c_0 and μ are given by Equations 56 and 58, respectively in the limit $h_2 \to \infty$, while $\gamma = \frac{1}{2}c_0 h_1(\rho_2/\rho_1)$. The derivation of Equation 60 is described by Kubota, Ko, and Dobbs [85] (also [86–87]). When $H \to \infty$, $\kappa \coth(\kappa H) \to |\kappa|$ and Equation 60 becomes the BDA equation, which has the solitary wave solution [83, 84].

$$A = a/1 + k^2(\theta - Vs)^2 \tag{61}$$

where $$V = \tfrac{1}{4}\mu a = \gamma k$$

When H is finite, the solitary wave solution of Equation 60 was obtained by Joseph [89] and describes a family of solitary waves which reduce to the KdV sech^2-profile as $H \to 0$ and to the BDA solitary wave (Equation 61) when $H \to \infty$. The evolution equation (Equation 60) has properties analogous to those of the KdV equation. In particular it has N-soliton solutions and is amenable to the inverse scattering transform technique [4]. Koop and Butler [90] and Segur and Hammack [91] have carried out experiments on solitary waves in two-layer fluids and compared their results with the KdV theory (Equation 57) and the deep fluid theory (Equation 60) described above. Generally they find the KdV theory is valid over a wide range of amplitudes and depth ratios h_1/h_2, (typically $|\eta|/h_1 < 0.2 - 0.3$ and $h_1/h_2 > 0.1$), while the deep fluid theory has only a limited range of validity.

Next we shall suppose that the stratification is continuous and characterized by a basic density profile $\rho_0(y)$ and a basic shear flow $u_0(y)$ in the wave direction. The fluid is inviscid, incompressible, and bounded below by the rigid plane $y = -h$; the undisturbed position of the free surface is $y = 0$. To derive the KdV equation for this system we again identify the small parameters α, the ratio of wave amplitude to the undisturbed depth, and ϵ, the ratio of the undisturbed depth to a typical wavelength. KdV scaling requires that $\alpha = \epsilon^2$. We again use the scaled variables (Equation 8) where c_0, the linear long wave speed, is defined in the following. In place of the expansion (Equation 9) we introduce the dependent variable $\eta(x, y, t)$ which is the vertical displacement of a fluid particle, and let

$$\eta = \alpha A(\theta, s)\phi(y) + 0(\alpha^2) \tag{62}$$

Here $\phi(y)$ is the vertical modal function, and satisfies the following boundary-value problem,

$$[\rho_0(c_0 - u_0)^2\phi_y]_y + \rho_0 N^2\phi = 0, \qquad \text{for } -h < y < 0 \tag{63a}$$

$$\phi = 0 \text{ at } y = -h \tag{63b}$$

$$\phi = \frac{(c_0 - u_0)^2}{g}\phi_y \text{ at } y = 0 \tag{63c}$$

Here N(y) is the Brunt-Vaisala frequency, and is defined by

$$N^2(y) = -g\rho_{0y}/\rho_0 \tag{64}$$

For a stable flow $\rho_{0y} \leq 0$ everywhere and hence $N^2(y) \geq 0$. The boundary-value problem (Equation 63) also serves to define the linear long wave speed c_0. In general there are an infinite number of modes, each satisfying Equation 63 with a distinct speed c_0. Hereafter we select just one of these modes and determine its evolution. We shall assume there are no critical layers, (i.e., c_0 lies outside the range of $u_0(y)$); for the modifications necessary when critical layers occur, see Maslowe and Redekopp [86] or Tung, Ko and Chang [88].

The KdV equation for this system has been derived by a number of authors [19, 73, 86–88]. The analogous theory for a compressible fluid is discussed by Grimshaw [92]. The result is the KdV equation (Equation 57) where now the coefficients μ and δ are given by

$$I\mu = 3\int_{-h}^{0} \rho_0(c_0 - u_0)^2\phi_y^3\,dy \tag{65a}$$

$$I\delta = \int_{-h}^{0} \rho_0(c_0 - u_0)^2\phi^2\,dy \tag{65b}$$

$$I = 2\int_{-h}^{0} \rho_0(c_0 - u_0)\phi_y^2\,dy \tag{65c}$$

The solitary wave solution is again given by Equation 59. Note that for right-going waves ($c_0 >$ max u_0) the coefficient δ is always positive, but the coefficient μ can take either sign. Thus solitary waves are waves of elevation or depression based on $\mu >$ or < 0. A useful approximation here is the Boussinesq approximation in which density variations are ignored except in $N^2(y)$; it is formally obtained by letting $g \to \infty$, $\rho_{0y} \to 0$ with $N^2(y)$ finite. We parameterize the Boussinesq approximation with the small parameter σ where $\sigma = \max(\rho_{0y}/\rho_0)$. As $\sigma \to 0$ it can be shown that

$$I\mu = \int_{-h}^{0} \rho_0\phi^3\{-2(N^2)_y - 3N^2u_{0y}/(c_0 - u_0) - (c_0 - u_0)^2[u_{0yy}/(c_0 - u_0)]_y\}\,dy + 0(\sigma) \tag{66}$$

Thus, in the absence of a shear flow (i.e. $u_0 = 0$), and for a near-surface pycnocline ($(N^2)_y > 0$), μ is negative for first-mode waves, which are thus waves of depression; that is, they face away from the free surface. For very weak stratification and very weak shear, $\mu \approx 0$, and it is necessary to include a cubic term in Equation 57. The subsequent development is analogous to that described above for the two-layer fluid, and is discussed by Miles [80] and Gear and Grimshaw [82].

When the horizontal waveguide is abutted by a deep fluid region within which the density is constant and the shear flow zero, it was shown by Benjamin [83] and Davis and Acrivos [84] that a different theory is needed. To be specific suppose that the waveguide retains a vertical scale h_1, determined by the near-surface structure of $\rho_0(y)$ and $u_0(y)$, but that the lower boundary is displaced to $y = -H/\epsilon$ (i.e. $H = \epsilon h$); other configurations are possible, but will not be considered here. If α is now defined as the ratio of h_1 to a typical wavelength, the appropriate balance between nonlinearity and dispersion is $\alpha = \epsilon$. The modal function $\phi(y)$ again satisfies Equation 63 with the proviso that the boundary condition at $y = -h$ is replaced by $\phi_y \to 0$ as $y \to -\infty$. The evolution equation has been derived by Maslowe and Redekopp [86], Grimshaw [73, 87] and Tung, Ko, and Chang [88]. The result is Equation 60 where now the coefficient γ is given by

$$I\gamma = (\rho_0 c_0^2 \phi^2)_{y \to -\infty} \tag{67}$$

and μ is again given by Equation 65 with the proviso that $-h$ is replaced by $-\infty$ in the lower terminal of the integrals.

Various extensions of the theory just described are possible, and are analogous to the corresponding extensions for free surface solitary waves. Thus, Gear and Grimshaw [82] have discussed the continuation of the KdV theory to second order in wave amplitude, and Grimshaw [93] has discussed the corresponding extension for the deep-fluid theory. Little is known about large-amplitude solitary waves, although some numerical results have been obtained by Tung, Chan, and Kubota [94]. Internal solitary wave interactions have been discussed by Gear and Grimshaw [95] for the case when the solitary waves belong to different vertical modes, and also for the case of the head-on collision. Strong interactions between internal solitary waves belonging to the same vertical mode can be discussed within the context of the KdV equation (Equation 57) or the deep-fluid equation Equation 60 for unidirectional waves. Strong oblique interactions are described by the solutions of the Kadomtsev-Petviashvili equation (Equation 35) or its deep-fluid counterpart. For internal solitary waves these two-dimensional equations require the addition of a term $\frac{1}{2} c_0 A_{ZZ}$ to the θ-derivative of the left-hand side of Equation 57 or 60 [77, 87] where $Z = \epsilon^2 z$ for the KdV-theory and $Z = \epsilon^{3/2} z$ for the deep-fluid theory.

Slowly varying unidirectional internal solitary waves are described by the analogue of Equation 43. With s and θ again given by Equation 42, the variable-coefficient KdV equation is [73, 87]

$$A_s + \frac{1}{2}\frac{\gamma_s}{\gamma} A + \frac{\mu}{c_0} A A_\theta + \frac{\delta}{c_0^3} A_{\theta\theta\theta} = 0 \tag{68}$$

where $\gamma = c_0^2 I$

Here μ, δ, and I are again given by Equation 65. The significance of γ is that γA^2 is proportional to the wave action flux in the X-direction. Slowly varying solitary waves are again described by Equation 44 where now $V = \frac{1}{3}\mu a/c_0 = 4\delta k^2/c_0^2$. The variation in amplitude is again determined by Equations 45 and 46, while the trailing shelf is now given by $A_1 = 3c_0 k_s/\mu k^2$ in place of Equation 48. The corresponding theory for the deep-fluid BDA equation has been discussed by Grimshaw [96]. The effect of viscous dissipation may be represented by the inclusion of a dissipative term in Equation 68 similar to the dissipative term in Equation 52. The contrasting case of fissioning of internal solitary waves has been discussed by Djordjevic and Redekopp [97]. Finally, we mention that the two-layer fluid model has been generalized by Grimshaw [98] to channels of slowly-varying nonrectangular cross-section, and provides a model for internal solitary wave propagation in lakes and fjords.

NOTATION

A	area
A_1, B_1	constants
a	distance
a, a_0	amplitude
b	parameter
c	wave speed
c_0	linear long wave speed
d_N	normalization constants
F	Froude number
F(A)	Fourier transform of A
f	constant
g	gravitational acceleration
h	level
k	characteristic length parameter
L(A)	operator defined by Equation 60
q	velocity at wave crest
r	distance
r_n	phase shift
s	parameter defined by Equation 8
T	surface tension
t	time
U_0	Ursell number
V	volume
X	function
x	horizontal coordinate
y	vertical coordinate
z	coordinate

Greek Symbols

α	ratio of wave amplitude to undisturbed depth
γ	wave velocity or displacement
δ	coefficient defined by Equation 58
ϵ	ratio of undisturbed depth to a typical wavelength
η	free-surface displacement
θ	phase variable
κ^{-1}	characteristic length
λ	eigenvalue
μ	coefficient defined by Equation 58
ν	kinematic viscosity
ρ	density
τ	time
υ	operator defined by Equation 53
ϕ	velocity potential
ψ_n	normalized eigenfunction
ω	expansion parameter

REFERENCES

1. Scott, A. C., Chu, F. Y. F., and McLaughlin, D. W., "The Soliton: A New Concept in Applied Science," *Proc. IEEE*, 61:1443–1483 (1973).
2. Miura, R. M., "The Korteweg-de Vries Equation: A Survey of Results," *SIAM REVIEW*, 18: 412–459 (1976).
3. Whitham, G. B., *Linear and Nonlinear Waves*, New York: Wiley (1974).
4. Ablowitz, M. J., and Segur, H., "Solitons and the Inverse Scattering Transform," *SIAM Studies in Applied Mathematics*, 4 (1981).
5. Miles, J. W., "Solitary Waves," *Ann. Rev. Fluid Mech.*, 12:11–43 (1980).
6. Russell, J. S., "Report on Waves," Rep. Meet. British Assoc. Adv. Sci., 14th, York (1844), pp 311–390, London: John Murray.
7. Boussinesq, M. J., "Theorie de l'intumescence liquide, appelée onde solitaire ou de translation, se propageant dans un canal rectangulaire," *Acad. Sci. Paris, Comptes Rendus*, 72:755–759 (1871).
8. Rayleigh, Lord, "On Waves," *Phil. Mag.*, T: 257–279 (1876).
9. Ursell F., "The Long-Wave Paradox in the Theory of Gravity Waves," *Proc. Camb. Phil. Soc.*, 49:685–694 (1953).
10. Korteweg, D. J., and de Vries, G. "On the Change of Form of Long Waves Advancing in a Rectangular Canal and on a New Type of Long Stationary Waves," *Phil. Mag.*, 39:422–443 (1895).
11. Gardner, C. S., et al., "Method for Solving the Korteweg-de Vries Equation," *Phys. Rev. Lett.*, 19:1095–1097 (1967).
12. Gardner, C. S., et al., "Korteweg-de Vries Equation and Generalizations. VI Methods for Exact Solution," *Comm. Pure Appl. Math.*, 27:97–133 (1974).
13. Hunter, J. K., and Vanden-Broeck, J. M., "Solitary and Periodic Gravity-Capillary Waves of Finite Amplitude," *J. Fluid Mech.*, 134:205–219 (1983).
14. Zabusky, N. J., and Kruskal, M. D., "Interaction of "Solitons" in a Collisionless Plasma and the Recurrence of Initial States," *Phys. Rev. Lett.*, 15:240–243 (1965).
15. Daily, J. W., and Stephan, S. C., Jr., "Characteristics of the Solitary Wave," *Trans. ASCE*, 118:575–587 (1953).
16. Daily, J. W., and Stephan, S. C., Jr., "The Solitary Wave," Proc. 3rd Conf. Coastal Eng., Cambridge Mass., October 1952 (1953), pp. 13–30.
17. Zabusky, N. J., and Galvin, C. J., "Shallow-Water Waves, the Korteweg-de Vries Equation and Solitons," *J. Fluid Mech.*, 47:811–824 (1971).
18. Hammack, J. L., and Segur, H., "The Korteweg-de Vries Equation and Water Waves. Part 2. Comparison with Experiments," *J. Fluid Mech.*, 65:289–314 (1974).
19. Benney, D. J., "Long Non-Linear Waves in Fluid Flows," *J. Mathematical Phys.*, 45:52–63 (1966).
20. Segur, H., "The Korteweg-de Vries Equation and Water Waves. Solutions of the Equation. Part 1," *J. Fluid Mech.*, 59:721–736 (1973).
21. Hammack, J. L., and Segur, H., "The Korteweg-de Vries Equation and Water Waves. Part 3. Oscillatory Waves," *J. Fluid Mech.*, 84:337–358 (1978).
22. Kadomstev, B. B., and Petviashvili, V. I., "On the Stability of Solitary Waves in Weakly Dispersing Media," *Sov. Phys. Doklady*, 15:539–541 (1970).
23. Tappert, F. D., and Zabusky, N. J., "Gradient-Induced Fission of Solitons," *Phys. Rev. Lett.*, 27:1774–1776 (1971).
24. Johnson, R. S., "Some Numerical Solutions of a Variable-Coefficient Korteweg-de Vries Equation (with Applications to Solitary Wave Development on a Shelf)," *J. Fluid Mech.*, 54:81–91 (1972).
25. Madsen, O. S., and Mei, C. C., "The Transformation of a Solitary Wave over an Uneven Bottom," *J. Fluid Mech.*, 39:781–791 (1969).
26. Byrne, R. J., "Field Occurrences of Induced Multiple Gravity Waves," *J. Geophys. Res.*, 74: 2590–2596 (1969).

27. Gallagher, B., "Some Qualitative Aspects of Nonlinear Wave Radiation in a Surf Zone," *Geophys. Fluid Dynamics*, 3:347–354 (1972).
28. Grimshaw, R., "The Solitary Wave in Water of Variable Depth. Part 2," *J. Fluid Mech.*, 46: 611–622 (1971).
29. Stokes, G. G., "The Outskirts of the Solitary Wave," *Math. and Phys. Papers*, V: 163, Cambridge Univ. Press (1905).
30. Fenton, J., "A Ninth-Order Solution for the Solitary Wave," *J. Fluid Mech.*, 53:257–271 (1972).
31. Longuet-Higgins, M. S., and Fenton, J. D., "On the Mass, Momentum, Energy and Circulation of a Solitary Wave. II," *Proc. R. Soc. London*, A340:471–493 (1974).
32. Schwarz, L. W., and Fenton, J. D., "Strongly Nonlinear Waves," *Ann. Rev. Fluid Mech.*, 14:39–60 (1982).
33. Byatt-Smith, J. G. B., and Longuet-Higgins, M. S., "On the Speed and Profile of Steep Solitary Waves," *Proc. R. Soc. London*, A350:175–189 (1976).
34. Hunter, J. K., and Vanden-Broeck, J. M., "Accurate Computations for Steep Solitary Waves," *J. Fluid Mech.*, 136:63–72 (1983).
35. Stokes, G. G., "On the theory of Oscillatory Waves," *Math. and Phys. Papers*, I:197–229, Cambridge, University Press (1880).
36. Williams, J. M., "Limiting Gravity Waves in Water of Finite Depth," *Phil. Trans. R. Soc. London*, A302:139–188 (1981).
37. Lavrent'ev, M. A., "A Contribution to the Theory of Long Waves," *C. R. (Doklady) Acad. Sci. URSS*, 41:275–277 (1943). [Reproduced in *Am. Math. Soc. Transl.*, 102:51–53 (1954)].
38. Lavrent'ev, M. A., *Variational Methods*, translated from Russian by J. R. M. Radok, Noordhoff (1960).
39. Friedrichs, K. O., and Hyers, D. H., "The Existence of Solitary Waves," *Comm. Pure Applied Math.*, 7:517–550 (1954).
40. Amick, C. J., and Toland, J. F., "On Solitary Water-Waves of Finite Amplitude," *Arch. Rat. Mech. Anal.*, 76:9–95 (1981).
41. Amick, C. J., and Toland, J. F., "On Periodic Water-Waves and their Convergence to Solitary Waves in the Long-Wave Limit," *Phil. Trans. R. Soc. London*, A303:633–609 (1981).
42. Amick, C. J., Fraenkel, L. E., and Toland, J. F., "On the Stokes Conjecture for a Wave of Extreme Form," *Acta Math.*, 148:193–214 (1982).
43. Miles, J. W., "Obliquely Interacting Solitary Waves," *J. Fluid Mech.*, 79:157–169 (1977).
44. Miles, J. W., "Resonantly Interacting Solitary Waves," *J. Fluid Mech.*, 79:171–179 (1977).
45. Su, C. H., and Mirie, R. M., "On Head-on Collisions Between Two Solitary Waves," *J. Fluid Mech.*, 98:509–525 (1980).
46. Mirie, R. M., and Su, C. H., "Collisions Between Two Solitary Waves. Part 2. A numerical study," *J. Fluid Mech.*, 115:475–492 (1982).
47. Fenton, J. D., and Rienecker, M. M., "A Fourier Method for Solving Nonlinear Water-Wave Problems: Application to Solitary-Wave Interactions," *J. Fluid Mech.*, 118:411–443 (1982).
48. Maxworthy, T., "Experiments on Collisions Between Solitary Waves," *J. Fluid Mech.*, 76:177–185 (1976).
49. Hirota, R., "Exact Solution of Korteweg-de Vries Equation for Multiple Collisions of Solutons," *Phys. Rev. Lett*, 27:1192–1194 (1971).
50. Weidman, P., and Maxworthy, T., "Experiments on Strong Interactions Between Solitary Waves," *J. Fluid Mech.*, 85:417–431 (1978).
51. Melville, W. K., "On the Mach Reflexion of a Solitary Wave," *J. Fluid Mech.*, 98:285–297 (1980).
52. Peregrine, D. H., "Long Waves in a Uniform Channel of Arbitrary Cross-Section," *J. Fluid Mech.*, 32:353–365 (1968).
53. Ostrovsky, L. A., and Pelinovsky, E. N., "Transformation of Surface Waves in Fluids of Variable Depth," *Izv. Atmosph. Ocean. Phys.*, 9:934–939 (1970).
54. Kakutani, T., "Effect of an Uneven Bottom on Gravity Waves," *J. Phys. Soc. Japan*, 30:272–276 (1971).
55. Johnson, R. S., "On the Development of a Solitary Wave Moving over an Uneven Bottom," *Proc. Camb. Phil. Soc.*, 73:183–203 (1973).

56. Johnson, R. S., "On the Asymptotic Solution of the Korteweg-de Vries Equation with Slowly Varying Coefficients," *J. Fluid Mech.*, 60:813–824 (1973).
57. Shuto, N., "Non-linear Long Waves in a Channel of Variables Section," *Coastal Eng. Jpn.*, 16:1–12 (1974).
58. Ostrovsky, L. A., and Pelinovsky, E. N., "Refraction of Nonlinear Sea Waves in a Beach Zone," *Izv. Atmosph. Ocean. Phys.*, II:67–74 (1975).
59. Miles, J. W., "Note on a Solitary Wave in a Slowly Varying Channel," *J. Fluid Mech.*, 80:149–152 (1977)
60. Grimshaw, R., "Slowly Varying Solitary Waves. I. Korteweg-de Vries Equation," *Proc. R. Soc. London*, A368:359–375 (1979).
61. Karpman, V. I., and Maslov, E. M., "Perturbation Theory for Solitons," *Zh. Exsp. Teor. Fiz.*, 73:537–559 (1977).
62. Kaup, D. J., and Newell, A. C., "Solitons as Particles, Oscillators, and in Slowly Changing Media: A Singular Perturbation Theory," *Proc. R. Soc. London*, A361:413–446 (1978).
63. Grimshaw, R., "The Solitary Wave in Water of Variable Depth," *J. Fluid Mech.*, 42:639–656 (1970).
64. Camfield, F. E., and Street, R. L., "Shoaling of Solitary Waves on Small Slopes," *J. Waterways Harb. Div., Proc. ASCE*, 95:1–22 (1969).
65. Chang, P., Melville, W. K., and Miles, J. W., "On the Evolution of a Solitary Wave in a Gradually Varying Channel," *J. Fluid Mech.*, 95:401–414 (1979).
66. Knickerbocker, C. J., and Newell, A. C., "Shelves and the Korteweg-de Vries Equation," *J. Fluid Mech.*, 98:803–818 (1980).
67. Miles, J. W., "On the Korteweg-de Vries Equation for a Gradually Varying Channel," *J. Fluid Mech.*, 91:181–190 (1979).
68. Miles, J. W., "An Axisymmetric Boussinesq Wave," *J. Fluid Mech.*, 84:181–191 (1978).
69. Johnson, R. S., "Water Waves and Korteweg-de Vries Equations," *J. Fluid Mech.*, 97:701–719 (1980).
70. Ko, K. and Kuehl, H. H., "Cylindrical and Spherical Korteweg-de Vries Solitary Waves," *Phys. Fluid*, 22:1343–1348 (1979).
71. Keulegan, G. H., "Gradual Damping of Solitary Waves," *J. Res. Natl. Bur. Stand.*, 40:487–498 (1948).
72. Miles, J. W., "Korteweg-de Vries Equation Modified by Viscosity," *Phys. Fluids*, 19:1063 (1976).
73. Grimshaw, R., "Solitary Waves in Density Stratified Fluids," in Nonlinear Deformation Waves, IUTAM Symposium Tallinn 1982, U. Nigul, J. Engelbrecht (Ed.), Springer: 431–447 (1983).
74. Farmer, D. M., "Observations of Long Nonlinear Internal Waves in a lake," *J. Phys. Ocean.*, 8:63–73 (1978).
75. Apel, J. R., "Satellite Sensing of Ocean Surface Dynamics," *Ann. Rev. Earth Planet. Sci.*, 8: 303–342 (1980).
76. Christie, D. R., Muirhead, K. J., and Hales, A. L., "On Solitary Waves in the Atmosphere," *J. Atmos. Sci.*, 35:805–825 (1978).
77. Redekopp, L. G., "Nonlinear Waves in Geophysics: Long Internal Waves," in *Lectures in Applied Mathematics, American Math. Soc.*, 20:59–78 (1983).
78. Keulegan, G. H., "Characteristics of Internal Solitary Waves," *J. Res. Natl. Bur. Stand.*, 51: 133–140 (1953).
79. Long, R. R., "Solitary Waves in One- and Two-Fluid Systems," *Tellus*, 8:460–471 (1956).
80. Miles, J. W., "On Internal Solitary Waves," *Tellus*, 31:456–462 (1979).
81. Kakutani, T., and Yamasaki, N., "Solitary Waves on a Two-Layer Fluid," *J. Phys. Soc. Japan*, 45:674–679 (1978).
82. Gear, J., and Grimshaw, R., "A Second-Order Theory for Solitary Waves in Shallow Fluids," *Phys. Fluids*, 26:14–29 (1983).
83. Benjamin, T. B., "Internal Waves of Permanent Form in Fluids of Great Depth," *J. Fluid Mech.*, 29:559–592 (1967).
84. Davis, R. E., and Acrivos, A., "Solitary Internal Waves in Deep Water," *J. Fluid Mech.*, 29: 593–607.

85. Kubota, T., Ko, D. R. S., and Dobbs, L., "Weakly-Nonlinear, Long Internal Gravity Waves in Stratified Fluids of Finite Depth," *J. Hydronautics*, 12:157–165 (1978).
86. Maslowe, S. A., and Redekopp, L. G., "Long Nonlinear Waves in Stratified Shear Flows," *J. Fluid Mech.*, 101:321–348 (1980).
87. Grimshaw, R., "Evolution Equations for Long Nonlinear Waves in Stratified Shear Flows," *Studies Appl. Math.*, 65:159–188 (1981).
88. Tung, K-K, Ko, D. R. S., and Chang, J. J., "Weakly Nonlinear Internal Waves in Shear," *Studies Appl. Math.*, 65:189–221 (1981).
89. Joseph, R. J., "Solitary Waves in a Finite Depth Fluid," *J. Phys. A: Math.* 10:L225–227 (1977).
90. Koop, C. G., and Butler, G., "An Investigation of Internal Solitary Waves in a Two-Fluid System," *J. Fluid Mech.*, 112:225–251 (1982).
91. Segur, H., and Hammack, J., "Solution Models of Long Internal Waves," *J. Fluid Mech.*, 118: 285–304 (1982).
92. Grimshaw, R., "Solitary Waves in a Compressible Fluid," *Pageoph.*, 119:780–797 (1980/81).
93. Grimshaw, R., "A Second-Order Theory for Solitary Waves in Deep Fluids," *Phys. Fluids*, 24:1611–1618 (1981).
94. Tung, K. K., Chan, T. F. and Kubota, T., "Large Amplitude Internal Waves of Permanent Form ," *Studies Appl. Math.*, 66:1–44 (1982).
95. Gear, J., and Grimshaw, R., "Weak and Strong Interactions Between Internal Solitary Waves," *Studies Appl. Math.*, (1984).
96. Grimshaw, R., "Slowly Varying Solitary Waves in Deep Fluids," *Proc. R. Soc. London*, A376: 319–332 (1981).
97. Djordjevic, V. C., and Redekopp, L. G., "The Fission and Disintegration of Internal Solitary Waves Moving Over Two-Dimensional Topography," *J. Phys. Ocean.*, 8:1016–1024 (1978).
98. Grimshaw, R. H. J., "Long Nonlinear Internal Waves in Channels of Arbitrary Cross-Section," *J. Fluid Mech.*, 86:415–431 (1978).

CHAPTER 2

STATISTICS OF DEEP WATER SURFACE WAVES

Bruce J. West

Center for Studies of Nonlinear Dynamics
(Affiliated with the University of California, San Diego)
La Jolla Institute
La Jolla, California USA

CONTENTS

INTRODUCTION, 26

STEADY-STATE LINEAR WAVES, 28
  Theory, 28
  Data, 31

STEADY-STATE NONLINEAR WAVES, 32

TRANSIENT LINEAR WAVE FIELD, 39

DYNAMIC NONLINEAR WAVE FIELD, 42

CONCLUSIONS, 44

NOTATION, 45

REFERENCES, 45

INTRODUCTION

One of the oldest problems in hydrodynamics is to describe the generation, evolution, and eventual dissipation of waves on the surface of water. In this article we attempt to sketch what is known about the statistical properties of wind-generated deep water waves in each of these three stages. In an evolving field of water waves, the statistics evolve coincidentally with the amplitudes and phases of the individual waves, and it is the evolution of the statistics that determines how the physical observables are transported in space and time. Herein we present the experimental evidence for the existence of non-Gaussian statistics for the water waves, review the arguments for both the linear and nonlinear steady-state wave fields giving rise to these statistics, and suggest a mechanism by which the transient state relaxes to the proper asymptotic statistical state.

The physical mechanisms coupling the air flow to the water surface must be understood in order to describe the evolution of the wind-generated spectrum of water waves on the ocean surface. Phillips [1] proposed a stochastic model of the excitation mechanism in which the pressure field at the fluid surface is assumed to fluctuate independently of the surface response. These incoherent fluctuations drive the surface at length and time scales which already exist in the pressure field spectrum. Miles [2] proposed a deterministic mechanism involving the modulation of the air flow by the vertical movement of the surface, resulting in the pressure field doing work on the surface in-phase with the surface response. These complementary mechanisms were later synthesized by Miles [3] into the Miles-Phillips model of inviscid resonant shear-flow instability. West and Seshadri [4] have recently extended this model by allowing the linear air-sea coupling parameter to fluctuate. The growth rates for long-wavelength gravity waves predicted by the model of West and Seshadri exceed those predicted by the Miles-Phillips model by an order of magnitude for some wavelengths, in close agreement with field data.

We are here interested in dynamic models which describe the evolution of the water surface driven by a fluctuating wind field from a state of rest to an asymptotic steady state. The dynamic models of Miles [2, 3], Phillips [1], and West and Seshadri [4] provide a description of the initial stages of wind-stimulated growth of water waves. In the gravity-capillary region of the spectrum the Miles-Phillips model does quite well in predicting the initial growth rates, and the model of West and Seshadri reduces to the Miles-Phillips model in that region. However, these models do not yield an asymptotic steady-state energy-spectral density. The steady-state observed in the data is presumed to be a consequence of the nonlinear interactions among the water waves, see e.g., Kitaigorodskii [5] or Phillips [6] for a qualitative discussion of this effect. West [7] obtained a solution to a set of dynamic equations including the average effect of the nonlinear interaction and does indeed obtain such a steady-state energy-spectral density for the wind generated gravity-capillary wave field.

In the transient regime where the wave-wave interactions can be important, the linear dynamic models are no longer adequate. In this region the nonlinear gravity wave field is often described using weak interaction theory. Such treatments expand the vertical displacement of the sea surface and the velocity potential at this surface in Fourier series. Waves on the surface are then described by linear mode amplitudes, and the nonlinear interactions among the waves appear as products of these amplitudes in systems of mode rate equations [8, 9]. The interactions among the modes fall into two categories:

1. Those that provide a rapid periodic interchange of energy (action).
2. Those that provide a slow unidirectional energy (action) flux. The latter interactions are called resonant whereas the former are nonresonant. Energy is supplied to this system by the turbulent air flow at the sea surface, but in most theoretical studies one usually restricts the wave dynamics to an isolated system evolving from a given initial state and ignores the wave generation question. Later we examine the wave generation problem as it relates to the question of water wave statistics. An approximation scheme is discussed by which the average wave-wave interaction can be incorporated into the linear evolution equation for wind-generated waves. Although the statistics of the wave amplitudes remain Rayleigh in this analysis, the mean-square wave amplitude is determined by a self-consistency condition.

The assumption that is invariably made in the study of the evolution of the water wave field is that the complex mode amplitude are Gaussian random variables. In the next section of this chapter we review the statistical properties of a linear wave field that was originally discussed in a water wave context by Longuet-Higgins [10]. These properties are assumed to be space and time independent, i.e., they are the steady-state statistical properties [8, 9]. Using the Gaussian assumption, the nonlinear mode rate equations are often replaced by a closed set of action transport equations, i.e., truncated hierarchical sets of low-order moments, even when the wave field is not homogeneous [11]. Analysis has shown that the steady-state solution of these transport equations yields *nonphysical* power spectral densities for the gravity wave field. Both Hasselmann [8] and Longuet-Higgins [12] demonstrate how the nonlinear interactions among a set of Gaussian mode amplitudes yields an action spectral density of the form $(\alpha + \beta\omega_{\mathbf{k}} + \gamma \cdot \mathbf{k})^{-1}$ where α, β, and γ are constants and the water wave has a linear frequency $\omega_{\mathbf{k}}$ and wave vector $\mathbf{k}$. If either β or γ is not zero, then the action density can become negative for some $\mathbf{k}$. If both of these quantities are in fact zero, but α is a nonzero constant, then the spectral density is uniform and the total action in the wave field is infinite for an unbounded spectrum.

The preceding argument forces the conclusion that (1) the wave amplitudes are Gaussian random variables, but the nonlinear dynamics are improperly represented by weak interaction theory, (2) the dynamics are properly represented by weak interaction theory but the statistics of the wave field are non-Gaussian, or (3) neither the statistics nor the nonlinear dynamics are adequately represented. We adopt the point of view that since the dynamics of the wave field have been derived from a Hamiltonian, i.e., Zakharov has shown that the gravity wave field on the surface of the ocean constitute a Hamiltonian system [13], it is unlikely that they are not correct. Further, we present later the evidence that the steady-state distribution of water waves that best describes laboratory data is non-Gaussian.

Thus we conclude that a new model to describe the evolution of the statistical properties of a field of isolated nonlinear deep water gravity waves is needed and not a new set of dynamic equations. West [31] has suggested such a model based on the time-scale separation between resonant and nonresonant interactions. This separation in a single group of waves gives rise to both a slow average evolution of the field (due to nonlinear resonances) and a jitter or rapid variation about this slow motion (due to nonresonant interactions). To describe these effects a two-field model of the gravity wave evolution process is proposed. The first field is one which interacts internally solely by means of nonresonant interactions and is coupled at most linearly to a second field. This second field interacts internally solely by means of resonant interactions and exchanges energy, action, and momentum with the first field. Thus the second field, which is the resonant test field (RTF), is embedded in a rapidly varying ambient and is described by means of a nonlinear Langevin equation as is discussed in a later section. Similar models have been used successfully in both plasma physics [14] and physical oceanography [15, 16].

STEADY-STATE LINEAR WAVE FIELD

Theory

The most well-studied physical model of a statistical wave field consists of a superposition of linear eigenmodes with mutually statistically independent wave amplitudes and phases. The physical model is that the wave components have been generated at different points on the sea surface and have propagated to the point of observation. The distance traveled is large compared to any correlation distance among the waves and the amplitudes are still sufficiently small that finite amplitude effects can be neglected. These restrictions will be relaxed in later sections. Consider the continuous random function $\zeta(x, t)$ defined on the spatial interval $-\infty \leq x \leq \infty$ and times $t \geq 0$ such that

$$\zeta(x, t) = \sum_{j=1}^{N} a_j \cos(\chi_j + \varphi_j) \tag{1}$$

where

$$\chi_j = k_j x - \omega_j t \tag{2}$$

and the φ_j's are a set of constants. In order for $\zeta(x, t)$ to represent a water wave field the frequency ω_j and wave number k_j are related through a dispersion relation, i.e.

$$\omega_j = \omega(k_j) \tag{3}$$

In the continuum limit the dispersion relation for linear waves on the ocean surface is given by

$$\omega(k) = (gk + \gamma k^3)^{1/2} \tag{4}$$

where g is the acceleration of gravity and γ is the kinematic surface tension [6]. The dynamic curve defined by the time-dependent function $\zeta(x, t)$ represents the ocean surface in this simple linear model. Longuet-Higgins used this model to describe the steady-state properties of water waves on the surface of the ocean [10].

Consider the complex function $Z(x, t)$ generated by the linear superposition of N gravity waves:

$$Z(x, t) = \sum_{j=1}^{N} a_j e^{i(\chi_j + \varphi_j)} \tag{5}$$

the real and imaginary parts being indicated by Z_R and Z_I, respectively. The real vertical displacement of the water surface is given by $Z_I(x, t)$ and the real velocity potential at the surface is proportional to $Z_R(x, t)$. The distribution of vertical displacements resulting from the superposition of N linear waves is that of $Z_I(x, t)$ and is determined by allowing the φ_j's to take on all values in the interval $(0, 2\pi)$ with equal probability. This assumption is often referred to as the random phase approximation. Since we assume that the phases are mutually statistically independent the correlation of the surface displacement between the phase point (x, t) and (x', t') is given by

$$C(x, x', t, t') = \langle \zeta(x, t)\zeta(x', t') \rangle = \sum_{j=1}^{N} \frac{1}{2} a_j^2 e^{i[k_j(x-x') - \omega_j(t-t')]} \tag{6}$$

where the brackets denote an average over an ensemble of realizations of phases. Equation 6 indicates that the process is spatially homogeneous and stationary, i.e., the correlation function depends only on $(x - x')$ and $(t - t')$ and not on the origin of the coordinate system or on the initial time. In the continuum limit the sum over modes is replaced by an integral over wave number and in any fixed interval $(k, k + dk)$ the continuous spectrum of the wave field $\psi(k)$ is given by

$$\psi(k)\,dk = \frac{1}{2} \sum_{k_j \epsilon k} a_j^2 \tag{7}$$

Inserting Equation 7 into the continuum limit of Equation 6 at $t = t'$ we obtain the Wiener-Khintchine relation between the spectrum and correlation function

$$C(x - x') = \int_0^\infty dk\, \psi(k) \cos k(x - x') \tag{8}$$

and by inverting the Fourier transform we obtain

$$\psi(k) = \frac{1}{\pi} \int_0^\infty dx\, C(x) \cos kx \tag{9}$$

To determine the steady-state distribution of wave amplitudes at the surface we first construct the joint probability density function $P(Z_R, Z_I)$. In general the probability density is given by the Fourier transform of the characteristic function $\varphi(\mathbf{K})$ defined by

$$\varphi(\mathbf{K}) \equiv \langle e^{i\mathbf{K} \cdot \mathbf{Z}} \rangle \tag{10}$$

The average in Equation 10 represented by brackets can be written [9, 17]

$$\varphi(\mathbf{K}) = \frac{1}{(2\pi)^N} \int_0^{2\pi} d\varphi_1 \cdots \int_0^{2\pi} d\varphi_N \exp i(K_R Z_R + K_I Z_I)$$

$$= \prod_{j=1}^{N} \frac{1}{2\pi} \int_0^{2\pi} d\varphi_j \exp\{ia_j[K_R \cos(\chi_j + \varphi_j) + K_I \sin(\chi_j + \varphi_j)]\} \tag{11}$$

which when integrated yields

$$\varphi(\mathbf{K}) = \prod_{j=1}^{N} J_0(|\mathbf{K}|a_j) \tag{12}$$

Here $J_0(*)$ is the zero-order Bessel function, and $|\mathbf{K}| = [K_R^2 + K_I^2]^{1/2}$. The inverse Fourier transform of Equation 12 then yields the joint probability density

$$P(Z_R, Z_I) = \frac{1}{(2\pi)^2} \int_{-\infty}^{\infty} dK_R \int_{-\infty}^{\infty} dK_I \prod_{j=1}^{N} J_0(|\mathbf{K}|a_j) \exp[-i\mathbf{K} \cdot \mathbf{Z}] \tag{13}$$

The contribution of the integrand of Equation 13 to the probability density is substantial for small $|\mathbf{K}|a_j$ and decreases with increasing $|\mathbf{K}|a_j$ because the oscillations in the J_0 Bessel function produce rapid inteference. Therefore we use the small slope approximation, i.e., $|\mathbf{K}|a_j \ll 1$, for all j, for $J_s(|\mathbf{K}|a_j)$: [9, 18]

$$J_{s-1}(|\mathbf{K}|a_j) \sim \frac{(\frac{1}{2}|\mathbf{K}|a_j)^{s-1}}{\Gamma(s)} \exp\{-K^2 a_j^2/4s\} \tag{14}$$

which when substituted into Equation 13 yields

$$P(Z_R, Z_I) \simeq \frac{1}{(2\pi)^2} \int_{-\infty}^{\infty} dK_R \int_{-\infty}^{\infty} dK_I \exp\left\{-\frac{K^2\Lambda^2}{4} - i\mathbf{K}\cdot\mathbf{Z}\right\} \tag{15}$$

with

$$\Lambda^2 \equiv \frac{1}{2}\sum_{j=1}^{N} a_j^2 \tag{16}$$

and Equation 15 becomes exact as $N \to \infty$. In this case we integrate Equation 15 to obtain

$$P(Z_R, Z_I) = \frac{1}{\pi\Lambda^2} \exp\{-(Z_R^2 + Z_I^2)/2\Lambda^2\} \tag{17}$$

and the distribution for the surface displacement $\zeta = (i/2)(Z - Z^*) = -Z_I$ is then given by

$$P(\zeta) = \int_{-\infty}^{\infty} P(Z_R, Z_I)\, dZ_R = \frac{1}{\sqrt{2\pi}\,\Lambda} \exp\{-\zeta^2/2\Lambda^2\} \tag{18}$$

The preceding arguments can be generalized to include a distribution of wave amplitude in the derivation of $P(Z_R, Z_I)$. If we assume that the wave amplitude a_j is described by a distribution function $P_0(a_j)$, then the quantity Λ^2 defined by Equation 16 becomes

$$\Lambda^2 = \frac{1}{2}\sum_{j=1}^{M} \int_0^{\infty} a_j^2 P_0(a_j)\, da_j = \frac{1}{2}\sum_{j=1}^{M} \langle a_j^2 \rangle \tag{19}$$

with the remainder of the argument remaining unchanged.

The distribution of the wave heights given by Equation 17 is determined from the condition

$$a(x, t)e^{i\varphi(x,t)} = Z(x, t) \tag{20}$$

so that $a^2 = Z_R^2 + Z_I^2$ and $dZ_R\, dZ_I = a\, da\, d\varphi$. For a narrow-band surface wave field we have

$$P(Z_R, Z_I)\, dZ_R\, dZ_I = P(a, \varphi)\, da\, d\varphi \tag{21}$$

so that using Equation 17 the distribution of the wave heights of a narrow band process is the Rayleigh distribution.

$$P(a) = \frac{a}{\Lambda^2} \exp[-a^2/2\Lambda^2] \tag{22}$$

as obtained by Longuet-Higgins [19] with a uniform distribution in angle

$$P(\varphi) = \frac{1}{2\pi} \tag{23}$$

From the Wiener-Khintchine relation Equation 8 we know that the mean square surface displacement is given by the integral over the spectrum, i.e.

$$\langle \zeta^2 \rangle = \int_{-\infty}^{\infty} dk\, \psi(k) \tag{24}$$

Further, by comparing Equations 16 and 18 we observe that

$$\Lambda^2 = \langle \zeta^2 \rangle \tag{25}$$

Thus Λ^2 is the mean square displacement of the water surface. In addition by direct integration we obtain

$$\Lambda^2 = \tfrac{1}{2}\overline{a^2} \tag{26}$$

where the overbar indicates an average with respect to the distribution Equation 22 itself. Thus Λ^2 is the mean square wave amplitude, which differs from Equation 25 except in the case of a linear wave field. For this reason Longuet-Higgins [20] stresses the use of Equation 26 for comparison with data.

Data

If we rewrite Equation 22 using Equation 26, then the probability that the amplitude of any given wave exceeds the value a_0 is given by

$$P(a > a_0) = \int_{a_0}^{\infty} e^{-a^2/\overline{a^2}} \frac{a\,da}{\frac{1}{2}\overline{a^2}} = e^{-a_0^2/\overline{a^2}} \tag{27}$$

It is remarkable that this distribution, constructed by Longuet-Higgins in 1952 has been found to agree so well with such a variety of field observations even when the observational conditions violate the assumptions under which the distribution was derived. For example you will recall that in the derivation of Equation 17 we assumed that the slope of each of the linear waves was sufficiently small so that no correlation between the individual waves could develop. The recent application of Equation 27 by Earle [21] to large hurricane-generated waves was successful even though the root mean square surface slope could exceed 0.1 in magnitude, clearly violating the small slope assumption.

Forristall [22] using waves from different records of storms in the Gulf of Mexico tested the prediction made by Equation 27, where instead of $\overline{a^2}$, he used $2\Lambda^2$ in the distribution

$$P(a > a_0) = e^{-a^2/2\Lambda^2} \tag{28}$$

with Λ^2 given by Equation 25. Normalizing the individual wave height to the variance of the record $\sqrt{2}\,\Lambda$, the fraction of the total number of waves exceeding a given value of $\sqrt{2}\,a_0/\Lambda$ is plotted as a triangle in Figure 1. The dashed curve in this figure is the Rayleigh distribution plotted from Equation 28 where Λ^2 is calculated using the experiment spectrum Equation 25. It is clear that Equation 28 over-predicts the probabilities of the highest waves, and the error increases in the tail region of the distribution where the probability is lowest. This tail region determines the extreme properties of the sea surface and can therefore be quite important for certain purposes, such as in establishing design specification for platforms.

Longuet-Higgins [20] countered Forristall's [22] analysis by pointing out that the latter had mistakenly referred to Equation 28 as "the" Rayleigh distribution, when in fact there is more than one possible such distribution. In particular Equation 27 not Equation 28 is the distribution first proposed by Longuet-Higgins [19] for use in the study of wave height distributions. It should be emphasized that Equations 27 and 28 are equivalent only when the surface wave field is linear, i.e., when Equation 26 is satisfied. Thus for applications of the Rayleigh distribution to data Equation 27 could be used with $\overline{a^2}$ fit by the data. Longuet-Higgins [20] chose the value $(\overline{a^2})^{1/2} = 0.925(2\Lambda^2)^{1/2}$ and the corresponding distribution is given by the solid curve in Figure 1. It is obvious that this latter curve fits the data much better than does Equation 28. Analysis indicates that finite-amplitude effects increase the mean square wave amplitude and can therefore not account for this improved agreement. Longuet-Higgins estimates the effect of free background "noise" in the spectrum, outside the dominant peak on a^2 and finds this can indeed account for the factor of 0.925 in the root mean square wave height [20].

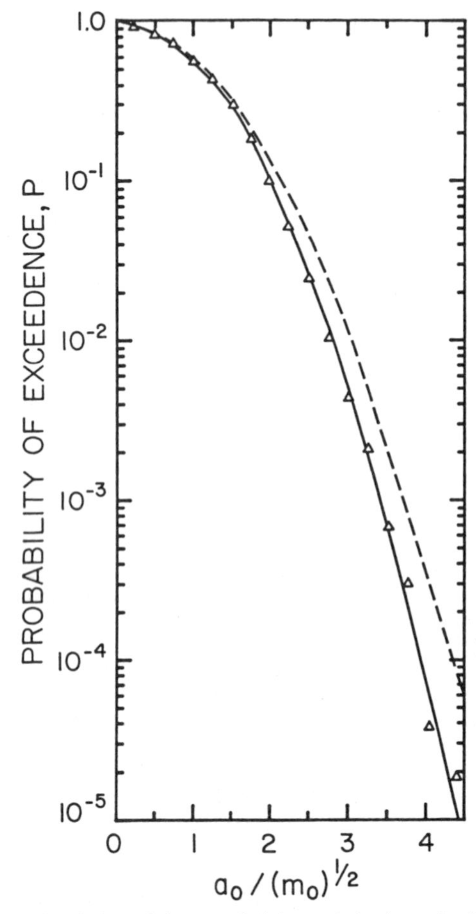

Figure 1. Probability of the wave heights exceeding a given value $2\alpha_0$. The triangles are the data points from Forristall [22]. The two curves are Rayleigh distributions; the solid one is calculated using Equation 27 and the dashed one using Equation 28.

Deviation of the wave height statistics from the Rayleigh form were observed in the laboratory experiments of Huang and Long [23]. In these experiments the wind was left on for a sufficiently long time that a steady-state spectrum of wind-generated waves was in evidence. Typical measured probability distributions are shown in Figure 2 for a number of wind speeds. The distribution becomes increasingly skewed with increasing wind speed. Huang and Long [23] point out that at high wind the skewness approaches unity and a hump appears on the positive side of the distribution curve indicating the occurrence of waves at a particular height with more than the expected frequency (see Figure 2C). They find that this hump is centered on $1.4\ (\langle\zeta^2\rangle)^{1/2}$, the location of the mean amplitude of the waves. This hump indicates that the amplitudes of the waves have a preferred range of height rather than being completely at random and that the spectra of even the wind-generated random waves are rather narrow-peaked. It should be emphasized that these laboratory data run counter to the claim of Kinsmann [24] who asserted that the distribution should approach a Gaussian with increasing wind speed and wave amplitude.

The deviation of the experimental data from Gaussianity suggests the need for a nonlinear analysis. In the next section we review the theoretical results of Longuet-Higgins' use of perturbation theory to calculate the Gram-Charlier approximation to the distribution of the surface displacements [25]. Here we merely note that the theoretical distribution gives an overall good fit to the experimental data. However there are a number of theoretical problems with this approximation which we will discuss in the next section.

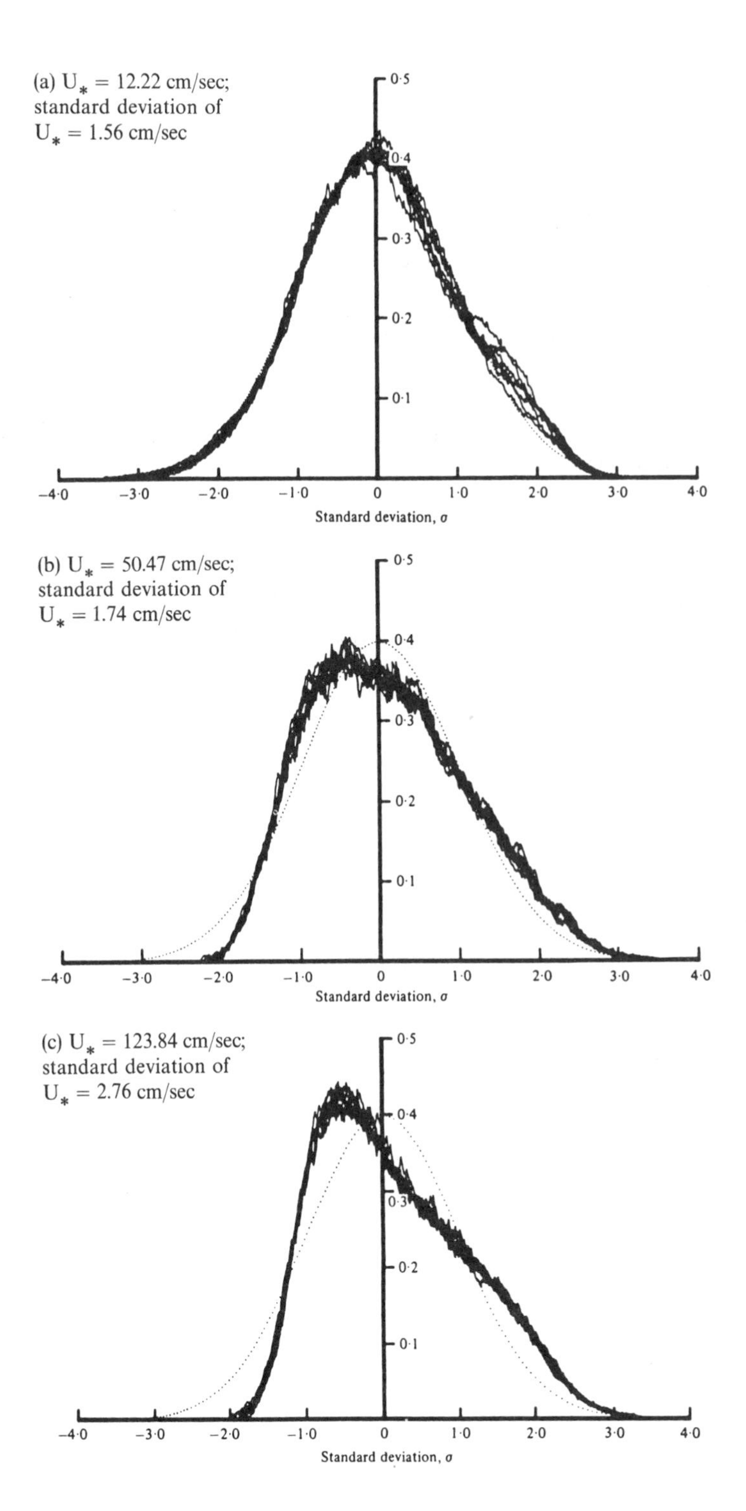

Figure 2. The normalized distribution of wave heights for wind friction U_*: Each case represents seven independent measurements under the same conditions. The heavy jagged curve is the data and the dotted curve is a normalized Gaussian distribution [23].

STEADY-STATE NONLINEAR WAVES

In the preceding section we had assumed that the surface displacement is the superposition of many wave components that have been generated at different points on the water surface by the wind and had propagated to the point of observation. The distance traveled is assumed larger than the correlation distance of the waves and the amplitudes are sufficiently small that wave-wave interactions can be neglected and the surface displacement can be regarded as the sum of a large number of independent contributions each with a statistically independent random phase. The central limit theorem then ensures that the statistics of the surface displacement is Gaussian (Equation 18) and the distribution of wave amplitudes is Rayleigh (Equation 22). When the slope of the sea surface is sufficiently large that the wave-wave interactions can no longer be neglected one must find corrections to these distributions.

Longuet-Higgins was the first to systematically investigate the effect of water wave nonlinearities on the statistical distribution of wave heights [25]. He proceeded on a speculation of Phillips [26] who pointed out that the coefficient of skewness in the wave height distribution is of the same order of magnitude as the slope of the surface displacement. In the following we present the barest sketch of the ideas used by Longuet-Higgins to determine the finite amplitude corrections to the linear analysis of the preceding section. Although still of some interest, these ideas have been superceded by those of Huang et al. [27] and Tayfun [28] which are discussed subsequently.

One way to evaluate the probability distribution for the nonlinear wave field is by means of the characteristic function introduced in the preceding section.

$$\varphi(K) = \int_{-\infty}^{\infty} P(\zeta) e^{iK\zeta}\, d\zeta \tag{29}$$

The characteristic function can be expressed as an infinite series of cumulants of the field variable:

$$\varphi(K) = \exp\left\{\sum_{n=1}^{\infty} \frac{(iK)^n}{n!} \lambda_n\right\} \tag{30}$$

or equivalently as an infinite series of moment of ζ:

$$\varphi(K) = \sum_{n=0}^{\infty} \frac{(iK)^n}{n!} \mu_n \tag{31}$$

The cumulants λ_n and moments μ_n are therefore related to each other, the lowest order terms being related as follow:

$$\begin{aligned} \lambda_1 &= 0 \\ \lambda_2 &= \mu_2 \\ \lambda_3 &= \mu_3 \\ \lambda_4 &= \mu_4 - 3\mu_2^2 \\ \lambda_5 &= \mu_5 - 10\mu_3\mu_2 \\ &\vdots \end{aligned} \tag{32}$$

and we have centered the moments with respect to the mean μ_1. For a Gaussian distribution all even moments can be expressed as products of the second moment, all odd moments vanish: $\mu_{2n} = (2n)!\mu_2^n/2^n n!$, $\mu_{2n+1} = 0$, $n \geq 1$, and all cumulants greater than $\lambda_2 = \mu_2$ vanish.

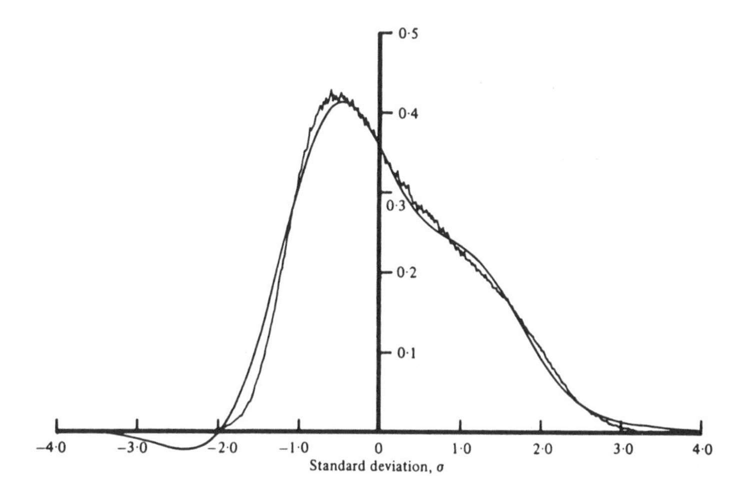

Figure 3. The distribution for a wind friction speed of 123.84 cm/sec (see Figure 2C) is fit by the Gram-Charlier expansion suggested by Longuet-Higgins. The data represents the average of seven independent measures [23].

The probability density, as mentioned, is given by the inverse Fourier transform of Equation 29.

$$P(\zeta) = \frac{1}{2\pi} \int_{-\infty}^{\infty} dK \exp\left\{ -iK\zeta + \sum_{n=1}^{\infty} \frac{1}{n!} (iK)^n \lambda_n \right\} \tag{33}$$

so that in terms of the normalized cumulants

$$\bar{\lambda}_n = \lambda_n / \lambda_2^{n/2} \tag{34}$$

we can, for small skewness λ_3, kurtosis λ_4, and higher cumulants, expand the exponential in Equation 33 to obtain the series expression for the probability density:

$$P(\zeta) = \frac{\exp[-\eta^2/2]}{(2\pi\bar{\lambda}_2)^{1/2}} \left\{ 1 + \sum_{n=3}^{\infty} C_n(\eta)\bar{\lambda}_n \right\} \tag{35}$$

where the first few coefficients are $C_3 = H_3(\eta)/6$, $C_4 = H_4(\eta)/24$, $C_5 = H_5(\eta)/120$; $H_n(\eta)$ is a Hermite polynomial of degree n and $\eta \equiv \zeta/\mu_2^{1/2}$. Equation 35 is the Gram-Charlier series first developed in the nonlinear water wave context by Longuet-Higgins [25]. Although the non-Gaussian distribution (Equation 35) has been shown to represent the data for surface displacements rahter well, it has a number of serious deficiencies. Firstly, this series has negative probability densities for some ranges of surface displacement-particularly in the case of waves with steep slopes. In Figure 3 the Gram-Charlier series as suggested by Longuet-Higgins is used to fit the data in Figure 2c. The region where $P(\zeta)$ as calculated using Equation 35 becomes negative is apparent. Since a probability density is always ≥ 0, Equation 35 cannot be interpreted as a probability density. Secondly, as emphasized by Huang et al. [27] both the skewness λ_3 and kurtosis or flatness, λ_4, values are needed to compute the distribution function, using Equation 35, neither of which is known *a priori*. These authors showed that λ_4 is necessary to fit the experimental data, and the amount of computation necessary to calculate λ_3 and λ_4 from first principles is prohibitive. Thus techniques yielding distributions other than the Gram-Charlier series are required.

Tayfun [28] has presented a simplified probabilistic model of nonlinear random waves which is potentially more useful than that of Longuet-Higgins. We describe both his ideas and recent extensions of them [27] in the following. The surface displacement satisfying Bernoulli's equation on the water surface and the free surface boundary condition can be represented to second order in the linear wave components are

$$\zeta(x, t) = \sum_{n=1}^{N} a_n \cos(\chi_n + \vartheta_n) + \sum_{n,m=1}^{N} \Gamma_{nm} a_n a_m \cos(\chi_n - \chi_m + \vartheta_n - \vartheta_m) + \sum_{n,m=1}^{N} \Gamma'_{nm} a_n a_m \sin(\chi_n - \chi_m + \vartheta_n - \vartheta_m) \quad (36)$$

where the coupling coefficients are obtained from the nonlinear equations of motion [6, 9, 28] and the a's, χ's, and ϑ's are the same as described earlier. Tayfun demonstrated that *if* the spectrum is narrow band *and* the distribution of the amplitude a(x, t) is Rayleigh *and* the phase is uniformly random, then Equation 36 can be replaced by the second-order Stokes wave:

$$\zeta(x, t) = a(x, t) \cos[\chi_0 + \vartheta(x, t)] + \tfrac{1}{2}k_0 a^2(x, t) \cos 2[\chi_0 + \vartheta(x, t)] + \mathcal{O}(k_0 \mu_0 \nu) \quad (37)$$

The parameters in (37) are given by the spectral moments

$$\mu_p = \int_0^\infty \omega^p S(\omega)\, d\omega \quad (38)$$

where the spectrum in wave number $\psi(\mathbf{k})$ and in frequency $S(\omega)$ are related by $\psi(k)\, dk = S(\omega)\, d\omega$ so that μ_0 and $\mu_1/\mu_0 = \omega_0$ represent the first-order variance and mean frequency, respectively. The spectrum is considered to be narrow band if

$$\nu^2 = \frac{\mu_2}{\mu_0 \omega_0^2} - 1 \ll 1. \quad (39)$$

Under this condition we can write $\chi_0 = k_0 x - \omega_0 t$ where $\omega_0^2 = g k_0$ so that the amplitude a(x, t) and phase functions $\vartheta(x, t)$, defined as in Equation 20 are slowly varying functions of space and/or time for a nonliner wave field.

If we normalize the surface displacement with the mean square wave amplitude $(\overline{a^2})^{1/2}$ and define $\hat{\zeta}(x, t) = \zeta(x, t)/(\overline{a^2})^{1/2}$, $\hat{a}(x, t) = a(x, t)/(\overline{a^2})^{1/2}$, then Equation 37 can be written as

$$\hat{\zeta}(x, t) = \hat{a}(x, t) \cos[\chi_0 + \vartheta(x, t)] + \tfrac{1}{2}(k_0^2 \overline{a^2})^{1/2} \hat{a}^2(x, t) \cos 2[\chi_0 + \vartheta(x, t)] + \mathcal{O}(k_0 \overline{a^2} \nu) \quad (40)$$

Equation 40 shows that the correction to the linear wave field is quadratic in the wave amplitude with a strength determined by the root mean square wave slope. This vertical asymmetry to the linear wave profile causes the wave crests to narrow and sharpen and the troughs to lengthen and become shallower. Further, although the linear surface elevation is Gaussian the nonlinear term in Equation 40 implies that the distribution of values of $\hat{\zeta}(x, t)$ is non-Gaussian.

Huang et al. [27] generalized the analysis of Tayfun [28] by directly considering the Stokes wave

$$\zeta(\mathbf{x}, t) = \tfrac{1}{2}k_0 a^2 + a \cos(\chi + \vartheta) + \tfrac{1}{2}k_0 a^2 \cos 2(\chi + \vartheta) + \tfrac{3}{8}k_0^2 a^3 \cos 3(\chi + \vartheta) + \cdots \quad (41)$$

to third order in the wave amplitude and assuming in an ad hoc manner that for a narrow band spectrum the amplitude and phases are slowly varying as was *proven* by Tayfun for Equation 40. They therefore go one order higher, retaining terms to third order in the wave amplitude and also including the "constant" term $\frac{1}{2}k_0 a^2(\mathbf{x}, t)$. Dynamically this second-order quantity is a consequence of wave-induced mass transport, i.e. the so-called Stokes drift, and although it is usually unimportant in deterministic studies it can be quite important in the determination of a true probability density. Recall the hump in the data mentioned earlier.

Here we follow closely the analysis of Huang et al. [27] and using $\cos 2\chi = 2\cos^2\chi - 1$, write the surface displacement as

$$\zeta = a\cos(\chi + \vartheta) + k_0 a^2 \cos^2(\chi + \vartheta) + \tfrac{3}{8}k_0^2 a^3 \cos 3(\chi + \vartheta) + \cdots \tag{42}$$

We calculate the variance σ^2 of the surface displacement ζ from its mean level $\bar{\zeta}$

$$\sigma^2 = \overline{\zeta^2} - \bar{\zeta}^2 \tag{43}$$

which using Equation 42 gives

$$\bar{\zeta} = \tfrac{1}{2}k_0\overline{a^2} \tag{44a}$$

$$\overline{\zeta^2} = \tfrac{1}{2}a^2 + \tfrac{3}{8}k_0^2\overline{a^4} = \tfrac{1}{2}\overline{a^2}(1 + \tfrac{3}{2}k_0^2\overline{a^2}) \tag{44b}$$

since ϑ is uniformly distributed, a is Rayleigh distributed and $\overline{a^4} = 2(\overline{a^2})^2$, so that

$$\sigma^2 = \tfrac{1}{2}\overline{a^2}(1 + k_0^2\overline{a^2}) \tag{44c}$$

We use σ to normalize the deviation of the surface from its average value (since this is not zero with the Stokes-drift term included in Equation 41) and define

$$\eta(\mathbf{x}, t) = [\zeta(\mathbf{x}, t) - \bar{\zeta}(\mathbf{x}, t)]/\sigma \tag{45}$$

The quantity of interest in the present analysis is the probability density $P(\eta)$, so we follow the earlier procedure and construct a two variable distribution function and integrate out one of the variables.

In terms of the complex quantity $Z(\mathbf{x}, t)$, introduced earlier, (cf. Equation 5) we write Equation 45 as

$$\eta = \frac{\hat{Z}_R}{N} + \frac{\hat{Z}_R^2}{N^2}k_0\sigma + \frac{3}{8}\frac{\hat{Z}_R^3 - 3\hat{Z}_R\hat{Z}_I^2}{N^3}k_0^2\sigma^2 - k_0\sigma \tag{46}$$

in which from Equation 44c

$$N \equiv \frac{\sigma}{(\overline{a^2}/2)^{1/2}} \simeq 1 + k_0^2\sigma^2 \tag{47}$$

we have used $\cos 3\chi = \cos^3\chi - 3\cos\chi\sin^2\chi$ and Z has been normalized with $(\overline{a^2}/2)^{1/2}$. An explicit expression for $\hat{Z}_R$ in terms of η can be obtained by inverting Equation 46 to any desired order in the small parameters $k_0\sigma$ and η. The resulting expression is an infinite series which to third order in η is

$$\begin{aligned}\hat{Z}_R &\simeq N\eta - N(\eta^2 - 1)k_0\sigma + N\left(\frac{13}{8}\eta^3 - 2\eta\right)k_0^2\sigma^2 + \frac{9}{8}\frac{\eta}{N}\hat{Z}_I^2k_0^2\sigma^2 \\ &= H_\eta + \frac{9}{8}\frac{\eta}{N}\hat{Z}_I^2k_0^2\sigma^2\end{aligned} \tag{48}$$

where we have introduced the function

$$H_\eta = N[\eta - k_0\sigma(\eta^2 - 1) + k_0^2\sigma^2(13\eta^3/8 - 2\eta)].$$

From Equation 17 we can write the probability density for the normalized quantity $\hat{Z}$ as

$$P(\hat{Z}_R, \hat{Z}_I) = \frac{1}{2\pi} \exp\left\{-\frac{1}{2}(\hat{Z}_R^2 + \hat{Z}_I^2)\right\} \tag{49}$$

which in terms of η can be written to $\mathcal{O}(k_0^2\sigma^2)$ and $\mathcal{O}(\eta^3)$ as

$$P(\eta, \hat{Z}_I) = \frac{1}{2\pi} \exp\left\{-\frac{1}{2}H_\eta^2 - \frac{1}{2}\left(1 + \frac{9}{4}\eta^2 k_0^2\sigma^2\right)\hat{Z}_I^2\right\} \tag{50}$$

and

$$P(\hat{Z}_R, \hat{Z}_I)\, d\hat{Z}_R\, d\hat{Z}_I = P(\eta, \hat{Z}_I) J\left(\frac{\hat{Z}_R, \hat{Z}_I}{\eta, \hat{Z}_I}\right) d\eta\, d\hat{Z}_I \tag{51}$$

where J(*) is the Jacobian of the transformation $(\hat{Z}_R, \hat{Z}_I) \to (\eta, \hat{Z}_I)$. The Jacobian to $\mathcal{O}(k_0^2\sigma^2)$ is

$$J\left(\frac{\hat{Z}_R, \hat{Z}_I}{\eta, \hat{Z}_I}\right) = \frac{\partial H_\eta}{\partial \eta} + \frac{9}{8} k_0^2\sigma^2 \frac{\hat{Z}_I^2}{N} \tag{52}$$

Since we are only interested in $P(\eta)$ we integrate Equation 51 over $\hat{Z}_I$ to obtain [27]

$$P(\eta) = \int_{-\infty}^{\infty} P(\eta, \hat{Z}_I) J\left(\frac{\hat{Z}_R, \hat{Z}_I}{\eta, \hat{Z}_I}\right) d\hat{Z}_I$$

$$= \frac{e^{-H_\eta^2/2}}{\sqrt{2\pi}} \left[\frac{\dfrac{\partial H_\eta}{\partial \eta}}{\left(1 + \dfrac{9}{4}k_0^2\sigma^2\eta^2\right)^{1/2}} + \frac{9}{8}\frac{k_0^2\sigma^2}{N} \frac{1}{\left(1 + \dfrac{9}{4}k_0^2\sigma^2\eta^2\right)^{3/2}}\right] \tag{53}$$

the non-Gaussian distribution for the surface displacement.

Equation 53 is a positive definite quantity so that unlike the Gram Charlier series it can be interpreted as a probability density throughout its parameter space. Another difference between the two expressions is that Equation 53 depends only on two parameters; the root mean square value of the surface elevation σ and the "significant slope" of the wave fields $\S = \sigma/\lambda (\lambda = 2\pi/k)$. In Figure 4 the probability density function $P(\eta)$ is graphed for various values of §. The probability density becomes increasingly skewed as the significant slope increases as was predicted by Phillips [26] and Longuet-Higgins [25] and observed in the laboratory by Huang and Long [23]. The hump that arises in the probability density was predicted using the Gram-Charlier distribution as noted earlier (cf. Figure 3). Its source using Equation 53 can be traced back to the "constant" term in the Stokes expansion of the surface elevation Equation 41. Thus the existence of this hump in all the laboratory data of Huang and Long (cf. Figure 2) and its association with the constant term in Equation 41, strongly suggest that Equation 53 is a proper description of the surface elevation density function in the steady state.

The preceding experimental and theoretical evidence that the surface elevation density function is non-Gaussian in the steady state is quite convincing. How and why the wind-generated wave field approaches this steady state still needs to be examined, however. This is done in part in the remaining sections.

TRANSIENT LINEAR WAVE FIELD

In the preceding discussions we did not address the question of the possible physical mechanisms that could give rise to random fluctuations in a wave field. Instead we assumed the existence of

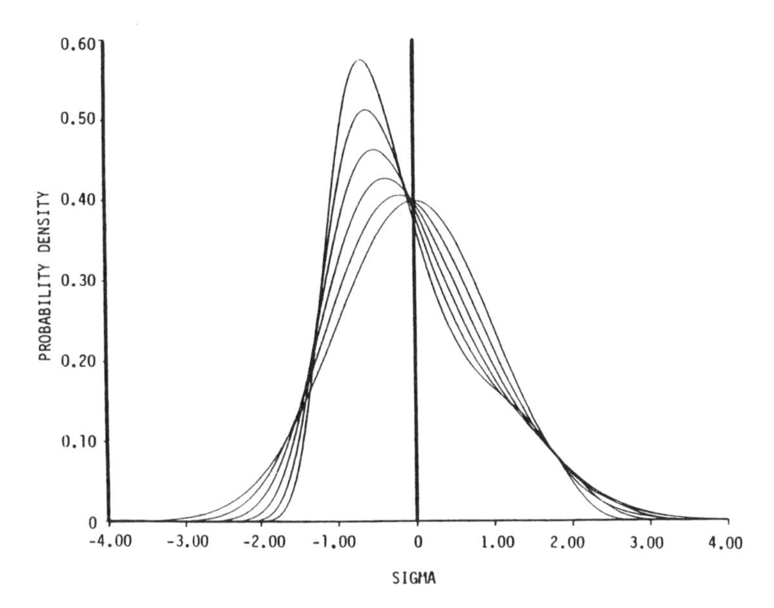

Figure 4. Surface elevation probability density function based on the third order Stokes wave model with bias for significant slope values ranging from 0 to 0.05 in steps from 0.01 [27].

linear eigenmodes with mutually statistically independent constant wave amplitudes and phases and then relied on the central limit theorem to determine the wave height statistics. This was done for both linear and nonlinear wave fields. A key feature of that analysis was the complete absence of dynamics. In this section we review some of the ideas that have been developed to describe the statistics of water waves in the transient region, i.e., the region of their generation and propagation prior to reaching the steady state discussed in the two prior sections.

Phillips [1] provided a dynamic model for the wind generation and initial evolution of water waves on the ocean surface. He considered the linear dynamic equations that describe the evolution of the mode amplitudes in Equations 5 and 36. In his dynamic equations the incoherent pressure fluctuations at the sea surface produced by the turbulent wind drive the linear surface waves. Concurrently Miles [2] studied a complimentary generation mechanism in which the wind field is modulated by the water waves. This modulation of the air flow leads to a positive energy flux from the wind to the water waves. The linear model having both the incoherent pressure fluctuations and the coherent wind modulation is the Miles-Phillips model of the air-sea interaction [3, 6]. The mode amplitudes in this model satisfy the linear equations of motion.

$$\dot{b}_n(t) + \mu_n b_n(t) = -f_n(t) \tag{54}$$

where in terms of the mode amplitudes used earlier we have

$$b_n(t) = \frac{1}{-i\mu_n} \dot{a}_n(t) - ia_n(t) \tag{55}$$

and $\mu_n(\equiv \nu k_n^2 - \gamma_n \omega_n + i\omega_1(n)$ and $\omega_1(n) = [\omega_n^2 - (\nu k_n^2 - \gamma_n \omega_n^2]^{1/2})$ is the complex eigenvalue of the linear dynamic equations. In the linear eigenvalue appearing in Equation (54) ν is the kinematic viscosity, $\omega_n = (gk_n + \gamma k_n^3)^{1/2}$ is the linear water wave frequency, and γ_n is the Miles air-sea coupling parameter. $p_n(t)$ is the Fourier mode amplitude of the turbulent pressure at the sea surface, so that Equation 55 describes a linear harmonic oscillator driven by incoherent fluctuations in the pressure field:

$$p_n(t) = \frac{-i\mu_n}{k_n} f_n(t) \tag{56}$$

and coherently driven by the average stress on the surface ($\gamma_n\omega_n > \nu k_n^2$) leading to an exponential growth on the n-th mode, on the average. A recent critique of the Miles-Phillips model has been given by Krasitaskii and Zaslavskii [29] in which they overcome a number of limitations of the earlier analyses.

The statistical properties of the solution to the harmonic oscillator equation (Equation 54) were first studied systematically by Wang and Uhlenbeck [30] in the case $\nu k_n^2 > \gamma_n\omega_n$, i.e., when the oscillator dissipates energy. For a dissipative harmonic oscillator the physical system will eventually reach a statistical steady state in which the energy supplied by the fluctuating force is balanced by the energy extracted in the form of heat by the dissipation. This balance implies that the fluctuations can be characterized by a single scalar parameter. In classical statistical mechanics this parameter is the temperature and the balance is the well known fluctuation-dissipation relation. Such an equilibrium does not arise in the present situation because the exponential growth ($\gamma_n\omega_n > \nu k_n^2$) does not lead to an asymptotic steady state of the wave field. That is to say that the deterministic solution to Equation 54, i.e., the solution with $p_n(t) = 0$, diverges in time so that there is no well defined value of the surface displacement $\zeta(x, t)$ as $t \to \infty$.

It is clear that after the surface mode amplitudes become finite the wave-wave interaction terms in the dynamic equations can no longer be neglected and the prior linear equations are an inadequate description of the wave field evolution process [12, 31]. The wave-wave coupling terms transfer energy from one spectral interval to another by both resonant and nonresonant interactions [25, 26]. The nonresonant energy transfer is periodic whereas the resonant transfer is unidirectional in wave number space. Therefore it is the resonance mechanism which provides the *net* energy flux among modes over finite times due to wave-wave interactions. Thus a nonlinear model which includes this energy transfer process will presumably circumvent the divergence imposed by Equation 54 as $t \to \infty$ for $\gamma_n\omega_n > \nu k_n^2$ and lead to an asymptotic statistical state. One must therefore include the effect of the nonlinear interaction in the dynamic description in order to achieve the observed steady state.

One way in which the nonlinear effect can be included in the context of a linear model is through a parameterization of the wave-wave interactions. In this approximation the coefficient $\nu k_n^2 - \gamma_n\omega_n$ is replaced by a quantity, λ_n say, having the property $\lambda_n < 0$ at early times before the other waves in the field have attained a significant amplitude, but $\lambda_n > 0$ at late time when other waves have nearly reached their steady-state levels. A technique which has had some success in achieving this objective is the method of statistical linearization. This technique has been applied by West [7] to the dynamic equations for a gravity capillary wave field at late times.

The full nonlinear equations of motion are written schematically as [6–9]

$$b_n(t) + \mu_n b_n(t) = -f_n(t) + T_n(\mathbf{b}) \tag{57}$$

where the function $T_n(\mathbf{b})$ represents the sum of both three- and four-wave interaction terms. The linearized equations of motion are then given by

$$\dot{b}_n(t) + \lambda_n b_n(t) = -f_n(t) \tag{58}$$

where λ_n is a variational parameter which reduces to μ_n when $T_n = 0$ (cf. Equation 54). It can be shown in a number of ways that because the system response (Equation 58) is linear, the surface mode has the same statistical properties as the fluctuations in the pressure field driving the surface [9]. In the simplest physically reasonable case $p_n(t)$ is assumed to be a zero-centered Gaussian process with the spectrum $\Phi(k_n)$ and to be both stationary in time and homogeneous in space:

$$\langle p_n(t)p_{n'}(t')\rangle = 2\Phi(k_n)\delta_{nn'}\varphi_n(t - t') \tag{59}$$

where the brackets denote an average over an ensemble of realizations of fluctuations in the pressure field. For simplicity the correlation function $\varphi_n(\tau)$ is often replaced by a delta function in time. In this case, for Real $\lambda_n > 0$, the surface response is also a Gaussian random process but with a correlation time dependent on the real part of λ_n. Recall that at late time λ_n includes the average effect of the wave-wave interactions. In this case the surface displacement spectral density is given by [7, 9]

$$\lim_{t \to \infty} \langle |b_n(t)|^2 \rangle = k_n^2 \Phi(k_n)/|\mu_n|^2 \text{ Real } \lambda_n \tag{60}$$

The mean square surface displacement is obtained by substituting the continuum limit of Equation 60 into Equation 20 to obtain

$$\langle \zeta^2 \rangle = \frac{1}{2} \int_{-\infty}^{\infty} dk \frac{\Phi(k)k^2}{\text{Real}[\lambda(k)]|\mu(k)|^2} \tag{61}$$

The parameter $\lambda(k)$ is then given by [9]

$$\lambda(k) = \nu k^2 - \gamma(k)\omega_k - \text{Real}\left[\frac{\langle b_k^* T_k(\mathbf{b}) \rangle_{ss}}{\langle |b_k|^2 \rangle_{ss}}\right] \tag{62}$$

where the brackets with an ss subscript denotes an ensemble average over the steady-state distribution of water waves. Thus in this model the mean square surface displacement depends directly on the spectrum of pressure fluctuations at the water surface and inversely on the Miles parameter $\gamma(k)$ and the projection of the nonlinear interaction T_n onto b_n^*.

The phase space equation of evolution for the probability density $P(\mathbf{b}, t|\mathbf{b}_0) \equiv P_t$ such that the dynamic variable $\mathbf{b}(t)$ is in the phase space interval $(\mathbf{b}, \mathbf{b} + d\mathbf{b})$ at time t t given the initial values $\mathbf{b}(0) = \mathbf{b}_0$ for a delta-correlated Gaussian pressure field is the Fokker-Planck equation. Restricting our considerations to waves propagating in the direction of the wind we obtain

$$\frac{\partial}{\partial t} P_t = \sum_n \left\{ \frac{\partial}{\partial b_n} [\lambda_n b_n] + cc + 2D_n \frac{\partial^2}{\partial b_n \partial b_n^*} \right\} P_t \tag{63}$$

where cc denotes the complex conjugate of the preceding terms and

$$D_n \equiv \frac{k_n^2}{|\mu_n|^2} \Phi(\mathbf{k}_n) \tag{64}$$

The steady-state solution to Equation 63 is obtained from $\partial P_t/\partial t = 0$ equation to be

$$P_{ss}(\mathbf{b}) = \prod_n \frac{\text{Real } \lambda_n}{\pi D_n} \exp\left\{ -\frac{\text{Real } \lambda_n}{D_n} |b_n|^2 \right\} \tag{65}$$

which is a Gaussian distribution in the complex mode amplitude b_n.

The argument used to construct the probability density from the characteristic function can be used here to obtain the distribution of wave amplitudes. Recalling that $b_n = a_n e^{i\phi_n}$ the characteristic function Equation 10 can be written

$$\varphi(\mathbf{K}) = \prod_n \int_0^{2\pi} \frac{d\varphi_n}{2\pi} \int_0^{\infty} \frac{2 \text{ Real } \lambda_n}{D_n} a_n \, da_n \exp\left\{ -\frac{\text{Real } \lambda_n}{D_n} a_n^2 \right\}$$
$$\exp\{iK_R a_n \cos(\chi_n + v_n) + iK_I a_n \sin(\chi_n + v_n)\}$$
$$= \exp\left\{ -\frac{1}{4}(K_R^2 + K_I^2)^{1/2} \sum_n \frac{D_n}{\text{Real } \lambda_n} \right\} \tag{66}$$

Fourier inverting Equation 66 (c.f. Equation 15) we obtain Equation 17 and consequentially the Rayleigh distribution (22) for the wave height distribution. In the present case however the parameter Λ^2 is given by

$$\Lambda^2 = \frac{1}{2}\sum_n \frac{D_n}{\text{Real } \lambda_n} \tag{67}$$

It is important to distinguish the use of Equation 65 from the usual Gaussian approximation in that Λ depends parametrically on the average wave-wave interaction (cf. Equation 62). Thus, to evaluate the steady-state energy-spectral density $\Psi(\mathbf{k})$ using Equation 65, i.e.

$$\Psi(\mathbf{k}) = \frac{1}{2}\int \langle |b_k|^2 \rangle_{ss} = \frac{1}{2}\int |b_k|^2 P_{ss}(\mathbf{b}) d^n\mathbf{b} \tag{68}$$

we obtain an explicit dependence on the average nonlinear interaction through Real $\lambda(k)$ in the distribution. Equation 67 actually constitutes a self-consistency condition for the average nonlinear interaction. West [7] has obtained the self-consistent solution numerically for a gravity-capillary water wave spectrum using an observed spectrum for the fluctuations in the wind field. The energy spectral density for the high frequency water wave was determined to be not too dissimilar from that observed [32], i.e., a predicted ω^{-4} spectrum rather than the observed ω^{-3} spectrum in this frequency interval.

DYNAMIC NONLINEAR WAVE FIELD

There have been relatively few attempts to systematically examine the theoretical properties of the gravity waves statistics in an *evolving* wave field. The first synthesis of statistics and dynamics was made in a related context by Prigogine [33]. He developed a phase space equation of evolution for the probability density of an ensemble of wave fields (not water waves) which relies on a random phase description for the initial state of the wave field. With this choice of the initial state, he shows that the Fokker-Planck equation describes the phase space evolution of an ensemble of waves. These arguments have been extended by Zaslavskii and Sagdeev [34] to the derivation of a Fokker-Planck-type equation for the probability density that does not require the random phase approximation on the initial state. They argue that the resonant triad interactions among the waves randomize the relative phases between them resulting in a loss of memory of the initial state of the system. Such an argument could be directly applied to gravity-capillary waves since the dominant energy exchange mechanism is a three wave resonance [35]. For gravity waves, however, the dominant interaction is a four wave resonance [25, 26] so that the lowest order interactions are (quasi-) periodic. This situation suggests that an alternative point of view be adopted for discussing the relation between the nonlinear interactions in, and the statistical properties of the gravity wave field. The resonant test field described in the "INTRODUCTION" is such an alternate viewpoint and is the one to which we now turn our attention.

The RTF model replaces the reversible equations of motion derived from the Hamiltonian description of the sea surface motion by an irreversible set of equations in which the nonresonant wave interactions are assumed to be well represented by a fluctuating flux of action and a dissipative current. Thus the source of the fluctuations in this model of the evolution of the sea surface is the nonlinear interactions among the surface wave modes. This is of course not the only source of fluctuations if the system is opened to the environment, but for the closed system model the highlights are sketched in the following where the equations of motion for the mode amplitudes are derived. In the present model there is a resulting modification in the strength of the self-interaction of the RTF waves which is shown to enable us to obtain an *exact* steady-state solution to the Fokker-Planck equation for the probability density. The steady-state probability density is found to be non-Gaussian and in fact closely resembles the phenomenological non-Gaussian distribution discussed earlier. With this distribution we calculate the energy spectral density for gravity (RTF) waves and find in a certain approximation a k^{-4} spectrum. The fact that the probability density obtained using this model provides an energy spectral density in essential agreement with a large body of data, suggests that there is some element of physical truth in the basic separation of effects into resonant and nonresonant interactions. Even without the approximation, i.e., when a k^{-4} spectrum is not obtained, the energy (action) spectral density is demonstrably finite, thereby improving on earlier theories [8–10, 12].

The resonant test field (RTF) model represents a broadband spectrum of surface waves by two distinct wave fields as described in the "INTRODUCTION" [31]. The ambient wave field with mode amplitudes $\{c_\nu(t)\}$ is assumed to be in a steady state with the wind field so that in the absence of the RTF waves it can be represented by a set of linear wave modes with random initial conditions. The second field has a discrete spectrum of waves $\{b_n(t)$ that interact resonantly with each other and couple linearly to the ambient waves. This latter wave field, the RTF, in the Markov approximation experience the ambient waves as a source of fluctuations. The mean value of these fluctuation determines the average coupling between the two wave fields.

Zakharov [13] proved that the wave field on the surface of deep water is a Hamiltonian system. Therefore we write the total Hamiltonian for an isolated system of water waves as [31]

$$H = H_R(\mathbf{b}) + H_A(\mathbf{c}) + H_{AR}(\mathbf{b}, \mathbf{c}) \tag{69}$$

The quantity H_R consists of the resonant test waves and can be written

$$H_R = \sum_n \omega_n b_n^* b_n + V_R(\mathbf{b}, \mathbf{b}^*) \tag{70}$$

where the nonlinear wave-wave interactions terms in Equation 57 is given by Hamilton's equation

$$T_n(\mathbf{b}) = -i \frac{\partial}{\partial b_n^*} V_R(\mathbf{b}, \mathbf{b}^*) \tag{71}$$

so that V_R is the nonlinear resonant interaction potential. The Hamiltonian for the ambient waves including the coupling between these waves and the RTF waves is

$$H_A + H_{AR} = \sum_\nu \omega_\nu [c_\nu + iG_\nu(\mathbf{b}, \mathbf{b}^*)][c_\nu^* - iG_\nu^*(\mathbf{b}, \mathbf{b}^*)] \tag{72}$$

where G_ν is a function desribing the modulation of the ambient wave field by the RTF waves and taken by West [31] to be linear in the b's.

Hamilton's equation of motion for $\{c_\nu\}$ can be solved exactly in terms of the initial conditions $\{c_\nu(0)\}$ and the RTF modes $\{b_n(t)\}$. Substituting this solution into the equations of motion $\dot{b}_n$, assuming that the initial conditions for the ambient wave field are specified by a canonical distribution in the Hamiltonian $H_A + H_{AR}$, leads in the Markov limit to the set of nonlinear stochastic differential equation

$$\dot{b}_n(t) + (\Gamma_n + i\omega_n) b_n(t) = \left(1 - i\frac{\Gamma_n}{\omega_n}\right) T_n(\mathbf{b}) + F_n(t) \tag{73}$$

West [31] has shown that under reasonable conditions $F_n(t)$ is a zero-centered Gaussian distribution delta-correlated in time. Thus the original Hamiltonian system which constituted a feedback system between two wave fields is projected onto a set of irreversible stochastic equations for the RTF waves. In addition to the zero-centered, delta-correlation Gaussian $F_n(t)$ and the dissipative flux of energy (action) to the ambient waves Real $\Gamma_n b_n$, there is a modification of the nonlinear interactions due to the back reaction of the ambient waves to the RTF waves, i.e., the $(\omega_n - i\Gamma_n)/\omega_n$ coefficient of the nonlinear term $T_n(\mathbf{b})$.

The statistical properties of the set $\{b_n(t)\}$ are uncertain due to the nonlinear interactions in Equation 73. However since $\langle F_n(t) F_{n'}(t') \rangle = 2D_n \delta_{nn'} \delta(t - t')$ we can use standard arguments [36, 37] to construct the Fokker-Planck equation equivalent to the nonlinear Langevin equation (Equation 73).

$$\frac{\partial}{\partial t} P_t = \sum_n \frac{\partial}{\partial b_n}\left[\left(1 - i\frac{\Gamma_n}{\omega_n}\right)\frac{\partial H_R}{\partial b_n^*} P_t\right] + cc + 2\sum_n D_n \frac{\partial^2 P_t}{\partial b_n\, \partial b_n^*} \tag{74}$$

An *exact* steady-state solution of Equation 74 exists [31]:

$$P_{ss}(\mathbf{b}) = Q^{-1} \exp\left\{-\sum_n \beta_n\left[\omega_n b_n^* b_n + \sum_{lmp} V_{mp}^{nl} b_n b_l b_m^* b_p^*\right]\right\} \tag{75}$$

where Q is the partition function, the coupling coefficients V_{mp}^{nl} are given by the four-wave interaction strengths in the Hamiltonian (Equation 70) and the $\{\beta_n\}$ are determined by the self-consistency condition

$$\frac{1}{\omega_n \beta_n} = \frac{D_n}{\text{Real}\, \Gamma_n} = \langle |b_n|^2 \rangle_{ss} + \frac{1}{\omega_n} \sum_{lmp}{}' V_{lm}^{np} \langle b_n^* b_p^* b_l b_m \rangle_{ss} \tag{76}$$

where the prime on the summation denotes the restriction on wave numbers $n + p = l + m$. Note that Equation 75 is non-Gaussian, in that there is a quartic correction of the exponential due to the nonlinear wave-wave interactions. In the absence of this term the distribution would be Gaussian in the b's. Additional analysis is required to directly compare Equation 75 with Equation 53 and with data.

At present we can use Equation 75 to calculate the mean square wave amplitude in the steady state. For a narrow spectrum of RTF waves we replace the wave-wave interaction strength by the diagonal term $V_{nn}^{nn} \equiv V_n \Phi(\Theta_n)$, and in the continuum limit we weight the integral over wave number by an element of volume in k-space, ie., αk^2 where α is a constant. Then the spectrum, by direct integration of Equation 75, can be expressed as

$$\Psi(\mathbf{k}) \propto \left(\frac{1}{k^{3-n/2} V_k}\right)^{\frac{1}{n/2-1}} \tag{77}$$

where n is the number of water waves involved in the resonant interaction. For deep water gravity waves $n = 4$, of course, and the interaction strength is $V_k \sim k^3$ so that Equation 77 yields

$$\Psi(\mathbf{k}) \propto \frac{1}{k^4} \tag{78}$$

for a bounded spectrum of deep water gravity waves. The spectrum (Equation 78) has also been obtained by Phillips [6] using a scaling argument and by West [31]. Equation 77 is quite general and applies to any thermodynamically open system dominated by a resonant interaction of order n. This is the first energy partitioning relation that depends explicitly on the strength of the nonlinear interaction and also the number of waves involved in the interaction. It is not an equipartitioning of energy as one expects when a single temperature characterizes the steady state, but rather each mode is weighted by the interaction strength as a consequence of the set of k-dependent "temperatures" (β_k). The generality of this result is discussed elsewhere.

CONCLUSIONS

In this brief article we have attempted to indicate that the statistics of water waves are at least as important as their deterministic dynamics. This has been done by examining the experimental data on the statistics of wind-generated water waves in their steady state and comparing these data with the predictions of both linear and nonlinear theory. The results obtained are summarized as follows:

1. Wave amplitudes in the steady state based on linear theory are Rayleigh distributed with the corresponding phases being uniform in the interval $(0, 2\pi)$. This distribution with minor adjustments has had remarkable success in fitting field data.

2. The Stokes expansion to third order, including the Stokes drift, is a good representation of the surface elevation in the statistical steady state. The linear wave amplitude in this expansion is Rayleigh distributed so that the surface elevation distribution for the wind-generated statistical steady state is non-Gaussian but is *not* of the Gram-Charlier form. The distribution has a more specialized functional dependence on the mean square surface elevation and the significant surface slope (c.f. Equation 53).
3. Statistical linearization has been used to generalize the linear theory of wind-generated water waves to the transient regime. This analysis is the first dynamic treatment of the relaxation of the wind-generated wave field to its asymptotic steady state. The energy spectrum in the steady state is determined directly by the spectrum of fluctuations in the pressure field at the water surface and inversely with the average wave-wave interaction. The calculated spectrum agress reasonably with observations.
4. The resonant test field (RTF) model is a statistical mechanical description of how an isolated field of water waves can generate fluctuations in the long wavelength waves. The steady-state distribution of complex mode amplitudes predicted by this model is not Gaussian and although not directly comparable with the distribution described in the third section, it does yield a k^{-4} energy spectral density for deep water gravity waves.

In this report I have avoided reference to the question of the evolution of the wind-generated energy spectrum as a function of fetch on the sea surface. I did this because, although interesting, I do not think these investigations at their present level of analysis have much to contribute to our understanding of water wave statistics. It is, however, worth pointing out that the majority of both theoretical and experimental studies have concluded that the nonlinear interaction among gravity waves is the dominant physical mechanism providing energy transfer from one spectral region to another [9, 38]. This is consistent with the present determination that these same wave-wave interactions lead to measurable non-Gaussian statistics for the surface elevation and to non-Rayleigh statistics for the wave amplitudes.

Acknowledgments

I wish to thank the Center for Studies of Nonlinear Dynamics, a division of the La Jolla Institute, for the independent research funds to support the writing of this paper. Also I would like to apologize to any scientists working in this area whose work was not suitably represented here.

NOTATION

a	wave amplitude
b_n	mode amplitude
g	acceleration of gravity
H	Hamiltonian, defined by Equation 69
J	Jacobian
J_0	zero-order Bessel function
k	wave vector
N	number of modes
P	probability density function
p_n	pressure field
Q	partition function
$S(\omega)$	frequency spectrum
T_n	function representing sum of both three- and four-wave interaction terms
t	time
x	distance
$Z(\mathbf{x}, t)$	complex displacement

Greek Symbols

α	constant
α, β, γ	constants
γ	kinematic surface tension
γ_n	Miles air-sea coupling parameter
$\zeta(x, t)$	displacement
Λ^2	mean square displacement
λ_n	cumulants
μ_n	moments
ν	kinematic viscosity
σ^2	variance
χ_j	parameter defined by Equation 2
$\psi(k)$	wave number spectrum

ψ_j constants
$\Phi(\kappa)$ pressure field spectrum
$\phi(k)$ characteristic function
ω frequency
ω_k linear water wave frequency

REFERENCES

1. Phillips, O. M., *J. Fluid Mech.*, 2:417 (1957).
2. Miles, J. W., *J. Fluid Mech.*, 3:185 (1957).
3. Miles, J. W., *J. Fluid Mech.*, 7:469 (1960).
4. West, B. J., and Seshadri, V., *J. Geophys. Res.*, 867:4293 (1981).
5. Kitaigorodskii, S. A., *The Physics of the Air-Sea Interaction*, Jerusalem, Israel Prog. Sci. Transl. (1973).
6. Phillips, O. M., *The Dynamics of the Upper Ocean 2nd Ed.*, Cambridge University Press (1977).
7. West, B. J., *J. Fluid Mech.*, 117:187 (1982); *J. Geophys. Res.*, 86:11073 (1981).
8. Hasselmann, K., *J. Fluid Mech.*, 15:172 (1963).
9. West, B. J., *Deep Water Gravity Waves*, Springer-Verlag, Berlin-Heidelberg-New York (1980).
10. Longuet-Higgins, M. S., *Proc. Symp. Appl. Math. Soc.*, 13:105 (1960).
11. Watson, K. M., and West, B. J., *J. Fluid Mech.*, 70:815 (1975).
12. Longuet-Higgins, M. S., *Proc. Roy. Soc. Lond.*, A 341:311 (1976).
13. Zakharov, V. Ye, *Prikl. Mekh. i. Teckhn. Fiz. No.*, 2:86 (1968).
14. Ichimaru, S., *Basic Principles of Plasma Physics, a Statistical Approach*, Benjamin, London (1973).
15. Meiss, J. D., Pomphrey, N., and Watson, K. M., *Proc. Nat. Acad. Sci. (USA)* 76:2109 (1979).
16. *Nonlinear Properties of Internal Waves*, B. J. West (Ed.)., Am. Ins. Phys. Proceed, 76 (1981).
17. Montroll, E. W., and West, B. J., in *Fluctuation Phenomena*, J. L. Lebowitz and E. W. Montroll, (Ed), North-Holland (1979).
18. Watson, G. N., *Quart. J. Math.*, 10:266 (1939).
19. Longuet-Higgins, M. S., *J. Mar. Res.*, 11:245 (1952).
20. Longuet-Higgins, M. S., *J. Geophys. Res.*, 85:1519 (1980).
21. Earle, M. D., *J. Geophys. Res.*, 80:377 (1975).
22. Forristall, G. Z., *J. Geophys. Res.*, 83:2353 (1978).
23. Huang, N. E., and Long, S. R., *J. Fluid Mech.*, 101:179 (1980).
24. Kinsman, B., Chesapeake Bay Ins., John Hopkins Univ. Tech. Rep. No. 19 (1960).
25. Longuet-Higgins, M. S., *J. Fluid Mech.*, 17:459 (1963).
26. Phillips, O. M., *J. Fluid Mech.*, 11:143 (1961).
27. Huang, N. E., et al, *J. Geophys. Res.*, 88:7597 (1983).
28. Tayfun, M. A., *J. Geophys. Res.*, 85:1548 (1980).
29. Krasitskii, V. P., and Zavlavskii, M. M., *Bound. Lay. Met.*, 14:199 (1978).
30. Wang, M. C., and Uhlenbeck, G. E., *Rev. Mod. Phys.*, 17:323 (1945).
31. West, B. J., *J. Fluid Mech.*, 132:417 (1983).
32. Lleonant, G. T., and Blackman, D. R., *J.* Fluid Mech., 87:455 (1980).
33. Prigogine, I., *Introduction to the Thermodynamics of Irreversible Processes*, Interscience, NY (1967).
34. Zaslavskii, Z. M., and Sagdeev. R. Z., *Soc. Phys. JETP*, 25 (4):718 (1967).
35. Crapper, G. D., *J. Fluid Mech.* (1957).
36. Lindenberg, K., et al., in *Probabilistic Analysis and Related Topics*, 3, A. T. Bharucna-Reid (Ed.) Academic Press, New York (1983).
37. van Kampen, N. G., *Stochastic Processes in Physics and Chemistry*, North-Holland, Amsterdam (1981).
38. Hasselmann, K., et al., 1973; "Measurement of Wind-Wave Growth and Swell Decay During the Joint North Sea Wave Project: (JONSWAP)" Deutshes Hydrographisches Aeitshrift, Supplement A8, 12.

CHAPTER 3

TURBULENCE STRUCTURE AND SCALAR DIFFUSION IN THERMALLY-STRATIFIED OPEN-CHANNEL FLOW

Satoru Komori

The National Institute for Environmental Studies
Ibaraki, Japan

CONTENTS

INTRODUCTION, 47

THEORETICAL WORK ON STRATIFIED SHEAR FLOW, 48
Second-Order Single-Point Model, 48
Spectral-Equation Model, 50

EXPERIMENTAL WORK ON STRATIFIED OPEN-CHANNEL FLOWS, 52

TURBULENCE STRUCTURE AND SCALAR DIFFUSION, 53
Typical Distributions of Mean Velocity and Temperature, 53
Turbulence Energy Budget and Its Transport Mechanism, 54
Buoyancy Effects on Scalar Diffusion, 58

DEPENDENCE OF TURBULENCE QUANTITIES ON RICHARDSON NUMBER AND COMPARISON WITH SPECTRAL-EQUATION MODEL, 61
Correlation of Turbulence Intensities and Fluxes with Ri, 61
Correlation of Eddy Diffusivities with Ri, 64

CONCLUDING REMARKS, 65

NOTATION, 67

REFERENCES, 69

INTRODUCTION

Turbulent stratified flows often occur on an environmental scale such as in natural streams, ocean currents, and the atmospheric boundary-layer flow. It is of great practical interest in the areas of air- and water-pollution to investigate buoyancy effects on turbulence structure and the scalar diffusion mechanism in stratified flows. Stratified flows are generally divided into two flow configurations; stably-stratified flows and unstably-stratified flows. In the case of stable stratification, the mean density gradient is negative in the upward gravitational direction, and the buoyancy force tends to decelerate eddies moving upward and downward. In unstably-stratified flow, the mean density gradient is positive and eddy motions are accelerated. However, actual stratified flows are not only predominated by such simple eddy motions but consist of complex turbulent flow structure and diffusion mechanisms.

A large number of field investigations have been conducted in the stratified lower regions of the atmospheric boundary layer flows [1–4]. Laboratory experiments also have been performed by many investigators (e.g., in wind tunnels [5–11]; in tilted pipe flow [12]; and in open channels [13–21]). Of these measurements, the field observations [1–4] and some of the laboratory experiments [7–9, 11] have been conducted in thermally stratified lower-boundary-layer flows generated by cooling or heating the wall or ground. In such stratified wall regions or atmospheric surface layers with large temperature and velocity gradients, the turbulence shear-production rate is strongly affected by buoyancy, so that organized turbulent motions (bursting phenomena) will be extremely changed. Contrary to such stratified flows, some of the wind-tunnel flows [5, 6, 10] and the open-channel flows [13–21] have both vertical temperature or concentration gradient and velocity gradient in the outer layer away from the wall.

Webster [5] and Young [6] have conducted experiments in stably stratified flows generated by heating the turbulence-producing grids in specially designed wind tunnels, and Piat and Hopfinger [10] have investigated the turbulence quantities in a developing stably stratified boundary-layer flow generated by discharging the warm air above the cold lower stream in a wind tunnel. Ellison and Turner [12], Schiller and Sayre [13], French [15, 16], and McCutcheon [17] also have conducted the experiments in developing stably stratified pipe or open-channel flows generated by injecting the saline water or heated water. Komori [18] and Komori et al. [19, 20, 21] have considered the turbulence structure and scalar diffusion in well-developed both stably and unstably stratified outer-layer flows generated by heating or cooling the top of the outer layer in an open channel.

These stratified outer-layer flows are considered to be relatively free of wall effects can be presumed to be close to a nominally homogeneous free shear flow in local equilibrium. Therefore, the buoyancy effects in the stratified outer-layer flows are different from those in the stratified lower-layer flows. The difference has been clarified both experimentally and theoretically. Launder [22] and Gibson and Launder [23] theoretically have shown that in the stratified surface-layer the ratio of height y to the Monin-Oboukhov length scale L becomes a significant parameter for representing the stability, whereas the local gradient Richardson number Ri becomes a significant parameter in a free shear flow in local equilibrium; energy production by the combined effects of mean shear and buoyancy being approximately balanced by viscous dissipation.

Experimentally, Fukui, Nakajima, and Ueda [11] have shown that the ratio y/L becomes a stability parameter in stratified lower-layer flow, whereas Webster [5], Young [6], Mizushina et al. [14], Komori [18], and Komori et al. [19, 20, 21] have confirmed that in stratified outer-layer flows eddy diffusivities and some turbulence quantities can be well correlated with the local gradient Richardson number Ri. Ueda, Mitsumoto, and Komori [24] have observed eddy diffusivities in the higher region of the atmospheric boundary layer, where ground effects are small, and also concluded that the stability dependencies of the observed diffusivities can well be expressed by the same function of Ri as those obtained in the stratified outer-layer open-channel flows by Mizushina et al. [14]. Furthermore, Komori [18] and Komori et al. [19, 20, 21] have applied a spectral equation model to stratified outer-layer flows in an open channel and have clarified the strong dependence of turbulence quantities on the local gradient Richardson number Ri.

This article focuses on the turbulence structure and scalar diffusion in the stratified outer-layer open-channel flows. Buoyancy effects on turbulence kinetic-energy budget and momentum, heat and mass transport will be discussed experimentally and theoretically. Since the results shown here are mainly quoted from a series of works by Komori [18] and Komori et al. [19, 20, 21], this article might be considered as a summary of their previous works.

THEORETICAL WORK ON STRATIFIED SHEAR FLOW

Statistical theories of turbulent transport have developed rapidly over the last twenty years. Of the theories, two types of turbulence models have generally been used to predict the buoyancy effects in stratified flows. One is a second-order single-point model based on the transport equations that apply to a single spatial point, and the other is a spectral-equation (two-point) model which involves two-point correlation or spectra. Brief review and classification of available types of turbulent models are given by Hill [25].

Second-Order Single-Point Model

Single-point models are based on the transport equations for the high-order moments of the turbulent quantities [26]. Recently, Launder [22], Gibson and Launder [23], Zeman and Lumley

[27] and Lumley, Zeman, and Siess [28] have applied second-order single-point models to stratified flows and have tried to estimate buoyancy effects on turbulent transport of momentum, heat, and mass. Zeman and Lumley [27] and Lumley, Zeman, and Siess [28] successfully applied an eddy-damped, quasi-Gaussian approximation to the equations for the third moments of turbulence quantities and predicted the second moments such as turbulence kinetic-energy, turbulent fluxes, etc. Their models, however, do not consider the stratified shear flow with velocity gradient. On the other hand, Launder [22] and Gibson and Launder [23] proposed a second-order single-point model based on approximated sets of momentum- and heat-transport equations for stratified shear flow. In thermally-stratified shear flow in local equilibrium, the transport equations of the Reynolds stress $\overline{u_i u_j}$ and heat flux $\overline{u_i \theta}$ are written as

$$\frac{D\overline{u_i u_j}}{Dt} = P_{ij} + G_{ij} - \frac{2}{3}\delta_{ij}\epsilon + \phi_{ij} \tag{1}$$

$$\frac{D\overline{u_i \theta}}{Dt} = -\overline{u_i u_k}\frac{\partial \bar{T}}{\partial x_k} + P_{i\theta} + G_{i\theta} + \phi_{i\theta} \tag{2}$$

where u_i = the velocity fluctuation in the i-direction
θ = the temperature fluctuation
T = the mean temperature
ϵ = the viscous dissipation

The quantities P_{ij}, G_{ij}, $P_{i\theta}$, and $G_{i\theta}$ represent the production rates of $\overline{u_i u_j}$ and $\overline{u_i \theta}$ due to the mean shear and buoyancy, respectively:

$$P_{ij} = -\left[\overline{u_i u_k}\frac{\partial \bar{U}_j}{\partial x_k} + \overline{u_j u_k}\frac{\partial \bar{U}_i}{\partial x_k}\right] \tag{3}$$

$$G_{ij} = -(\beta/\bar{T})[g_j\overline{u_i\theta} + g_i\overline{u_j\theta}] \tag{4}$$

$$P_{i\theta} = -\overline{u_k\theta}\frac{\partial \bar{U}_i}{\partial x_k} \tag{5}$$

$$G_{i\theta} = -(\beta g_i/\bar{T})\overline{\theta^2} \tag{6}$$

where g_i = the gravitational acceleration component in the i-direction
$\bar{U}_i$ = the mean velocity in the i-direction
β = the expansion coefficient

The symbols ϕ_{ij} and $\phi_{i\theta}$ denote the pressure-force term, that is, the pressure-strain and pressure temperature-gradient correlations:

$$\phi_{ij} = \overline{\frac{p}{\rho}\left(\frac{\partial u_i}{\partial x_j} + \frac{\partial u_j}{\partial x_i}\right)} \tag{7}$$

$$\phi_{i\theta} = \overline{\frac{p}{\rho}\frac{\partial \theta}{\partial x_i}} \tag{8}$$

In the preceding transport equations, calculating how to close these pressure correlations is a serious problem. Gibson and Launder [23] inferred the following simple linear model:

$$\phi_{ij} = \phi_{ij,1} + \phi_{ij,2} + \phi_{ij,3} \tag{9}$$

with

$$\phi_{ij,1} = -c_1(2\epsilon/\overline{q^2})(\overline{u_i u_j} - \tfrac{1}{3}\delta_{ij}\overline{q^2}) \tag{10}$$

$$\phi_{ij,2} = -c_2(P_{ij} - \tfrac{2}{3}\delta_{ij}P) \tag{11}$$

$$\phi_{ij,3} = -c_3(G_{ij} - \tfrac{2}{3}\delta_{ij}G) \tag{12}$$

where P and G are the production rates of turbulence energy due to the actions of mean shear and buoyancy:

$$P = -\overline{u_i u_k}(\partial \bar{U}_i/\partial x_k) \tag{13}$$

$$G = -(\beta g_i/\bar{T})\overline{u_i\theta} \tag{14}$$

Equation 9 is based on the assumption that the action of the fluctuating pressure field will be to make the turbulent velocity and the temperature fields more isotropic. In the case of the flow near the wall, the modified terms attributable to the wall effects are added to Equation 9. The optimum value of the constants c_1, c_2, and c_3 in Equations 10 to 12 can be determined from a comparison with the measurements. The pressure-temperature-gradient correlation term can also be modeled by the similar procedure. Finally, Equations 1 and 2 can be solved by using the prior closure assumptions and the transport equation for $\overline{\theta^2}$ in local equilibrium. The detailed calculations and the comparison with the experimental results are shown in the papers by Launder [22] and Gibson and Launder [23].

The single-point model is very useful for engineering purposes, but several empirical constants and approximated functions must be selected for a restricted flow in order for the predictions to agree with the measurements. This procedure may be a kind of parameter fitting and it is not easy to find out better closure functions.

Spectral-Equation Model

Of the multipoint models, the simplest model is a spectral-equation (two-point) model based on two-point correlations. Deissler [29, 30] considered spectral equations based on two-point correlations in homogeneous turbulence. The spectral equations could be solved easily under only two assumptions; the neglect of the inertial effects represented by higher-order correlations and the homogeneity of the flow. Deissler [31, 32] applied the spectral-equation model to homogeneous-stratified shear flow in the presence of a vertical body force due to uniform temperature and velocity gradients. Komori [18] and Komori et al. [19, 20, 21] also applied the spectral-equation model to the stratified outer-layer open-channel flow and compared the calculations with experimental results in detail.

These spectral-equation models are applicable to nominally homogeneous shear flow close to local equilibrium, and have an important advantage in the prediction of the pressure-force terms (pressure-strain or pressure-scalar-gradient correlations) which cannot be measured and must be closed by a sophisticated model in the case of the single-point models. Contrary to such advantage, the spectral models also have a disadvantage in that they are never applicable to stratified lower-layer flow near the wall.

The basic equations of the spectral-equation model are the continuity, Navier-Stokes, and energy equations for an incompressible stratified shear flow with constant temperature and velocity gradients in the vertical direction. If diffusive mass is imposed in the stratified shear flow, a mass-conservation equation with a constant concentration gradient is added to the basic equations. Under the assumptions of homogeneity and the neglect of triple correlations, two-point correlation equations can be transformed into the following spectral equations by taking the three-dimensional Fourier transforms of the two-point correlation. The pressure-velocity and the pressure-scalar correlations are obtained by taking the divergence of the equation of motion. The detailed procedures for analysis are described in Deissler's papers [32].

$$\underbrace{\frac{\partial E_{ij}}{\partial \tau}}_{I} = \underbrace{-\{\delta_{i1}E_{2j} + \delta_{j1}E_{i2}\}}_{II} \underbrace{+ [k_j - \delta_{j2}k_1\tau][2k_1E_{i2} - Ri(k_2 - k_1\tau)\tau_t E_{i\theta}]k^{-2} + [k_i - \delta_{i2}k_1\tau][2k_1E_{2j} - Ri(k_2 - k_1\tau)\tau_t E_{j\theta}]k^{-2}}_{III} \underbrace{+ \delta_{i2}Ri\tau_t E_{j\theta} + \delta_{j2}Ri\tau_t E_{i\theta}}_{IV} - \underbrace{\frac{2k^2E_{ij}}{\tau_t}}_{V} \tag{15}$$

$$\underbrace{\frac{\partial E_{i\theta}}{\partial \tau}}_{I} = \underbrace{-\left\{\frac{E_{i2}}{\tau_t} + \delta_{i1}E_{2\theta}\right\}}_{II} + \underbrace{[k_i - \delta_{i2}k_1\tau][2k_1E_{2\theta} - Ri(k_2 - k_1\tau)\tau_t E_{\theta\theta}]k^{-2}}_{III} \underbrace{+ \delta_{i2}Ri\tau_t E_{\theta\theta}}_{IV} - \underbrace{\left[1 + \frac{1}{Pr}\right]\frac{k^2E_{i\theta}}{\tau_t}}_{V} \tag{16}$$

$$\underbrace{\frac{\partial E_{\theta\theta}}{\partial \tau}}_{I} = -\underbrace{\frac{2E_{2\theta}}{\tau_t}}_{II} - \underbrace{\frac{2k^2E_{\theta\theta}}{Pr\tau_t}}_{V} \tag{17}$$

$$\underbrace{\frac{\partial E_{ic}}{\partial \tau}}_{I} = \underbrace{-\left\{\frac{E_{im}}{\tau_t} + \delta_{i1}E_{2c}\right\}}_{II} + \underbrace{[k_i - \delta_{i2}k_1\tau][2k_1E_{2c} - Ri(k_2 - k_1\tau)E_{\theta c}]k^{-2}}_{III} \underbrace{+ \delta_{i2}RiE_{\theta c}}_{IV} - \underbrace{\left[1 + \frac{1}{Sc}\right]\frac{k^2E_{ic}}{\tau_t}}_{V} \tag{18}$$

$$\underbrace{\frac{\partial E_{cc}}{\partial \tau}}_{I} = -\underbrace{\frac{2E_{mc}}{\tau_t}}_{II} - \underbrace{\frac{2k^2E_{cc}}{Sc\tau_t}}_{V} \tag{19}$$

$$\underbrace{\frac{\partial E_{\theta c}}{\partial \tau}}_{I} = \underbrace{-\{E_{m\theta} + E_{2c}\}}_{II} - \underbrace{\left[\frac{1}{Pr} + \frac{1}{Sc}\right]\frac{k^2E_{\theta c}}{\tau_t}}_{V} \tag{20}$$

where E_{ij}, $E_{i\theta}$, $E_{\theta\theta}$, E_{ic}, E_{cc}, and $E_{\theta c}$ are the dimensionless spectrum functions of $\overline{(u_i)_A(u_j)_B}$, $\overline{(u_i)_A(\theta)_B}$, $\overline{(\theta)_A(\theta)_B}$, $\overline{(u_i)_A(c)_B}$, $\overline{(c)_A(c)_B}$, and $\overline{(\theta)_A(c)_B}$ at two points A and B, respectively. The symbol k is the dimensionless wavenumber, k_i the dimensionless wavenumber component in the i-direction, Pr the Prandtl number, Sc the Schmidt number, τ the dimensionless time, τ_t the dimensionless form of the terminal time t_t, and c is the concentration fluctuation. The terms I to V in Equations 15 to

20 denote the change per unit of time (or, in a developing steady flow, advection), the mean field production by vertical velocity gradient, temperature gradient or concentration gradient, the pressure-force production, the buoyancy production, and the dissipation. In Equations 18 to 20, the values of subscript m in E_{im}, E_{mc}, and $E_{m\theta}$ are 1, 2, and 3 for longitudinal, vertical, and lateral concentration gradients respectively. For solving the spectral equations (Equations 15 to 20), the turbulence is assumed to be isotropic and temperature and concentration fluctuations are assumed to be absent at a nondimensional initial time $\tau = 0$. The conditions are satisfied by [29–32].

$$E_{ij} = (\delta_{ij}k^2 - k_i k_j)/12\pi^2 \qquad (\tau = 0) \tag{21}$$

$$E_{i\theta} = E_{\theta\theta} = E_{ic} = E_{cc} = E_{\theta c} = 0 \qquad (\tau = 0) \tag{22}$$

Substituting Equations 21 and 22 into Equations 15 through 20, the spectral functions can be numerically solved at the terminal time $\tau = \tau_t$ by using a Runge-Kutta method. Letting the distance between the two points A and B be equal to zero and integrating E_{ij}, $E_{i\theta}$, $E_{\theta\theta}$, E_{ic}, and E_{cc} over the surface of a sphere of radius $k = 0$ to ∞, the following time-averaged quantities are obtained

$$\overline{u_i u_j} = J_0 \int_0^\infty \int_0^\pi \int_0^{2\pi} E_{ij} k^2 \sin\psi \, d\phi \, d\psi \, dk/[\nu^{5/2}(t_t - t_0)^{5/2}] \tag{23}$$

$$\overline{u_i \theta} = J_0(\partial \bar{T}/\partial y) \int_0^\infty \int_0^\pi \int_0^{2\pi} E_{i\theta} k^2 \sin\psi \, d\phi \, d\psi \, dk/[\nu^{5/2}(t_t - t_0)^{3/2}] \tag{24}$$

$$\overline{\theta^2} = J_0(\partial \bar{T}/\partial y)^2 \int_0^\infty \int_0^\pi \int_0^{2\pi} E_{\theta\theta} k^2 \sin\psi \, d\phi \, d\psi \, dk/[\nu^{5/2}(t_t - t_0)^{1/2}] \tag{25}$$

$$\overline{u_i c} = J_0(\partial \bar{\Gamma}/\partial x_m) \int_0^\infty \int_0^\pi \int_0^{2\pi} E_{ic} k^2 \sin\psi \, d\phi \, d\psi \, dk/[\nu^{5/2}(t_t - t_0)^{3/2}] \tag{26}$$

$$\overline{c^2} = J_0(\partial \bar{\Gamma}/\partial x_m)^2 \int_0^\infty \int_0^\pi \int_0^{2\pi} E_{cc} k^2 \sin\psi \, d\phi \, d\psi \, dk/[\nu^{5/2}(t_t - t_0)^{1/2}] \tag{27}$$

where $\bar{\Gamma}$ is the mean concentration and J_0 is the constant that depends on initial conditions in the dimensional form of Equation 21.

Komori et al. [19, 20, 21] selected the parameter value of $\tau_t = 4.0$ from the comparison between the computed velocity correlation values at $Ri = 0$ and the measured ones in neutral open-channel flows, and also used a Prandtl number of 5 and a Schmidt number of 2,060 respectively from their experimental conditions. Futhermore, to take into account the significant buoyancy effect on the velocity gradient in unstable conditons, a corrected function of τ_t was used [20]. The calculation results will be shown later.

Higher-order three- or four-point models will be considered. In particular, the models will become useful for the predictions of the eddy motion in the higher-wavenumber range such as inertial subrange. However, such models have not yet been applied to the stratified shear flows because of the complication and numerous calculations. The basic concept of the higher-order model can be found in Deissler's papers [29, 33].

EXPERIMENTAL WORK ON STRATIFIED OPEN-CHANNEL FLOWS

As mentioned in the introduction, several experiments have been conducted in turbulent stratified open-channel flows in connection with the ecological problems due to the release of heated waste water from large power-generating plants into rivers and ocean. Schiller and Sayre [13] measured vertical temperature profiles in a developing stably-stratified flow and predicted the temperature profiles by solving a simplified convection-diffusion equation based on the parameters obtained from experimental tests.

Related experiments were carried out by French [15, 16] and McCutcheon [17] in a salt stratified flow. They compared the measured values of mean velocity and friction factors with those computed by using the Monin-Oboukhov analysis, and presented a better description of vertical mixing. However, these experiments consisted of only measurements of mean density and velocity by means of simple measuring instruments such as a thermister, a conductivity bridge, and a Pitot tube, and so could not clarify the detailed buoyancy effects on the turbulence structure. Also, the difference of stability parameters between stratified lower-layer flow and outer-layer flow have not been considered. Furthermore, since most of the experiments were conducted in the stably-stratified open-channel flows generated by discharging heated water or saline water into an ambient flow with the use of splitter plate at the entrance of the flume, artificial and disturbed initial conditions prevented the full development of the stratified flows at the downstream.

On the other hand, Komori [18] and Komori et al. [20, 21] measured several statistical turbulence quantities and clarified the turbulence structure in both stably- and unstably-stratified outer-layer open-channel flows close to fully-developed two-dimensional flows. In particular, the studies in unstably-stratified open-channel flow are restricted only to work by Komori [18] and Komori et al. [20]. To obtain stably-stratified flow, Komori [18] and Kormori et al. [21] mildly condensed saturated steam at 373 K on the free surface of the fully-developed neutral flow in a small rectangular flume 6.1 m long, 0.3 m wide and 0.06 m deep. For unstably-stratified flow, Komori [18] and Komori et al. [20] allowed water heated to a temperature between 303 and 343 K to evaporate from the free surface into the atmosphere and so to be cooled from above. In the flows, the flow depth δ was maintained at approximately 0.04 m, and the cross-sectional mean velocity $\bar{U}_{ave}$ ranged from 0.069 to 0.07 m/s. The Reynolds number Re based on the hydraulic radius ranged from 8,600 to 41,700, and the Froude number Fr was less than 1.0. The bulk Richardson number $\overline{Ri}$, which is an overall buoyancy parameter describing the whole flow, varied from -0.09 to 0.27. The details of the conditions are described in the papers by Komori [18] and Komori et al. [20, 21]. Under these flow conditions, Komori [18] and Komori et al. [20, 21] simultaneously measured instantaneous velocity and temperature by using a laser Doppler velocimeter (LDV) and a cold-film probe operated by a temperature bridge. Their measuring techniques and the accuracy of the velocity measured by LDV in nonisothermall flow are described by Mizushina et al. [34] and Komori and Ueda [35] in detail. From these measurements, they have calculated statistical turbulence quantities by a digital computer.

In addition, Komori et al. [19] have conducted the experiments of mass-diffusion from a vertical line source in the same stratified outer-layer open-channel flows. They fed a solution of methylene-blue into developed stratified flows, through a small line source mounted perpendicular to the floor of the flume, at a speed equal to the velocity of the ambient flow, and then simultaneously measured instantaneous concentration of the methylene-blue and lateral velocity at the downstream of the line source by using a small optical probe based on the principle of the light-absorption and the LDV respectively. From the fluctuations of the lateral velocity and concentration, they calculated the lateral eddy diffusivity of mass, and also calculated the vertical and longitudinal eddy diffusivities of heat from the temperature-velocity correlation measured by an LDV and a cold-film probe.

In the next section we shall show the experimental results obtained through the above experiments by Komori [18] and Komori et al. [19, 20, 21].

TURBULENCE STRUCTURE AND SCALAR DIFFUSION

Typical Distributions of Mean Velocity and Temperature

Figure 1 shows the typical distributions of the mean velocity and temperature measured by Komori et al. [20, 21] in well-developed stable, unstable, and neutral flows. In the figure, x is the streamwise direction from the inlet of the flume. Temperature gradients are established in the outer layer where the velocity gradients are comparatively small and constant, and the distributions of the mean velocity and temperature do not change noticeably in the flow direction. This means that both the stably and unstably stratified flows are close to a developed flow. In the outer layer of the stable flow, the velocity gradient becomes larger than that in the neutral flow, whereas in the unstable case it becomes smaller. Thus, the buoyancy affects the mean fields and will change the turbulence structure.

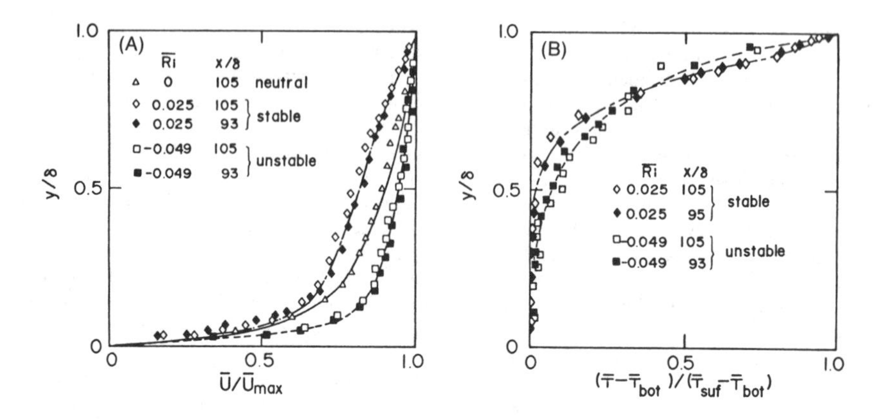

Figure 1. Typical distributions of the mean (A) velocity and (B) temperature in stable, unstable, and neutral flows [20, 21].

Turbulence Energy Budget and its Transport Mechanism

The turbulence kinetic-energy balance equation for two-dimensional stratified flow with high Reynolds number in which the work by the viscous shear stress of the turbulent motion can be neglected, is given by

$$\frac{1}{2}\frac{D\overline{q^2}}{Dt} = -\overline{uv}\frac{\partial \bar{U}}{\partial y} + \beta g\overline{v\theta} - \frac{\partial}{\partial y}\left[\overline{v\left(\frac{p}{\rho} + \frac{q^2}{2}\right)}\right] - \nu\overline{\left(\frac{\partial u_i}{\partial x_j} + \frac{\partial u_j}{\partial x_i}\right)\frac{\partial u_j}{\partial x_i}} \tag{28}$$

where $\overline{q^2}$ $(= \overline{u^2} + \overline{v^2} + \overline{w^2})$ is the turbulent kinetic energy. The first term on the right is the shear-production term, the second the buoyancy-production term, the third the diffusion term, and the last is the viscous-dissipation term. Each term in the prior equation can be made dimensionless by δ/u^{*3}. The dimensionless terms estimated by Komori et al. [20, 21] are plotted against y/δ in Figures 2A, B, C, and D for a neutral flow, a mildly stable flow, a strongly stable flow, and a strongly unstable flow respectively. In these diagrams, the energy dissipation is estimated from the normalized power spectral density of the streamwise velocity fluctuation by using the Kolmogoroff hypothesis for the inertial subrange. The diffusion of the turbulence energy is calculated as a residue of Equation 28, on the assumption that the flow is well developed and steady.

Under stable conditions, the shear production term is reduced significantly in the outer region. In a strongly stable flow, a small negative contribution is observed in the region of $y/\delta > 0.5$. This suggests that in strongly stable conditions momentum is transferred against the mean velocity gradient. The buoyancy-production term $\beta g\overline{v\theta}$ makes a negative contribution in a mildly-stable flow, but in a strongly-stable flow it makes a large positive contribution in the outer layer of $y/\delta > 0.5$. This means that the buoyancy works so as to reduce the turbulence kinetic energy in weak stratification, and that in strongly stable conditions turbulence energy is generated by upward heat transfer against the mean temperature gradient (a positive $\overline{v\theta}$). The contribution of the positive buoyancy production is larger than that of the negative shear production, and the countergradient heat transfer occurs in more weakly stable conditions than the countergradient momentum transfer. Komori et al. [21] confirmed that the countergradient momentum transfer is approximately compensated by molecular transfer, but the countergradient heat transfer greatly exceeds the molecular transfer. This suggests that in a strongly stable flow heat is supplied by advection from upstream and this heat is pumped up against the temperature gradient.

Komori et al. [21] have suggested that the advection of heat is attributable to the formation process of the stratified flow. Their stably stratified flow was obtained by steam condensation on

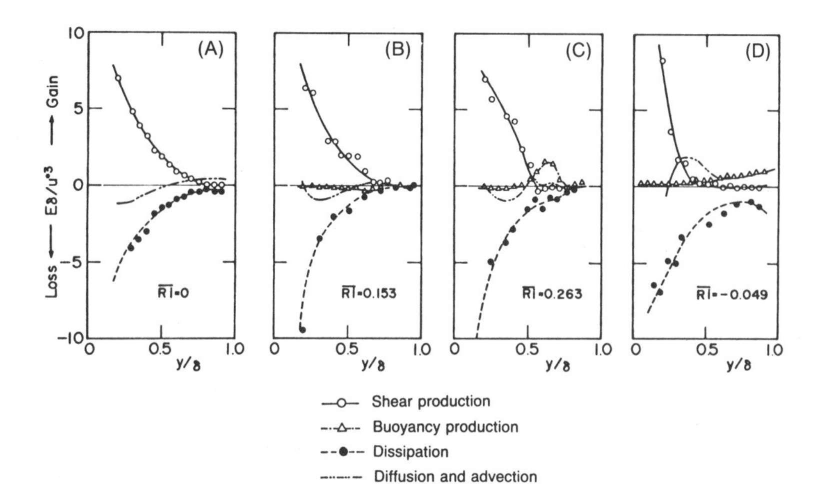

Figure 2. Dimensionless energy budget in stable, unstable, and neutral flows. Each term is nondimensionalized by δ/u^{*3}: (A) neutral flow; (B) a mildly-stable flow; (C) a strongly-stable flow; (D) a strongly-unstable flow. [20, 21].

the free surface of open-channel flow in the region where the flow had been a fully-developed flow under neutral conditions. In this flow configuration, the mean temperature gradient was first formed, and then temperature fluctuations were generated from the interactions of the mean temperature gradient with the existing velocity fluctuations. Thus, in strongly-stable flow with a large positive temperature gradient in the outer layer, very positively skewed temperature fluctuations may be generated and may be advected downstream. Similar flow configurations may also be formed in the case of other stratified open-channel or wind-tunnel flows.

Komori et al. [21] have observed very positively-skewed spikes of temperature fluctuation in a simultaneous recording of θ, v, and vθ (Figure 3A). The intermittent positive spikes of θ are accompanied by positive peaks of v, and the peaks result in the positive vθ-products shown by the arrows

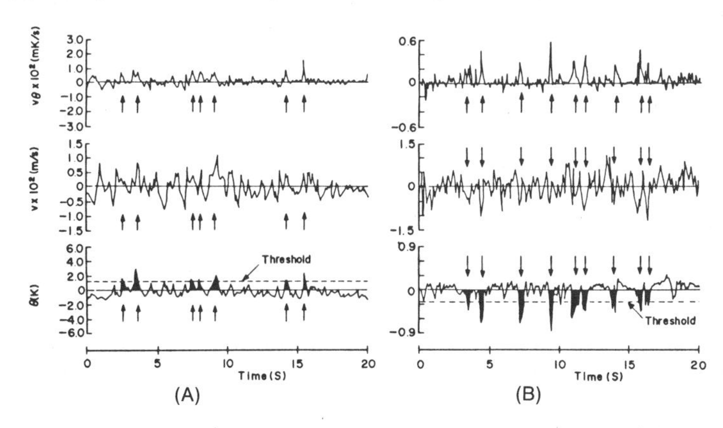

Figure 3. Simultaneous recordings of the instantaneous values of θ, v, and vθ in (A) a strongly stable flow (at $y/\delta = 0.61$ and $Ri = 0.91$) and in (B) a strongly unstable flow (at $y/\delta = 0.60$ and $Ri = -18.2$) [20, 21].

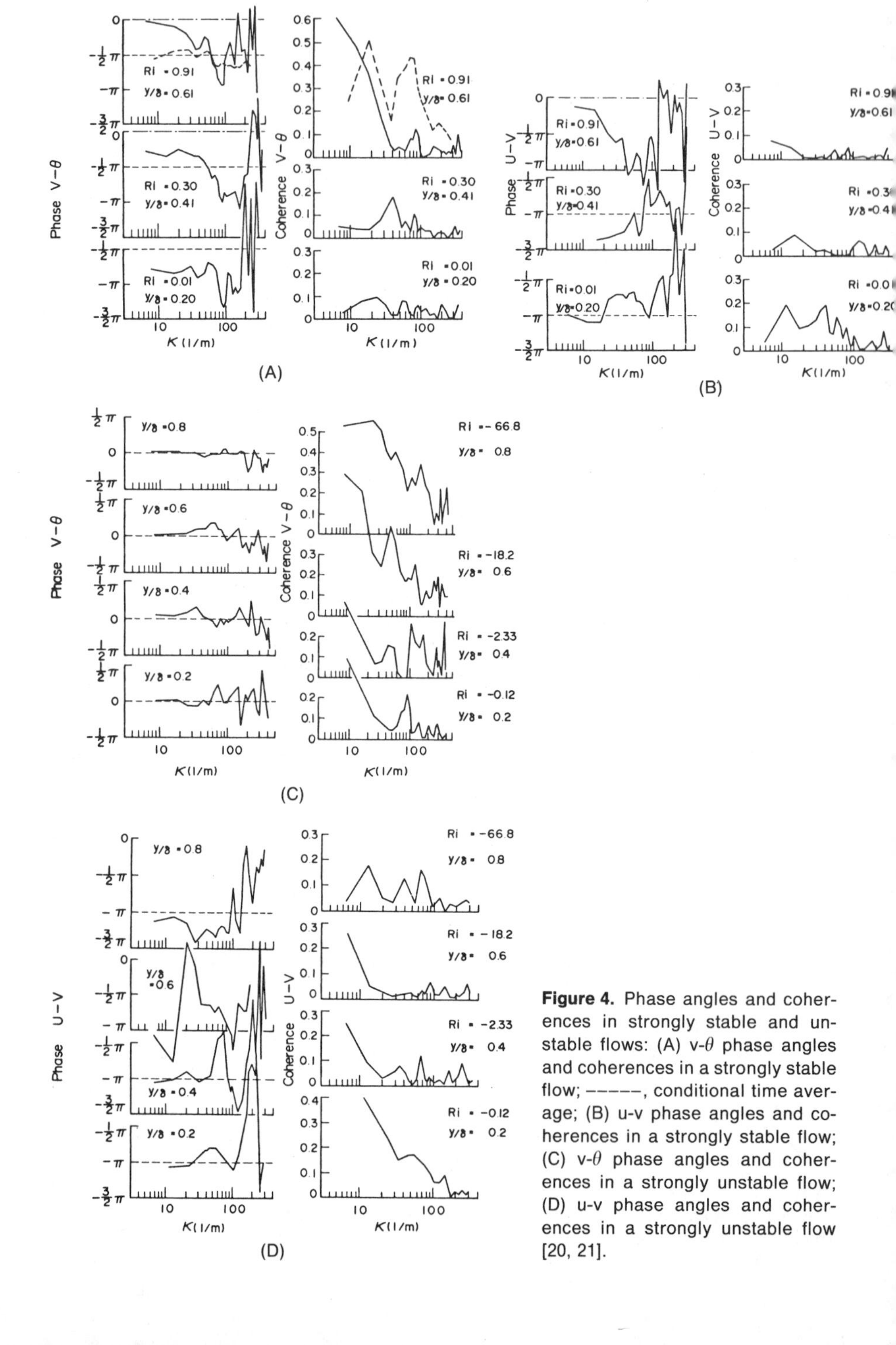

Figure 4. Phase angles and coherences in strongly stable and unstable flows: (A) v-θ phase angles and coherences in a strongly stable flow; -----, conditional time average; (B) u-v phase angles and coherences in a strongly stable flow; (C) v-θ phase angles and coherences in a strongly unstable flow; (D) u-v phase angles and coherences in a strongly unstable flow [20, 21].

in the figure. In order to estimate the contribution of these positive vθ-products to the total heat transfer, Komori et al. [21] set a threshold level with the magnitude of the r.m.s. value of θ in the θ-trace, as shown by a dashed line in Figure 3A, and calculated the heat flux due to the intermittent eddies with higher temperature fluctuations than the threshold level to the total turbulent heat flux. The results showed that the ratio attains about 0.75 in the region near $y/\delta = 0.6$, in spite of the small time fraction of the appearance of the positive spikes less than 0.29. This supports the assumption that intermittent upward motions of the advected eddies with the positive spikes of temperature fluctuation cause upward heat transfer against the temperature gradient.

The upward motion seems to be a kind of buoyancy-driven motion, and it may trigger off a breakdown of wavelike motion under strongly-stable conditions. Komori et al. [21] also confirmed the presumption from the velocity-temperature coherence-phase relationships. In the wave motions the v-θ coherence should be high and the phase angle should be equal to $\pm 1/2\pi$, while the turbulent shear flow has v-θ and u-v phase angles near $\pm\pi$ [3, 36].

Figures 4A and B show the v-θ and u-v phase angles and coherences as a function of wavenumber κ for three positions in a strongly stable flow. In the almost neutral region (e.g., $y/\delta = 0.2$ and $Ri = 0.01$) ordinary turbulent motions predominate, so that both v-θ and u-v phase angles are near $-\pi$ except for the range of high wavenumber. In stable stratification of $Ri = 0.3$ and $y/\delta = 0.41$, the v-θ phase angles approach $-1/2\pi$ and the coherences become somewhat higher in the wavenumber region of $\kappa < 50$. This phase-coherence relationships suggest that the turbulent motions become close to wavelike motions in stable conditions. In the wavelike motions, the u-v phase angles close to $-3/2\pi$ are observed in the lower-wavenumber region. In strongly stable stratification of $Ri = 0.91$ and $y/\delta = 0.61$, the v-θ phase angles approach zero in the lower-wavenumber region and coherences become extremely high. These high coherences without phase shift show the appearance of buoyancy-driven upward motion of the hot eddy.

In order to investigate the background motion during the absence of the buoyancy-driven motion, Komori et al. [21] eliminated the v- and θ-signals due to the buoyancy-driven motion by using a threshold level shown in Figure 3A and conditionally calculated the phase angles and coherences. The results show that the phase angles approach $-1/2\pi$ and the coherences become high even in strongly stable stratification of $Ri = 0.91$ and $y/\delta = 0.61$, as shown by a dashed line in Figure 4A. These suggest that also in strongly stable conditions background motion is wavelike motion and the wavelike motion is intermittently broken down by a buoyancy-driven motion of the hot eddy. The u-v phase angles exhibit the phase difference close to 0 in low-wavenumber space. This also suggests that the intermittent breakdown is caused by an upward motion of the accelerated hot eddy. Furthermore, these can be confirmed from the flow patterns obtained by the hydrogen bubble technique, as shown in Figures 5B and C. The flow pattern of strongly stably stratified flow clearly exhibits the existence of the wavelike motion in the outer layer (Figure 5B), compared with the neutral flow case (Figure 5A). Certainly, the breakdown of the wavelike motion is intermittently observed, as shown in Figure 5C.

In strongly unstable conditions, the buoyancy energy production makes a remarkable contribution to the turbulence energy balance in the outer region of $y/\delta > 0.4$–0.5, while the contribution of the shear production is very small (Figure 2D). This means that turbulent eddy motion is remarkably enhanced by buoyancy. In fact, its enhancement can be observed by violent streaks of hydrogen bubbles in the upper half region of the flow (Figure 5D) compared with the neutral flow (Figure 5A). To investigate the enhancement mechanism of turbulence eddy motion in unstable conditions, Komori et al. [20] examined some statistical characteristics of the turbulent fluctuations.

Simultaneous recordings of the instantaneous values of θ, v, and vθ at $Ri = -18.2$ and $y/\delta = 0.60$ in a strongly unstable flow are also shown in Figure 3B. Contrary to the stable case, intermittent negative spikes of θ are accompanied by negative values of v-fluctuation, and result in a positive vθ-product as noted by the arrows in the figure. The spikes in the θ- and vθ-signals are the most prominent feature in unstably-stratified flow cooled from above. Physically, this means that the downward motion of the eddies with negative temperature fluctuations predominates. Komori et al. [20] also estimated the contribution of the downward motion to the vertical heat transfer by using a threshold level with the negative magnitude of the r.m.s. value of θ shown by a dashed line in Figure 3B, as well as in the stably-stratified flow case. The calculation predicted that the contribution of the intermittent downward motion is significant and occupies more than 60 percent of

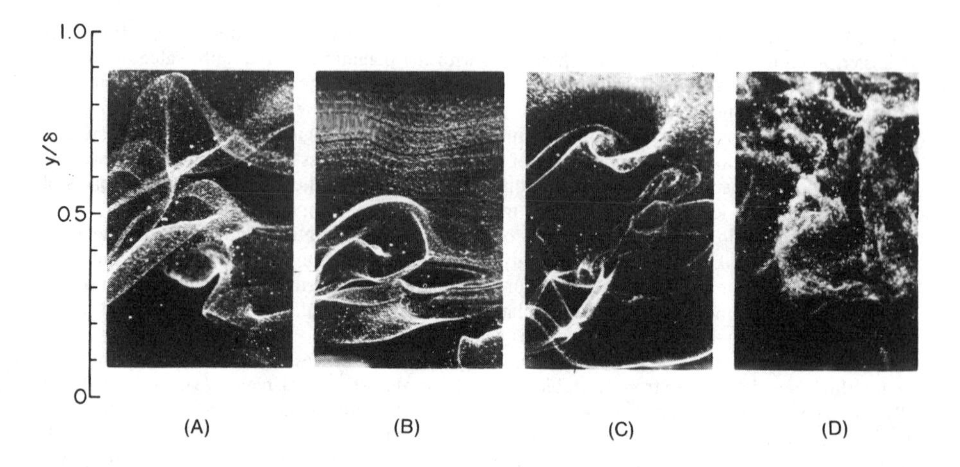

Figure 5. Flow pattern of neutral, stable, and unstable flows: (A) a neutral flow; (B) a strongly stable flow; (C) breakdown of wavelike motion in a strongly stable flow; (D) a strongly unstable flow [18, 20, 21].

the total flux. This means that the intermittent downward motion of the cold eddies are most significant in the buoyant convection cooled from above.

The buoyant convection can also be confirmed from the v-θ and u-v phase-coherence relationships shown in Figures 4C and D. As the instability increases, the region of zero v-θ phase shift extends to the higher-wavenumber region and ultimately occupies almost the whole wavenumber region in strongly unstable conditions of $\text{Ri} = -66.8$ and $y/\delta = 0.8$. The coherences also grow to higher levels with increasing instability (Figure 4C). This behavior means that not only the large-scale eddies but also small-scale eddies are affected by the buoyancy force, and that the upward and downward motions of the eddies are enhanced by buoyancy. Thus, in the entire wavenumber region, the correlation between v and θ increases. In contrast to the v-θ phase angles and coherences, the coherences of u and v become lower and the phase angles become scattered with increasing instability (Figure 4D). This behavior reflects the low correlation between the vertical and streamwise motion in unstable conditions. In the weakly unstable region of $y/\delta < 0.4$, the phase angles are near $-\pi$ except in the range of high wavenumber and the coherences are comparatively large. From this fact, it is found that ordinary turbulence predominates in the lower layer of flow with nearly neutral stratification, as well as in the stably stratified flow.

In Figure 2, the diffusion and advection term in the outer layer of $y/\delta > 0.5$ is relatively small under all conditions, and the buoyancy and shear productions are almost balanced by the viscous dissipation. However, in the range where both the shear and buoyancy productions are close to zero in stable conditions, the small diffusion and advection become comparable to or exceed the shear and buoyancy productions (Figures 2B and C). Thus, the stratified flows investigated by Komori et al. [20, 21] are not perfectly in local equilibrium, but they will be close to a free shear flow in local equilibrium.

Buoyancy Effects on Scalar Diffusion

To investigate the buoyancy effects on scalar diffusion, Komori et al. [19] estimated the eddy diffusivities of heat and mass by means of the simultaneous measurements of velocity and temperature or concentration in stratified conditions. Figures 6A, B, and C show the longitudinal eddy diffusivity of heat, vertical eddy diffusivity of heat, and lateral eddy diffusivity of mass in the non-

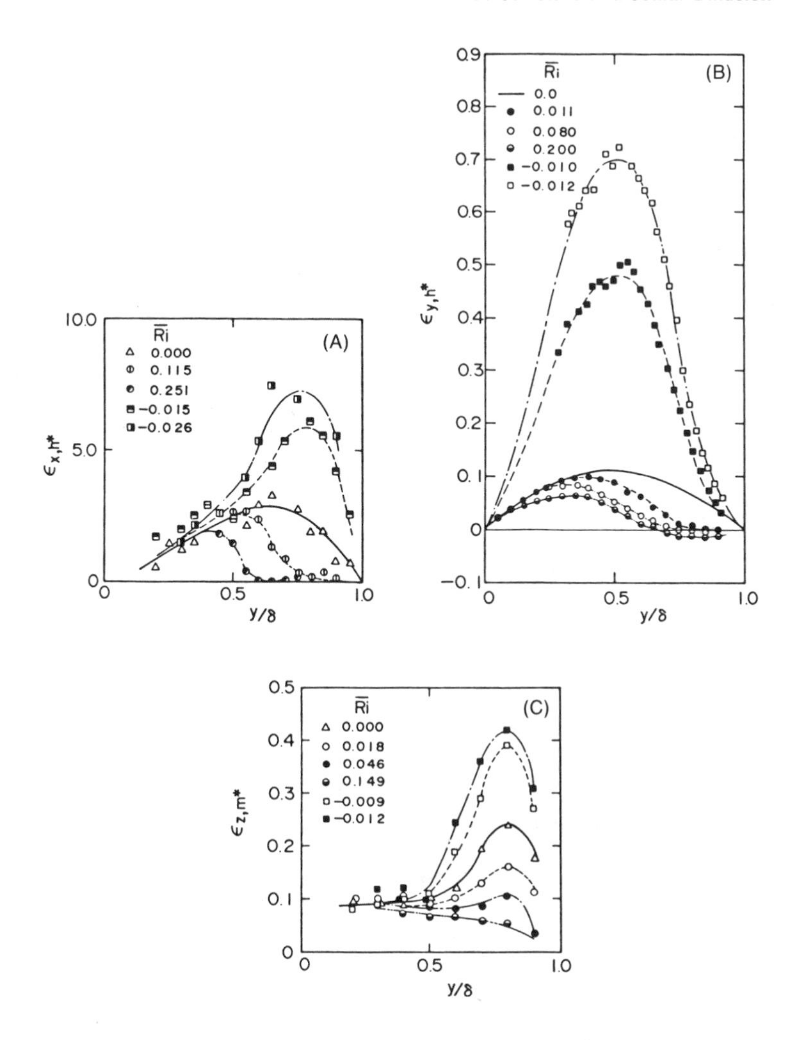

Figure 6. Distributions of dimensionless eddy diffusivities of scalar quantities: (A) longitudinal eddy diffusivity of heat; (B) vertical eddy diffusivity of heat; (C) lateral eddy diffusivity of mass [14, 19].

dimensional form, respectively. The dimensionless eddy diffusivities are defined as

$$\epsilon_{x,h}^* = \epsilon_{x,h}/\delta u^* = -\overline{u\theta}/[(\partial\overline{T}/\partial x)/\delta u^*] \tag{29}$$

$$\epsilon_{y,h}^* = \epsilon_{y,h}/\delta u^* = -\overline{v\theta}/[(\partial\overline{T}/\partial y)/\delta u^*] \tag{30}$$

$$\epsilon_{z,m}^* = \epsilon_{z,m}/\delta u^* = -\overline{wc}/[(\partial\overline{\Gamma}/\partial z)/\delta u^*] \tag{31}$$

where w is the velocity fluctuation in the lateral direction.

In the outer region under stable conditions of $y/\delta > 0.4$, the eddy diffusivities in all three directions decrease with increasing stability (increasing the bulk Richardson number $\overline{Ri}$). In stable flow, turbulent motion is organized into wavelike motion (see the preceding section), so that not only the vertical diffusion but also the lateral and longitudinal diffusion is suppressed. In strongly stable flow, the vertical eddy diffusivity $\epsilon^*_{y,h}$ becomes negative due to the countergradient transport of heat (see the preceding section).

Under unstable conditions, it is seen that all three eddy diffusivities components increase with increasing instability (increasing $-\overline{Ri}$). In the unstable case, vertical motion is remarkably enhanced by buoyancy. The extra energy of the vertical motion is redistributed in the longitudinal and lateral directions, so that the lateral and longitudinal motions are also enhanced (see the section, "Correlation of Turbulence Intensities and Fluxes with Ri"). Thus, the enhancement of the longitudinal and lateral eddy diffusivities may be attributed to those turbulence mechanisms. Indeed, Komori et al. [19] predicted the increase of the correlations between the longitudinal or lateral velocity and the scalar in unstable conditions.

The preceding behaviors of the eddy diffusivities show that the turbulent diffusion in all three directions is suppressed in stable conditions, whereas in unstable conditions it is enhanced..

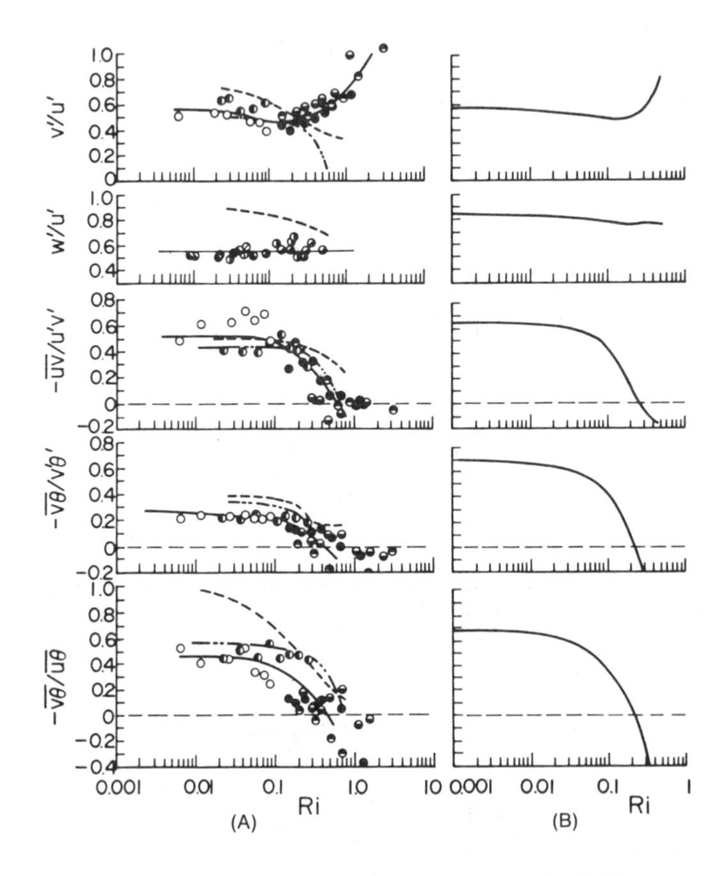

Figure 7. Correlation of turbulence quantities in stable conditions with Ri: (A) measured results: ————, best-fit curve of the data of Komori et al. [21]; – – – – – –, data of Webster [5]; –· ·–, data of Young [6]; (B) Predictions by the spectral-equation model [21].

DEPENDENCE OF TURBULENCE QUANTITIES ON RICHARDSON NUMBER AND COMPARISON WITH SPECTRAL-EQUATION MODEL

As mentioned in the introduction, the local gradient Richardson number Ri becomes a scaling parameter in stratified outer-layer flow close to a free shear flow in local equilibrium. Komori et al. [19, 20, 21] showed good correlations of turbulence quantities in the outer layer with Ri. Here, the correlation of turbulence quantities with Ri and the comparison with the spectral-equation model will be considered.

Correlation of Turbulence Intensities and Fluxes with Ri

Figures 7 and 8 show the correlations of the measured turbulence quantities with Ri and the comparison with the predictions of the spectral-equation model and other laboratory measurements [5, 6] in stable and unstable conditions, respectively. In these figures, only values measured in the outer layer of $0.4 < y/\delta < 0.75$ are adopted and correlated with Ri. For all turbulence quantities, it can be found that the theoretical predictions are in the qualitative agreement with the variations of the measured values against Ri.

In stable conditions, the ratio of the turbulence intensities (root-mean-square values), v'/u', decreases slightly as the stratification shifts from neutral to weakly-stable conditions and then increases as Ri becomes larger (Figure 7). Although these behaviors are different from Webster's and Young's measurements [5, 6], the fact that vertical motions are induced more than horizontal motions

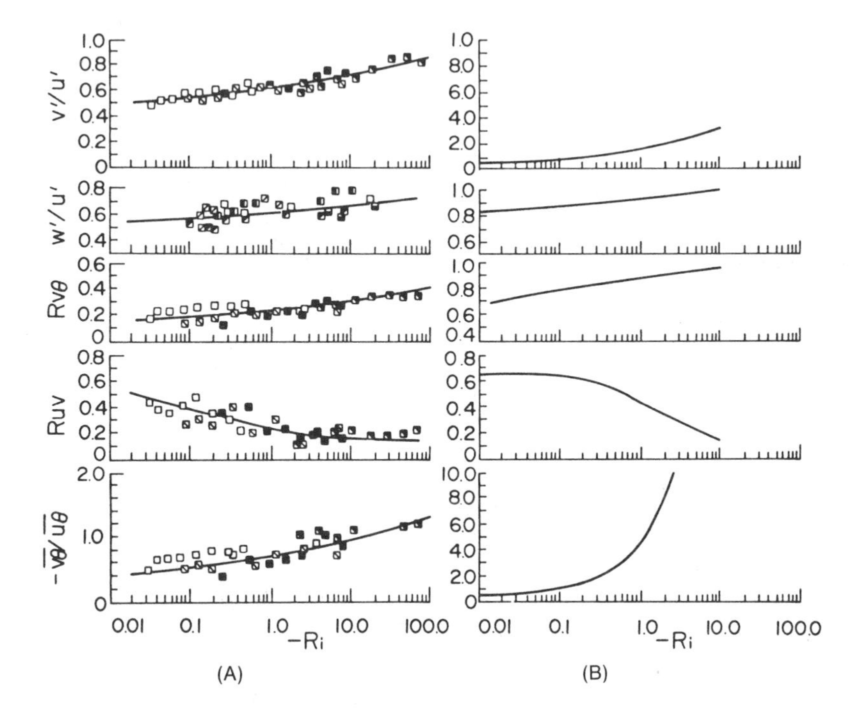

Figure 8. Correlation of turbulence quantities in unstable conditions with Ri: (A) measured results: ————, best-fit curve of the data of Komori et al. [20]; (B) predictions by the spectral-equation model [20].

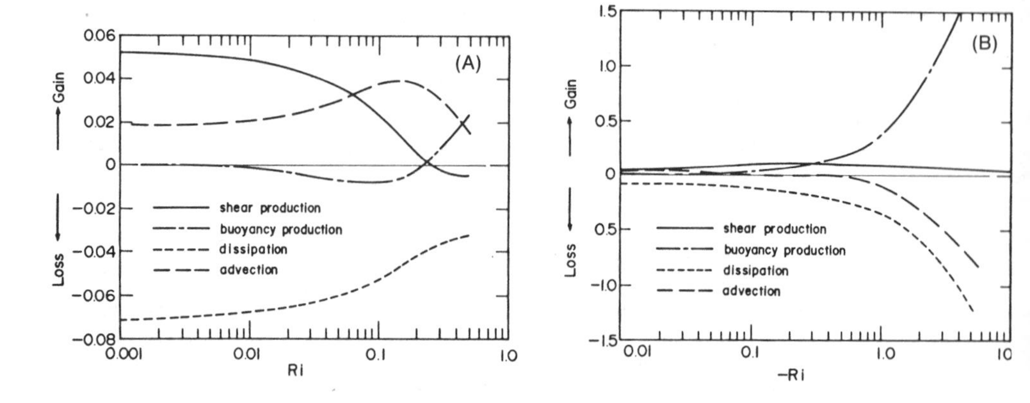

Figure 9. Turbulence energy budgets calculated from the spectral-equation model. Each term is nondimensionalized by $\nu^{5/2}(t_t - t_0)/[J_0(\partial\bar{U}/\partial y)]$: (A) in stable conditions; (B) in unstable conditions [20, 21].

under strongly stable conditions is very interesting and can be easily understood, since the buoyancy-production term, which comes from the vertical heat flux in the $\overline{v^2}$-transport equation, has a positive contribution to $\overline{v^2}$-production (see Figure 2C). The change of sign of $\overline{v\theta}$ is also predicted by the spectral-equation model (Figure 7B), and calculations of the terms in the turbulence kinetic-energy balance equation suggest that the contribution by buoyancy production becomes positive in the strongly stable condition (Figure 9A). The dissipation decreases with increasing stability and in strongly-stable conditions it is balanced by the positive buoyancy production. In contrast with v'/u', the ratio w'/u' is almost constant in the whole range of positive Ri (Figure 7).

In unstable conditions the ratios of v'/u' and w'/u' increase as $-$Ri becomes larger (Figure 8). This indicates that the vertical motion (v') is most strongly enhanced and second, the lateral fluctuation (w') is promoted. This noticeable enhancement of the vertical motion is due to the increase of the buoyancy-production term $\beta g\overline{v\theta}$ in the $\overline{v^2}$-transport equation with increasing instability (Figure 9B) and the extra energy by buoyancy production is probably redistributed to $\overline{w^2}$ and $\overline{u^2}$ through the pressure-force terms in the $\overline{u^2}$-, $\overline{v^2}$- and $\overline{w^2}$-transport equations. Komori et al. [20] have confirmed the energy transport process by the pressure-force terms from the predictions by the spectral-equation model. The results showed that the energy of $\overline{v^2}$ and $\overline{u^2}$ produced by buoyancy and shear is distributed to $\overline{w^2}$ in weakly unstable conditions, while in the strongly unstable range the energy transfer occurs from $\overline{v^2}$ to $\overline{w^2}$ and $\overline{u^2}$. In contrast to the increase of the buoyancy production, the variation of the shear production $-\overline{uv}(\partial\bar{U}/\partial y)$ against instability is small (Figure 9B). In unstably stratified conditions the turbulence motion is enhanced, but at the same time the velocity gradient is reduced by the promotion of the vertical turbulent mixing, as shown in Figure 1A, so that the total effect is very small for shear production. The theoretical predictions also show that in strongly-unstable conditions the shear production is very small and the large buoyancy production is balanced by the dissipation (Figure 9B).

The correlation coefficient of the Reynolds stress $R_{uv} = -\overline{uv}/u'v'$ decreases with increasing Ri and has small negative values in the strongly stable range of Ri > 0.7. (Figure 7). A similar behavior is seen in the distribution of the correlation coefficient of the vertical heat flux $R_{v\theta} = -\overline{v\theta}/v'\theta'$. That is, the vertical transfer of momentum and heat is suppressed in weakly stable conditions, and in the extremely stable range of Ri > 0.5–0.7 they occur against their mean gradients. It can also be seen that the negative value of $R_{v\theta}$ is larger than that of R_{uv} and the change of the sign of $R_{v\theta}$ occurs in more weakly stable conditions. The values of R_{uv} are in good agreement with the laboratory measurements by Young [6].

Webster's data [5] also show similar decreasing behavior with stability. The zero values of R_{uv} and $R_{v\theta}$ suggest that the turbulent motion approaches wavelike motion with increasing stability. The calculations of the correlation coefficients in stable conditions are in qualitative agreement with the measured values; in particular, the prediction of the negative correlation coefficients is significant (Figure 7B). This negative correlation supports the experimental results; in strongly-stable conditions buoyancy-driven motions appear, so that the phase angles decrease (Figure 4A). The calculations of the contribution terms in the transport equations of $-\overline{v\theta}$ and $-\overline{uv}$ in stable conditions are shown in Figure 10A. From the predictions, it is also found that the variations of $R_{v\theta}$ and R_{uv} are mainly

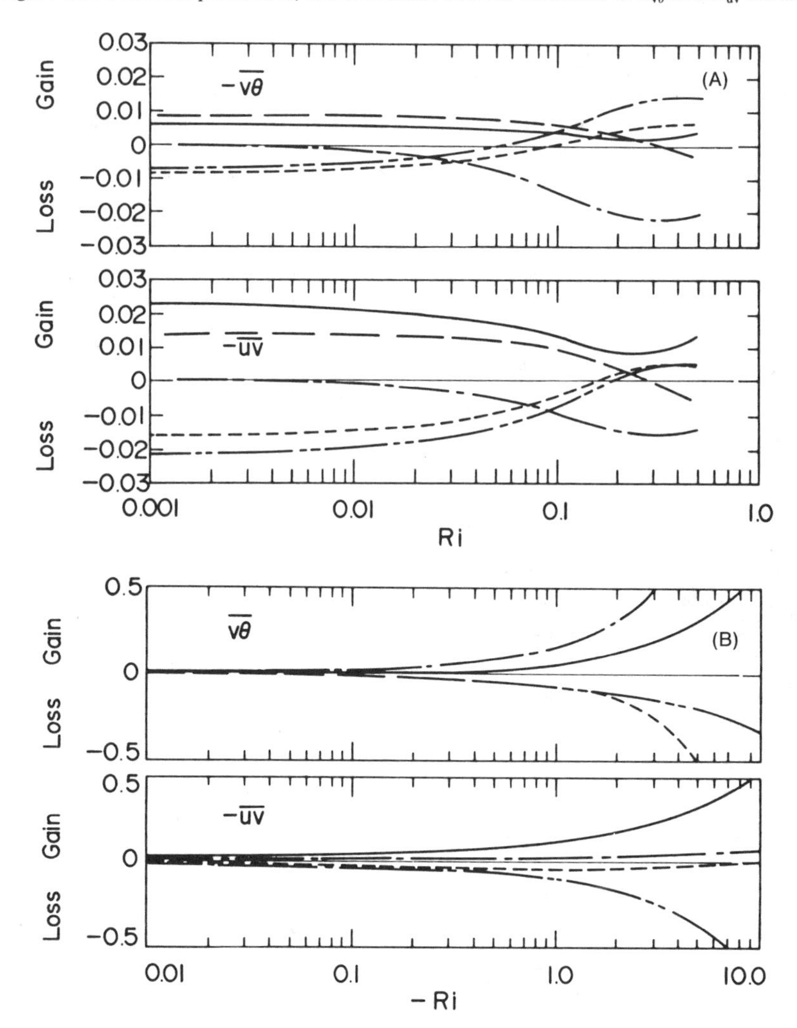

Figure 10. Budgets of vertical heat flux and shear stress calculated from the spectral-equation model. Contribution terms for the vertical heat flux and shear stress are nondimensionalized by $\nu^{5/2}(t_t - t_0)^{3/2}/[J_0(\partial\bar{U}/\partial y)(\partial\bar{T}\partial y)]$ and $\nu^{5/2}(t_t - t_0)^{7/2}/[J_0(\partial\bar{U}/\partial y)]$, respectively: (A) in stable conditions; (B) in unstable conditions; —, mean-field production; ———, buoyancy production; —·—·—, pressure-force production; ———, advection; ----, dissipation [20, 21].

due to the remarkable negative contributions of the buoyancy terms. The calculations also clarify that the large, negative contribution of the buoyancy term $\beta g \overline{\theta^2}$ in the transport equation of $-\overline{v\theta}$ is attributed to the advection term, while the negative contribution of the buoyancy term $\beta g \overline{u\theta}$ in the transport equation of $-\overline{uv}$ is attributed to both the advection and pressure-force production terms [21]. This supports the experimental presumption that the hot eddy is advected downstream and it is due to the countergradient transport (see the section, "Turbulence Energy Budget and its Transport Mechanism.").

In unstable conditions, $R_{v\theta}$ $(=\overline{v\theta}/v'\theta')$ increases with increasing instability, while R_{uv} $(=-\overline{uv}/u'v')$ decreases (Figure 8). The predictions of the contribution terms in the transport equations of $\overline{v\theta}$ and $-\overline{uv}$ by the spectral-equation model are shown in Figure 10B. From the theoretical predictions, it is found that the increase of $R_{v\theta}$ with instability is mainly due to the increase of the buoyancy term and that the pressure-force term works so as to reduce $R_{v\theta}$. The decrease of R_{uv} in strongly-unstable conditions is attributed to the negative contribution of the pressure-force term, while the influence of the buoyancy term is very small.

The ratio $-\overline{v\theta}/\overline{u\theta}$ of the vertical to the streamwise heat flux decreases rapidly with increasing stability and ultimately crosses the zero value (Figure 7). Except under strongly stable conditions, the behavior is in good agreement with Young's and Webster's data [5, 6]. In unstable conditions the ratio increases noticeably with increasing instability, and the vertical heat flux overcomes the streamwise flux in strongly unstable conditions (Figure 8). The predictions by the spectral-equation model are also in qualitative agreement with the measurements (Figures 7 and 8).

Correlation of Eddy Diffusivities with Ri

The measured values of the ratios of eddy diffusivities of scalar quantities under stratified conditions to those under neutral conditions, K_x $(=\epsilon^*_{x,h}/\epsilon^*_{x,h,0})$, K_y $(=\epsilon^*_{y,h}/\epsilon^*_{y,h,0})$ and K_z $(=\epsilon^*_{z,m}/\epsilon^*_{z,m,0})$, are shown by best-fit curves in Figure 11A against $|Ri|$.

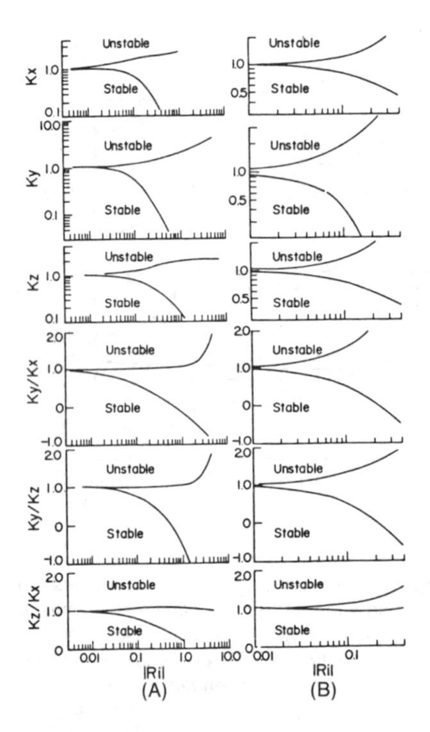

Figure 11. Correlation of eddy diffusivities of scalar quantities in stable and unstable conditions: (A) measured results: ————, best-fit curve of the data of Komori et al. [19]; (B) predictions by the spectral-equation model [19].

Under stable conditions, all the ratios decrease with increasing stability, whereas in unstable conditions, they increase with increasing instability. From these results it is seen that all three diffusivity components are suppressed under stable conditions and promoted under unstable conditions in comparison with neutral conditions.

To make more clear the difference between these buoyancy effects on the three diffusivities, the ratios K_y/K_x, K_y/K_z and K_z/K_x are also shown in Figure 11A. From the variations with Ri, it is seen that in stable conditions the vertical diffusion is most strongly suppressed, the lateral diffusion follows this, and the minimum effect lies with the longitudinal diffusion. This means that the buoyancy effects on the lateral and longitudinal diffusivities are rather weak in comparison with those on the vertical one. Under strongly-stable conditions, heat is transferred upward against the temperature gradient by the intermittent buoyancy driven-motion, that is, the vertical turbulent heat flux $\overline{v\theta}$ changes its sign (see the section "Turbulence Energy Budget and Its Transport Mechanism"). Thus, K_y becomes negative in the range of Ri > 0.5–0.7. However, such a countergradient scalar transfer occurs only in the vertical direction, not in the lateral or longitudinal direction. Therefore, the signs of K_z and K_x are not changed, and the ratios of K_y/K_x and K_y/K_z become negative in strongly-stable conditions. Under weakly-unstable conditions, all of the ratios are close to 1.0. From these values together with the variations of the eddy diffusivities of K_x, K_y and K_z, it is found that the enhancements of the rates of diffusion in the three directions are almost the same. However, in the strongly unstable range of $Ri < -1.0$, the vertical diffusion becomes extremely large. This is surely due to the dominant vertical motion driven by buoyancy (see the section, "Turbulence Energy Budget and Its Transport Mechanism").

The predictions of the ratios of eddy diffusivities by the spectral-equation model are shown in Figure 11B. The variations with Ri of the ratios of the predicted eddy diffusivities are in good qualitative agreement with the experimental results shown in Figure 11A. The enhancement mechanism of the vertical scalar diffusion has been discussed theoretically in the previous section, "Turbulence Energy Budget and Its Transport Mechanism."

To clarify the longitudinal and lateral scalar diffusion mechanisms, Komori et al. [19] have calculated the contribution terms in the spectral equations (Equation 18) for the longitudinal and lateral mass-fluxes $\overline{uc}$ and $\overline{wc}$. The main contribution terms for the lateral and longitudinal mass-transport are shown in Figures 12A and B, respectively. From these figures, it is seen that in stable conditions the suppression of the lateral diffusion is due both to the decrease of the mean-field production term and to the increase in the negative contribution of the pressure-force term.

For longitudinal diffusion, the pressure-force term works so as to promote mass-transfer in the strongly-stable range, whereas the mean-field term makes a negative contribution. Therefore, the suppression of the longitudinal diffusion is attributed mainly to the decrease of the mean-field production term. In unstable conditions, the positive contribution of the mean-field production term becomes large for both lateral and longitudinal mass-transport. The pressure-force term makes an important, positive contribution to the lateral transport, whereas for the longitudinal mass-flux, it works so as to reduce the longitudinal transport. From these results, it is understood that the mean-field production and pressure-force terms are important for both lateral and longitudinal transport of mass. Of the two terms, it is found that the contribution of the pressure-force term for the lateral mass-transport is contrary to that of the longitudinal mass-transport in both strongly stable and unstable conditions. This induces the difference between the stability dependencies of the lateral and longitudinal diffusion.

CONCLUDING REMARKS

This article has summarized a series of work conducted by Komori [18] and Komori et al. [19, 20, 21] and has shown conclusively the buoyancy effects on the turbulence structure and scalar diffusion in stratified outer-layer open-channel flows. The main conclusions that can be drawn from this article are that:

1. In stable conditions, fluctuating motions become close to wavelike motions, and turbulent heat and momentum transfer against the mean temperature and velocity gradients occurs in

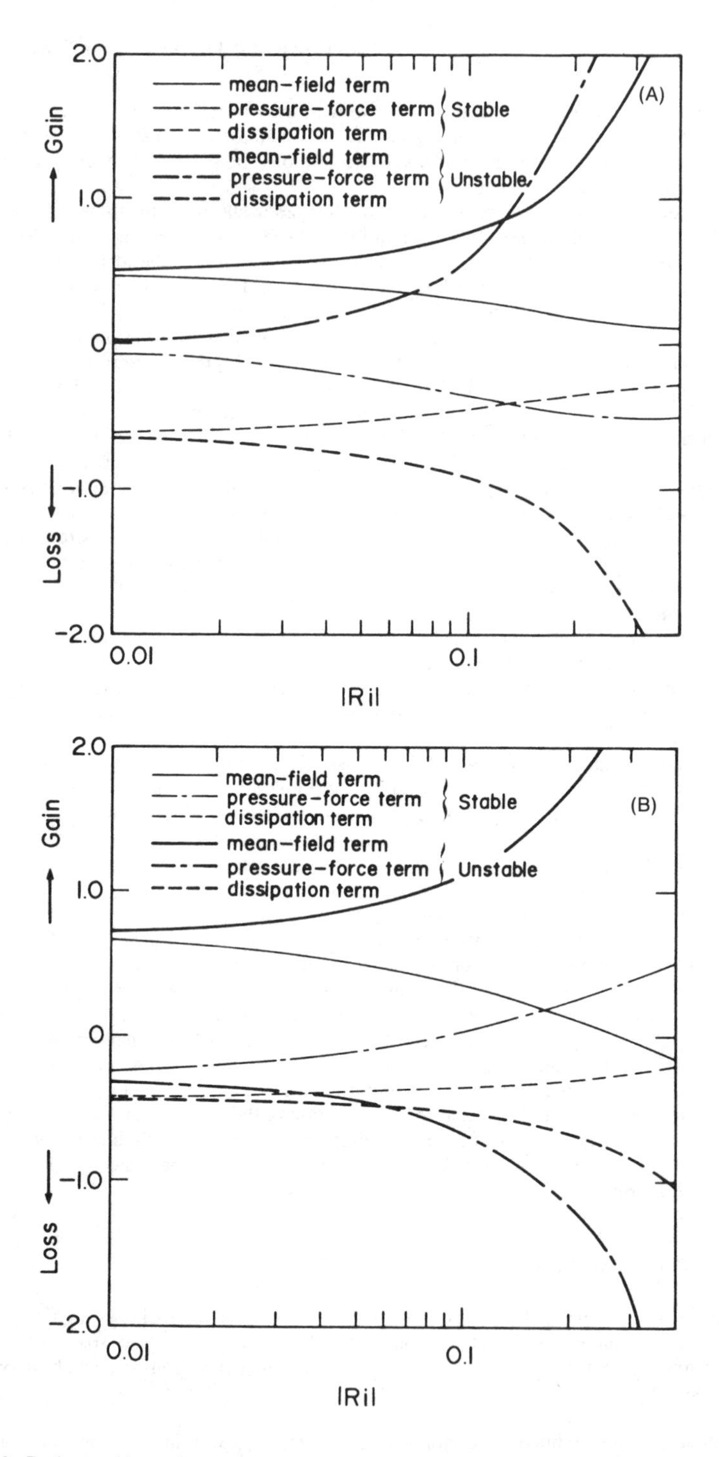

Figure 12. Budgets of lateral and longitudinal mass fluxes calculated from the spectral-equation model: (A) budget of the lateral mass flux nondimensionalized by $(\partial\bar{U}/\partial y)wc$ at $Ri = 0$; (B) budget of the longitudinal mass flux nondimensionalized by $(\partial\bar{U}/\partial y)uc$ at $Ri = 0$ [19].

strongly stable stratification. The countergradient transfer is due to the intermittent wave breakdown triggered by the buoyancy-driven motion of the advected eddies. In unstable conditions cooled from above, the intermittent downward motions of the cold eddies with large negative spikes of temperature fluctuation are pronounced and are the main contributors to the turbulent transport of heat in the vertical direction. With increasing instability, the upward and downward motions are enhanced and, in strongly unstable conditions, not only the large-scale eddies but also small-scale eddies are affected by the buoyancy-driven motions.

2. Longitudinal and lateral turbulent diffusion of scalar quantities, as well as the vertical diffusion is suppressed under stable conditions and is promoted under unstable conditions. Vertical diffusion is most strongly affected by buoyancy; lateral diffusion less so, and longitudinal diffusion is least dependent on buoyancy. The lateral and longitudinal diffusion depends mainly on the mean-field production and the pressure-force production, and the difference between buoyancy dependencies of the lateral and longitudinal diffusion is due to the contribution of the pressure-force production.
3. The stratified outer-layer flows in an open channel are close to local equilibrium, and there the local gradient Richardson number Ri becomes a significant parameter for representing the buoyancy effects.
4. The spectral-equation model is very useful for the predictions of the qualitative buoyancy effects on the turbulence structure and scalar diffusion in the stratified outer-layer away from the wall. In particular, the predictions of the pressure-force terms which are significant in stratified flows are striking points that any other model cannot depict.

The major future works emerging from this article are:

- Clarification of the detailed mechanisms of the wave-breakdown and the countergradient transport in strongly-stable conditions.
- Investigation of developing process of stratified flows.
- Investigation of buoyancy effects on the turbulence structure (bursting phenomena) and scalar diffusion in stratified lower-layer flows cooled or heated from the bottom floor of the flume.
- Theoretical development of higher-order models based on three- and four-point correlations.

Acknowledgments

The author would like to thank Dr. H. Ueda, The National Institute for Environmental Studies; for his continuing support in this work.

NOTATION

c concentration fluctuation kg/m^3

c_1, c_2, c_3 constants in Equations 10 to 12

D molecular diffusivity of mass, m^2/s

E_{ij} spectrum function of $\overline{(u_i)_A(u_j)_B}$ at two points A and B non-dimensionalized by $\nu\tau_t/[J_0(\partial\bar{U}/\partial y)]$

$E_{i\theta}$ spectrum function of $\overline{(u_i)_A(\theta)_B}$ non-dimensionalized by $\nu/[J_0(\partial\bar{T}/\partial y)]$

E_{ic} spectrum function of $\overline{(u_i)_A(c)_B}$ non-dimensionalized by $\nu/[J_0(\partial\bar{\Gamma}/\partial x_m)]$

$E_{\theta\theta}$ spectrum function of $\overline{(\theta)_A(\theta)_B}$ non-dimensionalized by $\nu(\partial\bar{U}/\partial y)/[\tau_t J_0(\partial\bar{T}/\partial y)^2]$

E_{cc} spectrum function of $\overline{(c)_A(c)_B}$ non-dimensionalized by $\nu(\partial\bar{U}/\partial y)/[\tau_t J_0(\partial\bar{\Gamma}/\partial y)^2]$

$E_{\theta c}$ spectrum function of $\overline{(\theta)_A(c)_B}$ non-dimensionalized by $\nu(\partial\bar{U}/\partial y)/[J_0(\partial\bar{T}/\partial y)(\partial\bar{\Gamma}/\partial x_m)]$

Fr Froude number, $Fr = \bar{U}_{ave}/(g\delta)^{1/2}$

f frequency, 1/s

g gravitational acceleration, m/s^2

g_i gravitational acceleration component in the i-direction, m/s^2

J_0 constant that depends on initial conditions in Equation 21, m^7/s^2

K Von Kármán constant

K_x ratio of the longitudinal eddy diffusivity of heat under stratified conditions to that under neutral conditions, $K_x = \epsilon^*_{x,h}/\epsilon^*_{x,h,0}$

K_y ratio of the vertical eddy diffusivity of heat under stratified conditions to that under neutral conditions, $K_y = \epsilon^*_{y,h}/\epsilon^*_{y,h,0}$

K_z ratio of the lateral eddy diffusivity of mass under stratified conditions to that under neutral conditions, $K_z = \epsilon^*_{z,m}/\epsilon^*_{z,m,0}$

k wavenumber nondimensionalized by $[\nu\tau_t/(\partial\bar{U}/\partial y)]^{1/2}$, $k = [k_1^2 + (k_2 - k_1\tau)^2 + k_3^2]^{1/2}$

ki dimensionless wavenumber component in the i-direction

L Monin-Obukhov length, $L = -K\beta g\overline{v\theta}/(\bar{T}u^{*3})$, m

Pr Prandtl number, $Pr = \nu/\alpha$

p pressure fluctuation, Pa

$\overline{q^2}$ turbulence kinetic energy, $\overline{q^2} = \overline{u^2} + \overline{v^2} + \overline{w^2}$, m^2/s^2

R hydraulic radius, $R = \delta W/(2\delta + W)$, m

Re Reynolds number, $Re = 4R\bar{U}_{ave}/\nu$

Ri local gradient Richardson number, $Ri = \beta g(\partial\bar{T}/\partial y)/(\partial\bar{U}/\partial y)^2$

$\overline{Ri}$ bulk Richardson number, $Ri = \beta gR(\bar{T}_{suf} - \bar{T}_{bot})/\bar{U}^2_{ave}$

R_{uv} correlation coefficient between u and v

$R_{v\theta}$ correlation coefficient between v and θ

Sc Schmidt number, $Sc = \nu/D$

$\bar{T}$ time-averaged (mean) temperature, °K

t time, s

t_0 initial time, s

t_t terminal time, s

$\bar{U}$ time-averaged (mean) velocity in the longitudinal direction, m/s

$\bar{U}_i$ time-averaged (mean) velocity in the i-direction, m/s

$\bar{U}_{ave}$ cross-sectional time-averaged (mean) velocity, m/s

u velocity fluctuation in the longitudinal direction, m/s

u_i velocity fluctuation in the i-direction, m/s

u^* friction velocity, m/s

v velocity fluctuation in the vertical direction, m/s

W width of the flume, m

w velocity fluctuation in the lateral direction, m/s

x space coordinate in the longitudinal direction, $x = x_1$, m

x_i, x_m space coordinate in the i- or m-direction, m

y space coordinate in the vertical direction, $y = x_2$, or the vertical distance from the bottom floor of the flume, m

z space coordinate in the lateral direction, $z = x_3$, m

Greek Symbols

α thermal diffusivity, m^2/s

β expansion coefficient, 1/K

$\bar{\Gamma}$ time-averaged (mean) concentration, kg/m^3

δ flow depth, m

δ_{ij} Kronecker delta

ϵ viscous-dissipation rate, m^2/s^3

$\epsilon^*_{x,h}$ longitudinal eddy diffusivity of heat nondimensionalized by δu^*

$\epsilon^*_{x,h,0}$ $\epsilon^*_{x,h}$ in neutral conditions

$\epsilon^*_{y,h}$ vertical eddy diffusivity of heat nondimensionalized by δu^*

$\epsilon^*_{y,h,0}$ $\epsilon^*_{y,h}$ in neutral conditions

$\epsilon^*_{z,m}$ lateral eddy diffusivity of mass nondimensionalized by δu^*

$\epsilon^*_{z,m,0}$ $\epsilon^*_{z,m}$ in neutral conditions

θ temperature fluctuation, °K

κ wavenumber, $\kappa = 2\pi f/\bar{U}$, 1/m

ν kinematic viscosity, m^2/s

ρ density, kg/m^3

τ dimensionless time, $\tau = (\partial\bar{U}/\partial y)(t - t_0)$

τ_t τ at $t = t_t$

ϕ dimensionless spherical coordinate in wavenumber space

ψ dimensionless spherical coordinate in wavenumber space

Superscript

— time-averaged

′ root mean squared

Subscript

A	at point A	suf	at the free surface
B	at point B	max	maximum
bot	at the bottom floor of the flume		

REFERENCES

1. Wyngaard, J. C., Coté, O. R., and Izumi, Y., "Local Free Convection, Similarity aand the Budgets of Shear Stress and Heat Flux," *J. Atmos. Sci.*, Vol. 28 (1971), pp. 1171–1182.
2. Haugen, D. A., Kaimal, J. C., and Bradley, E. F., "An Experimental Study of Reynolds Stress and Heat Flux in the Atmospheric Surface Layer," *Quart. J. Roy. Meteor. Soc.*, Vol. 97 (1971), pp. 168–180.
3. McBean, G. A., and Miyake, M., "Turbulent Transfer Mechanisms in the Atmospheric Surface Layer," *Quart. J. Roy. Meteor. Soc.*, Vol. 98 (1972) pp. 383–398.
4. Wyngaard, J. C., and Coté, O. R., "The Budgets of Turbulent Kinetic Energy and Temperature Variance in the Atmospheric Surface Layer," *J. Atmos. Sci.*, Vol. 28 (1971), pp. 190–201.
5. Webster, C. A. G., "An Experimental Study of Turbulence in a Density-stratified Shear Flow," *J. Fluid Mech.*, Vol. 19 (1964), pp. 221–245.
6. Young, S. T. B., "Turbulence Measurements in a Stably-Stratified Turbulent Shear Flow," Queen Mary Coll. Lond. Rep. QMC-EP 6018, 1975.
7. Arya, S. P. S., and Plate, E. J., "Modeling of the Stably Stratified Atmospheric Boundary Layer," *J. Atmos. Sci.*, Vol. 26 (1969) pp. 656–665.
8. Arya, S. P. S., "Buoyancy Effects in a Horizontal Flat-Plate Boundary Layer," *J. Fluid Mech.*, Vol. 68 (1975), pp. 321–343.
9. Schon, J. P., et al., "Experimental Study of Diffusion Processes in Unstable Stratified Boundary Layers," Vol. 18B. Frenkiel, F. N. and Munn, R. E., (Eds.) from *Turbulent Diiffusion in Environmental Pollution*, Academic Press, New York, 1974, pp. 265–272.
10. Piat, J. F., and Hopfinger, E. J., "A Boundary Layer Topped by a Density Interface," *J. Fluid Mech.*, Vol. 113 (1981), pp. 411–432.
11. Fukui, K., Nakajima, M., and Ueda, H., "A Laboratory Experiment on Momentum and Heat Transfer in the Stratified Surface Layer," *Quart. J. Roy. Meteor. Soc.*, Vol. 109 (1983), pp. 661–676.
12. Ellison, T. H., and Turner, J. S., "Mixing of Dense Fluid in a Turbulent Pipe Flow," *J. Fluid Mech.*, Vol. 8 (1960), pp. 514–544.
13. Schiller, E. J., and Sayre, W. W., "Vertical Temperature Profiles in Open-Channel," *J. ASCE*, HY6 (1975), pp. 749–761.
14. Mizushina, T., et al., "Buoyancy Effects on Eddy Diffusivities in Thermally Stratified Flow in an Open Channel," Vol. 1, MC16, Hooper, F. C. (Ed.) from *Heat Transfer 1978*, Hemisphere Publishing Corporation, Washington, 1978, pp. 91–96.
15. French, R. H., "Stratification and Open Channel Flow," *J. ASCE*, HY1 (1978), pp. 21–31.
16. French, R. H., "Transfer Coefficients in Stratified Channel Flow," HY9 (1979), pp. 1087–1100.
17. McCutcheon, S. C., "Vertical Velocity Profiles in Stratified Flows," HY8 (1981), pp. 971–988.
18. Komori, S., "Turbulence Structure in Stratified Flow," Ph.D. dissertation., Kyoto University, 1980.
19. Komori, S., et al., "Lateral and Longitudinal Turbulent Diffusion of Scalar Quantities in Thermally Stratified Flow in an Open Channel," Vol. 2, EN9, Grigull, U., Hahne, E., Stephan, K. and Straub, J. (Eds.) from *Heat Transfer 1982*, Hemisphere Publishing Corporation, Washington, 1982, pp. 431–436.
20. Komori, S., et al., "Turbulence Structure in Unstably-Stratified Open-Channel Flow," *Phys. Fluids*, Vol. 25, No. 9 (1982), pp. 1539–1546.
21. Komori, S., et al., "Turbulence Structure in Stably Stratified Open-Channel Flow," *J. Fluid Mech.*, Vol. 130 (1983), pp. 13–26.
22. Launder, B. E., "On the Effects of a Gravitational Field on the Turbulent Transport of Heat and Momentum," *J. Fluid Mech.*, Vol. 67 (1975), pp. 569–581.

23. Gibson, M. M., and Launder, B. E., "Ground Effects on Pressure Fluctuations in the Atmospheric Boundary Layer," *J. Fluid Mech.*, Vol. 86 (1978), pp. 491–511.
24. Ueda, H., Mistumoto, S., and Komori, S., "Buoyancy Effects on the Turbulent Transport Processes in the Lower Atmosphere," *Quart. J. Roy. Meteor. Soc.*, Vol. 107 (1981), pp. 561–578.
25. Hill, J. C., "Models for Turbulent Transport Processes," *Chem. Eng. Ed.*, Vol. 13 (1979), pp. 34–39.
26. Launder, B. E., and Spalding, D. B., *Lectures in Mathematical Models of Turbulence*, Academic Press, New York, 1972.
27. Zeman, O., and Lumley, J. L., "Modeling Buoyancy Driven Mixed Layers," *J. Atmos. Sci.*, Vol. 33 (1976) pp. 1974–1988.
28. Lumley, J. L., Zeman, O., and Siess, J., "The Influence of Buoyancy on Turbulent Transport," *J. Fluid Mech.*, Vol. 84 (1978), pp. 581–597.
29. Deissler, R. G., "On the Decay of Homogeneous Turbulence Before the Final Period," *Phys. Fluids*, Vol. 1 (1958), pp. 111–121.
30. Deissler, R. G., "Effects of Inhomogeneity and of Shear Flow in Weak Turbulent Fields," *Phys. Fluids*, Vol. 4 (1961), pp. 1187–1198.
31. Deissler, R. G., "Turbulence in the Presence of a Vertical Body Force and Temperature Gradient," *J. Geophys. Res.*, Vol. 67 (1962) pp. 3049–3062.
32. Deissler, R. G., "Growth Due to Buoyancy of Weak Homogeneous Turbulence with Shear," *z. Angew. Math. Phys.*, Vol. 22 (1971), pp. 267–274.
33. Deissler, R. G., "A Theory of Decaying Homogeneous Turbulence," *Phys. Fluids*, Vol. 3 (1960), pp. 176–187.
34. Mizushina, T., et al., "Application of Laser Doppler Velocimetry to Turbulence Measurement in Nonisothermal Flow," *Proc. Roy. Soc. Lond.*, A. 366 (1979), pp. 63–79.
35. Komori, S., and Ueda, H., "Turbulent Effects on the Chemical Reaction for a Jet in a Nonturbulent Stream and for a Plume in a Grid-generated Turbulence," *Phys. Fluids*, Vol. 27, No. 1 (1984), pp. 77–86.
36. Ueda, H., and Mizushina, T., "Turbulence Structure in the Inner Part of the Wall Region in a Fully Developed Turbulent Flow," Proc. 5th Symp. Turbulence in Liquid, Univ. Missouri-Rolla, 1977, pp. 357–366.

CHAPTER 4

FORM RESISTANCE IN OPEN-CHANNEL FLOW IN THE PRESENCE OF RIPPLES AND DUNES

M. I. Haque

The George Washington University
Civil, Mechanical and Environmental Engineering
Washington, DC

CONTENTS

INTRODUCTION, 71

AN OVERVIEW OF THE BEDFORM PHENOMENON, 72
Ripples, 76
Dunes, 76
Antidunes, 77

GENERAL REMARKS ON SURFACE AND PRESSURE DRAGS, 77

EMPIRICAL OR SEMI-ANALYTICAL STUDIES, 79
Einstein-Barbarossa's Bar-Resistance Curve, 79
Alam-Kennedy's Analysis, 84
Engelund's Analysis, 86

FORM RESISTANCE OF RIGID RIPPLES AND DUNES, 87
Haque-Mahood's Analysis, 87

FORM RESISTANCE OF NONRIGID RIPPLES AND DUNES, 93

NOTATION, 95

REFERENCES, 96

INTRODUCTION

A plane erodible bed is inherently unstable under the action of a flowing fluid. It deforms into various types of bed undulations, which offer increased resistance to flow. Bed features, such as ripples and dunes, play a significant role in the makeup of hydraulic resistance to flow in alluvial channels. Since most alluvial channels are wide and shallow, the major contribution to the resistance force which opposes the fluid acceleration comes from the channel bed. It is recognized that the total drag at the channel bed consists of two main parts:

1. The surface drag due to the grain roughness.
2. The form drag (pressure drag) due to the hydrodynamic forces generated by the macroscale bed features.

The laboratory studies [1–3] clearly demonstrate the fact that, in many cases, the form drag created by ripples and dunes achieves such a proportion that the surface drag due to grain roughness in comparison seems insignificant. In flume studies, more than a seven-fold increase in friction factor has been observed quite frequently, as the plane granular bed deforms into a train of ripples and dunes.

In alluvial hydraulics, the importance of form drag was recognized in the early fifties by Einstein and Barbarossa [4]. Since then, there have been many attempts to determine the friction factor in the presence of large roughness elements [5–10]. In view of the complexity of the turbulent flows, most of these studies have been empirical in nature, based largely on plausible hypotheses guided by dimensional analysis. These studies have furnished a vast storehouse of experimental data needed for engineering analysis and design. Despite the steadily increasing wealth of empirical data, no satisfactory theory on the form resistance in alluvial channels has emerged so far. The main difficulty seems to arise from the relatively large scatter of the experimental data in alluvial hydraulics, as compared to their counterparts in the rigid boundary hydraulics.

Any empirical inference obtained from a purely statistical regression analysis of these data is naturally subject to skepticism, unless some theoretical framework is available to guide and interpret the experimental observations. Since an ideal flow fails to shed any light on the resistance problem and the presence of viscosity combined with the complexity of the turbulent motion has offered unsurmountable difficulties, a rigorous mathematical solution of the resistance problem still defies our best analytical skills and taxes our computational facilities beyond their limits. This situation in the past forced the researchers, engaged in the problem of resistance to flow in the presence of large roughness elements, to follow closely the course set by their predecessors, who were dealing with the seemingly similar problem of resistance to flow through conduits with uniform surface texture. For one-dimensional flows, either through a conduit, or over a flat plate, with small roughness elements and uniform texture, we can find some theoretical justification for the Prandtl-Karman resistance laws. However, the extension of these laws to include a two- or a three-dimensional flow situation in the presence of large roughness elements (the maximum linear dimension of the roughness element comparable to the average flow depth) does not enjoy this privilege, and yet the analytical basis of these laws has set the pattern of subsequent empirical studies on resistance to flow in the presence of large-scale roughness elements to a great extent [2, 9].

Why should we consider the form and the grain roughness separately? Because they arise due to different reasons. Also, the form resistance plays a significant role in determining the intensity of flow field, while the skin friction is mainly instrumental in the transport of the bed material. Thus, the problem of identifying the two parts of the total friction factor occupies a central position in the mechanics of sediment transport.

In subsequent sections of this article, we first describe the phenomenon of bedform evolution and its influence on the hydraulic resistance in a qualitative manner. Then, some of the empirical, or semi-analytical, studies are examined with an objective of unifying the common elements. Finally, an analytical approach based on the fundamentals of hydrodynamics is presented and compared with the experimental studies and data.

AN OVERVIEW OF THE BEDFORM PHENOMENON

Much of our knowledge of bedforms and their interaction with the flow comes from the laboratory flume studies, pioneered by Gilbert [11] and later followed by Simons and Richardson [10], and others. Under the influence of fluid motion, an initially plane, erodible bed deforms successively into different types of bedforms as shown schematically in Figure 1, which is based on the observations reported by Simons and Richardson [10]. The sequence of bedforms shown in this figure follows, in general, an increasing magnitude of stream power, which equals the product of the average flow velocity with the average bed shear stress. As the flow velocity increases, a value is reached when the fluid exerts sufficient shearing stress at the bed surface to cause some sporadic movement of the individual sediment particles. This bed shear, when a few individual grain movements can be detected by the unaided eyes, is called the critical shear and it is generally denoted by τ_c. Its magnitude depends on the grain diameter, the specific weights of the fluid and the sediment particles and on the kinematic viscosity of the fluid. The critical shear stress for a given bed material can be predicted with a fair degree of accuracy by the Shield function [12, Chapter 6]. As the shear stress exerted by the fluid on the plane bed increases beyond this threshold of sediment movement, the plane bed changes into an undulating surface marking the first appearance of bedforms.

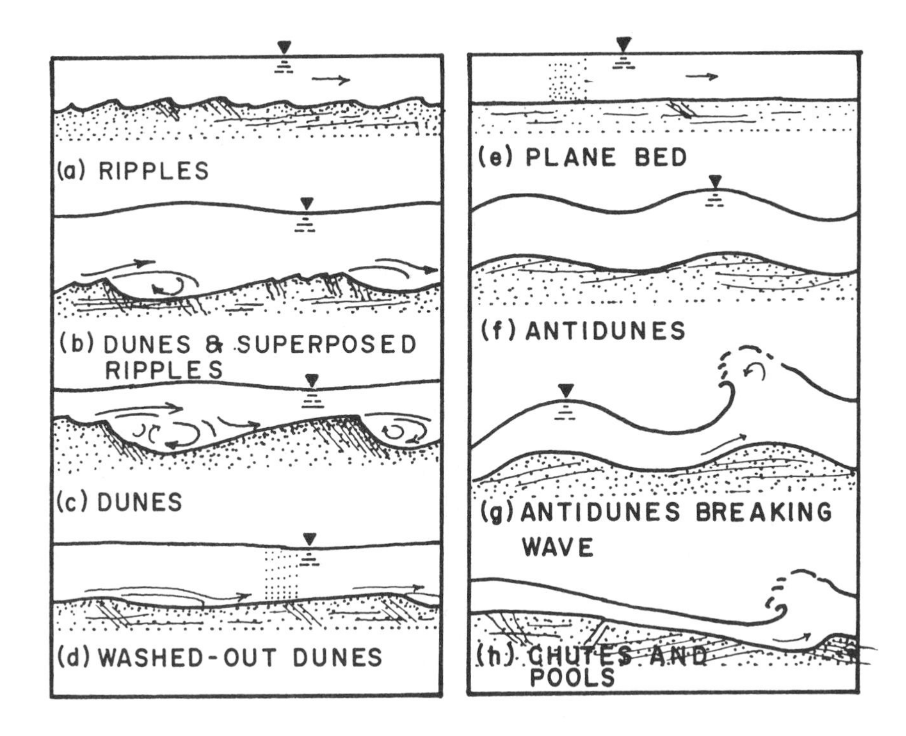

Figure 1. Forms of roughness configurations in alluvial channels [10].

The sequence in which these bedforms appear with an increasing flow intensity is as follows: ripples, dunes with superposed ripples, dunes, washed-out dunes, plane bed, antidunes and chutes and pools. For some combinations of the fluid and the sediment properties, the bedforms may not follow strictly the aforementioned sequence. For a complete description of the bedform phenomenon, the reader is suggested to refer to the original references [10, 11], or consult the standard textbooks on the subject [12–15]. The pertinent features of bedforms and their influence on the friction factor of flow are summarized in Table 1. In this table the bedforms are categorized into two groups based

Table 1
Flow Regime, Bedform and Friction Factor

Flow	Bedform	Friction Factor f''
Lower regime	Ripples	0.05–0.16
	Dunes with ripples	—
	Dunes	0.05–0.13
Transition	Washed-out dunes	
Upper regime	Plane bed	0.02–0.03
	Antidunes	0.02–0.07
	Chutes and pools	0.07–0.09

After Simons and Richardson [10].

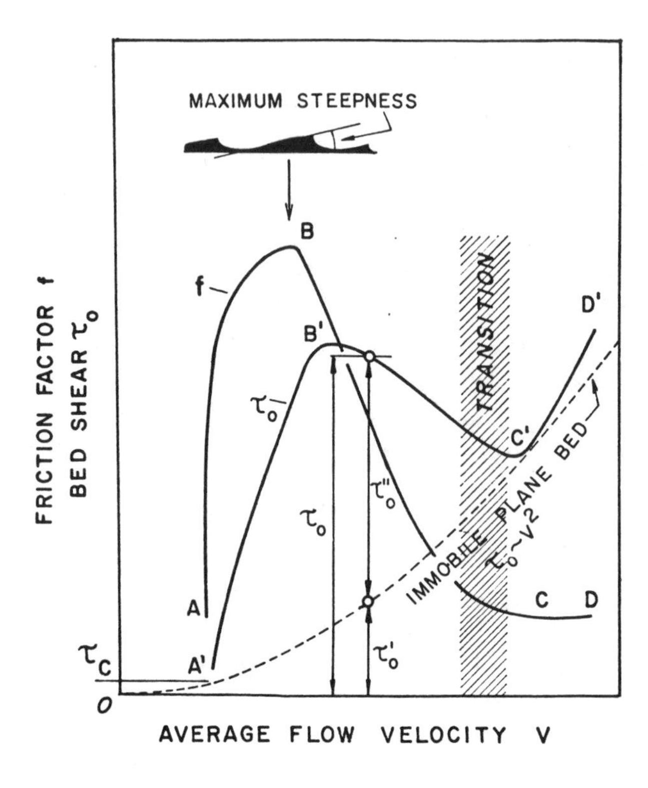

Figure 2. Changes in bed shear stress and friction factor due to increasing velocity [3].

on the flow conditions in which they thrive. The sedimentary ripples and dunes occur in tranquil flows in which the Froude number is less than unity. In contrast, the antidunes are found in rapid flows in which the Froude number exceeds one. From the viewpoint of flow resistance, as well as from the practical river engineering perspective, the lower-regime bedforms like ripples and dunes are the most important bed features. They occur more frequently in nature and their presence marks a significant rise in the friction factor.

Some intuitive grasp on the bedform phenomenon and its interaction with the hydraulic resistance can be achieved by examining Figure 2, which is adapted from Raudkivi's study [3] in a laboratory flume. In this figure, the abscissa represents the average flow velocity V through a rectangular flume, whose bottom is covered up to a uniform depth by noncohesive sand. The ordinate represents the corresponding bed shear (curve A'B'C'D'), or the variation in the total friction factor (curve ABCD). As the flow velocity increases beyond a certain value, the bed shear stress surpasses the critical shear τ_c and the plane erodible bed transforms into a train of ripples (points A and A' in Figure 2). The appearance of ripples is always marked by a rapid increase in the friction factor, because their presence increases the relative roughness of the bed and their peculiar shape (with a discontinuous surface gradient at the crest, Figure 3) causes the flow to separate. The segment AB in Figure 2 represents this sudden increase in the total friction factor due to emerging ripples. The segment A'B' represents the corresponding increase in the average bed shear τ_0. The dotted line in this figure represents the shear stress that would exist on a plane granular bed if the grain roughness were the only agent of surface drag. This shear stress varies as a quadratic function of the average

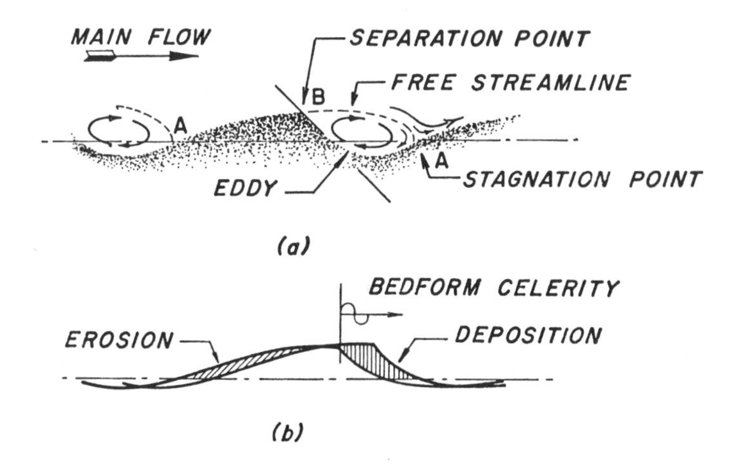

Figure 3. Definition sketch for ripples and dunes.

velocity. Using Darcy-Weisbach's friction factor f, this stress can be obtained from

$$\tau_0' = \frac{f}{4}\left(\frac{1}{2}\rho V^2\right) \quad \text{or} \quad f = 8\frac{(\tau_0'/\rho)}{V^2} \tag{1}$$

where τ_0' = the shear stress due to grain roughness
ρ = the mass density of the fluid
V = the average flow velocity

Equation 1 can also be regarded as the definition equation for Darcy-Weisbach's friction factor. For hydraulically rough boundaries, this friction factor depends only on the relative roughness k/R, where k denotes the roughness height of the granular material (generally, the mean diameter of the sediment particles) and R denotes the hydraulic radius. For wide channels the value of R is approximately equal to the average depth of flow. Henceforth, we shall assume the channel is wide unless stated otherwise. The solid curve A′B′C′D′ represents the average bed shear stress which is partly due to the surface drag and partly due to the pressure drag. The average bed shear stress is related to the slope of the energy grade line by the following equation

$$\tau_0 = \gamma R S_f \tag{2}$$

where γ = the specific weight of fluid
S_f = the slope of the energy grade line, also called the friction slope

For uniform flows in open channels, the friction slope is also equal to the average bed slope, or the average water surface slope. The component τ_0'' of the average bed shear stress which is due to the pressure drag on the bedform, is seldom obtained directly. It is obtained indirectly from

$$\tau_0'' = \tau_0 - \tau_0' \tag{3}$$

in which τ_0 is obtained experimentally using Equation 2 and τ_0' is calculated either from Equation 1 or from some other appropriate resistance formula such as the Manning-Strickler equation.

The point B or B′ in Figure 2 corresponds to the maximum steepness (the ratio of the bedform height to length) of ripples. As the flow velocity further increases, the ripples elongate and their steepness decreases. This phase of their development is represented by segments BC or B′C′ in this figure. As the flow further intensifies, the ripples or dunes are washed out and the bed changes into a plane granular bed with considerable sediment movement along its surface. This situation is represented by point C′ in Figure 2. With further increase in velocity, the bed changes into antidunes and other upper-regime bedforms. As long as the water surface does not become violent and the surface waves do not break, the resistance to flow in the case of antidunes is slightly higher than the resistance offered by an immobile plane bed (dotted curve in Figure 2).

The shaded area in Figure 2 represents the transition from the lower-regime bedforms to the upper-regime bedforms. The bedform phenomenon in this range is rather erratic and uncertain. The curves shown in Figure 2 are for a certain combination of hydraulic and sediment variables. For other combinations, qualitatively similar graphs exist. How do these graphs change quantitatively due to changes in the hydraulic and sediment parameters? The answer to this question, despite vigorous efforts in recent years, still eludes our analytical pursuit.

Due to the limited space, an in-depth discussion on the sedimentary bedforms in alluvial channels is not considered within the scope of this chapter. However, for those readers who are not familiar with the subject, a brief discussion on ripples, dunes, and antidunes is included for completeness.

Ripples

Ripples are small triangular-shaped elements whose upstream faces are slightly convex toward the flow and the downstream faces are steeply inclined at an angle which is roughly equal to the angle of repose of the bed material. The abrupt change in the surface gradient at the crest causes the flow to separate and form a leeward eddy (Figure 3A). Their length is less than 2.0 ft (0.6 m) and their height is less than 0.2 ft (0.06 m). In the case of ripples, the sediment transport rates are relatively small, with almost nothing being transported in suspension. The bed material is eroded systematically from the upstream face, transported over the crest by local velocities and selectively deposited along the downstream face. This pattern of erosion and deposition causes the bedforms to migrate slowly in the downstream direction (Figure 3B). As the bedforms move downstream, they show some random variations in shape and size, while maintaining their dominant profiles and identities over sufficient duration of time, or distance traveled. Despite their small size, ripples cause a larger friction factor than dunes. The ripple size is practically independent of the flow depth. Thus, their resistance to flow decreases as the flow depth increases, because of the diminishing relative roughness of the bedforms. According to some observers, the ripples are formed if a viscous sublayer is present when the critical shear stress is just surpassed.

Dunes

From a practical viewpoint, dunes are the most important bedforms due to their predominant presence in natural rivers and sand-bed channels. They are similar to ripples in shape, but their sizes are much larger. Their length ranges from 2.0 ft to thousands of feet, depending upon the scale of flow. Observations in prototype channels indicate that dunes develop in any size of the bed material, as long as the flow intensity is sufficient to cause transport of the bed material without exceeding a Froude number of unity. At lower flows, the upstream faces of dunes are often observed to carry small ripples (Figure 1B). These superimposed ripples disappear at larger flow velocities. The resistance to flow is large, but not as large as in the case of ripples. This is due to the fact that their steepness is smaller than the steepness of ripples. Although the relative roughness (the ratio of dune height to the average depth of flow) in the case of dunes is larger than in the case of ripples, the smaller steepness of dunes often reduces the overall friction factor. Since the dune height is affected by the flow depth, an increase in the depth of flow does not necessarily result in a decrease in the relative roughness. Sometimes, an increase in the flow depth causes enough increase in the steepness of dunes that a net gain in the friction factor is observed, despite a decrease in the relative roughness.

Just like ripples, dunes migrate slowly downstream. The sediment transport is larger in the case of dunes than it is in the case of ripples. Whereas, their presence invariably increases the friction factor, their disappearance particularly near the transition stage due to changes in the water temperature, causes a rapid decrease in the flow depth and, thus, poses some concern for the river navigation.

Antidunes

Antidunes are found in supercritical flows (Froude number larger than unity) only. Their shape resembles very closely a sinusoidal curve. In case of antidunes, water surface waves are always in phase with the bed waves. The amplitude of the water surface waves is about 1.2–2.0 times larger than the bedform height. The antidunes are highly unstable. They grow from a plane bed until their height becomes unstable. At this stage, the water surface either breaks like the sea surf and the antidunes dissipate, or it simply subsides. As the antidunes are formed, they migrate upstream, downstream, or remain stationary. It is their upstream movement which inspired Gilbert to designate them as antidunes. A detailed theoretical account can be found in Kennedy's work [16]. If antidunes do not break rapidly, the resistance to flow is about the same as offered by a plane, granular bed.

GENERAL REMARKS ON SURFACE AND PRESSURE DRAGS

Of the many mechanisms by which resistance to flow is caused, we shall consider only two: the normal and the tangential stresses acting at the solid boundaries. This restricts us to those circumstances where there are no spreading of gravity waves, no flow boundaries that deform under the action of fluid motion, and no sediment medium that diffuses profusely into the fluid continuum. Under these circumstances, the hydrodynamic behavior of ripples and dunes is very similar to the behavior of bluff rigid bodies immersed into a fluid stream. While considering the drag exerted by the fluid, it is helpful to regard the bedform geometry fixed, independently of the flow field; just as the shape of a rigid bluff body exists with no regard to the oncoming flow. The drag on the body depends only on the normal and tangential stresses acting over that part of the surface which comes in contact with the fluid. On each elementary area of the wetted surface, there acts a normal force, which is essentially due to pressure, and a tangential force which is due to fluid friction. Examining Figure 4A, we see that over a major portion of the body, the frictional force has a strong component in the direction of the main flow. Integration of this component over the entire wetted surface, yields

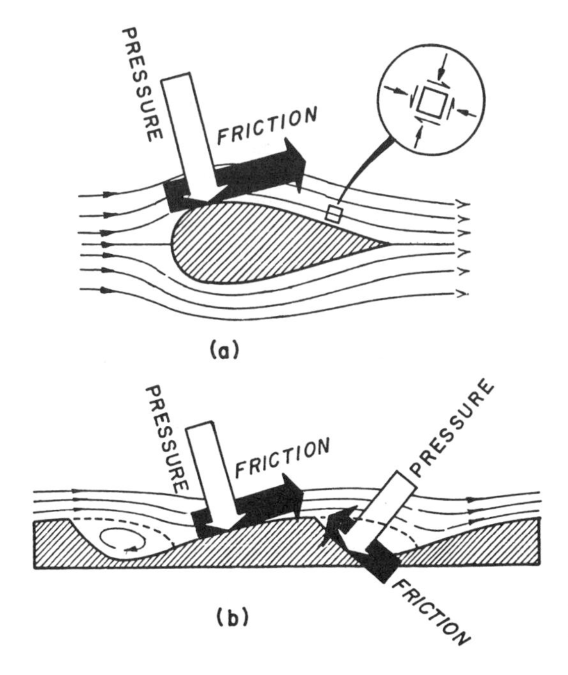

Figure 4. Surface forces on immersed bodies in a fluid stream.

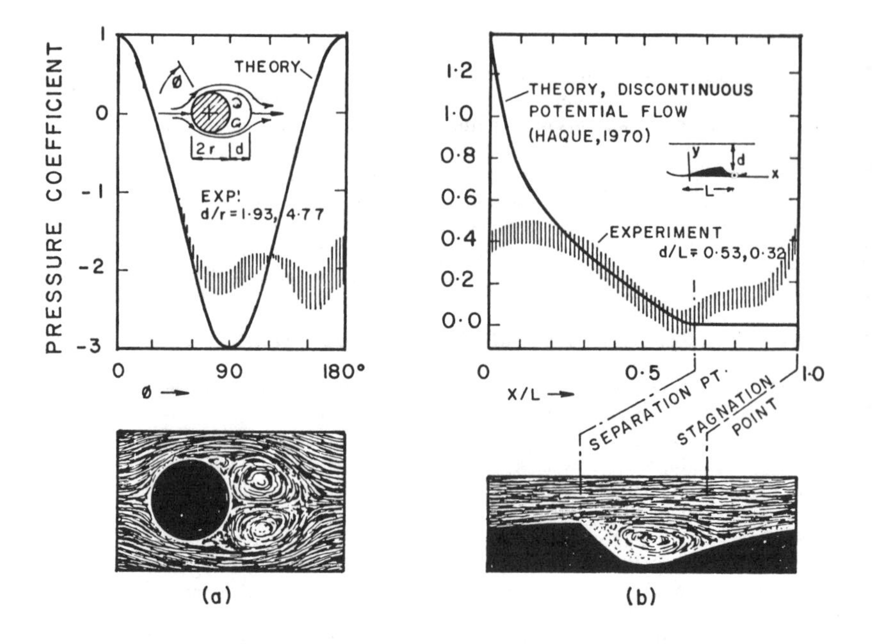

Figure 5. Theoretical and observed pressures: (A) around a cylinder; (B) along a ripple surface.

the surface drag (also called the skin friction, or drag due to the grain roughness). Whenever there is separation of flow, the frictional force on an elementary surface located inside the separation bubble, has a component which acts in a direction against the main flow. For bluff bodies, the surface drag per unit area, τ_0', never remains uniform over the entire surface; it varies considerably, depending upon the shape of the body and the separation of the boundary layer.

Now, consider the normal force acting on the wetted surface. With or without flow separation, the normal force over the anterior part of the body has a component which acts in the main flow direction. On the rear part, the normal force has a component which acts against the main flow (i.e., the fluid pressure tends to push the body against the oncoming flow). Integrated over the wetted surface, this component yields the pressure drag. When there is no separation of flow, the resultant force due to pressure on the anterior part almost equals the resultant force due to pressure on the rear part of the body. In this case, thus, there is no significant pressure drag on the body. Owing to the discontinuous nature of the surface gradient at the crest, there is always separation of flow in the lee of a ripple or a dune. This separation causes an imbalance between the resultant pressure forces acting on the upstream and the downstream faces, just as the separation of flow causes a decrease in pressure in the wake of a stationary cylinder in a fluid stream (see Figures 5A and B). The pressure drag depends both on the shape and the extent of the boundary-layer separation. In the case of bluff bodies with mild surface curvatures, the determination of the point of boundary-layer separation is tedious. On the other hand, for bodies like ripples and dunes with abrupt changes in the surface gradient, the point of separation is fixed and its location is known in advance. For such bodies, the location of the separation point does not depend on the Reynolds number and the pressure drag becomes simply a function of the body shape.

Let us, now, revert to the problem of surface drag, or to the more basic problem of shear stress distribution over the wetted surface. This shear stress depends on the shape of the body, the roughness texture of its surface, and on the extent of the boundary-layer separation. Now, even for the simple case of bodies with well-defined separation zones, the shear stress acting on a typical surface element depends on the shape of the body as well as on the roughness texture of the surface. Thus, in principle, the problem of surface drag in the presence of ripples and dunes is far more involved than the determination of the pressure drag. Hence, the notion of an average shear stress due to grain roughness only, is extremely tenuous, unless we pay due regard to the geometric shape of the bedforms.

EMPIRICAL OR SEMI-ANALYTICAL STUDIES

In this section, we examine some of the empirical studies on the form friction factor in alluvial channels. Only three studies have been selected for this purpose. This limited number has been deliberately chosen to identify any common elements and to construct a coherent picture depicting the basic physics of the phenomenon. As is often the case, a vast number of experimental observations lead to baffling information unless we begin to think and talk in terms of the fundamental principles of fluid motion.

Einstein-Barbarossa's Bar-Resistance Curve

The first significant contribution to the subject was made by Einstein and Barbarossa [4], who formally took into account the shear stresses associated with grain roughness and pressure drag separately. We shall review their analysis in some detail, and make it a point of departure for further discussions on this topic.

Let us consider the dynamic equilibrium of the fluid mass shown in Figure 6A. For uniform flows in a prismatic channel, this leads to the following equation

$$\gamma ALS - LP\tau_0 = 0 \tag{4}$$

where γ = the weight of fluid per unit volume
A = the cross-sectional area
L = the length of the channel segment
S = the channel slope
P = the wetted perimeter
τ_0 = the average bed shear stress which acts along a line parallel to the main flow

Equation 4 can also be written as

$$\gamma RS - \tau_0 = 0 \tag{5}$$

where $R \equiv A/P$, the so-called hydraulic radius, which is the only variable in this equation that contains information about the geometry of the channel. Since any function (or variable) can be expressed as a sum of two functions (or variables), it is mathematically permissible to decompose the cross-sectional area and the average bed shear stress into two parts

$$A = A' + A'' \tag{6}$$

$$\tau_0 = \tau_0' + \tau_0'' \tag{7}$$

Although it is possible, and often customary, to assign some physical meanings to each part of the cross-sectional area, it is not considered necessary to engage in such an effort. Regarding the two parts of the shear stress in Equation 7, they have the same meaning as described previously. Equation 7 is tantamount to saying that the bed shear stress consists of only two parts, one associated with the grain roughness and the other with the form roughness. Substituting Equations 6 and 7 into Equation 4 and dividing by LP, yields

$$\left(\frac{A'}{P} + \frac{A''}{P}\right)\gamma S - (\tau_0' + \tau_0'') = 0 \tag{8}$$

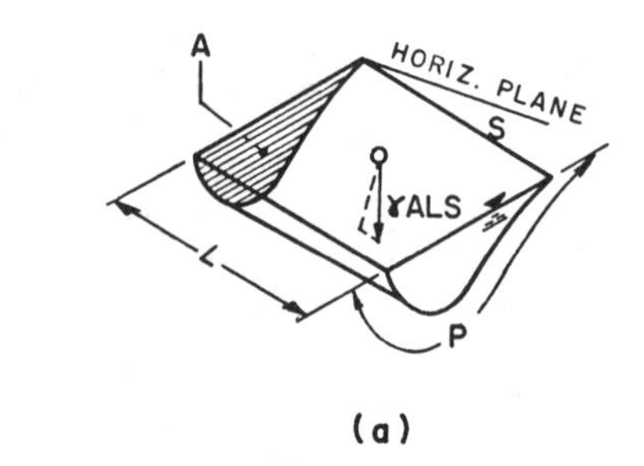

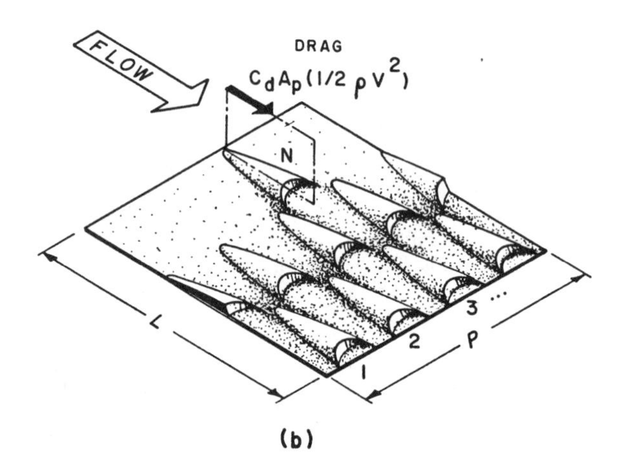

Figure 6. Definition sketches.

or

$$(R' + R'')\gamma S - (\tau_0' + \tau_0'') = 0 \tag{9}$$

where we have used the abbreviated notations R′, R″ for A′/P and A″/P, respectively. From Equations 9, 7 and 5, it immediately follows that

$$R = R' + R'' \tag{10}$$

Now, if we chose R′ (= A′/P) in such a way that

$$\gamma R'S = \tau_0' \tag{11}$$

then, it follows

$$\gamma R''S = \tau_0'' \tag{12}$$

The right-hand side of the preceding equation represents the "effective" shear stress due to the pressure forces acting on the bedforms. The pressure force on an individual bedform can be expressed as

$$D_p = C_d A_p(\tfrac{1}{2}\rho V^2) \tag{13}$$

where D_p = the pressure drag
C_d = the coefficient of drag
A_p = the projected area of the bedform normal to the main flow

The average shear stress due to pressure drag on a bed surface having an area LP (see Figure 6B), becomes

$$\tau_0'' = \frac{ND_p}{LP} \tag{14}$$

where N denotes the total number of bedforms. Combining Equation 14 with Equations 12 and 13 yields

$$\frac{(gR''S)^{1/2}}{V} = \sqrt{\frac{NC_dA_p}{2LP}} \tag{15}$$

The quantity $(gR''S)^{1/2}$ has the dimension of velocity and is generally denoted by u_*''. The function on the right-hand side of Equation 15 depends only on the geometry and total number of bedforms. For its determination, however, there is no analytical means available, so far. According to Einstein and Barbarossa, the bedform configuration depends in general on the sediment transport of the bed material, which in turn depends on the dimensionless variable

$$\psi = \frac{(\rho_s - \rho)D_{35}}{\rho R'S} \tag{16}$$

where ρ_s is the mass density of the sediment grains and D_{35} is the sediment size, such that, 35% of the sediment by weight is smaller. It is further assumed that a unique (?) functional relationship

$$\frac{u_*''}{V} = f(\psi) \tag{17}$$

exists in nature. This functional relationship was empirically determined using the river data obtained mostly from the Missouri River basin. The solid curve in Figure 7 represents this functional relation, graphically.

It is also assumed that the average shear stress due to grain roughness is given by Manning-Stricker's equation

$$\frac{V}{(\tau_0'/\rho)^{1/2}} = \frac{V}{(gR'S)^{1/2}} = 7.66\left(\frac{R'}{k_s}\right)^{1/6} \tag{18a}$$

where k_s is the grain roughness diameter in feet (according to Einstein and Barbarossa, $k_s = D_{65}$). This equation is recommended for hydraulically rough boundaries only. When grain roughness does not produce a hydraulically rough boundary, the following logarithmic equation can be used

$$\frac{V}{(\tau_0'/\rho)^{1/2}} = \frac{V}{(gR'S)^{1/2}} = 5.75\log\left(12.2\frac{R'}{k_s}x\right) \tag{18b}$$

where x is a correction factor taking into account the viscous effects.

The resistance problem can now be stated: given R (or depth for wide channels), S and sediment properties, find V, f′, f″. The solution to this problem can be found using the following steps:

1. Assume a value of R′ ($<$R) and compute V from Equation 18a or 18b and find ψ from Equation 16.

Alam-Kennedy's Analysis

The method developed by Alam and Kennedy rests on the notion of dividing the channel slope into two parts

$$S = S' + S'' \tag{19}$$

where S′ and S″ are the slopes required to overcome the surface drag due to the grain roughness and the form drag due to the pressure forces, respectively. From equilibrium Equation 5

$$\tau_0 = \gamma RS' + \gamma RS'' = \tau_0' + \tau_0'' \tag{20}$$

Also, from the definition of the Darcy-Weisbach friction factor, it follows from Equation 20 that

$$f = f' + f'' \tag{21}$$

where f′ and f″ are friction factors associated with the boundary shear stresses τ_0' and τ_0'', respectively. According to Alam and Kennedy the grain-associated friction factor f′ can be equated to the friction factor of a flat granular bed with sediment movement. To calculate the flat bed friction factor f_f, they used the previous work by Lovera and Kennedy [20] in which the following functional relationship was conjectured

$$f_f = \left(\frac{VR}{\nu}, \frac{R}{D}\right) \tag{22}$$

where ν denotes the kinematic viscosity, and D represents the mean grain diameter. Using laboratory and field data, they found the above relationship as shown in Figure 9.

Based on intuitive reasoning and reasonable assumptions, they started essentially with the following functional relationship for f″

$$f'' = f''\left(\frac{V}{\sqrt{gD}}, \frac{D}{R}, \frac{h}{L}, \frac{L}{D}\right) \tag{23}$$

in which h, L represent the height and the length of the bedform, and g is the acceleration due to gravity. The introduction of h and L into the list of the independent variables is intended to take into account the bedform geometry and its effects on f″. They further used the analytical results

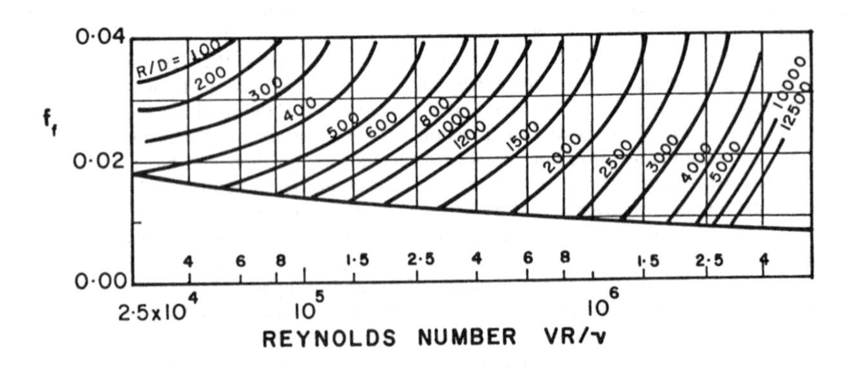

Figure 9. Flat-bed friction factor [20].

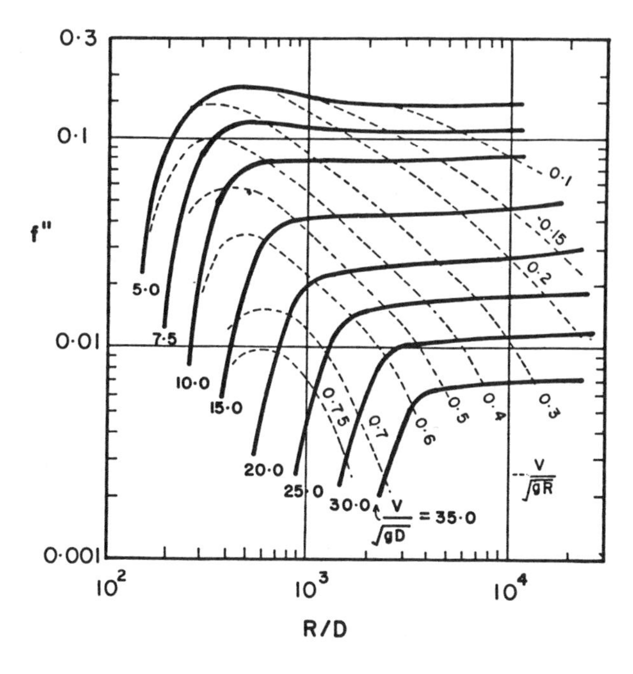

Figure 10. Form friction factors for sand-bed channels [21].

obtained previously by Kennedy [16] to eliminate the variables h/L and L/D. They finally proposed the following functional relationship

$$f'' = f''\left(\frac{V}{\sqrt{gD}}, \frac{D}{R}\right) \tag{24}$$

Using laboratory and field data, they found the preceding relationship as shown graphically in Figure 10. In this figure the abscissa represents the relative depth R/D and the ordinate gives the form friction factor f''. The curve parameters for the solid and dotted curves are $V/\sqrt{gD}$ and $V/\sqrt{gR}$, respectively.

The resistance problem can now be stated: given ν, D, R and S, find f', f'' and V. Using Alam and Kennedy's method, the answer to this problem can be found as follows

1. Assume a value of V and calculate $V/\sqrt{gD}$, R/D, and VR/ν
2. Obtain f_f and f'' from Figures 9 and 10.
3. Calculate V using $f = f_f + f''$ in the definition equation $V^2 = 8gRS/f$.
4. Compare the calculated V with the assumed value. Iterate if necessary.

A few comments are in order. In contrast to Einstein-Barbarossa's analysis, the present analysis suggests that at least two independent dimensionless variables are required to describe f''. As pointed out by Alam and Kennedy, in the region where the solid curves become almost flat, their analysis is equivalent to Einstein-Barbarossa's bar resistance curve since ψ is roughly proportional to gD/V^2. Most of the river data also falls in this region. In contrast, the laboratory data generally falls in the region where f'' varies with respect to both independent variables, R/D and $V/\sqrt{gD}$.

Since f_f represents the skin friction factor for mobile plane beds, it is certainly better to replace f' by f_f than to substitute the skin friction factor obtained from formulas for plane rigid boundaries. However, the problem of deciding f', or the more basic issue of deciphering a representation value of τ'_0 in the presence of bedforms, cannot be resolved easily without taking into account the geometry of large-scale roughness elements. These geometric effects are absent in the plane bed situation. The Kennedy's [16] theory on bedforms, and its use in eliminating the geometric variables, h and L, rest on the assumption of a sinusoidal shape for bedforms. Thus, strictly speaking, the simplified Equation 24 is more appropriate for antidunes.

Willis [22] using data from a large test channel found qualitative agreement with Alam and Kennedy's analysis. However, the dependence of f'' on the relative grain roughness was uncertain.

Engelund's Analysis

The method proposed by Engelund rests on his similarity hypothesis [23, 24]. Based on this hypothesis, he concluded that the dimensionless shear stress (also called Shield's parameter)

$$\tau_* = \frac{u_*^2}{(s-1)gD} = \frac{\tau_0}{\gamma(s-1)D} \tag{25}$$

is only a function of the grain associated dimensionless shear

$$\tau'_* = \frac{u'^2_*}{(s-1)gD} = \frac{\tau'_0}{\gamma(s-1)D} \tag{26}$$

where s = the specific gravity of the sediment particles
D = the mean diameter of sediment grains

Using flume data reported by Guy, Simons, and Richardson [25], he obtained the relationship between τ_* and τ'_*, as shown in Figure 11. In order to obtain the average shear stress τ'_0 due to grain roughness in the presence of bedforms, he proposed the following set of equations

$$\frac{V}{u'_*} = 6 + 2.5 \ln \frac{D'}{2.5D} \tag{27}$$

and

$$u'_* = \sqrt{gD'S} \tag{28}$$

where, according to Engelund, D′ is the thickness of the local boundary-layer near the bedform crest. For a given set of values of V and the mean diameter D, we can obtain u'_* or τ'_0 ($= \rho u'^2_*$) from the preceding two equations. And f' can be obtained from Equation 1.

Figure 11 can be viewed as a nondimensional version of Figure 2, if we replace abscissa V by τ'_* (instead of velocity, the shear stress due to skin friction can be taken as a measure of flow intensity) and change τ_0 to dimensionless shear τ_*. The thin line represents the situation, real or imaginary, in which the bed remains plane and does not develop any additional resistance other than the skin friction. The point $\tau_* = \tau'_* = 0.05$ represents the threshold of sediment movement as predicted by the Shield function for a rough turbulent regime. The discontinuity between the two curves represents the transition where the bedform behavior is uncertain. In this range, the bedforms are very sensitive to the variations in the physical properties of the fluid and the sediment.

Some verification of Engelund's analysis is given in Reference 15. However, the universality of the curves in Figure 11 rests heavily on his similarity hypothesis which has not been verified as yet. Data reported by Willis [22] deviate considerably, as shown by the shaded area in Figure 11.

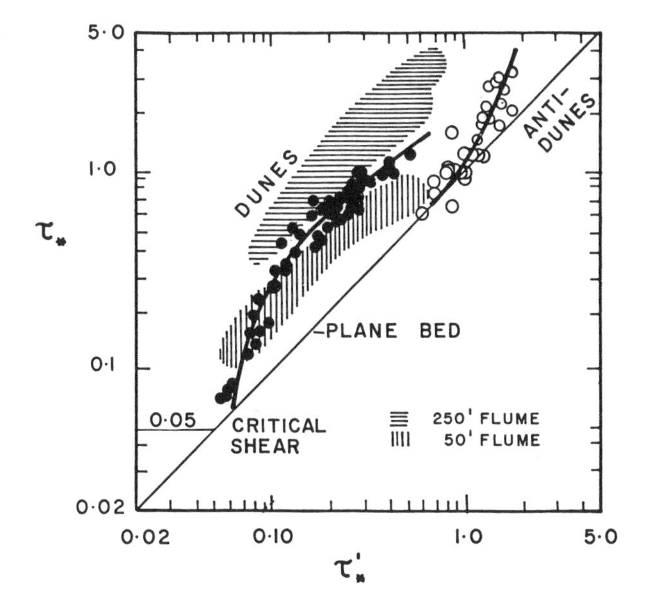

Figure 11. Relationship between dimensionless bed shear and dimensionless grain-associated shear [23].

Due to the restricted scope of the present article, only three studies have been reviewed in the preceding sections. Besides these studies, there have been other valuable contributions, notably by Raudkivi [3], Vanoni and Hwang [2], Simons and Richardson [10]. A good review can be found in Reference 15.

FORM RESISTANCE OF RIGID RIPPLES AND DUNES

In this section we shall discuss the form resistance of ripples and dunes which do not change their shape or size as they migrate downstream. It is assumed that the bedform geometry is known in advance, and it is not affected by the prevailing conditions. In this case, the form friction factor depends only on the geometry of the bedform, and on the relative depth of flow. Its magnitude can be obtained directly by integrating the pressure over a typical ripple or dune surface. The possibility of calculating the form friction factor by integrating the pressure has been suggested previously by Raudkivi [3], Mercer [27], Vanoni and Hwang [2], and others. However, the idea has only been recently carried out systematically by Haque and Mahmood [28]. We shall review their analysis in the following section and attempt to relate their findings with some of the empirical studies.

Haque-Mahmood's Analysis

Let us consider the flow taking place over an infinite sequence of bedforms of identical shape and uniform size. In this idealized situation, the roughness elements (bedforms) act independently and the form friction factor can be obtained by analyzing the flow over a typical element. Moreover, since there is no fundamental difference between ripples and dunes from the viewpoint of the form resistance, we shall not distinguish between them, unless necessary. A schematic representation of pressure variation along a two-dimensional ripple or dune surface of unit width normal to the plane of

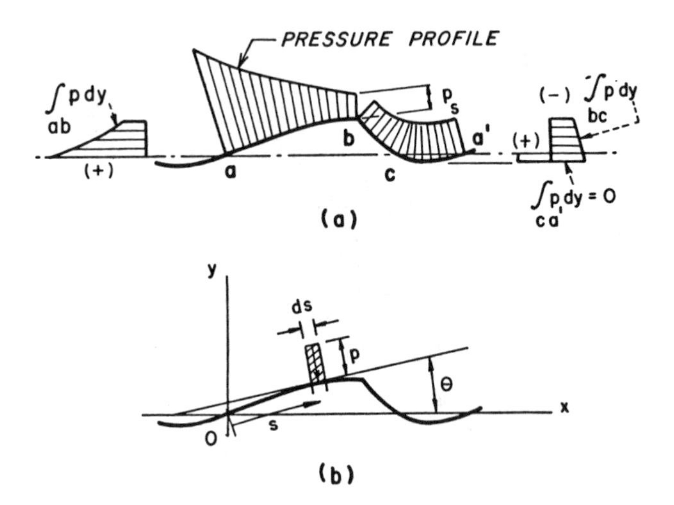

Figure 12. Schematic pressure variation on ripple surface.

paper is shown in Figure 12. By integrating the downstream component of the pressure force acting on an elementary surface ds, the drag force on a single bedform can be obtained as

$$D_p = \int_{abca'} p \sin\theta \, ds \tag{29}$$

where the integration is carried over the entire ripple surface abca'.

p = pressure

θ = the slope

s = the variable representing the arc length between the origin and a generic point (line) on the ripple surface, as shown in Figure 12.

Equation 29 can also be written as

$$D_p = \int_{ab} p\, dy + \int_{bc} p\, dy + \int_{ca'} p\, dy \tag{30}$$

in which the first, second, and third integrals represent the downstream components of forces acting, respectively, on different segments of the area whose traces in the two-dimensional space of Figure 12 are identified by ab, bc, and ca'. If we invoke Helmholtz-Kirchhoff's hypothesis of constant pressure in the separation zone in the absence of extraneous body forces, or a hydrostatic pressure distribution inside the eddy region under gravity, the last integral in Equation 30 can be dropped, and the form drag can be written as

$$D_p = \int_{ab} [(p - p_s) - \gamma(y_s - y)]\, dy \tag{31}$$

where p_s, y_s = the pressure and elevation, respectively, at the separation point, b. The preceding equation is valid both for viscous and inviscid flows. However, for inviscid flows, Bernoulli's equation can be used to replace the integrand by the kinematic pressure

$$D_p = \int_{ab} \tfrac{1}{2}\rho(U_s^2 - U^2)\, dy \tag{32}$$

where U = the magnitude of the local velocity and s refers to the separation point. The form friction factor can be obtained from the following equation

$$f'' = \frac{8(D_p/\rho L)}{V^2} \tag{33}$$

where V = the average flow velocity based on the average flow depth.

The integral in Equation 32 was obtained by analyzing the flow over bedforms of prescribed geometries. The bedform geometry was obtained previously by Mercer and Haque [29]. In their analysis, the bedform geometries were determined analytically on the basis of a two-dimensional discontinuous potential flow using the classical Helmholtz's method of free streamline. The details of analysis can be found elsewhere [29, 30]. We shall discuss here only the salient features of the theoretically admissible bedform profiles for ripples and dunes. These profiles were obtained under two requirements:

1. There should be an eddy region in the lee of each bedform.
2. The velocity gradient should be continuous at the separation point (point b in Figure 12).

The theoretical profiles meeting these requirements are shown in Figure 13. The abscissa and ordinate in this figure represent the normalized x and y coordinates defining the bedform geometry.

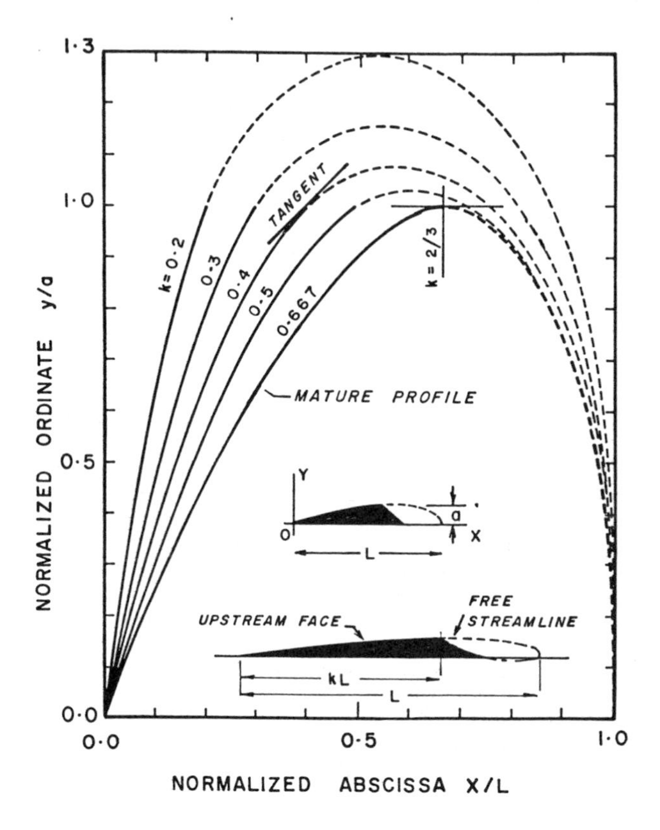

Figure 13. Theoretically admissible profiles for ripples and dunes.

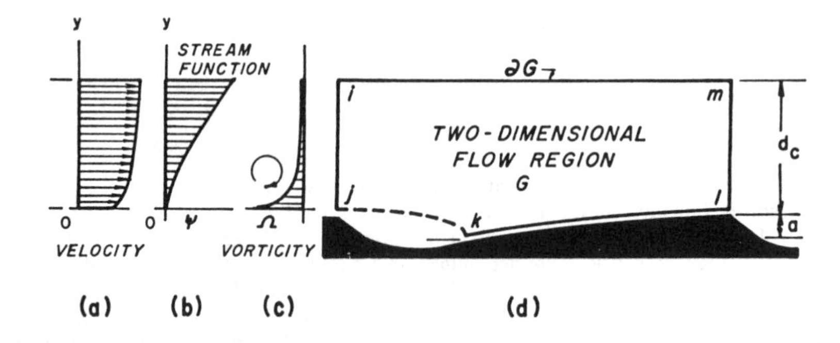

Figure 14. Flow region and boundary conditions.

The abscissa and ordinate are normalized, respectively, by the bedform length L and height a. The curve parameter $0 < k < 1$ represents the length of the upstream face relative to the bedform length (see Figure 13). By examining Figure 13, it is clear that there are an infinite number of theoretically admissible nondimensional profiles for ripples and dunes. However, the profile with $k = \frac{2}{3}$ is the only one whose surface gradient $dy/dx = 0$ at the separation point. As pointed out by Mercer and Haque [29], it is this profile which can theoretically maintain its height; all other profiles will either increase, or decrease their height with time, depending upon the case, whether k is less or more than $\frac{2}{3}$. The bedform profile corresponding to $k = \frac{2}{3}$ is referred to as the mature bedform profile. It was this mature profile whose geometry was used by Haque and Mahmood to analyze the flow in order to evaluate the integral in Equation 32. They obtained the pressure variations along the surfaces of bedforms by analyzing the flow over a number of bedforms with different steepness ratios a/L and different relative depths. The motion of fluid was described by an incompressible, rotational inviscid flow model

$$\frac{\partial^2\psi}{\partial x^2} + \frac{\partial^2\psi}{\partial y^2} = -\Omega(x, y) = -f(\psi) \text{ in G} \tag{34}$$

subject to boundary condition

$$\psi = \bar{\psi} \text{ on } \partial G \tag{35}$$

where ψ is the stream function, Ω is the vorticity, G and ∂G are the flow region and its boundary as shown typically in Figure 14. The vorticity was expressed as a function of ψ from the prescribed velocity boundary conditions at section ij (Figures 14B and C). The preceding boundary-value problem was solved numerically using the finite element method. The details of the procedure can be found in Reference 28. The results of a typical finite element analysis are shown in Figure 15. This figure corresponds to a bedform steepness $a/L = 0.06$ and a relative depth $d_c/a = 5$. The form friction factor f'' was obtained by numerically intergrating the pressure shown in Figure 15. This pressure corresponds to the integrand in Equation 32. As pointed out by Haque and Mahmood, the form friction factor depends only on the shape of the flow region G and not on its absolute size. Thus, for geometrically similar flow regions, the form friction factor is invariant. It also follows that the form friction factor is independent of the magnitude of the average flow velocity (for further details, see Reference 28). Since the geometry of the flow region G depends on two parameters only, for instance, a/L and a/d_c, the form friction factor must also depend on these two parameters only; i.e., $f'' = f''(a/L, a/d_c)$. This functional relation is shown in Figure 16, which was constructed from the results of the finite element analysis of the flow in a number of flow regions. By examining this figure, we notice that for small dunes ($L < 2d_c$), form friction factor can be expressed as

$$f'' = 4.9\left(\frac{a}{L}\right)^{1.477}\left(\frac{a}{d_c}\right)^{0.176} \simeq 5\left(\frac{a}{L}\right)^{3/2}\left(\frac{a}{d_c}\right)^{1/6} \tag{36}$$

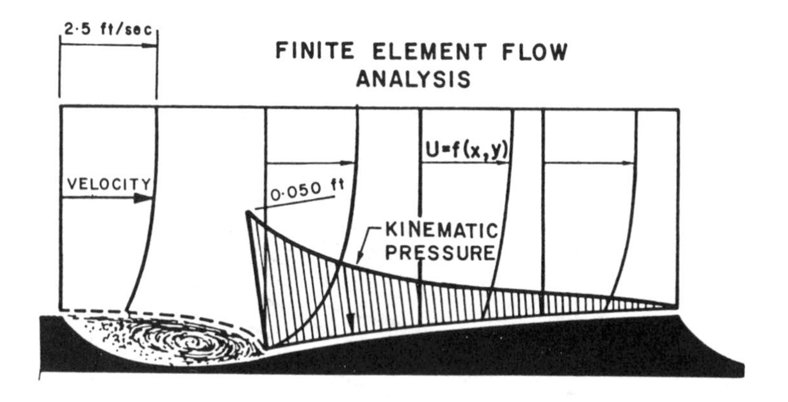

Figure 15. Typical finite element analysis of velocity profiles and pressure variation.

Thus, for small dunes ($L < 2d_c$), the form friction factor is a stronger function of the bedform steepness ratio a/L than it is of the relative roughness a/d_c. The same conclusion also follows by visually examining the curves in Figure 16.

The curves of Figure 16 are transformed into a set of new curves shown in Figure 17. In this figure, h denotes the crest-to-trough height of the bedform and d represents the average depth of flow. The transformation between Figures 16 and 17 is based on the relationship $a = 0.8h$. In Figure 17, the analytically obtained curves are compared with the experimentally obtained curves by Vanoni and Hwang [2], and with the Nikuradse equation for hydraulically rough surfaces.

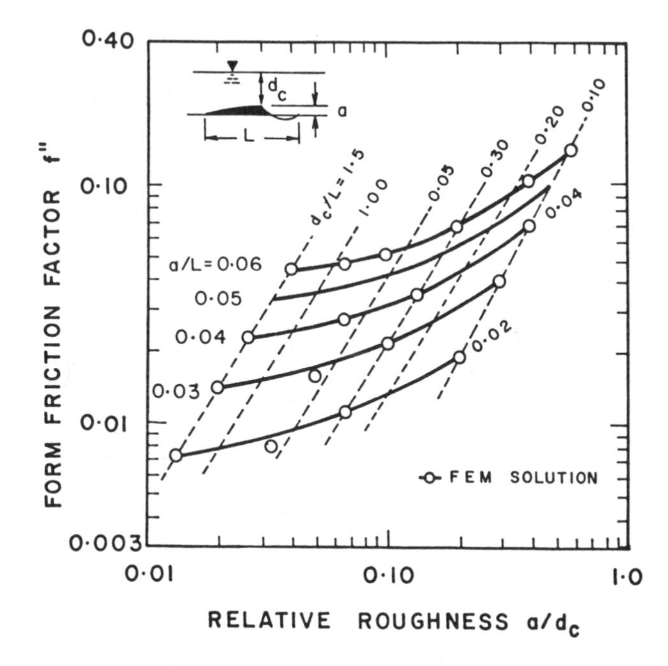

Figure 16. Relation among form friction factor, relative roughness and bedform steepness [28].

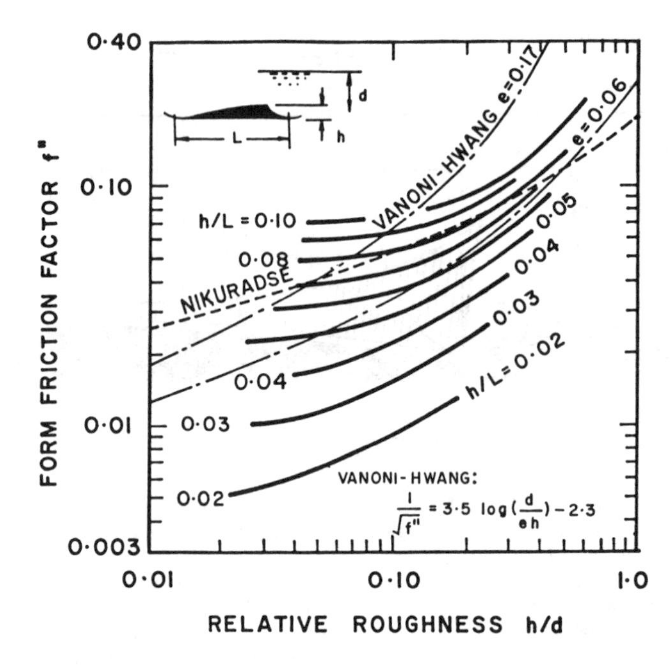

Figure 17. Comparison of analytical curves with Nikuradse's and Vanoni-Hwang's curves [28].

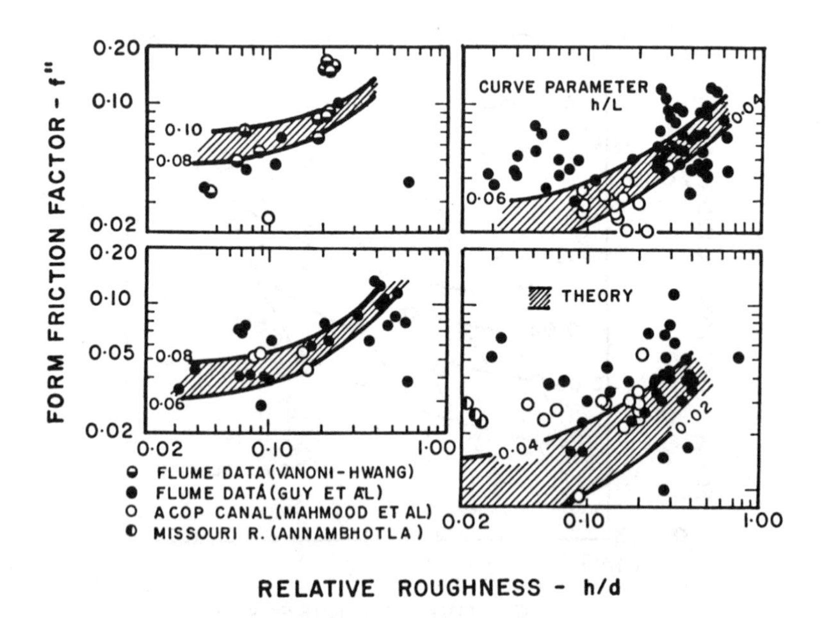

Figure 18. Comparison of the theoretical results with the observed data [28].

The analytical results are compared with the observed data in Figure 18. Each plot represents a certain range of the bedform steepness h/L. The results of larger steepness, $0.08 \leq h/L < 0.10$, are shown in the top-left area. It is evident from these figures that the trend of the theoretical curves is borne reasonably by the data. The larger scatter of the experimental data, for smaller steepness of bedforms can, in part, be attributed to the uncertainties involved in estimating the average grain roughness τ'_0, especially when the flow separates in the lee of the bedform. Since in this case, the form drag is comparable to any other sources of flow resistance, (e.g., surface waves, planform irregularities, etc.) any approximations or omissions, of these additional resistances will have noticeable errors in the estimated values of the observed f''.

FORM RESISTANCE OF NONRIGID RIPPLES AND DUNES

In this section, we shall discuss the phenomenon of form resistance in the presence of ripples and dunes whose geometry is affected by the prevailing flow conditions. Since the bedforms owe their existence to the motion of fluid, it is natural to expect that their geometry is, also, determined by the flow. As the bedforms evolve, their shape continually tends to acquire a limiting equilibrium profile, which does not change with time. When bedforms are in such a quasi-steady state, the sediment transported over their surfaces completely matches the sediment transport requirement for maintaining a steady profile.

As mentioned previously, out of all the theoretically admissible nondimensional bedform shapes shown in Figure 13, only the mature profile with $k = \frac{2}{3}$ can maintain a constant height a with time. Thus, only a mature profile can, if possible, represent a steady profile of a bedform which is in the limiting equilibrium. Neither the size nor the steepness of the mature bedform is unique. In fact, depending upon the values of the arbitrary parameters a and L (bedform height and length, Figure 13), there are infinite possibilities of selecting mature bedforms. Under a given depth of flow, and for an arbitrarily selected value of the bedform length, however, there is a unique mature profile with a unique bedform steepness a/L, which attains the limiting equilibrium. As pointed out by Haque and Mahmood [31], for a given relative depth d/L, the limiting bedform steepness (the steepness of a quasi-steady profile which is in limiting equilibrium) depends only on the nature of the relationship between the bed load and the local velocity, acting along the bed surface. If we assume that the bed load is proportional to the n-th power of this local velocity, then there is a unique relationship among the variables n, d/L, and h/L. This relationship is shown graphically in Figure 19. The abscissa and the ordinate in this figure represent the relative depth and the limiting bedform steepness, respectively. The curves in Figure 19 are obtained from an extensive analysis of the flow fields over different bedforms using the finite element formulation of an incompressible, rotational, inviscid flow model. The details of the theory of limiting steepness can be found in Reference [31]. The curves shown in this figure correspond to different assumed values of the dimensionless exponent n. The curve for $n = 6$ corresponds roughly to the Einstein-Brown bed load equation.

In the previous section, we have seen that the form friction factor for rigid bedforms depends on two independent dimensionless parameters, h/L and h/d. However, in the case of sedimentary ripples and dunes tending to acquire a limiting equilibrium profile, these two parameters are no more independent; for h, L, and d are now mutually related as shown in Figure 19. The form friction factor now depends on only one dimensionless parameter, say, h/d. This relationship is shown in Figure 20 by dotted lines for different assumed values of n, ranging from 3 to 8. These dotted curves (for $n = 3$, 6, and 8) correspond to the curves shown in Figure 19, and represent the constraint imposed on the previously independent parameters h/L and d/L, due to the limiting equilibrium requirements. In Figure 20, the family of curves shown by solid lines with the curve parameter h/L are the same curves shown in Figure 17 for the rigid bedforms with known geometry. The chain-dotted lines represent constant values of the ratio of the flow depth to the bedform length. In this figure the region extending toward the left represents smaller bedform lengths, relative to the flow depth. Likewise, the region extending toward the right, depicts larger bedforms compared to the depth of flow.

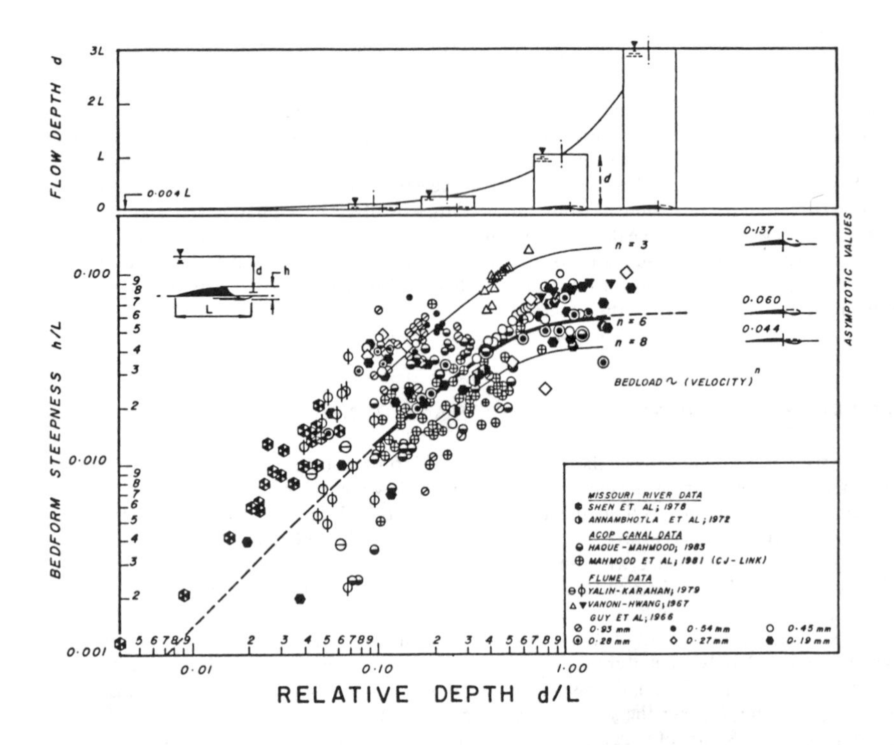

Figure 19. Constraint on geometric variables h/L and d/L for bedforms [31].

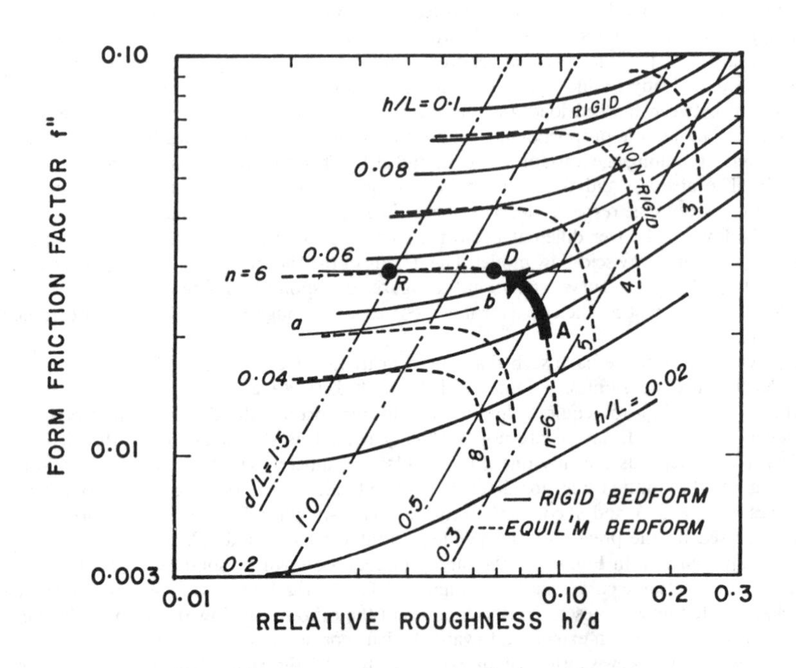

Figure 20. Bedform friction factor interaction diagram.

Let us now examine Figure 20 closely, because it promises some interesting theoretical explanations for the intriguing aspects of the bedforms and their influence on the resistance problem. To facilitate discussion, we shall assume that for sand-bed channels, f'' is roughly equal to f in the presence of lower-regime bedforms. If we now assume that $n = 6$ is a good representation in a power law for sediment transport along the bedform surface, then it follows from completely theoretical considerations that the friction factor for sand-bed channels with relatively small bedforms (length < depth) is roughly 0.03, as shown in the figure. For most natural rivers and sand-bed channels, the observed friction factor falls around this value. For a constant value of n, the friction factor f'' has a maximum value which occurs in the vicinity where the bedform lengths are roughly equal to the flow depths. Many researchers believe that the natural sand-bed channels operate around this state. In case of a uniform flow with a constant average slope, the friction factor is constant, and it is represented by a horizontal line in this figure. This line intersects the dotted curve at two points R and D as shown in this figure. Thus, in a uniform flow, there are two different lengths of bedforms, one corresponding to the point R and the other corresponding to the point D, which can exist in the channel without disturbing the uniformity of flow. The point R represents the smaller bedform, while the point D represents the larger bedform relative to the flow depth. For smaller bedforms (lengths less than the depth of flow), the dotted curves are almost horizontal, indicating that in a uniform flow a large variety of smaller bedforms with varying lengths can exist.

When a generic point A moves along a dotted curve, it represents the variations in the friction factor f'' due to changes in the geometry of a sedimentary bedform which continually maintains an equilibrium profile. When such a point moves along a solid curve, it describes the variation in f'' in the presence of rigid bedforms due to changes in the flow depth. In a region extending toward the left, a dotted curve tends to approach asymptotically a solid curve, as shown for the curve ab and the dotted curve $n = 7$. This implies that for small bedforms, an increase in the flow depth does not cause any change in the bedform steepness. Thus, f'' decreases in this case whenever an increase in the flow depth takes place. This is in agreement with the experimental observations reported in Reference 10 in regard to ripples. Now consider the larger bedforms (length greater than 2d). In this case, an increase in depth steepens the bedform and thus, results in larger f'', as shown by the arrow A. This, again, is in agreement with the experimental observations [10].

Figure 20 is similar in many respects to the figure presented by Alam and Kennedy. In fact, the abscissa in this figure becomes inversely proportional to the relative depth R/D if we assume that the bedform height bears a constant ratio to the grain diameter. Although Figure 20 explains the physics of the bedforms and their resistance behavior quite remarkably, there are, at present, no direct methods available to relate the exponent n to other pertinent flow and sediment variables. In this regard, the need for additional research could hardly be overemphasized.

Despite these comments, we hope that the analytical development presented in the foregoing will provide a useful theoretical framework for future studies.

Acknowledgment

This chapter is the outgrowth of several research studies undertaken by the author together with Prof. Khalid Mahmood at the George Washington University. To Dr. Albert G. Mercer, a special note of thanks is due for his guidance and encouragement in the early years. The financial support from the National Science Foundation is gratefully acknowledged.

NOTATION

A	cross-sectional area
A_p	projected area
a	distance
C_D	drag coefficient
D	mean grain diameter
D_p	pressure drag
D'	thickness of local boundary layer
D_{35}	sediment size
f	friction factor
f''	form friction
G	flow region

g	gravitational acceleration	S_f	slope of energy grade line
h	height	S	channel slope
k_s	grain roughness diameter	s	surface contour
L	length of channel segment	U	local velocity
N	number of bedforms	u''_*	friction velocity
P	wetted perimeter	V	flow velocity
p	pressure	x	correction factor for viscous effects
R	hydraulic radius	y_s	elevation

Greek Symbols

γ	specific weight	τ''_0	average bed shear stress
θ	slope	τ'_*	dimensionless shear
ν	kinematic viscosity	ψ	dimensionless variable defined by Equation 16
ρ	density	ψ	stream function
τ	shear stress	Ω	vorticity
τ_c	critical shear		

REFERENCES

1. Brooks, N. H., "Mechanics of Stream With Movable Beds of Fine Sands," *Transactions, ASCE*, Vol. 123 (1958), pp. 526–594
2. Vanoni, V. A., and Hwang, L., "Relationship Between Bed Forms and Friction in Streams," *Journal of Hydraulics Division ASCE*, Vol. 93, HY3 (May 1967), pp. 121–144.
3. Raudkivi, A. J., "Analysis of Resistance in Fluvial Channels," *Journal of Hydraulics Division*, Vol. 93, HY5 (September 1967), pp. 73–84.
4. Einstein, H. A., and Barbarossa, N., "River Channel Roughness," *Transactions, ASCE*, Vol. 117, No. 2528 (1952), pp. 1121–1146.
5. Laursen, E. M., "The Total Sediment Load of Streams," *Journal of Hydraulics Division, ASCE*, HY1 (Feb. 1958), pp. 1530–1536.
6. O'Loughlin, E. M., "Resistance To Flow Over Boundaries With Small Roughness Concentrations," Ph.D. dissertation presented to the University of Iowa, Iowa City, Iowa, Aug. 1965.
7. Raudkivi, A. J., "Study of Sediment Ripple Formation." *Journal of Hydraulics Division, ASCE*, Vol. 89, HY6 (Nov. 1963), pp. 15–33.
8. Roberson, J. A., and Chen, C. K., "Flow in Conduits With Low Roughness Concentration," *Journal of Hydraulics Division, ASCE*, Vol. 96, HY4, (April 1970), pp. 941–957.
9. Sayre, W. W., and Albertson, M. L., "Roughness Spacing In Rigid Open Channels," *Transactions, ASCE*, Vol. 128, Part I, (1963), pp. 343–372.
10. Simons, D. B., and Richardson, E. V., "Resistance To Flow In Alluvial Channels," Professional Paper 422J, U.S. Geological Survey, Washington, DC 1966.
11. Gilbert, G. K., "The Transport of Debris by Running Water," U.S. Geological Survey, Professional Paper 86, 1914.
12. Graf, W. H., *Hydraulics of Sediment Transport*, McGraw-Hill Inc., 1971.
13. Yalin, M. S., *Mechanics of Sediment Transport*, Pergamon Press, Ltd., First Edition, 1972.
14. Raudkivi, A. J., *Loose Boundary Hydraulics*, Pergamon Press, Ltd., First Edition, 1967.
15. Vanoni, V. A., (Ed.), *Sedimentation Engineering*, American Society of Civil Engineers, New York, 1975.
16. Kennedy, J. F., The Mechanics of Dunes and Antidunes in Erodible-bed Channels, *J. Fluid Mechanics*, Vol. 16, (1963) pp. 521–544.
17. Nordin, C. F., "Aspects of Flow Resistance and Sediment Transport, Rio Grande near Bernalillo, New Mexico," U.S. Geological Survey Water Supply Paper 1498-H, 1964.
18. Vanoni, V. A., and Brooks, N. H., "Laboratory Studies of the Roughness and Suspended Load of Alluvial Streams," Calif. Inst. Tech., MRD Sediment Serv. No. 11, 1957.

19. Garde, R. J., and Raju, K. G. R., "Resistance Relationships for Alluvial Channel Flow," *ASCE, Journal of Hydraulics Div.*, Vol. 92, HY4 (July 1966), pp. 77–100.
20. Lovera, F., and Kennedy, J. F., "Friction-Factors for Flat-Bed Flows in Sand Channels," *ASCE, Journal of Hydraulics Div.*, Vol. 95, HY4 (July 1969), pp. 1227–1234.
21. Alam, M. Z., and Kennedy, J. F., "Friction Factors for Flow in Sand-Bed Channels," *ASCE, Journal of Hydraulics Div.*, Vol. 95, HY6 (Nov. 1969), pp. 1973–1992.
22. Willis, J. C., "Flow Resistance in Large Test Channel," *ASCE, Journal of Hydraulic Engineering.* Vol. 109, No. 12 (Dec. 1983), pp. 1755–1770.
23. Engelund, F., "Hydraulic Resistance of Alluvial Streams," *ASCE, Journal of Hydraulics Div.*, Vol. 92, HY2 (March 1966), pp. 315–326.
24. Engelund, F., "Hydraulic Resistance of Alluvial Streams," *Closure, ASCE, Journal of Hydraulics Division* (July 1967), pp. 287–296.
25. Guy, H. P., Simons, D. B., and Richardson, E. V., "Summary of Alluvial Channel Data From Flume Experiments, 1956–61," Geological Survey Professional Paper No. 462-I, Washington, DC 1966.
26. Engelund, F., and Fredsoe, J., "Sediment Ripples and Dunes," *Annual Review of Fluid Mechanics*, M. V. Dyke (Ed.) Vol. 14 (1982).
27. Mercer, A. G., "Characteristics of Sand Ripples in Low Froude Number Flow ," Ph.D. dissertation, University of Minnesota, June 1964.
28. Haque, M. I., and Mahmood, K., "Analytical Determination of Form Friction Factor," *ASCE, Journal of Hydraulic Engineering*, Vol. 109, No. 4 (April 1983), pp. 590–610.
29. Mercer, A. G., and Haque, M. I., "Ripple Profiles Modeled Mathematically," *ASCE, Journal of Hydraulics Division*, Vol. 99, HY3 (March 1973), pp. 441–459.
30. Haque, M. I., "Analytically Determined Ripple Shapes," M. S. Thesis, Colorado State University, Fort Collins, Colorado, December 1970.
31. Haque, M. I., and Mahmood, K., "Analytical Study on Steepness of Ripples and Dunes," submitted for publication to *ASCE, Journal of Hydraulic Engineering*, December 1983.

CHAPTER 5

UNSTABLE TURBULENT CHANNEL FLOW

J. Berlamont

Laboratorium voor Hydraulica
Katholieke Universiteit Te Leuven
Heverlee, Belgium

CONTENTS

INTRODUCTION, 98

GOVERNING EQUATIONS, 98

STABILITY CRITERION, 101

FRICTION FACTOR, 109

FINITE AMPLITUDE ROLL WAVES, 112
Discontinuous Solution, 112
Cnoidal Wave Solution, 113
Approximate Solution from the Linearized Equations, 115
Higher-Order Theory, 115

NOTATION, 118

REFERENCES, 119

INTRODUCTION

In steep-sloped prismatic channels, with rectangular cross section, the theory of steady gradually varied turbulent flow is not applicable when the Froude number exceeds some critical value. Instead of obtaining uniform flow conditions some distance away from the channel inlet the flow becomes unstable: waves of various lengths, amplitudes and phase velocities appear in the channel. These waves, traveling downstream and occasionally taking over each other, are usually called roll waves [8, 13, 14, 21]. Roll waves were first described by Cornish [13]. A number of questions arise about unstable turbulent flow:

- What is the stability criterion? Which factors affect the change from stable to unstable flow?
- How is the friction factor affected by the flow instability
- Which are the characteristics of the roll waves, developing naturally: wave heights, celerity, and wave-length?

The basic equations are first presented.

GOVERNING EQUATIONS

The general two-dimensional Reynolds equations for turbulent flow in open channels, taking into account the vertical velocity and acceleration, are:

$$\frac{1}{\rho}\frac{\partial p}{\partial x} = -\left(\frac{\partial u}{\partial t} + u\frac{\partial u}{\partial x} + v\frac{\partial u}{\partial y}\right) + g\sin\theta + \frac{1}{\rho}\frac{\partial \tau_{xx}}{\partial x} + \frac{1}{\rho}\frac{\partial \tau_{xy}}{\partial y} \tag{1}$$

$$\frac{1}{\rho}\frac{\partial p}{\partial y} = -\left(\frac{\partial v}{\partial t} + u\frac{\partial v}{\partial x} + v\frac{\partial v}{\partial y}\right) - g\cos\theta + \frac{1}{\rho}\frac{\partial \tau_{yx}}{\partial x} + \frac{1}{\rho}\frac{\partial \tau_{yy}}{\partial y} \tag{2}$$

in which the x-axis is chosen along the channel bottom and the y-axis is perpendicular to it; and

where g = acceleration of gravity
$p(x, y, t)$ = pressure at point (x, y), at time t
$u(x, y, t)$ = velocity component in the x direction
$v(x, y, t)$ = velocity component in the y direction
θ = slope angle of the channel
ρ = mass density of the fluid
$\tau_{xx}, \tau_{xy}, \tau_{yy}, \tau_{yx}$ = Reynolds stresses

Integrating the continuity equation for an incompressible fluid

$$\frac{\partial u}{\partial x} + \frac{\partial v}{\partial y} = 0 \tag{3}$$

over the cross-sectional area, A, with the help of the kinematic surface condition gives the continuity equation for the mean flow

$$\frac{\partial(UH)}{\partial x} + \frac{\partial H}{\partial t} = 0 \tag{4}$$

where $U(x, t)$ = mean velocity over the cross section at x at time t
$H(x, t)$ = water depth at x at time t.

Equations 1 and 2 may be integrated over a cross section by assuming the following.

1. The flow is gradually varied, i.e., $\partial\tau_{xy}/\partial x = 0$.
2. $\tau_{xx} = \tau_{yy}$.
3. The expression for the mean shear stress on the channel walls is, even in unsteady flow:

$$(\tau_{xy})_{y=0} = \tau = \rho\frac{U^2}{8}f(U, H) \tag{5}$$

   in which f = the Darcy-Weisbach friction factor.
4. The velocity profiles are similar in all cross sections:

$$u(x, y, t) = F(z)U(x, t) \tag{6}$$

   in which $z = y/H$ and F is a shape function. This assumption is valid both in the case of laminar flow (parabolic profile) and turbulent flow (logarithmic profile). From $\int_H u(x, y, t)\,dH = UH$, it follows that $\int_0^1 F(z)\,dz = 1$.
5. Products of lower derivatives of U and H, with respect to x and t, are small as compared with the higher derivatives [25].

From Equations 3 and 6 it follows that

$$v = -\int_0^y \frac{\partial(FU)}{\partial x}dy = -\int_0^y\left(F\frac{\partial U}{\partial x} - \frac{U}{H}z\frac{\partial H}{\partial x}\frac{\partial F}{\partial z}\right)dy \tag{7}$$

Equation 2 becomes, with assumption 1:

$$\frac{\partial v}{\partial t} + u\frac{\partial v}{\partial x} + v\frac{\partial v}{\partial y} = -\frac{1}{\rho}\frac{\partial(p-\tau_{yy})}{\partial y} - g\cos\theta \tag{8}$$

Substituting Equation 7 into Equation 8, with the help of Equation 4 and assumption 5, one obtains, after integrating Equation 8 between y and H:

$$\frac{p-\tau_{yy}}{\rho} = g(H-y)\cos\theta + \frac{\partial^2 H}{\partial t^2}\int_y^H dy\int_0^z F\,dz + U\frac{\partial^2 H}{\partial x\,\partial t}\int_y^H dy\left(\int_0^z F\,dz + \int_0^z \frac{\partial F}{\partial z}z\,dz + F\int_0^z F\,dz\right) + U^2\frac{\partial^2 H}{\partial x^2}\int_y^H dy\left(F\int_0^z F\,dz + F\int_0^z \frac{\partial F}{\partial z}z\,dz\right) \tag{9}$$

After integrating Equation 1 over the cross section, A, substituting Equation 9, and taking assumption 2 into account, one obtains the momentum equation:

$$\frac{\partial U}{\partial t} + \alpha U\frac{\partial U}{\partial x} + (1-\alpha)\frac{U}{H}\frac{\partial H}{\partial t} + g\frac{\partial H}{\partial x}\cos\theta + \frac{HU^2}{3}\left(\frac{C_1}{U^2}\frac{\partial^3 H}{\partial x\,\partial t^2} + \frac{C_2}{U}\frac{\partial^3 H}{\partial x^2\,\partial t} + C_3\frac{\partial^3 H}{\partial x^3}\right) = g\sin\theta - \frac{\tau}{\rho R} \tag{10}$$

where R = the hydraulic radius
C_1, C_2, and C_3 = shape factors of the velocity profile, defined by

$$C_1 = 3\int_0^1 dz\int_z^1 dz\int_0^z F\,dz \tag{11}$$

$$C_2 = 3\int_0^1 dz\int_z^1 dz\left(\int_0^z F\,dz + \int_0^z \frac{\partial F}{\partial z}z\,dz + F\int_0^z F\,dz\right) \tag{12}$$

$$C_3 = 3\int_0^1 dz\int_z^1 F\,dz\left(\int_0^z F\,dz + \int_0^z \frac{\partial F}{\partial z}z\,dz\right) \tag{13}$$

$$\alpha = \frac{\int_0^1 u^2\,dz}{U^2} = \int_0^1 F^2\,dz = \text{the velocity-distribution coefficient.} \tag{14}$$

Equation 10 accounts both for the effect of the curvature of the stream lines (and thus of the vertical acceleration and the nonhydrostatic pressure distribution), and the nonuniformity of the velocity distribution.

Dropping the third derivatives, Equation 10 reduces to the "shallow water" equation. Equation 9 then reduces to the hydrostatic law.

On the contrary, assuming a uniform velocity distribution [$F(z) = 1$, $\alpha = 1$, $C_1 = C_3 = 1$, and $C_2 = 2$], but taking the higher derivatives into account, one finds the momentum equation, obtained by Keulegan [25] and Rodenhuis [33]. Equation 9 then reduces to:

$$\frac{p-\tau_{yy}}{\rho} = g(H-y)\cos\theta + U^2\frac{H}{2}(1-z^2)\left(\frac{1}{U^2}\frac{\partial^2 H}{\partial t^2} + \frac{2}{U}\frac{\partial^2 H}{\partial x\,\partial t} + \frac{\partial^2 H}{\partial x^2}\right) \tag{15}$$

Table 1
Typical Values of Constants C_1, C_2, and C_3

Velocity Profile	α	C_1	C_2	C_3
Parabolic	1.20	0.825	2.496	1.864
Uniform	1	1	2	1
Trapezoidal	1.03	0.977	2.068	1.094
	1.05	0.961	2.113	1.161
	1.10	0.921	2.227	1.341
Triangular	1.33	0.750	2.700	2.400

The coefficients C_1, C_2, and C_3 can be calculated for different values of α [6], if the turbulent velocity profile is approximated by a "trapezoidal" one (Table 1).

Assuming a logarithmic velocity profile, Iwasa has shown [19] that:

$$\alpha = 1 + 6.25\frac{\tau_0}{\rho U_0^2} = 1 + 0.781 f_0 \tag{16}$$

in which the subscript 0 refers to values in steady uniform flow.

STABILITY CRITERION

When searching for a stability criterion, one should find out under which conditions an infinitesimal small amplitude disturbance wave will grow or decay with time. Equations 4 and 10 can be linearized by putting: $U = U_0 + U'$ ($U' \ll U_0$); $H = H_0 + H'$ ($H' \ll H_0$); $R = R_0 + R'$ ($R' \ll R_0$); and $\tau = \tau_0 + \tau'$ ($\tau' \ll \tau_0$). Thus

$$\frac{\partial U'}{\partial t} + \alpha(U_0 + U')\frac{\partial U'}{\partial x} + (1 - \alpha)\frac{(U_0 + U')}{(H_0 + H')}\frac{\partial H'}{\partial t} + g\frac{\partial H'}{\partial x}\cos\theta$$
$$+ \frac{(H_0 + H')(U_0 + U')^2}{3}\left[\frac{C_1}{(U_0 + U')^2}\frac{\partial^3 H'}{\partial x\,\partial t^2} + \frac{C_2}{(U_0 + U')}\frac{\partial^3 H'}{\partial x^2\,\partial t} + C_3\frac{\partial^3 H'}{\partial x^3}\right]$$
$$= g\sin\theta - \frac{(\tau_0 + \tau')}{\rho(R_0 + R')} \tag{17}$$

$$\frac{\partial[(H_0 + H')(U_0 + U')]}{\partial x} + \frac{\partial H'}{\partial t} = 0 \tag{18}$$

Neglecting the product $U'H'$, Equation 22 becomes:

$$H_0\frac{\partial U'}{\partial x} + U_0\frac{\partial H'}{\partial x} + \frac{\partial H'}{\partial t} = 0 \tag{19}$$

Differentiating Equation 17 with respect to x, and neglecting products and powers of U' and H' greater than 1, one obtains, with the help of assumption 5 and Equation 19

$$\left(g\cos\theta - \alpha\frac{U_0^2}{H_0}\right)\frac{\partial^2 H'}{\partial x^2} - 2\alpha\frac{U_0}{H_0}\frac{\partial^2 H'}{\partial x\,\partial t} - \frac{1}{H_0}\frac{\partial^2 H'}{\partial t^2} + \frac{H_0 U_0^2}{3}\left(\frac{C_1}{U_0^2}\frac{\partial^4 H'}{\partial x^2\,\partial t^2} + \frac{C_2}{U_0}\frac{\partial^4 H'}{\partial x^3\,\partial t} + C_3\frac{\partial^4 H'}{\partial x^4}\right)$$
$$= -\frac{1}{\rho R_0}\frac{\partial \tau}{\partial x} + \frac{\tau_0}{\rho R_0^2}\frac{\partial R}{\partial x} \tag{20}$$

Applying a Taylor series to $\tau(U, H)$ and $R(H)$, and dropping the higher order terms, the right side of the equal sign in Equation 20 may be written as:

$$-\frac{1}{\rho R_0}\left(\frac{\partial\tau}{\partial U}\right)_0\frac{\partial U'}{\partial x}-\frac{1}{\rho R_0}\left(\frac{\partial\tau}{\partial H}\right)_0\frac{\partial H'}{\partial x}+\frac{\tau_0}{\rho R_0^2}\left(\frac{\partial R}{\partial H}\right)_0\frac{\partial H'}{\partial x}$$

By introducing Equation 23, one obtains:

$$=\frac{1}{\rho R_0}\left(\frac{\partial\tau}{\partial U}\right)_0\left(\frac{U_0}{H_0}\frac{\partial H'}{\partial x}+\frac{1}{H_0}\frac{\partial H'}{\partial t}\right)-\frac{1}{\rho R_0}\left(\frac{\partial\tau}{\partial H}\right)_0\frac{\partial H'}{\partial x}+\frac{\tau_0}{\rho R_0^2}\left(\frac{\partial R}{\partial H}\right)_0\frac{\partial H'}{\partial x}$$

$$=\left[\frac{U_0}{\rho H_0 R_0}\left(\frac{\partial\tau}{\partial U}\right)_0-\frac{1}{\rho R_0}\left(\frac{\partial\tau}{\partial H}\right)_0+\frac{\tau_0}{\rho R_0^2}\left(\frac{\partial R}{\partial H}\right)_0\right]\frac{\partial H'}{\partial x}+\frac{1}{\rho R_0 H_0}\left(\frac{\partial\tau}{\partial U}\right)_0\frac{\partial H'}{\partial t}$$

The linear differential equation, Equation 20, has a solution of the form:

$$H'=\eta\exp[\beta i(x-ct)] \tag{21}$$

where $\beta = 2\pi/\lambda$, the wave number;
λ = the wave length;
$c = c_r + ic_i$
c_r = the phase velocity
$i = \sqrt{-1}$
η = the infinitesimal small amplitude of the disturbance wave

Introducing Equation 21 in Equation 20, one obtains, when putting both the real and the imaginary part equal to zero [6]:

$$\left(\alpha-\frac{1}{F^2}\right)-2\alpha C_r-C_i^2+C_r^2+\frac{B^2}{3}[C_3-C_2C_r+C_1(C_r^2-C_i^2)]$$
$$-\frac{1}{B}\left(\frac{H_0}{R_0}\right)\left[\frac{1}{\rho U_0}\left(\frac{\partial\tau}{\partial U}\right)_0\right]C_i=0 \tag{22}$$

$$2C_i(C_r-\alpha)B+\frac{B^3}{3}(2C_1C_r-C_2)C_i-\left[\frac{1}{\rho U_0}\left(\frac{\partial\tau}{\partial U}\right)_0\right]\frac{H_0}{R_0}+\frac{H_0}{R_0}\left[\frac{1}{\rho}\frac{H_0}{U_0^2}\left(\frac{\partial\tau}{\partial H}\right)_0\right]$$
$$-\left(\frac{H_0}{R_0}\right)^2\left(\frac{\partial R}{\partial H}\right)_0\left(\frac{1}{\rho}\frac{\tau_0}{U_0^2}\right)+\frac{H_0}{R_0}\left[\frac{1}{\rho U_0}\left(\frac{\partial\tau}{\partial U}\right)_0\right]C_r=0 \tag{23}$$

where $F=\frac{U_0}{\sqrt{gH_0\cos\theta}}$, the Froude number

$C_i=\frac{c_i}{U_0}$, and $C_r=\frac{c_r}{U_0}$, the dimensionless wave celerity

$B=\beta H_0=\frac{2\pi}{\Lambda}$, the dimensionless wave number

$\Lambda=\frac{\lambda}{H_0}$, the dimensionless wave length and, with assumption 3:

$$\tau = \rho \frac{U^2}{8} f(U, H)$$

$$\left(\frac{\partial \tau}{\partial U}\right)_0 = \rho \frac{U_0^2}{8}\left(\frac{\partial f}{\partial U}\right)_0 + \rho \frac{U_0}{4} f_0$$

$$\left(\frac{\partial \tau}{\partial H}\right)_0 = \rho \frac{U_0^2}{8}\left(\frac{\partial f}{\partial H}\right)_0 \tag{24}$$

in which f_0 = value of f in uniform flow.

The friction losses are best taken into account by using the White-Colebrook-Thijsse formula [37]:

$$\frac{1}{\sqrt{f}} = -2.03 \log_{10}\left(\frac{\frac{k_s}{R}}{12.20} + \frac{3.03}{R\sqrt{f}}\right) \tag{25}$$

in which k_s/R = the equivalent relative roughness of the channel; R = the hydraulic radius; Re = 4 (UR/ν), the Reynolds number; and ν = the kinematic viscosity of the liquid.

In a smooth channel ($k_s = 0$), Equation 25 reduces to:

$$\frac{1}{\sqrt{f}} = 2.03 \log_{10} R\sqrt{f} - 0.977 \tag{26}$$

The expressions for $(\partial f/\partial U)_0$ and $(\partial f/\partial H)_0$ are summarized in Table 2 for different friction formulas.

Using the White-Colebrooke-Thijsse formula, a function ϕ is introduced:

$$\phi = \frac{2.671}{\left(\frac{\frac{k_s}{R}}{12.20} + \frac{3.03}{R\sqrt{f_0}}\right)R} \tag{27}$$

For hydraulic smooth surfaces:

$$\phi = 0.882\sqrt{f_0} \tag{28}$$

The stability criterion is obtained by putting $C_i = 0$ in Equations 22 and 23. It yields the critical Froude number, F_c, and the celerity, $C_{r,c}$, of an infinitesimal small amplitude disturbance wave at $F = F_c$. The flow will be unstable if $F > F_c$ ($C_i > 0$). Then the amplitude of the disturbance wave will increase with time.

Taking Equation 24 into account, one obtains:

$$\left(\alpha - \frac{1}{F_c^2}\right) - 2\alpha C_{r,c} + C_{r,c}^2 + \frac{B^2}{3}(C_3 - C_2 C_{r,c} + C_1 C_{r,c}^2) = 0 \tag{29}$$

$$\frac{U_0}{8}\left(\frac{\partial f}{\partial U}\right)_0 - \frac{H_0}{8}\left(\frac{\partial f}{\partial H}\right)_0 + \left[2 + \left(\frac{H_0}{R_0}\right)\left(\frac{\partial R}{\partial H}\right)_0\right]\frac{f_0}{8} = \left[\frac{U_0}{8}\left(\frac{\partial f}{\partial U}\right)_0 + \frac{f_0}{4}\right]C_{r,c} \tag{30}$$

Table 2
$(\partial f/\partial U)_0$ and $(\partial f/\partial H)_0$ for Different Friction Formulas

Friction Formula	$(\partial f/\partial U)_0$	$(\partial f/\partial H)_0$
Chézy: $f = \text{constant} = 8g/C^2$	0	0
Manning: $f = 8gn^2/R^{1/3}$	0	$-(1/3)(f_0/R_0)(\partial R/\partial H)_0$
White-Colebrook-Thijsse (Equation 31)	$(-\phi/1 + \phi)2(f_0/U_0)$	$(-\phi/1 + \phi)2(f_0/R_0)(\partial R/\partial H)_0$
Power law (Equation 37) $f = 8gR^{1-(1+b)(2/a)} \sin \theta^{1-2m/a} K^{2/a}$	0	$\left[1 - (1 + b)\frac{2}{a}\right]\frac{f_0}{R_0}(\partial R/\partial H)_0$
Laminar: $f = 64/R$	$-f_0/U_0$	$-f_0/R_0(\partial R/\partial H)_0$

To solve Equation 30 for $C_{r,c}$, a function M is defined:

$$M = 1 - R_0\left(\frac{\partial P}{\partial A}\right)_0 \tag{31}$$

in which P = the wetted perimeter.

Since for a rectangular section of width B:

$$P = B + \frac{2A}{B} \quad \text{and} \quad \left(\frac{\partial P}{\partial A}\right) = \frac{\partial\left(\frac{P}{B}\right)}{\partial H} = \frac{\partial\left(\frac{H}{R}\right)}{\partial H} = \frac{1}{R} - \frac{H}{R^2}\frac{\partial R}{\partial H}$$

Equation 31 then reduces to:

$$M = \frac{H_0}{R_0}\left(\frac{\partial R}{\partial H}\right)_0 = \frac{1}{1 + \dfrac{2H_0}{B}} \qquad 0 < M < 1 \tag{32}$$

For a channel of infinite width, M = 1. The value of $C_{r,c}$ depends upon the friction formula chosen and is computed from Equation 30 as given in Table 3.

The critical Froude number, F_c, can be obtained from Equation 29, by substituting the results of Table 3. One obtains, using the White-Colebrook formula:

$$\frac{1}{F_c^2} = \alpha\left(1 + \frac{C_3}{\alpha}\frac{B^2}{3}\right) - 2\alpha\left(1 + \frac{C_2}{2\alpha}\frac{B^2}{3}\right)\left[1 + \frac{M}{2}(3\phi + 1)\right] + \left[1 + \frac{M}{2}(3\phi + 1)\right]^2\left(1 + C_1\frac{B^2}{3}\right) \tag{33}$$

It is seen that the critical Froude number depends upon

1. The shape of the velocity profile (α, C_1, C_2, C_3).
2. The Reynolds number, R, and a parameter signifying the relative roughness, ϕ
3. A parameter representing the channel width, M
4. The dimensionless wave number, B.

For $B = 0$ (initial disturbance with infinite wave length), one finds:

$$F_c = \frac{1}{\left\{\alpha - 2\alpha\left[1 + \frac{M}{2}(3\phi + 1)\right] + \left[1 + \frac{M}{2}(3\phi + 1)\right]^2\right\}^{1/2}} \tag{34}$$

Table 3
Values of Critical Dimensionless Wave Celerity, $C_{r,c}$ for Different Friction Formulas (Rectangular Cross Section)

Friction Formula	$C_{r,c}$
Chézy	$1 + (M/2)$
Manning	$1 + (2/3)M$
White-Colebrook	$1 + (M/2)(3\phi + 1)$
Power law	$1 + (1/a)(1 + b)M$
Laminar	$1 + 2M$

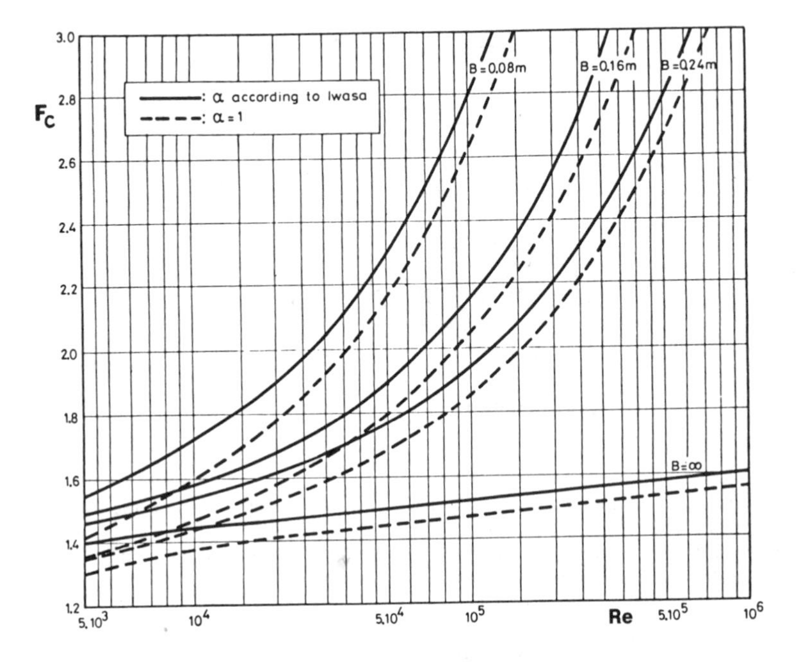

Figure 1. Critical Froude number, F_c, as function of Reynolds number, Re, and channel width, B, for Smooth Channel and $B = 0$.

Figure 1 shows the critical Froude number, F_c, as a function of R and the channel width, B, as computed with Equation 34 for a smooth channel and $B = 0$ and both for $\alpha = 1$ and α computed according to Equation 16 by Iwasa. For a given value of R, one obtains a marked increase of F_c for narrow channels, as compared with a channel of infinite width. It should be noted that, when a scale model is built at constant F, a stable flow in the model does not necessarily imply the same for the prototype.

For a particular value of B (e.g. B = 0.24 m (10 ft) in Figure 2) curves θ = constant and Q = constant can be drawn in the (Re, F) plane. It is seen from Figure 2 that for a given channel width, a minimum slope exists below which no flow instability occurs. For a given discharge, the flow turns from a stable to an unstable regime at a critical Froude number obtained by increasing the channel slope. For a fixed channel slope, instability may disappear for large values of the discharge Q (or Re).

Some experimental evidence for Figure 2 was provided by Berlamont [6]. Especially for the larger discharges and steeper slopes a minimum channel length is required for the development of roll waves [6, 17, 28], so that flows which should be unstable according to Equation 34 (Figures 1 and 2) may appear to be stable in relatively short flumes.

The effect of the channel roughness on F_c (through ϕ in Equation 34) is illustrated in Figure 3 for B = 0.24 m (10 ft). For a given Reynolds number F_c increases with the relative channel roughness.

The effect of $B \neq 0$ consists in a marked decrease of F_c if $B > 0.6$ or $\Lambda < 10$; this is for very short wave lengths. F_c tends to 0 for Λ tending to 0. Instabilities with very short wave lengths are nothing else but turbulence. F_c is seen to increase with R, α, and ks/R, and to decrease with increasing channel width B. Instability is more likely to occur in wide smooth channels, and at relatively moderate Reynolds numbers. The critical Froude number can also be calculated using an empirical friction law.

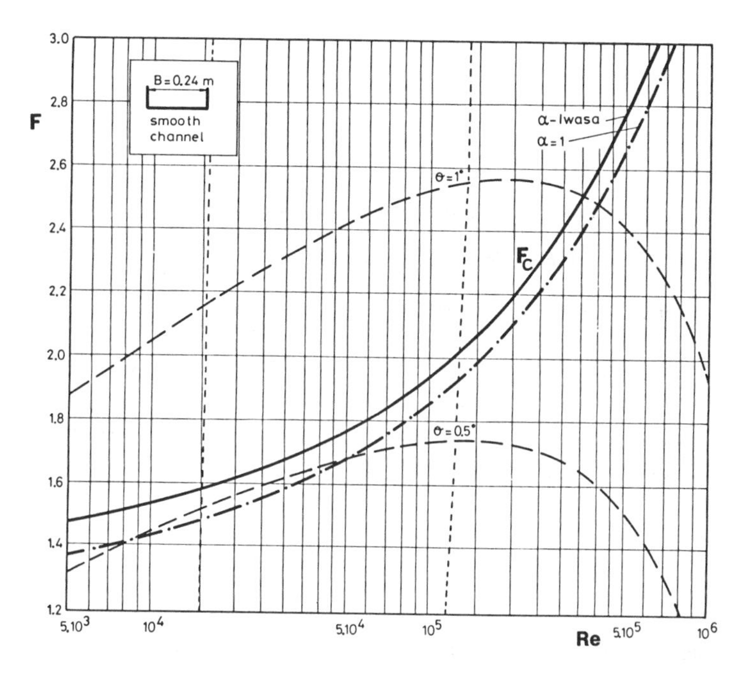

Figure 2. Computed values of Reynolds and Froude number for $B = 0.24$ m, $B = 0$, and smooth channel.

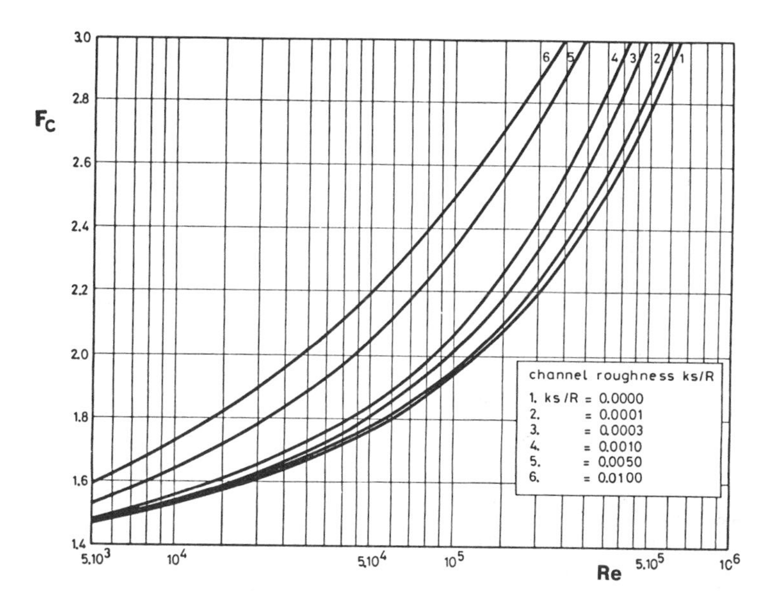

Figure 3. Influence of channel roughness on critical Froude number, F_c (α according to Iwasa).

Using the Chézy formula, one finds:

$$F_c = \frac{1}{\left[\alpha - 2\alpha\left(1 + \frac{M}{2}\right) + \left(1 + \frac{M}{2}\right)^2\right]^{1/2}} \quad (35)$$

If $\alpha = 1$, $F_c = 2/M$, and $F_c = 2$, for $M = 1$ ($B = \infty$), as found by Jeffreys [21], Dressler [14], Brock [8], and others.

$F_c = 2$ is also obtained from the White Colebrook formula with $\alpha = 1$, $M = 1$, and large values of ks/R and/or R.

With the Manning formula one obtains:

$$F_c = \frac{1}{[\alpha - 2\alpha(1 + \frac{2}{3}M) + (1 + \frac{2}{3}M)^2]^{1/2}} \quad (36)$$

If $\alpha = 1$, $F_c = \frac{3}{2}M$, and $F_c = 1.5$, for $M = 1$ ($B = \infty$), as found by Keulegan [24]. Figure 4 shows F_c as a function of α and M according to the Chézy or Manning formula. It is seen again that instability is more likely to occur in a wide section.

Using a power law

$$U_0^a = \frac{1}{K} R_0^{1+b} \sin \theta^m \quad (37)$$

for the calculation of the friction losses (Table 2) in Equation 29, one finds with $\alpha = 1$

$$F_c = \frac{a}{M(1 + b)} \quad (38)$$

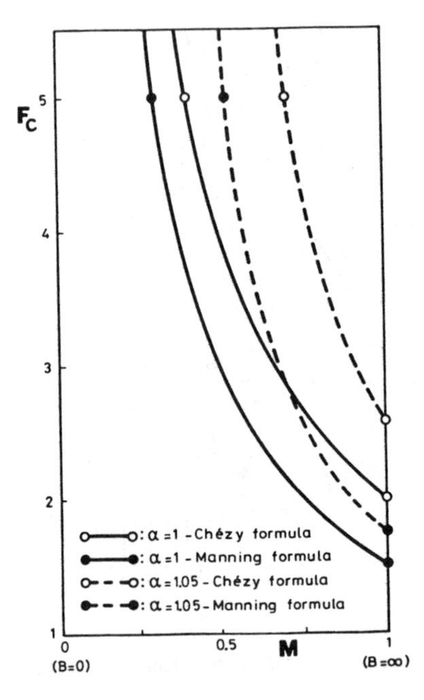

Figure 4. Critical Froude number, F_c, as function of M for different values of α according to resistance laws by Chézy and Manning.

Again, it is seen that F_c increases with decreasing channel width. The Vedernikov number [2, 12, 29, 40] is for a rectangular cross section defined by:

$$V = \frac{(1 + b)M}{a} \frac{1}{C_r - 1} \tag{39}$$

Since for large values of B [6] (see Equation 49)

$$C_r = \alpha + (\alpha^2 - \alpha + 1/F^2)^{1/2} \tag{40}$$

independent of the friction law chosen, and taking Equations 38 and 39 into account, the Vedernikov number for $\alpha = 1$ is

$$V = F/F_c$$

the flow becoming unstable when $V > 1$.

The following are some conclusions that can be made:

1. Since its derivation was general, Equation 29 indeed holds true for *laminar* flows in which $\alpha = 1.2$ (see Table 1 for parabolic velocity profile). One obtains:

$$F_c = \frac{1}{[1.2 - 2.4(1 + 2M) + (1 + 2M)^2]^{1/2}} \tag{41}$$

   in which $F_c = 0.577$ for $M = 1$ ($B = \infty$) or $R_c = 4 \operatorname{cotg} \theta$, and $C_{r,c} = 3$, as found by Ishihara [17].

   Yih [42] using a potential theory and the Orr-Sommerfeld equations found (not taking into account the capillarity effect): $F_c = 0.527$ and $C_{r,c} = 3$; and $R_c = 3.33 \operatorname{cotg} \theta$. The same result was obtained by Binnie [7] and Benjamin [3].

2. At very high Froude numbers $F > 9$ [26] air entrainment occurs, creating an other flow mode for which the presented theory is no longer valid. Using the White-Colebrook friction formula, the critical Froude number for a turbulent flow becoming unstable can be found to be a function of the shape of the velocity profile, the Reynolds number and the relative channel roughness, the wave length of the initial disturbance wave, and the channel width. For a given Reynolds number (discharge), flow instability is more likely to occur in wide rather than in narrow channels. When scale models are built at constant F of channels for unstable turbulent flow, it might occur that the flow in the prototype is unstable while the flow in the model remains stable.

FRICTION FACTOR

According to a dimensional analysis by Rouse [34], the friction factor, f, should be a function not only of the Reynolds number, the relative channel roughness, and the shape of the flow section, but also of the Froude number. Since the few indications of gravitational influence found in the literature [12, 30] were all obtained at Froude numbers well in excess of unity, Rouse assumed that the phenomenon was in some way connected with the surface instability involved in the formation of roll waves.

From experimental results obtained at Iowa [34, 35], and experimental evidence obtained by Nemec, Rouse found for smooth channels [34].

$$\frac{1}{\sqrt{f}} = 2.03 \log_{10} \frac{R\sqrt{f}}{\left(\frac{F}{F_c}\right)^{2/3}} - 0.99 \qquad \text{for } F > F_c \tag{42}$$

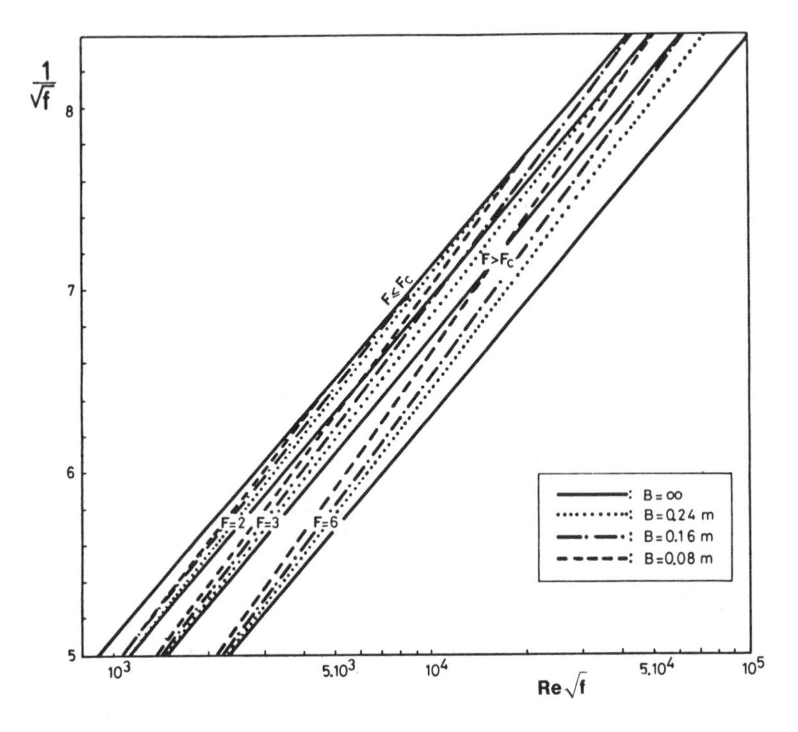

Figure 5. $1/\sqrt{f}$ as function of $R\sqrt{f}$ and Froude number, F, for different values of channel width, B.

If one puts $(F/F_c) = 1$ in Equation 42 when $F \leq F_c$, Equation 42 reduces to Equation 26 except for a small difference in the constant value, which agrees with the laws by Blasius, Thijsse [37] Keulegan [43], Straub [36], and others.

Chen [16], Powell [30], and others wrote

$$\frac{1}{\sqrt{f}} = a \log_{10} Re\sqrt{f} + b \tag{43}$$

in which a and b are some function of F and F_c for $F > F_c$ and constants for $F \leq F_c$.

Introducing the value of F_c (see Equation 34) into Equation 42 by Rouse, Berlamont found [6] that f is not only a function of Re and F, but also a marked function of the channel width. The value $1/\sqrt{f}$, as computed with Equations 34 and 42, is represented in Figure 5 as a function of $R\sqrt{f}$ and F for several channel widths. All stable flow situations ($F \leq F_c$) are situated on a straight line in the ($1/\sqrt{f}$, $R\sqrt{f}$) plane (see Equation 26). From Figure 5 it can be seen that the influence of F on f is not negligible.

Since for a smooth channel, and a given channel width, B, F_c is a function of R only (see Equation 34), Equation 42 can approximately by written as

$$\frac{1}{\sqrt{f}} = a(B) \log_{10} R\sqrt{f} + b(B) \log_{10} F + c(B) \tag{44}$$

A linear regression on computed results similar to those presented in Figure 5 yields, for the best straight line, according to Equation 44, the values of a, b, and c given in Table 4 valid for $0.014 < f < 0.040$ [6]. From Table 4 it can be seen that c is nearly independent of the channel width.

Table 4
Results of Linear Regression Analysis

B in meters (feet)	a	b	c	Correlation Factor, as a Percentage
0.08 (0.3)	2.32	−3.10	−0.89	99.75
0.16 (0.5)	2.24	−2.55	−0.93	99.51
0.24 (0.8)	2.21	−2.35	−0.91	99.51
0.40 (1.3)	2.17	−1.84	−0.99	99.64
1.00 (3.3)	2.12	−1.51	−0.98	99.80
∞	2.07	−1.34	−0.90	99.99
$F \leq F_c$	2.03	0	−0.99* −0.98†	100.00

* *See Equation 42*
† *See Equation 26*

Berlamont [6] showed that [44] is (at least qualitatively) supported by experimental results obtained by Jegorov [22], Tracy and Lester [38], Brock [8], and Van Heste [39].

To obtain the value of the friction factor in hydraulically rough channels at Froude numbers in excess of F_c, Equations 34 and 16 should be used together with the formula by Rouse [35]

$$\frac{1}{\sqrt{f}} = 2 \log_{10} \frac{\frac{R}{k_s}}{\delta\left(\frac{F}{F_c}\right)^{2/3}} - 0.82 \tag{45}$$

in which δ = the roughness concentration.

Experimental verification of the influence of the Froude number on the value of the friction factor, f, in hydraulically smooth channels, was obtained by Nemec [16, 34] Jegorov [22] Powell [30] Iwagaki [18] and Wakhlu [41]. Other experimenters [38] however, neither reported any influence of the Froude number, nor attributed the measured differences to errors of the measurements, [16].

Since the width of the various experimental channels [16, 18, 22, 30, 31, 34, 38, 41] varied between 0.03 m (1.2 in.) and 1.10 m (3.6 ft), Berlamont [6] suspected that the difference in channel width might partly account for the contradictory results on the influence of the Froude number on the friction factor.

In Figure 6 [6] the relative increase $(f - f_0)/f_0$ is shown as a function of Re and θ, in which f is computed by Equation 42 and f_0 by Equation 26. It is seen from Figure 6 that especially for large channels, the curves are very flat over a large interval of R numbers. This might explain the fact that some experimenters working in a rather wide channel and in a restricted interval of Reynolds numbers [16, 38] could not recognize the effect of the Froude number on the friction factor, which was, in the conditions of their experiments, of the order of the error on the measurements.

Both Tracy and Lester [38] and Brock [16] worked with a rather large channel (1.07 and 1.10 m respectively) and in an interval of Reynolds numbers in which $\Delta f/f_0$ is almost independent of R (see Figure 6). This might explain why they could not recognize the influence of the Froude number on the value of f, but instead, reported larger values for f_0 than those given by the Blasius or Thijsse formula. For example, for $R = 10^5$ Brock found $f_0 = 0.0202$, and Tracy and Lester found $f_0 = 0.0197$,

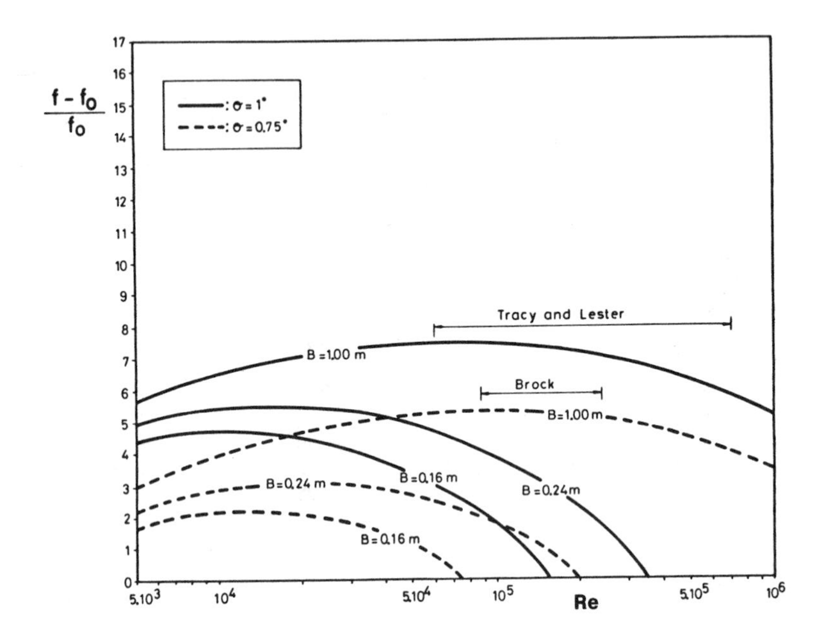

Figure 6. Relative Increase, $(f - f_0)/f_0$, as function of Reynolds number, Re, channel width, B, and slope angle θ.

while Equation 26 yields $f_0 = 0.0182$, which means an increase of 11% and 8%, respectively. These figures are in good agreement with the results presented in Figure 6 since reported channel slopes were less that 1°. Wakhlu, [41], although working in a large channel as well, (1 m), did find an influence of F on f, because he worked in a very large interval of R numbers ($2.5\ 10^2 - 1.5\ 10^5$). All other authors on the other hand, working in narrower channels, found an influence of F on f.

Introducing the critical Froude number in the formula by Rouse, the influence of the Froude number on the value of the friction factor for unstable turbulent flow in both smooth and rough channels can be calculated. The value of f increases with increasing Froude number (channel slope). Moreover, it is found that the coefficients in the formula relating f to R and F are functions of the channel width.

For large channels, the relative increase of f with respect to results obtained with the White-Colebrook formula, becomes almost independent of the Reynolds number, provided that $R > 4\ 10^4$.

FINITE AMPLITUDE ROLL WAVES

The equations describing roll waves, of course are the continuity Equation 4, the pressure Equation 9, and the momentum Equation 10. Different roll wave theories were presented, according to the simplifications introduced in the basic equations to make the integration possible.

Discontinuous Solution

As pointed out earlier, one obtains a first-order simulation of open channel flow by dropping the third derivatives in the momentum Equation 10, thereby neglecting the vertical acceleration

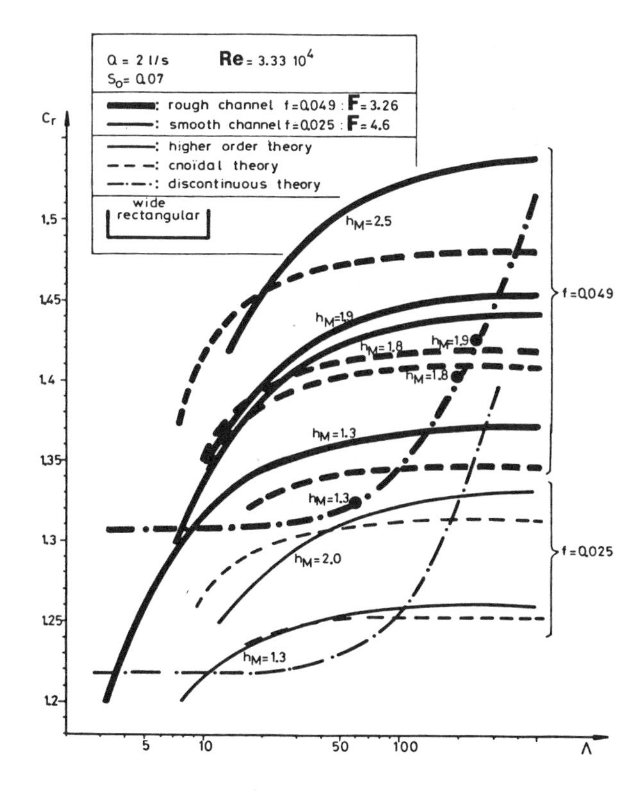

Figure 7. Different roll-wave solutions.

and assuming a hydrostatic pressure distribution. These so-called "shallow-water" equations do not give any finite amplitude periodical solution [14, 20]. Therefore Iwasa [19, 20], Dressler and Brock [8, 9, 14] constructed a roll wave solution by piecing together logarithmic solutions of the shallow-water equations by means of the hydraulic jump condition. This procedure was justified by the actual features of roll waves, which show a periodic pattern with a surge front [20]. When the Reynolds number and Froude number are given, one single possible roll wave length and corresponding phase velocity is found for each value of the maximum wave depth H_M (Figure 7). This seems to be at variance with experimental results: Brock [8] reported a standard deviation of the measured wave period λ/c_r of 50%. Other experiments [4] indicate that, at given R and F, a multitude of roll waves exists with a given maximum wave depth but different wave lengths and phase velocities.

Cnoidal Wave Solution

Dressler [14] advocated a higher-order approximation taking into account the vertical acceleration to find a periodical roll wave solution. Applying a perturbation method on the Euler equations, and neglecting resistance effects, Dressler [14] found a periodical, continuous, and symmetrical roll wave solution in terms of cnoidal waves (Korteweg and De Vries 1895).

$$h = h_M - a\ \mathrm{sn}^2(X\sqrt{3a}/2k) \tag{46}$$

where $h = H/H_0$, the dimensionless water depth

$h_M = H_M/H_0$, the dimensionless maximum wave depth

a = the dimensionless wave amplitude

$X = \zeta/H_0$, dimensionless ζ-coordinate

$\zeta = x - c_r t$

H_0 = normal depth

sn = the elliptical sine function [1]

Λ = the dimensionless wave length, $\lambda/H_0 = 4kF_1/(\sqrt{3a})$

$$a = k^2(h_M - 1)\frac{F_1}{F_1 - E_1} \tag{47}$$

$$b = (h_M - 1)\frac{2F_1 - 3E_1 - k^2F_1}{F_1 - E_1} \tag{48}$$

$F_1 = F(\pi/2, k)$, the elliptic integral of the 1° kind [1]

$E_1 = E(\pi/2, k)$, the elliptic integral of the 2° kind [1]

h_M and k are parameters.

k is related to the Froude number through $b + 1 = F^2(C_r - 1)^2$

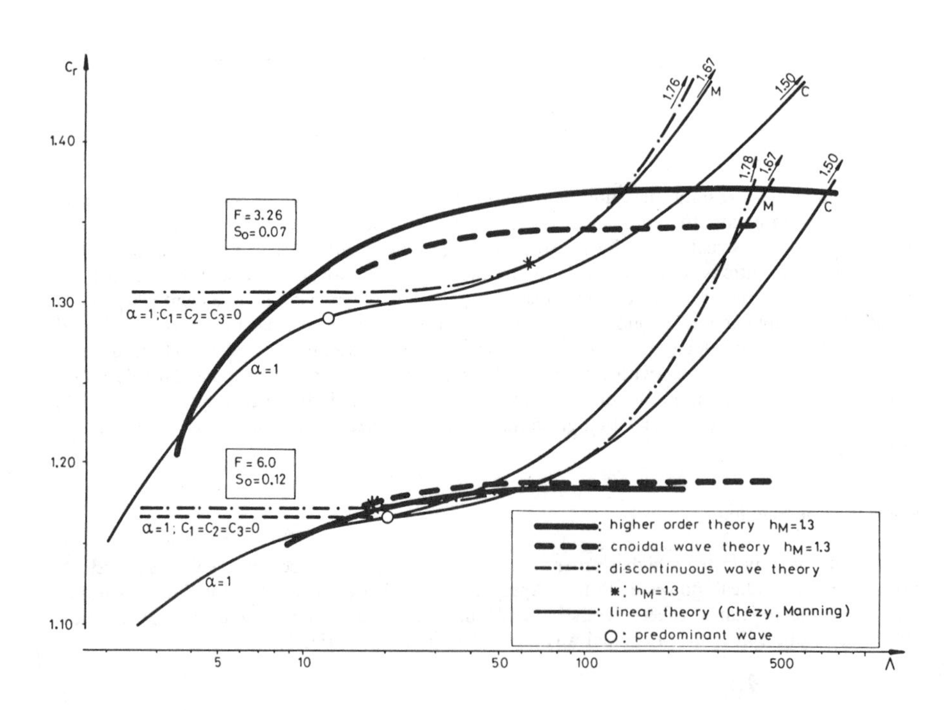

Figure 8. Different roll-wave solutions.

For each value of F and h_M one finds an infinity of possible wave lengths Λ and corresponding phase velocities (Figures 7 and 8), which are always greater than the fluid velocity: the roll wave spectrum (Λ, c_r-plane; Figures 7 and 8) is only dependent on F, independent of R. For each h_M a particular roll wave solution is found with infinite wave length Λ, but finite phase velocity c_r; this in fact is a solitary wave with

$$c_r = U_0 + \sqrt{gh_0}$$

In particular from Equation 46 a "limiting height" can be found for the roll waves by expressing the condition that the pressure gradient at the wave crest $(\partial p/\partial y)_{y=H}$, may not be positive, since otherwise the wave will break [25]. One finds [4]:

$$h_{M_{maximum}} = 3.28, \text{ independent of } \mathbf{F}$$

Dressler suggested that a more accurate solution could be found by taking into account friction in a more accurate way. One can show that the cnoidal wave solution is an approximation of the "higher-order solution" indeed [4].

Approximate Solution from the Linearized Equations

When considering small amplitude waves, Equations 4 and 10 can be linearized as shown earlier. These linearized equations have a solution of the form (Equation 25):

$$H' = \eta \exp(\beta i(x - ct)) \tag{21}$$

Assuming $C_i > 0$ in Equations 21, 22 and 23, one obtains a relation between the wavelength Λ of waves with a growing amplitude, their phase velocity c_r, F and c_i. Plotting this relation between c_r and Λ for a given F and different values of c_i, one obtains a "wave spectrum" as shown in Figure 8 (for a uniform velocity profile, and the friction losses computed either with the Chézy or Manning formula).

For each value of F one finds a *predominant* wave for which βc_i in Equation 25 is maximum. The waves with the fastest growing amplitude are expected to predominate in the channel, and eventually become finite amplitude roll waves. It should however be noted that the maximum of the function βc_i is not a sharp one, so that waves with wave lengths close to the *predominant* wave length, will grow at approximately the same rate.

For $\Lambda \to \infty$, C_r tends to $C_{r,c}$ (Table 3), and for $\Lambda \to 0$, and neglecting the higher order terms in Equations 22 and 23, one obtains:

$$\lim_{\Lambda \to 0} C_r = \alpha + (\alpha^2 - \alpha + 1/F^2)^{1/2} \tag{49}$$

This is the result obtained by Keulegan for short waves [25], and is very close to the result from the discontinuous theory for $\Lambda \to 0$ [4]. The linear theory does not give any information on the value of the maximum wave depth, since it deals with waves with amplitudes growing with time, and in fact, is only valid for small values of h_M.

Higher-Order Theory

Assuming $\alpha = 1$ in shallow water, the momentum equation (Equation 10) reads

$$\frac{\partial U}{\partial t} + g\cos\theta\frac{\partial H}{\partial x} + U\frac{\partial U}{\partial x} + \frac{HU^2}{3}\left(\frac{1}{U^2}\frac{\partial^3 H}{\partial x\,\partial t^2} + \frac{2}{U}\frac{\partial^3 H}{\partial x^2\,\partial t} + \frac{\partial^3 H}{\partial x\,\partial t^2}\right) = g\sin\theta - f\frac{U^2}{8H} \tag{50}$$

For a "permanent wave," whose shape and phase velocity do not change with time or position, the t-variable can be eliminated by introducing a new coordinate system (ζ, y), moving with the wave at the phase velocity c_r:

$$\zeta = x - c_r t \tag{51}$$

Introducing the dimensionless variables:

$$u = \frac{U}{U_0}$$

$$X = \frac{\zeta}{H_0}$$

$$h = \frac{H}{H_0} \tag{52}$$

$$C_r = \frac{c_r}{U_0}$$

$$F = \frac{U_0}{\sqrt{gH_0 \cos \theta}}$$

and assuming a uniform velocity distribution ($\alpha = 1$), Equations 4 and 10 can be written as:

$$(F^2(C_r - u)^2 - h)\frac{dh}{dX} - \frac{h^2}{3} F^2 \frac{d^3h}{dX^3}(C_r - u)^2 + tg\theta h - f\frac{F^2}{8} u^2 = 0 \tag{53}$$

$$\frac{\partial u}{\partial x} = (C_r - u)\frac{\partial h}{\partial X}\bigg/ h \tag{54}$$

The integration of Equation 54 yields:

$$(C_r - u)h = (C_r - u_{X=0})h_{X=0} = C' \tag{55}$$

in which C' is a constant and $u_{X=0}$ and $h_{X=0}$ are the values of u and h at $X = 0$.
Eliminating u from Equations 53 and 55 one obtains:

$$\frac{d^3h}{dX^3} = \frac{3}{(C'F)^2}\left[\left[\left(\frac{C'F}{h}\right)^2 - h\right]\frac{dh}{dX} + tg\theta h - f\frac{C_r^2F^2}{8} - \frac{f}{8}\frac{F^2C'^2}{h^2} + \frac{f}{4}\frac{C_rC'F^2}{h}\right] \tag{56}$$

Equation 9 then yields:

$$\frac{p}{\rho g \cos\theta} = (H - y) + \frac{(H^2 - y^2)}{2H}\frac{d^2H}{d\zeta^2}\left(\frac{c_r - U}{g\cos\theta}\right)^2 \tag{57}$$

showing that the pressure distribution is not hydrostatic, except at the inflexion point of the wave profile. The integration of the nonlinear differential Equation 56, was performed by Berlamont [4, 5, 6] using Hamming's modified predictor-corrector method [32].

For given values of the channel slope S_0, f, and Q, one can find a periodic slightly asymmetric roll-wave solution for a wave with a given maximum depth h_M and a given phase velocity C_r. $(\partial^2h/\partial X^2)_{X=0}$

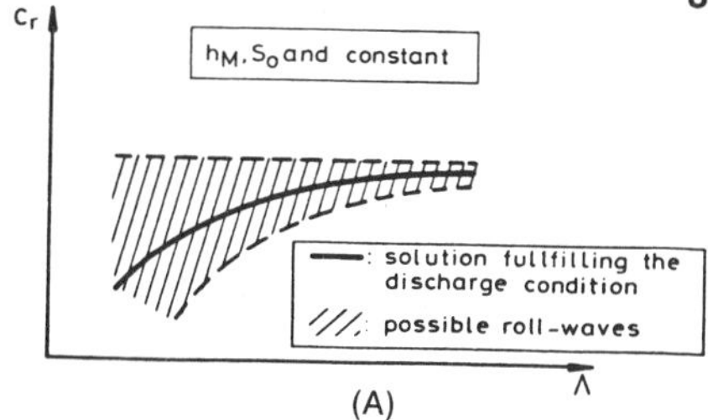

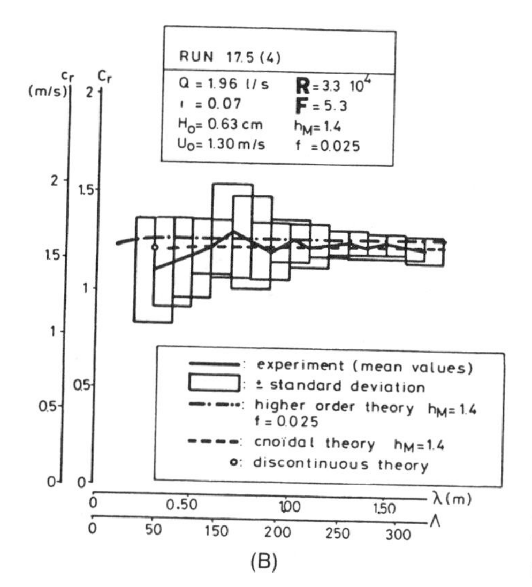

Figure 9. (A) All possible roll waves as compared with the solutions fulfilling the discharge condition. (B) Experimental results.

and C′ are determined by trial and error to satisfy both the periodicity condition, $(h)_{X=\Lambda} = h_M$, and the discharge condition

$$Q = 1/\lambda \int_0^\lambda UH\,dx \tag{58}$$

which states that the mean discharge over one wave length must equal the given discharge supplied to the channel. The latter assumption is not essential; apart from the solution obeying the discharge condition Equation 58, other waves can exist in the channel that do not convey individually the "correct" discharge, provided that the mean discharge over a wave train, i.e., different successive waves with different Λ and C_r, equals the given discharge. It can be shown [4] that the values of Λ and C_r of all possible roll waves (for constant h_M, S_0, and f) are situated within an almost triangular zone in the (Λ, C_r) plane (Figure 9).

The (C_r, Λ) values, obtained from Equation 56, taking into account the discharge condition, thus should be considered as mean values, around which a considerable scatter of the experimental results may be expected, especially at small Λ.

As an example, Figures 7 and 8 show the roll wave spectrum (C_r, Λ), as obtained from the higher-order theory for $Q = 0.002\ m^3s^{-1}$, $S_0 = 0.07$, $B = 0.24$ m, $f = 0.025$ and 0.049, and different values of h_M.

- For a given maximum wave depth the phase velocity C_r is seen to increase with Λ, tending to a finite terminal velocity for a wave with infinite wave length. The depth at the trough of such a wave tends to the normal depth. This "terminal velocity" increases with increasing h_M and *R*, and decreases

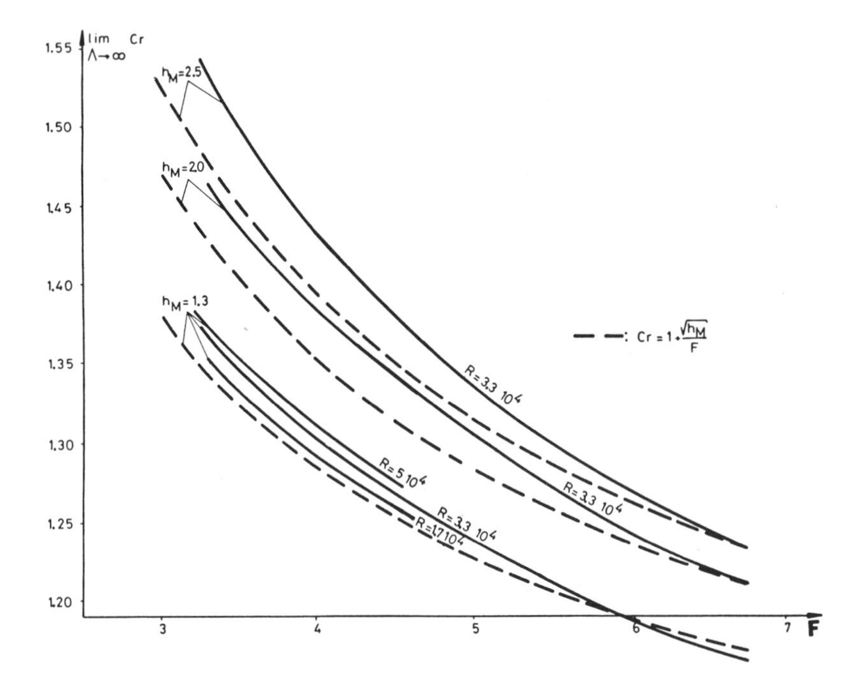

Figure 10. Terminal roll-wave velocity as obtained from the cnoidal wave theory.

with increasing F, approaching the celerity of a solitary wave

$$C_r = 1 + \sqrt{h_M}/F$$

as found from the cnoidal wave theory (Figure 10).

- For a fixed value of Λ, the phase velocity C_r increases with increasing h_M, f, and R, and with decreasing F and S_0. Due to the decrease of U_0 however, c_r decreases with increasing f or decreasing S_0.

Since for a given value of F and R many different roll waves will exist in the channel with different phase velocities C_r, a particular roll wave can be overtaken by another, faster wave, or itself overtake a slower wave. Then, several new roll waves of different lengths, depths, and celerities can be formed and eventually reach a steady state, represented in the (C_r, Λ) spectrum.

Berlamont [4] has compared experimental results with the higher order, the cnoidal, and the discontinuous solution and has found good agreement. Figures 7, 8, and 10 provide the basis for the cnoidal wave theory, which is a good approximation of the higher order theory, especially for small h_M, Λ, and R. The linearized solution is a good approximation for small h_M and Λ. The discontinuous roll wave solution yields only one of many possible (C_r, Λ) combinations for each h_M.

NOTATION

A	cross-sectional area
a	dimensionless wave amplitude
B	channel width
B	dimensionless wave number $(=\beta H_0)$

C Chézy roughness factor
C_r dimensionless wave celerity
$C_{r,c}$ critical dimensionless wave celerity
C_1, C_2, C_3 constants, the value of which depends upon F(z)
c_r wave celerity
F Froude number
F_c critical Froude number
F(z) shape function of the velocity profile
f Darcy-Weisbach friction factor
f_0 friction factor in uniform flow, according to White-Colebrook
g acceleration of gravity
H water depth
H_0 normal depth
H_M maximum wave depth
h dimensionless water depth ($=H/H_0$)
h_M dimensionless maximum water depth ($= H_M/H_0$)
i $\sqrt{-1}$
k_s/R equivalent relative channel roughness
M $1 - R_0(\partial P/\partial A)_0$
n Manning roughness factor
P wetted perimeter
p pressure at point (x, y), at time t
Q discharge
R hydraulic radius
R Reynolds number
S_0 channel slope
t time
U mean velocity over a cross section
u, v velocity components in the x direction and y direction
x dimensionless ζ-coordinate ($=\zeta/H_0$)
x, y rectangular Cartesian coordinates
z y/H

Greek Symbols

α velocity distribution coefficient
β wave number
δ roughness concentration
ζ $x - c_r t$
η infinitesimal small amplitude of the disturbance wave
θ slope angle of the channel
Λ dimensionless wave length ($=\lambda/H_0$)
λ wave length
ν kinematic viscosity of the liquid
ρ mass density of the fluid
τ bottom shear stress
$\tau_{xx}, \tau_{xy}, \tau_{yx}, \tau_{yy}$ Reynolds stresses
ϕ $(2.671/\{[(K_s/R)/12.20] + (3.03/R\sqrt{f_0})\}R)$

REFERENCES

1. Abramowitz, M., and Stegun, I. A., *Handbook of Mathematical Functions, Dover Publications*, New York, 1965.
2. Ali, K. M. H., "Roll Waves in Inclined Rectangular Open Channels," *International Symposium on Unsteady flow in Open Channels*, Newcastle-upon-Tyne, pp. X7-X9, April 12–15, 1976.
3. Benjamin, T. B., "Wave Formation in Laminar Flow Down an Inclined Plane," *Journal of Fluid Mechanics*, Vol. 2 (1957), p. 554.
4. Berlamont, J., "Roll Waves," thesis presented to the Ghent State University, at Ghent, Belgium, in partial fulfillment of the requirements for the degree of Doctor of Philosophy (in Dutch), 1975.
5. Berlamont, J., "Roll Waves in Inclined Rectangular Open Channels," *International Symposium on Unsteady Flow in Open Channels*, Newcastle-upon-Tyne, pp. 12–13 (BHRA), April 12–15, 1976.
6. Berlamont, J., and Vanderstappen, N., "Unstable Turbulent Flow in Open Channels," *Journal of the Hydraulics Division, ASCE*, Vol. 107 April 1981, pp. 427–449.
7. Binnie, A. M., "Instability in a Slightly Inclined Water Channel," *Journal of Fluid Mechanics*, Vol. 5 (1959), p. 561.

8. Brock, R. R., "Development of Roll Waves in Open Channels," W. M. Keck Laboratory of Hydraulics and Water Resources, California, Institute of Technology, Pasadena, California, Report by KH-R-16.
9. Brock, R. R., "Development of Roll Wave Trains in Open Channels," *Journal of the Hydraulics Division*, ASCE, HY 4 1969, (p. 1401).
10. Brock, R. R., "Periodic Permanent Roll Waves," *Journal of the Hydraulics Division*, ASCE, HY 12 (1970), p. 2565.
11. Brock, R. R., "Critical Analysis of Open Channel Resistance," *Journal of the Hydraulics Division*, ASCE, HY 2, Vol. 92 (1966), p. 395.
12. Chow, V. T., *Open Channel Hydraulics*, McGraw-Hill Book Company, N. Y., 1959.
13. Cornish, V., *Waves of the Sea and Other Water Waves*, The Open Court Publishing Company, La Sable, III, and T. Fisher Unwin, London, 1910.
14. Dressler, R. F., "Mathematical Solution of the Problem of Roll Waves in Inclined Open Channels," *Communications on Pure and Applied Mathematics*, N. Y., Vol. 2 (1949), pp. 149–194.
15. Engelund, F., "A Note on the Vedernikov's Criterion", *La Houille Blanche*, No. 8 (1965), p. 861.
16. Farell, C., Chen, C.-L., and Brock R., "Critical Analysis of Open Channel Resistance," by Hunter Rouse, *Journal of the Hydraulics Division, ASCE*, Vol. 92, HY 2, Proc. Paper 4708 (Mar. 1966), pp. 395–409.
17. Ishihara, T., et al., "On the Roll Wave Trains Appearing in the Water Flow on a Steep Slope Surface," *Memoirs*, Vol. 14, No. 2, Faculty of Engineering, Kyoto University, Japan (April 1952), p. 83.
18. Iwagaki, Y., "On the Laws of Resistance to Turbulent Flow in Open Smooth Channels," *Proceedings*, Second Japanese National Congress for Applied Mechanics; also *Applied Mechanics Review*, 2749 (1952), pp. 245–250.
19. Iwasa, Y., "The Criterion for Unstability of Steady Uniform Flows in Open Channels," *Memoirs*, Vol. 16, No. 4, Faculty of Engineering, Kyoto University, Japan (Oct. 1954), pp. 264–275.
20. Iwasa, Y., "Roll Waves in Inclined Rectangular Open Channels," *International Symposium on Unsteady Flow in Open Channels*, Newcastle-upon-Tyne, pp. X4–X5, Apr. 12–15, 1976.
21. Jeffreys, H., "The Flow of Water in an Inclined Channel of Rectangular Section," *Philosophical Magazine and Journal of Science*, London, England (1949), p. 793.
22. Jegorov, S. A., "Turbulente Uberwellenströmung", *Wasserkraft und Wasserwirtschaft*, Stuttgart, Germany, Vol. 35 (1940), pp. 55–57.
23. Keulegan, G. H., "Laws of Turbulent Flow in Open Channels," *Research Paper 1151, Journal of the National Bureau of Standands*, Vol. 21, (Dec. 1938), pp. 707–741.
24. Keulegan, G. H., and Patterson, G. W., "A Criterion for Instability of Flow in Steep Channels," *Transactions*, American Geophysical Union, Part II (1940), p. 594.
25. Keulegan, G. H., and Patterson, G. W., "Effect of Turbulence and Channel Slope on Translation Waves," *Journal of Research, National Bureau of Standards*, Vol. 30 (June 1943), p. 461.
26. Lai, K. K., "Roll Waves in Inclined Rectangular Open Channels," *International Symposium on Unsteady Flow in Open Channels*, Newcastle-upon-Tyne, pp. X4–X5, Apr. 12–15, 1976.
27. Laitone, E. V., "Limiting Conditions for Cnoidal and Stokes Waves," *Journal of Geophysical Research*, Vol. 67, No. 4 (April 1962), pp. 1555–1564.
28. Montuori, C., "Stability Aspects of Flow in Open Channels," by Francis F. Escoffier and Marden B. Boyd, *Journal of the Hydraulics Division, ASCE*, Vol. 89, HY 4 (July 1963), pp. 264–273.
29. Powell, R. W., "Vedernikov's Criterion for Ultra Rapid Flow," *Transactions, American Geophysical Union*, Vol. 29, No. 6, p. 882.
30. Powell, R. W., "Flow in a Channel of Definite Roughness," *Transactions, Journal of the Hydraulics Division, ASCE*. Vol. 3, No. 2276 (1946), pp. 531–554.
31. Powell, R. W., "Resistance to Flow in Smooth Channels," *Transactions, American Geophysical Union*, Vol. 30 Dec. 1949, pp. 875.
32. Ralston, A., and Wilf, H. S., *Mathematical Methods for Digital Computers*, John Wiley & Sons, New York, 1960.
33. Rodenhuis, G. S., "Difference Method for Higher-Order Equation of Flow," *Journal of the Hydraulics Division, ASCE*, Vol. 99, HY 3, (Mar. 1973), p. 471.

34. Rouse, H., Koloseus, H. J., and Davidian, J., "The Role of the Froude Number in Open Channel Resistance," *Journal of Hydraulic Research*, Vol. 1, No. 1 (1963), pp. 14–19.
35. Rouse, H., "Critical Analysis of Open-Channel Resistance," *Journal of the Hydraulics Division, ASCE*, Vol. 91, HY 4, Proc. Paper 4387, pp. 1–15, July 1965.
36. Straub, L. C., "Studies of the Transition Region between Laminar and Turbulent Flow in Open Channels," *Transactions, American Geophysical Union* (Aug. 1939), pp. 649–653.
37. "Friction Factors in Open Channels (Progress Report)," by the Task Force on Friction Factors in Open Channels of the Committee on Hydromechanics of the Hydraulics Division of the ASCE, E. Silberman, Chmn., *Journal of the Hydraulics Division, ASCE*, Vol. 89, HY 2, Proc. Paper 3464 (Mar. 1963), pp. 97–143.
38. Tracy, H. J., and Lester, C. M., "Resistance Coefficients and Velocity Distribution in Smooth Rectangular Channels," Water Supply Paper no. 1952-A., Department of the Interior, United States Geological Survey, Washington, D.C., pp. 1–17, 1961.
39. Vanheste, J. M., "Experimenteel Onderzoek naar de Invloed van het Froude Getal op de Weerstandscoëfficiënt f in Open Rechthoekige Kanalen," (Experimental Determination of the Influence of the Froude Number on the Friction Factor f in Open Rectangular Channels), thesis presented to the Ghent State University, at Ghent, Belgium, 1975.
40. Vedernikov, "Vedernikov's Criterion for Ultra Rapid Flow," *Transactions, American Geophysical Union*, Vol. 32, No. 4 (Aug. 1951), p. 603.
41. Wakhlu, O.N., "An Experimental Study of Thin-Sheet Flow over Inclined Surfaces," *Mitteilungen Heft 158*, Versuchanstalt für Wasserbau und Kulturtechnik, Universität Fredericiana, Karlsruhe, Germany, (1970), pp. 32–121.
42. Yih, C. S., "Stability of Liquid Flow down an Inclined Plane," *Physics of Fluids*, Vol. 6, No. 3 (Mar. 1963), p. 321.

CHAPTER 6

HYDRAULIC JUMPS AND INTERNAL FLOWS

John A. McCorquodale

Department of Civil Engineering
University of Windsor
Windsor, Ontario, Canada

CONTENTS

INTRODUCTION, 122

THE MACROSCOPIC APPROACH, 125
- General, 125
- Free Surface Hydraulic Jumps in Prismatic Channels, 127
- Hydraulic Jumps in Nonprismatic Channels, 139
- Internal Hydraulic Jumps, 150

INTERNAL FLOW IN THE HYDRAULIC JUMP, 153
- Velocity Distribution and Turbulence in the Classical Hydraulic Jump, 153
- Decay of Turbulence Downstream from a Stilling Basin, 158
- Pressure Fluctuations in the Hydraulic Jump, 159
- Internal Flow in Other Hydraulic Jumps, 159
- Analytical Approaches to Internal Flow, 160

CONCLUSIONS, 167

NOTATION, 168

REFERENCES, 168

INTRDOUCTION

The hydraulic jump is the flow phenomena associated with the abrupt transition of a supercritical (inertia-dominated) flow to a subcritical (gravity-dominated) flow. This transition is accompanied by a large conversion of kinetic energy to turbulent and potential energy. A prerequisite for a hydraulic jump is that there should be at least two fluids of different densities. The classical hydraulic jump, shown in Figure 1, involves the flow of water with air as the ambient fluid. The

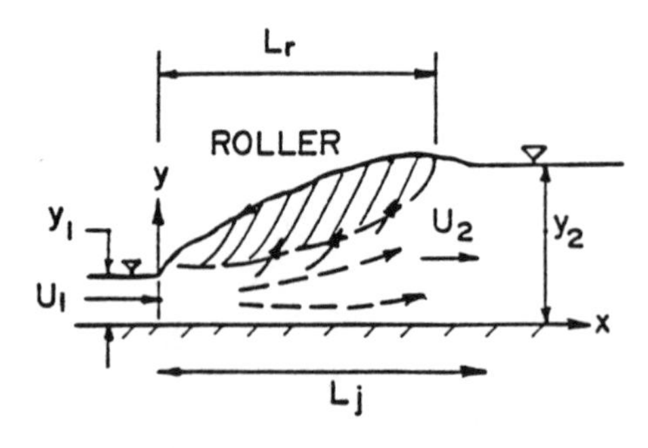

Figure 1. Defining diagram of the classical hydraulic jump.

internal hydraulic jump involves two liquids with different densities such as hot and cold water or two gases of different densities.

The importance of the inertia relative to the gravity forces is usually described by a Froude number, typically defined as

$$F = \frac{u}{\sqrt{g'D}} \tag{1}$$

where $g' = g\frac{\Delta\rho}{\rho}$

g = acceleration of gravity

ρ = density of active fluid

Δp = difference in density between the active and ambient fluid

u = fluid velocity relative to a coordinate system attached to the hydraulic jump

D = hydraulic mean depth; A/B

A = flow area

B = contact length between the active and ambient fluids perpendicular to u

The ambient fluid may be a liquid or a gas. If $F > 1$, the flow is supercritical and a hydraulic jump is possible. If $F < 1$, the flow is subcritical. A characteristic of hydraulic jumps is that active fluid tends to entrain the ambient fluid.

The hydraulic jump was known to Leonardo da Vinci nearly five hundred years ago, although the first study of the phenomenon was apparently made by Bidone [1] in 1818. The results of the analytical treatment by Belanger [2] in 1828 are still valid. Another important early treatment of the hydraulic jump was by Darcy and Bazin [3] in 1865. More recently, major contributions have been made by Bakhmeteff and Matzke [4] and Rajaratnam [5–11]. A classical paper on the internal flow characteristics of the hydraulic jump was published by Rouse, Siao, and Nagaratnam [12] in 1958. In 1964 an ASCE Task Force [13] on Energy Dissipators and Outlet Works listed almost 500 references.

A definition sketch of a typical hydraulic jump is shown in Figure 1. The important macroscopic parameters of the jump are:

y_1 = the initial depth

y_2 = the sequent depth

u_1 = initial mean velocity

u_2 = mean velocity at the exit of the jump

L_j = jump length

L_r = length of the roller

The hydraulic jump may be classified in several ways, as indicated in Figure 2. The full classification of any hydraulic jump involves specifying terms from several categories. Caution is advised in using and interpreting the term "free jump;" as indicated in Figure 2, this term is usually applied to the classical open-channel hydraulic jump in which there are no appurtenances and there is no submergence of the flow entering the jump.

The Froude number is one of the most important factors in determining the nature of the hydraulic jump. As the initial Froude number increases, the efficiency of the jump, as an energy dissipator, also increases. Figure 2 shows the types of jumps that can be expected at various Froude numbers [14, 15, 16].

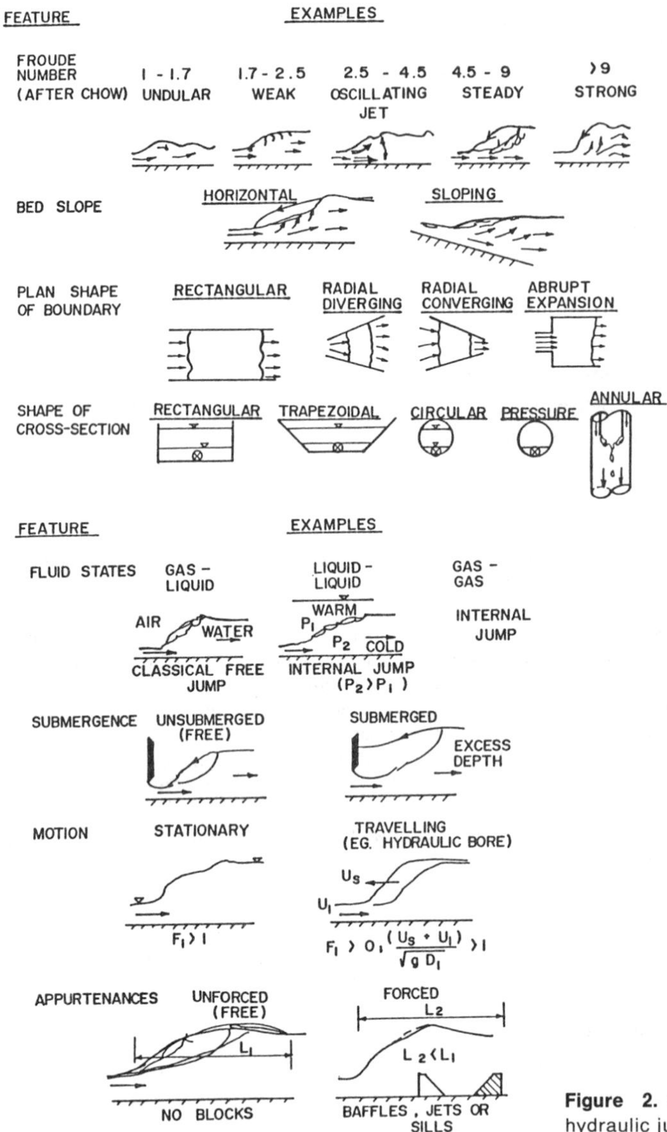

Figure 2. Classification of the hydraulic jump.

Hydraulic jumps are also classified according to the type of channel in which they occur (e.g.) prismatic channels (rectangular, trapezoidal, circular), nonprismatic (radially diverging, radially converging and curvilinear aprons) and horizontal and sloping beds).

The hydraulic jump is encountered in many engineering problems [17–24]. The most common application is the hydraulic jump stilling basin [17]. Other applications include.

1. Mixing of chemicals.
2. Entrainment of air.
3. Increasing pressure on channel aprons to reduce net uplift.

Internal hydraulic jumps have been identified in wastewater clarifiers [18–19], waste heat discharges, mixing of fresh and salt water, and sediment-laden flow into lakes or reservoirs.

The abrupt surges in open channels such as the hydraulic bore [16] are in fact traveling hydraulic jumps. Traveling jumps are also believed to be an important mechanism in the transition of gravity to pressure flow in closed conduits [21, 24].

Three approaches have been used in studying the hydraulic jump phenomenon:

1. Experimental studies using physical models.
2. The momentum and mass conservation principles applied on a macroscopic basis.
3. Solution of the differential equations for continuity and momentum using numerical integration techniques.

This chapter treats the subject of hydraulic jumps in two parts, namely, macroscopic aspects of the jumps (i.e. sequent depth ratios and length characteristics) and internal flow (i.e. velocity distributions and turbulence characteristics).

THE MACROSCOPIC APPROACH

General

Since the hydraulic jump takes place over a relatively short distance, of the order of five sequent depths, the transition is dominated by the initial momentum flux and pressure force due to the sequent depth. Boundary shear forces are secondary. Figure 3 shows an unsubmerged forced hydraulic jump in a radially diverging sloping channel. This will be used to illustrate the macroscopic approach.

Rouse, Siao, and Nagaratnam [12] have applied the Reynolds equation for turbulent flow in the form of

$$\frac{\partial(\bar{u}_i\bar{u}_j)}{\partial x_j} + \frac{\partial(\overline{u_i'u_j'})}{\partial x_j} = -\frac{1}{\rho}\frac{\partial \bar{p}}{\partial x_j} + X_i + \frac{\mu}{\rho}\frac{\partial^2 \bar{u}_i}{\partial x_j\,\partial x_i} \tag{2}$$

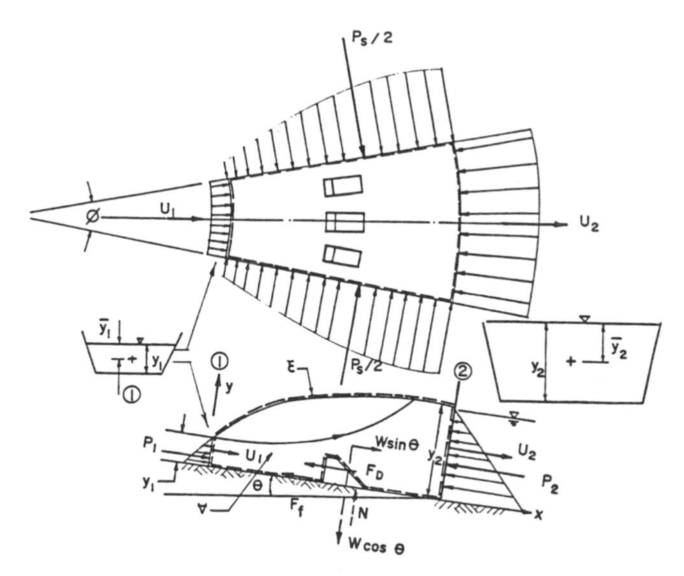

Figure 3. General defining diagram of a hydraulic jump in an open channel.

to the control volume $\bar{V}$ of a hydraulic jump,

in which $\bar{u}_i$ = local mean velocity in direction i
ρ = fluid density
$\bar{p}$ = local mean pressure
X_i = body force
μ = dynamic viscosity
u_i' = random component of $\bar{u}_i$

Using Green's theorem, Equation 2 can be integrated over the volume, $\bar{V}$, to obtain the following macroscopic equation:

$$\underbrace{\int_\xi \rho\bar{u}_i\bar{u}_j \frac{\partial x_j}{\partial n}\, d\xi}_{I} + \underbrace{\int_\xi \rho u_i' u_j' \frac{\partial x_j}{\partial n}\, d\xi}_{II} = \underbrace{-\int_\xi \bar{p} \frac{\partial x_i}{\partial n}\, d\xi}_{III} + \underbrace{\int_{\bar{V}} \rho X_i\, d\bar{V}}_{IV} + \underbrace{\int_\xi \mu \frac{\partial \bar{u}_i}{\partial x_j}\frac{\partial x_j}{\partial n}\, d\xi}_{V} \tag{3}$$

where n is the outwardly directed normal.
The terms represent the following:

I—The net flux of momentum through the boundary, ξ, due to the mean flow.
II—The net momentum transfer through the boundary, ξ, due to turbulence.
III—The resultant mean normal (pressure) force exerted on the fluid boundary, ξ.
IV—The net weight of the fluid within the control volume, $\bar{V}$.
V—The mean tangential force exerted on the boundary, ξ.

A macroscopic momentum equation is obtained if Equation 3 is applied to the control volume, $\bar{V}$, shown in Figure 3. Considering the streamwise, x-momentum, the following equation results:

$$\{\beta_2\rho u_2 Q - \beta_1\rho u_1 Q\} + \{\rho I_2 - \rho I_1\} = P_1 - P_2 + P_s \sin(\phi/2) - F_D + W \sin\theta - F_f \tag{4}$$

in which $\beta = \int_A \bar{u}^2\, dA/(Qu)$
u = average velocity over A
Q = total flow
A = flow area
$I = \int_A \bar{u}'^2\, dA$
P = pressure force = $\beta' g\rho A\bar{y}\cos\theta$
$\bar{y}$ = centroidal depth below the water surface measured perpendicular to the bed
P_s = total side force
F_D = form drag
W = weight of the water in volume $\bar{V}$
F_f = boundary shear force
θ = bed slope angle
ϕ = angle of divergence
β' = pressure correction factor

The mass conservation equation (the continuity equation) is

$$\frac{\partial \bar{u}_i}{\partial x_i} = 0 \tag{5}$$

or in the integrated form for the problem in Figure 3,

$$Q = u_1 A_1 = u_2 A_2 \tag{6}$$

Equations 4 and 6 can be solved simultaneously to determine the sequent conditions (u_2, y_2) for given initial conditions (u_1, y_1).

Free Surface Hydraulic Jumps in Prismatic Channels

The Classical Hydraulic Jump on a Horizontal Bed

Several terms in Equation 4 can be dropped for a hydraulic jump in a horizontal rectangular channel with no baffles. These are:

1. The side force $P_s \sin(\phi/2)$.
2. The drag force F_D.
3. The body force component $W \sin \theta$.

Furthermore, the turbulent intensity term $\rho(I_2 - I_1)$ and the bed shear force F_f are relatively small. The momentum and pressure correction factors β and β' assumed to be close to unity. With these simplifications Equation 4 reduces to

$$\frac{u_2^2 y_2}{g} + \frac{1}{2} y_2^2 = \frac{u_1^2 y_1}{g} + \frac{1}{2} y_1^2 \tag{7}$$

while the continuity equation becomes

$$y_2 u_2 = y_1 u_1 \tag{8}$$

The simultaneous solution of Equations 7 and 8 gives the well-known cubic equation in the sequent depth ratio ($y_o = y_2/y_1$),

$$y_o^3 - (2F_1^2 + 1)y_o + 2F_1^2 = 0 \tag{9}$$

which yields the classical Belanger solution,

$$y_o = \tfrac{1}{2}(\sqrt{1 + 8F_1^2} - 1) \tag{10}$$

Rajaratnam [9] showed that the effect of bed friction on y_o increases with increasing F_1, reaching a reduction of about 4% at $F_1 = 10$.

The Rectangular Hydraulic Jump on a Sloping Bed

If the bed of the hydraulic jump is sloped, the longitudinal component of the body force term in Equation 4 must be retained. Even if the appurtenances are excluded, the macroscopic momentum and continuity equations, by themselves, do not yield the sequent depth ratio since the body force depends on the length and surface profile of the jump. These are determined by the internal mechanics of the jump.

Chow [16] and Kindsvater [25] represented the body forces by

$$W = \tfrac{1}{2}K\gamma L_j(y_1 + y_2) \tag{11}$$

where K is a factor to account for the shape of the jump. Using the assumptions that $\beta = \beta'$ and $F_f \simeq 0$ and substituting Equations 6 and 11 into Equation 4 leads to

$$y_o^3 - K_o y_o^2 - \left(\frac{2F_1^2}{\cos\theta} + K + 1\right) y_o + \frac{2F_1^2}{\cos\theta} = 0 \tag{12}$$

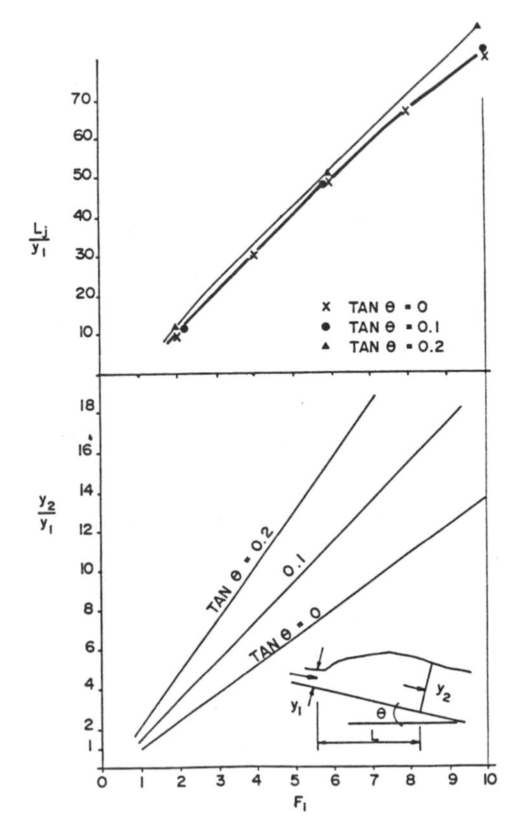

Figure 4. Jump length and sequent depth ratios for sloping rectangular channels [16].

where $K_o = (KL_j/y_1)\tan\theta = f(\theta, F_1)$

An approximate solution to Equation 12 was given by [9].

$$y_o = \frac{1}{2\cos\theta}(\sqrt{1 + 8G_1^2} - 1) \tag{13}$$

where $G_1 = F_1\Gamma_1$ (14)

$\Gamma_1 = 10^{1.55\theta^{rad}}$

Figure 4 compares the experimentally derived lengths and sequent depth ratios for horizontal and sloping hydraulic jumps. It is evident that the sequent depth increases sharply with increasing bed slope; however, the jump length in terms of y_1 is not greatly affected by θ.

A stilling basin with a sloping bed is sometimes used to accommodate uncertain or variable tailwater rating curves. In such cases, the stilling basin is designed to prevent sweep-out under the lowest tailwater levels. If higher tailwater levels are encountered, the start of the hydraulic jump will move upstream on the sloping apron. This arrangement is thought to give a more rapid reduction in the maximum velocity than would occur with a submerged hydraulic jump.

The Forced Hydraulic Jump in a Rectangular Channel

Another important case of Equation 4 is that involving appurtenances on the floor of the stilling basin. With the appropriate simplifications for a horizontal bed, the combination of Equations 4 and 6 gives

$$y_o^3 + \left[G\left(\frac{u_B}{u_1}\right)^2 F_1^2 - 2F_1^2 - 1\right] y_o + 2F_1^2 = 0 \tag{15}$$

where $G = S_B \dfrac{h_B^*}{y_1} C_D$ (16)

$$S_B = \frac{w_B}{s_B + w_B} = \text{blockage ratio} \tag{17}$$

$$h_B^* = y_B \text{ or } h_B \text{ whichever is less} \tag{18}$$

C_D = drag coefficient

s_B = baffle spacing

w_B = baffle width

h_B = baffle height

u_B = jet velocity at baffle

y_B = jet depth at baffle

The jet velocity at the baffle varies from u_1 to u_2 from the beginning to the end of the jump. McCorquodale and Regts [26] applied the momentum and continuity equations to estimate the expansion of the initial jet under the adverse pressure gradient of the forced hydraulic jump; the jet depth is

$$y_B = q/u_B \tag{19}$$

where $$\frac{u_B}{u_1} = 1 - \frac{y_o(x_B/y_1)}{\left(\dfrac{x_B}{y_1} + y_o\right)F_1^2}\left\{1 + \frac{1}{2}\left[\frac{y_o}{\dfrac{x_B}{y_1} + y_o}\right]\frac{x_B}{y_1}\right\} \tag{20}$$

x_B = distance from the initial section to the baffle

The determination of the drag coefficient on baffle blocks and sills in hydraulic jumps has been studied by Rajaratnam [6, 27], Harleman [28], Rand [29], Weide [30], Pillai and Unny [31], McCorquodale and Giratella [32], Narayanan [33], Tyagi et al. [34], and Karki [35]. Rajaratnam [9] represented the drag coefficient on a sill in a hydraulic jump as a function of the position of the wall from the start of the jump. He represented the drag force as

$$F_D = \tfrac{1}{2} C_d \rho u_1^2 h_B \tag{21}$$

where h_B = baffle height
$C_d = f(x/L_j)$

He found that C_d varied from about 0.6 at the start of the jump to about 0 at $x/L_j \simeq 0.8$; C_d then increased to about 0.12 for $x/L_j \geq 1.3$.

McCorquodale et al. [26, 32] attempted to define the drag coefficient, C_D, in terms of the baffle geometry. Thus, the drag force was defined, as in Equation 15, by

$$F_D = \tfrac{1}{2} C_D \rho u_B^2 A_B^* \tag{22}$$

where A_B^* = (area jet) $\cap$ (area baffle) (23)

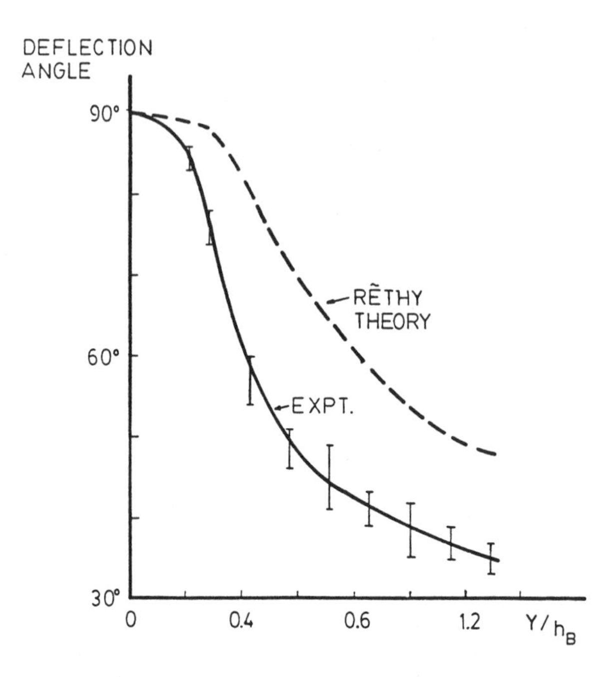

Figure 5. Experimentally and theoretically determined deflection angles for ventilated supercritical flow over a single row of baffle blocks [after Giratella (1969)].

In order to estimate an upper limit for C_D, McCorquodale et al. also studied the deflection of a supercritical jet by a single row of baffles. Figure 5 compares the experimental deflection angles with those estimated by conformal mapping. The decrease in the measured deflection angle is due to the effect of gravity, splash-back, air entrainment, and flow separation upstream of the baffle. The measured drag forces were the highest possible values since the wake was aerated.

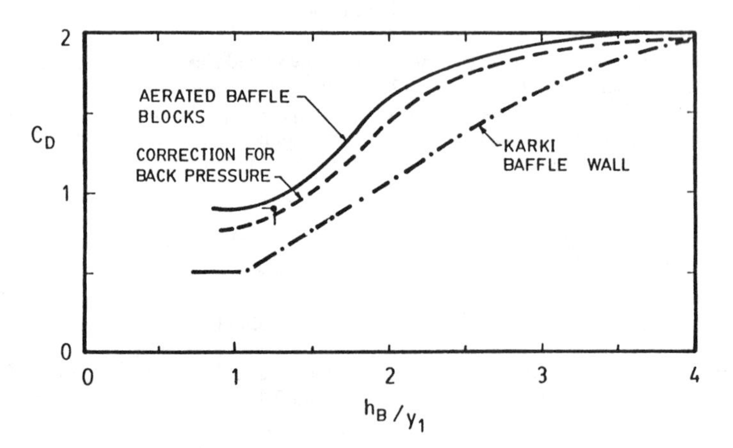

Figure 6. Drag coefficients on baffles in supercritical flow.

Figure 6 shows the C_D values obtained by Giratella. The author also carried out some experiments where a back pressure was allowed to develop. Figure 6 also shows the correction for this back pressure at a Froude number of approximately 5. The experimental values of C_D obtained by Karki [35] for an unventilated wall are also shown in Figure 6. The experimental conditions of Karki are probably closer to the hydraulic jump than those of the author. It was also noted by Giratella [32] that the upstream separation and associated vortex are stronger for a continuous wall than for a row of equally spaced baffle blocks. It is suggested that the Karki curve be used for C_D in the central portion of the jump while the higher values of C_D should be used to estimate the highest mean force on the baffles as well as for determining sweep-out conditions. In the subcritical flow in the downstream portion of the jump, the work of Rajaratnam [6] indicates that C_D varies from 0.6 to more than 2.

Two special cases of Equation 15 are of interest, i.e.

1. $x_B = 0$ and $u_B = u_1$; $y_B = y_1$
2. $x_B \simeq L_j$ and $u_B = u_2$; $y_B = y_2$

Case 1 corresponds to impending sweep-out and probably has a C_D very close to the value for a deflected jet. Figure 7 shows the sequent depth ratio, y_0, as a function of G and F_1 along with the specific example of: $h_B = 1.25y_1$; $h_B^*/y_1 = 1$; $S_B = 0.5$ and $C_D = 0.9$ which gives $G \simeq 0.45$.

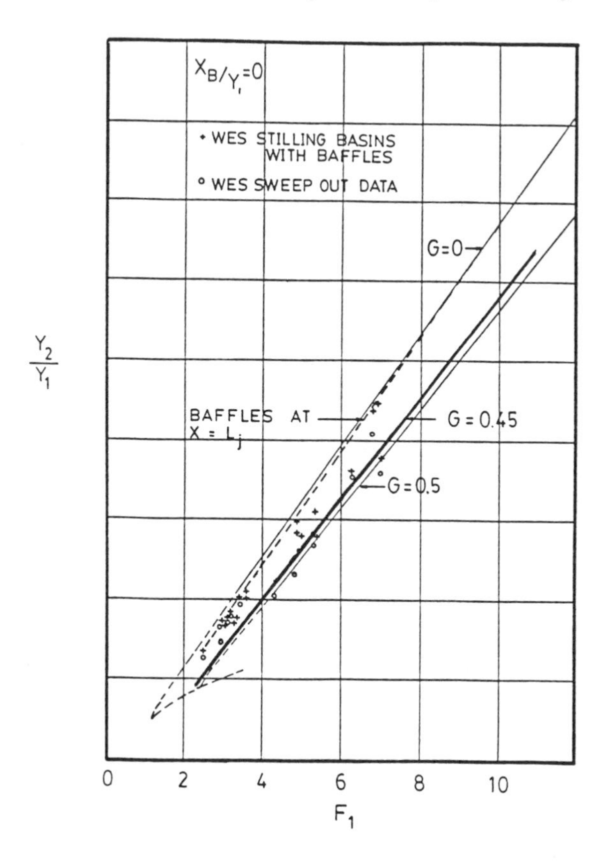

Figure 7. Sequent depth relationships for stilling basins with baffles [26] [WES (1969)].

In Case 2 the baffles are near the end of the jump and exert a relatively low drag force on the flow. Figure 7 shows the sequent depth ratio for $S_B = 0.5$, $C_D = 1$ and $(u_B/u_1) = (y_1/y_2)$ which gives $G \simeq 1/(2y_o)$. It is evident that baffle blocks or even a sill at the end of the jump are ineffective in substantially reducing the sequent depth; however, these appurtenances may be useful in deflecting the outflowing jet away from the bed. A large sill, such as that used in the USBR Stilling Basin III [36] may act as a weir to limit the minimum sequent depth to the sill height plus the critical depth at the sill. Forster and Skrinde [37] presented an analysis of the hydraulic jump with an abrupt rise at the end of the jump.

Figure 7 shows the Waterways Experimental Station, WES [38] data for stilling basins with baffles. The validity of Equation 15 and the magnitude of the C_D values in Figure 6 are supported by the agreement between the theoretical and WES experimental data.

Regts [39] used a dynamometer to measure forces in a forced hydraulic jump. He represented the mean drag force in dimensionless form as

$$\frac{F_D}{F_j} = \frac{F_b}{S_B \rho W y_1 u_1^2} \tag{24}$$

where $W = w_B + s_B$
F_b = drag on one baffle.

His results are compared with the theoretical solution [26] in Figure 8.

Narayanan and Schizas [33] studied the drag force on a baffle wall as well as the fluctuation in this force. They used the same definition for C_d as Rajaratnam. They found the rms value of the random component of C_d to be approximately 0.05; however, instantaneous fluctuations could be as high as seven times this value. Tyagi et al. [34] obtained similar results; however, they found the rms value of the random component of the fluctuating force to be about three times higher for $x/y_1 < 10$.

The standard baffle blocks [36] are subject to cavitation damage if the inflow velocity exceeds about 50 ft/sec. In order to overcome this problem, the U.S. Army Corps of Engineers [40] designed a streamlined baffle block with elliptical upstream curves. Harleman [28] studied these baffles and recorded similar drag forces to stepped baffles.

The Hydraulic Jump in Nonrectangular Channels

Occasionally site conditions (e.g. poor soil conditions) make it more economical to use sloping side walls in a stilling basin rather than the standard vertical walls. Such a stilling basin is referred to as a trapezoidal stilling basin. The angle of the side slopes is often determined by the inherent stability of the soil or rock underlying it.

Hydraulic jumps in circular sections often occur in sewers and in the tunnels of shaft spillways and in diversion tunnels.

Trapezoidal channels. Several researches, (e.g. [6, 41, 42]) have investigated the trapezoidal hydraulic jump. The basic analytical approach has been to apply Equations 4 and 6 neglecting the side force, body force, turbulence intensity, and friction terms and assuming $\beta = \beta' = 1$. An equation of the form

$$z^2 y_o^4 + z(2.5k + z)y_o^3 + (1.5k + z)(k + z)y_o^2 + \left[(1.5k + z)k - \frac{3(k + z)^2}{\left(\frac{k}{z} + 1\right)} F_\infty^2\right] y_o - 3(k + z)^2 F_\infty^2 = 0 \tag{25}$$

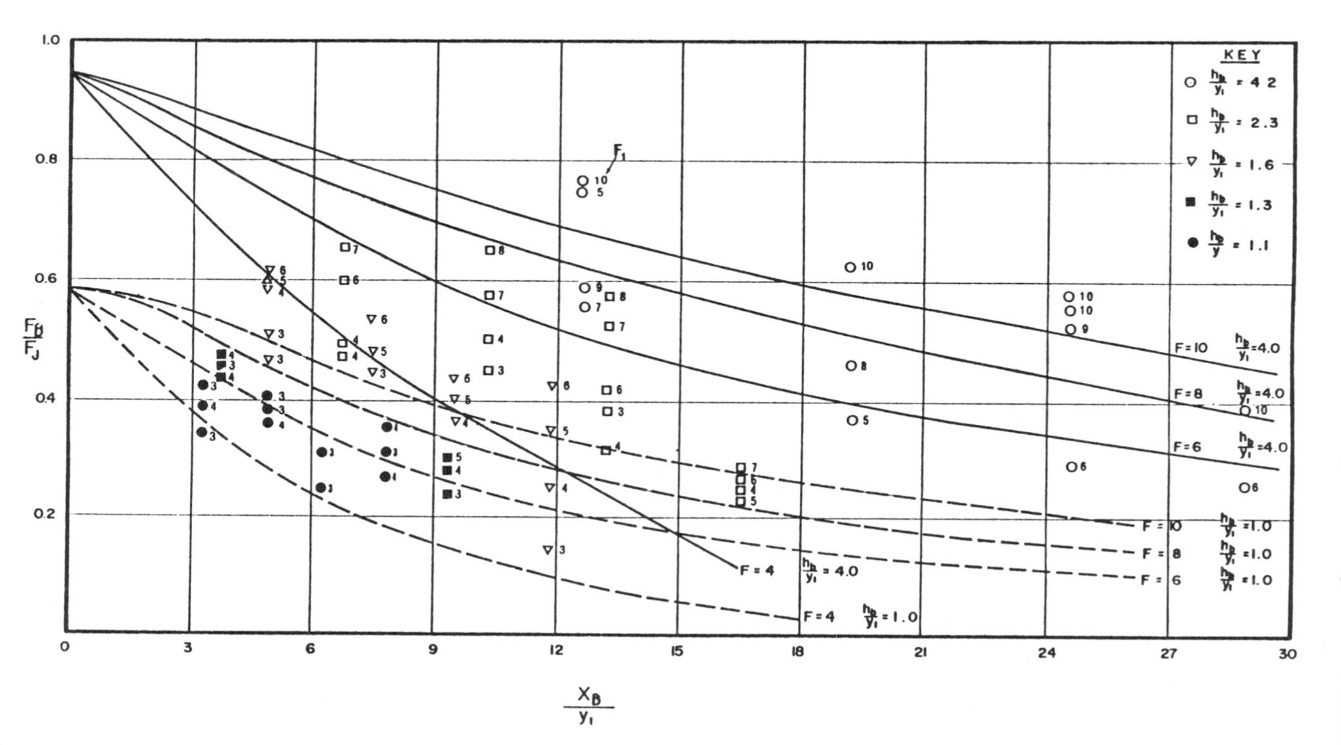

Figure 8. Comparison of predicted and measured mean baffle forces in a hydraulic jump [26, 39].

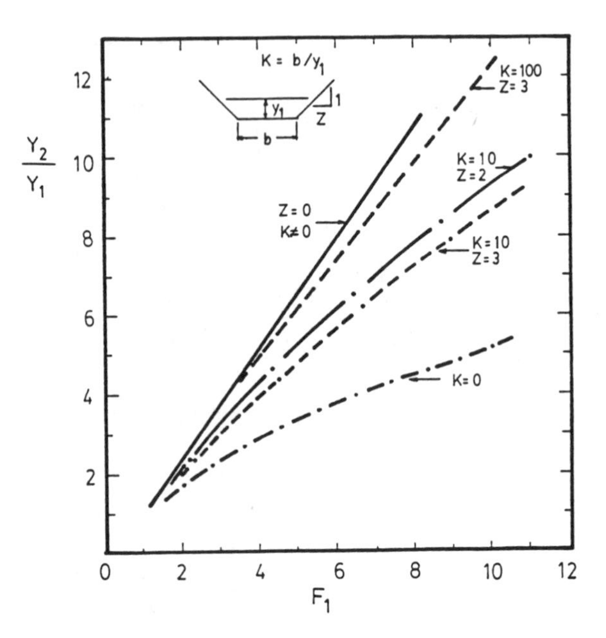

Figure 9. Sequent depth ratios for hydraulic jumps in nonrectangular prismatic channels (adapted from [9]).

in which $k = b/y_1$

$$z = \cot(\phi_s)$$

$$F_\infty = \frac{u_1}{\sqrt{gy_1}}$$

b = bottom width

$$F_1 = \frac{u_1}{\sqrt{gD}}$$

$$D = y_1(b + zy_1)/(b + 2zy_1)$$

is obtained.

Equation 25 reduces to the classical rectangular formulation for $z = 0$ and $k \neq 0$. The other extreme of Equation 25 is the jump in a triangular section which is obtained for $k = 0$ and $z \neq 0$, i.e.

$$y_o^4 + y_o^3 + y_o^2 - \tfrac{3}{2}F_1^2 y_o - \tfrac{3}{2}F_1^2 = 0 \tag{26}$$

Figure 9 illustrates the effect of depth-to-width ratio on the sequent depth ratio of a hydraulic jump in a trapezoidal channel.

One problem with the trapezoidal hydraulic jump is caused by oblique jets which emanate from the upstream corners of the jump and concentrate the outflow from the jump in the central portion of the channel as shown schematically in Figure 10. These jets are accompanied by counterrotating eddies which reduce the volume of the stilling basin that is available for energy dissipation. Furthermore, if the hydraulic jump occurs on a sloping trapezoidal section, upstream of the actual stilling

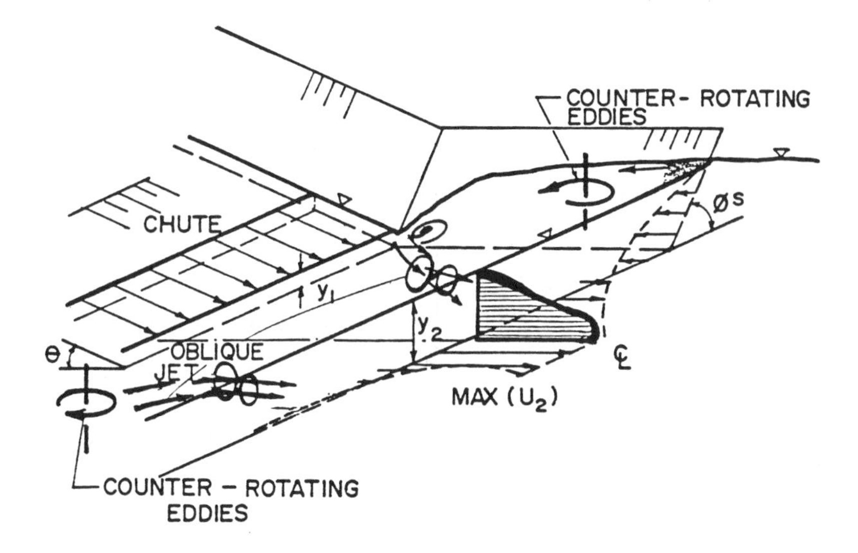

Figure 10. Schematic of trapezoidal hydraulic jump showing the oblique jets.

basin, the effect of the oblique jets and eddies may be even more severe as found by McCorquodale and Smith [20]. In these studies it was found that the undesirable flow concentration could be controlled by:

1. Directing a portion of the inflow onto the trapezoidal side slope.
2. Using a conical transition from the chute to the trapezoidal channel.
3. Installing guide vanes at the toe of the side slope.

The remedy, used in the McKeough Floodway Drop-Structure, is shown in Figure 11. This arrangement gave excellent energy dissipation over a wide range of tailwater depths (low and high submergence). It has recently been field tested and the field results appear to be in good agreement with the model predictions.

The normal trapezoidal hydraulic jump tends to be considerably longer than the corresponding rectangular jump [43]. This agrees with the author's observation [20] and appears to be related to the oblique jets and counterrotating eddies.

Circular channels. Thiruvengadan [44], Rajaratnam [9], Lane and Kindsvater [45], amongst others, investigated hydraulic jumps in circular channels. The research showed the applicability of the momentum approach. Figure 12 shows the sequent depth relationships obtained by momentum analysis for hydraulic jumps in horizontal circular channels.

Submerged Hydraulic Jump

Submerged hydraulic jumps in horizontal rectangular channels have been investigated by Rajaratnam and Govinda Rao [46]. A definition sketch of the submerged hydraulic jump is shown in Figure 13.

The submerged hydraulic jump is often encountered downstream from a sluice gate when the tailwater depth exceeds the sequent depth for the inflow. In such cases y_{2s} is known from the river rating curve and it is y_s that is unknown. It is important to be able to determine y_s because this has a direct bearing on the discharge from the gate.

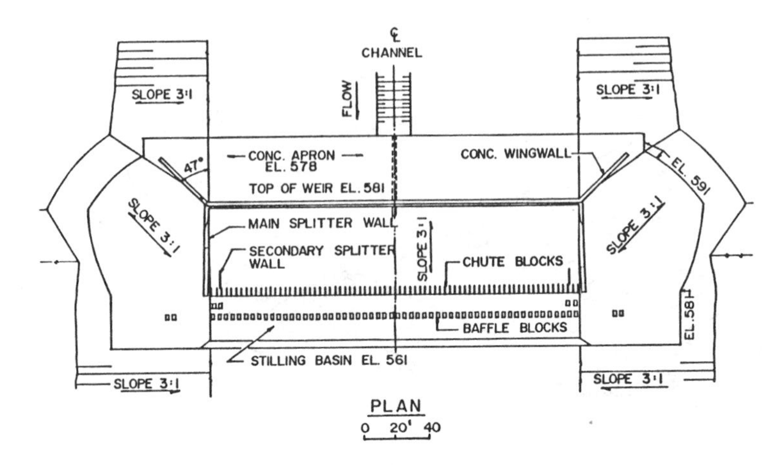

Figure 11. Modified trapezoidal stilling basin for the W. Darcy McKeough drop structure and stilling basin [67].

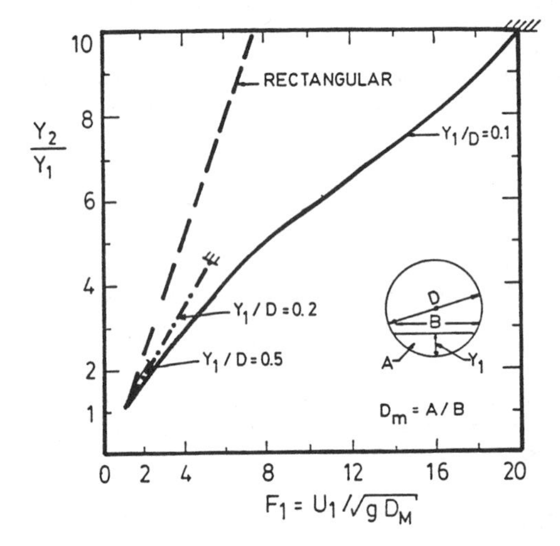

Figure 12. Sequent depth relationships for circular channels (adpated from [9]).

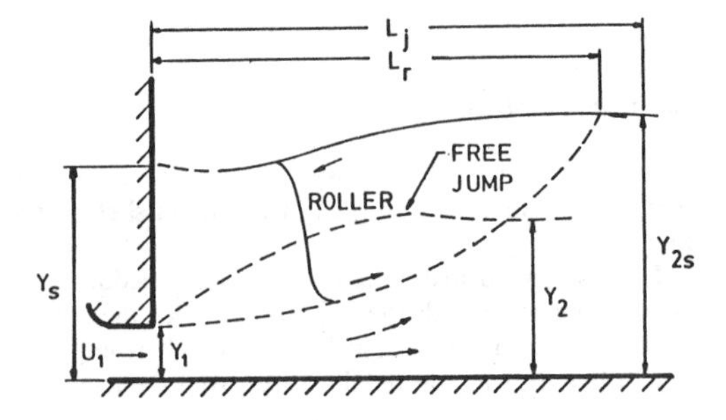

Figure 13. Definition sketch of submerged hydraulic jump.

The momentum method applied to the flow in Figure 13 [16] yields

$$\frac{y_s}{y_{2s}} = \sqrt{1 + 2F_2^2\left(1 - \frac{y_{2s}}{y_1}\right)} \tag{27}$$

where $F_2 = u_2/\sqrt{gy_{2s}}$

Rajaratnam [7–9] defined a submergence factor

$$S = \frac{y_{2s} - y_2}{y_2} \tag{28}$$

which he used for characterizing the submerged jump.

The Hydraulic Jump in Closed Conduits

Stationary hydraulic jump in closed conduits: Both steady and traveling hydraulic jumps are encountered in closed conduits, i.e., when a free surface flow changes abruptly to pressure flow. (See Figure 14.) Steady jumps may occur when a steep partially full conduit discharges through a submerged outlet or when the slope of the conduit changes from steep to mild or downstream of control gates in conduits. Earlier in this chapter, the momentum analysis was applied to hydraulic jumps in circular channels where tailwater did not pressurize the conduit. If the initial momentum is increased beyond this limit, the only way to create a hydraulic jump is to pressurize the pipe. Hamam [47] found the following simple momentum balance could be applied if the slope of the conduit was relatively flat (less than 1%):

$$A_1\left(\bar{y}_1 + \frac{u_1^2}{g}\right) = A_o\left(y_p + \frac{D}{2} + \frac{u_2^2}{g}\right) \tag{29}$$

or

$$y_p = \text{surcharge pressure head}$$
$$= \frac{A_1}{A_o}\left(\bar{y}_1 + \frac{u_1^2}{g}\right) - \frac{D}{2} - \frac{u_2^2}{g} \tag{30}$$

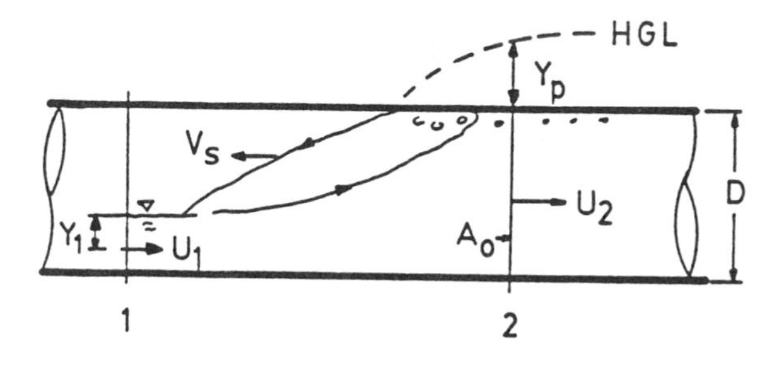

Figure 14. Definition sketch of a closed conduit hydraulic jump.

in which A_1 = initial flow area
$\bar{y}_1$ = centroid of A_1 from surface
$Q = u_1A_1 = u_2A_o$
A_o = full conduit area
D = diameter of conduit or full depth

An important characteristic of closed conduit flows is their capacity to remove air from the conduit. Kalinske and Robertson [48] suggested the following air demand equation:

$$\frac{Q_a}{Q_w} = 0.0066(F_r - 1)^{1.4} \tag{31}$$

where Q_a = air discharge
Q_w = water discharge
$F_r = u_1\sqrt{gy_1}$

The U.S. Army Corps of Engineers [49] suggested a design equation in the form

$$\frac{Q_a}{Q_w} = 0.03(F_r - 1)^{1.06} \tag{32}$$

as the upper envelope to the empirical data.

Traveling hydraulic jump in closed conduits. McCorquodale and Hamam [49–51, 21] studied the role of traveling hydraulic jumps in the transition of gravity to surcharge flow in circular and rectangular conduits. They found that a sudden surcharging anywhere in a conduit could produce a traveling hydraulic jump. These jumps traveled upstream, sometimes, at very high speeds (more than 10 m/s). Hamam showed that the momentum equation (Equation 29) could be applied to the hydraulic jump with a reference plane moving with the jump velocity, V_s, as given by

$$V_s = \frac{A_1u_1 - A_ou_2}{A_o - A_1} \tag{33}$$

in the upstream direction.

The traveling hydraulic jump in a closed conduit pushes a column of air in front of it. Experiments have shown [50, 21] that if this air cannot be readily exhausted, severe pressure transients can occur in the conduit.

Traveling Hydraulic Jumps in Open Channels

The subject of traveling hydraulic jumps in open channels is addressed in a number of standard references (e.g. [16, 52, 53]. The general approach as summarized by Chow is presented here. Chow considers four types of traveling hydraulic jumps as shown in Figure 15. Type B usually gives the most dramatic jumps. To analyze these jumps the classical hydraulic jump equations are applied to the relative velocity control volume as illustrated in Case B, Figure 15. This leads to a celerity

$$c = \sqrt{\frac{gy_2}{2y_1}(y_1 + y_2)} \tag{34}$$

and the surge velocities shown on Figure 15.

In the case of a sudden blockage of flow in a rectangular channel, the height of the surge

$$\Delta h = y_2 - y_1 = \frac{c}{g}\left(\frac{2y_1}{y_1 + y_2}\right)(V_1 - V_2) \tag{35}$$

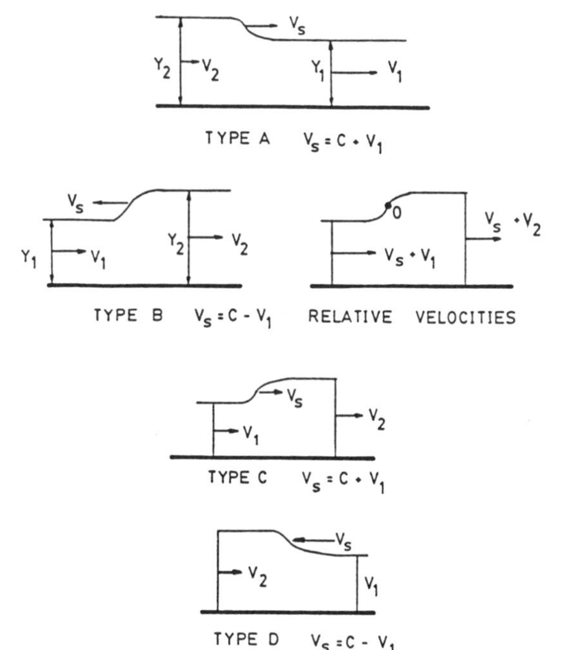

Figure 15. Chow's classification of traveling hydraulic jumps [16].

Hydraulic Jumps in Nonprismatic Channels

General

The two most distinctive nonprismatic channel stilling basins are the radially diverging stilling basin and the radially converging stilling basin [54]. Also classified as a nonprismatic stilling basin is the stilling basin with a relatively steep chute leading to a flatter apron in which the hydraulic jump occurs partially on the chute and partially on the apron. Hydraulic jumps also occur in flip buckets. Some examples of typical nonprismatic jumps are illustrated in Figure 16.

The Radially Diverging Hydraulic Jump on a Horizontal Bed

Figure 3 indicates the possible forces on a radially expanding jump. The significant difference between this jump and the classical rectangular jump is the longitudinal component of the side force, $P_s \sin \phi/2$. This force is a function of the jump length and the water surface profile, both of which are unknowns in the analysis. Therefore, some assumption must be made concerning the water surface profile and the overall length before applying the momentum equation (Equation 4). Gagnon [55] and Sadler [56] carried out a number of experiments to study the surface profiles in radial jumps, in the form of flow emanating radially from the center of a circle. Sadler estimated that the jump length L_j was $4(y_2 - y_1)$. Watson [57] also studied the circular jump (radial jump) caused by a vertical jet impinging on a horizontal plane. He ignored the side force and determined the following equation for y_2,

$$\frac{r_1 y_2^2 g a^2}{Q^2} = \frac{1}{\pi^2} - \frac{g y_2 a^4}{2Q^2} \tag{36}$$

in which a is diameter of the jet, and r_1 is radius to the start of the jump. Koloseus and Ahmad [58] studied the radial jump by assuming a straight line surface profile and introducing the radius

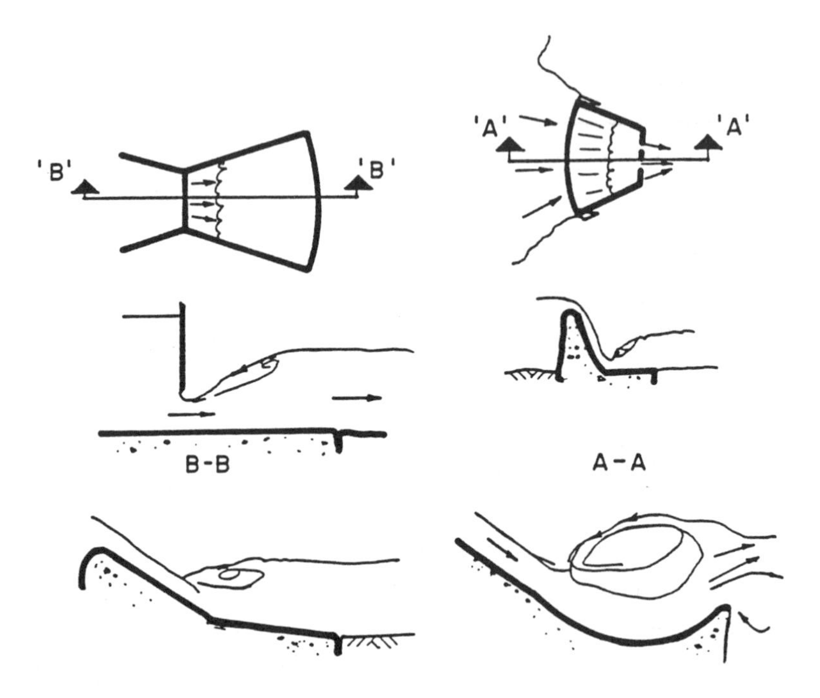

Figure 16. Typical nonprismatic hydraulic jumps.

ratio r_2/r_1 as a parameter which led to the equation,

$$y_o^3 - \frac{(r_o - 1)}{2r_o + 1} y_o^2 - \frac{(r_o + 6F_1^2 + 2)}{2r_o + 1} y_o + \frac{6F_1^2}{r_o(2r_o + 1)} = 0 \tag{37}$$

where $r_o = r_2/r_1$
$y_o = y_2/y_1$

Arbhabhirama and Abella [59] approximated the jump surface by a quarter ellipse and obtained the following relationship:

$$r_o y_o = \tfrac{1}{2}(\sqrt{1 + 8F_e^2} - 1) \tag{38}$$

where $F_e^2 = F_1^2 r_o + C_p'$ and C_p' is a correction for the side force.

Khalifa and McCorquodale [60] using experimental data, derived a parabolic equation for the water surface in the form

$$Y = \frac{y - y_1}{y_2 - y_1} = AR^2 + BR \tag{39}$$

$$A = 0.36 + 0.016F_1 + 0.155F_1F_p - 1.16F_p \tag{40}$$

$$B = 1.3 - 0.025F_1 - 0.146F_1F_p + 1.5F_p \tag{41}$$

$$F_p = \text{air content function} \simeq 0.034F_1 \tag{42}$$

where $y = (y - y_1)/(y_2 - y_1)$
$R = (r - r_1)/(r_2 - r_1)$

The development of the side force based on this effective water surface led to a cubic equation in y_o which was a function of the Froude number and r_o. The theoretical solution is shown on Figure 17 along with the empirical curves. The following empirical equations were also developed by Khalifa [60, 61],

$$y_o = 0.3r_o + 0.65F_1\left(1 + \frac{1}{r_o}\right) \tag{43}$$

and

$$\frac{L_j}{y_2} = 4.7 - 4.2/F_1 \qquad \text{for } 2 < F_1 < 9 \tag{44}$$

where $L_j = r_2 - r_1$

Equation 44 is compared with the U.S. Bureau of Reclamation jump length curve [15] in Figure 18. Khalifa also studied the air concentration in a radial jump. His results are compared with Rajaratnam's [9] results in Figure 19.

It was noted by Khalifa that the radial jump is subject to lateral instability. Scott-Moncrieff [62] observed a bistable behavior in a prototype diverging stilling basin. This instability depends on the F_1, ϕ, r_o and the initial conditions of the jump. The maximum angle of divergence with a radially curved inlet gate was given approximately by

$$\phi = 2\tan^{-1}\left(\frac{1}{2F_1^2}\right) \tag{45}$$

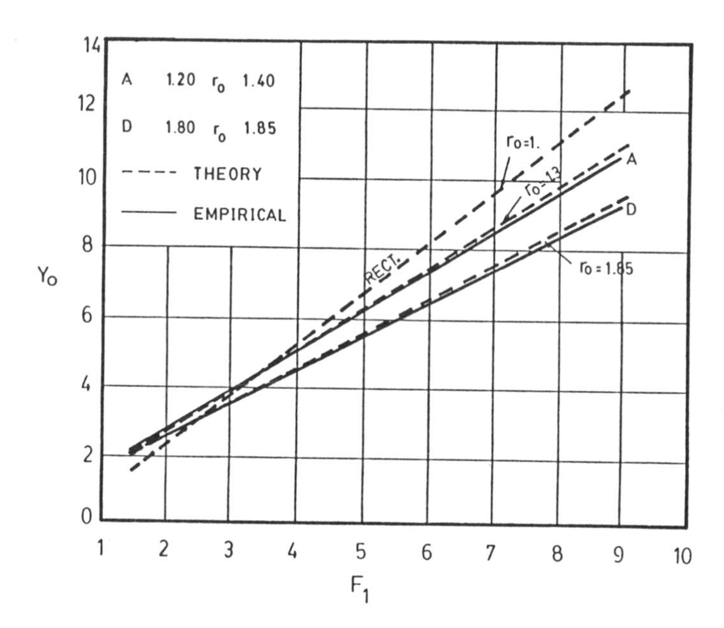

Figure 17. Sequent depth ratio for radial hydraulic jump—comparison of experimental and theoretical curves [61].

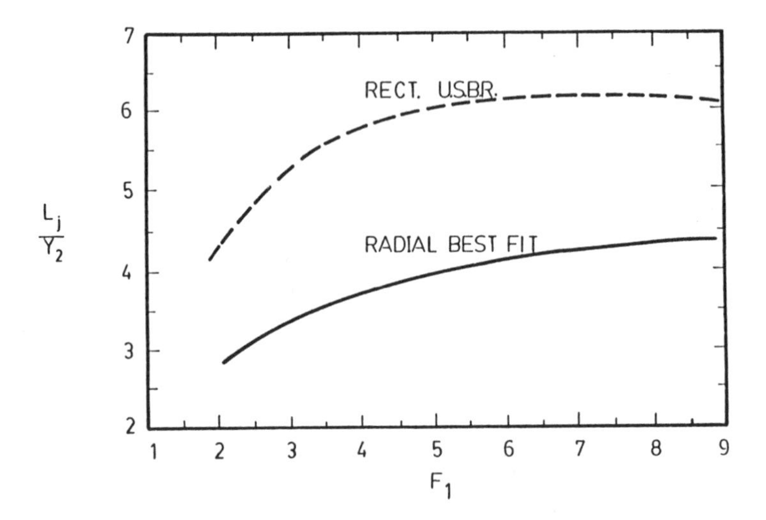

Figure 18. Comparison of the jump lengths for radial and rectangular hydraulic jumps [61].

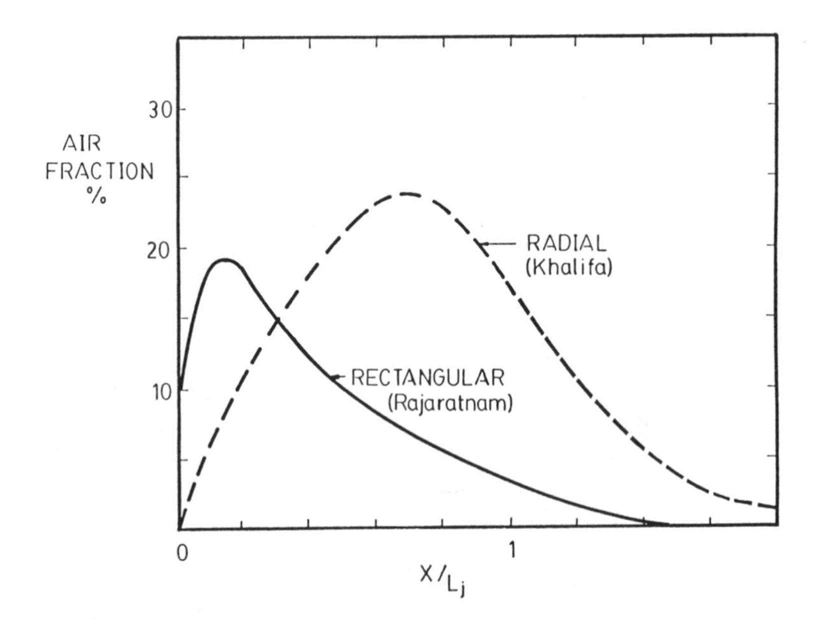

Figure 19. Longitudinal variation of air fraction in radial and rectangular hydraulic jumps [61, 9].

The radial hydraulic jump has been used as a basis for a number of stilling basin designs, namely a version of the St. Anthony Falls Stilling Basin [63] certain U.S. Corps of Engineers stilling basins, and for a number of critical flow flumes.

Radial flow stilling basins with large diverging angles were studied by Moore and Meshgin [64]. They proposed this type of stilling basin for use downstream from culverts.

The extreme case of a radially expanding stilling basin is the abrupt expansion. This jump was studied by Unny [65] who proposed an equation of the form

$$y_o = \frac{1}{2}(1 + K_1\lambda F_1^2)\left(\sqrt{1 + \frac{8F_1^2}{(1 + K_1\lambda F_1^2)^2}} - 1\right) \tag{46}$$

where $\lambda = y_1/b$
$K_1 = f(b/B)$
b = initial flow width
B = expanded flow width
$K_1 = 2.2$ for $B/b = 2$

The hydraulic jump at an abrupt expansion has also been studied by Herbrand [66] and Rajaratnam [10].

The Forced Radial Hydraulic Jump on a Horizontal Bed

The recently completed Darcy McKeough Dam near Wallaceburg, Ontario [67] utilizes a diverging stilling basin with baffle blocks and an end sill as shown schematically in Figure 20. McCorquodale [22] carried out a hydraulic study of this and alternate designs which did not include baffles. The radially expanding option with baffles was used because it permitted narrower and more economical gates to be used and at the same time distributed the outflow from the stilling

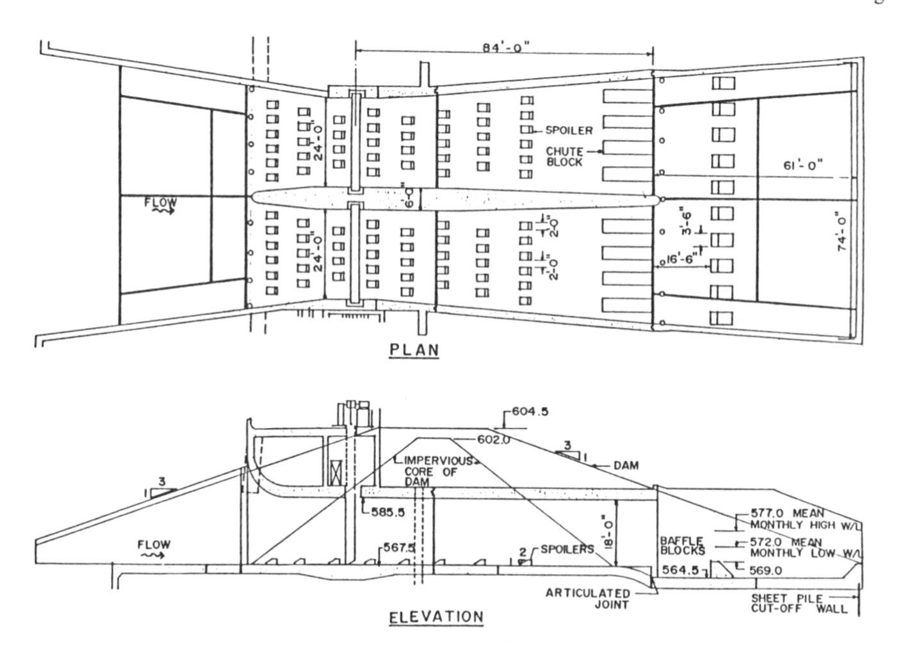

Figure 20. W. Darcy McKeough radial stilling basin [67].

basin over a greater width of the river than the width of the gates. The model tests indicated that the USBR [36] baffle blocks and end sill provided good energy dissipation, a high degree of protection against sweep-out, and a laterally stable outflow. In 1984 this structure experienced approximately the 1-in-20-year flood and performed as predicted by the model.

An analysis of the radial stilling basin with baffles was presented by Nettleton and McCorquodale [68]. An attempt was made to apply the macroscopic energy balance in order to determine the sequent depth; however, this approach was not totally successful because of the difficulty in determining the side force in the presence of the hump produced by the baffles. Figures 21A to E

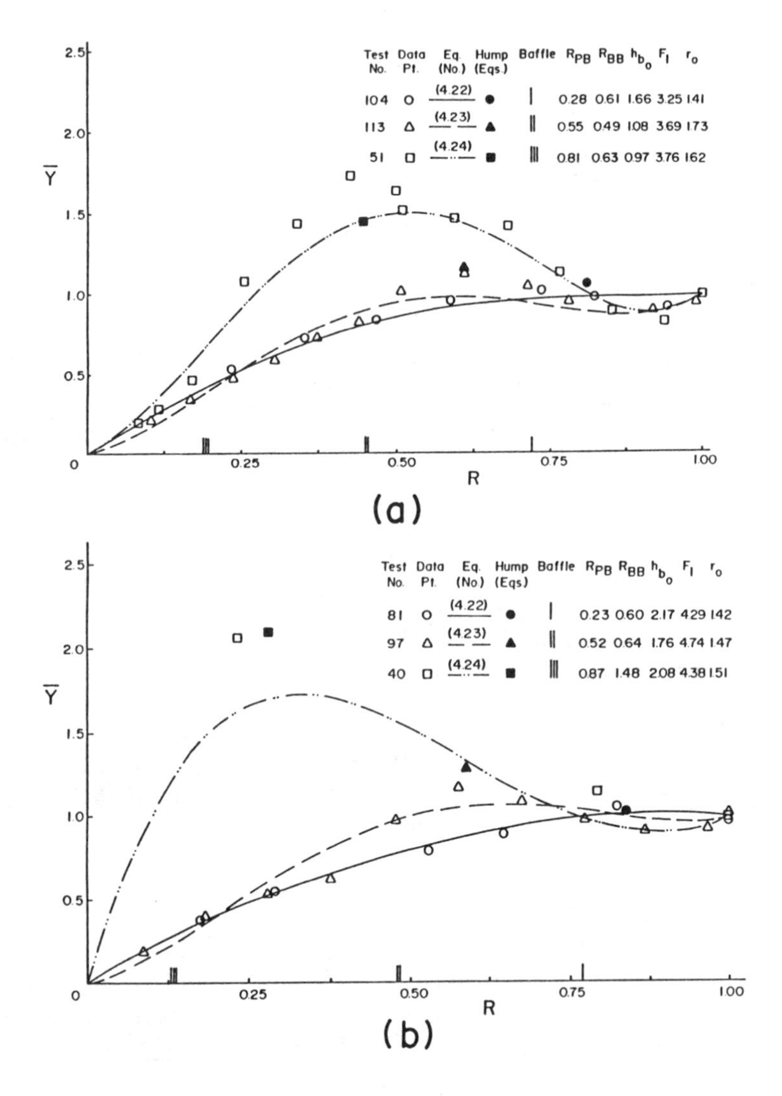

Figure 21. Water surface profiles for the forced radial hydraulic jump for various baffle positions and Froude numbers $\bar{y} = (y - y_1)/(y_2 - y_1)$; $R = (r - r_1)/(r_2 - r_1)$ [69]; (A) $3 < F_1 < 4$; (B) $4 < F_1 < 5$; (C) $5 < F_1 < 6$; (D) $6 < F_1 < 7$; (E) $7 < F_1 < 8$.

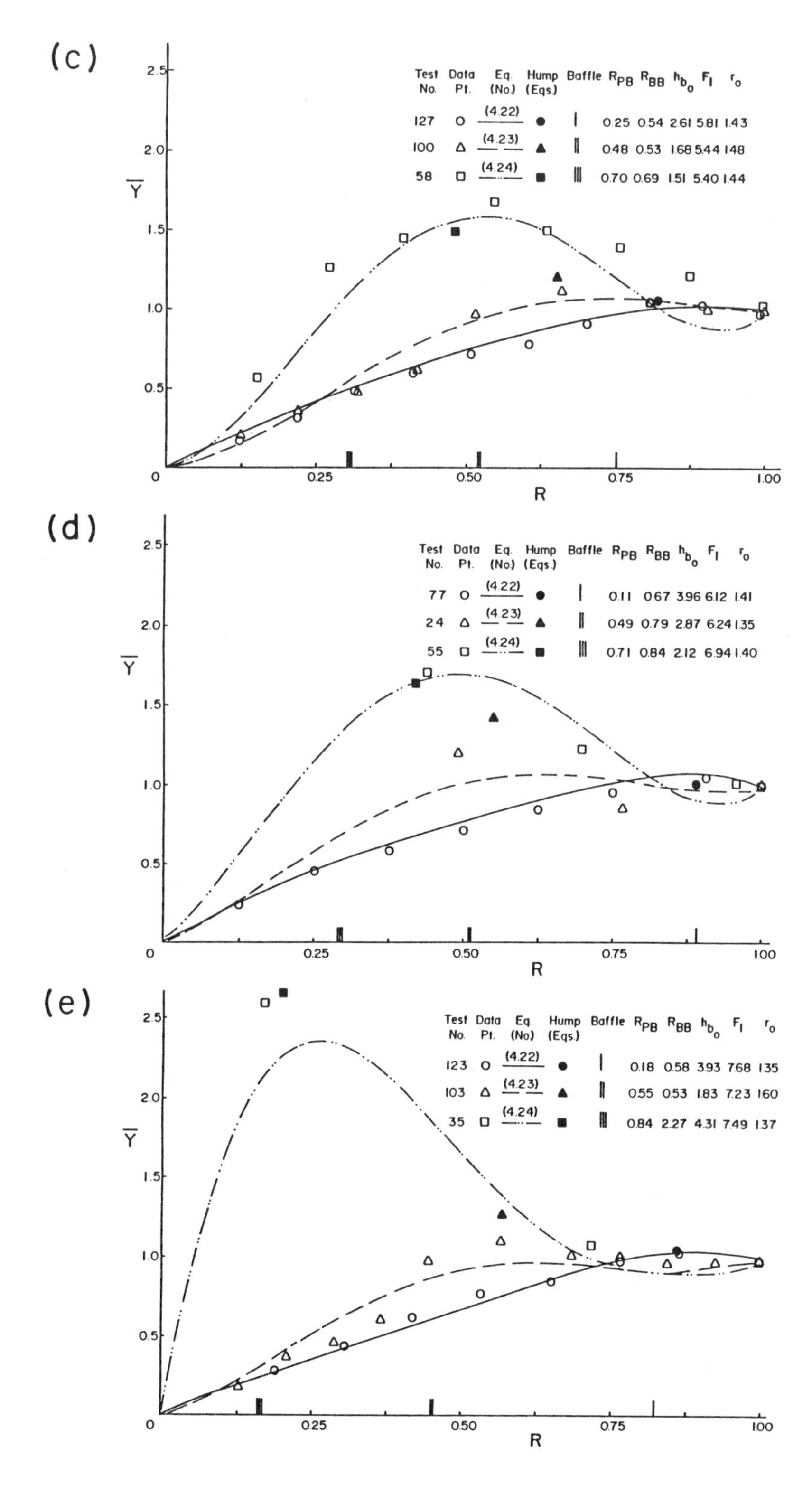

Figure 21. (Continued)

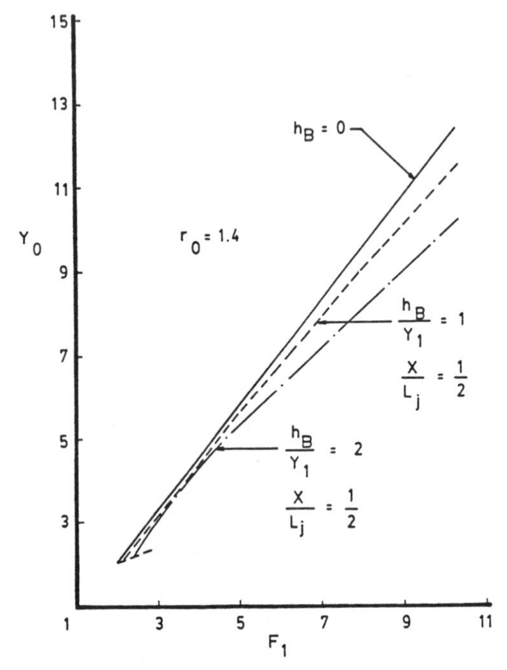

Figure 22. The suggested sequent depth ratio as a function of F_1 and h_B for $r_o = 1.4$.

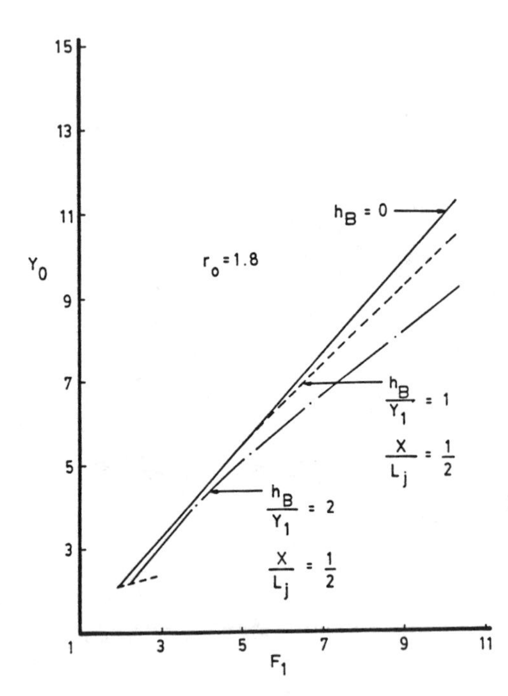

Figure 23. The suggested sequent depth ratio as a function of F_1 and h_B for $r_o = 1.8$.

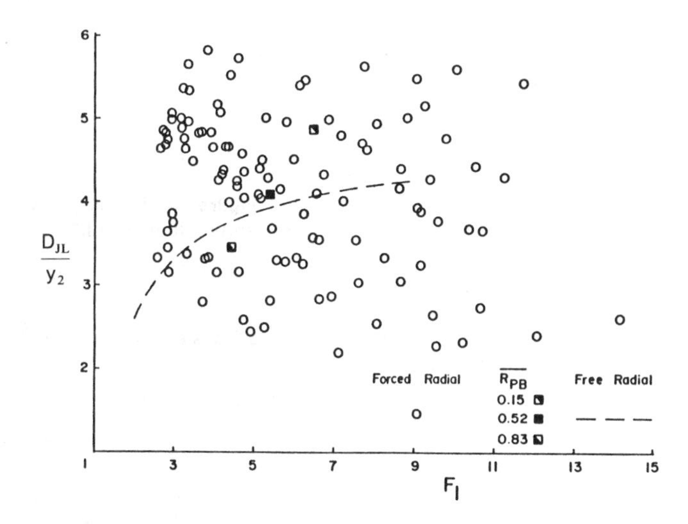

Figure 24. Estimated jump length as a function of F_1 (note D_{JL} = length to the second standing wave) [69].

illustrate the variability of the free surface profile with baffle position and Froude number. Figures 22 and 23 indicate typical theoretical sequent depth ratios. Nettleton [69] found a large degree of scatter in his measurements of the length of a forced radial jump. One of the problems in measuring the length of the jump involved the estimation of where the flow reaches the sequent depth because of secondary waves caused by the baffles. Figure 24 shows the scatter in the experimentally determined jump lengths. The work of Nettleton shows that the addition of baffles did not substantially decrease the jump length as has been observed when baffles are added to rectangular stilling basins.

Nettleton found that the outflow from a forced radial hydraulic jump became unstable for flare angle

$$\frac{\phi}{2} \geq \tan^{-1}\left(\frac{1}{1.5F_1}\right) \tag{47}$$

Typical lateral velocity profiles are shown in Figure 25. Lateral instability is evident at $F_1 = 7$.

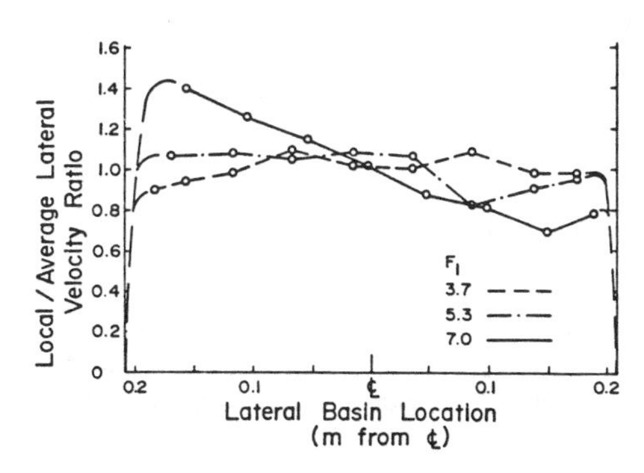

Figure 25. Lateral velocity profiles at the downstream end of a forced radial jump [67].

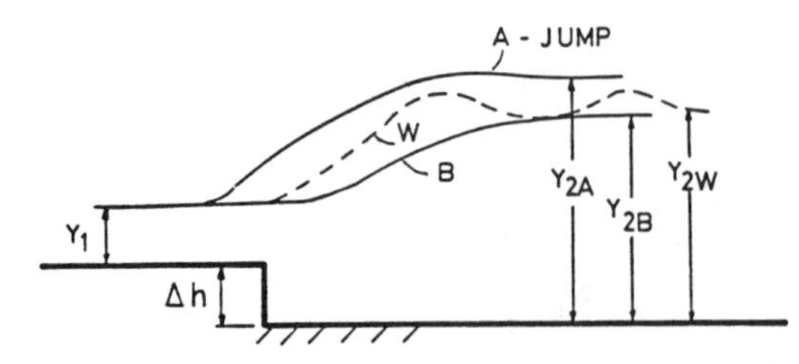

Figure 26. Forms of jumps at an abrupt drop [9, 16].

The Hydraulic Jump at an Abrupt Drop

The hydraulic jump at an abrupt drop can take the three forms shown in Figure 26. Hsu [70] developed the following equations: for Jump A

$$F_1^2 = \frac{1}{2}\left(\frac{y_o}{1 - y_o}\right)\left[1 - \left(y_o - \frac{\Delta h}{y_1}\right)\right] \tag{48}$$

where $y_o = y_{2A}/y_1$

and for Jump B

$$F_1^2 = \frac{1}{2}\left(\frac{y_o}{1 - y_o}\right)\left[\left(\frac{\Delta h}{y_1} + 1\right)^2 - y^2\right] \tag{49}$$

Rajaratnam and Ortiz [71] studied the velocity distribution in the W jump and the B jump. They found that the velocity decayed more rapidly in the W jump than in the B jump or in classical rectangular jump. They also found that there was no surface roller in the W jump.

The Hydraulic Jump at an Abrupt Rise

Forster and Skrinde [37] carried out analytical and experimental studies on the hydraulic jump at an abrupt rise. Their results are summarized on Figure 27.

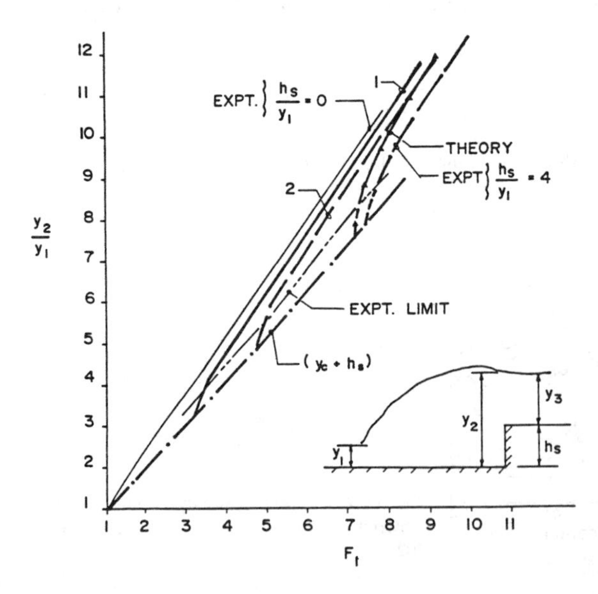

Figure 27. Sequent depth ratio at abrupt rises [37].

The abrupt rise is often required in a stilling basin when the tailwater depth in the river is insufficient to prevent sweep-out. The USBR standard end sill (Basin III) may fulfill the same purpose.

The Submerged Radial Hydraulic Jump

Khalifa and McCorquodale [24] carried out a study of the submerged radial hydraulic jump and derived the following equation for the submergence of the inflow.

$$\frac{y_s}{y_1} = -A_* + \sqrt{A_*^2 + r_o y_o^2 + 2F_1^2\left(\frac{1}{r_o y_o} - 1\right)} \tag{50}$$

in which $A_* = \frac{1}{2}y_o(r_o - 1)B$
$y_o = y_{2s}/y_1$
$r_o = r_2/r_1$
$B \simeq 1.04 - 0.074\sqrt{F_1}$

Abdel-Gawad and McCorquodale [72] extended the work of Khalifa [61] on the length of submerged radial jumps. The results are shown in Figure 28.

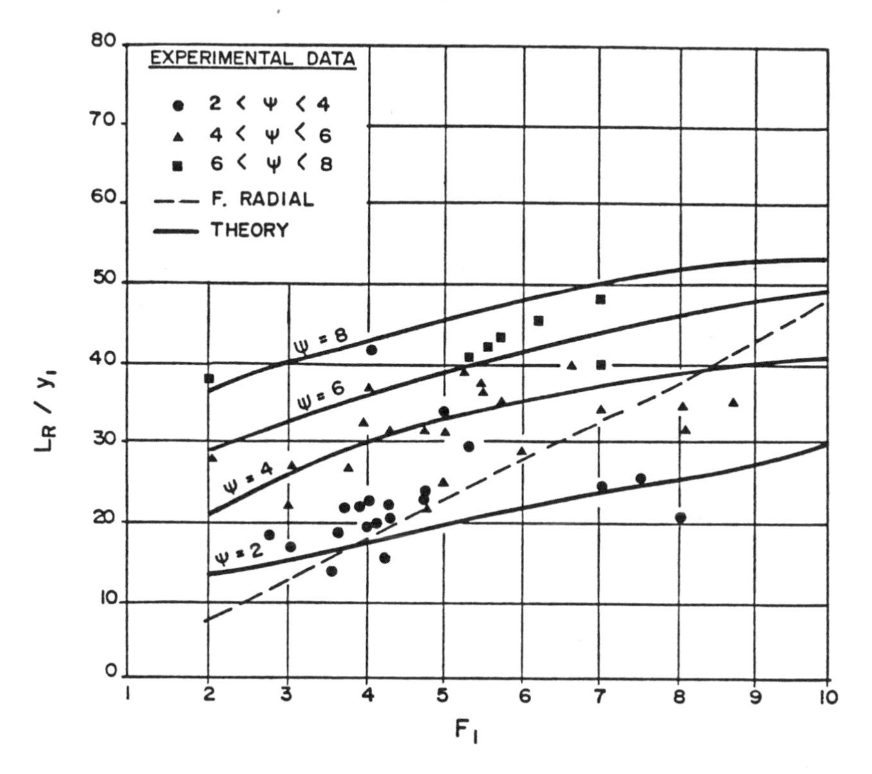

Figure 28. Comparison of measured and predicted jump lengths for the submerged radial hydraulic jump for different submergence conditions [72].

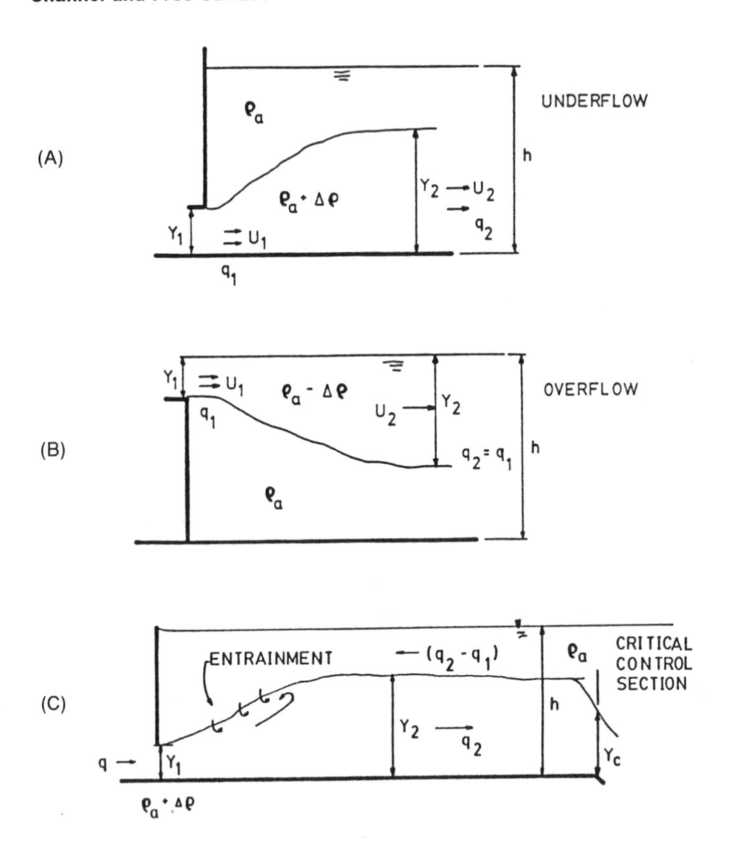

Figure 29. (A) Underflow internal hydraulic jump with immiscible fluids; (B) overflow internal hydraulic jump with immiscible fluids; (C) underflow internal hydraulic jump with dilution [73].

Internal Hydraulic Jumps

An internal hydraulic jump will occur when a heavy liquid flows under a lighter liquid, provided that

1. The densimetric Froude number is greater than 1
2. The sequent depth conditions are satisfied
3. The jump is stable

Figure 29A and B defines typical underflow and overflow internal hydraulic jumps with immiscible fluids. Internal hydraulic jumps can also occur when a light fluid flows at a supercritical velocity over a denser fluid. The more general case of an internal hydraulic jump was given by Baddour and Abbink [73] as illustrated in Figure 29C.

One of the earliest investigations of the internal hydraulic jump (immiscible fluids) was made by Yih and Guha [74]. They analyzed both the underflow and overflow jumps and derived an equation analogous to the classical hydraulic jump equation

$$\frac{y_2}{y_1} = \frac{1}{2}(\sqrt{1 + 8(F'_1)^2} - 1) \tag{51}$$

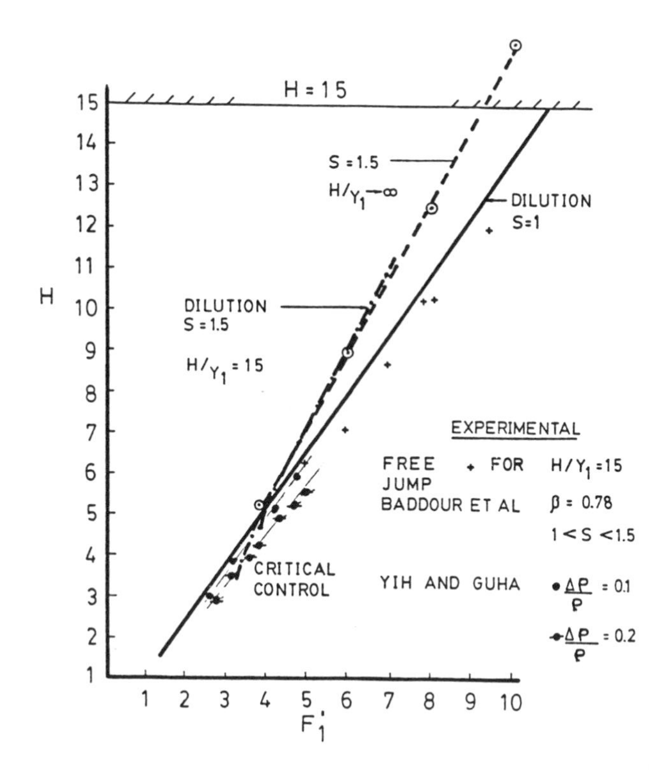

Figure 30. Internal hydraulic jump solutions with and without dilution [73, 74].

where $$F'_1 = u_1 \Big/ \sqrt{g \frac{\Delta\rho}{\rho_a} y_1} \qquad (52)$$

$\Delta\rho$ = the difference in density between the inflow and the ambient fluid.

Their theoretical (S = 1) and experimental results are given in Figure 30. It appears that the larger the density difference between the fluids the closer the agreement is between the experimental results and the theory.

Baddour et al. [73] carried out a number of experiments on internal hydraulic jumps in which a critical control section was located at the downstream end of his flume. He studied a wide range of Froude numbers and found the basic forms of jumps illustrated in Figure 31. They used a salt solution to create the heavy density currents with density differences of $0.004\rho_2$ to $0.03\rho_2$ with Reynolds numbers in the range of 800 to 2,100 based on the hydraulic radius of the dense fluid. They summarize the governing equations for the system shown in Figure 29C as follows:

Momentum:

$$2F_1'^2 S\left[\frac{S^2}{R} + \left(\frac{S-1}{H-R}\right)^2\left(\frac{H}{2} - R\right) - 1\right] = S - R^2 \qquad (53)$$

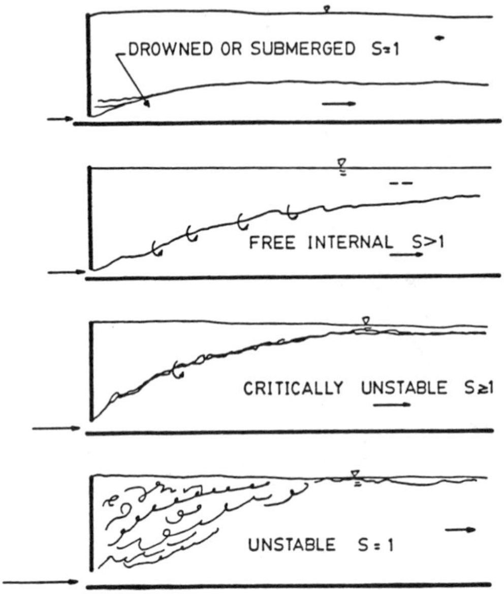

Figure 31. Basic forms of the internal hydraulic jump [73].

Energy:

$$R + \frac{1}{2} F_1'^2 S\left[\left(\frac{S}{R}\right)^2 - \left(\frac{S-1}{H-R}\right)^2\right] = Y_c + \frac{F_1'^2}{2\beta^2}\left[(SY_c)^2 - \left(\frac{S-1}{H-Y_c}\right)^2\right] \tag{54}$$

Conjugate depth:

$$O \leq R \leq H \tag{55}$$

Energy loss:

$$\Delta E_j \geq 0 \tag{56}$$

Subcritical region:

$$(F_1')^2 S\left[\frac{S^2}{R^3} + \frac{(S-1)^2}{(H-R)^3}\right] \leq 1 \tag{57}$$

where $R = y_2/y_1$
$S = q_2/q_1$
$H = h/y_1$
$Y_c = y_c/y_1$
β = contraction ratio at the critical control section

Figure 30 shows the effect of dilution on confined and unconfined internal hydraulic jumps as predicted by Baddour and Abbink. They also studied the conditions for the various jump types shown in Figure 31. For $\beta \to 1$, the instability depends on F_1' and H, e.g. for $H = 15$ a free internal jump occurs for $1 < F_1' \leq 11$ and for $H = 20$ the upper limit is 15.

The reader is also referred to the work of Baddour and Chu [75], Chu and Vanvari [76] Elison and Turner [77], Harleman [78], Jirka and Harleman [79], Koh [80], Long [81, 82], Stefan et al. [83], Kravenburg [84], French [85], Turner [86], Mehotra and Kelly [87], and Hayakawa [88].

THE INTERNAL FLOW IN THE HYDRAULIC JUMP

Velocity Distribution and Turbulence in the Classical Hydraulic Jump

Most of the research work on hydraulic jumps has dealt with their macroscopic behavior. The important parameters in these studies include the sequent depth ratio and the jump length which are required for stilling basin design. Unfortunately, the internal flow has received much less attention than it deserves.

There are a number of reasons for studying the internal flow in hydraulic jumps. Foremost is the problem of scaling-up physical model data to predict prototype behavior. Stilling basins are normally tested in small-scale models and the prototype data are determined by applying the Froude Law. The Froude Law assumes that inertia and gravity are the dominant independent forces acting on the stilling basin. Viscous forces and surface tension are assumed to be negligible. A complete mathematical model of the internal flow would permit the modeler to assess the possible scale effects in a physical model. Other effects such as the behavior of the air phase and the effect of the entrance boundary layer are generally not identical in the physical model and in the prototype. Other reasons for studying the internal flow are:

1. To assess the cavitation potential within the hydraulic jumps, which depends on the internal flow and turbulence pattern.
2. To estimate the forces on appurtenances in a stilling basin, which depend on the internal flow.
3. To assess the mixing and reaeration characteristics of a hydraulic jump.
4. To determine the size and extent of riprap protection downstream from a stilling basin.

Experimental studies are the only sound basis for a numerical model of a complex phenomenon like the hydraulic jump. A number of excellent experimental studies can be found in the literature.

Early investigation of the internal flow in hydraulic jumps was hampered by the problem of making accurate velocity measurements in the presence of entrained air. Studies by Rouse and his coworkers [12, 89, 90] used air models with solid boundaries that conformed to the mean configurations of actual hydraulic jumps. They applied hot-wire technology and obtained information about the mean flow, the turbulence intensities and the Reynolds stresses within the domain of the simulated jumps. The air model could not simulate all aspects of the hydraulic jump (e.g. the surface gravity waves, the entrainment of the ambient fluid, and certain aspects of the actual pressure field. In addition, the solid boundary at the surface introduced a shear stress which would be higher than the corresponding stress for the air-water interface of an actual jump. Nevertheless, their work was a breakthrough in the understanding of the internal mechanics of the hydraulic jump and has provided a framework for future research.

Their analytical treatment of the hydraulic jump was based on the Reynolds and continuity equations for the fluid (Equations 2 and 5). They integrated these equations over the flow domain to obtain the macroscopic Equations 3 and 6.

Furthermore, they developed a differential equation of work and energy in the form:

$$\rho \bar{u}_j \frac{\partial}{\partial x_j}\left(\frac{\bar{u}_T^2}{2} + \frac{\bar{u}_T'^2}{2}\right) + \overline{\rho u_j' \frac{\partial}{\partial x_j}\left(\frac{u_T'^2}{2}\right)} + \rho \frac{\partial}{\partial x_j}(\bar{u}_i u_i' u_j)$$
$$= -\bar{u}_j \frac{\partial \bar{p}}{\partial x_i} - \overline{u_i' \frac{\partial p'}{\partial x_i}} + \rho \bar{u}_i \bar{X}_i + \mu \bar{u}_i \frac{\partial^2 \bar{u}_i}{\partial x_j \partial x_i} + \overline{\mu u_1' \frac{\partial^2 u_i'}{\partial x_i \partial x_j}} \tag{58}$$

in which $\bar{u}_T$ = magnitude of the local mean velocity vector
u_T' = random component of u_T
p' = random component of the pressure p

This equation was segregated into two parts relating to the work energy relationships for the mean flow and turbulence, respectively, as follows:

$$\bar{u}_j \frac{\partial(\rho \bar{u}_T^2/2)}{\partial x_j} + \rho \bar{u}_i \frac{\partial \overline{u_i' u_j'}}{\partial x_j} = -\bar{u}_i \frac{\partial \bar{p}}{\partial x_i} + \rho \bar{u}_i \bar{X}_i + \mu \bar{u}_i \frac{\partial^2 \bar{u}_i}{\partial x_j \, \partial x_j} \tag{59}$$

$$\bar{u}_j \frac{\partial(\rho \overline{u_T'^2}/2)}{\partial x_j} + \overline{u_j' \frac{\partial(\rho u_T'^2/2)}{\partial x_j}} + \rho \overline{u_i' u_j'} \frac{\partial \bar{u}_i}{\partial x_j} = -\overline{u_i' \frac{\partial p'}{\partial x_i}} + \mu \overline{u_i' \frac{\partial^2 u_i'}{\partial x_j \, \partial x_j}} \tag{60}$$

Neglecting the initial turbulence and the free surface stresses and assuming hydrostatic conditions throughout, the following integrated energy equation was derived for transport of the mean flow kinetic energy:

$$\int_o^y \frac{\bar{u}_T^2}{2g} \bar{u} \, dy - \int_o^{y_1} \frac{\bar{u}_T^3}{2g} dy + \int_o^y \frac{\bar{u}\,\overline{u'^2} + \bar{v}\,\overline{u'v'}}{g} dy$$

net flux of K.E. due to mean motion — work of Reynolds stresses at surfaces (turb. energy flux)

$$-\frac{1}{g} \int_o^y \int_o^x \left[\overline{u'v'} \left(\frac{\partial \bar{u}}{\partial y} + \frac{\partial \bar{v}}{\partial x} \right) + (\overline{u'^2} - \overline{v'^2}) \frac{\partial \bar{u}}{\partial x} \right] dy \, dx$$

net work by Reynolds stresses from o to x (turb. energy flux)

$$= q y_1 - q y \quad + \frac{\mu}{\gamma} \int_o^y \left[2\bar{u} \frac{\partial \bar{u}}{\partial x} + \bar{v} \left(\frac{\partial \bar{u}}{\partial y} + \frac{\partial \bar{v}}{\partial x} \right) \right] dy$$

rate of work by pressure forces — rate of work by viscous forces (conservative)

$$-\frac{\mu}{\gamma} \int_o^y \int_o^x \left[4 \left(\frac{\partial \bar{u}}{\partial x} \right)^2 + \left(\frac{\partial \bar{u}}{\partial y} + \frac{\partial \bar{v}}{\partial x} \right)^2 \right] dy \, dx \tag{61}$$

dissipation rate of mean K.E. due to viscous stresses

A similar equation was obtained for the turbulent kinetic energy transport:

$$\int_o^y \frac{\overline{u_T'^2}}{2g} \bar{u} \, dy + \int_o^y \frac{\overline{u_T'^2 u'}}{2g} dy$$

(mean flow) (turb. diff.) flux of turbulent K.E.

$$+\frac{1}{g} \int_o^y \int_o^x \left[\overline{u'v'} \left(\frac{\partial \bar{u}}{\partial y} + \frac{\partial \bar{v}}{\partial x} \right) + (\overline{u'^2} - \overline{v'^2}) \frac{\partial \bar{u}}{\partial x} \right] dy \, dx$$

production rate of turbulent K.E.

$$= -\int_o^y \frac{\overline{p'u'}}{\gamma} dy \quad + \frac{\mu}{\gamma} \int_o^y \left[\frac{\partial}{\partial x} \left(\frac{\overline{u_T'^2}}{2} + \overline{u'^2} \right) + \frac{\partial \overline{u'v'}}{\partial y} \right] dy - \int_o^y \int_o^x \left[K \overline{\left(\frac{\partial u'}{\partial x} \right)^2} \right] dy \, dx \tag{62}$$

rate of work by fluctuating pressure on the D/s surface — rate of work by viscous stresses of turbulence on the D/s surface — rate of dissipation of turbulent K.E.

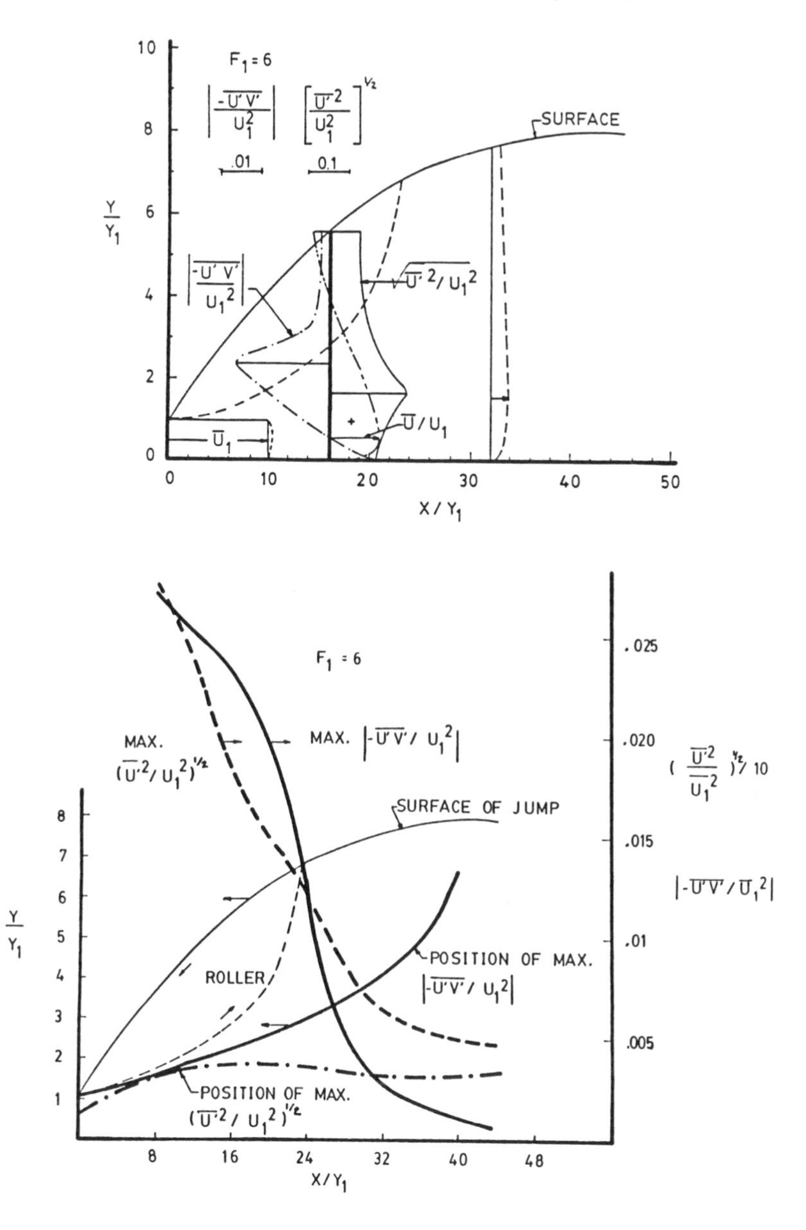

Figure 32. (A) Vertical distribution of velocity, turbulence intensity and turbulence shear stress at an $F_1 = 6$ [12]; (B) longitudinal variation of the maximum velocity, turbulence intensity and turbulence shear stress at $F_1 = 6$ [12].

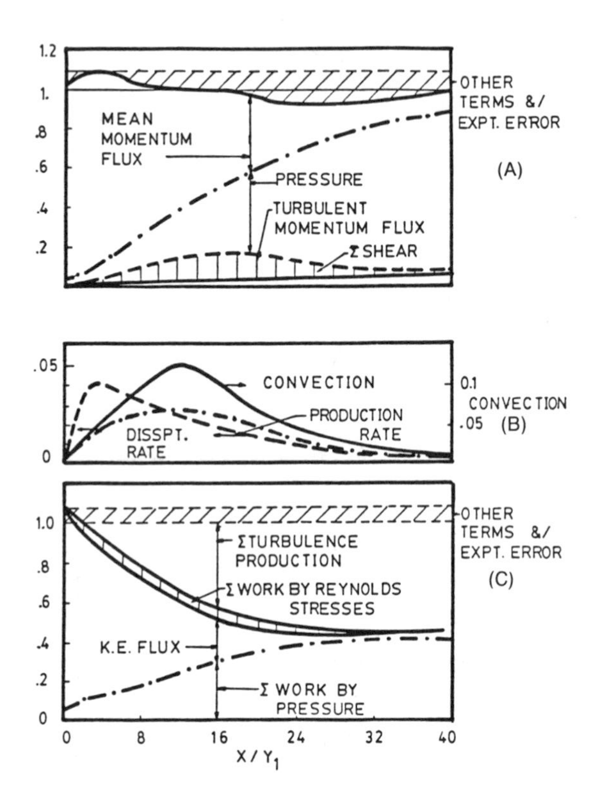

Figure 33. (A) Momentum balance for hydraulic jump at $F_1 = 6$; (B) convection, production, and dissipation rates for turbulent kinetic energy; (C) the energy balance in a hydraulic jump at $F_1 = 6$.

Rouse et al. presented experimental results for three values of F_1 (2, 4, and 6). Figure 32A shows the vertical distributions of mean velocity, turbulence intensity, and turbulent shear stress for the simulation of a jump at $F_1 = 6$. Figure 32B shows the longitudinal variation in the maximum velocity, turbulent shear stresses, and turbulence intensities in dimensionless form. Using Equation 2 integrated over intermediate ranges of x, a momentum balance was obtained as shown in Figure 33A. Similarly, Rouse investigated the mean energy balance as represented by Equation 61. Some of his experimental data are shown in Figure 33C. Rouse also presented experimental data on the rates of production, dissipation, and convection of turbulent kinetic energy. His results for $F_1 = 6$ are shown in Figure 33B.

Leutheusser and his coworkers [91, 92, 93, 94, 95] used a hot-film anemometer to measure the Reynolds stresses in actual hydraulic jumps. They investigated the effect of the initial boundary layer development on the internal and macroscopic flow of the jump. Their results for an undeveloped initial boundary layer are in general agreement with the air model data of Rouse et al. [12]. However, the results for the developed initial boundary layer are significantly different. Table 1 shows some typical data for Rouse et al. and Leutheusser et al. Leutheusser et al. [93] also found that the development of the initial boundary layer affected the entrainment of air. The undeveloped inflow resulted in deeper air entrainment in the upstream portion of the jump than the case of the developed inflow. Leutheusser et al. [92] also found that the tendency for flow separation on the bed and walls of the stilling basin is affected by the inflow boundary layer development. For example, with developed flow the range of potential separation was found to be $5 < x/y_2 < 7$ for

Table 1
Turbulence Characteristics in Hydraulic Jumps

Reference and Condition	F_1	Characteristic: $\max\left(\frac{\overline{U'^2}}{U_1^2}\right)^{1/2}$	Characteristic: $\max\left\lvert\frac{\overline{u'v'}}{U_1^2}\right\rvert$	Location: $\frac{x}{y_2} \simeq$	Location: $\frac{y}{y_2} \simeq$
Rouse (air)	2	0.20		1	0.6
			0.024	1	0.6
	4	0.24		1	0.25
			0.023	1	0.25
	6	0.28		1	0.2
			0.028	1	0.2
Leutheusser (water undeveloped)	2.85	0.23		1	0.3
			0.016		0.36
	6	0.32		1	0.15
			0.026	1	0.15
Leutheusser (water developed)	2.85	0.38		1	0.8
			0.1	1	0.85
	6	0.29		>1	0.6
			0.035	1	0.28

$F_1 = 4$ while for undeveloped flow this range was about $3 < x/y_2 < 7$. A similar delay in possible separation was noted at other values of F_1.

Rajaratnam [7, 9] has studied the velocity distribution, bed shear stress and air entrainment for various hydraulic jumps. He showed that the hydraulic jump could be studied as a wall jet under a strong adverse pressure gradient. He presented his data [9] in a dimensionless plot of $f(\eta) = \bar{u}/u_m$ versus $\eta = y/\delta_1$ where u_m is the maximum velocity and δ_1 is the depth to $\bar{u} = u_m/2$ in the outer layer (see Figure 34). The required values of δ_1 and u_m can be obtained from Figures 35 and 36.

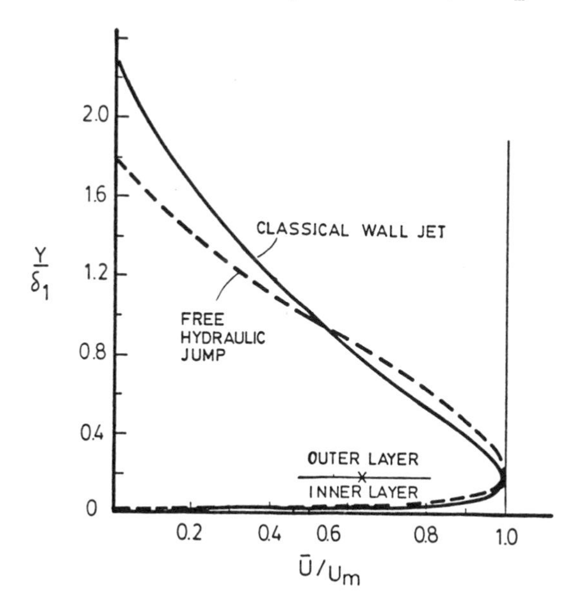

Figure 34. Comparison of the free jump velocity distribution with the wall jet distribution [9].

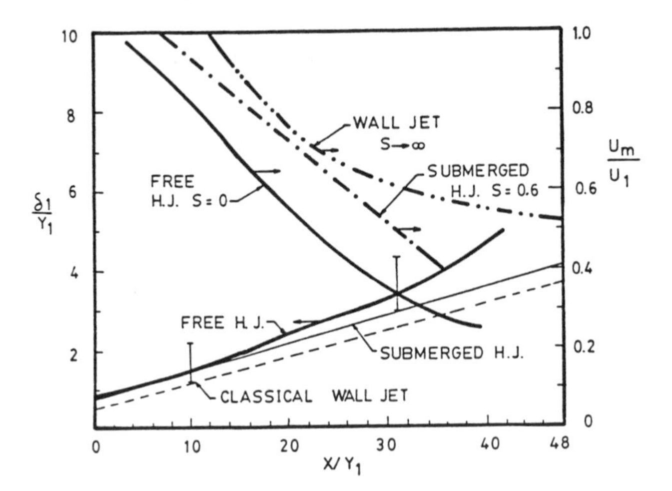

Figure 35. Longitudinal variation of the maximum velocity and boundary layer thickness of a hydraulic jump at $F_1 = 6$ [9].

Rajaratnam's [9] velocity decay curve is compared with that given by Rouse et al. [12] in Figure 36. The differences between the results cannot be explained by experimental error alone. Judging from the apparently longer potential core lengths in the Rouse study, it would appear that a more developed turbulent flow existed at the initial section in Rajaratnam's study.

Decay of Turbulence Downstream from a Stilling Basin

The decay of turbulence downstream from a stilling basin is important in the design of erosion protection. Lipay and Pustovoit [96] studied the decay of turbulence downstream from a stilling

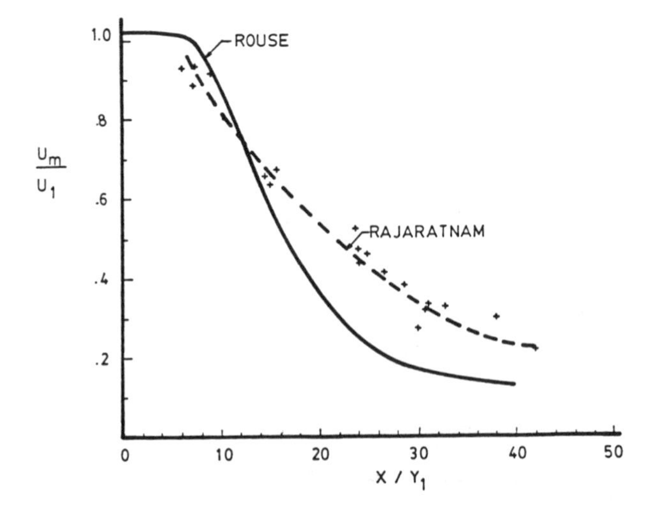

Figure 36. Longitudinal variation of the maximum velocity in a hydraulic jump at $F_1 = 6$ [9, 12].

basin. They summarized their results in an equation for u_A^*, the maximum instantaneous velocity near the bed,

$$\frac{u_A^*}{u_2} = 1.2 + \frac{0.2F_1}{1 + 0.07\left(\frac{x}{y_2}\right)^2} \tag{63}$$

where x = distance downstream of the basin

It is noted that between five and ten sequent depths are required for substantial decay of the excess turbulent velocity components in the outflow from a stilling basin.

Pressure Fluctuations in the Hydraulic Jump

Pressure fluctuations in the hydraulic jump have been studied by many researchers (e.g. [33, 97–103]). Typical pressure fluctuations at the bottom of free surface and open conduit hydraulic jumps are shown in Figure 37. The peak pressure fluctuations occur at about 25% of the jump length in both cases. The peak dimensionless pressure fluctuation is,

$$\left(\frac{\sqrt{\bar{p}'^2}}{\rho u_1^2}\right)_{max} \simeq 0.033 \text{ to } 0.04 \tag{64}$$

at $x \simeq 0.2L_r$ to $0.3L_r$. Figure 37 indicates that the pressure fluctuations in a closed conduit are slightly higher than with a free surface. Khader et al. [103] studied the hydraulic jumps with Froude number of approximately 6 and found that the constant in Equation 64 was approximately 0.042 at $x \simeq 10y_1$. He also determined the spectral density and autocorrelation functions at various positions on the bed of the jump. His analyses showed a peak in the spectral density function at

$$fy_1/u_1 \simeq 0.05 \tag{65}$$

where f is the frequency.

Internal Flow in Other Hydraulic Jumps

Rajaratnam [71, 104] studied the velocity field in hydraulic jumps at abrupt drops with W and B jumps. He noted that the maximum velocity in the B jump was deflected towards the bed while in the W jump it was deflected towards the surface.

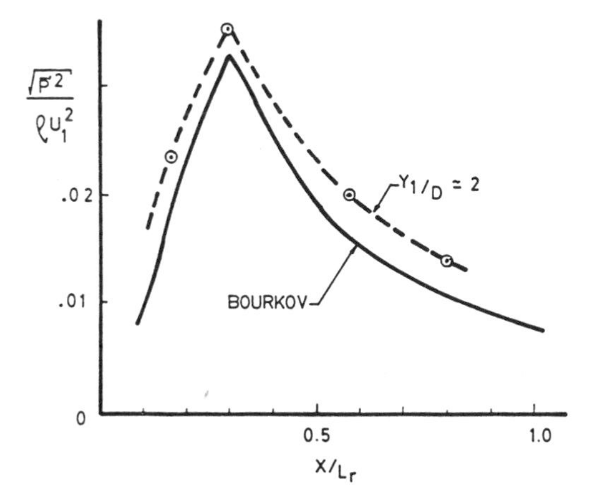

Figure 37. Typical pressure distributions under a hydraulic jump [98].

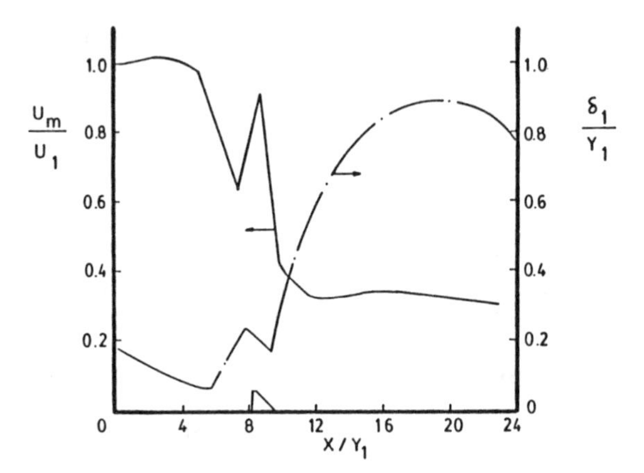

Figure 38. Decay of maximum velocity in a forced radial hydraulic jump [69].

Rajaratnam [9] showed that the velocity distribution in the rectangular submerged jump had the same form as a wall jet in the inner layer and was similar to the free jump in the outer layer. Abdel-Gawad and McCorquodale [72] came to similar conclusions about radial and submerged radial hydraulic jumps. Khalifa [61] carried out experimental and analytical studies of the internal flow in the radial hydraulic jump.

Nettleton and McCorquodale [68] measured the velocity field in the forced radial hydraulic jump. A typical velocity decay curve for this jump is given in Figure 38.

Analytical Approaches to Internal Flow

The Strip Integral Method was applied by Narayanan [105] to compute the flow patterns for the classical hydraulic jump. He used a power law relationship to represent the velocity distribution in the inner layer and a cosine function for the outer layer. McCorquodale and Khalifa [23] refined the work of Narayanan by:

1. Using the Gaussian velocity distribution suggested by Rajaratnam [104].
2. Incorporating the entrained air and its effect on the hydrostatic force and on the turbulent shear.
3. Using the kinematic boundary condition at the water surface.
4. Including the turbulence pressure.
5. Including the centrifugal effects.

A brief account of this method is presented here.

The Strip Integral Method is well established as a method of solving boundary-layer problems [106]. It has also been applied to other hydrodynamics and sedimentation problems in which a dominant flow direction can be identified. The method involves the selection of velocity shape functions with parameters which are functions of x (the dominant flow direction). These shape functions are inserted in the differential continuity and momentum equations which can then be partially integrated, over selected strips, to obtain a set of ordinary differential equations in the unknown parameters.

Based on the experimental observations of Rajaratnam [8] and Nagaratnam [106], the mean velocity distributions (Figure 39) in the hydraulic jump can be represented by:

$$\bar{u} = u_m \left(\frac{y}{\delta}\right)^{1/7}; \qquad 0 \leq y \leq \delta \tag{66}$$

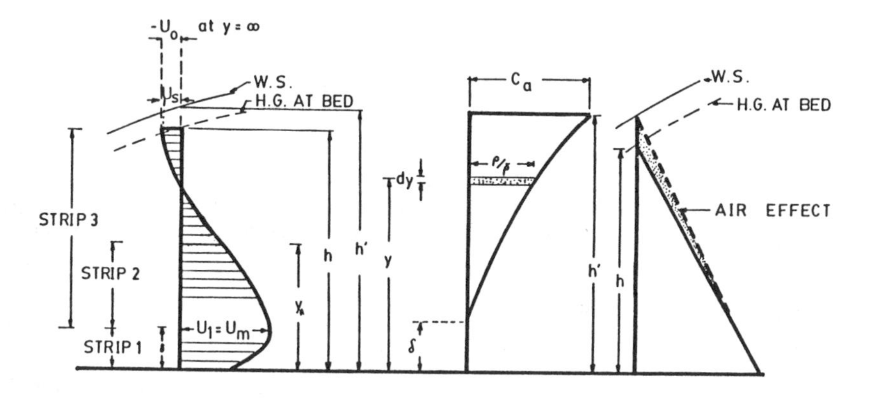

Figure 39. Definition sketch for strip integral method [23].

and

$$\bar{u} = u_o + u_t \exp(-4c[(y - \delta)/(h - \delta)]^2); \qquad \delta < y \leq h \tag{67}$$

where $\bar{u}$ = horizontal velocity
$u_t = (u_m - u_o)$
u_m = maximum horizontal velocity at x
δ = ordinate of the maximum velocity
u_o = horizontal velocity as $y \to \infty$
h = depth of flow
C = 0.693

The problem is to find u_m, δ, u_o, and h which are functions of x.
The momentum equation (Equation 2) is simplified to

$$\bar{u}\frac{\partial \bar{u}}{\partial x} + \bar{v}\frac{\partial \bar{u}}{\partial y} = -\frac{1}{\rho}\frac{\partial p_*}{\partial x} + \frac{1}{\rho}\frac{\partial}{\partial y}(\tau_b + \tau_t) \tag{68}$$

and

$$p_* = \gamma'(h' - y) + \rho(\overline{u'^2} - \overline{v'^2} + \overline{v_s'^2}) + \gamma c \tag{69}$$

where γ' = specific weight of air-water mixture above y
h' = depth to the surface of the air-water mixture
τ_b = laminar shear
τ_t = turbulent shear
$\overline{v_s'^2} = \overline{v'^2}$ at y = h
c = pressure head due to curvilinear flow

$$\simeq \frac{u_m^2(h - y)}{g\bar{R}} \tag{70}$$

$$\bar{R} \simeq 200 y_1 \tag{71}$$

The value of $(\overline{u'^2} - \overline{v'^2} + \overline{v_s'^2})$ is approximated by $\overline{u'^2}$.

The continuity equation

$$\frac{\partial \bar{u}}{\partial x} + \frac{\partial \bar{v}}{\partial y} = 0 \tag{72}$$

is also required for the solution.

The Strip Integral Method [105] is used to reduce Equations 68 and 72 to a set of ordinary differential equations which describe the longitudinal variation of the selected characteristics, u_m, δ, h, u_o, of the hydraulic jump. Four differential equations are required to obtain these unknowns.

The integral continuity equation is

$$\int_o^h \frac{\partial \bar{v}}{\partial y} dy = -\int_o^h \frac{\partial \bar{u}}{\partial x} dy = u_s \frac{dh}{dx} \tag{73}$$

in which u_s = horizontal component of the surface velocity; $u_s(dh/dx)$ is an approximation to the vertical velocity component at the effective jump surface, y = h.

Three other equations are obtained by integrating Equation 68 over the three strips shown on Figure 39:

$$\int_o^\delta \bar{u} \frac{\partial \bar{u}}{\partial x} dy + \int_o^\delta \bar{v} \frac{\partial \bar{u}}{\partial y} dy = -\frac{1}{\rho} \int_o^\delta \frac{\partial p_*}{\partial x} dy + \frac{1}{\rho} \int_o^\delta \frac{\partial \tau}{\partial y} dy \tag{74}$$

$$\int_\delta^{y_*} \bar{u} \frac{\partial \bar{u}}{\partial x} dy + \int_\delta^{y_*} \bar{v} \frac{\partial \bar{u}}{\partial y} dy = -\frac{1}{\rho} \int_\delta^{y_*} \frac{\partial p_*}{\partial x} dy + \frac{1}{\rho} \int_\delta^{y_*} \frac{\partial \tau}{\partial y} dy \tag{75}$$

$$\int_\delta^h \bar{u} \frac{\partial \bar{u}}{\partial x} dy + \int_\delta^h \bar{v} \frac{\partial \bar{u}}{\partial y} dy = -\frac{1}{\rho} \int_\delta^h \frac{\partial p_*}{\partial x} dy + \frac{1}{\rho} \int_\delta^h \frac{\partial \tau}{\partial y} dy \tag{76}$$

in which

$$y_* = \frac{1}{\sqrt{8C}}(h - \delta) + \delta \tag{77}$$

and

$$\tau = \tau_b + \tau_t \tag{78}$$

y_* is the ordinate at which maximum turbulent shear occurs.

Equations 74 to 78 introduce bed and turbulent shear at y_*. Zero shear is assumed at y = h. An expression suggested by Rajaratnam [7] was used for shear stress at the bed,

$$\tau_b = \frac{0.0424}{\left(\frac{u_m \delta}{\nu}\right)^{0.25}} \rho \frac{u_m^2}{2}; \qquad y = 0 \tag{79}$$

The Prandtl mixing length concept along with the free jet theory and Equation 67 were used to develop the following expression for the maximum turbulent shear:

$$\tau_t = \rho D_*^2 c \frac{u_t^2}{e}; \qquad y = y_* \tag{80}$$

where $D_* = 0.11$ which was obtained from the experimental air model data of Rouse [90] and agrees with the free jet theory [104].

The values of the average air fraction C_o along the jump can be estimated from the experimental data of Rajaratnam [9] which can be approximated by

$$C_o = 0.066F_1^{1.35}\frac{x}{L_j}; \qquad x < 8y_1 \tag{81}$$

$$C_o = 0.0115F_1^{1.35}\left(1 - \frac{x}{1.6}\right); \qquad x \geq 8y_1 \tag{82}$$

The experimental work of Rouse [90], Nagaratnam [106], Narayanan [101], Narasimhan and Bharagava [102] and Leutheusser et al. [91] shows that $\overline{u'^2}$ increases rapidly to a maximum at about $x \simeq 12y_1$ and thereafter decays. The following relationships are suggested for the turbulence "pressure" effect [107]:

$$\rho\overline{u'^2} = \rho\alpha\frac{u_1^2}{2}\left[K_1 + \frac{x(K_3 - K_1)}{12y_1}\right] \qquad \text{for } x \leq 12y_1 \tag{83}$$

$$\rho\overline{u'^2} = \rho\alpha\frac{u_1^2}{2}\left[\frac{1}{x}\right]^n \qquad \text{for } x > 12y_1 \tag{84}$$

in which $K_1 = 0.0067$
$K_3 = 0.0174$
$\alpha = 1.0$
$n = 1.0$

A set of four differential equations are obtained by substituting the velocities from Equations 66 and 67, and the shear functions from Equations 79 and 80 into the integral continuity and momentum equations, Equations 73 to 76; these equations can be expressed in the form of,

$$A_i\frac{dh}{dx} + B_i\frac{d\delta}{dx} + C_i\frac{du_o}{dx} + D_i\frac{du_m}{dx} = E_i \qquad (i = 1, 2, 3, 4) \tag{85}$$

Inverting Equation 85 gives,

$$\begin{Bmatrix} \frac{dh}{dx} \\ \frac{d\delta}{dx} \\ \frac{du_o}{dx} \\ \frac{du_m}{dx} \end{Bmatrix} = \begin{bmatrix} A_1 & B_1 & C_1 & D_1 \\ A_2 & B_2 & C_2 & D_2 \\ A_3 & B_3 & C_3 & D_3 \\ A_4 & B_4 & C_4 & D_4 \end{bmatrix}^{-1} \begin{Bmatrix} E_1 \\ E_2 \\ E_3 \\ E_4 \end{Bmatrix} \tag{86}$$

The coefficient matrix [A] and vector {E} are functions of h, δ, u_o and u_m (McCorquodale and Khalifa, 23).

A correction to the mixing length was made for the density gradient due to entrained air, viz.

$$\ell_p = \frac{\ell_m}{\sqrt{1 + \beta_* R_i}} \tag{87}$$

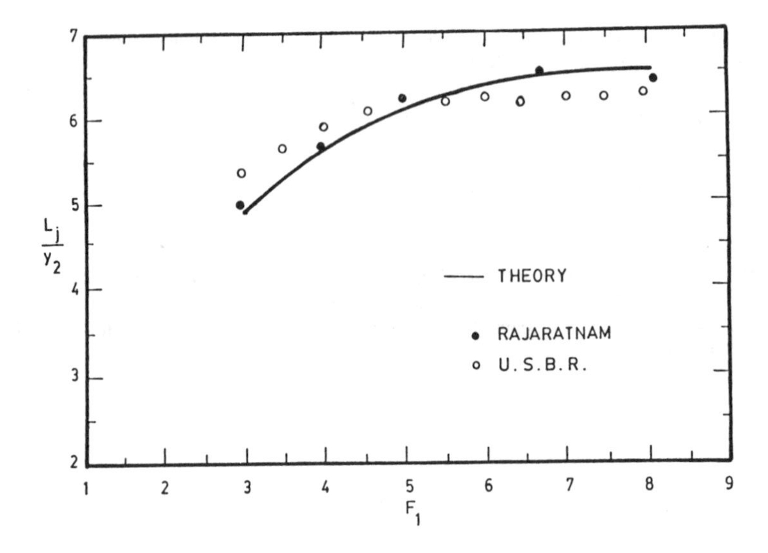

Figure 40. Comparison of computed and experimental jump lengths.

where R_i = the Richardson Number
$\beta_* \leq 160$ [108].

The bulking effect of the entrained air was also included where the air concentrations and distributions were estimated from the work of Rajaratnam [8, 9] and Leutheusser et al. [93].

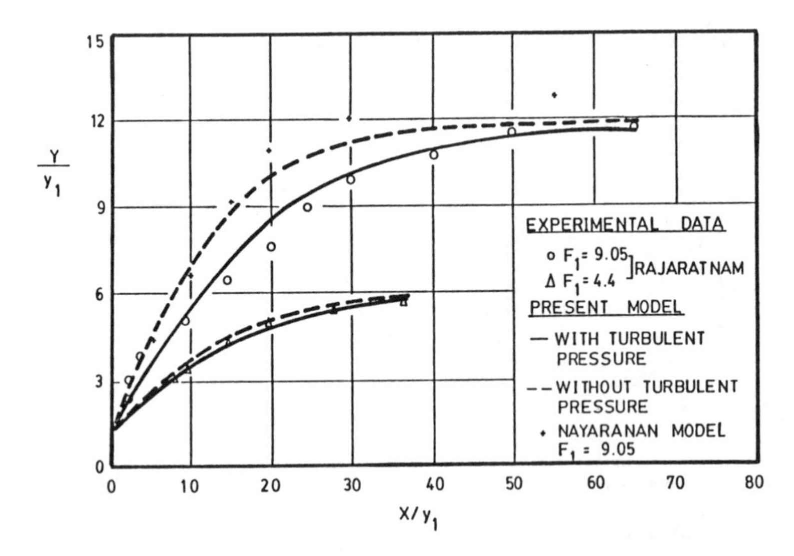

Figure 41. Effect of turbulence pressure on the profile of a hydraulic jump (after McCorquodale et al. (1983)).

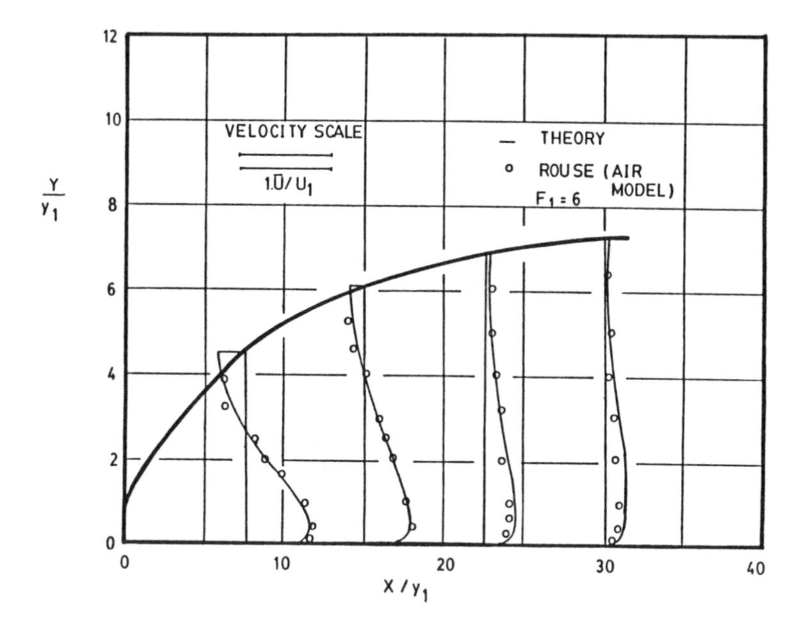

Figure 42. Comparison of the SIM computed velocity distributions with those measured by Rouse et al. (after McCorquodale *et al.* (1983)).

The solution was started with initial conditions at the end of the potential flow, $x \simeq 4y_1$. The macroscopic continuity and momentum equations were applied with the assumption that $\delta \simeq y_1/2$ and $u_m \approx u$, at $x = 4y_1$ and the requirement that the energy loss >0.

Figure 40 shows that the jump lengths of the SIM model and those given by Rajaratnam are very close; however, the predicted lengths are slightly lower than the USBR [109] for $F_1 = 3$ to 5 and slightly higher for $F_1 > 5$.

The model was run with and without the air effect. Except for the bulking effect of the air, the model indicated that air plays a minor role in determining the shape of the jump.

Figure 41 illustrates the effect of the turbulence "pressure" on the hydraulic jump at F_1 of 4 and 9.05. The model was run with and without the turbulence "pressure" term. Turbulence pressure had a much greater effect at $F_1 = 9.05$ than at $F_1 = 4$. The general effect was to lower the water surface near the beginning of the jump, although the length of the jump was not significantly changed. Equation 83 shows that the turbulence "pressure" relative to the hydrostatic pressure depends on F_1^2.

The relative importance of the centrifugal force also increases as F_1^2.

Figure 42 shows a reasonable agreement between the theoretical velocity distribution and the measured velocity distribution of Rouse et al. [12]. Figure 43 shows good agreement between predicted surface velocity and that of Rouse et al. while predicted maximum velocity falls between the measured value of Rouse et al. [12] and Rajaratnam [8, 9].

The Strip Integral Method has also been applied to the radial hydraulic jump by Khalifa [61] and to the radial submerged hydraulic jumps by Abdel-Gawad and McCorquodale [72]. Khalifa presented theoretical solutions for the radial jump. He showed that the roller length was very sensitive to the turbulent shear, i.e.

$$\frac{\partial L_r}{\partial(\tau/\bar{\tau})} \sim -y_2 \tag{88}$$

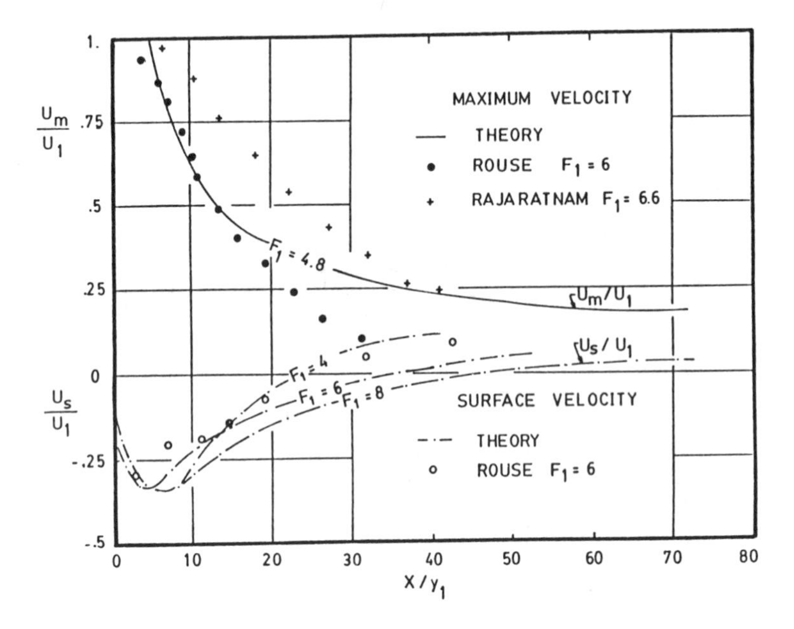

Figure 43. Comparison of predicted and measured surface and maximum velocities along the classical hydraulic jump [23].

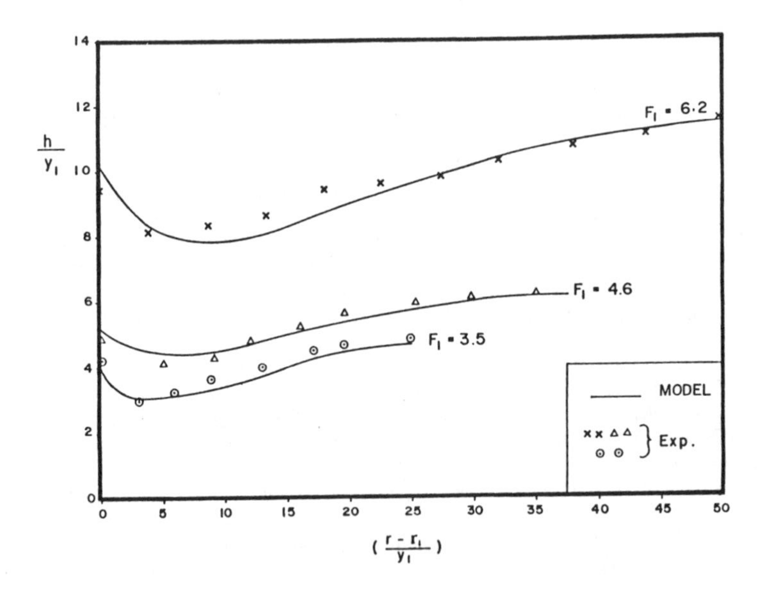

Figure 44. Predicted and measured surface profiles in a submerged radial hydraulic jump [72].

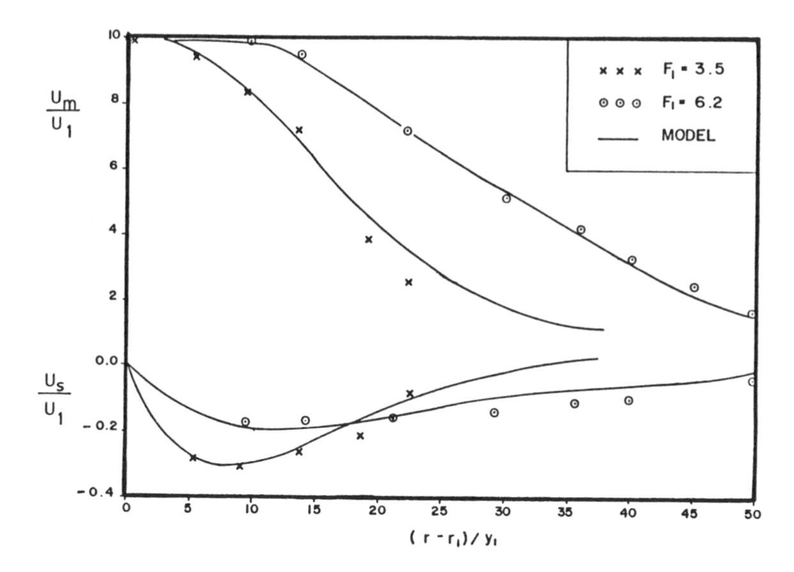

Figure 45. Longitudinal variation of the surface and maximum surface velocities in a submerged radial hydraulic jump [72].

Abdel-Gawad and McCorquodale [72] compared the results of a Strip Integral solution and an experimental study. Figure 44 shows the measured and predicted surface profiles in the submerged radial hydraulic jump. Figure 45 shows the longitudinal variation of surface and maximum velocities.

CONCLUSIONS

In Figure 2 the author indicated the wide variety of possible hydraulic jumps. It was shown that the sequent depth of the hydraulic jump can be adequately determined by applying a macroscopic momentum balance; this has been verified by numerous experimental studies. The length characteristics of the hydraulic jump depend on the internal flow and cannot be determined by the macroscopic approach. A considerable amount of experimental data are available for estimating jump lengths. An analytical approach referred to as the Strip Integral Method appears promising for the prediction of jump lengths and internal velocities in the hydraulic jump.

Some interesting areas for future research are:

- The application of higher order turbulence models such as the k-ε model [110].
- Investigation of the turbulence characteristics of internal radial and rectangular jumps.
- Simulation of flow in forced radial hydraulic jumps including the prediction of cavitation.

Additional references are given at the end of this chapter for further readings.

Acknowledgments

Some of the data presented in this chapter was developed with the help of my present and former graduate students. I wish to acknowledge the assistance of E. Regts, M. Giratella, C. Y. Li, A. Hannoura, A. Khalifa, M. Hamam, P. Nettleton, and S. Abdel-Gawad. In addition, the experience I gained while working on various projects for the following consulting firms was an asset to the writing of this chapter: H. G. Acres Ltd.; M. M. Dillon Ltd.; and Lafontaine, Cowie, Buratto, and Associates Ltd. Some of my research reported herein was supported by operating grants from the Natural Sciences and Engineering Research Council of Canada. Finally, I wish to thank Mrs. Judy Assef and my wife, Beth, for their assistance in preparing this manuscript.

NOTATION

- A flow area
- A_o conduit area
- B contact length between active and ambient fluids; also expanded flow depth
- b initial flow depth
- C_D drag coefficient
- c pressure head due to curvilinear shear; also celerity
- D hydraulic mean depth; diameter
- F Froude number
- F_f boundary shear force
- F_p air content function
- f frequency
- g acceleration of gravity
- g' acceleration term defined by Equation 1
- h_B baffle height
- h' depth to surface of air-water mixture
- I integral defined in Equation 4
- K shape factor
- L length
- P pressure force
- P_s total side force
- p' random component of pressure p
- Q volumetric flow
- R_i Richardson number
- r radius
- S submergence factor
- s_B baffle spacing
- u velocity
- u_B jet velocity at baffle
- u'_r random velocity component
- u'_i random component of $\overline{u_i}$
- $\overline{u_i}$ local mean velocity in direction i
- V volume
- V_s jump velocity
- W mass rate or body forces
- w_b baffle width
- X_i body force
- y depth
- y_B jet depth at baffle
- z coordinate

Greek Symbols

- β' pressure correction factor
- β^* constant in Equation 87
- Γ_1 parameter defined by Equation 14
- γ specific weight
- γ' specific weight of air-water mixture
- δ depth
- η fractional depth
- θ bed slope angle
- λ fractional flow depth
- μ dynamic viscosity
- ξ boundary layer
- ρ density
- τ shear
- ϕ angle of divergence

REFERENCES

1. Bidone, G., 1818. See Ven te Chow, 1959, p. 434.
2. Belanger, J. B., 1828. See Ven te Chow, 1959, p. 434.
3. Darcy, H., and Bazin, H., 1865, *Recherches Experimenteles Relatives Aux Remous et a La Propagation des Ondes*, Vol. II of *Recherches Hydraulic Academie des Sciences*, Paris.
4. Bakhmeteff, B. A., and Matzke, A. E., 1936, "The Hydraulic Jump in Terms of Dynamic Similarity," *Transactions, ASCE*, 101, Paper No. 1935, pp. 630–680.
5. Rajaratnam, N., "Profile Equation of the Hydraulic Jump," *Water Power*, 14 (1962), pp. 324–327.
6. Rajaratnam, N., "The Forced Hydraulic Jump," *Water Power*, 16 (1964) pp. 14–19, 61–65.
7. Rajaratnam, N., "The Hydraulic Jump as a Wall Jet," *Journal of the Hydraulics Division, ASCE*, 91: HY5, Proc. Paper 4482 (September 1965), pp. 107–132.

8. Rajaratnam, N., "Submerged Hydraulic Jump," *Journal of the Hydraulics Division, ASCE*, 91 (1965), pp. 71–96.
9. Rajaratnam, N., "Hydraulic Jumps," in *Advances in Hydroscience*, V. T. Chow, (Ed.), 4, Academic Press, New York, NY, 1965, pp. 198–280.
10. Rajaratnam, N., and Subramanya, K., "Hydraulic Jump Below Abrupt Symmetrical Expansions," *Journal of the Hydraulics Division, ASCE*, 94:HY2, (March 1968), pp. 481–503.
11. Rajaratnam, N., and Subramanya, K., "Profile of the Hydraulic Jump," *Journal of the Hydraulics Division, ASCE*, 94:HY3, Proc. Paper 5931, (May 1968), pp. 663–673.
12. Rouse, H., Siao, T. T., and Nagaratnam, R., "Turbulence Characteristics of the Hydraulic Jump," *Journal of the Hydraulics Division, ASCE*, 84:HY1, Proc. Paper 1528 (February 1958).
13. ASCE Task Force, "Energy Dissipators for Spillways and Outlet Works," *Journal of the Hydraulics Division, ASCE*, 90:HY1, Proc. Paper 3762 (January 1964).
14. Bradley, J. N., and Peterka, A. J., "The Hydraulic Design of Stilling Basins: Hydraulic Jumps on a Horizontal Apron (Basin I)," *Journal of the Hydraulics Division, ASCE*, 83:HY5 (1957), pp. 1–24.
15. U. S. Bureau of Reclamation, "Research Studies on Stilling Basins, Energy Dissipators and Associated Appurtenances," Hydraulic Laboratory Report No. Hyd-399.
16. Chow, V. T., *Open Channel Hydraulics*, McGraw-Hill Book Co., Inc., New York, 1959.
17. Elevatorski, E. A., *Hydraulic Energy Dissipators*, McGraw-Hill Book Co., Inc., New York, 1959.
18. McCorquodale, J. A., "Hydraulic Study of the Circular Clarifiers at the West Windsor Pollution Control Plant," for Lafontaine, Cowie, Buratto and Associates, Windsor, Canada, 1976.
19. McCorquodale, J. A., "Temperature Measurements in Circular Clarifiers," for City of Windsor, Canada, 1977.
20. McCorquodale, J. A., and Smith, A. F., "Hydraulic Model Study of the Wilkesport Drop-Structure and Trapezoidal Stilling Basin," Report to M. M. Dillon Limited, Windsor, Canada, 1977.
21. McCorquodale, J. A., and Hamam, M. A., "Modeling Surcharged Flow in Sewers, International Symposium on Urbar Hydrology, Hydraulics and Sediment Control," University of Kentucky, Lexington, KY (July 1983), pp. 331–338.
22. McCorquodale, J. A., "Hydraulic Studies of the W. Darcy McKeough Control Dam," IRI-15-61, for M. M. Dillon Limited, Windsor, Canada, 1983.
23. McCorquodale, J. A., and Khalifa, A., "Internal Flow in a Hydraulic Jump," *Journal of the Hydraulics Division, ASCE*, 106:HY3, (1983), pp. 355–367.
24. McCorquodale, J. A., and Khalifa, A., "Submerged Radial Hydraulic Jump," *Journal of the Hydraulics Division, ASCE*, 106:HY3, (March 1983), pp. 355–367.
25. Kindsvater, C. E., "The Hydraulic Jump in Sloping Channels," *Transactions, ASCE*, 109 pp. 1107–1154.
26. McCorquodale, J. A., and Regts, E. H., 1968, "A Theory for the Forced Hydraulic Jump," *Transactions of the Engineering Institute of Canada*, 11:C-1, May.
27. Rajaratnam, N., and Murahar, V., "A Contribution to Forced Hydraulic Jumps," *Journal of Hydraulic Research, IAHR*, 9:2 (1971).
28. Harleman, D. R. F., "Effect of Baffle Piers on Stilling Basin Performance," *Boston Society of Civil Engineers Journal*, 42:2 (April 1955).
29. Rand, W., "Flow Over a Vertical Sill in an Open Channel," *Journal of the Hydraulics Division, ASCE*, 91:HY4 (1965), pp. 97–121.
30. Weide, L., "The Effect of the Size and Spacing of Floor Blocks in the Control of the Hydraulic Jump," M. Sc Thesis, Colorado A and M, (1951).
31. Pillai, N. N., and Unny, T. E., "Shapes for Appurtenances in Stilling Basins," *Journal of the Hydraulics Division, ASCE* (November 1964).
32. McCorquodale, J. A., and Giratella, M., Supercritical Flow Over Sills, *Journal of the Hydraulics Division, ASCE*, 98:HY4 (1972), pp. 667–679.
33. Narayanan, R., and Schizas, L. S., "Force Fluctuations on Sill of Hydraulic Jump," *Journal of the Hydraulics Division, ASCE*, 106:HY4, (April 1980), pp. 589–599.
34. Tyagi, D. M., Prande, P. K., and Mittal, M. K., "Drag on Baffle Walls in Hydraulic Jump," *Journal of the Hydraulics Division, ASCE*, 104:HY4 (1978), pp. 515–525.

35. Karki, K. S., "Supercritical Flow Over Sills," *Journal of the Hydraulics Division, ASCE*, 102:HY10 (1976), pp. 1449–1459.
36. U.S. Department of Interior Bureau of Reclamation, *Design of Small Dams*, Second Edition (1974), pp. 442–446.
37. Forster, J. W., and Skrinde, R. A., "Control of the Hydraulic Jump by Sills," *Transactions, ASCE*, 115 (1950), pp. 973–1022.
38. Basco, D. R., Trends in Baffled, Hydraulic Jump Stilling Basin Designs of the Corps of Engineers Since 1947, Misc. Paper H-69-1, Waterway Experimental Station, Vicksburg, MS (1969).
39. Regts, E., "A Theoretical and Experimental Study of the Forced Hydraulic Jump," M.S. Thesis, University of Windsor, Windsor, Canada, 1967.
40. Berryhill, R. H., "Stilling Basin Experiences of Engineers," *Journal of the Hydraulics Division, ASCE*, 83:HY3 (June 1957).
41. Stevens, J. C., "The Hydraulic Jump in Standard Conduits," *Civil Engineering* (*NY*), 3 (1933), pp. 565–567.
42. Sandover, J. A., and Holmes, P., "The Hydraulic Jump in Trapezoidal Channels," *Water Power*, 14 (1962), pp. 445–449.
43. Posey, C. J., and Hsing, P. S., "Hydraulic Jump in Trapezoidal Channels," *Engineering News Record*, 106 (1938), pp. 797–798.
44. Thiruvengadam, A., "Hydraulic Jump in Circular Channels," *Water Power*, 13 (1961), pp. 496–497.
45. Lane, E. W., and Kindsvater, C. E., "Hydraulic Jump in Enclosed Conduits," *Engineering News Record*, 106 (1938), pp. 815–817.
46. Govinda Rao, N. S. and Rajaratnam, N., "The Submerged Hydraulic Jump," *Journal of the Hydraulics Division, ASCE*, 89:HY1, Proc. Paper 3404 (January 1963).
47. Hamam, M. A., *Transition of Gravity to Surcharged Flow in Sewers*, Ph.D. Dissertation, University of Windsor, Windsor, Canada (1982).
48. Kalinske, A. A., and Robertson, J. W., "Entrainment of Air in Flowing Water-Closed Conduct Flow," *Transactions, ASCE*, 108 (1943), pp. 1435–1447.
49. Hamam, M. A., and McCorquodale, J. A., "Transient Conditions in the Transition from Gravity to Surcharged Sewer Flow," *Canadian Journal of Civil Engineering*, 9:2 (1982), pp. 189–196.
50. Hamam, M. A., and McCorquodale, J. A., "Transition of Gravity to Surcharge Flow in Sewers," *Proceedings of the Eighth International Symposium on Urban Hydrology, Hydraulics and Sediment Control*, University of Kentucky, Lexington, KY (July 1981), pp. 173–177.
51. Hamam, M. A., and McCorquodale, J. A., "Surges in Sewers," *Porceedings of the CSCE* 1980 *Annual Conference*, Winnipeg, Manitoba, Canada (May 1980), pp. M/10: 1–14.
52. Rouse, H., *Hydraulic Engineering*, John Wiley & Sons, New York, NY (1949).
53. Henderson, F., *Open Channel Flow*, Macmillan Publishing Co., Inc., NY, 1966.
54. Whittington, R. B., and Ali, K. H. M., "Convergent Hydraulic Jumps," *Proceedings Institute of Civil Engineers*, 43 (1969), pp. 157–173.
55. Gagnon, A. R., "The Hydraulic Jump in Radial Flow," M.S. Thesis, MIT (May 1959).
56. Sadler, C., "Radial Free Surface Flow," M. S. Thesis, MIT (May 1963).
57. Watson, E. J., "The Radial Spread of a Liquid Jet Over a Horizontal Plane," *Journal of Fluid Mechanics*, 20, Part 3 (1964), pp. 481–499.
58. Koloseus, H. J., and Ahmad, D., "Circular Hydraulic Jump," *Journal of the Hydraulics Division, ASCE*, 95:HY1, Proc. Paper 6367 (January 1969), pp. 409–422.
59. Arbhabhirama, A., and Abella, A., "Hydraulic Jump Within Gradually Expanding Channel," *Journal of the Hydraulics Division, ASCE*, 97:HY1, Proc. Paper 7831 (1971), pp. 31–41.
60. Khalifa, A., and McCorquodale, J. A., "Radial Hydraulic Jump," *Journal of the Hydraulics Division, ASCE*, 105:HY9 (September 1979), pp. 1065–1078.
61. Khalifa, A., "Theoretical and Experimental Study of the Radial Hydraulic Jump," Ph.D. Thesis University of Windsor, Windsor, Canada (1980).
62. Scott-Moncrieff, A., "Behaviour of Water Jet in a Diverging Shallow Open Channel, Fifth Australian Conference on Hydraulics and Fluid Mechanics," Christ Church, New Zealand (December 1974), pp. 42–48.

63. Blaisdell, F. W., "The St. Anthony Falls Stilling Basin," U.S. Conservation Service, Report SCS-TP-79 (1949).
64. Moore, W. L., and Meshgin, K., "Adaptation of the Radial Energy Dissipator for Use With Circular or Box Culverts," Research Report 116-1, University of Texas at Austin (1970).
65. Meshgin, K., and Moore, W. L., "Design Aspects and Performance Characteristics of Radial Flow Energy Dissipators," Research Report 116-2F, Center for Highway Research, University of Texas at Austin (1970).
66. Herbrand, K., "The Spatial Hydraulic Jump," *Journal of Hydraulic Research, IAHR*, 11:3 (1973).
67. Meloche, L., and McCorquodale, J. A., "Hydraulic Design of Flood Control Structure with an In-Spillway Fishpass, International Conference on Hydraulic Design in Water Resources Engineering," Southampton, England (April 1984).
68. Nettleton, P. C., and McCorquodale, J. A., "Radial Stilling Basins with Baffles," *Proceedings 6th Canadian Hydrotechnical Conference*, Ottawa, Canada (1983), pp. 651–670.
69. Nettleton, P. C., *The Forced Radial Hydraulic Jump*, M.A.Sc. Thesis, Department of Civil Engineering, University of Windsor, Windsor, Canada (1983).
70. Hsu, E-Y., "Discussion" on Control of the Hydraulic Jump by Sills, *Transactions, ASCE*, 115 (1950), pp. 988–991.
71. Rajaratnam, N., and Ortiz, N., "Hydraulic Jumps and Waves at Abrupt Drops," *Journal of the Hydraulics Division, ASCE*, 103:4 (April 1977), pp. 381–394.
72. Abdel-Gawad, S. M., and McCorquodale, J. A., "Modelling Submerged Radial Hydraulic Jumps," *Proceedings Annual Conference of Canadian Society for Civil Engineering*, Halifax, Nova Scotia, Canada (1984).
73. Baddour, R. E., and Abbink, H., "Turbulent Underflow in a Short Channel of Limited Depth," *Journal of the Hydraulics Division, ASCE*, 109:6 (1983), pp. 722–740.
74. Yih, C. S., and Guha, C. R., "Hydraulic Jump in a Fluid System of Two Layers," *Tellus*, 7 (1955), pp. 358–366.
75. Baddour, R. E., and Chu, V. H., "Buoyant Surface Discharge on a Step and on a Sloping Bottom," Technical Report 75-2 (FML), Department of Civil Engineering and Applied Mechanics, McGill University, Montreal, Canada (1975).
76. Chu, V. H., and Vanvari, M. R., "Experimental Study of Turbulent Stratified Shearing Flow," *Journal of the Hydraulics Division, ASCE*, 102:HY6 (1976), pp. 691–706.
77. Ellison, T. W., and Turner, J. S., "Turbulent Entrainment in Stratified Flows," *Journal of Fluid Mechanics*, 6 (1959), pp. 423–448.
78. Harleman, D. R. F., "Stratified Flow," in *Handbook of Fluid Dynamics*, Ed. V. L. Streeter, McGraw-Hill Book Co., New York, 1960.
79. Jirka, G. H., and Harleman, D. R. F., "Stability and Mixing of a Vertical Plane Buoyang Jet in Confined Depth," *Journal of Fluid Mechanics*, 94 (1979), pp. 275–304.
80. Koh, R. C. Y., "Two-Dimensional Surface Warm Jets," *Journal of the Hydraulics Division, ASCE*, 97:HY6 (1971), pp. 819–836.
81. Long, R. R., "Some Aspects of the Flow of Stratified Fluids I," *Tellus*, 5 (1953), pp. 42–58.
82. Long, R. R., Some Aspects of the Flow of Stratified Fluids II, *Tellus*, 6 (1954), pp. 97–115.
83. Stefan, H., "Dilution of Buoyant Two-Dimensional Surface Discharges," *Journal of the Hydraulics Division, ASCE*, 98:HY1 (1972), pp. 71–86.
84. Kranenburg, C., "Internal Kinematic Waves and the Large-Scale Stability of Two-Layer Stratified Flow," *Journal of Hydraulic Research*, 21:4 (1983).
85. French, R. H., "Interfacial Stability in Channel Flow," *Journal of the Hydraulics Division, ASCE*, 105:HY8 (1979), pp. 955–967.
86. Turner, J. S., *Buoyancy Effects in Fluids*, Cambridge University Press, 1973.
87. Mehotra, S. C., and Kelly, R. E., "On the Question of Non-Uniqueness of Internal Hydraulic Jumps and Drops in a Two-Fluid System," International Symposium on Stratified Flows, Novosibirsk (1972).
88. Hayakawa, N., "Internal Hydraulic Jump in Co-Current Stratified Flow," *Journal of the Engineering Mechanics Division, ASCE*, 96:EM5 (1970), pp. 797–800.

89. Rouse, H., *Elementary Mechanics of Fluids*, Dover Publications Inc., New York, NY, 1946.
90. Rouse, H., and Simon, I., *History of Hydraulics*, Dover Publications, Inc., New York, NY, 1957.
91. Leutheusser, H. J., and Kartha, V. C., "Effects of Inflow Condition on Hydraulic Jump," *Journal of the Hydraulics Division, ASCE*, 98:HY8 (1972).
92. Leutheusser, H. J., and Alemu, S., "Flow Separation Under Hydraulic Jump, *Journal of Hydraulic Research, IAHR*, 17:3 (1979), pp. 193–206.
93. Resch, F., Leutheusser, H., and Alemu, S., "Bubbly Two-Phase Flow in an Hydraulic Jump," *Journal of the Hydraulics Division, ASCE*, 100:HY1, Proc. Paper 10297 (January 1974), pp. 137–150.
94. Resch, F. J., and Leutheusser, H. J., "Reynolds Stress Measurements in Hydraulic Jumps," *Journal of Hydraulic Research, IAHR*, 10:4 (1972), pp. 409–430.
95. Resch, F. J., and Leutheusser, H. J., "Research Studies on Stilling Basins, Energy Dissipators, and Associated Appurtenances," Hydraulic Laboratory Report No. Hyd-399, United States Bureau of Reclamation, Denver, CO, (June 1, 1955).
96. Lipay, I. E., and Pustovoit, V. F., *Proceedings 12th Congress, IAHR*, Fort Collins, CO (1967), pp. 362–369.
97. Vasiliev, O. F., and Bukreyev, V. I., "Statistical Characteristics of Pressure Fluctuations in the Region of Hydraulic Jump," *Proceedings 12th Congress IAHR*, 2 (1967).
98. Wisner, P., "Bottom Pressure Pulsations of the Closed Conduit and Open Channel Hydraulic Jump," *Proceedings 12th Congress, IAHR*, 2 (1967).
99. Bourkov, V. J., "Experimental Studies of Pressure Pulsations in the Hydraulic Jump (Russian)," V28, Gosenergoizet Leningrad and Moscow (January 1965).
100. King, D. L., "Analysis of Random Pressure Fluctuations in Stilling Basin," *Proceedings 12th Congress IAHR*, 2 (1967).
101. Narayanan, R., "Pressure Fluctuations Beneath Submerged Jump," *Journal of the Hydraulics Division, ASCE*, 104:HY9, Proc. Paper 14039 (September 1978), pp. 1331–1342.
102. Narasimhan, S., and Bharagava, P., "Pressure Fluctuations in Submerged Jump," *Journal of the Hydraulics Division, ASCE*, 102:HY3 (March 1976), pp. 339–350.
103. Khader, M. H. A., and Elango, K., "Turbulent Pressure Field Beneath a Hydraulic Jump," *Journal of Hydraulic Research, IAHR*, 12:4 (1974), pp. 469–489.
104. Rajaratnam, *Turbulent Jets*, Elsevier Scientific Co., New York (1976).
105. Narayanan, R., "Wall Jet Analogy to Hydraulic Jump," *Journal of the Hydraulics Division, ASCE*, 101:HY3, Proc. Paper 11172 (March 1975), pp. 347–360.
106A. Moses, H. L., "A Strip-Integral Method for Predicing the Behavior of the Turbulent Boundary Layers," *Proceedings, Computation of Turbulent Boundary Layers, AFOSR-IFP-Standford Conference* (1978).
106B. Nagaratnam, S., "The Mechanism of Energy Dissipation," M.S. Thesis, University of Iowa (June 1957).
107. Mehotra, C., "Length of Hydraulic Jump," *Journal of the Hydraulics Division, ASCE*, 162:HY7 (July 1976), pp. 1027–1033.
108. Odd, N., and Rodger, J., "Vertical Mixing in Stratified Tidal Flows," *Journal of the Hydraulics Division, ASCE*, 104: HY3, Proc. Paper 13599, (March 1978), pp. 337–351.
109. U.S. Army Corps of Engineers, *Engineering and Design, Structural Design of Spillways and Outlet Works*, Engineering Manual No. Em 1110-2-2400, 1964.
110. Rodi, W., *Turbulence Models and Their Application in Hydraulics*, Special Publication of IAHR, Delft, The Netherlands, 1980.
111. Advani, R. M., "Discussion of the Submerged Hydraulic Jump," *Journal of the Hydraulics Division, ASCE*, 89 (1963), pp. 147–150.
112. Arbhabhirama, A., and Wei-Chun Wan, "Characteristics of a Circular Jump in a Radial Wall Jet," *Journal of Hydraulic Research*, 13:3 (1975), pp. 239–262.
113. Argyropoulos, P. A., "General Solution of the Hydraulic Jump in Sloping Channels," *Journal of the Hydraulics Division, ASCE*, 88:HY4 (1962), pp. 61–75.
114. Bhowonik, Nani G., "Stilling Basin Design for Low Froude Number," *Journal of the Hydraulics Division, ASCE*, 101:HY2 (July 1975), pp. 901–915.

115. Blaisdell, F. W., "Development and Hydraulic Design, St. Anthony Falls Stilling Basin," *Proceedings, ASCE*, 73:2 (February 1947).
116. Bowers, C. E., and Tsai, F. Y., "Fluctuating Pressure in Spillway Stilling Basins," *Journal of the Hydraulics Division, ASCE*, 95:HY6 (November 1969).
117. Bradley, J. N., and Peterka, A. J., "Hydraulic Design of Stilling Basins: Stilling Basin With Sloping Apron (Basin V)," *Journal of the Hydraulics Division ASCE*, 85:HY5 (1957), pp. 1–32.
118. Diskin, M. H., "Hydraulic Jump in Trapezoidal Channels," *Water Power*, 13 (1961), pp. 12–17.
119. Hamam, M. A., "Surges in Storm Sewers," M.A.Sc. Thesis, University of Windsor, Windsor, Canada (1977).
120. Hinze, J. O., *Turbulence*, McGraw-Hill Book Co., Inc., New York, 1975.
121. Karki, K. S., Chander, S., and Malhotra, R. C., "Supercritical Flow Over Sills at Incipient Conditions," *Journal of the Hydraulics Division, ASCE*, 98:HY10 (October 1972), pp. 1753–1764.
122. McCorquodale, J. A., "Report on the Hydraulics of the Expanded West Windsor Pollution Control Plant," for Lafontaine, Cowie, Buratto and Associates, Windsor, Canada (1979).
123. McCorquodale, J. A., "Model Studies of the W. Darcy McKeough Dam and Diversion Control Studies," IRI-11-3, for M. M. Dillon Limited, Windsor, Canada (1978).
124. Mehotra, S. C., "Circular Jumps," *Journal of the Hydraulics Division, ASCE*, 100:HY8 (August 1974), pp. 1133–1140.
125. Meloche, L., McCorquodale, J. A., and Imam, E., "Hydraulic Characteristics of Flow Through Sluice Culverts," *Proceedings of the Annual Conference of the Canadian Society for Civil Engineering*, Montreal, Canada (June 1979), pp. 185–190.
126. Moore, W. L., and Morgan, C. W., "Hydraulic Jump at an Abrupt Drop," *Transactions, ASCE*, 124 (1959), pp. 507–524.
127. Rittal, W. F., and Hunt, B. W., "Internal Currents Resulting from Density Inflows in Stratified Reservoirs," Water Resources Research Report, University of Washington, Seattle, Washington, (1970).
128. Sarma, K. V. N., and Newnham, D. A., "Surface Profiles of Hydraulic Jump for Froude Number Less Than Four," *Water Power*, 25, London, England (April 1973), pp. 139–142.
129. Schlichting, *Boundary-Layer Theory*, McGraw-Hill Book Co., Inc., New York, NY, 1968.
130. Sharma, H. R., "Air Entrainment in High Head Gated Conduits," *Journal of the Hydraulics Division, ASCE*, 102:11 (November 1976), pp. 1629–1646.
131. Siao, T. T., Characteristics of Turbulence in an Air-Flow Model of the Hydraulic Jump," Ph.D. Dissertation, State University of Iowa (1954).
132. Sigalla, A., "Measurements of Skin Friction in a Plane Turbulent Wall Jet," *Journal of the Royal Aeronautical Society*, 62 (December 1958), pp. 874–877.
133. Silvester, R., "Hydraulic Jump in all Shapes of Horizontal Channels," *Journal of the Hydraulics Division, ASCE*, 90:HY1 (1964), pp. 23–55.
134. Wilson, E., and Turner, A., "Boundary Layer Effects on Hydraulic Jump Location," *Journal of the Hydraulics Division, ASCE*, 98:HY7 (July 1972), pp. 1127–1142.

CHAPTER 7

WAVE ATTENUATION IN OPEN CHANNEL FLOW

Angel N. Menendez

Iowa Institute of Hydraulic Research
University of Iowa
Iowa City, Iowa

Ruben Norscini

Laboratorio de Hidraulica Aplicada
Instituto Nacional de Ciencia y Tecnica Hidricas
Ezeiza, Argentina

CONTENTS

INTRODUCTION, 174

DEFINITION OF THE PROBLEM, 175
Theoretical Model, 175
Equations of Motion, 176

LINEAR ANALYSIS, 177
The Linear Approximation, 177
Conditions and Causes of Attenuation, 178
Spectrum of Shallow Water Waves, 182
Flow Instability, 184
Some Particular Solutions, 185

NONLINEAR ANALYSIS, 186
Nonlinear Phenomena, 186
Long Gravity Waves, 187
Dynamic-Gravity Waves, 191
Kinematic Waves, 192
Dynamic-Kinematic Waves, 193
Dynamic Waves, 194
Prismatic and Nonprismatic Channels, 197
Numerical Methods of Solution, 200

NOTATION, 201

REFERENCES, 201

INTRODUCTION

The understanding of wave propagation in open channels bears both practical and theoretical importance. Predicting the consequences of flood waves has long been recognized as an economical and social necessity. From a theoretical standpoint, this wave phenomenon represents an interesting combination of sometimes competing effects. These include attenuation and both frequency and amplitude dispersion, the latter being a nonlinear effect.

In the following, "attenuation" will be understood as the decrease of the wave crest while propagating, sometimes referred to as subsidence. It will be seen that, under certain conditions, the opposite effect (amplification) occurs. The material that will be presented is aimed to answer the two following questions:

1. When does attenuation occur?
2. Why does attenuation occur?

In other words, the conditions under which attenuation takes place will be established and the responsible mechanisms determined, within the limited current knowledge. A third question will be only partially addressed, namely

3. How much does a given wave attenuate?

In fact, the quantification of the attenuation effects can be obtained through analysis only for rather simplified cases, as general solutions are not available. However, this is not an insurmountable obstacle nowadays, as for any particular problem the solution can be obtained numerically (with varied levels of difficulty).

The differential equations which describe unsteady flow in open channels were formulated about a century ago by Barre de Saint Venant [1]. Important contributions to the analysis of the resulting wave phenomenon were made by Stoker [2] and Lighthill and Whitham [3]. The former studied long gravity waves based on the analogy with one-dimensional gas flow, and developed a method for calculating numerically flood wave propagation. Lighthill and Whitham, in turn, analyzed the flood wave phenomenon by introducing the concept of the kinematic wave. Furthermore, they established its relation with the general dynamic waves and, based on their theory, proposed a relatively simple semi-graphical method of calculation for the propagation of flood waves. Their results and ideas constitute an important part of this work.

DEFINITION OF THE PROBLEM

Theoretical Model

The set of hypothesis, over which the theoretical model for open channel flow is built, can be conveniently arranged according to how they restrict one of the three following factors: acting forces, channel geometry, and flow conditions. With this approach, the hypothesis can be expressed as follows:

1. Acting forces
   - Gravity is the driving force. Contributions such as wind forces, inertial forces due to the earth rotation (Coriolis), etc., are considered to be negligible.
   - The resistance to the motion, through boundary "friction" and turbulence generation, can be expressed using the same laws as for steady flow.

2. Channel geometry
   - The shape of the channel is arbitrary, but it does not change with time and does not present sudden variations (the latter restriction is related to the assumed flow conditions).
   - The (average) bottom slope is small.

3. Flow conditions
   - The flow is one-dimensional, i.e., it is unidirectional and the velocity is uniform throughout any cross section.
   - The acceleration components normal to the flow direction are negligible. Alternatively, this can be stated as requiring that the curvature and divergence of the streamlines are very small. As a consequence, the pressure distribution is hydrostatic and the free surface level is horizontal across each section.

The hypothesis presented under Factor 3 defines what is usually called the "shallow water" model. It can be shown that they are necessary conditions if the length scale of streamwise variation of the flow (typically, the wavelength) is much larger than the depth [4].

The relaxation of some of the hypotheses in Factors 1 and 2 does not greatly complicate the model. Regarding the flow conditions, however, only the relaxation of unidirectionality, allowing for two equally important (horizontal) velocity components, leaves the problem at a similar level of physical difficulty (the mathematical difficulty can be greatly increased). Changes in the remaining flow conditions lead into the so-called theory of nonlinear dispersive waves, in which the wavelength is now comparable to the depth. Typical examples are the cnoidal waves and their particular case, the solitary wave [4].

Presently, we will confine our attention to the shallow water waves, as defined by the preceding set of hypothesis.

Equations of Motion

Application of the principles of conservation of mass and momentum to the flow defined by the previous theoretical model lead, respectively, to the following equations [5]

$$B\frac{\partial h}{\partial t} + u\frac{\partial A}{\partial x} + A\frac{\partial u}{\partial x} = 0 \tag{1}$$

$$\frac{\partial u}{\partial t} + u\frac{\partial u}{\partial x} + g\frac{\partial h}{\partial x} = g\left(S_0 - \frac{u^2}{C^2R}\right) \tag{2}$$

where
- h = depth
- u = velocity
- A = cross-sectional area
- B = surface width
- S_0 = bottom slope
- R = hydraulic radius = A/P
- P = wetted perimeter
- g = acceleration of gravity
- x = space coordinates
- t = time coordinates
- C = Chezy coefficient

(see Figure 1)

The continuity equation, Equation 1, can be interpreted as establishing the balance between the following three "kinematic" mechanisms: the rate-of-rise (first term), the wedge storage (second term) and the prism storage (third term). In turn, the momentum equation, Equation 2, expresses the equi-

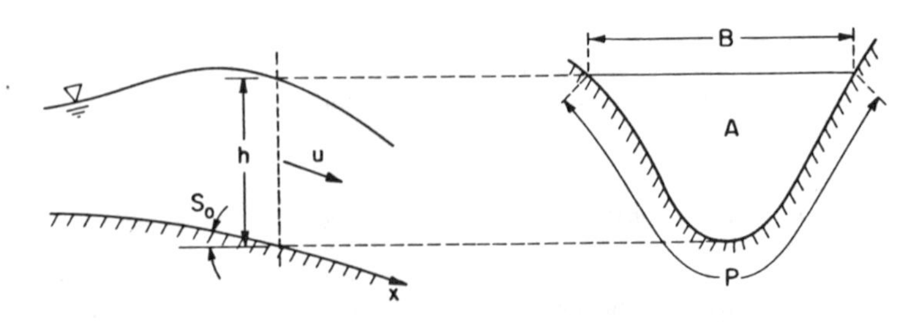

Figure 1. Flow variables.

librium between the following "dynamic" mechanisms: the inertia (first two terms on the left-hand side), the pressure gradient (third term on the left-hand side), the effective weight (first term on the right-hand side) and the hydraulic resistance (second term on the right-hand side).

For prismatic channels, $A = A(h)$ only. Using that $dA/dh = B$, Equation 1 is then more conveniently expressed as

$$B\left(\frac{\partial h}{\partial t} + u\frac{\partial h}{\partial x}\right) + A\frac{\partial u}{\partial x} = 0 \tag{3}$$

In the particular case of a rectangular prismatic channel, Equation 3 further simplifies to

$$\frac{\partial h}{\partial t} + u\frac{\partial h}{\partial x} + h\frac{\partial u}{\partial x} = 0 \tag{4}$$

which is independent of the channel width B. If this width becomes very large ($B \gg h$), an approximation often employed in theoretical analysis, $R \simeq h$ and the momentum equation, Equation 2, reduces to

$$\frac{\partial u}{\partial t} + u\frac{\partial u}{\partial x} + g\frac{\partial h}{\partial x} = g\left(S_0 - \frac{u^2}{C^2 h}\right) \tag{5}$$

The system of Equations 4 through 5 (or even further simplified versions) is the one which will be dealt with for the most part of this work. A special section will be devoted, however, to commenting on the effects of arbitrary cross-sectional shape and nonprismaticity.

LINEAR ANALYSIS

The Linear Approximation

The shallow water equations are a nonlinear system of differential equations. Interesting physical phenomena are associated with this nonlinearity. These are dealt with later. In this section a linear analysis is undertaken that will provide the basis for the global understanding of the wave phenomenon.

The linear approximation means mathematically that the wave contribution is a small perturbation over an otherwise undisturbed flow. In the case of a very wide rectangular channel with constant bottom slope, S_0, the undisturbed flow is a uniform regime characterized by constant values of depth, h_0, and velocity, u_0. Calling $h_1 = h - h_0$ and $u_1 = u - u_0$ to the wave contribution, the system of Equations 4 through 5 can be then transformed into

$$\frac{\partial h_1}{\partial t} + u_0\frac{\partial h_1}{\partial x} + h_0\frac{\partial u_1}{\partial x} = 0 \tag{6}$$

$$\frac{\partial u_1}{\partial t} + u_0\frac{\partial u_1}{\partial x} + g\frac{\partial h_1}{\partial x} = gS_0\left(\frac{h_1}{h_0} - 2\frac{u_1}{u_0}\right) \tag{7}$$

where the condition of uniformity of the undisturbed flow, namely $S_0 = u_0^2/C^2h_0$, has been used together with the assumption that C is a constant (the consequences of relaxing this last hypothesis will be mentioned later).

The main mathematical advantage of the system of Equations 6 through 7 over its nonlinear counterpart, Equations 4 through 5, is that the former provides explicit solutions under quite general initial and boundary conditions. Analysis of these solutions gives valuable physical insight into the phenomenon. In addition, they represent realistic approximations to the case when the wave amplitude is small relative to the depth. Some of these solutions are discussed later.

Another mathematical advantage of the linear system (related to the previous one) is that superposition of particular solutions is possible. Then, harmonic (Fourier) components can be studied separately, i.e., one can take

$$h_1 = \eta \exp[i(kx - \beta t)] \tag{8}$$

$$u_1 = \xi \exp[i(kx - \beta t - \delta)] \tag{9}$$

where k = wave number
η and ξ = initial amplitudes of the waves (real numbers)
δ = phase difference between them
β = propagation factor = $ck(1 + i\alpha/2\pi)$
c = phase velocity
α = growth rate of the wave expressed as a logarithmic decrement (α measures the relative change of amplitude after a time period)

This type of analysis is fruitful in providing the answers to the questions of when and why attenuation takes place, as will be shown later. These answers, of course, will depend on the wave number or, equivalently, the frequency (ck). The approach then leads naturally to the distinction of bands with different physical characteristics in the wave number spectrum. This is analyzed in a later section. It is also interesting to note that this methodology is the one used in the theory of linear hydrodynamic stability [6]. The connection with this theory is discussed later.

Conditions and Causes of Attenuation

When Equations 8 and 9 are introduced into Equations 6 and 7, one obtains

$$(ku_0 - \beta)\eta + kh_0(\xi e^{-i\delta}) = 0 \tag{10}$$

$$-g\left(\frac{S_0}{h_0} - ik\right)\eta + \left(2\frac{gS_0}{u_0} + iku_0 - i\beta\right)(\xi e^{-i\delta}) = 0 \tag{11}$$

Equations 10 and 11 constitutes a homogeneous system of linear algebraic equations in the unknowns η and $[\xi \exp(-i\delta)]$. For a nontrivial solution to exist, the determinant of the coefficient matrix must vanish. This condition provides a quadratic (complex) equation for β , from which two roots are obtained. They correspond to forward (c_1, α_1) and backward (c_2, α_2) waves, relative to the undisturbed flow. The solutions were provided by Ponce and Simons [7]. The results are

$$\hat{c}_1 \equiv \frac{c_1}{u_0} = 1 + c_r \tag{12}$$

$$\alpha_1 = -2\pi\frac{\zeta - E}{1 + c_r} \tag{13}$$

$$\hat{c}_2 \equiv \frac{c_2}{u_0} = 1 - c_r \tag{14}$$

$$\alpha_2 = -2\pi\frac{\zeta + E}{1 - c_r} \tag{15}$$

where

$$c_r = \left(\frac{C + A}{2}\right)^{1/2} \tag{16}$$

$$E = \left(\frac{C - A}{2}\right)^{1/2} \tag{17}$$

$$A = \frac{1}{F_o^2} - \zeta^2 \tag{18}$$

$$C = (A^2 + \zeta^2)^{1/2} \tag{19}$$

$$\zeta = \frac{1}{\sigma F_o^2} \tag{20}$$

$$F_o = \frac{u_0}{(gh_0)^{1/2}} \tag{21}$$

$$\sigma = \frac{kh_0}{S_0} \tag{22}$$

Note that the solutions finally depend only on two parameters: one characterizing the undisturbed flow, namely the Froude number, F_0, defined by Equation 21; another one characterizing the wave, namely the dimensionless wave number, σ, defined by Equation 22. Notice also that the phase velocity of the waves relative to the undisturbed flow is the same for both the forward and backward waves, and is given by Equation 16. If the complete solution is wanted, from Equation 10 one can obtain

$$\tan \delta = \frac{\alpha \hat{c}}{2\pi(1 - \hat{c})} \tag{23}$$

$$\hat{\Omega} \equiv \frac{\Omega h_0}{u_0} = \left[\left(\frac{\alpha \hat{c}}{2\pi}\right)^2 + (1 - \hat{c})^2\right]^{1/2} \tag{24}$$

where $\Omega \equiv \xi/\eta$. Introduction of Equations 12 and 13 (14 and 15) into Equations 23 and 24 gives the phase difference δ and the ratio of relative amplitudes $\hat{\Omega}$, respectively, for the forward (backward) wave.

Figure 2 shows the relative celerity, c_r, as a function of the wave number, σ, for differrent values of the Froude number, F_o. It is observed that it approaches its limiting values rather fast, specially for $F_o < 2$. These limiting values are

$$c_r \to 0.5 \qquad \text{for } \sigma \to 0 \tag{25}$$

$$c_r \to \frac{1}{F_o} \qquad \text{for } \sigma \to \infty \tag{26}$$

Note that, for $F_o = 2$, c_r remains constant. The logarithmic decrement, α_1, for the forward wave as a function of σ is presented in Figures 3A and B, for $F_o < 2$ and $F_o > 2$, respectively. For $F_o = 2$, α_1 is identically zero, i.e., the forward wave is neutral. It is observed that there is attenuation ($\alpha_1 < 0$) for $F_o < 2$ and amplification ($\alpha_1 > 0$) for $F_o > 2$. In the limits $\sigma \to 0$ and $\sigma \to \infty$, $\alpha_1 \to 0$, i.e., the forward wave again becomes neutral. The former limit corresponds to the so-called kinematic wave; the latter is a long gravity wave. Figures 4A and B, in turn, show the logarithmic decrement, α_2,

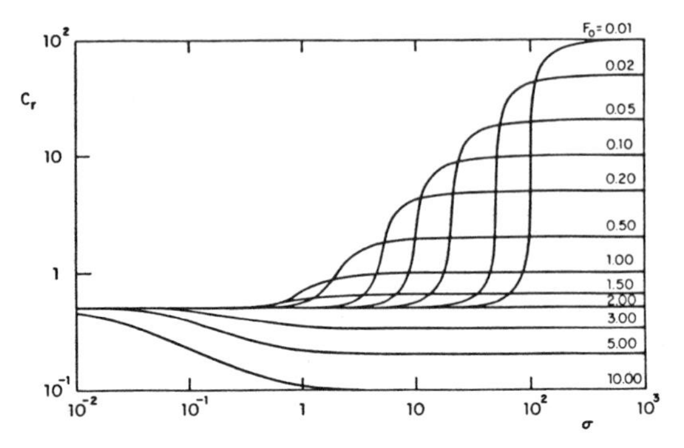

Figure 2. Relative celerity of the shallow water waves [9].

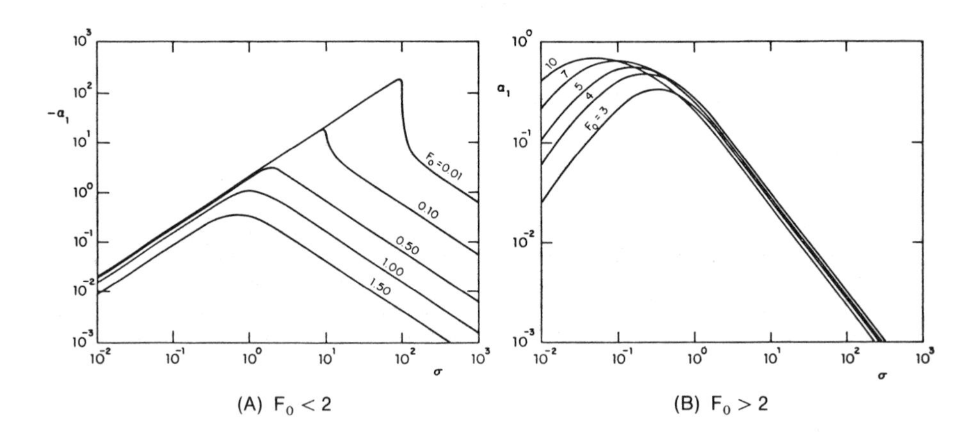

Figure 3. Attenuation for the forward shallow water waves.

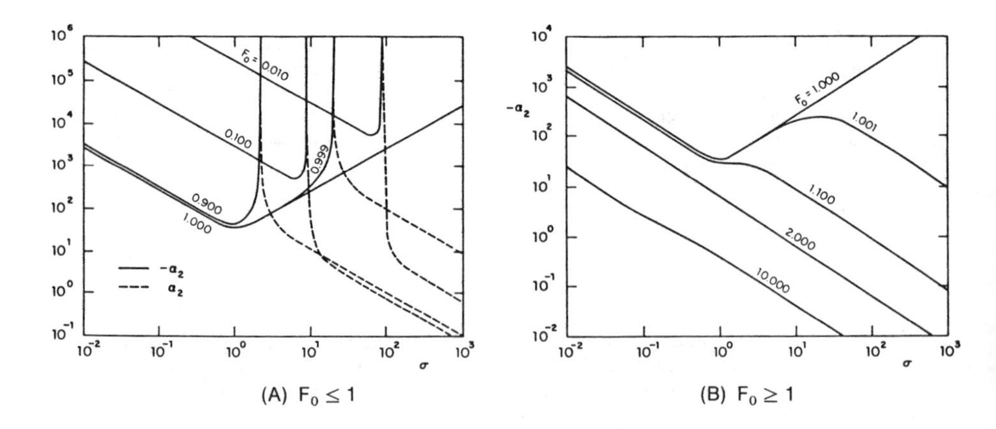

Figure 4. Attenuation for the backward shallow water waves.

for the backward wave for $F_o \leq 1$ and $F_o \geq 1$, respectively. The interpretation of the results is now more subtle. In fact, for $F_o < 1$ (subcritical flow) there is always a value of σ for which $c_r = 1$, i.e., the backward wave is stationary. This critical value, σ_c, is easily found to be

$$\sigma_c = \frac{1}{2F_o}\left(\frac{3}{1 - F_o^2}\right)^{1/2} \tag{27}$$

Note that for $F_o \to 1$, $\sigma_c \to \infty$. Now, for $\sigma \to \sigma_c$, $c_r \to 1$ and α_2 diverges, as seen from Equation 15. Furthermore, for $\sigma > \sigma_c$, $c_r > 1$ (see Figure 2) and α_2 changes sign (see Figure 4A). However, it must be realized that the attenuation is actually controlled by the product $(1 - c_r)\alpha_2$, which remains always finite and negative. It is then concluded that the backward wave attenuates for any value of F_o, if the results of Figure 4B are also taken into account. In the limit $\sigma \to 0$, $\alpha_2 \to -\infty$, i.e. there is no backward wave (for any F_o) in the "kinematic limit." On the other hand, when $\sigma \to \infty$, $(1 - c_r)\,\alpha_2 \to 0$, again corresponding to a (neutral) long gravity wave.

The previous discussion answers the question of when attenuation takes place, within the linear approximation. The issue of what the responsible mechanisms are, is addressed next. To clarify this matter, it is necessary to go back to Equations 10 and 11. From the former, the propagation factor can be expressed as

$$\beta = ku_0(1 + \hat{\Omega}e^{-i\delta}) \tag{28}$$

Equation 28 shows that (disregarding the trivial case $\hat{\Omega} = 0$), β is real, i.e., the wave is neutral, if, and only if, $\delta = 0$ or $\delta = \pi$, i.e., the depth and velocity-waves are in phase or in opposition of phase, respectively. Introducing now Equation 28 into Equation 11 yields

$$\begin{aligned} &2\hat{\Omega}e^{-i\delta} - i\sigma F_o^2\hat{\Omega}^2e^{-2i\delta} + i\sigma - 1 = 0 \\ &\quad R \quad - \quad I \quad + P - W = 0 \end{aligned} \tag{29}$$

where also a symbolic form of the equation has been written, with each term labeled according to the dynamic mechanism it represents, namely R = hydraulic resistance; I = inertia; P = pressure gradient; and W = effective weight. Note that R and W are independent of σ, while I and P are proportional to σ. These terms are represented qualitatively on the complex plane in Figure 5, taken from Menendez and Norscini [8]. Some interesting conclusions can be drawn from this figure and the general remarks following Equations 28 and 29. When $\sigma \to 0$, the contributions of I and P vanish, and the dynamic balance is established only between R and W. From Figure 5 it can be observed

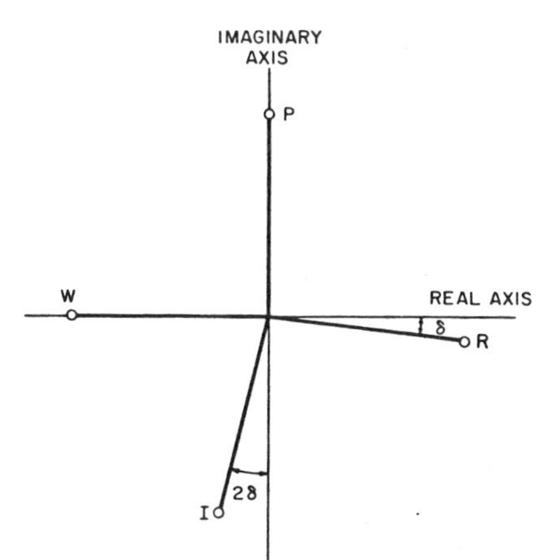

Figure 5. Representation of the terms of the momentum equation (Equation 29) on the complex plane [8].

that, in this case, δ must vanish; then, the resulting wave is neutral. This corresponds to the already mentioned kinematic wave. When $\sigma \to \infty$, the contributions of R and W are dwarfed by those of I and P. Figure 5 now shows that either $\delta = 0$ or $\delta = \pi$, resulting in the neutral forward or backward gravity wave, respectively. For σ finite, at least three mechanisms contribute to the dynamic balance. From Figure 5 it is further seen that a neutral (forward) wave with $\delta = 0$ can exist, for which the balance is set separately between I and P on one side and W and R on the other. It is apparent from Equation 29, that this corresponds to $F_o = 2$. It can further be shown that when $\delta > 0$ there is attenuation and vice versa, as demonstrated by Menendez and Norscini [9]. In fact, δ can be considered as a kinematic parameter which drives wave attenuation.

Spectrum of Shallow Water Waves

Concentrating the attention on the forward wave, the foregoing conclusions can be summarized in the sketch shown in Figure 6. A neutral wave zone is represented in the F_o-σ plane, with its associated dynamic balance. From previous results (see Figures 3A and B), it is known that there is attenuation to the left of the neutral wave zone (i.e. for $F_o < 2$) and amplification to its right ($F_o > 2$). Now, if a certain relative error $\epsilon(\epsilon < 1)$ is allowed, a "practical" neutral wave region can be defined according to the criterion $|\alpha| \leq \epsilon$. This includes the "theoretical" neutral wave zone (Figure 6) but spreads onto its neighborhoods. Menendez and Norscini [9], through an asymptotic analysis, provided its boundaries on the F_o-σ plane. Figure 7 shows the practical neutral wave zone for $\epsilon = 0.05$. With the asymptotic approach, the dynamic mechanism responsible for bringing attenuation (or amplification) when moving away from the neutral wave region can also be identified. This is also schematically represented in Figure 7. For example, pressure is the attenuation agent at the bottom left. Similarly, a neutral wave zone could be defined for the backward wave. This time, however, only an upper portion ($\sigma \gg 1$) would appear, limiting an attenuation region.

Another important property of traveling waves is frequency dispersion, i.e., the variation of the phase velocity with the wave number. From Figure 2 it is observed that shallow water waves are nondispersive (i.e., $dc_r/d\sigma = 0$) for $\sigma \to 0$ and $\sigma \to \infty$ for any F_o, and for $F_o = 2$ for any σ. In other words, the nondispersive wave zone coincides with the theoretical neutral wave region. Introducing the group velocity V, defined by

$$V = \frac{d}{dk}(ck) \tag{30}$$

a "practical" nondispersive wave zone can be defined according to the criterion $|(V - c)/c| \leq \epsilon$. Its boundaries can again be obtained through an asymptotic analysis [9]. Figure 7 also shows the practical dispersive wave region for $\epsilon = 0.05$. It is observed that the practical nondispersive wave zone completely contains the neutral wave region of the forward wave. Note that the dispersion characteristics are identical for the forward and backward waves (with the same wave number, σ).

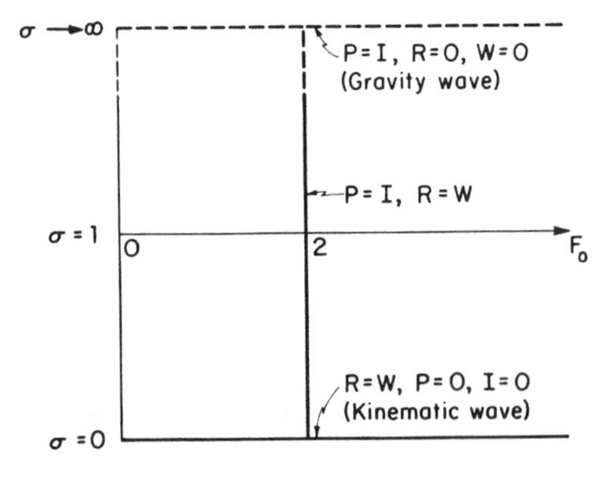

Figure 6. Neutral wave zone for the forward wave.

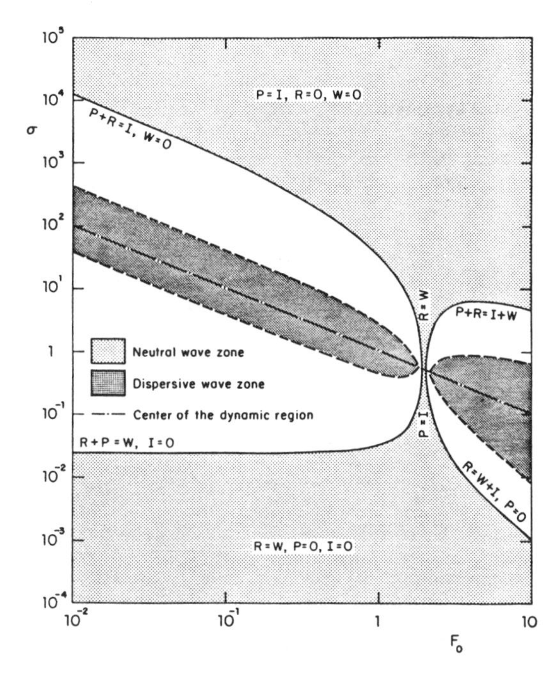

Figure 7. Practical neutral and dispersive wave zones for the forward wave, for $\epsilon = 0.05$ [9].

In the practical dispersive wave zone, the four dynamic mechanisms must contribute significantly to the momentum balance. This is also true in the region of intersection between the upper and lower parts of the practical neutral forward wave zone, around $F_o = 2$, though here the balance tends to be set separately between P and I on one side, and R and W on the other. This "dynamic region," where P, I, R, and W are all important, can be characterized, for every F_o, by the value of σ which corresponds to the point of inflexion of the curve c_r vs. σ (see Figure 2). This is given by [9]

$$\sigma = \frac{1}{F_o} \tag{31}$$

which is shown as a line in Figure 7.

From the foregoing discussion, the following description of the forward wave number spectrum can be given. For F_o beyond the neighborhood of $F_o = 2$, the spectrum is divided in five bands:

1. Two extreme bands ($k \ll S_0/h_0$ and $k \gg S_0/h_0$), which are neutral and nondispersive; they correspond, respectively, to the kinematic and long gravity waves, each one determined by the action of only two dynamic mechanisms.
2. A central band ($k \sim S_0/h_0$), which is nonneutral and dispersive; here, all the mechanisms are relevant; it corresponds to "dynamic" waves.
3. Two intermediate bands ($k \lesssim S_0/h_0$ and $k \gtrsim S_0/h_0$), which are nonneutral and nondispersive; they correspond, respectively, to "dynamic-kinematic" and "dynamic-gravity" waves; the dynamic balance is mainly established between two mechanisms, the remaining two giving small, but nonnegligible, contributions.

When $F_o \rightarrow 2$, the waves tend to be neutral and nondispersive throughout the spectrum, and kinematic and gravity waves coexist within the dynamic region.

These results are illustrated with the example presented in Table 1 [9], constructed using Figures 7 and 2. For the given particular flow conditions (S_0, h_0, u_0) and admitted relative error (ϵ), it comes

Table 1
Spectral Bands

Limit	σ	Wavelength (km)	c_r	c (m/s)	Period
Upper limit of lower neutral band (kinematic band)	0.025	7,600	0.50	1.4	64 days
Upper limit of lower nondispersive band	2.3	82	0.54	1.4	16 hr
Center of dynamic region	6.3	30	1.92	2.7	3.2 hr
Lower limit of upper nondispersive band	24	7.9	5.50	5.9	22 min
Lower limit of upper neutral band (gravity band)	620	0.30	5.97	6.4	48 sec

Note: $S_0 = 0.0001$; $h_0 = 3.05$ *m*; $u_0 = 0.91$ *m/s*; $\epsilon = 0.05$
From Menendez and Norscini [9].

out that: a kinematic wave model can be applied for waves of period greater than about 64 days, while a gravity wave model would require a period lower than 48 seconds. Attenuation will occur for intermediate values of the period. However, there will be still no dispersion for waves of periods between 48 seconds and 22 minutes or between 16 hours and 64 days. In fact, a diffusive wave model (including only R, W, and P) could be used for the latter range. The remaining band will show both attenuation and dispersion effects, which will attain a maximum for a period of about 3.2 hours.

Flow Instability

As already said, the mathematical procedure utilized in the present section to analyze the behavior of shallow water waves is the one used in linear hydrodynamic stability theory. From the latter point of view, the preceding results can be interpreted in a very simple way: the uniform flow regime is hydrodynamically stable for $F_o < 2$ and unstable for $F_o > 2$. Then, $F_o = 2$ is the critical Froude number for stability. This means that any small perturbation will be damped in the former case; for $F_o > 2$, instead, any perturbation traveling downstream, relative to the flow, will be amplified. Its rate of growth will be given by α_1 (Figure 3B) till the amplitude gets large enough for the linear theory to break down. This instability has been experimentally observed, and it leads to the formation of the so-called roll waves [10]. Jeffreys [11] was the first to establish this criterion for stability.

Consider now the consequences on the previous results of using a resistance law different from Chezy's. More specifically, assume that C depends on h. The results do not change qualitatively, but the critical value of the Froude number for stability and, consequently, the phase velocity in the kinematic limit, do. Assuming, for example, that $C^2 \sim h^n$, it is obtained that the critical F_o is

$$F_o = \frac{2}{1+n} \tag{32}$$

and that

$$c_r \rightarrow \frac{1+n}{2} \quad \text{for } \sigma \rightarrow 0 \tag{33}$$

which, of course, reduce to the previous values for $n = 0$. The well known Manning formula [5] corresponds to $n = \frac{1}{3}$. Then, $F_o = \frac{3}{2}$ is the critical value for stability (and $c_r \to 2/3$ in the kinematic limit), as first shown by Keulegan and Patterson [12]. Dressler and Pohle [13] extended the stability criterion to even more general resistance laws. They also demonstrated that the resistance law has to show a dependence on both the velocity and the depth for instability to occur. Equivalent stability criteria were obtained by Vedernikov [14] and Craya [15] using different approaches to the one presented here. Craya's criterion for instability can be written in the two following equivalent forms:

$$\frac{dq}{dh} > u + \sqrt{gh} \tag{34}$$

where $q = uh$, or

$$h\frac{du}{dh} > \sqrt{gh} \tag{35}$$

which are valid for any resistance law, and can be interpreted as requiring that the velocity of propagation of the kinematic wave must be larger than the velocity of the forward long gravity wave (see below for a general definition of the kinematic wave velocity). From this point of view, the instability can be thought, quoting the description given by Whitham [4], as arising from an "unresolvable competition between the two sets of waves."

Some Particular Solutions

The analysis of the previous sections allowed the identification of different types of shallow water waves, according to their wave number or, equivalently, their frequency. It is enlightening, now, to discuss a problem in which all kinds of waves are present, from kinematic to long gravity waves. A standard initial-value problem, typical for flood waves, will be considered: an initially uniform flow is disturbed by the passage of a flood wave, either in the form of a hump or of a smoothed step. The equations to be solved are Equations 6 and 7, with the following initial and boundary conditions

$$\begin{aligned} &\text{At } t = 0, \quad h_1 = 0, u_1 = 0 \quad \text{for } x > 0 \\ &\text{At } x = 0, \quad h_1 = f(t) \quad \text{for } t > 0 \end{aligned} \tag{36}$$

where f(t) is a given function of time. The problem is well posed provided $F_o < 1$, so the analysis is restricted to this range of Froude numbers. Lighthill and Whitham [3] solved it using the Heaviside calculus. The details will be omitted and only the major conclusions will be presented below.

Close to the wave-front, the solution is given by

$$h_1 \simeq \exp\left[-\frac{1}{2F_o}\frac{(2-F_o)}{(1+F_o)}\frac{S_0 x}{h_0}\right] f\left(t - \frac{x}{u_0\left[1+\frac{1}{F_o}\right]}\right) \tag{37}$$

Equation 37 shows that the waves traveling with a celerity $(1 + 1/F_o)u_0$, i.e., the forward long-gravity wave phase-velocity, attenuates exponentially. On the other hand, from an analysis of the solution far from the wave-front and for large times, i.e., $t \gg u_0/gS_0$, it can be shown that the "main disturbance" travels downstream with a celerity $(3/2)u_0$, i.e., the kinematic wave phase velocity. In

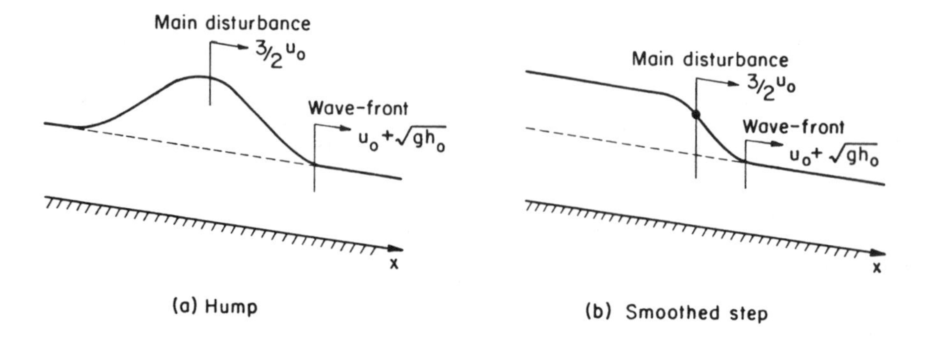

Figure 8. Particular solutions of the linearized equations.

the case of a hump, the main disturbance refers to the maximum amplitude, h_{1M}, and is approximately given by

$$h_{1M} \simeq \left(\frac{gS_0}{2\pi u_0 t}\right)^{1/2} \frac{3F_0}{(4 - F_0^2)^{1/2}} \int_0^\infty f(t)\, dt \tag{38}$$

Equation 38 shows that h_{1M} attenuates as $t^{-1/2}$; this is a diffusion effect. In the case of a smoothed step, the main disturbance is the maximum slope, and it is also damped as $t^{-1/2}$. Figure 8 illustrates the results. Interpreted in terms of the spectral components, the wave motion can be described as follows: the forerunner is a long gravity wave. It is followed by dynamic-gravity waves, which attenuate exponentially due to the action of the hydraulic resistance. The main disturbance, in turn, is a kinematic wave. It is surrounded by dynamic-kinematic waves, which diffuse due to the action of the pressure gradient. Between the wave-front and the main disturbance, dynamic waves propagate with intermediate celerities, driven by the simultaneous action of the four dynamic mechanisms.

NONLINEAR ANALYSIS

Nonlinear Phenomena

In a linear theory, the different spectral waves, i.e., harmonic waves with different wave numbers, propagate independently of each other. However, when the amplitude of the waves becomes comparable to the depth, the interaction between spectral waves is significant, and the linear theory breaks down. Moreover, even if the wave amplitudes remain small, the linear approximation may not be uniformly valid for large times, as cumulative nonlinear effects can become crucial [4].

One of the consequences of the breakdown of the linear approximation is that superposition of solutions is not valid anymore. Then, in a nonlinear analysis, each initial-value problem has to be studied separately. It is possible, however, to make a few general remarks about the nature of the main nonlinear phenomena. Nonlinearity produces what is called amplitude dispersion; this means that the individual waves have celerities that depend on the local conditions. For example, the dynamic-gravity waves, instead of traveling together with the velocity $(1 + 1/F_o)u_0 = u_0 + (gh_0)^{1/2}$, have celerities given by $u + (gh)^{1/2}$, where u and h are the local flow velocity and depth, respectively. Analogously, the dynamic-kinematic waves travel with $(3/2)u$, instead of $(3/2)u_0$.

As a consequence of amplitude dispersion, the general wave-form changes, producing what is called nonlinear distortion. Furthermore, some spectral waves can overtake others. Mathematically, this manifests as a multivaluation of the flow quantities at some point in space and time, which indicates wave breaking and the formation of a shock. A shock is a theoretical discontinuity in the flow quantities which evolves (i.e., propagates and attenuates) according to the principles of conservation of mass and momentum. The mathematical treatment of these discontinuous solutions requires the introduction of the weak solution concept [4]. Physically, a shock is a relatively abrupt transition within a distance considered negligible by the theory. This means that within that transition region there must be acting physical mechanisms not considered in the equations of motion. For example, when dealing with dynamic-gravity waves, distances of the order of h_0 (i.e., $\sigma \sim S_0^{-1}$) are neglected; then, any transition occurring within these distances will be treated as a shock by the theory. The same is true for dynamic-kinematic waves when the distances are of the order of h_0/S_0 (i.e., $\sigma \sim 1$).

One of the most interesting concepts arising from the linear analysis presented previously, is the distinction of spectral wavebands. Though they were defined according to the different physical characteristics of the spectral waves, the generalization of this concept to the nonlinear case is more easily done by using, as defining property, the relevant dynamic mechanisms. This approach is undertaken in the following sections, where the particularities of each one of the spectral bands are discussed, with emphasis on the associated nonlinear phenomena.

Long Gravity Waves

The defining characteristic of long gravity waves is that the effects of resistance and effective weight can be neglected. This means that Equation 5 reduces to

$$\frac{\partial u}{\partial t} + u\frac{\partial u}{\partial x} + g\frac{\partial h}{\partial x} = 0 \tag{39}$$

The system of Equations 4 and 39 lacks an explicit general solution. However, much insight into the nature of the solutions can be gained by recurring to the semi-graphical method of characteristics [16]. To this end, the above system is transformed, by simple algebraic manipulations, into the following equivalent set of equations [2]

$$\frac{\partial}{\partial t}(u + 2c) + (u + c)\frac{\partial}{\partial x}(u + 2c) = 0 \tag{40}$$

$$\frac{\partial}{\partial t}(u - 2c) + (u - c)\frac{\partial}{\partial x}(u - 2c) = 0 \tag{41}$$

where c is now defined as

$$c \equiv (gh)^{1/2} \tag{42}$$

Alternatively, Equations 40 and 41 can be expressed as requiring that

$$J_+ \equiv u + 2c = \text{const. along the curve } C_+\text{:}\ \frac{dx}{dt} = u + c \tag{43}$$

$$J_- \equiv u - 2c = \text{const. along the curve } C_-\text{:}\ \frac{dx}{dt} = u - c \tag{44}$$

The quantities J_+ and J_- are called Riemann invariants, and the curves C_+ and C_- on the x-t plane are the characteristic curves. According to Equations 43 and 44, then, the whole plane $x - t$ can be thought as covered by two families of characteristic curves. Along each curve of the $C_+(C_-)$ family, $J_+(J_-)$ is a constant, different, in general, for different curves of the family. In more physical terms, Equations 43 and 44 expresses that the "signal" $J_+(J_-)$ is transported with a celerity $c(-c)$ relative to the fluid. This shows that, in fact, this is a wave phenomenon and that it is composed of forward and backward waves; more specifically, these are hyperbolic waves [4].

Based on the fact that no source-like terms appear in Equations 4 and 39, and on the results of the linear analysis of the previous chapter, it is expected that long gravity waves are neutral. To

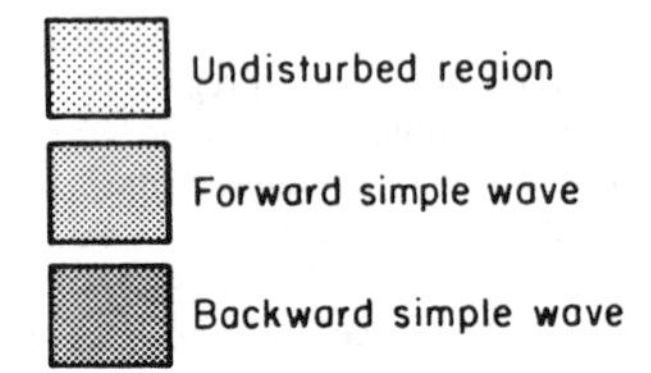

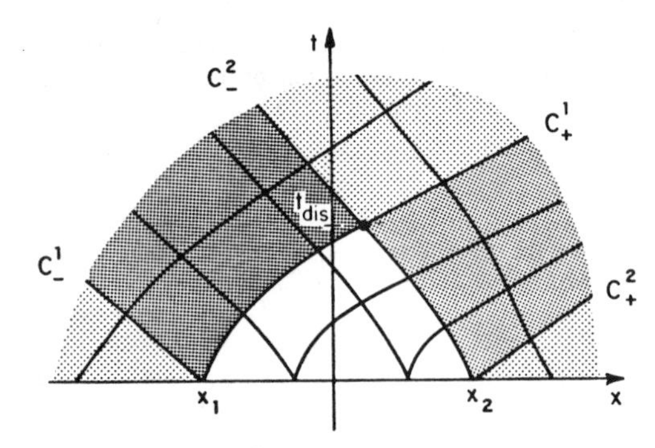

Figure 9. Characteristic curves for the evolution of a localized disturbance.

establish this unequivocally, it is first necessary to identify the most elementary (forward or backward) nonlinear wave. The method of characteristics is the appropriate tool for this analysis. Consider the following relatively general initial-value problem [17]: at $t = 0$ there is an undisturbed flow (u_0, h_0), except in the region $x_1 < x < x_2$ where an unknown (finite) perturbation exists. The general features about the evolution of the perturbation are sought. The situation is schematized on the x-t plane in Figure 9. A set of characteristic curves for each family is shown. Their precise shape is not known a priori (in fact, knowing the characteristic curves is equivalent to knowing the complete solution to the problem). However, significant conclusions can be drawn from the graph. First, note that all the C_+ curves ahead of C_+^2 (i.e., those with higher x-values for the same t-value) originate from the undisturbed region $x > x_2$. Then, from Equation 43 one gets

$$u + 2c = u_0 + 2c_0 \text{ ahead of } C_+^2 \tag{45}$$

where $c_0 = (gh_0)^{1/2}$. Analogously, it can be obtained that

$$u + 2c = u_0 + 2c_0 \text{ behind } C_+^1 \tag{46}$$

$$u - 2c = u_0 - 2c_0 \text{ ahead of } C_-^2 \tag{47}$$

$$u - 2c = u_0 - 2c_0 \text{ behind } C_-^1 \tag{48}$$

Now, realizing that the region ahead of C_+^2 is completely contained in the zone ahead of C_-^2, from Equations 45 and 47 one gets

$$u = u_0,\ c = c_0 \text{ ahead of } C_+^2 \tag{49}$$

which corresponds to the forward region still not disturbed by the wave. Similarly, Equations 46 and 48 yield

$$u = u_0,\ c = c_0 \text{ behind } C_-^1 \tag{50}$$

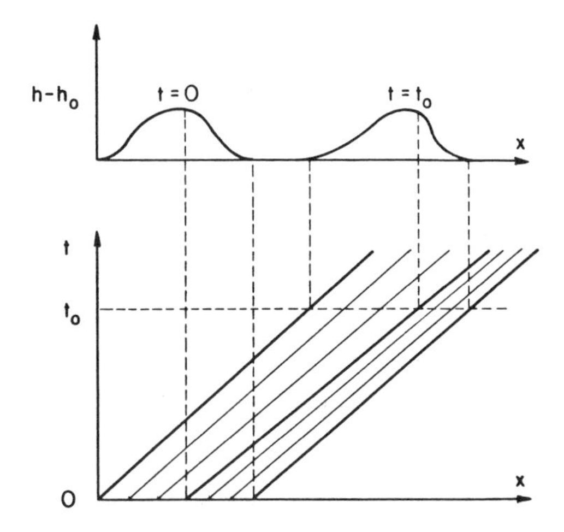

Figure 10. Amplitude dispersion in a forward simple wave.

corresponding to the backward undisturbed zone. Note that Equations 49 and 50 show that the perturbation spreads both forward and backwards onto the undisturbed region with a celerity c_0 relative to the fluid. The most interesting point, however, is that there exists a time $t = t_{dis}$ after which the regions behind C_+^1 and ahead of C_-^2 intersect. In this zone, from Equations 46 and 47, it holds again that $u = u_0$ and $c = c_0$, i.e. the flow is undisturbed. The conclusion is, then, that after a finite time any (finite) disturbance disentangles into a forward and backward waves. The nature of these elementary components or "simple waves" is studied next.

Consider the forward simple wave, i.e., the intersection of the regions ahead of C_-^2 and between C_+^1 and C_+^2. Along each one of the C_+ curves in this zone, Equation 43 is, of course, satisfied. Combining it with Equation 47, one readily obtains that

$$u = \text{const.},\ c = \text{const. along each } C_+ \text{ curve} \tag{51}$$

Equation 51 shows that the C_+ characteristics are straight lines in a forward simple wave (as $u + c =$ const. along each C_+). Most importantly, *there is no attentuation*, as both u and h are carried unchanged with the wave celerity c (relative to the fluid). Analogously, in a backward simple wave the C_- characteristics are straight lines, each one carrying constant values of u and h.

The phenomenon of amplitude dispersion, already mentioned in the previous section, stems clearly from the characteristic curve concept. It can easily be shown, for example, that in a forward simple wave larger depths tend to move away from smaller ones lying behind (diminishing the steepness of the wave in the rear), but closer to the smaller depths lying ahead (producing a steeper front) [2]. This is illustrated in Figure 10, both in the physical and characteristic planes (only the C_+ curves are represented for the sake of clarity). It is observed that the C_+ curves diverge from each other in the rear of the wave and converge to each other in the front. This means that there will be a time at which two characteristics will first intersect. It corresponds to the appearance of a point with an infinite slope in the front of the wave, and announces the imminence of shock formation. The theory of characteristics then breaks down locally, and the shock must be treated separately. The equations of motion governing this dynamic shock or "bore" are obtained using the mass and momentum conservation principles. They are [2]

$$U_s = \frac{u_H h_H - u_L h_L}{h_H - h_L} \tag{52}$$

$$(u_H - u_L)^2 = \frac{g}{2h_H h_L}(h_H - h_L)(h_H^2 - h_L^2) \tag{53}$$

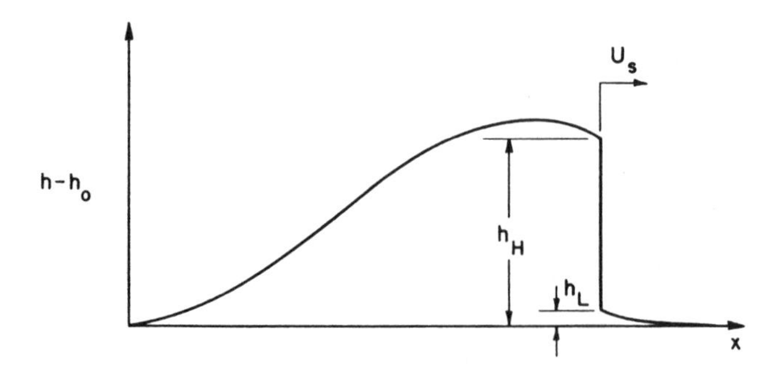

Figure 11. Theoretical representation of the hydraulic bore.

where U_s is the velocity of the bore and the subscripts H and L refer to conditions immediately to the higher and lower level sides, respectively (see Figure 11). Note that, according to Equation 53, the sign of $(u_H - u_L)$ remains undetermined. The energy principle has to be used for this purpose: fluid particles crossing the shock cannot gain mechanical energy [2]. This closes the problem of following the evolution of the bore. Regarding its structure, it could only be described recurring to the Navier-Stokes equations.

A typical example, for which the gravity wave model is particularly successful, is the description of the flow resulting from the sudden failure of a dam. It is specially so during the first stages of propagation of the wave, after its formation (during which vertical accelerations are significant). Figure 12 shows the solution for the wave profile [2]. Note that, as in the general initial-value problem discussed previously, there are two simple waves: a "depression wave" traveling backwards, and a bore propagating forward. But, unlike the general case, between them there exists a zone of uniform flow, instead of an undisturbed region. This is due to the crossing of the characteristic curves (leading to the existence of the bore), a possibility not considered in the general problem. Another flow dominated by gravity waves is that resulting from the operation of sluice gates, which can also be described in terms of simple waves [5]. More complicated situations require the use of numerical methods of integration of the differential equations.

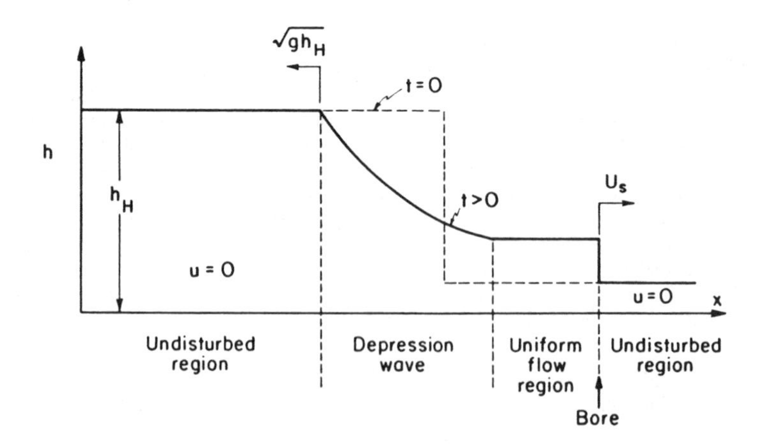

Figure 12. Theoretical solution of the dam failure problem.

Dynamic-Gravity Waves

The dynamic-gravity waves will be identified as those in which the effects of resistance and effective weight are small, though not negligible. A characteristic formulation of the equations of motion, Equations 4 and 5, leads now to the relations [2]

$$\frac{dJ_+}{dt} \equiv \frac{d}{dt}(u + 2c) = g\left(S_0 - \frac{u^2}{C^2h}\right) \text{ along } C_+: \frac{dx}{dt} = u + c \tag{54}$$

$$\frac{dJ_-}{dt} \equiv \frac{d}{dt}(u - 2c) = g\left(S_0 - \frac{u^2}{C^2h}\right) \text{ along } C_-: \frac{dx}{dt} = u - c \tag{55}$$

where c is still given by Equation 42. Equations 54 and 55 show that the wave properties J_+ and J_- change when moving along the corresponding characteristic curves, due to the effects of the weight and the resistance.

To determine whether these effects will result in attenuation or amplification, the flow in the neighborhood of the front of a flooding wave will be analyzed, where it is known from results found earlier that dynamic-gravity waves exist. Following Lighthill and Whitham [3], the flow quantities in this region can be expanded as

$$u = u_0 + \tau u_1(t) + \tau^2 u_2(t) + \cdots \tag{56}$$

$$h = h_0 + \tau h_1(t) + \tau^2 h_2(t) + \cdots \tag{57}$$

where $\tau \equiv t - x/[u_0 + \sqrt{gh_0}]$, and $\tau = 0$ is the wave-front. Introducing Equations 56 and 57 into Equations 4 and 5 the following equation is obtained [3]

$$\frac{dh_1}{dt} = \frac{3}{2h_0(1 + F_o)} h_1(h_1 - K) \tag{58}$$

where

$$K = \frac{gh_0S_0}{3u_0}(2 - F_o)(1 + F_o) \tag{59}$$

The last term of Equation 58 is the contribution of resistance and weight. As $h_1(0) > 0$ (this is a flooding wave), Equation 59 shows that this contribution is positive or negative depending on F_o being greater or smaller than 2, respectively. Then, the neutral behavior of a long gravity wave given by the first term on the right-hand side of Equation 58 will turn to amplification in a dynamic-gravity wave when $F_o > 2$ and to attenuation when $F_o < 2$, exactly as given by the linear analysis.

Equation 58 is also useful in studying shock formation. Note, first, that for long gravity waves (i.e., taking $S_0 = 0$), it predicts that h_1 will increase without limit. As $h_1 \sim \partial h_1/\partial x$, this means that a bore will eventually appear, in agreement with the discussion in the previous section. Now, for dynamic-gravity waves ($S_0 \neq 0$), a similar conclusion holds for $F_o > 2$, though in this case the bore will appear earlier as h_1 increases faster. Inversely, for $F_o < 2$ shock formation will be delayed. In fact, Equation 58 shows that if $h_1(0) < K$ no bore will appear at all. From the characteristic point of view, this means that the combined effect of resistance and weight will make the (C_+) characteristic curves to bend upwards and diverge from each other, avoiding their crossing [5]. If $F_o < 1$, the solution to Equation 58 is [3]

$$h_1(t) = \frac{Kh_1(0)e^{-bt}}{K - h_1(0)(1 - e^{-bt})} \tag{60}$$

where $b = (gS_0/2u_0)(2 - F_o)$, confirming the exponential decay predicted by the linear analysis. In fact, the linear expansion of Equation 60 is exactly Equation 37 on the wave-front. The same kind

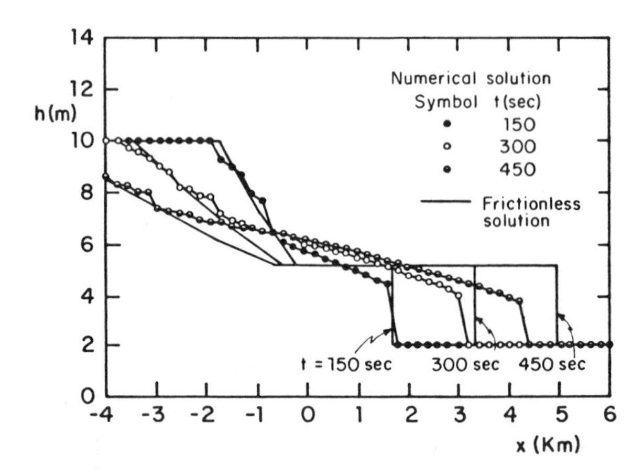

Figure 13. Effects of hydraulic resistance on the dam-break wave [19].

of analysis can be applied to the front of an upstream moving wave, which corresponds to the practical problem of the tidal bore. Also in this case the initial perturbation $[h_1(0)]$ has to exceed a critical value for the wave to break [18].

Once a bore is formed, its strength (measured by $h_H - h_L$) and celerity will be affected by resistance and weight. Figure 13 show the effect of resistance on the dam-break wave, obtained using a numerical method [19]. The region with larger velocities, close to the wave front, is the most affected. As intuitively expected, the strength and velocity of the bore are continually decreased by the action of the resistance.

Kinematic Waves

The kinematic waves result from the balance of resistance and effective weight, the remaining dynamic mechanisms being negligible. Then, Equation 5 can be simplified to the following algebraic expression

$$u = C(S_0 h)^{1/2} \tag{61}$$

Equation 61 shows that u and h are related univocally. Then, the last term in Equation 4 can be rewritten as

$$h\frac{\partial u}{\partial x} = \left(h\frac{du}{dh}\right)\frac{\partial h}{\partial x} \tag{62}$$

Using Equations 61 and 62, Equation 4 transforms into

$$\frac{\partial h}{\partial t} + \frac{3}{2}u\frac{\partial h}{\partial x} = 0 \tag{63}$$

Equation 63 is already in characteristic form. In fact, it can be restated as requiring that

$$h = \text{const. along the curve } \frac{dx}{dt} = \frac{3}{2}u \tag{64}$$

Then, in this case there exists only one family of characteristic curves, which correspond to a forward (hyperbolic) wave moving with celerity u/2 with respect to the fluid. It can be easily shown that u and the discharge (per unit width) $q = uh$ also satisfy Equation 63. As a consequence, the slope of the characteristics remains constant for each one of them (see Equation 64), i.e., they are straight lines. Equation 64 also shows that the values of h are carried unchanged by the waves, i.e., *they are neutral.* Note that the kinematic wave celerity is, according to Equation 62, given by $h\,du/dh$. This justifies the interpretation of Equation 35.

Amplitude dispersion occurs in kinematic waves very much as in long gravity waves: larger depths correspond (from Equation 61) to larger flow velocities and, then, to larger wave celerities. Thus, Figure 10 can also be used to illustrate qualitatively the phenomenon in this case (now, the drawn family of characteristic curves is the only one to exist). Again, amplitude dispersion leads to shock formation. The equation of motion of the resulting kinematic shock is simply Equation 52. Its structure can be described by the complete shallow water theory.

The kinematic wave model can be applied to study the propagation of (long) flood waves [3]. In this case, the kinematic shock, if present, may extend over a length of the order of 100 km.

Dynamic-Kinematic Waves

The dynamic-kinematic waves will be defined as those in which inertia and pressure gradient effects are small, though not negligible. No analysis has been reported that deals with the general case. Instead, the particular (important) case, in which only the pressure contribution is considered, has been studied in some depth. It corresponds to the condition $F_o \ll 1$ and it leads to the so-called diffusive waves.

If the inertia terms are dropped, Equation 5 reduces to

$$u = C\left[h\left(S_0 - \frac{\partial h}{\partial x}\right)\right]^{1/2} \tag{65}$$

Introducing Equation 65 into Equation 4, and considering that the pressure contribution is small ($\partial h/\partial x \ll S_0$), one obtains [3]

$$\frac{\partial h}{\partial t} + \frac{3}{2}u\frac{\partial h}{\partial x} = \frac{Ch^{3/2}}{2S_0^{1/2}}\frac{\partial^2 h}{\partial x^2} \tag{66}$$

or

$$\frac{dh}{dt} = \frac{Ch^{3/2}}{2S_0^{1/2}}\frac{\partial^2 h}{\partial x^2} \text{ along the curve } \frac{dx}{dt} = \frac{3}{2}u \tag{67}$$

Equation 67 shows that h varies slightly along the characteristic curves, at a rate which depends on the value of h on neighboring characteristics. For example, where h is a maximum, $\partial^2 h/\partial x^2 < 0$; then h decreases along the characteristics, i.e., *there will be attenuation.* Note that, on the contrary, if there is a minimum of h, it will tend to increase. This is a typical diffusion effect: the wave is being smoothed out. In fact, Equation 66 is a convection-diffusion equation, with a coefficient of diffusivity $Ch^{3/2}/2S_0^{1/2}$.

An interesting point, when dealing with diffusive waves, is the definition of the wave crest. Above, it has been demonstrated that what will be named the "instantaneous crest" (called "river crest" when treating flood waves), i.e., the point with $\partial h/\partial x = 0$, attenuates at a rate given by Equation 67. However, also a "local crest" can be identified at the point on the wave profile where the height is attaining a local maximum, i.e. that with $\partial h/\partial t = 0$. From Equation 4, this is equivalent to the condition $\partial q/\partial x = 0$ which, using Equation 65, leads to the relation

$$3\frac{\partial h}{\partial x}\left(S_0 - \frac{\partial h}{\partial x}\right) = h\frac{\partial^2 h}{\partial x^2} \tag{68}$$

that must be satisfied at the local crest. As this point cannot be far from the instantaneous crest (remember that in the kinematic waves both coincide), then $\partial^2 h/\partial x^2 < 0$. Equation 68 thus shows

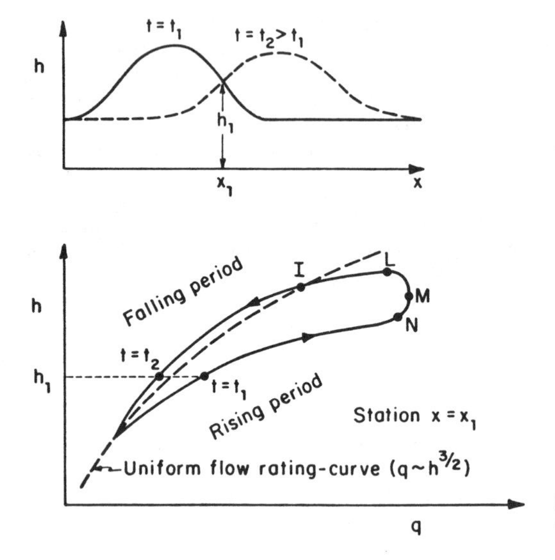

Figure 14. Rating curve for a flood wave.

that also $\partial h/\partial x < 0$, i.e., the local crest is downstream relative to the instantaneous one. This is in agreement with the linear theory, which shows that the u-wave leads the h-wave in a diffusive wave.

It turns out that the local crest has more engineering significance than the instantaneous one. Henderson [5] developed the following approximate expression to calculate the attenuation of the local crest:

$$\frac{dh}{dx} \simeq \frac{4}{27} \frac{1}{C^2 S_0^2} \frac{\partial^2 h}{\partial t^2} \tag{69}$$

Equation 69 is very convenient from a practical point of view, as the rate of attenuation can be calculated exclusively from local quantities.

The effect of the pressure gradient on shock formation is simple to understand. The diffusion effect counteracts the amplitude dispersion effect on the wave-front, thus producing a delay in the appearance of the kinematic shock. Once it is formed, it is expected that the pressure will contribute to its decay (both in strength and celerity).

The diffusive wave model is particularly successful in calculating the evolution of flood waves. It is relatively easy to obtain a qualitative picture of how the "stage" h and the discharge q will evolve locally in this case. From Equation 65, it is seen that, for a given value of h, there will be two values of u (or, equivalently, of q): a larger one during the rising period (as $\partial h/\partial x > 0$) and a smaller one during the falling period (∂h, $\partial x < 0$), as shown in Figure 14. The curve h-q ("rating curve") will be then a close loop, as shown in the figure. The points I and L identify the instantaneous and local crests, respectively. The point M, in turn, corresponds to the instant of maximum discharge, i.e., $\partial q/\partial t = 0$. Point N is also shown, at which the maximum flow velocity (q/h) is attained. Thus, at a given station, during the duration of the flood, it will first pass the maximum velocity, followed, in order, by the maximum discharge, the maximum stage, and the instantaneous crest.

Dynamic Waves

The dynamic waves are those in which the four dynamic mechanisms give significant contributions. Thye are described by Equations 4 and 5 or, in their equivalent characteristic form, by Equations 54 and 55. These last set of equations show that the details of the information is being carried by forward and backward long gravity wavelets, traveling with their typical celerity $c = \sqrt{gh}$ relative to the fluid.

It seems that the linear stability criteria can still be used with nonlinear waves to determine whether they will attenuate or amplify. But, as these waves are of finite amplitude, the criteria may give different answers at different heights. In the case of single-peaked waves, Di Silvio [20] proposed using the middle height of the wave as the adequate point to check for overall attenuation (or amplification). However, Jolly and Yevjevich [21], through numerical experiments, concluded that conditions at the wave peak determine the attenuation characteristics.

According to the linear theory, dynamic waves are highly attenuated and dispersed when $F_o < 2$ (and beyond the neighborhood of $F_o = 2$). This would indicate that these are short-term waves, in the sense that they would decay relatively fast, leaving the long-term kinematic or long gravity waves. However, nonlinearity, through amplitude dispersion, tends to counteract frequency dispersion in the wave front. In finite amplitude waves, the two mechanisms can, eventually, equilibrate each other producing longer-term or even neutral dynamic fronts.

A particular case, that of a fixed-shape forward dynamic front, is analyzed now in some detail to illustrate some of the general properties of dynamic waves. This is the kinematic shock, that, as explained earlier, is treated as a singularity in the kinematic wave theory. Its linear version was also discussed earlier. It was said there that the main disturbance, i.e., the maximum slope, moves as a kinematic wave (at least for $F_o < 1$). The existence of the kinematic shock can be proved by showing that there is a solution to Equations 4–5 in which both h and u depend only on the combination $x - U_s t \equiv X$, where U_s is the velocity of the wave form [3]. If the conditions upstream and downstream of the wave are identified by the subscripts H and L, respectively, U_s is given by Equation 52, for continuity reasons. In turn, from the momentum equation, Equation 5, and assuming that uniform-flow conditions exist upstream and downstream, the following expression is obtained [3]

$$\frac{dh}{dX} = -S_0 \frac{(h - h_L)(h_H - h)(h - H)}{h^3 - h_c^3} \tag{70}$$

where

$$H = \frac{h_H h_L}{(\sqrt{h_H} + \sqrt{h_L})^2} \tag{71}$$

$$h_c^3 = \frac{q_r^2}{g} \tag{72}$$

$$q_r = h(U_s - u) = h_L(U_s - u_L) \tag{73}$$

The quantity q_r is the (uniform) discharge measured from a reference system moving with velocity U_s, and h_c is the corresponding critical depth (i.e., the one at which transition from subcritical to supercritical flow occurs [5]). This quantity is related to the Froude number downstream of the wave, which will be called F_o, through

$$h_c^{3/2} = h_L h_H \frac{(h_H^{1/2} - h_L^{1/2})}{h_H - h_L} F_o \tag{74}$$

Three different cases are usually identified, depending on the value of F_o or, equivalently, of h_c. The first one corresponds to the condition $h_c < h_L$. Equation 70 shows that, in this situation, $dh/dX = 0$ at both ends, and at no other place (note, from Equation 71, that $H < h_L$ always). Therefore, the depth tends asymptotically to their limiting values both upstream and downstream. The profile is as shown in Figure 15A (the explicit solution is given in References 3 and 5). This is called the "monoclinal flood wave." A measure of the wave thickness is provided by the distance between locations upstream and downstream where the depth reaches the limiting values within a tolerance ϵ ($\ll 1$). This thickness is of the order of h_L/S_0, which, as already mentioned, is considered negligible in the kinematic wave theory.

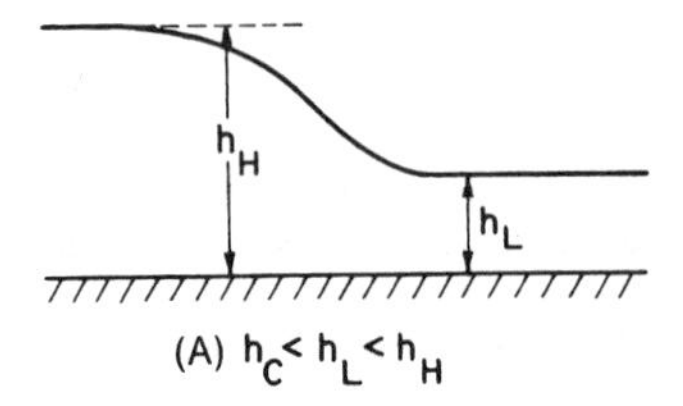

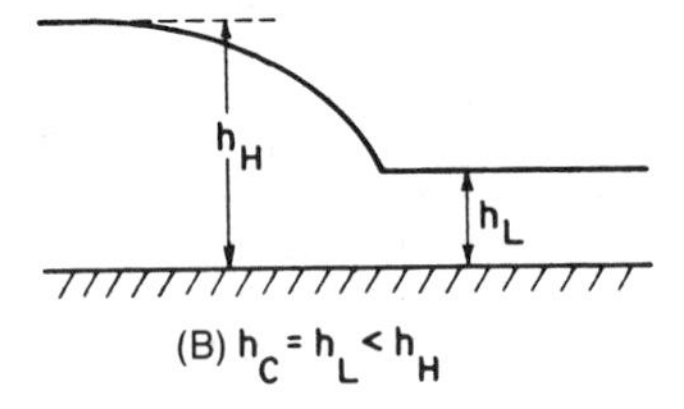

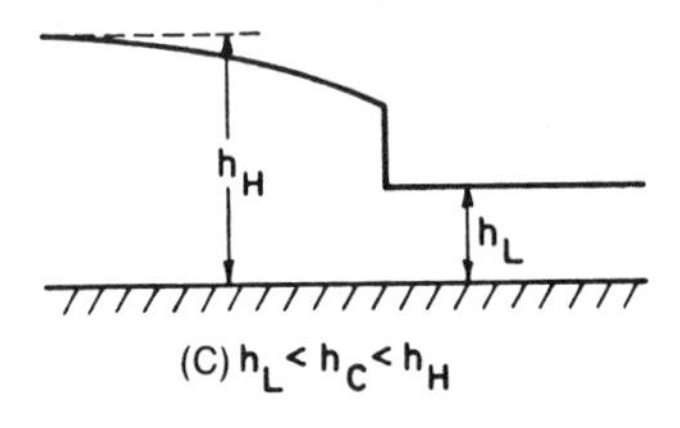

Figure 15. Structure of the kinematic shock.

The second case is when $h_c = h_L$. From Equation 70, it is now seen that dh/dX is finite at the downstream end, as shown in Figure 15B. The corresponding F_o, from Equation 74, is

$$F_o = \frac{\sqrt{(h_H h_L)} + h_L}{h_H} \tag{75}$$

which is always smaller than 2 ($F_o = 2$ for $h_H = h_L$). It can also be shown that

$$U_s = u_L + \sqrt{(gh_L)} \tag{76}$$

i.e., the wave form moves exactly with the same velocity as the long gravity wave forerunner. In addition, it can be obtained that

$$\left.\frac{\partial h}{\partial T}\right|_{\text{wave-front}} = K \tag{77}$$

where $T \equiv -X/U_s = t - x/U_s$ and K is given by Equation 59. Equation 77 shows, according to an earlier discussion, that the slope of the wave at the front, which is now the maximum slope (and then moves as a kinematic wave), has attained the critical value beyond which a dynamic shock will appear. In summary, there is a coalescense of the kinematic and long gravity waves which, in accordance with Craya's criterion, announces the imminence of bore formation.

The third case, finally, corresponds to $h_L < h_c < h_H$, and the profile is shown in Figure 15C. After the preceding discussion, it is not surprising to find a bore at the downstream end. The wave is now a combined kinematic-dynamic shock. Before h_c can reach h_H (in which case, the bore would separate two uniform flow states), F_o exceeds 2. Then, the flow becomes unstable, and a series of

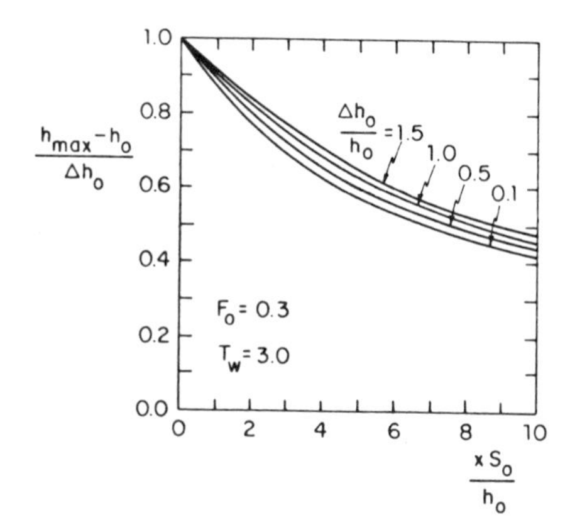

Figure 16. Attenuation of a sinusoidal hump for different initial amplitudes [22].

roll waves appears. It is interesting that the periodic roll-wave pattern is structurally very similar to the kinematic shock for $h_c > h_L$. In fact, it is constituted of smooth sections satisfying Equation 70, separated from each other by dynamic shocks [10].

If the dynamic wave form, instead of being a smoothed step, is a hump (a problem whose linear version was discussed earlier), nonlinear effects will again tend to steepen up the front, but they will smooth out the rear. At the front, a state of equilibrium can be attained, as before, between amplitude and frequency dispersion effects, giving rise to a kinematic shock. However, its strength will decay (as in the linear theory) as $t^{-1/2}$ [3]. Sridharan and Mohan Kumar [22] performed a numerical study of the evolution of a sinusoidal hump in subcritical flow. They found that the rate of attenuation of the peak is weakly dependent on the initial relative amplitude of the hump $\Delta h_0/h_0$, as shown in Figure 16. In this figure, the parameter T_w is a measure of the wave duration, and can be related to σ (Equation 22) as $T_w = 2\pi/\hat{c}_1\sigma$. The important conclusion is that results from the linear analysis will be relatively accurate (though they will overestimate the subsidence, as first pointed out by Supino [23]) even for amplitudes as high as 1.5 times the initial depth. They also confirmed the previous finding by Mozayeny and Song [24] that the decay is exponential with distance. However, they warned that the modification of the wave form while propagating may lead to the necessity of updating the value of the exponent. Di Silvio [20] made a semi-analytic study of the problem of the propagation of a single-peaked wave, assuming that the flood hydrograph at any section could be fitted by a triangular shape. He provided a set of nondimensional graphs from where to obtain the rate of attenuation and distortion of the wave. These results also show that, when the wave is sufficiently far from the breaking point, the peak discharge is only weakly dependent on the distortion. In this case, the attenuation in the discharge can be calculated through an analytical formula (see next section), whose accuracy was experimentally checked and found to be satisfactory even for nontriangular hydrograph forms.

Prismatic and Nonprismatic Channels

Up to now, the analysis (both linear and nonlinear) has been confined to an infinitely wide rectangular channel. The effects of arbitrary cross-sectional shapes both in uniform (prismatic) and nonuniform (nonprismatic) channels is briefly analyzed in the following, in the context of the more general nonlinear theory.

The equations of motion for a prismatic channel are Equations 3 and 2. If, following Escoffier [5], the "stage variable" w is introduced as

$$w \equiv \int_0^h \left(\frac{gB}{A}\right)^{1/2} dh \tag{78}$$

the equations of motion can be rewritten in characteristic form as

$$\frac{d}{dt}(u \pm w) = g\left(S_0 - \frac{u^2}{C^2R}\right) \text{ along the curves } \frac{dx}{dt} = u \pm c \tag{79}$$

where the characteristic celerity c is now

$$c = \left(\frac{gA}{B}\right)^{1/2} \tag{80}$$

Comparing Equation 79 with Equations 54 and 55, it is observed that the wave phenomenon occurring in prismatic channels is of the same kind as the one just analyzed, as expected. Then, the same qualitative conclusions must hold.

Note that, from Equation 79, in the particular case of long gravity waves, the quantities $u \pm w$ are invariants along the characteristic curves. At the other extreme, for kinematic waves, Equation 2 reduces to

$$u = C(S_0R)^{1/2} \tag{82}$$

which shows that u depends only on h. Then, Equation 3 can be rewritten as

$$\frac{\partial h}{\partial t} + \frac{dQ}{dA}\frac{\partial h}{\partial x} = 0 \tag{83}$$

which is formally identical to Equation 63, though now the kinematic velocity is given by dQ/dA, where Q = uA. For example, assuming that the variation of the wetted perimeter P and Chezy coefficient C with the depth can be neglected, one obtains dQ/dA = (3/2)u, as for infinitely wide rectangular channels.

In the numerical study of Sridharan and Mohan Kumar [22] on the propagation of a sinusoidal hump, referred to before, they also considered prismatic channels with rectangular and trapezoidal cross sections. Figure 17 shows the rate of attenuation for rectangular shapes, for different values of the relative width B/h_0. Significant effects of this parameter are observed, with larger subsidence occurring for narrower channels. Instead, the results are not very sensitive to the side slope in trapezoidal channels, except when they are very narrow.

Craya's instability criterion, Equation 34 or 35, is now generalized, respectively, to [15]

$$\frac{dQ}{dA} > u + c \tag{84}$$

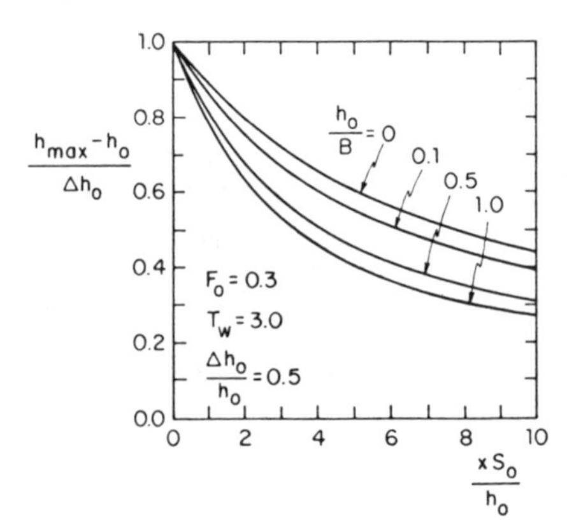

Figure 17. Attenuation of a sinusoidal hump for different channel widths [22].

or

$$A \frac{du}{dA} > c \tag{85}$$

which hold, as before, for any resistance law. Note that, according to the definition of Equation 78, Equation 85 can be rewritten simply as [5]

$$\frac{du}{dw} > 1 \tag{86}$$

A more practical expression can be derived from the foregoing equations if a resistance law in powers of u and R is assumed. Taking the classical Chezy's formula, and assuming that $C^2 \sim R^n/u^m$ one obtains

$$V \equiv F \sum > 1 \tag{87}$$

where F is the Froude number,

$$\sum \equiv \left(\frac{n+1}{m+2}\right)\left(1 - R\frac{dP}{dA}\right) \tag{88}$$

is a purely geometrical quantity and V is the "Vedernikov number."

As already mentioned, Di Silvio [20] found an approximate formula for the rate of attenuation of the peak discharge, assuming that the flood hydrograph can be fitted by a triangular shape. If conditions corresponding to the input hydrograph are identified with the subscript "0", Q_p is the peak discharge at a given downstream station, and t_p the propagation time of the peak, the formula is simply

$$\frac{Q_p}{Q_0} = \left(\frac{1}{1 + \dfrac{2\bar{a}t_p}{T'_0 T''_0}}\right)^{1/2} \tag{89}$$

where T′ and T″ are the rising and falling time periods, respectively, and $\bar{a}$ is the attenuation coefficient evaluated at the middle height of the wave, with the attenuation coefficient defined as

$$a \equiv \frac{4}{2+m} \frac{u}{gS_0} \left[\frac{1 - V^2}{(F + V)^2}\right] \tag{90}$$

When the channel is nonprismatic, the flow is governed by Equations 1 and 2. If, without any loose of generality, A is considered to depend on the variables h and x (and only implicitly on time through h), one can write

$$\begin{aligned} \left.\frac{\partial A}{\partial x}\right|_t &= \left.\frac{\partial A}{\partial h}\right|_x \left.\frac{\partial h}{\partial x}\right|_t + \left.\frac{\partial A}{\partial x}\right|_h \\ &= B\left(\frac{\partial h}{\partial x} + \gamma\right) \end{aligned} \tag{91}$$

where $\gamma \equiv (1/B)\, \partial A/\partial x|_h$ is a purely geometrical coefficient which measures the degree of divergence of the channel (at a constant depth). Using Equation 91, Equations 1 and 2 can be rewritten in characteristic form as

$$\frac{d}{dt}(u \pm w) = g\left(S_0 - \frac{u^2}{C^2R}\right) \pm \gamma u\left(\frac{g}{AB}\right)^{1/2} \text{ along } \frac{dx}{dt} = u \pm c \tag{92}$$

Equation 92 shows that nonprismaticity introduces a source-like effect similar to that of weight and resistance. For example, for a channel of diverging cross-section (i.e., $\gamma > 0$) it gives a negative (positive) contribution for the forward (backward) characteristic wavelets. Note that this holds even for long gravity waves. Using similar arguments, the kinematic wave equation now is

$$\frac{\partial h}{\partial t} + \frac{dQ}{dA}\frac{\partial h}{\partial x} = -\gamma u \tag{93}$$

which indicates attenuation (amplification) of the wave for diverging (converging) channels.

Similar qualitative effects to those introduced by nonprismaticity arise when discharge from tributaries or run-off is added to the flow, or subtracted by seepage through the bottom. This time, source-like terms appear in the continuity equation [5].

Numerical Methods of Solution

Although the analysis of the flow equations has provided a reasonable understanding of the phenomenon of wave propagation in open channels, only a few explicit solutions, for relatively simple situations, are available. Numerical methods of solution of the differential equations have been used to study more general cases. In fact, the practical interest in flow prediction has led to an important development of these numerical methods, which can deal with rather arbitrary flow situations. Some mention of their main features seems then appropriate.

An important property of numerical methods is that they introduce artificial attenuation and (frequency) dispersion. These nonphysical phenomena arise from the discretization of the differential equations. Numerical attenuation acts differently on different ranges of the wavelength spectrum. In practice, numerical attenuation is generally necessary in the low wavelength range of the spectrum (i.e., for wavelengths of the order of the spatial discretization interval) in order to obtain smooth solutions free of spurious short waves. Obtaining the maximum advantages and minimum disadvantages of numerical attenuation and dispersion, requires a careful election of the discretization intervals for each particular problem. Cunge, Holly, and Verwey [25] present a summary of the attenuation and dispersion characteristics of some practical numerical schemes.

Various methods have been devised to solve numerically the shallow water equations. The semi-graphical method of characteristics is particularly suited for computer calculations. Using a finite difference approximation, the characteristic mesh can be constructed step by step [26, 27, 28]. This procedure provides the solution (u, h) at the nodes. In order to obtain the profile at a given time or the hydrograph at a given location, some interpolation has to be made. One of the advantages of the method is that formation and evolution of bores can be handled accurately. However, it requires a relatively complex programming. A simplification of the preceding procedure consists in utilizing a fixed rectangular network as computational grid [29, 30]. Now, an interpolation is carried out at every time step, thus producing a loss of accuracy relative to the previous method.

Much more widespread is the use of "direct" methods, based on finite difference approximations of the original differential equations, both explicit [2, 25, 27, 29–31] and implicit [25, 27, 29, 30, 32]. Bores, if present, should now be treated separately. However, some of the methods allow a relatively accurate representation of bores by steep fronts that extend over a few grid points [33, 34]. Further steepening of the front is prevented by numerical diffusion.

Recently, a stochastic-deterministic method, originally proposed for gas dynamics [35], has been applied to open-channel flow [19, 36]. Its main advantages are that it is free of numerical diffusion and (using a fixed, rectangular net) can deal automatically with bores, these being represented as sharp discontinuities.

NOTATION

A	cross-sectional area	Q	volumetric flowrate
a	attenuation coefficient	q	discharge per unit width
B	surface width	R	hydraulic radius; also resistance
C	Chezy coefficient	S_0	bottom slope
c	phase velocity	T_w	parameter denoting wave duration
F_o	Froude number	t	time
g	acceleration of gravity	t_p	propagation time
h	depth	u	flow velocity
h_c	critical depth	V	Vedernikov number; also group velocity
I	inertia term	W	effective weight
J_+, J_-	Riemann invariants	x	space coordinates
k	wave number		
P	wetted perimeter; also pressure		

Greek Symbols

β	propagation factor	ϵ	relative error
α	growth rate of wave	η, ξ	initial amplitudes of waves
γ	geometrical coefficient	σ	dimensionless wave number
δ	phase difference	$\hat{\Omega}$	ratio of relative amplitudes

REFERENCES

1. Saint Venant, A. J. C. B. de, "Des diverses mainieres de poser les equations du movement varie des eaux courantes," *Annales des Ponts et Chaussees*, serites 6, Vol. 13 (1887).
2. Stoker, J. J., *Water Waves*, Wiley-Interscience, New York, 1957.
3. Lighthill, M. J., and Whitham, G. B., "On Kinematic Waves I, Flood Movement in Long Rivers," *Proceedings of the Royal Society of London*, Vol. A229 (May 1955), pp. 281–316.
4. Whitham, G. B., *Linear and Nonlinear Waves*, Wiley-Interscience, New York, 1974.
5. Henderson, F. M., *Open Channel Flow*, Macmillan Publishing Co., New York, 1966.
6. Lin, C. C., "The Theory of Hydrodynamic Stability," Cambridge University Press, London, 1966.
7. Ponce, V. M., and Simons, D. B., "Shallow Wave Propagation in Open Channel Flow," *Journal of the Hydraulics Division, ASCE*, Vol. 103, No. HY12 (Dec. 1977), pp. 1461–1476.
8. Menendez, A. N., and Norscini, R., discussion to "Nature of Wave Attenuation in Open Channel Flow," by V. M. Ponce, *Journal of Hydraulic Engineering, ASCE*, Vol. 109, No. 5 (May 1983), pp. 786–788.
9. Menendez, A. N., and Norscini, R., "Spectrum of Shallow Water Waves: An Analysis," *Journal of the Hydraulics Division, ASCE*, Vol. 108, No HYl (Jan. 1982), pp. 75–94.
10. Dressler, R. F., "Mathematical Solution of the Problem of Roll-Waves in Inclined Open Channels," *Commun. Pure Appl. Math.*, 2 (1949), pp. 149–194.
11. Jeffreys, H., "The Flow of Water in an Inclined Channel of Rectangular Section," *Philos. Mag.*, London, Sec. 6, Vol. 79 (1925), p. 793.
12. Keulegan, G. H., and Patterson, G. W., "A Criterion for Instability of Flow in Steep Channels," *Am. Geophys. Union Trans.*, Vol. 21 (1940), pp. 594–596.
13. Dressler, R. F., and Pohle, F. V., "Resistance Effects on Hydraulic Instability," *Commun. Pure Appl. Math.*, Vol. 6, No. 1 (1953), pp. 93–96.
14. Vedernikov, V. V., "Release Waves of the Real Liquid," *Transactions*, Unsteady Movement of the Water Current in an Open Channel, Akad. Nauk, USSR, 1947.
15. Craya, A., "The Criterion for the Possibility of Roll-Wave Formation," U.S. Natl. Bur. Standards Circ. 521 (1952), pp. 141–151.

16. Courant, R., and Hilbert, D., *Methods of Mathematical Physics*, Vol. 2, Wiley-Interscience, 1953.
17. Lighthill, M., *Waves in Fluids*, Cambridge University Press, London, 1978.
18. Abbot, M. R., "A Theory of the Propagation of Bores in Channels and Rivers," *Philos. Soc. (Cambridge) Proc.*, Vol. 52, No. 2 (1956), pp. 344–362.
19. Marshall, G., and Menendez, A. N., "Numerical Treatment of Nonconservation Forms of the Equations of Shallow Water Theory," *Journal of Computational Physics*, Vol. 44, No. 1 (Nov. 1981), pp. 167–188.
20. Di Silvio, G., "Flood Wave Modification along Prismatic Channels," *Journal of the Hydraulics Division, ASCE*, Vol. 95, No. HY5 (Sept. 1969), pp. 1589–1614.
21. Jolly, J. P., Yevjevich, V., "Stability of Single-Peaked Waves," Journal of Hydraulic Research, Vol. 12, No. 3 (1974), pp. 315–335.
22. Sridharan, K., and Mohan Kumar, M. S., "Parametric Study of Flood Wave Propagation," *Journal of the Hydraulics Division, ASCE*, Vol. 107, No. HY9 (Sept., 1981), pp. 1061–1076.
23. Supino, G., "Sur la propagation des ondes dans les canaux," *Rev. Gen. Hydraulique*, France, Vol. 5, No. 29 (1939), pp. 260–262.
24. Mozayeny, B., and Song, C. S., "Propagation of Flood Waves in Open Channels," *Journal of the Hydraulics Division, ASCE*, Vol. 95, No. HY3 (Mar. 1969), pp. 877–892.
25. Cunge, J. A., Holly, F. M., Jr., and Verwey, A., *Practical Aspects of Computational River Hydraulics*, Pitman Publishing Limited, 1980.
26. Fletcher, A. G., and Hamilton, W. S., "Flood Routing in an Irregular Channel," *Journal of the Engineering Mechanics Division, ASCE*, Vol. 93 (June 1965), pp. 45–62.
27. Liggett, J. A., and Woolhiser, D. A., "Difference Solutions of the Shallow Water Equations," *Journal of the Engineering Mechanics Division, ASCE*, Vol. 93, No. EM2 (Apr., 1967), pp. 39–71.
28. Sakkas, J. G., and Strelkoff, T., "Dimensionless Solution of Dam-Break Flood Waves," *Journal of the Hydraulics Division*, Vol. 102 (Feb. 1976), pp. 171–184.
29. Strelkoff, T., "Numerical Solution of Saint-Venant Equations," *Journal of the Hydraulics Division, ASCE*, Vol. 96 (Jan. 1970), pp. 223–253.
30. Price, R. K., "Comparison of Four Numerical Methods for Flood Routing," *Journal of the Hydraulics Division, ASCE*, Vol. 100 (July 1974), pp. 879–899.
31. Issacson, E., Stoker, J. J., and Troesch, B. A., "Numerical Solution of Flow Problems in Rivers," *Journal of the Hydraulics Division, ASCE*, Vol. 84, Paper 1810 (Oct. 1958), pp. 1–18.
32. Amein, M., and Chu, H. L., "Implicit Numerical Modeling of Unsteady Flows," *Journal of the Hydraulics Division, ASCE*, Vol. 101 (June 1975), pp. 717–732.
33. Terzidis, G., and Strelkoff, T., "Computations of Open Channel Surges and Shocks," *Journal of the Hydraulics Division*, Vol. 96 (Dec. 1970), pp. 2581–2610.
34. Chandhry, Y. M., Contractor, D. N., "Application of the Implicit Method to Surges in Open Channels," *Water Resources Research*, Vol. 9 (Dec. 1973), pp. 1605–1612.
35. Chorin, A. J., "Random Choice Solutions of Hyperbolic Systems," *Journal of Computational Physics*, Vol. 22 (1976), pp. 517–533.
36. Marshall, G., and Menendez, A. N., "Numerical Solution of the Dam Failure Problem by the Random Choice Method," *Advances in Water Resources*, Vol. 4 (Sept. 1981), pp. 125–133.

CHAPTER 8

STRAIGHT SEDIMENT STABLE CHANNELS

Mohammad Akram Gill

Institute of Hydraulic Research
The University of Iowa
Iowa City, Iowa USA

CONTENTS

INTRODUCTION, 203
Stability of Sediment Transporting Channels, 204

SEDIMENT-FREE STABLE CHANNELS, 205
Criterion of Bank Stability, 205
Shape Profile, 206
Design of Channel Cross Section, 207

SEDIMENT-TRANSPORTING STABLE CHANNELS, 208
Transverse Circulation of Sediments, 209
Simplification and Solution, 211
Geometric Properties of Stable Channels, 212
Comparison with Lane's Channels, 213
Design of Channel Cross Section, 213

SUSPENDED SEDIMENT TRANSPORT IN STABLE CHANNELS, 214

CONCLUSION, 217

NOTATION, 217

REFERENCES, 218

INTRODUCTION

Designing sediment-stable channels is an important problem confronted by hydraulic and irrigation engineers the world over, particularly in countries where irrigation is practiced on a large scale. Recently, link canals, which are earthen channels built mostly in noncohesive materials were designed and constructed in Pakistan. Some of these canals are very large, carrying discharges ranging between 10,000 and 20,000 cfs. They are interbasin channels connecting neighboring rivers with each other. They transfer water from one river to the other to ensure a year-round constant and stable supply of water in the irrigation canals.

These channels were required to be stable so that over a long period of time no substantial erosion and deposition of material would occur on the bed and banks. Excessive erosion can cause breaching of the canal banks, and deposition of suspended material necessarily leads to reduced capacity for carrying water flow. Both of these problems inevitably increase the cost of maintenance and lead to unsatisfactory operation. A sediment-stable canal is free of these undesirable problems. Lining the canal surface with impervious materials to avoid erosion is generally very expensive when the canals are very wide and long. The link canals in Pakistan are very wide with cross-sectional areas on the order of 5,000 ft^2 in some cases even greater and have a total length of several hundred miles.

The hydraulics of flow in these channels is indeed complex for many reasons. Such channels are required to carry a certain specified discharge of water which is charged with sediment load of suspended and bed material. Sediments are suspended in water and at the same time move in contact with the bed as bed load. The boundaries of the channel are thus in a state of constant flux. Bed and banks, although initially flat surfaces, may soon be covered with ripples and dunes. These (ripples etc.) are in effect large elements of surface roughness having a strong influence on the overall channel friction. While these macro-structures affect the sediment transporting capability of a channel, the size and spacing of these elements are in turn controlled by the capacity of the channel to transport sediment load.

There is a class of earthen channels which are not affected by the sediment transportation characteristics since they carry sediment-free water. Stability of these channels is defined as a state of equilibrium such that the particles on the surface of such channels are in a critical condition of movement. They do not actually move under the designed flow conditions, but if the bed stress exceeds even slightly the critical shear stress of the particles any where, the particles there will start moving. A channel shape which ensures that the particles everywhere on the channel surface are subjected to a resultant boundary stress which equals the critical shear stress is a stable profile. This class of channels is by far less important than the stable channels transporting sediment load, yet it is important from the viewpoint of analysis. Their shape and hydraulic characteristics were determined analytically some three to four decades ago, though similar success has not been achieved in the case of sediment-transporting channels. Some analytical results are presented herein later which constitute at least a starting point for further development of the sediment-transporting stable channels.

Stability of Sediment Transporting Channels

When the capacity for transporting sediment of a channel is equal to the time rate of sediment supply at its head, it is said to be a stable channel. Stable channels are also sometimes called regime channels, particularly in India, Pakistan, and other countries which at one time were under British influence. For want of comprehensive theoretical knowledge of the hydraulics of sediment-transporting channels with loose boundaries, the initial progress in their design was of necessity empirical. Even the empirical methods were developed slowly accompanied with a great deal of controversy and disputation. Kennedy [1] introduced the method of critical velocity which provided a control on the flow velocity only. Such a velocity which ensured a state of regime was called critical. Critical velocity was correlated with depth and a limiting velocity in turn delimited the value of depth. Increase in area was consequently achieved by increasing the width. This state of affairs was later improved by Lindley [2] who correlated width with depth thus providing a second relationship together with Kennedy's velocity formula. Lacey [3] achieved a successful synthesis of the various empirical methods and proposed a nearly complete set of empirical equations for stable channels. Heretofore sediment characteristics e.g. size and transport rate, were vaguely discussed but not included in the design equations. Lacey took care of this very important problem by introducing a "silt factor" in the design method. The silt factor by itself did not directly account for the size or quantity of sediment in movement; it was believed that the silt factor accounted for these characteristics somehow albeit cryptically.

The significance of Lacey's regime method can be judged from the fact that it is still popular in design offices and in research institutes in many countries after its formulation nearly half a century ago. The research engineers are interested to see if Lacey's method, which is considered to be unique by its supporters, can be brought in harmony with the basic rational laws of classical mechanics [4–7]. This effort in itself provided a good deal of impetus for theoretical development in the area of sediment transportation in open channels. The theoretical knowledge has not yet developed to a stage where it can provide a sure and foolproof law for designing the sediment stable channels.

Nonetheless, it has opened various possibilities for approaching these problems in a systematic manner thus reducing total dependence on gross and sometimes confusing empirical methods.

SEDIMENT-FREE STABLE CHANNELS

Criterion of Bank Stability

Consider a section of an open channel as in Figure 1. A typical sand particle P is subjected to a fluid tractive force $\tau_0(\pi D^2/4)$ in the flow direction and a gravitational force W sin α down the bank

where τ_0 = fluid boundary shear stress
D = average diameter of the bank material
W = submerged weight of the particle
α = inclination of the bank with the horizontal

The vector sum of these two forces is denoted by R_f. If the particle is to remain in a state of equilibrium, R_f should not exceed the force of static friction (F_f) which acts in the opposite direction to R_f. For the condition of critical equilibrium, $F_f = R_f$. Thus

$$[(\tau_0\pi D^2/4)^2 + (W \sin \alpha)^2]^{1/2} = W \cos \alpha \tan \phi \tag{1}$$

where ϕ = angle of repose of the bank material

The submerged weight W is given by

$$W = (\gamma_s - \gamma)\pi D^3/6 \tag{2}$$

where γ_s and γ = specific weights of sediment and water, respectively

It can be easily proved ($\tau_0 = \tau_c$ when $\alpha = 0$) that

$$W = (\tau_c \pi D^2)/(4 \tan \phi) \tag{3}$$

where τ_c = critical shear stress of the material

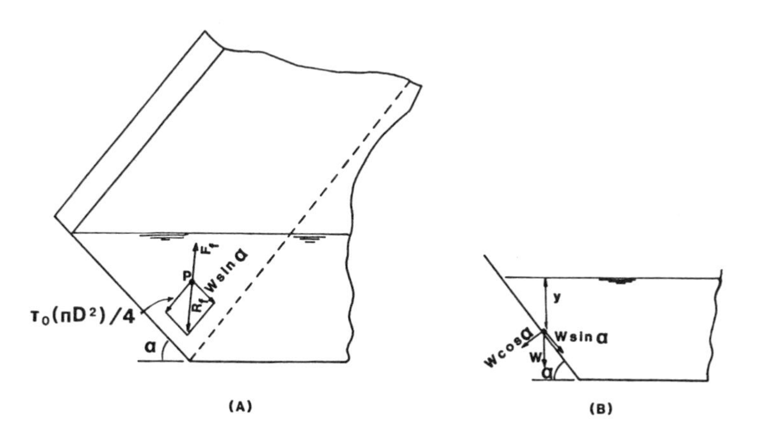

Figure 1. Schematic of forces acting on (A) a sediment particle or (B) particle on the bank of a channel.

Substituting W in terms of τ_c in Equation 1 and simplifying gives

$$\frac{\tau_0}{\tau_c} = [\cos^2 \alpha - (\sin^2 \alpha/\tan^2 \phi)]^{1/2} \tag{4}$$

Equation 4 can also be expressed in the following simplified manner.

$$\frac{\tau_0}{\tau_c} = [1 - (\sin^2 \alpha/\sin^2 \phi)]^{1/2} \tag{5}$$

Equation 4 or 5 expresses a very important criterion of bank stability. As long as the right-hand side of Equation 4 or 5 is less than or equal to τ_0/τ_c, the particles on the channel surface will remain stable i.e. they will not move. Equation 4 was first given by Lane and Carlson [8].

Shape Profile

A critically stable channel will have all the sediment particles on its surface in the state of threshold (critical) condition. To determine the cross-sectional shape of such a channel, Equation 4 applies to an elemental strip dx wide (Figure 2). The shear stress τ_0 equals γRS_0 where R is the hydraulic radius and S_0 is the longitudinal bed slope. In the case of an elemental strip that we are considering, $R = y \cos \alpha$. Therefore,

$$\tau_0 = \gamma y S_0 \cos \alpha \tag{6}$$

and

$$\tau_c = \gamma y_0 S_0 \tag{7}$$

where y = local depth
y_0 = maximum depth at the center of the channel due to symmetry

Equation 4 can now be written in a simplified manner as follows

$$dy/dx = \tan \phi[1 - (y/y_0)^2]^{1/2} \tag{8}$$

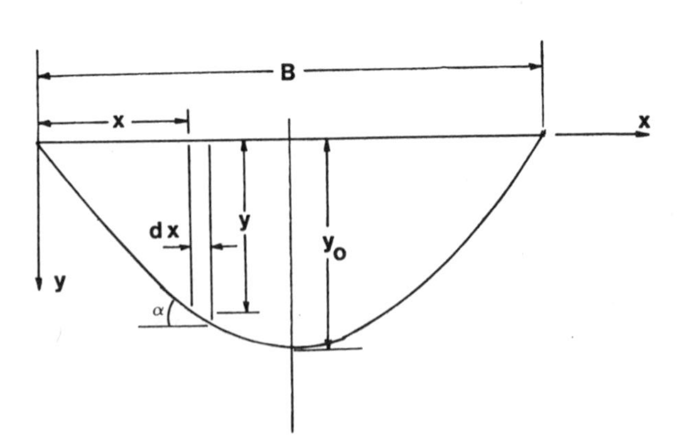

Figure 2. Schematic for determining the shape profile of a stable channel.

with the solution

$$y/y_0 = \sin(x \tan \phi / y_0) \tag{9}$$

evaluating the constant of integration using the boundary condition $x = 0$, $y = 0$. Equation 9 is a relationship between the geometric parameters and does not as such involve any hydraulic parameter such as Q, V, etc. It thus gives only the shape of the channel and does not ensure that it will provide sufficient cross-sectional area to carry a given design discharge. This point is discussed further later. Let us first determine the geometric properties of a channel which is described by Equation 9.

At $x = B/2$, $y = y_0$ so that

$$B = \pi y_0 / \tan \phi \tag{10}$$

Similarly, the cross-sectional area A is given by

$$A = 2y_0^2 / \tan \phi \tag{11}$$

The wetted perimeter P can be obtained from the following relationship

$$P = (2y_0/\sin \phi) \int_0^{\pi/2} (1 - \sin^2 \phi \sin^2 \beta)^{1/2} \, d\beta \tag{12}$$

where $\beta = x \tan \phi / y_0$

The integral in Equation 12 is a complete elliptic integral of the second kind and is usually denoted by E. Tables are available from which E can be read for a given value of β. Equation 12 is now written concisely as follows.

$$P = 2y_0 E / \sin \phi \tag{13}$$

Design of Channel Cross Section

The important restriction on the design of critically stable channels is that $\tau_0 = \tau_c$ everywhere on the channel surface. The critical shear stress is given by

$$\tau_c = f\gamma(S_s - 1)D \tag{14}$$

according to Shields formula. In Equation 14, S_s = specific gravity of the bed material, f = Shields parameter which is constant and equal to 0.056 when the particle Reynolds number [$Re = (\tau_c/\rho)^{1/2} D/\nu$] is greater than 400, Re = Reynolds number, and ρ and ν = density and kinematic viscosity of water respectively. In this range i.e. $Re \geq 400$, the bed material size is greater than $\frac{1}{4}$ inch (6.4 mm). For smaller values of Re, and hence finer bed material than $\frac{1}{4}$ inch diameter, f varies with Re but can however be determined from the Shields diagram for a given bed material size. Taking $f = 0.056$ and a typical value of $S_s = 2.6$, Equation 14 can be written as follows

$$D = 11RS_0 \tag{15}$$

using $\tau_c = \gamma RS_0$. Obviously for a given size of bed material D, RS_0 has a fixed value. A known value of S_0 will fix R and in turn the depth y_0. Now a hydraulic resistance formula such as Manning's formula is used noting that

$$n = 0.034D^{1/6} \tag{16}$$

where n = Manning's coefficient of bed roughness and D is in inches

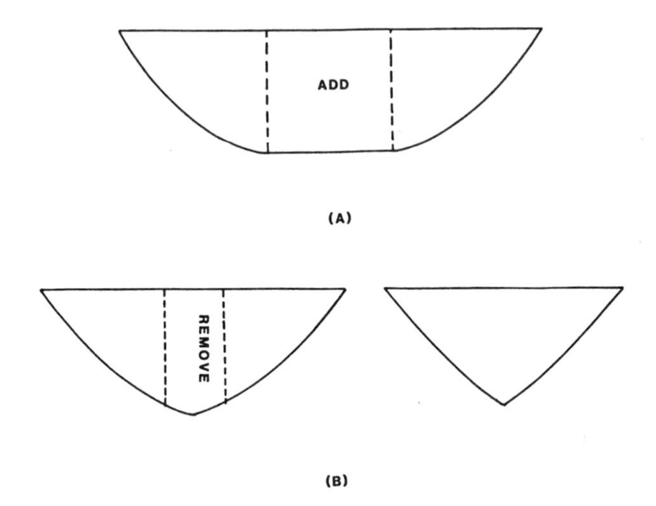

Figure 3. (A) Making up the required area by inserting a rectangular block; (B) reducing the area to a required value by removing a central block.

In the ft-sec units, Manning's formula reads as follows

$$V = 43.82R^{2/3}S_0^{1/2}/D^{1/6} \tag{17}$$

where V = flow velocity

Substituting for D from Equation 15 into Equation 17 gives

$$V = 29.4R^{1/2}S_0^{1/3} \tag{18}$$

which is the velocity formula for critically stable channels. This result is due to Henderson [5] and is valid for channels with average bed material size $D \geq \frac{1}{4}$ inch. It is possible to determine similar results for channels with $D < \frac{1}{4}$ inch using appropriate values of f from Shields diagram. It should be noted that

$$R = y_0 \cos \phi/E \tag{19}$$

For a typical design, the total discharge Q is known. Knowing or estimating the value of S_0 to suit the topography, R can be computed from Equation 15. Going to Equation 18 with known and computed values of S_0 and R, calculate V. Q/V gives the cross-sectional area A. This value of A will in most cases be different from that given by Equation 11. If the required value of A (=Q/V) is greater than A from Equation 11, make up the deficiency by providing a rectangular inset as in Figure 3A. If the required area is smaller, the design can be modified as in Figure 3B.

SEDIMENT-TRANSPORTING STABLE CHANNELS

A sediment-transporting stable channel is the one whose transporting capacity equals the rate at which sediment is admitted into it at its intake. The channel is thus adequate to convey a design discharge of water together with the imposed rate of sediment supply. In such a channel, the balance of net erosion of the channel surface and net deposition of material on the surface is almost nil over a long period of time. Stability however does not imply that the channel remains absolutely in its originally designed condition. Originally, the channel surface is plain; soon after operation it gets covered with moving ripples and dunes. Erosion and deposition of sediment occur continuously, but both of these effects cancel each other over a long stretch of channel and time.

Transverse Circulation of Sediments (Bank Erosion and Deposition of Suspended Sediments Over the Banks—Mechanism and Approximate Mathematical Formulation)

Transverse circulation of sediments is a normal feature of meandering channels in which strong secondary currents are generated due to centrifugal forces. In a straight channel, centrifugal forces are almost absent except in the curved bank regions. Transverse circulation is nevertheless set up due to deposition of suspended sediment on the bank surface (due mainly to a reduced capacity for transporting sediment in the bank region) which is eroded and transported inwards (towards the central region) because of gravity (Figure 4). In an approximate analysis, the effect of secondary currents (due to the bank curvature) can be ignored although it is not implied that such an effect does not exist. In order to ensure the stability of the channel, the rate of sediment transport towards the central region should be equal to the rate at which sediment is diffused towards the banks since there is no net transverse transport of sediment. The stability condition can also be expressed by specifying that the rate of deposition of sediments over the banks is equal to the lateral transport of bed load (eroded from the banks) towards the central region. Mathematically, these can be stated as follows:

$$G_L + g_L = 0 \tag{20}$$

$$d(G_L + g_L)/dx = 0 \tag{21}$$

where G_L = rate of lateral bed load transport, g_L = vertically integrated rate of lateral suspended transport both per unit length of the channel, and x = distance measured from one bank towards the other to a point to which G_L and g_L refer, Figure 4. Also

$$dG_L/dx = e - d \tag{22}$$

and

$$dg_L/dx = d - e \tag{23}$$

where e and d are rate of erosion and deposition per unit length of channel. Following earlier work of Parker [9]

$$g_L = -\epsilon_x \frac{d\psi}{dx} \tag{24}$$

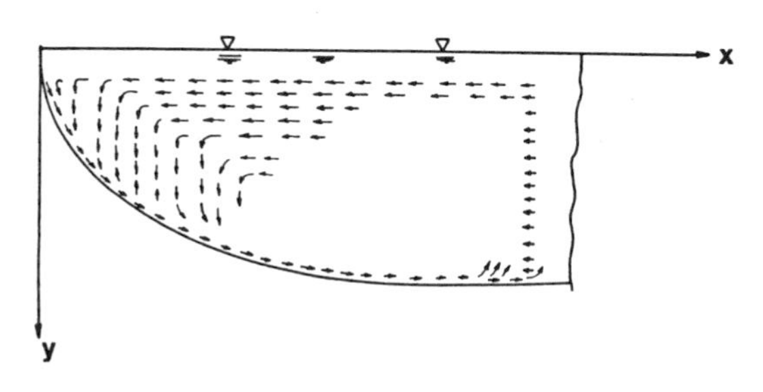

Figure 4. Schematic of transverse circulation of sediment in a stable channel section.

where

$$\psi = \int_0^y c(y)\,dy \qquad (25)$$

$c(y)$ = concentration of suspended sediment at an elevation y above the bed

ψ = vertically integrated sediment concentration

ϵ_x = coefficient of diffusivity or eddy viscosity in the transverse direction

Following the earlier work of Engleund [11], Parker gave an approximate result for the rate of deposition as follows

$$d = V_s^2\psi/\epsilon \qquad (26)$$

where V_s = terminal fall velocity of sediment of average size D
ϵ = diffusivity coefficient in the flow direction

Both ϵ and ϵ_x can be approximately specified by

$$\epsilon, \epsilon_x = a_1 U_* y_0 \qquad (27)$$

where a_1 = constant
U_* = shear velocity
y_0 = flow depth at the center of the channel

According to Parker [9], $a_1 = 0.077$ for ϵ and 0.13 for ϵ_x. Both ϵ and ϵ_x are assumed constant for a given flow condition herein.

For the lateral bed load transport, Parker used

$$G_L = (q_b/\mu)\,dy/dx \qquad (28)$$

a relationship which was initially proposed by Engleund. In Equation 28, q_b = unit rate of bed load transport in the main flow direction, and μ = Coulomb's coefficient of dynamic friction. A bed load formula of the following type is commonly used in sediment transportation studies,

$$q_b = a_2[gD^3(S_s - 1)]^{1/2}\left(\frac{\tau}{\tau_c} - 1\right)^{(m-1)} \qquad (29)$$

where a_2 and m = constants
τ = local bed shear stress
g = acceleration due to gravity
S_s = specific gravity of the bed material

For active mobile beds, Equation 29 can be simplified to

$$q_b = a_2[gD^3(S_s - 1)]^{1/2}\left(\frac{\tau}{\tau_c}\right)^{(m-1)} \qquad (30)$$

noting that $\tau/\tau_c \gg 1$. Using $\tau = \rho g y S_0$, assuming a wide and shallow channel,

$$G_L = (a_2/m\mu)[gD^3(S_s - 1)]^{1/2}\left(\frac{\tau_0}{\tau_c}\right)^{(m-1)} y_0 \frac{d(y/y_0)^m}{dx} \qquad (31)$$

where $\tau_0 = \rho g y_0 S_0$. Likewise, the lateral erosion rate is assumed to be given by

$$e = a_3[gD(S_s - 1)]^{1/2}\left(\frac{\tau}{\tau_c}\right)^n \tag{32}$$

after Parker [9] in which a_3 and n = constants.

All the mathematical relationships which are required for the determination of the shape of sediment-transporting stable channels have now been formulated. What remains is to simplify them and deduce a differential equation in terms of y and x and solve it for the shape profile as in the case of Lane's channels.

Simplification and Solution

Use the following ratios for developing dimensionless relationships:

$$y_* = (y/y_0)^m; \qquad \psi_* = \psi/\psi_0; \qquad x_* = x/B.$$

Equation 21 reduces to

$$(q_{b0}y_0/m\mu\psi_0\epsilon_x)d^2y_*/dx_*^2 = d^2\psi_*/dx_*^2 \tag{33}$$

where

$$q_{b0} = a_2[gD^3(S_s - 1)]^{1/2}\left(\frac{\tau_0}{\tau_c}\right)^{(m-1)} \tag{34}$$

For simplicity of notation, let $(q_{b0}y_0/m\mu\psi_0\epsilon_x) = \lambda$, so that

$$\lambda(d^2y_*/dx_*^2) = d^2\psi_*/dx_*^2 \tag{35}$$

Equation 23 leads to

$$\frac{d^2\psi_*}{dx_*^2} + \frac{B^2d_0}{\psi_0\epsilon_x}\left(\psi_* - \frac{e_0}{d_0}y_*^{n/m}\right) = 0 \tag{36}$$

where e_0 and d_0 are the values of e and d at x = B/2.

Boundary Conditions

$$\begin{aligned} x_* &= 0, & \psi_* &= 0, & y_* &= 0 \\ x_* &= 1, & \psi_* &= 0, & y_* &= 0 \\ x_* &= \tfrac{1}{2}, & \psi_* &= 1, & y_* &= 1 \end{aligned} \tag{37}$$

Solution

Solution of Equation 35 using Equation 37 gives

$$y_* = \psi_*, \qquad \lambda = 1 \tag{38}$$

Substitution of this result, Equation 38, in Equation 36 gives

$$d^2y_*/dx_*^2 + (B^2d_0/\psi_0\epsilon_x)[y_* - (e_0/d_0)y_*^{n/m}] \tag{39}$$

Equation 39 is similar to a result previously given by Parker [9]. Parker used $n = 3$ and $m = 4$ and proceeded to solve Equation 39 using a singular perturbation method. The final result was in the form of an integral. A simpler method is used here to solve Equation 39 which is nonlinear for all unequal values of n and m. Linearize Equation 39 as follows

$$d^2y_*/dx_*^2 + (B^2d_0/\psi_0\epsilon_x)[1 - (e_0/d_0)y_{*c}^{-(1-n/m)}]y_* = 0 \tag{40}$$

where $y_{*c}^{(1-n/m)}$ is a constant. With $n = 3$ and $m = 4$, $y_{*c}^{1/4} = y_c/y_0$. y_c is some characteristic value of y which can be taken equal to the hydraulic depth $= A/B$. For simplicity of notation, let $(B^2d_0/\psi_0\epsilon_x)[1 - (e_0/d_0)y_{*c}^{-(1-n/m)}] = \theta^2$, so that Equation 40 becomes

$$d^2y_*/dx_*^2 + \theta^2 y_* = 0 \tag{41}$$

Equation 41 can also be obtained using Equation 22 instead of Equation 23. Assuming θ^2 to be a positive quantity, solution of Equation 41 is obtained as follows

$$y_* = A_1 \sin(\theta x_*) + A_2 \cos(\theta x_*) \tag{42}$$

where A_1 and A_2 are constants. At $x_* = 0$, $y_* = 0$ which gives $A_2 = 0$. At $x_* = 1$, $y_* = 0$, so that $\theta = \pi$. At $x_* = \frac{1}{2}$, $y_* = 1$, so that $A_1 = 1$. Finally,

$$y_* = \sin(\pi x_*) \tag{43}$$

In terms of physical variables,

$$y = y_0[\sin(\pi x/B)]^{1/m} \tag{44}$$

Geometric Properties of Stable Channels

Using $m = 4$ after Parker, Equation 44 becomes

$$y = y_0[\sin(\pi x/B)]^{1/4} \tag{45}$$

The shape of a stable canal described by Equation 45 is shown in Figure 5. Expanding the sine term in series and performing term-by-term integration, the cross-sectional area of the channel is given by

$$A = 0.83By_0 \tag{46}$$

Hydraulic depth is thus given by

$$y_c = A/B = 0.83y_0 \tag{47}$$

Parker obtained a value of $y_c = 0.84y_0$ in his analysis which is quite close to that given in Equation 47. Approximate expressions for the wetted perimeter are given as

$$P = B[1 + 8.436(y_0/B)^2 - 313.52(y_0/B)^4], \qquad 0 < y_0/B < 0.15 \tag{48}$$

and

$$P = y_0[1.492 + 0.0253(B/y_0)^2 - 0.00004(B/y_0)^4] + B[0.80 + 2.012(y_0/B)^2 - 0.690(y_0/B)^4], \qquad 0.15 < y_0/B < 0.40 \tag{49}$$

The above formulas for P are adequate for almost all practical values of y_0/B. For instance, the

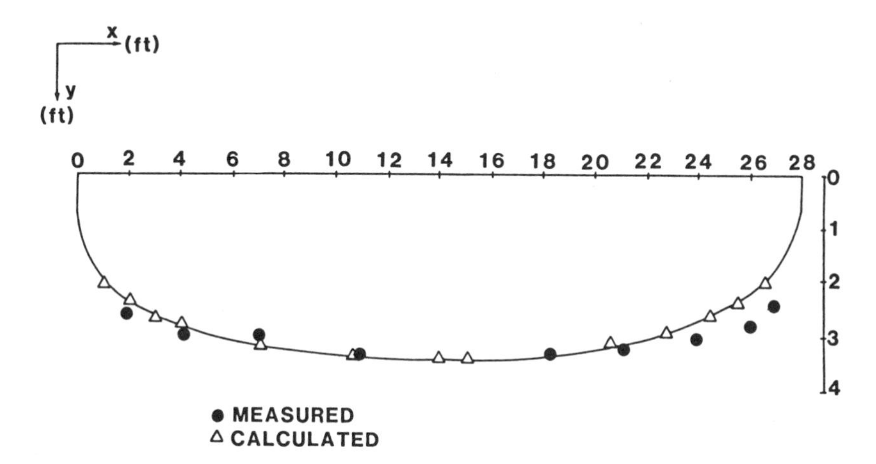

Figure 5. Verification of theory with measurement using Simons-Bender data. ● indicates measurements and △ the computed positions using Equation 45.

narrowest canal in the Simons-Bender [12] data had a value of $y_0/B = 0.23$. Values of y_0/B greater than 0.40 are seldom encountered in the practical design.

To verify the preceding approximate theoretical formulation, a typical canal profile is plotted using the Simons-Bender data in Figure 5 together with the theoretical profile. Agreement between theory and measurement is close and satisfying in view of a number of simplifying assumptions incorporated in the theory.

Comparison with Lane's Channels

Lane's stable channels correspond to a condition of incipient movement of particles on the channel surface. In these channels, $\tau/\tau_c = 1$ everywhere on the channel surface. Setting $m = 1$ in Equation 30 produces the same effect as setting $\tau/\tau_c = 1$. In a state of incipient movement, q_b should theoretically be zero as is correctly predicted by Equation 29. Equation 30 only approximates the condition of incipient movement. As $m \rightarrow 1$, bed load transport becomes increasingly small.

In view of the preceding considerations, it is natural to expect that the shape profile predicted by Equation 44 for $m = 1$ should correspond to the shape profile of Lane's channels. With $m = 1$, Equation 44 gives

$$y = y_0 \sin(\pi x/B) \tag{50}$$

Using Equation 10 in Equation 9, Equation 50 is obtained for Lane's channels. The expectation that the Lane channel should indeed be the limiting case of the stable sediment transporting channels is thus fulfilled. The approximate analysis presented here yields a unified theory of the stable channels.

Design of Channel Cross Section

The preceding analysis deals mainly with the shape profile of sediment-stable channels. If B and y_0 are known, the shape of such channels is completely determined. However, additional information is required for determining B and y_0 and indeed the longitudinal slope of the stable channels as well.

For this purpose, two main methods can be used. Firstly, purely empirical relationships like Lacey's regime equations are available which are known to produce satisfactory results at least in the Indo-Pakistani subcontinent where the method was first developed. Alternatively, a blend of semitheoretical and empirical formulas can be used [7] which is explained in the following. Equation 30 has already been introduced for estimating the bed load transport. The following equation which is of the Manning-Strickler type can be used for describing the hydraulic resistance of the channel.

$$V = a_4(y_0/D)^p S_0^q (g y_0)^{1/2} \tag{51}$$

In Equation 51, a_4, p, and q are constants. One more equation is needed to complete the design of stable channels. Lacey used his width equation for this purpose, which is as follows

$$P = a_5(Q)^{1/2} \tag{52}$$

where Q = design discharge
a_5 = constant

Equation 52 is an empirical relation but seems to have sound intuitive basis. It states that the channel wetted perimeter or width increases with the square root of the increasing discharge. Although it is arguable that the width is precisely a function of the square root of discharge, empirical evidence however strongly indicates that smaller canals tend to be deeper with respect to their width than the larger ones.

Alternatively, a functional relationship can be formulated as follows for the width-depth ratio of a canal

$$B/D = a_6(y_0/D)^r \tag{53}$$

where a_6 and r are positive constants. Equation 53 is identical to Equation 52 and is dimensionally homoegeneous. If a_6 and r are known with a reasonable accuracy, Equation 53 can be combined with Equations 30 and 51 to complete the design.

The fact that some relationship such as Equation 52 or 53 is applicable is supported by Gibson's comments [5]: "It is perhaps worth noticing in passing that what is in effect a distortion of scale (in a hydraulic model) is usual in nature. Since small streams flowing through alluvial ground have much steeper side slopes and gradients than large rivers of similar regime in similar ground. In a very large river such as the Mississippi, the Ganges, or the Irrawaddy, the maximum depth will rarely exceed 1:50 of the maximum width, while in a small stream in similar ground this ratio will seldom be less than 1:5."

In several of the link canals of Pakistan, the ratio B/y_0 is as high as 70 or even higher. The smallest value of this ratio in the Simons-Bender canal data is 4.32 corresponding to a discharge of 56.0 cfs and the largest is 23.39 for a discharge of 363.3 cfs.

SUSPENDED SEDIMENT TRANSPORT IN STABLE CHANNELS

According to Equation 38

$$\psi/\psi_0 = (y/y_0)^4 \tag{54}$$

The sediment transport per unit width, q_s, is given by

$$q_s = \int_0^y cu\, dy \tag{55}$$

where u = local velocity
$c = d\psi/dy = 4\psi_0 y^3/y_0^4$

The velocity distribution in a rough open channel is described by the logarithmic law as follows

$$u = U_*[8.48 + 2.5 \ln(y/\Delta)] \tag{56}$$

where Δ = characteristic height of bed roughness. Integrating Equation 56 into Equation 55 gives

$$q_s = \int_0^y (4\psi_0 U_*/y_0^4)[8.48y^3 + 2.5y^3 \ln(y/\Delta)]\, dy \tag{57}$$

Integration of Equation 57 and simplification leads to

$$q_s = (U_*\psi_0 y^4/y_0^4)[7.85 + 2.5 \ln(y/\Delta)] \tag{58}$$

The total discharge of suspended sediment, Q_s, is given by

$$Q_s = 2\int_0^{B/2} q_s\, dx \tag{59}$$

Using Equations 45 and 58 in Equation 59 gives

$$Q_s = 2U_*\psi_0 \int_0^{B/2} [7.85 \sin(\pi x/B) + 2.5 \ln(y_0/\Delta) \sin(\pi x/B) + 0.625 \sin(\pi x/B) \ln \sin(\pi x/B)]\, dx \tag{60}$$

To integrate Equation 60, U_* is assumed to be constant and equal to the shear velocity at the center of the channel i.e. U_{*0}. This assumption is reasonable since the local shear velocity does not change appreciably from its value at the center; any significant change that occurs is in the bank region. Recognizing that $\int_0^{\pi/2} \sin\theta \ln \sin\theta\, d\theta = \ln 2 - 1$, Equation 60 is solved to give

$$Q_s = (2U_{*0}\psi_0 B/\pi)[7.66 + 2.5 \ln(y_0/\Delta)] \tag{61}$$

It can be proved from Equation 56 that the average velocity V_0 in the central vertical is given by

$$V_0 = U_{*0}[6.0 + 2.5 \ln(y_0/\Delta)] \tag{62}$$

An approximate evaluation for the cross-sectional average velocity V for a stable channel gives effectively the same result as Equation 62. Thus replacing V_0 by V in Equation 62 and using it in Equation 61 leads to

$$Q_s = (2V\psi_0 B/\pi)(1.66 + U_{*0}/V) \tag{63}$$

The Darcy-Weisbach coefficient of friction f is defined by

$$f = 8(U_{*0}/V)^2 \tag{64}$$

so that

$$Q_s = (2V\psi_0 B/\pi)(1 + 0.587f^{1/2}) \tag{65}$$

Equation 65 can further be developed into a simpler and more meaningful form as follows. Use Equation 47 and simplify to rewrite Equation 65 as follows.

$$Q_s = (0.767Q\psi_0/y_0)(1 + 0.587f^{1/2}) \tag{66}$$

Obviously $\psi_0/y_0 = \bar{c}_0$, the average concentration of suspended sediment in the central vertical.

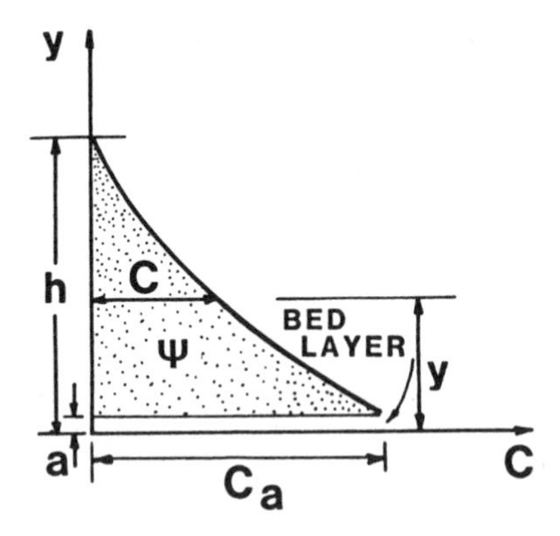

Figure 6. Schematic plot of c against y. Area under the curve gives ψ.

Denoting the average suspended transport, Q_s/Q, by c_A, Equation 66 then finally simplifies to

$$c_A = 0.767\bar{c}_0(1 + 0.587f^{1/2}) \tag{67}$$

The value of f in stable alluvial canals is usually in the range of 0.02 – 0.06. For f = 0.02, $c_A/c_0 = 0.831$ and for f = 0.06, $c_A/\bar{c}_0 = 0.877$. It is useful to know for practical purposes that the value of $c_A/\bar{c}_0$ is expected to lie in the narrow range of 0.83 to 0.88 in the stable alluvial canals.

Equation 67 is an important result; if $\bar{c}_0$ is known from measurement in the central vertical, c_A can be computed from which the total load Q_s is easily obtained. Indeed, measurement only in one vertical not necessarily the central one is required for computing Q_s. If for instance ψ is measured in any vertical, Equation 54 can be used to compute ψ_0 and $\bar{c}_0$. ψ can be measured either using a depth integrating sampler or by measuring local sediment concentration at a number of places in the given vertical. The values of c can then be plotted against y as in Figure 6 and the area under the curve gives ψ.

Purely theoretical methods can be used to compute Q_s. For instance, local concentration of suspended sediment can be computed [12] using

$$\frac{c}{c_a} = \left[\frac{a}{y}\left(\frac{h-y}{h-a}\right)\right]^{V_s/U_* k} \tag{68}$$

in which a = thickness of the bed layer
h = total depth in a vertical
$c = c_a$ at y = a
k = von Karman's universal constant
= 0.4 in clear water flows but varies in sediment charged flows

The value of c_a can be computed as the average bed load concentration in the bed layer of thickness a [$\simeq$2D as proposed by Einstein [13]]. Equation 68 can then be integrated over the depth to get ψ. Preceding steps from Equations 55 to 60 can be followed to compute Q_s.

Incidentally, Einstein's suggestion in respect of the bed layer thickness i.e. a = 2D, is suitable only for flat beds. When a bed is covered with moving ripples and dunes, the bed layer thickness will be considerably higher. The bed load transport in such cases, which are probably more frequent in nature than the flat bed cases, is largely due to movement of ripples and dunes. The height of these structures can be as large as one sixth to one third of the water depth [14, 15].

In principle, Equations 67 and 68 together can be used in place of Equations 52 and 53 for the design of stable channels. Average concentration of suspended transport can be estimated from measurements in the river near the canal intake. Using Equation 67, $\bar{c}_0$ can then be computed. By a trial-and-error method, y_0 and S_0 can be determined so that Equations 68, 51, and 30 are simultaneously satisfied. Equation 30 gives the bed load transport from which c_a can be computed which in turn is used in Equation 68. While this approach appears to be wholly feasible, some practical difficulties make its application to design still open to question at present. One of these difficulties is in respect to Equation 68. For clear-water flows, the value of k equals 0.4; in sediment-charged flows k is known to vary significantly but unpredictably over a wide range [12, 16, 17]. Although modifications of Equation 68 are available in literature; the status of such modifications is ad hoc to a large extent.

CONCLUSION

Sediment-stable channels appear to belong to a category whose shape is described by Equation 44. For sediment-free channels operating in a state of critical equilibrium, $m = 1$ and for actively mobile channels m is close to 4.0. Channels operating between these two limits presumably have different values of m between 1 and 4. More work remains to be done to ascertain appropriate values of m for different channels.

Computation shows that the average velocity in the central vertical of a stable channel is equal to the average cross-sectional velocity. A theoretical formula, Equation 67, is deduced for computing the total load of suspended sediment transport in a stable channel. This formulation can be refined in the future together with Equation 68 and a degree of empiricism can be reduced for the rational design of stable channels.

NOTATION

A cross-sectional area
A_1, A_2 constants
$a_1, a_2, \ldots, a_6$ constants
a thickness of bed layer
B channel width
c concentration of suspended sediment
$\bar{c}$ average value of c in a vertical
c_A cross-sectional average of the suspended sediment concentration $= Q_s/Q$
c_a concentration of the suspended sediment at the edge of the bed layer
$\bar{c}_0$ average concentration of the suspended sediment in the central vertical
D average diameter of the bed material
d rate of deposition of the suspended sediment
d_0 rate of deposition of the suspended sediment in the channel center $(x = B/2)$
E complete elliptical integral of the second kind
e rate of lateral erosion per unit length of channel
e_0 rate of lateral erosion at the channel center
f Shields' parameter; the Darcy-Weisbach coefficient of friction
G_L lateral transport of bed load per unit length of channel
g gravitational acceleration
g_L lateral transport of vertically integrated suspended sediment per unit channel length
h total depth of flow at any location
k von Karman's universal constant
n constant; Manning's coefficient of bed roughness
P wetted perimeter
p constant
q exponent
q_b unit rate of bed load transport
q_{b0} value of q_b at the channel center
q_s unit rate of suspended sediment
Q total discharge
Q_s total suspended sediment discharge
R hydraulic radius
Re Reynolds number
r exponent
S_0 longitudinal bed slope
S_s specific gravity of the sediment

U_* $[gyS_0 \cos\alpha]^{1/2}$; local shear velocity
U_{*0} $[gy_0S_0]^{1/2}$; shear velocity at the channel center
u local velocity
V average cross-sectional velocity
$\bar{V}_0$ average velocity in the central vertical
V_s terminal fall velocity of sediment
W submerged weight of a bed material particle
x coordinate in the transverse direction; distance measured from one bank to a reference point in the transverse direction
x_* x/B
y local depth; depth from the bed to a reference point in a vertical
y_c A/B; hydraulic depth
y_{*c} y_c/y_0
y_0 depth in the channel center
y_* $(y/y_0)^m$

Greek Symbols

α local inclination of a channel bank
β $x \tan\varphi/y_0$
Δ height of bed roughness
ϵ coefficient of apparent viscosity in the flow direction
ϵ_x coefficient of apparent viscosity in the x direction
ϕ angle of repose
γ specific weight of water
γ_s specific weight of sediment
λ $(q_{bo}y_0/m\mu\epsilon_x\psi_0)$
μ Coulomb's coefficient of dynamic friction
ν coefficient of kinematic viscosity
ρ density of water
θ an angle; $(B^2/\psi_0\epsilon_x) \times [d_0 - e_0 y_{*c}^{-(1-n/m)}]$
ψ $\int_0^y c\,dy$
ψ_0 value of ψ at the channel center $= \int_0^{y_0} c\,dy$
ψ_* ψ/ψ_0
τ local bed shear stress
τ_c critical shear stress
τ_0 bed shear stress at the channel center

REFERENCES

1. Kennedy, R. G., "The Prevention of Silting in Irrigation Canals," *Proc., Instn. of Civ. Engrs., London*, Vol. 119 (1895), pp. 281–290.
2. Lindley, E. S., "Regime Channels," *Proc., Punjab Engrg. Congress*, Vol. 7 (1919), pp. 63–74.
3. Lacey, G., "Stable Channels in Alluvium," *Proc., Instn. of Civ. Engrs., London*, Vol. 229 (1929–30), pp. 259–384.
4. Ackers, P., "Experiments on Small Streams in Alluvium," *Journal of the Hydraulics Division, ASCE*, Vol. 90, No. HY4 (July 1964), pp. 1–37.
5. Henderson, F. M., "Stability of Alluvial Channels," *Transactions, ASCE*, Vol. 128, Part I (1963), pp. 657–686.
6. Gill, M. A., "Rationalisation of Lacey's Regime Flow Equations," *Journal of the Hydraulics Division, ASCE*, Vol. 94, No. HY4 (July 1968), pp. 983–995.
7. Gill, M. A., "Discussion of 'A Contribution to Regime Theory Relating Principally to Channel Geometry,'" by D. I. H. Barr, M. K. Alam, and A. Nishat, *Proc., Instn, of Civ. Engrs.*, London, Part 2, Vol. 71 (Sept. 1981), pp. 957–961.
8. USBR, "Stable Channel Profiles," USBR Hydr. Lab. Reprt Hyd. 325 (Sept. 1951).
9. Parker, G., "Self Formed Straight Rivers with Equilibrium Banks and Mobile Bed. Part I. The Sand-Silt River," *Journal of Fluid Mechanics*, Vol. 89, Part I (1978), pp. 109–125.
10. Engleund, F., "Flow and Bed Topography in Channel Bends," *Journal of the Hydraulics Division*, ASCE, Vol. 100, No. HY11 (Nov. 1974), pp. 1631–48.
11. Simons, D. B., and Albertson, M. L., "Uniform Water Conveyance Channels in Alluvial Material," *Transactions, ASCE*, Vol. 128 (1963), pp. 65–105.

12. Vanoni, V. A., "Transportation of Suspended Sediment by Water," *Transactions, ASCE*, Vol. 111 (1964). pp. 67–133.
13. Einstein, H. A., "The Bed Load Function for Sediment Transportation in Open Channel flows," Technical Bulletin No. 1026, USDA, SCS, Washington, D.C. (Sept. 1950), pp. 1–71.
14. Yalin, M. S., "Geometrical Properties of Sand Waves," *Journal of the Hydraulics Division, ASCE*, Vol. 90, No. HY5, (Sept. 1964)
15. Gill, M. A., "Height of Sand Dunes in Open Channel Flows," *Journal of the Hydraulics Division, ASCE*, Vol. 97, No. HY12 (Dec. 1971), pp. 2067–73.
16. Einstein, H. A., and Farouk, M. Abdel-Aal, "Einstein Bed Load Function at High Sediment Rates," *Journal of the Hydraulics Division, ASCE*, Vol. 98, No. HY1 (Jan. 1972).
17. Gill, M. A., discussion of Ref. 16, *Journal of the Hydraulics Division, ASCE*, Vol. 98, No. HY10 (Oct. 1972), pp. 1888–89.
18. Colby, B. R., and Hembree, C. H., "Computation of Total Sediment Discharge, Niobrara River near Cody, Nebraska," USGS, Water Supply Paper 1357 (1955).

CHAPTER 9

TWO-DIMENSIONAL CHANNEL FLOWS OVER ROUGH SURFACES

Neil L. Coleman

Director
USDA Sedimentation Laboratory
Oxford, Mississippi USA

CONTENTS

INTRODUCTION, 220

THE TIME-MEAN VELOCITY DISTRIBUTION, 221

THE EFFECT OF ROUGHNESS ON THE VELOCITY DISTRIBUTION, 224

THE VELOCITY DEFECT LAW, 228

ROUGH-CHANNEL FLOW RESISTANCE, 230

SUMMARY, 233

NOTATION, 233

REFERENCES, 234

INTRODUCTION

In the general case of channel flow, both the sidewalls and the channel bed exert flow resistance, so that both velocity and shear stress can vary in streamwise, depthwise, and spanwise directions and the flow is three-dimensional. Many channels, however, have widths that are large relative to flow depth (large aspect ratio). As has been assumed by many investigators and verified by both Vanoni [1] and Tracy and Lester [2], a channel with a large aspect ratio has a region in the central portion of the flow where spanwise variation of velocity and shear stress is insignificant, and the flow can be treated as two-dimensional in nature. Furthermore, Pierce and Zimmerman [3] have shown that provided spanwise variations are not too large, even three-dimensional flows can be treated as two-dimensional flows to a good degree of approximation. Thus, particularly for considering the characteristics of velocity profiles, two-dimensional treatments have validity in many practical open channel flow situations.

Channel flows belong to the general class of bounded shear flows, as do pipe flows and flows around airplane components or ship hulls, and thus the treatment of channel flows by hydrodynamics rather than by hydraulics can be founded securely on the classical case of the developing boundary layer over a flat plate. Even natural river flows, with their contorted geometries and gross roughness elements in the form of propagating bed configurations, may ultimately succumb to this treatment.

Channel flows over rough surfaces are distinguished from flows over smooth surfaces by the need to defined a virtual origin for the velocity profile. The virtual origin is the plane, located somewhere between the tops and the bottoms of the roughness elements, where the local velocity is zero. In channel flows, as in wind tunnel flows [4], correct definition of the virtual origin is essential for correct interpretation of velocity profiles over rough surfaces.

Rough surface types encountered in natural channels include sand and gravel particle roughness, roughness of propagating bed forms, and roughness of channel bed vegetation and debris. Rough surface types encountered in man-made channels or channel-like structures can have many textures, ranging from that of finished concrete to isolated-element roughness like that of riveted steel plate, corrugated or mesh surfaces of different kinds, etc. Each of these rough surface types presents specific problems; an attempt to unify treatment of these problems using the concept of an "equivalent sand roughness" was started by Schlichting [5] and has continued to the present time [6].

Closely related to the type, concentration, and spacing of channel bed roughness elements is the Reynolds number at which the channel flow can be classed as fully hydrodynamically rough turbulent flow. The advent of hot-film anemometers has made possible the detailed study of turbulence in water channels as well as in wind tunnels [7, 8], although such studies are still laborious. In time, rough channel flows may be sufficiently understood to permit the calculation of the velocity profile, the turbulence distribution, and the expected head loss for every type of channel bed roughness, whether natural or man-made.

THE TIME-MEAN VELOCITY DISTRIBUTION

Figure 1 illustrates an ideal time-mean velocity profile U(y) in steady-state turbulent flow over a rough channel bed, where the roughness element height k is small relative to both the boundary layer thickness δ and the total flow depth y_t. As indicated, the velocity profile has a virtual origin where U is equal to zero and y, by definition, equals zero. The bed shear stress τ_0 may be assumed to act in the plane of the virtual origin. The maximum velocity U_m is by definition the velocity at the top of the boundary layer. Ideally, the free stream velocity everywhere equals U_m; in practice, channel flows often indicate a slight reduction in velocity between the top of the boundary layer and the water surface. This region can therefore be called a free stream by courtesy only. Equally often the boundary layer occupies the entire flow depth, so that δ equals y_t and U_m occurs at the water surface. Since the flow in Figure 1 is turbulent, in addition to the streamwise velocity component U(y), the orthogonal depthwise component V(y) also exists everywhere in the flow.

For steady flow, the momentum equation is

$$U\frac{\partial U}{\partial x} + V\frac{\partial U}{\partial y} = \frac{1}{\rho}\frac{\partial \tau}{\partial y} \tag{1}$$

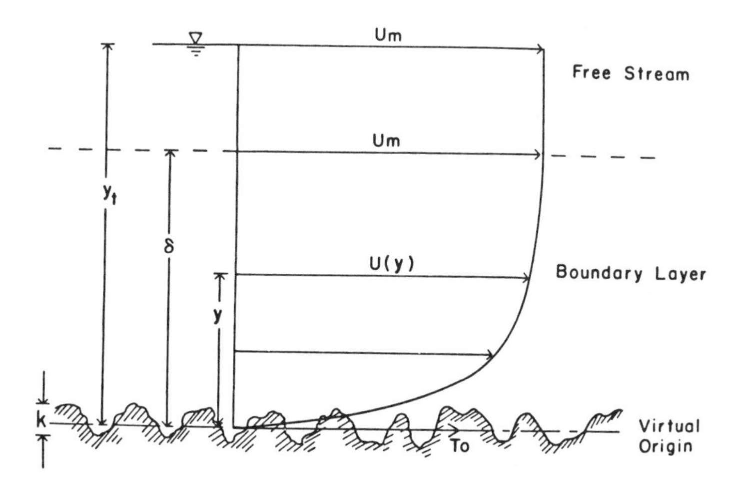

Figure 1. A definition sketch for the time-mean velocity profile in rough channel flow.

The shear stress equation is

$$\tau = \mu \frac{dU}{dy} - \rho uv \tag{2a}$$

$$\tau = \tau_{viscous} + \tau_{turbulent} \tag{2b}$$

and the continuity equation is

$$\frac{\partial U}{\partial x} + \frac{\partial V}{\partial y} = 0 \tag{3}$$

The boundary conditions are

$$\left.\begin{matrix} U = 0 \\ V = 0 \\ \tau = \tau_o \end{matrix}\right\} \text{at } y = 0 \tag{4}$$

and

$$\left.\begin{matrix} U = U_m \\ \tau = 0 \end{matrix}\right\} \text{at } y = \delta \tag{5}$$

Since U and V are zero at the virtual origin, it follows from the momentum equation that

$$\left.\frac{\partial \tau}{\partial y}\right]_{y=0} = 0 \tag{6}$$

The second partial derivative can also be shown to be zero by differentiating equation [1] to obtain

$$U \frac{\partial^2 U}{\partial x^2} + V \frac{\partial^2 U}{\partial y^2} + \frac{\partial U}{\partial y}\left(\frac{\partial U}{\partial x} + \frac{\partial V}{\partial y}\right) = \frac{1}{\rho} \frac{\partial^2 \tau}{\partial y^2} \tag{7}$$

introducing Equation 3, and evaluating again at U and V equal to zero. With both the first and second partial derivatives of τ shown to be zero at the virtual origin, it can be assumed that there is a small region near the channel bed where the shear stress is essentially constant and equal to τ_o.

In the constant-stress layer, U must be a function of fluid properties, channel bed shear stress, and distance from the virtual origin

$$U = \phi_1(\rho, \mu, \tau_o, y) \tag{8}$$

and, by dimensional analysis, the velocity profile is

$$\frac{U}{U_*} = \phi_1\left(\frac{U_* y}{\nu}\right) \tag{9}$$

where

$$U_* = \left(\frac{\tau_o}{\rho}\right)^{1/2} \tag{10}$$

is the channel bed shear velocity at the virtual origin, and

$$\nu = \frac{\mu}{\rho} \tag{11}$$

is the kinematic fluid velocity. The function ϕ_1 is the general "law of the wall" and is valid for all bounded shear flows including rough channel flows.

Very close to the virtual origin, where y, U, and V are very small, there is essentially no room for turbulence. The turbulent shear stress in Equations 2a and 2b can be assumed to be so small that viscosity dominates and

$$\frac{\tau_o}{\rho} = U_*^2 = \nu \frac{dU}{dy} \tag{12}$$

from which, integrating,

$$\frac{U}{U_*} = \frac{U_* y}{\nu} \tag{13}$$

This is the functional form of the law of the wall only in the viscous sublayer immediately proximate to the virtual origin. In rough channel flow, the viscous sublayer is easily perturbed by the wakes of individual roughness elements and becomes progressively thinner with increases in flow intensity. When the viscous sublayer thickness becomes infinitesimal at the virtual origin plane, the flow is said to be fully hydrodynamically rough [9].

A variety of phenomenological theories involving the concept of a mixing length or the related concept of an eddy viscosity have been used [10] to provide means of integrating Equation 2a to obtain a velocity profile for the flow outside the viscous sublayer. Millikan [11], however, has deduced the existence of a logarithmic velocity profile by a method free of any phenomenological assumptions. This argument proceeds from the fact that, outside the constant stress layer, the turbulent shear stress term in Equations 2a and 2b varies with distance from the virtual origin while the viscous shear stress can be neglected. The channel bed or other solid boundary may be considered to reduce the velocity in the boundary layer relative to the free stream velocity U_m. Thus, the velocity reduction ($U_m - U$) must be a function of the boundary layer thickness, the distance from the virtual origin, and the shear velocity at the virtual origin, or

$$U_m - U = \phi_2(\delta, y, U_*) \tag{14}$$

By dimensional analysis, the general velocity defect profile is

$$\frac{U_m - U}{U_*} = \phi_2\left(\frac{y}{\delta}\right) \tag{15}$$

As indicated in Figure 1, velocity profiles in boundary layer flows normally do not show any breaks or discontinuities that would indicate anything other than a smooth transition between the law of the wall and the velocity defect law. Therefore Millikan postulated that at least one point exists on the profile where the functions ϕ_1 and ϕ_2 are equal. If this is so, then, from Equations 9 and 15, the identity must be

$$\phi_1\left[\frac{U_*\delta}{\nu}\left(\frac{y}{\delta}\right)\right] = \frac{U_m}{U_*} - \phi_2\left(\frac{y}{\delta}\right) \tag{16}$$

Since (y/δ) appears as a multiplicative factor on the left side of Equation 16 and as an additive factor on the right side, the functions ϕ_1 and ϕ_2 must be logarithmic for the identity to hold.

Although deduced from very general considerations, Millikan's theory is quite specific in two regards. The first is that the law of the wall and the velocity defect law must be logarithmic at the point or locus of points where they overlap. The second is that there is absolutely no constraint on the defect law to be logarithmic where it does not overlap the law of the wall.

THE EFFECT OF ROUGHNESS ON THE VELOCITY DISTRIBUTION

In Figure 2 a velocity profile from water flow in a smooth channel (measurements by the author; hitherto unpublished) has been plotted in the law-of-the-wall coordinates from Equation 9. Also indicated are the (y/δ) values for each data point. From the concept of the constant-stress layer, the validity of the general law of the wall is restricted to a region close to the virtual origin; from Millikan's argument, the logarithmic part of the law of the wall is restricted to the region where it overlaps with the velocity defect law. This means that the logarithmic region must occur at small values of (y/δ). From this, it appears that the curve fitting the data is asymptotic near the channel bed to a function

$$\frac{U}{U_*} = N \ln \frac{U_* y}{\nu} + A \tag{17}$$

In this particular case, the data indicate that N is 2.5 while A is 5.4. Further, the data systematically deviate from Equation 17 with the maximum deviation $\Delta U_m/U_*$ at (y/δ) equal to one.

A velocity profile from flow in a rough-bed channel [12] is also plotted in Figure 2. This profile is also asymptotic to Equation 17, and N is 2.25 for this profile. As with the data from smooth-channel flow, the rough-channel flow data deviate systematically from Equation 17 with the maximum deviation $(\Delta U_m/U_*)$ at (y/δ) equal to one. The principal difference between the two profiles is that the logarithmic part of the rough-channel profile is displaced downward from the smooth-channel profile by a decrement $(\Delta U/U_*)$. For the two profiles in the figure, the value of this decrement is about 8.7, with some uncertainty being present because of the 10% difference displayed in values of N.

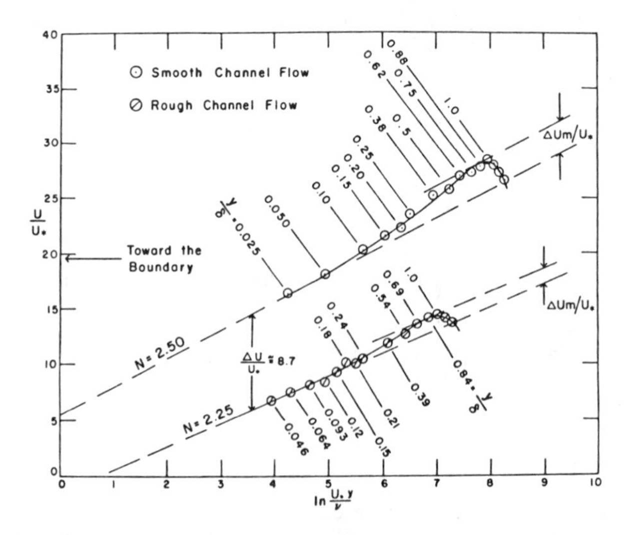

Figure 2. Smooth- and rough-channel flow velocity profiles in law-of-the-wall coordinates.

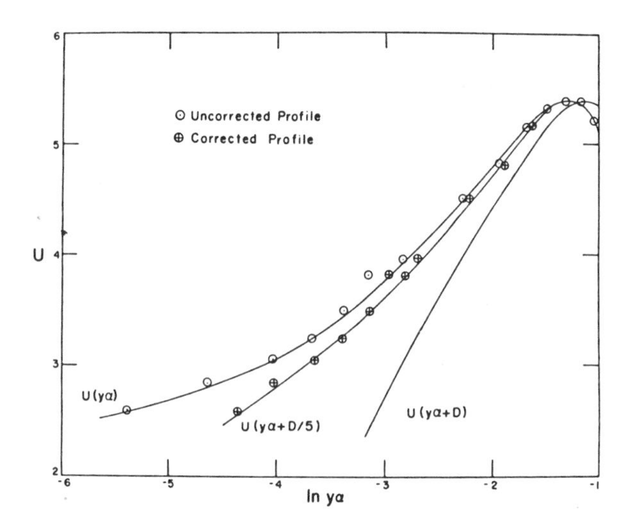

Figure 3. The Perry-Joubert [14] method of estimating the location of the virtual origin.

Concerning the logarithmic part of the law of the wall, plots like Figure 2 raise three questions:

1. How can the virtual origin in a rough-channel flow be defined so that rough-channel profiles can be compared with smooth-channel profiles?
2. How can the logarithmic profile slope N be represented as a function of channel roughness?
3. How can the velocity decrement ($\Delta U/U_*$) be represented as a function of channel roughness?

A practical answer to the first question appears to have been available for some time [13] before having been formally stated by Perry and Joubert [14]. This method takes advantage of the necessity for a logarithmic form of the law of the wall in the overlap region. It is illustrated in Figure 3, where the raw velocity data from the rough-channel profile in Figure 2 is plotted in simple coordinates of U against the logarithm of an apparent elevation y_α. For both this particular profile and others in the experimental series [12], local velocities were measured with a total head tube in flows over a bed of uniform close-packed spheres. With the virtual origin initially unknown, the working datum for measuring y_α for locating the total head tube was taken as the plane tangent to the tops of the spheres. For this particular case, the roughness height was taken as the sphere diameter D, and the curves $U(y_\alpha)$ and $U(y_\alpha + D)$ represent the velocity profile referenced to the top and the bottom of the sphere bed, respectively, on the logarithmic plot. The real virtual origin of the velocity profile, according to the Perry and Joubert method, will be indicated by trial and error, somewhere between these limits, by the appearance of a plot with a straight near-bed asymptote indicating the logarithmic region. For the particular velocity profile illustrated here the correction to be added to y_α to obtain y turns out to be about D/5. This method of estimating the virtual origin correction is applicable to all types and concentrations of channel bed roughness [9], so long as the roughness height is small relative to the boundary layer thickness and channel flow depth. The method is deceptive in that an erroneous virtual origin correction will be obtained if data points properly belonging in the wake region are included in the trial-and-error search for the near-bed logarithmic asymptote.

The logarithmic profile slope N is, in the classical phenomenological mixing length derivation of the logarithmic velocity profile [10], the reciprocal of the Karman coefficient κ; by convention this coefficient is discussed, rather than N itself. For the smooth-bed and rough-bed profiles in Figure 2, the values of κ are 0.40 and 0.44, respectively. The value of κ generally cited in fluid

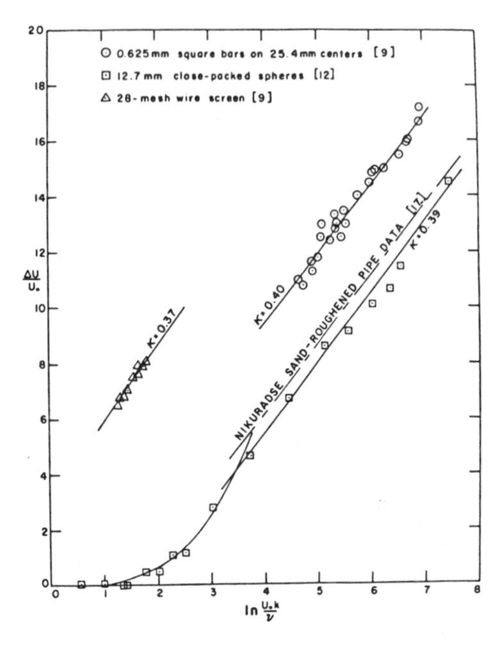

Figure 4. Some velocity reduction functions for common roughness types.

mechanics textbooks [15, 16] for air and water flows with no suspended particulate matter present is 0.41, although substantial experimental scatter around this value exists [15].

In Figures 2 and 3, the roughness height for the rough-bed velocity profile over a bed of close-packed spheres was taken as the sphere diameter D. For more complex bed roughness heights a single linear roughness measure is probably insufficient. Nonetheless, Hama [9] showed that a comprehensive treatment of roughness effects is possible provided that a single characteristic roughness measure k is defined for each family of geometrically similar roughness types. Thus, for both uniform spherical roughness and for uniform sand grains, the particle size D may be adopted as k; for rectangular slats or battens affixed to a channel bed the batten height may be adopted, etc.

With an appropriate virtual origin correction, rough channel profiles like that illustrated in Figure 2 can be plotted, and the displacements $\Delta U/U_*$ due to roughness can be determined. With an appropriate channel roughness k defined, $\Delta U/U_*$ can be plotted against a roughness Reynolds number U_*k/ν to answer questions 2 and 3. A plot like this for three different channel roughness types is shown in Figure 4. The channel roughness types included are 0.625 mm transverse square bars on 25.4 mm centers [9], 12.7 mm close-packed spheres [12], and 28-mesh wire screen [9]. The sphere roughness data were taken over a wide range of U_*k/ν. They show the gradual development of $\Delta U/U_*$ as the viscous sublayer thins, and also indicate that, for this roughness type, the flow becomes fully hydrodynamically rough at U_*k/ν near 33. For the other data sets, the flow was always in the fully rough range. Like the data sets included in Figure 4, flows in the fully rough range display a $\Delta U/U_*$ function in the form

$$\frac{\Delta U}{U_*} = \frac{1}{\kappa} \ln \frac{U_* k}{\nu} + B \tag{18}$$

For the data sets for the three roughness types included in Figure 4, and for other data sets over a considerable range of U_*k/ν, the slopes $(1/\kappa)$ of Equation 18, which are equivalent to N in Equation 17, remain within experimental error close to a general value of 2.5, or a general κ value of 0.40. From this, question 2 can be answered by concluding that, on the basis of present evidence, the Karman coefficient and hence the slope of the logarithmic velocity profile are unaffected by variations in roughness type or by variations in the roughness Reynolds number U_*k/ν as long as the flow is fully hydrodynamically rough.

Question 3 is also answered by plots like those in Figure 4. For the hydrodynamically rough case, Equation 18 defines the form of the function needed. The velocity decrement can be treated as an additive term on the smooth-boundary velocity profile Equation 17 so that

$$\frac{U}{U_*} = \frac{1}{\kappa} \ln \frac{U_* y}{\nu} + A - \frac{\Delta U}{U_*} \tag{19a}$$

from which

$$\frac{U}{U_*} = \frac{1}{\kappa} \ln \frac{U_* y}{\nu} + A - \frac{1}{\kappa} \ln \frac{U_* k}{\nu} - B \tag{19b}$$

or

$$\frac{U}{U_*} = \frac{1}{\kappa} \ln \frac{y}{k} + (A - B) \tag{20}$$

which is the common early elementary textbook form [18].

The broken line very near the sphere roughness line in Figure 4 is the $\Delta U/U_*$ function for pipes roughened with sand of various uniform particle sizes [17]. This rough pipe function has long been an engineering standard. It is customary engineering practice to define an equivalent sand roughness height k_s for complicated kinds of roughness with nonuniform actual roughness heights or nonuniform roughness element spacing [9, 20]. Where the sand particle diameter D_s equals the uniform roughness height k_s, the sand-roughened pipe velocity reduction function is

$$\frac{\Delta U}{U_*} = \frac{1}{\kappa} \ln \frac{U_* k_s}{\nu} - 3.6 \tag{21}$$

The equivalent sand roughness for any roughness for which a $\Delta U/U_*$ function like Equation 18 is known may be found by setting Equation 18, with the appropriate experimentally determined B value, equal to Equation 21 to obtain

$$\ln \frac{k_s}{k} = (B + 3.6)\kappa \tag{22a}$$

or

$$\frac{k_s}{k} = e^{(B + 3.6)\kappa} \tag{22b}$$

For the roughness types in Figure 4, the B values are −4.2, −0.4, and 3.0 for spheres, bars, and screen, respectively. Schlichting [5] and Keulegan [21] have given k_s values for a great many kinds of channel linings, including concrete, wood with and without battens, steel with spherical, conical, and angular roughness elements, and others. Kamphuis [6] has claimed that, for rock roughness of nonuniform size distribution, k_s is approximately $2D_{90}$ where D_{90} is the ninetieth percentile of the rock size distribution. The equivalent sand roughness concept is of great use in engineering practice, but is of little help in understanding the mechanics of flow in rough channels.

From Figure 4 it can be concluded that channel roughness types are unique in their effect on the velocity profile. However, a general velocity profile treatment may still be achieved using $\Delta U/U_*$ functions and U_*k/ν as a roughness Reynolds number, since this results in Equation 20, which is a universal equation for the logarithmic part of the law of the wall.

THE VELOCITY DEFECT LAW

Earlier, Millikan's [11] argument was used to deduce the existence of the inner (law of the wall) and outer (velocity defect) regions, with a logarithmic velocity profile region at small values of (y/δ), where the functions overlap. Figure 2 indicates what is meant by small (y/δ) values; generally the logarithmic region occupies the lower 10% to 20% of the boundary layer, with the velocity profile showing a systematic deviation from the logarithmic law of the wall in the upper part of the boundary layer. Incidentally, there is nothing rigid about the relative thickness of the logarithmic overlap region. For turbulent boundary layers on flat plates, both Landweber [22] and Clauser [23] have shown that, for decreasing values of the Reynolds number

$$R_\delta = \frac{U_m \delta}{\nu} \tag{23}$$

the outer region grows at the expense of the logarithmic region. Landweber [22] has in fact shown that, for an R_δ value less than about 1.5×10^5, the logarithmic overlap region disappears entirely, so that the velocity profile for the entire flow (except for any existing viscous sublayer) is of the outer region type.

From Figure 2 and from the foregoing discussion, the logarithmic law of the wall evvidently cannot be assumed a priori to predict the velocity U_m at δ; indeed, Figure 2 reveals that, if Equation 19a is evaluated at δ, it will give [Clauser, 24]

$$\frac{U(\delta)}{U_*} = \frac{U_m - \Delta U_m}{U_*} = \frac{1}{\kappa} \ln \frac{U_* \delta}{\nu} + A - \frac{\Delta U}{U_*} \tag{24}$$

where $(\Delta U_m/U_*)$ is a decrement equal to the maximum deviation found at $(y/\delta = 1)$ as in the figure. Equation 19a can be subtracted from equation 24 to produce a defect law in the coordinate system demanded by Millikan's [11] argument and given in Equation 15. This expression is

$$\frac{U_m - U}{U_*} = -\frac{1}{\kappa} \ln \frac{y}{\delta} + \frac{\Delta U_m}{U_*} \tag{25}$$

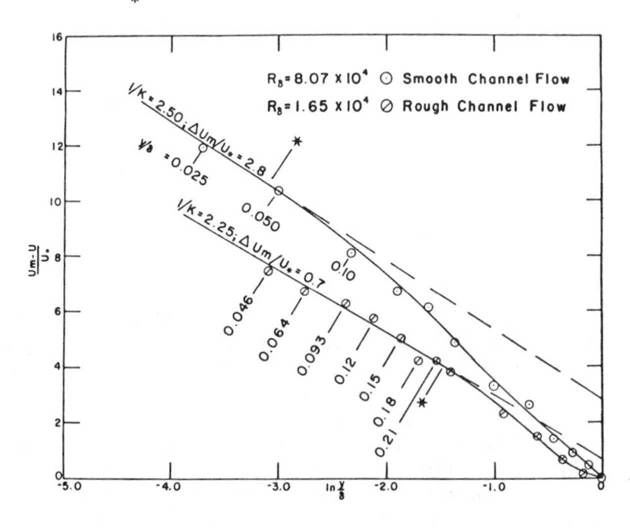

Figure 5. Smooth- and rough-channel flow velocity profiles plotted in velocity-defect coordinates.

In Equation 25 the roughness decrement $(\Delta U/U_*)$ is eliminated, so for a constant value of the intercept decrement $(\Delta U_m/U_*)$, which is the outer region deviation parameter, the velocity defect law given is universal for flows over both smooth and rough boundaries. This property of the velocity defect law has been demonstrated many times for flows over smooth and rough boundaries [9, 22, 23]. Differences in the velocity defect law from one situation to another must then be attributed entirely to difference in the intercept deficit.

Figure 5 is a velocity defect plot of the same two profiles plotted in Figure 2. In both cases, R_δ is on the order of a decade less than the 1.5×10^5 limit given by Landweber [22] for the disappearance of the logarithmic overlap region. Nonetheless, in both cases some logarithmic region evidently exists. The logarithmic region of the smooth channel flow extends over about 4% of the boundary layer, while the rough flow logarithmic region occupies about 21% of the boundary layer. The intercept deficits vary inversely with the logarithmic region thickness and the Reynolds number. This is qualitatively in accordance with the findings of Landweber [22] and Clauser [23] for boundary layer flows over flat plates, but illustrates the caution with which true boundary layer findings must be applied to the bounded shear flows found in channels. It must be emphasized that the differences in the defect plots for the smooth and rough channel cases in Figure 5 are not due to a general nonuniversality of the defect law, but rather are due strictly to the projection of a greater degree of roughness-generated turbulence into the outer region flow in the case of the rough channel.

From Figure 5, the asymptotes of a complete velocity defect law are the logarithmic defect law for $(y \ll \delta)$, and

$$\frac{U_m - U}{U_*} = 0 \tag{26}$$

at $y = \delta$. Between these asymptotes there must be a continuous function describing the systematic deviation of the actual velocity defect profile from that predicted by the logarithmic form. This function must have limits of zero at $(y/\delta = 0)$ and $\Delta U_m/U_*$ at $y/\delta = 1$. According to Cebeci and Bradshaw [25], the requirements for this function are fulfilled by any function $(\Pi/\kappa)\omega(y/\delta)$ where Π is a parameter proportional to the magnitude of the maximum deviation at $(y/\delta = 1)$. Such a function can be subtracted from Equation 25 to obtain

$$\frac{U_m - U}{U_*} = \frac{1}{\kappa}\ln\frac{y}{\delta} + \frac{\Delta U_m}{U_*} - \frac{\Pi}{\kappa}\omega\left(\frac{y}{\delta}\right) \tag{27a}$$

and this can be made consistent with the "wake law" of Coles [26] by setting $(\Delta U_m/U_*)$ equal to $(2\Pi/\kappa)$ so that Equation 27a becomes

$$\frac{U_m - U}{U_*} = \frac{1}{\kappa}\ln\frac{y}{\delta} + \frac{2\Pi}{\kappa} - \frac{\Pi}{\kappa}\omega\left(\frac{y}{\delta}\right) \tag{27b}$$

This is the complete velocity defect law; the equivalent complete law, in law-of-the-wall coordinates, is

$$\frac{U}{U_*} = \frac{1}{\kappa}\ln\frac{U_* y}{\nu} + A - \frac{\Delta U}{U_*} + \frac{\Pi}{\kappa}\omega\left(\frac{U_* y}{\nu}\frac{\nu}{U_*\delta}\right) \tag{28}$$

A variety of empirical forms for $\omega(y/\delta)$ have been proposed. The best-known are Cole's original function [26]

$$\omega\left(\frac{y}{\delta}\right) = 2\sin^2\left(\frac{\Pi}{2}\frac{y}{\delta}\right) \tag{29}$$

Table 1
Values of the Coles Wake Strength Coefficient and Related Parameters

Reference	Flow Type	Boundary Type	$\frac{k}{\delta}$	$\frac{U_* \delta}{\nu}$	π
1	Flume	0.47 mm sand	0.003	9,920	0.36
7	Flume	4.4 mm spheres	0.145	1,088	0.48
12	Rect. conduit	12.7 mm spheres	0.116	2,908	0
12	Rect. conduit	12.7 mm spheres	0.132	15,763	0
28	Flume	21.9 mm gravel	0.183	34,410	0
29	Flume	9.0 mm gravel	0.154	552	0.24

and that of Finley et al. [27]

$$\omega\left(\frac{y}{\delta}\right) = \frac{\kappa}{\Pi}\left(\frac{y}{\delta}\right)\left(1 - \frac{y}{\delta}\right) + 2\left(\frac{y}{\delta}\right)^2\left[3 - 2\left(\frac{y}{\delta}\right)\right] \tag{30}$$

Frequently, other functions fit measured data from channel flows as well or better than these, although for pure boundary layer flows the Coles function fits a vast data set. The parameter Π is known as the Coles wake strength coefficient; from its equivalency with $(\Delta U_m/U_*)$, it is a measure of the degree of deviation of the complete velocity profile equations (Equations 27b or 28) from their logarithmic near-boundary asymptotes.

Currently there is no body of theory about the function $\omega(y/\delta)$ or about Π. As yet, no systematic study of Π for water flows in rough channels has been made. Such flows are, however, part of the general class of bounded shear flows, and steady uniform two-dimensional channel flows are part of the class of equilibrium flows defined by Clauser [23]. For these flows, White [30] has indicated that, depending on the condition of the free-stream pressure gradient, Π can vary from zero to at least 4. Table 1 summarizes some of the few values of Π that have been established for specific rough channel or similar flow situations, together with values for some supposedly related flow parameters. The selection of these other parameters for inclusion in Table 1 was inspired by the parameters appearing in the different terms of Equations 27b and 28. On the very sparse information available, Π values for rough channels are within the bounds given by White [30].

ROUGH-CHANNEL FLOW RESISTANCE

The Darcy-Weisbach resistance coefficient f was originally employed as a universal nondimensional measure of the friction head loss (flow resistance) in pipes [31]. In this context, it was defined as the coefficient in the equation

$$\frac{h_f}{x} = \frac{f}{d}\frac{U_a^2}{2g} \tag{31}$$

The use of this equation was extended to noncircular conduits by substituting the conduit hydraulic radius H for the pipe diameter. The hydraulic radius H is the ratio of the conduit cross-sectional area to the wetted perimeter; since this has the relation $(d = 4H)$ for a pipe, Equation 31 becomes

$$\frac{h_f}{x} = \frac{f}{4H}\frac{U_a^2}{2g} \tag{32}$$

In Equations 31 and 32, the frictional head loss h_f per unit length is given in terms of the pipe diameter or conduit hydraulic radius and the average cross-sectional flow velocity U_a. The use of the Darcy-Weisbach coefficient has been extended, particularly in river mechanics [32], to use in wide rough-bedded channels where H approaches the total flow depth y_t as a limit, hence

$$\frac{h_f}{x} = \frac{f}{4y_t}\frac{U_a^2}{2g} \tag{33}$$

where the numerical constant 4 is conventionally retained to make numerical values of f for pipes, noncircular conduits, and wide channels at least consistent in order of magnitude. Solving Equation 33 for f gives

$$f = \frac{8gy_t h_f}{xU_a^2} = 8\frac{U_*^2}{U_a^2} \tag{34}$$

or

$$\left(\frac{8}{f}\right)^{1/2} = \frac{U_a}{U_*} \tag{35}$$

For a two-dimensional rough channel, U_a is defined as in Figure 6 and by

$$U_a = \frac{1}{y_t}\int_0^{y_t} U(y)\,dy \tag{36}$$

and Figure 6 gives by inspection the relations

$$y_a = ay_t \qquad a < 1 \tag{37a}$$

$$\delta = by_t \qquad b \le 1 \tag{37b}$$

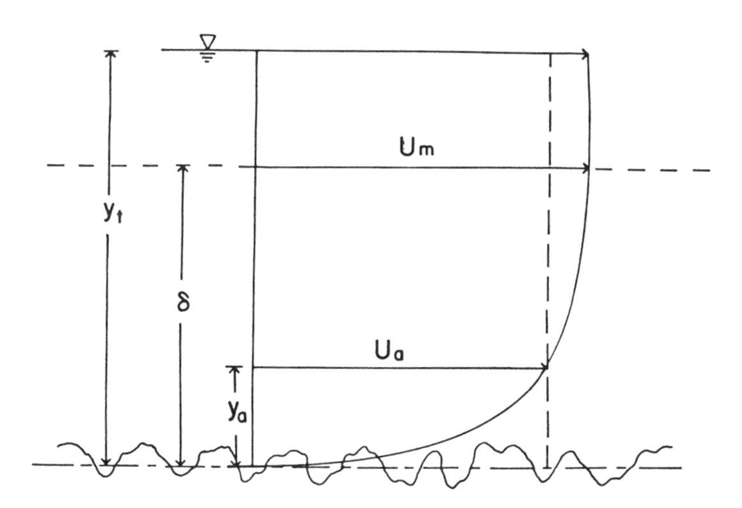

Figure 6. Definitions sketch for the average velocity integral and the relations between flow depth, boundary layer thickness, and the elevation of local velocity equal to average velocity.

which allows evaluating Equation 28 at (U_a, y_a) to obtain

$$\frac{U_a}{U_*} = \frac{1}{\kappa} \ln \frac{U_* y_t}{\nu} + \frac{1}{\kappa} \ln a + A - \frac{\Delta U}{U_*} + \frac{\Pi}{\kappa} \omega\left(\frac{a}{b}\right) \tag{38}$$

where the last term is constant for a given flow condition; that is,

$$\frac{\Pi}{\kappa} \omega\left(\frac{a}{b}\right) = \frac{a}{b} \frac{\Pi}{\kappa} \omega(1) = 2 \frac{a}{b} \frac{\Pi}{\kappa} \tag{39}$$

from the limit of ($2\Pi/\kappa$) at ($y/\delta = 1$) for the wake function, as described earlier. For fully hydrodynamically rough flow, $\Delta U/U_*$ is given by Equation 18. Introducing Equations 18, 35, and 39 into Equation 38 gives, after some manipulation,

$$\frac{1}{\sqrt{f}} = \frac{1}{\sqrt{8}\kappa} \ln \frac{y_t}{k} + \frac{1}{\sqrt{8}} \left[\frac{1}{\kappa} \ln a + A - B + 2 \frac{a}{b} \frac{\Pi}{\kappa} \right] \tag{40}$$

which is an expression for the Darcy-Weisbach coefficient in terms of the total flow depth y_t, the roughness height k, the roughness type constant B, and the ubiquitious wake strength coefficient Π.

With ($\kappa = 0.4$) for all types of roughness, the slope of the function for Equation 40 should be 0.88. Figure 7 is a plot of data collected by McQuivey [7] in flows over various flat particulate channel beds that might be expected to have similar B values. These data show reasonable agreement with the function

$$\frac{1}{\sqrt{f}} = 0.88 \ln \frac{y_t}{k} + 2.3 \tag{41}$$

where the constant 2.3 evidently is the mean value of the entire right-most term in Equation 40. Other roughness types may be expected to display flow resistance functions similar to Equation 41, but with different terminal constants, since the roughness type coefficient B will be different. For example, Bayazit [33] found, for flow over a bed of close-packed hemispheres 23 mm in diameter, that the

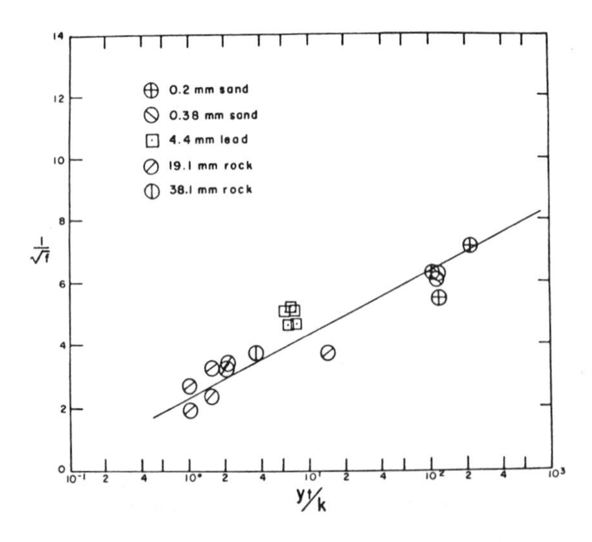

Figure 7. Resistance function for fully hydrodynamically rough flows over close-packed particulate channel beds.

resistance function was

$$\frac{1}{\sqrt{f}} = 0.85 \ln \frac{y_t}{k} + 0.74 \tag{42}$$

In Bayazit's experiments, k was taken as the hemisphere radius, which was, of course, the real height of the hemispheres when glued in close packing to the flat bottom plate of the experimental channel. In McQuivey's experiments, in which the channel beds were close-packed sand, lead shot, or rock, k was taken as the particle diameter. The importance of the roughness type constant B is thus demonstrated, since the resistance function constants for these two sets of results are quite different, although in plan view beds of close-packed hemispheres closely resemble beds of close-packed spheres or sediment particles.

SUMMARY

Two-dimensional flow in rough-bed channels has been treated here by applying principles first developed as part of modern boundary layer theory. This treatment is valid because channel flows belong to the general class of bounded shear flows. The application of boundary layer theory to rough-channel flows is very similar to boundary layer theory application to flows over smooth boundaries, provided that proper definition of a virtual origin for the time-mean velocity profile can be realized, and that the overall effect of roughness on the velocity profile is recognized. The effect of channel bed roughness on the velocity profile expressed in law-of-the-wall coordinates is an overall reduction or downward shift. Studies of this shift show that the Karman coefficient in rough-channel flow is independent of roughness element magnitude or type, at least for cases where the roughness elements are small relative to the overall flow depth.

Flow in a two-dimensional, rough-bed channel can be divided into near-bed and far-bed regions. In addition to a viscous sublayer immediately proximate to the bed, the near-bed region contains a relatively small zone in which the velocity profile is logarithmic. This logarithmic profile is the near-boundary asymptote toward which the velocity profile in the far-bed region tends. The far-bed region comprises most of the flow depth. In this region, the velocity profile is represented by a velocity defect function. The configuration of this function is extremely sensitive to the magnitude of a quantity called the wake-strength coefficient, which determines how rapidly the velocity profile approaches logarithmic form as the channel bed is approached.

The wake-strength coefficient appears on the basis of present best information to be Reynolds-number dependent, or in other words dependent on the intensity and intermittency of turbulence in the free stream and the upper part of the boundary layer. If the turbulence is relatively intermittent and of low intensity (low Reynolds numbers), the outer flow region is large relative to the inner logarithmic region. If the turbulence is not intermittent and is of relatively high intensity, the outer flow region is small relative to the inner logarithmic region. For flows including steady uniform channel flows, the wake strength coefficient may be expected to vary from zero to at least four.

For fully hydrodynamically rough conditions, flow resistance functions in two-dimensional rough-bed channels are similar in form, but are profoundly affected by roughness type, which evidently plays a major role in determining the absolute magnitude of the Darcy-Weisbach resistance coefficient.

NOTATION

A	intercept constant in the logarithmic velocity profile equation
B	intercept constant in the roughness decrement equation
D	bed particle diameter
H	hydraulic radius
N	slope of the logarithmic profile equation
R_δ	gross flow Reynolds number
U	local time-mean velocity; streamwise component
U_a	cross-sectional average flow velocity
U_m	maximum velocity; local velocity at δ

U_* shear velocity at the channel bed
V local time-mean velocity; depthwise component
a proportionality coefficient
b proportionality coefficient
d pipe diameter
f Darcy-Weisbach flow resistance coefficient
g gravity field strength
h_f frictional head loss
k roughness height
k_s equivalent sand roughness height
u mean value of the streamwise turbulent velocity fluctuation
v mean value of the depthwise turbulent velocity fluctuation
x streamwise coordinate
y depthwise coordinate
y_a elevation of average velocity
y_α apparent elevation
y_t total flow depth

Greek Symbols

δ boundary layer thickness; elevation of the top of the boundary layer
κ Karman coefficient
μ fluid dynamic viscosity
ν fluid kinematic viscosity
Π wake strength coefficient
ρ fluid mass density
τ local shear stress
τ_0 shear stress at the channel bed
ϕ functional symbol
ω functional symbol

REFERENCES

1. Vanoni, V. A., "Velocity Distribution in Open Channels," *Civil Engineering*, 11:6 (1941) 356–357.
2. Tracy, H. J., and Lester, C. M., "Resistance Coefficients and Velocity Distribution—Smooth Rectangular Channel," *Water Supply Paper 1592-A*, U.S. Geological Survey (1961) 30 pp.
3. Pierce, F. J., and Zimmerman, B. B., "Wall Shear Stress Inference From Two- and Three-Dimensional Turbulent Boundary Layer Velocity Profiles," *Jour. of Fluids Engineering*, Am. Soc. Mech. Engr., 95:1 (1973) 61–67.
4. Scottron, V. E., "Turbulent Boundary Layer Characteristics Over a Rough Surface in an Adverse Pressure Gradient." *Naval Ship Research and Development Center Report No. 2659* (1967) 154 pp.
5. Schlichting, H., "Experimental Investigation of the Problem of Surface Roughness," *NACA Technical Memorandum No. 823 (1937)*. Translation of Expermentelle Untersuchungen zum Rauhigkeits problem. *Ingenieur-Archiv*, III: 1 (1936) 1–34.
6. Kamphuis, J. W., "Determination of Sand Roughness for Fixed Beds," *Journal of Hydraulic Research*, 12:2 (1974) 193–203.
7. McQuivey, R. S., "Turbulence in a Hydrodynamically Rough and Smooth Open Channel Flow," Ph.D. Dissertation, Colorado State University; *U.S. Geological Survey Open-file Report*, (1967) 105 pp.
8. Blinco, P. H., and Partheniades, E. H., "Turbulence Characteristics in Free Surface Flows Over Smooth and Rough Boundaries," *Jour. of Hydraulic Research*, 9:1 (1971) 43–71.
9. Hama, F. R., "Boundary Layer Characteristics of Smooth and Rough Surfaces," *Trans. Soc. of Naval Architects and Marine Engr.* 62 (1954) 333–358.
10. von Karman, Th., "Mechanical Similitude and Turbulence," *NACA Technical Memorandum No. 611* (1930). Translation of Mechanische Aenlichkeit und Turbulenz. *Nachrichten von der Gesellschaft der Wissenschaften zu Goettingen.* I:5 (1930) 58–76.
11. Millikan, C. B., "A Critical Discussion of Turbulent Flows in Channels and Circular Tubes," *Proc. 5th International Congr. of Applied Mechanics, Cambridge, Mass.*, (1938) 386–392.
12. Coleman, N. L., "Bed Particle Reynolds Modeling for Fluid Drag," *Jour. of Hydraulic Research*, 17:2 (1979) 91–105.
13. Schubauer, G. B., and Tchen, C. M., *Turbulent Flow*, Princeton: Princeton University Press (1959) 123 pp.

14. Perry, A. E., and Joubert, P. N., "Rough-Wall Boundary Layers in Adverse Pressure Gradients," *Jour. of Fluid Mechanics*, 17 (1963) 193–211.
15. Tennekes, H., and Lumley, J. J., *A First Course in Turbulence*. Cambridge: the MIT Press, (1972) 300 pp.
16. White, F. M., *Fluid Mechanics*, New York: McGraw-Hill Book Co., Inc., (1979) 701 pp.
17. Nikuradse, J., "Laws of Flow in Rough Pipes," *NACA Technical Memorandum No. 1292* (1950). Translation of Stroemungsgesetze in rauhen Rohren. VDI-Forschungsheft 361. Beilage zu Forschung auf dem Gebiete des Ingenieurwesens, Ausgabe B, Band 4, (1933) 62 ss.
18. Rouse, H., *Elementary Mechanics of Fluids*, New York: John Wiley and Sons, Inc., (1946) 376 pp.
19. Kamphuis, J. W., "Determination of Sand Roughness for Fixed Beds," *Jour. of Hydraulic Research*, 12:2 (1974) 193–203.
20. Colebrook, C. F., "Turbulent Flow in Pipes, With Particular Reference to the Transition Region Between the Smooth and Rough Pipe Laws," *Jour. of the Institution of Civil Engineers*, 11 (1939) 133–145.
21. Keulegan, G. H., "Laws of Turbulent Flow in Open Channels," *Journal of Research of the National Bureau of Standards*, 21 (1938) 707–741.
22. Landweber, L., "The Frictional Resistance of Flat Plates in Zero Pressure Gradient," *Trans. Soc. of Naval Architects and Marine Engr.*, 61 (1953) 5–32.
23. Clauser, F. H., "The Turbulent Boundary Layer," *Advances in Applied Mech.*, 4 (1956) 1–51.
24. Clauser, F. H., "Turbulent Boundary Layers in Adverse Pressure Gradients," *Jour. Aeronautical Science*, 21 (1954) 91–108.
25. Cebeci, T., and Bradshaw, P., *Momentum Transfer in Boundary layers*, Washington: Hemisphere Publishing Corp., (1977) 391 pp.
26. Coles, D., "The Law of the Wake in the Turbulent Boundary Layer," *Jour. of Fluid Mech.*, 1 (1956) 191–226.
27. Finley, P. J., Khoo, C. P., and Chin, J. P., "Velocity Measurements in a Thin Turbulent Water Layer," *La Houille Blanche*, 21:6 (1966) 713–721.
28. Yalin, M. S., *Mechanics of Sediment Transport*, New York: Pergamon Press, (1972) 290 pp.
29. Grass, A. J., "Structural Features of Turbulent Flow over Smooth and Rough Boundaries," *Jour. of Fluid Mech.*, 50 (1971) 233–235.
30. White, F. M., *Viscous Fluid Flow*, New York: McGraw-Hill Book Co., Inc. (1974) 725 pp.
31. Rouse, H., *Elementary Mechanics of Fluids*, New York: John Wiley and Sons, Inc. (1946) 376 pp.
32. Bathurst, J. C., "Theoretical Aspects of Flow Resistance," in *Gravel-bed Rivers*, R. D. Hey, J. C. Bathurst, and C. R. Thorne (Eds.), New York: John Wiley and Sons, (1982) 875 pp.
33. Bayazit, M., "Free Surface Flow in a Channel of Large Relative Roughness," *Jour. of Hydraulic Research*, 14:2 (1976) 115–126.

CHAPTER 10

THREE-DIMENSIONAL DEEP-WATER WAVES

Ming-Yang SU

Naval Ocean Research and Development Activity
NSTL, Mississippi USA

CONTENTS

INTRODUCTION, 236

INSTABILITY, 237
Results of Experiments, 237
Comparisons with Theory, 243

BIFURCATIONS, 245
Structures of Skew Wave Patterns, 246
Structures of Symmetric Wave Patterns, 249
Interpretation and Comparison with Theory, 250

APPLICATIONS TO OCEAN SURFACE WAVES, 254
Short-Crestedness and Directional Spreading, 254
Wave Breaking in Deep Water, 255
Giant Waves, 255
Bubble Generation, 255

CONCLUDING REMARKS, 255

REFERENCES, 256

INTRODUCTION

Surface gravity waves, traditionally since Stokes [1], have been studied as a two-dimensional fluid phenomenon with two spatial coordinates: one pointed horizontally in the wave propagation direction and, the other pointing vertically in the earth's gravitational direction. The orbital motion of water particles associated with these waves is described in the two-dimensional coordinates; hence these waves are two-dimensional. (Such waves are often called, in literature, one-dimensional as well, apparently referring to their property of single propagational direction.) In order to avoid confusion in this chapter, we shall refer to such waves as two-dimensional (2-D) waves. Swells in lakes and oceans during windless or very light-wind conditions undoubtedly have inspired the traditional mathematical (2-D) idealization (or model). On the other hand, surface waves under active wind forces invariably exhibit random, short-crested appearances which require, in principle, two horizontal coordinates plus one vertical coordinate to describe them, thus these types of waves will be called three-dimensional (3-D) waves.

Since surface waves of small steepness have the remarkable property that they can pass each other without mutual interaction, and that they are dispersive (i.e. waves of different frequencies move at different phase speeds), the random, three-dimensional feature of wind-generated waves has been traditionally considered as a natural consequence of linearly superposition of many 2-D simple sinusoidal waves with different propagation directions, frequencies, and phases [2]. This linear mathematical model for random ocean waves does provide good first-order approximation for many statistical characteristics of real ocean waves. The usage of the linear model in both theoretical studies and engineering applications [3] is so prevailing in the last three decades, as to create a very strong impression on general readers on its scientific truth and general applicability.

During the same period when the linear 3-D wave models are being developed and applied, some other fundamentally important research on 2-D surface waves is being explored, concerned with more accurate computations of steep Stokes waves [4] and their instability. These investigations of nonlinear finite-amplitude 2-D gravity waves lead to surprisingly rich possible variations of surface waves, as described in a recent comprehensive review by Yuen and Lake [5]. It is along the same line of research on nonlinear wave dynamics that several new intrinsic 3-D properties of surface gravity waves on deep water have been most recently discovered, experimentally and theoretically. The main purpose of this chapter on 3-D deep-water waves is to describe these 3-D properties, and to point out their relevance to real ocean waves.

Briefly speaking, all of these 3-D properties have been initially evolved from a single train or packet of 2-D steep waves due to instabilities and bifurcations of the latter. These properties are intrinsically three-dimensional in nature in the sense that they are not produced by linear superposition of 2-D sinusoidal waves. The experimental observations and measurements of these 3-D waves are described first, which are followed by available theoretical intepretations. The aspects of instabilities are given in the following section and bifurcations in the section after that. Applications of these 3-D properties to actual ocean surface waves are then pointed out, and finally, some concluding remarks are made on this most recent development in the highly nonlinear, three-dimensional dynamics of surface gravity waves.

INSTABILITY

For a Stokes wave train of small (but finite) steepness, Benjamin and Feir [6] discovered a sideband instability. The long-time evolution of a two-dimensional wave train after the initial instability of Benjamin-Feir type was investigated experimentally by Lake et al. [7].

For finite-amplitude wave trains, a comprehensive, two-dimensional linear stability analysis was given by Longuet-Higgins [8, 9]. Concurrently, Longuet-Higgins and Cokelet [10–12] made accurate numerical computations of the evolution of steep two-dimensional wave trains. These computations show that the subharmonic instabilities ultimately cause every alternate crest to develop a fast-growing local instability that quickly leads to wave breaking.

McLean et al. [13] and McLean [14] showed theoretically that there are two main types of instabilities, designated as Types I and II respectively. Type I is the subharmonic instabilities, which occur roughly in the range $0 < ak < 0.38$. Type II is the much faster-growing instabilities that were first discovered by Longuet-Higgins [8, 9] in the two-dimensional case, when $ak \geq 0.40$.

We shall describe below the findings of experiments on the evolution of initially two-dimensional steep gravity-wave trains in deep water. For the values of steepness of wave trains in the range of $0.25 \leq ak \leq 0.34$, it is found that the subharmonic instablities cause alternate crests to develop a relatively fast-growing local instability, leading to breaking [15].

Results of Experiments

General Characteristics of Wave Evolution

We consider a representative example in which a wave train is generated with a plunger stroke of 5.1 cm, the basic frequency $f_o = 1.55$ Hz and the wavelength $\lambda_o = 65$ cm. The wave steepness $(ak)_o = 0.32$ measured at $x = 6.1$ m $(9\lambda_o)$. For reference, Figure 1 depicts the overall characteristics of the wave patterns in the evolution of the steep wave train.

Figure 2 shows the wavemaker and the wave patterns up to $x = 15$ m $(23\lambda_o)$. The first few waves are seen to contain small superharmonic disturbances, which grow in size and height. These perturbations are three-dimensional. The amplitudes of these perturbations reach their maxima at about

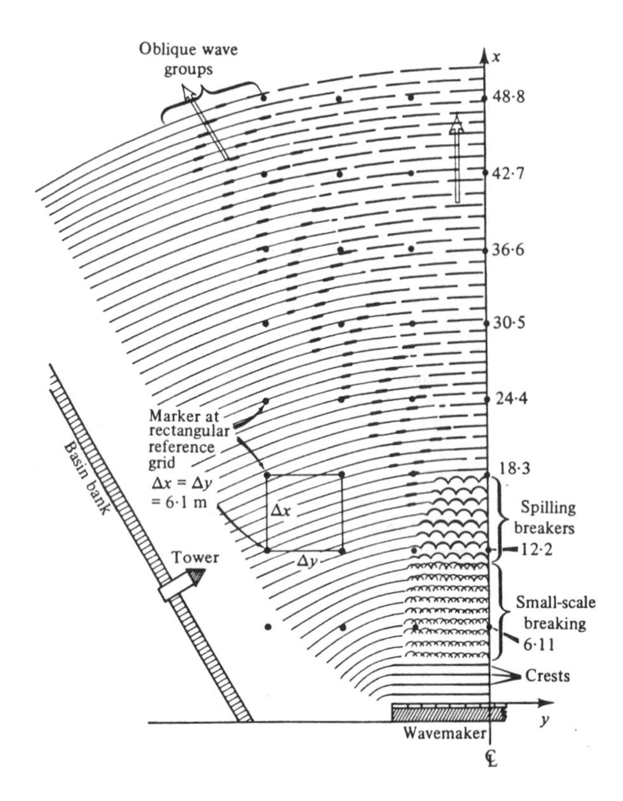

Figure 1. A sketch of overall characteristics of wave patterns for evolution of a steep wave train in the wave basin; $f_o = 1.55$ Hz, $(ak)_o = 0.32$.

Figure 2. The wavemaker and the wave patterns in the wave basin up to $x = 15.5$ m with $f_o = 1.55$ Hz, $(ak)_o = 0.32$. The picture is taken from a tower at 12 m above mean water surface. The expansion angle is $\theta = 17°$.

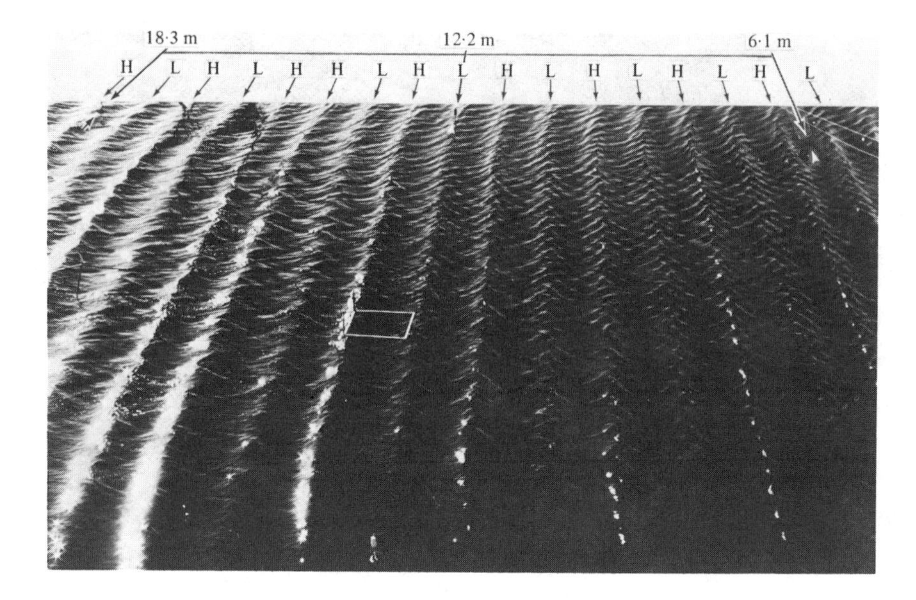

Figure 3. The same wave patterns as in Figure 2 but taken at a much lower sun angle to show two-dimensional subharmonic instabilities.

$10\lambda_o$ from the wavemaker, where some small-scale breaking is obvious on crests of the longer primary waves.

Figure 3 shows the same wave pattern as Figure 2, but the angle of incidence of the sunlight is lower; the presence of subharmonic modulation is more apparent under this condition. Starting from x = 6.1 m ($9\lambda_o$), we see that glitter created by superharmonic perturbations is accentuated at the higher wave crests, which alternate with lower, smoother crests for about ten wavelengths of the primary waves. The three-dimensional characteristics of subharmonic modulation can be clearly seen from about $10\lambda_o$ to $30\lambda_o$ of the wavemaker.

Figure 4 is a close-up view of the three-dimensional spilling breakers due to subharmonic instabilities. These breakers occur around $15 \text{ m} \leq x \leq 25 \text{ m}$.

In the next phase (Figure 5) the three-dimensionality of the patterns diminishes greatly, and long-crested waves dominate. We note that at the beginning of this phase a breaking wave appears on every third or fourth crest. Later the higher crests appear on the fifth, sixth, or seventh waves.

A new phenomenon appears to be related to the residues of the transition of the three-dimensional spilling breakers to the essentially two-dimensional wave forms. An example of the phenomenon is given in Figure 6, where a series of wave groups with about eight waves per group are shown propagating obliquely at an angle about 30° away from the basic wave direction; in the figure this phenomenon occurs along a strip extending from the lower right corner to the upper left. This feature persists for a distance of more than one hundred wavelengths.

Similar three-dimensional instabilities have also been observed in narrow wave tanks. For details, refer to Su et al. [15] and Melville [16].

The observations of the general qualitative characteristics of wave evolution just described can be presented more quantitatively by wave profiles measured at various distances away from the wavemaker. Figure 7 shows such wave profiles along the centerline of the wave tank from x = 6.1 m ($9\lambda_o$) to x = 106.7 m ($164\lambda_o$). The wave profile at x = 6.1 m ($9\lambda_o$) exhibits very slow and slight modulation

Figure 4. A close-up of three-dimensional spilling breakers in the wave basin around x = 15 m.

Figure 5. Wave patterns in the wave basin after the wave breaking.

Figure 6. Wave patterns at the final stage of the evolution starting in Figure 2. The oblique wave groups due to three-dimensional wave breaking are propagating from the lower right corner to the upper left corner of the photograph.

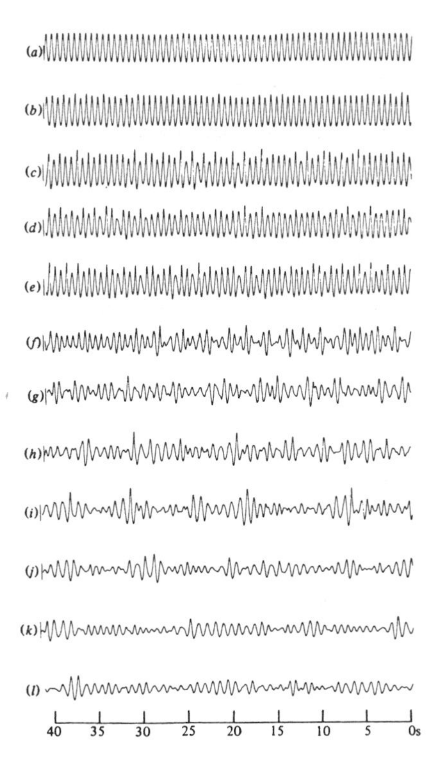

Figure 7. Wave profiles at various stages of evolution with $f_o = 1.55$ Hz, $(ak)_o = 0.318$; at locations from the wavemaker: (a) 6.1 m; (b) 12.2 m; (c) 15.25 m; (d) 18.6 m; (e) 24.4 m: (f) 30.5 m; (g) 36.6 m; (h) 48.8 m; (i) 61.0 m; (j) 76.25 m; (k) 91.5 m; (l) 106.75 m.

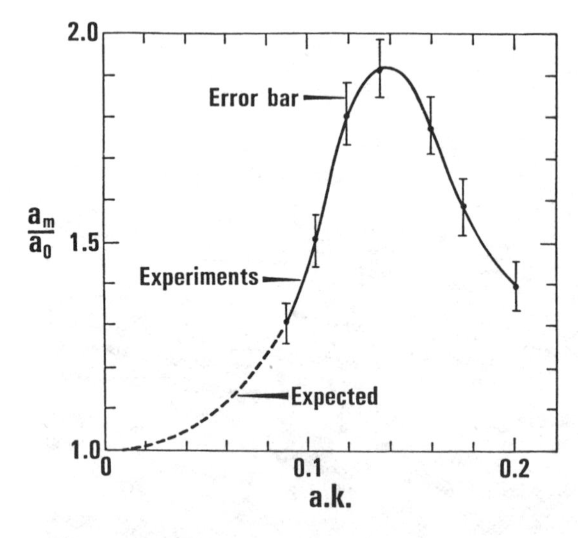

Figure 8. Amplification factors of wave amplitude with respect to the initial wave steepness a_ok_o.

with a periodicity of about six or more wavelengths. The wave profile at x = 12.2 m ($19\lambda_o$) shows more rapid modulations with obvious pairs of consecutive higher and lower waves. These types of subharmonic modulations intensify greatly in the next wave profile at x = 15.25 m ($24\lambda_o$), and continue to x = 24.4 m ($38\lambda_o$). These higher waves correspond to the three-dimensional spilling breakers.

The average steepness for these breakers (ak $\approx$ 0.42) is larger than the initial value (ak = 0.32), but is still smaller than that of the Stokes limiting waves (ak = 0.443).

From x = 30.5 m ($47\lambda_o$) to x = 36.6 m ($56\lambda_o$), wave breaking usually occurs every third or every fourth wave. From x = 48.8 m (75 λ_o) to 61.0 m ($94\lambda_o$), waves break less often. At this stage wave crests extend across the width of the wave tank. These breaking waves are nearly two-dimensional and are different in structure from the distinctly three-dimensional breakers occurring between x = 18.6 m ($29\lambda_o$) and x = 24.4 m ($38\lambda_o$).

As the wave train evolves further, near-breaking waves are observed only occasionally, and the distance between them also increases. Finally, the profiles at x = 91.5 m (141_o) and 106.75 m (164_o) exhibit modulational characteristics of a series of wave groups. The typical steepness of these waves is ak $\approx$ 0.15, which is only about half the initial steepness ak = 0.32 at x = 6.1 m ($9\lambda_o$). Furthermore, it can be noted from these wave profiles that the average wave period is longer than the period at x = 6.1 m. Detailed spectral analysis showed that the ratio of the initial frequency at x = 6.1 m, (f_0), to the final peak frequency at x = 106.75 m, (f_1), is $f_0/f_1 \approx \frac{4}{3}$.

Interactions of Two- and Three-Dimensional Instabilities

The experimental results of wave evolution are described here as if the two- and three-dimensional instabilities were not coexistent and not subject to interactions. These phenomena are coincident, although their growth rates differ according to a_ok_o. In fact, both types of instability influence each other during the evolution of wave trains and packets.

One particularly important effect of the interactions of the instabilities is found in the relation of initial steepness (a_ok_o) to relative growth of wave amplitude. Figure 8 shows the variation of the amplitude amplification factor, defined as (a_m/a_o), for $0.09 \leq a_ok_o \leq 0.20$, where a_m is the largest wave amplitude observed at the maximum modulation in each run of experiments at a fixed a_ok_o, with initial amplitude a_o. (a_m/a_o) is the relative maximum wave height growth in the wave evolution due to the Benjamin-Feir type instability.

A remarkable feature of the observed variation (Figure 8) is that (a_m/a_o) reaches a maximum value; $(a_m/a_o) \approx 1.9$ for $a_ok_o \approx 0.14$. At the stage of maximum amplitude of the wave envelopes $a_ok_o \times a_m/a_o = 0.27$ $(0.14 < a_ok_o < 0.2)$, i.e., the "effective steepness" (a_mk_o) reaches its maximum. This "effective steepness" puts the steepest waves into a range $(a_ok_o > 0.25)$ in which the three-dimensional instability limits the amplitudes of two-dimensional disturbances. The impact of the three-dimensional instability is clearly demonstrated in Figure 8. For wave packet modulations, Su [17] reported similar observations for $a_ok_o > 0.14$.

In short, the three-dimensional instability appears to be enhanced in the steeper waves by the presence of the two-dimensional instability. Breaking dissipation accompanying the evolution of the crescent-shaped patterns, which results from three-dimensional instability, appears to limit the amplitude attained by the two-dimensional instability.

Comparisons with Theory

A theoretical and numerical analysis by McLean et al. [13] and McLean [14] shows that there exists a new type (called Type II) of instabilties, which are predominantly three-dimensional, in contrast with the Benjamin-Feir (called Type I) instabilities, which are predominatly two-dimensional in nature. Type II are weaker than Type I when $ak \leq 0.3$, but become much stronger when $0.3 < ak < 0.44$. The two-dimensional extremely fast-growing instability when $ak > 0.40$, first discovered by Longuet-Higgins [8], is a special case of the Type II instability. Figure 9 shows two stability diagrams computed by McLean [14] for $ak = 0.30$ and 0.33 respectively. It should be

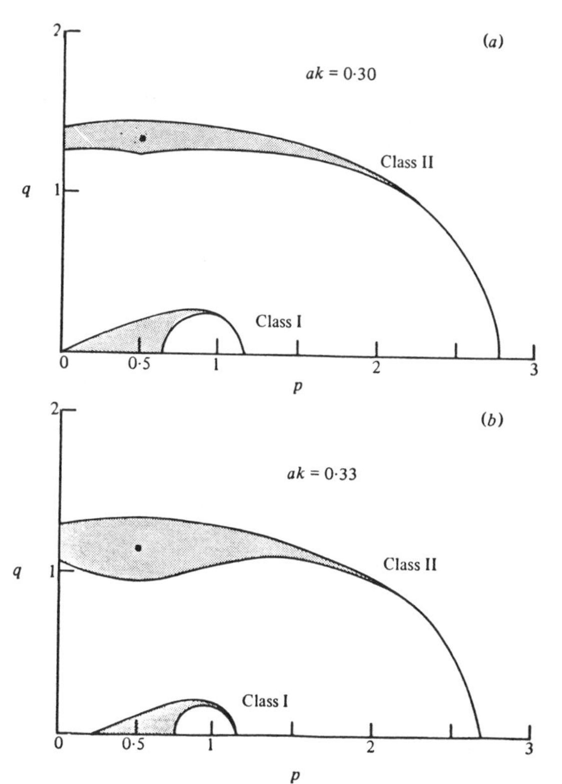

Figure 9. Instability regions of deep-water waves of finite amplitude. • labels the point of maximum instability [24] (Figure 2C and D) p and q are the perturbation wavenumbers along and transverse to the basic unperturbed two-dimensional waves.

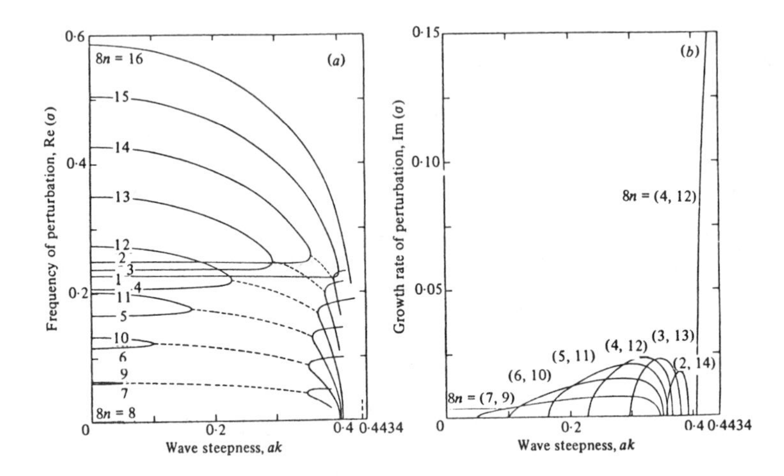

Figure 10. Frequencies (A) and growth rates (B) of subharmonic instabilities when m = 8 [9].

noted that the maximum growth rate for Type II instability occurs always at $p = 0.5$, i.e. with a perturbation wavelength equal to two unperturbed basic wavelengths. For $0.3 \leq ak \leq 0.33$, the corresponding q is $1.15 \leq q \leq 1.33$, i.e. the crestwise perturbation wavelength is slightly shorter than the basic wavelength. It seems to be most likely that the three-dimensional spilling breakers are due to the three-dimensional Type II instabilities discussed by McLean et al. [13] and McLean [14].

Longuet-Higgins [9] has given a two-dimensional linear-perturbation analysis for all normal modes of instabilities of arbitrary wave steepness for deep-water gravity waves. This analysis includes subharmonic perturbations, with wavenumbers less than the fundamental waves. We shall use his example of $m = 8$; i.e. the perturbations with a repetition distance of eight fundamental wavelengths as one of the theoretical results. Figure 10 is extracted from his paper to show the real part of perturbation frequencies and corresponding growth rates respectively. In Figure 10A a normal mode is denoted by $8n = 1, 2, 3, \ldots, 15, 16$ which refers to the number of perturbed wavelengths in a period of eight unperturbed waves. In Figure 10B a pair of normal modes that have the same growth rate is denoted by $8n = (r, s)$, with $r + s = 16$.

The mode of Type I subharmonic instability underlying the frequency downshifting of $f_o/f_1 \approx \frac{4}{3}$ is $8n = (4, 12)$ in Figure 10, which is unstable for $0.23 < ak < 0.36$, and which has a maximum growth rate at $ak = 0.32$. This prediction is qualitatively consistent with the experimental observations in which the alternate waves growth higher and lead to three-dimensional breaking. Subsequently, the $f_0/f_1 \approx \frac{4}{3}$ frequency downshifting is observed.

A typical example of the long-time evolution of an initially uniform steep wave train with $f_0 = 3.25$ Hz and $(ak)_0 = 0.23$ was given by Lake et al. [7]. Their Figures 5 and 6 show the evolution of wave profiles and corresponding power spectra. The ratio of the carrier frequencies for the initial and final stages in this example is about 1.3. That is to say, there is as much as 25% downshift in the peak frequencies. Lake et al. [7] and Yuen and Lake [18] have attributed the phenomenon of seemingly returning to a more or less uniform wave train to be a water-wave analogy of the Fermi-Pasta-Ulam recurrence phenomenon. On the other hand, no explanation for the definite frequency downshifting has been offered by those authors.

Experimental results by Su et al. [19] have not only confirmed earlier observations by Lake et al., but, more importantly, have established and identified the phenomenon of frequency downshifting as one of the most important characteristics of strong nonlinear interactions in the gravity waves in deep water.

BIFURCATIONS

Saffman and Yuen [20] have predicted the existence of two new types of three-dimensional permanent waveforms, resulting from bifurcations of a uniform two-dimensional Stokes wave on deep water. The first type of bifurcation produces a steady symmetric wave pattern propagating in the same direction as the Stokes waves. A computation of the waveforms for several configurations of the symmetric wave patterns is given by Mieron, Saffman, and Yuen [21]. The second type produces the steady skew wave patterns that propagate obliquely from the direction of the Stokes waves. We shall present experimental evidence for the existence and structures of the bifurcated symmetric and skew wave patterns. The skew wave patterns are found to occur when the steepness of the initial Stokes waves is $0.16 \lesssim a_o k_o \lesssim 0.18$, where a_o and k_o are, respectively, the wave amplitude and wavenumber of the initial Stokes waves.

The symmetric wave patterns are found to be most evident when

$$0.25 \lesssim a_o k_o \lesssim 0.34,$$

and are triggered by the three-dimensional instability. Three configurations of symmetric wave patterns with different subharmonic lengths from two to four basic wavelengths have been found.

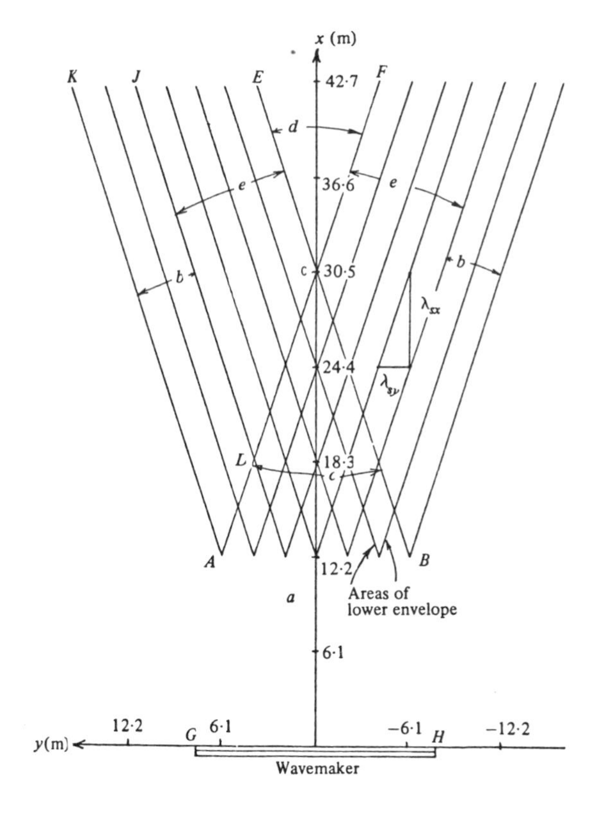

Figure 11. A sketch of the overall three-dimensional wave patterns observed in the experiments. (a) Stokes waves (ABHG); (b) skew wave patterns (ADJK); (c) interactions of skew wave patterns (ABC); (d) Benjamin-Feir modulations of Stokes waves (ECF); (e) Benjamin-Feir modulations of skew waves (JDCE).

Structures of Skew Wave Patterns

Since it is found that the skew wave patterns show up when $0.16 \lesssim a_o k_o \lesssim 0.18$, we shall use a typical example with $a_o k_o = 0.17$ to illustrate typical characteristics of the wave patterns observed in our experiments. For clarity in presenting the rather complicated three-dimensional wave patterns, the entire wave field is divided into the following five regimes:

1. Stokes waves.
2. Skew wave patterns.
3. Interactions of skew wave patterns.
4. Benjamin-Feir modulations of Stokes waves.
5. Benjamin-Feir modulations of skew waves.

Locations of these five regimes are shown in Figure 11. For the example with $a_o k_o = 0.17$, the other wave parameters used are $f_o = 1.24$ Hz, $a_o = 2.7$ cm, $\lambda_o = 0.95$ m and $c_o = 1.25$ m/s.

Stokes Waves

The first regime is composed of uniform two-dimensional waves which are generated by the mechanical wavemaker. The transient characteristics associated with the particular wave-making process die out exponentially. Thus, at the location $x = 6.1$ m, $y = 0$, where the initial reference waves are measured, the waves can be considered close to Stokes waves.

Skew Wave Patterns

Figure 12 shows a photographic image of four skew wave patterns that propagate to the right side of the x-axis. Figure 12B is an annotated line drawing corresponding to Figure 12A. The area with lighter tone has waves of lower wave heights than the area with darker tone. Figure 13 shows a clearer view of three-dimensional waveforms of the skew wave groups from a tower. The waves between two lighter tone areas (i.e. lower wave height) will be called a skew wave pattern, and each pattern propagates at an angle Ψ with respect to the direction (x-axis) of the initial Stokes wavetrain. The separation distances between two light-tone bands along the x- and y-axes will be called, respectively, the normal and the crestwise wavelengths of the skew wave patters λ_{xs} and λ_{ys} (Figure 11). The traveling velocity of the skew wave pattern normal to the Ψ direction will be called the pattern velocity C_{ps} of the skew wave pattern. From the experimental measurements for the initial wave steepness $0.16 \lesssim a_o k_o \lesssim 0.18$, it is found that:

$$15^\circ < \Psi < 20^\circ$$

$$2.5\lambda_o < \lambda_{ys} < 3.5\lambda_o$$

$$\tfrac{1}{35}C_o < C_{ps} < \tfrac{1}{65}C_o$$

Hence the propagation angle Ψ of the skew wave group varies only slightly, with a mean $\Psi = 18.5^\circ$, while the crestwise wavelength has a relatively larger variation. The pattern velocity of the skew wave patterns is seen to be about one-fiftieth of the phase velocity of the initial Stokes waves. Thus the skew wave patterns appear almost stationary to an observer standing on the wavemaker.

Interactions of Skew Wave Patterns

Referring to Figure 11 in the converging cone-shaped area defined by the triangle ABC, we can see a checkerboard pattern that results from interactions between two sets of the skew wave patterns

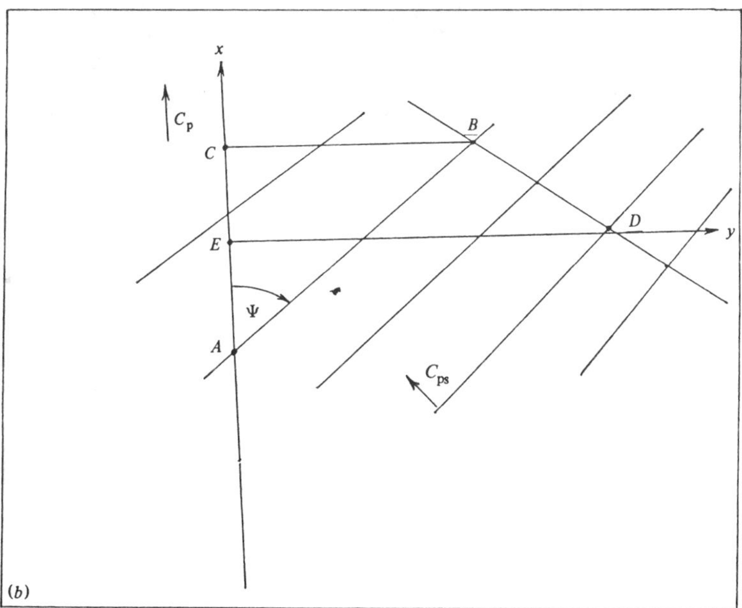

Figure 12. (A) Image of skew wave patterns with $a_o k_o = 0.17$, $f_o = 1.25$ Hz, $\lambda_o = 1.03$ m; (B) annotation of (A) for determining the propagating direction and λ_{sz} and λ_{sy}.

Figure 13. Close-up images of skew wave patterns for $a_o k_o = 0.17$.

which propagate in the different directions ($\pm\Psi$). This wave pattern is an actual example of a narrow-band two-dimensional wavenumber spectrum described by Longuet-Higgins [10] (Figure 1). We also observed that the wave heights in the center region of each diamond-shaped wave pattern are higher than both the wave height of the initial Stokes waves and the "pure" skew wave patterns. Furthermore, two sets of the skew wave paterns emerging after the mutual interactions seem to retain their pattern shape without obvious destructive interference.

Benjamin-Feir Modulations of Stokes Waves

Referring to the diverging cone-shaped (ECF in Figure 11) centered on the x-axis, and y $\gtrsim$ 36.6 m ($35\lambda_o$), we can see predominantly two-dimensional wave-envelope modulations in the direction of the original Stokes waves. These modulations are caused by the two-dimensional side-band instabilities [6, 11].

Figure 14. Three-dimensional symmetric waves of L_2 configuration observed in the wide wave basin; $f_o = 1.2$ Hz, $a_o k_o = 0.33$.

Benjamin-Feir Modulations of Skew Waves

Referring to Figures 11 and 12, in the region overlapped by the pure skew waves and the Benjamin-Feir modulations of Stokes waves, we can see a series of compact three-dimensional wave groups which are results of interactions between the above two characteristically different waveforms. These compact wave groups assume shapes of a skewed diamond. The normal wavelength is the same as for the Benjamin-Feir modulation of Stokes waves, while the crestwise wavelength is the same as for the "pure" skew wave patterns.

Structures of Symmetric Wave Patterns

For clarity of the description of observations which follow, we define the characteristic scales of the wave field. The wavetrain is initially composed of waves with length λ_o, amplitude a_o, wave number $k_o = 2\pi/\lambda_o$ and frequency f_o. The three-dimensional symmetric waves have crestwise wavelength λ_{BC} and their subharmonic periodicity is $L_m = m\lambda_o$ ($m = 2, 3, 4, \ldots$). The initial wave steepness is $a_o k_o$.

Figure 14 shows a typical example of crescent-shaped symmetric waves in the wide basin ($1 \times 100 \times 340$ m) for the conditions $f_o = 1.2$ Hz, $a_o k_o = 0.33$, $\lambda_o = 1.08$ m and $\lambda_{BC} = 0.915$ m.

Only three distinct spatial configurations of symmetric waves have been observed. The first and most frequent configuration is denoted as $L_2 = 2\lambda_o$, and is sketched in Figure 15, corresponding to Figure 14. Note that the crests are shifted by one-half of the width of the crescents ($1/2\lambda_{BC}$) on successive rows. I and II in Figure 12B are typical profiles observed at $y = 0$ and $y = 1/4\lambda_{BC}$, i.e. at crescent center and quarter-width respectively. The corresponding maxima (crests) and minima (troughs) occur at P_i ($i = 1, 2, \ldots, 8$). The trough at P_4 is deeper than the trough at P_2. The wave

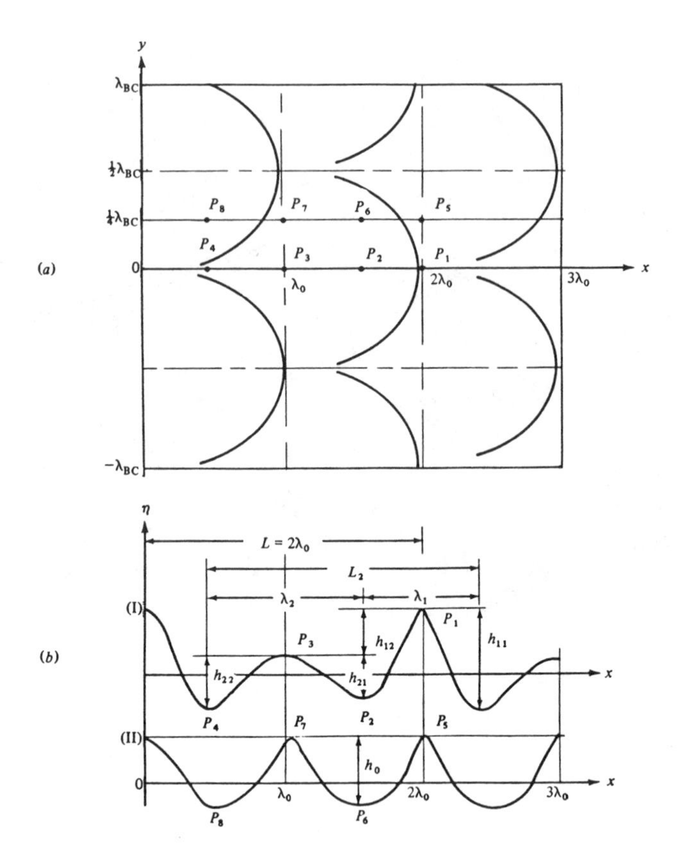

Figure 15. Definition sketch of L_2 configuration of symmetric waves: (A) top view of wave crests; (B) wave profiles along y = 0 (I) and along y = $1/4\lambda_{BC}$ (II).

profile II in Figure 15 exhibits less distortion than the profile I, and is similar to the form measured close to the wavemaker.

The second configuration of the symmetric waves characterized by the subharmonic scale $L_3 = 3\lambda_o$ is sketched in Figure 16. Figure 16A is the top view of the crests. The corresponding typical profiles for y = 0 and y = $1/2\lambda_{BC}$ are represented respectively by I and II in Figure 16B. Note that every third row of symmetric waves has a $1/2\lambda_{BC}$ crestwise shift with respect to the other two rows.

The third configuration $L_4 = 4\lambda_o$ occurs very infrequently and is described in Su [15]. Figure 17A and B present the surface displacement and local surface slope for the L_2 configuration of symmetric waves.

A rough estimate based on experimental measurements indicates that the L_2 configuration occurs more than 90% of the time, and that the L_3 and L_4 configurations have less than 10% and 1% incidence of occurrences, respectively.

Interpretation and Comparison with Theory

We shall now give an interpretation of the skew and symmetric waves by comparing the observations with the theoretical computations by Saffman and Yuen [20] and Meiron et al. [21], who

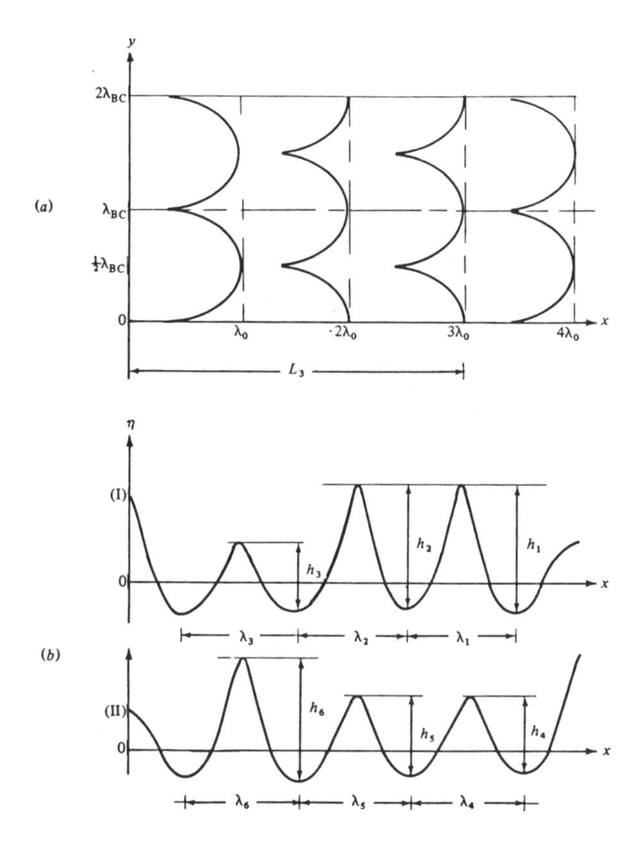

Figure 16. Definition sketch of L_2 configuration of symmetric waves: (A) top view of wave crests; (B) wave profiles along y = 0 (I) and along $1/4\lambda_{BC}$ (II).

predict the existence of such waves based on the Zakharov equation, which is valid for weakly nonlinear waves [22], and with the new type of three-dimensional instability discussed by McLean et al. [13] and McLean [14].

Three examples of the free surface corresponding to the bifurcated skew waves for $p = \lambda_o/\lambda_{sx} = 0.1$, $q = \lambda_o/\lambda_{sy} = 0.4$, and $b_1 = 0$, 0.2 and 0.4 are given by Lake and Yuen [5]. The example with $b_1 = 0$ corresponds to regular uniform Stokes waves, while the latter two examples correspond to bifurcated skew waves of increasing degree of bifurcation. The propagation direction for these cases is

$$\Psi = \arctan\frac{p}{q} = \arctan\frac{0.1}{0.4} = 14^\circ$$

The free surfaces resemble the skew wave patterns shown in Figure 12. Further, a critical wave steepness for the bifurcation for these examples (from Figure 1 of Saffman and Yuen [20]) is 0.20. This steepness is close to (but slightly larger than) the range of steepness $0.16 \lesssim a_o k_o \lesssim 0.18$ found to be most favorable for occurrence of the phenomenon.

We could improve the comparison by understanding a little better the computations by Saffman, Yuen, and Meiron. For this purpose, we have reproduced Figure 2 of Saffman and Yuen [20] in our Figure 18, which gives the critical bifurcation-wave steepness as a function of propagation

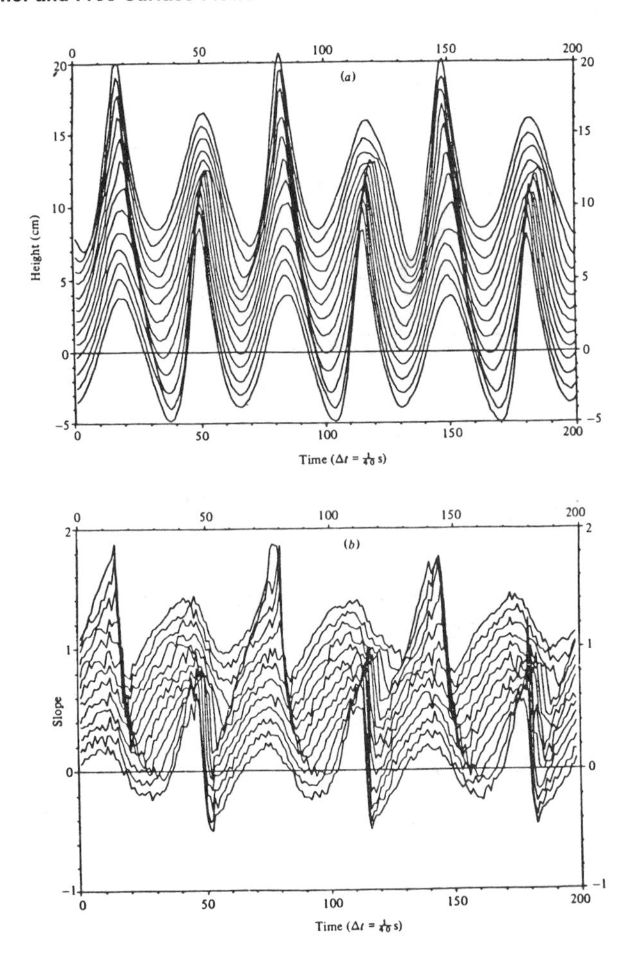

Figure 17. (A) Plots of wave profiles for L_2 configuration of symmetric waves; (B) plots of surface slopes corresponding to (A).

direction Ψ. Note that point A marked in Figure 18 corresponds to the example cited above. Also plotted are two shaded regions where the skew waves and symmetric waves have been predominantly observed in the experiments. For the skew waves:

$$15^\circ < \Psi < 20^\circ, \qquad 0.16 < ak < 0.18, \qquad 0.12 < p < 0.14, \qquad 0.3 < q < 0.5.$$

For the symmetric waves:

$$23^\circ < \Psi < 27^\circ, \qquad 0.25 < ak \leq 0.33, \qquad p = 0.5, \qquad 1.1 < q < 1.2.$$

The critical bifurcation-wave steepness for $p = 0.5$, 0.1 and 10^{-5} are all over-predicted for large ak; the values of steepness for $\Psi = 90^\circ$ (i.e. $q = 0$) as shown by points E, F, and G are well above the Stokes limiting steepness 0.443 (point H). On the other hand, the entire curve for $p = 0$ based on Whitham's theory [23], and the two points B and C for $q = 0$ based on the exact water-wave equation [13] are theoretically accurate. If we use as the necessary constraints the prior, known correct limiting points, and the general features of the curves by Saffman, Yuen, and Meiron for the critical bifurcation steepness, we might expect the curves for the critical bifurcation wave steepness based on the exact water-wave equation for $p = 0.5, 0.13$, and 0.10 to resemble the dashed lines sketched. These expected theoretical predictions then agree quantitatively with our experimental observations for both bifurcated skew waves and symmetric waves, as far as the critical wave steepness and the propagation direction are concerned.

The Saffman-Yuen-Meiron theory has built in it as assumption that the component of the phase velocity of the steady skew wave in the fundamental wave direction is the same as that of Stokes waves. In other words, these skew waves are phase-locked to the stokes waves. On the other hand, the observed skew waves patterns show only a small, yet finite, forward movement relative to the wavemaker at about one-fiftieth of the phase speed of Stokes waves.

Next, we consider the stability of the skew waves to infinitesimal perturbations, as demonstrated theoretically by Saffman, Yuen, and Meiron. What they have shown is that there is no change of stability with respect to infinitesimal disturbances near the bifurcation point. Since the Stokes waves are known to be unstable to subharmonic side-band instabilities of Benjamin-Feir type, the bifurcated skew waves should be subject to the same instabilities, as demonstrated clearly in our experiments.

The bifurcation theory (Saffman-Yuen-Meiron) for the skew waves (similarly for symmetric waves) does not provide information on the selection rules for the preferred values of p and q, which determine the spatial scales λ_{sx} and λ_{sy} of the skew wave pattern. The example in Saffman and Yuen [20] (Figure 4) for $p = 0.1$ and $q = 0.4$ cited earlier has been chosen based on some earlier experimental observations provided by our earlier investigation. Such selection rules may be related to the three-dimensional instability discovered theoretically by McLean et al. [24], using a linear perturbation analysis of the exact water-wave equation. According to this theory the most unstable mode occurs at $p = 0.5$ and $q \simeq 1.6$ for $0.1 < a_o k_o < 0.2$. The corresponding propagation direction of this mode is

$$\Psi m = \arctan \frac{0.5}{1.6} = 17.4^\circ$$

which is close to the preferred propagation direction of the skew wave patterns,

$$\Psi_s = \arctan \frac{1}{3} = 18.5^\circ$$

Figure 18 shows that the theoretical propagation direction of the skew waves is centered around 18.5° and is relatively insensitive to variation of wave steepness in the range $0.1 < ak < 0.20$. The three-dimensional unstable mode has a normal component of phase velocity equal to the Stokes waves. However, the crestwise wavelength of this unstable mode is only about one quarter ($0.4/1.5 = 1/4$) of the observed skew waves. Because of this discrepancy, we feel that the observed skew waves could not be explained satisfactorily by the just-mentioned theories.

An important difference between the Type I and II instabilities may be stressed. The perturbation frequency of the Type I is not small. The modulational envelope due to the subharmonic instabilities travels with the group velocity, which is roughly half the phase velocity. Hence each wave group displays about twice as many waves in a temporal record as in a spatial record (such as recorded on a photograph). In Type II instability, the perturbation frequency is small in both the two-dimensional case [9] and the three-dimensional case [13, 14]. Hence each wave group in the Type II instability travels with the basic phase velocity, and their variation in space looks roughly the same as their variation in a time record. So when in a time record we see alternately high and low crests, it is probably a Type II instability. But if we see wave groups with four or more waves, it is probably a Type I instability. Thus, it seems to be of no doubt that the three-dimensional Type II instability causes the L_2 configuration of symmetric waves.

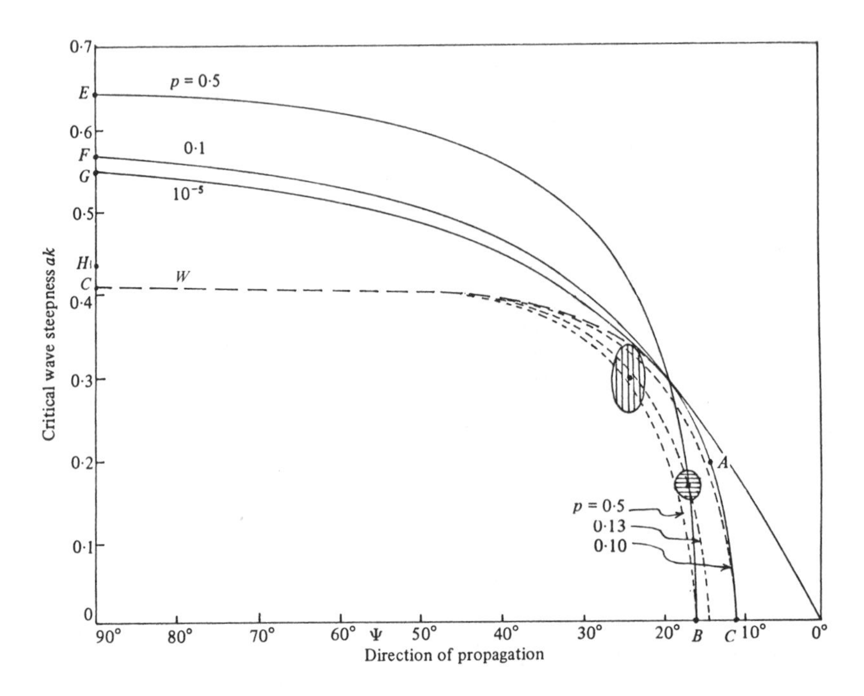

Figure 18. Critical bifurcation-wave steepness as a function of propagation angle; solid lines from Saffman and Yuen [20], based on Zakharov's equation; the curve marked W (long dashed line) represents the neutral stability curve obtained from Whitham's theory, which is exact for p, q approaching zero; short dashed lines show the expected "correct" theoretical prediction; horizontally shaded area shows the region of experimental data for the skew wave patterns.

APPLICATIONS TO OCEAN SURFACE WAVES

Short-Crestedness and Directional Spreading

Deep ocean waves, particularly in growing seas, are usually short crested; the crestlength is on the order of the wavelength. The short crests can be attributed to modulations produced by waves traveling in different directions, instabilities and bifurcations. Nature probably allows all of these phenomena a range of admixtures, but the laboratory experiments give some sharp focus on wave instabilities and bifurcations. The added acuity allows us to see that even small perturbations on waves of moderate ak create rapidly growing instabilities. The compounding of two- and three-dimensional instabilities, and bifurcations create short-crested waves.

Oceanic waves are not generated by a coherent source, nonetheless the perturbations will be present and the most unstable modes will amplify. These two-dimensional growth processes will modulate the waves to form groups containing waves steep enough to accelerate the growth of the three-dimensional instability and, finally, breaking. Weak resonannt interactions possibly complement wave instabilities. The laboratory results give clear evidence that bifurcations and three-dimensional instabilities lead to breaking waves and spreading of energy away from the primary wave direction.

In growing seas dominant waves may often have ak = 0.14 to 0.18, the range in which modulational instabilities grow rapidly, so it appears reasonable to surmise that the ocean wave instabilities lead to short-crestedness. The generation of short-crested waves spreads the directional power spectrum.

The experiments also show that three-dimensional wave breaking may be an additional source of directional spreading of wave energy.

Wave Breaking in Deep Water

We have been impressed by the apparent similarity of the spilling breakers seen in the experiment and white-capping deep sea breaking waves. In the laboratory we found that wave breaking is the result of three-dimensional instabilities and symmetric bifurcations. Analyses of storm waves records show that many waves have sufficient ak to trigger the sequence of these types of instabilities and bifurcations.

More indirect evidence of the physical similarity of these processes comes from the group properties of storm waves [17]; it appears that wave groups contribute to the most energetic bands of the spectra. The probabilities of occurrence of these groups exceeds that of a narrow-band Gaussian process. From this we surmise that modulational instabilities and bifurcations tend to reinforce phase locking of wave components; this also exists in ocean waves and leads to wave breaking. Donelan, Longuet-Higgins, and Turner [25] observed that oceanic wave breaking appears to happen most often at periods twice the dominant wave period. This is consistent with the experimental results of the three-dimensional breaking waves in the most frequent configuration of the symmetric bifurcation.

Giant Waves

We now add another example of evidence for strongly nonlinear wave evolution. This concerns the so-called "giant" waves encountered in the Aghulas Current and other regions of the oceans. These types of abnormally high waves have been described by Mallory [26] and Hamilton [27]. Results of our experiment that appear to bear on this phenomenon are the observations of rapid growth of wave height due to modulational instability. The giant waves appear to be in small groups that arise as a result of rapid changes in the wind or current field through which large amplitude swells arise. Swell entering the Aghulas Current from the south (with $ak = 0.08$) increases in steepness as it interacts with the opposing current. The increase in steepness to $ak = 0.12$ is sufficient to bring the swell into a range where rapid growth of modulational instabilities creates wave packets containing some large steep waves.

In the laboratory we found that the modulational instability could create waves with heights two times those of the initial waves. Intense modulation can also occur as a result of rapid and intense change of wind speed and direction in the presence of swells, such as in the case of the passage of a squall line [27]. The abrupt wind change rapidly generates waves that may increase the steepness of the initial wave field by the process of long-short wave interactions. These instabilities could be the sources of anomalous waves. Although the connections of the laboratory and oceanic processes are tenuous at this stage, the roles of nonlinear instabilities and bifurcations in ocean wave evolution seem to be important.

Bubble Generation

Bubbles are generated by wave breaking in storms. These bubbles are known to persist in the water much longer than the mean wave period. Experiments on bubble generation by the three-dimensional symmetric bifurcated waves [28] show that the three-dimensional wave breaking produce a string of vortices of bubble clouds with their axes nearly normal to the mean water surface. This new experimental finding may explain the longer lifetime of oceanic bubbles.

CONCLUDING REMARKS

The investigation on three-dimensional deep-water waves—their instabilities and bifurcations—is a new chapter of the long history of surface wave study, as evidenced by the citations in the preceding presentation. The problem of three-dimensional water waves is much more complicated than that of two-dimensional water waves, not only because of one additional dimension, but also because of the additional instability and nonlinear interactions. These added complexities increase substantially the difficulty in both experimental measurements and theoretical analyses. The properties described in this chapter are only the initial results obtained so far, much more detailed experiments and computations are needed to fully explore their complete characteristics and implications.

The study of three-dimensional waves is, at the same time, fundamentally important, since natural wind-generated waves in lakes and oceans are intrinsically three-dimensional. Particularly steep waves in storm seas with frequent wave breaking, which are most important considerations for marine operations, cannot be understood without considering nonlinear three-dimensional instabilities and bifurcations. Several examples cited earlier provide further applications.

The three-dimensional instability and bifurcation are not restricted to deep-water waves, but also exist in shallow-water waves. Actually, the Type II three-dimensional instability is enhanced as the water depth decreases. The interested reader is referred to papers by McLean [29] and Su et al. [30] for further details.

REFERENCES

1. Stokes, G. G., "On the Theory of Oscillatory Waves," *Trans. Camb. Phil. Soc.*, Vol. *8* (1847), pp. 441–55.
2. Pierson, W. J., Neumann, G., and James, R. W., "Observations and Forecasting Ocean Waves," U.S. Naval Oceanographic Office, H.O. Pub. No. 603 (1960).
3. Ochi, M. K., "Stochastic Analysis and Probabilitic Prediction of Random Waves," *Adv. in Hydroscience*, Vol. 13 (1982), pp. 217–375.
4. Schwartz, L. W., and Fenton, J. D., "Strongly Nonlinear Waves," *Ann. Rev. Fluid Mech.*, Vol. 14, pp. 39–60.
5. Yuen, H. C., and Lake, B. M., "Nonlinear Dynamics of Deep-Water Gravity Waves," *Advances in Applied Mechanics*, Vol. 22 (1982), pp. 67–229, Academic Press.
6. Benjamin, T. B., and Feir, J. E., "The Disintegration of Wave Trains on Deep Water, Part 1, Theory," *J. Fluid Mech.*, Vol. 27 (1967), pp. 417–430.
7. Lake, B. M., et al., "Nonlinear Deep-Water Waves: Theory and Experiment, Part 2, Evolution of a Continuous Wave Train," *J. Fluid Mech.*, Vol. 83 (1977), pp. 49–74.
8. Longuet-Higgins, M. S., "The Instabilities of Gravity Waves of Finite Amplitude in Deep Water, I. Superharmonics," *Proc. R. Soc. Lond.*, A 360 (1978a) pp. 417–458.
9. Longuet-Higgins, M. S., "The Instability of Gravity Waves of Finite Amplitude in Deep Water, II. Subharmonics," *Proc. R. Soc. Lond.*, A 360 (1978b) pp. 489–505.
10. Longuet-Higgins, M. S., and Cokelet, E. D., "The Deformation of Deep Surface Waves on Water, I. A Numerical Method of Computation," *Proc. R. Soc. Lond.*, A 350 (1976), pp. 1–36.
11. Longuet-Higgins, M. S., and Cokelet, E. D., "The Deformation of Deep Surface Waves on Water, II. Growth of Normal-Mode Instabilities," *Proc. R. Soc. Lond.*, A 364 (1976), pp. 1–36.
12. Longuet-Higgins, M. S., "On the Nonlinear Transfer of Energy in the Peak of a Gravity-Wave Spectrum: A Simplified Model," *Proc. R. Soc. Lond.*, A 347 (1976), pp. 331–328.
13. McLean, J. W., et al., Three-Dimensional Instability of Finite Amplitude Water Waves, *Phys. Rev. Lett.*, Vol. 46 (1981), pp. 817–820.
14. McLean, J. W., "Instabilities of Finite-Amplitude Water Waves," *J. Fluid Mech.*, Vol. 114 (1982a), pp. 315–330.
15. Su, M.-Y., "Three-Dimensional Deep-Water Waves, Part 1, Experimental Measurement of Skew and Symmetric Wave Patterns," *J. Fluid Mech.*, Vol. 124 (1982a) pp. 73–108.
16. Melville, J., "The Instability and Breaking of Deep-Water Waves," *J. Fluid Mech.*, Vol. 115 (1982), pp. 165–185.
17. Su, M.-Y., "Evolution of Groups of Gravity Waves with Moderate to High Steepness," *Phys. Fluids*, Vol. 25, No. 12 (1982b), pp. 2167–74.
18. Yuen, H. C., and Lake, B. M., "Instabilities of Waves on Deep Water," *Ann. Rev. Fluid Mech.*, Vol. 12 (1980), pp. 303–334.

19. Su, M.-Y., et al., "Experiments on Nonlinear Instabilities and Evolution of Steep Gravity-Wave Trains," *J. Fluid Mech.*, Vol. 124 (1982a) pp. 45–72.
20. Saffman, P. G., and Yuen, H. C., "A New Type of Three-Dimensional Deep-Water Waves of Permanent Form," *J. Fluid Mech.*, Vol. 101 (1980), pp. 797–808.
21. Meiron, D. I., Staffman, P. G., and Yuen, H. C., "Calculation of Steady Three-Dimensional Deep-Water Waves," *J. Fluid Mech.*, Vol. 124 (1982), pp. 109–121.
22. Zakharov, V. E., "Stability of Periodic Waves of Finite Amplitude on the Surface of a Deep Fluid," *J. Appl. Mech. Tech. Phys.*, Vol. 2 (1968), pp. 190–194.
23. Peregrine, D. H., and Thomas, G. P., "Finite Amplitude Deep Water Wave on Currents," *Phil. Trans. R. Soc. Lond.*, A 292 (1979), pp. 371–390.
24. McLean, J. W., "Instabilities of Finite-Amplitude Gravity Waves on Water of Finite Depth," *J. Fluid Mech.*, Vol. 114 (1982b), pp. 331–341.
25. Donelan, M., Longuet-Higgins, M. S., and Turner, J. S., "Periodicity in White Caps," *Nature*, Vol. 239 (1982), pp. 449–450.
26. Mallory, J. K., "Abnormal Waves on the South East Coast of South Africa," *Int. Hydrographic Rev.*, Vol. 51, No. 2 (1974), pp. 99–129.
27. Hamilton, G. D., "Buoy Capsizing Wave Conditions," *Mariners Weather Log*, Vol. 24, No. 3 (1980), pp. 165–173, D. Reidel Pub. Co.
28. Su, M.-Y., Green, A. W., and Bergin, M. T., "Experimental Studies of Surface Wave Breaking and Air Entrainment," *Gas Transfer at Water Surfaces*, W. Brutsaert and G. H. Jirka (Eds.), 1984, pp. 211–219.
29. Su, M.-Y., et al., "Experiments on Shallow-Water Grouping and Breaking," *Proc. First Int. Conf. on Meteorology and Air/Sea Interaction of the Coastal Zone*, The Hague, Netherlands, 1982b.
30. Phillips, O. M., *The Dynamics of the Upper Ocean*, Cambridge University Press, 1977.

CHAPTER 11

ESTIMATING PEAK FLOWS

B. P. Sangal

National Hydrology Research Institute
Ottawa, Canada

CONTENTS

INTRODUCTION, 259

EMPIRICAL FORMULAS, 259
Myers Formula, 260
Creager's Formula, 260
Rational Formula, 260

UNIT HYDROGRAPH APPROACH, 262
Unit Hydrograph Derived from Streamflow Records, 262
Synthetic Unit Hydrograph, 262
The S-Hydrograph, 265
Design Flood, 267

PEAKS FROM URBAN AREAS, 268
Rational Method, 268
Overland Flow Hydrograph Methods, 269
Hydrologic Simulation Models, 269

SNOWMELT FLOODS, 269
Snowmelt Equations, 269
Degree-Day Factor, 271
Snowmelt Flood Hydrograph, 271

HYDRAULIC METHODS, 272
Slope-Area Method, 272
Flow Through Contractions, 273
Flow Over Dams, 273

FLOOD ROUTING, 273
Modified Puls or Storage Indication Method, 273
The Muskingum Method, 274
Estimating Travel Time, 275

FLOOD FREQUENCY ANALYSIS, 276
Probability and Return Period, 276
Return Period and Risk, 277
Probability Distributions, 277
Frequency Factors, 281
Plotting Positions, 281
Single-Station Frequency Analysis, 282
Regional Frequency Analysis, 285

ESTIMATING PEAK FROM MEAN DAILY FLOWS, 286
Relation of Peak to Maximum Mean Daily Flow, 287
Suggested Approach, 290
Application of Method to Streams in Canada, 292
Application of Method to Streams in the United States, 299

ACKNOWLEDGEMENT, 307

APPENDIX A: IDENTIFICATION OF CANADIAN DATA USED IN FIGURE 11, 307

APPENDIX B: IDENTIFICATION OF ALL U.S. DATA EXCEPT HAWAII, 309

NOTATION, 311

REFERENCES, 312

INTRODUCTION

This chapter deals with one of the most important topics in engineering hydrology. Peak discharge data are needed in the design of hydraulic structures and other water planning and management projects such as floodplain mapping. Large sums of money are expended throughout the world on flood control measures every year. Such expenditures cannot be justified if design flows are not properly estimated. A record of past floods is an invaluable aid in making design flow estimates. However, such records are not always available, especially for project sites. Thus, various other methods involving physiographic and meteorologic data are used to gather needed data.

Systematic measurements of river flows have been made only for about 100 years although records of much longer period exist for some major rivers in the world. The network of recording gauges giving accurate data on peak flows has been expanding only recently. In many developing countries where the need for water resource development is greatest, this network is still meager and large sums of money are spent on projects designed on the basis of almost nonexistent data.

The title of this chapter is broad, and a detailed description of all the theoretical and practical methods currently available would be an enormous task and could constitute a handbook in itself. The purpose herein is to describe briefly some practical methods. Great emphasis has been placed on estimating peak from mean daily flows and results of an extensive analysis of Canadian and United States data have been presented. Other topics which have been covered include empirical formulas, the unit-hydrograph approach, peaks from urban areas, snowmelt floods, hydraulic methods, flood routing, and flood frequency analysis. All methods presented generally refer to those being followed in Canada and the United States. Methods being used in other countries have been omitted simply due to the author's lack of knowledge and experience in those techniques. Both English and metric systems of units have been used as necessary. The symbols and notation of the original publication have been retained. Although dual symbols sometimes had to be used for the same parameter, there is no chance of any confusion arising from this duplication.

EMPIRICAL FORMULAS

Many empirical formulas have been developed over the years to estimate flood peaks. The number of variables which affect flood flow is so large that it is almost impossible to consider them all. Most formulas consider the drainage area as the only variable and lump the effect of other variables into its coefficients and exponents. Detailed descriptions of these formulas are given in several publications [1–4]. The three main forms are

$$Q = CA^{n} \tag{1}$$

$$Q = CA^{mA^{-n}} \tag{2}$$

$$Q = CA[(a + bA)^{-m} + c] \tag{3}$$

where Q = peak flow rate
A = drainage area

C, a, b, c are coefficients and m and n are exponents.

Various countries have developed these formulas in different forms. The exponents and coefficients derived in one country or even in one region of the country are generally not applicable to other countries or other regions with different climatic and physiographic characteristics. The two most widely used formulas involving envelope curves of peak flows are due to Myers and Creager [5].

Myers Formula

The Myers formula as modified by Jarvis is

$$Q = 100pA^{0.5} \quad (4)$$

where Q = the peak discharge in cubic feet per second (cfs)
p = the percentage rating on the Myers scale
A = size of drainage area in sq. mi.

Creager's Formula

The Creager formula is

$$Q = 46CA^{0.894A^{-0.048}} \quad (5)$$

or its equivalent

$$q = 46CA^{(0.894A^{-0.048})-1} \quad (6)$$

where Q = the peak flow in cfs
A = the drainage area in sq. mi.
C = Creager's coefficient
q = peak flow in cfs per sq. mi.

A curve with $C = 100$ enveloped most of the floods considered by Creager.

Rational Formula

This is the most important of the empirical formulas and is given by

$$Q = ciA \quad (7)$$

where Q = peak flow in cfs
c = runoff coefficient
i = rainfall intensity (in./hr) of a storm whose duration is equal to the time of concentration of the basin
A = drainage area in acres

The formula takes advantage of the fact that one acre-inch per hour is very nearly equal to 1 cfs (1.008 cfs exactly) and assumes that for rainfall exceeding the time of concentration the rate of runoff equals the rate of rainfall reduced by an appropriate factor. The peak flow for any return period is determined by using a value of i having the corresponding frequency, assuming that the frequency of rainfall and the frequency of flood are the same. The success of the formula depends

Table 1
Values of Runoff Coefficient c

Topography and Vegetation	Open Sandy Loam	Clay and Silt Loam	Tight Clay
Woodland			
Flat (0%–5% slope)	0.10	0.30	0.40
Rolling (5%–10% slope)	0.25	0.35	0.50
Hilly (10%–30% slope)	0.30	0.50	0.60
Pasture			
Flat	0.10	0.30	0.40
Rolling	0.16	0.36	0.55
Hilly	0.22	0.42	0.60
Cultivated			
Flat	0.30	0.50	0.60
Rolling	0.40	0.60	0.70
Hilly	0.52	0.72	0.82
Urban areas	30% impervious	50% impervious	70% impervious
Flat	0.40	0.55	0.75
Rolling	0.50	0.65	0.80
Streets, roofs and other paved areas		0.70–0.95	

upon the accuracy with which the values of c, i, and the time of concentration are determined. The applicability of this formula is limited to small basins not exceeding 100 to 200 acres. A great deal of experience is required in determining the runoff coefficient c properly. Its value varies from as high as 0.95 for paved surfaces to as low as 0.1 for flat grassy fields or woodlands with highly permeable soil. Some representative values are given in Table I [6, 1].

The time of concentration T_c is defined as the time taken for storm runoff to travel from the most remote point of the basin to the point of interest. It is also the time of equilibrium at which the rate of runoff becomes equal to the rate of rainfall, assuming c = 1. Proper determination of the time of concentration is extremely important in applying the rational formula. Various formulas are in use for determining T_c. The most common one is the Kirpich formula [7]

$$T_c = 0.00013(L^2/S)^{0.385} \tag{8}$$

where T_c = time of concentration, hr
L = length of the longest watercourse, ft
S = slope of the stream (ft/ft) or the total fall divided by the total length

In its equivalent form used by the U.S. Soil Conservation Service (SCS) [8]

$$T_c = \left(\frac{11.9L^3}{H}\right)^{0.385} \tag{9}$$

where L = length of the longest watercourse, mi
H = total fall in ft

The reliability of this formula decreases as the size of the drainage area increases.

The value of rainfall intensity i is obtained from the intensity-duration-frequency curves for a nearby weather station. The design frequency has to be selected depending upon the type of project.

Despite its many deficiencies, the rational formula is still used in many urban design problems and estimating peaks from small basins. The quality of the results depends upon the experience of the designer.

UNIT HYDROGRAPH APPROACH

In 1932 L.K. Sherman [9] proposed the well-known theory of unit hydrographs. The unit hydrograph (originally named unitgraph) of a drainage basin is defined as a hydrograph of direct runoff resulting from 1 in. of effective rainfall generated uniformly over the basin area at a uniform rate during a specified period of time or duration. The unit hydrograph theory is based on the following assumptions:

1. The effective rainfall is uniformly distributed within its duration or specified period of time.
2. The effective rainfall is uniformly distributed over the whole basin.
3. The base or time duration of the hydrograph of direct runoff due to an effective rainfall of unit duration is constant.
4. The ordinates of direct runoff hydrographs of a common base time are directly proportional to the total amount of direct runoff represented by each hydrograph. This is also called the principle of linearity.
5. For a given drainage basin, the hydrograph of runoff due to a given period of rainfall reflects all the combined physical characteristics of the basin.

Under natural conditions, these assumptions cannot be satisfied perfectly. The patterns of rainfall and runoff, effect of physiographic characteristics of the basin, channel and basin storage, and man-made structures, all tend to distort the ideal conditions. However, the unit hydrograph has proved a very useful tool in engineering hydrology. If applied judiciously, it can give reliable results. Methods of its derivation and its use in estimating design floods are now discussed.

Unit Hydrograph Derived from Streamflow Records

A unit hydrograph can be derived from an observed hydrograph. In selecting a hydrograph for such an analysis, the assumptions involved in unit hydrograph theory should be carefully considered. A hydrograph resulting from an isolated, intense, short duration storm of nearly uniform distribution in space and time is most desirable. The observed hydrograph and the corresponding rainfall hyetograph averaged over the basin are plotted. The estimated base flow is subtracted from the total hydrograph and the hydrograph of only the surface runoff is obtained. This runoff is computed and an equal amount of effective rainfall is marked on the hyetograph, thus isolating the infiltration portion. Several parameters such as time base of hydrograph, time to peak, lag time defined as the time from the centroid of rainfall excess to the peak of hydrograph, etc. can be obtained from this graph. The ordinates of the hydrograph are divided by the total runoff in inches to obtain the ordinates of the unit hydrograph. This hydrograph corresponds to the duration of effective rainfall which is then designated as unit duration for the derived unit hydrograph. Sometimes it takes a long period of records to isolate a good hydrograph and corresponding rainfall. Therefore, a derived unit hydrograph is an extremely valuable tool. Sometimes, the unit hydrograph is expressed in dimensionless form with base length in units of time to peak and discharge as fractions of peak discharge.

Synthetic Unit Hydrograph

If the project site is ungauged or the recorded data are such that direct derivation of a unit hydrograph is not possible, then the unit hydrograph must be synthesized from the physiographic characteristics of the drainage basin. Two methods which are commonly used for this purpose are due to Snyder [10] and the United States Soil Conservation Service (SCS) [11]. These methods follow.

Snyder's Method

Snyder was the first to develop a procedure for synthesizing unit hydrographs. He analyzed a large number of hydrographs from drainage basins in the Appalachian Mountain region in the United States and developed the following equations:

$$t_p = C_t(LL_c)^{0.3} \tag{10}$$

$$t_r = t_p/5.5 \tag{11}$$

$$Q_p = (640C_pA)/t_p \tag{12}$$

$$T = 3 + 3t_p/24 \tag{13}$$

where t_p = lag time, hr
t_r = unit duration, hr
Q_p = unit peak, cfs
T = base length, days
C_t = lag coefficient varying from 1.8 to 2.2
C_p = peak coefficient varying from 0.56 to 0.69
L = length of main stream from outlet to divide, mi.
L_c = distance from the outlet to a point on the stream nearest the centroid of the basin, mi.
A = area of the watershed, sq. mi.

Equations 10 to 13 fully define a unit hydrograph for a duration t_r. For any other duration t_R the lag is

$$t_{pR} = t_p + \frac{t_R - t_r}{4} \tag{14}$$

This modified lag is used in Equations 12 and 13 to determine unit peak and base length.

SCS Method

The method of hydrograph synthesis presently used by the U.S. Soil Conservation Service and detailed in *Soil Conservation Service National Engineering Handbook* [11] is due to Mockus [12] developed in 1957. This method is also used by the U.S. Bureau of Reclamation, and the procedure is given in *Design of Small Dams* [8]. The method is based on a simple consideration that a hydrograph can be represented by a triangle.

Consider Figure 1. From this figure we have,

$$q_i = 2/(1 + H) \cdot Q/T_p \tag{15}$$

If H = 1.67 and A is the drainage area in sq. mi., then

$$q_p = 484AQ/(D/2 + L) \tag{16}$$

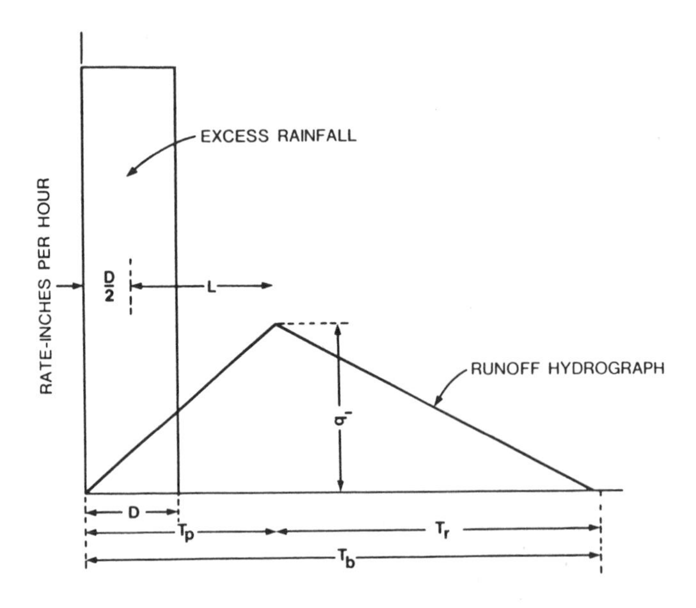

Figure 1. Schematic representation of unit graph.

where Q = total runoff, in.
q_i = peak rate, in./hr
T_p = time to peak, hr
T_r = time from peak to end of triangle, hr
$H = T_r/T_p$
q_p = peak rate, cfs
D = unit duration, hr
L = lag time, hr
T_b = base length, hr

The lag L is given by

$$L = 0.6T_c \tag{17}$$

where T_c is the time of concentration. It is computed by Equation 9, the Kirpich formula.

Channel Slope

Sometimes it has been found useful to establish a relationship between the observed value of lag and the parameters $LL_c/\sqrt{S}$ or $L/\sqrt{S}$ where L and L_c have the same meanings as given earlier and S is the slope of the main channel. This plots as a straight line on a log-log paper. Two methods of calculating S are presently used. These are the average slope method and the equivalent slope method.

In the average slope method, the channel slope is equal to the slope of a line which is drawn in such a way that the area under it is equal to the area under the stream profile [2]. If the profile is concave, then the starting point for the line is the outlet. However, if the profile is convex, then the starting point should be the upper end, otherwise the line will end at an elevation higher than that of the divide and give too steep a slope.

The equivalent slope method has been suggested by Taylor and Schwartz [13]. The equivalent slope is defined as the uniform slope which would produce the same travel time for the flow as that given by the actual profile. The channel is divided into n segments of equal length, and slope of each segment is computed as the difference in elevation divided by the length of the segment. The slope of the stream channel is given by

$$S = (n/\sum S_i^{-0.5})^2 \tag{18}$$

where S_i is the slope of the i-th segment. This method is considered better than the average slope method although very steep and very flat sections create problems in this method. A modification of the method is

$$S = (L/\sum L_i S_i^{-0.5})^2 \tag{19}$$

where L_i and S_i are the length and the slope of the i-th segment and L is the total stream length. As the channel profile is normally obtained from contour maps, slopes of the individual segments are easy to compute and very steep sections with short lengths have less effect on the slope.

If the stream drains into a large body of water, then the lower section can have a very flat slope for a long distance. This has the effect of considerably lowering the slope of the main channel in both the previously mentioned methods. In such a case, the whole flat section or at least the part likely to be affected by backwater under high flood conditions should be left out. The hydrograph of the upper section should be routed through this section.

The S-Hydrograph

If a unit hydrograph for duration t_0 hr is added to itself lagged by t_0, the resulting hydrograph represents the hydrograph for 2 in. of runoff in $2t_0$ hr. If the ordinates of this hydrograph are divided by 2, the result is a unit hydrograph for duration $2t_0$ hr. This example illustrates the ease with which a unit hydrograph of a short duration can be converted to a unit hydrograph of any integral multiple of the original duration. However, this technique cannot be used for the conversion of a long duration hydrograph to one of short duration. A general method of derivation applicable to unit hydrographs of any required duration is the S-hydrograph, or summation hydrograph method. It is also called the S-curve method and is as follows:

If a continuous effective rainfall at a rate of I in. per hour is considered, then the ultimate or equilibrium discharge from a watershed of A acres would eventually be equal to AI acre-inch per hour. This would occur after a time interval T_c called the time of concentration or the time of equilibrium of the basin. The hydrograph assumes the shape shown in Figure 2A. Such a hydrograph can be obtained by summation of a series of unit hydrographs of duration 1/I hr each lagged 1/I hr with respect to the preceding one. After the S-hydrograph is constructed, the unit hydrograph of any duration can be obtained. The S-hydrograph is advanced or offset by a time period equal to the desired duration of t_0 hr (Figure 2B). If this offset hydrograph is subtracted from the initial hydrograph, the difference between ordinates results in a hydrograph which contains a volume equal to I t_0 in. (Figure 2C). If each ordinate is divided by I t_0, then the resulting hydrograph will contain a volume equal to 1 in. due to a rainfall duration of t_0 hr or it would be a t_0-hr unit hydrograph (Figure 2D). The base length of this hydrograph will be equal to $T_c + t_0$ hr.

The principle of the S-hydrograph is further applied in Figure 3. Using a triangular unit hydrograph of 1 hr duration with base length of 8 hr, lag time 2.5 hr, time to peak equal to 3 hr and unit peak equal to 1 (cfs or m^3/sec.), an S-hydrograph is constructed which attains equilibrium flow of 4 (area of triangle divided by the unit duration) after a time interval of 7 hr, which is the time of concentration. By lagging this hydrograph 1, 4, and 7 hr, the original 1-hr and 4-hr and 7-hr unit hydrographs have been obtained. It may now be noted that the lag times of 4-hr and 7-hr hydrographs are 3 hr and 3.5 hr respectively. That is, the lag time continues to increase until a maximum of one-half the time of concentration is attained. No further increase is possible. It shows that lag time cannot exceed $0.5T_c$. A relationship between the lag time and the time of

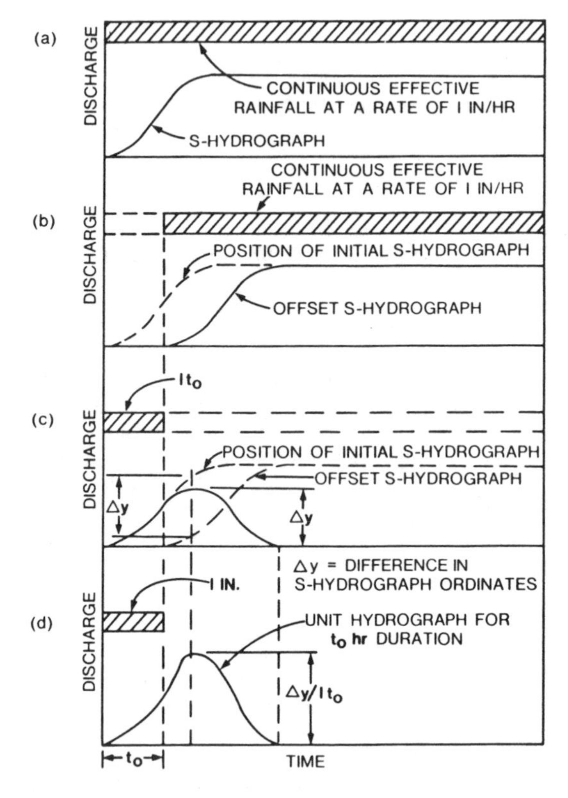

Figure 2. Derivation of unit hydrograph by the S-hydrograph method [1].

concentration can be obtained. From Figure 1,

$$T_c + D = (1 + H)(L + D/2) \tag{20}$$

or

$$L = (T_c + D)/(1 + H) - D/2 \tag{21}$$

If H = 1.67, then

$$L = 0.375T_c - 0.125D \tag{22}$$

If H = 1, then

$$L = 0.5T_c \tag{23}$$

Thus, only if the initial triangular hydrograph is an isosceles triangle will the constancy of lag time equal to $0.5T_c$ be maintained. In this case Chow's peak reduction factor [14] is equal to D/(0.5D + L)[15]. In all other hydrographs the lag time will continue to increase from an initial value given by Equation 21 to $0.5T_c$. The time of concentration is an invariable quantity and is a fundamental property of a drainage basin. The practical use of $L = 0.6T_c$ [8] is justified only if the computed value of T_c is considered low. This could as well be true for the Kirpich formula when applied to larger basins.

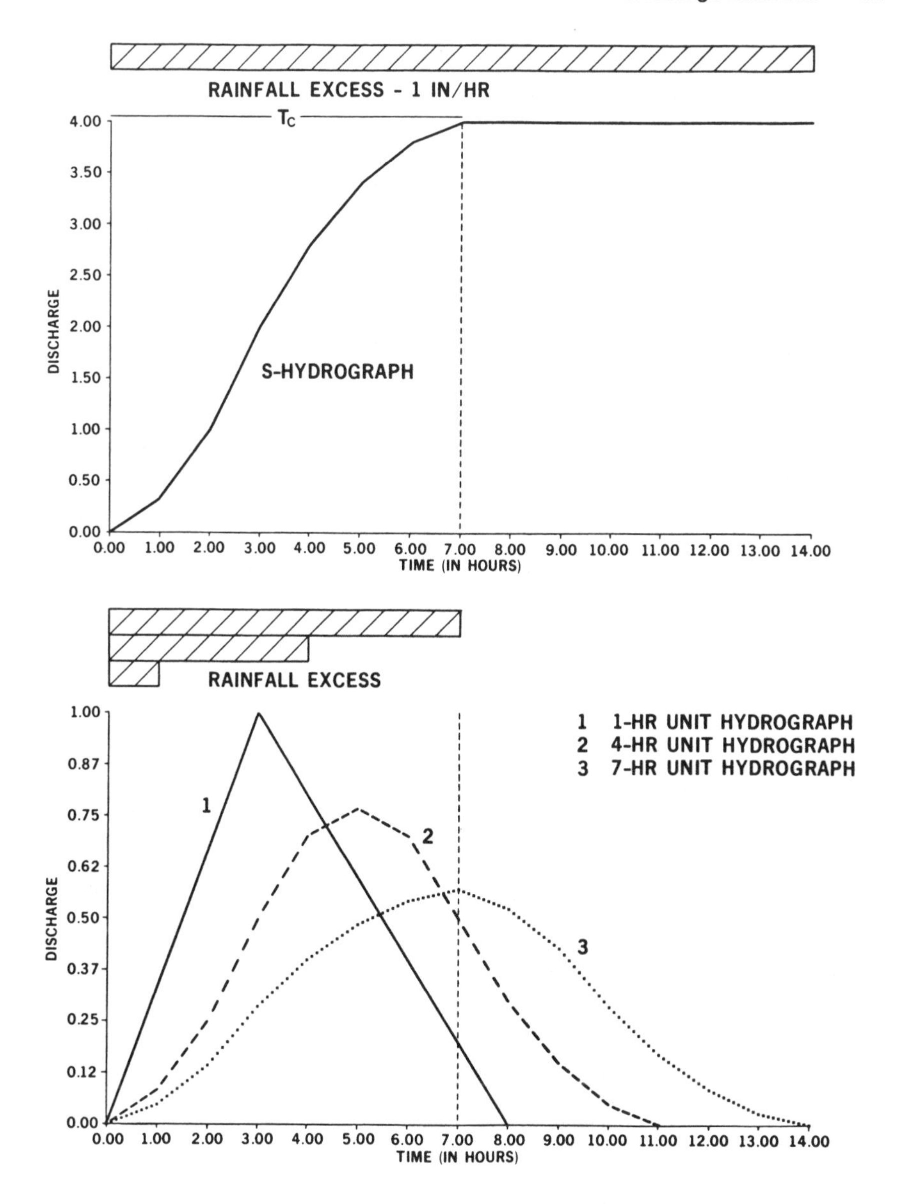

Figure 3. Application of the S-hydrograph method.

Design Flood

The ultimate use of a unit hydrograph is in estimating design flood for a water resource project. The process involves the selection of a design storm, estimation of runoff, and construction of a final design hydrograph. Small projects can be designed on the basis of a locally observed severe storm or storm of a given return period, say 100 yr or 50 yr. Large projects, however, require considerations such as the class of area to be protected, degree of acceptable risk, economic limitations, etc. For such projects, the U.S. Army Corps of Engineers and the U.S. Bureau of Reclamation [8] use two types of flood estimates:

1. Maximum probable flood
2. Standard project flood

Detailed discussion of these floods is provided by Dalrymple [1]

Maximum Probable Flood

The maximum probable flood (MPF) has been defined as the largest flood for which there is any reasonable chance of occurrence in a given region. This is a very large flood and is mainly used for the design of spillways of large reservoirs where failure could lead to great damage and loss of life. It represents the most severe combination of critical meteorologic and hydrologic conditions that are reasonably possible in the region. The flood hydrograph is synthesized using maximum probable precipitation estimates. These estimates have been prepared for the whole of the United States [8] and selected regions of Canada [6]. Detailed procedures are available for converting precipitation estimates to flood discharges. In the eastern United States, the point estimates of maximum probable precipitation for a duration of 6 hr vary from about 33 in. in the southern part to about 15 in. in the northern part. In the western United States due to orographic effects, the rainfall pattern is extremely complex. The point values for 6 hr storm vary from more than 10 in. in coastal and mountainous regions to less than 3 in. in desert areas.

Standard Project Flood

The standard project flood (SPF) represents the flood discharge that may be expected from the most severe combination of meteorologic and hydrologic conditions that are considered characteristics of the region, excluding extremely rare combinations. This flood is selected as design flood for the project where some small degree of risk can be accepted but an unusually high degree of protection is justified by hazards to life and high property values within the area to be protected.

The SPF may be defined as a flood hydrograph resulting from standard project storm. This storm represents the most severe rainfall considered reasonably characteristic of the region in which the drainage area is located. In those areas where snowmelt could be a factor, appropriate allowance is made for snowmelt runoff. Detailed procedures have been provided for determining this flood [16]. The standard project storm estimates vary from 40 to 60 percent of the maximum probable precipitation estimates. Approximately the same relation holds between MPF and SPF.

PEAKS FROM URBAN AREAS

Design flood estimates for urban areas are required for stormwater drainage and design of airports and expressways. The procedures involved are complex. For detailed study the reader is referred to several publications [17–20]. A comprehensive review of the topic is given by Jens and McPherson [1]. Briefly, the methods can be classified into three groups:

1. Rational method
2. Overland flow hydrograph methods
3. Hydrologic simulation models

Rational Method

This method makes use of the rational formula described earlier in this chapter. It is limited to estimation of peak only and cannot be used for projects where total volume of runoff and flood duration have to be considered. Considerable judgment is required in estimating the runoff coefficient c in this method.

Overland Flow Hydrograph Methods

Included in this group are methods which have been developed by some large metropolitan centers, such as Los Angeles [21] and Chicago [22, 23]. The basic relationships developed by Izzard [24] for overland flow have been used in these methods. A complete design hydrograph is obtained. Various types of data which are required in these methods include rainfall intensity-frequency-duration maps, overland flow characteristics of soil, rainfall-runoff relations, routing parameters, etc. The success of the methods depends upon the quality of these basic data.

Hydrologic Simulation Models

In recent years, due to computer facilities, hydrologic simulation models have been developed to a point where they provide an invaluable tool for the design of complex urban drainage problems. These models permit consideration of stormwater quality and quantity, effects of detention in various parts of the watershed, and many other variables and design options.

A hydrologic simulation model is a mathematical description of the response of a surface water system to the physical processes affecting the system. Two models which are frequently used in Canada and the United States are the Storm Water Management Model (SWMM) and the Illinois Urban Drainage Area Simulator (ILLUDAS).

SWMM was originally developed by the U.S. Environmental Protection Agency [25]. It is a deterministic model and can be easily transferred from one basin to another. Inputs to the model are rainfall intensity, watershed data, and data on the transporting and receiving system. This model has been modified in Canada and a user's manual has been prepared [26].

ILLUDAS is a simpler model and requires comparatively less data [27]. It simulates runoff from connected and unconnected paved areas and grassy fields. This model permits the analysis and design of both drainage and storage systems.

Several other models which are useful in urban design problems have been developed in many countries. A comparison of the capabilities of these models has been provided in a WMO publication [28].

SNOWMELT FLOODS

There are three considerations in regard to snowmelt floods:

1. Estimating the rate of snowmelt from snowpack
2. Estimating runoff from snowmelt
3. Derivation of flood hydrograph

The procedures that are currently used for solving these problems are as follows.

Snowmelt Equations

The process of snowmelt at a given point is a physical process. The most comprehensive study of this process has been carried out by the U.S. Army Corps of Engineers [29]. Equations have been derived for estimating snowmelt under different weather conditions and land cover:

1. During rainy periods

- For open ($<10\%$ cover) or partly forested (10%–60% cover) areas

$$M = (0.029 + 0.0084kv + 0.007P_r)(T_a - 32) + 0.09 \quad (24)$$

- For forested (60% to 80% cover) and heavily forested (>80% cover) areas

$$M = (0.074 + 0.007P_r)(T_a - 32) + 0.05 \quad (25)$$

2. During rain-free periods:

- For open (<10% cover) area

$$M = k'(0.00508I_i)(1 - a) + (1 - N)(0.0212T'_a - 0.84) + N(0.029T'_c) + k(0.0084v)(0.22T'_a + 0.78T'_d) \quad (26)$$

- For partly forested (10%–60% cover) area

$$M = k'(1 - F)(0.0040I_i)(1 - a) + k(0.0084v)(0.22T'_a + 0.78T'_d) + F(0.029T'_a) \quad (27)$$

- For forested (60% to 80% cover) area

$$M = k(0.0084v)(0.22T'_a + 0.78T'_d) + 0.029T'_a \quad (28)$$

- For heavily forested (>80% cover) area

$$M = 0.074(0.53T'_a + 0.47T'_d) \quad (29)$$

In Equations 24 through 29,

M = snowmelt rate (in./day)

T'_a = difference between air temperature at 10-ft. level and snow surface temperature, °F

T'_d = difference between dew-point temperature at 10-ft. level and snow surface temperature, °F

v = wind speed at 50-ft. level, mph

I_i = observed or estimated solar radiation on horizontal surface, langleys

a = observed or estimated average snow surface albedo

k' = basin short-wave radiation factor (between 0.9 and 1.1), depending on average exposure of open areas to short-wave radiation in comparison with an unshielded surface

F = estimated average basin forest canopy cover expressed as a decimal fraction

T'_c = difference between cloud base temperature and snow surface temperature, °F

N = estimated cloud cover expressed as a decimal fraction

k = basin convection-condensation melt factor

Pr = precipitation (in./day)

T_a = air temperature, °F

Equations 24 through 29 are quite elaborate and require extensive climatological data. Most often these data are not available. For such situations, equations based on temperature only have been suggested [30].

For open sites,

$$M = 0.06(T_{mean} - 24) \tag{30}$$

$$M = 0.04(T_{max} - 27) \tag{31}$$

For forest sites,

$$M = 0.05(T_{mean} - 32) \tag{32}$$

$$M = 0.04(T_{max} - 42) \tag{33}$$

These equations are applicable for T_{mean} in the range of 34° to 66°F and for T_{max} in the range of 44° to 76°F. T_{mean} and T_{max} are the mean daily and maximum daily temperatures.

Considerable discretion is required in the use of all these equations. They cannot be expected to apply in all regions. Adjustments to coefficients will be necessary to suit given conditions.

Degree-Day Factor

A degree-day represents the difference in temperatures between a reference temperature and a base temperature. The reference temperature can be either maximum daily or mean daily temperature. The base temperatures are selected by judgment and have varied from 24°F to 42°F (Equations 30–33). For uniformity, it is standard practice to consider mean daily temperature as the reference temperature and 32°F as the base temperature. Thus, a day with mean daily temperature of 52°F represents 20 degree-days. Degree-days as an index of heat are used in both snowmelt as well as in runoff computations. The ratio of snowmelt (or runoff) to the concurrent number of degree-days is called the degree-day factor. For point melt, this factor has been found to vary from 0.015 to 0.20 in./degree-day.

For estimating snowmelt flood, the point melt values must be converted into runoff. This has to be done in the same way as deriving a rainfall-runoff relation. Most often the derivation of these relationships is extremely complex and results are not always reliable. To avoid this problem, degree-day relationships are frequently established using streamflow directly. Daily runoff at a gauging station is divided by the concurrent number of degree-days averaged over the basin. This ratio is taken as the degree-day factor for runoff. This factor has been found to show a systematic variation with date during the melt season. This variation is in the form an S curve with the steep section lying in the middle of the season. The same type of curve is obtained if the accumulated runoff is plotted against accumulated number of degree-days starting from a convenient date at the beginning of melt season. Degree-day factors for runoff have been found to vary between 0.06 and 0.15 in./degree-day [3].

Snowmelt Flood Hydrograph

Estimation of floods from snowmelt is more complex than estimating rainfall floods. The principle of the unit hydrograph cannot be applied, as the basic assumptions underlying unit hydrograph theory are not satisfied by snowmelt. A very substantial part of snowmelt moves as subsurface and groundwater flow.

Garstka et al. [31] suggested a technique in which a fraction of snowmelt on a given day appeared as runoff for that day and the remaining departed as recession flow. Practical difficulties have been encountered in this technique, and thus it has not received wide acceptance.

Rockwood [32] used the method of phase routing in the Columbia basin studies. The total snowmelt runoff was separated into surface runoff and groundwater components. The two components were routed separately by using different storage constants. The storage times for each component were determined by trial and error by fitting synthesized hydrographs to recorded data. The method is laborious and requires extensive computer facilities.

HYDRAULIC METHODS

For ungauged areas it is sometimes possible to estimate flood peaks by indirect methods. These are called hydraulic methods and include the slope-area method, flow through contractions, and flow-over dams. The accuracy of these methods can be within 10% to 20% of the actual value with good measurements. Details of data collection and office procedures for these methods have been furnished by Benson and Dalrymple [33].

Slope-Area Method

This is a widely used method of calculating discharge in an open channel. The commonly used formula in this method is the Manning equation

$$Q = 1.49/nAR^{2/3}S^{1/2} \tag{34}$$

where Q = the discharge in cfs
A = the cross-sectional area in sq ft
R = the hydraulic radius (cross-sectional area divided by the wetted perimeter) in ft.
n = Manning's roughness coefficient
S = the slope of the energy gradient

The equation yields

$$Q/S^{1/2} = 1.49/nAR^{2/3} \tag{35}$$

The right-hand side of this equation contains only the terms related to physical characteristics of the cross section and is termed conveyance K. Its value can be computed. If the section is nonuniform, then K for each subsection is computed and values added to obtain total K for the whole section.

Difficulty is encountered in estimating the energy gradient S. If S_W is the slope of the water surface, then

$$S = S_W + (V_1^2 - V_2^2)/2gL \tag{36}$$

where V_1 and V_2 = velocities at the beginning and end of the reach, respectively
L = length of the reach
g = acceleration due to gravity

Table 2
Manning Roughness Coefficient for Natural Channels

Channel Condition	n, $(TL^{-1/3})$
Clean, straight, full stage, no pools	0.029
As above with weeds and stones	0.035
Winding, pools and shallows, clean	0.039
As above at low stages	0.047
Winding, pools and shallows, weeds and stones	0.042
As above, shallow stages, large stones	0.052
Sluggish, weedy, with deep pools	0.065
Very weedy and sluggish	0.112

Note: The above are average values. Variations of as much as 20% must be expected in individual cases.

As V_1 and V_2 are unknown, this equation must be solved by trial and error. Detailed description of the method is given by Chow [34]. Some selected values of the Manning n are given in Table 2.

Flow-Through Contractions

Reasonable estimates of flood flow can be made if accurate data on approach velocity, head loss, cross-sectional area, flow depth, and other hydraulic parameters can be obtained for a contracted opening such as a bridge. Equations in various forms can be employed. One such form is [4]

$$Q = CA_1A_2[2g(d_1 - d_2)/(A_1^2 - A_2^2)]^{1/2} \tag{37}$$

where Q = discharge
C = coefficient of discharge
A_1, A_2 = flow areas upstream and at the contracted section, respectively
d_1, d_2 = flow depths upstream and at the contracted section, respectively
g = acceleration due to gravity

Detailed application of this method is described by Matthai [35].

Flow-Over Dams

A dam has long been used as a means of estimating discharge. The equation is

$$Q = CLH^{3/2} \tag{38}$$

where Q = the discharge
C = a variable coefficient
L = the effective length of crest
H = the total head on crest, including velocity of approach head

Detailed computations for this method are given in hydraulics text books and other related publications [5, 8].

FLOOD ROUTING

Flood routing is defined as the method by which the hydrograph of a flood as it occurred at an upstream point is transferred to some point downstream. The intervening passage may be through a reservoir or the channel reach itself. Thus, two different types of procedures are considered, namely, reservoir routing and channel routing. For reservoir routing, generally the Modified Puls or the Storage Indication method is used and for channel routing, the Muskingum method is used. The former method can also be used for channel routing but the results are less accurate.

Modified Puls or Storage Indication Method

This method, as used in reservoir routing, is based upon the principle that the discharge from an uncontrolled reservoir can be expressed as a simple function of the water surface elevation. The known data on the reservoir are the elevation-storage curve and the elevation-discharge curve. The routing equation is the continuity equation which states that inflow minus outflow equals the change in storage in a given time interval. Expressed in symbolic form, the equation is

$$\tfrac{1}{2}(I_1 + I_2)\,\Delta t - \tfrac{1}{2}(O_1 + O_2)\,\Delta t = S_2 - S_1 \tag{39}$$

where Δt = the time interval
I_1, O_1, and S_1 = the inflow, outflow, and storage at the beginning of the interval
I_2, O_2, and S_2 = the inflow, outflow, and storage at the end of the interval

Taking all known values to the left, the equation becomes

$$\tfrac{1}{2}(I_1 + I_2)\,\Delta t + S_1 - \tfrac{1}{2}O_1\,\Delta t = S_2 + \tfrac{1}{2}O_2\,\Delta t \tag{40}$$

From the data, a storage indication curve $S + \frac{1}{2}O\,\Delta t$ vs. discharge is constructed. For a known value of initial outflow O_1, the quantity $S_1 - \frac{1}{2}O_1\,\Delta t$ is obtained as $(S_1 + \frac{1}{2}O_1\,\Delta t) - O_1\,\Delta t$. This quantity, when added to the average inflow during the interval gives $S_2 + \frac{1}{2}O_2\,\Delta t$. The outflow corresponding to this quantity is read from the storage indication curve. The computation is then repeated for the succeeding routing periods and the solution proceeds. The peak of the outflow hydrograph must fall on the recession limb of the inflow hydrograph. The method is known as the Modified Puls method because the original Puls method required the construction of $S\text{-}\frac{1}{2}O\,\Delta t$ vs. O curve also. The method was developed by L. G. Puls of the U.S. Army Corps of Engineers [36].

The Muskingum Method

This is the most common method of routing through a channel reach. It was developed in connection with studies of the Muskingum Conservancy District Flood Control Project of the U.S. Army Corps of Engineers in 1934–35 [37]. The routing equation in this method is:

$$S = K[xI + (1 - x)O] \tag{41}$$

where S = the storage
I and O = inflow and outflow
x = a factor expressing the relative importance of inflow and outflow in determining storage
K = a constant called the storage constant and has the dimension of time

If Equation 41 is substituted for S in Equation 39, the resulting equation becomes

$$O_2 = c_0 I_2 + c_1 I_1 + c_2 O_1 \tag{42}$$

where

$$c_0 = -(Kx - 0.5\,\Delta t)/(K - Kx + 0.5\,\Delta t) \tag{43}$$

$$c_1 = (Kx + 0.5\,\Delta t)/(K - Kx + 0.5\,\Delta t) \tag{44}$$

$$c_2 = (K - Kx - 0.5\,\Delta T)/(K - Kx + 0.5\,\Delta t) \tag{45}$$

with the result that $c_0 + c_1 + c_2 = 1$.

In order to determine the values of routing constants K and x, a linear relationship between S and the variable $xI + (1 - x)O$ must be established from the observed inflow and outflow hydrographs. A simple storage vs. outflow curve with $x = 0$ results in a loop. The best value of x is obtained when the loop converges to a straight line. The slope of this line gives the value of K. Once K and x are known, the coefficients c_0, c_1 and c_2 can be computed. The routing can now be carried out with Equation 42. For most streams, x is between 0 and 0.3 with a mean value near 0.2. The storage constant K is approximately equal to the travel time through the reach. Thus, if the recorded outflow hydrograph is not available, the values of K and x can be approximated. In channel routing, the peak of the outflow hydrograph falls outside the recession limb of the inflow hydrograph.

Estimating Travel Time

An estimate of travel time through a channel reach is required in flood routing studies as well as in flood forecasting and flood warning operations. If the reach is ungauged, then the time of travel is estimated from velocity estimates. If the reach is gauged and inflow and outflow hydrographs through the reach are available, then a simple technique of estimating this time is as follows:

1. Plot the inflow and outflow hydrographs together on the same time scale starting at the same time.
2. Take a time intercept on the inflow hydrograph corresponding to any discharge.
3. Make an equal intercept on the outflow hydrograph.
4. The time interval between the corresponding points on the rising limbs or the falling limbs is the travel time corresponding to the chosen discharge on the inflow hydrograph.
5. In the limit, the time interval between the peaks of the two hydrographs is the travel time corresponding to the peak flow on the inflow hydrograph.

The assumptions behind this technique are as follows:

1. For each flow on the inflow hydrograph, there is a corresponding flow, although different, on the outflow hydrograph.
2. The time interval between two equal flows on the rising and falling limbs of the inflow hydrograph is equal to the time interval between the corresponding equal flows on the rising and falling limbs of the outflow hydrograph.
3. The time difference between the corresponding points is the travel time corresponding to the chosen value of inflow.

These assumptions are not unrealistic and will hold if the local inflow in the reach is insignificant. The method is particularly useful if the peaks are not well-defined.

An example of this technique is provided in Figure 4. The inflow and outflow hydrographs represent actual data for the North Platte River between Bridgeport and Lisco, Nebraska [3].

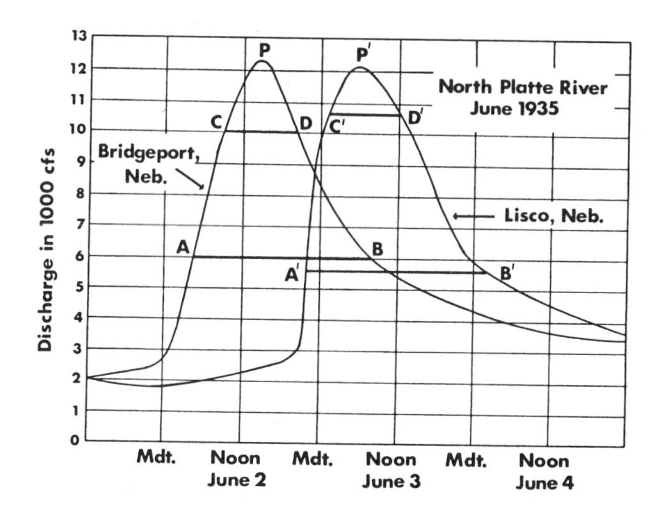

Figure 4. Estimating the time of travel through a channel reach.

The intercept AB on the inflow hydrograph at a discharge of 6,000 cfs corresponds to an equal intercept A'B' on the outflow hydrograph. The corresponding outflow is about 5,600 cfs. The travel time is equal to the interval AA' or BB' and is about 17.5 hours. Similarly, equal intercepts CD and C'D' corresponding to an inflow of 10,000 cfs give the travel time of 16 hours. The travel time between the two peaks PP' is also 16 hours. It may be noted that in this example travel time decreases from about 18.5 hours, corresponding to an inflow of 4,000 cfs, to 16 hours corresponding to the peak. That is, as the flow increases, the travel time decreases reaching an almost constant value at higher flows.

The author is not the originator of this technique. He had learned it during regular office work many years ago and does not know who first suggested it. To the best of his knowledge it has not been published.

FLOOD FREQUENCY ANALYSIS

Flood frequency analysis is perhaps the most important but most controversial topic in engineering hydrology. Several methods have been suggested and some have been standardized for office use but the general agreement on the best approach has not yet been reached. The following is a general discussion of various concepts and probability distributions commonly used in flood frequency analysis.

Probability and Return Period

The return period of an event is defined as the average number of years during which a given discharge rate will be equalled or exceeded. For example, a 100-yr flood will be expected to be equalled or exceeded 10 times in a long flood series of 1,000 years. In other words, a 100-yr flood has a one-percent chance of occurring in any one year. There is no implication that a flood of this magnitude would occur only once in 100 years. It could happen that there are 0, 1, 2, or more floods equal to or exceeding the 100-yr flood in a given flood series of 100 years but over a long period of time, the average will be satisfied (binomial distribution).

The probability of occurrence p of an event and its return period T are related by

$$T = 1/p \tag{46}$$

If q is the probability of nonoccurrence of an event, then

$$T = 1/(1 - q) \tag{47}$$

If J represents the probability that the event will occur at least once in the next n years, then

$$J = 1 - (1 - 1/T)^n \tag{48}$$

If n = T, then

$$J = 1 - e^{-1} = 0.63 \text{ (large n)} \tag{49}$$

That is, the probability that an event of return period T-yr will be equalled or exceeded in the next T years is 63%. In other words, if we make 100 series of 100 years each from a long flood series of 10,000 years, only 63 of these series will contain all the expected 100 floods equal to or exceeding the 100-yr flood and the remaining 37 will have no flood equal to the 100-yr flood. It shows the difficulty of estimating floods of high return periods even from a long series of recorded data.

Return Period and Risk

The most important question facing the hydrologist is: what is the risk of failure? For example, what is the probability that a project designed for 100-yr flood will be overtopped during its life of 25 years? With n = 25 and T = 100, Equation 48 gives J = 0.22. That is, there is 22% chance of failure. In order to estimate the return period for a given risk and the life of the project, we have, from Equation 48,

$$T = 1/[1 - (1 - J)^{1/n}] \tag{50}$$

Expanding the denominator and neglecting $J^2/2\,n^2$ and smaller terms, we have [38]

$$T = n/J - n/2 \text{ (approximately)} \tag{51}$$

Thus, a project with only 10% chance of failure and an economic life of 25 years should be designed for 238-yr flood. With further approximation, T is assumed to be equal to n/J only which Gumbel [39] called the "design quotient."

Probability Distributions

There are several probability distributions which are used in flood frequency analysis. The more important are:

1. Normal or Gaussian distribution
2. Extreme value Type I or Gumbel distribution
3. Pearson Type III distribution
4. Log-Pearson Type III distribution
5. Log-normal distribution
6. Three-parameter log-normal distribution

Details of these distributions are given in text books on probability and statistics and several hydrologic publications [1, 39–42]. The following is a brief description of these distributions.

Normal or Gaussian Distribution

This is a symmetrical, bell-shaped, and continuous distribution. Its probability density is

$$p(x) = \frac{1}{\sigma\sqrt{2\pi}}\, e^{-(x-\mu)^2/2\sigma^2} \tag{52}$$

where x = the variate
μ and σ = the mean and standard deviation respectively

The cumulative probability of a value being equal to or less than x or the distribution function is

$$P(x) = \frac{1}{\sigma\sqrt{2\pi}} \int_{-\infty}^{x} e^{-(x-\mu)^2/2\sigma^2}\, dx \tag{53}$$

This represents the area under the curve between $-\infty$ and x. Tables of these areas have been prepared and are available in text books and handbooks on statistics. The skewness of this distribution is zero. As the flood flows follow a skewed distribution, normal distribution is rarely used in flood frequency analysis. However, mathematical transformations of the original data have been done to render them amenable to analysis in accordance with this distribution. Occasionally, flood flows which have been routed through many lakes and swamps have been found to fit the normal distribution. Hazen [43], while studying the storage requirements for reservoirs, used this distribution and introduced the normal probability paper.

Extreme Value Type I or Gumbel Distribution

This is a skewed distribution. Gumbel [39] has described its theory and application in hydrology in great detail. This is a double exponential distribution. Its probability density is

$$p(x) = \alpha e^{-y-e^{-y}} \tag{54}$$

where $y = \alpha(x - \beta)$ with $-\infty < x < \infty$ and α and β are the scale and the location parameters. The distribution function is

$$P(x) = e^{-e^{-y}} \tag{55}$$

Using the method of moments, the parameters α and β can be obtained as

$$\alpha = 1.283/\sigma \tag{56}$$

and

$$\beta = \mu - 0.450\sigma \tag{57}$$

where μ and σ are the mean and the standard deviation. The coefficients of skew and kurtosis for this distribution are 1.139 and 5.4 respectively. The mean in this distribution occurs at a return period of 2.33 yr. This distribution has been used extensively in flood frequency analysis. The introduction of the Gumbel probability paper by Powell [44] further facilitated its use in practice.

Pearson Type III Distribution

The probability density of this distribution is

$$p(x) = \frac{1}{\alpha T(\beta)}\left(\frac{x - \gamma}{\alpha}\right)^{\beta - 1} e^{-(x-\gamma)/\alpha} \tag{58}$$

where α, β, and γ are the parameters and $T(\beta)$ is the gamma function. If the substitution $y = (x - \gamma)/\alpha$ is made, then the distribution reduces to one parameter gamma distribution.

Log-Pearson Type III Distribution

If the logarithms of a variable x are distributed according to Pearson Type III distribution, then the variable x is distributed as log-Pearson Type III distribution. Its probability density is given by

$$p(x) = \frac{1}{\alpha x T(\beta)}\left(\frac{y - \gamma}{\alpha}\right)^{\beta - 1} e^{-(y-\gamma)/\alpha} \tag{59}$$

where $y = \ln x$ and α, β, and γ are parameters and $T(\beta)$ is the gamma function.

Two-Parameter Log-normal Distribution

In this distribution the logarithm of the variate x is normally distributed. Its probability density is

$$p(x) = \frac{1}{x\sigma_y\sqrt{2\pi}} e^{-(y-\mu_y)^2/2\sigma_y^2} \tag{60}$$

where $y = \ln x$
μ_y and σ_y = mean and standard deviation of y, respectively

This is a skew distribution and has unlimited range in both directions. Chow [45] has derived the following relationships:

$$\mu_x = e^{\mu_y + \sigma_y^2/2} \tag{61}$$

$$\sigma_x = \mu_x(e^{\sigma_y^2} - 1)^{1/2} \tag{62}$$

$$M_x = e^{\mu_y} \tag{63}$$

$$g_x = 3v_x + v_x^3 \tag{64}$$

where μ_x = mean
σ_x = standard deviation
M_x = median
g_x = coefficient of skew
v_x = coefficient of variation

If the data series is such that its computed coefficient of skew is equal to 1.139 and its coefficient of variation is 0.364, it will satisfy Equation 64. Such a series will plot as a straight line on both the Gumbel probability paper as well as on the log-normal probability paper. The use of this distribution in flood frequency analysis was first suggested by Hazen [46, 47]. Whipple [48] designed the log-normal probability paper to linearize the plot.

Three-Parameter Log-normal Distribution

The three-parameter log-normal distribution represents the distribution in which the logarithm of the variable (x − a) is normally distributed. The variate x is the random variable, a is a parameter, and (x − a) is the reduced variable. This is a general distribution in which the logarithm of any linear function of x is normally distributed. For this distribution, the probability density is

$$p(x) = \frac{1}{(x-a)\sigma_y\sqrt{2\pi}} e^{-(y-\mu_y)^2/2\sigma_y{}^2} \tag{65}$$

where $y = \ln(x - a)$
μ_y and σ_y = mean and standard deviation of y, respectively
a = a parameter to be estimated

There are several methods of estimating a. The four among them are:

1. Method of moments
2. Maximum likelihood method
3. Median method
4. Graphical method

Method of moments. In this method, use is made of the property that the coefficients of skew of the original variate x and of the reduced variate (x − a) are the same, as the parameter a has no effect on this measure. Equation 64 thus gives

$$g_x = 3v_{(x-a)} + v_{(x-a)}^3 \tag{66}$$

The solution of this equation is

$$v_{(x-a)} = \frac{1 - w^{2/3}}{w^{1/3}} \tag{67}$$

where $w = \dfrac{-g_x + (g_x^2 + 4)^{1/2}}{2}$

An approximate but simpler solution suggested by Sangal [49] is

$$v_{(x-a)} = (-9 + \sqrt{81 + 12g_x^2})/2g_x \qquad (g_x > 0) \tag{68}$$

When v_{x-a} is known, then a can be calculated by

$$a = \bar{x} - \frac{s_x}{v_{(x-a)}} \tag{69}$$

Maximum likelihood method The method of maximum likelihood requires the solution of the equation [50]:

$$\sum_{i=1}^{n} \frac{1}{x_i - a} \left\{ \frac{1}{n} \sum_{i=1}^{n} \ln^2(x_i - a) - \left[\frac{1}{n} \sum_{i=1}^{n} \ln(x_i - a) \right]^2 - \frac{1}{n} \sum_{i=1}^{n} \ln(x_i - a) \right\} + \sum_{i=1}^{n} \frac{\ln(x_i - a)}{x_i - a} = 0 \tag{70}$$

This equation is difficult to solve manually but can be easily programmed for computer use.

Median method This method has been suggested by Sangal and Biswas [51]. In this method a is given by

$$a = z - \left[\frac{s_x^2}{2(\bar{x} - z)} \right] \tag{71}$$

where z, $\bar{x}$, and s_x are the median, mean, and the standard deviation of x. The median can be computed as the mean of the middle 1/5th of the data. Also, if the computed median is very close to or greater than the mean, then the median can be assumed to be equal to 99% of the mean.

Graphical method In this method, the series is plotted on the log-normal probability paper. If it plots concave downward, a will be negative. If the plot is a straight line, a will be zero. If the plot is concave upwards, a will be positive. For the curved plots, the value of a is so adjusted by trial and error that the curve for the variate (x − a) becomes a straight line. Sometimes, the values of a are plotted against the computed values of $g_{\log(x-a)}$. The value of a which very nearly reduces this skewness to zero is chosen for use.

The three-parameter log-normal is a very useful distribution. In a comparative study of several well-known distributions, Kite [41] found it to give extremely satisfactory results.

Frequency Factors

Chow [52] proposed a general equation for frequency analysis. The equation is

$$x = \bar{x} + K\sigma \tag{72}$$

where $\bar{x}$ and σ are the mean and standard deviation of the x series. The factor K is called the frequency factor and is a function of the recurrence interval and the type of probability distribution to be used in the analysis.

For the normal distribution, K is the same as the standard normal deviate whose values corresponding to the area under the normal probability curve have been widely tabulated in statistical textbooks and handbooks. These values plot as a straight line on the normal probability paper.

For the log-normal and the three-parameter log-normal distributions, the values of K corresponding to the normal distribution can be used if the variable x or the reduced variable (x − a) have been first logarithmically transformed. For the untransformed variable, the values of K can be obtained from table of frequency factors for log-normal distribution [1]. These values depend upon the probability as well as upon the coefficient of skewness. It may be mentioned that Hazen [47] used the three-parameter log-normal distribution rather indirectly with parameter a computed by the method of moments.

For the Gumbel distribution, the K values are given by

$$K = -\frac{\sqrt{6}}{\pi}\left[\gamma + \ln\ln\left(\frac{T}{T-1}\right)\right] \tag{73}$$

where $\gamma = 0.57721$, a Euler's constant
T = the return period

This formula is applicable to a sample of large size. For a small sample (< 100 years), a correction factor for the length of record has to be applied [53]. The values of K given by Equation 73 will plot as a straight line on the Gumbel probability paper.

Pearson Type III and log-Pearson Type III distributions have been extensively studied especially in the United States, as the latter distribution has been recommended for use throughout the country. Detailed procedures for its application including comprehensive tables of K factors have been prepared [54]. For computer use, approximate values of K can be obtained from the following formula [55] when skew coefficients are between 1.0 and −1.0.

$$K = \frac{2}{g_x}\left\{\left[\frac{g_x}{6}\left(t - \frac{g_x}{6}\right) + 1\right]^3 - 1\right\} \tag{74}$$

where g_x is the skew coefficient of the original data (Pearson Type III) or of their logarithms (log-Pearson Type III) and t is the standard normal deviate. Some selected values of K are given in Table 3.

Plotting Positions

A formula is needed to plot the data on the probability paper chosen for use. Many such formulas have been suggested over the years but no general agreement on the most suitable one appears to have been reached. In a recent study, Cunnane [56] points out the limitations of the Weibull formula which has been used in flood frequency analysis for many years. All formulas were found to have one shortcoming or the other which led him to suggest his own compromise formula. If an analysis

Table 3
Frequency Factors for Pearson Type III and Log-Pearson Type III Distributions

Coefficient of Skew	Probability of Exceedance, P (%)								
	99	95	80	50	20	5	1	0.1	0.01
-2.0	-3.605	-1.996	-0.609	0.307	0.777	0.949	0.999	0.999	1.000
-1.8	-3.499	-1.981	-0.643	0.282	0.799	1.020	1.087	1.107	1.111
-1.6	-3.388	-1.962	-0.675	0.254	0.817	1.093	1.197	1.238	1.247
-1.4	-3.271	-1.938	-0.705	0.225	0.832	1.168	1.318	1.394	1.418
-1.2	-3.149	-1.910	-0.733	0.195	0.844	1.243	1.449	1.577	1.628
-1.0	-3.023	-1.877	-0.758	0.164	0.852	1.317	1.588	1.786	1.884
-0.8	-2.891	-1.839	-0.780	0.132	0.856	1.389	1.733	2.017	2.184
-0.6	-2.755	-1.797	-0.800	0.099	0.857	1.458	1.880	2.268	2.525
-0.4	-2.615	-1.750	-0.816	0.067	0.855	1.524	2.029	2.533	2.899
-0.2	-2.472	-1.700	-0.830	0.033	0.850	1.586	2.178	2.808	3.299
0.0	-2.326	-1.645	-0.842	0.0	0.842	1.645	2.326	3.090	3.719
0.2	-2.178	-1.586	-0.850	-0.033	0.830	1.700	2.472	3.377	4.153
0.4	-2.029	-1.524	-0.855	-0.067	0.816	1.750	2.615	3.666	4.597
0.6	-1.880	-1.458	-0.857	-0.099	0.800	1.797	2.755	3.956	5.047
0.8	-1.733	-1.389	-0.856	-0.132	0.780	1.839	2.891	4.244	5.501
1.0	-1.588	-1.317	-0.852	-0.164	0.758	1.877	3.023	4.531	5.957
1.2	-1.449	-1.243	-0.844	-0.195	0.733	1.910	3.149	4.815	6.412
1.4	-1.318	-1.168	-0.832	-0.225	0.705	1.938	3.271	5.095	6.867
1.6	-1.197	-1.093	-0.817	-0.254	0.675	1.962	3.388	5.371	7.318
1.8	-1.087	-1.020	-0.799	-0.282	0.643	1.981	3.499	5.642	7.766
2.0	-0.990	-0.949	-0.777	-0.307	0.609	1.996	3.605	5.908	8.210

of all the formulas is done, a general formula for the return period can be written as [57]

$$T = \frac{n}{m - \gamma} \qquad (0 \leq \gamma \leq 1) \tag{75}$$

where n = total number of items in the series
m = order number if the series is arranged in descending order

The value of $\gamma = 0$ gives the most conservative estimates of return period and $\gamma = 1$, the least. The California formula is given by $\gamma = 0$ and the Hazen formula by $\gamma = \frac{1}{2}$. In other formulas, γ takes on the functional form of m and n. For example, $\gamma = m/(n + 1)$ for the Weibull formula. The Hazen and the Weibull formulas give identical results for $m = (n + 1)/2$. For $m < (n + 1)/2$, the Hazen formula gives higher values of return period and for $m > (n + 1)/2$, the Weibull formula gives higher values. All formulas give almost identical results after the first few values of m. Although the Weibull formula is still in common use, the Hazen formula also is simple and has considerable merit [58].

Single-Station Frequency Analysis

Flood frequency studies for a gauged station are required for design purposes or for general evaluation of the flood potential of a given region. Such studies have been carried out all over the world using different methods of analysis. For illustrative purposes, an example has been chosen from an earlier study [49]. The annual maximum mean daily flows of the Credit River near Cataract,

Ontario, Canada for 56 years from 1916 to 1971 inclusive are analyzed. The series is arranged in descending order and the return periods are computed from the Weibull formula $T = (n + 1)/m$. The original data, the ordered series, order numbers and return periods are given in Table 4.

Five frequency distributions have been applied:

1. Gumbel
2. Pearson Type III
3. Log-Pearson Type III
4. Log-normal
5. Three-parameter log-normal

The parameters were estimated using the following equations:

Mean

$$\bar{x} = \frac{1}{n} \sum_{i=1}^{n} x_i \tag{76}$$

Table 4
Frequency Analysis for Credit River near Cataract, Ontario, Canada

Year	Discharge (cfs)	Year	Discharge (cfs)	Order	Discharge (cfs)	Return Period (yr)	Order	Discharge (cfs)	Return Period (yr)
1916	1120	1944	685	1	2360	57	29	685	1.97
1917	1060	1945	402	2	2000	28.5	30	665	1.90
1918	1260	1946	820	3	1690	19	31	660	1.84
1919	2360	1947	850	4	1540	14.25	32	650	1.78
1920	1240	1948	1190	5	1470	11.4	33	638	1.73
1921	800	1949	638	6	1260	9.5	34	625	1.68
1922	765	1950	2000	7	1240	8.14	35	585	1.63
1923	1160	1951	953	8	1200	7.13	36	580	1.58
1924	940	1952	494	9	1190	6.33	37	580	1.54
1925	1690	1953	394	10	1190	5.70	38	571	1.50
1926	1120	1954	1030	11	1160	5.18	39	560	1.46
1927	1200	1955	571	12	1120	4.75	40	513	1.43
1928	765	1956	1470	13	1120	4.38	41	510	1.39
1929	1540	1957	705	14	1060	4.07	42	502	1.36
1930	560	1958	122	15	1030	3.80	43	501	1.33
1931	280	1959	665	16	953	3.56	44	498	1.30
1932	625	1960	494	17	940	3.35	45	494	1.27
1933	730	1961	510	18	850	3.17	46	494	1.24
1934	660	1962	502	19	820	3.0	47	473	1.21
1935	650	1963	585	20	800	2.85	48	464	1.19
1936	790	1964	226	21	790	2.71	49	423	1.16
1937	423	1965	755	22	765	2.59	50	402	1.14
1938	473	1966	284	23	765	2.48	51	394	1.12
1939	705	1967	501	24	755	2.38	52	284	1.10
1940	580	1968	513	25	730	2.28	53	280	1.08
1941	464	1969	580	26	705	2.19	54	266	1.06
1942	1190	1970	266	27	705	2.11	55	226	1.04
1943	685	1971	498	28	685	2.04	56	122	1.02

Standard deviation

$$s_x = \left[\frac{\sum_{i=1}^{n} (x_i - \bar{x})^2}{n - 1} \right]^{1/2} \tag{77}$$

Coefficient of skew

$$g_x = \left[\frac{n \sum_{i=1}^{n} (x_i - \bar{x})^3}{(n - 1)(n - 2)s_x^3} \right] \tag{78}$$

The preceding equations were extended to logarithmic distributions by substituting $\log x_i$ or $\log(x_i - a)$ for x_i where a is a parameter of the three-parameter log-normal distribution. The logarithms are to the base 10.

The parameters of the Gumbel distribution were obtained by the method of maximum likelihood [59]. The Pearson Type III coordinates were obtained from Equation 74. The parameter a of the three-parameter log-normal distribution was computed by:

1. The median method
2. The moment method
3. The maximum likelihood method.

For the median method, the median was computed as the mean of the 12 items from m = 23 to 34 inclusive and a was computed from Equation 71. For the moment method, $v_{(x-a)}$ was computed from Equation 68 and a was computed from Equation 69. For the maximum likelihood method, a was obtained by solving Equation 70. The results are presented in Tables 5 and 6. Table 6 suggests that the divergence among the various estimates is quite substantial at low probabilities (high return periods).

A plot of the data on the log-normal probability paper is given in Figure 5. Only the curves corresponding to the log-Pearson Type III (LP) and the three-parameter log-normal distribution (3-LN) fitted by the method of maximum likelihood have been drawn on the figure. Both curves are almost indistinguishable up to the return period of 20-yr. However, at higher return periods, a marked departure is apparent with LP giving somewhat lower estimates than 3-LN. A major difficulty with the LP distribution is the use of the coefficient of skewness. This parameter, whether of the original data or of their logarithms, is extremely unreliable. Although methods have been suggested to overcome this difficulty using regional generalized skew coefficients [54], it is doubtful whether such empirical techniques lead to a reliable solution. The question remains: can the coefficient of skew be regionalized? There is no definite answer as yet.

Regional Frequency Analysis

It is often necessary to obtain the frequency estimates for ungauged sites. For this purpose, the station frequency estimates are regionalized. The U.S. Geological Survey uses the index flood [60] method for regional analysis. First a homogeneous region is established according to a homogeneity test developed by Langbein. Within this homogeneous region, frequency data for single-station analyses are used to develop two relationships. The first expresses the ratio of flood of a given frequency to the mean annual flood or the 2.33-yr flood if the Gumbel distribution is used. The second expresses the mean annual flood as a function of drainage area. From these two curves, the flood of any desired frequency can be obtained for the ungauged site if the drainage area is known. Sometimes, more elaborate procedures using basin characteristics such as slope, land use, forest cover, area covered by lakes and swamps, mean annual runoff [49], etc. are used to develop relationships for the mean annual flood.

Regional frequency estimates are at best crude and cannot be used for design purposes without a thorough analysis of the physiographic and meteorologic characteristics of the basin.

Table 5
Frequency Computations for Credit River near Cataract, Ontario, Canada
(Drainage Area = 82 sq. mi.)

	Mean	Standard Deviation	Coefficient of Variation	Coefficient of Skew
x	795.32	434.32	0.546	1.413
log x	2.84009	0.23792	0.08377	-0.41482
Median Method, a = -198 median = 689.00 log (x-a)	2.96079	0.17783	0.06006	0.11379
Moment Method, a = -190 log (x-a)	2.95660	0.17951	0.06072	0.10074
Maximum Likelihood Method, a = -155 log (x-a)	2.93787	0.18736	0.06377	0.03882

Table 6
Comparative Frequency Results for Various Distributions

T	P	Gumbel	Pearson	Log-Pearson	Log-normal	3-Parameter Log-normal		
						Median	Moment	M.L.
1.01	99	135	219	164	193	154	155	162
1.05	95	268	286	264	281	267	268	270
1.11	90	350	343	336	342	342	339	340
1.25	80	461	436	442	436	449	449	448
2.00	50	722	698	718	691	715	714	711
2.33	42.92	788	772	792	764	786	785	782
5.00	20	1073	1098	1105	1097	1091	1091	1091
10.00	10	1306	1371	1356	1396	1346	1347	1351
20.00	5	1529	1633	1590	1704	1593	1595	1607
50.00	2	1818	1971	1882	2131	1920	1924	1947
100.00	1	2034	2223	2093	2474	2169	2176	2209
1,000.00	0.1	2749	3055	2751	3760	3039	3054	3131
10,000.00	0.01	3463	3890	3357	5310	3992	4019	4158

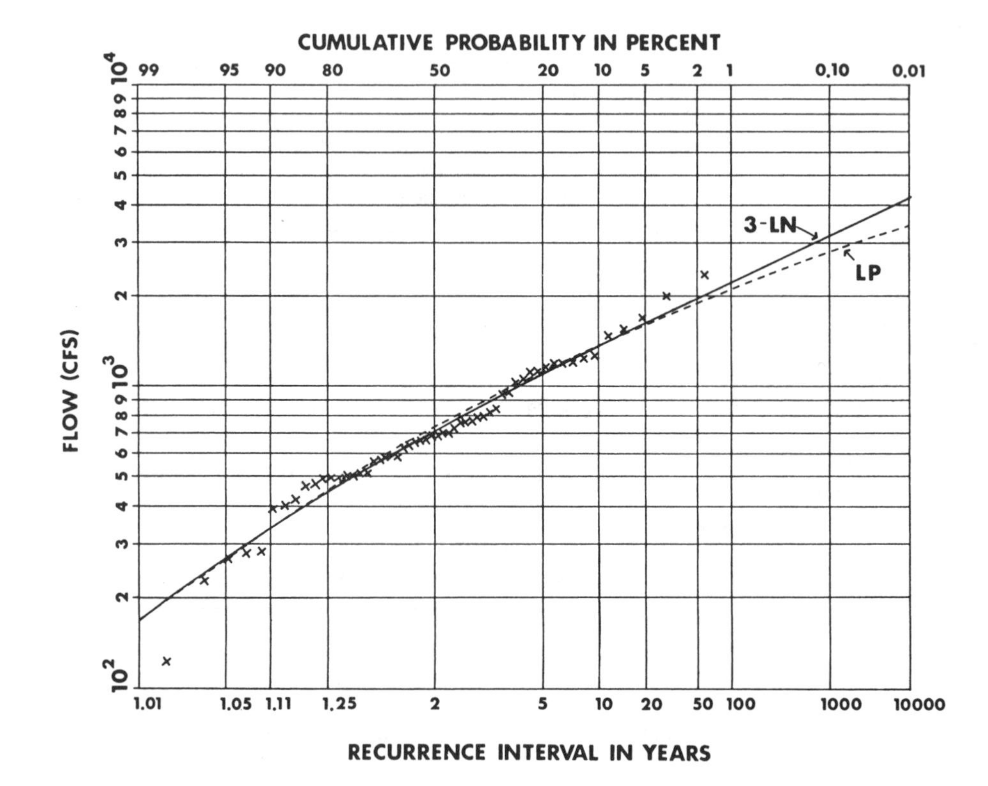

Figure 5. Flood frequency curves for the Credit River near Cataract, Ontario, Canada.

ESTIMATING PEAK FROM MEAN DAILY FLOWS

In hydrologic studies, it is often necessary to determine the peak flow at a given location. Discharge records, as usually published, provide only the mean daily flows, except for stations equipped with recording gauges for which the peak rates are also published. However, the data on peak flows are limited, and most often the investigations have to be based upon mean daily flows. These flows can vary considerably in relation to peaks, depending upon the flood and the basin characteristics. Thus, the design values obtained by the use of maximum mean daily flows instead of peak flows can be substantially in error, especially in the case of small watersheds.

Estimation of peak from mean daily flows has always been a problem in hydrology. The earliest study of this problem was carried out by Fuller [61]. He collected the available flood data of 24 river basins located in the eastern United States. The drainage areas of these basins varied from 1.18 sq. mi.–58,530 sq. mi. (3.06 km^2–151,592 km^2). He plotted the ratio of the excess of the peak over the maximum mean daily flow, to the maximum mean daily flow against the drainage area on a log-log paper and drew an average curve. This curve gave the relationship

$$Q(max) = Q(1 + 2A^{-0.3}) \tag{79}$$

in which Q(max) = peak flow, cfs

Q = maximum mean daily discharge, cfs

A = drainage area, sq. mi.

An examination of his (Fuller's) Plate XII [61] suggests that the relationship has considerable personal bias. Statistically, there is a very poor relationship between the drainage area and the ratio, [Q(max) – Q]/Q; the coefficient of determination is 0.45. The only reason that the formula has been so widely used is its simplicity and the lack of any alternative approach.

Another study was carried out by Jarvis et al. [62]. They collected a large amount of data on peak discharge, maximum calendar day, and maximum 24-hr flows but could not arrive at any definite result. They, however, suggested that a fair approximation of the peak could be made by plotting the mean daily discharges as bar ordinates, and sketching a hydrograph in such a manner that correct daily volumes were maintained. This method is subject to personal judgment and is not suitable for small watersheds.

Langbein [63] attempted another method. He derived a chart from reported data on peaks and corresponding mean daily flows. This chart is given in some hydrology textbooks [2–4]. The ratio of peak to maximum mean daily flow and the time of peak are shown as functions of the ratios of mean flow on the maximum day to the mean flow on the days immediately preceding and following the maximum day. This chart does not always give reliable results. Also, the maximum contour shown on it is 2.5, while much higher ratios of peak to maximum mean daily flow have been observed in practice.

The relation of peak to maximum mean daily flow is complex. Sangal [64] has examined this relationship. He has also developed a practical approach of estimating peak from mean daily flows. This approach and its application to a large number of streams in Canada and the United States are presented herein.

Relation of Peak to Maximum Mean Daily Flow

Consider the triangular hydrograph given in Figure 6. Day 0 is the measurement day of the peak. The preceding and succeeding days are shown with minus and plus signs, respectively. The areas on Day 0 prior to and after the peak are given by X and Y, respectively. The peak index is 0 if the peak occurs on the same day as the maximum mean daily flow. It is equal to −1 if the peak occurs on the preceding day.

The number of measurement days is equal to base length in days or exceeds it by one. Any part of base length less than 24 hr counts as one day. For instance, a hydrograph with a base length of 36 hr will be measured in either two or three days depending upon the location of peak.

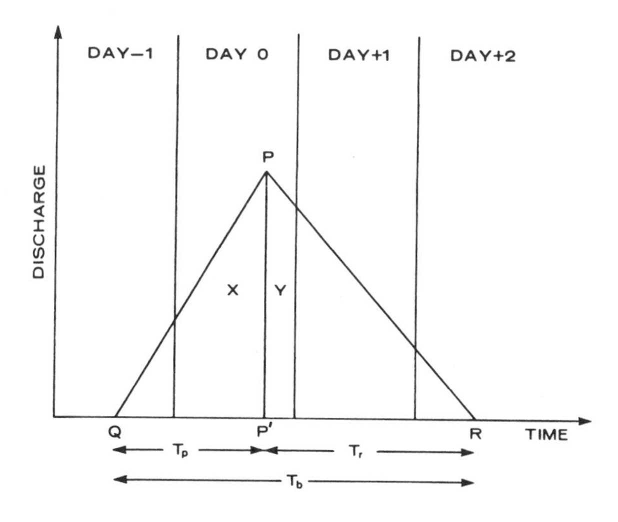

Figure 6. Schematic triangular hydrograph.

Table 7
Variation of QP/QD and other Hydrograph Properties with Time of Peak for a Triangular Hydrograph with T_b = 24 hr. and T_p/T_r = 0.6

Time of Peak (hrs)	A−1	X	Y	AO	A+1	QD	Peak Index	QP/QD
0	4.50	0	7.50	7.50	0	0.31	0	3.20
3	2.00	2.50	7.50	10.00	0	0.42	0	2.40
6	0.50	4.00	7.50	11.50	0	0.48	0	2.09
9	0	4.50	7.50	12.00	0	0.50	0	2.00
12	0	4.50	7.20	11.70	0.30	0.49	0	2.05
15	0	4.50	6.30	10.80	1.20	0.45	0	2.22
18	0	4.50	4.80	9.30	2.70	0.39	0	2.58
21	0	4.50	2.70	7.20	4.80	0.30	0	3.33
22:25	0	4.50	1.50	6.00	6.00	0.25	0	4.00
23	0	4.50	0.97	5.47	6.53	0.27	-1	3.67
24	0	4.50	0	4.50	7.50	0.31	-1	3.20

The ratio of peak, QP, to maximum mean daily flow, QD, depends upon three factors:

1. Base length
2. Shape of hydrograph, here defined as T_p/T_r
3. Time of peak

By varying these factors, the variation of the ratio, QP/QD, can be studied for a wide range of hydrographs. In the triangular hydrograph under consideration, the mean daily flow on a given day is obtained by dividing the area covered on that day by 24, where the time axis is in hours. Sample computations for a hydrograph with $T_p/T_r = 0.60$ ($T_r/T_p = 1.67$) and base length of 24 hr are given in Table 7. The total area of this hydrograph is 12 units (cubic feet per second-hours, or cubic meters per second-hours).

This table shows that, for the hydrograph under consideration, the ratio of peak to maximum mean daily flow is minimum when the peak is at 9 hr; the value is equal to 2. The maximum ratio occurs when the peak is at about 22:25 hr; the value is equal to 4. After this maximum is reached, the maximum mean daily flow occurs on the following day.

Figures 7 and 8 give the variation of the ratios of peak to maximum mean daily flow for sample hydrographs with T_p/T_r ratios of 1.0 and 0.6, respectively. The base lengths are 6, 12, 24, 48, and 72 hr. The arrows on Figure 8 indicate the positions of maximum ratios after which the maximum mean daily flow occurs on the following day and the ratio decreases. The following results can be derived from these analyses:

1. The maximum and minimum ratios of QP/QD are independent of T_p/T_r ratios. They depend only upon base length.
2. The maximum ratios of QP/QD for all T_p/T_r ratios are 16, 8, 4, 2, and 1.5 for base lengths of 6, 12, 24, 48, and 72 hr, respectively.
3. The minimum ratios of QP/QD for all T_p/T_r ratios are 8, 4, 2, 1.33, and 1.2 for the preceding base lengths, respectively. These ratios correspond to maximum 24 hr flow.

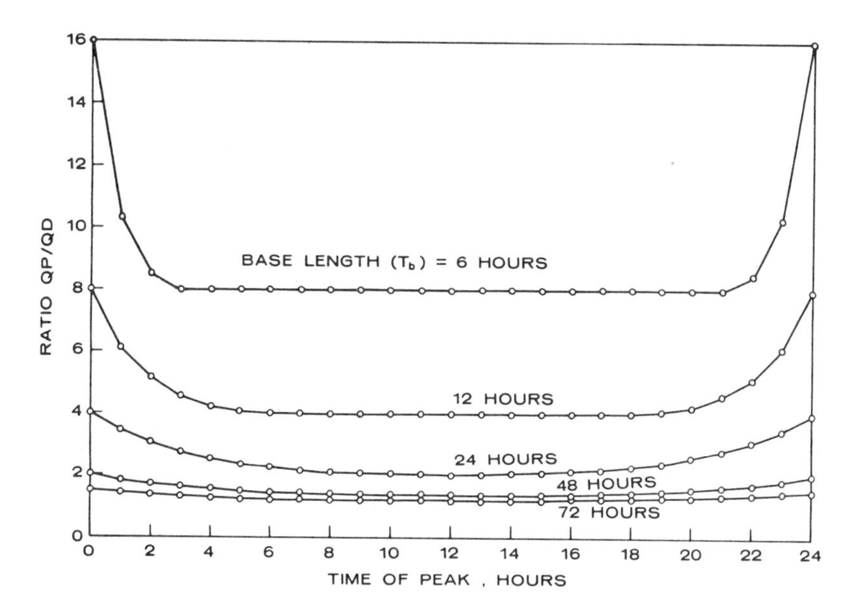

Figure 7. QP/QD versus time of peak when $T_p/T_r = 1.0$.

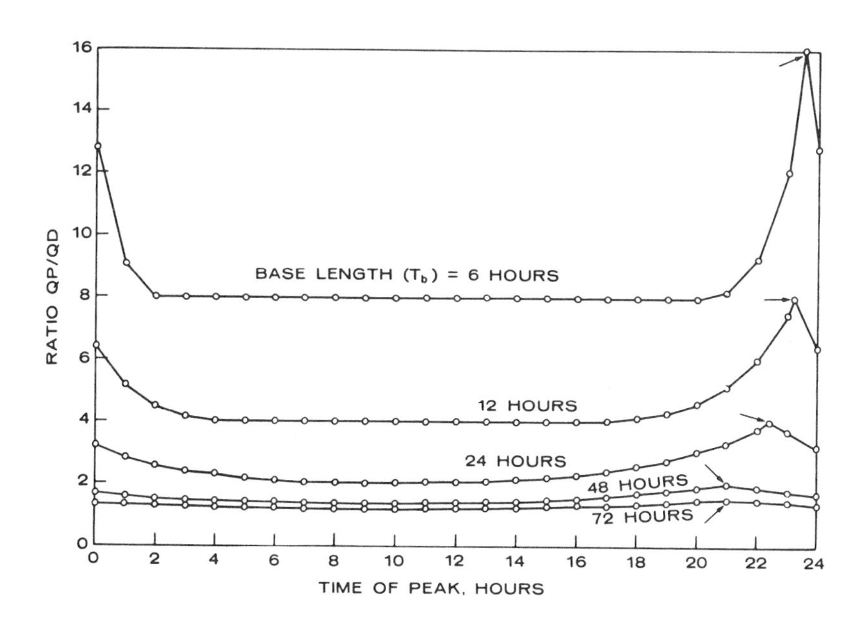

Figure 8. QP/QD versus time of peak when $T_p/T_r = 0.6$.

4. The times of occurrence of maximum and minimum ratios are different for hydrographs of different shapes.
5. There is only one time of peak for which the ratio is maximum. The minimum ratio can exist for peak occurring during a considerable length of the day.
6. If the base of the hydrograph is equal to or less than 24 hr, then the maximum 24-hr flow is obtained when the total hydrograph falls within the measurement day.
7. If the base of the hydrograph is greater than 24 hr, then the maximum 24-hr flow is obtained when the location of the peak is such that the flow at 0 hr is equal to flow at 24 hr.

The preceding analyses have been performed considering the initial flow in the stream as zero. Also, use of these results in estimating peak can be made only if the base of the hydrograph, its shape, and the time of peak are known. Most often these data are not available. To use the available data which consist only of the mean daily flows, a modified approach is needed. This approach is analyzed in the following.

Suggested Approach [64]

Consider the schematic diagram shown in Figure 9. Let Q2 be the maximum mean daily flow, with Q1 and Q3 being the mean daily flows on preceding and succeeding days, respectively. Let us assume that the actual hydrograph is such that Q1 can be extended forward inside the measurement day of Q2 through the time interval, α. Similarly, Q3 can be extended backward through the time interval, α'. It may be noted that α and α' can be negative as well. Assuming the average flow of $(Q1 + Q3)/2$ for the time interval of $1 - \alpha - \alpha'$ day, and a triangular hydrograph of height h superposed over this interval, we have

$$Q2 = \frac{Q1 + Q3}{2}(1 - \alpha - \alpha') + \alpha Q1 + \alpha' Q3 + (1 - \alpha - \alpha')\frac{h}{2} \tag{80}$$

or

$$h = \frac{2Q2 - Q1(1 + \alpha - \alpha') - Q3(1 + \alpha' - \alpha)}{(1 - \alpha - \alpha')} \tag{81}$$

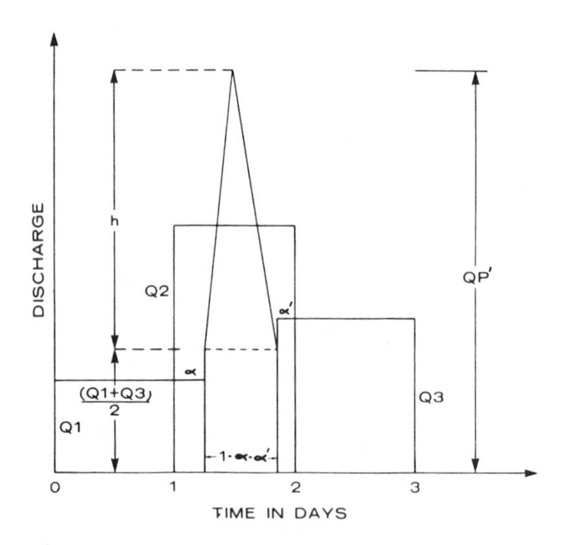

Figure 9. Schematic diagram of assumed flows.

The estimated peak, QP′, is equal to h + (Q1 + Q3)/2. Therefore,

$$QP' = \frac{Q1 + Q3}{2} + \frac{2Q2 - Q1(1 + \alpha - \alpha') - Q3(1 + \alpha' - \alpha)}{1 - \alpha - \alpha'} \tag{82}$$

This is the general equation which can be used for estimating the actual peak, QP. It can be simplified if α and α' are assumed to be equal. Thus

$$QP' = \frac{Q1 + Q3}{2} + \frac{2Q2 - Q1 - Q3}{1 - 2\alpha} \tag{83}$$

If $\alpha = \alpha' = 0$, then

$$QP' = \frac{4Q2 - Q1 - Q3}{2} \tag{84}$$

Equation 84 can be used for predicting the peak. If the predicted peak is denoted by QPP, then

$$QPP = \frac{4Q2 - Q1 - Q3}{2} \tag{85}$$

The upper limit of QPP from this formula is 2Q2 and the lower limit is Q2.

In Equation 83, the term $1 - 2\alpha$ can be defined as "base factor" and denoted by K. Substituting the actual peak QP for QP′, we have

$$K = \frac{4Q2 - 2Q1 - 2Q3}{2QP - Q1 - Q3} \tag{86}$$

Since $QP \geq Q2$, $K \leq 2$. Also, the minimum value of the numerator is zero as Q2 cannot be less than Q1 or Q3. Therefore, $0 \leq K \leq 2$.

If the ratio of the actual peak to the predicted peak is denoted by R, then

$$R = QP/QPP \tag{87}$$

Equations 85 and 86 now yield

$$KR = \frac{4Q2 - 2Q1 - 2Q3}{4Q2 - Q1 - Q3 - \dfrac{Q1 + Q3}{R}} \tag{88}$$

From this equation, we have the following:

$$\text{If } R > 1, \text{ then } KR < 1, \text{ and } K < 1 \tag{89}$$

$$\text{If } R = 1, \text{ then } KR = K = 1 \tag{90}$$

$$\text{If } R < 1, \text{ then } KR > 1, \text{ and } K > 1 \tag{91}$$

Thus, if KR (y-axis) is plotted against K (x-axis), all points will be enclosed in two right-angled triangles. The first one will have its end points at (0, 0), (0, 1), and (1, 1), and the second one at (1, 1), (2, 1), and (2, 2). The first triangle gives the zone of underprediction, and the second triangle gives the zone of overprediction. The line KR = K is the line of exact prediction. These plots have been prepared for a large number of streams in Canada and the United States and will be considered later. The limiting lines KR = 1 and KR = K correspond to the ratio (Q1 + Q3)/2 Q2 being equal to zero and one, respectively.

Importance of Base Factor K

The predicting Equation 85 is based upon K = 1. Peaks for which K > 1 will be overpredicted, and those for which K < 1 will be underpredicted. In practice, K is unknown. However, the values of K > 1 are not of much concern, as the degree of overprediction is limited by the condition that $K \leq 2$. The values of K < 1 are important. As K represents the base of a triangular hydrograph, its value can be estimated by the U.S. Soil Conservation Service method or any other method of synthesizing unit hydrograph. This value, expressed as a fraction of a day, can be substituted for $1 - 2\alpha$ in Equation 83 and the peak can be estimated. When peak data are available, values of K can be computed using Equation 86. If necessary, these values may be transferred to nearby streams of similar physical characteristics. Also, the historical values of K, together with those of QPP and R, may be useful for a stream if the recording gauge is to be discontinued or missing peaks have to be estimated. It may be emphasized that K is not a constant for a basin.

Characteristics of Q1, Q2, and Q3

When three-day flows are considered in defining a hydrograph, four cases are encountered:

1. Q1 and Q3 are both low in relation to Q2.
2. Q1 and Q3 are both high in relation to Q2.
3. Q1 is high and Q3 is low.
4. Q3 is high and Q1 is low.

The first two cases give a balanced hydrograph with respect to Q2, while the last two cases give an imbalanced hydrograph. In the case of a balanced hydrograph, the peak would generally occur during the daytime hours and on the same day as Q2. In the case of an imbalanced hydrograph, the peak would generally occur during nighttime hours and could occur on the same day as Q2 or on a different day. The formula as suggested here could underpredict the peak of an imbalanced hydrograph. This situation can be corrected by "balancing" the hydrograph. A small fraction, say 10%–15%, of Q1 or Q3, whichever is closer to Q2, can be taken and added to Q2. In this adjustment, Q1 + Q2, or Q3 + Q2 should remain the same. These adjusted values can then be used in Equation 85. This technique is not infallible and should only be applied with judgment.

The method as just presented is general and can be universally applied. Due to easy access to the data, a large number of streams in Canada and the United States have been analyzed according to this method and the results are now presented.

Application of Method to Streams in Canada

Canada is a large country with varying climate. The streamflow throughout the country is mostly due to snowmelt although sometimes thunderstorms and hurricanes cause heavy flooding locally. Table 8 gives the mean temperature and precipitation data for some selected stations in the country [65].

Figure 10 is a map of Canada showing provincial and territorial boundaries and 11 drainage divisions as adopted by the Water Survey of Canada in their station numbering system. The drainage division No. 11 is not numbered in this figure. It is the small division near the Canada-U.S. boundary in the provinces of Alberta and Saskatchewan and drains into the United States.

In order to apply the method throughout the country, the flow records of all gauging stations with recording gauges were retrieved from the Water Survey of Canada tapes. Computations were carried out on all those data which were found to be complete. Table 9 shows the computations for a sample station, Humber River near Cedar Mills, Ontario.

Table 8
Climate Data for Selected Stations in Canada (T, Temperature in °C; P, precipitation in mm)

Station		J	F	M	A	M	J	J	A	S	O	N	D	Year
Fort Nelson, B.C.	T	-23.2	-17.1	-9.2	1.2	9.7	14.5	16.7	14.8	8.8	1.2	-12.3	-20.7	-1.3
	P	26.4	24.4	24.9	21.6	37.6	64.3	74.7	55.6	38.6	25.7	26.7	25.9	446.4
Vancouver, B.C.	T	2.4	4.4	5.8	8.9	12.4	15.3	17.4	17.1	14.2	10.1	6.1	3.8	9.8
	P	147.3	116.6	93.7	61.0	47.5	45.2	29.7	37.1	61.2	122.2	141.2	165.4	1068.1
Dawson, Y.T.	T	-28.6	-23.0	-14.1	-1.8	7.8	13.9	15.5	12.7	6.4	-3.2	-16.5	-25.3	-4.7
	P	19.3	16.0	12.7	9.1	21.8	36.8	53.1	50.6	28.5	26.7	25.2	25.7	325.5
Yellowknife, N.W.T.	T	-28.6	-25.7	-18.6	-7.8	4.0	12.2	16.0	14.1	6.8	-1.2	-14.2	-23.8	-5.6
	P	13.7	12.2	11.7	10.2	14.0	17.3	33.3	36.3	28.2	30.7	23.9	18.5	250.0
Calgary, Alta.	T	-10.9	-7.4	-4.3	3.3	9.3	13.2	16.5	15.2	10.7	5.7	-2.6	-7.6	3.4
	P	17.0	19.8	20.3	29.5	49.8	91.7	68.3	55.9	35.3	18.8	16.0	14.7	437.1
Edmonton, Alta.	T	-14.7	-10.5	-5.4	4.0	10.9	14.7	17.5	15.9	10.9	5.4	-4.2	-10.7	2.8
	P	25.2	20.1	16.8	23.4	37.3	74.7	83.3	71.6	35.8	18.5	18.5	21.3	446.5
Regina, Sask.	T	-17.3	-14.3	-8.3	3.3	10.6	15.3	18.9	17.9	11.6	5.3	-5.2	-12.9	2.1
	P	18.0	17.3	18.3	23.4	40.9	82.6	57.9	49.8	36.3	19.1	18.0	16.3	397.9
Saskatoon, Sask.	T	-18.7	-15.1	-8.7	3.3	10.6	15.4	18.8	17.4	11.3	5.0	-5.8	-14.0	1.6
	P	18.3	18.0	16.8	20.6	34.0	57.4	53.1	45.2	33.0	19.1	18.8	18.3	352.6
Churchill, Man.	T	-27.6	-26.7	-20.3	-11.0	-2.3	6.1	12.0	11.5	5.7	-1.0	-11.9	-21.8	-7.3
	P	14.0	13.0	17.8	24.1	28.2	40.1	49.0	57.7	52.1	40.4	40.1	20.1	396.6
Winnipeg, Man.	T	-18.3	-15.7	-8.1	3.3	10.6	16.5	19.7	18.7	12.6	6.6	-4.4	-13.7	2.3
	P	23.6	19.1	26.2	37.3	57.2	80.3	80.3	73.7	52.6	34.8	27.2	22.9	535.2
Kenora, Ont.	T	-17.7	-14.7	-7.0	2.5	9.8	15.7	19.1	17.8	11.7	6.0	-4.6	-13.8	2.1
	P	30.0	26.9	33.3	42.9	57.7	86.6	99.1	82.6	72.6	37.6	42.7	35.3	647.3
Toronto, Ont.	T	-4.4	-3.8	0.6	7.6	13.2	19.2	21.8	21.1	17.0	11.2	4.8	-1.8	8.9
	P	62.5	56.6	65.5	67.3	72.9	63.0	80.8	67.3	61.2	61.5	67.3	64.0	789.9
Ottawa, Ont.	T	-10.9	-9.5	-3.1	5.6	12.4	18.2	20.7	19.3	14.6	8.7	1.4	-7.7	5.8
	P	59.9	56.9	61.0	67.6	70.1	72.6	81.3	81.5	78.7	65.8	78.5	77.0	850.9
Montreal, Que.	T	-8.9	-7.6	-1.4	6.7	13.6	19.1	21.6	20.4	15.8	10.1	2.9	-5.7	7.2
	P	79.5	71.4	75.2	77.0	74.9	87.1	93.0	91.7	86.6	79.0	92.7	90.9	999.0
Schefferville, Que.	T	-22.7	-21.1	-14.8	-6.9	0.9	8.5	12.6	10.8	5.6	-0.9	-8.7	-18.1	-4.6
	P	41.2	36.6	36.8	34.8	44.7	79.0	88.7	98.0	82.8	70.1	63.8	46.0	722.5
Fredericton, N.B.	T	-9.2	-8.5	-2.6	4.0	10.5	15.7	19.1	18.0	13.6	7.9	1.8	-6.3	5.3
	P	90.7	86.4	73.4	80.8	87.6	86.1	90.2	85.9	86.9	90.9	119.9	105.4	1084.2
Halifax, N.S.	T	-3.8	-4.2	-0.7	4.0	9.0	13.7	17.6	17.8	14.8	9.9	4.9	-1.1	6.8
	P	147.3	128.5	111.8	105.4	109.5	85.1	92.0	94.0	94.2	113.3	151.9	148.3	1381.3
Charlottetown, P.E.I.	T	-6.7	-7.2	-3.2	2.3	8.6	14.1	18.4	17.9	13.9	8.6	3.3	-3.6	5.6
	P	97.5	82.3	76.5	74.7	79.5	79.0	74.2	90.2	92.5	98.6	114.6	99.6	1059.2
Goose, Nfld.	T	-16.3	-14.4	-8.4	-1.8	4.9	11.1	15.8	14.5	9.8	3.2	-3.6	-12.3	0.2
	P	69.1	60.2	69.3	54.1	61.7	81.5	102.1	92.7	76.0	71.9	69.9	68.3	876.8
St. John's, Nfld.	T	-3.8	-4.2	-2.4	1.1	5.5	10.4	15.3	15.4	11.9	7.1	3.5	-1.3	4.9
	P	145.0	156.2	132.6	114.1	99.1	88.7	83.1	113.3	112.0	138.7	161.3	167.4	1511.5

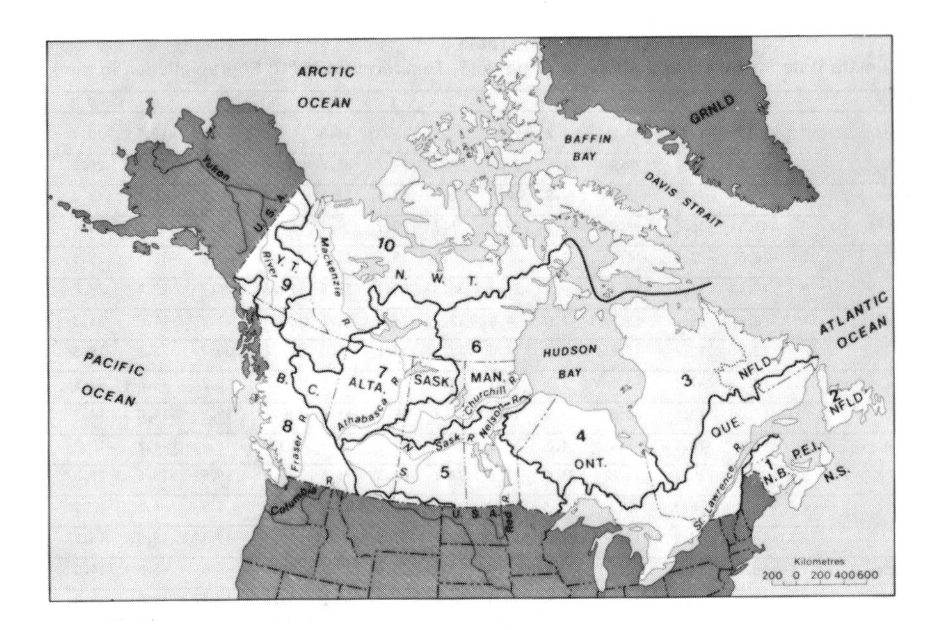

Figure 10. Map of Canada showing drainage divisions.

In this table, QP is the observed peak; Q1 is the flow on day preceding the maximum day; Q2 is flow on maximum day; Q3 is flow on day following the maximum day; QPP is the peak predicted by Equation 85; and K is the base factor computed by Equation 86. The peak index gives the day of peak in relation to the day of maximum mean daily flow. A positive value indicates that the peak occurred later than the maximum mean daily flow while a negative value indicates that the peak occurred earlier. A zero or blank indicates that the day of peak coincides with the day of maximum flow. The numbers give the respective number of days.

Statistical Analysis of Data Used

Gauging stations. A total of 2,410 stations equipped with recording gauges were used. All stations, regulated and unregulated, were analyzed. The drainage areas varied from less than 1 km^2 to more than 100,000 km^2. For some stations, the drainage areas were not available. Table 10 gives the distribution of these areas at specified intervals by drainage division.

Table 9
Computations for Sample Station (Discharges Are in m^3/sec)

Year	Date and Time of Peak	QP	Q1	Q2	Q3	Date of Q2	Peak Index	QP/Q2	Predicted Peak QPP	QP/QPP	K
1966	6/14, 18:00	8.86	0.963	4.59	3.82	6/14	0	1.93	6.79	1.31	0.68
1968	3/20, 23:30	38.5	13.6	26.6	18.4	3/19	1	1.45	37.2	1.03	0.94
1978	4/1, 17:43	13.9	8.95	9.40	4.67	4/2	-1	1.48	12.0	1.16	0.73

Table 10
Distribution of Drainage Areas (Canada)

Interval in km^2	Number of Stations (By Drainage Division) 1	2	3	4	5	6	7	8	9	10	11	Total Number of Stations
<1	3	5	0	0	17	3	0	0	0	0	0	28
1-100	48	128	0	0	93	0	35	99	1	9	3	416
100-500	58	172	0	1	212	8	41	146	3	11	20	671
500-1000	11	68	0	1	106	3	21	51	5	5	11	283
1000-5000	23	85	3	14	159	22	65	86	12	19	16	504
5000-100000	10	28	10	24	87	30	45	51	25	34	2	346
>100000	0	0	0	1	19	6	9	6	2	9	0	52
Unknown	3	11	0	1	47	1	5	30	0	9	3	110
Total	156	497	13	42	740	73	221	469	48	96	55	2410

Period of record with recording gauges. The data are up to and including the calendar year 1982. The data of discontinued stations were included in the analysis. The total number of stations was 2,410 with a total of 24,626 station-years of data. Table 11 gives the distribution of the period of recording gauges. About 61% of the gauges had records of 10 years or less.

Monthly percentage distribution of annual maximum peaks. Table 12 gives the monthly percentage distribution of annual maximum peaks. It shows that, for the country as a whole, about two-thirds of the peaks occur during the months of April, May, and June. The peaks in the southern part occur as early as March while in the north, they occur in July. As mentioned earlier, almost all these floods are due to snowmelt.

Time of peak and peak index. Theoretically, in the case of a simple triangular hydrograph, the peak can occur on the day of maximum mean daily flow or on the preceding day. It cannot occur on any other day. However, in practice, the peaks have been observed to occur on the day of maximum mean daily flow or a few days earlier or later. This is because of nontriangular shape of hydrograph. The pattern of rain and snowmelt, regulation, tributary inflow, ice jams, natural rounding of the peak, etc., all have an effect on the shape of the hydrograph. The peak index gives the day of the peak in relation to the day of maximum mean daily flow. Only the indices of −2, −1, 0, 1, and 2 have been considered. A peak reported to have occurred at an interval of three or more days was adjusted, by judgment, to correspond to one of these indices. The number of such cases was insignificantly small. Table 13 gives the distribution of the time of peak and the peak indices for 24,626 peaks.

Table 11
Distribution of Period of Recording Gauges (Canada)

Period (yr)	1–5	6–10	11–15	16–20	21–25	26–30	>30
Number of Stations	866	610	418	277	97	53	89

Table 12
Monthly Percentage Distribution of Annual Maximum Peaks (Canada)

Month	J	F	M	A	M	J	J	A	S	O	N	D
Percentage	1.80	2.48	8.86	23.90	20.17	21.56	8.08	3.31	1.89	2.52	2.24	3.19

This table suggests that about 79% of the peaks occur on the same day as the maximum mean daily flow. About 15% occur on the previous day and about 5% occur on the following day. The total number of peaks occurring at two-day intervals is about 1%. It is interesting to note that this distribution is almost identical to the distribution obtained for Ontario only with a much smaller data sample [66]. There is no clearly visible trend of the time of peak. About 46% of the peaks occur from midnight to noon and 54% occur from noon to midnight. Nonuniform distribution of gauges, a wide range of drainage areas, six time zones within the country and varying climate from coast to coast greatly affect the distribution of the time of peak. The peaks with indices −1 and +1 show a concentration during nighttime hours.

Table 13
Distribution of Time of Peak and Peak Indices (Canada)

	Peak Indices						
Time (hr)	−2	−1	0	1	2	Total	Percentage
0:00– 0:59	6	61	630	171	10	878	3.57
1:00– 1:59	6	17	887	188	6	1104	4.48
2:00– 2:59	3	12	813	127	8	963	3.91
3:00– 3:59	3	16	823	87	7	936	3.80
4:00– 4:59	3	10	766	45	3	827	3.36
5:00– 5:59	5	12	721	49	2	789	3.20
6:00– 6:59	5	9	842	47	2	905	3.67
7:00– 7:59	6	16	774	33	3	832	3.38
8:00– 8:59	8	29	902	34	4	977	3.97
9:00– 9:59	9	47	817	34	10	917	3.72
10:00–10:59	12	52	848	27	9	948	3.85
11:00–11:59	13	55	872	28	4	972	3.95
12:00–12:59	5	76	987	36	3	1107	4.50
13:00–13:59	7	85	819	19	10	940	3.82
14:00–14:59	15	78	834	25	10	962	3.91
15:00–15:59	13	125	862	22	5	1027	4.17
16:00–16:59	14	150	843	20	8	1035	4.20
17:00–17:59	13	179	839	26	13	1070	4.35
18:00–18:59	9	253	816	14	8	1100	4.47
19:00–19:59	14	283	728	21	6	1052	4.27
20:00–20:59	13	376	745	3	7	1144	4.65
21:00–21:59	14	435	619	13	7	1088	4.42
22:00–22:59	19	540	507	21	5	1092	4.43
23:00–23:59	22	648	427	18	4	1119	4.54
Missing Time	7	120	638	71	6	842	3.42
Total	244	3684	19359	1179	160	24626	100.00
Percentage	.99	14.96	78.61	4.79	.65	100.00	

Results and Analysis

Four parameters have been computed from the original data:

1. The ratio of peak to maximum mean daily flow, QP/Q2.
2. The predicted peak, QPP.
3. The ratio of observed peak to predicted peak, QP/QPP.
4. The base factor K.

Another relationship which has been studied is between KR and K.

Distribution of QP/Q2. This distribution is given in Table 14. About 52% of the values lie within the interval 1.0–1.1. This is due to predominance of snowmelt floods. Other factors such as the large size of the basin, routing through lakes and reservoirs, etc. also tend to lower this ratio. The higher ratios were mostly given by small urban streams.

Distribution of QP/QPP. The acid test of any predictive technique is provided by a comparison of the predicted values with the recorded data. This comparison is given in Table 15. About 61% of the values lie within ±10% of the actual values, and about 82% lie within ±20% of the actual values. No ratio is less than 0.6 although the theoretical minimum is 0.5. More than 66% of the ratios are less than 1.0, indicating the degree of overprediction. This table shows that the suggested technique can predict the peak with reasonable accuracy throughout Canada, especially for snowmelt floods. For rainfall floods occurring on small streams, smaller values of base factor K must be used. The value of K = 1 is too large for these streams.

Distribution of base factor K. Table 16 gives the distribution of base factor K. About 34% of the values are less than 1. The value of K = 0 is obtained when Q1 = Q2 = Q3 and K = 2 is obtained when QP = Q2.

Relationship between KR and K. A plot of KR vs. K clearly demonstrates the zones of undeprediction and overprediction and the line of exact prediction KR = K. As the number of data was large, a smaller sample of 977 stations was selected for this plot to avoid clutter. The stations were

Table 14
Distribution of QP/Q2 (Canada)

Interval	Number	Percentage of Total	Cumulative Percentage	Interval	Number	Percentage of Total	Cumulative Percentage
1.0–1.1	12866	52.25	52.25	2.0– 2.2	338	1.37	95.36
1.1–1.2	3790	15.39	67.64	2.2–2.4	260	1.06	96.42
1.2–1.3	2130	8.65	76.29	2.4–2.6	158	.64	97.06
1.3–1.4	1370	5.56	81.85	2.6–2.8	116	.47	97.53
1.4–1.5	879	3.57	85.42	2.8–3.0	98	.40	97.93
1.5–1.6	699	2.84	88.26	3.0–4.0	218	.89	98.81
1.6–1.7	472	1.92	90.17	4.0–5.0	99	.40	99.22
1.7–1.8	389	1.58	91.75	5.0–10.0	127	.52	99.73
1.8–1.9	307	1.25	93.00	> 10	66	.27	100.00
1.9–2.0	244	.99	93.99				

Table 15
Distribution of QP/QPP (Canada)

Interval	Number	Percentage of Total	Cumulative Percentage	Interval	Number	Percentage of Total	Cumulative Percentage
.6- .7	232	.94	.94	1.8- 1.9	81	.33	97.57
.7- .8	1474	5.99	6.93	1.9- 2.0	66	.27	97.84
.8- .9	3962	16.09	23.02	2.0- 2.2	101	.41	98.25
.9-1.0	10636	43.19	66.21	2.2- 2.4	94	.38	98.63
1.0-1.1	4348	17.66	83.86	2.4- 2.6	46	.19	98.81
1.1-1.2	1323	5.37	89.23	2.6- 2.8	42	.17	98.98
1.2-1.3	704	2.86	92.09	2.8- 3.0	35	.14	99.13
1.3-1.4	481	1.95	94.05	3.0- 4.0	86	.35	99.48
1.4-1.5	309	1.25	95.30	4.0- 5.0	34	.14	99.61
1.5-1.6	202	.82	96.12	5.0-10.0	64	.26	99.87
1.6-1.7	158	.64	96.76	> 10	31	.13	100.00
1.7-1.8	117	.48	97.24				

Table 16
Distribution of Base Factor K (Canada)

Interval	Number	Percentage of Total	Cumulative Percentage	Interval	Number	Percentage of Total	Cumulative Percentage
0.00	314	1.28	1.28	1.1-1.2	1934	7.85	49.61
0.0- .1	111	.45	1.73	1.2-1.3	1798	7.30	56.92
.1- .2	249	1.01	2.74	1.3-1.4	2009	8.16	65.07
.2- .3	396	1.61	4.35	1.4-1.5	1978	8.03	73.11
.3- .4	566	2.30	6.64	1.5-1.6	2123	8.62	81.73
.4- .5	695	2.82	9.47	1.6-1.7	1715	6.96	88.69
.5- .6	911	3.70	13.16	1.7-1.8	1332	5.41	94.10
.6- .7	1197	4.86	18.03	1.8-1.9	597	2.42	96.52
.7- .8	1181	4.80	22.82	1.9-2.0	141	.57	97.10
.8- .9	1309	5.32	28.14	2.00	715	2.90	100.00
.9-1.0	1451	5.89	34.03				
1.0-1.1	1904	7.73	41.76				

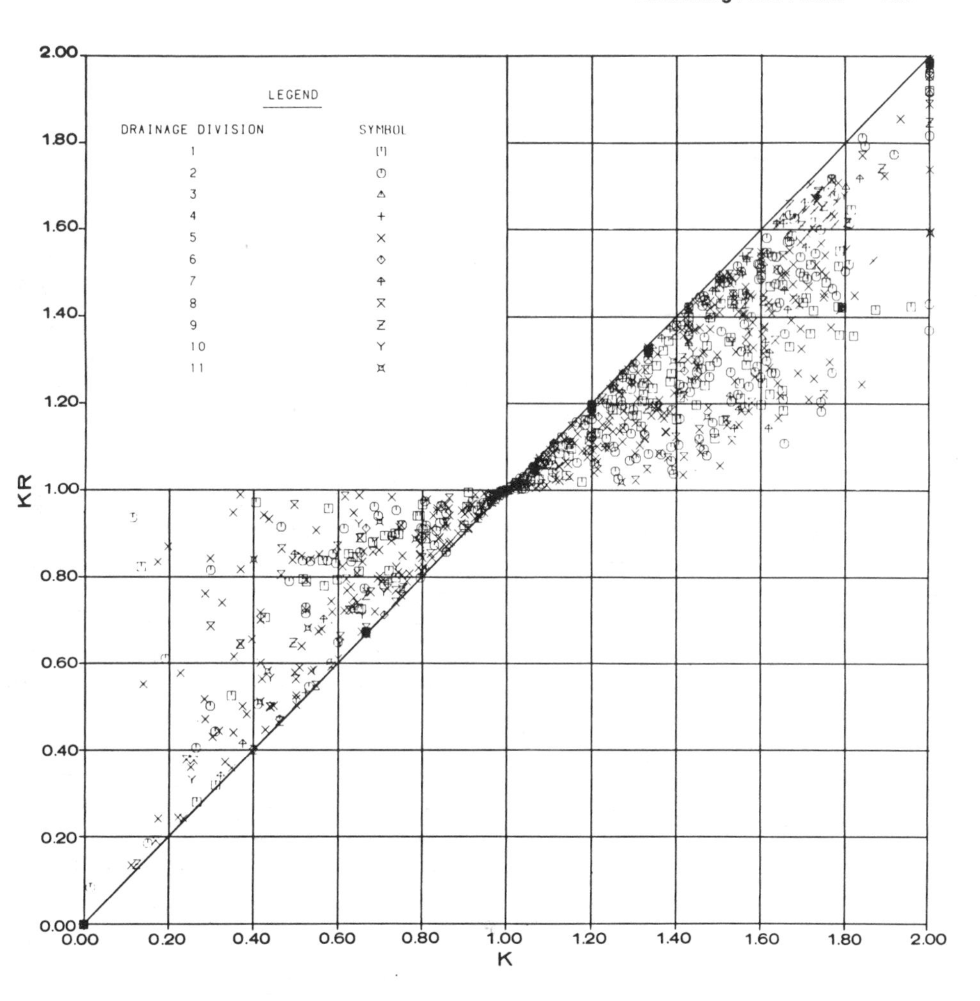

Figure 11. Relationship between KR and K (Canada).

spread almost uniformly throughout the country and covered a wide range of drainage areas. Only one year of data for each station was considered. This plot is presented in Figure 11. A different symbol has been used for each of the 11 divisions. The station numbers and years of peak used in this plot are identified in Appendix A. More detailed information about these stations can be obtained from the Water Survey of Canada Reference Index [67].

Application of Method to Streams in the United States

The method has been applied to flow data from all 50 states including Alaska and Hawaii. There are large variations in climatic conditions which affect the streamflow characteristics. A large proportion of the annual flow is due to snowmelt. However, often heavy flooding is caused by thunderstorms and hurricanes. There are desert conditions in a few western states but occasionally they too experience severe flooding due to sudden cloudbursts. Table 17 gives the mean monthly

Table 17
Climate Data for Selected Stations in the United States of America
(T, temperature °C;P, precipitation in mm)

Station		J	F	M	A	M	J	J	A	S	O	N	D	Year
Birmingham, Ala.	T	7.5	8.7	11.9	16.7	21.3	25.5	26.7	26.4	23.5	17.6	11.0	7.7	17.1
	P	128	134	152	114	87	102	131	123	85	75	90	128	1347
Anchorage, Alaska	T	-10.9	-7.8	-4.8	2.1	7.7	12.5	13.9	13.1	8.8	1.7	-5.4	-9.8	1.8
	P	20	18	13	11	13	25	47	65	64	47	26	24	374
Pheonix, Ariz.	T	10.4	12.5	15.8	20.4	25.0	29.8	32.9	31.7	29.1	22.3	15.1	11.4	21.4
	P	19	22	17	8	3	2	20	28	19	12	12	22	184
Little Rock, Ark.	T	4.8	6.9	11.0	16.9	21.4	26.1	27.7	27.4	23.5	17.3	9.7	5.5	16.5
	P	133	110	122	125	134	92	85	72	82	73	105	104	1237
San Francisco, Calif.	T	10.4	11.7	12.6	13.2	14.1	15.1	14.9	15.2	16.7	16.3	14.1	11.4	13.8
	P	116	93	74	37	16	4	tr	1	6	23	51	108	529
Denver, Colo.	T	-1.1	0.3	3.3	8.6	13.7	19.4	23.0	22.2	17.5	11.3	4.0	0.6	10.2
	P	14	18	31	54	69	37	39	33	29	26	18	12	380
New Haven, Conn.	T	-1.6	-1.6	2.83	7.83	13.7	18.8	21.8	21.0	17.6	11.8	6.1	0.0	9.83
	P	99	84	105	99	98	97	93	104	88	76	100	100	1143
Wilmington, Del.	T	0.7	0.9	5.8	11.0	17.1	22.1	24.4	23.2	20.0	13.4	7.5	1.7	12.3
	P	90	76	92	92	97	102	114	134	97	76	85	76	1130
Washington, D.C.	T	2.7	3.2	7.1	13.2	18.8	23.4	25.7	24.7	20.9	15.0	8.7	3.4	13.9
	P	77	63	82	80	105	82	105	124	97	78	72	71	1036
Miami, Fla.	T	19.4	19.9	21.4	23.4	25.3	27.1	27.7	27.9	27.4	25.4	22.4	20.1	23.9
	P	52	48	58	99	164	187	171	177	241	209	72	42	1520
Atlanta, Ga.	T	7.1	7.8	10.8	15.7	20.6	24.8	26.1	25.7	22.8	16.9	10.7	7.1	16.3
	P	113	115	136	114	80	97	120	91	83	62	75	111	1197
Honolulu, Hawaii	T	21.7	21.7	21.7	22.8	23.9	25.0	25.6	25.6	25.6	25.0	23.9	22.2	23.9
	P	94	109	97	58	48	28	33	38	38	48	107	104	803
Boise, Idaho	T	-1.9	1.1	5.1	9.9	14.3	18.2	23.7	22.3	17.3	11.4	3.9	0.1	10.4
	P	34	34	34	29	33	23	5	4	10	21	30	34	290
Chicago, Ill.	T	-3.3	-2.3	2.4	9.5	15.6	21.5	24.3	23.6	19.1	13.0	4.4	-1.6	10.5
	P	47	41	70	77	95	103	86	80	69	71	56	48	843
Indianapolis, Ind.	T	-1.6	-0.5	3.8	10.4	16.3	21.7	24.0	23.2	19.2	13.0	4.9	-0.5	11.2
	P	77	58	87	95	101	117	89	77	82	67	78	68	996
Concordia, Kans.	T	-2.2	0.5	5.3	12.2	17.3	23.3	26.7	25.8	20.8	14.3	5.6	-0.1	12.5
	P	15	22	34	57	92	107	86	81	62	44	29	15	644
Louisville, Ky.	T	1.6	2.9	7.6	13.3	18.5	23.4	25.5	24.5	21.2	14.8	7.6	2.7	13.6
	P	104	76	119	102	100	103	78	78	69	62	79	84	1194
New Orleans, La.	T	13.3	14.7	17.2	21.0	24.5	27.7	28.4	28.6	26.8	22.7	16.9	13.9	21.3
	P	121	106	167	138	138	141	180	163	148	93	102	116	1613
Portland, Maine	T	-5.7	-5.1	-0.3	5.8	11.7	16.7	20.1	19.3	14.8	9.2	3.4	-3.4	7.2
	P	111	97	110	95	87	81	73	61	89	81	106	98	1089

(Continued)

Table 17 (Continued)

Station		J	F	M	A	M	J	J	A	S	O	N	D	Year
Baltimore, Md.	T	1.6	2.1	6.2	12.3	18.0	22.5	24.9	23.9	20.1	13.9	7.5	2.1	12.9
	P	87	73	97	91	101	84	107	132	85	81	80	76	1094
Boston, Mass.	T	-2.8	-2.6	1.6	7.6	13.7	18.4	21.6	20.8	16.9	11.5	5.6	-1.1	9.3
	P	114	95	115	102	88	95	83	103	100	95	115	101	1206
Detroit, Mich.	T	-2.8	-2.7	1.6	8.4	14.7	20.7	23.3	22.4	18.1	12.1	4.7	-1.2	9.9
	P	52	53	61	76	90	72	72	73	62	67	56	53	787
Duluth, Minn.	T	-12.9	-11.5	-5.7	3.1	9.8	15.2	18.9	17.9	12.6	6.7	-2.9	-10.0	3.4
	P	29	24	41	60	84	108	90	97	73	55	45	29	735
Vicksburg, Miss.	T	9.4	11.0	14.2	18.7	23.0	26.5	27.7	27.6	24.8	19.7	13.4	10.2	18.8
	P	130	135	146	125	105	88	99	76	64	52	113	125	1258
St. Louis, Mo.	T	-0.1	1.8	6.2	13.0	18.7	24.2	26.4	25.4	21.1	14.9	6.7	1.6	13.3
	P	50	52	78	94	95	109	84	77	70	73	65	50	897
Billings, Mont.	T	-4.9	-3.5	0.7	7.5	13.2	17.6	22.9	21.6	15.8	10.0	2.3	-1.7	8.4
	P	14	15	27	33	48	65	23	23	30	28	16	15	337
Omaha, Nebr.	T	-5.4	-3.1	2.7	10.9	17.2	22.8	25.8	24.6	19.4	13.2	3.8	-2.1	10.8
	P	21	24	37	65	88	115	86	101	67	44	32	20	700
Reno, Nev.	T	-0.1	2.3	5.0	8.6	11.9	15.6	20.1	19.2	15.7	10.1	4.1	0.8	9.4
	P	30	26	17	14	13	9	7	4	6	13	14	27	180
Mount Washington, N.H.	T	-14.3	-14.7	-11.3	-5.0	1.7	7.2	9.5	8.7	5.0	-0.6	-6.5	-12.9	-2.8
	P	138	132	146	150	148	165	170	169	178	157	168	160	1881
Atlantic City, N.J.	T	1.6	1.5	5.1	10.6	16.3	21.1	23.9	23.2	19.6	14.0	8.2	2.6	12.3
	P	90	80	99	87	89	72	94	124	84	81	93	82	1075
Albuquerque, N. Mex.	T	1.7	4.4	7.9	13.2	18.4	23.8	25.8	24.8	21.4	14.7	6.7	2.8	13.8
	P	10	10	12	12	19	14	30	34	24	19	10	12	206
Buffalo, N.Y.	T	-4.7	-4.9	-0.8	6.1	12.4	18.2	21.0	20.2	16.3	10.4	3.7	-2.7	7.9
	P	72	69	82	76	75	65	65	77	80	76	91	76	904
Greensboro, N.C.	T	4.3	5.0	8.6	14.1	19.4	23.8	25.2	24.6	21.2	15.2	8.8	4.4	14.6
	P	86	84	94	87	84	88	122	117	93	69	68	80	1072
Williston, N. Dak.	T	-12.3	-10.3	-3.7	6.0	12.9	17.3	21.8	20.4	14.3	7.8	-1.9	-7.9	5.3
	P	14	12	18	24	36	84	48	38	28	19	15	13	349
Cleveland, Ohio	T	-2.4	-2.2	1.9	8.1	14.2	19.6	21.7	20.8	16.9	11.0	4.1	-1.4	9.3
	P	68	59	80	87	89	87	84	83	74	61	66	59	897
Tulsa, Okla.	T	2.9	5.1	9.2	15.2	19.9	25.2	27.9	27.8	23.5	17.7	9.5	4.8	15.7
	P	43	45	62	102	134	119	75	77	102	84	58	41	942
Portland, Oreg.	T	4.6	6.6	8.7	11.9	15.1	17.4	20.3	20.1	18.1	13.6	8.4	6.2	12.6
	P	161	124	121	62	52	43	10	18	44	99	153	188	1075
Harrisburg, Pa.	T	-0.1	0.3	4.6	11.0	17.1	21.8	24.3	23.1	19.1	13.2	6.6	0.8	11.8
	P	70	59	87	77	99	87	89	93	72	75	75	74	957
Providence, R.I.	T	-1.8	-1.9	2.7	7.8	13.8	18.7	21.7	20.8	17.1	11.5	5.9	-0.2	9.7
	P	95	72	91	86	77	81	78	92	81	72	95	88	1007

(Continued)

Table 17 (Continued)

Station		J	F	M	A	M	J	J	A	S	O	N	D	Year
Charleston, S.C.	T	10.2	10.8	13.7	17.9	22.2	25.7	26.7	26.5	24.2	19.0	13.3	10.0	18.3
	P	65	84	100	73	92	127	196	168	148	72	53	72	1250
Rapid City, S. Dak.	T	-5.6	-4.4	-0.5	6.9	13.2	18.3	23.2	22.2	16.4	10.0	1.7	-2.7	8.2
	P	9	12	26	42	68	78	45	31	24	20	10	8	373
Memphis, Tenn.	T	5.6	7.0	10.9	16.6	21.6	26.1	27.7	27.2	23.6	17.6	10.3	6.4	16.7
	P	154	119	129	118	107	93	90	75	72	69	111	125	1262
Dallas, Tex.	T	7.7	9.7	13.4	18.3	22.7	27.4	29.4	29.4	25.5	19.9	12.7	8.9	18.8
	P	59	65	72	102	123	82	49	49	72	69	69	68	879
Salt Lake City, Utah	T	-2.1	0.6	4.7	9.9	14.7	19.4	24.7	23.6	18.3	11.5	3.4	-0.2	10.7
	P	34	30	40	45	36	25	15	22	13	29	33	32	354
Burlington, Vt.	T	-7.7	-7.0	-1.6	6.2	13.2	18.7	21.4	20.1	15.5	9.2	2.7	-5.0	7.2
	P	50	45	54	67	76	89	98	86	84	75	67	54	845
Richmond, Va.	T	3.7	4.4	8.2	13.9	19.2	23.7	25.6	24.7	21.2	15.1	9.2	4.3	14.4
	P	88	74	87	80	94	95	142	141	93	76	77	75	1122
Seattle, Wash.	T	5.1	6.4	8.0	11.0	14.1	16.3	18.7	18.3	16.2	12.4	8.3	6.6	11.8
	P	132	99	84	50	40	36	16	19	42	83	127	138	866
Parkersburg, W. Va.	T	1.4	1.9	5.9	12.3	17.8	22.6	24.3	23.6	20.1	13.9	7.1	2.1	12.8
	P	85	72	90	83	94	108	104	96	69	52	60	72	985
Milwaukee, Wis.	T	-6.3	-5.3	-0.6	6.4	11.9	17.4	20.4	19.9	15.7	10.0	2.1	-4.1	7.3
	P	46	36	59	64	80	92	75	78	69	53	55	41	748
Lander, Wyo.	T	-7.1	-4.4	0.1	6.2	11.6	16.7	21.4	20.4	15.0	8.4	-0.6	-4.9	6.9
	P	12	18	29	62	67	35	20	12	26	31	23	11	346

and mean annual values of temperature and precipitation for selected stations in the country [68, 69].

Figure 12 is a map of the United States showing state boundaries and 16 drainage divisions used by the U.S. Geological Survey in their station numbering system. In view of the different physiographic and climatic conditions of the island of Hawaii, the data from that state have been analyzed separately. The analysis of the data from the contiguous United States and Alaska was carried out in one group. The data were collected from various publications of the U.S. Geological Survey.

Statistical Analysis of Data Used

Gauging stations. The number of gauging stations analyzed was 1,002. Only one year of record, although different years, for each station was used in the analysis. The station numbers and the years for which the records were used are identified in Appendix B. The distribution of drainage areas at specified intervals according to drainage divisions is given in Table 18. Approximately equal number of stations were selected from each state and covered a wide range of drainage areas.

Period of record. Only one year of record, although different years, was taken for each station. Thus, the total number of station-years analyzed was 1,002.

Monthly percentage distribution of annual maximum peaks. This distribution is given in Table 19. The major period of peak flow extends from March to June and reflects the period of snowmelt

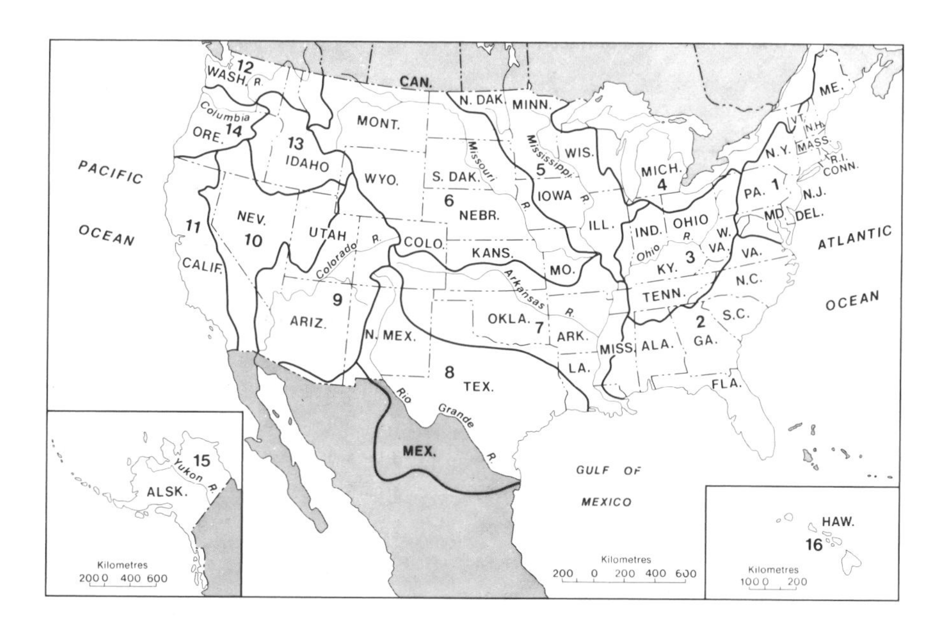

Figure 12. Map of U.S.A. showing drainage divisions.

Table 18
Distribution of Drainage Areas (USA)

Interval in km^2	Number of Stations (By Drainage Division)															Total Number of Stations
	1	2	3	4	5	6	7	8	9	10	11	12	13	14	15	
< 1	0	0	0	0	0	0	0	0	0	0	0	1	0	0	0	1
1-100	34	31	28	10	7	20	7	10	13	8	2	9	1	5	4	189
100-500	35	28	32	9	11	31	18	5	19	9	0	8	3	4	4	216
500-1000	28	28	31	7	19	20	16	5	15	2	4	6	4	6	3	194
1000-5000	27	33	29	11	16	28	15	4	15	5	5	11	0	4	5	208
5000-100000	23	33	24	9	13	24	7	6	14	3	0	11	0	2	4	173
>100000	0	0	4	1	2	4	4	0	5	0	0	0	0	1	0	21
TOTAL	147	153	148	47	68	127	67	30	81	27	11	46	8	22	20	1002

Table 19
Monthly Percentage Distribution of Annual Maximum Peaks (USA)

Month	J	F	M	A	M	J	J	A	S	O	N	D
Percentage	8.78	4.29	18.66	17.86	12.38	10.18	5.29	6.69	4.19	1.60	1.50	8.58

Table 20
Distribution of QP/Q2 (USA)

Interval	Number	Percentage of Total	Cumulative Percentage	Interval	Number	Percentage of Total	Cumulative Percentage
1.0–1.1	331	33.03	33.03	2.0– 2.2	24	2.40	81.44
1.1–1.2	141	14.07	47.11	2.2– 2.4	25	2.50	83.93
1.2–1.3	81	8.08	55.19	2.4– 2.6	22	2.20	86.13
1.3–1.4	67	6.69	61.88	2.6– 2.8	20	2.00	88.12
1.4–1.5	42	4.19	66.07	2.8– 3.0	15	1.50	89.62
1.5–1.6	43	4.29	70.36	3.0– 4.0	38	3.79	93.41
1.6–1.7	32	3.19	73.55	4.0– 5.0	12	1.20	94.61
1.7–1.8	20	2.00	75.55	5.0–10.0	36	3.59	98.20
1.8–1.9	21	2.10	77.64	>10	18	1.80	100.00
1.9–2.0	14	1.40	79.04				

in various regions. A large number of peaks occurs in other months also and is caused by rainfall floods.

Time of peak and peak indices. The time of peak was available for only 66 stations. It was too small a sample for any meaningful results. The indices of −2, −1, 0, 1, and 2 were given by 10, 143, 821, 27, and 1 peaks, respectively, with corresponding percentages of 1.0, 14.3, 81.9, 2.7, and 0.1.

Table 21
Distribution of QP/QPP (USA)

Interval	Number	Percentage of Total	Cumulative Percentage	Interval	Number	Percentage of Total	Cumulative Percentage
.6– .7	13	1.30	1.30	1.8– 1.9	4	.40	88.62
.7– .8	119	11.88	13.17	1.9– 2.0	13	1.30	89.92
.8– .9	178	17.76	30.94	2.0– 2.2	25	2.50	92.42
.9–1.0	272	27.15	58.08	2.2– 2.4	11	1.10	93.51
1.0–1.1	124	12.38	70.46	2.4– 2.6	4	.40	93.91
1.1–1.2	44	4.39	74.85	2.6– 2.8	8	.80	94.71
1.2–1.3	42	4.19	79.04	2.8– 3.0	2	.20	94.91
1.3–1.4	29	2.89	81.94	3.0– 4.0	19	1.90	96.81
1.4–1.5	18	1.80	83.73	4.0– 5.0	11	1.10	97.90
1.5–1.6	23	2.30	86.03	5.0–10.0	8	.80	98.70
1.6–1.7	10	1.00	87.03	> 10	13	1.30	100.00
1.7–1.8	12	1.20	88.22				

Results and Analysis

Distribution of QP/Q2. Table 20 gives this distribution. Almost one-third of the ratios lie between 1.0 and 1.1. The number of high ratios was found to be large with almost 20% exceeding a ratio of 2.0.

Distribution of QP/QPP. Table 21 gives the ratio of observed peak to predicted peak. About 40% of the values lie within ±10% and about 62% lie within ±20%. About 58% of the ratios are less than 1.0 indicating the degree of overprediction. About 42% of the peaks have been underpredicted. These were mostly rainfall floods occurring on small streams or even on some larger desert streams subject to incidence of flash floods. For these streams, a smaller value of base factor K must be used. The value of K = 1 is too large for them.

Distribution of base factor K. The distribution of base factor K is given in Table 22. About 42% of the values are less than 1, indicating the degree of underprediction of peaks. About 24% of the values are concentrated within the interval 1.4–1.7.

Relationship between KR and K. Figure 13 gives the plot of KR vs. K. Different symbols have been used for different drainage divisions. These indicate the flood potential of each division and the accuracy of predicted values. Lines with various values of R can be drawn on this figure to estimate the number of points lying above or below a given ratio of QP/QPP.

Analysis of Hawaiian Stations

Table 23 gives the analysis of 12 stations on the island of Hawaii. Six of these are on the island of Kauai and six on the island of Oahu. The drainage areas are small varying from about 4 km^2 to about 150 km^2. There are sharp peaks on these streams. The predicted values are generally

Table 22
Distribution of Base Factor K (USA)

Interval	Number	Percentage of Total	Cumulative Percentage	Interval	Number	Percentage of Total	Cumulative Percentage
0.00	6	.60	.60	1.1-1.2	57	5.69	53.39
0.0– .1	20	2.00	2.59	1.2-1.3	63	6.29	59.68
.1– .2	18	1.80	4.39	1.3-1.4	67	6.69	66.37
.2– .3	37	3.69	8.08	1.4-1.5	84	8.38	74.75
.3– .4	40	3.99	12.08	1.5-1.6	80	7.98	82.73
.4– .5	40	3.99	16.07	1.6-1.7	73	7.29	90.02
.5– .6	43	4.29	20.36	1.7-1.8	50	4.99	95.01
.6– .7	56	5.59	25.95	1.8-1.9	16	1.60	96.61
.7– .8	57	5.69	31.64	1.9-2.0	8	.80	97.41
.8– .9	45	4.49	36.13	2.00	26	2.59	100.00
.9-1.0	60	5.99	42.12				
1.0-1.1	56	5.59	47.70				

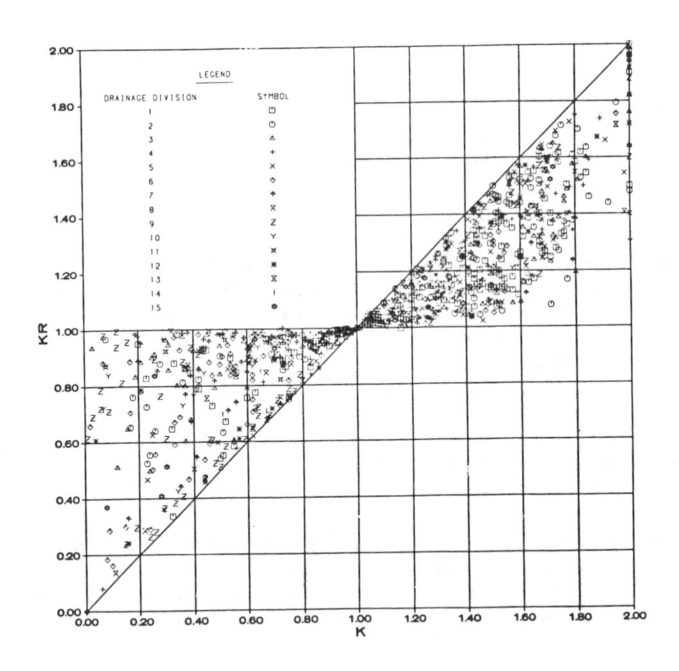

Figure 13. Relationship between KR and K (U.S.A.).

Table 23
Analysis of Hawaiian Data (Drainage Division 16) (Discharges Are in m^3/sec)

Station Number	Year	Date of Peak	QP	Q1	Q2	Q3	Date of Q2	Peak Index	QP/Q2	QPP	QP/QPP	K
					Island of Kauai							
16010000	1970	01/13	97.1	1.87	13.9	7.28	01/13		6.97	23.3	4.17	0.202
16013000	1970	01/13	15.9	3.26	3.68	.538	01/14	-1	4.31	5.47	2.90	0.256
16016000	1968	02/21	64.3	8.16	19.8	3.54	02/21		3.25	33.7	1.91	0.477
16031000	1970	01/13	535	61.7	66.0	9.34	01/14	-1	8.11	96.4	5.55	0.122
16036000	1970	01/13	111	10.5	18.3	2.58	01/14	-1	6.04	30.1	3.68	0.226
16071000	1970	01/13	134	14.2	26.9	5.58	01/14	-1	4.97	43.9	3.04	0.275
					Island of Oahu							
16200000	1970	07/25	92.0	1.05	10.0	3.74	07/25		9.18	17.7	5.21	0.170
16211000	1970	07/25	100	.181	11.2	1.98	07/25		8.92	21.4	4.68	0.205
16211600	1969	12/27	18.2	.031	2.55	.991	12/27		7.12	4.59	3.96	0.231
16212800	1969	12/24	40.8	2.38	2.97	1.30	12/25	-1	13.7	4.11	9.93	0.058
16213000	1970	07/25	96.6	9.93	15.9	1.44	07/26	-1	6.07	26.2	3.69	0.226
16216000	1970	07/25	436	56.6	57.5	2.38	07/26	-1	7.59	85.5	5.10	0.138

much lower than the observed values. The value of K = 1 is too large for these streams. The average value of K for the 12 streams analyzed is 0.22.

ACKNOWLEDGMENT

Thanks are due to Charles Mawby for collecting and analyzing all data reported in this chapter.

APPENDIX A: IDENTIFICATION OF CANADIAN DATA USED IN FIGURE 11

01AD002/82	01BU004/82	02AA002/78	02HJ002/82	023301 /73	05AB005/82	05BM004/82
01AD003/82	01BV005/71	02AC001/82	02HK004/82	023401 /72	05AB007/22	05BM008/82
01AD004/79	01CA003/82	02AC002/82	02HL001/82	023402 /72	05AB013/82	05BM014/82
01AE001/82	01CA004/78	02AE001/82	02KD004/82	023403 /73	05AB021/75	05BN012/82
01AF002/82	01CB002/81	02BA002/82	02KE002/48	023422 /72	05AB022/69	05CA001/72
01AF003/82	01CB003/81	02BB002/82	02OA028/82	023426 /74	05AB023/69	05CA004/82
01AF006/74	01CB004/82	02BB003/82	02OA030/82	023438 /72	05AB024/69	05CB001/82
01AF007/82	01CB005/82	02BC004/82	02OE018/82	023439 /72	05AB029/82	05CB004/82
01AG002/82	01CB006/82	02BF003/77	02OJ007/82	023601 /72	05AB030/82	05CC002/82
01AG003/82	01CC002/81	02BF004/82	02PD018/82	024003 /73	05AB038/82	05CC007/82
01AH002/82	01CC003/82	02BF005/82	02XA003/82	024007 /74	05AB039/82	05CD004/54
01AH003/82	01CD002/78	02CC008/82	02XA004/80	024007 /72	05AC003/82	05CE001/82
01AH005/82	01CD003/82	02CD002/82	02YA001/82	03NF001/80	05AC012/82	05CE006/81
01AJ001/82	01CE003/79	02CD004/82	02YC001/82	03NG001/82	05AC023/78	05CE013/69
01AJ003/82	01CE004/82	02CD005/82	02YD001/78	03OB002/70	05AC030/82	05CK004/82
01AJ004/82	01DA001/82	02CF005/82	02YD002/82	03OD003/71	05AD002/82	05DC007/68
01AJ007/78	01DC005/82	02CF008/82	02YF001/82	03OE001/82	05AD003/82	05DC008/72
01AJ009/79	01DC006/78	02CF100/74	02YJ001/82	03OE003/81	05AD005/82	05DC011/82
01AJ010/82	01DD004/82	02DD008/80	02YJ002/82	03PB001/70	05AD007/82	05DD004/82
01AJ011/82	01DE002/33	02DD010/81	02YK002/82	03PB002/82	05AD010/82	05DD005/82
01AK001/82	01DG003/82	02DD013/81	02YK003/66	03QC001/82	05AD016/82	05DE006/77
01AK002/67	01DG006/82	02DD015/82	02YK004/78	03QC002/81	05AD028/82	05EE004/79
01AK004/82	01DG018/82	02EA006/82	02YK005/81	030101 /73	05AD029/64	05EE007/82
01AK005/82	01DH002/72	02EA010/82	02YL001/82	030108 /72	05AD030/63	05EF001/82
01AK006/82	01DH003/82	02EA012/78	02YM001/82	030109 /72	05AD031/55	05EF004/82
01AK007/82	01DH004/82	02EB013/82	02YM003/82	030203 /72	05AD032/64	05EF005/82
01AK008/82	01DH005/82	02EB014/82	02YN002/82	030219 /74	05AD033/55	05EF006/82
01AL002/82	01DL001/82	02EC008/69	02YO005/82	030234 /74	05AD035/82	05EG004/82
01AL003/82	01DN004/82	02EC010/82	02YO006/82	030238 /72	05AE002/81	05EG005/82
01AL004/82	01DO001/82	02EC014/78	02YP001/82	030239 /74	05AE005/81	05EG006/82
01AL005/80	01DR001/82	02EC103/82	02YQ001/82	030273 /72	05AE006/82	05EG007/82
01AL008/82	01DR003/82	02ED003/82	02YR001/82	030304 /75	05AE009/50	05EG008/81
01AM001/82	01EA003/82	02ED005/82	02YR002/82	030305 /72	05AE016/78	05FE001/79
01AN001/82	01EC001/82	02ED007/82	02YR003/82	030316 /74	05AE017/25	05FE004/82
01AN002/82	01ED002/53	02ED100/82	02YS001/82	030401 /73	05AE027/82	05FE005/80
01AO009/82	01ED003/78	02FA001/82	02YS003/82	030408 /73	05AE031/36	05FF001/82
01AP002/82	01ED005/82	02FA002/79	02ZA001/82	030415 /74	05AE032/66	05GC007/82
01AP004/82	01ED007/82	02FB009/82	02ZA002/82	030903 /73	05AE037/81	05GF002/82
01AP006/82	01ED008/80	02FB010/82	02ZA003/82	030905 /73	05AE039/81	05GG001/82
01AQ001/82	01EE001/82	02FC002/82	02ZB001/82	030906 /74	05AE040/81	05GG007/82
01AQ002/82	01EE002/82	02FC012/82	02ZC001/69	030907 /74	05AG006/82	05HA003/82
01AQ006/76	01EE003/48	02FC013/79	02ZD001/66	030908 /75	05AH003/82	05HA072/62
01AR003/82	01EE004/82	02FC015/82	02ZD002/80	04AA001/73	05AH005/82	05HA075/67
01AR004/82	01EE005/82	02FD002/82	02ZE002/81	04AA004/82	05AH042/82	05HA076/82
01AR006/82	01EE006/82	02FE003/82	02ZF001/82	04AC005/81	05AH043/82	05HB001/75
01AR007/67	01EF001/82	02FE005/82	02ZG001/82	04AC007/82	05AH046/81	05HC005/82
01AR008/79	01EF003/79	02FE007/82	02ZG002/82	04AD002/74	05AJ001/82	05HE001/82
01BA001/79	01EG002/82	02FE008/82	02ZG003/82	04CA002/82	05AK001/81	05HF004/69
01BC001/82	01EH003/82	02FF002/82	02ZG004/82	04CB001/82	05BB001/82	05HF011/79
01BE001/82	01EJ001/82	02FF006/72	02ZH002/82	04CC001/75	05BC001/82	05HF012/73
01BJ001/82	01EJ004/82	02GA010/82	02ZJ001/82	04CD001/75	05BC008/77	05HF013/79
01BJ003/82	01EK001/82	02GA013/57	02ZK001/82	04DA001/82	05BD002/39	05HG001/81
01BJ004/81	01EK002/71	02GA014/82	02ZK002/82	04DA002/77	05BE004/82	05HG002/82
01BJ007/82	01EK003/82	02GA015/82	02ZL003/82	04DB001/82	05BE006/82	05HG020/82
01BJ009/82	01EN002/82	02GA016/82	02ZM006/82	04DC001/79	05BE008/82	05HH001/82
01BJ010/82	01EO001/82	02GA022/64	02ZM008/82	040201 /73	05BF013/82	05JA002/82
01BJ011/82	01EO003/78	02GA024/82	02ZM009/82	040212 /73	05BF018/82	05JB002/82
01BK004/74	01FB001/82	02GA030/82	02ZM010/82	041301 /74	05BG001/82	05JC001/74
01BK005/82	01FB003/82	02GA031/82	02ZM011/82	05AA004/82	05BH003/82	05JC007/82
01BK006/82	01FB005/78	02GA035/82	02ZN001/82	05AA006/75	05BH009/82	05JD004/82
01BL001/82	01FC002/82	02GA036/80	020401 /74	05AA008/81	05BH901/82	05JE001/82
01BL002/82	01FE001/77	02GB001/82	020501 /74	05AA011/82	05BJ001/82	05JE002/82
01BO001/81	010801 /72	02GC026/80	020602 /72	05AA013/81	05BJ004/82	05JE004/82
01BO002/82	010801 /74	02GD001/82	020802 /73	05AA022/82	05BJ010/82	05JE005/82
01BO003/82	010901 /74	02GD018/82	020901 /74	05AA023/82	05BJ011/79	05JE006/82
01BO004/82	011001 /72	02GE003/82	021405 /73	05AA024/82	05BL003/82	05JF001/82
01BP001/82	011201 /72	02GG007/82	021502 /73	05AA025/67	05BL009/82	05JF006/82
01BQ001/82	011507 /74	02HA014/82	021702 /74	05AA026/82	05BL013/82	05JF008/81
01BR001/82	011601 /72	02HB002/82	021702 /72	05AA027/82	05BL019/82	05JF011/82
01BS001/82	011601 /74	02HB004/80	022301 /74	05AA028/82	05BL020/63	05JF012/82
01BU002/82	013001 /72	02HB011/82	022504 /72	05AA030/82	05BL024/82	05JF014/82
01BU003/82	02AA001/79	02HF002/82	022703 /74	05AB002/82	05BM002/82	05JG001/82

APPENDIX A (Continued)

05JG007/82
05JG012/82
05JG015/82
05JH005/82
05JK004/82
05JL001/82
05JM010/82
05JM015/82
05KA001/82
05KB001/74
05KB003/82
05KB005/82
05KB006/82
05KB008/82
05KB010/74
05KB011/81
05KD001/58
05KD003/82
05KG002/82
05KG007/82
05KH007/82
05KH008/82
05KJ001/82
05LA003/82
05LA005/82
05LA006/82
05LB004/82
05LB007/81
05LB008/82
05LC001/82
05LC004/82
05LC005/81
05LD001/82
05LE004/82
05LE005/81
05LE006/82
05LE008/82
05LE009/76
05LE010/82
05LE011/82
05LF001/82
05LF002/82
05LG001/82
05LG003/77
05LG004/82
05LG005/77
05LG006/82
05LH005/82
05LJ005/81
05LJ007/82
05LJ010/81
05LJ011/81
05LJ012/82
05LJ015/81
05LJ019/81
05LJ021/81
05LJ022/79
05LJ024/75
05LJ024/75
05LJ025/82
05LJ026/76
05LJ027/82
05LJ030/75
05LJ031/81
05LJ032/81
05LJ033/79
05LJ034/74
05LJ035/75
05LJ040/81
05LJ043/81
05LJ045/81
05LJ048/81
05LL001/82
05LL002/81
05LL005/82
05LL007/81
05LL008/80
05LL009/82
05LL011/82
05LL013/82
05LL015/82
05LL017/81
05LL024/82
05LL027/81
05LM001/82
05LN003/82
05MA014/82
05MA021/82
05MB003/82
05MD004/82
05MD005/82
05MD007/80
05ME001/81
05ME003/82
05ME005/82
05ME006/81
05ME008/80
05MF001/82
05MF008/82
05MF018/82
05MG001/81
05MG004/82
05MG010/82
05MH001/72
05MH004/82
05MH005/80
05MH011/80
05MH013/82
05MJ001/82
05MJ003/82
05MJ003/82
05MJ007/78
05MJ009/78
05MJ010/78
05MJ011/81
05NA004/82
05NA005/82
05NB007/68
05NB009/82
05NB025/82
05NB026/65
05ND001/82
05ND004/82
05NF002/82
05NF007/82
05NF008/82
05NF012/80
05NG001/82
05NG003/82
05NG007/82
05NG012/82
05NG016/82
05NG020/82
05OA005/81
05OA006/82
05OA015/82
05OB006/81
05OB021/82
05OB023/82
05OB025/82
05OC001/82
05OC004/81
05OC008/79
05OC012/82
05OC019/80
05OD004/81
05OD029/82
05OE001/82
05OE002/82
05OE014/81
05OF006/81
05OF014/81
05OF015/82
05OF017/82
05OF018/78
05OG001/82
05OG005/82
05PA012/82
05PB014/82
05PB015/78
05PB018/80
05PC010/82
05PC010/82
05PC011/82
05PC018/82
05PD014/82
05PD015/82
05PD023/82
05PF064/61
05QC003/82
05QE009/82
05RA001/82
05RA002/82
05SA002/82
05SA004/82
05SB002/82
05SD004/81
05TB002/82
05TD001/82
05TE001/82
05UD004/82
05UE004/57
050409 /75
050701 /73
050701 /75
051001 /75
051002 /75
051003 /73
051004 /74
051005 /73
051006 /73
051006 /74
051007 /73
052803 /74
052806 /73
06AD006/82
06AF006/77
06AG001/82
06CA001/82
06CD002/82
06DA005/82
06EA002/81
06EA006/82
06FB001/81
06FB002/82
06FD001/79
06GB001/82
06HB002/82
06HD002/82
06JC002/82
06KC003/82
06LA001/82
06LC001/82
06LC002/82
06LC002/82
06LC003/82
06MA002/82
06MA005/82
06MB001/82
06NB002/80
06OA001/81
06OA002/81
06OA003/81
06OA004/81
06OA005/81
061901 /73
061905 /74
062101 /72
062101 /73
062102 /70
07AD002/82
07EA001/80
07EA002/81
07EA004/82
07EA005/81
07EB002/82
07EC002/82
07EC003/82
07EC004/82
07ED001/82
07ED003/82
07EE002/67
07EE009/82
07EE010/82
07EE011/82
07EF001/82
07FA001/82
07FA003/82
07FA004/82
07FA005/82
07FB001/82
07FB002/81
07FB003/82
07FB004/81
07FB005/82
07FB007/82
07FB008/82
07FB009/82
07FC001/82
07FC004/77
07FD001/82
07FD002/82
07FD007/82
07HA001/82
07HD001/67
07KA002/60
07KA002/60
07KC001/82
07LB002/82
07NB001/82
07OB001/77
07PA001/78
07QB002/79
07QC003/79
07QC004/78
07QD005/78
07RD001/82
07SA004/82
07SB009/82
07SB010/81
07TA001/82
07TB001/80
072301 /72
073502 /74
073801 /73
074902 /74
08AA001/82
08AA003/81
08AA008/82
08AA009/82
08AA010/81
08AB001/82
08BB001/82
08BB002/78
08CA002/82
08CB001/82
08CC001/82
08CF001/82
08CG003/82
08CG005/82
08CG006/82
08DA005/82
08DB010/82
08DB011/82
08DC006/82
08DD001/82
08EC004/82
08ED001/82
08EE008/82
08EE009/74
08EE012/82
08EE013/82
08EE018/79
08EE020/82
08EF004/71
08EG011/82
08EG012/82
08FB004/82
08FB004/82
08FB005/82
08FB009/82
08FE001/53
08FE003/82
08FF002/82
08GA026/28
08GA056/82
08GA057/80
08GA058/68
08GA060/82
08GA061/81
08GA062/74
08GA064/82
08GA065/81
08GA072/82
08GC003/30
08GC004/79
08GD004/81
08GD005/82
08GD007/82
08GD008/82
08GF001/28
08HA002/82
08HA010/81
08HA011/81
08HB003/61
08HB005/63
08HB006/82
08HB017/82
08HB030/78
08HB034/81
08HB041/82
08HB048/82
08HB062/78
08HB069/82
08HB074/82
08HC001/82
08HD003/70
08JA003/52
08JA010/52
08JA015/82
08JA017/82
08JB003/81
08JE005/81
08KG003/82
08KH010/82
08KH019/82
08LA013/82
08LA020/82
08LC039/82
08LE077/82
08LE077/82
08LE092/79
08LE093/79
08LE102/82
08LF049/74
08MC018/67
08MF005/82
08MF035/82
08MF040/82
08MH024/82
08NE058/82
080704 /75
080717 /75
080718 /75
09AA007/81
09AA009/73
09AA011/70
09AA012/82
09AB001/72
09AB008/82
09AB009/82
09AC001/82
09AC004/82
09AD001/82
09AD002/82
09AE001/82
09AF001/73
09AG001/82
09AH001/82
09AH003/82
09BA001/81
09BB001/82
09BC001/82
09BC002/74
09BC004/82
09CA002/82
09CA003/82
09CA004/82
09CB001/82
09CD001/82
09DA001/82
09DC002/79
09DC003/82
09DD002/73
09DD003/82
09DD004/82
09EA003/82
09EA004/82
09EB001/80
09FB002/82
09FC001/81
10AA001/82
10AA001/82
10AA002/82
10AB003/82
10AD002/82
10EA002/81
10EA003/81
10EB001/82
10EB002/82
10EC001/82
10EC002/82
10ED001/82
10ED002/81
10ED003/82
10ED004/82
10FA002/82
10FB001/78
10FB005/82
10FC001/82
10GB005/82
10GC001/82
10GC002/81
10GC003/82
10HA002/82
10HB003/82
10HC003/82
10JA002/80
10JC003/80
10JD001/81
10JD002/80
10JE001/82
10KA001/79
10KA005/78
10MD001/79
10MD002/82
10ND001/78
10ND003/80
10PC001/80
10QC002/80
10TC001/79
10TC002/79
10TE001/80
10TE002/82
10TE003/82
10TE004/82
10TE005/81
11AB001/82
11AB003/76
11AB076/66
11AB082/82
11AB088/66
11AB101/82
11AB103/82
11AB103/82
11AB112/76
11AB117/82
11AC008/45
11AC016/45
11AC025/82
11AC029/69
11AF002/51

APPENDIX B: IDENTIFICATION OF ALL U.S. DATA EXCEPT HAWAII

01010000/73
01010500/73
01011000/73
01013500/73
01015800/73
01016500/73
01017000/73
01017900/73
01018000/73
01019000/73
01021200/73
01022500/73
01023000/73
01024200/73
01030000/73
01031500/73
01031600/73
01033000/73
01034500/73
01036500/73
01038000/83
01046500/73
01049300/73
01052500/71
01053500/71
01054500/73
01059800/73
01064300/70
01064400/71
01064500/71
01064800/71
01065000/71
01072100/70
01092000/71
01094500/70
01096500/71
01100000/71
01100700/71
01104200/71
01111300/71
01111500/72
01112500/71
01116500/71
01118300/76
01119500/76
01120500/76
01121000/76
01122000/76
01122500/76
01123000/76
01124000/76
01126600/76
01127000/76
01127500/76
01128500/71
01129200/71
01129500/71
01138500/71
01144500/71
01151500/71
01184000/76
01184000/71
01186000/76
01187980/76
01199000/76
01200000/76
01200500/76
01208500/76
01303500/80
01304000/80
01304500/80
01305000/80
01377000/70
01377500/70
01378690/70
01380500/70
01381500/70
01383500/70
01384500/70
01387000/70
01387500/70
01388500/70
01389500/70
01391500/70

01402000/70
01405000/70
01417000/70
01428500/76
01429500/75
01431500/75
01434000/76
01439500/76
01440400/76
01446500/70
01446600/76
01446700/76
01447720/76
01448500/76
01449360/76
01449500/76
01451000/76
01453000/76
01457500/70
01465500/76
01467042/76
01470500/76
01471000/76
01474500/76
01477800/78
01478000/78
01478500/78
01480000/78
01480100/78
01481500/78
01483200/78
01483700/78
01485000/77
01485500/77
01487000/78
01496500/80
01500000/80
01500500/80
01502000/80
01502500/80
01503000/80
01505000/80
01509000/80
01509150/79
01512500/80
01515000/80
01520500/79
01521500/79
01526500/79
01527050/80
01529950/79
01578310/78
01589000/78
01594440/78
01595500/78
01598500/78
01601500/78
01603000/78
01608500/78
01610000/78
01613000/78
01614500/78
01636500/78
02011500/69
02012500/69
02013000/69
02014000/69
02016000/69
02016500/69
02017500/69
02018000/69
02018500/69
02019500/70
02020500/69
02021500/69
02022500/69
02030500/70
02034000/70
02034500/70
02036500/69
02037500/70
02038000/70
02038850/70
02039500/70

02040000/70
02065200/70
02075000/70
02075500/70
02077200/70
02077240/70
02079640/69
02080500/70
02081500/70
02081800/70
02082500/70
02083000/70
02084500/69
02084540/70
02087000/70
02088500/70
02089000/70
02091500/70
02096500/70
02096850/70
02097243/69
02098000/70
02100500/70
02102500/70
02107000/70
02110500/80
02110500/70
02113850/70
02116500/70
02129000/70
02129590/80
02130600/80
02130900/80
02131000/80
02131150/80
02131309/80
02132000/80
02135000/80
02135300/80
02135500/70
02146000/80
02147500/80
02148000/80
02148300/79
02156050/80
02156500/80
02157000/80
02160105/80
02160700/80
02165000/80
02175500/80
02176500/80
02177000/69
02187500/70
02193500/70
02202000/70
02203000/70
02205000/70
02206500/70
02211300/70
02212600/70
02213000/70
02213050/70
02213470/70
02214500/70
02217000/70
02217500/70
02218500/70
02219500/70
02220500/70
02222650/70
02223000/70
02226000/70
02226100/69
02314200/80
02314986/80
02315000/80
02315005/80
02315200/80
02315392/80
02315520/80
02317620/80
02317830/70
02319000/80

02319500/80
02321500/80
02324000/80
02326512/80
02326900/80
02327017/80
02327100/80
02329000/80
02329104/80
02329161/79
02358000/80
02365200/80
02370000/80
02400500/70
02404000/70
02408500/70
02411800/70
02414500/70
02415000/70
02417500/70
02418500/70
02420000/70
02424000/70
02429900/80
02429980/80
02430038/80
02430085/80
02430615/80
02431000/80
02433000/80
02435020/80
02439400/80
02441000/80
02443500/80
02448000/80
02472500/80
02473460/80
02474600/80
02475000/80
02475500/80
02476500/80
02478500/80
02479000/80
02479155/80
02488500/80
02489500/80
02490105/80
02490500/80
03013000/80
03014500/80
03086500/73
03092000/73
03093000/73
03094000/73
03102950/73
03110000/73
03111500/73
03115400/72
03117000/73
03117500/73
03124500/73
03125000/73
03127500/73
03129000/73
03130500/73
03131500/73
03136000/72
03137000/73
03139000/73
03140000/73
03140500/73
03178500/69
03179000/69
03180000/68
03180500/70
03181200/69
03183000/69
03183500/69
03184200/69
03186500/69
03187000/69
03189000/69
03189650/69
03190000/69

03191500/69
03192000/69
03193000/69
03193830/69
03194700/69
03197000/69
03198500/69
03199000/70
03201800/72
03206000/68
03207962/80
03207965/80
03208000/80
03209300/79
03210000/80
03213000/70
03215000/70
03216350/80
03216600/80
03216800/79
03231500/73
03234500/72
03237900/79
03248500/79
03250000/79
03250320/79
03252000/79
03270500/72
03274650/79
03274750/79
03274950/79
03276700/79
03277500/80
03278500/79
03282000/80
03283500/79
03284300/79
03285000/79
03289500/79
03291780/79
03294000/78
03294500/80
03295500/80
03298500/79
03302220/79
03302300/79
03302500/79
03302800/79
03303000/79
03303280/80
03303300/79
03311500/79
03322100/79
03322500/79
03322900/79
03324000/79
03324300/79
03326500/79
03327500/79
03328000/79
03329700/79
03335500/79
03336645/79
03336900/79
03337000/79
03339000/79
03340500/79
03342000/79
03343400/79
03344500/79
03345500/79
03354000/79
03360500/79
03378000/79
03378635/79
03380350/79
03380475/79
03381500/79
03414500/69
03416000/69
03417500/69
03418000/69
03420000/69
03421000/69

03422500/69
03425000/69
03426800/70
03431000/70
03431300/70
03433500/69
03435030/70
03465500/70
03467000/69
03469000/70
03470000/69
03484000/70
03487550/69
03490500/69
03496200/69
03497000/69
03498500/69
03538250/69
03559500/70
03574500/69
03575000/69
03575500/70
03576148/70
03576250/69
03576500/69
03577290/70
03592500/69
03597500/70
04001000/78
04016500/81
04018750/81
04019200/70
04024000/81
04024098/81
04024430/79
04025500/79
04027000/79
04031000/77
04031000/79
04031500/78
04031500/79
04033000/78
04034500/78
04037500/79
04040500/78
04043050/78
04044400/78
04046000/78
04056500/78
04058130/78
04058200/78
04059000/78
04059500/78
04061000/79
04061000/78
04061500/78
04062230/78
04062400/78
04063000/79
04063700/79
04065500/78
04066000/79
04067000/78
04069500/79
04071858/79
04079000/79
04079602/79
04084500/79
04085813/79
04087050/79
04087060/79
04101500/78
04116000/78
04119000/78
04122000/78
05014500/81
05017500/81
05020500/81
05051500/81
05054000/81
05056390/81
05057000/81
05058000/81
05062000/81

05067500/81
05089000/81
05089500/81
05092000/81
05100000/81
05267000/81
05275000/81
05278000/81
05287890/81
05304500/81
05316500/81
05340050/81
05340500/79
05344500/79
05345000/81
05373000/81
05376000/81
05376800/81
05378300/81
05385500/81
05408000/79
05413500/79
05443500/79
05476500/70
05480000/70
05481000/70
05481950/70
05482170/70
05482300/70
05483000/70
05485500/70
05486000/70
05486490/70
05487470/70
05487980/70
05488500/70
05489000/70
05491000/70
05494300/70
05495000/80
05496000/80
05497000/80
05501000/80
05502300/80
05503500/80
05503800/80
05505000/80
05525000/79
05525500/79
05528000/79
05528500/79
05531500/79
05533000/79
05536000/79
05536290/79
05543500/79
05552500/79
05587500/80
05587500/80
05591500/79
06014500/81
06019500/81
06024580/81
06024590/81
06026400/81
06035000/81
06036650/81
06040300/81
06050000/81
06062500/81
06090720/81
06115200/81
06136000/81
06137580/81
06154410/81
06231000/81
06235500/81
06311060/81
06311400/81
06313180/81
06334500/80
06335500/81
06336447/81
06336515/81

APPENDIX B (Continued)

06339180/81
06339300/81
06339490/81
06339500/81
06339900/81
06340000/81
06340528/81
06341800/81
06342040/81
06342230/81
06344600/81
06348000/81
06351000/81
06354860/80
06355500/80
06357800/80
06359500/80
06365300/81
06375600/81
06376300/81
06378300/81
06386000/81
06395000/81
06402000/80
06402500/79
06402600/80
06402600/80
06404000/80
06404998/79
06421500/80
06422500/80
06425500/80
06425750/81
06427500/81
06429500/80
06431500/80
06433000/80
06436700/80
06439300/80
06445980/80
06468170/81
06471898/80
06531850/81
06620000/81
06622700/81
06625000/81
06630000/81
06630330/81
06677100/70
06678000/70
06679000/70
06679500/70
06685000/70
06687000/70
06765000/70
06766000/70
06771500/70
06775500/70
06776500/70
06778000/69
06784800/70

06795000/70
06798000/70
06803000/70
06803520/70
06804000/70
06806500/70
06813000/80
06813500/80
06814000/80
06818000/80
06821150/80
06821190/80
06821280/80
06844900/80
06846500/80
06847900/80
06853500/79
06853500/79
06853800/80
06855800/79
06855900/80
06856000/80
06861000/80
06862700/80
06863500/80
06863900/80
06864050/80
06867000/80
06871500/80
06871900/80
06873700/80
06879650/80
06889100/80
06889120/80
06889580/80
06893300/80
06893500/80
06893670/80
06893793/80
06897500/80
06900000/80
06906200/80
06910230/80
06910800/80
07015000/80
07052100/80
07148350/70
07150500/70
07152000/70
07152500/70
07153000/70
07154500/80
07154500/70
07156900/69
07157960/70
07158150/70
07159000/70
07159200/70
07163000/70
07164500/70
07165550/70

07165700/70
07174600/70
07176500/70
07177000/70
07195000/70
07195800/70
07196900/70
07197000/70
07197000/70
07199000/80
07201420/80
07203000/80
07206000/80
07207000/80
07207500/80
07215500/80
07218000/80
07221500/80
07222500/80
07226500/80
07227100/80
07229300/70
07233500/79
07235000/79
07247000/70
07249400/70
07249500/70
07250000/69
07250550/70
07251500/70
07252000/70
07253000/70
07255500/70
07256500/70
07257000/70
07257500/70
07259500/70
07260000/70
07261500/70
07290000/80
07295100/80
07299200/79
07299300/79
07299670/79
07316000/79
07375000/80
07375500/80
07375800/80
07376000/80
07376500/80
07377000/80
08017200/79
08018730/79
08019500/79
08020960/79
08022070/79
08025307/79
08029500/79
08030500/79
08033300/79
08033300/79

08034500/79
08036500/79
08037050/79
08038000/79
08041500/79
08041700/79
08044000/79
08056500/79
08119500/79
08121000/79
08251500/80
08265000/80
08266000/80
08267500/80
08268700/80
08271000/80
08275300/80
08275600/80
08386000/80
08387600/80
08405150/80
09010500/79
09011000/79
09021000/79
09025000/79
09025400/79
09026500/79
09032000/79
09034500/79
09036000/79
09039000/79
09050700/79
09057500/79
09058000/79
09070000/79
09070500/79
09080400/79
09085000/79
09093500/79
09093700/79
09095000/79
09109000/79
09128000/79
09144250/79
09163500/80
09180000/80
09180920/80
09184000/80
09214500/81
09216000/81
09217000/81
09217000/81
09217000/80
09217900/80
09218500/81
09223000/81
09228500/81
09229500/80
09234500/79
09235600/80

09266500/80
09295000/80
09297000/80
09298000/80
09301200/80
09301500/80
09306405/80
09306430/80
09306800/80
09306870/80
09307300/80
09307900/80
09308500/80
09313000/80
09324500/80
09346400/80
09383200/70
09383400/70
09390500/70
09393500/70
09394500/70
09395900/70
09398000/70
09402500/70
09403000/70
09404040/70
09406000/70
09421500/79
09424000/70
09424000/79
09424450/70
09425500/65
09426000/70
09429600/78
09460150/70
09480000/70
09481500/70
09483000/70
09483100/70
09489100/70
09496000/70
09497800/70
09497980/70
09497980/70
10250800/79
10251300/79
10254050/78
10255700/79
10255800/79
10255810/79
10255850/79
10256000/79
10256500/79
10257600/79
10285700/79
10297500/70
10299100/69
10308800/69
10310400/70
10311000/70

10315500/70
10316500/70
10318500/70
10319500/70
10324500/70
10328000/70
10329500/70
10336600/70
10349300/70
10352500/70
10353000/70
11043000/79
11044000/79
11047200/79
11047500/79
11051500/79
11067890/79
11074000/79
11087020/79
11097500/79
11103000/79
11109600/79
12010000/81
12013500/81
12027500/80
12035450/81
12036650/81
12039220/81
12039500/80
12040500/80
12061500/80
12200500/80
12301300/81
12301550/81
12301933/81
12301933/81
12303000/80
12303100/80
12304500/80
12306500/80
12316800/80
12321500/80
12322000/80
12323000/81
12329500/81
12339450/81
12391400/81
12392000/80
12392300/80
12392895/80
12395000/80
12395500/81
12395500/80
12397100/80
12401500/81
12408500/81
12409000/81
12411000/80
12413000/80
12413140/80

12413150/80
12414500/80
12419000/81
12424000/81
12439300/81
12459000/80
12472300/81
12508779/80
12513000/81
13161500/80
13169500/80
13250600/80
13297330/80
13297355/80
13297597/80
13310700/80
13313000/80
14017000/81
14020000/70
14022500/70
14025000/70
14032000/70
14034350/80
14034500/70
14034800/70
14037500/70
14038530/70
14038530/70
14043560/70
14050500/70
14052000/70
14054000/70
14054500/69
14056500/70
14078000/70
14079500/70
14105700/70
14150000/70
14162500/70
14166000/70
15008000/71
15011500/71
15012000/71
15015600/70
15022000/70
15026000/71
15031000/71
15048000/71
15202000/71
15208000/71
15212000/71
15258000/71
15281000/71
15282000/71
15284000/70
15291000/71
15291200/71
15291500/71
15294500/71
15511000/71

NOTATION

a	coefficient of discharge; albedo; parameter of 3-parameter log-normal distribution
A	drainage area; cross-sectional area of channel
$A \pm N$	area of triangular hydrograph on day $\pm N$
b	coefficient of discharge
c	coefficient of discharge; runoff coefficient
C	coefficient of discharge; Creager's coefficient
C_p	peak discharge coefficient
C_t	lag time coefficient
D	unit duration
F	forest cover in decimal
g_x	coefficient of skewness of x
$g_{\log x}$	coefficient of skewness of log x
h	height of triangular hydrograph
H	total fall of stream; ratio T_r/T_p; total head
i	rainfall intensity (in./hr)
I	rainfall intensity (in./hr)
I_i	solar radiation (ly)
I_1	inflow at time t_1
I_2	inflow at time t_2
J	risk factor
k	basin constant for snowmelt
k'	radiation factor
K	channel conveyance; Muskingum constant; frequency factor; base factor
L	stream length; lag time; effective crest length
L_c	stream length from mouth to centroid of basin
L_i	length of i-th segment of stream
m	exponent in flood formulas; order number of items in flood series
M	daily snowmelt (in.)
M_x	median of x
n	exponent in flood formulas; number of channel segments; Manning's roughness coefficient; total number of items in flood series
N	cloud cover in decimal
O_1	outflow at time t_1
O_2	outflow at time t_2
p	percentage in Meyers formula; probability of occurrence of flood event
Pr	daily precipitation (in.)
$p(x)$	probability density function
$P(x)$	distribution function
q	probability of nonoccurrence of flood event; discharge per unit area
q_i	peak rate (in./hr.)
q_p	unit peak (cfs)
Q	peak discharge; total runoff
Q_p	peak flow
Q_{pR}	adjusted peak flow
$Q(max)$	peak flow
QP	recorded peak flow
QP'	theoretical peak flow
QPP	predicted peak flow
$Q1$	mean daily flow on day preceding maximum flow
$Q2$	maximum mean daily flow
$Q3$	mean daily flow on day following maximum flow
R	hydraulic radius; ratio QP/QPP
s_x	standard deviation of x
S	channel slope, channel storage; slope of the energy gradient line
S_i	slope of the i-th segment of stream
S_w	slope of water surface
S_1	storage at time t_1
S_2	storage at time t_2
t_p	lag time
t_r	unit duration
t_{pR}	adjusted lag time
t_R	unit duration other than t_r
t_0	unit duration
T	base length of unit hydrograph; return period
T_a	air temperature (°F)
T'_a	difference between air temperature and snow surface temperature (°F)
T_b	base length of unit hydrograph
T_c	time of concentration
T'_c	difference between cloud base temperature and snow surface temperature (°F)
T'_d	difference between dew point temperature and snow surface temperature (°F)
T_p	time to peak
T_r	recession time
T_{max}	maximum daily temperature (°F)
T_{mean}	mean daily temperature (°F)
v	wind velocity (mph)
v_x	coefficient of variation of x
x	variable
$\bar{x}$	mean of x

X partial area of triangular hydrograph

y natural logarithm of x

Y partial area of triangular hydrograph

z median of x

Greek Symbols

α parameter in probability distributions; extension of Q1 into Q2

α' extention of Q3 into Q2

β parameter in probability distribution

γ parameter in general plotting position formula; Fuler's constant

Δt time interval

μ mean

σ standard deviation

σ_x standard deviation of x

σ_y standard deviation of y

REFERENCES

1. Chow, Ven Te(Ed.), *Handbook of Applied Hydrology*, New York: McGraw-Hill Book Co., 1964.
2. Linsley, R. K., Kohler, M. A., and Paulhus, J. L. H., *Applied Hydrology*, New York: McGraw-Hill Book Co., 1949.
3. Linsley, R. K., Kohler, M. A., and Paulhus, J. L. H., *Hydrology for Engineers*, New York: McGraw-Hill Book Co., 1958.
4. Gray, D. M. (Ed.), *Handbook on the Principles of Hydrology*, Ottawa: National Research Council, 1970.
5. Creager, W. P., Justin, J. D., and Hinds, J., *Engineering for Dams in Three Volumes*, New York: John Wiley and Sons, Inc.; Chapman and Hall, Ltd., London, 1945.
6. Bruce, J. P., and Clark, R. H., *Introduction to Hydrometeorology*, Toronto: Pergamon Press, 1966.
7. Kirpich, Z. P., "Time of Concentration of Small Agricultural Watersheds," *Civil Eng.*, 10(6):362 (1940).
8. U.S. Department of the Interior. Design of Small Dams, Washington, D.C.: U.S. Govt. Printing Office, 1973.
9. Sherman, L. K., "Stream Flow from Rainfall by the Unit-graph Method." *Eng. News Rec.*, 108:501–505 (1932).
10. Synder, F. F., "Synthetic Unit-graphs," *Trans Am. Geophy. Union*, 19:447–454 (1938).
11. U.S. Department of Agriculture. "Soil Conservation Service National Engineering Handbook, Hydrology, Part 4," Washington, D.C.: U.S. Govt. Printing Office, 1972.
12. Mockus, Victor, "Use of Storm and Watershed Characteristics in Synthetic Hydrograph Analysis and Application," U.S. Soil Conservation Service (1957).
13. Taylor, A. B., and Schwartz, H. E., "Unit Hydrograph Lag and Peak Flow Related to Basin Characteristics," *Trans. Am. Geophys. Union*, 33:235–246 (1952).
14. Chow, V. T., "Hydrologic Determination of Waterway Areas for the Design of Drainage Structures in Small Drainage Basins," Univ. Ill. Eng. Expt. Sta. Bull. 462 (1962).
15. Sangal, B. P., "Discussion of Purpose and Performance of Peak Predictions by B. M. Reich and L. A. V. Heimstra," *Proc. Int. Hydrology Symp.*, Fort Collins, Colo., 2:378–380 (1967).
16. Standard Project Flood Determinations, U.S. Dept. of the Army Civil Works Eng. Bull., 52–8 (1952).
17. American Society of Civil Engineers, "Design and Construction of Sanitary and Storm Sewers," *Manual of Engineering Practice No. 37* (1960).
18. Fisheries and Environment Canada and Ontario Ministry of the Environment, "Manual of Urban Drainage Practice, Draft 3," Ottawa and Toronto (1977).

19. Roads and Transportation Association of Canada. "Drainage Manual, Vol. 1," Ottawa (1982).
20. Denver Regional Council of Governments, "Urban Storm Drainage Criteria Manual," Denver, Colo. (1969).
21. Hicks, W. I. "A Method of Computing Urban Runoff," *Trans. ASCE*, 109:1217–1253 (1944).
22. Keifer, C. J., and Chu, H. H., "Synthetic Storm Pattern for Drainage Design," *Proc. ASCE*, J. Hyd. Div., 83(4) Paper 1332, 1–25 (1957).
23. Tholin, A. L., and Keifer, C. J., "The Hydrology of Urban Runoff," *Trans. ASCE*, 125:1308–1379 (1960).
24. Izzard, C. F., "Hydraulics of Runoff from Developed Surfaces," *Proc. Highway Res. Board*, 26:129–150 (1946).
25. Metcalf and Eddy Inc., "Storm Water Management Model, Vol. 1," Final Report for the U.S. EPA, Contract No. 14-12-501, Water Pollution Control Research Series 11024 DOC 07/71 (1971).
26. Fisheries and Environment Canada and Ontario Ministry of the Environment, "Storm Water Management Model Study, Vol. III, Users Manual," Ottawa and Toronto (1976).
27. Terstriep, M. L., and Stall, J. B., "The Illinois Urban Area Simulator, ILLUDAS," Illinois State Water Survey Bulletin 58 (1974).
28. World Meteorological Organization. "Intercomparison of Conceptual Models used in Operational Hydrological Forecasting," Operational Hydrology Report No. 7, WMO No. 429 (1975).
29. U.S. Army Corps of Engineers, "Snow Hydrology," Summary Report of the Snow Investigations, (June 1956).
30. U.S. Army Corps of Engineers, "Runoff from Snowmelt," Engineering and Design Manuals, EM 1110-2-1406, (January 1960).
31. Garstka, W. U., et al., "Factors Affecting Snowmelt and Streamflow," U.S. Bureau of Reclamation and U.S. Forest Service (1958).
32. Rockwood, D. M., "Columbia Basin Streamflow Routing by Computer," *Proc. ASCE*, J. of Waterways and Harbours Div. 84 (WW5):1–15 (1958).
33. Benson, M. A., and Dalrymple, T., "General Field and Office Procedures for Indirect Discharge Measurements," U.S. Geological Survey Techniques of Water Resources Investigations, Book 3, Chapter A1, Washington, D.C.: U.S. Govt. Printing Office, 1968.
34. Chow, V. T., *Open Channel Hydraulics*, New York: McGraw-Hill Book Co., Inc., 1959.
35. Matthai, H. F., "Measurement of Peak Discharge at Width Contractions by Indirect Methods," U.S. Geological Survey Techniques of Water Resources Investigations, Book 3, Chapter A4, Washington, D.C., U.S. Govt. Printing Office, 1968.
36. Puls, L. G., "Flood Regulation of the Tenessee River," 70th Congress, 1st Session, M.D. 185, pt. 2, Appendix B (1928).
37. McCarthy, G. T., "The Unit Hydrograph and Flood Routing," paper presented at Conference of North Atlantic Div., U.S. Army Corps of Engineers (June 1938).
38. Kendall, G. R., "Statistical Analysis of Extreme Values," Proc. of Symposium No. 1, Spillway Design Floods, NRC, Ottawa 54–78 (1959).
39. Gumbel, E. J., *Statistics of Extremes*, New York: Columbia Univ. Press, 1960.
40. Yevjevich, V., "Probability and Statistics in Hydrology," Fort Collins, Colo.: Water Resources Publications 1972.
41. Kite, G. W., "Frequency and Risk Analyses in Hydrology," Inland Waters Directorate, Environment Canada, Ottawa (1976).
42. Natural Environment Research Council, Flood Studies Report in Five Volumes, London (1975).
43. Hazen, A., "Storage to be Provided in Impounding Reservoirs for Municipal Water Supply," *Trans. ASCE*, 77:1539–1669 (1914).
44. Powell, R. W., "A Simple Method of Estimating Flood Frequency," *Civil Eng.*, 13:105–107 (1943).
45. Chow, V. T., "The Log Probability Law and its Engineering Applications," *Proc. ASCE*, Paper No. 536, 80:1–25 (1954).
46. Hazen, A., "Discussion on Flood Flows by W. E. Fuller," *Trans. ASCE*, 77:626–632 (1914).
47. Hazen, Allen, *Flood Flows, A Study of Frequencies and Magnitudes* New York: John Wiley and Sons, Inc., 1930.

48. Whipple, G. C., "The Element of Chance in Sanitation," *J. Franklin Inst.* 182:37–59, 205–227 (1916).
49. Sangal, B. P., and Kallio, R. W., "Magnitude and Frequency of Floods in Southern Ontario," Technical Bulletin Series No. 99, Inland Waters Directorate, Fisheries and Environment Canada, Ottawa (1977).
50. Roche, M., *Hydrologie de Surface*, Paris: Gauthier-Villars, 1963.
51. Sangal, B. P., and Biswas, A. K. "The 3-Parameter Lognormal Distribution and Its Applications in Hydrology," *Water Resources Research*, 6(2):505–515 (1970).
52. Chow, V. T., "A General Formula for Hydrologic Frequency Analysis," *Trans Am. Geophys. Union*, 32:231–237 (1951).
53. Potter, W. D., "Simplification of the Gumbel Method for Computing Probability Curves," U.S. Dept. of Agriculture, Soil Conservation Service, SCS-TP-78 (May 1949).
54. United States Water Resources Council, "Guidelines for Determining Flood Flow Frequency," Bulletin #17B of the Hydrology Committee, Revised Sept. 1981. Washington, D.C.: U.S. Govt. Printing Office, 1981.
55. Wilson, E. B., and Hilferty, M. M., "The Distribution of Chi-Square," *Proc. National Academy of Science*, 17(12):684–688 (December 1931).
56. Cunnane, C., "Unbiased Plotting Positions—A Review," *J. of Hydrology*, 37:205–222 (1978).
57. Sangal, B. P., and Hill, H. M., "Reliability of Plotting Positions," *Proc. Int. Hyd. Symposium*, Fort Collins, Colo., 2:313–317 (1967).
58. Hald, A., *Statistical Theory with Engineering Applications*, New York: John Wiley and Sons, Inc., 1962, p. 131.
59. Panchang, G. M., "Improved Precision of Future High Floods," UNESCO Int. Symposium on Floods and their Computation, Leningrad, U.S.S.R., 51–59 (1967).
60. Dalrymple, T., "Flood Frequency Analyses, Manual of Hydrology, Pt. 3, Flood Flow Techniques," USGS Water Supply Paper 1543-A (1960).
61. Fuller, W. E., "Flood Flows," *Trans. ASCE*, Paper No. 1293, 77:564–617 (1914).
62. Jarvis, C. S., et al. "Floods in the United States, Magnitude and Frequency," USGS Water Supply Paper 771:90–96 (1936).
63. Langbein, W. B., "Peak Discharge from Daily Records," USGS Water Resources Bulletin (1944).
64. Sangal, B. P., "Practical Method of Estimating Peak Flow," *J. Hyd. Eng., ASCE*, 109(4):549–563 (1983).
65. Hare, F. K., and Thomas, M. K., Climate Canada, 2nd ed., Toronto: John Wiley and Sons Canada Ltd., 1979.
66. Sangal, B. P., "A Practical Method of Estimating Peak from Mean Daily Flows with Application to Streams in Ontario," National Hydrology Research Institute Paper No. 16, Inland Waters Directorate Tech. Bulletin No. 122, NHRI, Ottawa, Canada (1981).
67. Environment Canada. "Surface Water Data Reference Index," Inland Waters Directorate, Water Survey of Canada, Ottawa, Canada (1983).
68. Bryson, R. A., and Hare, F. K., *World Survey of Climatology, Vol. 11, Climates of North America*, New York: Elsevier Scientific Publishing Co., 1974.
69. U.S. Dept. of Commerce, NOAA, "Climates of the States, Vol. 1, Eastern States," New York: Water Information Centre, Inc., 1974.

SECTION II

MIXING PHENOMENA AND PRACTICES

CONTENTS

CHAPTER 12. HYDRODYNAMICS OF LAMINAR BUOYANT JETS 317

CHAPTER 13. IMPINGING JETS 349

CHAPTER 14. HYDRODYNAMICS OF CONFINED COAXIAL JETS 367

CHAPTER 15. TURBULENT MIXING AND DIFFUSION OF JETS 391

CHAPTER 16. HYDRODYNAMICS OF JETS IN CROSSFLOW 406

CHAPTER 17. MODELING TURBULENT JETS IN CROSSFLOW 430

CHAPTER 18. BATCHWISE JET MIXING IN TANKS 466

CHAPTER 19. STABILITY OF JETS IN LIQUID-LIQUID SYSTEMS 474

CHAPTER 20. MODELING TURBULENT JETS WITH VARIABLE DENSITY..... 511

CHAPTER 21. JET MIXING OF FLUIDS IN VESSELS........................ 544

CHAPTER 22. MIXING IN LOOP REACTORS.............................. 557

CHAPTER 23. VERTICAL CIRCULATION IN DENSITY-STRATIFIED RESERVOIRS 572

CHAPTER 24. ROLLOVER IN STRATIFIED HOLDING TANKS............... 637

CHAPTER 25. JET INJECTION FOR OPTIMUM PIPELINE MIXING........... 660

CHAPTER 26. SWIRL FLOW GENERATED BY TWISTED PIPES AND TAPE INSERTS.. 691

CHAPTER 27. MICROMIXING PHENOMENA IN STIRRED REACTORS........ 707

CHAPTER 28. BACKMIXING IN STIRRED VESSELS 772

CHAPTER 29. INDUSTRIAL MIXING EQUIPMENT.......................... 803

CHAPTER 30. SIMULATION OF MECHANICALLY-AGITATED LIQUID-LIQUID REACTION VESSELS.. 851

CHAPTER 31. SOLID AGITATION AND MIXING IN AQUEOUS SYSTEMS 886

CHAPTER 32. MIXING AND AGITATION OF VISCOUS FLUIDS 901

CHAPTER 33. USE OF HELICAL RIBBON BLENDERS FOR NON-NEWTONIAN MATERIALS.. 927

CHAPTER 12

HYDRODYNAMICS OF LAMINAR BUOYANT JETS

Yogesh Jaluria

Mechanical and Aerospace Engineering Department
Rutgers University
New Brunswick, New Jersey USA

CONTENTS

INTRODUCTION, 317

AXISYMMETRIC VERTICAL JETS, 319
Governing Equations, 319
Nonbuoyant Jet, 321
Axisymmetric Plume, 324
Buoyant Jet, 325

TWO-DIMENSIONAL VERTICAL JETS, 331
Governing Equations, 331
Nonbuoyant Jet, 332
Two-Dimensional Plume, 334
Buoyant Jet, 336

OTHER FLOWS, 339
Stratified Environment, 339
Other Orientations, 341
Transition to Turbulence, 342
Additional Effects, 343

SUMMARY, 345

NOTATION, 345

REFERENCES, 346

INTRODUCTION

In nature and in technology, we frequently encounter free boundary flows which are characterized by the absence of rigid boundaries. Among the most important of these is the buoyant jet, which is driven by an input of momentum and buoyancy in an extensive environment. Buoyant jets are important in meteorological, oceanographic, and environmental studies. The rejection of thermal energy and of chemical waste products to the atmosphere and to water bodies generally involves buoyant jets, such as the flow emerging from a cooling tower or a chimney. Similarly, the wake shed by a heated body possesses momentum as well as thermal buoyancy and may be treated as a buoyant jet. Heat extraction and energy storage systems, such as those employed in solar energy utilization, are also often concerned with the buoyant jet flows arising from heated discharges into extensive fluid regions.

The flow resulting from an input of momentum alone, with no associated buoyancy, is termed the nonbuoyant jet, or simply a jet. A considerable amount of work has been done on jets because of their importance in a wide variety of technological applications, ranging from jet flow in a combustion chamber to jets in aerospace devices. The flow arising from only a buoyancy input is termed a plume and represents the other extreme case of a buoyant jet. As a consequence of this, a buoyant jet is also often termed a forced plume, the two limiting circumstances of which are the jet and the buoyancy-driven plume. Work has also been done on the plume flow, mainly because of the practical importance of the natural convection wake generated by thermal sources such as heated bodies and fires. Far downstream of the location of discharge of the jet, the flow approaches the characteristics of a plume due to the continuing effect of the buoyancy force. Consequently, it is important to understand the basic nature of a plume flow resulting from a continuous input of buoyancy. Much less work has been done on buoyant jets, as compared to the effort directed at the nonbuoyant jet and the buoyancy-driven plume, mainly because of the complexity that results from the presence of both natural and forced convection mechanisms in the flow. Near the discharge location, the flow is dominated by the forced flow mechanisms, arising from the momentum input, and as the flow moves downstream, buoyancy effects increase, resulting in a buoyancy–dominated flow far downstream.

Most of the buoyant jet flows that arise in the environment are turbulent. Similarly, in technological applications; turbulent flows are generally encountered. However, the flow near the jet discharge if often laminar, though it may undergo transition to turbulence as the flow proceeds downstream. Also, in several practical problems, laminar flow is of interest because of the low flow rates that arise and the high viscosity of the fluid. Such circumstances arise, for instance, in sensible energy storage in solar energy systems, lubrication systems, experimental arrangements, and wakes generated by heated bodies. As a result of the greater interest in turbulent buoyant jets in practical problems, most studies have considered turbulent flow. However, as discussed, there are many problems in which laminar flow is of interest. It is important to determine the basic characteristics of laminar buoyant jets and to study the transition of the flow to turbulence. It is also necessary to consider the two limiting cases of buoyant jets and to determine the effect of buoyancy on the resulting flow.

Besides the preceding subdivisions of buoyant jets on the basis of whether they are laminar or turbulent and whether they are buoyant or nonbuoyant, several other classifications are possible. The jet may be discharged from a small circular hole which is generally idealized as a point source of momentum and buoyancy. The resulting flow is axisymmetric if the flow is discharged vertically. This represents the circumstance of an axisymmetric buoyant jet. Similarly, the jet may be discharged from a long slit, giving rise to a two-dimensional flow. The slit is generally idealized as a line source and the jet is termed as a plane, or two-dimensional, buoyant jet. Since the gravitational force affects the flow, through the buoyancy term, it is apparent that the orientation of the discharge, with respect to the vertical gravitational force, would be an important consideration. This leads to a classification of buoyant jets as vertical, inclined, or horizontal. In preceding discussion, we have considered the buoyancy force to be aligned with the direction of discharge for vertical jets. However, there are many circumstances in which the buoyancy force is in a direction opposite to that of the flow discharge. Such flows arise, for instance, in energy extraction systems where cold fluid is discharged back into the hot fluid after heat removal in a heat exchanger and in many natural processes. The jet is then said to be negatively buoyant.

Several other important considerations arise in the study of laminar buoyant jets. The ambient medium in which the jet is discharged may be of uniform density or may be stably stratified. Similarly, the medium may be quiescent or moving, as is the case for heat rejection into a river or into ambient air moving at a given speed. The buoyancy input may be due to a temperature or a concentration difference, the former circumstance being referred to as a thermally buoyant jet. Also, the stratification of the ambient medium may be due to a temperature variation, generally termed as thermal stratification, or due to a concentration variation. In some cases of practical interest, particularly in air and water pollution, combined thermal and material transport mechanisms may arise, resulting in combined buoyancy effects. Buoyant jets are sometimes discharged in the vicinity of rigid boundaries and the resulting flow may be affected by the presence of such surfaces. These effects are of particular interest in the discharge of buoyant jets into enclosed fluid regions.

This chapter discusses the basic nature of laminar buoyant jets, considering both axisymmetric and two-dimensional cases. The two limiting circumstances of the nonbuoyant jet and the buoyancy-induced plume are considered and the available analytical and experimental results on buoyant jets are presented. The effects of ambient medium stratification and motion are discussed. Also considered is the instability of the flow and its transition to turbulence. The effects of the input buoyancy and of the orientation of the flow, with respect to gravity, are considered. The behavior of a buoyant jet as the flow proceeds downstream from the location of discharge is discussed in detail. Various other effects mentioned are also outlined. The present status of available information on laminar buoyant jets is presented and the current and future trends discussed.

AXISYMMETRIC VERTICAL JETS

Among the earliest studies of the laminar axisymmetric, nonbuoyant, jet was that of Schlichting [1]. Using boundary-layer approximations, similarity solutions were found for the flow arising from a point source of momentum. A closed-form solution was presented. Experimental verification of the analytical results was obtained by Andrade and Tsien [2], who observed the flow optically by using suspended aluminum particles. Landau [3] obtained an exact solution of the full Navier-Stokes equations for a round laminar jet. The solution becomes equivalent to that obtained by employing the boundary-layer approximations for Reynolds numbers greater than 8, where the Reynolds number Re is based on the nozzle diameter, as defined later. Squire [4] presented an analysis similar to that of Landau [3] and also obtained the temperature distribution, neglecting the effects of buoyancy. At sufficiently large Reynolds numbers, the temperature solution of Squire [4] reduces to the boundary-layer results. We shall first outline the results obtained in these studies and then discuss the effect of buoyancy on the flow.

Governing Equations

Let us consider the governing equations for a vertical, axisymmetric, laminar thermally-buoyant jet, shown in Figure 1. We assume the flow to be incompressible and neglect the viscous dissipation and pressure work terms [5]. Then the following equations are obtained from the conservation of mass, momentum, and energy, for constant fluid properties except density, which must vary to give rise to buoyancy.

$$\frac{\partial \rho}{\partial \tau} + \bar{V} \cdot \nabla \rho + \rho \nabla \cdot \bar{V} = 0 \tag{1}$$

$$\rho \left[\frac{\partial \bar{V}}{\partial \tau} + (\bar{V} \cdot \nabla)\bar{V} \right] = \rho \bar{g} - \nabla p + \mu \nabla^2 \bar{V} \tag{2}$$

$$\rho C_p \left[\frac{\partial t}{\partial \tau} + (\bar{V} \cdot \nabla) t \right] = k \nabla^2 t \tag{3}$$

where $\bar{V}$ = the velocity vector in the flow
ρ = the fluid density
C_p = its speciific heat at constant pressure
k = the thermal conductivity
$\bar{g}$ = the gravitational acceleration
p = the local pressure
μ = the coefficient of viscosity of the fluid
t = the local temperature
τ = the time

These equations apply for a thermally-buoyant flow, which involves a consideration of the thermal transport mechanisms. Similarly, a buoyancy input due to a concentration difference may be considered [6, 7]. This circumstance is of particular interest in environmental pollution due to chemical waste disposal. The energy equation is then replaced by the conservation equation for

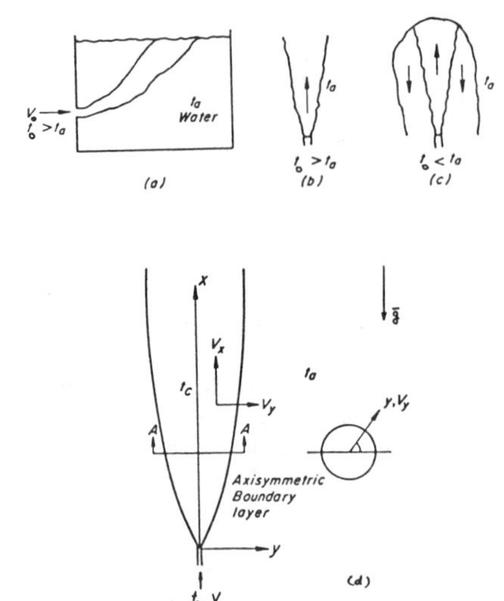

Figure 1. (A–C) Sketch of some common buoyant jet flows; (D) coordinate system for a vertical, laminar, axisymmetric, buoyant jet.

the chemical species. For a low concentration species diffusing in a two-component mixture, the mass conservation equation may be written, for constant fluid properties, as [5]

$$\frac{\partial C}{\partial \tau} + (\bar{V} \cdot \nabla)C = D\nabla^2 C \tag{4}$$

where C is the local concentration of the species and D the mass diffusivity. Several studies have considered this problem, as outlined later in this section.

The governing momentum equation has the body force term $\rho\bar{g}$ and the pressure force term $-\nabla p$. The pressure p may be subdivided into the motion pressure p_m due to the flow and the hydrostatic pressure p_h in the quiescent ambient medium. Then, the combination of the gravitational force and the pressure terms may be written as [8]

$$\rho\bar{g} - \nabla p = (\rho\bar{g} - \nabla p_h) - \nabla p_m = (\rho - \rho_a)\bar{g} - \nabla p_m \tag{5}$$

where ρ_a is the density in the ambient medium and $\nabla p_h = \rho_a\bar{g}$ for gravitational field. Therefore, $(\rho - \rho_a)\bar{g}$ is the buoyancy force term. Generally, the Boussinesq approximations are employed to simplify the governing equations [5]. Using these approximations, the continuity equation and the buoyancy term are obtained as

$$\nabla \cdot \bar{V} = 0 \tag{6}$$

$$\rho_a - \rho = \rho\beta(t - t_a), \text{ or, } \rho\beta^*(C - C_a) \tag{7}$$

where β is the coefficient of volumetric thermal expansion, β^* the corresponding coefficient of expansion due to concentration, and C_a the species concentration in the ambient medium. The two coefficients of expansion are defined by

$$\beta = -\frac{1}{\rho}\left(\frac{\partial \rho}{\partial t}\right)_{C,p} \qquad \beta^* = -\frac{1}{\rho}\left(\frac{\partial \rho}{\partial C}\right)_{t,p} \tag{8}$$

Since the governing equations for buoyancy input due to a concentration difference are identical to Equation 3, if $k/\rho C_p$ is replaced by D, the discussion here is largely directed at the thermally-buoyant jet, the results for the other case being obtained by employing the similarity between the two transport mechanisms.

The boundary-layer assumptions may be employed, with the Boussinesq approximations, to obtain the governing equations that would apply for a laminar, axisymmetric, buoyant jet. With these assumptions, the motion pressure term and the axial diffusion terms are neglected to yield the following equations for the steady, boundary-layer, thermally-buoyant flow shown in Figure 1.

$$\frac{\partial}{\partial x}(yV_x) + \frac{\partial}{\partial y}(yV_y) = 0 \tag{9}$$

$$V_x \frac{\partial V_x}{\partial x} + V_y \frac{\partial V_x}{\partial y} = g\beta(t - t_a) + \frac{\nu}{y}\frac{\partial}{\partial y}\left(y \frac{\partial V_x}{\partial y}\right) \tag{10}$$

$$V_x \frac{\partial t}{\partial x} + V_y \frac{\partial t}{\partial y} = \frac{\alpha}{y}\frac{\partial}{\partial y}\left(y \frac{\partial t}{\partial y}\right) \tag{11}$$

where V_x and V_y = the vertical and radial velocity components
$\nu\ (=\mu/\rho)$ = the kinetic viscosity of the fluid
$\alpha\ (=k/\rho C_p)$ = its thermal diffusivity
β = its coefficient of thermal expansion
g = the magnitude of gravitational acceleration
x = the vertical coordinate distance
y = the radial coordinate distance
t_a = the temperature of the quiescent, extensive ambient medium

The boundary conditions for the preceding equations may be written as

$$\begin{aligned} &\text{at } y = 0: \quad V_y = 0, \quad \frac{\partial V_x}{\partial y} = 0, \quad \frac{\partial t}{\partial y} = 0 \\ &\text{as } y \to \infty: \quad V_x \to 0, \quad \text{and } t \to t_a \end{aligned} \tag{12}$$

Therefore, the conditions at the centerline, y = 0, arise from symmetry and those far from the flow region, $y \to \infty$, from the conditions in the ambient medium. The centerline temperature is denoted by t_c and may be determined from energy conservation considerations, employing the energy input at the nozzle, x = 0. The problem is greatly simplified by restricting the analysis to large distances from the jet origin, as compared to the dimensions of the discharge opening. Then the buoyant jet is represented by a momentum input J_0 and a thermal energy input Q_0 at a point source located at the origin, x = 0. This allows a similarity analysis of the flow since a characteristic dimension does not arise in the problem [9].

Nonbuoyant Jet

For a nonbuoyant jet, or if the buoyancy effects are neglected, the momentum flux J(x) in the x direction is a constant, since there are no external forces acting on the flow. This gives

$$J(x) = \int_0^\infty 2\pi\rho V_x^2 y\, dy = J_0 = \text{constant} \tag{13}$$

In the presence of buoyancy, the momentum flux will not be a constant, but will increase if the buoyancy force acts in the direction of the flow. Also, if the buoyancy force term, $g\beta(t - t_a)$, in Equation 10 is neglected, the flow becomes independent of the temperature field. A stream function ψ may be defined to satisfy the countinuity equation as:

$$V_x = \frac{1}{y}\frac{\partial \psi}{\partial y}, \qquad \text{and } V_y = -\frac{1}{y}\frac{\partial \psi}{\partial x} \tag{14}$$

Similarity is obtained in the governing momentum equation if the following transformation is employed [1, 10, 11]

$$\eta = \frac{y}{\sqrt{K}x}, \qquad \psi = \nu x f(\eta) \tag{15}$$

where

$$K = \frac{16\pi}{3}\nu^2/(J_0/\rho) \tag{16}$$

Here

η = the similarity variable

f = the dimensionless stream function

J_0/ρ = termed the kinetic momentum

K = a constant

The relation between K and the input momentum J_0, as just given, is obtained by applying the similarity solution obtained to Eq. 13. The velocity components V_x and V_y are given by

$$V_x = \frac{\nu}{Kx}\frac{f'}{\eta}, \qquad V_y = \frac{\nu}{\sqrt{K}x}\left[f' - \frac{f}{\eta}\right] \tag{17}$$

Therefore, the boundary conditions are obtained in terms of the similarity variables as:

$$\begin{aligned} &\text{at } \eta = 0: \quad f = f' = 0 \\ &\text{as } \eta \to \infty: \quad f' \to \text{bounded} \end{aligned} \tag{18}$$

The governing momentum equation is obtained in terms of the similarity variables as

$$\left(f'' - \frac{f'}{\eta}\right)' = \frac{ff'}{\eta^2} - \frac{f'^2}{\eta} - \frac{ff''}{\eta} \tag{19}$$

where the primes denote differentiation with respect to η. The solution to this equation, with the prior boundary conditions, is given by Schlichting [1] as

$$f = \frac{\eta^2}{1 + \frac{1}{4}\eta^2} \tag{20}$$

which gives

$$\begin{aligned} V_x &= \frac{2\nu}{Kx}\frac{1}{(1 + \frac{1}{4}\eta^2)^2} \\ V_y &= \frac{\nu}{\sqrt{K}x}\frac{\eta - \frac{1}{4}\eta^3}{(1 + \frac{1}{4}\eta^2)^2} \end{aligned} \tag{21}$$

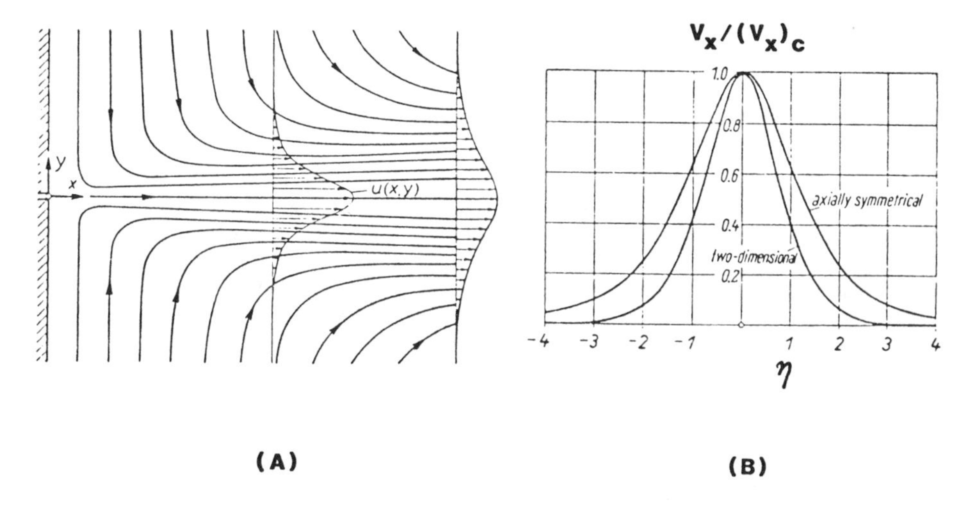

Figure 2. (A) Streamline pattern for an axisymmetric, nonbuoyant jet emerging from a circular opening in a wall; (B) velocity distribution in an axisymmetric and a two-dimensional jet [9].

Figure 2 shows the velocity distribution in an axisymmetric jet, along with that for a two-dimensional jet, discussed in the next section.

We may now consider the energy equation, Equation 11, employing the preceding similarity transformation along with the dimensionless temperature θ, defined as,

$$\theta = \frac{t - t_a}{t_c - t_a} \tag{22}$$

Similarity arises for the power-law variation of the centerline temperature excess, i.e., for $t_c - t_a = Nx^n$, where N and n are constant. The governing energy equation is obtained in terms of the similarity variables as

$$(\eta\theta' + \Pr f\theta)' = 0 \tag{23}$$

with the boundary conditions obtained from the preceding definition of θ as

$$\text{at } \eta = 0;\ \theta = 1, \text{ and, as } \eta \to \infty;\ \theta \to 0 \tag{24}$$

where Pr ($= \nu/\alpha$) is the Prandtl number. Squire [4] gave the temperature distribution in the flow as

$$\theta = (1 + \tfrac{1}{4}\eta^2)^{-2\Pr} \tag{25}$$

It may be noted that the symmetry condition of $\theta' = 0$ at $\eta = 0$ is satisfied by the preceding distribution. In fact, this condition may also be used as a boundary condition instead of $\theta \to 0$ as $\eta \to \infty$, which can be shown to be satisfied if the other two conditions at $\eta = 0$ are employed.

In the absence of viscous dissipation and any thermal sources in the flow, the convected thermal energy in the flow Q(x) must remain constant downstream. This condition may be written as

$$Q = \int_0^\infty 2\pi\rho C_p(t - t_a)V_x y\, dy = \text{constant} = Q_0 \tag{26}$$

Employing the similarity solutions, just given, the centerline temperature is, therefore, obtained from this equation as

$$t_c - t_a = \frac{(2\,Pr + 1)Q_0}{8\pi\mu C_p x} \tag{27}$$

which implies that $n = -1$ and $N = (2\,Pr + 1)Q_0/8\pi\mu C_p$ for an axisymmetric vertical jet in which the buoyancy effects are neglected. Therefore, the velocity and the temperature distributions in a nonbuoyant jet may be determined. These solutions would apply in the flow region, near the jet origin, where the buoyancy effects may be negligible. The extent of this region would depend on the input momentum and buoyancy. Farther downstream, the buoyancy effects would become important and the term due to buoyancy must be retained in the momentum equation. This results in a coupling between the flow and the thermal field, as discussed in the following.

Axisymmetric Plume

Before proceeding to the mixed-convection flow in a laminar, buoyant jet, let us consider the axisymmetric plume flow driven only by an input of thermal buoyancy. This flow is frequently observed in practical circumstances such as the natural convection flow due to heated electronic components, heaters, and fires. The velocity is zero at the source, and as the flow proceeds downstream, buoyancy force accelerates the flow to give rise to a boundary-layer flow far from the source. This problem has been considered in several studies, assuming a point source of buoyancy [12–15]. The wake due to a finite-sized source has also been studied and related to the axisymmetric plume arising from a point source [16]. The analysis of the point source plume and the characteristics relevant to the study of a buoyant jet are outlined here.

The governing boundary-layer equations for an axisymmetric point source plume are Equations 9 through 11. These may be solved by similarity methods, employing the following similarity variables

$$\eta = \frac{y}{x}(Gr_x)^{1/4}, \qquad \psi = \nu x f(\eta), \qquad \theta = \frac{t - t_a}{t_C - t_a} \tag{28}$$

where

$$Gr_x = \frac{g\beta x^3(t_c - t_a)}{\nu^2} \tag{29}$$

The Grashof number Gr_x is similar to the Reynolds number Re, which arises in forced flow, and represents a comparison between the buoyancy and the viscous forces. With this similarity transformation, the governing equations become

$$f''' + (f - 1)\left(\frac{f'}{\eta}\right)' - \left(\frac{1 + n}{2}\right)\frac{f'^2}{\eta} + \eta\theta = 0 \tag{30}$$

$$(\eta\theta')' + Pr(f\theta' - nf'\theta) = 0 \tag{31}$$

where n is the constant in the power-law variation of the centerline temperature excess, $t_c - t_a = Nx^n$. Since the convective heat flux Q(x) must remain constant downstream, in the absence of viscous dissipation and thermal sources, Equation 26 must be satisfied. If the preceding similarity variables are substituted in this equation, the result is

$$Q(x) = 2\pi\rho C_p \nu N x^{n+1} \int_0^\infty f'\theta\, d\eta = Q_0 \tag{32}$$

which implies that $n = -1$ for Q(x) to be independent of x. Also, the center-line temperature is given by

$$t_c - t_a = \frac{N}{x} = \frac{Q_0}{2\pi\rho C_p \nu I x}, \qquad \text{where } I = \int_0^\infty f'\theta \, d\eta \tag{33}$$

The integral I is computed from a solution of the coupled governing differential equations which are obtained, for $n = -1$, as

$$f''' + (f-1)\left(\frac{f'}{\eta}\right)' + \eta\theta = 0 \tag{34}$$

$$(\eta\theta' + \Pr f\theta)' = 0 \tag{35}$$

The energy equation is the same as that obtained earlier for the nonbuoyant jet. As seen later, it also applies for the buoyant jet flow. The momentum equation is quite similar to that for a jet (Equation 19), and the additional buoyancy term couples the flow with the temperature field. The velocity components V_x and V_y are given by

$$V_x = \frac{\nu}{x}\sqrt{Gr_x}\left(\frac{f'}{\eta}\right), \qquad \text{and } V_y = -\frac{\nu}{x}(Gr_x)^{1/4}\left(\frac{f}{\eta} - \frac{f'}{2}\right) \tag{36}$$

From these expressions, the boundary conditions for the preceding similarity equations may be obtained as

$$f(0) = f'(0) = 1 - \theta(0) = \theta'(0) = \frac{f'}{\eta}(\infty) = 0 \tag{37}$$

where the quantity within parentheses denotes the η location where the condition is applied. It can be shown that these five conditions are independent and that the sixth condition, $\theta(\infty) = 0$, may be obtained from these. Since only five conditions are needed for the fifth-order system of ordinary differential equations that govern the axisymmetric plume flow, Equations 34 and 35 may be solved with the boundary conditions given in Equation 37 to yield the velocity and temperature fields.

The centerline temperature can be obtained from Equation 33. The value of 1/(I Pr) is given by Fujii [13] for Pr values of 0.01, 0.7, 1.0, 2.0, 10, and ∞ as 0.759, 0.687, 0.667, 0.625, 0.561, and 0.5, respectively. Closed-form solutions for $\Pr = 1$ and 2 have also been given. Figure 3 shows the computed velocity and temperature profiles. The maximum velocity and temperature occur, with zero slope, at the centerline, as expected from the symmetry conditions, which give zero shear and adiabatic conditions there. The centerline velocity $(V_x)_c$ is given by

$$(V_x)_c = \sqrt{\frac{g\beta Q_0}{2\pi\rho C_p \nu I}}\left(\frac{f'}{\eta}\right)_0 \tag{38}$$

where $(f'/\eta)_0$ is the value at $\eta = 0$ in Figure 3. Therefore, the centerline velocity is predicted to remain constant downstream. This interesting result arises because of the accelerating effects of buoyancy being balanced by the decelerating effects of entrainment. Not much experimental work has been done on this flow. However, the basic trends are seen in flow arising above finite-sized sources [17].

Buoyant Jet

The two limiting cases of a nonbuoyant jet and an axisymmetric plume may be solved by similarity methods, as discussed previously, if a point source is assumed. However, in a buoyant jet, the buoyancy term in the momentum equation gives rise to a coupling between the flow and the

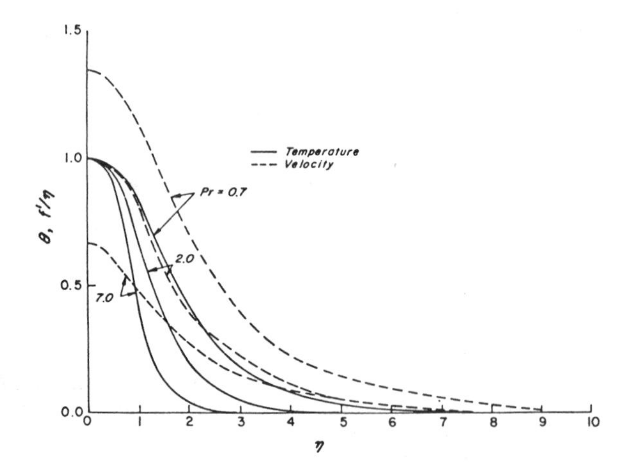

Figure 3. Computed velocity and temperature profiles in an axisymmetric plume arising from a point source of heat [15].

temperature field. The similarity formulation for a nonbuoyant jet does not yield similarity when this additional term is included. Similarly, the similarity analysis for a plume does not allow the inclusion of a momentum input at the source. This nonsimilar behavior is expected since the buoyancy effects increase downstream and become dominant far from the source. Near the source, these effects are small, as expected from the small values of the Grashof number Gr_x. Consequently, forced flow effects due to the momentum input dominate at small values of x. Therefore, a study of the buoyant jet flow involves a consideration of dominant forced-convection effects near the source and of dominant natural-convection effects far downstream, with an intermediate region where both mechanisms are of the same order of magnitude.

If we again consider the governing equations, Equation 9 through 11, and employ the similarity transformation previously given for the nonbuoyant jet, while retaining the buoyancy force term, an additional term appears in the momentum equation, which now becomes

$$\left(f'' - \frac{f'}{\eta}\right)' = \frac{ff'}{\eta^2} - \frac{f'^2}{\eta} - \frac{ff''}{\eta} - Gr_x K^2 \eta \theta \tag{39}$$

where the Grashof number Gr_x is defined in Equation 29. The energy equation remains unchanged. The additional term due to buoyancy is a function of height x and, therefore, similarity does not arise in the momentum equation for a laminar buoyant jet. Because of buoyancy, the momentum flux J(x) is not a constant, but increases downstream. The convected thermal energy Q(x), however, remains constant downstream, in the absence of thermal sources and viscous dissipation effects. The presence of the buoyancy term in the preceding equation also couples it with the energy equation, Equation 23, so that the two have to be solved simultaneously, as discussed earlier for the axisymmetric plume. However, similarity was obtained for the plume flow, resulting in coupled ordinary differential equations. The governing equations for the buoyant jet are nonsimilar and the solution is, therefore, more involved.

If the effect of thermal buoyancy on the flow is taken as small, the problem may be analyzed by a perturbation analysis, as done by Mollendorf and Gebhart [10] and by Schneider and Potsch [11]. As mentioned earlier, small buoyancy effects arise near the jet nozzle, since Gr_x is small in this region, being zero at the location of the source. As the flow proceeds downstream, Gr_x increases and, as seen from Equation 39, the buoyancy effects also become larger, restricting the validity of the perturbation analysis to a region close to the jet origin. Therefore, farther downstream, the governing partial differential equations would need to be considered, since the buoyancy effects would be of the same order of magnitude as the effects arising due to the input momentum. It is also evident that very far downstream, buoyancy effects will dominate and the effect of the input momentum may be treated as a perturbation on the buoyancy-driven plume. However, the flow may become turbulent before this occurs, leading to transport mechanisms which are obviously very different from those discussed here for laminar flow.

A perturbation analysis of the laminar buoyant jet may be carried out with a perturbation parameter $\epsilon(x)$ defined as

$$\epsilon(x) = K^2 \, Gr_x \tag{40}$$

For a uniform velocity distribution across a jet nozzle of diameter d, the input momentum J_0 is obtained from Equation 13 by integrating over the nozzle area. This gives

$$J_0 = \frac{\pi d^2}{4} \rho V_0^2 = \frac{\pi \rho \nu^2}{4} Re_d^2 \tag{41}$$

where

$$Re_d = \frac{V_0 d}{\nu} \tag{42}$$

V_0 = the uniform jet velocity at the nozzle

Similarly, other distributions may be considered. It can be shown that for a developed Poiseuille flow at the nozzle, $J_0 = \pi\rho\nu^2 Re_d^2/3$. The parameter K may be written in terms of Re_d as

$$K = 1/R^2, \text{ with, } R = Re_d/\gamma \tag{43}$$

where γ is a parameter that depends on the velocity profile at the nozzle. For a uniform velocity distribution at the nozzle, γ is $8/\sqrt{3}$ and for a parabolic distribution, it is 4. The perturbation analysis is carried out by defining the stream function f and the temperature θ as

$$f = f_0(\eta) + \epsilon(x) f_1(\eta) + \epsilon^2(x) f_2(\eta) + \cdots \tag{44}$$

$$\theta = \theta_0(\eta) + \epsilon(x)\theta_1(\theta) + \epsilon^2(x)\theta_2(\eta) + \cdots \tag{45}$$

The perturbation parameter $\epsilon(x)$ is Gr_x/R^4 and the zeroth-order solution, (f_0, θ_0), corresponds to the nonbuoyant case. Therefore, the boundary conditions for f_0 and θ_0 are those given in Equations 18 and 24. The first-order solution (f_1, θ_1) gives the effect of thermal buoyancy on the flow. The corresponding boundary conditions are

$$f_1(0) = f_1'(0) = \theta_1(0) = f_1'(\infty) = \theta_1(\infty) = 0 \tag{46}$$

and the velocity V_x is given by

$$V_x = \frac{\nu R^2}{x} [f_0'/\eta + \epsilon(x) f_1'/\eta + \cdots] \tag{47}$$

Numerical solutions for this problem were obtained by Mollendorf and Gebhart [10] for Prandtl number values of 6.7, 10, and 20. Figure 4 shows the computed velocity and temperature profiles at $Pr = 6.7$. The maximum effect on the velocity is found to be only about 7% for $\epsilon = 1$, with a still smaller effect on the temperature. A positive value of ϵ refers to the buoyant jet, with the buoyancy force in the same direction as the flow, and a negative value to a negatively-buoyant jet, with the

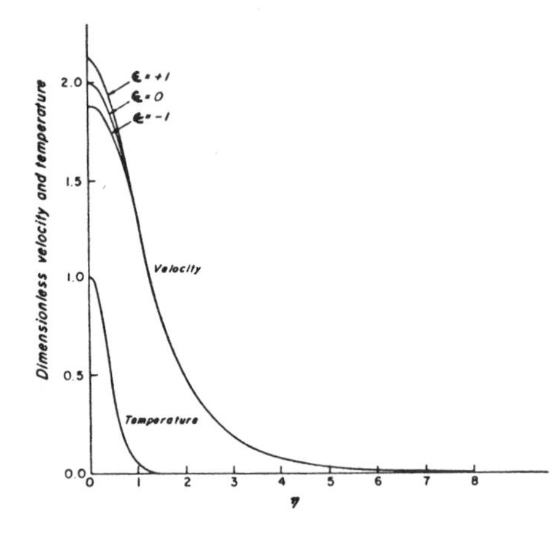

Figure 4. Computed velocity and temperature profiles in a vertical, laminar, axisymmetric jet for small thermal buoyancy effects, given by the perturbation parameter ϵ, at Pr = 6.7 [10].

buoyancy force opposing the flow. Therefore, an aiding buoyancy effect increases the velocity level and an opposing one decreases it, as expected. The effect was found to increase with decreasing Prandtl number, due to the increase in the relative thickness of the thermal boundary region. The buoyancy also perturbs the radial velocity component, V_y, as seen in Figure 5. The magnitude of the inward radial velocity is increased, particularly near the edge of the thermal boundary layer. This effect also results in a thinning of the axial velocity distribution.

The laminar buoyant jet flow with small buoyancy effects was also considered by Schneider and Potsch [11], who obtained the first approximation for the effect of buoyancy on the flow by the method of matched asymptotic expansions, for $\frac{1}{2} < \text{Pr} < \frac{3}{2}$. They showed that the analysis of Mollendorf and Gebhart [10] is not valid in a certain, outer, region of the jet if $\text{Pr} < \frac{3}{2}$ and that it breaks down if $\text{Pr} \leq \frac{1}{2}$. Figure 6 shows their results for Prandtl numbers between $\frac{1}{2}$ and $\frac{3}{2}$ and for nonbuoyant and positively and negatively buoyant jets. Expected trends are seen.

For relatively large buoyancy effects, which inevitably arise in regions far downstream, the perturbation analysis just given does not apply. The coupled boundary-layer equations need to be considered. These are generally solved numerically, employing various computational methods available. An approximate solution may, however, be obtained by employing entrainment models, such as those developed for free boundary turbulent flows [18]. Morton [19] developed the corresponding model for laminar jets, plumes, and wakes. The entrainment flux scale was obtained using order-of-magnitude arguments and the model was employed to study the ascent of laminar plumes in an isothermal or stably stratified environment. Laminar viscous jets were also considered and the solution was found to agree with the results obtained by Schlichting [1], discussed earlier. The jet radius was found to increase linearly with x and the centerline velocity to vary as 1/x.

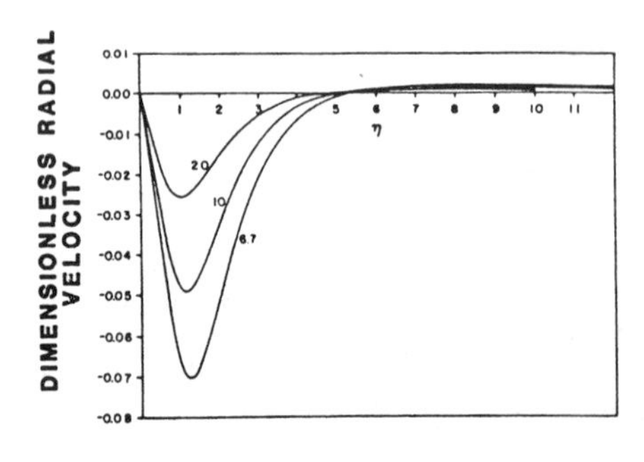

Figure 5. Radial component of velocity induced by thermal buoyancy in an axisymmetric jet [10].

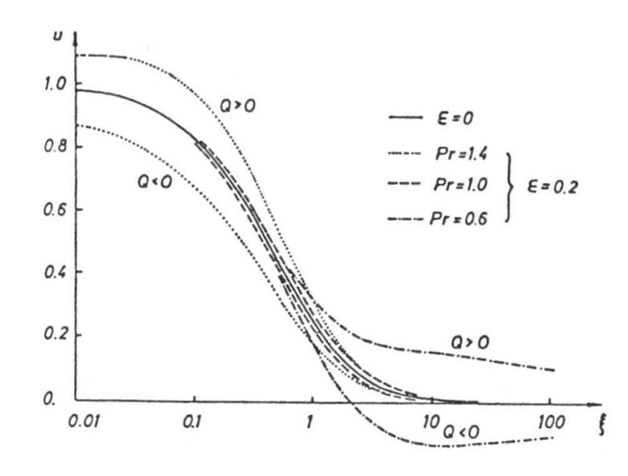

Figure 6. Velocity distribution in an axisymmetric buoyant, or negatively-buoyant, jet, as compared to that in a nonbuoyant jet [11].

The shape of the velocity and temperature profiles have to be assumed in such entrainment models and the entrainment coefficient E is determined experimentally. The profiles are assumed to be similar at all heights. Frequently, a Gaussian distribution, with the maximum at the axis and an exponential decay with the radial distance, y, is employed, since experimental evidence lends support to such a distribution [5, 6]. Thus, V_x and t may be taken as

$$V_x = (V_x)_c \exp(-y^2/b^2) \tag{48}$$

$$t - t_a = (t_c - t_a) \exp(-y^2/A^2b^2) \tag{49}$$

where $(V_x)_c$ = the centerline velocity
b = the characteristic halfwidth of the jet
A = the ratio of the spread of the temperature field to that of the velocity field

The mass conservation equation is given in terms of the entrainment model as

$$\int_0^\infty 2\pi\rho V_x(x, y)y\,dy = \begin{cases} 2\pi E\rho\nu x, & \text{for Pr} \geq 1 \\ 2\pi E\rho\alpha x, & \text{for Pr} \leq 1 \end{cases} \tag{50}$$

where E is the entrainment coefficient, whose value was found to be 4.0 for laminar viscous jets by Morton [19] and to be nearly 1.0 for laminar axisymmetric plumes by Wirtz and Chiu [20]. In general, it has to be determined experimentally. The momentum and energy conservation equations are

$$\frac{d}{dx}\int_0^\infty V_x^2 y\,dy = \int_0^\infty g\beta(t - t_a)y\,dy \tag{51}$$

$$\frac{d}{dx}\int_0^\infty V_x(t - t_a)y\,dy = 0 \tag{52}$$

With the assumed profiles, the preceding equations lead to ordinary differential equations. The input momentum and buoyancy provide the conditions at x = 0 and the equations may be solved analytically or numerically to yield the downstream temperature and velocity variation. Such integral entrainment models are approximate in nature and have been largely replaced by numerical techniques, which may be applied directly to the governing partial differential equations. However, the integral methods do provide some physical insight into the problem, without going into the complexity of a numerical solution. Also, in several practical problems, the approximate results provided by this analysis are of adequate accuracy.

Let us now consider the governing equations, Equations 9 through 11, for the boundary-layer buoyant jet flow far downstream of the jet origin. These equations may be nondimensionalized by employing the following dimensionless variables

$$\bar{V}_x = \frac{V_x}{V_0}, \qquad \bar{V}_y = \frac{V_y}{V_0}, \qquad X = \frac{x'}{x}, \qquad Y = \frac{y}{x}\sqrt{Re_x} \tag{53}$$

$$\theta = \frac{t - t_a}{t_0 - t_a}, \qquad Re_x = \frac{V_0 x}{\nu}$$

where the overbars denote dimensionless quantities, V_0 and t_0 are the velocity and temperature at the jet nozzle and x' is the local vertical coordinate distance, which varies from 0 to x. If the velocity and temperature distributions are nonuniform at the nozzle, the average or the maximum values may be employed. In any case, both V_0 and t_0 may be expressed in terms of the input momentum J_0 and thermal energy Q_0, as outlined earlier. The results obtained far downstream are independent of the diameter of the jet inlet and depend only on J_0 and Q_0. With the preceding nondimensionalization, the dimensionless boundary-layer equations are

$$\frac{\partial(Y\bar{V}_x)}{\partial X} + \frac{\partial(Y\bar{V}_y)}{\partial Y} = 0 \tag{54}$$

$$\bar{V}_x \frac{\partial \bar{V}_x}{\partial X} + \bar{V}_y \frac{\partial \bar{V}_x}{\partial Y} = \frac{Gr_x}{Re_x^2}\theta + \frac{1}{Y}\frac{\partial}{\partial Y}\left(Y \frac{\partial \bar{V}_x}{\partial Y}\right) \tag{55}$$

$$\bar{V}_x \frac{\partial \theta}{\partial X} + \bar{V}_y \frac{\partial \theta}{\partial Y} = \frac{1}{Pr}\frac{1}{Y}\frac{\partial}{\partial Y}\left(Y \frac{\partial \theta}{\partial Y}\right) \tag{56}$$

where

$$Gr_x = \frac{g\beta(t_0 - t_a)x^3}{\nu^2} \tag{57}$$

Therefore, Gr_x/Re_x^2 arises as a governing parameter, in addition to the Prandtl number Pr, and determines the effect of thermal buoyancy on the flow. For small values of Gr_x/Re_x^2, a perturbation solution may be obtained, as outlined earlier. The flow farther downstream has significant buoyancy effects and may be studied numerically, with the input thermal energy and momentum specified over a small finite region at $x = 0$. Himasekhar and Jaluria [21] considered this problem for flow in a thermally-stratified environment, employing finite-difference marching procedures for the parabolic energy and momentum equations. They considered both plumes and buoyant jets. Both flows were found to approach the characteristics of an axisymmetric plume, generated by a concentrated thermal source, far downstream. Therefore, thermal buoyancy dominates far from the jet origin and a thermal plume behavior results. Far downstream, the centerline velocity was found to approach a constant value, as predicted by the analysis for an axisymmetric plume, and the centerline temperature excess, $(t_c - t_a)$, to vary as $1/x$. The computed variation of the centerline velocity and temperature is shown in Figures 7 and 8. The parameter S refers to the thermal stratification in the ambient medium, an isothermal medium being given by $S = 0$. This aspect is considered later.

As is evident from the preceding discussion, not much work has been done on the laminar, axisymmetric buoyant jet. Since most jet flows encountered in practice are turbulent, much of the work has been directed at turbulent flows. The laminar jet also undergoes transition to turbulence downstream, as discussed later. Consequently, turbulent jets are of greater practical interest. However, as mentioned earlier, laminar jets are of interest in some practical problems and are also of importance in the region near the jet origin, where the discharge is often laminar. The laminar

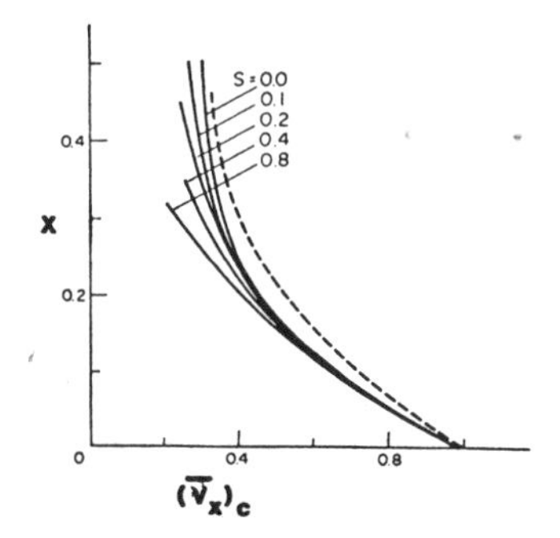

Figure 7. Downstream variation of the dimensionless centerline velocity $(\bar{V}_x)_c$ in a vertical, axisymmetric, buoyant jet at $Pr = 0.7$ and $Gr_x/Re_x^2 = 0.47$. The dashed curve is for $Gr_x/Re_x^2 = 0.94$ at $S = 0$ [21].

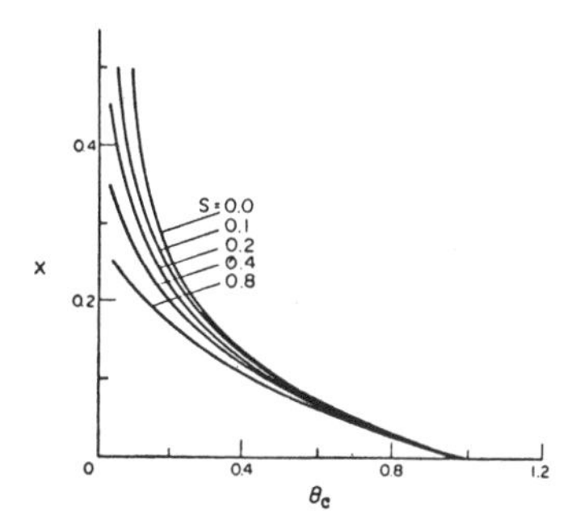

Figure 8. Downstream variation of the centerline temperature θ_c in an axisymmetric, buoyant jet at $Pr = 0.7$ and $Gr_x/Re_x^2 = 0.47$ [21].

analysis also provides inputs for a consideration of the process of transition to turbulent flow. After this detailed consideration of axisymmetric flows, we may now proceed to a discussion of two-dimensional buoyant jets.

TWO-DIMENSIONAL VERTICAL JETS

The two-dimensional, vertical, buoyant jet has received very little attention in the literature, because of the greater relevance of the axisymmetric jet to practical circumstances, such as flows emerging from chimneys and cooling towers. Also, a two-dimensional source is of finite length in practice and as the flow proceeds downstream, it tends to approach the characteristics of the axisymmetric flow because of entrainment from the sides [22]. However, in many applications, such as heat rejection, energy storage, and energy extraction systems, it is desirable to have a two-dimensional flow so as to utilize a greater portion of the fluid region for flow, than would be the case for a circular discharge [23]. A spreading out of the flow into a two-dimensional jet lowers the velocity level too, which is advantageous in practical problems, such as energy extraction from salt-gradient solar ponds [24]. Similarly, buoyancy-driven flows in enclosed spaces, such as those due to fire in a room, often lead to two-dimensional, buoyant, and negatively-buoyant, jets in the flow region [25].

The flow in a laminar, two-dimensional, nonbuoyant jet was first analyzed by Schlichting [1] and by Bickley [26], who presented a closed-form solution. The axial velocity was found to decay as $x^{-1/3}$ and the mass flow rate to increase as $x^{1/3}$, where x is the coordinate distance along the jet axis, measured from its origin. The measurements carried out by Andrade [27] corroborated the theoretical predictions, though the jet was found to twist and assume a non-two-dimensional form downstream. A review on laminar jets was presented by Pai [28]. We shall first discuss the nonbuoyant jet, followed by a consideration of the effect of buoyancy on the flow.

Governing Equations

Let us consider a jet emerging from a long, narrow slit, as shown in Figure 9. The following governing boundary-layer equations for a vertical, two-dimensional buoyant jet are obtained in a manner similar to that discussed for an axisymmetric buoyant jet, employing the Boussinesq approximations.

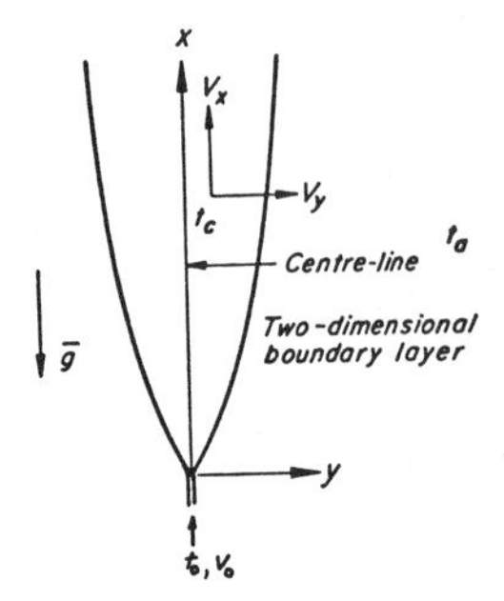

Figure 9. Coordinate system for a vertical, laminar, two-dimensional, buoyant jet.

$$\frac{\partial V_x}{\partial x} + \frac{\partial V_y}{\partial y} = 0 \tag{58}$$

$$V_x \frac{\partial V_x}{\partial x} + V_y \frac{\partial V_x}{\partial y} = \nu \frac{\partial^2 V_x}{\partial y^2} + g\beta(t - t_a) \tag{59}$$

$$V_x \frac{\partial t}{\partial x} + V_y \frac{\partial t}{\partial y} = \alpha \frac{\partial^2 t}{\partial y^2} \tag{60}$$

where the coordinate system is shown in Figure 9. The boundary conditions for the preceding equations are

$$\begin{aligned} &\text{at } y = 0: \quad V_y = 0, \quad \frac{\partial V_x}{\partial y} = 0, \quad \frac{\partial t}{\partial y} = 0 \\ &\text{as } y \to \infty: \quad V_x \to 0, \quad \text{and } t \to t_a \end{aligned} \tag{61}$$

Again, the centerline temperature t_c may be determined from the condition that the convected thermal energy in the flow is constant and equal to the heat input, per unit length of the source, Q_0, at the origin. The centerline temperature may, therefore, be used as a boundary condition, instead of the ambient condition, which can be shown to be satisfied if the remaining conditions are applied. The buoyant jet is, therefore, represented by a momentum input J_0 and a thermal energy input Q_0 at $x = 0$. By considering these inputs at a line source, it may be possible to solve the problem by similarity analysis.

Nonbuoyant Jet

For a nonbuoyant jet, the buoyancy term in the momentum equation is neglected, thus making the flow independent of the temperature distribution. A stream function ψ is defined as

$$V_x = \frac{\partial \psi}{\partial y}, \qquad \text{and } V_y = -\frac{\partial \psi}{\partial x} \tag{62}$$

Following the formulation of Schlichting [1], the following similarity transformation is employed

$$\eta = \frac{K_1}{3\nu^{1/2}} \frac{y}{x^{2/3}}, \qquad \psi = 2K_1 \nu^{1/2} x^{1/3} f(\eta) \tag{63}$$

where

$$K_1 = \left[\frac{9}{16\nu^{1/2}} (J_0/\rho) \right]^{1/3} \tag{64}$$

The relation between the constant K_1 and the input momentum J_0 is obtained by applying the resulting similarity solution to the condition of constant momentum flux J(x), per unit width, downstream. This may be stated as

$$J(x) = \int_{-\infty}^{\infty} \rho V_x^2 \, dy = J_0 = \text{constant} \tag{65}$$

The governing momentum equation is obtained in terms of the similarity variables as

$$f''' + 2ff'' + 2f'^2 = 0 \tag{66}$$

with the boundary conditions

$$f(0) = f''(0) = f'(\infty) = 0 \tag{67}$$

The preceding ordinary differential equation may be solved by integration, employing the given boundary conditions, to obtain the expression for the dimensionless stream function f as

$$f = \tanh \eta \tag{68}$$

The vertical velocity component V_x is obtained as

$$V_x = \tfrac{2}{3} K_1^2 x^{-1/3} (1 - \tanh^2 \eta) \tag{69}$$

If this distribution is substituted in Equation 65, the relation between K_1 and J_0, just given, is obtained. If the expression for K_1 is substituted in the preceding equation, V_x is obtained as

$$V_x = 0.4543 \left(\frac{J_0^2}{\rho^2 \nu x} \right)^{1/3} (1 - \tanh^2 \eta) \tag{70}$$

and the mass flow rate $\dot{m}$, per unit width, is given by

$$\dot{m} = 3.3019 \rho (J_0 \nu x/\rho)^{1/3} \tag{71}$$

Therefore, V_x varies as $x^{-1/3}$ and $\dot{m}$ as $x^{1/3}$. Figure 2 shows the velocity distribution for the two-dimensional nonbuoyant jet, along with that for the axisymmetric case.

In a manner similar to that discussed for an axisymmetric jet, the energy equation may be nondimensionalized by employing the preceding similarity transformation, along with the definition of θ given in Equation 22. Again, the convected thermal energy Q(x), per unit length in the third direction, remains constant as the flow moves downstream. Therefore,

$$Q(x) = \int_{-\infty}^{\infty} \rho C_p (t - t_a) V_x \, dy = Q_0 = \text{constant} \tag{72}$$

Similarity arises for the power-law variation of the centerline temperature excess, i.e., for $t_c - t_a = Nx^n$. If the velocity V_x, temperature excess $(t - t_a)$ and horizontal coordinate distance y are substituted in the preceding expression for Q(x), in terms of the similarity variables, it can be shown that $n = -\frac{1}{3}$ for Q(x) to be independent of x. It can also be shown that N varies as $Q_0/J_0^{1/3}$. The energy equation is obtained in similarity form as

$$(\theta' + 2 \, Pr \, f\theta)' = 0 \tag{73}$$

with the boundary conditions

$$1 - \theta(0) = \theta(\infty) = 0 \tag{74}$$

As before, the symmetry condition at the axis, $\theta'(0) = 0$, may also be used instead of the previously given second condition. Integrating the energy equation twice, with these boundary conditions, gives

$$\theta = \exp\left(-2\ \mathrm{Pr} \int_0^\eta \tanh \eta \, d\eta\right) = (\cosh \eta)^{-2\ \mathrm{Pr}} \tag{75}$$

Therefore, the velocity and temperature distributions for a nonbuoyant, two-dimensional jet may be obtained. These solutions will apply for negligible buoyancy effects, which would generally arise very near the origin of the jet.

Two-Dimensional Plume

The plane, or two-dimensional, plume arising from the input of thermal buoyancy has been studied analytically by several investigators [13, 14, 29–32]. The idealized problem is that of flow generated by a horizontal line thermal source and this circumstance has been studied by similarity methods. The similarity transformation employed is similar to that used for natural convection on a vertical plate and is given by

$$\eta = \frac{y}{x}\left(\frac{\mathrm{Gr}_x}{4}\right)^{1/4}, \qquad \psi = 4\nu\left(\frac{\mathrm{Gr}_x}{4}\right)^{1/4} f(\eta), \qquad \theta(\eta) = \frac{t - t_a}{t_c - t_a} \tag{76}$$

where $\mathrm{Gr}_x = \dfrac{g\beta x^3(t_c - t_a)}{\nu^2}$

Since there are no thermal sources in the flow and viscous dissipation is neglected, Equation 72 applies and may be employed to obtain n and N. The vertical velocity component V_x is given by

$$V_x = \frac{\partial \psi}{\partial y} = 4\nu\left(\frac{\mathrm{Gr}_x}{4}\right)^{1/4} \frac{1}{x}\left(\frac{\mathrm{Gr}_x}{4}\right)^{1/4} f'(\eta)$$

$$= \frac{2\nu}{x}(\mathrm{Gr}_x)^{1/2} f'(\eta) \tag{77}$$

Therefore,

$$Q(x) = \int_{-\infty}^{\infty} \rho C_p V_x (t - t_a)\, dy = \rho C_p \frac{2\nu}{x} \mathrm{Gr}_x^{1/2} N x^n \frac{x}{(\mathrm{Gr}_x/4)^{1/4}} \int_{-\infty}^{\infty} f'\theta \, d\eta$$

$$= 4\mu C_p N^{5/4}\left(\frac{g\beta}{4\nu^2}\right)^{1/4} x^{(5n+3)/4} \int_{-\infty}^{\infty} f'(\eta)\theta(\eta)\, d\eta = Q_0 \tag{78}$$

This indicates that $n = -\frac{3}{5}$ for a two-dimensional, laminar plume. If the integral is denoted by I, which is different in value from the similar integral obtained for the axisymmetric plume, the constant N is given by

$$N = \left(\frac{Q_0^4}{64 g\beta \rho^2 \mu^2 C_p^4 I^4}\right)^{1/5} \tag{79}$$

Table 1
Data of Gebhart et al.

Pr	f′(0)	I
0.01	0.9751	—
0.1	0.8408	3.090
0.7	0.6618	1.245
1.0	0.6265	1.053
2.0	0.5590	0.756
6.7	0.4480	0.407
10.0	0.4139	0.328
100.0	0.2505	—

From Gebhart et al. [31]

Table 1 lists the computed values of I and of f′(0), which gives the dimensionless centerline velocity, for various values of Pr.

The governing equations for the flow reduce to the following form when n is set equal to $\frac{-3}{5}$

$$f''' + \tfrac{12}{5}ff'' - \tfrac{4}{5}f'^2 + \theta = 0 \tag{80}$$

$$(\theta' + \tfrac{12}{5}\,\mathrm{Pr}\, f\theta)' = 0 \tag{81}$$

The five independent boundary conditions for this system of coupled ordinary differential equations are obtained as

$$\theta'(0) = f(0) = f''(0) = 1 - \theta(0) = f'(\infty) = 0 \tag{82}$$

It can be shown by integrating the energy equation that

$$\theta(\eta) = \theta(0)\exp\left[-\frac{12}{5}\,\mathrm{Pr}\int_0^\eta f(\eta)\,d\eta\right] \tag{83}$$

Since f is positive and becomes constant for large η, due to entrainment and also as seen from the boundary condition $f'(\infty) = 0$, the integral tends to infinity, giving $\theta(\infty) = 0$. Therefore, the preceding five conditions are adequate for the governing equations, which may be solved numerically. Figures 10 and 11 show the computed velocity and temperature profiles at various values of the Prandtl number. The velocity level decreases with increasing Pr, due to the increased viscous effect, and the thermal region becomes thinner, as expected.

For certain values of the Prandtl number, closed-form solutions may be obtained. At Pr = 2, for instance, Fujii [13] gives $\tilde{f}(\tilde{\eta})$ and $\tilde{\theta}(\tilde{\eta})$ as

$$\tilde{f} = \left(\frac{10a}{3}\right)^{1/3}\tanh\left[\left(\frac{3a}{10}\right)^{1/2}\tilde{\eta}\right] \tag{84}$$

$$\tilde{\theta} = \frac{4}{5}a^2\,\mathrm{Sech}^4\left[\left(\frac{3a}{10}\right)^{1/2}\tilde{\eta}\right] \tag{85}$$

where the tilde indicates a similarity transformation somewhat different from that just discussed and a is the nondimensional centerline vertical velocity component, being determined as 0.837 for Pr = 2. Experimental corroboration of the theoretical findings has been obtained by several studies

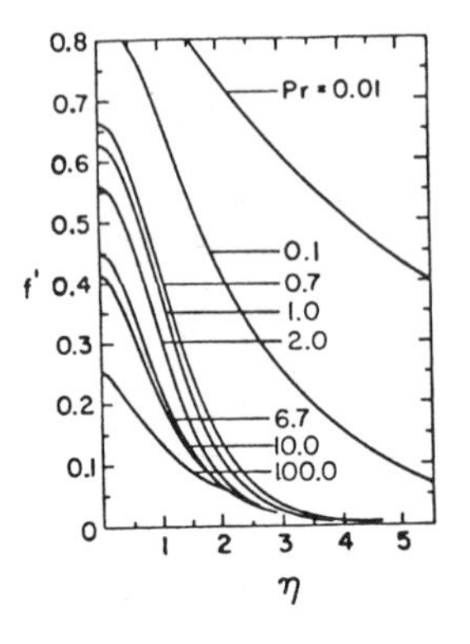

Figure 10. Computed velocity profiles in a two-dimensional, laminar plume arising from a heat source [31].

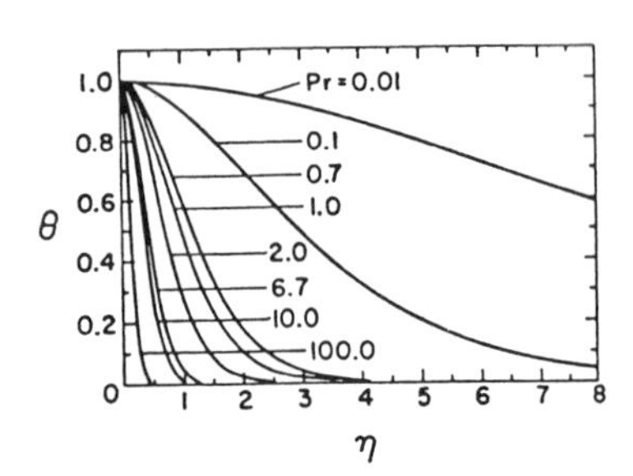

Figure 11. Computed temperature distribution in a two-dimensional, laminar plume [31].

[32–35]. Figure 12 shows the comparison between the theoretical and the experimental results for the temperature profiles, indicating a fairly good agreement. However, the centerline temperature has consistently been found to be lower than the predicted value. This difference may be attributed largely to the entrainment of fluid from below the line source in actual practice, whereas the boundary-layer analysis assumes no entrainment from below the origin. This would result in the analytical prediction of lower flow rates and higher temperatures in the plume. This consideration is also important in the boundary-layer analysis of a laminar, buoyant jet, as discussed later.

Buoyant Jet

Very little work has been done on the two-dimensional, laminar, buoyant, jet arising from an input of momentum and buoyancy. Near the jet origin, the buoyancy effects are expected to be small and the above treatment of a nonbuoyant jet may be employed. Buoyancy effects increase in magnitude downstream and would dominate far downstream. Therefore, the flow is expected to approach the characteristics of a two-dimensional plume far from the source. However, the flow may become turbulent before this occurs. Also, if a jet of finite length is considered, which is obviously the case in actual practice, the fluid entrainment at the edges causes the flow to become three-dimensional. Therefore, laminar, two-dimensional jets are generally of interest in a region close to the slit through which the jet is discharged. This region is often restricted to 20–50 times the width of the discharge slit. This flow is of interest in the study of buoyant wakes above long, heated bodies, such as fires and heated tubes, in heat rejection and energy storage applications, which employ heated two-dimensional discharges, and in the study of buoyancy-driven flows in enclosures with openings, such as rooms.

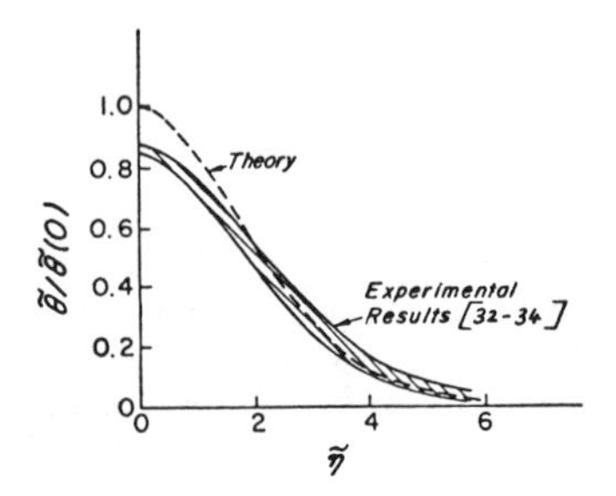

Figure 12. Comparison between theoretical and experimental results of the temperature profiles in a two-dimensional, laminar plume at Pr = 0.71 [32].

The equations governing the flow in a two-dimensional, buoyant jet may be nondimensionalized by employing the dimensionless variables given in Equation 53. The resulting boundary-layer equations are

$$\frac{\partial \bar{V}_x}{\partial X} + \frac{\partial \bar{V}_y}{\partial Y} = 0 \tag{86}$$

$$\bar{V}_x \frac{\partial \bar{V}_x}{\partial X} + \bar{V}_y \frac{\partial \bar{V}_x}{\partial Y} = \frac{Gr_x}{Re_x^2}\,\theta + \frac{\partial^2 \bar{V}_x}{\partial Y^2} \tag{87}$$

$$\bar{V}_x \frac{\partial \theta}{\partial X} + \bar{V}_y \frac{\partial \theta}{\partial Y} = \frac{1}{Pr}\frac{\partial^2 \theta}{\partial Y^2} \tag{88}$$

with the boundary conditions for $X > 0$ given by

$$\begin{aligned} &\text{at } Y = 0: \quad \bar{V}_y = \frac{\partial \bar{V}_x}{\partial Y} = \frac{\partial \theta}{\partial Y} = 0 \\ &\text{as } Y \to \infty: \quad \bar{V}_x \to 0, \qquad \theta \to 0 \end{aligned} \tag{89}$$

The input momentum J_0 and thermal energy Q_0, both being specified per unit length of the discharge slit, are given in terms of the velocity V_0 and temperature t_0, which are imposed over a small two-dimensional region at $x = 0$. As seen from the preceding equations, buoyancy effects are small near the origin and the problem may be studied by employing a perturbation analysis, with the mixed convection parameter Gr_x/Re_x^2 as a possible perturbation parameter. As the flow moves downstream, buoyancy effects increase, ultimately dominating the flow far from the source. Figures 13 and 14 show the centerline velocity and temperature for a two-dimensional, laminar, buoyant jet in an isothermal ($S = 0$) or a stably-stratified environment [36]. The flow was found to approach the characteristics of a two-dimensional plume, with respect to the profiles as well as the centerline temperature and velocity, at large x. The effect of stratification is discussed later.

For flow in the region near the source and for large slot dimensions, the effect of the size of the source is still present in the flow. Consequently, the slot width is a characteristic dimension in the problem. In this case, the governing equations may be nondimensionalized, employing the

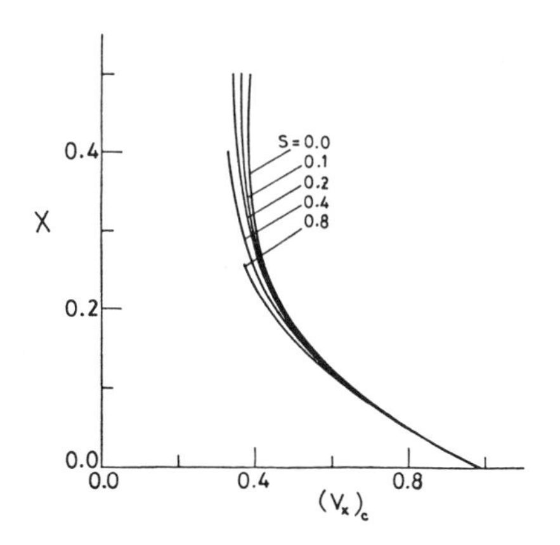

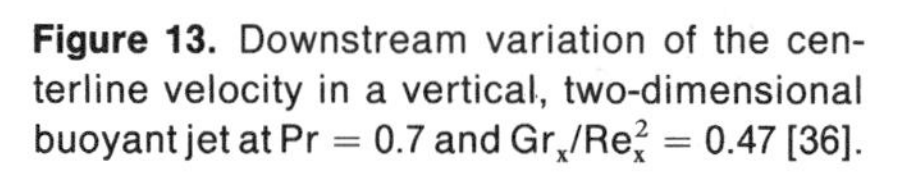
Figure 13. Downstream variation of the centerline velocity in a vertical, two-dimensional buoyant jet at $Pr = 0.7$ and $Gr_x/Re_x^2 = 0.47$ [36].

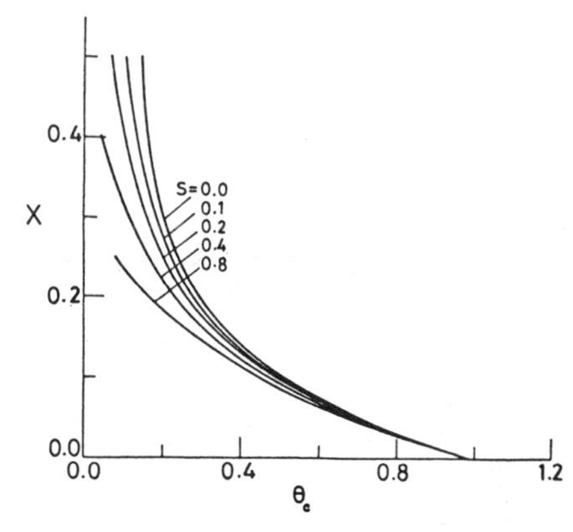

Figure 14. Downstream centerline temperature variation in a two-dimensional buoyant jet at $Pr = 0.7$ and $Gr_x/Re_x^2 = 0.47$ [36].

halfwidth d of the slot to obtain the dimensionless variables as

$$X = \frac{x}{d}, \qquad Y = \frac{y}{d}, \qquad \bar{V}_x = \frac{V_x}{V_0}, \qquad \bar{V}_y = \frac{V_y}{V_0} \tag{90}$$

$$\theta = \frac{t - t_a}{t_0 - t_a}, \qquad Re = \frac{V_0 d}{\nu} \tag{91}$$

The nondimensional boundary-layer equations are obtained as

$$\frac{\partial \bar{V}_x}{\partial X} + \frac{\partial \bar{V}_y}{\partial Y} = 0 \tag{92}$$

$$\bar{V}_x \frac{\partial \bar{V}_x}{\partial X} + \bar{V}_y \frac{\partial \bar{V}_x}{\partial Y} = \frac{1}{Re}\frac{\partial^2 \bar{V}_x}{\partial Y^2} + \frac{Gr}{Re^2}\theta \tag{93}$$

$$\bar{V}_x \frac{\partial \theta}{\partial X} + \bar{V}_y \frac{\partial \theta}{\partial Y} = \frac{1}{Re\,Pr}\frac{\partial^2 \theta}{\partial Y^2} \tag{94}$$

Therefore, the Reynolds number Re arises as an additional parameter, besides the mixed convection parameter Gr/Re^2 and the Prandtl number. The boundary conditions are the same as those given in Equation 89 for $X > 0$. At the origin, $X = 0$, the boundary conditions, for uniform velocity and temperature distributions, are

$$\begin{aligned} \text{at } X = 0: \quad & \bar{V}_x = \theta = 1.0, \qquad \text{for } Y \leq 1.0 \\ & \bar{V}_x = \theta = 0, \qquad \text{for } Y > 1.0 \end{aligned} \tag{95}$$

This problem may be solved numerically, employing finite-difference methods [37]. Figure 15 shows the computed downstream variation of the centerline velocity and temperature in a buoyant plane jet at various values of Re and Gr/Re^2, for $Pr = 0.7$. With increasing buoyancy effects, at a given Reynolds number, the velocity level increases, as expected. This results in a larger entrainment and, thus, a lower temperature level. For a nonbuoyant jet, the velocity decreases downstream, as

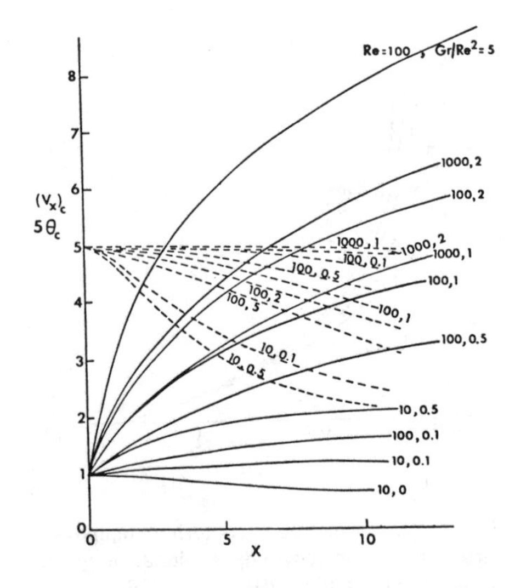

Figure 15. Centerline velocity and temperature variation in a vertical, two-dimensional, buoyant jet at $Pr = 0.7$ [37].

seen earlier from the similarity analysis. An increase in Re, for a fixed value of Gr/Re^2, also results in larger velocity due to a greater momentum input. The effect of this is to thin the flow and result in a larger temperature level downstream. Far downstream, the centerline temperature excess, θ_c, was found to decay as $x^{-3/5}$ and the velocity $(V_x)_c$ to increase as $x^{1/5}$, indicating the asymptotic approach to the thermal plume behavior at large x. The effect of thermal stratification and of the nonboundary-layer mechanisms near the jet origin are discussed later.

It is seen from the preceding discussion that a buoyant, two-dimensional jet behaves like a nonbuoyant jet near the origin for low values of the buoyancy input. Buoyancy effects increase in importance downstream, ultimately dominating the flow far from the jet discharge. In the mixed-convection region, where both the forced flow and the buoyancy force effects are of same order of magnitude, the problem may be solved as a boundary-layer flow, which is coupled with the energy equation through the buoyancy terrm. For large slot widths, particularly in regions close to the slot, the slot width arises as a characteristic dimension in the problem and must be considered in the analysis. Farther downstream and for small slot widths, the flow is independent of the slot dimension and depends only on the input momentum and thermal energy. Approximate results may be obtained from an integral formulation, with a suitable entrainment model, as discussed for the axisymmetric jet circumstance. Otherwise, one has to resort to numerical techniques to analyze the flow.

OTHER FLOWS

In the preceding discussion, we have considered the basic nature of vertical, laminar, buoyant jets. Both the axisymmetric and the two-dimensional circumstances have been discussed, while also considering the two limiting cases of the nonbuoyant jet and of the thermal plume. However, there are several other important considerations relevant to a study of buoyant jets. Among the most important of these are the condition of the ambient medium, concerning whether it is stratified and whether it is moving or quiescent, orientation of the jet with respect to gravity and transition of the flow to turbulence. Also of interest are flows with buoyancy input due to concentration differences and flows affected by the presence of neighboring surfaces and other free-boundary flows. In some flows, combined buoyancy effects, due to both temperature and concentration differences, are of interest, as is the case in environmental pollution studies. This section outlines the work done on several of these flow circumstances.

Stratified Environment

A substantial amount of effort has been directed at natural convection flows in an environment which is stably-stratified due to a density decrease with height [5, 6]. Such flows are of interest in many environmental problems, related to heat and material rejection into the atmosphere and into water bodies, and in engineering applications concerned with a heat input in enclosures, such as that due to a fire in a room. In most problems of practical interest, a stable thermal stratification is characterized by a temperature increase with height, for fluids which expand on heating, though the limiting case of adiabatic stratification is obtained with a slight decrease in temperature with height [39]. Because of the greater practical importance of turbulent flows, most studies have considered turbulent jets in stably-stratified media [6, 40]. However, some work has also been done on laminar jets, as outlined in the following.

Let us consider a vertical, axisymmetric, laminar, buoyant jet, which is governed by the dimensionless boundary-layer equations (Equations 54 through 56) with the nondimensionalization given in Equation 53. If a thermally-stratified environment is considered, with the ambient temperature $t_a(x)$ increasing with height x, an additional term arises in the energy equation (Equation 56) because of this variation. The resulting equation for a thermally-stratified medium is

$$\bar{V}_x \frac{\partial\theta}{\partial X} + \bar{V}_y \frac{\partial\theta}{\partial Y} + S\bar{V}_x = \frac{1}{Pr}\frac{1}{Y}\frac{\partial}{\partial Y}\left(Y\frac{\partial\theta}{\partial Y}\right) \tag{96}$$

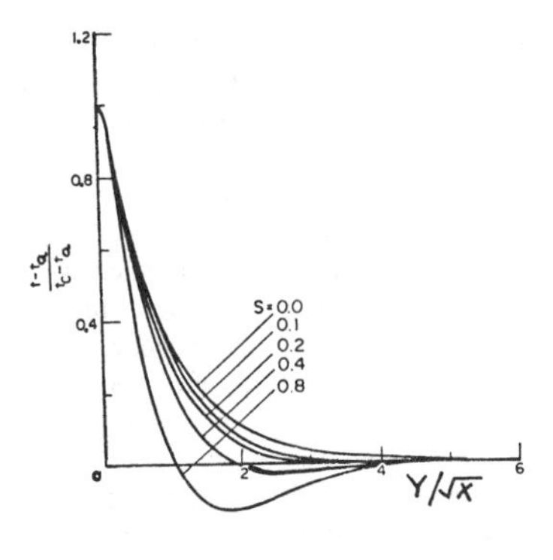

Figure 16. Temperature profiles in an axisymmetric jet rising in a thermally-stratified medium at Pr = 0.7, X = 0.2 and Gr_x/Re_x^2 = 0.47 for various stratification levels [21].

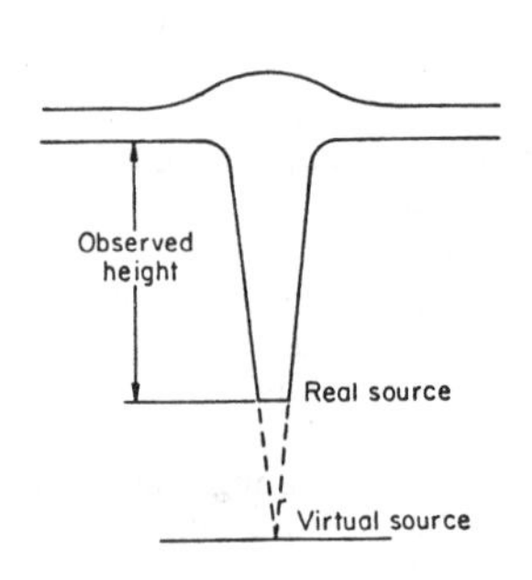

Figure 17. Sketch of plume or buoyant jet rise in a stably-stratified medium.

with

$$S = \frac{1}{\Delta t}\frac{dt_a}{dX}, \qquad \theta = \frac{t - t_a}{\Delta t}, \qquad \Delta t = t_0 - (t_a)_0 \tag{97}$$

where S is termed the stratification parameter and $(t_a)_0$ is the ambient temperature at x = 0. For an isothermal medium, S is zero and Δt is a constant. Therefore, S is proportional to the vertical ambient temperature gradient, being a constant for a linear variation. For this circumstance, Figures 7 and 8 show the downstream variation of the centerline velocity and temperature for various stratification levels.

It is seen from Figures 7 and 8 that the local centerline temperature excess, $t_c - t_a$, decreases more rapidly with height as S increases. This is expected, since a higher S implies a faster increase in t_a with x. It is also evident that negative buoyancy arises downstream, because of t_c becoming less than the local ambient temperature. This, in turn, would tend to decrease the velocity level, as seen in Figure 7. For the unstratified case, the centerline temperature never becomes less than the local ambient temperature and the centerline velocity approaches a constant value downstream, as predicted by the similarity analysis for an axisymmetric plume. Therefore, in a stratified environment, the flow rises to a finite height, comes to a stop, and then reverses in direction because of the downward buoyancy force. The temperature profiles also show local temperatures becoming less than the ambient temperature, even though the centerline temperature may be higher, as seen in Figure 16. The ambient temperature increases with height and the local temperature in the outer region of the flow is unable to catch up with this increase. This effect may also lead to reverse flow in the outer region of the jet [21]. Figure 17 shows qualitatively the behavior of a flow rising in a thermally-stratified medium.

A detailed experimental study on the flow of laminar, axisymmetric, buoyant, vertical jets in a stably-stratified environment was carried out by Tenner and Gebhart [7] using buoyancy input and stratification due to concentration differences. Employing fresh and salt water jets in a stably-stratified salt water environment, they studied the nature of the flow. The jet was found to give rise to a toroidal cell around itself, with the inner flow rising due to the input of momentum and buoyancy and the outer flow descending due to the negative buoyancy force. Under certain limiting conditions, a shroud, which tends to isolate the jet from the surrounding fluid, was found to arise.

Two-dimensional, laminar, buoyant jets and plumes rising in thermally-stratified media have also been studied [36, 37, 41]. Considering the formulation for a two-dimensional jet resulting from a line source of momentum and buoyancy, an additional term $S\bar{V}_x$ again arises on the left-hand side of the energy equation, Equation 88, with S defined, as before, by Equation 97. Figures 13 and 14

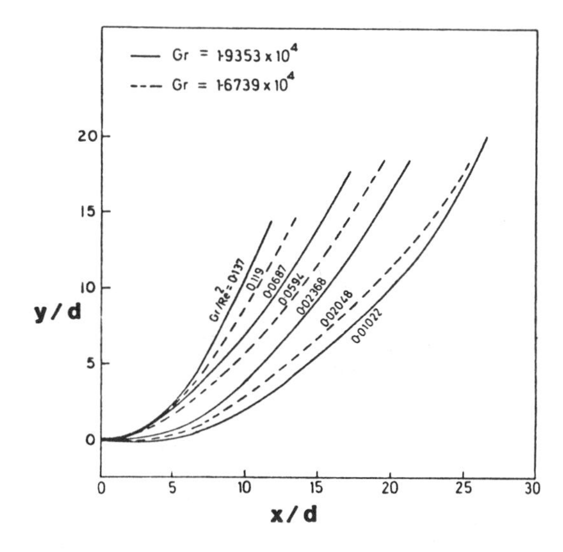

Figure 18. The trajectory of a circular buoyant jet discharged horizontally in water, Pr = 7.0 [47].

show the effect of thermal stratification on the centerline velocity and temperature. Trends similar to those discussed earlier for axisymmetric jets are observed. The flow in the region near a finite-sized slit has also been studied, employing the halfwidth d of the slit as a characteristic dimension. Numerical results have been presented by Jaluria [37], indicating the expected behavior of a buoyant jet ascending in a thermally-stratified medium. Integral models, with the entrainment assumption, as outlined earlier, may also be employed to obtain approximate results. Wirtz and Chiu [20] employed this method for axisymmetric plumes, generated by finite-sized sources, rising in thermally-stratified media.

Other Orientations

In many problems of practical interest, the jet discharge is not aligned with the vertical buoyancy force, but is at an inclination with it. Probably, the most important circumstance in this class of problems is the horizontal buoyant jet, which is of particular interest in heat rejection to water bodies [23, 42]. Another orientation which is of some importance is that of the buoyancy force opposing the input momentum. This circumstance is generally known as negatively-buoyant and would arise, for instance, in a jet of hot fluid discharged downwards. Such flows often arise in nature and in buoyancy-induced flow in enclosed spaces [6]. Much of the work on horizontal and negatively-buoyant jets has been done for turbulent flows [43–46].

Satyanarayana and Jaluria [47] have carried out a detailed experimental study of laminar, buoyant jets discharged at an inclination with the vertical buoyancy force. Axisymmetric, buoyant discharges were considered and a simple integral analysis was also carried out, indicating a fairly good agreement between the analytical and experimental results. The temperature distribution in the jet and the trajectory of the jet were determined. Figures 18 and 19 show the flow trajectory for jets discharged horizontally and at two downward inclinations with the horizontal. The temperature field was found to be largely dependent on the bouyancy parameter Gr/Re^2, where both the Grashof and Reynolds numbers are based on the nozzle diameter and the inlet temperature and velocity. The downward penetration of the jet for inclined downward discharges was also determined. The decay of the centerline temperature and the increase in the spread of the jet downstream were obtained. Several other thermal characteristics of the flow were studied and considered in terms of the underlying physical mechanisms.

Very little work has been done on laminar, buoyant jets discharged at an inclination with the buoyancy force. The main trends are similar to those observed in turbulent flow. However, the

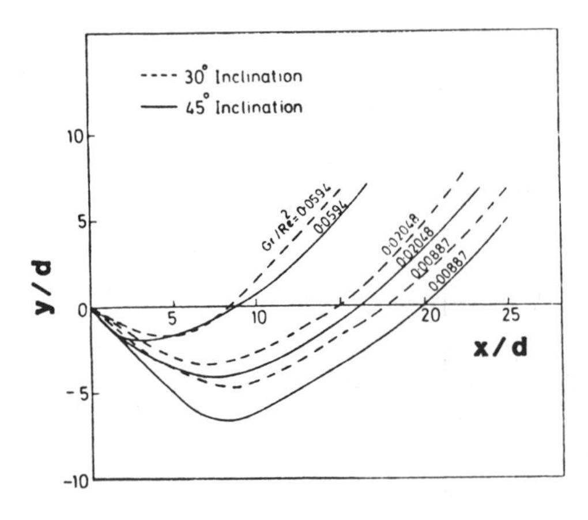

Figure 19. The trajectory of a circular buoyant jet discharged downwards, at two inclinations with the horizontal, in water, Pr = 7.0 [47].

underlying basic processes and the thermal characteristics of the flow are,obviously, quite different in laminar flows. There is a growing interest in some of these flows, mainly because of buoyancy-induced flows arising in electronic equipment and in various manufacturing processes, such as crystal growing. However, much needs to be done to provide information on the basic nature of the flow resulting from the combined effect of momentum and buoyancy input, when the two are not aligned with each other.

Transition to Turbulence

We are largely concerned with laminar buoyant jets. However, as the flow moves downstream, it becomes unstable to disturbances entering the flow and ultimately becomes turbulent. As mentioned earlier, turbulent buoyant jets have been studied in detail [6, 40, 48]. Similarly, the instability of buoyant and nonbuoyant jets and of thermal plumes has been investigated. Nonbuoyant jets have been studied in detail and the critical Reynolds number for the onset of instability has been determined [49]. The transition mechanisms have also been investigated. The instability of thermal plumes has been studied to determine the growth of disturbances, considering both symmetric and asymmetric disturbances [50]. The flow has been found to be less stable for the asymmetric mode. The transition to turbulence has been studied for many of these free boundary flows and the basic features have been found to be similar to other buoyancy-driven flows [5, 51].

A detailed experimental and numerical study of the stability of an axisymmetric, vertical, laminar, buoyant jet was carried out by Mollendorf and Gebhart [52], who also considered both symmetric and asymmetric disturbances. The computed critical Reynolds number Re for the onset of instability was found to be 9.4, where Re is based on the nozzle diameter. Initial instability was found to arise in the first asymmetric mode. The effect of buoyancy on the stability was studied, also considering the limiting case of an axisymmetric thermal plume for a Prandtl number of 2. Buoyancy was found to destabilize the flow. The laminar length of the jet was determined experimentally for varying amounts of thermal buoyancy and an empirical correlation was presented for determining the length over which the jet remains laminar. Disturbance growth characteristics were studied and presented in terms of amplification contours.

The measurements of Mollendorf and Gebhart [52] gave the laminar length of buoyant jets for various values of the input momentum and buoyancy. They suggested the following correlation for determining the laminar length L_ℓ

$$\frac{L_\ell}{d} = 568.5[(2\,Pr + 1)Gr/Re^3]^{3/8} \tag{98}$$

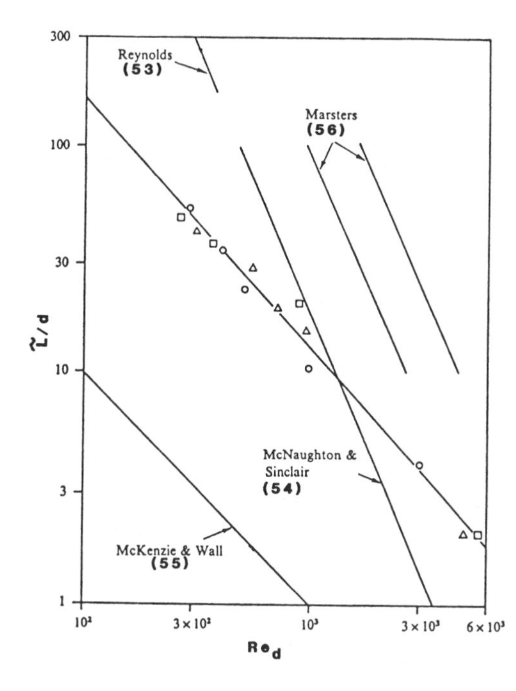

Figure 20. Experimental laminar length data for axisymmetric buoyant jets and correlations of the data of other investigators. ○, $\Delta t = 11.12°C$; △, $\Delta t = 16.67°C$; □, $\Delta t = 25°C$ [52].

where both Gr and Re are based on the nozzle diameter d and on the inlet velocity and temperature excess, over the ambient temperature. Figure 20 shows the results obtained by Mollendorf and Gebhart [52] along with those from other studies for small buoyancy effects [53–56]. Clearly, the laminar length is a strong function of the inlet conditions, particularly the Reynolds number. At large values of Re, the flow becomes turbulent immediately downstream of the nozzle. Of course, the flow may emerge as turbulent too, depending on the upstream conditions. For flow in a circular tube, leading up to the nozzle, a Reynolds number greater than 2,300 implies turbulent flow at the nozzle. Because of the onset of transition to turbulence downstream of the jet discharge, laminar flow analysis is generally restricted to low values of the Reynolds number and downstream distances less than about 50 times the nozzle diameter. Such a detailed study has not been done on two-dimensional jets but similar trends are expected.

Additional Effects

There are several other effects which may be considered in a study of laminar, buoyant jets. Though many of these additional effects are of considerable importance in practical applications, not much work has been done to understand the basic mechanisms involved. The discharge of a buoyant jet in a flowing medium, for instance, is of interest in environmental problems. Turbulent flows has been considered, as reviewed by List [40]. But laminar, buoyant jets in a moving ambient medium have not been studied. Similarly, flows arising due to combined buoyancy effects of thermal and mass diffusion have received little attention [5]. Again, these flows are often encountered in the atmosphere, particularly due to heat and material rejection, and in water bodies, such as the ocean and salt-gradient solar ponds. Most of these flows are turbulent and, therefore, work has been largely directed at turbulent flows [6].

As computed by Mollendorf and Gebhart [15] for a laminar axisymmetric plume, various flow circumstances arise, depending on the values of the Prandtl number Pr and the Schmidt number Sc, where $Sc = \nu/D$. Another important parameter that arises is $\tilde{N}$, which measures the relative

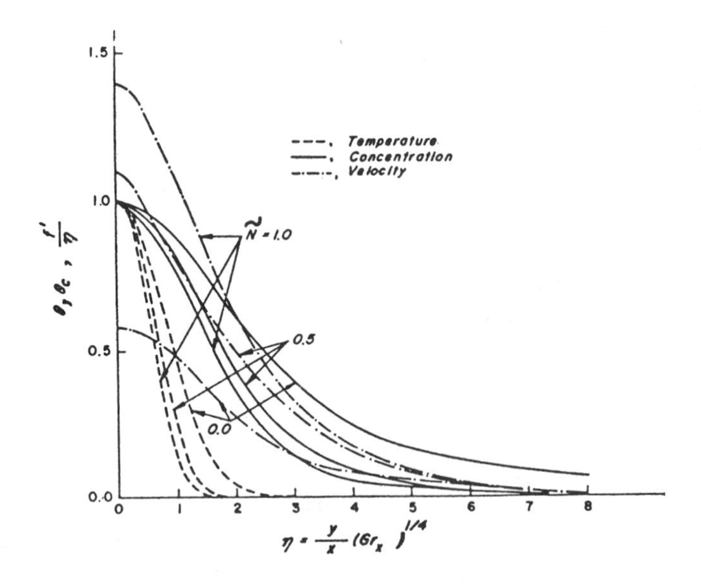

Figure 21. Temperature, concentration, and velocity distributions in an axisymmetric plume driven by buoyancy effects due to both heat and mass diffusion, for Pr = 7 and Sc = 1 [15].

importance of chemical and thermal diffusion in causing the density differences that give rise to the flow. This parameter is defined as

$$\tilde{N} = \frac{\beta^*(C_s - C_a)}{\beta(t_s - t_a)} \tag{99}$$

where C_s is the species concentration at the surface and C_a that in the ambient medium. Figure 21 shows the computed profiles for an axisymmetric plume, with θ_c defined as

$$\theta_c = \frac{C - C_a}{C_s - C_a} \tag{100}$$

Similar profiles are expected to arise for a laminar, buoyant jet driven by the input of momentum and of buoyancy arising from both temperature and concentration differences.

Another consideration of some importance is the effect of the boundary conditions adjacent to the jet origin on the flow. This aspect refers to the no-slip condition imposed by the presence of a wall which contains the opening through which the jet is discharged. Most analytical and numerical studies have considered boundary-layer flows, in which the entrainment condition at the origin does not affect the solution. However, nonboundary-layer effects may be important at low values of Re and in the region close to the jet origin. This question is similar to the leading edge effect in other buoyancy-driven flows [5]. The flow induced by jets and plumes has been studied by Schneider [57] and the no-slip conditions at the solid walls adjacent to the jet source are shown to be important even for large Reynolds numbers. Since, in actual practice, the nozzle or slit is bounded by solid walls, this becomes an important consideration. Then, the full, coupled, governing equations need to be solved to determine the flow in the immediate vicinity of the origin [37]. For low values of Re, the entrainment and thermal transport effects at the jet source are particularly important and need to be considered. Similarly, the flow may be affected by solid surfaces and other buoyancy-induced flows in the neighborhood. The flow, in such cases, is often substantially altered due to the presence of such boundary conditions and the full equations need to be solved to obtain the resulting flow [58].

SUMMARY

This review considers the flow in a laminar buoyant jet, driven by an input of momentum and of buoyancy due to temperature or concentration differences. The vertical, axisymmetric or two-dimensional jet, in which the momentum input is aligned with the buoyancy force, is of particular interest and has, therefore, received the most attention in the literature. The two limiting cases of a nonbuoyant jet and of a buoyant plume, arising due to an input of buoyancy alone, are important in the study of a buoyant jet. These two circumstances can be studied by similarity methods if a point or line source is assumed. Various analytical and experimental studies on these flows are discussed. Near the jet origin, buoyancy effects are generally small and a perturbation analysis may be carried out to determine the flow and the thermal field. The results obtained in such analyses are presented. Far downstream of the jet source, the flow is dominated by buoyancy effects and approaches the characteristics of a buoyancy-induced plume. These trends are considered in terms of available numerical and experimental results on buoyant jets.

The review also considers nonvertical flows, such as those resulting from horizontal and inclined discharges. The resulting jet trajectory is discussed. The rise of a laminar buoyant jet in a stably-stratified environment is considered. The flow rises to a finite height and the available results are discussed. The transition of the flow to turbulence is discussed and experimental results on the laminar length of the jet given. Several other considerations, particularly the effect of entrainment near the jet source, of multiple buoyancy input mechanisms, and of neighboring flows and surfaces on the resulting flow, are discussed. The basic processes underlying the flow in laminar, buoyant jets are considered in detail and the available information on these flows presented.

Acknowledgments

The author acknowledges the support of the National Science Foundation, through Grant No. MEA-82-14325, and of the U.S. Department of Commerce, through Grant No. NB83NADA4047, for some of the work reported here and for the preparation of this review.

NOTATION

A	ratio of spread of temperature field to that of velocity field
b	characteristic halfwidth of jet
C	local concentration
C_s	species concentration at surface
C_a	species concentration in the ambient medium
C_p	specific heat at constant pressure
D	mass diffusivity
d	diameter of jet
E	entrainment coefficient
f	dimensionless stream function
f', f'', f'''	similarity variables
Gr_x	Grashof number
$\bar{g}$	gravitational acceleration
I	integral defined by Equation 33
$J(x)$	momentum flux
K	constant
K_1	constant
k	thermal conductivity
L_ℓ	laminar jet length
m	mass flow rate per unit width
$\dot{m}$	mass flow rate
N, n	constants
$\tilde{N}$	parameter which measures relative importance of chemical and thermal diffusion in causing density differences
Pr	Prandtl number
P	local pressure
$Q(x)$	convective thermal energy in the flow
Re	Reynolds number
S	stratification parameter
t	local temperature
V_0	jet velocity at nozzle
V_x, V_y	vertical and radial velocity components
$\bar{V}$	velocity vector
x	vertical coordinate
y	radial coordinate

Greek Symbols

α	thermal diffusivity	η	similarity variable
β	coefficient of volumetric thermal expansion	θ	dimensionless temperature
		μ	viscosity
β^*	coefficient of expansion due to concentration	ν	kinematic viscosity
		ρ	density
γ	velocity dependent parameter, see Equation 43	τ	time
		ψ	stream function
$\epsilon(x)$	perturbation parameter		

REFERENCES

1. Schlichting, H., "Laminar Strahlenausbreitung." *ZAMM*, Vol. 13 (1933), pp. 260–263.
2. Andrade, E. N., and Tsien, H. S., "The Velocity Distribution in a Liquid-into-Liquid Jet," *Proc. Phys. Soc.*, Vol. 49, (1937), pp. 381–391.
3. Landau, L. D., and Lifshitz, E. M., *Fluid Mechanics*, Pergamon Press, N.Y., 1959, p. 86.
4. Squire, H. B., "The Round Laminar Jet," *Quart. J. Mech. Appl. Math.*, Vol. 4 (1951), pp. 321–329.
5. Jaluria, Y., *Natural Convection Heat and Mass Transfer*, Pergamon Press, Oxford, U.K., 1980.
6. Turner, J. S., *Buoyancy Effects in Fluids*, Cambridge Univ. Press, U.K., 1973.
7. Tenner, A. R., and Gebhart, B., "Laminar and Axisymmetric Vertical Jets in a Stably Stratified Environment," *Int. J. Heat Mass Transfer*, Vol. 14 (1971), pp. 2051–2062.
8. Gebhart, B., *Heat Transfer*, McGraw-Hill, N.Y., 2nd ed., 1971.
9. Schlichting, H., *Boundary Layer Theory*, McGraw-Hill, N.Y., 6th ed., 1968.
10. Mollendorf, J. C., and Gebhart, B., "Thermal Buoyancy in Round Laminar Vertical Jets," *Int. J. Heat Mass Transfer*, Vol. 16 (1973), pp. 735–745.
11. Schneider, W., and Potsch, K., "Weak Buoyancy in Laminar Vertical Jets," In *Rev. Dev. Theor. Exptl. Fluid Mech.*, U. Muller, K. G. Roesner, and B. Schmidt (Eds.), Springer, Berlin, 1979.
12. Yih, C. S., "Free Convection due to a Point Source of Heat," *Proc, 1st U.S. Nat. Cong. Appl. Mech.* (1951), pp. 941–947.
13. Fujii, T., "Theory of the Steady Laminar Natural Convection Above a Horizontal Line Heat Source and a Point Heat Source," *Int. J. Heat Mass Transfer*, Vol. 6 (1963), pp. 597–606.
14. Brand, R. S., and Lahey, F. J., "The Heated Laminar Vertical Jet," *J. Fluid Mech.*, Vol. 29 (1967), pp. 305–315.
15. Mollendorf, J. C., and Gebhart, B., "Axisymmetric Natural Convection Flows Resulting from the Combined Buoyancy Effects of Thermal and Mass Diffusion," *Proc. 5th Int. Heat Transfer Conf.*, Tokyo, Vol. 5 (1974), pp. 10–14.
16. Jaluria, Y., and Gebhart, B., "On the Buoyancy-Induced Flow Arising from a Heated Hemisphere," *Int. J. Heat Mass Transfer*, Vol. 18 (1975), pp. 415–431.
17. Jaluria, Y., "Natural Convection Flow Interaction Above a Heated Body." *Letters Heat Mass Transfer*, Vol. 3 (1976), pp. 457–466.
18. Morton, B. R., Taylor, G. I., and Turner, J. S., "Turbulent Gravitational Convection from Maintained and Instantaneous Sources," *Proc. Roy. Soc.*, Vol. A234 (1956), pp. 1–23.
19. Morton, B. R., "Entrainment Models for Laminar Jets, Plumes and Wakes," *Phys. Fluids*, Vol. 10 (1967), pp. 2120–2127.
20. Wirtz, R. A., and Chiu, C. M., "Laminar Thermal Plume Rise in a Thermally Stratified Environment," *Int. J. Heat Mass Transfer*, Vol. 17 (1974), pp. 323–329.
21. Himasekhar, K., and Jaluria, Y., "Laminar Buoyancy-Induced Axisymmetric Free Boundary Flows in a Thermally Stratified Medium," *Int. J. Heat Mass Transfer*, Vol. 25 (1982), pp. 213–221.
22. Bill, R. G., and Gebhart, B., "Transition of Plane Plumes," *Int. J. Heat Mass Transfer*, Vol. 18 (1975), p. 513.
23. Jaluria, Y., and Cha, C. K., "Heat Rejection to the Surface Layer of a Solar Pond," ASME *J. Heat Transfer*, Vol. 107 (1985), pp. 99–106.

24. Cha, C. K., and Jaluria, Y., "Recirculating Mixed Convection Flow for Energy Extraction," *Int. J. Heat Mass Transfer*, Vol. 27 (1984), pp. 1801–1812.
25. Jaluria, Y., and Steckler, K. D., "Wall Flow Due to Fire in a Room," *Proc. Fall Tech. Meeting, Eastern Sect. Combust. Inst.*, Atlantic City, N.J., Paper No. 47, (1982).
26. Bickley, W., "The Plane Jet," *Phil, Mag.*, Vol. 23 (1937), pp. 727–731.
27. Andrade, W. N., "The Velocity Distribution in a Liquid-into-Liquid Jet, Part 2: The Plane Jet," *Proc. Phys. Soc.*, Vol. 51 (1939), pp. 784–793.
28. Pai, S. I., *Fluid Dynamics of Jets*, Van Nostrand, N.Y., 1954.
29. Crane, L. J., "Thermal Convection From a Horizontal Wire," *Z. Angew. Math. Phys.*, Vol. 10 (1959), pp. 305–315.
30. Spalding, D. B., and Cruddace, R. G., "Theory of the Steady Laminar Buoyant Flow Above a Line Heat Source in a Fluid of Large Prandtl Number and Temperature-Dependent Viscosity," *Int. J. Heat Mass Transfer*, Vol. 3 (1961), pp. 55–59.
31. Gebhart, B., Pera, L., and Schorr, A. W., "Steady Laminar Natural Convection Plumes Above a Horizontal Line Heat Source," *Int. J. Heat Mass Transfer*, Vol. 13 (1970), pp. 161–171.
32. Fujii, T., Morioka, I., and Uehara, H., "Buoyant Plume Above a Horizontal Line Heat Source," Int. J. Heat Mass Transfer, Vol. 16 (1973), pp. 755–768.
33. Brodowicz, K., and Kierkus, W. T., "Experimental Investigation of Laminar Free-Convection Flow in Air Above Horizontal Wire with Constant Heat Flux," *Int. J. Heat Mass Transfer*, Vol. 9 (1966), pp. 81–94.
34. Forstrom, R. J., and Sparrow, E. M., "Experiments on the Buoyant Plume Above a Heated Horizontal Wire," *Int. J. Heat Mass Transfer*, Vol. 10 (1967), pp. 321–331.
35. Schorr, A. W., and Gebhart, B., "An Experimental Investigation of Natural Convection Wakes Above a Line Heat Source," *Int. J. Heat Mass Transfer*, Vol. 13 (1970), pp. 557–571.
36. Himasekhar, K., "An Analytical and Experimental Study of Laminar Free Boundary Flows in a Stratified Medium," M.S. thesis, Indian Inst. Tech., Kanpur, India, 1980.
37. Jaluria, Y., "Buoyant Plant Jets in Thermally Stratified Media," ASME Paper No. 82-WA/HT-57, 1982.
39. Jaluria, Y., "Vertical Natural Convection in an Adiabatically Stratified Medium," *Letters Heat Mass Transfer*, Vol. 2 (1975), pp. 151–158.
40. List, E. J., "Turbulent Jets and Plumes," *Ann. Rev. Fluid Mech.*, Vol. 14 (1982), pp. 189–212.
41. Jaluria, Y., and Himasekar, K., "Buoyancy-Induced Two-Dimensional Vertical Flows in a Thermally Stratified Environment," *Comp. Fluids*, Vol. 11 (1983), pp. 39–49.
42. Koh, R. C. Y., "Two-Dimensional Surface Warm Jets," *ASCE J. Hyd. Div.*, Vol. 97 (1971), pp. 819–836.
43. Madni, I. K., and Pletcher, R. H., "Buoyant Jets Discharging Nonvertically into a Uniform, Quiescent Ambient—A Finite Difference Analysis and Turbulence Modeling," *J. Heat Transfer*, Vol. 99 (1977), pp. 641–647.
44. Riester, J. B., Bajura, R. A., and Schwartz, S. H. "Effects of Water Temperature and Salt Concentration on the Characteristics of Horizontal Buoyant Submerged Jets," *J. Heat Transfer*, Vol. 102 (1980), pp. 557–562.
45. Turner, J. S., "Jets and Plumes with Negative Buoyancy," *J. Fluid Mech.*, Vol. 26 (1966), pp. 779–792.
46. Seban, R. A., Behnia, M. M., and Abreu, K. E., "Temperatures in a Heated Jet Discharged Downwards," *Int. J. Heat Mass Transfer*, Vol. 21 (1978), pp. 1453–1458.
47. Satyanarayana, S., and Jaluria, Y., "A Study of Laminar Buoyant Jets Discharged at an Inclination to the Vertical Buoyance Force," *Int. J. Heat Mass Transfer*, Vol. 25 (1982), pp. 1569–1577.
48. Chen, C. J., and Rodi, W. A., "Vertical Turbulent Buoyant Jets: A Review of Experimental Data," Pergamon Press, Oxford, U.K., 1979.
49. Rosenhead, L., *Laminar Boundary Layers*,Clarendon Press, Oxford, 1963, pp. 558–562.
50. Pera, L., and Gebhart, B., "On the Stability of Laminar Plumes: Some Numerical Solutions and Experiments," *Int. J. Heat Mass Transfer*, Vol. 14 (1971), pp. 975–984.
51. Gebhart, B., "Natural Convection Flows and Stability," *Adv. Heat Transfer*, Vol. 9 (1973), pp. 273–348.

52. Mollendorf, J. C., and Gebhart, B., "An Experimental and Numerical Study of the Viscous Stability of a Round Laminar Vertical Jet With and Without Thermal Buoyancy for Symmetric and Asymmetric Disturbances," *J.* Fluid Mech., Vol. 61 (1973), pp. 367–399.
53. Reynolds, A. J., "Observations of a Liquid-into-Liquid Jet," *J. Fluid Mech.*, Vol. 14 (1962), pp. 552—556.
54. McNaughton, K. J., and Sinclair, C. G., "Submerged Jets in Short Cylindrical Flow Vessels," *J. Fluid Mech.*, Vol. 25 (1966), pp. 367–375.
55. McKenzie, C. P., and Wall, D. B., "Transition from Laminar to Turbulence in Submerged and Bounded Jets," *Fluidics Quart.*, Vol. 4 (1968), p. 38.
56. Marsters, G. F., "Some Observations on the Transition to Turbulence in Small, Unconfined Free Jets," Queen's Univ., Kingston, Ont., Rep. No. 1–69, 1969.
57. Schneider, W., "Flow Induced by Jets and Plumes," *J. Fluid Mech.*, Vol. 108 (1981), pp. 55–65.
58. Pera, L., and Gebhart, B., "Laminar Plume Interactions." *J. Fluid Mech.*, Vol. 68 (1975), pp. 259–271.

CHAPTER 13

IMPINGING JETS

Nagy S. Nosseir

Department of Aerospace Engineering and
Engineering Mechanics
San Diego State University
California USA

CONTENTS

INTRODUCTION, 349

GENERAL CHARACTERISTICS, 350
The Free Jet, 351
The Impinging Region, 351
The Wall Jet, 354

RESONANCE PHENOMENON AT HIGH SPEEDS, 354
Two Branches of the Feedback Loop, 355
The Phase Lock, 358
Collective Interaction, 358

FAR-FIELD NOISE GENERATION, 360
The Mechanics of Noise Generation, 361

CONCLUDING REMARKS, 364

NOTATION, 364

REFERENCES, 365

INTRODUCTION

The impingement of a jet on a solid surface poses an interesting fluid mechanics problem that occurs in several engineering applications. For example, some short take-off and landing (STOL) aircrafts utilize an externally blown flap technique. Hot gases that exhaust from the engine at high speeds are deflected by direct impingement on the flaps to create extra lift during take-off and landing. Among other applications that involve jet impingement are VTOL aircrafts and ramjet combustors in the aerospace industry, and glass toughening, paper drying, and welding in chemical industry. Flow geometries which can be categorized under jet impingement include [1].

1. Two-dimensional, and axisymmetric jet-plate flow.
2. Jet-edge flow.
3. Internal cavity flow.
4. Jet-hole flow.

Although each flow has distinguishing features, all of them have common characteristics governed by similar mechanisms. In addition, under certain flow conditions all of these flows develop self-sustained oscillations. Such oscillations could result in fatigue of the impingement surface(s) due to a rise in unsteady loading, higher levels of heat transfer, and noise radiation.

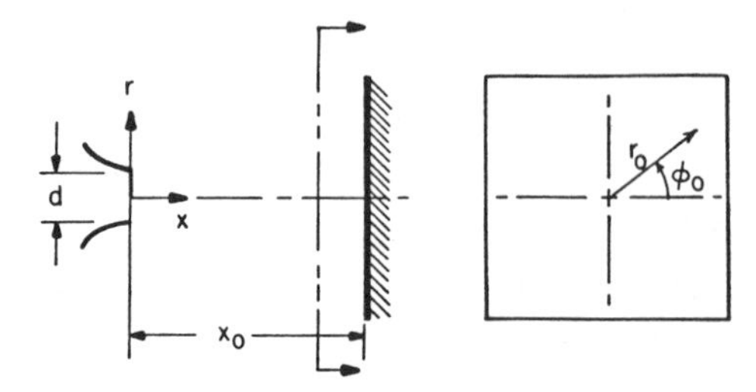

Figure 1. Schematic diagram of the axisymmetric impinging jet and the cylindrical coordinate system. Subscript 0 denotes the coordinates on the plate.

The main thrust of the present review will be focused on an axisymmetric, subsonic jet impinging on an infinitely large flat plate (Figure 1). Whenever applicable, the peculiarities of two-dimensional impinging jet will be addressed. Supersonic jet impingement and the heat transfer associated with hot jet impingement will not be included. (See References 2-4 for analyses of those cases.) An overview of the general characteristics of the flow field will be presented first. Then, the phenomenon of self-sustained oscillations in high-speed impinging jets will be examined. Finally, the noise radiated to the far-field will be addressed.

GENERAL CHARACTERISTICS

The flow field of an impinging jet can be viewed as a synthesis of the following flow modules:

1. A free turbulent jet upstream of the plate.
2. A stagnation flow near the plate (also referred to as the impinging region).
3. A wall jet.

A low-Reynolds number impinging jet (Figure 2) is used to describe the behavior of the flow in these modules. The jet is initially laminar and rolls-up into axisymmetric ring vortices one diameter downstream of the nozzle. These ring vortices are convected downstream in the same fashion as in a free jet [12]. They stretch outward as they approach the plate. The cores of the ring vortices reach the plate at locations $1.0 \leq r_0/d \leq 1.5$ depending on the nozzle-to-plate distance x_0. The impinging region extends $0.2x_0$ upstream of the plate [13, 14] and $0 < r_0/d < 3$ radially along the plate [15–17]. The flow develops into a wall jet downstream of the impinging region ($r_0/d > 3$). The following discussion will detail characteristics of the flow in each of these modules.

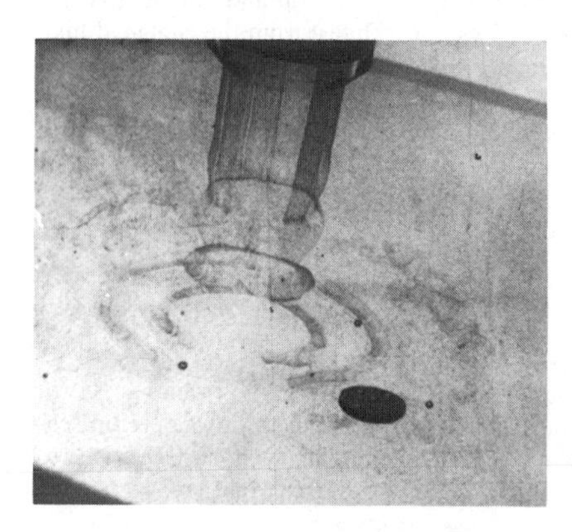

Figure 2. An oblique view showing the free jet module and the stretching of ring vortices in the impinging region near the plate. $Re = Ud/\nu = 5{,}000$, $x_0/d = 4$.

The Free Jet

The jet issuing from the nozzle has a potential core with a velocity U, surrounded by a shear layer in which the velocity drops to zero at its outer edge. The potential core extends to approximately six diameters downstream of the nozzle. The development of the shear layer downstream of the nozzle lip is initially dominated by a linear instability mechanism [18]. Axisymmetric waves (or two-dimensional waves in a two-dimensional jet) grow exponentially with downstream distance. The most amplified wave (also referred to as the natural instability waves of the jet) corresponds to a Strouhal number $St = f\theta/U = 0.016$, where θ is the momentum thickness at the nozzle lip. These waves roll-up into axisymmetric ring vortices (or two-dimensional vortices) which are convected downstream with an approximate speed $C_1 = 0.60U$. Amalgamation between these vortices, also referred to as "pairing" [19] in which two consecutive vortices roll over each other, causes the local shear layer to double its thickness. Since these amalgamations occur randomly in space the growth of the shear layer thickness with downstream distance is linear with a half-angle of $10°$ approximately. These vortices are usually referred to as large-scale coherent structures. The name reflects their deterministic nature (e.g. their shape, size, and convection motion), which can be determined within a relatively small standard deviation [20]. The successive amalgamations between these structures produce the classical random, small-scale turbulence [12]. Therefore, the turbulent jet is viewed here as double-structured in nature: large-scale coherent structures, and small-scale random structures.

The reader is referred to reviews by Ho and Huerre [21] and Husain and Hussain [22] for further discussions on the instability of free shear layers as well as long lists of references on the subject.

The Impinging Region

The impinging region extends upstream of the plate to a location where the mean properties of the flow deviate by 2% from what values the free jet would have had at the same location. This location (x_i) corresponds to [13]

$$(x_0 - x_i)/d = 1.2; \qquad x_0/d < 6.8$$

$$(x_0 - x_i)/d = 0.153(1 + x_0/d); \qquad x_0/d > 6.8$$

The shear layer in this region deflects outward and is under a stabilizing curvature effect. Reynolds stresses, among other turbulent quantities, decrease below their respective free jet values [23]. In effect, by forcing the shear layer to curve, the plate plays the role of a low-pass filter to the turbulent energy. Based on the longitudinal velocity spectra measured in the impinging region by Gutmark et al. [14] (Figure 3), a neutral scale ($k = 2.7\ m^{-1}$) is detected at which turbulent energy is neither augmented nor attenuated. At higher frequencies the turbulence is attenuated due to viscous dissipation, while at lower frequencies the turbulence is augmented due to vortex stretching as the plate is approached. Mathematically, the effect of shear layer curvature appears as an extra rate of strain, $e = \partial v/\partial x$, added to the $\partial U/\partial y$ term in the equations of motion of "thin" shear layers [24].

The centerline velocity decreases to zero as the flow approaches the stagnation point. However, the convection velocity of the coherent structures does not change significantly due to a balance between the retarding effect of the plate and the acceleration resulting from shear layer curvature. This is shown in Figure 4 by hydrogen bubble lines generated by a thin platinum wire stretched across the nozzle exit. The first line of bubbles downstream of the nozzle is horizontally straight indicating a uniform flow within the potential core. However, the line curves at the outer edge of the jet due to the retarded velocity in the shear layer. The distance between the bubble lines decreases as the flow approaches the plate which indicates that the flow is decelerating near the jet axis. The convex curvature toward the plate side of the two lines nearest to the plate indicates a slower speed near the jet axis as compared to the outer shear layer. This is further confirmed from x-t diagram of impinging ring vortices measured from frame-by-frame analysis of movies [17].

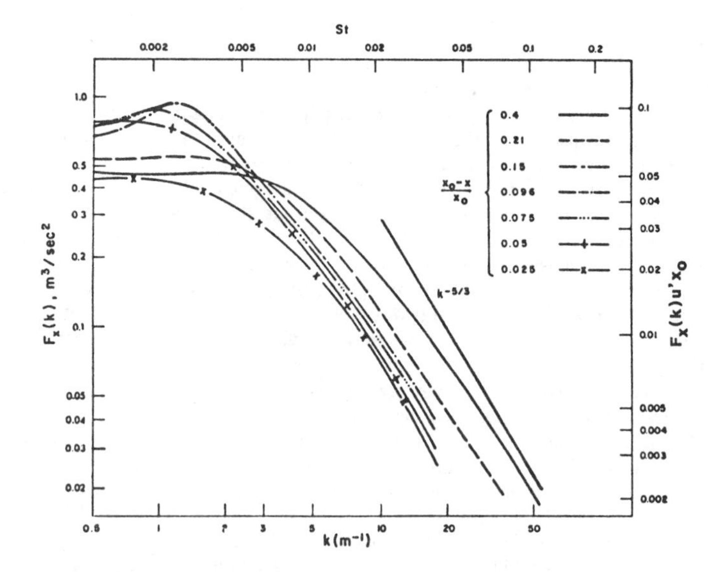

Figure 3. Spectral distribution of the logitudinal velocity fluctuations along a two-dimensional impinging jet's center plane [14].

Figure 4. Curvature of bubble lines indicating the velocity profiles during impingement. Re = 5,000, $x_0/d = 2$.

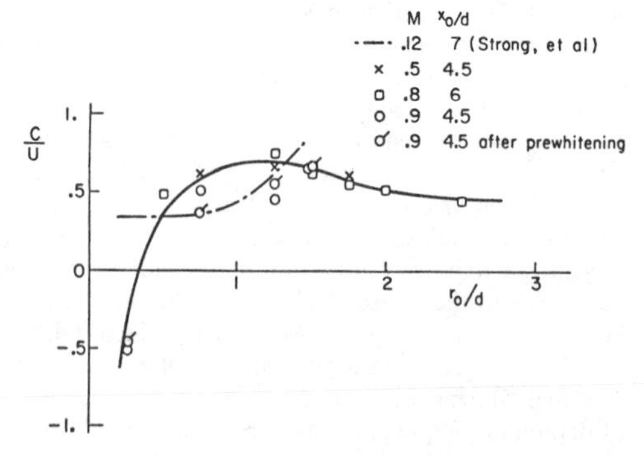

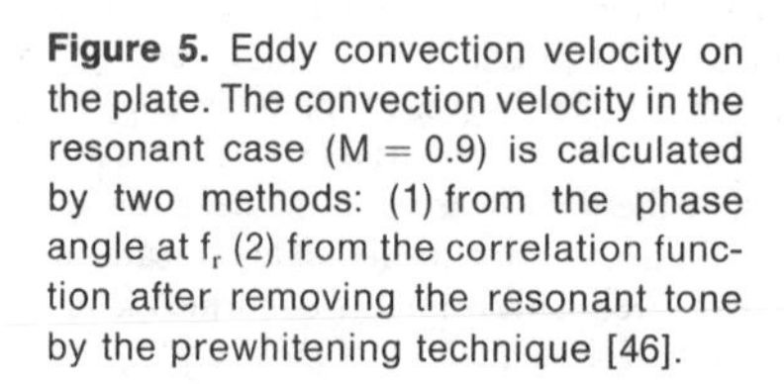

Figure 5. Eddy convection velocity on the plate. The convection velocity in the resonant case (M = 0.9) is calculated by two methods: (1) from the phase angle at f_r (2) from the correlation function after removing the resonant tone by the prewhitening technique [46].

Figure 6. Multiple exposure of hydrogen bubbles showing shear layer growth and ambient entrainment. Bubbles in the ambient region in Figure 4 display here pathlines of the ambient entrainment. Bubble lines appear clearer in the left-hand side of the jet due to their proximity to the source of light. These bubble lines indicate a high level of entrainment at the downstream end of the shear layer curvature near the plate.

The subsequent convection of the flow along the plate can be studied from space-time correlations of surface pressure fluctuations. The convection velocity based on the time delay of optimum correlation peaks is shown in Figure 5. From locations where ring vortices reach the plate ($1.0 \leq r_0/d \leq 1.5$), the convection velocity decreases from its free stream value of 0.6U asymptotically to its wall jet value. The figure also shows a negative value of C/U near the stagnation point ($r_0 = 0$) at the high speed case. It has been speculated that a flow reversal occurs there due to the so called "stagnation bubble" shown by the grease streak photographs of the impinging region on the plate by Donaldson and Snedeker [2].

The change in turbulent quantities (e.g. Reynolds stresses) from the free jet values upstream of the impinging region to those of the wall jet downstream is not monotonic. After initially decreasing due to the stabilizing curvature, the Reynolds stresses overshoot the values of the wall jet before finally decreasing [23]. The increase in turbulence is usually accompanied by an increase in ambient entrainment, which is indeed the case shown in Figure 6. Castro and Bradshaw [23] postulated that downstream of the curvature's end the large eddies are slower in reestablishing themselves to the values of the wall jet than other energy-containing eddies. An overshoot occurs since large eddies are instrumental in the transport of energy from the energy-containing eddies to the edge of the shear layer. A different explanation can be deduced from Didden and Ho's [25] experiment in which a dye having a color different from that of the ring vortices of the impinging jet is injected from the plate (Figure 7). A counter-rotating secondary vortex is generated by the impinging vortex. With respect to frames of reference fixed to the impinging vortex, the generation of the secondary vortex is caused by an unsteady separation of the boundary layer on the plate. The unsteady separation is induced by an adverse pressure gradient downstream of the core of the impinging vortex. The counter-rotating vortex-pair moves away from the wall due to the mutually induced

Figure 7. The counter-rotating vortex-pair in the impinging jet: the upper vortex is the impinging vortex ring and the lower is the secondary vortex generated by unsteady separation of the boundary layer. Courtesy of C. M. Ho.

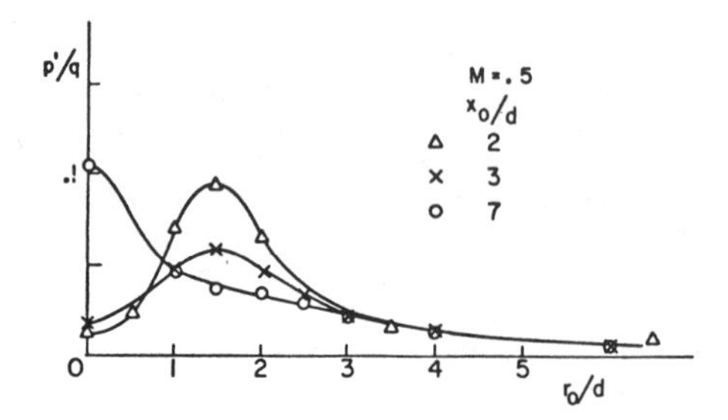

Figure 8. Radial variation of rms surface pressure fluctuations.

field on each other [26] The climax is reached when an "explosion-like" process occurs in which the two vortices breakdown into small-scale eddies, i.e. an increase in the turbulent energy.

The surface pressure distribution reflects footprints of the impinging turbulent structures. The radial distribution of the normalized surface pressure is shown in Figure 8 for different plate locations. For $x_0/d < 6.0$, the level of p'/q near the jet axis is small (2%) because of the impinging potential flow. A peak at $r_0/d = 1.5$ indicates the approximate location where ring vortices reach the plate. Beyond the length of the potential core, the peak of turbulent fluctuations in a free jet shifts toward the jet axis, and the peak of p'/q also moves to $r_0 = 0$ as in the case of $x_0/d = 7$. The maximum value of $(p'/q)_{r=0}$ is equal to 0.12 and is reached at $x_0/d = 7.5$ [15].

In the oblique jet impingement, the stagnation point becomes a stagnation streamline [27]. The shear layer curvature decreases near the downstream half of the plate. Accordingly, the magnitude of the low-frequency components of the surface pressure spectra decreases as compared to normal impingement [15]. The effect of changing the shape of the solid surface from that of a flat plate on the surface pressure distribution is investigated by Donalson and Snedeker [2], and on the turbulence characteristics in the stagnation region by Sadeh et al. [28, 29, 30]. The surface pressure distribution drops off more rapidly on a convex surface than on that of a flat plate, while on a concave surface it drops off less rapidly. Sadeh et al studied the flow impingement on a circular cylinder. Vorticity stretching is purported to be the mechanism triggering the formation of a coherent array of cross-vortex tubes in the stagnation region. Amplification of cross-vortices at scales larger than a neutral wavelength by stretching leads to amplification of the streamwise turbulence in the stagnation region.

The Wall Jet

For $r_0/d > 3.0$, Figure 8 indicates that p'/q is independent of the location of the plate which is a characteristic of the developed wall jet. In addition, p'/q does not change by changing the jet Mach number as was indicated from the measurements of Preisser and Block [31]. In this region, the flow is so dominated by the wall through viscous forces that it tends to "forget" its past history. For more discussion on wall jets the reader is referred to Glauert [32].

RESONANCE PHENOMENON AT HIGH SPEEDS

Generally surface pressure fluctuations have a broad spectrum (Figure 9) owing to the turbulent nature of the impinging jet at large Reynolds numbers. However, at high jet speeds ($M > 0.7$) and plate-to-nozzle distances of the order of the potential core length or less ($x_0/d < 7.5$), the pressure signals exhibit a large spectral peak (Figure 9). The energy increases in a relatively narrow frequency band, the resonance frequency (f_r), while it diminishes at all other frequencies. This phenomenon is accompanied by an audible screech tone as investigated by Wagner [33], Neuwerth [3], Ho and Nosseir [34], and Nosseir and Ho [35]. Another feature of the resonance phenomenon is that the dynamic loading on the surface increases by as much as 50% over the nonresonant jet [36].

A feedback mechanism was suggested by Powell [37] to explain the resonant phenomenon in jet-edge interaction (edge-tone). Powell's model is generally adopted in flows with self-sustained oscillations [1, 38]. Two wave trains, one traveling downstream and the other traveling upstream, are assumed to be phase locked at the nozzle lip and thus form a feedback loop.

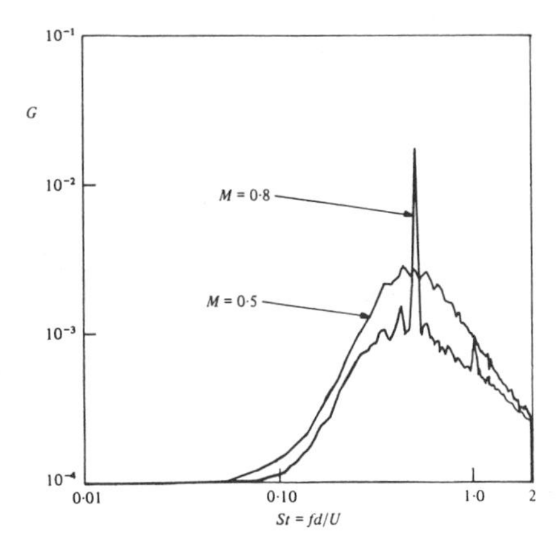

Figure 9. Normalized power spectra of surface pressure fluctuations ($x_0/d = 4$, $r_0/d = 1.5$). The curve M = 0.8 corresponds to resonance.

Two Branches of the Feedback Loop

The pressure at any point with a position vector $\vec{\xi}$ in the near field is the superposition of two waves and takes the form

$$p(t, \vec{\xi}) = a(\vec{\xi}) \exp i(2\pi f_r t - K_1 x + \nu_c) + b(\xi) \exp i(2\pi f_r t + \vec{K}_2 \cdot \vec{\xi}), \tag{1}$$

where the two waves are assumed for simplicity to be monochromatic plane waves having the same frequency f_r which is the resonant frequency. However, the wave numbers, $\vec{K}_1$ and $\vec{K}_2$, are different because of different phase velocities. The first term in Equation 1 represents a downstream-traveling pressure wave with velocity $C_1 = 2\pi f_r/K_1$. The second term is a wave traveling in the upstream direction at an angle θ_a to the jet axis with velocity

$$\vec{C}_2 = \frac{2\pi f_r}{K_2} \hat{\theta}_a$$

where $\hat{\theta}_a$ is a unit vector. The phase difference between the two waves at any location is

$$\nu = -K_1 x - \vec{K}_2 \cdot \vec{\xi} + \nu_c \tag{2}$$

where the constant phase shift ν_c can be evaluated by applying the boundary condition at the plate. Since the upstream-traveling wave is assumed to be generated by the downstream-traveling wave as it impinges on the plate, the phase difference between the two waves should be equal to zero there, i.e.

$$\nu = 0 \text{ at } \vec{\xi} = \vec{\xi}_s$$

where $\vec{\xi}_s$ is the position vector of an apparent sound source on the plate. Substituting in Equation 2, one obtains

$$\nu = K_1(x - x_0) + \vec{K}_2 \cdot (\vec{\xi}_s - \vec{\xi}) \tag{3}$$

The pressure in Equation 1 can be written in the simple form

$$p(t, \vec{\xi}) = A\alpha + B\beta \tag{4}$$

where A and B are the amplitudes of the two waves α and β.

The correlation between pressure signals is a good means to investigate waves propagating between the nozzle and the plate. The correlation between pressure signals at points x_i and x_j in the near field is

$$R_{i,j}(\tau) = A_iA_jR_{\alpha_i,\alpha_j}(\tau) + B_iB_jR_{\beta_i,\beta_j}(\tau) + A_iB_jR_{\alpha_i,\beta_j}(\tau) + A_jB_iR_{\beta_i,\alpha_j}(\tau) \tag{5}$$

The preceding equation shows that the correlation $R_{i,j}(\tau)$ of two signals, each of which is the sum of two stationary components, is the algebraic sum of the individual component correlations. The correlations of these components give four optimum time delays according to the following sequence:

$$R_{\alpha_i,\alpha_j}(\tau) \rightarrow \tau^{[1]} = (x_j - x_i)/C_1 \tag{6a}$$

$$R_{\beta_i,\beta_j}(\tau) \rightarrow \tau^{[2]} = -\vec{\xi}_{i,j} \cdot \hat{\theta}_a/C_2 \tag{6b}$$

$$R_{\alpha_i,\beta_j}(\tau) \rightarrow \tau^{[3]} = -\left(\frac{x_0 - x_j}{C_1} + \frac{\vec{\xi}_{i,s} \cdot \hat{\theta}_a}{C_2}\right) \tag{6c}$$

$$R_{\beta_i,\alpha_j}(\tau) \rightarrow \tau^{[4]} = \left(\frac{x_0 - x_1}{C_1} + \frac{\vec{\xi}_{j,s} \cdot \hat{\theta}_a}{C_2}\right). \tag{6d}$$

where $\vec{\xi}_{i,s}$ and $\vec{\xi}_{j,s}$ are vectors connecting positions i and j with the apparent sound source(s) on the plate.

The interpretation of the time delays in Equation 6 is as follows: $\tau^{[1]}$ is the time it takes the waves to travel downstream from x_i to $x_j \cdot \tau^{[2]}$ is the time needed for the upstream waves to travel from x_i to x_j and should therefore appear as a negative time delay in $R_{i,j}(\tau) \cdot \tau^{[3]}$ is the time required for the waves to travel downstream from x_j to the apparent source on the plate plus the time for the waves generated by the impingement to travel from the plate to $x_i \cdot \tau^{[4]}$ is the time for the waves to travel from x_i to the plate at a speed of C_1 plus the time to travel from the apparent source to x_j with a speed C_2.

For the case of $x_i = x_j$, the autocorrelation function $R_{i,i}(\tau)$ has peaks with time delays given by

$$\tau^{[1]} = \tau^{[2]} = 0, \tag{7a}$$

$$\tau^{[4]} = \tau^{[3]} = \frac{x_0 - x_i}{C_1} + \frac{\vec{\xi}_{i,s} \cdot \hat{\theta}_a}{C_2} \tag{7b}$$

Therefore, the extra peaks in the autocorrelation represent the time for waves to travel downstream from x_i to the plate and to return back from the plate to x_i.

The autocorrelations and cross-correlations of the signals from several pressure transducers placed along the streamwise direction are plotted in Figure 10. The vertical spacing of the correlation curves in the diagram are scaled to the streamwise distances between each microphone. Lines connecting all the peaks form a W-shape curve which indicates agreement between Equation 6 and the experimental values.

The convection velocities of the two waves are plotted in Figure 11 based on the time delays of peaks [1] and [2] of the measured correlations (e.g. Figure 10). The downstream waves are traveling with 0.62U which is the convection velocity of the coherent structures. The data is also in excellent agreement with the speed of the coherent structures measured by Neuwerth [3] from frame-by-frame analysis of movies of a similar high-speed impinging jet. On the other hand, the speed of the upstream-traveling waves is equal to the speed of sound (a_0). Ho and Nosseir [34] found that these waves are conical and propagate at an angle to the jet axis $\theta_a = 30^\circ$ approximately. When the convection speeds from Figure 11 were substituted in Equation 6C and D, Ho and Nosseir [34] found excellent agreement between the equations and the time delays of peaks [3] and [4] of the measured correlations.

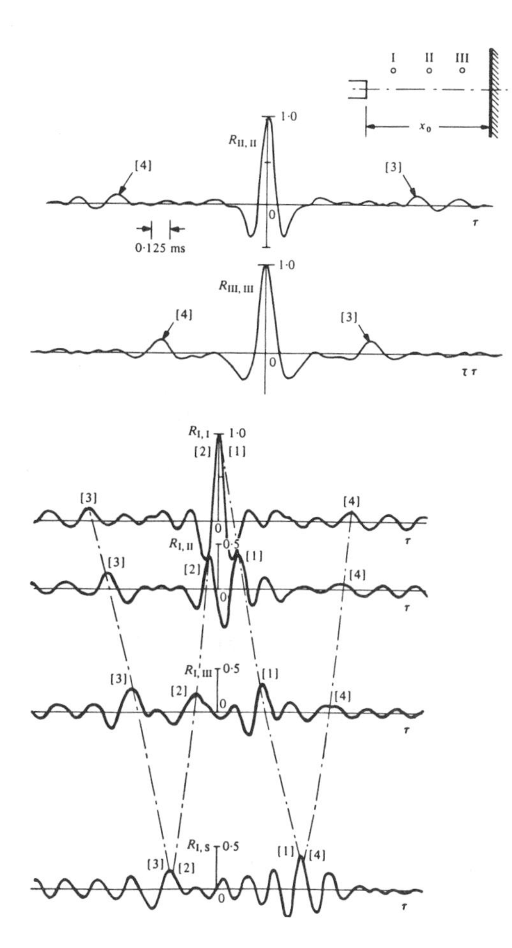

Figure 10. (A) Autocorrelations of near-field pressure signals (M = 0.8, $x_0/d = 7$). For microphone I, x/d = 1.09, r/d = 1.13; for II, x/d = 1.97, r/d = 1.3; for III, x/d = 3.25, r/d = 1.31. (B) Correlations of near-field pressure signals; (M = 0.8, $x_0/d = 5.5$, $r_s/d = 1$) indices refer to the sketch in Figure 10A.

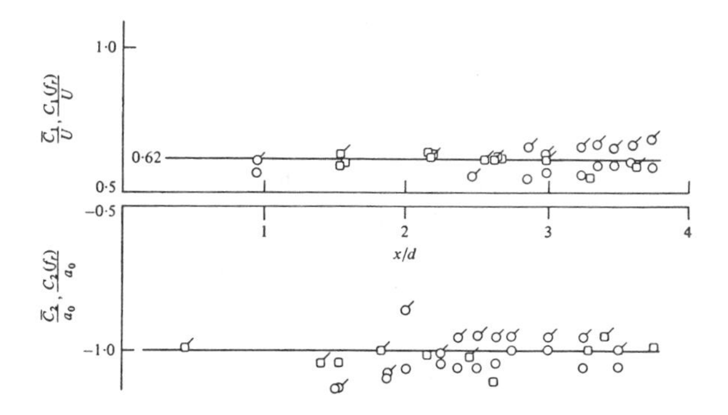

Figure 11. Propagation velocities of the downstream and the upstream waves: broadband velocity $\bar{C}_1$, $\bar{C}_2$ and phase velocity at the resonant frequency $C_1(f_r)$, $C_2(f_r)$. The straight line at $\bar{C}_1/U = 0.62$ follows Neuwerth [3].

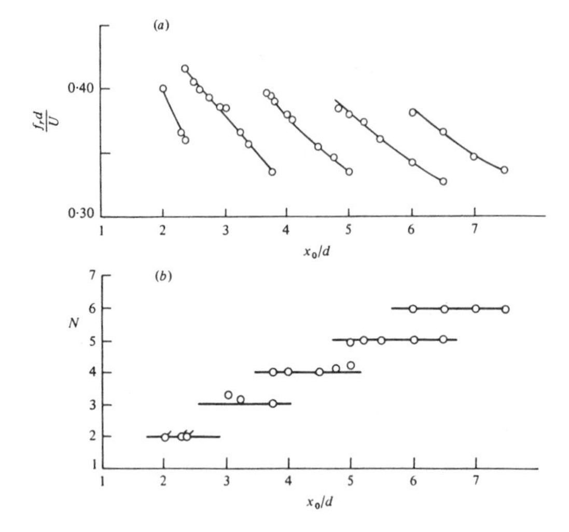

Figure 12. (A) Variations of the nondimensional resonant frequency parameter $f_r d/U$ with plate locations, $M = 0.9$; (B) resonant frequency stages, $N = x_0/\lambda_1 + x_0/\lambda_2$, O, measured; -ơ-, calculated from Equation 8 with f_r measured from a surface pressure transducer and $C_1/U = 0.62$.

The Phase Lock

The phase difference between the two waves, through direct measurements [17] or by using Equation 3, was found to be an integer multiple of 2π at the nozzle exit. The two waves remained in-phase despite changing the nozzle-to-plate distance. The frequency of the waves, i.e. the resonant frequency (f_r), changes (Figure 12) so as to maintain the phase lock between the waves at the nozzle exit. The variation of the resonant Strouhal number, $(St)_r = f_r d/U$, with x_0/d is shown in the figure. $(St)_r$ decreases with increasing x_0/d until it reaches a minimum value upon which it abruptly increases to a higher value. With further increase of x_0/d, the cycle is repeated. These frequency stages are typical of flows with self-sustained oscillations.

The phase lock between the two waves at the nozzle exit can be used to clarify the presence of the frequency stages. The phase lock requires an integer number of waves (N) to exist in the feedback loop. By definition

$$N = \frac{X_0}{\lambda_1} + \frac{X_0}{\lambda_2}, \tag{8}$$

where $\lambda_1 = C_1/f_r$ and $\lambda_2 = (C_2/\cos\theta_a)/f_r$ are the wavelengths of the downstream- and the upstream-traveling waves respectively. Equation 8 characterizes the most fundamental feature of flows with a feedback loop.

The measured f_r is plotted according to Equation 8 in Figure 12B. In each of the frequency stages, the number of waves, N, remains constant. As the nozzle-to-plate distance increases, the wavelength of both waves increases to preserve the phase lock at the nozzle exit. This results in a decrease in the resonant frequency. As the nozzle-to-plate distance is increased, the resonant frequency is decreased until a minimum value is attained. This minimum value is the lower limit of the frequency band which corresponds to the passage frequency of coherent structures [34]. A further increase in the nozzle-to-plate distance results in a frequency jump to a higher value and the number of waves in the feedback loop is increased by one.

Collective Interaction

The phase lock established at the nozzle exit is between two waves having a frequency in the range $0.3 < (St)_r < 0.4$. This frequency is an order of magnitude less than the natural instability

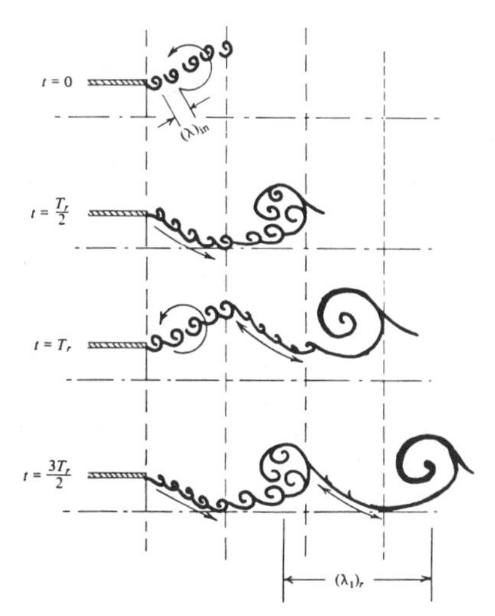

Figure 13. Collective interaction.

frequency (f_{in}) of the jet [34]. A mechanism must exist by which the frequency of the coherent structures would drop sharply from f_{in} to f_r downstream of the nozzle exit. The pairing process itself cannot produce such a frequency drop since it would take three to four pairings to decrease the frequency by a factor of 10, requiring a long distance not available in this case. However, a likely mechanism is described schematically in Figure 13, and is called the "collective interaction." The shear layer emerging from the nozzle is flapping due to the periodic forcing from the upstream-traveling waves. The shear layer goes through cycles of divergence ($t = 0, T_r, 2T_r, \ldots$) and convergence ($t = T_r/2, 3T_r/2, \ldots$), where T_r is the period of the resonant frequency. At $t = 0$, the small vortices rotate around each other under their induced field and coalesce into a large vortical structure. At $t = T_r/2$, the shear layer changes its orientation and the lower vortices move away from the upper vortices due to the mean shear. Therefore, this portion of the shear layer tends to stretch. The vortices are stable, and do not coalesce. These sketches show the evolution of the merging of multiple vortices under forcing. The passage frequency can drop by one order of magnitude within a short distance.

Collective interaction was confirmed further in the impinging jet [39] and in forced, two-dimensional shear layers [40, 41]. Figure 14 shows one of the cases in an experiment by Ho and Huang [40] in which they forced the shear layer at subharmonics of its natural instability frequency. The forcing frequency in the case shown is one order of magnitude lower than the instability frequency (f_{in}) similar to the case of the high-speed impinging jet. The picture displays the characteristic features of the collective interaction explained earlier, namely a sharp drop in the passage frequency of the coherent structures and a relatively large growth of the shear layer.

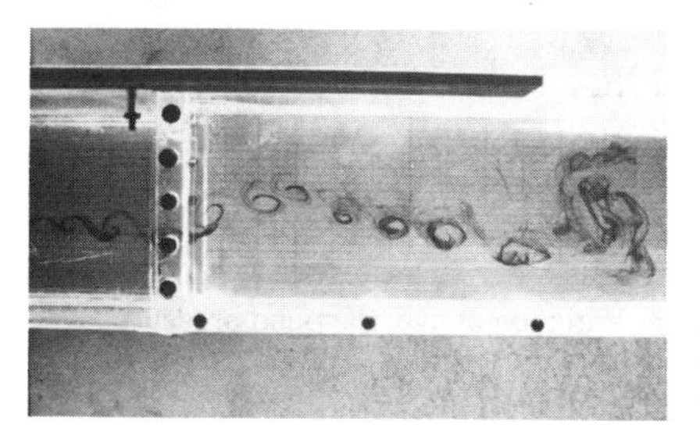

Figure 14. Collective interaction in a two-dimensional free shear layer, flow from left to right, forcing frequency $\ll$ natural instability frequency [40].

After the collective interaction the passage frequency remains constant downstream to the plate, i.e., there is no further amalgamation between the coherent structures. In the experiment of Ho and Huang [40], a high-amplitude subharmonic was needed for the coherent structures to merge. A long distance and a thick shear layer are required for the growth of the subharmonic. In an impinging jet neither of these requirements are met. Hence, no further merging appears after the collective interaction, and the passage frequency of the coherent structure after the collective interaction becomes the dominant resonant frequency.

In conclusion, the feedback loop is formed by the downstream-convected large coherent structures and by upstream-propagating pressure waves generated by the impingement of these structures on the plate. The upstream-propagating waves travel with the speed of sound in the quiescent medium. These waves force in-phase oscillations of the thin shear layer near the nozzle exit. The shear layer, oscillating at a frequency much lower than its intrinsic most unstable frequency, undergoes a collective interaction in which many small vortices merge together to form a large coherent structures. The collective interaction, therefore, generates large coherent structures at intervals phase-locked with the external forcing of the upstream-propagating waves causing the resonance. The sharp drop in the passage frequency of the vortices concurrent with the rapid growth of the shear layer within a short distance from the nozzle, by the collective interaction, makes the resonance independent of the initial conditions of the jet.

FAR-FIELD NOISE GENERATION

The noise generated by a turbulent flow in the presence of a solid surface can be expressed mathematically as the summation of a volume integral of quadupoles and a surface integral of dipoles [42]. The dipoles are produced by the pressure fluctuations exerted on the flow by the surface. Powell [43] applied this concept to the impinging jet replacing the plate by a mirror image of the jet. The volume integral over the image of the jet was found to be equal to the surface integral over the plate. Accordingly, the presence of the plate should double the noise levels which would have existed due to the free jet, i.e. a 6 db increase in noise levels.

The overall sound pressure level (OASPL) of an impinging jet is compared to that of a free jet for various jet speeds and plate locations in Figure 15. The OASPL changes approximately as U^8 similar to the free jet. However, the noise level of the impinging jet is greater than the noise level of the free jet by more than 6 db. The increase is even greater for smaller x_0/d. Similar results are reported by Marsh [44]. (The exceptionally high OASPL at the highest speed of the impinging jet is due to the resonance phenomenon discussed earlier.) The increase in noise persists in all directions as indicated by the directivity patterns in Figure 16. The data in these figures indicate that there is more to the role of the plate than merely reflecting the noise generated by the jet. Pan [45] proposed a theoretical method based on correlation techniques to separate the turbulent flow contribution to the noise from the contribution by surface pressure fluctuations. Near field measurements combined with surface pressure measurements are needed to determine the role of the plate in the generation of noise.

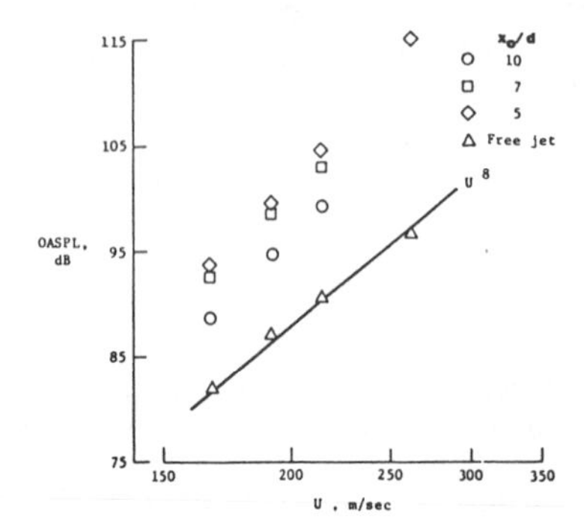

Figure 15. Variation of overall sound pressure level with jet velocity for various jet heights and for the free jet. h = 55° [16].

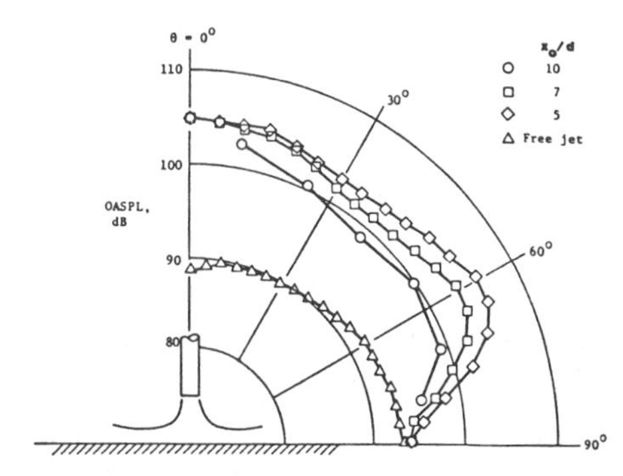

Figure 16. Directivity patterns. M = 0.70 [16].

The Mechanics of Noise Generation

The normalized spectra at $x_0/d = 5$ of Preisser [16] indicate that most of the increase in the noise of an impinging jet over a free jet occurs in the range $0.3 < St < 0.5$ which is the range of the passage frequency of the coherent structures near the end of the potential core. In addition, the cross-spectra of surface and far-field pressure signals [31] indicate that most of the noise is radiated from a region near the plate between one and three diameters from the stagnation point. These results trigger the question of what role the impinging coherent structures might play in the noise generation.

Noise Generated by the Impinging Coherent Structures

The correlation between the far-field pressure signal and both the near-field and the surface pressure signals are used to determine the propagation path(s) of the radiated pressure fluctuations. A few samples of the correlations are shown in Figure 17. The time delay in the figure is offset by

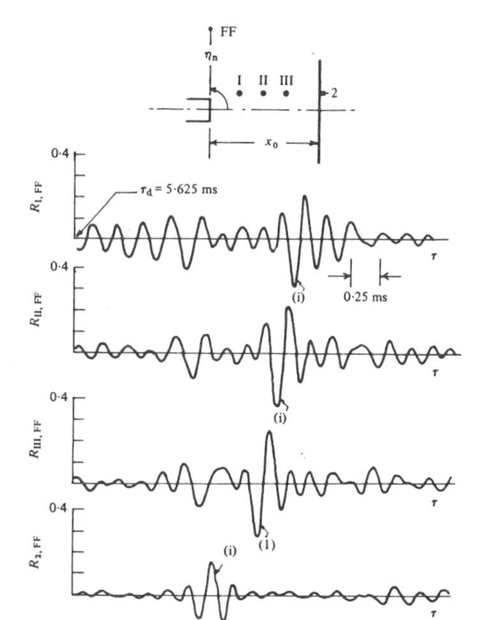

Figure 17. Cross-correlations with far-field microphone; point 2 is at $r_0/d = 1$ (M = 0.8, $x_0/d = 5.5$, $\eta_n/d = 90$).

mic	x/d	r/d
I	1.09	1.13
II	1.97	1.13
III	3.25	1.31

5.625 ms in order to present the main features of the correlations. The time delay of the maximum peak (i) between any microphone in the near field and the far-field microphone is greater than the ratio of the distance between the two microphones divided by the ambient speed of sound. This excludes the possibility of significant direct acoustic propagation between the two. The time delay decreases consistently as we correlate FF with further downstream locations in the near field. In fact, the time delay of peak (i) satisfies the relation

$$\tau^{(i)}_{j,FF} = \tau^{[1]}_{j,2} + \tau_{2,FF}; \qquad j = I, II, III \tag{9}$$

for different Mach numbers and plate locations [17]. The first term in the right-hand side of Equation 9, $\tau^{[1]}_{j,2}$ is the time delay of the maximum peak in the near-field correlations (Equation 6A) which represents the downstream convection of the coherent structures from point j to point 2 on the plate. The convection speed being 0.62U. The second term $\tau_{2,FF}$ is the propagation time of the pressure waves from position 2 on the plate to the far field with *the speed of sound.* As mentioned before, this is the region where the coherent structures are observed to "impinge" on the plate.

These experimental results indicate that the far-field noise is not dominated by pressure fluctuations generated by the coherent structures in the free jet module upstream of the plate. Rather, most of the noise is generated in the impinging region, and is due to the impingement of coherent structures on the plate. This is in disagreement with the passive role of the plate suggested by Powell [43]. Further confirmation of the former is offered by the conditional sampling technique [39] which found high peaks in the far-field signal phase locked with the impingement of the coherent structures on the plate.

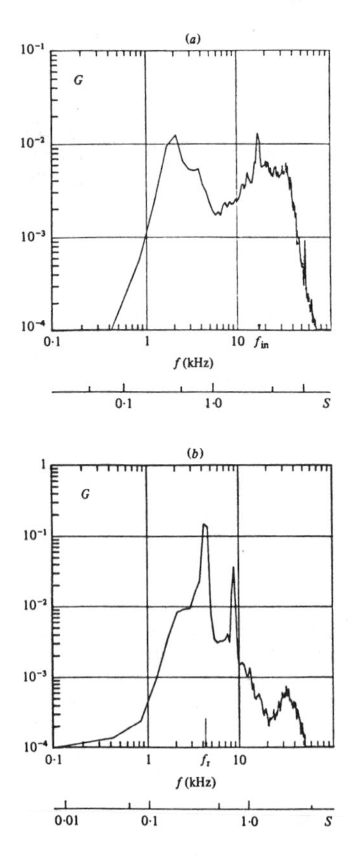

Figure 18. Far-field power spectra, $x_0/d = 4.5$: (A) M = 0.4; (B) M = 0.9.

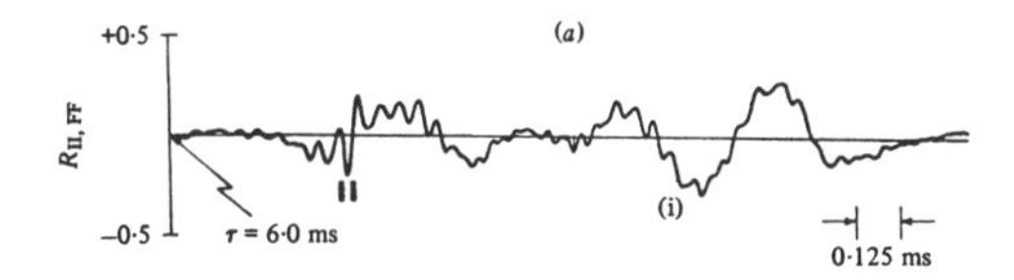

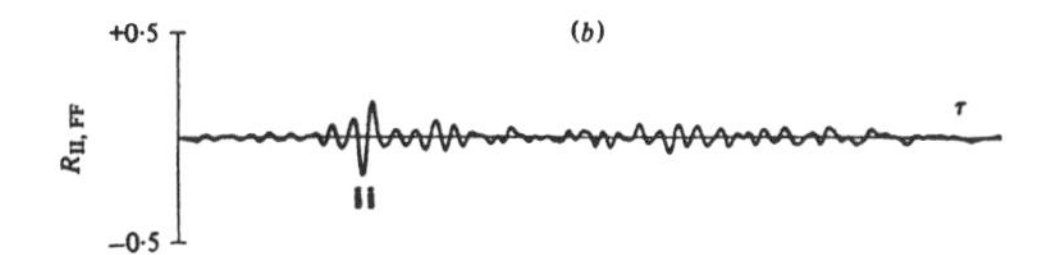

Figure 19. Cross-correlation between far-field and near-field signals ($M = 0.4$, $x_0/d = 4.5$, $x_{II/d} = 0.92$); (A) using raw signals: (B) using high-pass filtered signals (cut-off frequency = 8 KHz).

The power spectrum of the far-field pressure, according to the preceding results, is expected to peak at a low frequency band corresponding to the passage frequency of the coherent structures. However, the spectra (Figure 18) display two characteristic peaks which indicate that the impingement of the coherent structures on the plate is not the only mechanism for the generation of noise. The figure shows a spectral peak at a Strouhal number of 0.33, which corresponds to the noise generated by the impinging coherent structures. Another peak appears at a higher-frequency band centered on the jet instability frequency (f_{in}). The two peaks have the same level; however, since the energy is an integration over frequency, most of the noise energy is associated with the high-frequency instability waves of the jet for this low-Mach-number case. A similar distribution of energy (Figure 18B) is observed for a resonant jet at $M = 0.9$, except that in this case the large coherent structures prevail and a considerable amount of energy is concentrated at the resonant frequency and its first harmonic.

Noise Generated by the Instability Waves

The high-frequency instability waves, or small vortices, evolve shortly downstream of the nozzle into large-scale coherent structures by the collective interaction process. Therefore, the high-frequency noise associated with the instability waves cannot follow the same path discussed earlier because the waves lose their identity long before they reach the plate.

A cross-correlation function between the far-field microphone and a near-field microphone placed near the nozzle exit is shown in Figure 19A. A broad negative peak, peak (i), is associated with the noise generated by the impinging coherent structures. Another peak, denoted by (ii), emerges at an earlier time delay. In order to ensure that this peak corresponds to the radiated high-frequency noise, the pressure signals are filtered using a high-pass filter. The cut-off frequency of the filter is the 8 kHz frequency in the valley between the two spectral peaks in Figure 18A. The correlation of the filtered signals is shown in Figure 19B. Although peak (i) disappeared, peak (ii) survived. The noise radiation path, determined from the time delay of peak (ii), indicates direct acoustic radiation to the far-field.

The mechanism associated with such high-frequency noise-radiation is still unresolved. One possibility is an unsteady loading on the nozzle lip induced by the instability vortices, which would produce a dipole-type sound. Another possibility is the flapping of the thin shear layer due to the collective interaction, which would force the mean flow to undergo acceleration and decceleration. This would lead to oscillations of the mass flow rate, producing a monopole sound field. There is need for further experimentation to study the nature of such a mechanism.

In conclusion, Figure 20 schematically summarizes the noise generation mechanisms. Most of the noise is generated by the initial instability waves and the coherent structures. The noise generated by the coherent structures is caused by their impingement on the plate. The noise associated with the instability waves of the shear layer radiates directly from near the nozzle exit. The collective interaction confines the evolution of high-frequency instability waves into large coherent structures to a short distance, such that the two noise-generating mechanisms are spatially separated.

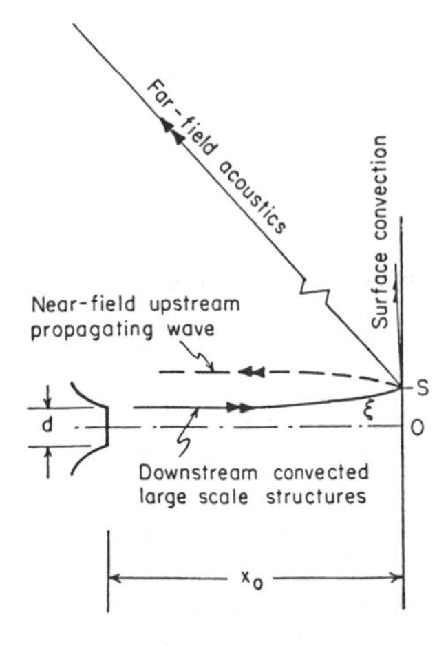

Figure 20. Summary of noise generation mechanisms in an impinging jet.

CONCLUDING REMARKS

This chapter has explained the "basic" mechanisms governing the behavior of impinging jets by consideration of mainly the experimental work done on the impingement of a turbulent jet on flat surfaces. We have discussed the characteristics of turbulence in the three modules of the impinging jet, namely the free jet, the stagnation flow, and the wall jet modules. The radial distribution of the surface pressure fluctuations reflects the foot prints of the turbulence structure in the stagnation flow module. The surface pressure fluctuations display a characteristic peak depending on the plate-to-nozzle distance, x_o. At a plate-to-nozzle distance larger than the potential core length ($x_o/d > 6$), the surface pressure fluctuations peak at the stagnation point ($r_0/d = 0$). At $x_o/d < 6$, the surface pressure fluctuations peak at $r_0/d = 1.5$ which is in the region where the large coherent structures approach the plate. For the latter case, self-sustained oscillations of the flow occur at high jet speeds ($M > 0.7$) due to the onset of resonance. The onset of resonance results in higher levels of surface pressure loading and noise radiation. A feedback loop of pressure waves between the nozzle and the plate is responsible for the onset of resonance. The feedback loop is formed by the downstream-traveling coherent structures and upstream-traveling pressure waves generated by the impingement of these structures on the plate. The upstream-traveling waves force the shear layer near the nozzle exit to generate new coherent structures at intervals phase-locked with the upstream-traveling waves through the collective interaction process.

Finally, it was shown that the far-field noise is generated by two different mechanisms and propagates along two different paths. The low-frequency noise is mainly generated by the impingement of the large coherent structures on the plate. The high-frequency noise is produced by the instability waves of the thin shear layer and radiates directly from the vicinity of the nozzle exit.

Acknowledgement

The author acknowledges the support of the U.S. Air Force Office of Scientific Research (Contract No. F49620 78 c 0060) and the U.S. Office of Naval Research (Contract No. N00014 84 k 0373) in the preparation of this chapter.

NOTATION

A, B	wave amplitudes	e	rate of strain
C_1	speed	f_r	resonance frequency
d	diameter	$\vec{K}$	wave number

k	constant	St	Strouhal number
M	Mach number	U	velocity
p'	pressure	x	coordinate
$R_{ij}(\tau)$	correlation function	y	coordinate
r	radius		

Greek Symbols

θ	momentum thickness at nozzle lip	$\vec{\xi}$	position vector
λ	wavelength	τ	delayed time
ν	phase difference between successive waves		

REFERENCES

1. Rockwell, D., and Naudascher, E., "Self-Sustained Oscillations of Impinging Free Shear Layers," *Ann. Rev. Fluid Mech.*, No. 11 (1979), pp. 395–409.
2. Donaldson, C. D., and Snedeker, R. S., "A Study of Free Jet impingement. Part 1. Mean Properties of Free and Impinging Jets," *J. Fluid Mech.*, Vol. 45, Part 2 (1971), pp. 281–319.
3. Neuwerth, G., "Acoustic Feedback Phenomena in Subsonic and Supersonic Free Jets that Strike a Perturbing Body," Ph.D. thesis, the Rhein-Westfalia University of Technology at Aachen, West Germany, 1973.
4. Donaldson, C. D., Snedeker, R. S., and Margolis, D. P., "A Study of Free Jet Impingement. Part 2. Free Jet Turbulent Structure and Impingement Heat Transfer," *J. Fluid Mech.*, Vol. 45, Part 3 (1971), pp. 477–512.
5. Seban, R. H., "Influence of Free Stream Turbulence on Local Heat Transfer from Cylinders," *J. Heat Transfer*, Vol. 82 (1960), p. 101.
6. Huang, G. C., "Investigation of Heat-Transfer Coefficients for Air Flow through Round Jets Impinging Normal to Heat-Transfer Surface," *Trans. ASME*, 85 (1963), p. 237.
7. Sutera. S. P., Maeder, P. F., and Kestin, J., "On the Sensitivity of Heat Transfer in the Stagnation-Point Boundary Layer to Freestream Vorticity," *J. Fluid Mech.*, Vol. 16, Part 4 (1963), pp. 497–520.
8. Gardon, R., and Akfirat, J. C., "Heat Transfer Characteristics of Impinging Two-Dimensional Air Jets," *ASME*, Paper no. 65-HT-20, 1965.
9. Sutera, S. P., "Vorticity Amplification in Stagnation-Point Flow and its Effect on Heat Transfer," *J. Fluid Mech.*, Vol. 21, Part 3 (1965), pp. 513–534.
10. Kestin, J., "The Effect of Freestream Turbulence in Heat Transfer Rates," *Advances in Heat Transfer*, Vol. 3, New York: Academic Press, 1966, pp. 1–32.
11. Traci, R. M., and Wilcox, D. C., "Freestream Turbulence Effects on Stagnation Point Heat Transfer," *AIAA J.*, Vol. 13, No. 7 (1975), pp. 890–896.
12. Browand, F. K., and Laufer, J., "The Role of Large Scale Structures in the Initial Development of Circular Jets," Proc. of 4th Biennial Symp. on Turbulence in Liquids, Univ. Missouri-Rolla, Sept. 1975.
13. Giralt- F., Chia, C., and Trass, O., "Characterization of the Impingement Region in an Axisymmetric Turbulent Jet," *Ind. Eng. Chem. Fund.*, Vol. 16, No. 1 (1977).
14. Gutmark, E., Wolfshtein, M., and Wyganski, I., "The Plane Turbulent Impinging Jet," Technion Report No. 226, Haifa, Israel. Also, *J. Fluid Mech.*, Vol. 88 (1978), pp. 737–756.
15. Strong, D. R., Siddon, T. E., and Chu, W. T., "Pressure Fluctuations on a Flat Plate with Oblique Jet Impingement," NASA CR-839, 1967.
16. Preisser, J. S., "Fluctuating Surface Pressure and Acoustic Radiation for Subsonic Normal Jet Impingement," NASA Tech. Paper 1361, 1979.
17. Nosseir, N. S., "On the Feedback Phenomenon and Noise Generation of an Impinging Jet," Ph.D. thesis, U. Southern California, U.S.A., 1979. Also, AFOSR, TR-79-1263.
18. Michalke, A., "Istabilitat eines Kompressiblen runden Freistrahls unter Berucksichtigung des Einflusses der Strahlgrenzschichtdicke," *Z. Flugwiss*, Vol. 19 (1972), pp. 319–32.

19. Winant, C. D., and Browand, F. K., "Vortex Pairing: the Mechanism of Turbulent Mixing Layer Growth at Moderate Reynolds Number," *J. Fluid Mech.*, Vol. 63 (1974), pp. 237–255.
20. Laufer, J., "New Trends in Experimental Turbulence Research," *Ann. Rev. Fluid Mech.*, No. 7 (1975), pp. 307–326.
21. Ho, C. M., and Huerre, P., "Perturbed Free Shear Layers," *Ann. Rev. Fluid Mech.*, No. 16 (1984).
22. Husain, Z. D., and Hussain, A. K. M. F., "Natural Instability of Free Shear Layers," *AIAA J.*, Vol. 21, No. 11 (1983), pp. 1512–1517.
23. Castro, J. P., and Bradshaw, P., "The Turbulence Structure of a Highly Curved Mixing Layer," *J. Fluid Mech.*, Vol. 73, Part 2, 1976, pp. 265–304.
24. Bradshaw, P., "Effects of Streamline Curvature on Turbulent Flow," *AGARDograph*, No. 169 (1973).
25. Didden, N., and Ho, C. M., "Unsteady Separation in the Boundary Layer Produced by an Impinging Jet," to be published.
26. Walker, J. D. A., "The Boundary Layer Due to Rectilinear Vortex," *Proc. Roy. Soc. Lond.*, A359 (1978), pp. 167–188.
27. Foss, J. F., and Kleis, S. J., "Mean Flow Characteristics for the Oblique Impingement of an Axisymmetric Jet," *AIAA J.*, Vol. 14 (1976), pp. 705–706.
28. Sadeh, W. Z., Sutera, S.P., and Maeder, P. F., "Analysis of Vorticity Amplification in the Flow Approaching a Two-Dimensional Stagnation Point," *Z. Angew. Math. Phys.*, Vol. 21 (1970), pp. 699–716.
29. Sadeh, W. Z., Sutera, S. P., and MAeder, P. F., "An Investigation of Vorticity Amplification in Stagnation Flow," *Z. Angew. Math. Phys.*, Vol. 21 (1970), pp. 717–742.
30. Sadeh, W. Z., Brauer, J., and Garrison, J. A., "Visualization Study of Vorticity Amplification in Stagnation Flow." Project Squid, TR CSU-1-PU, 1977.
31. Preisser, J. S., and Block, P. J. W., "An Experimental Study of the Aeroacoustics of a Subsonic Jet Impinging Normal to a Large Rigid Surface," AIAA Paper No. 76–520, 1976.
32. Glauert, M. B., "The Wall Jet," *J. Fluid Mech.*, Vol. 1, Part 6 (1956), pp. 625–643.
33. Wagner, F. R., "The Sound and Flow Field of an Axially Symmetric Free Jet Upon Impact on a Wall," *Zeit. fur Flugwiss* 19 (1971), pp. 30–44. Also, NASA TT F-13942, 1971.
34. Ho, C. M., and Nosseir, N. S., "Dynamics of an Impinging Jet. Part 1. The Feedback Phenomenon," *J. Fluid Mech.*, Vol. 105 (1981), pp. 119–142.
35. Nosseir, N. S., and Ho, C. M., "Dynamics of an Impinging Jet. Part 2. The Noise Generation," *J. Fluid Mech.*, Vol. 116 (1982), pp. 379–391.
36. Ho, C. M., and Nosseir, N. S., "Large Coherent Structures in an Impinging Jet," *Turbulent Shear Flows* 2, Berlin Heidelberg: Springer-Verlag (1980), p. 297.
37. Powell, A., "On the Edgetone," *J. Acoustic Soc. Am.*, Vol. 33 (1961), pp. 395–409.
38. Rockwell, D., "Oscillations of Impinging Shear Layers," *AIAA J.*, Vol. 21, No. 5 (1983), pp. 645–664.
39. Nosseir, N. S., and Ho, C. M., "Pressure Fields Generated by Instability Waves and Coherent Structures in an Impinging Jet," AIAA paper no. 80-0980, 1980.
40. Ho, C. M., and Huang, L. S., "Subharmonic and Vortex Merging in Mixing Layers," *J. Fluid Mech.*, Vol. 119 (1982), pp. 443–473.
41. Oster, D., Wygnanski, I., "The Forced Mixing Layer between Parallel Streams," *J. Fluid Mech.*, Vol. 123 (1982), pp. 91–130.
42. Curle, N., "The Influence of Solid Boundaries upon Aerodynamic Sound," *Proc. R. Soc. Lond.*, A231 (1955), pp. 505–514.
43. Powell, A., "Aerodynamic Noise and the Plane Boundary," *J. Acoust. Soc. Am.*, Vol. 32 (1960), pp. 982–990.
44. Marsh, A. H., "Noise Measurements around a Subsonic Air Jet Impinging on a Plane, Rigid Surface," *J. Acoust. Soc. Am.*, Vol. 33 (1961), pp. 1065–1066.
45. Pan, Y. S., "Cross-Correlation Methods for Studying Near- and Far-Field Noise Characteristics of Flow Surface Interactions," *J. Acoust. Soc. Am.*, Vol. 58 (1975).
46. Williams, K. C., and Purdy, K. R., "A Prewhitening Technique for Recording Acoustic Turbulent Flow Data," *Rev. Sci. Instrum.*, Vol. 41 (1970), pp. 1897–1899.

CHAPTER 14

HYDRODYNAMICS OF CONFINED COAXIAL JETS

Michael M. Gibson

Mechanical Engineering Dept.
Imperial College of Science & Technology
London, England

CONTENTS

INTRODUCTION, 367

PRESSURE RISE DUE TO MIXING, 369

SELF-PRESERVING DEVELOPMENT OF JETS, 371

DEVELOPMENT OF CONSTANT-PRESSURE JETS, 372

SIMILARITY CRITERIA FOR DUCTED JETS: THE CRAYA-CURTET NUMBER, 375

DEVELOPMENT OF CONFINED JETS IN OPEN DUCTS, 376

COAXIAL JETS IN ENCLOSED DUCTS, 381

CALCULATION OF CONFINED COAXIAL JETS, 384

NOTATION, 387

REFERENCES, 388

INTRODUCTION

Confined coaxial jets occur in a number of engineering devices. The aircraft ejector utilizes the momentum transfer from a high velocity stream to an entrained secondary flow to augment the thrust available for short take-off; the pressure rise produced by mixing has been used in the jet pump for over a century, and the occurrence of reversed flow regions surrounding an enclosed jet is important in combustion chambers to recirculate combustion products and stabilize the flame. Confined jets are also of interest because they exhibit a number of interacting basic turbulent flow phenomena, the details of which are not yet fully understood quantitatively: jet entrainment in an axial pressure gradient, interaction with a boundary layer, and the occurrence of flow separation and reattachment.

The main features that distinguish the flow in confined jets from free jet flow are the presence of an axial pressure gradient and the possibility of reversed flow. Generally the jet develops in a surrounding coaxial stream whose velocity decreases with distance from the nozzle. In a constant-area duct the pressure gradient is caused solely by the transfer of momentum from the primary to the entrained secondary stream. The total mass flow rate in the duct remains constant; the momentum flux varies with axial distance consistent with the changes in pressure. In a free jet in a constant-pressure environment the mass flow rate changes while the momentum flux remains constant. The pressure gradient generated by the ducted jet influences the spreading and entrainment rates and the relative strengths of the coaxial streams, and the dimensions of the nozzles and mixing duct

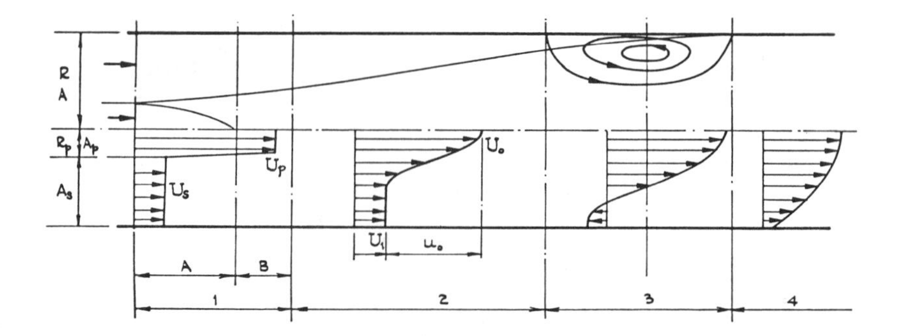

Figure 1. Ducted jet flow regimes showing notation.

determine the pressure variation and the flow pattern. Despite the apparent complexity of these patterns and the fundamental differences to free flow which have been noted, qualitative understanding and quantitative predictions may be obtained from a knowledge of simple free flow behavior.

Figure 1 illustrates the evolution of a confined jet which emerges with velocity U_p from a nozzle of radius R_p into a confining duct of radius R. The upstream end of the duct is open to the atmosphere from which a secondary coaxial stream is entrained with assumed uniform velocity U_s in the inlet plane. The arrangement is that of an idealized ejector or jet pump. If, for simplicity, the effects of the wall boundary layers are ignored, three principal flow regions may be identified [1].

1. A transition region in which the jet velocity distribution changes from uniformity in the plane of the nozzle to a nearly constant shape over a distance of some eight to twenty nozzle diameters. There are two sub-zones:
   - A region in which the potential core of the jet is eroded by an annular mixing layer (A in Figure 1).
   - The region from the end of the potential core to the point where the profiles become approximately self-preserving (B in Figure 1).

   The length of the potential core zone depends on the velocity ratio, the nozzle geometry, and the inlet plane flow conditions. For "clean" inlet conditions (a thin-lipped nozzle, uniform profiles, and low free-stream turbulence levels) the length of the potential core is approximately $(4 + 12\mu)$ nozzle diameters, where μ is the secondary/primary velocity ratio [2].
2. A region in which the external stream may be considered to be a uniform potential flow and the jet velocity and shear stress distributions may be assumed to be approximately self-preserving, that is, they may be scaled with single length and velocity scales. As the jet spreads it entrains fluid from the external stream rapidly enough to reduce the free-stream velocity and so to establish a positive axial pressure gradient. The structure of the flow in this region resembles in some ways that of a free jet evolving in a uniform-pressure environment.
3. A region of recirculating flow which will appear when the entire secondary stream is entrained before the primary jet spreads to the wall. Since the jet does not cease to entrain at this point the external velocity continues to decrease, fluid is entrained from downstream, and the external flow direction is reversed. The velocity profile in the jet changes shape and the flow outside it can no longer be considered to be irrotational because it consists of fluid recirculating through the jet itself. The recirculation zone may be considered to be approximately uniform in pressure up to the point where the jet approaches the wall. Further reductions in the mass flow rate of the entrained secondary stream in the open duct configuration increase the extent of the recirculation region until the secondary stream is cut off entirely and the primary jet is surrounded by a large toroidal recirculation in the vicinity of the nozzle. Under these conditions the flow ceases to resemble that in a free jet: the spreading rate is no longer even approximately constant as it is when the external flow is irrotational, and the flow pattern has more in common with that of a sudden pipe expansion, though there are some differences in detail.

4. A region downstream of the point at which the jet attaches to the wall. Considerable pressure gradients may be established in the upstream part of this region in the vicinity of the point of attachment; the velocity profile changes shape and the jet flow cannot be considered as even approximately self-preserving. Far downstream it reverts to fully developed pipe flow.

This simplified picture of a complex flow is further complicated by the presence of boundary layers on the duct walls and the annular wake from the primary nozzle. The boundary layers grow in an adverse pressure gradient which may be strong enough to produce local separation; even if this does not happen the boundary layer displacement effect and the interaction between the boundary layer and the jet may modify the flow pattern significantly. The initial jet-spreading rate and the length of region 1 is sensitive to the nozzle geometry and inlet flow conditions; vortex shedding from thick nozzle walls may accelerate the erosion of the potential core and enhance mixing with the entrained stream.

A distinction may be drawn between the "open-duct" flows of the type just described, which exemplify the flow in an ejector or in a combustor where the flame is stabilized on a central holder, and the "enclosed-duct" configuration typical of axially fired furnaces, where coaxial jets discharge into a sudden expansion. This situation has much in common with a sudden expansion in pipe flow but is more complicated. The turbulence generated in the high shear mixing zone of a free coaxial jet produces a higher initial spreading rate than that of a single round jet, and it is to be expected that the same effects will also influence the behavior of an enclosed coaxial jet. A further complication is added by the presence of outer jet swirl in practical combustion systems. The pressure field is then considerably modified to an extent that may result in a central recirculation for flame stabilization.

In this chapter the two types of coaxial jet flow in "open" and "enclosed" ducts are considered separately. The following section recapitulates the classical Borda analysis for the pressure rise in the duct extended for the particular case of an ejector or airmover. The next four sections are concerned mainly with the "open-duct" type with an induced coflowing secondary stream. The idea of similar or self-preserving flow is introduced in the third section and discussed in relation to free jets in the fourth section. These two sections form a necessary introduction to the sections following them in which the similarity criteria for ducted jets are presented and which cover the flow in open ducts of the ejector type. The last two sections deal with coaxial jets in enclosed ducts and, finally, an account of calculation methods for both types. The treatment is restricted, in the main, to uniform property flow of incompressible fluids.

PRESSURE RISE DUE TO MIXING

The ideal pressure rise due to the mixing of coaxial jets in a constant area duct may be determined from simple force-momentum considerations. If the shear stress at the wall of a duct of radius R is τ_w, the integral momentum and continuity equations are:

$$\frac{\partial}{\partial x}\int_0^R (p + \rho U^2)\, r\, dr + \tau_w R = 0 \tag{1}$$

$$\frac{\partial}{\partial x}\int_0^R \rho U r\, dr = 0 \tag{2}$$

If two incompressible, equal-density, fluid streams mix completely so that the velocities may be considered uniform and equal to $\bar{U}$, and the wall shear stress is neglected, Equations 1 and 2 reduce to:

$$p_1 A + \rho U_p^2 A_p + \rho U_s^2 A_s = pA + \rho \bar{U}^2 A \tag{3}$$

$$\rho U_p A_p + \rho U_s A_s = \rho \bar{U} A \tag{4}$$

in which subscripts p and s refer to the primary and secondary flows respectively, and unsubscripted quantities relate to the fully-mixed condition downstream. The pressure-rise coefficient is obtained by rearrangement of Equations 3 and 4 as:

$$\frac{p - p_1}{\frac{1}{2}\rho U_p^2} = \frac{2a}{[(a + 1)(\lambda_1 + 1)]^2} \tag{5}$$

where a is the area ratio A_s/A_p and λ_1 is the initial velocity ratio:

$$\lambda_1 = \frac{U_s}{U_p - U_s} \tag{6}$$

This result (Equation 5) provides a target figure for designers of jet pumps and ejectors, but because wall friction is neglected and complete mixing to a uniform profile is not achieved in practice, it overestimates the maximum pressure rise actually obtainable. The discrepancy is shown in Figure 2 [3], where the results are compared with wall pressure measurements for various values of λ_1, taken in a duct of area ratio a = 8.

The analysis can be taken a stage further for the particular case of an ejector drawing fluid from and discharging to a body of fluid at pressure p_∞. Then

$$p = p_\infty = p_1 + \tfrac{1}{2}\rho U_s^2 \tag{7}$$

and, consequently, with the aid of the continuity condition:

$$\lambda_1 = \sqrt{2a}/(a + 1) \tag{8}$$

It is seen that the effect of placing a shroud round a jet is potentially to increase substantially the static thrust available from the jet alone because reduced pressures act on the upstream surface of the inlet. The thrust-augmentation ratio is usually defined as the ratio of the static thrust of an ejector system to the thrust that would be obtained were the same primary mass flow to be expanded isentropically to the ambient pressure p_∞

$$\Phi = [1 + (a + 1)\lambda_1]^2/[(a + 1)(\lambda_1 + 1)\sqrt{2\lambda_1 + 1}] \tag{9}$$

Thus, typically, for a = 30, $\lambda_1 = 0.25$ from Equation 8, $\Phi = 1.61$. This figure for a constant-area duct can be very substantially increased by pressure recovery in a diffuser fitted at the exit of the mixing tube.

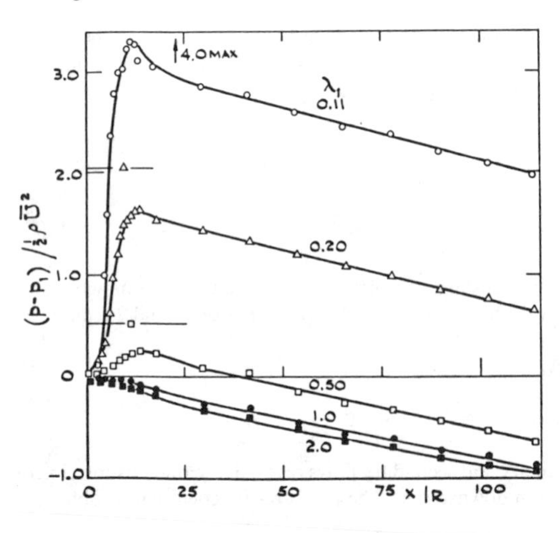

Figure 2. Pressure rise due to confined jet mixing for different velocity ratios. (From Razinsky and Brighton [3] by permission of the American Society of Mechanical Engineers.)

Current interest in the use of ejectors as thrust augmentors for V/STOL aircraft has resulted in a large body of literature in which the simple control-volume analysis has been extended to allow for the effects of nonuniform profiles, inlet losses, wall friction, and diffuser performance [4–7]. It is found that the pressure rise and thrust performance is highly sensitive to these factors and to the effects of compressibility and unequal density which are treated in References 8 through 10. Useful surveys of work on aircraft ejectors are contained in References 11 and 12.

SELF-PRESERVING DEVELOPMENT OF JETS

In a developing turbulent flow, the transverse distributions of mean velocity and other time-averaged quantities such as the Reynolds stresses change with distance downstream, but it is often found that the distributions retain the same functional forms, that is, they are self-preserving or self-similar. The analysis of thin shear layers is greatly simplified if the flow is self-preserving, or can be treated as such to a reasonable approximation, because the time-averaged partial-differential equations of motion can then be transformed into ordinary differential equations. The idea of self-preservation implies that the flow is in a condition of moving equilibrium, that it has essentially forgotten the conditions at its initiation, and that the local flow characteristics depend on only one or two simple parameters.

Subject to the usual thin shear layer approximations, the equation of motion for axisymmetric turbulent flow of an incompressible fluid at sufficiently high Reynolds numbers may be written as:

$$U\frac{\partial U}{\partial x} + V\frac{\partial U}{\partial r} = -\frac{1}{\rho}\frac{dp}{dx} - \frac{1}{r}\frac{\partial}{\partial r}(r\overline{uv}) \tag{10}$$

where $\overline{uv}$ is the time-averaged Reynolds shear stress and the streamwise normal stress gradients have been neglected. When there is an external potential flow of velocity U_1, the pressure gradient is given by:

$$\frac{1}{\rho}\frac{dp}{dx} = -U_1\frac{dU_1}{dx} \tag{11}$$

and the continuity equation is

$$\frac{\partial U}{\partial x} + \frac{1}{r}\frac{\partial}{\partial r}(rV) = 0 \tag{12}$$

The conditions under which the flow can be exactly self-preserving are established by substituting the following profile functions f and g in Equation 10.

$$U = U_1 + u_0 f(\eta)$$
$$\overline{uv} = u_0^2 g(\eta) \tag{13}$$

where $\eta = r/\delta$ and f and g are functions of η only; $\delta(x)$ is the flow width and $u_0(x)$ is the scaling velocity taken as the jet excess velocity $(U_0 - U_1)$ (Figure 1). The result of this substitution is the following ordinary differential equation with the profile function f as dependent variable [1, 13]:

$$\frac{1}{u_0}\frac{du_0}{dx}\left[\lambda f + f^2 - \frac{f'}{\eta}\int_0^\eta f\eta\, d\eta\right] + \frac{1}{U_1}\frac{dU_1}{dx}\left[\lambda f - \frac{1}{2}\lambda f'\eta\right]$$
$$-\frac{1}{\delta}\frac{d\delta}{dx}\left[-\lambda f'\eta - 2\frac{f'}{\eta}\int_0^\eta f\eta\, d\eta\right] = \frac{1}{\delta\eta}\frac{\partial}{\partial\eta}(\eta g) \tag{14}$$

where the primes denote differentiation with respect to η. For the flow to be self-preserving the coefficients must be independent of x, that is [13, 14]

$$\frac{d\delta}{dx} = \text{constant}$$

$$\lambda \equiv \frac{U_1}{u_0} = \text{constant}$$

$$\frac{\delta}{u_0}\frac{du_0}{dx} = \text{constant}$$

Two conditions are of interest in the present context: the short annular mixing layer surrounding the potential core of the jet and, more importantly, the developed jet downstream. It is evident that exact self-preservation is only possible in either case when $U_1 = 0$ or the streamwise pressure gradient is such that $\lambda = U_1/u_0 = \text{constant}$. Reference to the integral momentum equation for the flow in a free, unconfined, jet:

$$\frac{d}{dx}\int_0^\infty U(U - U_1)r\,dr + \frac{dU_1}{dx}\int_0^\infty (U - U_1)r\,dr = 0 \tag{15}$$

shows that in the latter case the self-preservation condition for the developed jet is:

$$\delta \propto x, \qquad u_0 \propto x^n$$

where $n = -\dfrac{I_2 + \lambda I_1}{I_2 + 1.5\lambda I_1}$

and the profile integrals I_1 and I_2 are defined by reference to the excess mass and momentum fluxes in the jet:

$$m_0 = 2\pi\rho\int_0^\infty (U - U_1)r\,dr = 2\pi\rho u_0\delta^2\int_0^\infty f\eta\,d\eta = 2\pi\rho u_0^2\delta I_1 \tag{16}$$

$$M_0 = 2\pi\rho\int_0^\infty U(U - U_1)r\,dr = 2\pi\rho u_0^2\delta^2\int_0^\infty f(f + \lambda)\eta\,d\eta = 2\pi\rho u_0^2\delta^2(I_2 + \lambda I_1) \tag{17}$$

The exponent n is a relatively weak function of λ. Measurements in round jets in stagnant surroundings or in a coflowing stream show that the mean velocity distributions can be tolerably well fitted by an error curve or by the cosine function:

$$f = \frac{U - U_1}{u_0} = \frac{1}{2}\left[1 + \cos\left(\frac{1}{2}\pi\eta\right)\right] \tag{18}$$

which, with δ chosen as the radius $r_{1/2}$ to the half-velocity point where $(U - U_1) = u_0/2$, gives $I_1 = 0.595$, $I_2 = 0.345$. n varies from -1 for $\lambda = 0$ to -0.81 at $\lambda = 0.5$. The first value corresponds to the case of a jet developing in stagnant surroundings. The jet in a coflowing constant velocity stream cannot be exactly self-preserving, nor save in exceptional circumstances, can the flow of confined jets, though in certain conditions approximate self-preservation can be assumed and the analysis greatly simplified as a result.

DEVELOPMENT OF CONSTANT-PRESSURE JETS

It will be useful first to consider the behavior of jets developing at constant pressure in still or moving surroundings when the effects of any confining duct are negligible. It has been stated that exact self-preserving development is only possible when the surroundings are at rest or the external pressure gradient is tailored so as to maintain a constant velocity ratio. It follows that a jet in a constant velocity stream cannot be exactly self-preserving; it may, however, under certain conditions, be approximately so. Dimensional arguments founded on the concept of a moving equilibrium, the "loss-of-memory" hypothesis [15–16], suggest that some way downstream of the jet origin the characteristics of the flow depend only on the excess momentum, M_0, and not on the conditions at the nozzle or the ratio of exit to free stream velocity. The experimental data are correlated when the jet width and distance from the origin are scaled by the momentum thickness defined by:

$$\theta = (M_0/\rho U_1^2)^{0.5} \tag{19}$$

When the jet develops at uniform pressure in a constant-velocity external stream θ is constant and given by:

$$\frac{\theta}{d} = \left[\frac{\pi}{4} \cdot \frac{\lambda_1 + 1}{\lambda_1^2}\right]^{1/2} \tag{20}$$

where d is the jet orifice diameter and λ_1 is, as before, the ratio of the external stream velocity to the initial excess velocity. Figure 3, adapted from Reference 14, shows free-jet data from different sources correlated with θ. The velocity decay data are reasonably well fitted by the expression:

$$\lambda \equiv U_1/u_0 = x/7.9\theta \tag{21}$$

in which the constant has been chosen to accord with the measured spreading rate of a round jet in still surroundings [14]: $r_{1/2} = 0.086x$. Also shown in Figure 6 is the jet width calculated from Equations 17 and 19, when λ is given by Equation 21 and δ is replaced by $r_{1/2}$,

$$\theta/r_{1/2} = [2\pi(I_2 + \lambda I_1)]^{1/2}/\lambda \tag{22}$$

with the profile integrals I_1 and I_2 evaluated from Equation 18 and assumed to be invariant to x. The good agreement with the measurements confirms the validity of this assumption and indicates that the mean velocity distribution retains approximately the same functional form. It is seen that the spreading rate is far from constant and may only be regarded as approximately so when the external stream velocity is low. This point is discussed in more detail in Reference 1, where it is

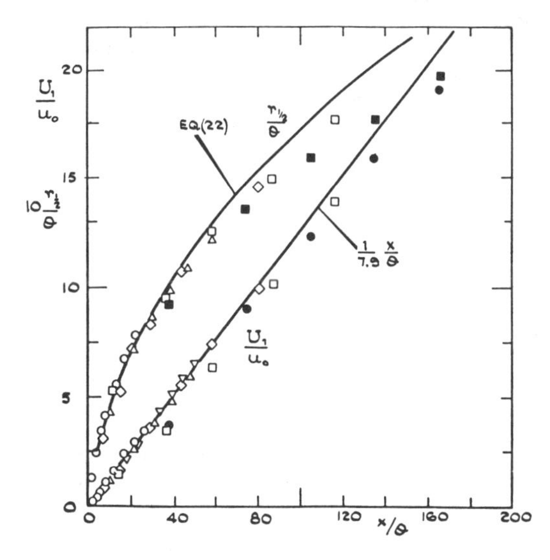

Figure 3. Growth and velocity decay of a round jet in a uniformly moving stream. (Data from various sources cited by Rodi [14], reproduced by permission of Academic Press Ltd., London.)

concluded that the use of the self-preservation hypothesis for the calculation of the mean field of these flows is limited to regions in which the velocity ratio λ does not exceed about one or two.

A number of methods for calculating constant-pressure jets are reported in the literature; these are relevant in the present context because they have been applied, with appropriate modifications, to the prediction of confined flows with pressure gradients and an external potential stream: region 2 of Figure 1. The principal difficulty in turbulent-flow calculation generally is that the time-averaged Reynolds equations do not form a closed set: information has been lost in the averaging process and a "closure" assumption is needed to express the turbulent Reynolds stresses in terms of known variables. There are numerous ways of closing the equation set at different levels of complexity. The equations themselves are converted into ordinary differential equations by means of similarity assumptions or solved directly by finite-difference techniques. A recent review of current methods is given in Reference 17.

Early calculations of the properties of free jets, reported in Reference 18, used the self-preservation concept to simplify the equations of motion which were then solved for the form of the mean velocity profile with the aid of the mixing length hypothesis for the eddy viscosity. An alternative and more generally applicable approach is the integral-profile method based on the observed invariance of the velocity profile function f to the external conditions. When the integrals I_1 and I_2 are evaluated for a preselected profile, the condition of overall momentum conservation expressed in Equation 22 relates local values of $r_{1/2}$ and λ. Another relationship is needed to determine the axial variation of these two quantities. It may be obtained from the differential Equation 14 in a number of ways.

Squire and Trouncer [19] used the cosine profile Equation 18 to integrate the momentum equation between the centerline and $r_{1/2}$, using the mixing length hypothesis to relate the shear stress to the local mean velocity gradient. They also dealt in a similar way with the mixing layer at the initial potential core. The method of Hill [1] consists of multiplying the momentum equation by y^2 and then integrating across the whole flow to form a moment-of-momentum integral equation which contains the integral of the shear stress distribution which is then evaluated from free ($\lambda = 0$) jet velocity measurements. The flow field is considered to be large enough for the jet core region to be replaced by a point source in the plane of the nozzle. Predictions of the mean field of round jets in moving streams obtained by this method agreed tolerably well with the measurements available at that time [20, 21] as long as the velocity ratio λ was not so large as to invalidate the assumption of approximate self-preservation. Although integral profile methods can give good results for the mean flow in jets, and have the merit of being relatively simple, they are limited in their range of applicability and by their inability to predict details of the turbulent field, such as are desired in combustion applications. The more complicated transport equation methods are more generally applicable and capable of providing more information. Typical results of calculations of this type for free shear flows are described in References 22 and 23 and for the free round jet in Reference 24. The relative performance of different integral-profile and transport equation methods in predicting the development of jets in coflowing streams is assessed in References 25 and 26.

In concluding this section reference should be made to some of the relevant experimental data from round free jets. A number of these [27–31] are reviewed in Reference 14; more recent measurements are reported in References 32 and 33, and a detailed discussion of the turbulence structure in jets is given in Reference 13. Earlier measurements made in round jets in moving streams [2, 20] have been widely used to verify theoretical hypotheses, but in the former case at least, the flow was influenced by the duct which guided the outer stream [14]. In References 20 and 33 the jets, which were heated, had a density lower than that of the surrounding stream. Entrainment measurements [34] reveal the rate of spread of a light jet in a heavy environment is greater than that of a jet in a fluid of the same density. The effects of the density difference are accounted for in correlating the data, and in an integral profile calculation method [1], by including the density ratio inside the square root in the formulation of the momentum thickness (Equation 20).

The approach of a free coaxial jet to a developed self-preserving state is generally faster than that of a single round jet. The investigations reported in References 35 and 36 for coaxial nozzles with thick separation walls show that the spread of the near-field mixing layer is strongly influenced by vortex shedding from the nozzle lip. These results may be compared with the near-field measurements of References 37 and 38 made in coaxial jets separated by thin walls.

SIMILARITY CRITERIA FOR DUCTED JETS: THE CRAYA-CURTET NUMBER

This section and the following one deal with ducted flows in which the outer coaxial stream is entrained by a central jet, the situation shown schematically in Figure 1. It has been observed that the main feature that distinguishes this flow from that in a free jet is that the confinement produces an adverse pressure gradient which affects the entrainment and spreading rates up to the point where the central jet attaches to the wall. For conditions where the central jet is approximately self-preserving in a coaxial potential flow, and provided that the entrance streams are clean with thin boundary layers and low free-stream turbulence level, the flow in a constant-area duct may be described in terms of a single similarity parameter. The following analysis analogous to the treatment of a jet in an unconfined free stream is due to Hill [1]. The total impulse per unit area of the duct is defined by

$$M = \frac{2}{A} \iint (p^* + \rho U^2)\, dA \tag{23}$$

where

$$p^* + \tfrac{1}{2}\rho U_s^2 = 0 \tag{24}$$

and the total mass flow rate per unit area is

$$m = \frac{1}{A} \iint \rho U\, dA = \overline{\rho U} \tag{25}$$

Consideration of the equation of motion combined with the assumption of approximate self-preservation then shows that the behavior of the confined jet can be expressed as a function of only two independent variables: x/D and ζ, the dimensionless ratio of the mass flow rate to the square root of the specific impulse. This quantity can be evaluated in the plane of the nozzle for uniform primary and secondary streams:

$$\zeta \equiv \frac{m}{(M\rho)^{0.5}} = \frac{\lambda_1 + A_p/A}{[\lambda_1^2 + 2(1 + \lambda_1)A_p/A]^{0.5}} \tag{26}$$

An alternative, but simply related, parameter which has been widely used is the Craya-Curtet number [39–42]:

$$Ct = \zeta[2/(1 - \zeta^2)]^{0.5} \tag{27}$$

The numerical value of ζ is normally between zero and unity, and of Ct between unity and infinity. A value of zero ζ signifies no mass flow in the duct; a value approaching unity means that the primary-to-duct-area ratio is negligible and that consequently the influence of the walls is insignificant.

Two values are of particular interest in describing the nature of the flow. Substitution of $\lambda_1 = 0$ in Equation 26 produces the value for zero secondary stream entrainment:

$$\zeta = (A_p/2A)^{0.5}; \qquad Ct = (A/A_p - 0.5)^{-0.5} \tag{28}$$

or Ct approximately equal to the nozzle/duct diameter ratio when this is small. Recirculation occurs when the dimensionless parameters fall below critical values variously reported as $\zeta < 0.45$, $Ct < 0.7$ [1], $Ct < 0.8$ [40], $Ct < 0.9$ [43].

The usefulness of the Craya-Curtet number or its equivalent in describing the nature of the flow in a constant-area duct is shown graphically in Figure 4, which has been reproduced from Reference 43. The diagrams show the observed streamline patterns, profiles of mean velocity, and the

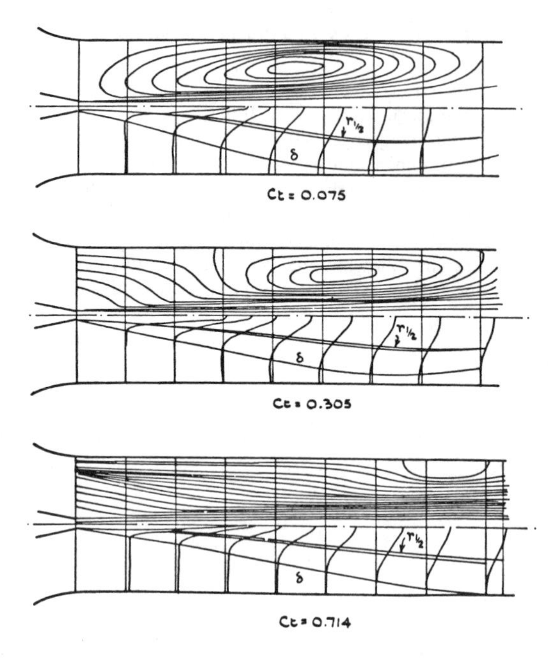

Figure 4. Streamlines, velocity profiles, and widths of jets mixing in a duct for different Craya-Curtet numbers. (From Barchilon and Curtet [43] by permission of the American Society of Mechanical Engineers.)

locations of the edge of the jet and the half-velocity radius for different values of Ct. The recirculation eddy, which first appears when Ct is slightly less than 0.976 (a flow condition which is not illustrated here) increases in extent as the relative mass flow rate of the entrained stream, and consequently the Craya-Curtet number, diminishes. At Ct = 0.075, consistent with the result given by Equation 28 when the diameter ratio of 6:80 is substituted, the secondary stream vanishes and the recirculation eddy fills the entire region between the nozzle and the point of jet attachment. It will be seen also that the width of the main jet, shown by the loci of $r_{1/2}$ and δ, grows nearly linearly with x when Ct is large, but is highly nonlinear in x when Ct is small.

The dimensionless similarity parameter Ct is sufficient by itself to describe the behavior of confined-jet flow only when the length of the initial region with the potential core is short compared with the total length of jet development, i.e., when the jet can reasonably be considered as emanating from a point source. When point-source behavior is not sufficiently approximated, as may be the case when the nozzle/duct diameter ratio is not sufficiently small, the diameter ratio must be introduced as a second parameter.

DEVELOPMENT OF CONFINED JETS IN OPEN DUCTS

Quantitative details of the development of a round jet in a confined coaxial stream are illustrated in Figures 5 through 8 which are reproduced from Reference 41. The "open-duct" type of experimental arrangement was that of Figure 1: the duct diameter was 19.7 cm, the primary nozzle diameter 0.635 cm, and the central jet flow velocity 130 m/s. Figure 5 shows the axial growth of the velocity half radius for different values of Ct consistent with the flow patterns illustrated in Figure 4. The spreading rate close to the nozzle exit increases with decreasing Ct (Figure 4 shows that for low Ct the rate decreases considerably downstream). For Ct > 0.7 approximately the growth is nearly linear; for values of Ct less than about 0.7 recirculation occurs and the growth increases and becomes noticeably nonlinear. Ct = 0.033 is the condition (Equation 28) for zero secondary stream induction when the central jet is surrounded by an extensive region of recirculation. The curve Ct = 0.673 is identical with that for a free jet in still surroundings: r = 0.084 x. The axial decay of the centerline excess velocity is shown in Figure 6 where values of u_0 have been

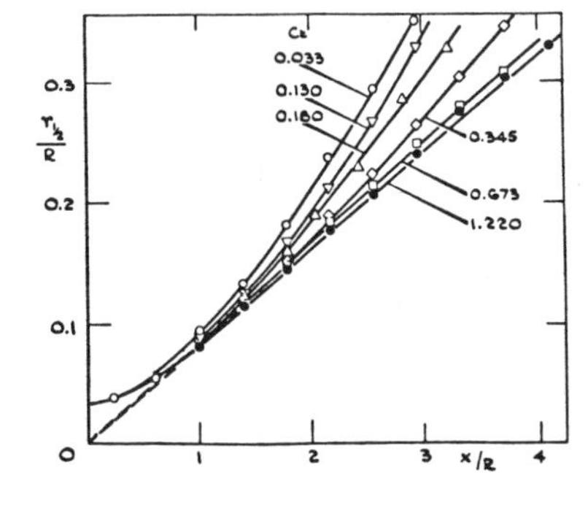

Figure 5. Axial growth of the half-velocity radius for different Craya-Curtet numbers. (From Becker et al. [4] by permission of the Academic Press Ltd., London.)

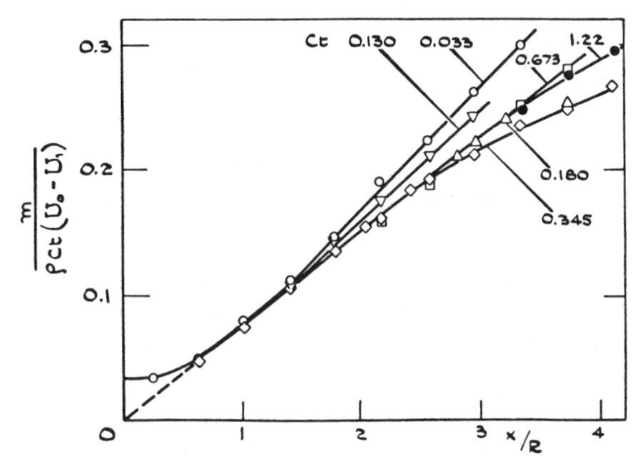

Figure 6. Axial decay of the excess velocity on the centerline. (From Becker et al. [41] by permission of the Academic Press Ltd., London.)

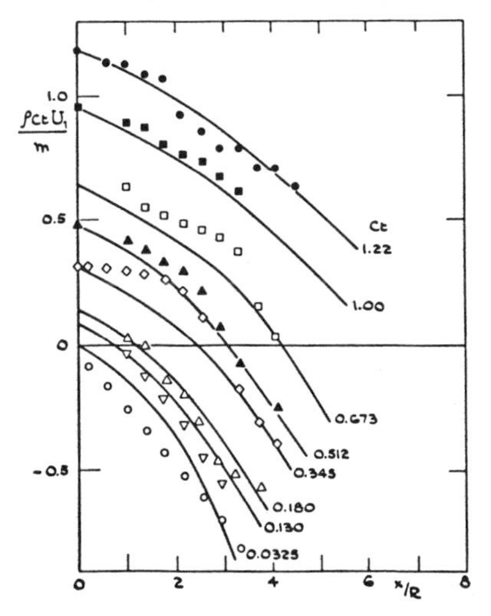

Figure 7. Axial variation of the velocity of the secondary stream. (From Becker et al. [41] by permission of the Academic Press Ltd., London.)

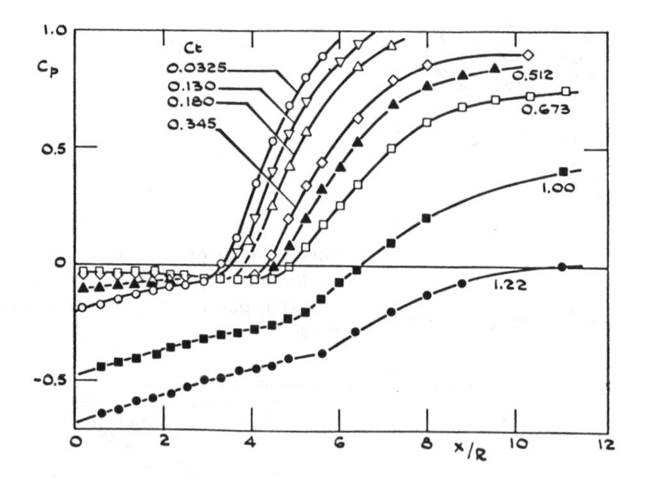

Figure 8. Axial variation of static pressure measured at the duct wall; $C_p = (p_w - p_1)Ct^2/\rho\bar{U}^2$. (From Becker et al. [41] by permission of the Academic Press Ltd., London.)

normalized by (mean duct velocity)/Ct. The curve for Ct = 0.673 is again identical with that for a free jet. The appearance of reversed flow is seen in Figure 7 which shows the axial variation of the secondary stream velocity: negative values occur for Craya-Curtet numbers less than about 0.7. The axial variation of static pressure measured at the wall, which is plotted in Figure 8, shows the presence of two distinct mixing regimes: the jet mixing zone and the succeeding zone of dying gradients in the mean velocity. The points at which the curves change shape may be identified approximately with the downstream locations where the secondary stream vanishes and entrainment ceases.

The main features of the recirculation eddy are shown schematically in Figure 9 which is reproduced from Reference 43. The extent of the eddy is marked by two zero-velocity points at the duct wall (N upstream and P downstream) and by a cross section containing the recirculation center or eye C. A recirculation flow can be defined by the integral of the negative velocities across

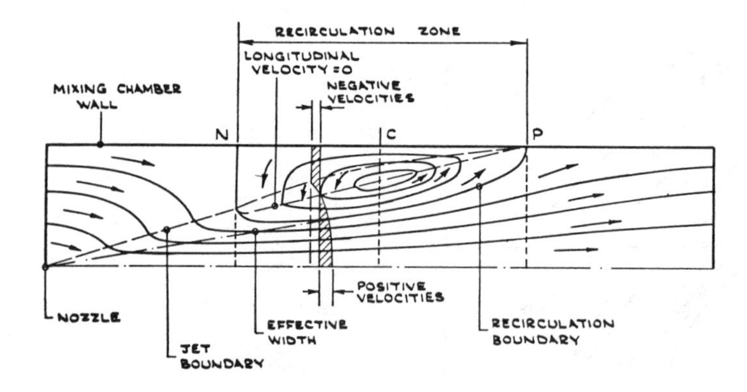

Figure 9. Details of the recirculation for a ducted jet. (From Barchilon and Curtet [43] by permission of the American Society of Mechanical Engineers.)

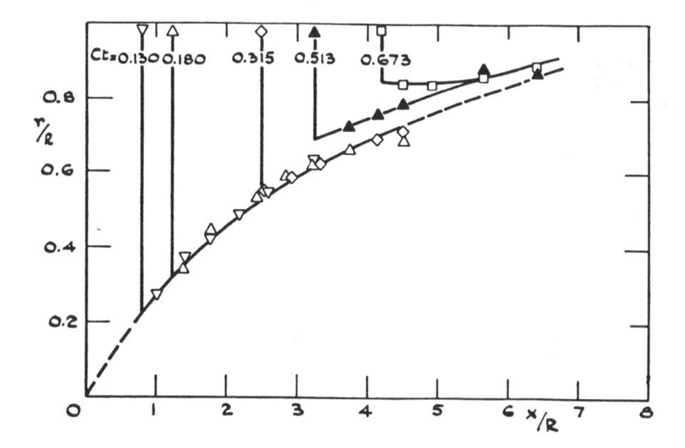

Figure 10. Mean boundaries of the recirculation on which U = 0. (From Becker et al. [41] by permission of Academic Press Ltd., London.)

the cross section. From zero at cross-section N, this recirculation flow rate increases to reach a maximum at cross-section C, then to fall to zero again at P. Quantitative details of the flow in the recirculation are given in References 41 and 43 for the flows of Figures 4 and 5 through 8, respectively. Figure 10 [41] shows the mean boundaries of the recirculatory flow defined as the surfaces on which the velocity is zero. The curves when extrapolated locate the downstream point P at approximately x/R = 8.5 for all values of Ct; this result, and the location of the upstream point N, are in broad agreement with the findings reported in Reference 43 with which they are compared for different values of Ct in Figure 11.

Detailed mean flow and turbulence measurements in confined jet flow over a range of conditions with small recirculation are reported in Reference 3. The axial pressure distribution for the case of a duct/nozzle diameter ratio of 3 have been reproduced in Figure 2; Figure 12 shows the development of the mean velocity profile with axial distance when $\lambda_1 = \frac{1}{9}$, Ct = 0.65. Profiles C and D show that the duct wall boundary layer separated in the adverse pressure gradient and the potential core has disappeared upstream of profile C, taken at x = 10d. The longitudinal turbulence intensity profiles measured at different axial locations and plotted in Figure 13 show high levels in the high shear region between the main jet and the entrained secondary flow and in the recirculation region. These levels remain high in the wall layer downstream of reattachment; in the central region the

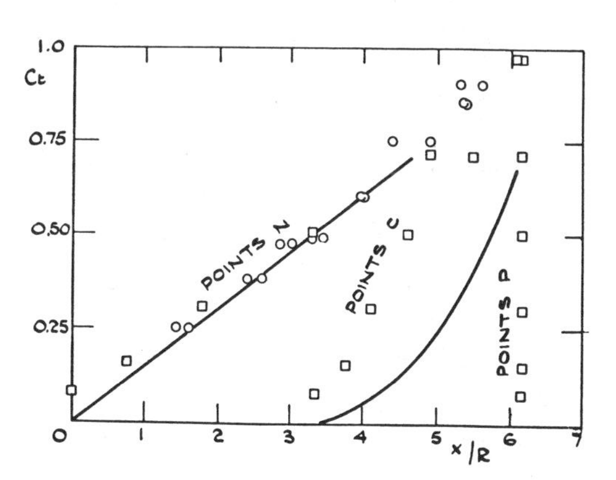

Figure 11. Location of points N, C, P of the recirculation eddy (Figure 9) as functions of the Craya-Curtet number. Points are data of Barchilon and Curtet [43]; lines deduced from measurements of Becker et al. [41].

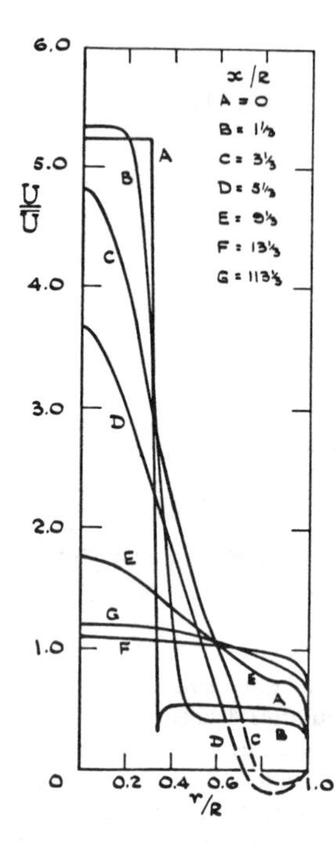

Figure 12. Mean velocity profiles measured in an open duct for Ct = 0.65. (From Razinsky and Brighton [3] by permission of the American Society of Mechanical Engineers.)

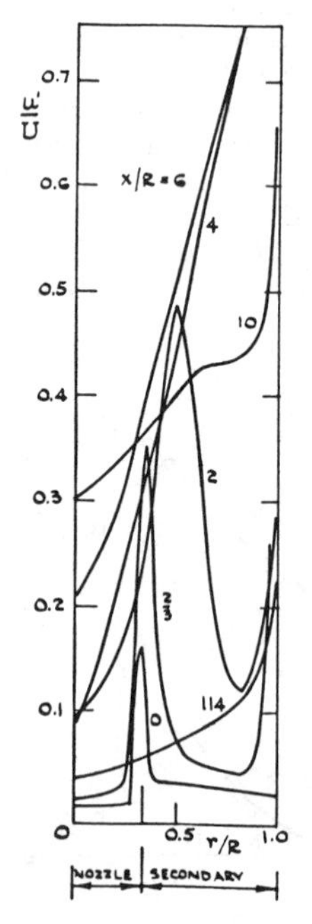

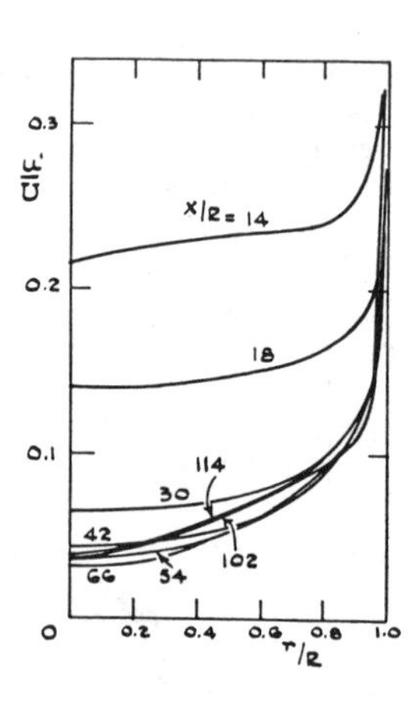

Figure 13. Evolution of longitudinal turbulent intensity distributions in open duct jet mixing for Ct = 0.65. (From Razinsky and Brighton [3] by permission of the American Society of Mechanical Engineers.)

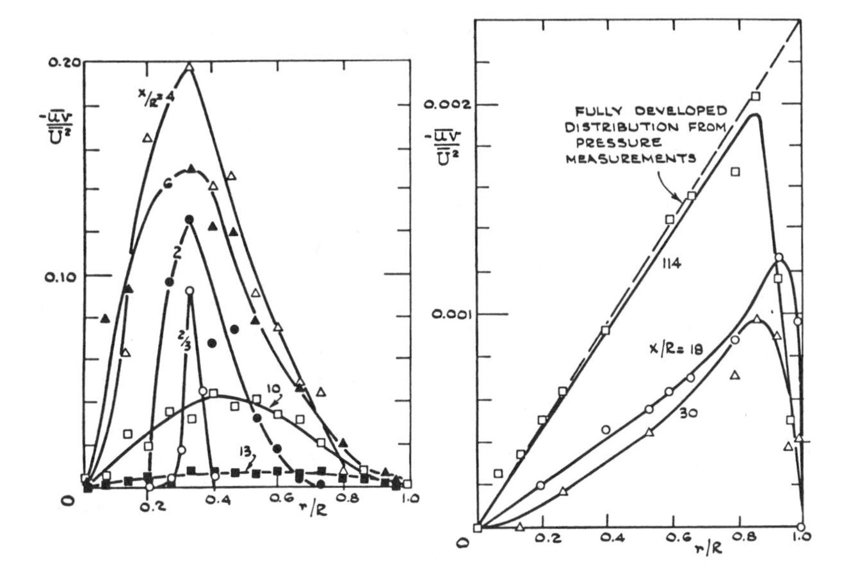

Figure 14. Evolution of turbulent shear stress distributions in open duct jet mixing for Ct = 0.65. (From Razinsky and Brighton [3] by permission of the American Society of Mechanical Engineering.)

decay to fully developed pipe flow levels is slow and arrived at for x > 42R. Profiles of the corresponding turbulent shear stresses are plotted in Figure 14. The highest levels are found in the high shear region where the jet mixes with the surrounding flow and these reach a peak just downstream of the potential core zone. The data also show the slow recovery to fully developed pipe flow. Turbulence measurements are also reported in Reference 44 from a ducted coaxial jet flow with the outer stream moving faster than the central jet.

More recent measurements are reported in Reference 45. These were made in an "open," diverging duct of a 5° included angle. In the plane of the nozzle the duct/nozzle diameter ratio was 10. The results were compared with those obtained in an equivalent constant-area duct. It was found that the divergence had an important effect on the flow especially when the inlet conditions led to recirculation which occurred for Ct = 1.1 compared to a value of 0.9 without divergence. The recirculation eddies extended further upstream, and had a larger volume flow, the pressure rise was steeper and higher, and the jet expanded faster than in the constant-area duct with the same initial conditions. Two points are particularly interesting:

1. The turbulence data did not confirm the similarity assumption in that the downstream evolution of the intensity depended on Ct and differed from that of the free jet, and the shear stress displayed a complete lack of similarity.
2. The measurements did not reveal much greater turbulence intensity than in a free jet, which could simply account for the larger entrainment and faster spreading rate observed in the confined flows.

COAXIAL JETS IN ENCLOSED DUCTS

This section deals with "enclosed-duct" flows in which secondary stream entrainment is physically prevented by the presence of a wall in the plane of the nozzle. The discharge of a coaxial jet without swirl into a sudden expansion is qualitatively similar to the discharge of a single jet, differing from it mainly in the changes in mixing and spreading rates due to enhanced turbulence activity resulting from high mean shear at the interface between the two coaxial streams. In unconfined flow, however, the individuality of the two streams is not maintained and at relatively short distances from the nozzle the dual jet assumes the character of a single, round jet. In combustion applications the velocity of the annular jet usually exceeds that of the center jet; if the momentum flux of the latter is small, a region of recirculating flow can appear on the axis [46].

The flow of a single jet into an enclosed duct, or in a sudden pipe expansion, has received much attention experimentally and analytically and it has become an important test case for flow calculation methods. A useful short survey and a general discussion of the data are given in Reference 47; the application of an integral-profile flow calculation method is described in Reference 48, and a more general method based on the numerical solution of modelled transport equations in Reference 49. In these more complex flows the similarity criteria which proved useful for data correlation and prediction of coflowing jets in open ducts are of limited value.

It has been noted in the foregoing discussion of free jets that the initial evolution of a coaxial jet is strongly influenced by the presence of a high shear mixing zone between the two jet streams on leaving the nozzle. The use of swirl in the outer jet further accelerates mixing and may produce a strong recirculation on the axis. A large and varied literature exists on the subject of flows of this type in combustion chambers, with experiments made in isothermal flow and with combustion. Work in this field to 1977 is summarized in Reference 50; subsequent work includes that of Whitelaw and his coworkers [35, 36, 51–53], who have examined the influence of the confining duct on the evolution of the mean flow and the turbulence structure by comparative measurements in free and confined coaxial jets, and Johnson and Bennet [54].

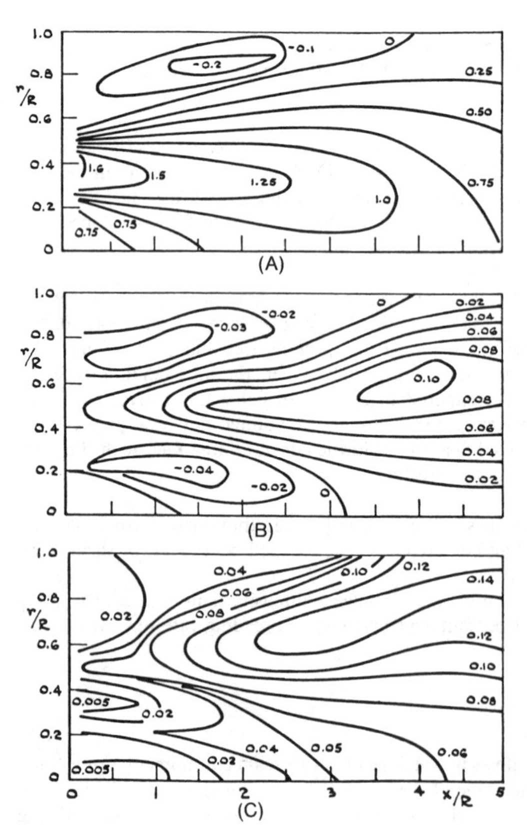

Figure 15. Contours of (A) axial mean velocity U m/s, (B) radial mean velocity V m/s, and (C) turbulent kinetic energy (m^2/s^2) in coaxial jet mixing in an enclosed duct configuration. (From Johnson and Bennett [54] by permission of United Technologies Inc.)

Although the main features of these flows emerge fairly clearly from the measurements, they are so complex and the details depend on numerous parameters that only a general description can be given. Figure 15 [54], shows contours of the axial and radial components of mean velocity and of turbulent kinetic energy, all obtained from laser-Doppler measurements in a water flow rig. The inner nozzle, annulus, and duct radii were 15.3, 29.5, and 61.0 mm respectively; the central jet and annulus flow velocities were initially 0.53 and 1.67 m/s. The inner/outer velocity ratio of about one-third is representative. The extent of the recirculation region is marked approximately by the $U = 0$ contour extending to four duct radii from the nozzle plane where the two streams have mixed to such a degree that the maximum velocity appears on the axis (the radial velocity contours showing inward flow upstream of the attachment point as fluid in the slow inner jet is accelerated). Downstream of the attachment only positive values of V are recorded as the combined flow fills the duct and the mean streamlines incline away from the axis. Peak values of the turbulent kinetic energy occur as expected in the vicinity of the mixing layer between the two jets and increase with distance downstream to a maximum near the attachment plane.

In the enclosed duct flow described in References 52 and 53 the inner and outer diameters of the coaxial nozzle were 16.1 and 21.6 mm, respectively and the duct diameter was 44.5 mm. Figure 16 [53] shows the effect of coaxial-jet mixing on the combined flow development. The comparison is made between the flow with an initial outer/inner velocity ratio of three and one where the velocities of the two streams were equal, thus resembling the flow of a single jet except for the effects on the near field of the annular wake from the separating wall. A second comparison shown in the figure is between the LDA measurements of Reference 53 and the earlier hot-wire data of Reference 52. The distributions of mean velocity on the axis show relatively rapid acceleration of the inner stream; the maximum value occurs at about 1.5 duct diameters from the nozzle, halfway to the combined jet attachment point at about 3.25 diameters, or relatively sooner than appears to be the case for the flow of Figure 15. The high shear of the coaxial jet produces substantially higher turbulence levels on the centerline when expressed in absolute terms; the intensity as a fraction of the local mean velocity is about the same in each case. It appears from these two sets of results that the extent of the recirculation is only marginally affected by the velocity ratio but that the negative velocities inside it are substantially greater when the coaxial jet velocities are initially unequal.

Figure 17 [53] shows the effects on this flow of swirl in the outer jet. As previously observed for the corresponding unconfined flow [36], the maximum mean velocity now occurs closer to the jet exit plane, but this peak is followed by a rapid decay to near zero value on the centerline as the

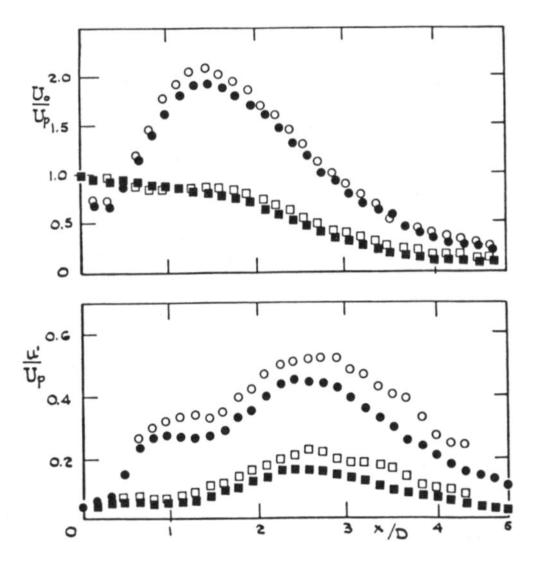

Figure 16. Axial variation of the decay of centerline velocity and longitudinal turbulent intensity in an enclosed coaxial jet flow. ○ ●: $U_s/U_p = 3.0$; □ ■: $U_s = U_{pN}$. Open symbols: LDA measurements; closed symbols: hot-wire data. (From Habib and Whitelaw [53] by permission of the American Society of Mechanical Engineers.)

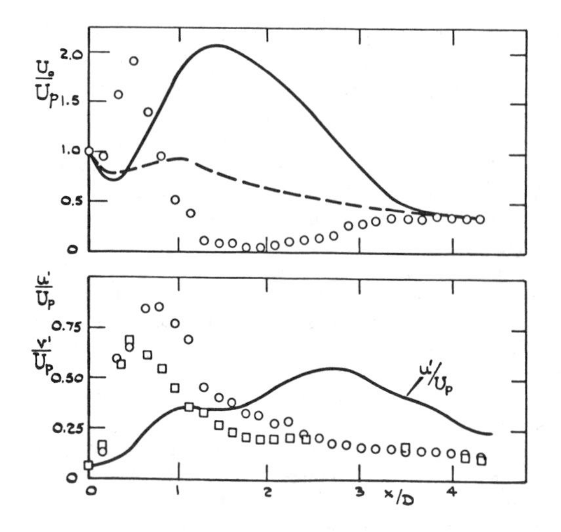

Figure 17. Effect of secondary stream swirl on the decay of mean velocity and turbulent intensities on the centerline of an enclosed coaxial jet flow. ○: u′; □: v′, w′; ———— u′ for zero swirl (Figure 16); — — — — swirling free jet. (From Habib and Whitelaw [53] by permission of the American Society of Mechanical Engineers.)

initial rate of spreading is increased. The axial distributions of the two components of fluctuating velocity are similar in form with the circumferential component slightly smaller; these distributions are qualitatively similar to those of the unconfined jet. The peak values are also located closer to the exit plane and are considerably increased by the high circumferential shear resulting from the swirl. Radial distributions of the two mean velocity components and turbulence quantities obtained in this flow are consistent with changes in the spreading rate. The profiles of axial velocity are nearly uniform at four duct diameters from the nozzle; in the near region the introduction of swirl produces rapid changes in profile shape so that the near zero value on the axis at about 1.5 duct diameters is accompanied by high velocities in the outer part of the flow. The distributions of swirl velocity show similar trends: an initially rapid change is succeeded by nearly solid body rotation for $x > 1.5D$.

CALCULATION OF CONFINED COAXIAL JETS

The simplest approach to the calculation of confined coaxial jets is through the integral momentum equation with assumed similar profiles for the mean velocity and Reynolds stresses. Brief reference has previously been made to the use of integral profile methods in connection with the development of free jets; the procedure for confined flow is essentially the same with the additional constraint imposed by the condition of continuity of mass flow in the duct and consequential flow development in a pressure gradient. The method of Hill [1] is representative and can be described briefly; it is appropriate for constant-area open-duct configurations with an entrained secondary stream and limited recirculation. The flow field is assumed to be large enough for the potential core region close to the nozzle to be replaced conceptually by a virtual source located approximately in the inlet plane. Three principal flow regions are identified as in Figure 1 and are treated differently. Where the secondary stream may be considered as a uniform potential flow in an axial pressure gradient, the integral momentum equation is solved by substituting jet similarity profile functions f and g (Equation 13) obtained from free jet data, the external conditions being determined from the condition of mass conservation and the use of Bernoulli's equation in the assumed potential flow. If a region of reversed flow appears, for the reasons discussed previously, the potential flow assumption for the external stream breaks down; the static pressure is then assumed to be constant throughout the recirculation, the jet similarity profiles being retained. In the third region, when the jet has spread to the walls of the duct, the assumption of approximate similarity is no longer valid as the profiles change shape. Use is then made of the gradient diffusion hypothesis [13] for the turbulent shear stress:

$$-\overline{uv} = \nu_t \, \partial U/\partial r \tag{29}$$

where the variation of the eddy viscosity ν_t is assumed to be similar to that in a self-preserving free jet.

The results of calculations [1] show that the axial pressure distributions in constant-area open ducts are predicted accurately for flows in which $0 < \zeta < 0.73$, where ζ is the similarity parameter defined by Equation 26, and that it is possible to have a region of recirculation for $\zeta < 0.45$, $Ct < 0.7$, a prediction in broad agreement with the observed behavior discussed earlier. For values of ζ greater than 0.6, however, the jet-spreading becomes decidedly nonlinear which is not in accord with the idea of approximate self-similarity or with the observed behavior. Comparison with the jet width measurements of Reference 41, which are reproduced in Figure 5, show that these are tolerably well predicted in the middle ranges of ζ. For $\zeta = 0.023$ ($Ct = 0.033$) the spread is under-estimated by about 8%, while for $\zeta = 0.65$ ($Ct = 1.22$), which is outside the range of linear growth, the discrepancy exceeds 20%. Similar calculations of the flow in converging-diverging open ducts are reported in Reference 55. The treatment of Razinsky and Brighton [56] differs only in detail from that just described, but two improvements are introduced: the inclusion of the short potential core region extends the range of the overall analysis to geometries of low duct/nozzle diameter ratios for which point-source behavior becomes increasingly unrealistic, and the growth of the duct wall boundary layer is also taken into account. The analysis due to Craya and Curtet [39, 40] is of fundamental interest in that it identifies the most common of the similitude parameters, Ct. The initial region is again neglected, implying point-source behavior, and the calculation entails the solution of a series of differential equations formed by multiplying the momentum equation by a general function and then integrating in the radial direction. However, Reference 40 deals mainly with the two dimensional flow between plane walls.

The limitations of integral-profile methods have been noted in foregoing sections; although capable of producing useful information for restricted classes of flow, they have now been largely superseded by finite-difference transport equation methods. A detailed description of these methods is inappropriate here and will not be attempted (see [17, 23, 57, 58] for this). The central problem is to "close" the time-averaged equations for turbulent flow by relating in some way the Reynolds stresses to known (mean flow) quantities in the calculation. The equations are then solved numerically over the flow field. Closure at the simplest, mean field level usually involves an eddy viscosity (Equation 29) which is then specified as a function of the flow dimensions, or by means of an assumed mixing-length distribution, or as is now most common, as a function of turbulence quantities such as the kinetic energy, k, and the energy dissipation rate, ϵ, which are themselves dependent variables of transport equations derived from the Navier-Stokes equations. Higher-order, turbulence, closures involve transport equations for the second moments: the Reynolds stresses and the turbulent heat fluxes.

Hendricks and Brighton [59] have used Spalding's [60] two-equation turbulence closure to calculate the effects of swirl on the development of a confined jet with an entrained secondary stream in a constant-area open duct. The eddy viscosity is here expressed in terms of the turbulent kinetic energy and vorticity. Figure 18 and 19 show typical results compared with the measurements reported in Reference 61. The effects of swirl are, as might be expected, to decrease the axial length required for mixing, to accelerate the decay of centerline velocity, and to locate the point of maximum pressure closer to the inlet plane; the calculations reproduce these trends reasonably well. The initial turbulence levels of the two streams were found to have significant effects on the velocity and pressure distributions. An interesting feature of these calculations is the relatively early use of a streamline curvature correction to the mixing length which is based on the buoyancy analogy [62]. It may be noted in passing that the calculation of swirling jets is fraught with difficulties which are described in Reference 24 and have yet to be satisfactorily overcome. A similar turbulence model has also been used by Pai et al. [63] to calculate the nonswirling isothermal coaxial flow in an open duct cement kiln configuration. Figure 20 [63] shows fair agreement between measured and calculated streamline patterns for a velocity ratio, λ, of 0.0226 and $Ct = 0.445$. The calculated flow development was found to be sensitive to the mean flow inlet conditions: the

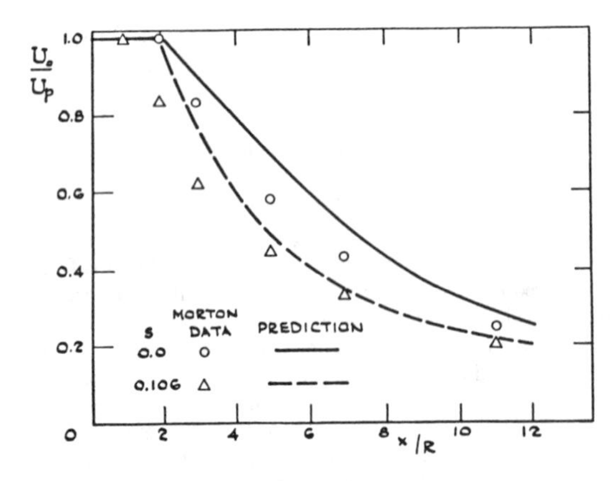

Figure 18. Measured and predicted effects of swirl on the decay of mean velocity on the centerline. Data of Morton [61]. (From Hendricks and Brighton [59] by permission of the American Society of Mechanical Engineers.)

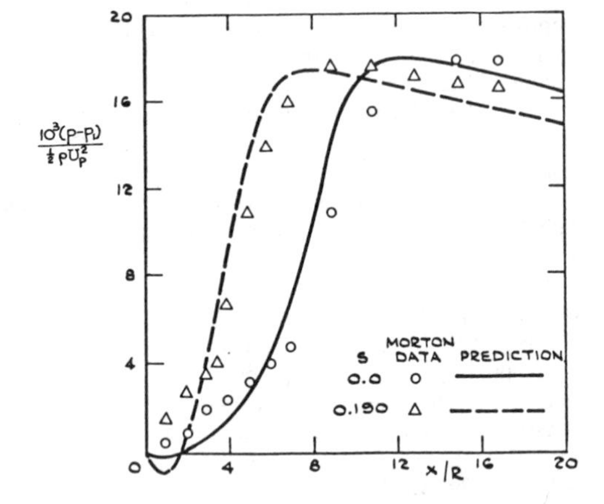

Figure 19. Measured and predicted effects of swirl on the axial pressure distribution. Data of Morton [61]. (From Hendricks and Brighton [59] by permission of the American Society of Mechanical Engineers.)

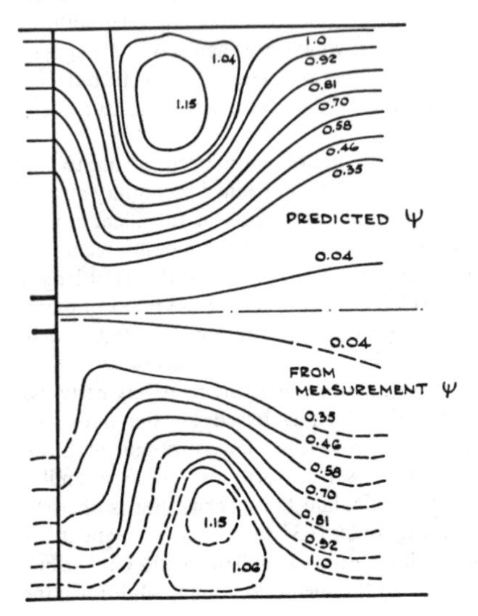

Figure 20. Comparison of measured and predicted streamline patterns in open duct jet mixing for Ct = 0.445. (From Pai et al. [63] by permission of the Institute of Energy, London.)

velocity profile assumed for the primary jet. On the whole the results appear to be consistent with the approximate theories [40, 64]. Similar results [49] have been obtained using a k-ϵ closure model.

The results of an early attempt to calculate the recirculatory flow with swirl in an enclosed duct are described in Reference 65. More recent and improved calculations, which may be taken as representative of current practice, are described in References 52, 53, and 66; the review by Lilley [50] cities a number of other works dealing mainly with reacting and swirling flows. The calculation of recirculating flows requires the numerical solution of elliptic equations, usually by iteration. Because this is an expensive process in terms of computer time and capacity, a coarse finite-difference grid is used and the number of differential equations to be solved is kept to a minimum. Simplifications of the turbulence model are justifiable when the main features of the flow are determined mainly by the pressure field rather than the turbulence structure, but it is difficult to eliminate numerical errors associated with the use of coarse grids. k-ϵ model calculations reported in Reference 52 are extended in Reference 53 to cover the effects of swirl in the secondary annular flow; the results are compared with the measurements described in the previous section. Solutions of modeled transport equations for the Reynolds stresses which are reported in Reference 66 show only marginal improvements on the mean-field closure methods.

NOTATION

- A duct cross-sectional area
- A_p nozzle cross-sectional area
- A_s $A - A_p$
- a area ratio A_s/A_p
- Ct Craya-Curtet similarity parameter, Equation 27
- D duct diameter 2R
- f velocity profile similarity function, Equation 13
- g shear stress profile similarity function, Equation 13
- I velocity profile integral, Equations 16 and 17
- M_0 excess momentum flux
- M total impulse per unit area, Equation 23
- m total mass flow rate per unit area, Equation 25
- P total pressure
- p static pressure
- R duct radius
- r radius
- $r_{1/2}$ radius to half-velocity point
- U axial component of mean velocity
- $\bar{U}$ average mean velocity in duct
- U_1 mean velocity of secondary stream
- U_p initial nozzle velocity
- U_0 mean velocity on duct axis
- U_s secondary stream velocity in plane of nozzle
- u' longitudinal component of fluctuating velocity
- u_0 excess mean velocity $U_0 - U_1$
- V radial component of mean velocity
- v' radial component of fluctuating velocity
- x axial distance

Greek Symbols

- δ jet width
- ζ similarity parameter, Equation 26
- η dimensionless radius r/δ
- θ jet momentum thickness, Equation 19
- λ velocity ratio $U_1/(U_0 - U_1)$
- μ velocity ratio U_s/U_p
- ρ fluid density
- τ_w shear stress at the wall
- Φ thrust augmentation ratio
- ψ stream function

Subscripts

- 1 conditions in the nozzle plane
- w conditions at the wall
- ∞ atmosphere or reference conditions

REFERENCES

1. Hill, P. G., "Turbulent Jets in Ducted Streams," *J. Fluid Mech.*, 22 (1965), pp. 161–186.
2. Forstall, W., and Shapiro, A. H., "Momentum and Mass Transfer in Coaxial Gas Jets," *Trans. ASME 72, J. Appl. Mech.* (1950), pp. 399–408.
3. Razinsky, E., and Brighton, J. A., "Confined Jet Mixing for Nonseparating Conditions," *Trans. ASME 93, J. Basic Engg.* (1971), pp. 333–347.
4. Quinn, B., "Recent Developments in Large Area Ratio Thrust Augmentors," AIAA Paper No. 72-1174 (1972).
5. Quinn, B., "Compact Ejector Thrust Augmentation," *J. Aircraft*, 10:8 (1973), pp. 481–486.
6. Fancher, R. B., "Low-Area Ratio, Thrust-Augmenting Ejectors," *J. Aircraft* 9:3 (1972), pp. 243–248.
7. Viets, H., "Thrust Augmenting Ejectors," 1968–9 von Karman Inst. Lecture Series published as Aerospace Research Lab. Rep. No. 75-0224, USAF Wright-Patterson Base, Ohio (1975).
8. Reid, J., "The Effect of a Cylindrical Shroud on the Performance of a Stationary Convergent Nozzle," ARC R & M 3320 (1962).
9. Nagaraja, K. S., Hammond, D. L., and Graetch, J. E., "One-Dimensional Compressible Ejector Flows," AIAA Paper No. 73-1184 (1973).
10. Quinn, B., "Ejector Performance at High Temperatures and Pressures," *J. Aircraft*, 13:12 (1976), pp. 948–954.
11. Quinn, B., "Thrust-Augmenting Ejectors: A Review of the Application of Jet Mechanics to V/STOL Aircraft Propulsion" Fluid Dynamics of Jets with Applications to V/STOL, AGARD CP, 308 (1982).
12. Porter, J. L. and Squyers, R. A. "A Summary/Overview of Ejector Augmentor Theory and Performance," Vought Corp. ATC Rep. R-91100-9CR-47 (1981).
13. Townsend, A. A., "The Structure of Turbulent Shear Flow," 2nd ed. Cambridge (1976).
14. Rodi, W., "A Review of Experimental Data of Uniform Density Free Turbulent Boundary Layers," *Studies in Convection*, Launder, B. E., (Ed.), 1 (1975), pp. 167–222.
15. Spalding, D. B., "Theory of the Rate of Spread of Confined Turbulent Premixed Flames," 7th Symposium (International) on Combustion, Butterworth, London (1958) pp. 595–603. cf. also "A Note on the Axially-Symmetrical Turbulent Jet in a Surrounding Stream," unpublished note, Imperial College (Dec 1958).
16. Bradbury, L. J. S., and Riley, J., "The Spread of a Turbulent Plane Jet Issuing into a Parallel Moving Airstream," *J. Fluid Mech.*, 27 (1967), pp. 381–394.
17. Bradshaw, P., Cebeci, T., and Whitelaw, J. H., "Engineering Calculation Methods for Turbulent Flow," Academic Press, London (1981).
18. Schlicting, H., "Boundary Layer Theory," 2nd ed., McGraw Hill (1968).
19. Squire, H. B., and Trouncer, J., "Round Jets in a General Stream," ARC R & M 1974 (1944).
20. Landis, F., and Shapiro, A. H., "The Turbulent Mixing of Coaxial Gas Jets," *Proc. Heat Transfer & Fluid Mechanics Inst.* (1951), pp. 133–146.
21. Pabst, O., "Die Ausbreitung Heisser Gasstrahlen in Bewegter Luft, I Teil-Versuche im Kerngebeit" Deutsche Luftfahrtforschung U.u.M. 8004 (1944) (Translated Central Air Documents Office).
22. Launder, B. E., et al., "Prediction of Free Shear Flows—A Comparison of the Performance of Six Turbulence Models," Proc. NASA-Langley Conference on Free Turbulent Shear Flows, NASA-SP-312 (1973).
23. Rodi, W., "Turbulence Models and their Applications in Hydraulics—A State of the Art Review," International Association for Hydraulic Research, Delft (1980).
24. Launder, B. E., and Morse, A., "Numerical Prediction of Axisymmetric Free Shear Flows with a Reynolds Stress Closure," Turbulent Shear Flows 1, Springer (1979).
25. Antonia, R. A., and Bilger, R. W., "The Prediction of the Axisymmetric Turbulent Jet Issuing into a Co-Flowing Stream," Aero. Quart. XXV (1974), pp. 69–80.
26. Madni, I. K., and Pletcher R. H., "Prediction of Turbulent Jets in Coflowing and Quiescent Ambients," *Trans. ASME 95, J. Fluids Engg.* (1975), pp. 558–564.

27. Wygnanski, I., and Fiedler, H. E., "Some Measurements in the Self-Preserving Jet," *J. Fluid Mech.*, 38 (1969), pp. 577–612.
28. Reichardt, H., "Zur Problematik der Turbulenten Strahlausbreitung in einer Grundstromung," Mitteilungen aus dem Max-Planck-Institut fur Stromungsforschung, No. 35 (1965).
29. Corrsin, S., and Uberoi, M. S., "Further Experiments on the Flow and Heat Transfer in a Heated Turbulent Air Jet," NACA TN 1865 (1949).
30. Maczynski, J. F. J., "A Round Jet in an Ambient Coaxial Stream," *J. Fluid Mech.*, 13 (1962), pp. 597–608.
31. Hinze, J. O., and Van der Hegge Zijnen, B. G., "Transfer of Heat and Matter in the Turbulent Mixing Zone of an Axially Symmetric Jet," *Appl. Sci. Res.*, A, 1 (1949), pp. 435–461.
32. Antonia, R. A., and Bilger, R. W., "An Experimental Investigation of an Axisymmetric Jet in a Coflowing Airstream," *J. Fluid. Mech.*, 61 (1973), pp. 805–822.
33. Antonia, R. A., and Bilger, R. W., "The Heated Round Jet in a Coflowing Stream," *AIAA J.* 14:11 (1976), pp. 1541–1547.
34. Ricou, F. P., and Spalding, D. B., "Measurements of Entrainment by Axisymmetrical Turbulent Jets," *J. Fluid Mech.*, 11 (1961), pp. 21–32.
35. Durao, D. F. G., and Whitelaw, J. H., "Turbulent Mixing in the Developing Region of Coaxial Jets," *Trans. ASME 95, J. Fluids Engg.* (1973), pp. 467–473.
36. Ribiero, M. M., and Whitelaw, J. H., "Coaxial Jets with and without Swirl," *J. Fluid Mech.*, 96 (1980), pp. 769–795.
37. Ko, N. W. M., and Kwan, A. S. H., "The Initial Region of Subsonic Coaxial Jets," *J. Fluid Mech.*, 73 (1976), pp. 305–332.
38. Ko, N. W. M., and Au, H., "Initial Region of Subsonic Coaxial Jets of High Mean-Velocity Ratio," *Trans. ASME 103, J. Fluids Engg.* (1981), pp. 335–338.
39. Craya, A., and Curtet, R., "On the Spreading of a Confined Jet," *C. R. Acad. Sci.*, Paris, 241 (1955), pp. 621–622.
40. Curtet, R., "Confined Jets and Recirculation Phenomena with Cold Air," *Comb. Flame*, 2:4 (1958), pp. 383–411.
41. Becker, H. A., Hottel, H. C., and Williams, G. C., "Mixing and Flow in Ducted Turbulent Jets," 9th Symposium (International) on Combustion, Academic Press, New York (1963), pp. 7–20.
42. Curtet, R., and Ricou, F. P., "On the Tendency to Self-Preservation in Axisymmetric Ducted Jets," *Trans ASME 86, J. Basic Engg.*, (1964), pp. 765–771.
43. Barchilon, M., and Curtet, R., "Some Details of the Structure of an Axisymmetric Confined Jet with Back-Flow," *Trans. ASME 86, J. Basic Engg.* (1964), pp. 777–787.
44. Kulik, R. A., Leithem, J. J., and Weinstein, H., "Turbulence Measurements in a Ducted Coaxial Flow," *AIAA J.*, 8:9 (1970), pp. 1694–1696.
45. Binder, G., and Khan, K., "Confined Jets in a Diverging Duct," Proc. 4th Turbulent Shear Flows Symposium, Karlsruhe (1983).
46. Owen, F. K., "Measurements and Observations of Turbulent Recirculating Jet Flows," *AIAA J.*, 14:11 (1976), pp. 1556–1562.
47. Johnston, J. P., "Internal Flows," Chap. 3 of *Turbulence*, Bradshaw, P. (Ed.), *Topics in Applied Physics 12*, Springer (1976).
48. Teyssandier, R. G., and Wilson, M. P., "An Analysis of Flow through Sudden Enlargements in Pipes," *J. Fluid Mech.*, 64 (1974), pp. 85–95.
49. Gosman, A. D., Khalil, E. E., and Whitelaw, J. H., "The Calculation of Two-Dimensional Turbulent Recirculating Flows," *Turbulent Shear Flows 1*, Springer (1979), pp. 237–255.
50. Lilley, D. G., "Swirl Flows in Combustion: A Review," *AIAA J.*, 15:8 (1977), pp. 1063–1078.
51. Ribiero, M. M., and Whitelaw, J. H., "Turbulent Mixing of Coaxial Jets with Particular Reference to the Near Exit Region," *Trans. ASME 98, J. Fluids Engg.* (1976), pp. 284–291.
52. Habib, M. A., and Whitelaw, J. H., "Velocity Characteristics of a Confined Coaxial Jet," *Trans. ASME 101, J. Fluids Engg.*, (1979), pp. 521–529.
53. Habib, M. A., and Whitelaw, J. H., "Measured Velocity Characteristics of Confined Coaxial Jets with and without Swirl," *Trans ASME 102, J. Fluids Engg.* (1980), pp. 47–53.

54. Johnson, B. V., and Bennett, J. C., "Mass and Momentum Transport Experiments with Confined Coaxial Jets," Proc. 4th Turbulent Shear Flows Symposium, Karlsruhe (1983).
55. Hill, P. G., "Incompressible Jet Mixing in Converging-Diverging Axisymmetric Ducts," *Trans ASME 89, J. Basic Engg.* (1967), pp. 210–220.
56. Razinsky, E., and Brighton, J. A., "A Theoretical Model for Nonseparated Mixing of a Confined Jet," *Trans ASME 94, J. Basic Engg.* (1972), pp. 551–556.
57. Launder, B. E., and Spalding, D. B., "Mathematical Models of Turbulence," Academic Press (1972).
58. Launder, B. E., "Stress Transport Closures—Into the Third Generation," *Turbulent Shear Flows 1*, Springer (1979), pp. 259–266.
59. Hendricks, C. J., and Brighton, J. A., "The Prediction of Swirl and Inlet Turbulence Kinetic Energy Effects on Confined Jet Mixing," *Trans. ASME 97, J. Fluids Engg.* (1975), pp. 51–59.
60. Spalding, D. B., "A Two-Equation Model of Turbulence," VDI-Forschungsheft, 549 (1972), pp. 5–16.
61. Morton, H. L., "Effect of Swirl on Turbulent Jets in Ducted Streams," MIT GTL Rep. 95 (1968).
62. Bradshaw, P., "Effects of Streamline Curvature on Turbulent Flow," AGARDograph No. 169 (1973).
63. Pai, B. R., Richter, W., and Lowes, T. M., "Flow and Mixing in Confined Axial Flows," *J. Inst Fuel*, 48 (1975), pp. 185–196.
64. Thring, M. W., and Newby, M. P., "Combustion Length of Enclosed Turbulent Jet Flames," 4th Symposium (International) on Combustion, Williams and Wilkins, Baltimore (1953), pp. 789–796.
65. Gibson, M. M., and Morgan, B. B., "Mathematical Model of Combustion of Solid Particles in a Turbulent Stream with Recirculation," *J. Inst. Fuel*, 43 (1970), pp. 517–523.
66. Habib, M. A., and Whitelaw, J. H., "Calculations of Confined Coaxial Jet Flows," Proc. 3rd Turbulent Shear Flows Symposium, Davis (1981).

CHAPTER 15

TURBULENT MIXING AND DIFFUSION OF JETS

N. Rajaratnam

Department of Civil Engineering
University of Alberta
Edmonton, Alberta, Canada

CONTENTS

INTRODUCTION, 391

PLANE JETS, 391

CIRCULAR JET, 396

SHEAR LAYERS, 398

JETS IN COFLOWING STREAMS, 400

WALL JETS, 402

NOTATION, 404

REFERENCES, 404

INTRODUCTION

This chapter presents a concise discussion on the mixing and diffusion of submerged turbulent jets, that is, jets of water (or air) diffusion in water (or air). This treatment also applies to the turbulent diffusion of submerged jets of any other Newtonian fluid. For a more extensive treatment of turbulent jets, see Reference 1.

PLANE JETS

Let us consider a (submerged) rectangular jet of water (or air) of thickness $2b_0$ and length B much larger than its thickness, discharging into a large body of water (or air) at rest as shown in Figure 1. The jet mixes with the surrounding fluid forming shear layers all around. Let us neglect the end regions and consider only the central region of this rather long jet and refer to this as the plane jet. The turbulence created in the shear layers penetrates outwards as well as inwards into the jet thereby reducing continuously the thickness of the central region of almost potential flow, known as the potential core. By the time the jet has traveled a distance of about 12 b_0 from the nozzle, the potential core would have vanished and beyond this section we have the so-called region of fully developed flow. The region from the nozzle to the end of the potential core is known as the flow development region. In this section, we limit our attention to only the fully developed flow region.

Let us look at some basic characteristics of the plane jet as observed from experiments before attempting to develop a general theoretical structure. Simple observations show that as the jet travels further from the nozzle, it grows in thickness; secondly it entrains the surrounding fluid and thirdly the velocity in the jet continuously decreases. With reference to the definition sketch in Figure 1, if u is the time-averaged velocity in the axial or x direction at any point (x, y), the typical

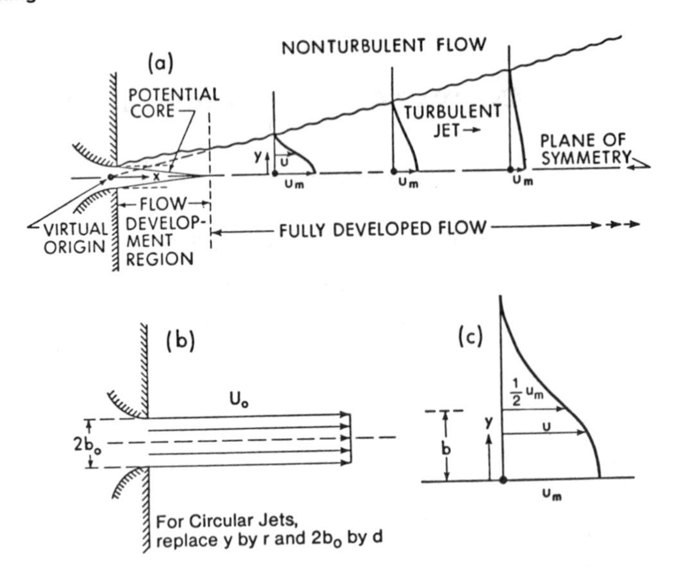

Figure 1. Plane and circular jets.

u(y) or velocity profiles at different x stations are shown in Figure 2A from the early observations of Foerthman [2], where $\bar{x}$ is the axial distance from the nozzle and x is that measured from a suitable virtual origin (to be discussed later). The velocity profiles at different sections appear to have the same shape and let us replot these profiles in a dimensionless from with u/u_m against y/b wherein u_m is the maximum value of u at any section occurring on the axis and b = y where u is equal to (say) $0.5u_m$. These two quantities can be referred to respectively as the velocity and the length scales. We find from Figure 2B that all these profiles are described by one general curve and the velocity profiles possessing this property are said to be "similar." This similarity property is possessed by many turbulent jets flows and is of great importance in developing general methods of solution.

The second interesting property is that the jet grows rather slowly; that is, its thickness at any section is small compared to the distance of this section from the nozzle. Hence, jets belong to the class of "slender flows" or boundary layer-type of flows and for such flows, some powerful approximations can be made to the full Reynolds equations of motion.

For the plane turbulent jet with a steady primary flow (u, v), the relevant Reynolds equations can be reduced to [7]:

$$u\frac{\partial u}{\partial x} + v\frac{\partial u}{\partial y} = \frac{1}{\rho}\frac{\partial \tau}{\partial y} \tag{1}$$

wherein most of the viscous stress terms have also been neglected and ρ is the mass density of the fluid. The shear stress τ is equal to the sum of τ_t the turbulent shear stress ($-\rho u'v'$, the primes denote fluctuating quantities) and the laminar shear stress $\tau_l = \mu(\partial u/\partial y)$ (μ being the coefficient of dynamic viscosity) and for most of the turbulent jet problems, $\tau_t \gg \tau_l$ and hence τ in Equation 1 can be considered as the turbulent shear stress itself. An exception to this rule will arise later when we consider turbulent wall jets, where the viscous shear stress will dominate over the turbulent shear stress as the wall is approached. The continuity equation reduces to

$$\frac{\partial u}{\partial x} + \frac{\partial v}{\partial y} = 0 \tag{2}$$

Equations 1 and 2 could be referred to as the equations of motion for the plane turbulent jet.

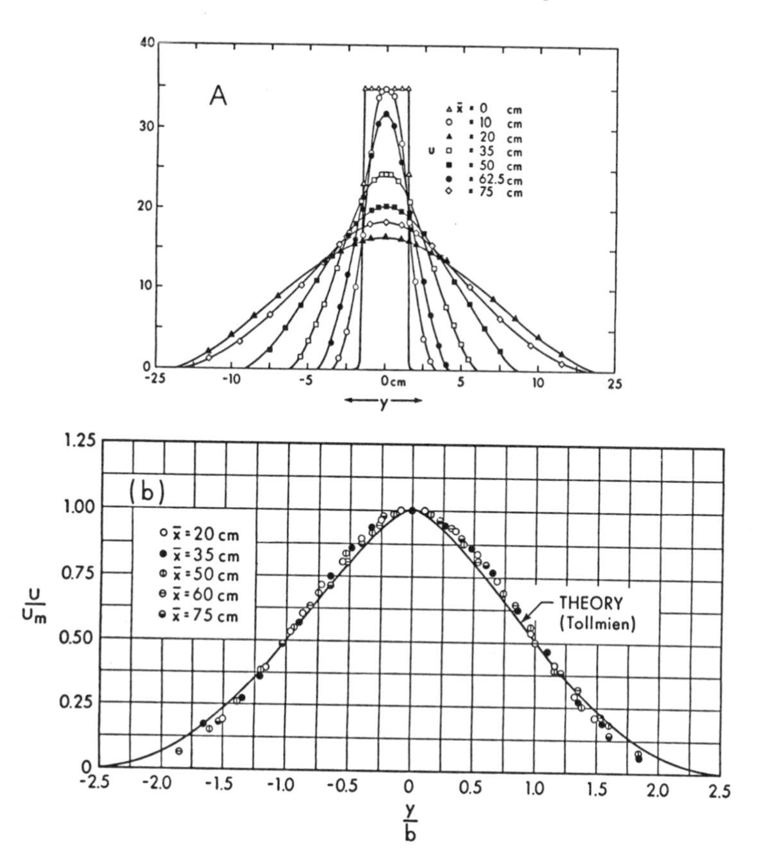

Figure 2. Velocity distribution in plane jets [2].

By integrating Equation 1, using Equation 2, we can obtain

$$\frac{d}{dx}\int_0^\infty \rho u^2\, dy = 0 \tag{3}$$

Equation 3 says that the axial momentum flux of the jet is invariant or remains constant and can be seen to be equal to the momentum flux M_0 from the nozzle, equal to $2b_0\rho U_0^2$.

We can integrate Equation 2 to obtain

$$\frac{d}{dx}\int_0^\infty u\, dy = -v_e \tag{4}$$

where v_e is the so-called entrainment velocity. Following Morton et al. (1956), we will now write

$$-v_e = \alpha_e u_m \tag{5}$$

where α_e is known as the entrainment coefficient. The structure of Equation 5 can be developed from the equations of motion assuming the similarity of v velocity profiles. Equation 5 expresses the rate at which the volumetric flux in the jet is increasing in the axial direction.

If we want to solve for the velocity field (u, v), since we have three unknowns u, v, and τ and only two equations, an additional equation is needed. Tollmien used the Prandtl mixing length hypothesis with the further assumption that the mixing length is constant at any x station and is proportional to the width of the jet whereas Goertler assumed a constant eddy diffusivity at any x station. Details of these solutions can be studied in Abramovich [3] or Rajaratnam [1]. There are a number of other closure models and for a discussion of some of these see Launder and Spalding [4]. In this analysis we are going to assume an empirical equation which describes the experimental observations very satisfactorily. The empirical equation adopted is

$$\frac{u}{u_m} = f(\eta) = \exp[-0.693\eta^2] \tag{6}$$

wherein $\eta = y/b$, b is the value of y where $u = u_m/2$ and u_m is the value of u on the axis of the jet.

Having assumed the form of the velocity profile, let us now predict the manner of variation of the velocity and length scales.

Let us further assume

$$\left.\begin{aligned} u_m &= k_1 x^p \\ b &= k_2 x^q \end{aligned}\right\} \tag{7}$$

where p and q are unknown exponents and k_1 and k_2 are the characteristic coefficients of the jet which do not depend on the variable x. Using the integral momentum and continuity equations, we can show the $q = 1$ and $p = -\frac{1}{2}$ (see [1]). Thus, for a plane turbulent jet

$$u_m \propto \frac{1}{\sqrt{x}} \qquad b \propto x \tag{8}$$

Using again the integral momentum and continuity equations, we can show that

$$k_1 = \sqrt{\frac{b_0}{k_2 F_2}}\, U_0 \tag{9}$$

$$k_2 = \frac{2\alpha_e}{F_1} \tag{10}$$

where $F_1 = \int_0^\infty f\, d\eta$ and $F_2 = \int_0^\infty f^2\, d\eta$. For the exponential profile suggested in Equation 6, $F_1 = 1.065$ and $F_2 = 0.753$. Amongst the three coefficients k_1, k_2, and α_e, at least one of them will have to be determined experimentally. It is convenient to evaluate k_2 from experimental observations and then k_1 and α_e can easily be evaluated.

If the jet carried a passive pollutant or tracer in it, we could study the mixing and spread of the tracer as shown in the following. If c is the time-averaged concentration of this pollutant at any point in the jet, in some suitable units (say mg/litre), then from the pollutant conservation equation, with the slender flow approximations, we can obtain the simplified equation

$$u\frac{\partial c}{\partial x} + v\frac{\partial c}{\partial y} = \epsilon_y \frac{\partial^2 c}{\partial y^2} \tag{11}$$

wherein the eddy diffusivity ϵ_y for the pollutant concentration has been assumed to be constant at any x station in the jet. Integrating Equation 11 across the jet from $y = 0$ to a large value, we can reduce Equation 11 to

$$\frac{d}{dx}\int_0^\infty uc\,dy = 0 \tag{12}$$

which says that the pollutant flux in the jet at any section remains constant and would be equal to the pollutant flux from the nozzle.

Experimental observations indicate that the pollutant concentration profiles at the different x stations are similar. That is

$$\frac{c}{c_m} = h(\eta) \tag{13}$$

where c_m is the concentration on the axis for a particular value of x. If we assume that

$$c_m = k_3 x^s \tag{14}$$

where s is the unknown exponent and k_3 is the x independent coefficient, from Equation 12, we can show that $s = -\frac{1}{2}$. Since the pollutant is known to spread slightly faster than the momentum in the jet, if $b_c = kb$ where $b_c = y$ where $c = \frac{1}{2}c_m$ and k is a constant, to be evaluated from experiments, we can show that

$$k_3 = c_0\sqrt{\frac{b_0 F_2}{k_2 F_3^2}} \tag{15}$$

where

$$F_3 = \int_0^\infty fh\,d\eta$$

The earliest experiments on the plane jet were performed by Foerthmann [2]. Since then, many other experiments on the plane jet have been done by Albertson et al. [5]. Zijnen [6, 7], Heskestad [8], and others. These measurements have shown conclusively that the u velocity profiles at different x sections are indeed similar in the fully developed region and the exponential equation (Equation 6) is indeed a good approximation to this similarity curve. The length scale b grows linearly with x, but the coefficient k_2 has been found to vary to some extent with the turbulence level and the degree of uniformity of the velocity profile at the nozzle. For practical purposes, k_2 could be given a value of 0.097 or even 0.1. For the concentration profiles, it appears that $k \simeq 1.47$ [7]. Hence, taking k_2 as equal to 0.097, $\alpha_e = 0.052$, $k_1 = 3.70\,U_0\sqrt{b_0}$ and $k_3 = 3.17c_0\sqrt{b_0}$. We can recast the equations for the scales of the plane jet as

$$b = 0.097x \tag{16}$$

$$\frac{u_m}{U_0} = \frac{3.70}{\sqrt{x/b_0}} \tag{17}$$

$$\frac{c_m}{c_0} = \frac{3.17}{\sqrt{x/b_0}} \tag{18}$$

In these equations, x is the axial distance measured from a virtual origin, from which the ideal jet of momentum flux of M_0 and zero thickness originates and this virtual origin appears to be located slightly behind the nozzle [9]. For practical purposes, unless $x/2b_0$ is small, we could consider $\bar{x}$ the distance from the nozzle to be essentially the same as x itself. If Q is the volumetric flux at any x station and if Q_0 is the flux from the nozzle (per unit length), we can show that

$$\frac{Q}{Q_0} = 0.44\sqrt{x/b_0} \tag{19}$$

Experimental observations show [7] that the eddy diffusivity $\epsilon \simeq 0.003xu_m$. Experimental observations [8] have also shown that the profiles of u'^2, v'^2, w'^2 and $u'v'$ in terms of u_m^2 at different sections are similar with the maximum values being respectively about 8, 3.5, 4.2, and 2 percent.

CIRCULAR JET

Let us next consider the diffusion of a (submerged) circular jet of diameter d with a velocity of U_0 in a large expanse of the same fluid at rest. As in the case of the plane jet, we will have a flow development region, in which we will observe an axisymmetric shear layer surrounding a continuously diminishing potential core, followed by the region of fully developed flow in which the maximum axial velocity u_m (occurring on the axis of the jet) decays continuously with the axial distance $\bar{x}$ from the nozzle. The jet grows in thickness but still behaves like a slender flow. In addition, the jet entrains the surrounding fluid thereby diluting any passive pollutant that it might transport.

If u and v are the time-averaged velocities in the axial (x) and radial (r) directions at any point (x, r) in the jet and if the primary flow in the jet is steady, the relevent Reynolds equations can be simplified, using the conditions of axisymmetry and slender flow to obtain [1]:

$$u\frac{\partial u}{\partial x} + v\frac{\partial u}{\partial r} = \frac{1}{\rho r}\frac{\partial r\tau}{\partial r} \tag{20}$$

where τ is the total shear stress at any point in the axial direction equal to the sum of the laminar and the turbulent components. For turbulent jets, if the nozzle Reynolds number is at least a few thousand, the viscous shear stress can be neglected and τ becomes equal to the turbulent shear stress itself. The continuity equation reduces to

$$\frac{\partial ru}{\partial x} + \frac{\partial rv}{\partial r} = 0 \tag{21}$$

Equations 20 and 21 could be referred to as the equations of motion for the circular turbulent jet.

Integrating Equation 20 with respect to r and using Equation 21, one obtains the result

$$\frac{d}{dx}\int_0^\infty 2\pi r\rho u^2\, dr = 0 \tag{22}$$

Equation 22 says that the axial momentum flux remains constant and can be seen to be equal to the momentum flux from the nozzle M_0 equal to $(\pi d^2/4)\rho U_0^2$.

If we next integrate the continuity equation, from $r = 0$ to $r = \bar{b}$ where $u \simeq 0$ (using the Liebnitz Rule) and using the entrainment hypothesis, we can obtain

$$\frac{d}{dx}\int_0^\infty r\, dr\, u = \alpha_e \bar{b} u_m \tag{23}$$

Experimental observations on the u velocity profiles at different x sections have shown that these velocity profiles possess the similarity property, and we can write

$$\frac{u}{u_m} = f\left(\frac{r}{b}\right) = f(\eta) \tag{24}$$

where for any x section u = the axial(x) velocity at radial distance of r
u_m = the value of u on the axis
b = a length scale, generally taken as equal to r where $u = \frac{1}{2}u_m$

The velocity profile similarity function $f(\eta)$ can be found by solving Equations 20 and 21 along with an auxiliary equation. For solutions of this kind using the mixing length idea and the eddy diffusivity concept, see Rajaratnam [1] and for solutions using other kinds of closure models, see Launder and Spalding [4]. In this work, we are going to assume the convenient exponential function described by Equation 6 which is very satisfactory for most purposes.

Following closely the treatment for plane jets, let us assume that

$$u_m = k_1 x^p \quad \text{and} \quad b = k_2 x^q \tag{25a}$$

where k_1 and k_2 are the x-independent coefficients and p and q are the unknown exponents. Using the integral momentum and continuity equations, we can show [1] $p = -1, q = 1, k_2 = \beta\alpha_e/F_4$ and

$$k_1 k_2 = \sqrt{\frac{M_0}{2\pi\rho F_5}} \tag{25b}$$

where $\beta = \bar{b}/b$, $\bar{b}$ being the value of r at the outer edges of the jet and $\beta \simeq 2.2 - 2.5$, $F_4 = \int_0^\infty \eta f \, d\eta$ and $F_5 = \int_0^\infty \eta f^2 \, d\eta$. Regarding these three unknowns, k_1, k_2, and α_e, if one of them (perhaps k_2) is determined experimentally, the other two can be evaluated using the expressions developed herein.

If the jet carries a passive pollutant, the pollutant conservation equation can be reduced to

$$u\frac{\partial c}{\partial x} + v\frac{\partial c}{\partial r} = \frac{1}{r}\frac{\partial}{\partial r}\left(r\epsilon_r \frac{\partial c}{\partial r}\right) \tag{26}$$

where c is the time-averaged pollutant concentration at any point and ϵ_r is the eddy diffusivity for pollutant transport. Equation 26 can be reduced to

$$\frac{d}{dx}\int_0^\infty r \, dr \, uc = 0 \tag{27}$$

which describes the constancy of the pollutant flux in the jet. Experimental observations have shown that the concentration profiles at different sections are similar [ie, $c/c_m = h(\eta)$] and that

$$\frac{c}{c_m} = \exp\left[-0.693\left(\frac{r}{kb}\right)^2\right] \tag{28}$$

where k appears to be a constant, equal to be about 1.17 [10]. If we assume that $c_m = k_3 x^s$, using Equation 27, we can show that $s = -1$ and

$$k_3 = \frac{P_0}{2\pi}\frac{1}{F_6 k_1 k_2^2} \tag{29}$$

where

$$F_6 = \int_0^\infty \eta f h \, d\eta = 0.722\frac{k^2}{1 + k^2}$$

and P_0 is the pollutant flux from the nozzle. If k is known, knowing k_1 and k_2, k_3 can be calculated.

Experiments on circular jets have been performed by Trupel [11], Hinze and Zijnen [10], Albertson et al. [5] and others. From these numerous studies, the u velocity profiles have been shown to be similar (see [1]) and the exponential equation is a good approximation to this similarity profile. The length scale has been found to grow linearly with the axial distance x, even though the value of k_2 has been found to vary to some extent with the turbulence level and nonuniformity of the velocity distribution in the jet at the nozzle. For practical purposes, a value of 0.097 has been suggested by Abramovich [3]. Strictly speaking, x is measured from a virtual origin, located behind the nozzle, but for reasonably large values of x/d, x could conveniently be measured from the nozzle itself.

Using the value of $k_2 = 0.097$, the velocity scale u_m in terms of the jet velocity U_0 at the nozzle is given by the equation

$$\frac{u_m}{U_0} = \frac{6.13}{x/d} \tag{30}$$

Using the value of 2.5 for β, $\alpha_e = 0.028$. The pollutant concentration scale c_m in terms of the concentration at the nozzle c_0 is given by the equation

$$\frac{c_m}{c_0} = \frac{5.34}{x/d} \tag{31}$$

If Q is the volumetric flux at any x section, we can show that

$$\frac{Q}{Q_0} = 0.33\frac{x}{d} \tag{32}$$

where Q_0 is the volumetric flux at the nozzle. At $x/d = 100$, $Q = 32Q_0$ where $31Q_0$ is the contribution from the entrainment. If E is the kinetic energy flux at any section and E_0 is that at the nozzle

$$\frac{E}{E_0} = \frac{4.1}{x/d} \tag{33}$$

Turbulence measurements in the circular jet have been made by Wygnanski and Fielder [12]. Profiles of $\overline{u'^2}$, $\overline{v'^2}$, $\overline{w'^2}$ and $\overline{u'v'}$ in terms of u_m^2 have been found to be similar, with the maximum values of the respective ratios equal to 8, 6, 6, and 1.7 percent.

A treatment of radial jets and three dimensional jets can be found in Rajaratnam [1].

SHEAR LAYERS

Having studied the fully developed plane jet in the first section, let us now consider the plane shear-layer problem. Consider a plane jet of large thickness and uniform velocity of U_0 flow over a stagnant mass of the same fluid as shown in Figure 3. The intense shear at the surface of velocity discontinuity induces turbulent mixing and as a result, the stagnant fluid is accelerated and a portion of the jet loses some momentum. The thickness of the layer affected by this momentum exchange is known as the shear layer and its thickness at any x station can be denoted by $\bar{b}$. In Figure 3, the ϕ_1 and ϕ_2 lines denote the edges of the shear layer. The velocity profile measurements by Liepmann and Laufer [13] have shown that the distribution of u/U_0 across the shear layer is similar, and here again, the exponential equation describes the similarity profile well.

The equations of motions for the plane shear layer are the same as Equations 1 and 2. Assuming that $u/U_0 = f(\eta)$ where $\eta = y/b$, b being a suitable length scale, and assuming that $\tau/\rho U_0^2 = g(\eta)$, using Equation 1 to obtain the relation for v as

$$v = \int_{y^*}^{y} \frac{\partial v}{\partial y}\,dy \tag{34}$$

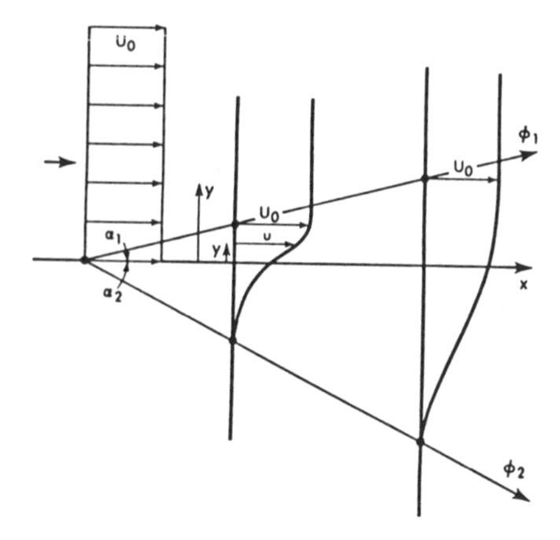

Figure 3. Plane shear layers.

where $y_* = y$ at the location of v being equal to zero, Equation 1 can be reduced to

$$g' = b'[\eta ff' - f'F_7] - b'h(x)f' \tag{35}$$

where the primes on g and f denote differentiation with respect to η and the prime on b denotes differentiation with respect to x, $F_7 = \int \eta f' \, d\eta$ and $h(x) = F_7(\eta_*)$. For equation 35 to be satisfied, $b \propto x$ and $y_* \propto x$.

Tollmien used the Prandtl mixing length hypothesis with the assumption that the mixing length $\ell \propto b$ and solved Equation 1 using Equation 2 to obtain the u velocity distribution in the shear layer, which indicated that on the $y = 0$ plane, $u = 0.68U_0$. For the solution to be complete, we need the experimental determination of just one parameter $a = [2(\ell/x)^2]^{1/3}$. Goertler employed the constant-eddy viscosity model and solved the general shear layer problem wherein a uniform stream with velocity of U_0 flows over a slower stream with a velocity of U_1 [1].

Experiments on the simple shear layer have been made by Liepmann and Laufer [13], Albertson et al. [5], Wygnansky and Fielder [14], and others. Using these results, we find that Tollmien's $a = 0.084$. With this value of a, for the plane turbulent shear layer, $b = 0.115x$, $\bar{b} = 0.263x$, and the angles α_1 and α_2 of the outer limits of the shear layer (see Figure 3) are equal to 4.8 and 9.5 degrees, respectively.

With the knowledge of plane shear layers, we can easily find the length of the potential core x_0 of plane jets as

$$x_0 = \frac{b_0}{\tan 4.8^0} = 11.91 b_0 \tag{36}$$

Combining the uniform velocity in the potential core and the velocity distribution in the shear layer, the characteristics of the flow development region of plane jets can be found [1].

If now we consider the general (or compound) shear layer which forms between a deep stream with uniform velocity of U_0 flowing tangentially over another deep stream with a smaller uniform velocity of U_1, the solutions of Tollmien and Goertler can be extended [1]. Using the experimental results of Watt [15], Miles and Shih [16], Yule [17, 18] and others, if Δb is the thickness of the shear layer bound by the lines of $(u - U_1)/(U_0 - U_1) = 0.1$ and 0.9, then

$$\frac{\Delta b}{0.165x} = \frac{\alpha - 1}{\alpha + 1} \tag{37}$$

where α is equal to the ratio of U_0 to U_1. Further, if α_1 and α_2 are the angles of penetration of the shear layer into the faster and slower streams respectively, the variation of α_1 and α_2 with

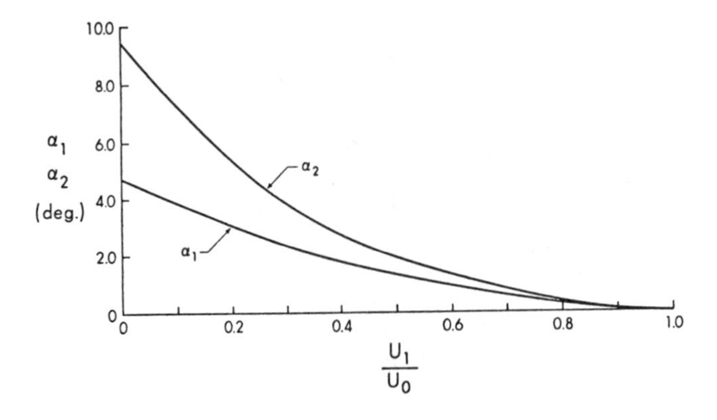

Figure 4. Spreading angles of compound shear layers.

$1/\alpha = U_1/U_0$ is shown in Figure 4. In Figure 4, it is seen that as $1/\alpha$ approaches unity, α_2 approaches α_1 and they both approach zero. For a similar analysis of axisymmetric shear layers see Rajaratnam [1].

JETS IN COFLOWING STREAMS

In the previous sections we studied the diffusion of plane and circular jets in a stagnant environment. If the jets are discharged into a moving stream in the same direction as the jet but with a velocity U_1 less than the jet velocity of U_0, we have the problem of the diffusion of a jet in a coflowing stream. We will discuss the diffusion of a plane jet in a coflowing stream of large extent in this section and for a similar treatment of circular jets, we refer the reader to Rajaratnam [1].

With reference to Figure 5, in the flow development region, we have a potential core with a (compound) shear layer on the top and bottom and using the results presented in the previous section, this region can be analyzed and in this section we will be primarily interested in the fully developed flow region. In this region, experimental observations have shown that if the flow in the jet is looked at with respect to the coflowing stream then the u velocity profiles possess the similarity

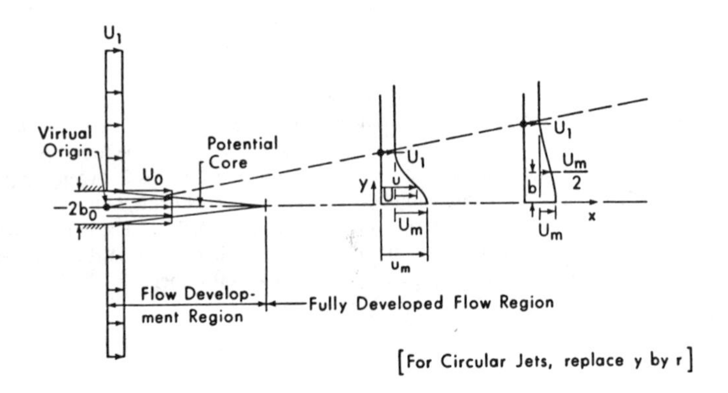

Figure 5. Jet in a coflowing stream.

property. That is, if $U = (u - U_1)$ and U_m is the maximum value of U equal to $(u_m - U_1)$, then $U/U_m = f(\eta)$ where $\eta = y/b$, b being equal to y where $U = \frac{1}{2}U_m$. It has also been observed that $f(\eta)$ is well described by the exponential relation discussed earlier. Then for predicting the structure of the mean flow in the jet immersed in a large coflowing stream, we will have to predict the manner of variation of U_m and b with the axial distance.

Equations 1 and 2 are the relevent equations and integrating Equation 1 using Equation 2, one can obtain

$$\frac{d}{dx}\int_0^\infty \rho u(u - U_1)\,dy = 0 \tag{38}$$

Equation 38 states that the excess momentum flux M_0 in the jet with respect to the free stream is preserved. Equation 38 can be rewritten as:

$$\frac{d}{dx}\left[b\frac{U_m}{U_1}\int_0^\infty f d\eta + b\left(\frac{U_m}{U_1}\right)^2\int_0^\infty f^2\,d\eta\right] = 0 \tag{39}$$

Now consider the so-called "strong jet case" where $U_m/U_1 \gg 1$. For this case, the first term in Equation 39 can be dropped as an approximation whereas for the "weak jet case" for which $U_m/U_1 \ll 1$, the second term in Equation 39 can be neglected leaving only the first term. If we assume that $U_m \propto x^p$ and $b \propto x^q$, then with these simplifications, for the strong jet case, $q + 2p = 0$, whereas for the weak jet case, $p + q = 0$. To get another equation for each one of these two cases, we can do a similarity analysis of the equations of motion. From such an analysis [1] it can be seen that for the strong jet case $q = 1$ and this leads to $p = -\frac{1}{2}$. For the weak jet case, $q - p = 1$. As a result, $p = -\frac{1}{2}$ and $q = \frac{1}{2}$. For the intermediate case where $U_m \sim U_1$, it does not appear to be possible to have a simple power law type of variation.

Using dimensional arguments, we can show that $U_m/U_1 = F\,(x/\theta)$ where F denotes a function and θ is a momentum thickness defined by the expression.

$$M_0 = 2\theta\rho U_1^2 \tag{40}$$

A number of methods have been proposed to predict the function F [1], and we would simply present a correlation developed by Pande and Rajaratnam [19] using the integral momentum and energy equation along with a hypothesis regarding the turbulence production in the transition region from the strong jet to the weak jet flows (see Figure 6). This correlation appears to describe the experimental data better than the other computational schemes. Once the velocity scale has been determined for a given problem, the growth of the length scale can be obtained using the integral momentum equation. Turbulence characteristics of the plane jet in a coflowing stream have been studied by Bradbury [20] and others.

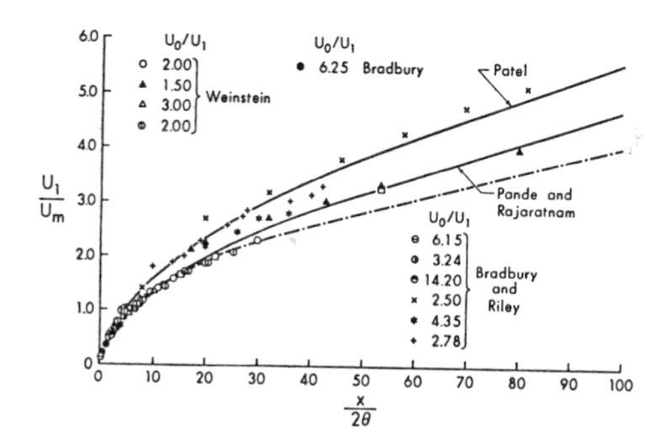

Figure 6. Velocity scale for plane jets in coflowing stream [19].

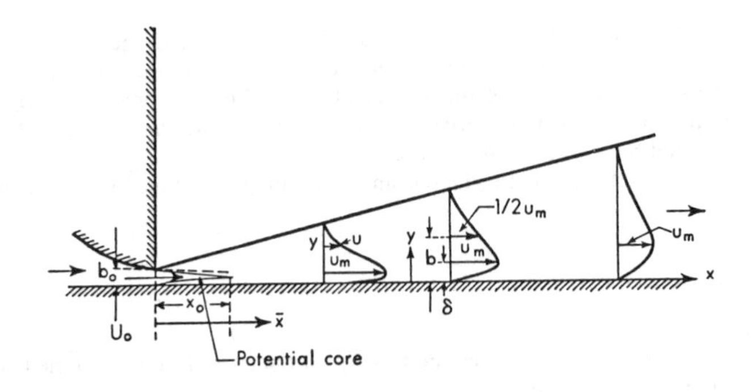

Figure 7. Plane wall jet.

WALL JETS

In the plane jet considered earlier, if the half-jet of thickness b_0 is discharged tangentially onto a smooth plate as shown in Figure 7, we have the problem of a plane wall jet. This plane wall jet will have a flow development region in which the boundary layer on the plate will also have to be considered. Considering now the fully developed region downstream of the end of the potential core, experimental observations have shown that (see Figure 8A) at any x section, as y increases from zero, the u velocity increases to a maximum of u_m at $y = \delta$, where δ is the thickness of the boundary layer and for y greater than δ, u decreases continuously from u_m to zero. If these profiles are plotted in a dimensionless form with u_m and b where b = y where $u = \frac{1}{2}u_m$ and $\partial u/\partial y < 0$ as the scales, as shown in Figure 8B, they are found to be similar.

The equations of motion are Equations 1 and 3 discussed before with one difference—τ must now be interpreted as the sum of the turbulent shear stress and the laminar shear stress, which dominates over the former as the boundary (or wall) is approached. Integrating Equation 1 with respect to y from y = 0 to y = ∞, we get

$$\frac{d}{dx}\int_0^\infty \rho u^2 \, dy = -\tau_0 \tag{41}$$

where τ_0 is the wall shear stress at any distance x. If τ_0 is neglected as a first approximation, Equation 41 reduces tp Equation 3. If $u_m \propto x^p$ and $b \propto x^q$, from Equation 3, $2p + q = 0$. A similarity analysis of Equation 1 with the neglect of the viscous term gives q = 1. The entrainment hypothesis would also give q = 1. Then $p = -\frac{1}{2}$. Then the variations will be the same as those for the plane jet. Dimensional considerations indicate that $\tau_0 \propto 1/x$.

Experiments on plane wall jets have been made by Foerthmann [2], Myers et al. [21], Schwarz and Cosart [22], and others. Based on these results, we could write

$$b = 0.068x \tag{42}$$

$$\frac{u_m}{U_0} = \frac{3.5}{\sqrt{x/b_0}} \tag{43}$$

$$\tau_0 = c_f \frac{\rho U_0^2}{2} \tag{44}$$

where

$$c_f = \frac{0.2}{(x/b_0)(U_0 b_{0/\nu})^{1/12}} \tag{45}$$

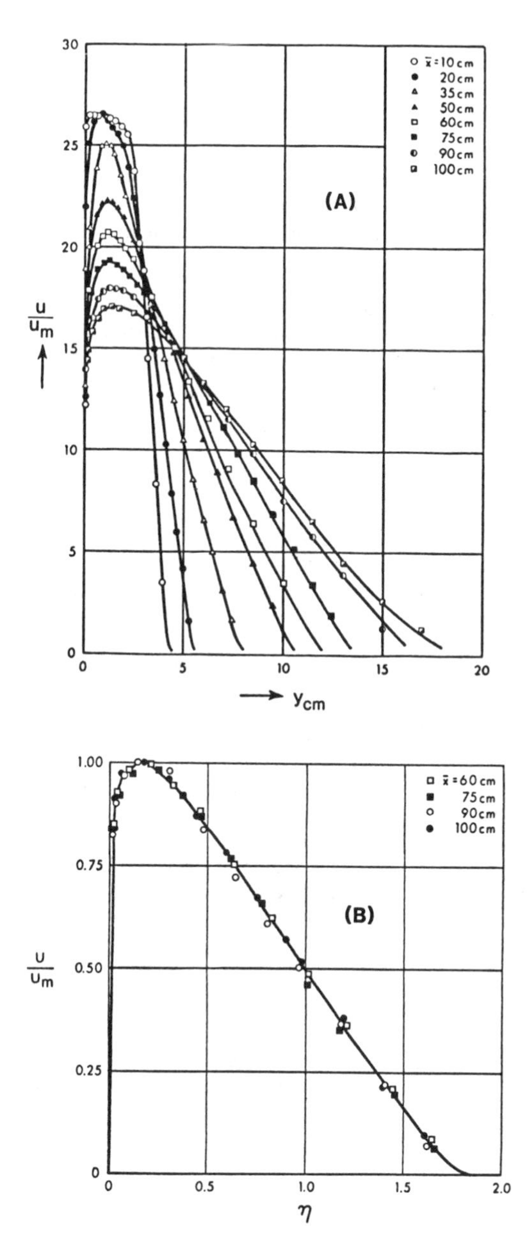

Figure 8. Velocity profiles in plane wall jets [2].

and ν is the coefficient of kinematic viscosity. Further, $\delta \simeq 0.16b$ and $\alpha_e = 0.035$. It can be noticed that the velocity scale for the plane wall jet is only slightly less than that for the plane (free) jet, but the length scale b for the wall jet is only about 0.7 times that of the free jet. In the boundary layer, the u velocity profiles can be described by power laws or a combination of the "law of the wall" and "defect law" of the turbulent boundary layers whereas in the free-mixing region above the boundary layer, the exponential equation is valid. Verhoff [23] has developed an empirical equation to describe the complete profile as

$$\frac{u}{u_m} = 1.48\eta^{1/7}[1 - \text{erf}(0.68\eta)] \tag{46}$$

where erf represents the error function. The length of the flow development region has been found to vary from $6.1b_0$ to $6.7b_0$ for $U_0 b_0/\eta$ from 10^4 to 10^5. Turbulence observations on plane wall jets have been made by Mathieu and Tailland [24]. The effect of wall roughness on plane wall jets has been studied by Rajaratnam [25].

A treatment of radial and cylindrical wall jets can be found in Rajaratnam [1]. For a discussion on the diffusion of three-dimensional wall jets see Rajaratnam [1] and Launder and Rodi [26].

If a submerged water jet is discharged tangentially at the water surface of a large pool, we call it a surface jet. Recent studies [27] have shown that surface jets resemble wall jets in many respects. For a treatment of surface jets, see Humphries and Rajaratnam [27].

NOTATION

A	area of cross section of nozzle
a	Tollmien coefficient
B	length of jet
b	length scale
b_0	half-thickness (width) of nozzle
c	time-averaged pollutant concentration
d	diameter of nozzle
e	suffix
F	definite integrals (with suffixes)
f	function (with suffixes), also suffix
g	function
h	function
k	ratio of length scales, with suffixes, denote characteristic coefficients
l	suffix to denote laminar flow
M_0	momentum flux at nozzle, also excess momentum flux
m	suffix to denote maximum value
p	exponent
Q	volumetric flux
q	exponent
r	radial distance
r_0	radius of nozzle
s	exponent
t	suffix to denote turbulent flow
U_0	jet velocity at nozzle
U_1	velocity of free stream
U	velocity relative to freestream
u	time averaged axial velocity at any point
v	time averaged velocity in y or r or z direction
v_e	entrainment velocity
x	axial distance from virtual origin
$\bar{x}$	axial distance from nozzle
y	coordinate distance

Greek Symbols

α	velocity ratio
α_e	entrainment coefficient
β	ratio of length scales
δ	boundary layer thickness
ε	eddy diffusivity
μ	coefficient of dynamic viscosity
ν	coefficient of kinematic viscosity
ρ	mass density
τ	shear stress

REFERENCES

1. Rajaratnam, N., *Turbulent Jets*, Elsevier Scientific Publishing Co., Amsterdam, The Netherlands 1976.
2. Foerthamnn, E., "Turbulent Jet Expansion," English Translation, A.C.A. TM-789, (original paper in German, 1934, Ing. Archiv., 5), 1934.
3. Abramovich, G. N., *The Theory of Turbulent Jets*, English translation published by M.I.T. Press, Massachusetts, U.S.A., 1963.

4. Launder, B. E., and Spalding, D. B., *Mathematical Models of Turbulence*, Academic Press, London and New York, 1972.
5. Albertson, M. L., et al., "Diffusion of Submerged Jets," *Trans. A.S.C.E.*, 115:639–697 (1950).
6. Zijnen, B. G., Van der Hegge, "Measurements of the Velocity Distribution in a Plane Turbulent Jet of Air," *Applied Sci. Research*, Sec. A, 7:256–276 (1958).
7. Zijnen, B. G., Van der Hegge, "Measurements at Turbulence in a Plane Jet of Air by the Diffusion Method by the Hot ire Method," *Applied Sci. Research*, Sec. A, 7:293–313 (1958).
8. Heskestad, G., "Hot-Wire Measurements in a Plane Turbulent Jet," *Trans. A.S.M.E., J. Applied Mechanics* (1965), pp. 1–14.
9. Flora, J. J., Jr., and V. W. Goldschmidt, "Virtual Origins of a Free Plane Turbulent Jet," *J.A.I.A.A.*, 7:2344–2346 (1969).
10. Hinze, J. O., and Zijnen, B. G., Van der Hegge, " Transfer of Heat and Matter in the Turbulent Mixing Zone of an Axially Symmetrical Jet," *J. Applied Sci. Research*, A1:435–461 (1949).
11. Schlichting, H., *Boundary Layer Theory*, (6th Edition), McGraw-Hill, New York (1968).
12. Wygnanski, I., and Fielder, H., "Some Measurements in the Self Preserving Jet," *J. Fluid Mechanics*, 38:577–612 (1969).
13. Liepmann, H. W., and Laufer, J., "Investigation of Free Turbulent Mixing," N.A.C.A., Technical Note 1257.
14. Wygnanski, I., and Fielder, H., "The Two-Dimensional Mixing Region," *J. Fluid Mechanics*, 41:327–361 (1970).
15. Watt, W. E., "The Velocity Temperature Mixing Layer," Report 6705, Department of Mechanical Engineering, University of Toronto.
16. Miles, J. B., and Shih, J. S., Similarity Parameter for Two-Stream Turbulent Jet Mixing Region, *J. A.I.A.A.*, 6:1429–1430 (1968).
17. Yule, A. J., "Two Dimensional Self-Preserving Turbulent Mixing Layers at Different Free Stream Velocity Ratios," Aeronautical Research Council, England, 1971.
18. Yule, A. J., "Spreading of Turbulent Mixing Layers," *J. A.I.A.A.*, 10:686–687 (1972).
19. Pande, B. B. L., and Rajaratnam, N., "Turbulent Jets in Coflowing Streams," *ASCE, J. of the Engineering Mechanics Div.*, 105 (EM6) (1979) pp. 1025–1038.
20. Bradbury, L. J. S., "The Structure of a Self-Preserving Turbulent Jet," *J. Fluid Mechanics*, 23:31–64 (1965).
21. Myers, G. E., Schauer, J. J., and Eustis, R. H., "The Plane Turbulent Wall Jet—Part I, Jet Development and Friction Factor." Technical Report 1, Department of Mechanical Engineering, Stanford University 1961 (Also published in Trans. A.S.M.E., *J. Basic Engineering*, (1963)).
22. Schwarz, W. H., and Cosart, W. P., "The Two-Dimensional Turbulent Wall Jet," *J. Fluid Mechanics*, 10:481–495 (1961).
23. Verhoff, A., "The Two-Dimensional Turbulent Wall Jet With and Without an External Stream," Report 626, Princeton University, (1963).
24. Mathieu, J., and Tailland, A., "Jet Parietal," *Compte Rend*, Academie des Sciences, Paris, 261:2282–2286 (1965).
25. Rajaratnam, N., "Plane Turbulent Wall Jets on Rough Boundaries," Technical Report, Department of Civil Engineering, University of Alberta, Edmonton, Canada (1965). (Also published in *Water Power*, England, 1967).
26. Launder, B. E., and Rodi, W., "The Turbulent Wall Jet," *Annual Rev. Fluid Mech.*, 15 (1983) pp. 429–459.
27. Rajaratnam, N., and Humphries, J. A., "Turbulent Non-Buoyant Surface Jets," *J. of Hydraulic Research, IAHR*, 1984, (in press).

CHAPTER 16

HYDRODYNAMICS OF JETS IN CROSSFLOW

J. A. Schetz

Department of Aerospace and Ocean Engineering
Virginia Polytechnic Institute & State University
Blacksburg, Virginia USA

CONTENTS

INTRODUCTION, 406

SINGLE JETS, 406
Low-Speed, Single-Phase Flows, 406
Transverse, Particle-Laden Jets, 412
Transverse Jets into Supersonic Flows: Gaseous Jets, 412
Transverse Jets into Supersonic Flows: Liquid Jets, 416
Trajectory Analyses, 417
Surface Pressure Distributions, 418

DUAL JETS, 419
Jet Trajectories, 419
Surface Pressures on a Flat Plate, 420
Surface Pressures on a Body of Revolution, 427

NOTATION, 427

REFERENCES, 427

INTRODUCTION

In this chapter, we are concerned with jet injection at large angles to a moving mainstream, which of course, produces strongly three-dimensional flowfields. This class of flows is often encountered in practice; smokestacks, some fuel injection systems, and sewage and cooling water outfalls can be cited as a few representative examples. One also can have two-phase cases. Further, one or both fluid streams may be supersonic. For essentially coaxial cases, that is not an important matter; however for transverse injection, supersonic flows require special treatment. Such cases are discussed in this chapter in separate sections. Lastly, cases where buoyancy forces are important are not discussed here.

SINGLE JETS

Low-Speed, Single-Phase Flows

The character of the flow for a case with 90-deg. injection can be seen in Figures 1 and 2 from Reference 1. The "kidney-shaped" nature of the jet as it is deflected and distorted by the cross stream is noteworthy. The simplest quantity of engineering interest is the gross penetration of the injected fluid into the mainstream. The trajectory of the center of the jet plume as a function of velocity ratio V_j/V_∞, and injection angle is shown in Figures 3 and 4 using data from various sources. The "theory" curves will be discussed later. For 90-deg. injection, trajectory results can be

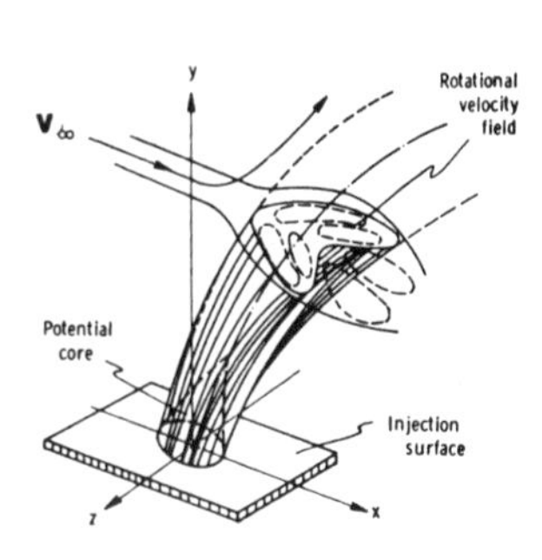

Figure 1. Schematic of transverse injection flowfield [1].

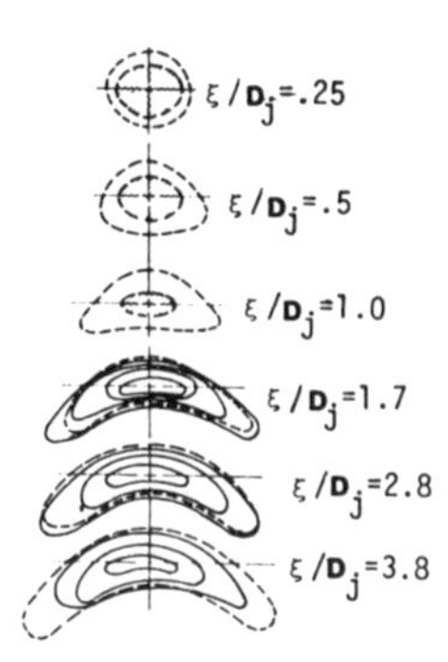

Figure 2. Cross-sectional pressure contours in a traverse jet with $V_j/V_\infty = 2.2$; solid and dashed lines correspond to constant total and static pressure and the shaded areas denote the potential core [1].

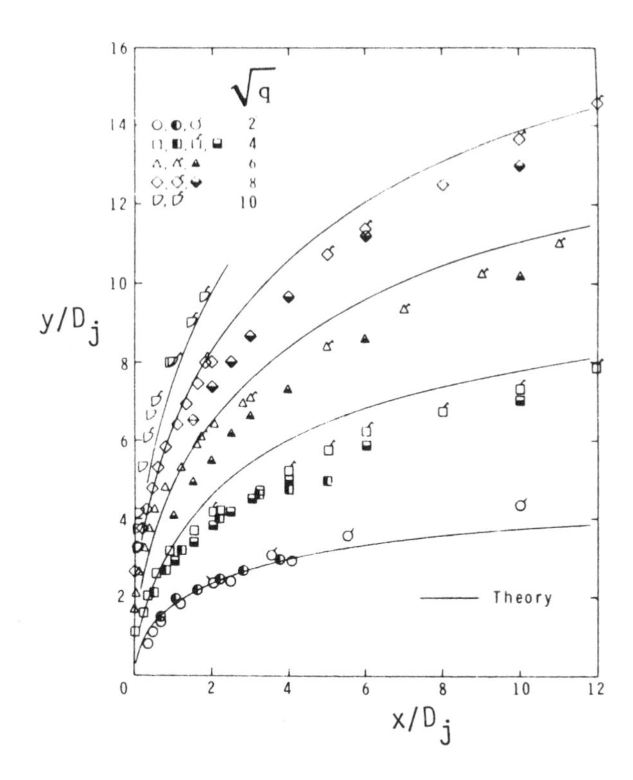

Figure 3. Trajectories for transverse ($\theta = 90$ deg.) air jets into air for various values of the parameter $\bar{q} = \rho_j V_j^2/\rho_\infty V_\infty^2$ [2].

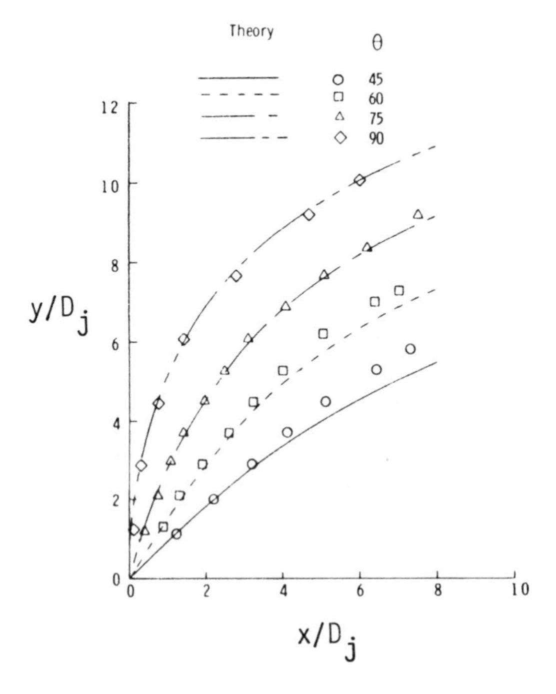

Figure 4. Trajectories of transverse jets for various injection angles with $\bar{q}^{-1/2} = 6.3$ [2].

correlated as in Figure 5. Here h is the vertical penetration, and ξ is the arc length along the jet centerline trajectory (see Figure 6). The next quantity of interest is the growth of the width of the mixing zone in the plane containing the trajectory and the direction perpendicular to that plane. These widths are denoted here as $\Delta\xi$ and Δz, and some results are reproduced as Figures 7 and 8.

Turning now to some details of the flow, we show the decay of the maximum velocity in the jet cross section in Figure 9. Free jet results are also shown for comparison. It is not surprising that

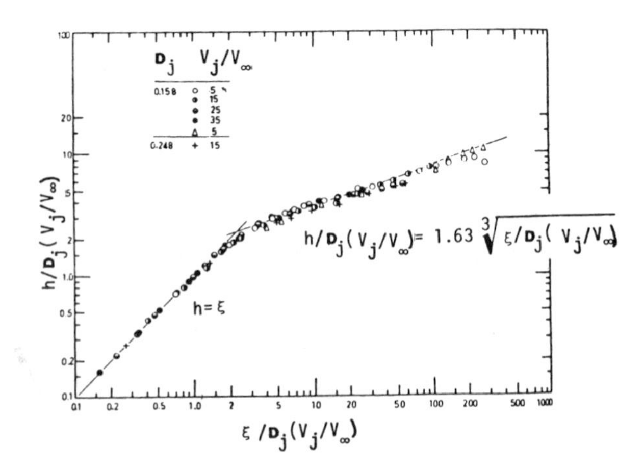

Figure 5. Correlation of the trajectory of the jet centerline along the arc length for transverse ($\theta = 90$ deg.) injection [3].

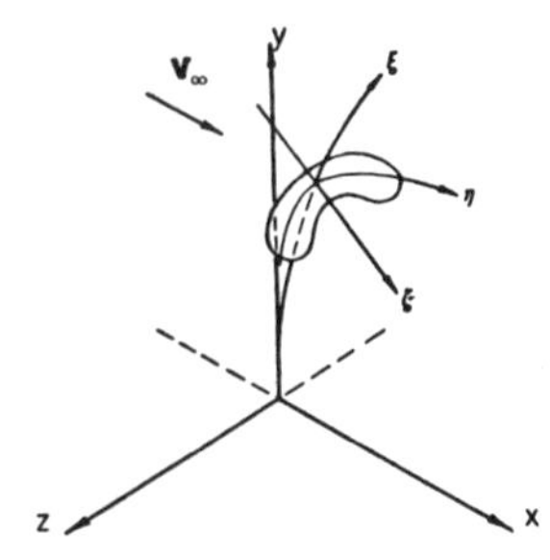

Figure 6. Coordinate system for transverse jets.

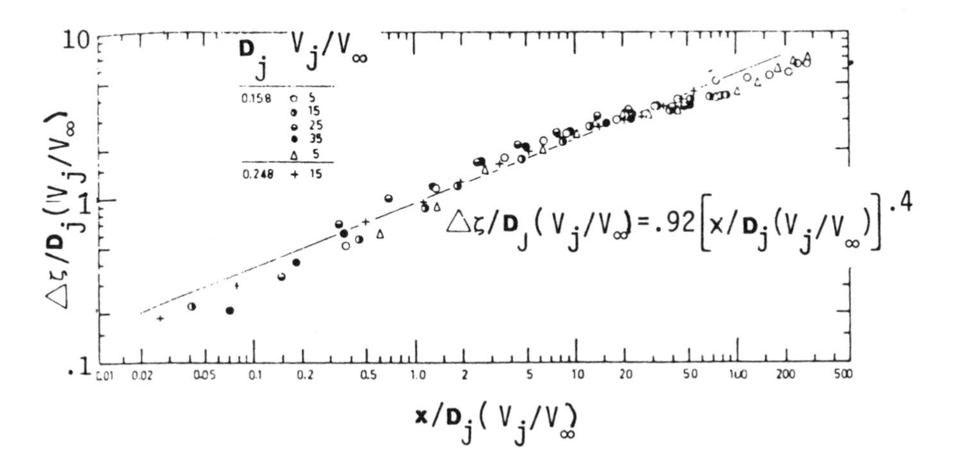

Figure 7. Spread in the ζ direction of a transverse jet [3].

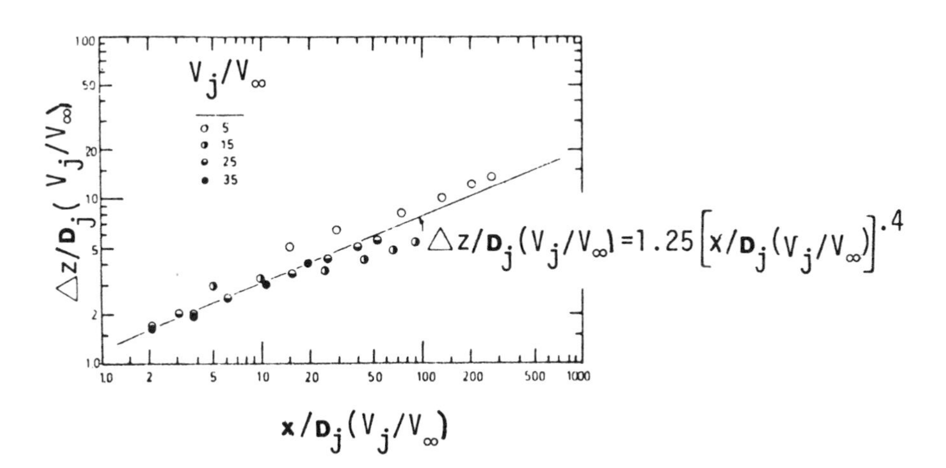

Figure 8. Spread in the z direction of a transverse jet [3].

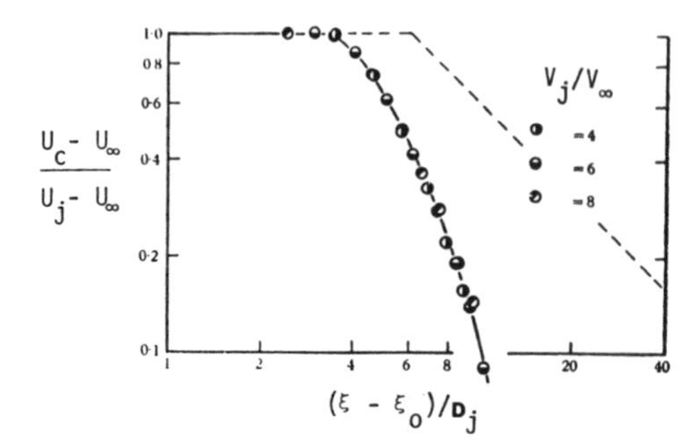

Figure 9. Variation of the centerline velocity along the jet trajectory measured from the virtual origin [4].

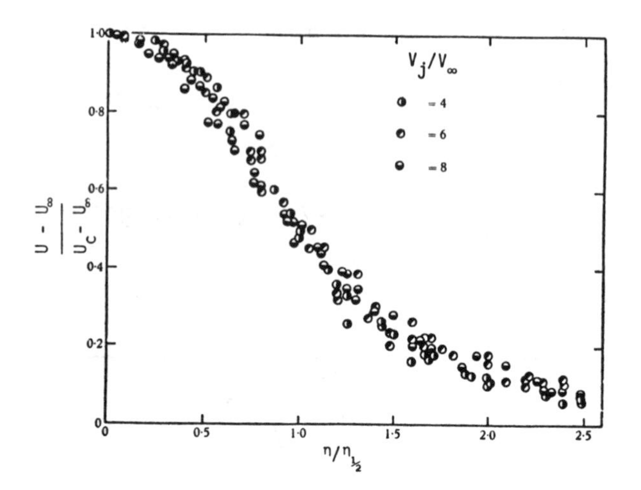

Figure 10. Transverse profiles across the jet [4].

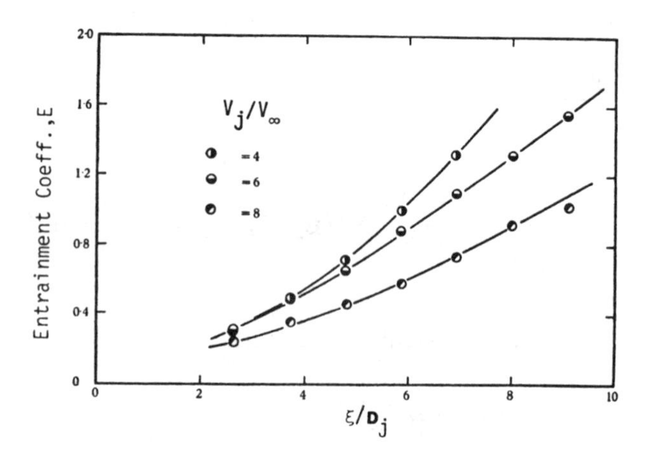

Figure 11. Entrainment coefficient as a function of arc length along the jet trajectory [4].

the transverse jet mixes faster, probably a direct result of the vortices induced in the jet plume. Velocity profiles across the plume are shown in Figure 10, where a near-similarity condition can be observed. Entrainment into the jet plume has been measured, and Figure 11 shows some results from Reference 4.

In Reference 5, velocity measurements in the jet are used to infer the properties of the vortices in the plume via back-calculation. The strength of the vortices as a function of injection angle, velocity ratio, and downsteam distance is shown here in Figure 12. It can be seen that the influence of injection angle is significant only for the larger velocity ratio. Information on vortex core size and spacing is also given in Reference 5.

The variation of the axial turbulence intensity along the jet centerline for three velocity ratios is shown in Figure 13. Also shown for comparison are some coaxial jet results from Reference 6. The turbulence is higher in the transverse jet case, which is consistent with the more rapid mixing of the mean flow noted earlier.

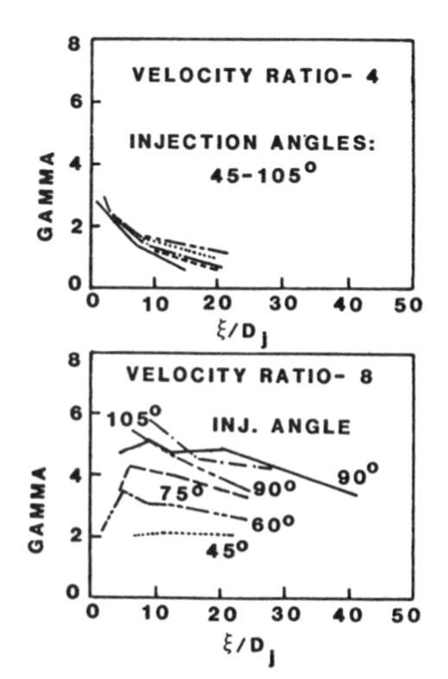

Figure 12. Strength of the vortices induced in transverse jets [5].

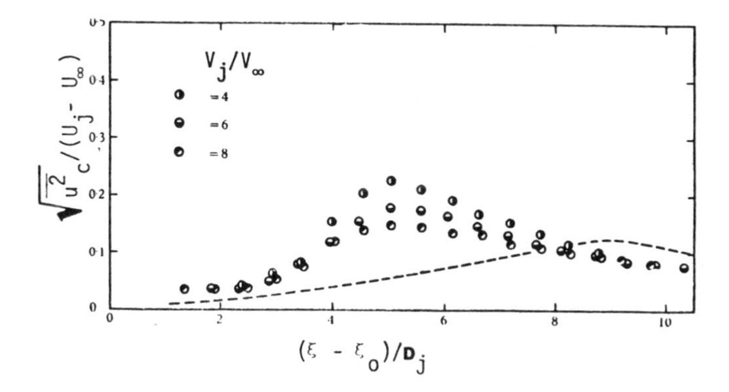

Figure 13. Turbulence intensity along the trajectory of a transverse jet [4].

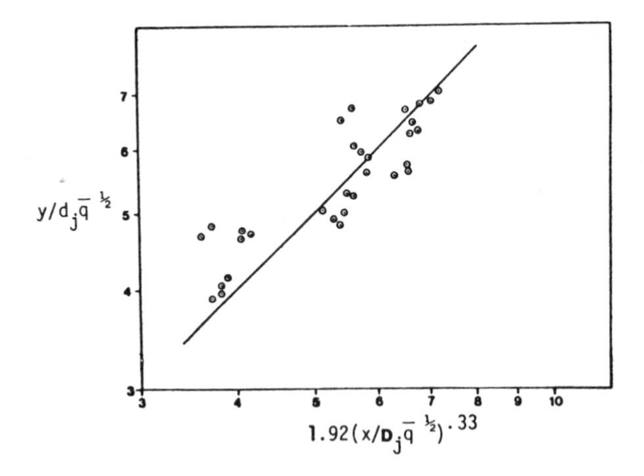

Figure 14. Correlation of plume centerline trajectory for a transverse particle-laden jet [7].

Transverse, Particle-Laden Jets

The situation of a particle-laden fluid jet injected transverse to a fluid stream is of interest because of the possible inertial effects of the particles. One might anticipate an appreciable "separation" of the injectant fluid and particle streams under the influence of the cross flow, at least for large particles.

The range of the variables that has been studied is very limited, and it is difficult at this time to arrive at firm general conclusions. In references 7 and 8, gas jets loaded with small particles (e.g., 15-μ silicate in Reference 7) were injected at 90 deg. to low-speed airstreams. The primary result is the trajectory of the center of mass of the solid phase under various conditions. Although the scatter in the data is large, the results of Reference 7 are correlated reasonably well (see Figure 14) by the equation

$$y/D_j = 1.92(\bar{q})^{0.335}(x/D_j)^{0.33} \tag{1}$$

where $\bar{q} \equiv \rho_j V_j^2/\rho_\infty V_\infty^2$

The data of Reference 8 indicate about 15% greater penetration. Both of these sets of data show a greater penetration (by roughly 20–30%) than single-phase cases at the same $\bar{q}$.

Transverse Jets into Supersonic Flows: Gaseous Jets

Transverse injection into supersonic streams is of engineering interest in several applications, including thrust vector control, fuel injection, and thermal protection,. Since such a jet presents an "obstruction" to the main supersonic flow, an "interaction shock" is generally produced, and the whole flow differs from low-speed cases in important ways. In Reference 9, the case of a sonic, underexpanded transverse jet was considered, and the resulting flowfield shown in Figure 15 was related to that for an underexpanded jet into quiescent surroundings. This relationship was accomplished through the notion of an "effective" back pressure, P_{eb}. It was then possible to use the experimental and theoretical results for the quiescent surroundings case to correlate data for the height (penetration) of the Mach disk in the transverse jet flow. This is shown in Figure 16 using $P_{eb} = 0.8P_2$, where P_2 is the static pressure behind a normal shock in the main flow. Similar

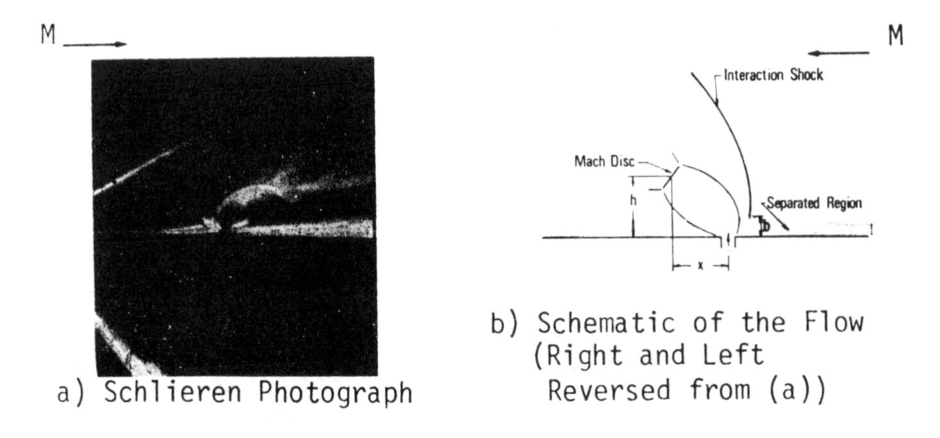

Figure 15. Flowfield observed for an underexpanded transverse gas jet into a supersonic stream: (A) Schlieren photograph; (B) schematic of the flow (right and left reversed from A) [9].

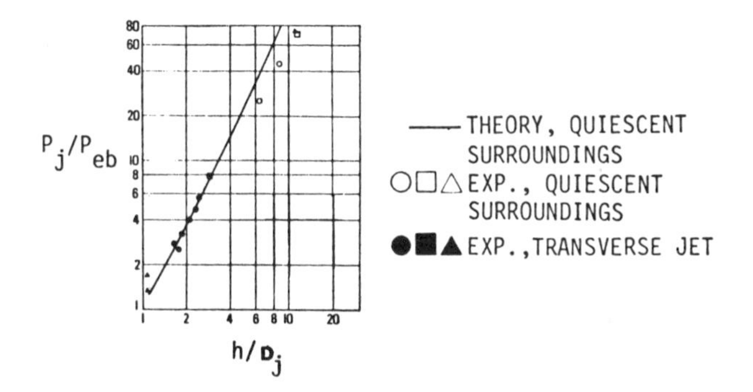

Figure 16. Correlation of the Mach disk height in an underexpanded transverse gas jet into a supersonic stream using theory and experiment for an underexpanded jet into quiescent surroundings [9].

results for supersonic, underexpanded transverse jets are given in Reference 10. The effect of the shape of injectors with the same cross-sectional area on the height of the Mach disk was studied in Reference 11 and found to be unimportant. From a practical viewpoint, these results together show that it is impractical to try to increase transverse gaseous penetration substantially into a supersonic crossflow by increasing injection pressure markedly, varying the shape of the injector, or employing supersonic injection.

In addition to the initial penetration of the transverse jet as given here in terms of the height of the Mach disk, the subsequent trajectory of the jet and mixing along that trajectory are also of interest. Some results for underexpanded, sonic, H_2 injection with $M = 2.7$ from Reference 11 are plotted in Figure 17.

The shape of the interaction shock is important in some applications, and some results are shown in Figure 18. The results of a simple analysis based upon an equivalent solid body obstruction are also shown, along with the less successful "blast-wave" model.

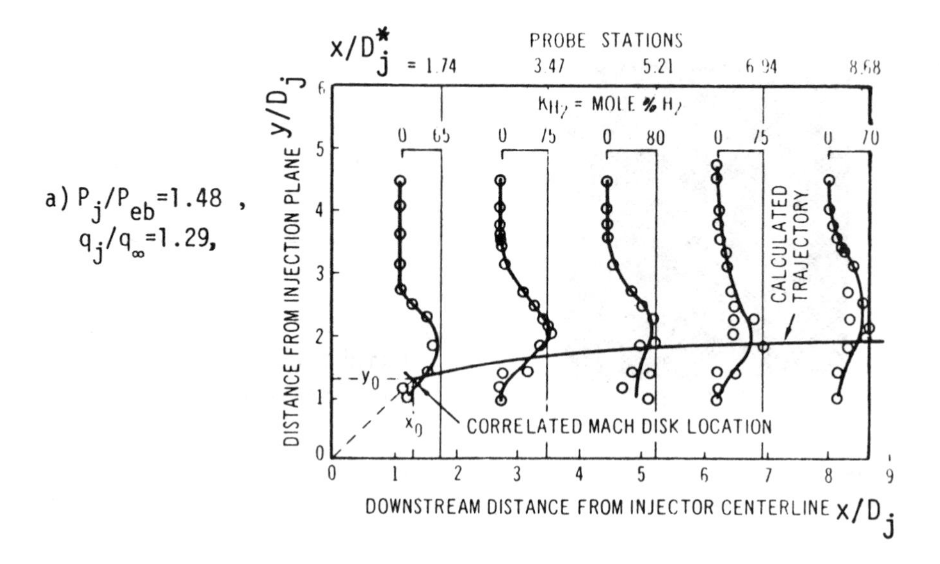

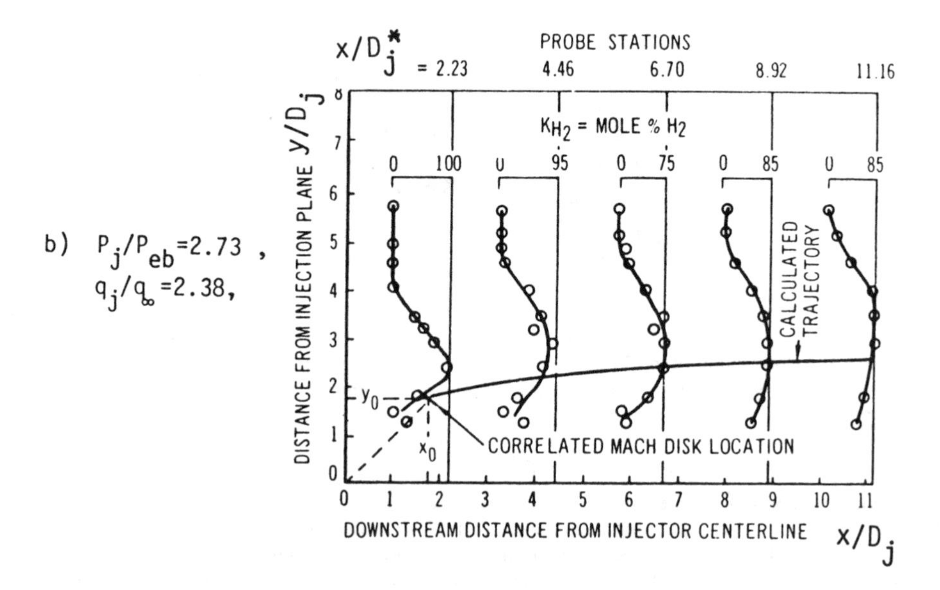

Figure 17. Comparison of prediction and experiment for an underexpanded transverse H_2 jet into $M_\infty = 2.7$ air [11].

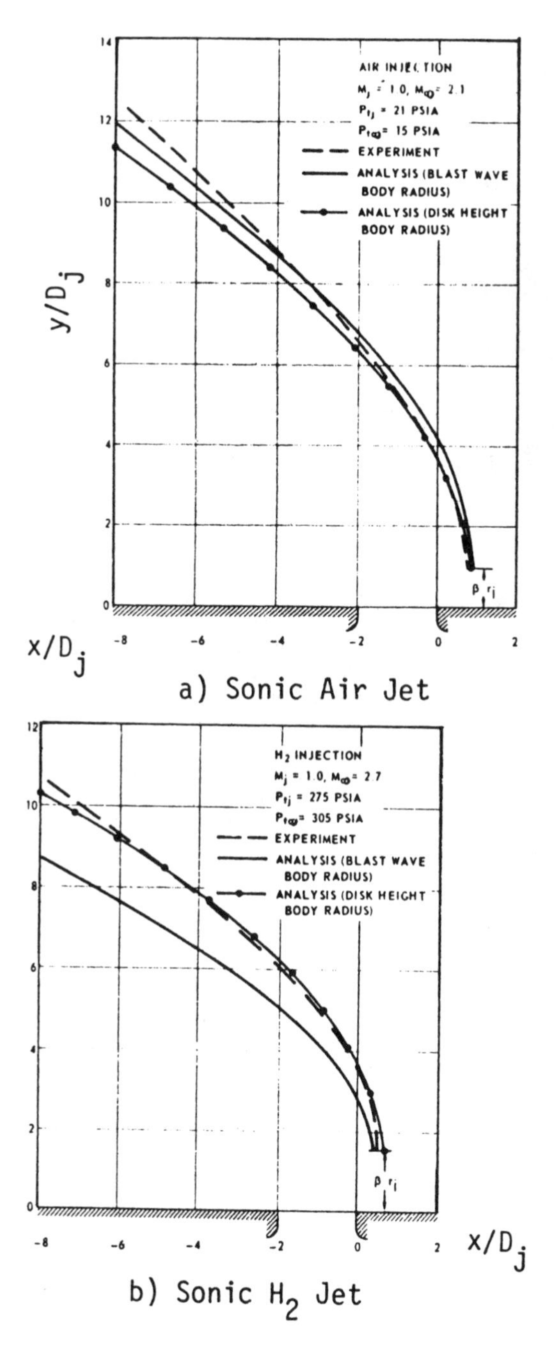

Figure 18. Comparison of predictions and experiment for the interaction shock shape of transverse gas jets into supersonic flow: (A) sonic air jet; (B) sonic H_2 jet [12].

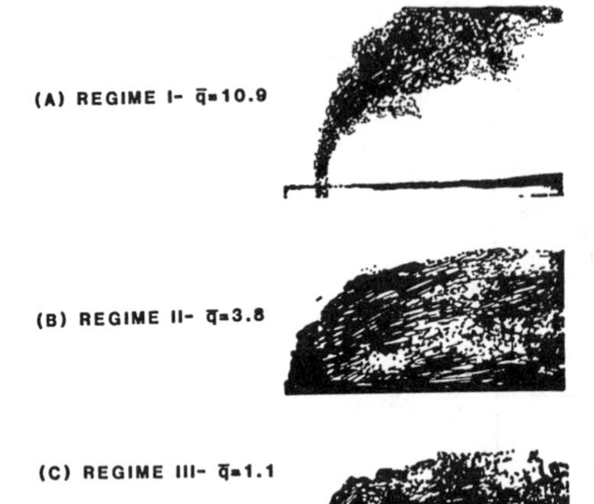

Figure 19. Spark photographs of transverse liquid jet injection into $M_\infty = 2.4$ air [13].

Transverse Jets into Supersonic Flows: Liquid Jets

This, of course, represents a two-phase flow, but the matters of primary interest here will be the gross penetration and behavior of the jet, not droplet processes. The interaction of the liquid jet column with the main supersonic stream is highly unsteady, and only high-speed (~20,000 pictures/s) motion pictures are really adequate to display all of the features of the flow. Some stop-action (10^{-6} s) stills are given in Figure 19 to show how the flow-field develops for various ranges of the important parameter $\bar{q}$.

Different workers have developed correlation formulas for the gross penetration of the liquid plume into the cross flow, including the effects of various parameters. Taking the influence of $\bar{q}$ from Reference 13, x/d from Reference 14, the aspect ratio d_f/d_s of the injector from Reference 15, and injection angle θ from new work, an overall correlation was developed in Reference 16. The

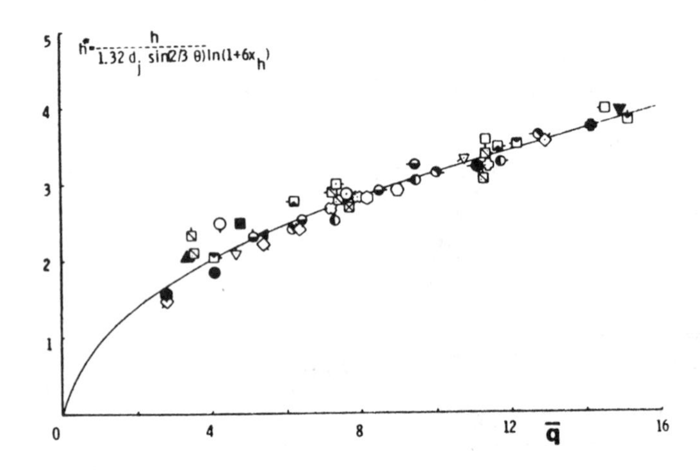

Figure 20. Correlation of liquid jet penetration into supersonic flow [16].

complete formula is

$$\frac{h}{D_j} = 1.32(\bar{q})^{1/2}C_d\left(\frac{d_{eq}}{d_f}\right)^2\left(\frac{d_f}{d_s}\right)^{0.46}\ln\left[1+6\left(\frac{x}{D_j}\right)\right]\sin\left(\frac{2\theta}{3}\right) \tag{2}$$

The adequacy of the correlation is indicated in Figure 20 for a number of varied experimental cases.

It should be noted that the liquid jet penetration results do show useful increases in penetration with a suitable change in injector shape and /or increased injection pressure. This is in contrast to gaseous jet results where the formation of the Mach disk in the jet flow plays a dominant role.

Information on the variation of droplet sizes in the plume as a function of location, $\bar{q}$ and physical properties is available in References 17 and 18. Some data for a case with a subsonic cross-flow age is given in Reference 19.

Trajectory Analyses

In this class of analysis, one is concerned only with predicting the penetration and spread of the jet plume. The details of the flow such as velocity profiles are not treated directly. A schematic of a model of this type is given here as Figure 21. Looking at an element of the jet column, the various overall forces are modeled, and the trajectory of the column may be predicted. One of the earliest such analyses was developed in Reference 1; however, the variation of normal momentum was neglected, and it proved necessary to use unrealistically high values of the drag coefficient acting on the jet column to obtain satisfactory results. That assumption was relaxed in Reference 20, and further refinements were introduced in Reference 2. The "theory" curves on Figures 1 and 2 are the result, and the rather good agreement with experiment can be noted.

It is a simple matter to add an energy equation to this type of analysis and thus treat heated or cooled jet cases. This was done in Reference 2, and the results are compared with experiment in Figure 22. The effects of a nonuniform crossflow can also be easily incorporated as in Reference 2, but no detailed experiments were available for comparison.

For the supersonic main flow cases, the only additional information required is suitable drag coefficient data at the normal Mach numbers of interest. If the jet flow is at least sonic and underexpanded, then the trajectory analysis is begun at the Mach disk location. The calculated trajectories on Figure 17 were obtained in that way using the analysis of Reference 20. It can be observed that there is little further penetration after the Mach disk. This is a result of the total pressure loss through the Mach disk (shock). The work was carried somewhat further in Reference 21, from which Figure 23 is taken. Quite good predictions of the gross behavior of the jet are clearly

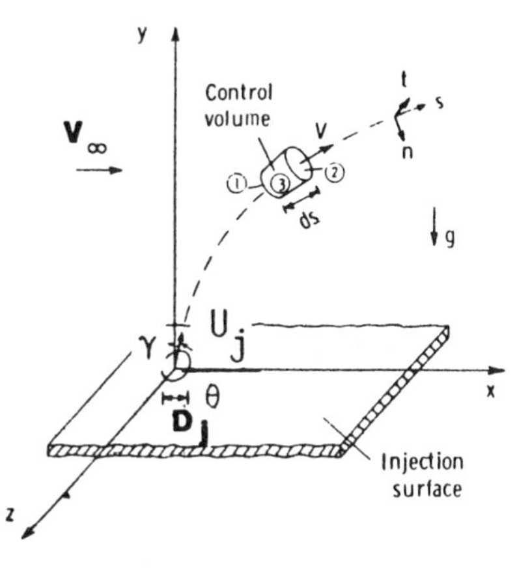

Figure 21. Flowfield model for transverse jet trajectory analysis.

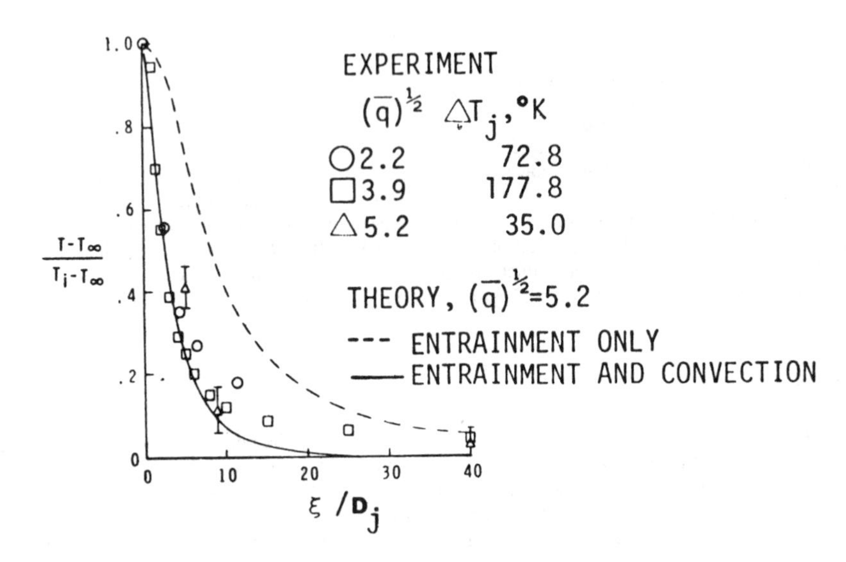

Figure 22. Comparison of prediction and experiment for temperature variation along the jet trajectory for transverse injection of a heated jet [2].

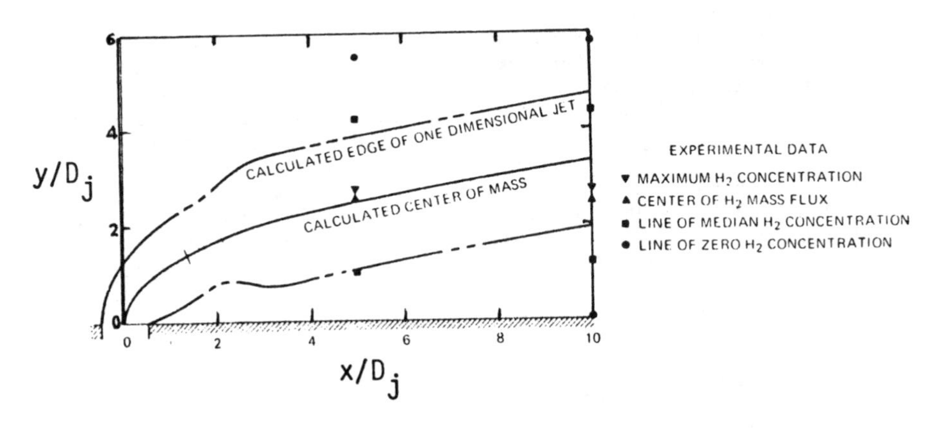

Figure 23. Comparison of prediction with the jet penetration analysis and experiment for a transverse H_2 jet into $M_\infty = 2.7$ air with $P_{0i}/P_0 = 1.6$.

obtainable with this level of analysis. For liquid jets into either low-speed or supersonic cross-flows, the breakup of the jet column influences penetration directly, and a simple trajectory analysis of the type discussed here is adequate only until the jet column "fractures."

Surface Pressure Distributions

One of the applications of transverse jets is to vertical and short take-off and landing (V/STOL) aircraft. In that application, the pressure field induced on adjacent surfaces is of particular importance. Thus, there have been a number of detailed experimental studies of that part of the flowfield covering many of the important variables and parameters (see References 22 through 32). Reviews of the early work can be found in References 33 and 34, and an up-to-date tabulation on the available information is contained in Reference 35. The jet generally induces negative (with respect to the freestream) pressures on the nearby surfaces, and this results in a net loss of lift on the body viewed as a whole. The longitudinal variation of the surface pressures is also important, since that determines the resulting pitching movement.

We will present data for single jets in comparison with that for dual jets in the next section.

DUAL JETS

Some of the applications of jets in a cross-flow involve dual-jet arrangements, either in-line or side-by-side. The mutual interference as a function of center-to-center spacing is the issue here. Until recently, few references (e.g., References 23, 28, and 36) existed. The work of References 37 and 38 has provided further information. Another item that has recently been identified as a prime candidate for further study is the behavior of a jet (or jets) injected from a body of revolution as opposed to the large flat plates usually considered. This is of obvious importance for VTOL aircraft with lifting jets in the fuselage. One can anticipate substantial transverse pressure "relief" around a cylindrical body. The only early work is Reference 39 which considered a case where $D_{jet}/D_{body} \ll 1$. That is not realistic for VTOL aircraft where $D_{jet}/D_{body} \approx \frac{1}{2}$ can be encountered. Reference 37 contains some data for that case. Another effect that has received little careful study until recent times is the effect of the angle of the jet with respect to the cross-flow. That is important because the transition to wingborne operation for VTOL aircraft is most commonly accompanied by a change in the angle of the jet thrust vector. There are few early investigations in the literature (see References 28, 31, and 40). Lastly, the interplay of the three items mentioned here over a range of the key parameter for all such flows, $R \equiv V_{jet}/V_{stream}$, is clearly of importance to the designer (see References 37 and 38).

Jet Trajectories

There is not a great deal of information on dual jet trajectories available. Figure 24 shows some measurements of the jet flowfield above the surface made with a yawhead probe. The intersection

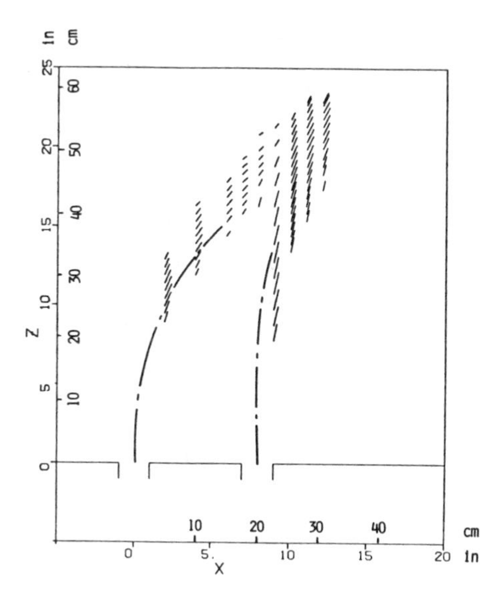

Figure 24. Comparison of trajectory prediction with analysis from Reference 42 and velocity vector data: 90 deg., R = 6.5, $S/D_j = 4$ from Ref. (37).

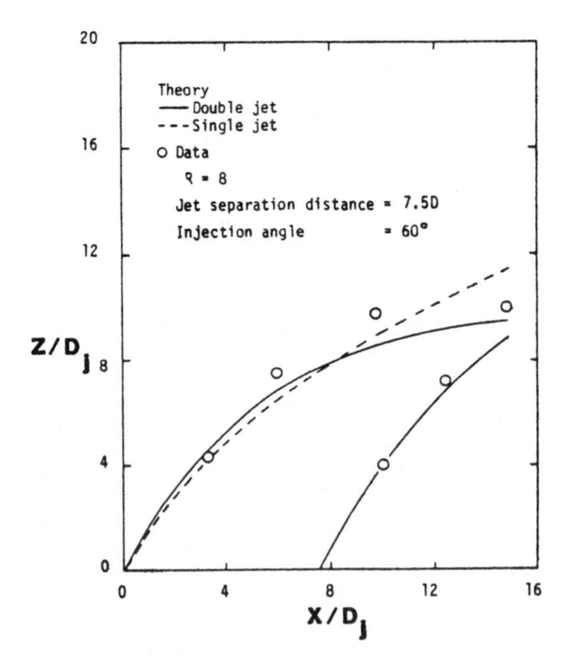

Figure 25. Jet centerline trajectories [36] and prediction [42].

region with two jets at R = 6.5 is displayed. One can observe that the rear is "sheltered" strongly by the front jet; the trajectory of the rear jet is nearly vertical until the intersection. Jet centerline measurements for another case are given in Figure 25. Other data is presented in Reference 41.

The simple trajectory analysis discussed earlier for single jets was extended to dual-jet arrangements in Reference 42. The key new item required was measurements of the drag of dual solid cylinders at various spacings. The ability of that simple analysis to predict the trajectory and dual jets is demonstrated in Figures 24 and 25. Obviously, the main features of the flow are accurately predicted. Other cases are considered in Reference 42 with equivalent results.

Surface Pressures on a Flat Plate

Longitudinal pressure distributions at selected lateral distances are plotted in Figure 26 for 90° injectors at R = 6 with a spacing of two diameters and the jets aligned one behind the other. The results for both jets are shown as circles, and those for a single jet only are stars.

Looking at the single jet results first, the expected pattern of negative ΔC_p's $[\Delta C_p \equiv (C_{p\text{jet on}} - C_{p\text{jet off}})]$ is evident. The pressure falls in the immediate vicinity of the jet, due largely to entrainment of external fluid into the jet.

The dual jet results show that the influence of the rear jet is less than that of the front jet at this close spacing. On the other hand, the presence of the rear jet seems to strengthen the influence of the front jet slightly.

The effects of the important parameter R can be seen in the isobar plots in Figure 27 A, B, and C. As R increases, the area of the surface influenced by the jets increases. This increase is mostly in terms of the areas with small to moderate negative values of ΔC_p (e.g., $0 \geq \Delta C_p \geq -1.0$). Since the area of influence increases with R, the total normal force also increases with R, but the increase

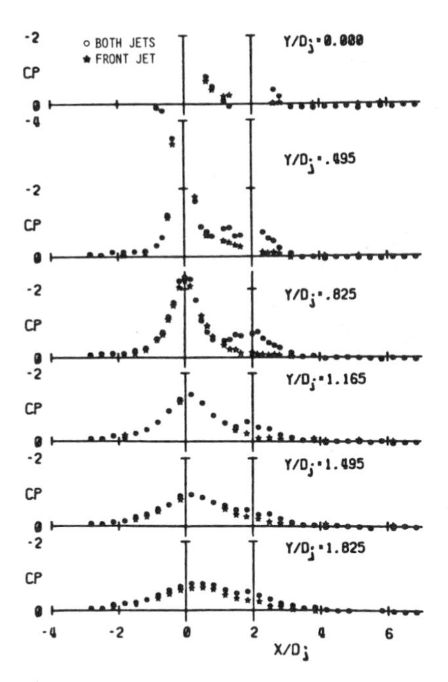

Figure 26. Surface pressure distribution ΔC_p on a flat plate (90 deg.): $R = 6$, $S/D_j = 2$ [38].

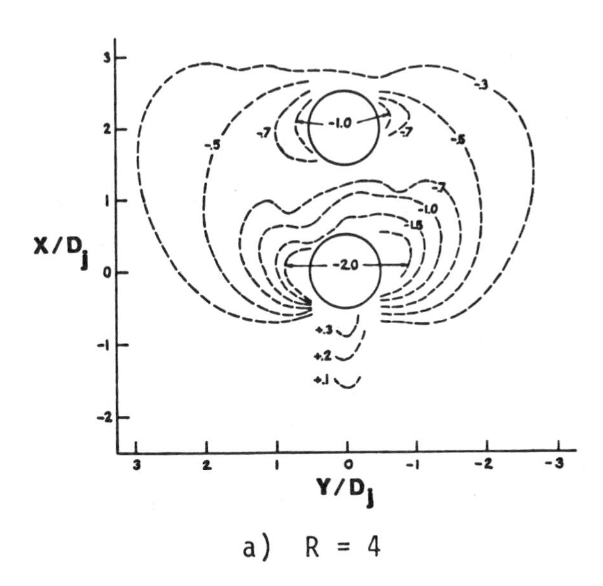

Figure 27. Isobar map of ΔC_p on a flat plate (both jets, 90 deg., $S/D_j = 2$, various R): (A) $R = 4$; (B) $R = 6$; (C) $R = 8$ [38].

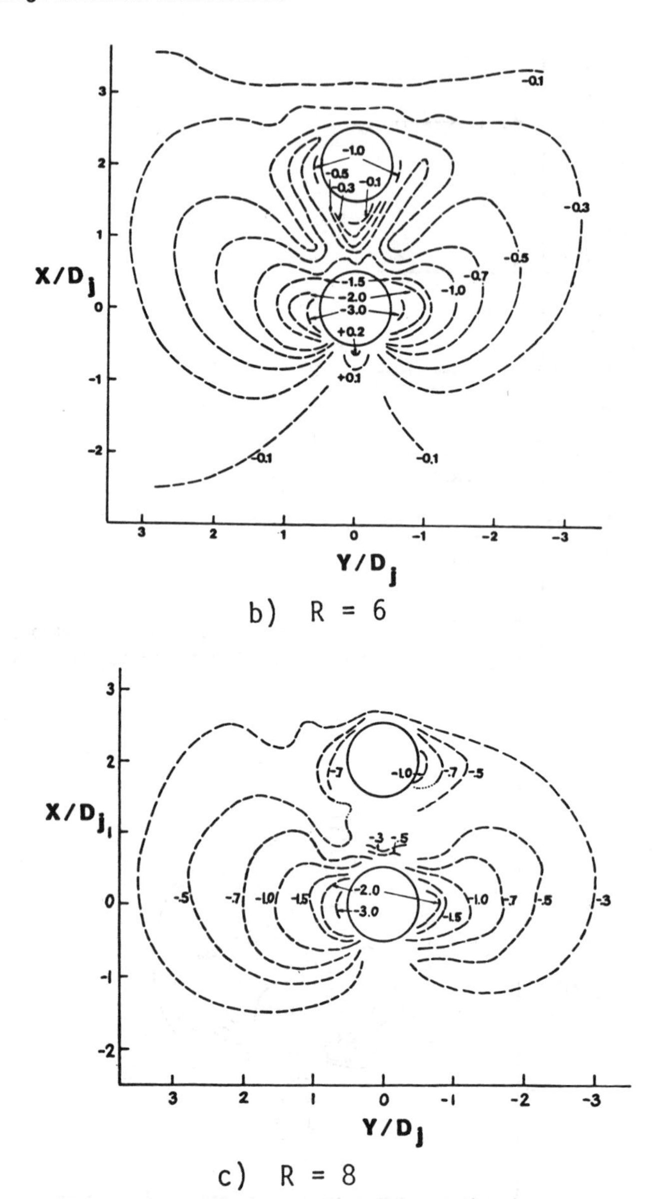

Figure 27 (B) and (C). (Continued)

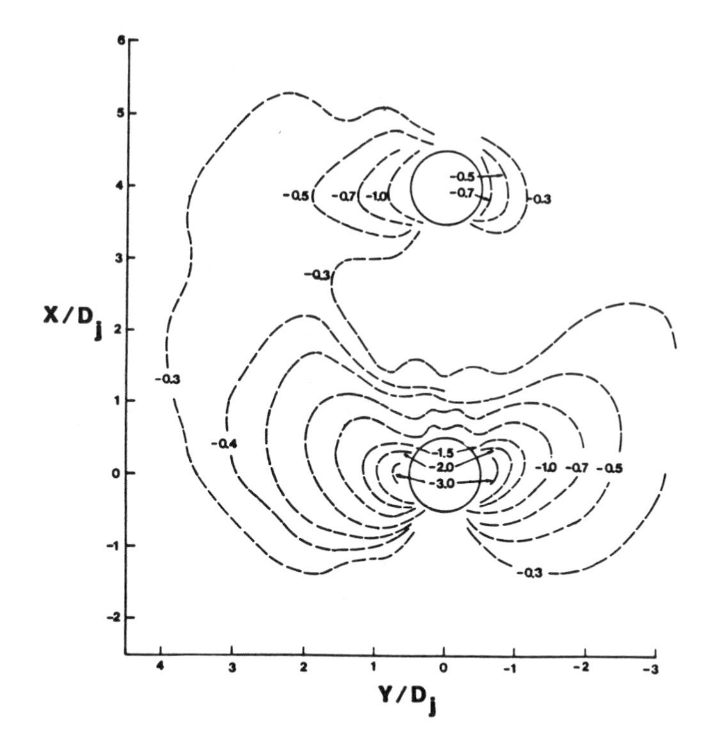

Figure 28. Isobar map of ΔC_p (both jets, 90 deg.): R = 6, S/D_j = 4 [38].

is slow, and the normal force normalized with the thrust of the jets actually decreases with increasing R. The effective center of force moves forward with increasing R. Estimates indicate that this center coincides with the center of the front jet at about R = 10. Lastly, the shape of the interaction region changes with increasing R. At low R, the isobars show asymmetrical lobes displaced in the downstream direction.

The effects of increasing the dimensionless spacing, S/D_j from 2 to 4 with 90° and R = 6 are shown in Figure 28. The rear jet is still sheltered by the front jet, but the influence of the front jet on the flow is reduced by the presence of the rear jet. Looking at the isobar patterns in Figure 28 and comparing with those in Figure 27B for the same case with $S/D_j = 2$, one can observe that the only overlap of the interaction regions of the two jets is now for small ΔC_p values only. Comparing isobar plots with the front jet only (not presented here due to space limitations) with those for both jets operating shows that the interaction of the two jets increases the surface area influenced by the front jet. The merging of the interaction regions was also found to be influenced by the velocity ratio, R. The merging is most pronounced at low R values. Also, the sheltering of the rear jet reduces the downstream distortion of its flow interaction area.

The influence of injection angle at R = 6 is shown in the next two figures. Figure 29 has results for a 75° angle, and Figure 30 has results for upstream injection at 105°, all at $S/D_j = 4$. The isobar plots in Figures 29 and 30 and that for 90° in Figure 28 show some interesting effects. The change from 75° to 90° produces only slight changes in the total interaction area influenced by the jets and, thus, in the normal force. However, the change from 90° to 105° leads to an increase in the total interaction area and, hence, also the normal force. Further, as the angle goes from 75° to 90° to 105°, the effective center of the interaction region moves forward. The data indicate that the

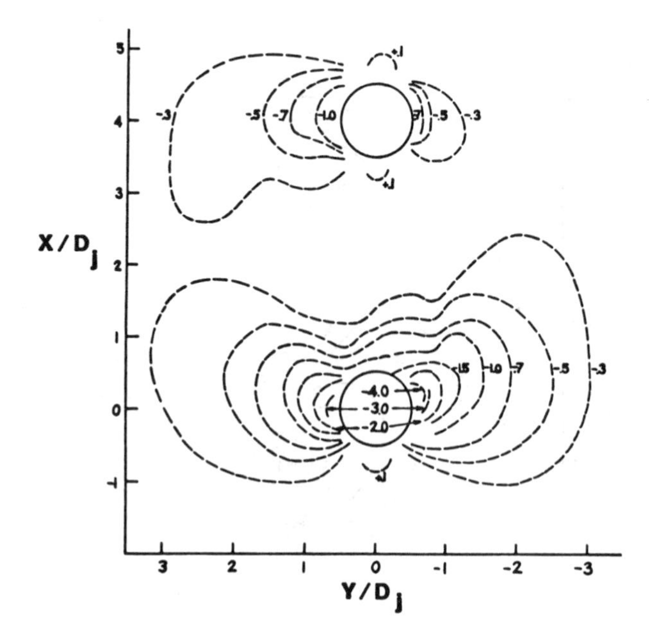

Figure 29. Isobar map of ΔC_p (both jets, 75 deg.): $R = 6$, $S/D_j = 4$ [38].

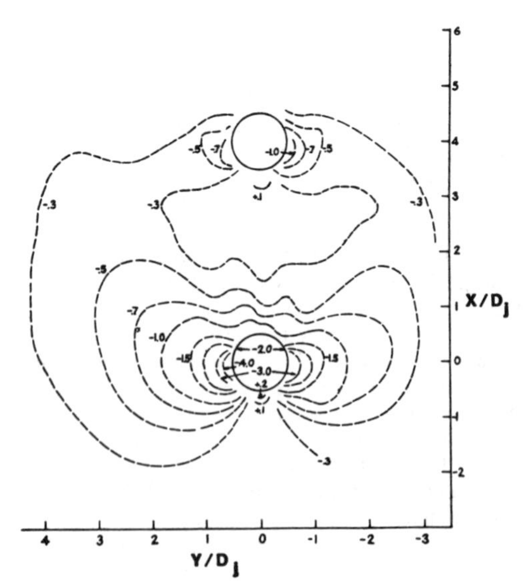

Figure 30. Isobar map of ΔC_p (both jets, 105 deg.): $R = 6$, $S/D_j = 4$ [38].

region ahead of the front jet is influenced somewhat more strongly by upstream angled injection. In addition, the sheltering of the rear jet by the front jet is stronger for 105° than for 90° or 75° injection.

The next series of figures present the results obtained with two jets in a side-by-side arrangement. Fewer parameters were varied in this configuration; all the data obtained are for the 90° injectors only. Figure 31A, B, and C shows isobar maps for $S/D_j = 2$ with R = 4, 6, and 8. As R is increased, there is a pronounced increase in the size of the interaction region around the jets. The interaction

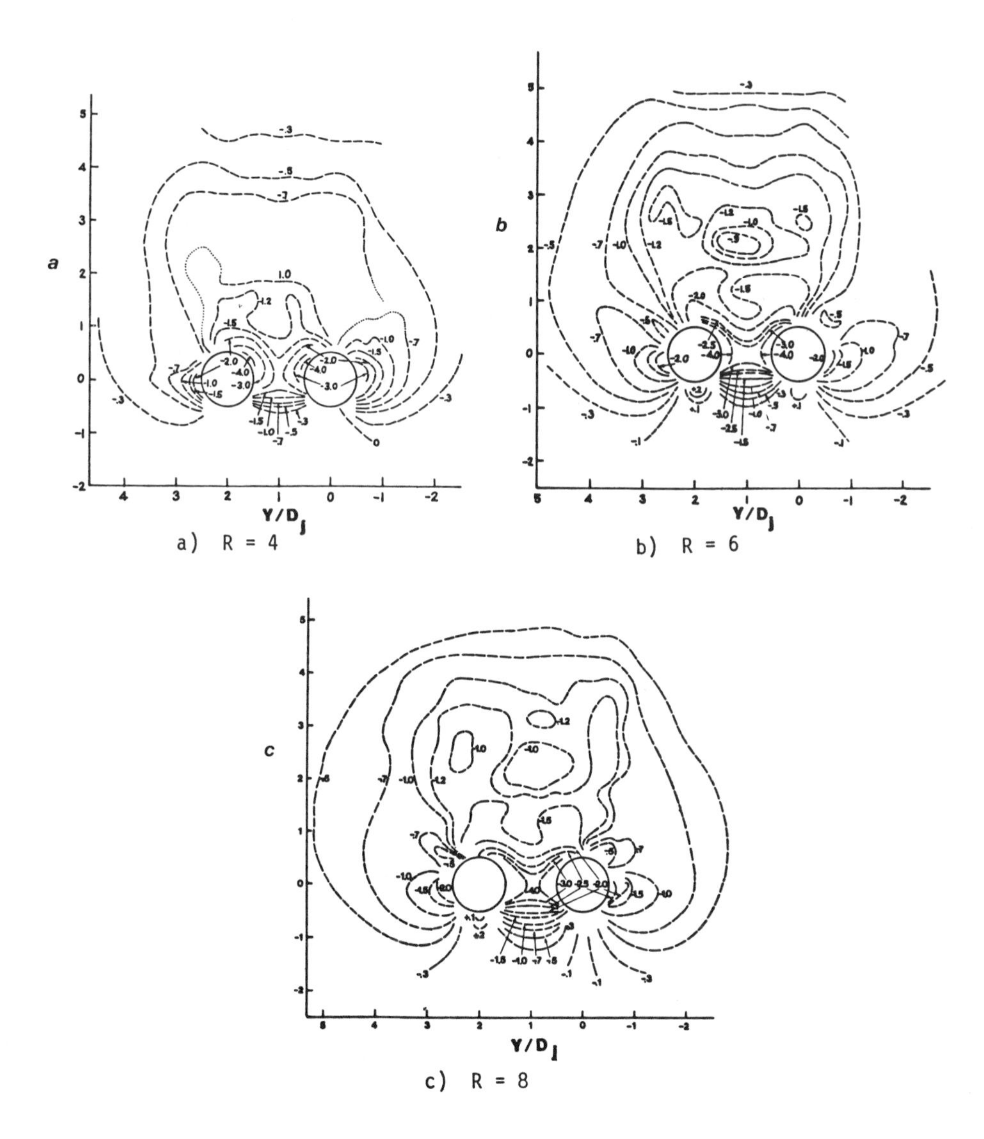

Figure 31. Isobar map of ΔC_p (both jets, 90 deg. $S/D_j = 2$, side-by-side, various R); (A) R = 4; (B) R = 6; (C) R = 8 [38].

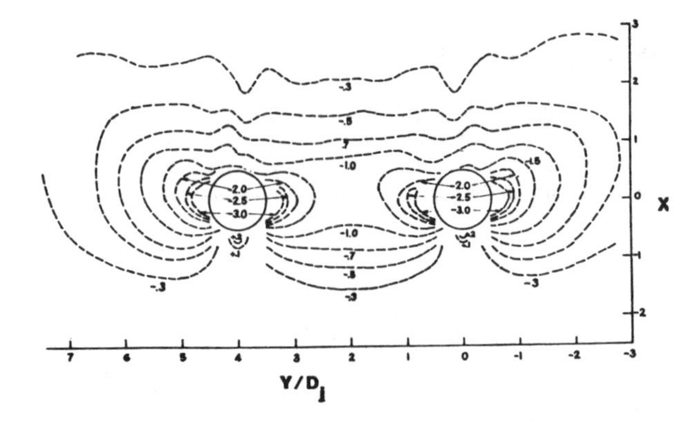

Figure 32. Isobar map of ΔC_p (both jets, 90 deg.): R = 6, S/D_j = 4, side-by-side [38].

normal force also increases, both as a result of the increase in area affected and increases (negative) in the ΔC_p values near the orifices. The effective center of the interaction force shifts upstream as R is increased. Taken altogether, these results indicate a strong increase in the interaction between the two jets as R is increased at this close spacing, $S/D_j = 2$. The large (negative) ΔC_p values in the region immediately behind the jets along the line of symmetry between them at the high R's is noteworthy. This is probably a result of the interaction between the pairs of counter-rotating vortices formed in each jet.

The influence of increasing S/D_j to 4 at R = 6 can be seen by comparing the results in Figure 32 to those given earlier for $S/D_j = 2$ at the same R in Figure 31B. The interaction of the jets is

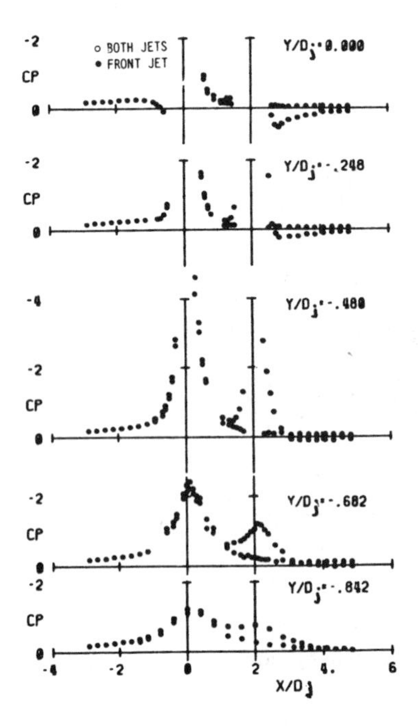

Figure 33. Surface pressure distribution on a body of revolution model (90 deg.): R = 7.7, S/D_j = 2 and contoured exit [37].

diminished significantly. There is much less overlap of the interaction regions of the two jets at $S/D_j = 4$. The overlap or merging of the interaction regions is R dependent; it is strongest at the higher R values. The interaction region of each jet on the "free" side (the side away from the other jet) is still always larger than for the single jet even at $S/D_j = 4$.

Surface Pressures on a Body of Revolution

Reference 37 gives surface pressure distributions on a body of revolution with $D_j/D_b \approx 1/2$. The results obtained for $R = 7.7$, 90° injection through nozzles with exits contoured to the body surface at a spacing of 2.0 diameters are shown in Figure 33 as open circles. The results obtained for injection from the front nozzle only with the exit of the rear nozzle carefully covered by tape are shown as the solid circles. The pattern is qualitatively similar to that found on flat plates (see, e.g., Figure 26 for $R = 6$). A more detailed comparison permits some interesting conclusions to be made. First, one finds that there are no positive ΔC_p values behind the second jet on the flat plate. Second, the peak values on the body of revolution are somewhat higher very near the nozzles. Third, by comparing values at $y/D_j = 0.8$, it can be seen that the peak values decay much faster with lateral distance on the body of revolution.

NOTATION

D	diameter	T	temperature
E	entrainment coefficient	U	velocity
h	vertical penetration	V	velocity
M_∞	Mach number	x	coordinate
P	pressure	y	distance

Greek Symbols

δ	width	ξ	arc length along the jet centerline trajectory
θ	injection angle	ρ	density

REFERENCES

1. Abramovich, G. N., *The Theory of Turbulent Jets*, MIT Press, Cambridge, Mass., 1960 (English edition).
2. Campbell, J. F., and Schetz, J. A., "Analysis of Injection of a Heated Turbulent Jet into a Cross Flow," NASA TR R-413, December 1973.
3. Pratte, B. D., and Baines, W. D., "Profiles of the Round Turbulent Jet in a Cross Flow," *Proceedings of ASCE, Journal of the Hydraulics Division*, November 1967, pp. 56–63.
4. Keffer, J. F., and Baines, W. D., "The Round Turbulent Jet in a Cross-Wind," *Journal of Fluid Mechanics*, Vol. 15, Pt. 4 (1963), pp. 481–496.
5. Krausche, D., Fearn, R. L., and Weston, R. P., "Round Jet in a Cross Flow: Influence of Injection Angle on Vortex Properties," *AIAA Journal*, Vol. 16 (June 1978), pp. 636–637.
6. Corrsin, S., and Uberoi, M., "Further Experiments on the Flow and Heat Transfer in a Heated Turbulent Air Jet," NACA TN 1895, 1949.
7. Salzman, R. N., and Schwartz, S. H., "Experimental Study of a Solid-Gas Jet Issuing into a Transverse Stream," *Journal of Fluids Engineering*, Vol. 100 (Sept. 1978), pp. 333–339.
8. Rudinger, G., "Some Aspects of Gas-Particle Jets in a Cross Flow," American Society of Mechanical Engineers, Paper 75-WA/HT-5, 1975.
9. Schetz, J. A., Hawkins, P. F., and Lehman, H., "The Structure of Highly Underexpanded Transverse Jets in a Supersonic Stream," *AIAA Journal*, Vol. 5 (May 1967), pp. 882–884.

10. Schetz, J. A., Weinraub, R., and Mahaffey, R., "Supersonic Transverse Jets in a Supersonic Stream," *AIAA Journal*, Vol. 6 (May 1968), pp. 933–934.
11. Orth, R. C., Schetz, J. A., and Billig, F. S., "The Interaction and Penetration of Gaseous Jets in Supersonic Flow," NASA CR-1386, July 1969.
12. Schetz, J. A., "Interaction Shock Shape for Transverse Injection," *Journal of Spacecraft and Rockets*, Vol. 7 (Feb. 1970), pp. 143–149.
13. Kush, E. A., Jr., and Schetz, J. A., "Liquid Jet Injection into a Supersonic Flow," *AIAA Journal*, Vol. 11 (Sept. 1973), pp. 1223–1224.
14. Yates, C. L., and Rice, J. L., "Liquid Jet Penetration," Applied Physics Lab., Johns Hopkins Univ., Research and Development Programs Quarterly Rept. U-RQR/69-2, 1969.
15. Joshi, P. B. and Schetz, J. A., "Effect of Injector Geometry on the Structure of a Liquid Jet Injected Normal to a Supersonic Airstream," *AIAA Journal*, Vol. 13 (Sept. 1975), pp. 1137–1138.
16. Baranovsky, S. I., and Schetz, J. A., "Effect of Injection Angle on Liquid Injection in Supersonic Flow," AIAA Journal, Vol. 18 (June 1980), pp. 625–629
17. Nejad, A. S., and Schetz, J. A., "Effects of Properties and Location in the Plume on Mean Droplet Diameter for Injection in a Supersonic Stream," *AIAA Journal*, Vol. 21, No. 7, pp. 956–961 (July 1983).
18. Nejad, A. S., and Schetz, J. A., "The Effects of Viscosity and Surface Tension of Liquid Injectants on the Structural Characteristics of the Plume in a Supersonic Airstream," Paper No. 82-0253, January 1982.
19. Schetz, J. A., Hewitt, P., and Situ, M., "Transverse Jet Break-up and Atomization with Rapid Vaporization along the Trajectory," AIAA Journal, Vol. 23, (April 1985), pp. 596–603.
20. Schetz, J. A., and Billig, F. S., "Penetration of Gaseous Jets Injected into a Supersonic Stream," *Journal of Space-craft and Rockets*, Vol 3, (Nov. 1966), pp. 1658–1665.
21. Billig, F. S., Orth, R. C., and Lasky, M., "A Unified Analysis of Gaseous Jet Penetration," *AIAA Journal*, Vol. 9 (June 1971), pp. 1048–1058.
22. Vogler, Raymond D., "Surface Pressure Distributions Induced on a Flat Plate by a Cold Air Jet Issuing Perpendicularly From the Plate and Normal to a Low-Speed Free-Stream Flow," NASA TN D-1629, 1963.
23. Vogler, Raymond, D., "Interference Effects of Single and Multiple Round or Slotted Jets on a VTOL Model in Transition," NASA TN D-2380, August 1964.
24. Bradbury, L. J. S., and Wood, M. N., "The Static Pressure Distributions Around a Circular Jet Exhausting Normally From a Plane Wall into an Airstream," C. P. No. 822, Brit, A. R.C., 1965.
25. Margason, Richard, J., "Jet-Induced Effects in Transition Flight," Conference on V/STOL and STOL Aircraft, NASA SP-116, 1966, pp. 177–189.
26. Gentry, Garl, L., and Margason, Richard, J., "Jet Induced Lift Losses on VTOL Configurations Hovering In and Out of Ground Effect," NASA TN D-3166, February 1966.
27. Soullier, A., "Testing at SI. MAS for Basic Investigation on Jet Interactions. Distributions of Pressures Around the Jet Orifice," NASA TTF-14066, April 1968.
28. Fricke, L. B., Wooler, P. T., and Ziegler, H., "A Wind Tunnel Investigation of Jets Exhausting Into a Cross-Flow," AFFDL-TR-70-154, Vols. I-IV, U. S. Air Force, December 1970.
29. Margason, Richard J., "Review of Propulsion-Indicated Effects on Aerodynamics of Jet/STOL Aircraft," NASA TN D-5617, February 1970.
30. Fearn, R. L., and Weston, R. P., "Induced Pressure Distribution of a Jet in a Crossflow," NASA TN D-7916, June 1975.
31. Taylor, P., "An Investigation of an Inclined Jet in a Crosswind," Aeronautical Quarterly, Vol. XXVIII, Part I, February 1977.
32. Kuhlman, J. M., Ousterhout, D. S., and Warcup, R. W., "Experimental Investigation of Effects of Jet Decay Rate on Jet-Induced Pressures on a Flat Plate: Tabulated Data," NASA CR-158990, November 1978.
33. Lee, C. C., "A Review of Research on the Interaction of a Jet With an External Stream," Tech. Note R-184, (Contract No. DA-01-021-AMC-11528 (z)), Res. Lab., Brown Engineering Co., Inc., March 1966. (Available from DDC as AD 630 294.)

34. Garner, Jack E., "A Review of Jet Efflux Studies Application to V/STOL Aircraft," AEDC-TR-67-163, U. S. Air Force, September 1967. (Available from DDC as AD 658 432.)
35. Perkins, S. C., Jr., and Mendenhall, M. R., "A Study of Real Jet Effects of the Surface Pressure Distribution Induced by a Jet in a Crossflow," NASA CR-166150, March 1981.
36. Ziegler, H., and Wooler, P. T., "Analysis of Stratified and Closely Spaced Jets Exhausting Into a Crossflow," NASA CR-132297, November 1973.
37. Schetz, J. A., Jakubowski, A. K., and Aoyagi, K., "Jet Trajectories and Surface Pressures Induced on a Body of Revolution with Various Dual Jet Configurations," *J. Aircraft*, Vol. 20, No. 11 (November 1983), pp. 975–982.
38. Schetz, J. A., Jakubowski, A. K., and Aoyagi, K., "Surface Pressures Induced on a Flat Plate with In-Line and Side-by-Side Dual Jet Configurations," AIAA Paper No. 83-1849, July 1983.
39. Ousterhout, D. S., "An Experimental Investigation of a Cold Jet Emitting from a Body of Revolution into a Supersonic Free Stream," NASA CR-2089, July 1972.
40. Margason, Richard J., "The Path of a Jet Directed at Large Angles to a Subsonic Freestream," NASA TN D-4919, November 1968.
41. Isaac, K. M., "Experimental and Analytical Investigation of Multiple Jets in a Cross-Flow," Ph.D. thesis, Virginia Polytechnic Institute and State University, 1982.
42. Isaac, K. M., and Schetz, J. A., "Analysis of Multiple Jets in a Cross-Flow," J. Fluids Engrg., vol. 104 (1983), pp. 489–492.

CHAPTER 17

MODELING TURBULENT JETS IN CROSSFLOW

A. O. Demuren

University of Karlsruhe
Karlsruhe, F. R. Germany

CONTENTS

INTRODUCTION, 430

ANALYSIS, 432
General Remarks, 432
Empirical Models, 434
Integral Models, 438
Numerical Models, 446

CONCLUDING REMARKS, 460

NOTATION, 461

REFERENCES, 462

INTRODUCTION

The flow of turbulent jets in cross flow is difficult to predict accurately. Needless to say, the flow has such important engineering and atmospheric applications that the development of better and more accurate mathematical models is likely to continue for a long time. Over the years, the modeling trends have moved from those based mainly on empirical findings, which were therefore limited in their range of applicability to numerical models of a more general nature. This development has been aided by the introduction in rapid succession of larger and faster computers over the last couple of decades. However, although numerical models show a great deal of promise for universality, they are still in their infancy, and none of them has yet been applied to (or is indeed capable of) predicting the whole range of turbulent jets in cross flow to a reasonable degree of accuracy. Fairly good prediction has been obtained in relatively few cases. Between the two extremes lie integral methods, which have been widely applied to jet flow problems; but these are semi-empirical in nature and require many assumptions which limit their range of validity.

In practical engineering applications, turbulent jets in cross flow are found in both confined and unconfined environments. Examples of confined jets in cross flow exist in:

- Internal cooling of turbine blades by air jets impinging on the leading edge and on both the pressure and suction surfaces, thus the latter jets operate in a cross flow composed of air from the leading edge and from upstream jets.
- Dilution air jets in combustion chambers of gas-turbine engines, where the jets are injected radially into the chamber, through discrete holes along its circumference, in order to stabilize the combustion process near the head, and to dilute the hot combustion products near the end. At the initial stage, the jets penetrate to the axis where they impinge on each other and some of the air flow is deflected upstream.

Other examples include Vertical and Short Take-Off and Landing (V/STOL) aircraft in transition from hover to forward flight, in which case, the jets from its engines impinge on the ground surface. Practical examples of turbulent jets in unconfined (semi-infinite) cross flow are even more

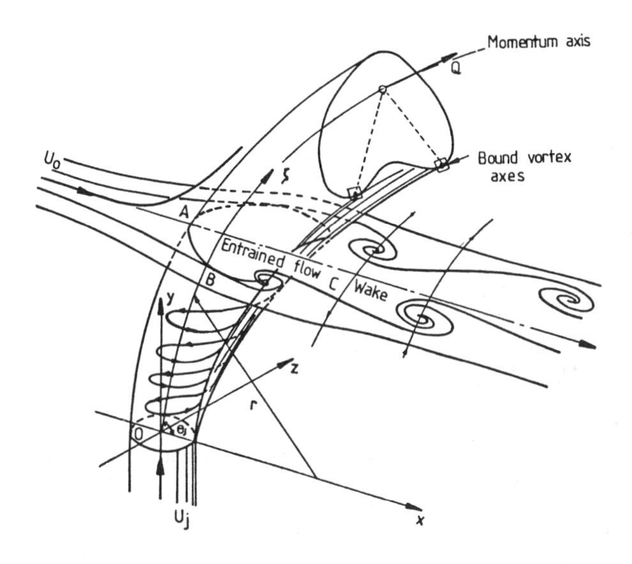

Figure 1. Flow field of a turbulent jet in cross flow [1].

numerous. These include flow situations resulting from the action of cross winds on effluents from cooling towers, chimney stacks, or flares from petrochemical plants; discharge of sewerage or waste heat into rivers or oceans; film cooling of turbine blades; the use of air curtains to prevent cold air from entering open spaces in industrial buildings. A similar flow situation may exist in the atmosphere when thermal plumes generated on the earth's surface rise to levels at which significant cross winds exist.

These practical flows are mostly quite complex, and in modeling them certain idealizations or simplifications are usually necessary. Thus, turbulent jets in unconfined cross flow may be studied by considering a flow configuration such as is depicted in Figure 1. This shows the evolution of a round jet issuing from a pipe into a nearly uniform cross flow, in the near field. The oncoming stream is obstructed by the jet in a way similar to that of a flow impinging on a rigid circular cylinder, but the boundaries of the jet are compliant and entraining. Further, the pressure gradient, induced across the jet, by the oncoming stream leads to its being bent over. The cross stream is itself accelerated along the jet boundary AB and the boundary interaction of the two flows leads to a periodical shedding of vortices behind the jet as in the von Karmàn-Benard street. Most of the vorticity issuing from the pipe is tilted and stretched by the flow, and bundles up into a pair of vortex tubes bound to the lee side of the jet. This gives the jet cross-section its characteristic kidney shape. The flow in the wake region is highly turbulent and a considerable entrainment of cross-stream fluid into the jet takes place through it. Beyond the near field, the jet is rapidly deflected, continously entraining cross-stream fluid, and far downstream its axis becomes nearly aligned with the cross-flow direction. However, the pair of counter-rotating vortices and the kidney shape are retained for a long distance downstream. Experimental studies [1–4] of this flow situation show that actual evolution of the near field is dependent on the geometry of the jet discharge hole, the jet-to-cross-flow velocity ratio, and the discharge angle. Further complexities arise when there are multiple jets, although there is experimental evidence [5–6] that these behave individually like single jets until just before merging. From the foregoing, it is evident that even for the idealized flow situation considered here, the flow field is still rather complex and it remains a difficult task for any mathematical model to simulate completely.

Figure 2 shows a typical (but again idealized) example of the confined turbulent jet in cross flow. Many features of the unconfined jet of Figure 1 may be recognized. Further, at higher jet-to-cross-stream velocity ratios the jet may impinge on the top surface, blocking the path of cross flow

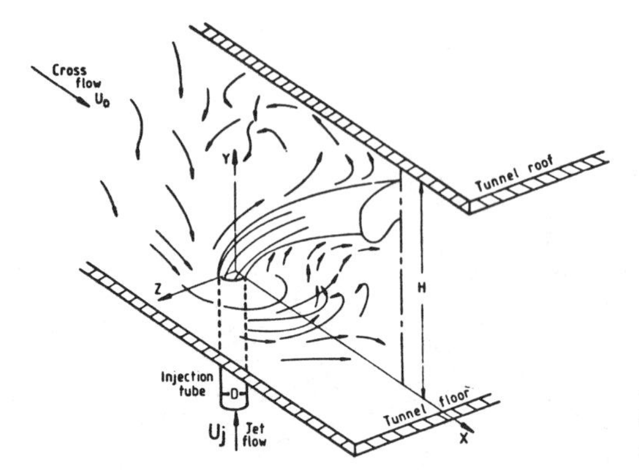

Figure 2. The turbulent jet in confined cross flow [8].

above the jet. If there is sufficient deceleration of the cross stream, a reverse flow region is formed near the top surface upstream of the jet [5, 7, 8].

On impingement of the jet with the top surface the characteristic kidney shape is destroyed (if it was at all able to develop) as the jet flattens against this surface and spreads in the transverse direction, with velocity maxima on either side of the plane of symmetry. At more moderate velocity ratios, the deflection of the jet and its bifurcation are only slightly pronounced by the effect of the confined cross flow. The flow of opposing jets in cross flow, studied experimentally by a number of investigators [5, 9, 10] also fall under this category. The symmetry plane between the jets seems to behave in much the same way as the opposite wall in confined jets. In the case of confined multiple jets in cross flow, it has been observed [5, 9] that, at low spacings, the jets rapidly coalesce laterally and the trend to two-dimensional flow may inhibit impingement by lowering the jet trajectory. Thus, there are certain similarities between the flow structure of turbulent jets in confined and unconfined cross flow, but also important differences. Smoke photographs by Kamotani and Greber [5] of confined single jets in cross flow at various discharge ratios and opposing channel wall spacings are reproduced in Figure 3. These give a good indication of the conditions under which confined jets are likely to behave significantly different from unconfined ones.

All the above experimental studies show that the main characterizing parameter for turbulent jets in cross flow is the jet-to-cross-flow velocity ratio R $(=U_j/U_0)$, or if the density of the jet is different from that of the cross flow, the momentum flux ratio, J $(=\rho_j U_j^2/\rho_0 U_0^2)$. In confined flows, the channel wall spacing H/D may be an important parameter, and in multiple jets, the jet spacings S/D together with their arrangement would be additional parameters. If the jet and cross-flow Reynolds numbers are sufficiently high, the flow development would be independent of the Reynolds number.

ANALYSIS

General Remarks

For the purpose of the analysis let us consider the plane of symmetry of a single round jet in cross flow. This is illustrated in Figure 4. The axis of the jet is usually defined [11] as the locus of the maximum velocity or total pressure in the plane of symmetry. The jet trajectory is usually referred to this line, as opposed to the center-line of the jet, which is mid-way between the inner and outer boundaries of the jet. The latter are often determined from photographs or flow visualization studies. Where these are not available the jet boundaries may also be derived from the velocity profiles. Where temperature profiles have been measured [12], they were found not to be coincident with the velocity profiles, so that the loci of the maximum values, and the half-widths were, in general, different.

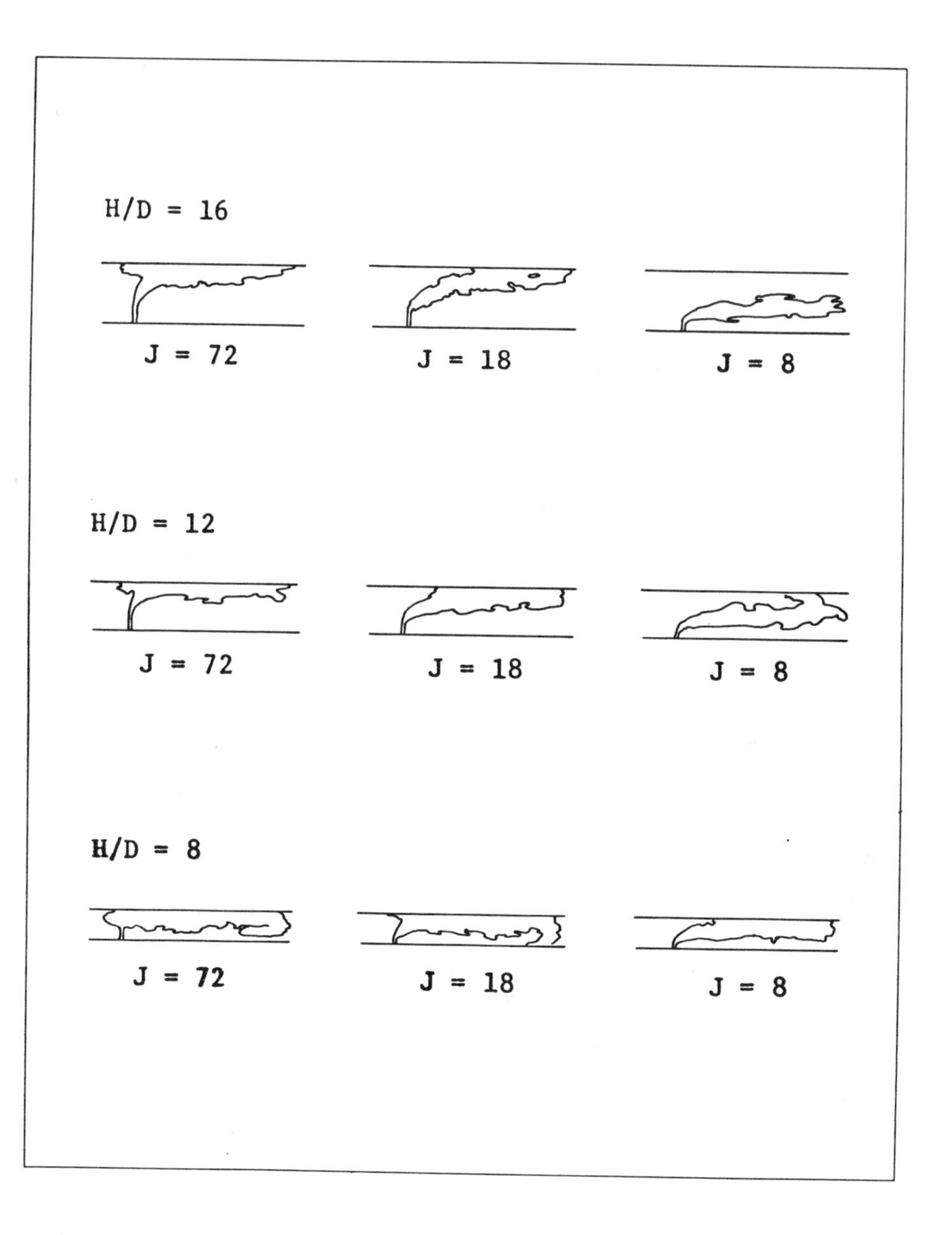

Figure 3. Tracings of smoke photographs of single jets in confined cross flow [5].

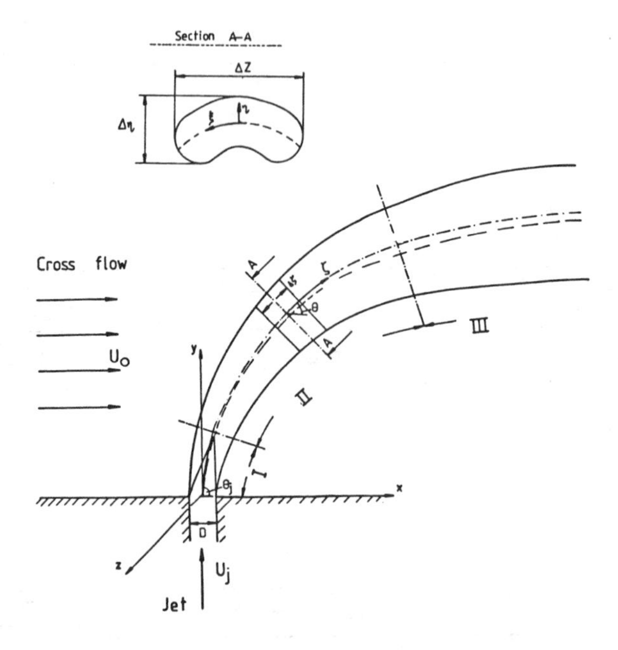

Figure 4. Description of coordinate systems.

Three regions of the jet are shown in Figure 4. The first zone is the potential core region within which the central portion of the jet remains relatively unaffected by the cross flow. At low R (less than about 4) it may be deflected slightly by the cross flow. Its length usually differs from that of free turbulent jets ($\sim$6D), depending on the jet-to-cross-flow velocity ratio R. From experimental data, Fan [13] proposed the relation $6.2De^{-3.3/R}$, which is also supported by the measurements of Pratte and Baines [14]. A simpler relation given by $6.4/(1 + 4.608/R)$ was proposed by Kamotani and Greber [12]. The two relations agree quite well for $R > 4$, but the relation of Fan decays much faster as R decreases. Thus, for $R = 1$, the former gives a potential core length of 0.23D, compared to 1.14D by the latter. The actual potential core length, at this velocity ratio, would be strongly dependent on the jet exit geometry and the cross-flow distribution.

The second region is the zone of maximum deflection or curvilinear zone where the jet experiences the most deflection. In the final region, the far-field, the jet axis approaches the cross-flow direction asymptotically. As in free jets, the velocity profiles in this region have been observed to be self-similar [15].

Models for predicting turbulent jets in cross flow may be divided into three broad classes, in ascending order of complexity, as empirical, integral, and numerical.

Empirical Models

Empirical models are the easiest means of predicting turbulent jets in cross flow. They depend largely on the correlation of experimental data, and the accuracy of the predictions may depend on the closeness of the particular problem of interest to the data base used for the correlation. Nevertheless, low cost and ease of use make them invaluable for first-order estimates or qualitative checks of results produced by more elaborate methods. Correlations exist in the literature, for many jet properties in the forms of equations or curves, the latter being the more common.

For the single circular jet injected normally to a cross flow the trajectory, based on maximum velocity (or total pressure), is usually given by an equation of the form [11, 12, 15] (for R in the range 1.4 ~ 50),

$$y/D = aR^{b}(x/D)^{c} \tag{1}$$

where, a has a value between 0.75 and 1.31, b varies between 0.74 and 1.0, and c between 0.33 and 0.39, depending on the experimental conditions. For example, Kamotani and Greber found that for R in the range 4–10, the trajectory was well correlated by

$$y/D = aR^{0.94}(x/D)^{0.36} \tag{2}$$

with a = 0.89, when the jet issued from a nozzle (flat initial profile) and a = 0.81 when it issued from a pipe (fully developed initial profile). This relation was also valid for heated jets with temperature excess of up to 180°C, ($\rho_j/\rho_0 = 0.625$), if R is replaced by $J^{1/2}$. However, the jet trajectory, based on the maximum temperature, was found to show a slight dependence on the density ratio as well and given by the relation

$$y/D = 0.73(\rho_j/\rho_0)^{0.11}J^{0.52}(x/D)^{0.29} \tag{3}$$

That the temperature trajectory usually lies below the velocity trajectory is confirmed by measurements of Ramsey and Goldstein [16] and numerical predictions by Demuren [17]. Kamotani and Greber [5] observed that these relations are also valid for jets in confined cross flow, even up to the point of contact, in cases with impingement on the opposing wall.

However, in plane jets discharged normally to a cross flow there is a marked influence of flow confinement on the trajectory. From experiments in the range of J from 8 to 72 and H/D from 12 to 44, the trajectory was found to be well correlated [5] by

$$y/D = 2.0J^{0.28}(x/D)^{0.5}(1 - e^{-0.7H/D}) \tag{4}$$

(for $y < 0.5H$). Note that the trajectory of the plane jet would be higher than that of a corresponding single jet, when the opposing wall is sufficiently far to be uninfluential. Trajectories for a row of jets in cross flow were found to decrease steeply from the plane jet ones, as jet spacings S/D were increased, to some minimum, after which they rose gradually, reaching nearly the single jet ones at a spacing ratio S/D of 12.

Correlations for predicting the trajectory of circular jets discharged at oblique angles θ into the cross flow are presented by Abramovich [18]. The first, due to Shandorov [19] has the form

$$x/D = J^{-1}(y/D)^{2.55} + \frac{y}{D}(1 + J^{-1})\cot\theta \tag{5}$$

and was derived from experiments of cold jets into hot air in the range of J from 2 to 22 and θ from 45° to 90°. The other relation is due to Ivanov [20], and may be expressed as

$$x/D = J^{-1.3}(y/D)^{3} + \frac{y}{D}\cot\theta \tag{6}$$

It was derived from experiments with J in the range from 12 to 1,000 and θ from 60° to 120°, and was also found to correlate correctly the trajectory of jets issuing from rectangular-sectioned (aspect ratios 1:5 and 5:1) orifices, if D was replaced by the corresponding hydraulic diameter. Abramovich pointed out that the two relations predict similar trajectories in the whole range of the experiments. Curves of trajectories of oblique jets in cross flow are also presented by Platten and Keffer [4], for the range of R from 4 to 8 and θ from 60° to 135°.

As we have discussed previously, the jet trajectory and the center-line (observed from photographs or other flow visualization techniques) do not usually coincide, so that different correlations would apply to the latter. Care should be taken when interpreting results presented in the literature, as the two terminologies are often mixed together. Pratte and Baines [4] found that for normal jets with R in the range of 5 to 35, the center-line was well correlated by the relation,

$$y/D = 2.05R^{0.72}(x/D)^{0.28} \tag{7}$$

(for x/D up to 1,000). Over a much smaller range of R (2 to 10) and x/D (< 10), Margason [21] found the relation

$$y/D = 1.59R^{0.67}(x/D)^{0.33} \tag{8}$$

to correlate to his experimental data. Curves of the center-line are also presented for oblique ($\theta = 30^\circ \sim 150^\circ$ and $R = 2 \sim 10$) circular jets in cross flow by the latter [21]. The former [14] give equations correlating the inner and outer boundaries of the jet as

$$y/D = 1.35R^{0.72}(x/D)^{0.28} \quad \text{and} \quad y/D = 2.63R^{0.72}(x/D)^{0.28} \tag{9}$$

respectively.

Although, empirical models have mostly been applied in the literature to predict jet trajectories, Rajaratnam and Gangadharaiah [22] have shown that velocity distributions and spreading rates in the far-field of a normal jet in cross flow can also be predicted with such a model. In this region the velocity profiles are self-similar and the well known exponental distribution

$$(U - U_0)/(U_m - U_0) = e^{-0.693n^2} \tag{10}$$

where n is the normalized distance $\eta/\eta_{1/2}$ or $\xi/\xi_{1/2}$, and U_m the axial or maximum velocity. The velocity scale $(U_m - U_0)$ and the length scales $\eta_{1/2}$, $\xi_{1/2}$ are obtained from empirical relations. Holdeman and Walker [23] devised an elaborate empirical model for predicting the temperature distribution downstream of a row of cold jets injected normal to a hot confined cross flow. The model assumes that by properly nondimensionalizing the vertical temperature profiles they can be expressed in self-similar form. The scaling parameters are expressed as functions of the independent variables J, S/D, H/D, X/H, and Z/S. Figure 5 shows the flow configuration investigated and some typical comparisons of model predictions with experimental data. The agreement is quite good in most cases, but this should be expected since the model was calibrated with the same experimental data base.

Another important parameter often required in jet modeling is the entrainment velocity U_e. This is the average velocity at which cross stream fluid is entrained into the jet. Keffer and Baines [24] proposed the relation

$$U_e = E(\bar{V} - U_0) \tag{11}$$

where $\bar{V}$ is the average jet velocity at the cross-section and E is an entrainment coefficient which is a function of R and ζ. Curves are presented to show the variation of E with ζ at R = 4, 6, 8, in their experiments. Kamotani and Greber [12] argued that the entrainment velocity should be related to the components of the velocity differences between jet and cross flow in the axial and transverse directions, thus U_e should be given by

$$U_e = E_1(V_{max} - U_0 \cos\theta) + E_2U_0 \sin\theta \tag{12}$$

where V_{max} is the axial velocity of the jet, θ is the local angle between the jet axis and the cross flow, and E_1 and E_2 are entrainment coefficients which are functions of R only. There was not enough data to derive reliable relations for E_1 and E_2, but they found that their measured entrainment ratios were well reproduced by Equation 12 if $E_1 = 0.070$ and $E_2 = 0.320$, for R = 4; $E_1 = 0.061$ and $E_2 = 0.240$, for R = 6; $E_1 = 0.067$ and $E_2 = 0.182$, for R = 8. This is somewhat supported by integral model calibration studies carried out by Carhart et al. [25]. They found that field data of cooling tower plumes ($R = 0.3 \sim 2.0$) were best represented using Equation 12, with $E_1 = 0.125$ and $E_2 = 0.575$.

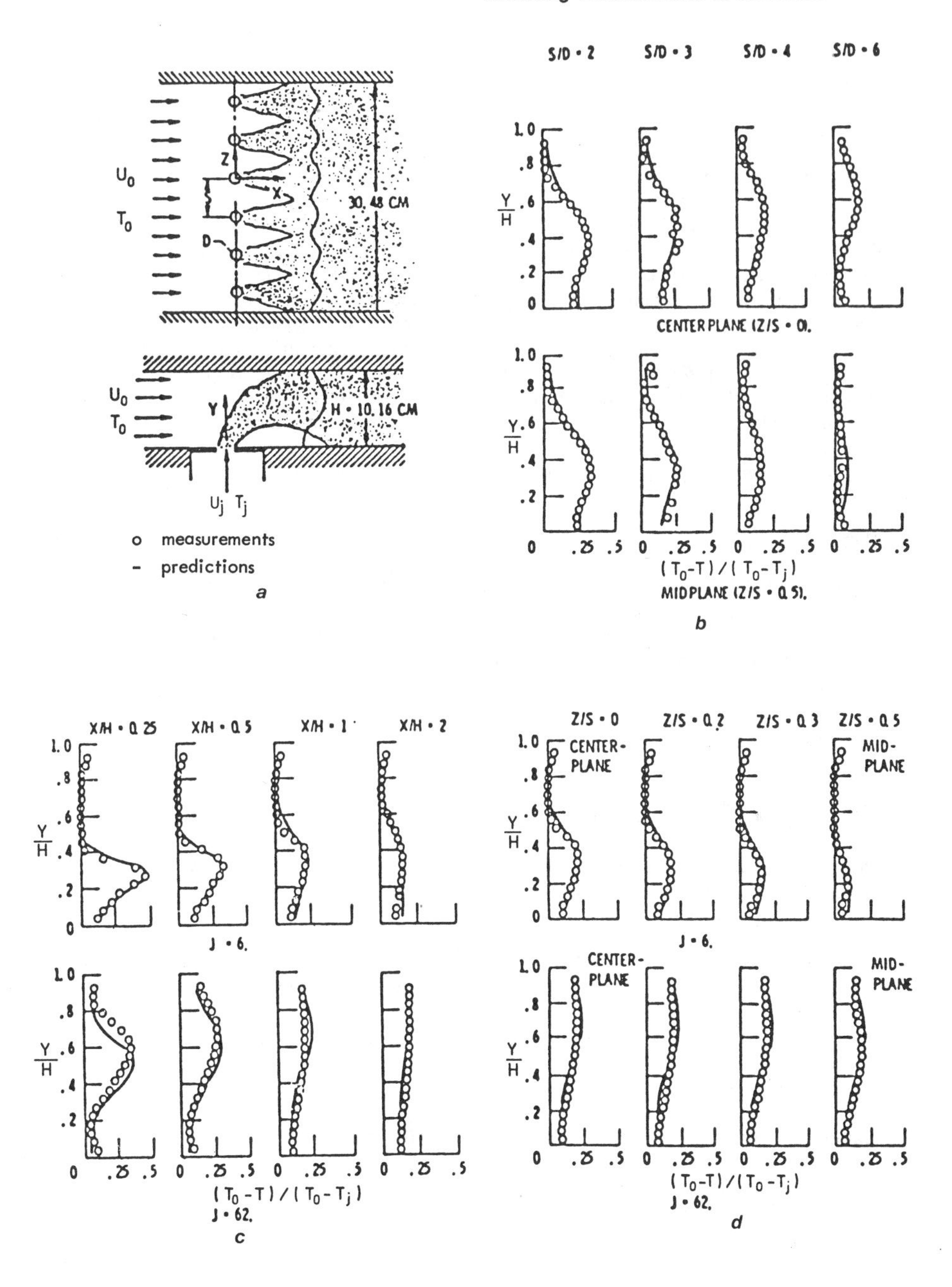

Figure 5. Flow configuration and empirical model predictions: (A) flow configurations; (B) influence of spacing ratio, x/H = 1, H/D = 8, J = 25; (C) profiles in center plane, s/D = 4, H/D = 8; (D) profiles in lateral plane, x/H = 1, s/D = 4, H/D = 8 [23].

Integral Models

Integral models are the first elaborate calculation procedures applied to predict the behavior of jets in cross flow. In these models integral equations are derived either by considering a balance of forces acting over an elementary control volume of the jet or by integrating in two special dimensions, the three-dimensional partial-differential equations governing the turbulent jet flow. In each case, the resulting equations are a set of ordinary differential equations which are solved analytically or numerically. Many physical phenomena such as pressure drag, entrainment of cross-stream fluid, and spreading rates are simulated by way of empirical relations. In order to obtain analytical solutions, it was often necessary to make some simplifying assumptions which could not always be justified on physical grounds. Although analytical solutions were employed in many of the earlier models, the later ones mostly required numerical solution of the final working equations. The approach of deriving the integral equations by integrating the governing partial-differential equations has the advantage over the control-volume one in that it is more transparent on the assumptions made and it affords more flexibility in dealing with complex trajectories. However, it involves tedious mathematical manipulations, for which reason most investigators have opted for the control-volume approach along lines first developed by Morton, Taylor, and Turner [26] for straight, vertical plumes. It appears that only Hirst [27] and Schatzmann [28] have presented models using the former method.

Earlier Models

Rajaratnam [11] presented a review of the main features of some of the earlier models. Volinsky and Abramovich [18] developed a model for predicting the trajectory of a jet in cross flow by assuming that the momentum of the jet in the direction normal to the cross flow was preserved and that the pressure difference across the jet was balanced by the centrifugal force due to its curvature. These led to the following equations:

$$A\rho\bar{V}^2 \sin\theta = \pi/4 D^2 \rho_j U_j^2 \sin\theta_j \tag{13}$$

$$C_D/2\rho_0 U_0^2 \sin^2\theta \, \Delta z \, d\zeta = \rho_0 A \frac{d\zeta \, \bar{V}^2}{r} \tag{14}$$

where C_D = a drag coefficient
A = the cross-sectional area of the jet
r = its radius of curvature

The kidney-shaped cross-section of the jet was approximated with an ellipse having a semi-major axis of 2.25D and a semi-minor axis of 0.45D in length. Abramovich further assumed that

$$\Delta z = 2.25D + 0.22\zeta, \text{ further simplified to,}$$

$$\Delta z = 2.25D + 0.22x \tag{15}$$

Equations 13 through 15 were combined to obtain the trajectory of the jet as

$$y/D = (39\Omega)^{1/2} \ln\left[\frac{10 + x/D + (x/D)^2 + 20(x/D) + 7\Omega \cot^2\theta_j}{10 + (7\Omega)^{1/2}\cot\theta_j}\right] \tag{16}$$

where $\Omega = \dfrac{1}{C_D} J \sin\theta_j$

Abramovich recommended a value of 3 for C_D, but much higher values have been found [11] necessary to obtain agreement with experimental data, in some cases. A somewhat similar model, by Crowe and Riesebieter [29], assumed that the axial momentum of the jet was preserved, which led to the equation:

$$A\rho V^2 = \pi/4D^2\rho_j U_j^2 \tag{17}$$

(in contrast to Equation 13).

Δz was assumed to be given by the following empirical relation:

$$\Delta z = D + 5.2J^{-0.29}\zeta \tag{18}$$

or, for low R (less than 5)

$$\Delta z = D + 5.2J^{-0.29}y \tag{19}$$

Combining Equations 14 and 17 with 18 or 19 gives the trajectory equations

$$x/D = \int_0^n \sinh\left[\frac{n + 5.2J^{-0.29}\int_0^n \frac{\zeta}{D}\,dn}{\pi/4(J/C_D)}\right] dn \tag{20}$$

or, for low R

$$x/D = \int_0^n \sinh\left[\frac{n + 2.6J^{-0.29}n^2}{\pi/4(J/C_D)}\right] dn \tag{21}$$

where $n = y/D$, and the drag coefficient C_D was given a value 1.5.

The third model, due to Vizel and Mostinskii [30], assumed that the momentum of the jet in the direction of the cross flow was enhanced by the drag force in this direction, and the jet momentum in the normal direction was preserved. Thus, the second equation was given by

$$A\rho\bar{V}^2 \cos\theta = \pi/4D^2\rho_j U_i^2 \cos\theta_i + C_D/2\rho_0 U_0^2 y\,\Delta z \tag{22}$$

Following Abramovich,

$$\Delta z = 2.25D + 0.11x \tag{23}$$

Equations 13, 22, and 23 were combined to yield the jet trajectory

$$y/D = 16.2[J/C_D \log(1 + 0.049x/D)]^{1/2} \tag{24}$$

with C_D given a value of 4.

Jet trajectories predicted by the three models (Equations 16, 21, and 24) are compared to experimental data of Shandorov [19], for $J = 4.75$ and 16.35, and $\theta_i = 90°$, in Figure 6. The model of Volinsky and Abramovich (with $C_D = 3$) showed good agreement with the low momentum ratio data, but predicted too high a trajectory for the data with $J = 16.35$. This may be due to the simplifying assumption made in Equation 15, where x was substituted for ζ. The larger J is, the smaller x will become in comparison to ζ, hence, the greater the error in underestimating Δz. This error may be offset by increasing C_D in Equation 14 as suggested by Rajaratnam. The model of Vizel and Mostinskii consistently underpredicted the jet trajectories. The model of Crowe and Riesebieter showed good agreement with the data for $J = 16.35$, but an underprediction for $J = 4.75$.

All the preceding models have neglected the effects of fluid entrainment and nonuniformity of the jet velocity profile on the jet momentum. A model proposed by Platten and Keffer [31] accounted for both effects. It assumed that the momentum of the jet in the direction normal to the cross flow was preserved and that in the direction of the cross flow was enhanced by the momentum of the entrained fluid. This model in turn neglected the effects of drag forces in deflecting the jet (see Rajaratnam [11] for a description of this model).

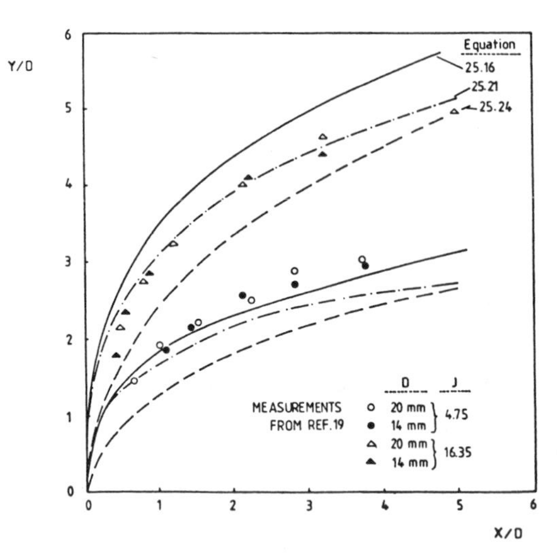

Figure 6. Comparison of trajectories predicted by three integral models [18, 29, 30] to experimental data [19].

Current Models

Most of the other integral models for predicting jets in cross flow, which derive their working equations through the control-volume approach, make assumptions that are related to the preceding in one way or the other, albeit with some refinements. However, unlike the prior models, which consider the effects of either drag forces or entrainment of ambient fluid in deflecting the jet, they mostly incorporate both effects. The main differences between these models lie in the representations of the drag forces, entrainment, cross-sectional shape of the jet, velocity profiles, and treatment of buoyancy effects or heat transfer, if present. Fan [13] developed an integral model for buoyant jets in cross flow. The cross-section of the jet was assumed circular with radius $\sqrt{2}b$, the excess velocity profile was assumed to be Gaussian $V - U_0 \cos\theta = (V_{max} - U_0 \cos\theta)e^{-\eta^2/b^2}$, and the entrainment velocity was assumed to be given by the magnitude of the vector difference between the axial jet velocity and the cross flow velocity. Thus,

$$U_e = E|\vec{V}_{max} - \vec{U}_0| = E[(V_{max} - U_0 \cos\theta)^2 + U_0^2 \sin^2\theta]^{1/2} \tag{25}$$

In general, the integral equations for such a model could be written as:

- Continuity

$$\frac{d}{d\zeta}(\rho A\bar{V}) = C\rho_0 U_e \tag{26}$$

- x-momentum

$$\frac{d}{d\zeta}(A\rho\lambda_v\bar{V}^2 \cos\theta) = C\rho_0 U_e U_0 + C_D/2\,\Delta z\,\rho_0 U_0^2 \sin^3\theta \tag{27}$$

- y-momentum

$$\frac{d}{d\zeta}(A\rho\lambda_v\bar{V}^2 \sin\theta) = -A(\rho - \rho_0)g + C_D/2\,\Delta z\,\rho_0 U_0 \sin^2\theta\cos\theta \tag{28}$$

- Scalar property ϕ

$$\frac{d}{d\zeta}(A\rho\lambda_\phi\bar{V}\phi) = 0 \tag{29}$$

where g = the acceleration due to gravity
ρ = the density of the jet fluid
A = the cross-sectional area of the jet
C = its circumference

λ_v is a momentum coefficient which depends on the assumed velocity profile and λ_ϕ is the equivalent coefficient for the scalar property, ϕ, which may be temperature, concentration, etc.

The first term on the right of Equation 28 is the buoyancy term, and in the same manner additional source or sink terms can be added to any of the equations if necessary. The entrainment equation is combined with Equations 26 through 29, which are then solved numerically to yield the jet trajectory and the distributions of all dependent variables. The particular form of the final working equations would depend on the profile assumptions and the equation of state through which the density ρ is calculated. In his calculations, Fan applied the Boussinesq approximation so that only the density variation in the buoyancy term needed to be considered, and the density difference $(\rho - \rho_0)$ was assumed proportional to the temperature difference $(T - T_0)$. The model of Fan, using Equation 25 for the entrainment velocity required different values of E and C_D to procure good agreement with each of his experimental data. To alleviate this problem, Abraham [32] proposed an entrainment model with two parts and two entrainment coefficients, in the manner of Equation 12. The entrainment velocity was given by:

$$U_e = E_{mom}(V_{max} - U_0 \cos\theta) + E_{th}U_0 \sin\theta \cos\theta \tag{30}$$

where $E_{mom} = 0.057$
$E_{th} = 0.50$

The first part represented the entrainment of a momentum jet in nearly stagnant ambient fluid, whereas the second the entrainment into thermals also moving through stagnant ambient fluid. Cos θ was introduced arbitrarily into the second part, to prevent it from contributing to entrainment when the jet is nearly perpendicular to the cross flow. E_{mom} and E_{th} were chosen, respectively, from experimental data for momentum jets and thermals in stagnant ambients. With this entrainment model and $C_D = 0.3$, Abraham obtained quite good predictions of the jet trajectory and axial concentration decay measured in Fan's experiment as is shown in Figure 7.

Equations 26 through 29 may be applied to plane jets in cross flow by substituting the appropriate relations for the area A and the circumference C. The entrainment model and the drag coefficient C_D would need to be calibrated with plane jet data.

Campbell and Schetz [33] proposed a rather elaborate model, which took into consideration the effects of drag forces, entrainment, buoyancy, axial pressure gradient, turbulent shear stress between jet and cross-stream fluid, and heat transfer due to forced convection from jet to cross-stream fluid. The integral momentum equations were derived in directions tangential and perpendicular to the jet's axis, and for the entrainment velocity the relation (Equation 11) due to Keffer and Baines [24] was assumed. The entrainment coefficient E was obtained by correlating their [24] data as

$$E = 0.2(\zeta/D)^{1.37}/(\bar{V}/U_0)^{0.6} \tag{31}$$

The pressure gradient term in the axial momentum equation was modeled by correlating results obtained from potential flow theory for flows impinging normally on a cylinder, and the turbulent shear stress term was calculated using Prandtl's eddy-viscosity (mixing length) hypothesis. In the thermal energy equation, (similar to Equation 29, with $\phi = T$) an additional sink term, representing forced convection losses from the jet to the cross-stream was introduced by way of empirical relations [34] developed for convective heat transfer from circular cylinders to a moving stream. Predictions with the model were compared with experimental data, showing relatively good

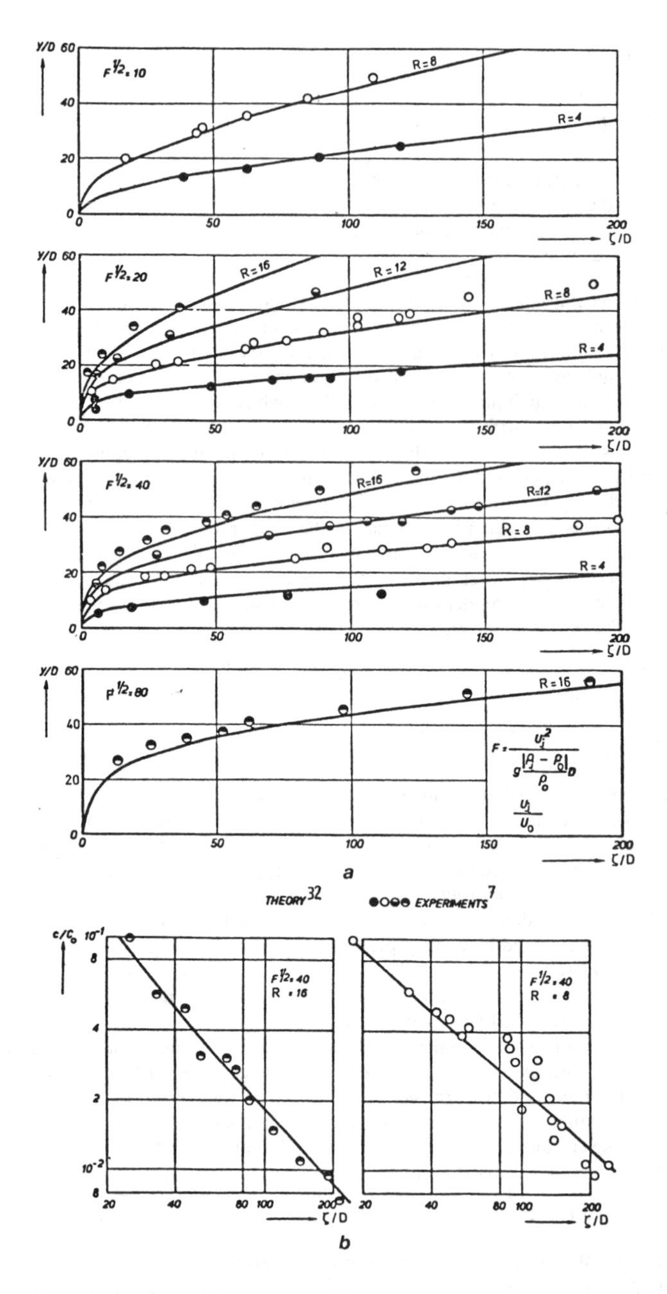

Figure 7. Prediction of jet trajectories and concentration decay: (A) trajectories; (B) concentration at axis of jet [32].

agreement. The model was extended by Isaac and Schetz [35] to predict trajectories of double jets arranged in a row or in tandem. The interactions between the jets were accounted for by modifying the drag coefficient C_D, based on experimental data [36] for interaction between circular cylinders. The model of Campbell and Schetz has demonstrated some ways in which integral methods could be extended to include more physics of the flow, however, it is not entirely convincing that such complicated empirical modeling of the physics was necessary in the particular flow situation studied. Much simpler models [7, 32] seem to have performed just as well in such flows.

Sucec and Bowley [37] proposed an integral model which differs from those just described in that no entrainment equation was specified; rather the rates of expansion of the jet boundaries were prescribed as empirical relations. This was following the approach of some earlier models [18, 29, 30]. The initial circular exit cross-section of the jet was approximated to a rectangle with sides $\xi_b:\eta_b$ of 4:1. The cross-section then grew at different rates in the two directions, which were prescribed as empirical relations.

η_b was given by,

$$\left.\begin{aligned}
&\frac{d\eta_b}{d\zeta} = (1.168 + 0.268U_0 \cos\theta/U_j) \times 0.27\left[\frac{U_j - U_0\cos\theta}{U_j - U_0\cos\theta}\right] \\
&\qquad\qquad\qquad\qquad \text{for } \eta_b < 2\int_0^{\zeta}\left[\frac{U_j - U_0\cos\theta}{U_j + U_0\cos\theta}\right]d\zeta \\
&\text{otherwise,} \\
&\frac{d\eta_b}{d\zeta} = 0.44\left[\frac{V - U_0\cos\theta}{V + U_0\cos\theta}\right]
\end{aligned}\right\} \tag{32}$$

ξ_b was given by:

$$\left.\begin{aligned}
&\frac{d\xi}{d\zeta} = 1.0, \qquad \text{for } 2 \le R \le 8 \\
&\frac{d\xi}{d\zeta} = 2.5 - 1.67\log R, \qquad \text{for } 8 \le R \le 16 \\
&\frac{d\xi}{d\zeta} = 0.86 - 0.3\log R, \qquad \text{for } 16 \le R \le 40 \\
&\frac{d\xi}{d\zeta} = 0.32, \qquad \text{for } 40 < R < \infty
\end{aligned}\right\} \tag{33}$$

Equations 32 and 33 were combined with the axial and transverse momentum equations (with, $C_D = 1.8$, for $R < 4.4$, and $C_D = 1$, for $R \ge 4.4$) to yield the set of working equations which were solved by a finite difference numerical method. Their model predictions of jet trajectory showed relatively good agreements with a wide range of experimental data, including some for oblique jets [4, 38] and some for jets in confined cross flow [7]. Predictions of trajectories from smokestack model studies showed good agreement for $R = 4$, deteriorating as R was decreased. It appears that for low R (~ 1), the interaction of the wake of the stack with the jet fluid may be of considerable importance. To take account of these effects in integral models, the entrainment or jet spreading relations would have to be calibrated with data exhibiting such interactions. Otherwise, some form of zonal modeling, such as that proposed by Chan, Liu, and Kennedy [39], may be employed. Carhart [25] et al. have developed a model, similar to Fan's [7] and Abraham's [32], for predicting cooling tower plumes with low R in the range of 0.3 to 2. This model was calibrated with 39 field and 6 laboratory data sets (for cooling tower configurations) of plume trajectory and dilution, and it was subsequently verified by comparison of predictions to further sets of measured data. The agreement with laboratory data was found to be fairly good and field data were reproduced satisfactorily, considering the uncertainty in the latter.

Other interesting applications of integral models are those of Stoy and Ben-Haim [7] for jets in confined cross flow with highly nonuniform velocity distributions, and those of Makihata and Miyai [40] for triple jets in a row, or triangular arrangement in a uniform cross flow. Figure 8 shows some typical results form the latter.

The models of Hirst [27] and Schatzmann [28] differ from the rest in the sense that the integral equations were derived by integrating the partial-differential equations governing the steady flow of turbulent jets in a cross-stream. The derivation was carried out on a curvilinear coordinate system, and by assuming axisymmetry and Gaussian profiles for the velocity excess, the partial differential equations could be integrated in the radial and circumferential directions to yield a set of ordinary differential equations much like Equations 26 through 29. Through the choice of coordinate system they were able to derive integral equations which were applicable to jets with two-, or three-dimensional trajectories: the other models were limited to jets with two-dimensional trajectories. The main difference between the models of Hirst [27] and Schatzmann [28] lies in the treatment of entrainment. The latter pointed out, correctly, that fluid entrainment into the jet led to an increase in the integrated excess mass velocity, rather than an increase in the total mass flux of the jet, as was assumed by the former [27] and in most of the other models. This point was illustrated with the case of a jet in a coflowing stream, where the increase in mass flux would only be equal to the entrained fluid if the free-stream velocity was zero or the jet width remained constant. However, this apparent inconsistency in most of the models is only of theoretical interest, since the entrainment models employed in them have mostly been developed from measurements of axial growth of the mass flux in the jet. Schatzmann, on the other hand, had derived the entrainment coefficient by integrating the integral form of the turbulent kinetic energy and combining the result with the continuity and axial momentum equations. The resulting equation for the entrainment coefficient had 4 constants which were determined empirically. In spite of its apparent sounder basis, the predictions with Schatzmann's model were not superior to those with the others. Integral models are semi-empirical in nature, and their applicability often depends not only on the assumptions made in deriving them, but also on the empirical relations used in specifying physical processes such as turbulent entrainment of cross-stream fluid into the jet, pressure effects, jet growth, etc. It appears that there are many combinations of these that would lead to a reasonable model. For simplicity, many of the models assume a circular cross-section and top hat profiles, and these do not even make them worse than the others.

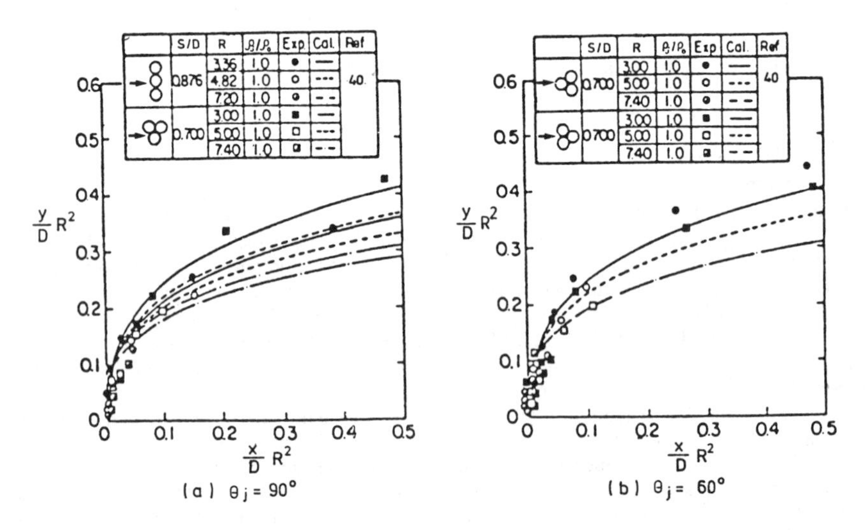

Figure 8. Prediction of trajectories of triple jets [40].

One of the main criticisms of integral models is the need to assume the shape of the jet cross-section and profile variations for the properties within the jet. They can, therefore, not be used to investigate certain important properties of the jet such as the development of the vortex structure and the resulting formation of the kidney-shaped cross-section, or the velocity or temperature distributions within the cross-section. Also, they can only be applied with much confidence to flow situations similar to those in which the experimental data, used in calibrating them were acquired. In spite of these criticisms, they are still the most widely used methods for predicting the trajectory of turbulent jets in cross flow, for practical purposes. The reason for this is their low computational cost, compared with that of currently used numerical models.

The first drawback was removed in a quasi-three-dimensional integral model developed by Adler and Baron [41], which did not assume the cross-sectional shape of the jet or similarity of velocity profiles but computed these. They succeeded in predicting the development of the characteristic kidney-shape cross section by considering the evolution of vortices distributed along the boundaries of the jet in a Lagrangian manner, using the ideas of Chen [42] and Strauber [43]. The cross-sectional area was assumed to grow at a rate which was the sum of the growth rate for a jet in stagnant fluid (similar to Equation 32) and the growth rate of the vortex pairs, based on the proposal of Tulin and Schwartz [44]. The velocity distribution in the zone of maximum deflection was assumed to be given by a family of profiles which were not constant but changed with axial distance until the third zone is reached, where the cross-sectional shape remained constant and the profiles became self-similar. The model resembles the other integral methods in all other respects. Figure 9 shows a typical computation of the development of the kidney-shaped cross section, and comparisons of

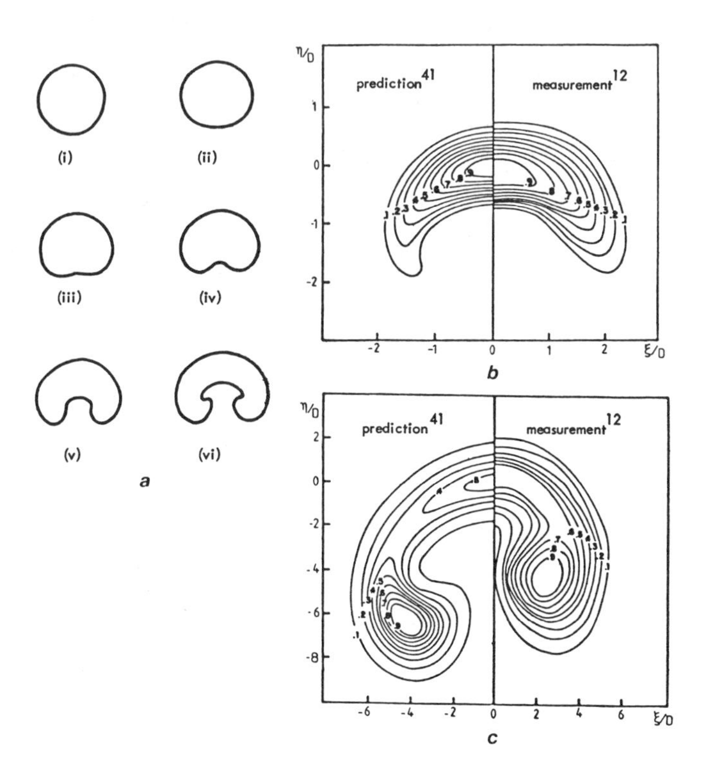

Figure 9. Prediction of jet cross section and velocity field: (A) computed development of cross-sectional shape; (B) velocity contours, $R = 3.91$, $\xi/D = 7$; (C) velocity contours, $R = 7.73$, $\xi/D = 23$ [41].

the predicted velocity contours with the measurements of Kamotani and Greber [12]. The agreement is seen to be remarkably good. This model must be considered quite promising, especially if it could be generalized to flow situations with nonuniform cross flow, confined cross flow and those with buoyancy effects.

Demuren [41] developed a model for predicting buoyant surface jets in cross flow, which can be classified as lying between integral and numerical models. Following integral models, the computations were carried out only within the jet and the effect of the cross-stream was transfered through its entrainment into the jet. The entrainment coefficient was also prescribed empirically. However, unlike integral methods the profiles of the two horizontal velocity components and the temperature did not have to be prescribed, but were obtained from the solution of two-dimensional partial-differential equations, for each of these fluid properties. The pressure gradients were obtained by solving a Poisson equation for pressure, derived with the aid of the continuity equation. An integral equation was also solved to determine the jet deflection. This model predicted trajectories and velocity and temperature decays which agreed fairly well with laboratory data, and the computational cost is only slightly higher than those of integral models. However, it is only applicable to flows with R greater than 8.

Numerical Models

Basic Features

Most of the numerical models which have been applied to predict the flow and heat transfer of turbulent jets in cross flow are of the finite-difference type. These solve the time-averaged partial-differential equations governing the turbulent transport of heat, mass or species concentration, using a finite-difference numerical procedure. Turbulence models are required for closure of the equations, and these mostly follow the eddy-viscosity concept. In the simplest form, the turbulent eddy-viscosity is prescribed as constant, while the more sophisticated models solve partial-differential equations for turbulent quantities, from which the eddy-viscosity distribution is obtained. The computational domain encompasses the whole region in which the influence of the jet is felt, or if need be, the whole flow field of the jet and the cross flow. No assumptions are required as to the evolution of the jet within the flow domain, but this is obtained as a result of the computations. It is only necessary to prescribe boundary conditions at the chosen computational boundaries.

Equations

The mean flow equations may be expressed for the steady-state in Cartesian coordinates as continuity

$$\frac{\partial \rho U}{\partial x} + \frac{\partial \rho V}{\partial y} + \frac{\partial \rho W}{\partial z} = 0 \tag{34}$$

momentum or energy conservation,

$$\frac{\partial \rho U\phi}{\partial x} + \frac{\partial \rho V\phi}{\partial y} + \frac{\partial \rho W\phi}{\partial z} = \frac{\partial}{\partial x}\left(\Gamma_\phi \frac{\partial \phi}{\partial x} - \rho\overline{u\varphi}\right) + \frac{\partial}{\partial y}\left(\Gamma_\phi \frac{\partial \phi}{\partial y} - \rho\overline{v\varphi}\right) + \frac{\partial}{\partial z}\left(\Gamma_\phi \frac{\partial \phi}{\partial z} - \rho\overline{w\varphi}\right) + S_\phi, \tag{35}$$

where ϕ may stand for any of the velocity components U, V, W, the temperature T, or a species concentration C. Γ_ϕ is the corresponding diffusion coefficient which is equal to the molecular viscosity μ when $\phi = U, V, W$ or the molecular thermal or concentration diffusivity with $\phi = T$

or C, respectively. S_ϕ is the corresponding source, and in the absence of internal energy generation through chemical reactions, etc.,

$$S_u = -\frac{\partial P}{\partial x}$$

$$S_v = -(\rho - \rho_r)g - \frac{\partial P}{\partial y} \tag{36}$$

$$S_w = -\frac{\partial P}{\partial z}$$

$$S_T = S_C = 0$$

Turbulence Model

$\overline{u\varphi}$ represents the turbulent stresses (if $\phi = U, V, W$) or the turbulent heat or concentration fluxes (when $\phi = T, C$). To achieve closure of Equations 34 and 35, the distributions of these quantities must be determined and this is the main purpose of the turbulence models. All the turbulence models follow the Boussinesq's eddy-viscosity concept which assumes that in analogy to viscous stresses in laminar flows the turbulent stresses are proportional to the velocity gradients. Thus, the turbulent (or Reynolds) stresses may be expressed as

$$-\rho\overline{u^2} = 2\mu_t\frac{\partial U}{\partial x} - \frac{2}{3}\rho k$$

$$-\rho\overline{v^2} = 2\mu_t\frac{\partial V}{\partial y} - \frac{2}{3}\rho k$$

$$-\rho\overline{w^2} = 2\mu_t\frac{\partial W}{\partial z} - \frac{2}{3}\rho k$$

$$-\rho\overline{uv} = -\rho\overline{vu} = \mu_t\left(\frac{\partial U}{\partial y} + \frac{\partial V}{\partial x}\right) \tag{37}$$

$$-\rho\overline{uw} = -\rho\overline{wu} = \mu_t\left(\frac{\partial U}{\partial z} + \frac{\partial W}{\partial x}\right)$$

$$-\rho\overline{vw} = -\rho\overline{wv} = \mu_t\left(\frac{\partial V}{\partial z} + \frac{\partial W}{\partial y}\right)$$

Similarly, the turbulent heat fluxes are expressed as:

$$-\rho\overline{uT} = \frac{\mu_t}{\sigma_T}\left(\frac{\partial T}{\partial x}\right)$$

$$-\rho\overline{vT} = \frac{\mu_t}{\sigma_T}\left(\frac{\partial T}{\partial y}\right) \tag{38}$$

$$-\rho\overline{wT} = \frac{\mu_t}{\sigma_T}\left(\frac{\partial T}{\partial z}\right)$$

(in which, for the concentration fluxes C is substituted for T). In Equations 37 and 38, μ_t is the turbulent eddy-viscosity, which, in contrast to the molecular viscosity μ, is not a fluid property but depends strongly on the state of turbulence, and its value usually varies considerably from point to point in the flow field. It is in the manner of determining the distribution of this quantity that the various turbulence models differ from each other. k is the turbulence kinetic energy and it is given by $k = \frac{1}{2}(u^2 + v^2 + w^2)$. σ_T and σ_c are, respectively, the turbulent Prandtl and Schmidt numbers whose values are empirically determined. A value between 0.5 and 1.0 is usually employed.

Apart from the model of Chien and Schetz [46], which prescribed a constant value for μ_t $(= 0.0256[U_j - U_0]D/2)$ all the other models obtained the distribution of μ_t through the so-called $k - \epsilon$ turbulent model, described in detail by Launder and Spalding [47]. According to this model,

$$\mu_t = \rho c_\mu k^2/\epsilon \tag{39}$$

where ϵ is the rate of dissipation of the turbulence kinetic energy and c_μ is an empirical constant. The distributions of k and ϵ are obtained from the solution of partial differential equations, which have the same form as equation 35, with

$$S_k = G - \rho\epsilon; \qquad S_\epsilon = c_1\rho\epsilon G/k - c_2\rho\epsilon^2/k \tag{40}$$

and the turbulent diffusion fluxes are obtained from Equation 38, by substituting k or ϵ for T. The turbulence model constants c_μ, c_1, c_2, σ_k, and σ_ϵ have the standard values [47], $c_\mu = 0.09$; $c_1 = 1.44$; $c_2 = 1.92$; $\sigma_k = 1.0$; and $\sigma_\epsilon = 1.3$. (These constants are empirical in origin, but the same values are employed in the calculation of a wide range of turbulent flows). G is the rate of generation of turbulence kinetic energy by the interaction of turbulent stresses with mean velocity gradients and is given by:

$$G = \mu_t\left[2\left(\frac{\partial U}{\partial x}\right)^2 + 2\left(\frac{\partial V}{\partial y}\right)^2 + 2\left(\frac{\partial W}{\partial z}\right)^2 + \left(\frac{\partial V}{\partial x} + \frac{\partial U}{\partial y}\right)^2 + \left(\frac{\partial W}{\partial y} + \frac{\partial V}{\partial z}\right)^2 + \left(\frac{\partial W}{\partial x} + \frac{\partial U}{\partial z}\right)^2\right] \tag{41}$$

Finite-difference Equations

Equations 34 through 41 now form a closed set which are solved by the finite-difference numerical procedure. In order to obtain the finite-difference equations, Equation 35 is discretized by integration over elementary control volumes for each grid node in the computational domain, or by Taylor series expansion. The former approach is sometimes called the finite-volume formulation and is much more popular than the latter. It is also fully conservative, in the sense that the outflow through one control-volume face is always equal to the inflow into the adjacent control volume. On integration the diffusion terms are usually represented by central differences.

Three types of differencing methods have been applied, in the literature to represent the control volume face value for the ϕ property in the convection terms. These are the upwind [46] the hybrid [8] (central/upwind) and the quadratic upstream weighted (QUICK) [17] differencing schemes. Figure 10 shows the approximations of $\phi_{i-1/2,j}$, due to each of these schemes. (A two-dimensional representation was employed, in the figure, for simplicity.) For the flow going from left to right (i.e., $U > 0$), the upwind scheme assumes that the control volume face value $\phi_{i-1/2,j}$ is equal to the value at the upstream node $\phi_{i-1,\cdot}$. The hybrid scheme assumes that $\phi_{i-1/2,j}$ is given by a linear interpolation (central difference) between the two nodal values $\phi_{i-1,j}$ and $\phi_{i,j}$, if the grid Peclet number $R_V(=[\rho U]_{i-1/2,j}\,\Delta x/[\Gamma_\phi + \mu_t/\sigma_\phi])$ is less than 2, and is given by the upwind difference scheme, if R is greater than 2. The QUICK scheme, determines $\phi_{i-1/2,j}$ from a quadratic interpolation of three nodal values; two upstream and one downstream of the face. The upwind scheme (or donor cell method) leads to a very stable numerical procedure, but is particularly prone to false or

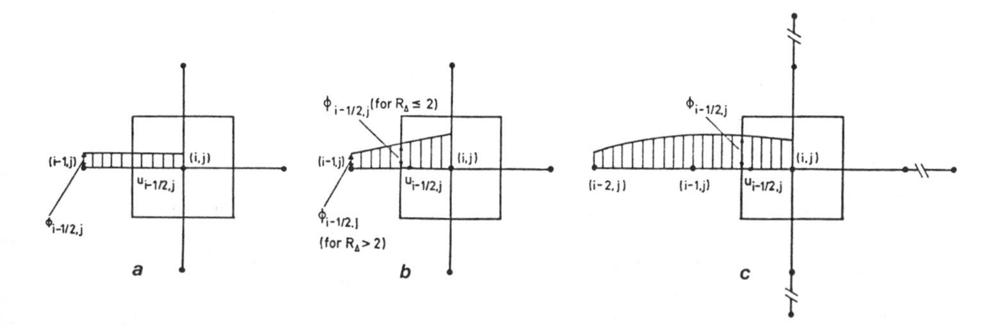

Figure 10. Methods of obtaining control-volume face values: (A) upwind scheme; (B) hybrid scheme; (C) QUICK scheme.

numerical diffusion [48], which may have adverse effect on the computed results when the velocity vectors are not aligned with the grid lines. The hybrid scheme suffers the same fate if $R_\Delta > 2$, since central differencing (which does not have false diffusion) tends to produce an unstable solution procedure if used when $R_\Delta > 2$. The QUICK scheme was proposed by Leonard [49] to take advantage of the stability of upwind differencing and the accuracy of quadratic interpolation. It should be noted that in three dimensions the value at any grid node is connected in the finite-difference equations to those at 12 neighboring nodes if the QUICK scheme is used but only to 6 if the upwind or hybrid scheme is used. Thus, the former is computationally somewhat more expensive.

Boundary Conditions

The boundary conditions applied to the equations depend on the particular problem. The various types of boundaries which may exist in these flow situations are inflow, outflow, wall, symmetry planes, and the free stream. At inflow boundaries, the values of the dependent variables are prescribed or deduced from experimental data. Except for the pressure, boundary conditions are not usually required at outflow planes. There the presumed level is either fixed or the gradient is prescribed, as appropriate. Along wall boundaries, no-slip conditions are usually applied at the walls, but to bridge the flow between the fully turbulent region and the viscous sublayer on the wall, the wall-function method described by Launder and Spalding [47] is employed to prescribe the variable values along the first grid nodes nearest to the walls. Zero normal velocity component and zero normal gradients for all other variables are usually prescribed along symmetry planes. Finally, known values are prescribed along free stream boundaries.

Solution Procedure

The various numerical models may be characterized through the manner in which the pressure gradient terms, Equation 36, were obtained. Chien and Schetz [46] again followed an entirely different approach from the others. The pressure gradient terms were eliminated from the momentum equations (Equation 35) by cross-differentiation. There then appeared, in these equations, terms involving gradients of the three vorticity components. The vorticity components were obtained by solving three extra partial differential equations, similar in form to equation 35. The other models invariably obtained the pressure distribution by solving a pressure-correction equation which was derived by combining the finite-difference form of the continuity equation (Equation 34) with those of the momentum equations (Equation 35). Thus, the solution procedure is a "guess and correct" one, similar to that of Chorin [50], Amsden and Harlow [51], and Patankar and Spalding [52].

Various Models

The first model in this group was due to Tatchel [53], who assumed that the flow was parabolic in the streamwise direction, i.e., flow information was transmitted only in the downstream direction. This enabled the solution for the whole field to be obtained by marching integration, plane-by-plane in the downstream direction. Only a single march was necessary and the dependent variables only needed to be stored in two dimensions. This solution scheme was economical, but the model could not be applied to jets in cross flow with a normal injection if R was greater than about 0.5. Even for $R = 0.1$, the computed results deviated considerably from measured data. Bergeles, Gosman, and Launder [54] proposed a model based on the work of Pratap and Spalding [55] in which only pressure effects could be transmitted in all directions; all the other effects could only be felt downstream. This was the so-called partially-parabolic procedure, and the pressure field was stored in three dimensions, but the other variables still required two-dimensional storage. The plane-by-plane marching integration process must now be repeated many times before a converged solution is obtained. The computational cost of this model was much higher than that with the parabolic method [53]. The model could only be applied to flow situations in which there was no flow reversal in the cross-flow direction, i.e., low R. The model was applied to flow situations occurring in film-cooling geometries, where the jets of cold air are required to remain attached to the wall so as to protect it from the hot cross flow. The predicted velocity distributions and film-cooling effectiveness showed very good agreement with their experimental data and those of Ramsey and Goldstein [16], when $R \leq 0.1$ for $\theta_j = 90°$, and $R \leq 0.3$ for $\theta_j = 30°$. The agreement was not so good, at higher values of R. This might have been due to the presence of some small regions of reverse flow which had to be suppressed in the solution. It should be mentioned that Bergeles et al. [54] found that it was necessary to introduce a nonisotropic eddy-viscosity distribution in the near-wall region to account for the observation that the normal components of the Reynolds stresses are reduced by the presence of a wall while the horizontal components are enhanced.

In all the other models, an elliptic procedure was employed, so that regions with flow reversal in the cross stream direction could be accommodated adequately. Patankar, Basu, and Alpay [56] applied a model, with a fully elliptic procedure, to predict the trajectory of jets issuing normally into a cross flow at moderate to high velocity ratios ($R = 2 \sim 10$). This procedure required three-dimensional storage for all dependent variables; hence, they were restricted to relatively coarse grid ($15 \times 15 \times 10$ in the x, y, z directions) calculations. Nevertheless, the predicted trajectories agreed very well with experimental [15, 16, 24] ones. However, the velocity profiles failed to show some of the features observed in the experiments of Ramsey and Goldstein [16], namely, the high velocity maximum in the longitudinal direction after the jet was bent over and the reverse flow in the wake region. Rodi and Srivatsa [57] developed a locally-elliptic calculation procedure in which only the small region with reverse flow was treated elliptically, the rest of the computational domain being considered partially-parabolic. Thus, three-dimensional storage was only required for the pressure in the whole computational domain, and for the other dependent variables in the small elliptic region. The computational time required by this procedure was comparable to that of the fully elliptic one, but the computer core storage memory was considerably smaller, and was not much more than that required in a partially-parabolic procedure. Thus, finer grid distributions could be employed, even for flows with reverse flow regions (high R). Their predictions of the velocity at concentration profiles showed good agreement with the experimental data of Tatchel [53], for $R = 0.1$ and 0.3, although, at the higher velocity ratio, the predicted concentration maxima were too low, in the near field. The latter may be due to the presence of false diffusion, resulting from their approximation of the convection terms with the hybrid differencing scheme.

Numerical models of the fully elliptic type have been applied by White [58] and Claus [59] to predict velocity and concentration distributions in single jets in unconfined cross flow at $R = 2.3$, investigated experimentally by Crabb, Durao and Whitelaw [60]. A similar model was also applied by Jones and McGuirk [8] to predict the experimental data of Kamotani and Greber [5] for single

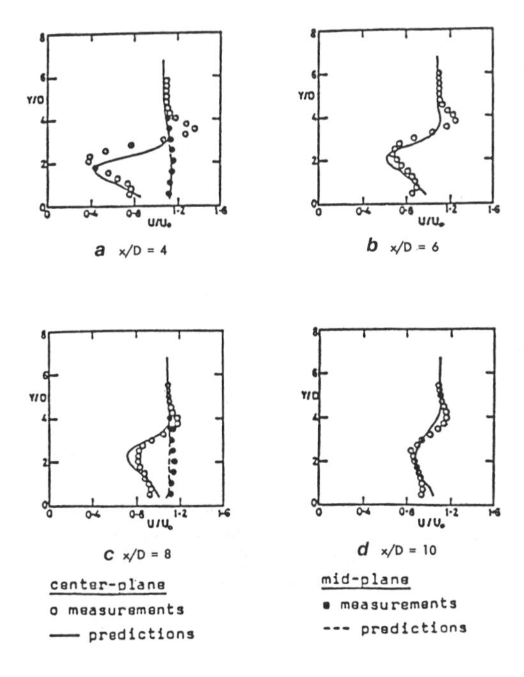

Figure 11. Predictions of a row of jets in confined cross flow, R = 2.3, H/D = 8, S/D = 4 [9].

jets in confined cross flow, and by Khan, McGuirk, and Whitelaw [9] to predict their own experiments for a row of jets in confined cross flow. All these models have approximated the convection terms with the hybrid differencing scheme and therefore suffered from false or numerical diffusion in the regions of the flow in which the cell Peclet numbers were higher than 2. The magnitudes of the false diffusion are often much higher than those of the turbulent diffusion, and this is the reason why some of the computed results show effects characteristic of excessive diffusion, such as velocity maxima which were much lower than observed experimentally (see Figure 11), or too rapid spreading of the temperature or concentration fields (see Figure 12). Another common feature of computations with these models was the difficulty in obtaining grid-independent solutions. Claus [59] found that even with a grid distribution of $90 \times 40 \times 22$, in the x, y, z directions, the results were still not grid independent, but continued to improve slightly with further grid refinement.

Demuren [17] extended the numerical model of Rodi and Srivatsa [57] by incorporating the option for approximating the convection term with the three-dimensional version of the QUICK differencing scheme. Calculations of the flow of a row of jets in nearly uniform cross flow at R = 1.96, investigated experimentally by Sugiyama and Usami [6], were then performed with both the original model (hybrid scheme) and the extended one (QUICK scheme). The jet spacing ratio in the experiments was S/D = 3, and this flow situation was chosen as a test case because, symmetry conditions along the center-plane of the jets and along the mid-plane between jets meant that computations only needed to be performed for a lateral extent of 1.5D. Hence, the flow in this direction could be resolved with relatively few grid points, and grid refinement was necessary in only the x and y directions. Some results of the grid-dependency tests are illustrated with the predicted vertical profiles of the streamwise velocity in Figure 13. In the hybrid scheme calculations, only the results for the finest grid distribution showed a significant velocity peak. This appeared to have been smeared out in the calculations with the other grids. Results with the QUICK scheme, on the other hand, showed velocity peaks for all grid distributions, and these increased in value with grid refinement. In fact, the velocity peak obtained with the coarsest grid of the latter was comparable to that with the finest grid of the former. The finest grid results of the QUICK scheme

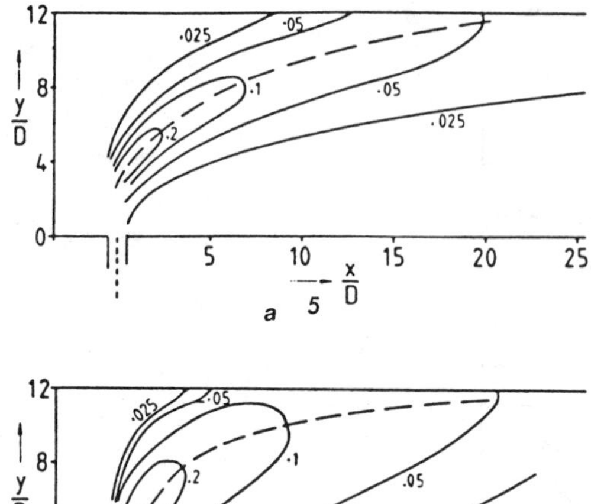

Figure 12. Prediction of normalized temperature contours for a jet in confined cross flow, J = 32, H/D = 12: (A) measurements; (B) predictions [8].

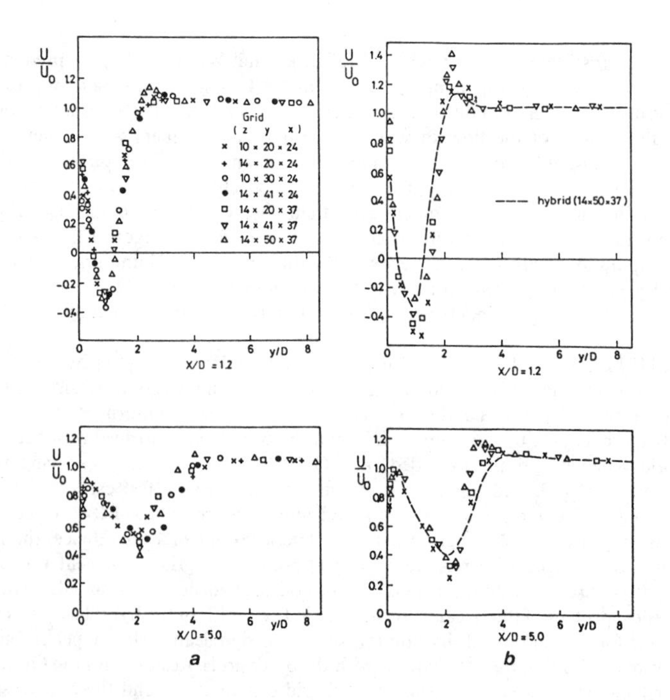

Figure 13. Test for grid dependency, R = 2.3, S/D = 3, z/D = 0: (A) hybrid scheme; (B) QUICK scheme.

may be considered "practically" grid independent. A calculation with a grid distribution of 37 × 70 × 14 in the x, y, z directions resulted in a maximum deviation (cf. 37 × 50 × 14) of less than 2% in the streamwise velocity. The computations with the latter grid and QUICK scheme required 400 iterations for convergence, and took about 16 hours CPU time on the Univac 1100/82 computer. 190k words of main core storage was required. The hybrid scheme calculations converged in 200 iterations and required about $7\frac{2}{3}$ hours CPU time. The coarsest grid computations required 1.42 hours CPU time.

The predicted normalized total pressure profiles are compared with experimental data [6] in Figure 14. The predictions with the QUICK scheme and with zero longitudinal velocity in the plane of the jet exit gave the right same pressure level as the experiments, but the profiles were shifted in the vertical direction by a distance of approximately $\frac{1}{4}$D. The reason for this was found in the experimental results of Andreopoulos [3] of the pipe flow issuing into a cross flow, at three velocity ratios, R = 0.5, 1.0, and 2.0. This showed that the flow inside the pipe was influenced by the cross flow to an extent which depended on R. At low R, the exit vertical velocity profile of the jet was highly distorted, and the cross flow intruded into the pipe near the upstream end. At R = 2.0 the distortion of the profile was only slight, and it was correctly reproduced by assuming uniform total pressure at the jet exit, in the calculations [17]. The component of the velocity in the direction of the cross flow measured in the plane of the jet exit was well correlated by the equation

$$U_{j,x} = 0.23U_0[1 - (2z/D)^2][1 - (2x/D + 1.2 \log R)^2] \tag{42}$$

for the three velocity ratios.

The application of this boundary condition brought the (QUICK) predictions in better correspondence with the measured data at x/D = 0.625 and 1.25. Further downstream, the size of the wake region was overpredicted. This was probably due to the inadequacy of the turbulence model. The flow in the wake has a highly complex turbulent structure and measurements of the Reynolds

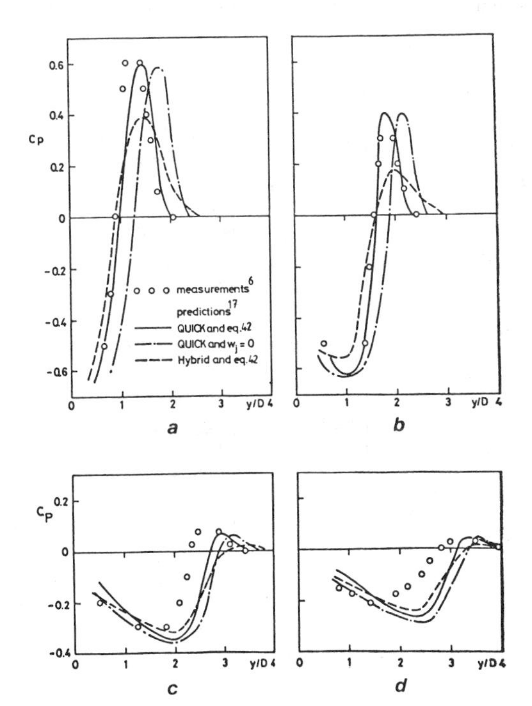

Figure 14. Prediction of normalized total pressure profiles, R = 2.3, S/D = 3, z/D = 0: (A) at x/D = 0.625; (B) at x/D = 1.25; (C) at x/D = 3.125; (D) at x/D = 6.25.

stresses by Crabb et al. [60] and Andreopoulos and Rodi [61, 62] have shown that the turbulence is both anisotropic and incongruous with the eddy viscosity concept. As may be expected, the hybrid scheme predicted too low values of total pressure peak indicative of numerical smearing.

False Diffusion

Demuren [63] derived equations for estimating the magnitudes of the false or numerical diffusivities in three perpendicular directions oriented with the velocity vector at any location in the flow. These are:

$$\Gamma_{f,\zeta} = \frac{QC_yC_z(C_y^2 + C_z^2)\,\Delta x\,\Delta y}{2(C_y^3\,\Delta y + C_z^3\,\Delta z)}$$

$$\Gamma_{f,\eta} = \frac{QC_x(C_y^2 + C_z^2)}{2[C_zC_x^3/\Delta z + C_yC_x^3/\Delta y + (C_z^2 + C_y^2)/\Delta x]} \tag{43}$$

$$\Gamma_{f,\xi} = \frac{Q}{2(C_z/\Delta z + C_y/\Delta y + C_x/\Delta x)}$$

Figure 15 shows contours of the ratio of the false to physical (effective) diffusivities computed on the basis of Equations 43. It is seen that there are large regions with false diffusivities much larger than the physical diffusivity in the flow. In some of these regions, the fluid property gradients are high and, thus, the false diffusion would be high, leading to numerical smearing. The false diffusion in the direction of the velocity vector, ξ, should not have much adverse effect on the results since property gradients, in this direction, may not be so large, and the convection is high. Further calculations [17] showed that when the magnitudes of $\Gamma_{f,\zeta}$ and $\Gamma_{f,\eta}$ are low compared with the

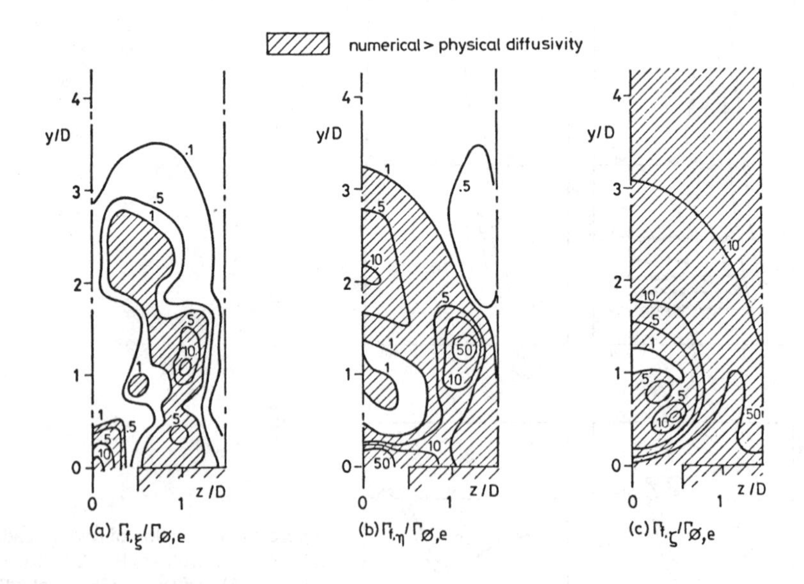

Figure 15. Contours of false to physical (molecular + turbulent) diffusivity at x/D = 1.25 (conditions as in Figure 14).

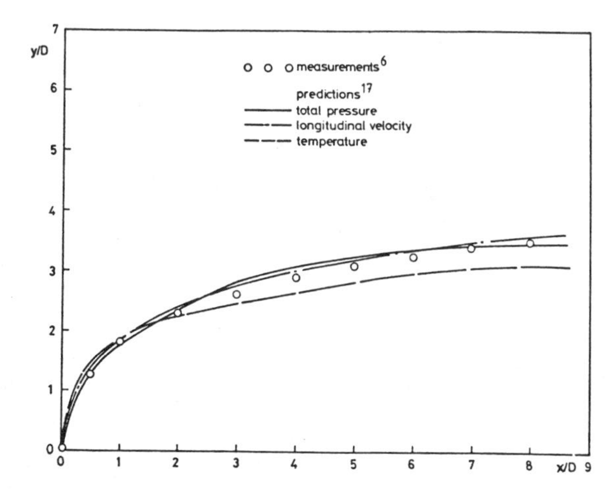

Figure 16. Comparison of jet trajectories (conditions as in Figure 14).

physical diffusion, false diffusion is unimportant, and the hybrid and QUICK schemes produce similar results. Model predictions (QUICK) of the jet trajectories are compared with the measured total pressure trajectory in Figure 16. The predicted velocity and total pressure trajectories are almost coincident and agree well with the measurements. The temperature trajectory lies somewhat below the others. It should be mentioned here, that, under certain conditions, the QUICK scheme like other higher-order methods suffers from lack of boundedness. The calculation of a neutral tracer under similar conditions as the results presented here showed at localized points overshoots of up to 9%. Although, higher-order methods have been shown to be necessary for increased accuracy, they are unlikely to find wide usage until their tendency to produce over- and undershoots can be cured. The FRAM (Filtering Remedy and Methodology) method of Chapman [64] may be one way to achieve this, and many investigators are now working along this or similar lines.

Further Results

The numerical model [17] presented previously was applied by Demuren and Rodi [65] to predict flow and heat transfer in film-cooling geometries, with the cross flow at various turbulence intensities. Figure 17 shows the flow configuration, and some comparisons of the predictions with measurements of Kadotani and Goldstein [66], in which the cross stream had a turbulence intensity level of 8.2% and a characteristic length scale of 0.33D, for case 1^A. The corresponding values for case 4^A were 4.8% and 1.2D. The predicted velocity distributions agreed quite well with the measurements, and the turbulence intensities showed qualitative agreement. The latter were some what under-predicted near the jet exit ($z/D \leq 0.5$). The predictions and measurements of the temperature contours agreed better for case 1^A than for case 4^A. The tendency of the jet to bifurcate as the cross-stream turbulence level was lowered was correctly reproduced, but not as strongly as in the measurements. These results show the importance of the cross-stream turbulence level and point to the need for a refined turbulence model. The proposals of Bergeles et al. [54] concerning the use of nonisotropic eddy-viscosity distribution were already incorporated in the presented calculations.

Demuren and Rodi [67] have performed some calculations of flow and concentration fields in cooling-tower configurations, with the locally-elliptic model [17, 57]. Both the flow around the tower and that of the jet issuing from its top were computed, but not the one inside the tower.

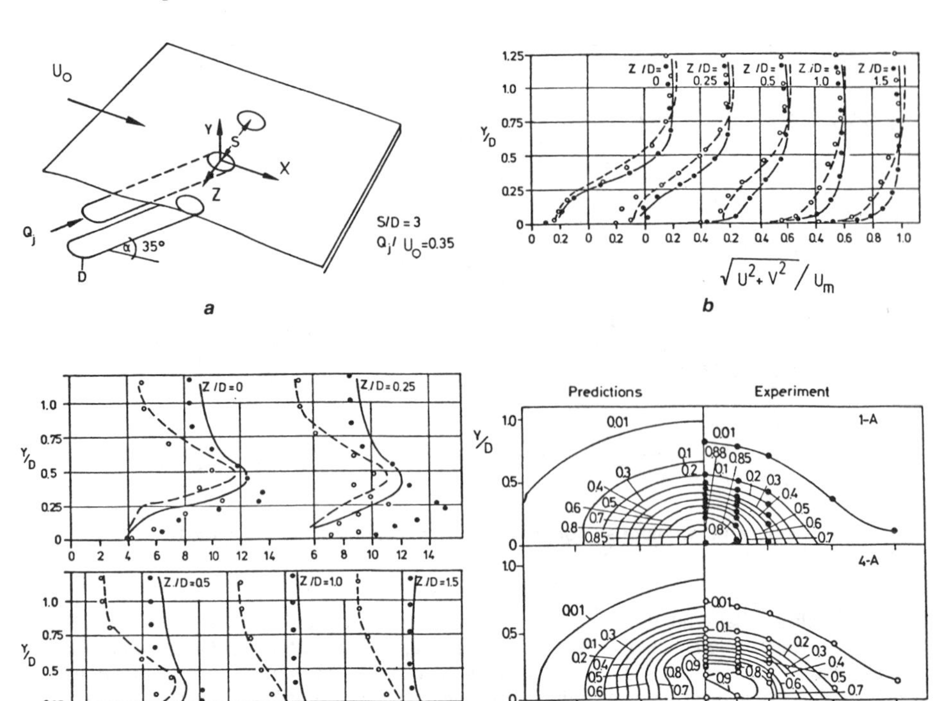

Figure 17. Flow configuration and comparison of predictions [65] with experimental data: (A) flow configuration; (B) velocity profiles; (C) turbulence intensity profiles; (D) normalized temperature contours [66].

Predictions were compared with laboratory data by Viollet [68] and Andreopoulos [69]. Figure 18 shows a comparison of predicted concentration contours with measured [68] data. (The calculations were performed with the hybrid scheme on a coarse grid (26 × 13 × 12) in the x, y, z directions and required 250 iterations for convergence and 12 minutes CPU time on the Siemens 7880 computer. The predictions show somewhat more spreading above the jet axis than the measurement, especially in the near field, but considering that this was only a coarse grid calculation, the agreement is satisfactory. Viollet [68] also performed some calculations of the experiments, but these employed a parabolic model (similar to Tatchel's [53]) with the initial conditions having to be prescribed from the measurements at a plane just downstream of the cooling tower. Thus, the bending over of the jet was not computed, and the method would only be useful in calculating flow situations in which the required experimental information was available to start it off.

Combustion Chamber Models

The actual realm of numerical models is in the calculation of flows in complex geometries, such as those inside combustion chambers. Theoretical and experimental investigations were performed by Green and Whitelaw [70] for two physical-model geometries of combustion chambers. Figure 19 shows the configuration and results of flow visualization studies in the two geometries. In both

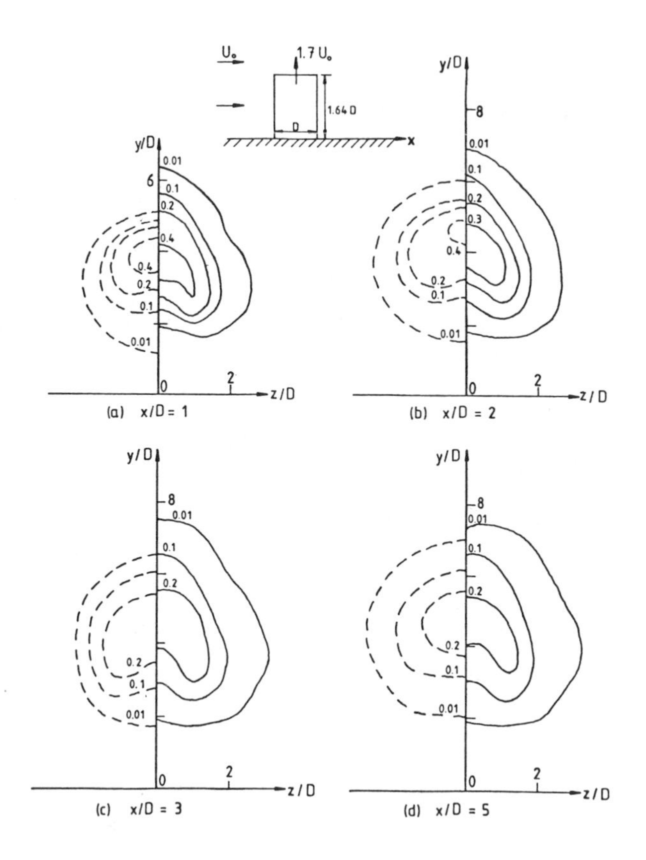

Figure 18. Comparison of normalized concentration contours. ———— predictions [67]; ------- experimental data [68].

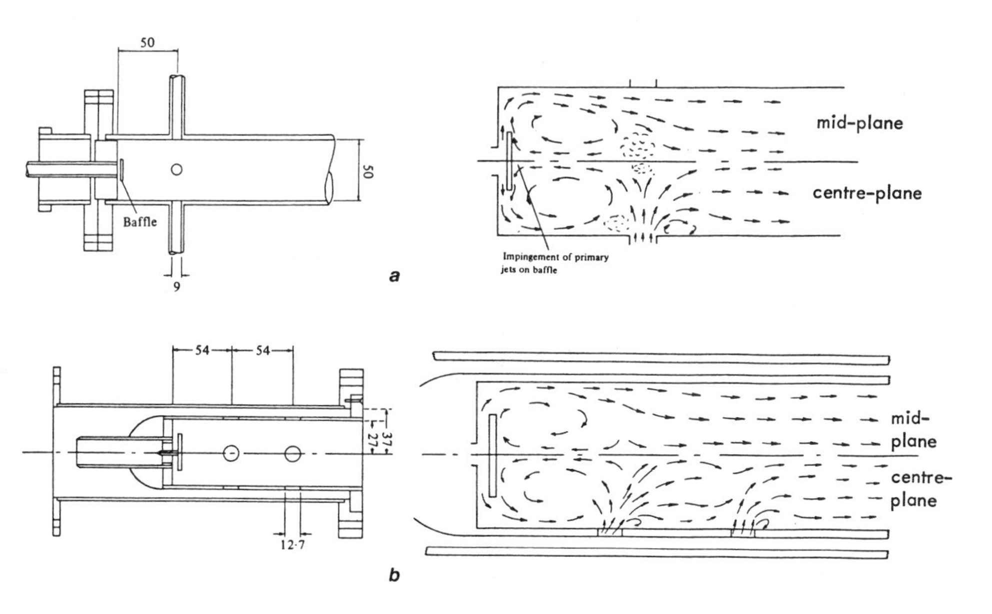

Figure 19. Geometries and flow patterns in gas-turbine combustion chambers [70]: (A) case 1; (B) case 2.

cases, an axial hole and baffle arrangement simulated a fuel-vaporizer inlet. In the first case, four jets were introduced into the chamber through pipes located symmetrically around its circumference to simulate, in a somewhat idealized manner, both the primary and the dilution jets. The second case had an annular arrangement through which the primary and secondary flows were supplied. These issued into the chamber through 2 sets of four holes, with each set arranged symmetrically around the circumference. The flow visualization studies showed that there were important differences between the two cases. The second case is of course closer to the practical situation. Figure 20 shows comparisons of predicted and measured longitudinal velocity contours for the first case,

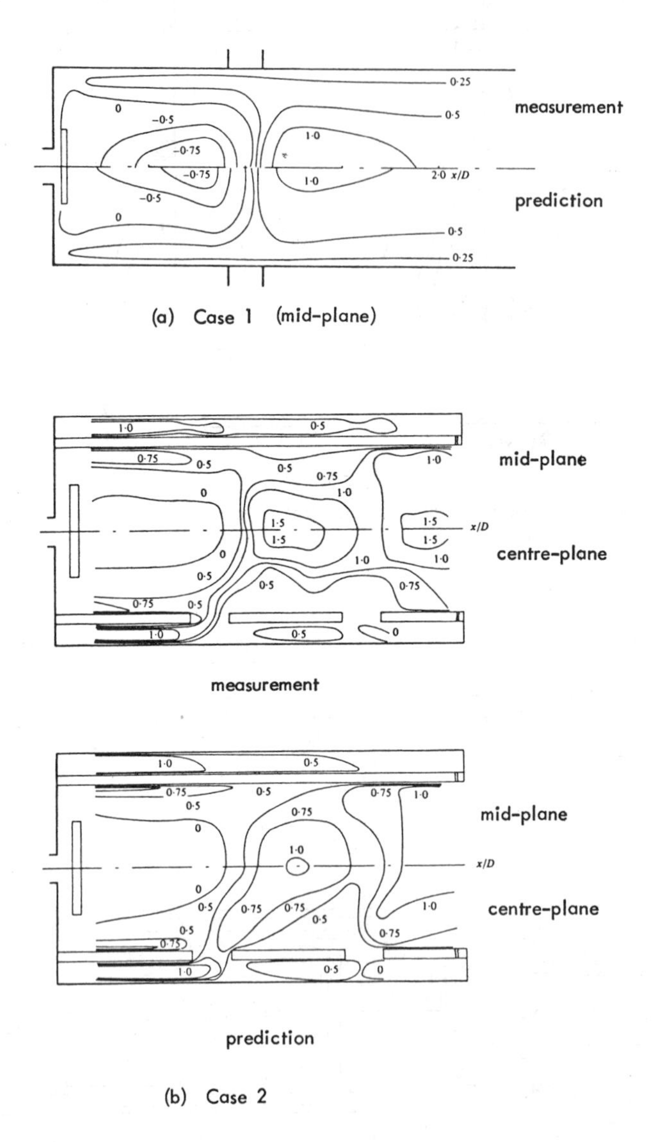

Figure 20. Comparison of predicted U-velocity fields with measurements [70].

along the mid-plane between the holes, and for the second case, both along the center plane through the holes and the mid-plane between them. The main features of the flow are adequately reproduced by the calculations, although the velocity maxima along the center line of the chamber were underpredicted. The discrepancy was up to about 40% in the second case. However, considering the complexity of these flow situations, the agreement is quite satisfactory, and certainly good enough for engineering applications. The calculations were performed with the same basic computer code as Jones and McGuirk [8], in 90° sectors, encompassing the space between adjacent mid-planes.

Computational grids of 10 × 19 × 27 in the circumferential, radial and axial directions were employed for case 1 and 10 × 20 × 30 for case 2, correspondingly. The number of iterations required for convergence were, respectively, 400 and 700. Serag-Eldin and Spalding [71] presented a numerical model for calculating the flow and heat transfer in a combustion chamber with swirl and combustion of natural gas. In addition to the hydrodynamic turbulence equations solved in the model just presented [70], partial differential equations were solved for the stagnation enthalpy, concentration, and the mean square of concentration fluctuations. The model was applied to predict their measurements of the temperature distribution in a combustion chamber in which the fuel and primary air flow were introduced through coaxial pipes at the center line of the chamber, with a 45° annular vane swirler acting on the air flow. Dilution air was introduced through six equally spaced pipes located around the circumference of the chamber. The predictions showed, in the worst case, qualitative agreement with the measurements, which is satisfactory for the very complex flow situation investigated.

Plane Jets

The plane jet in cross flow differs from the three-dimensional jet in cross flow in that it completely blocks the path of the cross-stream on its discharge side. The jet is itself more strongly deflected and where it cannot satisfy its entrainment requirement on the lee side a recirculating eddy is formed and the jet becomes attached to the downstream wall. The flow is elliptic, and it can in principle be solved with any of the three-dimensional numerical models already presented, since they all have an elliptic treatment in the plane normal to the streamwise direction. However, it is easier to use a two-dimensional method such as that employed by McGuirk and Rodi [72] to calculate full-depth side discharges into rectangular channel flow. The flow is nominally three-dimensional, but in the absence of significant buoyancy forces, three-dimensional effects are minimal [73], and plane jet calculations would be adequate. McGuirk and Rodi [72] accounted partially for the latter by employing a two-dimensional depth-averaged model in which the effects of non-uniformities over the depth could be lumped together in dispersion terms. Figure 21 shows the flow configuration and predicted streamlines. Predicted width of recirculation eddy and jet trajectory are also compared to experimental data [74, 75]. The agreement is reasonably good in both cases. The computational resources required for these calculations are much smaller, both in terms of core memory size and CPU time, than those for comparable three-dimensional ones. It is therefore of considerable advantage to use such a model, even when the jet is not completely two-dimensional, so long as the three-dimensional effects are only minor.

Future Developments

Numerical models have been applied to calculate fairly complex flow situations only since the last decade and half. The possibilities have been clearly demonstrated, but the accuracy and the computational costs are not yet satisfactory. A lot of development work is now going on in many groups to improve the situation. Numerical solution procedures are being developed which procure much faster convergence than the ones due to Chorin [50], Amsden and Harlow [51], or Patankar and Spalding [52] employed in most of the models presented here. One of such procedures is the PISO (Pressure Implicit Split Operator) algorithm proposed by Issa [76]. This algorithm has been found to procure convergence 4 to 6 times faster than the Patankar and Spalding method in many two-dimensional flow calculations. The present writer has obtained with it, in some three-dimensional calculations, convergence rates which are 2 to 3 times faster. On the accuracy side,

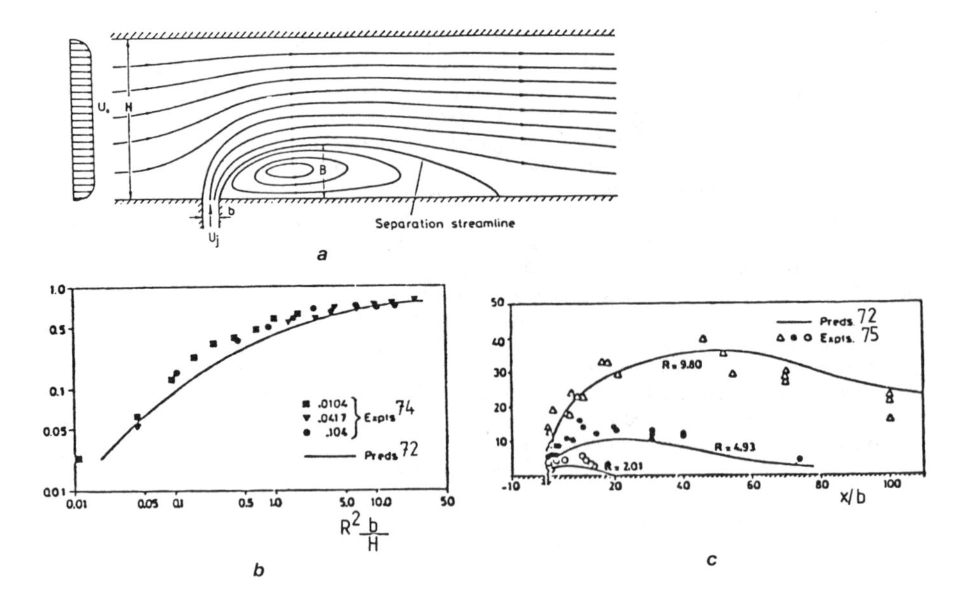

Figure 21. Prediction of plane jet in confined cross flow: (A) flow configuration and predicted streamlines; (B) recirculation-eddy width; (C) jet trajectory (locus of maximum temperature).

numerical methods are being developed, such as the finite-analytic method of Chen [77] and co-workers or finite element methods, [78] which are inherently more accúrate; in addition to the use of higher-order differencing methods in current models. Developments in the field of grid-generation techniques and body-fitted coordinate systems would improve the application of numerical model to flows in and around complex geometries. Improved turbulence modeling would also help to produce more accurate results. Coupled with these, the developments of larger and faster computers will make calculations with numerical models more commonplace.

CONCLUDING REMARKS

In the study of turbulent jets in cross flow empirical models offer quick and simple methods of obtaining first-order estimates and a qualitative picture of the jet trajectory, its extent, and the decay rates of its axial velocity or temperature. The main requirement for reasonable predictions is the use of correlation equations or curves derived from an experimental data base with similar characteristics as the problem of interest.

Integral models contain in simplified forms mathematical representations of the basic conservation laws, and are thus of wider applicability. Physical phenomena taking place in the flow are modeled with relations which are more or less empirical. Successful integral models have used a range of combinations of these. Again, integral models tend to perform well in flow situations similar to those in which they have been calibrated. They are not always simple to use, but they are computationally very cheap, and are thus the most widely used methods for the solution of practical problems.

Numerical models require the least assumptions or empirical input and are thus less problem dependent. They also offer the prospects of predicting flow phenomena under quite complex situations to which the other two model types cannot even be applied. Numerical models are attractive because, they can provide so much information on the flow properties as to make expensive experimental or field measurements superfluous, although they have not yet reached such a stage in their development. They are at present capable of providing qualitative results in most cases, and quantitatively accurate results in a few cases only. Their computational cost is still very high, but progress in the development of better algorithms, and larger and faster computers is bringing this down quite rapidly. Improved numerical techniques and turbulence models are also leading to more accurate predictions. Undoubtedly, these are the models of the future.

Acknowledgment

I wish to express my sincere gratitude to Frau Gisela Krause for the excellent typing of this paper.

NOTATION

B	width of recirculating eddy in plane jet
b	width of plane jet exit channel
C	concentration
C_D	drag coefficient
C_p	$(P_T - P_{T_0})/(P_{T_j} - P_{T_0})$
C_x, C_y, C_z	direction cosines of the velocity vector in the x, y, z directions
c_μ, c_1, c_2	constants in turbulence model
D	diameter of jet at exit
E	entrainment coefficient
e	exponential
F	densimetric Froude number
G	generation rate of k
g	acceleration due to gravity
H	distance between confining walls
J	jet-to-cross-flow momentum flux ratio
k	kinetic energy of turbulence
n	normalized distance from jet's axis
P	pressure
P_T	total pressure
Q	resultant velocity
Q_i	resultant jet velocity at exit
R	jet-to-cross-flow velocity ratio
R_Δ	cell Peclet number
r	radius of curvature of jet's axis
S	jet spacing
S_ϕ	source or sink of dependent variable ϕ
T	temperature
T_i	temperature of jet at exit
T_0	temperature of cross stream
U	longitudinal velocity
U_e	entrainment velocity
U_j	jet velocity at exit
$U_{j,x}$	longitudinal velocity in jet exit plane
U_m	maximum or axial velocity of jet in the far field
U_m	maximum value of U
$\tilde{u}$	root mean square of the fluctuating component of U
V	vertical velocity or jet velocity in the direction of the axis
V_{max}	maximum or axial velocity of jet in the near field
W	lateral velocity
x	longitudinal distance
y	vertical distance
z	lateral distance

Greek Symbols

Δ	cell width
ϵ	dissipation rate of k
Γ_f	false or numerical diffusivity
Γ_ϕ	molecular diffusion coefficient of ϕ
$\Gamma_{\phi,e}$	$(=\Gamma_\phi + \mu t/\sigma_\phi)$ effective diffusion coefficient of ϕ
ζ	axial distance
η, ξ	normal distances to jet's axis
λ_v	$(=\int_A V^2\, dA/(\bar{V}^2 A))$, momentum coefficient
λ_ϕ	$(=\int_A V\phi\, dA/(\bar{V}\bar{\phi}A))$, scalar property coefficient
μ	molecular viscosity
μ_t	turbulent eddy viscosity
ρ	density
ρ_r	reference density
σ_ϕ	Prandtl/Schmidt number for ϕ
ϕ, φ	dependent variable and its fluctuation
θ	angle between jet's axis and the horizontal direction
θ_j	angle of jet to the horizontal at exit

Subscripts

b	jet boundary	o	cross-stream value
j	jet value	1/2	half-width
m, max	maximum value		

Superscript

$\bar{\ }$	average value	$\vec{\ }$	vector

REFERENCES

1. Moussa, Z. M., Trischka, J. W., and Eskinazi, S., "The Near Field in the Mixing of a Round jet with a Cross-Stream," *Journ. Fluid Mechanics*, Vol. 80 (1977), pp. 49–80.
2. Foss, J., "Flow Visualization Studies of Jets in a Cross Flow," Sonderforschungsbereich 80, Universität Karlsruhe, Report No. SFB80/T/161, February 1980.
3. Andreopoulos, J., "Measurements in a Pipe Flow Issuing Perpendicular into a Cross Stream," *ASME Journ. Fluids Engineering*, Vol. 104 (1983), pp. 493–499.
4. Platten, J. L., and Keffer, J. F., "Deflected Turbulent Jet Flows," *ASME Journ. Applied Mech.* (Dec. 1971), pp. 756–758
5. Kamotani, Y., and Greber, I., "Experiments on Confined Turbulent Jets in Cross Flow," NASA Report, NASA CR-2392, March 1974.
6. Sugiyama, Y., and Usami, Y., "Experiments on the Flow in and around Jets Directed Normal to a Cross Flow," *Bulletin JSME*, Vol. 22 (1979), pp. 1736–1745.
7. Stoy, R. L., and Ben-Haim, Y., "Turbulent Jets in a longitudinal Cross Flow," *ASME Journ. of Fluids Engineering* (Dec. 1973), pp. 551–556.
8. Jones, W. P., and McGuirk, J. J., "Computation of a Round Turbulent Jet Discharging into a Confined Cross Flow," *Turbulent Shear Flows II*, Springer Verlag, New York, 1979, pp. 233–245
9. Khan, Z. A., McGuirk, J. J., and Whitelaw, J. H., "A Row of Jets in a Cross Flow," AGARD CP 308, paper 10. 1982.
10. Atkinson, K. N., Khan, Z. A., and Whitelaw, J. H., "Experimental Investigation of Opposed Jets Discharging Normally into a Cross-Stream," *Journ. Fluid Mechanics.*, Vol. 115 (1982), pp. 493–504
11. Rajaratnam, N., "Developments in Water Science 5," *Turbulent Jets*, Elsevier Scientific Publishing Company, New York, 1976.
12. Kamotani, Y., and Greber, I., "Experiments on a Turbulent Jet in a Cross Flow," *AIAA Journal*, Vol. 10, No. 11 (November 1972), pp. 1425–1429 (see also, NASA CR 72893, 1971).
13. Fan, L. N., "Turbulent Jets into Stratified or Flowing Ambient Fluids." Keck Laboratory of Hydraulics and Water Resources, California Institute of Technology, Report No. KH-R-15, 1967.
14. Pratte, B. D., and Baines, W. D., "Profiles of the Round Turbulent Jets in a Cross Flow," *Journ. Hydr. Div., Proc. ASCE*, HY6, Vol. 93 (1967), pp. 53–64.
15. Chassaing, P., et al., "Physical Characteristics of Subsonic Jets in a Cross-Stream," *Journ. of Fluid Mechanics*, Vol. 62 (1974), pp. 41–64.
16. Ramsey, J. W., and Goldstein, R. J., "Interaction of a Heated Jet with a Deflecting Stream," NASA CR 72613, 1972.
17. Demuren, A. O., "Numerical Calculations of Steady Three-Dimensional Turbulent Jets in Cross Flow," *Computer Methods in Applied Mechanics and Engineering*, Vol. 37 (1983), pp. 309–328.
18. Abramovich, G. N., *The Theory of Turbulent Jets*, Cambridge, MA, MIT Press, 1963.
19. Shandorov, G. S., "Flow from a Channel into Stationary and Moving Media," *Zh. Tekhn. Fiz.*, 37:1 (1957).
20. Ivanov, Yu. V., "Plane Jet in an External Cross Stream of Air," *Izv. Akad. Nauk Est. SSR*, 11:2 (1953).

21. Margason, R. J., "The Path of a Jet Directed at Large Angles to a subsonic Stream," NASA, TN.D.-4919, Langley Research Center, Hampton, Virginia, 1968.
22. Rajaratnam, N., and Gangadharaiah, T., "Scales for Circular Jets in Cross Flow," *Journ. Hydr. Div., ASCE*, HY4, Vol. 107 (1981), pp. 487–500.
23. Holdeman, J. D., and Walker, R. E., "Mixing of a Row of Jets with a Confined Cross Flow," *AIAA Journal*, Vol. 15, No. 2 (February 1977), pp. 243–249.
24. Keffer, J. F., and Baines, W. D., "The Round Turbulent Jet in a Cross Wind," *Journ. of Fluid Mechanics*, Vol. 15 (1963), pp. 481–497.
25. Carhart, R. A., et al., "Mathematical Model for Single-Source (Single-Tower) Cooling Tower Plume Dispersion," Electric Power Research Institute, Palo Alto, CA., CS-1683, Vol. 2, Project 906-1, Iterim Report, January 1981.
26. Morton, B. R., Taylor, S. G., and Turner, J. S. "Turbulent Gravitational Convection from Maintained and Instantaneous Sources," *Proc. Roy. Soc.* A 234 (1968), pp. 1–23.
27. Hirst, E. A., "Buoyant Jets with Three-Dimensional Trajectories," *Journ. Hydr. Div., Proc. ASCE*, HY 11, Vol. 98 (1972), pp. 1999–2014.
28. Schatzmann, M., "An Integral Model of Plume Rise," *Atmospheric Environment*, Vol. 13 (1979), pp. 721–731.
29. Crowe, C. T., and Riesebieter, H., "An Analytical and Experimental Study of Jet Deflection in a Cross Flow," Fluid Dynamics of Rotor and Fan-Supported Aircraft at Subsonic Speeds, AGARD Reprints, 1967.
30. Vizel, Y. M., and Mostinskii, I. L., "Deflection of a Jet injected into a Stream," *Fluid Dynamics*, Vol. 8 (1965).
31. Platten, J. L., and Keffer, J. F., "Entrainment in Deflected Axi-Symmetric Jets at Various Angles to the Stream," Univ. of Toronto, Report UTME, TP 6808, 1968.
32. Abraham, G., "The Flow of Round Buoyant Jets Issuing Vertically into Ambient Fluid Flowing in a Horizontal Direction," Delft Hydraulic Laboratory, Publ. No. 81 (1971).
33. Campbell J. F., and Schetz, J. A., "Flow Properties of Submerged Heated Effluents in a Waterway," *AIAA Journal*, Vol. 11, No. 2 (Feb. 1973), pp. 223–230.
34. Eckert, E. R. G., and Drake, R. M., Jr., *Heat and Mass Transfer*, 2nd ed., McGraw Hill, New York (1959), pp. 139, 239–243.
35. Isaac, K. M., and Schetz, J. A., "Analysis of Multiple Jets in a Cross Flow," *ASME, Journ. of Fluids Engineering*, Vol. 104 (Dec. 1982), pp. 489–492.
36. Hoerner, S. F., *Fluid-Dynamic Drag*, published by the author, 148 Busteed Drive, Midland Park, New Jersey, 07432, 1965.
37. Sucec, J., and Bowley, W. W., "Prediction of the Trajectory of a Turbulent Jet Injected into a Crossflowing Stream," *ASME, Journ. of Fluids Engineering*, Vol. 98 (Dec. 1976), pp. 667–673.
38. Braun, G. W., and McAllister, J. D., "Cross Wind Effect on Trajectory and Cross Sections of Turbulent Jets," in *Analysis of a Jet in a Subsonic Crosswind*, NASA SP-218 (1969), pp. 141–164.
39. Chan, D., et al., "Entrainment and Drag Forces of Deflected Jets," *ASCE, Journ. Hydr. Div.*, HY5, Vol. 102 (May 1976), pp. 615–635.
40. Makihata, T., and Miyai, Y., "Prediction of the Trajectory of Triple Jets in a Uniform Cross Flow," *ASME, Journ. of Fluids Engineering*, Vol. 105, (March 1983), pp. 91–97.
41. Adler, D., and Baron, A., "Prediction of a Three-Dimensional Jet in Cross Flow," *AIAA Journal*, Vol. 17, No. 2 (Feb. 1979), pp. 168–174.
42. Chen, C. L. H., "Aufrollung eines zylindrischen Strahles durch Querwind," doctoral dissertion, Univ. of Göttingen, Göttingen, W. Germany, 1942.
43. Strauber, M., "Berechnung von Strahlkonturen mit Hilfe eines Wirbelringmodells," *Zeitschrift für Flugwissenschaften*, Vol. 23 (Nov. 1975), pp. 394–400.
44. Schwartz, J., and Tulin, M. P., "Chimney Plumes in Natural and Stable Surrounding," *Atmospheric Environment*, Vol. 6 (Jan. 1972), pp. 19–35.
45. Demuren, A. O., "Prediction of Steady Surface-Layer Flows," Ph.D. thesis, University of London, 1979.
46. Chien, J. C., and Schetz, Y. A., "Numerical Solution of the Three-Dimensional Navier-Stokes Equations with Applications to Channel Flows and a Buoyant Jet in a Cross Flow," *ASME, Journ. of Applied Mechanics*, Sep. 1975, pp. 575–579.

47. Launder, B. E., and Spalding, D. B., "The Numerical Computation of Turbulent Flows," *Comp. Meths. Appl. Mech. Engrg.*, Vol. 3 (1980), pp. 293–312.
48. Roache, P. J., *Computational Fluid Dynamics*, Hermosa Publishers, Albuquerque, New Mexico, 1976.
49. Leonard, B. P., "A Stable and Accurate Convective Modelling Procedure based on Quadratic Upstream Interpolation," *Compt. Meths. Appl. Mech. Engrg.* 19, (1979), pp. 59–98.
50. Chorin, A. J., "Numerical Solution of the Navier-Stokes Equations," *Math. Comp.*, Vol. 22 (1968), pp. 745–762.
51. Amsden, A. A., and Harlow, F. H., "The SMAC Method," Los Alamos Scientific Lab. Report LA-4370, 1970.
52. Patankar, S. V., and Spalding, D. B., "A Calculation Procedure for Heat, Mass and Momentum Transfer in Three-Dimensional Parabolic Flows," *Int. J. Heat Mass Transfer*, Vol. 15 (1972), pp. 1787–1805.
53. Tatchel, D. G., "Convection Processes in Confined Three-Dimensional Boundary Layers," Ph.D. thesis, Univ. London, 1975.
54. Bergeles, G., Gosman, A. D., and Launder, B. E., "The Turbulent Jet in a Cross-Stream at Low Injection Rates: A Three-Dimensional Numerical Treatment," Report TF/78/3, Mech. Eng. Dept., Univ. California, Davis, 1978.
55. Pratap, V. S., and Spalding, D. B., "Fluid Flow and Heat Transfer in Three-Dimensional Duct Flows," *Int. J. Heat Mass Transfer*, Vol. 19 (1976), pp. 1183–1188.
56. Patankar, S. V., Basu, D. K., and Alpay, S. A., "Prediction of the Three-Dimensional Velocity Field of a Deflected Turbulent Jet," *ASME, J. Fluids Eng.*, Vol. 99 (1977), pp. 758–762.
57. Rodi, W., and Srivatsa, S. K., "A Locally Elliptic Calculation Procedure for Three-Dimensional Flows and its Application to a Jet in a Cross Flow," *Compt. Meth. Appl. Mech. Engrg.*, Vol. 23 (1980), pp. 67–83.
58. White, A. J., "The Predictions of the Flow and Heat Transfer in the Vicinity of a Jet in Cross Flow," ASME Paper 80-WA/HT-26, 1980.
59. Claus, R. W., "Analytical Calculation of a Single Jet in Cross Flow and Comparison with Experiment," Paper AIAA-83-0238, AIAA 21st Aerospace Sciences Meeting, Reno, Nevada, January 10–13, 1983.
60. Crabb, D., Durao, D. F. G., and Whitelaw, J. J., "Round Jet Normal to a Cross Flow," *ASME, Journal of Fluids Engineering*, Vol. 103, No. 1 (March 1981), pp. 142–153.
61. Andreopoulos, J., "Measurements in a low Momentum Jet into a Cross Flow," Proc. 3rd Turbulent Shear Flows Conference, Davis, CA, 1981.
62. Andreopoulos, J., and Rodi, W., "Experimental Investigation of Jets in a Cross Flow," *J. Fluid Mech.*, Vol. 138 (1984), pp. 93–127.
63. Demuren, A. O., "False Diffusion in Three-Dimensional Steady Flow Calculations," Report SFB80/T/224, Sonderforschungsbereich 80, Universität Karlsruhe, FRG, 1983.
64. Chapman, M., "FRAM-Nonlinear Damping Algorithms for the Continuity Equation," *Journ. of Computational Physics*, Vol. 44 (1981), pp. 84–103,
65. Demuren, A. O., and Rodi, W., "Three-Dimensional Calculation of Film Cooling by a Row of Jets," Proc. 5th GAMM Conference on Numerical Methods in Fluid Mechanics, Rome, October 1983.
66. Kadotani, K., and Goldstein, R. J., "On the Nature of Jets Entering a Turbulent Flow, Part A—Jet—Mainstream Interactions," Proc. 1977 Tokyo Joint Gas Turbine Congress, pp. 46–54.
67. Demuren, A. O., and Rodi, W., "Three-Dimensional Numerical Calculations of Flow and Plume Spreading Past Cooling Towers," Proc. Fourth IAHR Cooling Tower Workshop, Interlaken, Switzerland, 1984.
68. Viollet, M. P.-L., "Etude de Jets dans des Courants Traversiers et dans des Milieux Stratifies," Ph.D. thesis, Université Pièrre et Marie Curie, Paris, 1977.
69. Andreopoulos, J., "An Experimental Investigation of Cooling Tower Plumes, Part 1: Non-buoyant Plume in a Uniform Cross Flow," Report SFB80/E/214, Sonderforschungsbereich 80, University of Karlsruhe, 1982.
70. Green, A. S., and Whitelaw, J. H., "Isothermal Models of Gas-turbine Combustors," *J. Fluid Mech.*, Vol. 126 (1983), pp. 399–413.

71. Serag-Eldin, M. A. S., and Spalding, D. B., "Computation of Three-Dimensional Gas Turbine Combustion Chamber Flows," *ASME, Journ. of Engineering for Power*, Vol. 101, No. 2 (July 1979), pp. 327–336.
72. McGuirk, J. J., and Rodi, W., "A Depth-Averaged Mathematical Model for the Near Field of Side Discharges into Open-Channel Flow," *J. Fluid Mech.*, Vol. 86 (1978), p. 761–781.
73. Demuren, A. O., and Rodi, W., "Side Discharges into Open Channels: Mathematical Model," *ASCE, Journ. of Hydr. Engineering*, Vol. 109, No. 12 (Dec. 1983), pp. 1707–1722.
74. Mikhail, R., Chu, V. H., and Savage, S. B. "The Reattachment of a Two-Dimensional Turbulent Jet in a Confined Cross Flow," Proc. 16th IAHR Cong., Sao Paulo, Brazil, Vol. 3 (1975), pp. 414–419.
75. Carter, H. H., "A Preliminary Report on the Characteristics of a Heated Jet Discharged Horizontally into a Transverse Current. Part 1—Constant Depth," Tech. Rep. No. 61, Chesapeake Bay Inst., Johns Hopkins University, 1969.
76. Issa, R., "Solution of the Implicitly Discretized Fluid Flow Equations by Operator-Splitting," Report FS/82/15, Dept. of Mech. Eng., Imperial College, London, 1982.
77. Chen, C. J., and Li, P., "The Finite Analytic Method for Steady and Unsteady Heat Transfer Problems," ASME Paper 80-HT-86, ASME/AIChE National Heat Transfer Conference Orlando, Florida, July 27–30, 1980.
78. Baker, A. J. and Orzechowski, J. A., "An Assessment of Factors Affecting Prediction of Near-Field Development of a Subsonic VSTOL Jet in Cross-Flow," Naval Air Development Center, Warminster, PA, Report NADC-81177-60, 1982.

CHAPTER 18

BATCHWISE JET MIXING IN TANKS

P. Rice

Department of Chemical Engineering
University of Technology
Loughborough Leicestershire, England

CONTENTS

INTRODUCTION, 466

HISTORICAL DEVELOPMENT AND DESIGN FEATURES, 466

ALTERNATIVE DESIGNS, 468

PRINCIPLES OF JET MIXING, 469

NOTATION, 473

REFERENCES, 473

INTRODUCTION

Batchwise jet mixing is taking liquid from the tank and using a pump to reinject it as a fast moving stream. Revill [1] describes the principles of this method of mixing a liquid. The relative velocity between the jet and the bulk liquid creates a turbulent mixing layer at the jet boundary. This mixing layer grows in the direction of the jet flow, entraining and mixing the jet liquid with the bulk liquid.

HISTORICAL DEVELOPMENT AND DESIGN FEATURES

The original idea of jet mixing was conceived during the Second World War. Fossett and Prosser [2] and subsequently Fossett [3] published the original papers on the subject, in which they described an investigation using an inclined side-entry jet-mixer design for use in mixing TEL into high octane fuel in large-scale tanks up to 40 m (120 ft) in diameter. Subsequently Fossett [4] provided further information on the design of such mixers. Since their original work, jet mixing has received little attention, for although mixing is a very common process, there has been comparatively limited research and development on the mechanisms of the process. Commercially many companies still use the original Fossett design procedure. A typical application of this design is shown in Figure 1. Liquid is circulated from the tank and reinjected through a nozzle to form the jet. This

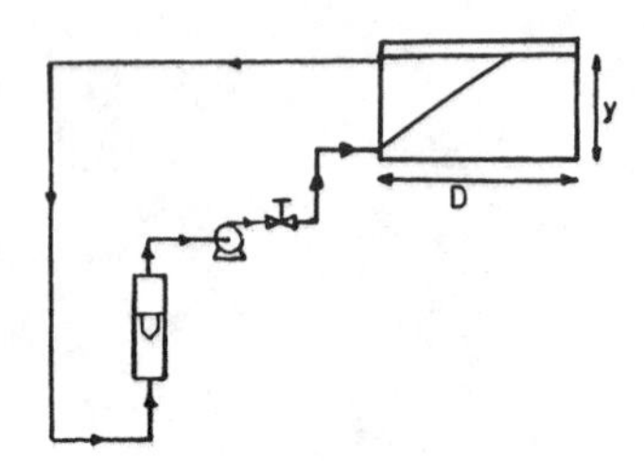

Figure 1. Typical Fossett and Prosser [2] design of jet mixer.

nozzle is positioned near the base of the tank inclined towards the liquid surface so that it can, if required, break the liquid surface two thirds of the way across the tank from the jet entry point. The important point to note is that the jet should be aimed diametrically so as not to produce a general solid rotation. The materials to be mixed may be added either to the circulating flow or directly into the tank. The flow patterns obtained are shown in Figure 2 for a vertical section through the jet and Figure 3 for a plan view of the tank.

The original correlation of Fossett and Prosser [2, 3] only included terms for tank diameter, jet diameter, and jet velocity

$$T = 9.0\frac{D^2}{Vd} \tag{1}$$

They specified however that the jet should just disturb the surface. They employed a conductivity technique to measure mixing time, but give no indication of the "degree of mixedness."

The time for the addition of their "tracer" represented a large fraction of the total mixing time. When the addition represents a very small fraction of the total mixing time Fossett [4] suggests that the constant in Equation 1 be 4.5. This equation is solely applicable to the turbulent jet regime and it suggests mixing time is independent of the jet Reynolds number.

In 1956 Fox and Gex [5] published results of an investigation of inclined side- entry jets covering both the laminar and turbulent jet regimes. They used a pH indicator technique to measure a "terminal" mixing time. Tanks up to 14 ft in diameter were used in their tests. Again no measure of mixedness is given for the data. Their results indicate that mixing time is dependent on jet Reynolds number; the dependence being strong in the laminar jet regime but less so in the turbulent jet regime.

For the laminar regime

$$T = C_1\frac{y^{0.5}D}{Re_j^{1.33}(Vd)^{0.67}g^{0.17}} \tag{2}$$

while for the turbulent regime

$$T = C_2\frac{y^{0.5}D}{Re_j^{0.17}(Vd)^{0.67}g^{0.17}} \tag{3}$$

In their results Fox and Gex [5] showed a lot of data scatter (24% average difference in the turbulent regime). This fact was noted by Okita and Oyama [6] who recorrelated a representative sample of the Fox and Gex data. They produced a mixing time equation of the same form as that

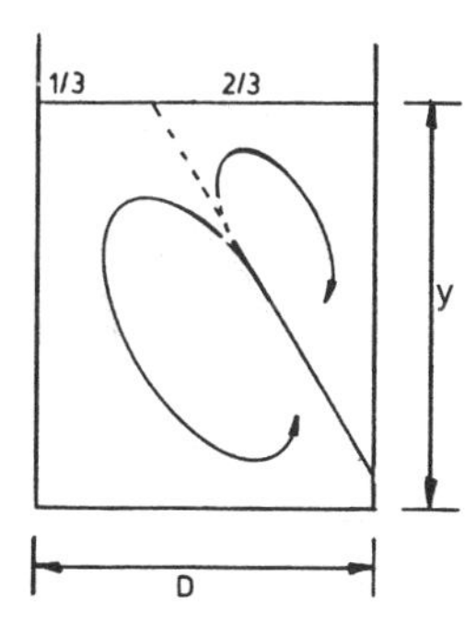

Figure 2. Vertical section through jet showing flow pattern for Fossett and Prosser design.

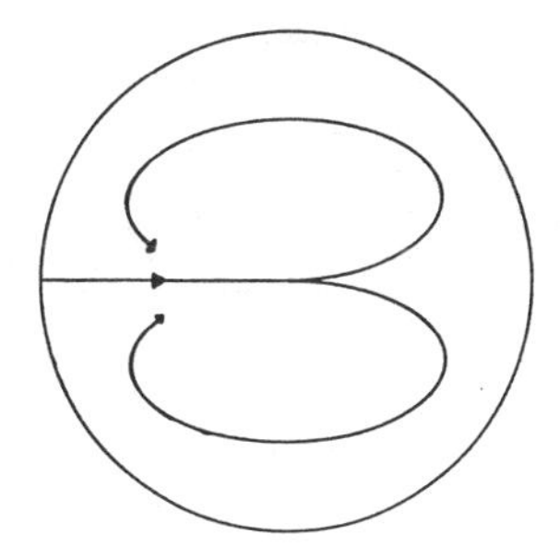

Figure 3. Plan view through jet showing flow pattern for Fossett and Prosser design.

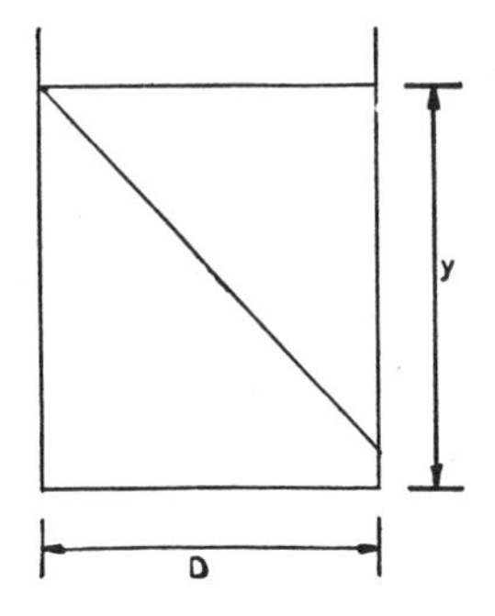

Figure 4. Cowdrey [8] design of jet mixer.

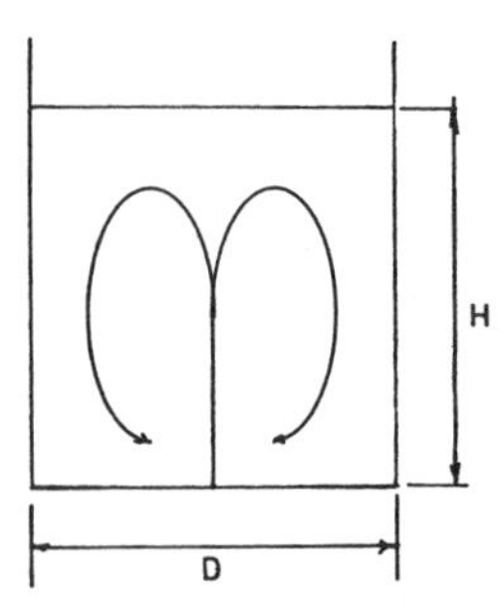

Figure 5. Hiby and Modigell [9] design of jet mixer.

of Fossett and Prosser, i.e., independent of jet Reynolds number

$$T = 2.6\frac{D^{1.5}y^{0.5}}{Vd}; \qquad Re_j > 7{,}000 \tag{4}$$

This equation, however, produces only a marginal improvement in the data scatter (20% average difference in the turbulent regime for their data sample). Variation in what was terminal mixing time could explain this scatter.

Further studies of inclined side-entry jets were carried out by Okita and Oyama [6] who used a conductivity method and Van de Vusse [7] who employed a density technique to measure mixing times. Okita and Oyama [6] obtained for their data the following mixing-time equation

$$T = \frac{5.5D^{1.5}y^{0.5}}{Vd}; \qquad 5 \times 10^3 < Re_j < 10^5 \tag{5}$$

while Van de Vusse [7] correlated his data with

$$T = 3.68\frac{D^2}{Vd} \tag{6}$$

Both conclude that mixing time is independent of jet Reynolds number. However no indication of the "degree of mixedness" is quoted in either paper.

In 1978 Coldrey [8] proposed a modified design for inclined side-entry jet mixing. He suggested utilizing the longest possible jet length. A typical design is shown in Figure 4. This longer jet length produces more effective mixing and therefore reduces mixing time. Coldrey formulated a mixing-time equation independent of the jet Reynolds number for the turbulent jet regime based on the theory that mixing time is inversely dependent on the amount of liquid entrained by the jet.

$$T = 3.54\frac{D^2yd}{XQ_j} \tag{7}$$

ALTERNATIVE DESIGNS

Investigations of alternative designs for jet mixing have been published in the past few years. Hiby and Modigell [9] tested a flat based cylindrical tank with an axial vertical discharging jet (Figure 5). They indicated that mixing time was dependent on a jet Reynolds number when the tank Reynolds number is less than 1×10^6. They were the first to introduce the concept of "degree of mixing." They used a conductivity technique with a trace output from which it was possible to estimate the time to any degree of mixing. They chose the time to 95% homogeneity, as their "standard" mixing time. In all previous reference, as already noted, the degree of mixing is not specified and it may be that the variability of the quoted correlations stems partly from this.

An analysis of jet mixing has been given by Revill [1] who derived an expression of the form

$$T = C\left\{\frac{D}{X}\right\}\left\{\frac{Dy}{Vd}\right\}$$

where C is dependent on the physical properties of the liquid and X is an "effective" jet mixing length which depends on the tank geometry.

PRINCIPLES OF JET MIXING

Lane and Rice have published a series of papers on jet mixing. In the first they examined the mixing performance of a jet mixer using a hemispherical bottom tank with an axial vertically discharging jet [10] (Figure 6). Subsequently they compared the performance of the Fossett and Cowdrey designs [11] and, also, the Hiby and Modigell design [13]. In their tests liquid was taken off from the top of the tank and injected at the bottom. They used a salt injection technique to measure the mixing time. A typical trace is shown in Figure 7. For their results, following Hiby and Modigell [9], they chose the time as that required to meet 95% total mixing as their "standard." Using dimensional analysis, Lane and Rice [10] derived a nondimensional mixing time as a function of jet Reynolds number similar to Fox and Gex. Their data indicated several regimes, which, they suggested, were related to the flow nature of the jet and the recirculating flow in the tank; that is, with a laminar jet the recirculating flow was laminar, and with the jet turbulent the

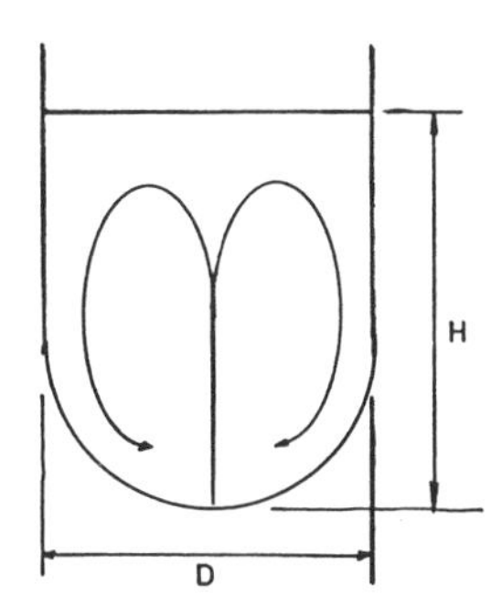

Figure 6. Lane and Rice [10] design of jet mixer.

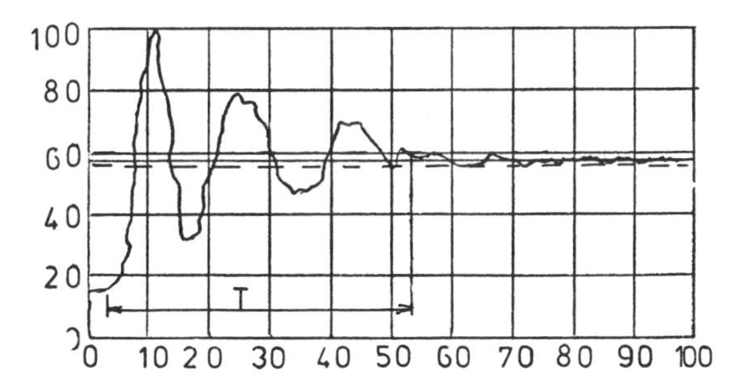

Figure 7. Typical trace from a pulse injection test of a batchwise jet mixer.

recirculating flow was laminar until with high jet velocities the recirculating flow became turbulent. The jet Reynolds number at which the change from one regime to the other was 2,000 and 100,000 respectively [10, 13].

The nondimensional mixing time relation they obtained for the hemispherical base and flat base designs were respectively

$$T = T\frac{(V \cdot d)^{0.5} g^{0.25}}{y^{0.5} D^{0.75}} \propto Re_j^{-0.15}; \qquad Re_j = 2{,}000 - 10^5$$

$$F = T\frac{(V \cdot d)^{0.65} g^{0.175}}{y^{0.5} D^{0.75}} \propto Re_j^{-1.30}; \qquad Re_j = 100 - 2{,}000$$

$$\left.\begin{array}{l} = 13.05 \text{ hemispherical base} \\ = 15.10 \text{ flat base} \end{array}\right\} Re_j > 10^5$$

Lane and Rice [11] in their comparison of the basic designs of side-entry jets concluded that the Cowdrey design was better than that of Fossett and Prosser. They used the salt concentration technique which they had already developed to measure mixing time. Again they used as the mixing time the time to achieve 95% complete homogeneity. Two tank sizes were used: 0.31-m and 0.573-m diameter. The Cowdrey design showed on average a 22% reduction in mixing time for both the laminar and turbulent jet regime (see Table 1) which Lane and Rice attribute to the jet having a longer jet length and so have a larger entrainment capacity hence a much greater mixing ability.

This result contradicts that of Okita and Oyama who stated that the angle of the jet did not influence the mixing time.

Lane and Rice [12] have also compared their design of a hemispherical base tank with that of a flat-based tank (as Hiby and Modigell). Both use a vertical discharging axial jet together with the Cowdrey design of a side-entry jet. Table 2 shows a comparison of these results. The design of Lane and Rice using the same diameter and either the same liquid depth or using the same volume as the flat-based cylindrical tank gives shorter mixing times than either that of Hiby and Modigell or Cowdrey. The Hiby and Modigell design is better than the Cowdrey design. Lane and Rice attribute this to the elimination of "dead" spaces within the tank.

Lane and Rice [13] using a dye injection with a cine photography technique, have investigated the jet behavior in a jet mixer. They examined how the jet spread and entrained the tank liquid. The jet angle decreased with increasing jet Reynolds number until it became turbulent when the jet angle was constant with a value of 22.5° approximately. The rate of increase in the entrained jet was related to the jet flow and the distance traveled by

$$\frac{Q_x}{Q_j} = 0.394\left\{\frac{x}{d}\right\}$$

in the turbulent jet regime. This indicated that the rate of entrainment was greater for the bounded jet than for a free jet.

Table 1
Comparison of Mixing Time Factors for the Fosset and Prosser Design and Cowdrey Design

		Laminar			Turbulent	
Re_j	500	1,000	2,000	5,000	10,000	20,000
Fosset & Prosser	236	92.6	36.4	30.4	26.4	23.8
Cowdrey	181	71.3	28.1	23.6	21.0	18.6
Comparison %*	23.3	23.0	22.8	22.4	21.9	21.9

* *Percentage decrease in mixing time factor (F) and therefore in mixing time shown by Coldrey design (all other conditions being the same).*

Table 2
A Comparison of Mixing Times for the Three Designs of Liquid Jet Mixers

		Laminar Jet Regime			Turbulent Jet Regime		
	$\mu = 10^{-5}\ m^2\ s^{-1}$	$\rho = 1{,}010\ kg\ m^{-3}$			$\mu = 10^{-6}\ m^2\ s^{-1}$	$\rho = 1{,}000\ kg\ m^{-3}$	
	$V\ ms^{-1}$	0.5	1.0	1.5	0.5	1.0	2.0
	(Re_j)	(500)	(1,000)	(1,500)	(5,000)	(10,000)	(20,000)
[1]	T(s)	4,245	1,071	471	563	313	175
[2]	T(s)	3,128	819	370	522	296	169
[3]i	T(s)	895	263	129	160	103	67
[3]ii	T(s)	980	288	141	175	113	73

Mixing system parameters $d = 10^{-2}$ m; D = 1 m
Note: [1] refers to Cowdrey design [y = m; volume = 0.785 m^3]
[2] refers to Hiby & Modigell design [y = 1 m; volume = 0.785 m^3]
[3]i refers to Lane & Rice design [y = 1 m; volume = 0.655 m^3]
[3]ii refers to Lane & Rice design [y = 1.2 m; volume = 0.785 m^3]

The effect of jet height relative to the tank base for side-entry jets has been examined by Maruyama, Ban, and Mizushina [14]. They used a jet which discharged diametrically but were able to alter the jet angle in the vertical plane. Another variable in their tests was the height for liquid off-take, which was always located diametrical to the jet but not necessarily at the same height.

They describe the flow pattern change as these variables were altered. They suggest the optimum geometrical conditions for minimum mixing time.

They nondimensionalized the mixing time using the mean residence time V/Q_j together with the jet diameter and the maximum jet length

$$T_N = \left\{\frac{T}{S/q}\right\}\left\{\frac{L}{d}\right\}$$

All of their tests are with tanks in which the nondimensional liquid height y/D is less than one. Their results for a horizontally discharging jet show a steady reduction in nondimensional mixing time when plotted against nondimensional jet height h/H up to a value of 0.25 and is then constant. This is for a liquid depth $y/D \approx 1$. For $y/D \approx \frac{1}{3}$ mixing time T_N was independent of h/H. The behavior between these two values of $y/D = \frac{1}{3}$ to 1 do not show a gradual change possibly due to the flow patterns changing.

For use in the brewing industry Denk et al. [15] examined the behavior of tank mixing using a tangential jet discharging horizontally. They showed that mixing was independent of jet velocity above a critical Reynolds number and that the mixing time measured to 95% homogeneity was proportional to the jet diameter and a complex function of jet height and tank liquid depth. Their results indicated that the flow pattern changed abruptly when the jet height was approximately 0.6 of the liquid depth. Interestingly they nondimensionalized the mixing time with the mean residence time S/q, h with H, H with D, and d with $S^{1/3}$.

The British Atomic Energy Authority has developed a pulse jet mixer. A sealed cylinder with the jet at the base is supported in the tank of the liquid to be mixed. Pressurized air is fed to the top of cylinder which expels the liquid through the jet. However the air is not allowed to pass through the jet. Suction is then applied to the cylinder refilling it with liquid. Several (odd number) cylinders are supported around the tank of liquid and these are "fired" in a random pattern. This system has been developed to keep particles suspended in the liquid. Since this is radioactive material, problems of servicing and sealing are minimized in contrast to mechanical devices. Jet stirring has been examined also by the British Atomic Energy Authority [Winfrith] to resuspend sediment in storage tanks.

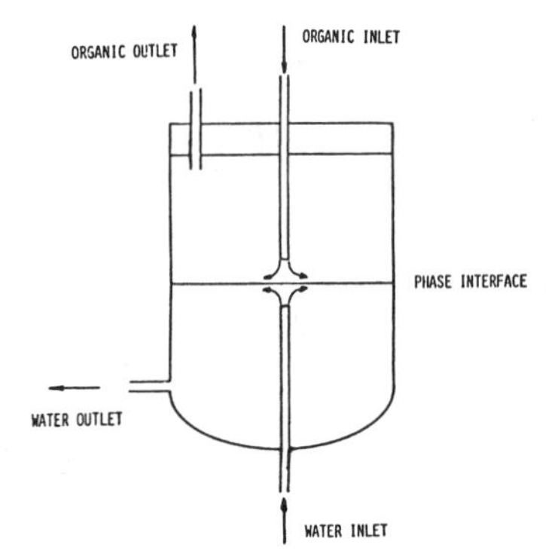

Figure 8. Blakeley et al. [17] arrangement for mass transfer tests with colliding axially opposed jets.

Although not directly concerned with batchwise mixing, five other references may be of some interest since a flow analysis is included in the work.

Blakeley et al. [16] used two axially opposed jets to study mass transfer between a water and an organic phase. The jets were arranged coaxially so that they impinged at the interface between the two phases. The arrangement is shown in Figure 8.

Clegg and his coworkers [17, 18] have studied continuous-flow jet mixing using a completely filled, closed, cylindrical vessel in which the jet was positioned off center (0.393D) with the take off also positioned off center (0.393D) but diametrically opposite the inlet in the closed ends of the tank (Figure 9).

Woods [19] studies were similar to Clegg and coworkers' but he positioned his jet and take off point axially in the tank (Figure 10).

Both Clegg and coworkers and Wood produced expressions describing the residence time distribution for their flow systems.

Sinclair and McNaughton [21, 22] have examined the continuous jet mixing for a similar arrangement to that of Wood. From an analysis of the fluid mechanics they produced expressions for residence time distributions. A range of jet Reynolds numbers covering both laminar and turbulent jets were used in their tests.

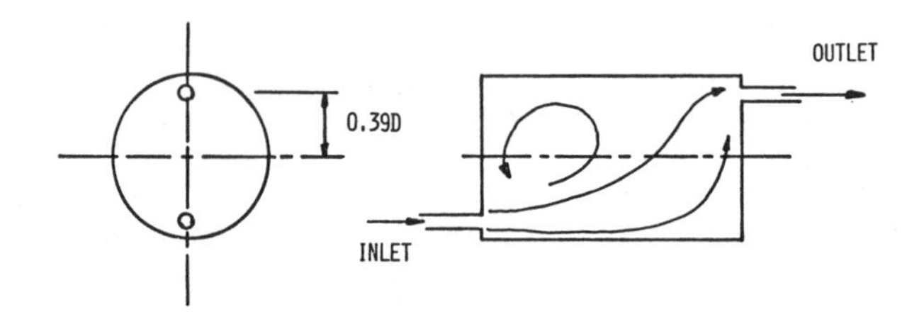

Figure 9. Clegg et al. [18, 19] continuous-jet stirred mixer.

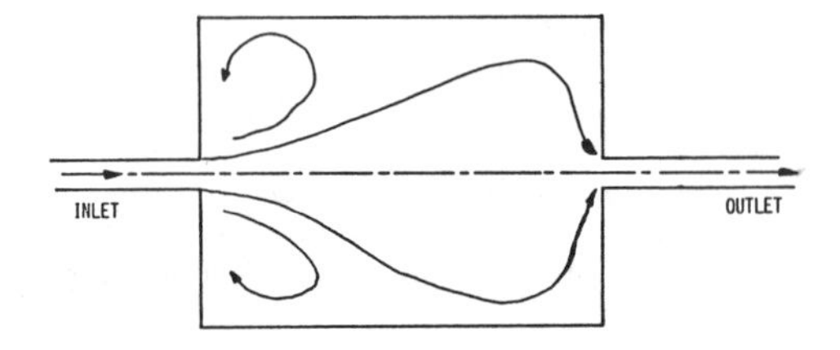

Figure 10. Arrangement of continuous-jet stirred mixer with axial inlet and outlet [20, 21, 22].

NOTATION

d	jet diameter	T_N	mixing time nondimensionalized
D	tank diameter	V	jet velocity
g	gravity	x	distance from injection point
H	injection height	X	effective jet mixing length
Q_J	jet volumetric flow rate at injection	y	liquid submergence (depth)
Q_X	jet volumetric flow rate at distance x	α	proportionality sign
Re_j	jet Reynolds number	S	tank volume
T	mixing time		

REFERENCES

1. Revill, B. K., "Jet Mixing," paper to I. Chem. E. course, Univ. of Bradford, England. (July, 1980).
2. Fossett, H., and Prosser, L. E., "The Application of Free Jets to the Mixing of Fluids in Bulk," *Proc. I. Mech. E.*, 160:224–251 (1949).
3. Fossett, H., "The Action of Free Jets in the Mixing of Fluids," *Trans. I. Chem. E.*, 29:322–332 (1951).
4. Fossett, H., "Some Observations on the Time Factor in Mixing Processes," in *Fluid Mechanics of Mixing*, Uram, E. and Goldschnudt, V. (Eds.), ASME Conf., 39–49 (1973).
5. Fox, E. A., and Gex, V. E., "Single-Phase Blending of Liquids," *A.I.Ch.E.J.*, 2:539–544 (1956).
6. Okita, N., and Oyama, Y., "Mixing Characteristics of Jet Mixing," *Kagaku Kogaku*, 27:252–259 (1963).
7. Van de Vusse, J. G., "Vergleichende ruhrversuche zum mischen loslicher flussigkeiten en einem 12,000 m^3 behalter," *Chemie. Ing. Tech.*, 31:583–587 (1959).
8. Cowdrey, P. W., "Jet Mixing," paper to I. Chem. E. course, Univ. of Bradford, England. July, 1978.
9. Hiby, J. W., and Modigell, M., "Experiments on Jet agitation," 6th CHISA Congress, Prague, 1978.
10. Lane, A. C. G., and Rice, P., "An Experimental Investigation of Liquid Jet Mixing Employing a Vertical Submerged Jet," *I. Chem. E. Symp. Series* 64, paper K1, (1981).
11. Lane, A. C. G., and Rice, P., "An Investigation of Liquid Jet Mixing Employing an Inclined Side Entry Jet," *Trans. I. Chem. E.*, 60:171–176 (1982).
12. Lane, A. C. G., and Rice, P., "Liquid Jet Mixing at High Jet Reynolds Numbers," 4th European Conf. on Mixing, 449–458 (1982).
13. Lane, A. C. G., and Rice, P., "Comparative Assessment of the Performance of the Three Designs for Liquid Jet Mixing," *Ind. Eng. Chem. Proc. Des. Dev.*, 21:650–653 (1982).
14. Lane, A. C. G., and Rice, P., "The Flow Characteristics of a Submerged Bounded Jet in a Closed System," *Trans. I. Chem. E.*, 60:245–248 (1982).
15. Maruyama, T., et al., "Jet Mixing of Fluids in Tanks," *J. Chem. Eng. Japan*, 95:342–348 (1982).
16. Denk, V., et al., "Beitrag zur Strahlmischung en einem behalter mit ausserer umwalzung," *Chemie. Ing. Tech.* 54:174–175 (1982).
17. Blakeley, D., et al., "Liquid-Liquid Mass Transfer in Systems Agitated by Submerged Jets," *Chem. Eng. Sci.*, 32:1457, 1463 (1977).
18. Stainthorp, F. P., and Clegg, G. T., "Fluid Stirring in Continuous Flow Systems," *Chem. Eng. Sci. Suppl.*, 32:167–172 (1965).
19. Clegg, G. T., and Coates, R., "A Flow Model for a Filled Cylindrical Vessel," *Chem. Eng. Sci.*, 22:1177–1183 (1967).
20. Wood, T., "Mixing Characteristics of a Bounded Turbulent Jet," *Chem. Eng. Sci.*, 23:783–789 (1968).
21. McNaughton, K. J., and Sinclair, C. G., "Submerged Jets in Short Cylindrical Flow Vessels," *J. Fluid Mechanics*, 25:367–375 (1966).
22. Sinclair, C. G., and McNaughton, K. J., "Residence Time Distributions in Jet Stirred Vessels with Linear Scale from 0.5 to 4 feet," *Can. J. Chem. Eng.*, 48:411–419 (1970).

CHAPTER 19

STABILITY OF JETS IN LIQUID-LIQUID SYSTEMS

Yoshiro Kitamura and **Teruo Takahashi**

Department of Industrial Chemistry
Okayama University
Okayama, Japan

CONTENTS

INTRODUCTION, 474

STABILITY THEORY OF LIQUID JET, 475
General Solution, 475
Limiting Solutions, 479

APPLICATION OF STABILITY THEORY TO BREAKUP OF JET, 480

BREAKUP OF LIQUID JET INJECTED INTO STAGNANT IMMISCIBLE LIQUID, 482
Breakup Phenomena and Breakup Length, 483
Jetting Velocity and Critical Velocity, 485
Correlation of Breakup Length, 488
Drop Formation from Jet, 491
Influence of Nozzle Length on Jet Breakup, 493
Effect of Mass Transfer on Jet Breakup, 494

BREAKUP OF JET IN CENTRIFUGAL FIELD, 495

DIRECT CONFIRMATION OF STABILITY ANALYSIS BY JET OF ZERO RELATIVE VELOCITY, 497
Breakup Length of Jet, 498
Drop Diameter, 500

EFFECT OF HYDRODYNAMIC RESISTANCE OF CONTINUOUS PHASE ON JET STABILITY, 500

STABILITY OF JET IN NON-NEWTONIAN-NEWTONIAN LIQUID SYSTEMS, 503
Breakup Pattern of Jet in Immiscible Non-Newtonian Liquid Systems, 504
Newtonian Jets in Non-Newtonian Liquids, 505
Non-Newtonian Jets in Newtonian Liquids, 506

NOTATION, 507

REFERENCES, 509

INTRODUCTION

The dispersion of liquids into immiscible liquids is widely used in many industrial processes such as liquid-liquid extraction, direct contact heat transfer, and emulsification. Although the dispersion in actual equipment is influenced by the complex flow and the subsequent shear field, the breakup of liquid column or jet into drops is one of most basic phenomena for understanding the dispersion in immiscible liquid-liquid systems.

When one liquid is injected into a second immiscible liquid through a nozzle or orifice, three principal mechanisms for drop formation are observed as flow rate increases, drop formation at nozzle, from jets, and from sprays. At moderate injection velocities, a laminar jet issues from the nozzle and breaks into drops in a regular pattern. Many investigators in chemical engineering have focussed their attention on the drop formation by liquid injection for the prediction of drop size in extractors. Drop formation at jetting condition is considered especially desirable for effective mass transfer operations.

On the other hand, the breakup of the liquid thread into droplets has been the object of study in the field of emulsion science, where high viscosity liquid was concerned. Although the physical conditions related to both fields are different from each other, drops are formed through the same mechanism. The stability theory of the liquid jet is an excellent model for the drop formation mechanism. Thus, knowledge of the hydrodynamic stability of a jet is vitally essential to understanding the dispersion of liquids into immiscible liquids.

STABILITY THEORY OF LIQUID JET

The stability and disintegration of liquid jets has been a phenomenon investigated in various scientific and technical fields for over one hundred and fifty years. Subsequent to earlier works [1, 2], Rayleigh [3] contributed the first quantitative description of jet stability. His analysis was concerned with the stability of an inviscid liquid jet in air; this corresponds to the case where surrounding fluids have no influence on stability. The stability of a viscous liquid jet in air was analyzed by Weber [4]. The extension of the analysis to immiscible liquid-liquid systems was pioneered by Tomotika [5, 6]. He analyzed the stability of a liquid column for general cases where both inertia and viscous force are significant and for two limiting cases where either inertia or the viscous force is negligible.

A historical survey of the jet stability is briefly summarized in Table 1. The major theoretical in vestigations were provided by Rayleigh, Weber, and Tomotika. Each investigation is a linearized analysis which assumes surface disturbances to be always small compared to jet radius. Recently, the effect of finite amplitude of disturbances on the jet stability were theoretically and experimentally investigated for liquid in air systems [7–12]. No one has studied the nonlinear effect on stability of the liquid-liquid jet.

The stability theory of the liquid jet in air is detailed by Levich [13] and Chandrasechar [14]. McCarthy and Molloy [15] reviewed the stability of jets mainly in air. No well-compiled review is available on the stability of immiscible liquid-liquid jets. Hence, Tomotika's stability analysis for the general case [5] is sumarized in this section. The solutions for limiting cases are subsequently described. Numerical solutions and the application of analysis to jet breakup are followed.

General Solution

We will consider a stationary cylindrical column of viscous liquid resting in an infinite mass of another viscous fluid. We assume the following: the fluid is imcompressible and Newtonian; there are no gross flows in both fluids; the motions of fluid are very small and symmetrical about the column axis. These assumptions simplify the equation of motion in the cylindrical coordinate (r, z) in terms of the Stokes's stream function as follows (because the squares and products of velocity components are negligibly small).

$$\left(E^2 - \frac{1}{\nu}\frac{\partial}{\partial t}\right)E^2\psi = 0 \tag{1}$$

where operator E^2 is expressed as

$$E^2 = \frac{\partial^2}{\partial r^2} - \frac{1}{r}\frac{\partial}{\partial r} + \frac{\partial^2}{\partial z^2} \tag{2}$$

Table 1
Historical Survey of Jet Stability

	System	Theoretical works	Experimental works	
			Jet Injected into Stagnant Air	
Liquid in air	Inviscid liquid	Rayleigh [3]	Haenlein [92]	
	Viscous liquid	Weber [4]	Tyler [21]	
			Grant and Middleman [20]	
	Aerodynamic effect	Weber [4]	Fenn and Middleman [77]	
			Sterling and Sleicher [67]	
	non-Newtonian liquid	Middleman [79]	Kroesser and Middleman [82]	
		Goldin et al. [81]	Goldin et al. [81]	
			Stationary column or jet of zero relative velocity	Jet injected into stagnant liquid
Liquid in immiscible liquid	Inviscid liquid in inviscid liquid	Tomotika [6]	Rumscheidt and Mason [26]	Meister and Scheele [49]
	Viscous liquid in viscous liquid	Tomotika [5]	Kitamura et al. [19]	Takahashi and Kitamura [17]
	Dynamic effect of surrounding liquid		Kitamura et al. [19]	
	Non-Newtonian systems	Lee et al. [90]	Kitamura and Takahashi [91]	Shirotsuka and Kawase [86]

A spectrum of infinitesimal disturbances of the following form is assumed to be initiated on the surface of liquid column.

$$\delta = \bar{\delta}_0 \exp(qt + ikz) \tag{3}$$

This is the surface wave whose wavelength is $2\pi/k$ and whose amplitude increases exponentially with time.

Assuming ψ to be proportional to $\exp(qt + ikz)$, we can solve Equation 1. The stream function ψ' inside the column is

$$\psi' = [A_1 r I_1(kr) + A_2 r I_1(k_D r)] \exp(qt + ikz) \tag{4}$$

The stream function ψ outside the column is

$$\psi = [B_1 r K_1(kr) + B_2 r K_1(k_C r)] \exp(qt + ikz) \tag{5}$$

where

$$k_D^2 = k^2 + q/\nu_D \tag{6}$$

$$k_C^2 = k^2 + q/\nu_C \tag{7}$$

$I_i(x)$ and $K_i(x)$ are modified Bessel functions of order i. A_1, A_2, B_1, and B_2 are arbitary constants to be determined by the following boundary conditions.

1. The velocity is continuous at the interface of both phases.

B.C.1: $v_r = v_r'$ at $r = a$;

$$\left.\frac{1}{r}\frac{\partial\psi}{\partial z}\right|_{r=a} = \left.\frac{1}{r}\frac{\partial\psi'}{\partial z}\right|_{r=a} \tag{8}$$

B.C.2: $v_z = v_z'$ at $r = a$;

$$\left.\frac{1}{r}\frac{\partial\psi}{\partial r}\right|_{r=a} = \left.\frac{1}{r}\frac{\partial\psi'}{\partial r}\right|_{r=a} \tag{9}$$

2. The shear stress is continuous at the interface.

B.C.3: $\tau_{rz} = \tau_{rz}'$ at $r = a$;

$$\mu_C E^2\psi\Big|_{r=a} = \mu_D E^2\psi'\Big|_{r=a} \tag{10}$$

3. The difference in the normal stress between the outside and inside of the column is due to the interfacial tension.

B.C.3:

$$(p_{rr} - p_{rr}')\Big|_{r=a} = \frac{\sigma(k^2a^2 - 1)}{a^3}\frac{ik}{q}\psi'\Big|_{r=a} \tag{11}$$

The constants in Equations 4 and 5 are determined from these boundary conditions. Eliminating these constants, we finally obtain the characteristic equation for the growth rate of a given disturbance.

$$\begin{vmatrix} I_1(x) & I_1(y') & K_1(x) & K_1(y) \\ xI_0(x) & y'I_0(y') & -xK_0(x) & -yK_0(y) \\ 2\hat{\mu}x^2I_1(x) & \hat{\mu}(x^2+y'^2)I_1(y') & 2x^2K_1(x) & (x^2+y^2)K_1(y) \\ h_1 & h_2 & h_3 & h_4 \end{vmatrix} = 0 \tag{12}$$

where

$$\left.\begin{aligned} h_1 &= -2\hat{\mu}x^2\dot{I}_1(x) - \hat{\mu}\frac{Q}{2Z_D}I_0(x) - \frac{\hat{\mu}}{QZ_D}(x^2-1)xI_1(x) \\ h_2 &= -2\hat{\mu}xy'\dot{I}_1(y') - \frac{\hat{\mu}}{QZ_D}(x^2-1)xI_1(y') \\ h_3 &= -2x^2\dot{K}_1(x) + \frac{\hat{\mu}}{\hat{\rho}}\frac{Q}{2Z_D}K_0(x) \end{aligned}\right\} \tag{13}$$

$$h_4 = -2xy\dot{K}_1(y)$$

$$Q = q\sqrt{2\rho_D a^3/\sigma} \tag{14}$$

$$x = ka, \qquad y = k_C a, \qquad y' = k_D a \tag{15}$$

$$\left.\begin{aligned} \hat{\mu} &= \mu_D/\mu_C, \qquad \hat{\rho} = \rho_D/\rho_C, \\ Z_D &= \mu_D/\sqrt{2\rho_D a\sigma} \end{aligned}\right\} \tag{16}$$

and $\dot{I}_i(x)$, $\dot{K}_i(x)$ are the first derivatives of $I_i(x)$, $K_i(x)$ with respect to x.

The relation between the growth rate and the wave number can be obtained by solving Equation 12. The equation has been numerically solved by Meister and Scheele [16] and Takahashi and Kitamura [17]. Figure 1 shows an example of numerical solutions. The numerical solutions indicate that the growth rate is positive in $0 < x < 1$ and reaches the maximum at $x = x^*$, which means as follows. Although a spectrum of disturbances corresponding to all possible wavelengths is initiated at $t = 0$ as a result of density and pressure fluctuations, the disturbance with the wavelength of $(2\pi a/x^*)$ grows most rapidly and consequently dominates the breakup of the liquid column to droplets.

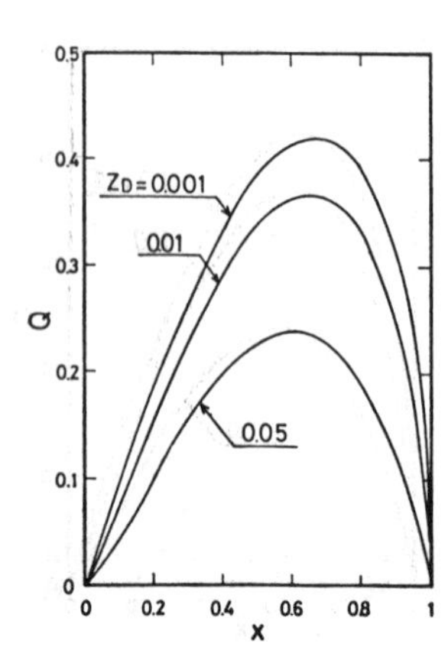

Figure 1. Dimensionless growth rate for viscous liquid column in viscous liquid. Numerical solution of Equation 12 for the case of $\rho_D/\rho_C = 1$ and $\mu_D/\mu_C = 1$.

Limiting Solutions

All of the limiting solutions for Newtonian fluids can be derived from the general solution just described. The limiting solutions for the case closely related to subsequent discussions are briefly summarized in this section.

Case 1: Viscous liquid column which is not affected by the surrounding fluid.
This case corresponds to a low velocity liquid jet in air. Substituting Equation 4 into Equations 10 and 11, we obtain the exact solution for the growth rate.

$$Q^2 \frac{xI_0(x)}{2I_1(x)} + 2QZ_Dx^2 \left\{ \frac{xI_0(x)}{I_1(x)} - 1 - \frac{2x^2}{x^2 + y'^2} \left[\frac{y'I_0(y')}{I_1(y')} - 1 \right] \right\} - (1 - x^2)x^2 \frac{y'^2 - x^2}{x^2 + y'^2} = 0 \tag{17}$$

Weber [4] reduced this complex equation to the following simple form, by approximating the terms of Bessel function for $x < 1$.

$$Q^2 - 6QZ_Dx^2 - (1 - x^2)x^2 = 0 \tag{18}$$

The dimensionless growth rate and wave number of the most rapidly growing disturbance are easily obtained from the approximate equation, respectively,

$$Q^* \equiv q^*/\sqrt{2\rho_D a^3 \sigma} = 1/[2(1 + 3Z_D)] \tag{19}$$

$$x^* \equiv k^*a = 1/\sqrt{2(1 + 3Z_D)} \tag{20}$$

Case 2: Inviscid liquid column in inviscid liquid.
When both phases are perfect fluids, the growth rate is given by the following equation [6, 18].

$$Q = \sqrt{2}\, F_1(x) \tag{21}$$

where $F_1(x)$ is a function of x and $\hat{\rho}$ and expressed as,

$$F_1(x) = \left\{ (1 - x^2)x \bigg/ \left[\frac{K_0(x)}{\hat{\rho} K_1(x)} + \frac{I_0(x)}{I_1(x)} \right] \right\}^{1/2} \tag{22}$$

Figure 2 shows the calculated value of F_1 as a function of x and $\hat{\rho}$. From the figure and Equation 21, it is clear that Q^* for this case is a function of $\hat{\rho}$.

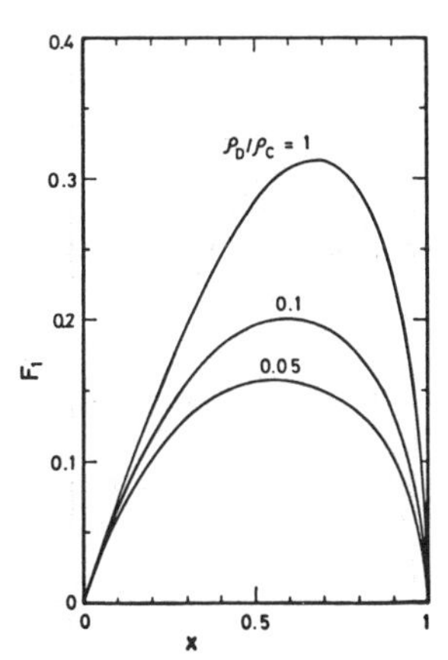

Figure 2. $F_1(x)$ defined by Equation 22; corresponding to dimensionless growth rate for inviscid liquid column in inviscid liquid.

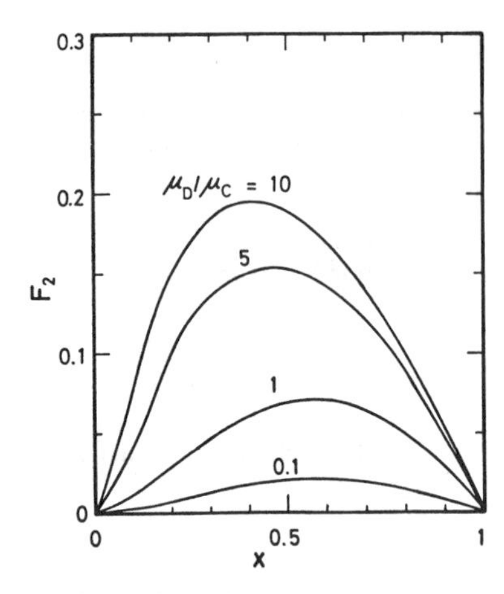

Figure 3. $F_2(x)$ defined by Equation 24, corresponding to dimensionless growth rate for high-viscosity liquid column in high-viscosity liquid.

Case 3: High viscosity liquid column in high viscosity liquid.
When viscosity of both phases is so high that the effect of inertia is insensible in comparison with that of viscous force, the general solution is simplified to the following explicit expression for growth rate [5].

$$q(2a\mu_D/\sigma) \equiv QZ_D = F_2(x) \tag{23}$$

where $F_2(x)$ is a function of x and $\hat{\mu}$ and given by

$$F_2(x) = (1 - x^2)\{I_1(x)\Delta_1 - [xI_0(x) - I_1(x)]\Delta_2\}/P(x) \tag{24}$$

with

$$\begin{aligned} P(x) = {} & [xI_0(x) - I_1(x)]\Delta_1 - [(x^2 + 1)I_1(x) - xI_0(x)]\Delta_2 \\ & - [xK_0(x) + K_1(x)]\Delta_3/\hat{\mu} - [(x^2 + 1)K_1(x) + xK_0(x)]\Delta_4/\hat{\mu} \end{aligned} \tag{25}$$

where Δ_i is a three-by-three determinant whose elements are given by striking out the i-th column of the following matrix.

$$\begin{pmatrix} I_1(x) & xI_0(x) - I_1(x) & K_1(x) & -xK_0(x) - K_1(x) \\ I_0(x) & I_0(x) + xI_1(x) & -K_0(x) & -K_0(x) + xK_1(x) \\ \hat{\mu}I_1(x) & \hat{\mu}I_0(x) & K_1(x) & -xK_0(x) \end{pmatrix} \tag{26}$$

The calculated value of F_2 is shown in Figure 3 which indicates the maximum growth rate for this case is a function of the viscosity ratio $\hat{\mu}$.

APPLICATION OF STABILITY THEORY TO BREAKUP OF JET

The stability analyses just described relate to the characteristics in jet breakup, breakup length and drop size, as follows. Assuming that jets break into drops when disturbances grow to be comparable to jet radius at t = T, the breakup time is given by

$$T = \ln(a/\bar{\delta}_0)/q^* \tag{27}$$

The breakup length ℓ relates to T by $T = \ell/u_D$ unless both jet radius and jet velocity do not change after injecting from nozzle. Thus, the breakup length is given by

$$\ell = \ln(a/\bar{\delta}_0)(u_D/q^*) \tag{28}$$

As the liquid column of one wavelength turns into one drop of corresponding volume, the diameter of drops formed from jets is given by

$$d = (1.5\pi D^3/x^*)^{1/3} \tag{29}$$

The maximum growth rate q^* of a disturbance and its corresponding wave number x^* are important for predicting the drop size from jets and the behavior of jets in immiscible liquid-liquid systems. Numerical solutions of the characteristic equation for growth rate in the general case indicate that both Q^* and x^* are functions of Z_D, $\hat{\rho}$ and $\hat{\mu}$. On the other hand, those for the case where the surrounding fluid has no influence are explicitly approximated by Equations 19 and 20. Utilizing the well-known approximation, the following expressions can be proposed to approximate Q^* and x^* for the immiscible liquid system where neither viscous force nor inertia is negligibly small.

$$Q^* = \phi_1/2(1 + 3Z_D) \tag{30}$$

$$x^* = \phi_2/\sqrt{2(1 + 3Z_D)} \tag{31}$$

where ϕ_1 and ϕ_2 are a correction factor for Q^* and x^*, respectively. Both factors can be estimated as a function of Z_D, $\hat{\mu}$, and $\hat{\rho}$ (since Z_C is expressed by them). The contribution of these parameters was assessed by numerical computations. Figures 4 and 5 show the correction factor ϕ_1 and ϕ_2, respectively. Both of them are plotted against $Z_D/\hat{\mu}$ for various values of $\hat{\mu}$ and for $\hat{\rho} = 1$. It is noted that numerical calculations support insensibility of the effect of the density ratio on both correction factors, and that the density ratio is close to unity in actual systems.

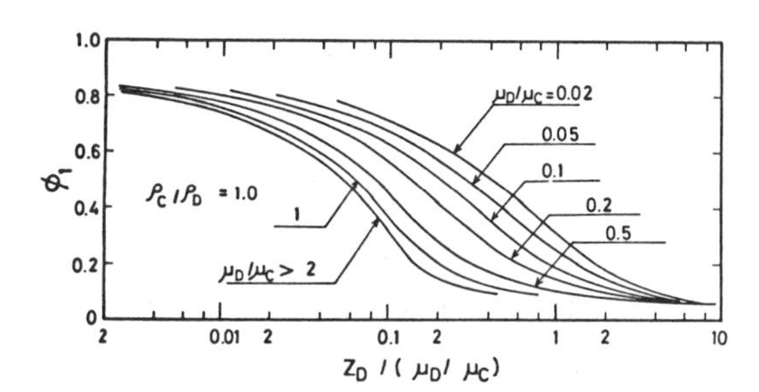

Figure 4. Correction factor ϕ_1 defined by Equation 30, which corresponds to maximum growth rate for general case.

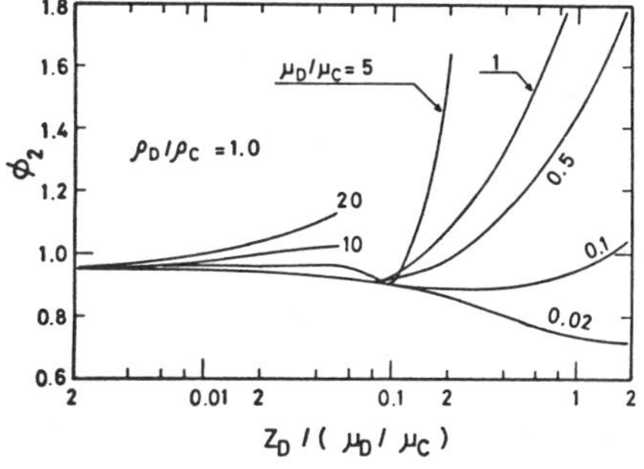

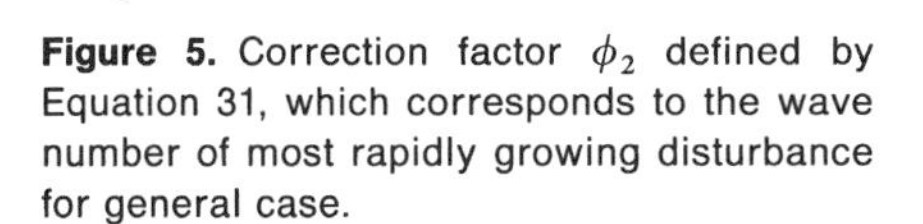

Figure 5. Correction factor ϕ_2 defined by Equation 31, which corresponds to the wave number of most rapidly growing disturbance for general case.

The correction factor ϕ_2 is almost independent of the density ratio and can be approximated as a constant 0.95 when $Z_D < 0.2$ and $0.1 < \mu < 20$, as shown in Figure 5. Thus, the wave number corresponding to maximum growth rate for immiscible liquid-liquid systems is approximated by

$$x^* = 0.95/\sqrt{2(1 + 3Z_D)} \quad (32)$$

Substitution of this expression into Equation 29 gives

$$d = 1.91\ D(1 + 3Z_D)^{1/6} \quad (33)$$

The preceding approximation is confirmed to agree very closely with the exact value calculated from Equations 12 and 29. The largest difference between them is only 0.8% [19].

All of the stability analyses mentioned before assume the stationary liquid column resting in surrounding fluid although the liquid jet injected into stagnant fluid has the velocity distribution and produces the secondary gross motion surrounding fluid. Thus in practice, jets are not always expected to satisfy the assumption. On the other hand, experimental breakup lengths of laminar jets in air are well correlated by Equation 28 [4, 20]. The size of drops formed from laminar jets in air agrees with the estimate from Equations 20 and 29, provided the contraction of the jet diameter is negligible [21–23]. These experimental results indicate that the stability analysis can be applied to liquid jets in air at lower velocities. In higher injection velocities, the drag by surrounding air plays an important role on the jet stability. The dynamic effect of surrounding fluid is discussed later.

In immiscible liquid-liquid systems, Tomotika [5, 24] afforded a reasonable explanation of Tayler's experimental work [25] on breakage of a cylindrical liquid thread resting in another liquid by use of the four-roller apparatus. Rumscheidt and Mason [26] confirmed that the wavelengths and growth rates of disturbances for stationary liquid threads have good agreement with Tomotika's limiting solution (Case 3: both phases are highly viscous).

Breakup data of jets in liquid-liquid systems, however, differ from those of liquid-in-air systems and those of a stationary column in the four-roller apparatus since the secondary motion induced by the injection cannot be negligible due to the comparable magnitude of the density and viscosity of both phases. So, the comparison of the stability analysis with breakup data of jets in liquid-liquid systems is not simple. To obtain an understanding of jet stability for immiscible liquid-liquid systems and breakup of liquid jets injected into stagnant liquids, we will examine the following:

- The application of stability theory to these data.
- The confirmation of jet stability by the experiment of zero relative motion.
- The effect of hydrodynamic resistance of the continuous phase on jet stability.
- The jet stability in non-Newtonian systems.

BREAKUP OF LIQUID JET INJECTED INTO STAGNANT IMMISCIBLE LIQUID

Early experimental studies on breakup of jets in liquid-liquid systems were conducted by Smith and Moss [27], Tyler and Watkin [28], and Merrington and Richardson [29]. Their objective was to develop the mercury jet as a cathode in the electrolysis of water. Some of their data were presumably complicated by the presence of soluble electrolytes and the electric field.

The second stage of progress was the investigation of drop formation in liquid-liquid spray columns. A number of experiments were caried out to estimate the volume of drops formed from orifices or nozzles [18, 30–40]. Several reviews are available on this problem [41–45]. In most of them, however, the attention is focused on the single-drop formations at low injection velocities.

Much attention was paid to the breakup phenomena of jets in immiscible liquid-liquid systems by Hayworth and Treybal [30], Keith and Hixson [31], Christiansen and Hixson [18], Fujinawa et al [46], Ranz [47], Ranz and Dreier [48], Meister and Scheele [49, 50], Takahashi and Kitamura [17, 51], and Skelland and Johnson [38]. Tomotika's stability analysis described before has been successfully applied to breakup data [17, 49, 51].

Recently, Bayerns and Laurence [52], Meister and Scheele [49], Sawistowski [53], Dzubur and Sawistowski [54], Burkholder and Berg [55], Skelland and Huang [56, 57], Coyle et al. [58], and

Nealson and Berg [59] have investigated the effect of mass transfer on jet breakup and drop formation at jetting conditions.

In the following sections experimental appearance of jet stability in liquid-liquid systems are comprehensively described mainly based on the author's works [17, 51, 60].

Breakup Phenomena and Breakup Length

When a liquid issues through a nozzle into a stagnant immiscible liquid, the pattern of drop formation depends on the injection velocity u_D as shown schematically in Figure 6. At low injection velocities a single drop is formed at the nozzle. As the velocity increases to reach a jetting velocity u_j, a laminar jet issues. The laminar jet seems to break by axisymmetric growing of surface disturbances. The breakup length of laminar jet increases with increasing u_D, and reaches a maximum at a critical velocity u_k. Above u_k, the breakup length decreases with increasing u_D and the jet seems to break primarily by the growth of asymmetric disturbances. The jet at velocities above u_k is called the turbulent jet.

On the other hand, drop sizes depend on the breakup pattern of jets. At low injection velocities, $u_D < u_j$, drops are uniform but relatively large. Above u_j, the drop size slightly decreases with increasing u_D and reaches to a minimum at u_M, the velocity corresponding to maximum interfacial area. The characteristic velocity u_M is almost equivalent to u_k in many systems [46]. Above u_M or u_k, the mean drop size slightly increases with increasing u_D and the distribution of drop sizes becomes broader because of irregular breakup due to turbulent shear. At velocities higher than 100 cm/s for the system of the figure, drops are formed by a spray and the drop sizes decrease again. Usually in this region, a cloud of dispersed droplets prevents photographing clear patterns of sprays when injecting jets into stagnant liquids.

Figure 7 shows the breakup lengths in several systems against injection velocities. The relation between ℓ and u_D varies from system to system. In most systems, the breakup lengths of laminar jets increase linearly with u_D at the velocity region above u_j. But data in n-buthanol-water systems is extremely different from other systems. Namely, the breakup length for this system increases

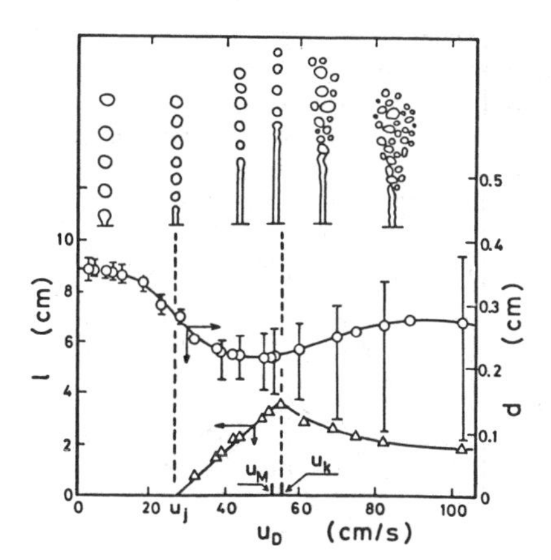

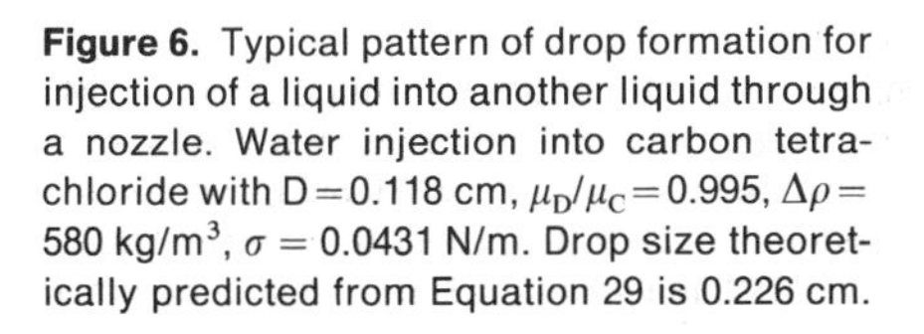

Figure 6. Typical pattern of drop formation for injection of a liquid into another liquid through a nozzle. Water injection into carbon tetrachloride with D = 0.118 cm, $\mu_D/\mu_C = 0.995$, $\Delta\rho = 580\ kg/m^3$, $\sigma = 0.0431$ N/m. Drop size theoretically predicted from Equation 29 is 0.226 cm.

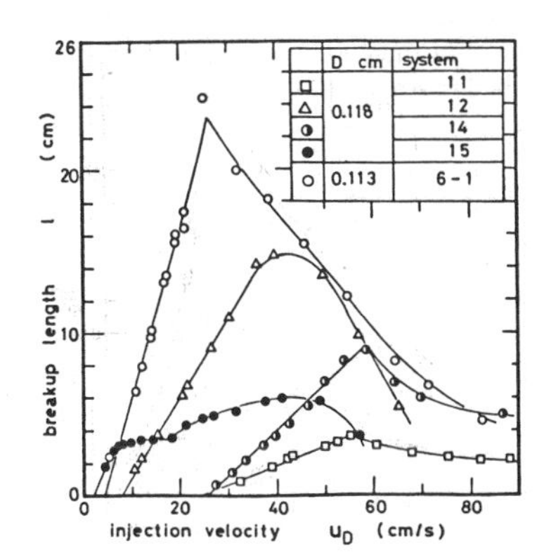

Figure 7. Breakup length of laminar jet in various systems. System 11—water-carbon tetrachloride; 12—water-furfural; 14—benzene + carbon tetrachloride-water; 15—n-buthanol-water; 6-1—water-n-buthanol [51].

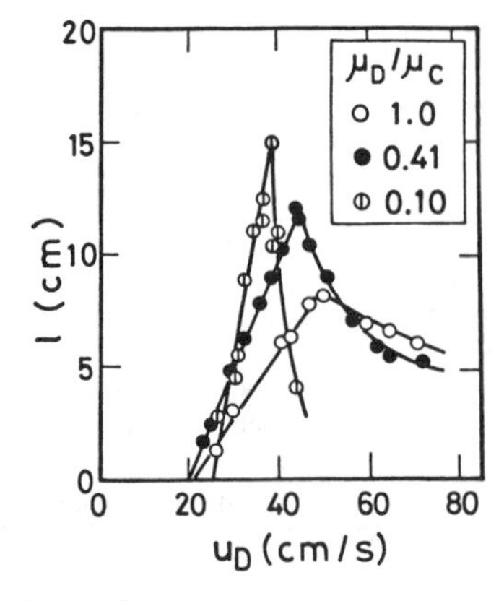

Figure 8. Effect of viscosity of continuous phase on breakup length. Water jet injected into the mixture of liquid paraffin and gasoline [17].

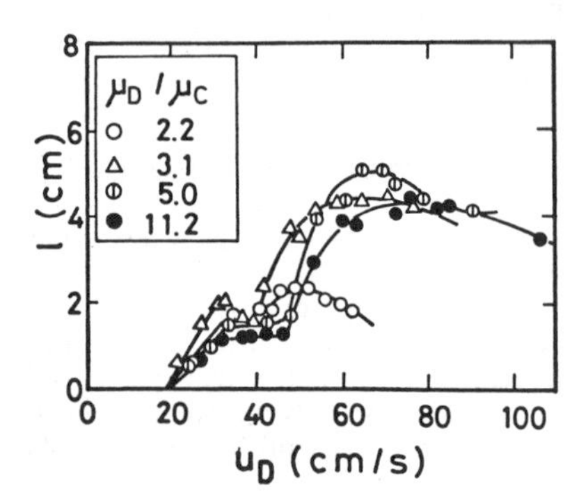

Figure 9. Effect of viscosity of dispersed phase on breakup length. For the case injecting water and aqueous solutions of starch syrup into gasoline, D = 0.160 cm [17].

linearly with the velocity in the initial region, reaching a plateau and then enlarging to attain a final maximum.

Takahashi and Kitamura [17, 51] found from breakup data from experiments in 33 systems that such a morphological difference in breakup curves mainly depends on the viscosity ratio of dispersed to continuous phase. The effect of viscosity ratio on breakup length are shown in Figures 8 and 9. Figure 8 shows breakup data for the case where continuous phases are more viscous than dispersed phases, while Figure 9 shows data for the systems where dispersed phases are more viscous than continuous phases.

For the systems of $\mu_D/\mu_C < 1$, the breakup length of the laminar jet is proportional to $(u_D - u_j)$ and increases with an increase of the viscosity of the continuous phase, as shown in Figure 8. Laminar jets in these systems are apparently disintegrated by growing axisymmetric disturbances as illustrated in Figure 10A. For the systems where dispersed phases are more viscous than continuous

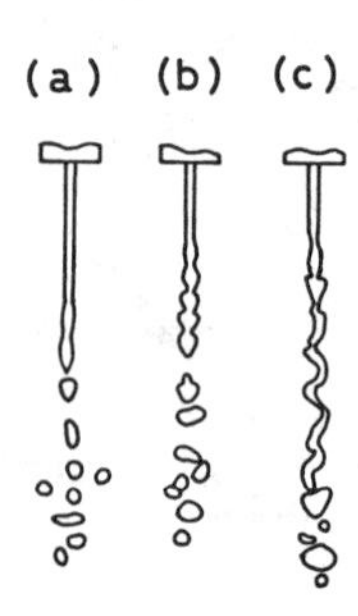

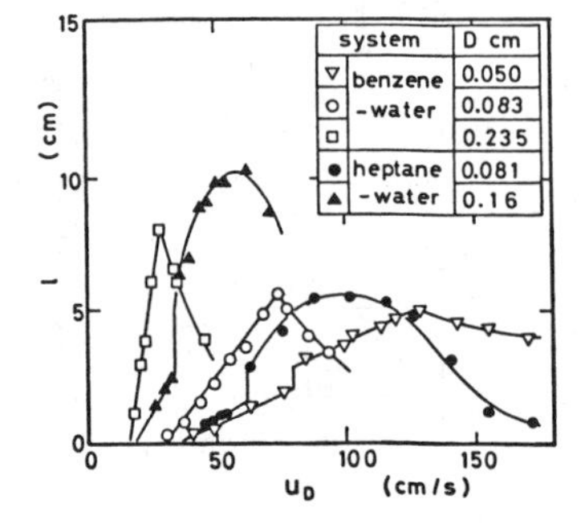

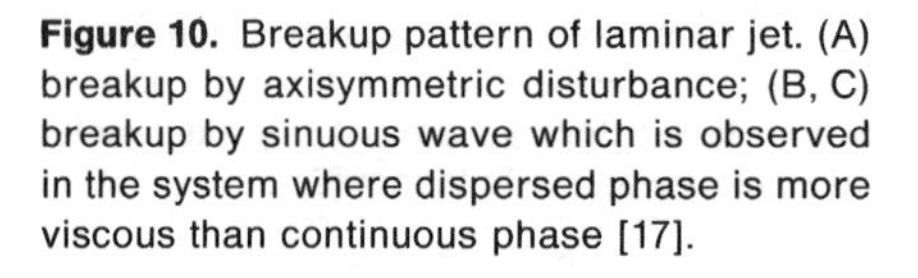

Figure 10. Breakup pattern of laminar jet. (A) breakup by axisymmetric disturbance; (B, C) breakup by sinuous wave which is observed in the system where dispersed phase is more viscous than continuous phase [17].

Figure 11. Abrupt lengthening of laminar jet observed in the system where dispersed phase is less viscous than continuous phase. Data for heptane-water system are from Meister and Scheele [49].

phases, however, breakup curves are nonlinear, and abrupt lengthening of the laminar jet is observed, as shown in Figure 9. When such a lengthening occurs, the breakup patterns illustrated in Figure 10B and C are observed. These breakup patterns differ apparently from those observed in lower velocities at which sinuous waves are not observed yet or those observed in the system of $\mu_D/\mu_C < 1$. Such sinuously wavy jets are formed by merging of drops with the jet, because the terminal velocity of the drop formed from the jet is lower than the jet velocity in these systems.

Meister and Scheele [49] reported such lengthening occurs in a system of $\mu_D/\mu_C < 1$. When breakup data in Figure 7 is elaborately examined, small lengthening is observed in the benzene + carbon tetrachloride-water systems (system 14) just before the critical velocity. Figure 11 shows breakup data for benzene injection into water as a function of liquid velocity and nozzle diameter. Data for a nozzle of 0.050 cm exhibit the significant lengthening in laminar breakup length at 85 cm/s. For the nozzles of larger diameters, such a lengthening hardly occurs because of the formation of larger drops and the decrease in the velocity region for laminar jets, as shown in Figure 11. In different systems, the lengthening is observed at lower velocities as exhibited by Meister and Scheele's data (heptane-water system) [49] illustrated in the figure.

Although the slight lengthening occurs in some systems, a linear relation passing through u_j can approximate laminar breakup lengths for the systems where the viscosity ratio of the dispersed-to-continuous phase is sufficiently low. The critical ratio of viscosity is estimated as 1.0 ~ 2.0 from the experiments for 33 systems although it depends slightly on the nozzle diameter and the density difference [17, 51].

Jetting Velocity and Critical Velocity

The prediction of jet behavior and drop sizes requires knowing the condition of transition from a single-drop formation to laminar jet and from laminar to turbulent jet because the drop size depends strongly on the breakup pattern of the jet.

Jetting Velocity

Fujinawa et al. [46] have first proposed a correlation of jetting velocity for immiscible liquid-liquid systems.

$$u_j = 4.4\sigma^{0.2}D^{-0.5} \tag{34}$$

The preceeding relation is valid only for cgs units.

Scheele and Meister [33] derived the following expression for jetting velocity by balancing the interfacial tension force, the kinetic force, and the excess pressure force necessary to sustain a spherical drop at nozzle tip.

$$We_j\ (\equiv \rho_D D u_j^2/\sigma) = 3[1 - (D/d_F)] \tag{35}$$

where We_j is the Weber number based on u_j. In the preceding equation, d_F is the diameter of the drop which would be formed at the liquid velocity u_j if a jet was not formed, and given by

$$d_F^3 = \frac{6F}{\pi}\left\{\frac{\pi\sigma D}{g\,\Delta\rho} + \frac{20\mu_C VD}{d_{FG}^2 g\,\Delta\rho} - \frac{4\rho_D V u_D}{3g\,\Delta\rho} + 4.5\left[\frac{V^2D^2\rho_D\sigma}{(g\,\Delta\rho)^2}\right]^{1/3}\right\} \tag{36}$$

where F is the Harkins-Brown correction factor [61]. For predicting the jetting velocity, Equations 35 and 36 have to be combined and to be iteratively calculated because the drop diameter is a function of the injection velocity. Scheele and Meister reported that the prediction from Equation 35 is in good agreement with experimental data obtained for 15 liquid-liquid systems.

To simplify the calculation from the preceding equations, Takahashi and Kitamura [17] approximated the drop diameter by

$$d_F^3 = 6\sigma D/\Delta\rho g \tag{37}$$

which is the diameter of drops suspended at the nozzle when $u_D \rightarrow 0$. Then, they proposed the following equation for predicting the jetting velocity.

$$We_j = 2.25[1 - (Bo/6)^{1/3}] \tag{38}$$

where Bo is the Bond number defined as

$$Bo = D^2 g\, \Delta\rho/\sigma \tag{39}$$

Figure 12 shows the comparison of Equation 38 with experimental data. Equation 38 is in good agreement with data of Fujinawa et al. [46], Scheele and Meister [33], and Takahashi and Kitamura [17, 51].

Lehrer [62] recently attempted to modify Scheele and Meister's prediction and proposed

$$u_j = \left\{\left(\frac{0.8\rho_D g D}{\sigma}\right)^2 + \left(\frac{3\sigma}{\rho_D D}\right)\left[1 + D\left(\frac{\Delta\rho g}{2\sigma}\right)^{1/2}\right]^{-1}\right\}^{1/2} - \frac{0.8\mu_D g D}{\sigma} \tag{40}$$

On the other hand, Tanazawa and Toyoda [63] correlated experimental jetting velocities for liquid-in-air systems by the following equation.

$$\sqrt{We_j} = 0.99\, Bo^{-1/4} + 0.45 - 0.57\, Bo^{1/8} \tag{41}$$

Kitamura [64] proposed the following empirical equation for liquid injection into stagnant air both from a stationary and from a rotating nozzle.

$$We_j = 0.78 \ln(6/Bo) + 0.35 \tag{42}$$

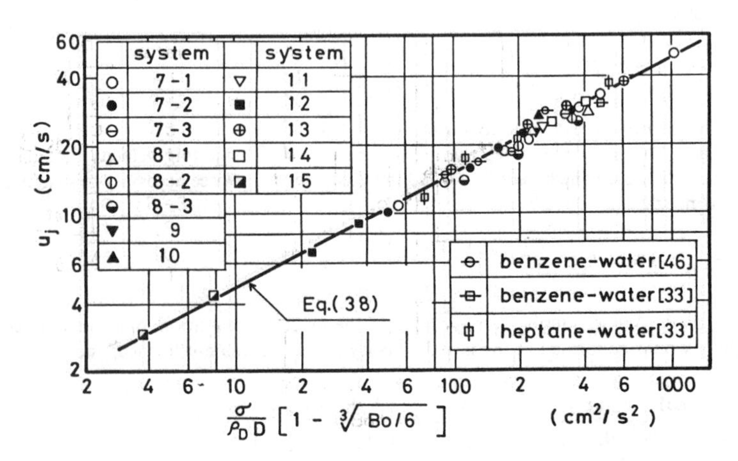

Figure 12. Comparison of experimental jetting velocities with Equation 38. Solid line shows Equation 38. System 7—kerosene-aq. glycerine sol.; 8—kerosene + paraffin-aq. glycerine sol.; 9—kerosene + paraffin-water; 10—kerosene-water; 13—benzene-water; other numbers are shown in Figure 7 [51].

For a centrifugal field Bo is defined as

$$Bo = D^2 \rho_D R \omega^2 / \sigma \tag{43}$$

where ω = angular velocity of rotating nozzle
R = rotating radius of nozzle

It must be noticed that these correlations for liquid-in-air systems consist of essentially the same dimensionless number as the correlation for liquid-liquid systems.

Critical Velocity from Laminar to Turbulent Jet

As described before, two characteristic velocities were investigated near the velocity region where the breakup length reaches a maximum; the critical velocity from laminar to turbulent jet u_k and the characteristic velocity at the maximum interfacial area u_M. The two characteristic velocities are almost equivalent in many systems [46]. In the system of low interfacial tensions, however, the minimum drop size occurs nearly at jetting velocity [35, 37, 64]. The details concerning this problem are discussed in the subsequent section.

Treyball [65] related the velocity u_M with system properties, using experimental data obtained by Christiansen and Hixson [18]. The result is put in the form

$$u_M = 2.69 \left(\frac{D_j}{D} \right) \left(\frac{\sigma / D_j}{0.5137 \rho_D + 0.4719 \rho_C} \right)^{1/2} \tag{44}$$

where the diameter ratio of nozzle to jet is given empirically by

$$D/D_j = 0.485\ Bo + 1, \qquad \text{for } Bo < 0.616 \tag{45}$$

and

$$D/D_j = 1.51 \sqrt{Bo} + 0.12, \qquad \text{for } Bo > 0.616 \tag{46}$$

The relation given by Equations 44 ~ 46 is widely used in subsequent investigations to predict the drop size at jetting conditions [36, 37, 38].

Fujinawa et al. [46] correlated the critical velocity u_k by

$$u_k^2 \rho_D D / \sigma = 19{,}000 Z_D^{1.34} (\mu_D / \mu_C)^{0.14} \tag{47}$$

where Z_D is the Ohnesorge number for the dispersed phase and is defined with nozzle diameter as (cf: Equation 16 is based on jet radius)

$$Z_D = \mu_D / \sqrt{\rho_D D \sigma} \tag{48}$$

A limitation to their study was the use of surfactants to vary interfacial tension and aqueous (vinyl alcohol) solution to change viscosity. The presence of surfactants gives rise to additional modifications of a complex nature, and some polymer solutions are non-Newtonian fluids.

Takahashi and Kitamura [17, 51] investigated experimentally the critical velocities for Newtonian liquid-liquid systems free from surfactants and proposed the following empirical equation.

$$Re_k\ (\equiv \rho_D D u_k / \mu_D) = 98 Z_C^{-0.11} Z_D^{-0.30} \tag{49}$$

where Re_k is the Reynolds number based on μ_k and Z_C is the Ohnesorge number for continuous phase, defined as

$$Z_C = \mu_C / \sqrt{\rho_C D \sigma} \tag{50}$$

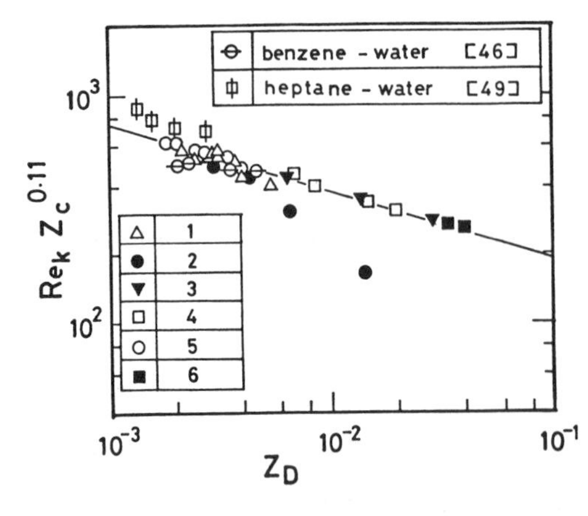

Figure 13. Correlation of critical velocity from laminar to turbulent jet. Solid line shows Equation 49. System 1—water-gasoline + paraffin; 2—aq. starch syrup sol.-gasoline; 3,4—aq. starch syrup sol.-gasoline + paraffin; 5—water-kerosene; 6—water-n-buthanol [17].

Figure 13 shows the comparison of experimental critical velocities with Equation 49. The experimental data exceedingly deviated from the equation are for the systems of $\mu_D/\mu_C = 5$; 11.2. In these systems, wavy jets are observed, and the breakup curves are as noted earlier (see Figures 9 and 10). Equation 49 is valid for the system where μ_D/μ_C is less than $1 \sim 2$.

On the other hand, the critical velocity for a liquid jet in air is correlated by Tanazawa and Toyoda [66] with

$$Re_k = 370 Z_D^{-0.318} \tag{51}$$

and by Grant and Middleman [20] with

$$Re_k = 325 Z_D^{-0.28} \tag{52}$$

The comparison of the preceding equations with Equation 49 indicates that the critical velocity from laminar to tubulent jets can be expressed in the similar form both in liquid-liquid and liquid-air systems although that in liquid-liquid systems is modified by Z_C.

Correlation of Breakup Length

As mentioned earlier the breakup length of a jet is given by Equation 28 based upon the stability analysis. Experimental breakup lengths of laminar jets in air are well correlated with the equation. However, application of this equation to liquid jets injected into stagnant liquids requires some considerations.

Primarily the difference between the average injection velocity at nozzle exit and the surface velocity of the jet cannot be neglected. Thus, the breakup length relates to breakup time T by $\ell = Tu_s$, where u_s is the substantial surface velocity of jet. The experimental facts that breakup lengths of liquid-liquid jets are proportional to $(u_D - u_j)$ suggests that u_s is presumably proportional to $(u_D - u_j)$. So the breakup time relates to ℓ by

$$\ell = K'T(u_D - u_j) \tag{53}$$

where K′ is a constant to be experimentally determined. Substitution of Equation 27 into the Equation 53 leads to

$$\ell = K(u_D - u_j)/q^* \tag{54}$$

where

$$K = K' \ln(a/\bar{\delta}_0) \tag{55}$$

The maximum growth rate of surface disturbances is given numerically in Figure 4, as mentioned before. For the purpose of correlating breakup length with jet velocity and physical properties by Equation 54, an explicit expression for the maximum growth rate is needed. Thus, the approximate solution is derived as follows [17].

The maximum growth rate of a viscous liquid jet on which the surrounding medium has no influence is given by Equation 19. We rewrite it in the dimensional form with new subscripts to prevent confusions.

$$\begin{aligned} q_G^* &= [\sqrt{8\rho_D a^3/\sigma} + 6\mu_D a/\sigma]^{-1} \\ &= [(1/q_I^*) + (1/q_V^*)]^{-1} \end{aligned} \tag{56}$$

where q_I^* and q_V^* is the maximum growth rate for the case where either a viscous force or inertia is negligible and written, respectively, as

$$q_I^* = \sqrt{\sigma/8\rho_D a^3} \tag{57}$$

$$q_V^* = \sigma/6\mu_D a \tag{58}$$

The prior relation between the maximum growth rate for the limiting cases can be presumably extended to liquid-liquid systems.

$$q_L^* = [(1/q_{II}^*) + (1/q_{VV}^*)]^{-1} \tag{59}$$

where q_L^* = the maximum growth rate of a viscous liquid jet in viscous liquid
q_{II}^* = the maximum growth rate for an inviscid liquid-inviscid liquid system
q_{VV}^* = the maximum growth rate for a high viscosity liquid-high viscosity liquid system

Recalling Equations 21 and 23 which gives the growth rate of these limiting cases, we rewrite them in dimensional form with subscript respectively as

$$q_{II} = \sqrt{\sigma/\rho_D a^3}\, F_1(x) \tag{60}$$

$$q_{VV} = (\sigma/2a\mu_D)F_1(x) \tag{61}$$

As described before, the maximum of $F_1(x)$ and $F_2(x)$ is given as a function of the density and viscosity ratio, respectively. We denote the maximum as f_1 and f_2, respectively. Thus the maximum growth rate is written as

$$q_{II}^* = \sqrt{\sigma/\rho_D a^3}\, f_1 \tag{62}$$

$$q_{VV}^* = (\sigma/2a\mu_D)f_2 \tag{63}$$

Substituting the preceding expressions into Equation 59 we obtain

$$q_L^* = [\sqrt{\rho_D a^3/\sigma}/f_1 + (2a\mu_D/\sigma)/f_2]^{-1} \tag{64}$$

Although Figure 2 indicates the dependence of f_1 on the density ratio, it saturates to $1/\sqrt{8}$ as $\hat{\rho}$ increases. For the usual liquid-liquid systems, furthermore, the density ratio is not so far from unity, so the dependence of f_1 on the ratio is negligible and f_1 can be approximated as

$$f_1 = 1/\sqrt{8} \tag{65}$$

Figure 3 indicates that f_2 is a function of the viscosity ratio. For $0.02 < \mu_D/\mu_C < 5$, f_2 can be approximated by

$$f_2 = 0.071(\mu_D/\mu_C)^{0.52} \tag{66}$$

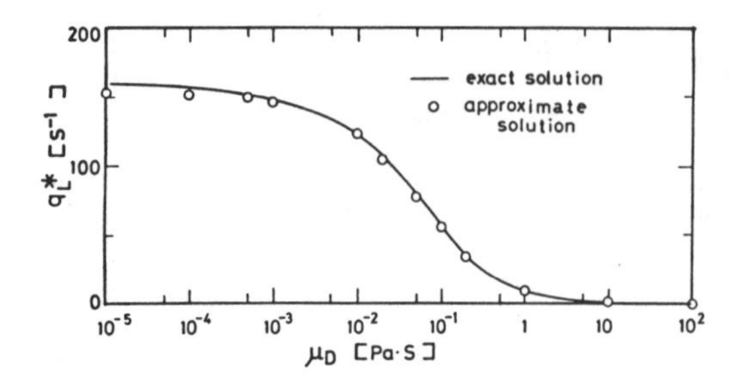

Figure 14. Comparison of approximate solution for maximum growth rate, Equation 64, with exact solution of Equation 12. Calculations are carried out for D = 0.1 cm, ρ_D = 1,000 kg/m^3, ρ_C = 870 kg/m^3, μ_C = 0.001 Pa s, σ = 0.03 N/m [17].

Figure 14 shows the comparison of Equation 64 (the approximate solution) with the numerically calculated value from Equation 12 (the exact solution). The figure illustrates that Equation 64 can satisfactorily approximate the maximum growth rate of a disturbance in the liquid-liquid systems. Combining Equations 54, 64, 65 and 66, we obtain

$$\frac{\ell}{D} = K\left[\sqrt{\frac{\rho_D D}{\sigma}} + 14.1\left(\frac{\mu_D}{\mu_C}\right)^{-0.52}\frac{\mu_D}{\sigma}\right](u_D - u_j) \tag{67}$$

The constant K which contains the initial amplitude of disturbances has to be determined from experimental data.

Figure 15 shows the correlation of K with the physical properties of the system. The constant K is put in the form.

$$K = 1.55\sqrt{Z_C}/Z_D \tag{68}$$

From the prior equations, the laminar breakup length of a liquid jet in an immiscible liquid is given by

$$\frac{\ell}{D} = 1.55\frac{\sqrt{Z_C}}{Z_D}\left[\sqrt{\frac{\rho_D D}{\sigma}} + 14.1\left(\frac{\mu_D}{\mu_C}\right) - 0.52\frac{\mu_D}{\sigma}\right](u_D - u_j) \tag{69}$$

The preceding semi-empirical equation cannot be applied to the system of a high viscosity ratio where ℓ is not proportional to $(u_D - u_j)$, as shown in Figure 9.

Meister and Scheele [49] related the stability analysis to breakup length by considering the dependence of the surface velocity on the distance from nozzle as follows. If the surface velocity u_s is not assumed to be equal to the nozzle velocity, instead of Equation 28, the breakup length is given by the more general form.

$$\int_0^{\ell}\frac{dz}{u_s} = \ln(a/\bar{\delta}_0)/q_L^* \tag{70}$$

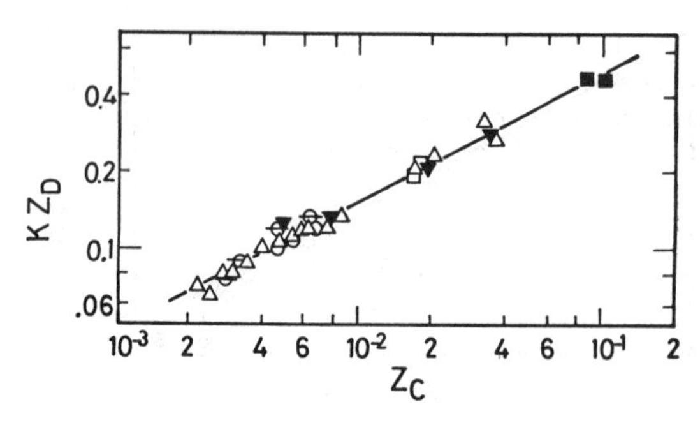

Figure 15. Correlation of coefficient K with Z_D and Z_C. Solid line shows Equation 68. Keys are shown in Figure 13 [17].

Using the approximate estimate of surface velocity, the above equation can be expressed as

$$\ell = \frac{1}{2q_L^*}\left[\left(\frac{D_j^2 u_s}{D^2}\right)\bigg|_{z=2.5D} + \left(\frac{D_j^2 u_s}{D^2}\right)\bigg|_{z=0}\right] \ln\left(\frac{D}{2\bar{\delta}_0}\right) \tag{71}$$

After the evaluation of u_s at a given distance, which requires additional equations and graphical numerical values, iterative calculations are necessary to predict the breakup length because u_s is a function of 1. Meister and Scheele described that Equation 71 is a satisfactory approximation.

Drop Formation from Jet

Stability analyses described in the first two sections assume that uniform drops are formed from a jet in a regular pattern, which predicts that the drop size is independent of jet velocity in the region of laminar jets. The drop size predicted from the stability theory is given by Equation 29 or 33. As shown in Figure 6, however, drops formed from the laminar jet ejected into stagnant liquids depend strongly on the injection velocity.

The comparison of experimental drop sizes with the prediction from Equation 29 are shown in Figure 16 (injection of benzene into the water phase), and in Figure 17 injection of water into furfural is shown (both phase are mutually saturated). Solid lines in these figures show drop sizes

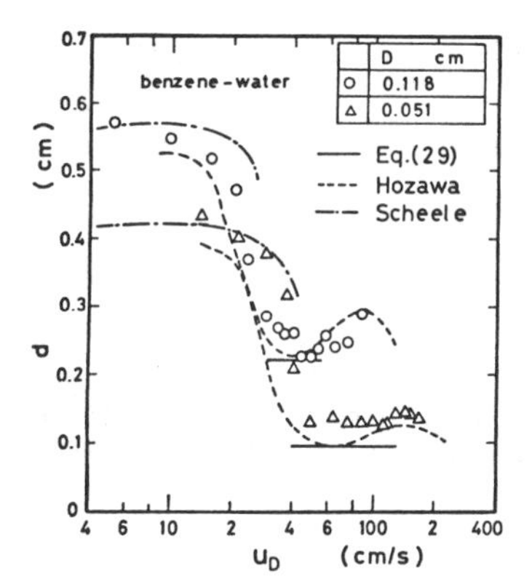

Figure 16. Diameter of drops formed by injection of benzene into stagnant water phase. ($\Delta\rho = 124$ kg/m^3, $\mu_D/\mu_C = 0.67$, $\sigma = 0.0302$ N/m.) Solid lines are calculated from Equation 29 or 33. Dotted curves show the prediction from Hozawa's chart (Figures 18 and 19). Broken curves calculated from Equation 36 [19].

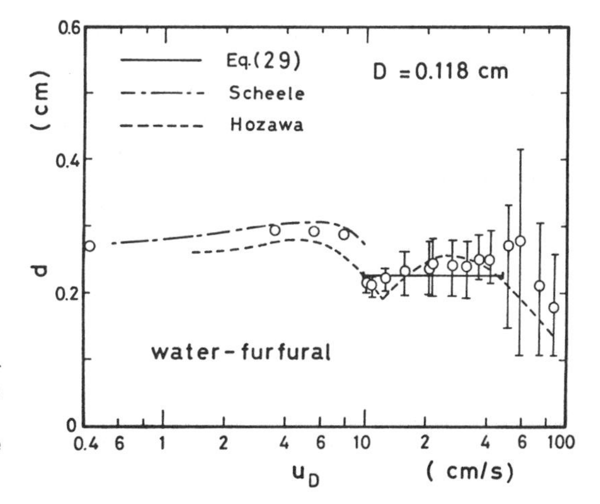

Figure 17. Drop sizes and their distribution for water-furfural system ($\Delta\rho = 137$ kg/m^2, $\mu_D/\mu_C = 0.69$, $\sigma = 0.0045$ N/m.) Estimated lines are the same as those in Figure 16.

calculated from Equation 29. In almost all experimental systems, the diameter of drops formed from a laminar jet reaches a minimum at u_M. In the systems of low interfacial tensions, however, the drop size at jetting velocity is minimum and increases with increasing injection velocity, as shown in Figure 17. The minimum drop sizes at jetting velocity were also observed in the system of low interfacial tensions by Perrut and Loutary [35] and Hozawa and Tadaki [37].

Christiansen and Hixson [18] and Fujinawa et al [46] compared the minimum drop size at u_M with Equation 29 and reported the satisfactory agreement. However, the largest disagreement of the author's data at u_M with the estimate from Equation 29 is about 40%.

Even at the velocity range for laminar jets, furthermore, the drop size extremely depends on the injection velocity. If the degree of this dependence is accounted, the discrepancy between theoretical and experimental drop size exceeds 100%. Two explanations are expected for such a big difference. One is the formation of velocity distribution in both phases and subsequent significance of surface velocity [49, 50]. The other is the effect of the hydrodynamic resistance of the surrounding liquid on the jet stability, discussed later.

Meister and Scheele [50] improved the drop formation model, by analyzing the dependence of surface velocity on the distance from the nozzle, and by introducing a complicated criterion for wave splitting. After sophiscated considerations, they modified Equation 29 to obtain

$$d = \left[1.5\pi/x^* M(u_s/u_a) \Big|_{\lambda/2} \right]^{1/3} (2a) \Big|_{\lambda/2} \tag{72}$$

where M is a criterion for wave splitting and is a function of R_c: $M = 1$ for $R_c < 2$; $M = 2$ for $2 < R_c < 4$; $M = 3$ for $4 < R_c < 8$ etc, and R_c is given as

$$R_c = \frac{(u_s/u_a)_{1-\lambda}(a^3)_{\lambda/2}}{(u_s/u_a)_{\lambda/2}(a^3)_{1-\lambda}} \tag{73}$$

Both the ratio of surface-to-average velocity of a jet u_s/u_a and the jet radius are given as a function of the axial distance by additional equations and graphs [49]. They reported that their analysis showed better agreement with experimental results than previous analyses, and the most significant success was attained especially for systems with a high continuous phase viscosity (heptane-glycerine system, $\mu_C = 0.515$ Pa s).

Furthermore, drops formed from a jet have distributions in size as shown in Figures 6 and 17. They also described that the disagreement of theoretical prediction with experimental data occurred in the system where the drop size distribution was significantly broad. The drop size distribution is caused by wave splitting, drop merging, interaction of the adjacent node of disturbances, and coalescence of drops as mentioned by Meister and Scheele [50]. A number of mechanisms for drop size distribution make the phenomena of drop formation from a jet more complicated, in addition to the presence of velocity distribution in both phases. So, a complete theoretical analysis for the drop formation from the jet injected into stagnant liquids is very difficult and far from practical use.

For practical purpose, much effort was made to correlate experimental drop sizes at jetting conditions [18, 30, 31, 37, 38]. There is still no satisfactory correlation appropriate in a wide range of liquid-liquid systems because of the complicated phenomena mentioned above. Among these works, Hozawa and Tadaki [37] proposed a simple but useful nomograph, applicable to relatively comprehensive liquid-liquid systems, as shown in Figure 18 for $u_D < u_M$, and in Figure 19 for $u_D > u_M$. In the nomograph, d_M is the minimum drop size at u_M which is given by Equation 44. The prediction from these figures is shown as a dotted line in Figures 16 and 17. The estimation from Scheele and Meister's Equation 36 are also shown in these figures for comparison. It must be reminded that Equation 36 can predict only at the velocity region of a single drop formation.

In practical equipment, the additional gross flow, the shape of the orifice, mass transfer, etc, influence the drop formation, and consequently mass and heat transfer rates. Thus, it is not important how accurate the correlation for predicting drop size is, but to what extent it is applicable.

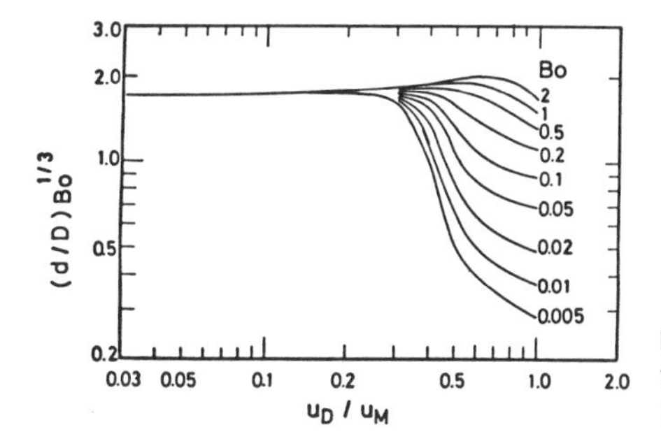

Figure 18. Prediction of drop size for the case where injection velocity is less than u_M. u_M is given by Equation 44 [36].

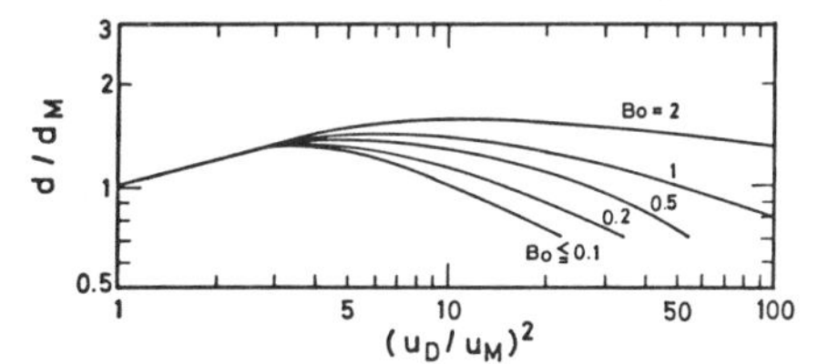

Figure 19. Prediction of drop size in the velocity region above u_M. Minimum drop size d_M is given by $2.0 \times D_j$, where the jet diameter D_j is calculated from Equation 45 or 46 [37].

Influence of Nozzle Length on Jet Breakup

Although the breakup of jets in liquid-liquid systems has been the object of many investigations as described before, most experiments were carried out by liquid injection through long nozzles. In practical equipment, liquids issue usually from thin orifices or sieve plates.

Kitamura and Takahashi [60] investigated systematically the effect of nozzle length on both breakup length and drop size. Figure 20 shows breakup lengths of jets injected from the nozzle whose length varies at a nearly constant diameter. The breakup length depends strongly on nozzle length as shown in the figure. As the nozzle becomes longer, the laminar breakup length increases to approach a constant for each diameter above a critical value of aspect ratio L/D.

The jetting velocities are independent of nozzle length. The dependence of critical velocity on nozzle length is not so significant as those for liquid jets in air [20, 60, 67].

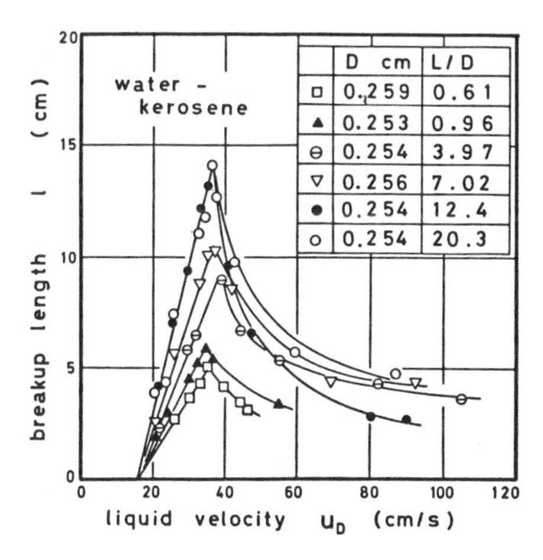

Figure 20. Effect of nozzle length on breakup length of water jet injected into kerosene [60].

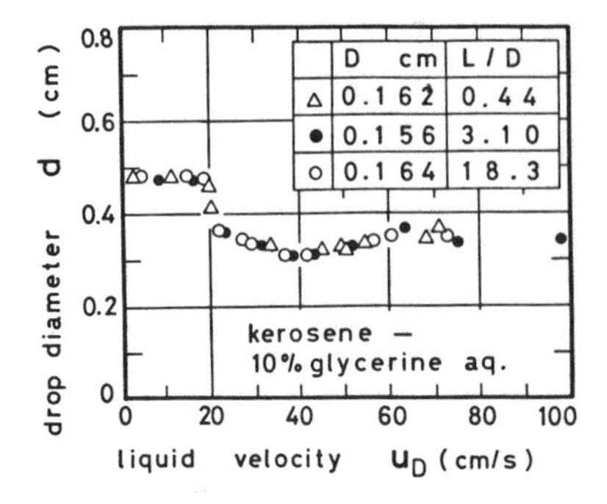

Figure 21. Effect of nozzle length on diameter of drops formed by injection of kerosene into 10% aqueous solution of glycerine [60].

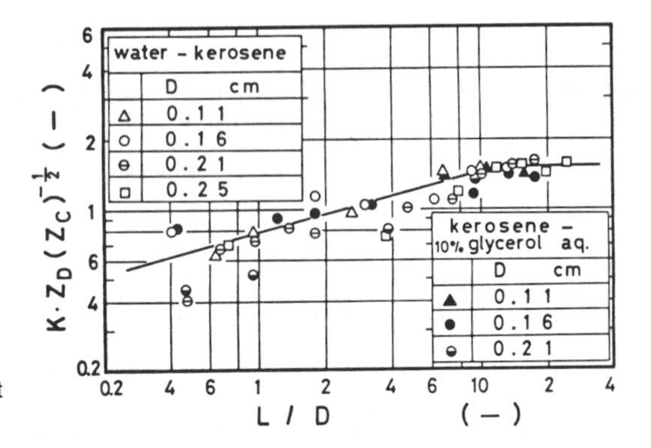

Figure 22. Dependence of coefficient K on aspect ratio of nozzle [60].

The effect of nozzle length on drop sizes is shown in Figure 21 which illustrates data for three principal mechanisms of drop formation: directly at nozzle exit, from laminar jets and from turbulent jets. In the whole region, drop sizes are apparently independent of nozzle length, while it greatly affects breakup lengths.

The effect of nozzle length on breakup length can be explained by the dependence of initial amplitude of disturbances upon it [60]. The disturbances on jet surface are initiated at the nozzle exit as a result of pressure fluctuations or eddies in the nozzle. The eddies are mainly produced at the nozzle entrance and dissipate passing through the nozzle. The initial amplitude of disturbances consequently becomes smaller as the nozzle is lengthened. Thus, the coefficient $\ln(a/\bar{\delta}_0)$, namely the breakup length, increases with an increase in nozzle length.

Figure 22 shows a correlation of the coefficient K which contains $\ln(a/\bar{\delta}_0)$ with the aspect ratio of the nozzle; see Equations 55 and 68. As L/D becomes longer, the ordinate in the figure increases and saturates to a constant beyond a critical aspect ratio (about 10 for the systems shown in the figure).

Effect of Mass Transfer on Jet Breakup

Industrial utilization of jet breakup is usually accompanied by mass transfer. Mass transfer across the jet interface introduces a twofold effect on the hydrodynamic stability of a jet. One is the decrease of interfacial tension due to the presence of a solute, which stabilizes the jet. The other is the Marangoni phenomena due to concentration and subsequent interfacial tension gradients, though there are some differences in the estimation as to its effect [54, 58, 68].

The breakup lengths of a jet under conditions of mass transfer were measured by Meister and Scheele [49], Dzuber and Sawistowski [54], Skelland and Huang [56, 57], and Coyle et al. [58].

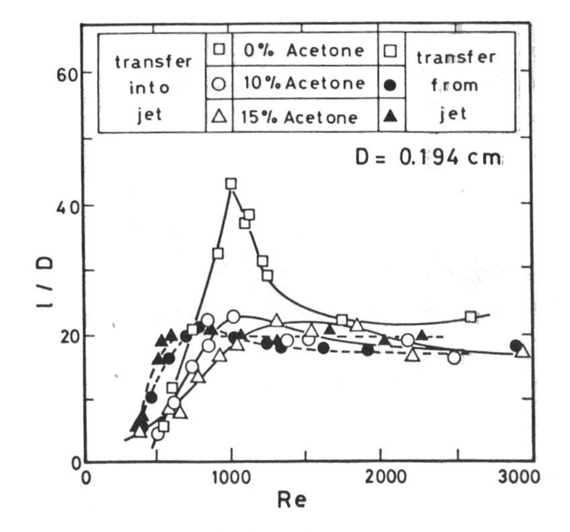

Figure 23. Effect of mass transfer on breakup length of jet: inward transfer—benzene-water + acetone; outward transfer—benzene + acetone-water. Reproduced with permission from *Chem. Eng. Sci.*, 36, B. W. Coyle, J.C. Berg and J. C. Niwa, "Liquid-Liquid Jet Breakup under Conditions of Relative Motion, Mass Transfer and Solute Adsorption," 1981, Pergamon Press Ltd. [58].

Although their results are not in essential agreement, the breakup length seems to depend on the direction of transfer, out of or into a jet, as shown in Figure 23.

Burkholder and Berg [55] developed the theory of hydrodynamic stability of liquid jets in immiscible liquids undergoing mass transfer. The presence of a tension-lowering solute affects in such a way where the radial concentration profile near the deformed interface produces axial variations in interfacial tension along the jet. They obtained the analytical results that the mass transfer effect is evaluated by the Marangoni number and it depends on the transfer direction. Transfer of a tension-lowering solute out of the jet (positive Marangoni number) destabilizes the jets, while transfer in the opposite direction is stabilizing only for smaller rates of mass transfer and becomes destabilizing as the rate increases.

It must be emphasized that the quantitative comparison of the theoretical with the experimental effects of mass transfer is prevented by the difficulty in estimating appropriate interfacial tension of nonequilibrium systems and by a limited number of experiments.

The direction of mass transfer also affects sizes of drops formed from jet [52–54, 56]. Bayens and Laurence [52] measured the drop size distribution and found that an increase in the driving force for mass transfer from a jet resulted in a substantial broadening of size distribution. They also found that the mean drop size increases with an increasing driving force for mass transfer from a jet although the mass transfer into a jet did not markedly change the drop size. Dżuber and Sawistowski [54] also found that the drop size for the outward transfer case was larger than for the equilibrated and the inward transfer case.

Recently the influence of chemical reaction on jet breakup was investigated by Nealson and Berg [59], who reported the jet was strongly destabilized by the presence of interfacial turbulence.

BREAKUP OF JET IN CENTRIFUGAL FIELD

The operating liquid velocity in ordinary extractors is controlled by the settling stage which is driven by the gravity force working through the density difference, and consequently large velocities are impossible. The application of a large centrifugal force makes high velocities and small extractor size possible. The essential characteristic of all centrifugal contactors such as the Podbielniak and Luwesta extractor is their short contact time and low hold-up. Such advantages are especially useful for handling labile or extensive materials in the pharmaceutical, fine chemicals, and nuclear industries [65, 69, 70, 71].

Knowledge of drop formation in the centrifugal field is important for designing the centrifugal extractor. Takahashi et al. [72] investigated the breakup of jets and drop formation in a centrifugal extractor.

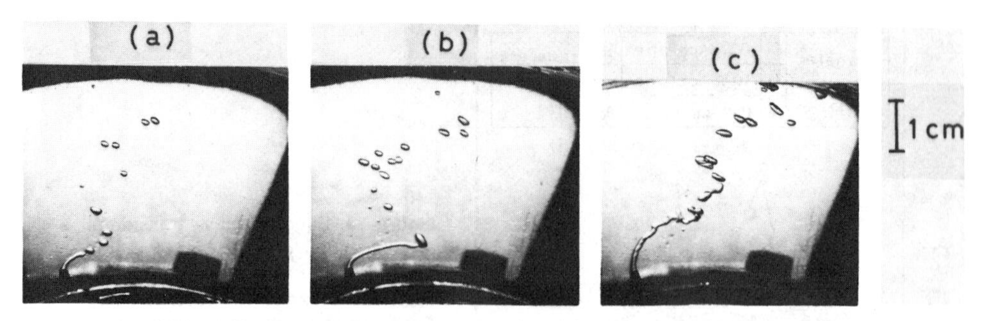

Figure 24. Breakup pattern of jet in centrifugal field. Water jet injected into mixture of kerosene and liquid paraffin at 800 r.p.m., D = 0.125 cm, R = 6.4 cm. a) $u_D = 6.3$ cm/s; b) $u_D = 25.5$ cm/s; c) $u_D = 83.3$ cm [72].

The principal mechanisms for drop formation in the centrifugal field are similar to those in a gravitational field, from the formation at a nozzle tip, from jets, and from sprays. Figure 24 shows a sequence of photographs of jet breakup in a centrifugal field: laminar jet (A, B) and turbulent jet (C). The jet is bent to the tangential direction because of the drag force by continuous phase.

The breakup lengths in the centrifugal field are shown in Figure 25. A dotted line in the figure is the breakup length for a system the same physical properties in the gravitational field calculated from Equation 69 which predicts about 20 cm in the maximum breakup length. The breakup lengths in the centrifugal field are considerably shorter than those in the gravitational field. The jetting velocity is also low.

Sizes of drops from the jet in the centrifugal field are shown as a function of injection and rotating velocity in Figure 26. The prediction for the system of the same physical properties in the gravitational field is presented as a dotted line which is estimated from Figures 18 and 19. A larger centrifugal force (45 ~ 140 times that of gravity in this case) makes the drops smaller as shown in the figure. Drop sizes at a constant centrifugal force increase with injection velocities to approach a constant; namely the minimum drop size is attained at jetting velocity.

The diameter of drops formed from a liquid-liquid jet in a centrifugal field are empirically correlated by [72]

$$d/D = 2.4\ We^{-0.14}\ Fr^{0.2} \tag{74}$$

where

$$We = \rho_D D u_D^2/\sigma \tag{75}$$

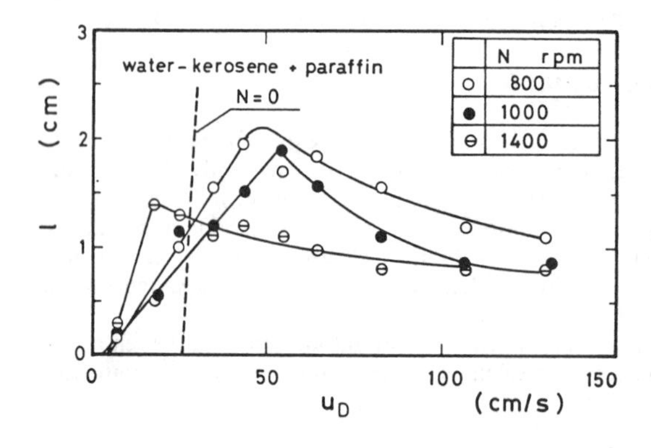

Figure 25. Breakup length of liquid-liquid jet in centrifugal field. Dotted lines are prediction for gravitational field, D = 0.125 cm, R = 6.4 cm [72].

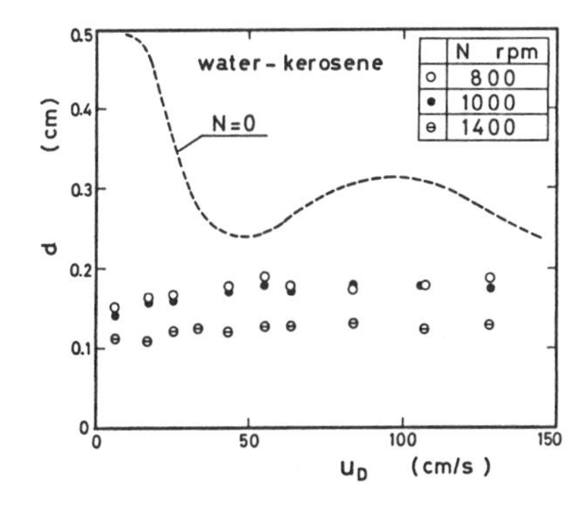

Figure 26. Diameter of drops for water-kerosene system in centrifugal field. Dotted curve shows prediction for gravitational field, D = 0.125 cm, R = 6.4 cm [72].

$$\mathrm{Fr} = u_D^2/R\omega^2 D \tag{76}$$

The preceding equation can be applicable for the case of

$$0.1 < \mathrm{We} < 50, \quad \text{and} \quad 0.002 < \mathrm{Fr} < 5$$

DIRECT CONFIRMATION OF STABILITY ANALYSIS BY JET OF ZERO RELATIVE VELOCITY

Although the knowledge of hydrodynamic stability of the liquid column is important for the prediction of breakup of liquid jets in immiscible liquids, complicated phenomena induced by the injection of a jet into stagnant liquids prevent the direct application of the stability theory to breakup data. Mason and co-workers [26, 73, 74] examined Tomotika's theory using a stationary liquid column produced in the four-roller apparatus which was first developed by Tayler [25], and also by using a binary liquid column inside a tube [75]. They confirmed that the wavelength of surface disturbances agrees with Tomotika's result. However, their examination was restricted to the limiting case where both jet and surrounding liquid were highly viscous.

Kitamura et al. [19] developed the experimental method using the zero relative velocity jet to confirm the stability analysis for the general case described in the first section (a viscous jet in a viscous liquid). Furthermore, the use of the zero relative velocity system provided useful information about both drag effect by surrounding liquids and non-Newtonian effects on the breakup of a liquid jet into drops.

The experimental apparatus for the jet of zero relative velocity is schematically shown in Figure 27. The test section is a glass tube of 1.52 cm I.D., in the axis of which a nozzle made from a stainless steel tube is fixed. The continuous phase flows in the test section. The jet issues from the nozzle

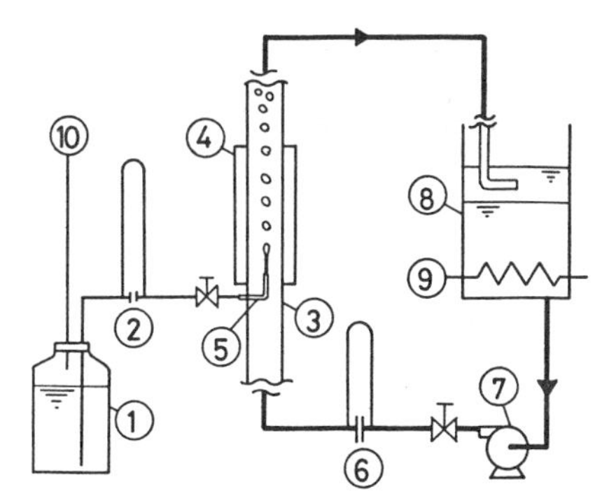

Figure 27. Experimental apparatus for jet of zero relative velocity: 1—liquid reservoir; 2—capillary flow meter; 3—test section (glass tube); 4—rectangular column to eliminate photographic distortion; 5—nozzle; 6—orifice flow meter; 7—pump; 8—liquid reservoir (Settler); 9—cooling water; 10—compressed air [19].

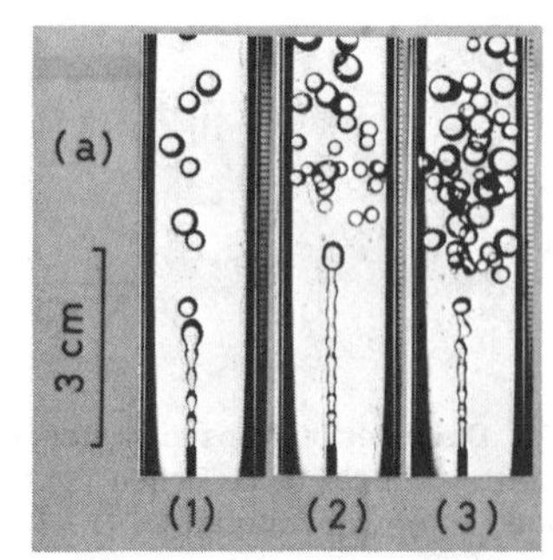

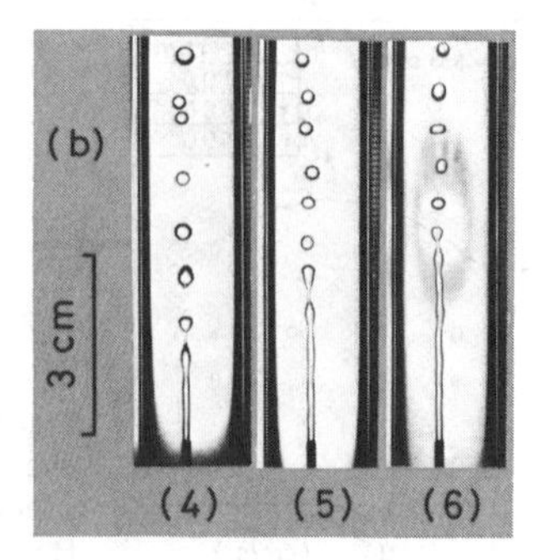

Figure 28. Breakup pattern of jet in immiscible liquid (liquid paraffin + kerosene-water system). Breakup data for this system are shown in Figure 30: (A) Jets injected into stagnant liquids. (1) $u_D = 53$ cm/s, (2) $u_D = 83$ cm/s, (3) $u_D = 113$ cm/s. (B) Jets injected at the same velocity as that of continuous phase ($U = 0$). (4) $u_D = 50$ cm/s, (5) $u_D = 80$ cm/s, (6) $u_D = 110$ cm/s [19].

in the same direction as the flow of continuous phase, and at the same velocity as the continuous phase. So the jet velocity relative to the continuous phase can be always maintained at zero

$$U = u_D - u_C = 0$$

Breakup of the jet injected into the stagnant continuous phase was examined in the same section as for the reference data, $u_C = 0$.

Breakup Length of Jet

A typical example of breakup pattern is shown in Figure 28, which contains two sequences of photographs; one is for jets, $u_C = 0$ and the other is for jets, $U = 0$. Surface waves under the condition of $u_C = 0$ seem to disagree with Equation 3 as shown in Figure 28A. Such irregular waves were observed even at a low injection velocity in some experimental systems for the condition of $u_C = 0$; especially in the systems of $\mu_D/\mu_C > 1$, as described in the preceding section. In contrast, for jets under the condition of $U = 0$, surface waves apparently satisfy the form of Equation 3; this is a basic assumption of stability analysis, as shown in Figure 28B.

The breakup of jets was researched for 14 systems where the viscosity ratio μ_D/μ_C varies from 0.11 to 18.2 and the interfacial tension ranged from 0.00133 to 0.056 N/m. Figures 29A and 30A show the relation of breakup length to jet velocity for a heptane-water and a liquid paraffin + kerosene-water system, respectively. Data for both conditions of $U = 0$ and $u_C = 0$ are plotted in these figures. For the case of $u_C = 0$, the viscosity ratio of both phases affects the pattern of breakup curves and furthermore the curve extrapolated from the data does not pass through the origin (as described earlier). On the contrary, the pattern of breakup curves for $U = 0$ is independent of the viscosity ratio and extrapolated curves pass through the origin, which means that the measured breakup length satisfies the relation of Equation 28. Breakup lengths satisfy such a relation in all systems experimented under the condition of $U = 0$.

These results and photographic observations indicate that the stability analysis can be applied to the jet for $U = 0$. The values of $\ln(a/\bar{\delta}_0)$ were calculated from experimental breakup lengths and the maximum growth rate q^* obtained by solving Equation 12 numerically. These values of $\ln(a/\bar{\delta}_0)$ vary from 2.7 to 17.1 are almost equivalent to those for liquid jets in air. While $\ln(a/\bar{\delta}_0)$ for jets in air is correlated as a function of the Ohnesorge number Z_D [20, 23], the coefficient for jets in immiscible liquids depends not only on Z_D but on Z_C. Figure 31 shows the correlation of $\ln(a/\bar{\delta}_0)$ for liquid-liquid systems. The data are fitted by the solid line with the following empirical equation.

$$\ln(a/\bar{\delta}_0) = 4.1 Z_C^{0.16} Z_D^{-0.3} \tag{77}$$

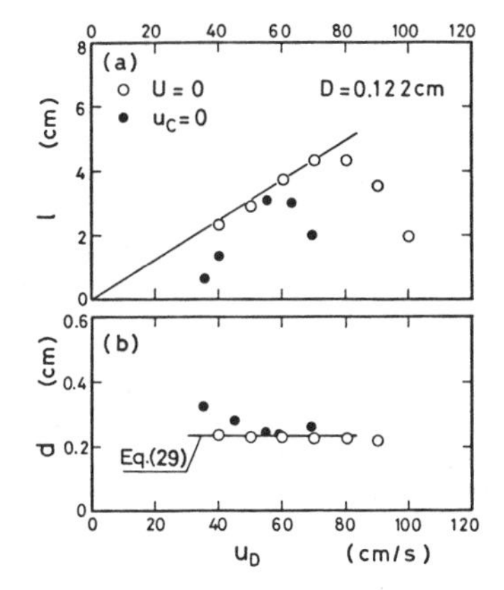

Figure 29. Breakup data for n-heptane-water system ($\mu_D/\mu_C = 0.49$): (A) breakup length of jets; (B) diameter of drops formed from jets. Solid line shows Equation 29 [19].

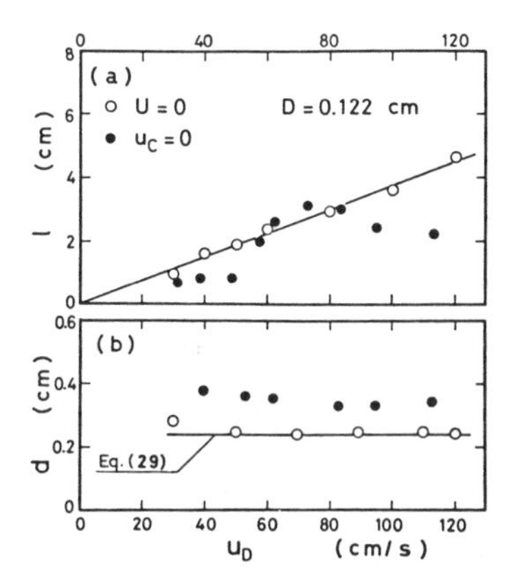

Figure 30. Breakup data for liquid paraffin + kerosene-water system ($\mu_D/\mu_C = 18.2$): (A) breakup length of jets; (B) diameter of drops formed from jets. Solid line shows Equation 29 [19].

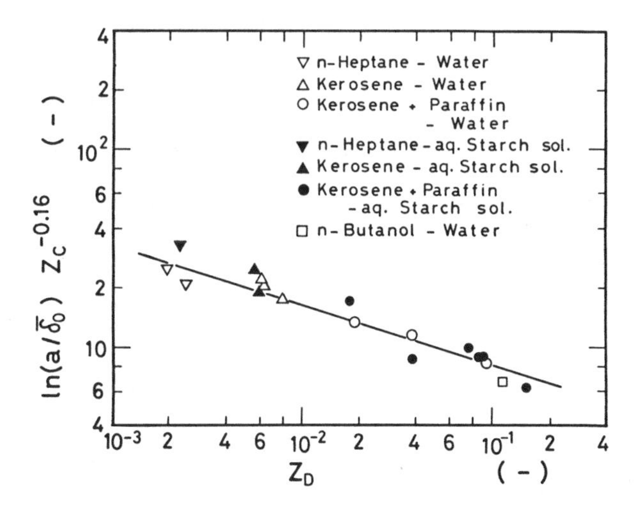

Figure 31. Correlation of $\ln(a/\bar{\delta}_0)$ for immiscible liquid-liquid systems with Ohnesorge numbers based on both phases [19].

The continuous phase flows were turbulent under many experimental conditions, especially in the systems of low viscosity continuous phase. However, Figure 28 (4) shows that the jet breakup seems not to be affected by the turbulence in the continuous phase flow, the Reynolds number of which is 7,500. The experimental breakup lengths are proportional to the injection velocity even at the conditions of a higher Reynolds number, as shown in Figures 29 and 30. Furthermore, the values of $\ln(a/\bar{\delta}_0)$ seem not to depend on the Reynolds number. These facts mean that the jet stability is little affected by turbulence in the continuous phase flow.

It should be discussed why the turbulence does not affect the jet stability. This problem is concerned not only with the validity of the experiment using a zero relative velocity jet, but is closely related in both experiments to the dynamic effect of continuous phase and non-Newtonian effect on the jet stability in immiscible liquid-liquid systems.

The turbulent stream contains eddies of very many different sizes differing from one another by an order of magnitude. Spectra of turbulent energy in fully developed pipe flow were experimentally obtained by Laufer [76]. His spectrum diagram indicates that a fraction of turbulent energy decreases as the turbulent frequency becomes higher or as the size of the eddies becomes smaller. For example, the turbulent energy becomes $1/10^4$ of the maximum energy whose wave number is 0.1, when the wave number of turbulence reaches 10.

On the other hand, the wave number of the most rapidly growing disturbances on the jet surface has the approximate dimension of 10 (Equation 32 indicates that the values of k^*a for the experimental conditions lie between 0.53 and 0.67). Thus the energy of turbulences, the wave number of which is equivalent to that of surface disturbances, is not great enough to amplify them by a reasonant interaction. Consequently, the turbulent flow of the continuous phase has little effect on the jet stability. In other words, the sizes of surface disturbances are much smaller than the sizes of the turbulent eddies which have enough energy to influence the growth rate of disturbances.

Drop Diameter

Experimental drop diameters at jetting conditions are shown in Figures 29B and 30B, which contain data under both conditions of $U = 0$ and $u_C = 0$. Data for $U = 0$ are apparently independent of injection velocity and constant for each system, while in almost all systems drop sizes for $u_C = 0$ depend on u_D. The solid line presenting the calculated values from Equations 12 and 29 is in good agreement with measured drop diameters for $U = 0$, whereas drop diameters for the case of $u_C = 0$ are larger than the estimation.

The drop diameter can be more easily predicted from Equation 33 which is confirmed to agree very closely with the exact solution from Equation 12 and 29 as described before. Drop sizes from jets of zero relative velocity in all experimental systems are independent of jet velocity, whereas drop sizes under the condition of $u_C = 0$ have a minimum at u_M. Experimental data for $U = 0$ agree with the theoretical diameter within 9% except for the system where the viscosity of both phases is rather high.

EFFECT OF HYDRODYNAMIC RESISTANCE OF CONTINUOUS PHASE ON JET STABILITY

As mentioned in the preceding sections, the behavior of jets injected into stagnant liquids is very different from that for zero relative motion systems. Such a difference in the behavior is caused either by the existence of a velocity distribution or the fact that the drag by surrounding liquids affects the stability of the jet. Which is the more reasonable cause of discrepancy between the theoretical and the experimental breakup data for immiscible liquid systems must then be determined. So it is estimated in this section how extensive the dynamic resistance of a continuous phase influences the stability of a liquid jet in immiscible liquids.

It has been pointed out for liquid jets in a gaseous medium that the dynamic effect of a surrounding medium plays an important role on the stability at higher jet velocities [4, 13, 67, 77, 78]. The qualitative picture of the aerodynamic effect on jet stability can be presented as follows: Let us assume a wave of finite magnitude on the surface of the jet. The dynamic pressure of gas over the crest of the wave is less than the average pressure, while on the contrary the pressure at the base of the wave is higher than the average. Then, the amplification of disturbances tends to accelerate as the relative velocity between the jet and the surrounding medium increases [13].

Weber [4] analyzed the aerodynamic effect on jet stability by assuming the surrounding air as potential flow; namely by accounting for only the form drag. He derived the following equation for the growth rate of disturbances.

$$Q^2 + 6QZ_Dx^2 - (1 - x^2)x^2 - We_A\, x^3K_0(x)/K_1(x) = 0 \qquad (78)$$

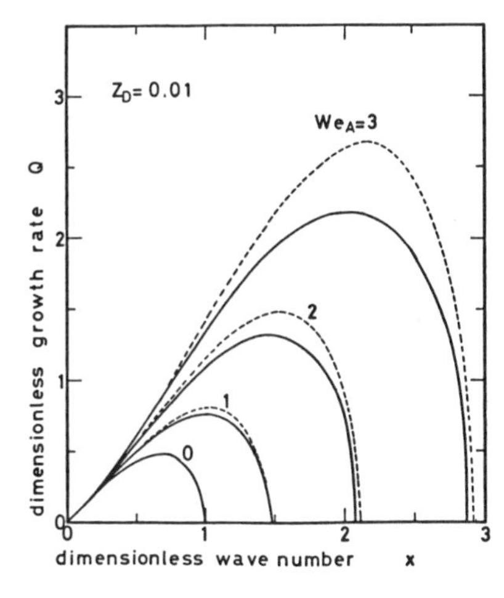

Figure 32. Aerodynamic effect on jet stability. An example of numerical solution. Solid line shows solution from exact equation. Dotted line shows Equation 78.

The preceding equation is obtained by the addition of the aerodynamic term to Equation 18. It must be noticed that Equation 18 is valid for $x < 1$. Thus, the exact equation for the growth rate of a given disturbance of a jet at higher injection velocities might be written an expression that can be added to the aerodynamic term of Equation 17 [13].

Figure 32 shows an example of aerodynamic effect on jet stability, where dotted curves are calculated from Equation 78 and solid curves are numerical solutions of the exact equation. As the Weber number based on air density increases, in any case, both the maximum growth rate and the corresponding wave number increase.

The quantitative comparison of Weber's theory with the experiment data has been carried out by Sterling and Sleicher [67]. They reported as follows: even though Weber's analysis overestimates the aerodynamic effect, it can be assessed as a function of We_A. They proposed the reduction of the aerodynamic term in the characteristic equation by replacing We_A by 0.175 We_A. Although no theoretical estimation is available for liquid-liquid systems, the effect of drag by the surrounding liquid on the stability of jets in immiscible liquids may be presumably assessed as a function of the ambient Weber number.

To estimate the effect of the dynamic resistance of the continuous phase, Kitamura et al. [19] examined the jet stability by the experiment where the relative jet velocity ($U = u_D - u_C$) was maintained at a constant and finite value.

Figure 33 shows the effect of relative velocity on breakup lengths for the case where jet velocities are higher than those of a continuous phase. Figure 34 shows data for the case where jet velocities

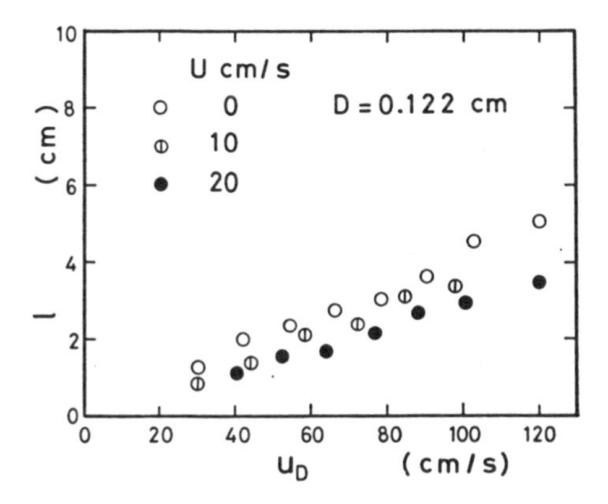

Figure 33. Dependence of breakup lengths on jet velocity relative to continuous phase for liquid paraffin + kerosene-water system ($\mu_D/\mu_C = 9.92$) [19].

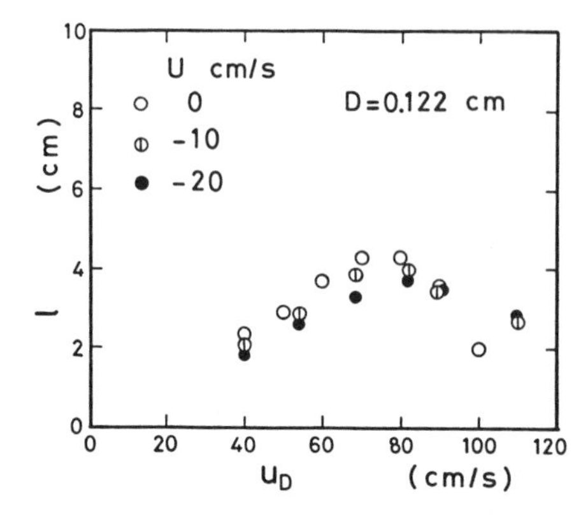

Figure 34. Dependence of breakup length on jet velocity relative to continuous phase for n-heptane-water system [19].

are lower than those of the continuous phase. The laminar breakup length decreases with an increase of the absolute value of the relative velocity. Experimental breakup curves for the case of finite relative velocities are linear, so the difference between average injection velocities and surface velocities of the jet is negligible. Thus the dependence of the breakup length on the relative velocity of the jet is clearly due to the drag effect of the continuous phase on the growth rate of the disturbances.

Figure 35 shows the correlation of the ratio of breakup length for the case where U is a finite constant to that for U = 0, corresponding to the inverse ratio of the maximum growth rate of disturbances, with the Weber number based on the density of the continuous phase and U. Weber's analytical results and Sterling and Sleicher's modified results are shown in the figure. At a lower Weber number, experimental data agree with Weber's estimation. When the Weber number increases the data deviate from the analytical results as in the case of data for liquid jets in air [67, 77].

Figure 35 indicates that the experimental drag effects on the growth rate have a tendency to approach a constant value as the ambient Weber number increases, which means that this effect is not so significant in immiscible liquid-liquid systems even though the ambient Weber numbers are much greater than those for liquid-air systems because of the higher order of density magnitude.

Figure 36 shows the effect of relative velocity on drop diameters. Data for $u_C = 0$ include the formation at nozzle tip, from laminar and turbulent jet. All data for finite relative velocities are

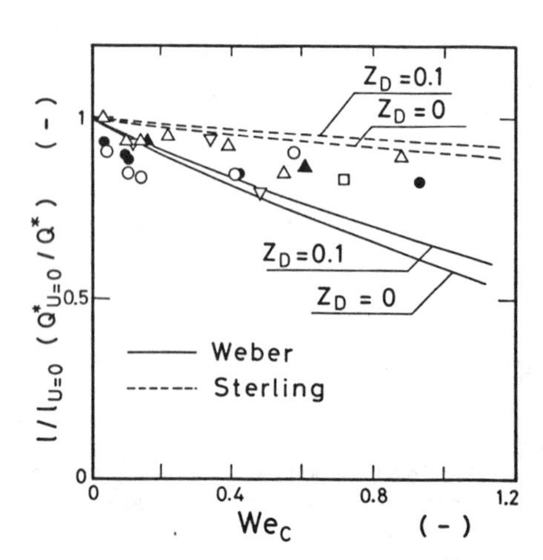

Figure 35. Correlation of effect of hydrodynamic resistance by continuous phase on maximum growth rate of disturbance with the Weber number based on continuous phase. Solid lines show estimation from Equation 78. Dotted lines are estimated from Sterling and Sleicher's modification [67]. Both estimations are valid only for liquid jet in air. Keys are shown in Figure 31 [19].

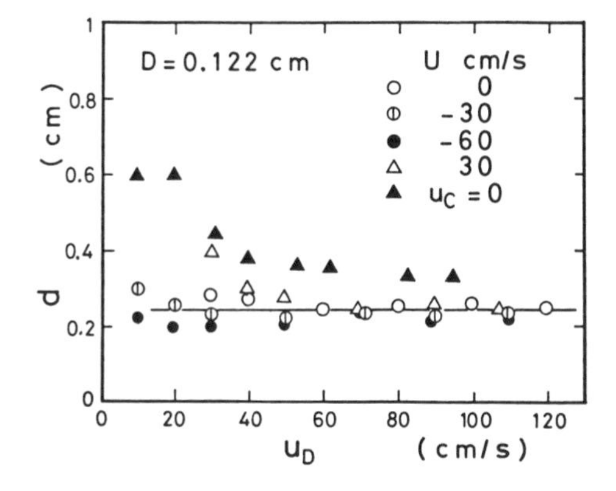

Figure 36. Dependence of drop diameter on jet velocity relative to continuous phase for liquid paraffin + kerosene-water system ($\mu_D/\mu_C = 18.2$) [19].

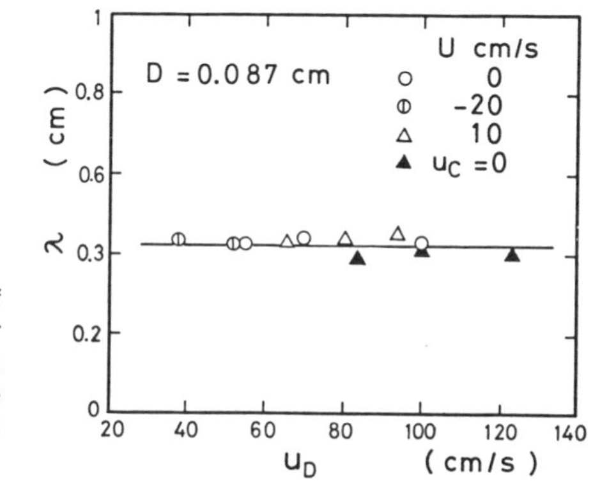

Figure 37. Dependence of wavelength of disturbances on relative velocities of jet for liquid paraffin + kerosene-aqueous solution of starch syrup ($\mu_D/\mu_C = 0.87$). Solid line shows the theoretical wavelength calculated from Equation 12 [19].

obtained under the condition of laminar jetting. The diameters of drops from laminar jets slightly decrease with decreasing relative velocity.

Figure 37 shows the effect of relative velocity on the wavelength of disturbances. Data indicate that the wavelength is independent of the relative velocity. Thus the dependence of drop diameter on relative velocity can be inferred to be due not to the hydrodynamic effect of ambient liquids on stability but to the velocity distribution or the subsequent change in jet diameter. It is remarked that the drag force of ambient liquids has little effect on the wavelength of disturbances, although their growth rate depends on it.

STABILITY OF JET IN NON-NEWTONIAN-NEWTONIAN LIQUID SYSTEMS

The non-Newtonian effect on the stability of a jet has been mainly analyzed for liquid-in-air systems. Middleman [79] first presented a stability analysis for a capillary jet in a medium which has no influence on stability based on a three-constant linearized Oldroyed fluid. The linear viscoelastic model used in the analysis obeys

$$\tau_{ij} + \lambda_1 \frac{\partial \tau_{ij}}{\partial t} = -\mu\left(e_{ij} + \lambda_2 \frac{\partial e_{ij}}{\partial t}\right) \tag{79}$$

Middleman showed that the characteristic equation for the growth rate of disturbances of a viscoelastic jet is identical to that for the Newtonian jet with μ replaced by μ_1.

$$\mu_1 = \mu \frac{1 + q\lambda_2}{1 + q\lambda_1} \tag{80}$$

where τ_{ij} = stress tensor
e_{ij} = rate of deformation tensor
λ_1 and λ_2 = viscoelastic constants
μ = Newtonian viscosity

Then, the ratio of breakup length of viscoelastic to Newtonian jets is obtained as

$$\frac{\ell(\text{for viscoelastic jet})}{\ell(\text{for Newtonian jet})} = 1 - \frac{(\lambda_1 - \lambda_2)u_D}{a} \frac{1}{\sqrt{We}(1 + 3Z_D)} \tag{81}$$

For real fluids whose behavior can be approximated by Equation 79, λ_1 is greater than λ_2 [80]. Hence the laminar jets of a linear viscoelastic fluid are less stable than Newtonian jets.

Goldin et al. [81] extended this analysis to a general viscoelastic material to show a lower stability as compared to a Newtonian fluid of the same zero shear viscosity. They also confirmed the analytical prediction by experiments in weakly elastic fluids. Breakup lengths of jets of viscoelastic liquid, of power law liquid, and of liquid having a yield stress were experimentally measured by Kroesser and Middleman [82], Lenczyk and Kiser [83], and Goldin et al. [84], respectively. It is confirmed in their experiments that breakup lengths of non-Newtonian jets in air are shorter than those of Newtonian jets.

Skelland and Raval [85] studied drop formation in immiscible non-Newtonian liquid systems at low injection velocities. Shirotsuka and Kawase [86, 87] investigated experimentally the breakup of jets in immiscible non-Newtonian liquid systems. Kawase and Shirotsuka [88] proposed an approximate method for predicting drop size from laminar jets. Even in Newtonian liquid-liquid systems, however, experimental breakup data for the jet injected into a stagnant liquid disagrees with the predictions from the stability theory because of the complicated behavior of the velocity distribution in both jets and the continuous phase. The behavior of jets injected into stagnant immiscible liquids depends enormously on the viscosity ratio of both phases as described in the preceding sections. Thus the non-Newtonian characteristics of jet breakup in previous works are hidden in complicated phenomena caused by the velocity distribution inside and outside the jet.

Lee et al. [89, 90] showed theoretically that a viscoelastic liquid column in immiscible liquids is less stable than a Newtonian column. They measured the wave number of most rapidly growing disturbances for a viscoelastic polymer solution (silicone oil systems) using a Tayler's four-roller device. The experimental results are in general agreement with Tomotika's solution for the limiting case (case 3 described earlier) using the zero shear viscosity.

Recently, Kitamura and Takahashi [91] applied the jet of a zero relative-velocity system to power-law non-Newtonian-fluid systems and successfully provided the useful information about the non-Newtonian effect on the jet breakup in immiscible liquid-liquid systems as follows.

Breakup Pattern of Jet in Immiscible Non-Newtonian Liquid Systems

The experimental apparatus shown in Figure 27 was used. The non-Newtonian effects on jet breakup were examined in both Newtonian-non-Newtonian and non-Newtonian-Newtonian systems. Carboxymethyl cellulose (CMC) aqueous solutions were used as the non-Newtonian continuous phase and polystyrene and styrene butadien rubber (SBR) solutions in xylene as the non-Newtonian dispersed phase. Viscosities of polymer solutions were determined in terms of a power-law relationship with a concentric cylinder rotational viscometer. Rheological parameters m and n are defined by

$$\tau_{rz} = -m|dv_z/dr|^{n-1}(dv_z/dr) \tag{82}$$

In either a non-Newtonian jet in Newtonian liquids or a Newtonian jet in non-Newtonian liquids, surface waves for jets under the condition of $U = 0$ ($u_D = u_C$) are apparently axisymmetric and seem to satisfy Equation 3—a basic assumption of stability analysis. There are no significant differences between the breakup pattern in both non-Newtonian-Newtonian systems and that in Newtonian-Newtonian systems. For the case of $u_C = 0$ (stagnant continuous phase), however, irregular waves which seem not to obey Equation 3 prevent the reasonable comparison between both breakup patterns in Newtonian and in non-Newtonian fluid systems.

Newtonian Jet in Non-Newtonian Liquids

Figure 38 shows the breakup lengths for kerosene + paraffin − CMC aqueous solution system. In this system the breakup length under the condition of $U = 0$ is almost proportional to jet velocity, which means the maximum growth rate of disturbance is constant. Then the breakup length satisfies Equation 28.

$$\ell = \ln(a/\bar{\delta}_0)(u_D/q^*) \tag{28}$$

For immiscible Newtonian liquid systems, the maximum growth rate q^* can be numerically calculated and the coefficient $\ln(a/\bar{\delta}_0)$ can be obtained from Equation 77.

The value of viscosity is necessary in the numerical calculation of breakup lengths. The apparent viscosity of non-Newtonian fluid depends on the velocity gradient. Instead of the actual velocity gradient that it is unable to know exactly, it is reasonable to use the average velocity gradient in a glass tube (see Figure 27). However, the velocity gradient in the center of a glass tube, where the jet exists, is very small. In the power-law model, the zero shear viscosity has no meaning so the velocity gradient of $1\ s^{-1}$ may be taken as another informative value, which has the advantage of using the rheological parameter m as the apparent viscosity. Actually the existence of the jet affects the velocity gradient in the center of a tube so the real velocity gradient depends on the flow condition and is expected to be between the two informative values mentioned before.

In Figure 38 a solid line is the breakup length calculated from Equation 28 using the average velocity gradient in the glass tube, and a dotted line is that calculated using the velocity gradient of $1\ s^{-1}$. Experimental data lie between both estimates, which means that the breakup length of Newtonian jets in non-Newtonian liquids agrees with the prediction from stability analysis as long as the apparent viscosity is exactly estimated.

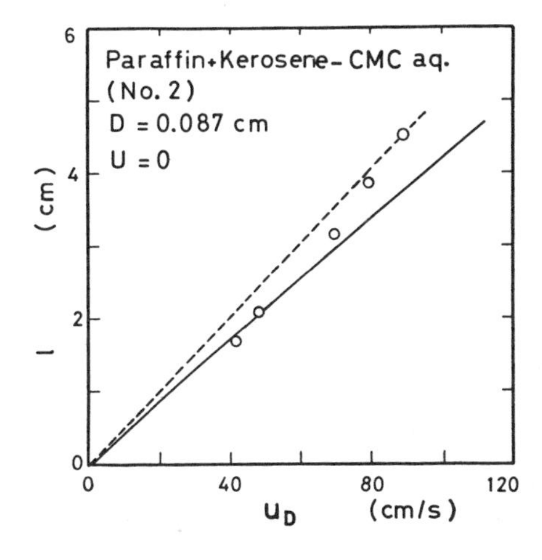

Figure 38. Breakup length of Newtonian jets in non-Newtonian liquid. Liquid paraffin + kerosene—0.12 wt% aqueous solution of CMC system: $n = 0.88$, $m = 0.0106\ kg/m\ s^{2-n}$. Solid line shows breakup length calculated from Equation 28 using the average gradient in glass tube to estimate apparent viscosity of continuous phase. Dotted line is calculated for the case that the velocity gradient is $1\ s^{-1}$ [91].

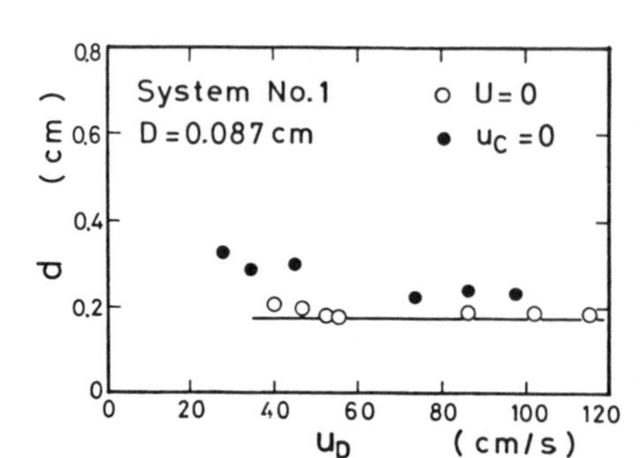

Figure 39. Diameter of drops formed from jets for n-heptane—0.25 wt% aqueous solution of CMC system: $n = 0.71$, $m = 0.028$ kg/m s^{2-n}. Solid line shows drop size calculated from Equation 33 [91].

Diameters of drops formed from Newtonian jets in non-Newtonian liquids under both conditions of $u_C = 0$ and $U = 0$ are shown in Figure 39. A solid line presents the drop diameter calculated from Equation 33 which does not contain any rheological properties of the continuous phase.

$$d = 1.91D(1 + 3Z_D)^{-1/6} \tag{33}$$

Experimental drop sizes for $U = 0$ are in good agreement with the estimate, while those for $u_C = 0$ are enormously larger than the prediction. Such a difference between the drop sizes for $U = 0$ and for $u_C = 0$ was observed even in Newtonian-Newtonian systems as mentioned before.

Non-Newtonian Jets in Newtonian Liquids

Figure 40 shows breakup lengths for two systems of nonNewtonian jets in Newtonian liquids. For the condition of $u_C = 0$, laminar breakup lengths are almost constant at low jet velocities and

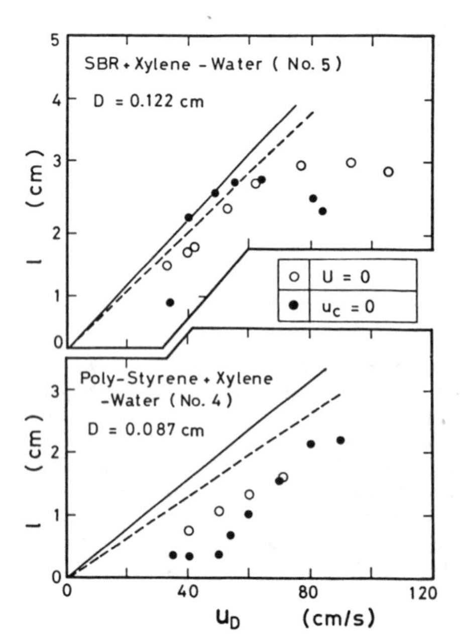

Figure 40. Breakup length of non-Newtonian jet in Newtonian liquid. SBR + xylene-water system: $n = 0.88$, $m = 0.0223$ kg/m s^{2-n}; polystyrene + xylene-water system: $n = 0.88$, $m = 0.0088$ kg/m s^{2-n}. Solid lines are calculated using the average velocity gradient in the nozzle to estimate apparent viscosity of jet. Dotted lines are calculated using the velocity gradient of 1 s^{-1} [91].

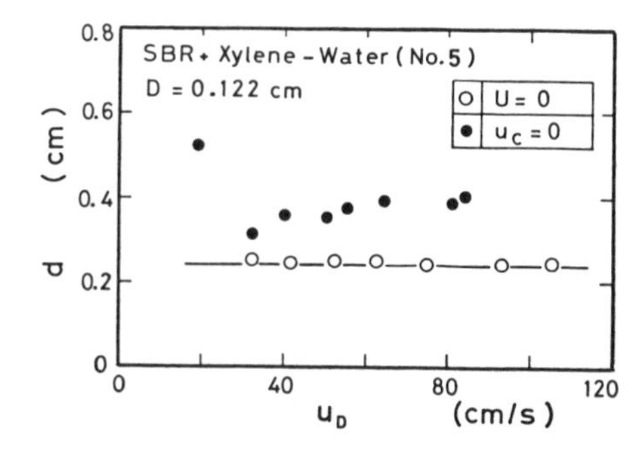

Figure 41. Diameter of drops from non-Newtonian jets for SBR + xylene-water system. Solid line shows drop sizes calculated from Equation 33 [91].

increases with increasing velocity above a critical velocity as shown typically in polystyrene + xylene − water system. Such nonlinear relations between ℓ and u_D are observed in the Newtonian-Newtonian systems where the viscosity of the dispersed phase is higher than that of the continuous phase. Linear curves of laminar breakup lengths are obtained under the condition of $U = 0$. The stability of non-Newtonian jets has to be discussed by comparing the analytical results with such data.

Solid lines in the figure are calculated from Equation 28. In the calculation the average velocity gradient in the nozzle is used to estimate the apparent viscosity of the dispersed phase. Dotted curves are informative breakup lengths calculated for the case where the rheological parameter m is used as the apparent viscosity (velocity gradient of $1\ s^{-1}$). Experimental breakup lengths are shorter than both estimates as shown in the figure, whereas data for a Newtonian jet in non-Newtonian systems are between both curves as shown in Figure 38. This indicates that non-Newtonian jets in Newtonian liquids are less stable than those in Newtonian systems. Such non-Newtonian effects in immiscible liquid systems are analogous to the stability of non-Newtonian jets in air [82, 83].

Figure 41 shows diameters of drops formed from non-Newtonian jets under both conditions of $u_C = 0$ and $U = 0$. The solid line is the diameter calculated from Equation 33. The modified ohnesorge number for power law fluids which is necessary for use in Equation 33 is given by

$$Z_D = \frac{2^{n-3}\left(3 + \frac{1}{n}\right)^n m}{u_D^{1-n}\sqrt{\rho_D D^{2n-1}\sigma}} \tag{83}$$

Experimental drop sizes for the condition of $U = 0$ are in good agreement with Equation 33 as shown in Figure 41. Although the laminar breakup lengths of non-Newtonian jets are shorter than those estimated from the stability analysis, the drop size agrees closely with the prediction from the analysis. The good agreement between experimental and theoretical drop size in non-Newtonian-Newtonian systems is partially due to the fact that the wavelength of the disturbance which grows most rapidly is insensitive to the viscosity of the jet [16].

NOTATION

a	radius of jet, m
Bo	Bond Number defined by Equation 39
D	diameter of nozzle, m
D_j	Diameter of jet given by Equations 45 and 46, m
d	diameter of drops, m
e_{ij}	rate of deformation tensor, s^{-1}
F	Harkins-Brown correction factor
Fr	Froude number defined by Equation 76
f_1, f_2	maximum of function $F_1(x)$, $F_2(x)$
g	gravitational acceleration, $m\ s^{-2}$
$I_i(x)$; $\dot{I}_i(x)$	modified Bessel function of 1st kind and of i-th order; 1st derivative with respect to x

i square of -1
$K_i(x)$; $\dot{K}_i(x)$ modified Bessel function of 2nd kind and of i-th order; 1st derivative with respect to x
k wave number $(2\pi/\lambda)$, m^{-1}
k_C; k_D modified wave number defined by Equation 6 and 7
L nozzle length, m
ℓ breakup length of jet, m
m rheology constant of power law model, $kg\ m^{-1}s^{n-2}$
n index of power law model
p_{rr} normal stress, Pa
Q dimensionless growth rate defined by Equation 14
q growth rate of disturbance, s^{-1}
q_L^* maximum growth rate for liquid-liquid jet, s^{-1}
q_G^* maximum growth rate for liquid jet in air, s^{-1}
q_I^* maximum growth rate for inviscid jet in air, s^{-1}
q_V^* maximum growth rate for highly viscous jet in air, s^{-1}
q_{II}^* maximum growth rate for inviscid-inviscid liquid systems, s^{-1}
q_{VV}^* maximum growth rate for highly viscous-highly viscous liquid system, s^{-1}
R rotating radius of nozzle; distance between nozzle exit and rotating axis, m
Re Reynolds number based on critical velocity $(=\rho_D D u_k/\mu_D)$
r radial coordinate, m
T breakup time, s
t time, s
U jet velocity relative to continuous phase $(= u_D - u_C)$, m/s
u_C velocity of continuous phase at center of tube, m/s
u_D injection velocity of jet, m/s
u_j jetting velocity, m/s
u_k critical velocity from laminar to tubulent jet, m/s
u_M characteristic velocity at minimum drop size, m/s
V flow rate of dispersed phase, m^3/s
v velocity component, m/s
We Weber number defined by Equation 75
We_A Weber number based on air density $(= \rho_A D U^2/2\sigma)$
We_C Weber number based on continuous phase $(= \rho_C D U^2/2\sigma)$
We_j Weber number based on jetting velocity $(= \rho_D D U_j^2/\sigma)$
x dimensionless wave number $(= ka)$
y, y' dimensionless number defined by Equation 15 $(= k_C a;\ k_D a)$
Z Ohnesorge number $(= \mu\sqrt{\rho D\sigma})$
z axial coordinate, m

Greek Symbols

δ disturbance on jet surface, m
$\bar{\delta}_0$ initial amplitude of disturbance, m
λ wavelength of disturbance, m
λ_1, λ_2 viscoelastic constants, s
μ viscosity of liquid, Pa s
$\hat{\mu}$ viscosity ratio $= \mu_D/\mu_C$
ν kinematic viscosity of liquid, $m^2\ s^{-1}$
ρ density of liquid, kg/m^3
$\hat{\rho}$ density ratio $= \rho_D/\rho_C$
σ interfacial tension, N/m
τ shear stress, Pa
ϕ_1, ϕ_2 correction factors defined by Equations 30 and 31
ψ stream function
ω angular velocity of nozzle, s^{-1}

Superscripts

* refers to disturbance that grows most rapidly
refers to inside of jet

Subscripts

A refers to air
C continuous phase
D dispersed phase
r radial component
z axial component

REFERENCES

1. Savart, F., *Ann. de Chimie*, 53:337(1833), cited by Lord Rayleigh, "Theory of Sound" *Vol. 2*, p. 362, Dover, New York, 1945.
2. Plateau, J., "Statique experimentale et théorique des liquides soumis aux seules forces moleculaires," Paris, 1873, cited by Lord Rayleigh, "Theory of Sound" *Vol. 2*, p. 363, Dover, New York, 1945.
3. Rayleigh, Lord, *Proc. London Math. Soc.*, 10:4 (1878). "Theory of Sound" Vol. 2, p. 351, Dover, New York, 1945.
4. Weber, C., and Angew, Z., *Math. Mech.*, 11:136 (1931).
5. Tomotika, S., *Proc. Roy. Soc.*, A150:322 (1935).
6. Tomotika, S., *Proc. Physico-Math. Soc.*, Japan, 18:550 (1936).
7. Donnelly, R. J., and Glaberson, W., *Proc. Roy. Soc*, A290:547 (1966).
8. Wang, D. P., *J. Fluid Mech.*, 34:299 (1968).
9. Yuen, M., *J. Fluid Mech.*, 33:151 (1968).
10. Rutland, D. F., and Jameson, G. J., *Chem. Eng. Sci.*, 25:1689 (1970).
11. Rutland, D. F., and Jameson, G. J., *J. Fluid Mech.*, 46:267 (1971).
12. Goedde, E. F., and Yuen, M. C., *J. Fluid Mech.*, 40:495 (1970).
13. Levich, V. G., "Physicochemical Hydrodynamics," p. 626, Prentice Hall Inc., Englwood Cliffs, N.J., 1962.
14. Chandrasechar, S., "Hydrodynamic and Hydromagnetic Stability," p. 515, Oxford University Press, London, 1961.
15. McCarthy, M. J., and Molloy, N. A., *Chem. Eng. Journal*, 7:1 (1974).
16. Meister, B. J., and Scheele, G. F., *A.I.Ch.E. Journal*, 13:682 (1967).
17. Takahashi, T., and Kitamura, Y., *Kagaku Kogaku*, 35:637 (1971).
18. Christiansen, R. M., and Hixson, A. N., *Ind. Eng. Chem.*, 49:1017 (1957).
19. Kitamura, Y., Mishima, H., and Takahashi, T., *Can. J. Chem. Eng.*, 60:723 (1982).
20. Grant, R. P., and Middleman, S., *A.I.Ch.E. Journal*, 12:669 (1966).
21. Tyler, E., *Phil. Mag.*, 16:504 (1933).
22. Takahashi, T., and Kitamura, Y., *Kagaku Kogaku*, 35:1229 (1971).
23. Takahashi, T., and Kitamura, Y., *Kagaku Kogaku*, 36:527 (1972).
24. Tomotika, S., *Proc. Roy. Soc.*, A153:302 (1936).
25. Taylor, G. I., *Proc. Roy. Soc.*, A146:501 (1934).
26. Rumscheidt, F. D., and Mason, S. G., *J. Colloid Sci.*, 17:260 (1962).
27. Smith, S. W., and Moss, H., *Proc. Roy. Soc.*, A93:373 (1919).
28. Tyler, E., and Watkin, F., *Phil. Mag.*, 14:849 (1932).
29. Merrington, A. C., and Richardson, E. G., *Proc. Phys. Soc.*, 59:1 (1947).
30. Hayworth, C. B., and Treybal, R. E., *Ind. Eng. Chem.*, 42:1174 (1950).
31. Keith, F. W., and Hixson, A. N., *Ind. Eng. Chem.*, 47:258 (1955).
32. Null, H. R., and Johnson, H. F., *A.I.Ch.E. Journal*, 4:273 (1958).
33. Scheele, G. F., and Meister, B. J., *A.I.Ch.E. Journal*, 14:9 (1968).
34. Chazal, L. E. M., and Ryan, J. T., *A.I.Ch.E. Journal*, 17:1226 (1971).
35. Perrut, M., and Loutaty, R., *Chem. Eng. Journal*, 3:287 (1972).
36. Hozawa, M., Tadaki, T., and Maeda, S., *Kagaku Kogaku*, 33:893 (1969).
37. Hozawa, M., and Tadaki, T., *Kagaku Kogaku*, 37:827 (1973).
38. Skelland, A. H. P., and Johnson, K. R., *Can. J. Chem. Eng.*, 52:732 (1974).
39. Rao, E. V. L. N., Kumar, R., and Kuloor, N. R., *Chem. Eng. Sci.*, 21:867 (1966).
40. Vedaiyan, S., et al., *A.I.Ch.E. Journal*, 18:161 (1972).
41. Kintner, R. C., "Advance in Chemical Engineering" *Vol. 4*, p. 51, Academic Press, New York, 1963.
42. Tavlarides, L. L., et al., *Ind. Eng. Chem.*, 62:6 (1970).
43. Heertjes, P. M., and De Nie, L. H., "Recent Advances in Liquid-Liquid Extraction," C. Hanson (Ed.), p. 367, Pergamon Press, Oxford, 1971.
44. Kumar, R., and Kuloor, N. R., "Advances in Chemical Engineering Vol. 8" p. 256, Academic Press, New York, 1970.

45. Clift, R., Grace, J. R., and Weber, M. E., "Bubbles, Drops and Particles," Academic Press, New York, 1978.
46. Fujinawa, K., Maruyama, T., and Nakaike, Y., *Kagaku Kogaku*, 21:194 (1957).
47. Ranz, W. E., *Can. J. Chem. Eng.*, 36:175 (1958).
48. Ranz, W. E., and Dreier, W. M., *Ind. Eng. Chem., Fundam.*, 3:53 (1964).
49. Meister, B. J., and Scheele, G. F., *A.I.Ch.E. Journal*, 15:689 (1969).
50. Meister, B. J., and Scheele, G. F., *A.I.Ch.E. Journal*, 15:700 (1969).
51. Takahashi, T., and Kitamura, Y., *Kagaku Kogaku*, 36:912 (1972).
52. Bayens, C. A., and Laurence, R. L., *Ind. Eng. Chem., Fundam.*, 7:521 (1968).
53. Sawistowski, H., *Chem. Ing. Tech.*, 45:1114 (1973).
54. Dżuber, I., and Sawistowski, H., *Proc. Solv. Ext. Conf.*, 1:379 (1974).
55. Burkholder, H. C., and Berg, J. C., *A.I.Ch.E. Journal*, 20:863 (1974).
56. Skelland, A. H. P., and Huang, Y. F., *A.I.Ch.E. Journal*, 23:701 (1977).
57. Skelland, A. H. P., and Huang, Y. F., *A.I.Ch.E. Journal*, 25:80 (1979).
58. Coyle, R. W., Berg, J. C., and Niwa, J. C., *Chem. Eng. Sci.*, 36:19 (1981).
59. Nealson, N. K., and Berg, J. C., *Chem. Eng. Sci.*, 37:1067 (1982).
60. Kitamura, Y., and Takahashi, T., *Proc. 1st Int. Conf. Liquid Atomization and Spray Systems*, 1 (1978).
61. Harkins, W. D., and Brown, F. E., *J. Am. Chem. Soc.*, 41:499 (1919).
62. Lehrer, I. H., *Ind. Eng. Chem., Process Des. Dev.*, 18:297 (1979).
63. Tanazawa, Y., and Toyoda, S., *Trans. J.S.M.E.*, 20:299 (1954).
64. Kitamura, Y., Doctoral dissertation, Kyoto Univ., 1975.
65. Treybal, R. E., "Liquid Extraction," 2nd ed., p. 467, p. 530, McGraw Hill, New York, 1963.
66. Tanazawa, Y., and Toyoda, S., *Trans. J.S.M.E.*, 20:306 (1954).
67. Sterling, A. M., and Sleicher, C. A., *J. Fluid Mech.*, 68:477 (1975).
68. Sawistowski, H., "Recent Advances in Liquid-Liquid Extraction," C. Hanson, (Ed.), p. 360, Pergamon Press, Oxford, 1971.
69. Logsdail, D. H., and Lowes, L., "Recent Advances in Liquid-Liquid Extraction," C. Hanson (Ed.), p. 139, Pergamon Press, Oxford, 1971.
70. Barson, N., and Beyer, G. H., *Chem. Eng. Prog.*, 49:243 (1953).
71. Jacobsen, F. M., and Beyer, G. H., *A.I.Ch.E. Journal*, 2:283 (1956).
72. Takahashi, T., Kitamura, Y., and Iwamoto, T., *Kagaku Kogaku Ronbunshu*, 3:632 (1977).
73. Rumscheidt, F. D., and Mason, S. G., *J. Colloid Sci.*, 16:238 (1961).
74. Mikami, T., Cox, R. G., and Mason, S. G., *Int. J. Multiphase Flow*, 2:113 (1975).
75. Mikami, T., and Mason, S. G., *Can. J. Chem. Eng.*, 53:372 (1975).
76. Laufer, J., NACA Tech. Report, No 1174 (1954).
77. Fenn, R., and Middleman, S., *A.I.Ch.E. Journal*, 15:379 (1969).
78. Chow, T. S., and Hermans, J. J., *Phys. Fluid*, 14:244 (1971).
79. Middleman, S., *Chem. Eng. Sci.*, 20:1037 (1965).
80. Oldroyd, J. G., *Proc. Roy. Soc.*, A200:523 (1950).
81. Goldin, M., et al., *Fluid Mech.*, 38:689 (1969).
82. Kroesser, F. W., and Middleman, S., *A.I.Ch.E. Journal*, 15:383 (1969).
83. Lenczyk, J. P., and Kiser, K. M., *A.I.Ch.E. Journal*, 17:826 (1971).
84. Godin, M., Pfeffer, R., and Shinnar, R., *Chem. Eng. J.*, 4:8 (1972).
85. Skelland, A. H. P., and Raval, V. K., *Can. J. Chem. Eng.*, 50:41 (1972).
86. Shirotsuka, T., and Kawase, Y., *Kagaku Kogaku Ronbunshu*, 1:219 (1975).
87. Shirotsuka, T., and Kawase, Y., *Kagaku Kogaku Ronbunshu*, 1:652 (1975).
88. Kawase, Y., and Shirotsuka, T., *Proc. 1st Int. Conf. Liquid Atomization and Spray Systems*, 21 (1978).
89. Lee, W. K., and Flumerfelt, R. W., *Int. J. Multiphase Flow*, 7:363 (1981).
90. Lee, W. K., Yu, K. L., and Flumerfelt, R. W., *Int. J. Multiphase Flow*, 7:385 (1981).
91. Kitamura, Y., and Takahashi, T., *Can. J. Chem. Eng.*, 60:732 (1982).
92. Haenlein, A., *Forshung auf dem Gebiete des Ingenieurwesens*, 2:139 (1931), NACA Tech Memo No. 659 (1932), cited by C. Weber, Z. Angew., *Math. Mech.*, 11:136 (1931).

CHAPTER 20

MODELING TURBULENT JETS WITH VARIABLE DENSITY

Kunio Kataoka

Department of Chemical Engineering
Kobe University
Rokkodai, Kobe, Japan

CONTENTS

INTRODUCTION, 511

OVERVIEW—GENERAL CHARACTER OF TURBULENT FREE JETS, 512

FUNDAMENTAL EQUATIONS, 516

EMPIRICAL MODELS OF JET DEVELOPMENT, 517
Core Length, 518
Centerline Decay, 520
Radial Spreading, 524
Radial Profile, 525

MODELS OF VARIABLE-DENSITY FREE TURBULENT MIXING, 527
Simple Explanation of Turbulent Mixing, 528
Warren and Donaldson-Gray Model, 529
Ferri-Kleinstein Model, 529
Schetz-Zelazny Model, 530
Other Modifying Models, 531

TURBULENT KINETIC ENERGY MODEL, 538

NOTATION, 541

REFERENCES, 542

INTRODUCTION

When a hot free jet is discharged from a nozzle into quiescent cold surroundings, the velocity and temperature on the jet axis decrease with increasing axial distance from the nozzle whereas the jet width continues to increase. It is of great importance in aerospace, chemical, and mechanical technology to predict local variations of velocity and temperature (and/or concentration) in nonisothermal (or compressible) turbulent free jets at various temperature levels. However, the great bulk of experimental data is for unheated or incompressible jets. The jet development is characterized basically by the width of the turbulent mixing and the decay rates of centerline properties.

When there is a considerable difference in fluid density between a jet and its ambient-receiving medium, the jet development is influenced greatly by the change in fluid density during the turbulent mixing process. For example, a decrease of jet density with respect to that of the ambient-receiving medium causes an increase in both the rates of centerline decay and radial spreading.

This chapter is concerned with modeling such variable-density turbulent-free jets. We begin with an overview of the general character of turbulent-free jets to be followed by a section on the

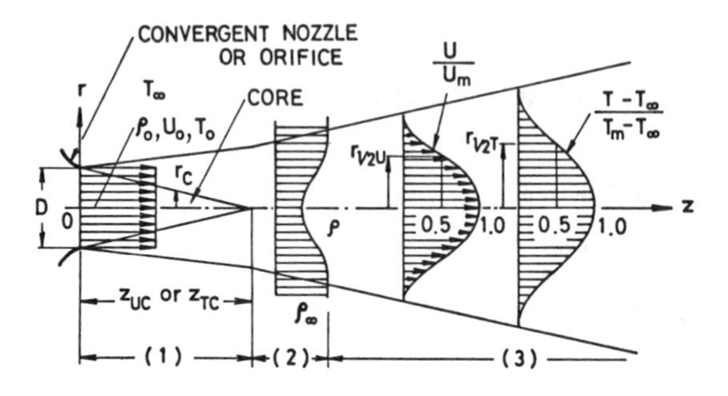

Figure 1. Schematic diagram of heated turbulent free jet and the coordinate system to be used: (1) potential core region, (2) transition region, and (3) fully-developed region.

derivation of the fundamental equations. The next section discusses empirical models of the development of nonisothermal free jets with variable density for their practical utilization. These models are useful for predicting the distributions of velocity, temperature, and/or concentration at any axial station where hot-wire anemometry does not work well owing to large temperature gradients. The subsequent section discusses semi-empirical models for prediction of the turbulent free mixing with variable density. Those models are based upon the gradient transport modeling of turbulent fluxes along the lines of Prandtl's incompressible turbulent diffusion model. Finally, the last section explains the turbulent kinetic energy model, although it is not widely used for variable-density jets.

OVERVIEW—GENERAL CHARACTER OF TURBULENT FREE JETS

A schematic diagram of a nonisothermal, compressible free jet is shown in Figure 1 with the coordinate system to be used.

The jet, in general, consists of three separate flow regions:

1. Potential core region.
2. Transition region.
3. Fully-developed region.

Close to the jet exit, there is a potential core region where the centerline velocity and temperature remain unchanged from their initial values. The core diminishes in width in a downstream direction due to the turbulent mixing which is initiated at the periphery of the jet. The core region extends to the point at which the turbulent mixing reaches the jet axis. The distance from the nozzle exit to this point is known as the core length. Downstream of the core region there is a short transition region where the velocity and temperature profiles have not been fully established in the cross section. Further downstream, the profiles become similar in shape, changing only in scale. In this sense, this region is called "fully-developed." The velocity distribution becomes similar at approximately six nozzle diameters downstream from the nozzle for the case of axisymmetric incompressible jets.*

Figure 2 is a photograph of an isothermal air jet visualized by a smoke-wire technique. The white lines indicate the streaklines of smoke. Ring-shaped vortices occur at about a nozzle diameter downstream of the jet exit owing to velocity difference between the jet and stationary surroundings. The inner jet region is affected by the outer intermittent vortex motion. The free jet becomes fully turbulent at approximately six nozzle diameters after the coherent vortex structure becomes invisible.

For the case of axisymmetric incompressible jets, the centerline decay is proportional to $1/z$ in the fully-developed region whereas the radial spreading is proportional to z. For the case of two-dimensional incompressible jets, the centerline decay is proportional to $1/\sqrt{z}$ whereas the lateral

* Strictly speaking, even in the fully-developed region, the turbulent properties have not yet been self-preserving up to fifteen to twenty nozzle diameters downstream from the jet exit.

Figure 2. Photograph of an isothermal air jet visualized by smoke-wire method: nozzle diameter D = 28 mm; $Re_0 = 2000$.

spreading is proportional to z. The jet width is characterized by the so-called half-radius. This is the radial distance from the jet axis, at which a jet property becomes just one-half of the centerline value.

Figures 3 and 4 show the centerline decay and radial spreading of axisymmetric gas jets observed by Kataoka, Shundoh, and Matsuo [1] at different jet-to-ambient temperature ratios. Their convergent nozzle has an exit diameter D = 10 mm and a contraction ratio 1/36. It can be seen from these figures that the higher-temperature jets have shorter core lengths and more rapid decay than the lower-temperature jets.

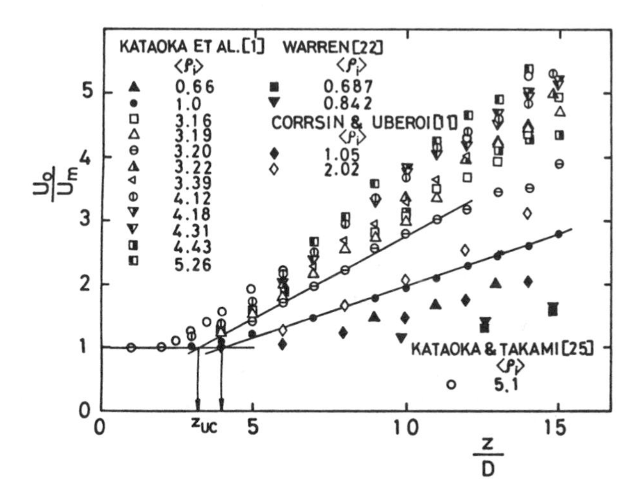

Figure 3. Axial decay of centerline velocity and definition of velocity core length [1].

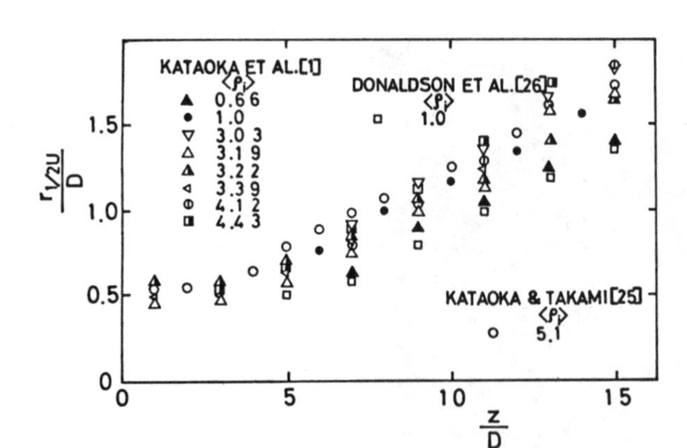

Figure 4. Axial variation of jet half-radius for velocity [1].

There has been unfortunately little work on turbulence modeling in variable-density nonisothermal jets. Prandtl's incompressible turbulent mixing model [2] gives an eddy diffusivity of the form

$$\epsilon_M = K_p b |U_m - U_\infty| \tag{1}$$

where b = the width of the mixing region
K_p = an empirical constant called the mixing rate parameter

For a single round jet exhausting into a stationary ambient (i.e. $U_\infty = 0$), $b \sim r_{1/2} \sim z$ and $U_m \sim 1/z$ in the fully-developed region. Hence Equation 1 suggests that ϵ_M remains constant over the whole fully-developed region.

The boundary layer theory [3] gives the following similar incompressible velocity distribution using the constant eddy diffusivity model:

$$\frac{U}{U_m} = \frac{1}{(1 + \xi^2/4)^2} \tag{2}$$

where $\xi = (3/16\pi)^{1/2} \dfrac{M_t^{1/2}}{\epsilon_M} \dfrac{r}{z}$

$$U_m = (3/8\pi) \frac{M_t}{\epsilon_M z}$$

$$M_t = 2\pi \int_0^\infty U^2 r \, dr$$

This equation is formally identical with that for the laminar jet if the eddy diffusivity is replaced by the kinematic viscosity.

In a similar manner, Hinze [4] treated a general case where a jet issues at a velocity U_0 from a circular orifice into a stream of uniform velocity U_∞. In the case $(U - U_\infty)/U_\infty \gg 1$, the spread of the jet is linear with $(z + a)$ and the velocity decreases with $(z + a)^{-1}$. The centerline velocity decay can be described by

$$\frac{U_0 - U_\infty}{U_m - U_\infty} = A \frac{z + a}{D} \tag{3}$$

where A = an empirical constant
a = a virtual origin

The similar distribution of the velocity difference is given by

$$\frac{U - U_\infty}{U_m - U_\infty} = \left[1 + \frac{U_m}{8\epsilon_M(z + a)} r^2\right]^{-2} \tag{4}$$

When $U_\infty = 0$, Equation 4 becomes identical with Equation 2. If the ratio of eddy diffusivities ϵ_H/ϵ_M is assumed constant, the similar distribution of temperature difference is given by

$$\frac{T - T_\infty}{T_m - T_\infty} = \left(\frac{U - U_\infty}{U_m - U_\infty}\right)^{\epsilon_H/\epsilon_M} \tag{5}$$

In the case $U - U_\infty/U_\infty \ll 1$, the spread of the jet is proportional to $(z + a)^{1/3}$ and the centerline velocity decreases at a rate of $(z + a)^{-2/3}$. If ϵ_M is assumed to depend only upon $(z + a)^{-1/3}$, the velocity distribution becomes of the form of the Gaussian error function:

$$\frac{U - U_\infty}{U_m - U_\infty} = \exp\left[-\frac{U_\infty}{6\epsilon_M(z + a)} r^2\right] \tag{6}$$

Using the experiments of an air jet issuing into stagnant air, Hinze and van der Hegge Zijnen [5] give the following relationships

$$\frac{U_0}{U_m} = 0.156 \frac{z + a}{D} \tag{7}$$

$$\frac{\epsilon_M}{U_0 D} = 0.013 \tag{8}$$

They also obtained an average value of $\epsilon_H/\epsilon_M \simeq 1.36$ from their experiments of heated air jets and found the ratio ϵ_D/ϵ_M to be equal to ϵ_H/ϵ_M from their measurements of 1% city gas in air.

Keagy and Weller [6] found that $\epsilon_D/\epsilon_M \simeq 1.38$ in their experiments of a nitrogen jet issuing into stagnant air. Ruden [7] and Reichardt [8] report similar air jet measurements. Forstall and Gaylord [9] and Kiser [10] made similar measurements in a submerged water jet by using sodium chloride as a tracer.

Corrsin and Uberoi [11] measured momentum and heat transport in two heated turbulent air jets. The initial temperature of the hotter jet was 300°C above the ambient temperature. They found that the potential core length decreased as the initial temperature was raised and that the hotter jet had more rapid decay and spreading than incompressible jets. In such variable density jets, we should consider the effect of density change on the turbulent mixing.

As the characteristic density parameter, we usually adopt the initial density ratio defined by

$$\langle \rho_i \rangle = \frac{\rho_\infty}{\rho_0} \tag{9}$$

This is equal to the reciprocal T_0/T_∞ of the initial temperature ratio between jet and ambient for the case of submerged nonisothermal gas jets.

Thring and Newby [12] and Sunavala, Hulse, and Thring [13] found that the axial decay of centerline properties should be correlated using the streamwise coordinate nondimensionalized by an equivalent diameter

$$D_{eq} = \frac{D}{\sqrt{\langle \rho_i \rangle}} \tag{10}$$

The main purpose of this chapter is how to model such turbulent variable-density jets taking into consideration the effect of density change.

FUNDAMENTAL EQUATIONS

It is very difficult to predict local variation of jet properties in turbulent variable-density jets because of a lack of experimental data. The boundary-layer equation for instantaneous quantities u, v, h_s can be applied to such an axisymmetric jet in the absence of pressure gradients:

$$\frac{\partial}{\partial z}(\rho u r) + \frac{\partial}{\partial r}(\rho v r) = 0 \tag{11}$$

$$\rho u \frac{\partial u}{\partial z} + \rho v \frac{\partial u}{\partial r} = \frac{1}{r}\frac{\partial}{\partial r}\left(r\mu \frac{\partial u}{\partial r}\right) \tag{12}$$

$$\rho u \frac{\partial h_s}{\partial z} + \rho v \frac{\partial h_s}{\partial r} = \frac{1}{r}\frac{\partial}{\partial r}\left(r \frac{\mu}{Pr}\frac{\partial h_s}{\partial r}\right) \tag{13}$$

Here the viscous dissipation and pressure work terms have been neglected in the last equation. The stagnation enthalpy h_s is the sum of local enthalpy and enthalpy rise due to adiabatic compression:

$$h_s = Cp\,T + \tfrac{1}{2}u^2 \tag{14}$$

First, our discussion in this section is restricted to low Mach number subsonic jets where the effect of compressibility on the velocity and enthalpy changes is neglected. The last equation then reduces to

$$\rho\, Cp\, u \frac{\partial T}{\partial z} + \rho\, Cp\, v \frac{\partial T}{\partial r} = \frac{1}{r}\frac{\partial}{\partial r}\left(r\kappa \frac{\partial T}{\partial r}\right) \tag{15}$$

For the case of turbulent jet, the molecular transport is unimportant compared with the turbulent transport. If all quantities are represented as the sum of a time-averaged and a fluctuating component, the time-averaged equations of continuity, momentum, and enthalpy become

$$\frac{\partial \bar{\rho}\bar{u}r}{\partial z} + \frac{\partial}{\partial r}(\bar{\rho}\bar{v}r + \overline{\rho' v'}r) = 0 \tag{16}$$

$$\bar{\rho}\bar{u}r \frac{\partial \bar{u}}{\partial z} + (\bar{\rho}\bar{v} + \overline{\rho' v'})r \frac{\partial \bar{u}}{\partial r} = -\frac{\partial}{\partial r}(r\bar{\rho}\overline{u'v'}) \tag{17}$$

$$\bar{\rho}\bar{u}\, Cp\, r \frac{\partial \bar{T}}{\partial z} + (\bar{\rho}\bar{v} + \overline{\rho' v'})\, Cp\, r \frac{\partial \bar{T}}{\partial r} = -\frac{\partial}{\partial r}(r\bar{\rho}\, Cp\, \overline{v'T'}) \tag{18}$$

In the derivation of those equations, the following assumptions were made:

$$\bar{v}\overline{\rho' u'} \ll \bar{u}\overline{\rho' v'} \quad \text{and} \quad \frac{\partial \overline{\rho' u'}}{\partial z} \ll \frac{\partial \overline{\rho' v'}}{\partial r}$$

The overbar and prime imply the time-averaged and fluctuating quantities, respectively. If we eliminate $\bar{\rho}\bar{v} + \overline{\rho' v'}$ by means of the equation of continuity, we obtain

$$r\bar{\rho}\bar{u} \frac{\partial \bar{u}}{\partial z} - \frac{\partial \bar{u}}{\partial r}\int_0^r \frac{\partial}{\partial z}(r\bar{\rho}\bar{u})\, dr = -\frac{\partial}{\partial r}(r\bar{\rho}\overline{u'v'}) \tag{19}$$

$$r\bar{\rho}\bar{u}\, Cp \frac{\partial \bar{T}}{\partial z} - Cp \frac{\partial \bar{T}}{\partial r}\int_0^r \frac{\partial}{\partial z}(r\bar{\rho}\bar{u})\, dr = -\frac{\partial}{\partial r}(r\bar{\rho}\, Cp\, \overline{v'T'}) \tag{20}$$

In a similar fashion, we obtain the equation of mass conservation given by

$$r\bar{\rho}\bar{u}\frac{\partial\bar{\gamma}_i}{\partial z} - \frac{\partial\bar{\gamma}_i}{\partial r}\int_0^r \frac{\partial}{\partial z}(r\bar{\rho}\bar{u})\,dr = -\frac{\partial}{\partial r}(r\bar{\rho}\overline{v'\gamma_i'}) \tag{21}$$

where γ_i is the mass fraction of i-component gas.

Those turbulent fluxes are too difficult to measure precisely by means of conventional experimental techniques. Usually they are related to local gradient of the time-averaged quantity by the expression

$$-\bar{\rho}\overline{u'v'} = \bar{\rho}\epsilon_M\frac{\partial\bar{u}}{\partial r} \tag{22}$$

$$-\bar{\rho}\,Cp\,\overline{v'T'} = \bar{\rho}\,Cp\,\epsilon_H\frac{\partial\bar{T}}{\partial r} \tag{23}$$

$$-\bar{\rho}\overline{v'\gamma_i'} = \bar{\rho}\epsilon_D\frac{\partial\bar{\gamma}_i}{\partial r} \tag{24}$$

Here ϵ_M, ϵ_H, and ϵ_D are known as the eddy diffusivities based upon Prandtl's mixing length theory [14]. For the case of a variable-density jet, we should treat those coefficients in connection with a local variation of the time-averaged fluid density.

For the case of turbulent compressible jets at high subsonic Mach numbers, the equation of enthalpy is given by

$$r\bar{\rho}\bar{u}\frac{\partial\bar{h}_s}{\partial z} - \frac{\partial\bar{h}_s}{\partial r}\int_0^r \frac{\partial}{\partial z}(r\bar{\rho}\bar{u})\,dr = -\frac{\partial}{\partial r}(r\bar{\rho}\overline{v'h_s'}) \tag{25}$$

Actually the velocity field is closely coupled to the temperature field. We can usually make an assumption that the following relation, called the Crocco integral [15], holds throughout the mixing region:

$$h_s = h + \tfrac{1}{2}u^2 = Au + B \tag{26}$$

For the case of a two-dimensional (plane) jet, the governing equations can be obtained in a similar manner:

$$\bar{\rho}\bar{u}\frac{\partial\bar{u}}{\partial z} - \frac{\partial\bar{u}}{\partial y}\int_0^y \frac{\partial}{\partial z}(\bar{\rho}\bar{u})\,dy = -\frac{\partial}{\partial y}(\bar{\rho}\overline{u'v'}) \tag{27}$$

$$\bar{\rho}\bar{u}\,Cp\frac{\partial\bar{T}}{\partial z} - Cp\frac{\partial\bar{T}}{\partial y}\int_0^y \frac{\partial}{\partial z}(\bar{\rho}\bar{u})\,dy = -\frac{\partial}{\partial y}(\bar{\rho}\,Cp\,\overline{v'T'}) \tag{28}$$

$$\bar{\rho}\bar{u}\frac{\partial\bar{\gamma}_i}{\partial z} - \frac{\partial\bar{\gamma}_i}{\partial y}\int_0^y \frac{\partial}{\partial z}(\bar{\rho}\bar{u})\,dy = -\frac{\partial}{\partial y}(\bar{\rho}\overline{v'\gamma_i'}) \tag{29}$$

In the following sections, for convenience, we use ρ, U, V, T, Γ, H in place of $\bar{\rho}$, $\bar{u}$, $\bar{v}$, $\bar{T}$, $\bar{\gamma}_i$, $\bar{h}$.

EMPIRICAL MODELS OF JET DEVELOPMENT

When a free jet of hot gas issues into a stagnant environment of cold gas, the turbulent mixing with density change promotes the rates of radial spreading and axial decay. If one has a jet of hydrogen exhausting into air, one can have similar variable-density turbulent jet. The practical calculation is often reduced to searching for the distribution of velocity and temperature along the jet axis and to determining the width and centerline quantities of the jet. The objective of this section is to consider a generalized model predicting the jet development over a wide range of the initial density ratio.

We should adopt $U_m - U_\infty$, $T_m - T_\infty$, and $r_{1/2U}$, $r_{1/2T}$ as the characteristic variables. As mentioned for the case of nonisothermal variable-density jets, the initial density ratio, i.e. $\langle \rho_i \rangle = \rho_\infty/\rho_0$ is the most appropriate parameter. In what follows, we discuss mainly how to establish a generalized model of variable-density jets at low Mach numbers. For distances not too far from the origin of the jets, the buoyancy effect may be small compared with the forced convection.

Core Length

The first step in analyzing the jet development is to determine the core length (or virtual origin) for standardizing the jet development length. This chapter adopts the core length in place of the virtual origin because the core length correlations are more useful from the practical viewpoint. It is one more reason that there is little work done systematically on the effect of initial density ratio on the virtual origin.

Kataoka, Shundoh, and Matsuo [1] measured local variation of the velocity and temperature in high temperature jets of the burned gas produced by premixed combustion of methane. They used an unheated CO_2 jet into air for the supplementary experiment. The initial density ratio ranged from 0.66 to 5.26.

Figure 3 shows the centerline behavior of axial velocity [1], nondimensionalized with respect to the nozzle exit value, for the case of heated jets into stagnant cold air. Here the reciprocal of axial velocity U_0/U_m is used to emphasize a linear decay.

The velocity core length z_{UC}, defined within the figure, tends to decrease with the initial density ratio raised. The jet development is influenced by the nonuniformity of the initial velocity and temperature profiles. Up to the present, it is too difficult to predict the decay and spreading rates of a jet issuing from an arbitrarily shaped exit. In this chapter, we adopt the experimental data obtained by convergent nozzles and orifices, the exit of which has almost uniform profiles of flow variables.

Figure 5 shows the centerline decay of the temperature difference [1]. By comparison between these two figures, it can be seen that heat is mixed faster than momentum. The temperature core

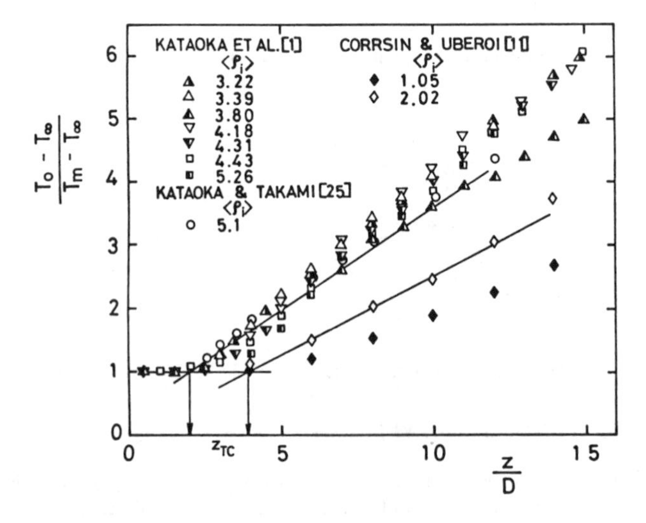

Figure 5. Axial decay of centerline temperature and definition of temperature core length [1].

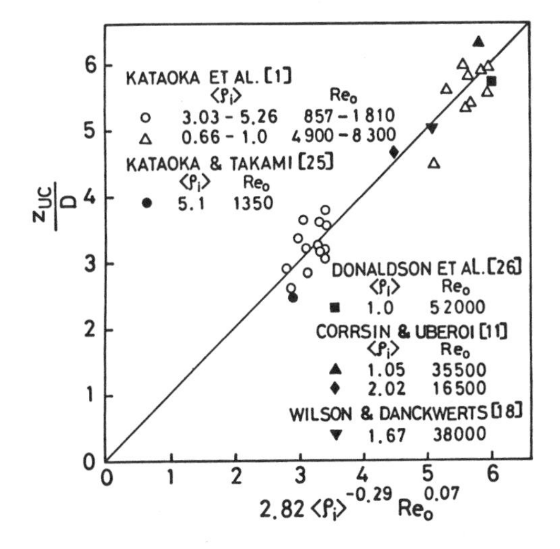

Figure 6. Correlation of velocity core length [1].

length z_{TC}, defined in Figure 5, is shorter than the velocity core length at the same temperature level.

As shown in Figures 6 and 7, the core lengths for velocity and temperature are well correlated in the range $0.66 \leq \langle\rho_i\rangle \leq 5.26$ by the following equations [1]:

$$\frac{z_{UC}}{D} = 2.82\langle\rho_i\rangle^{-0.29}\,\mathrm{Re}_0^{0.07} \tag{30}$$

$$\frac{z_{TC}}{D} = 3.80\langle\rho_i\rangle^{-0.45}\,\mathrm{Re}_0^{0.03} \tag{31}$$

where $\mathrm{Re}_0 = U_0 D\rho/\mu$ is the jet Reynolds number.

As the initial density ratio increases, the core lengths tend to decrease. The prior correlations were obtained in the range of $\mathrm{Re}_0 < 10^4$. The core lengths become insensitive to the jet Reynolds

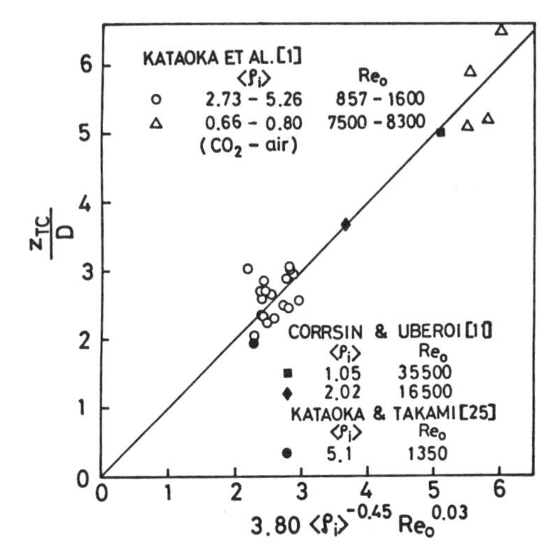

Figure 7. Correlation of temperature core length [1].

number, especially when Re_0 exceeds 10^5 at low Mach numbers. The length of the temperature core is shorter than that of the velocity core. It is also attributable to the preferential transport of heat over momentum.

At high subsonic Mach numbers, Witze [16] proposed the following relation of the velocity core length:

$$\frac{z_{UC}}{D} = 4.38(1 - 0.16M_j)^{-1}\langle\rho_i\rangle^{-0.28} \tag{32}$$

The density dependence is equal to that of Equation 30. His model gives good predictions of centerline behavior in the main region of most classes of free jets, including heterogeneous, nonisothermal and subsonic and properly expanded supersonic flows.

Centerline Decay

The second step is to obtain an appropriate length scale for the jet development length. The equivalent diameter concept introduced by Thring and Newby [12] is based upon the assumption that the jet development is governed by the total jet momentum. It can be considered that the total momentum and enthalpy fluxes are respectively conserved:

$$\int_0^\infty \rho U\, U\, 2\pi r\, dr \simeq \frac{\pi}{4}\rho_0 U_0^2 D^2 \tag{33}$$

$$\int_0^\infty \rho\, Cp(T - T_\infty)U\, 2\pi r\, dr \simeq \frac{\pi}{4}\rho_0\, Cp(T_0 - T_\infty)U_0 D^2 \tag{34}$$

At distances far downstream from the exit, the density in the jet approaches that of the ambient fluid. Hence, the integration of Equation 33 gives

$$\frac{U_0}{U_m} \sim \sqrt{\langle\rho_i\rangle}\,\frac{z}{D} \tag{35}$$

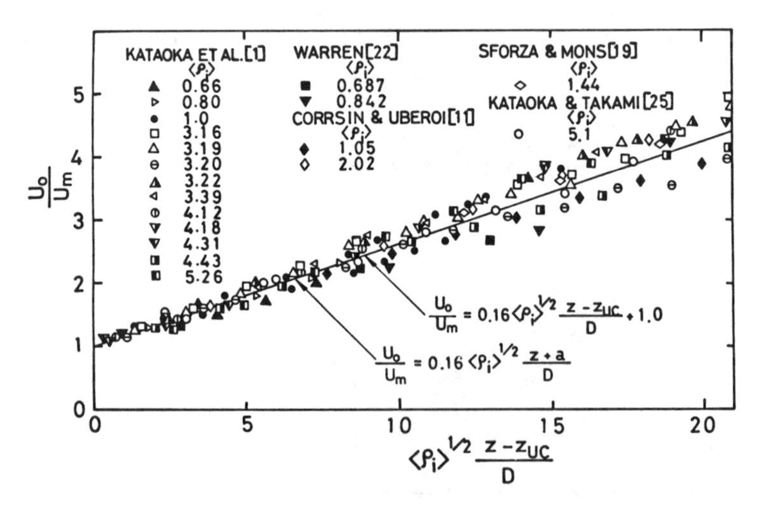

Figure 8. Correlation of centerline velocities [1].

Similarly

$$\frac{T_0 - T_\infty}{T_m - T_\infty} \sim \sqrt{\langle\rho_i\rangle}\,\frac{z}{D} \tag{36}$$

Kleinstein [17], Wilson and Danckwerts [18], and Sforza and Mons [19] also proposed $\sqrt{\langle\rho_i\rangle}z/D$ as the characteristic streamwise coordinate.

Kataoka, Shundoh, and Matsuo [1] proposed $\langle\rho_i\rangle^{1/2}(z - z_{UC})/D$ and $\langle\rho_i\rangle^{1/2}(z - z_{TC})/D$ as the most appropriate coordinates for the jet development from a practical viewpoint.

Figure 8 indicates that if the preceding streamwise coordinates are adopted, the normalized velocities follow one straight line in the fully-developed region [1]. The best-fit equation is given by

$$\frac{U_0}{U_m} = 0.16\langle\rho_i\rangle^{1/2}\,\frac{z - z_{UC}}{D} + 1.0 \tag{37}$$

The last constant 1.0 comes from the definition of the velocity core length that the initial velocity is maintained up to $z = z_{UC}$.

Similarly, as shown in Figure 9, the normalized temperatures follow another straight line [1]. The best-fit equation is given by

$$\frac{T_0 - T_\infty}{T_m - T_\infty} = 0.20\langle\rho_i\rangle^{1/2}\,\frac{z - z_{TC}}{D} + 1.0 \tag{38}$$

Sforza and Mons [19] measured the transport of momentum, heat, and material in three kinds of turbulent jets:

1. Isothermal air jet.
2. Isothermal air/CO_2 jet
3. Nonisothermal heated-air jet.

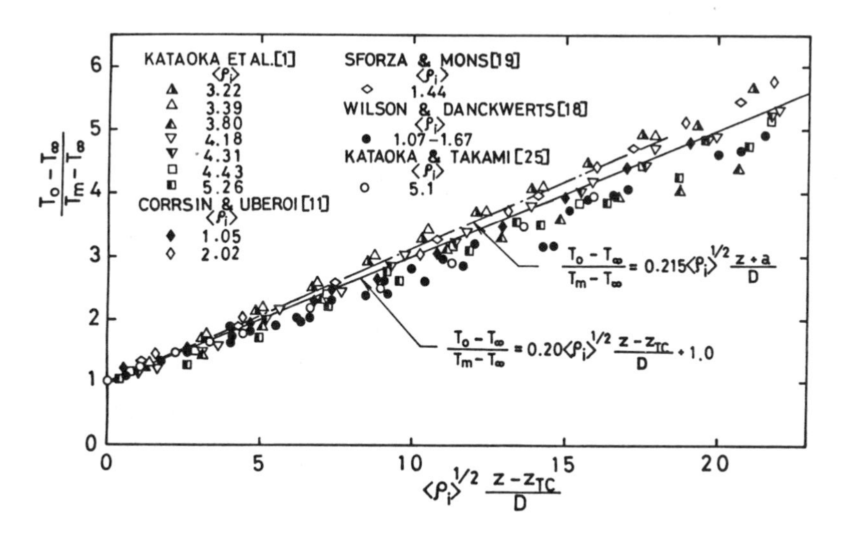

Figure 9. Correlation of centerline temperatures [1].

It can be considered that their highest initial density ratio was about 1.5 for a nonisothermal heated-air jet. The following correlations are given

$$\frac{U_0}{U_m} = 0.16\langle\rho_i\rangle^{1/2}\frac{z+a}{D} \qquad (39)^*$$

$$\frac{T_0 - T_\infty}{T_m - T_\infty} = 0.215\langle\rho_i\rangle^{1/2}\frac{z+a}{D} \qquad (40)^*$$

These rates of the centerline decay are the same as or very close to those of Equations 37 and 38. The virtual origin concept is often used as the counterpart of the core length. Unfortunately, no systematic investigation has been done about it. The decay rate of the centerline concentration can be considered to be equal to that of the centerline temperature.

Kleinstein [17] proposed theoretically another form of the axial decay law:

$$\frac{U_m}{U_0}, \frac{h_m - h_\infty}{h_0 - h_\infty}, \frac{\Gamma_m - \Gamma_\infty}{\Gamma_0 - \Gamma_\infty} = 1 - \exp\left[-\frac{1}{\kappa_1\sqrt{\langle\rho_i\rangle}\, z/r_0 - 0.70}\right] \qquad (41)$$

where $\kappa_1 = 0.074$, 0.102, and 0.104, respectively. At axial positions far downstream of the core region, this equation gives a similar form of the linear law:

$$\frac{U_0}{U_m}, \frac{h_0 - h_\infty}{h_m - h_\infty}, \frac{\Gamma_0 - \Gamma_\infty}{\Gamma_m - \Gamma_\infty} = \kappa_1\sqrt{\langle\rho_i\rangle}\,\frac{z}{r_0} - 0.70$$

Witze [16] improved Kleinstein's theory and proposed κ_1 as a function of the initial density ratio and jet Mach number:

Subsonic flow:

$$\kappa_1 = 0.08(1 - 0.16M_j)\langle\rho_i\rangle^{-0.22} \qquad (42)$$

Supersonic flow:

$$\kappa_1 = 0.063(M_j^2 - 1)^{-0.15} \qquad (43)$$

The last equation is applicable from the nozzle exit to the sonic point.

Generally, the centerline decay can be described by a linear law of the form

$$\frac{U_0}{U_m} = A_1\langle\rho_i\rangle^{1/2}\frac{z+a}{D} \quad \text{or} \quad A_1\langle\rho_i\rangle^{1/2}\frac{z - z_{UC}}{D} + 1.0 \qquad (44)$$

$$\frac{T_0 - T_\infty}{T_m - T_\infty}, \frac{\Gamma_0 - \Gamma_\infty}{\Gamma_m - \Gamma_\infty} = B_1\langle\rho_i\rangle^{1/2}\frac{z+a}{D} \quad \text{or} \quad B_1\langle\rho_i\rangle^{1/2}\frac{z - z_{TC}}{D} + 1.0 \qquad (45)$$

Table 1 lists the values found by different investigators for these constants A_1, B_1. The literature data, except for those of Sforza and Mons [19], Kataoka et al. [1], Wilson and Danckwerts [18], and Kleinstein [17], are those obtained under the constant density assumption, i.e. $\langle\rho_i\rangle \simeq 1$. The constants A_2, B_2 will be discussed in the next section.

---

* In their original paper [19], the virtual origin a is replaced by another notation $-x_0$.

Table 1
Constants in the Linear Decay and Spreading Laws

Investigator	Jet	A_1	B_1	A_2	B_2	Remarks*
Corrsin and Uberoi [11]	Air/air		0.278	0.084	0.107	Temperature
Sunavala, Hulse and Thring [13]	Gas/air		0.220		0.117	Temperature and concentration
Kristmanson and Danckwerts [20]	Water/water		0.210		0.112	Concentration
Kiser [10]	Water/water	0.164	0.200	0.0815	0.104	Concentration
Becker, Hottel and Williams [21]	Air/air		0.185		0.106	Concentration
Hinze and van der Hegge Zijnen [5]	Air/air	0.156	0.190	0.08	0.096	Temperature and Concentration
Keagy and Weller [6]	N_2/air				0.105	Concentration
Forstall and Gaylord [9]	Water/water			0.091	0.101	Concentration
Kleinstein [17]	Air/air	0.148	0.204			Corrsin and Uberoi's measurements
Wilson and Danckwerts [18]	Air/air	0.15	0.175		0.130	Nonisothermal jet
Sforza and Mons [19]	Air/air	0.16	0.215	0.089	0.108	Nonisothermal jet
Kataoka, Shundoh and Matsuo [1]	Gas/air	0.16	0.20	$0.09\langle\rho_i\rangle^{1/10}$	$0.08\langle\rho_i\rangle^{1/3}$	Nonisothermal jet

* *The remarks indicate from what scalar quantity data the constants B_1, B_2 were determined.*

Radial Spreading

From a viewpoint of dimensional analysis, the jet half-radius should be functions of physical axial length, but does not depend upon the initial density ratio. According to Sforza and Mons [19], the jet half-radii are independent of $\langle \rho_i \rangle$:

$$\frac{r_{1/2U}}{D} = 0.089 \frac{z + a}{D} \tag{46}$$

$$\frac{r_{1/2T}}{D} = 0.108 \frac{z + a}{D} \tag{47}$$

The spreading rate of concentration reported by Becker et at. [21] is very close to that of temperature. The preceding correlations can be utilized over a wide range of $\langle \rho_i \rangle$ if the virtual origin correlation is made available.

The jet half-radii for velocity and temperature measured by Kataoka et. al. [1] are shown in Figures 10 and 11. These figures suggest that the linearity of the half-radii with the axial coordinate is valid, but that the rates of linear spreading, especially $r_{1/2T}$, depend slightly upon the initial density ratio. This may be attributable to the fact that these correlations have been obtained from the experimental data covering a wide range of $\langle \rho_i \rangle$.

The best-fit equations are given by

$$\frac{r_{1/2U}}{D} = 0.09 \langle \rho_i \rangle^{1/10} \frac{z - z_{UC}}{D} + 0.5 \tag{48}$$

where $0.66 \le \langle \rho_i \rangle \le 5.26$

$$\frac{r_{1/2T}}{D} = 0.08 \langle \rho_i \rangle^{1/3} \frac{z - z_{TC}}{D} + 0.5 \tag{49}$$

where $2.73 \le \langle \rho_i \rangle \le 5.26$

The intercept with the ordinate in these figures should have a value of 0.5 from the definition of core lengths. The last equation, Equation 49, was obtained by using high-temperature burned gas jets.

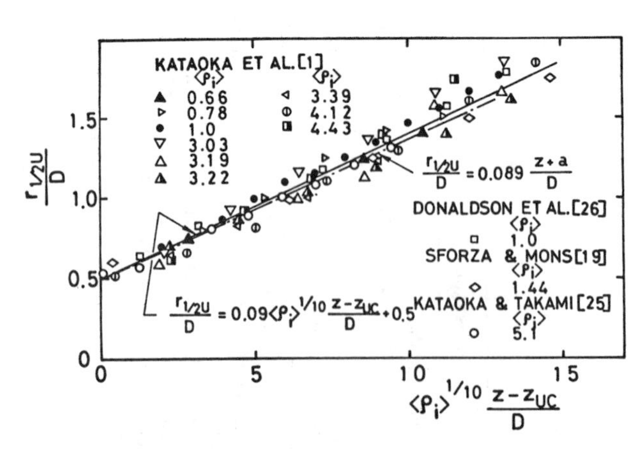

Figure 10. Correlation of jet half-radii for velocity [1].

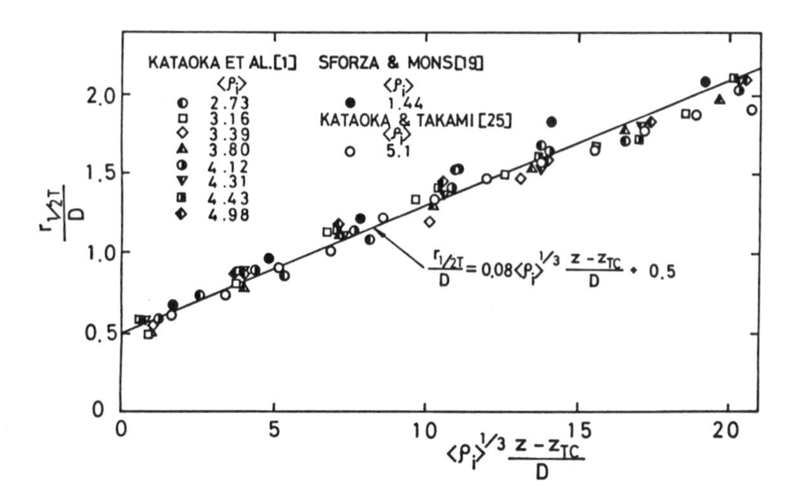

Figure 11. Correlation of jet half-radii for temperature [1].

These correlations indicate a preferential transport of heat and mass over momentum in the sense that at a given axial station, the centerline values of temperature and concentration are lower than that of momentum and their half-radii are larger.

Generally, the radial spreading can be described by a linear law of the form

$$\frac{r_{1/2U}}{D} = A_2 \frac{z+a}{D} \quad \text{or} \quad A_2 \frac{z - z_{UC}}{D} + 0.5 \tag{50}$$

$$\frac{r_{1/2T}}{D} = B_2 \frac{z+a}{D} \quad \text{or} \quad B_2 \frac{z - z_{TC}}{D} + 0.5 \tag{51}$$

Table 1 compares the values of these constants A_2, B_2 found by different investigators.

Radial Profile

The next step is to get the radial distribution functions of velocity and temperature at a given axial station.

As shown in Figure 12, the similarity in the radial distributions of velocity and temperature has been confirmed in the fully-developed region of highly heated gas jets [1]. The best-fit equations are given by

$$\left.\begin{aligned} \frac{U}{U_m} &= \left[1 + 0.414\left(\frac{r}{r_{1/2U}}\right)^2\right]^{-2} \\ \frac{T - T_\infty}{T_m - T_\infty} &= \left[1 + 0.414\left(\frac{r}{r_{1/2T}}\right)^2\right]^{-2} \end{aligned}\right\} \quad 0 \le r/r_{1/2} \le 1 \tag{52}$$

$$\left.\begin{aligned} \frac{U}{U_m} &= \exp\left[-0.693\left(\frac{r}{r_{1/2U}}\right)^2\right] \\ \frac{T - T_\infty}{T_m - T_\infty} &= \exp\left[-0.693\left(\frac{r}{r_{1/2T}}\right)^2\right] \end{aligned}\right\} \quad r/r_{1/2} > 1 \tag{53}$$

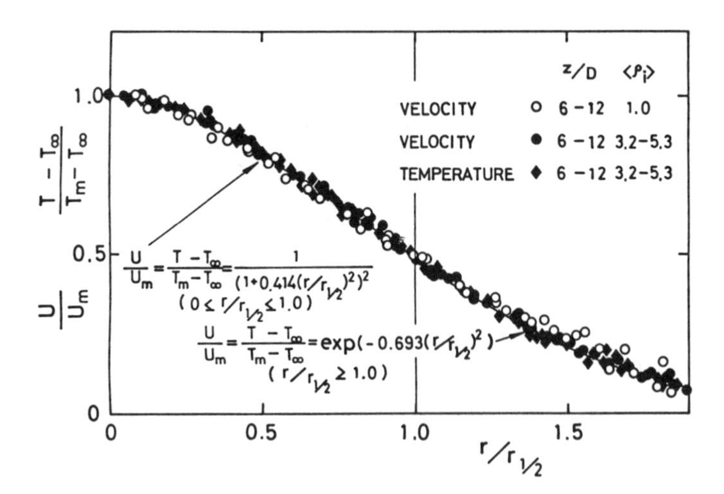

Figure 12. Radial distribution of the normalized velocity and temperature [1].

These equations are mathematically similar in form to those for incompressible jets. The only difference is that the effect of initial density ratio is taken into account in the jet half-radii and centerline quantities. The equations for the inner region (Equation 52) are essentially the same as those obtained under the assumption of constant incompressible eddy diffusivity by the boundary layer theory [3]. The equations for the outer region (Equation 53) are the same as Warren's Gaussian error function [22] to be mentioned below.

Warren [22] proposed a useful set of distribution functions for high-velocity compressible jets:

Initial region:

$$\frac{U}{U_0} = \exp\left[-(\ln 2)\frac{r^2 - r_C^2}{r_{1/2U}^2 - r_C^2}\right] \qquad r \geq r_C$$
$$= 1 \qquad r \leq r_c \tag{54}$$

Fully-developed region:

$$\frac{U}{U_m} = \exp\left[-(\ln 2)\left(\frac{r}{r_{1/2U}}\right)^2\right] \tag{55}$$

Here r_C refers to the radius to the outer edge of the core. The core-edge radius can be assumed to decrease with z:

$$\frac{r_C}{D} = \frac{1}{2}\left(1 - \frac{z}{z_{UC}}\right) \tag{56}$$

where $0 \leq z \leq z_{UC}$

If the initial density ratio and the jet Reynolds number (or the jet Mach number) are specified for axisymmetric variable-density turbulent jets, the jet development and distributions of velocity and temperature can be predicted at various temperature levels by the following steps:

1. Estimate the core lengths by Equations 30 to 32 or, if possible, estimate the virtual origins.
2. In the core region, $z \leq z_{UC}$ or z_{TC}, determine r_C by Equation 56 and calculate the radial profiles at a given axial station by Equation 54.
3. In the fully-developed region, $z \geq z_{UC}$ or z_{TC}, determine U_m, T_m, and $r_{1/2U}$, $r_{1/2T}$ using the given initial density ratio by Equations 37 to 45.
4. Calculate the radial profiles at a given axial station using Equations 52 to 55.

There is unfortunately little work on two-dimensional turbulent jets heated so highly that the effect of density change should be taken into consideration. Davies et al. [23] reported the spread of the temperature and velocity fields of a plane turbulent jet heated slightly, so they did not need a consideration as to the density effect.

MODELS OF VARIABLE-DENSITY FREE TURBULENT MIXING

The time-averaged boundary-layer equations for an axisymmetric turbulent free jet with variable density are

$$\frac{\partial}{\partial z}\rho U + \frac{1}{r}\frac{\partial}{\partial r}\rho r V = 0 \tag{57}$$

$$\rho U\frac{\partial U}{\partial z} + \rho V\frac{\partial U}{\partial r} = \frac{1}{r}\frac{\partial}{\partial r}\left(\rho\epsilon_M r\frac{\partial U}{\partial r}\right) \tag{58}$$

$$\rho\, Cp\, U\frac{\partial T}{\partial z} + \rho\, Cp\, V\frac{\partial T}{\partial r} = \frac{1}{r}\frac{\partial}{\partial r}\left(\rho\, Cp\, \epsilon_H r\frac{\partial T}{\partial r}\right) \tag{59}$$

$$\rho U\frac{\partial \Gamma}{\partial z} + \rho V\frac{\partial \Gamma}{\partial r} = \frac{1}{r}\frac{\partial}{\partial r}\left(\rho\epsilon_D r\frac{\partial \Gamma}{\partial r}\right) \tag{60}$$

All theoretical approaches are ineffective without the simplifying assumption of local variation in the eddy diffusivities with fluid density.

As mentioned earlier, Prandtl's incompressible turbulent mixing model [2] gives an eddy diffusivity of the form

$$\epsilon_M = K_p b|U_m - U_\infty| \tag{1}$$

Here the width of a turbulent mixing layer b can be taken as the jet half-radius $r_{1/2U}$. The velocity difference reduces to the centerline velocity U_m for a stationary environment ($U_\infty = 0$). Hence

$$\epsilon_M = K_p r_{1/2U} U_m \tag{61}$$

In the fully-developed region, $r_{1/2U} \sim z$ while $U_m \sim 1/z$. Hence ϵ_M remains constant over the whole fully-developed region of incompressible jets.

The eddy diffusivity for an incompressible round air jet measured by Hinze and van der Hegge Zijnen [5] gives

$$\epsilon_M = 0.013 U_0 D \tag{8}$$

This implies $K_p = 0.026$ in the fully-developed region. According to Peters [24], the K_p constant should be 0.014 in the initial region and 0.022 in the developed region. Most of the turbulent mixing models proposed for variable-density jets are directly or indirectly based upon this Prandtl's mixing-length model [2, 14].

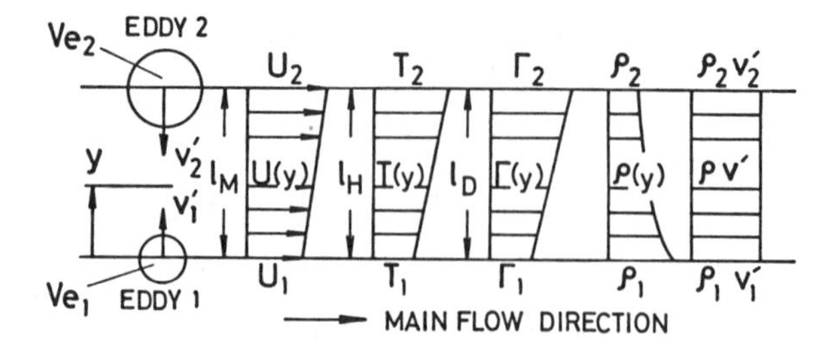

Figure 13. Schematic picture of variable-density turbulent mixing.

Simple Explanation of Turbulent Mixing

It is instructive to explain a simple model of compressible turbulent mixing [25]. Suppose two eddies are exchanged over the so-called mixing length (l_M, l_H, l_D) in a turbulent shear layer with a large density gradient in the transverse direction, as shown in Figure 13.

Assume that these eddies do not lose their original quantities (i.e. (U_1, T_1, Γ_1, ρ_1) and (U_2, T_2, Γ_2, ρ_2)) while they travel in the transverse direction. The volume Ve of the traveling eddies is inversely proportional to the local density of fluid

$$\rho \text{ Ve} = \text{constant} \tag{62}$$

It can be considered that the volume transported per unit area and unit time by the traveling eddies is proportional to the transverse component of the turbulent velocity v′. Hence

$$\rho_1 v'_1 = \rho_2 v'_2 = \rho v' = \text{constant} \tag{63}$$

everywhere in the transverse direction. The static pressure and heat capacity are assumed to be constant in the model consideration. The traveling eddies transport momentum, enthalpy, and mass undergoing the change in volume due to the temperature difference between two layers. Therefore, the turbulent fluxes of momentum, enthalpy, and mass may be expressed as

$$\begin{aligned} \tau_t &= \rho_2 U_2 v'_2 - \rho_1 U_1 v'_1 = \rho v'(U_2 - U_1) \\ &= \rho v' l_M \frac{dU}{dy} = \rho \epsilon_M \frac{dU}{dy} \end{aligned} \tag{64}$$

$$\begin{aligned} q_t &= \rho_2 \, Cp_2 \, T_2 v'_2 - \rho_1 \, Cp_1 \, T_1 v'_1 = \rho v' \, Cp(T_2 - T_1) \\ &= \rho \, Cp \, v' l_H \frac{dT}{dy} = \rho \, Cp \, \epsilon_H \frac{dT}{dy} \end{aligned} \tag{65}$$

$$\begin{aligned} j_t &= \rho_2 \Gamma_2 v'_2 - \rho_1 \Gamma_1 v'_1 = \rho v'(\Gamma_2 - \Gamma_1) \\ &= \rho v' l_D \frac{d\Gamma}{dy} = \rho \epsilon_D \frac{d\Gamma}{dy} \end{aligned} \tag{66}$$

These relations are the same in form as those for incompressible flows, but they suggest that $\rho\epsilon_M$, ρ Cp ϵ_H, $\rho\epsilon_D$ should be regarded as a new coefficient of the turbulent transport. They are called the dynamic eddy transfer coefficients for momentum, heat, and mass, respectively. Therefore, it is reasonable to make the following modification to the Prandtl's incompressible eddy diffusivity model (Equation 1):

$$\rho\epsilon_M = K_F b|\rho_m U_m - \rho_\infty U_\infty|$$

This is the Ferri et al. model [27, 28] to be discussed later.

Warren and Donaldson-Gray Model

Warren [22] and Donaldson and Gray [26] applied a compressibility correction to the mixing rate parameter K_p of the Prandtl model for high-velocity compressible jets. Warren performed an integral analysis assuming that the mixing rate parameter is a function of the initial Mach number. According to his correction factor, the turbulent flux of momentum can be written as

Initial region:

$$\tau_t = (0.0217 - 0.00345M_j)\rho(r_{1/2U} - r_C)U_0 \frac{\partial U}{\partial r} \tag{67}$$

Developed region:

$$\tau_t = (0.0217 - 0.00345M_j)\rho r_{1/2U} U_m \frac{\partial U}{\partial r} \tag{68}$$

Donaldson and Gray extended Warren's momentum integral method for the compressible turbulent jet mixing of two dissimilar gases at relatively high velocities. They considered that the maximum shear stress at a given axial station occurs at a radial position not too far from the jet half-radius $r_{1/2U}$. Assuming that K_p varies with z, they give a general relationship of the local mixing rate parameter with the local Mach number evaluated at the jet half-radius.

It should be noted that Warren used a sharp-edged nozzle but Donaldson and Gray used a blunt-edged nozzle. A jet issuing from a flat surface tends to decay more rapidly than the same jet issuing from a sharp-edged nozzle because of the difference in the entrainment of the surrounding fluid.

Peters [24] improved their correction factor for practical applicability. His version of the correction parameter is given by

$$\epsilon_M = [0.66 + 0.34\exp(-3.42M_{1/2}^2)]K_p b|U_m - U_\infty| \tag{69}$$

where K_p is the Prandtl's mixing rate parameter for incompressible flows.

Ferri-Kleinstein Model

Ferri, Libby, and Zakkay [27, 28] proposed a new form of the eddy viscosity modeled along the lines of Prandtl's incompressible model:

$$\rho\epsilon_M = K_F r_{1/2M}|\rho_m U_m - \rho_\infty U_\infty| \tag{70}$$

where K_F is Ferri's mixing rate parameter. The half-radius $r_{1/2M}$ is defined as the radius at which $\rho U = (\rho_m U_m + \rho_\infty U_\infty)/2$.

Assuming the eddy transfer coefficient $\rho\epsilon_M$ to be independent of the radial coordinate, they performed an analysis for compressible jet flow by linearizing the boundary-layer equation in the plane of the von Mises transformation. They recommend a value of 0.025 for their mixing rate parameter ($K_F = 0.025$).

The centerline value of the eddy viscosity in the fully-developed region of compressible jets is given by

$$\frac{(\rho\epsilon_M)_m}{\rho_0 U_0 D} = 0.0125\langle\rho_i\rangle^{1/2} \tag{71}$$

For a compressible jet exhausting into a moving coaxial stream, Alpinieri [29] gives

$$\frac{(\rho\epsilon_M)_m}{\rho_0 U_0 D} = 0.0125\frac{r_{1/2M}}{r_0}\left(\frac{\rho_\infty U_\infty}{\rho_0 U_0} + \frac{\rho_\infty U_\infty^2}{\rho_0 U_0^2}\right) \tag{72}$$

Kleinstein [17] performed an analysis along the lines of the Ferri et al. model using a similar formulation of the eddy viscosity:

$$\rho\epsilon_M = (\kappa_1/4) r_{1/2M}\rho_m U_m \tag{73}$$

As the result of analysis, the eddy viscosity is found to be a constant for a free jet exhausting into a quiescent environment:

$$\frac{(\rho\epsilon_M)_m}{\rho_0 U_0 r_0} = \frac{\kappa_1}{4}\langle\rho_i\rangle^{1/2} \tag{74}$$

The resulting expression for the centerline decay is of the form

$$\frac{U_m}{U_0} = 1 - \exp\left[-\frac{1}{\kappa_1\sqrt{\langle\rho_i\rangle}(z/r_0) - X_C}\right] \tag{75}$$

where X_C is the dimensionless core length:

$$X_C = \frac{z_{UC}}{r_0} = \frac{0.70}{\kappa_1\sqrt{\langle\rho_i\rangle}} \tag{76}$$

Using the heated jet mixing data of Corrsin and Uberoi [11], Kleinstein determined the universal constant to be 0.074. Hence

$$\frac{(\rho\epsilon_M)_m}{\rho_0 U_0 D} = 0.00925\langle\rho_i\rangle^{1/2} \tag{77}$$

His numerical coefficient seems to be smaller than Ferri's, (see Equation 71).

Schetz-Zelazny Model

Schetz [30] defined another form of the eddy viscosity model for a compressible jet mixing with a moving environment.

For compressible two-dimensional flow

$$\rho\epsilon_M = K_S\rho_\infty U_\infty\delta^* \tag{78}$$

where δ^* is the compressible displacement thickness defined as

$$\delta^* = \int_{-\infty}^{\infty}\left|1 - \frac{\rho U}{\rho_\infty U_\infty}\right| dy \tag{79}$$

For compressible axisymmetric flow

$$\rho\epsilon_M = \frac{K_S}{r_0}(\rho_\infty U_\infty \pi \delta_r^{*2}) \tag{80}$$

where

$$\delta_r^{*2} = \int_0^\infty \left|1 - \frac{\rho U}{\rho_\infty U_\infty}\right| 2r\,dr \tag{81}$$

He recommends a value of 0.018 for πK_S. That is

$$\rho\epsilon_M = \frac{0.036}{r_0}\int_0^\infty |\rho U - \rho_\infty U_\infty| r\,dr \tag{82}$$

This model predicts the eddy viscosity increasing too rapidly with increasing thickness of the mixing zone.

In order to make the Schetz model apply to both quiescent and coflowing axisymmetric flows, Zelazny [31] improved it by using the velocity half-radius $r_{1/2U}$ in place of r_o in the definition of displacement thickness. That is

$$\rho\epsilon_M = \frac{0.036}{r_{1/2U}}\int_0^\infty |\rho U - \rho_\infty U_\infty| r\,dr \tag{83}$$

The eddy viscosity given by Zelazny is in good agreement with Eggers' measurements of a Mach 2.2 quiescent air jet [32]. However, neither Schetz's nor Zelazny's model can describe well the data of a Freon jet into an air stream [33].

Other Modifying Models

Cohen-Guile Model

For variable-density mixing flows, Cohen [34] gives another correction factor to the Prandtl's constant mixing rate parameter:

$$\epsilon_M = K_P b[(1 + \rho_\infty/\rho_m)/2]^{0.8}|U_m - U_\infty| \tag{84}$$

Cohen and Guile [35] considered that the mixing of two streams in the far downstream region is dominated by the preturbulence levels before the two streams came into contact with each other.

In the region dominated by the preturbulence level of the two streams, Cohen [34] developed an eddy viscosity of the form

$$\epsilon_M = K_P b \frac{1 + (\rho_\infty/\rho_m)_1}{2} U_m \left(1 - \frac{U_\infty}{U_{m_1}}\right) \frac{1 + (\rho_\infty/\rho_m)_1}{1 + \rho_\infty/\rho_m} \frac{1 + (\rho_\infty U_\infty/\rho_m U_m)}{1 + (\rho_\infty U_\infty/\rho_m U_m)_1} \tag{85}$$

where the subscript 1 indicates the values evaluated at the station where the mixing is determined to be dominated by the preturbulence level. Cohen and Guile model consists of Equations 84 and 85. It gave a good prediction of the free mixing and combustion of high-velocity hydrogen jets into surrounding atmosphere.

Libby Model

Variable-density turbulent mixing can be analyzed by the incompressible flow equations through the use of suitable variable transformation. Libby [36] obtained a closed-form solution to the transformed equation for the compressible turbulent mixing problem in terms of offset circular probability functions.

For two-dimensional flow

$$\rho^2 \epsilon_M = \rho_m^2 \epsilon_M^* \tag{86}$$

For axisymmetric flow

$$r^2 \rho^2 \epsilon_M = 2\rho_m^2 \epsilon_M^* \int_0^r \left(\frac{\rho}{\rho_m}\right) r\, dr \tag{87}$$

where ϵ_M^* is the incompressible eddy viscosity and ρ_m is the centerline density. This model is too complex to use in compressible jet mixing analysis.

Tomich-Weger Model

The complexity of the core region requires special consideration in any jet mixing problems. Better prediction in the entire mixing region should take into account the effect of upstream history. The prediction schemes should begin at the point where the turbulent mixing is initiated.

Tomich and Weger [37] tried a modification of Kleinstein's model for high subsonic heated jets exhausting into quiescent cold air. In their numerical analysis, the following expressions for the eddy transfer coefficients were used:

$$\frac{\rho \epsilon_M}{\rho_m U_0 D} = K_T \langle \rho_i \rangle^{1/2} f\left(\frac{z}{D}\right) \tag{88}$$

where $f\left(\frac{z}{D}\right) = 0.2 \quad \text{for } z \leq z_{UC}$

$= 1.0 \quad \text{for } z \geq z_{UC}$

$$\frac{z_{UC}}{D} = 4.73 \langle \rho_i \rangle^{-1/2}$$

$$\frac{\rho\, Cp\, \epsilon_H}{\rho_m\, Cp\, U_0 D} = K_T \langle \rho_i \rangle^{1/2} f_T\left(\frac{z}{D}\right) Pr_t^{-1} \tag{89}$$

where $f_T\left(\frac{z}{D}\right) = 0.2 \text{ and } Pr_t = 1.0 \quad \text{for } z \leq z_{TC}$

$= 1.0 \text{ and } Pr_t = 0.715 \quad \text{for } z \geq z_{TC}$

$$\frac{z_{TC}}{D} = 3.43 \langle \rho_i \rangle^{-1/2}$$

Their mixing rate parameter, similar to Warren's [22], is given by

$$K_T = 0.00972 - 0.00751 M_j + 0.00298 M_j^2 \tag{90}$$

The boundary-layer equations were numerically solved by a finite-difference technique assuming that the eddy transfer coefficients are constant in the radial direction. The axial variation is taken into account as the axial variation of the centerline fluid density ρ_m. Their numerical solution was compared with the experiments of high Mach number subsonic jets of heated air into quiescent cold air. It should be noted that the specification of only two more initial properties suffices to get a solution of jet velocity and temperature characteristics for high velocity and temperature

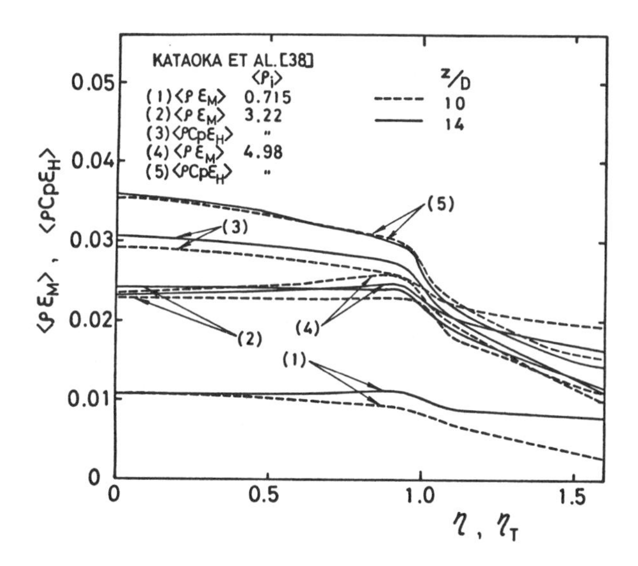

Figure 14. Radial distribution of eddy transfer coefficients [38].

subsonic jets; one is the jet Mach number M_j and the other the initial temperature or density ratio $\langle \rho_i \rangle$. For a given jet Mach number, an increase in the initial density ratio causes a more rapid axial decay of the centerline velocity and temperature. On the other hand, for a given initial density ratio, an increase in the jet Mach number causes a less rapid axial decay of the two properties.

Kataoka Model

Still we need some modifications in the radial distribution of the eddy transfer coefficients. Kataoka and Takami [25] and Kataoka, Matsuo and Shundoh [38] made measurements of radial and axial variations in the flowfield of highly heated gas jets into quiescent cold air. The Mach number was low enough (of the order of 0.1) and the initial density ratio ranged from 0.66 to 5.1. Their convergent nozzle has a diameter D = 10 mm and a contraction ratio 1/36.

Figure 14 shows the radial distribution of the eddy transfer coefficients for momentum and enthalpy. As in the case of incompressible jet, the eddy transfer coefficient for momentum $\rho\epsilon_M$ remains constant over the inner region ($0 \leq \eta \leq 1$), but decreases rapidly with radial distance in the outer region ($\eta \geq 1$). It is also noted that the eddy transfer coefficient for enthalpy ρ Cp ϵ_H decreases slightly in the radial direction even in the inner region. The eddy transfer coefficient for a chemical species remains constant as that for momentum, but the value is very close to that for enthalpy.

There are few experimental data available for comparison. Corrsin and Uberoi [11] measured precisely local variation of temperature and dynamic pressure in the flowfield of a heated air jet issuing into quiescent air at two initial density ratios; $\langle \rho_i \rangle = 1.05$ and 2.02. Figure 15 shows local variation of the eddy transfer coefficients that Kataoka et al. [38] calculated from Corrsin and Uberoi's experiments [11].

The dimensionless eddy transfer coefficient $\langle \rho\epsilon_M \rangle$ when $\langle \rho_i \rangle = 1.05$ has the same constant value over the inner region as the Prandtl's eddy diffusivity $\langle \epsilon_M \rangle$ for an incompressible jet. Here $\langle \rho\epsilon_M \rangle = \rho\epsilon_M/\rho_0 U_0 D$ and $\langle \epsilon_M \rangle = \epsilon_M/U_0 D$. The axial variation is small compared to the radial variation. It should be noted that $\langle \rho \text{ Cp } \epsilon_H \rangle$ decreases slightly radially in the inner region even when $\langle \rho_i \rangle = 1.05$. Here $\langle \rho \text{ Cp } \epsilon_H \rangle = \rho \text{ Cp } \epsilon_H/\rho_0 \text{ Cp } U_0 D$.

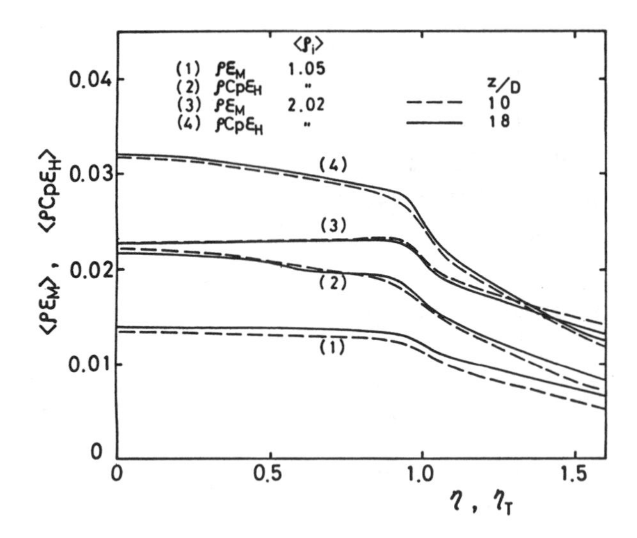

Figure 15. Radial distribution of eddy transfer coefficients calculated by Kataoka et al. [38] from the radial distribution of dynamic pressure and temperature measured by Corrsin and Uberoi [11].

Those coefficients on the jet axis are plotted against the initial density ratio in Figure 16. It is clear that those eddy transfer coefficients $\rho\epsilon_M$, ρ Cp ϵ_H, $\rho\epsilon_D$ increase in proportion to the square root of the initial density ratio. They can be expressed as

$$\langle\rho\epsilon_M\rangle_m = 0.0135\langle\rho_i\rangle^{1/2} \tag{91}$$

$$\langle\rho \text{ Cp } \epsilon_H\rangle_m = 0.0195\langle\rho_i\rangle^{1/2} \tag{92}$$

$$\langle\rho\epsilon_D\rangle_m = 0.0195\langle\rho_i\rangle^{1/2} \tag{93}$$

It should be noted that Equation 91 is almost the same as the Ferri et al. model. (see Equation 71). When $\langle\rho_i\rangle$ is unity, Equation 91 reduces to the Hinze and van der Hegge Zijnen eddy diffusivity relation, Equation 8, for incompressible jet.

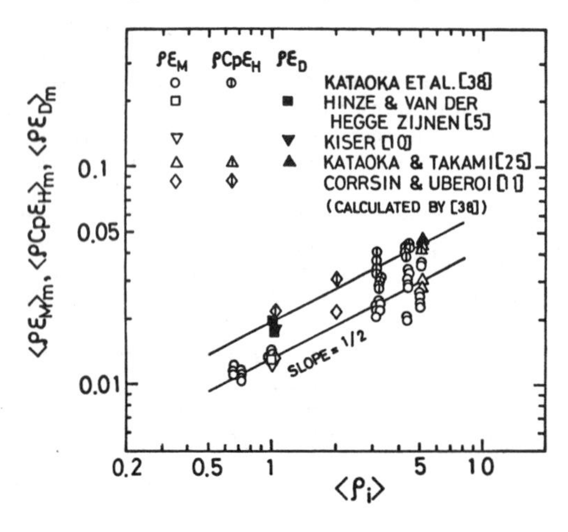

Figure 16. Variation of eddy transfer coefficients on jet axis with initial density ratio [38].

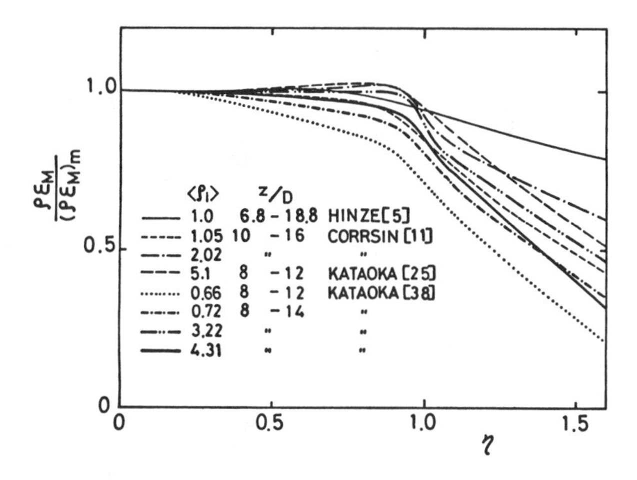

Figure 17. Radial distribution of the eddy transfer coefficients for momentum normalized with the centerline values [38].

Figures 17 and 18 show radial distribution of the eddy transfer coefficients normalized with the centerline values. These experimental lines are scattered slightly owing to experimental difficulties. If the universal function of the radial distribution is obtained, we can easily estimate local values of the eddy transfer coefficients with the aid of Equations 91 to 93. The Kataoka et al. model is practically useful over a wide range of initial density ratio ($0.66 \leq \langle \rho_i \rangle \leq 5.1$) to predict local values of the eddy transfer coefficients in the fully-developed region.

These models described up to here relate the eddy transfer coefficients to local gradients of the mean flow properties. The turbulent transport flux is determined only by local conditions without regard to the prior development of the flow.

Those models of eddy diffusivity and eddy transfer coefficient are summarized in Table 2.

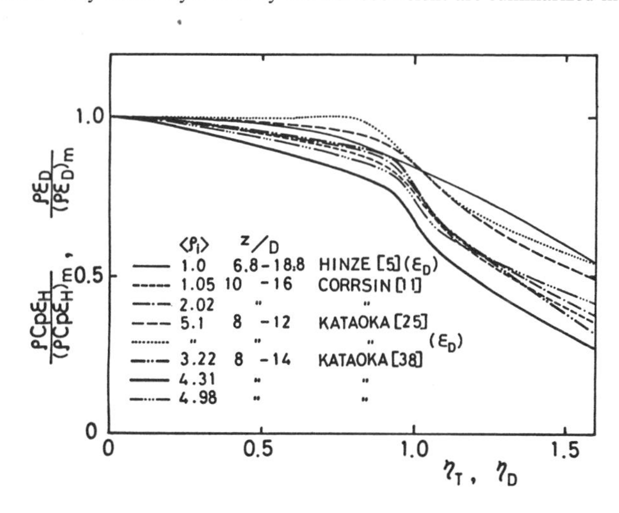

Figure 18. Radial distribution of the eddy transfer coefficients for temperature and concentration normalized with the centerline values [38].

Table 2
Eddy Diffusivity and Eddy Transfer Coefficient Models for Mixing Region of Jets

Investigator	Planar	Axisymmetric	Variable density	Expression	Remarks
Prandtl [14]	x	x		$\epsilon_M = 1^2 \frac{\partial U}{\partial y}$	Mixing length theory
Prandtl [2]	x	x		$\epsilon_M = K_P b \lvert U_{max} - U_{min} \rvert$	Velocity difference concept
Hinze and van der Hegge Zijnen [5]		x		$\epsilon_M = 0.013 U_0 D$ $\epsilon_H = 0.018 U_0 D$	Measurement (air jet)
Warren [22]		x	x	$\rho\epsilon_M = (0.0217 - 0.00345 M_j)\rho r^* U^*$ $r^* U^* = (r_{1/2U} - r_C) U_0 \quad \text{for } z \le z_{UC}$ $r^* U^* = r_{1/2U} U_m \quad \text{for } z \ge z_{UC}$	High velocity jet (sharp-edged nozzle)
Ferri et al. [27]		x	x	$\rho\epsilon_M = K_F r_{1/2M} \lvert (\rho U)_m - (\rho U)_\infty \rvert$	Variable-density model
Libby [36]	x	x	x	$\rho^2 \epsilon_M = \rho_m^2 \epsilon_M^* \quad \text{(planar)}$ $r^2 \rho^2 \epsilon_M = 2\rho_m^2 \epsilon_M^* \int_0^r (\rho/\rho_m) r\, dr \quad \text{(axisymmetric)}$	ϵ_M^*: constant density eddy diffusivity
Kiser [10]		x		$\epsilon_M = 0.0122 U_0 D$ $\epsilon_D = 0.0183 U_0 D$	Measurement (water jet)

Alpinieri [29]		x	x	$\langle \rho \epsilon_M \rangle = 0.0125(r_{1/2M}/r_o)\left(\frac{\rho_\infty U_\infty}{\rho_0 U_0} + \frac{\rho_\infty U_\infty^2}{\rho_0 U_0^2}\right)$	Extension of Ferri et al. model
Kleinstein [17]		x	x	$\rho \epsilon_M = (\kappa_1/4) r_{1/2M}(\rho U)_m$ $\langle \rho \epsilon_M \rangle = (\kappa_1/2)\langle \rho_i \rangle^{1/2}$	Extension of Ferri et al. model
Donaldson and Gray [26]		x	x	$\rho \epsilon_M = K \rho_{1/2U} r_{1/2U} U_m/2$ K: function of $M_{1/2}$	Modification of Warren's model
Tomich and Weger [37]		x	x	$\langle \rho \epsilon_M \rangle = K_T(\rho_m/\rho_0)\langle \rho_i \rangle^{1/2} f(z/D)$ $\langle \rho\ Cp\ \epsilon_H \rangle = K_T(\rho_m/\rho_0)\langle \rho_i \rangle^{1/2} f_T(z/D)\ Pr_t^{-1}$ $K_T = 0.00972 - 0.00751 M_j + 0.00298 M_j^2$	Modification of Warren's and Kleinstein's models
Cohen [34]	x	x	x	$\epsilon_M = K_P b[(1 + \rho_\infty/\rho_m)/2]^{0.8} \lvert U_m - U_\infty \rvert$	Mixing of two streams dominated by preturbulence
Schetz [30]	x	x	x	$\rho \epsilon_M = K_S \rho_\infty U_\infty \delta^*$ (planar) $\rho \epsilon_M = K_S(\rho_\infty U_\infty \pi \delta_r^{*2}/r_o)$ (axisymmetric)	δ^*, δ_r^*: compressible displacement thicknesses $\pi K_S = 0.018$
Peters [24]		x	x	$\rho \epsilon_M = [0.66 + 0.34 \exp(-3.42 M_{1/2}^2)] \times K_P b \lvert U_m - U_\infty \rvert$	Modification of Donaldson-Gray model
Zelazny [31]		x	x	$\rho \epsilon_M = K_S \rho_\infty U_\infty \pi \delta_r^{*2}/r_{1/2U}$	Modification of Schetz's model
Kataoka et al. [25, 38]		x	x	$\langle \rho \epsilon_M \rangle_m = 0.0135 \langle \rho_i \rangle^{1/2}$ $\langle \rho\ Cp\ \epsilon_H \rangle_m = 0.0195 \langle \rho_i \rangle^{1/2}$ $\langle \rho \epsilon_D \rangle_m = 0.0195 \langle \rho_i \rangle^{1/2}$	Extension of Ferri et al. and Kleinstein models Measurement of radial distribution

TURBULENT KINETIC ENERGY MODEL

There is no history effect incorporated in the eddy viscosity models. In order to get the history of turbulence into shear layer analysis, a suitable relation between turbulent shear stress and kinetic energy should be considered. A simple model for the turbulent shear stress was first suggested by Nevzgljadov [39] and discussed by Dryden [40].

This model was applied to the problem of turbulent free mixing of constant-density streams by Lee and Harsha [41, 42]. A linear relation between local turbulent shear stress and local turbulent kinetic energy is assumed:

$$\tau_t = a_1 \rho k \tag{94}$$

where a_1 is an empirical constant and $k = \overline{(u'^2 + v'^2 + w'^2)}/2$, the turbulent kinetic energy. They confirmed the linear relationship and recommended a value of 0.3 for the constant a_1 by correlation of the experimental measurements of various turbulent free-mixing flows [33, 41, 43, 44].

The local turbulent kinetic energy is always positive whereas the local shear stress changes in the same sign as the local velocity gradient. Therefore, the preceding equation should be modified as

$$\tau_t = a_1 \rho k \frac{dU}{dr} \Bigg/ \left|\frac{dU}{dr}\right| \tag{95}$$

An additional difficulty arises in this model when k still remains nonzero at the position where τ_t becomes zero with the local velocity gradient. To overcome this difficulty, the following restriction is added in the particular region:

$$\tau_t = a_1 \rho k \frac{dU}{dr} \Bigg/ \left|\frac{dU}{dr}\right|_{max} \tag{96}$$

where $\dfrac{dU}{dr} \simeq 0$

The governing equations for a free jet are given by

Axisymmetric flow:

$$\frac{\partial}{\partial z}(\rho U) + \frac{1}{r}\frac{\partial}{\partial r}(\rho r V) = 0 \tag{97}$$

$$\rho U \frac{\partial U}{\partial z} + \rho V \frac{\partial U}{\partial r} = \frac{1}{r}\frac{\partial}{\partial r}\left[r\mu_t \frac{\partial U}{\partial r}\right] \tag{98}$$

$$\rho U \frac{\partial k}{\partial z} + \rho V \frac{\partial k}{\partial r} = \frac{1}{r}\frac{\partial}{\partial r}\left[r \frac{\mu_t}{\sigma_t}\frac{\partial k}{\partial r}\right] + \mu_t\left(\frac{\partial U}{\partial r}\right)^2 - D_k \tag{99}$$

Two-dimensional flow:

$$\frac{\partial}{\partial z}(\rho U) + \frac{\partial}{\partial y}(\rho V) = 0 \tag{100}$$

$$\rho U \frac{\partial U}{\partial z} + \rho V \frac{\partial U}{\partial y} = \frac{\partial}{\partial y}\left[\mu_t \frac{\partial U}{\partial y}\right] \tag{101}$$

$$\rho U \frac{\partial k}{\partial z} + \rho V \frac{\partial k}{\partial y} = \frac{\partial}{\partial y}\left[\frac{\mu_t}{\sigma_t}\frac{\partial k}{\partial y}\right] + \mu_t\left(\frac{\partial U}{\partial y}\right)^2 - D_k \tag{102}$$

where σ_k is the ratio of the turbulent eddy viscosity to a coefficient for energy transfer (called the effective Schmidt number for turbulent kinetic energy).

According to Patankar and Spalding [45], the eddy viscosity is given by

$$\mu_t = l_k \rho k^{1/2} \tag{103}$$

The dissipation term D_k takes the form

$$D_k = a_2 \rho k^{3/2}/l_k \tag{104}$$

Here a_2 is an empirical constant and l_k is analogous to the mixing length for turbulent kinetic energy, which can be assumed to be equal to the width of the mixing layer.

The problem of incompressible free turbulent mixing can be considered by solving Equations 97 through 102 for U, V, k, τ_t, and μ_t. The calculation can be begun with defining the turbulent kinetic energy level k at the initial station. The turbulent kinetic energy model gives a much better description of the mixing process than the other models. However, it needs more starting conditions for input into the method of calculation. This additional input restricts the applicability of the model. Unfortunately, the experimental measurements for such initial conditions are not available for the case of nonisothermal turbulent jets with variable density.

Spalding [46] characterized the turbulence by three quantities k, W, and g for an axisymmetric incompressible free jet injected into a stagnant environment of different compositions. The quantity W can be regarded as the average value of the vorticity fluctuations having the dimensions of frequency squared. The quantity g is the average value of the square of the concentration fluctuations of injected fluid component. He adopted a single length scale l_k to influence the magnitudes of both the transport and the dissipation processes. The length scale is given by

$$l_k = \left(\frac{k}{W}\right)^{1/2} \tag{105}$$

The eddy viscosity is given by Equation 103.

In addition to Equations 97 to 99, the boundary-layer equation of mass conservation is

$$\rho U \frac{\partial f}{\partial z} + \rho V \frac{\partial f}{\partial r} = \frac{1}{r}\frac{\partial}{\partial r}\left[r \frac{\mu_t}{\sigma_f}\frac{\partial f}{\partial r}\right] \tag{106}$$

where f is the mass fraction of injected fluid component and σ_f the effective Schmidt number for concentration.

The equations for turbulence quantities are

$$\rho U \frac{\partial k}{\partial z} + \rho V \frac{\partial k}{\partial r} = \frac{1}{r}\frac{\partial}{\partial r}\left[r \frac{\mu_t}{\sigma_k}\frac{\partial k}{\partial r}\right] + \rho k^{1/2} l_k \left(\frac{\partial U}{\partial r}\right)^2 - C_D \rho \frac{k^{3/2}}{l_k} \tag{107}$$

$$\rho U \frac{\partial W}{\partial z} + \rho V \frac{\partial W}{\partial r} = \frac{1}{r}\frac{\partial}{\partial r}\left[r \frac{\mu_t}{\sigma_w}\frac{\partial W}{\partial r}\right] + C_1 \rho k^{1/2} l_k \left(\frac{\partial^2 U}{\partial r^2}\right)^2$$
$$+ C_3 \rho \frac{k^{1/2}}{l_k}\left(\frac{\partial U}{\partial r}\right)^2 - C_2 \rho \frac{k^{1/2}}{l_k} W \tag{108}$$

$$\rho U \frac{\partial g}{\partial z} + \rho V \frac{\partial g}{\partial r} = \frac{1}{r}\frac{\partial}{\partial r}\left[r \frac{\mu_t}{\sigma_g}\frac{\partial g}{\partial r}\right] + C_4 \rho k^{1/2} l_k \left(\frac{\partial f}{\partial r}\right)^2 - C_5 \rho \frac{k^{1/2}}{l_k} g \tag{109}$$

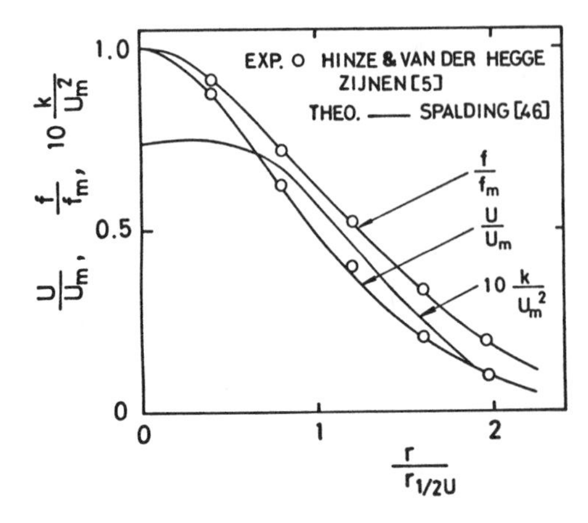

Figure 19. Comparison between theory and experiment of velocity, concentration, and kinetic energy in fully-developed region [46].

If the values of all the main fluid variables (U, f, k, W, g) are given at the initial station, the six parabolic differential equations can be solved simultaneously.

His values of constants recommended for this system are

$$C_D = 0.075, C_1 = 3.81, C_2 = 0.134, C_3 = 1.23, C_4 = 2.7, C_5 = 0.134$$

$$\sigma_f = 0.7, \sigma_k = 1.0, \sigma_w = 1.0, \text{ and } \sigma_g = 0.7$$

A general numerical procedure for solving the preceding parabolic differential equations is given in the book of Patankar and Spalding [45]. The predicted variations of τ_t, k, and g as well as those of U and f are in good agreement with the experimental measurements of many investigators [5, 21, 47] for incompressible turbulent jets. Some of his calculations are shown in Figures 19 and 20. Still much remains to be done for prediction of the development of heated jets.

Acknowledgment

The author would like to express his thanks to Y. Hirai and T. Harada for their assistance in the preparation of this chapter.

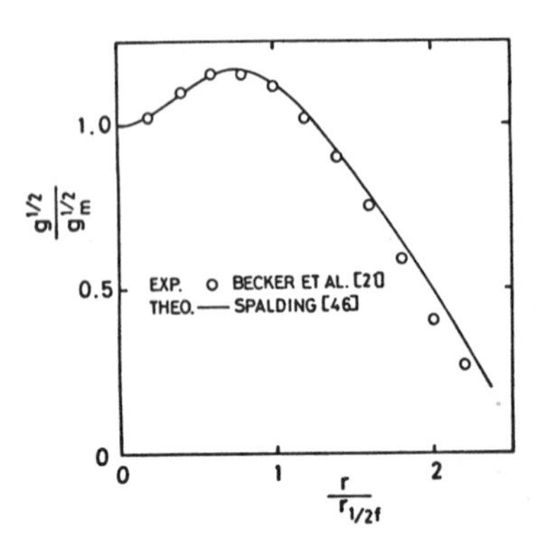

Figure 20. Comparison between theory and experiment of RMS concentration fluctuation in fully-developed region [46].

NOTATION

A	empirical constant (−)
A_1	empirical constant of centerline velocity decay (−)
A_2	empirical constant of velocity half-radius (−)
a	virtual origin (m)
a_1	empirical constant defined by Equation 94 (−)
a_2	empirical constant defined by Equation 104 (−)
B_1	empirical constant of centerline temperature decay (−)
B_2	empirical constant of temperature half-radius (−)
b	width of mixing region (m)
$C_1, \ldots, C_5$	constants of turbulence model (−)
C_D	constant for dissipation term of turbulence model (−)
C_p	heat capacity at constant pressure (J/kg K)
D	nozzle diameter (m)
D_k	dissipation of turbulent kinetic energy (kg/m s^3)
f	mass fraction of injected component (−)
g	square of the mass fraction fluctuations of injected component, time-averaged (−)
H	enthalpy time-averaged (J/kg)
h	enthalpy (J/kg)
j	mass flux (kmol/m^2 s)
K	mixing rate parameter (−)
k	turbulent kinetic energy (m^2/s^2)
l	mixing length (m)
M_j	jet Mach number (−)
M_t	total momentum flux (m^4/s^2)
Pr	Prandtl number (−)
q	heat flux (W/m^2)
Re	Reynolds number (−)
r	radial distance from jet axis (m)
r_o	nozzle radius (m)
T	temperature, instantaneous or time-averaged (K)
U	axial velocity time-averaged (m/s)
u	axial component of jet velocity (m/s)
V	radial velocity time-averaged (m/s)
v	radial component of jet velocity (m/s)
Ve	volume of traveling eddy (m^3)
W	property of local turbulence represented as the time-averaged value of the square of the vorticity fluctuations (1/s^2)
X_C	dimensionless core length (−)
y	lateral or transverse distance from jet axis (m)
z	axial distance from nozzle exit (m)

Greek Symbols

Γ	mass fraction time-averaged (−)
γ_i	mass fraction of i-component gas (−)
δ^*, δ^*_r	compressible displacement thickness defined by Equations 79 and 81 (m)
ϵ	eddy diffusivity (m^2/s)
η	dimensionless radial distance $r/r_{1/2}$ (−)
κ	thermal conductivity (W/m K)
κ_1	universal constant (−)
μ	fluid viscosity (kg/m s)
ξ	similarity variable (−)
ρ	fluid density, instantaneous or time-averaged (kg/m^3)
σ	effective Schmidt number (−)
τ	momentum flux (N/m^2)

Superscripts

′	fluctuation or RMS value of fluctuations

Subscripts

C	core
D	mass definition
eq	equivalent
F	Ferri model

f	mass fraction
g	quantity
H	heat
i	initial
k	turbulent kinetic energy
M	momentum
m	maximum or jet centerline
o	nozzle exit
P	Prandtl model
S	Schetz model
s	stagnation
T	temperature or Tomich model
t	turbulent
U	velocity
W	local turbulence W
1	at transverse distance y
2	at transverse distance y + dy
1/2	half-radius
∞	ambient

Overlines

— time-averaged

Brackets

$\langle\ \rangle$ nondimensionalized with respect to nozzle exit quantities

REFERENCES

1. Kataoka, K., Shundoh, H., and Matsuo, H., *J. Chem. Eng. Japan*, 15:17 (1982).
2. Prandtl, L., *Z. angew. Math. Mech.*, 22:241 (1942).
3. Schlichting, H., *Boundary Layer Theory*, 6th ed., Chapter 24, McGraw-Hill, New York, 1968.
4. Hinze, J. O., *Turbulence*, Chapter 6, McGraw-Hill, New York (1959).
5. Hinze, J. O., and van der Hegge Zijnen, B. G., *Appl. Sci. Res.*, 1A:435 (1949).
6. Keagy, W. R., and Weller, A. E., *Proc. Heat Transfer and Fluid Mechanics Inst.*, Berkeley, Calif., pp. 89 (1949).
7. Ruden, P., *Naturwissenschaften*, 21:375 (1933).
8. Reichardt, H., *Forsch. Gebiete Ingenieurw.*, No. 414 (1951).
9. Forstall, W., and Gaylord, E. W., *J. Appl. Mech.*, 22:161 (1955).
10. Kiser, K. M., *A.I.Ch.E. J.*, 9:386 (1963).
11. Corrsin, S., and Uberoi, M. S., *NACA TN*, No. 1865 (1949).
12. Thring, M. W., and Newby, M. P., *4th* Symposium on Combustion, Cambridge, Mass., pp. 789 (1952).
13. Sunavala, P. D., Hulse, C., and Thring, M. W., *Combustion and Flame*, 1:179 (1957).
14. Prandtl, L., *Z. angew. Math. Mech.*, 5:136 (1925).
15. Crocco, L., *Rend. Mat. Univ. Roma*, V2:138 (1941).
16. Witze, P. O., *A.I.A.A. J.*, 12:417 (1974).
17. Kleinstein, G., *J. Spacecraft Rockets*, 1:403 (1964).
18. Wilson, R. A. M., and Danckwerts, P. V., *Chem. Eng. Sci.*, 19:885 (1964).
19. Sforza, P. M., and Mons, R. F., *Int. J. Heat Mass Transfer*, 21:371 (1978).
20. Kristmanson, D., and Danckwerts, P. V., *Chem. Eng. Sci.*, 16:267 (1961).
21. Becker, H. A., Hottel, H. C., and Williams, G. C., *J. Fluid Mech.*, 30:285 (1967a).
22. Warren, W. R., *Ph.D. thesis* Princeton Univ., Princeton, N.J. (1957).
23. Davies, A. E., Keffer, J. F., and Baines, W. D., *Phys Fluids*, 18:770 (1975).
24. Peters, C. E., *TR-68-270*, Arnold Engineering Development Center, Tullahoma, Tenn. (1969).
25. Kataoka, K., and Takami, T., *A.I.Ch.E. J.*, 23:889 (1977).
26. Donaldson, C. dup. and Gray, K. E., *A.I.A.A. J.*, 4:2017 (1966).
27. Ferri, A., Libby, P. A., and Zakkay, V., *ARL 62-467*, Aeronautical Research Labs., Wright-Patterson Air Force Base, Ohio (1962).

28. Ferri, A., Libby, P. A., and Zakkay, V., *3rd ICAS Conference*, Spartan Books, Baltimore, Md. (1964).
29. Alpinieri, L. J., *A.I.A.A. J.*, 2:1560 (1964).
30. Schetz, J. A., *A.I.A.A. J.*, 6:2008 (1968).
31. Zelazny, S. W., *A.I.A.A. J.*, 9:2292 (1971).
32. Eggers, J. M., *NASA TN*, No. 3601 (1966).
33. Zawacki, T. S., and Weinstein, H., *NASA CR*, No. 959 (1968).
34. Cohen, L. S., *United Aircraft Research Laboratories Report*, G211709-1 (1968).
35. Cohen, L. S., and Guile, R. N., *NASA CR*, No. 1473 (1969).
36. Libby, P. A., *A.R.S. J.*, 32:388 (1962).
37. Tomich, J. F., and Weger, E., *A.I.Ch.E. J.*, 13:948 (1967).
38. Kataoka, K., Matsuo, H., and Shundoh, H., *J. Chem. Eng. Japan*, 15:255 (1982).
39. Nevzgljadov, V., *J. Phys. (USSR)*, 9:235 (1945).
40. Dryden, H. L., *Advances in Applied Mechanics*, Vol. 1. Academic Press, New York, pp. 1–40, 1948.
41. Lee, S. C., and Harsha, P. T., *A.I.A.A. J.*, 8:1026 (1970).
42. Harsha, P. T., and Lee, S. C., *A.I.A.A.J.*, 8:1508 (1970).
43. Heskestad, G., *J. Appl. Mech.*, 32:721 (1965).
44. Heskestad, G., *J. Appl. Mech.*, 33:417 (1966).
45. Patankar, S. V., and Spalding, D. B., *Heat and Mass Transfer in Boundary Layers*, Morgan-Grampian, London, 1967.
46. Spalding, D. B., *Chem. Eng. Sci.*, 26:95 (1971).
47. Wygnanski, I., and Fiedler, H. E., *Boeing Sci. Res. Lab. Doc.*, No. D1-82-0712 (1968).

CHAPTER 21

JET MIXING OF FLUIDS IN VESSELS

Toshiro Maruyama

Department of Chemical Engineering
Kyoto University
Kyoto, Japan

CONTENTS

INTRODUCTION, 544

RECIRCULATION TIME, 546

MIXING TIME, 548

MIXING BY USE OF HORIZONTAL JETS, 551

MIXING BY USE OF INCLINED JETS, 552

MIXING BY USE OF VERTICAL JETS, 554

CONCLUSIONS, 555

NOTATION, 555

REFERENCES, 556

INTRODUCTION

In modern chemical processing units, it is common practice for liquids in a tank to be circulated by drawing them through a pump and returning them to the tank through a pipe or nozzle for such purposes as homogenization of physical properties, prevention of stratification, prevention of deposition of suspended particles, and tank cleaning.

Figure 1 depicts diagrammatically the configuration and symbols of jet mixing in a tank. The situation chosen is jet injection through a nozzle attached to the side wall or the bottom wall of a cylindrical tank. The external recirculation system consists of a pipeline and a pump. The volume of the external system is negligibly small in comparison with the volume of the contents of the tank. For horizontal or inclined jet mixing the nozzle is directed across a diameter of the tank at an arbitrary height h_i and an arbitrary elevation angle θ (Figure 1A), and for vertical jet mixing the nozzle is vertically directed to the surface at an arbitrary radial position r (Figure 1B).

The correlating equations of the mixing time have been proposed by Fossett and Prosser [1], Fox and Gex [2], van de Vusse [3], and Okita and Oyama [4]. Fossett and Prosser [1] assumed that the momentum of the jet was preserved in the tank and that the jet diameter and jet axis length at the termination point of mixing were equal to the tank diameter. They measured concentrations electrically by a pair of electrodes in a tank (D = 152.4 cm, H = 25.7 cm) and correlated the mixing time with Equation 1 on the basis of the two assumptions.

$$t_M = 8D^2/\sqrt{qu} \qquad \text{for Re} \geqq 4{,}500 \tag{1}$$

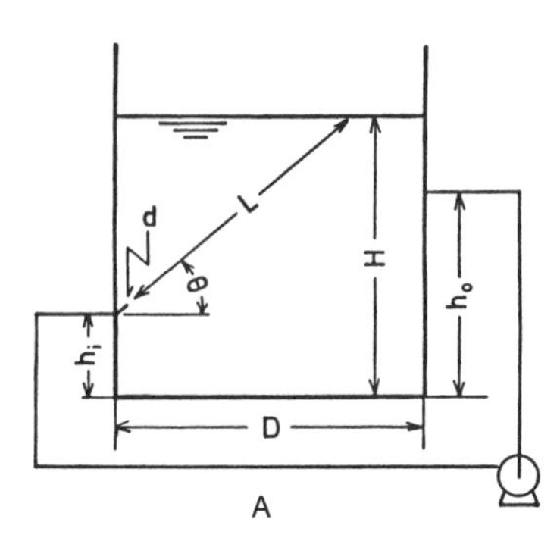

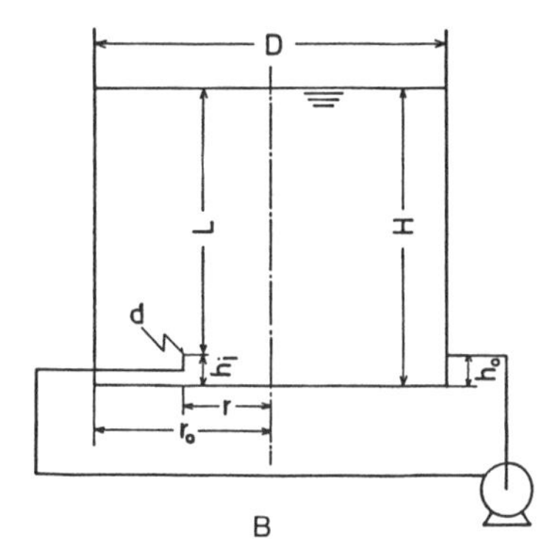

Figure 1. Configuration and symbols of jet mixing in tank: (A) horizontal ($\theta = 0$ rad) or inclined ($\theta = 0$ rad) jet mixing; (B) vertical jet mixing.

Furthermore, they made a number of tests on blending of hydrocarbons and confirmed the applicability of Equation 1 to some large tanks (D = 457 cm, H = 457 cm–D = 4,390 cm, H = 1,006 cm) by measuring the concentration of grab sampling at different levels of the tank. In addition, they indicated some important points in practical applications: such as gravity effects, the required power, tank shape, multiple jets, and stratification.

Their measurements of mixing time, however, comprise the period for introduction of the second component to the tank. The time taken to inject the second fluid occupied about half the mixing time. Therefore, the constant 8 in Equation 1 was replaced with 4 by a later investigator [3]. Rushton [5] compared the mixing time of Fossett and Prosser with that of a side-entering propeller mixer. Fox and Gex [2] measured the mixing time for a tank D = 30.5 cm by visual interpretation of endpoint and for a tank D = 152.4 cm by a concentration measurement. Their correlation of mixing time constitutes the dependences on gravity and Reynolds number. Van de Vusse [3] assumed that mixing was accomplished in the circulation time obtained by dividing the tank contents volume by the flow rate of the jet midway between nozzle and liquid surface.

$$t_M = 8.7D^2 \sin\theta/du \tag{2}$$

He measured the density difference in a tank (D = 3,600 cm, H < 1,180 cm) equipped with a nozzle of a fixed elevation angle ($\theta = 5\pi/36$ rad), and confirmed the validity of Equation 2. Okita and Oyama [4] obtained the mixing time for different sized tanks (D = 40 cm, 100 cm) by measuring concentration differences at two points with electrolyte conductivity probes placed in each tank. On the basis of a dimensional analysis of the mean circulation time, Okita and Oyama correlated the results by Equation 3:

$$t_M = 5.5(d/u)(D/d)^{1.5}(H/d)^{0.5} \qquad 5{,}000 < Re < 10^5 \tag{3}$$

In addition, they reported that the data of Fox and Gex [2] were also correlated well by Equation 3 if the constant 5.5 was replaced by 2.6.

In all the previously mentioned studies, the degree of mixing was defined qualitatively (e.g. terminal mixing or complete mixing), and consequently the mixing time has been determined from the time when the uniformity of composition (or color) in the specified sample size, within the precision of the instrument used (or of visual interpretation) is not further changed by additional mixing. Therefore, the mixing time is not sufficient to predict quantitatively the time required to achieve a desired degree of mixing. In addition, they did not describe the effect of jet location, i.e. the height and elevation angle of the nozzle, on efficient mixing because the correlation of mixing time is based on the concept of the mean circulation time.

Maruyama et al. [6, 7] discussed the mixing time and the circulation time on the basis of a recycle model of the tracer response and of the entrainment of the jet. An application of these models is shown by using the measured impulse response. In their experiments, a tracer which was trapped in the bypass line was instantaneously added to the system, already made steady in motion through regulation of the flow line by a three-way electromagnetic valve, without disturbing the steady flow rate and total contents of the system. Temporal changes in concentration were measured by using an electric conductivity meter. The mixing time was defined by the quantitative definition of the degree of mixing, and the circulation time was obtained from the period of a damping oscillation of the response curve. The optimum height and angle of nozzle for efficient mixing were determined from the measured mixing time and were discussed on the basis of the characteristics of the jet-induced circulation.

RECIRCULATION TIME

A recycle model for tracer response is applied to analyze mixing characteristics. When the Reynolds number of the jet is so large that the circulating flow in the tank dominates the mixing and the variance of circulation time is small, an output curve of the impulse response shows a damping oscillation. Khang and Levenspiel [8] proposed an approximate expression for the decaying amplitude of the damping oscillation.

$$A = 2\exp(-2\pi^2\sigma_c^2 t/t_c) \tag{4}$$

That is, a concentration deviation from a mixed mean value decreases exponentially in time and the time elapsed to a certain value of A depends on the mean circulation time t_c and the dimensionless variance of the circulation time, σ_c^2.

Figure 2 shows an example of the impulse response [6]. It is normalized by dividing by the final value. The abscissa is the elapsed time from the instant of tracer injection. In this case, the mean residence time is 934 s. An oscillation appears which damps and disappears within a short time compared to the mean residence time. The mean circulation time t_c is obtained from the mean period of the oscillation, i.e., mean time from peak to peak and from valley to valley in the oscillation. The measuring position has no measurable effect on the amplitude and period of oscillation, although in some cases high-frequency oscillations of small amplitude are added to the damping oscillation.

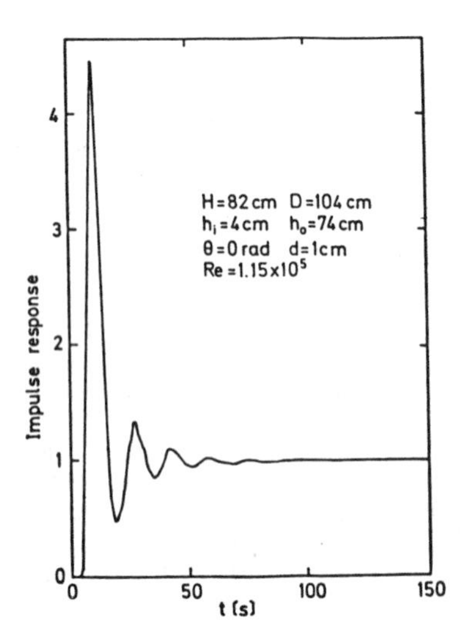

Figure 2. Typical example of impulse response.

Meanwhile, since the circulating flow is caused by the entrainment of the jet, the mean circulation time is expressed by the liquid volume in the tank and the flow rate of the jet at its termination point as:

$$t_c = V/q_j \tag{5}$$

If one assumes that the jet terminates at the point where the jet axis collides with the inner wall of the tank or intersects the liquid surface, and that the entrainment rate of the jet is constant, the flow rate at the termination point is expressed as:

$$q_j/q = kL/d \tag{6}$$

Then the mean circulation time t_c can be normalized by the mean residence time and the ratio d/L.

$$(t_c/t_R)/(d/L) = 1/k = \text{constant} \tag{7}$$

In fact, the experimental results of Maruyama et al. [6, 7] showed that the mean circulation time was independent of the jet injection position, and that as a representative length scale of tank the jet axis length L was more suitable than the diameter and depth of the tank.

Tables 1 to 3 show a dimensionless mean circulation time $(t_c/t_R)/(d/L)$ obtained from the mean period of the damping oscillation on the response curve. The value of $(t_c/t_R)/(d/L)$ ranges from 0.8

Table 1
Circulation Time for Horizontal Jet Mixing

D (cm)	H (cm)	L (cm)	h_i (cm)	d (cm)	$(t_C/t_R)/(d/L)$ (−)
104	104	90.7	4	1	1.57
104	104	90.7	4	1.8	1.67
104	104	90.7	4	0.5	1.71
104	104	90.7	44	1	1.46
104	104	90.7	44	0.5	1.24
104	93	98.3	4	1	1.54
104	82	90.7	4	1	1.59
104	73	98.3	4	1	1.76
104	61	98.3	4	1	1.88
104	52	90.7	4	1	1.88
104	43	98.3	4	1	2.08
56	56	53.7	4.38	1.8	1.63
56	56	53.7	4.38	1	1.60

Table 2
Circulation Time for Inclined Jet Mixing

θ (rad)	D (cm)	H (cm)	L (cm)	h_i (cm)	d (cm)	$(t_C/t_R)/(d/L)$ (−)
$7\pi/180$	104	53	90.7	4	1	1.61
$\pi/6$	104	104	105.3	24	1	1.22
$\pi/4$	104	104	104.3	24	1	1.52
$\pi/2$	104	104	91.0	4	1	1.26
$\pi/2$	104	104	70.3	24	1	1.03

Table 3
Mean Circulation Time for Vertical Jet Mixing

H/D	h_i (cm)	r/r_0	t_C (s)	$(t_C/t_R)/(d/L)$
1.0	13.5	0.96	16.7	1.28
1.0	13.5	0.90	17.2	1.32
1.0	13.5	0.84	18.3	1.40
1.0	13.5	0.81	15.1	1.16
0.8	13.5	0.96	19.7	1.43
0.8	13.5	0.90	18.1	1.31
0.6	13.5	0.96	11.9	0.81
1.0	0	0.96	16.9	1.51
1.0	0	0.90	15.4	1.38
1.0	0	0.80	15.8	1.41
0.8	0	0.96	14.6	1.26
0.8	0	0.90	15.3	1.32
0.8	0	0.80	16.3	1.41
0.8	0	0.70	18.3	1.58
0.6	0	0.96	14.3	1.24
0.6	0	0.90	14.8	1.29
0.6	0	0.80	14.3	1.24
0.4	0	0.96	21.5	1.78

to 2.1. Substituting these value into Equation 7 yields values of k ranging from 0.48 to 1.23, which are larger than those for the free jet ($k = 0.32$) obtained by Ricou and Spalding [9] from direct measurements of the entrainment rate.

MIXING TIME

The expressions of the mixing time, Equations 1 through 3, can be rewritten in terms of dimensionless mixing time which is similar to Equation 7.

$$(t_M/t_R)/(d/H) = 9 \qquad Re \geqq 4{,}500 \tag{8}$$

$$(t_M/t_R)/(d \sin\theta/H) = 8.7 \tag{9}$$

$$(t_M/t_R)/(d/\sqrt{DH}) = 5.5 \qquad 5{,}000 < Re < 10^5 \tag{10}$$

The constant in Equation 8 is replaced by 4.5 if the net mixing time is taken into account. In addition the constant in Equation 9 becomes 3.7 when the experimental value of van de Vusse [3], i.e. $\theta = 5\pi/36$ rad, is substituted into Equation 9. Thus the dimensionless mixing time falls into a narrow range of 2.6–5.5, including the measurements of Fox and Gex [2]. However, Equations 8 through 10 include different representative length scales of tank: i.e., liquid depth H in Equation 8, jet axis length $H/\sin\theta$ in Equation 9, and the geometric mean of H and D in Equation 10. None of the single length scales of the just mentioned ones represents the mixing time.

Substituting Equation 7 into Equation 4 yields the amplitude A as a function of both dimensionless time $(t/t_R)/(d/L)$ and the dimensionless variance of circulation time, σ_c^2.

$$A = 2\exp\{-2\pi^2\sigma_c^2 k(t/t_R)/(d/L)\} \tag{11}$$

The peak and valley values of the response curve were changed into the amplitude from the final value. Figure 3 shows a typical series of the decaying amplitude as a function of the dimensionless

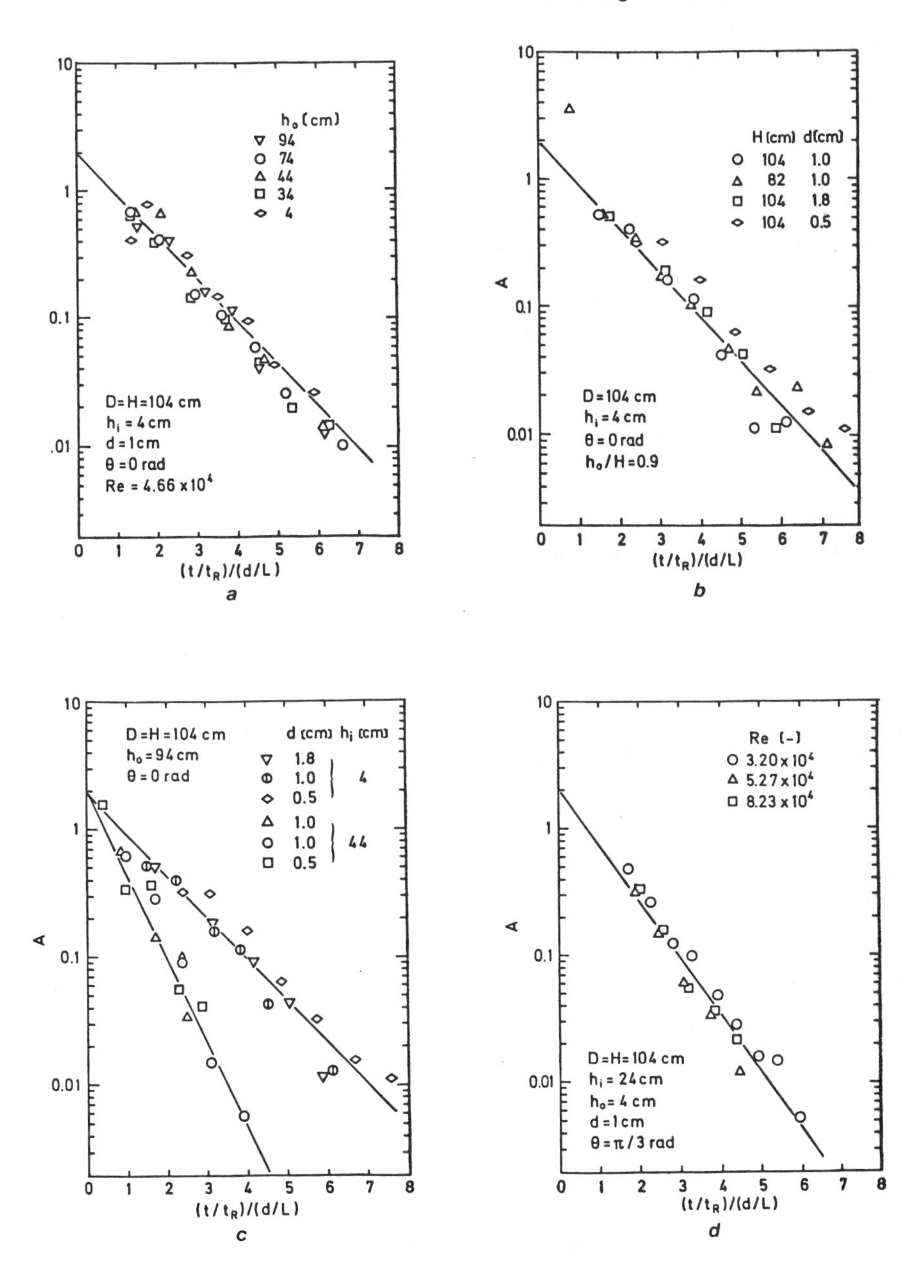

Figure 3. Amplitude A as a function of dimensionless time: (A) effect of suction height h_0; (B) effect of liquid depth H; (C) effect of nozzle height h_i; (D) $\theta = \pi/3$ rad.

time $(t/t_R)/(d/L)$. The decay of amplitude is independent of suction height h_0 and inner diameter of nozzle d. As shown by a solid line, measured amplitude can be correlated by the exponential function with the intercept 2 on the ordinate. That is, an expression of the recycle model, Equation 11 with σ_c^2 as a parameter, is a good approximation for the decaying amplitude except for the first few peak and valley values [8], and predicts how uniformity is approached. The difference in amplitude A shown in Figure 3C is not attributable to that of the mean circulation time.

At certain values of nozzle height, nozzle elevation angle, and liquid depth, the number of peaks and valleys on the response curve is too small to show the applicability of Equation 11 since the oscillation damps very rapidly or shows some deformations. At low values of Re where the circulation flow is not dominant to the mixing, no damping oscillation appears but the concentration decreases (after an overshoot) or increases monotonically to the mixed mean value. To compare the mixing characteristics for these various conditions, Maruyama et al. [6, 7] defined the mixing time as the time from the start of mixing to the time when the variation of measured concentration dropped to within $\pm 1\%$ of the mixed mean concentration. Therefore, the mixing time is obtained from the time at $A = 0.01$ when the response curve shows a precise damping oscillation. Because of a scatter of data for each set of system parameters, Maruyama et al. [6, 7] measured the mixing time more than three times. An average of the measurements was recorded as the final result.

One example of the Reynolds-number dependence of the mixing time is shown in Figure 4. Although at higher values of Re the dimensionless mixing time is independent of Re, at lower values of Re it becomes larger with decreasing Re. In addition, the response curve at $Re < 10^4$ is not characterized by damping oscillation. Depending on the relative height of the nozzle and suction pipe, the concentration increased or decreased (after an overshoot) monotonically to the mixed mean value. Thus, the dimensionless mixing time shows a complicated dependence on Re, h_i and h_0 in addition to its dependence on θ and H. That is, the mixing at the lower value of Re is not dominated by jet-induced circulation flow.

The transition-to-circulation-flow regime of mixing occurs at $1 \times 10^4 < Re < 3 \times 10^4$, where the dimensionless mixing time often shows the minimum value. At $Re \geqq 3 \times 10^4$ all the measured mixing times become independent of Re. This value of the lower bound of Re coincides with that associated with a free jet, i.e. $Re \geqq 2.5 \times 10^4$ where the entrainment rate of the free jet becomes independent of Re according to Ricou and Spalding [9]. On the other hand, the reported values [1, 4] of Re for the lower bound are small: Re = 4,500 with Equation 1, and Re = 5,000 with Equation 3. These smaller values are probably due to the fact that measurements were based on the qualitative definition of the degree of mixing without a detailed examination of the response curve.

The following discussion is based on the results in the circulation flow regime, i.e. $Re \geqq 3 \times 10^4$.

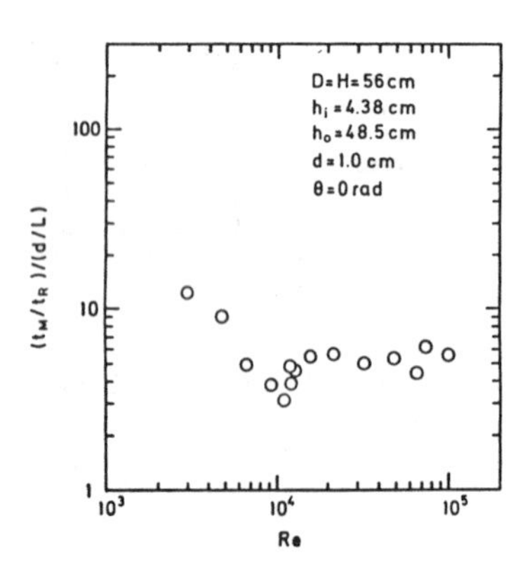

Figure 4. Dimensionless mixing time as a function of Reynolds number.

MIXING BY USE OF HORIZONTAL JETS [6]

Figure 5 shows the dimensionless mixing time $(t_M/t_R)/(d/L)$ at $H/D = 1$ plotted against dimensionless nozzle height h_i/H. The dimensionless mixing time $(t_M/t_R)/(d/L)$ is independent of the inner diameters of both nozzle and tank. It is constant, with a value of 3.5 at $h_i/H \geqq 0.25$, increasing to 7 with decreasing h_i/H at $h_i/H < 0.25$. The mixing time for $H/D < 1$ is shown in Figure 6. The results for different liquid depths show different dependence on h_i/D. The mixing time is least at $h_i/H = 0.5$. That is, there exists a range of optimum nozzle height for rapid mixing. The range becomes narrower with decreasing liquid depth. With variation in nozzle height from $h_i/H = 0.5$, meanwhile, the mixing time increases and at $h_i/H \leqq 0.05$ it reaches the maximum value, which is twice the minimum value. On the basis of the variance of circulation time, the dependence of mixing time on liquid depth can be qualitatively explained as follows.

Figure 7 schematically represents the jet-induced flow pattern in three dimensions. In general, an injected fluid develops a circular jet which induces three-dimensional circulations of larger variance of circulation time such as those in Figure 7A. In particular, the variance becomes large at the optimum nozzle height because a pair of strong circulations is formed at any cross section which includes the jet axis. On the other hand, a circular jet at a small value of h_i evolves into a wall jet along the vessel's bottom wall. According to Davis and Winarto [10], the spreading rate

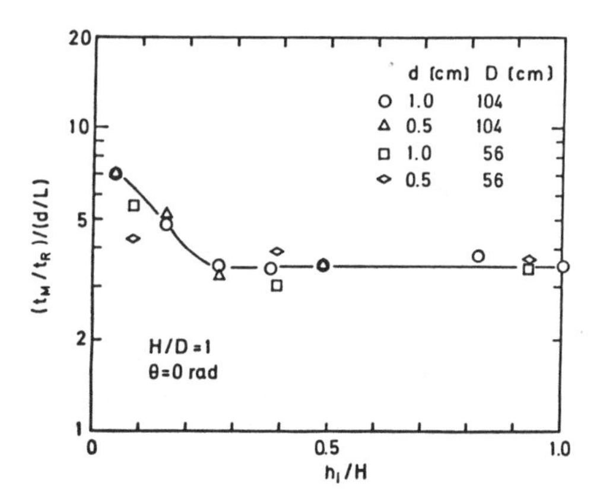

Figure 5. Dimensionless mixing time as a function of nozzle height for H/D = 1.

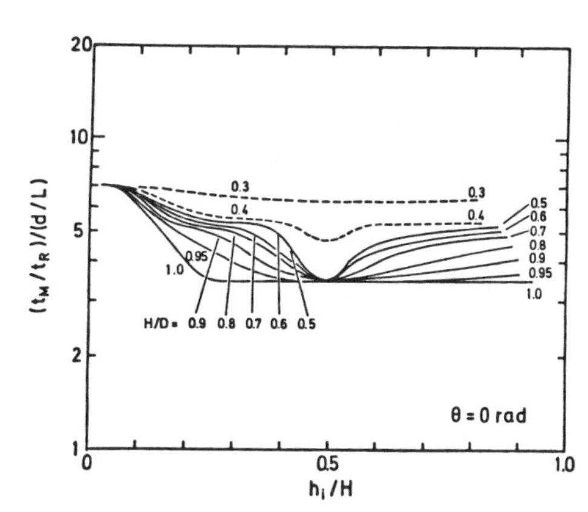

Figure 6. Dimensionless mixing time as a function of nozzle height and liquid depth.

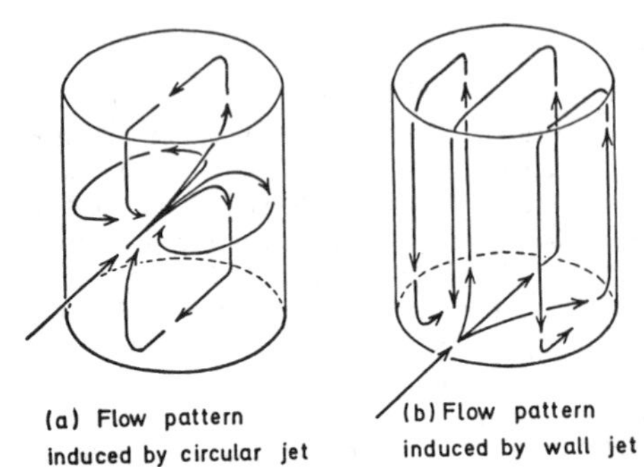

Figure 7. Qualitative sketch of flow pattern induced by jet.

of the wall jet is 8.5 times greater parallel to the wall than normal to the wall. Therefore, the wall jet spreads widely along the wall, inducing circulations which consist mainly of vertical two-dimensional loops as shown in Figure 7B. Hence the variance becomes small at the same mean circulation time. This is reflected by the large mixing time and by the precise damping oscillation on the response curve. From the results in Figure 6 it is obtained that the range of nozzle height where the wall effect appears in mixing time was $h_i/D \leqq 0.25$.

MIXING BY USE OF INCLINED JETS [6]

Figure 8 shows the dimensionless mixing time $(t_M/t_R)/(d/L)$ plotted against the elevation angle of the nozzle. Although the plot shows a scatter, it ranges from about 2.5 to about 7. Despite a wide range of jet axis length, the dimensionless mixing time and the dimensionless mean circulation time show the same ranges of value as those for $\theta = 0$ rad in which the jet axis length L is nearly equal to the tank diameter.

In Figure 8, the dimensionless mixing time becomes large at $\theta = \pi/2$ rad and at $\theta = 0$ rad and $h_i = 4$ cm. As explained in the preceding section, these larger values are due to the wall effect: i.e., the circular jet changes to a wall jet along the vessel's side wall, inducing circulations of small variances of circulation time. The range of the elevation angle affected by the wall is evident from results at $\theta < \pi/12$ rad where the mixing time at $h_i = 4$ cm differs from that at $h_i = 24$ cm. It is noted that the range of the elevation angle, i.e. $\tan(\pi/12) \leqq 0.27$, is equivalent to that of the nozzle height for $\theta = 0$ rad, i.e. $h_i/D \leqq 0.25$. Hence, the extent of the wall effect can be expressed in a

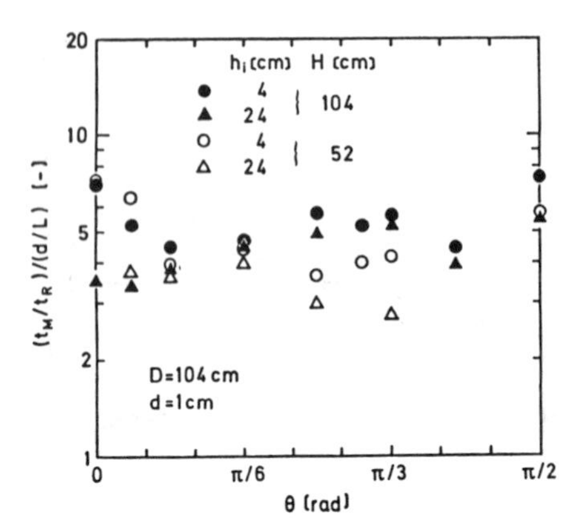

Figure 8. Dimensionless mixing time as a function of angular elevation.

unified manner by a cone in a tank as shown in Figure 9; the cone of vertical angle $\pi/6$ rad is concentric to the jet. So far as the cone does not contact the tank wall until the termination of the jet, the circular jet does not evolve into a wall jet and consequently the mixing time is not increased by the wall effect. It is worth noting that the vertical angle $\pi/6$ rad agrees with that associated with a free jet; according to Ricou and Spalding [9] the angle of exit orifice of $\pi/6$ rad is necessary for the orifice not to affect development of a free jet in direct measurement of the entrainment rate.

The dimensionless mixing time in Figure 8 comprises the dependence of a jet axis length L on nozzle elevation angle θ. To clarify the effect of elevation angle on the mixing time, the ratio of mixing time t_M to mean residence time t_R is shown in Figure 10. The large value at $\theta = 0$ rad corresponds to the maximum of the dimensionless mixing time, i.e. 7, and the small value at $\theta = 0$ rad corresponds to the minimum value, i.e. 3.5. In short, all results at $\theta = 0$ rad, if plotted, may be included between the two values. With the results for $h_i = 4$ cm, the ratio decreases with increasing θ and at $\theta = \pi/12$ rad the ratio shows the minimum value, which is nearly equal to that for $\theta = 0$ rad. Consequently, it is possible to decrease the mixing time to about the minimum value by giving the nozzle a sufficient elevation angle to exclude the wall effects. From a practical point of view, this fact is very useful since it is often the case that nozzle must be fixed at the side of the tank near the bottom for use irrespective of the level in the tank.

Meanwhile, the mean circulation time can be reduced by giving the nozzle an elevation angle and consequently making the jet axis length longer in the tank because the mean circulation time is inversely proportional to the jet axis length. This would reduce the mixing time if the variance

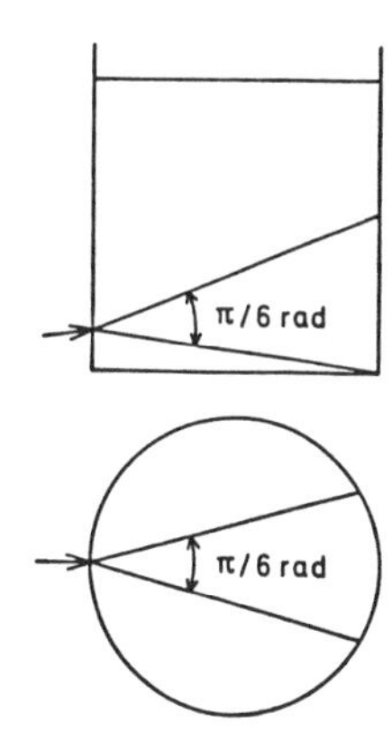

Figure 9. Cone of vertical angle $\pi/6$ rad in tank.

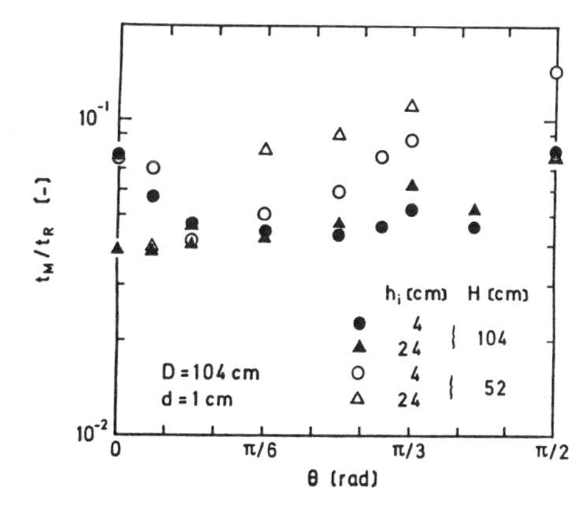

Figure 10. Ratio of mixing time to mean residence time as a function of angular elevation.

of the circulation does not decrease with increasing θ. Among the results in Figure 10, the largest value of L is given at the elevation angle $\theta = \pi/4$ rad at $h_i = 4$ cm and H = 104 cm; $\theta = \pi/6$ rad at $h_i = 24$ cm and H = 104 cm; $\theta = \pi/12$ rad at $h_i = 4$ cm, 24 cm and H = 52 cm. Evidently, the mixing times at these angles are small compared to the others but they are not smaller than the minimum value at $\theta = 0$ rad. That is, no reduction in mixing time from the minimum value for $\theta = 0$ rad is possible by tilting the nozzle upwards.

MIXING BY USE OF VERTICAL JETS [7]

Figure 11 shows the dimensionless mixing time plotted against the dimensionless radial position of a nozzle, r/r_0. The dimensionless mixing time is nearly constant, with a value of 3–6 at $r/r_0 \lesssim 0.5$, increasing with increasing r/r_0 at $r/r_0 \lesssim 0.5$. The increase can be interpreted by the wall effect described in a previous section. The extent of the wall effect is expressed in a unified manner by a cone in a tank as shown in Figure 12.

A nozzle position r/r_0 between 0 and 0.5 is recommended for rapid mixing irrespective of the level in the tank. When liquid depth is small, the horizontal or inclined jet mixing is preferable to the vertical jet mixing because the dimensionless mixing times are nearly equal to one another and consequently the mixing time is inversely proportional to the jet axis length L at the same values of d and t_R.

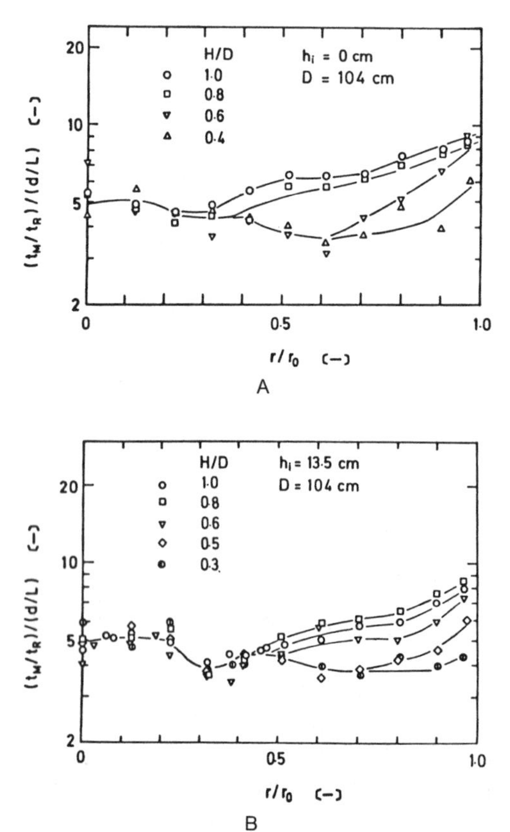

Figure 11. Dimensionless mixing time as a function of radial position of nozzle and liquid depth: (A) $h_i = 0$ cm; (B) $h_i = 13.5$ cm.

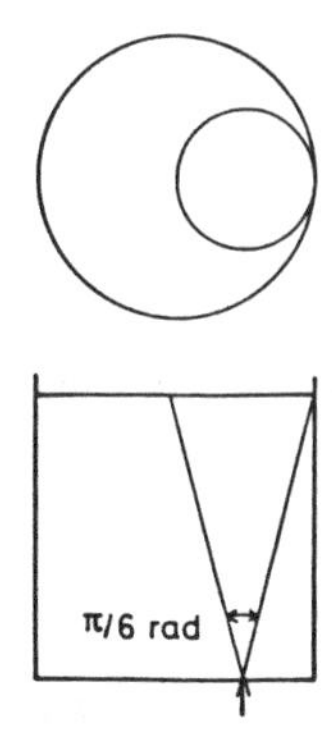

Figure 12. Cone of vertical angle $\pi/6$ rad in tank for vertical jet mixing.

CONCLUSIONS

In the circulation flow regime ($Re \geqq 3 \times 10^4$) of mixing the decaying amplitude of the output curve in the impulse response can be expressed by the recycle model. The mean circulation time obtained from the mean period of the damping oscillation can be normalized with a combined variable of the mean residence time in a tank and the ratio of nozzle diameter to jet axis length on the basis of the entrainment of a free jet. The dimensionless mean circulation time ranges from 0.8 to 2.1. The dimensionless mixing time depends on liquid depth, nozzle height, and nozzle elevation angle, and shows a value between 2.5 and 8.

When liquid depth is small, the horizontal or inclined jet mixing is preferable to the vertical jet case. As noted earlier the dimensionless mixing times are nearly equal to one another.

There exists an optimum geometric condition of nozzle for rapid mixing with a minimum value of dimensionless mixing time. For rapid vertical-jet mixing irrespective of the level in the tank, it is recommended to select the nozzle position r/r_0 between 0 and 0.5. For horizontal jet mixing the optimum nozzle depth ranges from the liquid surface level to three-quarters of the liquid depth when the liquid depth is equal to the tank diameter and is the mid-depth of liquid when the liquid depth is smaller than the tank diameter. If the nozzle height is within one-fourth of the tank diameter, the mixing time for horizontal jet becomes large because of the wall effect: i.e., the circular jet evolves into the wall jet, which induces circulations of small variance of circulation time. In this case, the mixing time can be decreased to about minimum value by giving the nozzle an elevation angle adequate to exclude the wall effect; a cone which is concentric to the jet and of a vertical angle $\pi/6$ rad should not contact the tank wall until the termination of the jet. The mixing time cannot be made less than the minimum value for a horizontal nozzle by tilting the nozzle upwards and consequently decreasing the mean circulation time.

NOTATION

A	normalized amplitude of response signal (−)
D	internal diameter of tank (cm)
d	internal diameter of nozzle (cm)
H	liquid depth in tank (cm)
h_i	height of nozzle from bottom of tank; see Figure 1 (cm)
h_0	height of suction pipe from bottom of tank; see Figure 1 (cm)
k	constant defined by Equation 3 (−)
L	length of axis of jet in tank; see Figure 1 (cm)
q	volumetric rate of flow through nozzle (cm^3/s)
q_j	volumetric flow rate of jet (cm^3/s)
Re	Reynolds number of jet ($=du/\nu$) (−)
r	radial position of nozzle; see Figure 1 (cm)
r_0	internal radius of tank; see Figure 1 (cm)
t	time (s)

t_C mean circulation time (s)
t_M mixing time (s)
t_R mean residence time ($=V/q$) (s)
u mean velocity of liquid through nozzle (cm/s)
V volume of liquid in tank (cm^3)

Greek Symbols

θ angular elevation of nozzle; see Figure 1 (rad)
ν kinematic viscosity of liquid (cm^2/s)
σ_c^2 dimensionless variance of circulation time (−)

REFERENCES

1. Fossett, H., and Prosser, L. E., *Proc. I. Mech. E.*, 160:224 (1949), idem: *Trans. Instn. Chem. Engrs.*, 29:322 (1951).
2. Fox, E. A., and Gex, V. E., *AIChE J.*, 2:539 (1956).
3. Van de Vusse, J. G., *Chemie-Ing.-Techn.*, 31:583 (1959).
4. Okita, N., and Oyama, Y., *Kagaku Kogaku*, 27:252 (1963).
5. Rushton, J. H., *Pet. Refiner*, 33:101 (1954).
6. Maruyama, T., Ban, Y., and Mizushina, T., *J. Chem. Eng. Japan*, 15:342 (1982).
7. Maruyama, T., Kamishima, N., and Mizushina, T., ibid, 17:120 (1984).
8. Khang, S. J., and Levenspiel, O., Chem. Eng. Sci., 31, 569 (1976).
9. Ricou, F. P., and Spalding, D. B., *J. Fluid Mech.*, 11:21 (1961).
10. Davis, M. R., and Winarto, H., *J. Fluid Mech.*, 101:201 (1980).

CHAPTER 22

MIXING IN LOOP REACTORS

Yasuhiro Murakami

Department of Chemical Engineering
Kyushu University, Japan

Tsutomu Hirose

Department of Industrial Chemistry
Kumamoto University, Japan

Shinichi Ono

Section of Production Technology
Shoei Chemical Industry Co., Ltd., Japan

CONTENTS

INTRODUCTION, 557

DISPERSION MODEL FOR RECIRCULATING FLOW, 559
 Batch Operation, 559
 Continuous Operation, 562

MIXING CHARACTERISTICS IN A LOOP REACTOR, 562
 Experimental Conditions, 563
 Data Processing, 564
 Peclet Number, 565

FLOW PROPERTIES IN A LOOP REACTOR, 567
 Background for Analysis, 567
 Discharge Characteristics, 568
 Discharge Pressure, 569
 Proposed Design Procedure, 570

NOTATION, 570

REFERENCES, 571

INTRODUCTION

A loop reactor is a reactor in which a recirculating channel and a fluid mover are combined into a loop. Various types as shown in Figure 1 belong to this category. An agitated vessel with a draft tube, as shown in Figure 1A and B [1], is a type of loop reactor, of which the former is a propeller driving type and the latter a jet driving type. A mammoth loop reactor shown in Figure 1C [1] is not equipped with a special fluid mover but fluid is circulated by the density difference caused by sparged gas. A type given in Figure 1D is a loop reactor proposed by Norwood [2] for slurry polymerization, and consists of a tubular loop with bends and an impeller type mover. The original I-shaped loop reactor can be modified into an L-shaped and a U-shaped loop reactor, as shown in Figure 2, of which the U-shape is the most similar to the commercial unit.

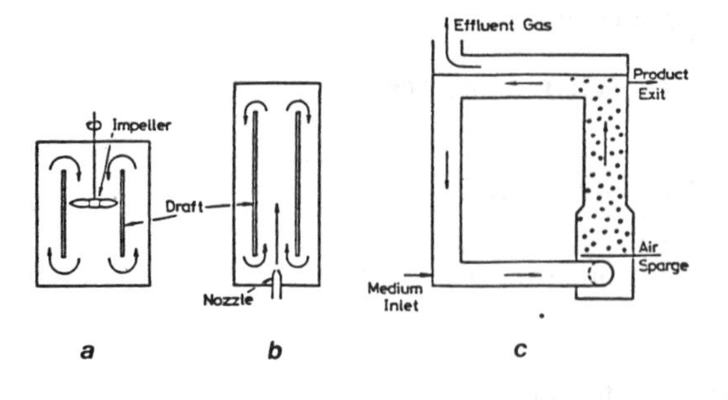

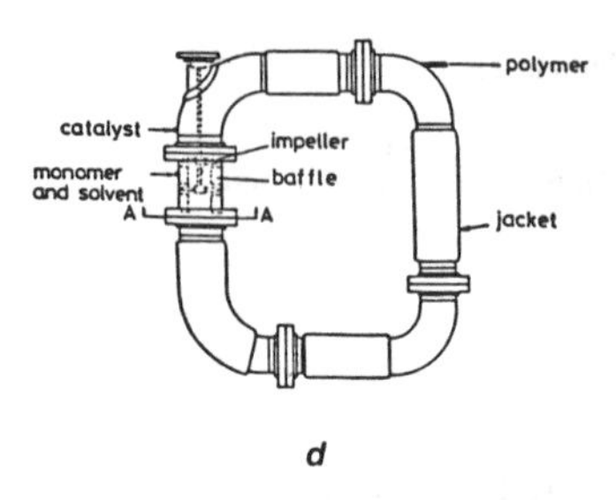

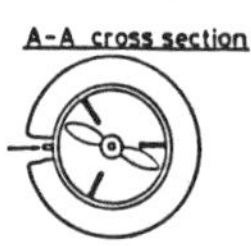

Figure 1. Schematic diagrams of various loop reactors: (A) propeller loop reactor; (B) jet loop reactor; (C) mammoth loop reactor; (D) loop reactor proposed by Norwood [2].

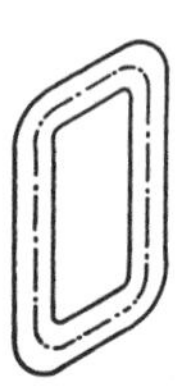
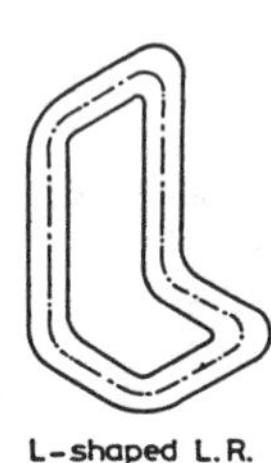
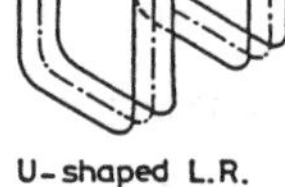

Figure 2. Skeletonized diagrams of I-, L-, and U-shaped reactors.

In this article special attention is paid to the last type, i.e. a tubular loop reactor, since it has been widely used for a large-scale production of homo- and copolymers of olefins in slurry and bulk (solventless) polymerizations. It allows high liquid velocity and thus a high heat transfer coefficient and large specific heat transfer area. Both of them may minimize the difficulties in the removal of polymerization heat generated at a high rate by the recently developed high activity catalyst. Besides this, a relatively uniform liquid velocity may avoid the problem of polymer deposition on the wall.

The Chemical Reaction Engineering Subdivision in the Society of Polymer Science, Japan, has given useful assessments on the production of polyethylene by the low-pressure process without deashing as shown in Figure 3 in which a loop reactor is involved as a main reactor. The report [3] of the process calculation for a production rate of 20,000 t/year stated that the 35% polymer slurry

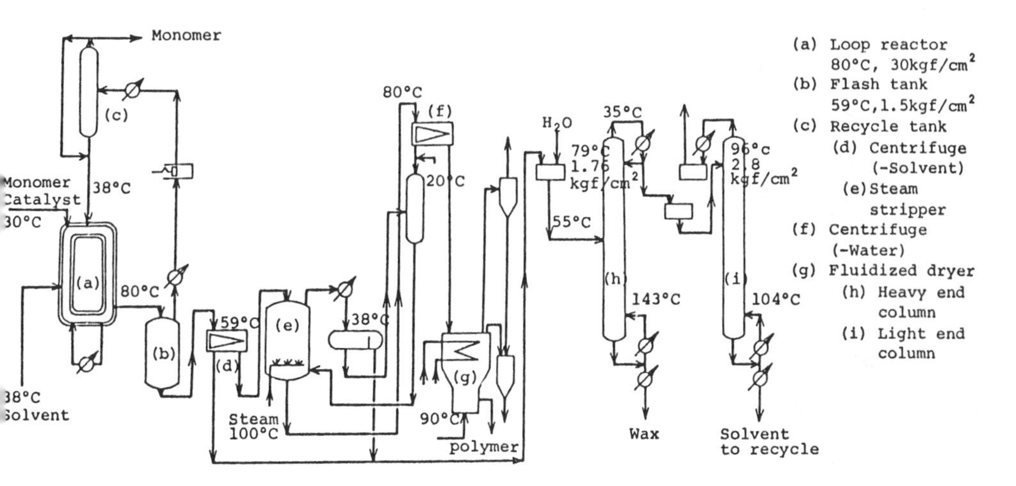

Figure 3. An example of the polyethylene process involving a loop reactor.

in a hexane solvent pressurized at 30 kgf/cm^2 and 80°C is supposed to be circulated at the velocity of 4 m/s in a 0.48 m i. d. and 126-m-long loop reactor with the volume of 23 m^3.

As has been often stated, axial mixing behavior in the reactor has a great significance on design and operation because the mixing characteristics influence yield and selectivity of products. Information on mixing together with findings of other flow phenomena is helpful to estimate the reactor performance.

In what follows, the mixing behavior of the loop reactor is described. Discussions deal with the mathematical formulation and the series solution of mixing in the loop reactor. Also information of mixing properties is given based on the experimental findings of the authors. The latter part of this chapter deals with flow behavior as related to mixing properties.

DISPERSION MODEL FOR RECIRCULATING FLOW

To visualize the axial mixing in the loop reactor, the dispersion model is applied in which some degree of axial mixing characterized by the dispersion coefficient D_a is superimposed on the plug flow. This model may be the most suitable for the loop reactor because of the absence of stagnant pocket and gross bypassing and of the small deviation from plug flow. The concentration response curve is discussed first for the batch operation and then for the continuous operation.

Batch Operation

A given amount of the tracer M is supposed to be injected in the delta functional impulse at x = 0 into a loop reactor of volume V and total loop length L in which the fluid circulates at the average velocity of U, as shown in Figure 4A. The local concentration of the tracer C = C(x, t) is described by the material balance equation based on the dispersion model,

$$\frac{\partial C}{\partial t} = D_a \frac{\partial^2 C}{\partial x^2} - U \frac{\partial C}{\partial x} \tag{1}$$

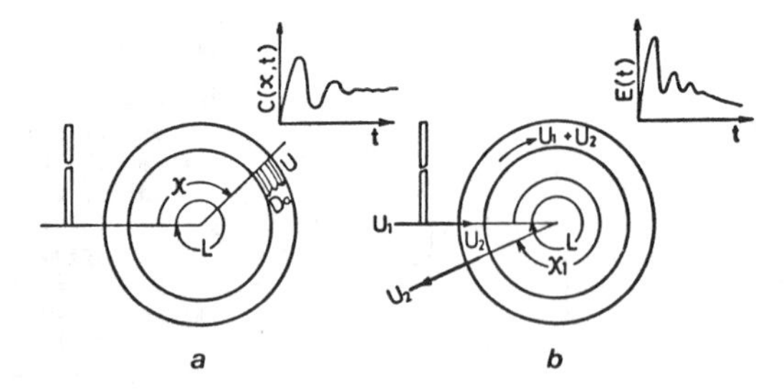

Figure 4. Schematic diagram of the dispersion model for loop reactors: (A) batch operation; (B) continuous operation.

The initial and boundary conditions are

$$C(x, 0) = \bar{C}\delta(x) \tag{2}$$

$$C(0, t) = C(L, t) \tag{3}$$

$$\frac{\partial C(0, t)}{\partial x} = \frac{\partial C(L, t)}{\partial x} \tag{4}$$

in which $\bar{C}$ is the average tracer concentration after the complete mixing, i.e. $\bar{C} = M/V$.

The solution is obtained [4, 5] as

$$\frac{C(x^*, t^*)}{\bar{C}} = \sqrt{\frac{Bo}{4\pi t^*}} \sum_{j=-\infty}^{\infty} \exp\left[-\frac{Bo}{4t^*}(j + x^* - t^*)^2\right] \tag{5}$$

in terms of dimensionless groups of

$$t^* = t/(L/U) \tag{6}$$

$$x^* = x/L \tag{7}$$

$$Bo = \frac{LU}{D_a} \tag{8}$$

by taking account of the periodic nature of the solution. The solution, Equation 5 satisfies also the normalization and the equilibrium conditions of

$$\int_0^L C(x, t)\, dx = \lim_{t \to \infty} C(x, t) = \bar{C} \tag{9}$$

Voncken et al. [6] and Blenke [1], respectively, applied the dispersion model to the agitated vessel and the jet loop reactor. They summed up the tracer response for an open vessel at every interval of loop length and obtained essentially the same result as Equation 5 except that the summation was taken only for positive j. Since the contribution of the terms with j less than 1 is negligibly small for Bo greater than 10, their results are practically identical with the present result.

The mixing behavior in recirculation flow is characterized by a single parameter of the Bodenstein number (Bo). The tracer concentration according to Equation 5 shows a unique oscillating decay for a given value of Bo (as shown in Figure 5, for example, at Bo = 300). According to Equation 5, the time t^* at which the k-th trough of the concentration appears approaches asymptotically the value of $x^* + k - 1$ with increasing k. For example, at Bo = 50, the difference between the true

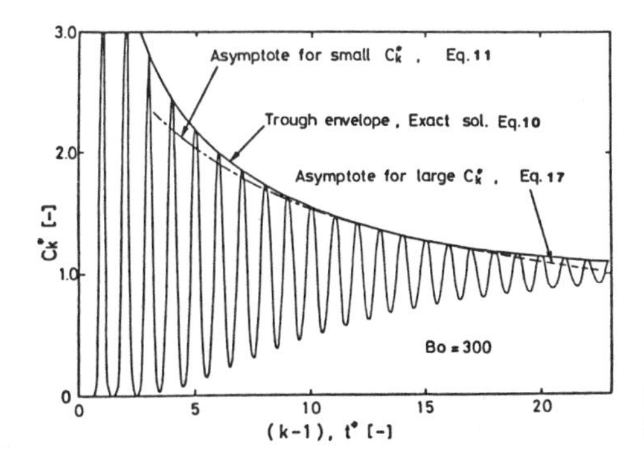

Figure 5. An example of the calculated response curve and trough envelope.

value t_k^* and asymptotic value $x^* + k - 1$ is only 0.02 at $k = 2$ and even less for large value of **k**. By substituting $t^* = x^* + k - 1$ in to Equation 5, the concentration at the k-th trough C_k is given approximately by

$$\frac{C_k}{\bar{C}} = \sqrt{\frac{Bo}{4\pi(k - 1 + x^*)}} \sum_{j=-\infty}^{\infty} \exp\left\{-\frac{Bo \cdot j^2}{4(k - 1 + x^*)}\right\} \quad \text{for } Bo \geq 50 \tag{10}$$

where $(j - k + 1)$ in the exponent has been newly replaced by j because of summation from $-\infty$ to ∞. Note that the terms with $j \neq 0$ in the summation vanish rapidly for large values of Bo/k, and Equation 10 can be approximated by the term with $j = 0$ alone.

$$\frac{C_k}{\bar{C}} = \sqrt{\frac{Bo}{4\pi(k - 1 + x^*)}} \tag{11}$$

to within 1% accuracy for $C/\bar{C} > 1.3$, as shown in Figure 5.

Equation 10 can be rewritten for the injection point ($x^* = 0$) through Jacobi's identity [7]

$$\frac{C_k}{\bar{C}} = \sqrt{z}\,\theta(z) \tag{12}$$

$$= \theta\left(\frac{1}{z}\right) \tag{13}$$

$$= 1 + 2 \sum_{j=1}^{\infty} \exp\left\{-\frac{4\pi^2(k - 1)}{Bo} j^2\right\} \tag{14}$$

in which

$$z = \frac{Bo}{4\pi(k - 1)} \tag{15}$$

and $\theta(z)$ is the theta function defined by

$$\theta(z) = \sum_{j=-\infty}^{\infty} \exp(-\pi j^2 z) \tag{16}$$

Equation 14 converges very rapidly for large values of k/Bo and can be approximated by the first term only

$$\frac{C_k}{\bar{C}} = 1 + 2\exp\left\{-\frac{4\pi^2(k-1)}{Bo}\right\} \tag{17}$$

to within 1% accuracy for $C_k/\bar{C} < 1.4$, as shown by the chain line in Figure 5.

The number of circulations $k - 1$ necessary to reach the given uniformity $(C_k/\bar{C} - 1)$ is derived readily from Equation 17, i.e.

$$k - 1 = -\frac{Bo}{4\pi^2}\ln\left\{\frac{1}{2}\left(\frac{C_k}{\bar{C}} - 1\right)\right\} \tag{18}$$

For example, it reduces to

$$k - 1 = 0.134\ Bo \tag{19}$$

for 1% deviation, i.e. $C_k/\bar{C} = 1.01$. This criterion can be applied to estimate the time required for the desired homogeneity.

Continuous Operation

The preceding results for the batch operation may be modified to obtain the residence time distribution E(t) for the continuous operation in the loop reactor. As shown in Figure 4B, the fresh fluid enters the reactor at $x = 0$ at the velocity of U_1 (reduced to the column cross-sectional area) to mix with the base circulating fluid of the velocity U_2 and leaves it at $x = x_1$ at the same velocity U_1. The total amount of the tracer decreases after each circulation by a factor of the ratio of flow rate $Y = U_2/(U_1 + U_2)$ since only the fraction Y of the total flux of the tracer remains to circulate in the reactor at the exit point $x = x_1$. Such a dilution effect may be taken into account by weighting a factor of Y^j in the summation of Equation 5. Thus, the tracer concentration at $x = x_1$ is described by

$$\frac{C}{\bar{C}} = \sqrt{\frac{Bo}{4\pi t^*}}\sum_{j=1}^{\infty}\left[Y^j \cdot \exp\left\{-\frac{Bo}{4t^2}(j + x^* - t^*)^2\right\}\right] \tag{20}$$

in which terms with negative j are dropped for simplicity. Since the dimensionless time t^* relative to the circulation time $(L/(U_1 + U_2))$ is related to the dimensionless time τ relative to the mean residense time (L/U_1) by the equation

$$t^* = \frac{\tau}{1 - Y} \tag{21}$$

the residense time distribution $E(\tau)$ for the continuous loop reactor is given by

$$E(\tau) = \sqrt{\frac{Bo(1-Y)}{4\pi\tau}} \cdot \sum_{j=1}^{\infty}\left[Y^j \exp\left\{-\frac{Bo(1-Y)}{4\tau}\left(j + x^* - \frac{\tau}{1-Y}\right)^2\right\}\right] \tag{22}$$

in which Bo is evaluated at $U = U_1 + U_2$.

MIXING CHARACTERISTICS IN A LOOP REACTOR

The mixing behavior in the loop reactor can be characterized by a single parameter of the Bodenstein number $Bo = UL/D_a$, as discussed in the previous chapter. The general trend of Bo will be discussed in what follows, based on the experimental findings obtained for various geometrical and operational variables in the authors' laboratory [4].

Experimental Conditions

The mixing behavior was measured in an I-shaped loop reactor, of which the main dimensions are given in Figure 6, under baffled and unbaffled conditions. Most experiments were carried out in a reactor with a tube diameter $d_t = 10$ cm, total length of loop $L = 28d_t$, and four baffle plates $W_b \times W_L = 0.15d_t \times 1.5d_t$ while some were carried out for smaller diameter ($d_t = 5$ cm), shorter length ($L = 22d_t$), or wider baffle plates ($W_b = 0.3d_t$) to find the geometrical effects. The impellers used were pitched paddle impellers with various pitched angles, an axial flow pump-type impeller and a marine screw, of which the detailed dimensions are given in Table 1 and Figure 7.

Measurements of axial dispersion in circulation flow in the loop reactor were carried out by an impulse response method. KCl solution used as tracer was injected from a syringe into the loop

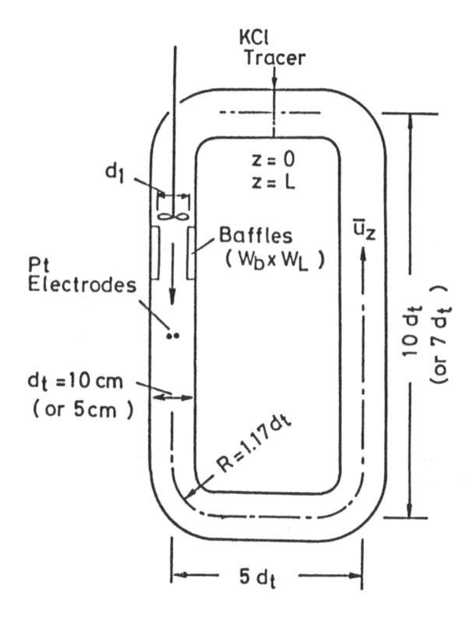

Figure 6. Loop geometry and notation.

Table 1
Geometries of Impellers Used and Legend for Figures 11 through 13

	d_t [cm]	d_1 [cm]	Pitched Paddles 20°	30°	40°	60°	Axial flow 30°	Marine Screw 33°
Baffled	10	9	—	⦶	—	—	◇	—
	10	8	□	○ ⍉	—	△	—	▽
	10	6	—	⊖	—	—	—	—
Unbaffled	10	8	—	⊘ ●	—	▲		
	5	4	—	—	■	—		

$L = 28d_t$ except for ⊘: $22d_t$
$W^b \times W_L = 0.15d_t \times 1.5d_t$ except for ⍉: $0.3d_t \times 1.5d_t$

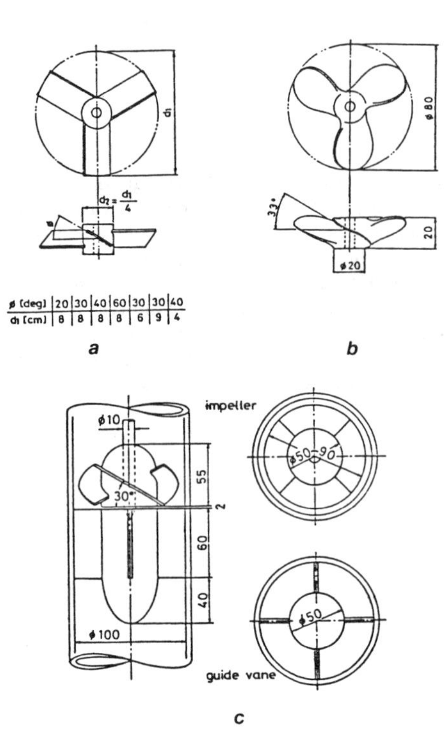

Figure 7. Schematic diagram and dimensions of impellers (unit: mm): (A) pitched blade paddle impellers; (B) marine screw impellers; (C) axial-flow pump impellers.

reactor at a section shown in Figure 6 and the conductivity of flowing liquid was monitored continuously by a set of electrodes located in a section $2.5d_t$ downstream from the impeller. The working fluid was tap water and the experimental range of pipe Reynolds number was $2.5 \times 10^3 - 2 \times 10^5$.

Data Processing

The usual method for determining Bo may be a curve-fitting of the whole response curve measured to that calculated from Equation 5. However, full curve-fitting is sometimes cumbersome in practice. Thus, the following simplified method is proposed to identify Bo. Equation 11 is rewritten

$$\left(\frac{\bar{C}}{C_k}\right)^2 = \frac{4\pi}{Bo}(k - 1 + x^*) \quad \text{for} \left(\frac{\bar{C}}{C_k}\right)^z < 0.6 \tag{23}$$

and the value of Bo can be simply determined from the slope ($= 4\pi/Bo$) in a plot of $(\bar{C}/C_k)^2$ against $(k - 1)$, as shown in Figure 8 for some examples of measured response curves. This method of determining Bo is applicable to the value of Bo greater than 50, a condition satisfied under most operations of the loop reactor.

Comparison of response curves between calculated and measured values is shown in Figure 9 in case of batch operation. An oscillating curve shows the response curve calculated from Equation 5 with Bo = 120 which was obtained by the simple plot proposed above. Points show the measured response curve obtained under the condition specified in the figure. In the range of small t^*, a little difference appears between the two response curves at the trough of the curve while in the range of $t^* > 3$ the curves show good agreement. Figure 10 is an example of the response curve

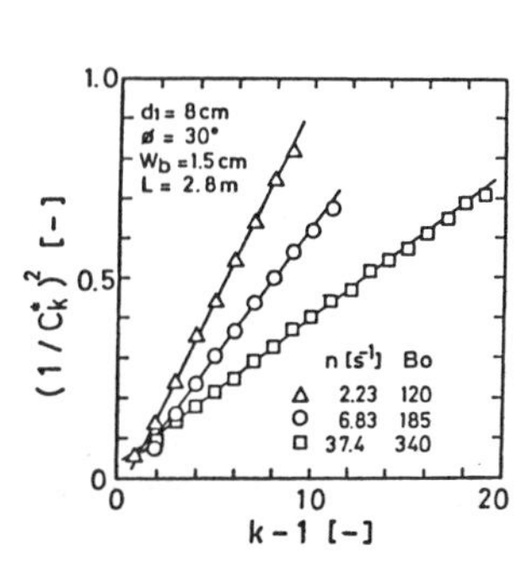

Figure 8. Typical plots for determination of the Bodenstein number.

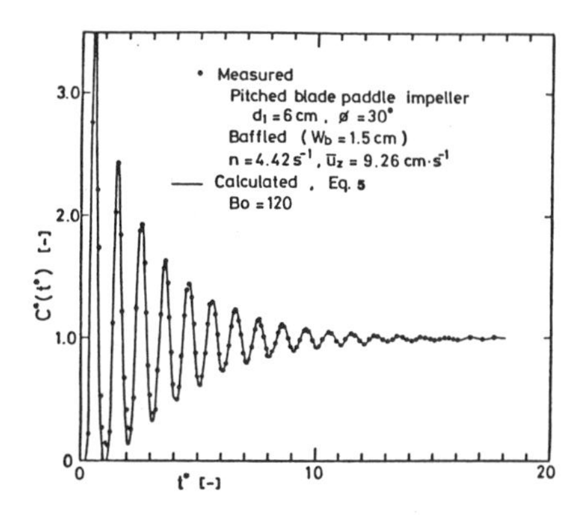

Figure 9. Comparison of response curves between calculated and measured.

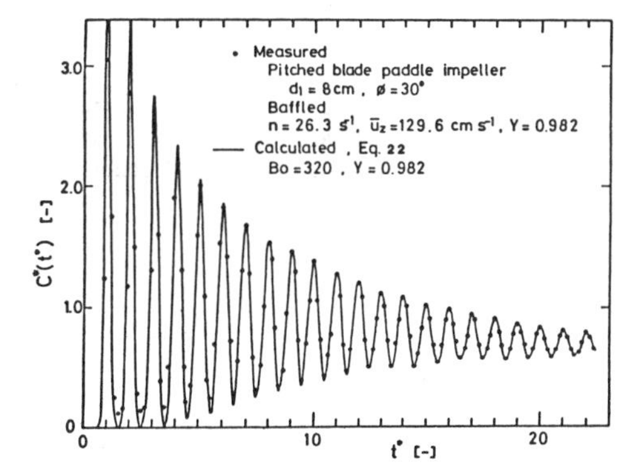

Figure 10. Typical example of measured residence time distribution compared with calculated distribution.

in the case of continuous operation. The experimental curve is well simulated by the solution to the dispersion model Equation 22 again. From these results, mixing behavior in the loop reactor can be successfully represented by Equation 5 (batch operation) and Equation 22 (continuous operation) and the proposed procedure to determine Bo is found to be adequate.

Peclet Number

The Bodenstein number obtained by the experiments was translated into the Peclet number defined as

$$Pe = d_t U/D_a = (d_t/L)Bo \tag{24}$$

The range of Pe obtained in this work is from 3 to 16. Some results with various pitched angles of an 8-cm impeller in a 10-cm loop reactor with four baffles are plotted against the mixing Reynolds number Re_M

$$Re_M = \rho n d_i^2/\mu \tag{25}$$

in Figure 11 in which is also plotted the circulation velocity normalized by the tip velocity of the impeller U/nd_i. The pitched angle ϕ influences Pe value to some extent in the same way that it

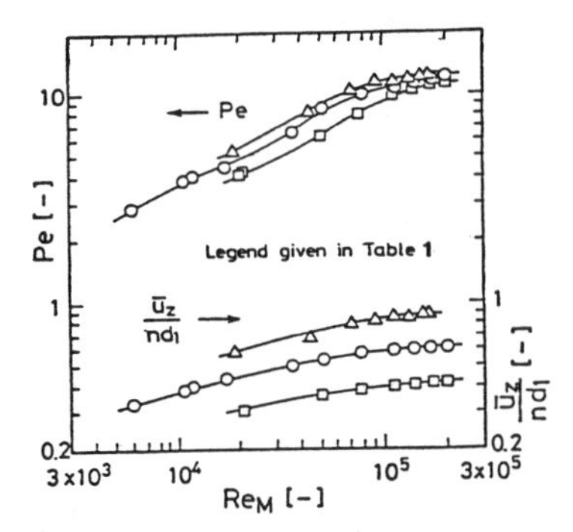

Figure 11. Dependence of Peclet number and average circulation velocity on mixing Reynolds number.

influences the value of U/nd_i. This fact implies that the circulation velocity, and therefore the pipe Reynolds number Re_p

$$Re_P = \rho d_t U/\mu \tag{26}$$

is a better correlating parameter than the mixing Reynolds number Re_M.

The same results are replotted against the pipe Reynolds number Re_p in Figure 12A, which shows that they are well correlated by a single curve independent of the pitched angle. The results with various impeller sizes and various impeller shapes in the same baffled condition are shown in Figure 12B and C, respectively, where the Peclet number is again found to be independent of

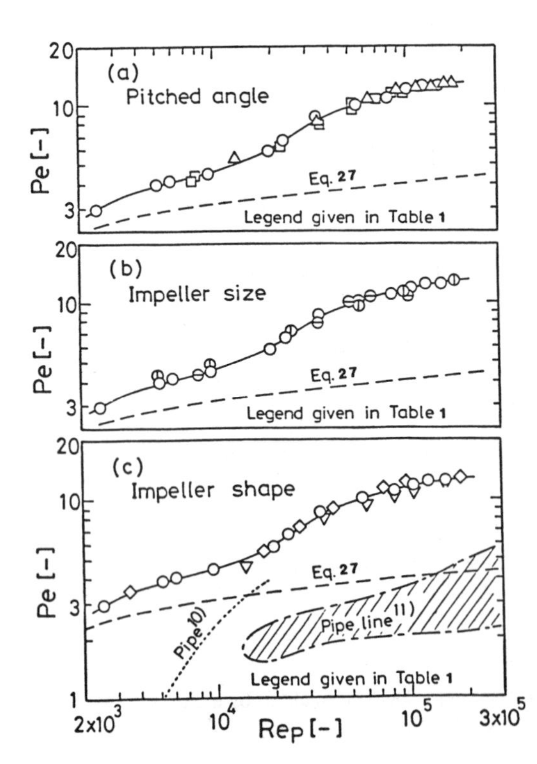

Figure 12. Effects of impeller geometries on Peclet number under baffled conditions.

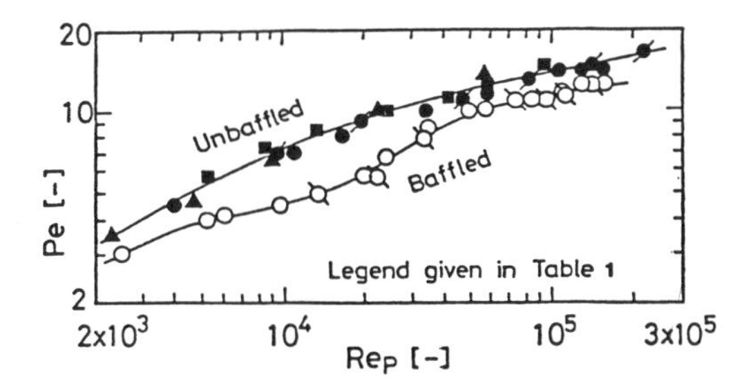

Figure 13. Comparison of Peclet number between baffled and unbaffled conditions.

both the size and shape of the impeller as well as the pitched angle mentioned in Figure 12A. Sato et al. [8] also observed flow patterns that correspond to this insensitive impeller geometry. They also noted that the dimensionless velocity distribution in the loop reactor is independent of the impeller geometry in the baffled condition. As shown in Figure 12, the Peclet number increases with the pipe Reynolds number and an inflection point appears around $Re_p = 2 \times 10^4$ on the Pe-Re_p diagram in the baffled condition. This value of Re_p corresponds to the critical Reynolds number for the bend used in this work.

The effect of baffles and their size is shown in Figure 13. When baffle width varies between 1.5 and 3 cm, no effect on Pe is detected. In the case of unbaffled conditions, however, the Peclet numbers are larger than those under baffled conditions and the inflection point disappears. Helical motion induced by the impeller survives far downstream in unbaffled conditions [8] and larger values of Pe may be ascribed to this helical motion.

The effects of loop length and pipe diameter are also examined briefly in the unbaffled condition. These parameters do not influence the Peclet number, as shown in Figure 13. This fact and the independence of impeller geometry imply a possible scaling-up criterion based on the pipe Reynolds number. Levenspiel [9] reviewed some results of experiments on axial mixing in pipe flow measured by some investigators [10, 11]. Those results were rearranged into the relation of Pe vs. Re_p as shown in Figure 12 to compare with the present results. The Peclet number in the loop reactor is larger than those obtained by other investigators and Taylor's equation [12].

$$Pe = 0.28/\sqrt{f} \qquad (27)$$

in pipe flow, probably because facilitation of radial mixing in bends or the impeller region decreases axial mixing in the loop reactor.

FLOW PROPERTIES IN A LOOP REACTOR

In the preceding chapter, it was shown that the Bodenstein number Bo characterizing the axial mixing is a unique function of the pipe Reynolds number $Re_p = d_t U \rho / \mu$. The next problem is selecting the reactor geometry and operational variables to get the desired pipe Reynolds number Re_p. This chapter deals with this topic as well as other flow properties such as power consumption and discharge efficiency.

Background for Analysis [13, 14]

The long range interaction between the impeller region and the tubular loop will decay rapidly, and each region can be considered to behave independently, just as in the conventional pump-tubing system. Thus, the effect of the loop geometry can be accounted through the single parameter of flow resistance or pressure loss, which is balanced by the discharge pressure of the impeller.

Taking this and turbulent flow into consideration, the discharge pressure Δp and power consumption P can be given, for a given impeller geometry, as a function of impeller speed n, impeller diameter d_i, fluid density ρ, and circulation velocity U.

$$\Delta p = \psi_1(n, d_i, \rho, U) \tag{28}$$

$$P = \psi_2(n, d_i, \rho, U) \tag{29}$$

This is reduced to

$$\beta = \Delta p/(\rho U^2/2) = \psi_3(q) \tag{30}$$

$$N_P = P/\rho n^3 d_i^5 = \psi_4(q) \tag{31}$$

in a dimensionless form, in which the parameter q is the discharge number and

$$q = U/nd_i \tag{32}$$

The parameter β is a friction loss factor and N_p is a power number. When the discharge efficiency η is defined as the ratio of the power transformed into the discharge pressure to the total power consumption, it is given as

$$\eta = \frac{\pi}{4} d_t^2 U \, \Delta p/P = \frac{\pi}{8}\left(\frac{d_t}{d_i}\right)^2 \frac{\beta q^3}{N_P} \tag{33}$$

Thus, all three parameters β, N_P, and η are expected to be functions of a single variable q for any loop geometry. The discharge number q is related to the pipe Reynolds number Re_P and the mixing Reynolds number Re_M as follows

$$q = \left(\frac{d_i}{d_t}\right) \frac{Re_P}{Re_M} \tag{34}$$

Thus, when the q value is found, Re_M and then impeller speed n can be readily evaluated for a given Re_P.

Discharge Characteristics

To determine the discharge characteristics according to the preceding analysis, the discharge pressure Δp, power consumption P, and circulation velocity U were measured by the authors. The flow resistance was controlled by inserting a throttle ring with different openings (m = 0, 0.4, 0.6, 0.8 and 1.0) and by changing the loop shape (I-shape, L-shape, and U-shape as shown in Figure 2) and the tube diameter d_t (5 and 10 cm). Pressure p and power consumption P, respectively, were measured by the conventional manometers and torque transducers. The circulation velocity U was measured by a pluse response method with a KCl tracer in which U was determined from time interval between successive concentration peaks.

A typical example for the pitched blade paddle with $\phi = 40°$ is shown in Figure 14. The characteristic curve obtained in I-, L-, and U-shaped loop reactors and in different-sized reactors coincides with each other satisfactorily. Thus, it is concluded that the impeller of the loop reactor has its own discharge characteristic curve, regardless of the loop geometry, as expected by the analysis in the preceding section.

The characteristic curve for other impellers is given in Figure 15. The efficiency η for the pitched blade paddle with $\phi = 20°$ tends to flatten at small values of discharge number q. The η value for the axial flow pump type is not too much higher than the 40°-pitched blade paddle in spite of a sophisticated configuration.

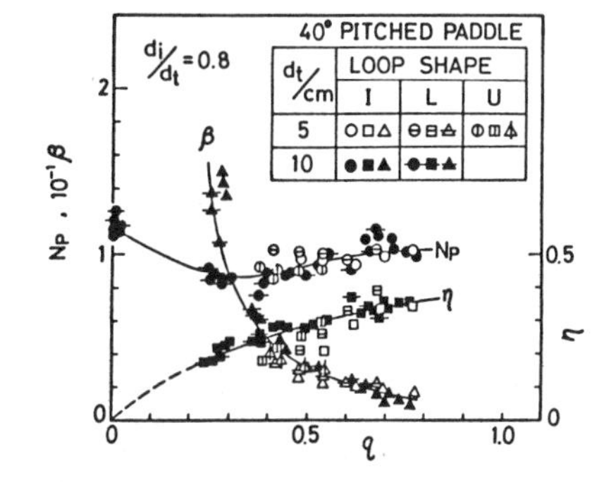

Figure 14. Discharge characteristics curve for the 40°-pitched blade paddle.

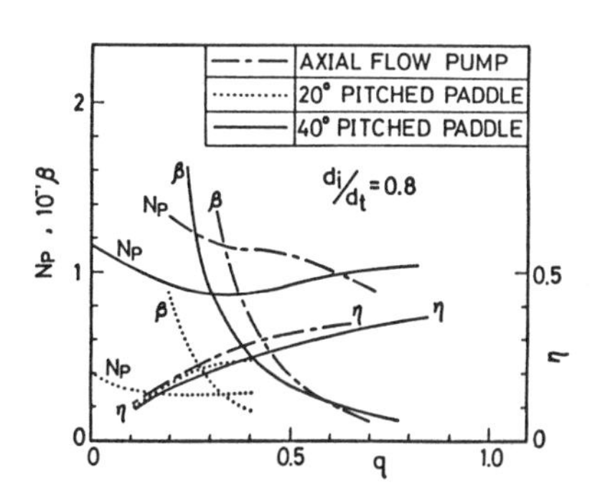

Figure 15. Comparison of discharge characteristic curve between various impellers.

Discharge Pressure

An example of measured pressure distribution is shown in Figure 16. Static pressure decreases gradually as fluid flows downstreams in the straight portion of the loop owing to the wall friction while it drops rapidly in bends. The discharge pressure Δp of the impeller, which is a jump-up in pressure at the impeller position, is counterbalanced by the total pressure loss.

The total pressure loss was separated into two contributions from the straight portion and bends, i.e.

$$\beta = \frac{\Delta p}{\frac{1}{2}\rho U^2} = 4f\frac{l_s}{d_t} + 4\sum\zeta \tag{35}$$

when the friction factor f and loss coefficient ζ was calculated by Drew's formula [15]

$$1/\sqrt{f} = 3.2\log(Re_P\sqrt{f}) + 1.2 \tag{36}$$

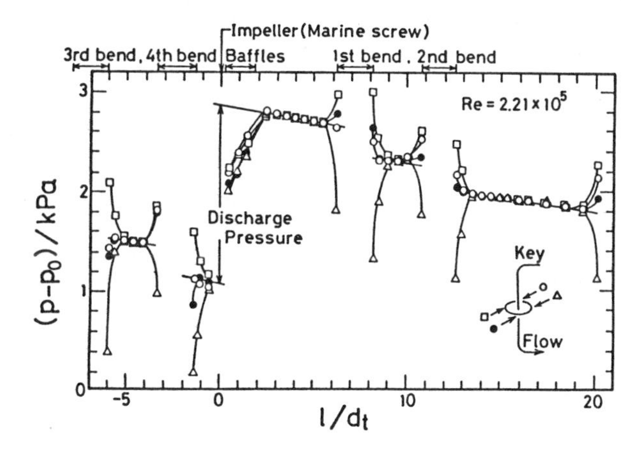

Figure 16. An example of pressure distribution in a 10-cm i.d. loop reactor of I shape with a marine screw.

and Ito's formula [16]

$$\zeta = 0.0241\alpha\gamma \, Re^{-0.17} (2R/d_t)^{0.84} \qquad (\alpha = 0.95 + 17.2(2R/d_t) \tag{37}$$

It is found that the experimental pressure loss can be well estimated by Equation (35).

Proposed Design Procedure

By use of the preceding discharge characteristic curve, the following procedure for the reactor design is proposed. Dimensions of the reactor and the required axial velocity U are first specified. The required discharge pressure Δp which is balanced by the pressure loss through the whole loop is calculated from Equation 35. Then, the corresponding value of η, N_p and q can be obtained from the value of β on the characteristic curve. Finally, required impeller speed n and resultant power consumption P are evaluated. When this procedure was applied to commercially operating units for olefin polymerization, the estimated values were very close to real ones.

Acknowledgments

This research has been supported in part by Grant-in-Aid for Scientific Research from the Ministry of Education of Japan (57850300 in 1982), for which the authors express their sincere thanks.

NOTATION

Bo	Bodenstein Number
C	tracer concentration (mol/m^3)
C_k	k-th peak of concentration (mol/m^3)
$\bar{C}$	concentration after complete mixing (mol/m^3)
D_a	axial dispersion coefficient (m^2/s)
d_i	impeller diameter (m)
d_t	diameter of loop reactor tube (m)
E	residence time distribution function
f	pipe friction factor
j	integer
k	integer
L	length of loop reactor (m)
l_s	length of straight portion of loop (m)
M	amount of tracer (mol)
m	fractional opening area
n	impeller speed (1/s)
N_p	power number
P	power consumption (W)
p	pressure (Pa)
Δp	discharge pressure or pressure loss (Pa)
Pe	Peclet number
q	discharge number
R	curvature radius of bend (m)
Re_M	mixing Reynolds number
Re_p	pipe Reynolds number
t	time (s)
t*	time relative to circulation time
U	average circulation velocity (m/s)
V	reactor volume (m^3)
x	axial distance (m)
x*	dimensionless axial distance
Y	ratio of recycle flow rate
z	variable defined by Equation 15

Greek Symbols

α	parameter defined by Equation 37
β	pressure loss coefficient defined by Equation 30
γ	curvature angle of bend (deg.)
ζ	loss coefficient in bend
η	discharge efficiency
θ	theta function
μ	viscosity (Pa·s)
ρ	density (kg/m^3)
τ	dimensionless residense time
ϕ	pitched angle of impeller (deg)

REFERENCES

1. Blenke, H., *Preprints for International Symposium on Mixing*, C-7, Faculte Polytechnique de Mons, 1978.
2. Norwood, F., Japan patent, Showa 37-10087, 1962.
3. Chemical Reaction Engineering Subdivison, The Society of Polymer Science, Japan, "Assesments on Polyolefin Production Processes," 1979.
4. Murakami, Y., et al., *Journal of Chemical Engineering of Japan*, Vol. 15, No. 2 (April 1982), pp. 121–125.
5. Takao, M., et al., *Chemical Engineering Science*, Vol. 37, No. 5 (May 1982), pp. 796–798.
6. Voncken, R. M., Holmes, D. B., and den Hartog, H. W., *Chemical Engineering Science*, Vol. 19 (1964), p.209.
7. Dym, H., and Mckean, H. P., *Fourier Series and Integrals*, Academic Press, 1972, p. 52.
8. Sato, Y., et al., *Journal of Chemical Engineering of Japan*, Vol. 12, No. 6 (Dec. 1979), pp. 448–453.
9. Levenspiel, O., *Industrial and Engineering Chemistry*, Vol. 50 (1958), p. 343.
10. Fowler, F. C., and Brown, G. G., *Transactions of AIChE*, Vol. 39 (1943), p. 491.
11. Hull, D. E., and Kent, J. W., *Industrial and Engineering Chemistry*, Vol. 44 (1952), p. 2745.
12. Taylor, G. I., Proceedings of the Royal Society, London, Vol. 223, Ser. A (1954), p. 446.
13. Murakami, Y., et al., *Industrial and Engineering Chemistry, Process Design and Development*, Vol. 21, No. 2 (1982), pp. 273–276.
14. Murakami, Y., et al., The 3rd Pacific Chemical Engineering Congress, Korea, 1983.
15. Drew. T. B., Koo, E. C., and McAdams, W. H., *Transactions of AIChE*, Vol. 28 (1936), p. 56.
16. Ito, H., *Journal of Japan Society of Mechanical Engineers*, Vol. 62 (1959), p. 1634.

CHAPTER 23

VERTICAL CIRCULATION IN DENSITY-STRATIFIED RESERVOIRS

Zygmunt Meyer

Water Engineering Institute
Technical University of Szczecin
Szczecin, Poland

CONTENTS

INTRODUCTION, 572

MATHEMATICAL MODELS FOR VERTICALLY PLANE MOTION, 574

VERTICAL CIRCULATION IN CASE OF WIND ACTION AND FLOW THROUGH RESERVOIR, 579
- Primary Stream Function Solution, 579
- Mechanism of Vertical Circulation, 585
- Vertical Circulation with Additional Horizontal Density Changes, 599
- Vertical Circulation in a Case of Vanishing Density Stratification, 602

WIND-GENERATED VERTICAL CIRCULATION IN STRATIFIED RESERVOIR, 605

AXISYMMETRIC VERTICAL CIRCULATION IN DENSITY-STRATIFIED RESERVOIR, 612

LABORATORY EXPERIMENTS, 622

PRACTICAL APPLICATIONS OF VERTICAL CIRCULATION MODELS, 626

CONCLUSIONS, 632

NOTATION, 633

REFERENCES, 635

INTRODUCTION

Vertical circulation in density-stratified reservoirs is of great importance for many practical purposes. Growing water demand requires more intensive reservoir management, as well as increased attention to water quality. These two factors have contributed to the need for research to estimate the internal structure of flow within a reservoir. From a practical point of view an important problem is to predict the flow pattern as a function of varying boundary conditions.

The vertical circulation problem is a part of a large group of flows called stratified flows or nonhomogeneous flows. The name comes from the fact that within the flowing stream, density differences exist. A characteristic feature of the stratified flow is that small density differences cause radical changes in flow patterns. The most practical problem within the group is that of selective withdrawal, defined [13, 28, 34] as an intake of water from only certain parts of the reservoir. Usually water is taken from the lower layers of the reservoir, and at the upper layers a circulation or stagnation region is created. The situation may be reversed, and water can be withdrawn from upper layers, with the circulation region arising at the bottom. This depends on the outlet location and discharge.

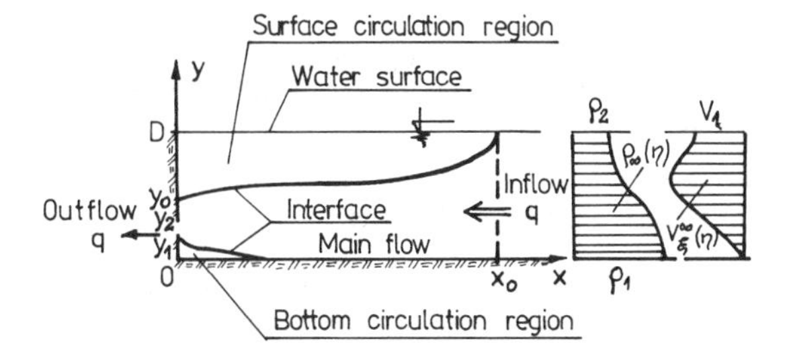

Figure 1. Vertical reservoir cross section with characteristic flow sub-areas: circulation regions and main flow area (courtesy of the American Society of Civil Engineers).

The streamline separating the circulation region from the main flow area is called the interface (Figure 1). Another example is wind-generated vertical circulation in a stratified reservoir. Full classification of the stratified flows is given by Harleman [13], Turner [28], and Yih [34].

In general vertical circulation is caused by the reservoir's inflow-outflow and wind action. The problem of flow patterns in a reservoir under natural conditions is very complicated and there are a great number of factors affecting the flow. However, two basic assumptions on this subject are commonly accepted: incompressible fluid, and small density stratification. The small density stratification means in practise relative density differences of order of several percent. To solve the problem of the vertical circulation, simplifications are usually made by taking factors of the strongest influence only. In the circulation flow investigations [2–3, 5–8, 12, 14–26, 29–34, 37] it is commonly accepted that the main factors are:

1. The mass budget (inflow, outflow, and changes of reservoir's volume).
2. Energy exchange (wind action and heat transfer).

The effect of wind action is waves and surface shear stress. Usually the associated effect of wind action upon waves—energy transfer across the surface—is neglected. Waves have a significant influence on equalizing water properties, especially at the upper layers.

Investigations concerning the circulation in reservoirs have proceeded in two main directions: one neglecting viscosity, and the other including viscosity.

The first group neglecting the effect of viscosity contains papers based on the potential flow theory [6, 14, 27, 31] and the theory of stratified flows developed by Long [18] and Yih [32–34]. If the potential flow assumption is imposed a circulation region consisting of a single eddy extending far upstream to infinity (with horizontal interface) is yielded. The stratified flow theory formed by Long [18] and Yih [32–34] takes into account vorticity and the momentum equation and as a result yields a circulation region consisting again of a single eddy but of limited extent. The extent depends on the flow parameters and the density stratification. Those models were developed mainly for the purpose of selective withdrawal.

The research of the second group (2–3, 5, 7–8, 16–17, 19–26, 29–30, 35–37) including the viscosity effect is based on Ekman's model. Among the papers special attention is given to those which solve the problem analytically. Koh [16] introduced a model based on the Navier-Stokes equation, using the small parameter method. The model solved a case of flow with linear density stratification, constant velocity far upstream, and no wind. He concluded that the circulation region can be formed of more than one eddy. Walesh and Monkmeyer [30] presented a model based on the theory of circulation. The authors estimated numerically the interface shape for the following case of flow: bottom withdrawal, point slot, and linear density stratification. From their results it follows that backward velocity occurs opposite to the main flow.

Recently Meyer (19–24) has put forward a model of vertical circulation in a density-stratified reservoir. The model takes into account such factors as: the different eddy viscosity in the x- and y-directions, wind action, nonlinear density, and velocity distribution at inflow (far upstream), additional horizontal density changes due to diffusion, and various boundary conditions. The Fourier series method used for solving the problem gives a very useful way for analysis of the existence and properties of the vertical circulation region.

The aim of the present work is to calculate analytically the flow pattern for the flow generated by the inflow-outflow, wind action, or both. Based on the calculated flow pattern the existence and properties of the circulation regions were examined for a rectangular and semi-infinite strip reservoir, in the case of vertically plane and axisymmetric flow.

MATHEMATICAL MODELS FOR VERTICALLY PLANE MOTION

For further analysis there has been assumed the following model of flow: steady, two-dimensional vertically plane motion of water in a density-stratified reservoir. Two cases of flow are considered:

- The flow generated by the mass exchange inflow-outflow and the wind action.
- The wind-generated circulation.

In a vertical cross section two reservoir shapes are taken into account:

- The rectangular shape.
- The semi-infinite strip shape.

The inflow to the reservoir may be formed by a group of slots located one above the other (rectangular shape reservoir), or with the full reservoir depth D (semi-infinite strip shape). The outflow is a vertical profile with a group of outlet slots. Within the reservoir the small density stratification exists. The water density at the surface is equal to ρ_2 and at the bottom ρ_1. The wind action is included by assuming the surface shear stress of water τ_w, equal to the wind shear stress τ_2. The basic system of Cartesian coordinates is directed: vertically upward, y and horizontally, x.

As a mathematical representation of the considered model of flow, the following set of equations was taken [1]:

- Momentum equations

$$\frac{dV_x}{dt} = -\frac{1}{\rho} \cdot \frac{\partial P}{\partial x} + \frac{\partial}{\partial x}(\tau_{xx}) + \frac{\partial}{\partial y}(\tau_{xy})$$

$$\frac{dV_y}{dt} = -g - \frac{1}{\rho}\frac{\partial P}{\partial y} + \frac{\partial}{\partial x}(\tau_{yx}) + \frac{\partial}{\partial y}(\tau_{yy}) \qquad (1)$$

- And because of nonhomogeneous fluid, mass conservation law

$$\frac{\partial}{\partial x}(\rho \cdot V_x) + \frac{\partial}{\partial y}(\rho \cdot V_y) = 0 \qquad (2)$$

In the above relationships $\tau_{i,j}$ are components of the turbulent shear stress tensor (Reynolds stress). From the tensor symmetry, we have

$$\tau_{x,y} = \tau_{yx}$$

and the tensor components are commonly expressed as

$$\tau_{xx} = \rho \cdot K_x \cdot \frac{\partial V_x}{\partial x}; \qquad \tau_{xy} = \rho \cdot K_y \frac{\partial V_x}{\partial y}$$

$$\tau_{yx} = \rho \cdot K_x \cdot \frac{\partial V_y}{\partial x}; \qquad \tau_{yy} = \rho \cdot K_y \frac{\partial V_y}{\partial y} \qquad (3)$$

Furthermore the stream function is introduced

$$V_x = \frac{\partial \psi}{\partial y}; \qquad V_y = -\frac{\partial \psi}{\partial x} \tag{4}$$

The basic solution of the investigated problem refers to the density distribution function assumed by Long [18] and Yih [32]

$$\rho = \rho(\psi) \tag{5}$$

This sort of density distribution means in practice, that the density along the chosen streamline remains constant. If

$\psi = \text{const}$, then $\rho = \text{const}$.

This distribution does not imply that the flow is nondiffusive. Such a density function means, that the diffusion of mass within each elementary volume is in balance. The above assumption (Equation 5) reduces the mass conservation equation to the common form

$$\frac{\partial V_x}{\partial x} + \frac{\partial V_y}{\partial y} = 0 \tag{6}$$

and it happens because

$$V_x \frac{\partial \rho}{\partial x} + V_y \frac{\partial \rho}{\partial y} = \frac{d\rho}{d\Psi}\left(V_x \frac{\partial \Psi}{\partial x} + V_y \frac{\partial \Psi}{\partial y}\right) = \frac{d\rho}{d\Psi}(-V_x \cdot V_y + V_x \cdot V_y) \equiv 0$$

Numerical estimations indicate that the case of small stratification formula [6] may be applied for practical calculation even if the density does not satisfy Equation 5. Taking the continuity equation 6, further analysis of the momentum relationships (Equation 1) can be made. After differentiations and combining terms we obtain

$$\frac{d}{dt}\left[\Omega + \frac{1}{\rho} \cdot \frac{d\rho}{d\Psi} \cdot \frac{1}{2} V^2\right] = \frac{g}{\rho} \cdot \frac{\partial \rho}{\partial x} + \frac{\partial^2}{\partial x\, \partial y}(\tau_{xx}) - \frac{\partial^2}{\partial x\, \partial y}(\tau_{yy}) + \frac{\partial^2}{\partial y^2}(\tau_{xy}) - \frac{\partial^2}{\partial x^2}(\tau_{yx}) \tag{7}$$

where Ω is the vorticity function defined as

$$\Omega = \frac{\partial V_x}{\partial y} - \frac{\partial V_y}{\partial x} \tag{8}$$

The exact derivation of the left-hand side terms of the relationship in Equation 7 was given by Long [18] and Yih [32]. From their analysis it comes out that

$$|\Omega| \gg \left|\frac{1}{2} \cdot \frac{1}{\rho} \cdot \frac{d\rho}{d\Psi} \cdot V^2\right| \tag{9}$$

And neglecting the term $\frac{1}{2} \cdot \frac{1}{\rho} \cdot \frac{d\rho}{d\Psi} \cdot V^2$ results in very small changes of the calculated stream function. The relative changes are of order $\left(\frac{\rho_1 - \rho_2}{\rho_1}\right)^{1/2}$. According to that conclusion, the above assumption was applied for further investigation, especially noting the small stratification.

For the convenience of further analysis and in order to make the solution more general, the following dimensionless variables, function and parameters were introduced:

$$\eta = \frac{y}{D}; \qquad \xi = \frac{x}{D}\sqrt{\frac{K_y}{K_x + K_y}} \qquad \psi = \frac{\Psi}{\Psi_0}; \qquad V_\xi = \frac{\partial \psi}{\partial \eta}; \qquad V_\eta = -\frac{\partial \psi}{\partial \xi} \tag{10}$$

$$\frac{\Psi_0}{\sqrt{K_y(K_x + K_y)}} = Re; \qquad \frac{\Psi_0^2}{g \cdot D^3} = Fr$$

thus (11)

$$V_x = \frac{\Psi_0}{D} \cdot V_\xi; \qquad V_y = \frac{\Psi_0}{D}\sqrt{\frac{K_y}{K_x + K_y}}$$

Parameters: Re denoting the Reynolds number and Fr Froude number, were chosen because of their similarity to the original numbers. In the engineering calculations, the densimetric Froude number is commonly used. Its connection with the original Froude number can be expressed as

$$Fr_1 = \left(Fr \cdot \frac{\rho}{\Delta\rho}\right)^{1/2} \tag{12}$$

Let us consider the components of the turbulent shear stress tensor appearing on the right-hand side of the relationship in Equation 7. Using the aforementioned dimensionless variables we can make the following approximation

$$\frac{\partial}{\partial y^2}(\tau_{xy}) \sim \frac{\Psi_0}{D^4} \cdot \rho \cdot K_y \frac{\partial^4 \psi}{\partial \eta^4}$$

$$\frac{\partial}{\partial x^2}(\tau_{yx}) \sim \frac{\Psi_0}{D^4} \cdot \rho \cdot \frac{K_y^2}{(K_x + K_y)^2} \cdot K_x \frac{\partial^4 \Psi}{\partial \xi^4}$$

As can be seen

$$\left|\frac{\partial^2}{\partial y^2}(\tau_{xy})\right| \gg \left|\frac{\partial^2}{\partial x^2}(\tau_{yx})\right| \tag{13}$$

This is so because at the prevalent part of the flow area the vertical changes of the stream function are many times greater than horizontal. Additionally, $K_x \gg K_y$ and Equation 13 does not depend on the density function (Equation 5). This may be considered as the "Prandtl assumption." Following simplifications in Equation 9 and Equation 13 the governing equation of motion becomes

$$\frac{d\Omega}{dt} = \frac{g}{\rho} \cdot \frac{\partial \rho}{\partial \xi} + \frac{\partial^2}{\partial x \, \partial y}\left[(K_x + K_y)\frac{\partial^2 \Psi}{\partial x \, \partial y}\right] + \frac{\partial^2}{\partial y^2}\left(K_y \frac{\partial^2 \Psi}{\partial y^2}\right) \tag{14}$$

Many of the authors, among them Bengtson [2], Liggett and Lee [17], Uzzell and Ozisik [29], Walesh and Monkmeyer [30], Young et al. [35], suggest that for practical calculation the eddy viscosity coefficients may be assumed as constants. This conclusion comes from theoretical and experimental investigations (including field measurements) and from the fact that the physical nature of these coefficients is a matter of continuous research. There does not exist one commonly accepted theory on this subject, though attempts are being made to develop one. However, if these coefficients are given functions, then the relationship in Equation 14 can be applied. Therefore we have

$$K_x, K_y = \text{constant} \tag{15}$$

Taking into account the preceding simplification in Equation 15 we obtain the following dimensionless equation of motion

$$\mathrm{Re}\cdot\frac{d\omega}{dt_*}=\frac{\mathrm{Re}}{\mathrm{Fr}}\cdot\frac{1}{\rho}\cdot\frac{\partial\rho}{\partial\xi}+\frac{\partial^2}{\partial\eta^2}(\nabla^2\psi) \tag{16}$$

where ω is the dimensionless vorticity function defined as

$$\omega=\frac{D^2}{\Psi_0}\cdot\Omega$$

and

$$t_*=t\cdot\frac{\Psi_0}{D^2}\sqrt{\frac{K_y}{K_x+K_y}}$$

where $\nabla^2\psi$ is the Laplace operator. As it comes from the relationship in Equation 11 the Reynolds number Re is based on the eddy viscosity coefficients and this value arises in dimensionless equation (Equation 16) as one of two constants. The values of the eddy viscosity in large water bodies are quite high. Consequently, the approximate Reynolds number Re, is of order $10\sim100$. It suggests that from the point of view of numerical calculation these are the low Reynolds number flows. Thus, the nonlinear terms in the governing equation (Equation 16) do not play an important role in momentum transfer. This has been confirmed in the research made by Findikakis and Street [11]. Another fact promoting this assumption is that we are considering the flow of very small velocities. Numerical computation that has been carried out indicates that velocities within the circulation region are of order of 0.1 to 0.01 of the average velocity V_0. Therefore in further analysis nonlinear terms are basically neglected, except for the case of vanishing density stratification. We have

$$\frac{d\omega}{dt_*}=0 \tag{17}$$

When density stratification disappears, the convective term

$$\mathrm{Re}\cdot\frac{d\omega}{dt_*}$$

and the bouyancy term

$$\frac{\mathrm{Re}}{\mathrm{Fr}}\cdot\frac{1}{\rho}\cdot\frac{\partial\rho}{\partial\xi}$$

are of the same order and this significantly affects the stream function. The aforementioned numerical computations made by the author lead to the conclusions that for practical calculation the nonlinear terms may be approximated by the following linear formulae [22]:

$$\frac{d\omega}{dt_*}=\alpha\frac{\partial^2\psi}{\partial\xi\,\partial\eta^2}-\beta\frac{\partial\psi}{\partial\xi} \tag{18}$$

where $\alpha,\beta=$ constants

To close the described mathematical model of the vertical circulation in a stratified reservoir the density distribution function in Equation 5 is needed. According to the earlier works: Long [18],

Walesh and Monkmeyer [30] and Yih [32–34] for the primary solution linear function was taken

$$\rho(\psi) = \rho_1 + \Delta\rho \cdot \psi$$

where

$$\Delta\rho = \rho_1 - \rho_2 \tag{19}$$

If the diffusion effects causing additional horizontal density changes cannot be neglected, another density distribution function is proposed [21]

$$\rho = \rho_1 + \Delta\rho \cdot \psi \cdot \mathrm{f}(\xi) \tag{20}$$

where $\mathrm{f}(\xi)$ is a given function representing additional horizontal changes. This function is subject to certain restrictions. The following requirement must be fulfilled

$$\lim_{\xi \to \infty} \frac{\mathrm{df}(\xi)}{\mathrm{d}\xi} = 0$$

and so

$$\lim_{\xi \to \infty} \rho = \rho_\infty(\eta) \tag{21}$$

The second restriction is that the additional horizontal density changes are not rapid and so it may be written

$$\frac{\partial\rho}{\partial\xi} = \Delta\rho \cdot \mathrm{f}(\xi) \cdot \frac{\psi}{\partial\xi} \tag{22}$$

Having the governing equation of motion, Equation 16, and the further conclusions in Equations 17, 18, 19 and 20, a set of three relationships for the three basic solutions has been formed:

- The primary stream function solution of the vertical circulation

$$\mathrm{A} \cdot \frac{\partial\psi}{\partial\xi} + \frac{\partial^2}{\partial\eta^2}(\nabla^2\psi) = 0 \tag{23}$$

where

$$\mathrm{A} = \frac{\Delta\rho}{\rho_1} \cdot \frac{\mathrm{Re}}{\mathrm{Fr}} = \frac{\Delta\rho \cdot \mathrm{g} \cdot \mathrm{D}^3}{\rho_1 \cdot \Psi_0 \cdot \sqrt{\mathrm{K_y(K_x + K_y)}}} \tag{24}$$

- The vertical circulation with the additional horizontal density changes

$$\mathrm{A} \cdot \mathrm{f}(\xi) \cdot \frac{\partial\psi}{\partial\xi} + \frac{\partial^2}{\partial\eta^2}(\nabla^2\psi) = 0 \tag{25}$$

where A remains the same as before in Equation 24,

- The vertical circulation in the case of vanishing density stratification

$$\mathrm{Re}\left(\alpha \frac{\partial^3\psi}{\partial\xi\,\partial\eta^2} - \beta \frac{\partial\psi}{\partial\xi}\right) = \mathrm{A} \cdot \mathrm{f}(\xi) \cdot \frac{\partial\psi}{\partial\xi} + \frac{\partial^2}{\partial\eta^2}(\nabla^2\psi) \tag{26}$$

where A is given by Equation 24 and Re is from Equation 11.

A detailed analysis of the whole aforementioned models is given by Meyer [19–23]. It should be emphasized, that from the final equations, Equations 24, 25 and 26 it becomes clear that in general, we are considering slow, creeping, density-stratified flow.

VERTICAL CIRCULATION IN CASE OF WIND ACTION AND FLOW THROUGH RESERVOIR

Primary Stream Function Solution

Wind-generated flow and subsequent mass exchange is most common under natural conditions. The basic solution, referred to as the primary stream function, applies to the case $\rho = \rho(\psi)$, hence we are looking for the solution of Equation 23. In this case the main system of Cartesian coordinates (x, y) is directed vertically upward at the outflow (y) and horizontally along the bottom (x). The inflow slots location is determined by edge coordinates $y_1'^{(m)}$ and $y_2'^{(m)}$, so the height is $h'_m = y_2'^{(m)} - y_1'^{(m)}$. Similarly, at the outflow the slot edge coordinates are $y_1^{(n)}$ and $y_2^{(n)}$, so the height is $h_n = y_2^{(n)} = y_1^{(n)}$. The flow through the m-th slot at the inflow is equal to q'_m, and at the n-th outflow slot q_n. Values q may be positive or negative depending on the direction of flow within the slot. It has been assumed that q is positive when its direction is opposite to x (Figure 2), otherwise it is negative. In the considered case the reference stream function Ψ_0 was chosen as the total net flow through the reservoir. We have

$$\Psi_0 = q = \sum_{n=1}^{n=N} q_n = \sum_{m=1}^{m=M} q'_m \tag{27}$$

The boundary conditions under which the solution is sought come from the earlier physical description of flow, and they are as follows (Figure 2):

when

$\eta = 1;\quad \psi = -1,\quad \rho = \rho_2,\quad \partial\psi/\partial\xi = 0$ and $\tau = \tau_2$

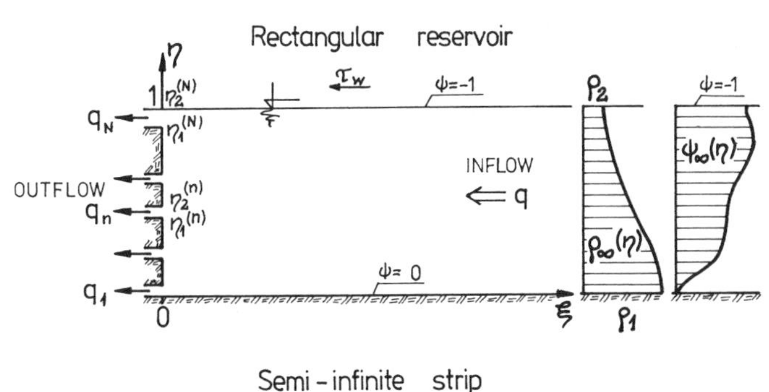

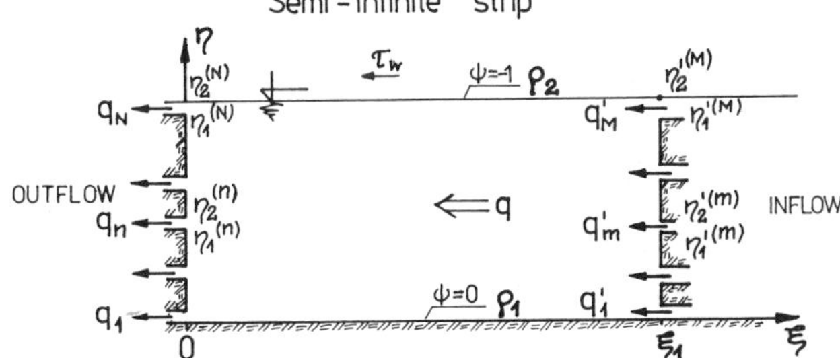

Figure 2. Definition sketch of flow area (courtesy of the American Society of Civil Engineers).

and when

$$\eta = 0; \qquad \psi = 0, \qquad \rho = \rho_1 \quad \text{and} \quad \partial\psi/\partial\xi = 0$$

At the outflow profile ($\xi = 0$) for the n-th slot and its vicinity [19, 20] the following criteria exist;

when $\eta_2^{(n)} \leq \eta \leq \eta_1^{(n+1)}$;

$$\psi = F_n = \text{const.} \quad \text{and} \quad \frac{\partial\psi}{\partial\eta} = 0$$

when $\eta_1^{(n)} < \eta < \eta_2^{(n)}$;

$$\psi = f_n(\eta)$$

and when $\eta_2^{(n-1)} \leq \eta \leq \eta_1^{(n)}$;

$$\psi = F_{n-1} = \text{const.} \quad \text{and} \quad \frac{\partial\psi}{\partial\eta} = 0 \tag{29}$$

where n = 1, 2, 3, . . . N. The functions F_n and f_n which appear in Equation 29 were estimated according to the previous research [19–21]:

$$F_n = -\frac{\sum_{i=1}^{i=n} q_i}{q} \tag{30}$$

and

$$f_n(\eta) = F_{n-1} + (F_n - F_{n-1}) \cdot \frac{\eta - \eta_1^{(n)}}{\eta_2^{(n)} - \eta_1^{(n)}} \tag{31}$$

At the inflow profile ($\xi = \xi_1$) for the m-th slot and its vicinity we have [19, 20]:

when $\eta_2'^{(m)} \leq \eta \leq \eta_1'^{(m+1)}$;

$$\psi = F'_m = \text{const.} \quad \text{and} \quad \frac{\partial\psi}{\partial\eta} = 0$$

when $\eta_2'^{(m)} < \eta < \eta_1'^{(m)}$;

$$\psi = f'_m(\eta)$$

and when $\eta_2'^{(m-1)} \leq \eta \leq \eta_1'^{(m)}$;

$$\psi = F'_{m-1} = \text{const.} \quad \text{and} \quad \frac{\partial\psi}{\partial\eta} = 0 \tag{32}$$

where m = 1, 2, 3, . . . M. By analogy to the inflow function, F'^m and $f'_m(\eta)$ are equal:

$$F'_m = -\frac{\sum_{j=1}^{j=m} q'_j}{q} \tag{33}$$

and

$$f'_m(\eta) = F'_{m-1} + (F'_m - F'_{m-1}) \cdot \frac{\eta - \eta_1'^{(m)}}{\eta_2'^{(m)} - \eta_1'^{(m)}} \tag{34}$$

The general solution of Equation 23 was obtained by using the Fourier series method and the preceding boundary conditions. It yields

$$\psi(\xi, \eta) = \psi_\infty(\eta) + \sum_{n=1}^{\infty} \{a_{1n} \cdot \exp(d_{1n} \cdot \xi) + a_{2n} \cdot \exp[-d_{2n}(\xi_1 - \xi)]\} \cdot \sin(\pi n \eta) \tag{35}$$

and

$$d_{1n} = \frac{A - \sqrt{A^2 + 4\pi^6 n^6}}{2\pi^2 n^2}; \qquad d_{2n} = \frac{A + \sqrt{A^2 + 4\pi^6 n^6}}{2\pi^2 n^2}$$

while

$$A = \frac{\Delta\rho}{\rho_1} \cdot \frac{g \cdot D^3}{q\sqrt{K_y(K_x + K_y)}} \tag{36}$$

Function $\psi_\infty(\eta)$ depends on the velocity distribution at the inflow (far upstream) [19–21].

$$\text{If } V_{\xi\infty}(\eta) = \text{const.} = -1; \qquad \text{then } \psi_\infty(\eta) = -\eta \tag{37}$$

and if $V_{\xi\infty}(\eta) = 3(V_1 + 2)\eta^2 - (V_1 + 3)\eta^2$ – parabolic distribution

then

$$\psi_\infty(\eta) = (V_1 + 2)\eta^3 - (V_1 + 3)\eta^2 \tag{38}$$

where $\quad V_1 = V_{\xi\infty}[1]$

The coefficients a_{1n} and a_{2n} are calculated from the boundary conditions at inflow and outflow:

$$a_{1n} = \frac{c_{1n} - c_{2n} \cdot \exp(-d_{2n} \cdot \xi_1) - b_n[1 - \exp(-d_{2n} \cdot \xi_1)]}{1 - \exp[-\xi_1 \cdot (d_{2n} - d_{1n})]} \tag{39}$$

$$a_{2n} = \frac{c_{2n} - c_{1n} \cdot \exp(d_{1n} \cdot \xi_1) - b_n[1 - \exp(d_{1n} \cdot \xi_1)]}{1 - \exp[-\xi_1 \cdot (d_{2n} - d_{1n})]} \tag{40}$$

Coefficients c_{1n}, c_{2n}, and b_n are the Fourier series coefficients, and they are evaluated from the boundary conditions. Assuming that $\psi(0, \eta)$ and $\psi(\xi_1, \eta)$ are given functions we have:

$$c_{1n} = 2 \cdot \int_0^1 \psi(0, \eta) \cdot \sin(\pi n \eta) \cdot d\eta$$

$$c_{2n} = 2 \cdot \int_0^1 \psi(\xi_1, \eta) \cdot \sin(\pi n \eta) \cdot d\eta$$

$$b_n = 2 \cdot \int_0^1 \psi_\infty(\eta) \cdot \sin(\pi n \eta) \cdot d\eta \tag{41}$$

Taking the boundary conditions of Equations 28 through 34 we obtain

$$c_{1n} = \frac{2 \cdot (-1)^n}{\pi \cdot n} - \frac{2}{\pi^2 n^2} \cdot \sum_{i=1}^{i=N} \left[\frac{q_i}{q} \cdot \frac{\sin(\pi n \eta_2^{(i)}) - \sin(\pi n \eta_1^{(i)})}{\eta_2^{(i)} - \eta_1^{(i)}} \right]$$

$$c_{2n} = \frac{2 \cdot (-1)^n}{\pi \cdot n} - \frac{2}{\pi^2 n^2} \cdot \sum_{j=1}^{j=M} \left[\frac{q'_j}{q} \cdot \frac{\sin(\pi n \eta_2'^{(j)}) - \sin(\pi n \eta_1'^{(j)})}{\eta_2'^{(j)} - \eta_1'^{(j)}} \right] \tag{42}$$

$$b_n = \frac{2(-1)^n}{\pi \cdot n} \tag{43}$$

for the case $V_{\xi\infty} = \text{const.} = -1$ and

$$b_n = \frac{2(-1)^n}{\pi n} + \frac{4}{\pi^3 n^3} [(V_1 + 3) + (2 \cdot V_1 + 3) \cdot (-1)^n] \tag{44}$$

for the parabolic function $V_{\xi\infty}$. In the case of an infinitely small slot, when

$$\eta_1^{(i)} = \eta_2^{(i)} = \eta_*^{(i)} \quad \text{and} \quad \eta_1'^{(j)} = \eta_2'^{(j)} = \eta_*'^{(j)} \tag{45}$$

the relationship in Equation 40 may be used, but the original terms in parenthesis must be replaced as follows

$$\lim_{\eta_2^{(i)} \to \eta_1^{(i)}} \frac{\sin(\pi n \eta_2^{(i)}) - \sin(\pi n \eta_1^{(i)})}{\eta_2^{(i)} - \eta_1^{(i)}} = \pi n \cdot \cos(\pi n \eta_*^{(i)}) \tag{46}$$

From Equation 39 it follows that other stream functions at the inflow and outflow may be assumed, provided they can be expanded in the Fourier series. The stream function $\psi_\infty(\eta)$ in Equations 35, 37, 38, and 41 may not be taken arbitrarily. It must satisfy the following equation [21]:

$$\frac{\partial^4}{\partial \eta^4} [\psi_\infty(\eta)] = 0 \tag{47}$$

The presented model can also be used in the case when permeable walls are at the inflow and outflow. Assuming that $V_\xi(0, \eta)$ and $V_\xi(\xi, \eta)$ are known functions, we may write

$$c_{1n} = \frac{2 \cdot (-1)^n}{\pi n} + \frac{2}{\pi n} \cdot \int_0^1 V_\xi(0, \eta) \cdot \cos(\pi n \eta) \cdot d\eta$$

$$c_{2n} = \frac{2 \cdot (-1)^n}{\pi n} + \frac{2}{\pi n} \cdot \int_0^1 V_\xi(\xi_1, \eta) \cdot \cos(\pi n \eta) \cdot d\eta \tag{48}$$

Function $\psi_\infty(\eta)$ and coefficients b_n remain the same as before. An important element of the flow pattern is velocity distribution. Velocity components can be evaluated from the stream function in Equation 35. Hence we have

$$V_\xi(\xi, \eta) = V_{\xi\infty}(\eta) + \sum_{n=1}^{n=\infty} \pi n \cdot \{d_{1n} \cdot \exp(d_{1n} \cdot \xi) + d_{2n} \cdot \exp[-d_{2n}(\xi_1 - \xi)]\} \cdot \cos(\pi n \eta)$$

$$V_\eta(\xi, \eta) = -\sum_{n=1}^{\infty} \{d_{1n} \cdot a_{1n} \cdot \exp(d_{1n} \cdot \xi) + a_{2n} \cdot d_{2n} \cdot \exp[-d_{2n}(\xi_1 - \xi)]\} \cdot \sin(\pi n \eta) \tag{49}$$

The final Equations 35 through 46 describe the primary stream function solutions of the vertical circulation in a stratified reservoir. Examples of the circulation in a rectangular reservoir are shown

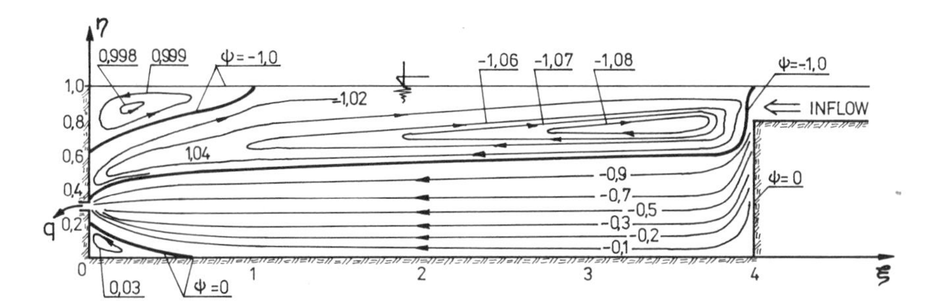

Figure 3. Vertical circulation in density-stratified reservoir (flow pattern in case of rectangular reservoir): A = 10,000; $\eta_1 = 0.28$; $\eta_2 = 0.32$; $\eta'_1 = 0.8$; $\eta'_2 = 1.0$; $\xi = 4.0$; $V_1 = -1.5$ (courtesy of the American Society of Civil Engineers).

in Figures 3 and 4. As can be seen in Figure 4, the model gives good results even in the case of short reservoir. However, one of the restrictions, Equation 13, needs the reservoir to be rather long in order to satisfy the "Prandtl assumption." In practice the reservoirs are very often of the semi-infinite strip shape. The case of semi-infinite strip shape can be calculated using the primary stream function for the rectangular reservoir. We have

$$\psi_1(\xi, \eta) = \lim_{\xi_1 \to \infty} \psi \tag{50}$$

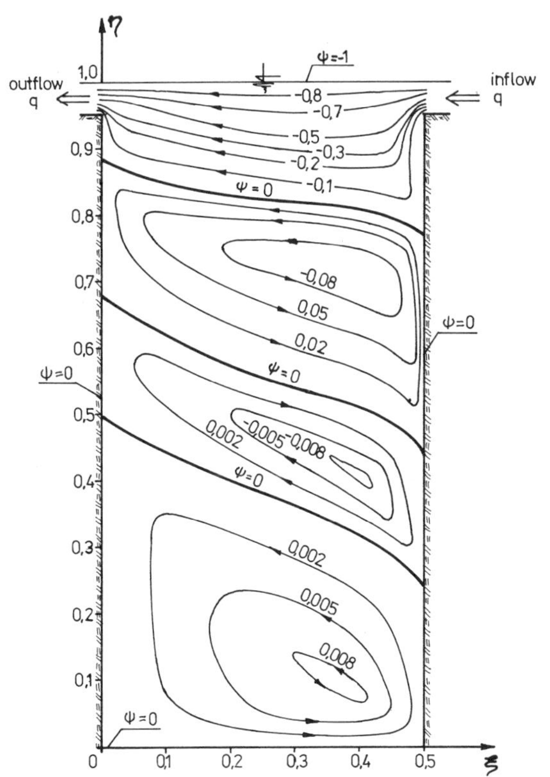

Figure 4. Vertical circulation in density-stratified reservoir (flow pattern in case of short rectangular reservoir): A = 10,000; $\eta_1 = 0.95$; $\eta_2 = 1.0$; $\eta'_1 = 0.95$; $\eta'_2 = 1.0$; $\xi = 0.5$; $V_1 = -1.5$ (courtesy of the American Society of Civil Engineers).

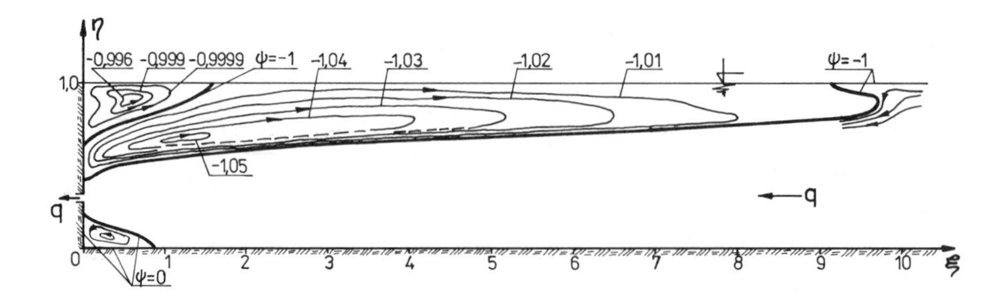

Figure 5. Vertical circulation in density-stratified reservoir (flow pattern in case of semi-infinite strip reservoir): A = 1,000; $\eta_1 = 0.28$; $\eta_2 = 0.32$; $V_1 = -2$ (courtesy of the American Society of Civil Engineers).

and so

$$\psi_1(\xi, \eta) = \psi_{\infty}(\eta) + \sum_{n=1}^{n=\infty} [(c_{1n} - b_n) \cdot \exp(d_{1n} \cdot \xi) \cdot \sin(\pi n \eta)] \tag{51}$$

The velocity components evaluated from the new stream function ψ_1 are as follows

$$V_\xi(\xi, \eta) = V_{\xi\infty}(\eta) + \sum_{n=1}^{\infty} [\pi n \cdot (c_{1n} - b_n) \cdot \exp(d_{1n} \cdot \xi) \cdot \cos(\pi n \eta)];$$

$$V_\eta(\xi, \eta) = -\sum_{n=1}^{\infty} [d_{1n} \cdot (c_{1n} - b_n) \cdot \exp(d_{1n} \cdot \xi) \cdot \sin(\pi n \eta)] \tag{52}$$

Examples of the flow pattern in the case of a semi-infinite strip reservoir are given in Figures 5 and 6. From these figures it becomes clear that the surface velocity V_1 plays an important role in

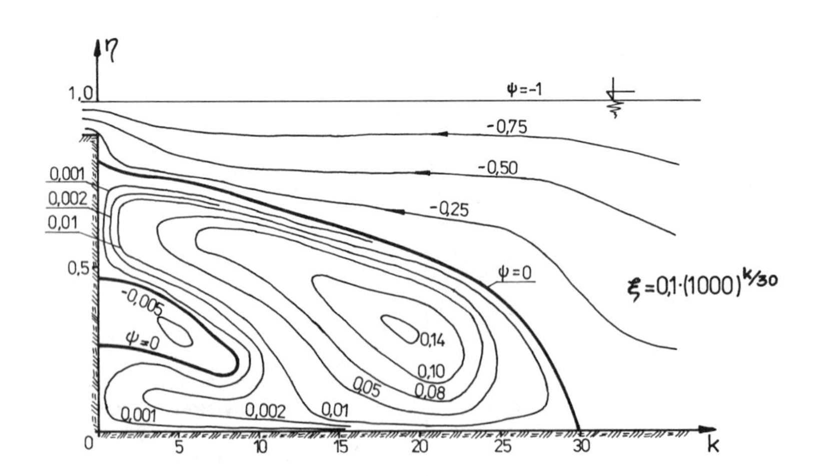

Figure 6. Vertical circulation in density-stratified reservoir (flow pattern in case of semi-infinite strip reservoir): A = 4,000; $\eta_1 = 0.9$; $\eta_2 = 1.0$; $V_1 = -2$ (courtesy of the American Society of Civil Engineers).

forming the circulation region. The increase of the surface velocity V_1 makes the circulation region smaller and finally leads to the case of a submerged circulation region. To close the analysis, a relation between the wind shear stress and the surface velocity V_1 was examined. In the case where

$$V_{\xi\infty}(\eta) = -1$$

we have

$$\tau_2 = 0 \tag{53}$$

that is flow without wind action. In the case of the parabolic distribution ($V_{\xi\infty}(\eta)$ from Equation 38) we have the following relationship

$$\tau_2 = 4 \cdot \frac{\rho_2 \cdot K_y \cdot q}{D^2} \cdot \left(V_1 + \frac{3}{2}\right) \tag{54}$$

It can be seen, that the surface shear stress τ_2 remains constant all along the water surface. From this formula it is shown that the parabolic type of $V_{\xi\infty}(\eta)$ may be applied in the case of no wind. In such a case we have

$$V_1 = -1.5 \tag{55}$$

Mechanism of Vertical Circulation

From the examples of the flow pattern shown in Figures 3, 4, 5, and 6 it follows that the circulation regions are the characteristic feature of the flow fields. Their shape and size depend on the flow parameters and the boundary conditions. The circulation regions are those places of vertical water mixing, and they are limited with their interfaces. Another characteristic feature of those regions is a number of closed internal eddies, existing within them and separated from each other by internal interfaces (Figure 7).

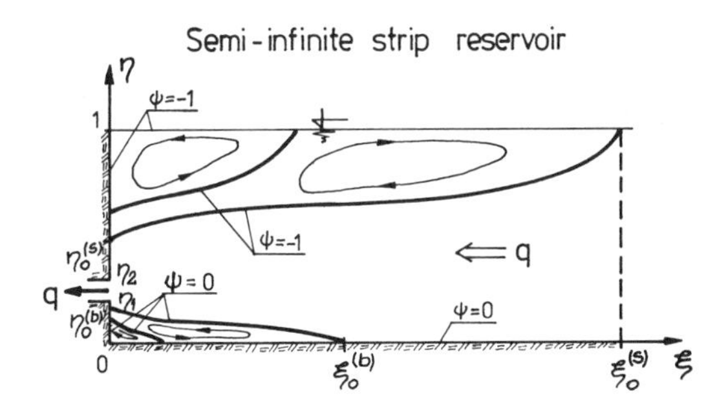

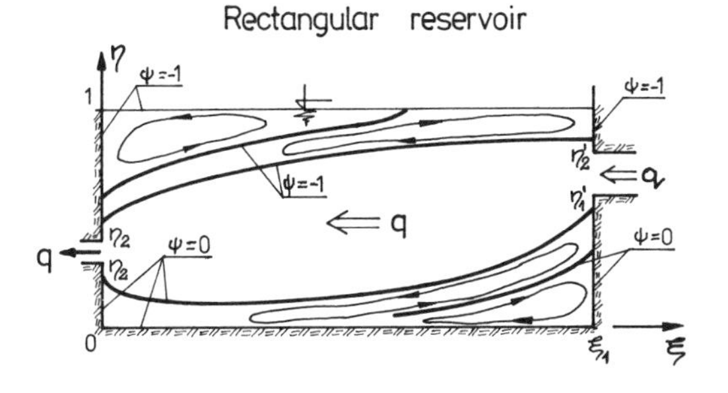

Figure 7. Definition sketch of typical shape of surface and bottom circulation regions (courtesy of the American Society of Civil Engineers).

The equations of the interface lines can be defined by using the earlier obtained stream functions in Equations 35 and 51. We have

- For the surface circulation region

$$\psi(\xi, \eta) = -1 \tag{56}$$

- For the bottom circulation region

$$\psi(\xi, \eta) = 0 \tag{57}$$

For the geometry of the circulation regions an important point is to specify their vertical and horizontal extents. These values can be estimated using the velocity distribution in Equations 49 and 52. For the surface circulation region we have

$$\begin{aligned} &V_\xi(\xi_0^{(s)}, 1) = 0 \\ &V_\eta(0, \eta_0^{(s)}) = 0 \\ &V_\eta(\xi_1, \eta_0^{(s)}) = 0 \end{aligned} \tag{58}$$

while for the bottom circulation region

$$\begin{aligned} &V_\xi(\xi_0^{(b)}, 0) = 0 \\ &V_\eta(0, \eta_0^{(b)}) = 0 \\ &V_\eta(\xi_1, \eta_0^{(b)}) = 0 \end{aligned} \tag{59}$$

Careful analysis of the above relationships indicates that each of them includes several solutions and that each solution corresponds to the one internal closed eddy. From the research carried out it follows that the semi-infinite strip reservoir with one outflow slots, gives the best opportunity for examining the mechanism of vertical circulation. This case causes the strongest changes of the circulation region when flow parameters vary. In the research of vertical circulation made by Long [18], Walesh and Monkmeyer [30], and Yih [32, 33] it was assumed, that far upstream we have $V_{\xi\infty}(\eta) = -1$. It refers to the case with no wind. In order to compare the present model and the previous results, this flow case was initially investigated. Following Equations 27 through 49 in the case of constant velocity far upstream, we have

$$\psi_1(\xi, \eta) = -\eta + \sum_{n=1}^{\infty} [a_{1n} \cdot \exp(d_{1n} \cdot \xi) \cdot \sin(\pi n \eta)] \tag{60}$$

$$V_\xi(\xi, \eta) = -1 + \sum_{n=1}^{\infty} [\pi n \cdot a_{1n} \cdot \exp(d_{1n} \cdot \xi) \cdot \cos(\pi n \eta)] \tag{61}$$

$$V_\eta(\xi, \eta) = -\sum_{n=1}^{\infty} [a_{1n} \cdot d_{1n} \cdot \exp(d_{1n} \cdot \xi) \cdot \sin(\pi n \eta)] \tag{62}$$

where

$$a_{1n} = -\frac{2}{\pi^2 n^2} \cdot \frac{\sin(\pi n \eta_2) - \sin(\pi n \eta_1)}{\eta_2 - \eta_1} \tag{63}$$

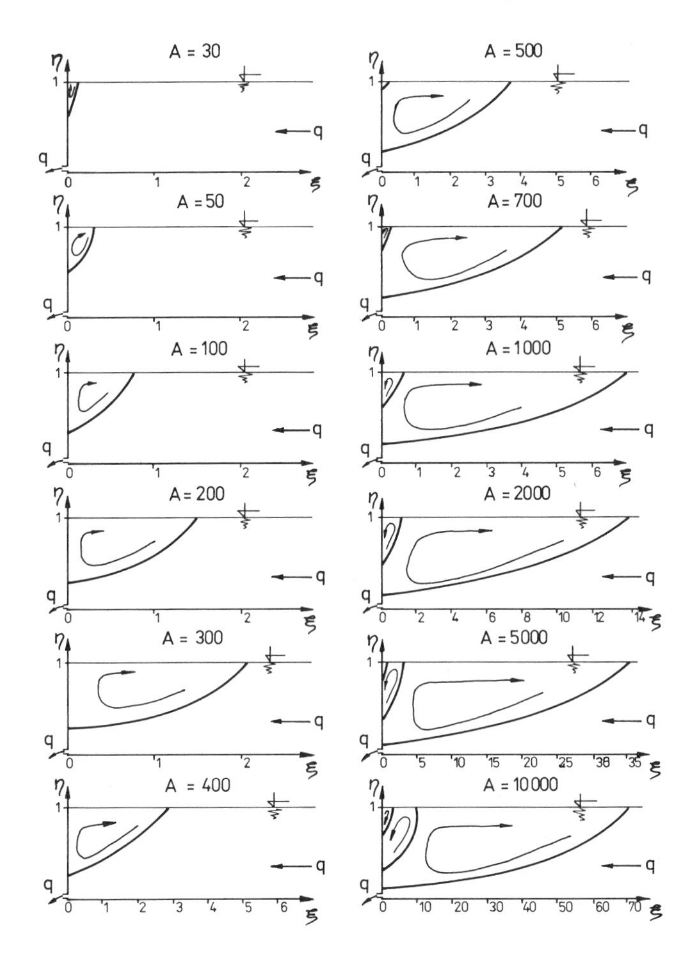

Figure 8. Mechanism of vertical circulation in case of constant velocity far upstream: $\eta_1 = 0$; $\eta_2 = 0.02$, (*Arch. Hydr.*, Vol. 3 (1979) courtesy of the Polish Academy of Science).

It can be seen that in this case the flow pattern depends only on two factors: parameter A and outflow location (η_2, η_1). Figure 8 shows the mechanism of vertical circulation in the case of constant velocity far upstream. From this figure it follows that the source of each newly-created eddy is in the corner of the outflow wall and water surface. The increase of A causes the growth of the circulation region and the number of internal closed eddies. The vertical and horizontal extent can be evaluated from Equations 58 through 61. Figure 9 shows the surface velocity distribution and the velocity changes along the outflow wall. In practical cases, especially in the laboratory experiments, the number of internal closed eddies can be concluded from the vertical distribution of $V_\xi(\xi, \eta)$ in a given profile. The idea of this conclusion is given in Figure 10. For the rough estimation of the horizontal extent of the surface circulation region in the considered case of flow, the following approximate formula may be used

$$\xi_0^{(s)} = \frac{A}{\pi^4} \cdot \ln[2\cos(\pi\eta_*)] \tag{64}$$

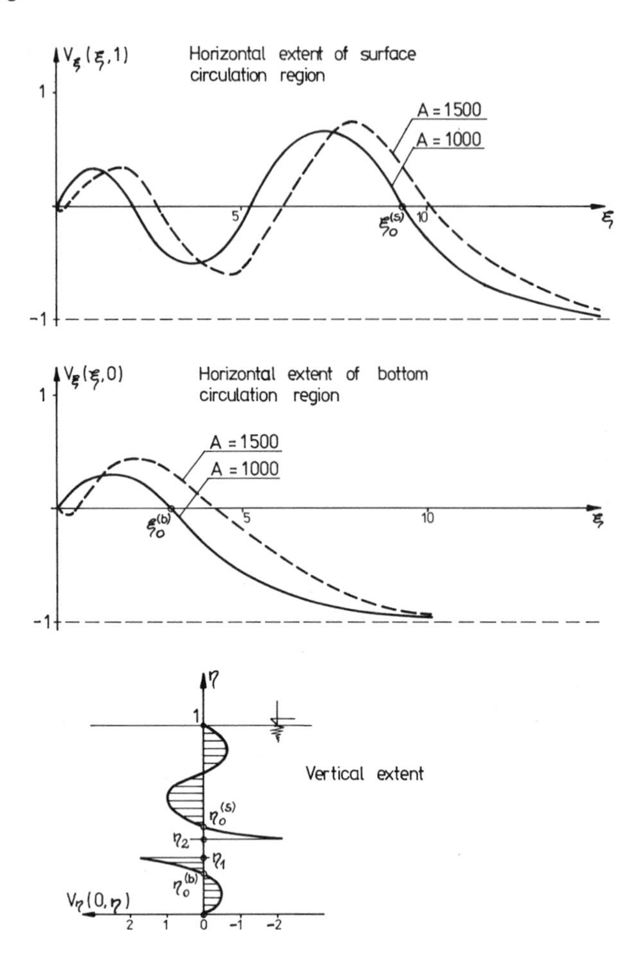

Figure 9. Estimation of horizontal and vertical extent of circulation regions from velocity distribution: $\eta_1 = 0.28$; $\eta_2 = 0.32$.

or coming back to the dimension values

$$\chi_0^{(s)} = \frac{D}{\pi^4} \cdot \frac{\Delta\rho}{\rho_1} \cdot \frac{q}{K_y} \cdot \frac{gD^3}{q^2} \cdot \ln[2 \cdot \cos(\pi\eta_*)] \tag{65}$$

These formulae give sufficient accuracy for practical calculation if

$$\eta_* \le 0.25 \quad \text{and} \quad A \ge 300 \tag{66}$$

From Equation 65 it follows that the horizontal extent $\chi_0^{(s)}$ does not depend on the eddy viscosity coefficient K_x; however both coefficients, K_x and K_y, were taken into account in the mathematical model.

The second case considered here is the flow with a parabolic velocity function $V_{\xi\infty}(\eta)$. As has been mentioned earlier, the flow pattern depends strongly on the surface velocity V_1 (Figures 3

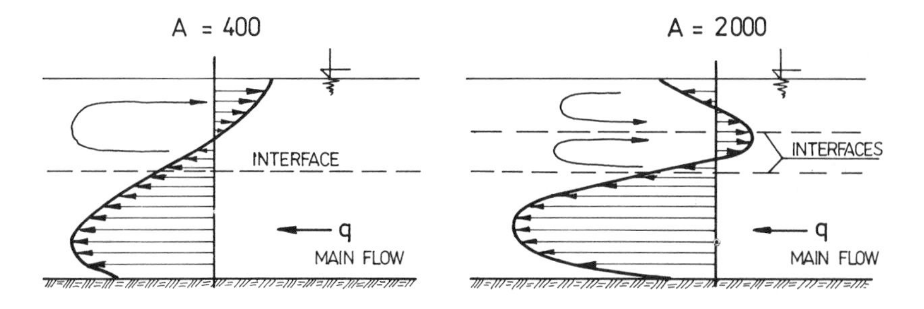

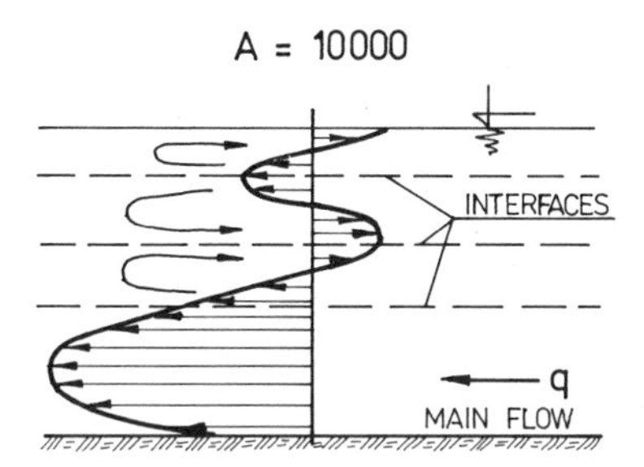

Figure 10. Dependence of number of closed internal eddies on vertical distribution $V_\xi(\xi, \eta)$, (*Arch. Hydr.*, Vol. 3 (1979), courtesy of the Polish Academy of Science).

through 6). Following Equations 27 through 49 we may write

$$\psi_1(\xi, \eta) = (V_1 + 2)\eta^3 - (V_1 + 3)\eta^2 + \sum_{n=1}^{\infty} [a_{1n} \cdot \exp(d_{1n} \cdot \xi) \cdot \sin(\pi n \eta)]$$

$$V_\xi(\xi, \eta) = 3(V_1 + 2)\eta^2 - 2(V_1 + 3)\eta + \sum_{n=1}^{\infty} [\pi n \cdot a_{1n} \cdot \exp(d_{1n} \cdot \xi) \cdot \cos(\pi n \eta)]$$

$$V_\eta(\xi, \eta) = -\sum_{n=1}^{\infty} [d_{1n} \cdot a_{1n} \cdot \exp(d_{1n} \cdot \xi) \cdot \sin(\pi n \eta)] \tag{67}$$

where

$$a_{1n} = -\frac{2}{\pi^2 n^2} \cdot \frac{\sin(\pi n \eta_2) - \sin(\pi n \eta_1)}{\eta_2 - \eta_1} - \frac{4}{\pi^3 n^3} \cdot [(V_1 + 3) + (2V_1 + 3) \cdot (-1)^n] \tag{68}$$

The solution described above indicates, that in this case we have three parameters influencing the flow pattern: parameter A, the outflow slot location (η_1, η_2), and the surface velocity $\mathbf{V}_1$. To formulate the mechanism of vertical circulation Equations 58 and 59 were examined. For the convenience of further calculation these relationships were rearranged in order to obtain functions of type

$$V_1 = V_1(A, \xi_0^{(s)})$$

$$V_1 = V_1(A, \eta_0^{(s)}) \tag{69}$$

This means that we are looking for a value V_1 that makes the surface velocity equal to zero at a certain distance $\xi_0^{(s)}$. Therefore we have

$$V_1(\xi_0^{(s)}) = \frac{\sum_{n=1}^{\infty} [c_{3n} \cdot \exp(d_{1n} \cdot \xi_0^{(s)})]}{1 - \sum_{n=1}^{\infty} [c_{4n} \cdot \exp(d_{1n} \cdot \xi_0^{(s)})]} \tag{70}$$

$$V_1(\eta_0^{(s)}) = \frac{\sum_{n=1}^{\infty} [c_{5n} \cdot \sin(\pi n \eta_0^{(s)})]}{\sum [c_{6n} \cdot \sin(\pi n \eta_0^{(s)})]} \tag{71}$$

where

$$C_{3n} = \frac{2 \cdot (-1)^n}{\pi \cdot n} \cdot \left[\frac{\sin(\pi n \eta_2) - \sin(\pi n \eta_1)}{\eta_2 - \eta_1} + 6 \frac{1 + (-1)^n}{\pi \cdot n} \right]$$

$$C_{4n} = \frac{4 \cdot (-1)^n}{\pi^2 \cdot n^2} \cdot [1 + 2 \cdot (-1)^n]$$

$$C_{5n} = -\frac{2 \cdot d_n}{\pi^2 n^2} \cdot \left[\frac{\sin(\pi n \eta_2) - \sin(\pi n \eta_1)}{\eta_2 - \eta_1} + 6 \frac{1 + (-1)^n}{\pi \cdot n} \right]$$

$$C_{6n} = \frac{4 \cdot dn}{\pi^3 n^3} [1 + 2 \cdot (-1)^n] \tag{72}$$

The analysis of Equation 71 indicates that this relationship is insensitive to variation in V_1. This means that for the given outflow slot location (η_1, η_2) and parameter A, a different V_1 does not change the value of $\eta_0^{(s)}$ significantly. Furthermore, it appears that this feature does not depend on the boundary conditions far upstream. This could explain why different authors carrying research on the critical conditions of the selective withdrawal, have obtained such a good agreement with experimental results since $\eta_0^{(s)}$ does not depend on the velocity distribution at inflow. A typical shape of $V_1(\eta_0^{(s)})$ is given in Figure 11, and the relation $\eta_0^{(s)} = \eta_0(A)$ is given in Figure 12. The evaluation of the mechanism of vertical circulation is more convenient by means of Equation

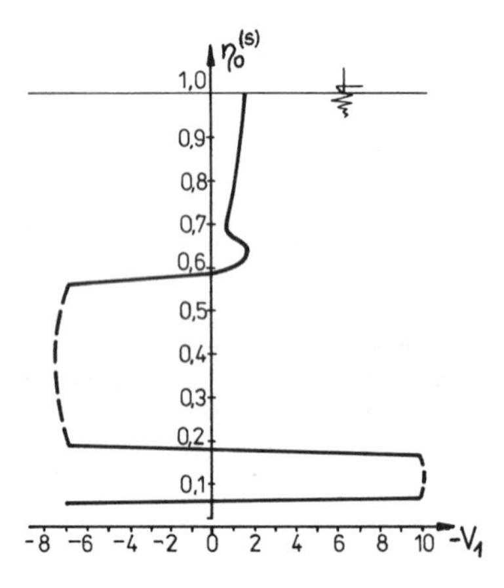

Figure 11. Plot of function $V_1(\eta_0^{(s)})$; $\eta_1 = 0$; $\eta_2 = 0.02$; A = 1,000 (*Hydr. Trans.*, Vol. 43, (1981), courtesy of the Polish Academy of Science).

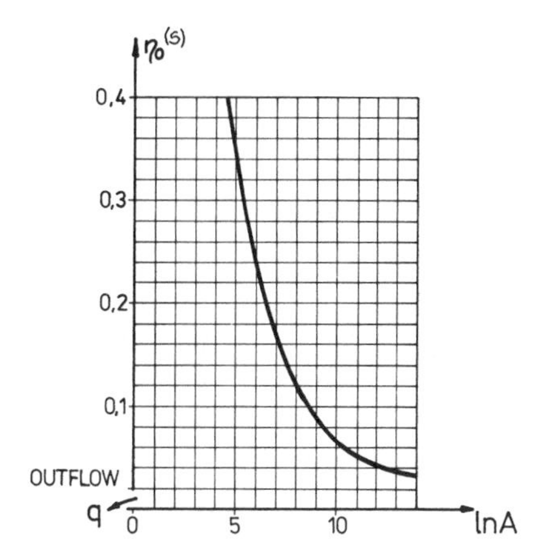

Figure 12. Curve $\eta_0^{(s)} = \eta_0^{(s)}(A)$, ($\eta_1 = 0$, $\eta_2 = 0.02$).

70. A typical curve of $V_1(\xi_0^{(s)})$ is given in Figure 13. In order to make the picture clearer the horizontal coordinate is given in the term of K, where

$$K = 10 \log(10 \cdot \xi_0)$$

and so

$$\xi_0 = 0.1 \cdot (1{,}000)^{K/30} \tag{73}$$

From this figure it follows that each arbitrarily chosen value, V_1 corresponds to several values of $\xi_0^{(s)}$. The curve $V_1(\xi_0^{(s)})$ has several extreme points (Figure 13) denoted as $(V_{01}, (\xi_{01})$, (V_{02}, ξ_{02}),

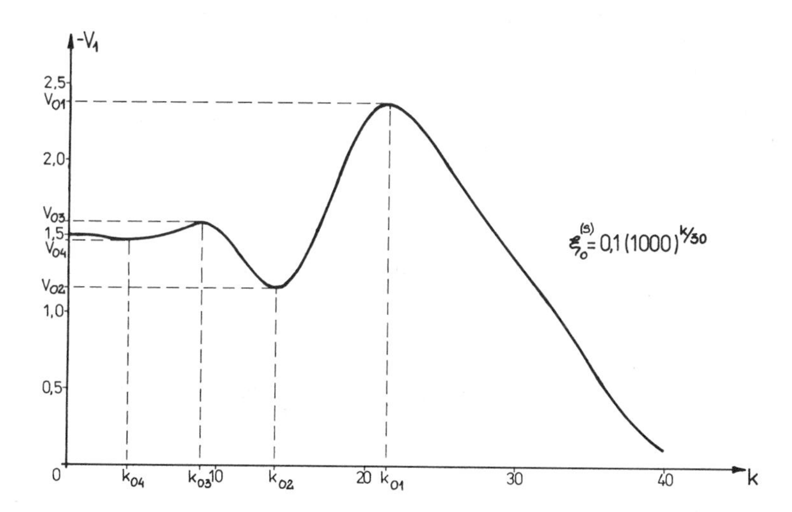

Figure 13. Typical curve $V_1(\xi_0^{(s)})$: $\eta_1 = 0.28$; $\eta_2 = 0.32$; A = 40,000, (*Hydr. Trans.*, Vol. 43 (1981), courtesy of the Polish Academy of Science).

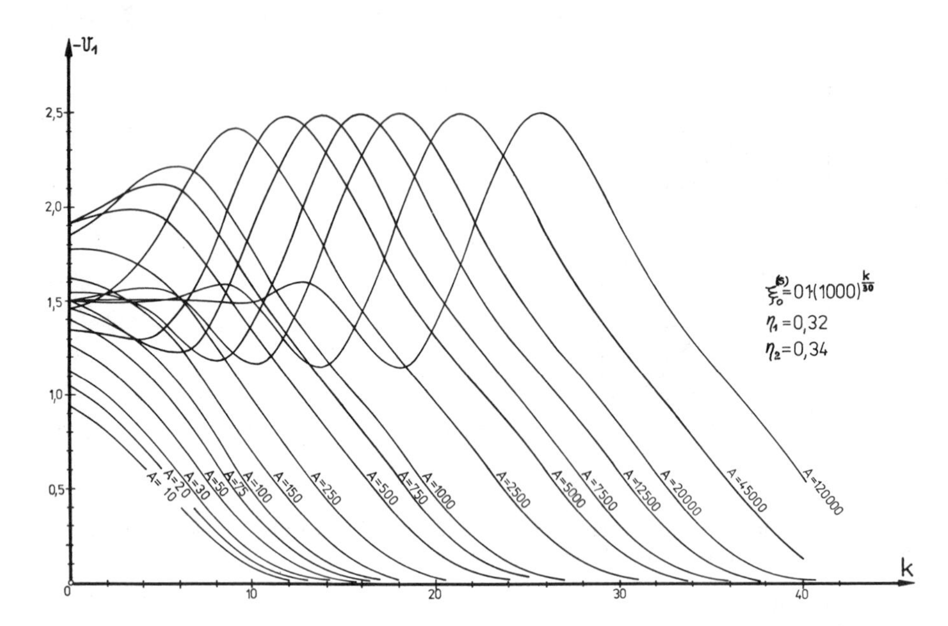

Figure 14. Influence of parameter A on shape of $V_1(\xi_0^{(s)})$ curve: $\eta_1 = 0.32$; $\eta_2 = 0.34$ (*Hydr. Trans.* Vol. 43 (1981), courtesy of the Polish Academy of Science).

(V_{03}, ξ_{03}), (V_{04}, ξ_{04}), etc. The number of those points depends on A. As an example, in Figure 14 the influence of A on the shape of the $V_1(\xi_0^{(s)})$ curve is shown. The points (V_{01}, ξ_{01}), $(V_{02}, \xi_{02}) \ldots$, play an important role in the interpretation of the vertical circulation. Based on these points the mechanism of the formation of the circulation regions can be explained. If

$$V_{02} < V_1 < 0 \tag{74}$$

(note: V_1, V_{01} and others are negative), then the circulation region consists of a single, fully developed eddy (the one that extends from the outflow wall to the water surface). If

$$V_{01} < V_1 < V_{02}$$

then the number of closed internal eddies within the circulation region varies with V_1 and can be determined from Figure 13. From that curve, the location of all the points where the internal interface intersects the surface, can be estimated. If

$$V_1 = -1.5$$

(in this case $\tau_2 = 0$; Equations 54 and 55), then at the corner of the outflow wall and the surface, a flow is formed of very slow velocities (of order $10^{-4} \cdot V_0$). If

$$V_1 < V_{01}$$

then the circulation region detaches from the water surface and sinks. The submerging of the circulation region does not depend on the value A. The process of detachment is shown in Figure

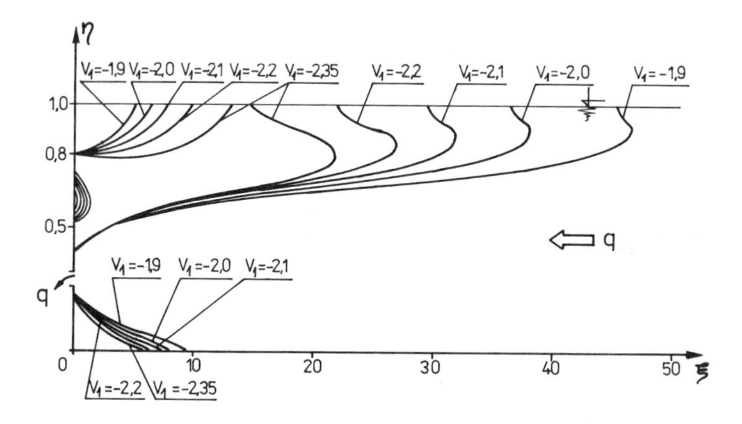

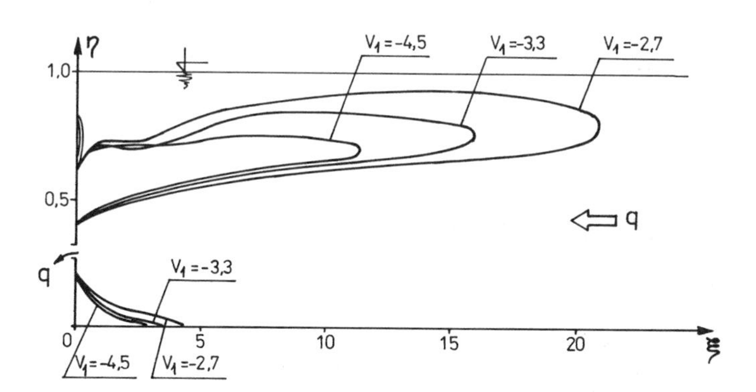

Figure 15. Influence of V_1 on shape of surface circulation region: $\eta_1 = 0.32$; $\eta_2 = 0.34$; $A = 40{,}000$ (courtesy of the American Society of Civil Engineers).

15, and the full development of the surface circulation region is given in Figure 16. The development of the bottom circulation region in the case of surface outflow is given in Figure 17.

For practical calculations an important thing is to estimate the horizontal extent of the circulation region. The analysis made by Meyer [20] indicated that if total length is considered the parameter U should be introduced as follows

$$U = \ln\left(\frac{A}{\pi^4 \cdot \xi_0^{(s)}}\right) \tag{75}$$

For such a parameter U, where

$$\xi_0^{(s)} > \xi_{01} \tag{76}$$

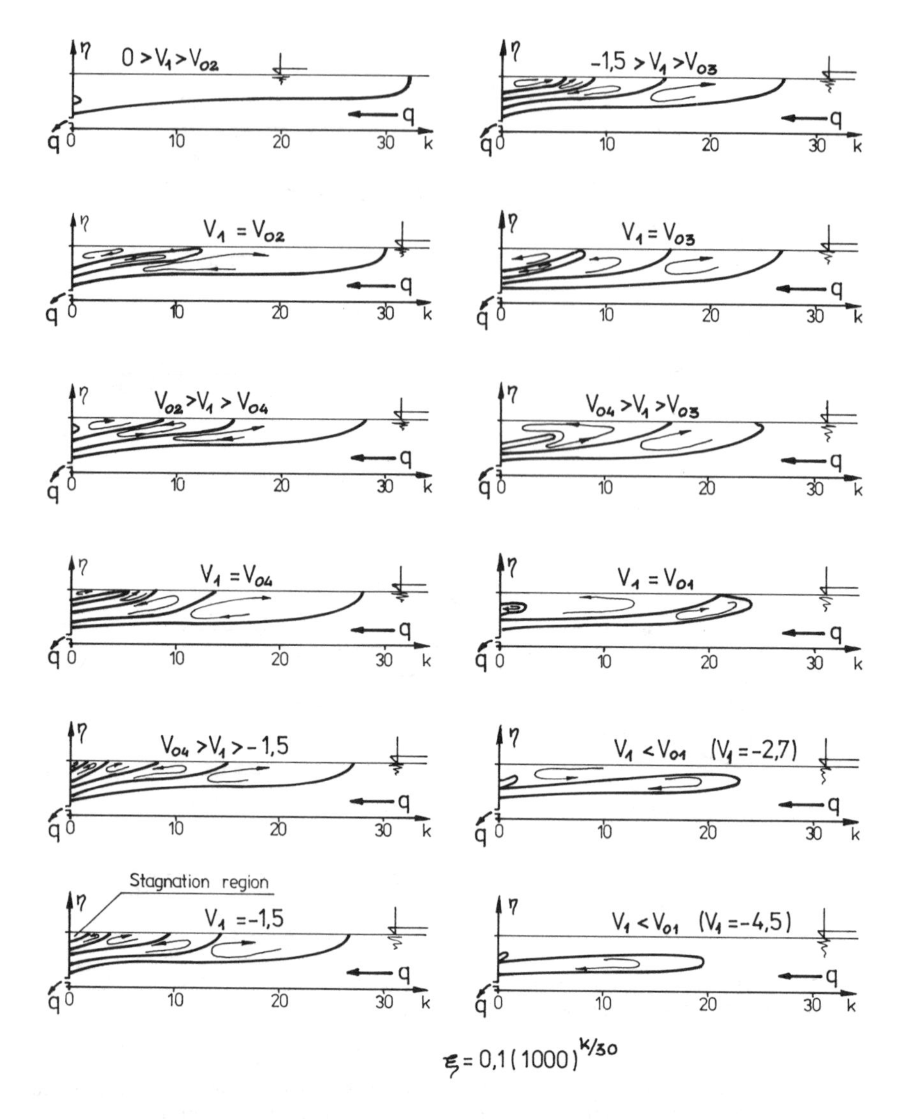

Figure 16. Mechanism of formation of surface circulation region in density-stratified reservoir: $\eta_1 = 0.32$; $\eta_2 = 0.34$; A = 38,000, (*Hydr. Trans.*, Vol. 43 (1981), courtesy of the Polish Academy of Science).

the set of curves $V_1(U)$ can be constructed. In Figure 18 the set of curves $V_1 = f(U)$ is shown. Each curve refers to the different outflow slot location. From these graphs it follows, that the total extent of the surface circulation region is equal to

$$\xi_0^{(s)} = \frac{A}{\pi^4} \cdot \exp[-U(V_1)] \tag{77}$$

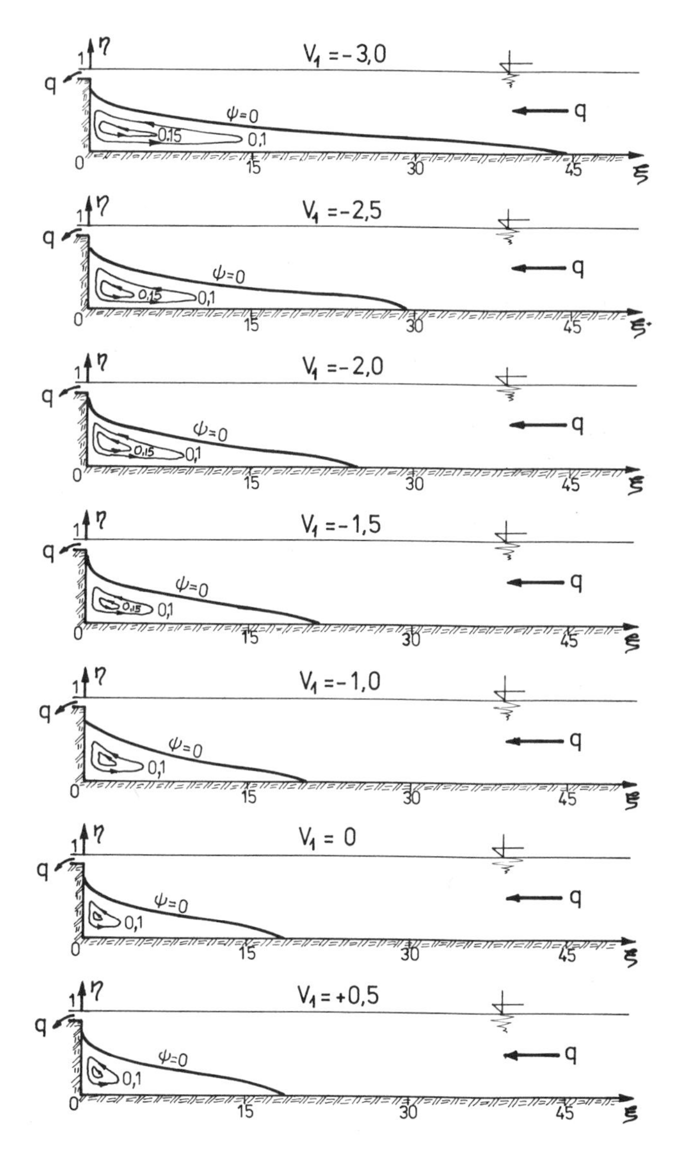

Figure 17. Development of bottom circulation region in case of surface outflow: $\eta_1 = 0.9$; $\eta_2 = 1.0$; A = 1,000 (courtesy of the American Society of Civil Engineers).

where $U(V_1)$ is the reverse function of $V_1 = f(U)$. And coming back to the dimension values we have

$$x_0^{(s)} = \frac{D}{\pi^4} \cdot \frac{\Delta\rho}{\rho_1} \cdot \frac{q}{K_y} \cdot \frac{gD^3}{q^2} \cdot \exp[-U(V_1)] \tag{78}$$

An interesting point is that similarly to the case with the constant velocity $V_{\xi\infty} = -1$ (Equation 65), the horizontal extent depends only on the vertical eddy viscosity coefficient K_y.

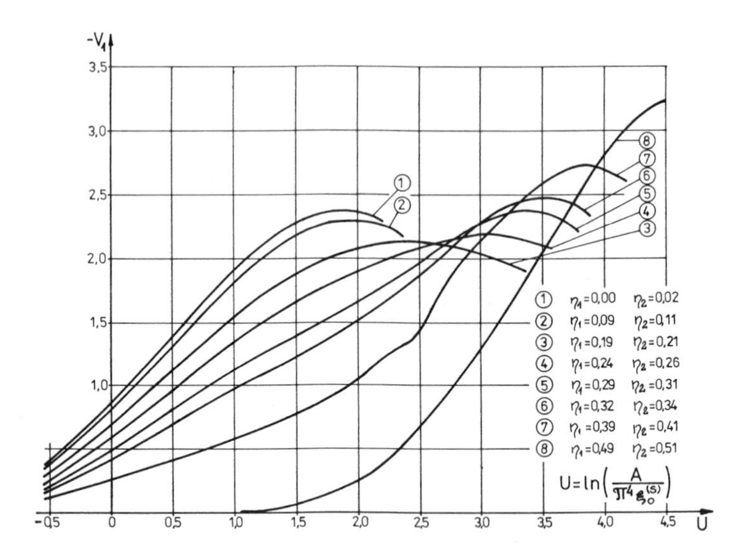

Figure 18. Curves $V_1(U)$ for various outflow location (*Hydr. Trans.*, Vol. 43 (1981), courtesy of the Polish Academy of Science).

In the case of a bottom circulation region produced by the surface outflow, the horizontal extent can be calculated similarly to the surface circulation region. Therefore we have

$$U_* = \ln\left[\frac{A}{\pi^4 \cdot \xi_0^{(b)}}\right]$$

and thus

$$\xi_0^{(b)} = \frac{A}{\pi^4} \cdot \exp[-U_*(V_1)] \tag{79}$$

The curve $U_*(V_1)$ is given in Figure 19.

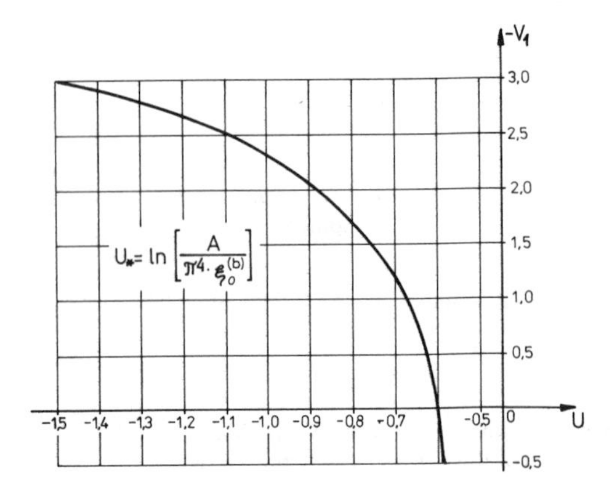

Figure 19. Curve $V_1(U_*)$ for bottom circulation region: $\eta_1 = 0.9$; $\eta_2 = 1.0$.

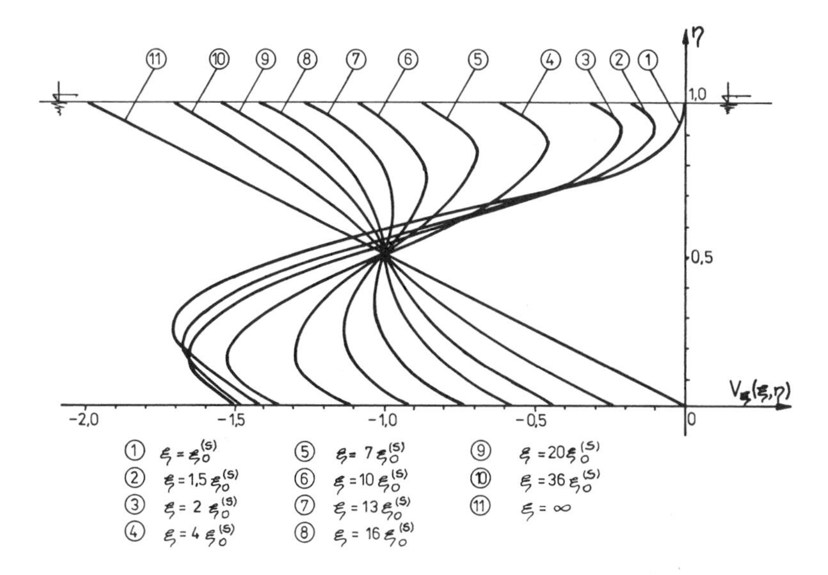

Figure 20. Vertical velocity profiles $V_\xi(\xi, \eta)$ for various distances from outflow (*Hydr. Trans.*, Vol. 43 (1981), courtesy of the Polish Academy of Science).

From the given analysis of the flow pattern it is shown that the parameter V_1 plays an essential role in the formation of the vertical circulation. A question arises as to how far away this value should be measured in order to obtain a relevant accuracy. The vertical distribution of $V_\xi(\xi, \eta)$ is given by Equation 67. When $\xi \to \infty$ then V_ξ tends to $V_{\xi\infty}$. In Figure 20 the vertical velocity profiles $V_\xi(\xi, \eta)$ are given at various distances from the outflow. It follows that the distance from where V_1 should be measured is quite long. From Figure 20 it is shown that a distance of about $36 \cdot \xi_0^{(s)}$ is needed in order to obtain a relative error less than 10%. In laboratory experiments such a distance means hundreds of meters. It suggests that the direct measurement of V_1 needs a very long model. Another characteristic feature of those velocity profiles is that a surface current of strong velocity arises in the upper layers. This phenomenon comes from the nature of the stratified flow. The terminal velocity distribution $V_{\xi\infty}(\eta)$ is shown in Figure 21. It can be seen that if $V_1 < -3$, then

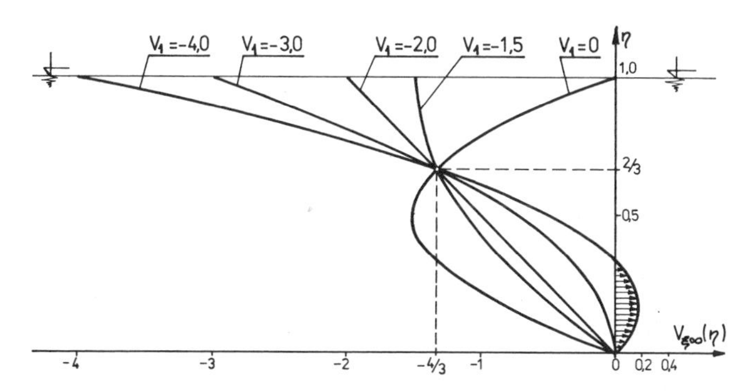

Figure 21. Terminal velocity distribution $V_{\xi\infty}(\eta)$ (*Hydr. Trans.*, Vol. 43 (1981), courtesy of the Polish Academy of Science).

at the bottom a backward current appears. This current is not due to the density stratification, but is produced by the momentum transfer phenomena. In Figure 22 the development of the circulation regions in the rectangular reservoir is given. It follows that the mechanism is similar to that for the semi-infinite strip reservoir.

In practice a question often arises about the maximum flow respecting the mechanism of the vertical circulation. Exceeding this maximum flow causes strong mass and energy exchanges through the interface and damages the cell structure of the circulation. The maximum flow is usually given in terms of the densimetric Froude number (Equation 12). An interesting subject is to relate the parameter A to the critical Froude number ($(Fr_1)_{cr}$). For the rough approximation we may carry out the following circulation: if we write

$$A = \frac{\Delta\rho}{\rho_1} \cdot \frac{gD^2}{V_0 \cdot K}$$

and moreover, if the value K is assumed to be

$$K = \kappa_0 \cdot V_0 \cdot D$$

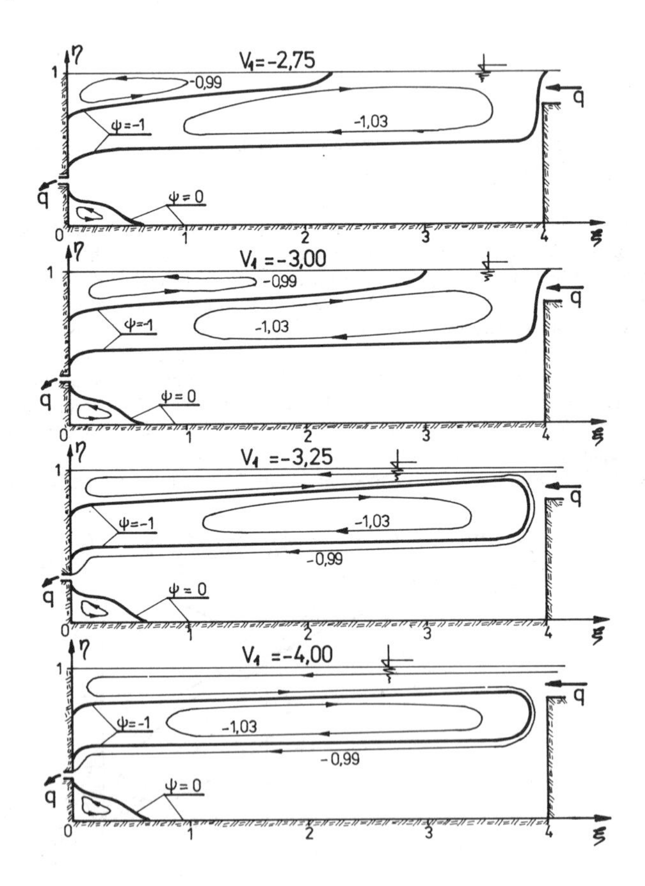

Figure 22. Wind influenced vertical circulation in rectangular reservoir: $\eta_1 = 0.28$; $\eta_2 = 0.32$; $\eta'_1 = 0.8$; $\eta'_2 = 1.0$; A = 10,000; $\xi_1 = 4.0$ (courtesy of the American Society of Civil Engineers).

(as it is often taken), where κ_0 = const. and it is of the order $\kappa_0 = 10^{-2} + 10^{-4}$, it becomes clear that

$$A = \frac{1}{\kappa_0} \cdot (Fr_1)^{-2}$$

And then the critical value of A can be defined as

$$A_{cr} = \frac{1}{\kappa_0} \cdot [(Fr_1)_{cr}]^{-2}$$

Under average conditions $A_{cr} = 1{,}000$.

Vertical Circulation with Additional Horizontal Density Changes

The primary solution described in the previous subsection assumes that density along the streamline is constant. Laboratory experiments made by the author [20] indicate that the additional horizontal density changes arise along the streamline. Those changes, although they are small, may play an important role in the formation of the circulation region. They have an especially strong influence in the case of a long reservoir (semi-infinite strip reservoir), when flow is generated by mass exchange and wind. The relevant relationship for this sort of flow is given by Equation 25. The solution of this equation is given by Mayer [21]. It is clear that the general solution of the problem may be expressed in terms of the Fourier series. Assuming boundary conditions to be the same as before, we obtain for the rectangular shape reservoir, the following solution [21]:

$$\psi(\xi, \eta) = \psi_{\infty}(\eta) + \sum_{n=1}^{\infty} \{a_{1n} \cdot \exp[D_{1n}(\xi)] + a_{2n} \cdot \exp[-D_{2n}(\xi_1) + D_{2n}(\xi)]\} \cdot \sin(\pi n \eta) \tag{80}$$

$$V_{\xi}(\xi, \eta) = V_{\xi\infty}(\eta) + \sum_{n=1}^{\infty} \pi n\{a_{1n} \cdot \exp[D_{1n}(\xi)] + a_{2n} \cdot \exp[-D_{2n}(\xi_1) + D_{2n}(\xi)]\} \cdot \cos(\pi n \eta) \tag{81}$$

$$V_{\eta}(\xi, \eta) = -\sum_{n=1}^{\infty} \{d_{1n} \cdot a_{1n} \cdot \exp[D_{1n}(\xi)] + d_{2n} \cdot a_{2n} \cdot \exp[-D_{2n}(\xi_1) + D_{2n}(\xi)]\} \cdot \sin(\pi n \eta) \tag{82}$$

where

$$D_{1n}(\xi) = \int_0^{\xi} d_{1n}(\xi) \cdot d\xi \qquad D_{2n}(\xi) = \int_0^{\xi} d_{2n}(\xi) \cdot d\xi \tag{83}$$

while in this case

$$d_{1n}(\xi) = \frac{A(\xi) - \sqrt{A^2(\xi) + 4\pi^6 n^6}}{2 \cdot \pi^2 n^2}; \qquad d_{2n}(\xi) = \frac{A(\xi) + \sqrt{A^2(\xi) + 4\pi^6 n^6}}{2 \cdot \pi^2 n^2} \tag{84}$$

and from Equation 25

$$A(\xi) = A \cdot f(\xi) \tag{85}$$

The value of A remains the same as before (Equation 36)

$$A = \frac{\Delta\rho}{\rho_1} \cdot \frac{g \cdot D^3}{q \cdot \sqrt{K_y(K_y + K_x)}}$$

Coefficients a_{1n} and a_{2n} in the present case should be evaluated from the following formulae:

$$a_{1n} = \frac{c_{1n} - c_{2n} \cdot \exp[-D_{2n}(\xi_1)] - b_n \cdot \{1 - \exp[-D_{2n}(\xi_1)]\}}{1 - \exp[-D_{2n}(\xi_1) + D_{1n}(\xi_1)]}$$

$$a_{2n} = \frac{c_{2n} - c_{1n} \cdot \exp[D_{1n}(\xi_1)] - b_n\{1 - \exp[D_{1n}(\xi_1)]\}}{1 - \exp[-D_{2n}(\xi_1) + D_{1n}(\xi_1)]} \tag{86}$$

Coefficients c_{1n}, c_{2n}, and b_n are the same as previously stated in Equations 41 through 48.

In the case of the semi-infinite strip reservoir the appropriate relationship describing the flow pattern results from:

$$\psi_1(\xi, \eta) = \lim_{\xi_1 \to \infty} \psi(\xi, \eta)$$

and they are as follows:

$$\psi_1(\xi, \eta) = \psi_\infty(\eta) + \sum_{n=1}^{\infty} \{(c_{1n} - b_n) \cdot \exp[D_{1n}(\xi)] \cdot \sin(\pi n \eta)\} \tag{87}$$

$$V_\xi(\xi, \eta) = V_{\xi\infty}(\eta) + \sum_{n=1}^{\infty} \{\pi n \cdot (c_{1n} - b_n) \cdot \exp[D_{1n}(\xi)] \cdot \cos(\pi n \eta)\} \tag{88}$$

$$V_\eta(\xi, \eta) = -\sum_{n=1}^{\infty} \{d_{1n} \cdot (c_{1n} - b_n) \cdot \exp[D_{1n}(\xi)] \cdot \sin(\pi n \eta)\} \tag{89}$$

Figure 23 shows a typical shape of the curve $V_1(\xi_0^{(s)})$ affected by the additional horizontal density changes. In this case the function $V_1(\xi_0^{(s)})$ was calculated from Equation 88. The additional horizontal density changes expressed by the function $f(\xi)$ are assumed to be of the type

$$f(\xi) = \begin{cases} 1 - \epsilon \cdot \dfrac{\xi}{\xi_0^{(s)}} & \text{for } 0 \leq \xi \leq \xi_0^{(s)} \\ 1 - \epsilon & \text{for } \xi > \xi_0^{(s)} \end{cases} \tag{90}$$

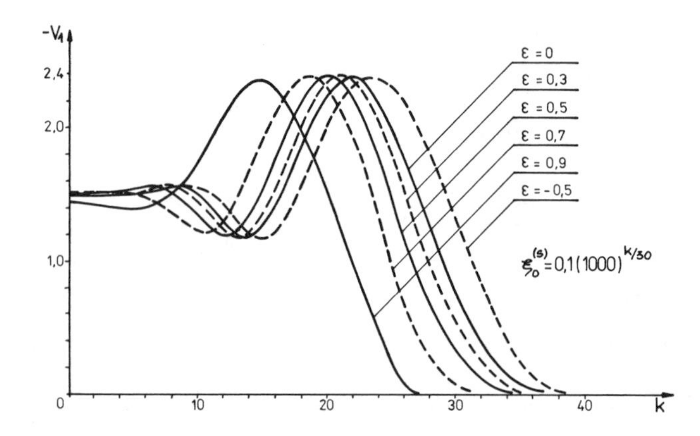

Figure 23. Typical curve $V_1(\xi_0^{(s)})$ for flow affected by additional horizontal density changes: $\eta_1 = 0$; $\eta_2 = 0.02$; A = 10,000.

and this means that

$$\rho_2 = \rho_2(\xi) = \rho_1 - [\rho_1 - \rho_2(0)] \cdot \left(1 - \epsilon^{\frac{\xi}{\xi_0^{(s)}}}\right) \tag{91}$$

From Figure 23 it follows, that when the density ρ_2 decreases towards upstream, then the horizontal extent $\xi_0^{(s)}$ increases, and if the density increases then $\xi_0^{(s)}$ decreases. Another conclusion coming from Figure 23 is that the additional horizontal density changes do not affect the aforementioned mechanism of the vertical circulation. Various circulation region shapes for the flow affected by the additional horizontal density changes are shown in Figure 24. Two f(ξ) functions were chosen for the computation

- $f(\xi) = 1$, no changes
- $f(\xi) = (0.2 \cdot \xi + 1) \cdot \exp(-0.1 \cdot \xi)$

In addition, the parameter V_1 was varied to account for wind effects. In order to clarify the picture the horizontal coordinate in Figure 24 is given in terms of $10 \log(10 \cdot \xi)$.

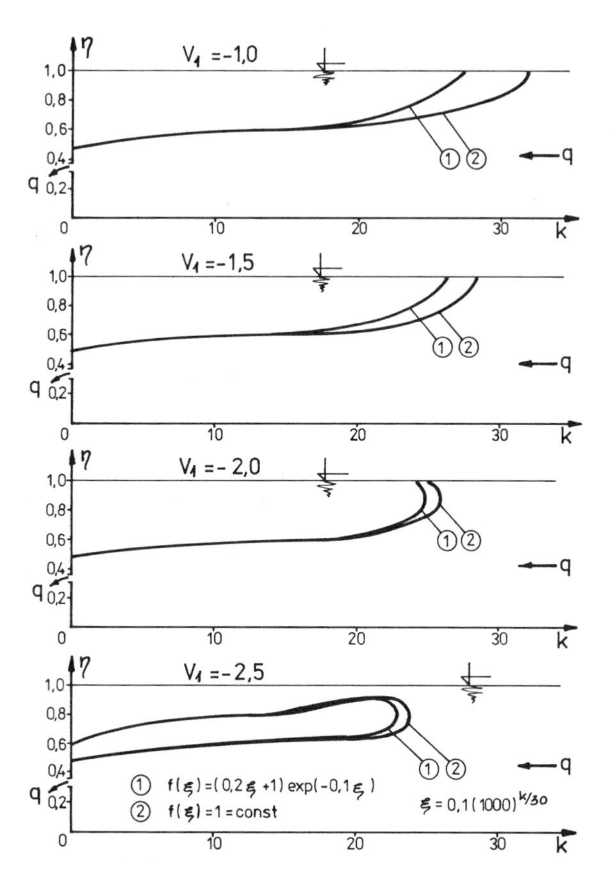

Figure 24. Wind influenced circulation region for case with additional horizontal density changes: $\eta_1 = 0.32$; $\eta_2 = 0.34$; A = 40,000 (courtesy of the American Society of Civil Engineers).

Vertical Circulation in a Case of Vanishing Density Stratification

It has been stated earlier that in the case of vanishing density stratification the convective terms and the buoyancy term in the governing equation of motion (Equation 14) are of the same order. The appropriate equation for solving the problem is the formulae in Equation 26. It turns out that because of the linear form of this equation, the Fourier series method can be applied [22]. The general solution of the flow pattern is of the same for as in the previous case of flow with the additional horizontal density changes (the set of Equations 80 through 89). The parameter A originally evaluated from Equation 36 should now be calculated as follows:

$$A = Re\left[\frac{f(\xi)}{Fr}\cdot\frac{\Delta\rho}{\rho_1} + \beta + \alpha\cdot\pi^2\cdot n^2\right] \tag{92}$$

An interesting matter is to examine how the parameters α and β change the mechanism of vertical circulation. From Equation 92 it can be noticed that if $\Delta\rho/\rho_1 \rightarrow 0$ then α and β takes the role of density stratification. If β is negative, then the circulation region is getting smaller because of the value A. The mechanism of vertical circulation in this case has been examined for the semi-infinite strip reservoir. The examination was based on the shape of the curve $V_1(\xi_0^{(s)})$, (Equation 58). Taking Equation 88 for the horizontal velocity component and function $\psi_\infty(\eta)$ from Equation 38, the appropriate formula for $V_1(\xi_0^{(s)})$ can be obtained. The detailed analysis of this problem is given by Meyer [22]. The relationship $V_1(\xi_0^{(s)})$ is of the form

$$V_1(\xi_0^{(s)}) = \frac{\sum_{n=1}^{\infty}\left\{(-1)^n\cdot\pi n\left[c_{1n} - \frac{2\cdot(-1)^n}{\pi n} - 12\cdot\frac{1+(-1)^n}{\pi^3 n^3}\right]\exp[D_{1n}(\xi_0^{(s)})]\right\}}{1 - \frac{4}{\pi^2}\cdot\sum_{n=1}^{\infty}\left\{\frac{2+(-1)^n}{n^2}\cdot\exp[D_{1n}(\xi_0^{(s)})]\right\}} \tag{93}$$

And in the case of one outflow slot it becomes

$$V_1(\xi_0^{(s)}) = \frac{\sum_{n=1}^{\infty}\left\{(-1)^n\cdot\frac{2}{\pi n}\left[\frac{\sin(\pi n\eta_2) - \sin(\pi n\eta_1)}{\eta_2 - \eta_1} + 6\frac{1+(-1)^n}{\pi\cdot n}\right]\exp[D_{1n}(\xi_0^{(s)})]\right\}}{1 - \frac{4}{\pi^2}\cdot\sum_{n=1}^{\infty}\left\{\frac{2+(-1)^n}{n^2}\cdot\exp[D_{1n}(\xi_0^{(s)})]\right\}} \tag{94}$$

As an example in Figure 25 various curves $V_1(\xi_0^{(s)})$ are given. From this figure it is shown that parameters α and β change the curve $V_1(\xi_0^{(s)})$ significantly. In the case of large Fr and β (lines 1, 2, 3, and 4), a circulation region is formed by a single eddy. It can also be seen from Figure 25, that line 5, is horizontal at a long distance (up to $k = 30$). It suggests that if $V_1 = -1.5$ (no wind), then the surface velocity $V_\xi(\xi, 1)$ is equal to zero on that distance, and a stagnation region can be formed. As one expects, parameters α and β should depend on the basic flow parameters. An approximation of them can be obtained in the following way. The parameter β can be estimated from the earlier obtained stream function $\psi_\infty(\eta)$. We have

$$\beta = \frac{\partial^3[\psi_\infty(\eta)]}{\partial\eta^3} = 6\cdot(V_1 + 2) \tag{95}$$

which in the practical cases is

$$-12 \leq \beta \leq 12 \tag{96}$$

And the parameter α can be evaluated from analysis of expanding the term $\exp[D_{1n}(\xi_0^{(s)})]$ in the power series. It suggests that α has to be of the order

$$\alpha = -\kappa_1\cdot(Fr_1\cdot Re)^{-2/3}$$

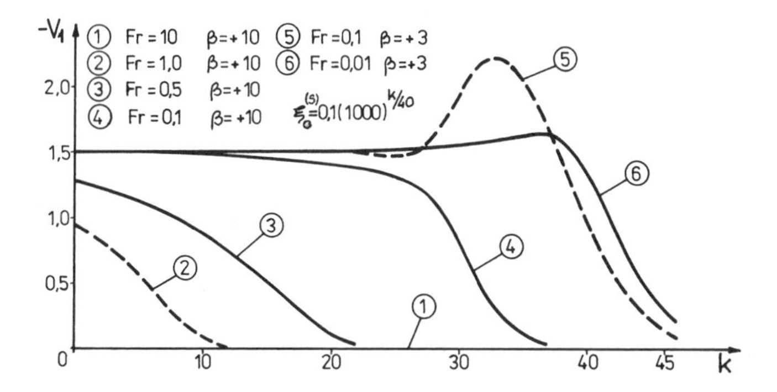

Figure 25. Plots $V_1(\xi_0^{(s)}, \beta, Fr)$ for Re = 1,000 and $\alpha = -1$, (*Arch. Hydr.*, Vol. 1–2 (1982), courtesy of the Polish Academy of Science).

where

$$0.5 \leq \kappa_1 \leq 2.0 \tag{97}$$

The coefficient κ_1 should be more accurately determined, but requires extensive laboratory experiments. The relation between Fr and Fr_1 is given by Equation 12. Taking the preceding assumptions the equation $V_1(\xi_0^{(s)})$ may now be written in the general form as

$$V_1 = V_1[\xi_0^{(s)}, \eta_1, \eta_2, D_{1n}(Re, Fr_1, f(\xi), V_1)] \tag{98}$$

In Figure 26 the shape of $V_1(\xi_0^{(s)})$ given by Equation 98 is shown. From this curve it is shown that the general mechanism of the vertical circulation has not been changed. However, it can be seen that for k = 15 when

$$-2.4 \leq V_1 \leq -1.2$$

the increase of V_1 causes a decrease in the extent of the internal closed eddy. That is a reverse situation from the previous typical shape given by Equation 70 and shown in Figure 13. From the numerical calculations it follows that the case of vanishing density stratification can now be defined as

$$\frac{1}{Fr_1^2} \leq 0.6 \cdot (V_1 + 2) \tag{99}$$

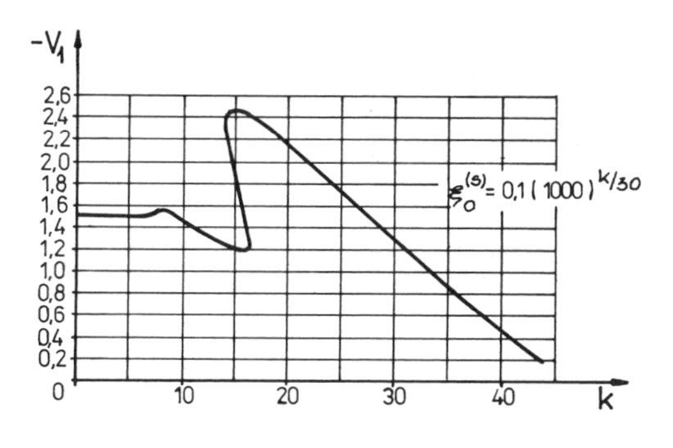

Figure 26. Curve $V_1(\xi_{(s)})$ Including $\beta(V_1)$ and $\alpha(Re, Fr_1)$, (*Arch. Hydr.*, Vol. 1–2 (1982), courtesy of the Polish Academy of Science).

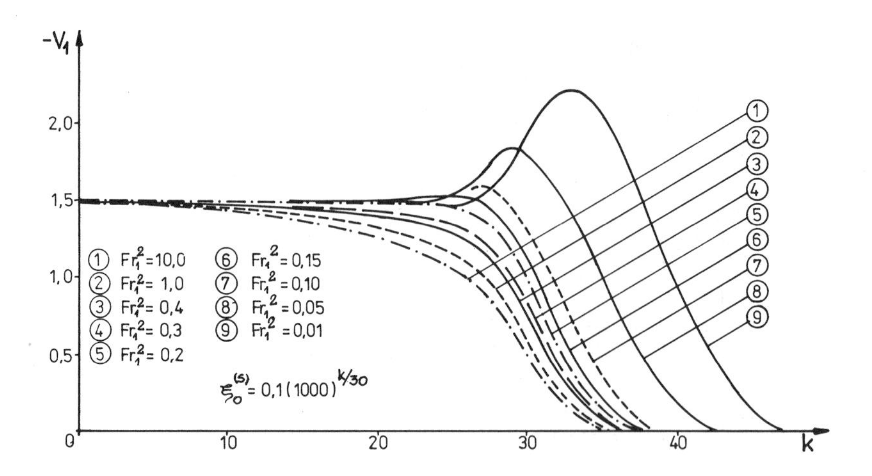

Figure 27. Set of $V_1(\xi_{(s)})$ curves for case of stagnation region, (*Arch. Hydr.*, Vol. 1–2, (1982), courtesy of the Polish Academy of Science).

As was stated earlier, the case with vanishing density stratification can lead to the formation of the stagnation region. And it happens when

$$V_1 = 1.5 \tag{100}$$

(no wind surface tension). Then from Equation 95 it follows

$$\beta = 3.0 \tag{101}$$

Assuming the relationship in Equation 92, we obtain

$$A(\xi) = \mathrm{Re}\left[\frac{f(\xi)}{Fr_1^{+2}} + 3 + \alpha \cdot \pi^2 n^2\right] \tag{102}$$

A set of curves $V_1(\xi_0^{(s)})$ including Equation 102 is shown in Figure 27. From this set it is seen that there exists one value of Fr_1 for which the curve is horizontal and without an extreme point. That

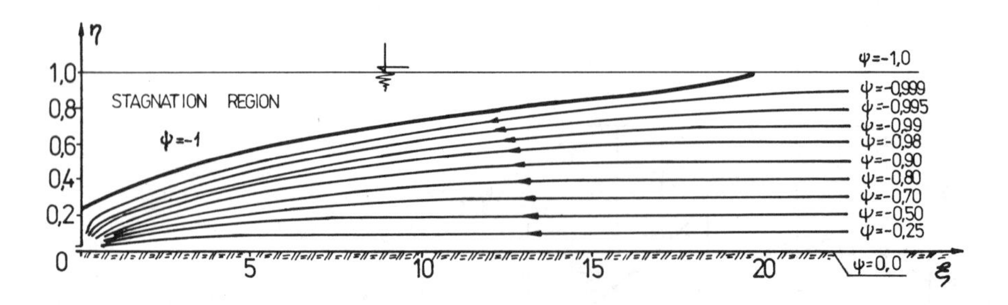

Figure 28. Flow pattern with stagnation region: $\alpha = -1.0$; Re = 1,000, (*Arch. Hydr.*, Vol. 1–2 (1982), courtesy of the Polish Academy of Science).

value of Fr_1 corresponds to the case of the stagnation region. And it means that on a long distance from the outflow wall the surface velocity is equal to zero, and that the internal eddies do not exist. That stagnation region is shown in Figure 28, and it denotes the area where

$$\psi(\xi, \eta) = \text{const.} = -1 \tag{103}$$

WIND-GENERATED VERTICAL CIRCULATION IN STRATIFIED RESERVOIR

Using the just-described model of vertical circulation, the case of wind-generated flow was investigated. In comparison to the previous cases (with inflow-outflow), the total flow through the reservoir is equal to zero. The general assumption is that the flow is induced solely by the wind surface tension. The problem of stratified circulation in a rectangular cavern was investigated by Young, Liggett, and Gallagher [35, 36]. The solution was obtained numerically by using a finite element scheme. The authors pointed out that the flow pattern consists of a series of separate eddies. They did not investigate a typical mechanism of circulation. The method presented here [21] is a solution of the flow case similar to that given by Young, Liggett, and Gallagher [35, 36]. Taking advantage that the solution presented here is expressed in a close analytical form, a mechanism of stratified circulation was formed.

The flow area in the considered case is given in Figure 29. The relevant equation of motion is the formula in Equation 26. The boundary conditions suiting the investigated problem are as follows (Figure 29):

$$\eta = 1; \quad \psi = 0, \quad \frac{\partial \psi}{\partial \xi} = 0 \quad \text{and} \quad \rho = \rho_2$$

$$\eta = 0; \quad \psi = 0, \quad \frac{\partial \psi}{\partial \xi} = 0 \quad \text{and} \quad \rho = \rho_1$$

$$\xi = 0; \quad \psi = 0, \quad \frac{\partial \psi}{\partial \eta} = 0 \quad \text{and} \quad \rho = \rho_0$$

$$\xi = \xi_1; \quad \psi = 0, \quad \frac{\partial \psi}{\partial \eta} = 0 \quad \text{and} \quad \rho = \rho_0 \tag{104}$$

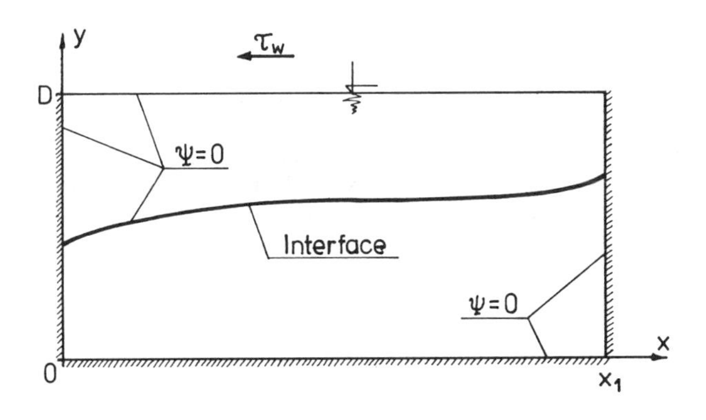

Figure 29. Definition sketch of wind vertical circulation in density-stratified reservoir.

The general solution of the considered problem may be expressed in the form:

$$\psi^{(1)}(\xi,\eta) = \psi^{(1)}_{\infty}(\eta) + \sum_{n=1}^{\infty} \{a^{(1)}_{1n} \cdot \exp[D^{(1)}_{1n}(\xi)] + a^{(1)}_{2n} \cdot \exp[-D^{(1)}_{2n}(\xi_1) + D^{(1)}_{2n}(\xi)]\} \cdot \sin(\pi n \eta) \tag{105}$$

where

$$D^{(1)}_{1n}(\xi) = D_{1n}(\xi); \qquad D^{(1)}_{2n}(\xi) = D_{2n}(\xi) \tag{106}$$

and they come from Equations 83 and 84, and

$$A(\xi) = \mathrm{Re}\left[\frac{1}{\rho_0} \cdot \frac{\rho_{max} - \rho_{min}}{\psi_{max} - \psi_{min}} \cdot \frac{f(\xi)}{\mathrm{Fr}} + \beta + \alpha \cdot \pi^2 n^2\right] \tag{107}$$

while Fr and Re are defined by Equation 11. The reference value of Ψ_0 in this case is as follows

$$\Psi_0 = V_s \cdot D \tag{108}$$

where V_s is the reference velocity related to the surface shear stress and it will be defined later. Value α remains the same as it was previously. The Fourier series coefficient $a^{(1)}_{1n}$ and $a^{(1)}_{2n}$ evaluated from the boundary conditions are

$$a^{(1)}_{1n} = -b^{(1)}_n \cdot \frac{1 - \exp[-D^{(1)}_{2n}(\xi_1)]}{1 - \exp[-D^{(1)}_{2n}(\xi_1) + D^{(1)}_{1n}(\xi_1)]}$$

$$a^{(1)}_{2n} = -b^{(1)}_n \cdot \frac{1 - \exp[D^{(1)}_{1n}(\xi_1)]}{1 - \exp[-D^{(1)}_{2n}(\xi_1) + D^{(1)}_{1n}(\xi_1)]} \tag{109}$$

where

$$b^{(1)}_n = \frac{4}{\pi^3 n^3}[1 + 2 \cdot (-1)^n] \tag{110}$$

and it follows the fact that [21]

$$\psi^{(1)}_{\infty}(\eta) = \eta^2(\eta - 1) \tag{111}$$

and then from Equation 95, $\beta = 4$. The density distribution appropriate to this case is given as [21]

$$\rho(\psi) = \rho_0 + \frac{\rho_{max} - \rho_{min}}{\psi_{max} - \psi_{min}} \cdot \psi \cdot f(\xi) \tag{112}$$

From Equation 107 it can be seen that the value A depends on ψ_{max} and ψ_{min} which are the result of calculation, and so an iterative procedure must be employed. The sign of A may be positive or negative depending on V_s, which correlates with the surface tension. To examine the mechanism of vertical circulation the numerical computations have been carried out in order to obtain various flow patterns. The calculations assume the primary solution i.e. $\alpha = 0$; $\beta = 0$; and $f(\xi) = 1$. In this case it can be proven that the change of wind direction does not alter the flow pattern, so we have

$$\psi^{(1)}(\xi, \eta, -\tau_2) = \psi^{(1)}(\xi_1 - \xi, \eta, \tau_2) \tag{113}$$

And it comes from the fact that

$$d_{1n}(-A) = -d_{2n}(A); \qquad d_{2n}(-A) = -d_{1n}(A);$$

$$a^{(1)}_{1n}(-A) = a^{(1)}_{2n}(A); \qquad a^{(1)}_{2n}(-A) = a^{(1)}_{1n}(A) \tag{114}$$

If f(ξ) is not equal to 1, then the condition in Equation 113 is satisfied only when

$$f(\xi) = f(\xi_1 - \xi) \tag{115}$$

In Figure 30 the mechanism of the vertical circulation in a rectangular reservoir is shown. From these figures it is shown that the increase of A intensifies the vertical circulation i.e. the number of the internal closed eddies (internal cells) grows up, and the difference between ψ_{max} and ψ_{min} decreases. This implies that the velocities of the circulating water are getting smaller. It can also

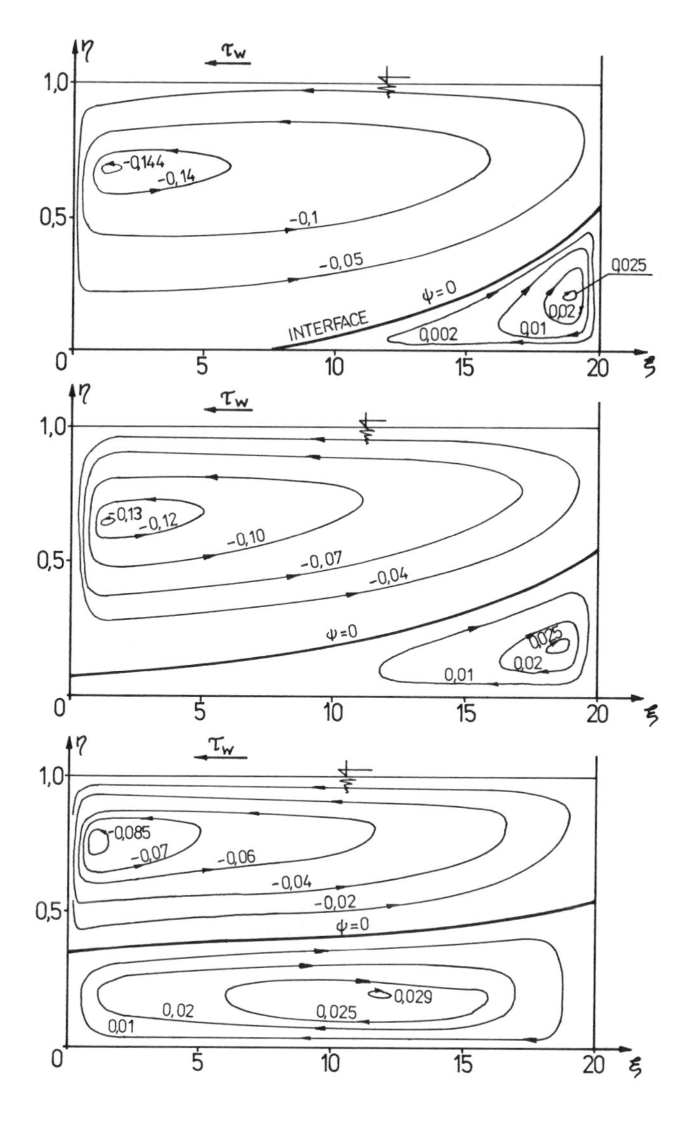

Figure 30. Mechanism of wind vertical circulation in rectangular stratified reservoir: $\xi_1 = 20$; A = −500; −1,000; −5,000; −10,000; −100,000; −200,000; −500,000.

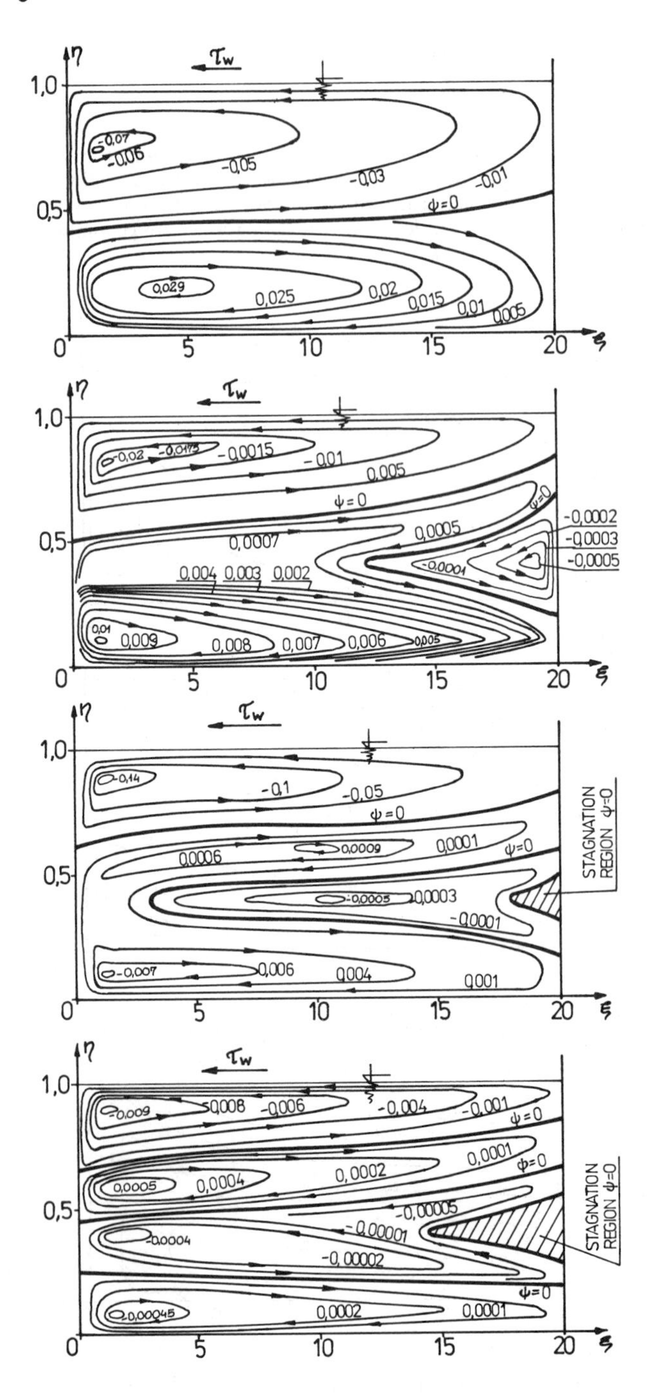

Figure 30. (Continued)

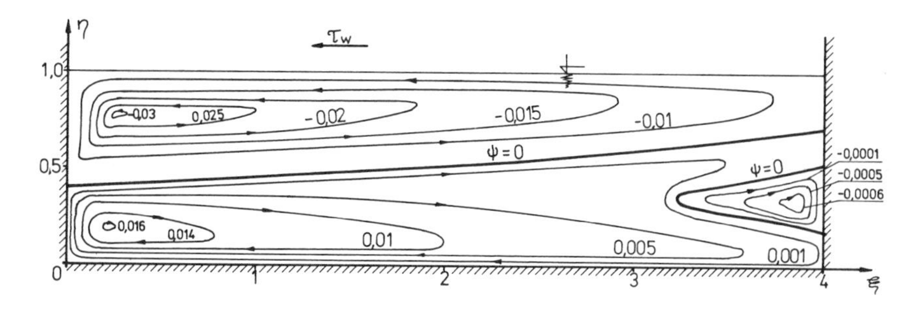

Figure 31. Wind vertical circulation in stratified short reservoir: $\xi_1 = 4$; A = −10,000.

be seen, that each newly created eddy starts from the reservoir wall at the leeside. If A exceeds a certain value, then instead of the subsequent eddy a stagnation region arises. The stagnation region denotes the area where

$$\psi^{(1)}(\xi, \eta) = 0 \tag{116}$$

The reservoir's length plays an important role in the wind vertical circulation. For the same value of A, the number of the internal eddies is higher in the short reservoir. An example of the vertical circulation flow pattern in the short reservoir is shown in Figure 31. It can clearly be seen that in this case the vertical circulation is developed much stronger than in the long reservoir (Figure 30). Figure 32 shows the shape of the internal eddies for various parameters α and β and for two values

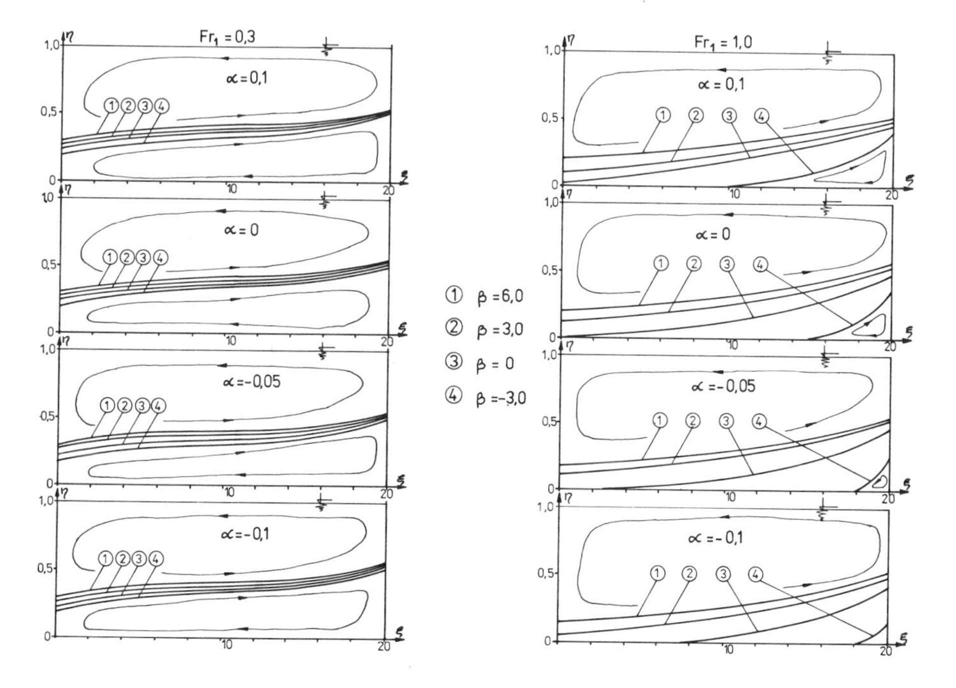

Figure 32. Influence of parameters α and β on shape of internal eddies: Re = 450; $\xi_1 = 20$.

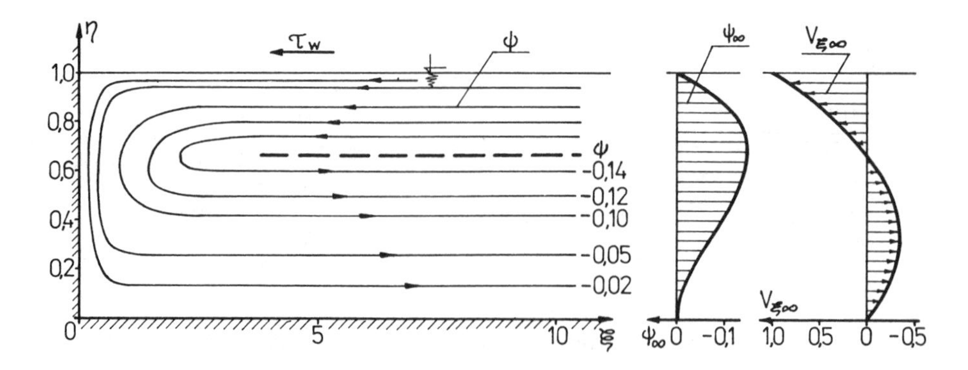

Figure 33. Wind vertical circulation in semi-infinite strip reservoir.

of Fr_1. From this picture it can be seen, that only in the case of a large Fr_1 is the influence significant. An asymptotic case may also be considered when the reservoir's length extends to infinity, i.e., the semi-infinite strip reservoir. The wind vertical circulation in such a case is defined as

$$\psi_1^{(1)}(\xi, \eta) = \lim_{\xi_1 \to \infty} \psi^{(1)}(\xi, \eta) \tag{117}$$

and it leaves

$$\psi_1^{(1)}(\xi, \eta) = \psi_\infty^{(1)}(\eta) + \sum_{n=1}^{\infty} \{a_{1n}^{(1)} \cdot \exp[D_{1n}(\xi)] \cdot \sin(\pi n \eta)\} \tag{118}$$

All the symbols were explained earlier. In Figure 33 the wind vertical circulation flow pattern is shown for the semi-infinite strip reservoir. It follows, that in this case the circulation consists of a single eddy extending upstream to infinity. Since the reference velocity V_s was introduced (Equation 108), it is the relation between this and the surface shear stress which becomes essential. Following the relationship for the primary solution (Equation 54) in the present case we can write, Meyer [21]

$$\tau_2 = -4\rho_0 \cdot K_y \cdot \frac{V_s}{D} = -\tau_w = \text{constant} \tag{119}$$

It should be emphasized that again the surface tension remains constant all along the water surface.

As it was pointed out earlier, to use the solution given by Equation 105 an iterative procedure must be employed. The initial value of A (or for rough estimation) may be taken from the graph $A = f(A^*, \xi_1)$ shown in Figure 34. For the convenience of further consideration the new parameter

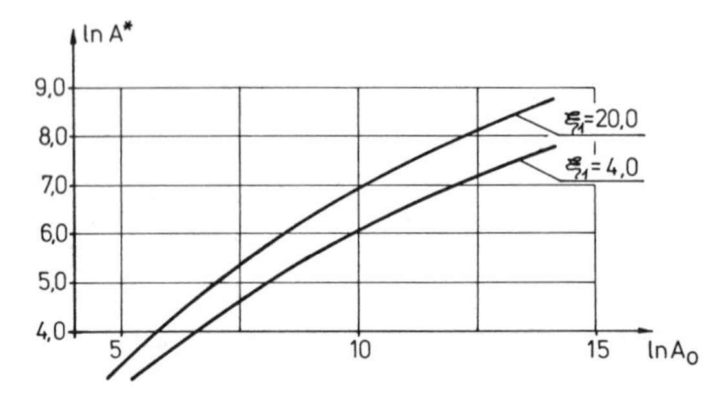

Figure 34. Plots $A = f(A^*, \xi_1)$.

A* was introduced

$$A^* = \frac{\rho_{max} - \rho_{min}}{\rho_0} \cdot \frac{g \cdot D^2}{V_s\sqrt{K_y(K_x + K_y)}} \tag{120}$$

and therefore

$$A^* = A \cdot [\psi_{max}(A, \xi_1) - \psi_{min}(A, \xi_1)] \tag{121}$$

or generally

$$A = f(A^*, \xi_1) \tag{122}$$

In practical calculations, the parameter A* can be estimated directly from field or laboratory experiments. For this value A* and given ξ_1, an appropriate value A can be chosen from Figure 34. When parameters α and β are involved, than the changes of ψ_{max} and ψ_{min} within the circulation regions are given in Figure 35.

The model presented here for the wind vertical circulation in a stratified reservoir is based on the assumption of density distribution (Equation 112). Deeper analysis implies that in certain areas this kind of density distribution does not satisfy the stably stratified flow condition $\partial\rho/\partial y = 0$. The reason for this is that the entire density gradient $\Delta\rho$ corresponds to small values of the stream function. This means that it is unlikely to keep the density constant along the streamline within the internal eddy. Turner [20] following Martin's research suggests that the effect of diffusion along the streamline may be neglected if

$$\frac{K_h}{\gamma}\left(\frac{Ri}{Re_0}\right)^{1/2} \leq 1 \tag{123}$$

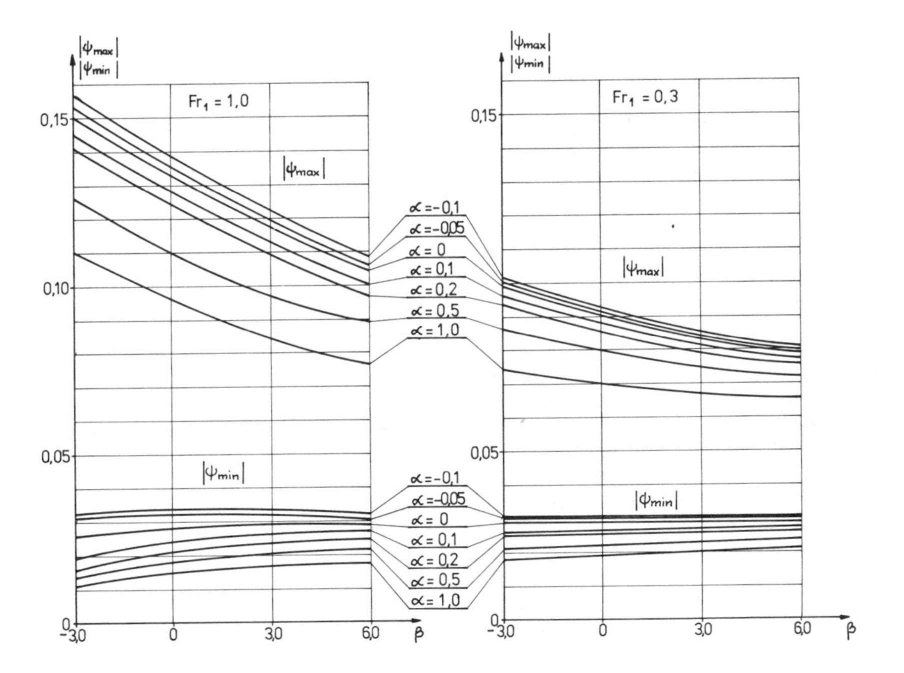

Figure 35. Values ψ max and ψ min for various α and β: Re = 450; $\xi_1 = 20$.

where in this case Re_0 is the original Reynolds number based on the kinematic viscosity and

$$Ri = -\frac{g \cdot \dfrac{\partial \rho}{\partial y}}{\rho \left(\dfrac{\partial V_x}{\partial y}\right)^2} \tag{124}$$

is the Richardson number. For practical calculations the condition in Equation 123 can be expressed in terms of minimum velocity, which must be kept in order to neglect the effects of diffusion. We have

$$V_{0\,min} = \sqrt{\frac{K_h^2}{\gamma} \cdot g \cdot \frac{\Delta\rho}{\rho_1}} \text{ and then } V_0 > V_{0\,min} \tag{125}$$

On the other hand, the model of the wind vertical circulation assumes Re from Equation 11 not to be too large, in order to neglect the nonlinear terms in the momentum equation (Equation 17). This needs to keep the velocity slow. We have

$$V_{0\,max} = \frac{Re_{max} \cdot \sqrt{K_y(K_x + K_y)}}{D} \text{ and then } V_0 < V_{0\,max} \tag{126}$$

There is a narrow gap of the velocity range where both requirements are satisfied. This limits the application of the described model.

AXISYMMETRIC VERTICAL CIRCULATION IN DENSITY-STRATIFIED RESERVOIR

The second largest group of the two-dimensional flows are axisymmetric flows. The flow pattern in this case cannot be estimated by using the models presented above, and hence a new model including the fact of axially-symmetric flow is needed. The case of axially-symmetric inflow to the deep intake from stratified ponds was calculated by Koh [16]. The solution was obtained by using the small parameter method, and it could be applied in the case of laminar flow. The solution does not allow distinction of the characteristic features of the circulation region. Recently an attempt has been made to include the turbulent viscosity [24]. Similarly to the case of vertically plane motion the following factors are of great importance for the flow pattern description: axially-symmetric motion, boundary conditions at the inflow and outflow, various eddy viscosity in radial and vertical directions, flow parameters, and density stratification.

For the purpose of this analysis it has been assumed that the flow is steady, stratified, and axisymmetric to the outflow slots situated on the vertical wall of the intake tower (Figure 36). In the vertical cross section the outflow and inflow are formed by the group of slots located above each other (Figure 2). It has been assumed, that the main flow is in the opposite direction to the r-axis. The basic system of the cylindrical coordinates is as follows: the axis of flow, vertically upward along the outflow tower axis (Figure 1) and the radial coordinate plane r-axis is the reservoir bottom surface. It has been assumed, that within the flowing stream, density stratification exists. The density of water at the surface is equal to ρ_2 and at the bottom ρ_1. The flow through reservoir is generated by mass exchange (inflow-outflow). Wind action was basically not taken into account; however, the discussion of its influence is given. Mass diffusion is included by assuming a certain form of the density distribution function. The wind action in the previous models (vertically plane motion) was included by assuming the wind shear stress, equal to the water shear stress at the surface. In the case considered here—axially symmetric flow—the direction of wind (in plane) is not parallel to the direction of the following stream. Because of the axisymmetric flow the angle between them varies with the streamline. Therefore, it is impossible to treat the wind surface shear stress as constant, along the whole streamline. In order to simplify the calculation and to obtain an analytical solution, the case with no wind was basically considered.

As a mathematical representation, the set of equations for the axisymmetric flow was taken, (see Batchelor [1]). Assuming the previous conclusion, the eddy viscosity coefficients in the horizontal

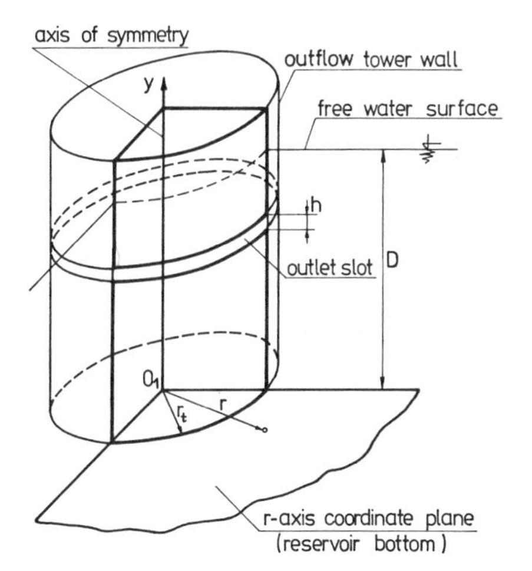

Figure 36. Location of outflow tower in reservoir (courtesy of the International Association for Hydraulic Research).

and vertical directions should be separated in the governing equation of motion. So an analysis of the Navier-Stokes equation was made. We have

$$\frac{dV_r}{dt} = X - \frac{1}{\rho}\cdot\frac{\partial P}{\partial r} + \gamma\cdot\frac{\partial}{\partial r}\left[\frac{1}{r}\frac{\partial}{\partial r}(r\cdot V_r)\right] + \gamma\frac{\partial^2 V_r}{\partial y^2};$$

$$\frac{dV_y}{dt} = Y - \frac{1}{\rho}\cdot\frac{\partial P}{\partial y} + \gamma\frac{1}{r}\cdot\frac{\partial}{\partial r}\left(r\cdot\frac{\partial V_y}{\partial r}\right) + \gamma\frac{\partial^2 V_y}{\partial y^2} \tag{127}$$

and following the previous consideration the mass conservation equation (Equations 5 and 6) is:

$$\frac{1}{r}\frac{\partial}{\partial r}\cdot(r\cdot V_r) + \frac{\partial V_y}{\partial y} = 0 \tag{128}$$

To adopt these equations to the case of turbulent flow additional friction forces caused by the turbulent velocity (fluctuation) components V_r' and V_y' are usually introduced. Including this in Equation 128 additional terms on the right-hand side arise

$$-\left[\frac{\partial}{\partial r}(\overline{V_r'^2}) + \frac{\overline{V_r'^2}}{r}\right] - \frac{\partial}{\partial y}\cdot(\overline{V_r'\cdot V_y'})$$

and

$$-\left[\frac{\partial}{\partial r}(\overline{V_y'\cdot V_r'}) + \frac{\overline{V_y'\cdot V_r'^2}}{r}\right] - \frac{\partial}{\partial y}(\overline{V_y'^2}) \tag{129}$$

The bar over the terms in Equation 129 marks mean values over a certain period of time, [1]. Instead of terms containing velocity fluctuation (Equation 129), eddy viscosity coefficients are commonly introduced. Thus

$$\overline{V_r'^2} = -K_{rr}^*\cdot\frac{\partial\bar{V}_r}{\partial r};\qquad \overline{V_r'\cdot V_y'} = -K_{ry}^*\cdot\frac{\partial\bar{V}_r}{\partial y}$$

$$\overline{V_y'\cdot V_r'} = -K_{yr}^*\frac{\partial\bar{V}_y}{\partial r};\qquad \overline{V_y'^2} = -K_{yy}^*\cdot\frac{\partial\bar{V}_y}{\partial y} \tag{130}$$

Symbols $\bar{V}_r$ and $\bar{V}_y$ denote the average value over a certain period of time, and in the further description they are omitted. To include the kinematic viscosity in the practical calculation, new turbulent coefficients are commonly introduced

$$K_{ij} = \gamma + K_{ij} \tag{131}$$

where i and j refer to x and y, respectively. Following the discussion in the previous models (Equations 14 and 15) it has been assumed that

$$K_{rr} = K_{yr} = K_r = \text{constant}$$

$$K_{ry} = K_{yy} = K_y = \text{constant} \tag{132}$$

The set of governing equations may be expressed as

$$\frac{dV_r}{dt} = -\frac{1}{\rho} \cdot \frac{\partial P}{\partial r} + K_r \cdot \frac{\partial}{\partial r}\left[\frac{1}{r} \cdot \frac{\partial}{\partial r}(rV_r)\right] + K_y \frac{\partial^2 V_r}{\partial y^2}$$

$$\frac{dV_y}{dt} = -g = \frac{1}{\rho} \cdot \frac{\partial P}{\partial y} + K_r \cdot \frac{1}{r} \cdot \frac{\partial}{\partial r}\left(r \frac{\partial V_y}{\partial r}\right) + K_y \frac{\partial^2 V_y}{\partial y^2} \tag{133}$$

$$\frac{1}{r}\frac{\partial}{\partial r}(r \cdot V_r) + \frac{\partial V_y}{\partial y} = 0 \tag{134}$$

Equation 133 incorporates the eddy viscosity in the horizontal and vertical directions separately. For further analysis, the Stokes stream function, Ψ, and vorticity function, Ω, were introduced. We have

$$V_r = \frac{1}{r}\frac{\partial \Psi}{\partial y}; \qquad V_y = -\frac{1}{r}\frac{\partial \Psi}{\partial r} \tag{135}$$

$$\Omega = \frac{\partial V_r}{\partial y} - \frac{\partial V_y}{\partial r} = \frac{1}{r}\left[\frac{\partial^2 \Psi}{\partial r^2} - \frac{1}{r} \cdot \frac{\partial \Psi}{\partial r} + \frac{\partial^2 \Psi}{\partial y^2}\right] \tag{136}$$

and after cross-differentiation of Equation 133 we obtained [24]:

$$\frac{d\Omega}{dt} = \frac{g}{\rho} \cdot \frac{\partial \rho}{\partial x} + K_x \frac{\partial}{\partial r}\left[\frac{1}{r} \cdot \frac{\partial}{\partial r}(\Omega \cdot r)\right] + K_y \cdot \frac{1}{r} \cdot \frac{\partial^2}{\partial y^2}(\Omega \cdot r) \tag{137}$$

As has been already stated in the considered case, the nonlinear terms play a limited role in the momentum transfer and so term $d\Omega/dt$ may be neglected [24]. Furthermore, a simplification adequate to the Prandtl assumption (Equation 13) was introduced. It leads to the following relationship

$$\frac{g}{\rho} \cdot \frac{\partial \rho}{\partial r} + (K_r + K_y) \cdot \frac{\partial}{\partial r}\left[\frac{1}{r} \cdot \frac{\partial^3 \Psi}{\partial r\, \partial y^2}\right] + K_y \cdot \frac{1}{r}\frac{\partial^4 \Psi}{\partial y^4} = 0 \tag{138}$$

In order to obtain a more general solution the following dimensionless coordinates and functions are introduced

$$\eta = \frac{y}{D}; \qquad \xi = \frac{r}{D} \cdot \sqrt{\frac{K_y}{K_r + K_y}} \tag{139}$$

$$\psi = \frac{\Psi}{\Psi_0}$$

where

$$\Psi_0 = q_1 = \frac{dQ}{d\varphi} \tag{140}$$

moreover

$$Re = \frac{\Psi_0}{K_y \cdot D}$$

and

$$Fr = \frac{\Psi_0^2}{g \cdot D^5} \tag{141}$$

As it can be seen q_1 is the flow per unit angle, and it is assumed to be constant. If the flow area consists of a full circle, there is

$$q_1 = \frac{Q}{2\pi}$$

where Q is the total flow through the reservoir. If $q_1(\varphi)$ is not constant, then the flow from each direction is different. And so the calculation should be done for every value of $q_1(\varphi)$ separately. The density distribution function is assumed to be of the form in Equation 19. Therefore we obtain

$$A_0 \cdot \xi \cdot \frac{\partial \psi}{\partial \xi} + \frac{\partial}{\partial \eta^2}\left[\xi \frac{\partial}{\partial \xi}\left(\frac{1}{\xi} \cdot \frac{\partial \psi}{\partial \xi}\right) + \frac{\partial^2 \psi}{\partial \eta^2}\right] = 0 \tag{142}$$

where

$$A_0 = \frac{\Delta\rho}{\rho_1} \cdot \frac{Re}{Fr} = \frac{\Delta\rho}{\rho_1} \cdot \frac{g \cdot D^4}{q_1 \cdot K_y} \tag{143}$$

The boundary conditions in the considered case are the same as they were in the previous models (Equations 27 through 34) with the following restrictions: $\tau_2 = 0$ and the outflow coordinate is ξ_t (instead of zero). The general solution of the problem is expressed in terms of the Fourier series as follows

$$\psi(\xi, \eta) = \psi_\infty(\eta) + \sum_{n=1}^{\infty} \{R_n[\delta(\xi)] \cdot \sin(\pi n \eta)\} \tag{144}$$

where

$$\delta(\xi) = \frac{A_0}{\pi^2 n^2} \cdot \frac{\xi^2}{2}$$

and

$$n = 1, 2, 3, \ldots$$

$$\psi_\infty(\eta) = -\eta \tag{145}$$

The function R_n appearing in Equation 144 is a solution of the Whittaker differential equation

$$\frac{d^2R_n}{d\delta^2} - \frac{dR_n}{d\delta} - \theta \cdot \frac{R_n}{\delta} = 0 \tag{146}$$

where

$$\theta = \frac{1}{2}\frac{\pi^4 n^4}{A_0} \tag{147}$$

The solution of the preceding differential equation is usually given in terms of the Whittaker functions, and they are defined by the confluent hypergeometric functions

$$R_n(\delta) = \exp(\tfrac{1}{2}\delta) \cdot [c_{1n} \cdot W_{\theta,1/2}(\delta) + c_{2n} \cdot W_{-\theta,1/2}(-\delta)] \tag{148}$$

where functions $W_{\theta,1/2}(\delta)$ and $W_{-\theta,1/2}(-\delta)$ are those mentioned previously. Coefficients c_{1n} and c_{2n} should be evaluated from the assumed boundary conditions. Practical calculation of the stream function indicates that the confluent hypergeometric series converge very slowly, making computer calculations a lengthy process. This fact limits the application of Equation 148, and it follows that another method sufficiently accurate for practical calculation is needed.

Numerical computation shows that for the practical calculation the governing Equation 142 may be reduced to the form [24]:

$$A_0 \cdot \xi \frac{\partial \psi}{\partial \xi} + \frac{\partial^2}{\partial \eta^2}(\nabla^2 \psi) = 0 \tag{149}$$

The same result can be achieved from the model of vertically plane motion by assuming V_0 as a variable. In the case of axially-symmetric flow we have (Figure 37):

$$V_0 = V_0(r) = \frac{Q}{\varphi \cdot r \cdot D} = \frac{q_1}{r \cdot D}$$

The parameter A in Equation 36 may now be evaluated as follows:

$$A = \frac{\Delta\rho}{\rho_1} \cdot \frac{r \cdot D}{q_1} \cdot \frac{g \cdot D^2}{\sqrt{K_y(K_y + K_r)}} = \frac{\Delta\rho}{\rho_1} \cdot \frac{g \cdot D^4}{q_1 \cdot K_y} \cdot \frac{r}{D} \cdot \sqrt{\frac{K_y}{K_r + K_y}} = A_0 \cdot \xi \tag{150}$$

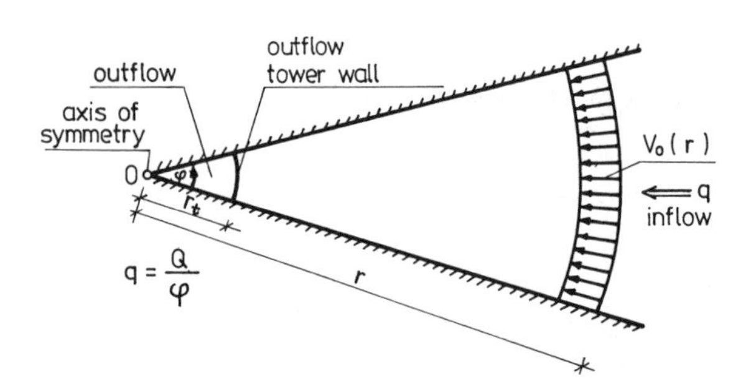

Figure 37. Definition sketch of reservoir in plane (courtesy of the International Association for Hydraulic Research).

and so the equation is reduced to the form in Equation 149. The solution of Equation 149 can be obtained from the previous models. Following the idea of including additional horizontal density changes (Equations 80 through 82) we have

$$A = A(\xi) = A_0 \cdot \xi \tag{151}$$

and so the solution can be expressed as

$$\psi^{(2)}(\xi, \eta) = -\eta + \sum_{n=1}^{\infty} \{a_{1n}^{(2)} \cdot \exp[D_{1n}^*(\xi)] + a_{2n}^{(2)} \cdot \exp[-D_{2n}^*(\xi_1) + D_{2n}^*(\xi)]\} \cdot \sin(\pi n \eta) \tag{152}$$

$$V_\xi(\xi, \eta) = -\frac{1}{\xi} + \frac{1}{\xi} \sum_{n=1}^{\infty} \{\pi n \cdot a_{1n}^{(2)} \cdot \exp[D_{1n}^*(\xi)] + \pi n a_{2n}^{(2)} \cdot \exp[-D_{2n}^*(\xi_1) + D_{2n}^*(\xi)]\} \cdot \cos(\pi n \eta) \tag{153}$$

$$V_\eta(\xi, \eta) = -\frac{1}{\xi} \sum_{n=1}^{\infty} \{d_{1n} \cdot a_{1n}^{(2)} \cdot \exp[D_{1n}^*(\xi)] + d_{2n} \cdot a_{2n}^* \cdot \exp[-D_{2n}^*(\xi_1) + D_{2n}^*(\xi)]\} \cdot \sin(\pi n \eta) \tag{154}$$

where in this case Equation 83 is of the form

$$D_{1n}^*(\xi) = \int_{\xi_t}^{\xi} d_{1n}(\xi) \cdot d\xi$$

$$D_{2n}^*(\xi) = \int_{\xi_t}^{\xi} d_{2n}(\xi) \cdot d\xi \tag{155}$$

The coefficients $d_{1n}(\xi)$ and $d_{2n}(\xi)$ remain the same as previously (Equation 84). The values of coefficients $a_{1n}^{(2)}$ and $a_{2n}^{(2)}$ should now be calculated as

$$a_{1n}^{(2)} = \frac{c_{1n} - c_{2n} \cdot \exp[-D_{2n}^*(\xi_1)] - b_n \cdot \{1 - \exp[-D_{2n}^*(\xi_1)]\}}{1 - \exp[-D_{2n}^*(\xi_1) + D_{1n}^*(\xi_1)]}$$

$$a_{2n}^{(2)} = \frac{c_{2n} - c_{1n} \cdot \exp[D_{1n}^*(\xi_1)] - b_n \cdot \{1 - \exp[D_{1n}^*(\xi_1)]\}}{1 - \exp[-D_{2n}^*(\xi_1) + D_{1n}^*(\xi_1)]} \tag{156}$$

The coefficients c_{1n} and c_{2n} should be taken from the primary solution i.e., Equations 41 and 42. The coefficient b_n, because of the no-wind case, should be taken from Equation 43.

In the case of a semi-infinite strip reservoir the solution in Equations 151 through 156 is reduced to the following form

$$\psi_1^{(2)}(\xi, \eta) = \lim_{\xi_1 \to \infty} \psi^{(2)}(\xi, \eta) \tag{157}$$

and so

$$\psi_1^{(2)}(\xi, \eta) = -\eta + \sum_{n=1}^{\infty} \{(c_{1n} - b_n) \cdot \exp[D_{1n}^*(\xi)] \cdot \sin(\pi n \eta)\} \tag{158}$$

$$V_\xi(\xi, \eta) = -\frac{1}{\xi} + \frac{1}{\xi} \cdot \sum_{n=1}^{\infty} \{\pi n(c_{1n} - b_n) \cdot \exp[D_{1n}^*(\xi)] \cdot \cos(\pi n \eta)\} \tag{159}$$

$$V_\eta(\xi, \eta) = -\frac{1}{\xi} \cdot \sum_{n=1}^{\infty} \{d_{1n} \cdot (c_{1n} - b_n) \cdot \exp[D_{1n}^*(\xi)] \cdot \sin(\pi n \eta)\} \tag{160}$$

Examples of the flow patterns corresponding to Equation 158 are given in Figure 38. From this figure it comes, that the flow pattern contains the circulation region and that its properties may be estimated from Equations 56 through 59 formed earlier. The mechanism of the vertical circulation remains the same as in the primary solution. Also for a large value of A_0 the horizontal extent $\xi_0^{(s)}$ is much longer than in the vertically plane motion. For a rough estimation of the horizontal extent of the surface circulation region, the following approximate formula may be used in the case of bottom outflow (Equation 159)

$$\xi_0^{(s)} = \frac{1}{\pi\sqrt{e}} \cdot f(\epsilon) \cdot \frac{\exp(\theta_1 \ln 2)}{\theta_1} \tag{161}$$

where

$$f(\epsilon) = (\epsilon + \sqrt{\epsilon^2 + 1}) \cdot \exp\left(\frac{\epsilon}{\epsilon + \sqrt{\epsilon^2 + 1}}\right) \tag{162}$$

and

$$\epsilon = \frac{\pi}{2} \cdot \xi_t \cdot \theta_1$$

$$\theta_1 = \frac{A_0}{\pi^4}$$

$$e = 2.7182 \tag{163}$$

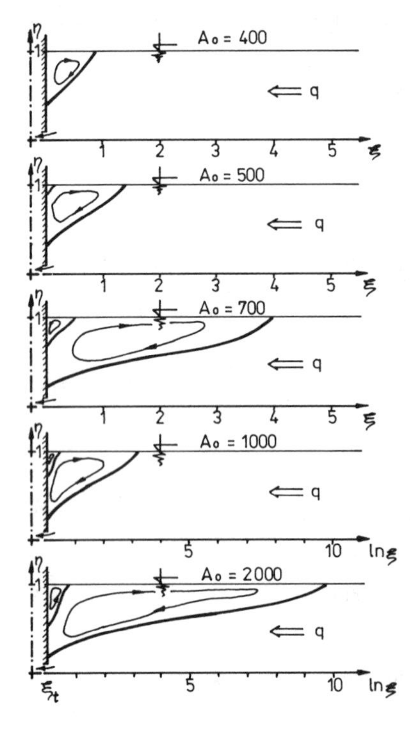

Figure 38. Mechanism of vertical circulation in axisymmetric stratified flow (courtesy of the International Association for Hydraulic Research).

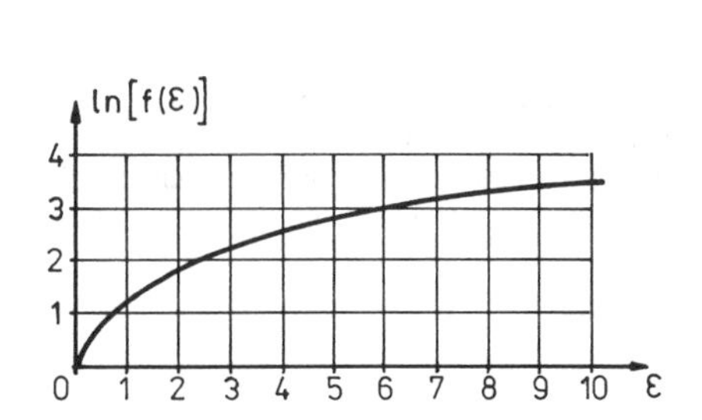

Figure 39. Curve f(ε) (courtesy of the International Association for Hydraulic Research).

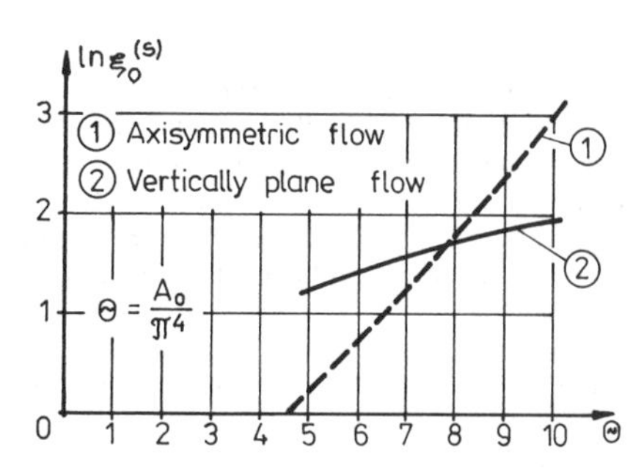

Figure 40. Horizontal extent of surface circulation region for axisymmetric and vertically plane motion (courtesy of the International Association for Hydraulic Research).

From a practical point of view a relation exists between the horizontal extents in the axisymmetric and the vertically plane motion, under corresponding conditions. For the vertically plane motion the horizontal extent of the surface circulation region is given by the following equation [19, 20]

$$\xi_0^{(s)} = \theta_1 \ln 2 \tag{164}$$

The functions $\xi_0^{(s)}$ given by Equations 161 and 164 for small ϵ are plotted in Figures 39 and 40. It can be seen that if $\theta_1 > 7$ then the horizontal extent for the axisymmetric motion is many times longer than it is for the vertically plane motion. From Equations 161 the role of the eddy viscosity coefficients K_r and K_y and the density stratification may be examined. Expressing Equation 161 in the dimensional form we obtain

$$r_0^{(s)} = \frac{\pi^3}{\sqrt{e}} \cdot D \cdot \frac{\rho_1}{\Delta\rho} \cdot \frac{q_1\sqrt{K_y(K_y + K_x)}}{g \cdot D^4} \cdot \exp\left(\frac{A_0}{\pi^4} \ln 2\right) \cdot f(\epsilon) \tag{165}$$

From this equation it follows, that in the case of axisymmetric flow both coefficients, K_r and K_y, are significant.

As it has been stated earlier, the method does not allow for examination of the influence of wind action in the axially-symmetric flow. For a rough estimation, a calculation may be made, using $\psi_\infty(\eta)$ from Equation 38 and b_n from Equation 44 and substituting into Equation 158. The effect of such a computation is shown in Figure 41. From that figure it follows that, just as in the case of

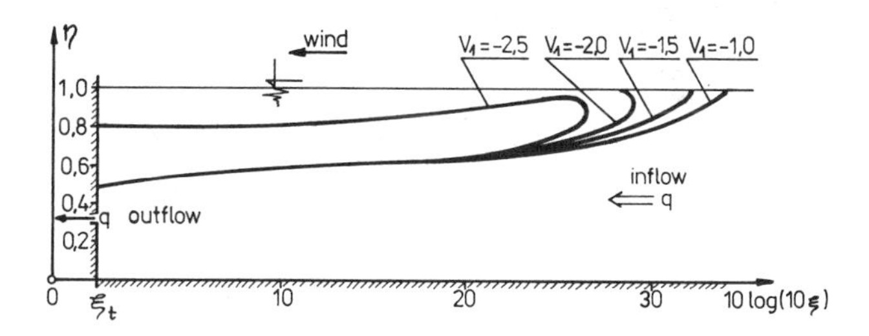

Figure 41. Effect of wind action in axisymmetric stratified flow, (courtesy of the International Association for Hydraulic Research).

vertically plane motion, if V_1 exceeds a certain value, then the flow pattern is reduced to the case of a submerged circulation region.

The method just described may be extended to the case of a gradually converging channel. The idea of such a flow is given in Figure 42. The solution is based on the assumption, that the two axes of flow are considered (located at O_1 and O_2). This is an approximate procedure because it assumes that each streamline (in plane) has a discontinuity point placed along the circle dividing the two areas of different axes. The new extended solution may be formed from Equations 84, 143, 155, and 158

- For the area of axis located at O_1, noting that

$$\xi = \xi^{(1)}$$

if

$$\xi_t^{(1)} \leq \xi^{(1)} \leq \xi_1^{(1)} \tag{166}$$

and substituting into Equation 158

$$D_{1n}^*(\xi) \rightarrow \exp\left[\int_{\xi_t^{(1)}}^{\xi^{(1)}} d_{1n}^{(1)}(\xi^{(1)}) \cdot d_\xi^{(1)}\right] \tag{167}$$

- For the area of axis located at O_2, noting that

$$\xi = \xi^{(2)}$$

if

$$\xi_t^{(2)} < \xi^{(2)} \tag{168}$$

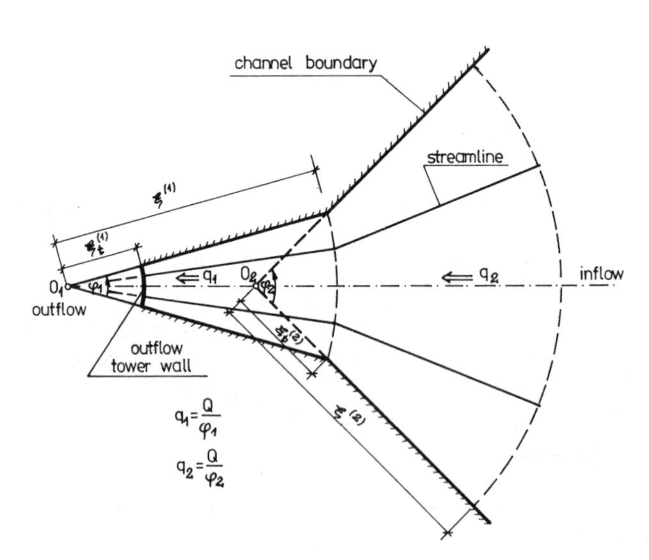

Figure 42. Geometry of axisymmetric flow in gradually converging channel (courtesy of the International Association for Hydraulic Research).

and substituting into Equation 158

$$D_{1n}^{*}(\xi) \rightarrow \exp\left[\int_{\xi_t^{(1)}}^{\xi_1^{(1)}} d_{1n}^{(1)}(\xi^{(1)}) \cdot d\xi^{(1)} + \int_{\xi_t^{(2)}}^{\xi^{(2)}} d_{1n}^{(2)}(\xi^{(2)}) \cdot d\xi^{(2)}\right] \tag{169}$$

where

$$d_{1n}^{(1)} = \frac{A^{(1)} - \sqrt{[A^{(1)}]^2 + 4\pi^6 n^6}}{2\pi^2 n^2}$$

$$d_{1n}^{\prime(2)} = \frac{A^{(2)} - \sqrt{[A^{(2)}]^2 + 4\pi^6 n^6}}{2\pi^2 n^2} \tag{170}$$

while

$$A^{(1)} = \frac{\Delta\rho}{\rho_1} \cdot \frac{g \cdot D^4}{q_1^{(1)} \cdot K_y} \cdot \xi^{(1)}$$

$$A^{(2)} = \frac{\Delta\rho}{\rho_1} \cdot \frac{g \cdot D^4}{q_1^{(2)} \cdot K_y} \cdot \xi^{(2)} \tag{171}$$

and

$$q_1^{(1)} = \frac{Q}{\varphi_1}$$

$$q_1^{(2)} = \frac{Q}{\varphi_2} \tag{172}$$

This procedure may be extended in the case of more than two axes of symmetry. Following that idea, a case with continuously converging channel was considered. The geometry of such a flow is given in Figure 43. The boundary of the flow area (in plane) is given by the curve u = u(x). Each

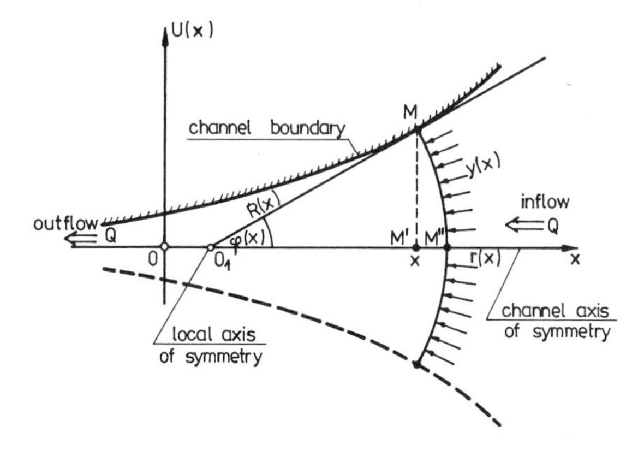

Figure 43. Geometry of axisymmetric flow in continuously converging channel (courtesy of the International Association for Hydraulic Research).

point on this curve corresponds to the value of the r-axis (measured along the x-axis), determined by the local axis of symmetry (Figure 43). We have

$$r(x) = x + \frac{u(x)}{u'(x)} \cdot \sqrt{1 + [u'(x)]^2} \tag{173}$$

where

$$u'(x) = \frac{du}{dx} \tag{174}$$

The parameter A in Equation 150 should now be calculated as

$$A = \frac{\Delta\rho}{\rho_1} \cdot \frac{g \cdot D^2}{V_0(x)} \cdot \frac{1}{\sqrt{K_y(K_r + K_y)}} \tag{175}$$

where

$$V_0(x) = \frac{Q}{2 \cdot R(x) \cdot D \cdot \varphi(x)}$$

and

$$R(x) = \frac{u(x)}{u'(x)} \cdot \sqrt{1 + [u'(x)]^2}$$

$$\varphi(x) = \text{arc tg}[u'(x)] \tag{176}$$

Generally, the value of A may be expressed as

$$A = A(x) = A[x(r)] = A(r) \tag{177}$$

Taking this value for A, the solution can be obtained by substituting into Equation 158

$$D^*_{1n}(\xi) \rightarrow \int_{r_1}^{r} d_{1n}[A(r)] \cdot \sqrt{\frac{K_y}{K_r + K_y}} \cdot \frac{dr}{D} \tag{178}$$

In the case of additional horizontal density changes, when the flow is axisymmetric the aforementioned solution may also be applied. Assuming the density distribution function to be from Equation 20, the solution can be constructed by substituting into Equations 151, 171, 175, and 177

$$A = A_0 \cdot \xi \cdot f(\xi) \tag{179}$$

where $f(\xi)$, as before, is the given function representing the additional density changes. That case is the most general solution presented here.

LABORATORY EXPERIMENTS

In order to test the mechanism of the vertical circulation, experimental investigations have been carried out. The main aim of the experiments was to examine whether the models presented here can be adopted to the case of thermal stratification [20]. Thermal stratification is the most common factor causing the density gradient in reservoir under natural conditions. The experiments were concerned with such things as the shape of interfaces, the shape and number of internal closed eddies within the circulation region, the presence of the surface currents, upstream conditions which were outside the circulation region, and the existence of backward bottom currents. Additionally, the formula for the horizontal extent of the surface circulation region (Equations 77 and 78) in the case of two-layered thermally stratified flow was verified. From a theoretical point of view the laboratory investigations refer to the primary solution case: steady, two-dimensional vertically plane motion in a semi-infinite reservoir, with one outflow slot, located at $y_1 = 0$ and $y_1 = \frac{1}{3}D$.

The experiments were carried out at the Imperial College of Science and Technology, University of London [20]. The hydraulic model was made in the form of a rectangular flume of the following dimensions: length—22 meters; width—0.05 meters, and height—0.07 meters. The front wall of the canal was made of perspex, to enable the observation of the circulation region. The back wall and the bottom were made of wood in order to decrease the heat exchange. A detailed description of the hydraulic model is given in the Meyer [20]. The heating equipment allowed the creation of various density distributions: from one of distinctly two-layers to one gradually changing. The temperature differences of the flowing water were: from 2°C (if $Q = 10^{-3}$ m^3/s) to 20°C (if $Q = 10^{-4}$ m^3/s). Because of its length, the hydraulic model did not permit the direct measurement of $\rho_\infty(\eta)$ and $V_{\xi\infty}(\eta)$, i.e., the density and velocity distribution far upstream. In the qualitative part of the investigations the mechanism of the formation of the surface circulation region was examined for various $\rho_\infty(\eta)$ and $V_{\xi\infty}(\eta)$. The second part of the laboratory experiments concerned verification of Equation 78 in the case of two-layered thermally stratified flow. In each run, the only unknown value V_1 was calculated indirectly, from a group of "l" measured velocities outside the circulation region.

$$V_\xi = V_\xi[\xi^{(l)}, \eta^{(l)}, A, V_1] \tag{180}$$

On the other hand from Equation 78:

$$A = \pi^4 \cdot \xi_0 \cdot \exp[U(V_1)] \tag{181}$$

The approximation of the parameters A and V_1 in each case was based on combining Equations 180 and 181 and using the least square method. Therefore

$$\sum \{V_\xi^{(l)} - V_\xi[\xi^{(l)}, \eta^{(l)}, A(V_1), V_1]\} = \text{minimum} \tag{182}$$

The numerically evaluated parameter A was compared with that obtained directly from Equation 36, which in this case, because of the laminar boundary effects (parabolic velocity distribution along the transverse direction) at the inflow, becomes

$$A = \frac{1}{3\sqrt{2}} \cdot \frac{V_0 \cdot D}{\nu(T_1)} \cdot \frac{g \cdot D}{V_0^2} \cdot \frac{\rho(T_1) - \rho(T_2)}{\rho(T_1)} \tag{183}$$

It leads to the conclusion, that in the case of two-layered thermally stratified flow an additional coefficient κ should appear when A is calculated from Equation 36 as follows

$$A = \frac{1}{\kappa} \cdot \frac{\Delta\rho}{\rho_1} \cdot \frac{Re}{Fr} \tag{184}$$

The values for κ calculated from the laboratory experiments are as follows: $\kappa = 1.86$ for the bottom outflow and $\kappa = 1.54$ when $y_1 = \frac{1}{3} \cdot D$. The results of the experiments concerning the horizontal extent of the surface circulation region in the two-layered thermally stratified flow are shown in Figure 44.

Some of the results concerning qualitative verification are shown in the following figures: Figure 45 shows the flow pattern based on the primary solution. The shape of internal closed eddies corresponds to the two measured velocity profiles. From numerical calculations it is shown that

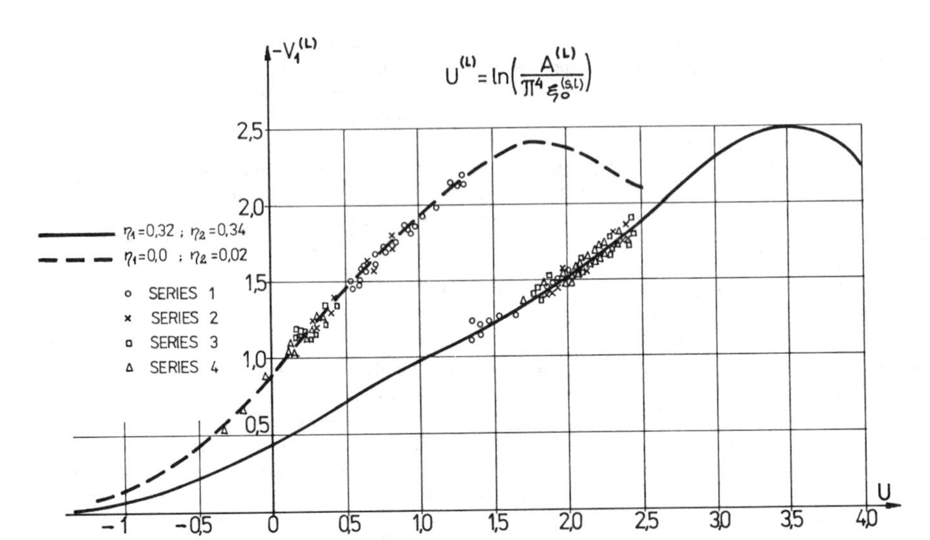

Figure 44. Results of laboratory experiments for horizontal extent $\xi_0^{(s)}$ in two-layered thermally stratified flow (*Hydr. Trans.*, Vol. 43 (1981), courtesy of the Polish Academy of Science).

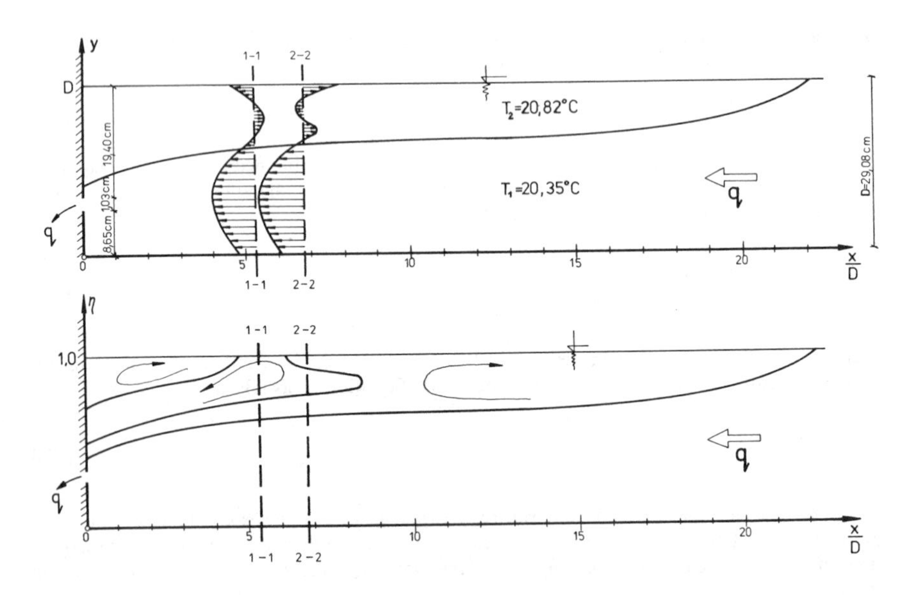

Figure 45. Vertical circulation region corresponding to measured velocity profiles (*Hydr. Trans.*, Vol. 43 (1981), courtesy of the Polish Academy of Science).

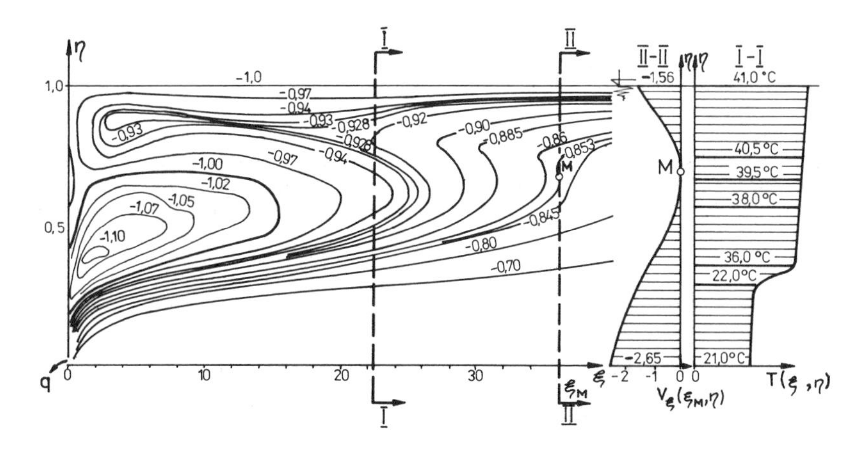

Figure 46. Submerged circulation region corresponding to measured velocity profile (*Hydr. Trans.*, Vol. 43 (1981), courtesy of the Polish Academy of Science).

A = 7,500 and $V_1 = -1.25$. The parameter A evaluated directly from Equation 183 is equal to 8,100. The case with a submerged circulation region is shown in Figure 46. The flow pattern performed here corresponds to the measured velocity profile at the point M. The point M is characteristic because $V_\xi = 0$, and it is easy to distinguish this profile during laboratory experiments. The numerical value of the parameter A in this case is equal to 10,500, and that calculated directly, 10,800.

Figure 47 shows the case of flow with the backward bottom current. Under laboratory conditions this case was produced as follows. The heating equipment was switched on for a short period of time and then switched off. The hot water was accumulated at the upper layers close to the outflow wall. Because of the heat exchange (with the flowing stream and with the air) the surface circulation

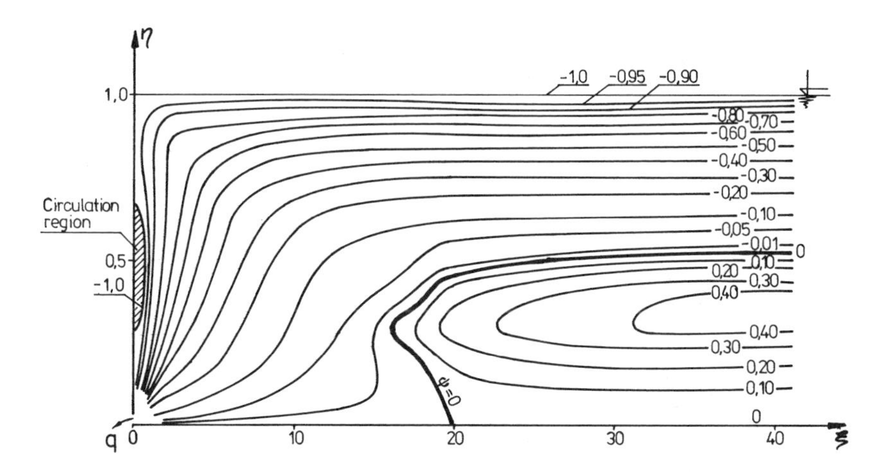

Figure 47. Submerged circulation region with backward bottom current, (*Hydr. Trans.*, Vol. 43 (1981), courtesy of the Polish Academy of Science).

region was getting cooler and cooler. When the density gradient was very small due to strong surface velocity, a short and submerged circulation region was created. In such a case the backward bottom current arises at the bottom layers. This phenomenon is due to the vertical momentum transfer, and it corresponds to the earlier mentioned flow, where $V_1 < -3$ (Figure 21). During laboratory experiments each case of the circulation region was visualized by using the dye. The internal closed eddies and the direction of flow inside them were marked by using the dye-streaks. In general, the flow patterns obtained analytically were in good agreement with the flow pattern observed in the laboratory. It should be emphasized that even in the case of very strong stratification (Figure 46), the increase of surface velocity V_1 submerges the circulation region. It seems to be independent from the density gradient. Full analysis of the laboratory experiments is given by Meyer [20].

PRACTICAL APPLICATIONS OF VERTICAL CIRCULATION MODELS

One of the most common applications of the vertical circulation in a stratified reservoir is the selective withdrawal problem, and the basic case of the selective withdrawal is the deep intake (Figure 48). Usually the shape of the main interface is calculated assuming that it is flat and horizontal. The essential point is to estimate the maximum intake discharge in order to avoid the inflow from the reservoir's epilimnion. The estimation usually neglects the viscosity effects, and the vertical heat transfer causing the additional horizontal density changes.

With regard to the surface circulation region the following practical conclusions can be formed:

1. For the given boundary conditions and the flow parameters, the shape of the circulation region is strictly defined.

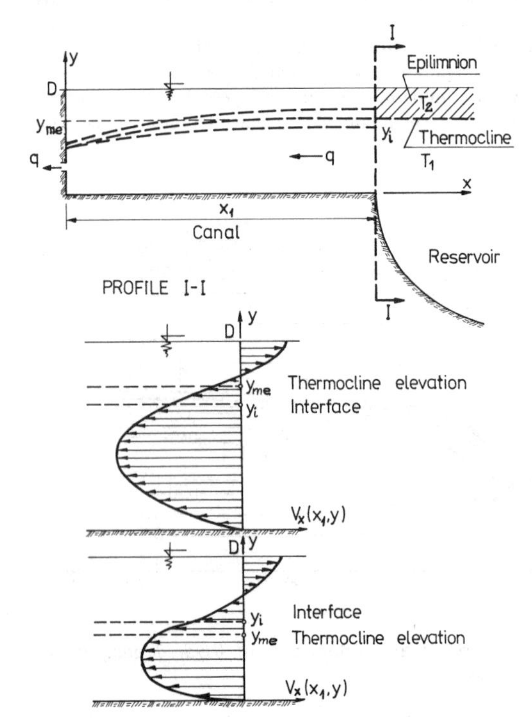

Figure 48. Selective withdrawal from canal supplied by stratified reservoir (courtesy of the Technical University of Szczecin).

2. Following the x-axis the main interface appears to tend to the water surface. The horizontal extent of the surface circulation region is limited and strictly defined.
3. The water quantity within the circulation region for the given flow conditions is constant. If an additional water volume is added to the circulation region (of the same temperature), it does not enlarge this region stably. The exceeding water volume will be discharged and the interface will return to its previous shape.
4. Within the circulation region there exists a number of closed internal eddies bounded by the internal interfaces. These eddies cause the vertical mixing of water.
5. The location of the outflow end of the main interface appears to be very stable. And it appears that its position does not depend strongly either on the wind action or on the density stratification far upstream.
6. The essential meaning in the vertical circulation mechanism has the surface velocity (wind action). Strong wind towards the intake submerges the circulation region and shortens its length. The additional horizontal density changes alter the horizontal extent of the surface circulation region, and so they have to be taken into account.
7. The density distribution at inflow seems to be insignificant for the aforementioned mechanism of vertical circulation. However, it affects the horizontal extent as well as the shape of the main interface.
8. When the parameter A is very large ($A \to \infty$), then the horizontal extent of the circulation region tends to infinity.

These conclusions make way for further analysis of the selective withdrawal from the canal, supplied by the stratified reservoir.

Figure 48, shows a typical shape of the interface within the canal linking the intake with the reservoir. The horizontal interface, may be assumed inside the reservoir since the value of parameter A in the reservoir is many times greater than in the canal. This is due to the large reservoir's dimensions and small velocity. So the criterion of withdrawing water may be defined as follows

- If $y_{me} > y_i$, then the water comes only from the area below the thermocline.
- If $y_{me} < y_i$, then the intake takes water also from the thermocline and epilimnion layers. In this description y_i means the curve of the main interface.

These conditions can also be expressed in terms of the stream function, which is the solution of the mathematical model.

- If $\psi(\xi_1, \eta_{me}) < -1$, then the intake takes water only from hypolimnion layer.
- If $\psi(\xi_1, \eta_{me}) > -1$, then the water comes also from the epilimnion layer.

The approximate quantity of the water coming from the epilimnion layer can be evaluated as

$$\frac{(\Delta q)_{epilimnion}}{q_{total}} = 1 + \psi(\xi_1, \eta_{me}) \tag{185}$$

And the rough estimation of the withdrawal water temperature gives

$$T_{intake} = T_1 + (T_2 - T_1) \cdot [1 + \psi(\xi_1, \eta_{me})] \tag{186}$$

The maximum intake outflow can now be defined as

$$q = q_{max} \qquad \text{when } \psi(\xi_1, \eta_{me}, A) = 1 \tag{187}$$

From this equation the value A can be determined. We have

$$A = A(\xi_1, \eta_{me}) \tag{188}$$

and furthermore combining with Equation 36 we obtain

$$q_{max} = \frac{\Delta\rho}{\rho_1} \cdot \frac{q \cdot D^3}{A(\xi_1, \eta_{me})\sqrt{K_y(K_x + K_y)}} \tag{189}$$

In fact the relationship expressing q_{max} is more complicated because the eddy viscosity coefficients depend on the velocity and flow. For the rough estimation the following approximate formula may be used [23]

$$q_{max} = \sqrt{\frac{\Delta\rho}{\rho_1} \cdot \frac{g \cdot D^3}{\kappa_0 \cdot A(\xi_1, \eta_{me})}} \tag{190}$$

where κ_0 is the same constant as before

$$\kappa_0 = 10^{-2} + 10^{-4}$$

The described procedure for calculating the inflow to the selective withdrawal can include the wind action and the flow with the additional horizontal density changes. The idea of such an influence in the considered cases is shown in Figure 49. From Figure 49B it follows that the density distribution function can also be estimated by introducing the surface density gradient $\Delta\rho_2$. Thus when $\psi = -1$

$$\rho_2(\psi, \xi) = \rho_1 - \Delta\rho \cdot f(\xi)$$

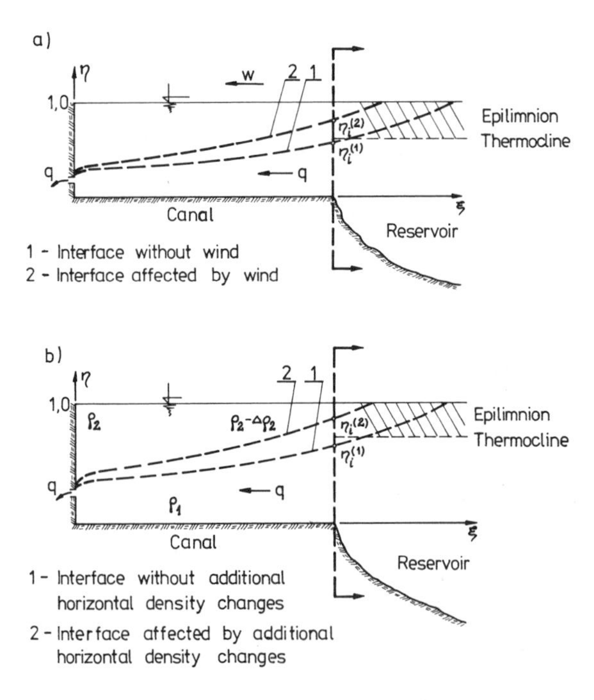

Figure 49. Wind effect on interface displacement in selective withdrawal canal (courtesy of the Technical University of Szczecin).

and

$$\rho_2(0) = \rho_1 - \Delta\rho \cdot f(0)$$

while

$$\rho_2(\xi_1) = \rho_1 - \Delta\rho \cdot f(\xi_1)$$

therefore

$$\rho(\xi) = \rho_1 - \Delta\rho_2 \cdot \frac{f(\xi)}{f(0) - f(\xi_1)} \tag{191}$$

To include certain factors affecting the flow, a proper mathematical model for estimation the stream function must be chosen.

In practice the problem often occurs to shorten the circulation region extent. This can be achieved by using a bottom obstacle. The scheme of the flow in such a case is given in Figure 50. The appropriate flow pattern can be calculated as follows:

- For area 1; $0 \leq \xi \leq \xi_2$

$$\psi(\xi, \eta) = \psi_\infty(\eta) + \sum_{n=1}^{\infty} [a_{1n} \cdot \exp(d_{1n}^{(1)} \cdot \xi) \cdot \sin(\pi n \eta)]$$

$$A^{(1)} = \frac{\Delta\rho}{\rho} \cdot \frac{q \cdot D^3}{q\sqrt{K_y(K_x + K_y)}}$$

$$d_{1n}^{(1)} = d_{1n}(A^{(1)}) \tag{192}$$

- For area 2; $\xi_2 < \xi < \xi_3$

$$\psi(\xi, \eta) = \psi_\infty(\eta) + \sum_{n=1}^{\infty} \left\{a_{1n} \cdot \exp[d_{1n}^{(1)} \cdot \xi_2 + d_{1n}^{(2)}(\xi - \xi_2)] \cdot \sin\left(\frac{\pi n \eta}{1 - \eta_P}\right)\right\}$$

$$A^{(2)} = A^{(1)} \cdot (1 - \eta_P)^3$$

$$d_{1n}^{(2)} = d_{1n}(A^{(2)}) \tag{193}$$

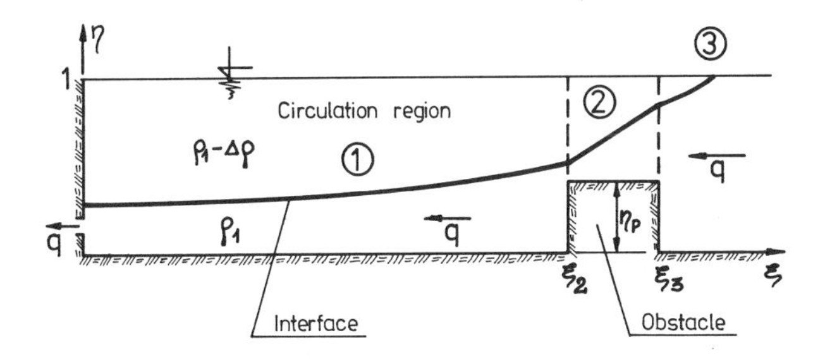

Figure 50. Surface circulation region limited by bottom obstacle (courtesy of the Technical University of Szczecin).

• For area 3; $\xi > \xi_3$

$$\psi(\xi, \eta) = \psi_\infty(\eta) + \sum_{n=1}^{\infty} \{a_{1n} \cdot \exp[d_{1n}^{(1)} \cdot \xi_2 + d_{1n}^{(2)} \cdot (\xi_3 - \xi_2) + d_{1n}^{(3)}(\xi - \xi_3)] \cdot \sin(\pi n \eta)\}$$

$$A^{(3)} = A^{(1)}$$

$$d_{1n}^{(3)} = d_{1n}(A^{(3)}) = d_{1n}^{(1)} \tag{194}$$

It can be seen that $A^{(2)} \gg A^{(1)}$ and it causes a rapid shortening of the circulation region. When the additional horizontal density changes are included then the original term $d_{1n} \cdot \xi$ should be replaced as follows

$$d_{1n} \cdot \xi \rightarrow D_{1n}(\xi) \tag{195}$$

and $D_{1n}(\xi)$ is determined by Equation 83. The same approach to the bottom obstacle problem can be used in the case of axisymmetric flow.

The mechanism of vertical circulation can also be applied for the case of simultaneous dam withdrawal and hot water discharge. The scheme of such a flow is given in Figure 51. The hot water discharge is equal to the flow over the weir. This sort of construction is taking advantage of the fact that the surface area close to the dam is working as the cooling region. From the point of view of practical calculation the extent of cooling surface is essential. In this case the stream function should be calculated as follows (Equations 35 through 44):

$$\psi(\xi, \eta) = \psi_\infty(\eta) + \sum_{n=1}^{\infty} [a_{1n} \cdot \exp(d_{1n} \cdot \xi) \cdot \sin(\pi n \eta)] \tag{196}$$

where

$$a_{1n} = c_{1n} - b_n$$

and taking symbols from Figure 51,

$$c_{1n} = -\frac{2}{\pi^2 n^2}(E_1 + E_2 + E_3)$$

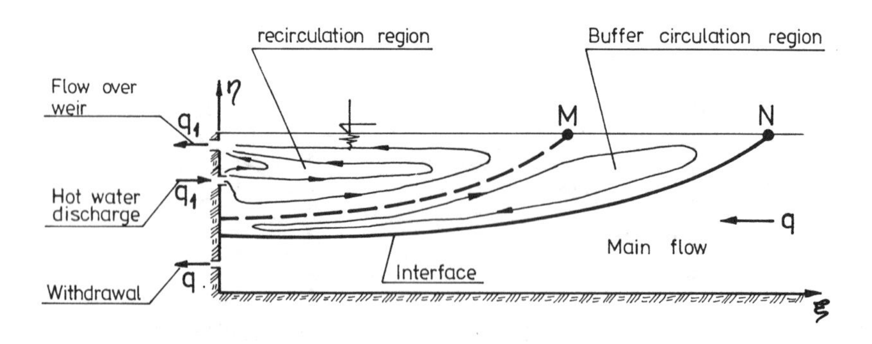

Figure 51. Dam selective withdrawal with hot water discharge (courtesy of the Technical University of Szczecin).

while

$$E_1 = \frac{\sin(\pi n \eta_2^{(1)}) - \sin(\pi n \eta_1^{(1)})}{\eta_2^{(1)} - \eta_1^{(1)}}$$

$$E_2 = -\frac{q_1}{q} \cdot \frac{\sin(\pi n \eta_2^{(2)}) - \sin(\pi n \eta_1^{(2)})}{\eta_2^{(2)} - \eta_1^{(2)}}$$

$$E_3 = \frac{q_1}{q} \cdot \frac{\sin(\pi n \eta_2^{(3)}) - \sin(\pi n \eta_1^{(3)})}{\eta_2^{(3)} - \eta_1^{(3)}} \tag{197}$$

The coefficients d_{1n} and the parameter A are given by Equation 36, and the values b_n by Equations 37 or 38. The detailed analysis of the flow pattern described by Equations 196 and 197 indicates, that the cooling surface does not cover the full circulation region, i.e., from the wall to point N. It appears that between the discharged circulating water and the main flow area a buffer region exists. This region does not take place in the cooling of water.

If the additional horizontal density changes exist, then the original term $d_{1n}\xi$ should be replaced as in Equation 195. The same idea can be used in the case of the axisymmetric flow.

It seems that the mechanism of vertical circulation can also be applied to certain cases of the vertical mixing in the dam reservoirs [9]. It appears that most open for this influence are the reservoirs of the reolimnical type. In this reservoir, full depth mixing occurs a couple of times a year while at other times density currents prevail. If the incoming stream (river) is cooler than the reservoir's water, then the main stream arises at the bottom and the upper layers the circulation region appears. The circulation region may consist of two or three separate cells. It appears that for the rough estimation of the flow pattern within such a reservoir, the ψ function earlier obtained may be used. The scheme of this flow is given in Figure 52. If the reservoir's slope is small, i.e., several percent, then the new extended solution can be constructed. The reservoir's depth is assumed to be variable:

$$D(x) = D - y_b(x) \tag{198}$$

And the new $\eta(\xi)$ coordinate is introduced

$$\eta(\xi) = \frac{\eta - y_b(x)}{D - y_b(x)} \tag{199}$$

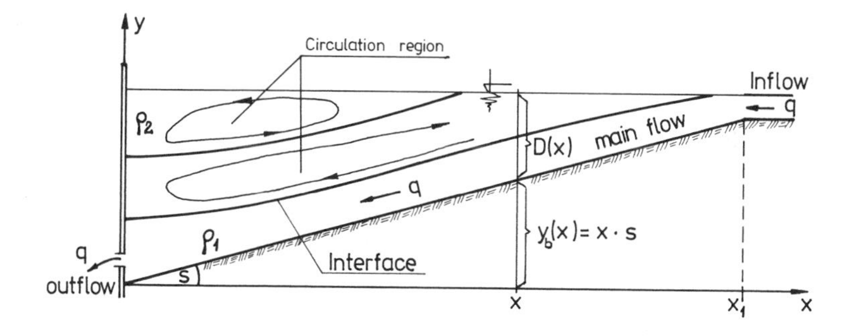

Figure 52. Vertical mixing in dam reservoir.

where $y_b(x)$ is the curve representing the reservoir's bottom. In this case the full reservoir's depth is penetrated if

$$0 < \eta(\xi) < 1$$

So we have

$$\psi(\xi, \eta) = \psi_{\infty}[\eta(\xi)] + \sum_{n=1}^{\infty} \left\{ a_{1n} \cdot \exp\left[\int_0^{\xi} d_{1n}(\xi) \cdot d\xi \right] \cdot \sin[\pi n \eta(\xi)] \right\} \tag{200}$$

where

$$d_{1n}(\xi) = d_1[A(\xi)]$$

and

$$A(\xi) = \frac{\Delta\rho}{\rho_1} \cdot \frac{g \cdot D^3(x)}{q\sqrt{K_y(K_y + K_x)}} \tag{201}$$

And if the reservoir's bottom is a straight line

$$y_b(x) = S \cdot x \tag{202}$$

where "s" is the bottom slope, then

$$D(x) = D - s \cdot x$$

or

$$D(\xi) = D(1 - S_1 \cdot \xi)$$

where

$$s_1 = s\sqrt{\frac{K_y + K_x}{K_y}}$$

and

$$A(\xi) = A \cdot (1 - s_1 \cdot \xi)^3 \tag{203}$$

The coefficients a_{1n} remain the same as before (Equations 39 through 42). The parameter A is defined in Equation 36; d_{1n} is given by Equation 36 and the function $\psi_{\infty}(\eta)$ by Equation 37 or 38. If the additional horizontal density changes exist, the solution can be obtained by putting A as variable $A(\xi)$. The same idea of the sloped reservoir can be applied in the case of axisymmetric flow.

CONCLUSIONS

1. Analytical models of vertical circulation in a density-stratified reservoir are presented in this work. The models include rectangular and semi-infinite strip reservoirs and two sorts of motion: the vertically plane and the axisymmetric flow. The investigations permit distinction of the characteristic mechanism of the vertical circulation. It appears that wind action plays a very important role in this mechanism. The models account for several distinctive features, which are characteristic stratified flows. They are as follows:

    - The important role of wind action in the vertical circulation.
    - The case of a submerged circulation region.

- The existence of internal closed eddies within the circulation region (cellular structure of the circulation region).
- The conditions of formation of the stagnation regions.
- The cellular structure of the wind vertical circulation.
- The appearance of surface current outside the circulation region.
- The influence of the additional horizontal density changes on the mechanism of vertical circulation.

Using the presented models several practical cases of stratified flow were considered:

- The selective withdrawal from the channel supplied by the stratified reservoir.
- The influence of bottom obstacle on the vertical circulation.
- The dam withdrawal with simultaneous hot water discharge.
- The vertical circulation in a sloped reservoir.

2. The flow patterns obtained for the case of wind vertical circulation, although in agreement with the previous results [35, 36] need deeper consideration. Laboratory investigations should be carried on in order to verify the mechanism of this circulation and to estimate to what extent of the parameter A the model may be used. Further analytical research should include mass diffusion instead of assuming the density distribution. From the laboratory experiments it follows, that the wall boundary layer effect may have some significance, especially at the corner of the side wall and outflow wall. It can affect producing vertical swirls.
3. Several problems remain unsolved and should be investigated in further research. They are as follows:

- The influence of varying eddy viscosity coefficients on the mechanism of vertical circulation.
- The unsteady vertical circulation problem particularly in reference to the unsteady wind action.
- The application of the presented models for the reservoirs with irregular geometry in plane.

The varying eddy viscosity coefficients do not seem to radically change the aforementioned mechanism of the circulation. And it results in making the shape of the interfaces different. If it corresponds to the slow density changes, the unsteady circulation problem can be estimated assuming the quasi-stationary flows and using the described models. The flow through the reservoir of irregular geometry in certain cases may be considered using the presented models. It needs to solve the plane problem first, and then the flow between the neighboring streamlines (in plane) can be treated by using the case of a continuously varying (converging or diverging) channel.

NOTATION

A, A, A_* dimensionless parameters including flow and density stratification

a_n, b_n, c_n Fourier's series coefficients

D depth of reservoir

D_{1n}, D_{2n} coefficients in Fourier series

$f(\xi)$ function representing horizontal density changes

F_n, F'_m constant values of stream function

$f_n(\eta), f'_m(\eta)$ stream-function distribution at outflow and inflow orifices

Fr_1 densimetric Froude number

Fr constant value equivalent to Froude number

g acceleration due to gravity

h_n height of n-th slot at outflow

h'_m height of m-th slot at inflow

K_h molecular diffusion coefficient of factor causing density changes

K_r eddy viscosity coefficient in radial horizontal plane, referred to axisymmetric flow

K_x, K_y eddy viscosity coefficient in x- and y-direction

P pressure

q total net flow in reservoir (per unit width)

q_1 net flow in reservoir (per unit angle) in case of axisymmetric flow
q_n flow in n-th outflow slot
q'_m flow in m-th inflow slot
Q total flow in axisymmetric motion
r_t radius of intake tower in axisymmetric flow
$r_0^{(s)}, y_0^{(s)}$ horizontal and vertical extent coordinates of surface circulation region
$r_0^{(b)}, y_0^{(b)}$ horizontal and vertical extent coordinates of bottom circulation region
r, y basic system of cylindrical coordinates in axisymmetric flow
R_n Fourier's series coefficient
Re constant value equivalent to Reynolds number
R_i Richardson number
s reservoir's bottom slope
T water temperature
T_1 bottom water temperature
T_2 surface water temperature
$u(x)$ function representing channel sideline
U parameter determining horizontal extent of surface circulation region
V velocity
V_0 depth averaged velocity in reservoir
V_x, V_y velocity components in x- and y-direction
V_r, V_y velocity components in case of axisymmetric flow, in r- and y-direction
V_ξ, V_η dimensionless velocity components in ξ- and η-direction
$V_{x\infty}, V_{\xi\infty}$ velocity distribution at inflow
V_s surface velocity at inflow (far upstream)
V_1 dimensionless surface velocity at inflow (far upstream)
$W_{\theta,1/2}(\delta)$ Whittaker's functions of parameters θ and $\frac{1}{2}$
x, y basic system of Cartesian coordinates
x_1 coordinate limiting reservoir's length
$x_0^{(b)}, y_0^{(b)}$ horizontal and vertical extent coordinates of bottom circulation region
$x_0^{(s)}, y_0^{(s)}$ horizontal and vertical extent coordinates of surface circulation region
y_{me} coordinate of thermocline (metalimnion) elevation
$y_1^{(n)}, \eta_1^{(n)}$ location of lower edge of n-th outflow slot
$y_2^{(n)}, \eta_2^{(n)}$ location of upper edge of n-th outflow slot
$y_1'^{(m)}, \eta_1'^{(m)}$ location of lower edge of m-th inflow slot
$y_2'^{(m)}\ \eta_2'^{(m)}$ location of upper edge of m-th inflow slot

Greek Symbols

α, β parameters representing nonlinear terms in momentum equation,
δ dimensionless variable
ϵ coefficient in horizontal extent equation in case of axisymmetric flow
η vertical dimensionless coordinate
η_* location of point outflow slot
θ dimensionless variable
κ coefficient from laboratory experiment correcting parameter A
ν kinematic viscosity of water
ξ, η dimensionless system of coordinates
$\xi_0^{(b)}, \eta_0^{(b)}$ dimensionless coordinates of horizontal and vertical extent of bottom circulation region
$\xi_0^{(s)}, \eta_0^{(s)}$ dimensionless coordinates of horizontal and vertical extent of surface circulation region
ξ_1 dimensionless coordinate of inflow location (reservoir's length)
ξ_t dimensionless coordinate of intake tower
ρ density of water
ρ_0 reference water density
ρ_1 bottom water density
ρ_2 surface water density
$\rho_\infty(\eta)$ density distribution at inflow
ρ_{max}, ρ_{min} maximum and minimum water density in reservoir
τ shear stress
τ_2 shear stress at water surface
τ_w wind shear stress
φ angle of convergence of channel in plane

Ψ stream function
Ψ_0 reference stream function
ψ dimensionless stream function
$\psi_\infty(\eta)$ dimensionless stream function at inflow
ψ_{max}, ψ_{min} maximum and minimum values of stream function within flow pattern
Ω vorticity function
ω dimensionless vorticity function

REFERENCES

1. Batchelor, G. K., "*An Introduction to Fluid Dynamics*," Cambridge University Press, 1970.
2. Bengtson, L., "Mathematical Models of Wind Induced Circulation in Lake," *Proceedings of the International Symposium on the Hydrology of Lakes*, IAHR, Publication No. 109, pp. 313–320, held July 23–27, 1973, at Otaniemi, Finland.
3. Blumberg, A. F., "Numerical Model For Estuarine Circulation," *Journal of the Hydraulics Division*, ASCE, Vol. 103, No. HY 3, Proc. Paper 12815, (Mar. 1977), pp. 295–310.
4. Bohan, J. P., and Grace, L. L., "Mechanics of Stratified Flow Through Orifices," *Journal of the Hydraulics Division*, ASCE, No. HY 12, (Dec. 1970), pp. 2401–2416.
5. Brooks, N. H., and Koh, R. C. Y., "Selective Withdrawal From Density-Stratified Reservoir," *Journal of the Hydraulics Division*, ASCE, Vol. 95. No. HY 4, Proc. Paper 6702 (July 1969), pp. 1369–1400.
6. Craya, A., "Recherches theoriques sur l'ecoulment de couches superposees de fluids de densite differents," *La Houille Blanche*, Januier-Fevrier, No. 1 (1949), pp. 45–55.
7. Csanady, G. T., "Wind Driven Summer Circulation in the Great Lakes," *Journal of the Geophysical Research*, Vol. 73, No. 8, (Apr. 1968).
8. Csanady, G. T., "Hydrodynamics of Large Lakes," *Annual Review of Fluid Mechanics*, Vol. 7 (1975), pp. 357–385.
9. Cyberski, J., and Cyberska, B., "Water Mixing in Different Types of Storage Reservoirs," *Proceedings of the International Symposium on the Hydrology of Lakes*, IAHR, Publication No. 109, pp. 374–379, held July 23–27, 1973, at Otaniemi, Finland.
10. Debler, W. R., "Stratified Flow Into a Line Sink," *Journal of the Engineering Mechanics Division*, ASCE, Vol. 85, No. EM 3 (July 1959), pp. 51–65.
11. Findikakis, A. N., and Street, R. L., "Vertical Structure of Internal Modes in Stratified Flows," *Journal of the Engineering Mechanics Division*, ASCE, Vol. 108, No. EM 4, Proc. Paper 17254, (Aug. 1982), pp. 583–595.
12. Findikakis, A. N., and Street, R. L., "Finite Element Simulation of Stratified Turbulent Flows," *Journal of the Hydraulics Division*, ASCE, Vol. 108, No. HY 8, Proc. Paper 17266, (Aug. 1982), pp. 904–920.
13. Harleman, D., "Stratified Flows Sect. 26," *Handbook of Fluid Dynamics*, V. L. Streeter (Ed. in-chief), McGraw-Hill Book Co. Inc., London, England, 1961, pp. 26.1–26.18.
14. Huber, D. G., "Irrotational Motion of Two Fluids Strata Towards a Line Sink," *Journal of the Engineering Mechanics Division*, ASCE, Vol. 86, No. EM 4 (Aug. 1960), pp. 71–88.
15. Kao, T. W., "Free-Streamline for Stratified Flow Into a Line Sink," *Journal of Fluid Mechanics*, Vol. 21 (1965), pp. 535–543.
16. Koh, R. C. Y., "Viscous Stratified Flow Towards a Sink," *Journal of Fluid Mechanics*, Vol. 24, Part 3 (Mar. 1966), pp. 555–575.
17. Liggett, J. A., and Lee, K. K., "Properties of Circulation in Stratified Flows," *Journal of the Hydraulics Division*, ASCE, Vol. 97, No. HY 1, Proc. Paper 7793, (Jan. 1971), pp. 15–29.
18. Long, R. R., "Some Aspects of the Flow of Stratified Fluids," Part I: A Theoretical Investigation, *TELLUS*, Vol. 5, No. 1 (Feb. 1953), pp. 42–58.
19. Meyer, Z., "Hydrauliczne warunki dopływu wody do ujecia w warunkach stratyfikacji gestości wody," *Archiwum Hydrotechnki*, Vol. XXVI, Part 3 (1979), pp. 371–396.
20. Meyer, Z., "Vertical Circulation in Reservoir Due to Selective Withdrawal," *Rozprawy Hydrotechniczne/Hydrotechnical Transactions*, Vol. 43 (1981), pp. 193–219.
21. Meyer, Z., "Vertical Circulation in Density-Stratified Reservoir," *Journal of the Hydraulics Division*, ASCE, Vol. 108, No. HY 7, Proc. Paper 17206, (July 1982), pp. 853–873.

22. Meyer, Z., "Pionowa cyrkulacja w zbiorniku w warunkach zanikąjacej stratyfikacji gęstości wody," *Archiwum Hydrotechniki*, Vol. XXIX, Part 1–2, 1982, pp. 33–52.
23. Meyer, Z., *Selektywne ujęcie wody-analiza hydraulicznych warunków dopływu*, Technical University of Szczecin Press, Vol. 201, SZCZECIN, 1982.
24. Meyer, Z., "Axisymmetric Vertical Circulation in a Stratified Reservoir," *Journal of Hydraulic Research*, Vol. 21, No. 2 (1983), pp. 133–152.
25. Oltesen Hanzen, N. E., "Effect of Wind Stress on Stratified Deep Lake," *Journal of the Hydraulics Division*, ASCE, Vol. 101, No. HY 8, Proc. Paper 11484, (Aug. 1975), pp. 1037–1052.
26. Plate, E. J., "Water Surface Velocities Induced by Wind Shear," *Journal of the Engineering Mechanics Division*, ASCE, Vol. 96, No. EM 3, Proc. Paper 7351 (June 1970), pp. 295–312.
27. Richardson, A. R., "Stationary Waves in Water," *Philosophical Magazine*, 86, Vol. 40, No. 235 (1920), pp. 97–110.
28. Turner, J. S., *Buoyancy Effects in Fluids*, Cambridge University Press, Cambridge, Mass., 1973.
29. Uzzell, J. C., and Ozisik, M. N., "Far Field Circulation Velocities in Shallow Lakes," *Journal of the Hydraulics Division*, ASCE, Vol. 103, No. HY 4, Proc. Paper 12873, (Apr. 1977), pp. 395–407.
30. Walesh, S. G., and Monkmeyer, P. L., "Bottom Withdrawal of Viscous Stratified Fluid," *Journal of the Hydraulics Division*, ASCE, Vol. 99, No. HY 9, Proc. Paper 10018, (Sept. 1973), pp. 1401–1419.
31. Yih, C. S., "On Stratified Flows in Gravitational Field," *TELLUS*, Vol. 9, No. 2 (1957), pp. 220–228.
32. Yih, C. S., "On the Flow of a Stratified Fluid," *Proceedings of the Third U.S. National Congress of Applied Mechanics* (1958), pp. 857–861.
33. Yih, C. S., "Exact Solutions of Steady Two-Dimensional Flow of a Stratified Fluid," *Journal Fluid Mechanics*, Vol. 9 (1960), pp. 101–124.
34. Yih, C. S., "*Dynamics of Non-Homogeneous Fluids*," The Macmillan Co., New York, N.Y., 1965.
35. Young, D. L., Liggett, J. A., and Gallagher, R. H., "Steady Stratified Circulation in Cavity," *Journal of the Engineering Mechanics Division*, ASCE, Vol. 102, No. EM 1, Proc. Paper 11903, (Feb. 1976), pp. 1–17.
36. Young, D. L., Liggett, J. A., and Gallagher, R. H., "Unsteady Stratified Circulation in Cavity," *Journal of the Engineering Mechanics Division*, ASCE, Vol. 102, No. EM 6, Proc. Paper 12643, (Dec. 1976), pp. 1009–1023.
37. Young, D. L., and Liggett, J. A., "Transient Finite Element Shallow Lake Circulation," *Journal of the Hydraulics Division*, ASCE, Vol. 103, No. HY 2, Proc. Paper 12741, (Feb. 1976), pp. 109–121.

CHAPTER 24

ROLLOVER IN STRATIFIED HOLDING TANKS

John W. Meader

Chemical Engineering Department
Worcester Polytechnic Institute
Worcester, Massachusetts USA

Janet Heestand

Cabot Corporation
Billerica Technical Center
Billerica, Massachusetts USA

CONTENTS

INTRODUCTION, 637

CELL FORMATION AND STABILITY, 638

BACKGROUND WORK, 639
Heat and Mass Transfer Rates, 639
Composition of LNG Mixtures, 641
Criteria for Rollover, 641
Mass Transfer Between Cells, 642
Conditions at the Vapor-Liquid Interface, 642
Material and Energy Balances on the Cells, 644
Application of the Balance and Rate Equations, 645
Results for the LaSpezia Incident, 646

RESULTS OF PARAMETRIC STUDIES, 651

APPENDIX A: DETAILS OF THE LASPEZIA INCIDENT, 655

APPENDIX B: DERIVATION FOR THE OVERALL HEAT TRANSFER COEFFICIENT AT THE INTERFACE BETWEEN TWO CELLS, 656

NOTATION, 656

REFERENCES, 659

INTRODUCTION

Rollover is the name given to a rapid mixing of strata (or cells) in a liquefied natural gas (LNG) storage tank because of the development of an unstable interface. Multiple, periodic rollovers can occur in peak-shaving storage tanks to which a continuous stream of LNG is being added. These rollovers can be preceded by nitrogen induced spontaneous stratifications. Extensive discussions of the rollover phenomena within peak-shaving tanks have been given by Chatterjee and Geist [1], for example. Chatterjee and co-workers have continued their efforts since then. A single, more intense rollover process may occur in a base-load LNG storage facility when fresh cargo of one composition and temperature is incorrectly added to a tank containing a heel of different composition and temperature. This chapter deals with the latter process only. Certain background details not included here can be found in the paper by Heestand, Shipman, and Meader [2].

Liquefied natural gas is stored in tanks at temperatures of about 115 K and pressures slightly above 1 atmosphere. Heat leaks, even in well-insulated tanks, cause a slow boil-off of the LNG, and this requires removal of some vapor to maintain constant pressure. During this "weathering" process the temperature and density of the LNG change. Also, the composition of the LNG changes because the small amount of nitrogen present is much more volatile than the methane and the heavier hydrocarbons are effectively nonvolatile.

Natural convection causes circulation of the LNG within the tank, maintaining a uniform liquid composition. Addition of new liquid, however, can result in the formation of strata of slightly different temperature and density. The energy and mass transfer between these cells during the incubation period may then lead to rollover. When the lower, more dense cell is also at a higher temperature, rollover causes the vapor pressure at the top of the LNG to rise very rapidly, and a rapid evolution of a large quantity of vapor results. Furthermore, the pressure in the tank may rise noticeably. At best, such a situation involves loss of valuable fuel. At worst, it could be dangerous.

This chapter presents a theoretical framework for rollover analysis. Quantitative computer results are given for the simulation of the only well-documented rollover incident generally known. This occurred at LaSpezia, Italy, and was described by Sarsten [3]. A parametric study has been performed with this computer program, and the results are reported here.

CELL FORMATION AND STABILITY

Stratification in an LNG storage tank can give rise to the circumstances shown in Figure 1. The liquid of the lower cell (cell 1) has a greater density than the liquid of the upper cell (cell 2). The individual cells may approach homogeneity because of the circulation patterns produced by heat leak into the tank bottom and sides. However, a quite stable interface can be produced between cells. For example, the initial densities for the LaSpezia incident were $\rho_1 = 541.118\ \mathrm{kg{\cdot}m^{-3}}$ (cargo) and $\rho_2 = 537.316\ \mathrm{kg{\cdot}m^{-3}}$ (heel), both calculated values. The interface presence and stability is a

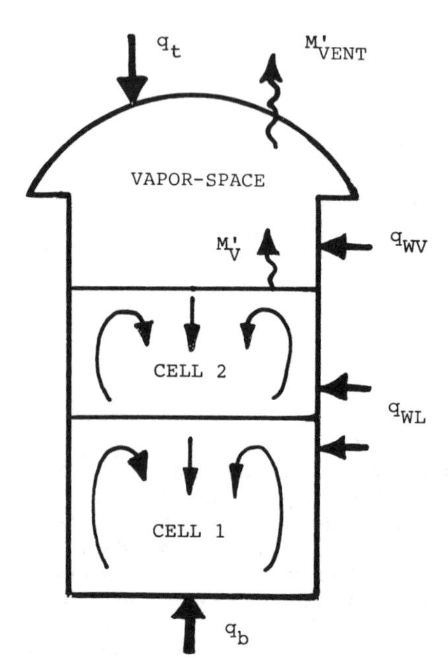

Figure 1. Stratification in an LNG tank showing two liquid cells with circulation and the vaporization and heat flux quantities.

hydrodynamic phenomenon, and it is not a matter of miscibility. The stability does not prevent heat and mass transfer between the cells. In fact, these transfer processes proceed and tend to equalize properties of cells 1 and 2 as time passes. An LNG storage tank may have more than two such cells; in that case cells are numbered i ($i = 1, 2, \ldots, N$).

As the densities of the two cell liquids approach equality two important considerations enter. The first of these is the hydrodynamic instability of an interface to a disturbance. Lamb [4] discusses this matter, and his analysis indicates that interfacial waves will grow in amplitude. The second consideration is that somewhere near the condition of equal densities the liquid rising within the boundary layer of the lower cell will possess enough momentum to punch through the interface and continue to the top of the upper cell. These two considerations explain the rapidity of the mixing process (or rollover) when the densities of adjacent cells become equal. When the lower cell liquid is at a higher temperature than the upper cell liquid the rapid mixing brings a rapid increase of vapor pressure at the vapor-liquid interface. Correspondingly, this causes a rapid evolution of vapor which must be removed from the tank to maintain constant pressure.

BACKGROUND WORK

Heat and Mass Transfer Rates

Rollover models have been developed by Chatterjee and Geist [1, 5] and their co-workers, Germeles [6] and Heestand, Shipman, and Meader [2]. The two former models included acceptance of the thermohaline experiments of Turner [7] as the basis for treating the heat and mass transfer between cells. Heestand et al. [2] attempted unsuccessfully to implement Turner's correlations. They are considered to be inapplicable for the LNG rollover analysis. The thermohaline expressions simply do not allow large enough mass transfer rates to occur at early times without unreasonable adjustments in the interface stability or in the initial conditions of the cells.

Turner [7] arranged two layers (each 12 cm thick) of different strength saline solutions in a cylindrical tank. The interface was made sharp by mechanically stirring each layer. Density differences were varied from 1 to 100 $kg \cdot m^{-3}$. Heat and mass transfer measurements were made by a combined thermistor and conductivity probe. Salinity and temperature of each layer were monitored as functions of time. The metallic bottom of the tank was uniformly heated by an electrical pad. Turner correlated his data with a stability parameter, $R = (\alpha\ \Delta S/\beta\ \Delta T)$, where $\alpha\ \Delta S$ is the density difference between layers created by salinity difference, ΔS, and $\beta\ \Delta T$ is the density difference because of the temperature difference. When R approaches unity the layers have essentially equal densities, the salinity and temperature effects being equal and opposite. Turner compared the measured heat transfer coefficients with those expected for turbulent natural convection between two heated plates with the same ΔT as in his experiments. The Rayleigh number was of the order of 10^8. The heat transfer results were correlated with the equation

$$(N_{Nu})_{THERMOHALINE} = f(R) \cdot [(N_{Nu})_{NAT.\ CONV.}] \tag{1}$$

where f(R) approached nearly ten when R approached unity. When R was two, f(R) was unity; and when R increased to seven, f(R) asymptotically decreased to about 0.1. Following Chandrasekhar [8], Turner chose

$$(N_{Nu})_{NAT.\ CONV.} = 0.085(N_{Ra})^{1/3} \tag{2}$$

Turner's mass transfer data were also correlated as a function of the stability parameter in the form

$$\begin{aligned}(k_x \cdot C_L/h_i) &= 1.85R^{-1} - 0.85 \qquad (1 \leq R \leq 2)\\ &= 0.15R^{-1} \qquad (R \geq 2)\end{aligned} \tag{3}$$

Huppert [9] represented Turner's thermohaline heat transfer coefficient in the form

$$h_i = 3.8R^{-2}h \tag{4}$$

where $h = 0.085k(g\beta\,\Delta T/\nu\kappa)^{1/3}$ is just Chandrasekhar's [8] form of the turbulent natural convection heat transfer coefficient between two parallel, horizontal flat plates. Huppert drew attention to the critical value of the stability parameter, $R_c = 2$, as being the condition for which the ratio of mass flux to heat flux for the thermohaline experiments becomes

$$(k_x \cdot C_L/h_i) \rightarrow 0.15R^{-1}.$$

Chatterjee and Geist [5] accepted Turner's thermohaline experiments because there were no heat and mass transfer data in the literature for LNG systems. They applied no correction to the heat flux correlation because the thermal diffusivities of methane and water differ by only about 10%. They applied a correction factor to the mass transfer correlation because of differences in diffusivity. The magnitude of $(k_x \cdot C_L/h_i)$ was double Turner's value because the Lewis number for the NaCl-water system is about twice that for LNG. For large values of the stability parameter they anticipated molecular mass transfer; for smaller values they expected eddy diffusion to become predominant.

Germeles [6] accepted Turner's and Huppert's correlations, but he adjusted them for the effect of a change in Lewis numbers between the saline solutions and LNG. He took the plateau value of $(R \cdot k_x \cdot C_L/h_i)_{\text{large } R}$ to be inversely proportional to the Lewis number. The critical stability parameter was accepted to be $R_c = 2$. Thus, he put

$$\begin{aligned}(k_x \cdot C_L/h_i) &= 1.80R^{-1} - 0.80 \qquad (1 \le R \le 2)\\ &= 0.20R^{-1} \qquad (R \ge 2)\end{aligned} \tag{5}$$

Heestand et al. [2] found that in attempts to use the thermohaline model of interfacial energy and mass transfer, with the initial conditions specified by Sarsten [3] for the LaSpezia incident predicted mass transfer rates were essentially negligible until the cell densities became nearly equal ($R \cong 1$). On this basis the time period needed to attain equal densities (beginning at the start of cargo transfer) depended wholly on the heat transfer rate and was approximately three times the reported period. This difference was far too large to be attributed to errors in the data. Furthermore, it seemed unreasonable that mass transfer was not playing an active role at an early stage of the incubation period. These considerations led Heestand et al. [2] to seek a more appropriate model for the transfer processes. They assumed that transfer of heat or mass from either cell to the interface is under fully turbulent conditions, that is, the transfer coefficients are related through the Reynolds' analogy: $h_i/(k_i \cdot C_L) = 1$. There is a turbulent convective film at each interface, one from each cell, and the motion is characterized by the Grashof number,

$$N_{Gr} = (gL^3\,\Delta\rho/\langle\rho\rangle\nu^2)$$

Note that there are contributions to $\Delta\rho$ both from temperature variations and from composition differences; only when composition is uniform does this expression reduce to the usual heat transfer Grashof number, $gL^3\beta\,\Delta T/\langle\rho\rangle\nu^2$. Heestand et al. believed that McAdams' correlation [10] for the heat transfer coefficient for a single, heated plate facing upward (or cooled plate facing downward) should apply to each convective film at the interface. However, they believed the Prandtl number played no role in the process in the LNG system. Accordingly, they removed the Prandtl number influence from the correlation, which was developed from data on air ($N_{Pr} = 0.7$). The modified correlation assumed the form

$$hL/k = 0.1243(N_{Gr})^{1/3} \tag{6}$$

If this expression gives the heat transfer coefficient for each convective film at an interface and the interfacial density is intermediate to the densities of the two bulk cells, then the overall coefficient of heat transfer at an interface would be $(2)^{-4/3}$ the value for a single, heated plate. (See Appendix B for a demonstration of this assertion). The heat transfer coefficient predicted for the cell to cell heat transfer from the modified McAdams correlation is then

$$h_i = 0.0493k(g\,\Delta\rho/\langle\rho\rangle\nu^2)^{1/3} \tag{7}$$

An alternative for the intercellular heat transfer coefficient is provided by the correlation of Globe and Dropkin [11] for transfer through the fluid placed between two horizontal plates and heated from below. Here the two convective films are already accounted for by the correlation:

$$\begin{aligned} hL/k &= 0.069(N_{Pr})^{0.407}(N_{Gr})^{1/3} \\ &= 0.0597(N_{Gr})^{1/3} \end{aligned} \tag{8}$$

The predicted heat transfer coefficient is then

$$h_i = 0.0597k(g\,\Delta\rho/\langle\rho\rangle\nu^2)^{1/3} \tag{9}$$

Heestand et al. [2] selected a coefficient value intermediate to those of the modified McAdams correlation and the modified Globe and Dropkin correlation in their final equation for predicting intercellular heat transfer. Thus, they selected the equation

$$h_i = 0.0553k(g\,\Delta\rho/\langle\rho\rangle\nu^2)^{1/3} \tag{10}$$

Composition of LNG Mixtures

Both Chatterjee and Geist [5] and Germeles [6] considered LNG to be a mixture of methane and nonvolatile heavy component. This simplification required changing the LaSpezia initial conditions as reported by Sarsten [3] to match temperatures and vapor pressures. These adjustments, described in detail by Germeles [6], have the effect of decreasing the time to rollover in both models. Chatterjee and Geist [1] considered LNG to contain methane, ethane, and nitrogen. This is an improvement because it recognizes that the presence of relatively small amounts of nitrogen increases both the vapor pressure and the density of the liquid. However, in order to match the temperatures, vapor pressures and densities given by Sarsten [3], three is not a sufficient number of species.

Heestand, et al. [2] examined the initial conditions for the LaSpezia incident. The reader should refer to Appendix A for a presentation of these details, and Table A-1 should be seen, in particular. From these data it is evident that temperature differences of the order of 5 K and liquid density differences of less than 0.7% are important. Thus, any approximation to the actual composition must permit matching of temperature and densities with an accuracy consistent with these data. For accurate modeling of the effects of boil-off on the composition good estimates of the vapor pressures are also required. Finally, even though nitrogen is a minor component of most LNG's, its effect on the vapor pressure and density is significant. Therefore, any approximation to the actual composition must include enough species to match liquid densities and vapor pressures at specified temperatures and heats of vaporization whether or not nitrogen is present. It was found that this can be accomplished by using methane, ethane, propane, n-butane, and nitrogen.

Criteria for Rollover

Chatterjee and Geist [5] used equality of temperature and composition as the criteria for rollover. However, there could be no sudden surge in boil-off if one mixes two cells of the same composition at the same temperature. The only reason Chatterjee and Geist [5] found an increase in boil off in their model was that it used small tolerances for equality of composition ($\Delta x(CH_4) < 0.002$) and ($\Delta T < 0.0278$ K). Properly, Germeles [6] used equality of cell densities as a criterion for rollover. He also discussed other inconsistencies in the Chatterjee and Geist [5] model. The later work of Chatterjee and Geist [1] did not directly address the issue of selection of criteria for rollover.

Heestand et al. [2] agreed with equal cell densities as the rollover criterion. They computed densities by interpolation within the tabulations of density, composition, and temperature prepared by Boyle [12].

Mass Transfer between Cells

Chatterjee and Geist [5] assumed there should be no net molar flux across the interface between liquid cells. Their later paper [1] did not state any change in position on this matter. Germeles [6] dealt with the mass transfer modeling on a mass-basis rather than a molar-basis; his model fixed the volumes of all cells except the uppermost one. Heestand et al. [2] took the interface stability between cells to be a purely hydrodynamic phenomenon. Hence, equimolar counter-diffusion was assumed to have been obtained.

Conditions at the Vapor-Liquid Interface

Huntley [13] conducted some experiments in a dewar flask partially filled with liquid nitrogen. He correlated bulk temperature and pressure rise (closed flask) with time, in order to evaluate the heat leak rate. The measured pressure rise was rather larger than expected from the temperature rise. In the course of seeking an explanation, he measured the vertical temperature profile and discovered the surface was 2.1°F (1.17°C) hotter than the bulk average liquid. The measured and calculated pressure rises agreed when surface temperature/vapor pressure was used instead of bulk temperature/vapor pressure. Stirring the liquid eliminated the temperature gradient and the differences in pressure rise. He also found that the surface temperature elevation and rate of pressure rise was increased by an increase of the relative heat flow to the interface. In answer to a question Huntley indicated that for a vented dewar . . . "the liquid below the surface was 0.3°F (0.17°C) warmer than the surface. The change in temperature occurred very near to the surface." There is good reason to expect similar behavior for a vented dewar flask and an LNG storage tank, at least a priori, although there is a great difference in their linear scales, surface-to-volume ratio, and so forth.

Hashemi and Wesson [14] proposed a mathematical model to predict the evaporation rate of a pure liquid. From experiments on water and other liquids they came to recognize a zone at the free surface, about two millimeters thick, across which heat transfer occurred to supply the latent heat of vaporization from the bulk liquid. They understood the surface temperature to be "extremely close to the equilibrium saturation temperature of the vapor phase," and that evaporation occurs "essentially on the liquid surface with no visible bubble formation." Their model assumed turbulent heat transfer between two horizontal, parallel plates with heat flux upwards. The model was applied to "LNG," without regard to composition and possible variations thereof. Their formulation for predicted boil-off rate was

$$M_V'' = 0.13(2)^{4/3}\frac{k}{\lambda}\cdot(\beta\cdot g/\nu\kappa)^{1/3}(T_L - T_S)^{4/3} \tag{11}$$

They admitted the possibility of radiant heat transfer to the surface layer "from the tank ceiling and unwetted walls."

Both the Chatterjee and Geist [5] and Germeles [6] models used the Hashemi-Wesson [14] description of boil-off rate as basis for predicting vaporization from the top layer. Chatterjee and Geist [5] assumed the vaporizing film to be at the same temperature as the uppermost cell. Germeles [6] used the ideal solution law and the Clausius-Clapeyron equation for describing the vapor-liquid equilibrium. Chatterjee and Geist [1] apparently used K-values for their three-component mixture, but insufficient details were given to be clear about how the vapor-liquid equilibrium was calculated. It is not known whether the K-values were constant or variable with temperature and/or composition. Models such as these predict a normal boil-off from the LaSpezia tank which disagree with the data by a factor of about two. This difference affects the temperature and density history of the top cell because the major portion of the energy of vaporization is supplied by that cell. Solutions containing nitrogen are quite non-ideal, and the presence of even small amounts of nitrogen exerts a critical role on the vaporization. For example, K-values for the LaSpezia rollover incident are approximately $K_1 = 1.4$ to 1.6 and $K_5 = 58$ to 66, respectively, for methane (liquid-phase mol fraction about 0.63) and nitrogen (liquid-phase mol fraction about 0.002). Heestand et al. [2] treated the vapor-liquid equilibrium by the method of Prausnitz and Chueh [15] as extended by Zudkevitch and Joffe [16], Joffe et al. [17], Soave [18], Hirata et al. [19], Kato et al. [20] and Valencia-Chávez [21]. In this method the Soave-Redlich-Kwong equation of state was applied to each species in the mixture and fugacity coefficients were evaluated as functions of temperature and composition. Direct equating of fugacities for each species then led to the vapor-liquid equilibrium state.

In order to maintain a proper accounting of the composition and temperature of the top cell and to describe "weathering" of the stored LNG, an accurate model of the composition and rate of evolution of the vapor is needed. Heestand et al. [2] assumed that the vapor evolved is in equilibrium with an arbitrarily thin film of liquid on the surface of the top cell. Vapor-liquid equilibrium was calculated as just described and a modified Hashemi-Wesson [14] model was chosen for the evolution rate, at least for the larger tanks of interest. Energy is transmitted to the film from the vapor space above it at rate q′ and energy and mass are exchanged between the film and the liquid below it. The concept is shown schematically in Figure 2. The appropriate energy and material balances are:

$$(M'' + M_V'')H_N - M''H_S - M_V''H_V + q' = 0 \tag{12}$$

$$(M'' + M_V'')x_N(j) - M''x_S(j) - M_V''y_S(j) = 0 \qquad (j = 1, 2, \ldots, 5) \tag{13}$$

M'' is the "Rayleigh flow," mass interchange between the bulk liquid and the film; there is no mass or energy accumulation in the film. In addition to these balance equations, the equilibrium between the surface liquid and the evolved vapor relates H_V and the $y_S(j)$ to H_N and the $x_S(j)$. The temperature of the film must be such that the vapor pressure matches the tank pressure. Then, the rate of vapor evolution must be

$$M_V'' = \left[\frac{M''(H_N - H_S) + q'}{(H_V - H_N)}\right] \tag{14}$$

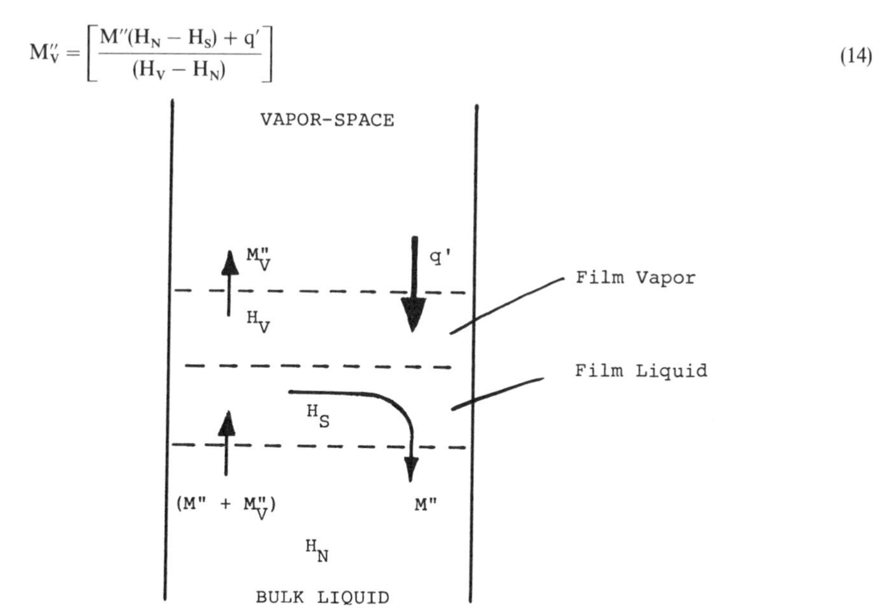

Figure 2. Boil-off from the interfacial film showing the circulatory Rayleigh flow between bulk liquid and film.

provided that M''_V, as calculated from this expression, is positive. A negative value would imply condensation, and the material and energy balances previously shown would not be correct for that case.

The Rayleigh flow was taken to be

$$M'' = 0.3276 \frac{k}{C_B} (g|\Delta\rho|/\nu\kappa\langle\rho\rangle)^{1/3} \tag{15}$$

after Hashemi and Wesson [14]. The absolute value of the density difference rather than the temperature difference has been used; and the numerical coefficient is that given by Hashemi and Wesson [14].

The value of q' in the vapor evolution rate expression is of some importance and difficult to estimate from first principles owing to the hydrodynamic stability of the mass in the vapor space, the venting of the vapor, and the transpiration of vapor from the surface of the liquid. The value of q' is not negligible, however, because the energy required for the normal boil-off rate is greater than the heat flux through the wetted side walls and the bottom of the tank. An energy balance on the vapor space, employing the heat fluxes q_t and q_{WV} (see below) and the normal boil-off rate, indicates that unless about 95% of the heat transmitted to the vapor space is transferred in turn to the liquid phase, the computed temperature of the vapor space becomes unreasonably high. Accordingly, Heestand et al. [2] assumed that 95% of the heat transmitted to the vapor space was transmitted to the liquid film of the top cell. The dynamics of the rollover are sensibly unaffected by a reduction of this percentage to 85%, and it is mainly the vapor space temperature history that is changed.

Material and Energy Balances on the Cells

The dynamic behavior of each cell of liquid and the vapor space is described by material and energy balances. These balances include cargo transfer and liquid recirculation. They are as follows:

Cell 1

$$\frac{d}{dt}[C_1L_1x_1(j)] = (M'_1/A)x'_1(j) - (M'_{R1}/A)x_1(j) + k_1[x_2(j) - x_1(j)] \tag{16}$$

$$\frac{d}{dt}[C_1L_1C_{L1}(T_1 - T_0)] = (M'_1/A)C_{C1}(T'_1 - T_0) + q_b + (q_{WL}\pi DL_1/A) + h_1(T_2 - T_1) \tag{17}$$

Cells i = 2 to N − 1

$$\frac{d}{dt}[C_iL_ix_i(j)] = k_{i-1}[x_{i-1}(j) - x_i(j)] + k_i[x_{i+1}(j) - x_i(j)] \tag{18}$$

$$\frac{d}{dt}[C_iL_iC_{Li}(T_i - T_0)] = (q_{WL}\pi DL_i/A) + h_{i-1}(T_{i-1} - T_i) + h_i(T_{i+1} - T_i) \tag{19}$$

Cell N (Top Cell)

$$\frac{d}{dt}[C_NL_Nx_N(j)] = (M'_Nx'_N(j)/A) - M''_Vy_S(j) + (M'_{RN}x_{RN}(j)/A) + k_{N-1}[x_{N-1}(j) - x_N(j)] \tag{20}$$

$$\frac{d}{dt}[C_NL_NC_{LN}(T_N - T_0)] = (q_{WL}\pi DL_N/A) + h_{N-1}(T_{N-1} - T_N) + f_Q(Q/A) - M''_VH_V + (M'_N/A)C_{CN}(T'_N - T_0) \tag{21}$$

where $Q = q_t A + q_{WV}\pi D L_{VS}$ and L_{VS} is the height of the vapor space, defined as $L_{VS} = L_{TANK} - \sum_{i=1}^{N} L_i$.

Vapor Space

$$\frac{d}{dt}[M_{VS}y_{VS}(j)] = M'_V y_S(j) + M'_F y_F(j) - M'_{VENT} y_{VS}(j) \tag{22}$$

$$\frac{d}{dt}\left\{\sum_{j=1}^{5} M_{VS}y_{VS}(j)[\lambda_j + C_V(j)(T_{VS} - T_0)]\right\} = \sum_{j=1}^{5} M'_V y_S(j)[\lambda_j + C_V(j)(T_S - T_0)] + \sum_{j=1}^{5} M'_F y_F(j)[\lambda_j + C_V(j)(T_F - T_0)] + (1 - f_Q)Q - \sum_{j=1}^{5} M'_{VENT} y_{VS}(j)[\lambda_j + C_V(j)(T_{VS} - T_0)] \tag{23}$$

An alternative form of the vapor space energy balance can be obtained in the following way. Multiply each species material balance by $[\lambda_j + C_V(j)(T_{VS} - T_0)]$, sum over species and subtract from the energy equation given above. This gives the following equation:

$$\sum_{j=1}^{5} M_{VS}Y_{VS}(j)C_V(j)\frac{dT_{VS}}{dt} = \sum_{j=1}^{5} M'_V Y_S(j)C_V(j)(T_S - T_{VS}) + \sum_{j=1}^{5} M'_F Y_F(j)C_V(j)(T_F - T_{VS}) + (1 - f_Q)Q \tag{24}$$

The average heat capacities of the vaporizing material, vapor-space material, and flash vapor, respectively, are

$$C_V = \sum_{j=1}^{5} Y_S(j)C_V(j)$$

$$C_{VS} = \sum_{j=1}^{5} Y_{VS}(j)C_V(j) \tag{25}$$

$$C_{VF} = \sum_{j=1}^{5} Y_F(j)C_V(j)$$

Hence

$$\frac{dT_{VS}}{dt} = \frac{M'_V C_V(T_S - T_{VS}) - M'_F C_{VF}(T_F - T_{VS}) + (1 - f_Q)Q}{M_{VS}C_{VS}} \tag{26}$$

Application of the Balance and Rate Equations

A computer program (ROLLO) was written to integrate the material and energy balance equations, given in the previous section, over time from given initial conditions within an LNG tank. Heat and mass transfer coefficients are calculated from the equations discussed in the first section. The rate of vaporization is predicted by the modified Hashemi and Wesson [14] expression as discussed earlier. Numerical integration of the balance equations is a fourth-order Runge-Kutta procedure. Cargo transfer, recirculation with or without flashing, and liquid withdrawal from the tank can be started and stopped at specified times. The program output is values of temperatures, compositions, densities, transfer coefficients, and so forth, at hourly intervals.

Table 1
Initial Conditions Used in ROLLO to Simulate the LaSpezia Incident

Components	Initial Heel	*Esso Brega* Cargo
Methane	63.62 (mol-%)	62.26 (mol-%)
Ethane	24.16	21.85
Propane	9.36	12.66
n-Butane	2.51	3.21
Nitrogen	0.35	0.02
	100.00	100.00
Vapor Pressure ($N \cdot m^{-2}$)	*3,923.0*	*16,280*
Temperature (K)	*114.355*	*118.997*
Density ($kg \cdot m^{-3}$)	*537.316 (calc'd)*	*541.118 (calc'd)*
Vapor-space Temperature (K)	*122.039 (assumed)*	
Initial Heel Depth (m)	*5.029*	
Vapor-space Height before filling (m)	*20.422*	
Vapor-space Composition	*95.0 mol-% methane and 5.0 mol-% nitrogen (assumed)*	

Heat-Leak Rates	*Location of Heat Leak*	*Initial Total Heat Flux*
$q_b = 20.82\ W\,m^{-2}$	*Tank bottom*	*39,390 W*
$q_{wL} = 6.94\ W\,m^{-2}$	*Tank walls to liquid*	*19,080 W (cargo)*
		5,390 W (heel)
$q_{WV} = 6.94\ W\,m^{-2}$	*Tank walls to vapor*	*4,250 W*
$q_t = 15.77\ W\,m^{-2}$	*Vapor dome*	*29,840 W*
		Total = 97,950 W

Results for the LaSpezia Incident

The initial conditions used for simulation of the LaSpezia incident are shown in Table 1, which should be compared with Table A-1 in Appendix A of this chapter. The latter tabulation shows the information given by Sarsten [3]. It can be seen that the butane-pentane components have been combined and termed "n-butane" in Table 1. The densities listed in this tabulation are computed values from the Boyle [12] data. They are about 0.8% smaller than the values reported by Sarsten [3]. The sort of agreement with Sarsten's observations that was found from the ROLLO program are fully described by Heestand et al. [2]. It is recommended that the reader refer to that paper, noting especially Figures 3 through 8. Those are representative predictions from ROLLO. To account for heat transfer and pumping power given to the cargo during transfer from the Esso Brega an increment of 0.278 K (0.5°F) was assumed. Thus, the delivery temperature to the tank was taken as 118.997 K, whereas the saturation temperature calculated from a vapor pressure of 16,280 $N \cdot m^{-2}$ is 118.719 K.

Table 2 presents the calculated histories during the cargo transfer period of the vapor-space temperature (T_{VS}), the equilibrium film temperature (T_S), the temperature difference between liquid in Cell 1 and liquid in Cell 2 (ΔT_{12}), the temperature difference between the uppermost cell and the vaporizing film (ΔT_{film}), the heat transfer coefficient between Cell 1 and Cell 2 liquids (h_1), the Rayleigh flow (3,600 M″) and the boil-off rate (3,600 $M_V''A$).

Table 3 extends the histories in Table 2 from the end of the cargo transfer period (13.0 hours) until 80 hours after the start of cargo transfer. The double entry at 30.15 hours represents conditions just prior to rollover and at the peak instant of rollover, respectively.

Table 2
Calculated Dynamics of the LaSpezia Incident (ROLLO)

Time (hours)	T_{VS} (K)	T_S (K)	ΔT_{12} (K)	ΔT_{film} (K)	h_1 ($W \cdot m^{-2} \cdot K^{-1}$)	Rayleigh flow ($kg\ mol \cdot hr^{-1} \cdot m^{-2}$)	Boil-off ($kg\ mol \cdot hr^{-1}$)
2.0	122.03	114.61	3.72	0.082	89.55	16.54	45.06
4.0	121.97	114.84	3.49	0.114	86.56	17.97	54.65
6.0	121.86	115.05	3.29	0.135	83.35	18.78	60.68
8.0	121.70	115.24	3.12	0.149	80.05	19.27	64.17
10.0	121.49	115.41	2.97	0.157	76.68	19.57	65.92
12.0	121.19	115.55	2.84	0.163	73.26	19.74	66.49
13.0	120.99	115.62	2.78	0.165	71.52	19.79	66.43

Cargo delivery completed at 13.0 hours.

Table 3
Calculated Dynamics of the LaSpezia Incident (ROLLO)

Time (hours)	T_{VS} (K)	T_S (K)	ΔT_{12} (K)	ΔT_{film} (K)	h_1 ($W \cdot m^{-2} \cdot K^{-1}$)	Rayleigh flow ($kg\ mol \cdot hr^{-1} \cdot m^{-2}$)	Boil-off ($kg\ mol \cdot hr^{-1}$)
14.0	120.77	115.68	2.70	0.165	69.25	19.82	66.49
16.0	120.41	115.80	2.55	0.164	64.65	19.79	65.81
18.0	120.13	115.89	2.42	0.161	59.96	19.69	64.35
20.0	119.93	115.97	2.32	0.156	55.11	19.52	62.33
22.0	119.79	116.04	2.23	0.150	49.97	19.31	59.84
24.0	119.70	116.10	2.16	0.142	44.36	19.03	56.91
26.0	119.65	116.15	2.10	0.133	37.88	18.69	53.46
28.0	119.65	116.19	2.07	0.122	29.53	18.24	49.25
30.0	119.70	116.22	2.05	0.105	11.47	17.52	43.31
30.15	119.70	116.22	2.05	0.103	3.76	17.42	42.608
30.15	119.70	116.81	—	1.113	—	36.49	634.27
32.0	117.78	116.81		0.984		35.03	539.90
36.0	117.28	116.81	—	0.767	—	32.28	391.74
40.0	117.35	116.82		0.612		29.99	293.77
44.0	117.48	116.82		0.498		28.05	226.91
48.0	117.63	116.83		0.413		26.40	180.05
52.0	117.80	116.83		0.347		24.99	146.41
56.0	117.98	116.84		0.297		23.78	121.77
60.0	118.17	116.84		0.257		22.75	103.41
64.0	118.37	116.84		0.226		21.86	89.50
68.0	118.57	116.85		0.200		21.09	78.82
72.0	118.78	116.85		0.180		20.43	70.52
76.0	118.98	116.86		0.164		19.87	64.00
80.0	119.18	116.86		0.150		19.38	58.83

Table 4
Calculated Dynamics of the LaSpezia Incident (ROLLO)

Time (hours)	$\Delta\rho_{12}$ ($kg \cdot m^{-3}$)	$\Delta\rho_{film}$ ($kg \cdot m^{-3}$)	K_1	K_5	ρ_{film} ($kg \cdot m^{-3}$)	x_1 (top cell)	x_1 (film)
2.0	2.855	0.208	1.303	58.63	537.87	0.6351	0.6348
4.0	2.579	0.268	1.328	59.35	538.25	0.6341	0.6338
6.0	2.303	0.307	1.351	60.01	538.60	0.6332	0.6329
8.0	2.041	0.333	1.372	60.58	538.92	0.6324	0.6320
10.0	1.795	0.349	1.390	61.12	539.21	0.6317	0.6312
12.0	1.565	0.360	1.407	61.58	539.48	0.6310	0.6305
13.0	1.457	0.363	1.414	61.78	539.60	0.6307	0.6302

Cargo delivery completed at 13.0 hours.

Table 4 presents the calculated histories during the cargo transfer period of the density difference between liquid in Cell 1 and liquid in Cell 2 ($\Delta\rho_{12}$), the density difference between liquid in the film and in the uppermost cell ($\Delta\rho_{film}$), the volatility of methane at the vaporizing film conditions (K_1), the volatility of nitrogen at the vaporizing film conditions (K_5), the density of the liquid film (ρ_{film}), the mol fraction of methane in the top cell liquid (x_1) and the mol fraction of methane in the liquid film (x_1).

Table 5 extends the histories in Table 4 from the end of the cargo transfer period (13.0 hours) until 80 hours after the start of cargo transfer. The double entry at 30.15 hours represents conditions just prior to rollover and at the peak instant of rollover, respectively.

Table 5
Calculated Dynamics of the LaSpezia Incident (ROLLO)

Time (hours)	$\Delta\rho_{12}$ ($kg \cdot m^{-3}$)	$\Delta\rho_{film}$ ($kg \cdot m^{-3}$)	K_1	K_5	ρ_{film} ($kg \cdot m^{-3}$)	x_1 (top cell)	x_1 (film)
14.0	1.322	0.365	1.421	61.99	539.72	0.6304	0.6299
16.0	1.076	0.364	1.434	62.35	539.93	0.6298	0.6293
18.0	0.859	0.359	1.445	62.67	540.11	0.6293	0.6288
20.0	0.667	0.351	1.455	62.95	540.27	0.6289	0.6284
22.0	0.497	0.340	1.463	63.19	540.41	0.6285	0.6280
24.0	0.348	0.326	1.470	63.36	540.53	0.6282	0.6277
26.0	0.217	0.309	1.476	63.54	540.63	0.6279	0.6274
28.0	0.103	0.287	1.481	63.67	540.70	0.6276	0.6272
30.0	0.006	0.255	1.484	63.77	540.76	0.6274	0.6270
30.15	0.000	0.250	1.484	63.76	540.76	0.6274	0.6270
30.15	—	2.330	1.557	66.08	542.85	—	0.6220
32.0		2.063	1.557	66.01	542.86		0.6220
36.0	—	1.616	1.558	66.10	542.87	—	0.6219
40.0		1.296	1.558	66.17	542.88		0.6219
44.0		1.061	1.559	66.07	542.89		0.6218
48.0		0.885	1.559	66.11	542.90		0.6218
52.0		0.751	1.560	66.14	542.91		0.6217
56.0		0.648	1.560	66.17	542.92		0.6217
60.0		0.567	1.561	66.19	542.93		0.6216
64.0		0.503	1.561	66.21	542.94		0.6215
68.0		0.452	1.562	66.23	542.95		0.6215
72.0		0.411	1.562	66.26	542.95		0.6214
76.0		0.378	1.563	66.29	542.96		0.6214
80.0		0.351	1.563	66.32	542.97		0.6213

Table 6
Calculated Dynamics of the LaSpezia Incident (ROLLO)

Time (hours)	x_1 (bottom cell)	x_5 (top cell)	x_5 (film)	x_5 (bottom cell)	Δx_1	Δx_5
2.0	0.6246	0.003189	0.002945	0.000671	2.77×10^{-4}	2.44×10^{-4}
4.0	0.6244	0.002909	0.002660	0.000630	3.34	2.49
6.0	0.6243	0.002652	0.002409	0.000600	3.79	2.43
8.0	0.6242	0.002419	0.002190	0.000573	4.13	2.29
10.0	0.6241	0.002211	0.001997	0.000549	4.38	2.14
12.0	0.6240	0.002026	0.001829	0.000527	4.55	1.97
13.0	0.6240	0.001941	0.001753	0.000517	4.62	1.88

Cargo delivery completed at 13.0 hours.

Table 6 presents the calculated histories during the cargo transfer period of mol fraction of methane in the bottom cell (x_1), the mol fraction of nitrogen in the top cell (x_5), the mol fraction of nitrogen in the liquid film (x_5), the mol fraction of nitrogen in the bottom cell (x_5), the difference of mol fractions of methane in the top cell and the film (Δx_1), and the difference of mol fractions of nitrogen in the top cell and the film (Δx_5).

Table 7
Calculated Dynamics of the LaSpezia Incident (ROLLO)

Time (hours)	x_1 (bottom cell)	x_5 (top cell)	x_5 (film)	x_5 (bottom cell)	Δx_1	Δx_5
14.0	0.6240	0.001862	0.001681	0.000531	4.69×10^{-4}	1.81×10^{-4}
16.0	0.6242	0.001721	0.001554	0.000554	4.80	1.67
18.0	0.6243	0.001600	0.001446	0.000574	4.83	1.54
20.0	0.6243	0.001496	0.001354	0.000589	4.81	1.42
22.0	0.6244	0.001406	0.001276	0.000602	4.75	1.30
24.0	0.6244	0.001330	0.001211	0.000612	4.65	1.19
26.0	0.6245	0.001265	0.001156	0.000620	4.51	1.09
28.0	0.6245	0.001211	0.001112	0.000626	4.30	9.9×10^{-5}
30.00	0.6245	0.001170	0.001081	0.000630	3.96	8.9
30.15	0.6245	0.001167	0.001080	0.000630	3.92	8.7
30.15	0.6252	—	0.000471	0.000750	3.15×10^{-3}	2.79×10^{-4}
32.0	0.6248		0.000470	0.000716	2.80	2.46
36.0	0.6242		0.000466	0.000659	2.21	1.93
40.0	0.6237		0.000462	0.000618	1.79	1.56
44.0	0.6233		0.000459	0.000586	1.48	1.27
48.0	0.6230		0.000455	0.000561	1.25	1.06
52.0	0.6228		0.000451	0.000542	1.07	9.1×10^{-5}
56.0	0.6226		0.000447	0.000526	9.4×10^{-5}	7.9
60.0	0.6224		0.000443	0.000512	8.35	6.9
64.0	0.6223		0.000439	0.000501	7.54	6.2
68.0	0.6222		0.000435	0.000491	6.88	5.6
72.0	0.6221		0.000431	0.000482	6.36	5.1
76.0	0.6220		0.000427	0.000475	5.95	4.8
80.0	0.6219		0.000423	0.000468	5.61	4.5

Table 7 extends the histories in Table 6 from the end of the cargo transfer period (13.0 hours) until 80 hours after the start of cargo transfer. The double entry at 30.15 hours represents conditions just prior to rollover and at the peak instant of rollover, respectively.

RESULTS OF PARAMETRIC STUDIES

Some understanding of the rollover process can be achieved with parametric studies. A listing of major variables which should be examined for such studies would include the following:

1. Changes of tank dimensions (but preserving dimensional similarity).
2. Changes in relative volumes of heel and cargo.
3. Decrease in heat leak rates (but preserving the distribution to the various zones).
4. Increase or decrease of the nitrogen content of the cargo.
5. Increase or decrease of the methane content of the cargo.
6. Increase or decrease of the cargo temperature.

Each of these cases was examined for its quantitative influence on rollover time, boil-off just prior to rollover, peak boil-off, Rayleigh flow just prior to rollover, and Rayleigh flow at peak boil-off. Tables 8 through 13 present the results calculated from the program ROLLO. The results are all presented as normalized ratios, using the standard values of the LaSpezia rollover as denominators. These pertinent LaSpezia values are:

$$\text{Rollover time} = 30.15 \text{ hours}$$

$$\text{Boil-off just prior to rollover} = 42.61 \text{ kg mol}\cdot\text{hr}^{-1}$$

$$\text{Peak boil-off} = 634.3 \text{ kg mol}\cdot\text{hr}^{-1}$$

$$\text{Rayleigh flow just prior to rollover} = 17.42 \text{ kg mol}\cdot\text{hr}^{-1}\cdot\text{m}^{-2}$$

$$\text{Rayleigh flow at peak boil-off} = 36.49 \text{ kg mol}\cdot\text{hr}^{-1}\cdot\text{m}^{-2}$$

It is interesting to compare the boil-off rates per unit area with the Rayleigh flows. The cross-sectional area of the LaSpezia tank is

$$A = \pi/4(D)^2 = \pi/4(49.07)^2 = 1{,}891.13 \text{ m}^2$$

Then,

$$\frac{(\text{boil-off rate})/(\text{tank area})}{(\text{Rayleigh flow})} = 0.00129 \qquad \text{just prior to rollover}$$

$$= 0.00919 \qquad \text{at peak boil-off}$$

Table 8 presents the influence of tank size on the time to reach rollover, boil-off just prior to rollover, peak boil-off, Rayleigh flow just prior to rollover, and Rayleigh flow at peak boil-off. Each quantity has been normalized by its corresponding LaSpezia value. Tank shape has been preserved by scaling all heights and diameter by the same factor. The same cargo pumping rate as that used at LaSpezia has been assumed.

Table 9 presents the influence of changes in the relative volumes of heel and cargo (in the LaSpezia tank) upon the same quantities tabulated in Table 8.

Table 10 presents the influence of the heat leak rates upon the same quantities tabulated in Table 8. In each case the heat leak rates to the various zones of the tank were changed by a uniform factor. The LaSpezia tank was taken to be the storage vessel.

Table 8
Influence of Tank Dimension Changes

Normalized Tank Dimensions	Normalized Rollover Time	Normalized Boil-off Prior to Rollover	Normalized Peak Boil-off	Normalized Rayleigh Flow Prior to Rollover	Normalized Rayleigh Flow at Peak Boil-off
0.50*	0.477	0.249	0.251	0.9989	1.0012
0.75†	0.725	0.562	0.564	0.9996	1.0008
1.0‡	1.000	1.000	1.000	1.000	1.000
1.50§	1.764	2.207	2.231	0.9946	0.9981

* *Tank volume decreased by a factor of 8.*
† *Tank volume decreased by a factor of* $\frac{64}{27} = 2.37$.
‡ *LaSpezia tank.*
§ *Tank volume increased by a factor of* $\frac{27}{8} = 3.375$.

Table 9
Influence of Relative Heel and Cargo Volumes

Heel Fraction of Final Liquid Volume	Normalized Rollover Time	Normalized Boil-off Prior to Rollover	Normalized Peak Boil-off	Normalized Rayleigh Flow Prior to Rollover	Normalized Rayleigh Flow at Peak Boil-off
0.10	0.509	1.024	1.430	1.0100	1.0985
0.234*	1.000	1.000	1.000	1.000	1.000
0.36	1.492	0.920	0.629	0.9697	0.8868
0.50	1.808	0.849	0.376	0.9392	0.7740

* *LaSpezia case.*

Table 10
Influence of Heat Leak Rate

Heat Leak Rate Relative to the LaSpezia Rate	Normalized Rollover Time	Normalized Boil-off Prior to Rollover	Normalized Peak Boil-off	Normalized Rayleigh Flow Prior to Rollover	Normalized Rayleigh Flow at Peak Boil-off
0.25	1.066	0.719	0.901	0.9616	0.9767
0.50	1.041	0.812	0.935	0.9745	0.9849
1.00 (LaSpezia)	1.000	1.000	1.000	1.000	1.000

Table 11 presents the influence of nitrogen content of the cargo upon the same quantities tabulated in Table 8. In each instance the mol fraction of nonvolatiles (of constant relative proportions) in the cargo was decreased by an amount equal to the increase of nitrogen mol fraction. The LaSpezia tank was taken to be the storage vessel.

Table 12 presents the influence of an increase or decrease of methane mol fraction, at constant nitrogen mol fraction, upon the same quantities tabulated in Table 8. In each case, as methane

Table 11
Influence of Cargo Nitrogen Content

Factor of Increase of Nitrogen mol Fraction in cargo	Cargo Density* ($kg \cdot m^{-3}$)	Normalized Rollover Time	Normalized Boil-off Prior to Rollover	Normalized Peak Boil-off	Normalized Rayleigh Flow Prior to Rollover	Normalized Rayleigh Flow at Peak Boil-off
0.5	541.126	1.017	0.974	0.962	0.9923	0.9909
1.0 (LaSpezia)	541.118	1.000	1.000	1.000	1.000	1.000
2.0	541.088	0.965	1.053	1.075	1.0153	1.0175
4.0	541.042	0.905	1.160	1.225	1.0443	1.0494
8.0	540.920	0.806	1.384	1.523	1.0973	1.1045
16.0	540.692	0.667	1.836	2.108	1.1844	1.1908

* *For constant methane mol fraction of 0.6226, the densities of cargos where nitrogen is traded for nonvolatiles (with constant proportions) can be represented by the linear expression*

$$\rho \cong 541.147 - 142x_5 \qquad [0.0001 \leqq x_5 \leqq 0.0032]$$

Note that the LaSpezia heel density = 537.316 $kg \cdot m^{-3}$.

Table 12
Influence of Changes in Methane Mol Fraction

Increment of Methane Mol fraction	Cargo Density* ($kg \cdot m^{-3}$)	Normalized Rollover Time	Normalized Boil-off Prior to Rollover	Normalized Peak Boil-off	Normalized Rayleigh Flow Prior to Rollover	Normalized Rayleigh Flow at peak Boil-off
−0.06	555.064	1.876	0.640	0.394	0.8518	0.7765
−0.03	548.242	1.491	0.830	0.631	0.9406	0.8835
0.00 (LaSpezia)	541.118	1.000	1.000	1.000	1.000	1.000
+0.03	533.775	Immediate mixing	—	—	—	—

* *For constant nitrogen mol fraction of 0.0002, the densities of cargos where methane is traded for nonvolatiles (with constant proportions) can be represented by the linear expression*

$$\rho \cong 685.831 - 232.43x_1 \qquad [0.5626 \leq x_1 \leq 0.6826]$$

Note that the LaSpezia heel density $= 537.316\ kg \cdot m^{-3}$

Table 13
Influence of Changes in Cargo Delivery Temperature

Incremental Change of Cargo Temperature (K)	Cargo Density ($kg \cdot m^{-3}$)	Normalized Rollover Time	Normalized Boil-off Prior to Rollover	Normalized Peak Boil-off	Normalized Rayleigh Flow Prior to Rollover	Normalized Rayleigh Flow at Peak Boil-off
−0.555	541.890	1.243	0.777	0.6439	0.9202	0.8937
0.000 (LaSpezia)	541.118	1.000	1.000	1.000	1.000	1.000
+0.555	540.349	0.803	1.234	1.441	1.0668	1.0985

For a cargo of the composition of that for the LaSpezia incident the density may be approximated, for a temperature change, by the linear expression

$$\rho \cong 541.118 - 1.388(\Delta T)\ kg \cdot m^{-3} \qquad [-0.555\ K \geq (\Delta T) \geq +0.555\ K]$$

mol fraction was increased the mol fractions of ethane, propane and n-butane were decreased without changing their relative proportions. The LaSpezia tank was taken to be the storage vessel.

Table 13 presents the influence of an increase or decrease by 0.555 K (1.0°F) of the cargo delivery temperature upon the same quantities tabulated in Table 8. (In each case shown in Table 13 the standard increment 0.278 K (0.5°F) was added to the saturation temperature 118.719 K to allow for pumping power and heat transfer during cargo transfer.) The LaSpezia tank was taken to be the storage vessel.

APPENDIX A: DETAILS OF THE LASPEZIA INCIDENT

Complete details of the LaSpezia incident were given by Sarsten [3]; only a brief summary is given here. Some quantitative details of the cargo and heel and the heat leak rates are presented in Table A-1. The storage tank had a 49.07-meter (161-foot) diameter and a 26.82-meter (88-foot) total height. It contained an initial LNG heel of 5.029 meters (16.5 feet) at 114.4 K (−253.83°F) under a gauge pressure of about 250 mm water. Cargo was pumped from the tanker *Esso Brega*, which had been sitting in the harbor for more than a month. During this holding period the cargo had become heavier and hotter. The ship's storage compartment gauge pressure was about 1,100 mm water, and the cargo temperature was 119 K (−245.48°F) at delivery time. The heavier, hotter, more volatile cargo was bottom-filled, and only minimal mixing of heel and cargo occurred. Cargo transfer required about thirteen hours, and approximately 57,200 kg of vapor was displaced from the tank. This was about 25% more than expected, based on the displacement volume of transferred liquid. Presumably, the 45,360 kg of vapor expected was returned by compressor to the ship. The disposition of the extra vapor was not stated, but it may have been vented to the atmosphere. During the approximately 18-hour period between completion of transfer and rollover, normal boil-off of about 15,900 kg was observed. During the rollover, which persisted for about an hour and a quarter, approximately 136,000 kg vapor was discharged and another 13,600 kg flowed out vents of a connected tank during a fifteen minute interval. An additional 49,900 kg vapor was vented in a two-hour period after the tank safety valves closed and the tank returned to a normal pressure of 250 mm water gauge. The usual boil-off rate was reported to be about 900 $kg \cdot hr^{-1}$; hence, the boil-off rate during rollover was about 120 times the normal.

Table A-1
Initial Conditions for LaSpezia Rollover

Components	Initial Heel	*Esso Brega* Cargo
Methane	63.62 mol%	62.26 mol%
Ethane	24.16	21.85
Propane	9.36	12.66
n-Butane	1.45	1.94
iso-Butane	0.90	1.20
n-Pentane	0.05	0.01
iso-Pentane	0.11	0.06
Nitrogen	0.35	0.02
	100.00	100.00
Temperature (K)	*114.356**	*118.994**
Liquid depth (m)	*5.029*	*17.831*
Density ($kg \cdot m^{-3}$)	*541.742*	*545.586*
Vapor pressure ($N \cdot m^{-2}$)	*3,923.0*	*16,280*
Vapor space height after filling	*3.962 meters*	

* *Temperatures were derived from the Boyle [12] correlation for the composition and vapor pressures reported by Sarsten [3].*

APPENDIX B: DERIVATION FOR THE OVERALL HEAT TRANSFER COEFFICIENT AT THE INTERFACE BETWEEN TWO CELLS

Consider the heat transfer between two neighboring LNG cells. Let the lower cell have temperature T_1 and density ρ_1. The upper cell has temperature T_2 and density ρ_2. The interfacial temperature and density are T_i and ρ_i, respectively. It is assumed that ρ_i is the arithmetic average of ρ_1 and ρ_2. Each cell has a convective film for heat and mass transfer. The heat transfer coefficient for each film follows an equation of the form suggested by McAdams' correlation [10]:

$$hL/k = 0.1243(N_{Gr})^{1/3} \tag{6}$$

or

$$h_M = 0.1243k[g(\Delta\rho)/\langle\rho\rangle v^2]^{1/3}$$

(where the subscript M simply means the heat transfer coefficient is obtained from the McAdams correlation). Thus,

$$h_1 = 0.1243k[g(\Delta\rho)_1/\rho_i v^2]^{1/3} \tag{27}$$

is the heat transfer coefficient from cell 1 to interface and

$$h_2 = 0.1243k[g(\Delta\rho)_2/\rho_i v^2]^{1/3} \tag{28}$$

is the heat transfer coefficient from interface to cell 2. Note that

$$(\Delta\rho)_1 = (\rho_1 - \rho_i) = 1/2(\rho_1 - \rho_2) = 1/2(\Delta\rho) \tag{29}$$

and

$$(\Delta\rho)_2 = (\rho_i - \rho_2) = 1/2(\rho_1 - \rho_2) = 1/2(\Delta\rho) \tag{30}$$

Thus,

$$\begin{aligned} h_1 = h_2 &= (1/2)^{1/3}(0.1243)k[g(\Delta\rho)/\langle\rho\rangle v^2]^{1/3} \\ &= (1/2)^{1/3}h_M \end{aligned} \tag{31}$$

The heat transfer rate from cell 1 to interface equals the heat transfer rate from interface to cell 2:

$$q'' = h_1(T_1 - T_i) = h_2(T_i - T_2) \tag{32}$$

Eliminating the unknown interfacial temperature,

$$q'' = (T_1 - T_2)/(1/h_1 + 1/h_2) = h_i(T_1 - T_2) \tag{33}$$

It is then easily shown that

$$h_i \equiv (1/2)^{4/3}h_M \tag{34}$$

NOTATION

- A cross-sectional area of LNG tank, m^2
- $\langle C\rangle$ averaged molar liquid concentration between two neighboring cells, $kg\ mol \cdot m^{-3}$
- C_i molar concentration of cell i liquid, $kg\ mol \cdot m^{-3}$
- C_L molar liquid heat capacity (taken to be a function of composition but in

dependent of temperature), $J \cdot kg\ mol^{-1} \cdot K^{-1}$

$C_L(j)$ molar liquid heat capacity of species j, $J \cdot kg\ mol^{-1} \cdot K^{-1}$

C_B molar heat capacity of bulk liquid in uppermost cell, $J \cdot kg\ mol^{-1} \cdot K^{-1}$

C_{Ci} molar heat capacity of cargo transfer liquid to cell i, $J \cdot kg\ mol^{-1} \cdot K^{-1}$

C_{LF} molar heat capacity of a liquid having the same composition as the flash vapor, $J \cdot kg\ mol^{-1} \cdot K^{-1}$

C_{Li} molar heat capacity of cell i liquid, $J \cdot kg\ mol^{-1} \cdot K^{-1}$

C_{LR} molar heat capacity of the residue liquid from flashing, $J \cdot kg\ mol^{-1} \cdot K^{-1}$

C_p mass heat capacity at constant pressure, $J \cdot kg^{-1} \cdot K^{-1}$

$C_V(j)$ molar vapor heat capacity of species j, $J \cdot kg\ mol^{-1} \cdot K^{-1}$

C_{VF} molar heat capacity of flash vapor, $J \cdot kg\ mol^{-1} \cdot K^{-1}$

C_V molar heat capacity of vaporizing vapor, $J \cdot kg\ mol^{-1} \cdot K^{-1}$

C_{VS} molar heat capacity of vapor-space material $J \cdot kg\ mol^{-1} \cdot K^{-1}$

D diffusion coefficient, $m^2 \cdot s^{-1}$; tank diameter, m

f_Q fraction of the total heat transfer to the vapor-space which is passed along to the vaporizing film, dimensionless

f_V fraction of the feed liquid which flashes, dimensionless

f( ) function of the quantity in parentheses

g acceleration of gravity, $9.80665\ m \cdot s^{-2}$

h turbulent natural convention heat transfer coefficient through fluid confined between two parallel plates held at different temperatures and with heat flow upwards, $W \cdot m^{-2} \cdot K^{-1}$

h_i heat transfer coefficient across the interface between cell i and cell (i + 1) liquids, $W \cdot m^{-2} \cdot K^{-1}$

H specific enthalpy of liquid or vapor, $J \cdot kg\ mol^{-1}$

H_N enthalpy of the uppermost cell liquid, $J \cdot kg\ mol^{-1}$

H_S enthalpy of liquid at vaporizing film, $J \cdot kg\ mol^{-1}$

H_V specific enthalpy of vapor evolved from the vaporizing film, $J \cdot kg\ mol^{-1}$

i index used to identify cell numbers (cells are numbered upwards from the bottom of the tank: $i = 1, 2, \ldots, N$)

j index used to identify species: $j = 1, 2, \ldots, 5$

1 = methane 4 = n-butane
2 = ethane 5 = nitrogen
3 = propane

k thermal conductivity of LNG, 0.185 $W \cdot m^{-1} \cdot K^{-1}$

k_i turbulent mass transfer coefficient for all species across the interface between cell i and cell (i + 1), $kg\ mol \cdot m^{-2} \cdot s^{-1} \cdot \Delta(\text{mol fraction})^{-1}$

k_x turbulent mass transfer coefficient for all species, $kg\ mol \cdot m^{-2} \cdot s^{-1} \cdot \Delta(\text{mol fraction})^{-1}$

K_1 y_1/x_1, volatility of methane at vaporizing film conditions, dimensionless

K_5 y_5/x_5, volatility of nitrogen at vaporizing film conditions, dimensionless

L height, m

L_i height of cell i liquid, m

L_{TANK} overall height of tank, m

L' molar flow-rate of recirculating liquid or cargo transfer to be flashed in the vaporspace, $kg\ mol \cdot s^{-1}$

M_{VS} total mols of vapor in the LNG tank vapor-space, kg mol

M'_i cargo transfer flow-rate to cell i, $kg\ mol \cdot s^{-1}$

M'_F molar rate of addition of flash vapor to vapor-space, $kg\ mol \cdot s^{-1}$

M'_{VENT} molar venting rate from the vapor-space, $kg\ mol \cdot s^{-1}$

M'_{R1} recirculation liquid flow-rate from cell 1, $kg\ mol \cdot s^{-1}$

M'_{RN} recirculation liquid flow-rate to cell N, $kg\ mol \cdot s^{-1}$

M'_V molar vaporization rate from the vaporizing film, $kg\ mol \cdot s^{-1}$

M'' Rayleigh flow-rate (circulation current between the vaporizing film and the bulk liquid in uppermost cell), $kg\ mol \cdot m^{-2} \cdot s^{-1}$

M''_V vaporization rate at the vapor-liquid interface, $kg\ mol \cdot m^{-2} \cdot s^{-1}$

N total number of stratified liquid cells in LNG tank (also designates the top liquid cell in tank), dimensionless

N_{Le} Lewis number ($N_{Le} = N_{Sc}/N_{Pr}$), dimensionless

N_{Nu} Nusselt number ($N_{Nu} = hL/k$), dimensionless

N_{Pr} Prandtl number ($N_{Pr} = C_p\mu/k$), dimensionless

N_{Ra} Rayleigh number ($N_{Ra} = \beta\,\Delta T\, gL^3/\nu\kappa$), dimensionless

N_{Sc} Schmidt number ($N_{Sc} = \mu/\rho D$), dimensionless

q_b heat leak rate into bottom of LNG tank, $W \cdot m^{-2}$

q_{WL} heat leak rate into liquid side-wall of LNG tank, $W \cdot m^{-2}$

q_{WV} heat leak rate into vapor side-wall of LNG tank, $W \cdot m^{-2}$

q_t heat leak rate into vapor dome of LNG tank, $W \cdot m^{-2}$

q' rate of heat transfer from vapor-space to vaporizing film, $W \cdot m^{-2}$

Q total heat transfer rate to vapor-space from surroundings, W

R stability parameter, $R = (\alpha\,\Delta S/\beta\,\Delta T)$, the ratio of the stabilizing influence of a difference in salinity, ΔS, to the destabilizing influence of a temperature difference, ΔT, dimensionless

R_c critical value of the stability parameter

t time, s or h

T temperature, K

T_L bulk liquid temperature of uppermost cell, K

T_i' temperature of cargo transfer liquid fed to cell i, K

T_F temperature of flash vapor added to vapor space, K

T_s interfacial temperature between uppermost cell and the vapor space, K

T_0 reference temperature for zero liquid-phase enthalpy for all species, 99.82 K

T_{VS} vapor-space temperature, K

x liquid-phase mol fractions, dimensionless

$x_N(j)$ mol fraction of species j in the bulk liquid of the uppermost cell, dimensionless

$x_i'(j)$ mol fraction of species j in liquid cargo transfer to cell i, dimensionless

$x_i(j)$ liquid-phase mol fraction of species j in cell i, dimensionless

$x_R(j)$ mol fraction of species j in the residue liquid from flashing, dimensionless

$x_{R1}(j)$ mol fraction of species j in the recirculation flow from cell 1, dimensionless

$x_{RN}(j)$ mol fraction of species j in the recirculation flow to cell N, dimensionless

$x_S(j)$ mol fraction of species j in the liquid-phase at the vaporizing film, dimensionless

y vapor-phase mol fractions, dimensionless

$y_F(j)$ mol fraction of species j in the flash vapor, dimensionless

$y_S(j)$ mol fraction in vapor-phase of species j in vaporizing film, dimensionless

$y_{VS}(j)$ mol fraction of species j in the vapor-space, dimensionless

Greek Symbols

α fractional density increase per unit salinity difference, $m^3 \cdot kg^{-1}$

β thermal expansion coefficient for liquid phase (a function of temperature and composition), K^{-1}

$$\beta = -\rho^{-1}(\partial\rho/\partial T)_x$$

ΔS salinity difference, $kg \cdot m^{-3}$

ΔT temperature difference between two neighboring cells or between bulk liquid and film, K

ΔT_{12} difference between the temperature of cell 1 and cell 2, K

ΔT^*_{film} difference between the temperature of the uppermost cell and the vaporizing film, K

Δx difference of liquid-phase mol fractions, dimensionless

Δx_1 difference between liquid-phase mol fractions of methane in uppermost cell and film, dimensionless

Δx_5 difference between liquid-phase mol fractions of nitrogen in uppermost cell and film, dimensionless

$\Delta\rho$ difference of liquid density for two neighboring cells or between bulk liquid and film, $kg \cdot m^{-3}$

$\Delta\rho_{12}$ difference between the liquid densities of cell 1 and cell 2, $kg \cdot m^{-3}$

$\Delta\rho_{film}$ difference between the liquid densities of the film and the uppermost cell, $kg \cdot m^{-3}$

κ thermal diffusivity of LNG ($\kappa = k\langle C\rangle C_L$), $1.267 \times 10^{-7}\ m^2 \cdot s^{-1}$

λ_F molar latent heat of vaporization of flash vapor, $J \cdot kg\ mol^{-1}$

$\lambda(j)$ latent heat of vaporization of species j at 99.82 K, $J \cdot kg^{-1}$

$\lambda(1) = 8.426\ MJ \cdot kg\ mol^{-1}$
$\lambda(2) = 17.39\ MJ \cdot kg\ mol^{-1}$
$\lambda(3) = 22.26\ MJ \cdot kg\ mol^{-1}$
$\lambda(4) = 26.31\ MJ \cdot kg\ mol^{-1}$
$\lambda(5) = 4.37\ MJ \cdot kg\ mol^{-1}$

μ viscosity of LNG, $kg \cdot m^{-1} \cdot s^{-1}$

ν kinematic viscosity of LNG, $2.787 \times 10^{-7}\ m^2 \cdot s^{-1}$

ρ_i density of cell i liquid, $kg \cdot m^{-3}$

$\langle \rho \rangle$ averaged density between two neighboring cells or between uppermost cell and the film, $kg \cdot m^{-3}$ [e.g., $\langle \rho \rangle = 1/2(\rho_i + \rho_{i+1})$]

ρ_N density of bulk liquid of the uppermost cell, $kg \cdot m^{-3}$

ρ_{FILM} density of liquid in the vaporizing film, $kg \cdot m^{-3}$

REFERENCES

1. Chatterjee, N., and Geist, J. M., "Spontaneous Stratification in LNG Tanks Containing Nitrogen," Paper 76-WA/PID-6, ASME Winter Annual Meeting, New York (December 5, 1976).
2. Heestand, Janet, Shipman, C. W., and Meader, J. W., "A Predictive Model for Rollover in Stratified LNG Tanks," *A.I. Ch.E. Journal*, 29(2):199–207 (1983).
3. Sarsten, J. A., "LNG Stratification and Rollover," *Pipeline and Gas Journal*, 199:37 (1972).
4. Lamb, H., *Hydrodynamics*, 6th Ed., Article 232, Dover Publications (1945).
5. Chatterjee, N., and Geist, J. M., "The Effects of Stratification on Boil-off Rates in LNG Tanks," *Pipeline and Gas Journal*, 199:40 (1972).
6. Germeles, A. E., "A Model for LNG Tank Rollover," *Advances in Cryogenic Engineering*, 21:326 K. D. Timmerhaus and D. H. Weitzel (Eds.), Plenum Press (1975).
7. Turner, J. S., "The Coupled Turbulent Transports of Salt and Heat across a Sharp Density Interface," *Intl. J. Heat Mass Transfer*, 8:759 (1965).
8. Chandrasekhar, S., *Hydrodynamic and Hydromagnetic Stability*, Figure 13, p. 68, Clarendon Press, Oxford (1961).
9. Huppert, H. E., "On the Stability of a Series of Double-Diffusive Layers," *Deep-Sea Research*, 18:1005 (1971).
10. McAdams, W. H., *Heat Transmission*, 3rd Edition, Chapter 7, McGraw-Hill Book Co., New York (1954).
11. Globe, S., and Dropkin, D., "Natural-Convection Heat Transfer in Liquids Confined by Two Horizontal Plates and Heated from Below," *Trans. ASME, J. Heat Trans.*, C81, 24 (1959).
12. Boyle, G. J., "Basic Data and Conversion Calculations for Use in the Measurement of Refrigerated Hydrocarbon Liquids," *J. Inst. Pet.*, 58:133 (May, 1972).
13. Huntley, S. C., "Temperature-Pressure-Time Relationships in a Closed Cryogenic Container," *Adv. in Cryogenic Eng.*, 3:342, Plenum Press (1957).
14. Hashemi, H. T., and Wesson, H. R., "Cut LNG Storage Costs," *Hydrocarbon Processing*, 117 (August 1971).
15. Prausnitz, J. M., and Chueh, P. L., *Computer Calculations for High Pressure Vapor-Liquid Equilibria*, Prentice-Hall, Inc. (1968).
16. Zudkevitch, D., and Joffe, J., "Correlation and Prediction of Vapor-Liquid Equilibria with the Redlich-Kwong Equation of State," *A.I.Ch.E. Journal*, 16(1):112 (1970).
17. Joffe, J., Schroeder, M., and Zudkevitch, D., "Vapor-Liquid Equilibria with the Redlich-Kwong Equation of State," *A.I.Ch.E. Journal*, 16(3):496 (1970).
18. Soave, G., "Equilibrium Constants from a Modified Redlich-Kwong Equation of State," *Chem. Eng. Sci.*, 27(6):1197 (1972).
19. Hirata, M., Ohe, S., and Nagahama, K., *Computer Aided Data Book of Vapor-Liquid Equilibria*, Kodansha Limited/Elsevier Scientific Publishing Co., Tokyo-New York (1975).
20. Kato, M. W., Chung, K., and Lu, B. C.-Y., "Binary Interaction Coefficients of the Redlich-Kwong Equation of State," *Chem. Eng. Sci.*, 31:733 (1976).
21. Valencia-Chávez, J. A., "The Effect of Composition on the Boiling Rates of Liquefied Natural Gas for Confined Spills on Water," ScD Thesis, *Chem. Eng.*, MIT (1978).

CHAPTER 25

JET INJECTION FOR OPTIMUM PIPELINE MIXING

L. J. Forney

School of Chemical Engineering
Georgia Institute of Technology
Atlanta, Georgia USA

CONTENTS

INTRODUCTION, 660

JET MIXER DESIGN, 661

MIXING CRITERIA, 662
Second Moment, 663
Geometrically Centered Jet, 663

THEORY, 664
Dimensional Analysis, 665
Trajectory Analysis, 665

PREVIOUS WORK, 667
Experiments, 667
Theories, 671

OPTIMUM MIXING CONDITIONS, 671
Optimum Jet-to-Pipe Diameter, 671
Optimum Mixer Geometry, 675
Effect of Reynolds Number, 680
Effect of Pipe Length, 682
Effect of Buoyancy, 683
Concentration Decay, 683
Effect of Secondary Currents, 684

POWER REQUIREMENTS, 685

INSTABILITIES, 687

LARGE VISCOSITY RATIOS, 688

NOTATION, 689

REFERENCES, 690

INTRODUCTION

Turbulence promotes most important chemical reactions, heat transfer operations, and mixing and combustion processes in industry. Effective use of turbulence increases reactant contact and decreases reaction times, which can significantly reduce the cost of producing many chemicals.

Good mixing is necessary to obtain profitable yields or to eliminate excessive corrosion in reactor or combustion chambers.

If the necessary contact time between fluids is short, it is common in many existing chemical process units to continuously mix two fluids in a pipeline with subsequent transport to other locations. Although the continuous mixing of two fluid streams can be achieved using a number of mixer geometries, many procedures such as the use of baffles or complex internal geometries will introduce excessive pressure drops and significantly increase the cost of the mixing device. An effective, simple method to mix two fluids within a pipeline is to introduce feed jets such that jet contact with pipe walls is minimized and mixing occurs rapidly within the turbulent core of the pipe.

Frequently, a pipeline mixer must be designed for a special processing requirement. Examples of such requirements include [1, 2]:

1. Rapid mixing of a small gas stream into a large gas stream with limited pipe length, where a low pressure drop is essential.
2. Mixing of a liquid that would flash if left in the pure state and heated to the mixture temperature.
3. Mixing of two highly reactive gases or liquids that could form undesirable products if not mixed rapidly.
4. Mixing of highly reactive liquids or gases (e.g., combustion) with geometries that minimize corrosion, scaling, and thermal shock to the walls.

In general, the distance necessary to achieve a desired degree of uniformity of concentration or temperature in a pipe by jet injection depends on the following quantities: ratio of jet-to-pipe diameter, geometry of the mixer, uniformity criterion, ratio of jet-to-pipe flow rates or velocities, ratio of specific gravities of the two feed streams, pipe or jet Reynolds number, surface roughness, and pipeline secondary currents. Additional information is available in the reviews of pipeline mixing with tees and other geometries prepared by Simpson [3] and Gray [4].

The selection of the best injection system for promoting rapid mixing in a specific application, as indicated above, normally requires consideration of a number of design constraints [5]:

1. Desired mixing ratio of jet-to-pipe flow rates.
2. Pipe length available for mixing.
3. Required degree of uniformity of the mixture.
4. Secondary current patterns within the pipeline upstream of the injection point.
5. Accessibility of the injection point.
6. Power requirements for the proposed mixer geometry.

JET MIXER DESIGN

There are a number of possible geometries for jet injection systems in pipelines. The choice among the potential designs may depend on the desired mixing characteristics or power limitations. The simplest jet injection design is a single jet oriented perpendicular to the pipe centerline ($\theta_j = 90°$) as illustrated in Figure 1. Variations of this design involve multiple jets around the pipe periphery or single or multiple oblique jets ($\theta_j \neq 90°$). Here, θ_j represents the angle subtended by the jet axis and the pipe wall.

Other possible jet mixing geometries involve the tangential injection of fluid relative to the pipe centerline as shown in Figure 2, where ϕ represents the angle subtended by the pipe diameter and the jet axis. The maximum value of $\phi = \phi_m$ is represented by the geometry on the right of Figure 2 where $\phi_m = 180°/\pi \sin^{-1}(1 - d/D)$.

In the present discussion the treatment will be limited to a single oblique jet ($\theta_j \neq 90°$) of the type illustrated in Figure 1 and to single or dual tangential jet injection as shown in Figure 2. In this context it should be noted that the geometry of Figure 1 for $\theta_j = 90°$ is the same design as the single centerline case ($\phi = 0°$) of Figure 2.

D
V
ρ_P
Z
ρ
u
2b
θ
X
ρ_J
u_J
JET
INLET
d

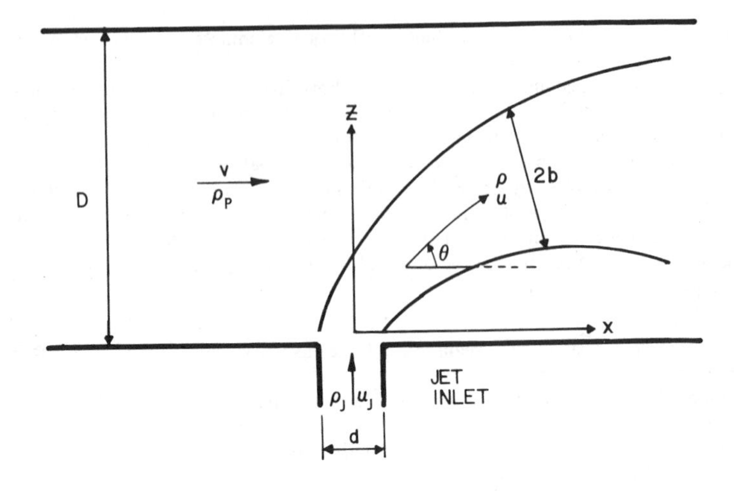

Figure 1. Schematic of a single jet mixer.

MIXING CRITERIA

There are a number of different parameters which can be used to characterize the degree of mixing or the uniformity of the concentration over the pipe cross section downstream from the injection point [7].

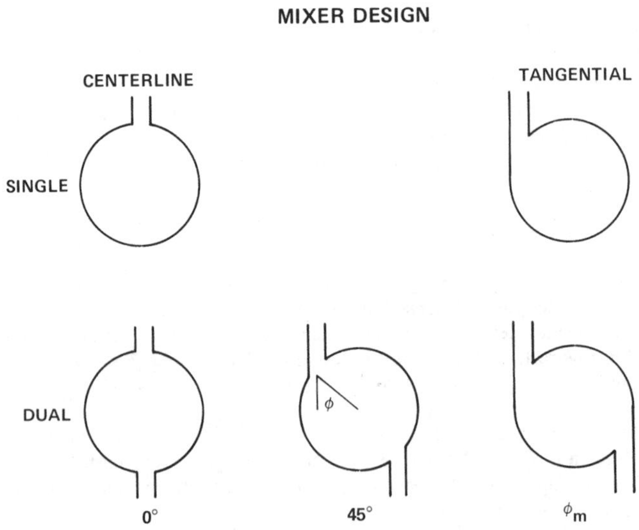

MIXER DESIGN
CENTERLINE
TANGENTIAL
SINGLE
DUAL
ϕ
0°
45°
ϕ_m

Figure 2. Schematic of single and dual jet mixers.

Second Moment

One useful method of characterizing the degree of mixing is to determine the second moment M of the concentration over the pipe cross section [5, 8]. Thus,

$$M = \frac{1}{A}\int_A \left(\frac{C}{\bar{C}} - 1\right)^2 dA \tag{1}$$

where $\bar{C}$ is the mean tracer concentration at a fixed distance downstream from the injection point. Since the standard deviation of the concentration $\sigma = M^{1/2}$, the mixing distance can be defined as the distance downstream from the injection point where a desired value of σ is achieved. Moreover, for a fixed mixer geometry and flow ratio, the mixing distance increases as the desired value of σ decreases.

When comparing the mixing characteristics defined by Equation 1 of different jet injection systems, it may be useful to normalize the second moment of the concentration M with respect to a reference value M_0 [8]. The parameter M_0 is defined as the second moment of the concentration over the cross section of the main and branch pipes before mixing with respect to the final bulk mean concentration $C_{ave} = q/(q + Q)$ where q is the volume flow rate of injected fluid and Q is the volume flow rate in the pipe. Thus, one obtains,

$$M_0 = \frac{(Q/q)^2(d/D)^2 + 1}{d^2/D^2 + 1} \tag{2}$$

Generally, one seeks a mixer geometry which minimizes the pipe length to achieve a desired degree of mixing or which minimizes the ratio M/M_0 for a fixed pipe length from the injection point [6, 9]. Problems can arise, however, when attempting to minimize M of Equation 1 to achieve optimum mixing, particularly when the ratio of jet-to-pipe diameter is small and the measurement point is less than ten pipe diameters from the injection point ($x/D \lesssim 10$). Recent data [6, 8, 13] which attempt to minimize M over a pipe length $x/D < 9$ suggest that optimum mixing is achieved in a short pipe length by impacting the jet against the opposite pipe wall between two and three pipe diameters from the injection point (i.e., the opposite pipe wall acts as a baffle). This would not provide optimum mixing over longer pipe lengths (measurement points $x/D > 10$) and could provide undesirable results if heat transfer, thermal shock, or corrosion to the pipe walls is a potential problem.

Geometrically Centered Jet

Another mixing criterion which is much simpler to measure in the laboratory and yet provides a basis for theoretical descriptions of optimum mixing is to center the jet on the pipe axis at a specified distance of at least two pipe diameters from the injection point. This technique is patterned after the pioneering work of Chilton and Genereaux [10] and has been developed in a number of recent publications [7, 11, 12]. The technique minimizes wall contact and correlates the optimum mixing data of a single 90° jet as determined by second moment measurements at distances $x/D > 10$. The principle of geometrically centered jets also reasonably correlates the data for optimum mixing at measurement cross sections closer to the injection point for large jet-to-pipe diameter ratios. The data, however, are significantly different from those determined using second moment measurements close to the jet for small jet-to-pipe diameter ratios $d/D < 0.13$ because of wall impaction as previously described. Examples of concentration profiles demonstrating both poor and good mixing with this criterion are shown in Figures 3 and 4.

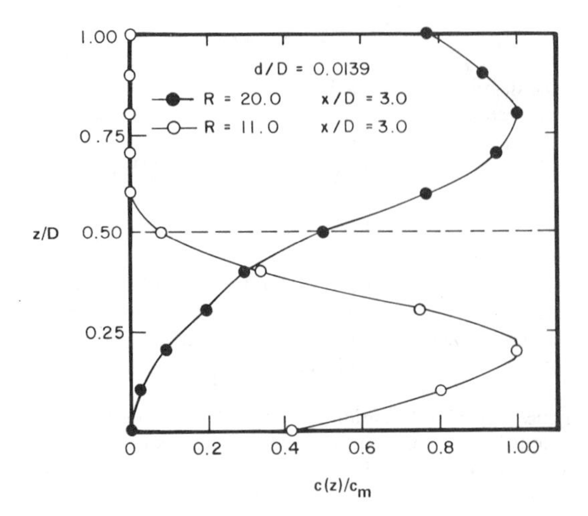

Figure 3. Concentration profiles for poor mixing (after [7]).

THEORY

The problem of predicting the optimum dimensions for jet injection in a pipe is a difficult one. The degree of mixing downstream from the point of injection is dominated by the properties of the jet near the point of injection. Near the inlet the jet entrains ambient pipe fluid and bends over in the cross flow as shown in Figure 1. Since this formally constitutes a three-dimensional turbulent shear flow commonly embedded in a fully developed turbulent pipe flow, it is unlikely that full analytical solutions based on first principles are possible.

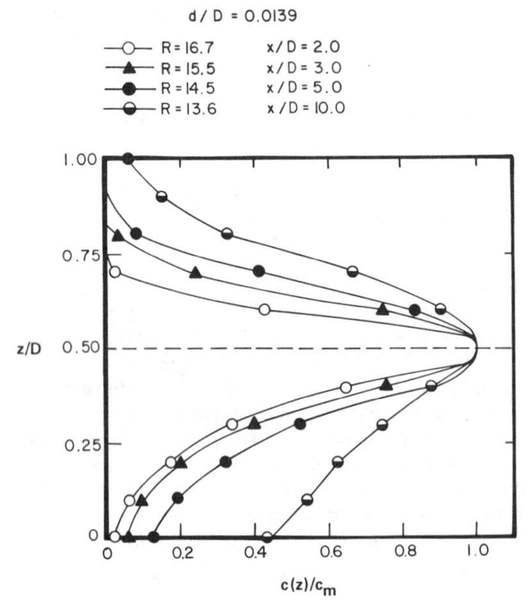

Figure 4. Concentration profiles for good mixing (after [7]).

Dimensional Analysis

From dimensional analysis it can be shown that the penetration, z, of a single nonbuoyant jet into a pipe as shown in Figure 1 can be written as

$$\frac{z}{D} = f\left(\frac{x}{D}, \frac{\ell_m}{D}, \frac{\ell_m}{d}, Re_p, Re_j, \theta_j\right) \tag{3}$$

where $\ell_m (= du_j/v)$ is the momentum length of the jet and Re_p, Re_j are the Reynolds numbers of the pipe and jet, respectively. Equation 3 can be simplified if Re_p and Re_j are large as discussed later, when experimental evidence suggests that the pipe and jet Reynolds numbers do not effect the mixing characteristics.

Since $\ell_m/D = Rd/D$, $\ell_m/d = R$ and $q/Q = R(d/D)^2$ in Equation 3 where $R = u_j/v$, it is convenient to write Equation 1 for design purposes for large Re_p, Re_j in the form

$$\frac{z}{D} = f\left(\frac{x}{D}, \frac{d}{D}, \frac{q}{Q}, \theta_j\right) \tag{4}$$

Moreover, if the ratio of jet-to-pipe diameter d/D is small, then the jet diameter d is not important in relation to the momentum length ℓ_m and the pipe diameter D [14] so that Equation 4 reduces to

$$\frac{z}{D} = f\left(\frac{x}{D}, \frac{qD}{Qd}, \theta_j\right) \tag{5}$$

Although Equations 3 through 5 express the jet penetration into the pipe, other mixing characteristics such as the ratio of the second moments M/M_0 of Equations 1 and 2 could be expressed as functions of the same dimensionless groups.

Since one normally wishes to optimize the mixing downstream from the injection point which implies centering the jet or minimizing the second moment of the concentration at a fixed distance downstream, Equations 4 and 5 reduce to

$$\frac{d}{D} = f\left(\frac{q}{Q}, \theta_j\right) \tag{6}$$

or for small d/D one obtains

$$\frac{qD}{Qd} = f(\theta_j) \tag{7}$$

Equation 7 states the principle that there is a unique ratio of jet-to-pipe momentum $(qD/Qd)^2$ for optimum mixing with small centerline jets [9, 15].

Trajectory Analysis

Considerable insight can be gained by analyzing the jet trajectory near the injection point for the simple geometry of Figure 1 [11]. Restricting the analysis to a right angle jet ($\theta_j = 90°$) and assuming uniform (top hat) profiles for jet properties, one obtains the following conservation expressions valid for small departures from the origin $\theta \simeq \theta_j$, x, z $\simeq 0$:

- Mass

$$\frac{d}{dz}(\pi b^2 u) = 2\pi b_j(\alpha u_j + \beta v) \tag{8}$$

- Tangential momentum

$$\frac{d}{dz}(\pi b^2 u^2) = 0 \tag{9}$$

- Normal momentum

$$\pi u_j^2 b_j^2 \frac{d\theta}{dz} = -v \frac{d}{dz}(\pi b^2 u) \tag{10}$$

Equations 8 to 10 are subject to the boundary conditions:

$$z = 0, \qquad u = u_j, \qquad b = b_j, \qquad \theta = \frac{\pi}{2}$$

Solution of Equations 8 to 10 to first order gives

$$\frac{b^2 u}{b_j^2 u_j} = 1 + 4\left(\alpha + \frac{\beta}{R}\right)\left(\frac{z}{d}\right) \tag{11}$$

$$u^2 b^2 = u_j^2 b_j^2 \tag{12}$$

$$\theta = \frac{\pi}{2} - 4\left(\alpha + \frac{\beta}{R}\right)\left(\frac{z}{\ell_m}\right) \tag{13}$$

Since $dx \simeq (\pi/2 - \theta)\, dz$ near the jet orifice, one obtains an approximate jet trajectory of the form

$$\frac{z}{\ell_m} = \left(\frac{1/2}{\alpha + \beta/R}\right)^{1/2}\left(\frac{x}{\ell_m}\right)^{1/2} \tag{14}$$

valid for $x/\ell_m < 1$. Normalizing the coordinates x, z of Equation 14 with respect to the pipe diameter D, one has

$$\frac{z}{D} = \left(\frac{1/2}{\alpha + \beta/R}\right)^{1/2}\left(\frac{\ell_m}{D}\right)^{1/2}\left(\frac{x}{D}\right)^{1/2} \tag{15}$$

One now assumes that optimum mixing is achieved when the jet is centered on the pipe axis at a fixed distance downstream from the injection point. For simplicity, imposing the boundary conditions $z/D = x/D = 0.5$, Equation 15 becomes

$$\frac{\ell_m}{D} = \alpha + \frac{\beta}{R}, \qquad R > 1 \tag{16}$$

where the entrainment parameters $\alpha = 0.11$ and $\beta = 0.6$ are universal constants. Since $q/Q = R(d/D)^2$, one obtains the mixing ratio

$$\frac{q}{Q} = \alpha/2\left(\frac{d}{D}\right)\left[1 + \left\{1 + \frac{4\beta}{\alpha^2}\left(\frac{d}{D}\right)\right\}^{1/2}\right] \tag{17}$$

which provides the useful limits

$$\frac{q}{Q} = 0.775\left(\frac{d}{D}\right)^{1.5}, \quad \frac{d}{D} \to 1 \tag{18}$$

and

$$\frac{q}{Q} = 0.11\left(\frac{d}{D}\right), \quad \frac{d}{D} \to 0 \tag{19}$$

Thus, the approximate analysis providing geometrically centered jets above predicts a simple scaling law for the limiting geometry of either large or small jet-to-pipe diameter ratios. Moreover, the form of the scaling laws is identical to the results of a dimensional analysis provided by Equations 6 and 7.

PREVIOUS WORK

The first systematic study of pipeline mixing by jet injection was conducted by Chilton and Genereaux [10], who used smoke visualization techniques to determine optimum mixing conditions at a glass tee. They concluded that right-angle configurations were effective for good mixing. Chilton and Genereaux also found that when the ratio of velocity of secondary to main flow was in the range of 2 to 3, satisfactory mixing was obtained in 2 to 3 pipe diameters. Narayan [16, 17] used quantitative methods to measure the degree of mixing of air-carbon dioxide feed streams in three pipeline mixers. Narayan, like Chilton and Genereaux, found it was possible to achieve quality mixing in a few diameters with perpendicular jet injection devices but that parallel flow devices required up to 250 pipe diameters. Holley [9, 15] compared standard deviations from measured tracer concentration distributions far downstream (7–120 pipe diameters) from the injection point. These results were compared to other mixer geometries with passive tracer sources near the wall or at the center of the pipe. For a limited range of small jet-to-pipe diameter ratios, Holley also concluded that complete mixing could be achieved in a short pipe length with right-angle jet configurations.

More recently, Forney [7, 11, 12, 18] and Maruyama et al. [6, 8, 13, 19] independently studied jet injection of fluid into a pipeline. Both studies concluded that the ratio of jet-to-pipe diameter versus jet-to-pipe velocity for optimum mixing was relatively independent of the measurement position down the pipe axis, provided the measurements were made at two or more pipe diameters downstream from the injection point. They also concluded that the mixing criterion was independent of the pipe Reynolds number for a sufficiently large Re_p.

The criterion for optimum mixing used by Forney was to center the jet and pipe axis at measurement points between two and 10 pipe diameters. Although their mixing criterion was different than that of Holley and Maruyama, who minimized the second moment of the distribution across the pipe, the measured results (dimensions for optimum mixing) of all three studies were identical within experimental scatter of the data for most conditions. The only exception was that the use for the second moment for the mixing criterion encouraged jet impaction on the opposite wall when d/D is small and the measurement point was close to the jet mixer as discussed earlier.

Experiments

Various mixing criteria have been used by experimentalists to determine conditions of optimum mixing for jet mixers of the type shown in Figures 1 and 2. Using either air or water as working fluids, the investigators have typically injected a chemical tracer into the fluid medium, measured the distribution of the tracer downstream from the injection point and compared the measurements to a given mixing criterion. Table 1 summarizes all of the previous experimental work including the range of parameters and mixing criteria used.

Table 1
Summary of Experiments

Jet Fluid	Pipe Fluid	Jet Diameter (cm)	Pipe Diameter (cm)	R (u_j/v)	Pipe Re_p	Jet Re_j	Measurement Point in Pipe Diameters	Measured Variable	Mixing Criterion	Reference
Air and $TiCl_4$	Air	0.68–2	4.45	1.5–3	4.3×10^3 -1.8×10^4	$> 1.8 \times 10^4$	2–3	Visual smoke conc.	Visual smoke uniformity	Chilton and Genereaux [10]
Aq. 0.5N HNO_3	Aq. 0.5N NaOH	0.635	0.635	1	1×10^4 -4×10^4	1×10^4 -4×10^4	6–7	Temp.	97% of final temp. rise	Swanson [32]
Aq. NaCl	Water	0.08 −0.32	15.24	6–24	6×10^4	$> 7.5 \times 10^3$	20–120	Elec. conductivity	Conc. stand. deviation	Ger and Holley [15]
Air and $TiCl_4$	Air	0.42 −1.5	5	1.5–3.3	4×10^3 -2×10^4	$> 1 \times 10^4$	2–3	Visual smoke conc.	Visual smoke uniformity	Winter [33]
Air and 1% CH_4	Air	0.16	6.35	2–7	2×10^3 -9×10^4	5×10^2 -2.3×10^4	2–5	CH_4 conc.	Max. conc. centered on pipe axis	Forney and Kwon [11]
Air 25°C	Air ~35°C	0.5 −1.3	5.1	3–4	1.6×10^4 -6.3×10^4	8.2×10^3 -2.3×10^4	2–10	Temp.	Temp. stand. deviation	Maruyama, Suzuki, and Mizushina [13]

Air 19% CO_2	Air	1.58	5.25	2.7	4.6×10^4	3.74×10^4	10	CO_2 conc.	Equal CO_2 conc. at pipe axis and periphery	Reed and Narayan [17]
Air and 0.3% CH_4	Air	0.1 –1.27	11.43	2.9 –28.3	1.3×10^4 -3.2×10^4	1.1×10^3 -7.2×10^3	2–10	CH_4 conc.	Max. conc. centered on pipe axis	Forney and Lee [7]
Aq. NaCl	Water	0.32	15.24	3 –10.3	3×10^4 -4×10^4	2.1×10^3 -7×10^3	7–47	Elec. conductivity	Conc. stand deviation	Fitzgerald and Holley [9]*
Air $>25°C$	Air $\sim 35°C$	0.5 –2.5	5.1	3–8	1.7×10^4	$>4 \times 10^3$	0.8	Temp.	Temp. stand. deviation	Maruyama, Mizushina and Watanabe [8]*
Air $>25°C$	Air $\sim 35°C$	0.6 –1.6	10	3–19	7×10^3 1.6×10^4	5.3×10^3	0.5–5	Temp.	Temp. stand. deviation	Maruyama, Mizushina and Shirasaki [13]
Air 25°C	Air $\sim 35°C$	0.42 –1.1	10	3–10	7×10^3 -1.5×10^4	5×10^3	0.5–5	Temp.	Temp. stand. deviation	Maruyama, Mizushina and Hayashiguchi [6]†
Air and 0.3% CH_4	Air	0.07 –0.635	11.43	4.7 –16	1.8×10^4 -1.1×10^5	4×10^3 -1.8×10^4	0.25–10	CH_4 conc.	Max. conc. centered on pipe axis	O'Leary and Forney [12]

* *oblique jets* ($30° \leq \theta_j \leq 150°$)
† *tangential jets* ($0° \leq \phi \leq \phi_m$)

Table 2
Predictions of Optimum Mixing

Prediction	Reference
Theories	
Single jet, $\theta_j = 90°$	
$\frac{q}{Q} = \frac{\alpha}{2}\left(\frac{d}{D}\right)\left[1 + \left\{1 + \frac{4\beta}{\alpha^2}\left(\frac{d}{D}\right)\right\}^{1/2}\right], \quad \frac{d}{D} < 1$	Forney and Kwon [11]
$\alpha = 0.11, \quad \beta = 0.6$	
$\frac{d}{D} = \frac{0.27}{Rf^2}, \quad R < 6 \quad \text{where } f(R) = \frac{0.17}{0.1 + 0.35/R^{1.25}}$	Forney and Lee [7]
$\frac{d}{D} = \frac{0.28}{Rg^{3/2}}, \quad R > 6 \quad g(R) = 0.83 + 0.2 \ln R$	
$Re_j > 9{,}000$	
Empirical Correlations	
Single jet, $\theta_j = 90°$	
$d/D = 8.77(q/Q), \quad q/Q < 0.0024$	Fitzgerald and Holley [9]
$d/D = 2.2(q/Q), \quad q/Q < 0.058$ $d/D = 0.78(q/Q)^{0.63}, \quad q/Q > 0.058$	Maruyama et al. [6]
$d/D = 8.54(q/Q), \quad q/Q < 0.0026$ $d/D = 0.81(q/Q)^{0.6}, \quad q/Q > 0.0026$ $Re_p > 40{,}000, \quad Re_j > 6{,}000$	O'Leary and Forney [12]
Oblique jet	
$d/D = 8.55[\cos(\theta_j - \pi/2)]^{1.3}(q/Q), \quad q/Q < 0.01, \quad \pi/2 \leq \theta_j \leq \frac{5\pi}{6}$	Fitzgerald and Holley [9]
$d/D = 0.78[\cos(\pi/2 - \theta_j)]^{0.26}(q/Q)^{0.63}, \quad q/Q > 0.058, \quad \theta_j < 90°$ $d/D = 0.78[\cos(\theta_j - \pi/2)]^{-0.26}(q/Q)^{0.63}, \quad q/Q > 0.058, \quad \theta_j > 90°$	Maruyama et al. [8]
Tangential jets	
$d/D = \frac{2.2}{n}(q/Q), \quad q/nQ < 0.058$ $d/D = \left[\frac{0.67}{n}(q/Q)\right]^{0.63}, \quad q/nQ > 0.058$	Maruyama et al. [6]

Theories

The analysis of Forney [7, 11] suggested a physical picture, although incomplete, which provided an elementary understanding of the mixing process and led to the development of useful scaling laws. The critical assumption in this work was that one could achieve rapid unretarded mixing of the jet and pipe fluids by centering the jet in the pipe at a fixed distance x_0 (ideally the Lagrangian integral length scale) down the pipe axis from the injection point. This criterion provided geometrically similar jet trajectories within the pipe. The reasoning behind this approach was that over the distance $0 \leq x \leq x_0$ jet-induced turbulence dominates the mixing process while ambient pipe turbulence dominates the mixing for $x > x_0$. A summary of these theories and additional empirical correlations are shown in Table 2 and many of these predictions are plotted in the following section.

OPTIMUM MIXING CONDITIONS

In practice, one needs to choose a mixer geometry and its dimensions to provide optimum mixing for a specified flow or mixing ratio q/Q. In this section, plots of experimental data and theoretical predictions are provided to aid in jet mixer design. In addition to providing optimum conditions, the potential effects of Reynolds number, pipe length, buoyancy, and secondary currents are also discussed. Although the present section is limited to the mixing of fluids with nearly equal viscosities, a brief discussion of the mixing of fluids with large viscosity ratios is presented later.

Optimum Jet-to-Pipe Diameter

Single 90° Jet

The largest research effort has been devoted to defining optimum conditions and mixer geometries for the single 90° jet which represents the simplest jet mixer geometry. Attempts to predict velocity, diameter, and flow ratios for optimum mixing conditions for a 90° jet (T-junction) are limited to several empirical correlations and the theoretical development of Forney [7, 11]. These predictions are plotted in Figure 5 subject to the limitations listed in Table 2. It can be noted from the expressions in Table 2 that the theory of Forney and Kwon [11] yields the limiting values for the mixing ratio $d/D = 1.29(q/Q)^{0.67}$ for $d/D \rightarrow 1.0$ and $d/D = 9.1(q/Q)$ for $d/D \rightarrow 0$. These limiting expressions compare favorably with the empirical results of $d/D = 0.78(q/Q)^{0.63}$ for $0.03 < d/D < 1.0$ ($q/Q > 0.058$) measured by Maruyama [6] and $d/D = 8.77(q/Q)$ for $d/D < 0.02$ ($q/Q < 0.0024$) measured by Holley [9].

In Figure 5 and Table 2, it is evident that the scaling laws for optimum mixing change form in the limiting case of either large or small diameter or flow ratios. This can be explained by noting that the scaling law for a jet trajectory in a crossflow changes from a $\frac{1}{2}$ to a $\frac{1}{3}$ power law with distance from the source [7]. Since the extent of the near-field $\frac{1}{2}$ power law decreases significantly with smaller jet diameters, one may expect jet mixing in a pipe to be dominated by the $\frac{1}{2}$ power law for large jet-to-diameter ratios and conversely the $\frac{1}{3}$ power law for small diameter ratios [7].

As indicated in Figure 5, all of the theories are in reasonable agreement with the exception of the empirical expression of Maruyama [6] for small diameter ratios or $d/D < 0.13$. Apparently, minimizing the second moment or standard deviation of the concentration near the mixer as done by Maruyama favors jet impaction on the opposite wall of the pipe as discussed earlier.

Figures 6 and 7 illustrate the measurements of optimum velocity ratios and flow ratios for a single 90° jet. The data of Maruyama ($d/D < 0.13$) which favors wall impaction has been excluded from Figures 6 and 7. The solid lines in Figures 6 and 7 are the empirical correlations of O'Leary and Forney [12], which clearly demonstrate the change in slope in the scaling laws at $d/D = 0.022$.

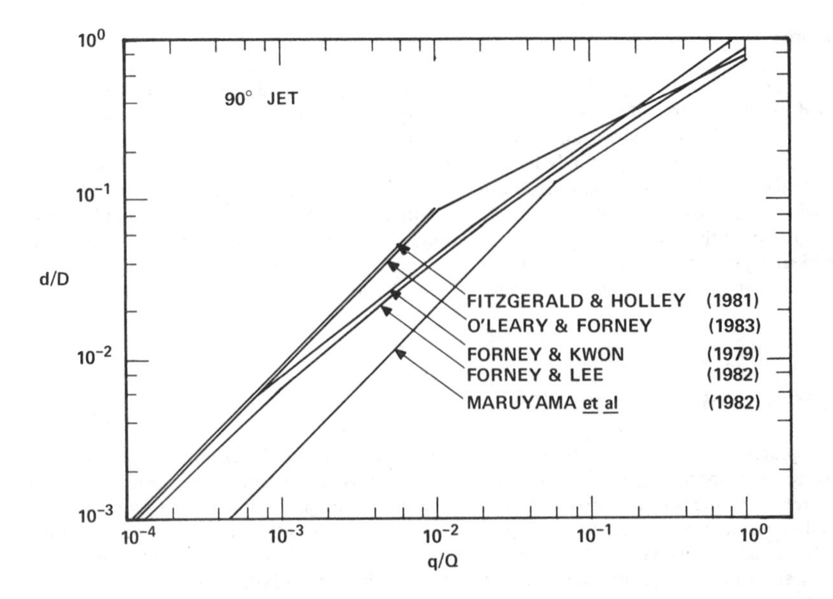

Figure 5. Jet-to-pipe diameter ratio versus volume flow ratio for optimum mixing.

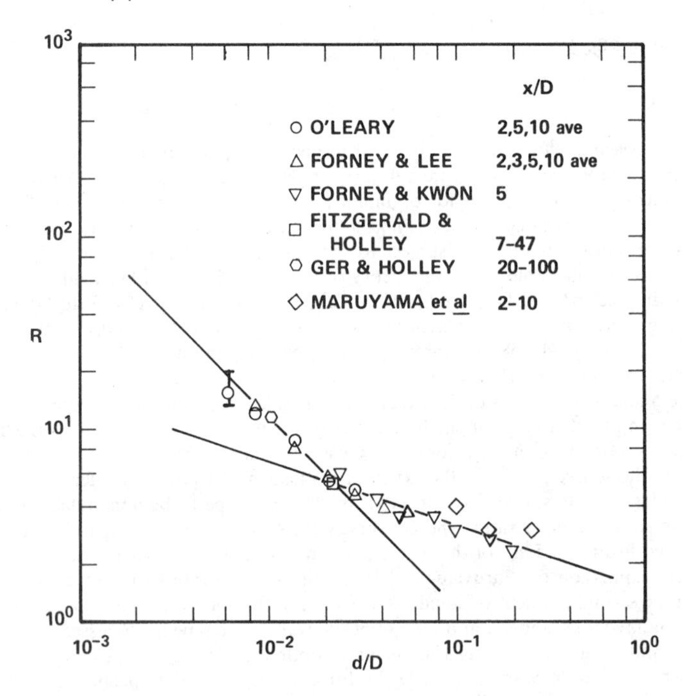

Figure 6. Velocity ratio versus diameter ratio for optimum mixing of a single 90° jet. Solid line is theory (after [12, 23]).

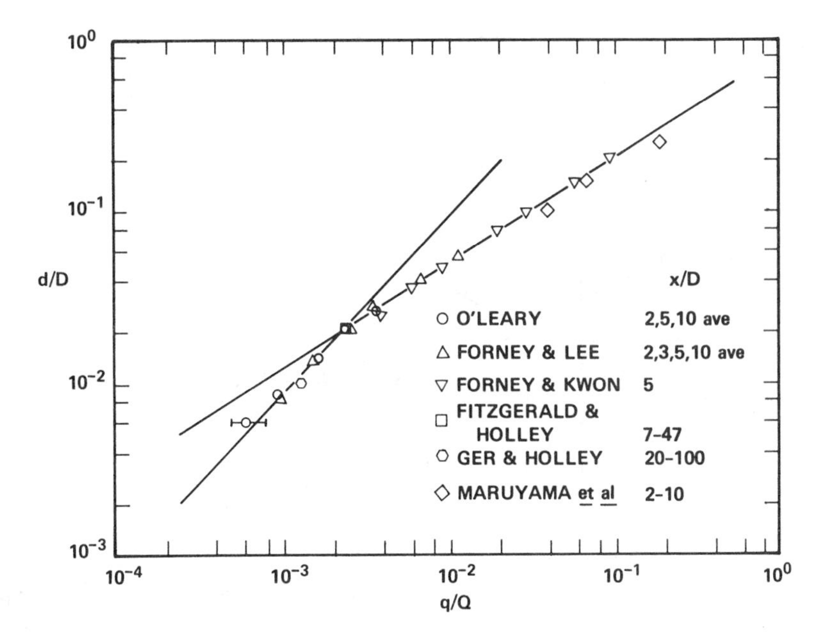

Figure 7. Diameter ratio versus volume flow ratio for optimum mixing of a single 90° jet. Solid line is theory (after [12, 23]).

Single Oblique Jet

If the jet is inclined such that the angle subtended by the jet axis and pipe wall $\theta_j > 90°$ (see Figure 1), mixing may be enhanced but the power requirements will increase. Holley has determined optimum conditions for small jet-to-pipe diameter ratios $d/D \leq 0.021$ for the angles $90° \leq \theta_j \leq 150°$ [5, 9]. For these conditions the scaling law reduces to Equation 7 and a universal curve can be determined for all diameter and flow ratios if the left ordinate is plotted as indicated in Figure 8. The right ordinate illustrates the mixing distance to reduce the standard deviation of the concentration to $\sigma = 0.03$. In Figure 8 on the right ordinate $\alpha_f = 0.048f^{1/2}$ Sc, where f is the pipe friction factor, Sc is the turbulent Schmidt number, and the solid line correlating the open square symbols is given in Table 2. It is clear that by using the jet injector oriented at $\theta_j = 150°$ (upstream injection) instead of $\theta_j = 90°$, a 35% decrease in mixing distance can be realized. However, the power requirements to sustain the mixing increase significantly. For example, in Figure 8, the momentum of the $\theta_j = 150°$ jet is six times that required for $\theta_j = 90°$.

Additional experimental data were taken by Maruyama [8] for large jet-to-pipe diameter ratios $d/D > 0.1$, and these data are plotted in Figure 9 for $30° \leq \theta_j \leq 150°$. The solid lines in Figure 9 are the expressions listed for oblique jets in Table 2.

It should be noted that a significant difference is apparent when comparing Figures 8 and 9 for $\theta_j > 90°$. Holley's results indicate that one must increase the momentum of the jet to achieve optimum mixing as θ_j increases beyond 90° while the data of Maruyama predict a decrease in momentum. Although both Maruyama and Holley used the same mixing criterion, Maruyama optimized mixing close to the jet ($x/D \leq 9$) while Holley measured the quality of mixing at much longer distances from the injection point.

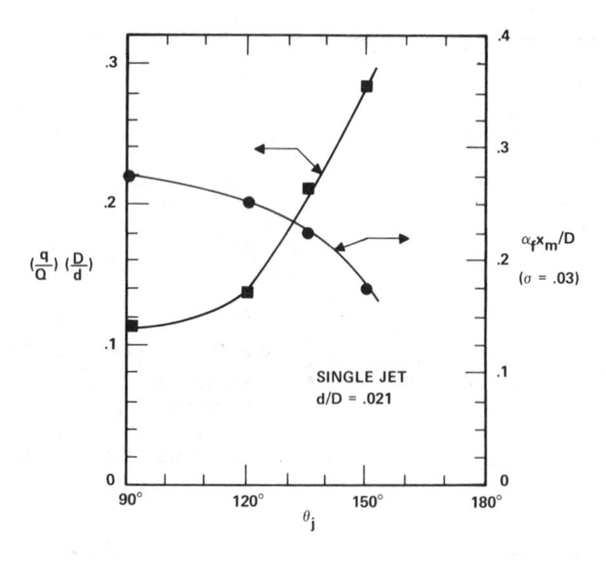

Figure 8. Optimum mixing conditions for a single oblique jet. Solid line is theory (after [5]).

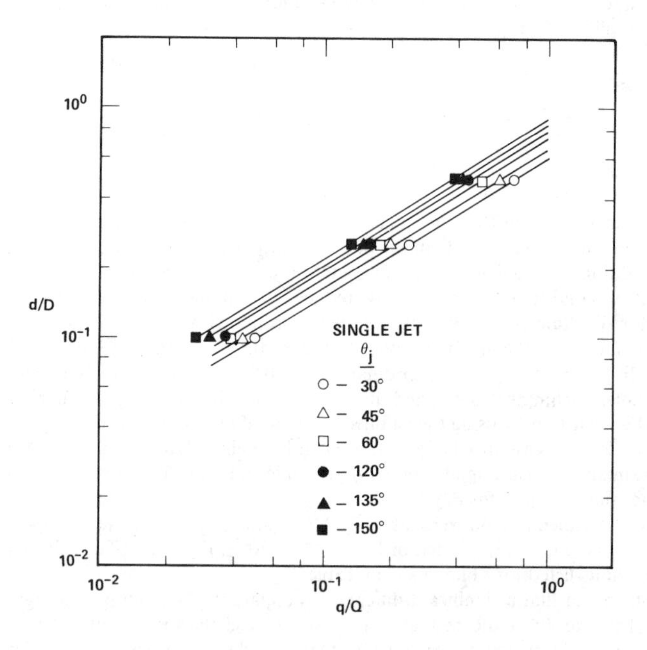

Figure 9. Diameter ratio versus volume flow ratio for optimum mixing of large oblique jet. Solid line is theory (after [8]).

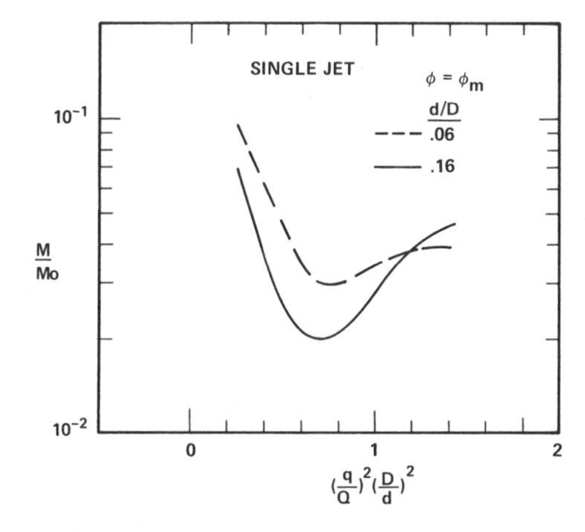

Figure 10. Concentration second moment versus momentum ratio for single tangential jet (after [19]).

Single and Dual Tangential Jets

Maruyama injected fluid into a pipeline in a tangential manner with single and dual jets as illustrated in Figure 2 [6, 19] for a range of angles $0° \leq \phi \leq \phi_m$. Maruyama points out that these geometries may be effective for the protection of wall surfaces during combustion or the mixing of multiphase flows, since the injected fluid is propelled along the walls and rotation is induced within the pipe.

For the case of a single tangential jet with the maximum angle $\phi = \phi_m$, Maruyama reduced a swirl number for the pipe to the ratio of the side-stream momentum to the pipe momentum $(qD/Qd)^2$ for large jet-to-pipe diameter ratios $d/D \geq 0.06$. A plot of the second moment of the concentration versus the momentum ratio as shown in Figure 10 illustrates that optimum mixing is achieved when

$$(q/Q)^2(D/d)^2 \simeq 0.7 \tag{20}$$

The optimum dimensions for single and dual tangential jet injection were also measured by Maruyama [6] for $d/D \geq 0.042$. These data are correlated with a single curve and plotted on Figure 11. The solid lines in Figure 11 are predicted by the equations for tangential jets listed in Table 2 where $n = 1$ for a single jet and $n = 2$ for dual jets.

Optimum Mixer Geometry

In the previous section optimum mixer dimensions are presented for each of the mixer geometries shown in Figures 1 and 2. In the present section the mixing performance of each of the geometries (i.e., oblique, tangential) is compared. Since an important design requirement is a desired mixing ratio q/Q, measurements of the second moment of the concentration have been replotted from the work of Holley [5, 9, 15] and Maruyama [6, 8, 19] and these results are compared for fixed q/Q. All of the experimental results are presented as plots of the concentration second moment M (or standard deviation $\sigma = M^{1/2}$) versus distance down the pipe axis from the jet. It should be noted that all of the data of Maruyama for both oblique and tangential jets are limited to large jet-to-pipe diameters $d/D \geq 0.042$. The only second moment data available for small diameter ratios $d/D = 0.021$ is the work of Holley [9] for single oblique jets.

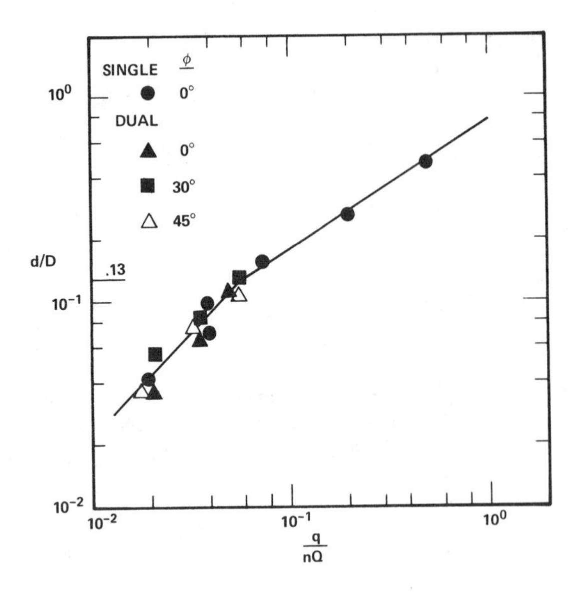

Figure 11. Optimum diameter ratio versus volume flow ratio for single and dual tangential jets. Solid line is theory (after [6]).

Single Oblique Jet

Figure 12 represents the decrease in the concentration standard deviation versus distance down the pipe axis for the optimum conditions presented in Figure 8. The results for four values of jet angle θ_j are compared with passive sources located at the pipe centerline and wall. It is clear that mixing performance is improved with jet injection and in particular by increasing θ_j.

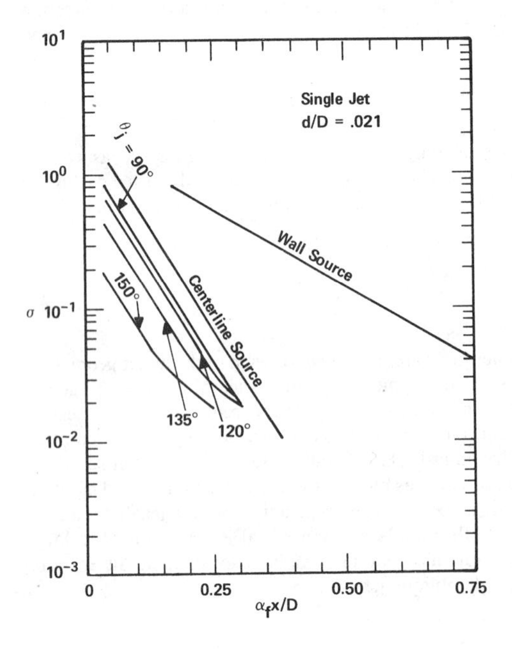

Figure 12. Minimum standard deviation versus distance from mixer for a small oblique jet (after [9]).

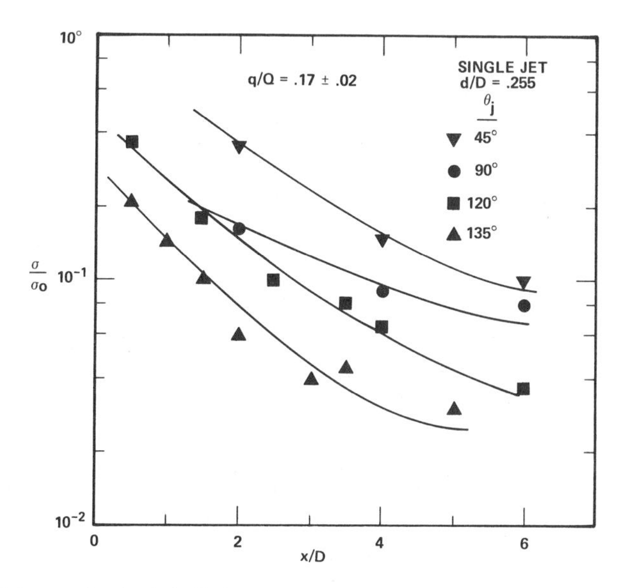

Figure 13. Minimum standard deviation versus distance from mixer for a large oblique jet (after [8]).

Figure 13 represents a normalized concentration standard deviation versus distance from the jet for large diameter ratios. As in Figure 12, the distance to mix is reduced by roughly a factor of two for larger θ_j.

Single and Dual Tangential Jets

A comparison of the optimum mixing performance for both single and dual tangential jets at a fixed mixing ratio $q/Q \sim 0.04$ follows. It should be noted that q, the total injected flow rate, represents the sum of flow rates for the dual jet geometries. Figures 14 through 17 were compiled by replotting the experimental data of Maruyama [6, 19] for large jet-to-pipe diameter ratios and comparing the mixing performance of each geometry of Figure 2 with its nearest neighbor moving clockwise beginning with the upper left.

Thus, Figure 14 demonstrates that a single jet orthogonal to the pipe axis ($\phi = 0°$) is superior to a jet aligned tangential to the pipe cross section ($\phi = \phi_m$) with roughly a 20% reduction in pipe length required for equal mixing quality. Similarly, comparing single and dual jet injection in Figure 15 for $\phi = \phi_m$ indicates that the use of two jets improves mixing quality. Likewise Figure 16 compares the performance of dual jet injection for varying ϕ. Clearly, injection nearly orthogonal to the pipe axis reduces the pipe length required to achieve equal concentration second moments. Finally, centerline injection is compared for both single and dual jet geometries in Figure 17. A dual centerline geometry reduced the mixing distance by only approximately 20% for a concentration second moment ratio of $M/M_0 = 0.01$. These latter conclusions were also reached by Holley [5, 9] for limited data taken with single and dual jets of small jet-to-pipe diameter ratios $d/D = 0.021$ as shown in Figure 25.

Thus, a careful study of Figures 14 through 17 demonstrates that tangential jet injection does little to improve the mixing quality. Moreover, the use of dual centerline jet injection only moderately

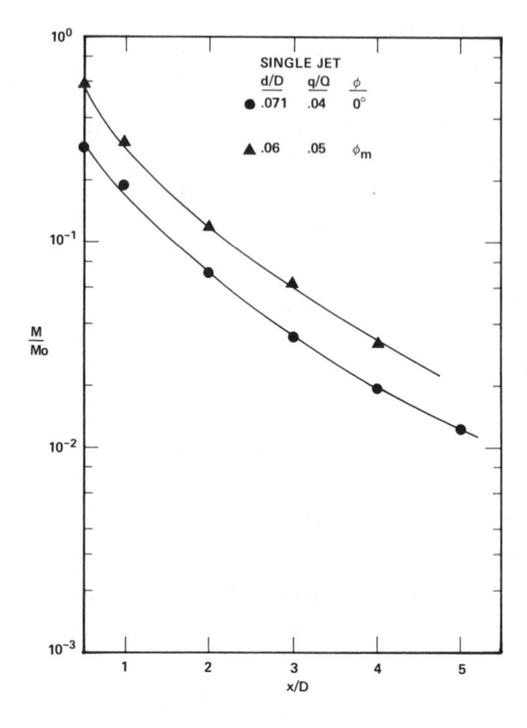

Figure 14. Minimum second moment versus distance from mixer for a single tangential jet (after [6, 19]).

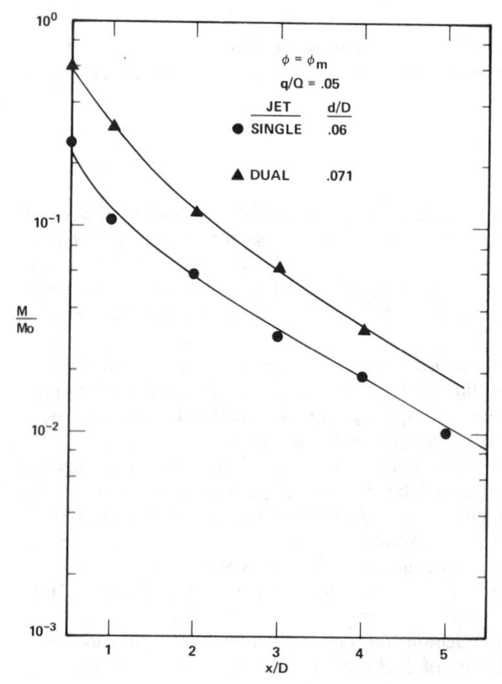

Figure 15. Minimum second moment versus distance from mixer for single and dual tangential jets (after [6, 19]).

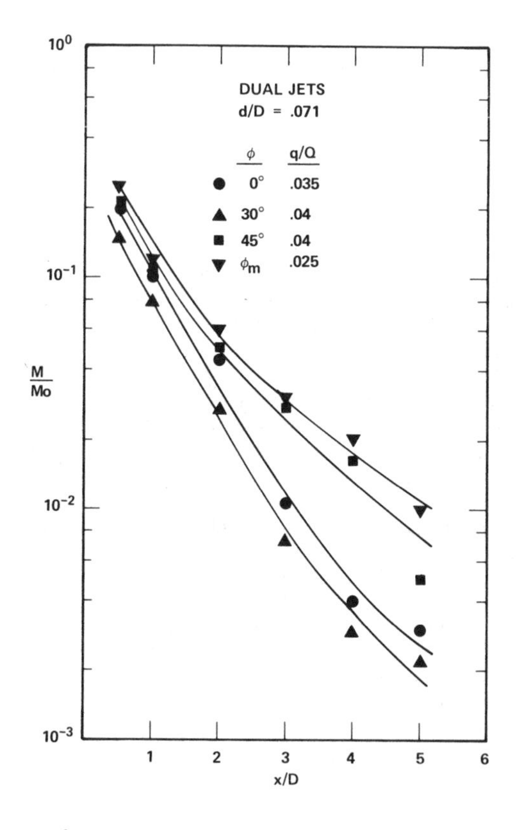

Figure 16. Minimum second moment versus distance from mixer for dual tangential jets (after [6]).

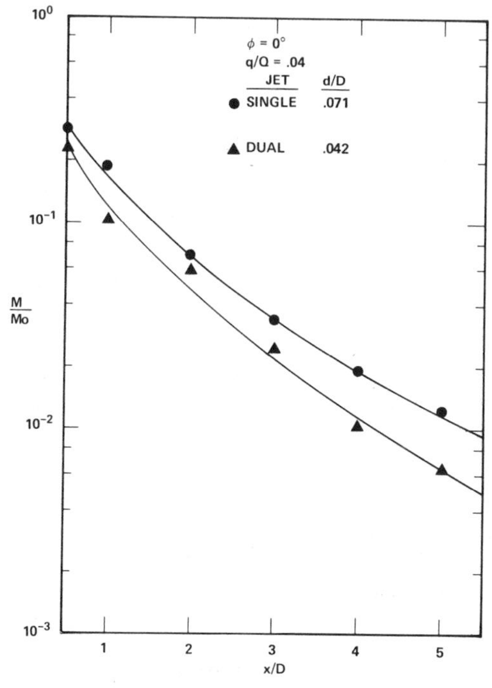

Figure 17. Minimum second moment versus distance from mixer for single and dual centerline jets (after [6]).

improves mixing quality but at the expense of greater power requirements and increased construction costs.

Effect of Reynolds Number

As demonstrated in Equation 3, a complete dimensional analysis of jet properties within a pipe includes both the pipe and jet Reynolds numbers. Currently, a thorough understanding of the possible influence of Reynolds number on jet mixing within pipes is lacking. For example, Yule [20] documented the development of transitional ring vortices in the mixing region of a normal turbulent jet injected into a stagnant fluid for values of Re_j down to 9,000. Recently, Sparrow and Kemink [21] observed the influence of pipe Reynolds number Re_p on the mixing at a tee. They found large variations in the circumferential average Nusselt numbers at short distances $x/D \leq 2$ from the mixing tee and $Re_p > 20{,}000$. Kadotani and Goldstein [22] also concluded that mainstream turbulence can influence thermal mixing by altering the mixing rate and changing the shape of the injected flow due to vortex formation.

O'Leary [12, 23] has attempted to characterize the effects of Reynolds number on optimum mixing conditions for $\theta_j = 90°$. With the optimum mixing criterion of a geometrically centered jet applied at 2, 5, and 10 pipe diameters downstream from the injection point, it was found that the velocity ratio R ($= u_j/v$) appeared to be a constant for $Re_p > 40{,}000$ and $Re_j > 6{,}000$ as shown in Figures 18, 19, and 20. In these figures, the R values have been normalized with respect to the average R from all data for which Re_p and Re_j were greater than 40,000 and 6,000, respectively. Figure 18 represents all the data of O'Leary, while Figure 19 illustrates the effect on R for small $Re_j < 6{,}000$ and for large $Re_p > 40{,}000$. In Figure 20 a sharp increase is shown in R for $Re_p < 40{,}000$ and $Re_j > 6{,}000$. Clearly, care must be taken when attempting to ensure optimum pipeline mixing at small Reynolds numbers.

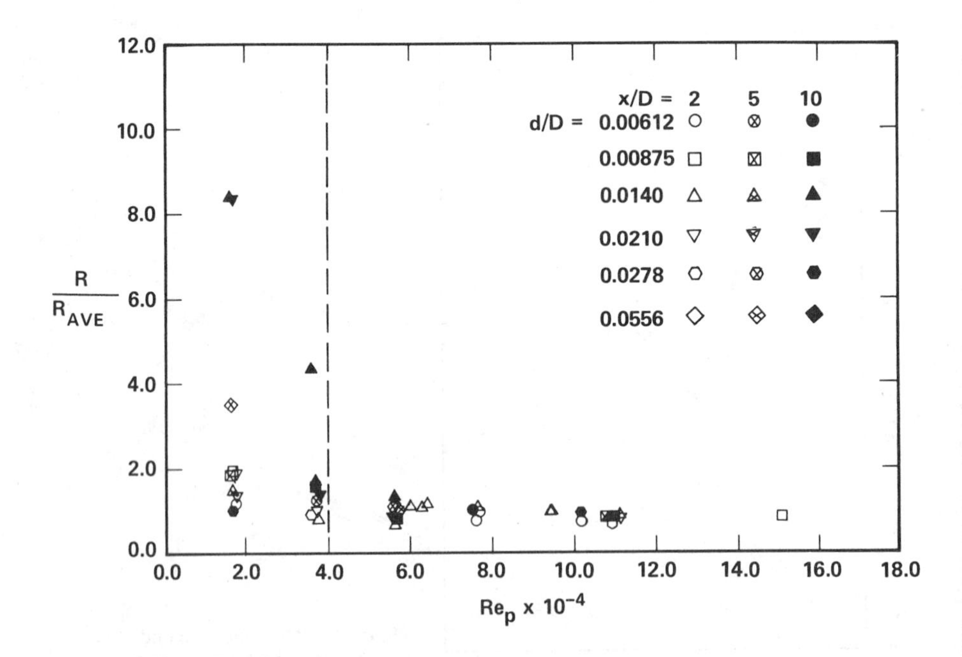

Figure 18. Optimum velocity ratio versus pipe Reynolds number (after [12, 23]).

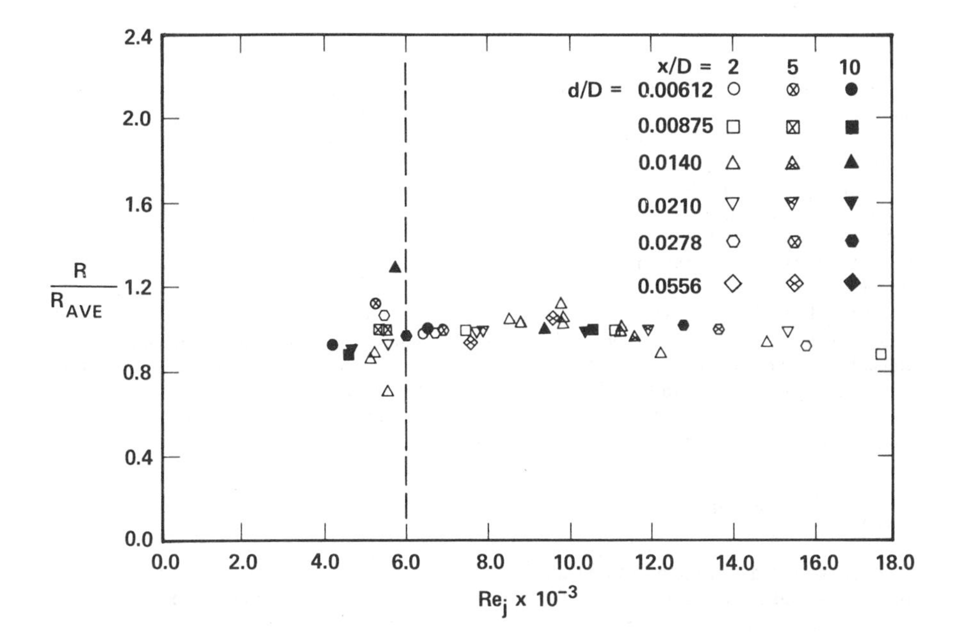

Figure 19. Optimum velocity ratio versus jet Reynolds number for $Re_p > 40{,}000$ (after [12, 23]).

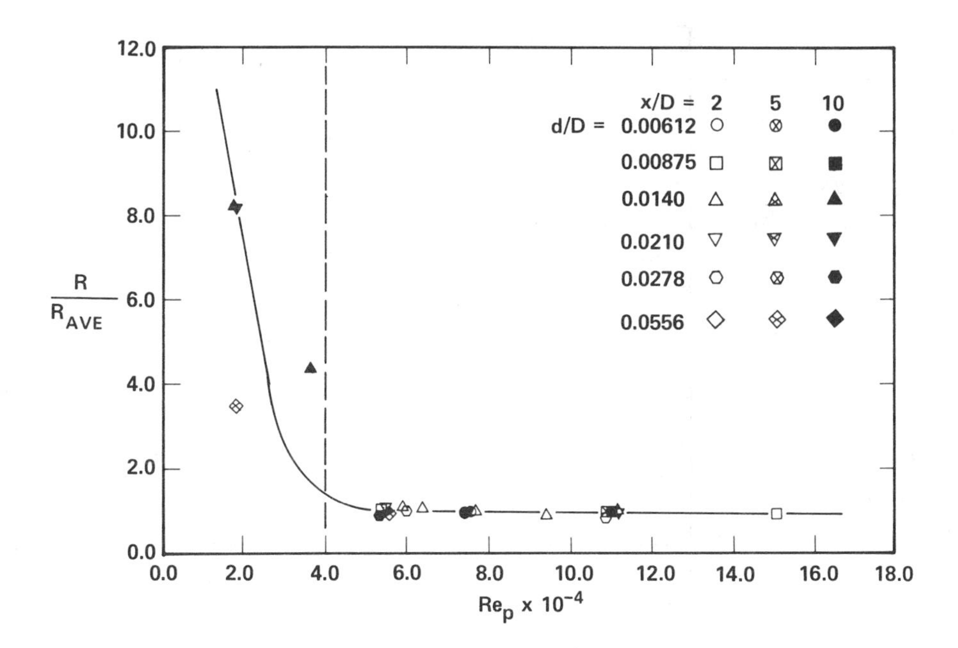

Figure 20. Optimum velocity ratio versus pipe Reynolds number for $Re_j > 6{,}000$ (after [12, 23]).

Effect of Pipe Length

Earlier discussions were concerned with the possible effect of pipe length from the jet to the measurement point in the determination of optimum mixing conditions for $\theta_j = 90°$. Figure 5 demonstrates that if one attempts to minimize the concentration second moment for fixed jet-to-pipe diameter, a significantly larger jet momentum, velocity, or volume flow rate is required when the available pipe length is short ($x/D \gtrsim 7$) compared to the results of identical measurements at longer distances ($x/D > 7$) from the injection point. To see this, compare the contour of Maruyama [6] with the predictions of Holley [9] in Figure 5 for $d/D < 0.1$. Both minimized the concentration second moment but the data of Maruyama near the jet promoted jet impaction on the opposite pipe wall.

O'Leary [12, 23] has measured the optimum mixing ratio q/Q for a range of pipe lengths $2 \leq x/D \leq 10$ and diameter ratios $0.006 < d/D < 0.028$ for geometrically centered jets. Excluding small Reynolds number data $Re_p < 40{,}000$ and $Re_j < 6{,}000$, and normalizing q/Q required for optimum mixing at each location with the average value for measurements at $x/D = 2$, 5 and 10, a graph of the increase in q/Q with decreasing distance from the injection point is shown in Figure 21. The solid line in Figure 21 is represented by the expression

$$\frac{(q/Q)x/D}{(q/Q)_{AVE}} = 1.4(x/D)^{-0.23}, \qquad 2 \leq x/D \leq 10 \tag{21}$$

It should be noted that the measurements in Figure 21 were made with small jet-to-pipe diameter ratios. It is unlikely that larger values of d/D would demonstrate as large a change and in any case the curve will level off for $x/D > 10$.

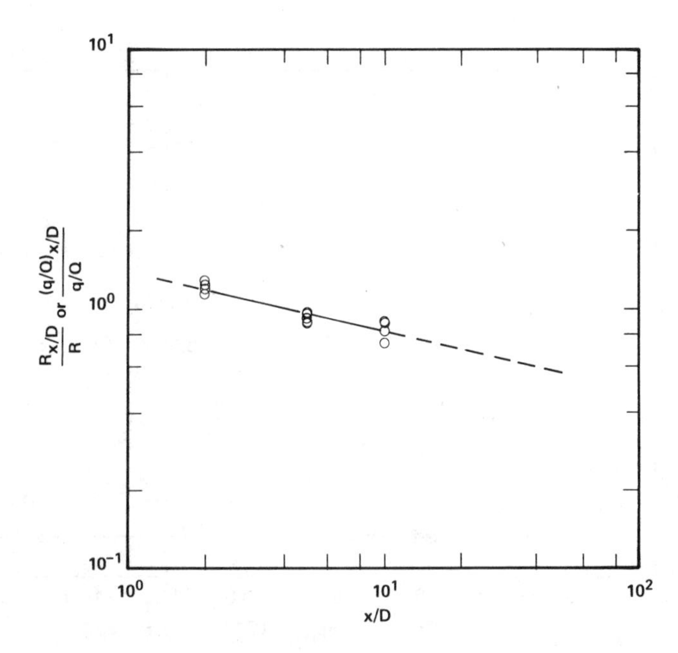

Figure 21. Variation of optimum conditions versus distance from mixer. Solid line is theory (after [12, 23]).

Effect of Buoyancy

The near field trajectory and growth of a buoyant jet in a crossflow are dominated by the initial flux of momentum at the jet exit. After the jet has entrained sufficient ambient fluid it will bend over in the crossflow [14]. At some point downstream depending on the magnitude of the buoyant flux in the jet, the buoyant forces will affect the rise and growth of the nearly horizontal plume. Since the degree of mixing at all locations within the pipe is strongly influenced by jet properties near the injection point (momentum dominated region), it takes a substantial flux of buoyancy to change the mixing quality.

O'Leary [23] introduced a 90% by volume mixture of helium into a pipe such that the Froude number covered the range $83 < Fr < \infty$. This reduced the air density to $\sim 17\%$ of ambient at standard conditions. The value of R required for effective mixing increased slightly and is correlated with the density ratio by the following equation:

$$R_b = R\left(\frac{\rho}{\rho_b}\right)^m \tag{22}$$

where $m = 0.09$.

Although these results agree reasonably well with the value of $m = 0.11$ reported by Nece and Littler [24], more data is necessary to establish a correlation between buoyancy and optimal mixing. However, the experimental data indicate that buoyancy can be neglected without introducing significant error for the range of conditions used by O'Leary. This conclusion is in agreement with previous investigations that predicted negligible effects of buoyancy for $Fr > 50$ [15, 25] from limited experimental results or $Fr > R$ [11] from an analysis of the equations of motion.

In conclusion, it is expected that the form of the scaling laws Equations 6 and 7 and all of the other formula presented would remain unchanged for unequal densities provided the Froude number of the jet is restricted to $Fr > R$.

Concentration Decay

In addition to characterizing the quality of mixing by a decay in the concentration second moment or standard deviation, it may be useful to predict the decrease in the maximum concentration within the jet for $\theta_j = 90°$. For a geometrically centered jet, maximum concentrations were recorded with distance from the injection point, and examples of these measurements are shown in Figure 22.

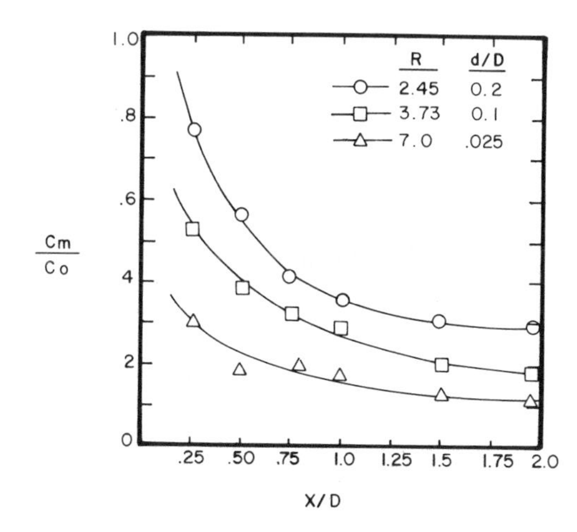

Figure 22. Concentration decay for optimum mixing conditions (after [7]).

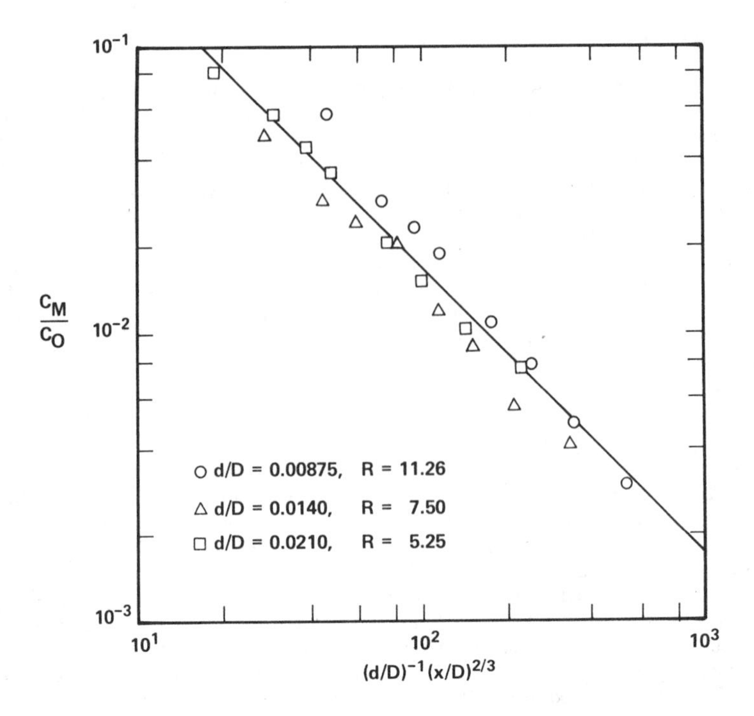

Figure 23. Concentration decay at optimum mixing conditions for small diameter ratios. Solid line is theory (after [12, 23]).

The analysis of Forney [7, 11, 12] provided a scaling law for small diameter ratios of the form

$$\frac{C_m}{C_0} = \frac{1.75(d/D)}{(x/D)^{2/3}}, \qquad d/D < 0.022 \tag{23}$$

where C_m is the maximum concentration within the jet and C_0 is the initial concentration at the jet exit. For large jet-to-pipe diameter ratios the maximum concentration can be predicted with the expression

$$\frac{C_m}{C_0} = \frac{0.7(d/D)^{0.34}}{(x/D)^{1/2}}, \qquad d/D > 0.022 \tag{24}$$

In both expressions the data were limited to values of Reynolds number $Re_p > 40{,}000$ and $Re_j > 6{,}000$. These data are plotted in Figures 23 and 24 and correlated with Equations 23 and 24, respectively, for both large and small jet-to-pipe diameter ratios.

Effect of Secondary Currents

Since secondary currents exist in many pipes, the flow may not be symmetrical with respect to the centerline of the pipe and the degree of mixing may be affected downstream from the mixer. Bends in the pipe, for example, will contribute to a secondary current or swirl downstream.

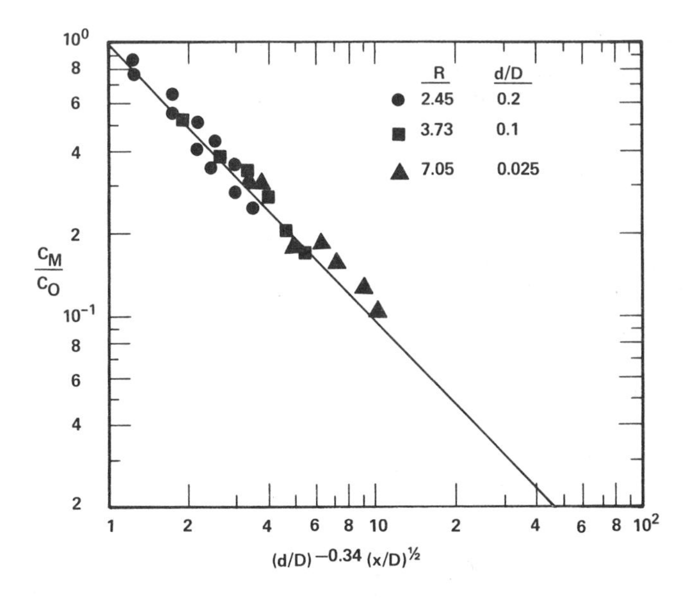

Figure 24. Concentration decay at optimum mixing conditions for large diameter ratios. Solid line is theory (after [12, 23]).

Limited experimental results on the possible influence of secondary current on mixing are provided by Holley [9]. Holley placed a three-blade fixed outboard motor propeller in the pipe 13.5 diameters upstream of the point of injection. The resulting secondary current consisted of a single cell swirl which was approximately symmetrical with respect to the pipe centerline and which caused a full rotation of the flow in ~8 pipe diameters. Both single and dual 90° jet injections were used to compare the mixing characteristics with and without the secondary currents.

A comparison of the single 90° jet with and without the secondary current as shown in Figure 25 indicates that the presence of the secondary current has little effect on the quality of mixing for the first 30 pipe diameters. Beyond this point, however, mixing was less effective with the secondary current. Apparently the single jet was deflected away from the pipe centerline by the current. On the other hand, comparison of the dual 90° jet mixer in Figure 25 indicates that the swirl decreased the mixing distance. One possible explanation for the dual jet mixer given by Holley [5] is as follows: With no secondary current, the two jets approached each other and merged about the pipe centerline. With the swirl in the flow, the jets were deflected in opposite directions so that they were spread out more uniformly across the pipe resulting in improved mixing.

POWER REQUIREMENTS

An important consideration is the power requirement for a pipeline mixer. A complete account of all power requirements for a specific mixing application should include the power required for the velocity head of the jet, the need to pump the jet against the internal pressure in high pressure pipelines, and the pump friction losses and efficiency. Since all of these contributions are unique to a particular mixer with the exception of the velocity head of the jet, the following discussion will focus on the latter.

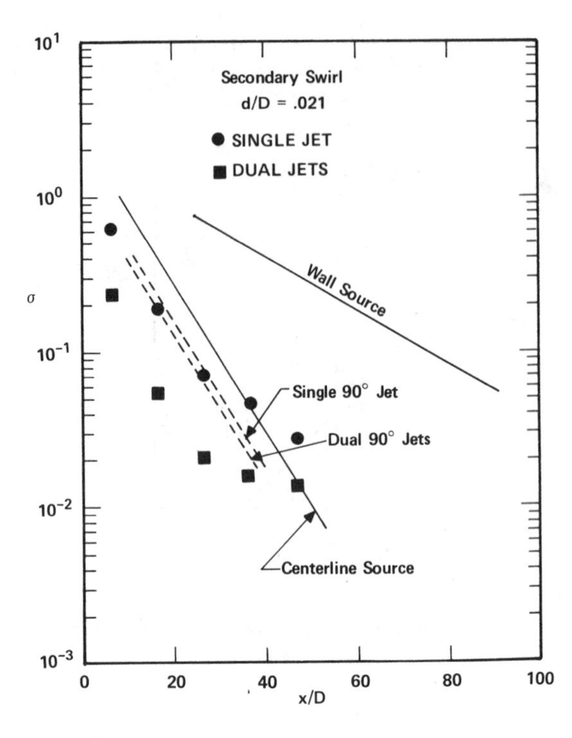

Figure 25. Minimum standard deviation versus distance from mixer showing single and dual centerline jets in a secondary swirl (after [9]).

The power requirement for supplying kinetic energy to the jets can be written in dimensionless form as [5]

$$\frac{P_j}{P_p} = \frac{n\gamma_j q_j \frac{u_j^2}{2}}{\gamma_p Q v^2/2} \tag{25}$$

where P = the power in kinetic energy of the jet or pipe flow
n = the number of jets
γ = the resistance coefficient of the flow in the pipe or jet
q_j = the flow in each jet

It is instructive to assume in Equation 25 that the ratio γ_j/γ_p is roughly a constant as suggested by Miller [26, 4] who found for a 90° jet mixer geometry that $5.71 \geq \gamma_i/\gamma_p \geq 4.9$ for a range in flow ratios of $9.2 < q/Q < 0.7$. Moreover, since one normally requires a fixed flow ratio $q/Q = nq_j/Q$, Equation 25 reduces to

$$\frac{P_j}{P_p} = \gamma_r\left(\frac{q}{Q}\right)R^2 \tag{26}$$

where $R = u_j/v$

$\gamma_r = \gamma_j/\gamma_p$

Restricting the analysis to 90° jets where $R \propto (d/D)^{-1}$ for $d/D < 0.022$ and $R \propto (d/D)^{-0.34}$ for $d/D > 0.022$ [12] one obtains

$$\frac{P_j}{P_p} \propto \left(\frac{q}{Q}\right)\left(\frac{D}{d}\right)^2, \qquad d/D < 0.022 \tag{27}$$

and

$$\frac{P_j}{P_p} \propto \left(\frac{q}{Q}\right)\left(\frac{D}{d}\right)^{0.68}, \qquad d/D > 0.022 \tag{28}$$

Since $q_j/Q \propto d/D$ for $d/D < 0.022$, $q_j/Q \propto (d/D)^{1.66}$ for $d/D > 0.022$, and $q_j/Q = (1/n)(q/Q)$, one obtains

$$\frac{P_j}{P_p} \propto \left(\frac{Q}{q}\right) n^2, \qquad d/D < 0.022 \tag{29}$$

and

$$\frac{P_j}{P_p} \propto \left(\frac{q}{Q}\right)^{2/3} n^{1/3}, \qquad d/D > 0.022 \tag{30}$$

Thus, the power requirements increase significantly with the number of jets for small jet-to-pipe diameter ratios but are considerably less dependent on n for larger d/D. It is also apparent from Equation 29 that an increase in the mixing ratio q/Q will increase the diameter ratio for optimum mixing resulting in a net decrease in power requirements. For other mixer geometries (e.g., oblique jets) the power requirements will increase above that predicted in Equations 29 and 30 since more momentum is required to project the jet to the centerline of the pipe.

INSTABILITIES

To achieve a uniform composition of a gas downstream from any pipeline mixer, the streams to be mixed must be controlled accurately. Since there is usually little backmixing in pipelines because of the relatively flat turbulent velocity profiles, small flow oscillations (possibly slugs) will be transmitted for long distances downstream [1, 27]. For example, Reed [7] states that pressure variations of one percent or more, having a period of one to two seconds, may exist in many gas systems. Gaube [28, 4] states that vortices can activate resonant pressure vibrations in a gas flowing in a pipe with an injected side stream. If the vortices shed from the injected gas stream have a frequency near the natural frequency of the gas column downstream of the mixer, large periodic pressure fluctuations will occur.

Narayan [16] modeled the mixing of two streams in a tee and the subsequent downstream flow with an expression for the energy balance of an isothermal compressible gas. It was demonstrated that small fluctuations in the upstream pressure could cause large changes in the relative flow rates of the two entering streams. These variations in the flow rates produced corresponding composition fluctuations in the downstream flow. Assuming a sinusoidal variation in upstream pressure with a period of approximately 0.5 sec and a peak-to-peak change of roughly 1%, a periodic change of 60% in the concentration could be expected 100 feet from the mixer.

Gray [4] and Simpson [27] state that large pressure drops in the feed streams leading to the mixer will dampen the flow or pressure fluctuations. Moreover, when the pressure drop downstream from the mixer is large, a change in downstream pressure may alter the flow ratio. In addition, a change in the flow rate or pressure of one feed stream will have less effect on the other feed stream flow rate when the after-mixer pressure drop is small. Simpson states that a pressure drop in the control values should be at least 30% of the total fractional pressure drop in the pipe mixer to dampen fluctuations.

LARGE VISCOSITY RATIOS

There are a number of industrial applications which require the mixing of small volumes of a relatively nonviscous fluid such as a dye, reactant or catalyst to a viscous reacting mixture. Several studies have been conducted on impingement mixing geometries which are similar to the dual centerline device of Figure 2 with the exception that all of the fluid originates within the two side tees and the upstream pipe segment of the mixer is closed. A review of impingement mixing of fluids is given by Etchells et al. [28].

Assuming that one or both of the inlet streams are of the same diameter as the pipe, a dimensional analysis for optimum mixing conditions would follow as indicated earlier. Limited experimental results [28, 29] indicate that the momentum ratio $(q/Q)^2(D/d)^2(\rho_j/\rho_p)$ and pipe Reynolds number Re_p play an important role in the quality of mixing downstream from the device. Since most of this work was conducted with side tees of equal diameters or $D \cong d$ and of equal viscosities such that $Re_p \cong Re_j$ the results are consistent with a dimensional analysis of the form

$$q/Q = f(Re_p) \tag{31}$$

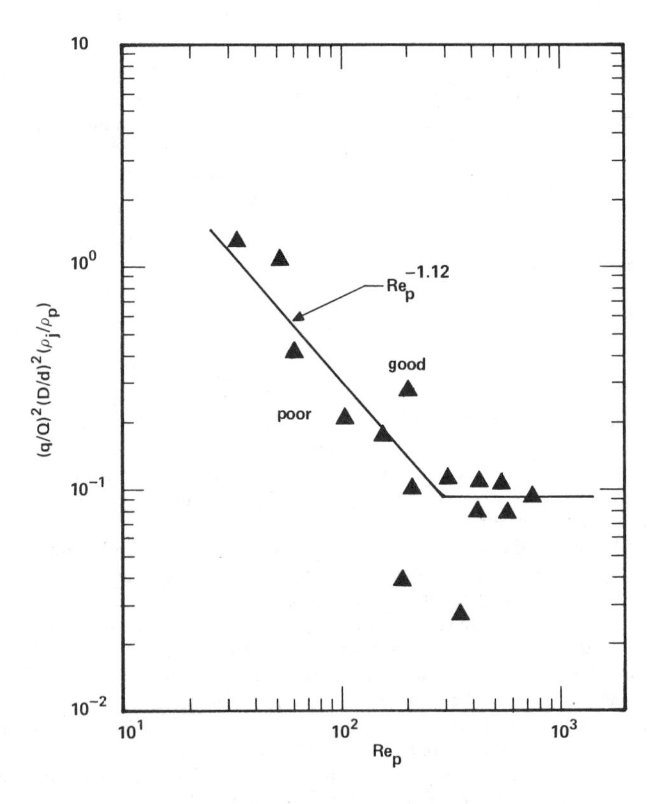

Figure 26. Momentum ratio versus large inlet stream Reynolds number for the mixing of viscous fluids (after [28]).

If, however, both side tees are not of the same diameter and the inlet streams are not of the same viscosity or density, a complete dimensional analysis for optimum mixing would yield

$$d/D = f\left(\frac{\rho_j q}{\rho_p Q}, Re_j, Re_p\right) \quad (32)$$

for d/D provided that either Re_j or Re_p or both are greater than 100 to insure turbulence. Here, Re_p represents the Reynolds number of the large diameter inlet. The restriction of $Re > 10^2$ on the Reynolds number is consistent with observations of the onset of fully developed turbulent jets in a crossflow [30, 31]. If d/D is small, then the diameter of the small inlet stream is not an important length and one obtains

$$\frac{\rho_j}{\rho_p}\left(\frac{q}{Q}\right)^2\left(\frac{D}{d}\right)^2 = f(Re_j, Re_p) \quad (33)$$

where Equations 31, and 33 become independent of Re_j or Re_p if either becomes large.

An example of experimental data for optimum mixing with an impingement device is shown in Figure 26 where $0.14 < d/D < 0.27$, $\rho_j/\rho_p = 0.77$ and $1.7 \times 10^{-3} < \mu_j/\mu_p < 6.3 \times 10^{-3}$. In this experiment the primary fluid was corn syrup and added water with a secondary fluid of dyed water in the smaller side tee inlet [28]. Since $Re_j \sim 10^4$ in this experiment, it is unlikely that Re_j would affect the mixing quality. Part of the scatter of the data in Figure 26, however, may be due to the failure to include the diameter ratio d/D as an independent parameter in the correlation. Clearly, more experimental data are necessary to verify Equations 32 and 33 for their application to the mixing of fluids with large viscosity ratios.

NOTATION

- A pipe cross sectional area (cm^2)
- b local jet radius (cm)
- b_j initial jet diameter (cm)
- C local concentration in pipe (gm/cm^3)
- $\bar{C}$ mean concentration over pipe cross-sectional area (gm/cm^3)
- C_M maximum concentration in jet at fixed x/D (gm/cm^3)
- C_0 initial jet concentration (gm/cm^3)
- d initial jet diameter (cm)
- D pipe diameter (cm)
- Fr jet Froude number, $u_j/(gd|\rho_j - \rho_p|/\rho_p)^{1/2}$
- f pipe friction factor
- g acceleration of gravity (cm/sec^2)
- ℓ_m momentum length, du_j/v (cm)
- m exponent
- M second moment of concentration
- M_0 initial second moment of concentration
- n number of jets
- P_j initial power in kinetic energy of jet flow (watts)
- P_p power in kinetic energy of pipe (watts)
- q total jet flow rate, $(\pi/4)n\, d^2u_j$ (cm^3/sec)
- q_j initial jet flow rate, $(\pi/4)\, d^2u_j$ (cm^3/sec)
- Q pipe flow rate, $(\pi/4)D^2v$ (cm^3/sec)
- $(q/Q)_{x/D}$ jet-to-pipe volume flow ratio at fixed x/D
- $(q/Q)_{AVE}$ average jet-to-pipe volume flow ratio for x/D = 2, 5, and 10
- R optimum velocity ration, u_j/v
- R_{AVE} average optimum velocity ratio for x/D = 2, 5, 10 and for $Re_p > 40{,}000$ and $Re_j > 6{,}000$
- $R_{x/D}$ optimum velocity ratio for fixed x/D
- R_b optimum velocity ratio for buoyant jet
- Re_j jet Reynolds number, $\rho_j u_j d/\mu_j$
- Re_p pipe Reynolds number, $\rho_p vD/\mu_p$
- Sc turbulent Schmidt number
- u_j initial jet velocity (cm/sec)
- v mean pipe velocity (cm/sec)
- x distance downstream from jet (cm)
- x_0 Lagrangian length scale in pipe (cm)
- z distance along pipe diameter from jet (cm)

Greek Symbols

θ local angle between jet and pipe axis
θ_j initial angle between jet and pipe axis
ϕ angle between jet axis and pipe diameter
ϕ_m maximum angle between jet axis and pipe diameter, $(180°/\pi) \sin^{-1}(1 - d/D)$
ρ_b density of buoyant jet (gm/cm^3)
ρ_j initial density of jet (gm/cm^3)
ρ_p pipe density (gm/cm^3)
μ_p pipe viscosity (gm/sec-cm)
μ_j initial jet viscosity (gm/sec-cm)
γ_j coefficient of resistance for jet
γ_p coefficient of resistance for pipe
γ_r ratio, γ_j/γ_p
α_f ratio, $0.048\ f^{1/2}/Sc$
α tangential entrainment parameter
β normal entrainment parameter
σ concentration standard deviation, $M^{1/2}$

REFERENCES

1. Simpson, L. L., *Chemical. Eng. Prog.*, 70:77 (1974).
2. McFarland, B. L., and Landy, D. G., *AIChE Symposium Series*, 76:351 (1980).
3. Simpson, L. L., Chapter VI in *Turbulence in Mixing Operations*, pp. 285–287, R. S. Brodkey (Ed.), Academic Press, N.Y. (1975).
4. Gray, L. B., chapter in *Mixing-Theory and Practice, III*, V. W. Uhl and J. B. Gray (Eds.), Academic Press, N.Y. (to be published).
5. Fitzgerald, S. D., and Holley, E. R., *Research Report No. 144*, Water Resources Center, University of Illinois at Urbana-Champaign, Urbana, Ill., Dec. (1979).
6. Maruyama, T., Hayashiguchi, S., and Mizushina, T., *Kagaku Kogaku Ronbunshu*, 8:327 (1982a).
7. Forney, L. J., and Lee, H. C., *AIChE J.*, 28:980 (1982).
8. Maruyama, T., Watanabe, F., and Mizushina, T., *International Chem. Eng.*, 22:287 (1982b).
9. Fitzgerald, S. D., and Holley, E. R., *J. Hydraulics Div., ASCE*, 107:1179 (1981).
10. Chilton, T. H., and Genereaux, R. P., *Trans. Am. Inst. Chem. Engrs.*, 25:103 (1930).
11. Forney, L. J., and Kwon, T. C., *AIChE J.*, 25:623 (1979).
12. O'Leary, C. D., and Forney, L. J., *I&EC Process Design and Development*, 24:332 (1985).
13. Maruyama, T., Suzuki, S., and Mizushina, T., *Intern. Chem. Eng.*, 21:205 (1981).
14. Wright, S. J., *J. Hydraulics Div., ASCE*, 103:499 (1977).
15. Ger, A. M., and Holley, E. R., *J. Hydraulics Div., ASCE*, 102:731 (1976).
16. Narayan, B. C., M.S. thesis, University of Tulsa (1971).
17. Reed, R. D., and Narayan, B. C., *Chem. Eng.*, 86:131 (1979).
18. Forney, L. J., *J. Hydraulics Div., ASCE*, 109:921 (1983).
19. Maruyama, T., Mizushina, T., and Shirasaki, Y., *Kagaku Kogaku Ronbunshu*, 7:215 (1981).
20. Yule, A. J., *J. Fluid Mechanics*, 89:433 (1978).
21. Sparrow, W. M., and Kemenk, R. G., *Intern. J. Heat Transfer*, 22:909 (1979).
22. Kadotani, K., and Goldstein, R. J., *Trans. ASME*, 101:466 (1979).
23. O'Leary, C. D., M.S. thesis, Dept. Chem. Eng., Georgia Institute of Technology (1982).
24. Nece, R. E., and Littler, J. D., *Tech. Report No. 34*, Charles Harris Hydraulic Lab., University of Washington, Seattle, Washington (1973).
25. Kamofani, Y., and Greber, I., *AIAA J.*, 10:1425 (1972).
26. Miller, D. S., *Internal Flow, A Guide to Losses in Pipe and Duct Systems*, British Hydrodynamics Research Assoc., Cranfield, Bedford, England (1971).
27. Gaube, E., *Chem. Eng. Tech.*, 51:14 (1979).
28. Etchells, A. W., Ford, W. N., and Horgan, J. P., Paper No. 60b, AIChE Annual Meeting, New Orleans, Louisiana (1981).
29. Tucker, C. L., and Suh, N. P., *Polymer Eng. and Science*, 20:875 (1980).
30. Hoult, D. P., and Weil, J. C., *Atmospheric Environment*, 6:513 (1972).
31. Fay, J. A., Escudier, M., and Hoult, D. P., *J. Air Pollution* Control Association, 20:391 (1970).
32. Swanson, W. M., Unpublished Data, Engg. Research Laboratory, E. I. du Pont de Nemours & Co., Wilmington, Delaware (1958).
33. Winter, D. D., M. S. Thesis, University of Illinois, Urbana, Illinois (1975).

CHAPTER 26

SWIRL FLOW GENERATED BY TWISTED PIPES AND TAPE INSERTS

K. Nandakumar and **J. H. Masliyah**

Department of Chemical Engineering
University of Alberta
Edmonton, Alberta
Canada

CONTENTS

INTRODUCTION, 691

GOVERNING EQUATIONS, 692

LAMINAR FLOW RESULTS, 696
Theoretical, 696
Experimental, 699

TURBULENT FLOW RESULTS, 701

CONCLUSION, 703

NOTATION, 703

REFERENCES, 703

INTRODUCTION

The primary motivation for using twisted pipes or twisted tape inserts in circular pipes is to enhance the convective heat and mass transfer processes. This enhancement is mainly due to the following two mechanisms. The changing curvature of the bounding wall forces the fluid elements to follow a curved path which in turn sets up a centrifugal force and induces a secondary flow, thus intensifying the convective transfer processes. This is the dominant in flow through coiled tubes as well. This has been reviewed elsewhere in greater detail [1, 2]. In addition to intensifying the convective transfer processes, a twisted pipe geometry also provides increased surface area per unit length of pipe. The importance of this effect in twisted tape inserts in circular pipes has been recognized, provided the contact between the tape insert and the tube is good, i.e. the contact resistance is not high enough to render this additional area ineffective for the convective transfer process. This has been termed the fin effect for heat transfer enhancement. However, both these beneficial effects are accompanied by an increase in the pressure drop. In order to evaluate the net benefit of using such systems, quantitative data are needed for the pressure drop and heat transfer coefficients. Such data have been gathered since the late 1950s for a number of systems. Detailed numerical predictions of flow and heat transfer in such configurations have started to appear since the early 1970s. The available theoretical and experimental results will be reviewed in this chapter.

The applications suggested in the literature for twisted tubes and tape inserts are quite varied indeed. They include steam condenser of marine power plants [3], gas-cooled nuclear reactor heat exchange systems [4], liquid-cooled nuclear reactor heat exchange systems [5], potassium boilers for an advanced Rankine cycle space power system [6], and transportation of slurry in pipelines with the use of helical vanes or half twisted tapes [7]. Twisted pipe configurations have also been used to model flow through flexible pipes with potential biomechanical applications [8]. The simplest models for flow through porous media treat the porous medium as a bundle of straight, cylindrical tubes. Recognizing the deficiency of such models, many investigators [9, 10] have studied the effect of periodic changes in cross-sectional areas along the axis of the pipe. Twisted tubes and tape inserts provide two more geometrical departures from the straight, cylindrical tube and could throw additional light on the effort needed to model flow in porous media. But the task of integrating the effects of these geometrical departures into a single consistent model for flow in porous media is a formidable one.

As there are two opposing factors to contend with in any augmentation method (viz. increased heat transfer rate and pressure drop), there is a clear need for a systematic method of evaluating the net benefit of using such configurations. Often the macroscopic results are presented as ratios of friction factors and Nusselt numbers of the augmented case to the straight tube or empty tube (without inserts) limit. Such results give a direct measure of the increase in the pressure drop and in the heat transfer over the straight tube limit. A better measure of comparison, particularly for a new system design, would be to compute the increase in Nusselt number for the same power requirement. In some cases the overriding factor may be the compactness of the equipment. In such cases the previous measures are of little use and one must seek a system with the highest transfer coefficient and the highest surface area per unit volume. For performance evaluation of an existing setup, it may be more appropriate to make the comparison based on equal pressure drop, as the head developed by an existing pump may be the limiting factor. Hence there is no single universally appropriate, evaluation technique. From among the many alternate methods available, the one that best fits a given situation must be selected. For additional discussion on the evaluation of augmentation techniques see Bergles [11]

GOVERNING EQUATIONS

The theoretical study of flow and heat transfer in twisted tubes and cylindrical tubes with twisted tape inserts is made difficult by the lack of an orthogonal coordinate system describing the geometry. A rotating frame of reference, sometimes called a helical coordinate system has been used with some success by Date [12], Todd [8] and Masliyah and Nandakumar [13, 14] in studying flow through twisted pipes. Some twisted tube geometries and a cartesian-helical coordinate system are shown in Figure 1. In the cartesian-helical system, the x-y plane undergoes a constant rate of rotation along the z-direction, covering a distance of 2H over a complete rotation of 2π radians. Denoting the fixed frame of reference as $(\bar{x}, \bar{y}, \bar{z})$ and the rotating frame of reference as (x, y, z), a point at (x, y) will trace a helical path in the fixed frame. The spatial coordinates of a point are transformed into the rotating coordinate as

$$\begin{bmatrix} x \\ y \\ z \end{bmatrix} = \begin{bmatrix} \cos\theta_z & \sin\theta_z & 0 \\ -\sin\theta_z & \cos\theta_z & 0 \\ 0 & 0 & 1 \end{bmatrix} \begin{bmatrix} \bar{x} \\ \bar{y} \\ \bar{z} \end{bmatrix} \tag{1}$$

where θ_z is the angle of rotation. For a constant twist rate it becomes, $\theta_z = \pi z/H$. The components of the velocity vector $\underline{v}$ in the helical coordinate system can be represented by

$$\underline{v} = \underline{i}_x v_x + \underline{i}_y v_y + \underline{i}_z v_z \tag{2}$$

where $(\underline{i}_x, \underline{i}_y, \underline{i}_z)$ are the unit vectors in the helical coordinate system. Note that $\underline{i}_z = \bar{\underline{i}}_z$, i.e. the z-direction is the same in both the reference frames. The components of the velocity vector are transformed as follows.

$$\bar{\underline{v}} = \begin{bmatrix} \cos\theta_z & -\sin\theta_z & 0 \\ \sin\theta & \cos\theta_z & 0 \\ 0 & 0 & 1 \end{bmatrix} \underline{v} \tag{3}$$

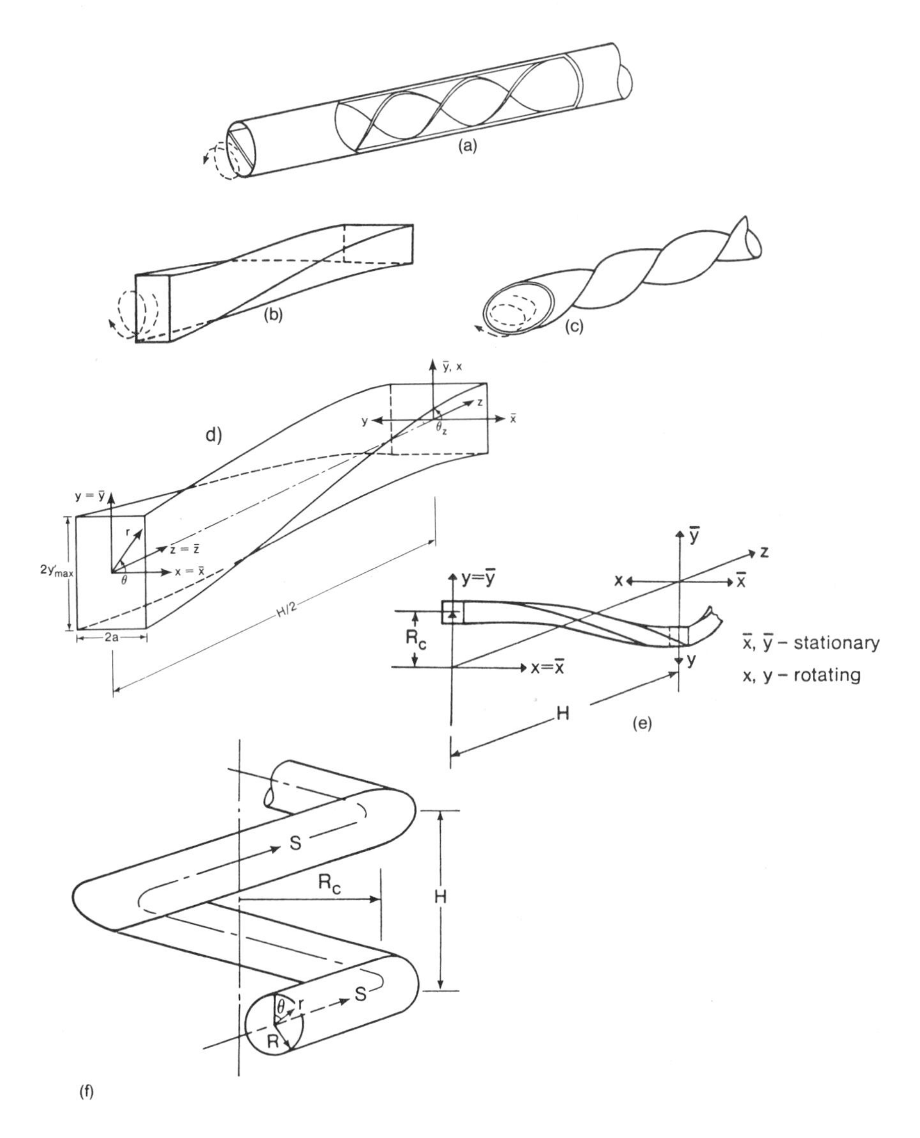

Figure 1. Useful geometrics and coordinate systems: (A) twisted tape insert; (B) twisted rectangular tube; (C) twisted elliptical tube; (D) cartesian-helical coordinate system; (E) nonorthogonal rotating coordinate system; (F) nonorthogonal helical coordinate (r, θ, s).

The derivatives of a scalar function f in the fixed reference frame are transformed as follows.

$$\begin{bmatrix} \dfrac{\partial f}{\partial \bar{x}} \\ \dfrac{\partial f}{\partial \bar{y}} \\ \dfrac{\partial f}{\partial \bar{z}} \end{bmatrix} = \begin{bmatrix} \cos\theta_z & -\sin\theta_z & 0 \\ \sin\theta_z & \cos\theta_z & 0 \\ \dfrac{\pi}{H} y & \dfrac{\pi}{H} x & 1 \end{bmatrix} \begin{bmatrix} \dfrac{\partial f}{\partial x} \\ \dfrac{\partial f}{\partial y} \\ \dfrac{\partial f}{\partial z} \end{bmatrix} \tag{4}$$

With the aid of the transformations defined in Equations 1–4, the momentum and energy equations can be obtained in the helical coordinate system. They follow for the case of a constant twist rate and an incompressible Newtonian fluid.

- *Continuity:*

$$\frac{\partial v_x}{\partial x} + \frac{\partial v_y}{\partial y} + \frac{\partial v_z}{\partial z} + \frac{\pi}{H}\left(y\frac{\partial v_z}{\partial x} - x\frac{\partial v_z}{\partial y}\right) = 0 \tag{5}$$

- *Axial velocity equation:*

$$\begin{aligned} &\rho\left[v_x\frac{\partial v_z}{\partial x} + v_y\frac{\partial v_z}{\partial y} + v_z\frac{\partial v_z}{\partial z} + \frac{\pi}{H}v_z\left\{y\frac{\partial v_z}{\partial x} - x\frac{\partial v_z}{\partial y}\right\}\right] \\ &\quad = -\frac{\partial p}{\partial z} - \frac{\pi}{H}\left\{y\frac{\partial p}{\partial x} - x\frac{\partial p}{\partial y}\right\} + \mu\left[\nabla^2_{x,y,z}v_z + 2\frac{\pi}{H}\left\{y\frac{\partial^2 v_z}{\partial x\,\partial z} - x\frac{\partial^2 v_z}{\partial y\,\partial z}\right\}\right. \\ &\quad\quad \left. + \left(\frac{\pi}{H}\right)^2\left\{y^2\frac{\partial^2 v_z}{\partial x^2} + x^2\frac{\partial^2 v_z}{\partial y^2} - x\frac{\partial v_z}{\partial x} - y\frac{\partial v_z}{\partial y} - 2xy\frac{\partial^2 v_z}{\partial x\,\partial y}\right\}\right] \end{aligned} \tag{6}$$

- *x-component of equation of motion:*

$$\begin{aligned} &\rho\left[v_x\frac{\partial v_x}{\partial x} + v_y\frac{\partial v_x}{\partial y} + v_z\frac{\partial v_x}{\partial z} + \frac{\pi}{H}v_z\left\{y\frac{\partial v_x}{\partial x} - x\frac{\partial v_x}{\partial y} - v_y\right\}\right] \\ &\quad = -\frac{\partial p}{\partial x} + \mu\left[\nabla^2_{x,y,z}v_x + \frac{2\pi}{H}\left\{y\frac{\partial^2 v_x}{\partial x\,\partial z} - x\frac{\partial^2 v_x}{\partial y\,\partial z} - \frac{\partial v_y}{\partial z}\right\}\right. \\ &\quad\quad \left. + \left(\frac{\pi}{H}\right)^2\left\{y^2\frac{\partial^2 v_x}{\partial x^2} + x^2\frac{\partial^2 v_x}{\partial y^2} - 2xy\frac{\partial^2 v_x}{\partial x\,\partial y} - y\frac{\partial v_x}{\partial y} - x\frac{\partial v_x}{\partial x} - v_x + 2x\frac{\partial v_y}{\partial y} - 2y\frac{\partial v_y}{\partial x}\right\}\right] \end{aligned} \tag{7}$$

- *y-component of equation of motion:*

$$\begin{aligned} &\rho\left[v_x\frac{\partial v_y}{\partial x} + v_y\frac{\partial v_y}{\partial y} + v_z\frac{\partial v_y}{\partial z} + \frac{\pi}{H}v_z\left[y\frac{\partial v_y}{\partial x} - x\frac{\partial v_y}{\partial y} + v_x\right\}\right] \\ &\quad = -\frac{\partial p}{\partial y} + \mu\left[\nabla^2_{x,y,z}v_y + \frac{2\pi}{H}\left\{y\frac{\partial^2 v_y}{\partial x\,\partial z} - x\frac{\partial^2 v_y}{\partial y\,\partial z} + \frac{\partial v_x}{\partial z}\right\}\right. \\ &\quad\quad \left. + \left(\frac{\pi}{H}\right)^2\left\{y^2\frac{\partial^2 v_y}{\partial x^2} + x\frac{\partial^2 v_y}{\partial y^2} - 2xy\frac{\partial^2 v_y}{\partial x\,\partial y} - y\frac{\partial v_y}{\partial y} - x\frac{\partial v_y}{\partial x} - v_y + 2y\frac{\partial v_x}{\partial x} - 2x\frac{\partial v_x}{\partial y}\right\}\right] \end{aligned} \tag{8}$$

- *Energy equation:*

$$\rho C_p\left[v_x\frac{\partial T}{\partial x}+v_y\frac{\partial T}{\partial y}+v_z\frac{\partial T}{\partial z}+\frac{\pi}{H}v_z\left\{y\frac{\partial T}{\partial x}-x\frac{\partial T}{\partial y}\right\}\right]$$
$$=k\left[\nabla^2_{x,y,z}T+\left(\frac{2\pi}{H}\right)\left\{y\frac{\partial^2 T}{\partial x\,\partial z}-x\frac{\partial^2 T}{\partial y\,\partial z}\right\}\right.$$
$$\left.+\left(\frac{\pi}{H}\right)^2\left\{y^2\frac{\partial^2 T}{\partial x^2}+x^2\frac{\partial^2 T}{\partial y^2}-2xy\frac{\partial^2 T}{\partial x\,\partial y}-x\frac{\partial T}{\partial x}-y\frac{\partial T}{\partial y}\right\}\right] \tag{9}$$

It is sometimes convenient to solve these equations in stream function-vorticity form. These equations, together with some auxiliary defining equations, can be found in Nandakumar and Masliyah [15]. A slightly less restrictive form of the equations, which makes allowance for a variable twist rate, can be found in Todd [8]. These equations are useful in studying flow through twisted rectangular tubes, as well as flow through helical coils of finite pitch. For example, if the flow domain does not include the z-axis (as illustrated in Figure 1E), then such a domain forms a helical coil of finite pitch. Details of this study can be found in Masliyah and Nandakumar [2]. An alternate form of the nonorthogonal helical coordinate system shown in Figure 1F has been used successfully by Manlapaz and Churchill [16, 17]. The effects of finite pitch on the flow through coiled tubes have also been examined by Wang [18] and Germano [19]. It should be pointed out that the use of a nonorthogonal rotating system of coordinates is not new in the study of laminar flows in helical geometries. Nebrensky et al. [20], Tung and Laurence [21], and Hami and Pittman [22] have made use of such coordinate systems in the study of flow in screw extruders.

The equations of motion in a rotating cylindrical coordinate system can be developed by a set of transformations similar to Equation 1, from a customary stationary system (r, z, θ) to a rotating system ($\bar{r}$, $\bar{z}$, $\bar{\theta}$).

$$r=\bar{r} \tag{10a}$$

$$z=\bar{z} \tag{10b}$$

$$\theta=\bar{\theta}+\frac{\pi z}{H} \tag{10c}$$

They are useful in studying flow through cylindrical pipes with twisted tape inserts. The equations of motion and energy, in their final form, can be found in Date [12] and Adjei [23] and will not be repeated here. Several additional useful equations for the heat transfer area of a twisted tape, the shear stress at a twisted pipe wall, and macroscopic force and energy balances have been developed in Nandakumar and Masliyah [15]. The surface area of a twisted tape of width d and twist rate H is given by,

$$A=\frac{dH}{2}\left[1+\left(\frac{\pi d}{H}\right)^2\right]^{1/2}+\frac{H^2}{2\pi}\ln\left\{\frac{\pi d}{H}+\left[1+\left(\frac{\pi d}{H}\right)^2\right]^{1/2}\right\} \tag{11}$$

The correct form of the viscous dissipation function* in the cartesian helical coordinate system for an incompressible Newtonian fluid is

$$\frac{\phi}{\mu}=2\left[\left(\frac{\partial v_x}{\partial x}\right)^2+\left(\frac{\partial v_y}{\partial y}\right)^2+\left(\frac{\delta v_z}{\delta z}\right)^2\right]+\left[\frac{\partial v_x}{\partial y}+\frac{\partial v_y}{\partial x}\right]^2+\left[\frac{\delta v_y}{\delta z}+\frac{\partial v_z}{\partial y}+\frac{\pi}{H}v_x\right]^2$$
$$+\left[\frac{\delta v_x}{\delta z}+\frac{\partial v_z}{\partial x}-\frac{\pi}{H}v_y\right]^2 \tag{12}$$

---

* Two of the terms are missing in the viscous dissipation function given in Nandakumar and Masliyah [15].

where $\frac{\delta}{\delta z}$ stands for $\frac{\partial}{\partial z} + \frac{\pi}{H}\left(y\frac{\partial}{\partial x} - x\frac{\partial}{\partial y}\right)$. The macroscopic results, either computed by solving the equation of motion for laminar flow or measured experimentally for turbulent flow, are generally presented in the form of friction factor-Nusselt number correlations. They are defined as follows.

$$f = d_h(dP/dz)/(2\rho\langle v_z\rangle^2) \tag{13}$$

$$Nu_h = hd_h/k \tag{14}$$

The equivalent diameter is defined as

$$d_h = 4 \times \text{area/wetted perimeter} \tag{15}$$

The available results for friction factor and Nusselt numbers in laminar and turbulent flows are reviewed next.

LAMINAR FLOW RESULTS

Theoretical

One of the earlier theoretical attempts to study the flow and heat transfer through tubes with twisted tape inserts was by Date [12]. He formulated and solved the equations of motion in a rotating-cylindrical coordinate system using a numerical method. Since then it has been reexamined by duPlessis and Kroger [24] for tubes with tape inserts. Such studies have been extended to twisted pipes of ellipical cross section by Todd [8] and for rectangular cross-sections by Masliyah and Nandakumar [13–15].

Date observed a fairly intense secondary flow and a strong coupling between the axial and secondary flow equations. However, in twisted pipes, such as in Figure 1B and C, axial flow is dominant in the core region and the secondary flow is strong only close to the pipe wall. In twisted tubes with large aspect ratios (i.e. ratio of two sides for a rectangle or the major-to-minor axis for anellipse, etc.) the core region is circular of the order of the smaller dimension and the secondary flow is intense only outside of this core.

As pointed out earlier, it is possible to take the flow domain to be a square or rectangle at a distance R_c away from the z-axis. In such a case a flow through a helical tube of finite pitch is simulated. The secondary flow in this case is intense uniformly over the entire flow domain. Several of these cases are illustrated in Figures 2 and 3 with the axial velocity profile and the ratio of the secondary velocity to axial velocity ($\sqrt{v_x^2 + v_y^2}/v_z$), respectively.

Figure 2A and B are for a twisted square tube. The axial velocity is essentially parabolic, and for tighter twist rates, it is essentially zero outside of the core. Figure 2C, D, and E show similar profiles for rectangular twisted tubes. Once again the flow is parabolic in the core and it is negligible outside this core for small twist rates, H = 2.5 (Figure 2D). However, for larger twist rates, H = 20 (Figure 2C and E), the axial flow outside the core is strongly influenced by centrifugal force and hence the profile shifts to one side. Figure 2F presents an extreme case (a helical tube with finite pitch) where there is no core flow, as the flow domain is outside of the z-axis. The centrifugal force is important over the entire domain and hence the axial velocity shows a shift over most of the flow domain. In this case the secondary velocity is important everywhere as shown in Figure 3F.

In twisted pipes then, if the pipe axis is coincident with the z-axis and if there is a high degree of symmetry (as in polygons of increasing sides) then the coupling between the axial and secondary flow becomes weaker progressively. In the limit of a circular pipe, the axial and secondary flow equations are uncoupled completely and a swirl component can be superimposed on the axial flow without affecting its Poisseuille character. On the otherhand, circular pipes with twisted tape inserts is analogous to helical pipes of very small radius of curvature and hence the coupling between the axial and secondary flow is very strong as observed by Date.

The friction factor data for pipes with tape inserts obtained by Hong and Bergles [25] and that for a helical tube of finite pitch obtained for Masliyah and Nandakumar [14] are given in Figures 4 and 5 respectively. Date observed that the effect of twist rate on friction factor was noticeable only at higher Reynolds number. This was later substantiated in the experiments of Hong and Bergles [25]. For helical pipes, Masliyah and Nandakumar found that the helical number defined as

$$\mathrm{He} = \mathrm{Re}/\sqrt{\mathrm{Rc}(1 + \mathrm{H}^2/\pi^2\mathrm{Rc}^2)} \tag{16}$$

was adequate to correlate the friction factor data. The correlation, developed according to the Churchill and Usagi [46] approach, is given in Table 1.

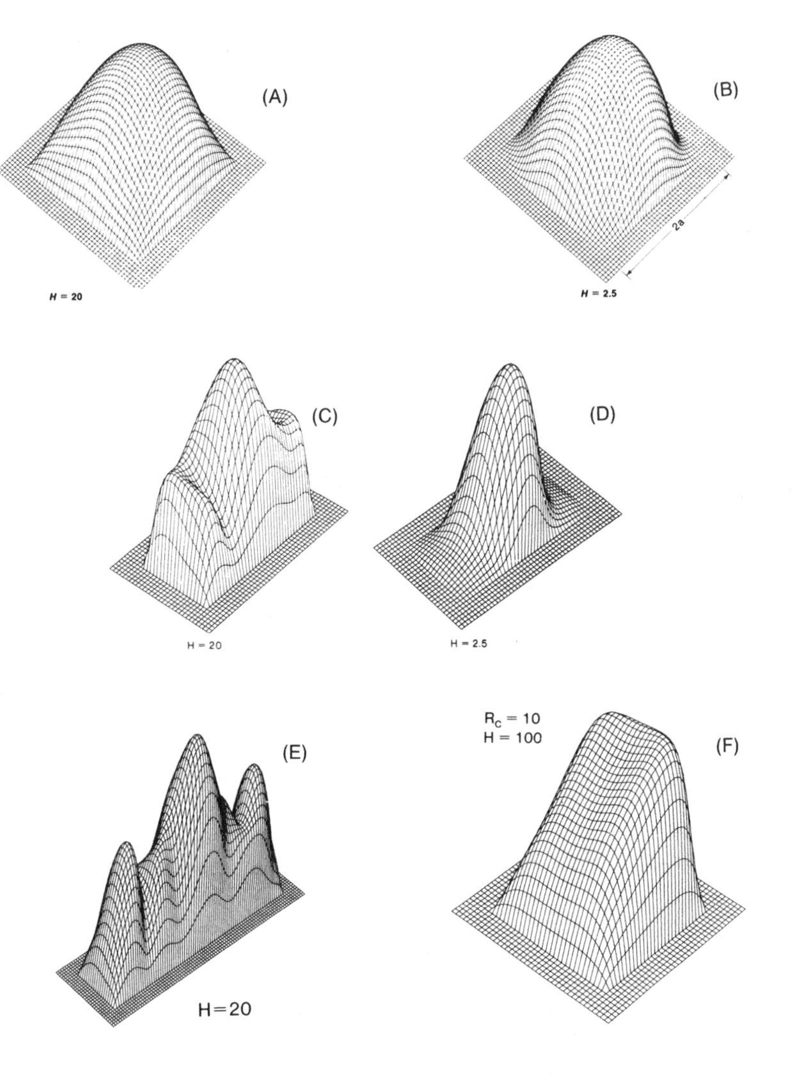

Figure 2. Axial velocity profiles in twisted pipes.

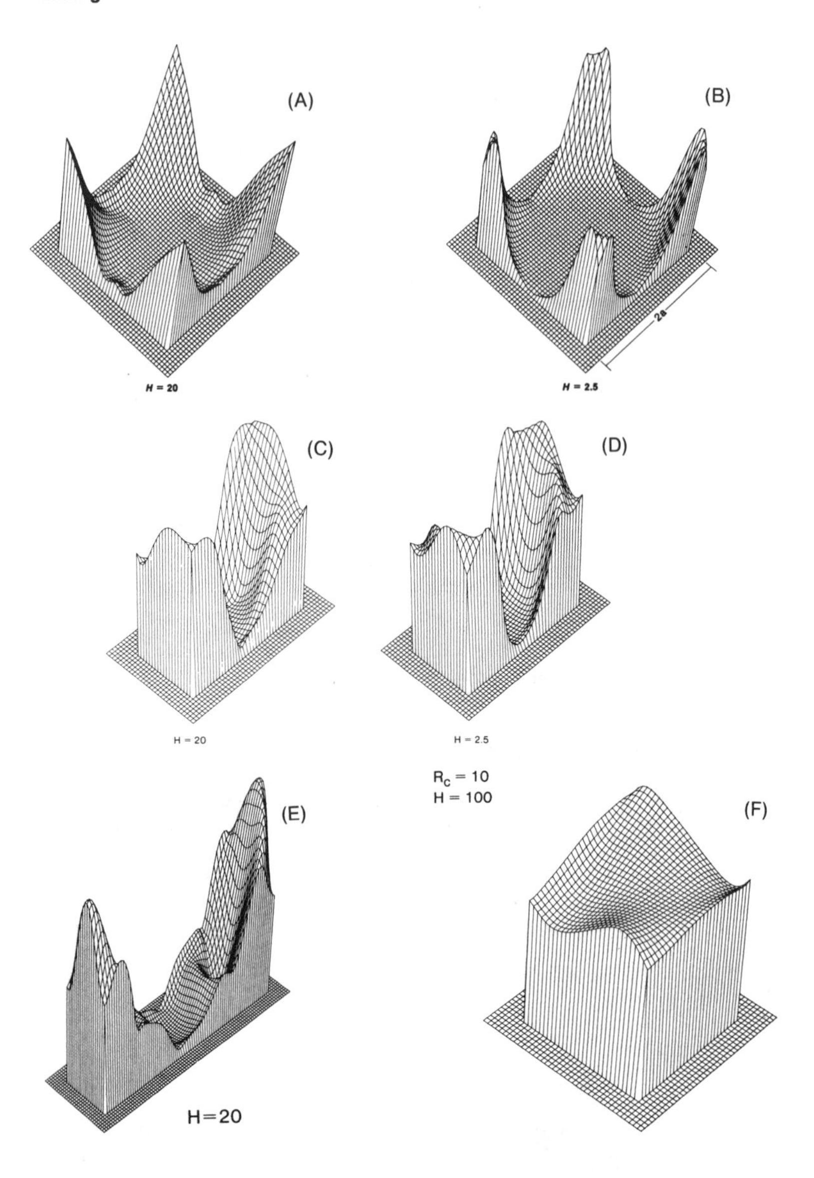

Figure 3. Ratio of secondary velocity to axial velocity in twisted pipes.

Experimental

Experimental measurement of macroscopic results such as friction factor and Nusselt number in laminar flow through a circular pipe with twisted tape inserts have been carried out by Hong and Bergles [25] and van Rooyen and Kroger [26]. Similar results for twisted pipes are not available. Clearly the former is more important, as existing heat exchangers can be more easily adapted with twisted tape inserts to augment heat transfer. The experiments of Hong and Bergles were conducted

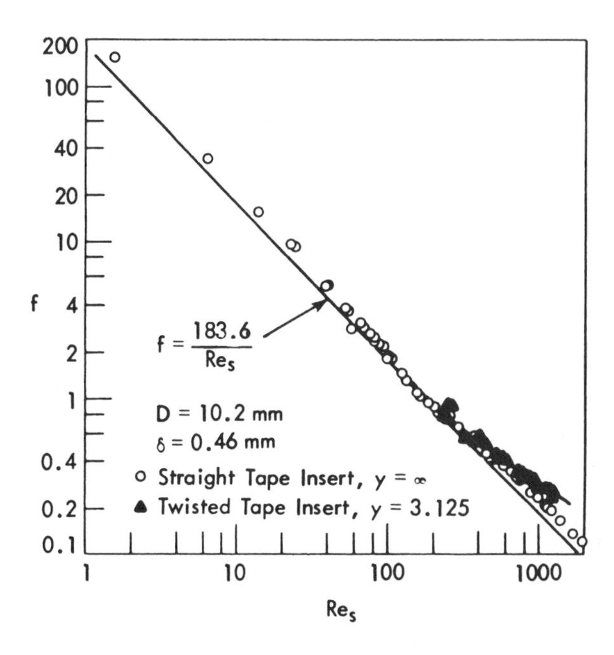

Figure 4. Friction factor for pipes with tape inserts. [25].

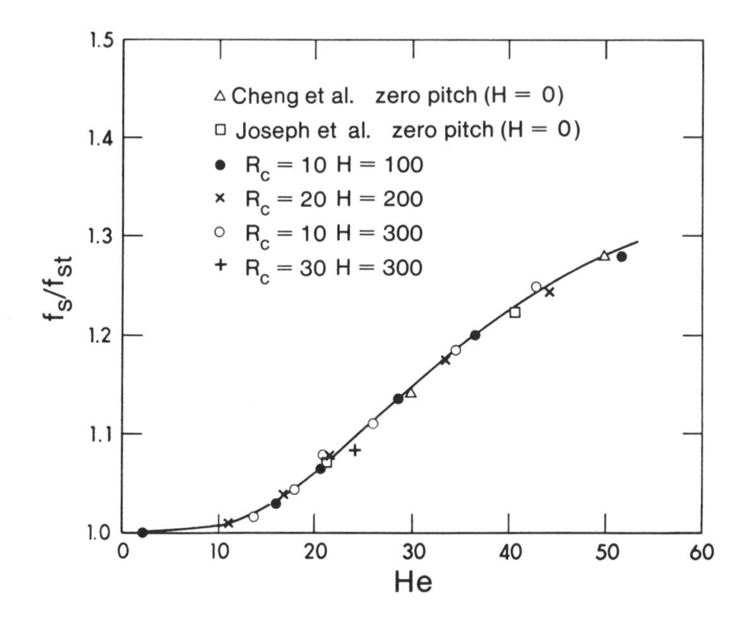

Figure 5. Friction factor in helical coils of finite pitch.

Table 1
Single-Phase Friction Factor Correlations for the Flow of Newtonian Fluids in Pipes with Twisted-Tape Inserts

Equation	Source	Remarks
1. $f_s/f_{st} = [1 + (0.546\ He^{0.219})^{20}]^{1/20}$	Masliyah and Nandakumar [14]	For helical coil of finite pitch, square cross-section laminar flow only based on side, a. He < 55
2. $f_s = f_{st} + (0.0525/y^{1.31})(Re_h/2{,}000)^{-n}$	Gambill and Bundy [32]	Isothermal, turbulent flow $2.5 \times 10^3 < Re_h < 9.5 \times 10^4$, $y > 2.3$
$n = 0.81 \exp[-1{,}700\epsilon/d_h]$		ϵ – absolute roughness
3. $f_s = \dfrac{29.5}{y\sqrt{Re}} + \dfrac{23.75}{Re} + 0.0088$	Seymour [31]	f and Re based on inside diameter, turbulent flow
4. $f_s/f_{st} = 2.75y^{-0.406}$ for non-isothermal case use the following $f_{s,\,iso} = f_s(\mu_b/\mu_w)^{0.35}(d_h/d)$	Lopina and Bergles [37]	Based on hydraulic diameter isothermal, turbulent flow, $y > 2.5$ Nonisothermal
5. $f_s = [0.046 + 2.1(2y - 0.5)^{-1.2}]\ Re_h^{-n}$ where $n = 0.2[1 + 1.7(2y)^{-0.5}]$	Smithberg and Landis [33]	$5 \times 10^3 < Re_h < 10^5$, $y > 1.8$

in an electrically heated tube in the fully developed region. They observe that the effect of tape twist on friction factor is evident only at large Re. This is an agreement with a similar theoretical prediction by Date. However the increase in Nusselt number is quite significant. The empirical correlation developed by Hong and Bergles is given in Table 2. They observe that the maximum increase in heat transfer coefficient is ten times the empty tube value while the corresponding pressure drop is less than four times the empty tube value. Turbulent flow results, to be given later, show a contrasting behavior. In the work by van Rooyen and Kroger the combined effect of an internally finned tube with twisted tape inserts is examined in laminar flow. Heat transfer enhancement of up to four times was observed for the same pumping power.

TURBULENT FLOW RESULTS

One of the earliest experimental studies on the effect of twisted tape insert was made by Colburn and King [27] for Reynolds numbers of up to 50,000. They observed a 50% increase in the heat transfer rate, but a four-fold increase in the pressure drop. Since then a number of investigators have carried out research on various aspects of single-phase flow and heat transfer in circular tubes with tape inserts.

Kreith and Margolis [28] measured heat transfer and friction coefficients for air and water and observed a four-fold increase in the heat transfer coefficient at the same mass flow rate. They also noticed the heat transfer coefficient depended on the temperature gradient. This is believed to be due to a density gradient in a centrifugal force field. Several of the subsequent correlations for the heat transfer coefficient, shown in Table 2, include this effect. They also identified the importance of other effects such as the roughness of the surface and the fin effect of the tape.

Gambill and Bundy [29], in a review of the data available up to that time, illustrated the wide diversity in experimental measurements and reiterated that surface roughness has a marked effect in twisted tape systems.

Seymour [30, 31] also examined the flow characteristics in tubes containing twisted tapes. In addition to the pressure losses, he measured velocity and pressure profiles. The correlation developed in that work is given in Table 1. This correlation departs significantly from others. Gambill and Bundy [32] developed some data under high heat flux conditions using ethylene glycol as the medium. They also examined the pool boiling and burnout characteristics. Smithberg and Landis [33] also measured the velocity profiles and the friction and heat transfer rates. They concluded that the velocity field is helicoidal and corresponds to a forced vortex in the core superposed on an essentially uniform axial flow. Thorsen and Landis [34] extended this work further to evaluate compressibility and buoyancy effects. They considered both heating and cooling of air for a Reynolds number up to 100,000. The temperature-dependent property effects are important under high heat transfer rates or with high wall-to-fluid temperature differences.

Kreith and Sonju [35] examined the decay of tape-induced turbulent swirl flow in pipes. They observed that the turbulent swirl decays to about 10%–20% of its initial intensity in a distance of about 50 pipe diameters. Backshall and Landis [36] measured the boundary-layer velocity distribution in turbulent flow and concluded that it is well represented by the universal logarithmic velocity profile. The correlations developed by Lopina and Bergles [37] take into account the fin effect and the buoyancy effect and are given in Tables 1 and 2. Subsequently Bergles et al. [39, 40] extended their work to include the effect of pipe roughness on flow and heat transfer characteristics. Kidd [40] conducted similar experiments using nitrogen at 14.6 atm as the working fluid over a Reynolds number range of 20,000 to 200,000. His heat transfer correlation is given in Table 2. More recently Klaczak [41] and Drizhyus et al. [42] have also measured the heat transfer coefficients. However, they have not compared their results wtih the earlier works.

Critical heat flux, boiling, and burnout data for water and ethylene glycol can be found in the work of Gambill and Bundy [32] and Gambil, Bundy and Wansbrough [5]. Nazmeev [43] has extended the heat transfer studies to a non-Newtonian fluid using aqueous solutions of sodium carboxymethyl cellulose of different concentrations. Pressure drop and flow characteristics for two-phase air-water flow inside pipes with twisted tapes were studied by Narasimhamurty and Varaprasad [61] for Reynolds numbers of up to 50,000. Their data have been correlated in terms of the Lockhart-Martinelli parameters.

Table 2
Heat Transfer Correlations for Flow in Pipes with Twisted-Tape Inserts

Equation	Source	Remarks
1. $Nu = 5.172[1 + 5.484 \times 10^{-3} Pr^{0.7} (Re/y)^{1.25}]^{0.5}$	Hong and Bergles [25]	Laminar flow, $y = H'/d'$ Re, Nu based on inside diameter fluid properties at bull temperature $13 < Re < 2{,}460$, $y > 2.45$
2. $Nu_h = 0.024\, Re_h^{0.8}\, Pr^{0.4} \left(\frac{T_w}{T_b}\right)^{-0.7} \left[1 + \left(\frac{L}{d}\right)^{-0.55}\right] \left(\frac{y}{y-1}\right)^{1.1}$	Kidd [40]	$\frac{T_w}{T_b} < 2.5$, $y > 2$ $2 \times 10^4 < Re < 2 \times 10^5$ properties based on bulk temperature
3. $Nu = F\{0.023(\alpha\, Re_h)^{0.8}\, Pr^{0.4} + 0.193[(Re_h/y)^2 (d_h/d) \beta\, \Delta T\, Pr]^{1/3}\}$ $\alpha = \sqrt{1 + \pi^2/4y^2}$	Lopina and Bergles [37]	Based on equivalent diameter. For cooling drop the 2nd additive term $10^4 < Re < 2 \times 10^5$, $y > 2.5$ ΔT = (wall temperature − fluid temperature)
4. $Nu = F\{8.8 \times 10^{-3} (\alpha\, Re_h)^{0.915}\, Pr^{0.4}$ $+ 0.193[(Re_h/y)^2 (d_h/d) \beta\, \Delta T\, Pr]^{1/3}\}$	Bergles et al. [38]	For rough tubes in turbulent flow
5. $Nu = 0.025\, Re^{0.84}\, Pr^{0.43} \left(\frac{\mu_b}{\mu_w}\right)^{0.06} \left(0.5 + \frac{8y^2}{\pi^2}\right)^{-0.11}$	Drizhyus et al. [42]	$2 < y < 10$, $6 \times 10^3 < Re < 6 \times 10^4$

CONCLUSION

Extensive experimental data are available on single-phase flow and heat transfer through tubes with twisted tape inserts. With the introduction of the rotating coordinate system by Date [12] for such systems, theoretical and numerical results are beginning to appear for a number of cases. Very limited results are available on the flow of non-Newtonian fluids and two-phase flow in such tubes. Experimental results in twisted pipes are also lacking at present. Further discussions can be found in References 47 through 66.

NOTATION

A	area of a twisted tape, Equation 11	L	length of tube
C_p	specific heat	Nu	Nusselt number, Equation 14
d	inside diameter of tube	p	pressure
d_h	hydraulic diameter, Equation 15	Pr	Prandtl number
f_s	Fanning friction factor, Equation 13, in the presence of swirl flow	r	radial coordinate
		R_c	radius of coiled tube
f_{st}	straight empty tube value	Re	Reynolds number
F	fin effect parameter, Equation 3 of Table 2	T_w	wall temperature
		T_b	bulk temperature
H	tape or tube pitch, length for 180° rotation	v_x, v_y, v_z	velocity components in rotating frame of reference
He	Helical number, Equation 16	y	dimensionless twist rate for tape insert
k	thermal conductivity		

Greek Symbols

θ_z	angle of rotation from the origin	β	coefficient of expansion
μ	viscosity	ϕ	viscous dissipation term
ρ	density		

Subscripts

b	bulk quantity	iso	isothermal condition
h	based on hydraulic diameter	$\langle \cdot \rangle$	average quantity
i	based on inside diameter		

REFERENCES

1. Berger, S. A., Talbot, L., and Yao, L.-S., "Flow in Curved Tubes," *Ann Rev. Fluid Mech.*, 15:461–512 (1983).
2. Nandakumar, K., and Masliyah, J. H., "Swirling Flow and Heat Transfer in Coiled and Twisted Pipes," to appear in *Advances in Transport Processes*, Vol. 4, Wiley Eastern, 1984.
3. Lustander, E. L., and Staub, F. W., "Development Contribution to Condenser Design, Desing and Operating Experiences," *INCO Power Conference*, North Carolina, USA, 1964.
4. Shiralkar, B., and Griffith, P., "The Effect of Swirl, Inlet Conditions, Flow Directions and Tube Diameter on the Heat Transfer to Fluids at Super Critical Pressure," *ASME Journal of Heat Transfer*, 92:465–474 (1970).
5. Gambill, W. R., Bundy, R. D., and Wansbrough, R. W., "Heat Transfer, Burnout, and Pressure Drop for Water in Swirl Flow Through Tubes With Internal Twisted Tapes," *Chemical Engineering Progress Symposium Series*, 57 *No* (32):127–137 (1961).

6. Peterson, J. R., Weltmann, R. N., and Gutstein, M. U., "Thermal Design Procedures for Space Rankine Cycle System Boilers," *IECEC Record Paper No. 689043*, 313–328 (1968).
7. Shook, C. A., and Sagar, S. K., "Turbulent Flow in Helically Ribbed Pipes," *Can J. Chem. Eng.*, 54:489–496 (1976).
8. Todd, L., "Some Comments on Steady, Laminar Flow Through Twisted Pipes," *J. of Engineering Mathematics*, 11:29–48 (1977).
9. Payatakes, A. C., Tien, C., and Turian, R. M., "A New Model for Granular Porous Media," *AIChE J.*, 19:67 (1973).
10. Deiber, J. A., and Schowalter, W. R., "Modelling the Flow of Viscoelastic Fluids Through Porous Media," *AIChE J.*, 27:912–920 (1981).
11. Bergles, A. E., "Augmentation Techniques for Low Reynolds Number in Tube Flow, in Low Reynolds Number Forced Convection in Channels and Bundles," *Proc. of NATO Advanced Study Institute*, July 13–24, 1981.
12. Date, A. W., "Prediction of Fully Developed Flow in a Tube Containing a Twisted-Tape," *Int. J. Heat Mass Transfer*, 17:845–859 (1974).
13. Masliyah, J. H., and Nandakumar, K., "Steady Laminar Flow Through Twisted Pipes: Fluid Flow in Square Tubes," *Trans. ASME, Journal of Heat Transfer*, 103:785–790 (1981).
14. Masliyah, J. H., and Nandakumar, K., "Steady Laminar Flow Through Twisted Pipes: Heat Transfer in Square Tubes," *Trans. ASME, Journal of Heat Transfer*, 103:791–796 (1981).
15. Nandakumar, K., and Masliyah, J. H., "Steady Laminar Flow Through Twisted Pipes: Fluid Flow and Heat Transfer in Rectangular Tubes," *Chem. Eng. Commun.*, 21:151–173 (1983).
16. Manlapaz, R. L., and Churchill, S. W., "Fully Developed Laminar Flow in a Helically Coiled Tube of Finite Pitch," *Chem. Eng. Commun.*, 7:57–78 (1980).
17. Manlapaz, R. L., and Churchill, S. W., "Fully Developed Laminar Convection from a Helical Coil," *Chem. Eng. Commun.*, 9:185–200 (1981).
18. Wang, C. Y., "On the Low-Reynolds-Number Flow in a Helical Pipe," *Journal of Fluid Mechanics*, 108:185–194 (1981).
19. Germano, M., "On the effect of Torsion on a Helical Pipe Flow," *Journal of Fluid Mehanics*, 125:1–8 (1982).
20. Nebrensky, J., Pittman, J. F. T., and Smith, J. M., "Flow and Heat Transfer in Screw Extruders: I. A. Variational Analysis in Helical Co-ordinates," *Polymer Engineering and Science*, 13:209–215 (1973).
21. Tung, T. T., and Laurence, R. L., "A Coordinate Frame for Helical Flows, *Polymer Science and Engineering*," 15:401–405 (1975).
22. Hami, M. L., and Pittman, J. F. T., "Finite Element Solutions for Flow in a Single-Screw Extruder, Including Curvature Effects," *Polymer Engineering and Science*, 20:339–348 (1980).
23. Adjei, D. A., "Numerical Prediction of Newtonian Fluid Flow and Heat Transfer in Twisted Horizontal Circular Tubes," Ph. D. Thesis, University of Saskatchewan (1981).
24. duPlessis, J. P., and Kroger, D. G., "Numerical Prediction of Laminar Flow with Heat Transfer in a Tube with Twisted Tape Inserts," *Numerical Methods in Laminar and Turbulent Flow*," C. Tayler et al., (Eds.) p. 775.
25. Hong, S. W., and Bergles, A. E., "Augmentation of Laminar Flow Heat Transfer in Tubes by Means of Twisted Tape Inserts," *Trans. ASME Journal of Heat Transfer*, 98:251-256 (1976).
26. van Rooyen, R. S., and Kroger, D. G., "Laminar Flow Heat Transfer in Internally Finned Tubes with Twisted Tape Inserts," *Proc. Sixth Int. Heat Transfer Conference*, Toronto, FC(a) 16, 577–581 (1978).
27. Colburn, A. P., "Heat Transfer and Pressure Drop in Empty, Baffled and Packed Tubes," *Ind. Eng. Chem.* 23:910–923 (1931).
28. Kreith, F., and Margolis, D., Heat Transfer and Friction in Turbulent Vortex Flow, *Applied Scientific Research*, A8:457–473 (1959).
29. Gambill, W. R., and Bundy, R. D., *ASME* 62-HT-42 (1962).
30. Seymour, E. V., "A Note on the Improvement in Performance Obtainable from Fitting Twisted Tape Turbulence Promoters to Tubular Heat Exchangers," *Trans. Instn. Chem. Engrs.*, 41:159–162 (1963).

31. Seymour, E. V., "Fluid Flow Through Tubes Containing Twisted Tapes," *The Engineer*, 222:634–642 (1966).
32. Gambill, W. R., and Bundy, R. D., "High-Flux Heat Transfer Characteristics of Pure Ethylene Glycol in Axial and Swirl Flow," *AIChE J.*, 9:55–59 (1963).
33. Smithberg, E., and F. Landis, "Friction and Forced Convection Heat Transfer Characteristics in Tubes with Twisted Tape Swirl Generators," *Trans. ASME J. of Heat Transfer*, 86:39–49 (1964).
34. Thorsen, R., and Landis, F., "Friction and Heat Transfer in Turbulent Swirl Flow Subjected to Large Transverse Temperature Gradients," *Trans. ASME J. of Heat Transfer*, 90:87–96 (1968).
35. Kreith, F., and Sonju, O. K., "The Decay of a Turbulent Swirl in a Pipe," *J. of Fluid Mechanics*, 22:257–271 (1965).
36. Backshall, R. G., and Landis, F., "The Boundary Layer Velocity Distribution in Turbulent Swirling Pipe Flow," *ASME J. of Basic Engineering*, 91:728–733 (1969).
37. Lopina, R. F., and Bergles, A. E., "Heat Transfer and Pressure Drop in Tape Generated Swirl Flow of Single-Phase Water," *Trans. ASME J. of Heat Transfer*, 91:434–441 (1969).
38. Bergles, A. E., Lee, R. A., and Mikic, B. B., "Heat Transfer in Rough Tubes with Tape Generated Swirl Flow," *ASME J. of Heat Transfer*, 91:443–445 (1969).
39. Bergles, A. E., "Survey and Evaluation of Techniques to Augment Convective Heat and Mass Transfer, Int. J. Heat Mass Transfer Ser.," *Progress in Heat and Mass Transfer*, 1:331–424 (1969).
40. Kidd, G. J., "Heat Transfer and Pressure Drop for Nitrogen Flowing in Tubes Containing Twisted Tapes," *AIChE J.*, 15:581–585 (1969).
41. Klaczak, A., "Heat Transfer in Tubes with Sprial and Helical Turbulators," *Trans. ASME J. of Heat Transfer*, 95:557–559 (1973).
42. Drizhyus, M. R., Shkema, R. K., and Shlanchyauskas, A. A., Heat Transfer in a Twisted Stream of Water in a Tube, *International Chemical Engineering*, 20:486–489 (1980).
43. Nazmeev, Yu, G., "Intensification of Convective Heat Exchange by Ribbon Swirlers in the Flow of Anomalously Viscous Liquids in Pipes, Translated from *Inzhenerno-Fizicheskii Zhurnal*, 37:239–244 (1979).
44. Date, A. W., "Some Measurements in Tubes Containing Twisted-Tape Swirl Generators," AE-RL-1157, *AB Atomenergi, Studsvik, Sweden* (1969).
45. Blatt, T. A., and Adt, R. R., Jr., "The Effects of Twisted Tape Swirl Generators on the Heat Transfer Rate and Pressure Drop of Boiling Freon II and Water," *ASME paper no. 63-WA-42*, 1963.
46. Churchill, S. W., and Usagi, R., "Generalized Expression for the Correlation of Rates of Heat Transfer and Other Phenomena," *AIChE J.*, 18:1121 (1972).
47. Danilore, I. B., and Keilin, V. E., "Heat Transfer and Hydraulic Resistance in Flow Along Tubes with Spiral Fins," *Int. Chem. Eng.*, 3:95–98 (1963).
48. Date, A. W., and Singham, J. R., "Numerical Prediction of Friction and Heat Transfer Characteristics of Fully Developed Laminar Flow in Tube Containing Twisted Tapes," *ASME paper no. 72-HT-17* (1972).
49. Date, A. W., "Flow in Tubes Containing Twisted Tapes, A Review of Single Phase Forced Convection Data," *Heating and Ventilation Engineering Journal*, Nov. 1973, 240–249.
50. Domansky, I. V., and Sokolov, V. N., "Heat Transfer in Turbulent Ascending Gas-Liquid Flow in a Vertical Tube with Twisted Tape Turbulence Promoter," *Heat Transfer-Soviet Research*, 8:70–75 (1976).
51. duPlessis, J. P, "The Velocity-Vorticity Procedure Applied to Developing Flow in Tubes with Twisted Tape Inserts," *Report no. TW 77-3*, Dept. of Applied Mathematics, Univ. of Stellenbosch, 1977.
52. Feinstein, L., and Lundberg, R. E., "Fluid Friction and Boiling Heat Transfer with Water in Tubes Containing Internally Twisted Tapes," *Stanford Res. Inst., RADC-TRR*-63-451, Defence Documentation Center A.D.
53. Gambill, W. R., "Subcooled Swirl Flow Boiling and Burnout with Electrically Heated Twisted Tapes and Zero Wall Flux," *Trans. ASME J. of Heat Transfer*, 87:342–348 (1965).

54. Gupta, R. K., and Raja Rao, M., "Heat Transfer and Friction Characteristics of Newtonian and Power Law Type non-Newtonian Fluids in Smooth and Spirally Corrugated Tubes," *AIChE Sym. Ser.*, 75:313 (1979).
55. Kidd, G. J., Jr., "The Heat Transfer and Pressure Drop Characteristics of Gas Flow Inside Spirally Corrugated Tubes," *Trans. ASME J. of Heat Transfer*, 92:513–519 (1970).
56. Klepper, O. H., "Heat Transfer Performance of Short Twisted Tapes," *AIChE J. Sym. Ser.*, 69:87–93 (1973).
57. Kreith, F., and Margolis, D., "Heat Transfer and Friction in Swirling Turbulent Flow," *Heat Transfer and Fluid Mechanics Institute*, Stanford University Press, 126–142, 1958.
58. Marner, W. J., and Bergles, A. E., "Augmentation of Tube Side Laminar Flow Heat Transfer by Means of Twisted Tape Inserts, Static-mixer Inserts, and Internally Finned Tubes," *Proc. Sixth Int. Heat Transfer Conf.*, Toronto, FC(a) 17:583–586 (1978).
59. Migai, V. R., "Intensification of Convective Heat Transfer in Channels by Using Artificial Turbulization of Flow," *Akad. Nauk. SSSR Izv. Energ. i. Trans.*, (6):123–131 (1965).
60. Migai, V. R., "Friction and Heat Transfer Inside Twisted Flow in a Pipe," *Akad, Nauk. SSSR Izv. Energ. i. Trans.*, (5):143–151 (1966).
61. Narasimhamurthy, G. S. R., and Varaprasad, S. S. R. K., "Effect of Turbulence Promoters on Two Phase Gas Liquid Flow in Horizontal Pipes," *Chem. Eng. Sci.*, 24:331–341 (1969).
62. Oullette, W. R., and Bejan, A., "Conservation of Available Work (Energy) by Using Promoters of Swirl Flow in Forced Convection of Heat Transfer," *Energy*, 5:587–596 (1980).
63. Poppendiek, H. F., and Gambill, W. R., "Helical, Forced-Flow Heat Transfer and Fluid Dynamics in Single and Two-phase Systems," Proc. *3rd. Int. Conf. on the Peaceful Uses of Atomic Energy*, United Nations, NY, 274–282 (1965).
64. Razgaitis, R., and Holman, J. P., "A Survey of Heat Transfer in Confined Swirl Flows," *Future Energy Production Systems*, Heat and Mass Transfer Processes, Vol. II, Academic Press, 831–866 (1976).
65. Royds, R., "Heat Transmission by Radiation, Conduction and Convection," 1st Ed., *Constable and Co., London*, 1921.
66. Sethumadhavan, and Raja Rao, M., "Turbulent Flow Heat Transfer and Fluid Friction in Helical-wire-coilinserted Tube," *Int. J. Heat Mass Transfer*, 26:1833–1845 (1983).

CHAPTER 27

MICROMIXING PHENOMENA IN STIRRED REACTORS

J. Villermaux

Laboratoire des Sciences du Genie Chimique, CNRS-ENSIC
Nancy, France

CONTENTS

INTRODUCTION, 708
What is Micromixing, 708

SINGLE-PARAMETER MODELS OF MICROMIXING, 710
Coalescence-Dispersion Model, 710
Interaction by Exchange with the Mean (IEM-Model), 716
Equivalence Between Coalescence-Dispersion and IEM Models, 717
Influence of Micromixing on Conversion and Yield of Chemical Reactions (Predictions of the IEM Model), 719

MIXING "EARLINESS" IN CONTINUOUS STIRRED REACTORS, 733
Entering and Leaving Environments, 733
Transfer from E.E. to L.E.: The "Segregation" Function, 734
Models Based on the "Segregation" Function, 735
Reactors with Two Unmixed Feedstreams, 737
Experimental Determination of Mixing Earliness Using Reactive Tracers, 737
Conclusion: Usefulness of Models Based on "Mixing Earliness", 737

SEGREGATION IN SINGLE-PHASE FLUIDS: MECHANISM OF MICROMIXING IN THE PHYSICAL SPACE, 738
Successive Stages of Micromixing, 738
Time and Space Microscales from Turbulence Theory, 738
Mechanisms for Micromixing in Stage 2, 740
Mechanism for Micromixing in Stage 3: Molecular Diffusion, 742
Micromixing and Chemical Reaction, 743
Experimental Determination of the State of Segregation by Physicochemical Methods, 748
Conclusion, 751

THE REAL STIRRED TANK, 751
Internal Circulation Patterns, 751
Distribution of Turbulence Parameters, 753
Cell Models, 755
Micromixing in a Nonuniform Stirred Tank, 755

PREDICTION AND SCALE-UP OF MICROMIXING PARAMETERS, 757
Dependence of Microscales and Time Constants on Stirring Speed and Reactor Size, 757
Scale-Up of the Stirring Power, 758
Example of Application, 759

EXAMPLES OF INFLUENCE OF MICROMIXING ON CHEMICAL REACTIONS, 759
Fast or Complex Liquid Phase Reactions, 759
Precipitation Reactions, 761
Polymerization Reactions, 761

Polycondensation Reactions, 764
Gas-Phase Reactions, 765
Thermal Segregation, 765

CONCLUSIONS, 766

NOTATION, 767

REFERENCES, 768

INTRODUCTION

Stirred reactors are basic devices for industrial operations such as mixing, blending, and chemical reaction. If good mixing at the macroscopic scale can be easily achieved, it is much more difficult to make sure that intimate mixing is effective on the molecular scale, and this may have a significant influence on the result of the operation, especially when fast and complex chemical reactions are involved. The aim of this chapter is to discuss micromixing phenomena from several viewpoints—definition, representation by suitable models, experimental study, tentative prediction and scale-up—and to see to what extent they may control the performance of reactors.

The scope of this review is limited to single (gas or liquid) phases in (mechanically or self-) stirred reactors of the batch, semi-batch, or continuous type.

Micromixing has been the subject of many academic investigations in the last twenty years. Unfortunately, many investigations have been little more than mathematical games that are over-sophisticated to represent very small effects which might be ignored in current practice. Therefore, it is important to clearly specify the conditions under which micromixing is a controlling factor and to show that in most cases its influence may be accounted for by very simple models.

What is Micromixing?

Let us consider an imperfect mixture in which the concentration C of a given component is not uniform and let p(C) be the local concentration distribution such that p(C) dC is the volume fraction of the mixture where the concentration is comprised between C and C + dC. The average concentration is

$$\bar{C} = \int_0^{C_{max}} Cp(C)\, dc \tag{1}$$

and the variance of the distribution is

$$\sigma^2 = \int_0^{C_{max}} (C - \bar{C})^2 p(C)\, dC \tag{2}$$

whereas the absolute deviation from the mean is $\Delta C = |C - \bar{C}|$. From these quantities, several indices may be defined to characterize the "quality of mixing" [1], namely

$$\delta = \overline{\Delta C}/\bar{C}, \qquad \delta_{max} = \Delta C_{max}/\bar{C}, \qquad \delta_\sigma = \sigma/\bar{C}$$

Let $\overline{\Delta C}_0$ be the value of $\overline{\Delta C}$ just before mixing. For instance, starting from two streams of reduced concentration 0 and 1,

$$\overline{\Delta C}_0 = 2\bar{C}(1 - \bar{C})$$

and

$$\sigma_0 = \bar{C}(1 - \bar{C})$$

Other indices are then

$$\Delta = \overline{\Delta C}/\overline{\Delta C_0}$$

or

$$I_s = \sigma^2/\sigma_0^2$$

the latter being called "intensity of segregation." $\sqrt{I_s}$ is also used.

If two species A and B are mixed, the definition of I_s is

$$I_s = -\frac{\overline{c_A c_B}}{\overline{C_{A0}} \cdot \overline{C_{B0}}} \tag{3}$$

where $c_A = C_A - \overline{C_A}$ and $c_B = C_B - \overline{C_B}$. The smaller all these quantities are, the better the quality of mixing. Corresponding degrees of homogeneity are $1 - \delta$, $1 - \delta_\sigma$, $1 - \sqrt{I_s}$, etc.

Mixing times may be defined as times required for δ, Δ, or I_s to fall down to some prescribed fraction (5%, 1%) of their values before mixing. In Lagrangian coordinates, a characteristic time-constant is also $-\sigma^2/(d\sigma^2/dt)$.

The size of segregated regions is characteristized by the scale of segregation, or concentration macroscale, defined from the autocorrelation function as

$$L_s = \int_0^\infty \frac{\overline{c(x)c(x + r)}}{\sigma^2}\, dr \tag{4}$$

where $c = C - \bar{C}$ at position x and r is the distance between two points.

In spite of the diversity of these indices, which are simultaneously used in the literature, it clearly appears that the value of the degree of mixing experimentally measured will depend on the spatial resolution of the probe used to estimate "local" concentrations C in the mixture. This leads directly to the concept of micromixing. The mixture appears as uniform if the scale of concentration gradients is smaller than the spatial resolution of the probe. Having made the choice of a scale of observation, we shall decide that nonuniformities larger than this scale are concerned with macromixing whereas phenomena occurring below this scale are concerned with micromixing. The distinction between macro- and micro-mixing is thus a matter of convention. Generally, it is admitted that micromixing takes place in the range between molecular dimensions up to a scale where nonuniformity of the fluid can be detected by usual macroscopic means of observation, let us say a few hundred micrometers in a liquid.

As far as mixing in physical space is considered, the fine texture of the fluid is characterized by the state of segregation. A totally segregated fluid is supposed to be made of small groups of molecules (called "aggregates" in the following) which keep their identity and do not mix with each other upon stirring. Such a fluid is called a macrofluid. On the opposite, a well micromixed fluid, also called a microfluid, consists of individual molecules which are free to move with respect to each other. This is the general picture everybody has in mind when thinking of a "fluid." Actually, partial segregation often exists in real fluids, which may be considered then as a mixture of macrofluid and microfluid, or as a collection of small aggregates exchanging material with each other.

Mixing may also be considered from the viewpoint of ages of fluid particles. In open reactors, where material is added to or withdrawn from the batch, all fluid particles have not the same age, and there exists an internal age distribution (IAD). A related concept is the Residence Time Distribution (RTD) in continuous reactors. It is well known, since the pioneering work of Danckwerts [2] and Zwietering [3] that the IAD or the RTD, which are concerned with macromixing, are not sufficient to characterize micromixing. There are for instance an infinite number of ways to arrange fluid particles of different ages with respect to each other within the same IAD (or RTD). In the limiting state of minimum mixedness, contact between particles which are to leave the reactor together occurs at the latest moment. Conversely, in the limiting state of maximum mixedness, this contact occurs at the earliest moment. This concept of "mixing earliness" is distinct from that of segregation, as is discussed in the following, but it is often described also within the frame of the micromixing processes.

SINGLE-PARAMETER MODELS OF MICROMIXING

The simplest description of micromixing in a stirred tank is provided by standard models relying on the following assumptions

- Perfect macromixing—there are no macroscopic gradients within the tank volume. In the case of a CSTR, the IAD and the RTD are respectively written:

$$I(\alpha) = (1/\tau)\exp(-\alpha/\tau) \quad \text{and} \quad E(t) = (1/\tau)\exp(-t/\tau) \tag{55}$$

- The intensity of micromixing is the same throughout the tank. Spatial distributions are ignored.
- Micromixing takes place via a single process or may be represented by one single parameter.
- The fluid is assumed to consist of small particles or "aggregates" having a size below the macroscopic scale and undergoing mass transfer with their environment while retaining their identity upon mixing.

Micromixing is actually involved in two processes:

- In the case of unmixed feed, it provides initial contacting between fluids of different composition. For instance, if both streams behave as macrofluids once in the tank, there is no mixing at all on the molecular scale (and eventually no chemical reaction) in spite of stirring.
- In the case of premixed feed, all components of the mixture are initially present in each aggregate, but micromixing controls whether they undergo further exchange with other aggregates or not. This influence is generally less obvious, except with chemical systems exhibiting a particular sensitivity to mixing effects.

The most popular models of this family are the coalescence-dispersion (C-D) model (or random-coalescence model), and the IEM model (Interaction by Exchange with the Mean). These are briefly described in the following.

Coalescence-Dispersion Model

This model [4] was initially proposed by Curl [5] to represent interaction between droplets in liquid-liquid suspensions. The basic assumptions of the models are:

- All aggregates have the same size.
- The probability of coalescence with a neighboring aggregate is the same for each aggregate and independent of time and of the chemical composition of the aggregate.
- Redispersion occurs immediately after coalescing and uniformization of concentration, producing again two aggregates of the same volume (Figure 1).

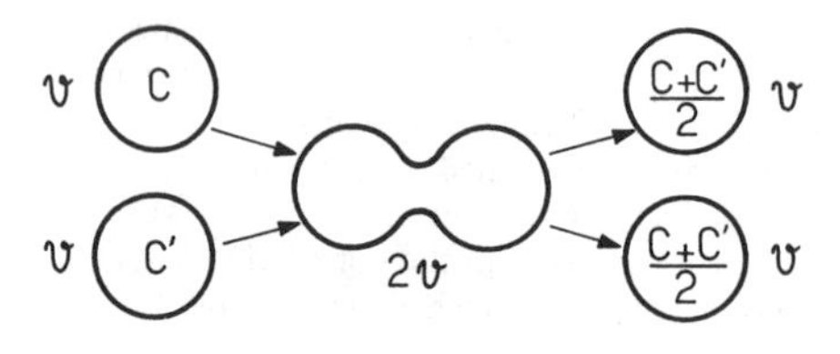

Figure 1. Mechanism for coalescence-dispersion in the C-D model.

- The frequency of coalescence-redispersion is ω, which means that each aggregate undergoes $\omega\,\Delta t$ C-D processes during time Δt.

This random coalescence process can first be represented by deterministic equations, which are presented below.

Batch Reactor Without Chemical Reaction

The tank volume is assumed to consist of n aggregates. The scale of concentration of a given solute is divided into m equal classes ($i = 1, 2, 3, \ldots, m$), each containing n(i) aggregates. The initial concentration distribution is characterized by $n_0(i)$. Interaction between aggregates proceeds with frequency ω. Let $n_k(i)$ be the number of aggregates in class i after k time increments Δt. The equation for the change of the distribution with time is then given by

$$\frac{n_{k+1}(i) - n_k(i)}{\omega\,\Delta t} = -n_k(i) + \sum_{j=1}^{m} \frac{n_k(j)n_k(2i-j)}{n} + \sum_{j=1}^{i-1} \frac{n_k(j)n_k(2i-j-1)}{n} + \sum_{j=1}^{i} \frac{n_k(j)n_k(2i-j+1)}{n} \tag{6}$$

Or, introducing the fraction $g_k(i) = n_k(i)/n$

$$\frac{g_{k+1}(i) - g_k(i)}{\Delta t} = \omega J_k \tag{7}$$

Where J_k is an interaction term defined by

$$J_k = \sum_{j=1}^{m} g_k(j)g_k(2i-j) + \sum_{j=1}^{i-1} g_k(j)g_k(2i-j-1) + \sum_{j=1}^{i} g_k(j)g_k(2i-j+1) - g_k(i) \tag{8}$$

which can also be written

$$J_k = \sum_{j=1}^{m} g_k(j)[g_k(2i-j) + \tfrac{1}{2}g_k(2i-j-1) + \tfrac{1}{2}g_k(2i-j+1)] - g_k(i) \tag{9}$$

If $\Delta t \to 0$, a differential equation is obtained for the continuous change of g(i) with time

$$\frac{dg(i)}{dt} = \omega J \tag{10}$$

J is given by Equation 8 or 9, index k being dropped. The concentration distribution at any time is thus given by the solution of m simultaneous differential equations (Equation 10) ($i = 1, 2, \ldots, m$). A simple example is shown in Figure 2 with $n = 100$ aggregates and $m = 10$ classes. Starting with two populations $C = 0$ and $C = C_0$, the simulation shows how C-D distributes the solute among aggregates, to end up with a uniform composition $C_0/2$.

CSTR Without Chemical Reaction

The description is the same as in a batch reactor, except that the contribution of incoming aggregates (distribution $g_E(i)$) has to be added and that the escape of aggregates at the reactor outlet has to be accounted for. The balance equations are then from Equations 7 and 10, respectively.

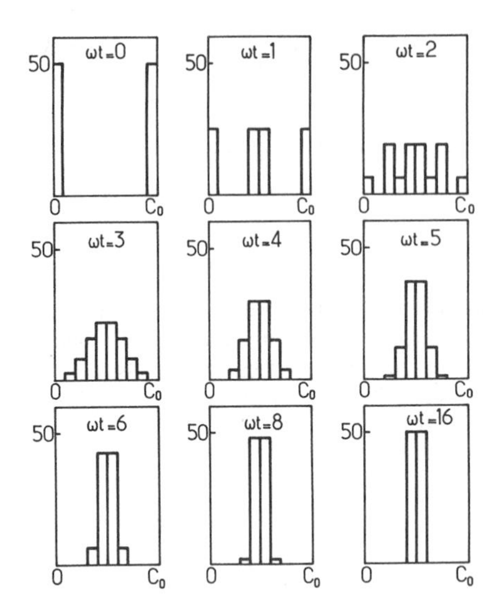

Figure 2. Distribution of concentration in aggregates as a function of time in a batch reactor under the influence of coalescence-dispersion: n = 100 aggregates; m = 10 classes.

$$\frac{g_{k+1}(i) - g_k(i)}{\Delta t} = \omega J_k + \frac{g_E(i) - g_k(i)}{\tau} \tag{11}$$

$$\frac{dg(i)}{dt} = \omega J + \frac{g_E(i) - g(i)}{\tau} \tag{12}$$

Distributions at steady state are obtained by equating the left-hand sides of these equations to zero. Reduction of time $\theta = t/\tau$ brings into evidence the interaction modulus $I = \omega\tau$, which is the micromixing parameter of the C-D model. I is the average number of C-D experienced by a particular aggregate during its stay in the reactor. Balance equations are then written

$$\frac{dg(i)}{d\theta} = IJ + g_E(i) - g(i) \tag{13}$$

or, at steady state

$$g(i) = g_E(i) + IJ \tag{14}$$

Figure 3 shows a simple example of a distribution obtained with different values of the interaction modulus by feeding a CSTR with two equal streams of concentration C = O and $C = C_0$ (n = 100 aggregates, m = 10 classes).

C-D in the Presence of a Chemical Reaction

It is assumed now that the chemical species in the aggregates is produced with rate $\mathscr{R}$ (amount of substance produced per unit volume per unit time). Then $n_{k+1}(i) - n_k(i)$ changes under the influence of interaction among aggregates but also of chemical reaction so that an additional term must be added in the balance. Let ΔC be the span of the concentration class and define the reaction term

$$R_k = \frac{g_k(i)\mathscr{R}_k(i) - g_k(i-1)\mathscr{R}_k(i-1)}{\Delta C} \tag{15}$$

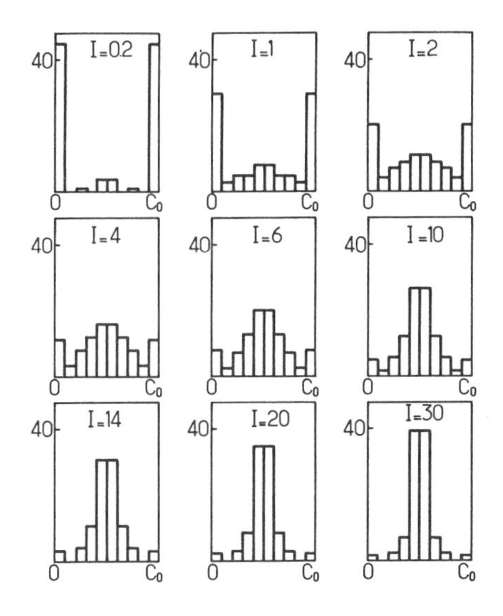

Figure 3. Distribution of concentration in aggregates as a function of the interaction modulus $I = \omega\tau$ in a CSTR under the influence of coalescence-dispersion: $n = 100$ aggregates; $m = 10$ classes.

where $\mathscr{R}(i)$ is the rate of reaction in class i (concentration C_i). It can be shown [4] that Equation 7 for $g_k(i)$ variation becomes

$$\frac{g_{k+1}(i) - g_k(i)}{\Delta t} = \omega J_k - R_k \tag{16}$$

or, with the continuous form (dropping index k):

$$\frac{dg(i)}{dt} = \omega J - R \tag{17}$$

in a batch reactor and

$$\frac{dg(i)}{dt} = \omega J - R + \frac{g_E(i) - g(i)}{\tau} \tag{18}$$

in a CSTR in the transient concentration regime. R is given by Equation 15. The equation for a CSTR at steady state is obtained by equating the left-hand side of Equation 18 to zero.

Example: Consider the case of one single species undergoing a n-th order decomposition with the rate $r = k_R C^n$. The reaction term is given by Equation 15, namely

$$R = \frac{-g(i)r(i) + g(i-1)r(i-1)}{\Delta C} \tag{19}$$

where $r(i) = k_R C(i)^n$. The concentration C(i) in class i may be taken equal to $i\,\Delta C$ or more accurately to $(i - \frac{1}{2})\,\Delta C$. In a well-micromixed fluid of initial concentration C_{A0}, $g(i) = 0$ if $i \neq i_{A0}$ and $g(i) = 1$ if $i = i_{A0}$ where i_{A0} is the integer part of $C_{A0}/\Delta C$.

Even to simulate a CSTR at steady state, it may be advantageous to use the unsteady-state Equation 18 and to start from any kind of distribution in the tank. The steady-state distribution is found asymptotically for long times. Such a numerical procedure is sometimes simpler than the direct solution of algebraic equations for steady state.

When several reacting species are simultaneously present in the aggregates, the formulation of the problem becomes more complex, as multivariable distributions must be introduced to calculate the reaction term. For instance, with two species A and B, two-dimensional quantities are involved, such as $n(i_A, i_B)$, the number of aggregates where the concentration is simultaneously in class i_A for A and in class i_B for B. The Monte-Carlo method described later is better adapted to the simulation of such systems.

Continuous Distributions: Curl's Equation

Instead of discrete distributions over m concentration classes (width ΔC), a continuous distribution $p(C)$ may be introduced as already defined earlier. It is easy to pass from the discrete formulation to the continuous distribution by observing that $g(i) \rightarrow p(C)\,dC$ and $\Delta C = C_0/m$ $(m \rightarrow \infty)$, where C_0 is the maximum concentration

$$\int_0^{C_0} p(C)\,dC = 1 \tag{20}$$

Setting $J = J'\,dC$ and $R = R'\,dC$, the interaction term and the reaction term become, respectively,

$$\begin{aligned} J' &= 2\int_0^{C_0} p(C')p(2C - C')\,dC' - p(C) = 4\int_0^{C} p(C')p(2C - C')\,dC' - p(C) \\ &= 4\int_0^{C} p(C - C'')p(C + C'')\,dC'' - p(C) \end{aligned} \tag{21}$$

$$R' = \frac{\partial(p\mathscr{R})}{\partial C} \tag{22}$$

Equation 18 is then written

$$\frac{\partial p}{\partial t} = \omega J' - R' + \frac{p_E - p}{\tau} \tag{23}$$

where $p(C, t)$ and $p_E(C, t)$ are the concentration distributions at time t in the reactor and in the feed, respectively. Equation 23 is known as the equation of Curl [5], who first established it. For a batch reactor, it reduces to

$$\frac{\partial p}{\partial t} = \omega J' - R' \tag{24}$$

and for a CSTR at steady state

$$p = p_E + IJ' + R' \tag{25}$$

Curl's equation is an integrodifferential partial derivative equation whose numerical solution may appear as difficult. In addition, it is limited to one single species. This is why, here also, the Monte-Carlo method is often preferable.

Moments of the Concentration Distribution

The concentration distribution among aggregates is an interesting feature of the fluid, but it is very difficult to measure, especially in transient regime. In practice, the distribution is often observed through its statistical moments about the origin, especially the average concentration and the variance of the distribution. Let the n-th-order moment of the distribution be

$$\mu_n = \int_0^{C_0} C^n p(C)\, dC \tag{26}$$

In a batch reactor without chemical reaction, Curl's equation is written

$$\frac{\partial p}{\omega\, \partial t} + p = 2 \int_0^{C_0} p(2C - C')p(C')\, dC' \tag{27}$$

Multiplying both sides by C^n and integrating over C, we obtain

$$\frac{d\mu_n}{\omega\, dt} + \mu_n = \frac{1}{2^n} \sum_{i=0}^{i=n} \frac{n!}{i!(n-i)!} \mu_{n-i}\mu_i \tag{28}$$

Taking into account $\mu_0 = 1$, it is found that $d\mu_1/dt = 0$ and that $\mu_1 = \bar{C}$ remains constant and equal to its initial value. More interesting, the variance $\sigma_B^2 = \mu_2 - \mu_1^2$ is found to decrease exponentially according to

$$\sigma_B^2(t) = \sigma_B^2(0) \exp(-\omega t/2) \tag{29}$$

The same treatment for a CSTR yields an equation similar to Equation 28.

$$\tau \frac{d\mu_n}{dt} = \mu_{En} - (1 + I)\mu_n + \frac{I}{2^n} \sum_{i=0}^{i=n} \frac{n!}{i!(n-i)!} \mu_{n-i}\mu_i \tag{30}$$

where μ_{En} is the n-th moment of the distribution in the feed. At steady state ($d\mu_n/dt = 0$), $\mu_1 = \mu_{E1} = \overline{C}_E$, the average concentration in the entering fluid and the variance of the distribution in the tank is

$$\sigma_s^2 = \frac{\sigma_E^2}{1 + I/2} \tag{31}$$

The larger the interaction modulus $I = \omega\tau$, the larger the reduction of the variance of the distribution under the influence of micromixing. It is interesting to notice that σ_s^2 results from σ_B^2 through the RTD

$$\sigma_s^2 = \frac{1}{\tau} \int_0^{\infty} \sigma_B^2(t) \exp(-t/\tau)\, dt \tag{32}$$

Simulation of Micromixing by the Method of Monte-Carlo

Instead of trying to write down balance equations for the concentration distribution and to solve them by deterministic methods, it is often more simple to rely on the random character of C-D processes and to simulate them using the method of Monte-Carlo. Substantially, this consists in picking up two aggregates at random in the population present in the reactor and causing them to coalesce, to equalize their concentrations, possibly to chemically react and then to redisperse them. For instance in a set of n aggregates, each containing a concentration C_i ($i = 1, 2, 3, \ldots, n$), a pair of aggregates (concentrations C_i and C_j) are chosen at random and the concentrations are replaced by $(C_i + C_j)/2$ in both aggregates. As there are $\omega n/2$ coalescences-redispersions per unit time in the reactor, this process is repeated $\omega n\, \Delta t/2$ times (supposed an integer) to obtain the

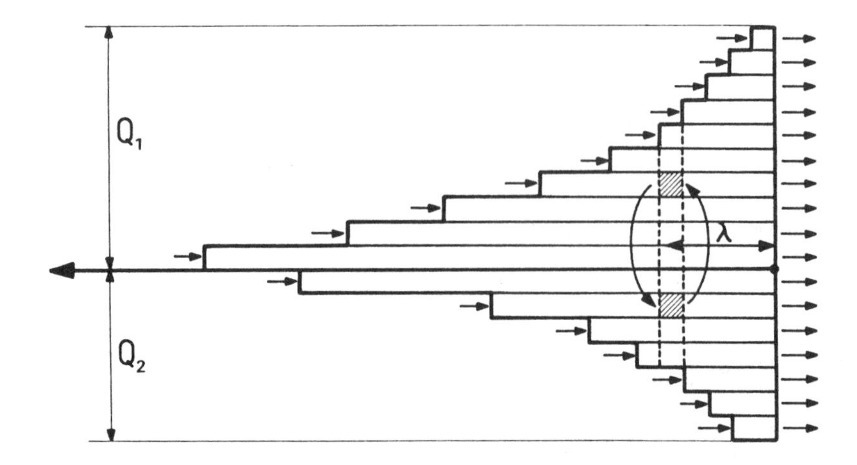

Figure 4. Simulation by the C-D model. Reactor with two unmixed feed-streams. Coalescence is allowed to occur between fluid elements having the same life expectancy λ.

concentration distribution at time $t + \Delta t$. In a CSTR, $n \Delta t/\tau$ aggregates have to be drawn at random during the same time Δt and replaced by entering aggregates having the proper inlet distribution. In addition, if a chemical reaction is taking place within the aggregates, the chemical species are allowed to react batchwise for the period of time between successive coalescences. Each coalescence initiates new initial conditions for the batch reaction. This is a brief outline of the method first proposed by Spielman and Levenspiel [6] and extensively used later by many authors [7, 8]. Of course, in order to avoid statistical noise, the number n of aggregates must be very large (on the order of several hundreds to several thousands of particles). The time increment Δt must also be properly chosen as a function of ω and n to avoid random fluctuations in the results. This requires large-capacity computers.

The method can be extended to stirred tanks with two unmixed feedstreams. Figure 4 shows a model of the kind proposed by Kattan and Adler [9] to simulate different intensities of micromixing in a reactor fed with separate streams of flow rates Q_1 and Q_2. The residence time distribution (which is here that of a perfect macromixer for both streams) is accounted for by the length of the parallel tubes in which the fluid is assumed to flow with the same velocity (bundle of parallel tubes, or BPT-model). All fluid elements on a given vertical have the same life expectancy λ and are thus allowed to coalesce. Micromixing is simulated by choosing two fluid elements at random (shaded blocks) and causing them to coalesce and redisperse with frequency ω. Simultaneously, the fluid flows from left to right and chemical reactions proceed along the streams in between coalescences. $\omega = 0$ corresponds to a macrofluid and $\omega = \infty$ to a microfluid. This method was later extended to any kind of RTD and exploited for simulating imperfect micromixing in various situations [10–12].

Interaction by Exchange with the Mean (IEM-Model)

The rigorous treatment of the C-D model leads to balance equations which are difficult to solve, owing to their integrodifferential character, or to statistical simulations requiring large computer capacities. This is essentially due to the complexity of the interaction term J. In order to overcome this difficulty, a much simpler model can be imagined where interaction between aggregates is accounted for by an equivalent mass transfer process. Mass exchange between aggregates is characterized by a micromixing time t_m which is nil in a macrofluid and infinite in a microfluid. The basic equation for the variation of the concentration C_j of a species A_j in an aggregate of age α is written

$$\frac{dC_j}{d\alpha} = \frac{\langle C_j \rangle - C_j}{t_m} + \mathscr{R}_j \tag{33}$$

$\langle C_j \rangle$ is the mean concentration of A_j in the neighboring aggregates with which interaction takes place and $\mathscr{R}_j$ the rate of production of A_j by chemical reaction. t_m may be a function of the aggregate's age, and $\langle C_j \rangle$ is defined by the condition that the net sum of all exchange fluxes over the whole population is equal to zero. Taking into account the exponential IAD, this leads to

$$\langle C_j \rangle = \frac{\int_0^\infty C_j t_m^{-1} \exp(-\alpha/\tau)\, d\alpha}{\int_0^\infty t_m^{-1} \exp(-\alpha/\tau)\, d\alpha} \tag{34}$$

Frequently, t_m is assumed to be constant, and $\langle C_j \rangle$ reduces to the average concentration $\overline{C_j}$ in the tank (and thence at the reactor outlet)

$$\overline{C_j} = \frac{1}{\tau} \int_0^\infty C_j \exp(-\alpha/\tau)\, d\alpha \tag{35}$$

The solution of the IEM model requires an iterative procedure as $C_j(\alpha)$ must be known to calculate $\langle C_j \rangle$, but this poses no numerical problem and the convergence is usually very fast.

The idea of the IEM model is not new. After Harada et al. [13], it was simultaneously proposed by Villermaux and Devillon [14] and Costa and Trevisoi [15], and later developed in a series of papers [16, 17]. The micromixing parameter is sometimes referred to as a mass transfer coefficient, which is the reciprocal of t_m in Equation 33.

Equivalence Between Coalescence-Dispersion and IEM Models

In order to show that the simplification introduced in the IEM model is valid, an equivalence may be sought with the more detailed C-D model. This can be done for instance by comparing the predictions of both models for the variances of concentration distributions in a batch reactor and in a CSTR fed with two unmixed populations of tracer of respective concentrations $C_{10} = C_0$ (fraction a) and $C_{20} = 0$ (fraction $1 - a$). The expressions obtained with the C-D model are given by Equation 29 and 31 where $\sigma_0^2 = \sigma_E^2 = C_0^2\, a(1 - a)$. With the IEM model, the corresponding expressions are obtained as follows: In a batch reactor, the average concentration remains equal to aC_0. The concentration in "rich" aggregates decreases according to

$$C_1/C_0 = a + (1 - a) \exp(-t/t_m) \tag{36}$$

whereas in the "poor" aggregates, it increases

$$C_2/C_0 = a(1 - \exp(-t/t_m)) \tag{37}$$

The variance of the whole population is thus

$$\sigma_B^2/C_0^2 = a(C_1/C_0 - a)^2 + (1 - a)(a - C_2/C_0)^2 = a(1 - a) \exp(-2t/t_m) \tag{38}$$

This is identical to Equation 29 provided that

$$\omega t_m = 4 \tag{39}$$

In a CSTR, Equation 36 and 37 represent the variation of concentration as a function of the aggregate's age. The concentration distribution within the tank is such that

$$\left.\begin{array}{ll} C > aC_0 & p_1(C_1)\,dC_1 = (a/\tau)\exp(-\alpha)\,d\alpha \\ C < aC_0 & p_2(C_2)\,dC_2 = ((1-a)/\tau)\exp(-\alpha)\,d\alpha \end{array}\right\} \tag{40}$$

Introducing dimensionless variables, $\gamma = C/C_0$ and the micromixing parameter $b = \tau/t_m$, the two branches of the distribution are obtained

$$\left.\begin{array}{ll} \gamma > a, & p_1(\gamma_1) = \dfrac{a}{b(1-a)^{1/b}}(\gamma_1 - a)^{(1-b)/b} \\ \gamma < a, & p_2(\gamma_2) = \dfrac{1-a}{ba^{1/b}}(a - \gamma_2)^{(1-b)/b} \end{array}\right\} \tag{41}$$

and the variance is calculated

$$\sigma_s^2/C_0^2 = \int_0^a (a-\gamma_2)^2 p_2(\gamma_2)\,d\gamma_2 + \int_a^1 (\gamma_1 - a)^2 p_1(\gamma_1)\,d\gamma_1$$

$$\sigma_s^2/C_o^2 = \frac{a(1-a)}{1+2b} \tag{42}$$

Comparing with Equation 31, the equivalence condition is found identical to Equation 39. In dimensionless form:

$$I = 4b$$

or

$$\omega\tau = 4\tau/t_m \tag{43}$$

However, the models are of different nature: with the C-D model, concentration is distributed in the whole range between 0 and C_0, whereas with the IEM model, there are only two values of C (Equations 36 and 37) at a given time. However, the IEM model yields predictions for chemical conversion which are in good agreement with those of the C-D model. For instance, in a CSTR fed with equal streams of A and B undergoing instantaneous reaction the IEM equations are ($\theta = \alpha/\tau$):

$$\left.\begin{array}{l} \dfrac{dC_A}{d\theta} = b(\overline{C_A} - C_A) + \mathscr{R}_A \\ \dfrac{dC_B}{d\theta} = b(\overline{C_B} - C_B) + \mathscr{R}_B \end{array}\right\} \tag{44}$$

with $\overline{C_A} = \overline{C_B}$ and $\mathscr{R}_A = \mathscr{R}_B = -r$ very large. By subtraction

$$\frac{d(C_A - C_B)}{d\theta} = -b(C_A - C_B) \tag{45}$$

but A and B cannot coexist in the same aggregate. For instance, in the A population, $C_B = 0$ and

$$C_A = C_{A0}\exp(-b\theta) \tag{46}$$

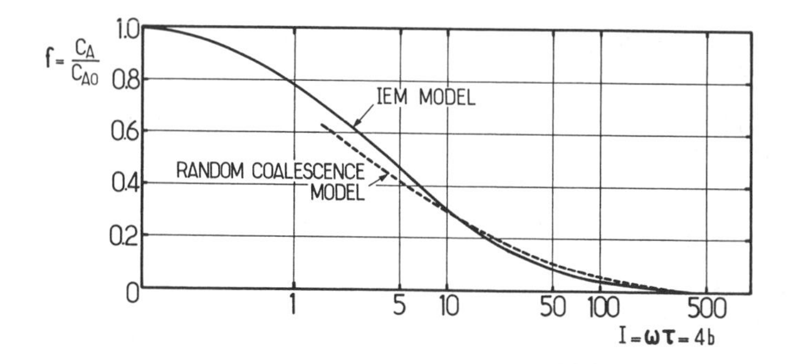

Figure 5. Equivalence between IEM and C-D models in the case of an instantaneous reaction.

Upon averaging

$$\overline{C_A} = C_{A0} \int_0^{\infty} \exp[-(b + 1)\theta] \, d\theta \tag{47}$$

and the residual concentration at the reactor outlet is

$$f_A = \overline{C_A}/C_{A0} = (1 + b)^{-1} \tag{48}$$

This very simple result is plotted in Figure 5 together with the prediction of the C-D model. The agreement is satisfactory, especially for $b > 1$. This is not surprising because $b < 1$ corresponds to very low coalescence frequencies and the statistical model becomes questionable. Many calculations performed with finite rate reactions have confirmed the relevance of equivalence conditions Equations 39 and 43 [18]. Owing to its simplicity and to the fact that it gives almost the same results as much more sophisticated models, the IEM model can thus be strongly recommended for chemical engineering calculations.

Influence of Micromixing on Conversion and Yield of Chemical Reactions (Predictions of the IEM Model)

A few examples of calculations that can be performed with the IEM model follow. The reader will easily generalize the method to cases of his own. Reduced variables will be used throughout. Figure 6 shows the notations when the stirred tank is fed with two streams (volumetric flow rates

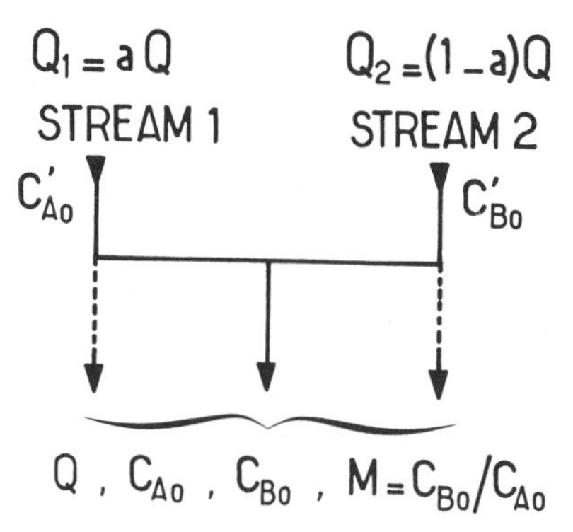

Figure 6. Notations for a CSTR fed with two unmixed streams.

Q_1 and Q_2) containing respectively A (concentration C'_{A0}) and B (concentration C'_{B0}). Two situations may be studied:

- *Premixed feed.* Both streams are supposed instantaneously micromixed before entering the reactor. The concentrations become $C_{A0} = aC'_{A0}$ and $C_{B0} = (1 - a)C'_{B0}$ where $a = Q_1/Q$ is the fraction of the total volumetric flow rate $Q = Q_1 + Q_2$ initially containing A.
- *Unmixed feed.* The streams are fed separately into the reactor but the reference is still the fictitious premixed stream.

In both cases, for a chemical reaction $A + B \rightarrow R$, the stoichiometric excess of B is $M = C_{B0}/C_{A0}$ and concentrations in the tank are normalized by C_{A0} (e.g., $f_A = C_A/C_{A0}, f_B = C_B/C_{A0}, f_R = C_R/C_{A0}$). For reaction of rate $r = kC_A^n C_B^n$ a characteristic reaction time $t_R = (kC_{A0}^{n-1})^{-1}$ may be defined, where $n = n_A + n_B$ is the global kinetic order. In a CSTR, the Damköhler number is then $Da = \tau/t_R = kC_{A0}^{n-1}\tau$. The case of a reaction with one single reactant $A \rightarrow$ products can be dealt with using the same equations and setting $M = 1$.

Micromixing parameters are either

$$b' = t_R/t_m \qquad \text{(all kinds of stirred tanks)}$$

or

$$b = \tau/t_m \qquad \text{(CSTR only)}$$

in the latter, $b = b' \cdot Da$.

b' has the advantage of being an intrinsic parameter which compares the relative fastness of chemical and micromixing processes.

Continuous Stirred Tank Reactor

Single reaction, premixed reactants. $A + B \rightarrow$ products, A and B initially premixed. There is no volume change upon reaction.

In reduced form, IEM equations are written

$$\left.\begin{aligned} &\frac{df_A}{d\theta} = -Da\, f_A^{n_A} f_B^{n_B} + b(\overline{f_A} - f_A) \\ &\text{where} \quad \theta = 0 \\ &\qquad\qquad f_A = 1 \\ &\qquad\qquad \overline{f_A} = \int_0^\infty f_A e^{-\theta}\, d\theta \end{aligned}\right\} \tag{49}$$

where $\theta = t/\tau$. Owing to stoichiometry

$$f_B = f_A + M - 1 \tag{50}$$

For one single reactant $A \rightarrow$ products, it suffices to set $M = 1$ and $f_A = f_B$ in Equation 49, $n = n_A + n_B$ being the reaction order. Figure 7A to E shows the variation of f_A as a function of Da for $n = 0$, 0.5, 1.5, and 2.

Several remarks can be made from these figures:

- The more different the reaction order from one, the larger the span between f_A for a microfluid and a macrofluid, at a given value of Da. This is especially noticeable for xero-order reactions. It is well known that conversion does not depend on micromixing for first-order reactions.

- Let f_{AM} and f_{AS} denote residual concentration in a microfluid and in a macrofluid, respectively. Then for $n > 1$, $f_{AM} < f_{AS}$ (conversion is higher in segregated flow) whereas for $n < 1$ the reverse conclusion is true.

Figure 7G and H shows the results for a Michaelis-Menten reaction, which may be considered as intermediate between 0 and 1st order. The rate of reaction is $r = k_m C_A/(C_A + k_m)$. Setting $da = k_m \tau / C_{A0}$ and $K = K_m/C_{A0}$, the first equation in Equation 49 is written

$$\frac{df_A}{d\theta} = -Da \frac{f_A}{f_A + K} + b(\overline{f_A} - f_A)$$

$K = 0$ corresponds to a zero-order reaction and $K \rightarrow \infty$ to a first-order reaction where micromixing effects no longer exist.

Figures 7A through I represent the residual concentration vs. the Damköhler number for a single reaction with premixed reactants ($M = 1$) or one single reactant A.

Single reaction, unmixed reactants. Two sets of equations have to be written, one for each population of aggregates, issuing from stream 1 and 2, respectively:

For a reaction $A + B \rightarrow$ products, of rate $r = kC_A^n C_B^n$, these equations are written, in reduced form,

$$\left.\begin{aligned}
&\frac{df_{A1}}{d\theta} = -Da\, f_{A1}^{n_A} f_{B1}^{n_B} + b(\overline{f_A} - f_{A1}) \\
&\text{where} \quad f_{A1} = \frac{1}{a} \text{ at } \theta = 0 \\
&\frac{df_{B1}}{d\theta} = -Da\, f_{A1}^{n_A} f_{B1}^{n_B} + b(\overline{f_B} - f_{B1}) \\
&\text{where} \quad f_{B1} = 0 \text{ at } \theta = 0 \\
&\frac{df_{A2}}{d\theta} = -Da\, f_{A2}^{n_A} f_{B2}^{n_B} + b(\overline{f_A} - f_{A2}) \\
&\text{where} \quad f_{A2} = 0 \text{ at } \theta = 0 \\
&\frac{df_{B2}}{d\theta} = -Da\, f_{A2}^{n_A} f_{B2}^{n_B} + b(\overline{f_B} - f_{B2}) \\
&\text{where} \quad f_{B2} = \frac{M}{1 - a} \text{ at } \theta = 0 \\
&\overline{f_A} = \int_0^\infty [a f_{A1} + (1 - a) f_{A2}] e^{-\theta}\, d\theta \\
&\overline{f_B} = \overline{f_A} + M - 1 \qquad \text{(stoichiometric constraint)}
\end{aligned}\right\} \tag{51}$$

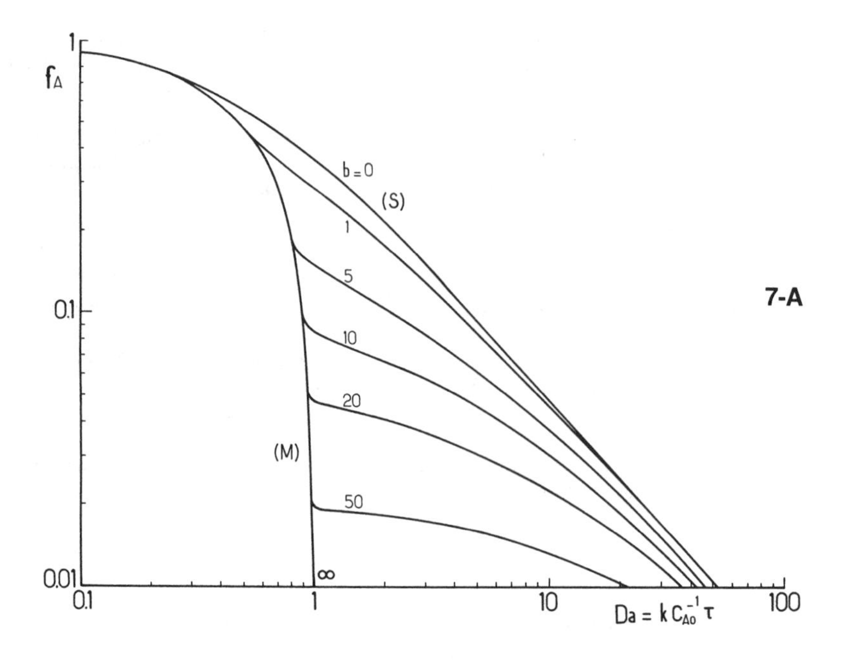

Figure 7A. n = 0 order reaction, logarithmic plot.

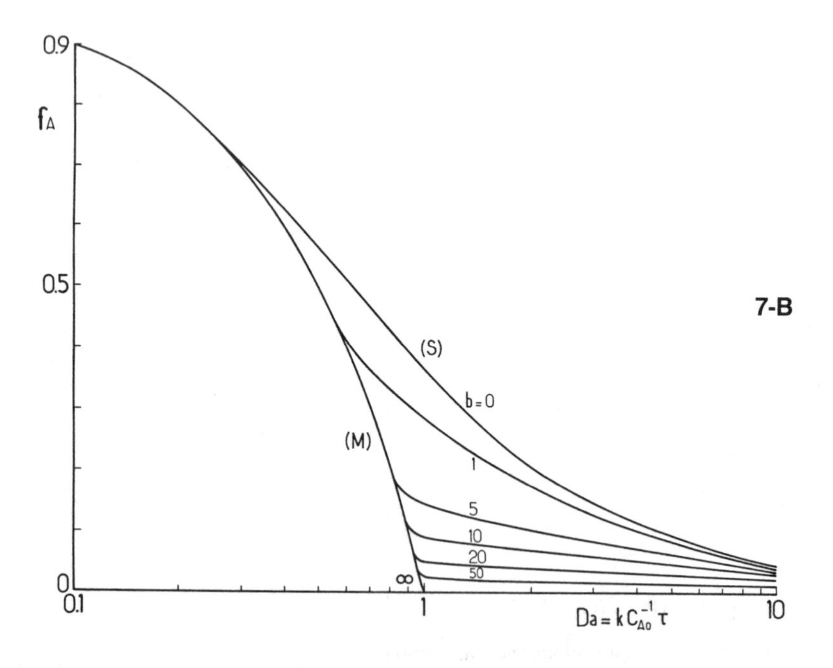

Figure 7B. n = 0 order reaction, semi-log plot.

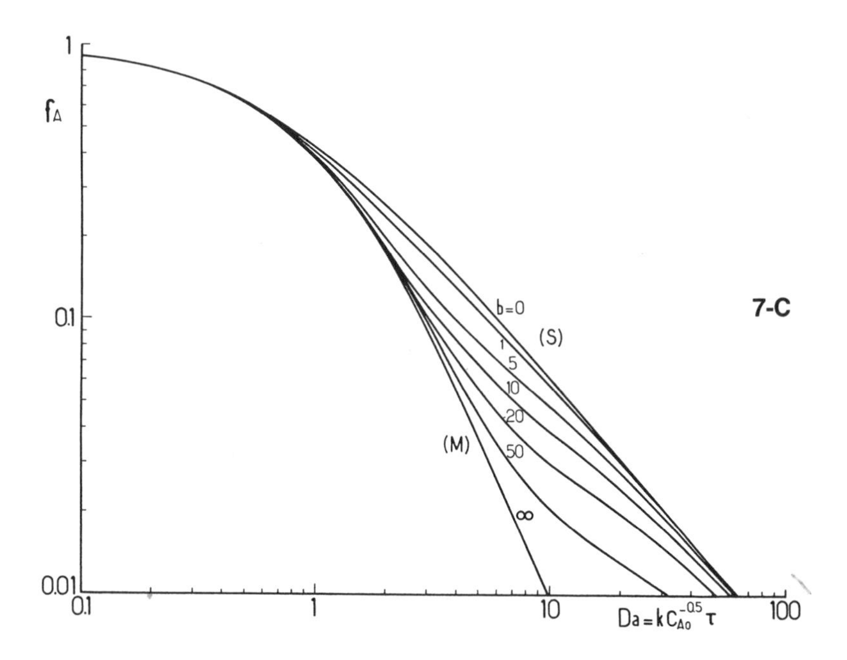

Figure 7C. n = 0.5 order reaction.

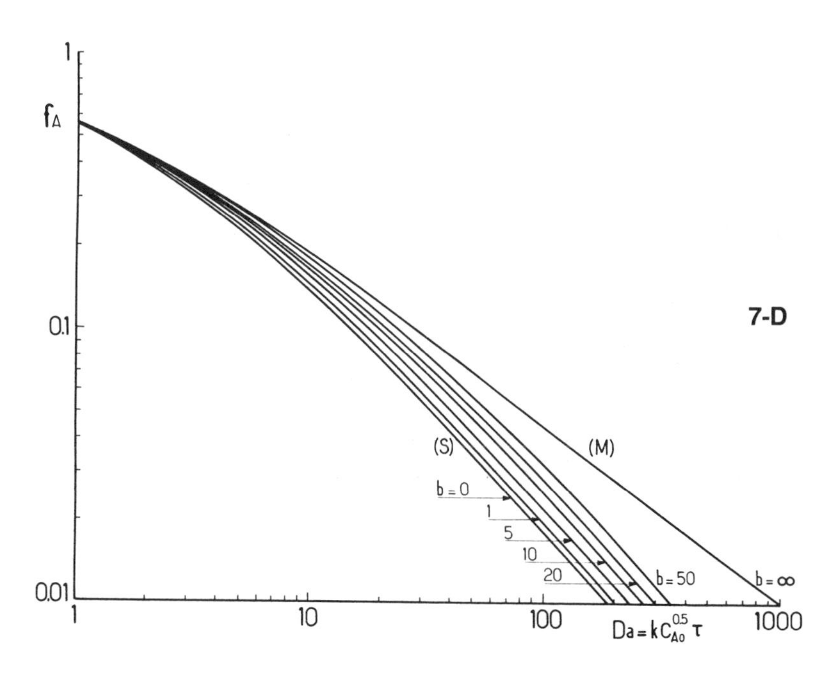

Figure 7D. n = 1.5 order reaction.

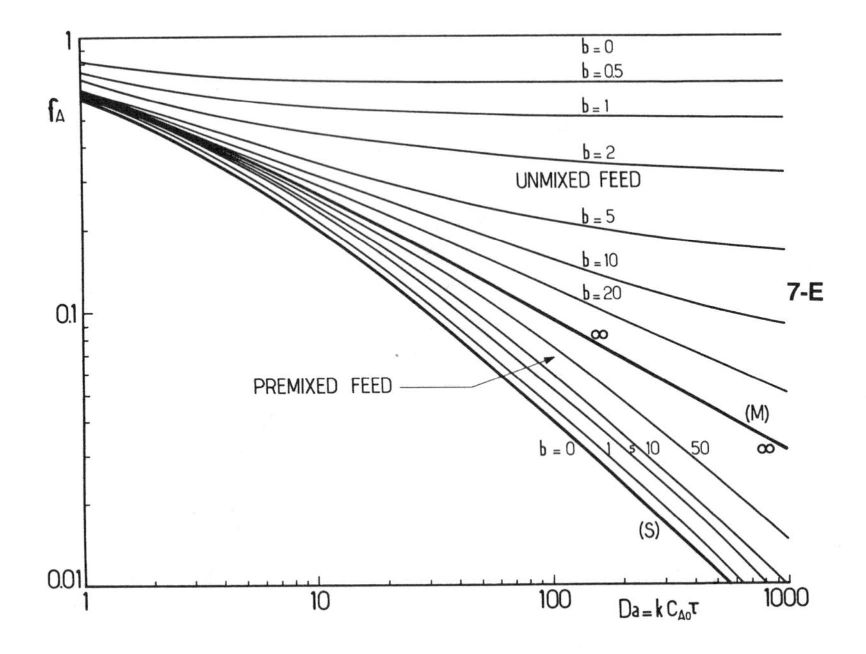

Figure 7E. n = 2 order reaction, logarithmic plot.

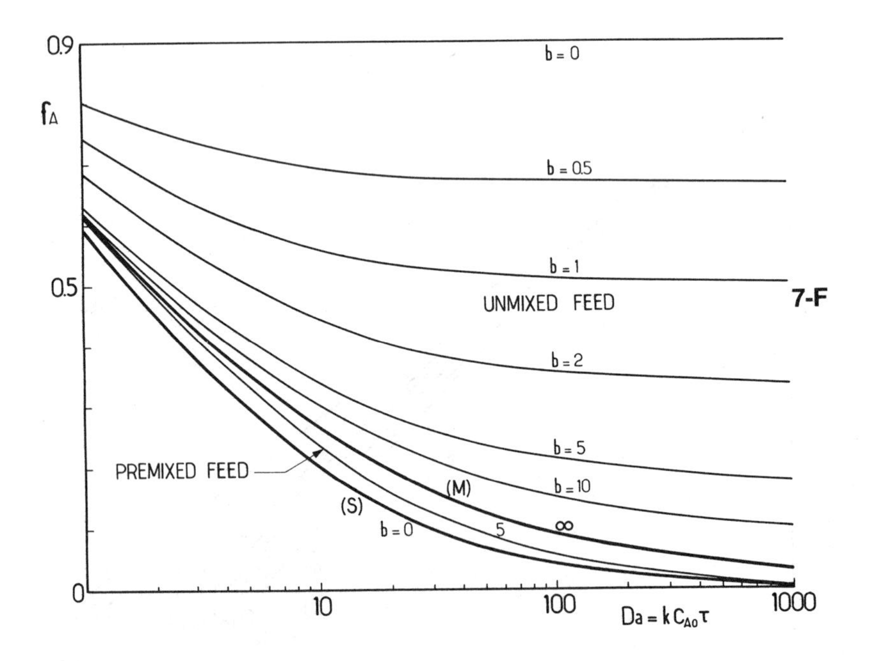

Figure 7F. n = 2 order reaction, semi-log plot.

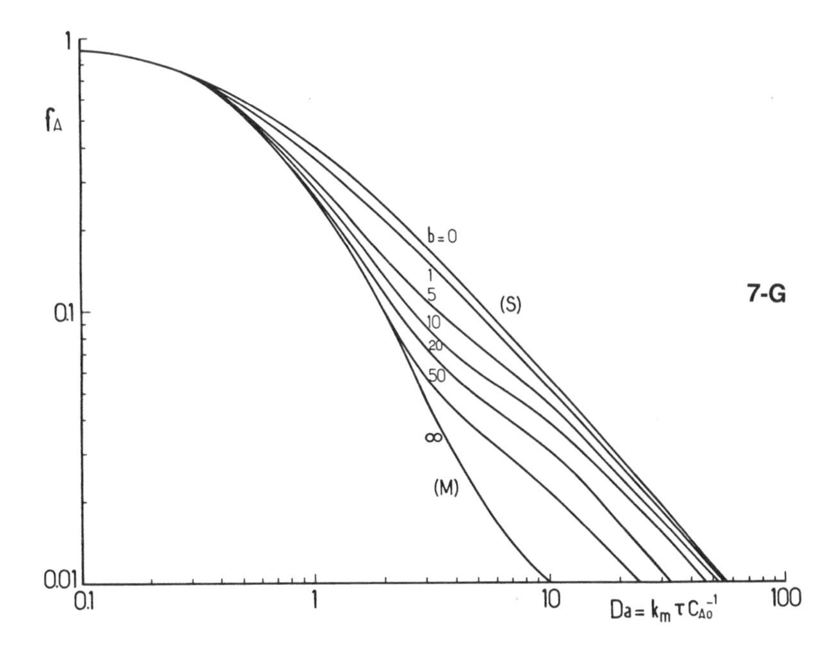

Figure 7G. Michaelis-Menten reaction, K = 0.1.

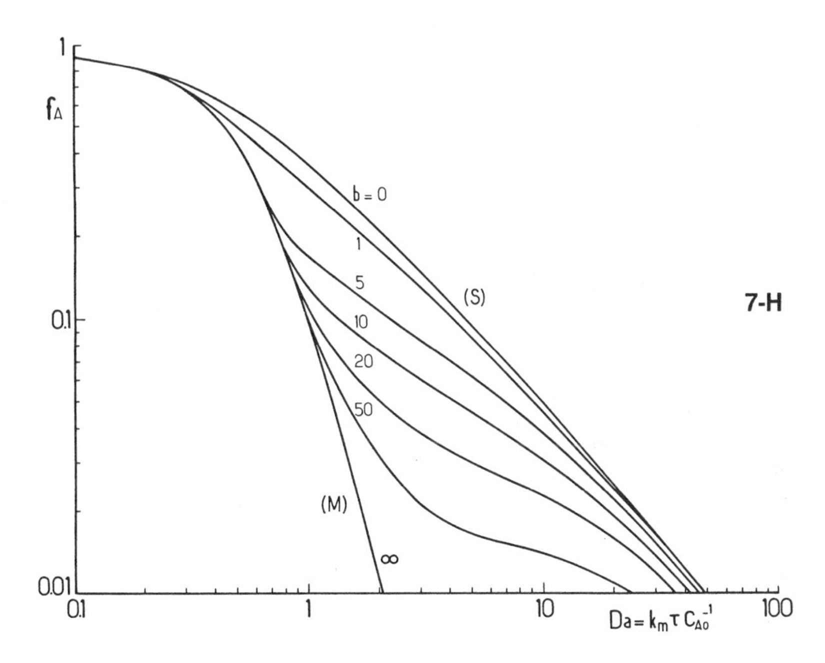

Figure 7H. Michaelis-Menten reaction, K = 0.01.

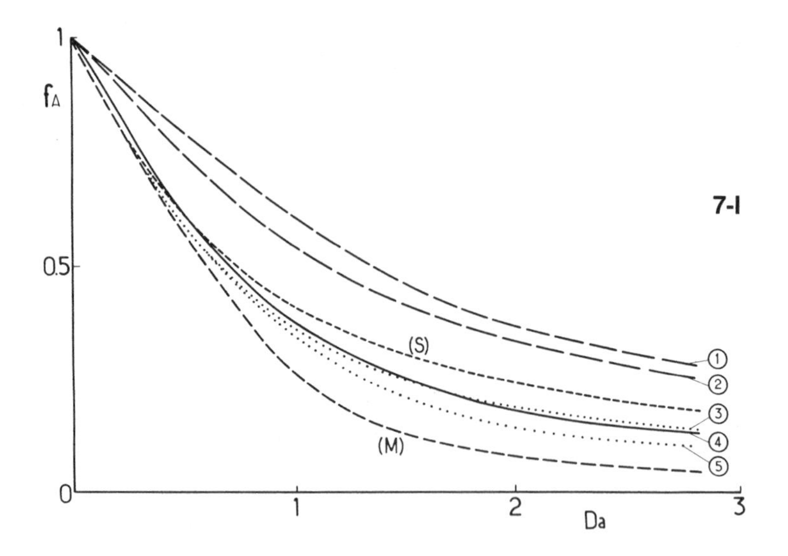

Figure 7I. Iodination of acetone

$Da = k\overline{C_A}\overline{C_H}\tau/C_0$, $K = k_m/C_0 = 0.1$, $a = 0.5$

where 1—$b' = 1$, unmixed feed, $b'_A = b'_H = b' = 1$
2—$b' = 1$, unmixed feed, $b'_A = b'_H = \infty$
3—$b' = 1$, premixed feed
4—$b' = 2$, unmixed feed $b'_A = b'_H = \infty$
5—$b' = 2$, premixed feed
S—segregated flow, premixed feed
M—well micromixed flow

As an example, Figure 7E and F show the case of a global second-order reaction $n_A = n_B = 1$ with equal feedstreams of A and B: $M = 1$, $a = 0.5$. If b (or b') = 0, no reaction occurs and $f_A = f_B = 1$ whatever the value of Da, whereas if b (or b') $\rightarrow 0$, the conversion tends to the well-micromixed case with premixed reactants; premixing is here achieved very quickly in the reactor itself. It must be noticed that for $n > 1$, there is no overlapping in the (f − Da) plane between conversion with premixed and unmixed reactants and variable values of t_m. This is no longer the case for orders $n < 1$, as the curve f(Da) for a microfluid is located below that for a macrofluid. A given value of $f_m < f < f_s$ may then be obtained either with premixed or unmixed reactants (but with two distinct values of b). An example of such a situation is presented in figure 7I. The reaction is the iodination of acetone, which obeys Michaelis-Menten kinetics [19]:

$$\text{Iodine} + \text{acetone} \xrightarrow{H^+,\,KI} \text{products}$$

$$r = \frac{kCC_AC_H}{C + k_m}$$

where C = iodine concentration
C_A = acetone concentration
C_H = acid concentration

The stirred reactor is fed with separate streams

1: Iodine + KI

2: Acetone (very large excess) + acid + KI

The equations are

$$\frac{df_1}{d\theta} = -Da \cdot \frac{f_1 f_{A1} f_{H1}}{f_1 + K} + b(\bar{f} - f_1)$$

where $f_1 = \frac{1}{a}$ at $\theta = 0$

$$\frac{df_2}{d\theta} = -Da \cdot \frac{f_2 f_{A2} f_{H2}}{f_2 + K} + b(\bar{f} - f_2)$$

where $f_2 = 0$ at $\theta = 0$

$$\bar{f} = \int_0^\infty [af_1 + (1 - a)f_2]e^{-\theta}\, d\theta$$

where $Da = k\overline{C_A}\,\overline{C_H}\tau/C_0$
$K = k_m/C_0$

The reduced concentrations of acetone f_A and acid f_H can be calculated by IEM equations taking into account the fact that they are (practically) not consumed by chemical reaction, but the rate of exchange with the mean environment of aggregates may be a priori different from that of iodine (b_A, $b_H \neq b$)

$$\frac{df_{Ai}}{d\theta} = b_A(1 - f_{Ai}) \quad (i = 1, 2)$$

$$f_{A1}(0) = 0$$

$$f_{A2}(0) = \frac{1}{1 - a}$$

$$\frac{df_{Hi}}{d\theta} = b_H(1 - f_{Hi}) \quad (i = 1, 2)$$

$$f_{H1}(0) = 0$$

$$f_{H2}(0) = \frac{1}{1 - a}$$

An example of results is reported in Figure 7I; the micromixing parameter is $b' = b/Da = t_R/t_m = C_0/(k\overline{C_A}\,\overline{C_H}t_m)$. Different values may be affected to t_m (to b′) according to the species. This shows the flexibility of the method and also illustrates the fact that closely related conversions may be obtained with unmixed feed (b′ = 2) or premixed feed (b′ = 1) in this special case.

Infinitely fast single reaction—unmixed reactants. A and B are fed separately into the reactor (volumetric flow rates proportional to a and 1 − a, respectively) and are assumed to react instantaneously according to the stoichiometry

$$\nu_A A + \nu_B B \longrightarrow \text{products}$$

as soon as they come into contact. The reaction extent is thus controlled entirely and only by micromixing. The stoichiometric parameter is here

$$M = \frac{C_{BO}}{C_{AO}} \cdot \frac{\nu_A}{\nu_B} = \frac{C'_{BO}}{C'_{AO}} \cdot \frac{(1-a)\nu_A}{a\nu_B} \tag{52}$$

The detailed treatment of this problem using the IEM model can be found elsewhere [4], with slightly different notations. The derivation is based on the same arguments as those exposed earlier for $\nu_A = \nu_B = 1$, a = 0.5. It leads to the following results

M > 1

$$\overline{f_A} = \frac{b}{b+1}\left(\frac{1}{b} - a(M-1)\left[1 - \left(\frac{M-1}{M-1+\frac{1}{a}}\right)^{1/b}\right]\right) \tag{53}$$

M = 1

$$\overline{f_A} = \frac{1}{b+1} \tag{54}$$

M < 1

$$\overline{f_A} = \frac{1}{b+1}\left(\frac{1}{b} + a(1-M)\left[1 + \frac{1-a}{a}\left(\frac{1-M}{1-M+\frac{M}{1-a}}\right)^{1/b}\right]\right) \tag{55}$$

When M > 1, A is the limiting reactant and the reaction extent is $X = 1 - \overline{f_A}$ but when M < 1, the limiting reactant is B and the reaction extent should be defined as $X = (1 - \overline{f_A})/M$ in order to be comprised between 0 and 1.

Consecutive competing reactions—premixed reactants. The simplest case for such reactions is

$$\left.\begin{aligned} &A + B \xrightarrow{k_1} R \\ &r_1 = k_1 C_A C_B \\ &R + B \xrightarrow{k_2} S \\ &r_2 = k_2 C_R C_B \end{aligned}\right\} \tag{56}$$

The wanted product may be R or S.

More complicated stoichiometries or kinetics may be involved, but the principles of utilization of the IEM model are the same as in the following. With the usual notations $f_A = C_A/C_{A0}$, $f_B =$

C_B/C_{A0}, $f_R = C_R/C_{A0}$, $M = C_{B0}/C_{A0}$, $K = k_2/k_1$, $Da = k_1 C_{A0}\tau$, the model equations are, for premixed reactants

$$\left.\begin{array}{l} \dfrac{df_A}{d\theta} = -Da\, f_A f_B + b(\overline{f_A} - f_A) \\ \text{where} \quad f_A = 1 \text{ at } \theta = 0 \\ \dfrac{df_B}{d\theta} = -Da \cdot f_B(f_A + Kf_R) + b(\overline{f_B} - f_B) \\ \text{where} \quad f_B = M \text{ for } \theta = 0 \\ \overline{f_A} = \displaystyle\int_0^\infty f_A e^{-\theta}\, d\theta \\ \overline{f_B} = \displaystyle\int_0^\infty f_B e^{-\theta}\, d\theta \end{array}\right\} \tag{57}$$

The reduced concentration of R is obtained from the stoichiometric balance, also valid for mean concentrations

$$f_R = 2(1 - f_A) + f_B - M, \qquad \theta = 0, \qquad f_R = 0 \tag{58}$$

Equation 57 assumes that the rate of exchange between aggregates is the same for A, B, and R (one single micromixing parameter). In a more refined version, all three species may be allowed to be exchanged with different micromixing times, giving rise to dimensionless parameters b_A, b_B, and b_R, each in their own balance equation. Equation 58 is then no longer valid in one particular aggregate but only holds true for average concentrations in the tank:

$$\overline{f_R} = 2(1 - \overline{f_A}) + \overline{f_B} - M \tag{59}$$

whereas a balance for R must be added to Equation 57, rewritten with b_A and b_B in the first two equations, namely

$$\frac{df_R}{d\theta} = Da\, f_B(f_A - Kf_R) + b_R(\overline{f_R} - f_R); \qquad \theta = 0, \qquad f_R = 0 \tag{60}$$

Consecutive competing reactions—unmixed reactants. Coming back to one single exchange parameter b, and assuming as usual two separate volumetric flow rates proportional to a (for A) and 1 − a (for B), the equations are

$$\left.\begin{array}{l} \dfrac{df_{A1}}{d\theta} = -Da\, f_{A1} f_{B1} + b(\overline{f_A} - f_{A1}) \\ \text{where} \quad f_{A1} = \dfrac{1}{a} \text{ at } \theta = 0 \\ \dfrac{df_{B1}}{d\theta} = -Da\, f_{B1}(f_{A1} + Kf_{R1}) + b(\overline{f_B} - f_{B1}) \\ \text{where} \quad f_{B1} = 0 \text{ at } \theta = 0 \\ \dfrac{df_{R1}}{d\theta} = Da\, f_{B1}(f_{A1} - Kf_{R1}) + b(\overline{f_R} - f_{R1}) \\ \text{where} \quad f_{R1} = 0 \text{ at } \theta = 0 \end{array}\right\} \tag{61}$$

An analogous system is written with subscripts "2", except for the initial conditions which are here $\theta = 0$, $f_{A2} = 0$, $f_{B2} = M/(1 - a)$, $f_{R2} = 0$.

The mean concentrations are defined as usual by

$$\left.\begin{aligned} \overline{f_A} &= \int_0^\infty [af_{A1} + (1 - a)f_{A2}]e^{-\theta}\, d\theta \\ \overline{f_B} &= \int_0^\infty [af_{B1} + (1 - a)f_{B2}]e^{-\theta}\, d\theta \\ \overline{f_R} &= 2(1 - \overline{f_A}) + \overline{f_B} - M \\ \overline{f_s} &= 1 - \overline{f_A} - \overline{f_R} \end{aligned}\right\} \tag{62}$$

When Da is large (K being kept constant), it may be advantageous to use the variable $\theta' = Da \cdot \theta$ and to introduce the micromixing parameter $b' = b/Da = 1/(kC_{A0} \cdot t_m)$ instead of b. The case of instantaneous reactions can be dealt with in a straightforward way, although the derivation is somewhat tedious. The calculation procedure of $\overline{f_R}$ can be found in Reference 17.

Equation 61 and similar ones for population 2 may also be used with different values of b for each species, as already mentioned.

Figure 8A and B show a few examples of results obtained with $K = 0.5$ and various values of M. In the $M = 1$ case, iso-Da, iso-b, and iso-b′ curves have been drawn both for premixed and unmixed reactants. The influence of segregation is especially noticeable on the selectivity of consecutive competing processes which may be used as test reactions to study the state of micromixing in stirred tanks as will be shown later.

Batch or Semi-Batch Reactor [9]

Semi-batch reactor with one single feedstream. Let Z characterize a volumetric property (for instance a concentration) produced with the chemical rate $\mathscr{R}$ in an incompressible fluid. If the fluid is poured out into a stirred tank with a volumetric flow rate Q(t), the balance for Z in the tank is written

$$\left.\begin{aligned} QZ_E &= \mathscr{R}V = \frac{d}{dt}(VZ) \\ V &= \int_0^t Q\, dt \end{aligned}\right\} \tag{63}$$

where Z_E is the inlet value of Z. However, this assumes that the fluid behaves as a microfluid. In a macrofluid, the variation of Z within an aggregate should be first determined by solving

$$\frac{dZ_B}{d\alpha} = R \tag{64}$$

where $\alpha = 0$

$Z_B = Z_E(t - \alpha)$

Then, the average value in the tank can be calculated according to

$$\bar{Z}(t) = \int_0^\infty Z_B(\alpha, t)\frac{Q(t - \alpha)}{V}\, d\alpha \tag{65}$$

$Q(t - \alpha)/V = I(\alpha, t)$ is nothing but the internal age distribution of aggregates in the tank at time t, which only depends on the history of Q (the tank is supposed to be empty at time $t = 0$).

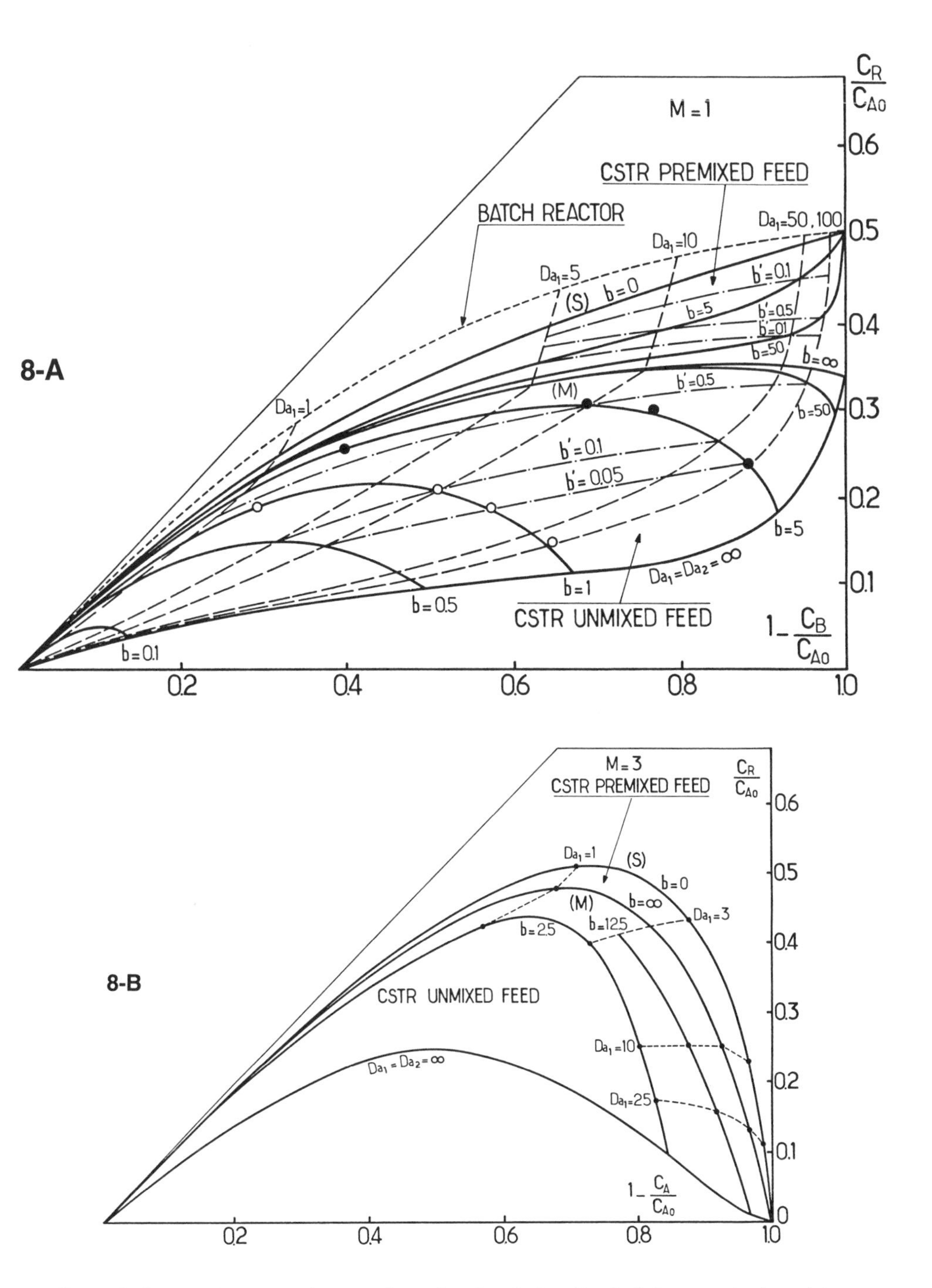

Figure 8. Consecutive-competing reactions: A + B = R, R + B = S. Influence of micromixing on the yield of R. (A) $K = k_2/k_1 = 0.5$, $M = C_{B0}/C_{A0} = 1$; (B) $K = k_2/k_1 = 0.2$, $M = C_{B0}/C_{A0} = 3$. Open and black circles are calculations by the C-D model showing excellent agreement with the IEM model.

The IEM model makes it possible to deal with partial segregation in the tank. Then Equation 64 is replaced by

$$\frac{dZ}{d\alpha} = \frac{\bar{Z} - Z}{t_m} + \mathscr{R}, \qquad \alpha = 0 \qquad Z = Z_E(t - \alpha) \tag{66}$$

and $\bar{Z}(t)$ is still calculated by Equation 65. Using these equations, the reader can verify that

- Conversion for first-order reactions does not depend on the state of segregation (same result using Equation 63 or 64 and 65.
- Conversion in a batch reactor does not depend on segregation either, as all aggregates have the same history. It suffices to set $Q(t) = V\,\delta(t)$ in Equation 65.

Semi-batch or batch reactor with unmixed feedstreams. As an example, the case of a semi-batch reactor fed with A and B (respective flow rates Q_1 and Q_2) will be considered. A and B are assumed to react with rate $r(C_A, C_B)$. According to the IEM model applied to both aggregate populations 1 and 2:

$$\left.\begin{aligned}
&\frac{\partial C_{A1}}{\partial \alpha} = \frac{\overline{C_A} - C_{A1}}{t_m} - r(C_{A1}, C_{B1}) \\
&\text{where} \quad \alpha = 0 \\
&\qquad C_{A1} = C'_{A0} \\
&\frac{\partial C_{B1}}{\partial \alpha} = \frac{\overline{C_B} - C_{B1}}{t_m} - r(C_{A1}, C_{B1}) \\
&\text{where} \quad \alpha = 0 \\
&\qquad C_{B1} = 0 \\
&\frac{\partial C_{A2}}{\partial \alpha} = \frac{\overline{C_A} - C_{A2}}{t_m} - r(C_{A2}, C_{B2}) \\
&\text{where} \quad \alpha = 0 \\
&\qquad C_{A2} = 0 \\
&\frac{\partial C_{B2}}{\partial \alpha} = \frac{\overline{C_B} - C_{B2}}{t_m} - r(C_{A2}, C_{B2}) \\
&\text{where} \quad \alpha = 0 \\
&\qquad C_{B2} = C'_{B0}
\end{aligned}\right\} \tag{67}$$

The mean concentrations are determined from the internal age distribution

$$\left.\begin{aligned}
V\overline{C_A} &= \int_0^t [C_{A1}(\alpha, t)Q_1(t - \alpha) + C_{A2}(\alpha, t)Q_2(t - \alpha)]\, d\alpha \\
V\overline{C_B} &= \int_0^t [C_{B1}(\alpha, t)Q_1(t - \alpha) + C_{B2}(\alpha, t)Q_2(t - \alpha)]\, d\alpha \\
V = V_1 + V_2 &= \int_0^t (Q_1 + Q_2)\, dt'
\end{aligned}\right\} \tag{68}$$

Notice the two time scales α and t in the numerical solution of these equations.

The cases of B poured out into A or A poured out into B, or A and B quickly mixed at $t = 0$ can be deduced from Equations 67 and 68. For instance if B is added to a volume V_1, of A,

$Q_1 = V_1\,\delta(t)$ and Equation 68 simply becomes

$$\left.\begin{aligned} V\overline{C_A} &= V_1 C_{A1}(t, t) + \int_0^t C_{A2}(\alpha, t) Q_2(t - \alpha)\, d\alpha \\ V\overline{C_B} &= V_1 C_{B1}(t, t) + \int_0^t C_{B2}(\alpha, t) Q_2(t - \alpha)\, d\alpha \\ V &= V_1 + \int_0^t Q_2\, dt' \end{aligned}\right\} \tag{69}$$

The case of a batch reactor is still simpler as there is only one time scale left, and $VC_A = V_1C_{A1} + V_2C_{A2}$ and one analogue for B.

MIXING "EARLINESS" IN CONTINUOUS STIRRED REACTORS

Entering and Leaving Environments

In a continuous reactor, the fluid is in a state of minimum mixedness (Min Mix) if mixing occurs at the latest moment: neighboring aggregates have the same age α. Conversely, the fluid is in a state of maximum mixedness (Max Mix) if aggregates of different ages are mixed as soon as they enter the reactor: mixing occurs at the earliest moment and neighboring aggregates have the same residual lifetime, or life expectancy λ. As fluid aggregates have different residence times $t_s = \alpha + \lambda$, they cannot spend their whole life in the reactor with neighbors having simultaneously the same α and the same λ. They start in an entering environment (E.E.) which is in a state of minimum mixedness (same ages) and gradually pass on to a leaving environment (L.E.) which is in a state of maximum mixedness (same residual lifetimes). In principle, this has nothing to do with the state of segregation of the fluid, which may behave either as a macrofluid or as a microfluid, whatever the state of age mixedness may be. However, many authors make the implicit assumption that the E.E. is totally segregated (macrofluid) and the L.E. is well micromixed (microfluid). The mechanism for micromixing is then confounded with that for passing from the E.E. to the L.E. But this assumption is not necessary, and partial segregation may be assumed to exist in both environments as will be seen below. In order to better visualize these phenomena, the "Bundle of Parallel Tubes" (BPT) model is helpful (Figure 9). The reactor volume is reorganized in the form of a bundle of small

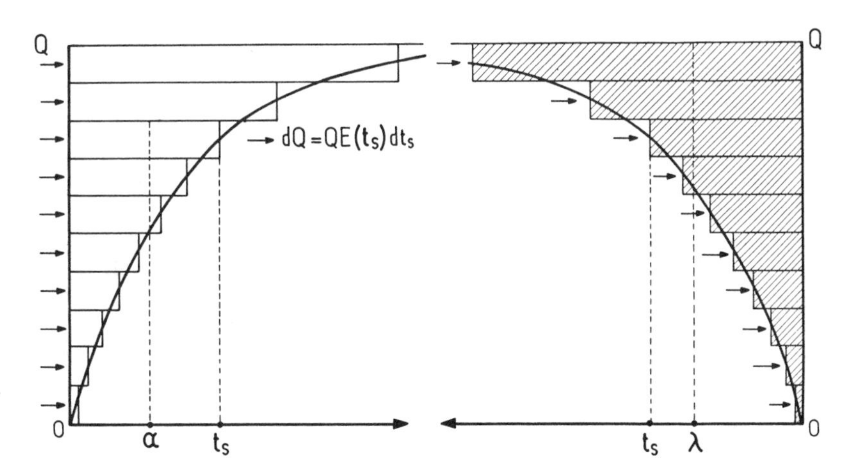

Figure 9. BPT (bundle of parallel tubes) model
left: minimum mixedness state
right: maximum mixedness state, mixing is achieved on a vertical line.

tubes of increasing length equal to the residence time t_s. The fluid flows at constant velocity and the elementary flow rate dQ in each tube is such that $dQ/Q = E(t_s)\,dt_s = (1/\tau)\exp(-t_s/\tau)$ in the case of a CSTR. The locus of the tubes extremity then pictures the familiar $F(t_s) = 1 - \exp(-t_s/\tau)$ curve. In Min Mix state, the tubes are piled up in such a way that aggregates of the same age are on the same vertical line. Conversely, in Max Mix state, the arrangement of tubes is symmetrical and aggregates with the same residual lifetime are on the same vertical (Figure 9).

Transfer from E.E. to L.E.: The "Segregation" Function

A lot of models have been (and are still) proposed in the literature to represent the transfer of fluid aggregates from the E.E. to the L.E. In spite of their apparent diversity, they are all more or less equivalent and special cases of the general model of Spencer et al. [20]. Along the axis of a small tube of the BPT model the fluid gradually passes from a Min Mix to a Max Mix state (Figure 10). The residence time in this particular tube lies between t_s and $t_s + dt_s$, and the flow rate is $dQ = QE(t_s)\,dt_s$. The flow transferred from the E.E. to the L.E. in the interval $d\lambda$ is

$$d^2Q = QA(\lambda, t_s)d\lambda\, dt_s \tag{70}$$

where $A(\lambda, t_s)$ is a mixing function also expressed by introducing $h(\lambda, t_s)$ and the RTD

$$A(\lambda, t_s) = h(\lambda, t_s)E(t_s) = h(\lambda, t_s)(1/\tau)\exp(-t_s/\tau) \tag{71}$$

For a given residual lifetime λ, the fraction of fluid in the E.E. is s (and the complementary fraction in the L.E. is 1 − s) such that (Figure 10)

$$s(\lambda, t_s) = s(0, t_s) + \int_0^\lambda h(\lambda, t_s)\, d\lambda \tag{72}$$

where one may have $s(t_s, t_s) \neq 1$ (tube inlet) and $s(0, t_s) \neq 0$ (tube outlet). $s(\lambda, t_s)$, also written $s(\lambda, \alpha)$ is known as the "segregation function" although this term is confusing (it is concerned with the decay of segregation only under the restricting assumption that E.E. = macrofluid, L.E. = microfluid); "segregation of ages function" would be a better term. In the particular tube previously con-

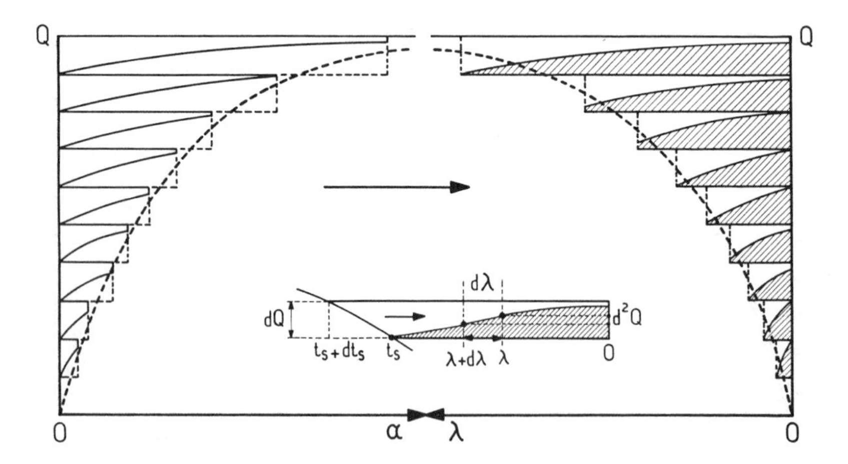

Figure 10. Model of Spencer and Leshaw. Transfer from the entering environment (left) to the leaving environment (right).

sidered, the flow rate of the Max Mix fluid is $dQ_{max} = Qg(\lambda)\, d\lambda$ with

$$g(\lambda) = \int_{\lambda}^{\infty} A(\lambda, t_s)\, dt_s = \int_{\lambda}^{\infty} h(\lambda, t_s)(1/\tau) \exp(-t_s/\tau)\, dt_s$$

In the whole reactor, the Max Mix flow rate of fluid of residual lifetime λ is thence $Q_{max}(\lambda) = Q \int_{\lambda}^{\infty} g(\lambda')\, d\lambda'$.

This model is represented in Figure 10. The fluid flows from left to right at constant velocity. It passes from the entering environment (unshaded tubes, minimum mixedness) to the leaving environment (shaded area, maximum mixedness).

The transfer between two environments may also be described as a Lagrangian process where a small band of volume $dV_{max} = Q_{max}\, d\lambda$ moves with constant velocity along the λ axis in the direction of decreasing λ. Spencer et al. [20] consider this volume as an "accumulator" whose dimensionless volume V is increasing according to

$$\frac{dv}{dt} = \int_{z=-\infty}^{t} A(-z, -t)\, dz \tag{73}$$

The accumulator leaves the reactor and discharges at time $t = 0$. The conversion may be calculated as follows: let $C_E(z)$ be the inlet concentration of a species entering the reactor at time z and produced with rate $\mathscr{R}$. In a segregated environment (here the E.E.), C varies according to the batch equation

$$\frac{dC}{d\alpha} = \mathscr{R} \tag{74}$$

where $\alpha = 0$
$C = C_E(z)$

whose solution is $C_B[C_E(z), \alpha]$

The accumulator receives at time t a flux resulting from all aggregates transferring from E.E. to L.E. Upon integration, the balance of the considered species in the accumulator is then written (L.E. = microfluid)

$$\frac{d(vC)}{dt} = \mathscr{R}v + \int_{z=-\infty}^{t} A(-z, -t)C_B[C_E(z), t - z)\, dz \tag{75}$$

This equation has to be integrated from $t = -\infty$ ($vC = 0$) up to $t = 0$. Such a Lagrangian formulation is interesting as it allows the representation of transient regimes by a succession of accumulators which are gradually filled during their transit in the reactor and are discharged when they leave it. This only requires to know $A(\lambda, t_s)$. In a CSTR, $A(\lambda, t_s)$ can be deduced from s through Equation 71 and $h(\lambda, t_s) = (\partial s/\partial \lambda)_{t_s}$. The "segregation" function $s(\lambda, t_s)$, or $s(\lambda, \alpha)$ thus entirely determines the micromixing process, apparently with much more flexibility than the single-parameter models presented earlier. Actually most models specifying the shape of $s(\lambda, \alpha)$ also involve one single parameter.

Models Based on the "Segregation" Function

Most models accounting for "intermediate micromixing" can be compared on the base of their "segregation" function. Some of them follow:

- α^*-model of Spencer et al. [20].

$s = 1$ for $\lambda > \alpha^* t_s$ (or $\alpha < (1 - \alpha^*)t_s$)

$s = 0$ for $\lambda < \alpha^* t_s$

- τ_D-model of Spencer et al. [20], which is also the series-model of Weinstein and Adler [21].

$$s = 1 \qquad \text{for } \alpha < \tau_D$$

$$s = 0 \qquad \text{for } \alpha > \tau_D$$

- Parallel-model of Weinstein and Adler [21].

$$s = 1 \qquad \text{for } t_s < \tau_p$$

$$s = 0 \qquad \text{for } t_s > \tau_p$$

- Model of Villermaux and Zoulalian [22]. s is only a function of t_s, for instance $s = \exp(-Kt_s)$. In this case, an alternative representation is possible where all the tubes of the EE and those of the L.E. have been piled up together (see Figure 11A).
- Model of Ng and Rippin [23]. s only depends on the age, for instance $s = \exp(-R_s\alpha)$. For representing the reduction of size of entering particles in an unmixed feedstream, Plasari et al. [24] have used a model of this kind (shrinking aggregate) where $s = (1 - \alpha/t_e)^3$.
- Model of Valderrama and Gordon [25, 26]. This is a two-parameter model (β, w)

$$s = 1 \qquad \text{for } \alpha < \alpha^* = -\ln(1 - \beta)$$

$$s = \frac{1 - \beta - w}{1 - \beta} \qquad \text{for } \alpha > \alpha^*$$

In all these models, the E.E. is assumed to consist of a macrofluid and the L.E. of a microfluid. A graphical representation of $s(\lambda, \alpha)$ and the corresponding pattern in the BPT model can be found in Reference 27, for any kinds of RTD.

Indeed, one can easily imagine an almost endless set of models of this kind without really bringing in anything new. Numerical simulations and theoretical arguments (based on the variance of concentrations or chemical conversion) lead to the conclusion that all single-parameter models yield approximately equivalent results if relationships between the micromixing parameters are established. For instance, generalizing Equation 39, it can be shown that a condition for equivalence

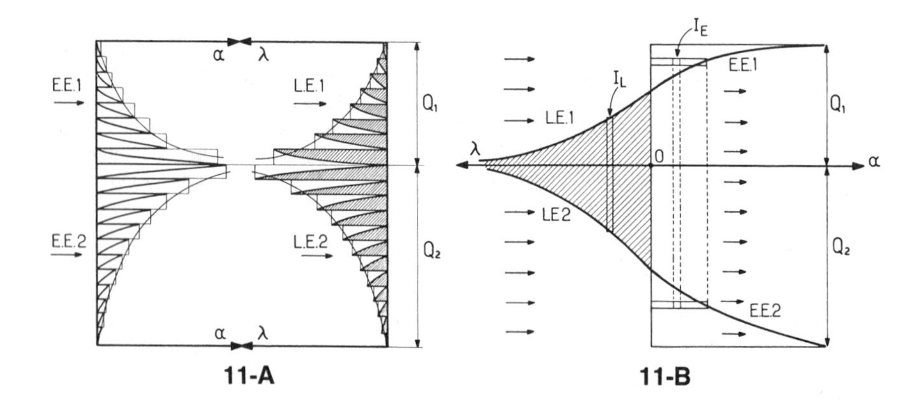

Figure 11. Extension of the model of Spencer and Leshaw to the case of a reactor with two feed-streams: (A) general case; (B) case where the segregation function depends only on residence time (Three-environment model).

between the C-D model (parameter $I = \omega\tau$), IEM model (parameter $b = \tau/t_m$), Ng and Rippin model (parameter $R_s\tau$), and shrinking-aggregate model [24] (parameter τ/t_e) is

$$\frac{\omega\tau}{2} = \frac{I}{2} = 2b = \frac{2\tau}{t_m} = R_s\tau = \frac{7\tau}{t_e} \tag{76}$$

Reactors with Two Unmixed Feedstreams

This case is perhaps more realistic for practical application. The BPT model offers a convenient picture of the reactor where both streams have their own RTD (here that of a perfect macromixer). The pattern of Figure 10 can easily be extended to the case of two separate feed streams by joining the corresponding entering and leaving environments side by side (Figure 11A). The assumption that the E.E. and L.E. consist respectively of macrofluids and microfluids may be dropped and partial segregation may be assumed to exist owing to interaction between aggregates, according to mechanisms which are discussed in the next section. Rules have also to be specified for these interactions. This problem has been thoroughly investigated in a series of papers by researchers of Exeter University [8, 10, 29, 60]. In the E.E., aggregates are generally allowed to interact if they have both the same age *and* the same residual lifetime. In the L.E., interaction takes place between aggregates having only the same residual lifetime. Interaction is often accounted for by the C-D model. For instance, a three-environment model has been proposed [12] (two E.E. and one L.E.) with the segregation function $s(\alpha) = \exp(-R_s\alpha)$. A generalization to four environments (two E.E. and two L.E.) was later studied [30]. A three-environment model where the segregation function only depends on t_s was also proposed [12]. In this case, one may use the representation of Figure 11B. The model depends on three parameters: two interaction moduli I_E and I_L for the E.E. and the L.E., respectively, and the "segregation" parameter $R_p\tau$ (the segregation function is $s = \exp(-R_pt_s)$. The simulation of the reactor is carried out by the Monte-Carlo method. With a little imagination, an almost endless suite of models of this kind can be conceived.

Experimental Determination of Mixing Earliness Using Reactive Tracers

Spencer et al. [20] proposed an interesting method to determine mixing earliness by using two tracers A and B which may react with each other. When the fluid is close to minimum mixedness, the best discrimination is obtained by injecting A and B in the form of two sharp pulses separated by a short interval. When the fluid is close to maximum mixedness, the recommended injection shape is a square pulse containing both A and B. The conversion in the outlet pulse is then strongly dependent on micromixing. This method was used to study imperfect mixing in a CSTR [82]. The reactive tracers were p-nitrophenyl acetate and sodium hydroxide which undergo a second-order reaction to produce p-nitrophenylate, an optically detectable product.

Conclusion: Usefulness of Models Based on "Mixing Earliness"

It has already been pointed out that all these phenomenological models are approximately equivalent to predict intermediate conversions or yields within the limits of minimum mixedness (segregated) flow and maximum mixedness (micromixed) flow. Equivalence between the set of micromixing parameters leading to comparable results may be established (e.g. Equation 76). The choice in favor of one or the other is often more a matter of computational convenience than of physical relevance. Actually, the major criticism which may be addressed to this family of models is that they do not rely directly on physical features of the process and treat mixing as a mathematical game in the age space. In addition, they ignore the spatial distribution of the micromixing intensity of the tank. This is why they should rather be considered as convenient means for correlating experimental results, but not as a sound base for scale-up, owing to the empirical character of the involved parameters.

SEGREGATION IN SINGLE-PHASE FLUIDS: MECHANISM OF MICROMIXING IN THE PHYSICAL SPACE

The definition of segregation, presented earlier involves the concept of "fluid aggregate," which may denote a blob, clump or particle of any shape (sphere, cylinder, slab, lamina), a group of molecules, or more generally any segregated region of the fluid that the mixing process has to destroy or disperse in order to achieve mixing on the molecular scale, where chemical reaction can occur. The aim of this section is to give a physical description amenable to quantitative prediction of segregation decay, relying on experimental parameters. Such a comprehensive description should also provide a physical interpretation for the phenomenological parameters of the models presented in the preceding sections.

Successive Stages of Micromixing

According to Beek and Miller [32], it is convenient to distinguish three successive stages in the mixing process.

1. Distribution of one fluid through the other and uniformization of average composition without decreasing local concentration variations.
2. Reduction of size of the regions of uniform composition and increase of contact between regions of different composition (fluid aggregates).
3. Mixing by molecular diffusion.

Stage 1 pertains to macromixing. Before reviewing models for the last two processes, it is helpful to recall the main results of the theory of turbulence in fluids.

Time and Space Microscales from Turbulence Theory

Turbulence theory provides a classical approach to mixing phenomena. Local velocities U and concentrations C are assumed to comprise an average nonfluctuating component and a fluctuating term:

$$U = \bar{U} + u$$
$$C = \bar{C} + c \tag{77}$$

Balance equations can be written for each species and a solution for the local uniformization of concentrations (with or without chemical reactions) may be sought in fixed (Eulerian) coordinates with respect to the tank. This is a difficult task because of the closure problem which arises owing to higher-order correlations between the fluctuating terms. This problem is discussed in other chapters and will not be dealt with here. Actually this is not the best way to come up with useful rules for practical design of stirred tanks. However, useful quantities may be defined, to characterize velocity and concentration fluctuations, based on the assumption of local homogeneous isotropic turbulence, which seems reasonable in most cases in stirred tanks. These quantities are recalled in Table 1. Most of them are deduced from spectral measurements which are easy for velocity fluctuations but less known for concentration fluctuations. There are two parallel families of characteristics pertaining to velocity fluctuations on one hand, and to concentration fluctuations on the other.

Macroscales L_f and L_s characterize large initial eddies, whose size is generally comparable to that of the impeller. It may be assumed that $L_f \approx L_s$. Upon mixing, the size of the eddies diminishes and energy, or segregation, is transferred to smaller structures. Taylor (λ_f) and Corrsin (λ_s) microscales are close to the maximum rate of dissipation of turbulent kinetic energy or segregation, respectively. The smallest size for turbulent eddies is given by the Kolmogorov microscale λ_K. Below this size, energy loss only occurs by viscous dissipation. Related microscales (λ_B, λ_C, see Table 1) have also

Table 1
Homogeneous Isotropic Turbulence

Velocity $U = \bar{U} + u$	Concentration $C = C + c$
Mean square fluctuation $u' = (\overline{u^2})^{1/2}$	$c' = (\overline{c^2})^{1/2}$
Autocorrelation $f(r) = \frac{\overline{u(x)u(x+r)}}{u'^2}$	$g(r) = \frac{\overline{c(x)c(x+r)}}{c'^2}$
Macroscale $L_f = \int_0^\infty f(r)\,dr$	$L_s = \int_0^\infty g(r)\,dr$
Kinetic energy of turbulent motion $q = \frac{3}{2}u'^2$	Segregation intensity $c'^2/c_0'^2$
Turbulent energy dissipation $\epsilon = -\frac{dq}{dt} = -\frac{3}{2}\frac{du'^2}{dt}$	Segregation dissipation $\epsilon_s = -\frac{dc'^2}{dt}$
Spectra: Three dimensional $E(k)$ One dimensional $E_1(k_1)$ $u'^2 = \frac{2}{3}\int_0^\infty E(k)\,dk = 2\int_0^\infty E_1(k_1)\,dk_1$ $\epsilon = 2\nu\int_0^\infty k^2E(k)\,dk = 30\nu\int_0^\infty k_1^2E_1(k_1)\,dk_1$	$E_s(k)$ $c'^2 = \int_0^\infty E(k)\,dk$ $\epsilon_s = 2\mathscr{D}\int_0^\infty k^2E_s(k)\,dk$
Taylor microscale $\lambda_f^2 = \frac{30\nu u'^2}{\epsilon} = \frac{10\int_0^\infty E(k_1)\,dk_1}{\int_0^\infty k_1^2E(k_1)\,dk_1}$	Corrsin microscale $\lambda_s^2 = \frac{12\mathscr{D}c'^2}{\epsilon_s} = \frac{6\int_0^\infty E_s(k)\,dk}{\int_0^\infty k^2E_s(k)\,dk}$
Taylor time constant $\tau_f = \left(-\frac{1}{u'^2}\frac{du'^2}{dt}\right)^{-1} = \frac{\lambda_f^2}{10\nu}$	Corrsin time constant $\tau_s = \left(-\frac{1}{c'^2}\frac{dc'^2}{dt}\right)^{-1} = \frac{\lambda_s^2}{12\mathscr{D}} \approx 2\left(\frac{L_s^2}{\epsilon}\right)^{1/3}$
$L_f = \frac{\eta E(0)}{u'^2} \sim \frac{q^{3/2}}{\epsilon}$	Small Sc: $\lambda_s^2/\lambda_f^2 = \mathscr{D}/\nu$, $\tau_s = \tau_f$ (Corrsin)
Viscous dissipation Kolmogorov microscale $\lambda_K = (\nu^3/\epsilon)^{1/4}$	Corrsin microscale $\lambda_c = (\mathscr{D}^3/\epsilon)^{1/4}$
Viscous dissipation time constant $\tau_k = (\nu/\epsilon)^{1/2}$	Batchelor microscale $\lambda_B = (\mathscr{D}^2\nu/\epsilon)^{1/4}$ $\tau_c = (\mathscr{D}/\epsilon)^{1/2}$

Table 2
Typical Values of the Time and Space Scales Given in Table 1*

	Case 1: $\epsilon = 1$ Watt kg^{-1} $L_s = 0.3$ m (large reactor)	Case 2: $\epsilon = 10$ Watts kg^{-1} $L_s = 0.1$ m (small reactor)
Kolmogorov λ_K	32 μm	18 μm
Corrsin λ_s	70 μm	16 μm
$\lambda_K^2/\mathscr{D} = \nu^{3/2}(\mathscr{D}\epsilon^{1/2})$	1 s	0.3 s
$\tau_s = 2(L_s^2/\epsilon)^{1/3}$	0.9 s	0.2 s
$t_\delta = \sqrt{2}\,\tau_K = (2\nu/\epsilon)^{1/2}$	1.4×10^{-3} s	0.45×10^{-3} s
$\lambda_c^2/\mathscr{D} = (\mathscr{D}/\epsilon)^{1/2}$	3.5×10^{-5} s	

* *Experimental parameters:* $\nu = 10^{-6}\ m^2\ s^{-1}$, $\mathscr{D} = 10^{-9}\ m^2\ s^{-1}$, $Sc = 10^3$

been introduced, taking into account concentration fluctuations. Typical values of these time and space scales are given in Table 2. A more detailed description of segregation decay in the frame of turbulence theory can be found in Volume 1 of this Encyclopedia.

It is not always obvious to establish a close link between all these quantities characterizing "eddies" in fixed (Eulerian) coordinates, or even in the spectral space, and the size or lifetime of fluid "aggregates" which are assumed to have a physical reality in convected (Lagrangian) coordinates. However the merit of turbulence theory is to give the dependence of time and space microscales upon physicochemical parameters, especially viscosity, diffusivity, and mechanical energy dissipation. Depending on the scale on which the micromixing process is considered, aggregates will be assumed to have the size and life duration of macro-eddies in the first stages of mixing, Taylor or Corrsin micro-eddies in the second stage, and Kolmogorov microstructures in the last one, but this is only a matter of convenience and the physical reality may be much more complex. When chemical reactions are involved, the characteristic reaction time may be compared with the different time scales to determine which stage of mixing will compete with the chemical consumption of reactants. Micromixing is not only concerned with the last stage (molecular diffusion in aggregates smaller than Kolmogorov microscale) but also with the second stage (mass transfer and shrinking of aggregates having roughly the size of Taylor/Corrsin microscale).

Mechanisms for Micromixing in Stage 2

Coming back to the description of micromixing in terms of fluid "aggregates" two different mechanisms may be involved to represent mixing in stage 2.

Purely Convective Mixing

Even in the absence of turbulence, mixing occurs under the influence of the convective velocity field. Fluid aggregates are stretched out and folded up, and the mixture finally exhibits a lamellar or striated structure characterized by a striation thickness δ. Ottino and coworkers [99–106] have established a general theory allowing the calculation of the rate of decrease of striation thickness once the velocity field is known; however, they use a mathematical formalism which may discourage practitioners unfamiliar with tensor analysis. They define a stretching time as

$$t_\delta \sim -\delta/(d\delta/dt) \tag{78}$$

In newtonian liquids, a lower bound for the stretching time is found to be

$$t_\delta > t_\delta^* = (2\nu/\epsilon_v)^{1/2} \tag{79}$$

where ϵ_v is the viscous dissipation per unit mass. For instance, for a filament stretching at constant velocity

$$\delta = \delta_0[1 + \alpha/t_\delta]^{-1} \tag{80}$$

In laminar shear flow, and for laminae initially normal to the direction of flow

$$\delta = \delta_0[+ (\alpha/t_\delta)^2]^{-1/2} \tag{81}$$

It may be seen from Table 2 that t_δ^* is usually small with respect to turbulence time-scales. Efficient mixing is achieved when the laminae are periodically reoriented with respect to the direction of stretching. When the striation thickness is small enough, molecular diffusion becomes effective to promote mixing between adjacent laminae (stage 3).

Erosive or Dispersive Mixing

Convective mixing assumes continuous motion and connectedness of material surfaces. Conversely one may think of a mixing mechanism that would gradually peel off smaller fragments from the segregated clumps by friction at their external surface. In the "shrinking aggregate" (SA) model [24], the peeling-off process is characterized by a mass-transfer coefficient h which is assumed to be expressed by the Calderbank-Moo Young correlation, applicable to small particles immersed in turbulent media

$$h\ell/\mathscr{D} = 0.13\ell(\epsilon/\nu^3)^{1/4}\,Sc^{1/3} \tag{82}$$

ℓ is the diameter of the shrinking aggregate. Actually, there is no fundamental reason to apply such a correlation, established for solid particles, to fluid aggregates. It is only a heuristic assumption which proved to be successful out of many others for interpreting experimental results [24, 26]. For spherical shape, this leads to a linear decrease of size as a function of age

$$\ell = \ell_0(1 - \alpha/t_e) \tag{83}$$

with the characteristic time constant

$$t_e = \frac{\ell_0\lambda_K}{0.26\mathscr{D}}\,Sc^{-1/3} \approx \frac{\ell_0\lambda_K}{2.6\mathscr{D}} \qquad \text{(in liquids)} \tag{84}$$

The dependence of the erosion time on experimental parameters is thus

$$t_e \sim \ell_0\nu^{5/12}\mathscr{D}^{-2/3}\varepsilon^{-1/4} \tag{85}$$

ℓ_0 is the initial size of aggregates, which has to be estimated (it has been found for instance that ℓ_0 was proportional to the velocity of the fluid at the feed pipe outlet [24], but this would deserve further study). The rate of energy dissipation comprises a contribution ϵ_1 from the kinetic energy of the entering fluid and a contribution ϵ_2 from mechanical agitation. In a conventional stirred tank, ϵ_2 has the usual expression $N_pN^3d^5/V$ where N_p is the power number and d is the stirrer diameter. From experience, it seems that ϵ_1 and ϵ_2 contribute to micromixing with different efficiencies

$$\epsilon = \eta_1\epsilon_1 + \eta_2\epsilon_2 \tag{86}$$

Quantitative expressions for ℓ_0, η_1, and η_2 depend on experimental conditions and are still unpredictable a priori.

Mixing by Coalescence-Dispersion (C-D)

This process has already been described earlier. The aggregates merge and redisperse at random with average frequency ω. The characteristic time-constant is $t_{CD} = 1/\omega$. This is mainly a phenomenological parameter, but it is generally accepted that t_{CD} is proportional to Corrsin's time constant τ_s in turbulent media.

Mechanism for Micromixing in Stage 3: Molecular Diffusion

All the processes mentioned above (decay of turbulent eddies, laminar stretching, erosion) give rise to a population of aggregates small enough for molecular diffusion to become effective: this is the ultimate and finally the only process really able to mix the components of a fluid on the molecular scale. The time constant for molecular diffusion is the diffusion time $t_D = \mu L^2/\mathscr{D}$ where μ is a shape factor and L the ratio of the volume to the external surface area of the aggregate. For instance

$$\left.\begin{aligned} \mu &= \frac{p+1}{p+3} \\ L &= \frac{R}{p+1} \\ t_D &= \frac{R^2}{(p+1)(p+3)} \end{aligned}\right\} \quad (87)$$

where p = 0 for slabs (laminae of thickness $\delta = 2R$)
p = 1 for cylinders (filaments of diameter 2R)
p = 1 for spheres (aggregates of diameter 2R)

The particle then behaves approximately as a first-order dynamic system of time-constant t_D with respect to mass transfer. The choice of the characteristic dimension $\ell = 2R$ depends on the kind of microstructure which is considered to exist when molecular diffusion becomes controlling, especially with respect to chemical reactions:

- In striated structures, $\ell = \delta$
- In aggregates close to the maximum of turbulent dissipation $\ell = \lambda_s$. Corrsin has established the corresponding expression for $t_D \sim \tau_s$:

$$\tau_s = \frac{\lambda_s^2}{12\mathscr{D}} \approx 2\left(\frac{L_s^2}{\epsilon}\right)^{1/3} + 0.5\left(\frac{\nu}{\epsilon}\right)^{1/2} \ln \frac{\nu}{\mathscr{D}} \quad (88)$$

  The first term is predominant in liquids and the second one in gases.
- In aggregates reaching the viscous dissipation zone where turbulence can no longer reduce the size. ℓ is then identified with one of the viscous dissipation microscales of Table 1, λ_K (Kolmogorov), λ_B (Batchelor), or λ_C (Corrsin). The most frequent assumption is that ultimate aggregates have the dimensions of a Kolmogorov microscale, undergoing subsequent stretching under the influence of viscous friction [41]:

$$\ell = \lambda_K(1 + t^2/t_\delta^2)^{-1/2} \quad (89)$$

In liquids, Table 2 shows that λ_K and λ_s lie between 10 and 100 μm and that $\lambda_K^2/\mathscr{D}$ and τ_s are in the same order of magnitude between 0.1 and 1 second. The other processes are much faster. The problem is to choose the adequate length scale ℓ (= 2R) when molecular diffusion is retained as the controlling micromixing mechanism. If one considers that the results of homogeneous isotropic turbulence theory are applicable, an experimental clue for the choice of the relevant microscale should be its dependence upon viscosity ν, molecular diffusivity $\mathscr{D}$, and the rate of turbulent energy dissipation ϵ (See Tables 1 and 2).

For instance, test experiments where chemical conversion is influenced by micromixing are performed by varying viscosity and mechanical energy dissipation. The dependence of characteristic microscales upon these parameters is deduced from a diffusion-mixing model (see the following discussions). However, the interpretation of such experiments is not obvious for several reasons:

- Homogeneous isotropic turbulence is only an idealized picture of reality. In addition, it is not straightforward to identify hypothetical "eddies," different for velocity and concentration, defined from spectral wavenumbers and described in Eulerian coordinates with aggregates having a physical existence in Lagrangian coordinates.
- It is difficult to make sure that test experiments are free from spurious effects (e.g., macromixing) and that they may be interpreted by invoking one single-stage for micromixing, namely molecular diffusion.
- Even in this case, assumptions have to be made concerning the shape of aggregates and the efficiency with which the mechanical energy input is actually used to produce turbulent kinetic energy.
- Shear viscosity and molecular diffusivities of reacting species may be affected in a very different way by additives used to change viscosity (for instance polymers may cause the viscosity to increase and let molecular diffusivity of small molecules unchanged).
- Spatial variations of ϵ have to be taken into account, as will be discussed later, and this is generally not obvious.

Therefore, the microscale ℓ for mixing by molecular diffusion should be considered in a first step as a phenomenological parameter and its dependence upon physicochemical parameters should be carefully checked in each specific case. Using fast consecutive-competitive reactions, Bourne and coworkers [41] have been able to interpret their results by assuming diffusion in stretching laminae (equation 89) having an initial size comparable to the Kolmogorov microscale but different conclusions are also reported in the literature and further research under different and perfectly controlled conditions is still required to come up with definitive conclusions.

Micromixing and Chemical Reaction

One Single-Stage Is Involved

The micromixing process may be characterized by one time-constant t_m depending on the mechanism involved. If chemical reactions are simultaneously occurring, the chemical process is generally characterized by some reaction time t_R (e.g. $t_R^{-1} = kC_0^{n-1}$ for an n-th order reaction). The key parameter for discussing which process is controlling is the ratio t_R/t_m. If t_R/t_m is small, the chemical reaction is faster than mixing and microscopic concentration gradients appear: the fluid becomes segregated. Conversely, when t_R/t_m is large, the fluid may be considered as well micromixed from the viewpoint of the chemical reaction. Simulations of simultaneous mixing and chemical reactions by the IEM-model or the C-D model have already been presented.

When molecular diffusion is involved, several authors (see Volume 1 and Reference 42) have studied R/D models in which partial derivative equations for transient diffusion with chemical reaction were solved numerically in fluid laminae or aggregates of various shapes. Actually, the competition between reaction and diffusion can also be simulated by the IEM-model. It suffices to identify t_m with the diffusion time $t_D = \mu L^2/\mathscr{D}$ given by Equation 87. In doing this, it must be reminded that different molecular diffusivities $\mathscr{D}_j$ (and thus different micromixing times t_{mj}) may be used for each species, as pointed out earlier. For instance, results obtained by both R/D and IEM models are compared in Figures 12 and 13. The IEM model is merely a lumped version of the distributed parameter R/D model. Numerical simulations prove that replacing concentration profiles in aggregates by average values does not change yield and conversion very much, even for

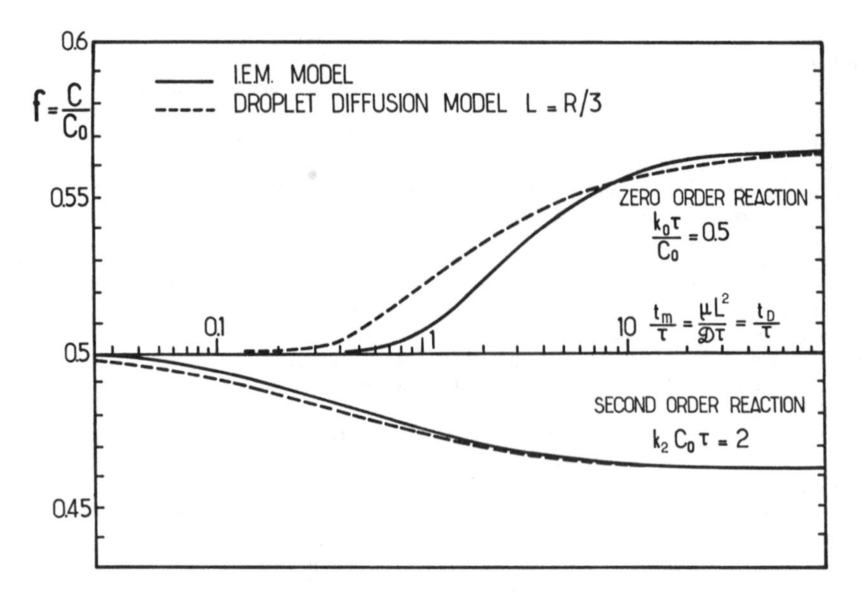

Figure 12. Equivalence between the reaction/diffusion model and the IEM model for a zero-order and a second-order reaction (f = 0.5 for perfect micromixing).

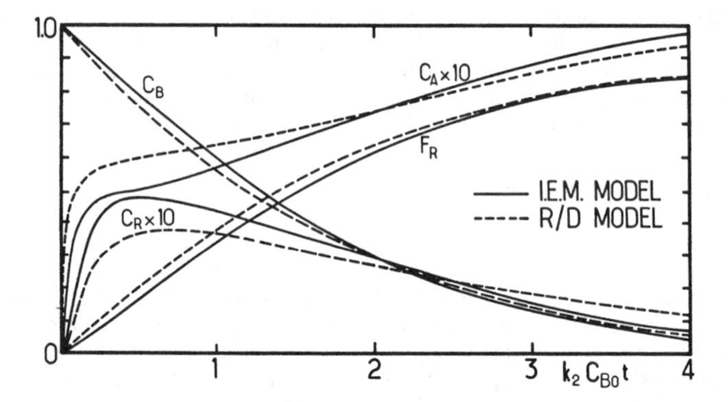

Figure 13. Equivalence between the reaction/diffusion and the IEM model for consecutive competing reactions $A + B \xrightarrow{k_1} R$, $R + B \xrightarrow{k_2} S$. Average concentrations vs. reduced time in a spherical particle immersed in a batch of constant composition ($C_A = 0.105$, $C_R = 0$), $C_{A0} = 0$, $C_{B0} = 1$, $C_{R0} = 0$. $k_2C_{B0}R^2/\mathcal{D} = 2$ for A and R. B cannot diffuse within the particle ($\mathcal{D} = 0$). $k_1/k_2 = 10$. F_R represents the total production of R (expressed as equivalent concentration in the particle). Equivalence is based on $t_m = t_D = \mu L^2/\mathcal{D}$; $\mu = 3/5$, $L = R/3$ (sphere).

rather "stiff systems" of fast reactions. Of course, very sensitive chemical systems can be found for which the two models will give different results; but this must be appreciated keeping in mind that both representations are ideal pictures of a very complex reality.

Therefore, in most current applications, a simple lumped parameter model should be preferred to more sophisticated distributed models. This saves much computer time and the difference in the

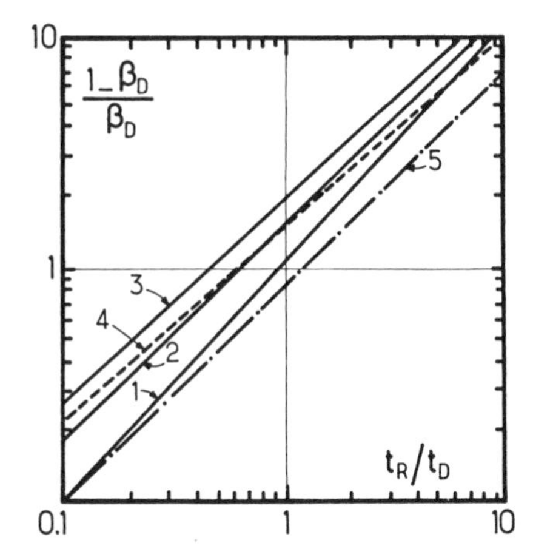

Figure 14. Microfluid/macrofluid volume ratio as a function of reaction/diffusion time ratio. Curves 1 to 4: simulations with the IEM model, $t_m = t_D$.
Curve 1: Second-order reaction $k_2C_{A0}\tau = 2$
Curve 2: Second-order reaction $k_2C_{A0}\tau = 5$
Curve 3: Second-order reaction $k_2C_{A0}\tau = 10$
Curve 4: Consecutive competing reactions, $C_{A0} = C_{B0}$, $k_2/k_1 = 0.5$, $t_R = 1/k_1C_{A0}$
Curve 5: Reaction and diffusion in a slab [43].

predictions is often not greater than that induced for instance by a change in the arbitrary assumptions about the aggregate shape.

An interesting property is revealed by numerical simulations via the IEM-model. The simplest way to represent partial segregation in a fluid is to consider that it consists of a mixture of macrofluid (fraction β) and microfluid (fraction $1 - \beta$). It comes out that the ratio $(1 - \beta)/\beta$ is always close to that of two characteristic times [43]. In the case of erosive mixing of two reactants in a CSTR (erosion controls mixing and the product of erosion is a microfluid), one finds

$$\frac{1-\beta_e}{\beta_e} \approx 3.5\frac{\tau}{t_e} \tag{90}$$

In the case of stoichiometric amounts of premixed reactants in a CSTR (Figure 14) an empirical correlation is found

$$\frac{1-\beta_D}{\beta_D} = a(t_R/t_m)^m \tag{91}$$

On the average $a \simeq 2$ and $m \simeq 0.8$. In the simplest case, $a = 1$ and $m = 1$.

The ratio of microfluid to macrofluid is thus roughly equal to that of reaction time to micromixing time–an interesting "rule of thumb" for the rapid estimation of segregation extent.

The equivalence condition $t_m = t_D$ completes the set of relationships presented in the previous section. Considering the segregation number $N_{seg} = t_D/\tau$ introduced by Nauman [42] and the Thiele modulus introduced by Zoulalian and Villermaux [43, 44] and by Bourne et al. [45], namely $M = kC_0\,\ell^2/\mathscr{D} \sim t_D/t_R$, a set of equivalent segregation parameters is presented in Table 3.

Several Stages of Micromixing Are Involved

More sophisticated mechanisms may exist, where chemical reaction takes place during successive stages of Beek and Miller, or combinations of these. Two examples of such mixed mechanisms follow:

Reaction occurring both during erosion and diffusion stages—mixed model of segregation decay. This happens when t_e, t_D, t_R are in the same order of magnitude.

Example: Reaction A + B → products, rate r, unmixed feedstreams of A and B in a CSTR (same notations as earlier). The fluid is considered as a mixture of fully segregated aggregates undergoing erosion (volume fraction β_e) and a population of smaller aggregates coming from erosion (volume

Table 3
Segregation Parameters*

	IEM model		Segregation Number	Thiele Modulus	Segregated Volume Fraction	$\frac{1-\beta}{\beta}$
	$b = \tau/t_M$	$b' = t_R/t_M$	$N_{seg} = 1/b$	$M \sim 1/b'$	β	
Macrofluid (total segregation)	0	0	∞	∞	1	0
Microfluid (molecular mixing)	∞	∞	0	0	0	∞

* $t_M = t_D$ *micromixing (diffusion) time.* t_R = *reaction time.*

fraction $1 - \beta_e$), and interacting in reaction diffusion regime with premixed feed. The fully segregated aggregates consist of fresh A (volume fraction $a\beta_e$, relative concentration $f_{Ae} = 1/a$) and fresh B (volume fraction $(1 - a)\beta_e$, relative concentration $f_{Be} = M/1 - a$). Erosion yields premixed aggregates as the result of a statistical loss of material from fresh aggregates of A and B. The population of small interacting aggregates is considered as a mixture of macrofluid (volume fraction β_D) and microfluid (volume fraction $1 - \beta_D$) so that the residual concentrations in this population are

$$f_{AD} = \beta_D f_{AS} + (1 - \beta_D) f_{AM} \tag{92}$$

$$f_{BD} = \beta_D f_{BS} + (1 - \beta_D) f_{BM} \tag{93}$$

where f_{AS}, f_{BS} (macrofluid) and f_{AM}, f_{BM} (microfluid) are calculated from usual mass balance equations for a CSTR with premixed feed as follows.

- Macrofluid

$$\left.\begin{aligned}
&f_{AS} = \frac{1}{\tau}\int_0^\infty f_{A,B}(\alpha)\exp(-\alpha/\tau)\,d\alpha \\
&\frac{df_{A,B}}{d\alpha} = -r \\
&\text{where}\quad \alpha = 0 \\
&\qquad f_{A,B} = 1 \\
&\qquad f_{BS} = \frac{1}{\tau}\int_0^\infty f_{B,B}(\alpha)\exp(-\alpha/\tau)\,d\alpha \\
&\qquad \frac{df_{B,B}}{d\alpha} = -r \\
&\text{where}\quad \alpha = 0 \\
&\qquad f_{B,B} = M
\end{aligned}\right\} \tag{94}$$

- Microfluid

$$\left.\begin{aligned} f_{AM} &= 1 - \tau r/C_{AO} \\ f_{BM} &= M - \tau r/C_{AO} \end{aligned}\right\} \tag{95}$$

Finally, the overall concentration at the reactor outlet is

$$\left.\begin{aligned} f_A &= \beta_e + (1 - \beta_e)[\beta_D f_{AS} + (1 - \beta_D)f_{AM}] \\ f_B &= M\beta_e + (1 - \beta_e)[\beta_D f_{BS} + (1 - \beta_D)f_{BM}] \end{aligned}\right\} \tag{96}$$

where β_e and β_D are approximately equal to

$$\beta_e = \frac{t_e/\tau}{3.5 + t_e/\tau} \qquad \text{(from Equation 90)} \tag{97}$$

and

$$\beta_D \approx \frac{1}{1 + t_R/t_D} = \frac{(t_D/\tau)Da}{1 + (t_D/\tau)Da} \tag{98}$$

where $Da = \tau/t_R$ is the Damköhler number.

These expressions suggest several remarks: Equations 92, 93 and 98 constitute an alternate but approximate way to calculate conversion with the IEM model without any iterative procedure, and this may be helpful for rapid estimations.

As β_e and β_D are independent parameters, f_A may take any value between 1 and the smallest of the two values f_{AS} or f_{AM}. In particular, if β_e is small enough, the conversion is the same as if the feedstreams were premixed: premixing is achieved by fast erosion. This was observed by Zoulalian and Villermaux [44] in the case of glycol diacetate hydrolysis for instance.

The existence of two micromixing stages in series was proven by several experimental observations: controlling erosion stage, followed by fast diffusion ($\beta_D \approx 0$) in the case of iodination of acetone [24]; competition between erosion and diffusion processes in the case of alkaline hydrolysis of nitromethane [46]. As the stirring speed is increased at a fixed space time, both t_e and t_D diminish (because ϵ increases) so that f_A first drops owing to better contacting of reactants and then augments because segregation decreases among the population of mixed aggregates. This is clearly shown in Figure 15.

Mixing relationships similar to Equation 96 and based on linear combination also hold for other additive properties. For instance, if an intermediate species is produced in the mixed aggregates only with the yield Y, then

$$Y = (1 - \beta_e)[\beta_D Y_s + (1 - \beta_D)Y_M] \tag{99}$$

where Y_s and Y_M are the yields calculated for a segregated and well micromixed CSTR respectively.

Reaction and diffusion in stretching aggregates. In order to represent reaction and simultaneous diffusion in lamellar structures (laminae) with a striation thickness $\delta(\alpha)$ function of age, some authors [35] introduced a "warped" time t_w defined by

$$\left.\begin{aligned} &\frac{dt_w}{d\alpha} = \frac{\mathscr{D}}{\delta^2} \\ &\text{or} \\ &t_w = \int_0^\alpha \frac{\mathscr{D}}{\delta^2(\alpha')} d\alpha' \end{aligned}\right\} \tag{100}$$

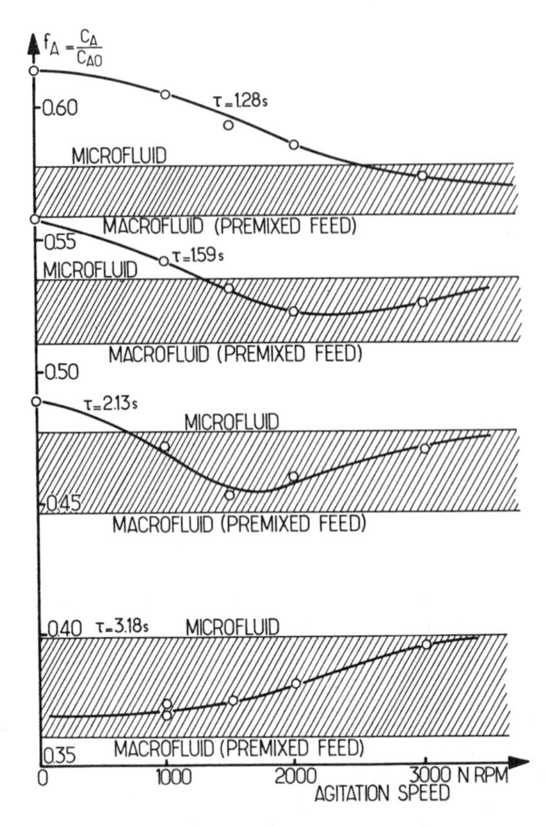

Figure 15. Alkaline hydrolysis of nitromethane. Experimental clue to the existence of two micromixing stages in series ($t_R = 1/kC_{A0} = 1.89$ s).

The equations for reaction/diffusion are then easier to solve in Lagrangian coordinates. In the same way, Bourne et al. [41, 47] extended their R/D model by assuming that the size of the aggregates would decrease according to Equation 89, which implies that a reduction of the striation thickness below the Kolmogorov microscale is conceivable. As already pointed out, the IEM-model makes it possible to account for such a variation of the aggregate's thickness. If the micromixing time is identified to a diffusion time, then

$$t_m(\alpha) = t_m(0) \cdot \delta^2(\alpha)/\delta^2(0) \tag{101}$$

without forgetting that the mean concentration $\langle C_j \rangle$ has to be defined properly, in such a way that the net exchange over the whole population is nil. An indication in favor of a possible stretching of aggregates is provided by in interpretation of Zoulalian's experiments [43, 44] in which exponent m in relationship Equation 91 is found larger than one (m = 1.4). This can be explained by assuming that the faster the reaction (the smaller t_R), the younger the aggregates in which the reaction takes place and consequently, the larger their size and the diffusion time $t_D = t_m$ so that $1 - \beta_D/\beta_D$ increases faster than $t_R/\bar{t}_m$ calculated on average conditions.

Experimental Determination of the State of Segregation by Physicochemical Methods

Use of Nonreactive Tracers

By mixing two solutions containing different amounts of tracer, microscopic concentration gradients exist, due to partial segregation of the fluid. If a local probe sensitive to microscopic and instantaneous values of the tracer concentration is available, it is possible to record the distribution function p(C). From the decay of the variance σ^2 of this distribution, one can deduce micromixing times or micromixing parameters as explained earlier (see Equations 29, 31, 38, and 42). However, this requires that the local probe has both a space and time-resolution smaller than segregation microscales and this condition is not easy to fulfill in usual fluids. Attempts have been made with conductometric microprobes [48] but the spatial resolution is not better than 100 μm and the time-response is about 5 ms, due to electrochemical limitations. This is not sufficient to study turbulent structures down to the Kolmogorov microscale (about 10 μm). Optical methods, and especially the use of fluorescent tracers [49] seem to be promising but still have to be developed. Major advances are awaited from experimental methods allowing us to obtain reliable spectra of concentration fluctuations in stirred tanks.

Chemical Methods

Although less direct, chemical methods are, at this moment, much more powerful than physical ones; they rely on the use of test chemical reactions which act as molecular probes for micromixing. However, partial segregation can only be studied through its influence on the conversion and yield of chemical reactions, and a model is required to extract micromixing parameters from experimental data.

Instantaneous reactions. If one assumes that the reactor is perfectly macromixed and that the state of segregation is uniform within the tank volume, the extent of an instantaneous reaction with unmixed reactants is directly related to the intensity of micromixing, which controls contacting of reactants. For instance if two reactants A and B, assumed to react instantaneously, are mixed in equal amounts into a batch reactor, the residual concentration as a function of time is, after the IEM model

$$f = \exp(-t/t_m) \tag{102}$$

In a CSTR

$$f = (1 + \tau/t_m)^{-1} \tag{103}$$

Corresponding relationships for nonstoichiometric mixtures are given by Equations 53, 54, and 55. This method was used in Reference 16 for the study of self-mixing in a liquid phase reactor. However, the previous assumptions are often too simple, and the correct interpretation of experimental data requires a more detailed analysis of hydrodynamic patterns at the microscopic level.

Finite rate reactions. The influence of micromixing on chemical conversion as predicted by various models has been described in the preceding sections. Therefore, after selection of a suitable test reaction, it suffices in principle to compare observed conversions or yields to model predictions in order to estimate micromixing parameters. In particular, if the fluid is considered as a simple mixture of microfluid and macrofluid (volume fraction β), the average residual concentration for a single reaction will be

$$f = \beta f_s + (1 - \beta) f_M \tag{104}$$

where f_s and f_M are values calculated from the kinetics and the space time under the assumption of total segregation and perfect micromixing, respectively. β can thus be determined from the measurement of f and further interpreted in terms of micromixing times (see Equations 90 or 91). A similar procedure may be applied to complex reactions, by measuring the yield of an intermediate product and applying Equation 99 to calculate β. More detailed models for the state of segregation may also be used, as the one presented earlier implying two parameters β_e and β_D, but means then have to be found to distinguish their respective influence.

Many experimental studies of this kind have been reported in the literature, but they present several drawbacks

- Segregration effects on conversion are often small and the kinetics have to be known with accuracy.

- The conversion has to be estimated on-line because the reaction continues after mixing of the reactants.
- The interpretation of data should lead both to a check on the postulated model and to the determination of micromixing parameters, and there is frequently some coupling between these two objectives.

Use of consecutive-competing reactions. Fast consecutive-competing reactions

$$A + B \xrightarrow{k_1} R \tag{1}$$

$$R + B \xrightarrow{k_2} S \tag{2}$$

with $k_1 \gg k_2$ and k_2 still large, are very interesting candidate test-reactions for studying the local state of segregation: if a small amount of B is mixed into a large excess of A, R is quasi-instantaneously formed. If the fluid is segregated, R stays at the contact of B and is immediately converted to S. If the fluid is well micromixed, R and B are dispersed within the whole volume but as the concentration of B is very small, R may subsist until the total consumption of B. Therefore, the amount of S formed is some kind of segregation index. A major advantage is that it keeps the memory of the micromixing process close to the point of injection and can be determined "off-line" by analyzing the tank content. Let $X_s = 2C_s/(2C_s + C_R)$ be the yield of S, then

$$\beta = \frac{X_S - X_{SM}}{1 - X_{SM}} \tag{105}$$

where X_{SM} is the yield one would observe in a well micromixed reactor. The method was extensively exploited by Bourne and coworkers, who enumerated the qualities of a good reaction for industrial tests:

- t_{R2} for the controlling reaction in the same order of magnitude as $t_D = t_m$.
- irreversible reaction, known mechanism.
- easy analysis and regeneration.
- safe and inexpressive chemicals.

No reaction proposed to date fulfills all these requirements. Azo-coupling of 1-naphtol (A) with diazotised sulphanilic acid (B) has been especially used. This and other possible reactions are discussed in Volume I of this Encyclopedia. A new reaction of this kind has recently been proposed [50]: the precipitation of barium sulphate complexed by EDTA (YH_4) in alkaline medium (reactant A) under the influence of an acid (reactant B). The stoichiometry can be written:

$$\left.\begin{aligned} A + B &\xrightarrow{k_1} R + W \quad (1) \\ nU + R + 2nB &\xrightarrow{k_2} nS + nT \quad (2) \end{aligned}\right\} \tag{106}$$

where
$A = (Ba^{2+}, Y^{4-})_n OH^-$
$B = H^+$
$R = (Ba^{2+}, Y^{4-})_n$
$W = H_2O$
$U = SO_4^{--}$
$S = BaSO_4$
$T = Y^{4-}, 2H^+$
$n = (Ba^{2+})_0/(OH^-)_0$

Reaction 1 is a quasi-instantaneous neutralization. Reaction 2 is limited by the rate of precipitation [51]

$$\left.\begin{aligned} r_2 &= k_2 C_R C_U (C_0)^{2/3} \\ k_2 &= 1.8 \times 10^{-2}\ \text{mol}^{-5/3}\ \text{m}^5\ \text{s}^{-1} \text{ at } 20°\text{C} \end{aligned}\right\} \tag{107}$$

C_0 is the concentration of available sites for the crystallisation of $BaSO_4$. This is the sum of the concentration of already precipitated sulphate in the medium (C_{SO}) and of the concentration of the reactant (Ba^{2+} or SO_4^{--}) in stoichiometric defect. $B = H^+$ is injected at a given point of the tank into an excess of A. $BaSO_4$ is easily measured after completion of the reaction by sampling and turbidimetry. A segregation index is then calculated according to

$$X_s = \frac{(2n + 1)C_s V}{n n_{BO}} \tag{108}$$

where n_{BO} is the number of moles injected and V the tank volume. n is the initial Ba^{2+}/OH^- ratio. The barium sulphate precipitation has many of the advantages cited above, but the reaction is not strictly instantaneous and still somewhat dependent on macromixing. Results obtained by this method are presented in the following.

A review of other work concerning the use of the chemical method for studying segregation can be found in Reference 27.

Conclusion

If the theory of homogeneous isotropic turbulence now provides a satisfactory picture of segregation decay (see Volume I of this Encyclopedia), there are still uncertainties in the final interpretation of mixing and chemical reaction in more complex hydrodynamic situations. Phenomenological models like the IEM-model are then helpful to account for interaction between aggregates, even by molecular diffusion. An important point is that in many cases, a single mechanism cannot represent the whole process and several simultaneous or successive stages of micromixing have to be invoked, each with their characteristic parameters: erosion, laminar stretching, coalescence-redispersion, molecular diffusion, etc . . . There is still a need for reliable test-reactions, usable in industrial reactors and obeying the criteria mentioned in the preceding section. Carefully designed experiments where such reactions would be carried out in perfectly controlled hydrodynamic conditions (no macromixing effects, high power input, control of viscosity) should make it possible to decide between existing theories.

THE REAL STIRRED TANK

Up to this point, the stirred tank has been considered as perfectly macromixed, concentration gradients within the tank have been ignored and micromixing has been characterized by average parameters without any spatial distribution. Actually, the real behavior of the fluid in the tank is much more complex, and this has to be taken into account for an accurate description of micromixing.

Internal Circulation Patterns

Using local probes and pulse injections of tracer, it is possible to determine the internal age distribution (IAD) at various points in the tank and to deduce from this information a representative model for the circulation pattern [52, 53]. Figure 16 shows an example of such a model [53] together with typical responses to a pulse injection obtained at different locations within the tank volume, clearly revealing the existence of short-circuits and delays. Other authors [54–56]

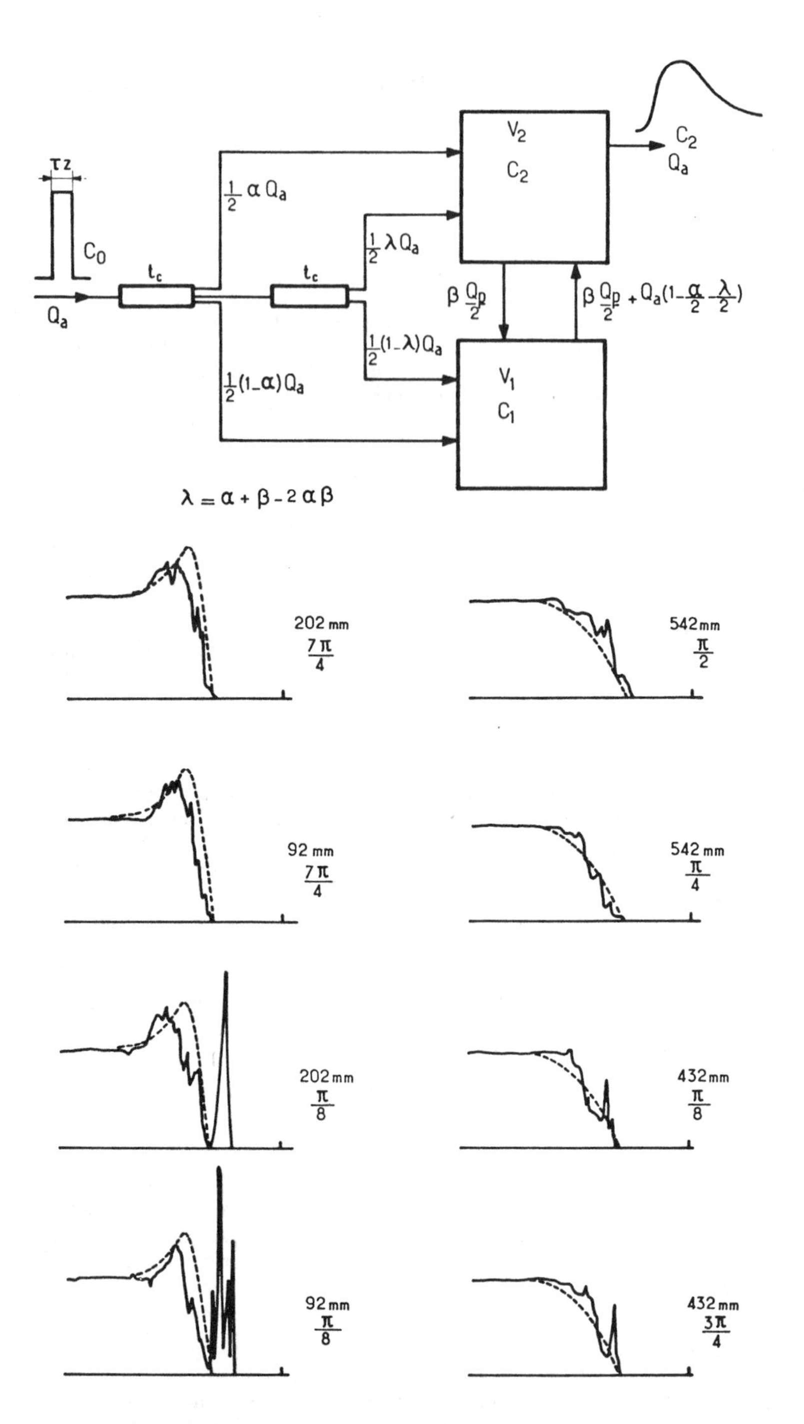

Figure 16. Model for representing internal age distribution in a stirred tank (top) and comparison between experimental (solid line) and predicted (broken line) responses to a pulse injection of tracer at different locations within the tank (bottom), showing short-circuits and delays. (see reference 53 for details).

studied circulation times in large stirred reactors (such as fermenters) by using radio flow-followers. They found that circulation times t_c were most of the time log-normally distributed

$$f(t_c) = \frac{1}{\sqrt{2\pi}\sigma t_c} \exp\left[-\left(\frac{\ln t_c - \mu}{\sigma\sqrt{2}}\right)^2\right] \tag{109a}$$

the mean value $\bar{t}_c$ and the natural variance s^2 are thus

$$\left.\begin{aligned} \bar{t}_c &= \exp(\mu + \sigma^2/2) \\ s^2 &= (\bar{t}_c)^2(\exp \sigma^2 - 1) \end{aligned}\right\} \tag{109b}$$

H being the height of liquid in the tank, these studies led to

$$\left.\begin{aligned} \bar{t}_c &\sim H/(Nd^3) \\ s &\sim H^{7/3}(Nd^3) \end{aligned}\right\} \tag{110}$$

The average circulation time is thus inversely proportional to the stirring speed N. More generally, $\bar{t}_c$ can be expressed from the power input per unit volume as

$$\bar{t}_c \sim (Nd)^2(P/V)^{-1} \tag{111}$$

A related quantity is the terminal mixing time θ_m, required for reducing concentration gradients down to a specified level. Many correlations for mixing time are available in the literature [57]. A comprehensive treatment of the problem was published by Khang and Levenspiel [58] on the basis of a recycle model. θ_m is defined as the time constant for the exponential decay of pseudo-periodic oscillations after a pulse-injection of tracer in a batch stirred reactor. When $Re > 10^4$, they obtain for turbines

$$N\theta_m(d/d_T)^{2.3} = 0.5 \approx 0.1P(\rho N^3 d^5)^{-1} \tag{112}$$

and for propellers

$$N\theta_m(d/d_T)^2 = 0.9 \approx 1.5P(\rho N^3 d^5)^{-1} \tag{113}$$

where d = the impeller diameter
d_T = the tank diameter
P = the power input

The terminal mixing time has also been related to the mean recirculation time. For instance in the work just cited

$$\left.\begin{aligned} &\theta_m/\bar{t}_c = A + B(s/\bar{t}_c)^2 \\ &s/\bar{t}_c > 0.8 \end{aligned}\right\} \tag{114}$$

Distribution of Turbulence Parameters

The fact that $\bar{t}_c$ and θ_m are not infinitely small shows that macroscopic uniformization of the tank contents is not instantaneous and that macromixing can never be considered as strictly perfect. Besides, measurements of velocity and concentration fluctuations have revealed that turbulence parameters were also distributed within the tank volume [49, 59–62]. The results of a typical investigation [48] concerning an aqueous medium in a semi-industrial (0.15 m^3) stirred tank are reported in Figure 17. The following ranges of variation were found.

- Turbulence intensity (with respect to turbine tip velocity) $5\% < u'/\pi Nd < 30\%$
- Turbulence macroscale $4 < L_f < 150$ mm
- Taylor microscale $1 < \lambda_f < 5$ mm
- Turbulent energy dissipation (with respect to the average value $\bar{\epsilon}$) $0.2 < \epsilon/\bar{\epsilon} < 2.5$
- Concentration fluctuations, determined by conductometry (for eddies > 100 μm) $2 \times 10^{-4} < c'/\bar{C} < 10^{-3}$

Similar results are reported in reference [59]. Generally, it is found that turbulence is not homogeneous immediately behind the impeller but that it may be considered as homogeneous elsewhere, especially in the recirculation streams. Upon integration of ϵ over the whole tank, one obtains nearly 90% of the mechanical power input, which proves that mechanical energy dissipation is mainly utilized to promote turbulence, and not to sustain the average velocity field, an interesting result for scaling-up.

Convenient models for predicting the distribution of turbulent parameters (See Table 1) are still lacking, even if attempts are made to solve turbulence equations with appropriate closure assumptions in the complex hydrodynamic pattern existing in a stirred tank.

The most popular models are the so-called "$k - \epsilon$" models, in which the mean flow momentum and continuity equations are linked to equations for k, the turbulent kinetic energy per unit mass (called $q = \frac{1}{2}u'^2$ in the present paper) and ϵ, the rate at which this energy is dissipated. Analogous

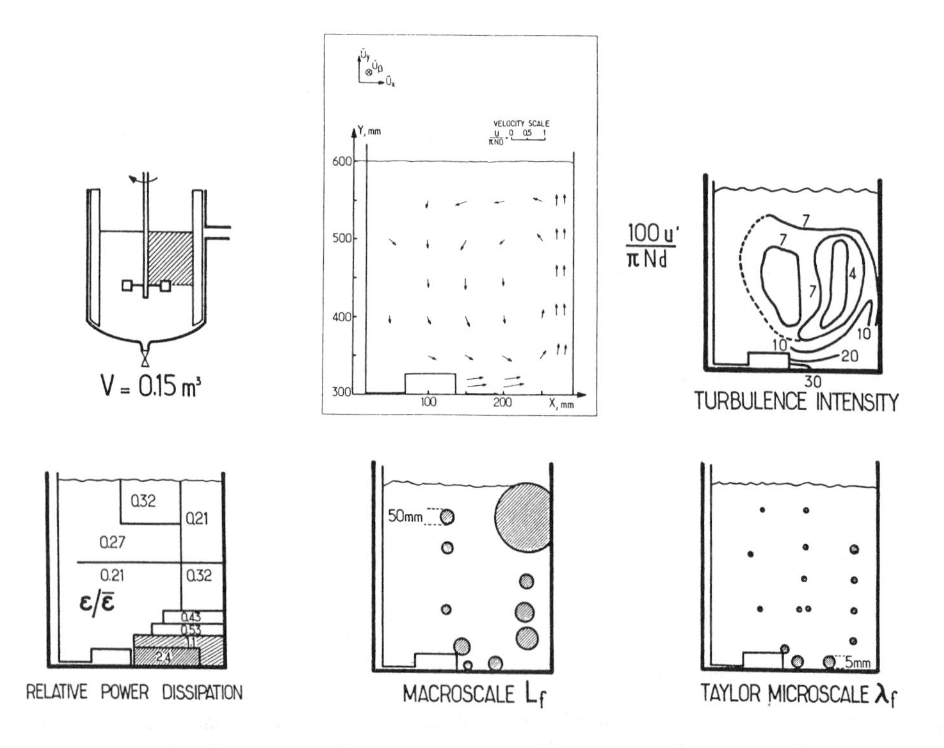

Figure 17. Distribution of average velocities, turbulence intensity, power dissipation, turbulent macro- and microscales in a semi-industrial stirred tank.

equations may be written for c'^2, the segregation and ϵ_s, the rate of segregation dissipation. Closure of equations is achieved by introducing effective turbulence exchange coefficients with appropriate constants. The reader will find a discussion of these problems in References 63–65 and in Volume I of this series. However, progress is still required to come up with universal and easy-to-implement models for standard design.

Cell Models

In order to predict chemical conversion in stirred tanks, Patterson and coworkers [63, 66–67] have proposed a model (HDM-model) in which the tank is divided into 30 mixing segments connected by specified flow rates Q_{ij} which are proportional to Nd^3. The turbulence level in each segment is characterized by the local values of L_s (proportional to d_T) and ϵ (proportional to d_T^2).

Mann and coworkers [68–70] also proposed a model where cells (or segments) are connected to each other according to the average flow pattern. Commutation according to a specified probability at each cell's outlet allows a stochastic path to be simulated. Such a model yields circulation time distribution in good agreement with experimental data [54–56].

All these models may be useful for the numerical simulation of turbulent mixing in the tank, but they require previous information about the hydrodynamic pattern.

Micromixing in a Nonuniform Stirred Tank

The problem of predicting chemical conversion in the presence of imperfect micromixing in a real stirred tank now appears as much more difficult than in the preceding sections where spatial uniformity was assumed. Actually, the characteristic micromixing times vary from point to point, owing to the spatial distribution of turbulence intensity and other physicochemical parameters. Although published works on this subject are still scarce, the following guidelines could be suggested:

- Selection of a macromixing model accounting for internal age distribution and/or circulation time distribution (i.e. the internal average hydrodynamic pattern).
- Estimation of local values of physicochemical and turbulence parameters along the trajectories or in the zones of the macromixing model.
- Calculation of characteristic micromixing times (or parameters) in the corresponding regions.
- Evaluation of conversion and/or yield by application of the micromixing models described in the preceding sections in the regions or along the trajectories of the macromixing model.
- Combination of the results according to this model in order to obtain overall conversion or yield.

An experimental proof of the spatial nonuniformity of the micromixing intensity in a stirred tank was provided by Barthole et al. [50], using the $BaSO_4$ precipitation method described earlier. Figure 18 shows the amount of precipitate obtained by injecting the same amount of acid at various places. The segregation index X_s is found to vary between 0.5% and 10.5%, i.e., a 20-fold variation. However, it is difficult to establish a direct link between these values and the distribution of turbulence parameters reported in Figure 17 because the precipitation time (Reaction 106) is in the same order of magnitude as the circulation time, which may be deduced from the average velocity pattern. A tentative model for interpreting these experiments is depicted in Figure 19. Roughly, the reactor is divided into two zones.

- Zone 1: turbine region, high turbulence intensity, small micromixing time t_{m1}.
- Zone 2: recirculation region, low turbulence intensity, large micromixing time t_{m2}.

After injection, the fluid starts a circulation loop and stays for t_1 in Zone 2. Then it passes through Zone 1 (residence time t_2) and through Zone 2 (residence time t_3 before closing the loop) etc . . . $t_1 + t_2 + t_3 = t_c$, the circulation time. By successive application of the IEM model

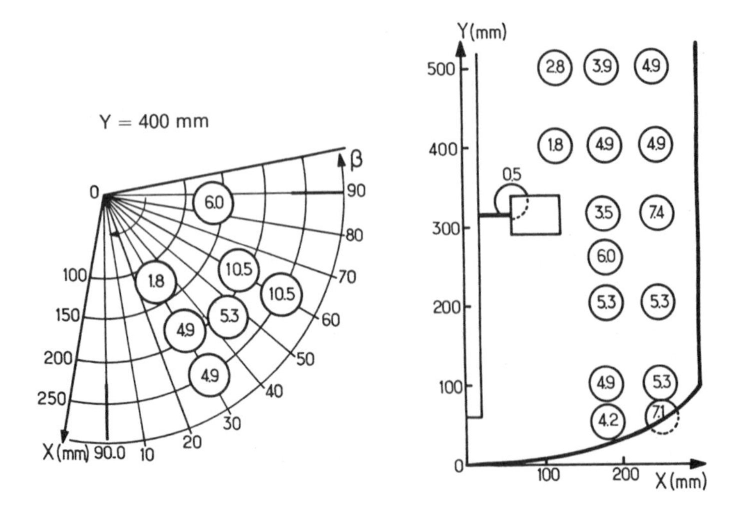

Figure 18. Distribution of the segregation index X_s (%) obtained by the $BaSO_4$ precipitation method in the tank of Figure 17.

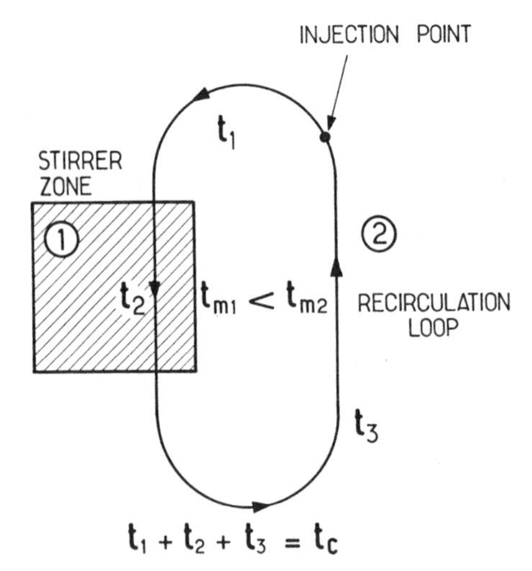

Figure 19. Model with two zones of different micromixing intensity for interpreting results of Figure 18.

in each zone, it is possible to calculate the final yield of precipitate. The micromixing times t_{m1} and t_{m2} can be calculated from the values of turbulence parameters in each zone. Clearly, the model can be complicated by adding more zones, taking into account the t_c-distribution etc... Such a model allows a first explanation of experimental data but work is still in progress to arrive at a definitive interpretation.

In conclusion to this section, it is obvious that a stirred tank is far from being the homogeneous and uniform system assumed in standard textbooks, but work is still required to provide designers with simple and reliable models taking into account the spatial heterogeneity of the micromixing intensity.

PREDICTION AND SCALE-UP OF MICROMIXING PARAMETERS

The scale-up of a stirred reactor in order to keep constant micromixing parameters and yield of chemical reactions controlled by micromixing has not been the subject of systematic investigations, as reliable and flexible test reactions to assess a scale-up strategy are not available.

Therefore, one can only suggest tentative relationships which ought to be carefully checked on experimental data.

The only reasonable basis available for establishing scale-up relationships is the theory of homogeneous isotropic turbulence presented earlier, although it is not applicable to all experimental situations, as already pointed out. However, in usual stirred tanks, this is not a too bad assumption outside the stirrer zone.

Dependence of Microscales and Time Constants on Stirring Speed and Reactor Size

Upon scale-up, reactors of different sizes are assumed to remain similar to each other so that their dimensions (height, diameter) are proportional to the stirrer diameter d. The only scaling factor is thus a length L proportional to d.

The following assumptions are made:

1. The results of the theory of homogeneous isotropic turbulence are applicable.
2. The fluid is Newtonian and incompressible.
3. The average velocity field is not distorted by scale-up, all average velocities in the tank being proportional to the stirrer tip velocity $\bar{U} = \pi Nd$.
4. The macroscale L_s and L_f are equal and proportional to d.
5. The turbulence intensity $u'/\bar{U}$ is independent of the turbulence Reynolds number $Re = u'L_f/\nu$ and roughly constant at a given position, whatever the reactor size.
6. The stirring power P is mainly dissipated under the form of turbulent kinetic energy and is used to a negligible extent for sustaining the average movement of the fluid.

In view of published experimental data, these assumptions are reasonable, at least for a first estimation. From these hypothesis and the results in Table 1, the exponents for scaling-up reported in Table 4 are obtained as follows:

From Assumptions 3 and 5, the scaling factor for the turbulent kinetic energy $q = \frac{3}{2}u'^2$ is N^2d^2.

From

$$\left.\begin{aligned} &Re_L = u'L_f/\nu \\ &\text{and} \\ &\lambda_f/L_f \approx Re_L^{-1/2} \end{aligned}\right\} \tag{115}$$

one obtains $Re_L \sim Nd^2\nu^{-1}$ and $\lambda_f \sim \nu^{1/2}N^{-1/2}$. The link between concentration and velocity fluctuations is made by

$$\lambda_s/\lambda_f = Sc^{-1/2} \tag{116}$$

So that $\lambda_s \sim \mathscr{D}^{1/2}N^{-1/2}$. The turbulent kinetic energy dissipation per unit mass is deduced either from $\epsilon \sim q^{3/2}/L_f$ or $\epsilon \sim \nu u'^2/\lambda_f^2$ and yields $\epsilon \sim N^3d^2$. Both dissipation time constants $\tau_f \sim \lambda_f^2/\nu$ and $\tau_s \sim \lambda_s^2/\mathscr{D}$ are then found to vary as N^{-1}. The scaling factors for viscous dissipation microscales and time constants are then easily deduced from the definitions and the expression of Kolmogorov microscale $\lambda_K = (\nu^3/\epsilon)^{1/4}$. In the case of the erosion time t_e (if the tentative expression (Equation 84) is accepted), an assumption has to be made about the dependence of ℓ_0. In Table 4, it is assumed that $\ell_0 \sim d$, but this could be discussed. The mean circulation time $\bar{t}_c$ together with τ_f and τ_s only depend on N^{-1}.

Table 4
Exponents for Scale-Up of Micromixing Parameters Under the Assumption of Homogeneous Isotropic Turbulence*

	ρ (a)	N (b)	$d \sim L$ (c)	ν (f)	$\mathscr{D}$ (g)
Average velocity $\bar{U}$	0	1	1	0	0
Macroscales $L_f = L_s$	0	0	1	0	0
Turbulent kinetic energy $q = 3/2u'^2$	0	2	2	0	0
Turbulent k.e. dissipation ϵ	0	3	2	0	0
Taylor microscale λ_f	0	$-\frac{1}{2}$	0	$\frac{1}{2}$	0
Corrsin microscale $\lambda_s = \lambda_f Sc^{-1/2}$	0	$-\frac{1}{2}$	0	0	$\frac{1}{2}$
Taylor time constant τ_f	0	-1	0	0	0
Corrsin time constant τ_s	0	-1	0	0	0
Kolmogorov microscale λ_K	0	$-\frac{3}{4}$	$-\frac{1}{2}$	$\frac{3}{4}$	0
Batchelor microscale λ_B	0	$-\frac{3}{4}$	$-\frac{1}{2}$	$\frac{1}{4}$	$\frac{1}{2}$
Corrsin microscale λ_C	0	$-\frac{3}{4}$	$-\frac{1}{2}$	0	$\frac{3}{4}$
Viscous dissipation times					
$\tau_K = (\nu/\epsilon)^{1/2} = \lambda_K^2/\nu$	0	$-\frac{3}{2}$	-1	$\frac{1}{2}$	0
$\tau_B = \tau_K Sc^{1/2}$	0	$-\frac{3}{2}$	-1	1	$-\frac{1}{2}$
$\tau_C = \tau_K Sc^{-1/2}$	0	$-\frac{3}{2}$	-1	0	$\frac{1}{2}$
$t_D \sim \lambda_K^2/\mathscr{D} = \tau_K \cdot Sc$	0	$-\frac{3}{2}$	-1	$\frac{3}{2}$	-1
Erosion time $t_e \sim \ell_0 \lambda_K Sc^{-1/3} \mathscr{D}^{-1}$	0	$-\frac{3}{4}$	1	$\frac{5}{12}$	$-\frac{2}{3}$
Power input P	1	3	5	0	0
Circulation time $\bar{t}_c \sim L/\bar{U}$	0	-1	0	0	0

* *Quantities in left column vary as $\rho^a N^b d^c \nu^f \mathscr{D}^g$. Exponents a, b, c, f, g are given in the table.*

Scale-Up of the Stirring Power

The stirring power depends on design parameters according to the familiar expression

$$P = N_p \rho N^3 d^5 \tag{117}$$

where N_p is the power number depending on the kind of design.

Let L be a characteristic linear dimension of the tank (homothetical design). From Equation 117 and the expressions in Table 4, the following rules may be deduced.

Quantity to Keep Constant	Scale-Up Rule	Micromixing Parameters which are Maintained
ϵ	$P \sim L^3$	$\lambda_K, \lambda_B, \lambda_c, \tau_K, \tau_B, \tau_c, t_D$
N	$P \sim L^5$	$\lambda_f, \lambda_s, \tau_f, \tau_s, \bar{t}_c$
ϵ/L^4	$P \sim L^7$	t_e

Keeping constant micromixing parameters for stage 3 (diffusion, viscous dissipation) is thus less demanding ($P \sim L^3$) than keeping them for stage 2 ($P \sim L^5$ or even L^7). It may be pointed out that the usual rule for scale-up at constant stirrer tip velocity (Nd = constant) leads to $P \sim L^2$. The average velocity field is then the same but micromixing characteristics are not preserved.

Example of Application

How is the 0.15 m^3 reactor described earlier ([48], Figure 17) scaled-up to a 15 m^3 stirred tank having the same geometrical shape?

1. *Linear scaling factor*. The volume ratio is $V_2/V_1 = 100$, therefore $L_2/L_1 = 100^{1/3} = 4.64$. The stirrer diameter is thus $d_2 = 4.64 \times 0.275 = 1.28$ m, and the tank diameter is $d_{T2} = 4.64 \times 0.58 = 2.69$ m.
2. *Preservation of ϵ, the dissipated power per unit mass* (The liquid is assumed to be the same). In the original design, $P_1 = 5\rho N_1^3 d_1^5$ and $N_1 = 1.6\ s^{-1}$ (96 RPM) in typical conditions. This corresponds to $P_1 = 32$ W. In order to preserve ϵ, the scale-up factor is $L^3 = 100$, which leads to $P_2 = 3.2$ kW. The stirring speed is then $N_2 = N_1(d_1/d_2)^{2/3} = 0.575\ s^{-1}$ (34.5 RPM).
3. *Preservation of N, the stirring speed*. The scale-up factor is then $L^5 = 100^{5/3} = 2{,}154$. The required power is then $P_2 = 0.032 \times 2{,}154 = 68.9$ kW.
4. *Preservation of Nd*. P is proportional to L^2 and $P_2 = 32 \times 100^{2/3} = 689$ W. The stirring speed is then $N_2 = N_1/4.64 = 0.345\ s^{-1}$ (20.7 RPM).
5. *Preservation of ϵ/L^4*. The scale-up factor for P is in that case $L^7 = 4.64 \times 10^4$ and this would lead to an unrealistic power requirement.

To sum up, the results are

Original design	V **0.15 m^3**	d_T **0.58 m**	d **0.275 m**	P **32 W**	N **1.6 s^{-1} (96 RPM)**
New design	15 m^3	2.69 m	1.28 m		
(2) ϵ = constant				3.2 kW	0.575 s^{-1} (34.5 RPM)
(3) N = constant				68.9 kW	1.6 s^{-1} (96 RPM)
(4) Nd = constant				689 kW	0.345 s^{-1} (20.7 RPM)

In Case 2 viscous dissipation microscales (Kolmogorov, Batchelor . . .) are preserved. In Case 3, turbulent dissipation microscales (Taylor, Corrsin) and circulation times are preserved. In Case 4, absolute mean velocities are preserved but micromixing parameters are not.

EXAMPLES OF INFLUENCE OF MICROMIXING ON CHEMICAL REACTIONS

As pointed out earlier, the key parameter for deciding whether micromixing has any influence on the conversion or selectivity of chemical reactions in a stirred tank is the ratio t_R/t_m where t_R and t_m are controlling time constants for the chemical reaction and the micromixing processes, respectively. Micromixing control is to be expected when this ratio is small (t_R/t_m roughly represents the microfluid/macrofluid volume ratio). Examples obtained either from experience or from simulations where micromixing effects have been brought to evidence follow.

Fast or Complex Liquid Phase Reactions

Such reactions have been mentioned previously, as they can be employed as test reactions for studying the state of segregation in reactors. They are also discussed in Volume I. The influence of micromixing on the selectivity of consecutive-competing reactions is especially noticeable in the case of the nitration of aromatic hydrocarbons [71], the bromation of 1-3-5 trimethoxybenzene [72], and the bromation of resorcin [73], where, owing to complex pH effects, the amount of 2–4 dibromoresorcin in the di-isomer may vary from 30% to 60% when the stirring speed passes from 0 to 360 RPM.

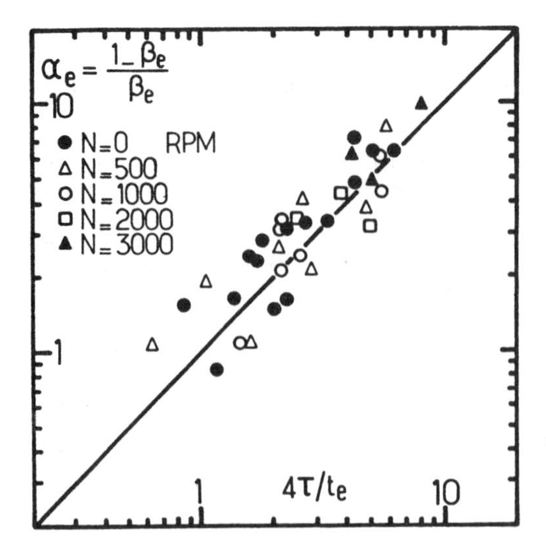

Figure 20. Iodation of acetone. Reaction mainly in the erosion stage. The quality of micromixing increases with the ratio τ/t_e.

The iodation of acetone was studied in a 196 cm³ CSTR with unmixed feed by Plasari et al. [19, 24, 43]. The reaction takes place essentially during Beek and Miller's stage 2 (erosion mechanism). Figure 20 shows that the quality of micromixing ($(1 - \beta_e)/\beta_e$) increases when the erosion time diminishes according to Equation 90.

Zoulalian and Villermaux [44, 43] studied the iodination of paracresol and the alkaline hydrolysis of glycoldiacetate (two consecutive competing reactions) and the alkaline hydrolysis of ethylacetate in an unbaffled 450 cm³ reactor stirred at the bottom. The micromixing stage 2 was very fast (there was no difference whether the feedstreams were unmixed or premixed) and the reactions mainly took place during stage 3, probably controlled by a mechanism of diffusion in stretching laminar eddies [43]. The average diffusion time was comprised between 2.4 s (glycol diacetate) corresponding

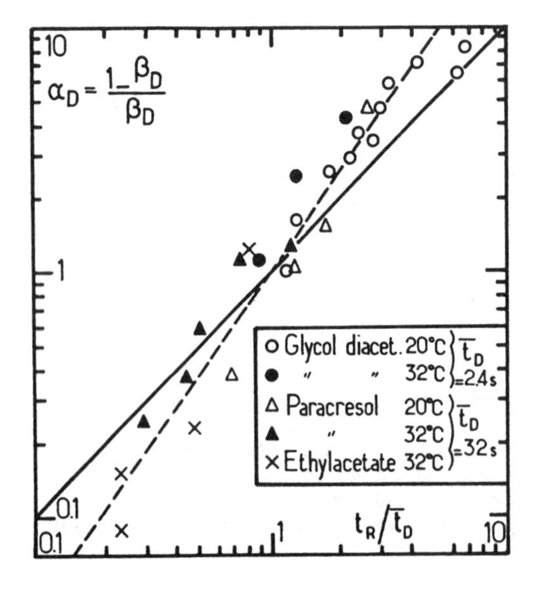

Figure 21. Reactions mainly in the diffusion stage with stretching aggregates. The quality of micromixing increases with the ratio $t_R/\bar{t}_D$, according to Figure 14.

to a striation thickness $\bar{\ell} \approx 50\ \mu m$ and 32 s (iodination of paracresol, ethylacetate) corresponding to $\bar{\ell} \approx 200\ \mu m$. Figure 21 shows experimental data plotted according to Equation 91. Using screens in the reactor, it was shown that the intensity of micromixing could be locally varied within the tank. Segregation could also be significantly reduced (a two-fold factor) in the presence of a 22 kHz-ultrasonic field.

Klein et al. [46] studied the alkaline hydrolysis of nitromethane in a standard 196 cm^3 CSTR with unmixed feed. As explained earlier, the reaction takes place both during stage 2 (erosion) and stage 3 (diffusion). Figure 22 shows the experimental results, compared with the predictions of a mixed model involving two time constants t_e and t_D (Equations 96–98). By fitting the free parameters, it was found, in S.I. units:

$$\left.\begin{aligned} t_e &= \frac{\nu^{5/12}}{3{,}700\mathscr{D}^{2/3}(\epsilon_1 + 15\epsilon_2)^{1/4}} \\ t_D &= \frac{2}{(\epsilon_1 + \epsilon_2)^{1/3}} \end{aligned}\right\} \tag{118}$$

indicating that mixing in stage 3 was probably controlled by turbulent dissipation (Corrsin time constant). But these results are concerned with a particular system and should not be generalized without any precaution.

Precipitation Reactions

Precipitation is a special case of crystallization produced by the apparition of a supersaturation resulting from the mixing of two reactants. Therefore, the rate of precipitation and the size of crystals may depend on micromixing intensity which influence contacting of reactants and subsequent processes of nucleation and growth.

This problem was studied by Pohorecki and Baldyga [83] in the frame of the theory of homogeneous turbulence but as pointed out earlier, macromixing effects must be carefully taken into account as they also control the local value of supersaturation. There is still little work published on this subject, but it may be anticipated that precipitation and control of the quality of crystals is one of the areas where micromixing theories will find industrial applications.

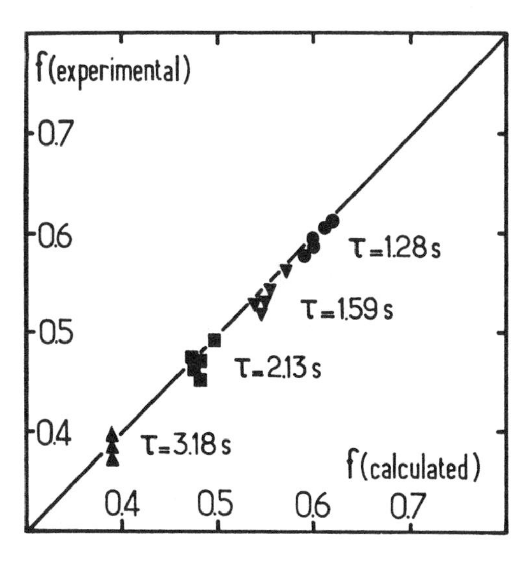

Figure 22. Alkaline hydrolysis of nitromethane. Comparison between experimental data and predictions of a model involving two micromixing stages in series.

Polymerization Reactions

Polymerization offers another industrial field where micromixing may play an important role [74–76]. Micromixing effects on free radical polymerization can be studied via numerical simulations based on the simple kinetic scheme:

$$\left.\begin{array}{lll} A \xrightarrow{k_{d,f}} 2R & & \text{initiation} \\ R + M \xrightarrow{k_p} R & & \text{propagation} \\ R + P \xrightarrow{k_{tp}} P + R & & \text{transfer to polymer} \\ R + R \xrightarrow{k_t} P + P & & \text{termination (disproportionation)} \end{array}\right\} \quad (119)$$

Figure 23 shows the chain length distribution at 66% conversion of the monomer in the case of a linear polymerization ($k_{tp} = 0$) in different reactors: batch reactor (B), segregated CSTR (S) and well micromixed CSTR (M). The shape of the chain length (molecular weight) distribution is strongly dependent on the state of segregation. In this special case, molecular mixing leads to a narrower distribution and to a lower dispersion index ($DI = \bar{M}_w/\bar{M}_n$). When transfer to polymer is involved, micromixing effects are still more pronounced. Figure 24 shows a plot of the dispersion index (or polydispersity) as a function of the monomer conversion X for segregated flow (S) and well micromixed flow (M). The dramatic influence of segregation can be noticed at high conversion. Moreover, an interesting effect is observed with diluted and slow-decomposing initiators, namely an inversion of the relative position of S and M curves when the transfer constant k_{tp} is increased. This doesn't happen with concentrated and fast-decomposing initiators.

These results were checked experimentally. In Reference 77 continuous polymerization of styrene was carried out in a CSTR and in cyclohexane solution in order to keep the viscosity low and constant. The dispersion index was measured as a function of space time and stirring speed. Limiting S and M curves were calculated from batch kinetics. Clear evidence for segregation effects can be seen on Figure 25 which shows that perfect micromixing may be very difficult to achieve, even

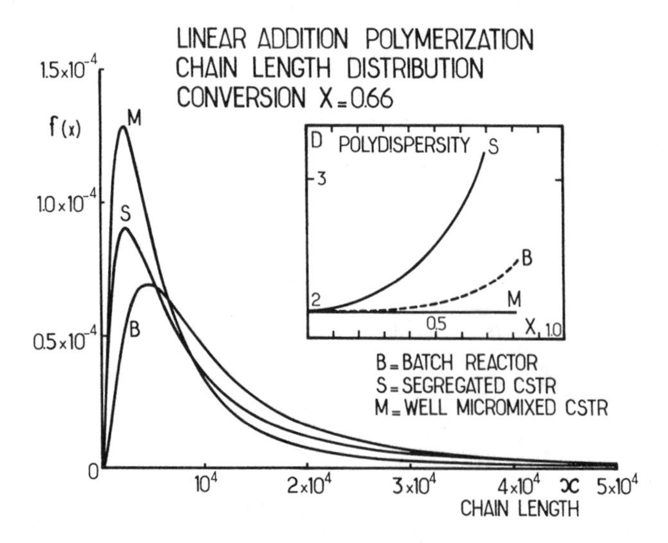

Figure 23. Influence of mixing on molecular weight (chain length) distribution. Linear free-radical polymerization.

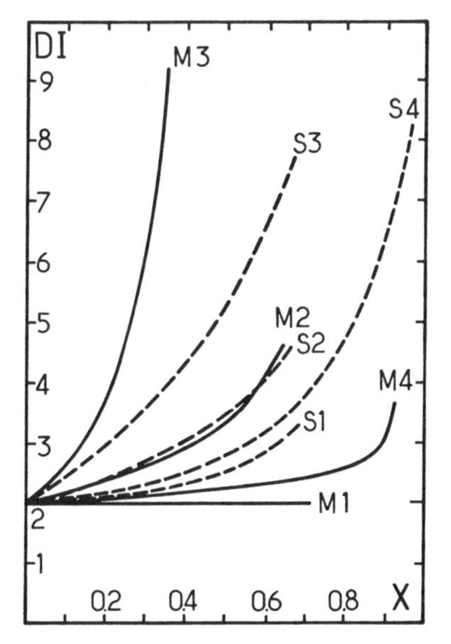

Figure 24. Influence of micromixing on dispersion index (DI) as a function of conversion (X): S = segregated flow, M = well micromixed flow.

Linear free-radical polymerization ($k_{tP} = 0$)

Curve 1: $k_p = 5 \times 10^7 \ \ell \ mol^{-1} \ hr^{-1}$
$k_t = 1.5 \times 10^{12} \ \ell \ mol^{-1} \ hr^{-1}$
$k_d = 3.3 \times 10^{-2} \ hr^{-1}$, $f = 0.5$,
$A_0 = 3 \times 10^{-3} \ mol \ \ell^{-1}$
$S_0 = 7.12 \ mol \ \ell^{-1}$ (solvent)
$M_0 = 3.56 \ mol \ \ell^{-1}$

Transfer to polymer

Curve 2: $k_{tp} = 3.5 \times 10^3 \ mol \ \ell^{-1} \ hr^{-1}$
Curve 3: $k_{tP} = 1.05 \times 10^4 \ mol \ \ell^{-1} \ hr^{-1}$
Curve 4: $k_{tP} = 1.05 \times 10^4 \ mol \ \ell \ hr^{-1}$ and
$A_0 = 3 \times 10^{-2} \ mol \ \ell^{-1}$
$k_d = 0.33 \ hr^{-1}$ (other parameters unchanged).

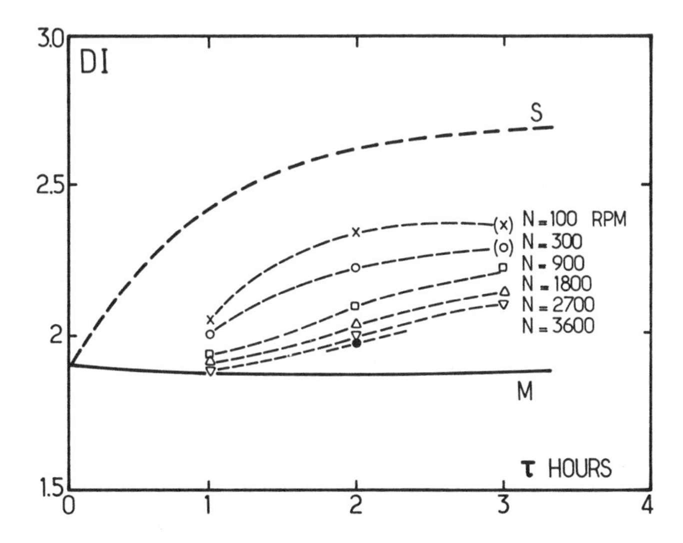

Figure 25. Effect of segregation on free radical polymerization of styrene in cyclohexane solution. Standard CSTR with 4 baffles and a 6 blade turbine V = 670 cm^3, T = 75°C. Dispersion index DI versus space time τ. Influence of agitation speed N. Curves S (segregated flow) and M (well micromixed flow) calculated from batch experiments. Initiator = PERKADOX 16.

$A_0 = 3.3 \times 10^{-2} \ mol \ \ell^{-1}$, $k_d = 5 \times 10^{-5} \ s^{-1}$, $f = 0.85$

$M_0 = 6.65 \ mol \ \ell^{-1}$, $S_0 = 2.22 \ mol \ \ell^{-1}$

with strong agitation and low viscosity. But such effects can only be observed with fast-decomposing initiators.

Besides these laboratory experiments, the analysis of industrial reactors may also reveal segregation effects, as for instance in high pressure free radical polymerization of ethylene where segregation of the initiator feedstream controls the reaction [78].

Polycondensation Reactions

Polycondensation is also an important industrial operation involved for instance in the manufacture of textile fibers (polyester or polyamide). The kinetic scheme is different from that of polymerization because of the evolution of a light product (e.g. water) and the existence of reversible reactions, exchange reactions, cyclizations etc.

A simple scheme for polyesterification can be written, for example [79]:

$$\left.\begin{array}{l} -\!\!-A + B-\!\!- \underset{k'}{\overset{k}{\rightleftarrows}} -\!\!-V-\!\!- + W \\ \text{esterification} \longrightarrow \longleftarrow \text{hydrolysis} \\ -\!\!-B + -\!\!-V-\!\!- \xrightarrow[\text{exchange}]{k_{eB}} -\!\!-V-\!\!- + B-\!\!- \end{array}\right\} \qquad (120)$$

Cyclization is neglected. X represents the conversion of terminal groups A and B, K is the equilibrium constant

$$K = \frac{Y_W X_e}{(1 - X_e)(r - X_e)} = \frac{k}{k'}$$

where r = the stoichiometric excess of initial B/A

Y_W/K = a characteristic parameter of the water content of the mixture (imposed for instance by the liquid vapor equilibrium)

$e_B = k_{eB}/K$ characterizes the relative importance of exchange reactions

Figure 26 shows the results of a simulation with the preceding simple model [79]. The dispersion index of linear macromolecules is plotted as a function of X for segregated flow (S) and well micro-

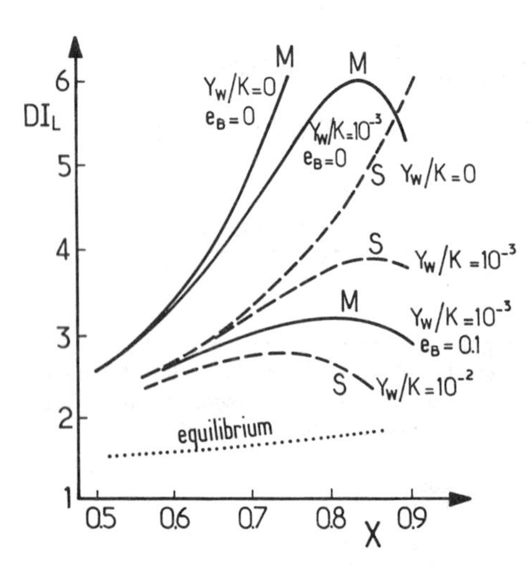

Figure 26. Effect of segregation on polycondensation. Dispersion index of linear macromolecules as a function of conversion of terminal groups.

mixed flow (M) in a CSTR. If Y_W/K and k_{eb}/k are both zero there is a noticeable influence of segregation at high conversion. However, it is difficult to achieve perfect elimination of water and it appears that either the presence of traces of water, or the existence of exchange reactions affect the polydispersity as much as segregation does. Nevertheless, micromixing should be here also considered as an important factor for the prediction of polymer quality in industrial reactors, especially at high conversion of terminal groups.

Gas-Phase Reactions

Up to this point, the present survey has mainly been concerned with liquid-phase reactors. What can be said about micromixing in gas-phase stirred reactors? In principle, the theoretical analysis previously exposed (e.g. the theory of homogeneous isotropic turbulence) is applicable, but the main difference is that the Schmidt number in a gas is close to 1, i.e. about 1,000 times smaller than in a liquid. As a consequence, molecular diffusion is much more efficient to wipe out concentration gradients, and it may be expected that in usual conditions, the gas behaves as a microfluid (the case of combustion reactions is particular and not dealt with in this survey).

This was checked in a reactor spontaneously stirred by gas jets, used for kinetic measurements at short residence times (a few seconds) [80]. The test reaction was the dehydrogenation of isobutane by iodine in the presence of oxygen

$$\left.\begin{aligned} i-C_4H_{10} + I_2 &\longrightarrow 2\,HI + i-C_4H_8 \\ 2\,HI + \tfrac{1}{2}O_2 &\longrightarrow I_2 + H_2O \end{aligned}\right\} \tag{121}$$

The reaction is pseudo-first order with respect to isobutane, but in the absence of oxygen it becomes inhibited by HI and obeys a complex nonlinear kinetics. It is thus possible to adjust the concentrations so that oxygen is used up in the reactor and the reaction then becomes very sensitive to segregation. Experiments were carried out in a 71 cm^2 jet-stirred reactor at 763 K (τ about 2 s). Under these conditions a 10% difference could be expected between conversion in a microfluid and in a macrofluid. Within experimental errors, it was shown that the reactor was perfectly micromixed, a result which was confirmed by the estimation of micromixing times (10^{-2} to 10^{-3} s).

Thermal Segregation

The preceding sections deal with isothermal systems. However, microscopic temperature gradients may arise in a partially segregated fluid upon mixing streams of different temperature, or by local release or absorption of heat by chemical reactions in the fluid aggregates. This phenomenon is especially important in combustion reactions and can be studied in the frame of turbulence theory. In a gas, the rates of decay of temperature and concentration segregation are roughly the same. This is not the case in a liquid, owing to the larger value of the Lewis number, as far as molecular diffusion controls ultimate mixing. The rate of decay of temperature segregation may then be 100 times faster than that of concentration. Consequently, in usual conditions, the assumption of local isothermicity in a liquid phase stirred tank is acceptable. Without treating the problem in great detail, it is interesting to mention the possibility to characterize thermal segregation by a chemical method. This method relies on the nonlinear dependence of the rate of reaction upon temperature.

For instance, one can use a first-order reaction giving a product which is easy to detect at the reactor outlet, and feed a CSTR with two streams (respective flow rates aQ and (1 − a)Q) at two different temperatures T_1 and T_2. If the reactor is adiabatic, the average temperature in the tank is $T = aT_1 + (1 - a)T_2$. The rate constant $k = A\exp(-E/RT)$ is assumed to be known. Let $Da = k(T)\tau$ be the Damköhler number.

In a well micromixed fluid, the extent of reaction is

$$X_M = \frac{Da}{1 + Da} \tag{122}$$

whereas in a totally segregated fluid, each aggregate keeps its initial temperature and

$$X_S = 1 - \frac{a}{1 + Da_1} - \frac{1 - a}{1 + Da_2} \tag{123}$$

$$\left.\begin{aligned} Da_1 &= Da \exp\left[\frac{E}{RT} \cdot \frac{(1 - a)\,\Delta T}{T + (1 - a)\,\Delta T}\right] \\ Da_2 &= Da \exp\left[-\frac{E}{RT} \cdot \frac{a\,\Delta T}{T - a\,\Delta T}\right] \\ \Delta T &= T_2 - T_1 \end{aligned}\right\} \tag{124}$$

owing to the nonlinearity of the Arrhenius relationship, $X_M \neq X_S$. An example of a possible reaction is the hydrolysis of sulfamic acid [81]:

$$\left.\begin{aligned} & NH_2SO_3H + H_2O \xrightarrow{k} SO_4^{--} + H^+ + NH_4^+ \\ & k = 1.986 \times 10^{13} \exp(-14{,}800/T)\ s^{-1} \end{aligned}\right\} \tag{125}$$

The extent of reaction can be easily determined by precipitation of $BaSO_4$ either "on-line" (the reactant being a mixture of sulfamic acid and barium chloride) or after sampling. X is proportional to the amount of precipitate $BaSO_4$. The reactor is preferably operated at small reaction extent (1 – 10%). For instance at 423 K and with a = 0.5, one obtains

Da	X_M%	X_S%	X_S/X_M	X_M%	X_S%	X_S/X_M
0.01	0.99	1.31	1.32	0.99	2.4	2.43
0.05	4.76	6.09	1.28	4.76	10.2	2.14
0.1	9.09	11.2	1.23	9.09	17.2	1.89
		$\Delta T = 20$ K			$\Delta T = 40$ K	

By application of the IEM model, it is easy to link the observed extent of reaction (comprised between X_M and X_S), namely $\beta = (X - X_M)/(X_S - X_M)$, to a thermal micromixing time t_m or to the dimensionless number $b' = kt_m$.

Such a chemical characterization of the temperature field heterogeneity in a stirred reactor is an interesting area to explore.

CONCLUSIONS

The major concepts explaining micromixing processes in stirred reactors are now well identified and progress has been made toward a unified theory allowing a priori predictions from the sole knowledge of physicochemical properties of the fluid and operating parameters of the reactor.

In continuous reactors, models for mixing earliness describing transfer between entering and leaving environments are superabundant. Models relying on a more physical description should be preferred. The reduction of segregation by interaction between fluid particles (aggregates) can be conveniently represented by simple models (exchange with the mean, coalescence-dispersion) but in most cases, several stages for mixing are involved, each with their own time constants, either in series or in parallel. In addition, there is a strong interaction between these micromixing processes and hydrodynamic patterns resulting from imperfect macromixing. In this respect, a global characterization of micromixing assuming that the tank content is uniform is not sufficient and more attention should be paid to internal age distributions and related quantities such as circulation-time distributions and local variation of mechanical energy dissipation.

In order to better understand the mechanism of the ultimate stage of mixing by molecular diffusion, turbulence theory offers one possible approach but carefully designed experiments (no macromixing effects, uniform power dissipation, controlled viscosity) and new chemical test reactions, easy to implement in industry, would be welcome.

The method of "characteristic times" is especially helpful for determining which processes are controlling. These are, for example, the sapce time τ for a continuous reactor; a characteristic time for internal macromixing pattern, (e.g., the average circulation time $\bar{t}_c$); one or several reaction times (e.g., $t_R = 1/kC_0^{n-1}$); and one or several characteristic micromixing times t_m (e.g., t_e (erosion) or $t_D \sim l^2/\mathscr{D}$, or $t_{cD} = 1/\omega$, or $t_\delta \sim (\nu/\epsilon)^{1/2}$, or $\tau_s \sim (L_s^2/\epsilon)^{1/3}$) depending on the appropriate controlling mechanism. A comparison between all these times allows the determination of the mixing regime, sometimes quantitatively. It was thus established that the microfluid/macrofluid volume ratio was nearly equal to t_R/t_m. Once the relevant combination of processes and the corresponding time constants have been determined, the IEM model may be recommended as a very simple and efficient tool for representing competition between interaction and chemical reaction.

A trend of current research is to combine the Eulerian approach of fluid mechanics and the Lagrangian and systems approach of chemical engineering. In particular, measurements of concentration fluctuations should be developed, both in presence and in absence of chemical reactions in order to obtain reliable spectral data. However, the final solution to micromixing problems should not be sought in turbulence theory alone, but rather in phenomenological interaction models, whose parameters could have a fundamental interpretation by this theory.

There are still important problems to be solved, concerning for instance the description and prediction of micromixing characteristics in large volume reactors or in non-Newtonian media, not to speak of multiphase reactors which are outside the scope of this survey. But above all, theory will progress in a direction useful to practitioners if more experimental data on realistic industrial situations are available to researchers. In the present status of our knowledge reliable methods exist for estimating the influence of micromixing on chemical reactions and for scaling-up laboratory- or bench-scale results to industrial reactors, at least in the simplest cases. More complex situations still require parameter fitting or experimental tests to calibrate theoretical models. But once this is done, existing models offer a flexible framework for correlating and extending results to new situations. The practical importance of mixing effects in operations such as polymerization, polycondensation, precipitation, crystallization, fast multi-step reactions, reactions in viscous media clearly proves that micromixing is no longer "a solution in search of a problem" [74].

NOTATION

b, b′	exponent
C	concentration
c′	concentration fluctuation
$\mathscr{D}$	molecular diffusion coefficient
Da	Damköhler number
d	stirrer diameter
E_s	spectral density
f_{AM}, f_{AS}	residual concentrations
I_s	segregation intensity parameter; see Equation 3
J_k	interaction term defined by Equation 8
K	equilibrium constant
k_R	reaction rate constant
L	volume to external area ratio
L_s	concentration fluctuation macroscale
ℓ	diameter of aggregate
M	Thiele modulus
m	exponent
N_p	power number
N_{seg}	segregation number
P	power
p(c)	local concentration distribution
Q	volumetric flowrate
q	turbulent kinetic energy
R_k	reaction term; see Equation 15
$\mathscr{R}_{(i)}$	rate of reaction in class i
r	reaction rate
Re	Reynolds number
Sc	Schmidt number
s	fraction of fluid in E.E.
t	time
t_δ	stretching time
U	local velocity
u′	velocity fluctuation

V volume
X reaction extent
x position
Y yield (concentration)
Z volumetric property

Greek Symbols

α age of aggregate
α^* segregation function
β volume fraction of macrofluid
δ index of mixing quality
ϵ rate of energy dissipation
η efficiency of energy dissipation
θ fractional time scale
λ life expectancy
λ_k Kolmogorov microscale
μ shape factor
μ_n n-th order moment of distribution
$\nu_{A,B}$ stoichiometric coefficients
σ^2 variance of distribution
τ time
ω frequency

REFERENCES

1. Hiby, J. W., "Definition and Measurement of the Degree of Mixing in Liquid Mixtures," *Internat. Chem. Eng.*, Vol. 21 (1981), 197–204.
2. Danckwerts, P. V., "The Effect of Incomplete Mixing on Homogeneous Reactions," *Chem. Eng. Sci.*, Vol. 8 (1958), pp. 93–99.
3. Zwietering, T. N., "The Degree of Mixing in Continuous Flow Systems," *Chem. Eng. Sci.*, Vol. 11 (1959), pp. 1–15.
4. Villermaux, J., "Drop Break-Up and Coalescence. Micromixing Effects in Liquid-Liquid Reactors, in *Multiphase Chemical Reactors*, Vol. I Fundamentals, A. E. Rodrigues, J. M. Calo and N. H. Sweed (Eds.), Nato Advanced Study Institutes Series E-N° 51, Sijthoff and Noordhoff, (1981), pp. 285–362.
5. Curl, R. L., "Dispersed Phase Mixing-Theory and Effects in Simple Reactors," *A.I.Ch.E. J.*, Vol. 9 (1963), p. 175.
6. Spielman, L. A., and Levenspiel, O., "A Monte-Carlo Treatment for Reacting and Coalescing Dispersed Phase Systems," *Chem. Eng. Sci.*, Vol. 20 (1965), pp. 247–254.
7. Kattan, A., and Adler, R. J., "A Conceptual Framework for Mixing in Continuous Chemical Reactors," *Chem. Eng. Sci.*, Vol. 27 (1972), pp. 1013–1028.
8. Treleaven, C. R., and Tobgy, A. H., "Monte-Carlo Methods of Simulating Micromixing in Chemical Reactors," *Chem. Eng. Sci.*, Vol. 27 (1972), pp. 1497–1513.
9. Villermaux, J., "Génie de la Réaction chimique. Conception et fonctionnement des réacteurs," *Technique et Documentation*, Lavoisier, Paris 1982.
10. Treleaven, C. R., and Tobgy, A. H., "Residence Times, Micromixing and Conversion in an Unpremixed Feed Reactor. I-Residence Time Measurements," *Chem. Eng. Sci.*, Vol. 27 (1972), pp. 1653–1668.
11. Treleaven, C. R., and Tobgy, A. H., "Residence Times, Micromixing and Conversion in an Unpremixed Feed Reactor. II-Chemical Reaction Measurements," *Chem. Eng. Sci.*, Vol. 28 (1973), pp. 413–425.
12. Ritchie, B. W., and Tobgy, A. H., "A Three-Environment Micromixing Model for Chemical Reactors with Arbitrary Separate Feedstreams," *The Chem. Eng. Journal*, Vol. 17 (1979), pp. 173–182.
13. Harada, M., et al., "Micromixing in a Continuous Flow Reactor (Coalescence and Redispersion Models)," *The Memoirs of the Faculty of Engineering*, Kyoto Univ., Vol. 24 (1962), p. 431.
14. Villermaux, J., and Devillon, J. C., "Représentation de la coalescence et de la redispersion des domaines de ségrégation dans un fluide per un modèle d'interaction phénoménologique," Proceed. 2nd Int. Symp. Chem. React. Engin. Amsterdam, (1972), pp. B 1–13.
15. Costa, P., and Trevisoi, C., "Reactions with Nonlinear Kinetics in Partially Segregated Fluids," *Chem. Eng. Sci.*, Vol. 27 (1972), p. 2041.

16. Aubry, C., and Villermaux, J., "Représentation du mélange imparfait de deux courants de réactif dans un réacteur agité continu," *Chem. Eng. Sci.*, Vol. 30 (1975), p. 457.
17. David, R., and Villermaux, J., "Micromixing Effects on Complex Reactions in a CSTR," *Chem. Eng. Sci.*, Vol. 30 (1975), p. 1309.
18. Ritchie, B. W., and Tobgy, A. H., "Mixing and Product Distribution with Series-Parallel Reactions in Stirred Tank and Distributed Feed Reactors," *Adv. Chem. Ser. Am. Chem. Soc.*, Vol. 133 (1974), pp. 376–392.
19. Plasari, E., David, R., and Villermaux, J., "L'iodation de l'acétone, une réaction-modèle de cinétique michaelienne pour la simulation de réacteurs chimiques et biochimiques," *Nouv. J. Chim.*, Vol. 1 (1977), p. 49.
20. Spencer, J. L., Lunt, R., and Leshaw, S. A., "Identification of Micromixing Mechanisms in Flow Reactors: Transient Inputs of Reactive Tracers," *Ind. Eng. Chem. Fundam.*, Vol. 19 (1980), pp. 135–141.
21. Weinstein, H., and Adler, J., "Micromixing Effects in Continuous Chemical Reactors," *Chem. Eng. Sci.*, Vol. 22 (1967), 65–75.
22. Villermaux, J., and Zoulalian, A., "Etat de mélange dans un réacteur continu. A propos d'un modèle de Weinstein et Adler," *Chem. Eng. Sci.*, Vol. 24 (1969), pp. 1413–1417.
23. Ng, D. Y. C., and Rippin, D. W. T., "The Effect of Incomplete Mixing on Conversion in Homogeneous Reactions," 3rd Europ. Symp. Chem. React. Eng. Amsterdam, 1964, Pergamon Press, Oxford, 1965, pp. 161–165.
24. Plasari, E., David, R., and Villermaux, J., "Micromixing Phenomena in Continuous Stirred Reactors Using a Michaelis-Menten Reaction in the Liquid Phase," A.C.S. Symp. Series, Chemical React. Eng. Houston, Vol. 65 (1978), pp. 126–139.
25. Valderrama, J. O., and Gordon, A. L., "Mixing Effects on Homogeneous p-Order Reactions. A Two-Parameter Model for Partial Segregation," *Chem. Eng. Sci.*, Vol. 34 (1979), pp. 1097–1103.
26. Ritchie, B. W., "Comments on Valderrama and Gordon," *Chem. Eng. Sci.*, Vol. 37 (1982), p. 800.
27. Villermaux, J., "Mixing in Chemical Reactors," ACS Symposium Series, Vol. 226 (1983), pp. 135–186.
28. Treleaven, C. R., and Tobgy, A. H., "Conversion in Reactors Having Separate Reactant Feed Streams. The State of Maximum Mixedness," *Chem. Eng. Sci.*, Vol. 26 (1971), pp. 1259–1269.
29. Ritchie, B. W., and Tobgy, A. H., "Mixing, Diffusion, and Chemical Reaction in an Unpremixed Feedstream Reactor," *The Canad, J. Chem. Eng.*, Vol. 55 (1977), pp. 480–483.
30. Mehta and Tarbel, "Four Environment Model of Mixing and Chemical Reaction," *A.I.Ch.E. J.*, Vol. 29 (1983), p. 320.
31. Ritchie, B. W., "Simulating the Effects of Mixing on the Performance of Unpremixed Flow Chemical Reactors," *The Canad. J. of Chem. Eng.*, Vol. 58 (1980), pp. 626–633.
32. Beek, J., Jr., and Miller, R. S., "Turbulent Transport in Chemical Reactors," *Chem. Eng. Prog. Symp. Series*, Vol. 55 (1959), pp. 23–28.
33. Ottino, J. M., Ranz, W. E., and Macosko, C. W., "A Lamellar Model for Analysis of Liquid-Liquid Mixing," *Chem. Eng. Sci.*, Vol. 34 (1979), pp. 877–890.
34. Ottino, J. M., "Lamellar Mixing Models for Structured Chemical Reactions and Their Relationship to Statistical Models: Macro and Micromixing and the Problem of Averages," *Chem. Eng. Sci.*, Vol. 35 (1980), pp. 1377–1391.
35. Ranz, W. E., "Applications of a Stretch Model to Mixing, Diffusion and Reaction in Laminar and Turbulent Flows," *A.I.Ch.E. J.*, Vol. 25 (1979), pp. 41–47.
36. Ottino, J. M., "Efficiency of Mixing from Data on Fast Reactions in Multi-jet Reactors and Stirred Tanks," *A.I.Ch.E. J.*, Vol. 27 (1981), pp. 184–192.
37. Ottino, J. M., Ranz, W. E., and Macosko, C. W., "A Framework for Description of Mechanical Mixing of Fluids," *A.I.Ch.E. J.*, Vol. 27 (1981), pp. 565–577.
38. Ranz, W. E., "Mixing and Fluid Dispersion of Viscous Liquids," *A.I.Ch.E. J.*, Vol. 28 (1982), pp. 91–96.
39. Ottino, J. M., "Description of Mixing with Diffusion and Reaction in Terms of the Concept of Material Surfaces," *J. Fluid. Mech.*, Vol. 114 (1982), pp. 82–103.

40. Chella, R., and Ottino, J. M., "Mixing, Diffusion, and Chemical Reaction in a Single Screw Extruder," ACS Symp. Series, Vol. 196 (1982), pp. 567.
41. Bourne, J. R., and Rohani, S., "Mixing and Fast Chemical Reaction VII Deforming Reaction Zone Model for the CSTR," *Chem. Eng. Sci.*, Vol. 38 (1983), pp. 911.
42. Nauman, E. B., "Tre Droplet Diffusion Model for Micromixing," *Chem. Eng. Sci.*, Vol. 30 (1975), pp. 1135–1140.
43. Villermaux, J., and David, R., "Recent Advances in the Understanding of Micromixing Phenomena in Stirred Reactors," *Chem. Eng. Comm.*, Vol. 21 (1983), pp. 105–122.
44. Zoulalian, A., and Villermaux, J., "Influence of Chemical Parameters on Micromixing in a Continuous Stirred Tank Reactor," *Adv. Chem. Series. Chem. React. Eng. II Evanston*, Vol. 133 (1974), pp. 348–361.
45. Bolzern, O., and Bourne, R. J., "Mixing and Fast Chemical Reaction, VI Extension of the Reaction Zone," *Chem. Eng. Sci.*, Vol. 38 (1983), pp. 999.
46. Klein, J. P., David, R., and Villermaux, J., "Interpretation of Experimental Liquid Phase Micromixing Phenomena in a Continuous Stirred Reactor with Short Residence Times," *Ind. Eng. Chem. Fundamentals*, Vol. 19 (1980), pp. 373–379.
47. Bourne, J. R., "The Characteristization of Micromixing Using Fast Multiple Reaction," *Chem. Eng. Comm.*, Vol. 16 (1982), pp. 79.
48. Barthole, J. P., et al., "Measurement of Mass Transfer Rates, Velocity and Concentration Fluctuations in an Industrial Stirred Tank," *Chem. Eng. Fundam.*, Vol. 1 (1982), p. 17–26.
49. Patterson, G. K., Bockelman, W., and Quigley, J., "Measurement of Mixing Effects on Local Reaction Conversion in Stirred Tanks," Proceedings of the Fourth Europ. Conference on Mixing, April 27–29, 1982, (BHRA Fluid Engineering), Paper J1, pp. 303–312.
50. Barthole, J. P., David, R., and Villermaux, J., "A New Chemical Method for the Study of Local Micromixing conditions in Industrial Stirred Tanks," ACS Symp. Series, Vol. 196 (1982), pp. 545.
51. Barthole, J. P., et al., "Cinetique macroscopique de la précipitation du sulfate de baryum en présence d'EDTA: une réaction chemique-test pour la caractérisation de la qualite du mélange dans les réacteurs industriels," *J. Chim. Phys.*, Vol. 79 (1982), p. 719.
52. Sasakura, T., et al., "Mixing Processes in a Stirred Vessel," *Internat. Chem. Eng.*, Vol. 20 (1980), pp. 251–258.
53. Rachez, D., David, R., and Villermaux, J., "Un nouveau modèle de circulation interne dans une cuve agitée de type industriel," *Entropie*, Vol. 72, No. 101 (1981), pp. 32–39.
54. Bryant, J., "The Characterization of Mixing in Fermenters," *Advances in Biochemical Engineering*, Vol. 5, Springer Verlag Berlin (1977), pp. 101–123.
55. Bryant, J., and Sadeghzadeh, S., "Circulation Rates in Stirred and Aerated Tanks," Proceed. Third Europ. Conf. on Mixing, April 4–6, 1979, York BHRA Fluid Engineering, paper F3, pp. 325–336.
56. Bryant, J., and Sadeghzdeh, S., "Terminal Mixing Times in Stirred Tanks," Proceed. Fourth Europ. Conf. on Mixing, April 27–29, 1982, York BHRA Fluid Engineering, paper B4, pp. 49–56.
57. Brennan, D. J., and Lehrer, I. H., "Impeller Mixing in Vessels Experimental Studies on the Influence of Some Parameters and Formulation of a General Mixing Time Equation," *Trans. I. Chem. Eng.*, Vol. 54 (1976), pp. 139–152.
58. Khang, S.J., and Levenspiel, O., "New Scale-Up and Design Method for Stirrer Agitated Batch Mixing Vessels," *Chem. Eng. Sci.*, Vol. 31 (1976), pp. 569–577.
59. Okamoto, Y., Nishikawa, N., and Hashimoto, K., "Energy Dissipation Rate Distribution in Mixing Vessels and its Effects on Liquid-Liquid Dispersion and Solid-Liquid Mass Transfer," *Internat. J. Chem. Eng.*, Vol. 21 (1981), pp. 88–94.
60. Van der Molen, K., and Van Maanen, H. R. E., "Laser-Doppler Measurements of the turbulent Flow in Stirred Vessels to Establish Scaling Rules," *Chem. Eng. Sci.*, Vol. 33 (1978), pp. 1161–1168.
61. Nishikawa, M., et al., "Turbulence Energy Spectra in Baffled Mixing Vessels," *J. Chem. Eng. Japan*, Vol. 9 (1976), pp. 489–494.

62. Fort, I., et al., "Turbulent Characteristics of the Velocity Field in a System with Turbine Impeller and Radial Baffles," *Collect. Czechoslov. Chem. Comm.*, Vol. 39 (1974), pp. 1810–1822.
63. Patterson, G. K., "Application of Turbulence Fundamentals to Reactor Modelling and Scale-Up," *Chem. Eng. Comm.*, Vol. 8, (1981), pp. 25–52.
64. Harvey, P. S., and Greaves, M., "Turbulent Flow in an Agitated Vessel. I: Predictive Model," *Trans. I. Chem. E.*, Vol. 60 (1982), p. 195.
65. Harvey, P. S., and Greaves, M., "Turbulent Flow in an Agitated Vessel. II: Numerical Solution and Model Predictions," Vol. 60 (1982), p. 201.
66. Canon, R. M., et al., "Turbulence Level Significance of the Coalescence-Dispersion Rate Parameter," *Chem. Eng. Sci.*, Vol. 32 (1977), pp. 1349–1352.
67. Waggoner, R. C., and Patterson, G. K., "Effect of Imperfect Mixing on the Performance and Control of Batch and Semibatch Reactors," ISA Transactions, Vol. 14 (1975), pp. 331–339.
68. Mann, R., Mavros, P. P., and Middleton, J. C., "A Structured Stochastic Flow Model for Interpreting Flow-Follower Data from a Stirred Vessel," *Trans. I. Chem. Eng.*, Vol. 59 (1981), pp. 271–278.
69. Mann, R., and Mavros, P., "Analysis of Unsteady Tracer Dispersion and Mixing in a Stirred Vessel," Proceedings of the Fourth Europ. Conference on Mixing, April 27–29, 1982, BHRA Fluid Engineering, Paper B3, 35–47.
70. Mann, R., "Gas-Liquid Contacting in Mixing Vessels," *I. Chem. E.*, Industrial research fellowship report, 1983.
71. Nabholtz, F., and Rys, P., "Chemical Selectivities Disguised by Mass Diffusion. IV: Mixing-Disguised Nitrations of Aromatic Compounds with Nitronium Salts," *Helvetica Chem. Acta*, Vol. 60 (1977), pp. 2937–2943.
72. Bourne, J. R., and Kozicki, F., "Mixing Effects During the Bromination of 1,3,5 Trimethoxybenzene," *Chem. Eng. Sci.*, Vol. 32 (1977), pp. 1538–1539.
73. Bourne, J. R., Rys, P., and Suter, K., "Mixing Effects in the Bromination of Resorcin," *Chem. Eng. Sci.*, Vol. 32 (1977), pp. 711–716.
74. Nauman, E. B., "Mixing in Polymer Reactors," *J. Macromol. Sci. Revs. Macromol. Chem.*, C10 (1974), pp. 75–112.
75. Gerrens, H., "Polymerization Reactors and Polyreactions. A Review," Proceed. 4th Int. Symp. Chem. React. Eng. Heidelberg, 1976, pp. 585–614.
76. Biesenberger, J. A., and Sebastian D. H., "Principles of Polymerization Engineering," Wiley, New York, 1983.
77. Sahm, P., "Effets de micromélange sur la polymerisation radicalaire du styrène en réacteur agité continu," thesis, Inst. Nat. Polyt. Lorraine, Nancy, 1978.
78. Villermaux, J., Pons, M., and Blavier, L., "Comparison of Partial Segregation Models for the determination of Kinetic constants in a High Pressure Polyethylene Reactor," to be presented ISCRE 8, Edinburgh, 1984, Inst. Chem, Eng. Symp. Series *87*, p. 553–560.
79. Costa, M. R., "Fondements de la modélisation des polycondensations linéaires réversibles en vue de la conception et de la conduite des procédés industriels," thesis, Inst. National Polytechnique de Lorraine, Nancy, 1983.
80. David, R., and Villermaux, J., "Détermination de l'état de micromélange dans un réacteur auto-agité en phase gazeuse à des temps de passage de l'ordre de la seconde," *J. Chim. Phys.*, Vol. 75 (1978), p. 656.
81. Berard, S., et al., "Une méthode de mesure de température par voie chimique," Entropie, 119, 1984, pp. 13–16.
82. Spencer, J. L., and Lunt, R. R., "Experimental Characterization of Mixing Mechanisms in Flow Reactors Using Reactive Tracers," *Ind. Eng. Chem. Fundam.*, Vol. 19 (1980), pp. 142–148.
83. Pohorecki, R., and Baldyga, J., "The Use of a New Model of Micromixing for Determination of Crystal Size in Precipitation," *Chem. Eng. Sci.*, Vol. 38 (1983), pp. 79–83.

CHAPTER 28

BACKMIXING IN STIRRED VESSELS

Koji Ando

Department of Chemical Engineering
Muroran Institute of Technology
Muroran, Japan

Takashi Fukuda

Government Industrial Development Laboratory
Sapporo, Japan

Kazuo Endoh

Department of Chemical Process Engineering
Faculty of Engineering
Hokkaido University
Sapporo, Japan

CONTENTS

INTRODUCTION, 773

BACKMIXED FLOW AND MIXING MODELS, 773
Axial Dispersion Model, 773
Cell Model, 776
Backflow Cell Model, 776
Parallel Combination Model, 779
Relationships Between Models, 780

BACKMIXING IN A STIRRED VESSEL WITH A SINGLE IMPELLER, 782
Mixing Time and Perfect Mixed Flow, 783
Mixing Scale in Complete Mixing, 785
Residence Time Distribution, 788

BACKMIXING IN STIRRED VESSELS WITH MULTIPLE IMPELLERS, 788
Flow Behavior of Liquid in Vessels and Mixing Models, 788
Degree of Backmixing in the Vessel with Multiple Impellers, 792

BACKMIXING IN HORIZONTAL STIRRED VESSEL, 795
Flow Behavior of Liquid in the Vessel, 796
Degree of Backmixing, 797
Backmixing in the Vessel with Perforated Partition Plate, 799

NOTATION, 799

REFERENCES, 800

INTRODUCTION

A stirred vessel is a multipurpose mixing device that facilitates the enhancement and acceleration of diffusion to achieve uniform concentrations in multicomponent mixtures, bubble dispersion, the disintegration of foreign droplets, chemical reactions, mass as well as heat transfer, and the development of particulate solids. It is a prevailing trend that mixing in batch operations is replaced by continuous flow systems. Fluid in a stirred vessel is forced into intensive movement rarely experienced in other equipment. It enables short mixing times while permitting longer mean residence times to achieve complete mixing in a continuous operation.

A number of explanations [1–7] for the mixing phenomena observed in a stirred vessel with a continuous flow system have been attempted, and they can be divided in two different groups. One uses macromixing which generates the RTD (residence time distribution) of the fluid element, and the other uses micromixing (a detailed description of this is given in other chapters) which is related to the level of molecular mixing.

Macromixing in a stirred vessel with continuous operation results in reductions in the effective driving force of the transport phenomena, thus reducing the volumetric efficiency of a device. Except when the rotational Reynolds number of impellers is very small, the stirred vessel establishes an approximate state of perfect mixed flow. In order to supplement the reductions in the transfer rate caused by the backmixing that accompanies the perfect mixed flow in the vessel, the following procedures are used:

1. Vessels connected in series,
2. A single vessel, which is longer than the diameter of the vessel, containing the multi-impellers,
3. A single vessel divided into multiple stages with stators.

The mixing in the vessel is related to the choices of control systems, as this plays the determining role in controlling the time required to achieve stabilized operation.

This text is an evaluation of backmixing in stirred vessels with characteristic values of the mixing models after a presentation of macromixing in a stirred vessel with a continuous flow system. The presentation is centered around experimental results obtained with mixing models, which are also detailed. A mathematical treatment of the recycle or circulation model (a description of this is given in other chapters) has been omitted even when it may appear mathematically interesting.

BACKMIXED FLOW AND MIXING MODELS

Generally, the flow state in the device was found to be between a perfect mixed flow and a plug flow. The backmixing flow can be considered a phenomenon where a part of the flow element moves in the opposite direction of the main flow. More accurately, the mixing of fluids in the vessel does not always depend on the backmixing. However, irrespective of the actual mechanism of the fluid mixing, the mixing characteristics can be explained by treating it as a case of nominal backmixing.

The deviation from plug flow in the device can be represented by the RTD function. In many instances, the flow within the vessel is too complicated to express simply, and models combining series of different, simplified flow patterns that can be treated mathematically have been proposed [6-11]. The following well-known mixing models for stirred vessels will be detailed here: axial dispersion model [12]; cell model [13]; backflow cell model [9]; and parallel combination model [14]. The recycle or circulation model [15, 16] that can be applied to stirred vessels with draft tubes is considered elsewhere in this volume.

Axial Dispersion Model

The axial dispersion model is also known as a longitudinal dispersion model, differential backmixing model, or diffusion model.

Peclet-Bodenstein Number, P_eB

This is the most frequently applied of the mixing models. In this model, fluid flow in a direction x, at an average uniform flow rate u, with the axial dispersion coefficient E_x is assumed to have a constant value in the flow direction. With the flow of a chemical component in a device which regulates the flow in the x direction at an average rate, u, the mass balance equation at a point, x, and time θ in the device becomes:

$$\frac{\partial c}{\partial \theta} = E_x \frac{\partial^2 c}{\partial x^2} - u \frac{\partial c}{\partial x} + \psi(c) \tag{1}$$

where C is the concentration of chemical component and $\psi(c)$ represents the reaction rate.

The RTD function of this model is given by Equation 4 which can be determined from $\psi(c) = 0$; the initial conditions Equation 2; and boundary conditions Equation 3 [12, 17]:

$$\left.\begin{array}{lll} \theta < 0, & \text{all } x: & c = 0 \\ \theta = 0, & : & c = \delta(x) \end{array}\right\} \tag{2}$$

$$\left.\begin{array}{lll} \theta > 0, & x = 0: & -E_x\left(\dfrac{dc}{dx}\right)_{x=0+} = u(c - c_{x=0+}) \\ \theta > 0, & x = L: & -E_x\left(\dfrac{dc}{dx}\right) = 0 \end{array}\right\} \tag{3}$$

$$E(\phi) = 2 \sum_{n=1}^{\infty} \frac{(-1)^{n+1}\mu_n 2 \exp(P_eB/2)}{(P_eB/2)^2 + P_eB + \mu_n{}^2} \exp\left(-\frac{(P_eB/2)^2 + \mu_n{}^2}{P_eB}\right)\phi \tag{4}$$

The dispersion, σ_d^2 of $E(\phi)$ is:

$$\sigma_d^2 = \frac{2}{(P_eB)^2}(P_eB - 1 + e^{-P_eB}) \tag{5}$$

where μ_n is the n-th positive root of $\cot \mu = (2\mu/P_eB - P_eB/2\mu)/2$, $P_eB = u \cdot L_T/E_x$, L_T is the overall length of vessel, $\phi = \theta/\theta_T$ and $\theta_T = L_T/u$. The function in Equation 4 appears complicated, but the relation between $E(\phi)$ and ϕ depends on a dimensionless number, P_eB, only. It is called the Peclet-Bodenstein number. With the diameter of the vessel, D_T, P_eB is the product of P_e ($= u \cdot D_T/E_x$) and B ($= L_T/D_T$), the product of the degree of mixing and a geometric factor. Figure 1 shows the relationship between the RTD function $E(\phi)$ and P_eB. When P_eB is zero, there is perfect mixing, and when P_eB increases enough, a plug flow is indicated. Since P_eB is proportional to the relative length of the system, L_T/D_T, it will approximate the plug flow with increasing in L_T/D_T.

For an axial dispersion coefficient, E_x, with a device of infinite length, the fluid moves at the mean flow rate, reaching the velocity of the state of normal distribution, as shown in Equation 6 [18], Equation 4 often approximates:

$$E(\phi) = \frac{1}{2\sqrt{\pi\phi/P_eB}} e^{-P_eB(1-\phi)^2/4\phi} \tag{6}$$

Equation 6 applies to cases of $P_eB \gg 1$.

Frequently a mixing degree is evaluated from the dispersion coefficient E_x, when a mixing phenomenon is approximated with a model other than the dispersion model.

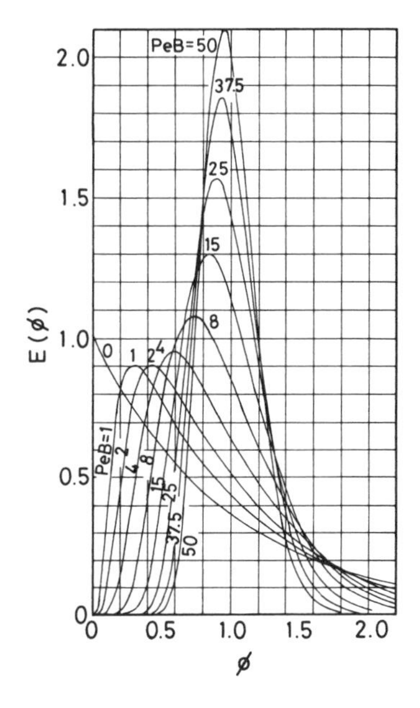

Figure 1. RTD function, $E(\phi)$ of axial dispersion model [9].

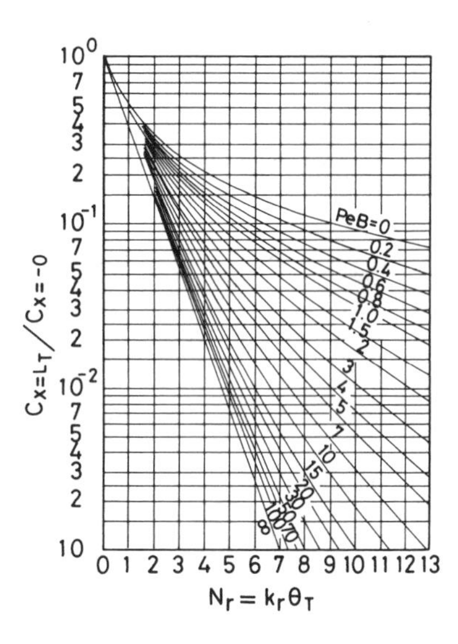

Figure 2. The ratio of concentration of outlet to that of inlet for first-order reaction by axial dispersion model [9].

Fractional Conversion of Reactant

When a reaction is in progress at a rate of $\psi(c)$ in a vessel with a continuous flow system, the following equation, derived from Equation 1, applies:

$$E_x \frac{d^2c}{dx^2} - u\frac{dc}{dx} + \psi(c) = 0 \tag{7}$$

For a first-order reaction with $\psi(c) = k_rC$, the solution [12, 19] subject to the boundary conditions of Equation (3) is

$$\left.\begin{aligned} \frac{c_{x=L_T}}{c_{x=-0}} &= \frac{4A\exp(P_eB/2)}{(1+A)^2\exp(A\cdot P_eB/2) - (1-A)^2\exp(-A\cdot P_eB/2)} \\ A &= \sqrt{1+4N_r/P_eB}, \qquad N_r = k_r\theta_T = \frac{k_rL_T}{u} = \frac{k_rV}{q} \end{aligned}\right\} \tag{8}$$

where, the term, $(c_{x=L_T}/c_{x=-0})$, represents the ratio of concentration of outlet to that of inlet and N_r is the number of reaction unit. Figure 2 shows the relations between $c_{x=L_T}/c_{x=-0}$ and P_eB, N_r. The fractional conversion of reactant, $1 - (c_{x=L_T}/c_{x=-0})$, is determined from Equation 8. Also for reactions that are not first-order [12, 20] $c_{x=L_T}/c_{x=-0}$ can be determined from Equations 7 and 3, when the order of $\psi(c)$ is known.

Cell Model

The cell model is also known as a tanks-in-series model.

Number of Cells, N

With the assumption that a vessel can be divided into N cells, with equal volumes and perfect mixing states, the RTD function, $E(\phi)$, for this model is as follows [21]:

$$E(\phi) = \frac{N}{(N-1)!}(N\phi)^{N-1}\exp(-N\phi) \tag{9}$$

where $\phi\ (=\theta/\theta_T)$ = dimensionless time
$\theta_T\ (= V/q)$ = mean residence time
V = the total volume of the vessel
q = the volumetric flow rate of main flow

The number of cells, $N = 1$, corresponds to a completely mixed flow and $N = \infty$ to a plug flow. With this model, N is the parameter that determines the degree of mixing of the flow.

Fractional Conversion of Reactant

For the ratio of concentration of outlet to that of inlet, c_{out}/c_{in}, the following formula may be used for first-order reactions, regardless of the flow mixing mechanism.

$$\left.\begin{aligned} \frac{c_{out}}{c_{in}} &= \int_0^\infty e^{-N_r\phi}E(\phi)\,d\phi \\ N_r &= k_r\theta_T = k_rL_T/u = k_rV/q \end{aligned}\right\} \tag{10}$$

Equation (10) is a Laplace transformation function of $E(\phi)$ with N_r as a parameter. A cell model gives the following formula by substituting Equation 9 for $E(\phi)$ in Equation 10:

$$\frac{c_{out}}{c_{in}} = \frac{1}{(1 + N_r/N)^N} \tag{11}$$

For plug flow at $N = \infty$, when N_r/N is replaced by y, it becomes:

$$\left(\frac{c_{out}}{c_{in}}\right)_{N=\infty} = \lim_{N\to\infty}\{(1+y)^{1/y}\}^{-N_r} = e^{-N_r} \tag{12}$$

Backflow Cell Model

Backflow Ratio, α

As shown in Figure 3A, the difference between the backflow cell model and the cell model lies in a backflow that runs upstream through the connecting points of the adjacent perfectly mixed cells. The flow rates supplied to the cell series and discharged from that are q. Between adjoining cells, there is a flow of $q + q'$ in the direction of the main flow and a backflow q' in the opposite direction of main flow. The ratio of the backflow to the main flow, $q'/q = \alpha$ is termed the backflow

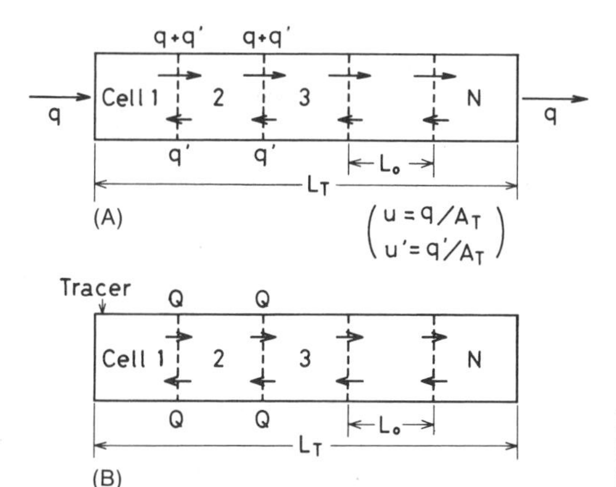

Figure 3. Backflow cell model: (A) Backflow cell model for q = q; (B) Backflow cell model for q → 0 or q = 0.

ratio $q'/(q + q) = \beta$ may also be used). When $\alpha = 0$, it represents the cell model. When the length of a single cell, L_0, approximates zero and the number of cells, $N \to \infty$, it becomes the axial dispersion model.

The mass balance with this model is expressed by the following formulas:

$$\frac{dc_1}{d\theta} = \frac{N_{q'}}{V}(-c_1 + c_2) + \frac{N_q}{V}(c_0 - c_1) + \psi(c_1) \tag{13}$$

$$\frac{dc_j}{d\theta} = \frac{N_{q'}}{V}(c_{j-1} - 2c_j + c_{j+1}) + \frac{N_q}{V}(c_{j-1} - c_j) + \psi(c_j) \tag{14}$$

$$\frac{dc_N}{d\theta} = \frac{N_{q'}}{V}(c_{N-1} - c_N) + \frac{N_q}{V}(c_{N-1} - c_N) + \psi(c_N) \tag{15}$$

The relationship between the RTD function, $E(\phi)$, and α for $N = 10$ is shown in Figure 4.

With this model the fractional conversion of reactant is often determined by converting measured α into P_eB, because the analysis of the RTD function is complicated. The details of this transformation are described later.

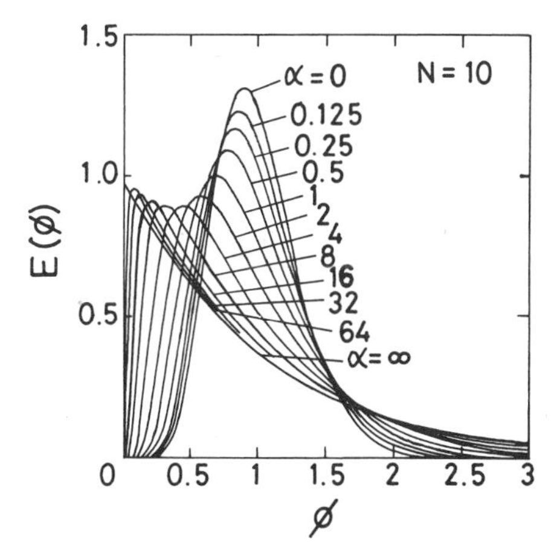

Figure 4. RTD function, $E(\phi)$ of backflow cell model for number of cells, $N = 10$, and various values of backflow ratio, α [22].

Exchange Rate of Fluid Between Cells, Q

Figure 3A shows the backflow cell model. Taking the critical range where the supply and discharge flow, q, become very small and approach zero as the batch operation, the mixing model is as shown in Figure 3B. In the model, Q is defined as an exchange rate of liquid through the connection part of adjoining cells. Then, the exchange rate, Q, governs the mixing process. Thus, when the flow q does not influence the backflow q′, the condition $q \approx 0$ or $q \ll q'$, it is considered that the backflow q′ is equivalent to Q. The exchange rate Q can be obtained experimentally.

With the mixing model in Figure 3B, the mass balance of tracer added to cell 1 can be expressed by:

$$\text{Cell 1:}\quad \frac{V}{N}\frac{dc_1}{d\theta} = Q(c_2 - c_1) \tag{16}$$

$$\text{Cell j:}\quad \frac{V}{N}\frac{dc_j}{d\theta} = Q(c_{j-1} - 2c_j + c_{j+1}), \qquad 2 \le j \le N \tag{17}$$

$$\text{Cell N:}\quad \frac{V}{N}\frac{dc_N}{d\theta} = Q(c_{N-1} - c_N) \tag{18}$$

The dimensionless concentration, C_j is given as c_jV/M, where V is the total volume of fluid in the vessel and M is the total amount of tracer material. The concentration C_j is given by Equation 19 by solving Equations 16 through 18 under the initial condition as $\theta = 0$; $C_1 = N$ and $C_j = C_N = 0$ [23, 24]:

$$C_j = 1 - \sum_{K=1}^{N-1} \frac{N \sin\psi_K[\sin(N-1+j)\psi_K - \sin(N-j)\psi_K]}{N\cos\psi_K \cos N\psi_K - N\cos N\psi_K - \sin\psi_K \sin N\psi_K} \exp(-S_K \cdot \phi_M) \tag{19}$$

where $S_K = 2N(1 - \cos\psi_K)$

$$\psi_K = \frac{K}{N}\pi$$

$$\phi_M = \frac{Q}{V}\theta$$

The concentration C_N in cell N is determined from Equation 19 as:

$$C_N = 1 - \sum_{K=1}^{N-1} \frac{N \sin^2\psi_K}{N\cos\psi_K \cos N\psi_K - N\cos N\psi_K - \sin\psi_K \sin N\psi_K} \exp(-S_K \cdot \phi_M) \tag{20}$$

Further, the concentration differences between cell 1 and N, $\Delta C_N \equiv (1/N)(C_1 - C_N)$, are given:

$$N = 2:\quad \Delta C_2 = \frac{1}{2}(C_1 - C_2) = \exp(-4\phi_M) \tag{21}$$

$$N = 3:\quad \Delta C_3 = \frac{1}{3}(C_1 - C_3) = \exp(-3\phi_M) \tag{22}$$

$$N = 4:\quad \Delta C_4 = \frac{1}{4}(C_1 - C_4) = \frac{2+\sqrt{2}}{4}\exp\{-4(2-\sqrt{2})\phi_M\} + \frac{2-\sqrt{2}}{4}\exp\{-4(2+\sqrt{2})\phi_M\} \tag{23}$$

$$N = 5: \quad \Delta C_5 = \frac{1}{5}(C_1 - C_5) = \frac{5+\sqrt{5}}{10}\exp\left\{-\frac{5(3-\sqrt{5})}{2}\phi_M\right\} + \frac{5-\sqrt{5}}{10}\exp\left\{-\frac{5(3+\sqrt{5})}{2}\phi_M\right\} \tag{24}$$

$$N = 6: \quad \Delta C_6 = \frac{1}{6}(C_1 - C_6) = \frac{2+\sqrt{3}}{6}\exp\{-6(2-\sqrt{3})\phi_M\} + \frac{2-\sqrt{3}}{6}\exp\{-6(2+\sqrt{3})\phi_M\} + \frac{1}{3}\exp(-12\phi_M) \tag{25}$$

The concentration, ΔC_N for each vessel of $N = 4 \sim 6$ is approximated by the first term of the right side of Equations 23 through 25 at large ϕ_M, respectively. For $N = 5$, the relationships between ΔC_5, the values of the first term of Equation 24, ΔC_5^*, and ϕ_M are shown in Figure 5. In the case of $\phi_M > 0.3$, ΔC_5 can be adequately approximated by the first term of Equation 24. Thus, the Q value of the exchange flow can be accurately determined from the slope of the measured values of ΔC_N plotted with time θ on a semilog scale [25–28]. The exchange flow Q can also be determined from the C_N–ϕ_M curve that gives the most favorable fit when measured values and Equation 20 are compared [23, 29–31].

Parallel Combination Model

Fraction of Plug Flow, ϕ°

When the curve for the internal age distribution function, $R(\phi)$, of a stirred vessel with multiple impellers is plotted on a semilog scale, as in Figure 6, it almost coincides with 1 at a smaller ϕ. When ϕ becomes larger than ϕ°, the approximation is often depicted with a line dropping to the right. The value of the abscissa where the rightward descending line and $R(\phi) = 1.0$ cross is defined as ϕ°. Fraction ϕ° corresponds to the fraction of flow passing through the vessel as plug flow, and the remaining part $(1 - \phi^\circ)$ shows the fraction of flow as complete mixing. ϕ° is called the fraction of plug flow [7, 14, 32].

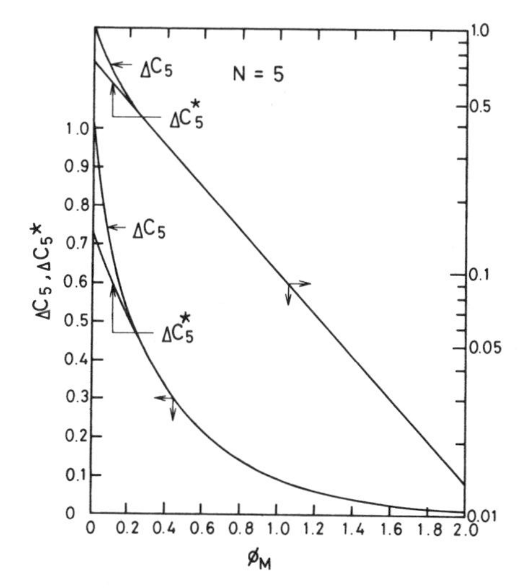

Figure 5. Comparison of ΔC_5 with ΔC_5^*.

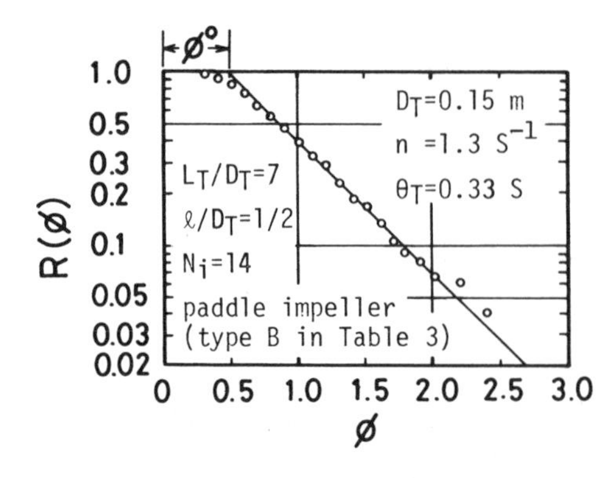

Figure 6. An example of internal age distribution $R(\phi)$ for a stirred vessel with multiple paddle impellers [14].

Fractional Conversion of Reactant

When the vessel represented by this model is used as a reactor in a continuous flow system, it is considered that the reaction proceeds separately in the region of plug flow and of completely mixed flow in the vessel. The ratio c_{out}/c_{in} is given [14]:

$$\frac{c_{out}}{c_{in}} = \phi^{\circ} e^{-k_r \theta_T/(2-\phi^{\circ})} + \frac{(1-\phi^{\circ})^2}{(1-\phi^{\circ}) + k_r \theta_T} \tag{26}$$

The ratio, c_{out}/c_{in}, can be calculated from ϕ° and $k_r\theta_T$.

The model is characterized by offering a simple formula to estimate the fractional conversion of reactant, as a fraction of the flow rates between the two extreme conditions—the plug flow and the completely mixed flow.

Relationships Between Models

Relationship Between Axial Dispersion Model and Backflow Cell Model

The parameter, P_eB, of the mixing characteristics of the axial dispersion model and the mixing characteristics of the backflow cell model, N and α, are related as follows [33]:

$$\underbrace{\frac{1}{P_eB}}_{\text{axial dispersion model}} = \underbrace{\underset{\substack{\text{intra-stage}\\ \text{backmixing}}}{\frac{1}{2N}} + \underset{\substack{\text{interstage}\\ \text{backmixing}}}{\frac{\alpha}{N}}}_{\text{backflow cell model}} \tag{27}$$

The right side of Equation 27 is made up of two terms: the first term is the effect of the intra-stage backmixing and the second term of the inter-stage backmixing [34]. These two effects added together give the total backmixing effect ($1/P_eB$). In the case of the inter-stage backmixing term, $\alpha/N = 0$, Equation 27 becomes Equation 28.

$$\frac{1}{P_eB} = \frac{1}{2N} \tag{28}$$

This formula represents the relationship between the axial dispersion model and cell model.

The relationship in Equation 27 can be elaborated as follows [8]: For $\psi(c) = 0$, with the mass balance equation Equation (1) based on the axial dispersion model, it can be expressed by P_eB, dimensionless time ϕ (= θ/θ_T, $\theta_T = V/q$), and dimensionless distance χ (= x/L_T) as follows:

$$\frac{dc}{d\phi} = -\frac{dc}{dx} + \frac{1}{P_eB}\frac{d^2c}{dx^2} \tag{29}$$

Similarly, when $\psi(c) = 0$ with mass balance equation Equation 13, based on the backflow cell model, the following equation is obtained with α and ϕ.

$$\frac{1}{N}\frac{dc_j}{d\phi} = c_{j-1} - c_j + \alpha(c_{j-1} - 2c_j + c_{j+1}) \tag{30}$$

The mathematical relationship [35] between the two models may be found by relating the concentration in adjacent stages by means of a Taylors Series:

$$\left.\begin{aligned} c_{j+1} &\approx c_j + \frac{1}{N}\frac{dc}{dx} + \frac{1}{2N^2}\frac{d^2c}{dx^2} + \cdots \\ c_{j-1} &\approx c_j - \frac{1}{N}\frac{dc}{dx} + \frac{1}{2N^2}\frac{d^2c}{dx^2} - \cdots \end{aligned}\right\} \tag{31}$$

Substitution of the series given by Equation 31 into Equation 30, neglecting third and higher order terms, and omitting the suffix from c produces:

$$\frac{dc}{d\phi} = -\frac{dc}{dx} + \left(\frac{\alpha}{N} + \frac{1}{2N}\right)\frac{d^2c}{dx^2} \tag{32}$$

Comparing Equation 29 and Equation 32 gives Equation 27 which expresses the relationship between the two models.

The relationship between the models, given by Equation 27, is only approximate. Comparing the differential and finite difference equations terms, $1/N^3$ (d^3c/dx^3) and higher terms in the Taylors Series were disregarded [36]. Thus the difference between the two models is of the order of $1/N^3$ per stage or $1/N^2$ for N stages [8].

Some calculated impulse response curves are shown in Figure 7 for six cells with various backflow rates. Impulse response curves are also indicated for the axial dispersion model with the Peclet-Bodenstein number, P_eB, equal to $1/2N + \alpha/N$. In this figure a solid line for the backflow cell model and a broken line for the axial dispersion model coinside at $\alpha = \infty$, indicating also that difference between the 2 curves becomes high with decreases in α. Thus the agreement becomes better for higher N and α values [37].

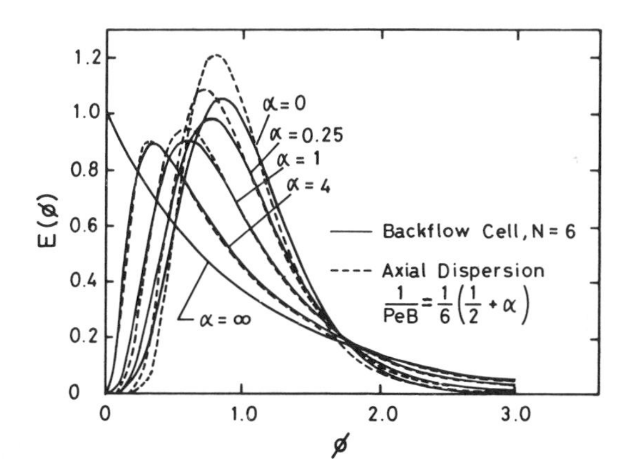

Figure 7. Comparison of RTD function $E(\phi)$ of backflow cell model for $N = 6$ with $E(\phi)$ of axial dispersion model with $1/PeB$ equal to $\{1/(2 \times 6) + \alpha/6\}$ [37].

Several equations different from Equation 27 have been made for the relationship between P_eB and α, N [33, 38–40].

$$\frac{1}{P_eB} = \frac{1}{2(N-1)} + \frac{\alpha}{N-1} \tag{33}$$

$$\frac{1}{P_eB} = \frac{1}{2(N-1)\{1+(1/2N)\}} + \frac{\alpha}{N-(1/2)} \tag{34a}$$

$$\approx \frac{1}{2N-1} + \frac{\alpha}{N}, \qquad N \gg 1 \tag{34b}$$

$$\frac{1}{P_eB} = \frac{1}{2(N-1)\{1-(1/N)\}} + \frac{\alpha}{N+1-r} \tag{35a}$$

$$\approx \frac{N}{2(N-1)^2} + \frac{\alpha}{N}, \qquad r = 1 \tag{35b}$$

For a finite number of stages, Latian [38] adopts Equation 33 for both physical transients and a homogeneous first-order reaction. The term 2(N − 1) is based upon Kramers and Alberdas' treatment [41] of the cell model. Equations 34a and 34b is an empirical equation obtained by setting $\sigma_a^2 = \sigma_b^2$, where σ_a^2 and σ_b^2 are variances of the axial dispersion model's RTD and the backflow cell models, respectively [33]. It is established for a wide range of N and α values. Equations 35a and 35b shows the relations between the two models when a reaction proceeds with reaction order r in the system [33]. It is established within the range of $\alpha \geq 0$, $N \geq 2$ and $r = 1/2, 1, 2$, with an accuracy better than $\pm 10\%$ [42]. Equations 33 through 35 correspond to Equation 27 for higher N.

Relationship Between the Axial Dispersion Model and the Parallel Combination Model

Figure 8 [32] depicts the relationship between P_eB and fraction of plug flow determined from the internal age distribution function [12, 19], based on the axial dispersion model.

BACKMIXING IN A STIRRED VESSEL WITH A SINGLE IMPELLER

The stirring operation extends the residence time distribution (RTD) of the flow in a continuous flow system. The operation may sometimes be used merely for expanding the RTD. On the contrary, smaller RTD's are often desirable, according to circumstances. With a single impeller, the mixing in the vessel does not attain a minimum RTD, but is an approximation to complete mixing. Thus, deviation from complete mixing may become significant, depending on the purpose of a stirred

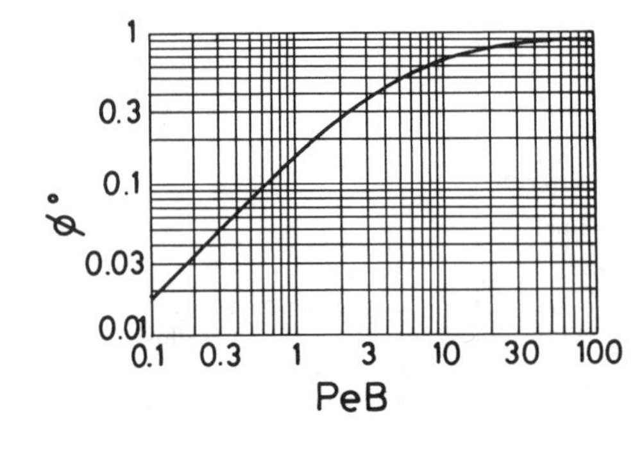

Figure 8. Correlation between fraction of plug flow, ϕ° and Peclet-Bodenstein number, P_eB [32].

vessel. In this case, the relationship between the extent of the deviation and the operational stirring conditions is described.

Mixing Time and Perfect Mixed Flow

The mixing state of a liquid in the stirred vessel with a single impeller and a continuous flow system can be evaluated from the relation between the mixing time, θ_M, of the batch operation and the mean residence time, θ_T, of the continuous flow operation in the vessel. If $\theta_M/\theta_T < 0.1$, it can be justified as a case of perfect mixing [43].

Generally, Equation 36 is established between θ_M and the rotational speed n of the impeller [44].

$$\theta_M \cdot n = \text{const} \tag{36}$$

$\theta_M \cdot n$ is called the dimensionless mixing time and its reciprocal is the dimensionless mixing velocity. Figure 9 shows the relation between the dimensionless mixing time and the rotational Reynolds

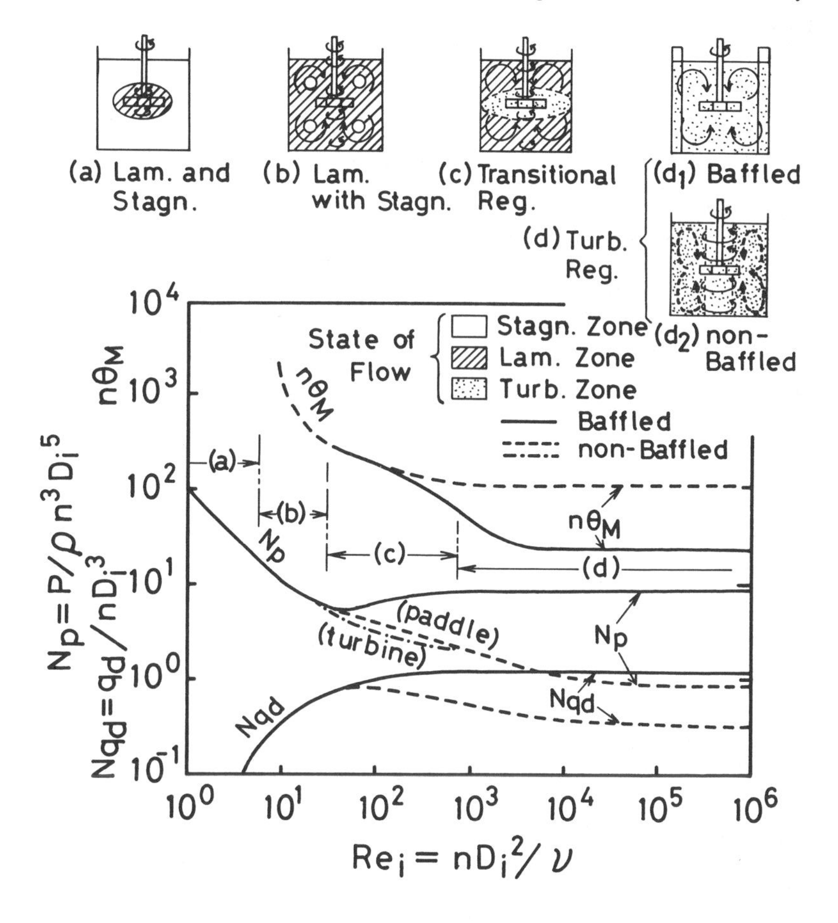

Figure 9. Correlation of characteristic flow with dimensionless mixing time $n \cdot \theta_M$, power number N_p, and discharged-flow rate number N_{qd} for an eight-bladed paddle and turbine impeller [45]. Geometry of stirred vessel is shown as Case a of Table 1.

Table 1
Geometry of Stirred Vessels

						Impeller			
	L_T/D_T	L_s/D_T	D_s/D_T	ℓ/D	W_c/D_T	N_i	Type*	D_i/D_T	W_i/D_i
a	1	—	—	—	$\frac{3}{40}$	1	P, T	$\frac{1}{2}$	$\frac{1}{5}$
b	4	—	—	1	$\frac{1}{10}$	4	T***	$\frac{1}{3}$	$\frac{1}{5}$
c		$\frac{1}{2}$	$\frac{1}{2}$	$\frac{1}{2}$	—		RD	$\frac{1}{2}$	—
d		1	$\frac{1}{3}$	1	$\frac{1}{10}$, —		T***	$\frac{1}{3}$	$\frac{1}{5}$
e	$1 \sim 4$	—	—	$\frac{2}{3} \sim 2$	$\frac{1}{10}$		T	$\frac{9}{10}$	$\frac{1}{5}$
f	2	1	—	1	$\frac{1}{10}$	2	T	$\frac{9}{10}$	$\frac{1}{5}$

* P: paddle; T: turbine; RD: disk.
** H/D_T.
*** Oldshue and Ruston type impeller; $D_i:W_b:W_i:D_d = 20:5:4:15$.

number, Re_i, comparing with liquid flow patterns. Geometry of equipment is shown as case of Table 1. The configuration of vessels and impellers were shown in Figures 10 and 11. In the figure, the power number, N_p, and the discharge flow rate number, N_{qd}, are expressed in a similar manner. There is little difference between the flow patterns obtained by a paddle impeller and a turbine. Since the turbine impeller is equipped with a disc, the degree of mixing in the volumes above and below the impeller is lower than for paddle impellers. When the flow is turbulent, the state appears differently, due to the existence of a baffle plate. In the figure, the solid line is the strong main flow and the broken line is the secondary flow. When a baffle plate is inserted, the horizontally rotating flow in the direction of the impellers rotation decreases, and the circulation flow returning along the rotational axis from the tips of the impeller becomes the main flow. As a result, $\theta_M \cdot n$ becomes smaller while N_{qd}, which correspond to the discharge flow, increases, as does the power

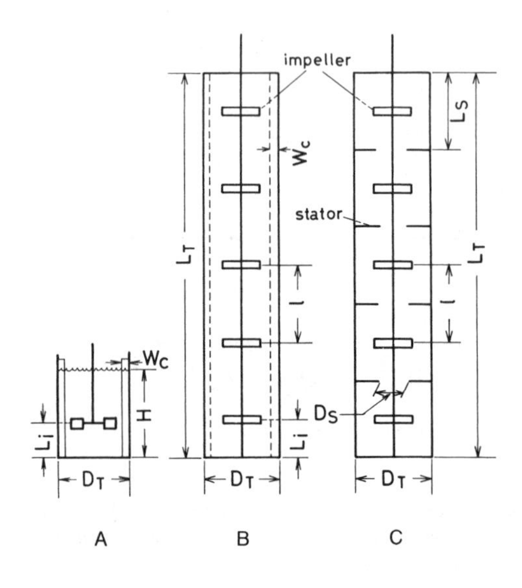

Figure 10. Geometry of stirred vessel: (A) vessel with single impeller; (B) vessel with multiple impellers; (C) RDC and Mixco column.

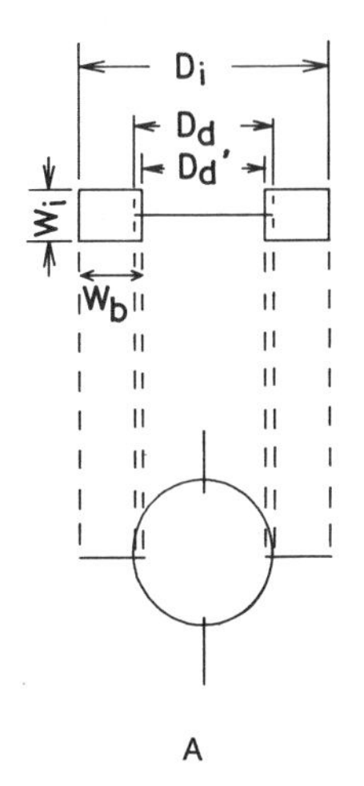

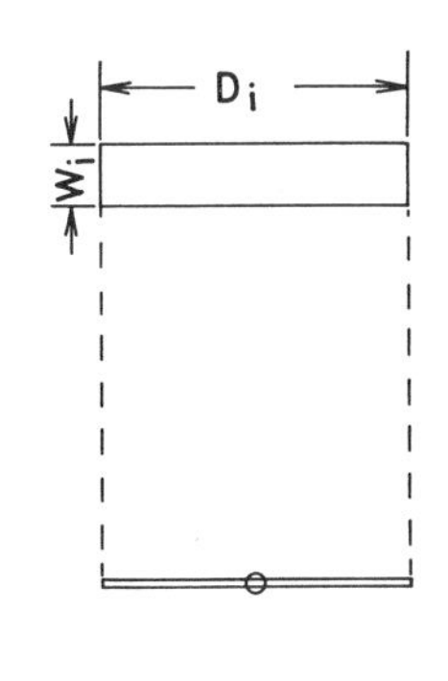

Figure 11. Geometry of impeller: (A) turbine impeller; (B) paddle impeller.

consumption, N_p. Without a baffle plate, the system shows decreases in the stirring effect due to the development of forced vortex flow. When the flow state is in laminar, the baffle plate displays limited effects, and causes some defects such as dead liquid space behind the plate.

Using a conductometric method, Kamiwano et al. [46] measured the dimensionless mixing time of uniform liquids in the perfect turbulence range of batch mixing systems, and expressed the results with the following formula:

$$\frac{1}{n\theta_M} = 0.092\left\{\left(\frac{D_i}{D_T}\right)^3 N_{qd} + 0.21\left(\frac{D_i}{D_T}\right)\left(\frac{N_p}{N_{qd}}\right)^{1/2}\right\}(1 - e^{-13(D_i/D_T)^2}) \tag{37}$$

It is considered that the liquid circulation and the turbulence caused by the rotating impeller greatly contribute to the mixing velocity. In Equation 37 the first term inside the { } on the right side corresponds to the circulation and the 2nd term of it to the turbulence. The exponential term of Equation 37 indicates the influence of the impeller diameter, and below $D_i/D_T \approx 0.4$ there are rapid decreases in the mixing velocity [43]. There are formulas devised by Sato et al. [47] and Nagase et al. [48], to estimate the discharge flow rate number N_{qd} ($= qd/n \cdot D_i^3$), in practical applications. Power number, N_p, is also found elsewhere [7].

Figure 12 [49] shows $n \cdot \theta_M$ of impellers used as for low and high viscosity liquids. In the figure, the reported values of $n \cdot \theta_M$ are inversely proportioned to $\gamma \cdot N_p \cdot Re_i$. From Figure 12:

$$\left.\begin{aligned} (n \cdot \theta_M)(\gamma \cdot N_p \cdot Re_i) &= \frac{P_v \cdot \theta_M}{\mu \cdot n} = 1.3 \times 10^4 \\ \gamma &= 4D_i/(\pi \cdot D_T^2 \cdot H) \end{aligned}\right\} \tag{38}$$

where P_v = power per unit volume of liquid
H = depth of liquid
N_p = power number
Re_i = rotational Reynolds number

The term $P_v \cdot \theta_M$ of the formula is the total energy required for mixing. Equation 38 indicates that a lower rotational speed is more economical.

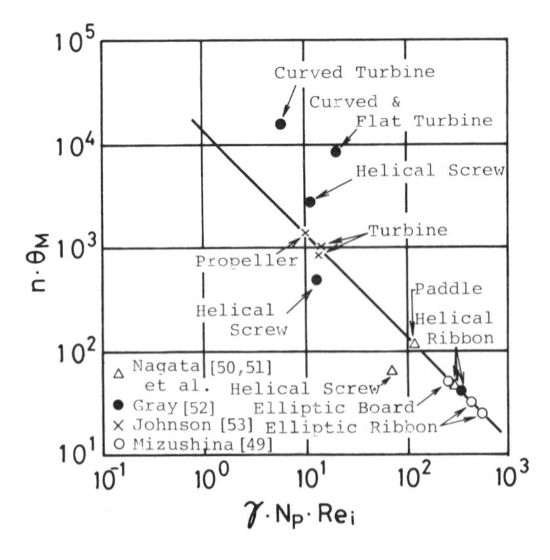

Figure 12. Relationship between power and mixing time [49].

Mixing Scale in Complete Mixing

As stated earlier, the mixing characteristics of the stirred vessel with a single impeller can often be approximated with a perfectly mixed flow. This is adequate, when considering the macromixing based on the RTD, in usual continuous flow systems. However, when the reaction velocity or the flow viscosity is high, or when the reaction is in progress within the droplet in heterogeneous system, the mixing scale must be considered [5–8].

Below is a discussion of the operation of the stirred vessel in Figure 11 [6, 54]. In the case of Figure 11A, the incoming flow is assumed to disperse into molecules immediately after flowing into the vessel. If stirring is adequate, the concentration in a vessel is perfectly uniform, and the mass balance is expressed by the formula:

$$V \frac{dc_{out}}{d\theta} = q(c_{in} - c_{out}) - V\psi(c_{out}) \tag{39}$$

In the case in Figure 13B it is assumed that the incoming fluid is dispersed in the vessel uniformly as aggregates and further that the molecules which compose the aggregate behave jointly within the aggregate until the aggregate flows out of the vessel.

The RTD function related to Figure 13A is given by substituting $\psi(c_{out}) = 0$ into Equation 39:

$$E(\theta) = \frac{1}{\theta_T} e^{-(\theta/\theta_T)} \tag{40}$$

This indicates a perfect mixed flow of microfluid. For Figure 13B the individual aggregate in the vessel is not composed of the same component, even though the mixing is satisfactory. However, a good stirring gives the same RTD as that of Figure 13A, is expressed by Equation 40[55]. The $E(\theta)$ observed by the ordinary tracer response technique is not affected by the scale of mixing in the flow. In the case of Figure 13B, the flow behavior is termed segregated flow with exponential RTD (macrofluid), distinct from the perfect mixed flow of the microfluid.

$$\frac{(c_{in} - c_{out})}{\psi(c_{out})} = \frac{V}{q} = \theta_T \tag{41}$$

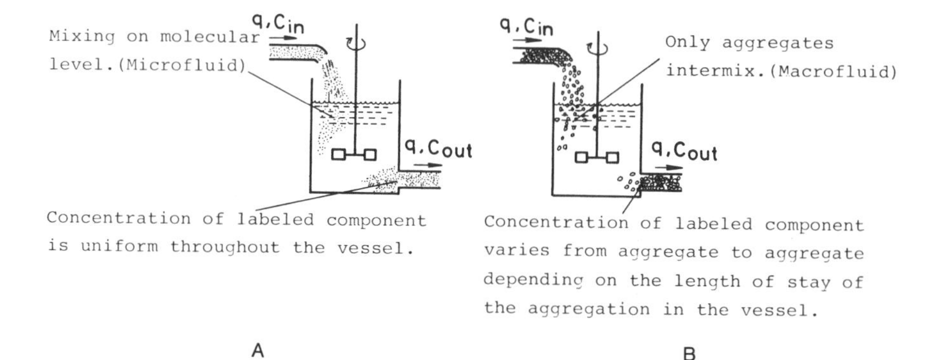

Figure 13. Behavior of microfluid and macrofluid in vessel [6]: (A) perfect mixed flow of microfluid; (B) segregated flow with exponential RTD.

However, this equation cannot be applied to the segregated flow with exponential RTD. In the complete mixing vessel, there are a number of components in the aggregates of the vessel and the outflow includes such aggregates. The component of an aggregate is a function of time spent in the vessel. When the RTD of the outgoing flow is expressed as $E(\theta)$ and representing the concentration of the labeled component with $c(\theta)$, the mean concentration of the component in the outgoing flow, $\bar{c}_{out}$, is given by:

$$\bar{c}_{out} = \int_0^\infty c(\theta) \cdot E(\theta)\, d\theta \tag{42}$$

When the concentration of a component of individual aggregates is considered to be determined unilaterally by the time spent in the vessel, $c(\theta)$ is the same as for the reaction in batch operation and can be obtained by integrating the rate equation. For the r-th order of reaction, indicated by $\psi(c) = k_r c^r$, the function $c(\theta)$ is given by:

$$\left.\begin{aligned} r = 1: &\quad c(\theta) = c_0 e^{-k_r \theta} \\ r \neq 1: &\quad c(\theta) = \left[\frac{c_0^{r-1}}{1 + (r-1)k_r c_0^{n-1} \cdot \theta}\right]^{1/(r-1)} \end{aligned}\right\} \tag{43}$$

When Equations 43 and 40 are substituted into Equation 42, for the segregated flow with exponential RTD, the concentration of the labeled component in the outgoing flow under the steady state is expressed as in Table 2. Figure 12 is a graphic representation of this concentration. For $r = 1$, the concentration corresponds to the case of perfect mixed flow.

Table 2 was determined from the critical state of the mixing scale. Mixing on a molecular scale is not necessarily achieved in the device, and an aggregate may occasionally disintegrate. In view of this, the actual conversion of reactant is thought to lie between the two extreme ranges shown in Table 2.

The mixing strength, measured by standard physical methods is related to the RTD function, $E(\theta)$, while for $c(\theta)$ it can be measured by chemical methods. The mixing related to $E(\theta)$ is called macromixing and that related to $c(\theta)$ is known as the micromixing [54].

Numerous mixing models have been presented for use in analyzing the micromixing phenomena [1–6, 56]. The mixing time is significant to evaluate the uniformity of fluid. Whether or not the

Table 2
Effect of Mixing Scale on Conversion [9]

Reaction Order r	Rate Equation $-\frac{dc}{d\theta} = k_r c^r$	Number of Reaction Unit N_r	Conversion; $(1 - X_A) = \bar{c}_{out}/c_{in}$		
			(i) Batch Reaction	(ii) Perfect Mixed Flow of Microfluid	(iii) Segregated Flow with Exponential RTD
0	$-\frac{dc}{d\theta} = k_0$	$\frac{k_0\theta_T}{c_{in}}$	$1 - N_0$	$1 - N_0$	$1 - N_0(1 - e^{-1/N_0})$
1	$-\frac{dc}{d\theta} = k_1 c$	$k_1\theta_T$	e^{-N_1}	$\frac{1}{1 + N_1}$	$\frac{1}{1 + N_1}$
2	$-\frac{dc}{d\theta} = k_2 c^2$	$k_2\theta_T c_{in}$	$\frac{1}{1 + N_2}$	$\left[\left(\frac{1}{2N_2}\right)^2 + \frac{1}{N_2}\right]^{1/2} - \frac{1}{2N_2}$	$-\frac{e^{1/N_2}}{N_2} E_i\left(-\frac{1}{N_2}\right)$

$$E_i(-x) = \int_x^{\infty} \frac{e^{-Z}}{Z} dZ$$

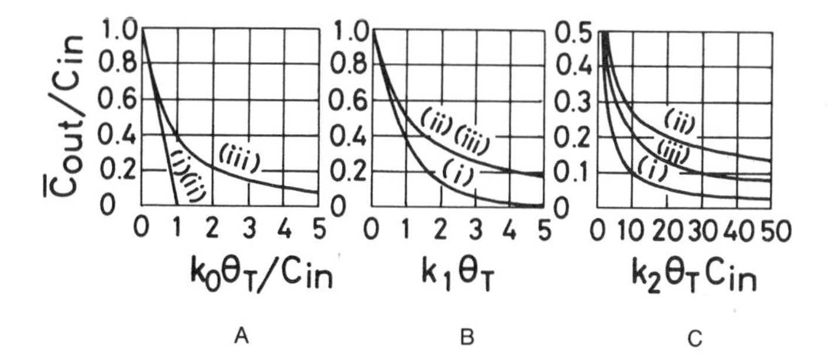

Figure 14. Comparison of conversion c_{out}/c_{in} in micro and macrofluid for some reaction [54]: (A) zero-order reaction; (B) first-order reaction; (C) second-order reaction.
(i) Batch.
(ii) Perfect mixed flow of microfluid, corresponding to Figure 13A.
(iii) Segregated flow with exponential RTD, corresponding to Figure 13B.

RTD function can be approximated at the perfect mixed flow is, as stated earlier, profoundly dependent on the mixing time. The mixing time is also related to the mixing parameters of the micromixing model [57]. In this case, a standard to determine the completion of mixing is very important in measurements of the mixing time. Here, macromixing in relation to the RTD function is evaluated. Details of the micromixing process are described later.

Residence Time Distribution

The RTD (Residence Time Distribution) of liquid in stirred vessel with a single impeller in general operations, is represented by $E(\theta) = (1/\theta_T)\exp(-\theta_T)$. However, since new liquid flows in during the mixing, it cannot be assumed that the liquid in the vessel is kept uniform. The deviation in uniformity is dependent not only on the relation between the mixing time and the mean residence time, but also on the geometric configuration of the vessel, the relative positions of the inlet and outlet, the impellers, and so forth. In order to express these relations quantitatively, it is necessary to understand clearly the mixing process in the vessel. The analysis is made with the RTD related to the mixing process within the vessel [2, 5–7, 58–66].

BACKMIXING IN STIRRED VESSELS WITH MULTIPLE IMPELLERS

Flow Behavior of Liquid in Vessels and Mixing Models

Baffled Vessel with Turbine Impellers

The following presents some studies that have been made on simplified mixing models that adequately identify the mixing state of a liquid in baffled vessels with turbine impellers.

Figure 15 is an example of a measured RTD curve for a baffled vessel equipped with four turbine impellers with six blades. For comparison, the calculated values based on the completely mixed flow, the backflow cell model, and the axial dispersion model are also plotted in the figure. The calculated values of the completely mixed flow is widely apart from that of the measured RTD. However, the values of both the backflow cell model and the axial dispersion model are located in the immediate vicinity of the measured RTD curve, and it cannot be determined which one fits better. It may be inferred that when a mixing state in a vessel is approximated by a mixing model,

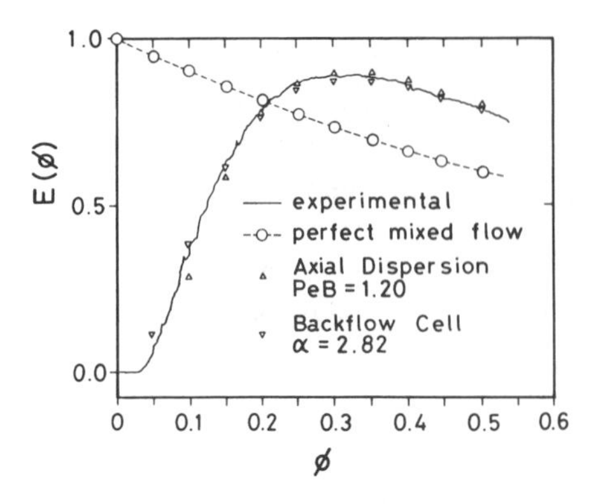

Figure 15. Comparison of the experimental RTD with calculated RTD function $E(\phi)$ based on the perfect mixed flow, axial dispersion model, and backflow cell model [30]. Turbines with 6 blades, $Re_i = 5420$, $Re_f = 449$. Geometry of stirred vessel is shown as Case b of Table 1.

the measured RTD and the RTD of the mixing model agree with each other. The reverse is not necessarily true.

Figure 16A shows the flow patterns of the liquid in the vicinity of two adjoining impellers in a stirred vessel. The liquid in the vicinity of one impeller is pushed against the wall of the vessel by the centrifugal force exerted by the rotating blades. Upon reaching the wall, it is diverged up and

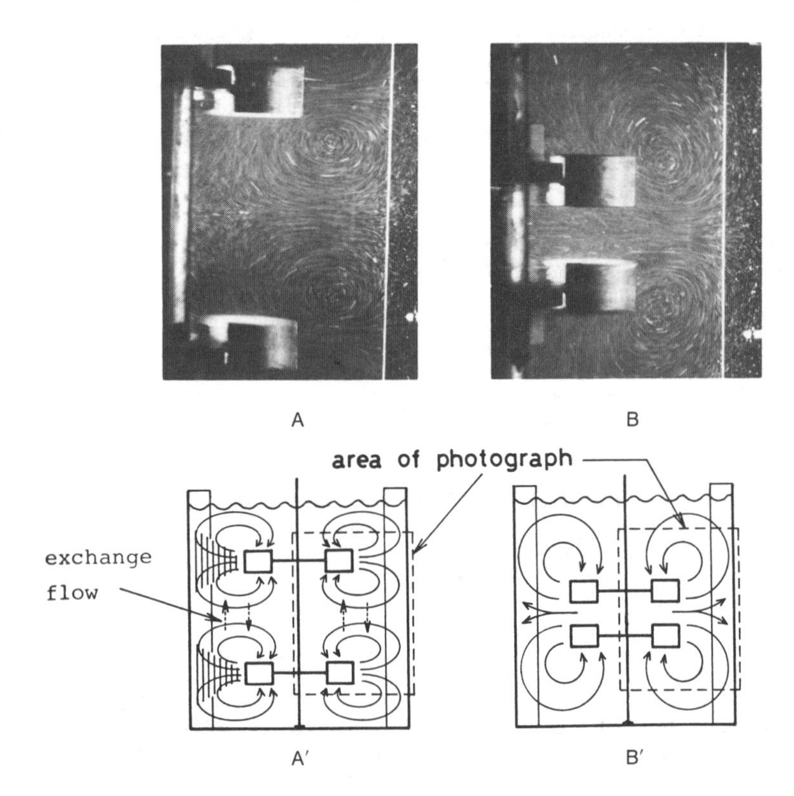

Figure 16. Illustration of flow patterns. Dual impeller $D_i/D_T = 0.5$: (A, A′) $\ell/D_T = 0.5$; (B, B′) $\ell/D_T = 0.2$ (Oldshue and Ruston type impellers with 6 blades).

down before being sucked along the stirring axis, eventually resulting in a circulating flow. Each circulation flow formed by the two impellers meets between the impellers and returns to its own blade. Thus, the liquid exchange flows occur across the horizontal section between the center of two impellers where the two adjoining circulating flows meet. Hence, the liquid exchange flow regulates the mixing rate [65], and the flow pattern in Figure 16A indicates that the backflow cell model is adequate.

When the distance between impellers becomes smaller, two adjacent impellers act like a single impeller with a wide blade, and so the partitioning effect decreases. As a result, the mixing rate in the entire vessel is increased [67]. Figure 16B shows such flow with one set of double impellers. The partitioning effect is lost when the impeller distance ratios are $\ell/D_T < 0.3$ for turbine blades [59] and $\ell/D_T < 0.5$ for paddle blades [67].

Figure 17 shows the profile of the tracer concentration in the vessel when a minute volume of tracer is regularly supplied around the top outlet of the stirred vessel with multiple turbine impellers in a continuous flow system. In the figure, a normalized concentration is used to show the concentration profile, with 1 (one) for the plane of the upper turbine and 0 (zero) for that of the lower turbine. The concentration range is clearly far from uniform as is required for perfect mixed flow. However, an outer zone of nearly uniform concentration exists for every turbine. This zone is indicated by the shaded area in Figure 16A'. And it accounts for approximately 30% of the total volume. This supports the applicability of the backflow cell model, but the concentration in the center core of the liquid between adjacent turbines varies, and this matches with the axial dispersion model [30].

Nonbaffled Vessel with Paddle Impellers

With paddle impellers and no baffle plate, the flow pattern is different from that of the baffled vessel with turbine impellers. Most of the liquid supplied to the top of the vessel reaches the bottom outlet after a comparably short period of time, with the dimensionless time value, $\phi\ (= \theta/\theta_T) = 0.05 \sim 0.10$. Around the stirring axis of the vessel, there is a forced vortex flow, and within the vortex flow there is a liquid flow toward the main flow. The impellers mix this liquid with the

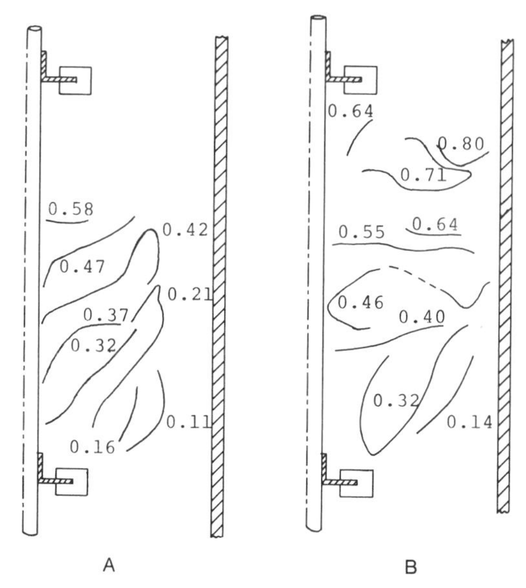

Figure 17. Profiles of normalized concentration inside the vessel [30]: (A) six-blade turbines. $Re_i = 5640$, $Re_f = 200$; (B) four-blade turbines. $Re_i = 5870$, $Re_f = 198$. Geometry of stirred vessel is shown as Case b of Table 1.

surrounding liquid [14]. It is not simple to approximate the mixing phenomena in this apparatus, while Nagata et al. [14, 32] have attempted the application of a parallel combination model.

The flow state of a nonbaffled vessel with turbine impellers approaches that of baffled vessel with turbine impellers, because in the nonbaffled vessel the liquid flow toward the main flow through the forced vortex flow is repressed by the disc plate of the impellers.

The flow state of a baffled vessel with paddle impellers approaches to that of a baffled vessel with turbine impellers, because in the baffled vessel with paddle impellers the circulating flow created by the impeller increases.

Rotating Disc Column and Mixco Column

In order to repress the backmixing in the main flow direction, the stator-rings are attached to the equipment such as the rotating disc column and the Mixco column, between the impellers of those columns.

Figure 18 shows the concentration profiles with RDC and Mixco columns determined with the same method as in Figure 17. A region of nonuniform concentration in the RDC and Mixco columns exists only above and below the perforated partitions. Each volume of the respective regions in the RDC and Mixco columns accounts for 10% and for 30% of the total volume, respectively. From this, it is understood that the backflow cell model explains the mixing state in the apparatus

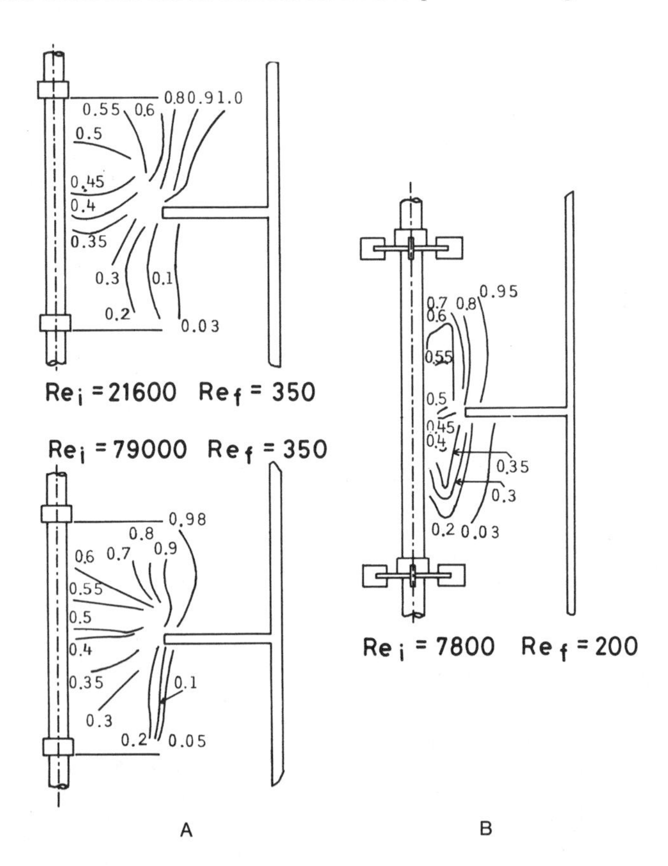

Figure 18. Profiles of normalized concentration inside the vessel [29]: (A) RDC (B) Mixco column without baffles. Geometry of stirred vessel is shown as Case c and d of Table 1.

better than the axial dispersion model. The measured RTD agree better with that of the backflow cell model than with the axial dispersion model at small rotational Reynolds number Re_i [29]. As shown in Figure 18A, the line indicating concentration value of 0.5 veers toward the upper impeller for concentrations at comparably small rotational Reynolds numbers, as $Re_i = 21{,}600$. When the rotational Reynolds number is larger, $Re_i = 79{,}000$, the fields of concentration both in the upper and lower sections are broadly symmetrical. When the flow Reynolds number, Re_f, is constant, the field of concentration with small Re_i is affected by Re_f. Thus RTD is also expected to be affected by Re_f.

Degree of Backmixing in the Vessel with Multiple Impellers

Baffled Vessel with Turbine Impellers

Regarding the Reynolds number of passing liquids, Re_f, Figure 19 shows the relations between the dimensionless backflow, q'/nD_i^3 of the baffled vessel with turbine impellers and the rotational Reynolds number, Re_i. For $Re_i > 8 \times 10^3$, q'/nD_i^3 indicates a constant value, and with $Re_i < 8 \times 10^3$, q'/nD_i^3 is affected by the viscosity. Over the entire field of Re_i, no influence from Re_f is observed [30]. Under ordinary operational conditions, circulating flow driven by the turbine impeller and the backflow q' would be larger than the passing liquid flow, so that Re_f has no effect on q'/nD_i^3.

From the experimentally obtained relations in Figure 19 and Equation 44, the values of a dimensionless backflow rate for turbine impellers with an arbitrary number of blades, n_b, can be calculated [30].

$$\left(\frac{q'}{nD_i^3}\right) = \left(\frac{q'}{nD_i^3}\right)_{n_b=6} \times \left(\frac{n_b}{6}\right)^{0.27} \tag{44}$$

Nonbaffled Vessel with Paddle Impellers

The following fraction of plug flow, ϕ°, was obtained after analyzing the degree of backmixing, applying the parallel combination model related to the nonbaffled vessel with flat paddle impellers [14, 32].

$$\phi^\circ = K_1\left[\frac{L_T u}{\nu}\bigg/\frac{nD_i^2}{\nu}\right]^{2/3}, \qquad \phi^\circ < 0.4 \tag{45}$$

In the preceding equation, when arranged as $L_T u/\nu = (D_T u/\nu) \times (L_T/D_T)$, $D_T u/\nu$ (= Re_f) expresses the Reynolds number for a passing liquid, and nD_i^2/ν (= Re_i) represents the rotational Reynolds number. The preceding equation shows that for geometrically similar vessels, at a constant ratio of Re_f to Re_i, values of ϕ° are equivalent. In addition, if Re_f/Re_i is constant, ϕ° becomes larger with increases in L_T/D_T.

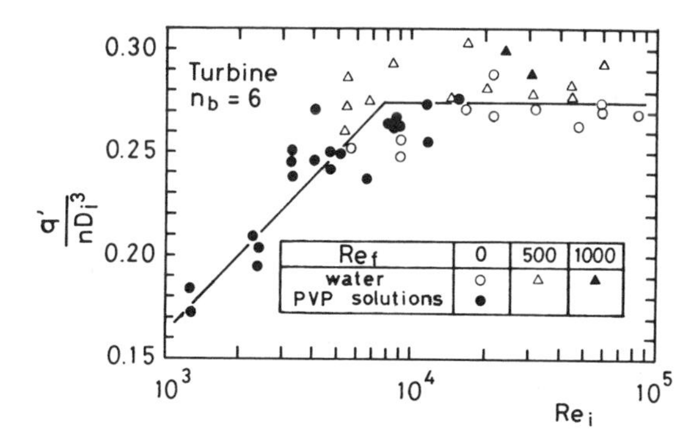

Figure 19. Effect of Reynolds numbers Re_i and Re_f on dimensionless backflow, q'/nD_i^3, for six-bladed turbines [30]. Geometry of vessel is shown in Case b of Table 1.

Table 3
Geometry of Paddle Impeller [32]

Type	n_b	ℓ/D_T	D_i/D_T	W_i/D_i	K_1
A	2	1	$\frac{1}{2}$	$\frac{1}{5}$	1.6
B	2	$\frac{1}{2}$	$\frac{1}{2}$	$\frac{1}{10}$	1.9
B′	2	variable	$\frac{1}{2}$	$\frac{1}{10}$	Fig. 18
C	2	$\frac{1}{3}$	$\frac{1}{2}$	$\frac{1}{15}$	2.45

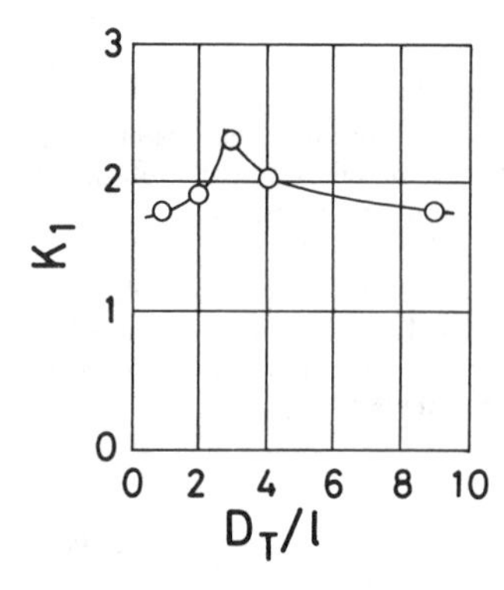

Figure 20. Effect of the distance between paddle impellers on K_1 [32].

K_1 of the preceding equation is constant related to the configuration of the impellers, as shown in Table 3. Figure 20 shows the relations between the inverse of the impeller distance ratio, D_T/ℓ and K_1, for paddle impellers of type B′ indicated in Table 3. With a smaller range of D_T/ℓ, K_1 becomes larger when D_T/ℓ and the number of impellers, N_i, increases. Conversely, in the range of $D_T/\ell > 3$, K_1 decreases with an increase in D_T/ℓ. The partitioning effects between the impellers decrease because the multiple impellers are changed to act like a single impeller as ℓ becomes smaller.

Rotating Disc Column and Mixco Column

The relationships between the backflow of RDC and the baffled and nonbaffled Mixco columns equipped with turbine impellers and Reynolds number Re_i are shown in Figure 21. With $Re_i < 10^5$, q'/nD_i^3* gets affected by Re_f. The effects of Re_i and Re_f on the values of q'/nD_i^3 become smaller with increases in Re_i.

For backmixing of RDC and Mixco columns, a number of empirical equations have been presented [25, 68–80]. Typical equations are as follows:

$$\frac{u'}{nD_i} = 1.7 \times 10^{-2} N_p^{1/3}(D_T/L_0)^{1/2}(D_s/D_T)^2 \tag{46}$$

$$\alpha\left(=\frac{u'}{u}\right) = 0.0098\left[K_2\frac{nD_i}{u}\left(\frac{D_i^2 A_h}{D_T L_s A_T}\right)^{1/2}\right]^{1.24} \tag{47}$$

$$\text{* RDC:} \quad \frac{q'}{nD_i^3} = \frac{u' \times (\pi/4)D_T^2}{nD_i(D_T/2)^2} \approx 3.14\frac{u'}{nD_i}$$

$$\text{M.C.:} \quad \frac{q'}{nD_i^3} = \frac{u' \times (\pi/4)D_T^2}{nD_i(D_T/3)^2} \approx 7.07\frac{u'}{nD_i}$$

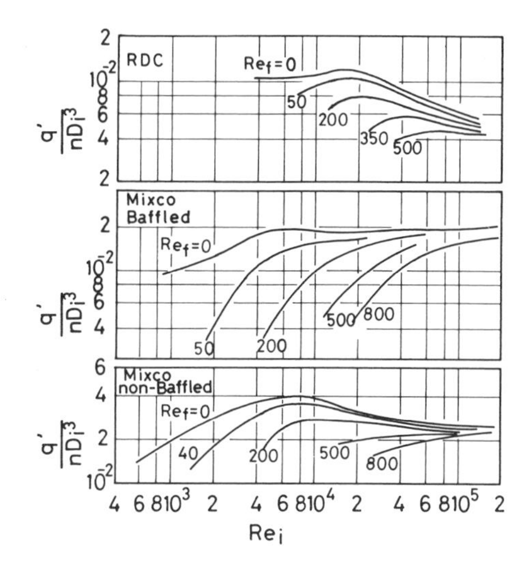

Figure 21. Effect of Reynolds number Re_i and Re_f on dimensionless backflow q'/nD_i^3 for RDC and Mixco column [23, 29]. Geometry of vessel is shown as Case c and d of Table 1.

Equation 46 designates that the backflow rate is affected by the geometric configuration of vessels and by the power number, N_p [25]. As good data for N_p is available [7, 43], the backflow rate can easily be calculated. N_p of RDC is shown in Figure 22.

Equation 46 contains backflow rate (Q/A_T) computed from the exchange flow Q determined through batch operations. Where the exchange rate of fluid between cells, Q, has already been explained. It represents that the backflow rate u′ measured on a continuous flow system and Q/A_T determined with a batch system are regarded as being the same. Thus, when $Re_i > 10^5$ or Re_f is small, Equation 46 is effective.

The perforation ratio of the stator, $(D_s/D_T)^2$, affecting the backmixing shown in Equation 46, can be applied to the baffled Mixco column indicated in Figure 21 [31].

Equation 47 is the simplest empirical equation which accurately summarizes the various reported data [68]. The impeller correction factor K_2 of Equation 47 takes into account differences in impeller types, and for six-blade turbines of the Rushton-Oldshue type, K_2 is unity. For other

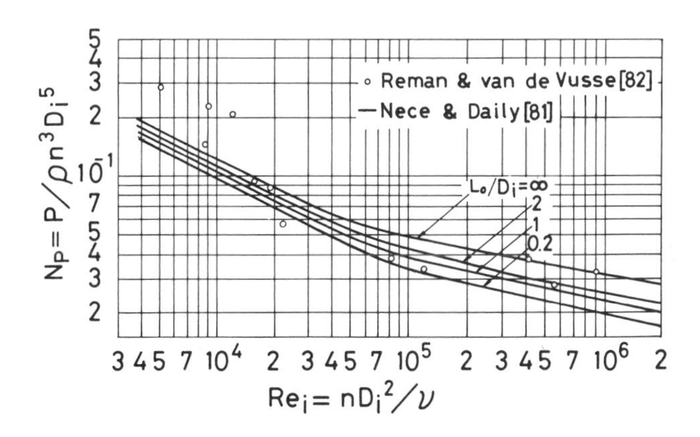

Figure 22. Power number, N_p for RDC [25].

configurations, K_2 is taken as the ratio of the power number, N_p, of the impeller used to that of the six-blade impeller. The N_p is approximately proportional to the square root of the number of blades, n_b [68]. It was experimentally confirmed [68] that Equation 47 cannot be established at $u = 0$, but is validated within the range $\alpha = 0.5 \sim 20$, $K_2(nD_i/u) \times (D_i^2A_h/D_TL_sA_T) = 3 \sim 500$.

The preceding is satisfied with a single-phase flow. Equation 46 and 47 are valid for a two-phase continuous flow system, such as liquid extraction, by using the following equation [25, 68].

$$\begin{aligned} u &= u_c/\epsilon_c \\ q &= q_c/\epsilon_c \\ u' &= u_c'/\epsilon_c' \\ \nu &= \mu_c/\rho_c \\ \rho &= \rho_c\epsilon_c \end{aligned} \tag{48}$$

In the above equation, subscript c indicates the continuous phase c and ϵ_c indicates the volume fraction of phase c. The axial dispersion coefficient, E_{xc}, is determined by substituting u′ obtained from Equation 46 and/or 47 into the following equation which is obtained from Equation (34b) and $L_T = NL_O$.

$$E_{xc} = \frac{u_cL_O}{2 - (1/N)} + \epsilon_c \cdot u'L_O \tag{49}$$

Note that L_O can replace L_s.

There are reports [75, 76] holding that u_c' would become greater due to the continuous phase accompanied by dispersed liquid droplets. However, correlations of backmixing for a two-phase system, including the preceding phenomena have not yet been established.

BACKMIXING IN HORIZONTAL STIRRED VESSEL

A horizontal stirred vessel effectively mixes gas and liquid with a rotating impeller installed on a horizontal shaft at the gas-liquid interface. (See Figures 11 and 23).

The apparatus has a large contact area for the volume [83, 84], and it is possible to control flow rates of gas and liquid separately. Most research reports on horizontal stirred vessels are concerned with the relationship between the gas-liquid contact [84–89], power consumption [90–93], type of impeller [83, 87], and operating conditions.

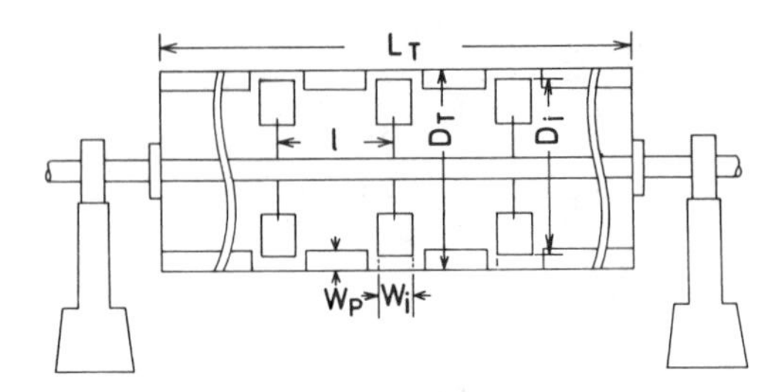

Figure 23. Schematic diagram and geometry of horizontal stirred vessel.

Flow Behavior of Liquid in the Vessel

Colored liquid tracer was injected for a short period in either the right or left cell, separated by the rotating impeller which was set in the center of a batch horizontal stirred vessel. Then, the cell where the tracer was injected shows a uniform color, before the coloring spreads to the entire vessel [26].

As a result, the mixing of the liquids can be approximated by the model shown in Figure 24B. The liquid in a horizontal stirred vessel is divided into two equal parts by the impeller. The liquid in the cells is mixed by the circulating flow of each cell, which contributes to an approximation of complete mixing. Moreover, the exchange flow rate, Q_i, between the cells controls the mixing throughout the whole vessel.

In the case of a continuous flow system, this mixing model forms a two-stage backflow cell model with a backflow rate, q_i', as shown in Figure 24A. In the case of the vessel with multiple impellers, it becomes a backflow cell model with $N = N_i + 1$ cells, as indicated in Figure 3A and 3B. N_i represents the number of impellers.

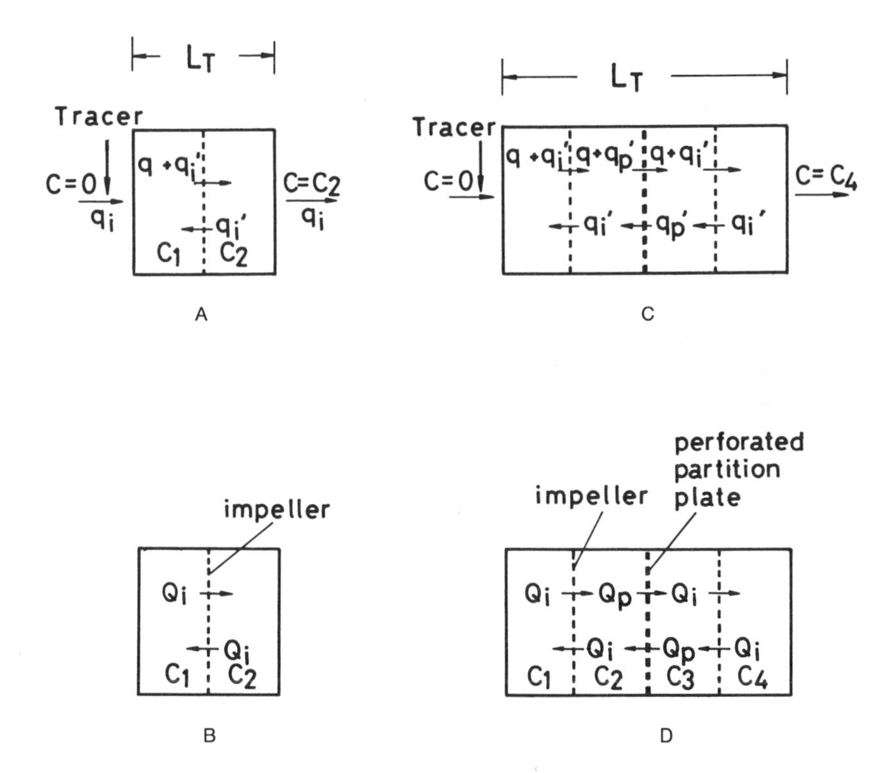

Figure 24. Mixing models for horizontal stirred vessels: (A) model of backflow cell separated by an impeller; (B) model of mixing in batch vessel; (C) model of backflow cell separated by impellers and a perforated partition plate; (D) model of mixing in batch vessel separated by impellers and a perforated partition plate.

Degree of Backmixing

The experimental relationship between the backflow rate q_i' and the supplied flow q is shown in Figure 25. The relationship between q_i' and q is expressed by:

$$q_i' = q_{io}' - q(1/2) \tag{50}$$

where, q_{io}' is the value of q_i' at $q = 0$. Since $q_{io}' = Q_i$ is established [28], the backflow ratio α is:

$$\alpha = \frac{q_i'}{q} = \frac{Q_i}{q} - \frac{1}{2} \tag{51}$$

where, Q_i is the exchange rate of liquid in the batch operation. For the impeller shown as Case e of Table 1, Q_i would be determined from the equation containing the power number, N_p, for one impeller [27, 94]:

$$\frac{Q_i}{n \cdot D_T^3 \cdot \epsilon} = 2.0 \times 10^{-2}\left(\frac{N_p}{\epsilon}\right)^{0.79} \tag{52}$$

where D_T = the diameter of the vessel
n = the rotational velocity of the impeller
ϵ = the ratio of liquid volume to the vessel volume

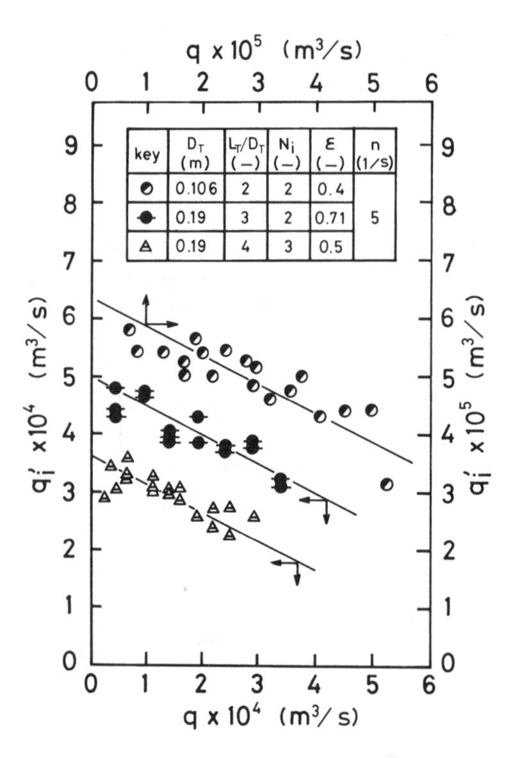

Figure 25. Experimental correlation of volumetric backflow rate q_i' with volumetric flow rate of main stream q [95].

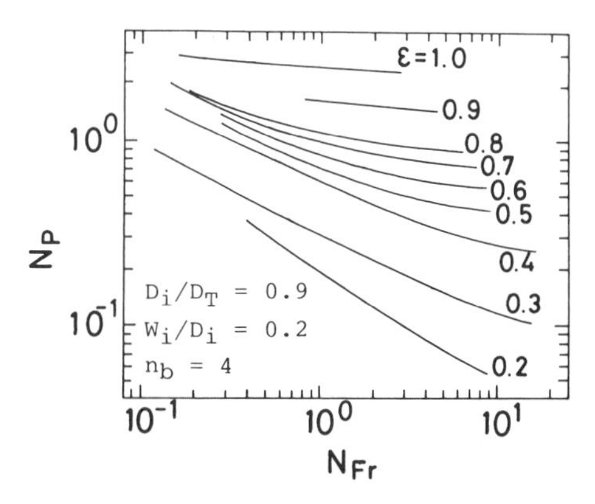

Figure 26. Power number for one impeller, N_p, of horizontal stirred vessel [91]. Geometry of vessel is shown as Case e of Table 1.

N_p is obtained from Figure 26.

The conditions establishing this model are:

1. The exchange flow rate is not affected by the action of adjacent impellers
2. Perfect mixing in each cell is assumed.

The conditions 1 and 2 would experimentally be expressed by [95]:

$$\frac{2}{3} < \frac{\ell}{D_T} < 2 \tag{53}$$

where ℓ = the distance between adjacent impellers

Backmixing in the Vessel with Perforated Partition Plate

When a horizontal stirred vessel is used in an actual gas-liquid contact operation [96], the counter-backmixing method with a perforated partition plate is effective. The liquid mixing model for a horizontal stirred vessel equipped with a perforated partition plate between the two impellers is expressed in Figure 24C and D. The backflow ratio related to the partitioning plate can be expressed [28] by:

$$\alpha_p = \frac{u'_p}{q} = \frac{Q_p}{q} - \frac{1}{2} \tag{54}$$

where Q_p is the liquid exchange flow rate at the partition plate in the batch system (Figure 24D). For the vessels shown as case f of Table 1, Q_p is given [28] by:

$$\frac{Q_p}{Q_i} = \frac{2_\gamma^{0.73}}{1 + 0.4(\delta/d)^{0.5}} \tag{55}$$

where γ = the opening ratio of the perforated plate
δ = the thickness of the perforated plate
d = the hole diameter of perforated plate

NOTATION

- A_h free passage area in a partition plate between compartments ($= \pi D_S^2/4$)
- A_T cross-sectional area of vessel ($= \pi D_T^2/4$)
- C dimensionless concentration
- c concentration
- c_0 concentration at time $\theta = 0$
- D_i diameter of impeller
- D_S diameter of hole in stator
- D_T diameter of vessel
- d hole diameter of perforated partition plate
- E_x axial dispersion coefficient
- $E(\phi)$ residence time distribution function (RTD function)
- g acceleration of gravity
- j cell (stage) number
- k_r reaction rate constant
- L_0 length of one compartment ($= L_T/N$)
- L_S distance between adjacent partition plate
- L_T total length of vessel
- ℓ distance between adjacent impellers
- N number of cells (or stage) or n-th cell
- N_i number of impellers
- N_p power number of impeller ($= P/\rho n^3 D_i^5$)
- N_{Fr} Froude number ($= n^2 D_i/g$)
- N_{qd} discharged-flow rate number ($= q_d/nD_i^3$)
- N_r number of reaction unit
- n rotational speed of impeller
- n_b number of blades
- P_eB Peclet-Bodenstein number ($= u \cdot L_T/E_x$)
- P_v power per unit volume of liquid
- Q volumetric exchange rate of liquid
- Q_i volumetric exchange rate of liquid at impeller
- Q_p volumetric exchange rate of liquid at partition plate
- q volumetric flow rate of main stream
- q_d volumetric discharge flow rate from the tip of impeller blades
- q' volumetric backflow rate
- Re_f flow Reynolds number ($= q/\nu D_T$)
- Re_i rotational Reynolds number ($= nD_i^2/\nu$)
- $R(\phi)$ internal age distribution function
- r reaction order
- u superficial velocity of main stream
- u' superficial velocity of backflow
- u_i' backflow rate of liquid due to impeller
- u_p' backflow rate of liquid through perforated partition plate
- V vessel volume
- V_ℓ volume of liquid in horizontal stirred vessel
- X x/L_T: dimensionless length
- X_A fractional convertion of reactant A ($= 1 - c_{out}/c_{in}$)
- Z liquid depth

Greek Symbols

- α backflow ratio ($= q'/q$)
- γ volume fraction ($= 4 \cdot D_i^3/D_T^2 H$)
- δ thickness of perforated partition plate
- ϵ ratio of volume of liquid to volume of horizontal stirred vessel ($= V_\ell/V$)
- ϵ_c volume fraction of phase c
- θ time
- θ_M mixing time
- θ_T mean residence time ($= L_T/u = V/q$)
- μ viscosity
- ν kinematic viscosity
- ρ density
- σ^2 normalized variance of $E(\phi)$
- ϕ° fraction of piston flow
- ϕ dimensionless time ($= \theta/\theta_T$)
- ϕ_M dimensionless time ($= (Q/V \cdot \theta)$
- $\psi(c)$ volumetric rate of reaction

Subscripts

- in incoming
- out outgoing
- x direction of x
- j j-th cell (or stage)
- c continuous phase
- d axial dispersion model
- i impeller
- p perforated partition plate

REFERENCES

1. Nauman, E. B., *Chem. Eng. Commun.*, 8:53 (1981).
2. Nauman, E. B., and Buffham, B. A., *Mixing in Continuous Flow Systems*, Wiley (Interscience), New York, 1983.
3. Takao, M., and Murakami, Y., *Kagaku Kōgaku*, 47:642 (1983).
4. Villermanx, J., and David, R., *Preprint of the 2nd World Congress of Chem. Eng.*, Montreal, 397 (1981).
5. Oldshue, J. Y., *Fluid Mixing Technology*, McGraw-Hill, New York, 1983.
6. Levenspiel, O., *Chemical Rection Engineering: An Introduction to the Design of Chemical Reactors*, Wiley (Interscience), New York, 1962.
7. Nagata, S., *Mixing; Principles and Applications*, Wiley, New York, 1975.
8. Mecklenburgh, J. C., and Hartland S., *The Theory of Backmixing*, Wiley (Interscience), New York, 1975.
9. Miyauchi, T., *Ryūkeisōsa to Kongōtokusei*, Zoku. Shin Kagaku Kōgaku Kōza, Nikkan Kōgyo Shinbunsha, Tokyo, 1960.
10. Shah, Y. T., Stiegel, G. J., and Sharma, M. M., *AIChE J.*, 24:369 (1978).
11. Wen, C. Y., and Fan, L. T., *Models for Flow Systems and Chemical Reactors*, Chemical Processing and Engineering Monograph Series, Dekkar, New York, 1975.
12. Danckwerts, P. V., *Chem. Eng. Sci.*, 2:1 (1953).
13. MacMullin, R. B., and Weber, M., *Trans. Am. Inst. Chem. Engrs.*, 31:409 (1935).
14. Nagata, S., et al., *Kagaku Kōgaku*, 17:387 (1953).
15. Oliver, E. D., and Watson, C. C. *AIChE J.*, 2:18 (1956).
16. Weber, A. P. *Chem. Eng. Progr.*, 49:26 (1953).
17. Yagi, S., and Miyauchi, T., *Kagaku Kōgaku*, 17:382 (1953).
18. Lebenspiel, O., and Smith, W. K., *Chem. Eng. Sci.*, 6:227 (1957).
19. Yagai, S., and Miyauchi, T., *Kagaku Kōgaku*, 19:507 (1955).
20. Lebenspiel, O., *Ind. Eng. Chem.*, 51:4131 (1959).
21. Ham, A., and Coe, H. S., *Chem. Met. Eng.*, 19:663 (1918).
22. Kats, M. B., and Genin, L. S., *Khim. Prom.* 42:50(1966), *Internat. Chem. Eng.*, 7:246 (1967).
23. Lelli, U., Magelli, F., and Sama, C., *Chem. Eng. Sci.*, 27:1109 (1972).
24. Sawinsky, J., and Hunek, J., *Trans. Instn. Chem. Eng.*, 59:64 (1981).
25. Miyauchi, T., Mitsutake, H., and Harase, I., *AIChE J.*, 12:508 (1966).
26. Ando, K., Hara, H., and Endoh, K., *Kagaku Kōgaku*, 35:805 (1971).
27. Ando, K., Fukuda, T., and Endoh, K., *Kagaku Kōgaku*, 38:460 (1974).
28. Ando, K., et al. *AIChE J.*, 27:599 (1981).
29. Lelli, U., Magelli, F., and Pasquali, G., *Chem. Eng. Sci.*, 31:253 (1976).
30. Fajner, D., Magelli, F., and Pasquali, G., *Chem. Eng. Commun.*, 17:285 (1982).
31. Magelli, F., Pasquali, G., and Lelli, U., *Chem. Eng. Sci.*, 37:141 (1982).
32. Nagata, S., Eguchi, W., Kasai, H., and Morino, I., *Kagaku Kōgaku*, 21:784 (1957).
33. Miyauchi, T., and Vermeulen, T., *Ind. Eng. Chem. Fund.*, 2:305 (1963).
34. Hartland, S., and Mecklenburgh, J. C., *Chem. Eng. Sci.*, 23:186 (1968).
35. Hartland, S., and Mecklenburgh, J. C., *Chem. Eng. Sci.*, 21:1209 (1966).
36. J. Mecklenburgh, C., and Hartland, S., *Chem. Eng. Sci.*, 23:81 (1968).
37. Roemer, M. H., and Durbin, L. D., *Ind. Eng. Chem. Fund.*, 6:120 (1967).
38. Latinen, G. A., and Stockton, F. D., *AIChE St. Paul, Minn.*, Sept. (1959).
39. Nishiwaki, A., and Kato, Y., *Kagaku Kōgaku Ronbunshu*, 2:530 (1976).
40. Gutoff, E. B., *AIChE J.*, 12:472 (1966).
41. Kramers, H., and Alberda, G., *Chem. Eng. Sci.*, 2:178 (1953).
42. Miyauchi, T., *Kagaku Kōgaku*, 28:615 (1964).
43. Yano, T., et al., *Kagaku Kōgaku Benran, CH* 18, *4th ed.*, S, Maeda (Ed.), Maruzen, Tokyo, 1978.
44. Kramers, H., Baars, G. M., and Knoll, W. H., *Chem. Eng. Sci.*, 2:35 (1953).
45. Yamamoto, K., in *Kakuhansōchi no Sekkei to Sōsa; Kakuhan riron no Kiso*, Kagaku Kōgyosha, Tokyo, 1970.
46. Kamiwano, M., Yamamoto, K., and Nagata, S., *Kagaku Kōgaku*, 31:365 (1967).

47. Sato, T., and Taniyama, I., *Kagaku Kōgaku,* 29:153 (1965).
48. Nagase, Y., Goto, S., and Yoshida, T., *Kagaku Kōgaku,* 38:684 (1974).
49. Mizushina, T., et al., *Kagaku Kōgaku,* 34:1205 (1970).
50. Nagata, S., Yanagimoto, M., and Yokoyama, T., *Kagaku Kōgaku,* 21:278 (1957).
51. Nagata, S., et al., *Kagaku Kōgaku,* 21:708 (1957).
52. Gray, J. B., *Chem. Eng. Progr.,* 59:55 (1963).
53. Johnson, R. T., *Ind. Eng. Chem. Process and Develop.,* 6:340 (1967).
54. Eguchi, W., and Kubota H., in *Shokubai Kōgaku Koza; Shokubai Sōchi oyobi Sekkei* H. Kobayashi, (Ed.) Vol. 3, Chijinshokan, Tokyo, 1965.
55. MacMullin, R. B., and Weber, M., *Trans. AIChE,* 31:409 (1935).
56. Angst, W., Bourne, J. R., and Sharma, R. N., *Chem Eng. Sci.,* 37:585 (1982).
57. Takao, M., et al., *Kagaku Kōgaku,* 45:588 (1981).
58. Gray, J. B., in *Mixing; Flow Patterns, Fluid Velocities, and Mixing in Agitated Vessels,* V. W. Uhl and J. B. Gray (Eds.) Vol. I, Academic Press, New York, 1966.
59. Sato, K., *Kagaku Kōgaku,* 33:320 (1969).
60. Gutoff, E. B., *AIChE J.,* 6:347 (1960).
61. van de Vusse, J. G., *Chem. Eng. Sci.,* 17:507 (1962).
62. Garceau, K., Cloutier, L., and Cholette, A., *Can. J. Chem. Eng.,* 46:82, 88 (1968).
63. Takamatsu, T., and Sawada, T., *Kagaku Kōgaku,* 30:1025 (1966).
64. Inoue, I., and Sato, K., *Kagaku Kōgaku,* 29:518 (1965).
65. Sato T., and Taniyama, I., *Kagaku Kōgaku,* 29:38 (1965).
66. Nauman, E. B., and Buffham, B. A., *Chem. Eng. Sci.,* 32:1233 (1977).
67. Takeda, K., et al. *Kagaku Kōgaku,* 32:376 (1968).
68. Haug, H. F., *AIChE J.,* 17:585 (1971).
69. Strand, C. P., Olney, R. B., and Ackerman, G. H., *AIChE J.,* 8:252 (1962).
70. Westerterp, K. R., and Landsman, P., *Chem. Eng. Sci.,* 17:373 (1963).
71. Stemerding, Ir. Sr., Lumb, E. C., and Lips, J., *Chemie Ing. Techn.,* 35:844 (1963).
72. Stainthorp, F. P., and Sudall, N., *Trans. Instn. Chem. Engrs.,* 42:198 (1964).
73. Gutoff, E. B., *AIChE J.,* 11:712 (1965).
74. Bibaud, R. E., and Treybal, R. E., *AIChE J.,* 12:472 (1966).
75. Sullivan, G. A., and Treybal, R. E., *Chem. Eng. J.,* 1:303 (1970).
76. Ingham, J., *Trans. Instn. Chem. Engrs.,* 50:372 (1972).
77. Bruin, S., *Trans. Instn. Chem. Engrs.,* 51:355 (1973).
78. Murakami, A., and Misonou, A., *Kagaku Kōgaku Ronbunshu,* 2:321 (1976). *Internat. Chem. Eng.,* 18:22 (1978).
79. Wolf, K. H., *Chem. Tech.,* 31:553 (1979).
80. Wolf, K. H., *Chem. Tech.,* 32:65 (1980).
81. Nece, R. E., and Daily, J. W., *Trans. ASME J. Basic Sci.,* D 82:562 (1960).
82. Reman, G. H., and van de Vusse, J. G., in *Liquid Extraction,* McGraw-Hill, New York, 1963.
83. Ganz, S. N., and Lokshin, M. A., *Zh. Prikl. Khim.,* 31:191 (1958).
84. Ando, K., Tabo, H., and Endoh, K., *J. Chem. Eng. Japan,* 5:193 (1972).
85. Yamane, T., and Yoshida, F., *J. Chem. Eng. Japan,* 5:381 (1972).
86. Tamaki, Y., Harada, E., and Ito, S., *Kagaku Kōgaku,* 38:601 (1974).
87. Sasaki, E., *J. Chem. Soc. Japan, Ind. Chem. Section,* 74:799 (1971).
88. Takeuchi, T., Ando, K., and Osa, T., *J. Japan Petroleum Inst.,* 19:1022 (1976).
89. Fukuda, T., et al. *J. Chem. Eng. Japan,* 13:298 (1980).
90. Ando, K., Hara, H., and Endoh, K., *Kagaku Kōgaku,* 35:466 (1971). *Internat. Chem. Eng.,* 11:736 (1971).
91. Ando, K., and Endoh, K., *Kagaku Kōgaku,* 36:1151 (1972).
92. Tamaki, and Y., and Ito, S., *Kagaku Kōgaku,* 37:725 (1973).
93. Ando, K., et al. *Kagaku Kōgaku Ronbunshu,* 4:154 (1978).
94. Ando, K., Fukuda, T., and Endoh, K., in *Ryūtaikongō,* Kagaku Kōgaku Symposium Series 6, Society of Chem. Eng., Japan, Tokyo, 1984.
95. Fukuda, T., doctor thesis, Hokkaido Univ. (1981).
96. Fukuda, T., et al. *Japan Kōgyō Yōsui,* No. 219:19 (1976).

CHAPTER 29

INDUSTRIAL MIXING EQUIPMENT

J. Y. Oldshue

Vice President of Mixing Technology
Mixing Equipment Co., Inc.
Rochester, New York

CONTENTS

INTRODUCTION AND MIXING TERMINOLOGY, 803
Impeller Types, 805
Impeller Function, 811
Tank Design, 814
Baffle Design, 814
Mixing Operations, 817

SCALE-UP OF FLUID MIXING DEVICES, 818
General Principles of Impeller Fluid Mechanics, 819
Change in Scale-Up Relationships, 829

CURRENT PRACTICE IN MIXER SCALE-UP TECHNIQUES, 830
Exponential Scale-Up Relationships, 832
Fluid Motion on Scale-Up, 833
Scale-Up Concepts, 833
Some General Principles of Pilot Planting, 835
Gas-Liquid Mass Transfer Example, 837
Fermentation Pilot Plant Example, 840
Zinc Purification Process Example, 847
Flocculation Example, 848

NOTATION, 850

REFERENCES, 850

INTRODUCTION AND MIXING TERMINOLOGY

When designing a commercial-scale system, agitation equipment often contributes a major role in optimizing the total process. It is essential to determine which mixing parameters are significant, in terms of meeting the overall process objectives, then properly incorporate them into a commercial-scale design. Substantial economic benefits can be achieved by using current technology to assure that a specific process is optimized with respect to agitation.

The effect of agitation on the process is established by evaluating the following:

1. Process design requirements.
2. Impeller power response characteristics.
3. Agitator mechanical design requirements.
4. Vessel and structural requirements.

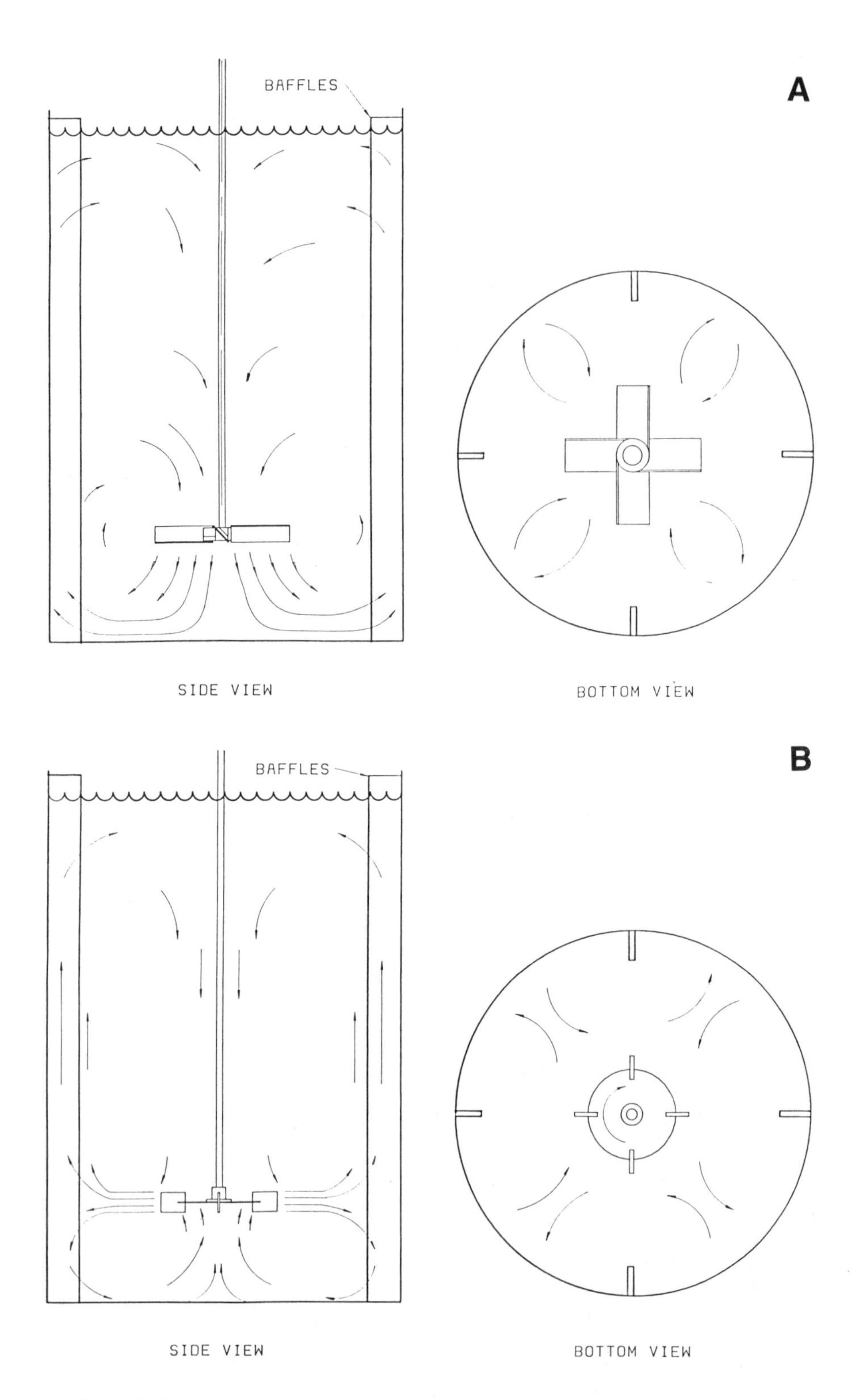

Figure 1. Flow patterns produced by (A) axial flow and (B) radial flow impellers [27].

Predicting the power consumed by an impeller in a specific process environment is completely independent of achieving a desired process result. Hence, the wide spectrum of flow and fluid shear characteristics available from a number of possible impeller configurations must be tailored to the process requirement.

The application of mixers to flow-controlled processes such as blending, heat transfer, or solids suspension uses well-established technology. This technology has been expanded for applications requiring specific mass transfer characteristics and for complex reactor systems. In many cases, it is necessary to conduct pilot-plant work to develop data for process optimization, mixer optimization, and mixer scale-up. Pilot work will ensure that the full-scale equipment is *process effective*, and feasible, from both an economic and equipment design standpoint.

Impeller Types

Mixing impellers are usually classified as two basic types—axial flow impellers and radial flow impellers. As the name implies, an axial flow impeller produces flow along the impeller axis, parallel to the impeller shaft. A radial flow impeller produces flow along its radius. Figure 1 illustrates the flow patterns developed by each type.

Since there is virtually no limit to the number of possible impeller designs, industry has developed the basic configurations shown in Table 1. Although impeller design may vary slightly from one manufacturer to another, the impellers shown are representative of those in common use. The designations R and A refer to radial and axial flow impellers respectively and the numerical part of each designation is arbitrary. Let's examine the properties of the basic impellers shown in the table.

Characteristics of Radial Flow Impellers

The R-1 impeller is used in applications requiring a high degree of turbulence or relatively high fluid shear rates. It is especially useful in gas-liquid contacting. Because of the flat, circular disc to which the blades are attached, gas introduced below the impeller is directed along a path of maximum liquid contact. This is in contrast to the direct upward vertical route and minimum contact that would result if the disc were removed.

The highest fluid shear rate of the basic designs is produced by the R-2 impeller. At a given power level, higher speeds are required for the R-2 impeller because of its low power number (refer to the section on scale-up). This also means smaller gear boxes are required for speed reduction, because of the lower torque developed, compared to other impellers at the same power level.

The R-3 impeller is a contoured, two-bladed device (also called an "anchor" impeller) typically used for higher viscosity applications, up to 40,000 or 50,000 cP. Power consumption varies directly as H/D, the blade height to diameter ratio. Some applications require "wipers" which are attached to the blade and fitted between the impeller outside diameter and the vessel inside diameter. Thus, the radial clearance between impeller and tank wall is essentially reduced to zero, resulting in a considerable increase in power draw. Wiping the vessel wall is particularly important for heat transfer. The anchor impeller is especially useful for blending and heat transfer when the fluid viscosity is between 5,000 and 50,000 cP. Below 5,000 cP there is not enough viscous drag at the tank wall to promote pumping, resulting in fluid swirl.

Characteristics of Axial Flow Impellers

The A-1 or marine impeller is primarily used with portable (Figure 2) and fixed-mounted (Figure 3) mixers in the $\frac{1}{4}$ to 3 horsepower range. Propeller sizes for this power range are approximately 3 to 15 inches in diameter. Propellers produce a down-pumping action toward the tank bottom. The A-1 impeller has a pitch ratio of 1.5. This means it would generate a plug or cylinder of fluid 1.5 times the impeller diameter in length for each impeller revolution, when operated in a theoretical environment with no slippage and at 100% efficiency. Other marine-type propellers are also in use where the pitch ratio is 1.0 (also referred to as "square pitch").

Table 1
Basic Impeller Designs in Common Use

Number	Name	Description	
R-1	Flat blade	Vertical blades bolted to support disk	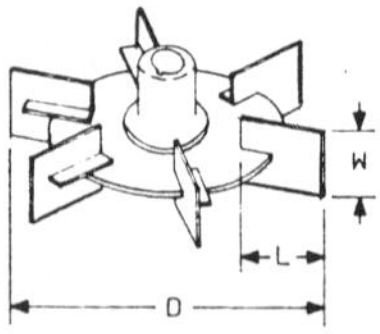 L=1/4D W=1/5D DISC DIA.=2/3D
R-2	Bar turbine	Six blades bolted/welded to top and bottom of support disk	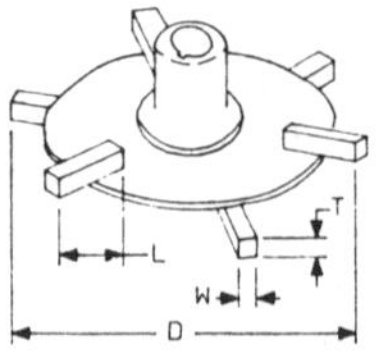 L=1/4D W=1/20D T=1/20D DISC DIA.=2/3D
R-3	Anchor	Two blades with or without cross arm	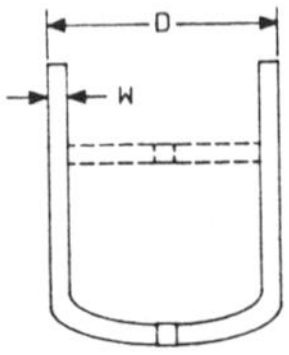 W=1/10D
A-1	Propeller three blades	Constant pitch, skewed-back blades	 1.5 PITCH RATIO
A-2	Axial flow four blades	Constant angle at 45°	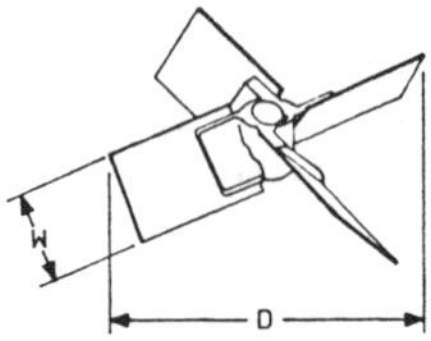 W=1/5D ∡=45°
A-3	Axial flow three blades	Variable blade angle, near constant pitch	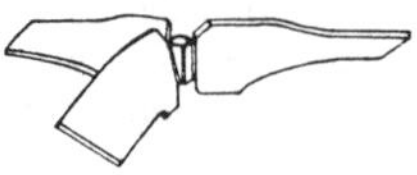 BLADE ANGLE AND WIDTH DECREASES HUB TO TIP
A-4	Double spiral	Two helical flights, pitch = $\frac{1}{2}D_0$	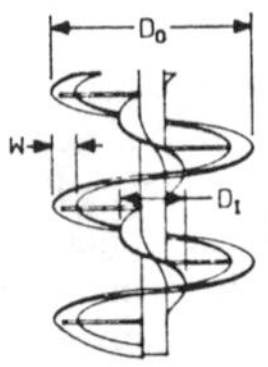 D_0=(OUTER) D_I=1/3D_0 W=1/6D_0

From Oldshue [27].

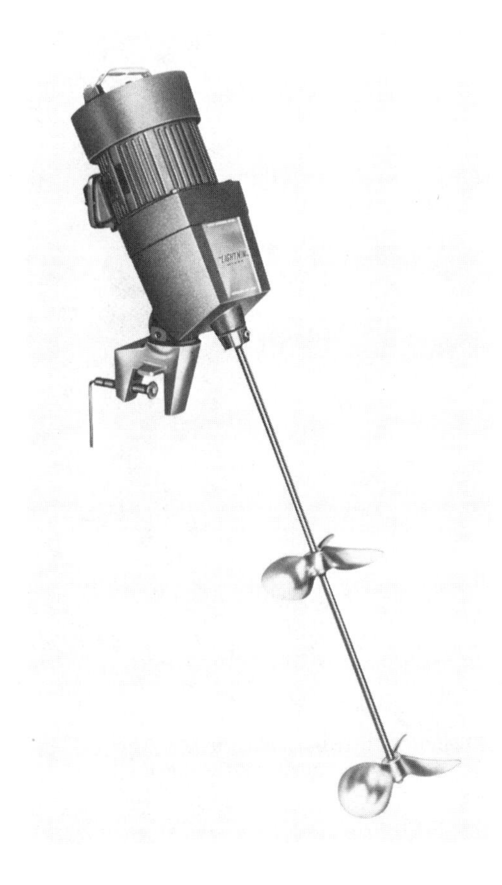

Figure 2. Portable mixer (courtesy of Mixing Equipment Co., Inc.).

A-1 impellers are also used on side-entering mixers from 1 through 75 horsepower. Impeller diameters range from approximately 10 to 33 inches. Side-entering mixers are mounted on a tank flange with the impeller shaft positioned at an angle to the tank centerline (Figure 4). Correct positioning is important for optimum process results (Figure 5).

The A-2 impeller is a 45° pitched-blade turbine normally supplied with four blades. It is available for the entire range of top-entering mixers, from 1 to 500 horsepower (Figure 6). For closed-tank installations, or open tanks with small openings, the blades are bolted to the hub, which allows impeller assembly inside a vessel having a small opening. Where tank openings are large enough, the impeller may also be supplied as an allwelded construction. A-2 impellers are normally used for flow-controlled applications.

The A-3 impeller produces greater flow and less fluid shear than the A-2 impeller. It was developed to perform as an A-1 impeller but without the physical limitations of the latter, which include excessive weight (which limits shaft design) higher cost, and a fixed, one-piece, cast construction. The pumping surface of the A-3 impeller was developed to approximate the blade angles of the A-1 impeller. Thus, near the hub, the blade width is wider and the blade angle steeper than at the tip. In progressing from hub to tip, blade width and blade angle decrease, providing a nearly constant pitch across the blade. This produces an almost uniform velocity profile across the entire discharge area.

Figure 3. Fixed-mounted mixer (courtesy of Mixing Equipment Co., Inc.).

Figure 4. Typical side-entering mixer (courtesy of Mixing Equipment Co., Inc.).

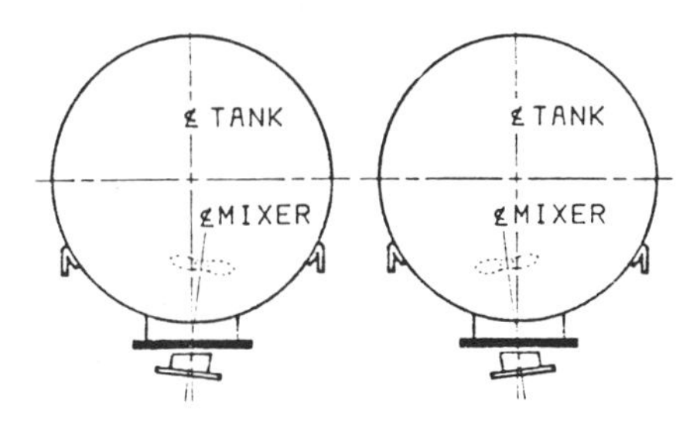

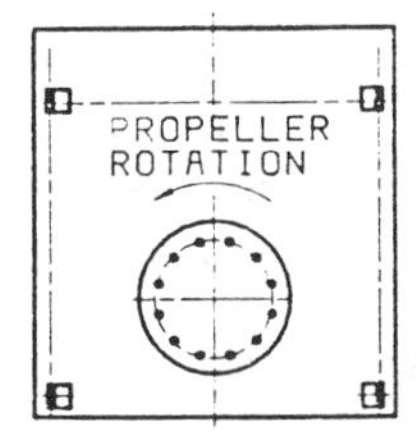

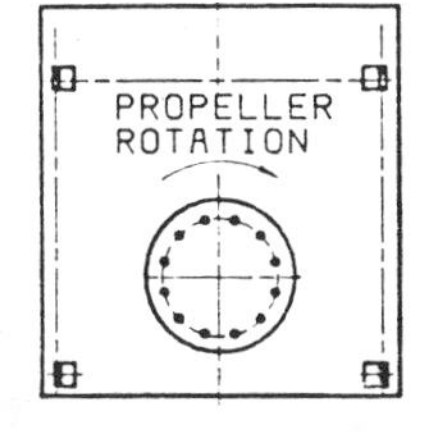

Figure 5. Correct positioning for side-entering mixers [27].

Figure 6. Typical top-entering mixer (courtesy of Mixing Equipment Co., Inc.).

The A-4 double helix impeller is used in highly viscous fluids exceeding the maximum allowable viscosity for anchor impellers. It is constructed with an inner flight which pumps downward and an outer flight which pumps upward. It is applied in the 1- to 250-motor horsepower range at speeds from 5.5 to 45 rpm and diameters from 20 to 120 inches. The relatively low speeds required by the helix means higher torque at a given power level, thus larger speed reducers.

High-Efficiency Impellers

Recently, efficient use of applied power has become more important than ever before. With energy shortages, increasingly higher energy costs and a worldwide concern with energy conservation, equipment suppliers have concentrated their efforts on producing highly flow-efficient impeller designs.

New analytical tools, hitherto unavailable, have been applied in analyzing impeller design characteristics and their relationships to energy efficiency. For example, the laser has been used to measure fluid velocities and fluid velocity components which were previously undetectable due to

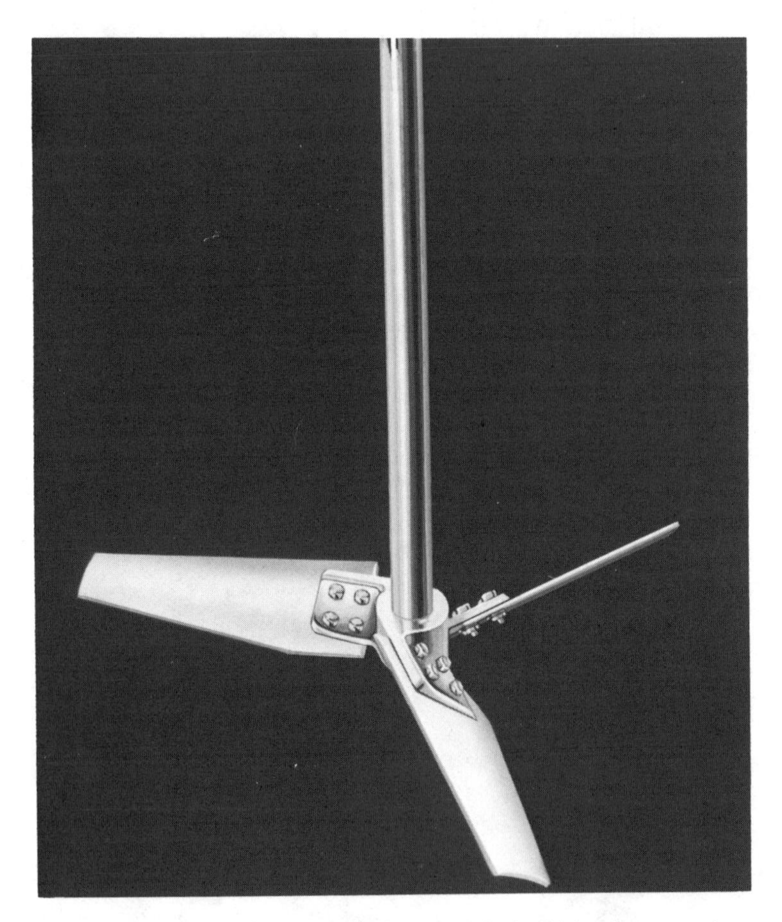

Figure 7. High-efficiency axial-flow impeller. Optimized design higher flow per unit power than traditional axial-flow impellers (courtesy of Mixing Equipment Co., Inc.).

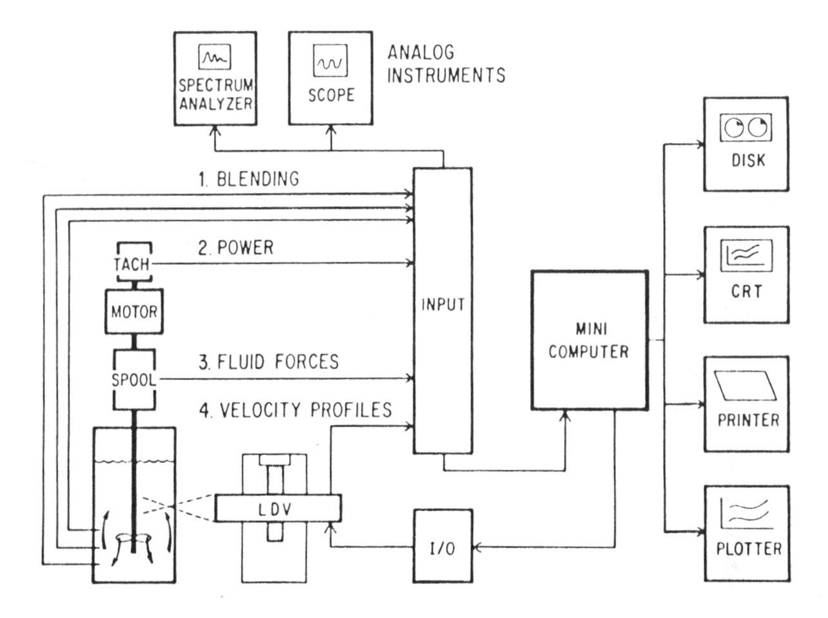

Figure 8. Schematic of integrated laboratory with computer-controlled laser anemometer and analog devices for measuring the mechanical properties of the shaft and impeller.

the limitations of other analytical equipment. As a result, a new generation of impeller designs with unique characteristics is being developed (Figure 7).

Furthermore, an integrated laboratory (Figure 8) using a computer-controlled laser velocimeter and analog devices for measuring mechanical properties of the shaft and impeller has been used to develop totally new systems. Weetman and Salzman [1] described the technique in which the fluid in the vessel is scanned by intersecting laser beams. Signals are received by the computer which plots velocity vectors, integrates them to determine flow through a given envelope, then differentiates them to determine average and maximum fluid shear rates (refer to the section on scale-up). Simultaneously, the equipment measures and displays bending moments, energy input, and average and transient torque. Figure 9 illustrates the velocity vectors of an impeller running in water. These results, obtained from a laser-doppler velocimeter, represent the time-average flow. The velocity vector lines indicate both magnitude and direction. The laser technique allows the study of many different impeller designs in a very short period of time. The net result has been optimized impeller designs in terms of flow efficiency and power consumption.

Impeller Function

The function of any mixing impeller is to provide flow, head, and shear characteristics consistent with the demands of a given operation or process. The majority of industrial mixer applications are more sensitive to flow than impeller head. (In mixing technology, the terms *shear* and *head* may be used interchangeably). Rotating mixing impellers transmit mixing energy in both forms (flow and head). The portion devoted to either of these two major components of mixing is controlled by impeller design.

The many different impeller designs which have evolved over the years can be classified in a spectrum according to the type of flow they produce (Figure 10). Furthermore, a second impeller spectrum shows the classes of processes related to specific impeller types (Figure 11) which we call

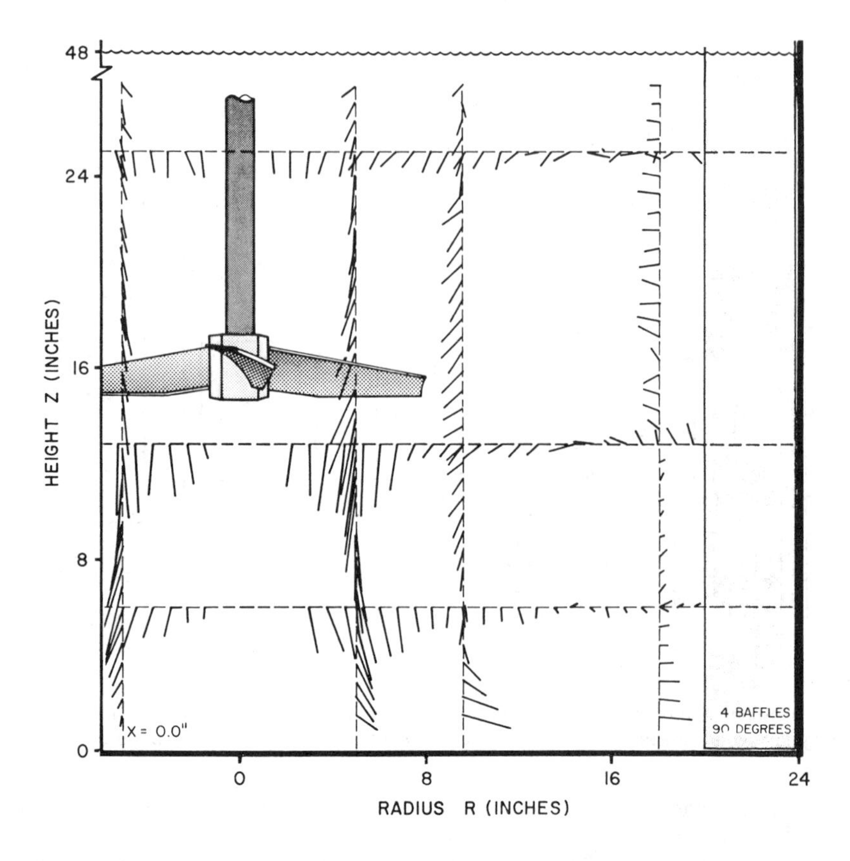

Figure 9. Computer-generated display showing velocity vectors for an impeller running in water.

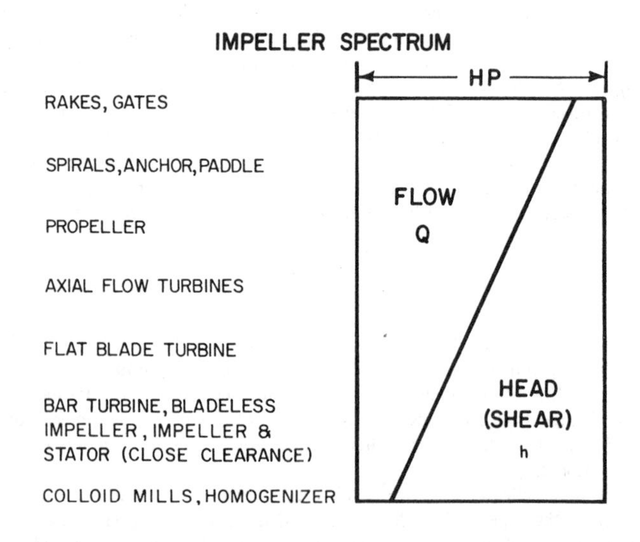

Figure 10. The impeller spectrum. Power is split between head and flow development. The correct impeller can be selected if the specific process requirements are known.

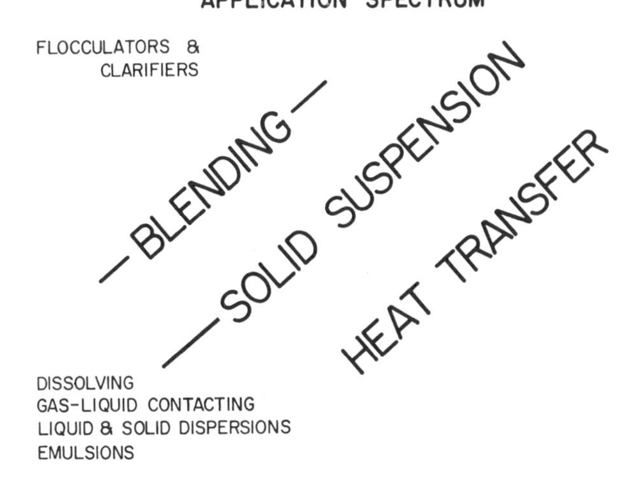

Figure 11. Classes of processes are related to specific impeller types.

the application spectrum. We also associate specific impeller types with equipment size, power, retention time, flow, and shear as shown in Figure 12.

Three factors are most important in designing a mixing impeller:

1. Impeller size (diameter) relative to vessel size.
2. Impeller speed, RPM.
3. Impeller geometry (number, width and angle of blades, pitch, etc.).

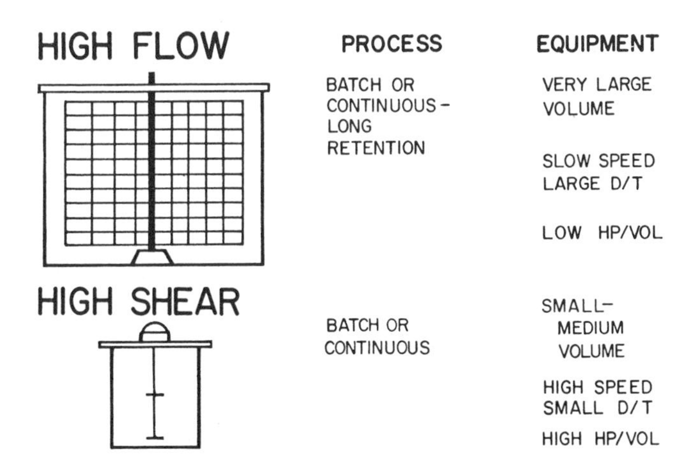

Figure 12. Associating impeller types with equipment size, power, retention time, flow, and fluid shear.

These factors are interrelated. All of them are necessary to define the mixer input to the system. In general, very large, low-speed impellers, which essentially span the entire tank, are used in viscous or hard-to-move fluids. Here, a direct "pushing" action is required in order to ensure the entire tank contents are subjected to a positive blending action. The other end of the spectrum uses thin-bladed, small diameter impellers at very high speeds (tip speeds of 5 or 6 thousand feet per minute, for example, for paint pigment dispersion and overall circulation.)

Comparing impellers for a given process application is usually based on the power input requirement. Blending very viscous materials is usually accomplished with a large, low-speed impeller but it could also be done with a small high-speed impeller at a *considerable* additional expenditure of power. The small, high-speed impeller produces flow with high shear which is not required for viscous blending, thus energy would be wasted. The reverse is also true. For example, a large impeller primarily designed for flow production would waste an enormous amount of power in trying to meet the shear rate level required for pigment dispersion.

Both flow and shear rate are required in every operation, but the proper *balance* of the two is necessary for the best impeller design. Thus a large impeller produces the low shear rate level required for blending and the high mass flow necessary for circulation, and a smaller dispersing impeller provides the high-intensity fluid shear for dispersion, plus sufficient flow, in order to circulate all of the material through the high-intensity zone of the impeller.

Flow Patterns

The first demand on the impeller is flow production. The flow pattern produced is a function of both the impeller and vessel designs. A rotating impeller with vertical blades produces radial flow by centrifugal force. Inclining the blades at an angle to the horizontal will produce axial flow components, useful for drawing flow from the liquid surface or directing flow against the bottom of the vessel for solids suspension.

A rotary flow pattern (swirling) has little radial or vertical flow components. But, if vertical wall baffles are installed, they produce radial components, producing vertical streams for blending. The resulting discontinuities in the flow produce the highest overall shear rate in the system.

Tank Design

It is not always possible to optimize the mixer design for a given process if the tank design is fixed. Tank geometry plays a very important role in the final mixer design. In some cases it might be less costly in the long run to begin with the flexibility of considering various tank designs rather than designing a mixer around a fixed tank geometry.

In most applications, the best tank design results in a "square batch." This refers to the tank diameter being equal to the static liquid depth ($Z/T = 1.0$). A square batch allows an optimum distribution of power and optimum power input in most cases. However, if conditions do dictate a design other than a square batch, the minimum normal operating level should be about one-third of the tank diameter. Where $Z/T > 1.0$, multiple mixing impellers are usually required.

Tank bottoms are either flat or dished but conical and spherical bottoms are also used. A dish-bottom tank is normally preferred but flat-bottom tanks may also be used. However in solids suspension operations, it may be more economical to provide a small fillet in a flat-bottomed tank. Often, power savings will accrue as a result of fillet formation (Figure 13).

Baffle Design

Baffles are normally flat, vertical plates equally spaced, and fixed to the wall of the vessel (Figure 14). Without baffles, swirling and vortexing may result with very little blending accomplished. An unbaffled fluid has virtually no vertical currents or velocity gradients, both of which are necessary for most mixing processes. Furthermore, vortex formation invariably produces undesirable air entrainment.

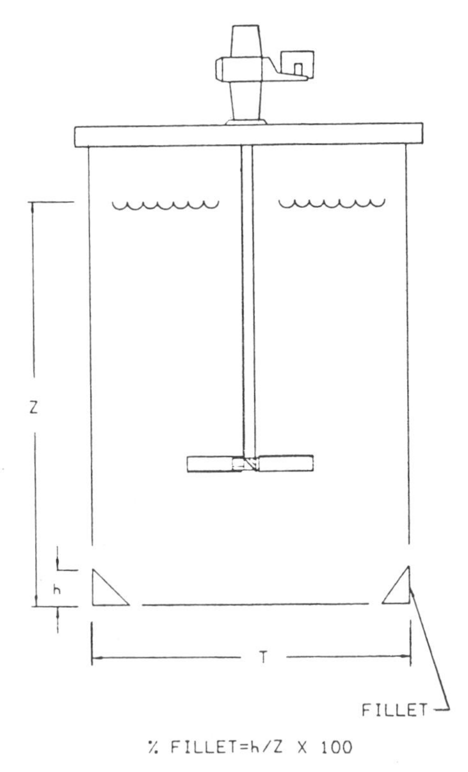

Figure 13. A fillet may reduce power requirements for solids suspensions [27].

Proper baffling produces a flow pattern which is carried throughout the batch, assuring adequate mixing. However, *excessive baffling* localizes the mixing and reduces mass flow, resulting in poor overall performance. Therefore a given baffle design cannot be used indiscriminantly.

The correct baffle design is largely a function of process viscosity (Figure 15). Viscous drag increases as fluid viscosity increases, reducing the baffle width required for proper loading and flow development. In Figure 15, baffle width is shown as a function of tank diameter, T, and as a percent of *normal* baffle width. Normal or *standard* baffles are those most frequently used for lower viscosity

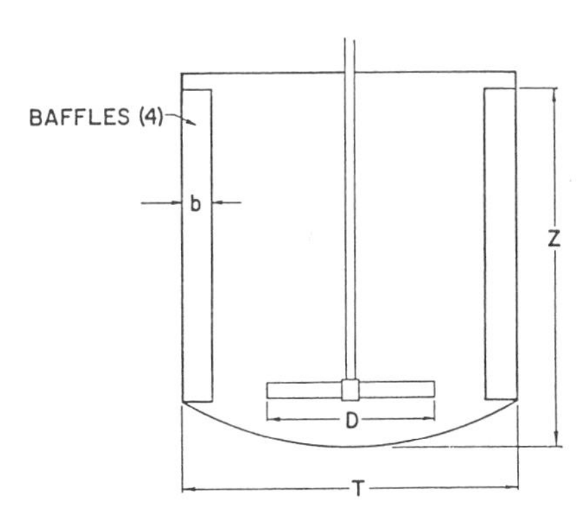

Figure 14. Typical baffle configuration in cylindrical vessels.

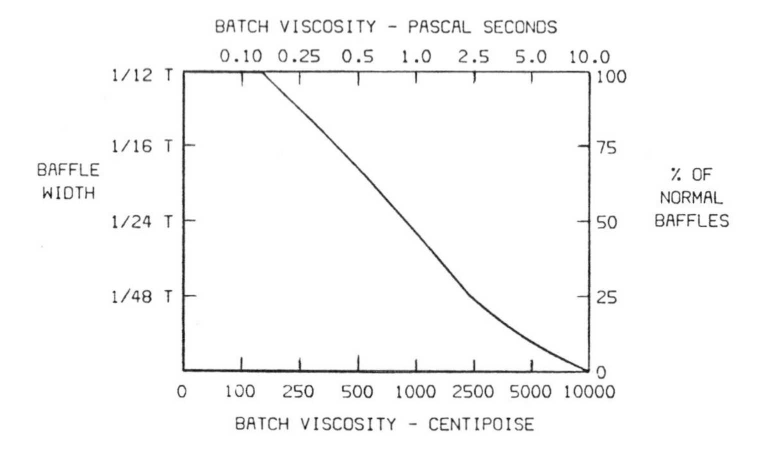

Figure 15. Baffle width is a function of fluid viscosity [27].

applications. They are vertical rectangular plates, four in number, equally spaced and one twelfth the tank diameter in width. Offsetting standard baffles a third of the baffle width from the vessel wall for low viscosity fluids, or half the baffle width for higher viscosity fluids, permit the fluid flow to scour the area behind the baffles. This is especially important when product carry-over from one batch to the next may be detrimental.

In some cases, baffles are not required at all. For example, in using portable mixers (Figure 16) or side-entering mixers (Figure 5) developing a good mixing flow pattern is solely a function of properly positioning the mixer. In viscous fluids (viscosities > 50,000 cP) baffles are usually not required since the overall blending flow patterns and vertical flow currents can be obtained without them. Here, the viscous drag of the fluid acts as a dynamic baffle. In highly viscous systems, the presence of baffles, internal tank fittings, and impeller stabilizing fins may actually impede the

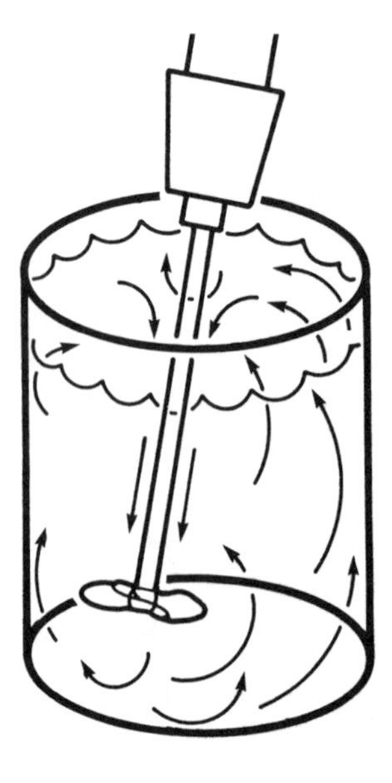

Figure 16. Correct mounting position for portable mixer which allows mixing without the need for baffles.

mixing process. Conversely, baffles *are* required for high intensity, low viscosity mixing to develop localized turbulence for high shear stresses.

In a rectangular tank, the corners of the tank act as baffles to disrupt swirling, producing desirable blending currents in lower viscosity fluids. Because of this natural baffling effect, highly viscous materials should *not* be mixed in a rectangular tank.

Mixing Operations

Economical equipment design and selection for any mixing operation depends on specifying the desired process result. Whenever possible, future process modifications should also be considered which will allow added capacity and flexibility. Many of the common mixing requirements are briefly stated in the following paragraphs.

Blending

When combining suspensions or solutions to form a continuous uniform material, the final viscosity of the blended materials and the inlet stream viscosities determine the mixing intensity required. Flow throughout the system is the primary objective, which requires propellers or axial turbine impellers for low-viscosity systems (under 500 cP). For higher viscosities, the impeller size increases for a process-effective, economical selection.

Solids Suspension

Solids must be kept in suspension to provide uniform storage feed and to prevent solids from settling. Fluid flow in the system is the primary requirement. Fluid velocities must be maintained in excess of the settling rate of the solids. For nonpacking, slowly settling solids, intermittent mixer operation may be adequate for long-term storage. A typical specification for solids suspension includes uniformity within a specified time, for resuspension, or maintaining uniform solids concentration.

Solids Dispersion

Wetting solids to form a suspension in a fluid is called solids dispersion. Rapid wetting of solids requires vigorous surface flow. It includes reducing the size of agglomerated particles. High intensity mixing (high fluid shear rates) may be required to further reduce particle size, accompanied by adequate flow capacity to recirculate the fluid through the high intensity mixing zone of the impeller. Specifications include wetting solids within a specified time and dispersing to a desired particle size within a specified time.

Heat Transfer

In order to increase the heat transfer rate, fluid is circulated over a heat transfer surface (e.g., a coil or jacket) to keep the surface from fouling and to avoid localized overheating or overcooling. Heat may also be transferred by directly dispersing steam to maintain a uniform batch temperature. A typical specification includes heating or cooling over a specified temperature range in a prescribed time. The viscosity of the fluid film at the heat transfer surface must also be stated. For direct steam addition, the maximum steam rates must be known.

Fluid Motion

In storage systems maintaining fluid motion in every part of the batch prevents stagnation, limits surface evaporation and eliminates surface crusting. A statement indicating the process objective, flow, viscosity, and temperature characteristics must be included in the specification.

Figure 17. Static-type inline mixing device. Mixer internals remain stationary while flow passes through alternately pitched mixing chambers.

Continuous Processing

A continuous supply of uniform product is required for high production rates. Improved process control and savings in direct labor can be attained by translating batch systems to continuous operations. In a continuous system, major savings accrue by eliminating both between-cycle cleanings and filling and emptying times required in a batch system. Materials must be held long enough in a continuous system to complete the operation. Frequently, staged operations for retention time control are necessary. Some continuous operations may also be achieved by mixing in a pipeline. Recently, mixing devices have been developed in which process fluids or suspensions are easily mixed by bolting a static type mixer into the line. The power required for mixing is derived from the energy supplied by a process pump, gravity, or pressure source (Figure 17).

SCALE-UP OF FLUID MIXING DEVICES

To view scale-up of a fluid mixing process objectively requires us to realize that a large vessel is not identical to a small vessel with respect to most of the mixing parameters and fluid mechanics parameters which may be involved in determining process results, (Figures 18 and 19).

To scale up the production rate from a small mixing vessel, either a large number of small vessels, or a few larger vessels can be used. Figure 18 illustrates that a large number of small vessels, identical to the original vessel, will maintain all the mixing parameters constant.

Figure 19 shows that as we increase vessel size we have greater distances to traverse in the fluid; some scale ratios increase linearly and some increase with the square or the cube of linear dimensions. A combination of the fluid mechanics criteria considered for Figure 19 is not the same for a large number of small vessels as in Figure 18.

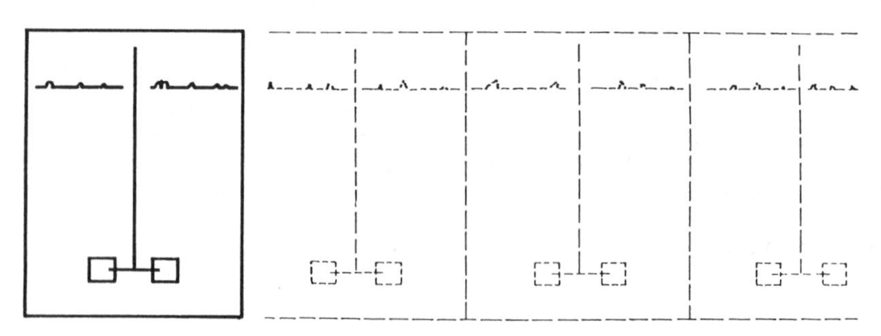

Figure 18. Schematic of pilot-scale tank, illustrating identical mixing parameters when using a large number of small tanks.

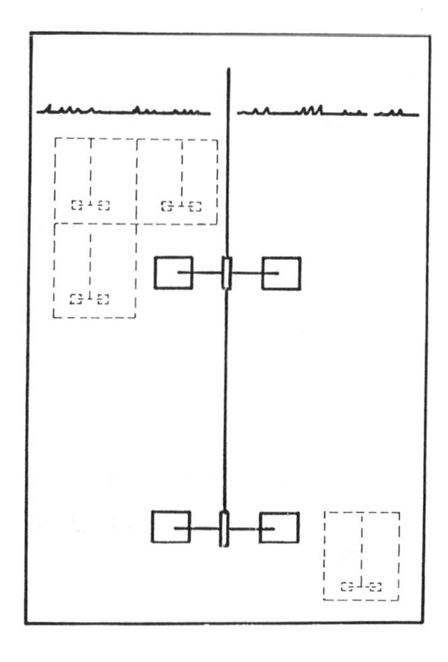

Figure 19. Full-scale mixer tank, showing that individual mixing parameter values in a large tank are usually different from corresponding values in small tanks.

This means we must have an appreciation of the ways in which the larger vessel is different from a number of smaller vessels, which requires an understanding of some of the basic principles of fluid mixing in the role of process performance.

Ultimately, we must determine a mixer selection for the larger vessel, for which we must specify the power, type of impeller(s), impeller diameter(s) speed, and many other parameters which may affect the process. Normally there will be more than one mixer which will satisfy the process requirement, and these will differ in capital and operating costs. The process performance may be expressed in quantitative, nonambiguous terms, such as a mass transfer rate (e.g., lb. mols of component A per hour to be transferred in the fluid under specified operating conditions). On the other hand, the process may be affected by many different characteristics of the fluid mixer, some of which may not be quantitatively definable.

For example, in making a polymer successfully, the polymer product specification may include specific tensile strength, brittleness, dielectric constant, yield stress, and particle size distribution in the powder form. Since the role of mixing may not be completely understood as it effects these parameters, the process description may be fairly general relating to the uniformity of various components and/or the physical dispersion or suspension of the chemical components of the system. This makes a qualitative understanding of the differences between the pilot plant and the full-scale plant, and their possible effects, much more important.

Batch-versus-continuous operation is another area of important concern since the pilot plant may be run batchwise, while the full-scale plant may be either batch or continuous.

General Principles of Impeller Fluid Mechanics

The power supplied to the mixer impeller shaft appears as heat in the batch at the rate of 1 Joule/sec/watt. The power in the fluid stream can be expressed as a product of the pumping capacity and the velocity head. In SI units, pumping capacity is normally expressed in kilograms per second, and the energy in the fluid velocity is expressed as Newton-meters per kilogram or Joules/kilogram. In the English system, pumping capacity is usually expressed in lb. per hr. and velocity head, $V^2/2_g$, is expressed in feet. The height of a fluid of the same density, having the same equivalent potential energy as the velocity of the flowing stream, can be calculated.

Since the moving fluid in the mixing tank is not confined in a channel, there is considerable arbitrariness in determining flow from the impeller, therefore in calculating the velocity head from the measured power.

Actually, these concepts are more useful in a relative form to predict changes from one scale to another or from one impeller and vessel geometry to another. However, there is seldom a need to calculate, with high precision, the actual pumping capacities and velocity heads in the streams discharging from the impeller.

The result of this discussion can be expressed in the impeller power consumption equation:

$$P = QH \quad (1)$$

where Q = the pumping capacity of the impeller, kg/s
H = the velocity energy of the flowing fluid, Nm/kg

There are two other relationships which are helpful in a general discussion of fluid mechanics in the mixing tank.

In the medium and low viscosity regions, flow varies directly with the impeller speed and with the cube of the impeller diameter as shown in the following equation:

$$Q = KND^3 \quad (2)$$

where K = a different constant for each impeller type
N = impeller rotational speed
D = impeller diameter

Figure 20 shows the relationship between the power number, Np, and the Reynolds number, N_{Re}. The power number vs Reynolds number curve can be divided into three regions—laminar, transitional, and turbulent flow. In the turbulent region the power number is constant for a given

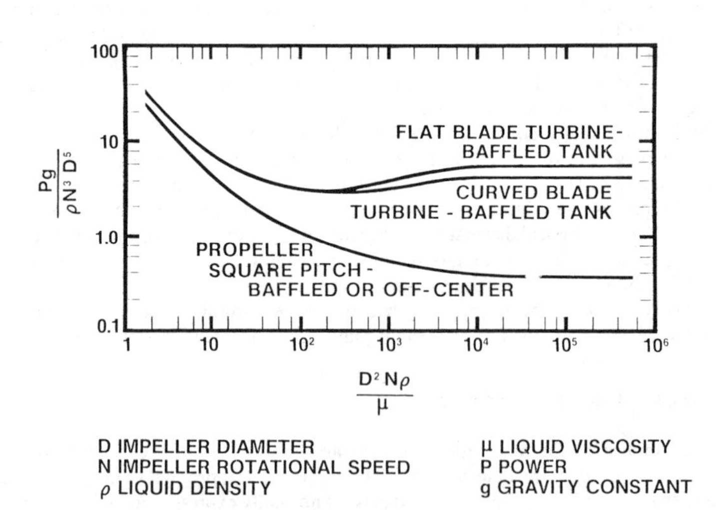

Figure 20. Power-number/Reynolds-number curve for three different impellers.

impeller. We can express the power consumption of any impeller as

$$P = K'N^3D^5 \tag{3}$$

We will use this relationship for the next interpretation, although it is obvious that the exponent on N can be anywhere from 2.0 to 3.2, and the exponent on D can be anywhere from 3.0 to 5.2. Using these three equations, we can determine at constant power, that the flow-to-head ratio is proportional to diameter to the 8/3 exponent.

$$(Q/H)_P \propto (D)^{8/3} \tag{4}$$

In the transition and laminar regions, exponents vary, such that Q/H at constant power can vary between $D^{7/3}$ and $D^{9/3}$.

We can use the exponent 8/3 as being a reasonable approximation for a wide variety of fluid viscosities.

This shows if we want to provide large pumping capacities at relatively low velocity heads we should use large diameter impellers at a given power level, which necessitates a low speed. Conversely, small diameter impellers at high speeds are required at a given power level if we need to produce high impeller heads to satisfy the process requirement.

The velocity head is related to the fluid shear rate of the impeller. As velocity head increases, fluid shear rate also increases. The functional form of the relationship is somewhat complex, so it is better to think of Equation 4 as indicating that the flow-to-fluid shear rate is proportional to D/T to a positive exponent, where T is the vessel diameter.

All of these equations hold true for a given impeller type, but become much more complicated and approximate when trying to compare different impeller types such as radial and axial flow turbines or propellers.

Fluid Shear Rates

If the discharge velocities from a radial flow turbine blade are measured, we get the velocity profile shown in Figure 21. The slope at any point is a measure of fluid shear rate at that point. The fluid shear rate is a velocity gradient and has the units of reciprocal time. For example, a velocity in meters per second, divided by distance in meters, gives us a shear rate of reciprocal seconds.

A key consideration is that the fluid shear rate from the impeller must be multiplied by the viscosity in the fluid to obtain the shear stress which actually promotes the mixing process. Even at low viscosities, e.g., from 1 to 5 cP., the shear stress resulting from a given impeller at a constant shear rate varies by a factor of 5.

At any point in the fluid in the mixing vessel there can be various degrees of velocity fluctuation. If the flow is fully turbulent, the velocity versus time at a point is as represented in Figure 22. The average velocity can be used to obtain the shear rate between adjacent layers of fluid, or the velocity fluctuations can be used to estimate the intensity and scale of the high frequency shear rates in the

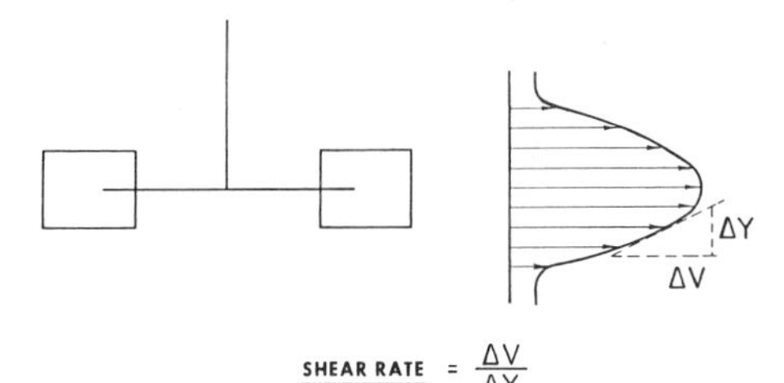

Figure 21. Typical velocity patterns from the blades of a radial flow turbine showing the shear rate, $\Delta v/\Delta y$.

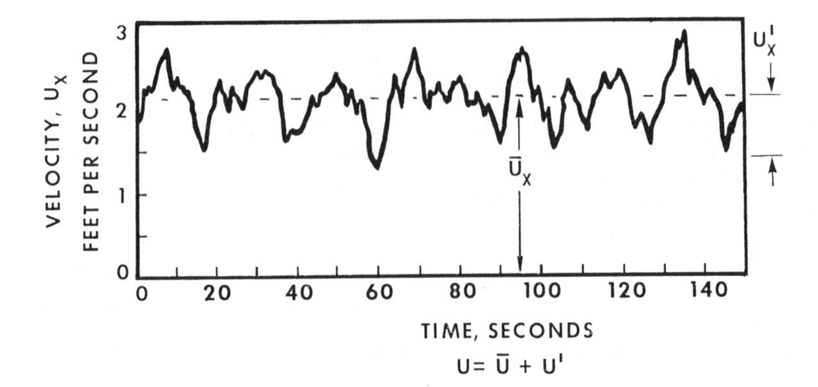

Figure 22. Velocity measurement at a point with time, showing velocity fluctuations and average velocity.

velocity fluctuations. The fluid shear rate between the average velocities represented by Figure 21 is a very important factor in macro-scale processes involving large particles. By comparison, fluid shear rates corresponding to the velocity fluctuations are inherent in micro-scale processes. Velocity fluctuations are often expressed as root-mean-square (RMS) values.

There are several different interpretations relative to macro-scale and micro-scale mixing in small volumes. It has been variously estimated that somewhere in the vicinity of 100–500 microns, the transition between macro-scale and micro-scale mixing occurs.

Probably 70% to 80% of industrial mixing applications are more sensitive to the pumping capacity of the impeller in the fluid (flow-controlled applications) than to various fluid shear rates. In flow-controlled applications, we need not be concerned about shear rates and their distribution. However, some processes do require a complete understanding of the fluid shear rate effects. In those cases it is necessary to consider, for macro-scale shear rates, at least four different points in the vessel, as shown in Table 2, and for micro-scale mixing, the RMS values in several different regions in the fluid.

Table 2
Types of Shear Rates Encountered in the Impeller Zone and Bulk Fluid*

AVERAGE POINT VELOCITY
MAX. IMP. ZONE SHEAR RATE
AVE. IMP. ZONE SHEAR RATE
AVE. TANK ZONE SHEAR RATE
MIN. TANK ZONE SHEAR RATE
RMS VELOCITY FLUCTUATIONS
$\sqrt{\overline{(u')^2}}$

* *Definition of RMS velocity fluctuations.*

Table 3
Three Dimensionless Groups Relating Four Fluid Forces*

FORCE RATIOS

$$\frac{F_I}{F_v} = N_{Re} = \frac{ND^2\rho}{\mu}$$

$$\frac{F_I}{F_g} = N_{Fr} = \frac{N^2D}{g}$$

$$\frac{F_I}{F_\sigma} = N_{We} = \frac{N^2D^3\rho}{\sigma}$$

* *F_i—inertia; F_v—viscosity; F_g—gravity; F_σ—surface tension.*

Fluid Forces

Dynamic similarity—In terms of dynamic similarity there are four fluid forces in a mixing tank which should be considered (Table 3). One is the inertia force applied by the mixer, F_I; the other three are opposing forces resisting the accomplishment of the process. They are the viscous force, F_v, gravitational force F_g, and surface tension force, F_σ. The ratio of inertia force applied by the mixer to the opposing force of viscosity is the Reynolds number N_{Re}; to the opposing force of the gravity, the Froude number, N_{Fr}, and to the opposing force of surface tension the Weber number, N_{We}.

There is another group of dimensionless ratios (Table 4), in which the inertia forces of both the model and prototype must be equal to the ratios of the viscous forces, the gravitational forces, and the surface tension forces for dynamic similarity. However, using the same fluid in both the pilot plant and full-scale plant can only satisfy two of these relationships. If we use the inertia force of the mixer as one of them, we can then work with either the opposing forces of viscosity or gravity or surface tension.

In many problems in civil and mechanical engineering calculating fluid forces is the ultimate goal of experimentation, and it is possible to use different fluids to model several of these

Table 4
Definition of Dynamic Similarity*

HYDRAULIC SIMILITUDE

GEOMETRIC $\frac{X_M}{X_P} = R_X$

DYNAMIC $\frac{(F_I)_M}{(F_I)_P} = \frac{(F_\mu)_M}{(F_\mu)_P} = \frac{(F_g)_M}{(F_g)_P} = \frac{(F_\sigma)_M}{(F_\sigma)_P} = R_F$

* *The ratios of four fluid forces (inertia, viscosity, gravity, surface tension) are related for a model (M) and prototype (P).*

Table 5
Relationship of Impeller Power Characteristics on Scale-Up*

IMPELLER POWER CHARACTERISTICS

$$P = f\left[D, N, \rho, \mu, g, T\right]$$

$$\frac{\text{APPLIED FORCE}}{\text{FLUID ACCELERATION}} = f\left[\frac{\text{APPLIED FORCE}}{\text{RESISTING FORCE}}\right]$$

$$\frac{Pg}{\rho N^3 D^5} = \left[\frac{ND^2 \rho}{\mu}\right]^x \left[\frac{N^2 D}{g}\right]^y \left[\frac{D}{T}\right]^z$$

* *Power number is related to Reynolds number, Froude number, and D/T ratio.*

dimensionless ratios. However in a mixing system, that luxury is not available; usually the same process fluid must be used in the pilot plant and full-scale plant.

Using the dimensionless groups technique produces some examples which give excellent correlation. For example, referring to Table 5, it can be estimated that the power drawn by a mixer is a function of the variables listed. The basic premise can be established that the inertia force divided by the fluid acceleration, (the power number), should be a function of the ratio of inertia force to viscous force, which is the Reynolds number. When the data are plotted (shown as smoothed curves in Figure 20) we get excellent correlations with all mixing impellers, vessel sizes, and fluid properties.

Using a dimensionless group to describe a process, we can, for example, relate the heat transfer coefficient to the thermal conductivity of the fluid with a length term to complete the dimensionless number (Table 6). This correlates very well with the Reynolds number as shown in Figure 23.

Similarly with blend time (Table 7) a dimensionless blend number, θN, is seen to be a function of the Reynolds number, shown in Figure 24.

However, in approaching the ordinary mixing process, of which several thousand are handled each year by mixer manufacturers, there is no easy way to express the process result in the form of a dimensionless group. Thus, dynamic similarity and dimensional analysis do not usually apply to mixing processes.

Table 6
Relationship for Dimensionless Group Correlation of Heat Transfer*

APPLICATION OF HYDRAULIC SIMILARITY TO HEAT TRANSFER

$$h = f\left[N, D, \rho, \mu, C_p, k, d\right]$$

$$\frac{\text{PROCESS RESULT}}{\text{SYSTEM CONDUCTIVITY}} = f\left[\frac{\text{APPLIED FORCE}}{\text{RESISTING FORCE}}\right]$$

$$\frac{hd}{k} = \left[\frac{N D^2 \rho}{\mu}\right]^x$$

* *Includes Nusselt number, Reynolds number, Prandtl number, and ratio of impeller diameter to the tube diameter of the heat transfer surface.*

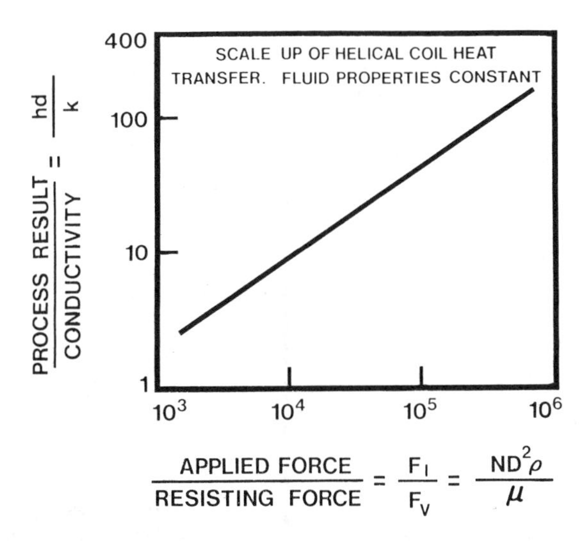

Figure 23. Heat transfer process dimensionless group, hd/k, as a function of Reynolds number.

Table 7
Relationship for Blend Number vs. Reynolds Number and D/T Ratio

APPLICATION OF HYDRAULIC SIMILARITY TO BLENDING

$$\theta = f\,[N, D, \rho, \mu, T]$$

$$\frac{\text{RESULT}}{\text{SYSTEM CONDUCTIVITY}} = f\left[\frac{\text{APPLIED FORCE}}{\text{RESISTING FORCE}}\right]$$

$$\theta N \propto \left[\frac{ND^2\rho}{\mu}\right]^X \left[\frac{D}{T}\right]^Z$$

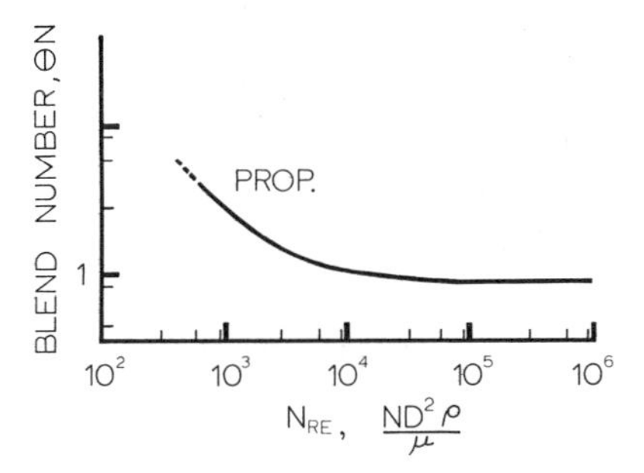

Figure 24. Blend number, θN, as a function of Reynolds number.

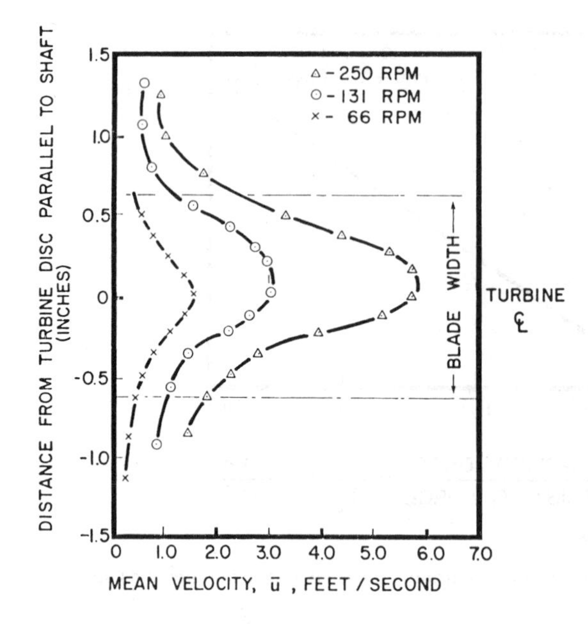

Figure 25. Experimental data showing mean velocity at various positions above and below the impeller centerline.

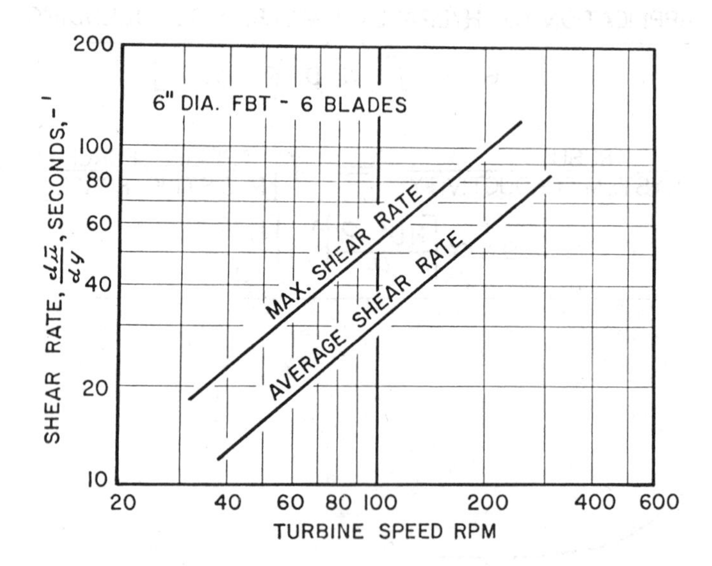

Figure 26. Correlation of maximum impeller zone shear rate and average impeller zone shear rate with impeller speed (smoothed data).

The procedure used for mixing processes is to first obtain data on one scale which is extensive enough to give some indication of the controlling factor or factors in the process. Thus, a particular relationship can be ratioed to the expected full-scale performance relationship.

Fluid Shear Rate Relationships

From the velocity fluctuation in Figure 22, the mean velocities at each point are calculated and plotted as a function of the velocity probe positions above and below the impeller centerline for a radial flow, flat blade turbine. The results are given in the curves shown in Figure 25. Taking the slope at any point gives the fluid shear rate, and we can calculate the average and maximum fluid shear rates in the impeller zone.

Using data for three different impeller speeds shows that the maximum and average fluid shear rates are functions of impeller speed (Figure 26).

In repeating this series of studies on different impeller sizes, the most significant finding is that the impeller zone *average* fluid shear rate *does not* change with impeller diameter at the same speed, but the *maximum* fluid shear rate *does* change with impeller diameter at constant speed as shown in Figure 27. This leads to the concept shown in Figure 28, which illustrates that the maximum fluid shear rate increases on scale-up while the average fluid shear rate decreases on scale-up. This means there are a greater variety of fluid shear rates in large tanks than in small tanks, which accounts for one of the differences in the overall performance on different scales.

From the root-mean-square velocity fluctuations a plot of RMS as a function of position above and below the impeller centerline is shown in Figure 29. The ratio of the RMS value to the average velocity at a point can be calculated, producing the curve shown in Figure 30.

This shows that the turbulent intensity is about 50% of the mean velocity in the impeller zone. This is a very high level of velocity fluctuation, compared to most pipeline fluctuations which are

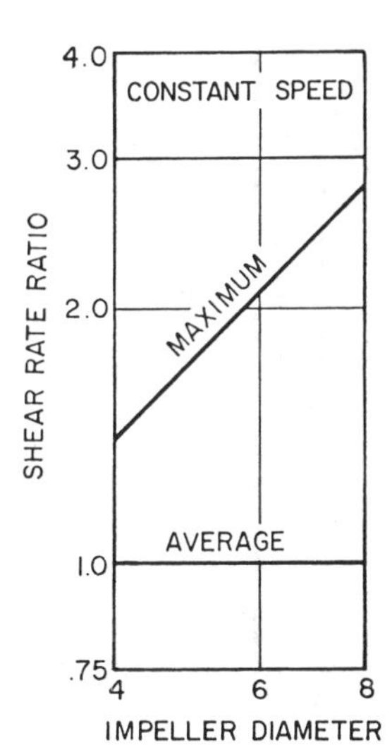

Figure 27. Correlation of maximum shear rate vs. Impeller diameter and average shear rate versus impeller diameter at constant speed.

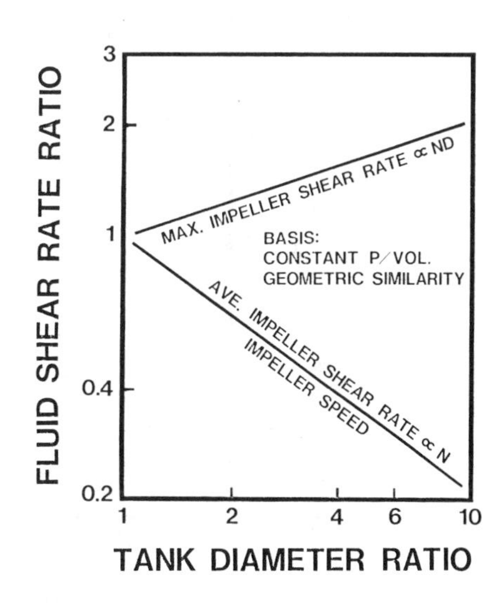

Figure 28. Correlation of fluid shear rates around a mixing impeller as a function of the ratio of tank diameters.

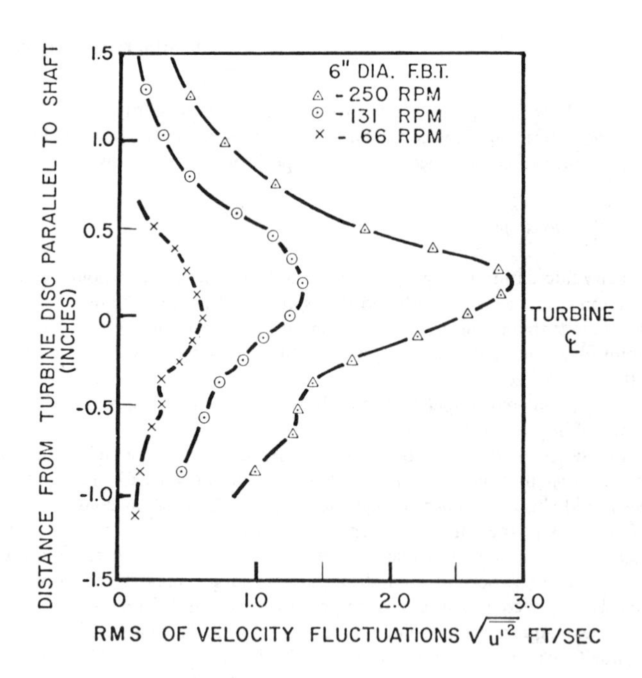

Figure 29. Experimental data for the root mean square velocity fluctuation, above and below the imepller centerline.

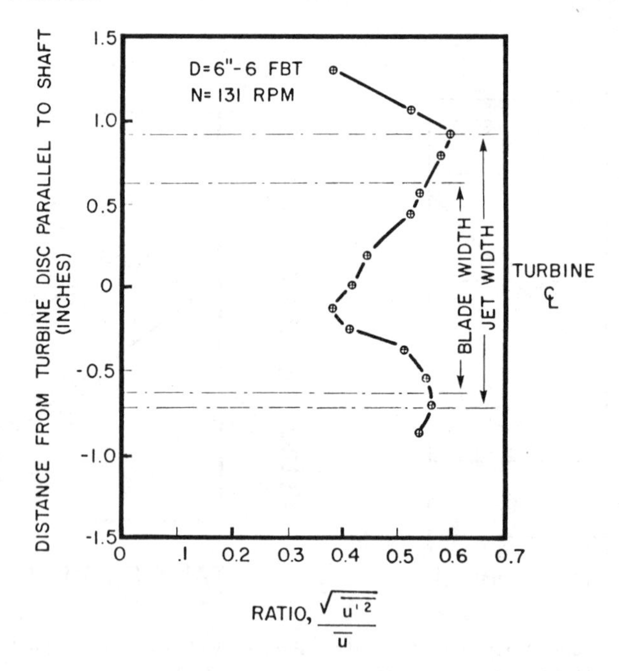

Figure 30. Experimental data for the ratio of velocity fluctuation and mean velocity at a point as a function of position around impeller discharge zone.

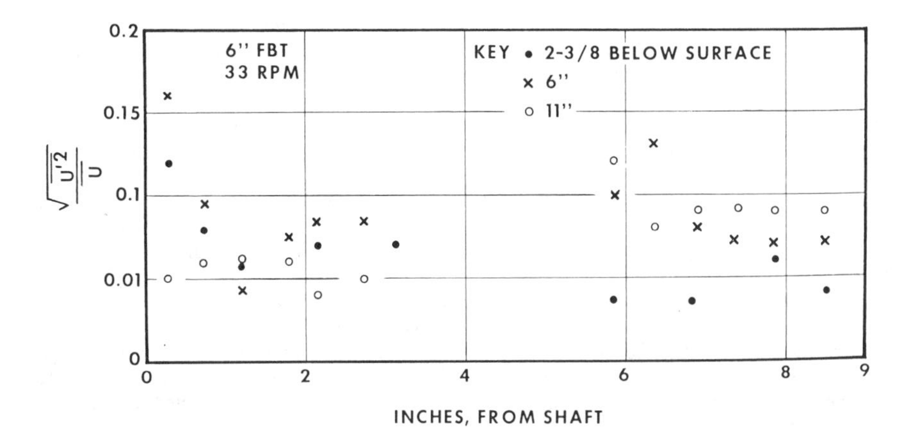

Figure 31. Ratio of RMS value to average velocity as a function of probe position at different locations in the mixing tank, excluding the immediate impeller zone.

on the order of 5% to 15%. Figure 31 shows data from other parts of the tank indicating that the velocity fluctuations are on the order of 5% to 15% in the rest of the batch.

Figure 28 pertains to macro-scale shearing effects and the particle sizes and fluid clumps associated with that mechanism. Figure 30 refers to the micro-scale shearing effects which are mainly a function of the power dissipation per unit volume.

Change in Scale-Up Relationships

Table 8 shows how scale-up factors change from a pilot-scale vessel of 20 gallons capacity to a full-scale vessel of 2,500 gallons capacity, a five-fold linear scale ratio. This assumes a vessel in which a constant D/T ratio is maintained, and the same fluid is used in both vessels. (Z/T is the ratio of static fluid depth to tank diameter.) These are calculated values, assuming four different parameters are held constant as described in the following:

Table 8
Properties of a Fluid Mixer on Scale-Up

PROPERTY	PILOT SCALE 20 GALLONS	PLANT SCALE 2500 GALLONS			
P	1.0	125	3125	25	0.2
P/VOL.	1.0	[1.0]	25	0.2	0.0016
N	1.0	0.34	1.0	0.2	0.04
D	1.0	5.0	5.0	5.0	5.0
Q	1.0	42.5	125	25	5.0
Q/VOL.	1.0	0.34	[1.0]	0.2	0.04
ND	1.0	1.7	5.0	[1.0]	0.2
$\frac{ND^2\rho}{\mu}$	1.0	8.5	25.0	5.0	[1.0]

On the left side of Table 8 are recorded many of the parameters involved in mixing processes; power, power per volume, speed, diameter, flow, flow per volume, tip speed, and Reynolds number. There could also be many other parameters considered. Each is assigned a value of 1 in Column 2, to see how it changes with four different scale-up calculations in Columns 3, 4, 5, and 6. In Column 3, P/vol has been held constant and everything else changes. Similarly, in Column 4 flow per unit volume is held constant; in Columns 5 and 6, tip speed and Reynolds number are respectively held constant.

The general observation is that on scale-up, the ratios of the different parameters change and there is no way to keep all the factors constant when a given scale-up parameter is held constant. These are correlating parameters and it is necessary to determine how they change on scale-up, to successfully satisfy the process.

Thus, a basic question must be asked—is it possible that for every conceivable mixing process there is a constant scale-up parameter which can be used for design?

There may very well be a constant scale-up parameter for a given mixing process. For example, impeller tip speed might be a constant on scale-up which would then determine all the other variables. But if a constant scale-up parameter exists, it may not be easily detected. In general, a practical scale-up usually lies between constant P/ vol as a conservative estimate and constant tip speed as a very unconservative estimate.

CURRENT PRACTICE IN MIXER SCALE-UP TECHNIQUES

Scale-up involves the translation of data on a small scale to production size equipment. Mixing involves many different kinds of operations, and there is no single scale-up rule that applies to all of them. This section summarizes the differences in the many physical and mass transfer mixing parameters exhibited in large tanks and small tanks. It reviews some of the rationale for using different scale-up concepts for various kinds of mixing operations.

In particular, we will look carefully at geometric similarity, dynamic similarity, and dimensionless groups, since they inherently represent the basic tools at our disposal. Their applicability is a key factor in reviewing current practice.

The overall concepts of mixer design are presented in Table 9. Process design involves choosing any two of the three variables of power, P, speed, N, and diameter, D. In considering the process, we have correlations, which describe the mixer's process performance. Those correlations involve

Table 9
Elements of Mixer Design

I PROCESS DESIGN	1 - FLUID MECHANICS OF IMPELLERS
	2 - FLUID REGIME REQUIRED BY PROCESS
	3 - SCALE UP; HYDRAULIC SIMILARITY
II IMPELLER POWER CHARACTERISTICS	4 - RELATE IMPELLER HP, SPEED & DIAMETER
III MECHANICAL DESIGN	5 - IMPELLERS
	6 - SHAFTS
	7 - DRIVE ASSEMBLY

Table 10
Mixing Process Classification

PHYSICAL PROCESSING	APPLICATION CLASSES	CHEMICAL PROCESSING
SUSPENSION	LIQUID—SOLID	DISSOLVING
DISPERSION	LIQUID—GAS	ABSORPTION
EMULSIONS	IMMISCIBLE LIQUIDS	EXTRACTION
BLENDING	MISCIBLE LIQUIDS	REACTION
PUMPING	FLUID MOTION	HEAT TRANSFER

fluid properties, tank geometry, and only two of the three variables, P, N, or D because the third variable is fixed for us by the power correlation (Figure 19) which is independent of process performance. The third variable can be calculated , but it doesn't enter into the process correlation. It is usually preferable to use power and diameter as the two parameters for process correlation.

Considering what is required by a process in the way of a fluid flow pattern, we examine the mechanical hardware combinations involved and select that which provides the optimum cost for the application. As stated earlier, there are many possible mixer selections for a given application. The best selection is the optimum in terms of overall system economics.

Table 10 lists five basic application classes which can be distinguished by uniformity criteria on the left side (Physical Processing) and mass transfer, chemical reaction, or diffusion processes on the right side (Chemical Processing). Designing for solids suspension can be quite different from designing for solids dissolving. The requirements for gas dispersion may be quite different from those for gas absorption, etc. Thus, there are 10 separate mixing technologies. Each of these has its own scale-up rules and principles, associated application know-how and experience. We will look at some examples and see how to prepare a pilot-plant program. The overall process result may be composed of many component steps. Our job is to analyze the process in a way that makes those steps meaningful, yet manageable, in terms of present day technology.

We will first discuss scale-up principles, and then determine how to make specific test runs to determine the controlling factors in order to apply the correct scale-up principle to a given process. There is no shortage of scale-up principles or possibilities, but we must apply the right one to the right circumstance. Demonstration plants do not normally yield much useful pilot-plant data. There is usually only one impeller diameter and speed setting for the mixer in a demonstration plant, which limits the number of variables we can study.

We can't study a fluid regime in the laboratory unless we have a process result in mind; nor can we decide that a flow pattern is adequate or inadequate by simply observing or measuring it unless we have a process objective.

Even then, we can't always make an evaluation unless we have installation size and process economics in mind. For example, are we concerned about a 50 gallon tank in which the applied power level is insignificant, or a 50,000 gallon tank where power and capital costs are very important?

Thus, we require an economic basis to evaluate the mixer variables. For example, in the waste treating industry, to estimate the cost of power over 20 years, mixers are currently evaluated at approximately $5,000 per HP. This means treatment plants are willing to pay $5,000 today to save one HP in the future over a 20 year period. In a chemical plant, where a 1- to 3-year evaluation normally applies, the values are about $500 to $1,000 per HP, which plant operators are willing to pay to save 1HP over the appropriate time period. Therefore the economic basis is extremely important when considering the effects of mixer variables in the laboratory.

If we have a successful pilot-plant process and we want to duplicate each individual mixing parameter in the pilot-plant vessel upon scale-up, we would have to use a number of vessels of exactly the same size, because once we go to a larger vessel, the design parameters change. In a larger vessel, for example, blend time is normally longer, there is a higher maximum fluid shear rate, a lower average fluid shear rate, and a greater variety of shear rates [2].

Fortunately, many processes are not all that sensitive to these parameters, and the full-scale tank behaves very much as expected. It may take a little longer to complete the blending or otherwise satisfy the process, but by and large, it may be a wholly satisfactory operation. But when we examine in detail what is happening at the molecular level, or to fluid shear rates or pumping capacities, we find that the larger tank behaves quite differently from the smaller tank. This provides the challenge of deciding which factors are of major importance and which are not.

For convenience, we used constant power per unit volume and geometric similarity in Figure 28. It is very difficult to bring these curves together as we scale-up, but we can tip the whole ratio curve. To do this, we have to use nongeometric similarity. Geometric similarity controls no mixing variable whatsoever. To control mixing variables, we must use a nongeometric approach to cause several variables to behave as we wish. However in doing so, we must suffer the consequences of the behavior of other variables. We can't maintain all of the variables in a desired ratio, even with nongeometric similarity, but we can vary 2 or 3 of them as desired.

Considering why a small tank doesn't model a large production tank, the small tank has a relatively high pumping capacity, therefore allowing a relatively short blend time. It has too low a maximum shear rate to give the shear effects we see in the large tank. So, in order to duplicate those two parameters in the small tank, we must make the impeller blade narrower or the impeller diameter smaller, or both, and increase the speed. These adjustments reduce the pumping rate and increase the fluid shear rate to values similar to those in the large tank. However, doing so doesn't necessarily model the other parameters in the large tank [3]. If other parameters have to be controlled, we may have to run different pilot plant geometries to duplicate, step by step, the series of parameters in question. This does not imply, however, that nongeometric similarity can duplicate every mixing variable. But, there are many examples where a pilot plant can be sensitized to behave as a full-scale plant by using nongeometric similarity techniques.

Exponential Scale-Up Relationships

Generalized relationships can be demonstrated, similar to those in Figure 32. This shows a variety of published scale-up relationships for liquid-solid suspensions. There are almost as many scale-up proposals as there are investigators. Figure 32 is taken from the article by Einenkel, W. D., [8] which references many other articles, [10–18] particularly from the European literature. Another investigator, Herringe [21], found that scale-up is a function of particle size. Data averaged for a variety of particle sizes will show scale-up parameters between the extremes found for small particles and large particles.

Articles by Raetzen et al. [20] and Connelly and Winter [26] suggest several exponential scale-up relationships for various types of processes.

All exponential relationships must be used with caution in practice, for several reasons:

1. They are usually based on geometric similarity, which does not always allow for increasing and decreasing relevant parameters as need be on scale-up.
2. They do not usually relate to important, but somewhat complicated, processes such as polymerization, crystallization and fermentation.
3. Often they do not account for such scale-up effects as differences in droplet or bubble sizes or linear velocities of immiscible fluids.
4. They do not always apply to macro-scale and micro-scale shear rate relationships which are important in many kinds of mixing processes.

In the remainder of this article, we will provide concepts and principles which can be used to apply qualitative techniques to existing problems. Normally, estimates must be made for several

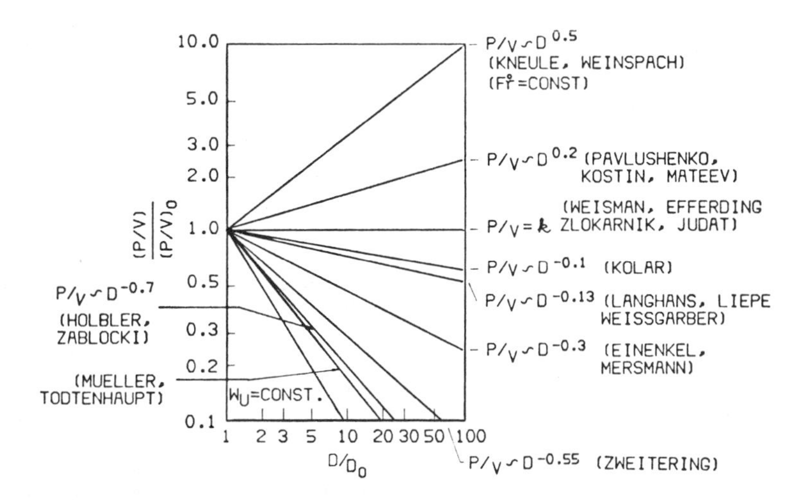

Figure 32. Illustration of the change of power per unit volume on scale-up from many different investigators for solid suspension [27].

important scale-up parameters. The final choice for scale-up must be a conservative, yet practical, selection. The result will be full-scale equipment with realistic performance, but not an excessively conservative design.

Fluid Motion on Scale-Up

Table 8 shows that impeller flow per unit volume decreases on scale-up. Middleton [22] used a radio transmitter sealed in an 8-mm diameter plastic capsule as a tracer particle and placed a fixed radio antenna around the impeller zone. He recorded each time the capsule passed by the antenna, in order to determine how particles circulated throughout the tank. He discovered as tank size increased, particles tend to develop small, independent mixing zones.

A large tank requires a longer blend time than that predicted by impeller pumping per unit volume. Table 8 shows at constant power per unit volume, where the scale ratio is 5:1, the impeller pumping capacity per unit volume in the large tank is $\frac{1}{3}$ of the value in the small tank. The circulation time turns out to be more like $\frac{1}{6}$, not $\frac{1}{3}$, of the time in the small-scale tank.

In a 50 gallon tank there was a single flow pattern. A large tank (1,000 gallons) behaved as two, five, or as many as 10 tanks in series, depending upon the impeller speed. It was found that in a small tank, the ratio of impeller flow per unit volume closely predicted the circulation time. Therefore in the 50 gallon tank, if we divide the batch volume by the pumping capacity, we can predict circulation time. The larger the tank size, the longer the circulation time than that calculated from V/Q.

Furthermore, in a gas-liquid system, an increase in gas rate increases circulation time and also increases the standard deviation of the blend time.

Scale-Up Concepts

There is a minimum-size pilot plant for heterogeneous processes. Referring to Figure 33, the maximum shear rate at the jet boundary gives a value of 10 sec^{-1}. Across $\frac{1}{8}$ of a centimeter, the fluid shear rate is 9.5 sec^{-1}, across a $\frac{1}{4}$ centimeter 7 sec^{-1}, across half the impeller 5 sec^{-1}, and across the entire impeller, the fluid shear rate is 0. Thus, a particle on the order of a centimeter or

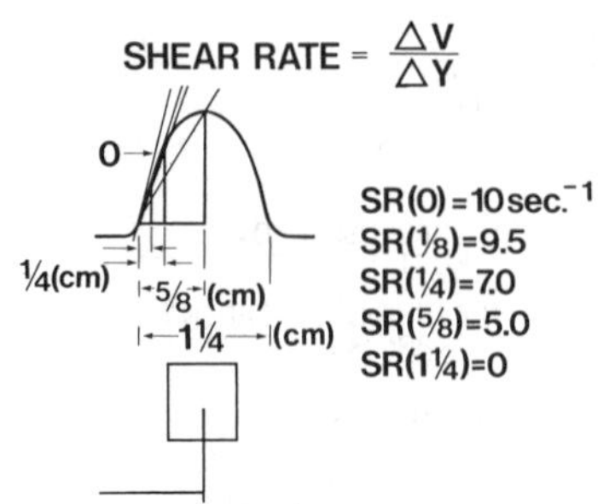

Figure 33. Various shear rates around a radial flow impeller blade.

two in size will "see" a fluid shear rate of essentially zero, while a micron-size particle will experience a fluid shear rate of 10 sec^{-1}. Thus, it is important to maintain appropriate impeller blade dimensions with respect to the particle sizes in the system.

The physical dimension of the impeller blades must be determined with respect to the particle size. On the plant scale, particle size is almost always smaller than the impeller blade width. In the pilot plant, in general, the impeller blade width should be at least two or three times larger than the largest particle to maintain a similar shear rate relationship on scale-up.

In an homogeneous chemical reaction, we could use a thimble-sized pilot vessel since there is no problem in scaling homogeneous chemical reactions with respect to shear rate. But if we have a gas bubble that is $\frac{1}{2}$ in. diameter for example, a dispersion using a $\frac{1}{4}$-in. blade height is markedly different from that obtained using a 1-in. blade height. These extremes represent different mechanisms of fluid shear rate. That doesn't mean we can't run the experiment with a narrow blade. We can obviously run the experiment no matter what size equipment we have, but the results do not follow any logical scale-up relationship when the impeller is out of proportion with respect to particle size. For example, in paper pulp, we need a tank almost 30 in. in diameter with a correspondingly larger impeller (compared with using a smaller vessel size) to keep the ratio of fiber length and blade proportions within reason to obtain meaningful data for scale-up.

The major goal of a pilot-plant or full-scale plant study is to determine the ways the process responds to changes in mixing variables. We can then describe and evaluate the important process parameters, provided sufficient data are available. Figure 34 shows the effect of a change in power on process results for several different application classes. Power is most conveniently changed by changing the speed of the impeller, holding the diameter constant. When we do this, we change the flow rate and the fluid shear rate and if increasing both of these does not cause a change in process result, there is little likelihood that other variables will. It is always possible, however, that an increase in flow rate may improve the process, while an increase in shear rate may hinder it. As a result, they may neutralize each other with little effect of power on the process. To be absolutely sure, a study must be made with a different impeller diameter to determine if there was just a peculiarity of the particular combination used in the first experiment.

In Figure 34, if the exponent on the slope is relatively high (0.8), it normally means a mass transfer process is involved. Gas-liquid mass transfer is typically the most sensitive process to mixer power, but liquid-liquid mass transfer can also be involved. However, liquid-solid mass transfer usually correlates at a much lower slope. If the slope is zero it is often caused by a chemical reaction that is controlling. The jagged line shown on the left side of the curve (A) indicates the power levels below those required to provide a satisfactory blend time. Here, process results may be quite erratic. If the slopes are in the middle range, between 0.1 and 0.4, the controlling mechanism may be illusive, requiring further experimentation.

Power per unit volume is a useful scale-up parameter. For some types of chemical reactions it is a constant because the micro-scale mixing required is less dependent on shear rate and more dependent on power dissipation, so equal power per unit volume is an appropriate criterion for chemical reactions. However, the power per unit volume parameter is not always constant on scale-

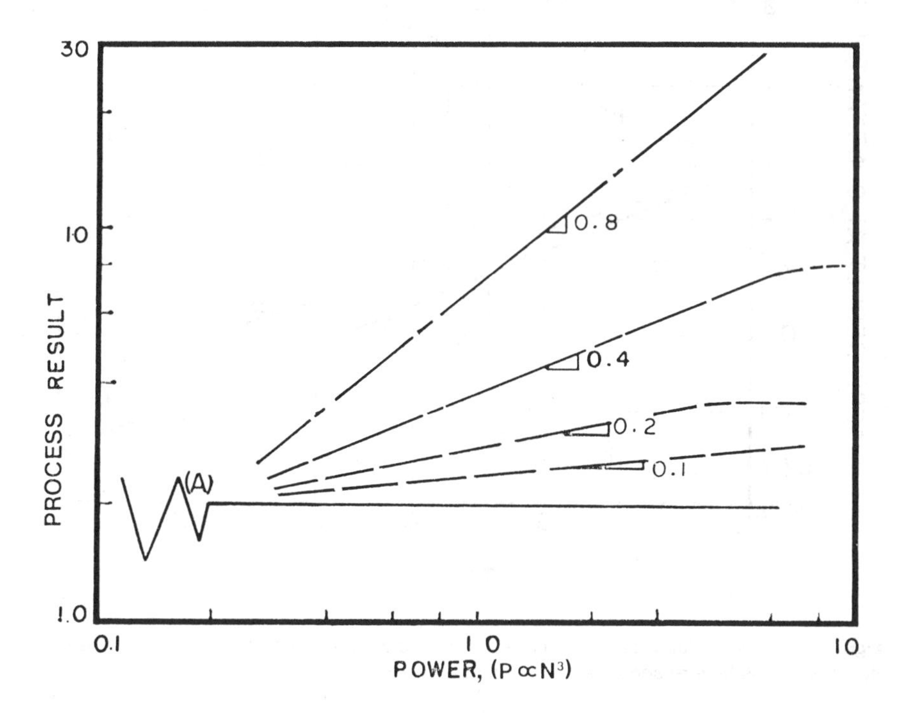

Figure 34. Slope of process results vs. mixer power level is an indication of the controlling factor in the process.

up. (Figure 35). In blending and slurry suspensions, power per unit volume often decreases on scale-up. But if we want to maintain equal blend time on both scales, power per unit volume increases with the square of the tank diameter.

For a gas-liquid-solid process, a simple measurement of the effect of gas rate separates the liquid-solid controlled process from the gas-liquid controlled process. The mass transfer coefficient of a liquid-solid step is unaffected by changing gas rates; so if there's a very low slope in the process result curve (curve A, Figure 36) liquid-solid mass transfer controls. Conversely, a high slope (Curve B, Figure 36) indicates gas-liquid mass transfer is controlling.

Some General Principles of Pilot Planting

When conducting pilot-plant tests, we normally begin with a fixed impeller diameter and tank diameter and vary the speed. Increased speed raises the power level, which increases both flow and fluid shear. If speed doesn't affect the process result, then there is little chance that any other mixing variable will. If we want to determine whether flow or fluid shear is controlling, we study the effect of D/T ratio. We run the experiment in a fixed tank diameter and vary the impeller diameter. This determines which D/T gives the best result.

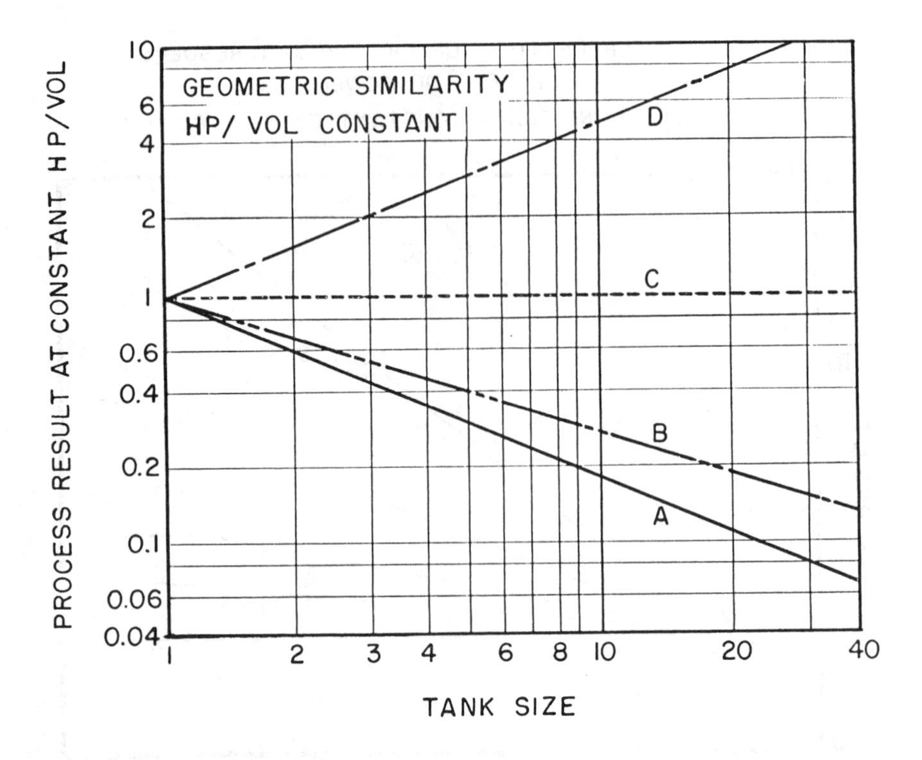

Figure 35. Power/volume can be a constant on scale-up, but it typically either increases or decreases. It is better-used as a correlating parameter.

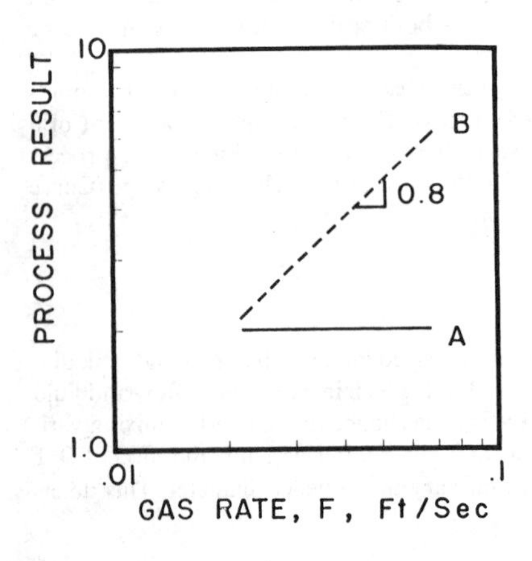

Figure 36. Two extremes of possible effects when gas rate is changed in a combination gas-liquid-solid slurry process. Curve A represents liquid-solid-controlled mass transfer, and Curve B represents gas-liquid-controlled mass transfer.

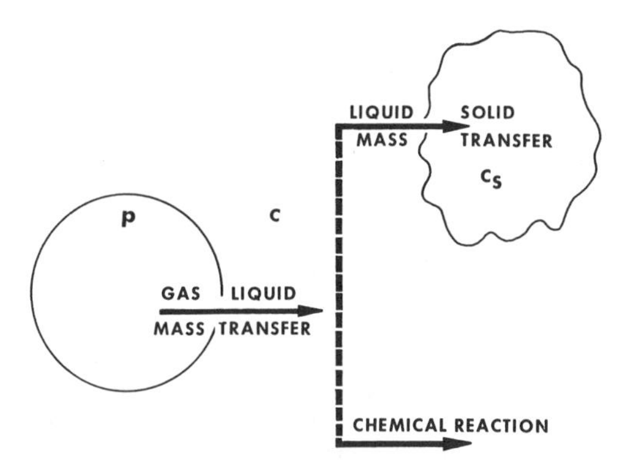

Figure 37. Gas-liquid mass transfer is usually in conjunction with a liquid-solid mass transfer step or a chemical reaction.

To determine the effect of fluid shear rate, we have to systematically vary the blade width in a series of experiments to evaluate micro-scale mixing action separately from macro-scale mixing. This is more time consuming, but in doing so, valuable information can be obtained about the process.

Mixer manufacturers do not routinely run pilot-plant tests on every application they analyze. Correlations and data are already available to size perhaps 95% of the applications; many operations can therefore be sized from those existing correlations.

For some processes where we *know* the scale-up effect, we still need a quantitative value from a small-scale test. In those cases, all we need is data from one tank size. Examples of this are gas-liquid mass transfer and chemical reactions. Of course, if the process is so new that little is known about the controlling factors, then it's probably best to use two tank sizes and initially assume a scale-up relationship empirically. If the initial assumption is correct, further testing will not be required. If the initial assumption is incorrect, we can eliminate parameters by following the procedures previously outlined.

Gas-Liquid Mass Transfer Example

The controlling mechanism in a typical gas-liquid mass transfer process is frequently not limited to gas-liquid dispersion. It may involve liquids and solids such as a biological process step in a fermentation or a chemical reaction (Figure 37). Here the procedure is to measure the mass transfer rate of the process, then estimate the gas-liquid concentration driving force and calculate a mass transfer coefficient (Figure 38). Mass transfer *coefficients* are scaled up, *not mass transfer rates.* We translate the coefficient as correlated with power level, gas rates, and tank size as in Figure 39. In the larger tank, we use the pressure and concentrations to obtain the gas-liquid driving force. The rate is known which gives us the mass transfer coefficient, K_Ga. If we calculate the superficial gas velocity F (gas flow rate, ft^3/sec., divided by tank cross-sectional area, ft^2), we can determine power per volume from Figure 39. Once again, the key consideration is that scale-up is based on the mass transfer coefficient, not the mass transfer rate. In other words, the mixer can only affect the mass transfer coefficient. Figure 40 shows a process curve for a specific tank size and various gas rates.

Figure 41 shows the sensitivity of gas-liquid mass transfer to D/T ratio, which is related to the ratio of flow to fluid shear rate. At low mixer power, (left-hand side) the gas overpowers the mixer and controls the flow pattern. A mixer-controlled regime occurs (center) when the mixer power is at least two or three times higher than the isothermal expansion power of the gas. At the right, the power level is so high that it makes no real difference which D/T we use.

For gas-controlled conditions, we want a relatively large D/T. (The optimum D/T is in the shaded band). For mixer controlled conditions, we want a small D/T. But in the case of fermentation, the

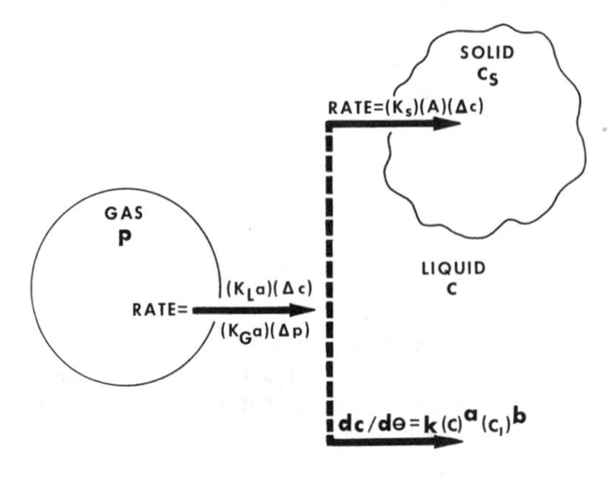

Figure 38. For scale-up purposes, the mass transfer coefficient is used rather than the mass transfer rate.

high shear rates produced by the optimum D/T ratios of 0.15 to 0.2 would be detrimental to the biological organisms; therefore high shear rates are undesirable. D/T ratios from 0.35 to 0.45 are necessary to provide blending and heat transfer and to prevent physical shear damage to the organism. They are not the most effective D/T ratios for gas-liquid mass transfer, however.

In waste-treatment plants, aeration basins are so large that it is impractical to use large impellers (large D/T ratios). Economics dictate, to some extent, how closely the optimum designs can be approached. This illustrates that we cannot, as a rule, optimize every individual mixing parameter in the process.

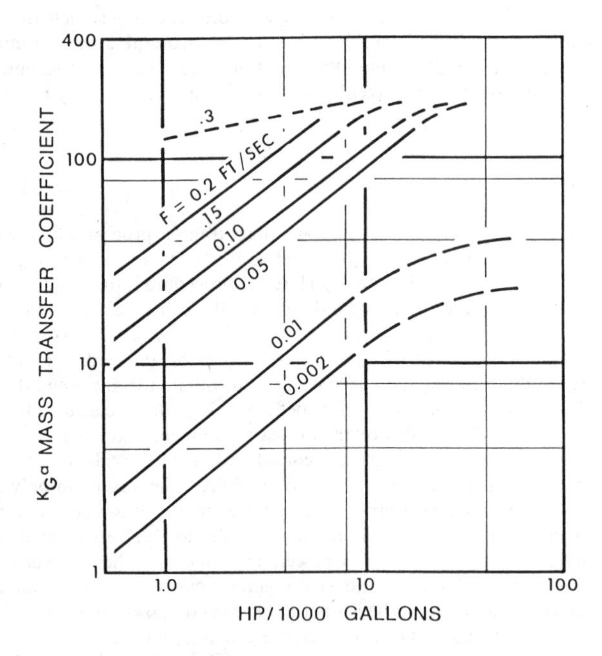

Figure 39. Typical correlation of gas-liquid mass transfer coefficient vs. power and superficial gas velocity.

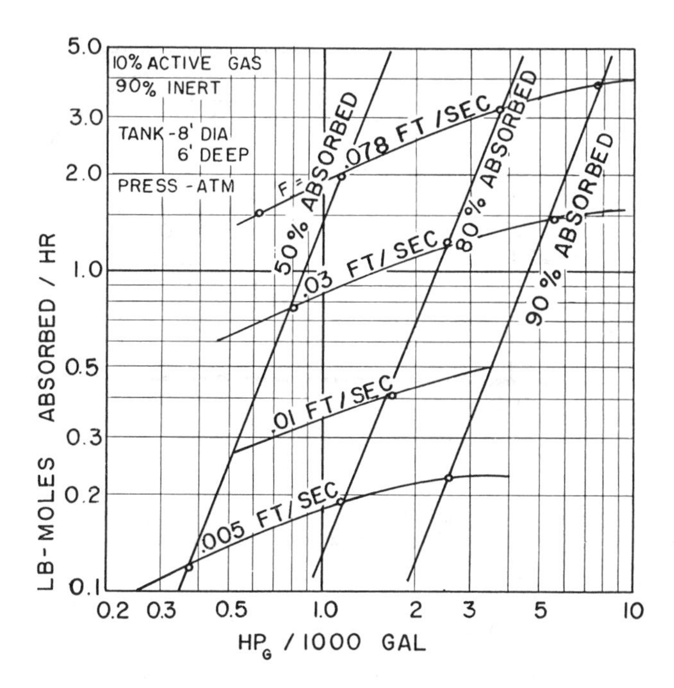

Figure 40. Typical correlation for an actual vessel and process fluid for various gas concentrations. This is derived from a mass transfer coefficient by using the actual driving forces for the process.

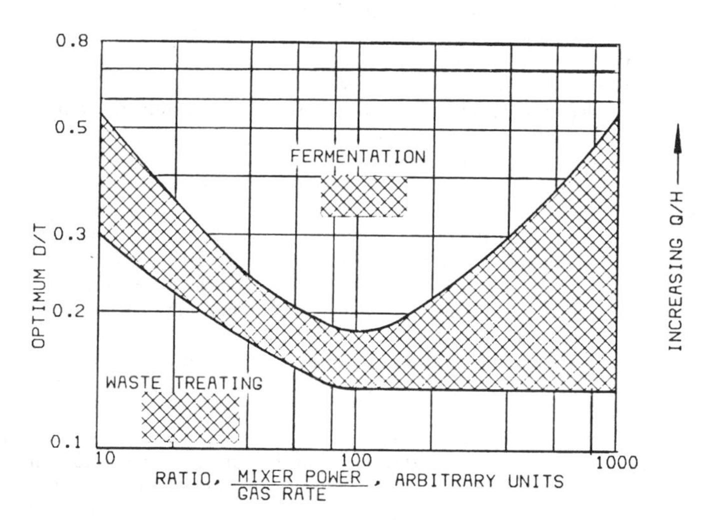

Figure 41. Relationship showing optimum D/T at various combinations of mixer HP to gas rate. On the left is gas controlled. In the center is mixer controlled and on the right, mixer HP is in excess of gas HP.

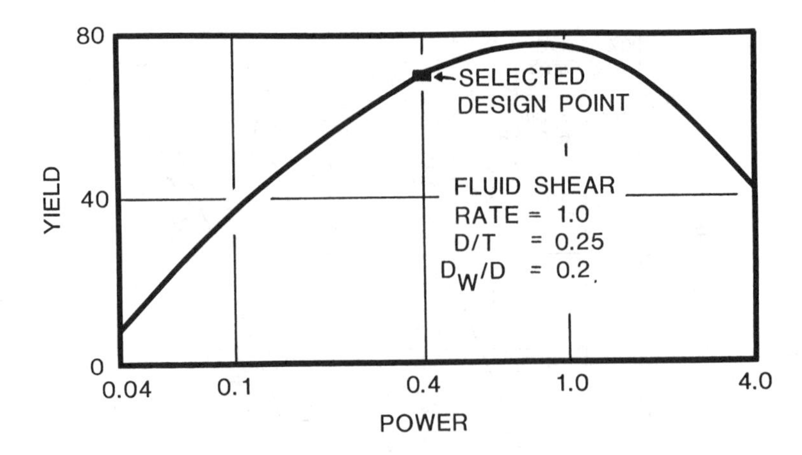

Figure 42. Typical data from fermentation process showing that as power increases, the yield after a 5-day batch fermentation reaches a maximum, then decreases.

Fermentation Pilot Plant Example

This is an example of a batch fermentation process, in which we begin our tests with a D/T ratio of 1/4, and a standard blade-width-to-impeller-diameter ratio. We assume this impeller gives us a shear rate of 1 relative to other impellers we'll be using later. We find the yield at the end of five days increases with increased power and then begins to taper off (Figure 42). We suspect that the shear rate is too high for the organisms to tolerate. To be on the safe side, we select a reasonable yield and see what happens when we go through the scale-up procedure.

For the pilot plant, we record the ratios of the variables involved. In fermentation we're interested in the gas-liquid mass transfer rate, which supplies oxygen to the organism. We're also interested in the shear rate in the impeller zone because we suspect it may be a key variable.

The first scale-up calculation (Table 11 Column B) uses geometric similarity with the same D/T ratio as the pilot plant, although the plant-scale tank is relatively taller, since the Z/T ratio is approximately 2 compared to 1 in the pilot tank. For equal gas-liquid mass transfer and geometric similarity, the tip speed relating to the maximum shear rate has increased approximately 80% in the proposed plant-scale tank. Assuming another mixer design, with a larger D/T ratio of 0.38, we must run at a lower speed to get the same mass transfer rate (Column C). This produces a maximum shear rate in the fluid about 40% higher than it was in the pilot plant. The most expensive part of a mixer is the gearbox which is related to the torque on the impeller shaft. So, by cutting the speed in half, we have doubled the torque, which means a much more expensive mixer, almost twice the cost of the mixer in Column B.

In Column D, using a 0.6D/T ratio, we can reduce the shear rate to what it was in the pilot plant. But by doing so, we need four times the torque, which will cost almost four times the design in Column B.

Let's return to the pilot plant to see if either mixer B or C will be satisfactory. Using an impeller $\frac{1}{3}$ the blade width of the original impeller (Dw/D = $\frac{1}{3}$ at the same D/T, the speed is higher and we develop a 40% higher shear rate than the original impeller (Figures 42 and 43). We find that when we run the experiment, we get increased yield with increasing power at low power levels, but at the design point the curve starts to flatten out. This indicates the higher shear fluid rate is causing the lower yield. Based on these data, mixer C is a good choice.

If we reduce blade width further,by another factor of $\frac{1}{3}$ (Mixer B) the blade width is smaller than the gas bubbles in the system. Since blade width should be larger than the particles we are trying

Table 11
Relationship Between Process Performance and Cost for Three Different D/T Ratios Selected for Full-Scale Mixers

	PILOT	PLANT		
	A	B	C	D
T.	1.2	7.0	7.0	7.0
P.	1.0	340	340	340
N	1.0	.31	0.16	.075
D/T	.25	.25	0.38	0.6
NO. of IMP.	1	2	2	2
Z.	1.2	12.0	12.0	12.0
F.	F	10 F	10 F	10 F
L-S. MTR	1.0	1.0	1.0	1.0
G-L. MTR	1.0	1.0	1.0	1.0
MAX. I.Z. SHEAR RATE	1.0	1.8	1.4	1.0
TORQUE	---	1.0	2.0	4.1
COST ($)		10,000	18,000	35,000

to disperse, there's no way to determine for certain if mixer B is satisfactory on this scale. Evaluation of mixer B therefore would only be possible in a larger pilot-plant tank, where blade proportions would be suitable.

Typical Fermentation Specifications

There are four ways in which mixers are often specified when considering larger units for a fermentation plant. In this context, we will consider either a larger tank, with a suitable mixer, or improving the productivity in the existing tank by installing a different combination of mixer

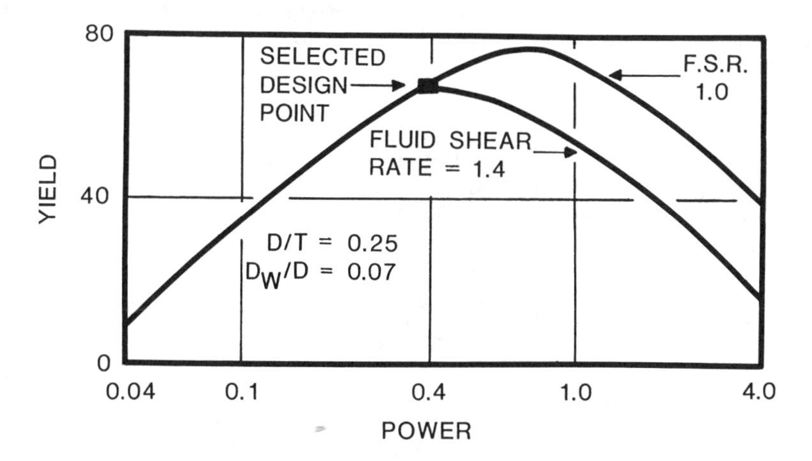

Figure 43. A second impeller type, having a shear rate 1.4 × original impeller, shows a decrease in yield at a lower power level.

horsepower and gas rate:

1. A change in productivity requirements based on production data from the existing plant fermenter.
2. A new production capacity based on pilot-plant studies.
3. Specifying the agitator based on the sulfite absorption rate in aqueous sodium sulfite solution (an accepted practice).
4. Specifying the oxygen uptake rate in the actual broth for the new system.

The power input from the gas increases on scale-up because there is a greater head pressure on the system. There is also an increased gas velocity.

The power level for the mixer may be reduced since the energy from the gas passing through the broth is necessarily higher in order to maintain the required mass transfer coefficient, K_Ga.

However, this changes the relative mixer power level compared to the power applied by the gas, and it changes other mass transfer rates, such as the liquid-solid mass transfer rate. But the capacity for blending with changes in the mixer power level is not affected in the same way as the gas-liquid mass transfer coefficient.

Scale-up Based on Data from an Existing Production Tank

If data are available from fermentation in a production-size tank, scale-up may be accomplished by increasing in a relative proportion the mass transfer, blending, and shear rate requirements for the full-scale system. For example, it may be determined that the new production system requires a new mass transfer rate which is a certain percentage of the existing mass transfer rate. Furthermore, there may be specifications on maximum or average shear rates, and a desire to consider changes in blend and circulation times. In addition, there may also be concern for the relative change in the CO_2 stripping efficiency in the revised system.

At this point, any tank size and shape may be considered. To illustrate the principles involved in gas-liquid mass transfer Figures 37 and 38 give the three different mass transfer and reaction steps common to fermentation. The mass transfer rate must be divided by a suitable driving force, which yields the mass transfer coefficient required. The mass transfer coefficient is then scaled to the larger tank, and it is normally related to superficial gas velocity raised to an exponent, power per unit volume raised to an exponent, and to geometric variables such as the D/T ratio.

A thorough analysis considers every proposed tank shape, the gas rate range required, the gas phase mass transfer driving force, and the required k_Ga to meet those conditions. Reference is made to data on the mixer for the conditions specified, to obtain the correct mixer power level for each gas rate and D/T ratio.

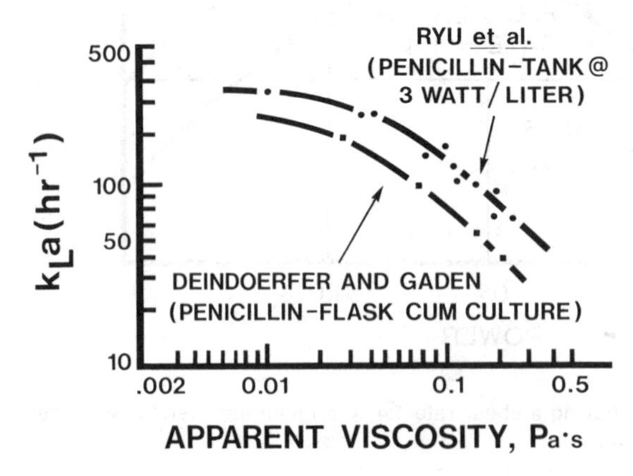

Figure 44. Illustration that the mass transfer coefficient decreases with the viscosity of the medium in the fermentation.

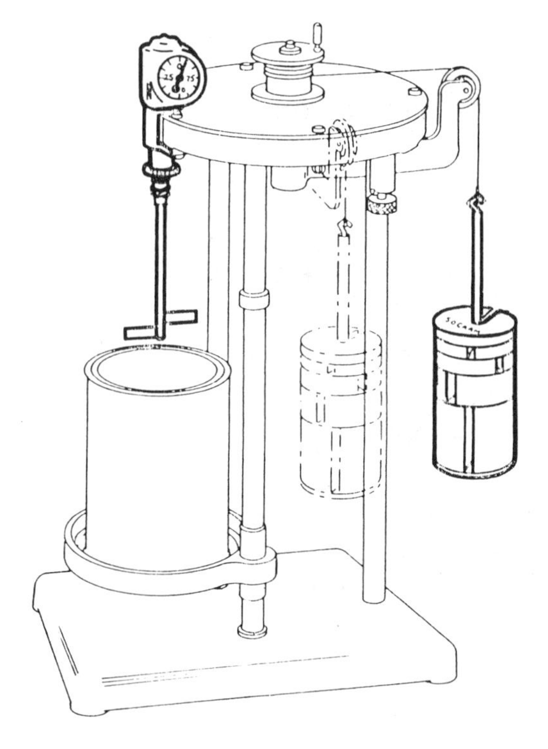

Figure 45. Stormer viscosimeter used for measuring viscosity at known shear rates.

The role of viscosity must also be considered. Figure 44 gives typical data for the effect of viscosity on the mass transfer coefficient [5]. Data must be obtained with a viscosimeter which mixes while it measures viscosity. Figure 45 shows the Stormer viscosimeter which is a device used to measure viscosity under mixing conditions at known shear rates.

For the proposed tank size, estimates should be made of the shear rate profile within the system. Since viscosity is a function of shear rate, we can estimate the viscosity throughout the tank, and estimate shear stress from the product of viscosity and shear rate. Performance estimates can then be determined for the proposed tank and compared with the known performance of the existing production tank.

Data Based on Pilot Plant Work

It is possible to model the biological fermentation process from a fluid mechanics standpoint, even though the impeller is not geometrically appropriate with respect to bubble size in the gas-liquid mass transfer step. Thus, data from one scale pilot plant may be usable for one or two of the fermentation mass transfer steps and/or chemical reaction steps, where impeller dimensions *are* appropriate, but may not be suitable for analysis of other mass transfer steps where impeller dimensions are *not* appropriate. The decision is then based on how suitable the resulting data are for any steps which are not properly modeled in the pilot plant.

Ideally, data should be taken during the course of the fermentation and should include gas rate, gas absorption, dissolved oxygen level, dissolved carbon dioxide level, yield of desired product and any other parameters which may influence the decision for the overall process. Figure 46 shows a typical set of data.

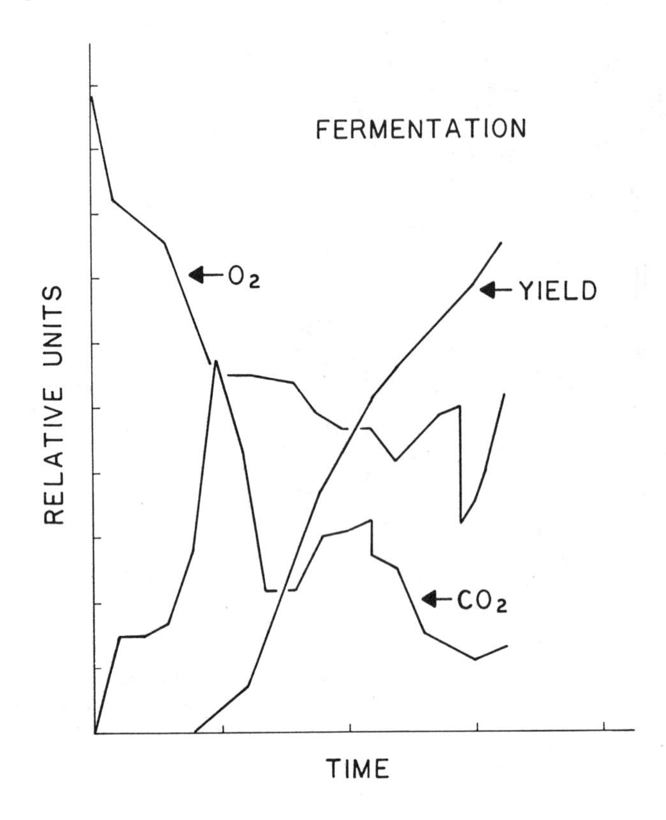

Figure 46. Typical relationship of dissolved oxygen, dissolved CO_2, and fermentation yield vs. time for batch fermentation.

If the pilot plant is to duplicate certain properties of fluid mixing, it may be necessary to use nongeometric impellers and tank geometries to duplicate mixing performance. In general, geometric similarity does not normally control mixing scale-up properties.

It may also not be possible to duplicate all of the desired variables in each run, therefore a series of runs may be required changing various relationships systematically, from which a synthesis of the overall results is obtained.

The linear superficial gas velocity is particularly important. It is defined as the ratio of the volumetric flow rate of the gas to the cross-sectional area of the vessel (e.g., $ft^3/sec{:}ft^2 = ft/sec$). Duplicating the anticipated plant-scale linear superficial gas velocity in the pilot plant will closely duplicate foam generation. Enough vessel head space must be available in the pilot vessel to adequately control the foam. Foam level is usually related to the square root of tank diameter on scale-up or scale-down.

If it is desired to duplicate maximum impeller zone shear rates on a small scale, there may be mechanical problems related to allowable shaft speed, shaft deflection, critical speed and mechanical seal limitations. This means careful consideration must be given to the allowable mixer operation limits.

Figure 47 shows what often happens in the pilot plant in terms of correlating the mass transfer coefficient, K_Ga, with power and gas rate. These data are translated to a suitable relationship for full scale. At the higher superficial gas velocity, the power level may be reduced in the full scale to maintain the same mass transfer coefficient. This is shown by shifting from the dashed rectangle

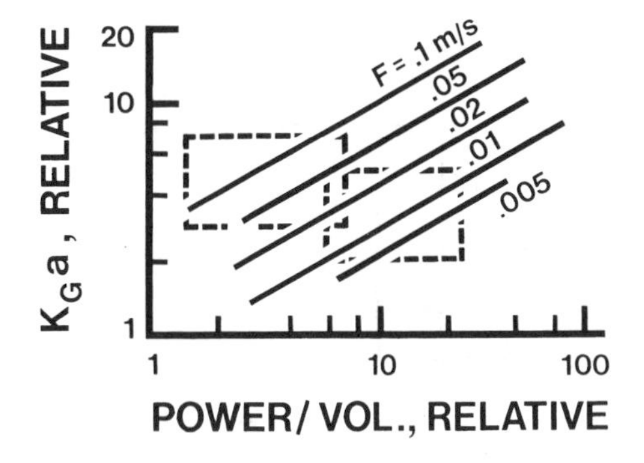

Figure 47. Correlation of K_Ga mixer power per unit volume and gas flow in pilot plant and plant size units.

on the right in Figure 47 (pilot scale) to the dashed rectangle on the left (plant scale). Doing so, changes the ratio of the mixer power level to gas power level in the system, and changes the blend time, flow pattern, and foaming characteristics. It can also markedly affect the liquid-solid mass transfer rate (Figure 48) if that is part of the process.

In all cases, a suitable mass transfer driving force must be used. Figure 49 illustrates a typical case for fermentation processes and shows a marked difference between the average driving force and the exit gas driving force. In a larger fermenter, experience shows that gas concentrations are essentially step-wise stage functions, and a log-mean driving force is applied. Figure 50 represents a small laboratory fermenter with a Z/T ratio of 1. In this case, depending on the power level, the suitability of the exit gas concentration driving force compared to the log-mean driving force must be evaluated.

In the waste-treating industry, it is quite common to run an unsteady-state reaeration test in which the fluid is stripped of oxygen, air is introduced with the mixer operating, and the increase in the dissolved oxygen level is monitored until the fluid is saturated with dissolved oxygen and no further mass transfer occurs. At that point, the bulk dissolved oxygen level is usually between the saturation value at the surface and the saturation value at the bottom of the tank (Figure 51). This means, for steady state, there must be enough absorption at the bottom and enough stripping at the surface, and an atypical mass transfer situation results compared to actual waste treatment production or fermentation. Therefore, running experimental tests and basing calculations on a particular driving force may give a marked difference from the way the mixer will operate in actual practice.

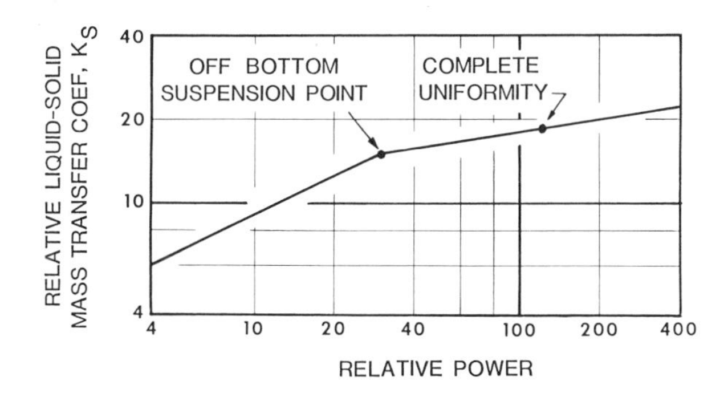

Figure 48. In liquid-solid systems, the mass transfer coefficient, K_s, increases markedly with power until the off-botton suspension point is reached. Thereafter, the increase in K_s is at a much lower rate.

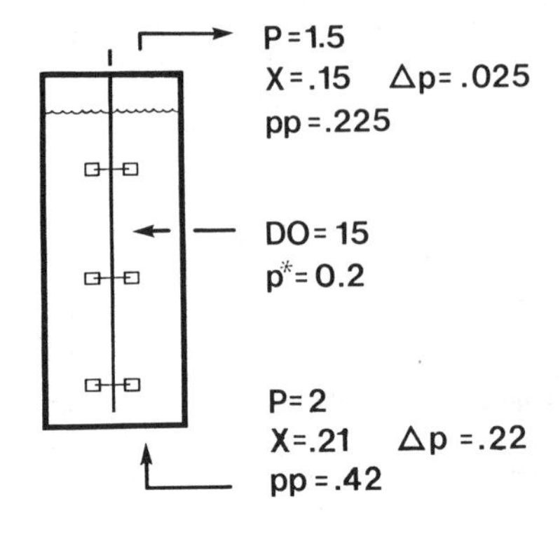

Figure 49. On a large-scale fermenter, the height and mixing intensity is such that a log-mean driving force is more suitable than an exit driving force.

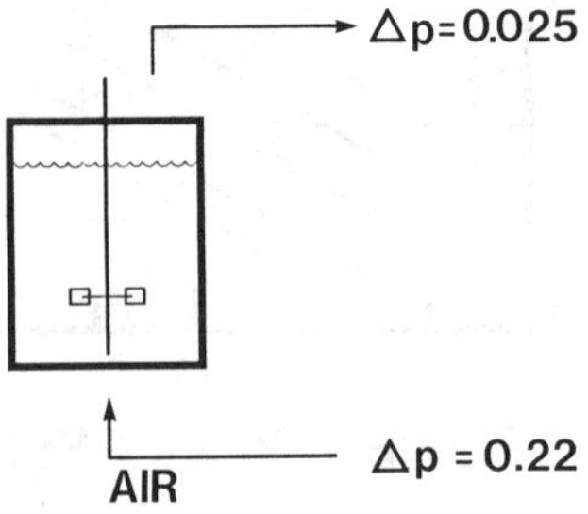

Δp=0.025

AIR

Δp = 0.22

Figure 50. A gas concentration driving force for a well-mixed batch fermenter, indicating that a choice must be made between an average driving force and an exit driving force.

Sulfite Oxidation Data

Mass transfer data have been obtained using excess sodium sulfite in tap water, containing suitable catalysts to keep the dissolved oxygen level at zero. Results have been obtained for both small and large size fermentation tanks on this basis. Data are collected when the tank is completely free of anti-foams which may be residual from the fermentation process. Anti-foams and other trace organics can cause marked differences in the mass transfer coefficients.

If we have a relationship between the sulfite oxidation number (an industry performance standard) and the performance required in the fermenter, it is a perfectly valid way to specify equipment. Tests can then be run to determine the resulting overall mass transfer rate.

Oxygen Uptake Rate in the Broth

If it is desired to relate fermenter performance to oxygen uptake rate in the fermentation broth, this can be specified along with suitable gas rates, and the mixer design can be estimated. But we must have the link between the mass transfer specification and the actual performance of the fermenter [6, 7].

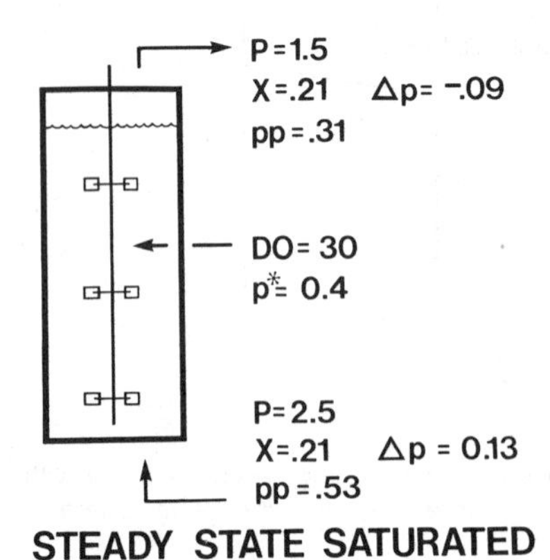

Figure 51. In an unsteady-state reaeration test, process equilibrium shows the top of the tank has a negative driving force while the bottom of the tank has a positive driving force.

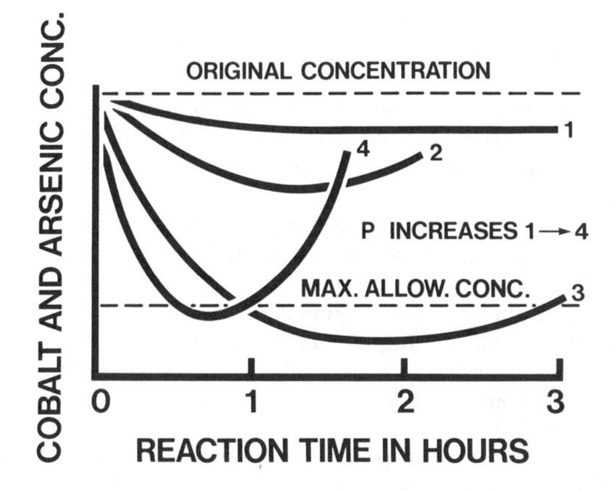

Figure 52. Illustration of four different power levels in the zinc purification process described in the text.

If performance is based on pilot-plant data, then the effects of different shear rates, blend times, and viscosities on both the mass transfer relationship and the resulting fermentation, must be considered.

Zinc Purification Process Example

From a study made in conjunction with American Zinc [23], an illustration of the interaction of the many variables in a zinc purification process is shown in Figure 52. It was desired to retain the cadmium concentration at a high level in the purification process, while removing both cobalt and arsenic. The procedure involved adding zinc dust to a solution from a leach system which produced a sudden drop in cadmium, as well as cobalt and arsenic concentrations. After sufficient time the cadmium was regenerated, returning to an acceptable recovery level.

It was desired to see whether mechanical mixing would affect the reaction time, the cadmium recovery, and the rejection of the arsenic and cobalt.

Figure 53 shows the results in a 1,016-mm-diameter pilot-plant study. In this study, for each run there was a typical curve as shown. At the optimum power level there was adequate recovery of the desired cobalt. On the same scale, Figure 53 shows that the 355-mm diameter impeller was much more effective than the 230-mm impeller, indicating the importance of flow-to-shear ratios.

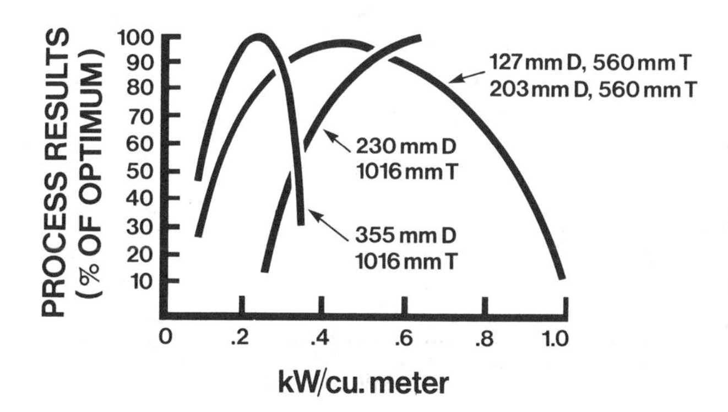

Figure 53. Percent of optimum process result is a function of power and D/T in the zinc purification process.

Data were also obtained in a 560-mm diameter tank. While similar process profiles were obtained with power and time, the two impeller diameters tested (127-mm and 203-mm) produced the same result.

In interpreting the data from the small-scale, 560-mm diameter tank, there was so much extra pumping capacity and such a short blend time (compared with the 1,016-mm-diameter tank and the full-scale tank) that any reduction in the pumping capacity due to the 127-mm-diameter impeller was not sufficient to affect the process result.

However, on the 1,016-mm-diameter tank scale, a reduction in the pumping capacity and an increase in fluid shear rate did affect the process result. Thus, it should be expected that D/T is an important ratio in scaling up to a full-size system.

If it were desired to obtain this kind of information on the 560-mm-diameter scale only, it would have been necessary to change the impeller blade-width-to-diameter ratio. This would markedly reduce the pumping capacity to the probable range in the full-scale plant. It would indicate whether pumping capacity on full scale would be a serious detriment to the process performance.

Plant-scale data indicated, by using suitable D/T ratios, that performance was predicted from the overall scale-up effect in the 1,016-mm diameter tank, when accounting for the optimum D/T ratio needed to satisfy the process requirement.

Flocculation Example

Flocculation is the aggregation and compacting of coagulated particles into large assemblages called *floc particles.* It is an especially important process in water purification and in waste treat-

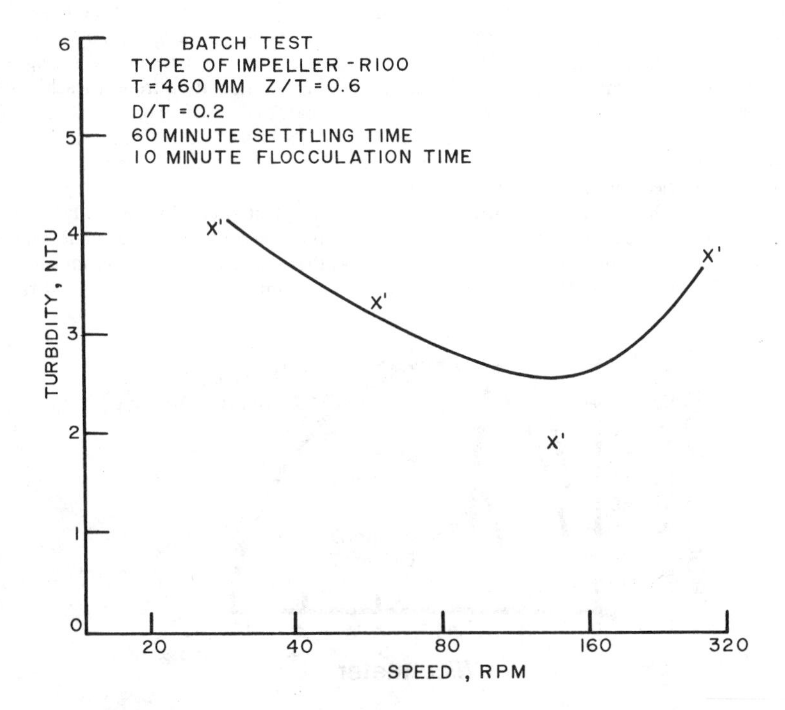

Figure 54. Data from batch flocculation pilot plant in 460-mm-diameter tank showing minimum turbidity is obtained at an optimum speed and therefore at an optimum shear rate.

ment, where pollution control regulations require a solids-free effluent. Minimum turbidity is a measure of optimum flocculator performance.

Oldshue and Mady [24, 25] derived scale-up correlations for two different flocculating impellers. Figure 54 shows an optimum impeller speed (therefore an optimum fluid shear rate) is required for minimum turbidity. Floc particle size increases with increasing fluid shear rate but decreases with the resulting shear stress, therefore an optimum speed is required for every system. The data in Figure 54 were obtained from a 460-mm pilot tank. Upon scale-up, optimum performance is expressed as a G factor, which is defined as the square root of power per unit volume divided by viscosity, at the optimum impeller speed:

$$G = \text{average fluid shear rate} = k\sqrt{\frac{PN}{\mu}} \tag{5}$$

Figure 55 shows the relationship of the G factor to tank diameter at constant D/T, indicating a decrease in power per unit volume as tank size increases. Figure 56 illustrates that for the axial flow flocculating impeller as D/T increased the G factor increased and turbidity (NTU units) decreased. However, performance of the paddle impeller was fairly insensitive to D/T.

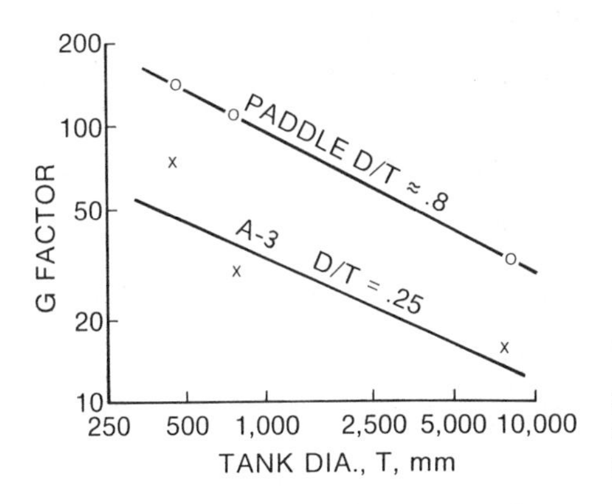

Figure 55. Results of flocculation experiments at optimum turbidity for various tank sizes.

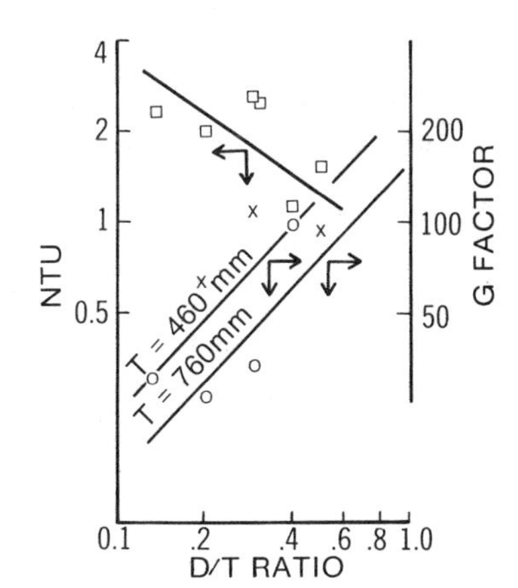

Figure 56. NTU (nephalometric turbidity units) at various D/T ratios for axial flow flocculation impellers showing that the G factor is a function of D/T.

NOTATION

P	power	N_{Fr}	Froude number
N	impeller speed	N_{We}	Weber number
D	impeller diameter	K_Ga	gas-liquid mass transfer coefficient
Z	liquid level	K_La	liquid-liquid mass transfer coefficient
T	tank diameter	K_S	liquid-solid mass transfer coefficient
μ	viscosity	DO	Dissolved oxygen
σ	surface tension	P*	equilibrium partial pressure in gas phase corresponding to dissolved solute concentration
g	gravitational constant		
N_{Re}	Reynolds number		

REFERENCES

1. Weetman, R. J., and Salzman, R. N., "Impact of Side Flour on a Mixing Impeller," *Chem. Eng. Prog.* (June 1981), pp. 71–75.
2. Oldshue, J. Y., "Spectrum of Fluid Shear Rates in a Mixing Vessel," CHEMECA '70, Australia (Butterworth, 1970).
3. Oldshue, J. Y., "Fermentation Mixing Scale-up Techniques," *Biotechnology and Bioengineering*, VIII (1966), pp. 3–24.
4. Oldshue, J. Y., and Gretton, A. T., *Chemical Engineering Progress*, Vol. 50 (1954), p. 615.
5. Dienderfer, F. H., and Gaden, E. L., *Applied Microbiology*, Vol. 3 (1955), p. 253.
6. Oldshue, J. Y., Coyle, C. K., et al., "Fluid Mixing Variables in the Optimization of Fermentation Production," *Process Biochemistry*, Vol. 13, No. 11 (1977) (Published Wheatland Journals Ltd., Waterford, Herts, England.
7. Ryu, D. Y., and Oldshue, J. Y., "A Reassessment of Mixing Cost in Fermentation Processes," *Biotechnology and Bioengineering*, XIX (1977), pp 621–629.
8. Levenspiel, Octave, and Kang, Soon J., *Chemical Engineering*, Vol. 83 (1976), p. 141.
9. Einekel, W. D., *German Chemical Engineering*, Vol. 3 (1980), pp. 118–124.
10. Kneule, F., and Weinspach, P. M., *Verfahrenstechnik* (*Mainz*) Vol. 1, Nr. 12, (1967) pp. 531–540.
11. Zwietering, Th. N., *Chemical Engineering Science*, Vol. 8 (1958), pp. 244–253.
12. Pavlushenko, J. S., and Kostin, N. M., *J. Appl. Chem.*, USSR, Vol. 30 (1957), pp. 1235–1243.
13. Kolar, V., *Coll. Czech, Chem. Commun.*, Vol 26 (1961), pp. 613–627.
14. Weisman, J., and Efferding, L. E., *AIChE J.*, Vol. 6, No.3 (1960), pp. 419–426.
15. Einenkel, W. D., and Mersmann, A., *Verfahrenstechnik* (*Mainz*), Vol. 11, No. 2 (1977), pp. 90–94.
16. Kotzek, R., et al., *Mitt. Inst. Chemieanl.* Vol. 9, No. 2 (1969), pp. 53–58.
17. Zlokarnik, M., and Judat, H., *Chem-Ing-Tech.*, Vol. 41, No. 23 (1969), pp. 1270–1273.
18. Hobler, T., and Zablocki, *J., Chem. Tech.* (*Leipzig*), Vol. 18, No. 11 (1966), pp. 650–652.
19. Muller, W., and Todtenhaupt, E. K., *Aufbereit. Tech.*, Vol. 13, No. 1 (1972), pp. 38–42.
20. Rautzen, R. R., Corpstein, R. R., and Dickey, D. S., *Chemical Engineering* (October 25, 1976), pp. 119—126.
21. Herringe, R. A., Proceedings 3rd European Conference on Mixing, York, England, April 1979, Vol. I, pp. 199–216 Paper D1.
22. Middleton, J. A., Proceedings 3rd European Conference on Mixing, York, England, Vol 1, p. 15 (1979) Paper A2.
23. Carpenter, R. K., and Painter, L. A. (Presented 1955 Annual Meeting AIMME) February 1955.
24. Oldshue, J. Y., and Mady, O. B., *Chemical Engineering Progress*, Vol. 74, (August 1978), p. 103.
25. Oldshue, J. Y., and Mady, O. B., *Chemical Engineering Progress*, Vol. 75 (May 1979), p. 72.
26. Connelly, J. R., and Winter, R. L., *Chemical Engineering Progress*, Vol. 65, No. 8 (1969), p.70.
27. Oldshue, J. Y., *Fluid Mixing Technology*, Chemical Engineering/McGraw-Hill Publications Co. (1983).

CHAPTER 30

SIMULATION OF MECHANICALLY-AGITATED LIQUID-LIQUID REACTION VESSELS

William Resnick

Department of Chemical Engineering
Technion-Israel Institute of Technology
Haifa, Israel

CONTENTS

INTRODUCTION, 851

FACTORS AFFECTING DISPERSION GEOMETRY AND BEHAVIOR, 852
Flow Patterns and Velocity Fields, 852
Behavior of Drops in Dispersions, 853
Interphase Heat and Mass Transfer, 853

CHEMICAL REACTIONS IN DISPERSIONS, 854

MODELING OF LIQUID-LIQUID REACTORS, 856
Noninteracting Models, 856
Dynamic Noninteraction Simulation, 857
Simulation of Two-Phase Liquid-Liquid Reactor, 857
Dynamic Simulation of Agitated Liquid-Liquid Reactor, 863

INTERACTION MODELS, 867
Circulation-Interaction Model, 867
Development of the Circulation-Interaction Model, 868

POPULATION BALANCE TECHNIQUES, 877
Monte Carlo Simulation, 879
Monte Carlo Simulations and Population Balances, 881

NOTATION, 881

REFERENCES, 882

INTRODUCTION

The experimental study of agitation is generally done in laboratory or pilot-scale equipment. The intention is, however, that the resulting information and its analysis be used for the design of larger apparatus which are required to meet given process requirements. The engineer would like to design the apparatus based on knowledge of basic hydrodynamic parameters and molecular properties and from these to derive the required gometry and operating parameters. A mechanically agitated mixing system is complex with mutual interactions taking place among many elements. A few of these elements can be mentioned—impeller and vessel geometry, fluid properties, hydraulic characteristics, fluid shear, power and pumping rates. The system is sufficiently complicated when a single liquid is considered. It becomes even more complicated when a liquid-liquid system is considered in which, in addition, a reaction is taking place.

The successful transfer from laboratory or pilot-plant scale to commercial scale requires appropriate attention to the relevant scale-up criteria. In the absence of an adequate understanding of the governing phenomena these criteria are usually determined empirically. The appropriate scale-up criteria can be determined, however, only when sufficient understanding of the physical phenomena occurring is available to permit modeling or simulation of the operation. In this paper several procedures for the simulation of mechanically agitated liquid-liquid reactors will be discussed.

FACTORS AFFECTING DISPERSION GEOMETRY AND BEHAVIOR

The description of an agitated liquid-liquid system requires knowledge of dispersion geometry and behavior. The phenomena, parameters, and properties affecting dispersion geometry and behavior not only are numerous and complex but also are interrelated. As an example of some of these interrelationships consider the size of drops in a liquid-liquid dispersion. The drop size distribution at "steady state" is determined by a dynamic equilibrium between drop breakup and coalescence occurring in the dispersion. Any change in the factors which affect coalescence and breakup rates, such as holdup, will cause a change in the drop size distribution. Local holdup itself depends upon a variety of factors such as flow rates, equipment geometry and settling and slip velocities. The latter two factors depend not only upon drop size but also upon among other things phase viscosities, interfacial tension, and electrical properties, which in turn are among the factors which affect drop size.

Some of the information that is immediately relevant to the simulation of mechanically agitated liquid-liquid reactors will now be reviewed. This review will consider dispersion geometry and behavior, heat and mass transfer in dispersions, and chemical reactions in dispersions.

Flow Patterns and Velocity Fields

Flow patterns and velocity fields in vessels equipped with turbine impellers and containing one- or two-phase liquid systems have been described and reviewed by a number of investigators [1–5]. A simplified image of the flow pattern and circulation loops in batch operation is shown in Figure 1. Inflow and outflow in continuous operation affect the flow pattern to a relatively small extent at the usual ratios of feed rate to tank volume.

The amount of fluid discharged from a rotating impeller depends on impeller diameter, impeller speed, the geometry of the equipment, and physical properties of the fluids. Wolf and Manning [6] and Cooper and Wolf [7, 8] developed analytical expressions for the pumping capacity of turbine impellers and propellors and tested them experimentally. They suggest the following correlation for the impeller pumping capacity

$$F = K_t \cdot I_s \cdot D_I^2 \tag{1}$$

Many other investigators measured pumping capacity average velocities at several locations along the circulation loops and circulation times. The local continuous phase velocity was found

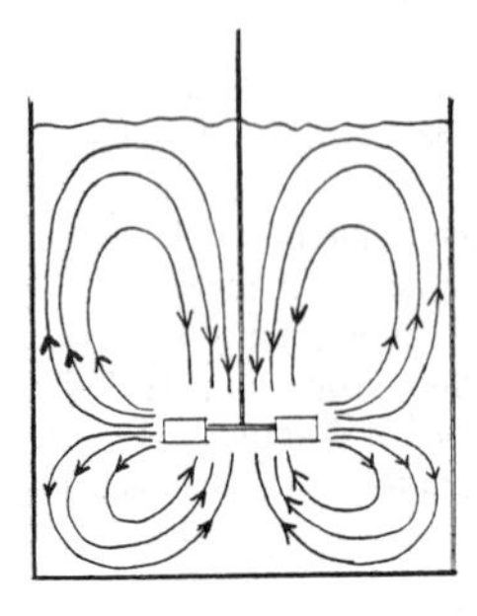

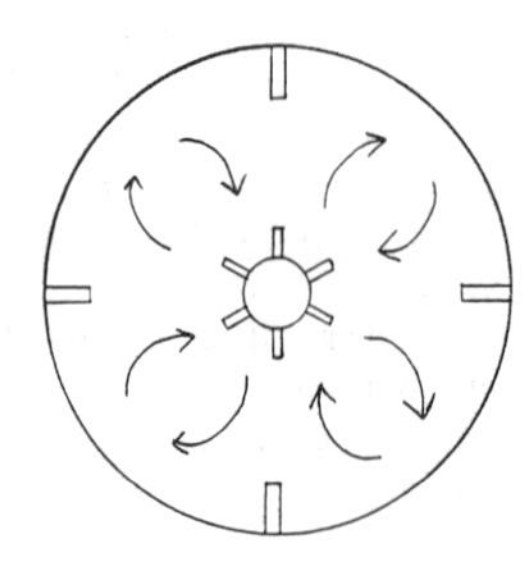

Figure 1. Flow pattern and circulation loops in a mechanically agitated vessel.

to be directly proportional to the impeller speed. Impeller speeds required for liquid-liquid dispersions have been reported [14, 15].

Behavior of Drops in Dispersions

A number of drop phenomena and the influences of these phenomena on the behavior of liquid-liquid dispersions have been reviewed [17–21]. The phenomena include drop formation, shape, deformation and breakup, internal circulation, oscillation, coagulation, coalescence, surface renewal, boundary layers, double electric layers, surfactant layers, and the like. Studies of relative velocities between drops and the continuous phase [1, 3, 22, 23] are also of immediate relevance to an understanding of dispersed-phase reactors. An important conclusion from these works is that for regions of strong turbulence, as in many industrial dispersions, the slip velocity of the drops relative to the continuous-phase, the internal circulation, and the shape oscillation are negligible, and consequently, convective heat and mass transfer are negligible, the most important factors being diffusion and coalescence and breakup.

The drop size distribution at steady state in an agitated liquid-liquid dispersion reflects a state of dynamic equilibrium between breakup and coalescence phenomena. For a given set of conditions, drops above a certain size will almost certainly break up because they become very deformed while drops below a certain size will, with a high probability, coalesce upon collision with other drops because of interfacial tension.

The characteristics and intensities of turbulence are different in different regions of the vessel and turbulence and local power dissipation per unit volume are strongest in the vicinity of the impeller. As a result, different phenomena take place in different regions of the mixer:

1. In the high-turbulence region near the impeller, drop breakup is the dominant phenomenon.
2. In regions remote from the impeller, little drop breakup occurs, and a certain amount of coalescence occurs [24–26].

The dependence of drop size on location in a stirred tank was studied by several investigators [9, 13, 18, 24–25, 33–35]. The main findings confirm that the drop size is smallest in the impeller discharge region and that the dependence of drop size on location is strong when the coalescence rate is high when compared with the circulation rate. Knowledge of coalescence behavior in the dispersion is important for design of reactors and the analysis of reactions in agitated dispersions, because the extent of interaction between drops in a dispersion has a marked influence on conversion and temperature in the reactor.

The dynamic response of local drop diameter to changes in phase composition and temperatures has been measured [29] and the importance of considering dynamic variations in interfacial area for proper description of reactor behavior has been pointed out [30]. An analysis of the observability of agitated liquid-liquid reacting systems shows that on-line information about changes in drop size sometimes is important for the estimation of variables which are difficult to measure [31].

Interphase Heat and Mass Transfer

This subject has attracted much research attention, experimental as well as theoretical. A discussion of some of the results with the greatest relevance for this work follows. In dispersions with small surfactant-contaminated drops, the dominant mechanisms for heat and mass transfer inside the drops are thermal conduction and molecular diffusion, respectively. Outside the drops, convection is dominant, but the main resistance to heat and mass transfer remains the resistance inside the drop. A large proportion of the total mass transfer occurs shortly after the formation of a new interface [14, 32]. Therefore, the rates of coalescence and redispersion and any phenomena influencing them will markedly affect mass transfer rates [13, 27, 33–34]. Thus, there is a mutual, complex dependence between mass transfer rates and coalescence redispersion rates, and hence, indirectly, between mass transfer rates and dispersion geometry as represented by drop size and holdup.

A wide range of interfacial mass and heat transfer correlations have been presented for liquid-liquid dispersions in stirred tanks. A number of researches that include many individual and total mass and heat transfer correlations have been reviewed [18, 20]. An accepted correlation for the individual mass transfer coefficient in the continuous phase is

$$Sh_C = K_1 \cdot (Re)^{\beta}(Sc)_C^{\gamma} \tag{2}$$

The density and viscosity for the calculation of the Re number are expressed by mean values [35]. The mass transfer coefficient in the dispersed phase when the governing mechanism for mass transfer in the droplets is molecular diffusion is

$$k_D = \frac{4\pi^2 \, Dif_D}{6 \cdot \phi \cdot D_{32}} \tag{3}$$

In a review of simultaneous interface heat and mass transfer [36] the following analogy between mass and heat transfer coefficients was suggested

$$\frac{h/(C_p \cdot \rho)}{k} = \left[\frac{Tcon/(Cp \cdot \rho)}{Dif}\right]^n \tag{4}$$

The value of the exponent ranges between 0.5 and 1 and depends upon the transport mechanisms and phase properties.

Holland and Chapman [37] and Oldshue [38] review many experimental correlations of heat transfer coefficients between fluids in agitated vessels and coolant. The correlations are of the following form

$$Nu_C = K_2 \cdot Re^{\beta} \cdot Pr_C^{\gamma}(\mu_w/\mu_b)^{\delta} \tag{5}$$

and the physical properties that appear in the various equations and correlations cited above are functions of the appropriate phase temperature.

CHEMICAL REACTIONS IN DISPERSIONS

Liquid-liquid reactions are of importance in a number of processing systems and many chemical reactions of industrial importance are carried out in mechanically-agitated liquid-liquid systems. The nitration and halogenation of aliphatic and aromatic hydrocarbons generally are performed in liquid-liquid reactors. Tonnage quantities of PVC are produced by emulsion polymerization in an aqueous medium in mechanically-agitated reactors. Metal chelation reactions are of importance in hydrometallurgical processes and are applied industrially in the extraction of tonnage quantities of copper. Phase transfer catalysis [39, 40], developed fairly recently, represents a technique in which reactions are conducted in aqueous-organic two-phase systems and are catalyzed by an ammonium or phosphonium salt. Phase transfer catalysis has been used as a preparative technique but may find industrial applications.

The design of this important class of reactors requires that dispersion behavior with respect to hydrodynamics as represented by factors such as drop size, breakup, coalescence, and dispersed-phase holdup along with interphase mass and energy transfer be coupled with reaction kinetics. Thus, microscopic considerations must be invoked when dealing with interphase transfer and kinetics while macroscopic conditions are invoked when dealing with drop size, surface area, holdup, breakup, and coalescence.

Some of the results of a small fraction of the large number of works devoted to liquid-liquid reactor behavior and characteristics will be cited to give an indication of the complexity of such systems and of the care which must be exercised in the interpretations of observations.

In their studies of liquid-liquid reactor behavior Schmitz and Amundson [41] and Luss and Amundson [42] analyzed stability and transient behavior in two-phase reactors. They show that for certain operating conditions multiple steady states exist and that the steady-state obtained depends on the initial conditions. Luss and Amundson in their analysis of completely segregated systems show that steady states may be obtained in which the temperature of some of the drops greatly exceeds that of the other drops. Barnea, Hoffer, and Resnick [43, 44] show that "hot spots" could develop and that overshooting and undershooting responses could be obtained in response to step changes in some operating conditions.

Cox and Strachan [45–47] and Chapman, Cox, and Strachan [48] performed experimental work on two-phase nitrations and presented solubility data, reaction rate constants, and interfacial heat transfer coefficients for these systems. They show that whereas the nitration rate of chlorobenzene is governed by the reaction kinetics, in the case of toluene the mass transfer rate has a marked effect on the overall rate. With respect to the kinetics of toluene nitration they observed a transition from first to zero kinetics order with respect to toluene.

The importance of metal chelation reactions in hydrometallurgical processing has generated much activity in the study of these reactions [49–58]. The hydrometallurgical recovery of copper involves the leaching of copper from the ore by sulfuric acid solutions followed by the extraction of the copper from the aqueous acid by chelation agents in an organic solvent. The overall chelation reaction can be written as

$$[Cu^{2+}] + 2\,HR \rightleftharpoons CuR_2 + 2[H^+]$$

where both the chelating agent, HR, and the complexed copper, CuR_2, are in the organic phase, whereas the bracketed ions are in the aqueous phase.

As in other situations where transport and reaction kinetics are involved, three general cases can be observed depending on the relative rates of the transport and kinetic processes.

1. Fast reactions with the result that the transport is rate-controlling and chemical equilibrium is achieved at the interface.
2. Slow reactions in which the chemical kinetics are the rate-controlling factor and transport rate is unimportant.
3. The intermediate case in which the mass transfer rate and the reaction rate are of the same order of magnitude.

In their study of the copper-LIX-64 system (LIX-64 is a β-hydroxy omide plus a low concentration of an α-hydroxy oxime and kerosene) Whewell, Hughes, and Hansen [50] found that the rate was proportional to the interfacial area. Flett et al. [51] found that the rate-controlling step in the extraction by alkylated 8-hydroxy quinoline was the formation of the neutral chelate at the interface. Hughes, Sergeant, and Whewell [58] point out, however, that the reported activation energies for the chelation reaction suggest that diffusion may be rate controlling. They note that an interfacial organic diffusion process must occur and that this may be of significance to the rate equation. Part of the uncertainties concerning the significant and rate-controlling mechanisms in this chelating reaction is undoubtedly due to the different experiment techniques used in the laboratory kinetic studies. Tavlerides [59] recently reviewed some of the laboratory reactors used for kinetic studies and listed the criteria for a suitable reactor for successful and significant liquid-liquid kinetic studies.

The rate and equilibrium processes in phase transfer catalysis have not yet been studied thoroughly but this technique can provide a fertile medium for the study of liquid-liquid reacting systems. The steps involved in a alkylation reaction as presented by Dehmlow [40] are shown in Figure 2. Ion pairs are indicated in this figure as bracketed ionic compounds. Thus, the ammonium salt migrates back and forth between the phases and transfers the hydroxide ion as an ammonium hydroxide ion pair. The ammonium hydroxide pair is converted into an ion pair in the organic phase with the substrate which is then alkylated. Scale-up of phase transfer catalytic reactions by rational modeling and simulation procedures will require the results of many studies of factors such as mechanistic aspects, kinetic and transport phenomena, and diffusion and equilibrium consideration. Kinkel and Tomlinson [60] developed a segmented flow/phase splitter assembly which is shown to be suitable for the study of fast reactions occurring in phase-transfer catalysis two-phase systems.

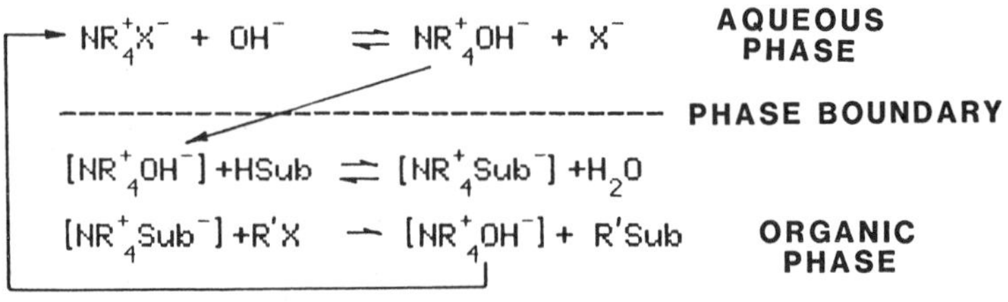

$$NR_4^+X^- + OH^- \rightleftharpoons NR_4^+OH^- + X^-$$

AQUEOUS PHASE

PHASE BOUNDARY

$$[NR_4^+OH^-] + HSub \rightleftharpoons [NR_4^+Sub^-] + H_2O$$

$$[NR_4^+Sub^-] + R'X \rightarrow [NR_4^+OH^-] + R'Sub$$

ORGANIC PHASE

Figure 2. Alkylation by phase transfer catalysis [40].

MODELING OF LIQUID-LIQUID REACTORS

From this brief review of some of the numerous and complex phenomena which take place in an agitated liquid-liquid dispersion it should be apparent that the problem of scale-up of liquid-liquid reactors is neither a simple nor straightforward one. To assist in understanding and analyzing the phenomena and, in addition, to permit scale-up of liquid-liquid reactors it is necessary to resort to modeling and subsequent simulation of the reactor behavior with the aid of the model. A variety of models has been proposed and they may be subdivided into two general categories. One category includes those models, termed noninteraction models, which do not concern themselves with micromixing phenomena in either phase. The other category comprises those models which do deal in one way or another with micromixing, generally in the dispersed phase. The latter are termed coalescence and dispersion models or interaction models. In the following sections one model from each of the categories will be presented in some detail whereas other models will be discussed only briefly.

Noninteracting Models

The simplest noninteracting models are those which are based on effective interfacial surface areas coupled with modeling of mass transfer and reaction while not considering dispersed phase interactions. The surface area can be estimated with information on drop diameter and holdup obtained from experimental measurements as described earlier with scale-up being based on some criteria such as power input per unit volume. Two-film, penetration, or surface renewal models are invoked for mass transfer with reaction [46, 48, 61–63, 65]. The simpler models in this category do not consider that drop breakup and coalescence occur, that the drops are not of uniform size nor that energy and mass transfer rates and interfacial fluxes are affected by the drop-size distribution. Drop-size distribution and residence time distribution can be taken into account by models which are based, generally, on the concept of spherical cells. This concept was used by Resnick and Gal-Or [65–67] for the case of equal-sized spherical bubbles in gas-liquid dispersions and extended and generalized by Tavlerides and Gal-Or [68] to systems consisting of drops, bubbles or solid particles with size and residence time distributions. For the liquid-liquid case the spherical-cell model assumes that the continuous phase is divided into a number of fluid cells, the number of cells being equal to the number of drops, generally assumed to be constant with time, in the dispersed phase. Each drop is then assigned a cell or element of continuous phase which accompanies it throughout its lifetime in the system.

Expressions are then developed for mass transfer and reaction for the system of the spherical drop embedded in its spherical cell of continous phase. These cell models, although they can account for drop-size and residence time distributions, cannot account for drop breakup and coalescence. It is highly unlikely that a drop or dispersed-phase element would retain its identity during its lifetime in the vessel without experiencing breakup events and subsequent coalescence of daughter droplets with daughter droplets produced in a different breakup event.

Dynamic Noninteraction Simulation

In this section the noninteraction model developed by Barnea, Hoffer, and Resnick [30, 44] will be presented and discussed in some detail. This model permits not only steady-state but also dynamic simulation of a mechanically agitated liquid-liquid reactor.

Dynamic responses of dispersed-phase holdup, of local average drop diameter, and of local specific interfacial area to disturbances in a number of system variables as well as the effect of these variables on the steady-state dispersion geometry have been reported [29, 69]. It was shown that phase composition has an effect upon the local dispersion geometry. In addition, although the observed dynamic responses could generally be represented by linear, overdamped systems with dead time, the responses to large disturbances were nonlinear and, in some cases, over- and undershooting and oscillatory responses were observed.

In order to find out whether and when it may be necessary to take dynamic variations in the interfacial area into account in liquid-liquid reactor simulation and to permit a study of the possible implications in control system design, Barnea et al. [30] developed a relatively simple model system in which the assumption of infinite breakup and coalescence rates was made. This assumption was mathematically convenient but still permitted the extraction of important information which might be valid also for systems with finite coalescence and breakup rates.

Simulation of Two-Phase Liquid-Liquid Reactor

The assumption of infinite breakup and coalescence rates implied a stirred tank of perfectly mixed phases. Thus, the assumption was made that each phase represented a perfectly mixed cell. The interfacial area and the dispersed-phase volume are calculable on the basis of dispersed-phase holdup and average drop diameter. The model development, however, treats the average diameter and interfacial area correctly by demanding that the appropriate average diameter be used. Thus, the interfacial area and dispersed-phase volume are represented by the Sauter or volume-surface mean diameter, D_{32}, which is the appropriate average diameter for use when transport considerations, i.e., the interfacial area, must be considered along with reaction and conversions, i.e., the volume or mass of the reacting system. The implied assumption is that the mass and energy fluxes are independent of drop diameter. In the model the continuous phase (C) and the dispersed phase (D) are fed continuously to the tank at rates F_{CO} and F_{DO}, respectively. The phases' properties and volume are assumed to be independent of changes in solute concentrations that result from chemical conversion or mass transfer. A single, exothermic, irreversible reaction is assumed: A → product, which takes palce in both phases. Assuming reaction kinetics of order α, the following differential equations describe the material and energy balances of the phases:

$$\frac{dC_{AD}}{dt} = \frac{F_{DO}}{V_t \cdot \phi}(C_{ADO} - C_{AC}) - \frac{6 \cdot K_A}{D_{32}}(\bar{D}C_{AD} - C_{AC}) - k_{OD}e^{-(Q_D/T_D)}C_{AD}^{\alpha} \tag{6}$$

$$\frac{dT_D}{dt} = \frac{F_{DO}}{V_t \cdot \phi}(T_{DO} - T_D) - \frac{6 \cdot h_t}{\rho_D Cp_D D_{32}}(T_D - T_C) + \frac{(-\Delta H)}{\rho_D Cp_D} k_{OD}e^{-(Q_D/T_D)}C_{AD}^{\alpha} - \frac{U_D \cdot s_D}{\rho_D Cp_D V_t \cdot \phi}(T_D - T_w) \tag{7}$$

$$\frac{dC_{AC}}{dt} = \frac{F_{CO}}{V_t(1 - \phi)}(C_{ACO} - C_{AC}) + \frac{6 \cdot K_A \cdot \phi}{(1 - \phi)D_{32}}(\bar{D}C_{AD} - C_{AC}) - k_{OC}e^{-(Q_C/T_C)}C_{AC}^{\alpha} \tag{8}$$

$$\frac{dT_C}{dt} = \frac{F_{CO}}{(1 - \phi)V_t}(T_{CO} - T_C) + \frac{6 \cdot h_t \cdot \phi}{\rho_C Cp_C(1 - \phi)D_{32}}(T_D - T_C) + \frac{(-\Delta H)}{\rho_C Cp_C} k_{OC}e^{-(Q_C/T_C)} \cdot C_{AC}^{\alpha} - \frac{U_C \cdot s_C}{-(\rho_C Cp_C V_t)(1 - \phi)}(T_C - T_w) \tag{9}$$

The physico-chemical properties that might have a major effect on conversion and reactor temperature are:

1. The distribution coefficient between the phases ($\bar{D}$), which defines the phase in which reactant A is more soluble.
2. The activation energy of A in each phase (Q), which defines the more reactive phase.

The relative values of these two properties should influence the behavior of the reactor and the magnitude of the effect of changes in mean drop diameter. To determine the range of values of these variables for which dynamic variations in the interfacial area would have a significant effect, the sensitivity of reactant conversion to reactant solubility and reactivity characteristics was studied as a function of steady-state values of D_{32}. Assuming a first-order reaction, $\alpha = 1$, and that the mean residence times for both phases, τ_D and τ_C are equal, the steady-state mass balance equations yield, for $\tau_D = \tau_C = \tau$,

$$C_{AC} = \frac{C_{ACO}/\tau + KA_C \cdot D \cdot C_{ADO}/(\tau \cdot TC_D)}{TC_C - KA_C \cdot KA_D \cdot \bar{D}/TC_D} \tag{10}$$

$$C_{AD} = \frac{C_{ADO}/\tau - KA_D \cdot C_{AC}}{TC_D} \tag{11}$$

where $KA_D = \dfrac{6 \cdot K_A}{D_{32}}$

$$KA_C = \frac{6 \cdot K_A \cdot \phi}{D_{32}(1 - \phi)}$$

$$TC_D = \frac{1}{\tau} + KA_D \cdot \bar{D} + k_{OD}e^{-(Q_D/T_D)}$$

$$TC_C = \frac{1}{\tau} + KA_C + k_{OC}e^{-(Q_C/T_C)}$$

The conversion in the two-phase system is determined as

$$Ex = 1 - \frac{F_D \cdot C_{AD} + F_C \cdot C_{AC}}{F_{DO} \cdot C_{ADO} + F_{CO} \cdot C_{ACO}} \tag{12}$$

The effect of reactivity expressed as Q_C/T_C and Q_D/T_D was examined for each of the three possible cases for the solubility of the reactant, A:

1. $\bar{D} > 1$, i.e., A is more soluble in the continuous phase.
2. $\bar{D} < 1$, i.e., A is more soluble in the dispersed phase.
3. $D = 1$, A is equally soluble in both phases.

and the conversions for these combinations was calculated as a function of D_{32}.

The following values for the parameters were used: $\tau_D = \tau_C = 500$ sec; $F_{DO} = 10$ cm^3/sec; $F_{CO} = 40$ cm^3/sec; $\phi = 0.2$; $K_{OD} = K_{OC} = 10^9$ 1/sec; $K_A = 0.0001$ cm/sec; and some typical results are presented in Figures 3 and 4.

Figures 3A and 4A describe a system with a constant and relatively high value of Q_C/T_C, i.e. the continuous-phase reactivity is small. The system conversion is shown vs. Q_D/T_D for two values of the distribution coefficient $\bar{D}$, at a drop diameter, D_{32}, of 0.01 cm. Two regions in which the conversion is independent of Q_D/T_D are observed: values of Q_D/T_D that are low as compared to Q_C/T_C and one at high values of Q_D/T_D (low reactivity). In the intermediate region a strong dependence

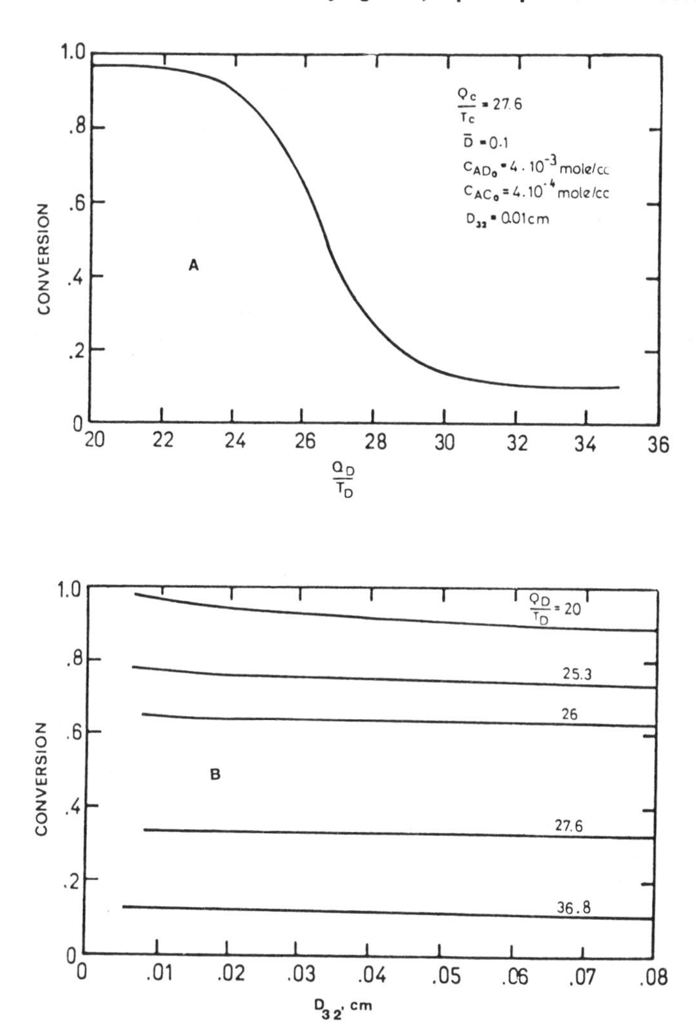

Figure 3. (A) Conversion vs. dispersed phase reactivity. (B) Conversion vs. mean drop diameter. Reactant A is more soluble in the dispersed phase [30].

of conversion upon Q_D/T_D is observed. Favorable solubility characteristics of the reactant in the dispersed-phase results in strong dependence of conversion on Q_D/T_D, whereas favorable solubility characteristics in the continuous phase result in strong dependence of conversion on Q_C/T_C.

The behavior of the system shown in Figure 4A, for example, in which reactant A has favorable solubility characteristics in the continuous phase ($\bar{D} = 10$), can be explained as follows: in the high conversion region at small values of Q_D/T_D, the rate of reaction in the dispersed phase is so rapid that it is controlled by the mass transfer rate of A from the continuous phase. The continuous phase is much less reactive but it contains most of the reactant. The mass transfer rate is relatively constant since the concentration of A in the dispersed phase is almost constant and approaches zero. In the region of high Q_D/T_D where the conversion is low and independent of Q_D/T_D, most of the reaction takes place in the continuous phase and at a relatively low rate. The reaction rate

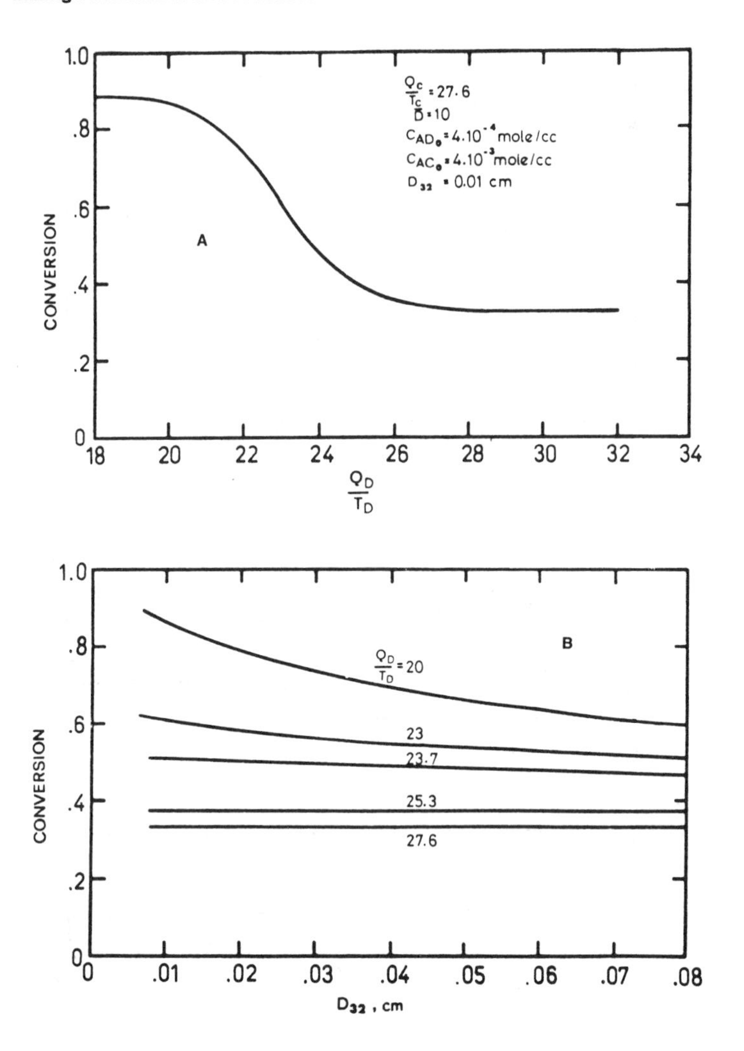

Figure 4. (A) Conversion vs. dispersed phase reactivity. (B) Conversion vs. mean drop diameter. Reactant A is more soluble in the continuous phase.

in the dispersed phase is so low at the high values of Q_D/T_D that the situation could be described as a reaction in a homogeneous phase (the continuous phase in the present case) with $Q_C/T_C = 27.6$ and $V_C = V_t(1 - \phi)$.

The effect of mean drop diameter (D_{32}) on steady-state conversion for each case previously mentioned is shown in Figures 3B and 4B in which Q_D/T_D appears as the parameter. The most prominent dependence of conversion on drop diameter as shown in Figure 4B occurs at high reactivity in the dispersed phase, i.e., at low magnitudes of Q_D/T_D. In this zone the extent of reaction is entirely controlled by the mass transfer rate from the continuous phase, which contains most of the reactant, to the dispersed phase in which the reactant conversion occurs. For $Q_D/T_D = 20$, changes of D_{32} from 0.008 cm to 0.08 cm cause a significant decrease in conversion from 0.885 to 0.585 whereas for $Q_D/T_D = 24 - 33$, the same changes in D_{32} do not affect conversion at all.

Table 1
Data for Figures 5 through 7

$C_{ADO} = 4 \times 10^{-4}$ mole/cm^3	$K_A = 0.0001$ cm/sec
$C_{ACO} = 4 \times 10^{-3}$ mole/cm^3	$h_1 = 10K_A/\rho_C Cp_C$ cm/sec
$D = 10$	$U_D \times s_D = 5$ cal/(sec)(°C)
$\tau_D = \tau_C = 500$ sec	$U_C \times s_C = 150$ cal/(sec)(°C)
$T_{DO} = T_{CO} = T_w = 300$ K	$\rho_D = 0.8$ g/cm^3
$\phi = 0.2$	$\rho_C = 1$ g/cm^3
$k_{OD} = 10^{10}$ 1/sec	$C_{PD} = 0.75$ cal/(g)(K)
$k_{OC} = 10^9$ 1/sec	$C_{PC} = 1$ cal/(g)(°J)
$\Delta H = -30{,}000$ cal/mole	

The charactertistic dependence of system conversion on Q_D/T_D as demonstrated in Figures 3A and 4A disappears if the continuous phase has favorable solubility characteristics and Q_C/T_C is relatively small. For the same parameters used for the case described in Figure 4 but using $Q_C/T_C = 23.7$ instead of 27.6, a constant conversion of 0.96 was obtained over the entire range of Q_D/T_D examined. The phase with the larger holdup has better dissolution characteristics as well as high reactivity, therefore, changes in the dispersed-phase reactivity and drop diameter will not affect system conversion.

A similar parameter study in which the effects of reactivity, expressed as Q, and temperature were examined was made by solving the simultaneous equations (Equations 6 through 9). Qualitative results similar to those previously described were obtained. Values of the parameter used in this part of the study are shown in Table 1.

The results of the numerical solution of the differential equations are presented graphically in Figures 5 and 7. The changes in system temperatures are governed by the changes in reaction rate because the heat transfer coefficients and inlet temperatures are constant parameters in this example. It can be seen that higher Q_D results in a smaller difference between phase temperatures and in many cases a physical equilibrium is achieved in phase temperatures. The temperature dependence on drop diameter is governed essentially by the dependence of conversion on drop diameter. A comparison between T_D and T_C changes at several mean drop diameters points out that the difference between phase temperatures increases as the drop diameter increases, i.e. the heat removal ability of the continuous phase becomes smaller.

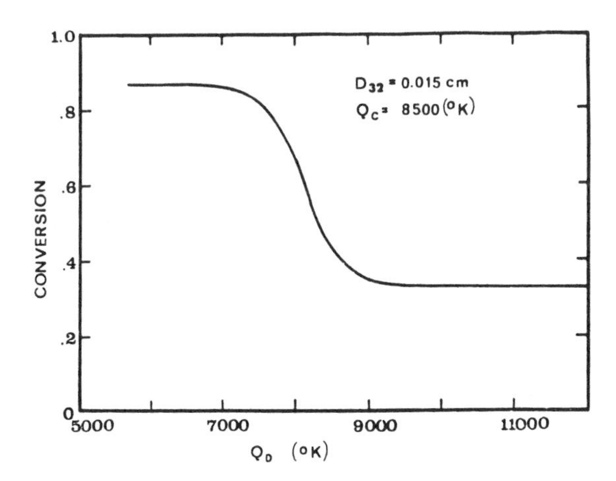

Figure 5. Conversion vs. Q_D. Reactant A is more soluble in the continuous phase.

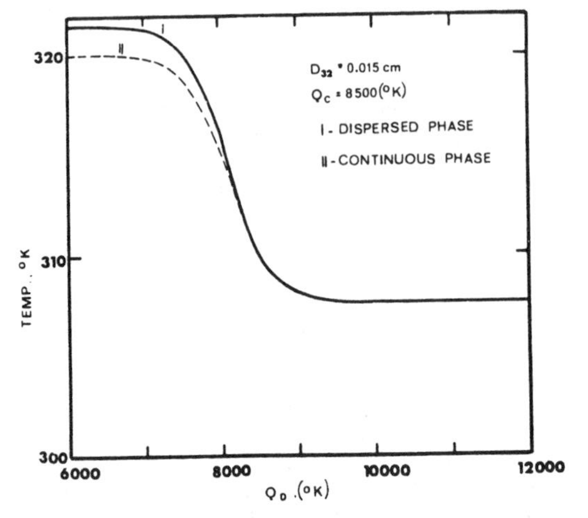

Figure 6. Phases temperatures vs. Q_D.

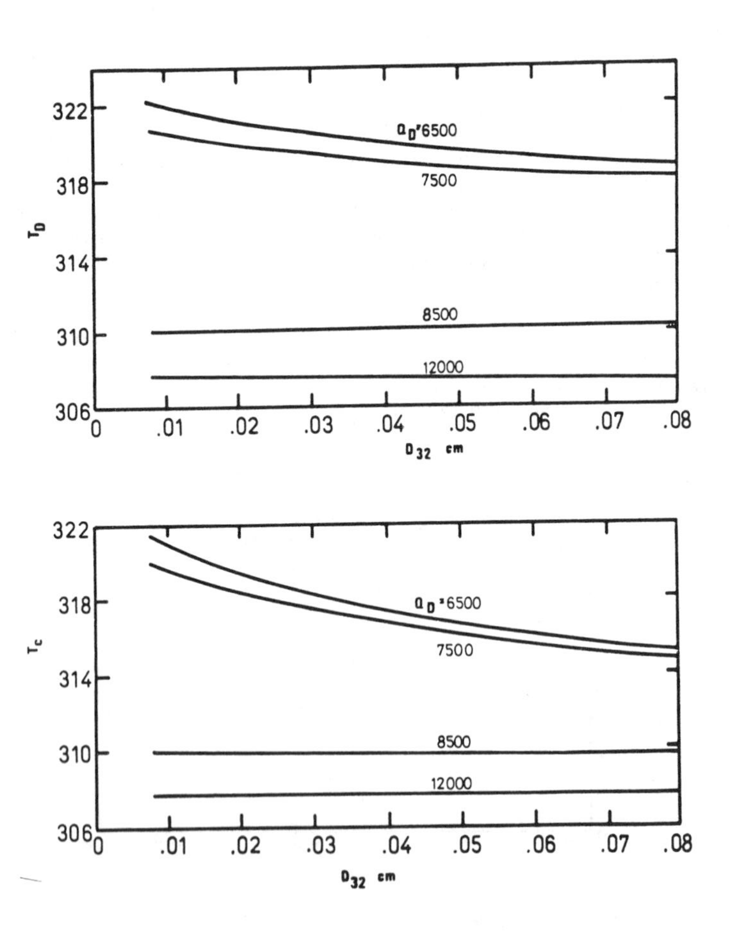

Figure 7. Phases temperature vs. mean drop diameter.

In summary, the simulation showed that the average drop diameter affected conversion at the steady state in all cases in which interphase mass transfer is the rate-determining factor in the chemical reaction. The conversion dependence on drop diameter was more marked when the less reactive phase had the favorable solubility characteristics. The conversion dependence on drop diameter is more prominent if the reactive phase has the larger holdup. In exothermic systems with a strong dependence of conversion on drop diameter a strong dependence of the same direction on temperature is found. The larger the drop diameter, the larger is the difference between phase temperatures.

Dynamic Simulation of Agitated Liquid-Liquid Reactor

The model Equations 5 through 9 were coupled with the appropriate additional dynamic and algebraic constitutive equations in order to perform dynamic simulations of reactor behavior. Appropriate dynamic equations based on experimental drop behavior data [29, 69] were developed to describe the dynamic dependence of drop diameter on dispersion concentration, temperature, and impeller speed. Algebraic constitutive equations were based on reported correlations among the transfer coefficients and state variables, operating conditions, tank geometry, and physical properties of the fluids.

The object of the dynamic simulations by Barnea et al. [30] was to determine if and when it is necessary to measure dynamic changes in drop diameter for proper description of the process dynamics. The nature of the approach to the steady state from several initial conditions, the system response to changes in some operating conditions around the steady state, and the system response to sudden changes in state variables around the steady state were examined for three different drop behavior assumptions:

1. Constant drop diameter during the transient response. The drop diameter was assumed to correspond to the physical conditions existing at the initial state prior to the change in the operating conditions.
2. The drop diameter changes immediately as a result of the imposed changes in the process conditions, i.e., an algebraic dependence between drop diameter and the system variables C_{AC}, T_C, and I_S was assumed.
3. A dynamic drop diameter dependence on system variables was assumed.

The earlier parameter study showed that the most marked effect of drop diameter changes on a reactor performance was obtained in the cases where the reaction rate was mass transfer limited. Similar conditions were, therefore, selected for the dynamic simulations [53].

Approach to steady-state from start up. The system was assumed to be at hydrodynamic and thermal equilibrium with no reacting component A present and system temperature of 310 K. At time zero, system start-up was initiated by feeding continuous and dispersed phases at the appropriate ratio in accordance with the assumed ϕ. The two phases, at 300 K, contained reactant A at concentrations of C_{ACO} and C_{ADO} in continuous and dispersed phases, respectively.

The dynamic changes in reactant concentration and continuous-phase temperature while the system approaches its new steady state are presented in Figure 8 for the three cases assumed for drop size dependence.

When the drop diameter was assumed constant or related algebraically to the variables, the concentration in the continuous phase increased quickly to a maximum value, different in each case. Later, the concentration decreased slowly to the steady state. The continuous phase temperature decreased in these cases to a minimum and then increased to steady-state value. When the drop diameter had an assumed dynamic dependence, both the reactant concentraton in the continuous phase and its temperature have an oscillatory overdamped response and the concentration response is similar to that of a second-order under-damped system.

The response shown in Figure 8 can be explained as follows: In the system of concern the reactant A was assumed to be more soluble in the continuous phase. Although the reaction occurs

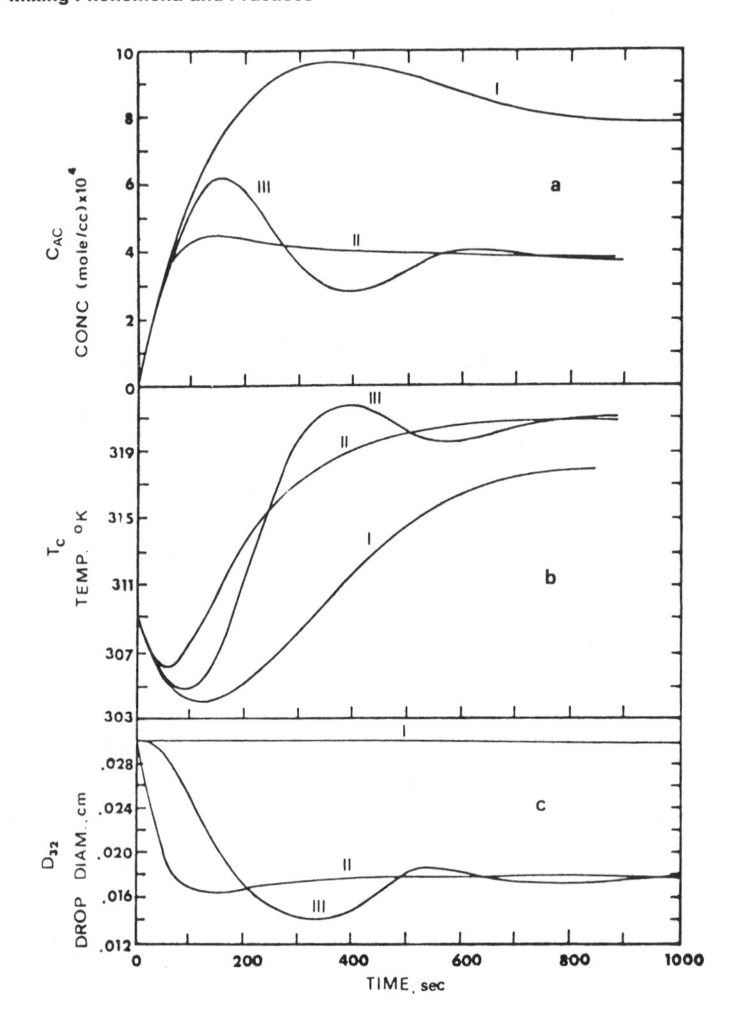

Figure 8. System startup. I—constant drop diameter, II—algebraic dependence of drop diameter on C_{AC} I_S; III—dynamic dependence of drop diameter on $C_{AC}T_C$, I_S.

in both phases the rate of reaction in the dispersed phase is much faster. Thus, most of the reactant disappearance from the continuous phase is caused by mass transfer to the dispersed phase. The dynamic response of concentration and temperature in each phase is determined, the practice, by the relation between the rate of addition of reactant and heat to the system and their rate of disappearance or removal by reaction or transport phenomena. When the drop diameter is constant, as in case 1, it is observed that the concentration of A in the continuous phase increases over a relatively long period before a slow decrease to the new steady-state value. The mass transfer rate to the dispersed phase is smaller than for cases 2 and 3 because of relatively small and constant interfacial area which results in a slower reaction in the dispersed phase and therefore also in a small mass transfer driving force. Only after 360 sec does the rate of A disappearance exceed its addition rate from the inlet stream and a slow decrease in A concentration is subsequently observed leading to a steady state with a relatively high reactant concentration. For this case the overall steady-state conversion of A to product was 0.8.

For case 2 in which the drop diameter is assumed to be algebraically related to the variables, the maximum in A concentration is achieved in 100 sec with this maximum being lower than that observed in case 1. The increase in A concentration immediately yields a rapid decrease in drop diameter and a concomitant increase in interfacial area, (Figure 8C), hence the rate of mass transfer to the dispersed phase increases, the reaction rate in the dispersed phase is accelerated, and this causes a larger concentration gradient between the two phases which stimulates the mass transfer rate even more. In addition, the fast reaction causes an increase in the dispersed-phase temperature which effects a rapid rise in the continuous-phase temperature. This rise in temperature, in turn, accelerates the reaction rate in the continuous phase and contributes to A removal from this phase. Therefore, after the relatively low maximum in concentration is achieved, a subsequent decrease toward steady state is observed. For this case a high overall conversion of A of 0.905 is attained.

When drop diameter is assumed to be dynamically dependent on the variables, case 3, the concentration response is oscillatory. The initial increase in this case was continuous during a longer period and achieved a higher maximum than in case 2. It can be seen from Figure 8C that the change in drop diameter is not immediate and after a dead time of approximately 20 sec a second-order under-damped response begins. This slow decrease in drop diameter results in a slower increase in the mass transfer rate than in case 2; therefore, a higher maximum of A concentration in the continuous phase is achieved. This maximum in concentration, however, results after a time lag in a greater decrease in drop diameter because of the dynamic nature of the drop diameter response. The decrease, in turn, causes an additional increase in the mass transfer rate from the continuous to the dispersed phase and to a decrease in A concentration in the continuous phase, which yields, again after a time lag, an increase in drop diameter and decrease in the mass transfer rate and the cycle repeats until a damped steady state is reached.

Similar considerations apply to the temperature behavior. Every change in concentration causes a change in reaction rate in each phase and also in the heat generation rate which, in turn, affects drop diameter, reactant concentration, and temperature in each phase.

Dynamic response of state variables to step changes in operating conditions. The dynamic responses of some state variables to step changes in several operating conditions were studied but only the transient response to a step change in impeller speed will be discussed here. The initial condition for dynamic response calculations was the steady-state achieved during start-up with a drop assumed to have had a dynamic or algebraic dependence, cases 2 and 3 of the preceding section. For each input variable the effect of drop behavior with cases 1, 2, and 3 dependence was tested.

Figure 9 shows the dynamic response of several state variables of the system to a step change in impeller speed. A change in impeller speed affects the interfacial heat and mass coefficient and the rate of heat transfer to coolant. Therefore, temperature and concentration in continuous phase are affected even for the case of constant drop diameter, (Figure 9A and B). A positive step change in impeller speed results in enhanced heat transfer from the dispersed to continuous phase but, on the other hand, increases the rate of heat removal from the continuous phase to the coolant. In addition, a smaller reaction rate and less reaction-heat-generation results which also contributes to the decrease in phase temperatures. In Case II a step change in impeller speed results in an immediate and significant decrease in the drop diameter (Figure 9C). The interfacial mass transfer rate increases and a fast decrease in A concentration in the continuous phase results. This decrease in concentration causes, on the other hand, an increase in drop diameter and a decrease in the mass transfer rate from the continuous to the dispersed phase and, therefore, the concentration of A in the continuous phase increases. These opposite phenomena result in a minimum in reactant A concentration followed by a moderate rate of increase to the steady-state value.

The continuous phase temperature response in Case II is characterized by a rapid increase followed by a slower decrease to the new steady state (Figure 9B). The decrease in drop diameter and the increase in transfer coefficients cause a higher heat transfer rate from the dispersed to the continuous phase. In addition, the mass transfer rate from the continuous to dispersed phase increases, therefore the rate of heat-generation due to reaction in the dispersed phase also increases.

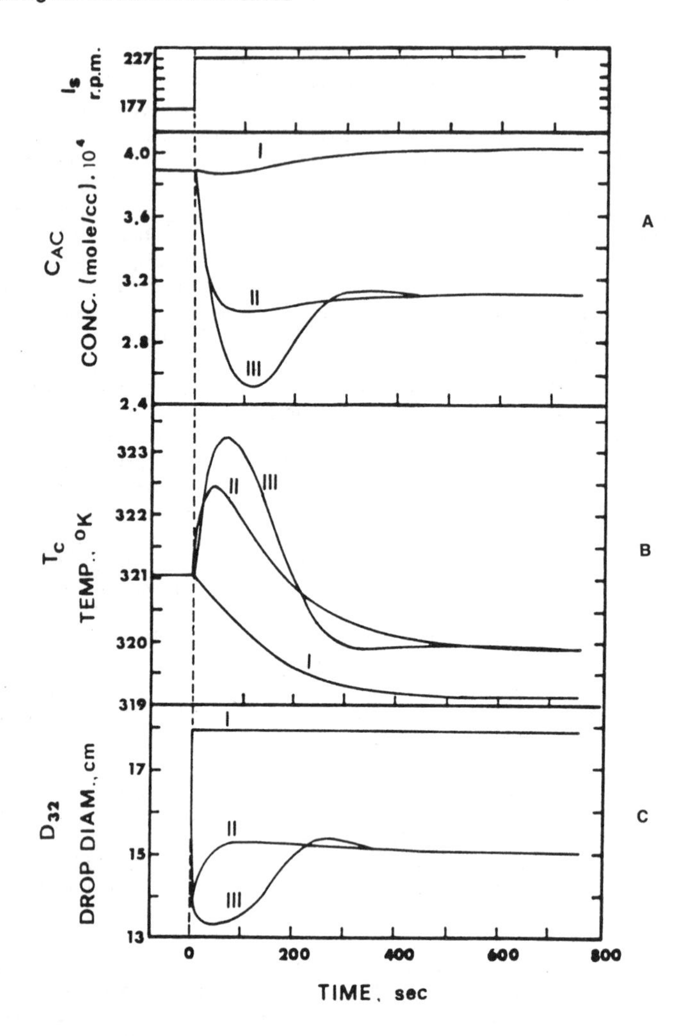

Figure 9. System response to step changes in I_S.

The rate of heat removal, however, simultaneously increases (u_C increases) and causes a decrease in phase temperature, reaction rate, and heat generation. The initial decrease in reactant concentration in the continuous phase causes an increase in drop diameter (Figure 9C) and a decrease in heat transfer rate from the dispersed to continuous phase. All these opposing factors cause a maximum in temperature followed by a considerable decrease later on toward a new steady-state temperature.

When the drop diameter has a dynamic dependence, a sharp decrease in reactant concentration in the continuous phase is observed after a short dead time of three to four seconds. The concentration reaches a lower minimum value than that observed for Case II. When the diameter is algebrically dependent, immediately after the decrease in drop diameter caused by the increase in impeller speed, there is an increase in diameter caused by the decrease in the concentration of A in the continuous phase (Figure 9C). When, on the other hand, drop diameter changes dynamically, the increase in diameter that results from a decrease in C_{AC} is noted after a dynamic lag. The drop diameter, therefore, reaches lower values and remains at these values for a longer time than in Case II.

Later on the increase that occurs in drop diameter slows the mass transfer rate and increases A concentration in the continuous phase. The increase in the rate of heat transfer to coolant operates in the same direction. Again, this increase in A concentration causes after a dynamic lag a decrease in drop diameter and in A concentration. The dynamic lag in diameter response causes, therefore, and oscillatory response in reactant concentration in the continuous phase.

The prominent maximum and minimum obtained in temperature response can be explained by similar considerations.

These simulations represented an attempt to determine if and when it is necessary to consider the dynamic behavior of drops in the dynamic description of liquid-liquid reacting systems. The importance of using dynamic relations between drop diameter and system variables for proper description of the transient response was shown by the dynamic simulations based on the mathematical model and experimental drop dynamics. Although sometimes the differences in the steady state obtained were not considerable, there usually were marked differences throughout the dynamic response. It was found that in most cases in which there is a dynamic dependence of drop size on phase composition, phase temperature, and impeller speed was assumed. The results showed oscillatory behavior of concentration and temperature with time. The oscillatory character of the transient response was a result of the dynamic lag between drop size and system variables. Large time constants in the dynamic response of drop size make the oscillatory behavior more pronounced. It is clear, therefore, that the dynamic changes of interfacial area must be considered in developing the controller equations for such systems (A, B).

INTERACTION MODELS

Interaction or coalescence and dispersion models attempt to take into account the fact that the dispersed phase undergoes breakup and coalescence in the agitated vessel and that macromixing as well as micromixing phenomena take place within and between the continuous and dispersed phases. The three types of coalescence and dispersion models to be discussed are a recently developed circulation-interaction model, population balance models, and Monte Carlo simulation models. The circulation-interaction model will be considered in some detail, and the remaining categories will be discussed only briefly.

Circulation-Interaction Model

This model proposed by Rietema [70] takes into account the general flow pattern in a baffled, liquid-liquid agitated reactor and also the observation that drop sizes are smallest near the impeller discharge region and larger far from it. The model assumes that a particle of dispersed phase passes through the impeller region several times during its stay in the vessel and that only drop breakup occurs in the impeller region, thus the drops leaving the impeller are small. It further assumes that only coalescence can occur on the circulation path from the impeller through the vessel and back to the impeller, thus, drops reentering the impeller region are larger than when they left it. This circulation-interaction model is reasonably tractable.

Reitema [70] suggests that two parameters are needed to handle this model. The first parameter is

$$\lambda_1 = \left(\frac{\text{maximum drop diameter}}{\text{minimum drop diameter}}\right)^3$$

which is the average number of small drops that coalesce during circulation to give one big drop. This parameter together with the circulation time gives an estimate of the coalescence rate.

The second parameter is

$$\lambda_2 = \frac{\text{average residence time}}{\text{circulation time}}$$

which indicates the average number of times that a dispersed phase element passes through the impeller zone.

The dependence of drop size on location in a stirred tank is large when the coalescence rate is high as compared to the circulation rate, i.e. at high values of λ_1. This condition is obtained at high values of holdup, low values of interfacial tension, or when some phenomenon which favors coalescence is present. In systems in which the circulation time is short and only few drops have time to coalesce along the loop, drop size is almost independent of location.

Barnea et al. [43] proposed a deterministic mathematical model for the Rietema circulation-interaction approach. Simulations performed with this model allowed the prediction of dynamic and steady-state behavior of two-phase reactors.

Development of the Circulation-Interaction Model

According to the basic assumptions of the circulation-interaction model the dispersed phase can be described as a plug-flow loop in which the drop diameter increases with distance from the impeller . By assuming that the impeller is located in the vessel center and that the circulation loops are symmetrical in shape, the model was confined to one circulation loop as shown in Figure 10.

It was assumed that all of the continuous phase circulates through an outer loop which is in contact with the dispersed phase circulating through an inner loop. In this dispersed-phase loop it was assumed, in accordance with Rietema [70] that all breakup that occurred took place in the impeller discharge zone and that only coalescence occurred along the rest of the loop. The loop was assumed to be made up of a series of equal-volume cells of dispersed phase of volume V_D in contact with cells of continuous phase (Figure 10). The number of cells is N and each cell of volume V_D/N is assumed to perfectly mixed. The equal-volume assumption means that the dispersed-phase interfacial area per cell changes along the loop in accordance with drop-size, assumed to be uniform in each cell, and that the interfacial area is largest in the impeller cell.

The division of the dispersed phase into several cells is a reasonably good approximation to the physical situation. One can expect, for example, that along the circulation loop there will be droplet groups with uniform drop diameter and uniform physical properties. The number of cells and the circulation capacity must meet the actual residence time distribution characteristics of the phases,

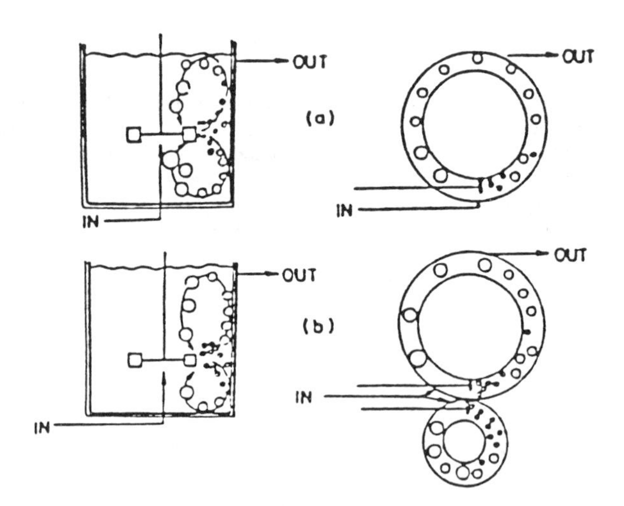

Figure 10. Circulation interaction model-physical aspect and schematic representation: (A) one loop (all loops equal); (B) two unequal loops.

and, in addition, be consistent with the drop diameter dispersed phase temperature, and concentration profiles.

Although the continuous phase is usually described as a perfectly mixed vessel with homogeneous properties the degree of homogeneity of the continuous phase depends on the recirculation flow which results from the impeller pumping action and on the local turbulence. In the circulation-interaction deterministic model the continuous phase of volume V_C is described as a series of W equal-volume perfectly mixed cells, with $W < N$ (Figure 10). The volume of each cell equals V_C/W. For convenience it is assumed that the ratio of N/W will be an integer, L.

Mixing in the dispersed phase will be enhanced by a high rate of interaction between drops. This condition will be achieved in the model when coalescence rates are high and circulation time is small as compared to the residence time. An additional contribution to the homogenization in the dispersed-phase properties will follow also from heat and mass transfer to the continuous phase.

The fact that the continuous phase has better mixing properties than the dispersed phase is expressed in the model by $W < N$. Any change in the continuous phase temperature or composition will, therefore, be felt simultaneously by several cells of the dispersed phase. As a result, plug flow is not the only means by which information is transferred from cell-to-cell in the dispersed phase. This fact decreases to a great extent the effect of the transportation lag when local changes in the dispersed-phase properties occur.

The model, shown schematically in Figure 11, assumes one circulation loop whose volume is assumed equal to that of the entire vessel, i. e., it contains all of the dispersed and continuous phases. Cells are numbered so that Cell Number One of the dispersed phase and Cell Number One of the continuous phase are both located in the impeller discharge zone.

The model development assumed a stirred tank with the continuous phase (C) and the dispersed phase (D) fed continuously at rates F_{CO} and F_{DO}, respectively. Although the simulation results to be discussed later are for the case where both the dispersed and continuous phases enter the vessel in the same region, the model can deal with the case when the two phases enter and leave the

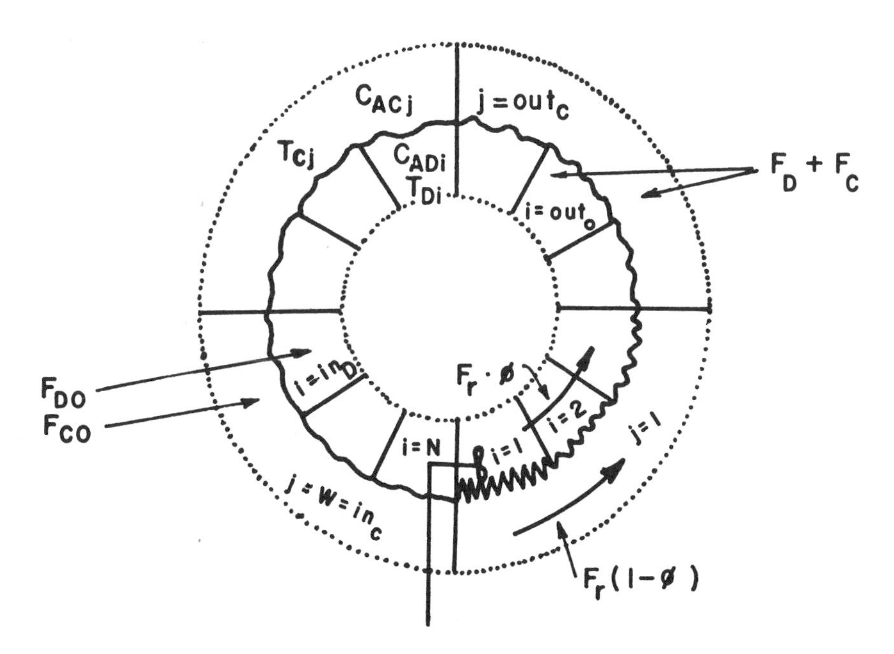

Figure 11. Schematic representation of the circulation-interaction model—one loop.

vessel at different locations. In addition, the model permits the number of dispersed and continuous phase cells along the loop to be varied.

The equations developed assumed a single, exothermic, irreversible reaction, A → Product, which takes place in each phase. Heat and mass transfer occur between the phases, and the vessel is cooled by a coolant system. The properties and volume of each phase are assumed to be constant and independent of changes in solute concentration that result from chemical conversions or mass transfer.

The state variables considered in the model [43, 44] were temperature and concentration in both phases, drop diameter, and dispersed-phase holdup. Input variables included inlet temperature, concentration and flow rate of each phase, coolant temperature, and impeller speed as well as the effect of adventitious contaminants on the dispersed-phase characteristics. The model not only permits changes in the location of the reactor feed inlet and product outlet locations but also calculates concentration and temperature profiles along the circulation path in each phase.

This circulation-interaction model [43, 44] results in a set of ordinary, nonlinear differential equations and a set of linear algebraic equations. The interdependence among the state variables is highly complex, the system is nonisothermal, there is a complex flow of "information" from the continuous phase to different regions of the dispersed phase, and "information" also flows within each phase by internal circulation. In addition there is a mutual dependence between the physical properties of the phases and the state variables of the system. The complexity of the equations is such that it is difficult if not impossible to predict system behavior by inspection of the model equations.

Simulation with the circulation-interaction model. Barnea et al. [43, 44] performed steady-state and unsteady-state simulations for two cases. Although for both cases the rate of conversion in the dispersed phase was taken to be higher than in the continuous phase, different reactant solubility characteristics were assumed. For Case A the reactant was assumed to have better solubility in the dispersed than in the continuous phase whereas for Case B the reactant was taken to have poorer solubility in the dispersed rather than in the continuous phase. Values of the parameters and data used in the examples were reasonable and realistic, although they did not refer to any given, specific system. The steady-state profiles calculated with these values permitted a study of reactor characteristics and provided the bases for the dynamic study of the reactor.

Steady-state simulation. Temperature and reactant concentration profiles along the circulation loop at steady state are of significant interest. The nature and shape of this profile will be affected by the physical and kinetic properties of the system, interfacial heat and mass transfer, location of the feed streams, the extent of mixing in the dispersed phase, the circulation flow rate, the impeller Reynolds number, and the drop size profile. To obtain the steady-state profiles it is necessary to solve a set of nonlinear algebraic equations for which the possibility of multiple steady-states exists. In a complex system described by a large number of equations, it is impossible, in most cases, to predict the number of steady states. For the studies performed the steady state of interest around which the system transient behavior was later studied was found by numerical integration of the differential equations from reasonable initial conditions. Steady state was assumed to have been reached when the system variables showed no further change with time.

In Case A a sharp temperature profile along the circulation loop might arise in this case of an exothermic reaction. This phenomenon indeed occurred and was especially marked when the reactant feed entered the circulation loop upstream of the impeller, this being the zone of large drops and low transfer coefficients. The effect of the feed location on the "hot" regions in the dispersed phase was studied and can be observed in Table 2. When the dispersed phase is fed into the circulation loop close to the impeller cell, a hot spot still results, but the total "hot" volume in the dispersed phase is reduced. When the dispersed phase is fed into the impeller discharge zone (a zone of small droplets where only break-up occurs), an appreciable reduction in the hot spot temperature occurs and a moderate temperature profile is obtained over the entire loop. This is shown as Case c in Table 2.

In addition to the marked temperature profile observed in the steady state for Case A a significant concentration profile was also obtained. The concentration profile will be determined not only by

Table 2
Steady-State Concentrations in Circulation Loop, Effect of Feed Location (Example A)

Case a Conversion 0.959 Dispersed Phase				Case b Conversion 0.959 Dispersed Phase				Case c Conversion 0.958 Dispersed Phase			
Cell No.	Temp. (°K)	Conc. of A (mole/cc)	Ave. Dia (cm)	Cell No.	Temp. (°K)	Conc. of A (mole/cc)	Ave. Dia (cm)	Cell No.	Temp. (°K)	Conc. of A (mole/cc)	Ave. Dia
1	340.53	$6.45 \cdot 10^{-6}$	0.0215	1	344.05	$9.55 \cdot 10^{-6}$	0.0215	1**	344.20	$50.17 \cdot 10^{-6}$	0.0215
2	339.63	$4.65 \cdot 10^{-6}$	0.0223	2	341.65	$4.72 \cdot 10^{-6}$	0.0223	2	343.27	$11.55 \cdot 10^{-6}$	0.0223
3	339.48	$2.30 \cdot 10^{-6}$	0.0231	3	341.00	$2.11 \cdot 10^{-6}$	0.0231	3	342.60	$3.21 \cdot 10^{-6}$	0.0231
4	339.25	$1.75 \cdot 10^{-6}$	0.0240	4	340.40	$1.60 \cdot 10^{-6}$	0.0240	4	341.70	$1.71 \cdot 10^{-6}$	0.0240
5***	339.06	$1.61 \cdot 10^{-6}$	0.0248	5	339.93	$1.49 \cdot 10^{-6}$	0.0248	5	340.94	$1.45 \cdot 10^{-6}$	0.0248
6	338.91	$1.55 \cdot 10^{-6}$	0.0256	6***	339.58	$1.46 \cdot 10^{-6}$	0.0256	6***	340.37	$1.40 \cdot 10^{-6}$	0.0256
7	338.80	$1.51 \cdot 10^{-6}$	0.0265	7	339.30	$1.44 \cdot 10^{-6}$	0.0265	7	339.92	$1.40 \cdot 10^{-6}$	0.0264
8**	352.42	$27.22 \cdot 10^{-6}$	0.0273	8	339.10	$1.42 \cdot 10^{-6}$	0.0273	8	339.58	$1.39 \cdot 10^{-6}$	0.0273
9	350.97	$3.47 \cdot 10^{-6}$	0.0281	9	338.93	$1.40 \cdot 10^{-6}$	0.0281	9	339.33	$1.38 \cdot 10^{-6}$	0.0281
10	348.49	$1.02 \cdot 10^{-6}$	0.0289	10	338.81	$1.37 \cdot 10^{-6}$	0.0289	10	339.13	$1.36 \cdot 10^{-6}$	0.0289
11	346.43	$0.78 \cdot 10^{-6}$	0.0298	11	338.72	$1.35 \cdot 10^{-6}$	0.0298	11	338.98	$1.34 \cdot 10^{-6}$	0.0297
12	344.82	$0.81 \cdot 10^{-6}$	0.0306	12**	352.76	$26.58 \cdot 10^{-6}$	0.0306	12	338.88	$1.32 \cdot 10^{-6}$	0.0306
Continuous Phase				**Continuous Phase**				**Continuous Phase**			
1	337.03	$56.83 \cdot 10^{-6}$	—	1	337.00	$57.05 \cdot 10^{-6}$	—	1	337.03	$58.15 \cdot 10^{-6}$	—

* *Dispersed phase hold-up—0.2*
** *Feed cell*
*** *Outlet cell*

the feed location, but also by the mass transfer of reactant A from the continuous phase which is the less reactive phase. When the reactant was fed well upstream of the impeller the highest dispersed-phase concentration was obtained in the feed cell and the concentration decreased markedly because of the significant conversion which occurs in the feed and adjacent cells until the impeller zone is reached. At Cell 1, the impeller discharge cell, an increase in concentration is obtained as a result of the improved mass transfer from the continuous phase. From this cell until the feed cell, a moderate decrease in reactant concentration was seen along the circulation loop. In Case B in which the reactor conversion will be limited by the mass transfer, hot spots were not obtained along the circulation loop but sharp concentration profiles were obtained.

Information about the concentration profile and its behavior is important for control purposes. It should be apparent from these concentration profile results that the measurement equipment used for conversion control must be appropriately located in the vessel and that the measurement should be made in a region close to the outlet stream. In addition, if sampling is to be used for control purposes the sampling should be made in defined regions in the vessel, preferably in those regions where, for example, the reactant concentration value is relatively high.

The effect of the number of continuous-phase cells was also studied. Two-cases were considered for the continuous phase: perfectly-mixed, W = 1, and a 4-cell continuous-phase loop, W = 4. For both cases moderate and almost identical temperature profiles were observed. On the other hand,

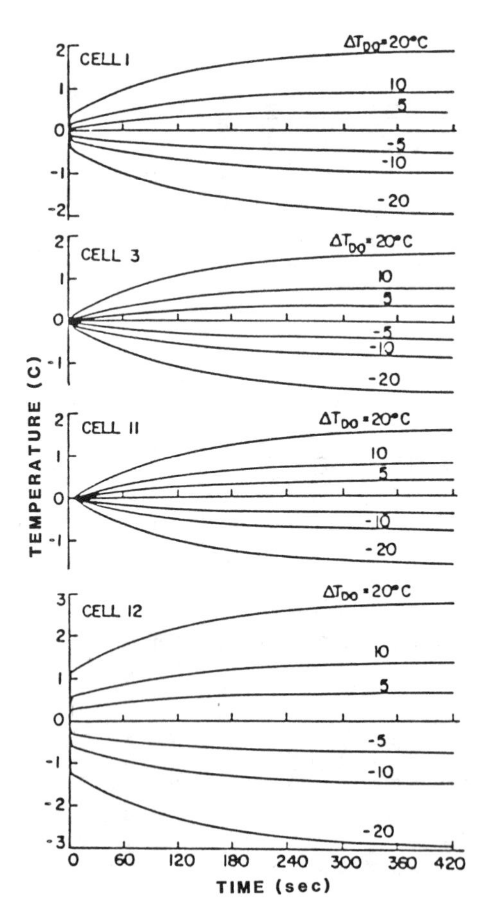

Figure 12. Response of temperature (presented as deviation) of the dispersed phase cells to step changes in T_{DO} (Example A).

there is a significant concentration profile, where the highest concentration is in the impeller discharge zone and the concentration decreases with distance from the impeller although a slight jump in concentration is obtained in the feed cell. A slight increase in conversion is, however, observed as the number of cells in the continuous phase increases. An increase in the number of continuous-phase cells in the model is equivalent to a decrease in the continuous-phase back-mixing (the impeller speed was kept constant). In practice, vessel design that would be such as to minimize continuous phase mixing might, in certain cases, result in improved conversion.

Dynamic simulations. The dynamic responses were studied [43, 44] for the two cases described earlier with the steady-state simulations being used as the initial conditions for the dynamic simulations being used as the initial conditions for the dynamic simulations. The transient responses of the system were obtained for individual step changes in each of the operating variables and also for simultaneous random changes in these variables.

The response of the system variables to positive and negative step changes in the feed temperature and reactant concentration for the dispersed phase described in Table 2 are shown in Figures 12–15.

The dynamic responses to the feed temperature of concentration and temperature along the loop may be expressed approximately as responses of a first-order linear system. A decrease in gain, an increase in the time constant, and an increasing dead time appear as the distance from the feed cell increases along the circulation loop.

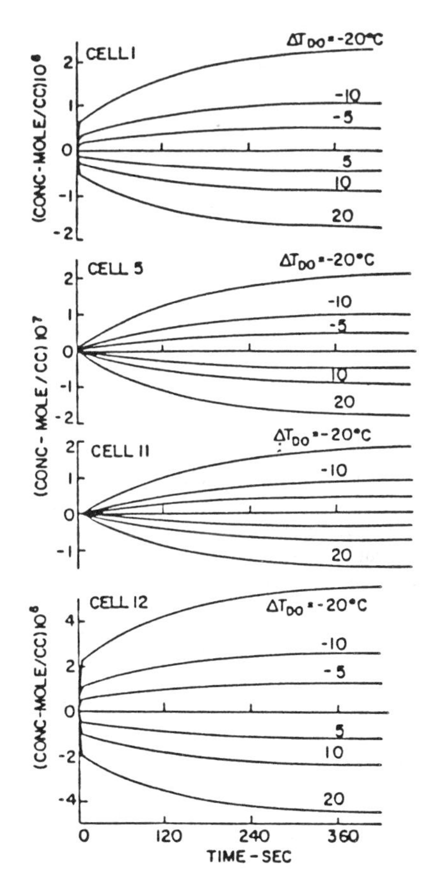

Figure 13. Response of concentration of A (presented as deviation) in the dispersed phase cells to step changes in T_{DO} (Example A).

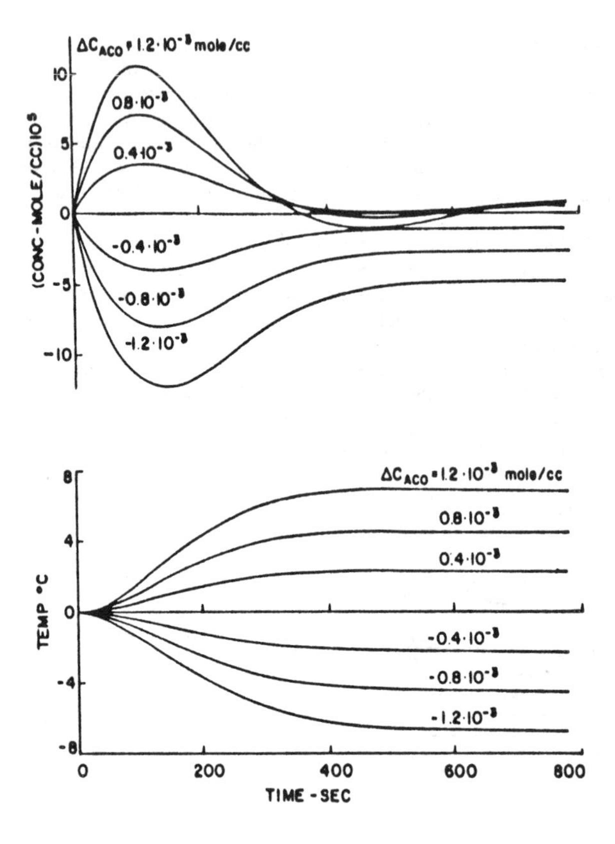

Figure 14. Response of concentration of A and temperature (presented as deviation) in the continuous phase to step changes in C_{ACO} (Example B).

The existence of dead time points to the important role that the flow term plays in this case in which most of the reactant is supplied to the system by the dispersed-phase feed stream and most of the conversion occurs in the feed and adjacent cells. Step changes in the dispersed-phase feed temperature will cause, therefore, an immediate change in the dispersed-phase feed zone temperature (Figure 12, Cell 12). The exothermic reaction and the high dependence of conversion on temperature will also cause fast and marked changes in reactant concentration in this zone (Figure 13). As the distance from the feed cell increases the local forcing function felt by each cell as compared to the feed cell is attenuated, modified in shape by dynamic lag, and its time of appearance delayed by transportation lag. Relatively low gains and long time constants are, therefore, obtained.

Concentration and temperature responses in the continuous and dispersed phases to step changes in reactant concentration in the continuous phase feed stream are shown in Figures 14 and 15 for Case B. Reactant concentration response in the continuous phase is not symmetric (Figure 14). During the first stages of the response a negative change in feed concentration of the continuous phase causes a decrease in reactant concentration in that phase while later on an increase in reactant concentration occurs after a marked undershoot. Positive input changes in C_{ACO} cause an oscillatory response that is more marked than in the case of a negative input. Addition of reactant to the continuous phase cause a marked increase in reactant concentration, but, simultaneously, the rate of reaction and mass transfer rate to the dispersed phase increase and a decrease in reactant con-

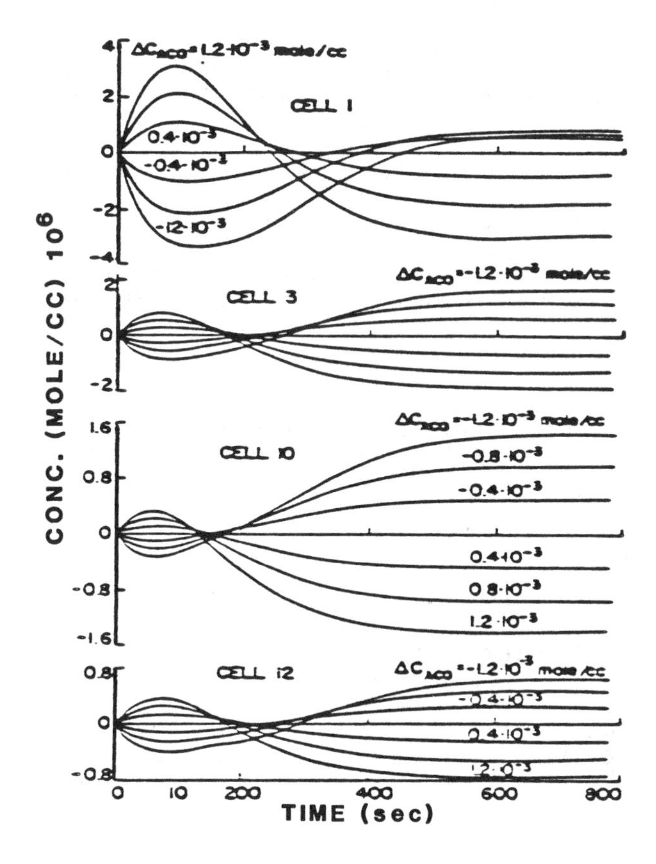

Figure 15. Response of concentration (presented as deviation) of the dispersed phase cells to step change in C_{ACO} (Example B).

centration occurs. This decrease causes, after a dynamic lag, an increase in drop diameter and a decrease in the mass transfer rate.

In the dispersed phase cells an inverse response is obtained in reactant concentration (Figure 15). The response resembles a linear system in a small range of input changes whereas for a larger range the principle of superposition is not fulfilled. The inverse response phenomena is most obvious at the impeller zone.

Temperature responses of the continuous phase and the various regions of the dispersed phase are similar. The simulation indicates that because of the similarity in the temperature responses of the two phases, temperature sensors can be located in the continuous phase where the measurements are more easily made even if the control objective is the dispersed-phase temperature. The dead time obtained in the temperature responses might cause problems in case of temperature control by regulating C_{ACO} but the large steady-state gains might be of assistance for good control.

Response to random changes in operating conditions. The circulation-interaction model permits the simulation of the system and its response to simultaneous changes in a number of operating conditions. In an actual process random changes in a number of operating conditions might occur rather than changes in only one. For example, changes in the phases feed rates, in feed temperature and in concentration, and in coolant temperature can occur and unexpected contaminants with surfactant-type behavior can enter the system. The frequency of appearance of each disturbance and its amplitude around the original steady state can be fitted into the model in accordance with what might be expected in an actual process. Barnea et al. [71] studied the response to simultaneous random changes in each of the operating conditions. The periods of each disturbance appearance and the maximum range of change around the steady-state value were held within reasonable limits and the actual magnitude of each change was calculated from random numbers with rectangular distribution.

As an example the temperature responses in several zones of the dispersed phase to simultaneous random changes in operating conditions are shown in Figure 16. The most marked changes in temperatures around the steady state of concern occur in the feed cell of the dispersed phase. The changes in temperatures range between -8 to $+5°C$ and might be problematic in the hot zone of

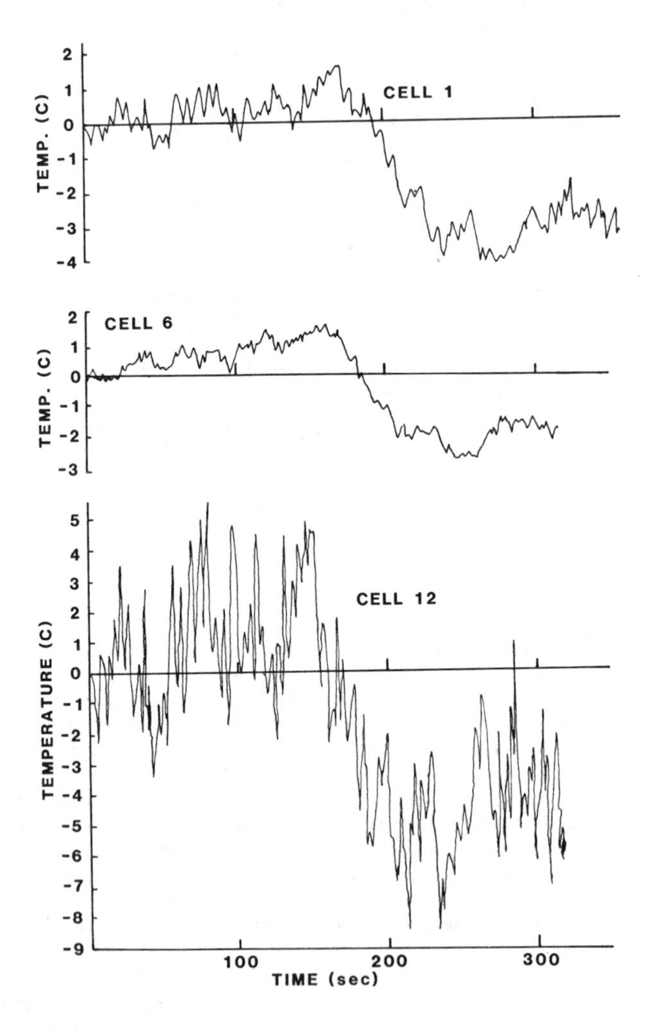

Figure 16. Response of temperature (presented as deviation) of the dispersed phase cells to random changes in operating conditions (Example A).

the reactor. The changes occurring in this zone result from the relatively high values of concentration and temperature in the feed zone of the dispersed phase and also from the fact that a major part of the simultaneous disturbances are given in the feed rates and properties, i.e., the most effective forcing functions are in the feed zone. With increasing distance from the feed zone the deviations from steady-state decrease significantly and the frequency of change is smaller.

The effectiveness of control loops proposed to maintain desired conditions can be checked investigated with the circulation-interaction model by imposing on it the expected random changes in operating conditions.

Simulation in control system design. The simulation results obtained from the circulation-interaction model were examined for their implications with respect to control system design. For example, the hottest spot in the reactor responds to changes in T_{CO} only after a dead time, whereas the response in the impeller zone is not only in the same direction as the response in the hot zone but also is immediate. Thus, if the control objective is the hot spot in the dispersed phase by changes in T_{CO} the simulation indicates that one should locate the measuring point in the impeller zone where the responses are fastest rather than in the hot spot zone.

If regulation of the continuous-phase feed temperature is to be used for the control of temperature and concentration in the reactor it is important that the controller includes information about the dependence between the gains and T_{CO}. Numerical results can be used to develop the gain functions which depend on the magnitude of change of the controlled variable.

The linear transfer function for the response to a forcing function can be obtained by process identification methods. Barnea et al. [71] obtained a matrix of simple linear transfer functions by applying these methods to the numerical responses obtained for small forcing functions. Each transfer function of the matrix gives a linear approximation of the dynamic relationship in some regions of the reactor between a state variable, such as temperature or concentration, and an external forcing function, such as coolant temperature or inlet feed concentration. Thus, these matrices give a map of gains and time constants in different regions of the dispersed phase and in the continuous phase. Matrices were presented, for the responses of the state variables of temperature and concentration to each of the operating variables studied in the simulation work.

Such matrices are useful tools for correct pairing between controlled and manipulated variables, in which both gain and speed of dynamic response have to be considered. This matrix helps decision making as to which variables would be desirable as control variables and which, because of large dead time or inverse response, would be undesirable.

POPULATION BALANCE TECHNIQUES

The concept of a population balance is useful in the description of dispersed or multiparticle behavior. Behnken et al. [72] generalized the concept of size in the population balance to mean any part of the particle distribution which advances or retreats with time according to an appropriate physical law. In their particular case the particle "size" was the amount of reactant material on a catalyst particle. Population balances arise in the description of the behavior of a population of particles from information available or which can be generated about the behavior of a single particle. The equations which result generally are integro-differential, and numerical techniques are usually required for their solution.

The population balance approach is particularly useful for the description of agitated liquid-liquid dispersions because it permits the breakup and coalescence processes to be considered in terms of the operating variables and parameters. The theory not only permits the prediction of the drop size distribution but also, at least conceptually, permits the description of interface mass and heat transfer with reaction in either one or both of the liquid phases.

The general framework of population balances for the analysis of a multivariate particle number density was presented by Hulburt and Katz [23]. The phase space over which the number density is defined is characterized by three external spatial coordinates and m internal coordinates. The internal coordinates permit the quantitative description of the state of the individual particle, such as its mass, temperature, concentration, etc. The number density continuity equation in particle phase space is shown to be [72–75]:

$$\frac{\partial n}{\partial t} + \nabla \cdot V_e n + \nabla \cdot V_i n - B + D = 0 \tag{13}$$

where V_e is the external or spatial particle velocity and V_i is the velocity of the particle in internal phase space. The two final terms on the right-hand side represent birth and death terms, i.e., the number of drops produced or destroyed in the phase space. Equation 13 is the most general form of the population balance and assumes that there are enough particles to approximate a continuum and that the particle trajectories in particle phase space are identical.

Other forms of the population balance include the macro-distributed balance which is useful for a spatially homogeneous distribution. In such cases the system can be described knowing only the distribution of drops in the internal phase space and the external spatial coordinates are of little interest. The spatial-averaged population balance equation can then be written as

$$\frac{\partial n}{\partial t} + \nabla \cdot V_i n + n d(\log V)/dt - B + D + \sum_k (Q_k n_k / V) = 0 \tag{14}$$

The summation terms represent the flow rate Q_k of streams k into the system where n_k is the number density of stream k.

For engineering purposes it is often sufficient to represent the particle distribution by some average or total quantities. For the case of liquid-liquid reactors it may be sufficient to know the average drop size, surface area, or reactant concentration for design purposes. For these cases a knowledge of the complete particle distribution is unnecessary and the population balance equation can be transformed so as to average the distribution with respect to the internal coordinate properties. By this transformation of the population balance in terms of moments of the distribution, the dimensionality of the equation is reduced as well as the magnitude of the computation required. This macro-moment balance equation can also be useful for stability transient studies of particle distributions.

Valentas and Amundson [76] employed population balance equations in their study of the effect of drop size-age distribution on the performance of continuous dispersed-phase reactors. The size and age of the dispersed-phase drops was expressed in terms of an integro-differential equation and numerical examples were presented to illustrate the effect of dispersed-phase interactions for simple as well as complex reaction schemes. Segregated and mixed reactor models were examined as these establish the operating limits for a real reactor whose performance will be somewhat between these two extremes. They showed that drop breakup influenced conversion when mass transfer is involved even for first-order systems.

Coulaloglou and Tavlerides [77] proposed phenomenological models to describe drop breakup and coalescence in an agitated liquid-liquid dispersion which were used to develop the appropriate breakup and coalescence rate functions. These rate functions were used in the population balance equations describing drop interactions in the continuous-flow vessel. The values for the parameters which appear in their models were evaluated by comparison with experimental data obtained by the authors on drop size distributions and mixing frequencies. Favorable agreement was obtained between experimental observation and the model simulations. Although the values obtained for the model compared favorably with the authors' observations they differed somewhat from the results obtained by investigators who studied other systems. This would tend to indicate that not all of the phenomena active in drop breakup and coalescence were accounted for in the phenomenological models used by Coulaloglou and Tavlerides in their population balance equations.

A detailed simulation of a dispersed-phase reaction system which can be cited is that of Min and Ray [78] who developed a model for the simulation of a batch emulsion-polymerization reactor. The model consists of particle size-distribution balances, individual particle balances, aqueous phase balances, and general material and energy balances. The aqueous phase, particle, and overall material and energy balances are ordinary differential or algebraic equations. The population balance equations, however, are coupled, multivariate, partial differential-difference equations.

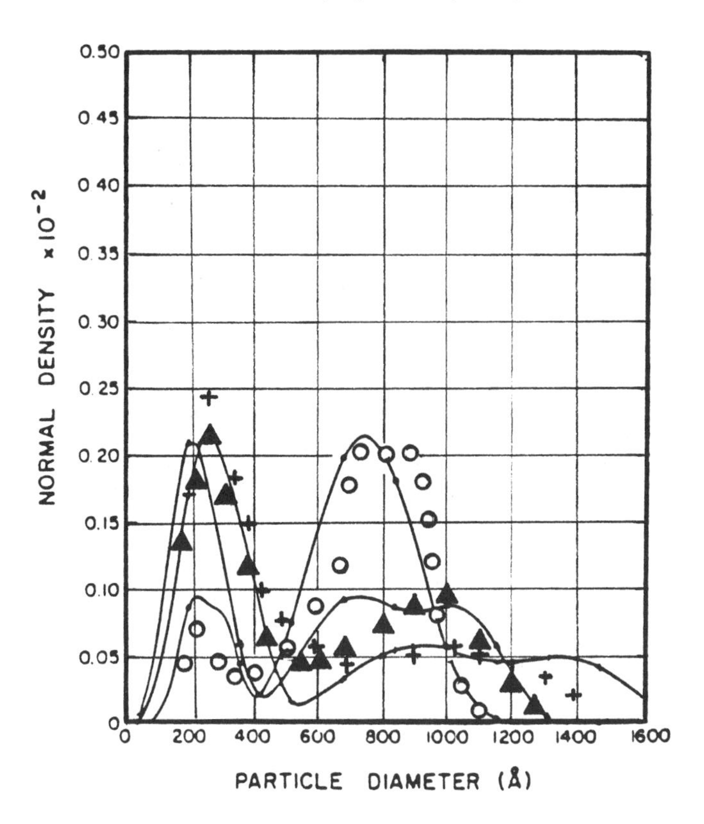

Figure 17. Particle size distributions for various initial polymerization initiator concentrations. Comparison between model and experimental data [78].

In their work Min and Ray developed a computational algorithm with some simplification of the population balance equation which permitted reasonable computational times. They compared their computer simulation results for methyl methacrylate with experimental data. Some final polymer particle size distributions as obtained by computer simulation are compared in Figure 17 with experimental results for a range of initiator concentrations. The comparison between simulated and experimental results for the total rate of polymerization as a function of batch time was equally good.

After obtaining these favorable comparisons between model and experiment the authors exercised their model to predict the effect of particle size on the properties of the polymer produced. The results showed that the polymerization rate and molecular weight of the polymer are strongly dependent on particle size. Thus, the results suggest that without chain-transfer control of molecular weight the polymer molecular weight can be strongly influenced by the shape of the particle size distribution.

Monte-Carlo Simulation

The computational effort required to solve the integro-differential equations arising from the population balance approach becomes formidable as the dimensionality of the dispersed phase increases. The Monte Carlo simulation models for reacting, dispersed-phase systems offers a procedure for bypassing these difficulties.

Monte-Carlo simulation methods require that a probabalistic model be built of the system under investigation. They are, at least in principle, relatively easy to carry out and can be applied to systems that are too complex or too large to solve by other numerical methods. The digital simulation moves from one distinct state to another where the state of the model at any time is represented by the values of the set of variables. The model state is caused to change by an event; thus, whenever an event occurs the values of the variables representing the model are changed. For the case of an agitated liquid-liquid reactor it is obvious that the model cannot include all of the drop particles because of the extremely large number of particles in the system. Drop diameters in such systems generally are smaller than 1 mm, thus, in the usual range of dispersed-phase holdup, the drop density in the dispersion will exceed 10^6 drops per liter. The model study, therefore, must be based on a sample population whose number must be large enough to represent the entire drop population but small enough to permit the system to be simulated without excessive computational effort.

Spielman and Levenspiel [79] were the first to apply the Monte-Carlo method to these dispersed-phase systems. They used the same physical model as Curl [80] who had used the population balance method to study the effect of coalescence and redispersion on a zero-order reaction. They studied the case of a single reactant disappearing not only in accordance with a zero-order rate law but also by a second-order rate law. They studied, in addition, the case of two reactants disappearing by a second-order rate expression.

The physical model [79, 80] assumed a large, constant number of equal-sized drops which coalesce two at a time and immediately redisperse to form two identical drops. Drop coalescence occurs at a constant rate and the probability of coalescence is the same for all drops. Reaction is assumed to occur only in the drops and the concentration of reactant is assumed to be uniform throughout the drop. The continuous phase is homogeneous and chemically inert.

In the simulation the drops are pictured as entering and leaving the reactor one at a time for the single-reactant case and as pairs for the two-reactant case, where one drop of the pair contains one reactant and the second drop contains the other reactant. This entering drop or pair of drops displace a randomly-selected drop or drop-pair which leaves the reactor. Within the reactor coalescence, redispersion, and reaction take place. The simulation clock is advanced by a time, T, which is the average time between entering drops or pairs of drops. During this time randomly selected drops undergo binary coalescence and redispersion, the number of coalescence events being in accordance with the drop coalescence rate. During the time increment T the concentration of each of the drops changes in accordance with the appropriate reaction-rate expression. The chemical conversion can be calculated by averaging on the compositions obtained in all of the drops. The authors prepared performance charts in which the performance of a reactor with a finite coalescence rate was compared with that to be expected for an infinite coalescence rate. No attempt was made in the work to relate the drop coalescence-rate parameter to the operating conditions nor to the system properties.

Zeitlin and Tavlerides [81] extended the technique by considering fluid-fluid interactions and the hydrodynamic effects in a fully baffled, turbulently agitated system. They assumed, although only binary coalescences occur, that the drops have a nonuniform size distribution. Breakup is assumed to result in the formation of two drops of randomly selected nonuniform size. The model simulates the system by a two-dimensional slice which comprises a grid system with a continuous-phase cell in each square of the grid. Each continuous cell can contain two, one, or no dispersed-phase particles. The particles now move through the grid system in accordance with the flow pattern selected, coalescence can occur in accordance with a coalescence efficiency if two particles arrive at a cell simultaneously, and breakup can occur in accordance with a breakage efficiency. Although the authors developed coalescence and breakup functions and calculated drop-size distributions, they did not include chemical reactions in their simulation study.

These Monte-Carlo studies used a periodic scan technique in which the system was examined for a fixed time increment for the occurrence of an event. The implicit assumption made was that the particles retain their identities during this period. The smooth changes in particle states were obtained by solving the appropriate algebraic and differential equations governing the particle state. The length of time over which the particle retains its identity is uncertain and will be a randomly varying quantity among the various particles. It would appear reasonable, therefore, to use a time interval for simulation over which the particles present at the beginning of the increment remain intact. This time has been referred to as the quiescence interval [74, 75] and provides a different basis for advancing the simulation clock than does the periodic scan technique. In using this method the occurrence of events is observed for the entire population and the simulation clock is advanced according to the time of event occurrence rather than at a predetermined time increment as in the periodic scan method.

Shah et al. [82] used the interval-of-quiescence method to simulate the Curl model for a liquid-liquid continuous flow agitated reactor [79]. They concluded that the method is computationally superior to the algorithm based on periodic scan as applied by Spielman and Levenspiel [79] to the same model. Bapat et al. [83] simulated a liquid-liquid continuous-flow stirred-tank extractor by this method. The model predicted the dispersed-phase drop-size distribution and extractor performance efficiency for a single solute. The procedure can be extended, however, to the case of multicomponent systems as well as to chemical reactions.

Monte-Carlo Simulations and Population Balances

Both Monte-Carlo simulations and population balances of dispersed-phase systems utilize information about single-particle behavior to provide a description of the behavior of the entire population of particles. Ramkrishna [84] inquired into the precise connection between simulation methods and population balances. The population balance equation represents the average behavior of a population and the number density function appearing in the equation is the expected population density. Higher-order product density functions are associated with the calculation of fluctuations about the average values.

These expectations or averages are calculated based on a master probability density function—the Janossy density function. Using Ramkrishna's notation the probability that the population at time consists of ν particles with one particle in each of the volume dV_i in multidimensional space located about x_i, $i = 1, 2, \ldots, \nu$ is given by

$$J(\bar{x}_1, \bar{x}_2, \ldots, x_\nu; t)\, dV_1\, dV_2 \cdots dV_\nu \tag{15}$$

subject to, if the particles are constrained in volume, V,

$$\sum_\nu \frac{1}{\nu!} \int_V dV \int_V dV_{\nu-1} \cdots \int_V dV_1\, J_\nu(\bar{x}_1, \bar{x}_2, \ldots, x_\nu; t) = 1 \tag{16}$$

The author points out that the population balance equation is obtained by averaging this master density equation whereas the simulation procedures efficiently evaluate the solutions to the master density equation by elimination of low probability events. The solution of the Janossy density equation displays the combinatorial complexity that is characteristic of the evolution of a dispersed phase or particulate system with time. The Monte-Carlo simulation represents an efficient way of obtaining numerical values for such combinatorially complex analytical solutions.

NOTATION

A	reactant
C	concentration
C_p	heat capacity
$\bar{D}$	distribution coefficient
D_I	impeller diameter
Dif	diffusivity of reactant A
D_T	vessel diameter
D_{32}	mean drop diameter
E_a	activation energy
Ex	conversion, defined in Equation 12
F	volumetric flow rate
F_0	total volumetric flow rate
F_r	total recirculation flow rate
h	individual heat transfer coefficient
h_t	overall heat transfer coefficient between the two phases

ΔH	heat of reaction
I_S	impeller speed
k	individual mass transfer coefficient
k_0	frequency factor in Arrhenius' law
K_t	constant in Equation 1
K_A	overall mass transfer coefficient between the two phases
KA_C, KA_D	variables defined in Equation 10 and 11
N	number of dispersed phase cells
Nu	Nusselt number
Pr	Prandtl number
Q	E_a/R_g
R_g	gas constant
Re	impeller Reynolds number ($D_I^2 \cdot I_S\rho/\mu$)
s	heat transfer area to coolant
Sc	continuous-phase Schmidt number ($\mu_C/\rho_C \cdot Dif_C$)
Sh	continuous-phase Sherwood number ($k_C \cdot D_T/Dif_C$)
T	temperature
TC_C, TC_D	variables defined in Equations 10 and 11
Tcon	thermal conductivity
T_w	coolant temperature
t	time
u_C	heat transfer coefficient between wall and continuous phase
V	volume
V_t	tank volume
W	number of continuous phase cells
We	Weber number

Greek Symbols

α	reaction order
μ	viscosity
μ_b	bulk viscosity
μ_w	viscosity at the wall
ρ	density
τ	average residence time
ϕ	holdup

Subscripts

A	component A
C	continuous phase
D	dispersed phase
O	feed stream

REFERENCES

1. Rushton, J. H., *AIChE Inst. Chem. Engrs. Symp. Ser.*, 10:3 (1965).
2. Rushton, J. H., and Oldshue, J. Y., "Mixing of Liquids," *Chem. Eng. Prog. Symp. Ser.*, 55, (25):181 (1959).
3. Schwartzberg, H. G., and Treybal, R. E., "Fluid and Particle Motion in Turbulent Stirred Tanks," *Ind. Eng. Chem. Fundtl. Quart.* 7:7 (1968).
4. Oldshue, J. Y., "Mixing," *Ind. Eng. Chem.*, 62 (11):44 (1970).
5. Oldshue, J. Y., *Fluid Mixing Technology*, New York: McGraw-Hill Publishing Company, 1983, pp. 12–22, 27–34.
6. Wolf, D., and Manning, F. S., "Impact Tube Measurements of Flow Patterns, Velocity Profiles and Pumping Capacities in Mixing Vessels," *Canad. J. Chem. Eng.*, 44:137 (1966).
7. Cooper, R. G., and Wolf, D., "Pumping in Stirred Tanks—Theory and Application," *Canad. Jour. Chem. Eng.*, 45:197 (1967).
8. Cooper, R. G., and Wolf, D., "Velocity Profiles and Pumping Capacities for Turbine Type Propellers," *Canad. J. Chem. Eng.*, 46:84 (1968).
9. Mlynek, Y., and Resnick, W., "Drop Sizes in an Agitated Liquid-Liquid System," *AIChE J.*, 18:122 (1972).
10. Kim, W. J., and Manning, F. S., "Turbulence Energy and Intensity Spectra in a Baffled, Stirred Vessel," *AIChE J.*, 10:747 (1964).
11. Holmes, D. B., Voncken, R. M., and Dekker, J. A., "Fluid Flow in Turbine-stirred Baffled Tanks-I. Circulation Flow," *Chem. Engng. Sci.*, 19:202 (1964).

12. Voncken, R. M., Rotte, J. W., and Houten, Th., *AIChE-Inst. Chem. Engrs. Symp. Ser.*, 10:24-32 (1965).
13. Mok, Y. I., and Treybal, R. E., "Continuous-Phase Mass Transfer Coefficients for Liquid Extraction in Agitated Vessels-II," *AIChE J.*, 17:916 (1972).
14. Kintner, R. C., "Drop Phenomena Affecting Liquid Extraction," in *Advances in Chemical Engineering*, Volume 4, New York: Academic Press, 1963, pp. 52–95.
15. Skelland, A. H. P., and Seksaria, R., "Minimum Impeller Speeds for Liquid-Liquid Dispersion in Baffled Vessels," *Ind. Eng. Chem. Process Des. Dev.*, 17:58 (1978).
16. Skelland, A. H. P., and Lee, J. M., "Agitator Speeds in Baffled Vessels for Uniform Liquid-Liquid Dispersions," *Ind. Eng. Chem. Process Des. Dev.*, 17:473 (1978).
17. Treybal, R. E., *Liquid Extraction*, 2nd ed., 1963, New York: McGraw-Hill Publishing Company.
18. Heertjes, P. M., and De Nie, L. H., "Mass Transfer to Drops," in *Recent Advances in Liquid-Liquid Extraction*, C. Hanson (Ed.) Oxford: Pergamon Press, 1971, Chapter 10.
19. Jeffreys, G. V., and Davies, G. A., "Coalescence of Liquid Droplets and Liquid Dispersions," in *Recent Advances in Liquid-Liquid Extraction*, C. Hanson (Ed.) Oxford: Pergamon Press, 1971, Chapter 14.
20. Tavlerides, L. L., et al., "Bubble and Drop Phenomena," *Ind. Eng. Chem.*, 62, (11):6 (1970).
21. Tavlerides, L. L., and Stamatoudis, M., "Analysis of Interphase Reactions and Mass Transfer in Liquid-Liquid Dispersions," in *Advances in Chemical Engineering*, Volume 11, New York: Academic Press, 1981, pp. 199–273.
22. Gal-Or, B., and Waslo, S., "Hydrodynamics of an Ensemble of Drops (or Bubbles) in the Presence or Absence of Surfactants," *Chem. Engng. Sci.*, 23:1431 (1968).
23. Thorsen, G., Stordalen, R. M., and Terjesen, S. G., "On the Terminal Velocity of Circulating and Oscillating Liquid Drops," *Chem. Engng. Sci.*, 23:413 (1968).
24. Sprow, F. B., "Drop Size Distributions in Strongly Coalescing Agitated Liquid-Liquid Systems," *AIChE J.*, 13:995 (1967).
25. Vanderveen, J. H., "Coalescence and Dispersion Rates in Agitated Liquid-Liquid Systems," Lawrence Radiation Lab. Report UCRL-8733, Univ. California, Berkeley, 1960.
26. Vermeulen, T., Williams, G. M., and Langlois, G. E., "Interfacial Area in Liquid-Liquid and Gas-Liquid Agitation," *Chem. Eng. Prog.*, 51 (2):85F–94F (1955).
27. Schindler, H. D., and Treybal, R. E., "Continuous-Phase Mass-Transfer Coefficients for Liquid Extraction in Agitated Vessels," *AIChE J.*, 14:790 (1968).
28. Konno, M., Aoki, M., and Saito, S., "Scale Effect on Breakup Process in Liquid-Liquid Agitated Tanks," *J. Chem. Engng. Japan*, 16 (4):312 (1983).
29. Hoffer, M. S., and Resnick, W., "Study of Agitated Liquid/Liquid Dispersions-I. Dynamic Response of Dispersion Geometry to Changes in Composition and Temperature," *Trans. Inst. Chem. Engrs.*, 57:1 (1979).
30. Barnea, D., Hoffer, M. S., and Resnic, W., "Dynamic Behavior of an Agitated Two-Phase Reactor with Dynamic Variations in Drop Diameter-I Numerical Simulations," *Chem. Engng. Sci.*, 33:205 (1978).
31. Barnea, D., Hoffer, M. S., and Resnick, W., "Dynamic Behavior of an Agitated Two-Phase Reactor with Dynamic Variations in Drop Diameter-II. Observability Using Drop Diameter Information," *Chem. Engng. Sci.* 33:219 (1978).
32. Sawistowski, H., and Goldz, G. E., "The Effect of Interfacial Phenomena on Mass-Transfer Rates in Liquid-Liquid Extractors," *Trans. Inst. Chem. Engrs.*, 41:174 (1963).
33. Rajan, S. M., and Heideger, W. J., "Drop Formation Mass Transfer," *AIChE J.*, 17:202 (1971).
34. Skelland, A. H. P., and Minhas, S. S., "Dispersed Phase Mass Transfer During Drop Formation and Coalescence in Liquid-Liquid Extraction," *AIChE J.*, 17:1316 (1971).
35. Treybal, R. E., "Estimation of the Stage Efficiency of Simple, Agitated Vessels Used in Mixer-Settler Extractors," *AIChE J.*, 4:202 (1958).
36. von Berg, R., "Simultaneous Heat and Mass Transfer," in *Recent Advances in Liquid-Lquid Extraction*, C. Hanson (Ed.), Oxford: Pergamon Press, 1971, Chapter 11.
37. Holland, F., and Chapman, F. S., *Liquid Mixing and Processing in Stirred Tanks*, New York: Reinhold Publishing Company, 1966.

38. Oldshue, J. Y., *Fluid Mixing Technology*, New York: McGraw-Hill Publishing Company, 1983, Chapter 14.
39. Harriott, A. W., and Picker, D., "Phase Transfer Catalysis: An Evaluation of Catalysts," *J. Am. Chem. Soc.*, 97:2345 (1975).
40. Dehmlow, E. V., "Phase Transfer Catalysis," *CHEMTECH*, 5 (4):210 (1975).
41. Schmitz, R. A., and Amundson, N. R., "An Analysis of Chemical Reactor Stability and Control-VI," *Chem. Engng. Sci.*, 18:415 (1963).
42. Luss, D., and Amundson, N. R., "An Analysis of Chemical Reactor Stability and Control-XIII," *Chem. Engng. Sci.*, 22:267 (1967).
43. Barnea, D., Hoffer, M. S., and Resnick, W., "Development of a Dynamic Circulation-Interaction Model for Mechanically Agitated, Liquid-Liquid Reactors," *Chem. Engng. Sci.*, 34:901 (1979).
44. Barnea, D., "Dynamic Behavior of Mechanically Agitated, Liquid-Liquid Reactors and Its Applications to Process Control," dissertation for the D.Sc. degree, Israel Inst. of Tech. Haifa, 1976.
45. Cox, P. R., and Strachan, A. N., "Two Phase Nitration of Chlorobenzene," *Chem. Engng. Sci.*, 26:1013 (1971).
46. Cox, P. R., and Strachan, A. N., "Two Phase Nitration of Toluene-I," *Chem. Engng. Sci.*, 27:457 (1972).
47. Cox. P. R., and Strachan, A. N., "Two Phase Nitration of Toluene-II," *Chem Engng. J.*, 4:253 (1972).
48. Chapman, J. W., Cox, P. R., and Strachan, A. N., "Two Phase Nitration of Toluene-III," *Chem. Engng. Sci.*, 29:1247 (1974).
49. Spink, D. R., and Okuhara, D. N., "Comparative Equilibrium and Kinetics of an Alkylated Hydroxy Quinoline and a β-hydroxy Oxime for the Extraction of Copper," *Met Trans. S.*, (1974) p. 1935.
50. Whewell, R. J., Hughes, M. A., and Hanson, C., "Kinetics of the Solvent Extraction of Copper II with LIX Reagents—Single Drop Experiments," *J. Inorg. Nucl Chem.*, 37:2303 (1975).
51. Flett, D. S., et al., "The Extraction of Copper by an Alkylated 8-hydroxy Quinoline," *J. Inorg. Nucl. Chem.*, 37:1967 (1975).
52. Atwood, R. L., Thatcher, D. N., and Miller, J. D., "Kinetics of Copper Extraction from Nitrate Solutions by LIX 64N," *Met Trans.*, B, 6B:465 (1975).
53. Miller, J. D., and Atwood, R. L., "Discussion of the Kinetics of Copper Solvent Extraction with Hydroxy Oximes," *J. Inorg. Nucl. Chem.*, 39:701 (1977).
54. Whewell, R. J., Hughes, M. A., and Hanson, C., "Kinetics of the Solvent Extraction of Copper II with LIX Reagents—the Effect of LIX 64," *J. Inorg. Nucl. Chem.*, 38:2071 (1976).
55. Flett, D. S., Melling, J., and Spink, D. R.. "Reply to the Discussion on the Discussion on the Kinetics of Copper Solvent Extraction with Hydroxy Oximes," *J. Inorg. Nucl. Chem.*, 39:701 (1977).
56. Fletts, D. S., Okuhara, D. N., and Spink, D. R., "Solvent Extraction of Copper by Hydroxy Oximes," *J. Inorg. Nucl. Chem.*, 35:2471 (1973).
57. Flett, D. S., Cox, M., and Heels, J. D., "Kinetics of Nickel Extraction by α-Hydroxy Oxime/Carboxylic Acid Mixtures," *J. Inorg. Nucl. Chem.*, 37:2533 (1975).
58. Hughes, M. A., Sergeant, H. C., and Whewell, R. J., "Aspects of the Diffusion of Copper in Solutions Encountered During Solvent Extraction with Hydroxyoximes—II. Organic Phases," *J. Inorg. Nucl. Chem.*, 41:1603 (1979).
59. Tavlerides, L. L., "Modeling and Scaleup of Dispersed Phase Liquid-Liquid Reactors," *Chem. Eng. Commun.*, 8:133 (1981).
60. Kinkel, J. F. M., and Tomlinson, E., "Phase-Transfer Catalysis in a Segmented Flow Assembly. Study of Transfer and Reaction Rates," *Sepn. Sci. & Tech.*, 18 (9):857 (1983).
61. Kuo, C. H., and Huang, C. J., "Liquid Phase Mass Transfer with Complex Chemical Reaction," *AIChE J.*, 16:493 (1970).
62. Mhaskar, R. D., and Sharma, M. M., "Extraction with Reaction in Both Phases," *Chem. Engng. Sci.*, 30:811 (1975).
63. Merchuk, J. C., and Farina, I. H., "Simultaneous Diffusion and Chemical Reaction in Two-Phase Systems," *Chem. Engng. Sci.*, 31:645 (1976).

64. Sharma, M. M., and Nanda, A. K., "Extraction with Second Order Reaction," *Trans. Inst. Chem. Engrs.*, 46:T44 (1968).
65. Resnick, W., and Gal-Or, B., "Gas-Liquid Dispersions," in *Advances in Chemical Engineering*, Volume 7, New York: Academic Press, 1968, p. 295.
66. Gal-Or, B., and Resnick, W., "Mass Transfer from Gas Bubbles in an Agitated Vessel," *Chem. Engng. Sci.*, 19:653 (1964).
67. Gal-Or, B., and Resnick, W., "Gas Residence Time in Agitated Gas-Liquid Contactor: Experimental Test of Mass Transfer Model," *Ind. Eng. Chem. Proc. Des. Dev.*, 5:15 (1966).
68. Tavlerides, L. L., and Gal-Or, B., "A General Analysis of Multicomponent Mass Transfer with Simultaneous Reversible Chemical Reactions in Multiphase Systems," *Chem. Engng. Sci.*, 24:533 (1969).
69. Hoffer, M. S., and Resnick, W., "Study of Agitated Liquid/Liquid Dispersions-Part II: Dependence of Steady-State Dispersion Geometry on Phase Composition and Location," *Trans. Inst.* Chem. Engrs., 57:8 (1979).
70. Rietema, K., "Segregation in Liquid-Liquid Dispersions and its Effect on Chemical Reactions," in *Advances in Chemical Engineering*, Volume 5, New York: Academic Press, 1964, p. 237.
71. Barnea, D., Hoffer, M. S., and Resnick, W., "Simulation of Mechanically-Agitated Liquid-Liquid Reactor Behavior and Some Control Implications," *Chem. Engng. Sci.*, 38:182 (1983).
72. Behnken, D. W., Horowitz, J., and Katz, S., "Particle Growth Processes," *Ind. Eng. Chem. Fund.*, 2:212 (1963).
73. Hulburt, H. M., and Katz, S., "Some Problems in Particle Technology—A Statistical Mechanical Evaluation," *Chem. Engng. Sci.*, 19:555 (1964).
74. Randolph, A. D., "A Population Balance for Countable Entities," *Can. J. Chem. Engng.*, 42:280 (1964).
75. Randolph, A. D., and Larson, M. A., *Theory of Particulate Processes*, New York: Academic Press, 1971, Chapter 3.
76. Valentes, K. J., and Amundson, N. R., "Influence of Droplet Size—Age Distribution on Rate Processes in Dispersed-Phase Systems," *Ind. Eng. Chem.* 7 (1):66 (1968).
77. Coulaloglou, C. A., and Tavlerides, L. L., "Description of Interaction Processes in Agitated Liquid-Liquid Dispersions," *Chem. Engng. Sci.*, 32:1289 (1977).
78. Min, K. W., and Ray, W. H., "Computer Simulation of Batch Emulsion Polymerization Reactors through a Detailed Mathematical Model," *J. Appl. Poly. Sci.*, 22:89 (1978).
79. Spielman, L. A., and Levenspiel, O., "A Monte Carlo Treatment for Reacting and Coalescing Dispersed Phase Systems," *Chem. Engng. Sci.*, 20:247 (1965).
80. Curl, R. L., "Dispersed Phase Mixing: I. Theory and Effects in Simple Reactors," *AIChE J.*, 9:175 (1963).
81. Zeitlin, M. A., and Tavlerides, L. L., "Fluid-Fluid Interactions and Hydrodynamics in Agitated Dispersions: A Simulation Model" *Can. J. Chem. Engng.*, 50:207 (1972).
82. Shah, B. H., Ramkrishna, D., and Borwanker, J. D., "Simulation of Particulate Systems Using the Concept of Interval of Quiescence," *AIChE J.*, 23:897 (1977).
83. Bapat, P. M., Tavlerides, L. L., and Smith, G. W., "Monte Carlo Simulation of Mass Transfer in Liquid-Liquid Dispersions," *Chem. Engng. Sci.*, 38:2003 (1983).
84. Ramkrishna, D., "Analysis of Population Balance—IV: The Precise Connection Between Monte Carlo Simulation and Population Balances," *Chem. Engng. Sci.*, 36:1203 (1981).

CHAPTER 31

SOLID AGITATION AND MIXING IN AQUEOUS SYSTEMS

R. Conti and **A. Gianetto**

Dipartimento di Scienza dei Materiali e Ingegneria Chimica
Politecnico di Torino, Italy

CONTENTS

INTRODUCTION, 886

ZWIETERING'S DIMENSIONAL ANALYSIS APPROACH TO COMPLETE SUSPENSION, 888

DEVELOPMENTS OF ZWIETERING'S APPROACH, 888

BALDI ET AL.'S THEORETICAL APPROACH, 891

FURTHER DEVELOPMENTS OF THE ZWIETERING/BALDI APPROACH, 893

OTHER STUDIES, 893

CONCENTRATION PROFILES, 893

SCALING-UP, 896

RESUSPENSION OF THE SOLIDS SETTLED AFTER STOPPING THE STIRRER, 897

SOLIDS SUSPENSION AND MASS TRANSFER, 897

NOTATION, 899

REFERENCES, 900

INTRODUCTION

Several chemical engineering operations of considerable industrial importance, such as dissolution, leaching, crystallization, and liquid/solid reaction, involve the suspension of solid particles in a liquid phase. Suspension is commonly attained in mechanically stirred vessels where some idealized states can be recognized as a function of the geometry of the system, the stirrer speed, the properties of the solid and liquid, and then of the power input, namely:

- *Homogeneous suspension.* In this condition the solid particles concentration and size distribution are uniform throughout the vessel. Obviously the practical attainability of a really homogeneous suspension is rather low since the nearly horizontal flow patterns which characterize the upper layers of the liquid are unable to withstand the settling tendency of the solid particles. Other zones not easy to control are those around the stirrer and immediately before and behind the baffles. Consequently concentration gradients are almost always present and a "degree of homogeneity" or a "quality of distribution" is often supplied on the basis of the concentration profiles of the solid particles in the whole vessel.

- *Complete suspension.* In this condition, also named "complete off-bottom suspension", the solid particles are just fully suspended; that is, solid particles neither remain at rest on the vessel bottom nor roll around on it. In reality it is not easy to recognize this state (the identification is based on the visual observation of the vessel bottom and it is often difficult to determine when the solid particles leave off simply moving on the vessel bottom and are truly suspended, i.e., show a vertical upward velocity component) and very often reference is made to the "one-second criterion" which states that a "complete" suspension can be considered achieved when no solid particle remains in a fixed position on the vessel bottom for more than one second.

  The approach of the one-second criterion to the complete suspension concept can be accepted since also in this condition the total surface area of the solid particles is available for processing.

  Interest in this condition is also justified by the consideration that at stirrer speeds above the critical one corresponding to the one-second criterion, the increase of the rate of several transfer processes is moderate and that, on the other hand, the power required for a fairly homogeneous suspension may be a good deal higher than that spent in this condition. At any rate in what follows every reference to the complete suspension state is in reality made to the state corresponding to the one-second criterion.

- *Sufficient suspension.* The most evident defect of the approaches which make reference to the complete suspension condition is that they frequently disregard what happens in the upper part of the vessel. In fact, as a rule from which only very small solid particles or solids with a density very near to the one of the liquid phase escape, the critical stirrer speed for complete suspension is not sufficient to lift the solid particles up to the upper layers of liquid. If the solids concentration is high enough, it is also possible to perceive at some height of the vessel a rough boundary of solids which separates the upper layers of liquid without solid particles from the lower, main part in which the solid particles are suspended.

  This approach (the "layer height criterion") has suggested the introduction of a new critical situation named "sufficient suspension," corresponding to a lifting of the solid particles up to 90% of the liquid height. In this condition (which is commonly attained at higher stirrer speeds than complete suspension) a great part of the volume occupied by the liquid is characterized by the presence of the solid particles and the concentration profiles are reasonably satisfactory since, even if they are not uniform (obviously), often they show more or less wide zones of pseudo-homogeneity, i.e., zones in which the solid particles concentration is the same as the average one in the vessel. This can be useful in continuous apparatuses for the withdrawal of the suspension.

  A mere widening of the sufficient suspension concept is "uniform or total suspension," which corresponds to a lifting of the solid particles up to 100% of the liquid height. In this condition the concentration profiles are frequently satisfactorily uniform (i.e., very near to those for a homogeneous suspension), if the obvious defects directly below the surface of the liquid are neglected.

- *Other suspension conditions.* Some authors have endeavored to find suspension conditions easy to be acknowledged instrumentally. So, for instance, Kolář [1] measured with a selenium photocell the light passing through the suspension at various distances from the bottom of the vessel; he considered homogeneous a suspension characterized by the same horizontal light transmission at one fourth and three fourths of the liquid height.

  A similar technique using a silicon photocell and a helium-neon laser as a source of light (instead of Kolář's microscope lamp) has been utilized by Bohnet and Niesmak [2] to measure the solids distribution throughout the vessel; they defined the solid particles completely suspended when the average solids concentration in the vessel thus determined was equal to the real one. Other authors have used measurements of pressure [3] and conductivity [4]. The attempt of rearrangement by Gates, Morton, and Fondy [5] which inserted the suspension conditions in a ten-degree scale of agitation also deserves mentioning.

Complete suspension is still the most interesting condition (at the least for its meaning of threshold condition) of this variety of situations; therefore this chapter will be mainly devoted to it.

ZWIETERING'S DIMENSIONAL ANALYSIS APPROACH TO COMPLETE SUSPENSION

The work of Zwietering [6] deserves peculiar attention since most probably, in spite of its age (it was published in 1958), it is still the widest investigation of the complete-suspension condition, and the correlation suggested for the calculation of the critical stirrer speed is one of the more reliable. Moreover this work has somehow affected a great part of the following work on the subject.

According to Zwietering the critical stirrer speed for complete suspension N_{js} is related to the other parameters by the equation

$$N_{js} = \underline{S}\,\frac{\nu_L^{0.1} D_P^{0.2}\left(g\dfrac{\Delta\rho}{\rho_L}\right)^{0.45} X^{0.13}}{D^{0.85}} \tag{1}$$

where ν_L and ρ_L = kinematic viscosity and the density of the liquid
D_p = solid particles size
g = acceleration due to gravity
$\Delta\rho$ = density difference between solid and liquid
D = stirrer diameter
X = solid concentration (weight per weight of liquid × 100)

The geometric parameters of the system taken into consideration, i.e., type of the stirrer, stirrer-clearance-to vessel-diameter ratio (C/T, see Figure 1), stirrer-diameter to vessel-diameter ratio (D/T), contribute to define the value of the dimensionless constant s. More exactly for each type of stirrer considered (six-bladed disc turbine, vaned disc turbine, two-bladed paddle, propeller (all the vessels were flat bottomed and fully baffled) the values of s were plotted versus T/D with T/C as a parameter. The origin of Equation 1 is dimensional analysis and therefore this calculation mode is purely empirical, but it has a very effective support in the extensive experimental work carried out by Zwietering who investigated the following ranges of the affecting parameters:

$0.31 < \mu_L < 9.3\ (10^{-3}\ kg\ m^{-1}\ s^{-1})$, where μ_L = liquid viscosity

$125 < D_p < 850\ (10^{-6}\ m)$

$720 < \Delta\rho < 1{,}810\ (kg\ m^{-3})$

$0.5 < X < 20$

$0.05 < D < 0.22\ (m)$

In these ranges the use of Equation 1 is rather safe. The only serious doubt concerning the disc turbine whose behavior was, according to Zwietering, independent of the clearance from the vessel bottom.

DEVELOPMENTS OF ZWIETERING'S APPROACH

Even though questioned as a suitable stirrer for suspending solid particles [1], the disc turbine has been extensively investigated mainly because of its ability to dissipate a lot of power. This characteristic was judged interesting in various processes in which solid particles suspension was accompanied by mass transfer phenomena. So as a part of a wider work related to mass transfer, Nienow [7] reexamined the behavior of the disc turbine with particular reference to the clearance.

It is well known that the clearance of a radial flow stirrer heavily affects the hydrodynamics in the vessel: when the stirrer is set rather far from the vessel bottom the classical double-eight figure

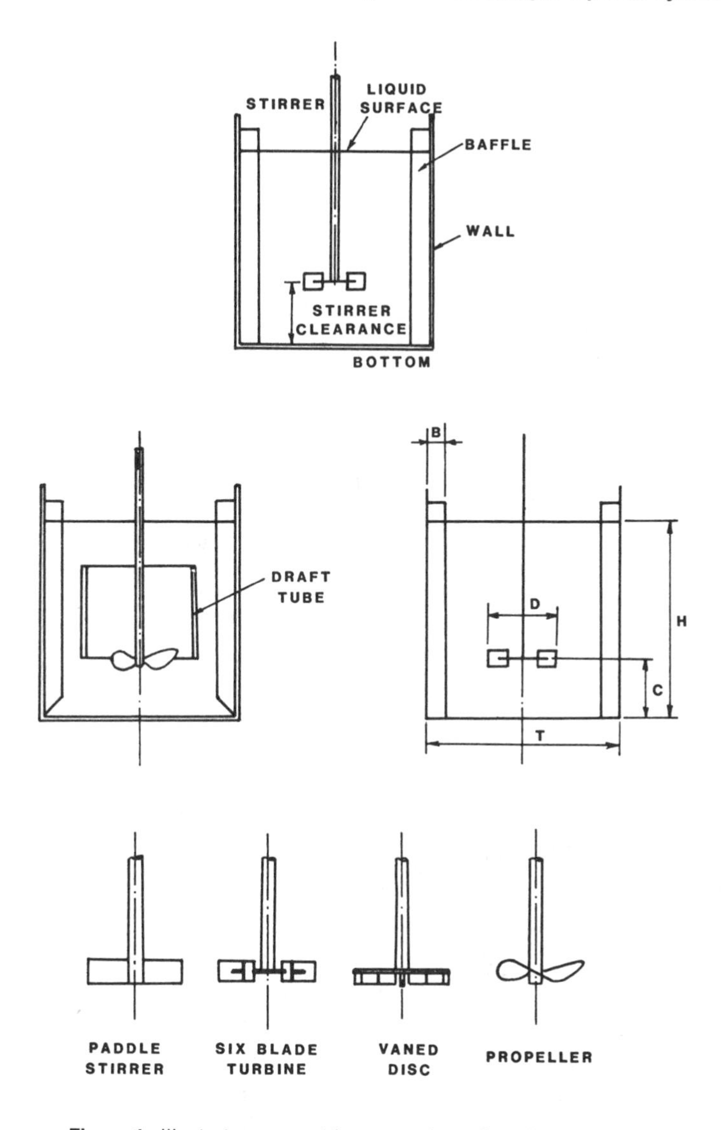

Figure 1. Illustrates geometric parameters of a mixing system.

persists in the vessel (Figure 2A). Owing to the presence of the lower vortices the bulk flow near the vessel bottom is directed from the peripheral zone near the wall to the central zone beneath the stirrer. It is in this zone that the solid particles tend to collect before being suspended again.

When the stirrer is close to the bottom the lower vortices are absent (Figure 2B), and the bulk flow near the vessel bottom is directed rather horizontally from the stirrer to the peripheral zone where the solid particles tend to accumulate; in this case the presence of baffles (number, size, shape) can have a remarkable importance.

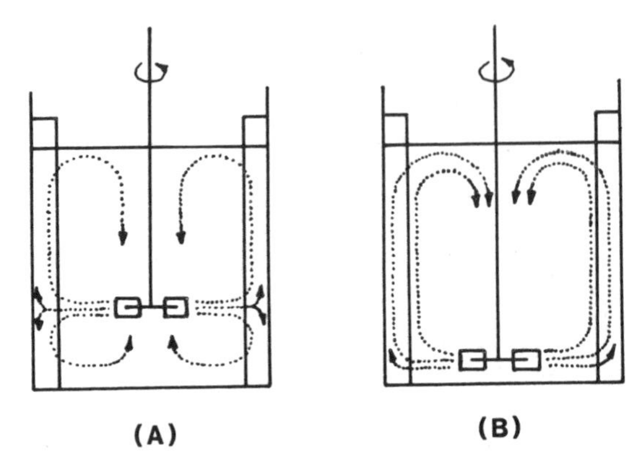

Figure 2. (A) Fluid circulation when stirrer is positioned far above vessel bottom; (B) fluid circulation when stirrer is positioned close to vessel bottom.

In this situation the result obtained by Zwietering was rather surprising, and in effect Nienow showed that while the reliability of Equation 1 was substantial, the value of s was a function of the clearance, always decreasing when the clearance decreased (Figure 3).

The influence of the stirrer clearance has been later on confirmed by other authors; Conti et al. [8] showed that the correlation between the minimum stirrer speed for complete suspension and C/T clearly feels the effects of the hydrodynamics in the vessel (Figure 4) as the power spent in the vessel does.

This means that the calculation method based on Equation 1 is very likely still improvable; this notwithstanding the use of the values of s obtainable from Figure 3 makes it considerably valid.

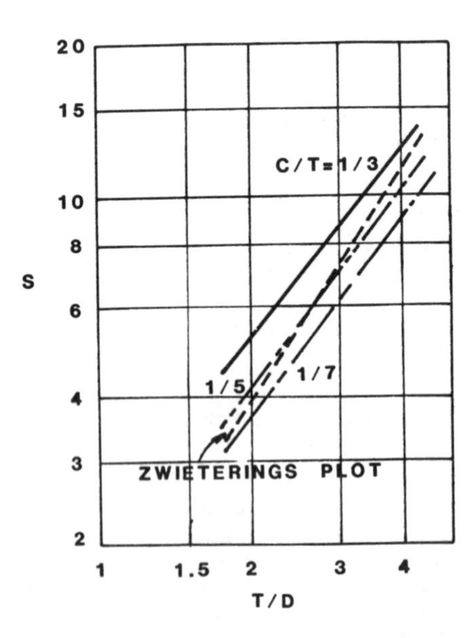

Figure 3. Correlation for suspension reported by Nienow [7].

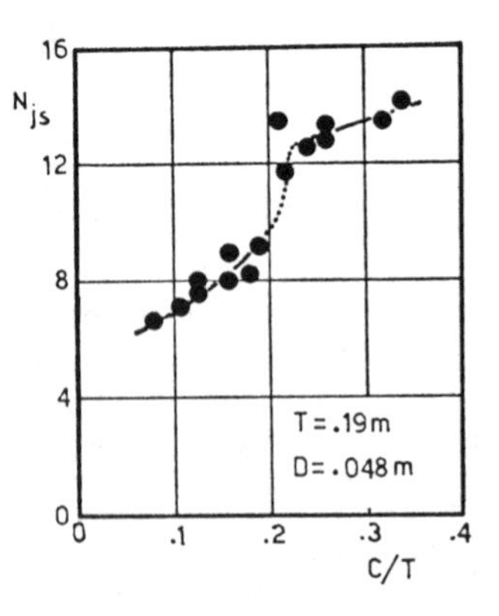

Figure 4. N_{js} vs. C/T reported by Conti et al. [8].

BALDI ET AL.'s THEORETICAL APPROACH

So far few theoretical models have been proposed to describe the attainment of the complete suspension state. Kolář [1, 9] attempted to explain the conservation of a "homogeneous suspension" (defined on the basis of measurers done at one fourth and three fourths of the liquid height) assuming that in the turbulent field generated by the stirrer the eddies of the same size of the particles have a velocity equal to the terminal settling velocity of the particles.

The weak point of Kolář's approach is that in a highly turbulent fluid the settling velocity of a particle is different from the one in a still fluid [10]. Narayanan et al. [11] considered the velocity of the vortices formed in the vessel owing to mechanical agitation with a radial flow stirrer and compared it with the minimum velocity of the fluid needed to suspend the particles. It is surely a valid approach, but in the development of the model the pick-up velocity was determined with a balance of the vertical forces acting on particles with some heavy assumptions on the drag coefficient value, on the solid wettability and on the slip velocity between solid particles and liquid. Moreover the peculiarity of the hydrodynamic situation near the vessel bottom was neglected so that the model of Narayana et al. seems somewhat improvable even if already able to give satisfactory results.

Most probably the soundest endeavour to explain theoretically the mechanism of complete suspension is due to Baldi et al. [12, 13]. According to this model the suspension of the solid particles is due to turbulent disturbances rather than to the whole velocity field over the vessel bottom. More exactly the particles are lifted from the vessel bottom, where they are momentarily at rest, when they are hit by turbulent eddies which have a scale of the order of, or proportional to, the particle size. Supposing the height at which the particles are lifted is of the order of, or proportional to, the particle size too, an energy balance can be made on the basis that the potential energy acquired by the particles is proportional to the kinetic energy lost by the eddies.

Besides, supposing that the decay of turbulence from the generation zone (the stirrer) to the vessel bottom can be treated in analogy with the decay of turbulence downstream from a grid, the following correlation can be written:

$$\frac{\left(g\dfrac{\Delta\rho}{\rho_L}\right)^{0.5} T D_p^{1/6}}{N_{js}\, Po_{js}^{1/3}\, D^{5/3}} = f\left(Re^*; \frac{D}{T}; \frac{C}{D}\right) \qquad (2)$$

In it Po_{js} is the power number at the complete suspension state

$$\left(Po_{js} = \frac{P_{js}}{\rho_L N_{js}^3 D^5}\right.$$

with P_{js} = power dissipated by the stirrer) and Re* a modified Reynolds number

$$\left(Re^* = \frac{\rho_L N_{js} D^3}{\mu_L T}\right)$$

f is a function symbol.

The nature of the function depends on the geometry of the system and must be determined experimentally. For instance for an eight-bladed disc turbine and C/D = 1, the dimensionless group on Equation 2 (that henceforth will be termed Z) is proportional to $(Re^*)^{0.2}$ so that:

$$N_{js} \propto \frac{\mu_L^{0.17} D_p^{0.14} (g\,\Delta\rho)^{0.42} T}{\rho_L^{0.58} D^{1.89}\, Po_{js}^{0.28}} \qquad (3)$$

Taking into consideration that Po is nearly constant for turbulent systems, the resemblance between Equation 3 and Equation 1 is good indeed. In reality the link between Z and Re* changes with the geometry of the system and therefore also the values of the exponents in Equation 3 change;

Table 1

$$N_{js} \propto \mu_L^a D_p^b (\Delta\rho)^c \rho_L^d X^e D^f \left(\frac{T}{D}\right)^g$$

Radial Stirrers: iA = i – bladed disc turbine; jB = j – bladed paddle stirrer

Author	Ref.	Stirrer	a	b	c	d	e	f	g	Notes on Clearance
Zwietering	6	6A	0.1	0.2	0.45	−0.55	0.13	−0.85	1.5*	Any C/T
Zwietering	6	2B	0.1	0.2	0.45	−0.55	0.13	−0.85	1.2†	Effect of the clearance on the constant of proportionality
Nienow	7	6A	—	0.21	0.43	—	0.12	−0.91‡	1.3‡	Effect of the clearance on the constant of proportionality
Narayanan et al.	11	8B	—	0.5	0.5		0.22	−1	$\left\{\frac{(T/D)^2}{[(2T/D)-1]}\right\}^{**}$	C/T = 1/2
Baldi et al.	12	8A	0.17	0.14	0.42	−0.58	0.125	−0.89	1	C/D = 1
Baldi et al.	12	8A	0	0.17	0.5	−0.5	0.15	−0.67	1	C/D = 1/2
Rieger and Ditl	14	6A–6B	0	0	0.5	−0.5	—	−0.5	1.05	Effect of the clearance on the constant of proportionality
Chapman et al.	17	6A	0	0.15	0.40	—	—	−0.76§	1.69§	C/T = 1/4

* $s = 1.45\left(\frac{T}{D}\right)^{1.5}$ *from graph in Reference 6.*

† $s \propto \left(\frac{T}{D}\right)^{1.2}$ *from graph in Reference 6.*

‡ $s = \left(1 + 3.25\frac{C}{T}\right)\left(\frac{T}{D}\right)^{1.3}$ *from graph in Reference 7.*

§ *Deduced from* $N_{js} \propto D^{-2.45}$ *(T = constant) and* $N_{js} \propto T^{-0.76}$ *(D/T = constant).*

** Group containing T/D.

nonetheless they remain very close to those suggested by Zwietering. It is worth noticing that the link between Z and Re* is also affected by the value of C/D; this means that the influence of the stirrer clearance on N_{js} is rather complex as confirmed by Conti et al. [8]. It deserves to be noticed also that rather recently Rieger and Ditl [14] have denied the influence of C/D on the exponents in the correlation of N_{js} with the other parameters; in effect their data regarding disc turbines refer only to values of C/D equal to 2/3 or lower. The model of Baldi et al. does not consider the influence of the solid concentration X on N_{js}; some experimental runs demonstrated that Z was proportional to $X^{-0.15}$ in full agreement with the results of Zwietering [6] and Nienow [7]. In Table 1 some experimental results are summarized.

FURTHER DEVELOPMENTS OF THE ZWIETERING/BALDI APPROACH

The fitness of Equation 2 to interpret the behavior of flat-bottomed vessels with draft tubes, which showed good ability to suspend solid particles when fitted out with propellers [15, 16], was also tested by Baldi and Conti [13]. Instead of propellers they used angled-blade-disc turbines which in preliminary trials gave excellent results.

In effect these systems proved to be very efficient, but while with solid particles larger than 2.10^{-4} m Z was definitely proportional to $(Re^*)^{0.05}$, with particles smaller than 2.10^{-4} m an open cluster of data points was obtained with a very poor possibility of correlation. This suggested that the Baldi et al. model had a lack of generality, showing evident defects when applied to stirrers producing an axial component of flow.

This limitation has been confirmed lately by Chapman et al. [17] in an exhaustive report in which even if the Baldi et al. model is judged the best one so far, Zwietering's correlation is still suggested as the soundest basis for design purposes.

Since the mixed-flow stirrers (like the angled-blade turbines) were not considered by Zwietering, a help for the prediction of N_{js} can be found also in the above quoted report of Champan et al. (6-angled-blade-disc turbine) and in the report of Nienow and Miles [18] (4-blade 45°-pitch turbine). This last stirrer showed N_{js} experimental values somewhat close to those predictable for a three-blade marine propeller using Zwietering's calculation method; since these mixed-flow stirrers are normally easier to make and cheaper than propellers, the attempt of Baldi and Conti seems to deserve further attention.

OTHER STUDIES

Of course several other studies on solid particles suspension in stirred vessels have been carried out hitherto and for the most part of remarkable value.

Among these studies it is right to include the ones of the "German School." A basic review of the equations suggested by some German researchers for calculating the minimum stirrer speed necessary for a complete suspension of the solid phase is presented in the previously cited report of Bohnet and Niesmak [2]. A comparison between the stirrer speeds measured by Bohnet and Niesmak and those calculated using the previously mentioned equations bore rather unsatisfactory results, but it has to be taken into consideration that the systems examined by Bohnet and Niesmak were sometimes somewhat different from those utilized to derive the equations; in particular they used a flat-bottomed vessel whereas most of the equations were related to vessels with dished bottoms. Owing to the well-known industrial interest for this shape of bottom, these studies should be the object of further consideration. The equations for calculating the minimum stirrer speed considered by Bohnet and Niesmak in their work [2] have been inserted by Oldshue [19].

CONCENTRATION PROFILES

The concentration profiles of the solid particles suspended in stirred vessels, worked in conditions different from the one of homogeneous suspension, have been considered mainly with reference to continuous steady-state operations.

Years ago the problem of the continuous removal of the suspension from a stirred vessel was examined by Rushton [20] in terms of isokinetic withdrawal conditions, but shortly afterwards Oldshue [21, 22] called attention to the expediency to know the solid particles distribution in the vessel, especially when the solid particles were of nonuniform size.

Rather recently the problem has been faced again by several authors. Baldi et al. [23] have shown that at the complete suspension condition the solid concentration profiles are heavily affected by the particle size (Figure 5), but are frequently characterized by the presence of parts in which the concentration is rather constant. In these parts the value of the concentration may be very close to the average one in the vessel and then they are true zones of pseudo-homogeneity: if a withdrawal tube is placed within their limits the use of the vessel as a continuous-flow device (i.e. the maintenance of a steady solid concentration) should be easy enough. But often the value of the concentration in these parts is different from the average one; in these cases the withdrawal tube should be placed at the level at which the concentration profile intersects the co-ordinate corresponding to the average concentration in the vessel (value 1 in Figure 5). This solution is not satisfactory, as it is difficult to scale-up. It is at any rate inapplicable when the solid particles are not uniformly sized. In this case, as noted by Oldshue [22], it is not possible to obtain steady-state conditions, as the composition in the vessel is different from the discharge (and feed) compositions. (See also [19] where rather high solid concentrations are considered, too.)

Since it was related to continuous flow vessels the work of Baldi et al. [23] was carried out using withdrawal tubes in the wall of the vessel: so the concentration they measured might be a little different from the real one inside the vessel, but was useful for their aims. In order to get data reflecting the real situation inside the vessel Bohnet and Niesmak [2] determined the solids concentration

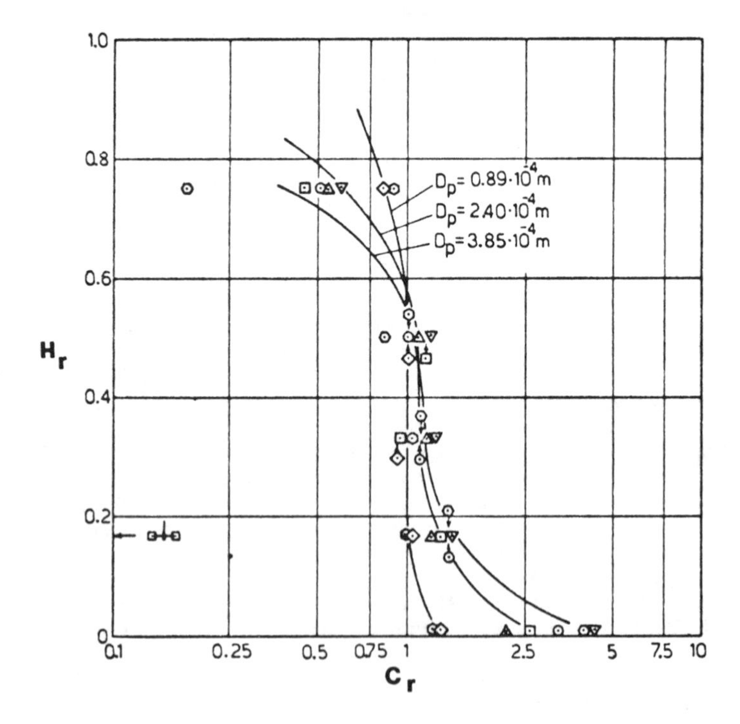

Figure 5. Solid concentration profiles for various particle sizes and mean concentrations (from Baldi, et al. [23]). H_r = off bottom distance to liquid height ratio; C_r = solid concentration related to average concentration in vessel.

measuring the intensity of light attenuation when passing through the suspension. At any rate their results are not comparable with those of Baldi et al. since they used as stirrers pitched-blade turbines and propellers (downward thrust) instead of radial turbines.

This notwithstanding the results of Bohnet and Niesmak confirm once more the deep influence of both the particles size and the density difference between solid and liquid on the quality of distribution of the solid particles, but above all confirm that to improve this quality much more power might be necessary. Figure 6, in which σ represents the relative standard deviation:

$$\sigma = \sqrt{\frac{1}{i}\sum_{1}^{i}\left(\frac{c_i}{c_0} - 1\right)^2} \tag{4}$$

where c_i is the local solid concentration (i gives the level considered) and c_0 is assumed, shows that with the used system geometry no quality improvement can be achieved raising the stirrer speed above 1,000 rpm.

Lately Niesmak [24] by measuring the attenuation of γ-rays emitted by a Cs-137 generator has confirmed that also the total solid concentration affects the quality of solid particles distribution; the mechanisms of this influence are not yet completely understood. At present chemical engineers have at their disposal no equation able to describe entirely the solid particles concentration profiles starting from the system parameters; a model suggested by Tojo and Miyanami [25] fits rather well the data of Bohnet and Niesmak with the exception of the lower part of the vessel where solid particles tend to accumulate. For this reason the model still needs some improvement.

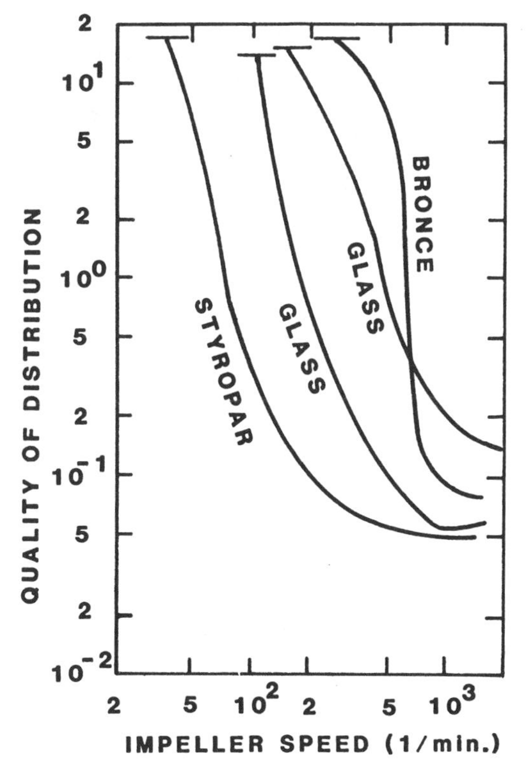

Figure 6. Plot of quality of distribution vs. impeller speed for different solid materials (propeller, downward thrust, D = 100mm, C = 50mm) [2].
* terminal settling velocity 10.81×10^{-2} m/s
ϕ terminal settling velocity 2.24×10^{-2} m/s

SCALING-UP

According to Zwietering's approach if geometrical similarity is maintained between the small, experimental apparatus and the large, technical one (i.e. s of Equation 1 has the same value in both cases) and the properties of the solid and of the liquid are the same, the scaling-up criterion is simply:

$$N_{js}D^{0.85} = \text{constant} \tag{5}$$

For turbulent systems the power number Po is rather constant and consequently P_{js} is proportional to $N_{js}^3D^5$. Then $P_{js} \propto D^{2.45}$ and, for constant D/T,

$$\left(\frac{P}{V}\right)_{js} \propto T^{-0.55} \tag{6}$$

where $(P/V)_{js}$ represents the power spent per unit volume at the complete suspension condition. Hence a sizable reduction in $(P/V)_{js}$ should be considered when scaling-up. This result has been only partially confirmed by the following research. Baldi et al. [12] showed that for radial stirrers the correlation between Z and Re* (see Equation 2) was considerably affected by the value of C/D and while for C/D = 1 Equation 3 was derived, which corroborates the result of Zwietering, for C/D = 0.5Z resulted practically independent of Re* and therefore the power spent per unit volume was almost unaffected by the scale of the system ($N_{js} \propto T^{-0.67}$).

Recently Chapman et al. [17] using a very wide range of vessel sizes (from T = 0.29 m to T = 1.83 m) showed that the correlation

$$N_{js} \propto T^{-0.76} \tag{7}$$

could be regarded as suitable both for disc turbines and for mixed-flow impellers, but they observed a remarkable scatter in terms of power dissipation per unit volume. This scatter was attributed to minor geometrical differences of the vessels used in their work. This opinion can be shared, and bearing in mind also the review of Niesmak [24] which has found in literature exponents of T in Equation 6 ranging from −1 to +0.5, the first scaling-up suggestions seem to be that only correlations obtained from exactly similar geometries can be used and that in any case when the scale ratio is rather high a bit of prudence is advisable. In a recent report Kneule [26] has endeavored to justify the wide range of values derived from experimental work for the exponent of T in Equation 6. He has found that when the lifting of the solid particles takes place in the presence of accumulations of solids on the bottom of the vessel the exponent of T can be also positive while when the previously mentioned accumulations are absent the value of the exponent is included between −1 and 0.

This conclusion is very interesting, but in order to be truly useful in scaling-up it should be completed with research on its connections with the geometry of the stirred vessels. At any rate the work of Kneule seems very valid and not only for having drawn the attention to the influence of the distribution of the solids over scaling-up, but also for having insinuated the idea that the exponent of T in Equation 6 might be a function of T. To sum up, a lot of research work has still to be done before truly safe scaling-up rules can be at disposal for any stirring system. This notwithstanding, some scaling-up suggestions can be considered rather sound if used for fully similar apparatuses, sparge pipes, and other minor geometrical details included. Among these can be enclosed for instance the Chapman et al. criterion:

$$N_{js} \propto T^{-0.76}$$

i.e.

$$\left(\frac{P}{V}\right)_{js} \propto T^{-0.28}$$

for disc turbines and mixed-flow impellers.

RESUSPENSION OF THE SOLIDS SETTLED AFTER STOPPING THE STIRRER

Restarting the stirrer can be a very difficult problem if the solid particles concentration is rather high. In this case, in fact, the solid particles bed can even completely cover the stirrer, and the torque required to set it in motion can exceed some times that at steady state [27, 28]. Besides some slow chemical reaction may bind together the solid particles hardening the problem. Kipke has suggested the following equation to correlate the starting power requirement P_s with the steady-state power consumption P:

$$\frac{P_s}{P} \propto \frac{\rho_{ss} - \rho_L}{\rho_{su}} g \frac{y_0}{PoN^2D^2} \tag{8}$$

where ρ_{ss}, ρ_L, and ρ_{su} = densities of the sediment (with voids filled with the liquid), of the liquid, and of the homogeneous suspension, respectively

y_0 = immersion depth

D = stirrer diameter

N = stirrer speed and P_0 the power number

Po = power number

At the complete suspension condition, if the Chapman et al. scaling-up criterion (Equation 7) is applicable, Equation 8 gives:

$$\frac{\left(\frac{P_s}{P}\right)_{js,\,large}}{\left(\frac{P_s}{P}\right)_{js,\,small}} = \left(\frac{T_{large}}{T_{small}}\right)^{0.52} \tag{9}$$

This relationship indicates that the ratio of restarting to steady-state power consumption significantly increases when the scale of the apparatus increases and that, consequently, in large-scale vessels it can be impossible to restart without additional devices. Unfortunately Kipke himself admits that his approach is still improvable; restarting will be therefore another important item for forthcoming research on solid-liquid stirred vessels.

SOLID SUSPENSION AND MASS TRANSFER

The stirring of a solid-liquid system and then the suspension of particles are important not only to mantain in the apparatus steady and controlled hydrodynamic conditions but also to accomplish some other more complex operation that is in general connected with solid-liquid mass transfer phenomena.

In an interesting review of mass transfer kinetics in such systems Aussenac et al. [29] attain as a first rough conclusion that when the solid particles are completely suspended the solid-liquid mass transfer coefficient k_s shows no significant differences from one stirring vessel to another and is rather near to the value characteristic of a single sphere at its terminal velocity of fall [30]. The value of k_s shows a scatter less than one order of magnitude.

Nevertheless the design economy is related with many other parameters like the power consumption or the investment for the apparatus. Both these last factors must be choosen in order to get operative the vessel by attaining an adequate particle suspension.

In Figure 7 [31] $Sh \cdot Sc^{-1/3}$ is reported as a function of the power input group $(\bar{\epsilon}D_p^4/\nu^3)$ introduced on the basis of Kolmogoroff theory [32]; $\bar{\epsilon}$ represents the power consumption per unit of mass, Sh and Sc the numbers of Sherwood and Schmidt, respectively. In Figure 8 [31] the k_s enhancement at stirrer speeds higher than the one which just causes the complete particle suspension is reported as a function of the stirrer speed itself.

The increase of the solid-liquid mass transfer coefficient with the power spent is evident, but it is also evident that this increase may be very expensive: to double the mass transfer rate it is necessary to double the stirrer speed, i.e., to multiply by eight the power spent.

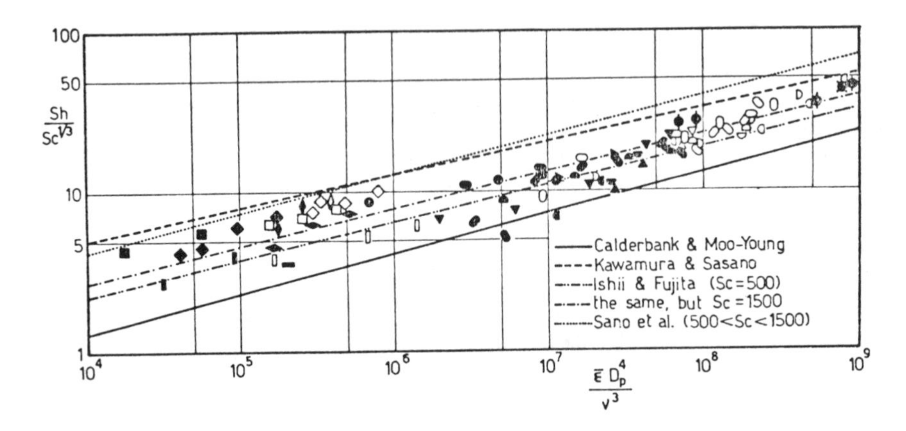

T $.10^3$ [m]	$\frac{D}{T}$	$\frac{C}{T}$	D_p $.10^3$ [m]			symbols $N \leq N_{js}$	symbols $N > N_{js}$
128	0.36	0.33	3.6	benzoic acid cylindrical particles	4 flat blades impeller		
128	0.56	0.33	3.6				
190	0.24	0.33	1.1				
190	0.38	0.33	1.1				
190	0.24	0.17	2.4			.17	j.s. only
190	0.24	0.22	2.4			.22	
190	0.24	0.33	2.4			.33	
190	0.24	0.50	2.4			.50	
190	0.24	0.17	3.6				
190	0.24	0.22	3.6				
190	0.24	0.33	3.6				
190	0.24	0.50	3.6				
190	0.30	0.33	3.6				
190	0.38	0.33	3.6				
190	0.24	0.17	6.1			.17	j.s. only
190	0.24	0.22	6.1			.22	
190	0.24	0.33	6.1			.33	
190	0.24	0.50	6.1			.50	
228	0.25	0.33	2.4				j.s. only
228	0.19	0.33	3.6				
228	0.25	0.33	3.6				
228	0.32	0.33	3.6				
228	0.25	0.33	6.1				j.s. only
190	0.33	0.33	3.6		6 flat blades turbine		
190	0.33	0.15	0.92	ion exch. resins			
190	0.33	0.20	0.92				
190	0.33	0.33	0.92				
190	0.33	0.50	0.92				

Flat-bottomed vessels with 4 wall baffles.

Figure 7. Correlation of Sh $Sc^{-1/3}$ with $\bar{\epsilon}D_p^4/\nu^3$ as reported by Conti and Sicardi [31]. Data symbols defined in the table.

In this situation it may be more profitable to increase the mass transfer rate by increasing the solid surface area, i.e., using a higher solid concentration; N_{js} in fact varies with the solid concentration following the previously mentioned $N_{js} \propto X^e$ with e included between 0.12 and 0.22. So to double the solid surface area it should be sufficient to multiply by 1.5, or 2 at the most, the power spent.

This is only an example of the utility of the research on solid particles suspension.

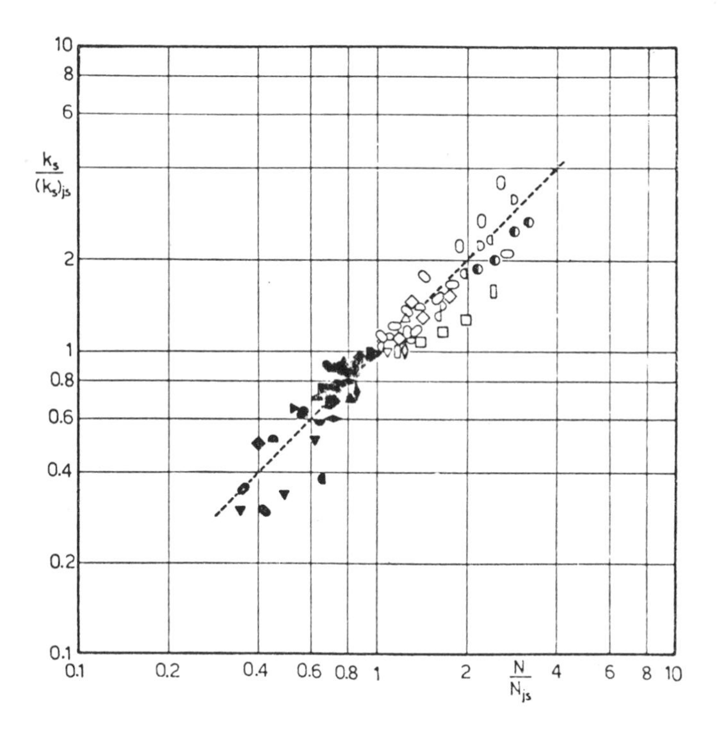

Figure 8. Correlation of the K_s enhancement with the ratio of the actual stirrer speed to the speed corresponding to complete suspension (from Conti and Sicardi [31]).

NOTATION

C	stirrer clearance
c_i	local solids concentration
D	stirrer diameter
D_p	solid particle size
g	acceleration due to gravity
k_s	mass transfer coefficient
N	stirrer speed
P	power dissipated by stirrer
P_s	starting power
Po	power number
Re	Reynolds number
Re*	modified Reynolds number
Sc	Schmidt number
Sh	Sherwood number
s	dimensionless constant
T	tank diameter
V	volume
X	solid concentration
y_0	immersion depth
Z	dimensionless group defined by Equation 2

Greek Symbols

$\bar{\epsilon}$	power consumption per unit mass
μ	viscosity
ν	kinematic viscosity
ρ	density
σ	relative standard deviation

REFERENCES

1. Kolář, V., *Collection Czechoslov. Chem. Commun.*, 26:613 (1961).
2. Bohnet, M., and Niesmak, G., *Ger. Chem. Eng.*, 3:57 (1980).
3. Weisman, J., and Efferding, L. E., *AIChE J.*, 6:419 (1960).
4. Musil, L., and Vek, J., *Chem. Eng. Sci.*, 33:1123 (1978).
5. Gates, L. E., Morton, J. R., and Fondy, P. L., *Chem. Eng.*, (May 24, 1976), p. 144.
6. Zwietering, T. N., *Chem. Eng. Sci.*, 8:224 (1958).
7. Nienow, A. W., *Chem. Eng. Sci.*, 23:1453 (1968).
8. Conti, R., Sicardi, S., and Specchia, V., *The. Chem. Eng. J.*, 22:247 (1981).
9. Kolář, V., *Collection Czechoslov. Chem. Commun.*, 32:526 (1967).
10. Schwartzberg, H. G., and Treybal, R. E., *I.E.C. Fundamentals*, 7:6 (1968).
11. Narayanan, S., et al., *Chem. Eng. Sci.*, 24:223 (1969).
12. Baldi, G., Conti, R., and Alaria, E., *Chem. Eng. Sci.*, 33:21 (1978).
13. Conti, R., and Baldi, G., Proc. Int. Symp. on Mixing, Mons, March 1978, paper B2.
14. Riegel, F., Ditl, P., Proc. 4th Europ. Conf. on Mixing, Leeuwenhorst, 1982, BHRA Fluid Engineering, 263.
15. Aeschbach, S. and Bourne, J. R., *The Chem. Eng. J.*, 4:234 (1972).
16. Bourne, J. R., and Sharma, R. N., *The Chem. Eng. J.*, 8:243 (1974).
17. Chapman, C. M., et al., *Chem. Eng. Res. Des.*, 61:71 (1983).
18. Nienow, A. W., and Miles, D., *The Chem. Eng. J.*, 15:13 (1978).
19. Oldshue, J. Y., *Fluid Mixing Technology*, McGraw-Hill Pub. Co., New York (1983).
20. Rushton, J. H., AIChE-IChemE Symposium Series No. 10 19:3 (1965).
21. Oldshue, J. Y., *Ind. Eng. Chem.*, 61 (a):79 (1969).
22. Oldshue, J. Y., Report presented at First Pacific Area Chem. Eng. Congress, Kyoto, 1972.
23. Baldi, G., Conti R., and Gianetto, A., *AIChE J.*, 27:1017 (1981).
24. Niesmak, G., *Chem.-Ing.-Tech.*, 55:318 (1983).
25. Tojo, K., and Miyanami, K., *Ind. Eng. Chem. Fundam.*, 21:214 (1982).
26. Kneule, F., *Chem.-Ing.-Tech.*, 55:275 (1983).
27. Kipke, K., *Ger. Chem. Eng.*, 6:119 (1983).
28. Kipke, K., *Ger. Chem. Eng.*, 6:264 (1983).
29. Aussenac, D., et al., "Solid Liquid Mass Transfer in Stirred Vessels," to be published.
30. Aussenac, D., Alran, C., and Couderc, J. P., Proc. 4th Europ. Conf. on Mixing, Leeuwenhorst, 1982, *BHRA Fluid Engineering*, p. 417.
31. Conti, R., and Sicardi, S., *Chem. Eng. Commun.*, 14:91 (1982).
32. Brian, P. L. T., Hales, H. B., Sherwood, T. K., *AIChE J.*, 15:727 (1969).

CHAPTER 32

MIXING AND AGITATION OF VISCOUS FLUIDS

Shozaburo Saito, Kunio Arai, Koji Takahashi and **Masafumi Kuriyama**

Department of Chemical Engineering
Tohoku University
Sendai, Japan

CONTENTS

INTRODUCTION, 901

FLOW PATTERNS IN AGITATED VESSELS, 902
Flow Patterns in Anchor Agitated Vessels, 902
Flow Patterns in Helical Ribbon Agitated Vessels, 903

POWER CONSUMPTION OF AGITATED VESSELS, 906
Shear Characteristics, 906
Derivation of Power Correlation for Both Anchor and Helical Ribbon Impellers, 907

MIXING OF VISCOUS NEWTONIAN LIQUIDS IN AGITATED VESSELS, 912
Liquid Crystal Method, 912
Mixing Patterns in Agitated Vessels, 913
Relation Between Mixing Time and Impeller Geometries, 921
Mixing Performance, 924

NOTATION, 924

REFERENCES, 925

INTRODUCTION

The mixing and agitation of viscous liquids is frequently encountered in the chemical, food, and pharmaceutical industries. The mixing of low viscous liquids is usually carried out in a turbulent flow region and is easily accomplished by bulk flow and turbulent diffusion. On the other hand, the mixing of highly viscous liquids requires extremely high power input if the turbulent flow is attained and is usually operated in a laminar flow region. In this case, the mixing proceeds mainly in the high-shear field and the overall mixing is controlled by the exchange of fluid elements between the high- and low-shear fields resulting from the bulk flow. The flow pattern produced in the mixing equipment may be easily affected by small changes of the geometries, therefore, a careful selection of the geometrical configuration of the equipment is important.

For mixing of highly viscous liquids, there are several types of equipment available, such as an agitated vessel, extruder, kneader, and motionless mixer. Of this equipment, an agitated vessel is most commonly used. The eight types of impellers most frequently used for the mixing of highly viscous liquids in industrial applications are illustrated in Figure 1. These impellers provide close clearance between blades and vessel wall where the fluid is submitted to very high shearing. Impellers 1 to 4 produce mainly tangential flow and from 5 to 8 the axial flow adding to the tangential one. The typical impellers in these two groups are an anchor and a helical ribbon impeller, respectively.

In this chapter, we introduce our recent studies for mixing and agitation of viscous Newtonian liquids in agitated vessels equipped with various types of anchor and helical ribbon impellers.

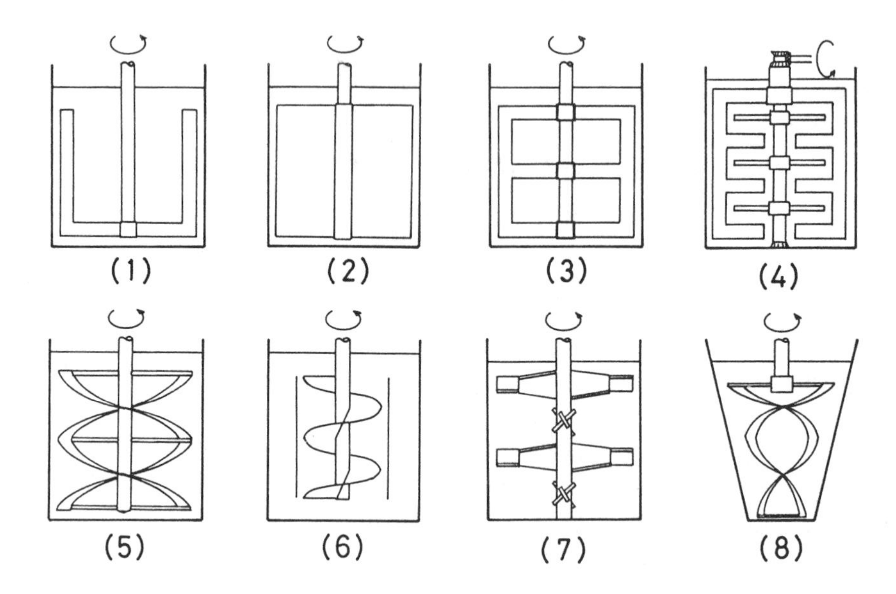

Figure 1. Impellers used for mixing highly viscous liquids: (1) anchor; (2) paddle; (3) gate; (4) double motion paddle; (5) helical ribbon; (6) helical screw; (7) MIG; (8) helicone.

FLOW PATTERNS IN AGITATED VESSELS

Detailed information on velocity distributions in an agitated vessel is very helpful to an understanding of mixing performance, since convective transport tends to control the overall mixing of highly viscous liquids. Velocity distributions for highly viscous liquids have been investigated experimentally by many authors in various mixer geometries [4, 20, 27, 28]. However, the quantitative expression for the flow patterns over the entire flow field seems not to have been presented. In order to obtain quantitative information, numerical study may be useful. However, some simplification of a flow field is required to solve a flow problem numerically. The application of numerical method is consequently restricted to an agitated vessel of simple geometry; such as an anchor agitated vessel. The authors have solved numerically two-dimensional Newtonian flow problems in the horizontal plane of anchor agitated vessels [17]. For agitated vessels of complicated geometries, such as a helical ribbon agitated vessel, the experimental study is necessary to obtain useful information on velocity distributions. The authors have developed a surface-fitting method for the velocity data scattered on a given surface by using a bicubic B-spline function [2]. Using this method, three dimensional velocity distributions of helical ribbon agitated vessels have been obtained [29].

In this section, these velocity distributions in anchor and helical ribbon agitated vessels are shown and effects of detailed geometrical variables on flow patterns are discussed.

Flow Patterns in Anchor Agitated Vessels

Fluid motion in an anchor agitated vessel is characterized by flow in a horizontal plane induced by the impeller. Detailed experimental studies on flow around the anchor impeller blades have been made by Peters and Smith [27, 28]. They reported the streamline patterns and velocity distributions in a horizontal plane of the vessel which were obtained by using tracer particles. On the other hand, numerical study on the flow has been made by the authors [17]. A two-dimensional

Newtonian flow problem was solved by using a new iteration method for the determination of the boundary values of stream function, and good agreements of calculated distributions of velocity with measured ones were obtained. The variations of streamline pattern and of tangential and radial velocity distributions with increasing Reynolds numbers are shown in Figure 2. In this figure, the streamlines are represented in a coordinate system fixed on the impeller and the velocity distributions of that are fixed on the vessel wall.

With increasing Reynolds numbers, the symmetry evidence of the flow with respect to the blade is gradually lost. In case of low Reynolds numbers of Re = 1 or 10, the small drag of the rotating impeller blades creates the stagnant zone, the region enclosed by the streamline of $\psi^* = 0$, which is observed near the shaft at an angle of about 90° from the blades. At Re = 50, although this stagnant zone disappears, a region where the liquid moves with the blade is seen behind the blade. The existence of this region has been pointed out also by Peters and Smith. This region is not very large in this case of Re = 50; however, it can be easily expected that this region becomes larger with an increasing Reynolds number. Peters and Smith have indicated that the axial velocity in this region will affect the liquid mixing in an anchor agitated vessel.

Flow Patterns in Helical Ribbon Agitated Vessels

It is with difficulty that the numerical method is applied to the flow in agitated vessels of complicated geometries. The authors have developed a surface-fitting method for the velocity data scattered on a given surface with the use of a bicubic B-spline function [2] and obtained the quantitative expression of the velocity distributions throughout a helical ribbon agitated vessel [29].

Surface-Fitting Method for the Velocity Data in an Agitated Vessel

From the photographs of the motion of a small particle in an agitated vessel, a great number of path lines are obtained by using a cubic polynominal spline:

$$r = F_r(t), \qquad \theta = F_\theta(t), \qquad z = F_z(t) \tag{1}$$

The velocity components can be obtained by differentiating Equation 1 with respect to time t:

$$V_r = F_r'(t), \qquad V_\theta = rF_\theta'(t), \qquad V_z = F_z'(t) \tag{2}$$

where the functional forms of $F_i(t)$ and $F_i'(t)$ are given in the literature [1, 14].

The parameter vectors for all path lines of Equation 1 are stored on a file in a computer with information on the region where each observed path line exists. Through the use of this data file and Equations 1 and 2, the coordinates of all intersections of the path lines and a given plane as well as the velocity components at the intersections can be calculated and recorded in a file as the velocity data scattered on the plane, which must be correlated. Then the least square surface fitting is carried out for these data. Detailed information on this surface-fitting method are given in the literature [2].

Velocity Distributions in Helical Ribbon Agitated Vessels

Velocity distributions in helical ribbon agitated vessels [29] with different clearances between blades and wall are shown in Figure 3 and the geometrical configuration of these impellers is shown in Figure 4. These velocities are normalized by dividing by πdN, which is the speed of the outertip of the ribbon. The impellers are rotated in a direction as the liquid flows upwards at the blade. These velocity distributions are closely connected with the mixing patterns and are discussed later.

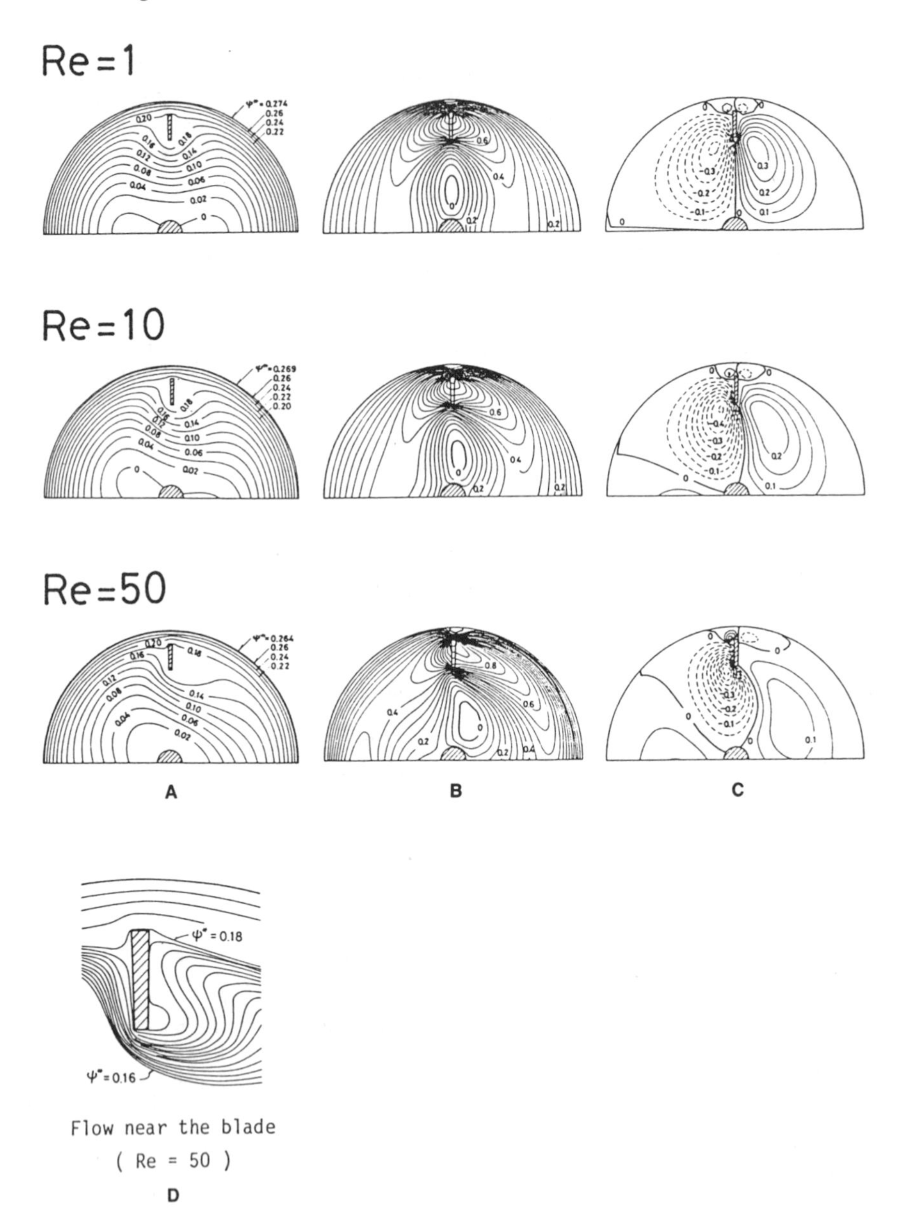

Figure 2. Effects of Reynolds number on flow pattern in an anchor agitated vessel (w/D = 0.1, c/D = 0.05): (A) stream line; (B) tangential velocity distribution; (C) radial velocity distribution; (D) flow near the blade (Re = 50).

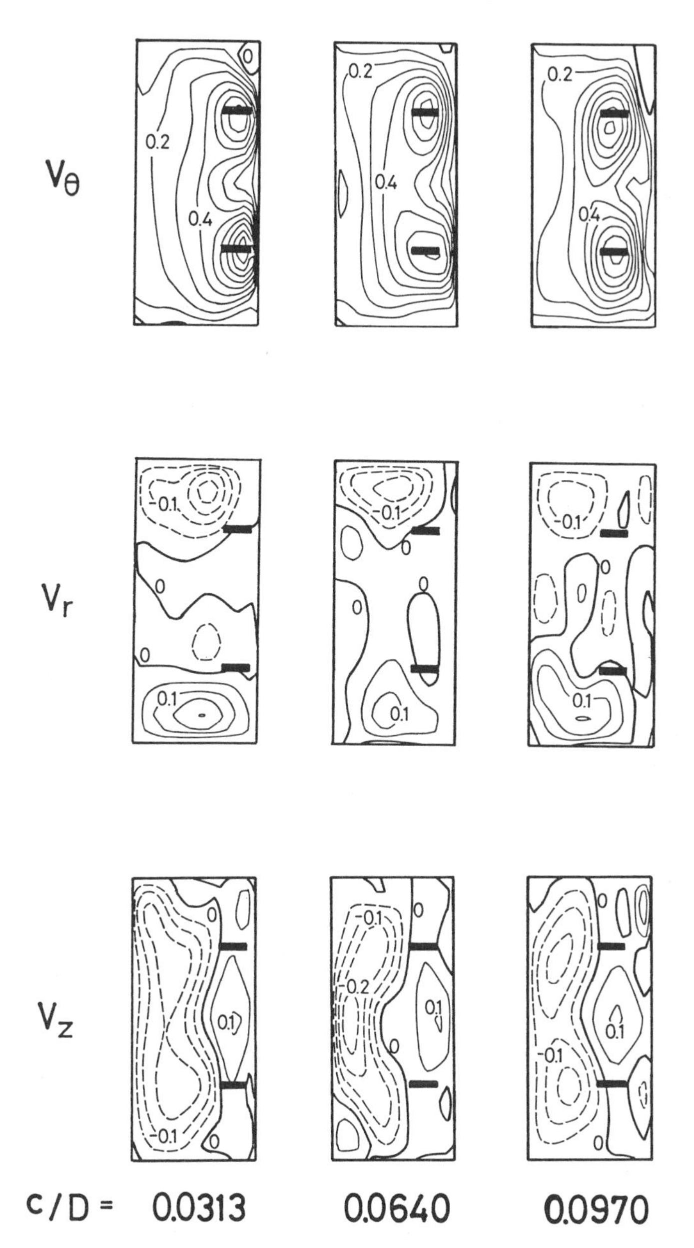

Figure 3. Velocity distributions in helical ribbon-agitated vessels with different clearances: (A) tangential velocity distribution; (B) radial velocity distribution; (C) axial velocity distribution.

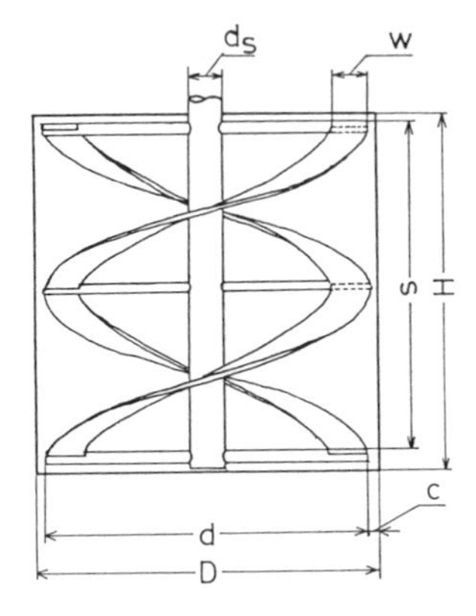

Figure 4. Geometrical configuration of helical ribbon impellers.

POWER CONSUMPTION OF AGITATED VESSELS [30, 31]

For the design of mixing equipment, it is necessary to predict its power consumption. In the laminar flow region, power consumption increases proportionally with liquid viscosity and the following relation has been confirmed experimentally [5, 6, 8, 12, 22, 25].

$$N_P \cdot Re = C_1 \tag{3}$$

C_1 is a function of the geometrical variables. The power correlations [3, 5, 8, 9, 12, 22] published previously for close-clearance impellers, such as anchor and helical ribbon impellers, are relatively limited. Most of them are empirical and are restricted to particular impellers.

In order to correlate the power consumption with geometrical variables, it may be useful to adopt a physical model which can represented the flow system in agitated vessels approximately. The authors have carried out power consumption measurements for various types of anchor and helical ribbon impellers and have proposed a power correlation for both types of impellers on the basis of a physical model [30, 31].

In this section, basic information of the shear characteristics is discussed and a power correlation is presented.

Shear Characteristics

Power consumption is characterized by the shear rate in the tangential direction. Figure 5 shows the distributions of normalized tangential shear rate, relative to the coordinate system fixed on the impeller blade in the horizontal plane of an anchor agitated vessel. This figure was obtained from the numerical analysis presented earlier. In this figure, it is evident that a high shear rate appears near the inner and the outer edges of the blade. However, the shear rate becomes suddenly small in the region apart from the blade. Although the pattern of the distribution changes with an increasing Reynolds number, the degree of nonuniformity in normalized shear rate in the flow field is relatively unaffected by Reynolds number.

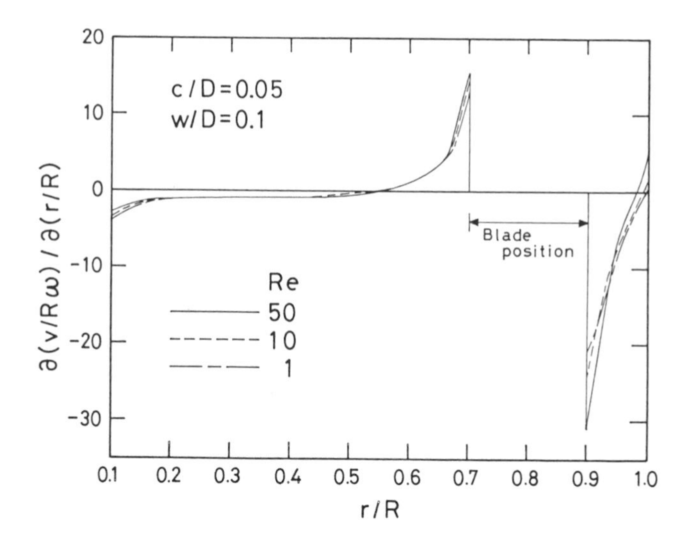

Figure 5. Tangential shear rate distributions in anchor-agitated vessels.

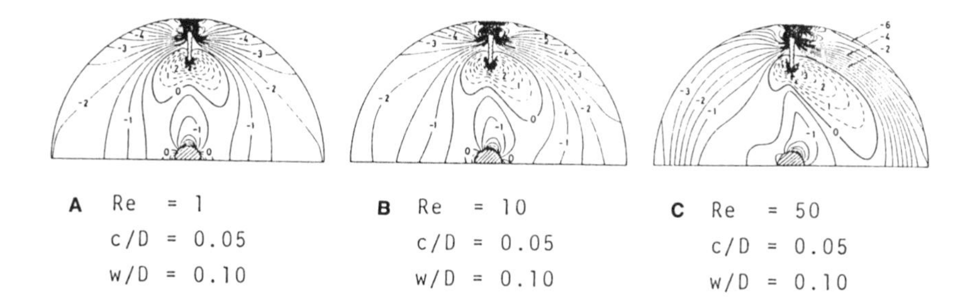

Figure 6. Shear rate distributions in anchor-agitated vessels along radius at blade position.

The detailed distributions along the radius of the blade position are shown in Figure 6. The distribution curves are similar to those reported by Peters and Smith [27, 28]. From this figure, it can be seen that the differences among the normalized curves at each Reynolds number is enhanced at the regions near the solid boundaries, especially at the clearance blades and wall.

From this the following physical model of flow in the clearance region will be effective in correlating power consumption.

Derivation of Power Correlation for Both Anchor and Helical Ribbon Impellers

Figure 7 shows the geometrical configurations of anchor and helical ribbon impellers used for correlation of power consumption and Table 1 summarizes the geometrical variables and experimental values of C_1 obtained by the authors.

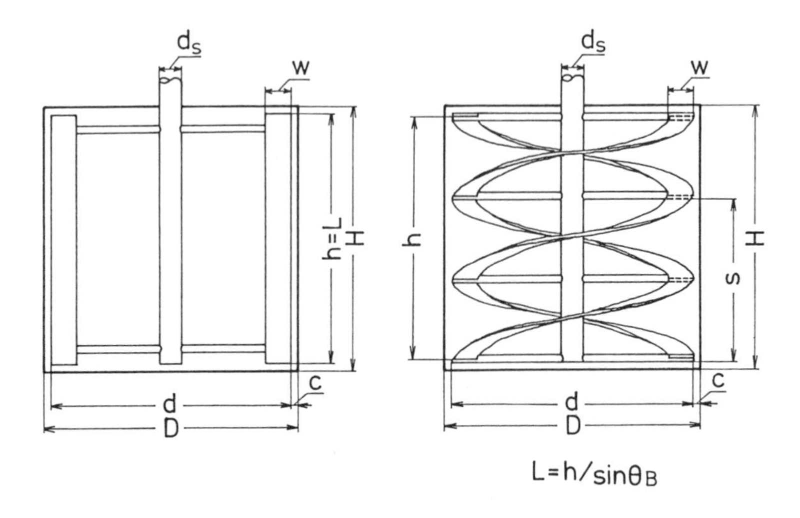

Figure 7. Geometrical configurations of anchor and helical ribbon impellers.

Table 1
Geometrical Variables and Measured Power Consumption for Anchor and Helical Ribbon Impellers

Geometry No.	n_p (−)	n_a (−)	d (mm)	c/D (−)	s/D (−)	L (mm)	w (mm)	C_1 (−)
Anchor impellers								
AC1	2	2	115.2	0.0500	∞	115.0	13.0	173.3
AC2	2	2	121.6	0.0250	∞	115.0	13.0	202.5
AC3	2	2	124.8	0.0125	∞	115.0	13.0	268.7
AC4	2	2	125.4	0.0100	∞	115.0	13.0	301.7
AC5	2	2	126.7	0.0050	∞	115.0	13.0	366.6
Helical ribbon impellers								
SH1	1	3	120.0	0.0313	0.980	397.1	13.0	171.9
SH2	1	3	114.0	0.0547	0.980	379.3	13.0	148.4
SH3	1	3	103.9	0.0943	0.980	349.4	13.0	152.1
SH4	1	4	113.2	0.0577	0.649	547.9	13.0	178.2
SH5	1	5	114.6	0.0522	0.488	731.1	13.0	196.8
SH6	1	3	116.4	0.0453	1.30	301.4	13.0	151.1
SH7	1	3	115.7	0.0480	1.95	220.6	13.0	148.4
DH1	2	3	120.0	0.0313	0.980	397.2	13.0	351.1
DH2	2	3	112.4	0.0609	0.980	374.6	13.0	290.6
DH3	2	3	102.8	0.0984	0.980	346.3	13.0	275.3
DH4	2	4	112.9	0.0590	0.649	546.5	13.0	356.6
DH5	2	5	113.8	0.0555	0.488	725.9	13.0	425.1
DH6	2	3	116.0	0.0469	1.30	300.6	13.0	291.0
DH7	2	3	116.1	0.0465	1.95	221.1	13.0	272.5
DH8	2	3	116.4	0.0454	0.980	386.5	9.75	294.1
DH9	2	3	116.8	0.0437	0.980	387.6	16.3	332.7
DH10	2	3	117.0	0.0428	0.980	388.2	19.5	339.4
DH11	2	3	116.6	0.0444	0.980	387.1	26.0	372.4
TH1	3	3	116.4	0.0453	1.30	301.4	13.0	448.1
TH2	3	3	115.9	0.0473	1.95	220.8	13.0	416.3
QH1	4	3	116.0	0.0469	1.95	221.0	13.0	491.5

$D = H = 128.0$ (mm) $d_a = 3.0$ (mm) $d_s = 12.0$ (mm)

Power Correlation for Anchor Impellers

Fluid flow around an anchor blade in an agitated vessel is similar to that around a flat plate moving at a low speed in viscous liquid bounded by a plate as shown in Figure 8. The drag D_r applied by the liquid to the plate per unit length can be represented by the following equation derived by Takaishi [33] on the basis of Oseen's linearized equations of motion.

$$D_r = 8\pi\mu U \frac{1}{2\ln(4 + 8c/w) - 1} \tag{4}$$

If the curvature of the vessel wall is negligible, Equation 4 is applicable to power consumption for anchor blade. The torque T_M produced by the rotation of the impeller is defined as follows:

$$T_M = D_r \cdot L \cdot \frac{d}{2} \cdot n_p$$

$$= 8\pi\mu U\, dL \frac{1}{2\ln(4 + 8c/w) - 1} \tag{5}$$

Consquently, we obtain the power P.

$$P = \omega T_M = 2\pi N \cdot 8\pi\mu U\, dL \frac{1}{2\ln(4 + 8c/W) - 1} \tag{6}$$

The velocity U is equal to that of the blade tip πdN. Equation 6 can be rewritten as follows:

$$N_{Pb} \cdot Re = \frac{16\pi^3}{2\ln(4 + 8c/w) - 1} \cdot \frac{L}{d} \tag{7}$$

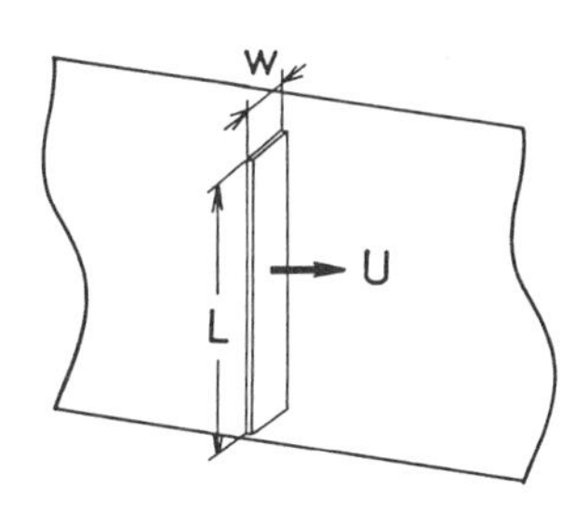

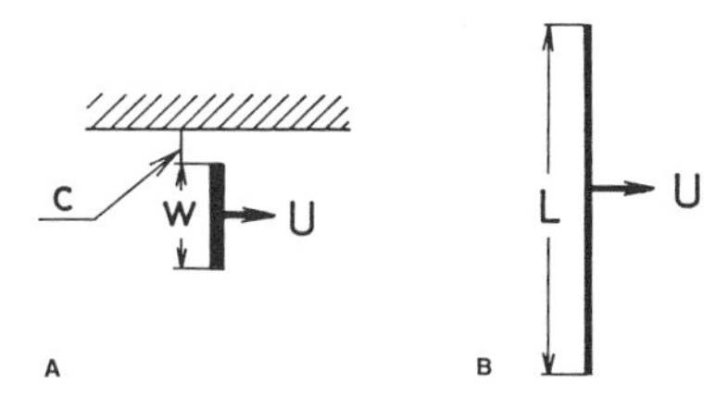

Figure 8. Schematic diagram of anchor blade motion: (A) top view of anchor blade; (B) side view of anchor blade.

Furthermore, the power consumption of an arm can be derived from the drag flow analysis and can be described by the following equation:

$$N_{Pa} \cdot Re = 2.08\pi^3 n_a n_p \left(\frac{d_a}{d}\right)^{0.15} \left(\frac{r_{bi}}{d}\right)^{3.15} \tag{8}$$

Thus the total power for the impeller is given by

$$\begin{aligned} N_P \cdot Re &= N_{Pb} \cdot Re + N_{Pa} \cdot Re \\ &= \frac{16\pi^3}{2\ln(4 + 8c/w) - 1} \cdot \frac{L}{d} + 2.08\pi^3 n_a n_p \left(\frac{d_a}{d}\right)^{0.15} \left(\frac{r_{bi}}{d}\right)^{3.15} \end{aligned} \tag{9}$$

This equation is a theoretical power correlation for anchor impellers.

This correlation agrees approximately with the experimental data at large c/D, but the discrepancy between Equation 9 and the experimental results increases with a decrease in clearance between the blades and wall. Therefore, an empirical factor of the clearance was introduced in Equation 9 with the help of the experimental data. The result is given by

$$\begin{aligned} N_P \cdot Re = {} & \frac{16\pi^3}{2\ln(4 + 8c/w) - 1} \left(\frac{L}{d}\right) f(c/D) \\ & + 2.08\pi^3 n_a n_p \left(\frac{d_a}{d}\right)^{0.15} \left(\frac{r_{bi}}{d}\right)^{3.15} \end{aligned} \tag{10}$$

where $f(c/D) = 1 + 0.00539(c/D)^{-0.876}$

Extension of Power Correlation to Helical Ribbon Impellers

An anchor impeller is considered a variation of the helical ribbon impeller which has its blades at a right angle to the direction of motion. The power correlation for helical ribbon impellers is related to that for anchor impellers by blade angle. We also considered a plate with an arbitrary

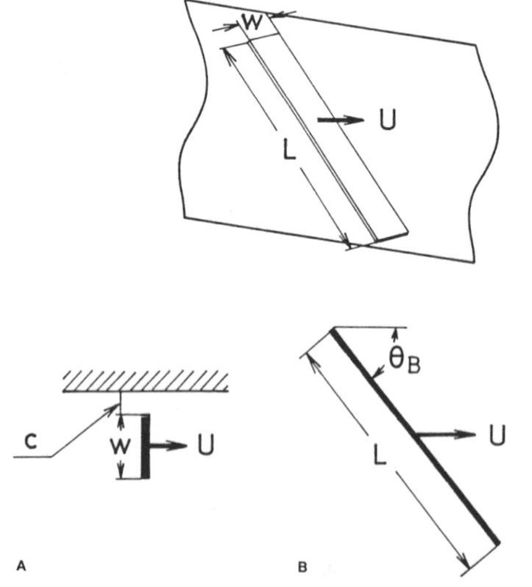

Figure 9. Schematic diagram of helical ribbon blade motion: (A) top view of helical ribbon blade; (B) development of side view of helical ribbon blade.

Table 2
Power Consumption for Helical Ribbon Impellers With and Without a Free Surface

Geometry No.	N_p Re	
	With a Free Surface	Without a Free Surface
DH1	347.1	351.1
DH2	285.6	290.6
DH3	271.1	275.3

angle moving at low speed in viscous liquid as shown in Figure 9. This motion is similar to that of the helical ribbon impeller. In this case, the drag experienced by the plate will be mostly dependent on the force normal to the platte. Therefore, power correlation for helical ribbon impellers was obtained by modifying Equation 10 in terms of blade angle on the basis of the experimental data. Furthermore, from the experimental results, it can be concluded that power consumption increases nearly proportionally to the number of blades while the effect of blade width on power consumption is expressed by the theoretical term in Equation 8. Thus the correlation is given by:

$$N_P \cdot Re = \frac{16\pi^3}{2\ln(4 + 8c/w) - 1}\left(\frac{L}{d}\right) f(c/D)(\sin\theta_B)^{0.555}\frac{n_P}{2} + 2.08\pi^3 n_a n_p \left(\frac{d_a}{d}\right)^{0.15}\left(\frac{r_{bi}}{d}\right)^{3.15} \tag{11}$$

The authors carried out power consumption measurements also for agitated vessels with a free surface, but the differences between the agitated vessels with and without a free surface were hardly noticeable. The results are summarized in Table 2. The effects of the lid wall on power consumption seem to be negligible.

Equation 11 is compared with the experimental data [11–13, 15, 22, 34] in Figure 10. The agreement between measured and calculated values was satisfactory except in works by Nagata et al.

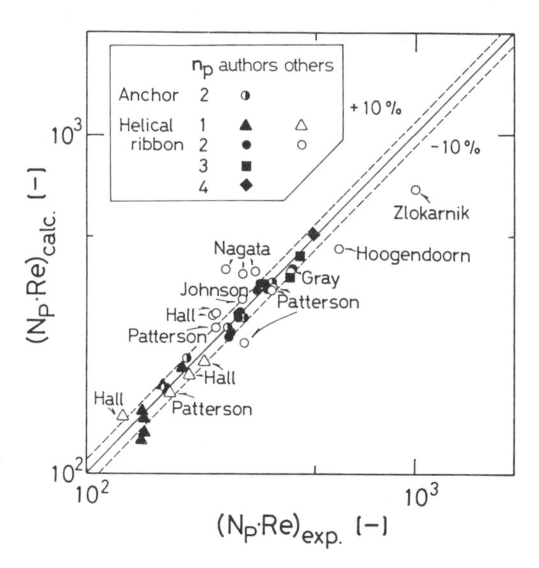

Figure 10. Power correlation for both anchor and helical ribbon impellers.

and Zlokarnik. This correlation is valid for the following ranges of geometrical variables: $n_p = 1$ to 4, $0.019 \leqq c/D \leqq 0.13$, $s/D \leqq 0.49$ and $0.076 \leqq w/D \leqq 0.20$ in the laminar flow region of $Re = 0.1$ to 10.

MIXING OF VISCOUS NEWTONIAN LIQUIDS IN AGITATED VESSELS [32]

The molecular diffusion is very small and there is no turbulent diffusion, when the mixing of highly viscous liquids is carried out in a laminar flow region. In this case, we must make an effort to decrease the diffusion length by stretching and subdividing the lump into small pieces. From Figures 3A and 5 the highshear field where the deformation of the liquid takes place is limited to the clearance between blades and wall. Therefore, the renewal flow rate of the liquid in this region may control the overall mixing. In other words, the mixing appears to be controlled by the flow pattern, especially the axial and radial flows. The velocity components in axial and radial directions are easily affected by small changes of impeller geometrical variables as shown in Figures 3B and C. Several investigators [7, 21, 23, 28] have reported the effects of geometrical variables on mixing time, but the relation between mixing pattern and geometrical variables has not been throughly established. In the literature, the coloration method [10, 11, 18, 23] and decoloration method [7, 13, 23, 26] are commonly used for measurement of mixing time. However, these methods are unsuitable for observation of a three-dimensional mixing process in an agitated vessel, because of the overlap of the colored liquids in the vessel.

The authors have developed the method using a tracer liquid containing small capsules of liquid crystal. This method, which is called the liquid crystal method, solves the overlap problem because only the liquid crystal capsules in the lighted zone can be observed. Using this method, the authors have observed mixing patterns and measured mixing times of anchor and helical ribbon impellers of different geometrical variables [32].

Liquid Crystal Method

The experimental apparatus is shown in Figure 11. The geometrical configurations of anchor and helical ribbon impellers are already shown in Figure 7. Details of impeller geometries are summarized in Table 3.

The agitated vessel was filled with the test liquid. A small amount of tracer liquid containing capsules of liquid crystal was put into the top of the vessel near the shaft. The mixing of tracer

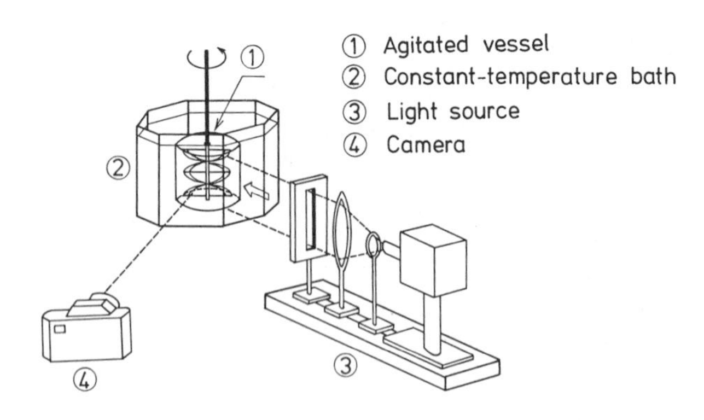

Figure 11. Sketch of experimental apparatus for liquid crystal method.

Table 3
Geometrical Variables for Anchor and Helical Ribbon Impellers used for Observation of Mixing Patterns

Geometry No.	n_p (−)	n_a (−)	d (mm)	c/D (−)	s/D (−)	L (mm)	w (mm)
Anchor impellers							
AC1	2	2	95.6	0.0220	∞	90.0	10.0
AC2	2	2	84.6	0.0770	∞	90.0	10.0
Helical ribbon impellers							
SH1	1	3	96.2	0.0190	0.901	315.0	10.0
SH2	1	3	87.9	0.0606	0.901	290.1	10.0
SH3	1	3	81.8	0.0910	0.901	272.0	10.0
DH1	2	3	96.2	0.0191	0.901	315.0	10.0
DH2	2	3	88.8	0.0558	0.901	293.0	10.0
DH3	2	3	82.7	0.0865	0.901	274.7	10.0
DH4	2	4	90.1	0.0496	0.599	434.8	10.0
DH5	2	5	89.7	0.0515	0.450	570.2	10.0
DH6	2	3	88.8	0.0599	1.20	227.7	10.0
DH7	2	3	88.0	0.0598	1.80	165.1	10.0
DH8	2	3	89.7	0.0507	0.901	296.0	7.50
DH9	2	3	89.8	0.0508	0.901	296.0	12.6
DH10	2	3	90.2	0.0489	0.901	297.1	15.2
DH11	2	3	91.1	0.0447	0.901	299.6	20.0

$D = H = 100.0$ (mm) $h = 90.0$ (mm) $d_s/D = 0.10$ (−)

with the mother liquid was visualized in the vertical plane illuminated by a plane light beam and recorded photographically.

Mixing time was defined as the time required until the detection of flow lines of tracer liquid ceased to be observable to the naked eye.

Mixing Patterns in Agitated Vessels

Mixing Patterns for Anchor Agitated Vessels

Figure 12 shows the illuminated vertical planes of an anchor agitated vessel. Figures 13 and 14 show photographs of mixing patterns in these vertical planes in anchor agitated vessels with different clearances. In each figure, N_r represents the number of impeller rotations counted from the start of mixing. The tracer liquid injected into the top of the vessel near the shaft flows downwards rotating along the gourd-shaped stream line of $\psi^* = 0$ to 0.02 in Figure 2. Figure 15 shows this downward flow region schematically. From the photographs at $N_r = 100$ in Figures 13 and 14, the influence of this flow is shown clearly and this flow tends to produce an unmixed zone which is shown in Figure 15. Next, the tracer flows towards the wall in the middle of the vessel, near $\theta = 0°$. The photographs at $N_r = 200$ show that the tracer flows upwards in the clearance region for the impeller of wide clearance or in the region near the inner edge of blades for the impeller of narrow clearance. It is also shown that the tracer liquid hardly flows into the lower region of the vessel for the impeller of wide clearance, but does flow gradually for that of narrow clearance. These observation results can be explained by considering both the primary tangential flow due to the rotating

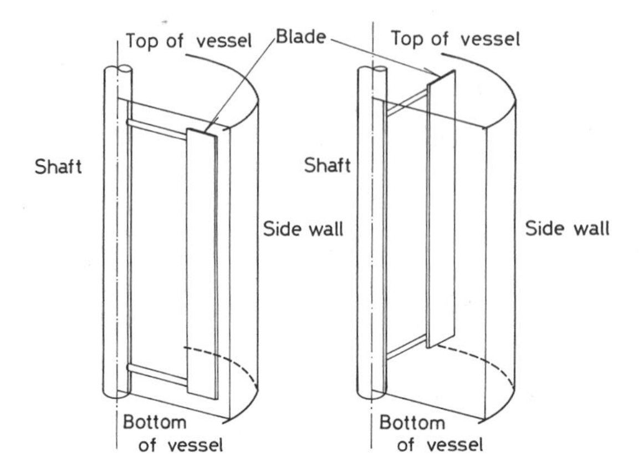

Figure 12. Sketch of illuminated vertical plane of an anchor-agitated vessel.

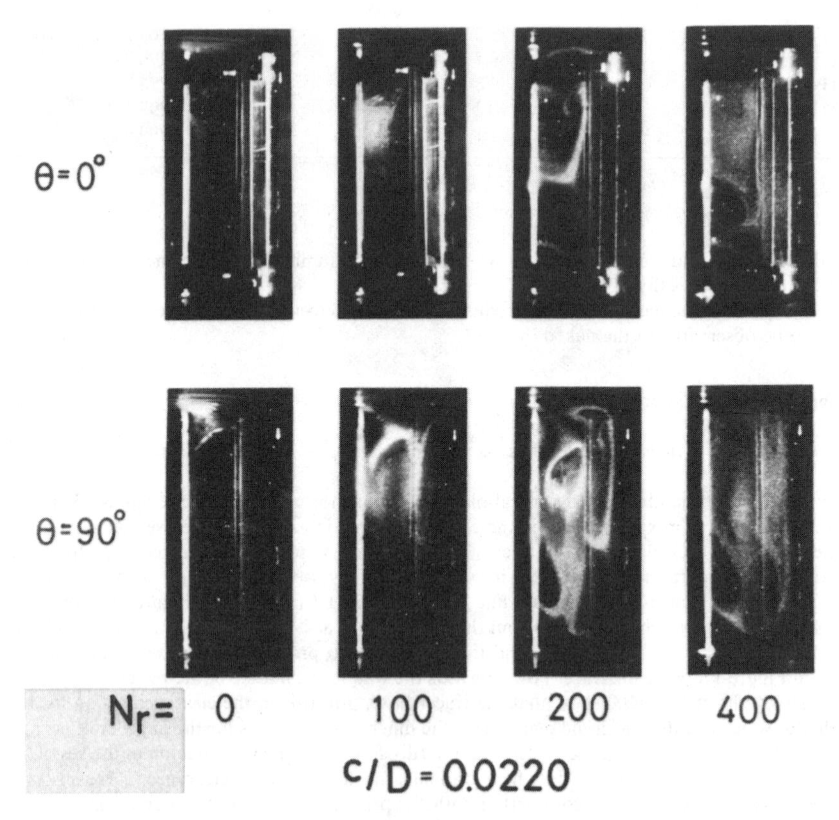

Figure 13. Mixing patterns in an anchor-agitated vessel with narrow clearance impeller.

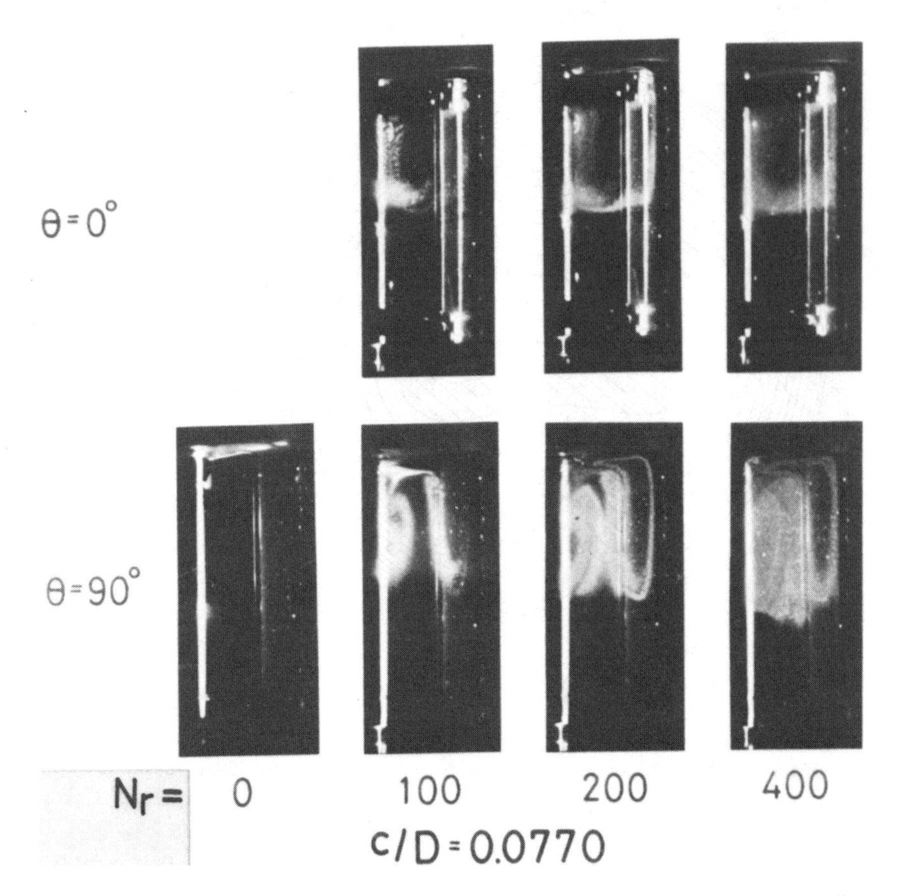

Figure 14. Mixing patterns in an anchor-agitated vessel with wide clearance impeller.

action of the blades, and the secondary flow due to the change of centrifugal force in the axial direction. However, even after 400 impeller revolutions, unmixed zones were observed in both anchor agitated vessels.

Mixing Patterns for Helical Ribbon Agitated Vessels

Figure 16 shows a cross-section of the primary circulation flow in the illuminated vertical plane of a helical ribbon agitated vessel. Figures 17 and 19 through 21 show photographs of mixing patterns in this cross-section in agitated vessels with different geometrical variables. The patterns of the primary circulation flows are almost the same as mentioned by previous investigators [4, 7, 23] and the mixing patterns are approximately similar in spite of the different impeller geometries. Namely, as observed by Carreau et al., [7] shear mixing proceeds mainly in the clearance between blades and vessel wall, and the well-mixed liquid flowing out of the clearance is easily distributed

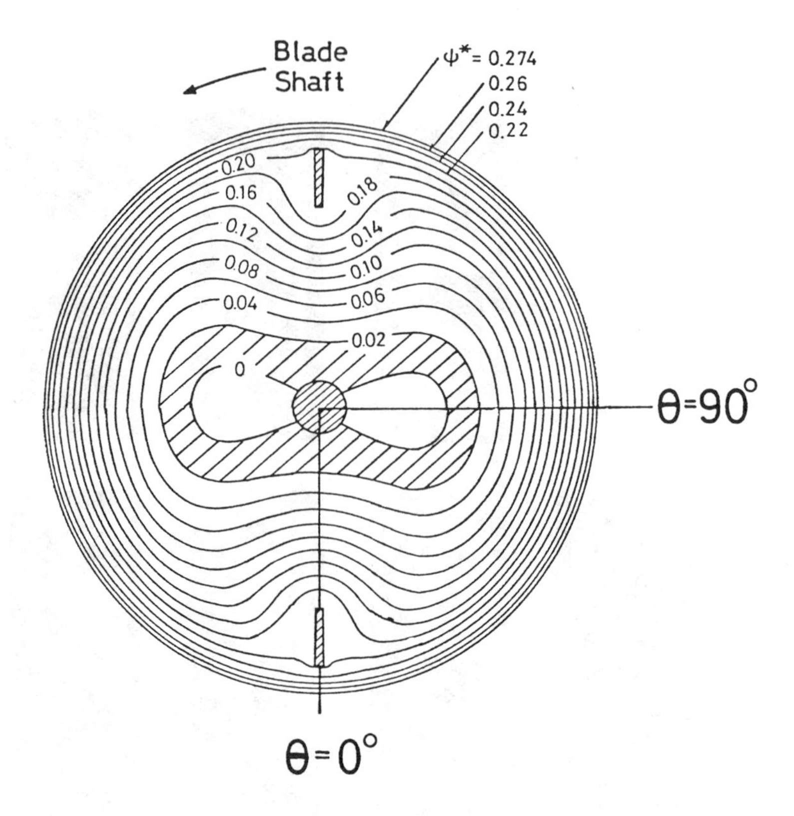

Figure 15. Top view of downward-flow region in an anchor-agitated vessel.

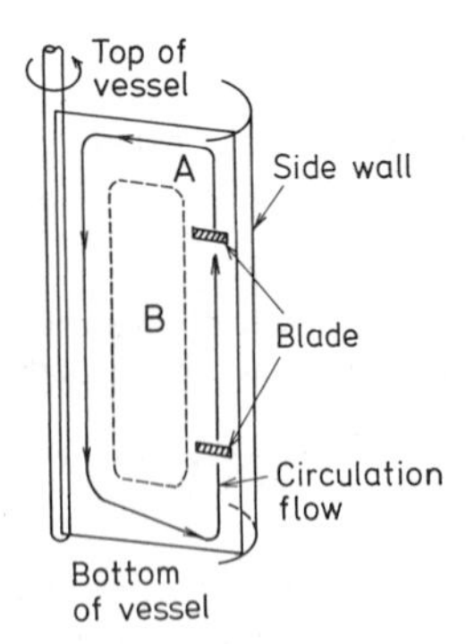

Figure 16. Sketch of illuminated vertical plane of a helical ribbon-agitated vessel.

to the circulation path between the blades by the primary flow. This well-mixed zone is denoted by A in Figure 16. The tracer in this region, however, can permeate only gradually into the inner circulation flow situated in the region inside the inner edge of the blade, denoted by B in Figure 16, and the exchange flow between Regions A and B is commonly observed to be a controlling step in the mixing process.

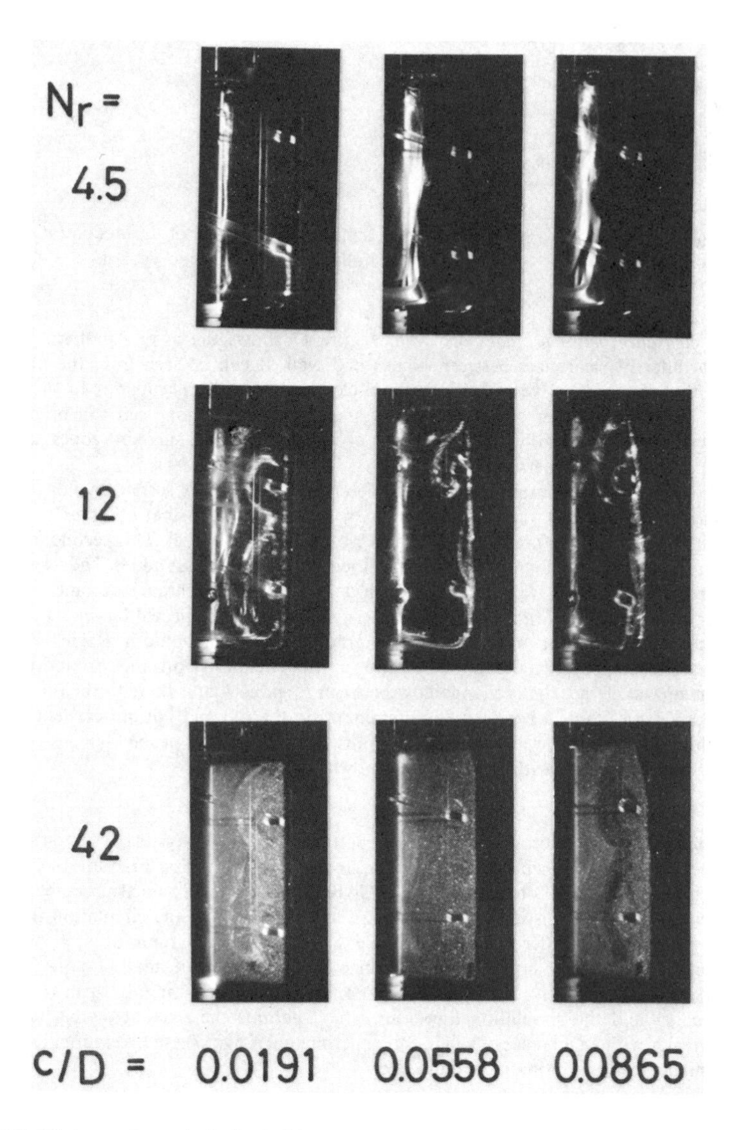

Figure 17. Mixing patterns in helical ribbon-agitated vessels with different clearance impellers ($n_n = 2$, $s/D = 0.901$, $w/D = 0.100$).

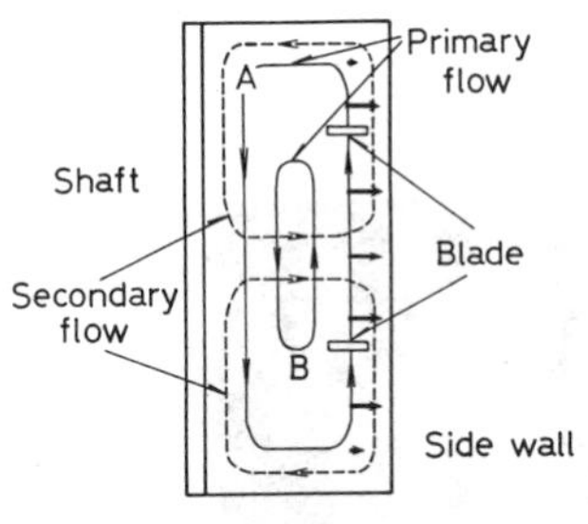

Figure 18. Schematic diagram of the secondary flow produced in a helical ribbon-agitated vessel.

Effect of clearance between blades and wall. Figure 17 shows the mixing patterns in agitated vessels with different clearances between blades and wall. It can be seen from the photographs at $N_r = 42$ that the agitated vessel with medium-clearance impeller is superior to the others for the mixing. With the impeller of narrowest clearance, an unmixed zone can still be observed in Region B. On the other hand, with the impeller of widest clearance, stagnant zones are detected at both top and bottom corners in Region A.

These phenomena may be explained by the difference in the effect of secondary flow on primary flow. As shown in Figure 18, the secondary flow, which is similar to that observed in an anchor agitated vessel, must also be created in a helical ribbon agitated vessel. This secondary flow will deform the primary flow to enhance the exchange flow between Regions A and B. The secondary flow will be depressed by the side wall, an effect which decreases as the clearance becomes wider. With the impeller of narrowest clearance, the flow pattern is not so much affected by the secondary flow and thus the liquid in Region A exchanges only gradually with the liquid in Region B. With the medium clearance impeller, moderate deformation of the stream in both the top and the bottom regions seems to accelerate the exchange flow between Regions A and B. With the impeller of the widest clearance, the stable secondary flow dominates in the top and bottom corners and closed loops develop. These considerations can be supported by the results of the changes of axial and radial velocity distributions with clearances as shown in Figure 3.

Effect of impeller pitch. Photographs of mixing patterns in agitated vessels of different pitches are shown in Figure 19. As shown in the photographs at $N_r = 42$, mixing time for the impeller of $s/D = 0.90$ is shortest. The photographs at $N_r = 9$ show that the axial velocity becomes greater as the s/D ratio increases in this experimental range, as expected from the circulation flow models proposed by Bourne and Butler [4] and Carreau et al. [7]. The photographs at $N_r = 14$ show that the deformation of the tracer liquid in the clearance for the impeller of small pitch is greater than for the impeller of large pitch. Namely, the shear rate, which affects the mixing in the clearance, becomes smaller and the circulation flow rate, which permits the renewal of the liquid in the clearance, greater with an increase of s/D in this experimental range. These two mutually competing effects on mixing lead an optimum in pitch size.

Effect of blade width. Photographs of mixing patterns in agitated vessels of different blade widths are shown in Figure 20. The photographs at $N_r = 42$ show that the impeller of medium blade width is superior to the others. As shown in the photographs at $N_r = 9$, the circulation flow becomes stronger as the blade widens. But if blade width exceeds a certain limit, the circulation flow rate decreases because the area of the upward flow becomes too great compared with that of the downward flow. Then the circulation flow rate, which permits the distribution of liquid in the clearance, has a maximum value at the certain ratio of w/D.

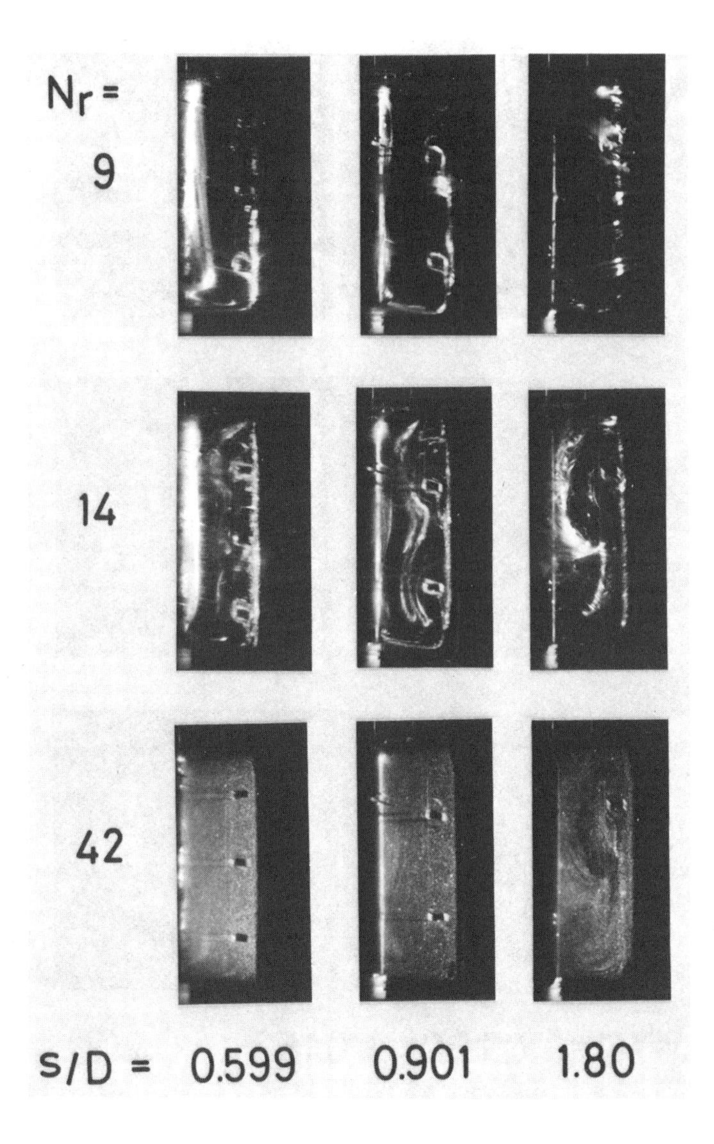

Figure 19. Mixing patterns in helical ribbon-agitated vessels with different pitch impellers ($n_p = 2$, $c/D = 0.058$, $w/D = 0.100$).

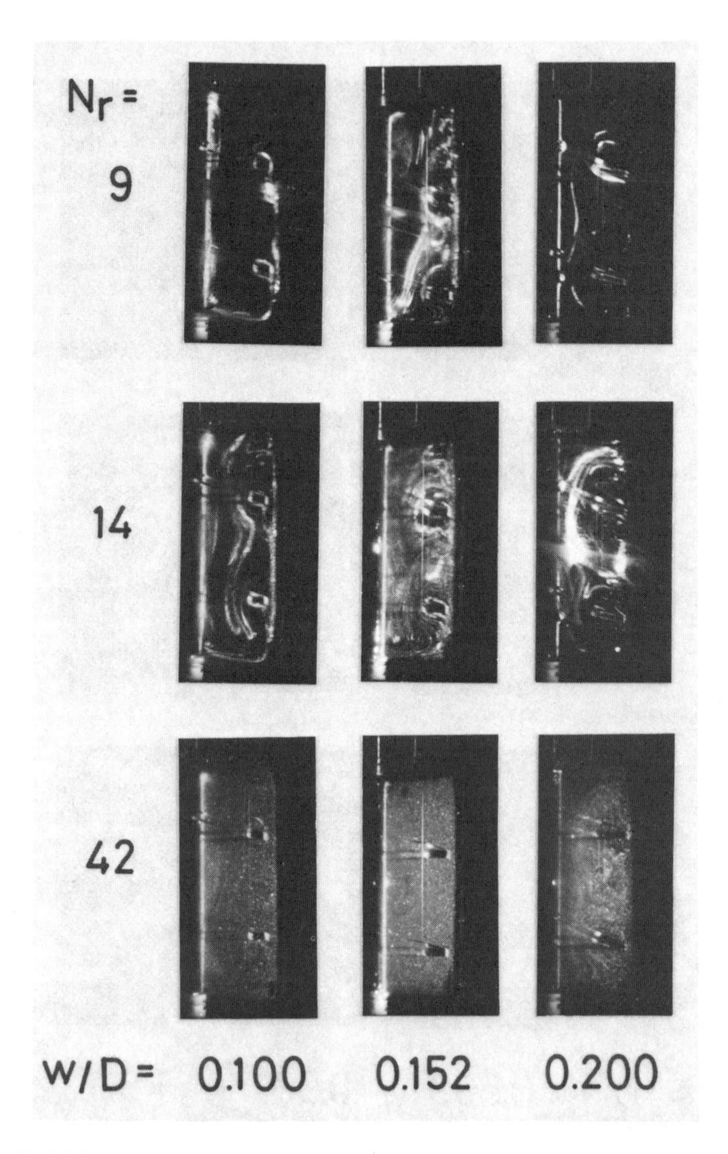

Figure 20. Mixing patterns in helical ribbon-agitated vessels with different blade width impellers ($n_p = 2$, $c/D = 0.050$, $s/D = 0.901$).

Effect of number of blades. As shown in Figure 21, even after 500 impeller revolutions, a poorly mixed zone was observed inside the blade for all single helical ribbon impellers. In single helical ribbon agitated vessels, the secondary flow is weak, due to the long periodic passage of the blade, and thus the overall flow is controlled by the primary flow. Therefore, the double helical ribbon impeller is more suitable than the single one for mixing.

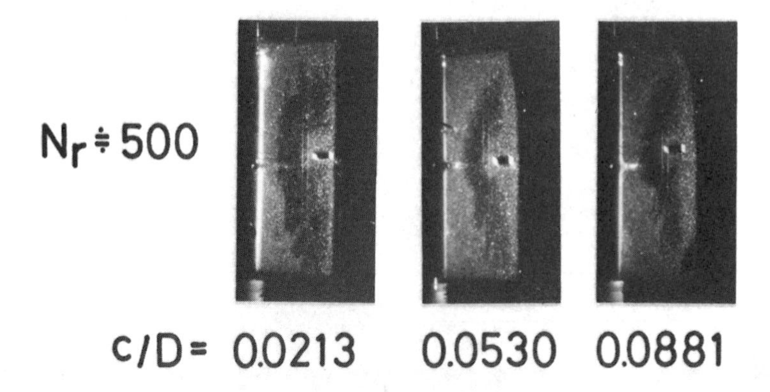

Figure 21. Mixing patterns in helical ribbon-agitated vessels with different clearance impellers ($n_p = 1$, $s/D = 0.901$, $w/D = 0.100$).

Relation Between Mixing Time and Impeller Geometries

As discussed in the previous section, the anchor impeller and the single helical ribbon impeller are unsuitable for mixing. Therefore, our discussion is limited to the mixing time for the double helical ribbon impeller.

For the double helical ribbon impeller, it is suggested that the relation between the mixing time, t_m, and the rotational speed of the impeller, N, can be written as follows in the laminar flow region [7, 11, 13, 15, 19, 21, 23].

$$N \cdot t_m = C_2 \quad (12)$$

where C_2 is a geometrical constant.

Mixing times measured by the liquid crystal method for 11 different impellers are summarized in Table 4. The relation between C_2 and each of three geometrical ratios, c/D, s/D, and w/D, is shown in Figures 22, 23, or 24. The shaded area represents the range where unmixed zones were observed even after 500 impeller revolutions.

As mentioned in the previous section, there are optimum values for the three geometrical ratios in the mixing process. The optimum values of c/D, s/D, and w/D obtained by the authors are 0.060, 0.90, and 0.15 respectively. These values differ from those suggested by Nagata et al. [21, 23], by the decoloration method, i.e. 0.025, 0.95, and 0.10.

Correlation of Mixing Time

Shear mixing of the liquid proceeds mainly in the clearance between blades and vessel wall where the shear rate is high [7]. But the renewal flow rate of the liquid in the clearance, which is assumed to be nearly equal to the circulation flow rate, and the exchange flow rate between the Regions A and B in Figure 18, due to the combined effect of the primary and secondary flows, also play important roles in the uniformity of concentration in an agitated vessel, as suggested in the previous section. On the basis of this consideration, mixing time is expressed as a function of magnitude of

Table 4
Correlations of Circulation Flow Rate and Mixing Time

Geometry No.	$(Q_1/Nd^3)_{exp.}$	$(Q_1/Nd^3)_{calc.}$	$(N \cdot t_m)_{exp.}$	$(N \cdot t_m)_{calc.}$	$(N_p \cdot Re)_{calc.}$
DH1	0.108	0.110	239	236	406.2
DH2	0.112	0.119	53.8	53.1	288.9
DH3	0.123	0.117	∞	592	256.6
DH4	0.0732	0.0867	56.4	60.0	352.6
DH5	0.0708	0.0678	68.5	79.0	392.7
DH6	0.113	0.136	245	69.4	259.4
DH7	0.120	0.132	∞	82.4	222.6
DH8	0.104	0.0795	106	122	273.6
DH9	0.150	0.143	42.3	43.5	315.7
DH10	0.185	0.155	33.0	33.9	334.6
DH11	0.137	0.150	56.2	33.4	365.6

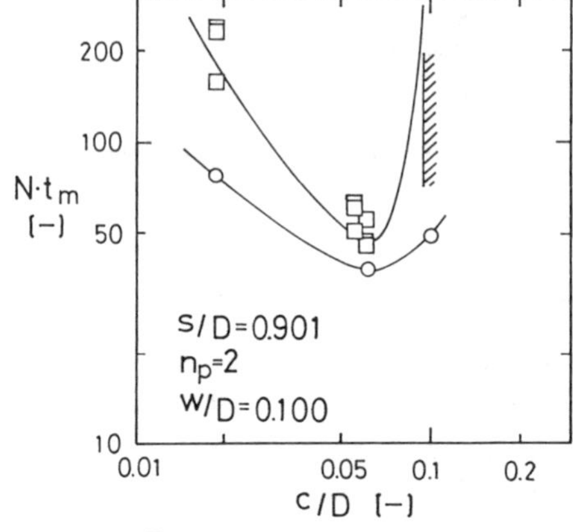

Figure 22. Measured mixing time of helical ribbon-agitated vessels with different clearance impellers.

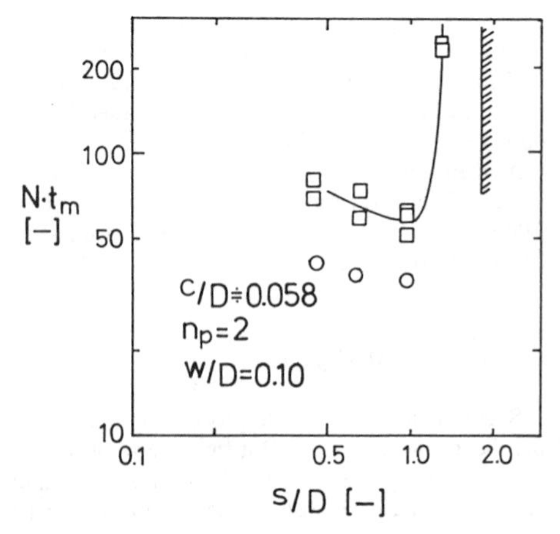

Figure 23. Measured mixing time of helical ribbon-agitated vessels with different pitch impellers.

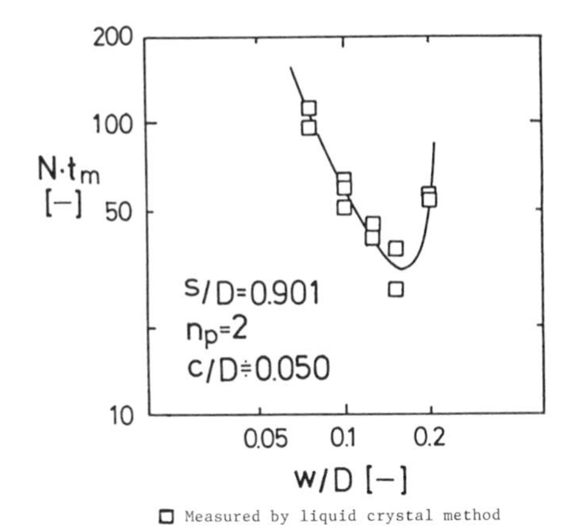

Figure 24. Measured mixing time of helical ribbon-agitated vessels with different blade width impellers.

shear rate in the clearance, M, circulation flow rate, Q_1, and exchange flow rate, Q_e. That is:

$$\frac{1}{Nt_m} = f(M, Q_1, Q_e) \tag{13}$$

Equation 13 is assumed to be given by the following equation:

$$\frac{1}{Nt_m} = M \frac{Q_1}{Nd^3} \frac{Q_e}{Nd^3} \tag{14}$$

In this equation, M and Q_e/Nd^3 are given as follows:

$$M = \frac{1}{N}\sqrt{\frac{P_v}{\mu}} = N_P \cdot Re \cdot \left(\frac{d}{D}\right)^3 \cdot \frac{4}{\pi} \tag{15}$$

$$\frac{Q_e}{Nd^3} = \alpha(Q_2/Q_1 - \beta)^2 + \gamma \tag{16}$$

where

$$\frac{Q_2}{Nd^3} = a \frac{\pi}{256}\left(\frac{D}{d}\right)^5 \left(\frac{\kappa}{\kappa - 1/\kappa}\right)^2 \left(\frac{1}{2\kappa^2} - \frac{\kappa^2}{2} + 2 \ln \kappa\right) \times \left\{1 - \kappa^4 - \frac{(1-\kappa^4)^2}{\ln(1/\kappa)}\right\} \tag{17}$$

$$\kappa = 1 - \frac{2w/D}{\ln[\{(D/d) - (1 - 2w/d)\}/(D/d - 1)]} \tag{18}$$

$$a = \frac{h}{d}\frac{w}{s}\sqrt{\pi^2 + (s/d)^2}\, n_p \tag{19}$$

$$\alpha = -12800, \qquad \beta = 0.000909, \qquad \gamma = 0.0114 \tag{20}$$

$N_P \cdot Re$ can be calculated from Equation 11. The values of Q_1/Nd^3 for the impellers used are shown in Table 4 and for those of different geometry can be estimated from a correlation presented by the authors. Details of derivation are given in the literature [32].

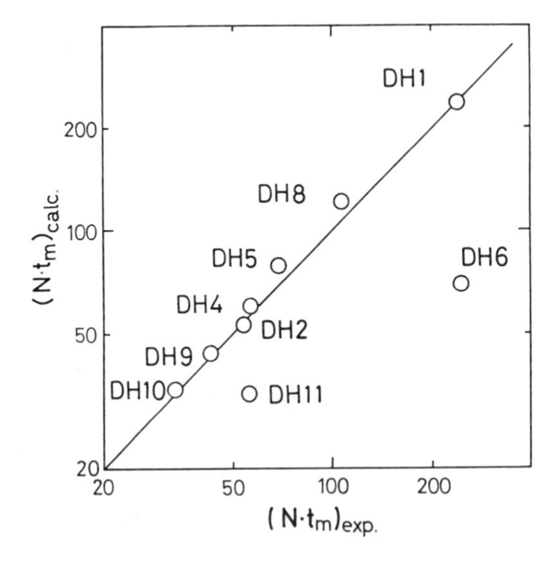

Figure 25. Correlation of mixing time for helical ribbon-agitated vessels.

The calculated values of $N \cdot t_m$ are summarized in Table 4. Figure 25 shows the comparison of calculated and experimental values. The agreement is satisfactory except for DH6 and DH11 impellers. Therefore, this correlation is applicable to the following ranges of each geometrical variable of $n_p = 2$, $0.019 \le c/D \le 0.60$, $0.45 \le s/D \le 1.0$ and $0.075 \le w/D \le 0.16$ in the laminar flow region Re = 0.1 to 10.

Mixing Performance

To compare the mixing performances of the agitated vessels, several criteria have been proposed [13, 20, 24, 25]. However, the results, listed in Table 4, indicate that a small variation in the impeller dimensions has a great effect on mixing time, but a small effect on power consumption. Thus, the mixing time mainly governs the selection of geometrical configuration of the impeller.

NOTATION

a	dimensionless area defined by Equation 19 (–)
C_1, C_2	geometrical constant in Equation 3 or 12 (–)
c	clearance between blades and vessel wall (m)
D	vessel diameter (m)
D_r	drag ($N \cdot m^{-1}$)
d	impeller diameter (m)
d_a	arm diameter (m)
d_s	shaft diameter (m)
H	height of vessel (m)
h	height of impeller (m)
L	length of blade ($= h/\sin \theta_B$) (m)
M	magnitude of shear rate (–)
N	rotational speed of impeller (s^{-1})
N_p	power number of impeller ($= P/\rho d^5 N^3$) (–)
N_{Pa}	power number of an arm (–)
N_{Pb}	power number of blade (–)
N_r	number of impeller revolutions counted from start of mixing (–)
n_a	number of arms on each blade (–)
n_p	number of blade (–)
P	power consumption (W)
P_v	power consumption per unit volume ($W \text{-} m^{-3}$)
Q_1	circulation flow rate ($m^3 \text{-} s^{-1}$)
Q_2	secondary flow rate ($m^3 \text{-} s^{-1}$)
Q_e	exchange flow rate ($m^3\ s^{-1}$)
R	vessel radius (–)
Re	Reynolds number ($= d^2 N \rho/\mu$) (–)

r	radial position (m)	U	velocity of uniform flow ($m\text{-}s^{-1}$)
r_{bi}	radial position of inner edge of blade (m)	V_r, V_θ, V_z	dimensionless radial, tangential or axial velocity (−)
s	impeller pitch (m)	w	blade width (m)
T_M	torque acting on blade (N·m)	z	axial position (m)
t_m	mixing time (s)		

Greek Symbols

α, β, γ	constants in Equation 16 (−)	μ	viscosity (Pa-s)
θ	angular position (rad)	ρ	density ($kg\text{-}m^{-3}$)
θ_B	blade angle (rad)	ψ^*	dimensionless stream function (−)
κ	dimensionless diameter defined by Equation 18 (−)	ω	angular velocity of impeller ($rad\text{-}s^{-1}$)

REFERENCES

1. Ahlberg, J. H., Nilson, E. N., and Walsh, J. L., *The Theory of Spline and Their Applications*, New York and London: Academic Press, 1967.
2. Arai, K., Takahashi, K., and Saito, S., "Correlation of Velocity Profiles in an Anchor Agitated Vessel Using a Bicubic B-Spline Function," *J. Chem. Eng. Japan*, Vol. 15 (1982), pp. 383–389.
3. Beckner, J. L., and Smith, J. M., "Anchor-Agitated Systems: Power Input with Newtonian and Pseudo-Plastic Fluids," *Trans. Inst. Chem. Engrs.*, Vol. 44 (1966), pp. T224–T236.
4. Bourne, J. R., and Butler, H., "An Analysis of the Flow Produced by Helical Ribbon Impellers," *Trans. Inst. Chem. Engrs.*, Vol. 47 (1969), pp. T11–T17.
5. Bourne, J. R., and Butler, H., "Power Consumption of Helical Ribbon Impellers in Viscous Liquids," *Trans. Inst. Chem. Engrs.*, Vol. 47, 1969, pp. T263–T270.
6. Calderbank, P. H., and Moo-Young, M. B., "The Power Characteristics of Agitators for the Mixing of Newtonian and Non-Newtonian Fluids," *Trans. Inst. Chem. Engrs.*, Vol. 39 (1961), pp. 337–347.
7. Carreau, P. J., Patterson, I., and Yap, C. Y., "Mixing of Viscoelastic Fluids with Helical-Ribbon Agitators. I-Mixing Time and Flow Patterns," *Can. J. Chem. Eng.*, Vol. 54, 1976, pp. 135–142.
8. Chavan, V. V., Jhaveri, A. S., and Ulbrecht, J., "Power Consumption for Mixing of Inelastic Non-Newtonian Fluids by Helical Screw Agitators," *Trans. Inst. Chem. Engrs.*, Vol. 50 (1972), pp. 147–155.
9. Chavan, V. V., and Ulbrecht, J., "Internal Circulation in Vessels Agitated by Screw Impellers," *Chem. Eng. J.*, Vol. 6 (1973), pp. 213–223.
10. Coyle, C.K., "Mixing in Viscous Liquids." *AIChE J.*, Vol. 16 (1970), pp. 903–906.
11. Gray, J. B., "Batch Mixing of Viscous Liquids," *Chem. Eng. Progr.*, Vol. 59 (1963), pp. 55–59.
12. Hall, K. R., and Godfrey, J. C., "Power Consumption by Helical Ribbon Impellers," *Trans. Inst. Chem. Engrs.*, Vol. 48 (1970), pp. T201–T208.
13. Hoogendoorn, C. J., and Den Hartog, A. P., "Model Studies on Mixers in the Viscous Flow Region," *Chem. Eng. Sci.*, Vol. 22 (1967), pp. 1689–1699.
14. Ichida, K., and Yoshimoto, F., *Spline Functions and Applications*, Tokyo: Kyoiku Shuppan Co. Ltd., 1979.
15. Johnson, R. T., "Batch Mixing of Viscous Liquids." *Ind. Eng. Chem. Process Des. Dev.*, Vol. 6 (1967), pp. 340–345.
16. Kuriyama, M., et al., "Heat Transfer and Temperature Distributions in an Agitated Tank Equipped with Helical Ribbon Impeller," *J. Chem. Eng. Japan*, Vol. 14 (1981), pp. 323–330.
17. Kuriyama, M., et al., "Numerical Solution for the Flow of Highly Viscous Fluid in Agitated Vessel with Anchor Impeller," *AIChE J.*, Vol. 28 (1982), pp. 385–391.
18. Lee, R. E., Finch, C. R., and Wooledge, J. D., "Mixing of High Viscosity Newtonian and Non-Newtonian Fluids," *Ind. Eng. Chem.*, Vol. 49 (1957), pp. 1849–1854.

19. Moo-Young, M., Tichar, K., and Dullien, F. A. L., "The Blending Efficiencies of Some Impellers in Batch Mixing," *AIChE J.*, Vol. 18 (1972), pp. 178–182.
20. Murakami, Y., et al. "Evaluation of Performance of Mixing Apparatus for High Viscosity Fluids," *J. Chem. Eng. Japan*, Vol. 5 (1972), pp. 297–303.
21. Nagata, S., *Mixing*, Tokyo: A Halsted Press, Kodansha, 1976.
22. Nagata, S., et al. "Power Consumption of Helical Mixer for the Mixing of Highly Viscous Liquid," *Kagaku Kōgaku*, Vol. 34 (1970), pp. 1115–1117.
23. Nagata, S., Yanagimoto, M., and Yokoyama, T., "A Study on the Mixing of High-Viscosity Liquid," *Kagaku Kōgaku*, Vol. 21 (1957), pp. 278–286.
24. Novak, V., and Rieger, F., "Homogenization with Helical Screw Agitators," *Trans. Inst. Chem. Engrs.*, Vol. 47 (1969), pp. T335–T340.
25. Novak, V., and Rieger, F., "Homogenization Efficiency of Helical Ribbon and Anchor Agitators," *Chem. Eng. J.*, Vol. 9 (1975), pp. 63–70.
26. Oya, H., and Miyauchi, T., "Mixing and Diffusion in High Viscous Fluid," *Kagaku Kōgaku*, Vol. 30 (1966), pp. 915–922.
27. Peters, D. C., and Smith, J. M., "Fluid Flow in the Region of Anchor Agitator Blades," *Trans. Inst. Chem. Engrs.*, Vol. 45 (1967), pp. T360–T366.
28. Peters, D. C., and Smith, J. M., "Mixing in Anchor Agitated Vessels," *Can. J. Chem. Eng.*, Vol. 47 (1969), pp. 268–271.
29. Takahashi, K., "Mixing and Heat Transfer in an Agitated Vessel Operated in Laminar Flow Region," *Doctor Thesis*, Tohoku Univ., 1981.
30. Takahashi, K., Arai, K., and Saito, S., "Power Correlation for Anchor and Helical Ribbon Impellers in Highly Viscous Liquids," *J. Chem. Eng. Japan*, Vol. 13 (1980), pp. 147–150.
31. Takahashi, K., Arai, K., and Saito, S., "An Extended Power Correlation for Anchor and Helical Ribbon Impellers," *J. Chem. Eng. Japan*, Vol. 15 (1982), pp. 77–79.
32. Takahashi, K., et al., "Effects of Geometrical Variables of Helical Ribbon Impellers on Mixing of Highly Viscous Newtonian Liquids," *J. Chem. Eng. Japan*, Vol. 15 (1982), pp. 217–224.
33. Takaishi, Y., "The Wall-Effect upon the Forces Experienced by an Elliptic Cylinder in a Viscous Liquid," *J. Phys. Soc. Jpn.*, Vol. 13 (1958), pp. 496–506.
34. Zlokarnik, M., "Eigung von Rühren zum Homogenisieren von Flüssigkeitsgemischen," *Chem.-Ing.-Tech.*, Vol. 39 (1967), pp. 539–548.

CHAPTER 33

USE OF HELICAL RIBBON BLENDERS FOR NON-NEWTONIAN MATERIALS

Shozaburo Saito, Kunio Arai, Koji Takahashi and Masafumi Kuriyama

Department of Chemical Engineering
Tohoku University
Sendai, Japan

CONTENTS

INTRODUCTION, 927

FLOW PATTERNS, 928
Velocity Distributions, 928
Circulation Flow Rates, 931

POWER CONSUMPTION, 931
Power Correlations Using a Modified Reynolds Number, 932
Power Correlations Using an Apparent Reynolds Number, 932

HEAT TRANSFER TO PSEUDOPLASTIC LIQUIDS IN A HELICAL RIBBON AGITATED VESSEL, 935
Relation Between Temperature Distribution and Flow Patterns, 935
Model for Heat Transfer at the Vessel Wall, 939
Correlation of Heat Transfer Coefficients, 941
Estimation of Temperature Distribution Based on Flow Model, 942

MIXING OF PSEUDOPLASTIC LIQUIDS, 944
Relation Between Mixing Time and Liquid Properties, 944

NOTATION, 946

REFERENCES, 947

INTRODUCTION

Most viscous fluids handled in the chemical industry are non-Newtonian liquids. The agitation of these liquids in an agitated vessel requires high power input and is usually operated in a laminar flow region. In this region, the states of nonuniformities of concentration and temperature are easily created in the vessel. Therefore, the studies for agitation of non-Newtonian liquids are very important. However, little information is available concerning the agitation of non-Newtonian liquids.

Four typical flow-curves are shown in Figure 1. A dilatant liquid increases viscosity with increasing applied shear rate, which is termed "shear-thickening". Alternately, a pseudoplastic liquid decreases viscosity with increasing shear rate, which is termed "shear-thinning." If the thinning effect is very strong, the liquid is termed plastic liquid. The limiting case of a plastic substance is one which requires a finite yield stress before it begins to flow. The majority of non-Newtonian liquids belong to the category of pseudoplastic liquid.

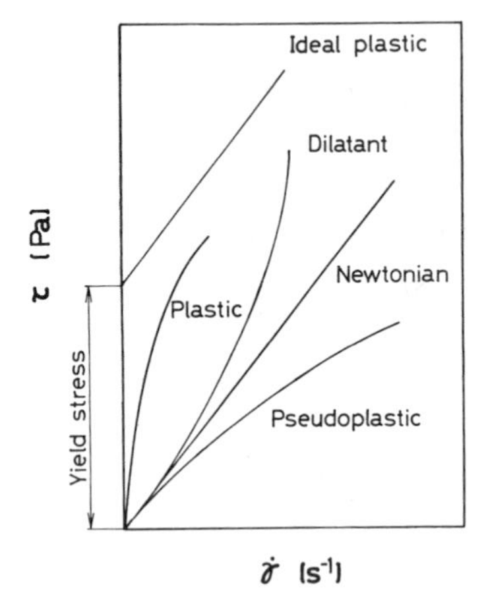

Figure 1. Typical flow-curves.

For the agitation of non-Newtonian liquids, the submitting of the liquid to high-shearing and the exchange flow between high- and low-shearing fields are necessary to eliminate stagnant zones and to accelerate heat exchange. One of the best impellers for this purpose is a helical ribbon impeller.

From this point of view, we introduce our recent studies for agitation of pseudoplastic liquids in helical ribbon-agitated vessels.

FLOW PATTERNS

A knowledge of the flow patterns in an agitated vessel is very helpful in understanding the mixing and heat transfer. For Newtonian liquids, flow patterns in a helical ribbon-agitated vessel have been investigated by many workers [2, 5, 17]. However, the information on the flow patterns produced in non-Newtonian liquids by helical ribbon impellers are very limited [2, 5]. Moreover, no detailed information exists for the effects of liquid properties on flow patterns produced by helical ribbon impellers with different geometrical variables.

The authors have obtained the quantitative expression of the velocity distributions for Newtonian and pseudoplastic liquids in helical ribbon-agitated vessels with three different clearances between blades and wall [13].

Velocity Distributions

Velocity distributions for Newtonian and pseudo-plastic liquids in helical ribbon-agitated vessels with three different clearances are shown in Figure 2. These velocities are normalized by dividing by πNd, which is the speed of the outer-tip of the ribbon. The flow behaviors of test liquids are expressed by the following power-law model.

$$\tau = K\dot{\gamma}^{n} \tag{1}$$

where K and n are consistency and flow-behavior indexes, respectively, and their values are shown in Table 1. The geometrical variables of impellers used are shown in Figure 3.

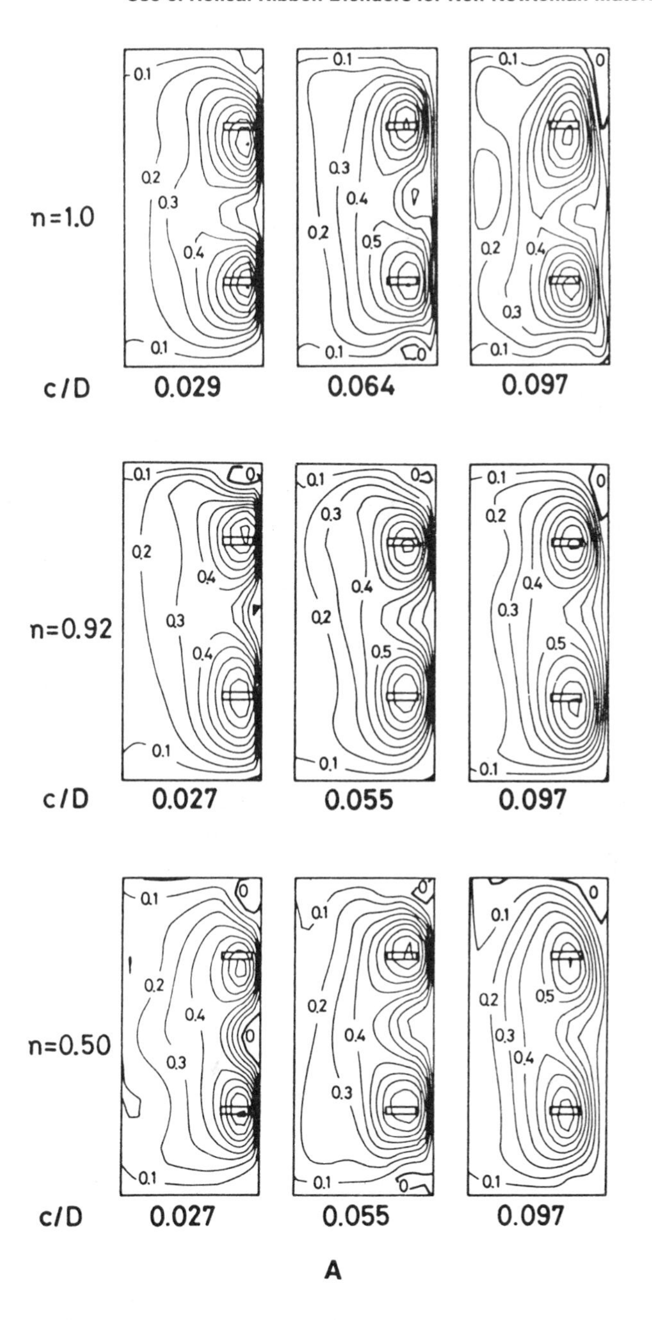

Figure 2. Velocity distributions: (A) tangential velocity; (B) axial velocity.

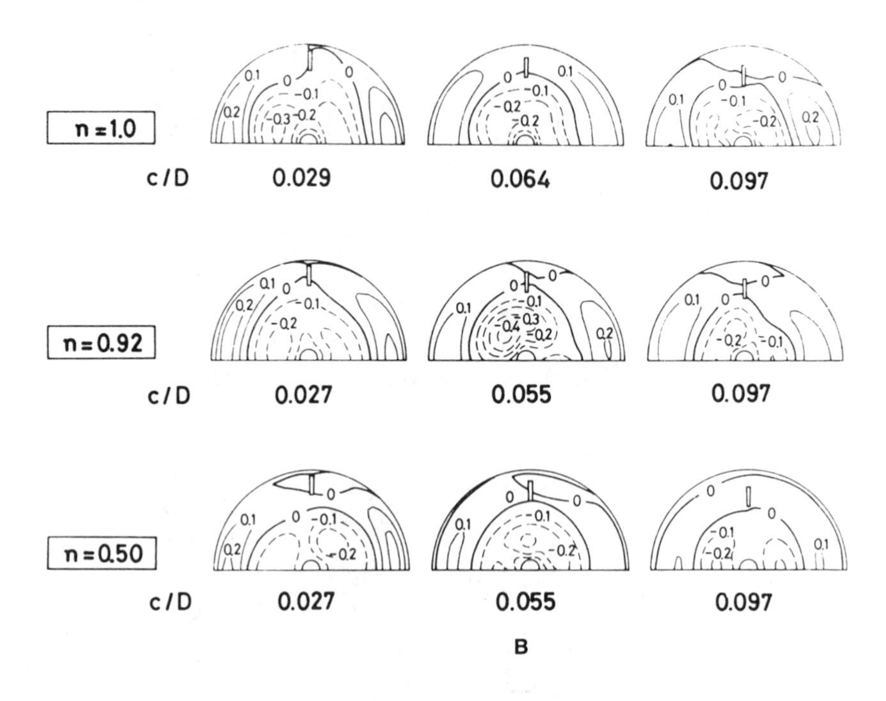

Figure 2. (Continued)

Table 1
Properties of Test Liquids

Liquid		Properties
Newtonian liquid		$\mu = 2.0$ Pa-s
		$\rho = 1.3 \times 10^3$ Kg/m^3
Pseudoplastic liquids*	1	K $= 1.84 \times 10^{-1}$ N-s^n/m^2
		n = 0.915
		$\rho = 1.0 \times 10^3$ kg/m^3
	2	K $= 1.97 \times 10$ N-s^n/m^2
		n = 0.502
		$\rho = 1.0 \times 10^3$ kg/m^3

* *Aqueous solutions of carboxymethyl cellulose (CMC).*

The velocity distributions are approximately similar in spite of the different properties of liquids and the different clearances. In the central core region between shaft and blades, the liquids rotate with an angular velocity which is only slightly smaller than that of the impeller itself. The high shear is observed in the clearance between blades and wall. In the region between the blades, tangential velocity decreases as the flow-behavior index becomes smaller. The axial velocity is reduced in the case of the pseudo-plastic liquid of the smallest flow-behavior index.

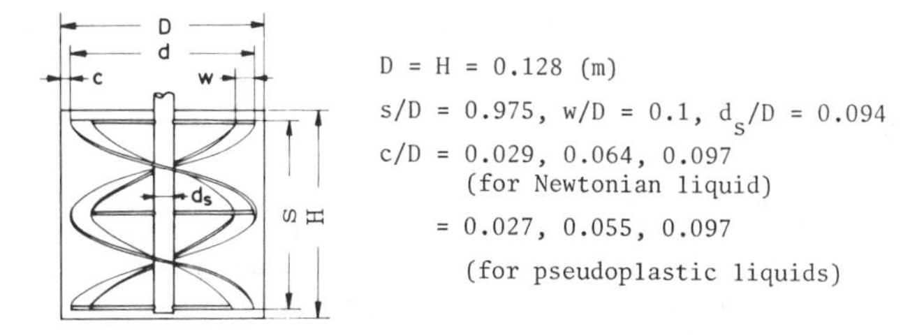

Figure 3. Geometrical configurations of helical ribbon-agitated vessel.

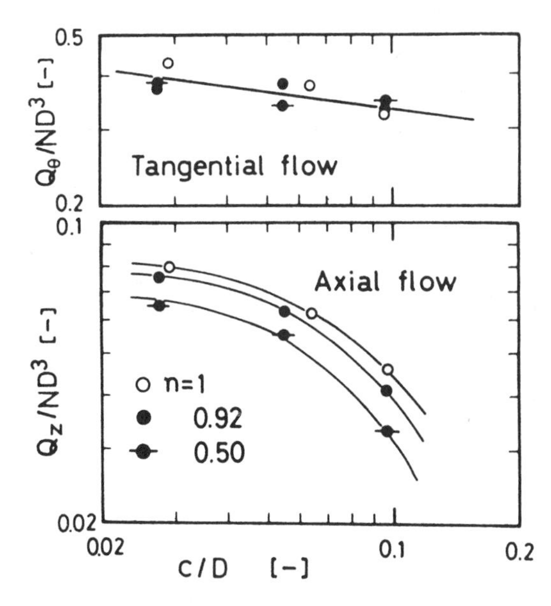

Figure 4. Circulation flow rates.

Circulation Flow Rates

The tangential and axial circulation flow rates, obtained by integration of the previous velocity distributions, are shown in Figure 4.

The tangential flow rates are almost the same in spite of the different liquid properties. On the other hand, the axial flow rate decreases as the flow-behavior index decreases. In the laminar flow region, both mixing and heat transfer are mainly controlled by the axial flow rate. Therefore, a shear-thinning property may have a negative influence upon these operations.

POWER CONSUMPTION

Power consumption is generally described by a relationship between the power number and the Reynolds number. For Newtonian liquids, the relation can be represented by the following equation as suggested by many investigators [3, 8, 19, 24, 26]:

$$N_P \cdot Re = C_1 \quad (2)$$

where C_1 is a geometrical constant. For non-Newtonian liquids, several investigators have obtained a similar relation [3, 6, 7, 8, 19, 25, 26]:

$$N_p \cdot Re' = C_1' \tag{3}$$

where C_1' is a function of the flow-behavior index adding to the geometrical variables. In this equation, Re′ represents the modified Reynolds number, Re_m, [3, 6, 7, 26] or apparent Reynolds number, Re_a [8, 25]. The power correlations published previously do not agree with each other and are not available for the different geometries of helical ribbon impellers.

Power Correlations Using a Modified Reynolds Number

For the liquids which can be described by the power-law model, the modified Reynolds number is expressed as:

$$Re_m = d^2N^{2-n}\rho/K \tag{4}$$

The power correlations using this modified Reynolds number have already been proposed by several investigators as follows:

Bourne and Butler [3]:

$$N_p \cdot Re_m = \pi^2\left(\frac{h}{d}\right)\left(\frac{D}{d}\right)^2\left[\frac{4\pi}{n\{(D/d)^{2/n} - 1\}}\right]^n \tag{5}$$

Chavan and Ulbrecht [6, 7]:

$$N_p \cdot Re_m = 2.5\pi a\left(\frac{de}{d}\right)\lambda^2\left\{\frac{4\pi}{n(\lambda^{2/n} - 1)}\right\}^n \tag{6}$$

where $a = \pi(h/d)(w/d)/(s/d)$

$$\lambda = D/de \tag{7}$$

$$\frac{de}{d} = \frac{D}{d} - \frac{2(w/d)}{\ln\left\{\frac{(D/d) - \{1 - 2(w/d)\}}{(D/d) - 1}\right\}}$$

Sawinsky, Havas, and Deák [25]:

$$N_p \cdot Re_m = \exp\{(n - 1)(4.2d/D - 0.5)\} \times \{19(L/d)(D/c)^{0.45}\} \tag{8}$$

These equations can be converted to the correlations for Newtonian liquids by substituting 1 for n and μ for K.

Power Correlations Using an Apparent Reynolds Number

Basic Concept

The simple and common method used very frequently for power correlation in the agitation of non-Newtonian liquids was proposed by Metzner and Otto [14]. They assumed that the average level of shear rate in an agitation system was proportional to the rotational speed of the impeller:

$$\dot{\gamma}_{av.} = k_M \cdot N \tag{9}$$

where k_M will depend on geometrical variables and flow-behavior index. The apparent viscosity is evaluated from viscometric measurement by using the following equation if k_M in Equation 9 is given in advance.

$$\mu_a = K(\dot{\gamma}_{av})^{n-1} \tag{10}$$

Using an apparent Reynolds number based on μ_a and the power correlation established previously for Newtonian liquids, the power consumption in pseudoplastic liquids can be calculated.

The values of k_M are usually determined from the power measurement at a specified rotational speed of the impeller for a pseudoplastic liquid and a power correlation for Newtonian liquid in a given agitated vessel. This is used to estimate the power consumption at any rotational speed of the impeller. However if a correlation of k_M with geometrical variables of agitated vessels and the flow-behavior index is given, this method will be applicable to a general estimation of power consumption in pseudoplastic liquids.

Correlation of k_M

The values of k_M determined experimentally by several investigators and the geometries of helical ribbon-agitated vessels used are tabulated in Table 2. Shown in the previous section [3, 6, 7, 26] are proposed power correlations for Newtonian and non-Newtonian liquids. Relationships between the modified Reynolds number, Re_m, and the apparent Reynolds number, Re_a, can be expressed as follows:

$$Re_a = Re_m/k_M^{n-1} \tag{11}$$

Combining Equation 11 and their power correlations for Newtonian and non-Newtonian liquids, the following equations for K_M are obtained:

Bourne and Butler [3]:

$$k_M = 4\pi\{[n\{(D/d)^{2/n} - 1\}]^n/\{(D/d)^2 - 1\}\}^{1/(1-n)} \tag{12}$$

Chavan and Ulbrecht [6, 7]:

$$k_M = 4\pi\{[n(\lambda^{2/n} - 1)]^n/(\lambda^2 - 1)\}^{1/(1-n)} \tag{13}$$

Sawinsky, Haves, and Deák [26]:

$$k_M = \exp\{4.2(d/D) - 0.5\} \tag{14}$$

Equation 12 or Equation 13 shows that k_M is a function of flow-behavior index in addition to geometrical variables, while Equation 14 shows that k_M depends only on geometrical variables.

Table 2
Experimental Values of k_M

Worker	n_p	c/D	s/D	w/D	k_M
Nagata et al. [21]	2	0.025	0.95	0.10	30.0
Rieger and Novak [24]	2	0.025	0.95	0.095	36.73 ± 1.45
Hall and Godfrey [8]	2	0.044	0.912	0.106	27.0

However, values of k_M calculated from Equations 12 and 13 for the agitated vessel of d/D = 0.9 and w/D = 0.1 vary slightly from 59.86 to 59.75 and from 31.69 to 31.48, respectively, as n increases from 0.5 to 0.99. It is, therefore, concluded that k_M is independent of n.

Figure 5 shows the calculated values of k_M from these equations and those determined from power measurements of Newtonian and pseudoplastic liquids for an agitated vessel with w/D = 0.1 and s/D = 0.9. Only Equation 14 agrees with the experiments. However, this equation involves only the influence of the clearance of k_M.

The authors proposed the following relation [13] between k_M and the power correlation for Newtonian liquids on the basis of a model of two coaxial cylinders.

$$k_M = (N_p \cdot Re)\frac{d}{\pi^2 H} \tag{15}$$

where the value of $N_p \cdot Re$ for Newtonian liquids is constant for the given geometries of an agitated vessel. This equation principally provides power estimation for pseudoplastic liquids in any close-clearance-type agitated vessel if its power correlation for Newtonian liquids is established. The authors presented the following power correlation for Newtonian liquids applicable to various geometries of anchor- and helical ribbon-agitated vessels.

$$N_p \cdot Re = \frac{16\pi^3}{2\ln(4 + 8c/w) - 1}\left(\frac{L}{d}\right) f(c/D)(\sin\theta_B)^{0.555}\frac{n_p}{2} + 2.08\pi^3 n_a n_p \left(\frac{d_a}{d}\right)^{0.15}\left(\frac{d/2 - w}{d}\right)^{3.15}$$

where

$$f(c/D) = 1 + 0.00539(c/D)^{-0.876} \tag{16}$$

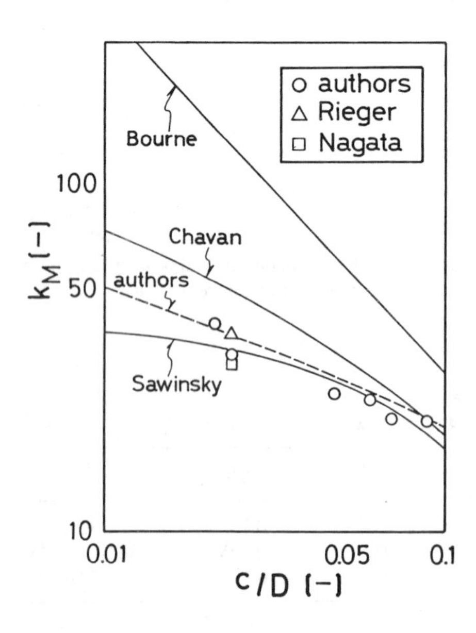

Figure 5. Relationship between k_M and c/D.

Table 3
Summary of Calculated and Experimental Values of k_M

n_p [−]	n_a [−]	c/D [−]	s/D [−]	w/D [−]	N_P Re [−]	$k_{M,calc.}$ [−]	$k^*_{M,exp.}$ [−]
2	3	0.0227	0.914	0.102	390	37.0	37.0
2	3	0.0623	0.966	0.102	284	25.2	23.1
2	3	0.097	0.956	0.102	260	21.1	18.9
2	4	0.0583	0.643	0.102	354	31.5	34.7
2	5	0.0502	0.452	0.102	406	36.5	41.7
2	3	0.0404	1.26	0.102	295	27.1	27.2
2	3	0.0457	1.90	0.102	249	22.9	23.7

* *Obtained for an agitated vessel of* $D = H = 128$ *mm,* $d_a = 3$ *mm,* $d_s = 12$ *mm, and* $h = 125$ *mm.*

The broken line in Figure 5 shows the calculated result from Equations 15 and 16. It can be seen from this figure that excellent agreement is obtained. The same results are obtained for various types of anchor- and helical ribbon-agitated vessels as shown in Table 3.

HEAT TRANSFER TO PSEUDOPLASTIC LIQUIDS IN A HELICAL RIBBON-AGITATED VESSEL

The main purpose of most studies on non-Newtonian heat transfer has been to extend the correlation of heat transfer coefficients at the vessel wall obtained for Newtonian liquids to non-Newtonian heat transfer. However, even for Newtonian heat transfer, many correlations are different from each other.

Some examples of the previous correlations of heat transfer coefficients [1, 9, 10, 15, 16, 22, 23] are summarized in Table 4 and are illustrated in Figure 6. It is clearly seen that there are considerable differences in their correlations, especially in the laminar flow region. This seems to result from the existence of temperature distribution in bulk in laminar flow region.

In case of laminar flow, a degree of non-uniformity in temperature exists even with agitation by a helical ribbon impeller, and in this case it is necessary to clarify the relation between heat transfer and fluid flow in vessels in order to obtain an accurate correlation of heat transfer coefficients and to predict the temperature distributions.

Relation Between Temperature Distribution and Flow Patterns

Patterns of temperature distribution in a helical ribbon-agitated vessel is mainly controlled by the flow pattern in the axial direction. It is, therefore, useful to investigate the growth process of temperature distribution in the vertical plane of the vessel.

The authors [11] visualized temperature distributions in the vertical plane of a helical ribbon-agitated vessel during unsteady heat-transfer operations by using liquid crystals. In this method of temperature visualization, temperature distribution is observed as color distribution. A sketch of the primary circulation flow due to the pumping action of the ribbon and its cross view in the vertical plane are shown in Figure 7.

Figure 8 shows some examples of the photographs displaying the change in color of liquid crystal according to temperature and photographs of color distributions for a Newtonian liquid. (The patterns of temperature distributions in Newtonian and pseudoplastic liquids are essentially almost similar to each other.) In Figure 8B and C, a thin layer with a steep temperature gradient is observed near the vessel wall. It is also found that the color change in the axial direction in bulk is not very considerable. This means that the temperature distribution in bulk is mainly controlled by axial

Table 4
Summary of Heat Transfer Correlations for Helical Ribbon-Agitated Vessel

Worker	Correlation	Re range	c/D	s/D	n_p	Type of Test	Heating or Cooling	Year
Mizushina et al. [16]	$Nu = 1.10\ Re^{1/2}\ Pr^{1/3}\ \mu_r^{-0.14}$ $Nu = 0.52\ Re^{2/3}\ Pr^{1/3}\ \mu_r^{-0.14}$	$10 \sim 100$ $100 \sim 10^4$	0.025	0.95	2	Quasi-steady	C	1970
Nagata et al. [22]	$Nu = 4.2\ Re^{1/3}\ Pr^{1/3}\ \mu_r^{-0.20}$ $Nu = 0.42\ Re^{2/3}\ Pr^{1/3}\ \mu_r^{-0.14}$	$1.5 \sim 10^3$ $10^3 \sim 10^5$	0.033	0.93	2	Unsteady	H & C	1972
Mitsuishi and Miyairi [15]	$Nu = 0.78\ Re^{1/3}\ Pr^{1/3}\ \mu_r^{-0.14}$ $Nu = 0.53\ Re^{1/2}\ Pr^{1/3}\ \mu_r^{-0.14}$ $Nu = 0.23\ Re^{2/3}\ Pr^{1/3}\ \mu_r^{-0.14}$	$1.5 \sim 10$ $10 \sim 180$ $180 \sim 4 \times 10^3$	0.025	0.95	2	Unsteady	H	1973
Nishikawa et al. [23]	$Nu = 1.75\ Re^{1/3}\ Pr^{1/3}\ \mu_r^{-0.20} \times ((D - d)/D)^{-1/3}$ $Nu = 0.52\ Re^{2/3}\ Pr^{1/3}\ \mu_r^{-0.14}$	$0.1 \sim Re_c$ $Re_c = 39\ D/(D - d)$ $Re_c \sim 10^6$	0.027, 0.034 0.060, 0.10	0.93	2	Steady	H & C	1975
Heim [9, 10]	$Nu = 0.628\ Re^{1/2}\ Pr^{1/3}\ \mu_r^{-1/7}$ $Nu = 0.146\ Re^{3/4}\ Pr^{1/3}\ \mu_r^{-1/7}$	$150 \sim 600$ $2.5 \times 10^3 \sim 10^4$	0.067	0.83	1	Unsteady	H	1980
Blasinski and Kuncewicz [1]	$Nu = 0.248\ Re^{0.5}\ Pr^{0.33}\ \mu_r^{-0.14} \times (c/D)^{-0.22}(s/d)^{-0.28}$ $Nu = 0.238\ Re^{0.67}\ Pr^{0.33}\ \mu_r^{-0.14} \times (s/d)^{-0.25}$	Laminar Turbulent	0.17, 0.033 0.050	0.33, 0.42 0.57, 0.83	2	Unsteady	H	1981

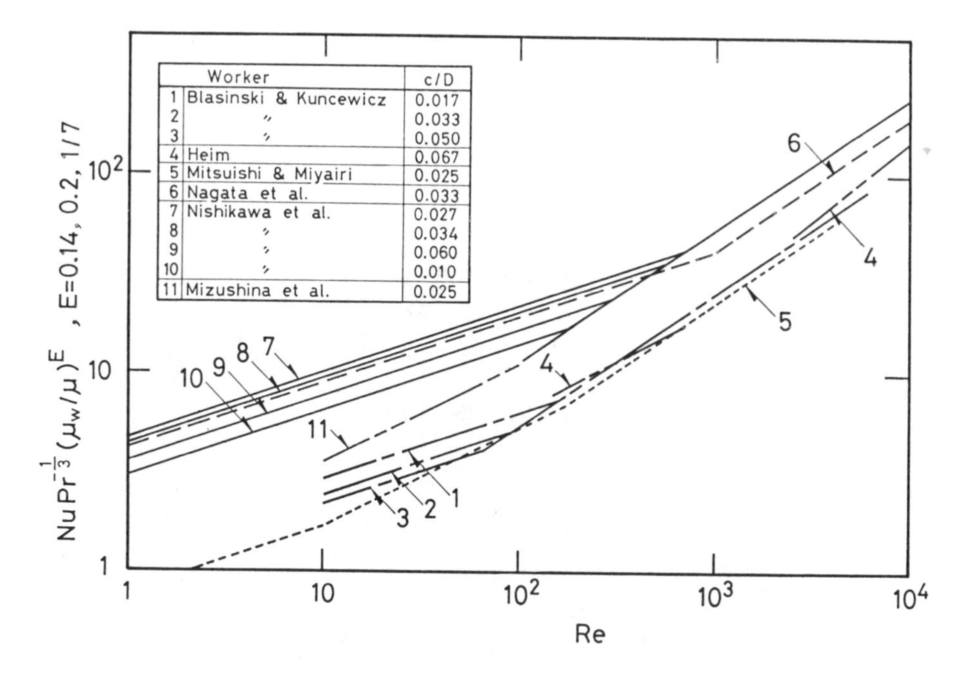

Figure 6. Illustrations of heat transfer correlations.

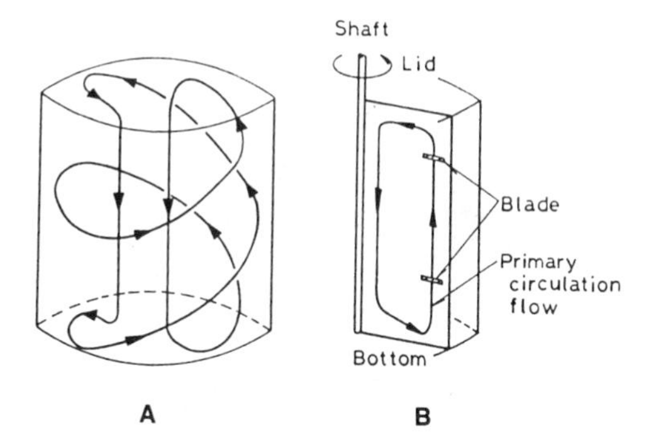

Figure 7. Primary circulation flow: (A) flow pattern reported by Nagata et al. [18]; (B) view of section illuminated by plane light beam.

flow. The patterns of these temperature distributions can be explained by the circulation flow shown in Figure 7.

Figure 9 shows examples of temperature distributions in horizontal planes of the vessel during steady-state heat-transfer operations. This figure represents the isothermal lines which were obtained from cooling experiments for a flow-through agitation system. Some degree of non uniformity in temperature exists in the bulk near the shaft even in the region of high rotational impeller speed. It is also noted that in each case the outlet temperature is lower than that in bulk. These facts can also be explained by the axial flow pattern in the vessel.

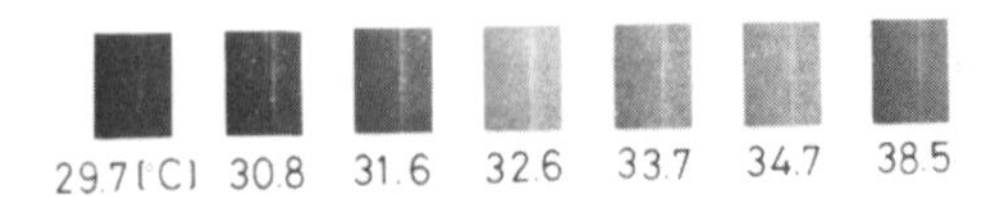

A

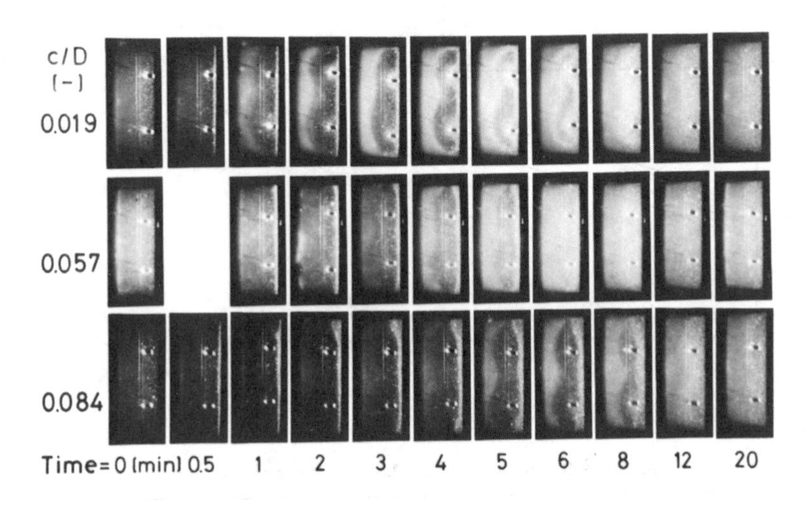

B

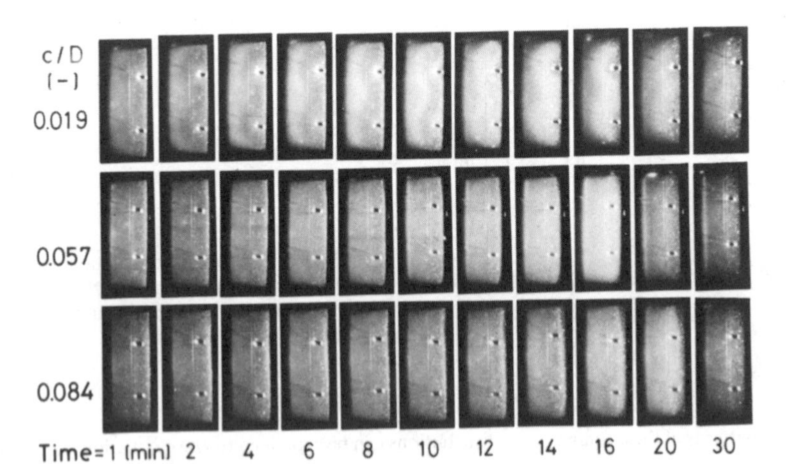

C

Figure 8. Visualization of temperature distributions: (A) changes in color of liquid crystal; (B) temperature distributions in unsteady heating experiments (N = 15 rpm., initial temp. = 30°C, temp. of heating water = 38°C); (C) temperature distributions in unsteady cooling experiments (N = 15 rpm., initial temp. = 38°C, temp. of cooling water = 30°C).

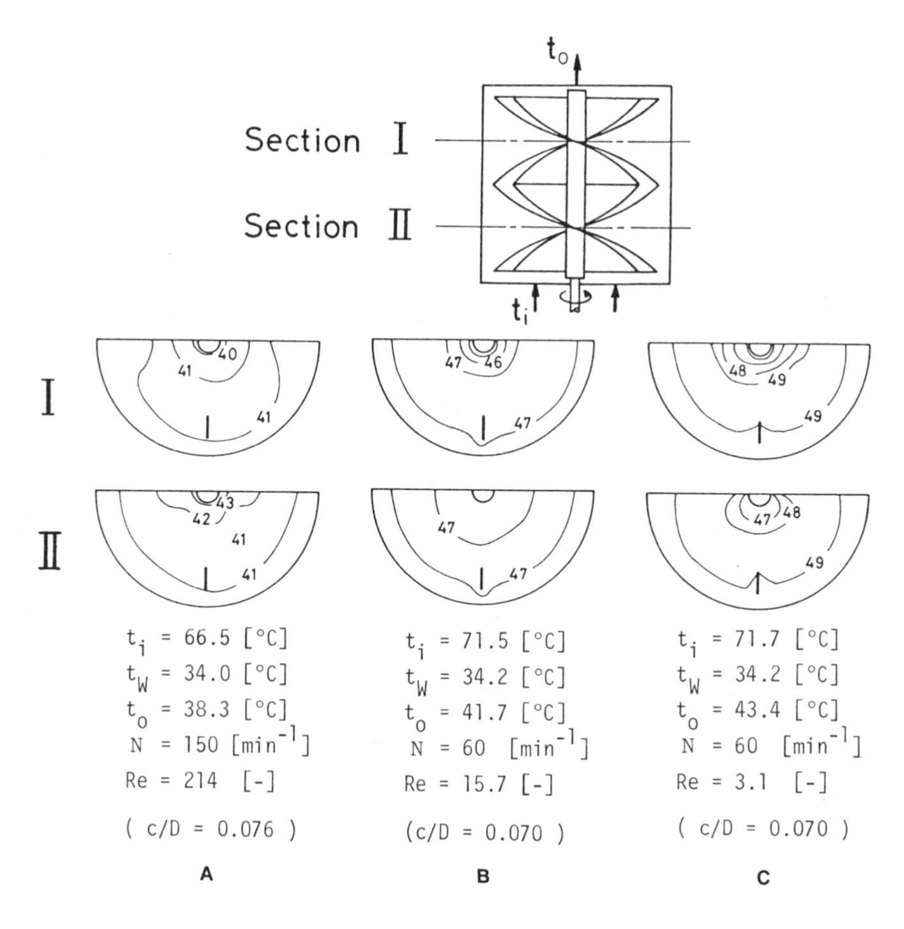

Figure 9. Typical temperature distributions in horizontal plane of $z/D = 0.5$, obtained from cooling experiments: (A) corn syrup solution ($n = 1$); (B) 3 wt% CMC solution ($n = 0.7$); (C) 5 wt% CMC solution ($n = 0.5$).

From the facts just described, it can be concluded that the mechanism of heat transfer in a helical ribbon-agitated vessel consists of the following two processes:

1. Heat transfer at the vessel wall.
2. Convective heat transfer by the circulation flow, which controls both the temperature distribution in the bulk and the renewal of the liquid in the thermal boundary layer at the wall.

Therefore, the introduction of a model for heat transfer at the wall and a flow model realistic enough to express the axial flow pattern will be effective to represent quantitatively the thermal mixing process in the vessel.

Model for Heat Transfer at the Vessel Wall

A model for heat transfer at the vessel wall [12] is shown in Figure 10, which represents the thermal boundary layer at the wall. The following assumptions are made: The boundary layer

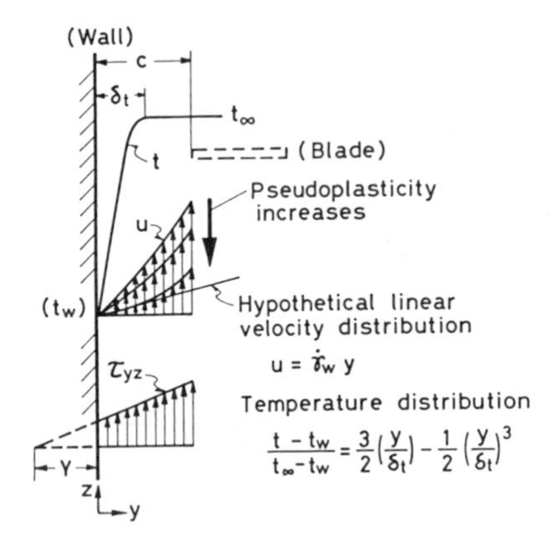

Figure 10. Description of thermal boundary layer.

develops from the bottom to the top along the wall and the distributions of velocity and temperature in the layer are approximated by linear and cubic forms as shown in Figure 10. Using these distributions with the heat-balance equation in the boundary layer [27], the following relation is obtained.

$$Nu = \frac{hD}{\lambda} \propto \left\{ Re\ Pr\ \frac{D^2 \dot{\gamma}_W}{d^2 N} \right\}^{1/3} \tag{17}$$

In this equation, $\dot{\gamma}_W$ is the shear rate at the wall, which will be affected by geometries of the system and flow-behavior of the agitated liquid.

The following relation is assumed for Newtonian liquids.

$$(\dot{\gamma}w)_N \propto sN/c \tag{18}$$

In addition, the effect of non-Newtonian behavior of the liquid is evaluated by the introduction of the following correction term, F,

$$F = (\dot{\gamma}_W)_{nN}/(\dot{\gamma}_W)_N \tag{19}$$

where subscripts N and nN represent Newtonian and non-Newtonian liquids. By using Equations 18 and 19, Equation 17 is rewritten as follows:

$$Nu = C^* \left\{ Re\ Pr\ \frac{D^2 s}{d^2 c} F \right\}^{1/3} \mu_r^{-0.2} \tag{20}$$

where the viscosity correction term, $\mu_r^{-0.2}$, is introduced empirically following previous investigations, and C^* is a proportional constant which should be determined experimentally. The correction term, F, is evaluated as follows.

It is assumed that the same axial pressure gradients may appear in spite of liquid pseudoplasticity under the same agiatation conditions if the liquids have the same values of the representative viscosity of non-Newtonian viscosity or Newtonian viscosity. According to this assumption, the axial flow in the clearance region may be approximated by the flow along a flat plate under the condition of a constant pressure gradient, $\Delta P/\Delta z$, as shown in Figure 10. The governing equation for the flow of pseudoplastic liquid is as follows:

$$\frac{d}{dy}(\tau_{yz}) = \frac{\Delta P}{\Delta z} \tag{21}$$

Since this equation provides a linear distribution of shear rate, τ_{yz}, as shown in Figure 10, the following boundary condition [28] is adopted for the convenience of calculation:

$$\tau_{yz} = 0 \quad \text{at} \quad y = -Y \tag{22}$$

where Y represents the distance from the wall to the hypothetical position of $\tau_{yz} = 0$ as shown in the figure.

Solving Equation 21 for shear rate, the following equation is obtained.

$$\dot{\gamma} = \frac{du}{dy} = \left\{ \frac{\Delta P/\Delta z}{K} (y + Y) \right\}^{1/n} \tag{23}$$

Assuming that the average shear rate at the position of the blade-tip is independent of liquid pseudoplasticity, the pressure gradient, $\Delta P/\Delta z$, can be described by Y as follows:

$$\Delta P/\Delta z = K/(Y + c) \tag{24}$$

From Equations 23 and 24, the correction term, F, can be derived as follows:

$$F = \frac{(\dot{\gamma}_W)_{nN}}{(\dot{\gamma}_W)_N} = \left(1 + \frac{c}{Y}\right)^{(n-1)/n} \tag{25}$$

Although the distance, Y, seems to depend not only on the clearance but also on the flow-behavior of the liquid, it is assumed that the former is the dominant factor affecting Y and that the latter can be ignored. Equation 25 will be rewritten as follows:

$$F = \left\{ 1 + \left(\frac{c/D}{\epsilon_1} \right)^{\epsilon_2} \right\}^{(n-1)/n} \tag{26}$$

In this equation, ϵ_1 and ϵ_2 are model parameters and are determined so that Equation 20 may best represent the experimental data.

Correlation of Heat Transfer Coefficients

To correlate heat transfer coefficients according to Equations 20 and 26, it is necessary to take appropriately the characteristic temperature (bulk temperature) and the representative value of non-Newtonian viscosity. The temperature in the region where impeller blades pass is adopted as the characteristic temperature, t_∞, according to the heat transfer model shown in Figure 10 and therefore the heat transfer coefficient is defined by the temperature difference, $|t_\infty - t_W|$, which is obtained experimentally. In addition, the apparent viscosity used in the power correlation is employed to calculate the Reynolds number, Prandtl number, and viscosity correction term.

The correlation results for Newtonian and pseudoplastic liquids are shown in Figure 11, in which all the experimental data have been obtained for a helical ribbon-agitated vessel of $D = H = 0.16$ m, $s/D = 0.90$, and $w/D = 0.10$. All the data are correlated by the following equation:

$$Nu = 0.64 \left\{ Re\, Pr \frac{D^2 s}{d^2 c} F \right\}^{1/3} \mu_r^{-0.2} \tag{27}$$

where $$F = \left\{ 1 + \left(\frac{c/D}{0.046} \right)^4 \right\}^{\frac{n-1}{n}}$$

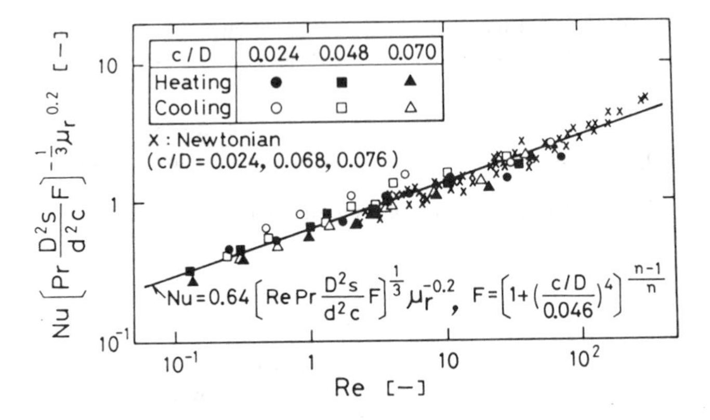

Figure 11. Correlation of heat transfer coefficients.

Estimation of Temperature Distributions Based on Flow Model [11, 12]

Figure 12 shows the flow model for a flow-through agitation system, which has inlet and outlet flows. This figure shows the construction of simplified flow paths in the vertical plane of a helical ribbon-agitated vessel. In this figure Q_1 to Q_5 represent the volumetric flow rates, and t_1 to t_4 and t_∞ represent the temperature in each flow path. Q represents inlet flow rate, and t_i and t_o inlet and outlet temperatures, respectively. Since in this model heat transfer through the wall results in a

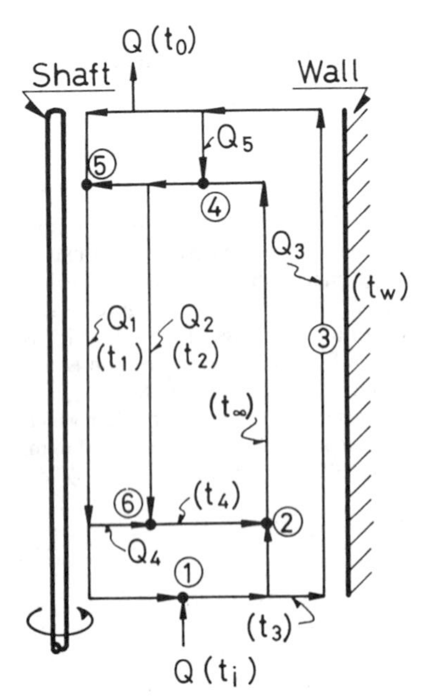

Figure 12. Flow model for flow-through agitation system.

temperature change in the liquid in Path 3, the steady-state heat balances at Locations 1 through 6 can be expressed as follows.

$$
\begin{aligned}
&1.\ Qt_i + (Q_1 - Q_4)t_1 - (Q + Q_1 - Q_4)t_3 = 0 \\
&2.\ (Q + Q_1 - Q_3 - Q_4)t_3 + (Q_2 + Q_4)t_4 - (Q + Q_1 + Q_2 - Q_3)t_\infty = 0 \\
&3.\ \rho C_p Q_3(t_3 - t_o) = \psi \\
&4.\ Q_5 t_o + (Q + Q_1 + Q_2 - Q_3)t_\infty - (Q + Q_1 + Q_2 - Q_3 + Q_5)t_2 = 0 \\
&5.\ (Q + Q_1 - Q_3 + Q_5)t_2 + (Q_3 - Q - Q_5)t_o - Q_1 t_1 = 0 \\
&6.\ Q_4 t_1 + Q_2 t_2 - (Q_2 + Q_4)t_4 = 0
\end{aligned}
\tag{28}
$$

In Equation 28, liquid density, ρ, and specific heat, C_p, are known and flow rate, Q, and inlet temperature, t_i, are specified as operational conditions. Rate of heat transfer at the vessel wall, ψ, is also obtained from Equation 27. Thus, the temperatures, t_1 to t_4, t_∞ and t_o can be determined through solutions to the preceding equations when the flow rates, Q_1 to Q_5 are given.

It has been found that good agreements between calculated and observed temperatures are obtained when the following relations are used in estimating the flow rate in each flow path:

$$Q_1 = \frac{\beta_1 Q_z}{1 + \beta_1} \tag{29}$$

$$Q_2 = \frac{Q_z}{1 + \beta_1} \tag{30}$$

$$Q_3 = \frac{\beta_2 G(Q_1 - Q_4 + Q)}{1 + \beta_2 G} \tag{31}$$

$$Q_4 = \frac{Q_1}{1 + \beta_3 (Q_1 + Q)/Q_1} \tag{32}$$

$$Q_5 = \frac{Q_3}{1 + \beta_4 (Q_3 + Q)/Q_3} \tag{33}$$

where $\beta_1 = 0.10$
$\beta_2 = 8.0 \times 10^2\ m^{-1}$
$\beta_3 = 2.0$
$\beta_4 = 1.0$

Q_z in Equations 29 and 30 is the overall flow rate of axial circulation flow and is equal to the sum of Q_1 and Q_2. The values of Q_z can be evaluated by interpolating the experimental data shown in Figure 4. G in Equation 31 is the characteristic length proportional to the average thickness of the thermal boundary layer and is given by the following equation which is derived from the model for heat transfer at the wall:

$$G = \sqrt[3]{\alpha c D/(sNF)} \tag{34}$$

The model analysis described here is an attempt to treat systematically both the heat transfer at the wall and the temperature distribution in bulk.

MIXING OF PSEUDOPLASTIC LIQUIDS

There are few investigations [4, 5, 20] about the mixing time of non-Newtonian liquids in a helical ribbon-agitated vessel in the laminar flow region, and considerable confusion exists about the effect of non-Newtonian properties on the mixing time. Namely, several investigators [4, 5] reported that the mixing time increases drastically as the liquid shear-thinning property increases, while others [20] reported that the mixing time is almost independent of the liquid properties.

The authors have carried out the observation of mixing patterns and the measurement of mixing times for Newtonian and pseudoplastic liquids in helical ribbon-agitated vessels by means of a liquid crystal method. The properties of test liquids are summarized in Table 5 and the geometrical variables of impellers are listed in Table 6.

Relation Between Mixing Time and Liquid Properties

The time required for the tracer liquid to complete one circuit was measured. The results are shown in Figure 13. The circulation time for each impeller decreases as the flow-behavior index becomes smaller.

Figure 14 shows the relation between the dimensionless mixing time, $N \cdot t_m$, and the apparent Reynolds number, Re_a. With the widest clearance impeller, stagnant zones are detected at both the top and bottom corner near the wall. It is clear from this figure that the dimensionless mixing time for pseudoplastic liquids, as well as that for Newtonian liquids, is constant in the region of a low apparent Reynolds number as reported by several investigators [4, 5, 20]. That is:

$$N \cdot t_m = C_2 \tag{35}$$

where C_2 is a function of liquid properties and geometrical variables.

Figure 15 shows the relation between the dimensionless mixing time and the clearance ratio, c/D. For pseudoplastic liquids, the mixing time becomes longer as the flow-behavior index becomes smaller with each impeller. With the narrowest clearance impeller which produces a strongly isolated zone inside the inner edge of the blades, the mixing time for Newtonian liquids is longer than those for pseudoplastic liquids. On the other hand, the medium-clearance impeller produces a

Table 5
Properties of Test Liquids

Liquid	K ($N \cdot S^n/m^2$)	n (−1)
3 wt% HEC* soln	7.82×10^{-1}–11.9×10^{-1}	0.710–0.763
5 wt% HEC* soln	3.02–4.61	0.619–0.685

* *Hydroxyethyl cellulose.*

Table 6
Geometrical Variables of Helical Ribbon-Agitators

No.	c/D (−)	s/D (−)
1	0.0212	0.926
2	0.0606	0.892
3	0.0902	0.912

$D = H = 100$ mm, $w/D = 0.10$, $n_p = 2$

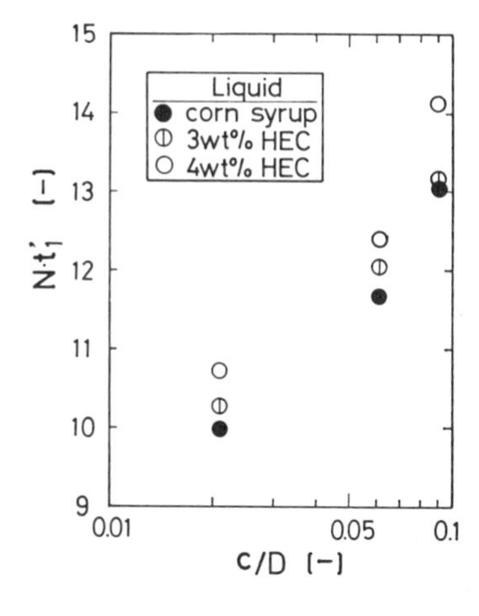

Figure 13. Relationship between nondimensional circulation time and clearance ratio.

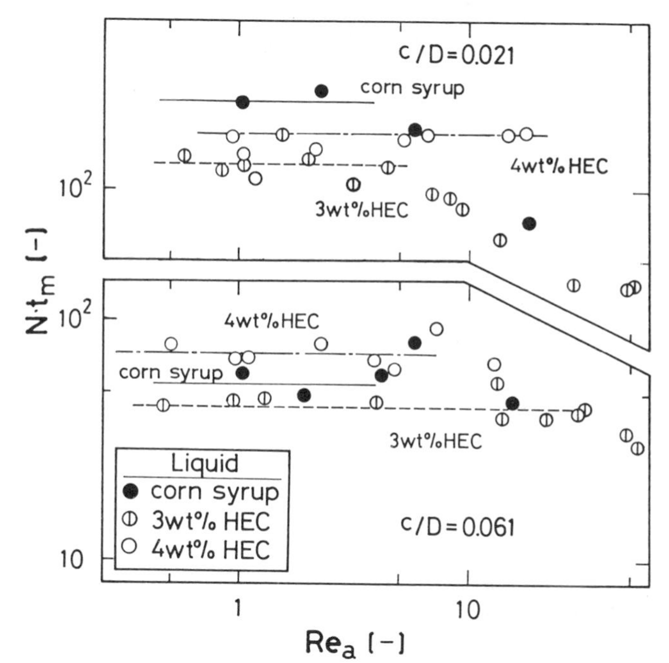

Figure 14. Relationship between nondimensional mixing time and apparent Reynolds number.

rather weakly isolated zone, and the mixing times for Newtonian and pseudoplastic liquids are almost independent of liquid properties. These results may indicate that the mixing of pseudoplastic liquids is less affected by the geometries of agitated vessels than that of Newtonian liquids. However, more quantitative information is required in order for detailed discussion. In conclusion, the optimum geometry for mixing pseudoplastic liquids is almost the same as for that of Newtonian liquids.

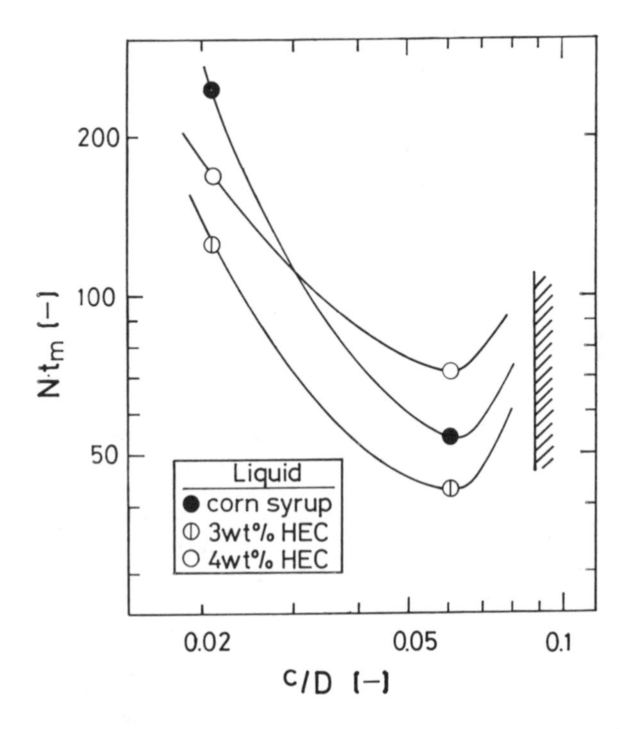

Figure 15. Relationship between nondimensional mixing time and clearance ratio.

NOTATION

c	clearance between blades and vessel wall (m)
C_p	specific heat (J/(kg·K))
c_1, C_2, C^*	geometrical constants in Equations 2, 35 and 20, respectively (−)
D	vessel diameter (m)
d	impeller diameter (m)
d_a	arm diameter (m)
d_S	shaft diameter (m)
F	correction term defined by Equation 25 (−)
G	characteristic length defined by Equation 34 (m)
H	height of vessel (m)
h	heat transfer coefficient (W/(m^2·k)) or height of impeller (m)
L	length of blade (= h/sin θ_B) (m)
k	fluid consistency coefficient (N·s^n/m^2)
k_M	proportional constant in Equation 9 (−)
N	rotational speed of impeller (s^{-1})
N_p	power number (= $P/\rho d^5N^3$)(−)
n	flow-behavior index (−)
n_a	number of arms for each impeller (−)
n_p	number of blade (−)
P	power consumption (W)
Pr	Prandtl number ($C_p\mu/\lambda$ or $C_p\mu_a/\lambda$) (−)
$\Delta P/\Delta z$	axial pressure gradient (Pa/m)
Q	inlet flow rate (m^3/s)
Q_1, Q_2, Q_3, Q_4, Q_5	flow rate in each flow path shown in Figure 12 (m^3/s)
Q_z	axial circulation flow rate (m^3/s)
Q_θ	tangential flow rate (m^3/s)
Re	Reynolds number (= $d^2N\rho/\mu$) (−)

Re_a apparent Reynolds number ($= d^2N\rho/\mu_a$) (−)
Re_m modified Reynolds number ($= d^2N^{2-n}\rho/K$) (−)
s impeller pitch (m)
t temperature (°C)
t_1, t_2, t_3, t_4 temperature in each flow path shown in Figure 12 (°C)
t'_1 time required for tracer liquid to complete one circuit (s)
t_i, t_o inlet and outlet temperatures, respectively (°C)
t_W wall temperature (°C)
t_∞ characteristic temperature (°C)
u axial velocity shown in Figure 10 (m/s)
w blade width (m)
y y-coordinate shown in Figure 10 (m)
Y distance from vessel wall shown in Figure 10 (m)
z axial distance from vessel bottom (m)

Greek Symbols

α thermal diffusivity ($\lambda/\rho C_p$) (m^2/s)
$\beta_1, \beta_3, \beta_4$ adjustable constants (−)
β_2 adjustable constant (m^{-1})
$\dot{\gamma}$ shear rate (s^{-1})
$\dot{\gamma}_{av}$. average shear rate (s^{-1})
$\dot{\gamma}_W$ shear rate at vessel wall (s^{-1})
δ_t thickness of thermal boundary layer (m)
ϵ_1, ϵ_2 adjustable constants (−)
θ_B blade angle (−)
λ thermal conductivity (W/(m·K))
μ viscosity (Pa · s)
μ_a apparent viscosity (Pa · s)
μ_r viscosity correction term (μ_W/μ or μ_W/μ_a) (−)
ρ density (kg/m^3)
τ shear stress (Pa)
τ_{yz} shear stress shown in Figure 10 (Pa)
Ψ rate of heat transfer (W)

Subscripts

N Newtonian liquid
nN non-Newtonian liquid

REFERENCES

1. Blasinski, H., and Kuncewicz, C., "Heat Transfer During the Mixing of Pseudoplastic Fluids with Ribbon Agitators," *Int. Chem. Eng.*, Vol. 21 (1981), pp. 679–683.
2. Bourne, J.R., and Butler, H., "An Analysis of the Flow Produced by Helical Ribbon Impellers," *Trans. Int. Chem. Engrs.*, Vol. 47 (1969), pp. T11–T17.
3. Bourne, J. R., and Butler, H., "Power Consumption of Helical Ribbon Impellers," *Trans. Inst. Chem. Engrs.*, Vol. 47 (1969), pp. T267–T270.
4. Carreau, P.J., "Mixing of Rehological Complex Fluids," *Proceedings of Second World Congress of Chemical Engineering, Montreal*, Canada (1981), pp. 379–382.
5. Carreau, P.J., Patterson, I., and Yap, C. Y., "Mixing of Viscoelastic Fluids with Helical-Ribbon Agitators. I-Mixing Time and Flow Patterns," *Can. J. Chem. Eng.*, Vol. 54 (1976), pp. 135–142.
6. Chavan, V. V., and Ulbrecht, J., "Power Correlation for Helical Ribbon Impellers in Inelastic Non-Newtonian Fluids," *Chem. Eng. J.*, Vol. 3 (1972), pp. 308–311.
7. Chavan, V. V., and Ulbrecht, J., "Power Correlation for Close-Clearance Helical Impellers in Non-Newtonian Liquids," *Ind. Eng. Process Des. Develop.*, Vol. 12 (1973), pp. 472–476.
8. Hall, K. R., and Godfrey, J. C., "Power Consumption by Helical Ribbon Impellers," *Trans. Inst. Chem. Engrs.*, Vol. 48 (1970), pp. T201–T208.
9. Heim, A., "Model of Momentum and Heat Transfer in Mixers with Close-Clearance Agitators," *Int. Chem. Eng.*, Vol. 20 (1980), pp. 271–278.
10. Heim, A. "Verification of a Model of Momentum and Heat Transfer in Mixers with Close-Clearance Agitators.," *Int. Chem. Eng.*, Vol. 20 (1980), pp. 279–288.

11. Kuriyama, M., et al., "Heat Transfer and Temperature Distributions in an Agitated Tank Equipped with Helical Ribbon Impeller," *J. Chem. Eng. Japan*, Vol. 14 (1980), pp. 323–330.
12. Kuriyama, M., Arai, K., and Saito, S., "Mechanism of Heat Transfer to Pseudoplastic Fluids in an Agitated Tank with Helical Ribbon Impeller," *J. Chem. Eng. Japan*, Vol. 16 (1983), pp. 489–494.
13. Kuriyama, M., "Fluid Flow and Heat Transfer Characteristics in Agitated Vessels for Highly Viscous Liquids," doctoral dissertation, Tohoku Univ., 1983.
14. Metzner, A. B., and Otto, R. E., "Agitation of Non-Newtonian Fluids," *AIChE J.*, Vol. 3 (1957), pp. 3–10.
15. Mitsuishi, N., and Miyairi, Y., "Heat Transfer to Non-Newtonian Fluids in an Agitated Vessel," *J. Chem. Eng. Japan*, Vol. 6 (1973), pp. 415–420.
16. Mizushina, T., et al., "Performance of Eliptic Board and Eliptic Ribbon Agitators," *Kagaku Kōgaku*, Vol. 34 (1970), pp. 1213–1219.
17. Murakami, Y., et al., "Evaluation of Performance of Mixing Apparatus for High Viscosity Fluids," *J. Chem. Eng. Japan*, Vol. 5 (1972), pp. 297–303.
18. Nagata, S., Yanagimoto, S., and Yokoyama, T., "A Study on the Mixing of High-Viscosity Liquid," *Kagaku Kōgaku*, Vol. 21 (1957), pp. 278–286.
19. Nagata, S., et al., "Power Consumption of Helical Mixer for the Mixing of Highly Viscous Liquid," *Kagaku Kōgaku*, Vol. 34 (1970), pp. 1115–1117.
20. Nagata, S., et al., "Mixing of Highly Viscous Liquids," *Kagaku Kōgaku*, Vol. 35 (1971), pp. 794–800.
21. Nagata, S., et al., "Power Consumption of Mixing Impellers in Pseudoplastic Liquids," *J. Chem. Eng. Japan*, Vol. 4 (1971), pp. 72–76.
22. Nagata, S., Nishikawa, M., and Kayama, T., "Heat Transfer to Vessel Wall by Helical Ribbon Impeller in Highly Viscous Liquids," *J. Chem. Eng. Japan* (1972), pp. 83–84.
23. Nishikawa, M., Kamata, N., Nagata, S. "Heat Transfer for Highly Viscous Liquids in Mixing Vessel," *Kagaku Kogaku Ronbunshu*, Vol. 1 (1975), pp. 446–471.
24. Novak, V., and Rieger, F., "Homogenization Efficiency of Helical Ribbon and Anchor Agitators," *Chem. Eng. J.*, Vol. 9 (1975), pp. 63–70.
25. Rieger, F., and Novak, V., "Power Consumption of Agitators in Highly Viscous Non-Newtonian Liquids," *Trans. Inst. Chem. Eng.*, Vol. 51 (1973), pp. 105–111
26. Sawinsky, J., Havas, G., and Deak, A., "Power Requirement of Anchor and Helical Ribbon Impellers for the Case of Agitating Newtonian and Pseudoplastic Liquids," *Chem. Eng. Sci.*, Vol. 31 (1976), pp. 507–509.
27. Schlichting, H., *Boundary Layer Theory*, 6th ed., MacGraw-Hill, 1968, p. 291.
28. Tomita, Y., *Rehology*, Corona Publishing Co. Ltd., 1975, p. 262.

SECTION III

FLUID TRANSPORT EQUIPMENT

CONTENTS

CHAPTER 34. PUMP CLASSIFICATIONS AND DESIGN FEATURES........... 951

CHAPTER 35. SYSTEM ANALYSIS FOR PUMPING EQUIPMENT SELECTION .. 1001

CHAPTER 36. DESIGN AND OPERATION OF MULTISTAGE CENTRIFUGAL PUMPS... 1038

CHAPTER 37. OSCILLATING DISPLACEMENT PUMPS 1057

CHAPTER 38. CAVITATION AND EROSION: MONITORING AND CORRELATING METHODS 1119

CHAPTER 39. SIZING OF CENTRIFUGAL PUMPS AND PIPING 1138

CHAPTER 40. FLUID DYNAMICS OF INDUCERS 1152

CHAPTER 41. HYDRODYNAMICS OF OUTFLOWS FROM VESSELS 1187

CHAPTER 42. DESIGN FEATURES OF FANS, BLOWERS, AND COMPRESSORS . 1208

CHAPTER 43. ANALYSIS OF AXIAL FLOW TURBINES.................... 1335

CHAPTER 44. GUIDELINES FOR EFFICIENCY SCALING PROCESS OF HYDRAULIC TURBOMACHINES WITH DIFFERENT TECHNICAL ROUGHNESSES OF FLOW PASSAGES 1355

CHAPTER 45. STABILIZING TURBOMACHINERY WITH PRESSURE DAM BEARINGS.. 1375

CHAPTER 46. FLUID DYNAMICS AND DESIGN OF GAS CENTRIFUGES...... 1393

CHAPTER 47. RUPTURE DISC SIZING AND SELECTION.................. 1433

CHAPTER 34

PUMP CLASSIFICATION AND DESIGN FEATURES

N. P. Cheremisinoff

Exxon Chemical Co.
Linden, New Jersey USA

CONTENTS

INTRODUCTION, 951

PUMP TERMINOLOGY, 952

CENTRIFUGAL PUMPS, 957
Theoretical Design Equations, 966

POSITIVE DISPLACEMENT PUMPS, 977

MOTIVE FLUID PUMPS, 989

PUMP SELECTION, 994

NOTATION, 999

REFERENCES, 999

INTRODUCTION

This chapter presents general information on pumps. Major classes of pumps along with their operating principles are described. Selection criteria and typical data on operating characteristics and service factors are included. The chapter is mainly an overview of data and information needed for specifying pumps for different process applications, as well as a review of pump technology principles and nomenclature. Specific details on sizing, operation, and maintenance of various types of pumps, as well as on overall process pipe network design, are presented in subsequent chapters of this volume.

The major pump types used in process plant operations are centrifugal, axial, regenerative turbine, reciprocating, metering and rotary. These classes are grouped under dynamic pumps and positive displacement pumps.

Dynamic pumps include centrifugal and axial types. They operate by developing a high liquid velocity, which is converted to pressure in a diffusing flow passage. In general, they are lower in efficiency than the positive displacement types. They do operate at relatively high speeds, thus providing high flow rates in relation to the physical size of the pump.

Positive displacement pumps operate by forcing a fixed volume of fluid from the inlet pressure section of the pump into the pump's discharge zone. This is performed intermittently with reciprocating pumps. In the case of rotary screw and gear pumps, the action is continuous. This pump category operates at lower rotating speeds than dynamic pumps. Positive displacement pumps also tend to be physically larger than equal-capacity dynamic pumps.

PUMP TERMINOLOGY

There are four characteristics descriptive of all pumps, namely:

1. Capacity—the quantity of liquid discharged per unit time.
2. Head H (m)—the energy supplied to the liquid per unit weight obtained by dividing the increase in pressure by the liquid's specific weight. This specific energy is determined from the Bernoulli equation. Head can be described as the height to which 1 kg of discharged liquid can be lifted by the energy supplied by a pump and therefore, it does not depend on the specific weight $\gamma(kg_f/m^3)$ or density $\rho(kg/m^3)$ of liquid to be pumped.
3. Power, $N(kg_f\text{-}m/sec)$—the energy per unit time consumed by a pump. Power is equal to the product of specific energy, H, and the mass flow rate γQ:

$$N = \gamma QH = \rho gQH \tag{1}$$

All machines have an effective power, N_e, which is larger than N due to inefficiencies. The relative value of N_e is evaluated in terms of the machine's efficiency η_p:

$$Ne = \frac{N}{\eta_P} = \frac{\rho gQH}{\eta_P} \tag{2}$$

4. Overall efficiency, η—the ratio of useful hydraulic work performed to the actual work input. This characterizes the perfection of the design and the pump's performance. The value of η reflects the relative power losses in the pump and is expressed as follows:

$$\eta = \eta_v \eta_h \eta_m \tag{3}$$

where η_v is the volumetric efficiency defined as the ratio of liquid actually pumped to that which is theoretically discharged. That is, it indicates the degree of losses (or slippage). In practice, slip should not exceed 5%. η_h is the hydraulic efficiency defined as the ratio of the actual head pumped to the theoretical head:

$$\eta_h = \frac{H}{H + \text{hydraulic losses}} \tag{4}$$

Hydraulic losses are those head losses in the suction and discharge sections of a pump. In the suction end, these losses consist of:

- Velocity head.
- Entrance head.
- Friction head in the suction line.
- Losses in bends and losses in suction valves.

Discharge line losses include losses in the discharge valves, velocity head, and friction in the discharge piping.

η_m is the mechanical efficiency defining the relation between the indicated pump horsepower and the actual power input from the drive. It characterizes mechanical losses in the pump (for example, in bearings, stuffing boxes, etc.).

The overall efficiency, η, depends on the pump design and varies from 50% for small pumps to about 90% for large units. The power consumed by a motor (defined as the nominal power of a motor), N_m, exceeds the brake power by mechanical losses incurred in transmission and in the motor itself. These losses are included in Equation 3 in terms of the efficiencies of transmission (η_{tr})

Table 1
Typical Values of Coefficient β as a Function of Motor Power

N_A (kW):	less than 1	1–5	5–50	greater than 50
β:	2–1.5	1.5–1.2	1.2–1.15	1.1

and the motor, (η_m):

$$N_m = \frac{N_e}{\eta_m \eta_{tr}} = \frac{N}{\eta_e \eta_m \eta_{tr}} \tag{5}$$

The product $\eta_e \eta_m \eta_{tr}$ is the total efficiency of a pump defined as the ratio of hydraulic power to the motor's nominal power:

$$\eta = \frac{N}{N_m} = \eta_p \eta_{tr} \eta_{mot} \tag{6}$$

From Equations 3 and 6, the total efficiency can be expressed by the product of five values:

$$\eta = \eta_v \eta_h \eta_m \eta_{tr} \eta_{mot} \tag{7}$$

The actual power of the pump's motor, N_A, is based on the energy required to overcome the fluid's inertia at startup, in order to avoid overloading the unit.

$$N_A = \beta N_m \tag{8}$$

Coefficient β is determined from the size of the motor and typical values are given in Table 1.

Head, H., characterizes the excessive energy, $\ell = gH$, added to 1 kg of liquid in a pump. Figure 1 illustrates a simple flow system which can be described by the Bernoulli equation as follows:

At suction (sections 1-1 and 1′-1′, Figure 1),

$$\frac{P_1}{\rho g} + \frac{w_1^2}{2g} = H_s + \frac{w_s^2}{2g} + \frac{P_s}{\rho g} + h_{\ell s} \tag{9}$$

At discharge (sections 1′-1′ and 2-2, Figure 1)

$$\frac{P_d}{\rho g} + \frac{w_d^2}{2g} = H_d + \frac{w_2^2}{2g} + \frac{P_2}{\rho g} + h_{\ell d} \tag{10}$$

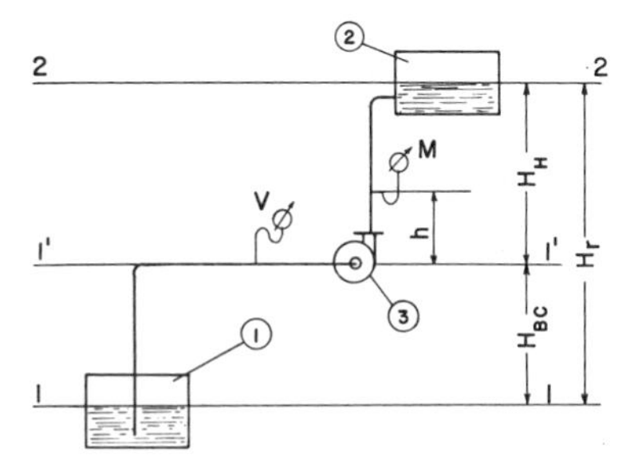

Figure 1. Illustrative pumping network: 1—tank, 2—tank, 3—pump, M—manometer, V—vacuum gauge.

where w_1 and w_2 = liquid velocities in tanks 1 and 2 , respectively
w_s and w_d = liquid velocities in the suction and discharge pump nozzles
$h_{\ell s}$ and $h_{\ell d}$ = head losses in the suction and discharge pipings, respectively.

Assuming velocities $w_1 = 0$ and $w_2 = 0$, the total dynamic head, H, of a pump is the difference between discharge head, H_d, and suction head in the nozzles:

$$H = \frac{P_d - P_s}{\rho g} \tag{11}$$

From Equations 9 and 10:

$$H = \frac{P_2 - P_1}{\rho g} + \frac{w_s^2 - w_d^2}{2g} + H_d + H_s + h_{\ell d} + h_{\ell s} \tag{12}$$

If the suction and discharge nozzles of a pump have the same diameter, then $w_d = w_s$ and $H_t = H_d + H_s$; $h_{\ell d} + h_{\ell s} = h_{\ell t}$. Equation 12 then simplifies to:

$$H = H_T + \frac{P_2 - P_1}{\rho g} + h_{\ell t} \tag{13}$$

Equation 13 states that the total head is expended in:

1. Lifting the liquid to height H_T.
2. Overcoming the pressure difference in tanks 1 and 2.
3. Overcoming the hydraulic resistances in the suction and discharge pipings.

If $P_1 = P_2$, this expression reduces to:

$$H = H_T = h_{\ell,t} \tag{14}$$

For a horizontal section of piping, $H_T = 0$:

$$H = \frac{P_2 - P_1}{\rho g} + h_{\ell t} \tag{15}$$

In the case in which P_1 and P_2 and $H_T = 0$:

$$H = h_{\ell t} \tag{16}$$

The total head of the operating pump may be evaluated from measurements obtained from the pressure, P_m, and vacuum gauges, P_v.

Total suction head can be obtained from the measurement h_{sg} on the gauge on the pump's suction nozzle (corrected to the pump centerline and converted to feet of liquid), plus the barometer reading in feet of liquid and the velocity head, h_{vs} (ft), at the point of gauge attachment:

$$H_s = H_{sg} + \text{atm} + H_{vs} \tag{17}$$

When the static pressure at the suction flange is less than atmospheric, the measurement obtained from a vacuum gauge replaces H_{sg} in Equation 17 and is assigned a negative sign.

Total discharge head, H_d, is obtained from a gauge, H_{dg}, at the pump's discharge flange (corrected to the pump centerline and converted to feet of liquid), plus the barometer reading and the velocity head, H_{vg}, at the point of gauge attachment:

$$H_d = H_{dg} + atm + H_{vg} \tag{18}$$

Again, if the discharge gauge pressure is below atmospheric, the vacuum gauge measurement replaces H_{dg} with a negative sign. Before installation, it is possible to estimate the total discharge head from the static discharge, H_{sd}, and discharge friction, H_{fd}, as follows:

$$H_d = H_{sd} + H_{fd} \tag{19}$$

Static suction head, H_{ss}, is defined as the vertical distance between the free level of the source of supply and the pump centerline, plus the absolute pressure at this level (converted to feet of liquid).

Total static head, H_{ts}, is the difference between discharge and suction static heads. The suction generated by a pump is derived from a pressure difference between the suction source (P_1) and the pump (P_s) or is due to the action of the head difference $P_1/\rho g - P_s/\rho g$. Suction height is:

$$H_s = \frac{P_1}{\rho g} - \left(\frac{P_s}{\rho g} + \frac{w_s^2 - w_1^2}{2g} + h_{\ell s}\right) \tag{20a}$$

or

$$H_s = \frac{P_1}{\rho g} - \left(\frac{P_s}{\rho g} + \frac{w_s^2}{2g} + h_{\ell s}\right) \tag{20b}$$

because $w_1 \simeq 0$. These expressions illustrate that the suction head increases with P_1, and decreases with increasing w_s and $h_{\ell s}$.

If liquid is pumped from an open tank, i.e., $P_1 = P_a$ where P_a corresponds to atmospheric, the suction pressure, P_s, must exceed pressure P_t (the pressure of the saturated vapor of the liquid at the pumping temperature ($P_s > P_t$)), otherwise the fluid begins to boil. When the pumped liquid vaporizes, the suction head goes to zero at the limit and flow stops. Consequently,

$$H_s \leq \frac{P_a}{\rho g} - \left(\frac{P_t}{\rho g} + \frac{w_s^2}{2g} + h_{\ell s}\right) \tag{21}$$

Hence, the suction head is a function of atmospheric pressure, fluid velocity and density, temperature, the liquid's vapor pressure, and the hydraulic resistance of the suction piping. When pumping from an open vessel, the suction head cannot exceed the head of flowing liquid, which corresponds to atmospheric pressure (the value of which depends on the height of the pump installation above a specified datum, normally sea level). Thus, for example, if 20°C water is pumped, the suction head cannot exceed 10 m at sea level. If the same pumping system is used at an elevation of 2,000 m, the suction head cannot exceed 8.1 m, which corresponds to the atmospheric pressure, in m of water column.

At temperatures approaching the boiling point of the liquid the suction head becomes zero; that is:

$$H_s = 0 \quad \text{at} \quad \frac{P_a}{\rho g} = \frac{P_1}{\rho g} + \frac{w_s^2}{2g} + h_{\ell s}$$

In this situation, the pump must be installed below the suction line to provide a back liquid. This method also is used for pumping high viscosity liquids. In addition to evaluating the friction head and local resistance losses, inertia losses (for piston pumps in particular), H_i, and the effect of cavitation (for centrifugal pumps), h_k, must be accounted for in the overall suction head.

Head losses due to overcoming inertia forces, H_i (in piston pumps), may be estimated by an expression that relates the pressure acting on the piston to the inertia force of a liquid column moving in the suction piping:

$$H_i = \frac{6}{5}\frac{\ell}{g}\frac{f}{f_1}\frac{u^2}{r} \quad (22)$$

where ℓ = height of liquid column in the piping (for pumps having a gas chamber, the distance between pump centerline and the liquid level in the chamber)

g = acceleration due to gravity

f, f_1 = cross-sectional areas of the piston and piping, respectively

u = circumferential crank velocity

r = crank radius

Cavitation in centrifugal pumps arises from high velocities or handling hot liquids under conditions of vaporization. It is a phenomenon caused by the formation and collapse of vapor cavities existing in a flowing liquid. Vapor cavities can form at any point in the fluid at which the local pressure approaches that of the liquid vapor pressure at operating temperatures. At these positions, a portion of the liquid vaporizes to form bubbles or cavities of vapor. Low-pressure zones are generated in several ways:

1. By a local increase in velocity resulting in eddies or vortices near the boundary contours.
2. By rapid vibration of the boundary.
3. By separating or parting of the liquid due to "water hammer."
4. By an overall reduction in static pressure.

Collapse of the bubbles begins when they move into regions where the local pressure exceeds the liquid's vapor pressure. This often results in objectionable noise and vibration, as well as extensive erosion or pitting of the system's walls. Even more important, cavitation results in a decrease in pumping performance and efficiency. The cavitation number σ is used to correlate performance and is defined as the ratio of the net static pressure available to collapse a bubble and the dynamic pressure initiating bubble formation.

$$\sigma_c = \frac{P - P_v}{\rho w^2/2g_c} \quad (23)$$

where P = static pressure (absolute) in the undisturbed flow, lb-force/sq ft
P_v = liquid vapor pressure (absolute), lb-force/sq ft
ρ = liquid density, (lb/cu ft)
w = free-stream velocity of the liquid
g_c = constant (32.17 (lb)(ft)/(lb-force)(sec^2))

The value of this dimensionless group at conditions of incipient cavitation, σ_c, depends on the pump's geometry. Typical values are reported by Perry and Chilton [1], Karassik et al. [2], Church [3], and Cheremisinoff [4, 5] and in Chapter 39.

A correction factor for head is:

$$\tilde{H}_c = 0.019\frac{(Qn^2)^{2/3}}{H} \quad (24)$$

where Q = pump capacity (m^3/sec)
n = number of revolutions (sec^{-1})
H = head of the pump

Table 2
Typical Suction Head Limits to Avoid Cavitation*

Temperature (°C)	10	20	30	40	50	60	65
Suction head, (m)	6	5	4	3	2	1	0

* *Based on water flow.*

In practice, to avoid cavitation the suction head for pumping liquids with physical properties close to those of water should exceed the values in Table 2.

We shall now describe the major pump types.

CENTRIFUGAL PUMPS

Centrifugal pumps are used widely throughout the process and allied industries because of their simplicity in design, low initial cost and maintenance, and versatility in application. This type of pump is available in a wide range of sizes—in capacities ranging from a few gpm up to 100,000 gpm and with discharge heads from a few feet up to several thousand psi. A major design feature is the impeller, which is a series of radial vanes of various shapes and curvatures rotating within a circular casing (see Figure 2).

The liquid from suction piping (1) enters at the axis of a rotating impeller (2) into the pump chamber (3) and is thrown outwards by centrifugal action against the blades, (4). The impeller's high speed of rotation causes the liquid to acquire kinetic energy. A pressure difference between the suction and discharge sides of the pump is produced by the conversion of kinetic energy of the liquid flow into pressure energy in the discharge piping (5). A reduction in pressure occurs at the entrance of the impeller, and the liquid is fed continuously into the pump from a supply tank. Without filling the pump chamber with liquid, the impeller cannot produce an adequate pressure difference, which is necessary for lifting liquid in the suction line. An example of a heavy-duty end suction centrifugal pump used to pump large quantities of non-aggressive liquids is shown in Figure 3. These are employed for general service applications in industry, irrigation and municipal water supply.

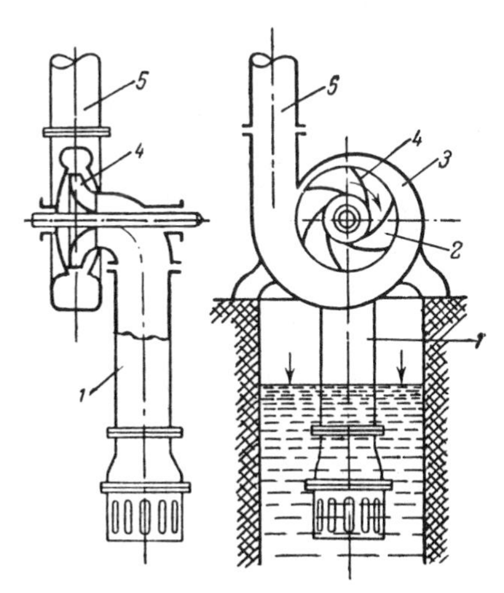

Figure 2. Centrifugal pump scheme: 1—suction piping, 2—impeller, 3—casing, 4—vanes, 5—delivery piping.

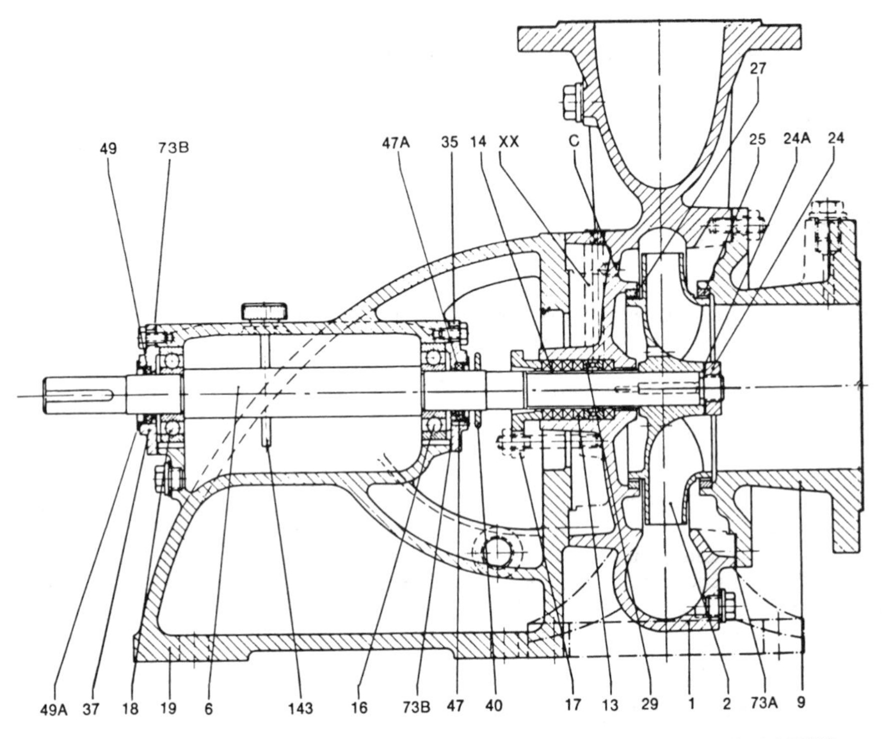

XX—SEALING LIQUID DUCT—EXTERNAL SOURCE
C—SEALING LIQUID DUCT—INTERNAL SOURCE

(A)

PART NO.	PART NAME	STD. MATERIAL	BRONZE FITTED
1	CASING	CAST IRON	CAST IRON
2	IMPELLER	CAST IRON	BRONZE
6	SHAFT	CARBON STEEL	CARBON STEEL
9	SUCTION COVER	CAST IRON	CAST IRON
13	PACKING	ASBESTOS	ASBESTOS
14	SHAFT SLEEVE	CAST IRON	BRONZE
16	BALL BEARING—INBOARD	STEEL	STEEL
17	GLAND	CAST IRON	CAST IRON
18	BALL BEARING—OUTBOARD	STEEL	STEEL
19	FRAME	CAST IRON	CAST IRON
24	IMPELLER NUT	STEEL	STEEL
24A	LOCKING PLATE	STEEL	STEEL
25	WEAR RING—SUC. COVER	CAST IRON	BRONZE
27	WEAR RING—STUFF. BOX	CAST IRON	BRONZE
29	LANTERN RING	CAST IRON	CAST IRON
32	IMPELLER KEY	STEEL	STEEL
35	BEARING COVER—INBOARD	CAST IRON	CAST IRON
37	BEARING COVER—OUTBOARD	CAST IRON	CAST IRON
40	DEFLECTOR	STEEL	STEEL
47	OIL SEAL—INBOARD	FELT	FELT
47A	CAP—INBOARD SEAL	STEEL	STEEL
49	OIL SEAL—OUTBOARD	FELT	FELT
49A	CAP—OUTBOARD SEAL	STEEL	STEEL
73A	GASKET—COVER	ASBESTOS	ASBESTOS
73B	GASKETS—BEAR. COVER	PAPER	PAPER
143	OIL GAUGE	STEEL	STEEL

CAST IRON PARTS ARE ASTM A-48-35.

Figure 3. (A) Horizontal, end-suction centrifugal pump; (B) cross-sectional drawing of the pump. (Courtesy Carver Pump Co.)

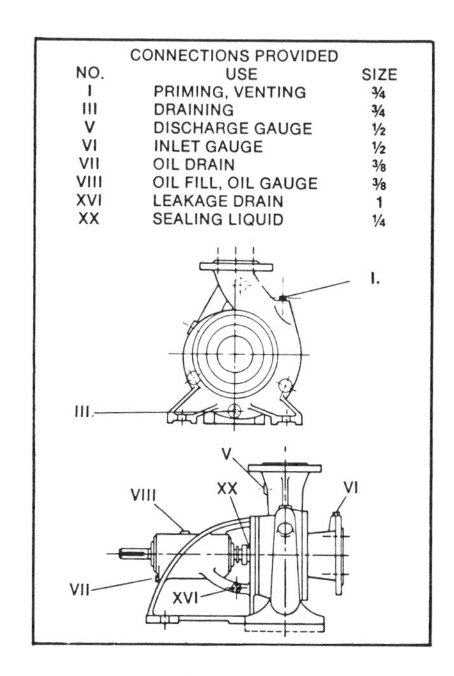

CONNECTIONS PROVIDED

NO.	USE	SIZE
I	PRIMING, VENTING	¾
III	DRAINING	¾
V	DISCHARGE GAUGE	½
VI	INLET GAUGE	½
VII	OIL DRAIN	⅜
VIII	OIL FILL, OIL GAUGE	⅜
XVI	LEAKAGE DRAIN	1
XX	SEALING LIQUID	¼

(B)

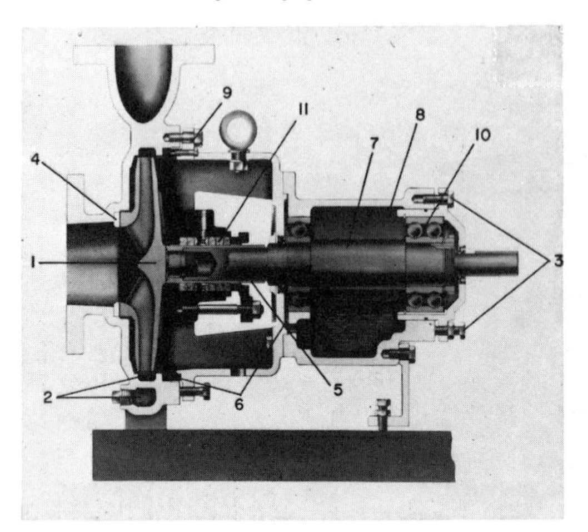

Figure 4. Cross-section of a centrifugal pump used in handling process chemicals. (Courtesy Ingersoll-Rand Co.)

Most centrifugal pumps are not self-priming and, hence, cannot evacuate vapor from the suction line; hence, liquid can flow into the pump casing without external assistance. The impellers on centrifugal pumps are specially designed for efficient pumping and are not operated at high enough tip speeds to convert them into vapor compressors. The differential head that the pump impeller can deliver is the same on the vapor as on the liquid. However, the equivalent differential pressure rise capability is typically much lower with vapor. To prime a centrifugal pump, both the suction line and pump casing must be filled with liquid. When the suction source is at positive pressure or is positioned above the pump, priming is accomplished by opening the suction valve and venting the trapped vapor from a valve connection on the pump casing or discharge line. Liquid then flows into the suction line and pump casing to displace the escaping vapor.

Centrifugal pumps are used in a multitude of applications throughout the chemical industry. Designs may consist of an open impeller system mounted on an externally adjustable shaft for handling clear liquids, slurries, or liquids with suspended solids. Also, they can be closed impellers for pumping clear liquids or light slurries. Figure 4 shows a cross section of one manufacturer's chemical process pump. The specific features of this pump are given in Table 3.

This particular design utilizes a dual volute. All pump volutes are designed to generate uniform radial thrust on the impeller shaft and bearings when operating at the best efficiency point on the pump curve (described in the next subsection). As a result, there is minimum radial thrust on the pump components. However, when a pump is not operating at its best efficiency point, the casing design no longer balances the hydraulic loads and radial thrust increases. A dual volute incorporates a flow splitter in the casing, which directs the liquid into two separate paths through the casing. The contour of the flow splitter follows the contour of the casing wall 180° opposite. Both are approximately equidistant from the center of the impeller; thus, the radial thrust loads acting on the impeller are balanced and greatly reduced.

Most chemical pumps are built with casings cast in alloys. Casings frequently are foot supported or bearing bracket supported rather than centerline supported. Chemical pumps are available in a wide range of operating conditions but most often are limited to low to moderate flows. Another chemical process pump for high-pressure, high-temperature service is shown in Figure 5. This particular unit has hydraulic coverage to 9,500 gpm, 700 ft of total dynamic head, and 570 psig discharge pressure. A cross-sectional view of this pump is shown in Figure 5B; its major features are listed in Table 4.

Many centrifugal pumps have single casings; that is, a single wall between the liquid under discharge pressure and the atmosphere. Double casings are used in horizontal, multistage, high-pressure pumps and in vertical pumps. In the former, a heavy barrel-shaped casing surrounds the stack of stage diaphragms. The stack of diaphragms makes up the inner casing, while the barrel forms the outer casing. This type of arrangement is used most often in boiler feed pumps.

Table 3
Features of the Chemical Process Pump

Feature	Advantage	Benefit
1. Open impellers or closed impellers.	Allows selection of proper impeller for individual needs. Open impellers handle slurries and particles. Closed impeller handles clear liquid and smaller particles at higher efficiencies.	Increased reliability; less maintenance downtime; lowest power costs for specific application.
2. Dual volute.	a. Reduced radial thrust. b. Reduced shaft deflection at face of stuffing box.	Increased bearing life; less downtime; lower maintenance. Extended seal and packing life; less downtime; lower maintenance costs.
3. External axial impeller adjustment.	Allows field setting of impeller clearances to compensate for wear, thus restoring high efficiencies without overhaul.	Less downtime; lower maintenance costs; reduced power costs.
4. Replaceable casing ring for closed impellers and casing shroud plate for open impellers.	Provides for quick renewal of casing surface with inexpensive flat plate or ring. No need to replace a cast part (casing, stuffing box cover, or cast wear plate).	Lower maintenance costs; increased availability of replacement parts.
5. Replaceable hook-type sleeve with O-ring seal between shaft sleeve and impeller.	Assures accurate seal setting. Protects shaft and impeller threads from contamination by pumped liquid.	Higher reliability; less downtime; lower maintenance costs.
6. Dry rabbet fit construction.	Provides for accurate, positive alignment and reduced possibility of crevice corrosion.	Increased reliability, less downtime; lower maintenance costs.
7. Heavy duty shaft system (thick shaft, optimum bearing spans, short impeller overhang).	Reduced shaft deflection.	Extends seal and packing life; less downtime and lower maintenance costs.
8. Three bearing cradles for all sizes.	Maximum parts interchangeability, reduced spare parts stock levels.	Lower maintenance costs; increased parts availability.
9. Studded casing	Reduced possibility of stripping casing threads during assembly and disassembly.	Lower maintenance repair costs; less downtime.
10. Double row thrust bearings.	Enables the bearing to carry high thrust loads at all suction pressures and operating conditions.	Increased bearing life; reduced downtime and lower maintenance costs.
11. Versatile stuffing box designed for packing and all seal types.	Provides capability to change from packing to seals without changing the stuffing box cover. All seals (balanced, unbalanced, outside, and tandem) can be interchanged without changing the stuffing box cover.	Greater interchangeability; lower repair costs.

Courtesy Ingersoll-Rand Co., 253 East Washington Ave., Washington, NJ.

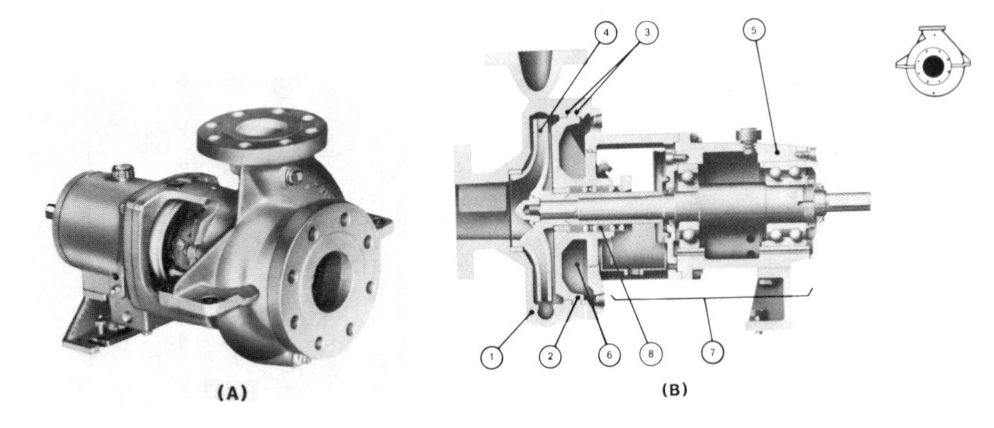

Figure 5. (A) High-pressure, high-temperature service chemical process pump; (B) cross-sectional view (refer to Table 4 for a listing of major features). (Courtesy Carver Pump Co.)

Table 4
Features of the Chemical Process Pump

1. *Pump casing*
   Back pull-out, self-venting top centerline discharge, vortex suppressing guide vane in suction nozzle, $\frac{1}{8}''$ corrosion allowance, rugged integral casted centerline supports to allow thermal or pressure expansion without causing misalignment and shaft deflection.
2. *Positive displacement*
   Positively and permanently achieved by full-circle registered fit on all mating parts. All such fits are away from liquid being pumped, preventing crevice corrosion.
3. *Fully confined gaskets*
   On wet and dry side of casing cover as well as between impeller nut, impeller and shaft sleeve provide safety against leakage.
4. *Enclosed impeller*
   For high efficiency and low NPSH, with back vanes for axial thrust balancing and low stuffing box pressure, keeping erosive impurities out of shaft seal area, keyed to shaft for positive fastening. Positioned by acorn-type impeller nut with Heli-Coil lock insert, it cannot come loose under reserve rotation.
5. *Bearing frame*
   Five sizes for 27 models from $1\frac{1}{2}$ in. to 12 in. discharge. Designed to carry maximum load with under 0.02 in. shaft deflection and at least 2-year bearing life.
6. *Built-in casing heating or cooling jacket*
   Increases the application scope without need of added costs for optional or extra parts. Special intensive cooling of stuffing box only is also available.
7. *Stuffing box*
   Can be adjusted to particular application requirements. Replaceable shaft sleeve is provided with stuffing box packing or mechanical seal, providing maximum adaptability for various seal designs. Step design shaft sleeve simplifies accurate seal setting for long seal life.

Courtesy Carver Pump Co., 2415 Park Ave., P.O. Box 389, Muscatine, IA.

Figure 6. Cross-section of a single-stage, double suction pump. (Courtesy Ingersoll-Rand Co.)

Casings may be joined on the same plane as the shaft axis (called axially split) or perpendicular to the shaft (called radially split). Axially split horizontal pumps most commonly are referred to as "horizontally split." Figure 6 shows a single-stage, double-suction, horizontally split case centrifugal pump.

Radial split horizontal pumps are commonly called "vertically split." Radial joining is used on horizontal overhung pumps to allow ready removal of the rotor-and-bearing bracket assembly for maintenance. This design configuration also is employed on high-pressure multistage pumps because of the structural problems associated with bolting together the halves of axially split casings exposed to high internal pressure.

The term *single-stage overhung* refers to the impeller mounting/support arrangement. The casings for these designs are supported at the centerline. Two shaft bearings are mounted close together in the same bearing bracket, with the impeller cantilevered or overhung beyond them. Normally, this type configuration utilizes top suction and discharge flanges, wearing rings both on the front and back of the impeller and casing, a single-suction closed impeller, and a single stuffing box fitted with a mechanical seal. These pumps are well suited to high-temperature operations and can be used for handling flammable liquids.

A two-stage overhung pump is a modification of the single-stage process pump and is capable of higher head. Usually the stuffing box pressure is approximately halfway between suction and discharge pressures. Figure 7 shows a cross section of a two-stage, horizontally split pump.

Multistage centrifugal pumps generally are used for generating higher heads (pressures) than can be obtained by single-stage pumps. These pumps are available for pressures as high as 3,000 lb/sq in^2 at capacities greater than 3,000 gpm. The operation of this type pump is illustrated in Figure 8. As shown, these designs have impellers (A) in one aggregate casing (B), which are located in series on one shaft (C). Liquid discharged from the first impeller enters (D) through the offtake in the second impeller, where it acquires additional energy, from the second impeller through the offtake in the third impeller, and so on. Thus, multistage pumps may be thought of as several single-stage pumps on one shaft, with the flow in series. Hence, the total head developed is the head of one impeller multiplied by the number of impellers (usually designs do not exceed five impellers). See Chapter 37 for a discussion of multistage centrifugal pumps.

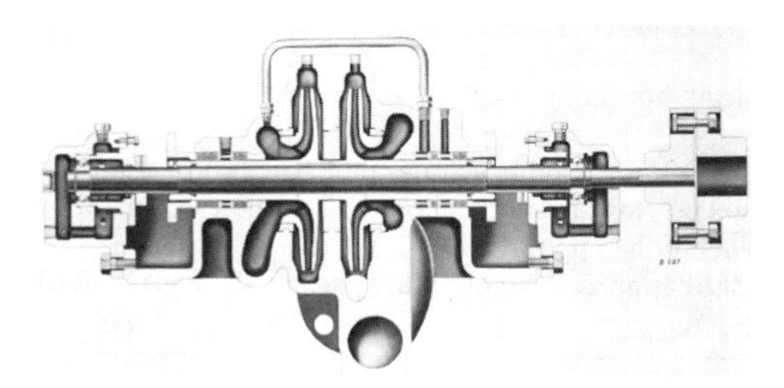

Figure 7. Cross-section of a two-stage, horizontally split pump. (Courtesy Ingersoll-Rand Co.)

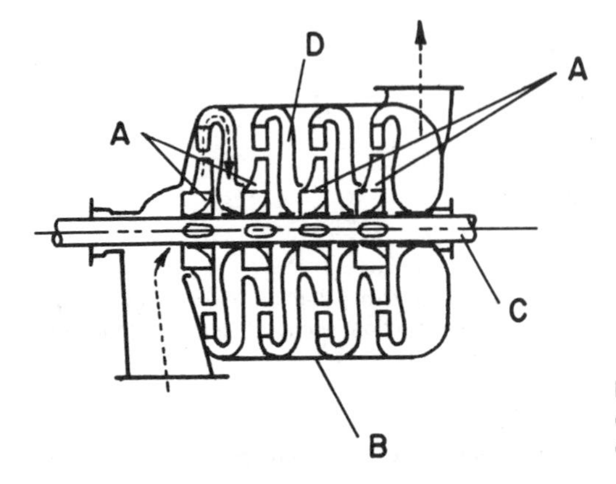

Figure 8. Operation of a multistage centrifugal pump: A—impeller; B—casing; C—shaft; D—offtake.

Further details and examples of multistage centrifugal pumps are shown in Figures 9 through 12. Figure 9 is a cross section of a multistage horizontally split pump. (Refer to Table 5 for an explanation of the features.) This design incorporates horizontally split channel rings and center bushings, so the rotor is easy to assemble, inspect, and balance dynamically. Figure 10 shows the same pump used to produce a high-pressure stream of water to loosen uranium at a mine. Multi-

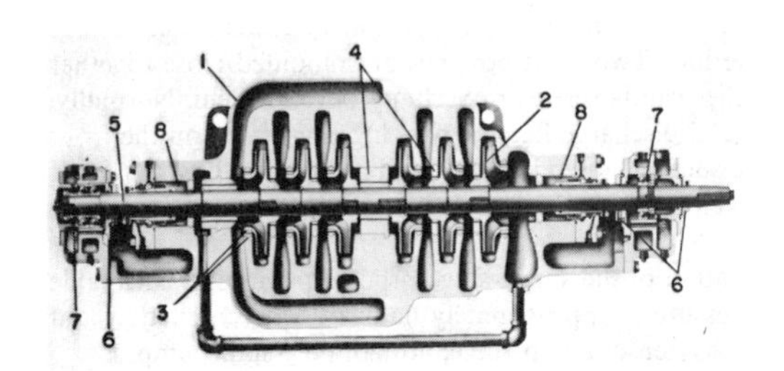

Figure 9. Cross-section of a multistage, horizontally split pump. (Courtesy Ingersoll-Rand Co.)

Table 5
Features of the Multistage Horizontally Split Pump

1. Heavy-walled, dual-volute casting allows for high working pressures and nozzle pipeloads. Capnuts are seated on the top half of the axially split casing to simplify assembly and disassembly.
2. Dynamically balanced opposed impellers of one-piece construction are shrunk on the shaft and keyed. Shaft is stepped by 5/1000 in at each impeller fit to facilitate removal.
3. Renewable casing rings and impeller rings control interstage leakage.
4. Channel rings, including center channel ring, are split horizontally to facilitate replacement and simplify dynamic balancing of rotor without dismantling.
5. Large-diameter shaft and minimum bearing span reduce deflection for longer bearing, mechanical seal and wear ring life.
6. Labyrinth flanges at each end of the bearing housing protect the lubrication against contamination.
7. Standard ring oil-lubricated bearings assure complete oil penetration into the bearings without foaming, for increased bearing life. Optional bearing arrangements and lubrication systems allow a "customized" fit to meet the requirements of the application.
8. Large stuffing boxes, integrally cast with the casing, can handle either packing or single, tandem or double-mechanical seals.

Courtesy Ingersoll-Rand Co., 253 East Washington Ave., Washington, NJ.

Figure 10. Multistage, horizontally split pumps used to produce a high-pressure stream of water to loosen uranium at a Colorado mine. (Courtesy Ingersoll-Rand Co.)

stage pumps are employed in a multitude of processing applications. Examples include hydrocarbon processing and refining, boiler feed operations, descaling operations, mine dewatering and hydraulic power recovery in which excess plant energy is recovered to drive other equipment.

Figure 11 shows a high-pressure, horizontal shaft, vertically split multistage pump. The ring section stage casing construction provides a wide range of discharge pressures through the addition of individual stages. Its principal function is high-pressure boosting of clear, non-aggressive liquids.

Vertical pumps are another orientation used widely throughout the process industries. In this type, a vertical cylinder buried in the ground houses the pumping element. Suction liquid enters the outer cylinder and flows to the bottom and then up through the pumping element stages. The diaphragms of the stages in the pumping element comprise the inner casing. As *inline pumps*, the casings are designed to be bolted directly to the piping, much like a valve. Two basic configurations of inline pumps are coupled and close-coupled. Service life and maintenance requirements for both are about the same. Figure 12 shows the discharge pressure measured at the discharge nozzle above the floor plate, the velocity head at the same location, and the vertical distance from the centerline of the pressure gauge to the liquid surface in the pump. These pumps generally are equipped with tail pipes, which allow pumping down below normal liquid level. The pump must be capable of operating under pumpdown conditions without allowing the throttle bushing to run dry.

Figure 11. High-pressure, horizontal-shaft, vertically split, multistage centrifugal pump. The maximum capacity and head of this pump are 2,200 gpm and 825 ft, respectively, with a maximum discharge pressure of 400 psi. (Courtesy Carver Pump Co.)

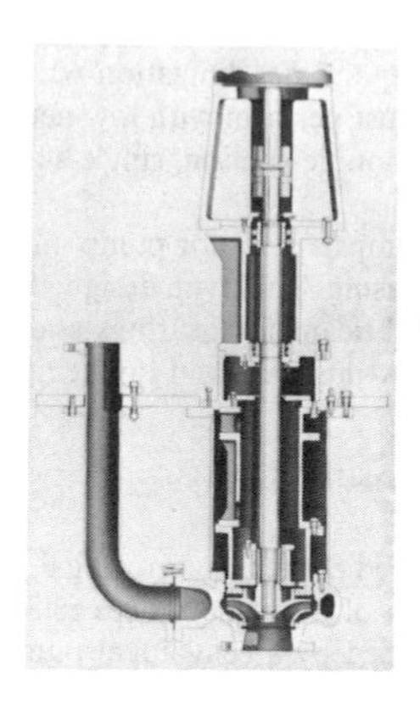

Figure 12. A top pullout vertical pump. This unit is suitable for continuous operation in ambient temperatures up to 50°C. (Courtesy Carver Pump Co.)

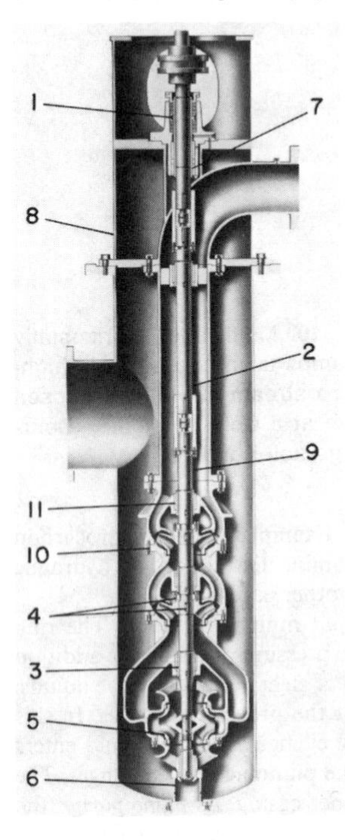

Figure 13. Cross-section of a double-suction vertical pump. (Courtesy Ingersoll-Rand Co.)

Vertical multistage pumps can have 24 or more stages. High specific-speed impellers often are used. The first stage is usually at the bottom of the assembly, below grade. These pumps require a large number of close-running clearances and, thus, are sensitive to damage by solids and by dry or two-phase flow conditions. This type of pump is employed in a broad range of applications. Examples include: fossil power plants where they are used for condensate service in large power generating plants; nuclear power plants where they are used in condensate and feedwater heater drain service; and desalinization whose operating facilities require large-capacity pumping equipment that must perform with low net positive suction head (NPSH) available. Figure 13 shows the details of a double-suction, single-stage vertical pump. The features of this pump are explained in Table 6.

"Cam" pumps are motor pump units with the rotating rotor and impeller housed entirely within a pressure casing. This type design eliminates the need for a stuffing box. The pumped fluid serves both as a lubricant for bearings and as a coolant for the motor. Designs typically are limited to low-flow, low-pressure, and low-temperature services.

Theoretical Design Equations

As described earlier, in centrifugal pumps the liquid flows along the surface of the impeller vanes while the tip of the vane moves relative to the casing of the pump. An expression of the virtual head developed by a centrifugal pump can be derived by assuming a path followed by the liquid volume as it passes through the pump in relation to a stationary impeller, with the fluid having the same relative velocity as an actual rotating impeller. Figure 14 defines the system under consideration.

c_1 and c_2 are the vector sums of the relative and tangential velocities of the fluid entering and exiting the impeller, respectively. The relative velocity components along the vanes are w_1 and w_2, and the tangential components (tangent to the circumference of rotation) are u_1 and u_2. Further,

Table 6
Features of Double Suction Vertical Pump

1. Multi-element stuffing box utilizes an advanced combination of bearing, serrated sleeve and throttle bushings, and a vent to suction. Vents vapor that collects in stuffing box increases packing and bearing life.
2. Shaft surface is ground after machining to assure shaft trueness for long bearing life and low vibration levels.
3. Hardened stainless steel shaft sleeves under carbon bearings prevent premature opening of clearances and premature replacement of shaft.
4. Keys and lock collars properly position impeller and transmit torque. This prevents impellers from breaking loose during operation.
5. Double-suction first-stage impeller reduces NPSH and meets Hydraulic Institute standards for suction-specific speed. This means a shorter, more rigid pump and reduces the possibility of cavitation over a wide operating range.
6. Self-seating snubbers on long settings eliminate column movement and prevent excessive wear on bearings during low-flow operations or system upset.
7. Replaceable, hardened stainless steel shaft sleeve, keyed and sealed with an o-ring, eliminates the need to replace the complete upper shaft due to wear.
8. Discharge head's fundamental frequency and reed critical frequency of motor are both analyzed to eliminate vibration problems caused by sympathetic resonance.
9. Large shafts, with low shaft stress levels, mean less shaft whip, longer bearing and ring life.
10. Gaskets in intermediate stages seal against interstage leakage and prevent premature failure of flange surface due to "wire drawing."
11. Bearings are available in carbon or bronze throughout, have inherent self-lubricating properties, last longer, and are more durable in two-phase liquid operations.

Courtesy Ingersoll-Rand Co., 253 East Washington Ave., Washington, NJ.

a reference datum is defined to be the surface of the impeller in Figure 14. From an energy balance for the fluid passing through the impeller at $Z_1 = Z_2$:

$$\frac{P_1}{\rho g} + \frac{w_1^2}{2g} = \frac{P_2}{\rho g} + \frac{w_2^2}{2g} \tag{25}$$

When the impeller rotates, the liquid obtains an additional energy, "A," which is derived from the work of centrifugal force along the path $r_2 - r_1$. Hence

$$\frac{P_1}{\rho g} + \frac{w_1^2}{2g} = \frac{P_2}{\rho g} + \frac{w_2^2}{2g} - A \tag{26}$$

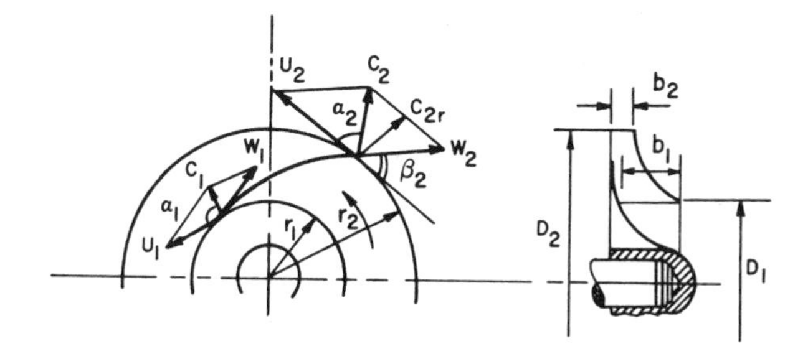

Figure 14. Defines system for the derivation of equations for centrifugal machines.

The centrifugal force, c, acting on the liquid particle of mass, m, is

$$c = m\omega^2 r = \frac{G}{g}\omega^2 r \tag{27}$$

where G = weight of fluid particle
ω = angular velocity
r = moving radius of particle rotation

The work, A_G, derived from the centrifugal force by displacement of this fluid particle along the path $r_2 - r_1$ is determined as follows:

$$A_G = \int_{r_1}^{r_2} \frac{G}{g}\omega^2 r\, dr = \frac{G\omega^2}{2g}(r_2^2 - r_1^2) = \frac{G}{g}\left(\frac{u_2^2 - u_1^2}{2}\right) \tag{28}$$

The specific work per unit weight of liquid is equal to the specific energy obtained by the fluid in the pump:

$$A = \frac{u_2^2 - u_1^2}{2g} \tag{29}$$

Combining Equations 29 and 28

$$\frac{P_2 - P_1}{\rho g} = \frac{w_1^2 - w_2^2}{2g} + \frac{u_2^2 - u_1^2}{2g} \tag{30}$$

The heads of liquid at the inlet and outlet from the pump are

$$H_1 = \frac{P_1}{\rho g} + \frac{c_1^2}{2g} \tag{31a}$$

$$H_2 = \frac{P_2}{\rho g} + \frac{c_2^2}{2g} \tag{31b}$$

Hence, the head of the pump is equal to the difference of heads between the pump's inlet and its outlet:

$$H_T = H_1 - H_2 = \frac{P_2 - P_1}{\rho g} + \frac{c_2^2 - c_1^2}{2g} \tag{32}$$

Substituting $(P_2 - P_1)/\rho g$ from Equation 30 into Equation 32 gives:

$$H_T = \frac{w_1^2 - w_2^2}{2g} + \frac{u_2^2 - u_1^2}{2g} + \frac{c_2^2 - c_1^2}{2g} \tag{33}$$

And from Figure 14 we have

$$w_1^2 = u_1^2 + c_1^2 - 2u_1c_1 \cos\alpha_1 \tag{34a}$$

$$w_2^2 = u_2^2 + c_2^2 - 2u_2c_2 \cos\alpha_2 \tag{34b}$$

Substituting this last set of expressions into Equation 33 results in an expression for the virtual head of a centrifugal pump:

$$H_T = \frac{u_2 c_2 \cos \alpha_2 - u_1 c_1 \cos \alpha_1}{g} \tag{35}$$

This equation represents the theoretical maximum head for a specified set of operating conditions. Note that the liquid entering the pump usually moves along the impeller in the radial direction. In this case, the angle between the absolute velocity value of the liquid entering the impeller and the tangential velocity is $\alpha_1 = 90°$, which corresponds to the liquid entering the impeller without any shock. Equation 35 simplifies to

$$H_T = \frac{u_2 c_2 \cos \alpha_2}{g} \tag{36}$$

From Figure 14

$$c_2 \cos \alpha_2 = u_2 - w_2 \cos \beta_2$$

Hence,

$$H_T = \frac{u_2^2}{g}\left(1 - \frac{w_2}{u_2} \cos \beta_2\right) \tag{37}$$

From the width of the impeller, b, the length of the circumference, $2\pi r_2$, and the cross section of the flow leaving the impeller, $2\pi r_2 b$, the quantity of liquid being pumped is

$$V = 2\pi r^2 b w_2 \sin \beta_2 \tag{38}$$

Hence

$$w_2 \cos \beta_2 = \frac{V}{2\pi r_2 \tan \beta_2} \tag{39}$$

Substituting Equation 39 into Equation 37 results in the following relationship between the head, H, and the volumetric rate of flow through the pump, V:

$$H = \frac{u_2^2}{g} - \frac{V}{g(2\pi r_2 b) \tan \beta_2} \tag{40}$$

For a given speed of rotation there is a linear relation between the head developed and the rate of flow. This is illustrated for different vane configurations in Figure 15.

If the outlet vane angles are inclined backwards, β_2 is less than 90°; hence, $\tan \beta$ is positive, and therefore the head decreases as the throughput increases (curve A, Figure 15). If β_2 is greater than 90°, i.e., the outlet vane is inclined forwards, the head increases at higher throughputs (curve B, Figure 15). Radial vanes provide a constant head (curve C, Figure 15). When the flow is zero ($V = 0$), then regardless of the vane angle our head expression is

$$H_T = \frac{u_2^2}{g} \tag{41}$$

Figure 15 shows that the maximum theoretical head achievable in a centrifugal pump is with vanes curved in the direction of rotation of the impellers, whereas the minimum occurs with vanes curved

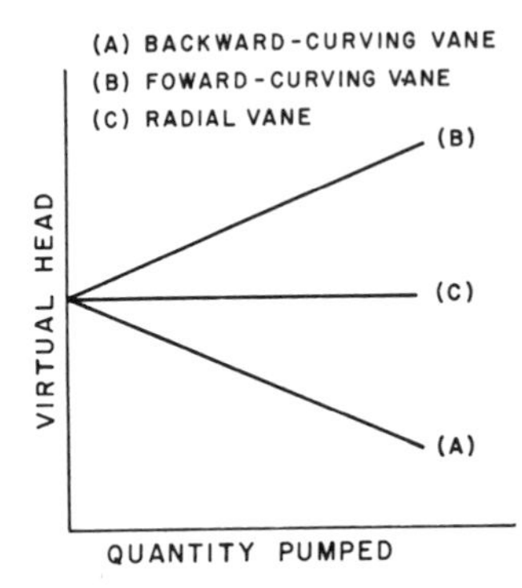

Figure 15. Plot of virtual head versus flow capacity for different outlet vane angles.

in the opposite direction. However, pumps are fabricated with angles $\beta_2 < 90°$ because an increase in β increases hydraulic losses and decreases the hydraulic pump efficiency.

The actual head is always less than virtual head for the following reasons:

1. The fluid circulating in the spaces between the vanes forms eddies.
2. Frictional losses occurring in the suction port, the impeller, and the discharge nozzle increase with pump speed and liquid viscosity.
3. Losses occur as the liquid is discharged from the impeller. The vane angles are correct only for the designed head and throughput. Deviation from these conditions causes an increase in the losses due to turbulence.
4. Leakage reduces the head developed, especially at low discharge rates.

The actual head is equal to

$$H = H_T \eta_h \epsilon \tag{42}$$

where η_h is the hydraulic efficiency (with typical values between 0.8 and 0.95), and ϵ is a coefficient that accounts for the number of vanes, ($\epsilon = 0.6$–0.8).

Referring back to Figure 14, the throughput, Q, corresponds to the liquid discharge through the channels between the impeller vanes having widths b_1 and b_2:

$$Q = b_1(\pi D_1 - \delta z')c_{1,r} = b_2(\pi D_2 - \delta z')c_{2,r} \tag{43}$$

where δ = vane thickness
z' = number of vanes
b_1, b_2 = width of impeller at the internal and external circumferences, respectively
$c_{1,r}, c_{2,r}$ = radial components of absolute velocities at the inlet and outlet of the impeller ($c_{1,r} = c_1$)

The output and head of a centrifugal pump depend on the number of revolutions per unit of time made by the impeller. From Equation 43, throughput is directly proportional to the radial component of the absolute velocity at the exit from the impeller, i.e., $Q \propto c_{2,r}$. If the number of revolutions changes from n_1 to n_2 (thus changing from Q_1 to Q_2), the trajectories of the motion of liquid particles remain unaltered and the velocity parallelograms at any corresponding points will be geometrically similar, as illustrated in Figure 16. Consequently,

$$\frac{Q_1}{Q_2} = \frac{c'_{2r}}{c''_{2r}} = \frac{u'_2}{u''_2} = \frac{\pi D_2 n_1}{\pi D_2 n_2} = \frac{n_1}{n_2} \tag{44}$$

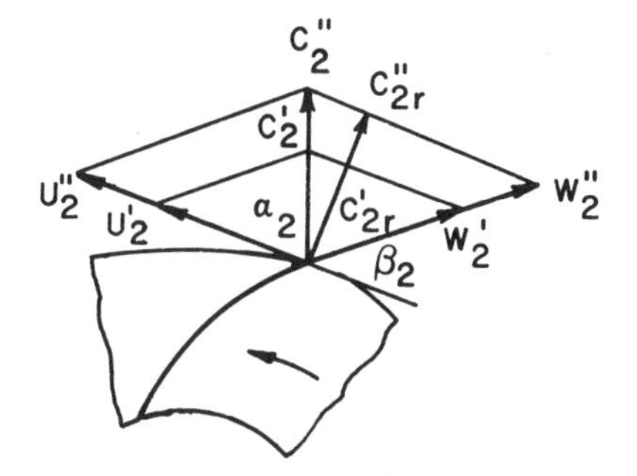

Figure 16. Shows how the similarity of the velocity parallelograms is maintained when the number of impeller revolutions is changed.

From Equation 37, the head is proportional to the square of circumferential velocity, i.e.

$$\frac{H_1}{H_2} = \left(\frac{u_2'}{u_2''}\right)^2 = \left(\frac{n_1}{n_2}\right)^2 \quad (45)$$

The power developed by the pump is proportional to the product of volumetric flow rate, Q, and head, H. From Equations 2, 44, and 45, we have

$$\frac{N_1}{N_2} = \left(\frac{n_1}{n_2}\right)^3 \quad (46)$$

Equations 44 through 46 are the *equations of proportionality* for centrifugal machines which state the following pump laws:

1. A change in the number of impeller revolutions (from n_1 to n_2) causes the pump throughput to change in a manner that is directly proportional.
2. The heads of the two systems are proportional to the number of revolutions raised to the second power.
3. The powers developed by pumps are proportional to the number of revolutions to the third power.

In a practical sense, these relationships do not hold rigorously. The proportionality between pump parameters is not strictly maintained when the number of revolutions is changed by more than a factor of two.

The preceding theoretical treatment assumes the absence of friction, flow turbulence, and various mechanical losses. In real pumps, these complications exist and influence pump performance. Now we shall examine commercial pump behavior, with attention given to backward-curved vanes. This type impeller configuration is used most extensively. (For discussions on pump performance with forward-curved vanes, see Metzner [6] and Cheremisinoff [4].) Chapters 35 and 36 treat losses and methods of sizing actual pumps; however, a brief introduction to these subjects is warranted here.

The main parameters that reduce the virtual head of a centrifugal pump (using backward-curved vanes) to its developed head are summarized in Figure 17. The dashed line represents the "virtual head," and thus defines the theoretical characteristics of this type of impeller, ($\beta_2 < 90°$). Physically, the virtual head line represents an infinite number of vanes and only infinitesimal amounts of liquid exist between them. In an actual pump, finite volumes of liquid exist between the vanes giving rise to inertia effects. Consequently, the portions of liquid near the impeller periphery have a greater velocity than inside the impeller. This creates circulation across the space between the vanes. Hence, inertial effects cause the virtual head curve in Figure 17 to shift downward.

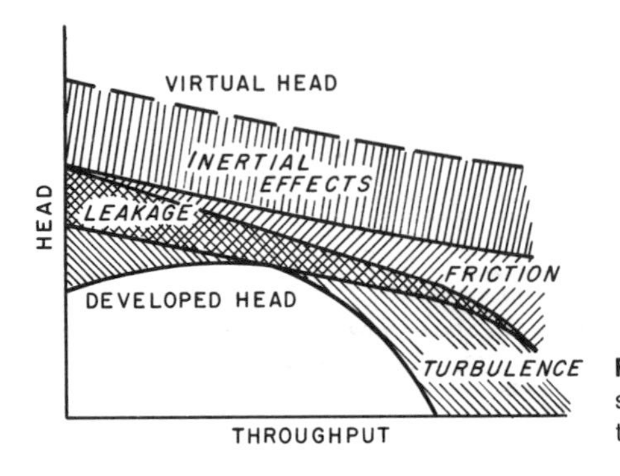

Figure 17. The head versus throughput plot summarizes the major factors that reduce virtual head.

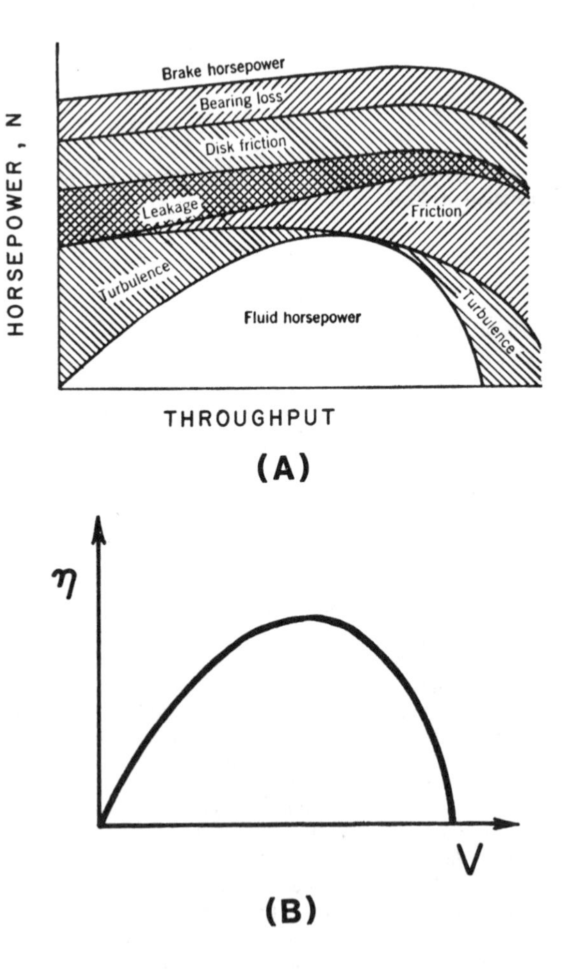

Figure 18. (A) Plot of pump horsepower vs. throughput, summarizing major factors affecting pump power; (B) typical curve of mechanical efficiency.

We shall assume that the liquid velocity through the pump increases while the same number of impeller rotations remains fixed. This can be accomplished by increasing a valve opening head of the pump, then the head in this case will decrease because of an increase in frictional resistance. These losses are compensated for partially by a decrease in leakage, which at low throughputs is larger than at a high throughput. As shown in Figure 17, this counterbalancing effect is not that significant. The last factor noted in Figure 17 responsible for decreasing head is turbulence or eddy formation on the vanes. All these factors result in the "developed head curve," which approaches zero on attaining a maximum throughput.

For the ideal pump, a maximum volumetric throughput corresponding to zero head, $H = 0$, exists for a constant, n. Conversely, a maximum head exists for the case of no flow. In real pumps, however, there is always some head at the maximum flow and some small flow at the pump's maximum head. The behavior of real pumps between these extremes is best explained by performance curves. Note that the product of the developed head (in units of pressure) and volumetric flow rate represents the power absorbed by the pumped fluid. Because the head approaches zero at the maximum flow rate, power first increases from zero, i.e., at $V = 0$, to a maximum and then decreases to zero at a maximum volumetric flow rate. This is illustrated by the pump horsepower versus throughput curve in Figure 18.

The power required to drive the pump is also used to overcome all the losses in the system and to supply the energy added to the liquid. Losses include frictional losses at the impeller as well as turbulent losses; the disk friction (or energy required to rotate the impeller in the fluid); leakage from the periphery back to the eye of the impeller; and mechanical friction losses in various pump components, such as bearings, stuffing boxes and wearing rings. The sum of all these power consumption items produces the final brake horsepower curve shown in Figure 18. As shown, brake horsepower is required even when the volumetric flow rate is zero. With increasing flow rate, brake horsepower increases even when the head is zero; in this case, the flow rate will be at a maximum.

An additional cross plot can be obtained from this last figure by dividing the fluid horsepower by the brake horsepower values (the definition of mechanical efficiency of a pump) and plotting efficiency versus throughput, as shown in Figure 18B. All three performance curves head (H-V) (Figure 17), power (N-V) (Figure 18A), and efficiency (η-V) (Figure 18B), are combined into a single diagram (Figure 19A or B). Figure 19A and B are illustrative only, and manufacturers should always be consulted for specific performance data.

The plot shown in Figure 19A contains several curves corresponding to different impeller diameters for a specific type of machine. Also shown are several lines of constant brake horsepower and constant efficiency, whose paths could be predicted from Figure 18A and B. Such a diagram provides information on the characteristics of a pump for a definite head at a specified liquid flow rate. By specifying coordinates, H, V, we can interpolate among the curves of brake horsepower, impeller diameter, and efficiency to obtain all characteristic values of a pump under consideration.

It is important also to evaluate the influence of the number of impeller revolutions on pump performance. From Equation 40 we note that an increase in the number of rotations, n, is accompanied by an increase in head at a constant flow rate. This is illustrated graphically in Figure 20, showing the influence of number of impeller rotations on head (H-V), brake horsepower (N-V), and efficiency (η-V).

Higher viscosity translates to higher resistance to flow and consequently greater frictional losses. Hence, we may expect a decrease in the head, an increase in brake horsepower, and a decrease in efficiency. Typical curves are shown in Figure 21.

In selecting a pump, it is necessary to consider the entire pump system's characteristics, i.e., the arrangement of piping, fittings, and equipment through which liquids flow. The characteristics of a pumping system express the relationship between flow rate, Q, and head, H, needed for liquid displacement through a given arrangement. Head, H, is the sum of the geometric height, H_g, and head losses, h (see Equation 14). For $V_{sec} = WS$ and $h_\ell = \sum \xi(w^2/2g)$ and denoting $V_{sec} = Q$, we determine that the head losses are proportional to the square of the flow rate:

$$h_\ell = KQ^2 \tag{47}$$

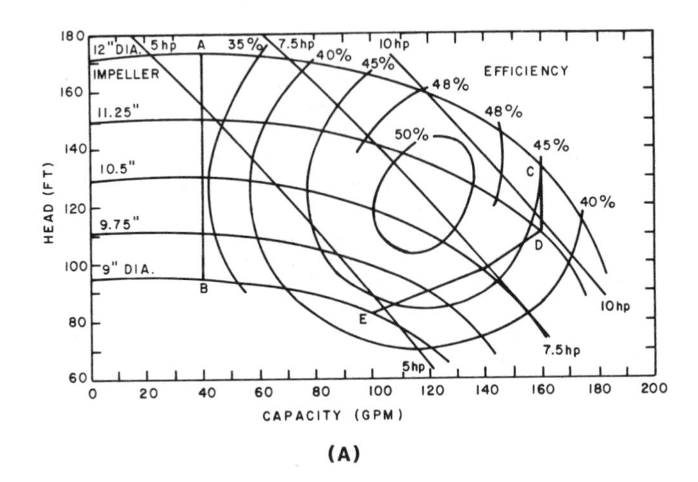

(A)

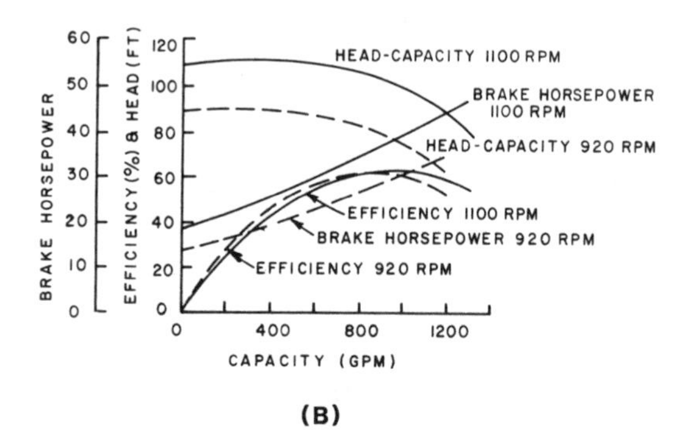

(B)

Figure 19. (A) Total characteristics of a centrifugal pump; (B) plots of head, efficiency, and brake horsepower as functions of capacity.

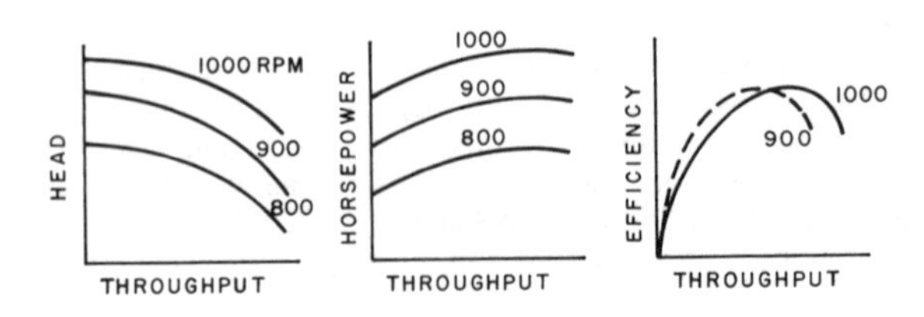

Figure 20. The influence of the number of impeller rotations on pump performance.

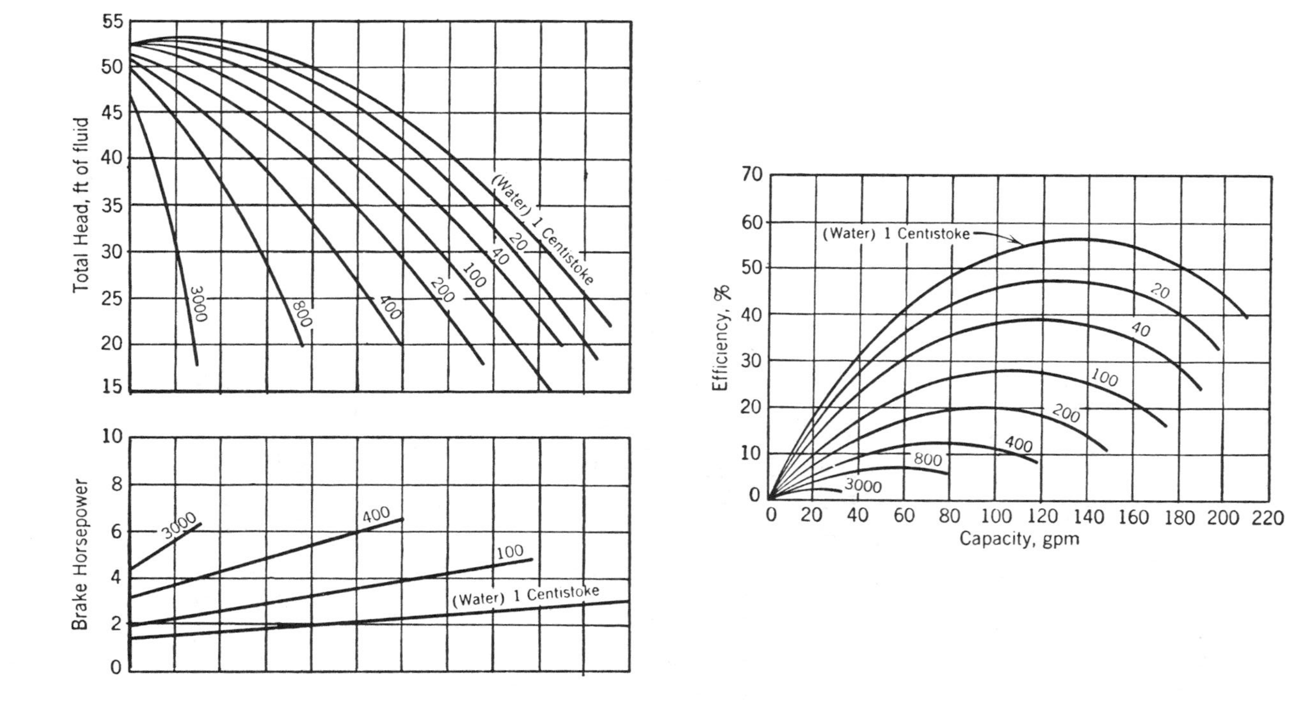

Figure 21. Characteristic curves of a typical centrifugal pump for liquids of different viscosities.

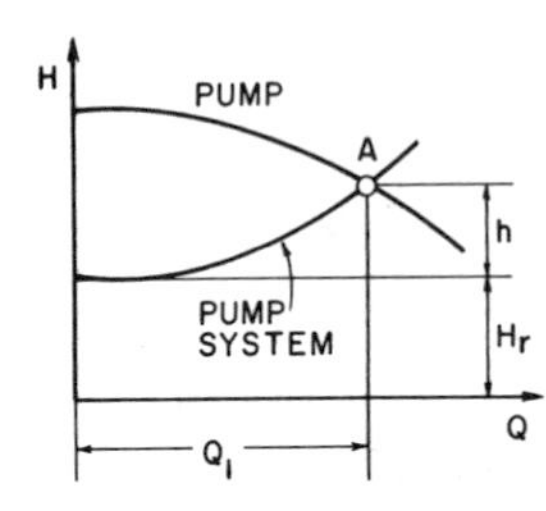

Figure 22. Pump and pump system characteristics represented on a single plot.

where K is a coefficient of proportionality. Hence, the characteristics of a pump system may be expressed by the following parabolic expression:

$$H = H_g + KQ^2 \tag{48}$$

The characteristics of both the pump system and the pump can be represented on a common plot, as shown in Figure 22. Point A (the intersection of both characteristics curves) represents the *operating point* and corresponds to the maximum capacity of the pump, Q, while operating for a pump system. If a higher capacity is required, it is necessary either to increase the number of motor rotations or to change to a larger pump. Increased capacity also can be achieved by decreasing the hydraulic resistance of the pump system, h_ℓ. In this case, the operating point is displaced along the pump characteristics towards the right. Hence, a pump should be selected so that the operating point corresponds to a desired head and capacity.

The high-speed coefficient (also referred to as the specific number of revolutions, n_s) is the number of revolutions of geometrically similar models of an impeller, which at the same efficiency and capacity of 0.075 m^3/sec has a head of 1 m. The high speed is a basic characteristic of a series of similar pumps having equal angles, α_2 and β_2, and coefficients, ϵ and η_h. The high-speed coefficient is expressed by the following relationship:

$$n_s = \frac{3.65n\sqrt{Q}}{\sqrt[4]{H^3}} \tag{49}$$

where n = the number of rotations (min^{-1})
Q = the maximum pump capacity (m^3/sec)
H = the total pump head (m)

From Equation 49, the high-speed coefficient, n_s, increases with increasing capacity and decreasing head. Therefore, low-speed impellers generally are used for obtaining higher heads at low capacities and high-speed impellers are used for creating high capacities at low heads. The impellers are divided into three groups, depending on the value of the high-speed coefficient. Ranges of values are summarized in Table 7.

Additional discussions on centrifugal pumps are given by Perry [1], Karassik [2], Stepanoff [3], Church [3], Cheremisinoff [4, 5], Hicks and Edwards [9], Brown [10] and Karassik and Carter [11].

Table 7
Range of Values for the High-Speed Pump Coefficient

Type Speed	n_s
Low	40–80
Normal	80–150
High-Speed	150–300

POSITIVE DISPLACEMENT PUMPS

Positive displacement pumps operate on the principle of forcing a fixed volume of liquid from the inlet pressure zone into the discharge zone of the pump. The two basic types of positive displacement pumps are *reciprocating* and *rotary*. Whereas the total dynamic head developed by a centrifugal pump is determined uniquely by the speed at which the impeller rotates, a positive displacement pump ideally will produce whatever head is impressed on it by the restrictions to flow on the discharge side.

Reciprocating pumps produce pulsating flow, developed high shutoff or stalling pressure, display constant capacity when motor driven and are subject to vapor binding at low NPSH (net positive suction head) conditions. There are three classes of reciprocating pumps: piston pumps, plunger pumps, and diaphragm pumps.

The piston pump consists of a cylinder with a reciprocating piston connected to a rod that passes through a gland at the end of the cylinder. The basic design is illustrated in Figure 23A and the operating principle is outlined in Figure 23B. The liquid suction and delivery in the pump are derived from the reciprocating motion of the piston (1) in the pump cylinder (2) of Figure 23B. When the piston moves towards the right, a vacuum develops in the closed space between the head (3) and the piston. Due to the pressure gradient between the suction tank and cylinder, the liquid is lifted through the suction piping and enters the cylinder via the suction valve (4). The delivery valve (5) is closed during the piston's motion to the right as it is subjected under liquid pressure from the suction piping. When the piston moves towards the left, the pressure generated in the cylinder closes one valve (4) and opens another (5). The liquid passing through the delivery valve enters into the discharge piping and then into the tank. Thus, both liquid suction and delivery by a single-acting piston pump vary with time because of the periodic motion of the piston. The volume delivery starts from zero at the instant the piston begins to move forward, reaches a maximum when it is fully accelerated at approximately the mid-point of its stroke, and then gradually falls off to zero. There will be a short time interval during the return stroke when liquid fills the cylinder and the delivery remains at zero. This is illustrated by the plot shown in Figure 24.

The piston shown in Figure 23B is driven by a crankshaft assembly (6), which converts the rotary motion of the drive wheel to the back-and-forth linear movement of the piston rod. Piston pumps are classified as single-acting and double-acting. In a single-acting piston pump, a single revolution of the crankshaft provides suction and delivery; in a double-acting pump, two strokes of a

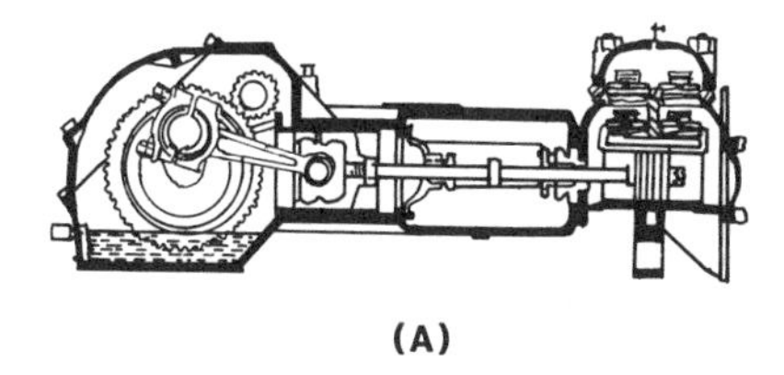

(A)

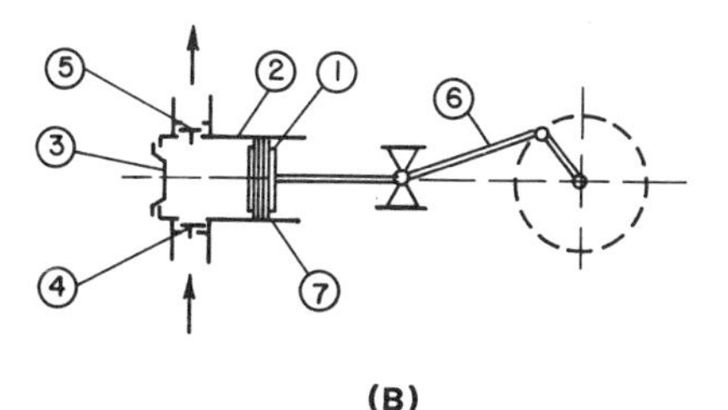

(B)

Figure 23. (A) Basic design of the piston pump; (B) operating principle for a horizontal, single-acting piston pump: 1—piston; 2—cylinder; 3—head of cylinder; 4—suction valve; 5—delivery valve; 6—crank and connecting rod assembly; 7—seal rings.

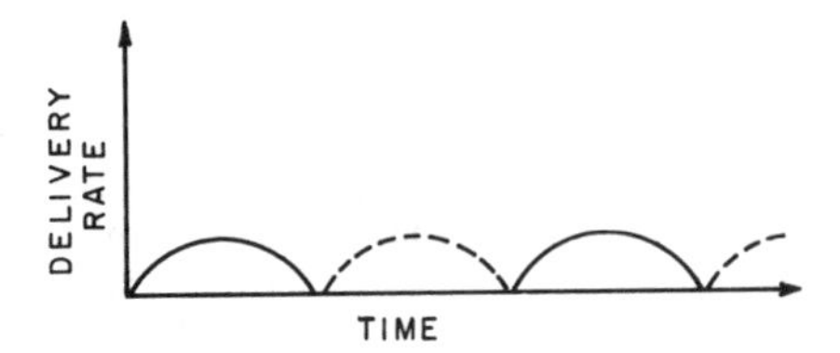

Figure 24. Liquid delivery rate from a simplex pump.

piston effect delivery/suction. As noted in Figure 23B, the main working component of a piston pump is the piston itself, consisting of disks and seal rings (7) housed inside a polished cylinder.

A *plunger pump* differs from a piston pump in that it has a plunger reciprocating through packing glands, causing the displacement of liquid from the cylinder. In this type pump, considerable radial clearance exists between the plunger and cylinder walls. Plunger pumps usually are thought of as single-acting in the sense that only one end of the plunger is used for pumping liquid. Figure 25 provides details of the operating principle of a single-acting, horizontal plunger pump. As shown, a piston serves the role of the plunger (1), reciprocating in the cylinder (2) and sealed with a packing gland (3). The internal cylinder surface of a plunger-type pump does not require as high a degree of finishing as does a piston pump. Also, leakage is readily minimized by tightening or replacing the packing gland. Plunger pumps are well suited to handling suspensions and viscous fluids because of their large radial clearances and ability to generate high pressures.

A more even delivery is achieved with piston and plunger pumps when operation is converted to double-acting. The horizontal, double-acting plunger pump (Figure 26) may be considered to

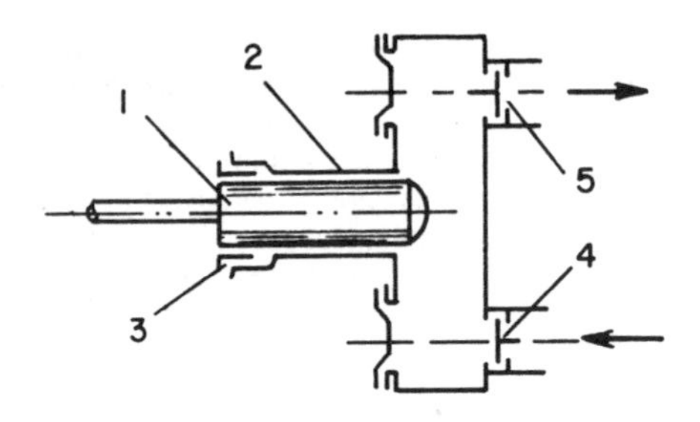

Figure 25. Operating principle behind a horizontal, single-acting plunger pump. 1—plunger; 2—cylinder; 3—packing gland; 4—suction valve; 5—delivery valve.

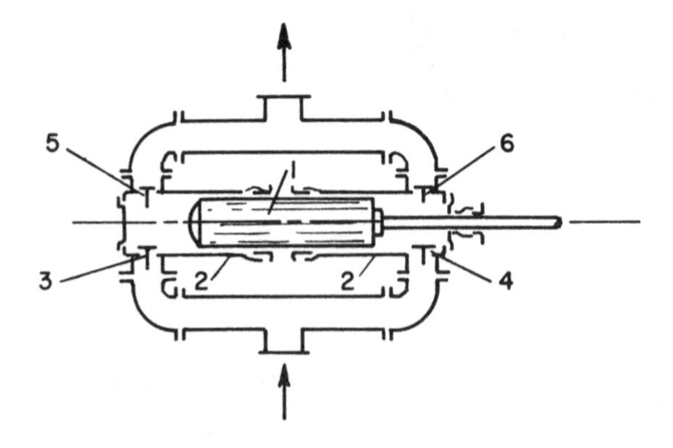

Figure 26. Operating principle behind a horizontal, double-acting plunger pump: 1—plunger; 2—cylinders; 3, 4—suction valves; 5, 6—delivery valves.

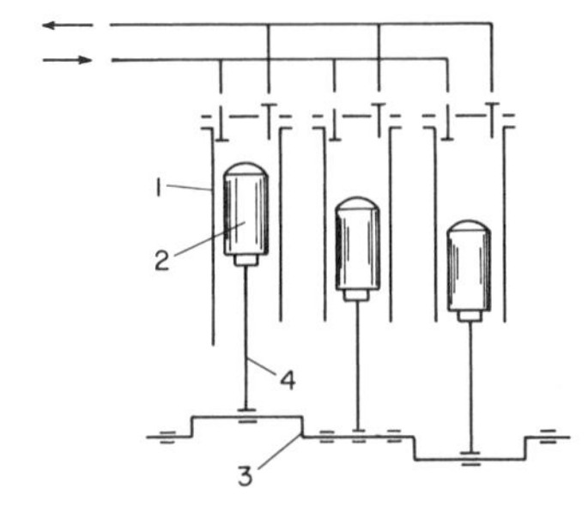

Figure 27. Operating principle of a triplex pump: 1— cylinders; 2—plungers; 3—crankshaft; 4—connecting rods.

be an aggregate of two single-acting pumps, having four valves (two suction and two delivery). When the plunger (1) moves toward the right of Figure 26, liquid enters into the cylinder (2) through the suction valve (3), while liquid passes through the delivery valve (6) into the discharge piping. During the reverse stroke of the plunger, suction occurs in the right side of the cylinder through the suction valve (4) and delivery takes place through the valve (5). Hence, with a double-acting pump, suction and delivery takes place during each piston stroke. Consequently, the capacity of these pumps is greater and delivery is smoother than with single-acting designs.

The operation of a three-plunger pump (triplex pump) is illustrated in Figure 27. These designs are widely used because of the pulsation-free flow rate they provide, as well as for their smooth torque at maximum stroke frequencies, which can exceed 1,500 strokes per minute (spm). The triplex pumps are single-acting tripled pumps having cranks located 120° from each other. The total triplex-pump delivery is the sum of the deliveries of the individual single-acting pumps. For one rotation of the crankshaft the liquid is suctioned and delivered three times. Figure 28 shows a cutaway view of a triplex plunger pump with an integrated gear drive. Figure 29 shows an actual triplex plunger pump.

Piston and plunger pumps may be actuated directly by steam-driven pistons or by rotating crankshafts through a cross head. The direct-acting steam pump consists of a steam cylinder end in line with a liquid cylinder, with a straight rod connection between the steam and pump pistons or plunger.

Direct-acting steam pumps are available as simplex (one steam and liquid cylinder) and duplex (dual side by side) units. Duplex units are employed in large-capacity services and for reducing the flow pulsations below that of the simplex. Dual pumps are designed with an interconnecting steam valve linkage arrangement so that one side pumps when the other side reaches the end of its stroke. Steam pumps consist of a rod and piston design and are double acting; that is, each side pumps on every stroke. Consequently, a duplex pump will have four pumping strokes per cycle.

Power pumps convert rotary motion to low-speed reciprocating motion via speed-reduction gearing, a crankshaft, connecting rods and cross-heads. Plungers or pistons are driven by the cross-head drives. Rod and piston construction, similar to that in duplex, double-acting steam

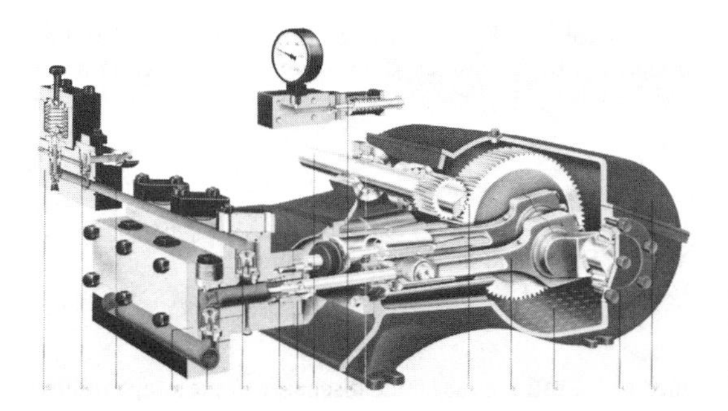

Figure 28. Cutaway view of a triplex plunger pump with integrated gear drive. (Courtesy American LEWA Inc.)

Figure 29. A triplex plunger pump with variable-speed drive. (Courtesy American LEWA Inc.)

pumps, is used by the liquid ends of the low-pressure, higher-capacity units. The higher-pressure units are normally single-acting plungers. This latter style generally employs three (triplex) plungers. Three or more plungers substantially reduce flow pulsations relative to simplex and even duplex pumps.

In general, the effective flow rate of reciprocating pumps decreases as viscosity increases because the speed must be reduced. High viscosity also leads to a reduction in pump efficiency. In contrast to centrifugal pumps, the differential pressure generated by reciprocating pumps is independent of fluid density. Rather, it is dependent entirely on the magnitude of force exerted on the piston.

Reciprocating pumps are used most often for sludge and slurry services, particularly where other pump types are inoperable or troublesome. Maintenance in such services tends to be high because of valve, cylinder, rod, and packing wear.

The theoretical delivery of a piston pump is equal to the total volume swept out by the piston in the cylinder times the number of strokes of the piston per unit time. Thus, the theoretical delivery in a single-action pump is

$$Q_{th} = F \cdot S \cdot n \tag{50}$$

where F = piston cross-sectional area
S = stroke length
n = number of crankshaft revolutions (or number of double strokes)

In the double-acting pump, there are two suctions and two deliveries per crankshaft revolution. When the piston moves to the right of Figure 26, the liquid volume sucked from the left side is equal to FS and that delivered from the right side is (F − f)S, where f is the rod cross-sectional area. When the piston moves to the left, the volume FS is delivered from the left side into the delivery piping; from the right side, the liquid volume (F − f)S is sucked in from the suction line.

Consequently, for n rotations of the crankshaft the theoretical delivery of a double-acting pump will be

$$Q_{th} = FSn + (F - f)Sn = (2F - f)Sn \tag{51}$$

The actual delivery may be less than the theoretical value because of leakage past the piston and valves or because of inertia in the valves. The actual delivery of the pump is

$$Q = Q_{th}\eta_v \tag{52}$$

where η_v is the volumetric efficiency, defined as the ratio of the actual discharge to the swept volume.

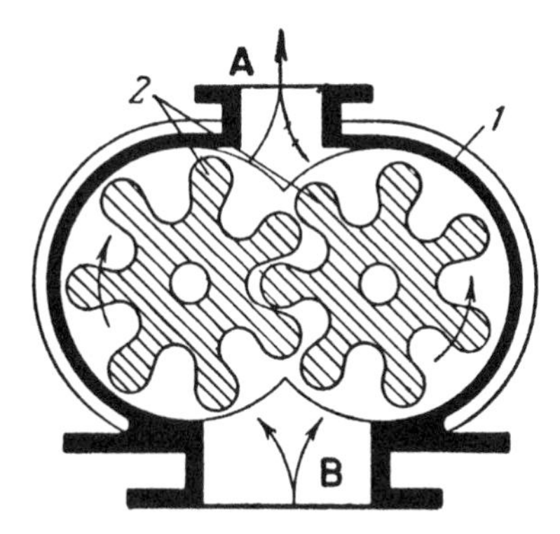

Figure 30. Operation of a gear-type rotary pump: 1—casing; 2—gear wheels.

The volumetric efficiency of large pumps is typically 0.97–0.99; for medium flow rates (Q = 20–300 m^3/hr), $\eta_v = 0.9$–0.95; and for small flow rates, $\eta_v = 0.85$–0.9.

In some cases, however, the actual delivery may be greater than the theoretical value because of fluid momentum in the delivery line and sluggishness in the delivery valve operation resulting in continued flow.

Rotary pumps combine the rotating movement of the working parts with the positive displacement of the fluid. There are a variety of rotary pumps, but all operate on basically the same principle.

The rotating parts move in relation to the casing to create a space that first enlarges, drawing in the fluid in the suction line, is sealed, and then reduces in volume, forcing the fluid through the discharge port at a higher pressure. The rotating elements of the pump generate a reduced pressure in the suction line, thereby allowing the external pressure to force liquid into the pump. The capacity of a rotary pump is a function of its size and speed of rotation. This type of pumping equipment provides near constant deliveries in comparison to the fluctuating flows of reciprocating pumps. The main reason for selecting rotary pumps over centrifugals is to take advantage of their high-viscosity handling capability. In addition, rotaries are simple in design and efficient in handling flow conditions that generally are considered too low for economic application of centrifugals. Rotary pumps operate in moderate pressure ranges and have small to medium capacities.

Rotary pumps may be divided into five main types according to the character of the rotating parts: gears, screw, lobe, cam, and vane. There is, however, considerable overlap among these types.

The simplest rotary pump is the *external gear pump*, illustrated in Figure 30. Two gear wheels (2) operate inside the casing (1), which provides a snug fit to effectively seal the spaces between each pair of adjacent teeth. One of the gear wheels is driven by the driver and the other rotates in mesh, as indicated by the arrows. As the spaces between the teeth of the gear wheel pass the suction opening "A," liquid is impounded between them, carried around the casing to discharge opening "B" and then delivered outside through the opening. The straight teeth in gear wheel pumps produce pulsations in the delivery with a frequency equivalent to the product of the number of teeth on both gear wheels and the speed of rotation. Pulsation can be eliminated by the use of gear wheels having *helical* teeth with a particular angle.

Gear pumps handle liquids of a very high viscosity but cannot be used with suspensions. Generally, spacings between gear teeth are too close to handle solids without suffering significant erosion. The rotating parts in the casing create a space that draws in the liquid from the suction line. As the parts rotate, the liquid is trapped between them and the pump casing, forcing the liquid out through the delivery side of the pump at the necessary higher pressure. Typical performance characteristics of external gear pumps are shown in Figure 31.

Internal gear pumps are of two principal types. Both kinds employ a modified spur gear, which rotates inside a larger gear rotating around an axis that is parallel to that of the spur gear and displaced somewhat to mesh snugly at one point on the periphery. Figure 32 shows one internal gear pump design whose two gears maintain a continuous series of sliding seals with each tooth.

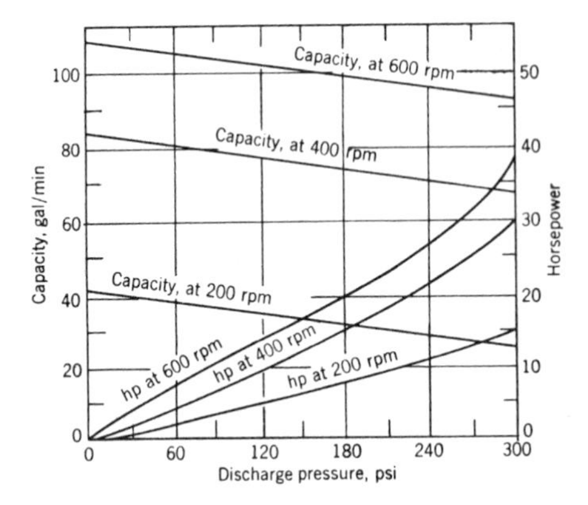

Figure 31. Typical performance characteristics of an external gear pump.

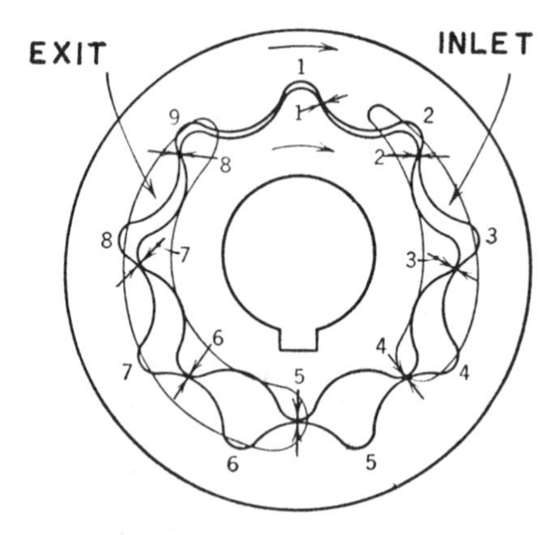

Figure 32. An internal gear pump with sliding seal.

This action forms pockets of fluid entirely between the gears. This type of pump can be equipped with any number of teeth in the spur gear (provided one more socket is included in the ring gear).

Screw pumps are modified helical gear pumps. They may have one, two, or as many as three screws turning along the pump axis, with liquid flowing between the screw threads and the casing. Both single-rotor and multiple-rotor screw pumps are commercially available. Pumping action is accomplished by progressing cavities, which advance the fluid along the rotating screw from inlet to outlet. This axial flow pattern minimizes vibration, producing a rather smooth flow.

A single-rotor screw pump is referred to as a progressive capacity pump and is used for handling slurries with relatively large particles. This type of pump is composed of a rotor that revolves within a stator, executing a compound movement. The rotor revolves about its axis while the axis itself travels in a circular path. Hence, the rotor is a true helical screw, and the stator consists of a double internal helical thread. With each complete revolution of the rotor, the eccentric movement allows the rotor to contact the entire surface of the stator. Voids between the rotor and stator contain entrapped fluid, which is moved continously toward the pump outlet. This pumping action provides continuous flow at low, smooth and uniform rates. The action minimizes fracturing of particles as well as abrasion damage to the pump. Single-screw pumps are used extensively in the food processing and chemical industries for handling solid/liquid mixtures that are abrasive or require gentle handling of the solids.

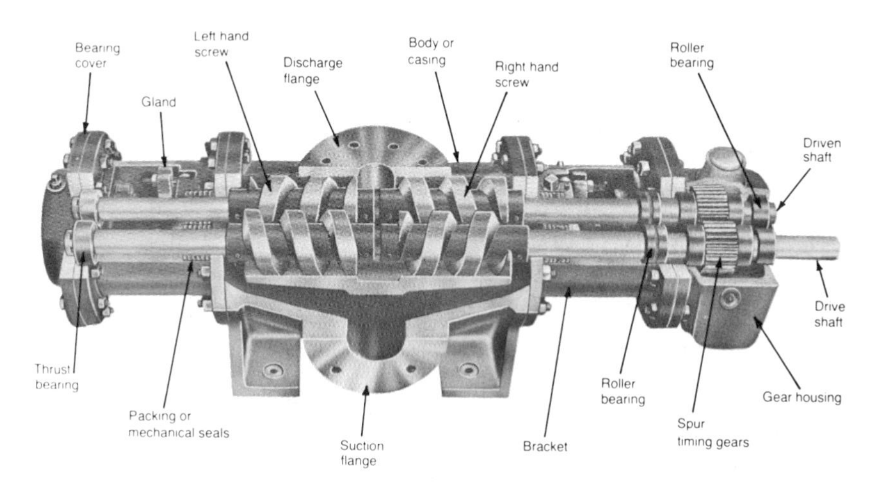

Figure 33. Cross-section of a twin-screw pump, showing main features. (Courtesy Worthington Pump Inc.)

Figure 33 shows a twin-screw pump consisting of two sets of screws that rotate and mesh in an accurately bored casing. Relatively tight operating clearances are maintained between the screws. This tight clearance usually is maintained by a pair of timing gears mounted on the shafts. The gears also transmit power from the drive shaft to the driven shaft. The body consists of a casing with two precision-machined bores, which house the rotating screws. Fluid passes from the inlet into the pumping chamber and then to the discharge zone.

Some screw pumps are operated without timing gears. In these units, a driver is used to turn the screws directly. The designs of twin-screw pump bodies and screw assembly arrangements are interrelated. Figure 34 illustrates typical body designs.

The operation of a three-rotor-screw pump is shown in Figure 35. The center rotor (1) is the driving member, while the other two (2) are driven. Liquid enters at the end of the rotors, is sealed between the rotors and casing, and forced smoothly to the delivery end.

The mechanical displacement of liquid from inlet to outlet is generated by trapping a slug of fluid in the helical cavity (referred to as a "positive lock"), created by the meshing of the screws.

Two- or three-lobe pumps are similar in design to gear pumps (Figure 36), but gear wheels are replaced either by two or three lobes, which are driven separately through an external gearing mechanism. This arrangement makes it possible to avoid actual contact of the lobes with each other. Wear on the lobes and casing can be minimized by maintaining a small clearance between them. The clearances are a few thousandths of an inch, sufficient to reduce friction and wear but maintain minimum leakage from the delivery to the suction side. The characteristics of the lobe pumps are generally similar to those of the gear pumps.

Cam pumps consist of an eccentrically mounted circular rotor that sweeps a circle whose radius is the sum of the radius of the rotor and the eccentricity. Figure 37 shows such a pump utilizing a plunger valve. The circular cam is fixed rigidly to the shaft and is housed inside the rotor ring, which is free to rotate about the cam. The plunger slides freely through a slide pin to act as a discharge valve. The plunger may be replaced by any form of vane, which seals the suction line from the discharge line. In general, contacts between surfaces are almost free of friction and wear except for the cam inside the rotor.

As the cam rotates it expels liquid from the space ahead of it and sucks in liquid behind it. The characteristics again are similar to those of a gear pump.

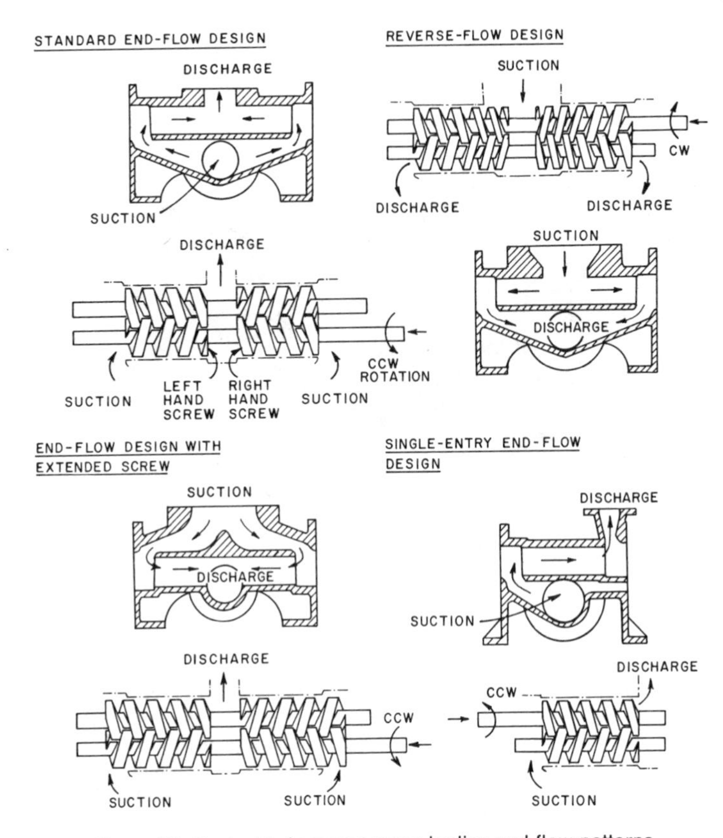

Figure 34. Typical twin-screw pump bodies and flow patterns.

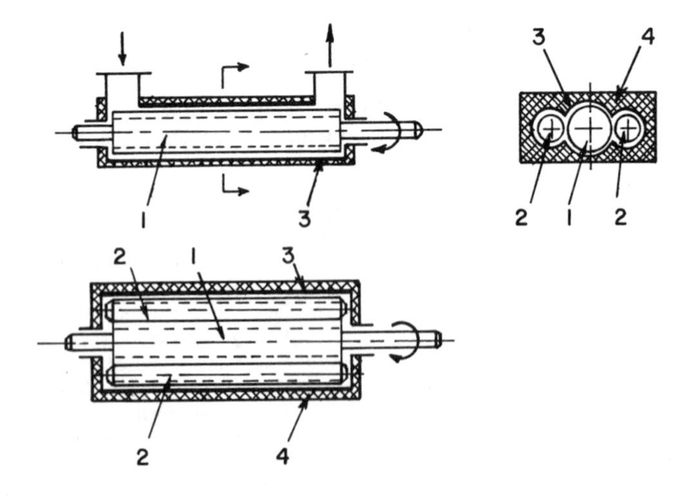

Figure 35. Operating scheme of a three-rotor screw pump: 1—driving screw; 2—driven screws; 3—sleeve; 4—casing.

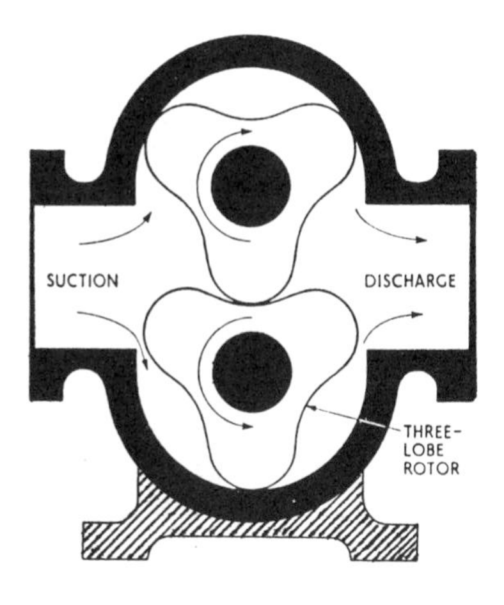

Figure 36. Sectional diagram of a three-lobe pump.

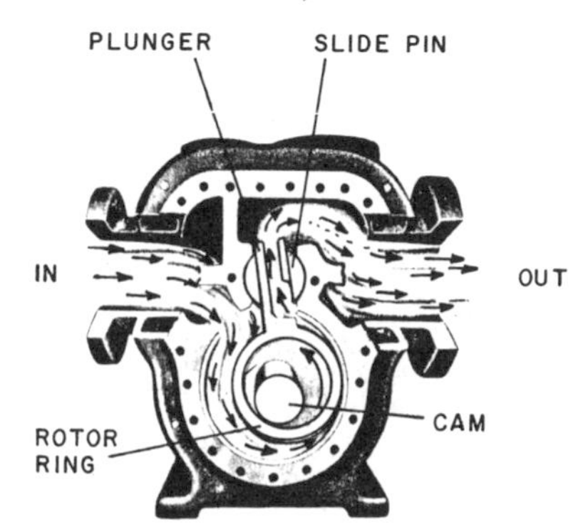

Figure 37. Cross-sectional view of a cam pump.

Vane pumps (Figure 38) include a massive cylinder (1), which carries rectangular vanes in a series of slots arranged at intervals around the curved surface of the rotor located eccentric to the casing (2). The tip of the vanes (3) is thrown outward by centrifugal force. As the cylinder rotates, the space behind a vane enlarges when it moves from the suction nozzle (5) to the vertical axis of the pump, resulting in the formation of a vacuum in space (4), thus drawing in liquid. At this point in the cycle liquid is trapped between the vanes. When the vane moves from the vertical axis in the direction of rotation, the volume of the chamber (space 4) decreases, and the liquid eventually is forced out through the discharge nozzle (6). Because of wear, the vane dimensions change; however, this is compensated for somewhat until the seal is broken. At that time new vanes must be inserted.

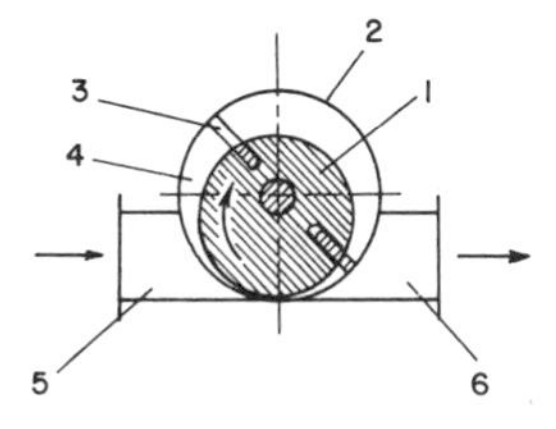

Figure 38. A sliding vane-pump: 1—rotor; 2—casing; 3—vanes; 4—working space; 5—suction nozzle; 6—delivery nozzle.

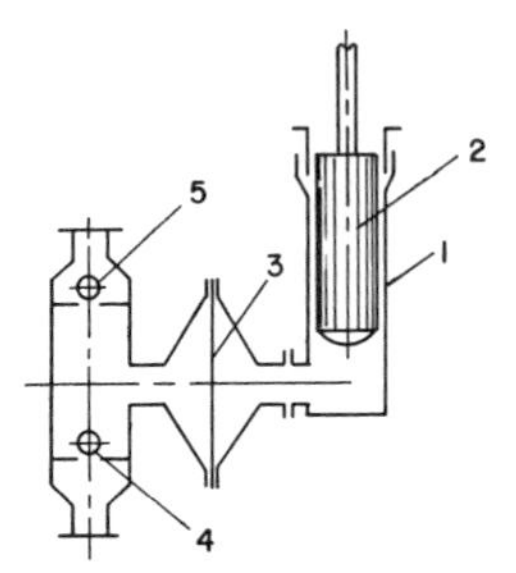

Figure 39. Operating scheme for a diaphragm pump: 1—cylinder; 2—plunger; 3—diaphragm; 4—suction valve; 5—delivery valve.

A *diaphragm pump* is a special type of positive displacement pump that operates by the periodic movement of a flexible diaphragm. It has the advantages of no stuffing boxes and high tolerance to abrasive slurries and chemically aggressive liquids. The operation is illuatrated in Figure 39. The plunger (1) operates in a cylinder (2) in which a non-corrosive liquid is displaced. The movement of the fluid is transmitted by means of the flexible diaphragm (3) made from soft rubber or a special steel. When the plunger moves upward, the diaphragm is bent to the right and liquid is sucked into the pump through the globe vale (4). When the plunger moves downward, the diaphragm is bent to the left and liquid is discharged to the delivery piping through the delivery valve (5). The parts of the pump that are in contact with the liquid to be pumped are protected against corrosion with corrosion- and erosion-resistant materials. Figure 40 shows a high-pressure diaphragm pump for handling large volumes of process chemicals. See Chapter 38 for a discussion of these types of pumps.

New and improved fully sealed diaphragm and bellows pumps have been developed to minimize leakage. Types range from micrometering pumps in the milliliter size to large diaphragm pumps having drives of several hundred kilowatts for high-pressure processes. These designs often achieve economic viability through reduction of maintenance and decontamination costs, and improvement of process reliability.

Operating limits and the advantages of various constructions are described in Table 8. Critical, toxic or abrasive media require the use of leakfree metering or production pumps, with a realistic upper size limit of about 300 kW.

A widely used construction utilizes a mechanical diaphragm drive (usually limited to less than 85 psi). The metering rate usually is limited to less than 200–500 charges/hr because the diaphragm loading becomes unfavorable with increasing diameter. The stroking rate also is restricted to less than 150 strokes/min because of acceleration shocks that occur at part-stroke settings with the widely used cam-and-spring-return or magnetic-drive units.

Figure 40. Diaphragm pump in horizontally opposed arrangement for high pressures up to 350 bar. (Courtesy American LEWA Inc.)

Table 8
Operating Limits and Applications of Various Metering Pumps

Pump Type	Limits	Application	Material of Wetted Parts
Low-pressure diaphragm pump with direct mechanical drive	<6–10 bar (85–140 psi), <500/h	Metering, pumping	PVC or austenitic stainless steel, elastomer diaphragm
Low-pressure bellows pump with direct mechanical drive	<5 bar (70 psi)	Metering, pumping	Glass; PTFE bellows
High-pressure micro-diaphragm metering pump with hydraulic drive	<700 bar (9,940 psi), <10/h	Metering	Acid-resistant steel
Diaphragm pump with hydraulic compression of tubular member	<50 bar (710 psi)	Metering, pumping	Acid-resistant steel; elastomer diaphragm and tube
Diaphragm pump with hydraulic drive	<350 bar (5,000 psi) for PTFE diaphragm <3,000 bar (42,600 psi) for metal diaphragm	Metering, pumping	PVC, PTFE, titanium acid-resistant steel

For a given membrane geometry, diaphragm life depends on stroke length, pressure, temperature, and compatibility with the fluid being processed. The detailed design is empirical, based on extensive fatigue trials. Diaphragm life easily can reach 3,000 hours under permissible maximum conditions and will considerably exceed this at lower pressures, shorter strokes, etc.

Diaphragm failure can be signaled by a float switch that responds to the presence of process fluid in the space behind the diaphragm. Another failure detection method involves two mechanically coupled diaphragms. A rise in pressure in the space between them indicates diaphragm failure and will cause simple sensors to trigger a warning signal.

Depending on the diaphragm geometry, the pumping characteristic is usually not quite linear. Metering accuracy is thus related to the tolerances of the operating conditions. The metering rate in relation to stroke frequency is linear.

Bellows-type metering pumps (Figure 41) are almost exclusively constructed of glass and polytetrafluorethylene (PTFE) components. This limits their pressure level to less than 5 bar (70 psi) because of the glass parts. The output capacity is virtually unlimited because there is no problem in producing fatigue-free bellows with diameters of several hundred millimeters.

Bellows require more care than do diaphragms during forming and manufacture, but normally achieve lives of 5,000–10,000 hours at maximum load. Because of radial stiffness of the bellows, the metering characteristic curve is purely linear and only slightly pressure dependent.

As in the case of the diaphragm pump, bellows failure is monitored via level or pressure sensors in the chamber below the bellows. This chamber is sealed from the drive by an auxiliary packing.

Plunger displacement by means of hydraulic fluid offers the benefit of uniform diaphragm support. The loading of the diaphragm then becomes solely the result of elastic deformation and not of resistance to pressure forces. Strain is reduced to a minimum, and operation becomes possible at very high pressures. Pressure is limited only by diaphragm fatigue strength under compression.

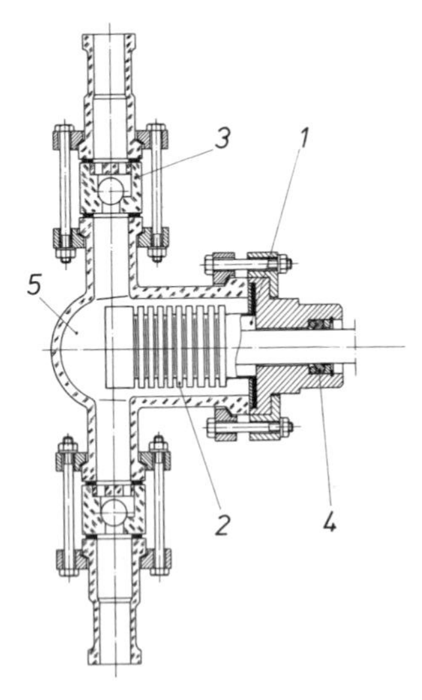

Figure 41. A bellows-type metering pump: 1—rupture chamber; 2—bellows; 3—check valve; 4—packed seal; 5—pump chamber.

Hydraulically driven pumps allow greater diaphragm displacements, but their operating pressure is restricted to about 5,000 psi and 120°C maximum by the fatigue limit under compression for PTFE and by clamping effects.

The largest number of applications can be met by compact diaphragm metering pumps. Stroke-adjustable drives operate the piston via a connecting link, the piston sliding in a liner tight enough to provide a seal. The lubrication and hydraulic systems are conventional. Plunger movement displaces the diaphragm, which can deflect between two perforated support plates. Topping up the space between plunger and diaphragm (the hydraulic space) is controlled at negative pressure by a vacuum-replenishing valve when the diaphragm has reached the lower support plate.

The hydraulic system also contains a relief valve for protection against excess pressure. This sometimes can take the place of an external safety valve that otherwise would be in contact with the pumped fluid. The upper support plate, together with this relief valve, protects the diaphragm against excessive deflection under certain process conditions. An automatic vent valve in the highest point of the hydraulic system provides a degassed hydraulic medium and thus, a faultless delivery characteristic. Compact diaphragm pumps with common oil systems generally require the use of sandwich diaphragms.

The pump is ideally suited to slurries of all types and cleaning-in-place in the food industry. Topping-up the hydraulic system is controlled via a gate that is diaphragm-position dependent and by a relief-and-snifter valve. The continuous vent can be seen at the highest point.

For services beyond the practical limits of PTFE, *metal diaphragms* are used made of a dead-parallel cold-rolled sheet. Pumps with metal diaphragms are especially suited to:

1. Pressures above 5,000 psi and temperatures up to 200°C (in special cases, 400°C maximum).
2. Microdiaphragm dosing pumps because of the low compressibility of diaphragm and clamping.
3. Applications where all-metal construction is mandatory, usually to ensure radiation resistance.

Because of the lower elasticity of metal diaphragms, the same stroke-displacement volumes require larger-diaphragm diameters and larger pump heads. For larger outputs, diaphragm pumps with metal diaphragms are more costly to manufacture than are pumps with nonmetallic diaphragms. In addition, because of the risk of notch formation and because of their lower deflection capability, thin metal diaphragms are not suitable for the unsupported type of diaphragm.

Important design considerations are an even-flow distribution through the support perforations and the prevention of local adhesion. Because of diaphragm durability and adhesion avoidance, the dimensioning of the perforations and the profiling of the diaphragm support surface cannot be readily calculated. They depend on experience gained through successful construction and installation of high-pressure diaphragm pumps.

In micrometering pumps for high pressures, the displacement volumes and the diaphragm pump head dimensions are very small. Therefore, the design parameters have a different rank of importance than for high-output pumps. The pump cavity is designed for minimum dead volume and maximum rigidity. For improved degassing, oil circulation in the hydraulic system has proved to be a satisfactory concept. This is achieved by plunger displacement, aided by two nonreturn valves.

Hydraulic-linkage pipes (also referred to as "pendulum" or "remote-head" systems) serve to remove high temperatures and other troublesome and dangerous influences from the drive, making it feasible to extend the field of application of diaphragm pumps (e.g., to 400°C). A familiar application is in the radioactively heated zone of atomic fuel recovery plants, where extreme demands are made for safety.

Loss in suction pressure (largely influenced by the suction valve) is much the same for diaphragm and plunger pumps. In particular, types having unrestricted forward chambers experience no additional internal pressure losses on the process fluid side. However, at high stroking frequencies, in excess of 300 strokes/min, it is essential that internal flow patterns be optimized.

Some types of diaphragm pumps require minimum suction pressures of 7–42 psi. These include pumps with metal diaphragms for larger output and faster stroking rates. Metal diaphragms with their lower stiffness are more sensitive to cavitation than are the heavier PTFE diaphragms.

Cavitation can be minimized by proper layout of the pipe system and pressure conditions. Upsets in metering accuracy and performance owing to cavitation can be tremendous. Delayed compression, with cavitation present, leads to sizable pressure and loading shocks.

Because of elastic influences, the instantaneous delivery flow is not in phase with plunger displacement. Hence, the delivery often starts jerkily at the end of the compression phase (final plunger speed) with a corresponding shock wave as the result. This shock problem, which occurs in all high-pressure pumps, can be overcome by shock dampers that have been sized in line with pulsation theories. Further discussions are given in Chapter 38.

MOTIVE FLUID PUMPS

In addition to mechanically displaced pumping systems, there are a class of pumps which operate on the principle of liquid displacement by a secondary fluid. This class includes jet pumps, acid eggs, hydraulic rams, and airlifts.

A *jet pump* (Figure 42) takes advantage of the momentum of a high-velocity secondary fluid stream (steam or water), referred to as the "working fluid," to impart momentum to the fluid to be pumped. In this operation, both streams are actually mixed. The working fluid (I) enters at a high velocity from the nozzle (1) through the mixing chamber (2) into the diffuser (3), thus entraining (because of surface friction) the liquid to be pumped (II). In the narrowest section of the diffuser

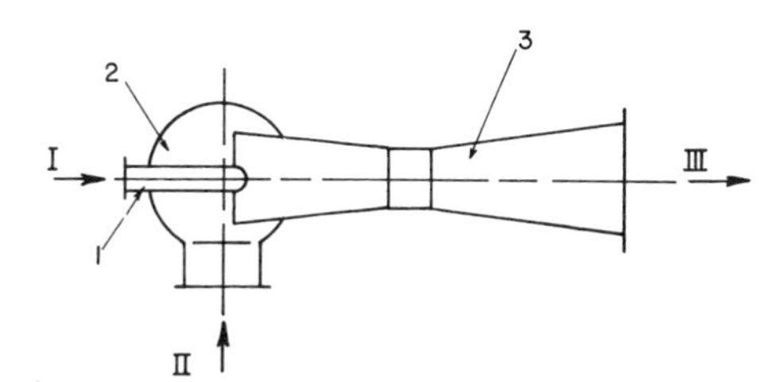

Figure 42. Operating principle of a jet pump.

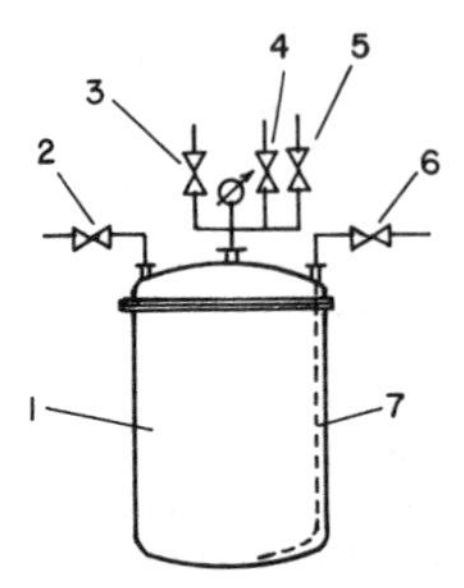

Figure 43. Operation of an acid-egg pump: 1—container; 2, 6—valves; 7—outlet pipe.

(having a geometry similar to a venturi), the velocity of the mixture (working and pumped fluids) reaches a maximum value, and the static pressure of the flow becomes minimum according to Bernoulli's equation. The pressure drop in the mixing chamber and the diffuser provides the delivery of liquid (II) in the mixing chamber from the suction line. In the expanding section of the diffuser the flow velocity is decreased and the kinetic energy of the flow converted into potential pressure energy. Hence, the liquid under pressure enters the delivery line.

The *acid egg pump* or *blowcase pump* (Figure 43) consists of a horizontal or vertical container (1) filled with the liquid to be pumped by means of gas pressure. In pumping liquids using kinetic, potential, or pressure energy as a motive force, the egg operates by displacement of one fluid by another. As with the jet pump, this design has no moving parts and can be operated simply. Liquid enters the egg from the feed pipe through the open valve (2); normally the vent valve (3) is open if filling is done under atmospheric pressure. Liquid also may enter through the valve (4) if filling is to be done under vacuum. In displacing the liquid, some valves (2–4) are closed and valves (6) on the discharge pipe (7) on the line of compressed gas (5) are opened. After the egg is discharged, some valves (5, 6) are closed and one valve (3) is opened to connect the egg to the atmosphere. The acid egg is useful despite its low efficiency (typically 10%–20%), in cases in which conventional pumps are cost prohibitive in handling corrosive or erosion-causing liquids. Its low efficiency is due to its intermittent action and the fact that at the end of each cycle compressed gas must be vented to the atmosphere without recovering work. Often these pumps require close manual operation or excessive instrumentation.

The *hydraulic ram*, (Figure 44), utilizes the kinetic energy of a moving column of liquid (usually water) to raise a portion of the pumping liquid to a higher pressure or elevation. If the liquid flowing in the supply line, or fall pipe, is stopped abruptly by the closure of the escape check valves, the static pressure at the valve suddenly increases due to the stream's conversion of kinetic energy. When the delivery valve opens, some of the liquid flows into the delivery system. As the energy of the liquid in the supply line is absorbed, the static pressure at the base decreases. When

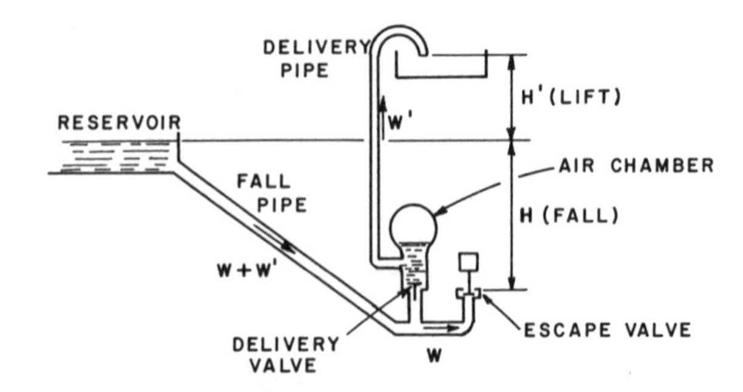

Figure 44. Operating principle of the hydraulic ram.

the escape valve opens the delivery valve will close, thus causing the flow in the column to increase. This causes the motive fraction of the fluid to flow out to waste through the escape valve until the velocity becomes great enough to pick up and seal this valve. The cycle is repeated with a frequency as low as 15 or as high as 200 times per minute.

The efficiency of a properly designed hydraulic ram may be as high as 90%, where efficiency is defined as

$$\text{Efficiency } (\%) = \frac{W'h'}{Wh} \times 100 \tag{53}$$

where W' = mass of liquid delivered
W = mass of motive liquid exhausted through the escape valve
h = the fall
h' = the lift

Thus, the essential components of this pump are the moving fluid column, the escape and delivery valves, and the air chamber. The efficiency and capacity strongly depend on the design of these elements.

The important factors of the moving fluid column are the mass of motive liquid and the friction of the flow of the motive liquid. These factors can be expressed as the ratio of the length of the fall pipe to the height of fall. For a vertical fall pipe an optimum ratio would contain too small a mass of liquid. For a relatively flat fall pipe, the ratio would have too great a friction loss. The optimum value thus varies with the lift.

Valve design for hydraulic rams depends on the weight and the length of the stroke. As a rule of thumb, efficiency varies inversely with the length of the stroke and the weight of the valve, whereas capacity varies directly with the weight of the valve and the length of the stroke. Decreasing the weight of the valve decreases the length of each cycle.

An air surge chamber is necessary to eliminate intermittent flow in the delivery line. In practice, the volume of the surge chamber is designed to be approximately the volume of the delivery pipe. To maintain a supply of air to the vessel, a small check valve (designed to open inward from the atmosphere) is installed just below the delivery valve. At the end of the delivery cycle, a small quantity of air is inspired and carried upward at the start of the next delivery.

Conventional rams may be modified to permit the pumping of one liquid by another. This can be done by replacing the delivery check valve with either a piston or a diaphragm, separating the two fluids. Brown (10) notes that such an arrangement allows the pumping of a clean fluid by a dirty liquid without contamination.

The *airlift pump* (Figure 45) is a device for raising liquid by means of compressed air. It consists of a pipe (1) for introducing compressed air and a mixer (2) for creating a gas-liquid mixture having a lower density than the liquid alone. The two-phase mixture thus rises up the pipe (3) because of

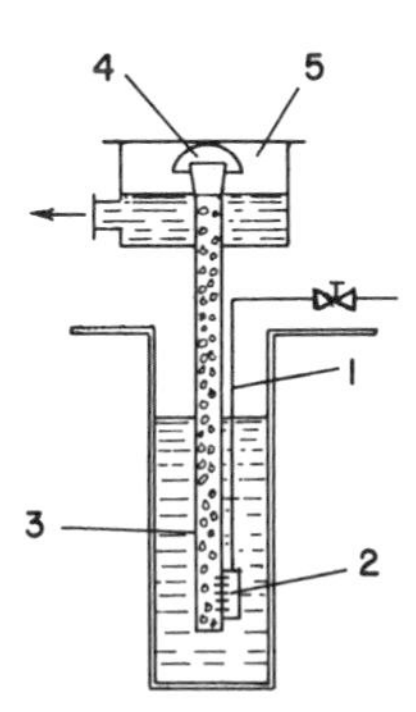

Figure 45. Operation of an airlift pump: 1—pipe for delivery of compressed air; 2—mixer.

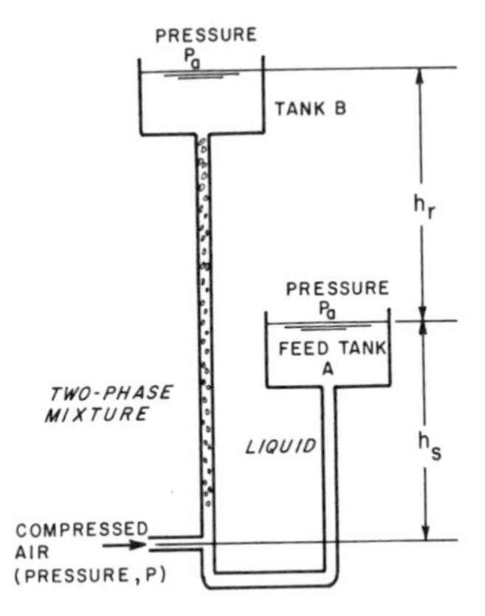

Figure 46. Example of a lift pump operation.

the lower density and gas expansion. At the exit of the pipe (3) the mixture flows around the baffler (4), and the liquid, after separation from air, enters the container (5).

The airlift pump can be applied to transferring liquid from feed tank A to reservoir B at an elevation of h_r above the feed tank liquid surface in Figure 46. The total work performed on the liquid by a mass, "m," of air is Mgh_r, where M is the mass of liquid raised. Assuming that the air enters the lift line at pressure P, then the work done by the gas under isothermal expansion to atmospheric is

$$\tilde{W} = P_a v_a m \ln(P/P_a)$$

where v_a = specific volume of air at atmospheric pressure

From the preceding, we may write the efficiency of the pump as follows:

$$\eta = \frac{Mgh_r}{mP_a v_a \ln(P/P_a)} \tag{54}$$

or rearranging this expression, the mass of air required to pump a unit mass of liquid is

$$\frac{m}{M} = \frac{gh_r}{\eta P_a v_a \ln(P/P_a)} \tag{55}$$

Coulson and Richardson [12] note that when the density of the two-phase mixture in the riser is not significantly different than the feed liquid, the pump will deliver infinitely slowly at 100% efficiency ($\eta = 1$).

Under this condition, the pressure at the point of injection of the compressed air equals the sum of the atmospheric pressure and the pressure due to the column of liquid of height h_s, i.e., the vertical distance between the feed tank level and the air inlet point. Then the mass of air required to pump a unit of mass of liquid for a perfectly efficient pump is

$$\left(\frac{m}{M}\right)_{\eta=1} = \frac{h_r g}{P_a v_a \ln\left(\frac{h_s + h_a}{h_a}\right)} \tag{56}$$

This expression is derived by noting that $P_a = h_a \rho g$, and $P = (h_a + h_s)\rho g$, where ρ is the liquid density (note that the derivation has ignored slip between the air and liquid). Equation 56 represents the minimum air requirement as frictional losses have been ignored in the derivation.

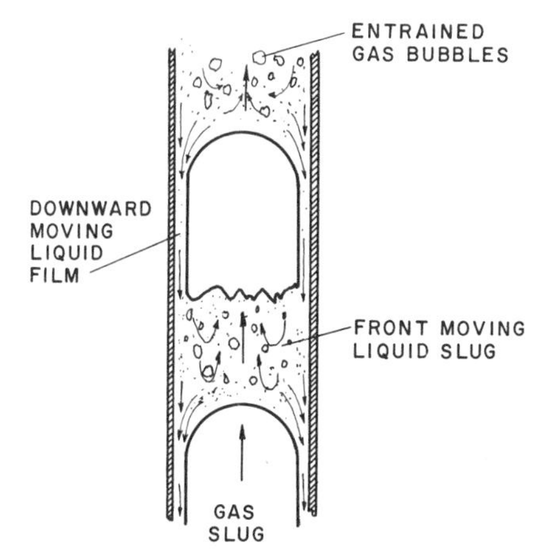

Figure 47. Slug flow in a vertical line.

The actual flow regime of the two-phase mixture in the riser depends on the design of the injection point and the size of the air bubbles. The size of the bubbles influences their velocity of rise relative to the liquid. With very small bubbles, the relative velocity is minimized; however, small bubbles are difficult to form at the rates needed for operation. Generally, bubble coalescence occurs near the injection point and, subsequently, in the upper portions of the riser. Usually coalescence is complete within a few pipe diameters of the injection point, and hence, the dispersion over most of the riser is not greatly different than that produced by introducing large air bubbles. Very often, particularly in small lines (under 3 in. in diameter), a slug flow regime develops, as shown in Figure 47. The air slug is almost bullet shaped and occupies only the central core of the pipe cross section, while a film of liquid flows downward at the walls. The gas slugs essentially push a volume of gas-entrained liquid up the pipe. As the liquid slug moves upward, it is continuously draining to the slug below. The pump functions properly only if the rate of drainage is less than the upward rate of transfer. Four sources lead to inefficient pumping with airlift pumps:

1. The injection point (footpiece).
2. The pump discharge.
3. Friction losses in the riser line.
4. The air supply line.

The gas feed point is a critical part of the design of airlift pumps. The air normally is introduced to the riser through an arrangement of orifices. Two typical feed point arrangements, or footpieces, are illustrated in Figure 48. Energy losses occur because of the sudden expansion of the gas flow and as a result of friction at the orifice walls. In addition, the gas is accelerated as it passes through the footpiece because the total volumetric gas flow in the lift line is increased by the addition of the air). To minimize slugging and accelerate the liquid as much as possible, the air should be injected to form relatively small bubbles over a considerable length of the pipe.

Losses from the discharge of the pump arise because the kinetic energy of the fluid cannot be converted effectively into pressure energy. This is attributed to the rapid fluctuations in the flow.

Friction losses in the riser generally are quite large due to the high degree of turbulence and unsteady slug flow. The relative velocity of the upward-moving air to the liquid promotes high turbulence, resulting in an increase in the flow of air required to maintain a desired fluid density in the riser. To minimize losses, an even feeding and distribution of air bubbles across the flow area is desirable.

Finally, minor losses in the air supply lines contribute to lower pump operating efficiency. Airlifts generally are more efficient than other pump types that employ compressed air directly, especially

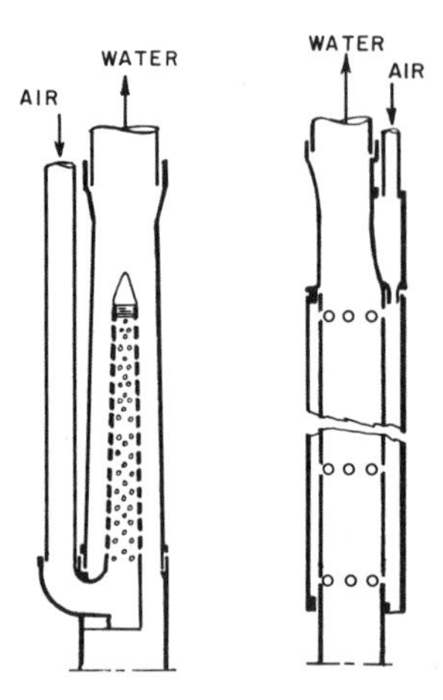

Figure 48. Two designs for gas injection into airlift pumps.

as the air is allowed to expand almost to atmospheric pressure when contacting the liquid. No valves are used other than those in the air supply lines and the nonreturn valve in the suction line. This type pump is well suited to handling liquids containing high concentrations of suspended solids. Both the pipe and the footpiece can be made of metal, stoneware, or glass, according to the material to be handled and the duty required of the pump. They are used extensively in the oil industry and for circulating nitric acid in absorption equipment.

Their primary disadvantage is that to obtain a high efficiency the air must be introduced at a considerable depth below the liquid feed point. In addition, it often is difficult to predict accurately the total energy losses for such a system during design. Additional discussions of secondary fluid displacement pumps are given in References 12–17.

PUMP SELECTION

Pump selection requires understanding the principles of mechanics and physics affecting the pumping system and the fluid. Pump efficiency is strongly dependent on the behavior of the fluid. Principles of operations defining head and flow relationships must be understood clearly before the system performance can be evaluated accurately.

General performance characteristics of commercially available pumps can be obtained from manufacturers. Selection is made on the basis of desired capacity and head, which are calculated in accordance with overall piping system layout. The electric motor for a pump is chosen from brake horsepower determined from the equations given earlier.

The overall procedure in choosing a pump for a particular application is as follows:

1. Obtain information on the physical and chemical properties of the liquid at the intended operating conditions, i.e., specific gravity, viscosity, vapor pressure, corrosiveness, toxicity, etc.
2. Lay-out the piping system on paper, defining major flow resistances in the system. Calculate total heads for the system.
3. Establish the capacity requirements in terms of a range. That is, define normal average capacity needs as well as system lows and peak flow required. If possible, estimate how long pumps will have to operate at peak loads.
4. Based on the preceding information, the class and type of pump can be selected. A more detailed specification can be made from examination of the manufacturer's literature.

Of the pumps described in this chapter, centrifugals are the most versatile and widely used throughout the chemical process industries.

The following advantages of centrifugal pumps should be remembered when comparing different pump classes for an application:

1. They are simple in construction and, as a general rule, less expensive than many positive displacement types. They are available in a wide range of materials.
2. They do not require valves for their operation.
3. They operate at high speeds (4,000 rpm or higher) and, therefore, can be coupled directly to an electric motor. In general, higher speeds typically mean smaller pumps and motors for a given duty.
4. They give steady deliveries.
5. Depending on the application, maintenance costs are lower than for any other type of pump.
6. They are typically smaller than other pumps of equal capacity. Therefore, they can be made into a sealed unit with the driving motor and directly immersed in the suction tank.
7. Liquids having relatively high concentrations of suspended solids can be handled.

At the same time, centrifugal pumps have several disadvantages, the primary ones being as follows:

1. Single-stage pumps cannot develop high pressures. Multistage pumps will develop greater heads but are much more expensive and cannot be readily constructed from corrosion-resistant materials without significantly higher costs due to their greater complexity. As a general rule, it is better to use very high speeds to reduce the number of stages required.
2. High-efficiency operation is usually only obtained over a limited range of conditions. This is especially true for turbine pumps.
3. The vast majority of centrifugal pumps commercially available are not self priming.
4. A nonreturn valve must be installed in the delivery or suction line or the liquid will run back into the suction tank when the unit is not running.
5. Centrifugal pumps have problems handling highly viscous materials. They typically operate at greatly reduced efficiencies.

For pumping applications requiring relatively small capacities and high heads (e.g., 50–10,000 or more atm) piston pumps are recommended. These pumps are well suited in these ranges to pumping liquids of high viscosity that are flammable and of an explosive nature (steam pumps). In addition, they make excellent metering pumps.

Screw pumps are best suited for handling high-viscosity liquids, fuels, and petroleum products. These pumps are used for capacities up to 300 m^3/hr and pressures up to 175 atm at speeds of rotation up to 3,000/min. The advantages of screw pumps are their high speed, compactness, and quiet operation. Pump capacity is practically independent of pressure, and efficiency is rather high (in the range of 0.75 to 0.80). The field of application of single-screw pumps is restricted by capacity up to 3.6–7 m^3/hr and pressures up to 10–25 atm. Their cost and maintenance are similar to those of centrifugal pumps of low capacity operating under pressures up to 3–5 atm. Screw pumps are considerably more economical when their delivery pressures exceed 10 atm. Single-screw pumps are employed in handling dirty and aggressive liquids, solutions, and high-viscosity polymer solutions.

Sliding-vane pumps are used for the displacement of clean liquids (without solid particles) at moderate capacities and heads.

Gear pumps are best suited to pumping viscous liquids without solid particles at low delivery rates (not higher than 5–6 m^3/min) and high pressures up to 100–150 atm.

Jet pumps are typically employed in operations in which pumping requirements are intermittent, an inexpensive standby unit is desirable or corrosion is important. They are used for the displacement of low-viscosity clean liquids at low delivery rates up to 40 m^3/hr and relatively high heads (up to 250 m). Efficiencies are typically low ($\eta = 20$–50%).

Acid-egg pumps and airlifts are used in industries in which moving and friction parts are highly undesirable.

Further discussions of various pump types, operating limitations, and applications are given in the references cited at the end of this chapter. Subsequent chapters in this volume (in particular Chapters 35–39) cover selection criteria in further detail. Tables 9 and 10 summarize construction

Table 9
Major Pump Types and Construction Styles

Pump Type And Construction Style	Distinguishing Construction Characteristics	Usual Orientation	Usual No. of Stages	Relative Maintenance Requirement	Comments
Dynamic					Capacity varies with head.
Centrifugal					Low to medium specific speed.
Horizontal					
Single-stage overhung, process type	Impeller cantilevered beyond bearings.	Horizontal	1	Low	Most common style used in process services.
Two-stage overhung, process type	2 impellers cantilevered beyond bearings.	"	2	Low	For heads above single-stage capability.
Single-stage impeller between bearings	Impeller between bearings; casing radially or axially split.	"	1	Low	For high flows to 330 m head.
Chemical	Casting patterns designed with thin sections for high cost alloys; small sizes.	"	1	Medium	Low pressure and temperature ratings.
Slurry	Large flow passages, erosion control features.	"	1	High	Low speed; adjustable axial clearance.
Canned	Pump and motor enclosed in pressure shell; no stuffing box.	"	1	Low	Low head-capacity limits for models used in chemical services.
Multistaged, horizontally split casing	Nozzles usually in bottom half of casing.	"	Multi	Low	For moderate temperature-pressure ratings.
Multistage, barrel type	Outer casing confines inner stack of diaphragms.	"	Multi	Low	For high temperature-pressure ratings.
Vertical					
Single-stage process type	Vertical orientation.	Vertical	1	Low	Style used primarily to exploit low NPSH requirement.

Multistage, process type	Many stages, low head/stage.	"	Multi	Medium	High head capability, low NPSH requirement.
In-line	Arranged for in-line installation, like a valve	"	1	Low	Allows low cost installation, simplified piping systems.
High speed	Speeds to 380 rps, head to 1,770 m	"	1	Medium	Attractive cost for high head/low flow.
Sump	Casing immersed in sump for installation convenience and priming ease.	"	1	Low	Low cost installation.
Multistage deep well	Very long shafts	"	Multi	Medium	Water well service with driver at grade.
Axial (propeller)	Propeller-shaped impeller, usually large size.	Vertical	1	Low	A few applications in chemical plants and refineries.
Turbine (regenerative)	Fluted impeller; flow-path-like screw around periphery.	Horizontal	1,2	Med. to High	Low flow-high head performance. Capacity virtually independent of head.
Positive Displacement					
Reciprocating					
Piston, plunger	Slow speeds; valves, cylinders, stuffing boxes subject to wear.	Horizontal	1	High	Driven by steam engine cylinders or motors through crankcases.
Metering	Small units wtih precision flow control system	"	1	Medium	Diaphragm and packed plunger types.
Diaphragm	No stuffing box; can be pneumatically or hydraulically actuated.	"	1	High	Used for chemical slurries; diaphragms prone to failure.
Rotary					
Screw	1, 2 or 3 screw rotors	"	1	Medium	For high viscosity, high flow high pressure.
Gear	Intermeshing gear wheels	"	1	Medium	For high viscosity, moderate pressure, moderate flow.

Table 10
Typical Operating Performances Data of Pumps

Pump Type/Style	Solids Tolerance	Capacity (dm³/s)	Capacity (gph) U.S.	Max. Head (m)	Max. Head (ft)	Typical NPSHR (m)	Typical NPSHR (ft)	Max. Kinematic Viscosity (mm²/s)	Max. Kinematic Viscosity (in.²/s)	Efficiency (%)	Max. Pumping Temperature (°C)	Max. Pumping Temperature (°F)
Centrifugal												
Horizontal												
Single-stage overhung	MH	1 ~ 320	950 ~ 3×10^5	150	492	2 ~ 6	6.6 ~ 20	650	1.01	20 ~ 80	455	851
2-Stage overhung	MH	1 ~ 75	950 ~ 7.1×10^4	425	1394	2 ~ 6.7	6.6 ~ 22	430	0.67	20 ~ 75	455	851
Single-stage impeller between bearings	MH	1 ~ 2,500	950 ~ 2.4×10^6	335	1099	2 ~ 7.6	6.6 ~ 25	650	1.01	30 ~ 90	205 ~ 455	401 ~ 851
Chemical	MH	65	6.2×10^4	73	239	1.2 ~ 6	3.9 ~ 20	650	1.01	20 ~ 75	20.5	401
Slurry	H	65	6.2×10^4	120	394	1.5 ~ 7.6	4.9 ~ 25	650	1.01	20 ~ 80	455	851
Canned	L	0.1 ~ 1,250	95 ~ 1.2×10^6	1,500	4922	2 ~ 6	6.6 ~ 20	430	0.67	20 ~ 70	540	1,004
Multi., horiz. split	M	1 ~ 700	950 ~ 6.7×10^5	1,675	5495	2 ~ 6	6.6 ~ 20	430	0.67	65 ~ 90	205 ~ 260	401 ~ 500
Multi. barrel type	M	1 ~ 550	950 ~ 5.2×10^5	1,675	5495	2 ~ 6	6.6 ~ 20	430	0.67	40 ~ 75	455	851
Vertical												
Single-stage process	M	1 ~ 650	950 ~ 6.2×10^5	245	804	0.3 ~ 6	1 ~ 20	650	1.01	20 ~ 85	345	653
Multistage	M	1 ~ 5000	950 ~ 4.8×10^6	1,830	6004	0.3 ~ 6	1 ~ 20	430	0.67	25 ~ 90	260	500
In-line	M	1 ~ 750	950 ~ 7.1×10^5	215	705	2 ~ 6	1 ~ 20	430	0.67	20 ~ 80	260	500
High speed	L	0.3 ~ 25	285 ~ 2.4×10^4	1,770	5807	2.4 ~ 12	7.9 ~ 39.8	109	0.17	10 ~ 50	260	500
Sump	MH	1 ~ 45	950 ~ 4.3×10^4	60	197	0.3 ~ 6.7	1 ~ 22	430	0.67	40 ~ 75	—	—
Multi. deep well	M	0.3 ~ 25	285 ~ 2.4×10^4	1,830	6004	0.3 ~ 6	1 ~ 20	430	0.67	30 ~ 75	205	401
Axial (propeller)	H	1 ~ 6500	950 ~ 6.2×10^6	12	39.4	~ 2	6.6	650	1.01	65 ~ 85	65	149
Turbine (regenerative)	M	0.1 ~ 125	95 ~ 1.2×10^5	760	2493	2 ~ 2.5	6.6 ~ 8.2	109	0.17	55 ~ 85	120	248
Positive Displacement												
Reciprocating				(kPa)	(PSI)							
Piston, plunger	M	1 ~ 650	950 ~ 6.2×10^5	345,000	50,038	3.7	12	1,100	1.71	55 ~ 85	290	554
Metering	L	0 ~ 1	0 ~ 950	517,000	74,985	4.6	15.1	1,100	1.71	~ 20	290	554
Diaphragm	L	0.1 ~ 6	95 ~ 5.7×10^3	34,500	5,004	3.7	12.1	750	1.16	~ 20	260	500
Rotary								(SSU)	(SSU)			
Screw	M	0.1 ~ 125	95 ~ 1.2×10^5	20,700	3002	~ 3	~ 9.8	150×10^6	150×10^6	50 ~ 80	260	500
Gear	M	0.1 ~ 320	95 ~ 3.0×10^5	3,400	493	~ 3	~ 9.8	150×10^6	150×10^6	50 ~ 80	345	653

MH—moderately high; H—high; M—medium; L—low.

characteristics and typical operating ranges of various pumps, respectively. Operating performance data reported in Table 10 should be considered as approximate for comparisons between pump types. The manufacturer's performance data should always be used when selecting and sizing pumps for a specific application.

NOTATION

A	energy from centrifugal force, J
atm	atmospheric head or pressure, ft or atm
b	impeller width, m
c	centrifugal force, N; see Equation 26
c_1, c_2	vector sums of relative and tangential velocities, m/sec
f, F	area, m^2
G	weight of fluid particle, kg
g	acceleration due to gravity, m/sec^2
H	head, m
$\tilde{H}_c$	cavitation correction number; refer to Equation 24
H_d	discharge head, ft or m
H_{fd}	discharge friction head, ft or m
H_{ss}	static suction head, ft or m
H_{ts}	total static head, ft or m
h_ℓ	head loss, m
h_{vs}	velocity head, ft or m
K	proportionality coefficient in Equation 47
l	height of liquid column, ft or m
m, M	mass, kg or lb
N	power, kW
N_e	effective power, kW
n	number of revolutions
n_s	specific number of revolutions
P	pressure, N/m^2
Q	flow capacity, m^3/sec
r	crank radius, m
S	stroke length, m
t	temperature, °C
u	circumferential crank velocity, m/sec
V	volume of liquid pumped, m^3
W	mass flow, lb/hr
$\tilde{W}$	work, Joules
w	average fluid velocity, m/sec
z	height, m
Z'	number of vanes

Greek Symbols

α	angle, °
β	motor efficiency coefficient; see Equation 8
β_2	angle defined in Figure 15,
γ	specific weight, kg_f/m^3
δ	vane thickness, mm
ϵ	coefficient accounting for number of vanes (refer to Figure 42)
ξ	kinetic energy correction term
η_h	hydraulic efficiency
η_m	mechanical efficiency
η_p	pump efficiency
η_{tr}	transmission efficiency
η_v	volumetric efficiency
μ	viscosity, poise
v	specific volume, m^3/kg
ρ	density, kg/m^3
δ_c	cavitation number; refer to Equation 23
ω	angular velocity, m/sec

REFERENCES

1. Perry, R. H., and Chilton C. H., (Eds), *Chemical Engineers' Handbook*, 5th edition, McGraw-Hill Book Co., New York, 1973.
2. Karassik, I. J., et al. *Pump Handbook*, John Wiley & Sons Inc., New York, 1976.
3. Church, A. H., *Centrifugal Pumps and Blowers*, John Wiley & Sons, Inc., New York, 1944.
4. Cheremisinoff, N. P., *Fluid Flow: Pumps, Pipes, and Channels*, Ann Arbor Science Pub., Ann Arbor, MI, 1981.
5. Azbel, D. S., and Cheremisinoff, N. P., *Fluid Mechanics and Unit Operations*, Ann Arbor Science Pub., Ann Arbor, MI, 1983.
6. Metzner, A. B., *Handbook of Fluid Dynamics*, McGraw-Hill Book Co., New York, 1961.

7. Stepanoff, A. J., *Pumps and Blowers*, John Wiley & Sons Inc., New York, 1965.
8. Hicks, T. G., *Pump Selection and Application*, McGraw-Hill Book Co., New York, 1957.
9. Hicks, T. G., and Edwards, T. W., *Pump Application Engineering*, McGraw-Hill Book Co., New York, 1971.
10. Brown, G. G., *Unit Operations*, John Wiley & Sons Inc., New York, 1950.
11. Karassik, I. J., and Carter, R., *Centrifugal Pump Design and Selection*, R. P. Worthington Corp., Harrison, NJ, 1981.
12. Coulson, J. M., and Richardson, J. F., *Chemical Engineering*, Pergamon Press Inc., Elmsford, NY, 1962.
13. Foust, A. S., et al., *Principles of Unit Operations*, 2nd edition, John Wiley & Sons Inc., New York, 1980.
14. Datta, R. L., "Studies for the Design of Gas Lift Pumps," *J. Imp. Coll. Chem. Eng. Soc.*, 4 (157) (1948).
15. Bergelin, O. P., "Flow of Gas-Liquid Mixtures," *Chem. Engr.*, 56(5):104 (1949).
16. Simonin, R. F., "Workings of an Air Lift Water Pump," *Comp. Rend. Acad. Sci.*, 233:465 (1951).
17. Bonnington, S. T., and King, A. L., "Jet Pumps and Ejectors," British Hydraulic Research Association: Fluid Engineering, Cranfield, England, 1972.
18. Kirk, R. E., and Othmer, D., *Encyclopedia of Chemical Technology*, 2nd ed., John Wiley & Sons Inc., New York, 1963.
19. Swindin, N., *The Modern Theory and Practice of Pumping*, Ernest Benn Ltd., London, 1924.
20. Taylor, I., *Chem. Engr. Prog.*, 46(637), (1950).
21. Tetlow, N., *Trans. Inst. Chem. Engr.*, 28(63), (1950).

CHAPTER 35

SYSTEM ANALYSIS FOR PUMPING EQUIPMENT SELECTION

Technical Staff of Peerless Pump
a Sterling Company
Montebello, California USA

CONTENTS

INTRODUCTION, 1002

CHARACTERISTICS OF CENTRIFUGAL PUMPS, 1002

NET POSITIVE SUCTION HEAD (NPSH), 1006

PUMP CHARACTERISTICS, 1008
Head-Flow Curve, 1008
Performance Curves, 1009

PUMP AFFINITY LAWS, 1013

PUMP SYSTEM CHARACTERISTICS, 1015
Pipe Deterioration, 1017
Pipe Friction, 1017

SYSTEM HEAD CURVES AND COMPONENTS, 1019

SYSTEM OPERATION AND CONTROL, 1020
Total System Head Alteration, 1022
Total Available Head Alteration, 1023

VARIABLE-SPEED PUMPS, 1025
Functions, 1025
Types, 1025
Location, 1025
Sensing Multiple Branch Systems, 1027

TOTAL SYSTEM EVALUATION, 1028

PROJECTING PUMP HORSEPOWER, 1030
Constant Speed, 1030
Variable Speed, 1030

EQUIPMENT POWER COMPARISONS AND TOTAL ENERGY PROJECTION, 1034

NOTATION, 1036

REFERENCES, 1037

INTRODUCTION

"Turn us on and we'll come running" is a phrase used by an investor-owned water utility in advertising their service. The faucet which we open to water the lawn or wash our hands is one end of a very large and complex piping system; at the other end is a pump. Many people are aware of a pump only when it does not function; others must be concerned about it from the time the piping system is conceived, through its design, construction, and operational life. By considering the pump early in the system design and applying energy evaluation procedures, long-term operating cost benefits can result.

This chapter has been prepared to guide readers at all levels of experience through a basic understanding of centrifugal pumps, pump characteristics, and system operation and control. These topics form the foundation for a total system evaluation.

Energy evaluation procedures are applicable to all systems, existing or proposed, large or small. In many systems the procedures will show that a reduction in operating power requirements can be achieved. Whether the reduction is large or small, you must determine its true significance.

We have all been challenged to use our available energy more efficiently. This chapter will serve as a working tool to assist you in meeting that challenge.

CHARACTERISTICS OF CENTRIFUGAL PUMPS

The selection of a centrifugal pump for an energy-efficient pumping system requires an understanding of the principles of mechanics and physics that can affect the pumping system and the pumped liquid. The efficiency of a centrifugal pump is also dependent on the behavior of the liquids being pumped. The principles of centrifugal pump operation that govern head and flow must be understood clearly before pump performance can be evaluated accurately.

The behavior of a fluid depends on its state—liquid or gas. Liquids and gases offer little resistance to changes in form. Typically, fluids such as water and air have no permanent shape and readily flow to take the shape of the containing enclosure when even a slight shear loading is imposed. Factors affecting behavior of fluids include:

- Viscosity
- Specific gravity
- Vapor pressure

Viscosity is the resistance of a fluid to shear motion—its internal friction. The molecules of a liquid have an attraction for each other. They resist movement and repositioning relative to each other. This resistance to flow is expressed as the viscosity of the liquid. Dynamic viscosity also can

Table 1
Specific Gravity Values for Water at Selected Temperatures

Water Temperature		Water Specific Gravity	Effect of Specific Gravity on Energy Input Constant Flow	
(°C)	(°F)		(kW)	(bhp)
4	39.2	1.0	74.6	100
60	140	0.983	73.3	98.3
100	212	0.958	71.5	95.8
125	257	0.939	70.0	93.9
150	302	0.917	68.4	91.7

Table 2
Effect of Specific Gravity on Pump Head

Specific Gravity	Effect of Specific Gravity on . . .			
	Linear Head		Gauge Pressure	
	(ft)	(m)	(psi)	(bar)
0.75	246	75	79.9	5.51
1.00	246	75	106.5	7.34
1.20	246	75	127.8	8.81

be defined as the ratio of shearing stress to the rate of deformation. The viscosity of a liquid varies directly with temperature; therefore, viscosity is always stated at a specific temperature.

Liquid *viscosity* is very important in analyzing the movement of liquids through pumps, piping, and valves. A change in viscosity alters liquid handling characteristics in a system; more or less energy then may be required to perform the same amount of work. In a centrifugal pump, an increase in viscosity reduces the pressure energy (head) produced while increasing the rate of energy input. In piping systems, a liquid with a high viscosity has a high energy gradient against which a pump must work, and more power is required than for pumping low-viscosity liquids.

Specific gravity is the ratio of the density of one substance to that of a reference substance at a specified temperature. Water at 4°C is used as the reference for solids and liquids. Air generally is used as the reference for gases. The specific gravity of a liquid affects the input energy requirements, or brake horsepower (bhp), of centrifugal pumps. Brake horsepower varies directly with the specific gravity of the liquid pumped. For example, water at 4°C has a specific gravity of 1.0. Table 1 includes some specific gravity values for water at selected temperatures.

Specific gravity affects the liquid mass but not the head developed by a centrifugal pump, as shown in Table 2. The specific gravity also affects the energy required to move the liquid and therefore must be used in determining the pump's horsepower requirement.

Vapor pressure is the pressure at which a pure liquid can exist in equilibrium with its vapor at a specified temperature. Fluids at temperatures greater than their specified (critical) temperatures will exist as single-phase liquids (vapors), with no distinction between gas and liquid phases. At less than the critical temperature, two fluid phases can coexist; the denser fluid phase exists as a liquid and the less dense phase as a vapor. At a specific temperature, the liquid phase is stable at pressures exceeding the vapor pressure and the gas phase is stable at pressures less than the vapor pressure.

For a fluid to exist in a liquid state its surface pressure must be equal to, or greater than, the vapor pressure at the prevailing temperature. For example, water has a vapor pressure of 0.1781 psia at 10°C and 14.69 psia at 100°C. The vapor pressure of a volatile liquid (such as ether, alcohol, or propane) is considerably higher than that of water at the same temperature; consequently, much higher pressures must be applied to maintain volatile materials in their liquid states. The surface pressure of a liquid must be greater than its vapor pressure for satisfactory operation of a centrifugal pump.

Centrifugal pump characteristics remain constant unless an outside influence causes a change in operating conditions. Three conditions can alter pump performance:

1. Changes in impeller or casing geometry.
2. Increased internal pumping losses caused by wear.
3. Variation of liquid properties.

For example, if the impeller passages become impacted with debris, the head-flow relationship will be reduced. Similarly, performance will decline if mechanical wear increases the clearance between the rotating and stationary parts of the pump.

Except for specially designed pumps, most centrifugal pumps can handle liquids containing approximately 3%–4% of gas (by volume) without an adverse effect on performance. An excess of gas will reduce the flow of liquid through the pump, and under certain conditions, flow will cease, setting up a condition that may damage the pump.

The function of a pump is to move liquids by imparting pressure energy (head) to the liquid. The ability of the pump to perform its function is based on the law, which states that energy cannot be created or destroyed, but can only be converted in form. A pump converts mechanical energy into pressure energy. Part of the converted energy is required to overcome inertia and move the liquid; most of the remaining energy is stored in the liquid as elevated pressure, which can be used to perform useful work outside the pump. A centrifugal pump is basically a velocity machine designed around its impeller. The interaction between the impeller and its casing produces the characteristics of head or pressure energy.

The *developed head* is a function of the difference in velocity between the impeller vane diameter at the entrance and the impeller vane diameter at the exit. The expression of theoretical head can be related to the law of a falling body:

$$H = \frac{u^2}{2g} \tag{1}$$

where H = height or head (ft)
u = velocity of moving body (fps)
g = acceleration of gravity (32.2 ft/s^2)

When the height of fall is known (for example, H = 100 ft), the terminal velocity can be determined (in this case u = 80.3 fps). Conversely, if the direction of motion is reversed, a liquid exiting through an impeller vane tip at a velocity of 80.3 fps reaches a velocity of 0 fps at 100 ft above the impeller tip (Figure 1).

When the developed head required is known, the theoretical impeller diameter for any pump at any rotational speed can be determined from the equation for peripheral velocity of a round rotating body:

$$D = \frac{(229.2)u}{N} \tag{2}$$

where D = unknown impeller diameter (in.)
u = velocity derived from $u = \sqrt{2gH}$ (fps)
N = rotational speed of the pump (rpm)

If pump head is 100 ft and rotational speed is 1,750 rpm, the formula indicates that a 10.52-in. theoretical diameter impeller is required.

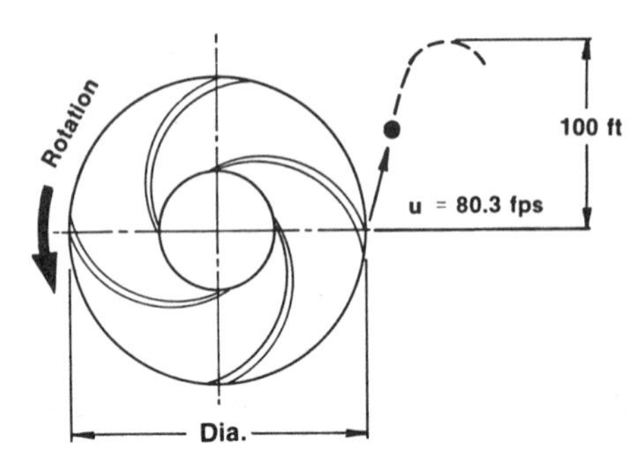

Figure 1. Example to determine theoretical impeller diameter.

The function of a centrifugal pump is to transport liquids by transferring and converting mechanical energy (foot-pounds of torque) from a rotating impeller into pressure energy (head). By the transfer of this energy to a liquid the liquid can perform work, move through pipes and fittings, rise to a higher elevation or increase the pressure level. Because a centrifugal pump is a velocity machine, the amount of mechanical energy per unit of fluid weight transferred to the liquid depends on the peripheral velocity of the impeller, regardless of fluid density. This energy per unit of weight is defined as pump head and is expressed in feet of liquid. If the effect of liquid viscosity is ignored, the head developed by a given pump impeller at a given speed and flow rate remains constant for all liquids.

Head is the vertical height of a column of liquid. The pressure this liquid column exerts on the base surface depends on the specific gravity of the liquid. A 10-ft column of liquid with a specific gravity of 0.5 exerts only 2.16 psi; a 10-ft column of liquid with a specific gravity of 1.0 exerts 4.33 psi.

The formula for converting feet of head to pressure is

$$H = \frac{P \times 2.31}{\gamma} \tag{3}$$

where P = pressure (psi)
γ = specific gravity

The proper selection of a centrifugal pump requires that pressure be converted to feet of head. If not, a pump that is incapable of imparting the required energy to a liquid may be installed before the inadequacy is discovered. The heads required to produce a pressure of 100 psi for three liquids of different specific gravities are given in Table 3.

From this analysis a single-stage pump selected for pumping a liquid with a specific gravity of 1.2 will have the lowest peripheral velocity at the impeller tip, while the pump for a liquid with 0.75 specific gravity will have the highest peripheral velocity.

The volumetric flow rate is a function of the peripheral velocity of the impeller and the cross-sectional areas in the impeller and its casing; the larger the passage area, the greater the flow rate. Consequently, the physical size of the pump increases with higher flow requirements for a given operating speed. The liquid flow rate is directly proportional to the area of the pump passages and can be expressed as

$$Q = \frac{u \times A}{0.321} \tag{4}$$

where Q = flow (gpm)
u = velocity (fps)
A = area (in.^{2})

Table 3
Heads Required to Produce 100 psi Pressure for Different Liquids

Head equivalent of 100 psi for liquids of different specific gravity	
Specific Gravity	Head (ft of liquid)
0.75	308
1.0	231
1.2	193

For example, a centrifugal pump having a 15-fps liquid velocity at the discharge flange can pump approximately 330 gpm through a 3-in.-diameter opening. If the flow requirements are increased to approximately 1,300 gpm and the velocity remains at 15 fps, a 6-in.-diameter opening will be required. Liquid flow is also directly proportional to the rotational speed that produces the velocity. The four preceding equations establish two relationships:

1. Head is directly proportional to the square of the liquid velocity.
2. Flow is directly proportional to the peripheral velocity of the impeller.

Cavitation in a centrifugal pump can be a serious problem. Liquid pressure is reduced as the liquid flows from the inlet of the pump to the entrance to the impeller vanes. If this pressure drop reduces the absolute pressure on the liquid to a value equal to or less than its vapor pressure, the liquid will change to a gas and form vapor bubbles. The vapor bubbles will collapse when the fluid enters the high-pressure zones of the impeller passages.

This collapse is called cavitation and results in a concentrated transfer of energy, which creates local forces. These high-energy forces can destroy metal surfaces; very brittle materials are subject to the greatest damage. In addition to causing severe mechanical damage, cavitation also causes a loss of head and reduces pump efficiency. Cavitation will also produce noise.

If cavitation is to be prevented, a centrifugal pump must be provided with liquid under an absolute pressure that exceeds the combined vapor pressure and friction loss of the liquid between the inlet of the pump and the entrance to the impeller vanes. Chapter 38 discusses cavitation in detail.

NET POSITIVE SUCTION HEAD (NPSH)

NPSHA (net positive suction head available) is the absolute pressure of the liquid at the inlet of the pump. NPSHA, a function of the elevation, temperature, and pressure of the liquid, is expressed in units of absolute pressure (psia). Any variation of these three liquid characteristics will change the NPSHA. An accurate determination of NPSHA is critical for any centrifugal pump application.

The net positive suction head required (NPSHR) by a specific centrifugal pump remains unchanged for a given head, flow, rotational speed, and impeller diameter, but changes with wear and liquids.

Specific speed is a correlation of pump flow, head, and speed at optimum efficiency. It classifies pump impellers with respect to their geometric similarity. Specific speed is usually expressed as

$$N_s = \frac{N\sqrt{Q}}{H^{3/4}} \tag{5}$$

where N_s = pump specific speed
Q = flow at optimum efficiency (gpm)

The specific speed of an impeller is defined as the revolutions per minute at which a geometrically similar impeller would run if it were of a size that would discharge 1 gpm against a head of 1 foot. Specific speed is indicative of the impeller's shape and characteristics.

Centrifugal pumps are divided into three classes: radial flow, mixed flow, and axial flow.

There is a continuous change from the radial flow impeller (which develops head principally by the action of centrifugal force) to the axial flow impeller (which develops most of its head by the propelling or lifting action of the vanes on the liquid). Typically, centrifugal pumps also can be categorized by physical characteristics relating to the specific speed range of the design (see Figure 2). Once the values for head and capacity become established for a given application, the pump's specific speed range can be determined and specified to ensure the selection of a pump with optimal operating efficiency.

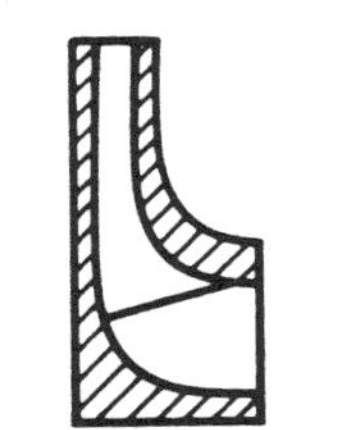

Single suction, horizontal or vertical centrifugal pumps with narrow port impellers have low capacities and deliver high heads. Specific speed range is 500 to 1000.

Single or multistage, double (illustrated) and single suction, volute and diffuser design centrifugal pumps deliver medium capacity and medium heads. Specific speed range is 1000 to 2000.

Single or multistage, and single (illustrated) and double suction, Francis-type impellers, operating in volute or diffuser type casings, produce medium to high capacity at medium to low speeds. Specific speed range is 2000 to 4000.

Single or multistage single suction centrifugal pumps, in volute or diffuser-type casings with mixed flow impellers, deliver high capacity at low head. Specific speed range is 4000 to 6000.

Single or multistage pumps with mixed flow and propeller-type impellers have very high capacities and deliver very low heads. Specific speed range is 6000 to 10,000.

Figure 2. Centrifugal pump configurations and specific speed ranges.

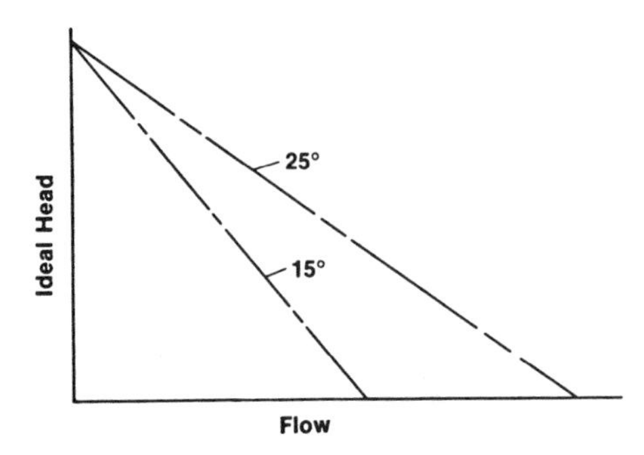

Figure 3. Illustrates the ideal head curve.

PUMP CHARACTERISTICS

Differing pump hydraulic characteristics will result in one pump being better suited for a given application than another. This section examines head-flow curves, pump performance curves, variations in curve shape as a function of specific speed and pump affinity laws.

Head-Flow Curve

The head-flow curve of an ideal pump with ideal (frictionless) fluid is a straight line whose slope from zero flow to maximum flow varies with the impeller exit vane angle. For example, a given impeller and casing combination with an impeller vane exit angle of 25° will have a greater maximum flow than a similar impeller with a 15° vane exit angle (Figure 3). However, the actual head-flow characteristic of a centrifugal pump is not an ideal straight line. Its shape is altered by friction, leakage, and shock losses that occur in impeller and casing passages.

Friction losses in a centrifugal pump are proportional to the surface roughness and the wetted areas of the impeller and casing. Leakage losses result from the flow of liquid between the clearance of rotating and stationary parts, such as impeller-to-case wear ring clearances. Shock losses occur as the liquid enters the impeller entrance vanes and as the liquid flows from the impeller into the casing. These internal losses characteristically reduce pump performance from the ideal to the actual total head-flow curve shown in Figure 4. The flow at which the sum of all these losses is the least determines the point of maximum efficiency.

Mechanical losses in bearings, packings, and mechanical seals further reduce pump efficiency. Although mechanical losses may be calculated, the results are generally not accurate; actual pump performance can be determined only by testing.

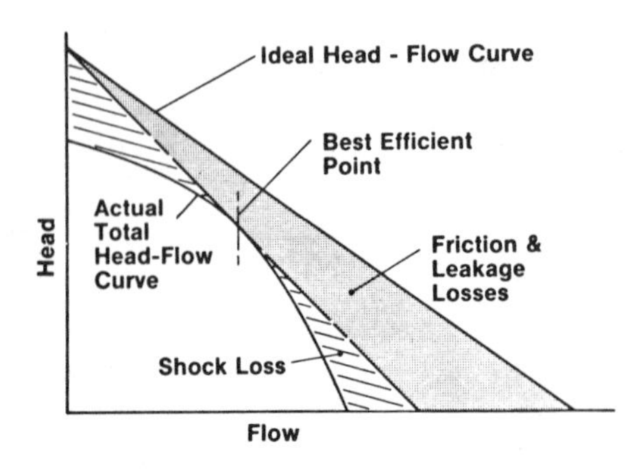

Figure 4. Illustrates deviation from the ideal head-flow curve due to internal losses.

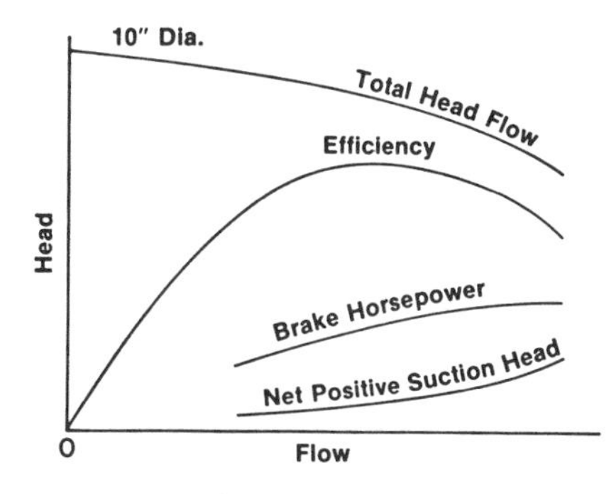

Figure 5. Pump performance curve for a single impeller diameter.

A centrifugal pump is designed around its impeller. The function of the casing is to collect the liquid leaving the impeller and convert its kinetic energy into pressure energy. A case is designed in conjunction with an impeller to achieve the most efficient match for conversion of liquid velocity to pressure energy. The interaction between the impeller and its casing for a given pump determines the pump's unique performance characteristics.

Performance Curves

Pump performance curves depict the total head developed by the pump, the brake horsepower required to drive it, the derived efficiency and the net positive suction head required over a range of flows at a constant speed. Pump performance can be shown as a single line curve depicting one impeller diameter (Figure 5) or as multiple curves for the performance of several impeller diameters in one casing (Figure 6).

The performance characteristics of pumps are classified by discharge specific speed and have the approximate curve shapes shown in Figure 7. According to shape, centrifugal pump head-capacity and horsepower curves are classified as follows:

Drooping head characteristics are those in which the head at zero flow is less than the head developed at some greater flow (Figure 8).

Continually rising head characteristics are those in which the head rises continuously as the flow is reduced to zero (Figure 9).

Steep head characteristics are those in which the head rises steeply and continuously as the flow is reduced. Curve steepness is only a relative term because there is no defined value of curve slope for comparison (Figure 10).

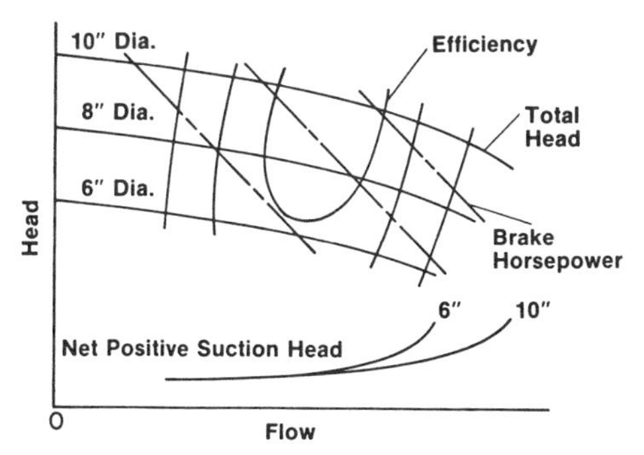

Figure 6. Pump performance curves for several impeller diameters.

BEP is the reference to the flow point at which the best pump efficiency occurs.

All specific speed values are in English units.

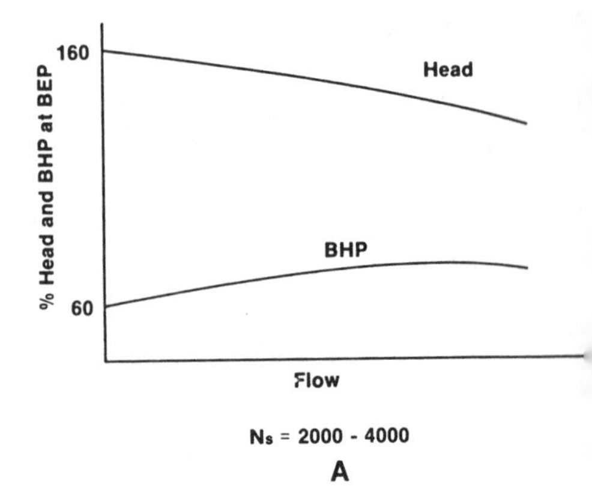

N_s = 2000 - 4000

A

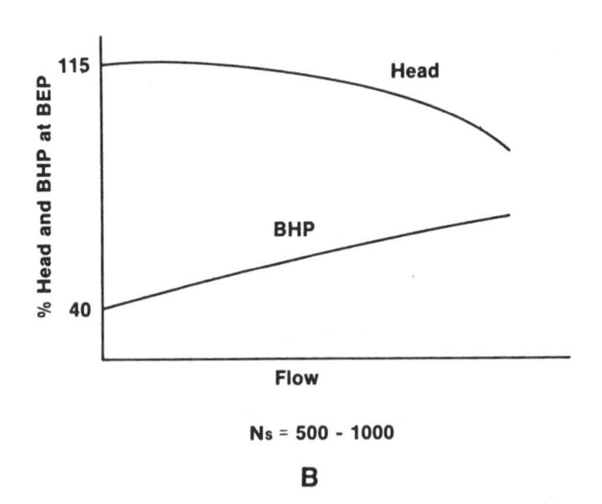

N_s = 500 - 1000

B

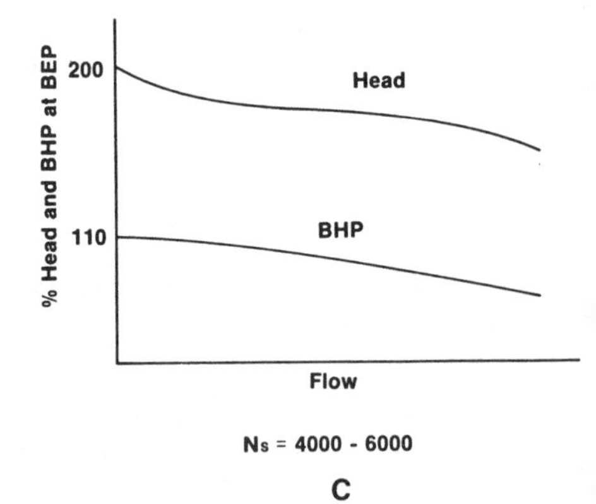

N_s = 4000 - 6000

C

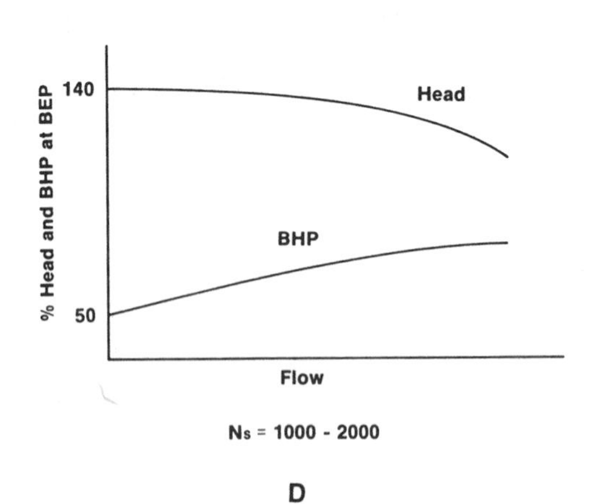

N_s = 1000 - 2000

D

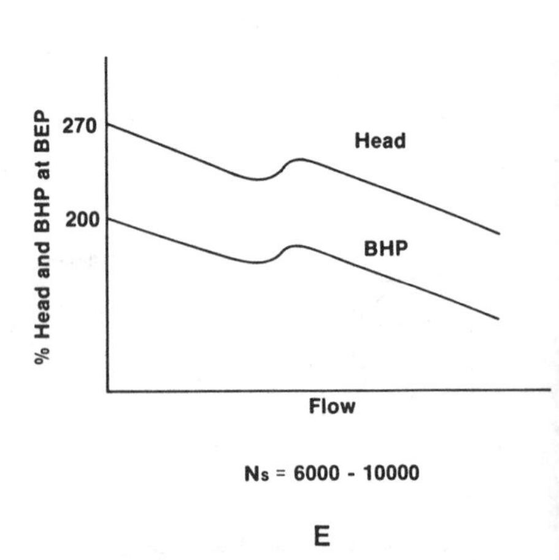

N_s = 6000 - 10000

E

Figure 7. Performance curve shapes for centrifugal pumps having different specific speeds.

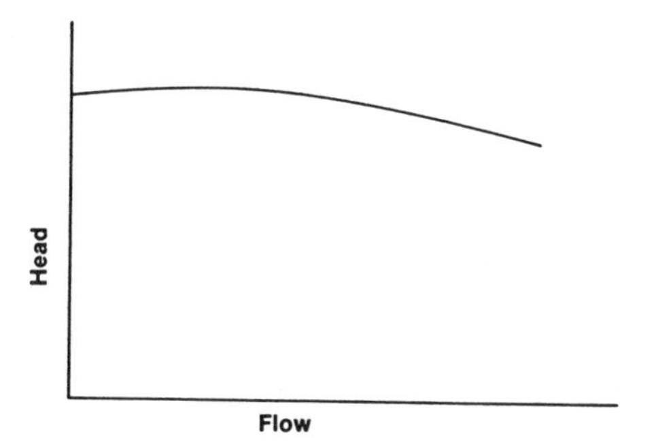

Figure 8. Drooping head curve.

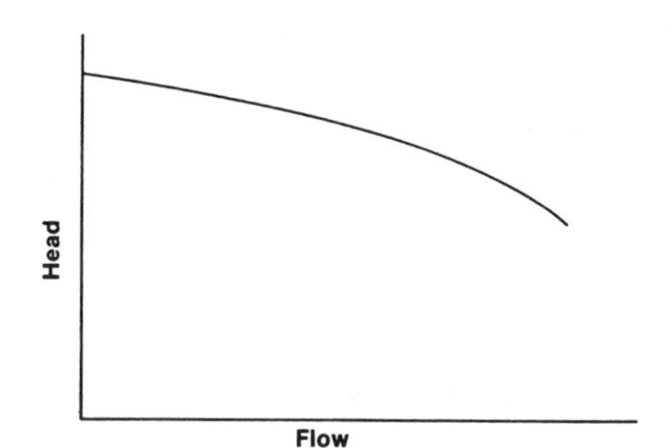

Figure 9. Continuously rising head curve.

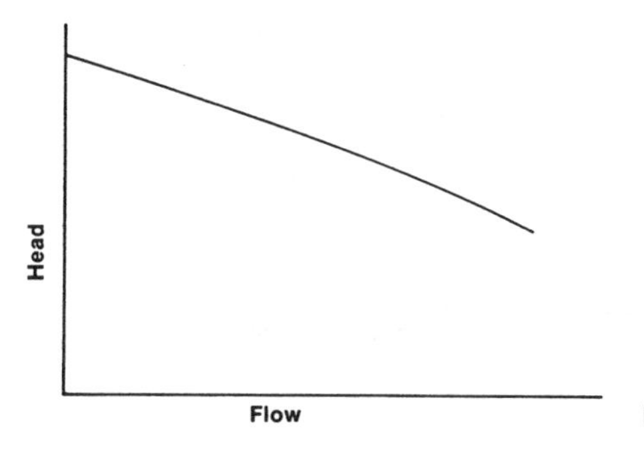

Figure 10. Steep head curve.

Flat head characteristics are those in which the head rises only slightly as the flow is reduced. As with steepness, the magnitude of flatness is a relative term (Figure 11).

Discontinuous head characteristics are those in which a given head is developed by the pump at more than one flow rate. Many pumps in the high specific speed ranges have this characteristic (Figure 12).

Continually increasing horsepower characteristics are those of low specific speed pumps where the horsepower increases at flows greater than the best efficiency point (BEP) and decreases at flows to the left of BEP (Figure 13).

Peaking horsepower characteristics are those of medium specific speed pumps where maximum horsepower occurs in the BEP range and decreases at all other values of flow (Figure 14).

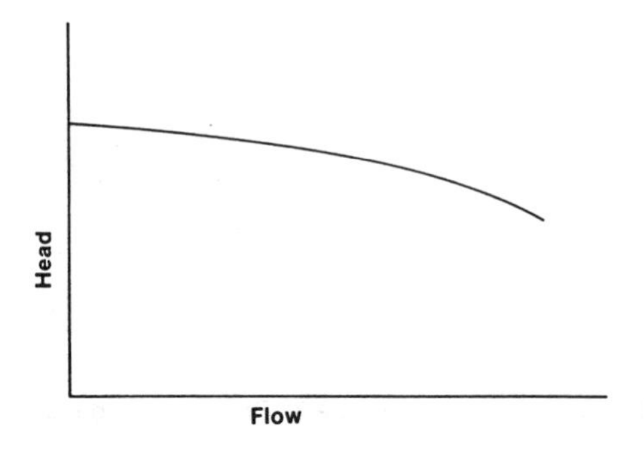

Figure 11. Flat head curve.

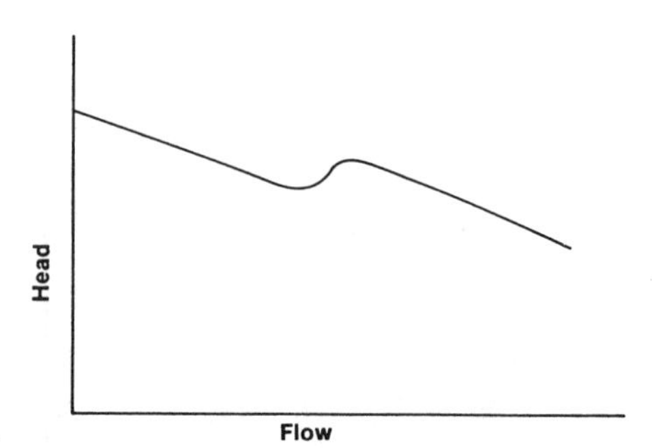

Figure 12. Discontinuous head curve.

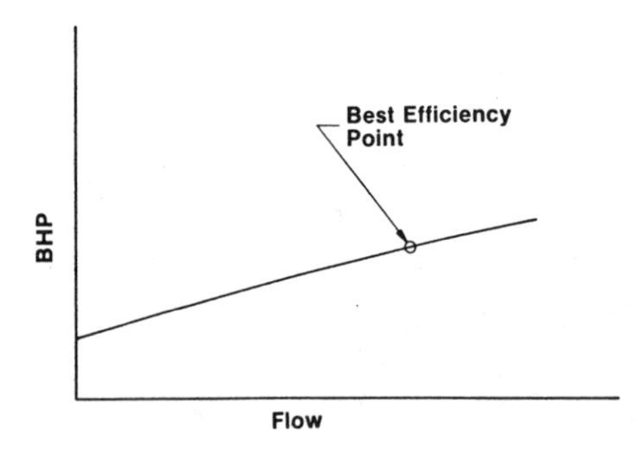

Figure 13. Illustrates continually increasing horsepower characteristics.

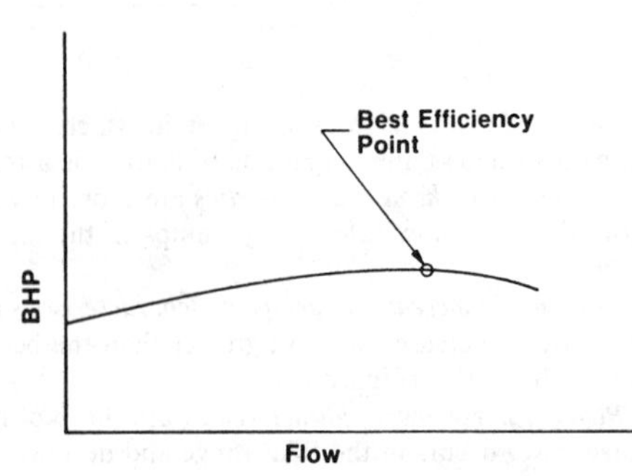

Figure 14. Illustrates peaking horsepower requirements.

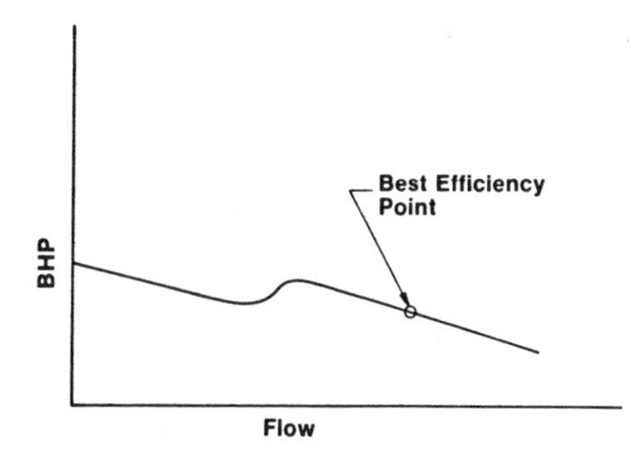

Figure 15. Illustrates discontinuous horsepower characteristics.

Discontinuous horsepower characteristics are those of high specific speed pumps where horsepower increases at flows less than BEP and decreases with flow to the right of BEP. The discontinuity usually occurs at flows less than BEP (Figure 15).

PUMP AFFINITY LAWS

The pump affinity laws state that flow is proportional to impeller peripheral velocity and head proportional to the square of the peripheral velocity. Specifically,

- Flow, Q, will vary directly with the ratio of change of speed, N, or impeller diameter, D:

$$\frac{Q_2}{Q_1} = \frac{N_2}{N_1} \quad \text{or} \quad \frac{Q_2}{Q_1} = \frac{D_2}{D_1} \tag{6}$$

- Head, H, will vary as the square of the ratio of speeds or impeller diameters:

$$\frac{H_2}{H_1} = \left(\frac{N_2}{N_1}\right)^2 \quad \text{or} \quad \frac{H_2}{H_1} = \left(\frac{D_2}{D_1}\right)^2 \tag{7}$$

- The pump required horsepower (BHP) will vary as the cube of the ratio of speeds or impeller diameters:

$$\frac{BHP_2}{BHP_1} = \left(\frac{N_2}{N_1}\right)^3 \quad \text{or} \quad \frac{BHP_2}{BHP_1} = \left(\frac{D_2}{D_1}\right)^3 \tag{8}$$

A basic premise throughout these analyses is that pump efficiency remains constant for speed or impeller diameter changes. For example, the data in Table 4 show a projection of pump performance from 1,760 rpm to 1,450 rpm. The subscript 2 is used for the calculated performance and subscript 1 for the values selected from the 1,760-rpm curve:

$$Q_2 = \left(\frac{N_2}{N_1}\right) Q_1 \tag{9}$$

$$H_2 = \left(\frac{N_2}{N_1}\right)^2 H_1 \tag{10}$$

$$BHP_2 = \left(\frac{N_2}{N_1}\right)^3 BHP_1 \tag{11}$$

Table 4
Calculated and Actual Pump Performance Data

Tabulated Performance Data from 1760-rpm Curve				Calculated Performance Data for 1450-rpm Operation			
Q_1 (gpm)	H_1 (ft)	Eff_1	BHP_1	Q_2 (gpm)	H_2 (ft)	Eff_2	BHP_2
1000	184	0.61	76.2	824	125	0.61	42.6
1500	175	0.76	87.2	1236	119	0.76	48.9
2000	166	0.84	99.8	1648	113	0.84	56.0
2500	151	0.86	110.8	2060	103	0.86	62.3
3000	128	0.82	118.3	2478	87	0.82	66.4
3250	110	0.73	123.7	2678	75	0.73	69.5

A similar table can be developed for changing impeller diameter. In projecting pump performance, care should be taken to project from the speed or diameter that is closest to the calculated value. Changes in both speed and diameter may alter the efficiencies taken from the originating pump performance curve.

Pump performance curves are developed as a means of conveniently viewing the complete characteristics of a given pump and selecting a pump that will successfully meet system requirements. The pump is a source of pressure energy and, within its limits of design and operating speed, has the capability of providing the liquid in a system with the potential to do work.

Pressure energy capability is dependent on the velocity the impeller imparts to a liquid. For a given pump, this velocity can be changed by varying the impeller diameter (Figure 16) or by changing the operating speed. Therefore, by controlling impeller angular velocity a centrifugal pump can be used as a variable energy source to satisfy pumping systems having variable pressure energy requirements.

When system requirements dictate that the centrifugal pump *must* have particular curve shape characteristics, the approximate discharge specific speed index range may be used to define the general requirement for the specified pump.

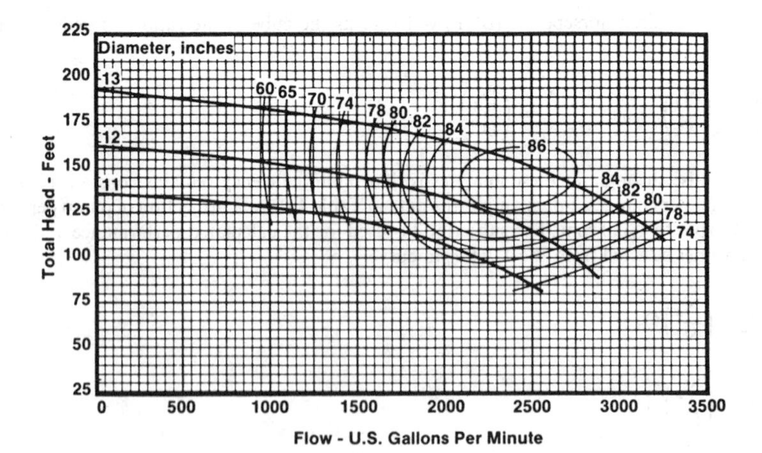

Figure 16. Pump head curves as a function of impeller diameter.

PUMP SYSTEM CHARACTERISTICS

A pumping system is the arrangement of pipe, fittings, and equipment through which liquids flow. Only a few pump applications are such that a pumping unit can operate alone, one example being a pump lifting water from a well into an irrigation ditch. The ditch acts as a conduit distributing the water by gravity flow. Pumps are required when liquids need to be transported from lower to higher elevations, moved over long distances, distributed in grids, or circulated in loop or pressurized systems to perform work.

Mechanically, systems can be classified by path of liquid flow as follows:

1. *Nonreturn systems*, in which all the liquid is discharged from the system.
2. *Return systems* in which (a) none of the liquid is discharged from the system; and (b) some of the liquid is discharged from the system.

Functions of systems can be classified by the characteristics of work performed as follows:

1. Thermal exchange, in which some form of thermal exchange is conducted for the purpose of satisfying a design condition.
2. Removal/delivery, in which the system is designed to remove the liquid from, or deliver it to, some point to satisfy a specific design service.

The following lists illustrate typical pump application names classified by system function:

1. Thermal exchange applications

- Chilled water
- Cold well
- Condenser water
- Cooling tower
- Heat recovery
- Hot well
- Mill roll cooling
- Plant circulating water
- Plant cooling water
- Spray pond
- Strip mill quench

2. Removal/delivery applications

- Ash sluice
- Boiler feed
- Condensate
- Domestic water
- Effluent
- Filter backwash
- Flood irrigation
- High service
- Low service
- Municipal booster
- Raw water
- River intake
- Sewage ejector
- Sprinkler irrigation
- Storm water

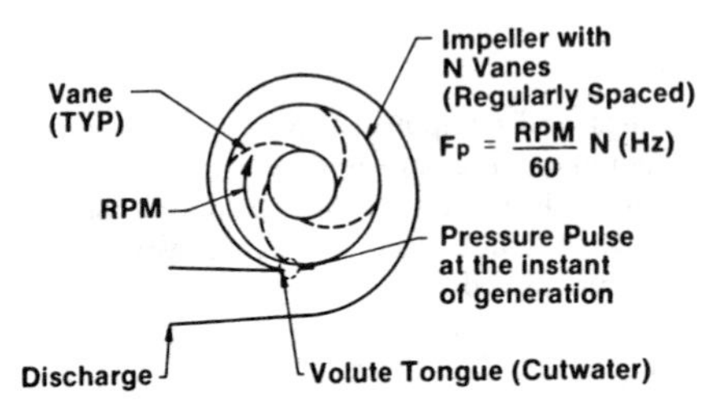

Figure 17. Illustrates noise problem in pumps generated by vanes in an impeller passing the volute tongue.

Items that affect the cost and operation of piping systems include noise, water hammer, pipe aging, and friction loss.

Flow noise, or "hiss," contains all sound frequencies at an evenly distributed sound level called "white noise." This noise is generated by turbulent flow through system pipes and increases with velocity; or it is generated by cavitation in the system. Flow noise occurs in pumps, valves, elbows, tees, or wherever flow changes direction or velocity.

Conduction of noise proceeds through the liquid, as well as through the pipe material, and is usually controlled by flexible fittings, sound insulating pipe supports, and pipe insulation.

Pipe resonance can result in objectional sound in a piping system. Resonance is a single tone sound magnified and emitted by a length of piping. It occurs when the natural frequency of a length of piping is matched by the frequency of some regular energy source such as the vanes in an impeller passing the volute tongue (cutwater) as shown in Figure 17. An illustration of this type of resonance is shown in Figure 18. Other regular energy sources, such as the rotating elements in a bearing or the power supply frequency, also can cause pipe resonance. Pipe resonance can be corrected by increasing or decreasing the resonating length or weight of pipe under the direction of an acoustical consultant.

Water hammer is the result of a strong pressure wave in a liquid caused by an abrupt change in flow rate. As an illustration, for the maximum possible instantaneous head increase above the normal head due to a water hammer pressure wave the following expression, known as Joukowsky's law, applies:

$$H_{wh} = \frac{Cu}{g} \tag{12}$$

where C = velocity of sound in the liquid (ft/s), e.g., C for 15.6°C water is 4,820 ft/s.
u = normal velocity in the conduit before closing valve (ft/s)
g = acceleration due to gravity (32.2 ft/s^2)

Assuming $u = 15$ ft/s, H_{wh} for 15.6°C water is 2,245 ft, or a pressure surge of 972 psi. This is the maximum possible pressure rise by instantaneous closing of the valve and may be more than the system can withstand.

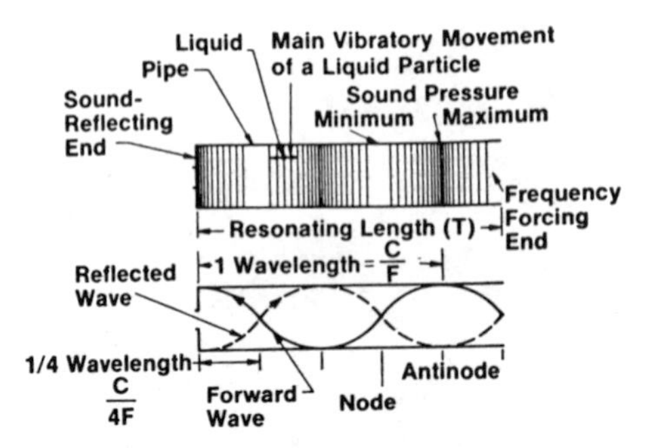

Figure 18. Illustrates noise generated by pipe resonance.

Pipe Deterioration

Some deterioration of the inside surfaces of a piping system normally occurs with age. This deterioration may result from corrosion, deposits, or a combination of both, which impede flow by increasing the relative roughness ratio, ϵ/D

where ϵ = height of protrusion on a side in feet or inches
D = inside diameter of pipe in feet or inches

Table 5 contains some representative values for the height of the internal surface protrusions.

For old material, the values listed may be increased by 100 or more times due to corrosion or deposits. Past experience is generally the most accurate guide to determine the extent of pipe deterioration caused by aging.

Table 5
Representative Values for the Height of Internal Surface Protrusions in Pipes

Material (New)	ϵ(ft)	ϵ(in.)
Copper or Brass Tubing	5×10^{-6}	4.2×10^{-7}
Steel Pipe	1.5×10^{-4}	1.3×10^{-5}
Galvanized Steel	5×10^{-4}	4.2×10^{-5}
Concrete	4×10^{-3}	3.3×10^{-4}

Pipe Friction

Pipe friction is resistance to flow and results in a loss of head, which is expressed as friction head H_f, in feet of liquid. Methods of calculating friction head have been discussed in previous chapters and only supplementary notes are given. First, proper viscosity conversion is sometimes confusing in evaluating friction losses. If absolute viscosity is expressed in centipoise (cp), the following conversions should be used:

$$2.089 \times 10^{-5} \times \text{cp} = \text{lb-s/ft}^2$$

which, in turn, can be converted into kinematic viscosity (ft²/s) by dividing by the mass density (lb-s²/ft⁴).

If kinetic viscosity is expressed in centistokes (cs), use the following conversion:

$$1.076 \times 10^{-5} \times \text{cs} = \text{ft}^2\text{/s}$$

If viscosity is expressed in SSU, there is no distinction between the absolute and the kinetic designations. Convert SSU to kinematic viscosity as follows:

For SSU larger than 100:

$$1.076 \times 10^{-3} \times \left(0.0022\ \text{SSU} - \frac{1.35}{\text{SSU}}\right) = \text{ft}^2\text{/s}$$

For SSU equal to 100 or less:

$$1.076 \times 10^{-3} \times \left(0.00226\ \text{SSU} - \frac{1.95}{\text{SSU}}\right) = \text{ft}^2\text{/s}$$

The following examples illustrate these conversions:

Example 1: Given—40°C water with kinematic viscosity of 1.58 cs. Convert viscosity into ft^2/s units.

Solution:

$$1.076 \times 10^{-5} \times 1.58 \text{ cs} = 1.70 \times 10^{-5} \text{ ft}^2/\text{s}$$

Example 2: Given—ethylene glycol at 21.1°C with a viscosity of 88.4 SSU. Convert viscosity into ft^2/s units.

Solution:

$$1.076 \times 10^{-3} \times \left(0.00226 \times 88.4 - \frac{1.95}{88.4}\right) = 1.912 \times 10^{-4} \text{ ft}^2/\text{s}$$

Example 3: Given—ethyl alcohol at 20°C with specific weight of 49.4 lb/ft^3 and absolute viscosity of 1.2 cp. Convert viscosity into ft^2/s units.

Solution:

$$2.089 \times 10^{-5} \times 1.2 = 2.507 \times 10^{-5} \text{ lb-s/ft}^2$$

$$\text{Mass density} = \frac{49.4 \text{ lb/ft}^3}{32.2 \text{ ft/s}^2} = 1.5342 \frac{\text{lb-s}^2}{\text{ft}^4}$$

$$\frac{2.507 \times 10 \text{ lb-s/ft}^2}{1.5342 \text{ lb-s}^2/\text{ft}^4} = 1.63 \times 10^{-5} \text{ ft}^2/\text{s}$$

Once proper viscosities have been determined, friction factors can be obtained.

Through many years, simplified variations of the Darcy-Weisbach formula have been devised for limited areas of application. The accuracy of such variations is supported by successful use and allows the user to bypass the diagram interpretation and calculation necessary with the basic formula. The use of specific formula variations outside their known areas of successful application may lead to inaccurate projections of friction head.

This simplified derivation of the Hazen-Williams formula can be used to calculate friction head for liquids having a kinematic viscosity of 1.1 cs. (Water at 60°F has a viscosity of 1.13 cs.)

$$\text{Friction head, } H_f, = 10.45\left(\frac{\text{gpm}}{C}\right)^{1.85}\left(\frac{L}{D^{4.87}}\right) \tag{13}$$

where gpm = gallons per minute
C = pipe roughness factor ranging from 60 to 160
D = inside diameter of pipe (in.)
L = length of pipe (ft)

The Hazen-Williams formula is generally used for cast iron pipes of 3-inch and larger diameter. As pipes deteriorate, the roughness factor, "C," decreases. This decrease in C depends on the pipe material, pipe linings, pipe age, and the characteristics of the liquid.

For liquids with viscosities other than 1.1 cs, methods more accurate than the Hazen-Williams formula should be used in determining friction head. For example, friction head computed with the Hazen-Williams formula will increase by 20% at a viscosity 1.8 cs and decrease by 20% at a viscosity

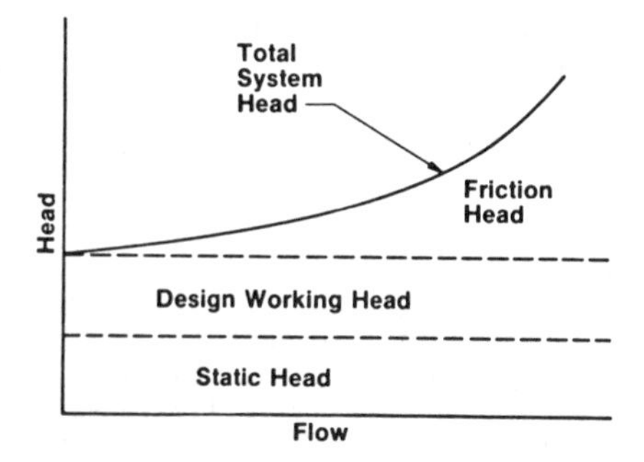

Figure 19. Illustrates total system head curve components.

of 0.29 cs. In addition, friction loss will occur in fittings such as elbows, tees, and valves. Many handbooks and manuals provide means for determining these losses. Manufacturer's data must be used to determine friction loss for mechanical equipment installed in the system.

SYSTEM HEAD CURVES AND COMPONENTS

A system head curve is the graphic representation of the head required at all flows to satisfy the system function. Regardless of the mechanical configuration, function, or means of control, the system head curve or system head band is used to define total head versus flow for any piping system. System head curve development is a combination of calculations and the designer's "best feel" for the variable conditions.

The three components that make up total system head are static head, design working head, and friction head. These components are defined as follows:

1. *Static head* is the vertical difference in height between the point of entry to the system and the highest point of discharge.
2. *Design working head* is the head that must be available in the system at a specified location to satisfy design requirements.
3. *Friction head* is the head required by the system to overcome the resistance to flow in pipes, valves, fittings, and mechanical equipment.

Total system head at a specified flow rate is the sum of static head, design working head, and friction head. The values of static head and design working head may be zero. Figure 19 illustrates a total system head curve made up of the friction head and the design working head and static head.

For most systems, the total system head requirements are best illustrated by a band formed by two total system head curves. This band of system requirements is a result of variable factors that affect total system head calculations. The following list of variables should be considered in projecting system requirements.

Static head will vary as a result of change in the elevation of the highest point of discharge of the system.

Design working head is usually treated as a constant component of total system head.

Friction head at any specified flow will vary as a result of:

- Method of calculation or source of tabulated data.
- Change in viscosity resulting from a change in liquid temperature.
- Deterioration of the piping system.
- Load distribution.
- Systems differences between design and "as-built."
- Manufacturing tolerances of mechanical equipment.
- Accumulation of solids in the system.

Most systems are not completely defined by a single line system head curve. Those variables that are applicable determine the band of maximum and minimum system head requirements. The band between maximum and minimum head requirements at any given flow will be significant in some systems and not in others. For all systems, interaction of total head available and total head required must be evaluated to achieve satisfactory operation.

SYSTEM OPERATION AND CONTROL

For a system to operate at design flow, total head available to the system must be equal to the total system head required. For proper evaluation of system operation at flow rates other than design, a comparison of total available head and system required head must be made based on the means selected to control system flow. The reference point used to determine total available head must be the same as that used to determine total system head required. The point of entry to the system, which is also the pump discharge connection, is used in this text as the reference point.

Total available head at the point of entry to the system is the sum of static head and pump head.

Static head available is the vertical difference in height between the point of entry to the system and the liquid supply, minus any friction head in the supply pipe and fittings. Static head may be relatively constant. The decrease in static head with increasing flow is the result of friction head in the supply pipe. In some applications static head will vary, and is normally illustrated by a band of maximum and minimum values.

When the liquid supply is below the point of entry to the system, as shown in Figure 20, static head available is a negative value.

The difference in elevation between the fluid supply and the point of entry to the system must be developed by the pump. Therefore, pump head must include static head and friction head in the supply pipe and fittings.

In systems in which the liquid supply is from a pressurized main, the pressure must be expressed in feet of liquid. The equivalent height of the liquid above the system entry point is static head available. In Figure 21, with a 100 psig pressure at point A and a liquid with a specific gravity of 1.0, the static head available will be 231 feet minus elevation height, X. In return piping systems, in which none of the working fluid is discharged from the system, static head available is zero.

Pump head is the total head developed by a pump. Its value at any given flow rate must be obtained from the pump manufacturer's performance curve.

A curve of total available head versus flow is developed by adding static head and pump head at several values of flow. The total available head curve is slightly steeper than the pump head curve due to the slope of the static head curve. To evaluate the operation of a system, the total system head curve or curves and the total available head curve are plotted on a common graph. The two curves will intersect at maximum flow for that system. As shown in Figure 22, system flow greater than 100% cannot occur since total system head required exceeds the total available head.

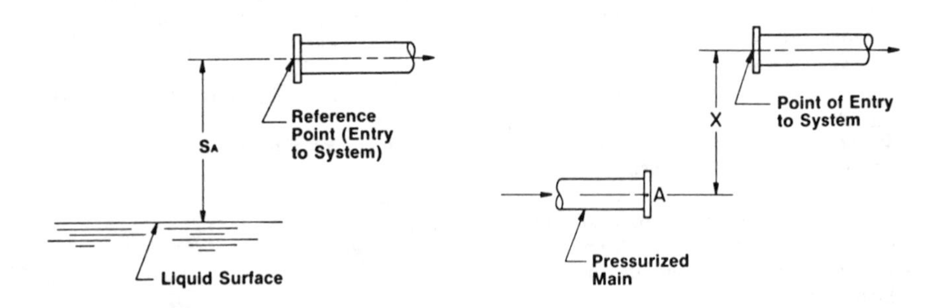

Figure 20. Illustrates a case in which the static head available is a negative curve.

Figure 21. Example problem to determine static head available.

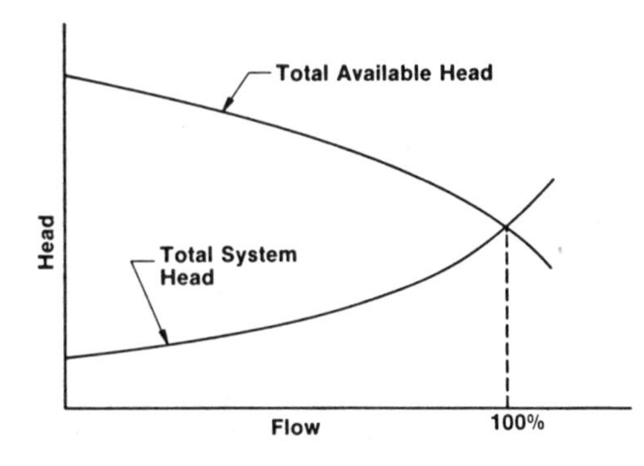

Figure 22. Shows that the total system head required exceeds the total available head above 100% flow.

Consider a system when the total available head must be represented by two curves forming a band as shown in Figure 23A. This is a common situation that is a result of maximum and minimum values of static head available. It is now possible for system flow to exceed design flow of 100% since the pump selection was based on the minimum value of static head available at 100% flow.

Figure 23B illustrates a system with a band of total system head and a band of total available head. At any given time, system operation can occur anywhere within the area A-B-D-C-A. While

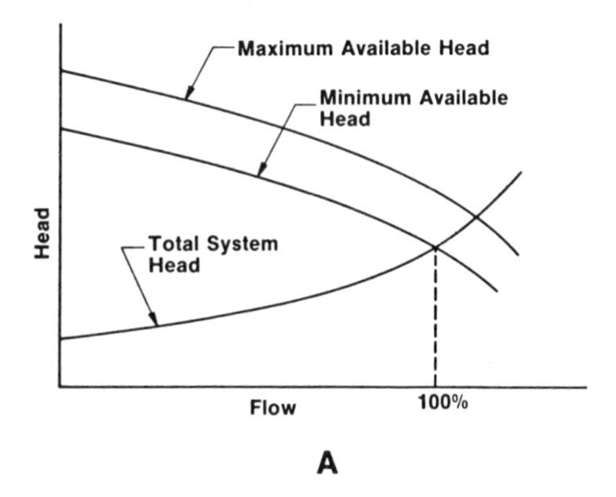

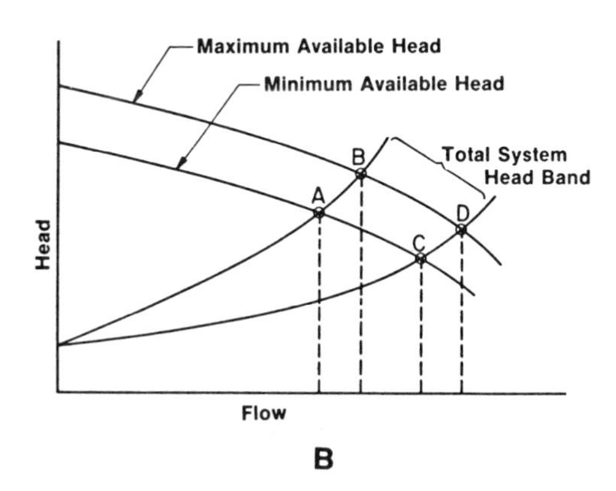

Figure 23. Illustrates interpretation of flow head curves.

the pump is selected for proper operation at design flow, operation at flows other than design may result in problems such as noise, cavitation, low pump efficiency, pump or system damage, or poor system function. To avoid these problems, should they occur, some means of system control must be employed.

Control of system operation is achieved by altering total system head, total available head, or both. Each method has advantages and disadvantages depending on the system, its design function and the characteristics of total available head. Methods of control should be selected that best satisfy the operation and cost objectives of the system.

Total System Head Alteration

Control of system operation by altering total system head is acomplished with valves that change the friction head component of total system head. Two methods of valve control are bypassing and throttling, each having a different effect on system head.

Valve bypassing usually is accomplished with valves that have three ports: one for entrance of liquid and two that determine the liquid flow path, as shown in Figure 24A. Movement of the valve mechanism reduces the cross-sectional area of one port and simultaneously increases the cross-sectional area of the other port. This action causes increased head through one flow path and reduced head through the other; however, the head required at the entrance to the valve remains relatively constant. Figure 24B illustrates how flow remains constant at the entrance of the bypass value as flows change in the functional portion of the system. The bypass valves may be automatic or

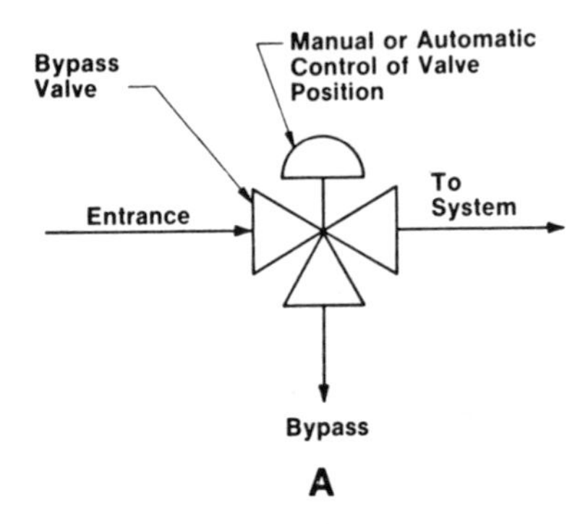

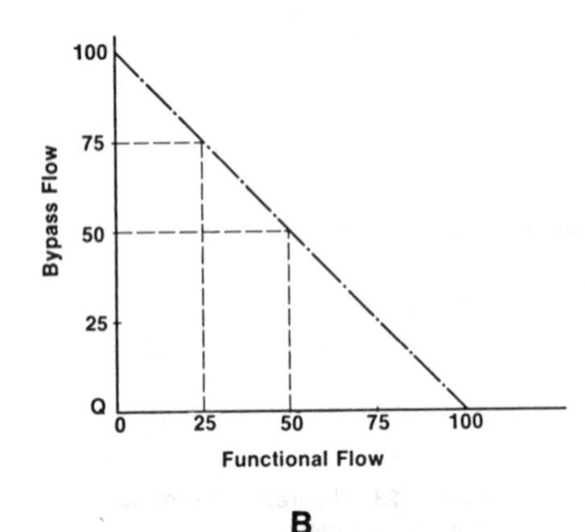

Figure 24. Illustrates valve bypassing. Flow remains constant at the entrance of the bypass valve as flows change in the functional portion of the system.

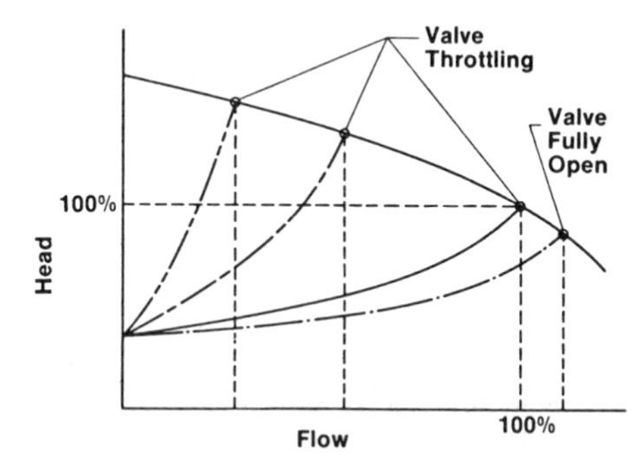

Figure 25. Illustrates valve throttling effect on head curve.

manual and can be located anywhere in the system. Regardless of the type of operation or location, the end result of valve bypass control is constant total flow from the pump for any given value of total available head.

Valve throttling control of system operation usually is accomplished with valves that have two ports: one for the entrance and one for the exit of liquid. Movement of the valve mechanism changes the cross-sectional area of the valve port causing friction head to increase to a greater value than the full open valve friction head. From fully open to fully closed position, valve throttling produces a series of system head curves as shown in Figure 25.

System friction head will vary depending on relative valve opening, and resulting system flow will occur at the point of intersection between the system head curve and the total available head curve.

Total Available Head Alteration

Control of system operation by altering the pump head portion of total available head can be accomplished by changing pump impeller diameter, selecting pumps for series operation or parallel operation, or by changing the pump operating speed. Each method of control will have a different effect on system operation.

Change of impeller diameter will alter the pump head component of total available head. In systems in which constant speed pump impeller diameters have been selected for the *calculated* system design head, and the *actual* system head is less than that calculated, system operating efficiency usually can be improved by machining the impeller diameter to develop the actual operating head of the system. All aspects of pump and system characteristics explained in the preceding sections should be reviewed before considering changing impeller diameter to ensure satisfactory pump and system operation.

In *series pumping*, the pump head component of total available head is the sum of the heads developed by each pump at any given flow. Each pump must be selected to operate satisfactorily at system design flow. Pumps operated in series can be referred to as "pressure additive," illustrated in Figure 26. With one pump operating, system flow will occur at point A. With both pumps operating, system flow will occur at point B, which is system design flow. In this mode of operation, both pumps are developing equal head at full system flow.

In *parallel pumping*, the pump component of total available head is identical for each pump and the system flow is divided among the number of pumps operating in the system. The flows produced by individual pumps can represent any percentage of total system flow. Where series pumping is described as "pressure additive," parallel pumping is described as "flow additive." When operating in parallel, pumps will always develop an identical head value at whatever their equivalent flow rate is for that developed head, and the sum of their capacities will equal system flow. Figure 27 illustrates parallel pumping in which each pump develops 50% of total flow at 100% head. With one pump operating, system flow will occur at point A; with both pumps in operation, flow will occur at point B.

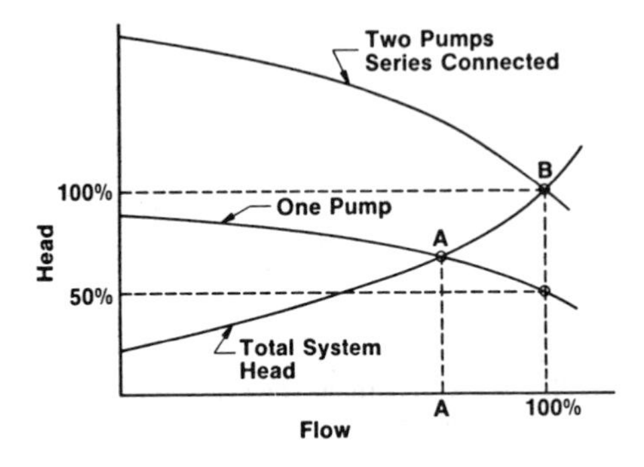

Figure 26. Effect of series pumping on the head curve.

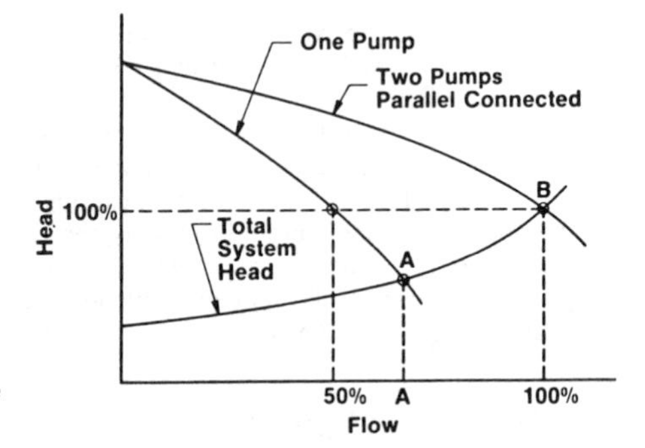

Figure 27. Effect of parallel pumping on the head curve.

Changing pump speed will alter the pump head component of the total available head. The performance characteristics of a pump at various speeds can be calculated by use of the affinity laws and will result in a pump head-flow curve for each operating speed, as shown in Figure 28A. The application of variable-speed pumps in systems can result in system operation at the intersecting

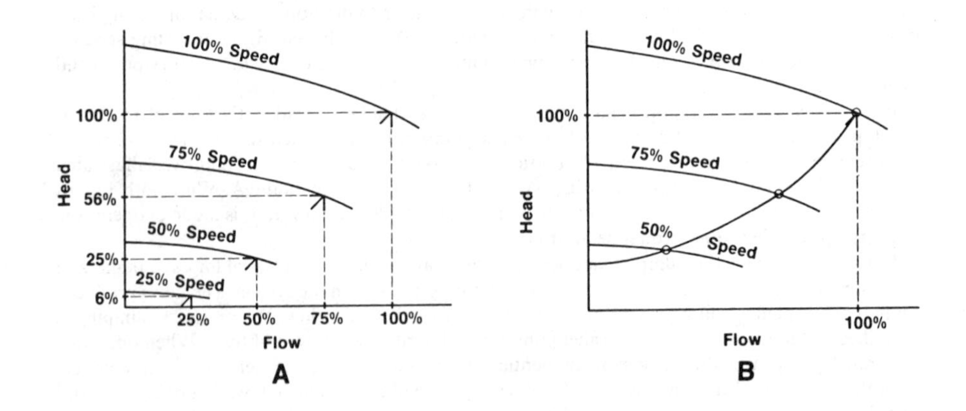

Figure 28. Effect of changing pump speed on head curve.

points shown in Figure 28B. The number of intersection points will be dependent on the number of pump speed changes.

There are several devices available to vary pump operating speeds. Some cause fixed increments of speed change, such as multiple-speed motors, while others operate at an infinite number of speeds within the design limits of the device. With drives that have fixed increments of speed change, a fixed number of intersection points occur between the pump head curves and the system head curve. Pump drives that have an infinite number of operating speeds can produce an infinite number of intersecting points on the system head curve.

VARIABLE-SPEED PUMPS

When variable speed pumps are employed as a means of system control, the system sensing method that will control the pump speed must be determined. This is necessary because the total available head at the point of entry to the system is a function of the type and location of the monitor.

Functions

The sensing equipment must perform three functions:

1. Monitor the system variable.
2. Compare the value of the system variable to the required value (set point).
3. Cause the variable-speed pump to restore the variable to the required value.

System values typically monitored include level, flow, pressure, differential pressure, temperature, and differential temperature.

Types

Sensing equipment may be as simple as a piece of tubing connecting the piping system to the variable-speed drive or as complex as electronic telemetry equipment that monitors variables many miles from the pump.

Location

The total operation of the system must be evaluated at the location where the system variable is being monitored.

When equipment in the system causes a change in flow, the path of pump operating points will not follow the unaltered system head curve. An infinite number of new system head curves will occur when system components cause flow change. For example, in a system that requires a fixed design working head, the path of pump and system operating points can be predicted once the location of the system monitoring device is established. This path of operation will be controlled by the location of the monitor and will affect the operating speeds and power requirements of the pump.

Figure 29A illustrates a simple system in which flow change is caused by a valve near the end of the system. The system head curve for this example is shown in Figure 29B and includes static head (S_1) and friction head (H_f) between points A and C, with the valve in a fixed position.

A variable-speed pump with a pressure monitor at the point of entry to the system, A, will produce a constant total available head at A for any system flow. The setting for the pressure monitor is determined from the following formula:

$$P_1 = P_2 + H_f + S \quad (14)$$

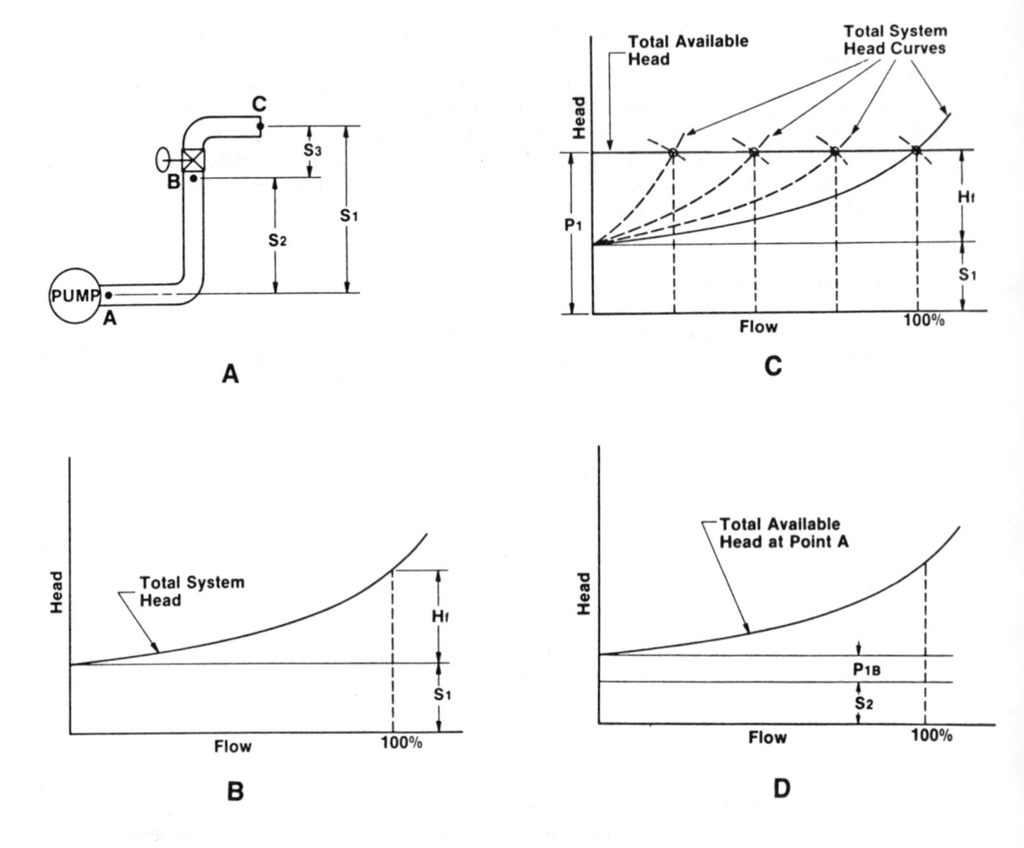

Figure 29. Pumping problem example.

where P_1 = total available head, in feet, to be maintained at the pressure monitor

P_2 = design working head, in feet

H_f = full flow friction head, in feet, between the monitor and the location in the system where P_2 is to be maintained, or from the monitor to the end of the system (A to C) when P_2 is zero

S = static head, in feet, between the monitor and the location in the system where P_2 is to be maintained, or from the monitor to the highest point of discharge; static head, S, is applicable only in systems employing pressure monitors

In Figure 29A, P_2 is zero and $P_{1A} = H_f + S_1$.

As system flow is throttled, a series of new system head curves will be produced, as shown in Figure 29C. System flow will occur at those points where the system head curves intersect the available head curve. At any flow of less than 100% there is more head available than required, and the excess is consumed as friction head across the throttling valve. Now consider the same system with the pressure monitor located in Figure 29A at point B. The system has not changed, and the total system head required is still as shown in Figure 29B. The pressure to be maintained at the monitor is determined from the same formula:

$$P_1 = P_2 + H_f + S$$

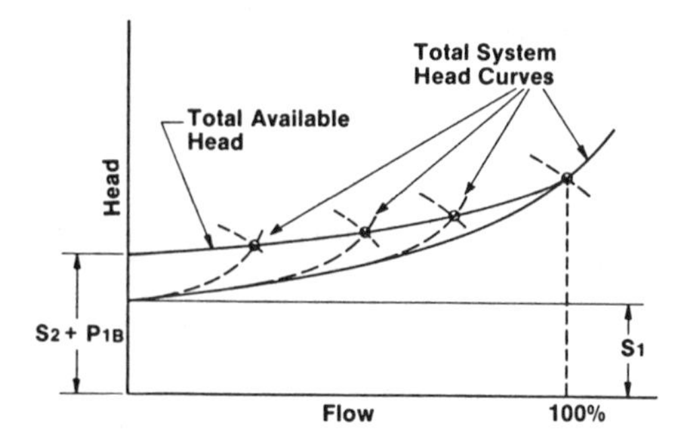

Figure 30. Series of head curves generated for example problem.

Design working head, P_2, is zero, and static head is equal to S_3. H_f is the full flow friction from B to C. The pressure setting is $P_{1B} = H_f + S_3$. The total head that must be available at point A to satisfy the monitor at point B will be equal to the pressure monitor setting, P_{1B}, plus the static head, S_2, plus the friction head from A to B. This is illustrated in Figure 29D.

As system flow is throttled, a series of system head curves will be produced, as shown in Figure 30. System flow will occur at those points where the system head curves intersect the total available head curve. By comparing Figures 29C and 30, the throttling losses are less with the pressure monitor at B than they are with the pressure monitor at A for specific values of flow. This can result in lower pump operating speed, lower horsepower required, and lower operating cost.

Sensing Multiple Branch Systems

In parallel piping systems with two or more branches, a single pressure monitor must be set to satisfy the branch with the maximum calculated friction head. In some instances, load variation in the system can cause the pressure in nonmonitored branches to drop below acceptable limits, even though the monitored branch is satisfied. This type of problem can be resolved by analyzing the various branches based on the location and setting of a single pressure monitor. Additional pressure monitors can be added to those branches that experience the low pressure condition to restore system operation to acceptable levels.

Evaluation of the total available head band and the total system head band over the expected range of system flow is necessary to make decisions that will affect the cost and operation of a pumping system. The decisions are based on problem statements and evaluation of the possible solutions with respect to the cost and functional objectives of the system design. The following questions outline some of the considerations that should be reviewed to assure satisfactory operation of a pumping system:

1. If the system is throttled for reduced system flow, does the maximum total available head exceed the system working pressure? If so, the alternatives include:

   - Increase system working pressure.
   - Select pump with a "flatter" head capacity characteristic.
   - Consider variable-speed pump drive.
   - Add bypass valve or pressure-reducing valve to system.

2. Does pump horsepower increase with decreasing flow? If it does, the alternatives include:

   - Add bypass valve to system.
   - Consider variable-speed pump drive.
   - Select several lower specific speed pumps to operate in parallel.

3. Will operation at reduced flows occur for significant periods of time? If so, pump efficiency will be reduced and the service life of pump bearings, packing, mechanical seals and close

clearance wearing rings usually will be shortened, and the alternatives include:

- Select smaller pump for the reduced flow operation.
- Add bypass valve to system.
- Consider variable-speed pump drive.

4. Will actual system head at design flow be considerably less than the calculated value? If so, the system flow will exceed design flow, and the alternatives include:

   - Reduce the pump impeller diameter.
   - Install valve to throttle system flow.
   - Consider variable-speed pump drive.

5. In systems planned for two or more pumps operating in parallel, have all the operating possibilities been considered? For example, one pump alone can operate at higher flow than its rating, as selected for parallel operation. For the higher flow operation the pump will require more net positive suction head and may be subject to undesirable hydraulic loads, motor overload, increased noise level, and shortened life.

These are only a few of the many problems that can be encountered in the course of designing a pumping system. In many instances, several solutions can provide equally satisfactory results if operating power requirements are not considered. If design objectives include minimum power consumption, an energy evaluation of the pumping system must be made.

TOTAL SYSTEM EVALUATION

Total system evaluation involves developing information about the system and its pumping equipment necessary to satisfy a system's design requirements, concluding with a projection of total power requirements. To perform the evaluation, system head and total available head must be established. These two components are combined with pump power requirements and system load profile to project total system operating costs.

Load profile expresses the measure of work executed in a system compared to a unit of time. Work performed has a direct relationship to flow; therefore, load expressed as flow provides a common base for any system energy requirement projection. System load profile can be illustrated with a curve as shown in Figure 31; however, it is generally easier to use the flow/time relationship in tabular form, as illustrated in Table 6. The tabular format organizes total operational time at specific flow values and simplifies identifying high flow/time concentration areas, which should be the focal point for preliminary pumping equipment selection.

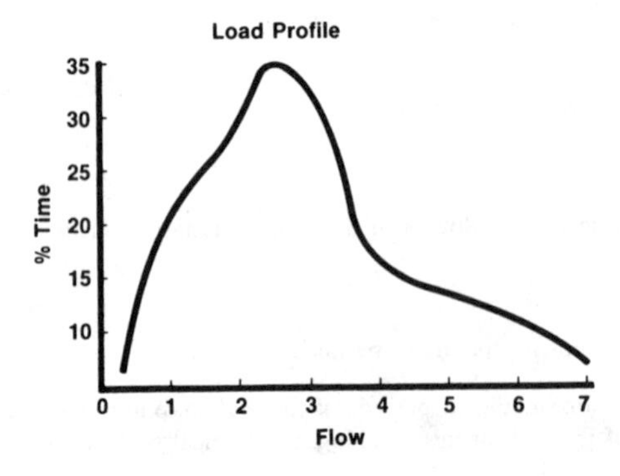

Figure 31. System load profile plot.

Table 6
Tabulated Flow/Time Data

Flow (gpm)	Hours	Flow (gpm)	Hours
280	87	4480	788
840	1138	5600	613
1680	1841	6440	350
2520	2629	7000	87
3640	1227	---	8760

Figure 32 illustrates how the flow/time relationship can be combined with a system head curve to aid in pump selection procedures. In this example, a significant amount of system operational time will occur in the flow ranges of 20% to 30% and 60% to 75% of the total design flow. Pumping equipment should be selected to provide maximum operating efficiencies in these flow ranges. One possible pump equipment selection for this system would include a single pump selected for the system head and flow requirements in the 20% to 30% range, as well as two identically sized pumps to share the total system capacity equally at full system flow, but selected so that each will be in its best efficiency range when operated independently in the 60% to 75% system flow range (Figure 32).

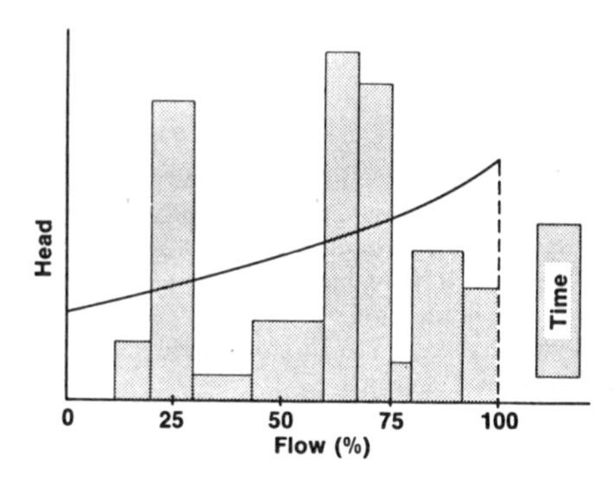

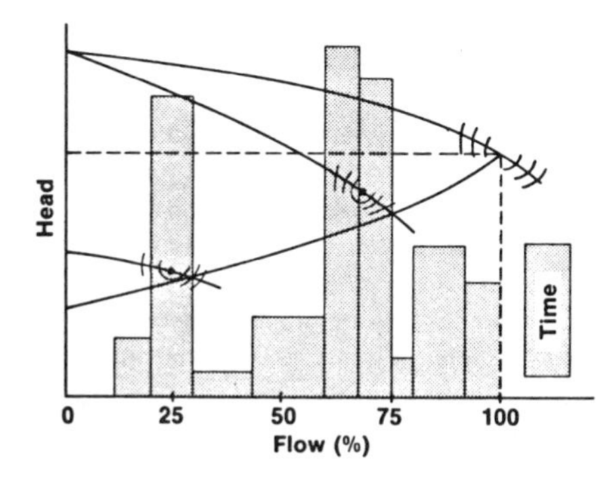

Figure 32. Illustrates how the flow/time relationship can be combined with a system head curve.

The time base for computing load requirements should be of sufficient length to incorporate all variables affecting system operation. The minimum practical load profile time length is one complete cycle, from minimum to maximum flow requirements. A twelve-month time base for total system load evaluation generally encompasses all production, load, and climatic variables for a system and can be used to project system operating requirements and costs. System load profile is the central component of any energy evaluation process.

In projecting total system energy requirements, total available head must be reviewed. If static head available is variable with respect to system flow, time, or both, a profile of these relationships should be made. This projection should represent the magnitude of change compared to the length of time the change is expected to occur, thus developing a correlation with system load profile. If the accuracy of these data is questionable, a band of minimum to maximum values should be established from which a total system power requirement band can be developed.

In projecting total power requirements for any liquid handling system, pump horsepowers at all significant flow rates between minimum and maximum conditions must be developed. Significant flow rates can be determined by reviewing the system load profile. To determine power inputs at various flow rates, it is necessary to know whether the pumping equipment shares the total system flow or head equally or unequally, and whether the pumping equipment operates at fixed speed, infinitely variable speed, or a combination of both. The system monitoring means also must be known when infinitely variable-speed pumps are employed.

When pumps are combined to share total system flow or head equally, the system power requirement at any flow rate can be computed based on individual pump horsepower. This definition is compatible with any range or quantity of pumps employed to satisfy all system flow requirements. The energy input will be governed by the number of pumps necessary to satisfy those requirements. In this case, the flows or heads of individual pumps combine equally to satisfy total system requirements and the total power required at any flow rate is the sum of the power required by the individual pumps.

When unequal size pumps are combined to satisfy system flow or head requirements (i.e., unequal percentage of total required), special consideration must be given to the pump operating sequence to ensure that maximum system efficiency is achieved.

Initial pump selections should be compared to the system head band and system load profile to establish a basic operating sequence for significant system flow values. Horsepower computa tions for the individual pumps are then made and recorded on a data form. To ensure that the selected order of pump operation results in the minimum power input necessary to satisfy the system requirements, rearrange the pump operating sequence at individual system flow rates, compute new pump horsepower requirements, and record this information.

This procedure will illustrate that for any system flow rate there is a single combination of pumps that requires minimum power input to satisfy the system requirements.

PROJECTING PUMP HORSEPOWER

Constant Speed

To compute constant speed pump horsepower for parallel, series, equal or unequal combinations, all pump characteristics necessary are available on the manufacturers' performance curves. For pump operation with fixed-speed change increments, the characteristics of flow, head, and brake horsepower can be calculated from the manufacturers' curves using the affinity laws. Therefore, power requirements for all constant-speed pumps and pumps of fixed-speed increments can be projected from available data.

Variable Speed

To compute variable-speed pump horsepower, it is first necessary to determine system monitor type and location. Type and location will determine head available, and head available will deter-

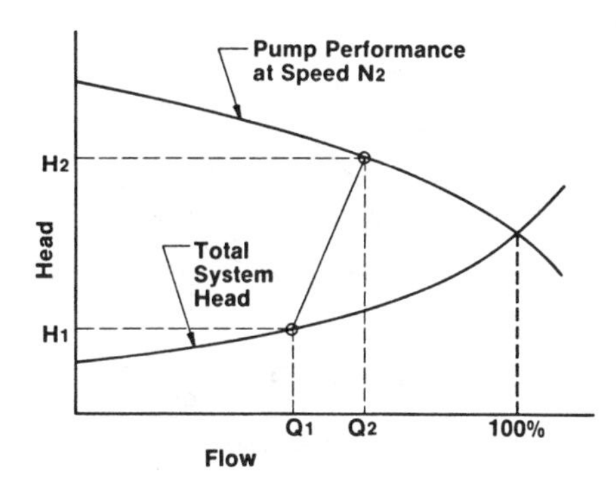

Figure 33. Shows relation of pump affinity law to total system head curve.

mine power required. Unfortunately, there is no simple mathematical relationship that can be developed for computing the interacting characteristics of systems and variable-speed pumps. The derivation of the affinity laws,

$$Q_2 = Q_1 \sqrt{\frac{H_2}{H_1}}, \tag{15}$$

is helpful in computing performance for individual variable-speed pumps. In this formula, Q_1-H_1 represents any single value of flow and head taken from the total system head curve. Q_2-H_2 represents a point on a pump total head curve of known operating speed that will satisfy the formula. The relationship of these values is shown in Figure 33. When the formula has been satisfied, the pump operating speed necessary to produce system conditions Q_1-H_1 can be calculated.

In using the proposed formula, values known include system flow, Q_1, system head, H_1, and full pump operating speed, N_2. Values unknown are the equivalent pump flow, Q_2, equivalent pump head, H_2, and the unknown speed, N_1, at which the pump must operate to produce the required system flow and head. When making variable-speed pump brake horsepower calculations, it is necessary to know the equivalent point on the full-speed characteristic curve that equates to the required system flow and head, since the efficiency at the equivalent point will reoccur at the reduced flow and head condition. Variable-speed pump brake horsepower is computed from the following formula:

$$BHP_1 = \frac{Q_1 \times H_1 \times \gamma}{3960 \times \eta_2} \tag{16}$$

where η_2 is the efficiency from the full-speed pump curve at Q_2, H_2 expressed as a decimal value, Q_1 is U.S. gpm, and H_1 is feet of head. The reduced pump operating speed necessary to satisfy the system flow, and head can be projected from the affinity law derivation:

$$N_1 = \left(\frac{Q_1}{Q_2}\right) N_2 \tag{17}$$

This value is used when computing brake horsepower requirements for variable-speed equipment where drive efficiency is a function of output-to-input-speed ratio. The expected range of pump operating speed should be determined and compared by the pump manufacturer to the critical speed of the pump. If operating speed occurs at or near pump critical speed, it will be necessary to modify the pump or to select a different pump for the application.

Determining the precise Q_2-H_2 point on the pump curve usually will require reiterative calculation. The triangulation procedure illustrated in Figure 34 locates an intercept point on the pump

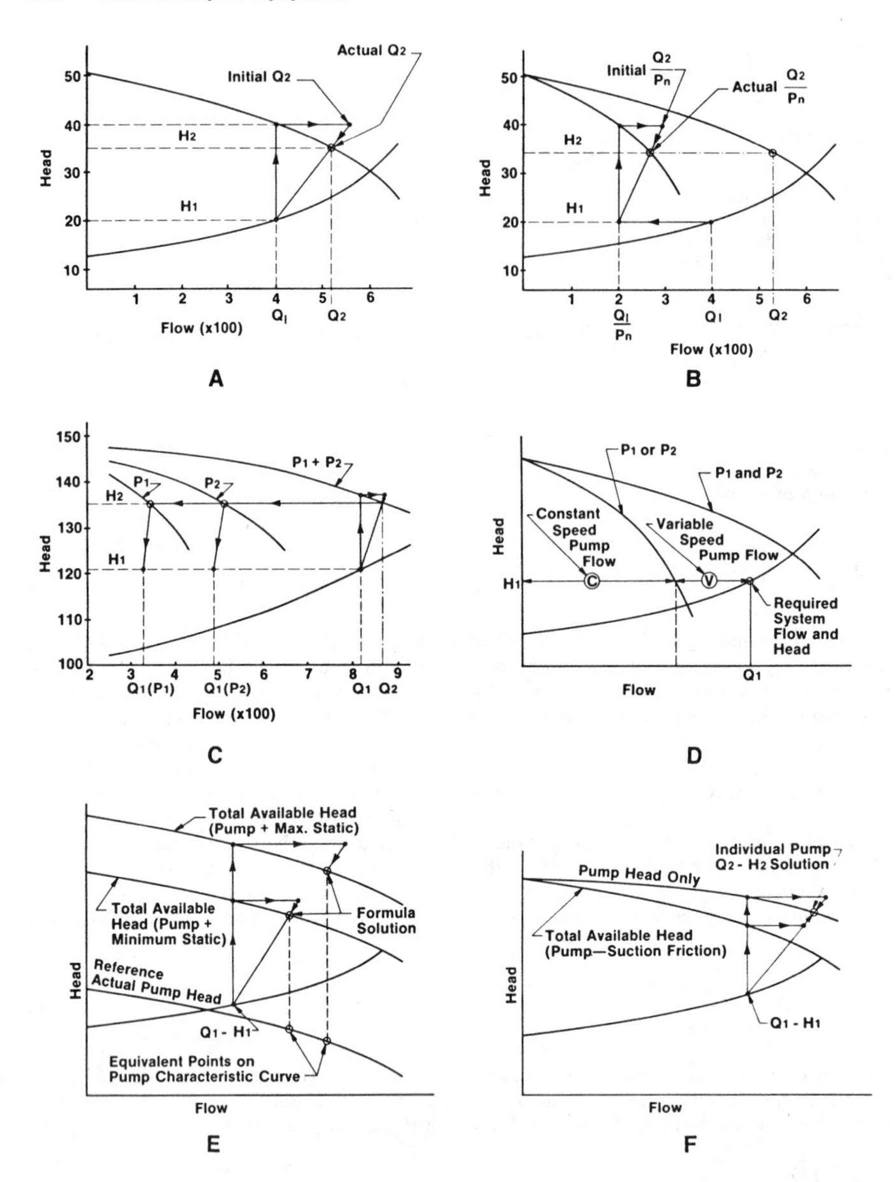

Figure 34. Triangulation procedure to determine actual flow-head solution for pumps.

curve that closely approximates the actual flow and head solution, Q_2-H_2. The accuracy of that approximated point is affected by the relative magnitude of H_1, H_2, and the slopes of the pump and system head curves. To satisfy the formula, and thus determine the point on the pump curve of the known speed that produces the required system values Q_1-H_1 at reduced speed, it is first necessary to calculate the initial Q_2-H_2 condition as a reference point. This is done by entering known system conditions Q_1-H_1 and an initial value of H_2 in the formula. The initial value of

H_2 is taken from the known pump performance curve at system flow, Q_1. By solving the formula for the initial value of Q_2, a reference point for initial Q_2-H_2 is established. By connecting the initial (reference) values of Q_2-H_2 with required system condition Q_1-H_1, an intersection point with the known pump curve is established. This intersection point becomes the approximate Q_2-H_2 value that will, at reduced operating speed, produce system conditions Q_1-H_1.

In a parallel pumping system where system flow is shared equally by all pumps, the procedure is identical to that for a single pump system, except that system flow, Q_1, must be divided by the number of pumps, P_n, operating. Figure 34B illustrates the triangulation process for a system incorporating two equally sized pumps operating in parallel.

When pumps selected to share unequal portions of total system flow are operating, the individual pump solution formula must be modified to $Q_1 = Q_2\sqrt{H_1/H_2}$. This is necessary since the flow produced by each pump for a selected system flow, Q_1, is unknown. To determine the flow and head developed by unequal size variable-speed pumps when they are both contributing to total system flow, it is first necessary to solve the basic triangulation process for system flow, Q_1, on the combined pump characteristic curve (Figure 34C). The intersection point that occurs on the combined pump curve establishes the H_2 value for the individual pumps. Use of the formula $Q_1 = Q_2\sqrt{H_1/H_2}$ determines individual pump flow, $Q_1(P_1)$ and $Q_2(P_2)$, for the system operation at Q_1. System power requirements at flow Q_1 are determined by adding the horsepower requirements of each pump.

When a variable-speed pump is combined with a constant-speed pump, the constant-speed pump will always produce flow at the point where the developed head matches the required system head. When system flow requirements are such that the constant-speed and variable-speed pumps must be operated in parallel, flow produced by the variable-speed pump will be the difference between required system flow, Q_1, and constant-speed pump flow at system head, H_1.

Figure 34D illustrates a two-pump system in which both pumps are of equal size—one constant speed and one variable speed. With a required system flow of Q_1 and head of H_1, the two pumps will develop unequal percentages of the total required flow. As system flow changes, the percentage of flow contribution by each pump will change.

To determine total system power requirements under these circumstances, it is necessary to exercise the basic triangulation procedure for the variable-speed pump based on the known capacity it will develop at the required head of the system. The horsepower of the variable-speed pump is calculated and added to the horsepower required by the constant-speed pump at its developed head and flow.

Projecting variable-speed pump performance under conditions of variable static available head requires the formula solution to be equated to the actual developed head of the pump. Use of the normal projection formula is sufficient, and the equivalent point on the pump head curve will result from projecting the actual Q_2,-H_2 solution and subtracting the static head. This will determine accurately the equivalent point on the pump head curve that equates to the system required head and flow. Figure 34E illustrates the effect of variable available static head.

When friction head in the supply pipe and fittings significantly reduces pump head, it is necessary to project the triangulation solution to the pump head curve only. This is illustrated in Figure 34F. The projection values can be computed from either the total available head curve or the pump curve, but the intersection solution always must be related to the equivalent point on the individual pump head curve.

The accuracy of the triangulation procedure will depend to a great extent on the accuracy of pump and system curve data used.

In the employment of variable-speed drives whose efficiency is a function of output to input speed, system operating efficiency can be improved when major system operating conditions require pump speed changes below the nominal speeds available from standard, multiple-speed ac induction motors. Careful comparison of the operating speed range to the system load profile should be made. If the time/load relationship is of sufficient magnitude, motor speed can be reduced to improve the variable-speed drive efficiency by altering output to input speed ratios.

Numerous contingencies must be considered in the analysis of this type of application to ensure that all mechanical and hydraulic considerations are taken into account.

EQUIPMENT POWER COMPARISONS AND TOTAL ENERGY PROJECTION

Accurate comparisons of the various kinds of pumping equipment under consideration for a given application make it necessary to establish a common point of reference. That reference point is the power that the prime mover must supply at its output shaft:

1. *Constant speed.* To project the total system power required for a constant-speed pump, divide the pump horsepower by the prime mover efficiency.
2. *Variable speed.* The horsepower at the output shaft of the prime mover for a pump driven by a variable-speed device is derived by dividing the pump brake horsepower by both the prime mover efficiency and the variable-speed device efficiency. This applies whether the variable-speed device is between the pump and the prime mover or ahead of the prime mover.

Do not be misled by the lower net efficiency of the equipment which includes the variable-speed device. Via pump operation which is closely aligned to system needs, significantly valuable operating cost savings can be obtained, particularly in systems that have variable loads. The greater initial cost of the variable-speed equipment often can be amortized in a relatively short time, after which a net reduction in overall system operating costs can be obtained.

To project the total energy requirements for any system with any combination of pumping equipment and to ensure that all answers are projected from a common base, a structured procedure for the analysis should be followed; such a procedure is outlined in Figure 35. Each item in the procedure will have a significant impact on the total energy requirement of a specified pump-equipment combination and system operating procedure.

To make comparisons of dissimilar equipment or to evaluate the impact of system operation change, it is necessary to follow the evaluation procedure for each specified condition. The individual steps of the procedure are defined as follows:

Step 1. At the point of entry to the evaluation procedure, it is necessary to have established and assembled all information relative to the system design requirements. An accurate projection of system flow and head requirements, as illustrated by a system head curve or band, is necessary at this stage of evaluation.

Step 2. The load profile of the system should be reviewed to make preliminary pump equipment selections and to establish the evaluation base for the system under consideration. If future system conditions such as expansion, etc., will significantly alter the load profile, these considerations should be confirmed at this time.

Step 3. Pump selection evaluation. Beyond the normal mechanical and hydraulic criteria associated with pump selection for a specific application, the selection procedure should encompass the primary operating efficiency at the point of selection and the amount of efficiency deviation that will be caused by system flow variation. These values then should be compared to the operating time at specified flow rates to ensure that the selection offers the maximum potential operating efficiency through the range of maximum flow/time concentration.

Step 4. Preliminary decisions must be made relative to the control means that will be employed to satisfy system flow and head requirements. The decision to control system required head, total available head, or both, will be a significant factor in the total energy requirements projected for the system being evaluated.

Step 5. This stage of the evaluation requires a projection of the pumping equipment power requirements over the entire system operating flow range. The pump required power can be organized in tabular or graphical form, as best suits the user. In tabular form, the power requirements of the pumping system, compared to required system flow, should be of sufficient quantity to allow the user an analysis of all significant operating areas, as dictated by the flow/time profile.

Step 6. The load profile is used to compute the energy requirements of the selected pumping equipment and system operating procedures. When compared to the power projection in Step 5,

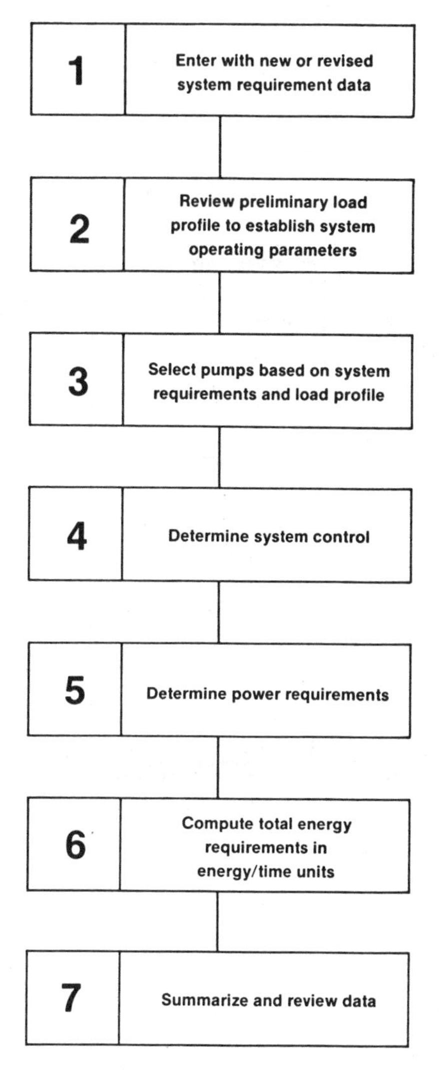

Figure 35. Summary of procedure to project total energy requirements.

the load profile will determine the total energy requirements of the system, thus concluding the evaluation in energy time units (brake horsepower hours, kilowatt hours, etc.).

Step 7. Determination of the total energy required to satisfy all projected system operating conditions is the basis from which all other considerations and comparisons can be made. This is the reviewing and decision-making phase of the total system evaluation procedure. If the energy summary is adequate, the evaluation procedures are concluded; if not, the evaluation can be repeated with new or modified data to determine what effect the change will have.

When the total system energy answer has been determined, normal decision-making processes to select that combination of pumping equipment and system operating procedures that most effectively satisfy the design goals of the system can begin. Total system evaluation ensures that all projections, comparisons, and decisions are based on as much factual information as possible.

POWER COMPARISON DATA

FROM:
INDIVIDUAL ____________

COMPANY ____________

ADDRESS ____________
(Street) (City) (State) (Zip)

TELEPHONE ( ) ____________

SYSTEM FLOW
Maximum ________ GPM
Minimum ________ GPM
Liquid ________
Specific Gravity ________
Viscosity ________

SYSTEM HEAD
Constant Pressure
Static ________ (FT)
Delivery ________ (FT)
Friction ________ (FT)
Total ________ (FT)

PUMP SUCTION SUPPLY
☐ Constant
☐ Variable
Minimum ________ (FT)
Maximum ________ (FT)

LOAD PROFILE—ESTIMATED OR MEASURED

GPM									
Hours									

(Fill in as many conditions as applicable.)

APPLICATION, SERVICE OR PROCESS BEING CONSIDERED:
____
____
(Please provide simple piping sketch.)

PRESENT ELECTRICAL COST:
________ ¢ Per KWH

TYPE OF COMPARISON TO BE MADE
(check one)
☐ Peerless variable speed drives with existing pumps _to_ existing constant speed pumps.
☐ Peerless variable speed drives with new Peerless pumps _to_ existing constant speed pumps.
☐ Peerless variable speed drives with new Peerless pumps _to_ proposed constant speed pumps.

PUMPS PROPOSED FOR/OR EXISTING IN SYSTEM
Quantity ________
Design Capacity ________ (GPM EACH)
Design Head ________ (FT EACH)
Driver ________ HP ________ RPM ________
(Please provide performance curve for each pump)

Figure 36. Power comparison data sheet for pump energy comparison evaluation.

When comparisons of various combinations of pump equipment and system operating procedures are desired, total system evaluation allows the comparisons to be made on an equal basis—the requirements of the system. Total system evaluation is a structured, fundamental process for determining the pump and system combination that best suits the cost-benefit-functionality requirements of the user.

Figure 36 is a recommended data sheet for recording the information for the energy analysis.

Further discussions and illustrative examples on pump sizing can be found in the references listed at the end of this chapter.

NOTATION

A	cross-sectional area	N_s	pump specific speed
BEP	best efficiency point	NPSH	net positive suction head
BHP	brake horsepower	P	pressure
C	discharge coefficient for pipe flow	Q	volumetric flow rate
c	velocity of sound in liquid	S	static head
D	impeller diameter	u	velocity
g	acceleration of gravity	γ	specific gravity
H	head	ϵ	height of roughness element
L	length	η	pump efficiency
N	rotational speed		

REFERENCES

1. Cheremisinoff, N. P., *Fluid Flow: Pumps, Pipes and Channels*, Ann Arbor Science Publishers, Ann Arbor, MI, 1981.
2. Azhel, D., and Cheremisinoff, N. P., *Fluid Mechanics and Unit Operations*, Ann Arbor Science Publishers, Ann Arbor, MI, 1983.
3. Cheremisinoff, N. P., and Gupta, R., (Eds.), *Handbook of Fluids in Motion*, Ann Arbor Science Publishers, Ann Arbor, MI, 1983.
4. Cheremisinoff, N. P., *Fluid Flow Pocket Handbook*, Gulf Publishing Co., Houston, TX, 1984.

CHAPTER 36

DESIGN AND OPERATION OF MULTISTAGE CENTRIFUGAL PUMPS

H. Miyashiro and **M. Kondo**

Mechanical Engineering Research Laboratory and
Tsuchiura Works
Hitachi, Ltd., Japan

CONTENTS

INTRODUCTION, 1038

DETERMINATION OF NUMBER OF STAGES, 1039

DESIGN OF IMPELLER, 1041
Number of Vanes and Outlet Vane Angle, 1041
Impeller Outlet, 1041
Impeller Inlet, 1041
Meridional Profile, 1045
Configuration of Vanes, 1045

DESIGN OF CASING, DIFFUSER AND RETURN CHANNEL, 1045
Suction Casing, 1045
Diffuser and Return Channel, 1046
Discharge Casing, 1048

CONSTRUCTION OF PUMPS, 1048
Axial Thrust, 1049
Shaft Seal, 1052

VIBRATIONS, 1052

PROBLEMS IN OPERATION, 1053
Surging, 1053
Operation in Low Capacity Range, 1054
Control of Capacity, 1054

NOTATION, 1055

REFERENCES, 1055

INTRODUCTION

Multistage centrifugal pumps are usually designed on the basis of empirical data.

At first the number of stages is determined from the given rated point with consideration for specific speed and cavitation. The next step is the design of hydraulic components, i.e., the impeller, the stage that consists of diffuser and return channel, and the casing. To get a good performance, flow and losses in the hydraulic components are calculated and estimated, respectively.

Axial hydraulic thrust, shaft seal, and critical speed must be taken into account. Operation in a small-capacity range and control of capacity are important problems.

In this chapter specific energy, y, (the energy per unit mass of liquid, $y = gH$) is used along with conventional head.

DETERMINATION OF NUMBER OF STAGES

The type of pump and the number of stages must be determined by examining which type of pump is suitable to the given rated point (i.e., rated head, H_R, and rated capacity, Q_R) and whether the pump is of single-stage or multistage.

Figure 1 is used to determine the type of pump. In the case of a multistage pump, the number of stages is determined with consideration for pump speed, n, and a relation between specific speed, n_s, and head, H, per stage shown in Figure 2. Table 1 shows the synchronous speed of an AC motor. Pump efficiency and cavitation performance must be taken into account. For this purpose net positive suction head, NPSH, or net positive suction energy, NPSE, is selected as a parameter in Figure 2. Specific speed, n_s, is defined by Equation 1.

$$n_s = n \frac{Q^{1/2}}{H^{3/4}} \tag{1}$$

where n = pump speed in rpm
Q = capacity per impeller eye in gpm
H = head per stage in ft

Dimensionless specific speed N_s is defined by Equation 2.

$$N_s = 2\pi n \frac{Q^{1/2}}{(gH)^{3/4}} \tag{2}$$

where units of time and length must be second and foot or meter, respectively, i.e., n is rps, Q is ft^3/s or m^3/s, g is ft/s^2 or m/s^2, and H is ft or m. Conventional specific speed is not dimensionless.

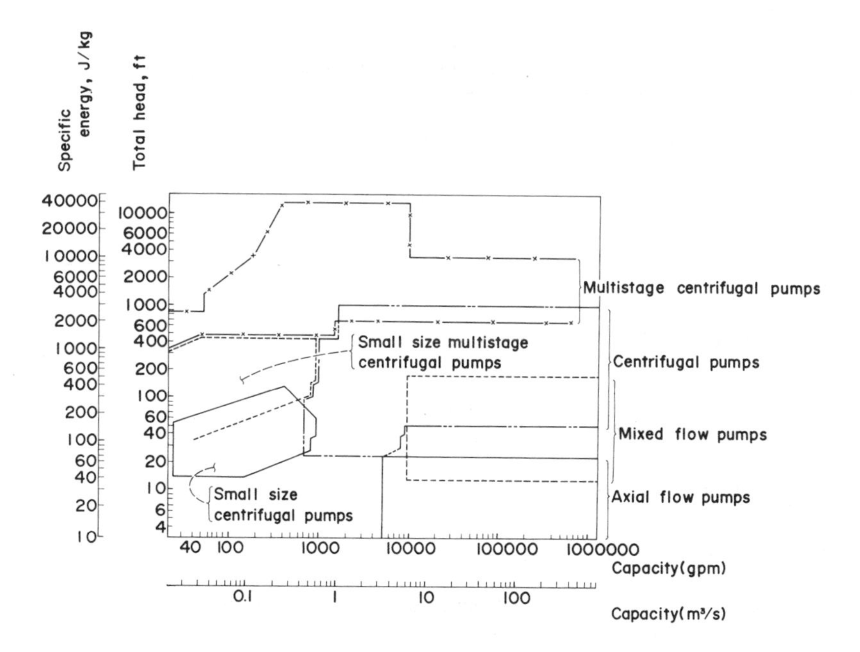

Figure 1. Determination of type of pumps.

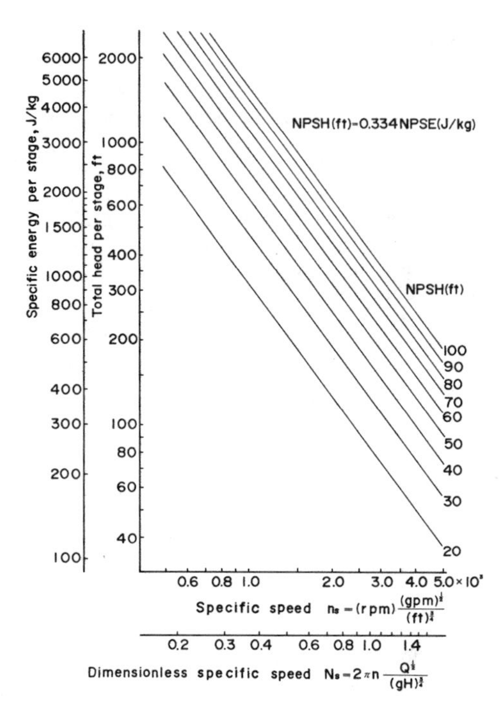

Figure 2. Relation between specific speed and head per stage.

Table 1
Synchronous Speed of AC Motors

Number of Poles	50 Hz (rpm)	60 Hz (rpm)
2	3,000	3,600
4	1,500	1,800
6	1,000	1,200

Conversion between various specific speeds is shown as follows.

$$n_s\,(\text{rpm, m}^3\text{/min, m}) = \frac{1}{6.67} \times n_s\,(\text{rpm, gpm, ft})$$

$$N_s\,(\text{dimensionless}) = \frac{1}{2{,}734} \times n_s\,(\text{rpm, gpm, ft})$$

The number of stages is determined with the following procedure.

1. *Estimate of n_s.* In case of centrifugal pumps n_s is usually selected between 600 and 2,000 with consideration for efficiency and construction of the pump.
2. *Determination of head per stage.* Head H per stage is determined from the values of n_s and available NPSH with the aid of Figure 2.

3. *Determination of n_s.* The value of n_s is calculated by Equation 1. If this value is different from the estimated value, the procedure is repeated till both values agree.

DESIGN OF IMPELLLER

Figure 3 shows a design procedure of the impeller.

Number of Vanes and Outlet Vane Angle

Number of vanes, z, and outlet vane angle, β_2, are estimated. Values of z and β_2 are selected usually between 2 and 9 and between 15 and 30 degrees, respectively, since they have a great influence on the stability of head-capacity curve, as shown in Figure 4 [1, 2].

When the flow toward the impeller has prewhirl, the head-capacity curve is steeper than the flow without prewhirl, then a larger outlet angle can be selected, as shown in Figure 5.

Impeller Outlet

Empirical values of speed constant K_{u2} and capacity constant K_{m2} are assumed.

$$K_{u2} = u_2/\sqrt{2gH} \tag{3}$$

$$K_{m2} = c_{m2}/\sqrt{2gH} \tag{4}$$

where u_2 = peripheral velocity, ft/s
c_{m2} = meridional component of absolute velocity of flow, ft/s

Figure 6 [3] shows a relation between n_s and K_{u2} for β_2 of 22.5 degrees. From the value of K_{u2} obtained from Figure 6, u_2 ft/s and the mean outlet impeller diameter D_{2m} in. are calculated by using Equations 5 and 6, respectively.

$$u_2 = K_{u2}\sqrt{2gH} \tag{5}$$

$$D_{2m} = \frac{229}{n} K_{u2}\sqrt{2gH} \tag{6}$$

where n is pump speed in rpm.

Similarly c_{m2} ft/s and impeller outlet width b_2 in are calculated.

$$c_{m2} = K_{m2}\sqrt{2gH} \tag{7}$$

$$b_2 = \frac{0.321Q}{(D_{2m}\pi - zs_{u2})c_{m2}} \tag{8}$$

where D_{2m} in. is mean outlet diameter of the impeller and s_{u2} in. is vane peripheral thickness of the impeller shown in Figure 7 [4].

Impeller Inlet

At first the configuration of the hub is determined with consideration for shaft torque and construction.

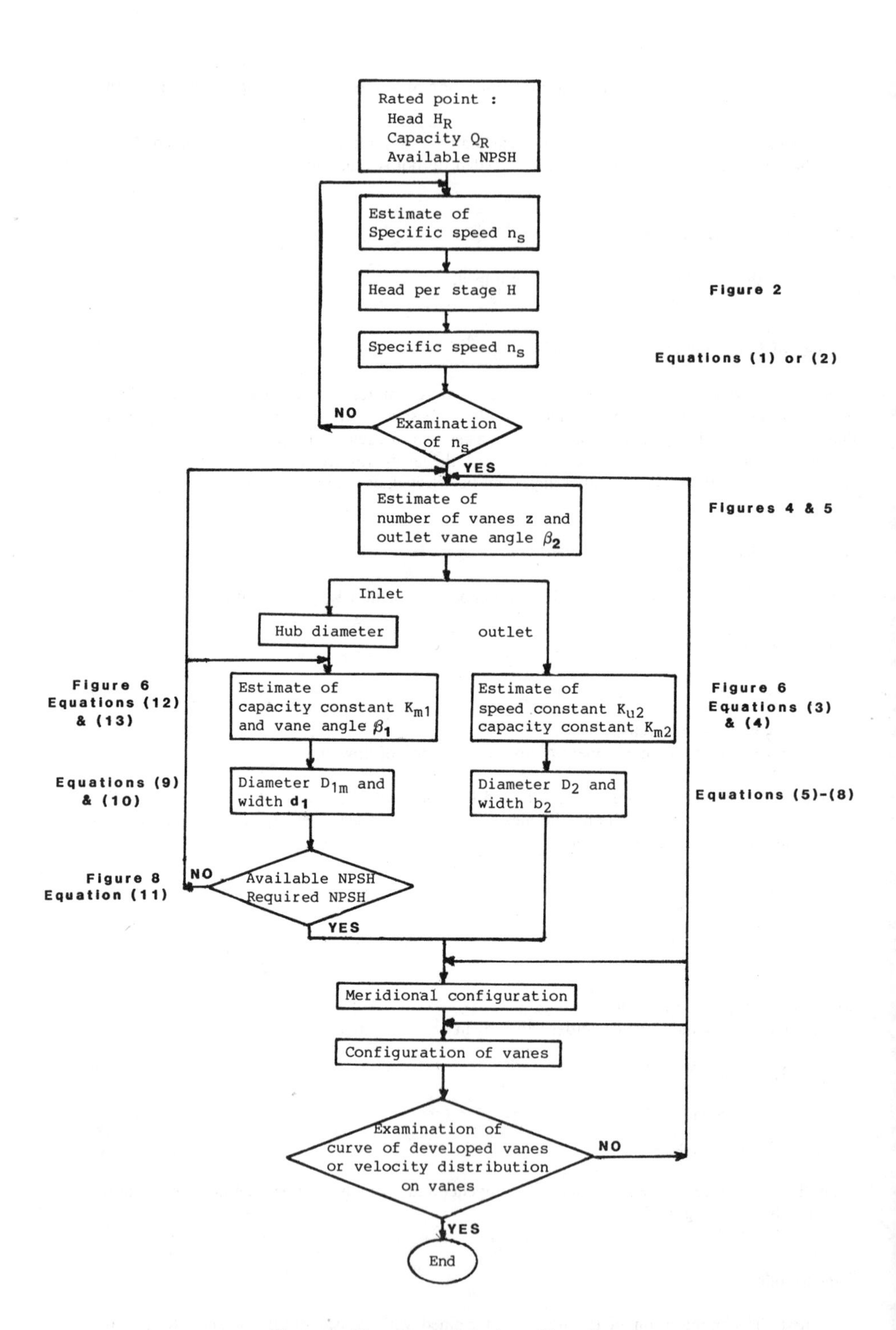

Figure 3. Flow chart of design of impeller.

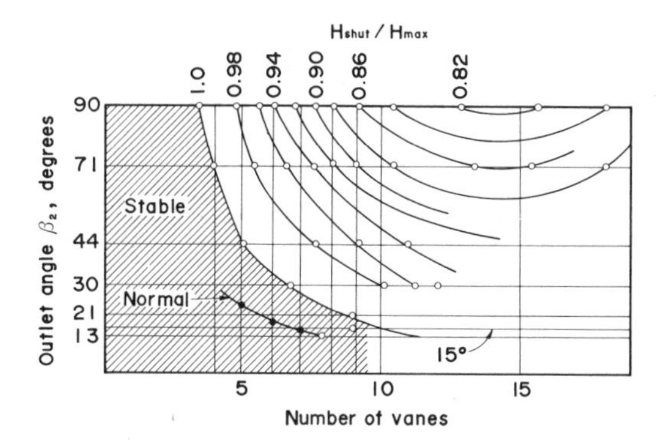

Figure 4. Conditions for a stable head-capacity curve [1, 2].

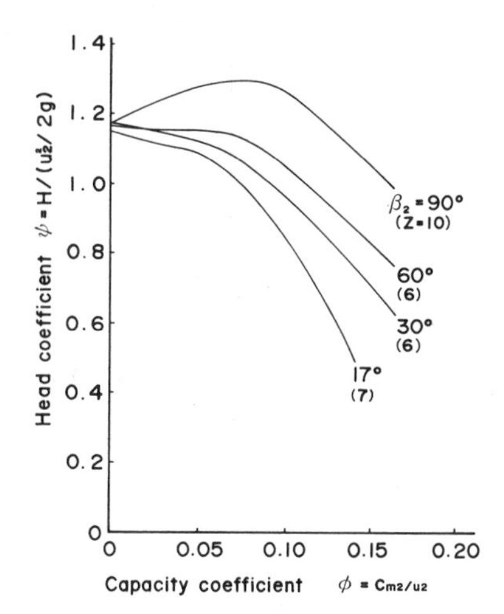

Figure 5. Head-capacity curve with prewhirl.

The value of K_{m1} is obtained from Figure 6, and the mean inlet diameter of the impeller D_{1m} in. is calculated from Equations 9 and 10.

$$c_{m1} = K_{m1}\sqrt{2gH} \tag{9}$$

$$D_{1m} = \frac{0.321Q}{\pi d_1 c_{m1}} \tag{10}$$

where d_1 in. is impeller inlet width.

Required NPSH is estimated by Equation 11.

$$\text{NPSHR} = \lambda_1 \frac{w_1^2}{2g} + \lambda_2 \frac{c_1^2}{2g} \tag{11}$$

where w_1 and c_1 are relative and absolute velocities at the inlet respectively, and λ_1 and λ_2 are dimensionless numbers obtained empirically. The first term of the right side of Equation 11 increases with the impeller eye diameter. On the other hand the second term decreases. Therefore,

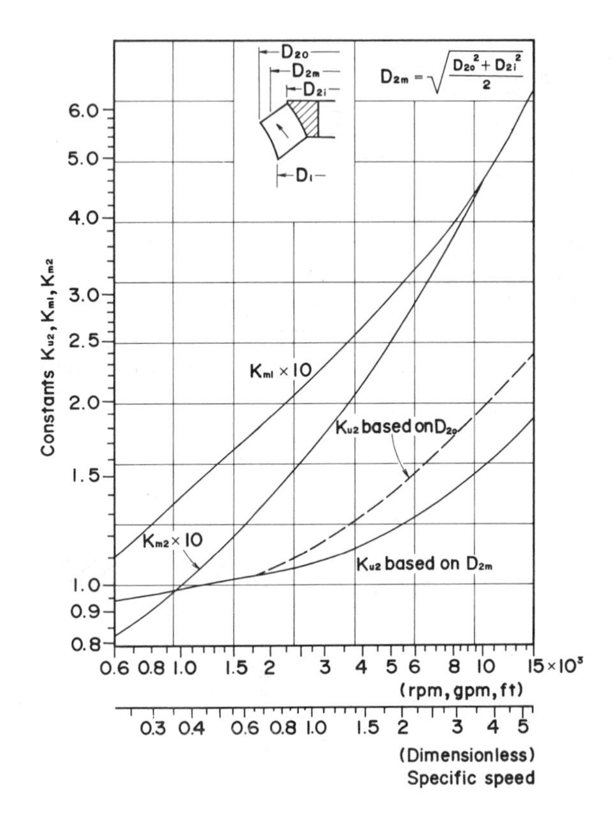

Figure 6. Impeller constants [3].

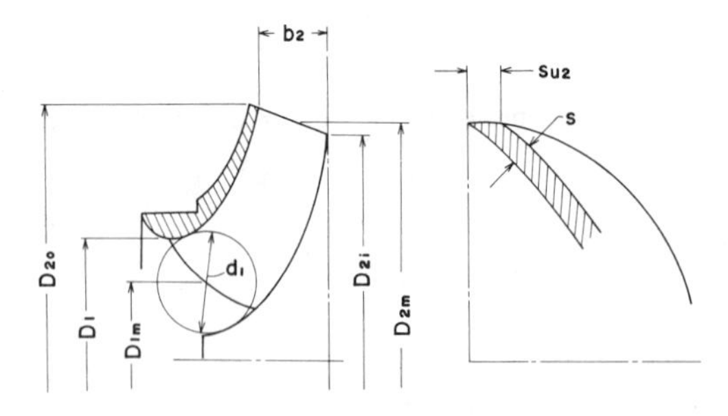

Figure 7. Inlet and Outlet of Impeller [4].

the required NSPH becomes minimum at a diameter of the impeller eye. This condition expressed by Equation 12 is determined only by inlet vane angle β_1 [5].

$$\tan(\beta_1)_{opt} = \sqrt{\tfrac{1}{2}\lambda_1/(\lambda_1 + \lambda_2)} \tag{12}$$

This relation is shown in Figure 8. The value of β_1 at the impeller front shroud is selected to satisfy Equation 12. When a positive prewhirl exists at the impeller inlet, c_1 and w_1 are larger and smaller,

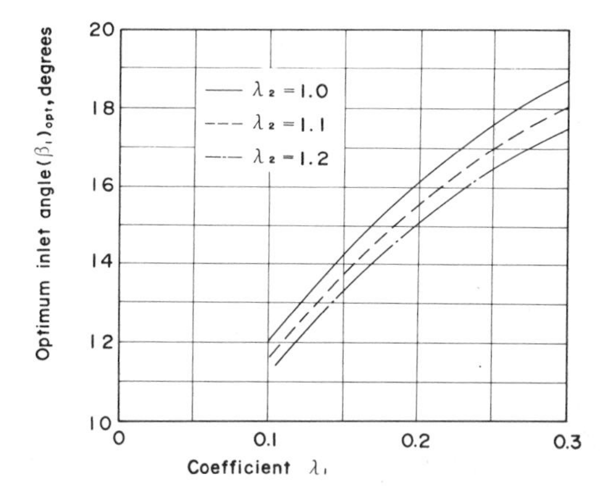

Figure 8. Optimum inlet angle of impeller vanes.

respectively, than c_1 and w_1 without prewhirl, and peripheral component c_{u1} of the inlet absolute velocity at the optimum condition of the NPSHR is expressed by Equation 13.

$$(c_{u1})_{opt} = \lambda_1 u_1/(\lambda_1 + \lambda_2) \tag{13}$$

When a double suction impeller is chosen as the first stage, a suction casing of volute type is often used. This casing induces prewhirl that enables the impeller to decrease the required NPSH. Equation 13 is used for estimation of an adequate prewhirl.

At last the NPSHR at D_1 shown in Figure 7 is calculated by Equation 11. It must be confirmed that the required NPSH is lower than the available NPSH.

Meridional Profile

The meridional profile is determined as the area perpendicular to the flow direction changes gradually from the impeller inlet to outlet.

Configuration of Vanes

The configuration of vanes must have a smooth curve from the inlet angle β_1 to the outlet one β_2. To draw the curve, the method of error triangles is usually used: Stream lines on the vane are developed on a plane as successive small triangles. Recently the numerical calculation of the velocity distribution on the vane is used to examine whether the deceleration rate of the relative velocity is not too high [6–9].

Various losses, i.e. friction, flow separation, disk friction and leakages through wearing rings, stage rings, shaft seal, and balancing devices, are estimated. If the velocity distribution is not good, the design procedure restarts from Step 1.

The design procedure of the pump is carried out effectively by the computer aided interactive method, which is supported by database and graphics.

DESIGN OF CASING, DIFFUSER, AND RETURN CHANNEL

Suction Casing

Two types of the suction casing are usually used—Concentric and volute.

An example of the concentric type is shown in Figure 9 [10]. Configurations a and b are used wih baffles shown in the figure. Configuration c is not suitable for a uniform velocity distribution around the impeller eye.

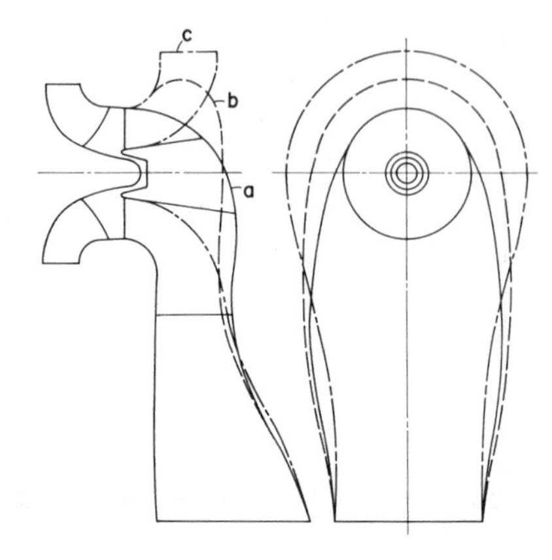

Figure 9. Concentric suction casing [10].

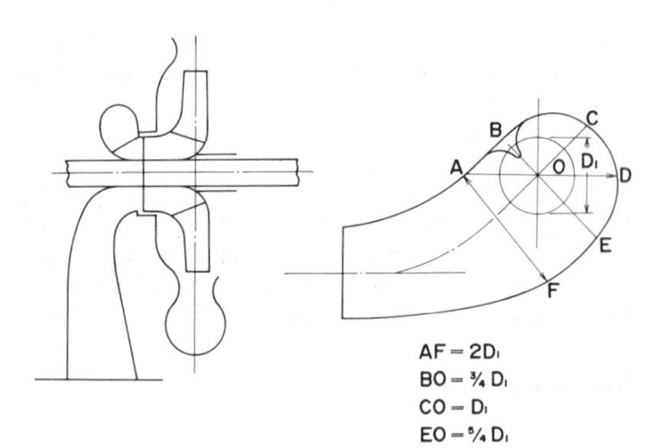

Figure 10. Volute suction casing [11].

Suction casing of volute type shown in Figure 10 [11] is used, in order to attain a uniform velocity distribution with prewhirl at the impeller inlet. The area of section AF is some 50 percent or more larger than that of the impeller eye. Location and shape of the baffle has a great influence on the velocity distribution.

Diffuser and Return Channel

The velocity at the impeller outlet must be converted into pressure effectively by a vaned diffuser or a volute diffuser. The return channel must introduce the flow from the diffuser to the inlet of the next stage impeller without causing large losses.

The width b_3 of the vaned diffuser is determined by Equation 14 [12].

$$b_3 \approx 1.1 b_2 \qquad (14)$$

It must be noted that the relative position of the impellers and diffusers has a great influence on the pump efficiency and the shape of the head-capacity curve near the shut-off. The best efficiency is obtained when the front shroud is lined up with the diffuser wall. The shut-off head tends to rise at the same time [13].

The inlet diameter D_3 of the diffuser is determined by Figure 11 [14].

Number z_g of diffuser vanes must have no common measure with the number z of the impeller vanes to avoid the severe pressure fluctuation caused by interference between impeller vanes and diffusers.

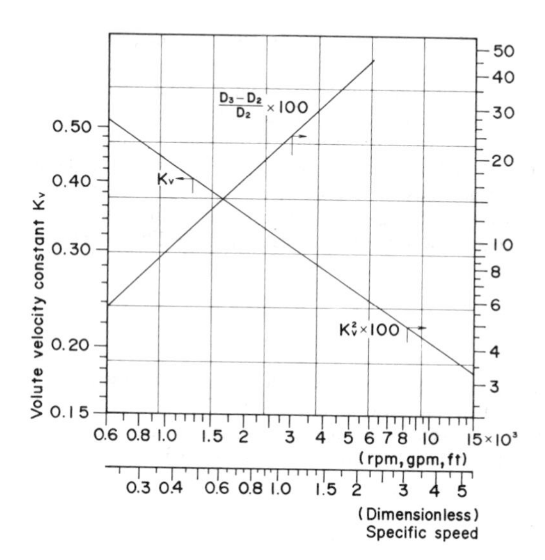

Figure 11. Volute constants [14].

Inlet angle α_3 of the vaned diffuser is determined from the flow continuity and the following assumption: The flow from the impeller outlet to the diffuser follows the free vortex motion, i.e., $rc_u = r_2c_{u2}$. Flow contraction by the diffuser vane thickness is taken into account. Less value of α_3 is often chosen, in order to avoid the drooping head curve at low-capacity range.

In order to reduce hydraulic losses, a diffuser flow passage with a small hydraulic radius (for example, a square section) must be chosen. For a rectangular diffuser section formed between two parallel walls the optimum divergence angle is about 11 degrees. An example is shown in Figure 12 [15], the dimension is expressed by Equations 15 through 17 [12, 13, 15].

$$\ell = 4a \tag{15}$$

$$b/a = 1.6 \tag{16}$$

$$D_4 = (1.35 \sim 1.6)D_2 \tag{17}$$

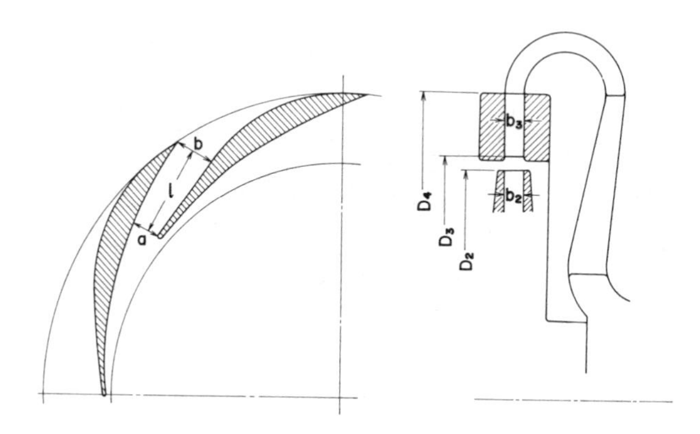

Figure 12. Diffuser [15].

The flow from the diffuser goes through a U turn, which has usually no vanes, into a return channel. At the diffuser outlet the flow deviates from the outlet angle α_4 of the diffuser, i.e., slip occurs. The number of return vanes z_R is usually different from z and z_g. Inlet angle α_5 of the return channel is determined with consideration for the flow contraction by the vane thickness. A larger value of α_5 is usually chosen because of the friction in U turn. In the return channel the flow usually has the equal relative velocity or is gradually accelerated. Velocity and pressure distributions in the U turn and return channel are experimentally investigated [16, 17]. Outlet angle α_6 of the return channel is 90 degrees, i.e. radial, or about 95 degrees with consideration for slip.

Numerical calculation method, for instance FEM, is applied to the flow in the diffuser and return channel near the best efficiency point as well as in the impeller.

Discharge Casing

The flow from the last stage goes through the volute or concentric casing to the discharge pipe. The following method for the design of the volute casing is based on the assumption that the average velocity in each sectional area of the casing is equal. The average velocity v_v is obtained by Equation 18 and Figure 11 [14], which shows the volute velocity constant K_v.

$$v_v = K_v\sqrt{2gH} \tag{18}$$

CONSTRUCTION OF PUMPS

The following types of construction are used for multistage centrifugal pumps:

1. Pumps with horizontally split casing. Each component of pumps is easily removed.
2. Pumps with stages assembled together by strong bolts (Figure 13). They are more suited to higher pressure than pumps of the former type.
3. Pumps with double casings (Figure 14). They are most suitable to high pressure. The outer barrel casing sustains high pressure, and the inner casing that needs not sustain high pressure is easy to manufacture.

The balancing axial thrust and shaft seal must be examined. Hydraulic radial thrust may be neglected.

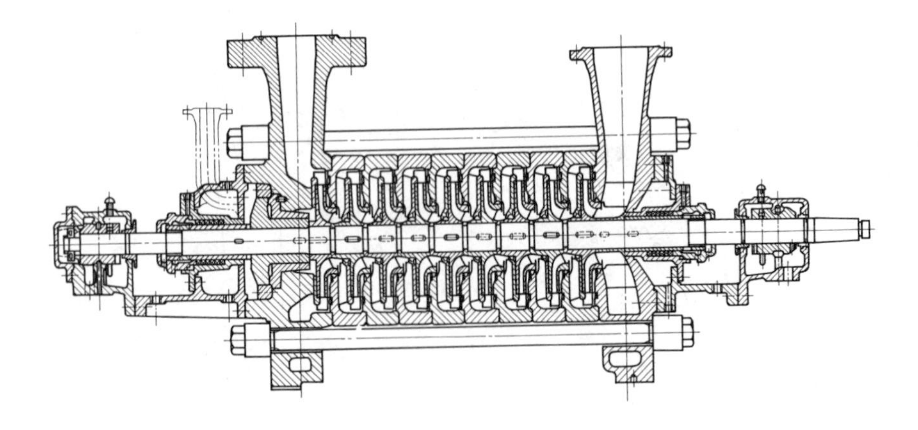

Figure 13. Pump with stages assembled by bolts.

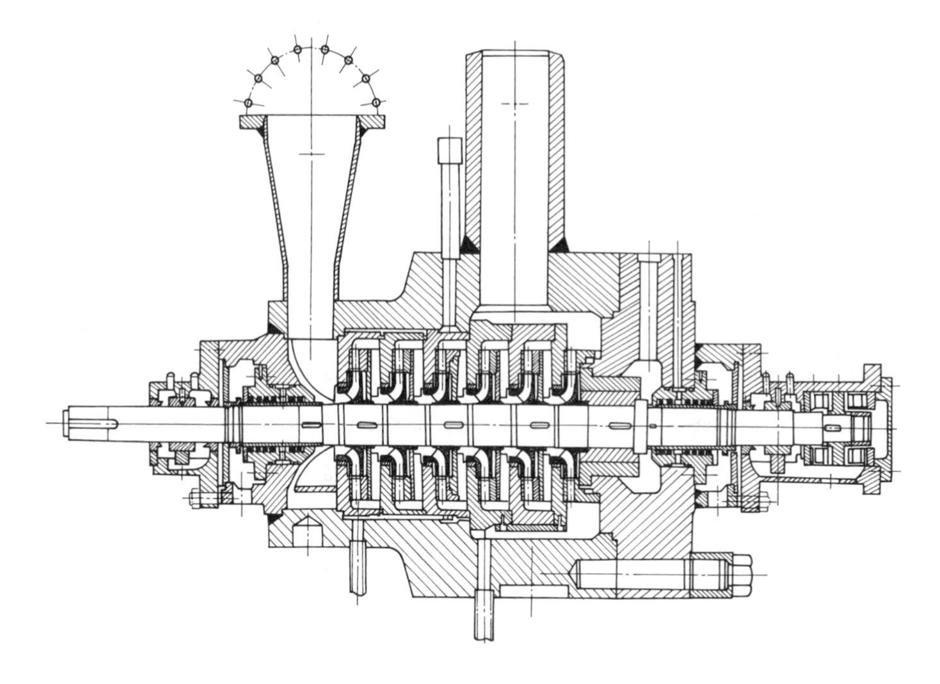

Figure 14. Pump with double casings.

Axial Thrust

The hydraulic force which acts on the back shroud of each impeller is not equal to that on the front shroud, as shown in Figure 15 [18]. The unbalancing thrust T per impeller is calculated by Equation 19 or Equation 20 on the following assumptions:

1. The average velocity v_v in the vaned diffuser is calculated by Equation 18 and Figure 11.
2. The liquid between the impeller shrouds and the stage walls rotates with one half of the angular velocity of the impeller.

$$T = \frac{A_1 - A_s}{144}\gamma\left[H(1 - K_v^2) - \frac{1}{4}\frac{u_2^2 - u_r^2}{2g}\right], \text{lb} \tag{19}$$

where A_1 and A_s = areas, in.^{2}

γ = specific weight, lb/ft^3

H = total head per stage, ft

u_2 and u_r = peripheral velocity at impeller outlet and that at impeller ring, respectively, ft/s

g = acceleration due to gravity, ft/s^2

$$T = (A_1 - A_s)\rho\left[gH(1 - K_v^2) - \frac{1}{4}\frac{u_2^2 - u_r^2}{2}\right] \tag{20}$$

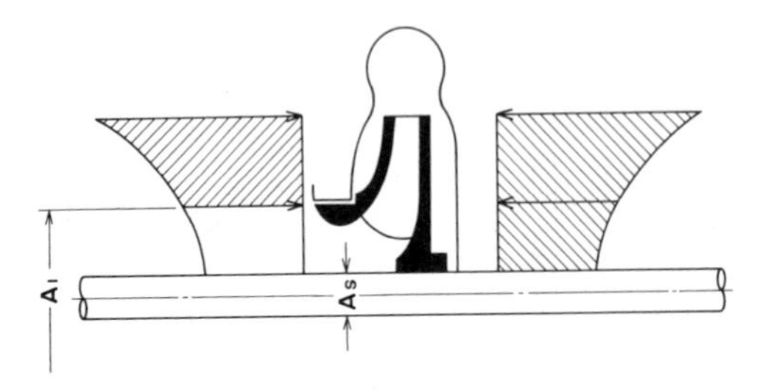

Figure 15. Hydraulic axial thrust [18].

in N (N is an SI unit in kg/s^2, where the following units are used: A_1 and A_s, m^2; density, ρ, kg/m^3; H, m; u_2 and u_r, m/s; and g, m/s^2).

In order to estimate the axial thrust more precisely, the leakage through the impeller ring clearance and other factors are taken into account [19, 20].

When the flow changes its direction from the impeller inlet to outlet, the change in momentum of the flow causes an opposed axial thrust. When the flow makes a 90 degree turn in the meridional profile of the impeller, the opposed thrust F per impeller is calculated by Equation 21 or Equation 22.

$$F = 6.92 \times 10^{-5} \gamma Q v_{m0} \tag{21}$$

in lb, where the following units are used: γ, lb/ft^3; Q, gpm; and meridional velocity through the impeller eye, v_{m0}, ft/s.

$$F = \rho \frac{Q}{60} v_{m0} \tag{22}$$

in N, where ρ, kg/m^3; Q, m^3/min; and v_{m0}, m/s are used.

The following methods are used to balance the axial thrust:

1. *Opposed impellers.* Half the impellers are arranged opposite to the other impellers, as shown in Figure 16 [21]. The impellers must be arranged with the following considerations.

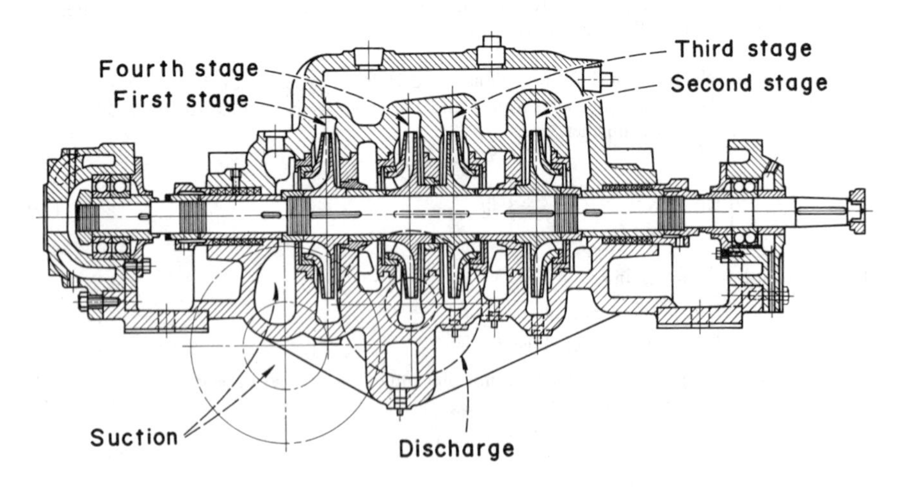

Figure 16. Impeller arrangement for balancing axial thrust [21].

- The pump construction must be not too complicated.
- The earlier-stage impellers are arranged at right and left ends, in order to reduce the leakage through shaft seal.
- When the diameter of the shaft is not constant, axial thrust is caused by shaft shoulders and impeller hubs [22].

2. *Balancing devices.* When all impellers are arranged in the same direction, the balancing disk shown in Figure 17 [23] or the balancing drum shown in Figure 18 [24] is used. High pressure from the last-stage acts on these balancing devices opposite to the direction of the axial thrust.

   The balancing disk operates automatically on the following principle [25]. When the axial thrust tends to move the disk to the left, the axial clearance between the disk C and the stationary face B is reduced. This reduces the pressure in the balancing chamber that is connected through an orifice to the first-stage suction. This reduction and the full pump pressure move the disk to the right. When the thrust tends to move the disk to the right, the increased pressure in the chamber and the reduced pump pressure move the disk to the left. Only a simple thrust bearing is needed, in order to avoid contact of the disk with face B, when the pump does not develop full pressure at starting and stopping.

   Since the balancing drum does not balance the axial thrust completely over the entire operating range, a thrust bearing is also employed to sustain any residual thrust.

   Since the clearance between the balancing devices and the stationary parts is very small, the pumped liquid must be clear.

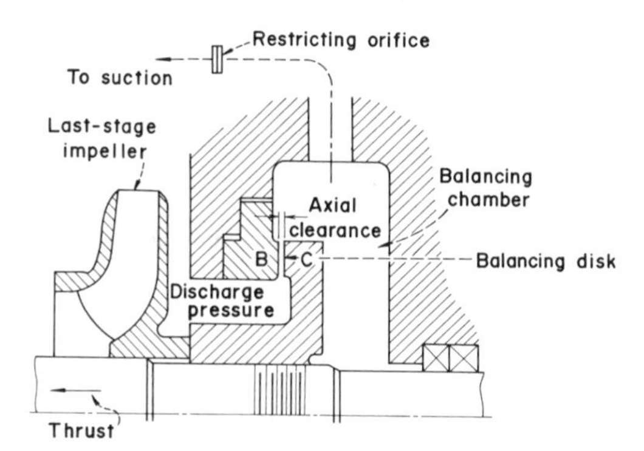

Figure 17. Balancing disk arrangement [23].

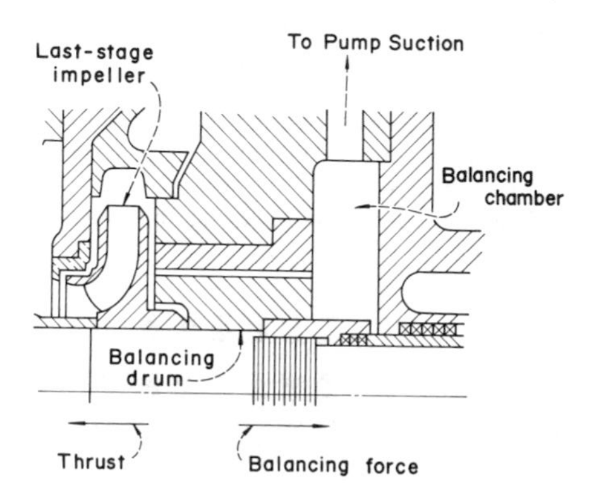

Figure 18. Balancing drum arrangement [24].

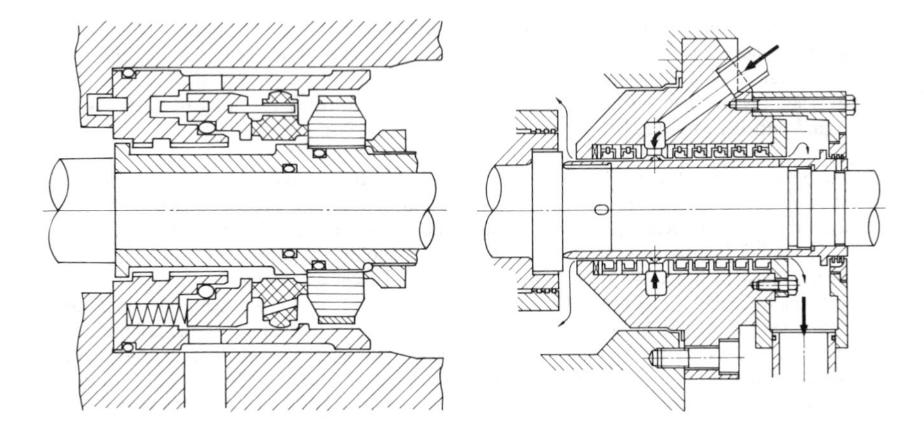

Figure 19. Mechanical seal arrangement.

Figure 20. Floating rings arrangement.

Shaft Seal

The following types of shaft seal are used.

- Stuffing box shown in Figure 16.
- Mechanical seal shown in Figure 19.
- Floating rings shown in Figure 20.

VIBRATIONS

The critical speed of the pump rotor is calculated usually on the following assumptions (Figure 21).

1. The rotating part consists of a shaft, disks equivalent to impellers, and bearings.
2. Bearings are equivalent to springs and oil dampers.

The rotating part is divided into several elements shown in Figure 22, where d_i, d_O, E, ℓ, and ρ are inner diameter, outer diameter, modulus of longitudinal elasticity, length and density of shaft; D_i, D_O, I_d, I_p, and M are inner diameter, outer diameter, moment of inertia around disk diameter, moment of inertia around axis, and mass of disk; and k and c are spring constant and damping constant, respectively. The empirical value of the damping coefficient is usually used. The spring constant of the bearings is estimated as follows.

- Bearing supported rigidly—Value of spring constant may be assumed as infinity.
- Journal bearing—Value of spring constant is usually selected between $(6 \sim 600) \times 10^4$ lb/in., i.e. $10 \sim 10^3$ MN/m.
- Ball bearing—Value of spring constant is usually between $(6 \sim 60) \times 10^4$ lb/in., i.e. $10 \sim 10^2$ MN/m.

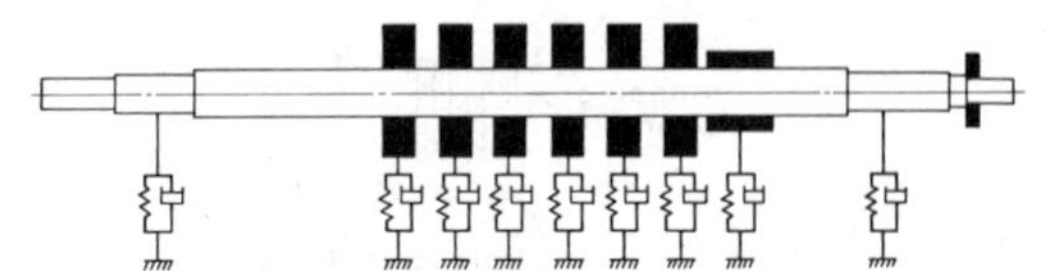

Figure 21. Modeling of rotor for vibration analysis.

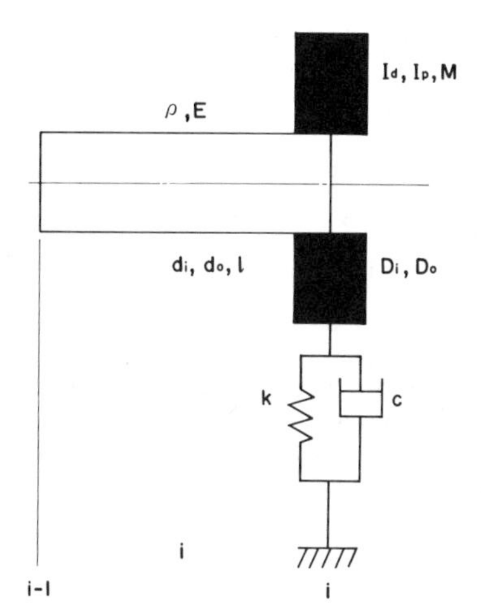

Figure 22. Part of rotor for vibration analysis.

Self-excited vibrations must be examined. Narrow clearances in multistage centrifugal pumps, for instance, interstage rings, act as a kind of journal bearings, and sometimes cause a self-excited vibration like an oil whip [26].

PROBLEMS IN OPERATION

Surging

When pumps are operated in low-capacity range, severe vibrations, i.e., surging, sometimes occur. The surging never occurs when a pump has a stable head-capacity curve, i.e., the head is monotonically increasing toward the shut-off point. Relations between factors in the impeller design and the stable curve were described earlier. When a head-capacity curve has a drooping range, it does not always mean that surging occurs. When the inclination of the head-capacity curve is smaller than that of the system head curve at point A, as shown in Figure 23, surging does not occur. When the inclination of the head-capacity curve is larger at point B, as shown in Figure 24, surging occurs.

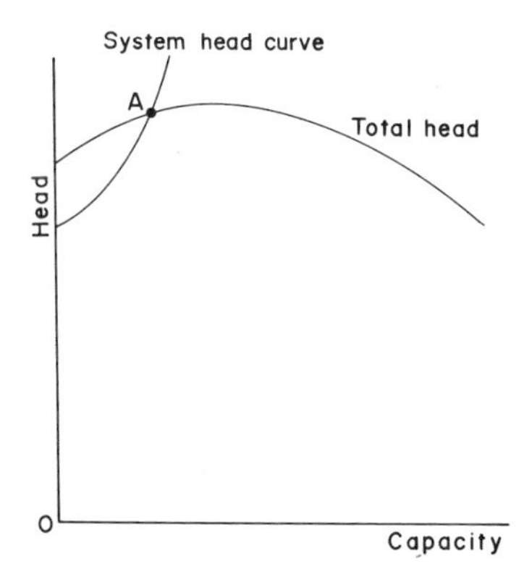

Figure 23. Stable operating point.

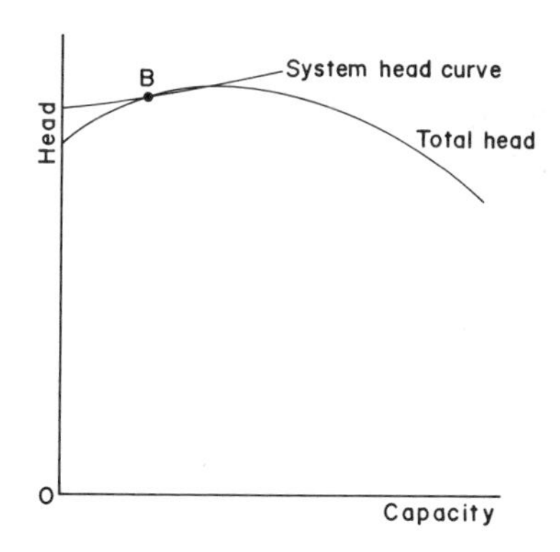

Figure 24. Unstable operating point.

Operation at point B is unstable: When the flow increases slightly there, the pump head becomes higher than the system head, then the liquid in the pipe accelerates, and the flow further increases. When the flow decreases slightly at point B, the pump head becomes lower than the system head, and the flow further decreases.

Surging is prevented in two ways:

1. Pumps are designed to have a stable head-capacity curve.
2. Resistances in the pump system, for instance, the discharge valve and the orifice flow meter, are attached near the pump.

The following phenomenum is sometimes experienced. A pump system with a discharge valve attached close to a pump is stable, as shown in Figure 23. When the same valve is attached to the end of the discharge pipe, the system becomes unstable, as shown in Figure 24.

At the pipe arrangement attention must be paid to avoid air pockets in the pipeline. Air reserved in the pipe acts as an air chamber and sometimes causes the surging.

Operation in Low Capacity Range

The long-time operation of pumps at the shut-off condition or at a very low-capacity range must be avoided. Under this operating condition vibration caused by pressure fluctuation is large and the power input consumed raises the liquid temperature. Recirculation of capacity to the suction side prevents the pump from operating at low-capacity range.

The minimum flow settled for boiler feed pumps is determined to hold back the maximum temperature rise of 20°C (36°F). Unsteady cavitation of the vortex type sometimes occurs at the inlet of the first-stage impeller by a reverse flow in a low-capacity range, as shown schematically in Figures 25 and 26 [27].

Control of Capacity

There are two methods to control the capacity: Valve control, as shown in Figure 27 and speed control, as shown in Figure 28. The former cost is lower than the latter one. The former, however, consumes more power input than the latter, as shown in Figure 27.

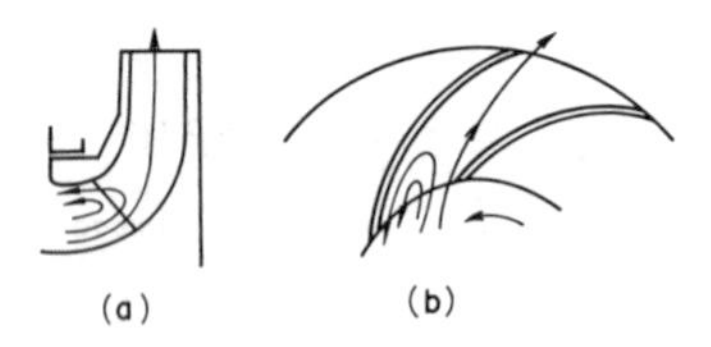

Figure 25. Reverse flow in low capacity range [27].

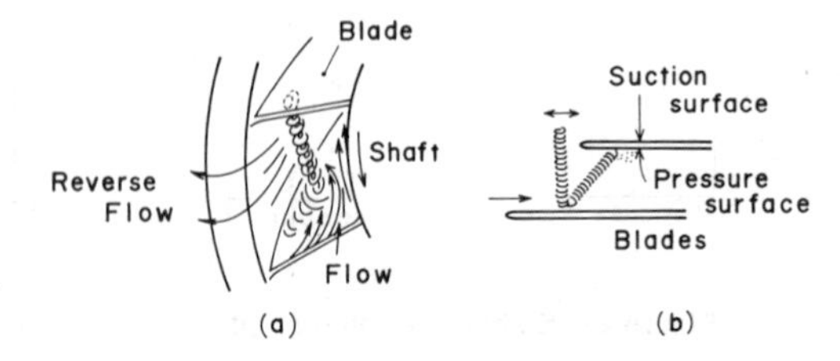

Figure 26. Vortex cavitation in low capacity range [27].

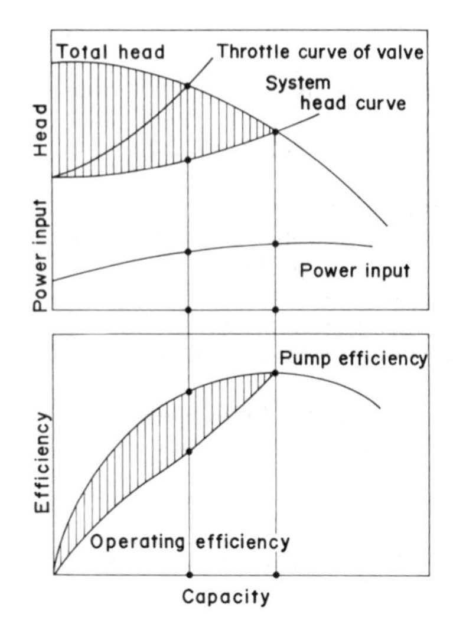

Figure 27. Capacity control by valve.

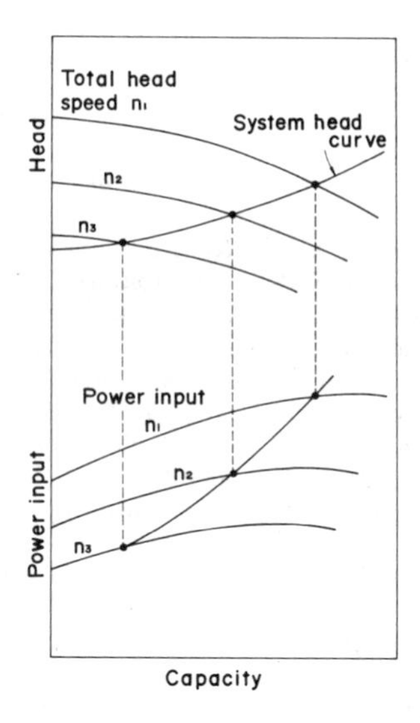

Figure 28. Capacity control by variable speed.

NOTATION

A	area
b_2	impeller outlet width
c_m	meridional component of absolute velocity of flow
c	absolute velocity of flow; damping constant
D_{2m}	impeller diameter
D_3	inlet diameter of diffuser
d	diameter
E	modulus of longitudinal elasticity
F	thrust
g	acceleration of gravity
H	head per stage
I	moment of inertia
K_v	volute velocity constant
K_m	capacity constant
K_u	speed constant
k	spring constant
ℓ	length
M	mass
N_s	dimensionless specific speed
NPSE	net positive suction energy
NPSH	net positive suction head
n	pump speed
n_s	specific speed
Q	capacity (volumetric flow rate)
r	radius
s_{u2}	vane peripheral thickness
T	unbalancing thrust
u	peripheral velocity
v_v	average velocity in volute casing
w	relative velocity of flow
z	number of vanes

Greek Symbols

α	angle
β	vane angle
γ	specific weight
ρ	density

REFERENCES

1. Pfleiderer, C., "Erfahrungen und Fortschritte in der Berechnung der Kreiselpumpen," *VDI-Z.*, Bd. 82, Nr. 9 (1938), S. 265.

2. Stepanoff, A. J., *Centrifugal and Axial Flow Pumps*, 2nd ed., New York: John Wiley, 1957, p. 297.
3. Stepanoff, A. J., *Centrifugal and Axial Flow Pumps*, 2nd ed., New York: John Wiley, 1957, p. 79.
4. Stepanoff, A. J., *Centrifugal and Axial Flow Pumps*, 2nd ed., New York: John Wiley, 1957, p. 81.
5. Pfleiderer, C., *Die Kreiselpumpen*, 5. Aufl. Berlin: Springer, 1961, S. 190.
6. Katsanis, T., "Use of Arbitrary Quasi-Orthogonals for Calculating Flow Distribution in a Turbomachine," *Trans. ASME, Series A*, Vol. 88, No. 2 (1966), pp. 197–202.
7. Senoo, Y., and Nakase, Y., "A Blade Theory of an Impeller with an Arbitrary Surface of Revolution," *Trans. ASME, Series A*, Vol. 93, No. 4 (1971), pp. 454–460.
8. Senoo, Y., and Nakase, Y., "An Analysis of Flow Through a Mixed Flow Impeller," *Trans. ASME, Series A*, Vol. 94, No. 1 (1972), pp. 43–50.
9. Krimerman, Y., and Adler, D., "The Complete Three-Dimensional Calculation of the Compressible Flow Field in Turbo Impellers," *Journal Mechanical Engineering Science*, Vol. 20, No. 3 (1978), pp. 149–158.
10. Krisam, F., "Der Einfluss der Leitvorrichtung auf die Kennlinien von Kreiselpumpen," *VDI-Z.*, Bd. 94, Nr. 11/12, 1952, S. 322.
11. Stepanoff, A. J., *Centrifugal and Axial Flow Pumps*, 2nd ed., New York: John Wiley, 1957, p. 110.
12. Stepanoff, A. J., *Centrifugal and Axial Flow Pumps*, 2nd ed., New York: John Wiley, 1957, p. 126.
13. Stepanoff, A. J., *Centrifugal and Axial Flow Pumps*, 2nd ed., New York: John Wiley, 1957, p. 128.
14. Stepanoff, A. J., *Centrifugal and Axial Flow Pumps*, 2nd ed., New York: John Wiley, 1957, p. 113.
15. Stepanoff, A. J., *Centrifugal and Axial Flow Pumps*, 2nd ed., New York: John Wiley, 1957, p. 127.
16. Akaike, S., and Toyokura, T., "Flow in Interstage Return Bend of Centrifugal Turbomachinery," *Proc. of the 6th Conference on Fluid Machinery*, Vol. 1, Budapest, 1979, pp. 11–20.
17. Ellis, G. O., "Crossover Systems Between the Stages of Centrifugal Compressors," *Journal of Basic Engineering, Trans. ASME* (1960), pp. 155–168.
18. Stepanoff, A. J., *Centrifugal and Axial Flow Pumps*, 2nd ed., New York: John Wiley, 1957, p. 207.
19. Kurokawa, J., and Toyokura, T., "Study on Axial Thrust of Radial Flow Turbomachinery," *Proc. of the Second International JSME Symposium Fluid Machinery and Fluidics*, Vol. 2, Tokyo, 1972, pp. 31–40.
20. Iino, T., Sato, H., and Miyashiro, H., "Hydraulic Axial Thrust in Multistage Centrifugal Pumps," *Journal of Fluids Engineering, Trans. ASME*, Vol. 102 (1980), pp. 64–69.
21. Carter, R., Karassik, I. J., and Wright, F. F., *Pump Questions and Answers*, New York: McGraw-Hill, 1949, p. 43.
22. Stepanoff, A. J., *Centrifugal and Axial Flow Pumps*, 2nd ed., New York: John Wiley, 1957, pp. 212–215.
23. Carter, R., Karassik, I. J., and Wright, F. F., *Pump Questions and Answers*, New York: McGraw-Hill, 1949, p. 44.
24. Carter, R., Karassik, I. J., and Wright, F. F., *Pump Questions and Answers*, New York: McGraw-Hill, 1949, p. 42.
25. Stepanoff, A. J., *Centrifugal and Axial Flow Pumps*, 2nd ed., New York: John Wiley, 1957, p. 209.
26. Black, H. F., "Calculation of Forced Whirling and Stability of Centrifugal Pump Rotor Systems," *Journal of Engineering for Industry, Trans. ASME*, Vol. 96, No. 3 (1974), pp. 1076–1084.
27. Okamura, T., and Miyashiro, H., "Cavitation in Centrifugal Pumps Operating at Low Capacities," *Polyphase Flow in Turbomachinery, ASME*, 1978, p. 250.

CHAPTER 37

OSCILLATING DISPLACEMENT PUMPS

Gerhard Vetter

Erlangen University
West Germany

Ludwig Hering

American Lewa Inc.
Natick, Massachusetts USA

CONTENTS

INTRODUCTION, 1057

PRINCIPLES OF OPERATION AND TYPES OF DESIGN, 1058

BASICS, 1062
- Kinematics of Displacement, 1062
- Flow Rate, the Pumping Process and its Characteristics, 1066
- Metering Errors, 1068
- Displacement Kinematics in Real Terms, 1070
- Torque, Output, Efficiency, 1071
- Specific Gravity, Cavitation, Gas Bubbles, Elasticities, 1072
- Efficiency Level, 1073
- Suction Process, Entry-Pressure Loss, Cavitation, 1076

DESIGN OF OSCILLATING DISPLACEMENT PUMPS, 1078
- Volumetric Metering Pumps, 1078
- Diaphragm Pumps for Heterogeneous Fluids, 1094
- In-Line Piston Pumps, 1098

INSTALLATION OF OSCILLATING DISPLACEMENT PUMPS, 1105
- Pressure Fluctuations without Damping, 1106
- Pulsation Damping, 1110
- Type of Installation, 1115

NOTATION, 1117

REFERENCES, 1117

INTRODUCTION

Oscillating displacement pumps are among the oldest types of pumps. They owe their high popularity to their very high rate of efficiency as well as to their volumetric metering applications. Although in the last century, these pumps have been displaced by the lighter and less expensive centrifugal pump, new types of applications have been created for displacement pumps in the fields of high pressure engineering, process technology, and performance hydraulics—applications which lead to strong and new impulses of development.

Table 1
Design Types of Oscillating Displacement Pumps as per Area of Application

Area of Application	Design Type
Water supply	Deep-well piston pumps Pressure-booster piston pumps
Crude oil and natural gas production	Deep-hole reciprocating piston pumps Flush pumps
Power supply	Boiler feed pumps with steam and electric drives
Lubrication engineering	Piston-type lubricating pumps
Automotive engineering	Fuel injection pumps
Performance hydraulics	Axial piston pumps Radial piston pumps In-line piston pumps
Medical technology	Heart pumps, infusion pumps
Homogenizing engineering	Homogenizing pumps in in-line piston design
Production engineering	High pressure purification pumps Water jet torching pumps
Process and chemical engineering	Piston-type metering pumps Diaphragm-type pumps Filling pumps Discharge pumps

This large area of application produced a multitude of designs (see Table 1). This chapter will concentrate mainly on the explanation of oscillating displacement pumps for application in process technology and production engineering.

PRINCIPLES OF OPERATION AND TYPES OF DESIGN

Most types of design can be traced back to the principle of operation as shown in Figure 1. A piston (1) is driven in an oscillating mode by means of drive mechanism (2) at a defined frequency n. In accordance with cross-section A_k and stroke length h, the piston displaces the stroke volume $V_h = A_k \times h$. The working space (3) is intermittently sealed by the self-acting suction and pressure valves (4 and 5) as well as by the piston gland (6). After the effective seals are operating almost clearance-free, the leakage flows occurring during the pumping process are, as a rule, extremely insignificant. Oscillating displacement pumps always exhibit for the single cylinder a time-dependent

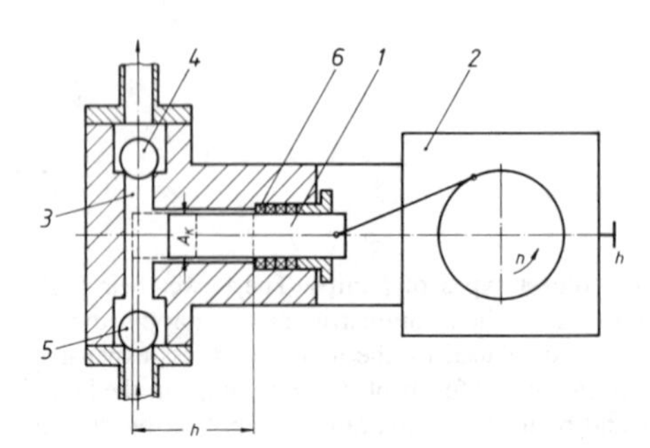

Figure 1. Principles of operation for oscillating displacement pumps.

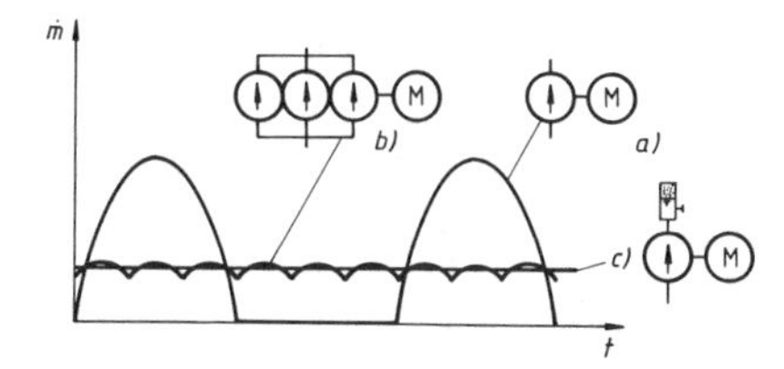

Figure 2. Time-dependent progression of flow rate: (A) single-cylinder pump; (B) Multi-cyclinder pump; (C) Pump with ideal damper system.

pulsating flow rate which can be smoothed out only by superimposing several cylinders or damping systems (Figure 2).

The principle of operation based on an intermittent displacement mode and the working space being at high density will, in contrast to other types of pumps (Figure 3), create rigid pressure characteristics as well as an outstanding level of efficiency.

Table 2 explains the limitations to the areas of application. There is no pump type that can achieve such a spread of applications. The main application oscillating displacement pumps have is in cases of high pressures and smaller flow rates. The hydraulic output for metering pumps achieves a range of up to 100 kW approximately and, for transfer pumps, up to and over 1,000 kW.

Because of the large variety of applications, the design types are of a diverse nature. The majority of displacer drives (Table 3) have their origin in the direct-thrust crank gear unit. In the field of metering pump manufacture the drive mechanisms are mostly stroke-adjustable. Transfer pumps are produced in a multi-cylinder design for the purpose of pulsation equalization (in-line piston pumps). With respect to the large number of extensively disseminated types of displacer drives, Table 3 will, in addition, outline the spring-action cam system as well as the magnetic and pneumatic linear drives.

The stroke-adjustable spring-action cam system is extremely popular in metering pump manufacturing, especially for small performance rates. At part stroke, this type of stroke adjustment causes unsteady and jerky kinematics, which, limits the operation to smaller flow rates and low-stroke frequencies. There is a similar effect from all operating principles that in a hydraulically stroke-adjustable mode transmit a given piston motion onto lost motion systems by means of adjustable overflow openings.

The magnetic linear drive has significance as a displacer drive for metering and diaphragm pumps of very small output rates. The kinematics of displacement are determined by the law of motion of the magnetic drive; in places, there exists a nearly uniform motion. Magnetic linear drives are adjustable as to stroke and frequency of strokes. They are easy to incorporate into control and regulating circuits. Pneumatic linear drives, also known to exist with a direct action of the high-pressure gas on diaphragms, are used to fulfill special application requirements such as protection against explosion.

More than 90% of the application possibilities are covered by the design types enumerated. In addition, we would like to point out special designs with a pulsation-poor flow rate which serve particular applications in the chemical industry and analytical measurement technology [1].

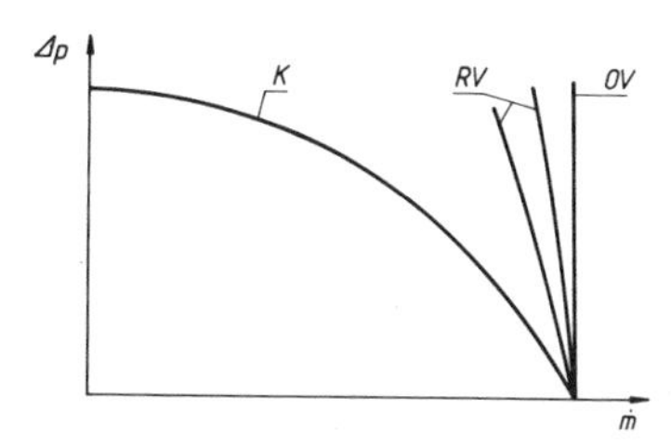

Figure 3. Schematics of pressure and flow-rate characteristics of the different pump types (K—centrifugal pumps; RV—rotating displacement pumps; OV—oscillating displacement pumps).

Table 2
Fluids Suitable for Pumping by Oscillating Displacement Pumps

Properties	Examples	Limits	Lay-out and Sizing Characteristics
Flow rate	The entire field of process technology	A few ml/hour up to many m^3/hour	Within the ml/h range, application of special micro-metering pumps; large flow-ins require high stroke frequencies (up to 350 strokes per minute), double-acting pump heads or multi-cylinder arrangement.
Pressure	Pumping from vacuum; initiator metering against 3,500 bar in the production of polyethylene	Vacuum up to 3,500 bar	Design of pump chamber (working space) and of valve to be in accordance.
Specific gravity	From volatile media to mercury	None	Spring-loaded valves at high specific gravity; leakproof metering pumps at low specific gravity.
Corrosion	Acids and solutions	None	All kinds of materials; diaphragm pumps.
Toxicity, dangerousness	Hydrocyanic acid, chlorine, bromine, plutonium salts, nitrotoluene, sodium	None	Diaphragm pumps.
Suspensions, slurries	Coal slurry, catalyst, TiO_2, ceramic slip, contact suspensions	Grain size Sedimentation	Wear-resistent materials; flow-favorable design; pumps without stuffing box (diaphragm pumps).
Viscosity	Resins, pastes	Normal $<50{,}000$ cP valve operation, cavitation	Low stroke frequency; large cross-sections of flow; valves to be spring loaded or externally controlled; valveless gate control.
Temperature	Liquid oxygen, hot melts	$-270°$ to $+500°C$	Insulation, cooling, heating.

Designation	Principle of Operation	Application	Displacement Kinematics	Adjustment	Displacement System
stroke-adjustable quasi-harmonic drive (one-cylinder type)	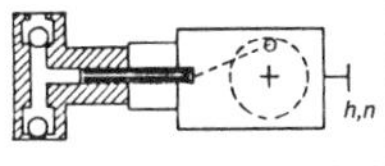 h,n	Metering	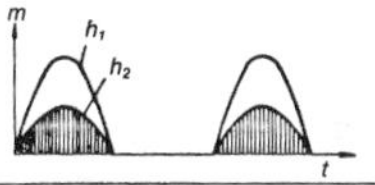 m, h_1, h_2, t	h $n < 350 \frac{1}{min}$	K FM MH SH
Direct-thrust crank gear drive (multi-cylinder type)	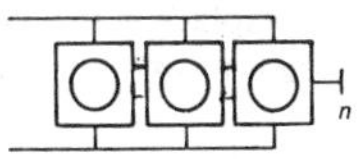 n	Pumping	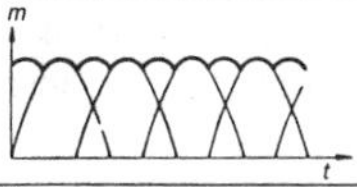 m, t	$n < 1000 \frac{1}{min}$	K (MH) (SH)
Spring-action cam drive	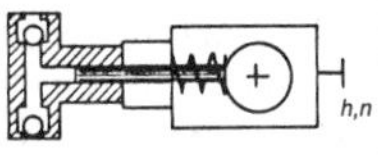 h,n	Metering	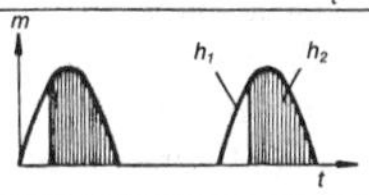 m, h_1, h_2, t	h $n < 150 \frac{1}{min}$	K FM MH MM
Magnetic linear drive	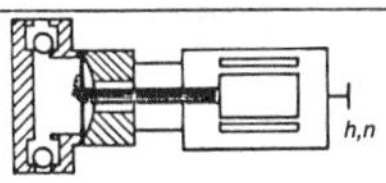 h,n	Metering	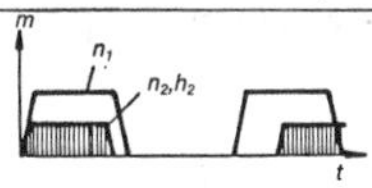 m, n_1, n_2,h_2, t	h $n < 120 \frac{1}{min}$	MM MH
Pneumatic linear drive	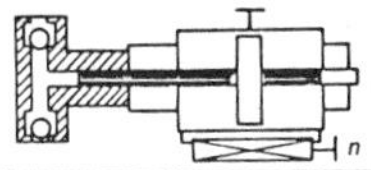 n	Pumping	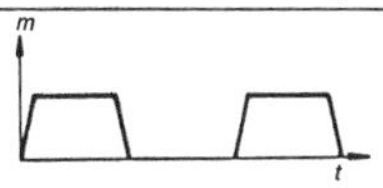 m, t	$h < 120 \frac{1}{min}$	K MM

With regard to the displacement systems (Table 4), there is a distinction between piston-displacement systems with a directly moving sealing of the fluid towards the outside area and displacement systems with diaphragms, bellows, or tubes that are leakproof and do not require moving seals for the fluid. Hydraulically driven diaphragms are suitable for higher pressures because of the hydraulic support. Mechanically driven diapragms and bellows experience the pumping pressure directly in the form of stress, and therefore they are limited to low pressures.

Table 4
Overview of Displacer Systems

Designation	Area of Application
1. Piston (K)	Pressure up to 3,500 bar; temperature −270 up to +500°C; suitable for all flanges; leakage and maintenance of stuffing box unavoidable.
2. Piston-diaphragm; hydraulic (MH)	Pressure up to 3,500 bar; center of gravity up to 700 bar; temperature <250°C; free of leakage—suitable for dangerous, toxic, and abrasive fluids, for suspensions, pastes, and slurries
3. Piston-diaphragm with tube; hydraulic (SH)	Pressure up to 50 bar; temperature <150°C; free of leakage—suitable for certain slurries at higher output
4. Diaphragm; mechanical (MM)	Pressure <10 bar; temperature <150°C; for smaller outputs and subordinate applications; leakproof design at good price.
5. Bellows; mechanical (FM)	Pressure <5 bar; temperature <120°C; mostly in glass/PTFE design.

The design types are represented in the first two lines of Table 4 compose approximately 80% of the possibilities of application.

BASICS

Kinematics of Displacement

The time-bound progression of the stroke motion of the displacement system determines the kinematics of displacement. Most displacement drives produce a more or less harmonic motion of displacement.

All systems that can be traced back to the direct-thrust crank gear unit (for instance, Figures 4A, B) are characterized by the rod ratio $\lambda_s = r/\ell$, r being the crank radius and ℓ the rod length. Other systems (for instance, Figures 4C, D, E, circular cams and cross-type cranks) are characterized by the ratio of $\lambda_s = r/1 = 0$. By means of crank angle $\psi = \omega t$, where ω denotes angular speed, you will find the following equations for piston speed and acceleration:

$$v_k = r \cdot \omega \cdot \left(\sin \omega t + \frac{\lambda_s}{2} \cdot \sin \omega t \right) \quad (1)$$

$$b_k = r \cdot \omega^2 \cdot (\cos \omega t + \lambda_s \cdot \cos 2\omega t) \quad (2)$$

In this respect, the crank radius corresponds to the half-stroke length h = 2r, and the angular speed is computed as $\omega = \pi h/30$, in speed $\min^{-1}$.

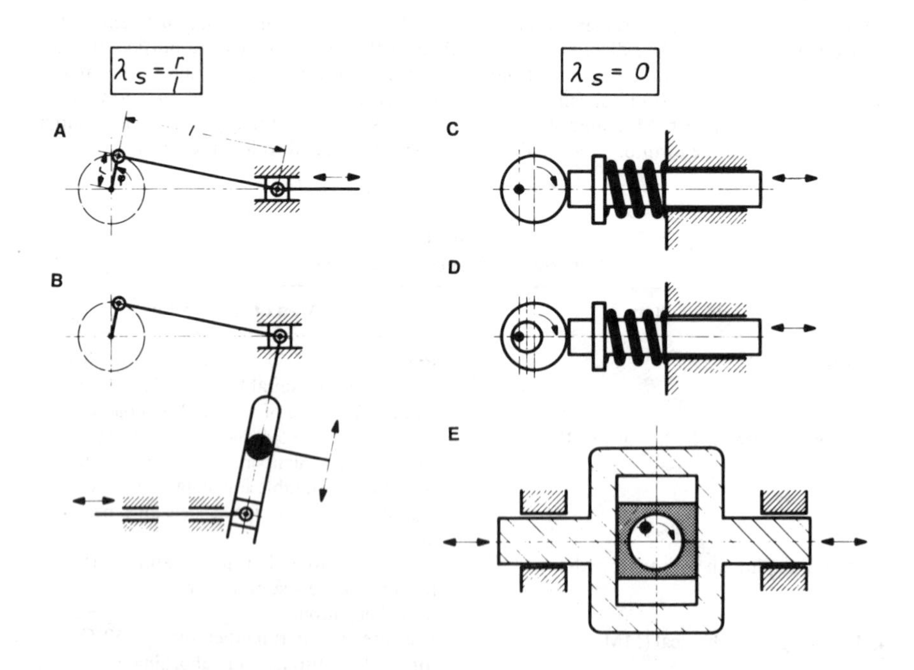

Figure 4. Drive gear systems with harmonic and quasi-harmonic kinematics: (A, B, D) with stroke adjustment; (C, E) without stroke adjustment.

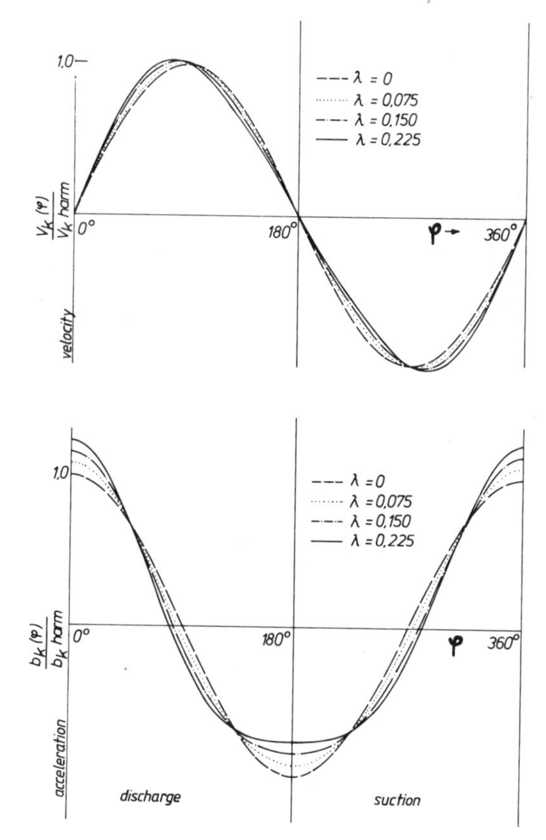

Figure 5. Kinematics in dependence on rod ratio.

At $\lambda_s = 0$, one finds harmonic kinematics as follows:

$$v_k = r \cdot \omega \cdot \sin \omega t \tag{3}$$

$$b_k = r \cdot \omega \cdot \cos \omega t \tag{4}$$

Figure 5 shows the difference of Equations 1 through 4 as a ratio for different λ. While the speed rates hardly differ from each other, the acceleration values of the direct-thrust crank gear unit are strongly unsymmetric in extreme positions. Ordinary values of the λ ratios rest between 0 and 0.2. In stroke adjustable displacement systems, the values of V_k and b_k also vary in accordance with the stroke length $h = 2r$. In this regard, the stroke kinematics, as a rule, basically remain as is (for examples, see Figures 6, 7, and 8).

With the frequently used stroke-adjustable spring-action cam drive mechanism as per Figure 9, the partial stroke is actuated by the stop limitation of the piston rod movement. This causes the stroke kinematics to become unsteady at part stroke, entraining consequences which will still be explained (refer to Table 4, Line 3). The displacement motion is transmitted on to the piping circuit whereby a completely rigid fluid and piping system are assumed as per the following conditions

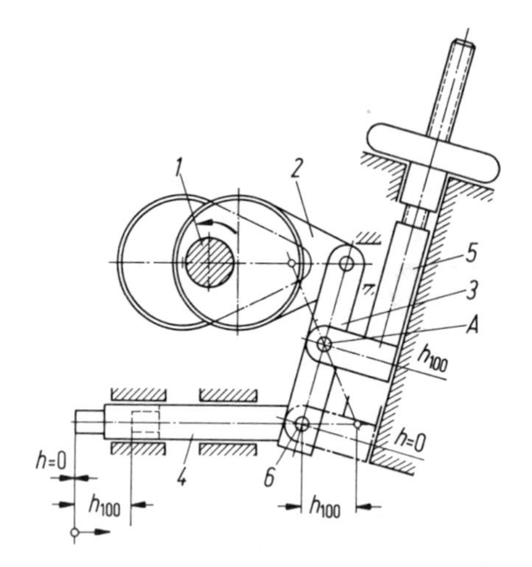

zero stroke position
≙ upper dead point

Figure 6. Rocker-arm system.

of continuity:

$$\omega(t) = v_k(t) \cdot \frac{A_k}{A} \tag{5}$$

$$b(t) = b_k(t) \cdot \frac{A_k}{A} \tag{6}$$

Pressure variations are produced in the piping system due to friction and mass acceleration and/or delay, factors that will be discussed later. The time-bound progression of the mass flow rate of oscil-

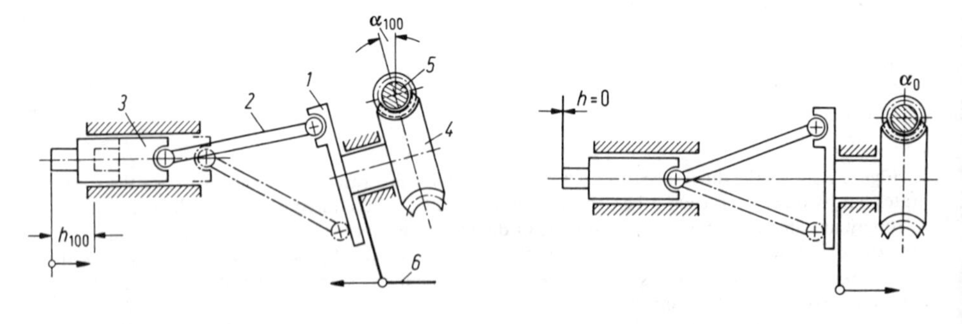

zero stroke position
≙ upper dead point

Figure 7. Rotary crank system.

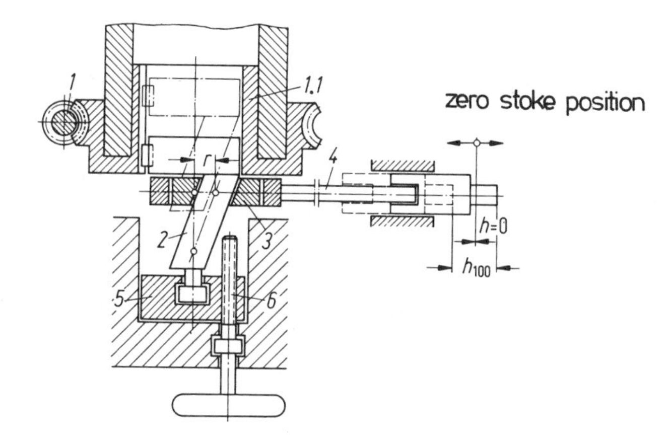

Figure 8. Eccentric adjustment system (angular type crank).

lating displacement pumps is of a pulsating nature which is demonstrated by Figure 10, exhibiting the example of the single-cylinder and triple-cylinder pump for the identical average volume flow rate Q_m at a volumetric efficiency of $\eta_v = 1$.

Thus, a smoothing of the mass flow is achieved by evenly superimposing several cylinders whereby, once again, a λ_s influence does exist (Figure 11). A multi-cylinder arrangement is the rule for discharge pumps (mostly triplex or quintuplex). In contrast, single-cylinder pumps are also frequently used for metering pumps. This explains why transfer pumps exhibit a stroke frequency of up to 1,000 min^{-1} and metering pumps rarely more than 300 min^{-1}.

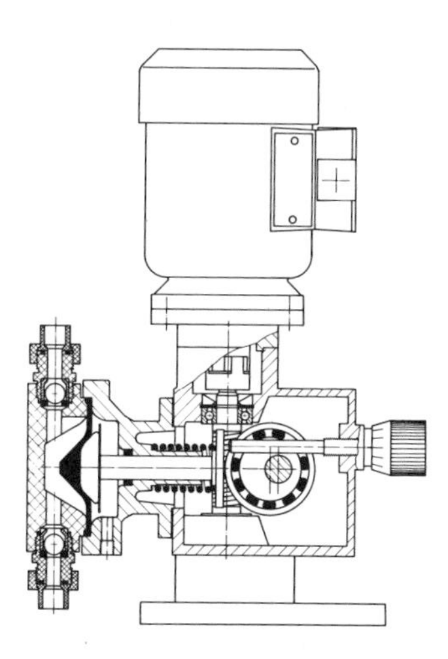

Figure 9. Diaphragm pump with spring-action cam drive gear.

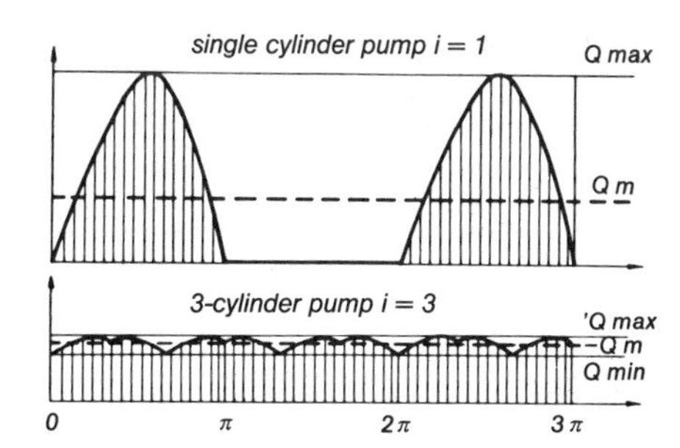

Figure 10. Progression of instantaneous flow in single and triple-cylinder pumps.

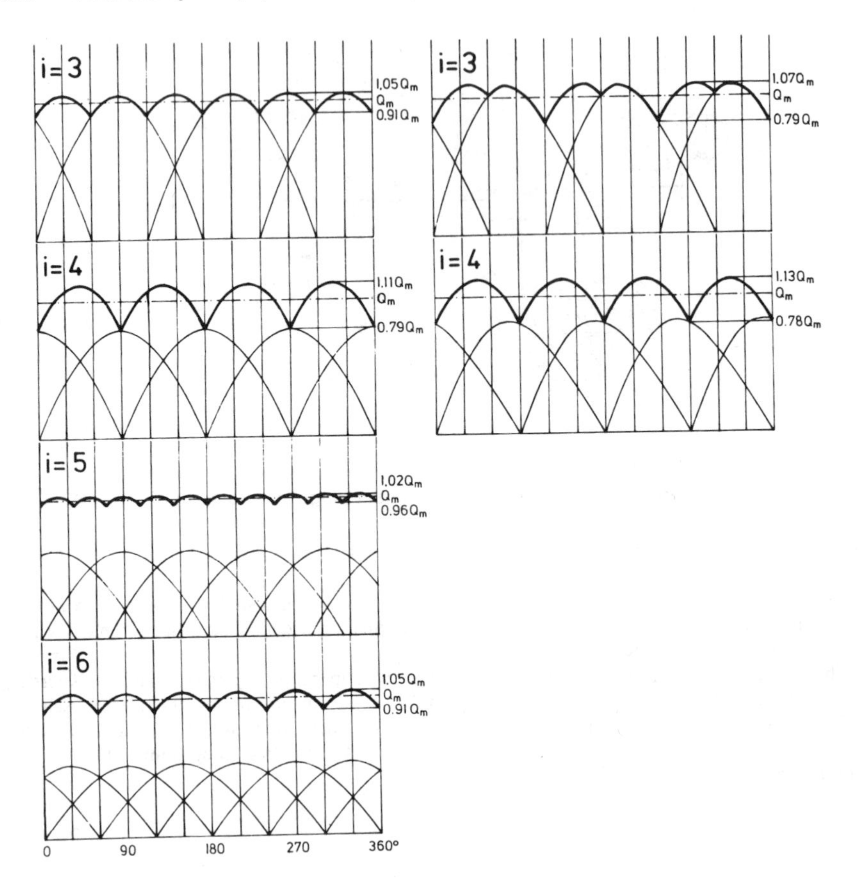

Figure 11. Progression of instantaneous flow at $\lambda_s = 0$ (LH) and $\lambda_s = 0.225$ (RH).

Flow Rate, the Pumping Process and its Characteristics

The flow rate of oscillating displacement pumps amounts to [2]:

$$\dot{m} = \rho \cdot i \cdot V_h \cdot \eta_V(E_p, E_F) \tag{7}$$

where ρ = specific gravity of the fluid
i = the number of cylinders
n = stroke frequency

The stroke volume V_h is calculated by the use of cross-section $A_K = (\pi d_K^2)/4$ and stroke length h to become $V_h = A_K \cdot h$.

For diaphragms and bellows directly driven by means of con rods (see Table 4, Lines 4 and 5), it may become necessary to determine the stroke volume V_h by means of experimentation because a simple connection to a diameter variable does not exist.

In Equation 7 all variables are known from the pump and fluid data. It is only the volumetric efficiency η_v representing a correction variable for the theoretical operating mode which is still unknown. To this effect, we should refer to the sequence of the pumping process shown in Figure 12.

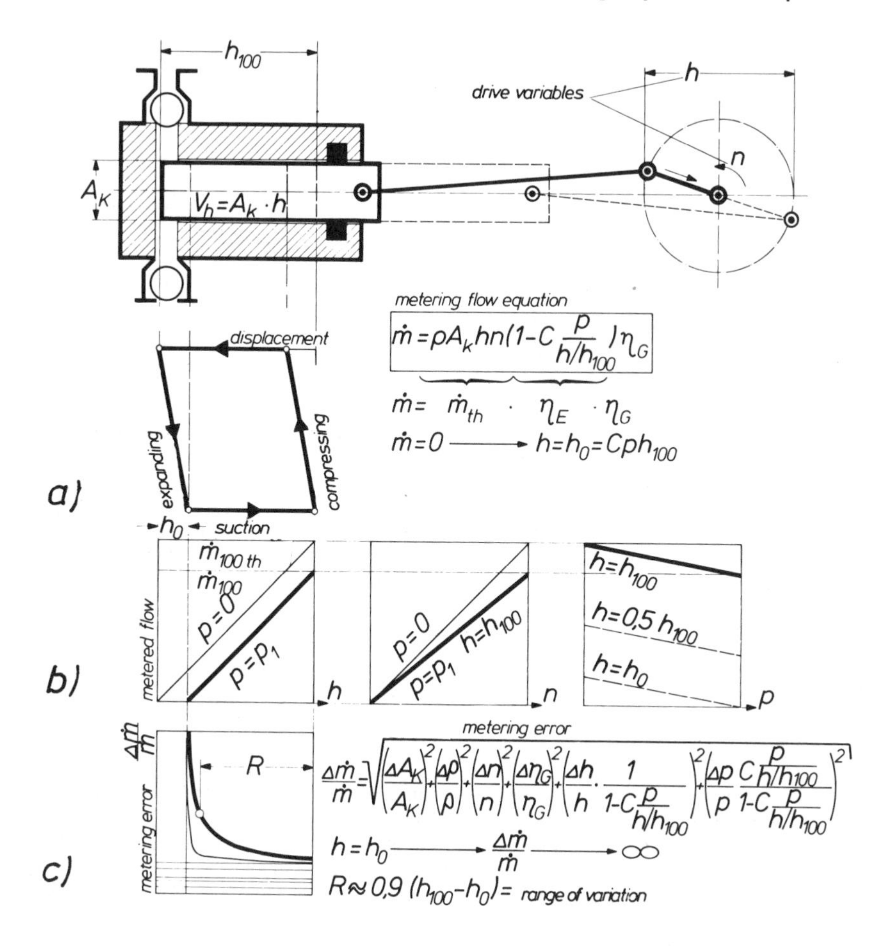

Figure 12. Pumping process.

After each fluid has become somewhat compressible, the pressure increase along the compression line will require a certain piston travel. Then the fluid will be expelled. It is along the expansion line that the return expansion of the so-called dead space will take place. The pump will begin suction in real terms only after the piston travel h_0 has been executed. The stroke motion following then corresponds to the effective suction stroke. This means that a pump the stroke length of which amounts to h_0 will compress just only under the conditions selected, but will not pump.

If one determines the influence the various variable values have on the flow rate for the general case of the stroke-adjustable pump, then with h/h100 being the stroke-adjustment ratio, the result [2] will be:

$$\dot{m} = \rho \cdot A_K \cdot h \cdot i \cdot \eta \left(1 - C \frac{P}{h/h100}\right) \cdot \eta_G \tag{8}$$

In this, the volumetric efficiency η_v is composed of elasticity level η_E and efficiency η_G as follows:

$$\eta_v = \eta_E \cdot \eta_G = \left(1 - C \frac{P}{h/h100}\right) \cdot \eta_G \tag{9}$$

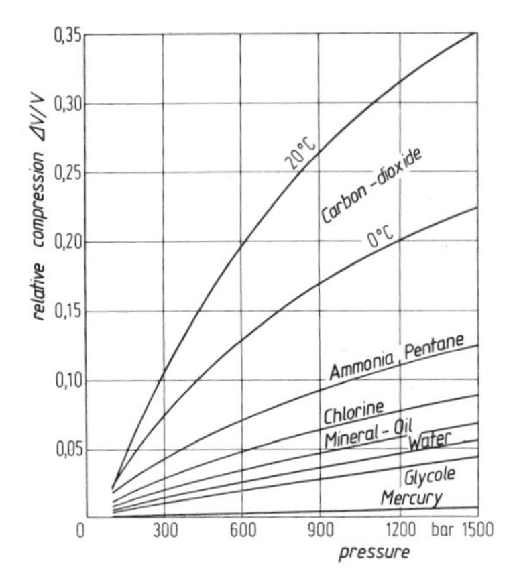

Figure 13. Relative volume compression of fluids.

The elasticity level can be calculated and takes into consideration (summed up in the constant C) the fluid elasticity and the working space elasticity κ or λ respectively, as well as the dead space V_T as per the following equations:

$$C = \frac{V_T}{V_h} \cdot \kappa + \lambda \tag{10}$$

$$\kappa = \frac{1}{\rho} \cdot \frac{\Delta V}{V} \tag{11}$$

The working space elasticity has an effect on the walls as well as on the glands. The relative volume compression rate of fluids can be gathered from Figure 13. The efficiency η_G takes into account the leakages such as those occurring on valves and piston glands. The variable can be determined only through experimentation, but it is usually close to 1. For stroke nonadjustable pumps (discharge pumps) $h/h_{100} = 1$. For stroke-adjustable metering pumps, the metered flow-in becomes zero at h_0/h_{100} (see Figure 11).

The shape of the essential characteristics is directly deducible from Equation 8. The flow rate and the stroke length progress in a linear mode. At $h = h_0$ the characteristic line will bisect the abscissa. The stroke length $h = h_0$ is necessary to overcome all elasticities at a given pressure. The flow rate progresses together with the stroke frequency in a linear and proportional mode.

Typical characteristics of metering pumps gained from measurements for particular pump types are summarized in Figure 14. The influence of the pressure is conspicuous, especially in diaphragm pumps having a hydraulic diaphragm drive. In mechanical diaphragm drives, non-linearities can mostly be detected.

Metering Errors in Volumetric Oscillating Metering Pumps

The relative fluctuation of the mass flow $\Delta\dot{m}/\dot{m}$ is a measurement for the metering error. It arises due to the fluctuation of the individual influence variables $\Delta n/n$, $\Delta p/p$, $\Delta h/h$, etc. If, on the basis of this one determines the fluctuation as per the method of the smallest error squares, one will find the relationship as shown in Figure 11. One recognizes that at $h = h_0$ the denominator expression $1 - C \times P/(h/h_{100}) = 0$ and thus becomes $\Delta\dot{m}/\dot{m} = \infty$. In stroke-adjustable metering pumps, therefore, the range of variation is available only as $R \approx 0.9(h_{100} - h_0)$.

Based on various published measurements [1, 3] Figure 15 impressively corroborates the findings made. The metering error increases asymptotically at $h = h_0$. Since the relationship of metering flow

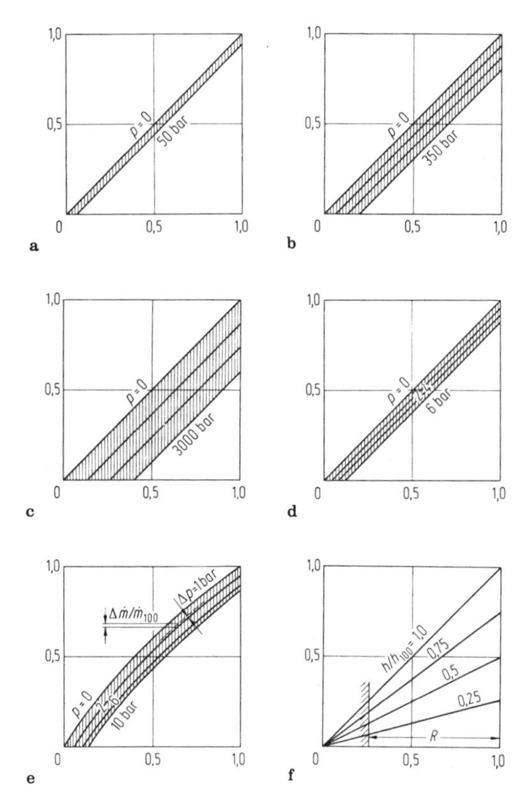

Figure 14. Characteristics of different metering pumps: (A) Piston-type metering pump (K); (B) Diaphragm-type metering pump (MH); (C) Diaphragm-type metering pump (MH); (D) Bellows-type metering pump (FH); (E) Diaphragm-type metering pump (MM); (F) Stroke-frequency adjustment at constant stroke length.

(flow-in) is proportional to the stroke frequency, the variation range can be further extended via the stroke frequency if the stroke length variation range does not suffice.

For the purpose of achieving a good metering accuracy, the factors η_E and η_G should approach the value 1 as closely as possible so the fluctuation rates of the marginal conditions will have a less strong effect. This is why manufacturers design volumetric metering pumps with dead spaces, non-elastic working spaces, and precise sealing valves.

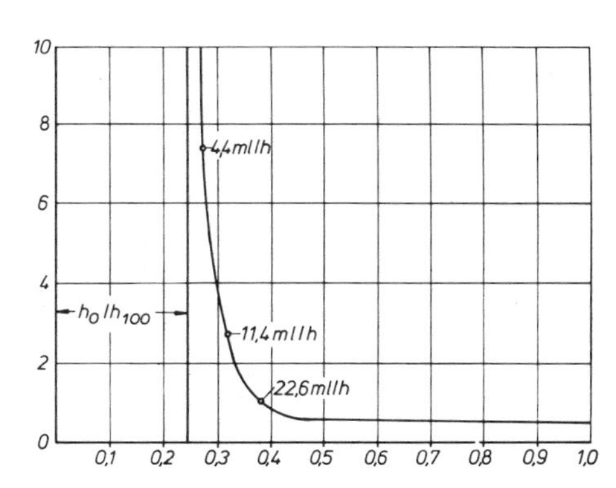

Figure 15. Metering error measured on a micro-metering pump.

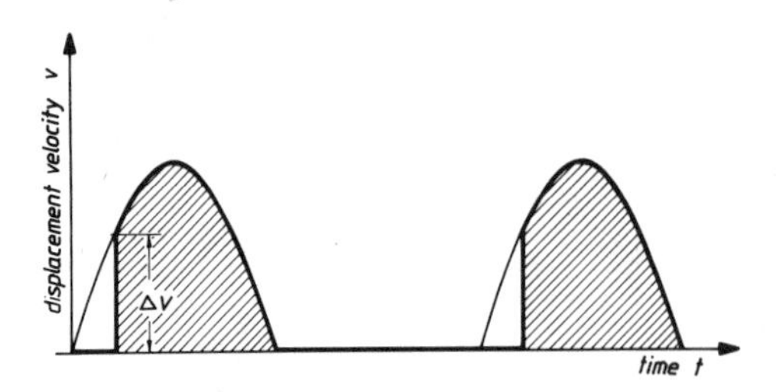

Figure 16. Displacement rate in real terms at a volumetric efficiency of $g_v < 1$.

Displacement Kinematics in Real Terms

The displacement kinematics represented in Figure 11 apply only to $\eta_V = 1$ [4]. Since $\eta < 1$ is always the case, the displacement action will only begin after the compression phase whereby, as per Figure 16, a part of the progression existent at missing elasticity will be cut off. This action will create considerably larger irregularities (Figure 17). In multi-cylinder pumps the instantaneous flow rate can indicate even zero positions (see Figure 17; i = 3; $\eta_V = 0.7$). The velocity bounce ΔV leads to pressure jolts which must be taken into account when the layout of the piping is designed. This

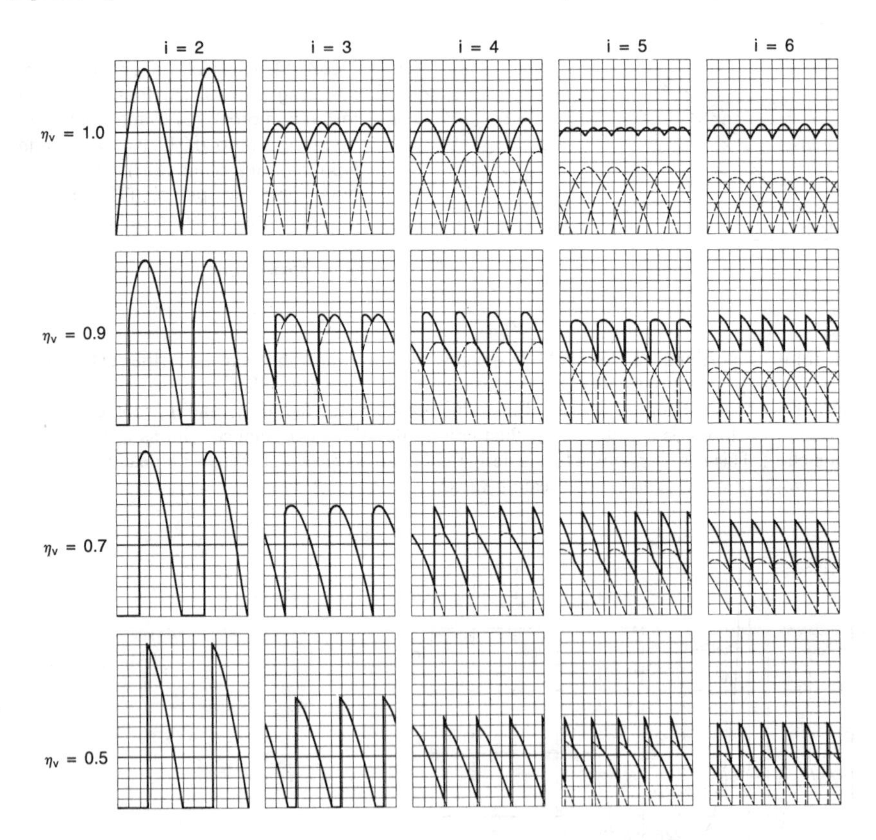

Figure 17. Displacement velocity in dependence on number of cylinders and volumetric efficiency ($\lambda_s = 0.225$).

means that in order to achieve favorable kinematics, you must aim at bringing the volumetric efficiency as closely to 1 as possible.

It is herewith pointed out that the lead characteristic of the displacement progression at part stroke is also exhibited by all pumps equipped with spring-action cam drive mechanism or with similar systems.

Torque, Power Requirement, Efficiency

As with all displacement pumps, the required torque is speed-independent, which means the power requirement is speed-proportional. The hydraulic output (rated output) is:

$$P_N = \dot{m} \cdot \frac{\Delta p}{\rho} \tag{12}$$

The required drive power P taking into account the total efficiency η is:

$$P_A = \frac{P_N}{\eta} \tag{13}$$

Due to the mostly pulsating flow rate, there will also arise time-bound fluctuations for torque T and power P.

The fluctuations are particularly large for single-cylinder and double-cylinder pumps. Figure 18 shows in a schematic form the ratios for $\eta_V = 1$ and $\eta_V < 1$. At $\eta_V = 1$ the torque fluctuates from

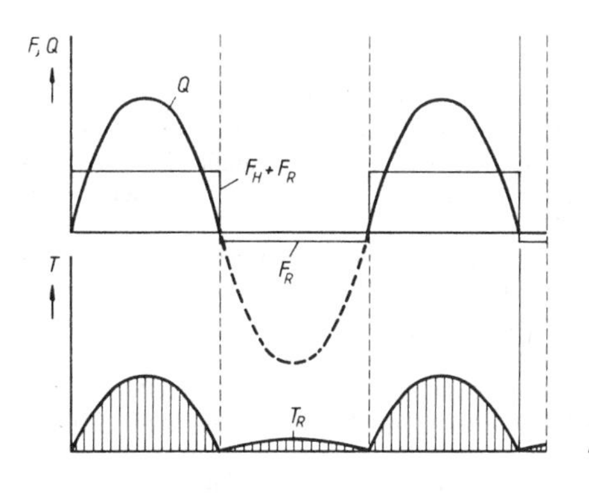

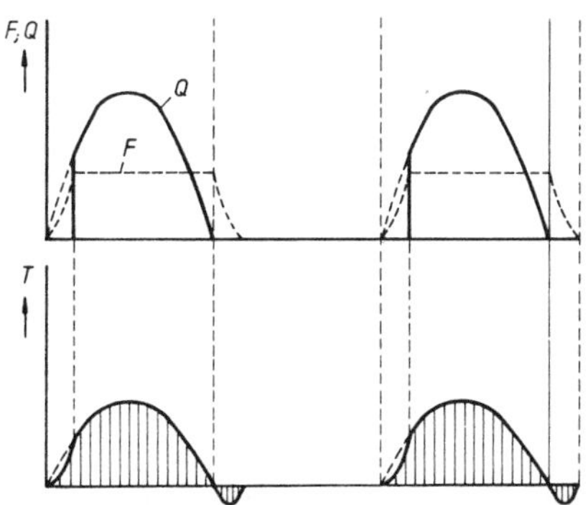

Figure 18. Torque progression of single-cylinder pump (Q—instantaneous volume flow; F_H—hydraulic piston rod load; F_R—stuffing box friction): (A) $\eta_V = 1.0$; (B) $\eta_V 1$, $F_R = 0$.

0 to T_{max}. During the suction stroke, the piston friction (piston gland) produces a small torque T_R of the identical sign. At beginning suction stroke at $\eta_V < 1$, the return expansion of the compressed liquid causes a sign change of the torque, which in gear drives with play will cause dynamic compression stress. This effect is demonstrated for $T_R = 0$ in Figure 18B. By mounting an additional centrifugal mass (flywheel), this unfavorable load condition can be prevented. Multicylinder pumps exhibit mostly a smoother torque action. As per Figure 17 there are, however, also operating conditions during which the torque will return to zero. However, sign changes at $i > 2$ will, as a rule, not be found.

For the rating and layout of electric motors, gear units, and shaft connections, it is imperative that the type of dynamic load be drawn into consideration. Where electric motors are involved, attention must be directed to the start-up conditions, the equivalent heating while the torque is pulsating, and possible fluctuations in the electronics. Gear units and shaft connections must be able to withstand, free from wear, the oscillating load. As a rule, jolt factors for the purpose of oversizing are applied, but they must be of an experience-based nature. Joint connections of shafts require free-from-play seats to avoid frictional corrosion.

The efficiency = P_N/P_A (useful output P_N, absorbed output P_A) can in smaller pumps produce unfavorable values in the range of 0.5 to 0.7 due to the relatively high portion of the frictional output P_R (piston gland and gear unit). Furthermore, in single and double-cylinder pumps, due to the strong fluctuation of P_A, the efficiency can only be accurately determined by integrating cycle $P_A(t)$ over time.

In contrast to centrifugal pumps, oscillating pumps of larger output capacities ($P_N > 1$ kW) exhibit excellent efficiencies due to the infinitely small internal leakage losses, so that it frequently pays to establish a cost-benefit analysis. Total efficiency rates from 0.85 to 0.92 are the rule for larger oscillating discharge pumps.

Specific Gravity, Cavitation, Gas Bubbles, Elasticities

Oscillating displacement pumps operate on a volumetric basis. The pumped mass flow $\dot{m}$ is directly influenced by the density. In particular, in fluids hovering closely at the boiling point, the density varies with pressure and temperature (Figure 19). Density as well as compressibility vary in a particularly strong mode when gas or vapor bubbles form. A penetration of air during the suction process must be avoided by blocking the piston gland.

The gas evolution or evaporation (cavitation) of the liquid pumped in the pump chamber (working space) ensues when a certain pressure is reached (evolution and/or vapor pressure). Cavitation will lead to an insufficient fluid charge of the cylinder as well as to pressure jolts with the formation of noise due to an impetuous collapse of the bubbles. The consequences are damage to the pump or facility as well as catastrophic metering errors in metering pumps. Figure 20 depicts cavitations in the indicator diagram. If steam bubbles form (see figure 20B) the pressure boost occurs in a delayed mode because the bubbles must first be collapsed.

We will discuss later measures to be taken to prevent cavitation and explain the dimensioning of piping systems for oscillating displacement pumps.

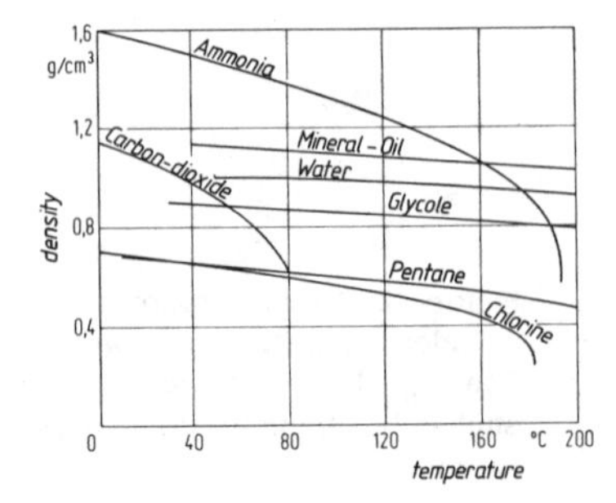

Figure 19. Specific gravity of fluids.

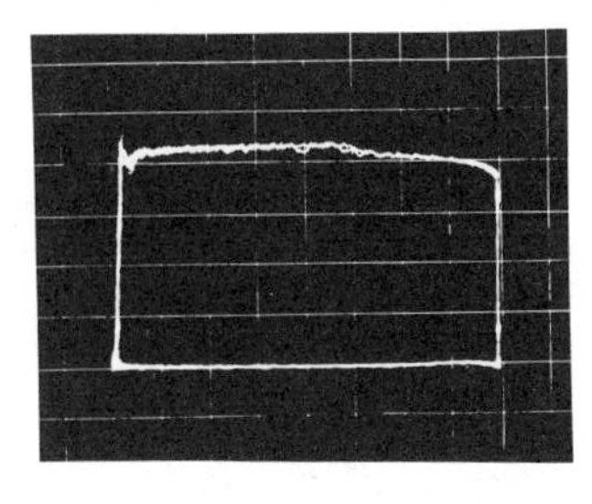

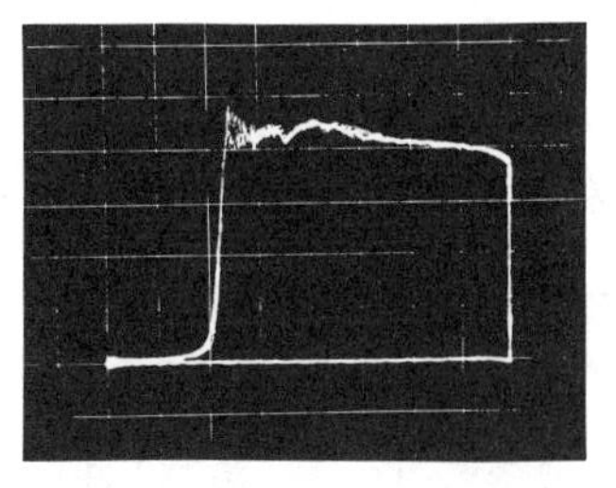

Figure 20. Influence of cavitation on fluid charge of pump's working space: (A) without cavitation; (B) with cavitation.

Efficiency Level

The efficiency level (η_G) (see Equation 8) takes into account the leakage losses at pump valves and piston glands (flexible seals). All efforts are directed to bringing η_G as closely as possible to a value of 1. Let us briefly look at the individual influences.

Leakage Losses at Movable Seals

As a rule, these losses are negligibly small. However, in extreme cases (such as poor maintenance, too much wear, small stroke volume) the efficiency can be lowered markedly. Piston seals, packing rings, or lip rings are predominantly used, as they are adjustable to an almost leakproof extent. In contrast, split seals require a certain leak flow that can no longer be neglected.

Leakage Losses at Pump Valves

An overview of typical valve designs is provided in Figure 21. Ball valves are suited for small nominal widths (up to about 15 mm). Spring-loading is used to compensate for unfavorable geodetic conditions or where viscous fluids are involved. For larger nominal widths and higher stroke frequencies, the use is concentrated on ball- and plate-type valves with a light spring load. The closing motion over time (t_0 = opening; t_s = closing) is determined by the stroke frequency, valve weight,

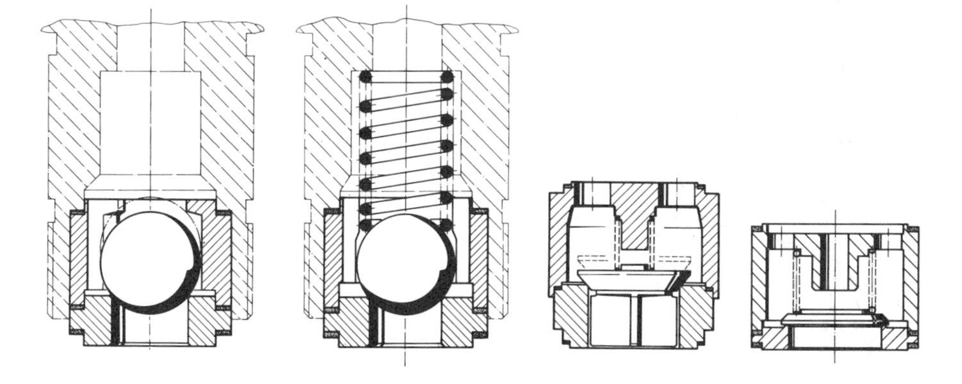

Figure 21. Typical pump valves.

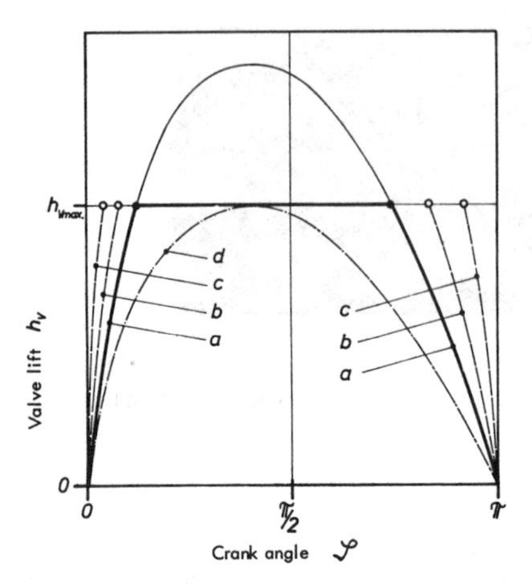

Figure 22. Time-dependent progression of valve motion.

spring load, and pumped liquid (viscosity). The optimal situation would be a valve which opens to its maximum capacity and closes without lag at a small closing velocity v_s (Figure 22, curve d). The closing energy $E_s = \frac{1}{2}m_v \cdot v_s^2$ is small in light-weight valves (valve mass m_v small) and at a low closing velocity. Material stress and wear increase rapidly at a certain closing energy which is also the main reason for the run-out noises ("beat limit").

Valve closing lag. The delayed closing action produces genuine "internal leakage" (reflux) because a liquid plug (see Figure 23A, which shows this detail for the pressure valve in a simplified mode) at stroke and end returns into the cylinder due to the delayed closing motion. This causes a fluid charge loss which is visible in the indicator diagram (see Figure 23B).

Figure 24 will explain the measured results for the efficiency η_G of a defined valve type in dependence on stroke frequency n, viscosity ν, and spring load.

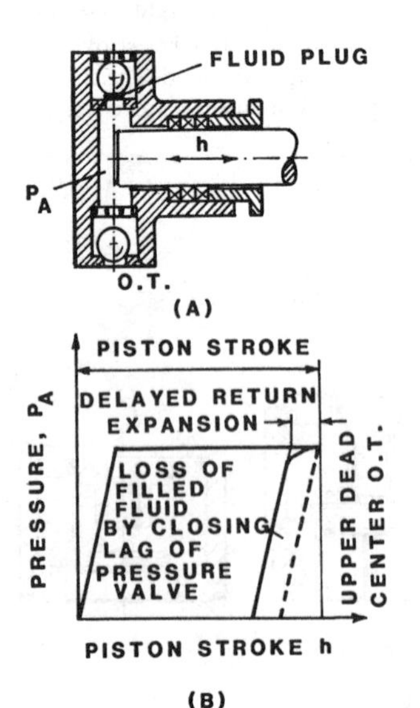

Figure 23. Effect of valve-closing lag on cylinder fluid charge.

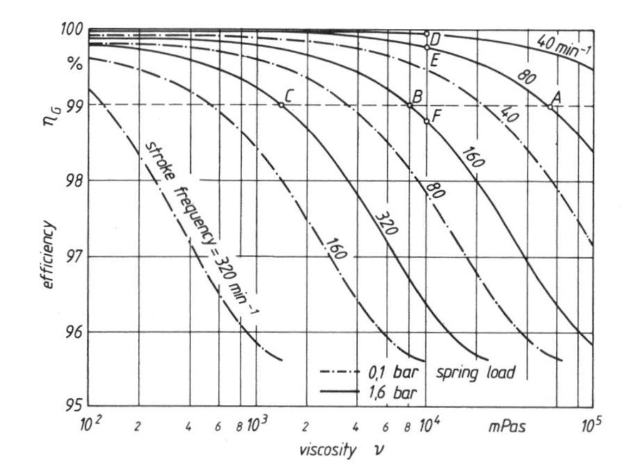

Figure 24. Quality efficiency in dependence on viscosity, spring load, and stroke frequency.

Two examples have been selected here: at $\eta_G = 0.99$ (very good value!) and stronger spring load, the permissible viscosity ν in the stroke frequency range of 80 to 320 min^{-1} changes from more than 5×10^4 to about 10^3 m Pa s (points ABC). At constant viscosity (points DEF), the uppermost stroke frequency, at an otherwise identical valve situation for $\eta_G = 0.99$ is calculated to result in approximately 160 l/min. If you have at your disposal diagrams for a valve as shown in Figure 24, an optimization of the operating range is thus easily possible. Deviating rheological properties of the medium again produce other relations.

Valve-closing lags are also created by the friction of the movable valve parts in the guide. Of favorable design are guideless types (for instance, plate valves with spring suspension). A good through-flow must be assured for suspensions, an objective which can be reached by the use of special suspension valves. In this case, one dispenses entirely with the valve guide; the flow channel has no edges or corners whatsoever which would otherwise encourage sedimentation. This design, however, is a compromise between an exact valve motion and a "choke-free" operation, which as a rule, means a somewhat lower efficiency (Figure 25).

Valve leakage and valve wear. Reflux losses are also generated by the valve not being tight in a closed state. The causes can be wear of the tight surfaces or insufficient bearing contact created by particles stuck in between. In most instances, there are no difficulities where homogeneous media (that is, without abrasive particles) are involved to keep the closing energy within limits and to find materials which withstand pressure, temperature, and corrosion conditions. If there are hard and abrasive particles in the fluid to be pumped (such as coal, quartz, and catalyst suspensions), they

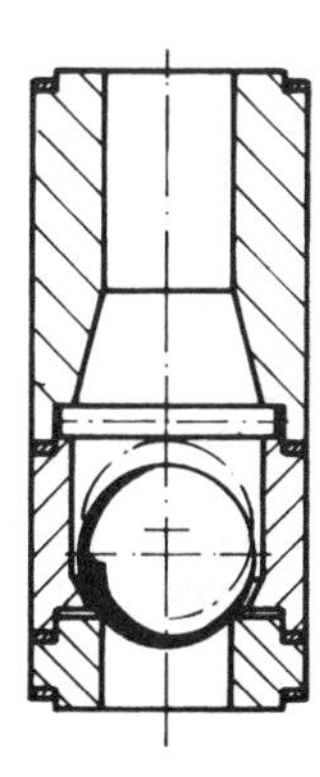

Figure 25. Suspension valve.

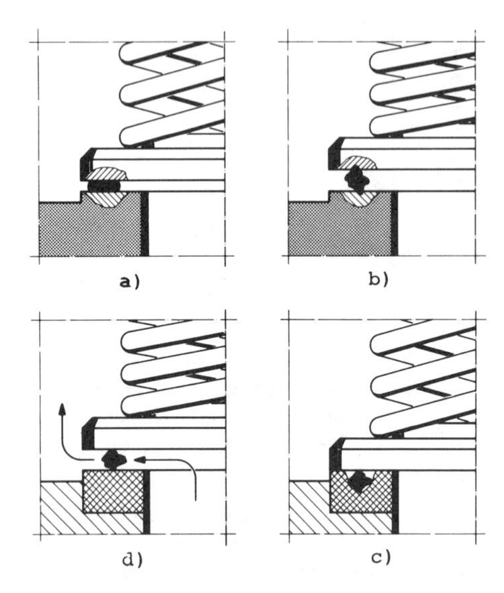

Figure 26. Abrasive particles in pump valves.

will succeed in getting between the valve closing element and seat during the closing action under a high rate of squeeze. If the modules are of a considerably harder nature, the wear will then be small or equal to zero (Figure 26A). In most cases, similar orders of magnitude are given for the hardness of modules and particles so that one will find oneself in a medium wear position. Under the high rate of squeeze, the particle penetrates somewhat into the module and thus, more or less, produces wear. Where brittle materials are involved, the squeeze across the edges generated by the inclined contact is also considered dangerous. Hard module materials such as carbide metals, stellite, and ceramics are brittle, and, therefore, confine the stroke frequency (closing energy). As a rule, the most suitable items for abrasive suspensions are hard and yet sufficiently tough materials as well as light-weight valve-closing parts having a small spring load. Since the wear rate increases with the stroke frequency (closing energy), it is advisable to layout and size the system on the basis of a reduced stroke frequency. Whenever pressure, temperature, and fluid permit, elastomeris-type valve seats provide an effective way to avoid wear. Because of the elastic penetration of the particles during the closing action (Figure 26C) and the "catapulting out" during the closing phase (Figure 26D), the squeeze rate is lower.

Suction Process, Entry-Pressure Loss, Cavitation

An important parameter for oscillating displacement pumps is the so-called entry-pressure loss, for it determines in a dominating mode the required NPSH value of the pump [4]. For an explanation, let us have a look at the suction process (Figure 27) of a single pump head (Figure 27B). If the piston starts at the upper dead center OT to execute the suction stroke, the return expansion from pressure P_{DF} (pressure at discharge flange, point a) to pressure P_{SF} (pressure at suction flange, point b) will first have to ensue. This requires a finite piston travel and/or a crank angle $\Delta\varphi$. Since the suction action and/or the initial coupling of the fluid onto the piston movement can only commence here, the fluid will be coupled in a jolty manner with a finite speed jump of Δv_k 2. This causes a sharp low-pressure peak of up to $P_{A1\,min}$ and an aperiodic high-frequency oscillation. This oscillation phase passes at point d over into the flow phase, which due to the superposition of frictional and mass pressures, reaches its pressure minimum at $P_{A2\,min}$.

At the end phase of the suction stroke, the pressure rises due to the mass delay via the average suction-flange pressure.

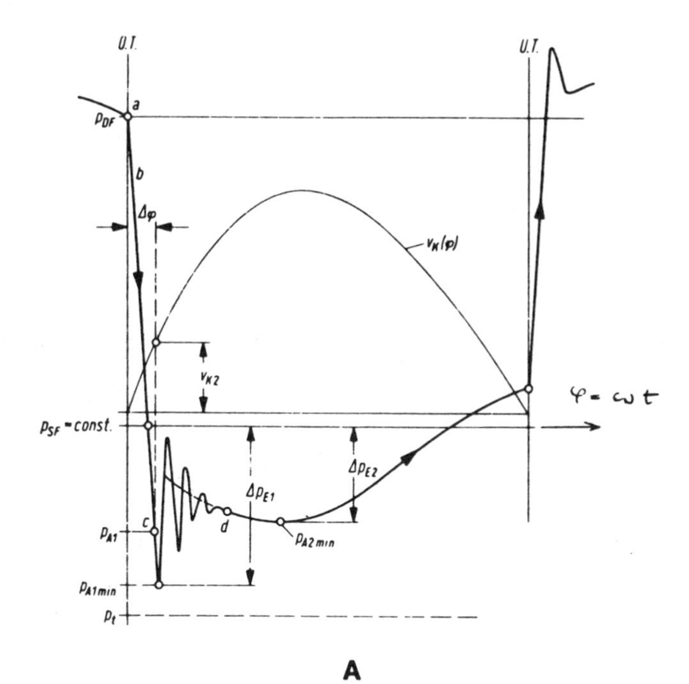

A

B

Figure 27. Suction process.

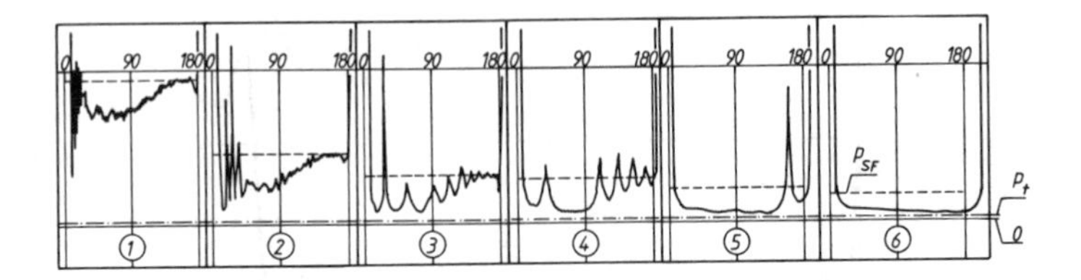

Figure 28. Pressure progression in working space under different cavitation conditions.

With respect to the danger of cavitation, the oscillation phase does not, based on past experience, play any essential role. Even if cavitation occured locally at this point, it would collapse again during the further progression. In this regard, a glance at the measured results as per Figure 28 will be informative. At stage 1 the average suction-flange pressure P_{SF} (broken horizontal line) is far away from the steam pressure p_t (dot-dash horizontal line). Already at stage 2 at a suction-flange pressure level closer to the steam pressure, it is quite obvious that the steam pressure is locally reached within the range of the oscillation phase, which is quite evident from the strong pressure peaks and the altered frequency.

It is only at stage 3 that one spots the steam bubbles in the proximity of the suction valve in a pump head that has been designed to be transparent, however, these bubbles will already have collapsed again by the end of the suction stroke. One can, however, recognize that the steam pressure will obviously be reached within the range of the flow phase. From this behavior it was deduced to define the entry-pressure loss, which is of principal importance for the cavitation, in the form of $\Delta P_E = P_{SF} - P_{A2\,min}$. This means that any occurrence of cavitation during the coupling phase is considered a less-disturbing factor. Incidentally, it is only from stage 4 on that any cavitation in the indicator diagram (Figure 20) can be recognized.

The entry-pressure loss ΔP_E can be, by the way, rather reliably determined from the static through-flow measurements at pump valves, as measurements of this type have confirmed this fact [5]. However, it is recommended to give preference to dynamic tests because these would also provide a better coverage with regard to the influences from mechanical friction. Efficient pump manufacturers have also available measurement results for Newtonian and non-Newtonian fluids of all viscosity ranges.

The NPSH comparison, customary in centrifugal pump manufacturing, requires extreme caution with respect to oscillating displacement pumps because most of the pressure influences are time-dependent variable functions. The entry-pressure loss ΔP_E is such a variable. It is therefore recommended that absolute pressures be taken into account with regard to a pulsating pump, and to waive the NPSH concept. If somebody wants to draw the NPSH concept into consideration anyway, he should take into account that NPSHr and NPSHa are time-bound variables.

DESIGN OF OSCILLATING DISPLACEMENT PUMPS

From the multitude of design types available we will select volumetric metering pumps, diaphragm pumps for heterogeneous fluids, and in-line piston pumps. This selection will explain the essential technical details, but it will not be complete.

Volumetric Metering Pumps

All displacement pumps have a feature in common—a defined volume is displaced per revolution or stroke. Oscillating piston, diaphragm and bellows pumps are the best units to fulfull the high requirements expected from a reproducible and long-time stable metering flow and from the disturbance-variable-insensitivity characteristics. The effect of the impervious state of the working space of the pump is that at the lower viscosity ratings the internal and external leakage losses are negligibly small. Generally, when speaking of metering pumps, one always means "oscillating displacement pumps."

Rotating displacement pumps with gears, eccentric worm, rotary piston, or roller hose used as displacers are only in a very limited sense suitable for metering purposes because the clearance losses can only be kept within limits at low pressure and high viscosity.

Almost all free-flowing fluids can be metered by the oscillating metering pumps, but their design features must be adapted to the fluid conditions. Table 2 will provide an overview of those fluids broken down into various viewpoints and layout directions. There is no other metering process which is similarly flexible and universally applicable. However, for the appplication to be successful, some experimentation will be necessary. The economical area of application is concentrated on metered flow-ins of below 10 m^3/h and on pressure ratings of over 5 to 20 bar. As shown by the comparison of the capital equipment expenditure, the economical incentive increases with decreasing flow-in and rising pressure. It is known that with metering flows of below 1 m^3/h the competing flow measuring processes will become relatively costly, increasingly inaccurate, and more prone to disturbances. The operating principle as per Figure 1 fulfills the basic functions of the metering process: measuring, pumping, adjusting. The displacer (for instance, the piston in Figure 1) displaces the stroke volume V_h per each stroke. The controlled variables—stroke length h and stroke frequency n—can be adjusted independently from each other. Once the displacement ensues intermittently, the metering flow will start to pulsate.

Some remarks with regard to the concept of "measuring": Volumetric metering systems do not measure in the physical sense. They will turn into a high-grade "measuring device substitute," but only if one constrains the disturbance variables by controlling the marginal conditions so that the volumetric principle can act quantitatively in an accurate mode. Metering pumps are designed to have greatly different displacement drives and displacment systems. The most important displacement drives are listed in Table 3.

The displacement system per se is exposed to the chemical and physical properties of the fluid to be pumped. This is why a report on a particular multitude of designs and types would be in order. Table 4 will only provide an overview of the main groups which all can more or less be attached to the different displacement drives as indicated in Table 3.

The piston-type displacement systems with direct piston gland or front-mounted hydraulically driven diaphragms (also bellows or tube) distinguish themselves by being able to transfer on a best-possible basis the precisely defined stroke volume of the piston on to the pumped liquid. The elasticities of the working space can be minimized. There is a trend to use leakproof displacement systems [6–8], which must be used with toxic media.

Displacement systems with mechanically driven diaphragms or bellows are actually a compromise between expenditure and precision. The elastic displacers are directly exposed to the pumping pressure difference, the consequences being disturbing elasticity influences and corresponding mechanical stresses. The extensive application of this simple design is limited to low pressures, smaller outputs, and mostly a much inferior metering accuracy.

It is only in extremely rare cases that valveless control-piston displacement systems can be applicable for lubricous and viscous fluids. The advantage rests with the small entry losses. The split seal of the piston will become a problem when contaminations or wear are involved [9].

From the large multitude of designs a few particularly frequently used types will be selected. The user of metering pumps must have an excellent sense of judgment with regard to design details—to a higher degree than for other metering processes. For the universal type of metering pumps it is customary to build up their structure on the basis of the building-block system. Suitable metering pumps can be assembled "tailormade" from series-built components. The essential component groups such as drive gear units, pump heads, stroke adjustment, add-on equipment, and drives are depicted in Figure 29.

Drive Gear Units, Drives, Actuators

Stroke-adjustable drive gear units with quasi-harmonic kinematics enjoy a great popularity among users. Figures 6 to 8 show some examples. In the rocker-arm system (Figure 6), fulcrum A

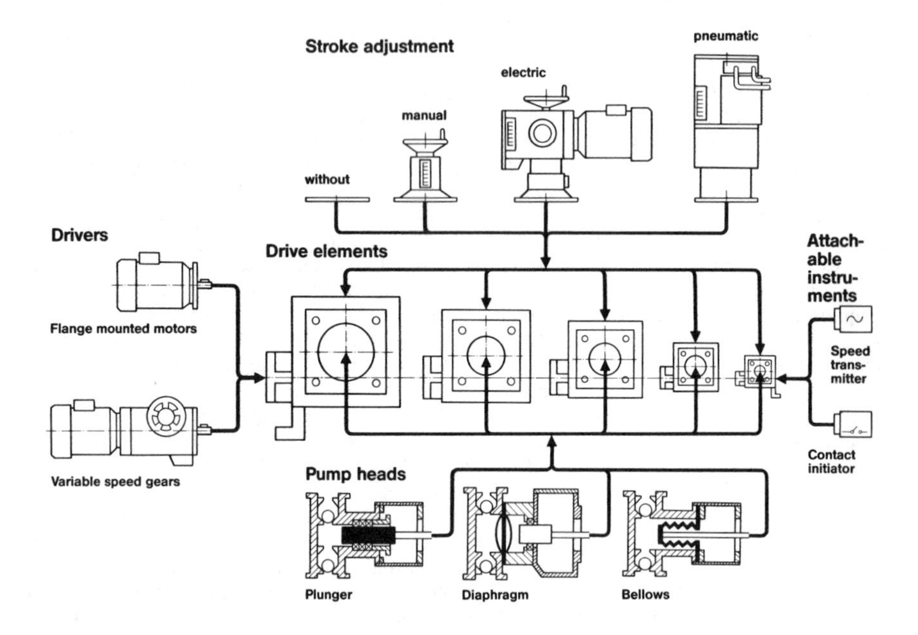

Figure 29. Modular building-block system for metering pumps.

of the rocker arm (3) actuated by the eccentric shaft (1) can be adjusted. The upper dead-center position of the piston and thus the clearance space V_T remain the same for all stroke adjustments. This is also the case in the crank system (Figure 7) in which the swash plate (1) with the worm wheel (4) is swivelled about the worm (5). The angular crank system (Figure 8) exhibits a constant center position. The angular crank (2) is axially moved in relation to eccentric (3). Figure 30 provides us with the impression of a design for a stroke-adjustable eccentric drive unit as a component of a diaphragm metering pump. The worm gear unit (1) turns the hollow shaft (2) which

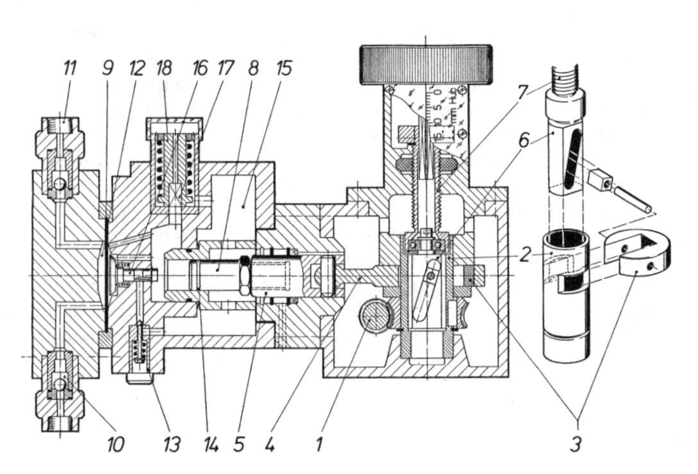

Figure 30. Diaphragm metering pump with stroke-adjustable eccentric drive gear (LEWA).

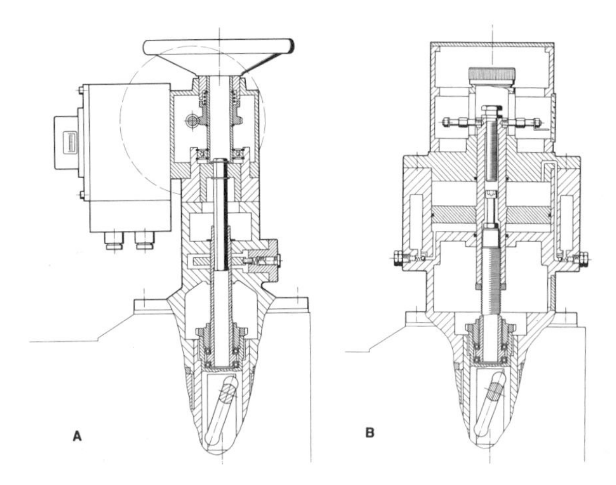

Figure 31. Actuators for stroke adjustment (LH—electric; RH—pneumatic).

itself is coupled with the eccentric disk (3). An axial movement of the spindle with the angular guide (7) makes it possible to adjust the eccentricity and thus the stroke of the piston rod and connecting rod (4, 5). Most drive gear units cover an output range of 0.1 to 100 kW.

The stroke can be adjusted manually or by means of electric or pneumatic actuators. Drive is triggered by using the conventional current, voltage, resistance, pulse, or pneumatic signals. Facilities endangered by explosions mostly prefer pneumatic actuators although explosion-proof electric actuators are also available (Figure 31). In most cases, A.C. motors are used as drives. Where stroke-frequency adjustment is required, semi-conductor-controlled A.C. and D.C. motors are increasingly gaining wider acceptance over the varidrives. This is because a variation of time and range are more favorable for the applications intended. Also here, all possibilities to protect from explosions are given.

Figures 32 to 34 are representative of typical designs for metering pumps. The horizontal in-line design (Figures 32 and 33) provides the advantage of an organized arrangement and good accessibility. In contrast, the vertical tower design requires little space (Figure 34). In contrast to the quasi-harmonic kinematics (for all stroke adjustments) of the drive gear unit principles shown, the spring-action cam drive system is marked by a jerky displacement motion at part stroke (Table 3). Because of these jerky kinematics, its application is mainly limited to low-priced diaphragm pumps of small outputs for subordinate purposes (Figure 9). In micro-metering pumps, this jerky motion is, for small stroke lengths, useful to exactly actuate the valves (Figure 35) [1]. Furthermore, the freedom-from-play as achieved by the recuperating spring is also important.

Figure 36 explains the example of the direct-digital proportional coupling of a solenoid-type diaphragm metering pump. This very compact metering unit is extensively used for injecting drinking water systems with small chemical, flows.

Figure 32. Multiple piston-type metering pump in horizontal in-line construction (speed-variable drive) (LEWA).

Figure 33. Multiple disphragm-type metering pump in horizontal in-line construction (pneumatic stroke adjustment) (LEWA).

Figure 34. Multiple piston-type metering pump in vertical multi-level (tower) construction (pulsation damper on suction side) (Bran and Lübbe).

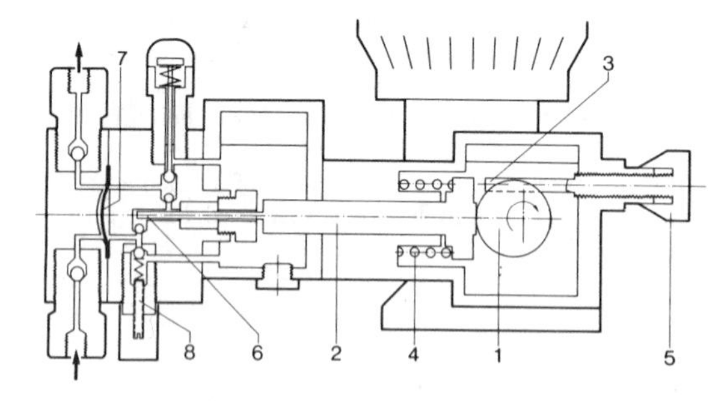

1 Eccentric cam
2 Piston rod
3 Adjustable stop
4 Return spring
5 Adjusting screw with micrometer scale
6 Displacer
7 Metal diaphragm
8 Pressure relief valve

Figure 35. Diaphragm-type micrometering pump with spring-action cam drive gear (LEWA).

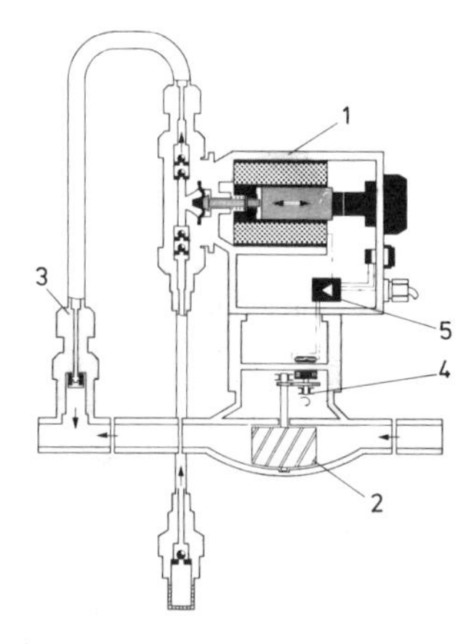

Figure 36. Solenoid metering pump for proportional coupling to flow meter (JESCO): (1) solenoid metering pump; (2) flow meter; (3) inoculation point; (4) contact actuator; (5) controls.

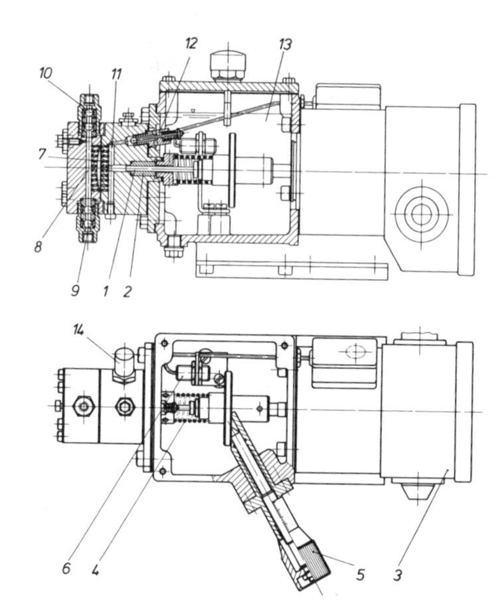

Figure 37. Explosion-proof solenoid diaphragm-type metering pump (LEWA).

The solenoid-type diaphragm metering pump (Figure 37), which is explosion proof and operates with an oil-flooded electromagnet is applicable for areas where high pressures are prevalent. The solenoid (3) drives, controlled by an electronic circuitry, the piston (1) of a diaphragm pump head with hydraulic diaphragm actuation. The piston stroke is variable by means of stroke adjustment (5). A well-known application for such high-pressure diaphragm pumps is found in natural gas odorization. Solenoid-type diaphragm meteing pumps with direct diaphragm linkage and explosion-proof capabilities are also available.

With respect to linear drives, the pneumatic linear drive in addition to the magnetic drive is playing a larger role. As a rule, the reason for their application is in explosion protection (compressed-air drive or missing electrical power; for instance, natural gas drive) (see Figure 38).

The hydraulic linear drive, especially that one with two superimposed displacers, plays a certain role in high-pressure metering to achieve pulsation-free flow rates [10].

For very small flow-ins (range preferably below 1,000 ml/h), superposition drives having the most different functions [1] are in application. They are capable of controlling the superposition of two

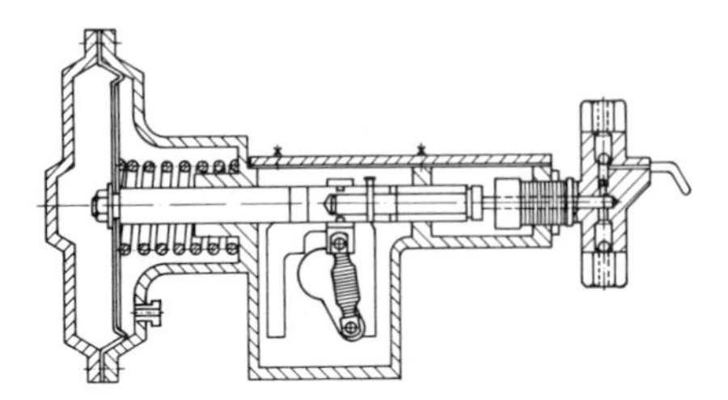

Figure 38. Piston-type metering pump with pneumatic linear drive (Texsteam).

displacers in the direction of freedom from pulsation. Figure 39 shows details of a superposition-drive equipped with special cams. One can recognize the two pump heads whose pistons are driven by specially shaped cams. The pump is adjusted by a varispeed motor. For the purpose of compensating the elasticity influences of the pumped fluid, a more or less large clearance stroke can be adjusted via the handwheel. As a rule, the clearance stroke is set automatically by the actuator via a pressure-dependent controller. In this mode, even though it is expensive, one obtains a proper and even flow-in. This expenditure is only justified if the units are slated for special applications in research and analytical technology. These fields have a great number of problems to be solved about which we cannot report at this point.

Pump Heads

Pump heads are usually separted from the drive train by brackets so that harmful fluid influences can be kept away. When selecting materials, the problems of corrosion, strength, and wear must be taken into account. Rust and acid resisting steels, titanium, tantalum, high-alloy special

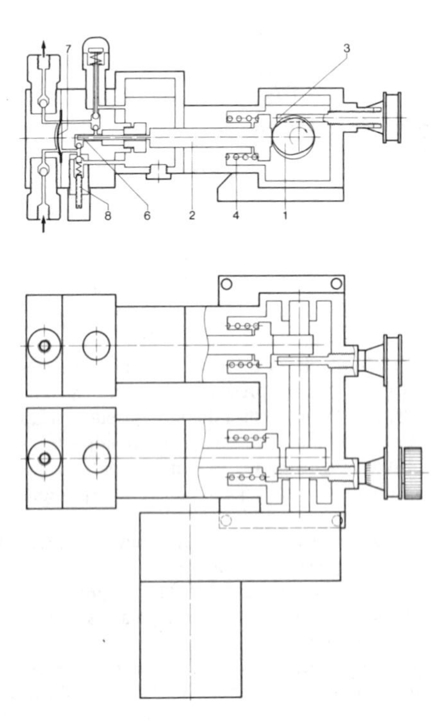

Figure 39. Twin-cylinder superposition displacement system with diaphragm pump heads.

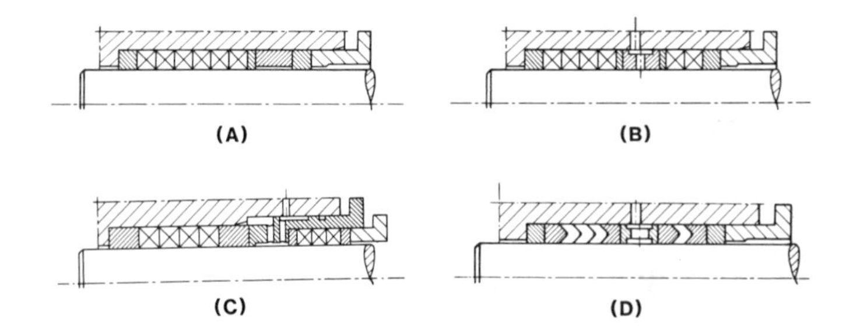

Figure 40. Piston glands: (A) packing rings; (B) packing rings with flush lantern; (C) sleeve and collar rings with flush lantern; (D) double-stage packing-type stuffing box with flush system.

steels, plastics, ceramics, and glass are being used. The design of maximum pressure pump heads for pressures up to 3,500 bar raises difficult strength problems [11] which can only be solved by a notch-poor design and autofrettage.

The application of *piston-type pump heads* is mostly limited to fluids which are allowed to exit to the outside in small leakage quantities. Pistons and gland-types require a great deal of experience for the purpose of lay-out and have high maintenance requirements. Extremely hard piston materials such as aluminum oxide, carbide metal, and stellite cladding prevent wear phenomena that would otherwise quickly lead to leakage problems of the piston gland. Pistons are sealed by means of differently formed sleeves and collars or rough-pressed packing rings (Figure 40). Sleeves are preferred over packing rings as they are able to seal pressure-independently without having to be tightened from the outside. Packing rings require outside tightening, but are more rugged and sturdy with respect to damage or wear. Figures 41 and 42 demonstrate design examples of conventional piston-type pump heads. The type with controlled valves is suitable for inhomogeneous media and particularly high viscosity. It requires a provision to pick up the valve control motions from the drive gear unit (Figure 43). A direct valve control with pneumatic, hydraulic, or electromagnetic actuation is also popular.

For very high pressures, the area of the T-bore where the axes of piston bore, suction head, and discharge channels meet, must be autofrettaged and is subject to special treatment with regard to local smoothness of the surface so that fatigue fractures due to internal dynamic pressure stresses can be prevented. Furthermore, the zone subject to maximum stress will be designed to be easily interchangeable—as a T-piece in Figure 44.

Environmental regulations force us to prevent any dangerous leakages from happening. This is the reason why *leakproof pump heads* should be used for toxic and dangerous fluids. Metering pumps, which are maintenance-free and do not have a stuffing box, are also advantageous for

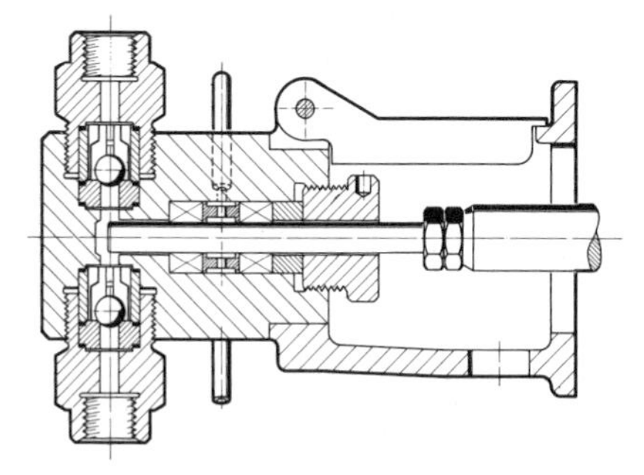

Figure 41. Metallic piston pump head (LEWA).

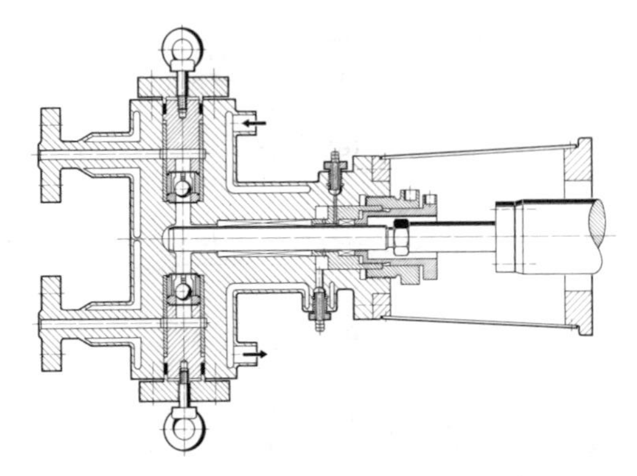

Figure 42. Cross-head type piston pump head with heating system (LEWA).

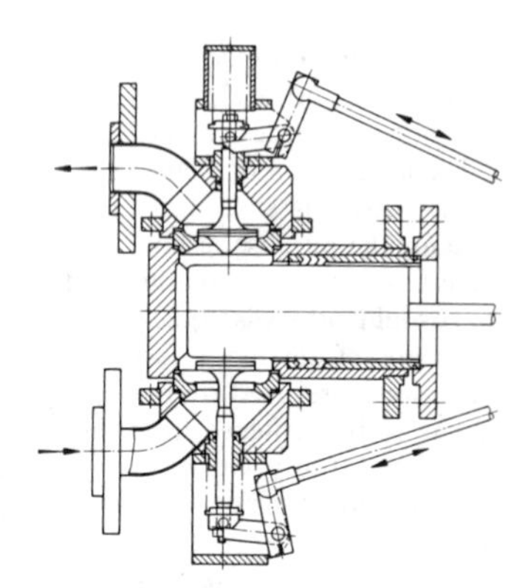

Figure 43. Forcibly controlled pump valves (Bran & Lübbe).

abrasive fluids, and those media not easily sealable due to their low viscosity [6]. The simplest design of a diaphragm metering pump with a mechanically linked diaphragm as per Figure 9 can only be used for low pressures and small flow rates. Elastomers only should be considered as a material for the diaphragm. Due to the pressure load of the diaphragm, its life is limited to approximately 3,000 hours. Diaphragm ruptures can be signalled by a sensor to be mounted to the plug bore hole (7) (Figure 14, characteristic line e). The bellows of bellows-type metering pumps are also mechanically linked (Figure 45). These pumps are mostly built with PTFE bellows and glass

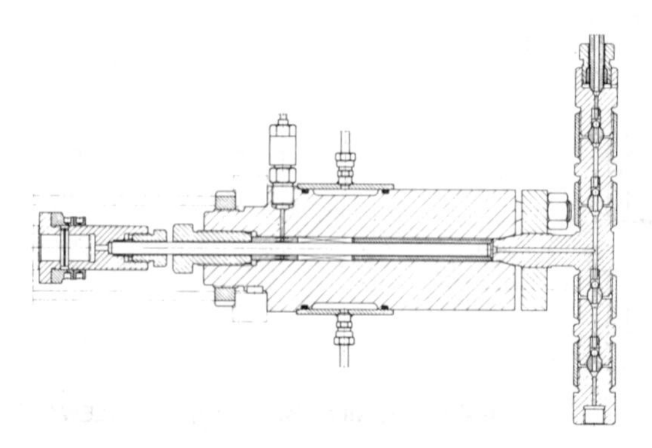

Figure 44. Maximum-pressure piston pump head with capacity of up to 3,500 bar (Uhde).

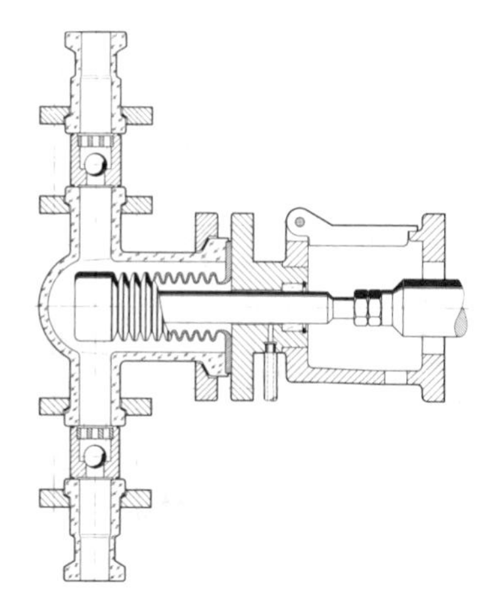

Figure 45. Bellows-type pump head (LEWA).

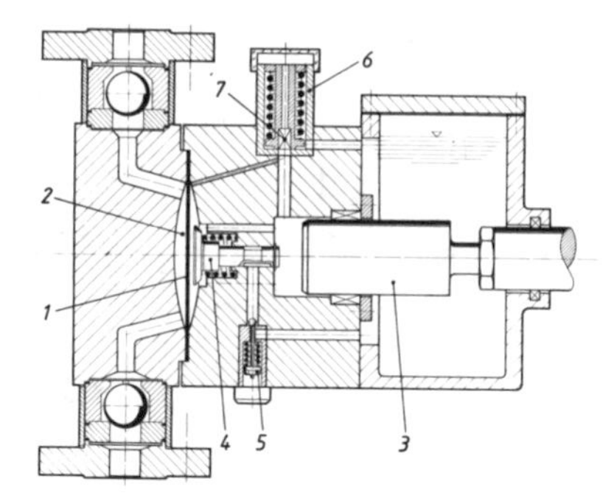

Figure 46. Diaphragm-type pump head with gate control (LEWA).

cylinders for special applications (Figure 14, characteristic line d); the pressures are limited to values below 10 bar—similar to the diaphragm pumps in Figure 9.

Best characteristics and maximum diaphragm life (up to 15,000 hours) are provided by diaphragm pumps equipped with hydraulic diaphragm drive (Figure 14, characteristic lines b + e). Diaphragm pumps with a PTFE diaphragm and gate control (Figure 46) are a particularly characteristic and broadly applicable design. Diaphragm (1) is freely oscillating on the fluid side in working space (2). The appropriate filling with hydraulic oil between the piston (3) and the diaphragm (1) is carried out by a pulse gate (4) together with a replenishing valve (5). At the extreme upper position, a gas venting valve (7) is located in combination with a pressure relief valve (6), which operates in the hydraulic system. Such metering pumps are produced for pressures of up to 350 bar in the entire output range (Figure 47).

There is still another series of successful diaphragm-control principles which permit the diaphragm to oscillate freely, but they are too complicated to discuss here. Because of the material strength, metallic diaphragms rather than those made of PTFE are still frequently used for pressures of more than 350 bar, although there are developments afoot to widen this limit to accommodate higher pressures. Figure 35 has already shown a diaphragm-type micro-metering pump with a metallic diaphragm in which the diaphragm operates between two contact faces (plain and concave). The hydraulic system is checked by means of a replenishing valve (9) and an overflow

Figure 47. Diaphragm-type process pump for toxic fluids and high pressures (200 kW drive output) (LEWA).

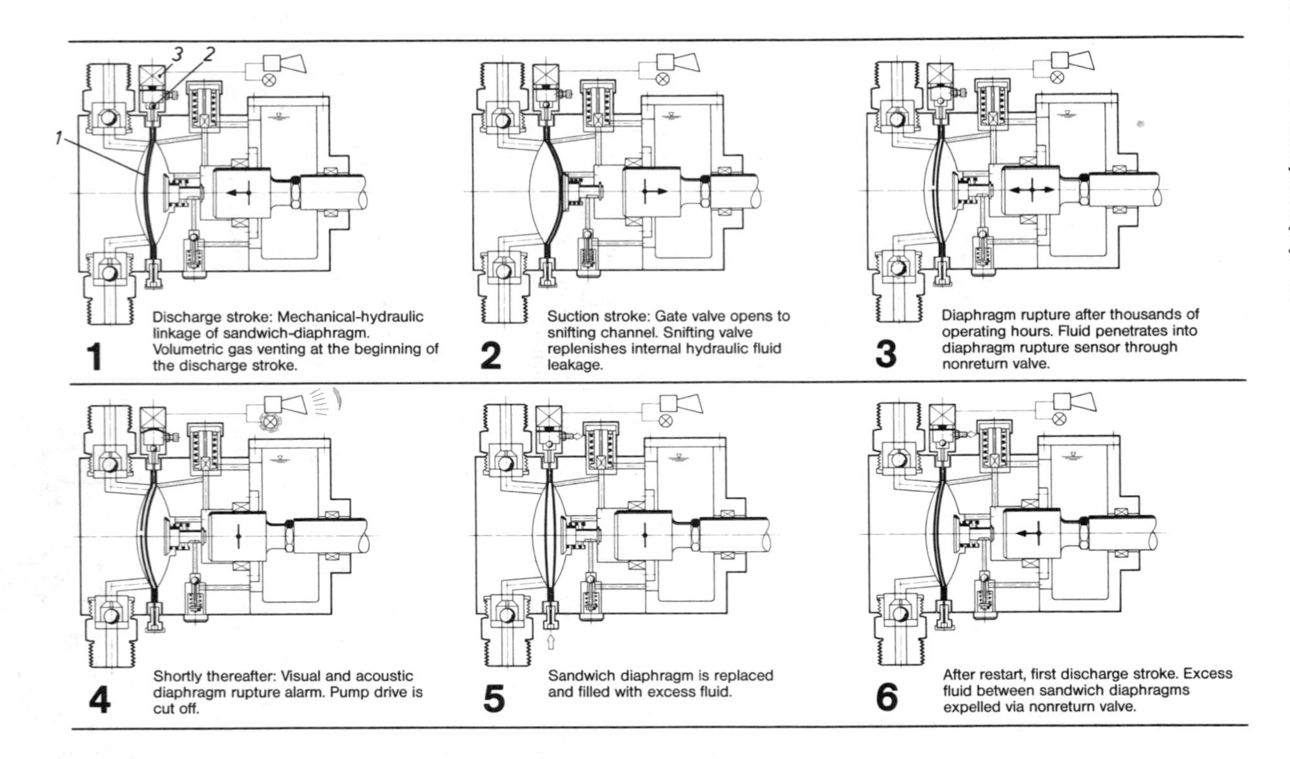

Figure 48. Diaphragm-rupture signalling by means of sandwich diaphragm (LEWA).

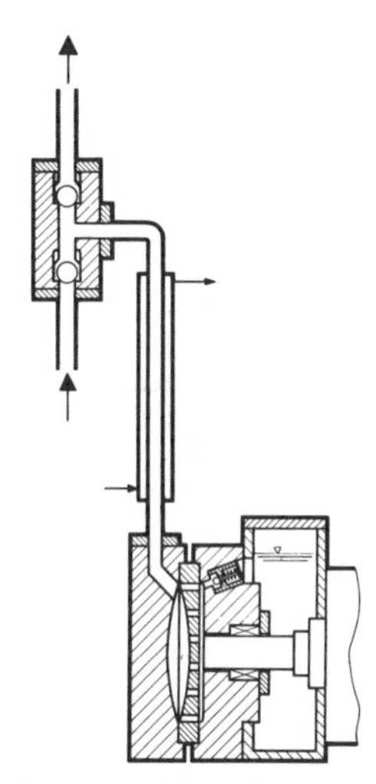

Figure 49. Remote head design.

valve (8). Leakage of hydraulic oil through the piston gland is automatically compensated for by the replenishing valve because the diaphragm will then gain access to the rear contact face, and the arising low pressure will cause the replenishing system to open. As a special design for micrometering, Figure 35 depicts check valves in the working space which serve to produce an unequivocal circulation direction and, thus, a gas flushing effect. Diaphragm pumps with metallic diaphragms are built for pressures of up to 3,000 bar and for all output variables. Dependent on material and type of design, the life of the diaphragms is anywhere from 5,000 to 15,000 hours. Metallic diaphragms are sensitive and as a rule do not reach the lower limit of the life figures mentioned.

A simple signal or alarm method for a diaphragm rupture is to use a sandwich diaphragm with a pressure sensor (Figure 48) [12]. During the discharge stroke the sandwich diaphragm (1) is positive-locking whereas in the suction stroke it is hydraulically coupled by an oil film. The interspace is linked with a pressure sensor (3) via a check valve (2). The sensor responds when one of the two outer diaphragm layers rupture. Another diaphragm rupture signal method operates on the basis of a vacuum-coupling of the diaphragm layers.

Occasionally, metering pumps with a "remote head" are required where special conditions and difficult fluids are prevalent. For such applications, the diaphragm drive system is separated from the valve head via a piping system. Where special conditions demand it, suspensions at high temperature and radioactive and explosive fluids can be controlled. The arrangement of the "hydraulic linkage" can be of a very different nature (Figure 49).

Metering Pumps for Control and Regulating Systems

Oscillating metering pumps are suited for the automation of manufacturing processes because they possess the following outstanding features:

1. The metering flow $\dot{m}$ can be adjusted by two independent setting variables—stroke length h and stroke frequency n.
2. The juxtaposition of several metering elements (drive gear units) by direct coupling allows simple recipe metering.
3. The flow-in consists of a series of individual batches (stroke volume), a fact which can be used for batch-metering processes by counting the strokes.

In the field of chemical process engineering, the proportional metering process and the matter-value control by means of metering pumps play a special role.

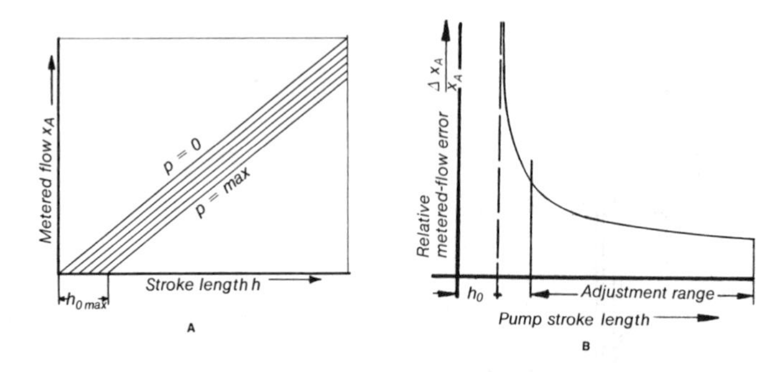

Figure 50. Stroke length as manipulated variable.

Proportional metering. Its task is to proportionally allocate substances to a main flow (such as waste water flow and precipitant metering). Main flows can be gas, liquid, or solid matter flows in pipelines, chutes, or other systems of conveyance. A quantitative metering action is expected from the metering pump. Depending on the variable to be controlled, the following points are to be observed:

1. *Stroke length as a manipulated variable* (*Figure 50*). At higher pressure rates the fact must be considered that the metering process begins only at stroke length h_0 (zero point compensation). Furthermore, the variation range is, at a given accuracy of the proportional allocation, determined by the progression of the metered error in dependence on the stroke length (Figure 50B). A desired variation range can possibly be attained by the application of two metering pumps of very different sizes. The stroke can be adjusted by means of electric or pneumatic actuators (Figure 31).

   Electric actuators include reversing electric motors (Figure 31A—run/standstill; B—run). Initial drive is actuated through a three-step controller. The feedback can be completed by means of ohmic or inductive transmitters.

   Pneumatic actuators contain a position-control circuit whose set point is preindicated by a standardized pneumatic signal emanating from the flow sensor (for instance, 0.2 to 1.0 bar).
2. *Stroke frequency as a manipulated variable.* Since the speed n (stroke frequency) exerts influence on the flow-in $\dot{m}$ in a linear *and* proportional mode (line cuts through point of origin), no zero point correction whatsoever needs consideration. However, attention should be directed (Figure 51) to the varying "break ratio" which can create mixing problems. If necessary, a combined stroke-length/stroke-frequency control circuit will be required.

   The RPM adjustment can be completed by means of speed-variable gear units via actuators (control range mostly < 10:1) or variable electric motors (D.C., A.C. with frequency conversion) which have a large variation range (20:1 up to 100:1) (Figure 52).

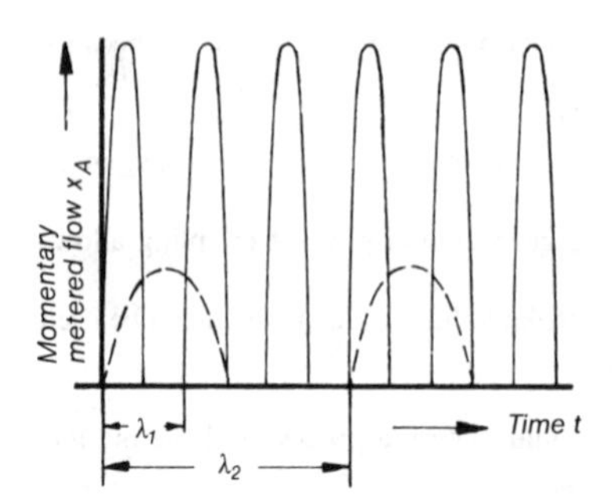

Figure 51. Break ratio at variable stroke frequency.

Figure 52. Diaphragm-type metering pump with D.C. motor drive (LEWA).

It is only the solenoid metering pump which allows direct-digital drive by a pulse sequence. For smaller flow-in rates, the result is an extremely economical process of proportional metering. Figures 53 to 56 depict some typical design examples suited for proportional metering.

Control of substance variables by means of metering pumps. The task is to control physical or chemical variables (pH-value, cloudiness, pressure, specific gravity etc.) by allocation, through metering, of substances into a process.

The metering pump is only a variable attenuator in a control circuit, and its quantitative accuracy is not directly required, but its responsiveness is. Stroke length or stroke frequency serve as manipulate variables; occasionally, both do at the same time. Attention must be directed to the shape of

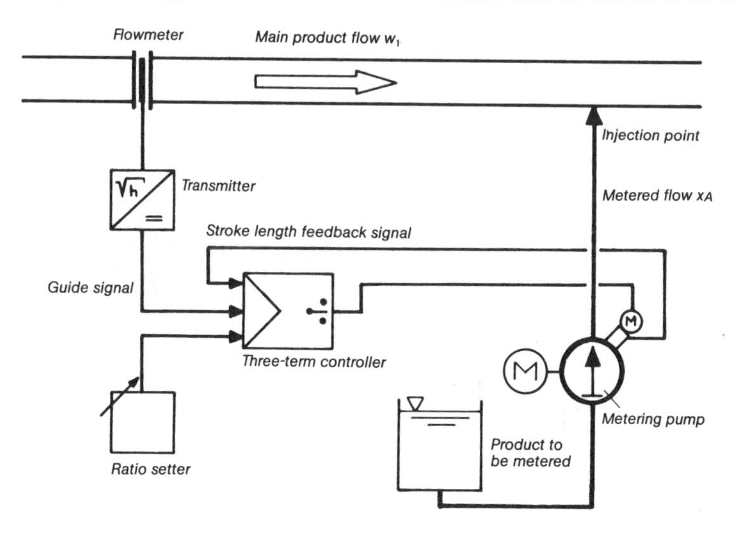

Figure 53. Proportional metering of one component by stroke adjustment.

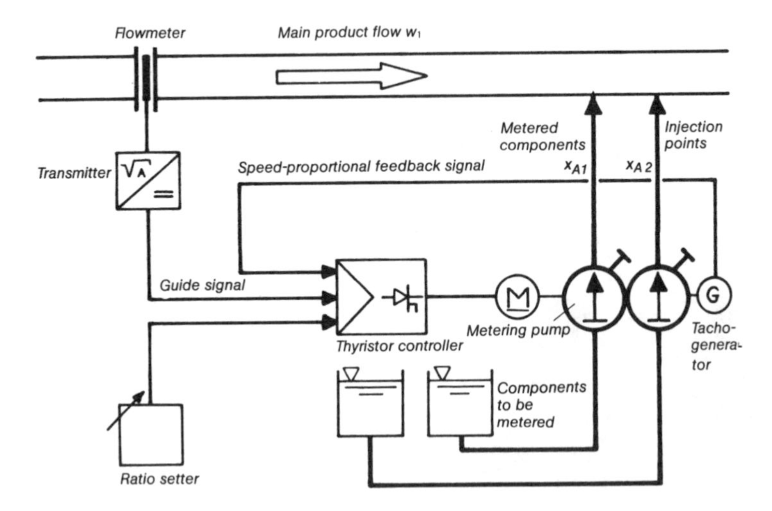

Figure 54. Proportional metering of two components via stroke frequency.

the characteristic with regard to the variation range. Depending on the time constant of the variation range, three-point controllers as well as three-point step controllers are well suited for the electric stroke actuators. The latter exhibit a switch-on behavior which is dependent on the variable of the deviation (Figure 57). Some typical examples are shown in Figures 58 to 60.

In the control of the pH-level (Figure 58), the speed variation is a disturbance variable compensation which will compensate immediately for fluctuations of the main flow. The flow control (Figure 59) of the metering pump has the job of leveling any disturbance.

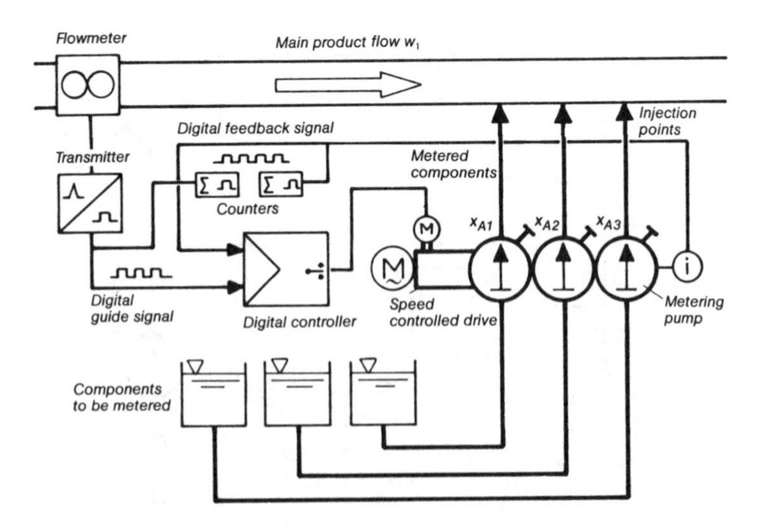

Figure 55. Digital proportional metering of several components via stroke frequency.

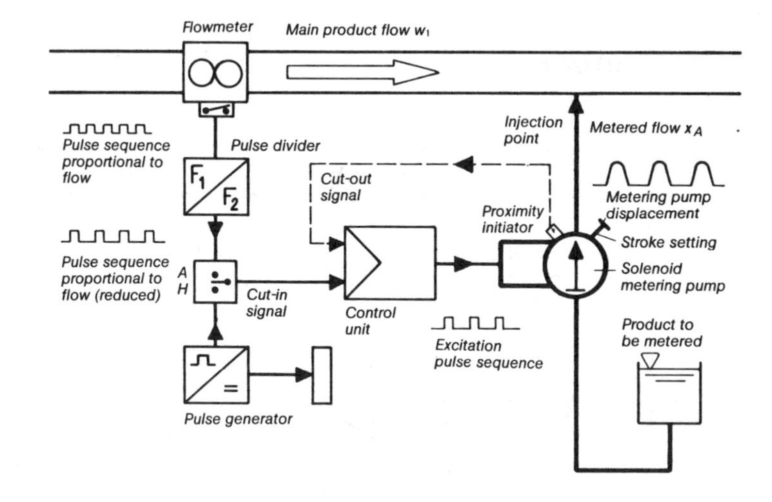

Figure 56. Direct-digital proportional metering with a solenoid pump.

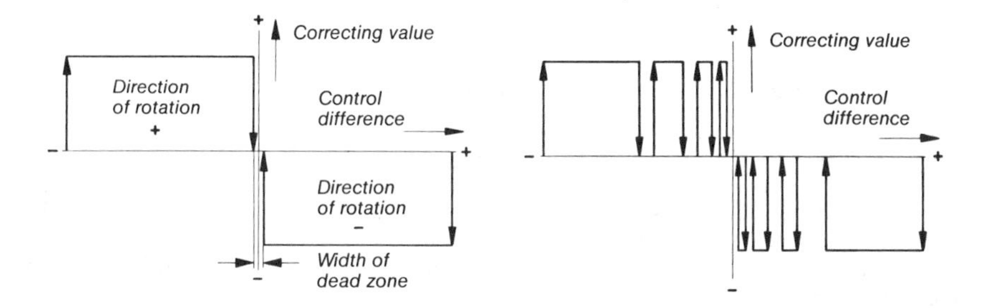

Figure 57. Switch-on behavior of three-level controllers (LH) and three-level step controllers (RH).

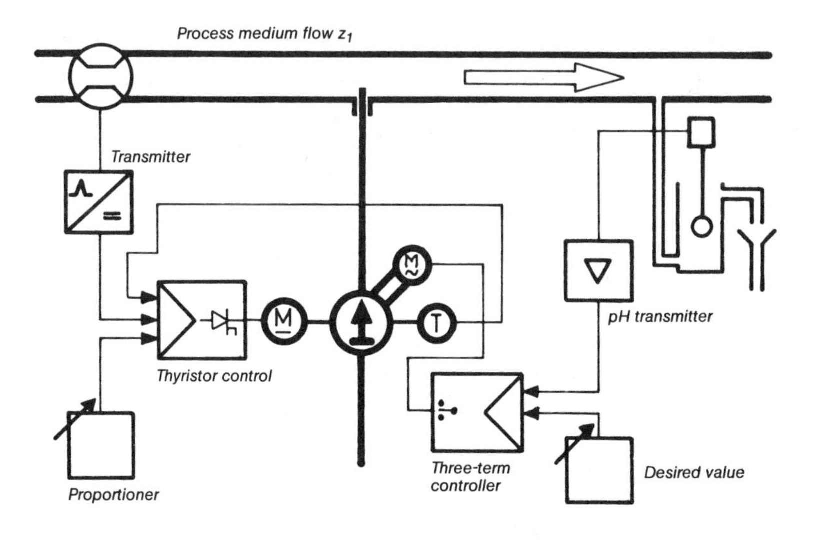

Figure 58. PH-control with disturbance variable compensation.

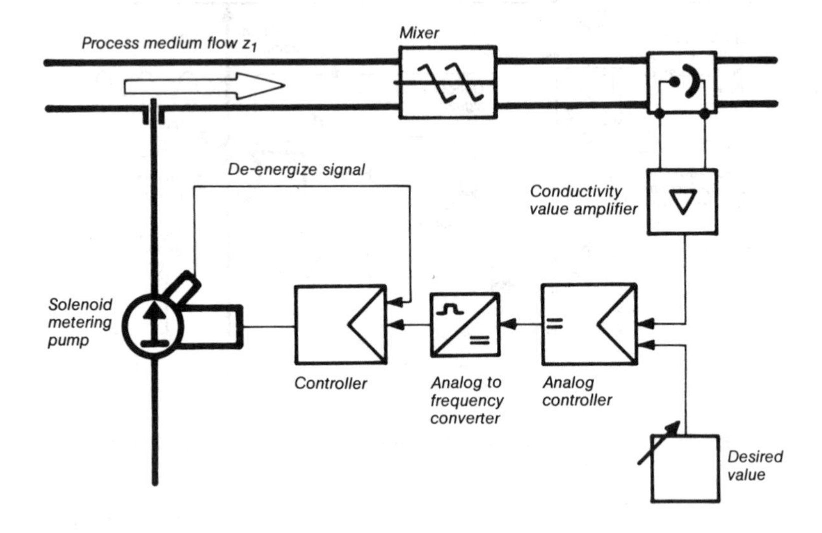

Figure 59. Conductivity control by means of solenoid pumps.

Diaphragm Pumps for Heterogeneous Fluids

The described metering pumps as a diaphragm design are used for all fluids, no matter whether homogeneous or heterogeneous. In general, we are dealing here with small- to medium-sized flow rates, and the building block design of the metering pumps definitely accomodates the applications as required in the process technology field.

For the purpose of pumping free-flowing, but mostly viscous sludge, slurries, or suspensions, frequently containing coarser particles (up to 5 mm), the application of rugged and maintenance-friendly types of diaphragm pumps has proved to be a success. Areas of application are the following:

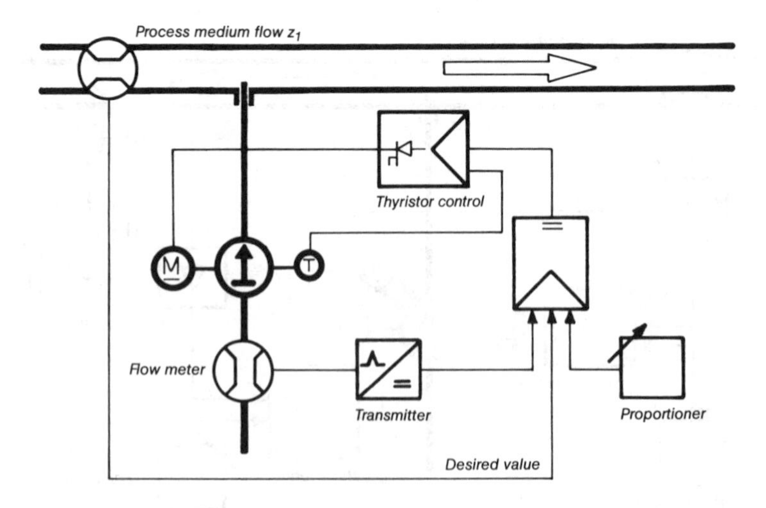

Figure 60. Flow control (verification) of metering pumps.

conveyance by pipeline of mineral, ore, and coal slurries; disposal of waste and decantation slurries, pressure filtration as well as the feed-in of heterogeneous components into chemical processes. The pumps have to be designed for high hydraulic performance, easy maintainability, easy cleaning, sufficient sturdiness, low wear rates, and long life.

The application of piston pumps for suspensions is only possible when soft and not too abrasive particles are involved or when intensive flushing of the piston gland into the working space of the pump is permissible and economical (as described later). Experience shows that diaphragm pumps without a piston gland that is touched by the fluid are, for heterogeneous fluids, operating much more reliably and with less maintenance.

Diaphragm Pumps Equipped with Direct-Thrust Crank Gear Units

The direct mechanical diaphragm drive, as it is occasionally used in dirty-water pumps for lower pressures and intermittent operation, is not suited for higher pressures. It is only the *hydraulic diaphragm drive* which provides a satisfactory life at high pressure rates due to little diaphragm stress. The operating principles as shown in Figure 61 [13, 14] are authoritative for this application. Diaphragm pumps (Figure 61A) are being built for larger outputs (200 m^3/h, 150 bar). However, the diaphragm covers are expensive because of the large cover forces. The hose-type displacer adapter unit has the following advantageous points (Figure 61B):

- Sedimentation is avoided due to the smooth piping system.
- The deflection area at the clamp-in site is located on the hydraulics side and particles cannot get stuck or abrade.

The ball-shaped execution of the tube-type diaphragm pump heads (Figure 61C) diminishes the expense used to dominate the cover forces at high pressure. However, the "ball diaphragm" is dependent on the tension-strain deformation, a fact which only similar-to-rubber elastomer materials permit. In contrast, plastomers such as PTFE can also be used for the predominant bending stress of the diaphragms and externally impinged hoses.

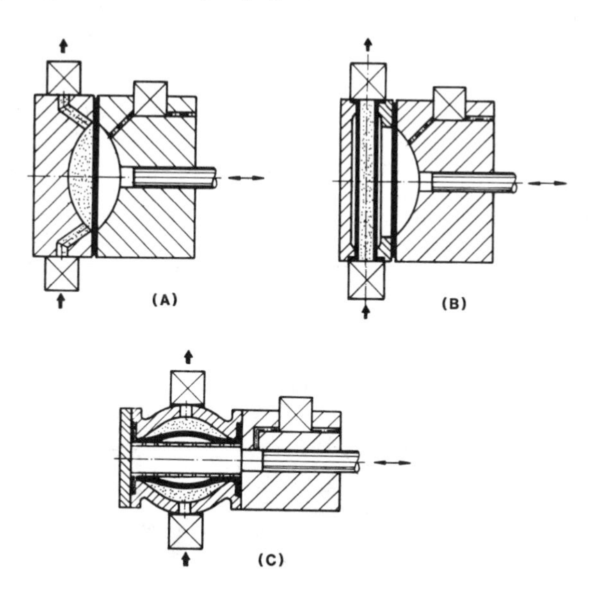

Figure 61. Schematics of operating principles of diaphragm pumps.

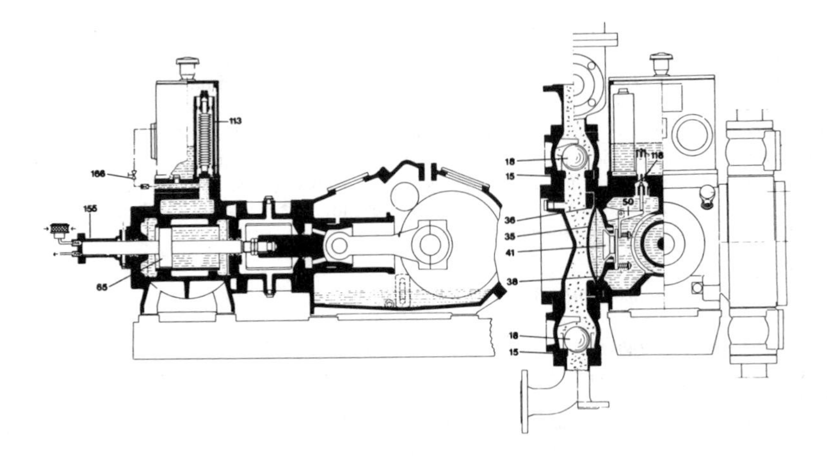

Figure 62. Two-cylinder diaphragm pump (FELUWA).

Figures 62 and 63 demonstrate the design of the drive gear unit. The displacer piston operates in a single or double-acting mode, and the diaphragm pump heads are often arranged in a lateral fashion. The gang arrangement of diaphragm pump heads of large diameters requires a corresponding distance between the drive gear axes. Two-cylinder double-acting pumps equipped with four diaphragm pump heads suited for very large output rates have made their successful entry into the market place (Figure 63).

In a similar manner as demonstrated in Figure 46, the hydraulic system is controlled by a diaphragm-layer scanner via a gate system (Figure 62). The hydraulic system assures an automatic leakage replacement (items 50 and 54) via a diaphragm-layer control circuit (item 41) as well as a relief pressure checking and continuous-venting system (items 105 to 112).

Another method of diaphragm layer control is provided by an inductive scanner which senses the diaphragm limit positions. In these cases, the preferred hydraulic fluids to be applied are clear water or a water-oil emulsion.

Furthermore, filter press pumps are using special control systems that limit the pressure when the ultimate filtering pressure is reached, keep it at the set height, and reduce the flow rate of the pump. A diaphragm rupture in diaphragm-type slurry pumps can mostly be signaled by the use of conductive and capacitive electrodes.

Diaphragm pumps for use on sludge generally operate at a low stroke frequency ($n < 150\ min^{-1}$) because this is a requirement to prevent any wear to the pump valves. To be able to overcome valve wear and to avoid clogging and stop-up special design provisions are required. Self-acting ball

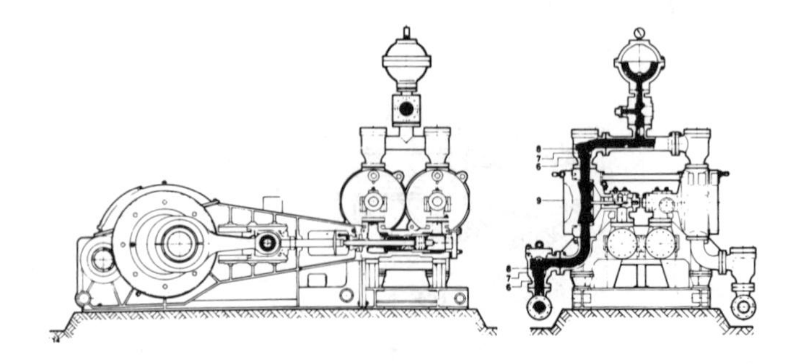

Figure 63. Large four-cylinder diaphragm pump (GEHO).

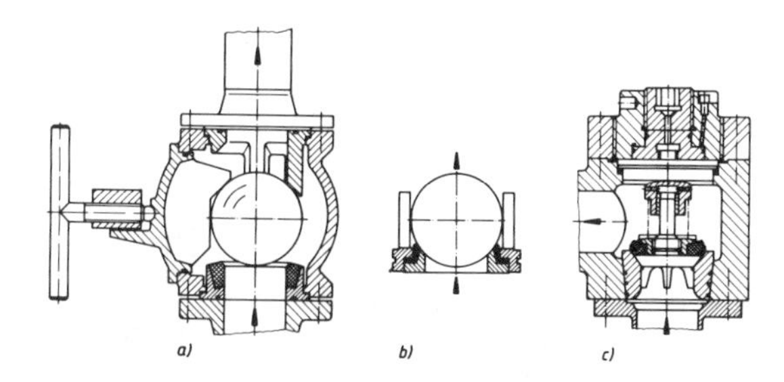

Figure 64. Typical valve designs.

valves that exhibit optimal properties and characteristics with respect to guide of the flow rate, wear and producibility are preferred. The balls are made from hardened steel, surface-coated with wear-resistent materials, from carbide metal and plastic-coated. Ball valves are also suitable for very large flow rates (nominal width > 100 mm) because of the lower weight of the movable valve part.

Typical valve designs have been assembled in Figure 64. The ball valve (Figure 64A) operates with a vulcanized rubber valve seat. The valve body can be opened up. Figure 64B explains the details of a ball valve equipped with an interchangeable elastomer lip ring. The sealing function (lip ring) and the contact function for power transmission (metallic valve seat) are separated, which gives the advantage of a much lower stress on the elastomer part. With respect to the ball valve as in Figure 64C, the same basic principle is applied; however, here the elastomer lip ring is also movable. When the material is selected, the corrosion and abrasion factors must be drawn into consideration.

Pneumatic Diaphgram Pumps [15]

Gas-driven oscillating displacement pumps are known from various special cases of application (Figure 38). In natural gas fields, for instance, one uses natural gas for driving if and when electricity as the driving energy is not available on-site. In the field of process technology, directly air-driven diaphragm pumps are used for subordinate purposes of all kinds. The pump is leakproof, easily installable, transportable, explosion-proof, and easily controllable via the air pressure. Air consumption is high, and the efficiency bad with a rating of 10% to 20%, which for continuous operation, makes a cost-benefit analysis necessary.

Figure 65 demonstrates the operating principles. Both diaphragms (5) are directly connected by means of a control rod (3) which passes the control valve (1). Stroke reversal happens in the extreme

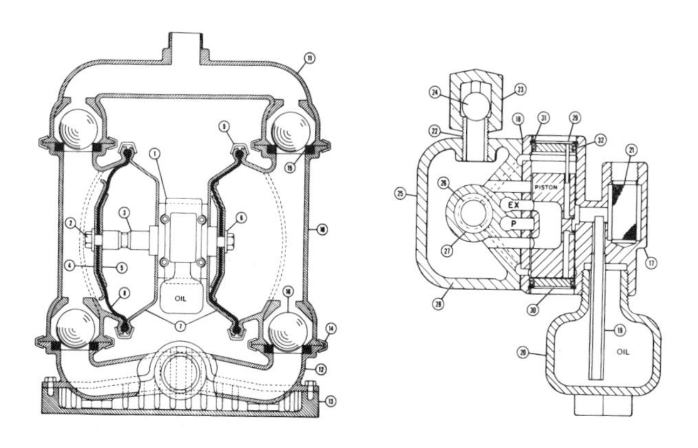

Figure 65. Diaphragm pump with direct pneumatic drive: schematic structure of pump (LH); pneumatic reversing valve (RH) (Wilden).

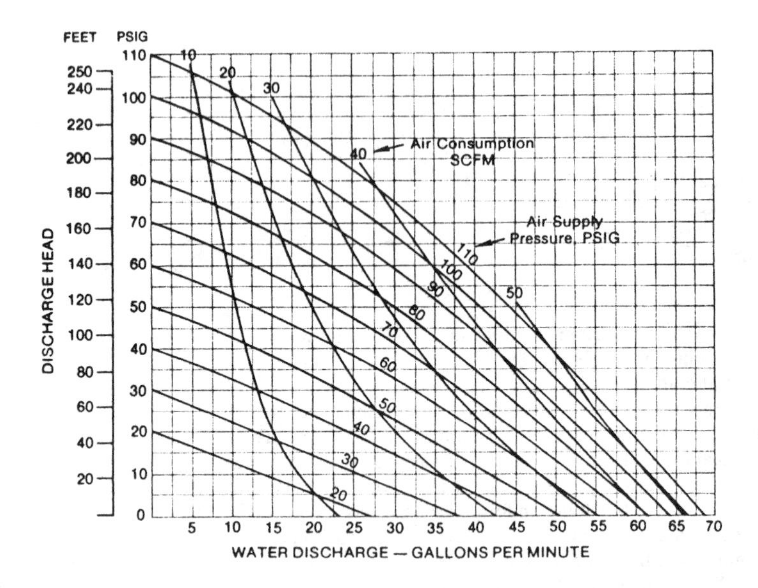

Figure 66. Characteristics of a pneumatic diaphragm pump.

position. The two-cylinder pump achieves a total stroke frequency of 140 strokes per minute—the frequency being adjustable up to a standstill by means of a compressed-air throttle. Figure 66 reproduces a typical flow rate/discharge pressure together with an air-consumption curve. At zero flow rate (ordinates), the air and discharge pressure agree. With increasing stroke frequency (= flow rate), the discharge pressure drops off, because, obviously, the throttling action being produced in the control valve is too large. The bad efficiency can be read easily at the comparison between the air and flow rate.

In-Line Piston Pumps

Drive Gear Unit and Drive

The triple-crank drive gear unit in the horizontal position with a double-bearing crankshaft is the predominant type used. It allows for very compact design and construction over a very wide size range. Normally, the crankshaft runs in antifriction bearings. The connecting rod bearings are mostly slide bearings; in rarer cases they are of the antifriction type. The crosshead guide always runs in slide bearings. In very large pumps the connecting rod bolt bearings are designed to accept antifriction bearings.

The triplex-crank drive gear unit (Figure 67) is a particularly compact design with antifriction bearings for the crankshaft as well as slide bearings for the con rod, con rod bolt, and crosshead. It is built for outputs of up to 1,000 kW approximately, (rod force approximately 30 tons, stroke length just under 300 mm). It goes without saying that drive gear units of larger outputs require forced oil lubrication. In addition to the horizontal position triple-crank arrangement, horizontal position five-crank drive units with four-bearing crankshafts are built for smaller output rates. The advantage of this design rests mainly in a pulsation-poor flow rate.

The vertical arrangement of drive gear units (Figure 68) saves space. It is being built for the entire upper performance range. In designs where the piston gland is located on the top, a favorable feature

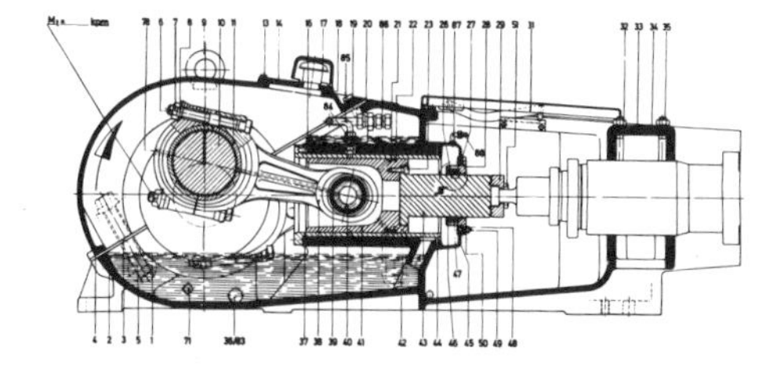

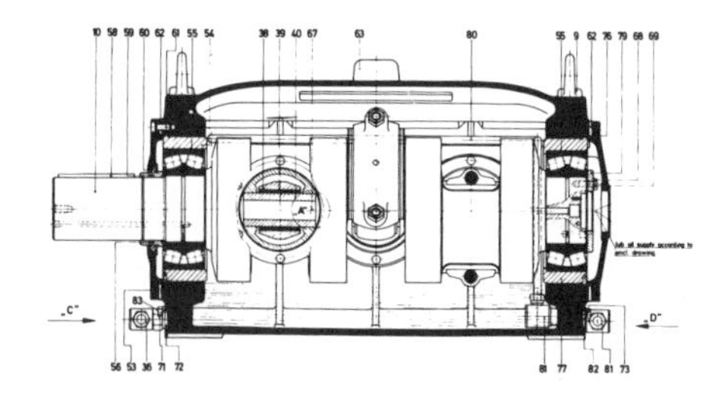

Figure 67. Horizontal triple-crank drive unit for large outputs (URACA).

is accessibility for replacement of wear parts. Vertical designs can also feature triple- or quintuple-crank arrangements.

Opposed-arrangement drive gear units (Figure 69) are suitable for mounting pump heads on two sides and are almost exclusively used for diaphragm pumps (see also Figure 4E).

For reasons of maximum possible piston speed in the piston gland (friction) and the flow velocity in the pump valves (valve entry-pressure loss) for process pumps, the stroke frequency for in-line piston pumps is limited to a range of 150 to 500/min. Speed rates that allow a direct motor coupling can only be found in smaller output rates and homogeneous, but somewhat lubricious media. In most cases, internal or external gear units are required (Figure 70). For particularly compact arrangements the drive motor with a belt drive is arranged above the drive gear unit block. With regard to generation of noise, the belt drive has, by the way, remarkable advantages over geared drives.

Figure 68. Vertical three-cylinder piston pump (Worthington).

Figure 69. Two-cylinder diaphragm pump with opposed drive unit arrangement (LEWA).

Varidrives for speed rate control are considered suitable units for smaller outputs (< 50 kW). Other than that, thyristor-controlled D.C. motors, A.C. motors with an electronic frequency converter, and the hydraulic torque converter are commanding the field of drives. As a rule, a decision is made after consideration of the required technical parameters such as protection against explosion, efficiency, variation range, and cost-benefit analysis. For very large units, the direct drive by means of steam turbine or combustion engine offers advantages—depending on local conditions and possibilities.

Parts in Contact with Fluids (Pump Head)

In single-cylinder metering pumps, one refers to parts in contact with fluids as the "pump head," whereas in in-line piston pumps the term "fluid part" has gained a greater prevalence. A good overview of design types of fluid parts (16) is shown in Figures 71A to F. The single drop-forged pump head represented in Figure 71A are made from carbon steel and used for noncorrosive fluids. Header pipes arranged on the suction and discharge pressure sides serve as feed-in and delivery lines. The valves are accessible, in pairs, from above. The piston gland is located in the pump head. Any exchange would require dismounting the pump head.

For critical fluid conditions, a design as per Figure 71B is suited in which the individual pump heads with stuffing box are bolted to the header pipe block in an easily dismountable and inter-

Figure 70. Triple-piston pump with gear unit and hydro-converter for speed-variation (URACA).

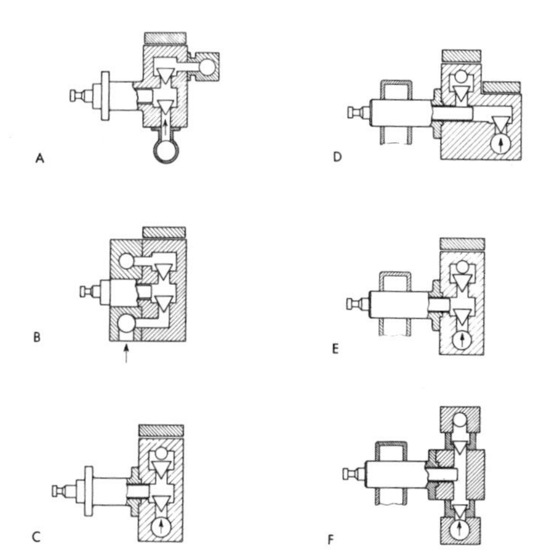

Figure 71. Overview of fluid parts on discharge pumps.

changeable mode. The block design for all three cylinders as per Figure 71C is suitable for lower pressures and noncorrosive liquids. The risk of a cylinder crack, which would destroy the solid block, must be negligibly small. The stuffing box housings are mounted on the cylinder as single units or as a block and are easily removable.

Individually accessible valves and easily interchangeable stuffing box housings are characteristic features for the valve block as shown in Figure 71D. When repairs to the stuffing boxes and valves have to be done, the header pipe system will not have to be disassembled. The design as per Figure 71E corresponds with Figure 71C, but both valves can be removed as one cartridge toward the top, and easily dismountable stuffing box bodies are provided for each cylinder. For corrosive fluids, for which cylinder cracks as well as expensive high-alloy materials cannot yet be excluded because of the pressures involved, the cylinder block is split into three individual pieces which are connected to joint headers on the suction and discharge sides (Figure 71F).

Where difficult fluids are involved, this type of design very much facilitates the exchange of individual parts, a fact which can be of corresponding economical significance for larger machines. Figure 70 shows how the pump bodies are mounted and the joint pressure headers are attached. The design of horizontally arranged suction valves offers installation advantages and reduces the entry-pressure loss (Figure 72). Where high pressures and unfavorable fluid influences are involved, the T-shaped bore holes of pump cylinders are creating considerable problems due to the pulsating internal pressure. It would then be a favorable situation if the endangered parts were easily interchangeable (Figure 73). The coaxial arrangement of the pump valves will lead to a considerable reduction of the stresses (Figure 74).

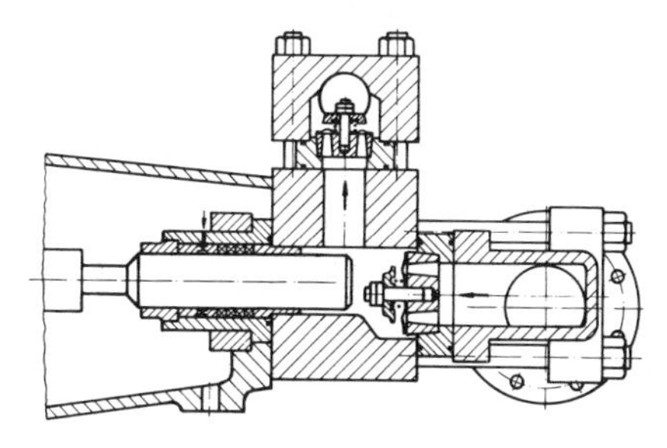

Figure 72. Horizontal installation of suction valve (Ingersoll-Rand).

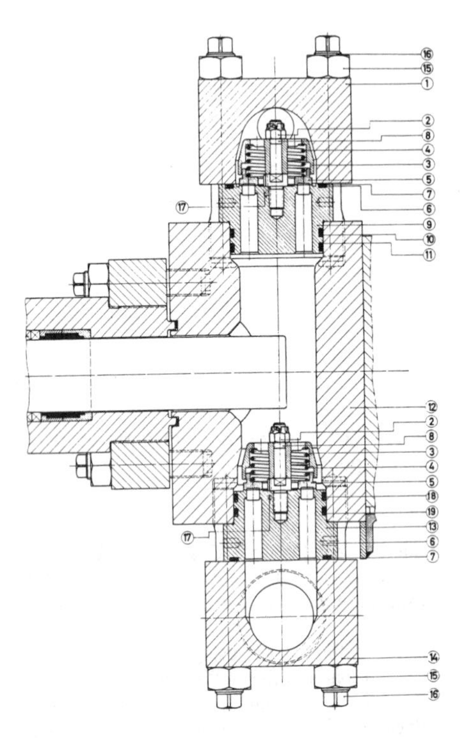

Figure 73. Structuring of fluid part from individual components at high pressure (URACA).

Piston Glands and Pump Valves

The piston glands and pump valves for metering pumps illustrated in Figure 40 do not differ from each other in principle, but rather in details from the designs for in-line piston pumps whose performance range is also larger. To complete the overview, the essential differences are outlined herewith. Lip rings with spring-tension adjusting features are used for noncorrosive and, in particular, for somewhat lubricating media (Figure 75). Figure 76 depicts the piston gland for a larger process pump with packing rings and locking seal having a separate tensioning feature. The various flush points (14) can be connected to the different flush circuits. Similar designs are used for suspensions, but with a volumetric injection flushing system right through a throttle gap located before the seal. By the arrangement of sedimentation spaces, the particles can be kept away from the piston seal (Figure 77) [17]. Also piston pump heads designed as "remote heads" are known [18].

In most cases, the packings are made of PTFE-containing braid-like tissue. As to the life of the piston glands, there are a number of influence variables such as the piston hardness grade, surface smoothness quality, lubrication, stuffing box structure, ring material, piston guide, and piston suspension. Their life will fluctuate (also dependent on maintenance) within the same type of seal and

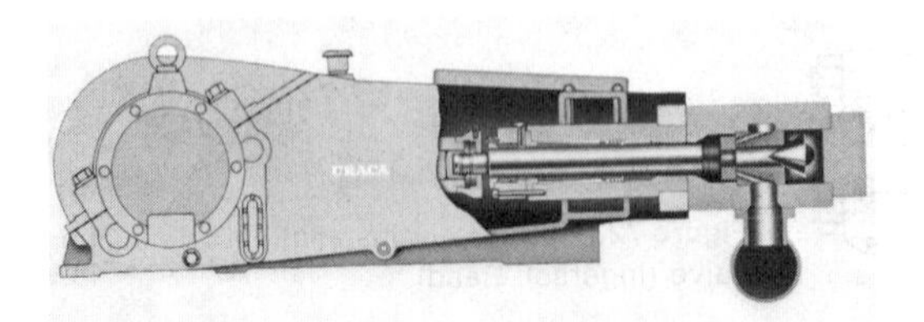

Figure 74. Coaxial arrangement of pump valves (URACA).

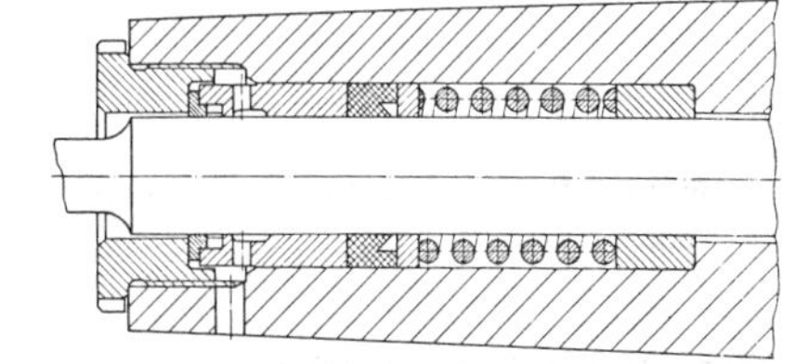

Figure 75. Piston gland with spring-loaded lip ring.

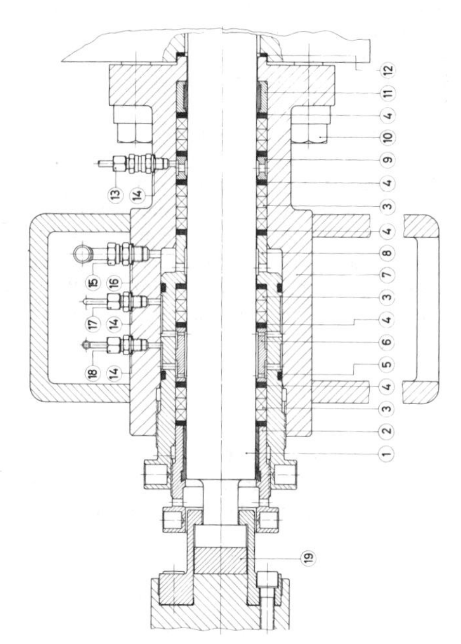

Figure 76. Piston gland of a process pump with various flush and lube points (URACA).

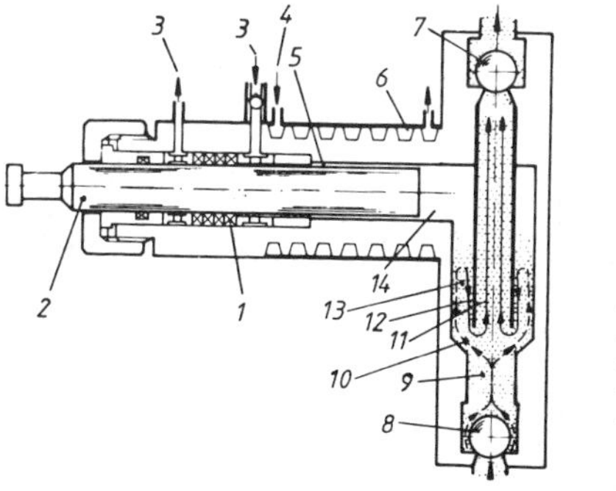

1—Plunger packing
2—Plunger
3—Leakage—drainage or flushing oil
4—Cooling water
5—Annular clearance
6—Cooling
7—Discharge valve
8—Suction valve
9—Suspension
10—Suction flow
11—Discharge flow
12—Separation pipe
13—Separation chamber
14—Chamber

Figure 77. Piston pump head for suspensions sedimentation area and injection flushing.

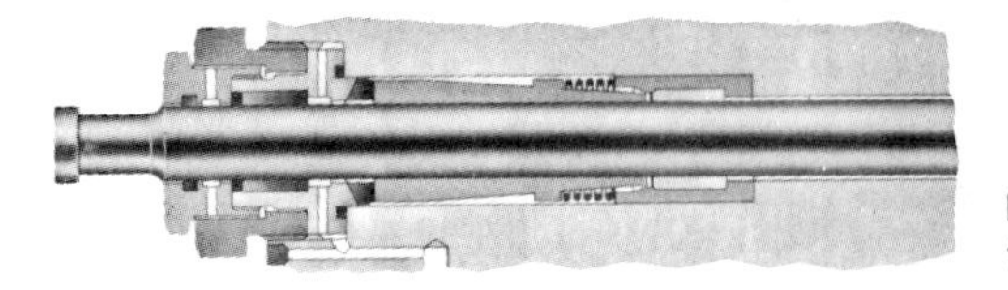

Figure 78. Split seal for oscillating pistons (URACA).

will, under favorable conditions, reach the order of 5,000 hours. Efforts to develop oscillating split seals are becoming increasingly successful [19]. In these cases, advantage is taken of the centering and sealing actions of a split which tends to widen in the direction of the pressure. To a certain extent, split seals (Figure 78) reach an amazingly long life; however, in other cases, one cannot yet dominate over them. However, more and more experience is being gained and it is expected that a very efficient sealing system will crystallize from these efforts (perhaps analogous to the mechanical seal in centrifugal pumps). The unavoidable split leakage in terms of a few percentage points of the flow rate is mostly more than compensated for in the total efficiency by the much lower friction. Dirt particles, surface destruction, and material configuration do present problems.

The pump valves of in-line piston pumps are frequently designed as insertable sets which are dismountable through the top (Figures 79 and 80). The lighter the movable part (for instance, a cup-shaped cone, Figure 79), the higher is the potential stroke frequency. Ball valves in suspensions permit only small stroke frequencies (<150 stroke per minute). Larger output requires suitable double-flow plate valves (Figure 73), as they allow a larger opening area at the smallest mounting space. The detailing of the plate guide is important (see guideless plate valves, Figure 20). When selecting the type of material used for pump valves, corrosion and abrasion factors must be taken into account.

Thick-Matter Piston Pumps

The diaphragm-type slurry pumps described are not suitable for media of very high consistency such as concentrated sewage sludge, "compact" pasty media, and coarse grained admixtures (in the range of several centimeters!). Such pumping tasks, apart from treatment of construction materials,

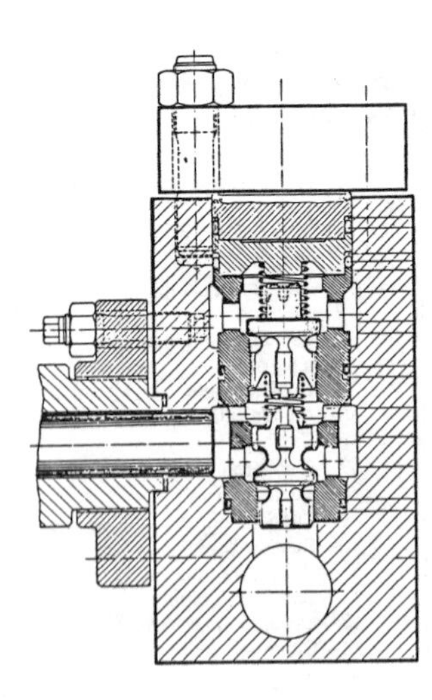

Figure 79. Fluid part with ball valves (URACA).

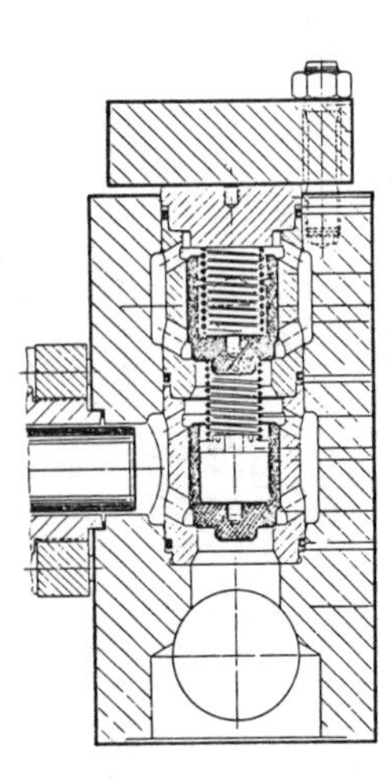

Figure 80. Fluid parts with cup-shaped plate-type valves (URACA).

Figure 81. "Compact" fluids.

also occur in the disposal of garbage (dumping) and the removal and recycling of waste matter (waste incineration) as well as in other areas of chemical process technology (coal upgrading).

The problems arising from pumping "compact" media having coarse admixtures (Figure 81) rest with the cylinder fluid charge, valve control, and control over wear. One interesting approach is shown in [21] Figure 82 for use of a feed screw, consisting of a hydraulically controlled pipe distributor as a valve (large opening area) and a hydraulically actuated long-stroke piston (stroke of 1 meter) of low piston speed.

With the help of such pumps one is able to deal with capacities of up to and over 30 m^3/h against approximately 100 bar, and if the materials have been appropriately selected, there will be a long life for the modules. For free-flowing media as, for instance, mortar, the same pump system with slow-running, self-acting ball valves has been successfully used (Figure 83). Piston pumps used on thick matter are also built as double-acting multicylinder in-line units.

INSTALLATION OF OSCILLATING DISPLACEMENT PUMPS

Because of the flow rate fluctuating periodically in a time-dependent mode, the lay-out and sizing operation of the installation for oscillating displacement pumps presents some peculiarities [21]. The displacement characteristics together with the piping system produce pressure fluctuations.

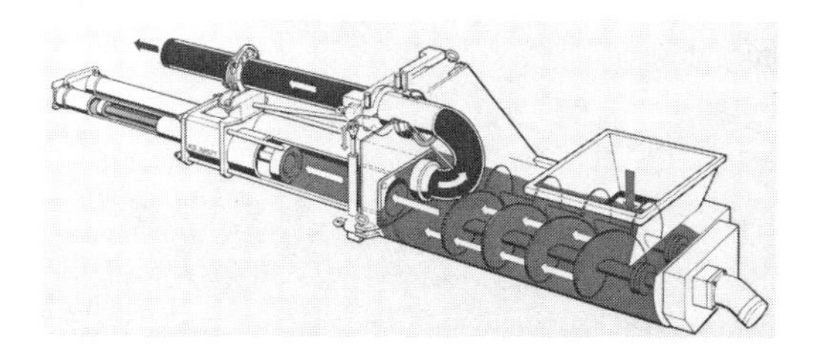

Figure 82. Thick-matter piston pump with hydraulic drive and pipe distributor valve (Putzmeister).

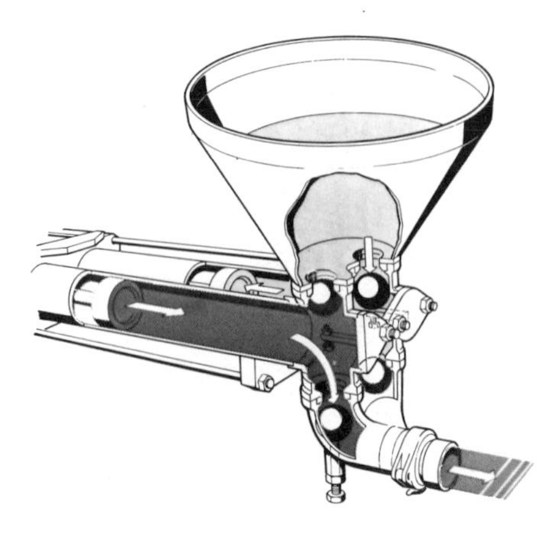

Figure 83. Piston pump for slurries with self-acting valves (Putzmeister).

For many cases, for calculation purposes, it suffices to consider the fluid as incompressible. However, the most accurate method of calculation starts out from the compressible fluids and is based on the wave theory [21–23].

Pressure Fluctuations without Damping

Incompressible Fluid

In this case, the fluid in the pipeline moves exactly as suggested by the displacement characteristics. The pressure fluctuations are casued by friction (Δp_r) and mass acceleration (Δp_m). For Newtonian fluids, the following equation is applicable:

$$\Delta p_r(t) = \left(\lambda_R \cdot \frac{L}{d} + \sum \zeta\right) \cdot \frac{\rho}{2} \cdot \omega^2(t) \tag{14}$$

where λ_R = pipe coefficient of friction
L = pipe length
d = I.D. of pipe
ζ = drag coefficient

The flow velocity $\omega(t)$ results from the piston speed v_K, the continuity equation with pipe cross-section A_R and piston cross-section A_K:

$$\omega(t) = v_K(t) \cdot \frac{A_K}{A_R} \tag{15}$$

The mass pressure follows from the Newtonian law:

$$\Delta p_m = \rho \cdot L \cdot b(t) \tag{16}$$

with

$$b(t) = b_K(t) \cdot \frac{A_K}{A_D} \tag{17}$$

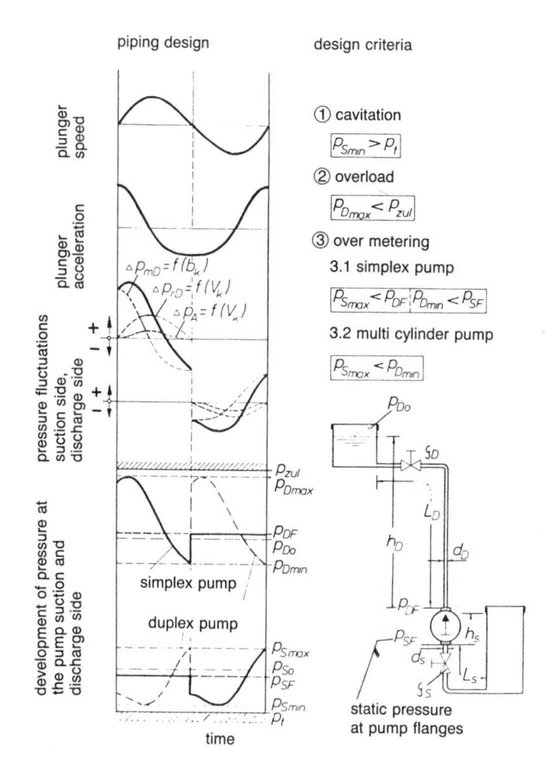

Figure 84. Pressure fluctuations in piping systems—suction and discharge sides.

Predominant friction pressures are also the entry and exit-pressure losses of the pump head Δp_E and Δp_A. The superposition of the individual pressure functions is shown by Figure 84. The total pressure fluctuation must fulfill certain criteria as marked in Figure 84. The superposition of the maximum values for frictional and mass pressure variation as per Equations 14 and 16 can occur, nearly square-like, for almost all applications:

$$\Delta p_{max} = \sqrt{\Delta p_{m\ max}^2 + \Delta p_{r\ max}^2} \tag{18a}$$

Here, equi-phased portions of Δp_m and/or Δp_r will have to be added in a scalar mode. The sizing and rating criteria as depicted in Figure 85 are once again shown for the suction and discharge stroke in this illustration [25].

Compressible Fluid

In many practical cases of application, the methods of calculation valid for incompressible fluids are also sufficiently accurate for real fluids. Under certain marginal conditions, however, the low compressibility of real fluids has the effect that the pressure progression measured at the pump flanges looks entirely different than expected according to the calculation with incompressible fluids.

Various effects are responsible for this situation:

1. The volumetric efficiency η_v distinctly deviates from value 1 ($\eta_v < 1$) at higher pump pressures, which is, more or less, always the case when high pressures and compressible fluids are involved (for example, liquid gases) (Figure 17).
2. Therefore, discharge and suction strokes begin with a jolt which is transmitted to the liquid column on the discharge and suction pressure sides and progresses in it at high velocity.

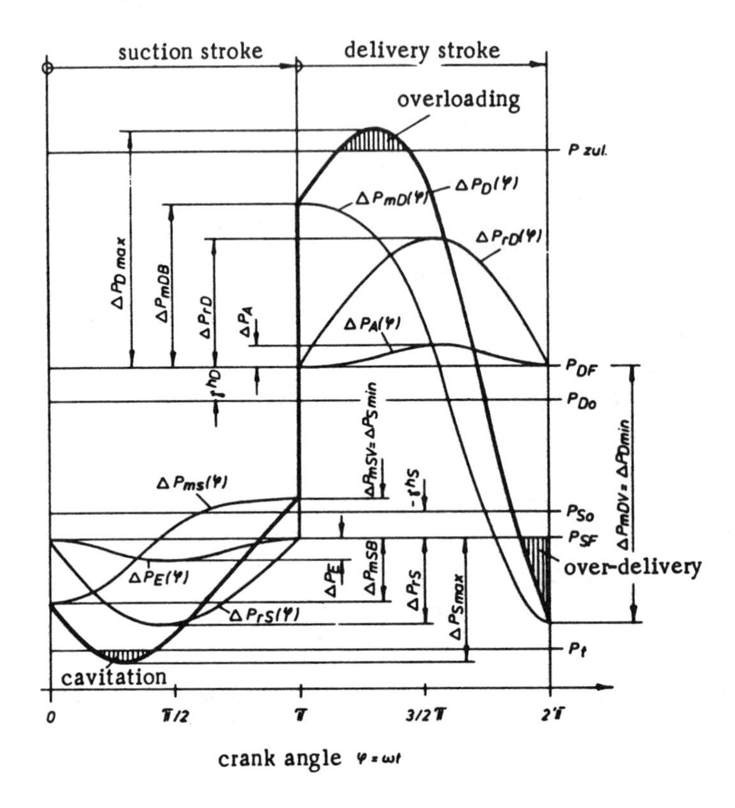

Figure 85. Pressure fluctuations due to displacement characteristics.

3. Because of its elasticity, the liquid does not follow the law of motion forced upon by the displacer along the entire length of the piping system.

The effects mentioned are basically always prevalent in real fluids. Their influence on the pressure progression is all the more strong, the higher the pumping pressure and the compressibility of the pumped liquid, and the longer the pipelines, and the higher the frequencies of the displacment kinematics are.

In the physical sense, a pipe system with an oscillating displacement pump represents an elastic continuum with defined properties upon which a jolt-like-commencing periodic law of motion is being forced. Principally, we are dealing here with a separately excited oscillatable pattern.

The calculation of the oscillation processes is complicated [26]. For the purpose of determining the extreme values, it is useful to superimpose the coupling jolt by the incompressibly calculated pressure progression. As can be determined from the displacement kinematics, the jolt-like coupling (see Figure 16) occurs with the speed jump Δv. The maximum pressure during the coupling action below the first self-generated frequency of the pipe system is calculated from the Joukowsky jolt as follows:

$$\Delta p = \Delta v \cdot p \cdot a \qquad (18b)$$

where a is the speed of sound of the fluid in the pipe system.

Measurement results are given by Reference 21. Figure 86 portrays the measured pressure progression for a defined pipeline geometry and design of pump at an extremely low volumetric

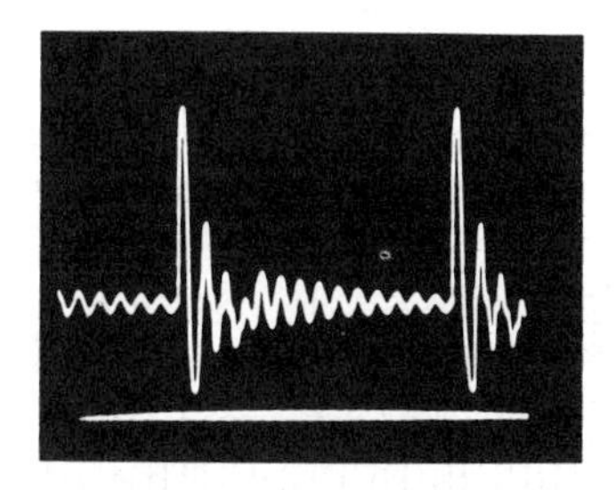

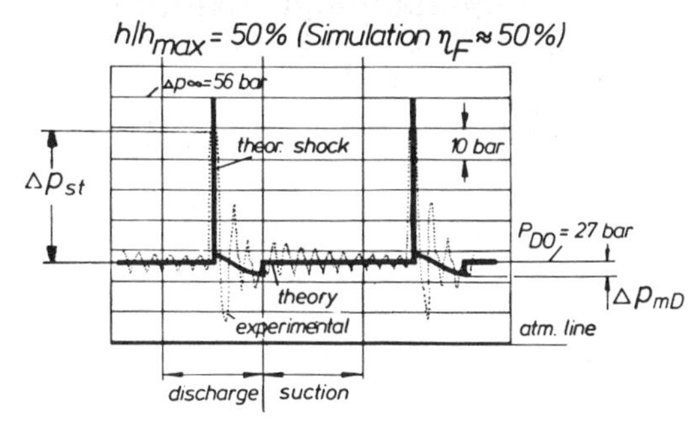

Figure 86. Pressure oscillations in a discharge pressure pipeline—volumetric efficiency $\eta_v = 50\%$.

efficiency $\eta_v = 0.5$ as it can occur in metering pumps for high pressure within the part-stroke range, in comparison to the incompressible theory and Joukowsky jolt. One recognizes that the method suggested reflects reality quite properly.

It is more accurate to carry out an exact calculation by using numerical arithmetical methods [26]. From the example shown in Figure 87, it is recognizable that the rate of accuracy will be excellent only when suitable software is applied. Figure 87 also intimates the pressure progression resulting from the incompressible theory. When severely jolt-affected excitation and extended pipe systems are involved, the deviations with respect to reality are very large.

Without generating a big calculation job, the frequency of the arising fluctuations can easily be determined from the existing reflection conditions where n is the eigenvalue figure (a continuum has an infinite number of self-frequencies (eigenvalues)) and f the self-frequency:

Open pipe end:

$$f = \frac{8}{2L} \cdot \left(n - \frac{1}{2}\right) \tag{19}$$

$$f_1 = \frac{a}{4L} \quad \text{(for n = 1, first self-frequency)} \tag{20}$$

Closed pipe end: (borderline case)

$$f = \frac{a}{2L} \cdot n \tag{21}$$

$$f_1 = \frac{a}{2L} \quad \text{(for n = 1, first self-frequency)} \tag{22}$$

In most cases of installations of the process technology field it is the "open pipe end" that applies.

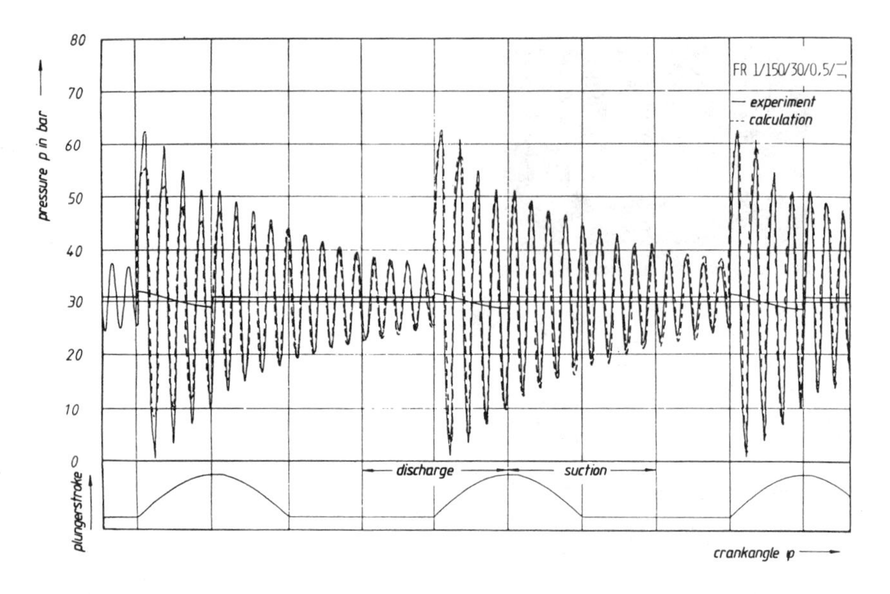

Figure 87. Pressure oscillations in a discharge pressure pipeline—volumetric efficiency $\eta_v = 50\%$.

Pulsation Damping

If an installation does not at least fulfill the criteria mentioned earlier under "Incompressible Fluid" with reference to cavitation, overload, and excess pumping, steps to dampen the pulsation will have to be taken. For a given pump there is the possibility of mounting absorption and reflection-type pulsation dampers (Figure 88). As mentioned, the continuum (Figure 88A) has an infinite number of self-frequencies (for example, Equation 19: $n = 1$ to ∞). If an absorption damper, that is, a resilient additional volume (for instance, gas) is installed, the system is then generally considered, with sufficient approximation, to be a spring/mass system. In general, the fluid is then looked upon as being rigid [27]. In contrast, reflection dampers base their actions on the wave

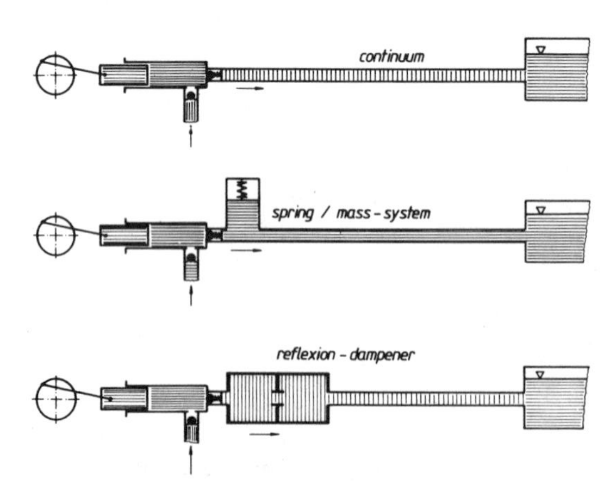

Figure 88. Pulsation damping: (A) continuum; (B) absorption damper; (C) reflection damper.

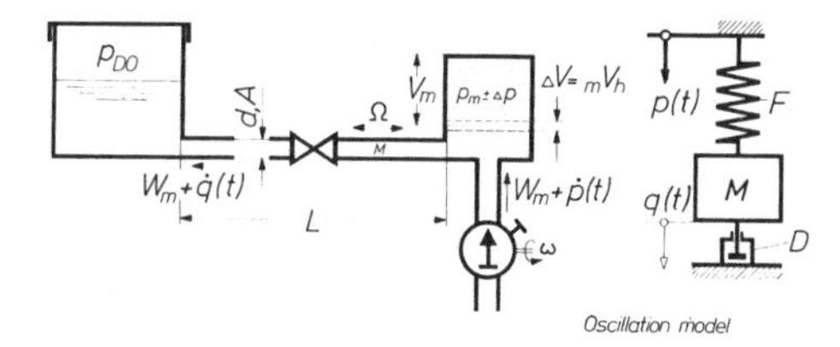

Figure 89. Pattern of oscillation system of a pipe system with absorption damper.

propagation of pressure waves caused by the elasticities. By using a suitable geometry with impedance jumps, a local multiple reflection of the waves is achieved—with dissipation [28, 29].

Absorption Pulsation Dampers

By using the pattern of the damped, forced oscillation (Figure 89), the propagation of the excitement function p(t) of the pump can be determined in accordance with damper q(t) [27]. The oscillatable system has a self-frequency Ω

$$\Omega = \sqrt{\frac{\kappa \cdot p_m \cdot A}{\rho \cdot L \cdot V_m}} \tag{23}$$

where κ = the adiabatic exponent of the damping gas
p_m = the mean damper pressure
A = the pipeline cross-section
V_m = the mean damping volume (gas volume under pressure p_m)

One must always make sure that the excitation frequency ω is situated far above the self-frequency (Figure 90). The friction damping action of system D results approximately from the following

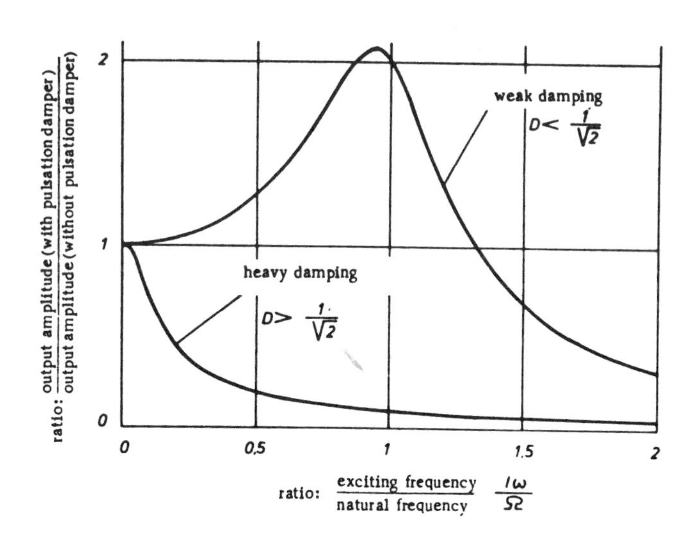

Figure 90. Amplitude reinforcement by means of resonance.

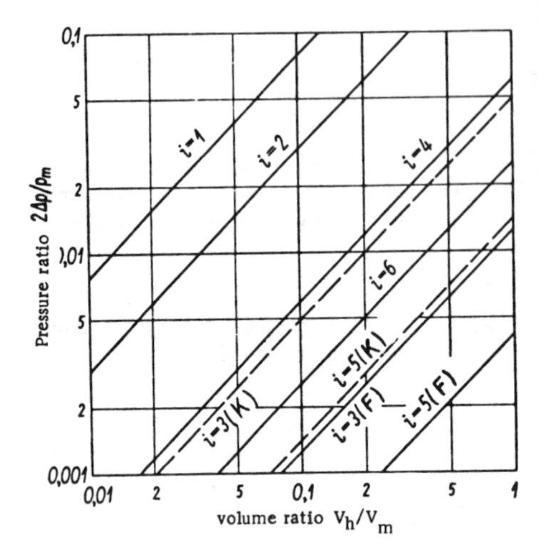

Figure 91. Rating/sizing of absorption pulsation dampers (K: $\lambda \approx 0.2$ F: $\lambda = 0$).

equation (see Equation 17 for designations):

$$D = \left(\frac{L \cdot \lambda_R}{d} + \zeta\right) \cdot \frac{w_m}{2\Omega L} \tag{24}$$

The relative pressure fluctuation $\Delta p/p_m$ can be calculated for $D > 0.5 \sim 0.7$ or $\Omega \gg \omega$ as follows:

$$\frac{\Delta p}{p_m} = \frac{\kappa \cdot m}{2} \cdot \frac{V_h}{V_m} \tag{25}$$

Δp is half of the pressure amplitude around the mean pressure p_m. Variable $m \cdot V_h$ is determined on the basis of the number of cylinders and the rod ratio provided the volumetric efficiency is $\eta_v > 0.8$.

The term $2\Delta p/p_m$ can be taken directly from the details of Figure 91. It must always be remembered that V_h means the stroke volume of the single cylinder and V_m the mean damping volume under pressure.

If unfavorable values are found for Ω or D, a parameter variation must be effected by means of a concerted effort. If D is too small, an additional choke valve must, for instance, be installed.

Designs of absorption dampers are shown in Figures 92 through 95. The mounting of such equipment is practical with a provision for through-flow (Figure 92C). The important factor in dampers that have a direct gas/fluid contact is that the gas volume is retained, which, to a certain extent, is diminished on a continuous basis by diffusion. There are possibilities for manual or automatic replenishment (Figure 93A and B).

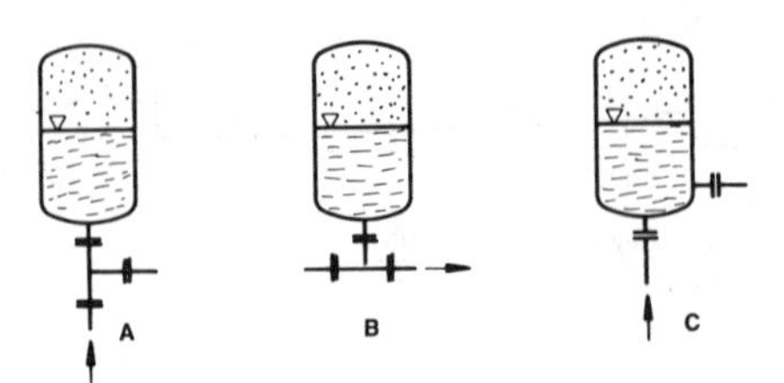

Figure 92. Mounting arrangements.

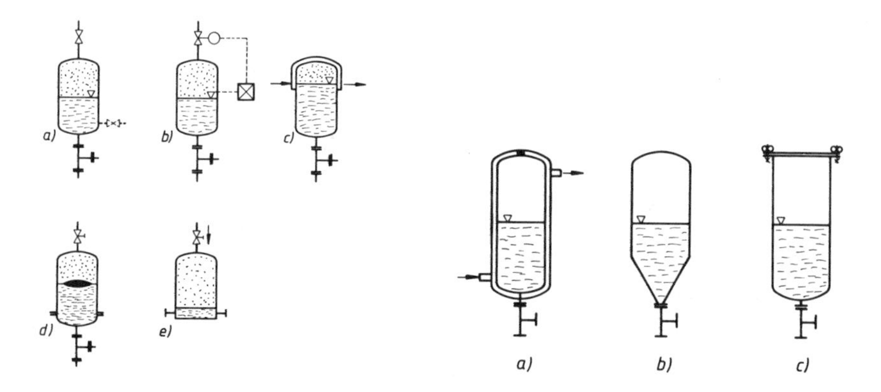

Figure 93. Gas supply line.

Figure 94. Types of construction.

In rare cases, the local heating feature (Figure 93C) serves to produce a gas volume. It is seldom that floats are installed to reduce the exchange area (Figure 93D). If the fluid brings with it small gas quantities or if a continuous gas flow can be piped in, the design as per Figure 93E would be the most appropriate one.

The designs of pulsation dampers that are required are those equipped with a heating feature (Figure 94A) for suspensions (Figure 94B) and with openings for cleaning purposes (Figure 94C).

The pulsation dampers must be checked for their effectiveness by verifying the pressure fluctuation (Figure 95A) or by optical observation (transparent design as per (Figure 95B and D). Damper designs with diaphragms for gas-liquid separation are shown by Figures 96 and 97. The operating principle is based on the pre-charge action (Figure 96) to a rate of approximately 70% of the mean discharge pressure (Figure 96A). A damping volume is then automatically produced under operating pressure (Figure 96B).

It must be noted that the fluid can dwell for a long time in absorption pulsation dampers. Therefore, one must be sure during any application that the fluid will not thus experience any harmful changes. Figure 98 depicts a pump with a damper automatically monitored by the suction and discharge system.

Reflection Damper (Resonator)

The usual design types are single- or twin-chamber resonators (Figure 99) which can also be supplied with a ball shape. Since the computation is rather complicated [29], we limit ourselves to some general explanations.

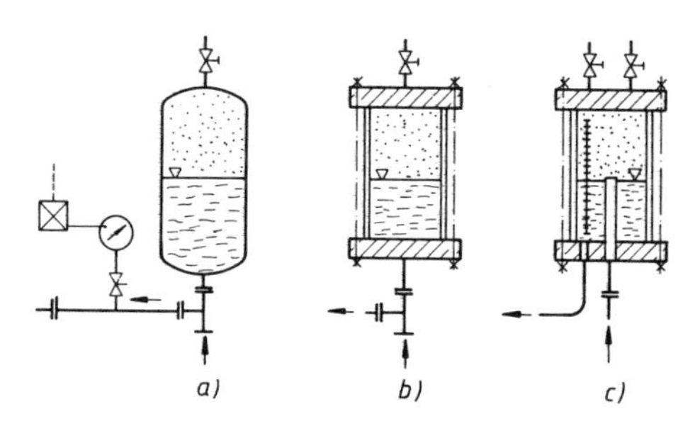

Figure 95. Gas fill verification.

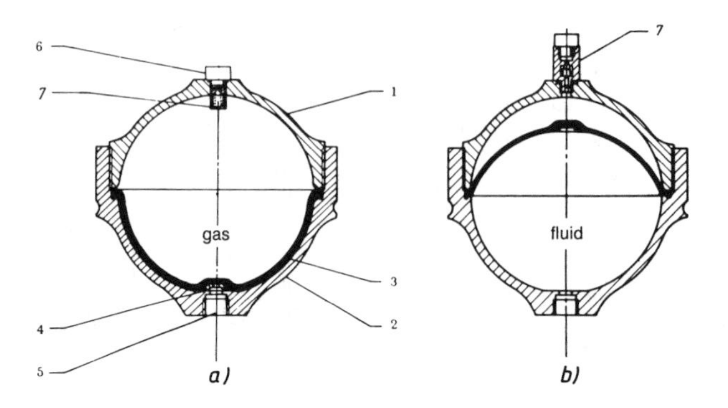

Figure 96. Pulsation damper with separation diaphragm.

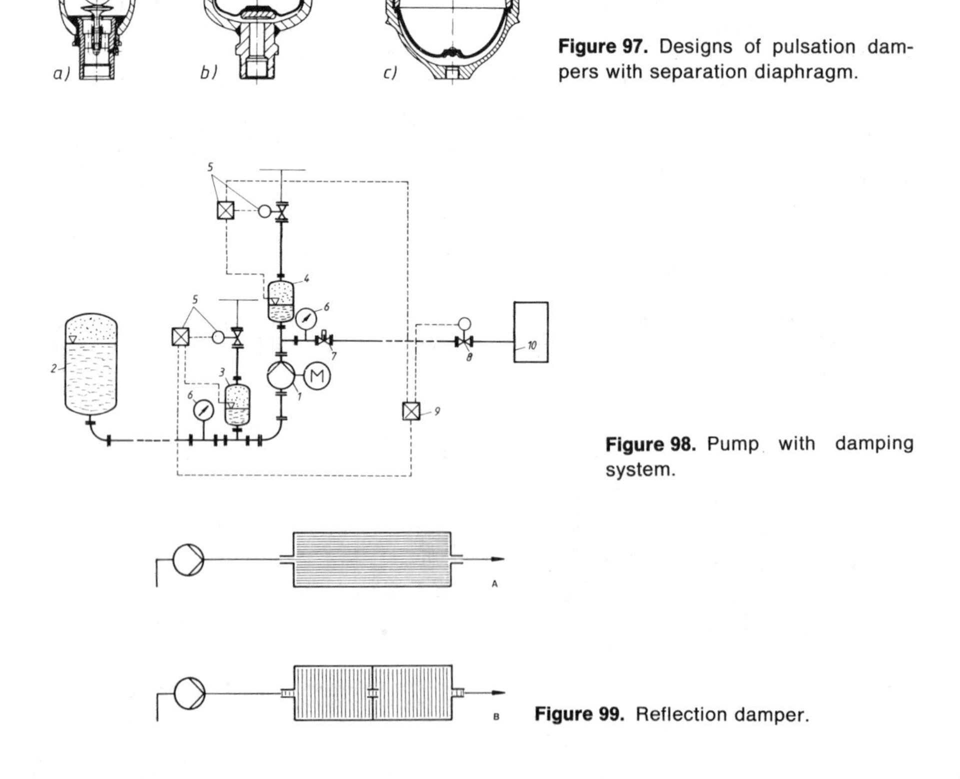

Figure 97. Designs of pulsation dampers with separation diaphragm.

Figure 98. Pump with damping system.

Figure 99. Reflection damper.

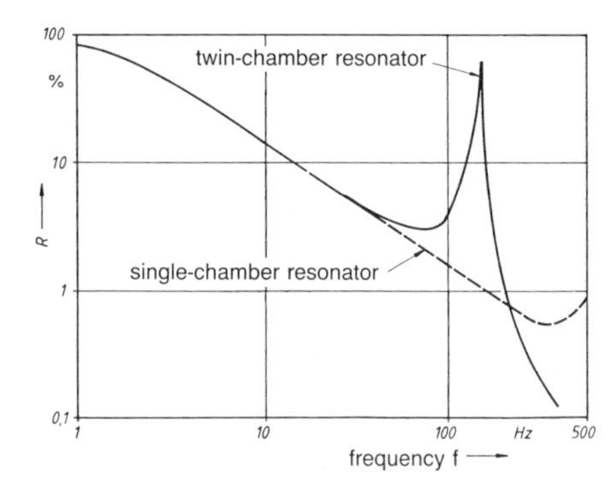

Figure 100. Damping curve of reflection dampers: (A) single-chamber system; (B) twin-chamber system.

The reflection damper destroys oscillations by means of a multiple reflection in the internal area of the damper—dependent on its frequency. Each resonator distinguishes itself by a characteristic damping curve which portrays the residual pulsation in dependence on the frequency.

The curve purports which proportion of an occurring pressure oscillation amplitude of a defined frequency will leave the resonator again. The higher the frequency rate (Figure 100), the lower the residual pulsation.

This means that resonators, in contrast to absorption dampers, do preferably damp high-frequency pressure oscillations. They are therefore preferred for use on multi-cylinder pumps, but hardly for single-cylinder pumps. Each resonator has a characteristic transmission frequency (Figure 100). When sizing the layout of the system, the frequency content of the excitation spectrum will have to be determined by way of a harmonic analysis in order to know where the transmission frequency is allowed to be.

Reflection dampers have important advantages: They have good through-flow characteristics and do not require any maintenance. Unfortunately, their ability to dampen is mostly limited to high-frequency oscillations. A combination of absorption and reflection dampers is frequently a practical approach.

Type of Installation [25]

In general, oscillating displacement pumps are installed as per the details shown in Figure 101. Filter, pulsation damper, and safety module are important components. Some hints of installation can be taken from the details of the illustrations of Table 5.

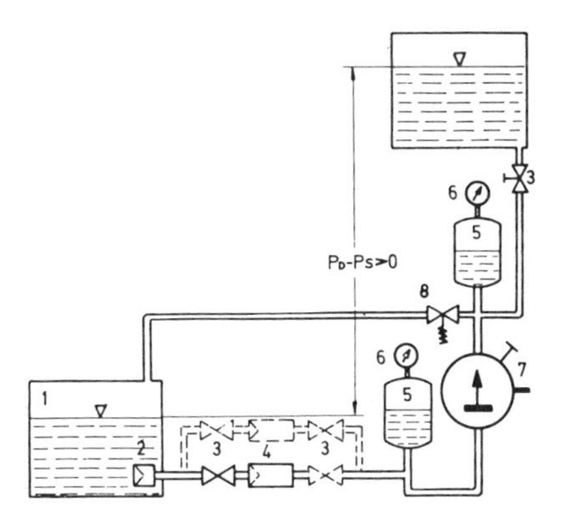

Figure 101. Installation of oscillating displacement pumps.

Table 5
Installation Hints

Arrangement	Action	Alternate Action	Problem
Wrong: suction lift too great	Reduce suction lift	Increase positive pressure at suction flange	Cavitation on suction side: short pipelines; smaller suction lift; if needed, produce positive pressure.
Wrong: too long and too narrow; air will be trapped in vertical loop.	Correct: keep line short: increase diameter	If long, narrow pipe is unavoidable, fit pulsation damper.	1. Bad venting: mount pipe system—to be self-venting. 2. Pressure drop too large: larger pipeline cross-section; shorter line; pulsation damper.
Float-controlled valve			Suction from pressure-conducting pipe system: if necessary, separate hydraulically.
Wrong: $P_S > P_D$	Pressure retaining valve	Overflow	Overdelivery: create positive pressure difference between pressure and suction sides; geodetically correct arrangement or pressure retaining valve or "artificial" pipe riser.
			Batch metering: prevention of second drip; controlled shut-off device; appropriately sized nozzle with defined overflow behavior.
1. Suction vessel 2. Agitator 3. Suction strainer 4. Metering pump 5. Flush valve			Suspension metering: good suspension homogenizing; suction strainer to prevent entry of coarse-grained solids; sufficient flow velocity and suitable pipe layout; flushing feature; avoidance of sedimentation onto pressure valve by shifting of pipeline axis.
Heating	Cooling		Metering of boiling fluids: create additional pressure by superheating vapor volume or additional supercooling by cooler; objective: reduce distance to vapor pressure.

NOTATION

A	area	ℓ	length
a	speed of sound	$\dot{m}$	mass flow rate
b_k	piston acceleration	n	speed
C	constant	P	power
D	friction damping action	P_A	pressure
d	diameter	Q	volumetric flow rate
f	frequency	r	crank radius
h	half-stroke length	T	torque
i	number of cylinders	t	time
L	length		

Greek Symbols

ξ	drag coefficient	v_k	piston speed
η	efficiency	ρ	density
κ	parameter defined by Equation 10	ψ	crank angle
λ_R	pipe coefficient of friction	ω	angular speed or excitation velocity
λ_s	rod radius		

REFERENCES

1. Vetter G., Fritsch, H., and Lange R., "Dosierung im Millimeterbereich gegen hohe Drücke," (Metering against high pressures in the millimeter range), *Verfahrenstechnik*, 12 (1978) (*Process Technology*).
2. Vetter, G., Fritsch, H., and Müller, A., "Einflüsse auf die Dosiergenauigkeit oszillierender Verdrängerpumpen" (Influences on the metering accuracy of oscillating displacement pumps), *Aufbereitungstechnik* (1974), S. 1–12, (mag. *Process Engineering* (1974), pp. 1–12).
3. Vetter, G., "Genauigkeit von Dosierkolbenpumpen" (Accuracy of piston-type metering pumps), CIT (1963), (p. 267 ff.).
4. Vetter, G., and Fritsch, H., "Die Auslegung der Rohrleitung für Dosierpumpen" (Layout and sizing of pipe systems for metering pumps), 3-R International (1981) (p. 328 ff.).
5. Vetter, G., and Fritsch, H., "Untersuchungen zur Kavitation bei oszillierenden Verdrängerpumpen" (Examinations with regard to cavitation in oscillating displacement pumps). CIT (1969) (p. 271 ff.).
6. Vetter, G., Fritsch, H., and Müller, A., "Leckfreie oszillierende Dosier-und Förderpumpen für Prozessbetrieb" (Leakproof oscillating metering and discharge pumps for process operation), CIT (1978) (p. 433 ff.).
7. O'Keffe, W., "Metering pumps for power plants," *Power* (1983), p. 1 ff.
8. Bristol, J. M., "Diaphragm metering pumps," *Chemical Engineering* 1981, p. 124 ff.
9. Orlita, F., "Ventillose Dosierkolbenpumpen" (Valveless piston-type metering pumps), Lecture at Karlsruhe 1979 Pump Meeting.
10. Körner, P., and Köpl, M., "Höchstdruckkolbenpumpen für Drücke über 3,000 bar" (Maximum pressure piston pumps for pressures of more than 3,000 bar), Lecture at Karlsruhe 1978 Pump Meeting.
11. Körner, P., and Wüstenberg, D., "Festigkeitsuntersuchungen von autofrettierten T-Stücken" (Strength examinations of autofrettaged T-pieces), CIT MS086, (1974).
12. Vetter, G., DBP 1.800.018 (German Federal Patent 1.800.018).
13. "Neuartige Schlauch-Membran-Kolbenpumpen" (New tube-type diaphragm piston pumps), *Export-Market*, Vogel Publishing, Vol. 24 (1973).
14. "Piston-Membrane Pumps; Contamination-sensitive fluids," Cahners Publishing Company Inc., *Design News*, (Sept. 1977).

15. Heitz, E., "Membranpumpen für Schlämme, Pasten, Pulver" (Diaphragm pumps for muds, pastes, powders), Die Chemische Produktion, Juli–Aug. 1982, S. 54 ff., (*Chemical Production*, (July–Aug., 1982) p. 54 ff.).
16. URACA Information TI 9-1/70.
17. Dettinger, W. W., "Eigenschaften von Kohlemaischepumpen für Hydrieranlagen" (Properties of coal mash pumps for hydrogenation plants), CIT (1982) (pp. 500–501.)
18. Vetter, G., "Die Förderung und Dosierung von Suspensionen gegen hohen Druck" (Pumping and metering of suspensions against high pressure), Verlag GVC-Hochdruckverfahrenstechnik, 01.04.81, Freiburg, (Publishers: GVC High-Pressure Process Technology, 1 Apr 81, Freiburg).
19. Dettinger, W. W., "Dichtungsreibung und Dichtungsleckage" (Seal friction and seal leakage), *Chemie-Anlagen* + Verfahren, 1979, Edition 5 (*Chemical Facilities and Processes*).
20. Putzmeister "Pumpen für das Schlimmste" (Pumps for the worst applications). IP 502-2.
21. Vetter, G., and Fritsch, H., "Auslegung der Rohrleitungen für Dosierpumpen" (Layout and sizing of pipe systems for metering pumps), 3R-International (1981), pp. 328–336.
22. Fritsch, H., "Druckpulsations-und Resonanzerscheinungen in Rohrleitungen oszillierender Verdrängerpumpen" (Pressure pulsation and resonance aspects in pipelines of oscillating displacement pumps), 3R-International (1982), (pp. 99–105).
23. Kästner, A., "Beitrag zur Beurteilung der instationären Strömung in den Druckleitungen von Kolbenarbeitsmaschinen bei Flüssigkeitsförderung; Dissertation Universität Stuttgart 1979," (Contribution to the assessment of instationary flows in the pressure lines of piston machines during fluid pumping; Dissertation at Stuttgart University in 1979).
24. von Nimitz, N. W., "Massnahmen gegen Pulsationen und Schwingungen bei der Auslegung von Kolbenverdichter-und Kolbenpumpenanlagen" (Measures against pulsations and oscillations in the layout and sizing of piston-compression and piston-pump facilities), 3R-International 21 (1982), (p. 507 ff.).
25. LEWA Information D10-011d, 1976.
26. Vetter, G., Kellner, A., Grimm, E., "Druckschwigungen durch oszillierende Verdrängerpumpen—Rechnung und Experiment" (Pressure oscillations by oscillating displacement pumps—
Calculation and Experiment), Lecture at Karlsruhe Pump Meeting in Sept. 1984.
27. Vetter, G., Fritsch, H., "Auslegung von Pulsationsdämpfern für oszillierende Verdrängerpumpen" (Rating and sizing of pulsation dampers for oscillating displacement pumps), CIT (1970), (p. 609 ff.).
28. Dettinger, W., "Druckschwankungen und ihre Dämpfung bei Kolbenpumpen" (Pressure fluctuations and their damping in piston pumps), CAV (1981), (p.17 ff.).
29. Dillmann, G., "Resonatoren zur Pulsationsdämpfung bei Kolbenpumpen" (Resonators for pulsation damping in piston pumps), CAV (1978), revised edition 6 (1981).

CHAPTER 38

CAVITATION AND EROSION: MONITORING AND CORRELATING METHODS

F. G. Hammitt

Department of Mechanical Engineering and Applied Mechanics
University of Michigan
Ann Arbor, Michigan USA

CONTENTS

INTRODUCTION AND GENERAL BACKGROUND, 1119
Nature of Cavitation, 1119
Effects of Cavitation, 1120
Scale Effects, 1121

EROSION-DAMAGE EFFECTS, 1124
General Background, 1124
Material Resistance Parameters, 1125

VIBRATIONS AND CAVITATION, 1126
Machinery Vibrations, 1126
Vibration-Induced Cavitation and Test Devices, 1126

CAVITATION NOISE, 1127

CAUSES AND PREDICTION CAVITATION AND EROSION, 1127
Cavitation Flow-Regime Prediction, 1127
Cavitation Damage Prediction: State-of-the-Art, 1128

OCCURRENCE OF CAVITATION, 1134
General, 1134
Rotating Devices, 1134
Stationary Flow Devices, 1135

SUMMARY, 1135

REFERENCES, 1135

INTRODUCTION AND GENERAL BACKGROUND

Nature of Cavitation

Cavitation* is an important problem in many liquid handling devices where low pressure regions exist. It is similar to local "boiling," providing bubbles and vapor regions in the liquid, due to local

* Word was coined by R. E. Froude, but the phenomenon was much earlier suggested by Euler (1754) [1].

pressure reduction below vapor pressure rather than increased temperature. Two important consequences are:

1. Damage to surrounding structure from bubble collapse.
2. Deterioration of machine performance (decrease of output or efficiency) due to modification of flow stream-lines around vapor regions.

Basic questions are

1. Will cavitation occur?
2. If it is unavoidable, can a device still perform its function acceptably?

Bubble growth and collapse are at a nominal rate when caused by gas diffusion or gradual pressure change. However, they may be explosive if caused by vaporization effects. Collapse occurs "implosively" if pressure is increased rapidly (i.e., position change or vibration effects) with small gas content. Such collapses can damage nearby structure.

"Cavitation" implies certain ideas, e.g.:

1. Cavitation occurs only in liquid, not gas, flows.
2. It results from pressure reduction to pressures near vapor. If pressure is maintained for a sufficient time below a "critical" pressure, depending on liquid flow conditions, cavitation will occur. If not, it will not.
3. Cavitation involves the appearance and disappearance of cavities (or "bubbles") in a liquid. Since the bubbles are "empty," they can play no part in the phenomenon, which then depends entirely on liquid behavior. This model is not entirely valid for initial nucleation and final collapse.
4. It may occur for liquids in motion or at rest, since pressure oscillations may also be due to surface vibration or acoustic radiation.
5. It may occur in a body of liquid or along walls.
6. Cavitation erosion occurs at the point of bubble collapse rather than at inception. In some cases these points coincide.

Many excellent reviews of the very extensive literature exist [1–5], but there are only two comprehensive books in English devoted entirely to it [6, 7]. Several foreign language books also exist, at least one monograph in English [8], and several handbook chapters [9–11]. It is also treated in portions of several fluid flow texts [12–14].

Effects of Cavitation

Major effects, according to their assumed importance follow.

Performance Effects

A substantial quantity of cavitation ("degree" or extent) in flow passages will alter streamlines and thus machine performance. The effect is minimal for conditions near inception (nucleation), since the few bubbles then present will not affect overall flow. For extensive (well-developed) cavitation, a large portion of the low-pressure region may become vapor-filled. The velocity in the remainder of the passage is then increased, and the pressure further decreased, giving added evaporation. Distortion of the flow pattern and energy losses caused by the two-phase regime can then cause a sudden decrease in overall machine performance (head, efficiency, etc). Figure 1 shows the performance of a typical centrifugal pump for three flow coefficients as functions of inlet suppression pressure, and (Figure 2) typical impeller inlet passage flow patterns with cavitation.

The most common nondimensional parameter to correlate with cavitation performance is the "cavitation number," K (or cavitation sigma, σ[6–8, 12, 14].

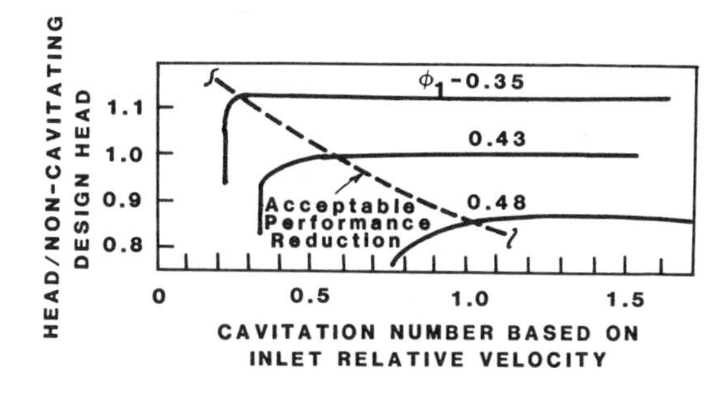

Figure 1. Effect of cavitation on performance of centrifugal pumps (from C. Blom, "Development of Hydraulic Design for the Grand Coulee Pumps," *Trans. ASME*, Vol. 72, (1950), p. 53).

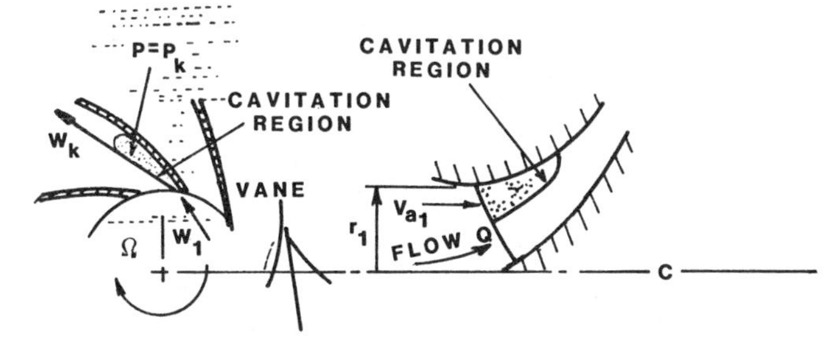

Figure 2. Sketch showing probable location of cavitation in a centrifugal pump [12].

Then

$$K = \sigma = p_{sv}/\rho V^2/2 = NPSH/V^2/2$$

where V = reference velocity
ρ = liquid density
p_{sv} = suppression pressure
NPSH = net positive suction head.

$$(p_{sv} = \rho \text{ NPSH}) = p - p_v$$

where p_v = vapor pressure

Another important parameter applied especially to pumps is "suction specific speed," S. The larger is S, the more probable is cavitation. In English units, for $S > 8{,}000$ in centrifugal pumps, [15] the possibility of cavitation is implied. S is analogous to "specific speed," N_s, except NPSH is substituted for total head rise. S reflects pump inlet conditions rather than the entire pump performance, assuring similarity for inlets. $S = N\,Q^{1/2}NPSH^{-3/4}$
where N = rotating speed (RPM)

Q = flowrate (GPM)

S and N_s are not nondimensional unless converted to consistent units, which can be done.

Scale Effects

General

Because of so-called "scale effects" [6–8, 16, 17], conventional similarity parameters often do not predict test results well. Reasons for discrepancies are not well understood. However, divergence

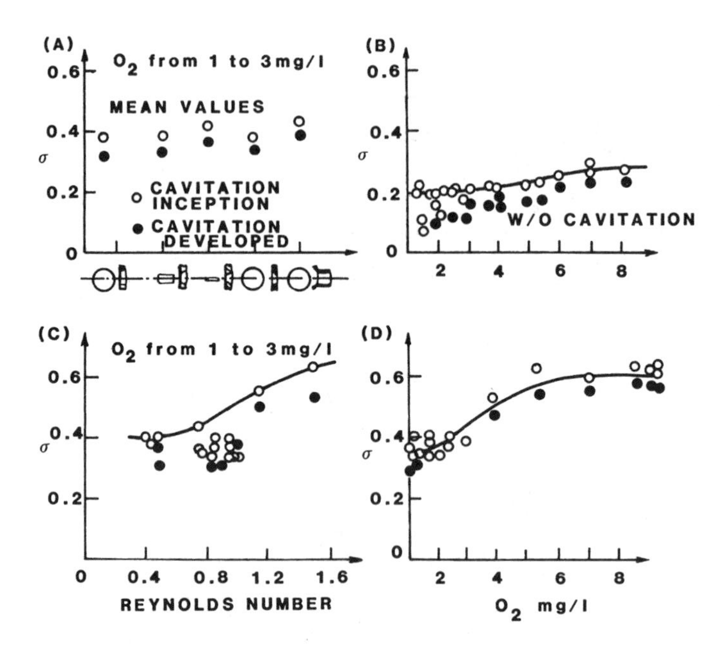

Figure 3. Influence on inception sigma of Reynolds number and oxygen content for various orifice shapes. Curve A shows various orifice shapes, and curves B, C, and D are for circular orifices. (Tests reported J. Duport, SOGREAH, Grenoble, France.)

from classical scaling laws which are laws pertinent to single-phase flow, may be due to the great complexity and lack of understanding concerning any multiphase flow, and cavitating flows in particular. Scale effects occur for velocity, size, p_{sv}, and fluid property and temperature changes. The latter two, termed "thermodynamic effects," are discussed later.

Figure 3 [7] shows effects on inception for well-developed cavitation numbers, σ, for the cases of water orifice flows, and as functions of Reynolds number, Re, and air content. The Re effect is from changes in velocity and diameter. Inception σ increases by $\sim 50\%$ for increase of Re by ~ 4 X, but σ for well-developed cavitation is little affected.

Figure 4 [7] shows velocity effects on inception σ, σ_{inc} for water or sodium in a venturi with various gases. For well-deareated water, σ_{inc} increases substantially for increased velocity, consistent with Figure 3 for orifices. For high gas content, it first decreases strongly, then increases for a continued velocity increase. Curves converge at high velocity. The gas content effects are expected, but not those for velocity. Tests with sodium gave similar results [7]. Results for centrifugal pumps are often similar. Optimum performance (maximum S) may occur at intermediate speeds, and S often then decreases for higher speeds, for any liquid.

Figure 5 shows data from a low specific speed pump [7] for mercury and water. Strong S increase (two fold) occurred for four fold RPM increase, with little difference between mercury and water, when plotted against RPM rather than Re. Thus Re was not a good correlating parameter. Data are available separating effects of velocity and diameter is not available.

Thermodynamic Effects

Substantial "thermodynamic" scale effects for both σ and damage exist [6, 7, 17–19] for changes in fluid properties or temperature. The primary mechanism is the large increase in vapor density with temperature for any fluid. Considering water as an example, for "cold" (room temperature)

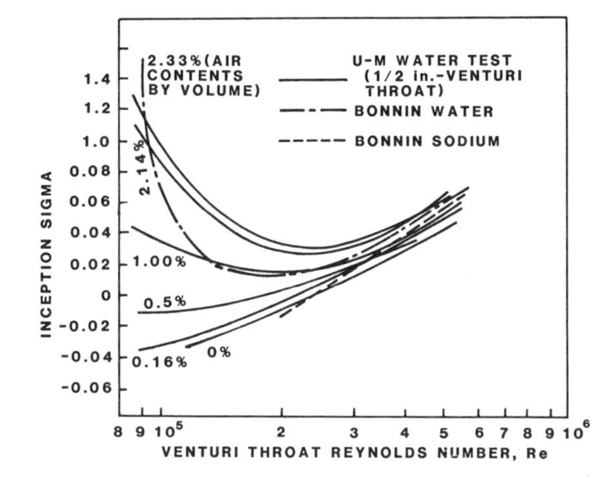

Figure 4. Venturi cavitation inception sigma, σ, vs. Reynolds number for water and molten sodium. (Water tests at University of Michigan. Sodium tests at Elec. de France. J. Bonnin.) Air content effect shown for water. Note the inception Sigma is defined as $\sigma = (p_t - p_v)/\rho V^2/2$, where p_t = throat pressure.

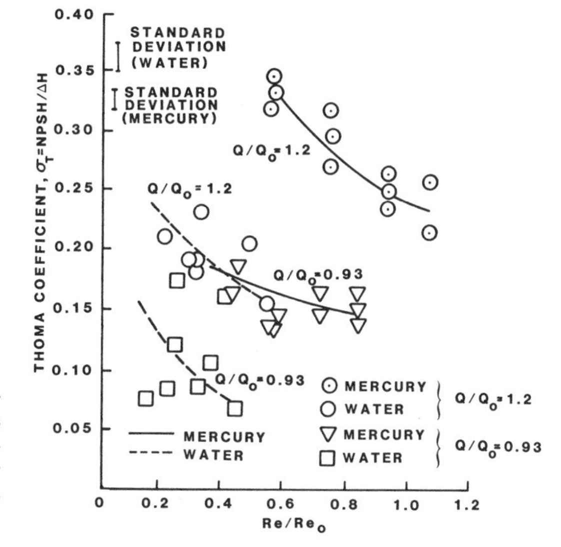

Figure 5. Thoma cavitation parameter vs. normalized Reynolds number for low specific speed centrifugal pump (Univ. Mich. tests) for mercury and water). Re_0 for pump design condition.

water, vapor density, and bubble thermal content are negligible. Bubble growth and collapse are then inertially dominated, as in the classical Rayleigh analysis [20], and heat transfer restraints are negligible. For "hot" water (>100°C, e.g.), vapor density is increased forty fold and the thermal content of the bubble is then increased by a similar ratio. Rapid growth or collapse is then restrained by heat transfer as well as inertial effects. Both are reduced compared to the Rayleigh model. Hot-water cavitation is thus very similar to subcooled boiling. Boiling is thus not damaging due to the "thermodynamic effects" and the lack of high pressures with boiling to drive collapse.

Similar effects occur comparing "cold" water with such fluids as cryogenics, petroleum products, chemical process fluids, freons, liquid metals, etc. Equivalent temperatures depend on physical properties, e.g., low for cryogenics, high for liquid metals, and intermediate for petroleum products, freons, etc. Cavitation effects on performance and erosion, as compared with "cold" water, are strongly reduced. Thus 200°C water may not be damaging at all. Experimental data [7] confirms this hypothesis. Very rough estimates can be obtained by using equivalent values of Stepanoff's B-factor [7, 18, 19].

EROSION DAMAGE EFFECTS

General Background

Cavitation damage has been important since the early 1900s, when first seen on British torpedo-boat propellers [6, 7]. The cause was high propeller speeds when turbines replaced reciprocating engine drives. The classical work of Rayleigh [20] showed that the spherical collapse of vapor cavities in "ideal" liquids caused "infinite" pressures and velocities at the site of collapse. Later analyses [6, 7] showed that the Rayleigh model is not entirely valid, especially concerning spherical symmetry near a wall, but it still shows correctly damage potential. The point of damage is thus that of collapse rather than inception.

In the past 30 years, studies have been made to study spherical bubble collapse in "real" liquids [21, 22], and nonspherical collapse [23, 24] near the wall, in pressure gradients [24], and with viscosity [24]. Effects in real fluids are shown in Figure 6 [7, 21]. Bubble shapes from high-speed movies, for collapse near wall [7, 25] are shown in Figure 7.

It is generally agreed that damage is due to both liquid "microjet" impact for nonsymmetrical collapse (Figure 7) and shock waves from "rebounding" bubbles [7, 22], which grow again after collapse due primarily to compressed internal gas and vapor. Such rebounds have been often

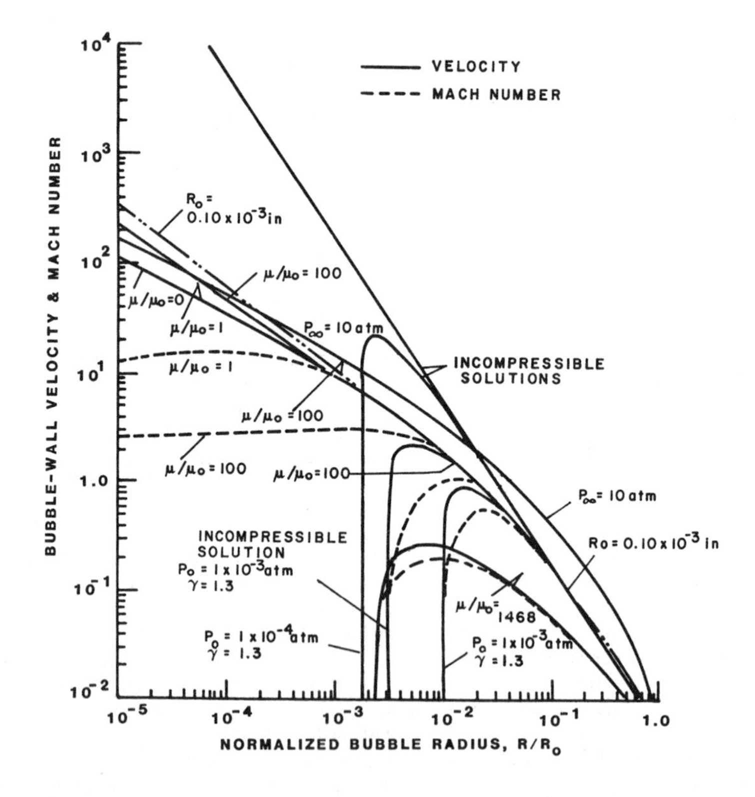

Figure 6. Spherical bubble collapse velocity and Mach number vs. normalized bubble radius. Compressibility, viscosity, and surface tension effects. 1 in. = 25.4 mm. Parameters unless otherwise marked are: viscosity $\mu/\mu_0 = 1$, surface tension $\sigma/\sigma = 1$, (σ_0, μ_p denote water properties), $P_\infty = 1$ atm, $P_0 = 0$, $R_0 = 50 \times 1^{-3}$ in.

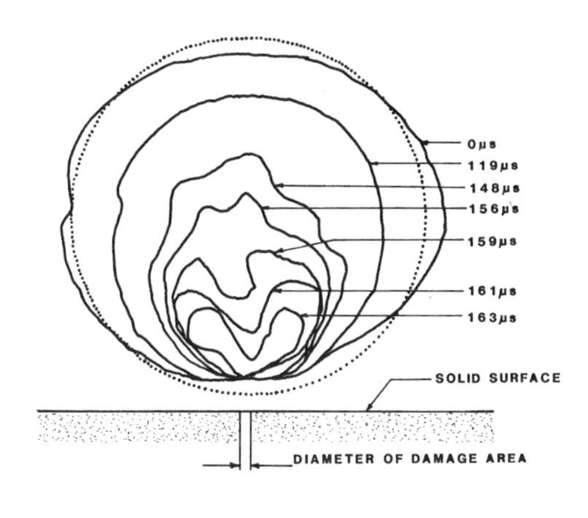

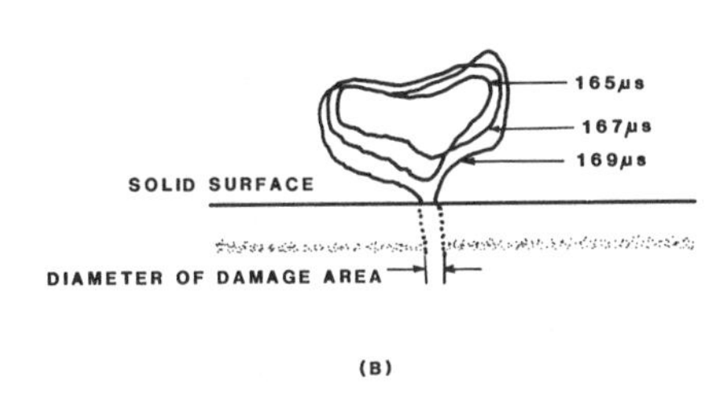

Figure 7. Venturi bubble collapse near wall of single (spark-generated) originally spherical bubble. High-speed cinematography, C. L. Kling [7, 25]. (A) initial portion bubble collapse (B) final portion bubble collapse showing microject.

observed in water tunnel tests [7, 25, e.g.]. This disagrees with the Rayleigh model [6, 7, 20], which predicts damage from shock waves from spherically-symmetric collapse. High-speed movies [6, 7, 25] show that spherical collapses do not occur near walls. This is confirmed by computer studies [7, 23, 24]. The potential intensity of such mechanisms can explain damage for most materials.

Material Resistance Parameters

All materials (even the hardest and strongest) can be damaged by cavitation. However, resilient materials such as elastomers may have good resistance, since their surface resilience may prevent microjet formation [7]. Still in most cases, harder materials are more resistant. This effect may be reduced if corrosion occurs. The combination of corrosion with "mechanical" cavitation can produce much more damage than each acting alone.

The best, most available, and easiest to apply index for cavitation resistance is then hardness (Brinell, or other), and it has been so used traditionally [6, 7, 9, 10]. It appears, however [7, 9, 10], that better correlations are often found with more complex parameters. One example is ultimate resilience (UR), for which theoretical backing and dimensional consistency exist, since UR is failure energy per volume for brittle fracture which is typical of cavitation failure.

$$UR \cong TS^2/2E$$

where TS = tensile strength
E = elastic modulus

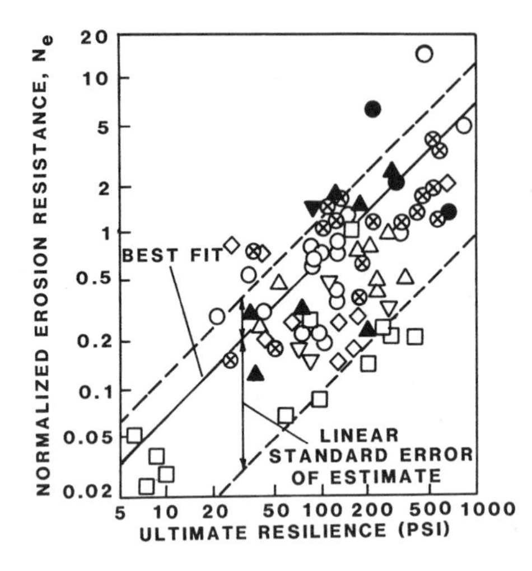

Figure 8. Normalized erosion resistance vs. ultimate resilience (UR). Data from previous Hammitt [7] and Heymann [7, 26] correlations. Cavitation and liquid impact data: $UR = TS^2/2E$.

Then

$$MDPR^{-1} \propto UR^n$$

where MDPR = mean depth of penetration rate or volume loss rate/exposed area

n must be determined experimentally, but $n \cong 1$ is expected. Data show that $n \cong 1$ is often the best-fit exponent [7, 9, 10, 26]. Since BHN $\propto$ TS for metals, $n = 1$ corresponds to a hardness exponent of 2, whereas 1.89 [9, 10, 26] has been reported. The discrepancy is not significant. Figure 8 shows data for a large group of materials, combining cavitation and liquid impact data. Standard deviation is ~three fold typical of fluid erosion data. Since liquid impact and cavitation are very similar, data-fits for metals may be applied to both.

The discrepancy from best-fit curves for some materials is approximately ten fold, due to micro-characteristics not reflected by conventional mechanical properties. Stellite 6-B, e.g., is much more resistant than expected from either BHN or UR. Micro-structures are important, and conventional mechanical properties, due partially to their relatively low loading rates, are not fully pertinent, since damage mechanism occurs in a few μs. Little confidence can be placed in damage rate correlations for untested materials. The large data scatter is not surprising, since the range of resistances, $MDPR^{-1}$, is 10^3–10^4 between the most and least resistant.

VIBRATIONS AND CAVITATION

Machinery Vibrations

Major machinery vibrations, sometimes damaging to structure, can occur. These involve macro- rather than the micro-loading for usual erosion. They are due to large and varying forces on the structure due to the collapse of large cavities, or conceivably smaller bubbles in phase. Resultant large noise emission can also prevent satisfactory machine operation. Little published data exists on these effects.

Vibration-Induced Cavitation and Test Devices

In some cases (diesel engine wet cylinder liners, e.g.). vibration-induced cavitation is the important damage mechanism [7, 27, 28].

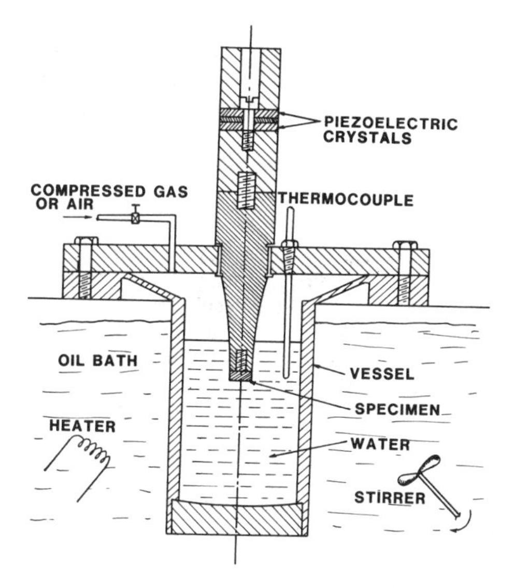

Figure 9. University of Michigan vibratory cavitation erosion test facility, 20 kHz, 14.3 mm (0.562 in.) specimen diameter. Amplitude to 3 miles (75 μm).

Vibratory cavitation also provides an important test device to measure affects of material properties and fluid conditions [6, 7, 29, 30]. This essentially zero-flow device is the only cavitation-erosion test method yet standardized [29, 30] by ASTM. It consists of a "vibratory horn" to which damage specimens (buttons) are attached (Figure 9). The Michigan unit (Figure 9), can be used over large temperature and pressure ranges, though most are for room temperature and pressure only.

The "vibratory" test and others such as "rotating disc," Venturi, etc., provide highly accelerated erosion so that corrosion is suppressed compared to "mechanical" cavitation. Thus results cannot be well applied for prediction of field damage rates, which are usually much smaller. This is a major difficulty in present state-of-art technology.

CAVITATION NOISE

Noise inevitably accompanies cavitation and may be important in various naval applications, such as in detection avoidance in submarines. It also limits transmitted power of sonar transducers. Cavitation produces essentially "white" noise out to very high frequencies (order of 10 MHz), since the damaging portion of bubble collapse often require only ~ 1, μ_s [7, 23–25]. Noise studies are necessarily limited by instrumentation, so that the maximum frequency is unknown. Most noise is from collapse rather inception. Thus little noise occurs for boiling where violent collapse is absent.

Hydrophones can be used to signal inception and also to correlate noise with damage. A low-frequency filter is useful to attenuate ordinary flow and machine noise, which do not usually have the very high frequency components of cavitation. A cut-off frequency of ~ 40 kHz would be suitable.

CAUSES AND PREDICTION OF CAVITATION AND EROSION

Cavitation Flow-Regime Prediction (Performance Effects)

For stationary devices, inception and development (i.e., degree or condition) are best correlated by the cavitation number, K (or sigma, σ) which is the ratio of total suppression pressure, P_{sv} and kinetic pressure, $\rho V^2/2$. Alternatively, the ratio between net positive suction head, NPSH and kinetic head, $V^2/2$. This is usually, but not always, referred to as upstream values or some point of cavitation within the system. Increased σ means less cavitation.

For rotating machines other parameters are used, e.g., suction specific speed, S, (already discussed) and Thoma parameter, σ_T, i.e., NPSH/ΔH, where ΔH is total pump head. The Thoma parameter (1930s) predates S (1940s). It is less successful in predicting cavitation since it depends on total performance rather than on that of the low-pressure region only. It is thus affected by more variables, such as conventional specific speed, N_s.

Performance "Scale Effects" and Modeling

Flow modeling, using similarity parameters already discussed, may be very imprecise due to substantial "scale effects," involving "scale" (size), velocity, pressure, and temperature ("thermodynamic effects"). The mechanisms are only partially understood and the details are too complex to present here. The mechanisms can best be evaluated from papers listed, elsewhere [6, 7, eg]. Figures 3–5 are typical. If not for "scale effects," inception σ and S would be constant for a given machine. In fact, they usually vary substantially (Figures 3–5).

Impurities and Gas Content

The greatest effects are upon inception, σ, through pressure thresholds for nucleation. For zero particulate and gas content and perfect wetting, cavitation and boiling would not usually occur, since the tensile strength of pure liquids would prevent nucleation. That they do occur is thus proof of a "stress-raising" mechanism such that small under-pressures do cause cavitation. Nucleation is assumed to be due unwetted acute angle micro-crevices in which gas cavities do exist. Nucleation occurs in the liquid from micro-particles with such nonwetted micro-crevices. This is the "Harvey mechanism" proposed in the 1940s [6, 7, 31].

The Harvey mechanism (or some alternative) is needed to explain the stable source of "nuclei", which is in fact observed.

Small bubbles would quickly dissolve, since the pressure within is greater than for surrounding liquid due to surface tension. Larger bubbles would be removed by buoyancy or centrifugal effects. Other mechanisms such as a particle shell around "nuclei" [7], or combined effects of solution and vaporization [32] have been suggested. However, is still assumed that the Harvey mechanism is of primary importance in most cases. Wetted particles have no effect, and nonwetted particles have an effect similar to gas bubbles.

Dissolved gas has little effect, since solution proceeds slowly. Thus only entrained gas is important, but measurement thereof is difficult, and no standard method exists. Presumably, entrained gas content increases with total gas content. Laser-light scattering seems the best approach for its measurement [7, 33] but is complex. Entrained gas is typically only a small part of the total, so its measurement alone may be misleading. Total gas can be measured easily (Van Slyke apparatus, e.g. [6, 7]. Suitable automated versions of this and other types are available.

Cavitation Damage Prediction: State-of-the-Art

General Background

Cavitation damage prediction until recently was seldom attempted. Rather, cavitation was to be avoided, or if that were not possible, performance of similar machines applied to new designs to assure suitable operation. Damage prediction is then essentially empirical, but some attempts at prediction are explained later. Damage can often be repaired quite easily (welding in new material, for example); however, in new applications, such as nuclear power-plants, rocket pumps, etc., past procedures may not be suitable. Thus more precise information is now needed, and improved prediction capability is essential.

"Zero" damage is impossible to guarantee with present technology, unless p_{sv} is set prohibitively high. Though NPSH vs head curves can be measured accurately, the NPSH value necessary to eliminate bubble collapse entirely is usually not known [7]. It may be 10 times (Figure 10) that

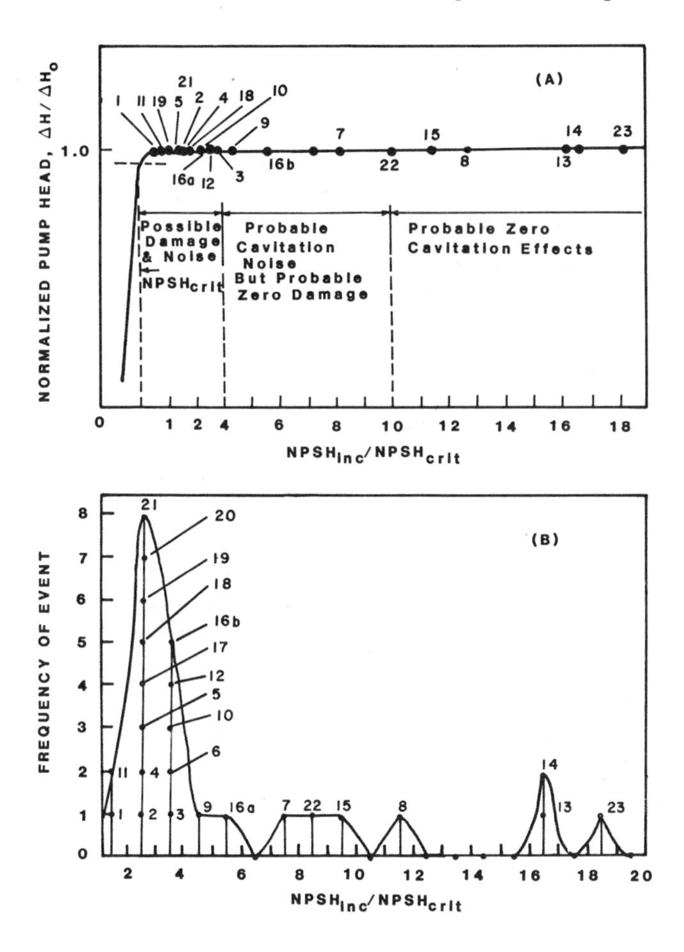

Figure 10. Normalized pump head vs. NPSH for actual inception (acoustic or visual) compared to head fall-off and frequency of events (A) normalized pump head vs. NPSH. (note ΔH_0 = design head; $NPSH_{inc}$ = actual inception, $NPSH_{crit}$ = head fall-off point); (B) frequency of event vs. ratio of inception to critical NPSH.

for the usual inception point (three % head drop-off). In most cases, needed NPSH for "zero cavitation" is two to four times "inception" value, and much greater in some cases [7, e.g.]. Accurate prediction of ratio is not yet possible.

Damage Prediction Methods

Several quite imprecise methods exist. These are discussed in order of probable utility.

Characteristic erosion curve and incubation period. Cavitation (or liquid impact) erosion vs. time often follows a "characteristic" (S-shaped) curve [7, 9, 10] (Figure 11). An "incubation period," IP, (damage rate very small) is followed by a zone of increasing rate, then a maximum rate period ($MDPR_{max}$). This may persist for some time, depending on other parameters. The rate then usually decreases, sometimes reaching an eventual "steady-state" value. However, second or third maxima

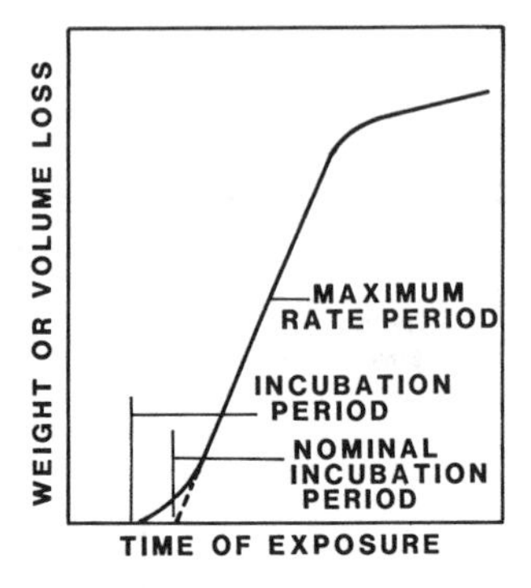

Figure 11. Characteristic cavitation of liquid impact "S-shaped" erosion curve.

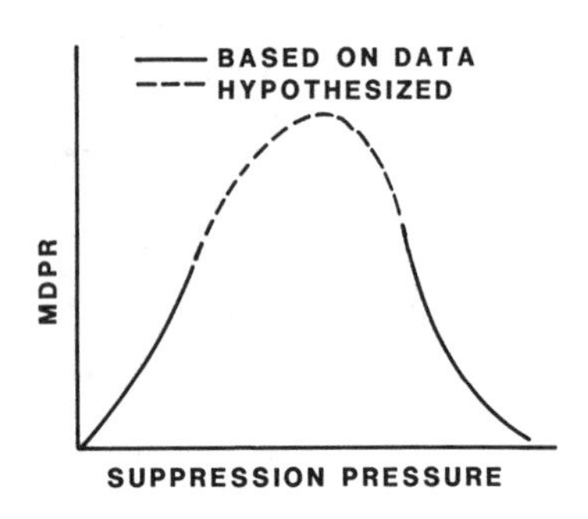

Figure 12. Cavitation erosion rate (volume or weight loss) vs. suppression pressure, p_{sv} ($= p - p_v$) or net positive suction head, NPSH, Note: erosion rate is MDPR (mean depth of penetration rate, or volume loss rate per exposed rate).

sometimes occur [7, 9, 10], if the test is continued long enough. Use of the "characteristic curve" for prediction is a rough approximation at best, unless it can be verified experimentally for a given case. An approximate relation, $MDPR_{max}^{-1} \propto IP^n$ exists [7, 9, 10], where n ranges from ~ 0.7 to ~ 1.2 [7]. It depends on test parameters, type of device, material, etc. Recent tests (vibratory and venturi) [34] for five materials (carbon steel, cast iron, aluminum alloys, and SS-316), showed $n = 0.94$ (± 0.1). If the unity exponent were approximately valid in general, the method's utility would be increased.

Tests to determine IP, based on the appearance of significant pitting or other erosion, may be feasible in prototype machines, using soft inserts (AL-1100-0, e.g.) in the expected damage region. Lab tests, such as vibratory, could compare IP and the remainder of erosion curve, including $MDPR_{max}$, for soft insert and prototype material. Assuming that the material-resistance ratio between lab and prototype units is reasonably valid, IP and $MDPR_{max}^{-1}$ for the prototype and the time of occurence can be predicted. The "characteristic curve" can be measured in the lab and assumed valid for the prototype.

Scale and "thermodynamic effects." Several important "scale effects" determine scaling between model and prototype for both erosion and performance. Probably the most important concerns velocity, V. Many tests [6, 7, 9, 10] show a strong damage increase with increased V. Assuming $MDPR \propto V^n$, usually $n \sim 6$ [6, 7, 9, 10]. However, n obviously depends on the relation between p_{sv} in collapse region and V. This relation then depends on geometry, and other test parameters. In fact P_{sv} in the collapse region for fully developed cavitation may not increase with V. Damage increase with V would then be minimal. The range of n reported in literature is usually $0 < n < 10$, but even negative values have been reported [7]. Thus the damage-exponent model is in general not valid. Nevertheless increased velocity can produce catastrophic damage, where little existed before. Thus damage tests at prototype velocity are vital.

Suppression pressure, p_{sv} effects. A maximum-damage P_{sv} exists [7, 10] (Figure 12) for all machines, including vibratory, between low values (low collapse pressure, as in boiling) and high (cavitation completely suppressed). Conventional machines operate on the right side of the curve (Figure 12). Thus increased p_{sv} usually reduces damage, but not always. It always reduces the extent of cavitation. Maximum damage often occurs near inception and then decreases for reduced NPSH. Vibratory tests are typically on the left-hand side of the curve. Damage increases about as p_{sv}^2 (7, 10).

Diameter (size) effects. Little information is available, though it appears [7, 16] that diameter-damage exponents are in the range of 3–4 for total volume loss, giving an MDPR exponent of 1–2.

Thermodynamic (temperature) effects [6, 7, 10, 18, 19]. A maximum damage temperature exists for any fluid in a given test (Figure 13). Damage decreases markedly for higher temperatures (thermodynamic effects), and less so for lower. The high temperature effects are best correlated by Stepanoff's B-factor. Figure 14 shows damage predictions with modified B for various fluids. Figure 13 shows that damage in water may not be important for T $\gtrsim$ 200 C. Of course, temperature also affects both material properties and corrosion rates. A second increase at even higher temperatures occurred in tests for both water and sodium, due primarily to increased corrosion effects.

Gas content and impurity effects. Entrained gas facilitates nucleation, increasing bubble size and number. Thus more gas (entrained or total) will increase erosion through this effect. However, increased gas in bubbles weakens collapse (Figure 6) [7, 21] causing a "cushioning effect." Air injection to the collapsed region has been thus used [6, 7, 10, 16, 35] to reduce damage. While information is scanty, Figure 15 shows the presumed overall effects [7, 10, 16]. A gas content of

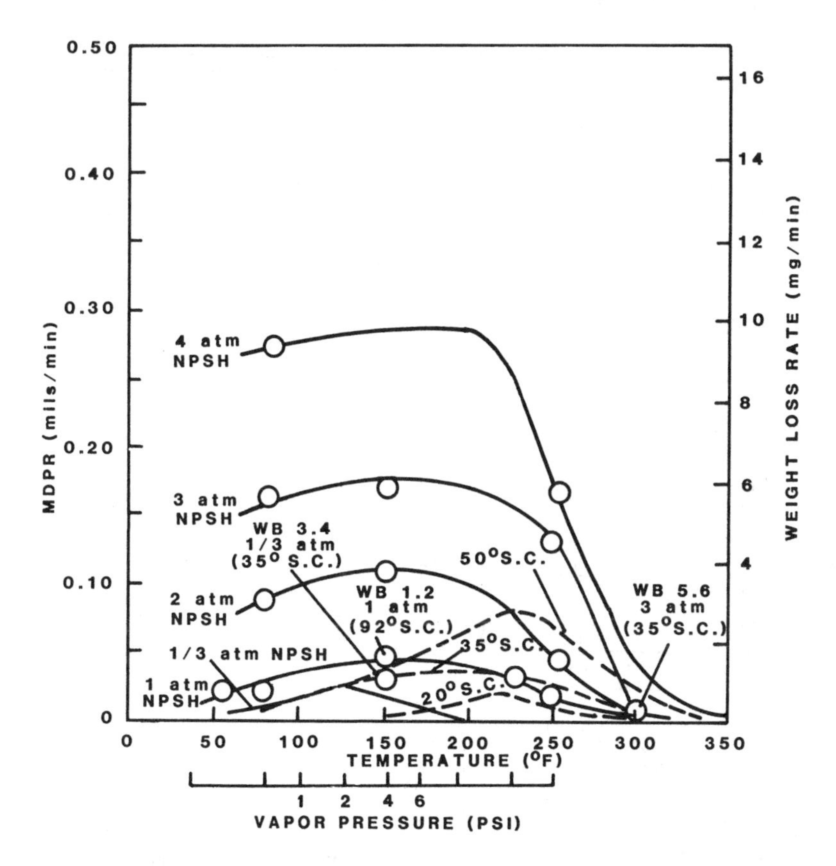

Figure 13. Temperature (thermodynamic) effect on erosion rate in water vibratory test [7, 10]. Maximum MDPR (mean depth of penetration rate) and weight loss rate vs. temperature and vapor pressure for bearing brass (SAE-660) at various static NPSH values (1, 2, 3, and 4 bar). Curves of constant subcooling (S.C.) also shown. Note: 1 mil = 25.4 μm; 14.7 psi = 1 bar.

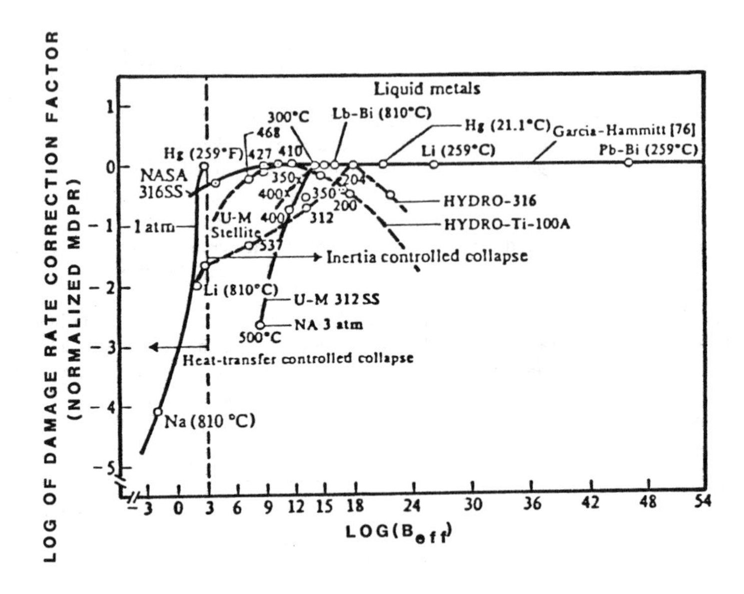

Figure 14. Effect of modified Stepanoff thermodynamic parameter, B_{eff}, on cavitation erosion rate for various liquid metals and test materials, showing strong damage decrease at high temperature.

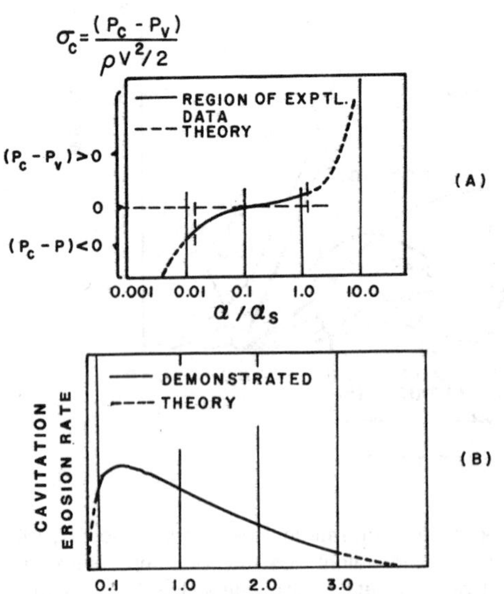

Figure 15. Air (or gas) content effects upon cavitation inception sigma and erosion rate—measured and hypothetical trends. P_c = pressure at point of cavitation inception, α_s = saturated gas content (1 atm), α = actual gas content.

near zero should reduce damage by suppressing inception through increased liquid tensile strength. Moderate gas content reduces damage through the "cushioning" effect (more so for high gas). Tests, verifying the high content effect especially, do exist [7, 10, 35].

Gaseous or particulate additives could increase corrosion and have other effects, and entrained particulates could strongly increase damage. As an extreme case, cavitating slurries can be very damaging as compared to the liquid alone. Examples are dredge pump sand-water slurries and coal or ore-bearing pipelines. Such a pipeline itself and the associated pumps, valves, or fittings sometimes present major erosion problems. It may then be required to design for "zero" cavitation. Little pertinent data exist.

A milder case for hydroelectric plant component is the particulate erosion in sand-bearing rivers, e.g., the Yellow River in China, where the particulate content is high [36]. Recent vibratory test of particulate and chemical additives on 1018 carbon steel [36] showed a damage increase $\sim 50\%$ over tap water.

Predictions from noise measurements. Overall noise (or that in a set frequency band) has been used successfully (1, 7, 10, 37–39) to predict and correlate damage for both pumps and Venturis. For pumps, maximum damage and noise often occur near inception (as defined by first measurable performance effects). Application of this method seems limited to actual geometry tested, since given overall noise (or noise in a fixed frequency band) could result either from numerous collapses of nondamaging intensity, or from a few much stronger collapses of damaging intensity.

An approach of more general applicability, tried in several labs including the author's with success [10, 40, 41, 42] is to count and quantify individual collapses, using suitable micro-transducers.

An approximately linear relation (Figure 16 and 17) between the measured acoustic collapse energy and the venturi erosion rate energy was found for various materials. The ratio of these was called the cavitation erosion efficiency, η_{cav} and is typically of order 10^{-10} [40, 41, e.g.).

Since the damaging portion of bubble collapse typically lasts only few μs, and the loaded area is often few μm, a microprobe of adequate smallness and high response rate is critical. State-of-art

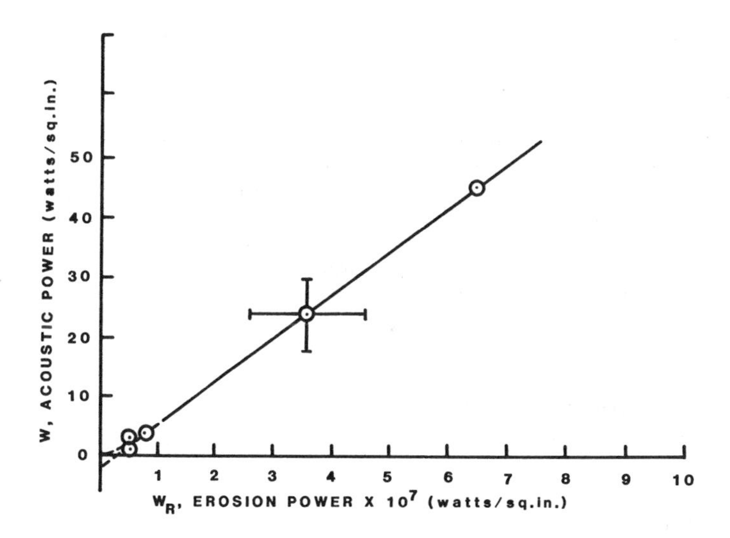

Figure 16. Bubble collapse acoustic power vs. measured erosion power for soft (1100-0) aluminum in Venturi [7, 10, 40]. "Erosion power" is product of material failure energy rate per volume ("ultimate resilience," UR) and eroded volume/exposed area, ("mean depth of penetration rate," MDPR).

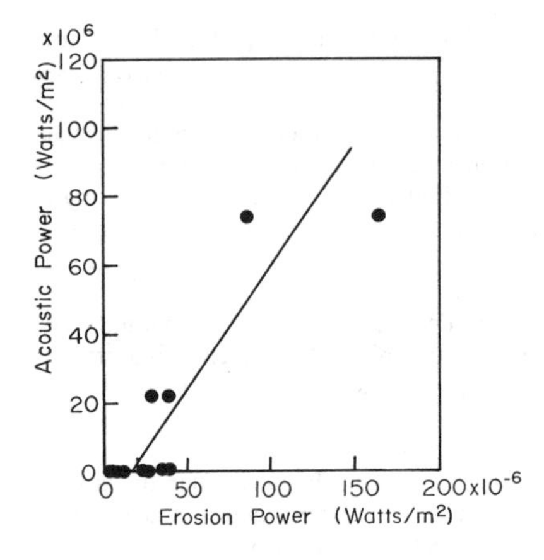

Figure 17. Relationship between acoustic power and erosion power for 1100-0 aluminum.

technology is 5 mm and 0.1 MHz. A 10 MHz and even smaller probe is needed but is not presently available. The lack of adequate probes is partially responsible for low values of η_{cav}.

OCCURRENCE OF CAVITATION

General

Cavitation can occur in any liquid-handling machine when p_{sv} is low and velocity sufficiently high. There is then no danger of cavitation in many cases. Important cases, rotating and stationary, where cavitation is a problem, are now described.

Rotating Devices

Centrifugal and Axial-Flow Pumps

Pumps liable to cavitation problems are those in large powerplants (fossil or nuclear), liquid-propellant rocket pumps, aircraft fuel pumps, and those driving jet-propelled ships, for example. In some cases, liquids other than water are involved. Nuclear applications are molten sodium pumps (LMFBR) and rocket pumps often use cryogenics.

"Cavitating inducers" (axial-flow) are often used to moderate effects for the main stage. These may involve "super cavitation," where the collapse region is downstream of the blades. Such designs are also used for high-speed propellors.

Hydraulic Turbines

Cavitation is a major limitation for most hydraulic turbines, affecting siting of the turbine and rotating speed to obtain adequate σ. Thus larger and more expensive units than otherwise required may be needed.

Marine Propellors

Design is limited by need to avoid prohibitive cavitation from both performance and erosion viewpoints. Corrosion, combined with "mechanical" cavitation, is often important. Special materials (relatively inexpensive) have been developed (propellor bronze, for example) and used. Cavitation at increased depth (pressure) such as for torpedoes and submarines, can cause increased damage.

Stationary Flow Devices

1. General—Erosion and performance effects occur, in many components, as discussed below. Effects can generally be again correlated by σ.
2. Pipe fittings (elbows, tees, etc.) and relatively sharp pipe or tubing bends.
3. Flow measuring devices as orifices, nozzles, and venturis. Cavitation is usually of the separated-flow type, so that while noise and performance effects exist, damage is usually limited, since bubbles collapse away from the surface. Figure 3 shows typical performance effects.
4. Valves—These are similar to flow-measuring devices with regard to cavitation, but damage is often important, since the collapse point depends on valve geometry.
5. Hydrofoils—Performance effects are important, and damage may be also. Correlation is with σ, velocity, and depth of subsequence.

SUMMARY

Cavitation as a whole has been examined, and special emphasis placed on damage effects. Methods of prediction, especially for damage, have been examined, including correlation with acoustic output. Effects on structure and bearings must also be considered. Performance can be correlated by σ.

REFERENCES

1. Arndt, R. E. A., "Cavitation in Fluid Machinery and Hydraulic Structures," *Ann. Rev. Fluid Mech.*, Vol. 13 (1981), pp. 273–328, Annual Reviews Inc. See also L. Euler, "More Complete Theory of Machines Driven by Hydraulic Reaction," (in French), *Historie de l'Academie Royale des Sciences et Belles Lettres, Classe de Philosophie Experimentale*, 227–295, *Mem 10*, 1754, Berlin, 1765.
2. Acosta, A. J., "Cavitation and Fluid Machinery," *Cavitation* (1974), pp. 383–396, *Proc. Conf. Heriot-Watt Univ.*, Edinburgh, Scotland, Sept. 1974, Inst. Mech. Engrs., London.
3. Acosta, A. J., and Parkin, B. R., "Cavitation Inception—A Selective Review," *J. Ship Res.*, Vol. 19 (1975), pp. 193–205.
4. Thiruvengadam, A., *Handbook of Cavitation Erosion*, Tech. Rep. 7301–1, Hydronautics, Inc., Laurel,Md., 1974.
5. Robertson, J. M., and Wislicenus, G. F., (Eds.), *Cavitation State of Knowledge*, ASME, N.Y., 1969.
6. Knapp, R. T., Daily, J. W., and Hammitt, F. G., *Cavitation*, 1970, McGraw-Hill; N.Y.
7. Hammitt, F. G., *Cavitation and Multiphase Flow Phenomena*, Adv. Book Series, McGraw-Hill, 1980.
8. Pearsall, I. S., *Cavitation*, Mills & Boone, London, 1972.
9. Hammitt, F. G., and Heymann, F. J., "Liquid-Erosion Failures," *Metals Handbook*, Vol. 10, 8th ed., pp. 160–167, Am. Soc. Metals, Metals Park, Ohio, 1975.
10. Hammitt, F. G., "Cavitation and Liquid Impact Erosion," *Wear Control Handbook*, M. B. Peterson and W. O. Winer (Eds.), ASME, 1980, pp. 161–230.
11. Eisenberg, P., "Mechanics of Cavitation," *Handbook of Fluid Dynamics*, V. L. Streeter (Ed.) N.Y., McGraw-Hill, 1961, pp. 12.2–12.24.
12. Sabersky, R. H., Acosta, A. J., and Hauptmann, E. G. *Fluid Flow*, 1971, MacMillan Co.
13. Streeter, V. L., *Fluid Mechanics*, McGraw-Hill, 1951 p. 71.
14. Wislicenus, G. F., *Fluid Mechanics of Turbomachinery*, Dover Pub. Co., N.Y., 1965.
15. Marks, L. S., (Ed.) *Mechanical Engineers' Handbook*, McGraw-Hill, N.Y.
16. Hammitt, F. G., "Effects of Gas Content upon Cavitation Inception, Performance and Damage," *J. Hyd. Res.* (LAHR), Vol. 10, No. 3 (1972), pp. 259–290.
17. Bonnin, J., et al., "Survey of Present Knowledge on Cavitation in Liquids Other than Cold Water (Thermodynamic Effect)," *J. Hyd. Res.* (*LAHR*), Vol. 19, No. 4 (1981), pp. 277–305.

18. Stahl, H. A., and Stepanoff, A. J., "Thermodynamic Aspects of Cavitating Centrifugal Pumps," *Trans. ASME*, Vol. 78 (1956), pp. 169–193.
19. Stepanoff, A. J., "Cavitation Properties of Liquids," *Trans. ASME, J. Engr. Power*, Vol. 83 (1961), pp. 195–200.
20. Rayleigh Lord, (John William Strutt), "On the Pressure Developed in a Liquid During the Collapse of a Spherical Cavity," *Phil. Mag.*, Vol. 34 (Aug. 1917), pp. 94–98.
21. Ivany, R. D., and Hammitt, F. G., "Cavitation Bubble Collapse in Viscous Compressible Liquids,"*Trans. ASME, J. Basic Engr.*, Vol. 87, D, (1965), pp. 977–985.
22. Hickling, R., and Plesset, M. S., "Collapse and Rebound of a Spherical Bubble in Water," *Phys. Fluids*, Vol. 7 (1964), pp. 7–14.
23. Plesset, M. S., and Chapman, R. B., "Collapse of an Initially Spherical Cavity in Neighborhood of a Solid Boundary," *J. Fluid Mech.*, Vol. 47, No. 2, (May 1971), p.283.
24. Mitchell, T. M., and Hammitt, F. G., "Asymmetric Cavitation Bubble Collapse," *Trans. ASME, J. Fluids Engr.*, Vol. 95, No. 1 (March 1973), pp. 29–37.
25. Kling, C. L., and Hammitt, F. G., "A Photographic Study of Spark Induced Cavitation Bubble Collapse," *Trans. ASME, J. Basic Engr.*, Vol. 94, No. 4 (1972), pp. 825–833.
26. Heymann, F. J., "Erosion by Cavitation, Liquid Impingement, and Solid Impingement," Westinghouse Electric Engr. Rept. E-1460, Mar. 15, 1968; see also F. J. Heymann, "Toward Quantitative Prediction of Liquid Impact Erosion," *ASTM STP*-474 (1969), pp. 212–248.
27. Zhou, Y. K., He, J. G., and Hammitt, F. G., "Cavitation Erosion of Diesel Engine Wet Cylinder Liners," *Wear*, Vol. 76, No. 3 (March 1982), pp. 321–328; also same authors, "Cavitation Erosion of Cast Iron Diesel Engine Liners," *Wear*, Vol. 76, No. 3 (1982), pp. 329–335.
28. Speller, F. N., and LaQue, F. L., "Water Side Deterioration of Diesel Engine Cylinder Liners," *Corrosion*, Vol. 6 (July 1950) pp. 209–215.
29. Hammitt, F. G., et al., "ASTM Round-Robin Test with Vibratory Cavitation and Liquid Impact Facilities of 6061-T-6511 Aluminum Alloy, 316 Stainless Steel, and Commercially Pure Nickel," *ASTM Mat. Res. and Stds*, Vol. 10, No. 10 (1970), pp. 16–36.
30. *ASTM Standard Method*, "Standard Method of Vibratory Cavitation Erosion Test," ANSI/ ASTM G32–77, 1977.
31. Harvey, E. N., McElroy, W. D., and Whitely, A. H., "On Cavitation Formation in Water," *J. Appl. Phys.*, Vol. 18, No. 2 (1947), pp. 162—172.
32. Cha, Y. S., "On Bubble Cavitation and Dissolution," *Intl. J. Heat and Mass Transfer*" (1983); also C. S.Cha, "On Equilibrium of Cavitation Nuclei in Liquid-Gas Solutions," *Trans. ASME, J. Fluids Engr.*, Vol. 103 (1981), pp. 425–430.
33. Keller, A., "The Influence of Cavitation Nucleus Spectrum on Inception, Investigated with a Scattered Light Counting Method," *Trans. ASME, J. Basic Engr.*, Vol. 94, No. 4 (1972), pp. 917–925; also A. Keller, F. G. Hammitt and E. Yilmaz, "Comparative Measurements by Scattered Light and Coulter-Counter Method for Cavitation Nuclei Spectra," *1974 ASME Cavitation and Polyphase Flow Forum*, 16–18; also (same authors) *J. Acoust. Soc. Amer.* (Feb. 1976), pp. 324–333.
34. He, J. G., and Hammitt, F. G., "Velocity Exponent and Sigma for Venturi Cavitation Erosion," *Wear*, Vol. 80, No. 1 (1982), pp. 43–58; also F. G. Hammitt, J. G. He, et al. "Cavitation Erosion of Ferrous and Aluminum Alloys in Vibratory and Venturi Facilities," *1981 Cavitation and Polyphase Flow Forum, ASME*, June 1981, pp. 24–27.
35. Rasmussen, R. E. H., "Some Experiments on Cavitation Erosion in Water Mixed with Air," *Proc. 1955 NPL Symp. on Cavitation in Hydrodynamics*, Paper 20, H. M. Stationary Office, London, 1956.
36. Zhou, Y. K., and Hammitt, F. G., "Vibratory Cavitation Erosion of Aqueous Solutions," *Wear*, Vol. 87, No. 3 (1983), pp. 163–171.
37. Lush, P. A., and Hutton, S. P., "The Relation Between Cavitation Intensity and Noise in a Venturi-type Section," *Proc. Intl. Conf. on Pump and Turbine Design*, Glasgow, Sept. 1976, I. Mech. Engr., London.
38. Varga, J. J., and Sebestyen, Gy., "Determination of Hydrodynamic Cavitation Intensity by Noise Measurement," *Proc. 2nd Intl. Conf., JSME Fluid Machinery and Fluidics*, Sept. 1972, 285–292, Tokyo; also J. J. Varga, et al., *La Houille Blanche*, No. 2, 1969, pp. 137–149.

39. Pearsall, I. S., and McNulty, P. J., "Comparisons of Cavitation Noise with Erosion," *1968 ASME Cavitation Forum* (June 1968), pp. 6–7.
40. De, M. K., and Hammitt, F. G., "New Method for Monitoring and Correlating Cavitation Noise to Erosion Capability," *Trans. ASME, J. Fluids Engr.*, Dec. 1983; also same authors, "Instrument System for Monitoring Cavitation Noise," *J. Phys. E: Sci. Instruments* (U.K.), *15*, 1982, pp. 741–744.
41. Hattori, S., "An Application of Bubble Collapse Pulse Height Spectra to Venturi Cavitation Erosion of 1100-0 Aluminum," DRDA Report No. UMICH 014571-69-I, July 1983, submitted *ASME J. Fluids Engr.*
42. Wang, Zhou, and Hammitt, ASME Cavitation Forum, February 1984, p. 18.

CHAPTER 39

SIZING OF CENTRIFUGAL PUMPS AND PIPING

Robert Kern

Hoffmann-La Roche, Inc.
Nutley, New Jersey USA

CONTENTS

PUMP PIPING, 1138
Suction, 1138
Discharge, 1139

DISCHARGE CHARACTERISTICS OF PUMPS, 1139
Performance Curves, 1139
Viscous Flow, 1142
Two-Phase Flow, 1143

TOTAL HEAD CALCULATIONS, 1144

DESIGN EXAMPLE, 1145
Suction Line, 1146
Discharge Line, 1147
Total Head, 1148

ECONOMY OF PUMP PIPING, 1150

NOTATION, 1150

REFERENCES, 1151

PUMP PIPING

Suction

Centrifugal pumps require vapor-free liquid flow in the suction line at the entrance to the impeller vanes to operate satisfactorily. A centrifugal pump cannot pump vapor-liquid mixtures.

When pumping saturated liquid, the minimum pressure in the suction line and the pump impeller should not drop below the vapor pressure of the liquid at the flowing temperature. System design and graphic piping design should satisfy this essential principle.

For vapor-free flow in the suction line pump manufacturers require a positive suction pressure called net positive suction head, $NPSH_r$.

Available net positive suction head, $NPSH_a$, is the potential energy of an installation, expressed in feet of liquid head, and can be calculated from:

$$(NPSH_a)(\rho)/144 = P$$

where ρ = density of liquid at flowing temperature, lb/ft^3

P = pressure above vapor pressure at the centerline of the suction flange, psi

If we let $\rho = S\rho_w$, where ρ_w = density of water at 60°F, 62.37 lb/ft^3, and S = specific gravity of liquid at flowing temperature, then:

$$NPSH_a = 2.31P/S$$

For reliable operation it is essential that the available suction pressure should exceed the manufacturers requirements:

$$NPSA_a > NPSH_r$$

Size of the suction pipe is usually one or two sizes larger than that of the pump's suction nozzle. A suction pipe smaller than the pump's suction nozzle should never be used. Total suction line losses can be estimated between 1 to 3 ft of liquid head. Liquid velocities can be in the range of 1 to 5 ft/s; for pumping viscous liquids, 0.5 to 3 ft/s.

Discharge

Economy of both capital outlay and operating cost is the essential requirement when designing the discharge piping for a pump. This contrasts with the need for reliable operation when sizing the suction piping.

If the centrifugal pump receives saturated liquid on the suction side, the liquid becomes well subcooled in the discharge line due to the increased pressure. This is one reason why filters, orifice runs, control valves, exchangers, and other flow restrictors are placed on the discharge side of process pumps.

Economical sizes for discharge pipe up to about 12-in. diameter can be readily selected. For larger diameters, a more detailed cost comparison should be made in order to choose between alternative discharge pipe sizes.

DISCHARGE CHARACTERISTICS OF PUMPS

By reviewing typical pump data and knowing only the flow rate, a reasonable design for the hydraulic system can be estimated, regardless of discharge-pressure requirements of the pumped liquid.

Performance Curves

Figure 1 shows a composite rating chart for a series of standard chemical process pumps and the head-capacity curve for a specific pump. The performance curves for this pump show that it is suitable for handling flows from 150 to 300 gpm (close to the maximum efficiency points). Additional data are:

Flow, gpm	150 to 300
Impeller diameter, in.	6 to 10
Power, hp.	3 to 10
Efficiency, %	~58 to ~67
Total head, ft. water	30 to 90

All these data were obtained from Figure 1B for the pump having a 4-in. suction nozzle and a 3-in. discharge nozzle. The total head represents a differential pressure of 13 to 39 psi (when pumping water) between suction and discharge flanges.

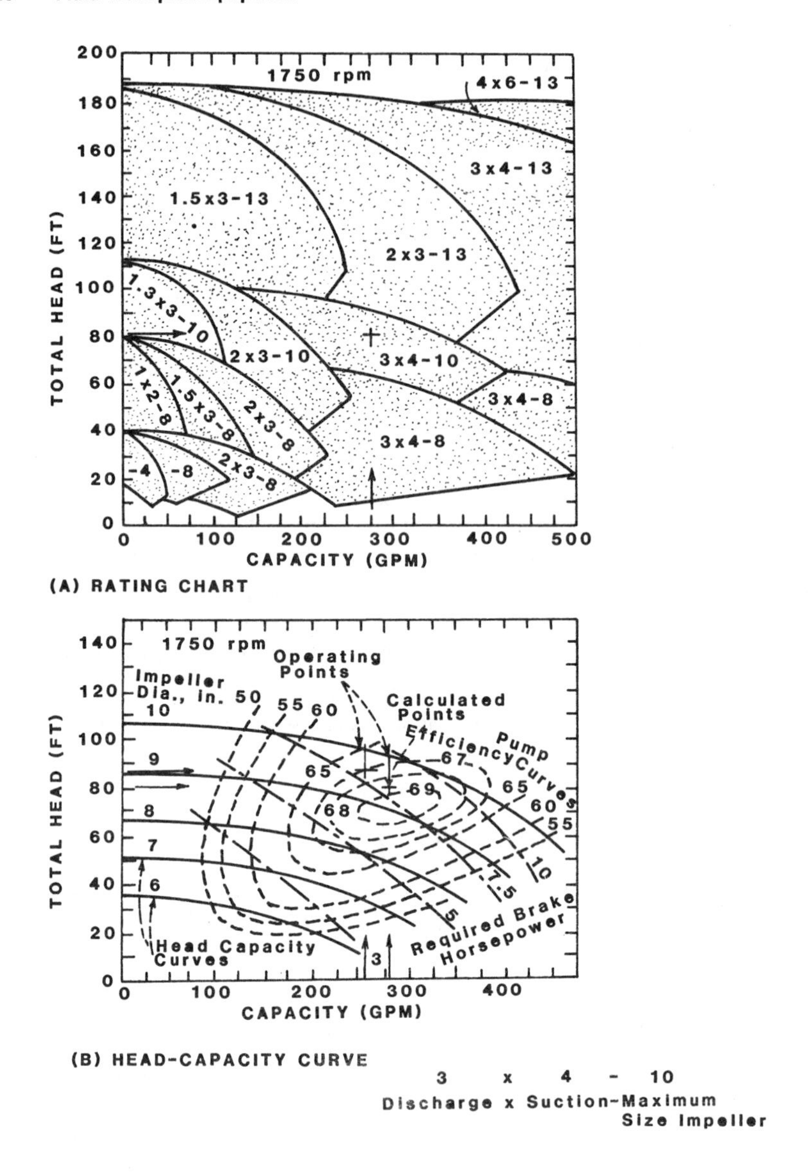

Figure 1. Performance characteristics of centrifugal pumps.

For a flow of about 250 gpm, the suction-pipe size can be 6 in., the discharge pipe 4 in. The orifice run can be 4 or 6 in., the control-valve size 3 to 4 in.; block valves and strainer should be line size. With these data, detailed calculations can be made, the pump total-head requirement determined, and impeller diameter and motor size selected.

A centrifugal pump is a dynamic machine that can perform only on points on the head-capacity curve. Consequently, the sum of piping and components resistances during operation must fall exactly on the head-capacity curve. The relationship between the pump's head-capacity curve and pipe-system resistance is shown in Figure 2A. Resistance of the throttling valve should not be included as part of the system resistance. The gap between the head-capacity curve and the system-resistance curve is available for throttling resistance (i.e., control-valve ΔP). While pipe resistance increases with higher flow rate, throttling resistance decreases. Or, at higher flow rates, the throttling valve has to open and pass larger flows with less resistance.

According to the Hydraulic Institute Standards [1], the velocity-head difference between pump nozzles should be included in the head-capacity curves. Some of the velocity energy converts to pressure energy if the discharge pipe is larger than the size of the pump's discharge nozzle. When computing required total head, velocity head often does not enter into the calculations, as in the total-head calculation for the example in this chapter.

In general, the velocity head is usually much less than 10% of the total head and is frequently neglected. For high-head pumps, the velocity head is negligible but for low-head pumps, it can be a large portion of the total head.

In a great number of publications, the head-capacity and system-resistance curves are reproduced as shown in Figure 2B.

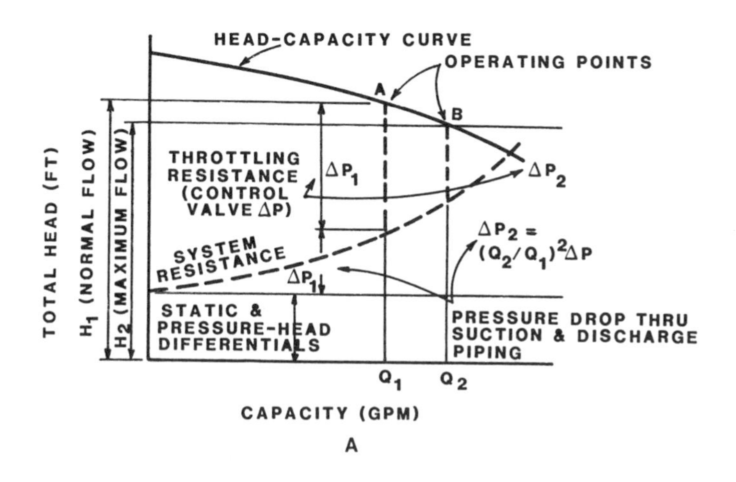

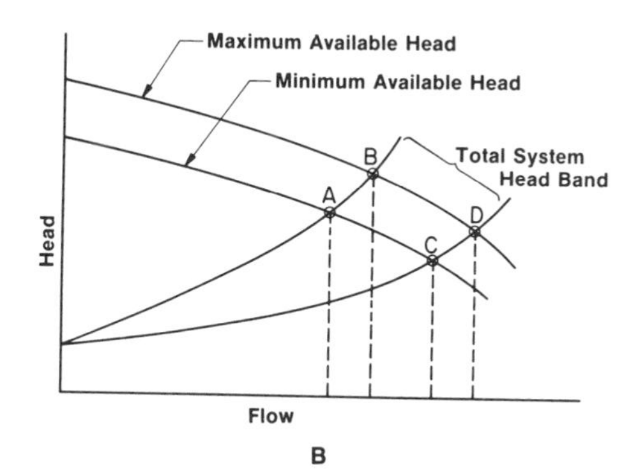

Figure 2. Head-capacity and system requirement curves.

The curves in the figure suggest the when the discharge is throttled, the system-resistance curve moves to another position. Not so! Only one system-resistance curve exists for a given discharge line. The varying ΔP of the control valve brings the system resistance up to a point on the head-capacity curve.

Head-capacity curves are drawn for pumping water. However, a centrifugal pump with a given impeller, speed, and size will develop the same head in feet, no matter what the liquid or specific gravity—providing viscosities are normal. The static-head pressure at the pump nozzles will be higher with a heavier liquid, and the horsepower requirement will also increase. Brake-horsepower requirements for pumping water are superimposed on the head-capacity curve (see Figure 1B). Multiply the horsepower required to pump water by the specific gravity of the liquid being pumped to obtain actual brake horsepower. Efficiency curves are also superimposed on the head-capacity curves (Figure 1B). The ideal, normal operating point is in the maximum efficiency range.

When calculating total-head requirements at alternative capacities, it is useful to know the operating points on the head-capcaity curve, so that the available control-valve pressure differentials can be more closely estimated.

If a pump has not been selected, we can still estimate the increase in total head for a given pump when capacity decreases by X% from a given capacity. This increase in total head can range from 0.5 to 1.0 multiplied by X%. A high value for this increase (selected from the range 0.5 to 1.0), say 0.85X%, is justified for three reasons:

1. The pump will not be undersized.
2. The head-capacity curve for a worn impeller will have a more pronounced decline with increasing flow as compared with a new. impeller.
3. A somewhat increased control-valve ΔP will provide a more desirable plug position at a maximum flow.

Viscous Flow

A reduction in head, capacity, and efficiency, and an increase in horsepower, will occur when pumping viscous liquids instead of water. (NPSH requirement does not change.) Figure 3 shows a performance-correction chart for conventional, single-stage, centrifugal pumps handling viscous Newtonian liquids for capacities between 10 to 100 gpm. A similar chart is available for 100 to 10,000 gpm in the *Hydraulic Institute Standards* [1].

These charts should not be used for multiple-impeller, mixed-flow, and axial-flow pumps, or for pumps with special hydraulic design. Manufacturers can supply viscosity-correction charts for their pumps.

To use Figure 3, we select a pump from the manufacturer's head-capacity curves at optimum efficiency, and note the head, capacity, and efficiency. Enter Figure 3 with the same capacity. Intersect the head line, in feet, and move horizontally left or right to viscosity; then vertically up to intersect the efficiency-correction factor, C_E, capacity-correction factor, C_Q, and head-correction factor, C_H. An example is drawn in Figure 3.

The corrected performance characteristics for viscous flow will be:

$$Q_w \times C_Q = Q_{vis}$$

$$H_w \times C_\theta = H_{vis}$$

$$E_w \times C_E = E_{vis}$$

where Q, H, and E are capacity, total head, and efficiency, respectively, and where subscript w stands for water, and subscript vis stands for viscous liquid.

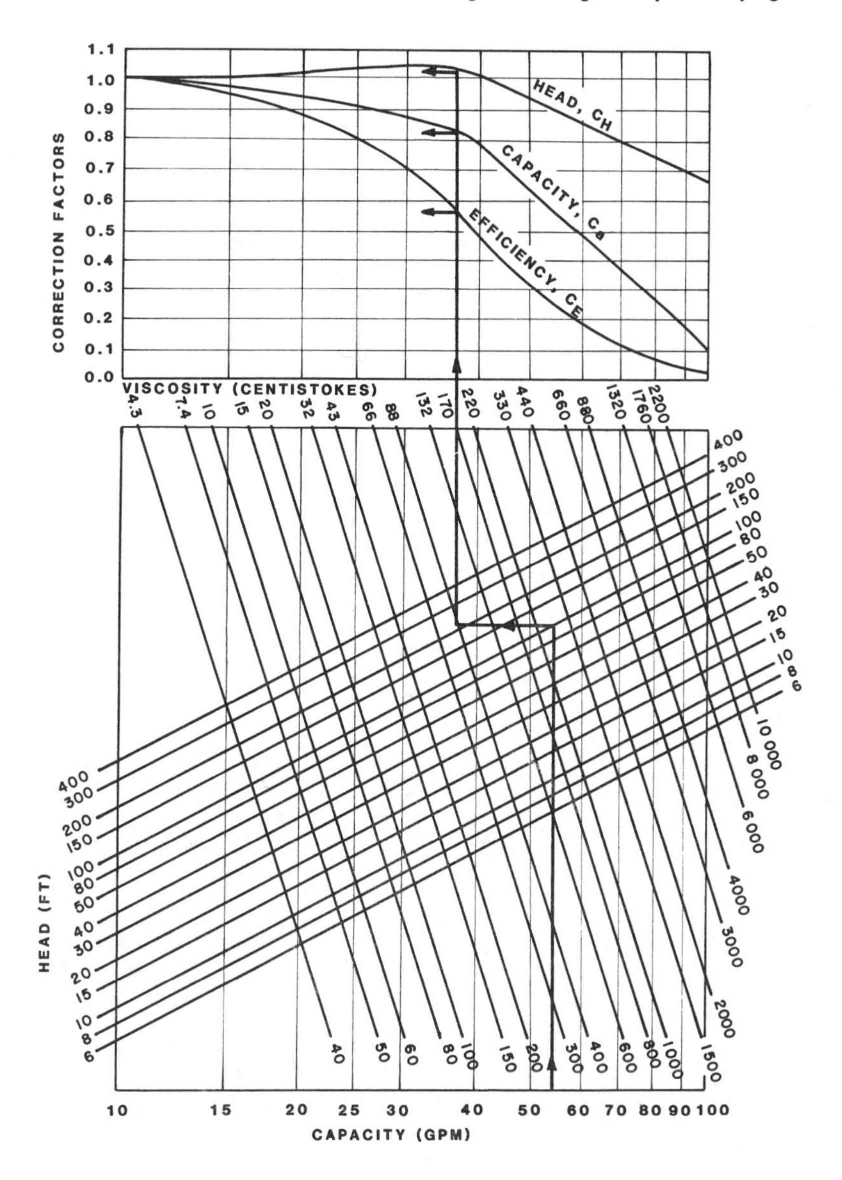

Figure 3. Correction factors for viscous Newtonian liquids flowing through single-stage centrifugal pumps.

Two-Phase Flow

If the pressure drops below the vapor pressure of the liquid in the discharge line (for example, after a heat exchanger or a control valve), vaporization will occur. Friction resistance in the pipe increases considerably with the increased vapor content of the liquid. This section of the discharge pipe should be calculated for two-phase flow. If two-phase flow is expected in a pipeline between two pieces of process equipment, capital cost and operating cost can be reduced by locating this equipment side by side. Generally, lines with two-phase flow should be short.

TOTAL-HEAD CALCULATIONS

We can use the data plotted in Figure 4 for selecting a suitable size for the discharge piping of a centrifugal pump. The shaded zones in Figure 4 will help in selecting economical sizes for alloy or carbon-steel piping in the case of electric-motor or steam-turbine drives.

In general, economical sizes for discharge pipes can be easily recognized if we list the unit losses for several adjacent pipe sizes. For example, the unit losses in a Schedule 40 pipe for a 300-gpm flow, as obtained from Figure 4, yield:

Pipe Size, In.	ΔP_{100}, Psi/100 Ft
3	8
4	1.9
6	0.23

In this case, the 4-in. size is reasonable for the discharge pipe. The unit loss of a 3-in. pipe will, most probably, increase utility cost to an extent that cannot be compensated for by lower capital

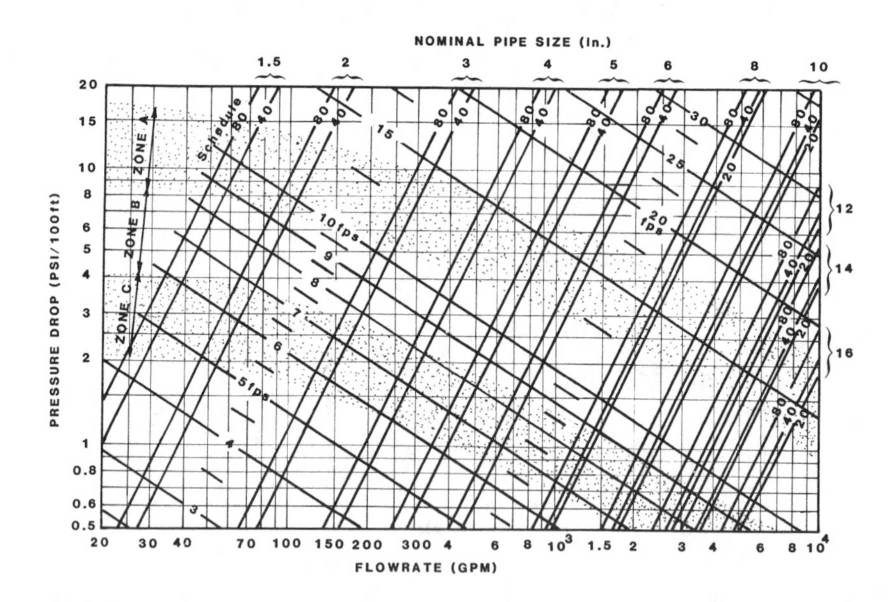

Figure 4. Unit losses in new, commercial steel pipe. *Note*: Economical sizes for discharge pipe:
Zone A: Alloy-Steel pipe, 2-yr payout
Zone B: Carbon-Steel pipe, electric-motor drive or 2-yr. payout. Also, alloy-steel pipe, long payout.
Zone C: Carbon-Steel pipe, turbine drive, or long payout

Note: Pressure drop values are for S = 1 and total turbulence. Multiply ΔP by specific gravity and correction for friction factor: f (actual)/f (total turbulence).

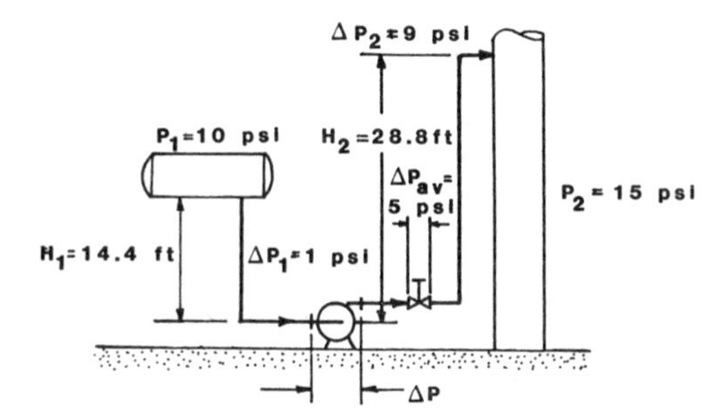

Figure 5. System example.

costs of the pipeline and components. And the cost of a possibly larger pump and motor has not been taken into account. The unit loss of a 6-in. pipe is far too low for an average process pump. Where the discharge line is very long, a 6-in. line might be reasonable, such as the cooling-water supply header to a distant point.

An example will best illustrate the concepts for calculating the pump's differential pressure:

A pump moves a liquid having a density, ρ, of 50 lb/ft^3 from an overhead drum to an elevated point on a column, as sketched in Figure 5.

Pressure in the drum, P_1, is 10 psi, and pressure in the column, P_2, is 15 psi. Pressure drop across the control valve, ΔP_{cv}, equals 5 psi; and the friction losses in the suction and discharge lines are $\Delta p_1 = 1$ psi and $\Delta p_2 = 9$ psi, respectively. Let us calculate the required total head for the pump, i.e., the differential pressure, ΔP, between the suction and discharge flanges.

Both suction head and discharge head consist of four components: static head, pressure head, velocity head, and piping and components resistance. Velocity-head differences normally do not have to be calculated because they are taken into account by U.S. pump manufacturers in their published head-capacity curves. Pressure and static head in a system usually remain constant with changing capacities. Suction-pipe, discharge-pipe, and components resistances vary with flow rate (see Figure 2). The analysis, therefore, becomes:

	Suction Psi	Discharge Psi
Vessel pressure	$P_1 = 10$	$P_2 = 15$
Static-head pressure	$P_{H1} = 5$	$P_{H2} = 10$
Friction loss	$\Delta p_1 = -1$	$\Delta p_2 = 9$
Total	14	34

We now subtract the total suction pressure from the discharge line backpressure to find the pump's differential pressure excluding the control valve: 34 − 14 = 20 psi. Choosing a control-valve ΔP_{cv} of 5 psi, we must add this to the pump's differential pressure of 20 psi to obtain the total differential pressure across the pump flanges as 25 psi, or expressed as head of liquid (the required total head):

$$H = (25)(144)/50 = 72 \text{ ft}$$

DESIGN EXAMPLE

A centrifugal pump having a 4-in. suction nozzle and a 3-in. discharge nozzle will handle a gas oil at a normal flow rate of 250 gpm through a piping and components system, as drawn in Figure 6. Allowing for a safety factor of 1.1, we find that the maximum flow rate will be 1.1(250) = 275 gpm. Specific gravity and density are: $S_{60} = 1.18$, and $\rho_{60} = 73.6$ lb/ft^3, respectively. At the flowing temperature of 555°F, $S = 1.04$ and $\rho = 64.87$ lb/ft^3. Viscosity of the gas oil is 0.6 cp. There is a flow control valve in the discharge line.

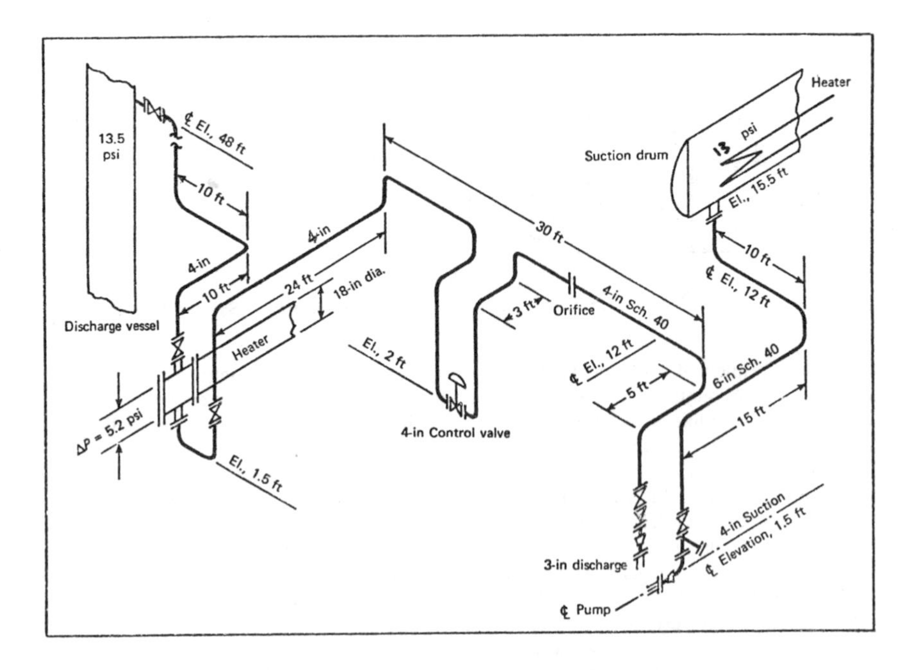

Figure 6. Example for suction and discharge pipe sizing and total head calculations.

Let us calculate the total head on the pump when it is expected to operate at a normal flow rate and at a maximum flow rate.

We will begin the analysis by first determining the loss in the suction line and then that in the discharge line. Pertinent data for the Schedule 40 pipe are:

	Suction Line	Discharge Line
Nominal size, in.	6	4
Inside dia., d, in.	6.065	4.026
d^5	8,206	1,058

Suction-Line

We find the loss in the suction line by initially calculating the Reynolds number from:

$$N_{Re} = 50.6(Q/d)(\rho/\mu)$$

$$N_{Re} = 50.6(250/6.065)(64.87/0.6) = 225{,}500$$

The friction factor, f, for this Reynolds number equals 0.0175 (from charts in Volume 1.). By substituting in the following equation, we obtain the unit loss:

$$\Delta P_{100} = 1.35fSQ^2/d^5$$

$$\Delta P_{100} = 1.35(0.0175)(1.04)(250)^2/8{,}206$$

$$\Delta P_{100} = 0.19 \text{ psi}/100 \text{ ft}$$

We then find the equivalent length for the suction line and its fittings from data in Volume 1.

Pipe length	39 ft
5 Short-radius elbows	75 ft
1 Reducer	4 ft
1 Strainer	30 ft
1 Gate valve	6.5 ft
1 Inlet to pipe	18 ft
Total equivalent length, L,	172.5 ft

Hence, the overall loss for the line and its fittings at the normal flow of 250 gpm becomes;

$$\Delta P = \Delta P_{100}(L/100)$$

$$\Delta P = 0.19(172.5/100) = 0.33 \text{ psi at normal flow}$$

$$\Delta P = 0.33(250/275)^2 = 0.4 \text{ psi at maximum flow}$$

To find the pressure at the suction nozzle, we calculate the static-head pressure: (14)(64.87)/144 = 6.3 psi, and add this to the vessel pressure of 13 psi to get 19.3 psi. Since the pipe-friction loss at normal flow is 0.33 psi, the pressure at the suction nozzle becomes 19.3 − 0.33 = 18.97 psi. At maximum flow, the pressure at the suction nozzle is 19.3 − 0.4, or 18.9 psi.

Discharge-Line

We now perform similar computations to find the loss in the discharge line. Therefore, for the discharge line:

$$N_{Re} = 50.6(250/4.026)(64.87/0.6) = 340{,}000$$

The friction factor, f, for this Reynolds number is 0.0178. We now use Figure 4 to find the unit loss, ΔP_{100}, as 1.32 psi for totally turbulent conditions at the normal flow of 250 gpm. Since the flow for this Reynolds number is in the transitional region, we must correct the unit loss as follows:

$$\Delta P_{100} = 1.32(1.04)(0.0178/0.0165) = 1.48 \text{ psi/100 ft}$$

We then find the equivalent length for the discharge and its fittings from data (see Volume 1).

Pipe length	156 ft
20 Short-radius elbows	210 ft
4 Gate valves	18 ft
1 Reducer	3 ft
2 Exits	40 ft
1 Inlet	10 ft
Total equivalent length, L	437 ft

Hence, the overall loss for the line and its fittings at the normal flow of 250 gpm becomes:

$$\Delta P = \Delta P_{100}(L/100)$$

$$\Delta P = 1.48(437/100) = 6.47 \text{ psi}$$

And at maximum flow,

$$\Delta P = 6.47(275/250)^2 = 7.83 \text{ psi}$$

Total Head

A detailed, specific, and systematic procedure is shown in Figure 7 for calculating the total head on the pump when it operates at a normal flow rate and at the maximum flow rate.

For the calculations presented in Figure 7, the following steps are recommended:

Lines 1 to 6 can be worked out simultaneously for normal and maximum flow. Lines 1 and 2: *discharge-vessel pressure* and *discharge static head* do not change with alternative flow capacities. Lines 3, 4, and 5: *orifice ΔP*, *pipe-friction loss*, and *equipment-friction loss* increase with capacity. Hence:

$$\Delta P_{max} = (\text{Pump Safety Factor})^2 \, \Delta P_{normal}$$

Line 3: *orifice pressure drop* depends on the manometer deflection and on the orifice diameter/pipe diameter ratio.

Line	Variable		Normal Flow psi	Maximum Flow psi
1.	Discharge-vessel pressure		13.5	13.5
2.	Discharge static head		20.94	20.94
3.	Orifice ΔP_o	1.52		
4.	Discharge pipe-friction loss	6.47		
5.	Exchanger ΔP	5.2		
	[a]	+		
6.	Discharge-system resistance	↳	13.19×1.1² =	15.96
7.	Subtotals		47.63 [b]	50.40
8.	Control valve ΔP		10.71 [d]	+ 5.12 [c]
9.	Discharge pressure		+ 58.34	55.52
10.	Suction-nozzle pressure		+ 18.97	− 18.90
11.	Pump differential pressure		39.37	36.62
12.	Required total head, (144 × ΔP)/P	[e]	87.4 ft	81.3 ft
13.	Total head from head-capacity curve		95 ft	92 ft
14.	Total-head safety margin		7.6 ft	10.7 ft

[a] Other equipment resistance.

[b] Pump safety factor, s.f. = 1.1

[c] With C_{vc}/C_v = 1.

[d] With C_{vc}/C_v = 0.5 to 0.8

[e] ΔH (normal flow) = ΔH (maximum flow) × 1.075
= 81.3 × 1.075 = 87.4

Figure 7. Calculations for total head of pump.

The manometer deflection for an orifice at a 250-gpm flow will be:

$$h_w = [0.176Q\sqrt{S}/(d_1^2\beta^2C)]^2$$

$$h_w = [0.176(250)\sqrt{1.04}/(16.21 \times 0.339)]^2$$

$$h_w = 80.7 \text{ in. or } 6.73 \text{ ft}$$

For this deflection, we can use a 100- or 125-in.-long manometer. The orifice ΔP_0 will be:

$$\Delta P_0 = 6.73(62.37)/144 = 2.92 \text{ psi}$$

With $\beta = 0.7$, the permanent loss will be 52% of the orifice pressure differential.

$$\Delta P_0 \text{ (loss)} = 0.52(2.92) = 1.52 \text{ psi}$$

Line 4: *Friction loss in the discharge* (*and suction*) *line* has been previously computed as 6.47 psi. Line 5: *Pressure drop through the exchanger* (and other equipment) can be obtained from the manufacturers. Line 6 is the sum of Lines 3 to 5 at the normal flow rate. For a 10% greater flow and a pump safety factor of 1.1, resistance of the discharge line will increase by $(1.1)^2$. Line 7 is the subtotal of Lines 1, 2, and 6 at normal flow and at maximum flow.

At this point, we will continue the calculations at the maximum flow rate for reasons that will become evident.

Line 8: For the control valve to operate in an optimum range at normal flow, we usually consider the valve plug in a fully open position at maximum flow. This also gives a minimum pressure drop through the control valve. A 4-in. single-seat control valve has a valve coefficient $C_v = 124$. And, with $C_{vc}/C_v = 1$:

$$\Delta P_{(min)} = \left[\frac{Q}{(C_{vc}/C_v)C_v}\right]^2 S$$

$$\Delta P_{(min)} = \left[\frac{275}{1 \times 124}\right]^2 1.04 = 5.12 \text{ psi}$$

Line 9 is the required discharge pressure, including control-valve ΔP at maximum flow. Suction-nozzle pressure has been previously computed. By deducting the suction-inlet pressure from the required discharge pressure, we obtain the pump's differential pressure at maximum flow (Line 11). This is converted to the equivalent head (Line 12) at maximum flow by using the previously determined valve of 36.62 psi from Line 11:

$$\frac{144(\Delta P)}{\rho} = \frac{144(36.62)}{64.87} = 81.3 \text{ ft}$$

Let us now summarize some of these results. The pressure needed at the pump-discharge nozzle to overcome backpressure in the discharge line is the sum (Line 9 in Figure 7) of actual pressure in the discharge vessel, static-head lift up to the terminating nozzle (or liquid level in discharge drum), control-valve ΔP, and total discharge-pipe and equipment resistances. Pump differential pressure (Line 11) equals discharge pressure (Line 9) minus suction-nozzle pressure (Line 10).

We can now estimate the total head at normal flow rate. Total head will increase by an amount ranging from 0.5 to 1 of the percentage decrease in capacity. In this example, there is a 10% decrease in capacity. Hence, for a single-impeller pump, we will assume an increase in the total head of about 7.5% (i.e., 0.75 of 10%). The computed value is on Line 12 for the normal flow. Calculated values for the total head for this example are shown in Figure 1B.

Suction-nozzle pressure at normal flow is on Line 10. Line 9 = Line 10 + Line 11.

The available pressure differential at normal flow for control-valve sizing (Line 8) equals discharge pressure (Line 9) minus the line backpressure without the control valve (Line 7). This ΔP should give a control-valve coefficient falling within the recommended ranges of $C_{vc}/C_v = 0.5$ to 0.8 for equal-percentage contoured plugs:

$$C_{vc} = Q\sqrt{S}/\sqrt{\Delta P} = 250(\sqrt{1.04})/\sqrt{10.71}) = 77.9$$

For the selected 4-in. control valve, $C_{vc}/C_v = 77.9/124$, or 0.63, which is acceptable.

For a normal flow of 250 gpm and 87.4 ft total head on the one hand, and a maximum flow of 275 gpm and 81.3 ft on the other, we can now select the pump, as shown in Figure 1. Impeller diameter for the selected pump is 10 in., and a standard motor of 10 bhp is required. The motor will work with a better than 65% efficiency.

The calculated total-head points fall between the 9- and 10-in. impellers. The pump will operate at 95-ft total head at 250 gpm, and at 92-ft at 275 gpm. The extra head (7.6 and 10.7 ft, respectively, here) provides a safety margin compensating for inaccuracies in the assessment of the flow-properties and line data. These additional pressure differentials can be absorbed by the control valve. Or, the block valve in the discharge line can be slightly closed to bring the operating point up to the head-capacity curve. Also, the motor will be able to drive the pump when the liquid is colder and specific gravities are greater than at operating conditions.

If a pump has not been selected and the head-capacity curve is not available, a safety margin of 5% to 15% can be estimated and added to the required total head of Line 12.

ECONOMY OF PUMP PIPING

As the data in Figure 1B show, several sizes of impellers can be placed in one pump case. The cost difference between impeller sizes is negligible. Motor sizes are usually well determined. A difference in the cost of the pump and motor occurs for pumps falling in adjacent envelopes of the composite rating chart. In some borderline cases, it may be more economical to redesign the discharge piping for lower pressure drop rather than to invest in a larger pump and motor.

For economy in utility cost, the pump should work at its highest efficiency. High pump efficiency results in minimum horsepower input, and minimum wear and maintenance. High-efficiency pumps last longer, are quieter, and vibrate less than low-efficiency ones.

Small pumps should not be oversized. The total of oversized small motors in a plant can add to substantial waste in energy usage.

For pipe diameters above 12 in., more than one size may be selected initially because capacity increments in large-size pipe are very close. Piping costs, of course, increase with diameter, while utility costs decrease because of smaller pipe and components resistances. The best size can be determined by adding the total cost of utilities over the period of capital payout to the cost of the mechanical and electrical installation. The lowest total cost calculated for a 2-, 5-, or 10-yr amortization will give the most economical design.

A detailed investigation for the most economical pipe size is justified if line sizes are large, pipe configurations are long or complicated, or if the piping material is expensive. Pipe friction must contribute a major portion of the discharge pressure—otherwise, there will be little difference in total heads between alternative designs. Actual vessel pressures and static liquid heights usually cannot be altered.

NOTATION

C_v	control-valve coefficient	Δp	pressure loss, psi
C_{vc}	calculated control-valve coefficient	ΔP_{100}	unit loss, psi/100 ft
d	internal pipe diameter, in.	P	pressure, psi
f	friction factor	ΔP	pressure difference
L	equivalent pipe length	Q	flow rate, gpm
N_{Re}	Reynolds number	S	specific gravity at flowing temperature

Greek Symbols

- β ratio of orifice bore to inside pipe diameter
- $\beta^2 C$ orifice capacity coefficient (if $\beta = 0.7$, $\beta^2 C = 0.339$)
- μ viscosity, cp
- ρ density of liquid at flowing temperature, lb/ft^3

REFERENCES

1. Hydraulic Institute Standards, Hydraulic Institute Cleveland, Ohio.
2. Marischen, J. P., "Critical Centrifugal Pump Information," Ampco Metal Inc., Milwaukee, Wisconsin.
3. "Cameron Pump Manual," Ingersell-Rand Co. Phillipsburg, N.J.
4. Kern, R., "Pump Suction Piping," *Chemical Engineering* (April 28, 1975), p. 119.

CHAPTER 40

FLUID DYNAMICS OF INDUCERS

B. Lakshminarayana

Department of Aerospace Engineering
Pennsylvania State University
University Park, PA

CONTENTS

INTRODUCTION, 1152

INDUCERS, 1153

TECHNOLOGY OVERVIEW AND NATURE OF FLOW FIELD, 1153

ANALYSIS OF THE INDUCER FLOW FIELD, 1155
 Analyses Based on Simplified Radial Equilibrium Analysis and Euler's Equation, 1156
 Meridional Flow Solution Including Viscous Effects, 1158
 Viscid Analysis Based on the Empirical Loss Coefficient, 1160
 Approximate Viscid Analysis Based on Shear Pumping Effect, 1162
 Three-Dimensional Inviscid Analysis (Exact), 1163
 Three-Dimensional Viscid Analysis Based on Empirical Wall Shear Stress, 1165

INDUCER OVERALL PERFORMANCE (NONCAVITATING), 1166
 Flat-Plate Inducers Tested in Water and Cryogenic Liquids, 1166
 Tandem and Hubless Inducers, 1169

INDUCERS TESTED IN AIR, 1170
 Measurements Downstream of the Aerodynamically Designed and Flat-Plate Inducers, 1171
 Measurements Inside the Blade Row, 1176
 Measurements Inside the Blade Passage of a Three-Bladed Inducer Using Hot-Wire Probes, 1176
 Boundary Layer and Related Phenomena on Inducer Blades and Passages, 1177

CONCLUDING REMARKS, 1179

NOTATION, 1181

REFERENCES, 1182

INTRODUCTION

The objective of this chapter is to review the experimental and analytical investigations on the fluid dynamic aspects of noncavitating inducers. Various analyses available for the prediction of the flow field, starting from the simple radial equilibrium analysis to the numerical solution of exact three-dimensional inviscid and viscid flow equations, are reviewed. The experimental data on the overall inducer performance, rotor blade passage measurements, and boundary layer on rotor blades are critically examined. Some of the available data have been reinterpreted to provide suitable guidance to the designer. The review and the concluding remarks include a global view of the state of the art and applicability of the research to design, and some viewpoints on the analysis and design of inducers.

INDUCERS

Inducers are employed in rocket pump feed systems, water jet propulsion, centrifugal impeller, and various other applications. The aerospace industry has made substantial contributions to the current state of the art of inducers used in rocket pump feed systems [1–3]. The inducer technology has found new applications in the commercial market, such as in high-speed ships (in excess of 50 knots) [4, 5], and auxiliary power units for aircraft [6]. Inducers are also used in centrifugal impellers and aircraft fuel feed systems.

The turbopumps used in liquid rocket feed systems operate at high speeds so as to minimize weight and size of the system—an important design criteria for an aerospace vehicle. Conventional pumps cavitate at suction-specific speeds (SS) in excess of 8,000. The need to increase the speed led to the development of a cavitation-resistant inducer, which is essentially an axial-flow pump with high-solidity blades used in front of the main pump. The value of the suction specific speed (SS) can be improved by fitting the inducer blade to the pump, thus decreasing the pump size for application in water jet propulsion of ships, hydrofoil craft, chemical pumps, and rocket turbopumps.

The inducer usually contains fewer blades (usually 3 to 4) than conventional pumps. Long and narrow passages provide the time and space for the collapse of the cavitation bubbles and for the gradual addition of energy. The blades in most practical inducers used in rocket pumps wrap around almost 360 deg. The purpose of the inducer is to pressurize the flow sufficiently to enable the main pump to operate satisfactorily. The physical reasoning for the selection of such unconventional pumps is described by Acosta [7]. Inducers have operated successfully at suction specific (SS) speeds in excess of 70,000. Various types of inducers used in rocket applications are described in Reference 8 and shown in Figure 1. A design and performance summary of typical rocket pump inducers is tabulated in Reference 8.

The major characteristic features of the inducer are low flow coefficient (0.05 to 0.2), large stagger angle (70 to 85 deg.), and high-solidity blades (few blades of very long chord length as opposed to short chords of a many-bladed compressor rotor) of little or no camber. The resulting configuration, even though beneficial from the point of view of cavitation, results in highly viscous, turbulent, and almost invariably, fully developed flow inside these passages. This results in a large departure from idealized and design fluid dynamic performance. The inducers for pumps are designed to keep the cavity thickness to a minimum and to provide part of the pressure rise (as high as 6,000 psi in space shuttle engines). Extreme compactness and light weight are required to obtain the maximum payload in the rocket vehicle. The light weight is achieved by high rotational speeds. The low pressure oxidizer turbopump inducer used in the space shuttle main rocket engine (470,000 lb thrust each) is shown in Figure 2. The turbomachinery (oxidizer, fuel pumps, and turbines) in the space shuttle engine has the highest power-to-weight ratio (75,840 hp with a weight of 760 lb) of any turbomachinery known. Details of this engine can be found in References 2 and 3.

In view of these cavitation requirements, the blades are designed to be very thin and the passages very long. Jakobsen [8] provided the most comprehensive review of the rocket pump inducer design, with specific details on the blading and passage design, including hub size and shape, inlet diameter and contour, blade profile, leading edge, sweep, cant, lead, thickness of blades, solidity, and other aspects. Details of the design can be found in References 2–16. Ross and Benerian [14] provided the analytical base for the design of inducers.

TECHNOLOGY OVERVIEW AND NATURE OF FLOW FIELD

The investigations reviewed [1–73] here are mainly concerned with the effect of viscosity, not the effects of cavitation. The papers dealing with cavitation and cavitation-induced instability are listed under "Uncited References" [75–113] at the end of the References.

In any research involving an understanding of the complex flow phenomena in a real environment, one has to resort to a "building block approach" and study the phenomena individually first and collectively later. This is the only approach that will lead to eventual solution and improvement of the flow in inducers. Inducer flows are dominated by viscous and turbulence effects and cavitation. The regions in which the primary or inviscid flow exists are very small. Therefore, the group at

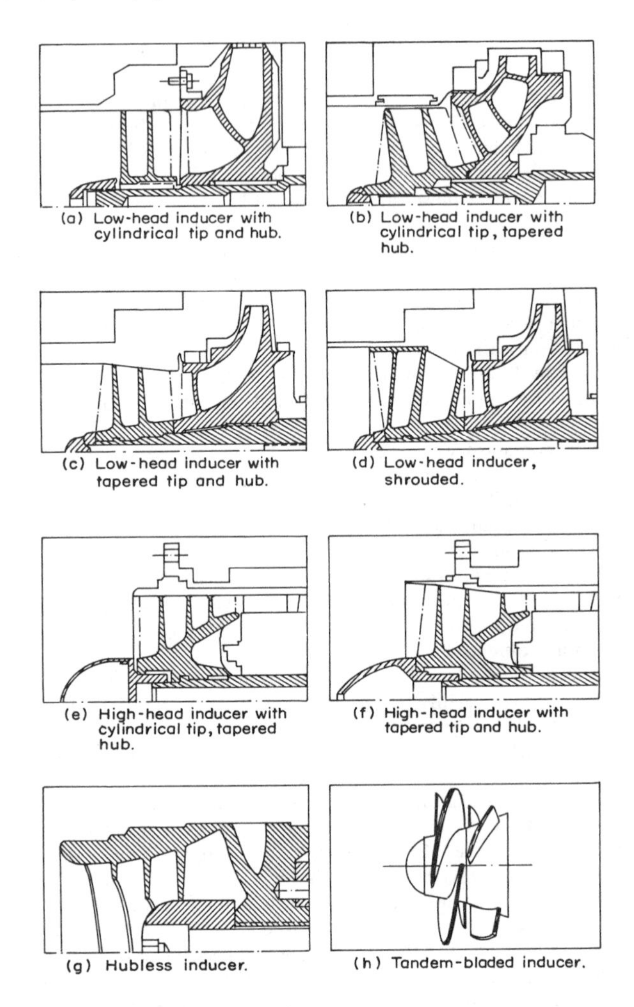

Figure 1. Basic inducer types (updated version of Reference 1).

Figure 2. Photograph of the low pressure oxidizer turbopump inducer impeller used in space shuttle main engine $\phi = 0.07$, SS = 70,000. (Courtesy: Rocketdyne Div. of Rockwell Corp.)

Pennsylvania State University studied one aspect of this complex machinery, viscous and turbulent flow, in great detail. A thorough understanding of the viscous and turbulent effects only (in the absence of cavitation) will eventually lead to better design (e.g., viscous design), a more systematic approach, and an understanding of this complex flow. The objective of this review is to review all aspects of this viscous flow phenomenon so that the designers and analysts in industry can include these effects in a systematic manner. This will also lay the ground work for the future investigation of the flow field inside the rotor passages of a cavitating inducer.

The experimental aspects of cavitation in inducers are not reviewed in this paper as the information available on this topic is related mainly to the overall performance and visualization of the cavitation phenomena [75–95]. The work on unsteady cavitation and stability of inducers has been reviewed by Greitzer [113]. The analytical part of the cavitating flow is incorporated later in this paper. Hopefully, this review will stimulate a systematic investigation of the research, such as the Penn State effort, to study the detailed flow phenomena in cavitating inducers. The availability of non-intrusive measurement technique (laser doppler velocimeter) would make such a study an attractive one at the present time. This review, as well as Penn State's work on inducers, will then form important ground work for such a study.

A review of the experimental data acquired inside the passage and at the exit of the rotor, tested with air and water, is covered. Various analytical, numerical analyses available for the prediction, of the flow field are described. Attempts are made to emphasize the significance and application of these results to inducer design. Advantages of tandem and hubless inducers are also briefly discussed. There has been no state-of-the-art review on fluid dynamic aspects of inducers, and therefore, a review of this field is long overdue. This is the major objective of the chapter.

The mechanism by which the stagnation-static-pressure rise takes place in inducers is due to a curious combination of inviscid turning effects and shear forces. While the former has a dominant influence in conventional machinery such as compressors and pumps, the shear forces dominate the flow in inducers, at least near the tip. The viscous and turbulence effects in these blade rows are not confined to thin regions at the blades and in the end zones, but extend over the entire cross section of the flow. It is hoped that the data, analysis, and correlations reviewed in this paper will serve to establish analytical methods for the analysis and modeling of turbomachinery flows dominated by viscous, turbulent, and secondary flow effects and under conditions where the real fluid effects are at least as important as the ideal effects which form the basis of the existing design.

The boundary layers that develop on the blades of rotating machinery such as compressors, pumps, inducer, propellers, and turbines are not two-dimensional. The rotation of the blade produces spanwise flow, resulting in a skewed or three-dimensional boundary layer. The extent of three-dimensionality depends on the angular velocity, flow coefficient, space/chord ratio, aspect ratio, stagger angle of the blade, etc. A basic understanding of the effects of centrifugal and Coriolis forces on the boundary-layer behavior is essential for the improved analysis and performance of such rotating fluid machinery. The radial flow inside the boundary layer, when encountered by the annulus wall, produces a complex flow near the tip, resulting in radially inward flow. These and other interaction effects result in extremely complex boundary-layer characteristics, as shown in Figure 3. It is one of the additional objectives of this paper to review the state of the art of the understanding of boundary-layer phenomena on these inducer blades.

ANALYSIS OF THE INDUCER FLOW FIELD

Many of the analyses carried out for the prediction of the inducer flow field are equally applicable to both cavitating and noncavitating flow; hence, a brief review of the two-phase flow analysis is included, even though this chapter primarily is restricted to noncavitating flows. The reader is referred to books by Horlock [44], Wislicenus [45], and Vavra [46] for several techniques available for the prediction of inviscid flow fields in axial-flow turbomachinery. These techniques are equally applicable in the computation of inviscid flow fields in inducers. A word of caution is in order. The inducer flow field is dominated by viscous and turbulence effects, and therefore, the inviscid theories are at best qualitative.

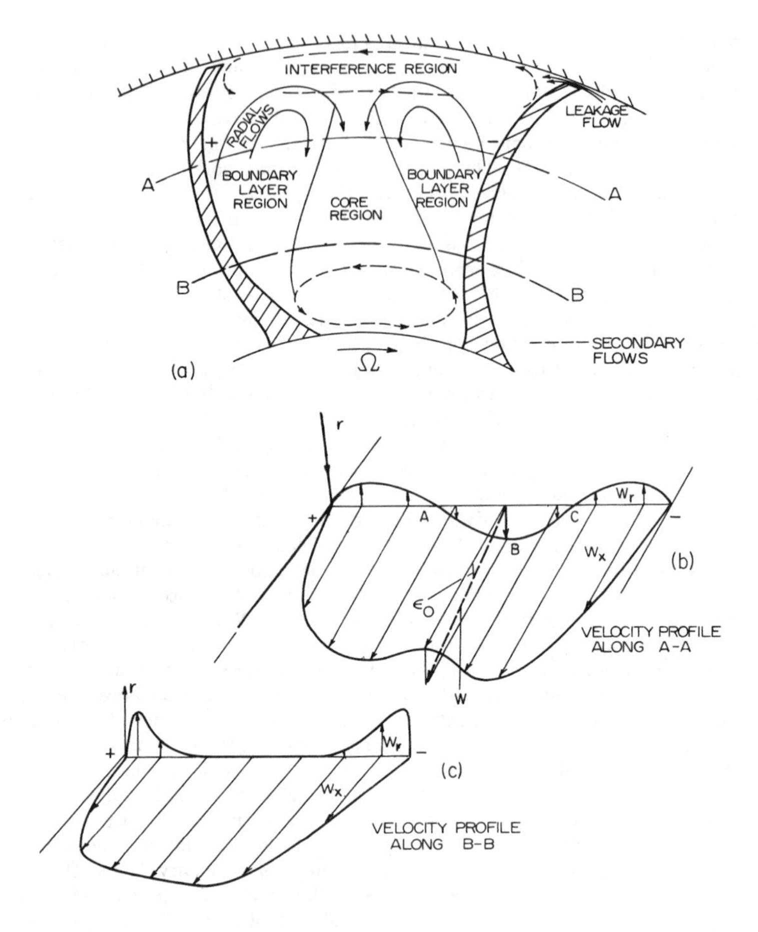

Figure 3. Nature of flows in inducers.

The analyses available can be classified as:

1. Analyses based on simplified radial equilibrium ($V_r = W_r = 0$) and Euler's equation.
2. Meridional flow solution including viscous effects.
3. Approximate viscid analysis based on shear pumping effect.
4. Three-dimensional inviscid analyses (exact).
5. Three-dimensional viscid analysis.

Analyses Based on Simplified Radial Equilibrium Analysis and Euler's Equation

In this section, the analyses based on one- or two-dimensional blade-to-blade solution, coupled with the complementary solution based on the simplified radial equilibrium ($V_r = W_r = 0$) analyses, are described.

Earlier analyses [14, 48–55] were mainly concerned with the calculation of the pressure head rise characteristics using one-dimensional equations. The analyses are mainly based on the Euler's equation, with the tangential velocity computed from the inlet and outlet blade angles. Mirolyubov [48] allowed for the losses in the channel through an empirical loss based on the friction loss in a pipe. Similar analysis was carried out by Dumov [49]. They both conducted a series of experiments (in water and fuel) with space/chord ratios varying from 0.70 to 1.5 and correlated these results and the analysis to prove that the ratio of actual to theoretical head rise is given by (for the range of values of $k = 0.2$ to 0.8):

$$\frac{H_A}{H_i} = \frac{1}{0.4 + (S/C)}\left(1 - \frac{k}{(S/C)}\right) \tag{1}$$

where $k = \bar{V}_{m2}/\Omega r \tan\beta$ is the meridional velocity
β = the helix angle

The equation is valid for a flat-plate helical inducer ($r \tan\beta$ = const).

The theoretical head rise was calculated using a simplified radial equilibrium equation ($\partial p/\partial r = \rho V_\theta^2/r$) and the Euler's equation ($H = UV_\theta/g$). The theoretical head is predicted to be:

$$H_i = \frac{\Omega^2}{g}\left(\frac{\Omega L}{2\pi V_{z/m}} - 1\right)\left[k_1 + \left(1 - \frac{2\pi \bar{V}_{zm}}{\Omega L}\right)k_2\right] \tag{2}$$

where k_1 and k_2 are functions of r_h, r_t, and L. The measured data shows good agreement with Equation 1. It should be remarked here that the theory is based on two-dimensional flows, while the correlation (Equation 1) based on the experimental data includes the three-dimensional effects. The good agreement between predictions and data may be somewhat fortuitous.

Some of the early work in this area is due to Rains [51]. His analysis is valid for a flat-plate helical inducer ($r \tan\beta$ = const) with a constant hub/tip ratio, without any inlet swirl. The outlet flow angle is assumed to be equal to the blade angle, and the flow is assumed to be in simple radial equilibrium ($V_r = W_r = 0$). The axial velocity predicted from this analysis is given by:

$$\frac{V_z}{U_t} = \frac{R^2 \tan\beta_t + C}{R^2 + \tan^2\beta_t}$$

The continuity equation is used to evaluate the constant C. Two expressions are given for the head rise coefficient (ψ_E), one based on the area-averaged values of $\psi(r)$ and the other based on the flow-averaged quantity. The agreement between the experiments of Newoz [53] and Rains is found to be poor at off-design conditions where hub flow shows the tendency to separate.

Analyses similar to Rain's have been carried out by Huppert [47] and Montgomery [55]. The latter included the viscous effects (through constant as well as radially varying loss correlations) and provided a comprehensive set of predictions for various inducers tested at the NASA Lewis Research Center [17–20]. The analysis [55] included inducers with varying hub/tip ratios, radially constant loss correlations, and radially varying loss correlations. The analysis is valid for a flat-plate helical inducer. Unlike Rain's [51], Montgomery's analysis included the effects of the deviation angle.

The simplified radial equilibrium equation, in combination with the integral form of the continuity equation, is used to derive an expression for the radial distribution of the axial velocity at various off-design conditions for the four-bladed inducer used in Penn State's investigation [31]. It is an extension of Huppert's [47] analysis for a flat-plate inducer, where $r \tan\beta$ = constant. This has been extended in Reference 52 for an inducer based on aerodynamic design [31], where r^2 and β is nearly constant. The axial velocity distribution for this case is given by:

$$\frac{V_{z_2}}{V_{z_m}} = \frac{1}{Y}\left[Y_m + \frac{\tan\beta_t}{\phi}\frac{V_{z_1}}{V_{z_m}}\left(R^2 - R_m^2\right)\right] \tag{3}$$

$$\frac{\bar{V}_{z_1}}{V_{z_m}} = \frac{2KY_m \int_{R_h}^{1.0} (R/Y)\,dR}{1 - 2\tan\beta_t/\phi\left[\int_{R_h}^{1.0} (R^3/Y)\,dR - R_m^2 \int_{R_h}^{1.0} (R/Y)\,dR\right]} \tag{4}$$

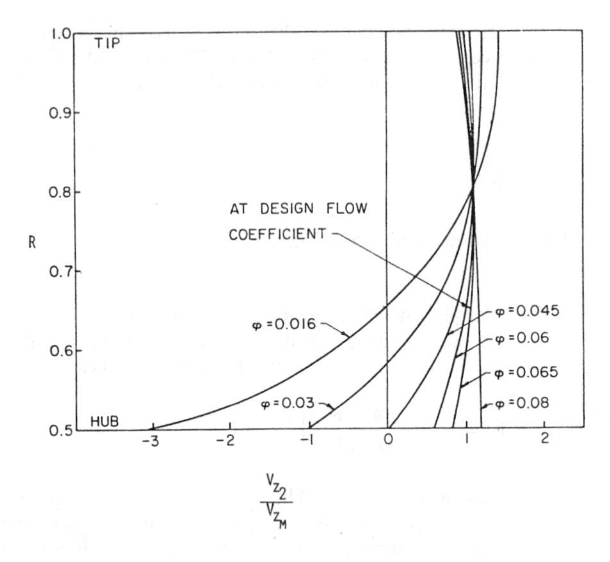

Figure 4. Predicted radial variation of outlet axial velocity for the Penn State four-bladed inducer [52, 71].

where $\phi = \bar{V}_{z_1}/\Omega r_1$, $Y = (\tan^2 \beta_t + R^4)^{3/4}$

$Y_m = (\tan^2 \beta_t + R_m^4)^{3/4}$,

$r_m = \sqrt{(r_t^2 + r_h^2)/2}$

k is a blockage factor to allow for blade thickness and boundary-layer growth, and V_{z_m} is calculated using Equation 4, which is then substituted in Equation 3 to predict the axial velocity at various radii. The prediction for the four-blade inducer, shown in Figure 4, confirms the results from flow visualization and measurements reported later. It should be recognized that the above analysis does not take into account the viscous effects, hence, the tendency for the backflow to occur near the hub at the design condition is not predicted. The tendency for flow reversal at the root at flow coefficients lower than the design value is, however, correctly predicted.

Meridional Flow Solution ($V_r = W_r \neq 0$) Including Viscous Effects

Cooper [56] included losses as well as two-phase flow (cavitation) effects and utilized approximate momentum equations to solve the flow field in an inducer with an annulus of varying area. The viscous effects are included through a loss term in the streamwise momentum equation. The momentum equation is written in the meridional (m) and quasi-normal directions shown in the insert of Figure 5. The basic assumptions made are as follows:

1. The fluid is assumed to be either liquid or a variable-density homogeneous two-phase medium.
2. Fluid flows in annuli (Figure 5) formed by stream surfaces of revolution generated by revolving meridional streamlines about the axis of rotation.
3. Average flow conditions exist as the mid-passage. These are determined from the meridional flow solutions, which are then utilized in the blade-to-blade solution.
4. The relative flow follows the blade mean line, and the exit deviation angles are based on specified values.
5. The blade-to-blade relative velocity variation is linear.

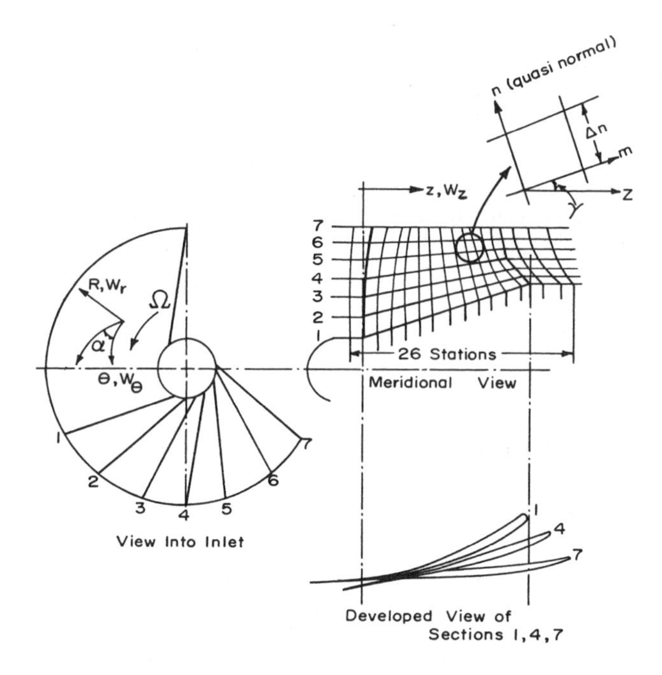

Figure 5. Inducer geometry, coordinate system, and grid for numerical analysis.

The meridional streamlines and the quasi-normals are derived using the method developed by Katsanis [54]. The approximate equations employed in the m, n, and θ directions are given, respectively, by:

$$\frac{dP}{\rho} = d\left(\frac{\Omega^2 r^2}{2}\right) - d\left(\frac{W^2}{2}\right) - dL \tag{5}$$

$$\frac{dP}{dn} = \frac{\rho V_\theta^2 \cos\gamma}{r} \tag{6}$$

$$\frac{(p_p - p_s)}{(\theta_p - \theta_s)} = \rho V_m \frac{d}{dm}(rV_\theta) \tag{7}$$

$$dL = f\frac{x}{d_h}\frac{W^2}{2} - \zeta d\left(\frac{W^2}{2}\right) \tag{8}$$

$$\rho = \begin{cases} \rho_f & p \geq p_{sat} \\ \rho_f\left[\dfrac{1}{1 + (B/\rho_f)(p_{sat} - P)}\right] & p < p_{sat} \end{cases} \tag{9}$$

where γ = the meridional angle
dL = the work done by the friction force
f = the friction coefficient
ζ = the diffusion loss factor (due to the pressure gradient and bubble collapse)
B = a fluid thermodynamic vaporization constant [56]

Equations 5 through 9 and the continuity equation are solved interactively for pressure, the overall pressure coefficient, and efficiency. The simplifying assumptions made (especially for the blade-to-blade solution) are serious limitations on the application of the analysis. The overall performance (ψ and η) can be predicted reasonably accurately provided the loss correlation (including its radial variation) is accurate. The analysis described in the next section is based on the three-dimensional form of the radial equilibrium equation and a more accurate loss model.

Davis et al. [57] and Cool et al. [59] utilized a channel approach for the blade-to-blade solution (including cavity bubbles) and the following radial equilibrium equation with streamline curvature in the meridional plane:

$$\frac{1}{\rho}\frac{\partial p}{\partial r} = \frac{V_\theta^2}{r} - V_m^2 \cos^2\gamma \frac{\partial \gamma}{\partial z} - V_m \sin\gamma \cos\gamma \frac{\partial V_m}{\partial z}$$

where V_m is the meridional velocity. The boundary layer growth is calculated from a two-dimensional momentum integral equation. This analysis is at best qualitative in view of the approximations used in the blade-to-blade solution (both viscous and inviscid). The analysis was mainly directed towards computation of aerodynamic loads for the stress and vibration analysis of the blade [58].

Viscid Analysis Based on the Empirical Loss Coefficient

An approximate analysis of the inducer flow, based on an empirically derived friction-loss coefficient, was carried out by Lakshminarayana [31]. The salient points of this analysis are now discussed.

Inducer data collected from various sources (NASA, M.I.T., TRW, etc.) indicate that the measured friction losses are several times higher than those of an equivalent stationary inducer channel, especially near the tip. The friction losses in an inducer are strongly dependent on the rotation factor (inverse of flow coefficient). A new friction-loss coefficient applicable to inducers operating in the range of flow coefficients $\phi = 0.065$ to 0.2 is defined and derived. The frictional losses estimated from this newly defined friction-loss coefficient agree well with the measured values for the Penn State inducer.

The loss coefficient is given by:

$$\psi_{loss} = \frac{2gH_{loss}}{U_t^2} = \lambda_R \frac{R_{ht}}{\phi} \frac{1}{R_N^{1/4}} \frac{C}{d_h} \left(\frac{\bar{W}}{U_t}\right)^2 \tag{10}$$

where λ_R is a function of radius, as shown in Figure 6.

A circumferentially averaged radial equilibrium equation is used to predict the relative and absolute tangential velocities. The analysis is based on suitable assumptions for the radial and main-flow velocity profiles (based on the existing three-dimensional turbulent boundary-layer data available) and loss coefficients just discussed. The circumferentially averaged radial equilibrium equation valid for inducers operating at low flow coefficients is given by:

$$\frac{-0.275}{\sin^2\beta} \bar{W}_\theta \left(\tan\alpha \frac{\partial \bar{W}_\theta}{\partial x} + \bar{W}_\theta \sec^2\epsilon \frac{\partial \epsilon}{\partial x}\right) + 1.015\bar{W}_\theta \frac{\partial \bar{W}_\theta}{\partial r} + 1.015 \frac{\bar{W}_\theta^2}{r} - 2\Omega\bar{W}_\theta + \frac{\partial}{\partial r}\left(\frac{\lambda_R}{R_N^{1/4}} \frac{R_{ht}}{\phi} \frac{C}{d_h} \frac{\bar{W}^2}{2}\right) = 0 \tag{11}$$

This equation was solved with the assumption that W_θ varies linearly with x (chord-wise distance) and by using experimental values for ϵ. The pressure rise coefficient was computed from the predicted $\bar{W}_\theta$. This predicted and measured value of ψ_t for the Penn State inducer (four-bladed inducer) is shown in Figure 7. Predicted values for three- and two-bladed inducers are shown in References 33 and 27, respectively. The losses and efficiences predicted from the empirical loss

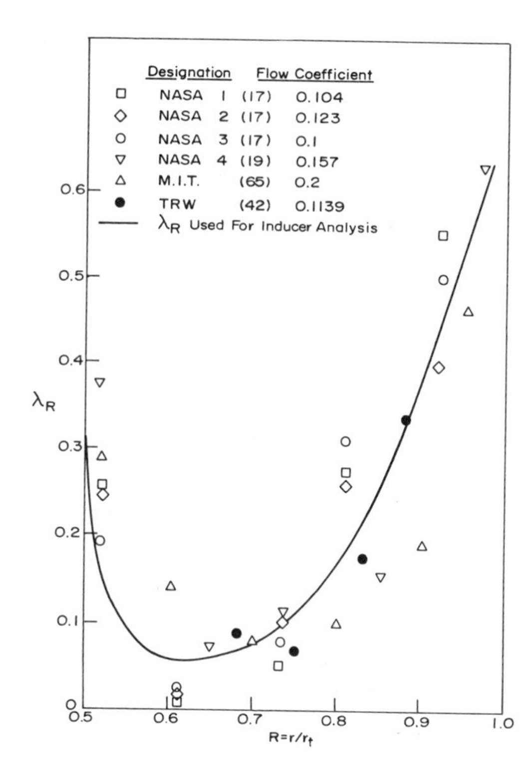

Figure 6. Radial variation of the modified friction loss coefficient λ_R for various inducers calculated from Equation 10 [31].

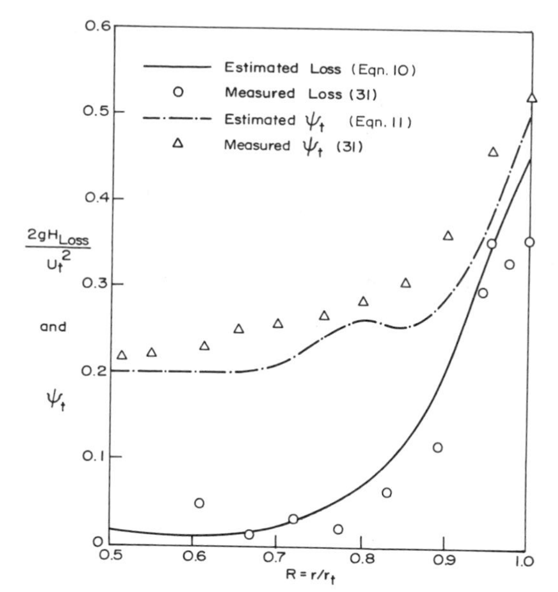

Figure 7. Estimated and measured friction losses and stagnation pressure coefficient for the four-bladed inducer [31].

correlation are also in good agreement (Figure 7). Hence, it is evident that if the frictional effects are known either empirically or analytically, the flow properties at the exit of the inducer can be predicted quite accurately. The very purpose of the boundary-layer investigation on the helical blade and a channel described later is to provide the most important missing link in the frictional effects, which are otherwise based on empiricism, and hence not universally valid.

The axial velocities predicted, using the continuity and axial momentum equations with assumed radial velocity profiles and the derived tangential velocity distribution, agree qualitatively with the value measured from a stationary probe at the exit of a four-bladed inducer [31].

Approximate Viscid Analysis Based on Shear Pumping Effect

The investigations carried out at Penn State [23–41] have revealed that the flow field inside these rotating passages is highly complex and three-dimensional in nature. The flow is highly turbulent and viscous both inside and outside of the rotor. Even though the inducers are designed as axial flow machinery, the radial velocities generated by viscous, Coriolis, and centrifugal forces inside the blade boundary layers and found to be of the same order of magnitude as the axial velocity. The energy exchange process is complicated in terms of physical phenomena as well as mathematical description.

The conventional design and analysis procedures used in other types of pumps or compressors have to be supplemented by some innovative technique to include major viscous effects (both direct and indirect) responsible for the energy addition. The pump has to be treated as a mixed-flow pump. The axial velocity has the same effect (inviscid) as in the conventional pump. The pressure rise due to this effect is caused by incidence and camber and both static- and stagnation-head increases through the blade row.

The other contribution to the energy addition comes from the radial component of velocity arising from the Coriolis and centrifugal forces in the boundary layers. These boundary layers are found to cover the entire passage, and hence the radial velocity occurs across the entire passage width. Therefore, the effect of radial velocity on head rise may be significant. The radial velocity provides a vehicle for momentum exchange between the fluid and the rotor via the shear stress. This phenomenon will be referred to here as the "shear pumping effect." Because of the shear stresses exerted on the flow, it is accelerated tangentially during the radial path, thus providing an additional head rise. It should be recognized here that this phenomenon involves no potential flow effect or flow turning. Even though explored in connection with multiple-disk shear pumps or compressors, it has never been recognized as a source of pressure rise in an axial flow inducer which has very long and narrow passages. The major purpose of the analysis carried out by Lakshminarayana [41] was to explore this effect analytically and provide for the first time an estimate of the "direct" effect of shear stresses in providing a head rise. In this analysis, the flow is assumed to be an incompressible single phase. Blade camber and normal shear stresses as well as pressure gradients normal to the blade inside the blade boundary layer are neglected. The passage is assumed to have fully developed flow.

The equations of motion in a rotating coordinate system are integrated normal to the blade, with suitable assumptions for the velocity profiles and the skin-friction coefficient, to derive the following equation [41]:

$$\frac{d\bar{W}}{dr} - 2.2\Omega - 0.254\Omega\frac{\sin\beta}{\epsilon^2} + \frac{\bar{W}}{r}\left[0.127r\frac{\partial\beta}{\partial r}\tan\beta + \sin^2\beta + 0.259\frac{\sin\beta}{\epsilon^2} - 0.254 + \frac{0.075}{\epsilon\cos\beta}N\left(\frac{W_e r}{\nu}\right)^{-1/5}\right] = 0 \qquad (12)$$

where W is the passage-averaged velocity given by:

$$\bar{W} = \frac{1}{\delta}\int_0^\delta u\,dy = \frac{7}{8}W_e$$

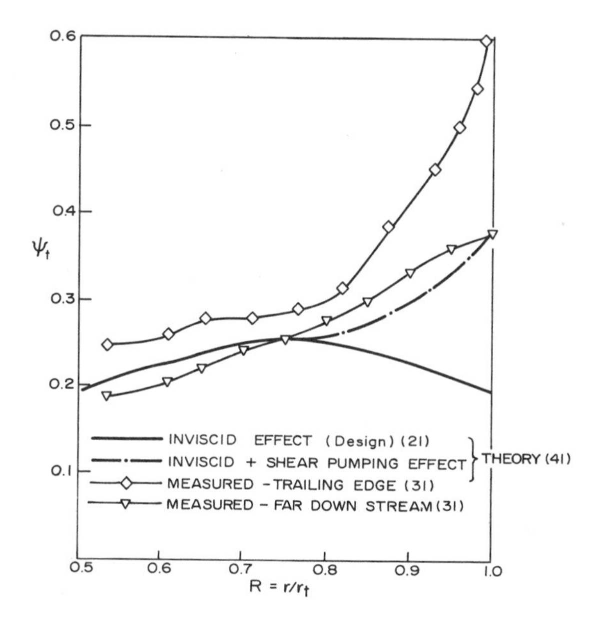

Figure 8. Predicted and measured radial variation of absolute stagnation-pressure rise coefficient (ψ_t) for the Penn State inducer [41].

Equation 12 is a first-order nonlinear differential equation and can be solved from known values of $\epsilon(r)$, Ω, $\beta(r)$, and ν.

The preceding expression provides an estimate for the reduction in relative velocity due to the shear pumping effect alone. It can be used to calculate the stagnation-pressure rise due to this shear pumping effect. An estimate of the total-pressure-rise coefficient can be made by the addition of the head rise due to inviscid and shear pumping effects. The predicted pressure-rise coefficient for the four-bladed inducer with aerodynamically designed blades operating at $\phi = 0.065$ is shown compared with the measured values in Figure 8. It is evident that the steep head rise characteristics observed are partly due to the shear pumping effect. The agreement between the theory and the experiment, shown in Figure 8, is only reasonable. The discrepancy may well be due to the three-dimensional inviscid effects, which are not accounted for in the "mean streamline method," used in the design, as well as leakage and secondary flows that exist in this region.

Three-Dimensional Inviscid Analysis (Exact)

A thorough knowledge of all significant inviscid effects (blade blockage, flow turning, finite hub/tip ratio, etc.) and viscid effects (boundary-layer growth, energy dissipation, etc.) is essential in the accurate prediction of the flow in all turbomachinery. Relevant to this, the availability of modern computers with large storage capacities and fast computation times greatly enhances the possibility of numerically solving the complete equations of motion. One of the early investigations in this area was made by Cooper and Bosch [42] and Bosch et al. [43] for the three-dimensional inviscid flow through axial flow inducers. Application and extension of this method of analysis was carried out in References 30, 33, 34, and 37. Modifications to the Cooper-Bosch method have been attempted to reduce the convergence time of the solution and to provide a viscid solution capability based on the empirically determined blade skin-friction coefficients. A method of initializing the blade flow parameters as input to the Cooper-Bosch method has also been attempted in a search for a faster convergence to the solution. A new technique for satisfying the Kutta condition at the trailing edge has been incorporated into the program.

This method developed by Cooper and Bosch [42] solves the momentum and continuity equations iteratively in three dimensions for a finite grid of points representing the channel between the blades. The nonlinear partial differential equations governing the flow in a rotating cylindrical system r, θ, z are:

r momentum:

$$\frac{g_0}{\rho}\frac{\partial p}{\partial r} + W_r \frac{\partial W_r}{\partial r} + \frac{W_\theta}{r}\frac{\partial W_r}{\partial \theta} + W_z \frac{\partial W_r}{\partial z} - \frac{1}{r}(W_\theta + r\Omega)^2 + F_r = 0 \tag{13}$$

θ momentum:

$$\frac{g_0}{\rho}\frac{\partial p}{\partial \theta} + W_r \frac{\partial W_\theta}{\partial r} + \frac{W_\theta}{r} + \frac{W_\theta}{r}\frac{\partial W_\theta}{\partial \theta} + W_z \frac{\partial W_\theta}{\partial z} + \frac{W_r W_\theta}{r} + 2W_r\Omega + F_\theta = 0 \tag{14}$$

z momentum:

$$\frac{g_0}{\rho}\frac{\partial p}{\partial z} + W_r \frac{\partial W_z}{\partial r} + \frac{W_\theta}{r}\frac{\partial W_z}{\partial \theta} + W_z \frac{\partial W_z}{\partial z} + F_z = 0 \tag{15}$$

Continuity:

$$\frac{W_r}{r}\frac{\partial W_r}{\partial \theta} + \frac{1}{r}\frac{\partial W_\theta}{\partial \theta} + \frac{\partial W_z}{\partial z} + \frac{1}{\rho}\left(W_r \frac{\partial \rho}{\partial r} + \frac{W_\theta}{r}\frac{\partial \rho}{\partial \theta} + W_z \frac{\partial \rho}{\partial z}\right) = 0 \tag{16}$$

where W_z, W_θ, and W_r are relative velocities in the axial, tangential, and radial directions, respectively. F_r, F_θ, and F_z are the components of the viscous forces and are zero for the inviscid case considered in this section. The boundary condition to be satisfied on the hub, annulus walls, and blade surfaces is $W_R \cdot n = 0$, where n is the direction normal to the channel boundaries and W_R is the total relative velocity. In the Cooper-Bosch method, the preceding equations are rearranged to give residuals, which are reduced to zero by a relaxation procedure. The total residual (RT) of one relaxation cycle is calculated by:

$$RT = \sum_{i=1}^{IMAX} \sum_{j=1}^{JMAX} \sum_{k=1}^{KMAX} [(R1)^2 + (R2)^2 + (R^3)^2 + (R4)^2]_{i,j,k} \tag{17}$$

where R1, R2, R3, and R4 are the residuals calculated for the three momentum equations (Equations 13–15) and the continuity equation (Equation 16), and IMAX, JMAX, and KMAX are the number of grid stations in the radial, tangential, and axial directions, respectively, which are used in the numerical analysis.

The initial estimation for the velocity and static pressure distribution throughout the inducer flow passage is calculated by the two-dimensional Douglas-Neumann program for a cascade, described in detail in Reference 60. This solution was modified to include three-dimensional effects, approximately, using the solution of Lakshminarayana and White [61] for the effects of converging or diverging ducts on the airfoil performance.

Since the extension of the stagnation stream surfaces downstream have been constructed to be uniformly periodic with a spacing of $2\pi/N$ (N being the number of blades), the values of velocity and pressure at the downstream tangential channel, boundaries should be equal. This condition is applied at the blade trailing edge after each iteration cycle. If the pressure and suction surface parameters differ from each other at the trailing-edge grid point, the average value is used in the residual calculations. Changes made to the original Cooper-Bosch program are given in Appendix B of Reference 34 and are concentrated in subroutine "MAIN." This program was run for the three-bladed configuration tested at Penn State. Some of the results and their interpretation are discussed in a later section.

Three-Dimensional Viscid Analysis Based on Empirical Wall Shear Stress

A method incorporating some of the major effects of viscosity was presented in References 34 and 37. In this formulation, the major viscous terms in Equations 13 through 16 are retained and modeled. The expressions for F_r, F_θ, and F_z are given by:

$$F_r = -\frac{1}{\rho}\left[\frac{\partial \tau_{r\theta}}{r\partial\theta} + \frac{\partial \tau_{rz}}{\partial z} + \frac{\partial \sigma_{rr}}{\partial r} + \frac{(\sigma_{rr} - \sigma_{\theta\theta})}{r}\right]$$

$$F_\theta = -\frac{1}{\rho}\left[\frac{\partial \sigma_{\theta\theta}}{r\partial\theta} + \frac{\partial \tau_{\theta z}}{\partial z} + \frac{\partial \tau_{\theta r}}{\partial r} + \frac{2}{r}\tau_{\theta r}\right]$$

$$F_z = -\frac{1}{\rho}\left[\frac{\partial \tau_{z\theta}}{r\partial\theta} + \frac{\partial \sigma_{zz}}{\partial z} + \frac{\partial \tau_{rz}}{\partial r} + \frac{\tau_{rz}}{r}\right]$$

where τ = the shear stress
σ = the normal stress

For example, $\tau_{r\theta}$ is the shear stress in the r direction in a plane perpendicular to θ. σ_{rr} is the normal stress in a plane normal to the r direction.

Since the stagger angle is very large, these viscous terms can be approximated by retaining the dominant terms as well as neglecting the normal shear stress, resulting in:

$$F_r = -\frac{1}{\rho}\frac{\partial \tau_{rz}}{\partial z}$$

$$F_\theta = -\frac{1}{\rho}\frac{\partial \tau_{\theta z}}{z}$$

$$F_z = -\frac{1}{\rho}\frac{\partial \tau_{z\theta}}{r\partial\theta}$$

The distribution of shear stress is assumed to be linear across the flow passage from the pressure surface to the suction surface. The values of wall shear stresses are assumed to be known from the empirical skin-friction coefficient (C_f) data for a four-blade flat-plate helical channel presented in Reference 38. Details of incorporating this into the main program are discussed in References 34 and 37.

An additional requirement placed on the viscid analysis is to satisfy the viscid boundary condition, which requires that all components of velocity are zero at the blade surface.

The geometry and the grid system used are shown in Figure 5. The comparison of blade static pressure is given in Reference 37. Inviscid and viscid analysis results for the total relative velocity distribution near the trailing edge for a three-bladed inducer are plotted in Figure 9. Magnitudes of W near the tip are comparable to those found experimentally, whereas near the hub, the inviscid velocities are significantly lower than the predicted results. The viscid analysis prediction for the total relative velocity distribution is also shown in Figure 9. It is a striking departure from the inviscid analysis distribution, especially near the tip where the viscous effects are known to be appreciable. The viscid analysis also provides crude approximations for the suction and pressure surface boundary layers. A large velocity deficiency near the tip is predicted at approximately 50 percent passage width and agrees favorably with the experimental W profiles plotted in Figure 9.

The prediction of radial velocity is at best qualitative. The reasons for this are discussed in References 34 and 37.

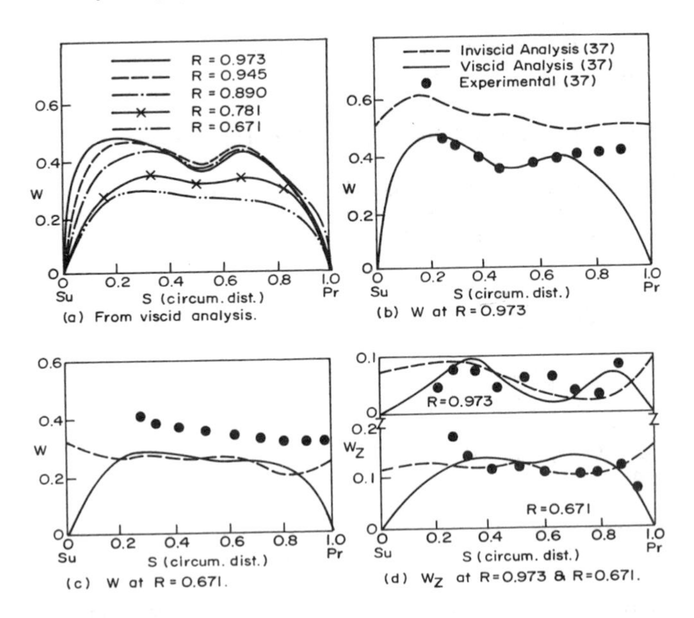

Figure 9. Comparison between theory and experiment of the blade-to-blade distribution of total relative velocity and axial velocity near the trailing edge for the Penn State inducer [37].

The viscid analysis is approximate due to the various assumptions and simplifications made. In particular, the viscid boundary conditions imposed on the solution are rather drastic, since the grid geometry spacing used in the exact analysis is relatively large. More tangential grid stations would be needed, especially close to the blade surface, to better define the shape of the blade boundary layer.

INDUCER OVERALL PERFORMANCE (NONCAVITATING)

Flat-Plate Inducers Tested in Water and Cryogenic Liquids

For the experimental aspects of the flow, many authors have studied the overall inducer performance with different configurations and inlet angles, and in various fluids such as water, liquid hydrogen, or nitrogen, under a wide range of flow parameters [4, 5, 7, 12, 14, 17–20, 42, 43, 48, 49, 53, 56, 57, 62–68, 72, 73]. Cavitating inducer performance has been studied by investigators listed in Uncited References. However, in most cases these studies are very specific and deal with cavitation performance, overall efficiency, and radial distribution of flow properties rather than with a general and basic investigation of the flow phenomena in inducers. The cavitating and noncavitating performances of 84-, 81-, and 78-deg. inducers under different flow coefficients have been studied by Acosta [7]. Acosta observed a deterioration in the radial distribution of the axial velocity and head rise at the exit for $\phi = 0.07$ under a noncavitating condition, with a backflow near the hub and a sharp positive gradient in head rise near the tip. These results are similar to those obtained by many investigators. Acosta attributed the large departure from design values based on the simplified radial equilibrium equation (SRE) to strong three-dimensional and viscous effects. Acosta also noticed a decrease in efficiency at large solidity, which is confirmed by the results obtained for the four- and three-bladed inducers tested at Penn State [27]. This effect can be attributed to the influence of blade blockage on the flow characteristics and an increase in viscous and turbulent mixing losses due to a decrease in channel width.

Mullan [65] tested two inducers, one designed on the basis of radial equilibrium consideration [11] and the other with a constant pitch angle (simple helix). Both inducers were tested at various flow coefficients. The inducer designed on the basis of radial equilibrium equations had superior performance. The axial velocities were found to be low at the hub and high at the tip even at design condition, and this trend is reversed at flow coefficients higher than design. At flow coefficients lower than design, however, the performance deteriorates with the presence of the back-flow at $\phi = 0.17$ and 0.14. Steep head rise near the tip is observed at a low coefficient ($\phi = 0.12$). The peak efficiences of the inducers were 80 percent and 72 percent, for the aerodynamically designed inducer and the flat-plate inducer, respectively.

R. F. Soltis et al. [19] also measured steep head rise near the blade tip of a noncavitating 78-deg. axial inducer under various flow coefficients. Using the simplified radial equilibrium equation which relates the static head rise gradient and the tangential velocity

$$\frac{1}{\rho}\frac{\partial p}{\partial r} = \frac{V_\theta^2}{r}$$

they derived the outlet axial velocity profile using experimental values for the total pressure and outlet flow angles. However, at their station of measurement, located about 1/5 diameter downstream of the rotor, the radial velocities are likely to be very small and the flow nearly axisymmetric, since the wake diffusion in such inducers is very rapid.

This analysis serves to establish that the flow is axisymmetric at small axial distances downstream of the trailing edge, which may correspond to large distances along the streamline path (since the streamlines follow tightly wound helical paths). However, the existence of large head rise near the tip cannot be accounted for in this analysis.

Three flat-plate inducers of 78, 80.6, and 84 deg. were tested, respectively, by Soltis et al. [19], Sandercock et al. [17], and Anderson et al. [20] at the NASA Lewis Research Center. The rotors had three blades with a tip diameter of 127 mm, hub/tip of 0.5, and a solidity ranging from approximately 1.8 for the 78-deg. inducer to 3.0 for the 86-deg. inducer. The investigation was carried out in the Center's water tunnel, powered by a 3,000 hp variable-frequency motor. The rotor was tested at various speeds. Details of this tunnel are given in Reference 67. They carried out overall performance measurements at both noncavitating and cavitating conditions, including a radial survey at the exit. The following properties were derived from these measurements: pressure rise coefficient, efficiency, axial velocity at the exit, deviation angle, loss coefficient, and ideal head rise coefficient. In addition, flow visualization was carried out to determine the extent of caviation. The radial distribution of outlet parameters for the 80.6-deg. inducer [17] is shown in Figure 10. The performance is similar to those observed by Mullan [65], who reported a deterioration in exit velocity distribution with a decrease in the flow coefficient. The overall pressure (or head) rise and efficiency showed marked improvement at the lowest flow coefficient ($\phi = 0.107$) tested. This is an improvement over the data reported by Mullan [65] at low flow coefficients. The noncavitating performance showed nearly linear variation of the head rise coefficient with the flow coefficient with a maximum efficiency of 82 percent. A very sharp increase in the loss coefficient towards the tip region was observed at all flow coefficients tested. Other conclusions are similar to those of Mullan [65] described earlier.

Overall noncavitating performances of the inducers with 78-, 80.6-, and 84-deg. stagger angles at the tip are shown in Figure 11 for the range of rotor speeds from 9,000 to 15,000 rpm. The solid symbols on the performance curves for each inducer locate the operating conditions at which zero axial velocities are first observed at the hub outlet measuring stations. The slope of the ψ-ϕ curve increases as the stagger angle is increased. The NASA Lewis Research Center personnel have thus demonstrated successful operation of the inducer at very low flow coefficients.

The inducers tested by other investigations [4, 5, 7, 12, 14, 42, 43, 53, 57, 62–68, 72, 73] showed trends similar to those measured by the NASA Lewis group [17, 19, 20] and hence will not be described here.

Oshima [64] tested two-bladed inducers with two different tip clearances (0.5 mm and 10 mm), a stagger angle of 75 deg. at the tip, and a tip diameter of 206 mm, and found that the suction performance deteriorates with an increase in tip clearance height when the eye diameter of the

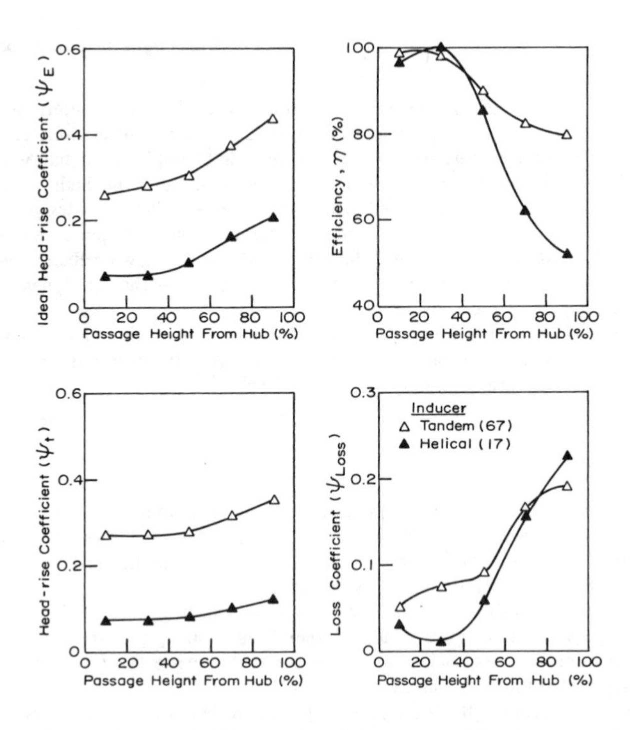

Figure 10. Comparison of blade element parameters for tandem-bladed inducer and 80.6-deg. helical inducer [67].

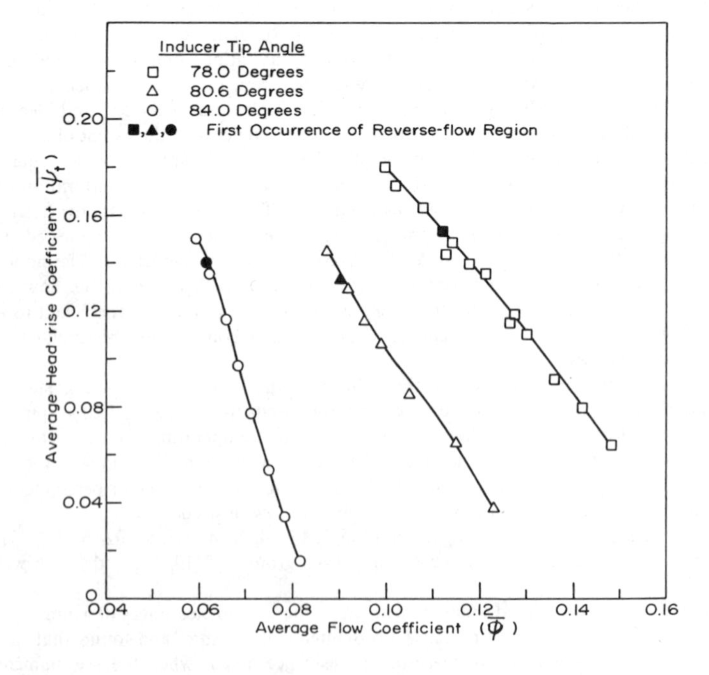

Figure 11. Comparison of overall performance of 78-, 80.6-, and 84-deg. helix angle inducers [20].

main impeller is nearly equal to the normal inducer diameter. Other configurations tested showed that the behavior is less sensitive to the tip clearance.

The main conclusions of the various studies just described are:

1. The overall head rise coefficient increases when the operating flow coefficient decreases. The head rise coefficient is very high near the tip (two to three times the design values) and nearly uniform from the hub to mid-radius.
2. The total head rise coefficient increases when the solidity of the blades is decreased in the practical range of solidities used in inducers.
3. The radial distribution of outlet velocity tends to deteriorate when the flow coefficient is decreased. At low flow coefficients and for most inducer configurations, there is a large positive radial gradient in the exit axial velocity with a backflow near the hub. This is dictated by the radial equilibrium as described earlier. Another probable cause of this backflow is the large redistribution of the flow as it leaves the trailing edge, the radial velocity decaying from positive values inside the blade passage to zero as the flow becomes axisymmetric downstream.

The radial loss distribution shows substantially higher losses from mid-radius to the tip. The efficiency is maximum near the hub and reaches very low values near the tip of the blade.

Tandem and Hubless Inducers

Soltis et al. [62, 67] tried to improve the performance of flat-plate inducers by employing a tandem rotor, as shown in Figure 1. In this approach, the high blade loading is distributed between two or more closely coupled blade sections so that the blade surface boundary-layer separation is prevented on any of the individual blades. The comparison between the flat-plate inducer and the tendem inducer tested by Soltis et al. [62, 67] is shown in Figure 10. The average flow coefficient ($\bar{\phi}$) and head rise coefficient ($\bar{\psi}$) of the flat-plate inducer were 0.107 and 0.085, respectively. These values for the tendem rotor were 0.109 and 0.303, respectively. The comparison indicates that the efficiency and head rise are generally higher and the losses lower near the tip for a tandem rotor. Substantial improvement near the tip section (η and ψ_{loss}) indicates that the tandem rotor prevents a large accumulation of low-energy fluid in the tip region. The tandem rotor tested by Soltis et al. [62, 67] also maintained great efficiency over a large portion of the flow range.

Another concept which has been tried to improve the performance of these inducers is to utilize a hubless inducer, shown schematically in Figure 1. The entire shroud-blade assembly rotates. The advantages of this design configuration are:

1. The elimination of tip vortices.
2. The centrifuging effect on the cavitation bubbles, causing them to cluster around the center of the inducer where they collapse without causing matrial damage.
3. The achievement of high suction speed by employing high sweepback in the vanes.

This concept is originally due to Jekat [15, 73] and has been explored further by Miller and Gross [18] and Lindley and Martinson [72]. Jekat observed all the advantages mentioned earlier. The tip vortex cavitation was eliminated, and the inlet vane cavitation was reduced by large sweep back in the vanes. The predicted centrifuging of the cavitation bubble and its collapse (without causing material damage) was confirmed. The peak efficiences measured by Miller and Gross [18] indicate that their value did not exceed 65 percent, while a flat-plate inducer (with hub) tested by Sandercock et al. [17] has a peak efficiency of 82 percent, and the corresponding tandem inducer [67] has a peak efficiency of 85 percent. Even though hubless inducers have better cavitation characteristics, their poor overall efficiency makes them less attractive for commercial application. Lindley and Martinson [72] concluded from a full-scale test that the cavitation performance of a hubless inducer is not significantly better than the conventional inducer.

Since no information is available on the exit flow properties, such as the radial distribution of the axial and tangential velocities and the pressure rise characteristics, it is difficult to conclude from the available data that the hubless inducer has better overall performance and flow characteristics. It is not clear how the core flow near the center would affect the performance of the main pump.

INDUCERS TESTED IN AIR

The experimental investigation of the flow phenomena in rocket pump inducers carried out at the Pennsylvania State University is summarized here. All the experimental investigations were conducted with air as the test medium. Details can be found in References 21 through 41 and a summary in Reference 71. Rocketdyne [74] tested pumps with air as the working fluid and found that a very good correlation exists between air test data and the liquid test data. While the characteristic form of the inducer is dictated by cavitation requirements, the flow is subjected to major effects of viscosity and turbulence in the long and narrow passages. The investigations reported here are concerned with the effects of viscosity, not the effects of cavitation.

One of the major objectives of this investigation was to understand the flow phenomena in these inducers through flow visualization and conventional and hot-wire probe measurement inside and at the exit of the blade passage, and to provide analytical methods for the prediction of flow through inducers. To achieve the objectives just mentioned, the experimental work was carried out using four-, three-, and two-bladed inducers with aerodynamically designed blades. The blades were identical in all these cases. Wislicenus and Lakshminarayana [21] made the first attempt to design inducers on the basis of rational aerodynamic concepts. Extensive data is available for this inducer.

The basic research and boundary-layer investigation was carried out using a helical flat plate (of the same dimensions as the inducer blades tested), and a flat-plate helical inducer (four-bladed). The flow measurements were carried out using triaxial hot-wire probes and pitot probes. Details of the measurement technique are given in Reference 69. Detailed mean and turbulence flow fields inside the passage as well as at the exit of the rotor were derived from these measurements.

The boundary layer, end-wall, and other passage data revealed the extremely complex nature of the flow, in particular the major effects of viscosity present across the entire passage. A summary of the configurations used, the nature of measurements taken, and the references where they are reported is tabulated in Table 1. The inducer characteristics and the design of the inducer are described in References 21 and 31. Both the published and the unpublished results from the Penn State program are reviewed here.

A visualization study of the flow through the four-bladed inducer configuration is reported in References 28 and 52. The flow was found to be highly three-dimensional, with appreciable radial

Table 1
Summary of Inducers Tested in Air at the Pennsylvania State University
(Tip Dia, = 0.915 m, Re = 6.6×10^5)

Configuration	Design	C^* in.	(C^*/S^*)	β_1^* degrees	(β_1)hub degrees	ϕ	Measurements and analysis carried out	Refs. where reported
Four-bladed inducer	Cambered blades-free vortex design	83	2.86	86°15′ -	75°30′	0.065	Flow visualization inside the passage; exit, entry, and passage flow measured with a pitot probe. Approximate analyses.	[21,22,23,27, 28,31,32,41]
Four-bladed inducer	Flat plate	95	3.4	84°	78°	0.04 to 0.067 and open throttle	Exit flow measurement; extensive hot-wire measurement at 1600 locations inside th- passage to define mean velocity, turbulence intensity, and stress; blade static-pressure and skin-friction stress. Boundary-layer analysis.	[26,30,38,39, 40]
Three-bladed inducer	Cambered blades	83	2.14	86°15′	75°30′	0.065	Exit and entry flow using both hot-wire and pitot probe; hot-wire probe measurements inside the passage. Numerical solution of inviscid and viscid equations.	[26,27,30,32, 33,34,35,36, 37,41]
Two-bladed inducer	Cambered blades	83	2.14	86°15′	75°30′	0.065	Exit and entry flow measurements.	[27]
Single blade	Flat plate	95		84°	78°	open throttle	Mostly three-dimensional turbulent boundary-layer measurements.	[24,25,26,29]

*Values at the tip

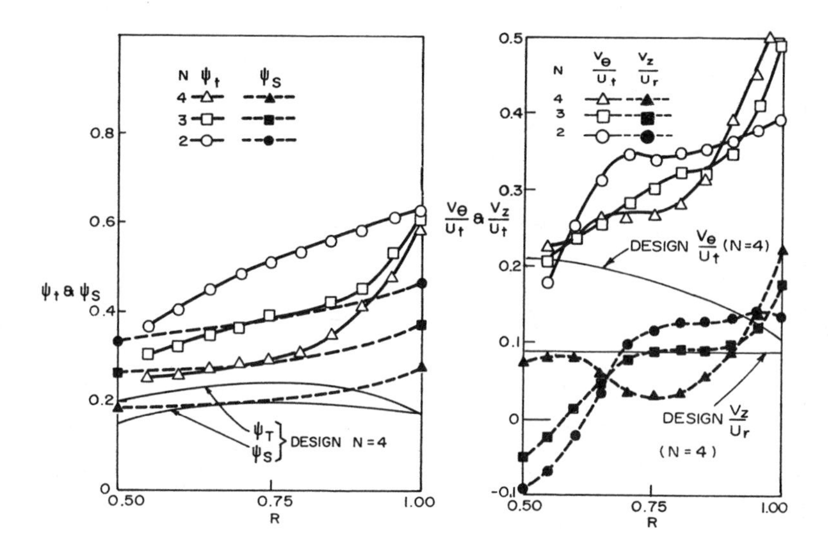

Figure 12. Radial variation of flow properties for four-, three-, and two-bladed inducers.

velocity throughout the passage. At or near the design flow coefficient, no backflow is observed upstream of the inducer. A separated region of the flow exists near the hub at the discharge of the inducer. The extent of the backflow increases considerably both at inlet and at exit, for flow coefficients lower than the design value. The radial velocities within the blade passage appear to be quite strong at all radii. The radial movement inside the blade boundary layer, when encountered by the annulus wall, tends to deflect towards the mid-passage and then radially inward. These radial flows exist near the outer radius (mid-radius to tip) and are found to be large (Figure 3).

Measurements Downstream of the Aerodynamically Designed and Flat-Plate Inducers

The flow measurement carried out at several stations downstream of the four-, three-, and two-bladed inducer blade rows tested at the same flow coefficient are reported in References 22, 23, 25–28 and 30–41. The major conclusions from the four-bladed inducer results are presented here.

The test inducer, designed approximately for uniform head distribution over its discharge area, actually produces a very nonuniform head near the tip, as shown in Figure 12. The axial velocity distribution shows maximum departure from design, as shown in Figure 12. The velocity near the hub shows a tendency to separate, with maximum values occurring near the tip. Measurements very close to the trailing edge reveal a different trend, with minimum velocity occurring at mid-radius [27]. The measured tangential velocity (absolute) is substantially higher than the design, especially from the mid-radius to the tip (Figure 12).

The loss coefficient (ζ) calculated from the measured Euler head coefficient ($\psi_E = 2UV_\theta/U_t^2$) and measured pressure coefficient (ψ_t) is plotted in Figure 13. The losses are much higher than for any other conventional turbomachinery, especially in the outer radii. The hydraulic efficiency, $\eta = \psi_t/\psi_E$, plotted in Figure 13, shows near-normal efficiencies (90 percent) near the root and 50 percent efficiency near the tip for the four-bladed inducer. The mass-averaged efficiency given by:

$$\bar{\eta} = \int_{r_h}^{r_t} \eta V_z r \, dr \Big/ \int_{r_h}^{r_t} V_z r \, dr \tag{18}$$

is found to be 64 percent of the four-bladed inducer, 76 percent for the three-bladed inducer, and 88.5 percent for the two-bladed inducer.

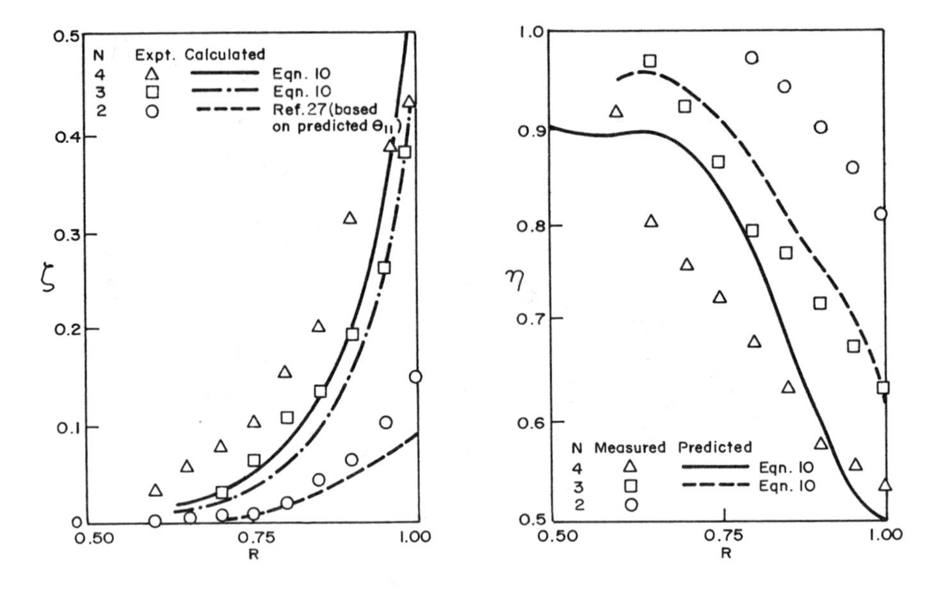

Figure 13. Calculated and measured distribution of loss coefficient (ζ) and efficiency (η) for Penn State inducers.

Effect of Solidity (or Number of Blades) on Overall Performances

Major conclusions derived by comparison of the measurements at the exit of the two-, three-, and four-bladed inducer are as follows (Figures 12 and 13).

The results indicate that the performance, both pressure rise and efficiency, improve continuously with a decrease in solidity (at the same flow coefficient ϕ). The two-bladed inducer shows the best performance, even though the backflow region near the hub at the exit is increased with a decrease in solidity. The static-pressure (or head) rise increases continuously, at all radii, with the decrease in solidity (Figure 12).

The radii distribution of the stagnation-pressure rise coefficient indicates that strong real fluid effects are evident at all solidities. ψ_t increases continuously with the decrease in solidity. The frictional effects play a dual role here. The energy exchange or head rise is the desirable part. Associated with friction is the energy dissipation of the fluid flowing through the pump. These losses arise from (in addition to secondary and leakage flow near the annulus and hub walls):

- Skin friction on the blade surface.
- Interaction between pressure surface and suction surface boundary layers, especially near the outer wall.

The radial movement inside the blade boundary layer, when encountered by the annulus wall, tends to deflect towards midpassage and then radially inward. Intense turbulence and mixing in this region give rise to considerable flow loss. Both of these losses are reduced when the spacing is increased, even though the effect of spacing on energy transfer is not too appreciable. The net effect is a larger pressure rise, at all the radii, as the blade spacing is increased. It can also be seen in Figure 12 that a steep rise in ψ_t (towards the tip), observed in the case of three- and four-bladed inducers, is substantially reduced for N = 2, thus indicating minimization of the interaction effects near the tip as the solidity is decreased. An observed increase in ψ_t, with the decrease in solidity, is a result of major significance, indicating that the frictional losses are reduced substantially with the increased spacing, even though energy transfer due to inviscid and shear pumping effects show no appreciable change with the spacing. The mass-averaged stagnation-pressure (or head) rise coefficient, defined by:

$$\bar{\psi}_t = \int_{r_h}^{r_t} \psi_t V_z r \, dr \Bigg/ \int_{r_h}^{r_t} V_z r \, dr \tag{19}$$

and shown in Figure 14, indicates that ψ_t increases continuously with a decrease in blade number at all the radii.

The axial velocity profiles are found to be similar, qualitatively, downstream of all the inducers tested, as shown in Figure 12. The steep rise in axial velocity toward the tip observed in the case of three- and four-bladed inducers is absent in the case of the two-bladed configuration, but the extent of the separated zone (or back-flow region) near the hub increases continuously with the decrease in solidity. Thus, the axial velocity profile deteriorates with the decrease in the solidity.

The steep gradient in ζ that exists near the tip for $N = 4$ and 3 is reduced drastically for the two-bladed case (Figure 13). There is a drastic reduction in loss between a three- and two-bladed inducer, thus indicating that the boundary layers in a two-bladed inducer do not cover the entire passage. The hydraulic efficiency for the inducer can be derived from $\eta = \psi_t/\psi_E$, where ψ_E is the ideal or Euler head rise coefficient (Figure 13). This clearly indicates that the efficiency improves, at all the radii, as the solidity is decreased. It is surprising that the local efficiencies improve as much as 30 to 50 percent when the solidity is halved. Even though efficiency is not a major consideration in the selection of inducers for the present-day spacecraft, it would assume added importance when space travel becomes a commercial venture.

The mass-averaged efficiency derived from Equation 18 is shown plotted in Figure 14 and compared with those of Acosta [7], Iura [70], and Soltis et al. [19]; the latter measurements correspond to noncavitating conditions. An appreciable decrease in $\bar{\eta}$ is observed with the increase in solidity, the trend being the same for all of the inducers shown. A physical reasoning for the observed trend in the variation of overall efficiency ($\bar{\eta}$) with σ is as follows: consider inducers operating as the same flow and blade parameters, except for the change in blade spacing S. Since the average loss coefficient (ζ) is proportional to C/d_h or C/S, the solidity, the overall efficiency is:

$$\eta = 1 - \frac{k\sigma}{\psi_e}$$

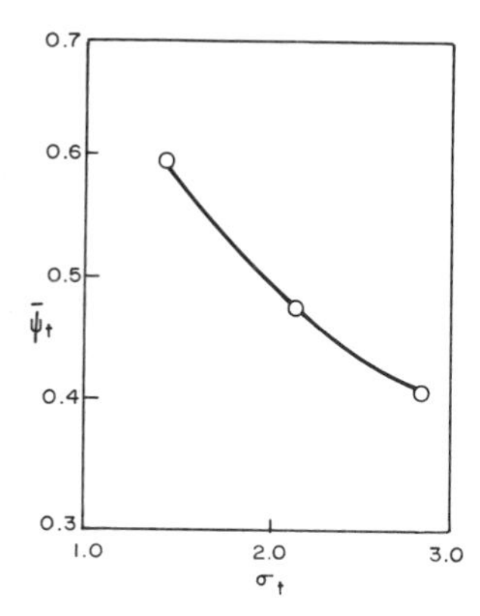

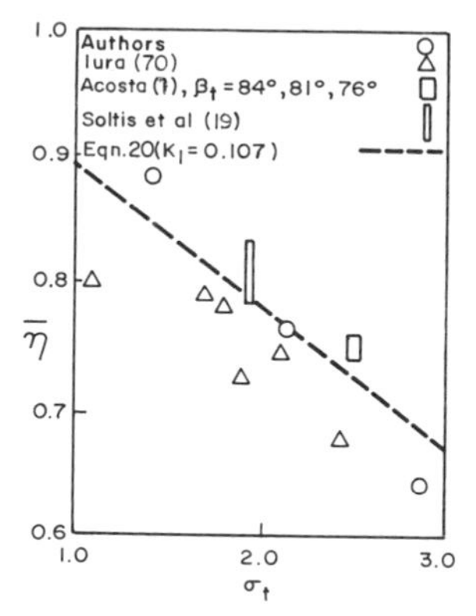

Figure 14. Variation of mass-averaged stagnation pressure rise coefficient ($\bar{\psi}_t$) and ($\bar{\eta}$) with solidity for Penn State inducers.

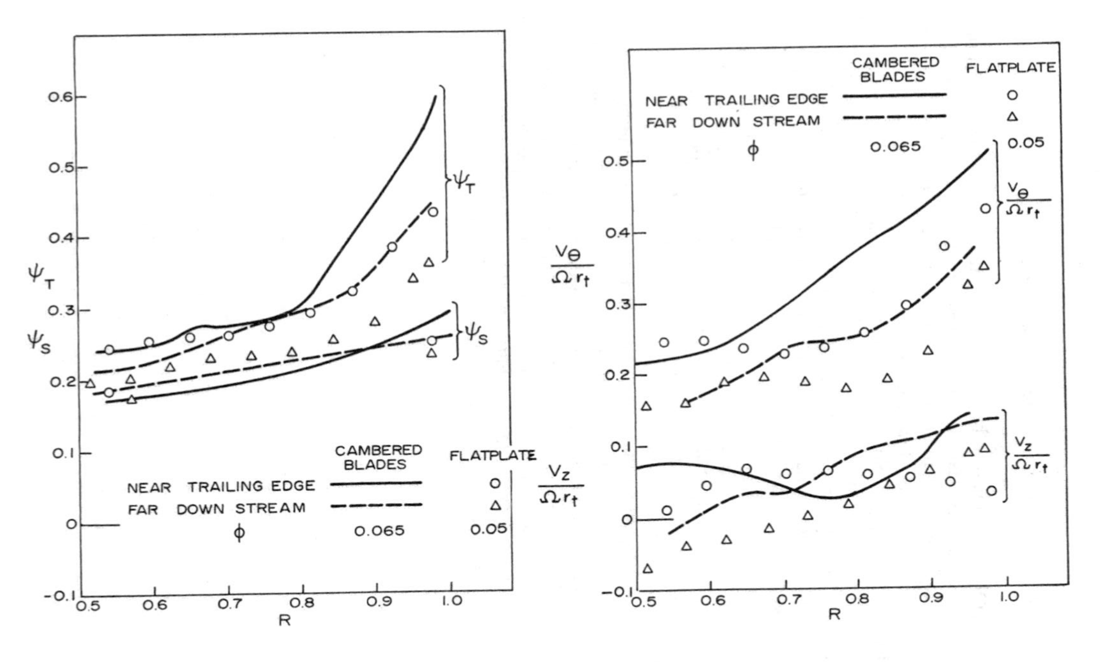

Figure 15. Radial variation of flow properties for the flat-plate inducer and cambered-bladed inducers tested at Penn State [41].

where k is a constant. Since ψ_E, the average Euler head rise coefficient, is nearly the same for all the inducers tested in this program (Figure 14), overall efficiency ($\bar{\eta}$) can be expressed as:

$$\eta = 1 - k_1 \sigma \tag{20}$$

The values of $\bar{\eta}$ derived from Equation 20 with $k_1 = 0.107$ agree reasonably well with the measured efficiencies (Figure 14). This adequately explains the decrease in efficiency for increased solidity.

Flat Plate Inducer

A flat-plate inducer, whose details are shown in Table 1 and described in Reference 39, was tested in air. The stagnation-pressure rise for the flat-plate inducer exhibits characteristics which are very similar to an aerodynamically designed inducer. A steep rise in stagnation pressure was observed near the tip even in the case of the flat-plate inducer with an open throttle (no inviscid effects). This provides a convincing argument that the tip of the inducer blade behaves as a shear pump, where the pressure or head rise is brought about entirely by the viscous effects. The absolute tangential velocity distribution [71], even in the absence of inviscid turning effects, is very similar to the distribution observed for the aerodynamically designed inducer.

For the purpose of comparison, the flat-plate inducer data at $\phi = 0.05$ and the data from the aerodynamically designed inducer ($\phi = 0.065$) are chosen. This choice is based on nearly identical pressure rise across the inducers from hub to tip [71, 41]. The static-pressure and stagnation-pressure rise coefficients, plotted in Figure 15, indicate nearly similar characteristics, even though the stagnation-pressure rise near the tip is about 50 percent lower for the flat-plate inducer near the trailing edge. From mid-radius to the tip, the aerodynamically designed inducer shows a much steeper gradient in the head rise coefficient. The major cause of this is probably due to the "shear pumping effect."

The radial distribution of the absolute tangential velocity and the axial velocity are also compared in Figure 15. The tangential velocity follows the same trend as the stagnation head rise coefficient, but the major discrepancy occurs in the axial velocity distribution. Near the trailing-edge location, the flow in the flat-plate inducer shows a tendency to separate near the hub, while the flow in the aerodynamically designed inducer does not. The axial velocity near the tip for the aerodynamically designed inducer is higher, a consequence of higher radial velocity [41]. Both inducers show the same trend far downstream, with a larger backflow region observed for the flat-plate inducer.

Blade-to-Blade Distribution of Flow Properties at the Exit of the Inducer

Pitot probes, wedge probes, and three-sensor hot-wire probes (with ensemble-averaging technique) were utilized at the exit of the inducers to drive flow properties at the exit. Details of these measurements are given in References 22, 33 and 35. These results are summarized in Reference 71. The relative stagnation-pressure coefficient (ψ_R) is found to be nearly uniform across the passage near the hub and mid-radius. The measurements near the tip indicate considerable boundary-layer thickening. This confirms the presence of severe radial flows and, hence, the blade boundary-layer transport towards the tip.

The radial velocities obtained from the hot-wire measurements are found to be of the same order of magnitude as the axial velocities throughout the flow passage. The tubulence-intensity components in the radial, axial, tangential, and relative directions (based on local mean-velocity component) are found to be about 75 percent, 45 percent, 20 percent, and 15 percent, respectively. The turbulence intensities are found to be nearly uniform in the entire flow passage at the exit. These measurements reveal the highly turbulent nature of the flow in inducers.

The distribution of axial, tangential, and relative velocities are similar for both the four- and the three-bladed inducers even though the tangential velocities are higher for the three-bladed inducers.

Measurements Inside the Blade Row

Some of the blade static pressures and the limiting streamline angles for the four-bladed inducers are reported in References 28 and 32, respectively. The most extensive blade-pressure distribution was carried out for the three-bladed inducer and is reported in References 34, 36, and 37.

The ammonia trace technique provides a satisfactory method for determining blade limiting streamline angles on the inducer rotor blade. The blade limiting streamline angle measurements provide several observations which have either been noted in previous investigations or have been found from other experimental results quoted in References 25 through 41. These include: an increase in ϵ from the leading edge to the trailing edge, indicating the existence of higher radial velocities as the flow proceeds downstream within the inducer channel; higher values of ϵ near the hub indicating higher radial velocities in this region; negative values of ϵ at the pressure surface tip (up to 45 percent chord) indicating radially inward flow due to the presence of the annulus-wall boundary-layer scraping effect; and values of α_w (arc tan ϵ) greater than 90 deg. near the hub trailing edge indicating the existence of a backflow region in this area.

The magnitude of blade static pressure distributions are found to be considerably higher than the design values.

In order to understand the flow behavior inside the inducer passages, experimental investigation of the relative flow (blade-to-blade) inside the inducer passages was undertaken using rotating probes and the pressure-transfer device. Pressure-probe measurements inside the four-bladed inducer are reported in References 22, 23, and 32, and for the three-bladed inducer in References 25, 26, 30, 32, 33, and 37. The flow was surveyed in one passage and at several radial locations. Some of the major conclusions derived from measurements in four-bladed inducers as presented in References 22, 23, and 32 are now described.

1. The measurement of relative flow near the trailing edge reveals the presence of a loss core located slightly inward from the tip. The mid-passage at this radius is found to have minimum relative stagnation pressure and, hence, maximum loss. A substantial portion of total flow losses occurs near the leading edge. The losses near the tip are found to be about five times those at other radii (see Figure 11 in Reference 32).
2. The blade boundary layers are found to be quite thin near the hub. The boundary-layer thickness near the mid-radius is about 25 percent of the blade passage, being thicker on the suction surface than on the blade pressure surface. A conventional boundary-layer type of profile exists from hub to mid-radius; whereas, the velocity profile near the tip (mid-radius to tip) is of conventional type near the blade surfaces and "wake" type near the mid-passage. Qualitative reasons for the existence of this type of profile are discussed in References 32.
3. Near the trailing-edge station, the passage-averaged values of relative velocity have a maximum value at mid-radius, being considerably lower than the design values from mid-radius to tip. The diffusion of relative flow is caused mainly by the inviscid turning effects from the hub to mid-radius; whereas, the diffusion from mid-radius to the tip is dominantly influenced by the viscid and "shear pumping" effect.

The following conclusions are derived by comparing the relative flow measurements taken in the three-bladed inducer with those of the four-bladed configuration [32, 33]. Measurements taken near the leading edge show marked reduction in boundary-layer growth, flow losses, and radial inward velocity in the case of the three-bladed inducer. The losses near the tip are nearly halved from those of the four-bladed inducer, and the "wake" type of profile observed near the mid-passage of the tip disappears in the case of the four-bladed inducer, as shown in Figure 16. There is appreciable improvement in hub and wall static-pressure distributions in the case of the three-bladed inducer.

Measurements Inside the Blade Passage of a Three-Bladed Inducer Using Hot-Wire Probes

Detailed measurements of mean velocity and turbulent flow field inside the three-bladed inducer were carried out using the rotating hot-wire technique described in References 30, 34, 36, and 37.

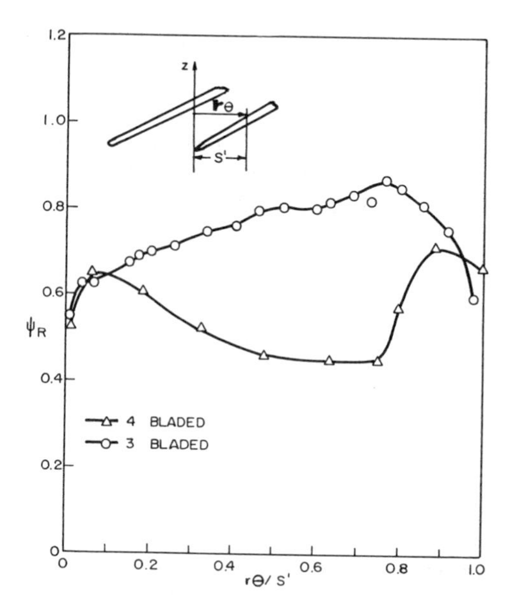

Figure 16. Stagnation pressure coefficient of the relative flow at 1/3 blade chord from the leading edge for R = 0.975 for four- and three-bladed inducers.

The flow was measured at various radii and across the blade at axial stations approximately at 1/3 blade chord and trailing edge.

Total relative velocity profiles, derived from the rotating triaxial hot-wire measurements, indicate a substantial velocity deficiency near the tip at mid-passage, which expands significantly as the flow proceeds downstream toward the inducer trailing edge (Figure 9). The higher values of axial velocity near the hub, shown in Figure 9, indicate the significant effect of blade blockage within the inducer flow passage. A slight backflow was found to exist at the extreme tip location at 1/3 blade chord from the leading edge and can be attributed to the annulus-wall boundary-layer scraping effect identified in the limiting streamline angle (ϵ) measurements. The measured magnitudes of radial velocity are found to be the same order as those of axial velocity within the inducer passage. The large values of radial velocity confirm the highly three-dimensional characteristics of the inducer flow and emphasize the necessity for a three-dimensional theory for accurate flow analysis. Turbulence levels within the blade passage, indicated from the experimental results, are generally high near the tip regions. The radial component of turbulence intensities appears to have the largest magnitudes, reaching values of up to 24 percent in the mixing region.

Boundary Layer and Related Phenomena on Inducer Blades and Passages

A systematic study of the boundary layer on a rotating blade and inside a rotating channel was undertaken by the Penn State group. In the first phase of the program, the boundary layer on a simpler configuration, namely, a rotating helical blade of large chord length enclosed in an annulus, was studied [24, 26, 29]. In the absence of chordwise pressure gradients, the effect of rotation on the boundary layer can be discerned.

The interference due to adjoining blades ("channel effect") that exists in turbomachinery passages is not simulated in the preceding model. With this as an objective, studies of the influence of the rotation and "channel effects" as well as pressure gradient effects were carried out in a rotating channel, with the geometry of a four-bladed flat-plate inducer [30, 38–40]. In both of these experiments, air was employed as the test medium.

The following conclusions may be drawn from the theoretical analytical investigations carried out on a single rotation helical blade [24, 29]:

1. The solution of the momentum integral equations predict, accurately, the boundary-layer growth and the limiting streamline angle ϵ at various locations on the rotating-blade surface.
2. About 100 skin-friction measurements carried out at various radial and tangential locations of the blade and at several rotational speeds indicate that the skin-friction coefficient on a rotating blade can be represented by:

$$\frac{c_f}{c_{f0}} = 1 + 0.85R_0^{1/2}$$

where $R_0 = \epsilon\theta$, the theoretically derived rotation number (ratio of the Coriolis force to the inertial force inside the boundary layer), and c_{f0} = skin-friction coefficient for a flat plate at zero pressure gradient. A similar expression is given for the developed region of the flow.

Boundary Layer Development in a Flat-Plate Channel

In the axial flow inducer, the interaction between pressure surface and suction surface boundary layers would result in an extremely complex flow, especially near the outer half of the blade span. These effects were studied in a helical flat-plate inducer specifically designed and fully instrumented for this purpose. Details of the program and the Reynolds equation valid for this flow are given in Reference 25, and velocity-profile analysis and details of instrumentation in Reference 26. Details on the velocity profiles carried out with mild and strong pressure gradients, as well as detailed analyses, are described in References 38–40.

A boundary-layer flow analysis was carried out using the momentum integral technique, which employs three-dimensional equations of motion in the rotating coordinate system [38–40]. The measurements of the boundary-layer characteristic (three components of velocity, turbulence intensities, Reynolds stresses, skin-friction coefficient, and limiting streamline angle) were carried out utilizing the conventional and triaxial hot-wire probes rotating with the rotor inside the passage [38, 40].

The inducer was tested with open throttle [38, 39] and at $\phi = 0.05$ [39, 40]. The momentum thickness and the limiting streamline angle on the leading surface predicted from a three-dimensional momentum integral analysis is shown in Figure 17. The momentum thickness increases rapidly near the leading edge, reaching an asymptotic value at nearly 1/3 chord from the leading edge. It starts to decrease near the trailing edge. As expected, the momentum thickness increases toward the tip. The predictions are quite good, except near the blade tip. The limiting streamline angle, shown plotted in Figure 17, indicates that the radial velocities are large in the laminar region and lower in the turbulent region. The limiting streamline angle decreases as the outer wall is approached. The predictions from the momentum integral analysis are quite good.

The skin-friction coefficient for the rotating channel is found to vary with three important three-dimensional boundary-layer parameters: θ_{11}, ϵ, and H. The proposed correlation for the skin-friction coefficient, based on the limited data in the rotating channel with mild pressure gradients, is:

$$C_f = 0.172R_{\theta_{11}}^{-0.268}10^{-0.678H}[1 + 0.52\sqrt{\epsilon_w(x - x_t)/C}]$$

The term in the brackets is due to three-dimensional effects. Variation with $R_{\theta_{11}}$ and H is found to be the same as that for a stationary channel.

The streamwise and radial velocity profiles inside the four-bladed flat-plate inducer at $\phi = 0.05$ measured by a triaxial hot wire probe are shown in Figure 18. The boundary layers are thin near the hub (R = 0.565) and show a substantial increase as the tip is approached. As the flow proceeds downstream the hub boundary layer is thrown outwards, resulting in a complex flow near the tip. Turbulent shear stress measurements [40] show that in three-dimensional rotating turbulent layers,

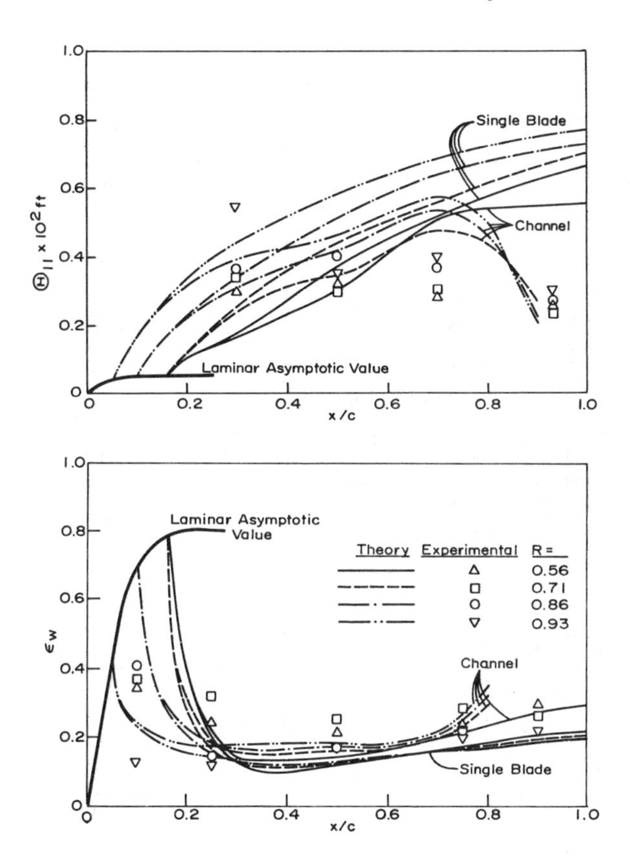

Figure 17. Predicted and measured variations of momentum thickness and limiting streamline angle on the leading surface of the Penn State flat-plate inducer rotor (open throttle) [38].

all three of the velocity cross-correlations are of the same order of magnitude inside the boundary layer.

CONCLUDING REMARKS

The major significance of the research carried out hitherto is that it has provided a good understanding of the flow in inducers where the real fluid effects are at least equally important to the ideal fluid effects which form the basis of existing design methods of turbomachinery. It is hoped that the results reviewed will help establish a theoretical model for the eventual analysis and design of turbomachinery dominated not only by viscous and turbulence effects inside the turbomachinery passages, but also by the secondary flow and other interaction effects near the end-wall regions. Techniques have been developed for the solution of three-dimensional inviscid equations of turbomachinery. It is anticipated that the analytical and experimental investigations reviewed here will lead to the eventual development of a three-dimensional viscid program that will include all the dominant viscous and turbulence terms. The Penn State research program has provided a large amount of accurate data for the boundary layer on rotating blades and in rotating channels. These could be utilized by investigators presently involved in the development of computer codes.

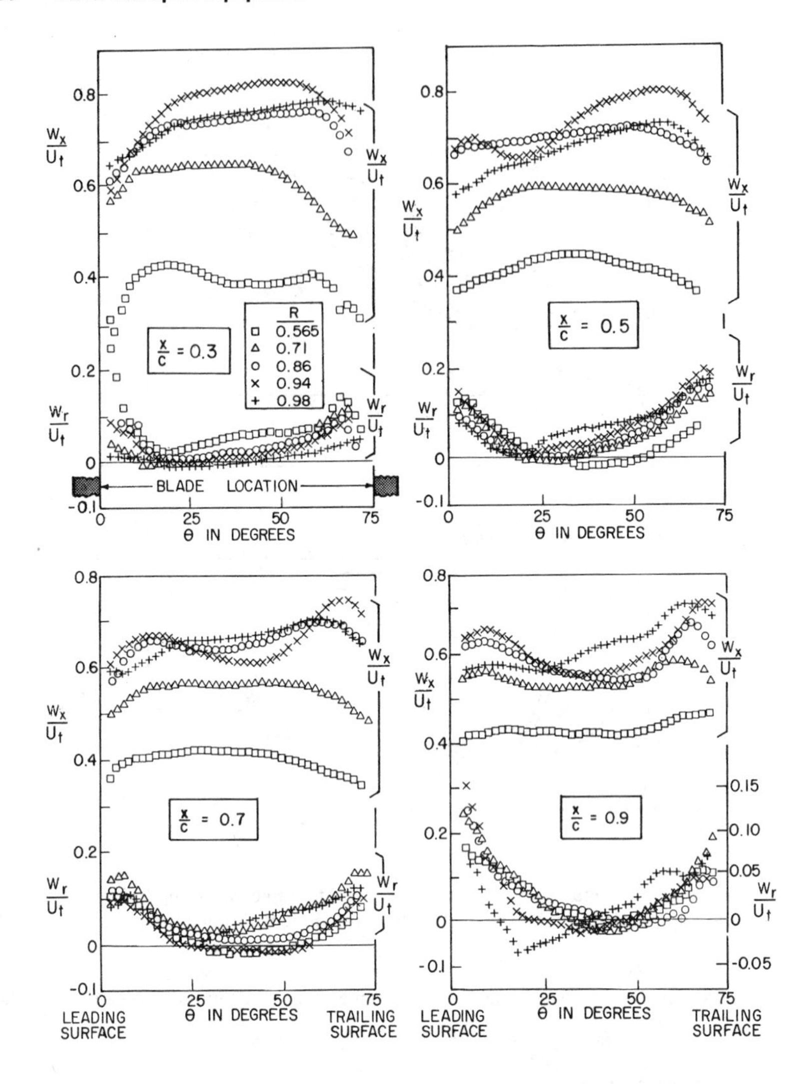

Figure 18. Blade-to-blade variation of streamwise and radial velocity profile for the Penn State flat-plate inducer at $\phi = 0.05$ [40].

The radial velocities are of the same order of magnitude as the axial velocities; hence, inclusion of these as well as the dominant turbelence terms is essential for the accurate prediction of the flow. The inner part of the rotor (hub to mid-radius) could be designed from inviscid considerations, but the design of the outer part (mid-radius to tip) should take into account the "drag or shear" pumping effect. The semiempirical friction loss coefficient developed and used for the prediction of head rise characteristics, even though unsatisfactory in view of our better insight into the loss mechanism, could serve as a useful tool in the preliminary design or analysis of inducers. The performance deteriorates with an increase in the number of blades. These results provide a design criteria in the selection of solidity in inducers.

Acknowledgments

The author wishes to acknowledge the National Aeronautics and Space Administration for sponsoring (NASA Grant NGL 39-009-007) the research carried out at The Pennsylvania State University on inducer flows. Comments, criticism, and encouragement by Werner R. Britisch of the NASA Lewis Research Center and George F. Wislicenus of The Pennsylvania State University are gratefully acknowledged.

Special thanks is also given to the American Society of Mechanical Engineers for granting permission of reprint. This chapter first appeared in *Journal of Fluids Engineering*, vol. 104 (Dec. 1982).

NOTATION

C	chord length
C_f	skin-friction coefficient ($2\tau_0/\rho W_e^2$)
d_h	hydraulic mean diameter
H	stagnation head rise (in absolute flow)
H_E	head rise derived from Euler's equation (UV_θ/g)
H_{loss}	$H_E - H_m$
H_m	measured stagnation head rise (in absolute flow)
H_{sv}	net positive suction head, ft
h	static head
L	lead of the helix
N	number of blades
p, P_0	static and stagnation pressure
P_{OR}	stagnation pressure of the relative flow
Q	flow rate, gpm
R	r/r_t
Re	Reynolds number based on blade tip speed ($U_t r_t/v$)
R_{ht}	hub/tip ratio
R_N	Reynolds number, $\bar{W}d_h/\nu$
r, θ, z	radial, tangential, and axial coordinates
SS	suction specific speed,
S	blade spacing
U	blade speed
V, ν	resultant and fluctuating components of absolute velocity
V_r, V_θ, V_z	radial, tangential, and axial components of absolute velocity
W, w	resultant and fluctuating components of relative velocity
W_r, W_θ, W_z	radial, tangential, and axial components of relative velocity
x, y	coordinates parallel to and perpendicular to blade camber line and lying on a cylindrical surface
x_t	location of transition point
ϵ	angle between limiting streamline and x-direction
β	relative air angle measured from axial direction
λ_R	friction coefficient for a rotating channel
ν	kinematic viscosity
ρ	density
ϕ	flow coefficient $W_z/\Omega r_t$
ψ_E	$2gH_E/U_t^2$
ψ_{loss}	$2gH_{loss}/U_t^2$
ψ_S	static head rise coefficient, $2gH/U_t^2$
ψ_t	stagnation headrise coefficient, $2gH/U_t^2$
ψ_R	stagnation head coefficient of relative flow, $2P_{OR}/\rho U_t^2$
Ω	angular velocity
η	hydraulic efficiency, ψ_t/ψ_E
ζ	$\psi_E - \psi_t$
θ_{11}	momentum thickness in chordwise direction
$R\theta_{11}$	$W_e\theta_{11}/\nu$
σ	solidity (C/S)
τ	shear stress

Subscripts

h, t	hub, tip
m	values at mid-passage, free stream
r, θ, z	components along r, θ, z directions
x, y	components along x, y directions
1, 2	inlet, outlet
e	edge of the boundary layer

Superscript

– mass-averaged over the entire passage/circumferential average

REFERENCES

1. Ross, C. C., "Some Historical Notes on Turbopump Systems for Liquid Rockets," ASME Paper 64WA/FE22, 1964.
2. Rothe, K., "Turbopump Configuration Selection for the Space Shuttle Main Engine," ASME Paper 74-FE-33, 1974.
3. Johnson, J., and Colbo, H., "Update of the Space Shuttle Main Engine (SSME)," AIAA Paper 78-1001, 1978.
4. Susuda, S., and Kitamura, N., "Experimental Studies on Centrifugal Pump with Inducer for Water Jetted Propulsion," Inst. Mech. Engrs. (London), Paper C165/74, 1974.
5. Arcand, L., "The Performance of Two Axial Flor Water Jet Pumps," ASME Symp. on Pumping Machinery for Marine Propulsion, Mar. 1968.
6. Luscher, W. P., "Commercial Application of Rocket Turbopump Technology," NASA SP 5030 (Symposium on Technology Status and Trends), Apr. 1965, pp. 21–35.
7. Acosta, A. J., "An Experimental Study of Cavitating Inducers," Proc. Second Symposium on Naval Hydrodynamics, Washington, DC, Aug. 1958, pp. 553–557.
8. Jacobsen, J. K., "Liquid Rocket Engine Turbopump Inducers," Series on NASA Space Vehicle Design Criteria, NASA SP8052, May 1971.
9. King, A. L., "References on Inducers for Rotodynamic Pumps," British Hydromechanics Research Association Report B1B24, Sept. 1967.
10. Wislicenus, G. F., "Preliminary Design of Turbopumps and Related Machinery," NASA Reference Publication (RP) (to be published 1982).
11. Stenning, A. H., "The Design of Axial Inducers for Turbopumps," M.I.T. Gas Turbine Lab Rept. 44, Feb. 1958.
12. Wright, M. K., "Design Comments and Experimental Results for Cavitation Resistant Inducers Up to 40,000 Suction Specific Space," *Trans. ASME* (Apr. 1964), pp. 176–180.
13. Janigro, A., and Ferrini, F., "Recent Progress in Pump Research," Von Karman Inst. Lecture Series 61, 1973.
14. Ross, C. C., and Benerian, G., "Some Aspects of High Suction Speed Pump Inducers," *Trans. ASME* (Nov. 1956), pp. 1715–1721.
15. Jekat, W. K., "The Worthington Inducer," NASA Final Report on Contract NAS8-2680, 1964.
16. Farquahr, T., and Lindley, B. K., "Hydraulic Design of the M-1 Liquid Hydrogen Turbopump," NASA CR 54822, 1966.
17. Sandercock, D. M., Soltis, R. F., and Anderseon, D. A., "Cavitation and Non-Cavitation Performance of 80.6° Flat Plate Helical Inducer at Three Rotational Speeds," NASA Technical Note D-1439, Nov. 1962.
18. Miller, C. D., and Gross, L. A., "A Performance Investigation of an Eight-Inch Hubless Pump Inducer in Water and Liquid Nitrogen," NASA TND 3807, Mar. 1967.
19. Soltis, R. F., Anderson, D. A., and Sandercock, D. M., "Investigation of the Performance of a 78° Flat Plate Helical Inducers," NASA Technical Note D-1170, Mar. 1962.
20. Anderson, D. A., Soltis, R. F., and Sandercoeck, D. M., "Performance of an 84° Flat Plate Helical Inducer and Comparison with Performance of a Similar 78° and 80.6° Inducers," NASA TND-2553, Dec. 1964.
21. Wislicenus, G. F., and Lakshiminarayana, B., "Design of a Test Inducer," NASA CR-67129, 1965.
22. McCafferty, H. G., "Errors in Measuring the Fluctuating Flow at the Discharge of an Inducer," M.S. thesis, Dept. of Aerospace Engineering, The Pennsylvania State University, 1967.
23. Lakshminarayana, B., "Investigations and Analysis of Flow Phenomena of Secondary Motions in Axial Flow Inducers," NASA CR-103291, June 1969, pp. 1–23.

24. Jabbari, A., "Turbulent Boundary Layer Characteristics on a Rotating Helical Blade," M.S. thesis, Dept. of Aerospace Engineering, The Pennsylvania State University, 1969 (also as NASA CR-105649, NASA Star Index N69-36288, Sept. 1969).
25. Poncet, A., Yamaoka, H., and Lakshminarayana, B., "Investigations and Analysis of Flow Phenomena of Secondary Motions in Axial Flow Inducers," NASA CR-107267, July 1970, pp. 1–115.
26. Yamaoka, H., Lakshminarayana, B., and Anand, A. K., "Investigations and Analysis of Flow Phenomena of Secondary Motions in Axial Flow Inducers," Report to NASA, July 1971, 97 pp.
27. Lakshminarayana, B., and Anand, A. K., "Solidity Effects in Axial Flow Inducers," Proc. Second International J.S.M.E. Conf. on Fluid Machinery and Fluidics, Tokyo, Sept. 1972, pp. 157–166.
28. Lakshminarayana, B., "Visualization Study of Flow in Axial Flow Inducers," *ASME Journal of Basic Engineering* (Dec. 1972), pp. 777–787.
29. Lakshminarayana, B., Jabbari, A., and Yamaoka, H., "Turbulent Boundary Layer on a Rotating Helical Blade," *Journal of Fluid Mechanics*, Vol. 51 Part 3 (1972), pp. 545–569.
30. Anand, A. K., et al., "Investigation of Boundary Layer and Turbulence Characteristics Inside the Passages of an Axial Flow Inducer," NASA CR-121248, July 1973, pp. 1–220.
31. Lakshminarayana, B., "Experimental and Analytical Investigation of Flow Through a Rocket Pump Inducer," Fluid Mechanics, Design and Acoustics of Turbomachinery, NASA SP304, Part II, 1974, pp. 690–731.
32. Lakshminarayana, B., "Three Dimensional Flow in Rocket Pump Inducers, Part 1: Measured Flow Field Inside the Rotating Blade Passage and at the Exit," *ASME Journal of Fluids Engineering*, Vol. 95 (Dec. 1973), pp. 567–578.
33. Poncet, A., and Lakshminarayana, B., "Investigations of Three Dimensional Flow Characteristics in a Three-Bladed Rocket Pump Inducer," NASA CR-2290, 1973.
34. Gorton, C. A., and Lakshminarayana, B., "Analytical and Experimental Study of the Three-Dimensional Mean Flow and Turbulence Characteristics Inside the Passage of an Axial Flow Inducer," NASA CR3333, 1980, pp. 1–170.
35. Lakshminarayana, B., and Poncet, A., "A Method of Measuring Three-Dimensional Wakes in Turbomachinery," ASME Journal of Fluids Engineering, Vol. 96, No. 2 (June 1974), pp. 87–91 (discussion in *Journal of Fluids Engineering*, Vol. 97, No. 4, (Dec. 1975), pp. 627–628).
36. Gorton, C. A., and Lakshminarayana, B., "A Method of Measuring the Three Dimensional Mean Flow and Turbulence Characteristics Inside a Rotating Turbomachinery Passage," *ASME Journal of Engineering for Power*, Vol. 98, No. 2 (Apr. 1976), pp. 137–146.
37. Lakshminarayana, B., and Gorton, C. A., "Three-Dimensional Flow Field in Rocket Pump Inducers, Part 2: Three Dimensional Viscid Flow Analysis and Hot Wire Data on Three Dimensional Mean Flow and Turbulence Inside the Rotor Passage," ASME Journal of Fluids Engineering, Vol. 99, No. 1 (Mar. 1977), pp. 176–186.
38. Anand, A. K., and Lakshminarayana, B., "Three Dimensional Turbulent Boundary Layer in a Rotating Channel," *ASME Journal of Fluids Engineering*, Vol. 97 (June 1975), pp. 197–210.
39. Anand, A. K., and Lakshminarayana, B., "An Experimental and Theoretical Investigation of Three Dimensional Turbulent Boundary Layer Inside the Passage of a Turbomachinery Roto," NASA CR2888, 1977.
40. Anand, A. K., and Lakshminarayana, B., "An Experimental Study of Three Dimensional Boundary Layers and Turbulence Characteristics Inside a Rotating Channel," *ASME Journal of Engineering for Power*, Vol. 100, No. 4 (1978), pp. 676–690.
41. Lakshminarayana, B., "On the Shear Pumping Effect in Rocket Pump Inducers," book chapter in *Pump—Analysis, Design and Application*, Vol. 1, Worthington Pump Inc., 1978, pp. 49–68.
42. Cooper, P., and Bosch, H., "Three-Dimensional Analysis of Inducer Flow," NASA Report CR-54836, TRW ER-6673A, Feb. 1966.
43. Bosch, H. B., Cooper, P., and Stoermer, W. P., "Advanced Inducer Study," TRW ER-5288, May 1963.
44. Horlock, J. H., *Axial Flow Compressors*, Butterworth, Inc., 1958.
45. Wislicenus, G. F., *Fluid Mechanics of Turbomachinery*, Dover, Vol. II, 1965, pp. 646–683.

46. Vavra, M. H., *Aerothermodynamics of Turbomachines*, Wiley, 1960.
47. Huppert, M. C., et al., "Some Cavitation Problems in Rocket Propellant Pumps," unpublished report, Rocketdyne.
48. Mirolyubov, I. V., "Calculations of the Characteristics of Axial Flow Force Pumps," Aviattsionnaya Teknika, No. 1, 1959, pp. 81–88 (English translation AFSC, FTDTT63-418).
49. Dumov, V. I., "Calculation of Pressure Head Characteristics of Axial Helical Impellers," Taploenergetika, Vol. 11, 1962, pp. 23–27 (English translation AFSC-FTD-TT63-407).
50. Roelks, R. J., "Analytical Investigation of Three Turbopump Feed Systems Suitable for High Pressure Hydrogen Oxygen Rocket Engine Applications," NASA TN D-2974, Aug. 1965.
51. Rains, D. A., "Head Flow Characteristics of Axial Flow Inducers," *Jet Propulsion* (Aug. 1958), pp. 557–558.
52. Lakshminarayana, B., and Wislicenus, G. F., "Investigations and Analysis of Flow Phenomena in Axial Flow Inducers," Report to NASA, Aug. 1965.
53. Newoj, H. J., "Cavitation Studies in Axial Inducers," Professional Degree Thesis, Calif. Inst. 1956.
54. Katsanis, T., "Use of Arbitrary Quasi-Orthogonals for Calculating Flow Distribution in the Meridional Plane of a Turbomachine," NASA TN D-2546, 1964.
55. Montgomery, J. C., "Analytical Performance Characteristics and Outlet Flow Constant and Variable Lead Helical Inducers for Cryogenic Pumps," NASA TN D-583, March 1961.
56. Cooper, P., "Analysis of Single and Two-Phase Flows in Turbopump Inducers," *ASME Journal of Engineering for Power*, Vol. 89 (1967).
57. Davis, R. E., Coons, L. L., and Scheer, D. D., "Internal Streamline Flow Analysis for Turbopump Inducers Under Cavitating and Non-Cavitating Conditions," AIAA Paper 70–629, 1970.
58. Cool, L. L., et al., "Study of Inducer Load and Stress," NASA CR 72712, Vol. 1 (1970), Vol. 2 (1972), Vol. 3 (1972).
59. Barten, H. J., Scheurenbrand, J. A., and Scheer, D. D., "Stress and Vibration Analysis of Inducer Blades Using Finite Element Technique," AIAA Paper 70–630, 1970.
60. Giesing, J. P., "Extension of the Douglas-Neumann Program to Problems of Lifting, Infinite Cascades," Douglas Aircraft Division Report LB-31653, July 1964.
61. Lakshminarayana, B., and White, M. T., "Airfoil in a Contracting or Diverging Stream," *J. Aircraft*, Vol. 9, No. 5 (May 1972), pp. 354–360.
62. Soltis, R. F., Urasek, D. L., and Miller, M. J., "Overall Performance of a Tandem Bladed Inducer Tested in Water," NASA TN D-5134, May 1969.
63. Moore, R. D., and Meng, P. R., "Comparison of Non Cavitation Performance for 78°, 80.6°, and 84° Helical Inducers Operated in Hydrogen," NASA TN D-6361, May 1971.
64. Oshima, M., "The Effect of Inducer Tip Clearance on Suction Performance," *Bull. JSME*, Vol. 13, No. 58 (1970), pp. 555–581.
65. Mullan, P. J., "An Investigation of Cavitating Inducers for Turbopumps," Gas Turbine Laboratory, M.I.T., Report No. 53, May 1959.
66. Osborn, W. M., "Investigation of a Liquid-Fluorine Inducer and Main-Stage Pump Combination Designed for a Suction Specific Speed of 20,000," NASA TM X-1070, Mar. 1965.
67. Soltis, R. F., Urasek, D. C., and Miller, M. J., "Blade Element Performance of a Tandem-Bladed Inducer Tested in Water," NASA Technical Note D-5562, Nov. 1969.
68. Crouse, J. E., Montgomery, J. C., and Soltis, R. F., "Investigation of the Performance of an Axial Flow Pump Designed by the Blade Element Theory—Design and Overall Performance," NASA TN D-5915, 1961.
69. Lakshminarayana, B., "Techniques for the Measurement of Rotor Flow in Turbomachinery," Proc. AS, E Symposium on Measurement Methods in Rotating Components of Turbomachinery, edited by B. Lakshminarayana and P. Runstadler, Jr., 1980 (published in *ASME Journal of Engineering for Power* (April 1981).
70. Iura, T., discussion of Reference 6, p. 554.
71. Lakshminarayana, B., "Analytical and Experimental Study of Flow Phenomena in Non-Cavitating Rocket Pump Inducers," NASA CR 3471, Oct. 1981.

72. Lindley, B. K., and Martinson, A. R., "An Evaluation of a Hubless Inducer and a Full Flow Hydraulic Turbine Driven Inducer Boost Pump," NASA CR 72995, 1971.
73. Jekat, "A New Approach to the Reduction of Pump Cavitation, The Hubless Inducer," *ASME Journal of Basic Engineering*, Vol. 89 (1967), pp. 125–139.
74. King, J. A., "Testing Pumps in Air," *ASME Journal of Engineering for Power* (1968) (ASME Paper No. 67-WA/FE-4).

Uncited References

Cavitation Performance

Also see References 4–20, 43, 53, 55–58, 63–67, 73 for cavitating aspects of inducers.

75. Stripling, L. B., and Acosta, A. J., "Cavitation in Turbopumps, Part 1," *ASME Journal of Basic Engineering*, Vol. 84, No. 3, (Sept. 1962), p. 326.
76. Stripling, L. B., "Cavitation in Turbopumps, Part 2," *ASME Journal of Basic Engineering*, Vol. 84, No. 3 (Sept. 1962), p. 329.
77. Jekat, W. K., "Reynolds Number and Incidence-Angle Effects on Inducer Cavitation," ASME Paper 66-WA/FE-31, 1966.
78. Oshima, M., "A Study on Suction Performance of a Centrifugal Pump with an Inducer," *Bulletin, JSME*, Vol. 10, No. 42 (1967), pp. 959–965.
79. Lewis, G. W., and Tysl, E. R., "Cavitation Performance of an 83° Helical Inducer Operated in Liquid Hydrogen," NASA TM X-419, Mar. 1961.
80. Jakobsen, J. K., "On the Mechanism of Head Breakdown in Cavitating Inducers," *ASME Journal of Basic Engineering*, Vol. 86 (June 1964), pp. 291–304.
81. Bisell, W. R., Wong, G. S., and Winstead, T. W., "An Analysis of Two Phase Flow in LH_2 Pumps for O_2/H_2 Rocket Engines," AIAA Paper 69–549, 1969.
82. Deeprose, W. M., et al., "Cavitation Noise, Flow Noise, and Erosion," Proc. Conf. and Cavitation, Inst. Mech. Engrs. London, 1974.
83. Moore, R. D., and Meng, P. R., "Cavitation Performance of Line Mounted 80.6° Helical Inducer in Hydrogen," NASA TM X-1854, Aug. 1969.
84. Meng, P. R., and Moore, R. D., "Hydrogen Cavitation Performance of 80.6° Helical Inducer with Blunt Leading Edges," NASA TM X-2022, July 1970.
85. Meng, P. R., and Moore, R. D., "Hydrogen Cavitation Performance of 80.6° Helical Inducer Mounted in Line with Stationary Center Body," NASA TM X-1935, Jan. 1970.
86. Kovich, G., "Experimental and Predicted Cavitation Performance of 80.6° Helical Inducer in High Temperature Water," NASA TN D-6809, May 1972.
87. Kovich, G., "Comparison of Predicted and Experimental Cavitation Performance of 84° Helical Inducer in Water and Hydrogen," NASA TN D-7016, Dec. 1970.
88. Ball, C. L., Meng, P. R., and Reid, L., "Cavitation Performance of an 84° Helical Pump Inducer Operated in 37° and 42° R Liquid Hydrogen," NASA TN D-7016, Dec. 1970.
89. Moore, R. D., and Meng, P. R., "Thermodynamic Effects of Cavitation of an 80.6° Helical Inducer Operated in Hydrogen," NASA TN D-5614, 1970.
90. Meng, P. R., "Change in Inducer Net Positive Suction Head Requirement with Flow Coefficient in Low Temperature Hydrogen (27.9° to 36.6°R)," NASA TN D-4423, 1968.
91. Meng, P. R., and Moore, R. D., "Cavitation Performance of 80.6° Helical Inducer in Liquid Hydrogen," NASA TM X-1808, 1969.
92. Mang, P. R., and Connelly, R. E., "Investigation of Effects of Simulated Nuclear Radiation on Inducer Performance in Liquid Hydrogen," NASA TM X-1359, 1967.
93. Moore, R. D., and Meng, P. K., "Effect of Blade Loading Edge Thickness on Cavitation Performance of 80.6° Helical Inducer in Hydrogen," NASA TN D-6855, 1970.
94. Sutton, M., "Improving the Cavitation Performance of Centrifugal Pumps with Helical Inducer," Brit. Hydromechanics Res. Assoc., TN 814, 1964.

95. Jacobsen, J. K., "Computer Program to Calculate Cavity on Inducer Blades," NASA Tech. Brief NAR 53852, 1968.

Unsteady Cavitation, Instability and Oscillations

96. Kim, J. H., and Acosta, A. J., "Unsteady Flow in Cavitating Turbopumps," *ASME Journal of Fluids Engineering*, Vol. 97, (1975), pp. 413–418.
97. Rubin, S., "Longitudinal Instability of Liquid Rockets Due to Propulsion Feedback (POGO)," *Journal Spacecraft and Rockets*, Vol. 3, No. 8 (1966), pp. 1188–1195.
98. Sack, L. E., and Nottage, H. B., "System Oscillations Associated with Cavitating Inducers," *ASME Journal of Basic Engineering*, Vol. 87, No. 4 (1965), pp. 917–925.
99. Natanzon, M. S., et al., "Experimental Investigation of Cavitation Induced Oscillations in Helical Inducers," *Fluid Mech.-Soviet Research*, Vol. 3 (1974), pp. 38–45.
100. Brennen, C., and Acosta, A. J., "Theoretical Quasi-Static Analysis of Cavitation Compliance in Turbopumps," *Journal Spacecraft and Rockets*, Vol. 10, No. 3 (1973), pp. 175–180.
101. Ghahremani, F. G., and Rubin, S., "Empirical Evaluation of Pump Inlet Compliance," Final Report No. ATR-73 (7257)-1, Aerospace Corp., E1 Segundo, CA, July 1972.
102. Kolesnikov, K. S., and Kinelev, V. G., "Mathematical Model of Cavitation Phenomena in Helico Centrifugal Pumps," *Soviet Aeronautics*, Vol. 16, No. 4 (1973), pp. 64–68.
103. Brennen, C., Acosta, A. J., "The Dynamic Transfer Function for a Cavitation Inducer," *ASME Journal of Fluids Engineering*, Vol. 98 (1976), pp. 182–191.
104. Jackson, E. D., "Summary: Study of Pump Discharge Oscillations," Rocketdyne Report R-6693-1, Oct. 1966.
105. Wong, G. S., MacGregor, C., and Hoshide, R. K., "Suppression of Cavitation and Unstable Flow in Throttled Pumps," *Journal Spacecraft*, Vol. 2, No. 1 (Jan.-Feb. 1965), pp. 73–80.
106. Kamijo, K., Shimura, T., and Watanabe, M., "An Experimental Investigation of Cavitating Inducer Instability," ASME Paper 77-WA/FE-44, 1977.
107. Ng, S. L., "Dynamic Response of Cavitating Turbomachines," Ph.D. thesis, Cal. Tech., 1976.
108. Young, W. E., et al., "Study of Cavitating Inducer Instabilities," NASA CR-123939, 1972.
109. Badowski, H. R., "An Explanation for Inducer Instability in Cavitating Inducer," ASME Forum on Cavitation, 1969.
110. Watanabe, T., and Kawata, Y., "Research on the Oscillation in Cavitating Inducer," *IAHR*, 1979, pp. 265–276.
111. Jackson, E. D., "Study of Pump Discharge Oscillations," (Final Report) NASA CR 80153, 1968.
112. Kamijo, K., Shimura, T., and Watanabe, M., "A Visual Observation of Cavitating Inducer Instability," National Aerospace Laboratory (Japan) Report NAL TR 598T, May 1980.
113. Greitzer, E., "Stability of Pumping Systems," *ASME Journal of Fluids Engineering*, Vol. 103, No. 2 (June 1981).

CHAPTER 41

HYDRODYNAMICS OF OUTFLOWS FROM VESSELS

J. Kubie

Central Electricity Generating Board
Health and Safety Department
London, England

CONTENTS

INTRODUCTION, 1187

BASIC PRINCIPLES, 1189
- Introductory Remarks, 1189
- Vessel-Controlled Regime, 1189
- Onset of Choking of the Outlet Orifice, 1190
- Outlet-Controlled Regime, 1191
- Variation of the Liquid Level Along the Vessel, 1192
- Vessel-Controlled and Outlet-Controlled Regimes, 1193
- Drawdown of the Supernatent Phase, 1194

APPLICATIONS, 1194
- Horizontal Cylindrical Vessels, 1194
- Vertical Cylindrical Vessels, 1197
- Design of the Outlet and the Downcomer, 1200
- Effect of the Liquid Temperature, 1203
- Effect of the Liquid Inflow, 1204
- Effect of the Density of the Supernatent Phase, 1204
- Effect of Mixed or Dissolved Gases, 1204
- Effect of Vortices, 1205
- Effect of Adverse Pressure Gradient, 1205
- Transient Flow, 1205
- Vessel Level Control, 1206

NOTATION, 1206

REFERENCES, 1207

INTRODUCTION

Large vessels are commonly used in many engineering applications. They are typically employed:

- As holding or surge tanks in process engineering plants.
- As drums of direct contact regenerative feed heaters used extensively in the low pressure part of the feed heating train of large turbo-generator units.
- As collectors used to gather and re-deploy spilt liquids, as in containment sumps of pressurized water reactors.

A diagram of a simple vessel to be analyzed in this chapter is shown in Figure 1. It will be assumed throughout that the vessels are only partially full, i.e., that there is a free surface in the vessel. The reason for making this assumption is that when there is no free surface in the vessel, i.e., when the

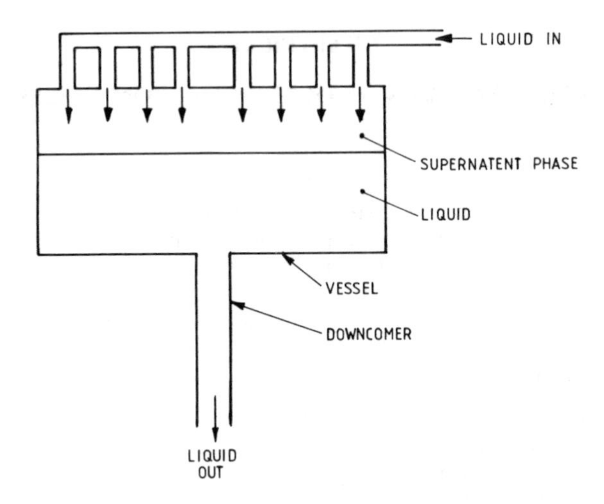

Figure 1. A diagram of a typical vessel considered in this chapter.

vessel is full, the vessel can be considered as just another flow obstruction in the pipework, and its resistance may be estimated accordingly.

Three major problems will be discussed in this chapter. The first one is the variation of the liquid level in the vessel with the liquid throughflow. This problem is considered because the liquid level is one of the more important design parameters. The level must be known for the positioning of pipe connections, level switches, etc., and, additionally, because it is related to the overall liquid inventory used in assessing the suitability of the design.

The second problem is related to the onset of drawdown, or gross entrainment, of the supernatent phase, since it is usually required that in their normal operation surge tanks do not permit the drawdown of supernatent gas, vapor, or liquid into their bottom offtake or downcomer [1]. This aspect is closely related to the problem of selected withdrawal from vessels or reservoirs [2].

The final problem considered here is concerned with the performance of the vessel and its outlet in the pipework system. For example, the outlet should be sufficiently large so as not to provide the controlling resistance to the flow, which could potentially result in pumps downstream of the vessel being starved of flow.

The vessels considered here are large in the sense that viscous and surface tension forces are negligible in comparison with inertia and gravitational forces. The parameters which can influence the hydrodynamic characteristics of the vessels can be classified as follows:

1. The size and the shape of the vessel outlet.
2. The size, the shape, and the orientation of the vessel.
3. The various flow resistances within the vessel and in the downcomer.
4. The temperature of the liquid relative to its saturation temperature with respect to the pressure above the liquid surface in the vessel.
5. The method of supply of the liquid into the vessel.
6. The density of the supernatent phase relative to the density of the liquid.
7. Real-life complications, such as the concentration of mixed or dissolved gases, the presence of vortices, adverse pressure gradient across the outlet, and transient flow.

Some of the problems to be investigated are quite complex, but where possible simple theoretical models describing the basic physical principles will be presented. Thus, initially, we will concentrate on steady-state vortex free outflow. In the following section we will demonstrate all the important concepts on a simple model of one-dimensional flow. This will be followed by application of the basic concepts to real problems.

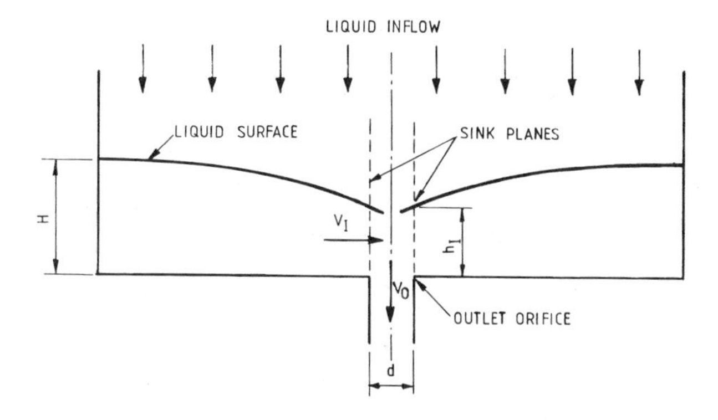

Figure 2. A diagram of a vessel considered in the analysis of one-dimensional flow.

BASIC PRINCIPLES

Introductory Remarks

The basic principles applicable to this problem will be demonstrated, and the fundamental equations derived, for a simple vessel shown in Figure 2. It will be assumed that the flow in the vessel is one-dimensional towards the outlet, and that the liquid is supplied uniformly over the surface between the walls and the outlet (i.e., the liquid is not supplied over the outlet). It is further assumed that the density of the supernatent phase is negligible compared with the density of the liquid, and that the flow has reached its steady state.

It will be shown that there are two regimes of outflow. For low liquid flow rates the liquid flows freely through the outlet, which thus provides negligible resistance to the flow. The liquid level is only necessary for supplying the liquid towards the outlet, and depends only on the design of the vessel. This is the vessel-controlled regime. For high liquid flow rates the outlet provides the controlling resistance to the flow, and the liquid level is necessary to overcome it, and thus depends on the design of the outlet. This is the outlet-controlled regime. Both regimes are discussed in the following sections, which follow the analysis of References 3 and 1.

Vessel-Controlled Regime

It is assumed that for flows in this regime the outlet is sufficiently large, so that it does not interfere with the flow. As shown in Figure 2 the outlet can thus be substituted by two vertical planes which provide uniform sinks for the incoming liquid. Because the vessel is symmetric about the outlet, we will concentrate on just one half. Application of the momentum equation between the end wall and the sink plane leads to the following equation

$$\frac{2Q_H^2}{gw^2} = h_I(H^2 - h_I^2) \tag{1}$$

where Q_H = the liquid flow rate to one half of the vessel (equal to one half of the total flow rate Q)
H = the liquid level at the end wall of the vessel
h_I = the liquid level at the outlet
w = the width of the vessel
g = the gravitational acceleration

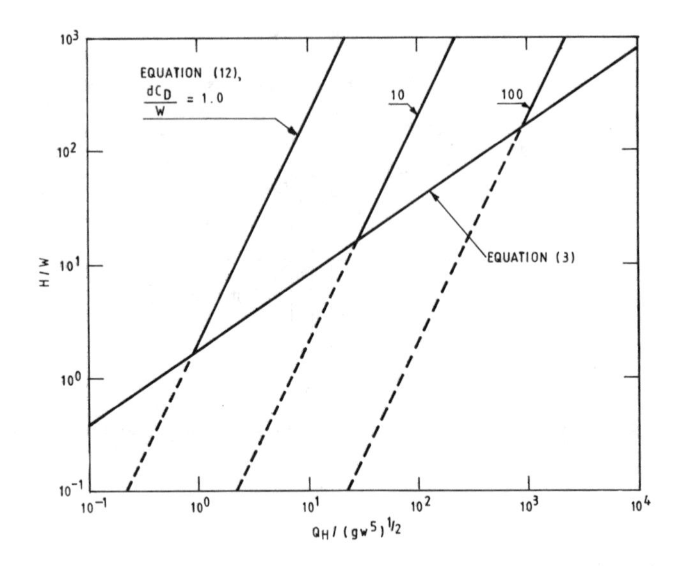

Figure 3. Variation of the liquid level at the end of the vessel with the flow rate for the case of one-dimensional flow.

The level h_I is obtained by assuming that the flow at the outlet is critical, i.e., that for a given H the flow is at its maximum. Thus Equation 1 is differentiated with respect to h_I and the following relationships can be established

$$\frac{h_I}{H} = 3^{-0.5} \tag{2}$$

$$\frac{H}{w} = 3^{0.5}\left[\frac{Q_H}{(gw^5)^{0.5}}\right]^{2/3} \tag{3}$$

Equation 3 gives the variation of the liquid level H with the liquid flow rate Q_H. Equation 3 is plotted in Figure 3.

Onset of Choking of the Outlet Orifice

The results of the previous section are only applicable if the approximation of vertical sink over the outlet is satisfied, and this will be the case if the outlet is self-venting and provides negligible resistance to the flow. Eventually, however, the outlet will not be able to discharge all the liquid freely, and it will start to choke. Application of the Bernoulli's equation over the outlet (see Figure 2) yields

$$V_I^2 + 2gh_I + 2\frac{\Delta p}{\rho} = V_O^2 + C_V V_O^2 \tag{4}$$

where V_I = the liquid velocity at the outlet plane
V_O = the liquid velocity at the outlet orifice
$C_V V_O^2$ = the flow losses between the outlet plane and the outlet orifice
ρ = the liquid density

Δp = the difference between the static pressure above the liquid surface in the vessel, p_V, and the static pressure below the outlet orifice, p_0.

For self-venting flow the pressures above the liquid surface and below the outlet are equal and thus the pressure difference Δp is zero. Hence the pressure difference Δp is also zero at the onset of choking and can be eliminated from Equation 4. Next, the flow rate across the outlet plane can be expressed as

$$Q_H = V_l h_l w \tag{5}$$

and once the choking starts (i.e., when the outlet is just full), the flow rate through the outlet can be expressed as

$$Q_H = 0.5 C_C w d V_0 \tag{6}$$

where d = the width of the outlet
C_C = the coefficient of contraction of the outlet orifice

Equations 5 and 6 can be substituted into Equation 4 with $\Delta p = 0$ to obtain

$$\frac{Q_H^2}{g h_l^3 w^2} = \frac{2C_D^2}{(2h_l/d)^2 - C_D^2} \tag{7}$$

where C_D, the outlet discharge coefficient, is given as

$$C_D = C_C/(1 + C_V)^{0.5} \tag{8}$$

Since at the onset of choking the level of Equation 7 and the level of Equations 2 and 3 are identical, the conditions for the onset of choking are determined as

$$\frac{H^C}{w} = 1.5\frac{dC_D}{w} \tag{9}$$

$$\frac{Q_H^C}{(gw^5)^{0.5}} = 0.75^{0.75}\left(\frac{dC_D}{w}\right)^{1.5} \tag{10}$$

The outlet will be choked for $Q_H > Q_H^C$ or $H > H^C$.

Outlet-Controlled Regime

For liquid flow rates above the critical flow rate Q_H^C, when the outlet orifice is choked, the relationship between the liquid level and the liquid flow rate will depend on the detailed arrangement of the outlet orifice, its downcomer, and the vessel itself. For simplicity, it is assumed here that the static pressure above the liquid surface in the vessel and the static pressure just below the outlet orifice are equal. (This will be obtained, for example, if the downcomer is short and if the downcomer and the top of the vessel are both opened to atmosphere.) Thus in this case the driving mechanism for the flow through the outlet is only the liquid level above the outlet, and Equation 7 will be applicable.

In order to relate the variation of the liquid level at the end of the vessel H to the liquid flow rate Q_H, the relationship between H and h_l in this flow regime must be established. This is done by simultaneously solving Equations 1 and 7, and using Equations 9 and 10 to obtain

$$\left(\frac{Q_H}{Q_H^C}\right)^{4/3} = \frac{3 + (H/h_l)^2}{108^{1/3}[(H/h_l)^2 - 1]^{1/3}} \tag{11}$$

Equation 11 has several solutions, but since in this case the flow in the vessel is controlled by a downstream obstruction, only tranquil flow (for which $h_I > H/3^{0.5}$) is considered. Equation 11 then demonstrates that as Q_H/Q_H^C increases, the level h_I approaches the level H and the liquid surface in the vessel becomes practically horizontal.

This conclusion, together with the observation that in this flow regime $H > H^C$, can be used to approximate Equation 7 by

$$\frac{H}{w} = 2\left(\frac{w}{dC_D}\right)^2\left[\frac{Q_H}{(gw^5)^{0.5}}\right]^2 \tag{12}$$

It can be shown that for $Q_H > 1.5Q_H^C$ Equation 12 is accurate to within 2%. Equation 12 is also plotted in Figure 3.

Variation of the Liquid Level Along the Vessel

The equations just presented only give the liquid level either at the end of the vessel or at the outlet plane. However, in order to estimate the volume of the liquid in the vessel or to interpret the reading of a liquid level gauge placed anywhere in the vessel, variation of the liquid level along the vessel must be known [4]. The variation of the level along the vessel is due to:

- Acceleration of the liquid supplied to the vessel.
- Friction in the vessel.

Since the latter effect is small the influence of friction is neglected, and only level changes due to acceleration are considered.

It is assumed that the liquid is supplied uniformly to the vessel, but only in the region indicated in Figure 4. Since there is no liquid acceleration between the end of the vessel and station 1, and between station 2 and the outlet plane, the liquid levels there are uniform. Thus $H = h_1$ and $h_2 = h_I$.

To obtain the level variation between stations 1 and 2, we consider the momentum balance between stations X and 2 in Figure 4, also noting that the conditions at station 2 and the conditions at the outlet plane are identical; thus

$$\frac{2Q_H^2}{gw^2h_I} - \frac{2Q_X^2}{gw^2h_X} = h_X^2 - h_I^2 \tag{13}$$

where Q_X = the liquid inflow between stations 1 and X
h_X = the liquid level at station X,

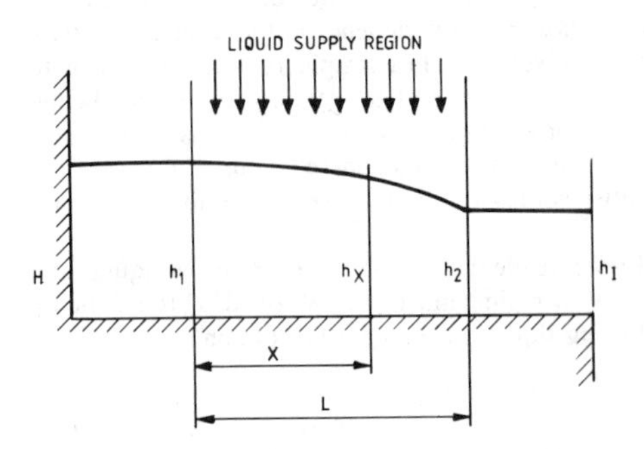

Figure 4. Liquid inflow considered in the analysis of variation of the liquid level along the vessel.

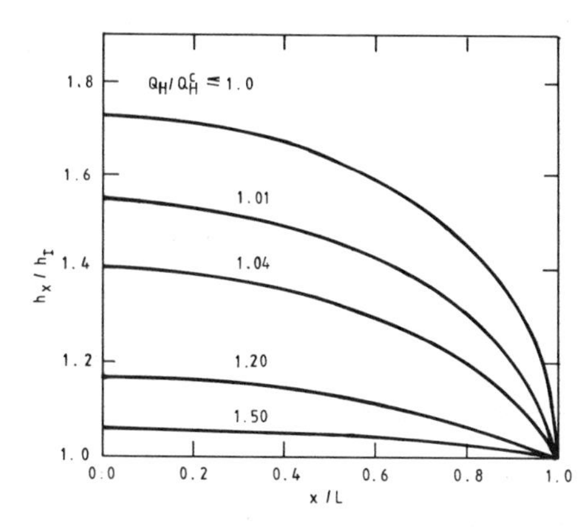

Figure 5. Theoretical results for the variation of the liquid level along the vessel for the case of one-dimensional flow.

and x and L are shown in Figure 4. Since the liquid is supplied to the vessel uniformly, $Q_X = xQ_H/L$ and Equation 13 can be rewritten as

$$1 - \frac{h_I}{h_X}\left(\frac{x}{L}\right)^2 = \frac{(h_X/h_I)^2 - 1}{(H/h_I)^2 - 1} \tag{14}$$

In order to use Equation 14 the ratio H/h_I must be known. For flows below the critical flow Q_H^C this ratio is given by Equation 2, and for flows above the critical flow this ratio is given by Equation 11. The theoretical results for the variation of the liquid level along the vessel are plotted in Figure 5 for various ratios Q_H/Q_H^C.

Vessel-Controlled and Outlet-Controlled Regimes

The analysis in the previous sections demonstrates that for liquid flows below the critical flow Q_H^C the liquid level in the vessel is independent of the size of the outlet, and the level rises slowly with the flow rate, but depends appreciably on the position in the vessel. This implies that in this flow regime the liquid level is relatively stable and that the total liquid volume in the vessel does not depend strongly on the liquid throughflow.

However, for flow rates above the critical flow the liquid level in the vessel depends strongly on the size of the outlet and rises rapidly with the flow rate, but tends to become more uniform along the vessel.

It should be stressed that Equation 12 for the level variation with the flow rate in the outlet-controlled flow regime is based on the assumption that the pressure difference Δp over the outlet is zero (or at least negligible). Such conditions can obtain for either the case discussed earlier, but they also become applicable for flows of liquids saturated with respect to their pressure from vessels with unobstructed downcomers, when the pressures p_V and p_0 both tend to the saturation pressure, and are thus identical. This then implies that for flows of saturated liquids the liquid level in the vessel rises first in accordance with Equation 3 and then in accordance with Equation 12.

On the other hand for flows of subcooled liquids through vessels with long unobstructed downcomers (see Figure 6) there is a potential for pressure recovery downstream of vena contracta, which may result in the development of suction pressure just below the outlet. This obviously is not possible for vessel-controlled flows when the outlet is self-venting, but it can occur for outlet-controlled flows when the outlet orifice is choked and can thus support considerable difference between the static pressure above the liquid surface in the vessel p_V and the static pressure below the outlet orifice p_0. In such a situation the outlet will behave as an effective sink, able to discharge freely all the liquid supplied to it even for liquid flows above the critical flow Q_H^C. The level in the vessel is then controlled by the process of supplying the liquid towards the outlet, i.e.,

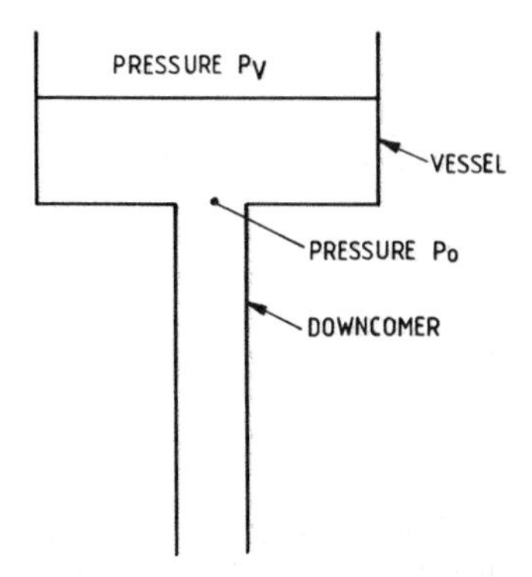

Figure 6. A diagram of a vessel with a long unobstructed downcomer.

by Equation 3. Hence it follows that in such a case the transition to the outlet-controlled flow regime is not necessarily manifested by a rapid increase in the liquid level in the vessel, i.e., by Equation 12.

Drawdown of the Supernatent Phase

For liquid flows below the critical flow Q_H^C, when the flow through the outlet is self-venting, the drawdown (or entrainment) of the supernatent phase is negligible.

For liquid flows above the critical flow the outlet is choked and can support a considerable difference between pressures p_V and p_0. Thus, as discussed earlier, it may be possible to create conditions in which the pressure p_0 is lower than the pressure p_V. In such situations the outlet will appear as an effective sink which is able to discharge freely all the liquid supplied to it. A pump in the drain is an example of a mechanism which can create such conditions.

As discussed earlier the maximum liquid flow which can be supplied from the vessel towards the outlet is obtained when the conditions at the sink plane are critical, i.e., as given by Equation 3. This is then the maximum liquid flow that the drain pump can demand. If the pump demands more liquid flow than the vessel can supply towards the outlet, the supernatent phase must complement this flow and will be entrained. Thus Equation 3 also gives the minimum liquid levels in the vessel which prevent the drawdown of the supernatent phase.

However, there is one complication which must be considered. The treatment of the outflow problem in the vessel-controlled regime is similar to the treatment of liquid flow ever a broad-crested weir [3], and it is well known that for the liquid flow over a broad-crested weir the liquid velocity reaches its critical value a short distance upstream of the crest [5] (or the sink plane). This phenomenon does not influence the results in this case, but, as discussed later, it has a significant influence on the prediction of drawdown from axisymmetric vessels with bottom outlets.

APPLICATIONS

Horizontal Cylindrical Vessels

Flow regimes

This section considers steady-state liquid outflow from horizontal cylindrical vessels with a vortex-free circular outlet in the center of their bottom surface. Such an arrangement is shown in Figure 7, which also shows that the liquid is supplied uniformly over the liquid surface in the vessel.

As shown in the various theoretical studies [1, 4, 6], solutions of the various problems can be only obtained numerically. However, for $H/R_V < 1$, where R_V is the radius of the cylindrical vessel,

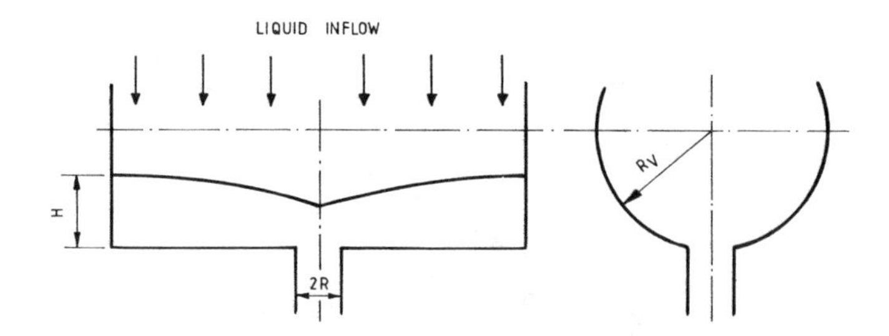

Figure 7. A diagram of a horizontal cylindrical vessel with a single outlet.

the numerical solutions can be usually substituted by simple formulae which are accurate to within 5%.

It is shown in Reference 4 that the total liquid flow rate Q^C required for the onset of choking of the outlet orifice is given as

$$\frac{Q^C}{(gR^5)^{0.5}} = 5C_D^{4/3}\left(\frac{R}{R_V}\right)^{1/6} \tag{15}$$

where R is the radius of the outlet and C_D, the coefficient of discharge, is typically about 0.6 [10].

For total liquid flow rates below the critical flow rate Q^C the outlet is self-venting and the level H at the end of the vessel is related to the liquid flow rate Q by [1, 6]:

$$\frac{H}{R_V} = 0.88\left[\frac{Q}{(gR_V^5)^{0.5}}\right]^{0.51} \tag{16}$$

It should be noted that in this flow regime the liquid level does not depend on the size of the outlet orifice. Equation 16 was confirmed experimentally [1, 4, 6] in large vessels. Beij [7] conducted experiments in semi-circular gutters and found that

$$\frac{H}{R_V} = 0.98\left[\frac{Q}{(gR_V^5)^{0.5}}\right]^{0.5} \tag{17}$$

The levels of Equation 17 are about 11% higher than the levels of Equation 16, perhaps because frictional resistance was of some importance in Beij's experiments.

As pointed out earlier for liquid flow rates above the critical flowrate Q^C, when the outlet is choked, the relationship between the liquid level and the liquid flow rate will depend on the detailed arrangements of the outlet orifice, its downcomer, and the vessel itself. If it is assumed, as above that the pressure p_V is equal to the pressure p_0, the liquid level is related to the liquid flow rate by [4]

$$\frac{H}{R_V} = \frac{0.051}{C_D^2}\,\phi\left(\frac{R_V}{R}\right)^4\left[\frac{Q}{(gR_V^5)^{0.5}}\right]^2 \tag{18}$$

where the correction factor ϕ is given as

$$\phi = [1 + 1.75(Q^C/Q)^6]^{0.4} \tag{19}$$

As expected, for high liquid flow rates when $Q \gg Q^C$, the correction factor approaches unity and can be eliminated from Equation 18. It should be noted that in this flow regime the liquid level depends strongly on the size of the outlet orifice.

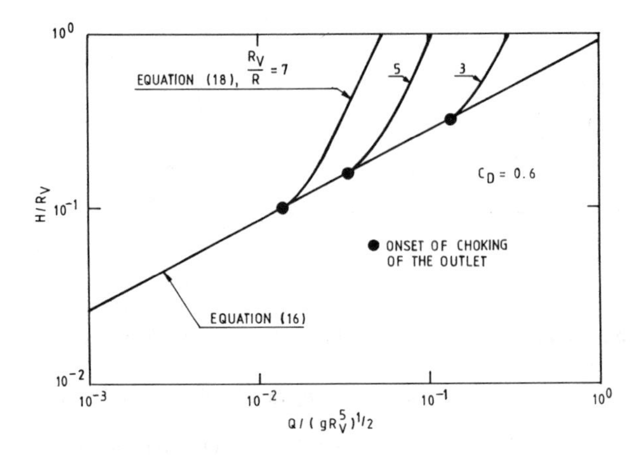

Figure 8. Variation of the liquid level at the end of the vessel with the flow rate for the case of a horizontal cylindrical vessel with an unobstructed downcomer.

The levels of Equations 16 and 18 are compared in Figure 8, which shows that in the vessel-controlled regime ($Q < Q^C$) the level rises slowly with the flow rate, but that in the outlet-controlled regimed ($Q > Q^C$) the level rises rapidly with the flow rate.

Variation of the Level in the Vessel

As argued earlier, the liquid level along the vessel can vary only in regions of liquid acceleration, which in this case can take place only in the liquid supply region, as shown in Figure 4. It is shown in [4] that in the liquid supply region the variation of the liquid level along the vessel is given as

$$\left(\frac{h_X}{h_I}\right)^{2.5} = 1 + 1.75\psi^6\left[1 - \left(\frac{X}{L}\right)^2\left(\frac{h_I}{h_X}\right)^{1.5}\right] \tag{20}$$

where for

$$\begin{aligned} &Q \leq Q^C \qquad \psi = 1.0 \\ &Q > Q^C \qquad \psi = Q^C/Q \end{aligned} \tag{21}$$

The theoretical results of Equations 20 and 21 are plotted in Figure 9, which shows that in the vessel-controlled regime the liquid level depends appreciably on the position in the vessel, but that in the outlet-controlled regime the level tends to become more uniform.

Drawdown of the Supernatent Phase

As discussed earlier for liquid flow rates above the critical flow rate drawdown of the supernatent phase will be obtained, unless the liquid level is kept sufficiently high. The minimum liquid level in the vessel required to prevent the drawdown of the supernatent phase is given by Equation 16, but the level should be suitably increased for long vessels where friction effects could be of importance.

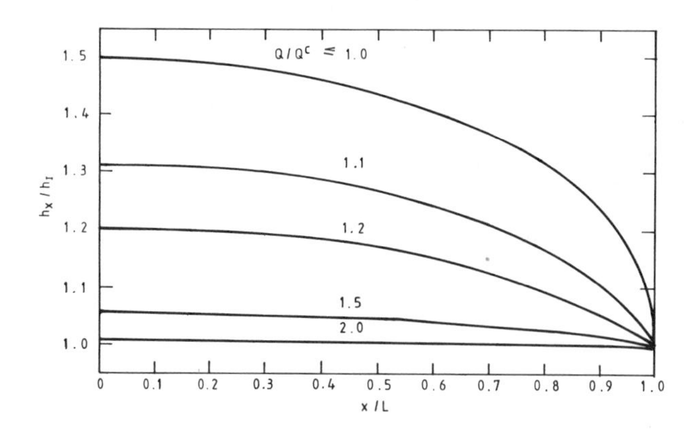

Figure 9. Theoretical results for the variation of the liquid level along the vessel for the case of a horizontal cylindrical vessel.

Effect of the Downcomer

It should be noted that it has been implicity assumed in this section that the downcomer is unobstructed and freely discharging, and thus does not provide the controlling resistance to the flow. If this is not the case the liquid levels in the vessel could be dominated by the downstream resistance and could thus be much higher than the levels just discussed. The influence of flow resistances downstream of the vessel outlet is discussed later.

Vertical Cylindrical Vessels

Flow Regimes

This section considers steady-state liquid outflow from vertical cylindrical vesels with a vortex-free circular outlet in the center of their bottom surface. It is also assumed that the liquid is supplied to the vessel uniformly over the liquid surface, but only between the outer wall and the outlet (see Figure 10).

As shown in Reference 3 the total liquid flow rate Q^C required for the onset of choking of the outlet orifice is given as

$$\frac{Q^C}{(gR^5)^{0.5}} = 5C_D^{1.5} \tag{22}$$

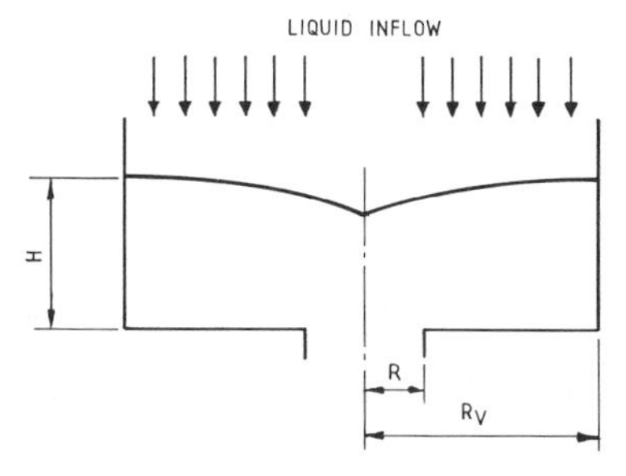

Figure 10. A diagram of a vertical cylindrical vessel with a single outlet.

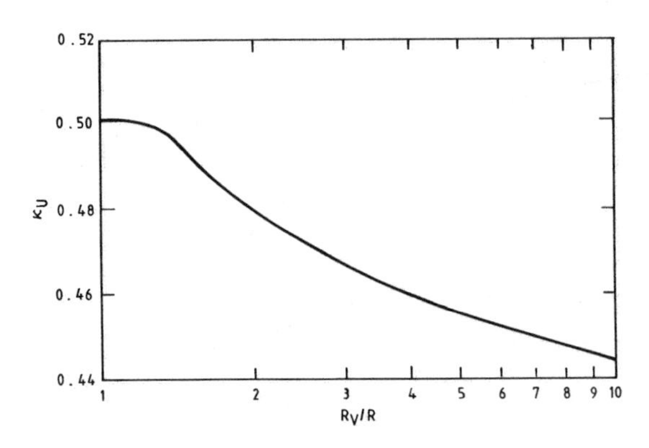

Figure 11. A plot of the parameter κ_U versus the ratio R_V/R.

Furthermore, for a vessel with $R_V/R > 3$; the critical level at the wall H^C required for the onset of choking of the outlet orifice is given as

$$\frac{H^C}{R} = 1.3C_D \tag{23}$$

Since it can be shown that in the present case $C_D \simeq 0.6$ [10], the critical value of the liquid level is estimated as about 0.8R, which is in the experimental range observed by Simpson [8] and McDuffie [9].

For liquid flow rates below the critical flow rate Q^C, when the outlet is self-venting, the level H is related to the liquid flow rate Q by [3]:

$$\frac{H}{R} = \kappa_U \left[\frac{Q}{(gR^5)^{0.5}}\right]^{2/3} \tag{24}$$

where the parameter κ_U is a function of the ratio R_V/R. The numerical values of κ_U are shown in Figure 11, from which it can be seen that for $R_V/R > 3$ the asymptotic value

$$\kappa_U = 0.44 \tag{25}$$

can be used with the maximum error of less than 10%. Using the energy based Francis equation [11] the parameter κ_U can be calculated as

$$\kappa_U = 0.41 \tag{26}$$

The preceding results are in reasonable agreement with available experimental evidence.

For liquid flow rates above the critical flow rate Q^C, when the outlet is choked, the detailed arrangements must be known. However if it is assumed that the pressure p_V is equal to the pressure p_0, the liquid level is related to the liquid flow rate by

$$\frac{H}{R} = \frac{0.051}{C_D^2}\left[\frac{Q}{(gR^5)^{0.5}}\right]^2 \tag{27}$$

which for $Q/Q^C > 1.5$ is accurate to within 2%.

The levels of Equation 24 with $\kappa_U = 0.44$ and of Equation 27 are compared in Figure 12, which demonstrates once more that in the vessel-controlled regime ($Q < Q^C$) the level rises slowly with the flow rate, but that in the outlet-controlled regime ($Q > Q^C$) the level rises rapidly with the flow rate.

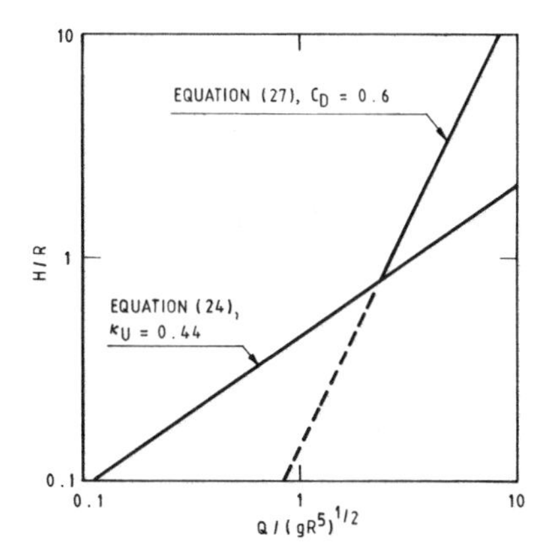

Figure 12. Variation of the liquid level at the wall of the vessel with the flow rate for the case of a vertical cylindrical vessel with an unobstructed downcomer.

Variation of the Level in the Vessel

It can be shown that in the vessel-controlled flow regime the liquid level depends appreciably on the position in the vessel, but that in the outlet-controlled flow regime the level tends to become more uniform. This aspect is discussed further in Reference 3.

Drawdown of the Supernatent Phase

Several studies of drawdown have been published. As discussed earlier drawdown of the supernatent phase can be a problem for liquid flow rates above the critical flow rate Q^C. The most recent study [3] gives Equation 24 with $\kappa_U = 0.44$ for the minimum liquid level required to prevent the drawdown. However, a study by Lubin and Springer [12] suggests that the minimum liquid level required to prevent drawdown is given as

$$\frac{H}{R} = 0.69\left[\frac{Q}{(gR^5)^{0.5}}\right]^{0.4} \tag{28}$$

whereas the work of Harleman et al. [2] suggests that Equation 28 could be used only if the constant were 0.82 instead of the theoretically determined 0.69.

Figure 13 shows the minimum liquid level required to prevent drawdown obtained in the following investigations:

1. Equation 24 with $\kappa_U = 0.44$.
2. The theoretical and experimental results of Lubin and Springer [12] for air-water flow.
3. The experimentally correlated results of Harleman et al. [2].
4. The experimental results of McDuffie [9].

It is clear from Figure 13 that the experimental data of Lubin and Springer [12], obtained in a transient situation by allowing the liquid level in the vessel to fall until the effective start of drawdown was observed, follow their own theoretical results extremely well. Furthermore, when McDuffie [9] carried out his experiments in a similar manner his experimental results also followed the theoretical prediction of Reference 12 reasonably closely.

When, however, McDuffie's [9] experimental arrangements followed those of Harleman et al. [2], i.e. when the liquid level in the vessel was kept constant by recirculating the liquid and the onset of drawdown was established by increasing the liquid flow, his experimental results followed the semi-empirical prediction of Reference 2.

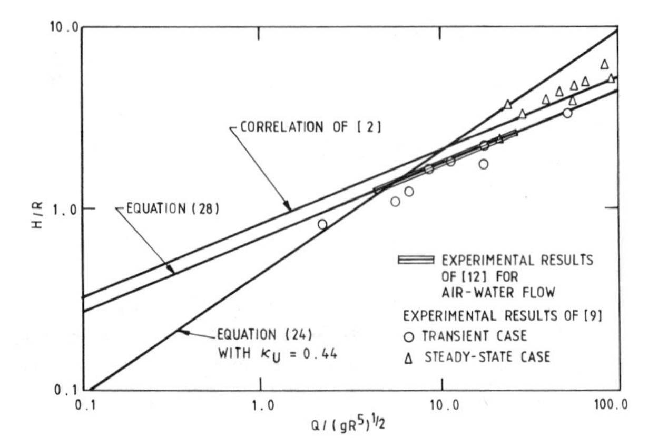

Figure 13. A comparison of the results of various studies on the minimum liquid level required to prevent the drawdown of the supernatent phase.

Hence it appears that in a transient situation, when the liquid level is allowed to fall, flow losses can be neglected and Equation 28 describes well the minimum liquid level required to prevent drawdown. When, however, the liquid is being recirculated to keep the level constant, a correction for flow losses must be taken and the equation of Harleman et al. [2] gives the minimum liquid level required to prevent drawdown.

Finally, Figure 13 demonstrates that the theoretical results of Equation 24 with $\kappa_U = 0.44$ give the upper bound for the minimum liquid level required to prevent drawdown. This is also confirmed in Reference 3 where it is shown that Equations 24 and 28 are related by the radial position where the liquid velocity in the vessel reaches its critical value. Equation 24 is consistent with the liquid reaching its critical velocity at the edge of the outlet and Equation 28 is consistent with the liquid reaching its critical velocity a short distance upstream of the edge of the outlet.

Effect of the Downcomer

It should be noted that it has been assumed in this section that the downcomer does not provide the controlling resistance to the flow. The influence of the downcomer resistances downstream of the outlet orifice is discussed in the next section.

Design of the Outlet and the Downcomer

Central Outlets

As pointed out previously for liquid flow rates below the critical flow rate Q^C the liquid level rises slowly with the liquid flow rate. This then implies that in certain cases it is preferable for the vessel to operate in the vessel-controlled flow regime, since in this regime the liquid level is relatively stable and the total liquid volume in the vessel (the liquid inventory) does not depend strongly on the liquid flow rate.

As shown earlier, for the vessel to operate in the vessel-controlled flow regime the maximum liquid flow rate likely to be encountered in operation must be lower than the critical liquid flow rate. Equations 15 and 22 can then be used for sizing the outlet orifice.

However, it may not be practicable to design the outlet in accordance with the above requirements. Instead, it may be preferable to use multiple outlets and combine them in a junction downstream of the vessel. For example, Figure 14 shows a horizontal cylindrical vessel with twin outlets

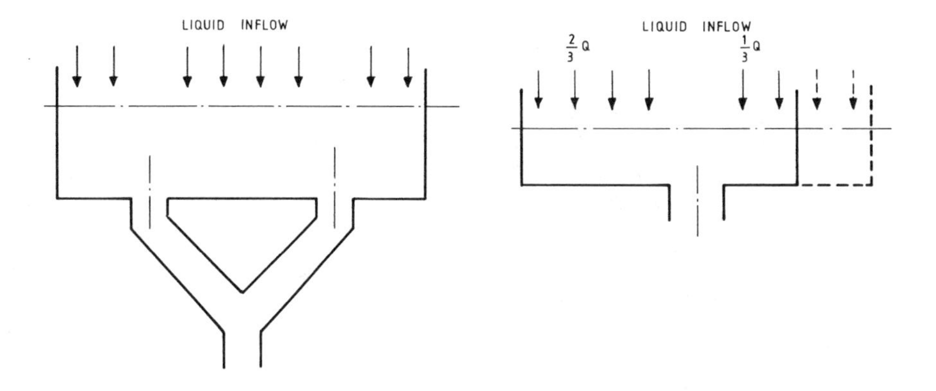

Figure 14. A diagram of a horizontal cylindrical vessel with twin outlets.

Figure 15. A diagram of a horizontal cylindrical vessel with a single off-centre outlet.

combined in a Y-junction to form a single downcomer. Since in this case the arrangement is symmetric about the central plane, the vessel can be assumed to consist of two vessels, each with a central outlet. Hence the vessel and the outlets can be sized using equations given earlier but with the flow rate equal, clearly, to only one half of the total flow rate.

Off-Center Outlets

As already mentioned, for vessel-controlled flow regimes the liquid level depends appreciably on the position in the vessel. In the case of an off-center outlet its position must be taken into account when the liquid levels are calculated. Consider, for example, the horizontal cylindrical vessel shown in Figure 15, where two thirds of the total liquid flow approach the outlet from one side and one third of the flow from the other side. The end liquid level on the side of the higher flow is calculated from Equation 16 but with $Q = 4Q'/3$, where Q' is the total flow through the vessel, since the outlet could be considered central if the vessel were increased in size as indicated in Figure 15. It should be noted that this procedure can only be used on the side of the vessel which supplies the greater proportion of the flow. The end liquid level on the other side of the vessel is somewhere between the level calculated by the above approach for this side (i.e., with $Q = 2Q'/3$) and the end level on the side of the higher flow. Finally, it should be also noted that Equation 15 can be used to predict the critical total liquid flow rate for the onset of choking of the outlet orifice also in the case of an off-center outlet.

Design of the Downcomer Pipework

As previously discussed, in order to improve the vessel discharge large outlets should be used. For economy, however, the diameter of the downcomer pipework should be as small as practicable and this may require a reduction in the downcomer diameter somewhere below the vessel. When such a system is being designed, care must be taken to ensure that the transition to the smaller diameter pipe does not provide the controlling resistance to the flow, since in such a case the outlet orifice is not used effectively [13]. The following method can be used to design the downcomer pipework for the flow of saturated liquids.

For the steady-state flow of saturated liquids from vessels with unobstructed downcomers the liquid level in the vessel rises first in accordance with the equations of the vessel-controlled flow regime and then in accordance with equations of the outlet-controlled flow regime. Figure 16 shows

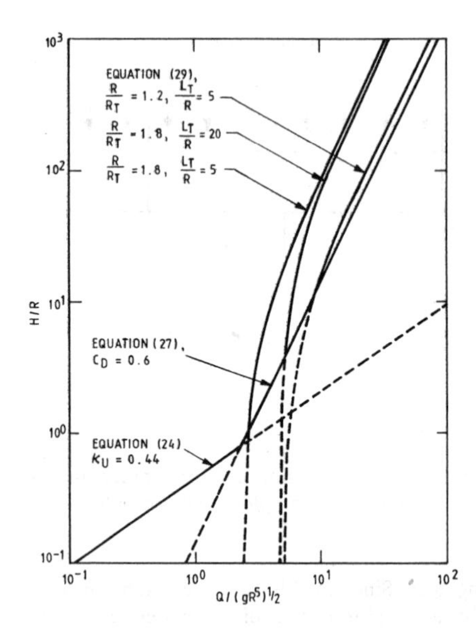

Figure 16. A plot of the liquid level as a function of the flow rate for the case of a vertical cylindrical vessel with (i) an unobstructed downcomer (Equations 24 and 27), and (ii) a downcomer with flow obstructions (Equation 29).

the liquid level H as a function of the flow rate Q for the case of a vertical cylindrical vessel with a central outlet and an unobstructed downcomer. The levels indicated in Figure 16 are the minimum (natural draining) levels which can be obtained in such a situation. However, it should be noted that the levels could be higher if any downstream obstruction became controlling and too restrictive, since the liquid would then fill the downcomer and the liquid level in the vessel would rise above the minimum levels.

Consider the downcomer arrangement shown in Figure 17, which shows a large outlet orifice with a radius R which reduces to a smaller pipe with a radius R_T at a vertical distance L_T below the vessel. As discussed in Reference 13 the level H can be obtained as

$$\frac{H}{R} = \frac{0.051}{C_D^2}\left[\frac{Q}{(gR^5)^{0.5}}\right]^2\left[C_D^2 C_V + \frac{C_D^2}{C_{DT}^2}\left(\frac{R}{R_T}\right)^4\right] - \frac{L_T}{R} \tag{29}$$

where C_{DT} is the discharge coefficient at the transition from the larger to the smaller pipe. Typically, the numerical values of C_D, C_V, and C_{DT} are respectively 0.6, 0.5, and 0.8 [14], and for these values the results of Equation 29 are also included in Figure 16 for various values of R/R_T and L_T/R.

The results of Equation 29 are physically meaningful only if the levels of Equation 29 are higher than the liquid levels which are obtained in the outlet-controlled flow regime, and in these situations

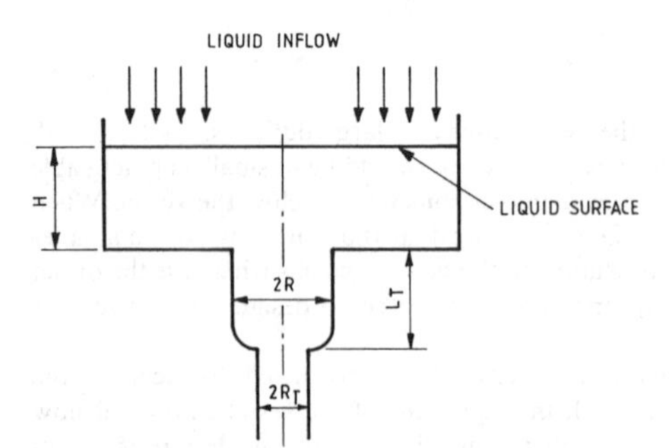

Figure 17. A diagram of a downcomer with a reduction in its diameter.

the transition from the larger to the smaller pipe provides the controlling resistance to the flow. It can be seen from Figure 16 that the design objective of eliminating the influence of the transition can be achieved by either increasing the radius R_T or by increasing the distance L_T.

Effect of the Liquid Temperature

Temperature Effects

For outlet-controlled flows through vessels with long downcomers there is a potential for pressure recovery downstream of vena contracta, which may result in the development of suction pressure just below the outlet. As shown in Reference 13 the liquid level in the vessel is then given as

$$\frac{H}{R} = \frac{0.051}{C_D^2}\left[\frac{Q}{(gR^5)^{0.5}}\right]^2 - \frac{p_V - p_0}{R\rho g} \tag{30}$$

although it must be stressed that this level cannot be lower than the minimum liquid level determined by the equations of the vessel-controlled regime.

It should be noted that there are limitations on the minimum value of p_0. If this pressure reaches the saturation value p_S corresponding to the liquid temperature at the outlet from the vessel (which is assumed to be the same as the liquid temperature in the vessel itself), vapor starts to be generated by flashing. This vapor cannot be rapidly removed from the downcomer and thus a further decrease of p_0 is not possible. The minimum liquid level, which is obtained for $p_0 = p_S$, is then given by

$$\frac{H}{R} = \frac{0.051}{C_D^2}\left[\frac{Q}{(gR^5)^{0.5}}\right]^2 - \frac{p_V - p_S}{R\rho g} \tag{31}$$

Hence when the temperature of the liquid is low relative to its saturation temperature with respect to the pressure above the liquid surface in the vessel, i.e., when the liquid is highly subcooled, the pressure difference $p_V - p_S$ is large and the level in the vessel is then determined, as previously discussed, by the equations of the vessel-controlled regime. If, however, the temperature of the liquid approaches the saturation temperature, the pressure difference $p_V - p_S$ is low and the liquid level in the vessel is only just below the level of Equation 27. This also implies that if the liquid in the vessel is at uniform temperature and the pressure in the vessel p_V equals the saturation pressure at this temperature, p_S, the second term on the right hand side of Equation 31 becomes negligible and the level of Equation 31 is then equal to the level of Equation 27.

This observation can be reinterpreted by noting that for flows of saturated liquids only the conditions above the outlet determine the maximum discharge from the vessel. Hence the driving head in the outlet-controlled flow regime is only the liquid level in the vessel and any downcomer has a negligible influence on the maximum discharge from the vessel—it can only decrease it.

Thus any increase in the liquid temperature which does not affect appreciably the pressure difference $p_V - p_S$ is of secondary importance, since the liquid level and the liquid flow are governed by the equations of the vessel-controlled flow regime and are thus independent of the pressure difference $p_V - p_S$. The only parameters which can be affected by an increase in temperature in this range are the viscosity and the surface tension of the liquid whose influence, as discussed in the Introduction, is negligible. Thus, an increase in the liquid temperature becomes important only when the temperature approaches the saturation temperature corresponding to the pressure p_V.

Implications for Modeling

The preceding discussion has an important bearing on the modeling of outflow problems. There are many instances where flows of saturated liquids occur in practice, such as in regenerative feed heaters or in containment sumps of nuclear reactors during certain postulated accidents. These

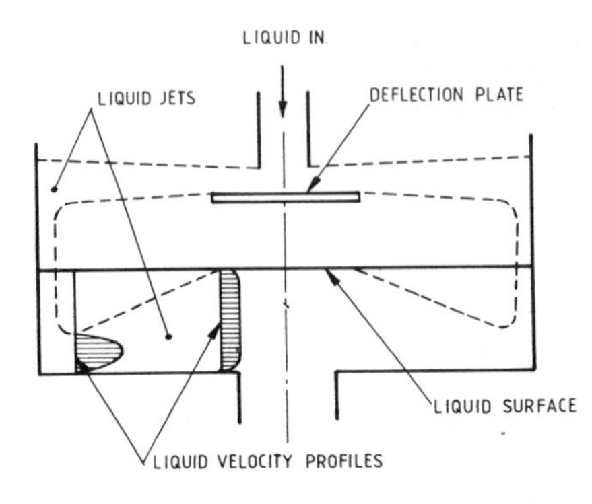

Figure 18. A diagram of an arrangement with a nonuniform liquid velocity profile.

real systems are complex and thus must be investigated experimentally. Because working with saturated liquids is costly and time-consuming, air-water analogues are used instead, and they can be effective in reproducing the flow of saturated liquids. However, as previously discussed, care must be taken to ensure that where flow separation and/or vapor flashing is likely the geometry of the analogue is such that suction pressures are prevented from occurring by suitable venting arrangements [13].

Effect of the Liquid Inflow

The method by which the liquid is introduced into the vessel can have an important influence on the flow characteristics of the vessel. For example, Gardner and Crow [1] investigated the influence of a jet of liquid aiding the flow towards the outlet in a horizontal cylindrical vessel. They showed that in the vessel-controlled flow regime such a jet considerably decreases the liquid levels. Similarly, it was shown [3] that in a vertical cylindrical vessel the method of liquid supply to the vessel can also have a significant influence on the liquid level. Finally, it can be argued that the method of liquid supply to the vessel is responsible for the disagreement between theoretical results and the experimental data reported by Cranfield [15].

It should be also noted that in deriving the various theoretical formulae it was assumed that the liquid velocity profile at any given cross-section was uniform. In certain circumstances, such as those shown in Figure 18, the liquid velocity profile can be highly nonuniform, and this can, once again, influence the liquid level in the vessel.

It then follows that in any experimental investigation care should be taken to model the inflow characteristics as closely as possible.

Effect of the Density of the Supernatent Phase

So far it has been assumed that the density of the supernatent phase ρ_S is negligible. If it is not negligible, the equations applicable to the vessel-controlled flow regime and to the onset of choking of the outlet orifice should be modified by substituting $g' = g(1 - \rho_S/\rho)$ for g. The modification of the equations applicable to the outlet-controlled flow regime is not that straightforward, since it will depend on the detailed arrangement of the outlet orifice and its downcomer. These cases must be examined individually.

Effect of Mixed or Dissolved Gases

The presence of gases, mixed or dissolved, in the liquid in the vessel promotes the formation of vortices and thus has an indirect effect on the flow characteristics of the vessel. This aspect is discussed in the following section.

The direct influence of gases mixed with the liquid in the vessel is examined in Reference 6 where it is shown that high gas concentrations can have an appreciable effect on the liquid level and the discharge performance of the vessel.

Effect of Vortices

The effect of vortices can be dramatic. The rotating liquid can open a core of the supernatent phase that will propagate towards the outlet, and the supernatent phase will be entrained even though the liquid level may be well above the minimum liquid level required to prevent the draw-down. Additionally, the liquid level in the vessel will become unsteady and could fluctuate with large amplitudes.

The formation of vortices is difficult to predict because the forces that initiate them are generally weak [8]. The initiation and the prevention of vortices is extensively discussed in the literature [16, 17], where it is suggested that tangential liquid inlets into the vessel are frequently responsible for the initiation of vortices. Whatever the causes, vortices can be easily eliminated with vortex breakers installed over the vessel outlet [8].

Effect of Adverse Pressure Gradient

In several industrial applications the liquid is sometimes required to drain against an adverse pressure gradient. This problem, which is sometimes seen in power plant deaerators, was recently examined by Cranfield [18], who observed three stable and three unstable flow regimes, and presented techniques for their identification.

Transient Flow

Transient effects may have a considerable influence on the discharge performance of vessels. This is demonstrated here on a case of outlet-controlled steady-state discharge from a vessel with a long downcomer (Figure 6) affected by pressure fluctuations above the liquid level in the vessel.

The maximum discharge from such a vessel can be obtained from Equation 31 as

$$\frac{Q}{(gR^5)^{0.5}} = 4.4C_D\left[\frac{H}{R} + \frac{p_V - p_S}{R\rho g}\right]^{0.5} \tag{32}$$

As discussed earlier, for subcooled liquids the pressure difference $p_V - p_S$ is large and thus small fluctuations in the pressure p_V do not have appreciable influence on the maximum discharge from the vessel. However, as also discussed earlier, the situation is different if the liquid is saturated, since in that case $p_V \simeq p_S$, and small fluctuations in the pressure p_V can considerably influence the maximum discharge from the vessel.

As an example, consider a power plant deaerator discharging saturated water and operating in steady state, so that $p_V \simeq p_S$. If the pressure in the vessel p_V then suddenly decreases, due either to pressure fluctuations in the bled steam line or to dumping of colder water into the vessel through the steam space, the second term in the brackets of Equation 32 becomes negative and the discharge from the vessel decreases. This means that in a transient situation the discharge capacity of the vessel outlet can be appreciably cut, which could result in the water column in the downcomer to be broken and thus in the loss of the pump suction head and possible cavitational damage to the pump.

The preceding discussion demonstrates that the maximum discharge from the vessel is considerably more sensitive to pressure fluctuations in the case of saturated liquids than in the case of subcooled liquids. The reason is that in the case of saturated liquids the pressure fluctuations are associated only with the head given by the liquid level in the vessel and not, as is the case with subcooled liquids, with the total head given by the liquid level in the vessel *and* the length of the downcomer (up to a certain limit depending on the saturation pressure p_S). Since the downcomer is usually substantially longer than the liquid level in the vessel, the pressure fluctuations are disproportionately more important in the former case. Thus anticipated pressure transients should be taken into account when sizing vessels for saturated liquids.

Vessel Level Control

As discussed earlier the liquid levels obtained in a vessel when the flow obstructions downstream of the vessel are negligible are the minimum liquid levels for that vessel and its outlet. If, however, the flow obstructions downstream of the vessel are too restrictive the flow will fill the downcomer and the level in the vessel will rise above the minimum level. The final level will then depend on the flow characteristic of the obstructions [13].

It is shown in Reference [13] that if the liquid level in the vessel is to be controlled by a downstream device, such as a downstream valve and a downstream pump, it is important for the range of the liquid levels covered by the controller to be above the minimum level. Otherwise, at some time, a situation can develop in which the pump will demand more flow than can be supplied by the vessel, liquid separation will occur in the downcomer and the suction head will be lost.

Acknowledgment

This work is published by permission of the Central Electricity Generating Board.

NOTATION

C_C coefficient of contraction
C_D discharge coefficient
C_{DT} discharge coefficient at the transition
C_V velocity loss coefficient
d width of the outlet
g gravitational acceleration
g' $(1 - \rho_S/\rho)g$, corrected gravitational acceleration
h_I liquid level at the edge of the outlet
h_X liquid level at station X
h_1 liquid level at station 1
h_2 liquid level at station 2
H liquid level at the end wall of the vessel
L length of the liquid supply region
L_T vertical distance between the vessel and the transition
p_0 static pressure below the outlet orifice
p_S saturation pressure
p_V static pressure in the vessel
Δp $p_V - p_0$, pressure difference
Q total volumetric liquid flow rate
Q_H volumetric liquid flow rate to one half of the vessel
Q_X volumetric liquid flow rate at station X
R radius of the outlet orifice
R_T radius of the smaller downcomer pipe
R_V radius of the vessel
V_I mean liquid velocity at the edge of the outlet
V_0 mean liquid velocity in the outlet
w width of the vessel
x coordinate

Greek Symbols

κ_U constant in Equation 24
ρ liquid density
ρ_S density of the supernatent phase
ϕ correction factor given by Equation 19
ψ factor given by Equation 21

Superscript

C condition for the onset of choking of the outlet

REFERENCES

1. Gardner, G. C., and Crow, I. G., *Chem. Eng. Sci.*, 26:211 (1971).
2. Harleman, D. R. F., Morgan, R. L., and Purple, R. A., *Int. Association for Hydraulic Research, 8th Congress, Montreal*, p. 10-C-1 (1959).
3. Kubie, J., *Proc. Instn. Mech. Engrs*, 198A:113 (1984).
4. Kubie, J., *Proc. Instn. Mech. Engrs.*, 196:259 (1982).
5. Henderson, F. M., *Open Channel Flow*, Macmillan, London (1966).
6. Kubie, J., and Oates, H. S., *Trans. ASME-J. Fluids Eng.*, 102:324 (1980).
7. Beij, K. H., *Bur. Stand. J. Res.*, 12:193 (1934).
8. Simpson, L. L., *Chem. Eng.*, 75:192 (1968).
9. McDuffie, N. G., *AIChE J.*, 23:37 (1977).
10. Rouse, H., and Abul-Fetouh, A.-H., *Trans. ASME–J. Appl. Mech.*, 71:421 (1950).
11. Souders, M., et al., *Ind. Eng. Chem.*, 30:86 (1938).
12. Lubin, B. T., and Springer, G. S., *J. Fluid Mech.*, 29:385 (1967).
13. Kubie, J., Rowe, M., and Jones, E. W., *Steam Turbines for the 1980's, I. Mech. E. London*, 141 (1979).
14. Robertson, J. M., *Hydrodynamics in Theory and Application*, Prentice-Hall, Englewood Cliffs, NJ (1965).
15. Cranfield, R. R., *J. Mech. Eng. Sci.*, 23:77 (1981).
16. Swainston, M. J. C., *Proc. Instn. Mech. Engrs.*, 190:671 (1977).
17. Rajagopal, H. Y., *Power Engineering*, 85:52 (1981).
18. Cranfield, R. R., *Proc. Instn. Mech. Engrs.*, 196:301 (1982).

CHAPTER 42

DESIGN FEATURES OF FANS, BLOWERS, AND COMPRESSORS

N. P. Cheremisinoff

Exxon Chemical Company
Linden, New Jersey USA

P. N. Cheremisinoff

Department of Civil and Environmental Engineering
Newark, New Jersey USA

CONTENTS

FAN ENGINEERING, 1208
General Information and Terminology, 1208
Fan Performance and Design Basics, 1212
Radial-Flow Fans, 1221
Axial-Flow Fans, 1227
Controllable Pitch Fans and System Design, 1233
Air-Conditioning Systems, 1249
Noise, Vibration, and Mechanical Considerations, 1257
Overall Design Considerations, 1280

COMPRESSION MACHINES, 1294
General Information, 1294
Thermodynamic Principles, 1297
Piston Compressors, 1303
Rotary Compressors, 1320
Blowers, 1325
Centrifugal Compressors, 1328

NOTATION, 1331

REFERENCES, 1333

FAN ENGINEERING

General Information and Terminology

Introduction

Pumps, compressors, blowers and fans all belong to the same family of machines called *turbo-machinery* or *rotating equipment*. Pumps can handle only incompressible fluids, viz. liquids, whereas compressors, blowers, and fans handle compressible fluids such as air and other gases. In this first section a review of fundamental concepts and definitions is presented.

The word fan applies to a rotating device for delivering or exhausting a quantity of air or other gas with little change in pressure, by means of an impeller using centrifugal or propeller action. Fans are termed "compressors" when they operate at sufficiently high speeds to compress the gas materially. For purposes of testing and rating compressors the ASME Power Test Code includes in the compressor class all devices producing over a seven-percent pressure rise. The term "fan"

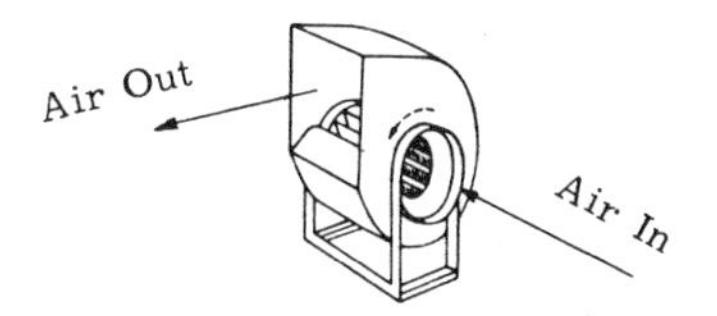

Figure 1. Shows a radial flow fan.

is loosely used to refer only to the impeller, or it may mean the impeller, casing, and any other components that may be considered part of the unit, such as the fan shaft, guide vanes, inlet cone, outlet, and diffuser. Frequently the impeller is referred to as the fan wheel or rotor, and the fan casing as housing, shell, volute, scroll, or diffuser.

Geometric similarity. Two fans are geometrically similar when all dimensions of the second fan are scaled up (or down) in equal proportion to give a fan of the same shape but a different size compared to the first one. It is common manufacturing practice when a shape of a fan has been obtained which will give desirable operating characteristics (for example, high peak efficiency), to construct a series of geometrically similar fans to handle different requirements of pressure and rates of flow. All the fans of this series, within wide limits of size and speed, will have the same peak efficiency and the same type of operating characteristics. The actual pressures, rates of flow, speeds, and horsepower requirements of the different fans are related by the fan laws. These laws are explained later.

The amount of noise produced by similar fans can be calculated by a similar set of relationships if the fans are operating in similar surroundings. These noise laws are also explained later in this chapter.

Major types of fans. Radial-flow fans are fans in which the motion of the fluid over the impeller occurs in a substantially radial direction. These fans are frequently called "centrifugal" fans because the compression of the gas occurs mainly through centrifugal action. "Paddle wheel," "multiblade," "squirrel cage," and "sirocco" are terms often applied to specific types of radial-flow fans. (See Figure 1.)

Axial-flow fans are fans in which the motion of the fluid over the impeller is parallel to the impeller shaft. Fans of this type are frequently termed "propeller" fans, although this term is more accurately restricted to one type of axial-flow fan. The most common axial-flow fan is the familiar office cooling fan. (See Figure 2.)

The term *mixed-flow fan* is sometimes applied to radial-flow fans which receive their energy from the axial direction and whose blade shape gives the fluid an axial component of velocity while passing through the blades. The term is also sometimes applied to axial-flow fans in which the fluid passing through the impeller has a large radial component of velocity.

Pressure. Three pressure terms are worth noting: static, velocity, and total pressures.

In fan calculations, the *static pressure*, or static head, at a given point in a moving gas is defined by the height of the column of a fluid, usually water, which the pressure at that point will just sustain. Figure 3A shows how to obtain a measurement of static pressure.

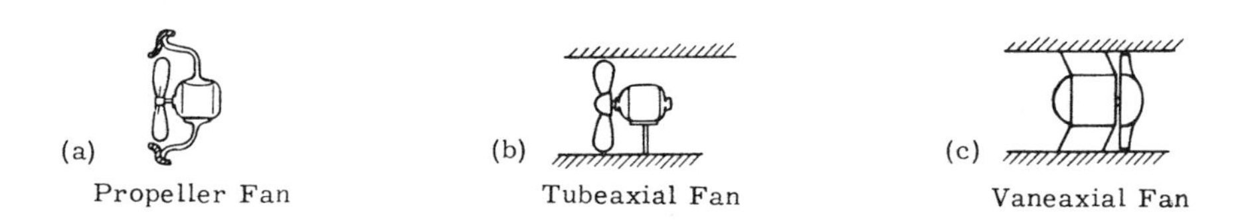

Figure 2. Shows axial flow fans.

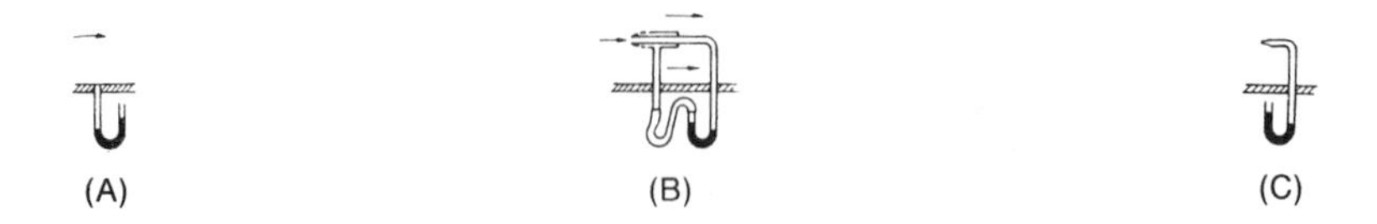

Figure 3. Illustrates common methods of pressure measurements using manometers.

The velocity pressure or velocity head, at a given point in a moving gas is defined as the pressure corresponding to the kinetic energy of the gas moving at velocity, V. It is obtained by converting the velocity at that point into equivalent pressure (expressed in feet of the gas flowing) by means of the expression $\Delta p_v^* = V^2/2g$ from Bernoulli's equation. For purposes of comparison, in fan calculations this pressure is usually converted into inches of water by means of the expression $\Delta p_v^* = (V'/4{,}000)^2$, provided the gas is air at close to atmospheric pressure and at a temperature of 70°F. A special pitot tube for measuring velocity pressure is shown in Figure 3B.

The total pressure at a given point in a moving fluid is equal to the sum of the static and velocity pressures at that point. It is sometimes called impact pressure or head, because it is measured with a simple impact tube pointed against the direction of flow of the fluid as in Figure 3C.

Efficiency. The efficiency of a fan is defined as the ratio of useful power output to power input. The useful power output, termed the air horsepower, is directly proportional to the product of the rate of air flow and the increase in air pressure. Since the gas has a total pressure which is composed of static and velocity components, efficiency is commonly specified in terms of static efficiency or total efficiency.

In cases where static pressure is desired, there is little worth in total efficiency values. However, where velocity pressure is important in the service application, total efficiency is more important than static efficiency. Static efficiency is more commonly used because it is easier to measure and because the static efficiency characteristics of some fans are more rigorously defined than the total efficiency characteristics.

The static and total efficiency are given by

Static efficiency

$$\eta_s = \frac{0.000157Q\ \Delta p_s^*}{P_H} = \frac{0.118\ \Delta p_s^*\ Q}{P_w} \tag{1a}$$

Total efficiency

$$\eta_t = \frac{0.000157Q\ \Delta p_t^*}{P_H} = \frac{0.118\ \Delta p_t^* Q}{P_w} \tag{1b}$$

where P_H and P_w = actual power input to the fan in horsepower and watts respectively, Δp^* is expressed in inches of water, and Q = flow in cfm.

Performance coefficients. From dimensional analysis, the pressure developed by a fan system can be expressed as:

$$\Delta p = \frac{\rho u_2^2}{g} f\left[\left(\frac{Q}{u_2 D_2^2}\right)\left(\frac{u_2 D_2}{u}\right)\left(\frac{u_1}{\frac{\gamma p_1 g}{\rho}}\right)\left(\frac{L_1}{D_2}\right)\left(\frac{L_2}{D_2}\right)\left(\frac{L_n}{D_2}\right)\right] \tag{2}$$

where L_n/D_2 is the ratio of a linear dimension of the fan (L_n) to the fan tip diameter (D_2). Defining the following notation:

Pressure coefficient

$$\psi = \frac{\Delta p}{\frac{\rho u_2^2}{2g}}$$

Flow coefficient

$$\phi = \frac{Q}{u'_2 A_2} = \frac{V'_{2m}}{u'_2}$$

Mach number

$$M_1 = \frac{u}{\sqrt{\frac{\gamma p_1 g}{\rho}}}$$

Reynolds number

$$Re = \frac{u_2 D_2}{\nu}$$

Shape factor

$$S_n = \frac{L_n}{D_2}$$

The pressure coefficient can then be rewritten as:

$$\psi = f(\phi, Re, M_1, S_1, S_2, \ldots, S_n)$$

For many applications ψ will be independent of the Reynolds and Mach numbers (Re, M_1). Thus for a given geometric fan family ($S_1, S_2, \ldots, S_n$) ψ is dependent only on

$$\psi = f(\phi)$$

Under these conditions, there is a one-to-one relationship between the flow coefficient and the pressure coefficient. (See Figure 4.)

The relation

$$\psi = f(\phi) \quad \text{or} \quad \frac{\Delta p}{\frac{\rho u_2^2}{2g}} = f\left[\frac{Q}{u'_2 A_2}\right]$$

also incorporates all of the fan laws. As such, it is often convenient to plot fan data in terms of ψ vs. ϕ and η (static efficiency). Plots of this type are useful for:

1. Prorating known fan performance to different rpm, scale, and fluid mediums within the limits of independence on Reynolds number or Mach number.
2. Comparing the performance of geometrically different fans, run at different conditions.
3. Correlating test information.
4. Identifying Reynolds number and Mach number effects.

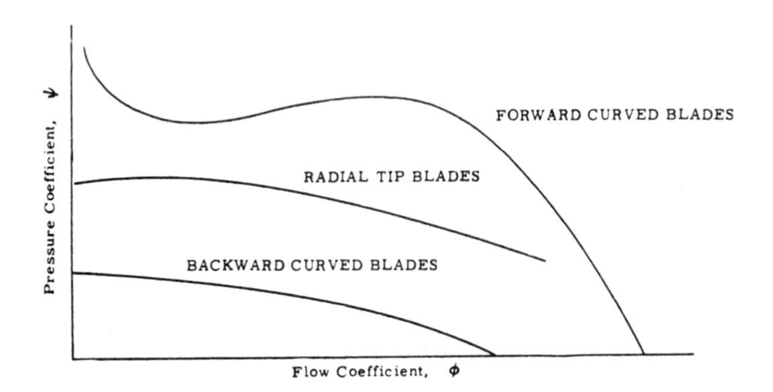

Figure 4. Correlation of pressure coefficient and flow coefficients.

Pressure coefficient ψ can be interpreted as the pressure developed in proportion to the dynamic head of the blade tip.

The flow coefficient ϕ is the ratio of the average velocity (normal to the rotor discharge area) to the rotor tip speed.

For a given operating point, the relation $\psi = f(\phi)$ implies that with ϕ, ψ, η fixed:

- Pressure $\sim$ rpm^2
  $\sim$ density
- cfm $\sim$ rpm, independent of density

Also if a single fan is moving air through a fixed square law resistance, the fan will perform at the same operating point regardless of rpm.

Examples of coefficient plots for typical fans are given later along with examples of curves for radial and for axial fans.

For problems in fan selection it is often more convenient to use plots of specific diameter and efficiency vs. specific speed rather than ψ vs. ϕ plots.

Fan Performance and Design Basics

In correlating the effect of Reynolds number with fan performance, a general trend of increasing efficiency with an increase in Reynolds number is observed. To a lesser extent ψ is dependent in the same manner. Both η and ψ approach a plateau at Re $\simeq 10^6$ with minor dependence in the region of Re $\simeq 10^5$ to 10^6. Performance decreases at lower Reynolds numbers such that the efficiency at Re $\simeq 10^4$ may only be 50% to 60% of that at Re $\simeq 10^6$.

The Reynolds number for fan rotors should be determined from the hydraulic diameter of the flow channel formed by the fan blades and the relative velocity of the fluid at the mean rotor diameter. In general, for axial-flow fans, inlet flow is axial and the velocity triangle takes the form of Figure 5, where

$$v_{\infty} = \sqrt{V_1^2 + \left(\frac{2u_2 - V_{2t}}{2}\right)^2} \tag{3}$$

However, for simplicity the Reynolds number as calculated on the characteristic performance curves is of the form

$$Re = \frac{u_2 D_2}{\nu}$$

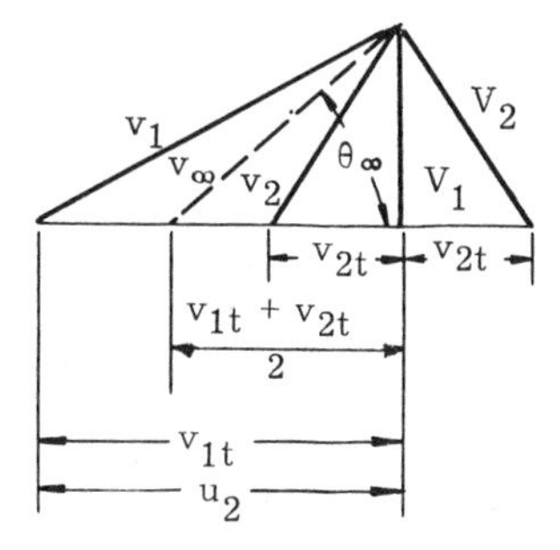

Figure 5. Velocity vectors for air flow in an axial flow fan.

As such, this Reynolds number for any one fan design cannot be compared with the Reynolds number of another design. Its value is restricted to a comparison of fan performance among fans prorated in terms of dimension and speed from one design.

The Mach number expresses the compressibility effects in fluid flow. Physically, it is the ratio of the local velocity to the local speed of sound. In compressors the relative velocity is always highest at the blade inlet so the Mach number is most appropriately evaluated at this location.

$$M = v/(\gamma p_1 g/\rho_1)^{0.5} \tag{4}$$

Pressure coefficient ψ and efficiency for any one design are essentially independent of Mach number at $M < 0.5$. Performance deteriorates progressively at higher M and falls off very sharply as M approaches unity. The dependence of η on M above 0.5 is markedly influenced by the geometric design. By careful shaping, rotors have been run successfully at transonic ($0.8 < M < 1.2$) and supersonic ($M > 1.0$) inlets.

Another performance coefficient is *specific speed*, N_s, defined as the speed of a geometrically similar fan, forcing one cubic foot of air per minute against a head of one inch of water.

For standard air,

$$N_s = \frac{nQ^{1/2}}{\Delta p^{*3/4}} \tag{5}$$

If the density is other than 0.075 lb/ft^3,

$$N_s = \frac{nQ^{1/2}}{\Delta p^{*3/4}}\left(\frac{\rho}{0.075}\right)^{3/4} \tag{6}$$

Also in terms of fan coefficients

$$N_s = 8.05 \times 10^4 \left[\frac{A_2^{1/2}}{D_2}\right]\left[\frac{\phi^{1/2}}{\psi^{3/4}}\right] \tag{7}$$

Specific diameter is a dimensionless coefficient used in selecting a fan. It is similar to specific speed in that one of the two main selection parameters, namely, size and rpm, is eliminated. When used in fan selection, specific diameter is based on D_c, a diameter which characterizes the fan-diffuser package.

For standard air,

$$d_s = \frac{D_c\,\Delta p^{*1/4}}{Q^{1/2}} \tag{8}$$

If the density is other than 0.075 lb/ft³,

$$d_s = D_c \frac{\Delta p^{*1/4}}{Q^{1/2}} \left(\frac{0.075}{\rho}\right)^{1/4} \tag{9}$$

In terms of fan characteristics

$$d_s = 1.58 \times 10^{-2} \frac{D_c}{A^{1/2}} \frac{\psi^{1/4}}{\phi^{1/2}} \tag{10}$$

The system into which the fan discharges fixes the pressure (head), against which the fan operates. For a given section the pressure drop will vary with the square of the flow. However, different sections result in different drops for the same flow. Long pipes will have more drop than short pipes. Pipes with a large area will have less drop than those with a small area. Each part of the system has a certain resistance to flow, and its pressure drop may be expressed as

$$\Delta p^* \propto Q^2 R$$

or any section may be likened to an equivalent orifice, with no vena contracta, whose area, A_0, is called the effective area, and the drop may be expressed as

$$\Delta p^* \propto \frac{Q^2}{A_0^2}$$

It is evident that if the same units are used for Q and Δp^* in the two equations

$$R = \frac{1}{A_0^2}$$

At this point, a general review of centrifugal machine fundamentals is warranted. Most *centrifugal machines* have a housing with an inlet and an outlet. Inside the housing is a wheel or an impeller which rotates and imparts kinetic energy to the fluid. The fluid comes in at a pressure, P_1, and leaves the housing at a higher pressure, P_2. These machines will, therefore, always generate a constant *pressure differential* when operating at the same flow.

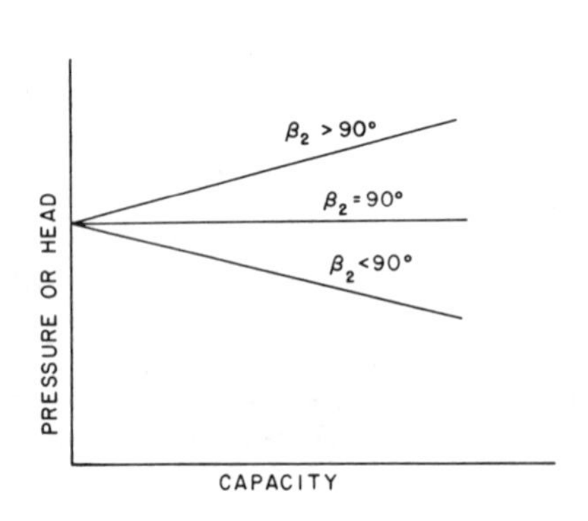

Figure 6. Shows theoretical pressure-volume curves for different blade discharge angles.

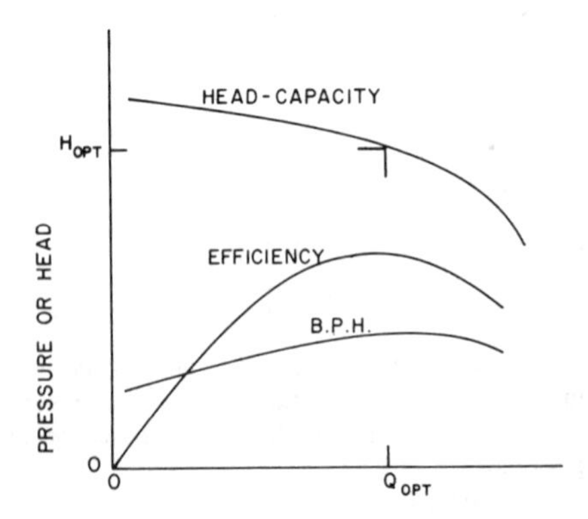

Figure 7. Shows typical head-capacity curve for centrifugal machines with maximum efficiency at optimum flow.

Theoretically, the pressure-volume relationship is linear when the discharge angle on the blades is 90 degrees. The ideal pressure-volume lines for discharge angles less than and greater than 90 degrees are as shown in Figure 6. In actual practice, however, this is never the case. The reason for this is that there are certain losses within the housing: disk friction (wheel friction), blade inefficiency, circulation within the blades, etc.

Kinetic energy is generated by the rotary motion of the impeller and is imparted to the fluid moving through the machine. Each machine is designed to produce a certain pressure rise or head at a given capacity with a minimum loss, or that is, with a maximum efficiency, as shown in Figure 7.

To understand how the head is generated by an impeller, consider Figure 8. Figure 8A shows a typical sectional elevation view of the hydraulic path of an impeller, whereas Figure 8B shows the same impeller in plan view with inlet and discharge velocity triangles.

Assuming that in time dt, there is a mass dm of a very thin layer of fluid leaving the impeller and at the same time an equal amount of mass of fluid entering the impeller. Let T denote the moment of external forces. Then T is given by the following equation:

$$T = dm/dt(r_2 c_2 \cos \alpha_2 - r_1 c_1 \cos \alpha_1) \tag{11}$$

and when the term (dm/dt) is applied to all the fluid in all the blades, it becomes $Q\rho/g$

where Q = the flow through the machine in cubic feet per second
ρ = the specific weight in pounds per cubic feet
r_2 = the outer radius of the impeller
r_1 = the inner radius of the impeller
c_2 = the absolute velocity of the fluid at the discharge
c_1 = the absolute velocity of the fluid at the inlet
$\alpha_{2,1}$ = the angle between the absolute and the peripheral velocities at the exit and the inlet, respectively
u_2 = the peripheral velocity at the exit
u_1 = the peripheral velocity at the inlet

Substituting for (dm/dt), in Equation 11, gives:

$$T = Q(\rho/g)(r_2 c_2 \cos \alpha_2 - r_1 c_1 \cos \alpha_1) \tag{12}$$

Multiplying both sides by w, the angular velocity, then, Tw equals power denoted by P, and substitute: $u_2 = w \cdot r_2$; $u_1 = w \cdot r_1$; and $c_{u2} = c_2 \cos \alpha_2$, $c_{u1} = c_1 \cos \alpha_1$. The Equation 12 becomes:

$$P = Q(\rho/g)(u_2 c_{u2} - u_1 c_{u1}) \tag{13}$$

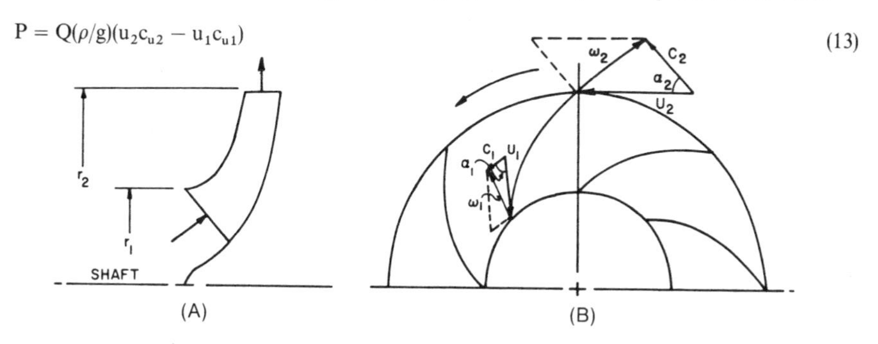

Figure 8. Shows layout of impeller: (A) illustrates elevation view of impeller; (B) plan view of blade layout. Inlet and outlet velocity triangles are shown.

Since $P = Q\rho H_i$ where H_i = the ideal head, we get

$$H_i = (u_2 c_{u2} - u_1 c_{u1})/g \tag{14}$$

No losses are considered in deriving Equation 14, which is called Euler's head or the theoretical head equation.

The actual process that a gas or fluid undergoes in an impeller is approximated by the polytropic process, which is

$$PV^n = \text{constant} \tag{15}$$

From thermodynamics the *polytropic head equation* is:

$$H_p = \frac{Z_{av} R T_1 \cdot n}{(n-1)} \left[(P_2/P_1)^{(n-1)/n} - 1\right] \tag{16}$$

where H_p = polytropic head, ft-lb_f/lbm
R = gas constant, ft-lb_f/lbm-°R
n = polytropic exponent, dimensionless
Z_{av} = average compressibility factor, $(Z_1 + Z_2)/2$
T_1 = absolute inlet temperature, °R
P_2 = absolute discharge pressure, psia
P_1 = absolute suction pressure, psia

Once the polytropic head is known, it is relatively easy to calculate the theoretical or actual brake horsepower:

$$\text{thp} = wH_p/33{,}000 \tag{17}$$

$$\text{bhp} = wH_p/33{,}000\eta_p \tag{18}$$

where thp and bhp = theoretical and brake horsepowers, respectively
w = flow in lbs per minute
η_p = polytropic efficiency in decimal

The theoretical or actual *horsepower* can also be computed without knowing the polytropic head. The head equation is useful in centrifugal and axial machinery because the number of stages or wheels required to obtain the desired head can be computed. For reciprocating compressors the head need not be known. There compression is achieved in a cylinder, and if the pressure ratios are quite high, additional cylinders are simply added until the desired pressure ratio is achieved. Thus to calculate horsepower, without the head, all that is needed is the inlet volume flow (icfm—inlet cubic feet per minute or also sometimes called the acfm—actual cubic feet per minute). This is the flow in the machine at inlet temperature and pressure conditions. If the process flow is given only in terms of the pounds per hour, pounds per minute or moles per hour, use the following equations to obtain the icfm and then the horsepower:

$$w = (\text{moles/hr}) \times (\text{MW})/60 \text{ lb per minute}$$

$$\text{scfm} = (\text{moles/hr}) \times (379/60)$$

(Note: Each gas occupies the 379 cubic feet volume at standard conditions) (19)

$$\text{scfm} = (w/\text{MW}) \times 379$$

$$\text{icfm} = \text{scfm} \times (14.7P_1) \times (T_1/520)$$

where MW is the molecular weight, and 14.7 and 520°R are the standard conditions. Blower and the fan manufacturers, however, use 68°F as the standard temperature, instead of the 60°F. Denoting V_1 as the inlet cfm or the acfm, the horsepower is given as follows:

$$thp = \frac{P_1 V_1}{229.17}(n/(n-1))[(P_2/P_1)^{(n-1)/n} - 1] \tag{20}$$

The next step is to divide the thp by the polytropic efficiency to the bhp. In fans, however, the horsepower is calculated in a different manner:

$$bhp = 0.000157 \times acfm \times SP/SE \tag{21}$$

or

$$bhp = acfm \times SP/6356 \times SE \tag{22}$$

With fans the discharge temperature is not an important consideration since fans are a very low pressure ratio machine. In blowers and compressors the discharge temperature can be very high, and therefore *intercooling* (between stages) or *aftercooling* (i.e., cooling of the gas after it leaves the machine housing) is usually required. To calculate the *discharge temperature*, the following equations are used:

$$T_2 = T_1(P_2/P_1)^{(n-1)/n} \tag{23}$$

As is quite evident from Equation 23, the discharge temperature varies linearly with the inlet temperature (T_1), but more so with the pressure ratio (P_2/P_1). For this reason intercoolers and aftercoolers are used in multistage centrifugal compressors, or reciprocating compressors. The discharge temperature can also be computed from:

$$T_2 = T_1 + H_p/[Z_{av}R(n/(n-1))] \tag{24}$$

Impeller Designs

There are basically three different designs used in the centrifugal machines (see Figure 9). In the *radial design*, the fluid enters the impeller along the shaft, but leaves in a direction perpendicular to the shaft. In the mixed flow design, the fluid enters along the shaft as before, but leaves the impeller at some angle. In the axial design, however, the fluid enters and leaves parallel to the shaft. The radial blades are used in low specific speed range, mixed flow blades are used in the medium range, and axial impellers or the propeller design are used in the high specific speed range.

As noted earlier, specific speed (Equation 5) establishes machine geometry. Machines having the same specific speed are geometrically similar and therefore the performance of a new machine can be predicted. Specific speed, N_s, is defined as the speed which will produce one foot of head at a flow of one cfm.

The relationship between the specific speed and the *specific diameter* is shown in Figure 10. The higher the specific speed is, the higher the flow is and smaller the pressure or head and vice versa. Figure 10A covers a wide range of turbomachinery, viz. centrifugal and positive displacement pumps

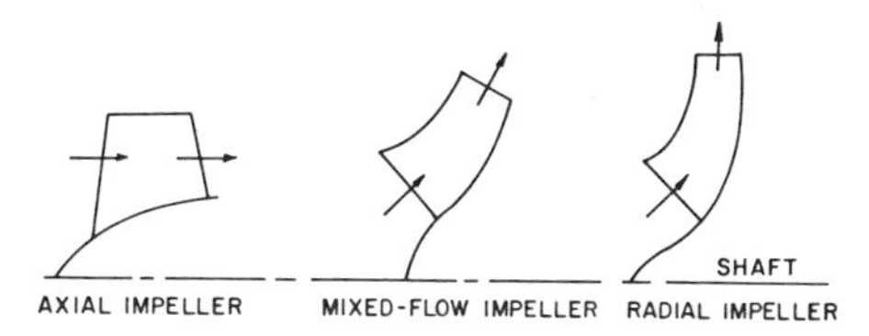

Figure 9. Shows different impeller types.

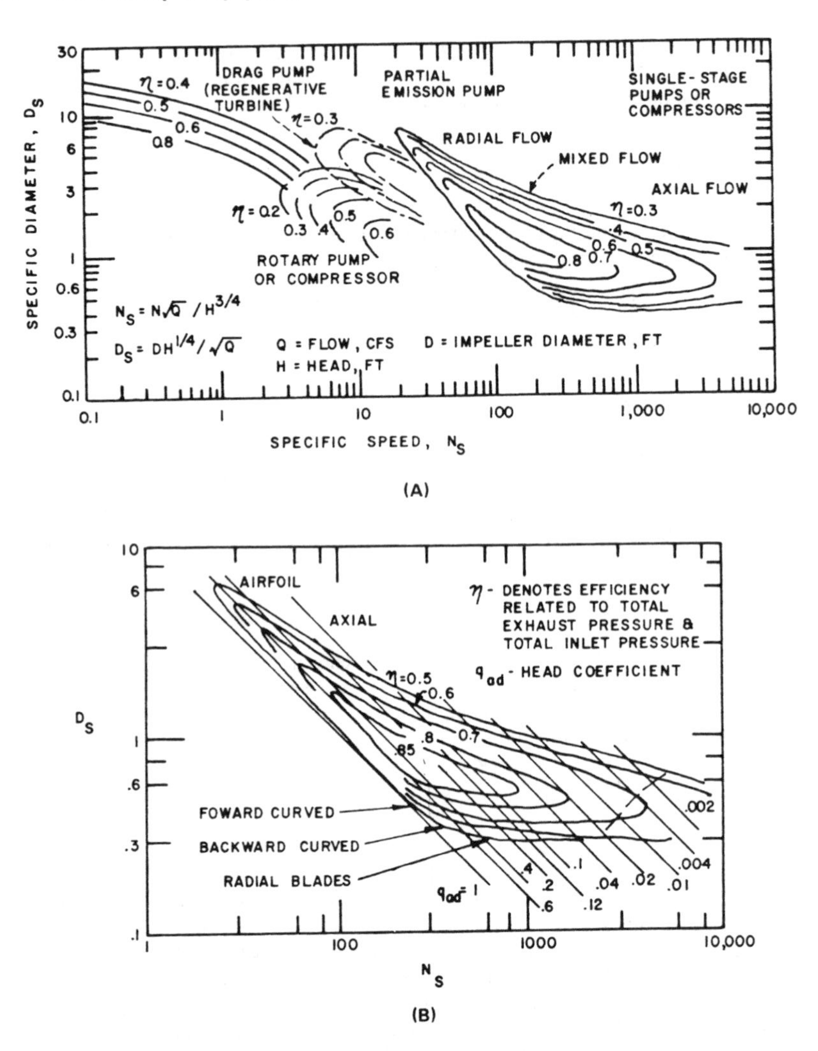

Figure 10. (A) N_s–D_s diagram for single-stage pumps and compressors; (B) N_s–D_s curves for single-stage pumps and low pressure blowers and fans.

and compressors. Figure 10B, however, covers just low pressure, single-stage, pumps and blowers or fans. Figure 10C is a modified version of the N_s – D_s diagram developed by the fan industry for their convenience.

The following example illustrates the use of Figure 10B and C for fans.

Example

It is required to have 54,000 inlet cfm at 30 inches of wg. Determine the optimum wheel diameter for the best efficiency for an airfoil type blade. Also determine the optimum speed.

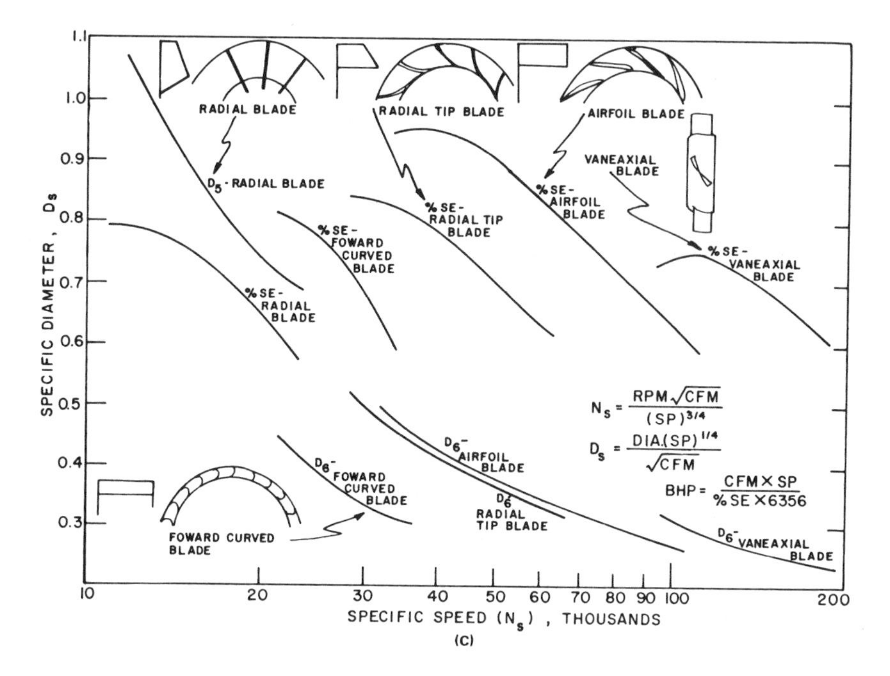

Figure 10. (Continued)

Solution. (Using curves in Figure 10B)

$$Q = 54{,}000 \text{ cfm}$$
$$= 900 \text{ ft}^3/\text{sec}$$

$$H = 30 \cdot (69.3)$$
$$= 2{,}079 \text{ ft}$$

For an airfoil blade, for $N_s = 200$, the efficiency is at a maximum. This gives $D_s = 0.82$ and

$$D = D_s \cdot \sqrt{Q}/H^{1/4}$$
$$= 0.82 \times 30/6.75$$
$$= 3.644 \text{ ft or } 43.7 \text{ inches}$$

To find the speed

$$N = N_s \cdot H^{0.75}/\sqrt{Q}$$
$$= 200 \times 307.9/30$$
$$= 2{,}052.6 \text{ rpm}$$

Another approach would be to use Figure 10C:

$$N_s = N \cdot \sqrt{Q}/P_s^{0.75}$$
$$= 2{,}052.6 \times 232.38/12.82$$
$$= 37{,}206$$

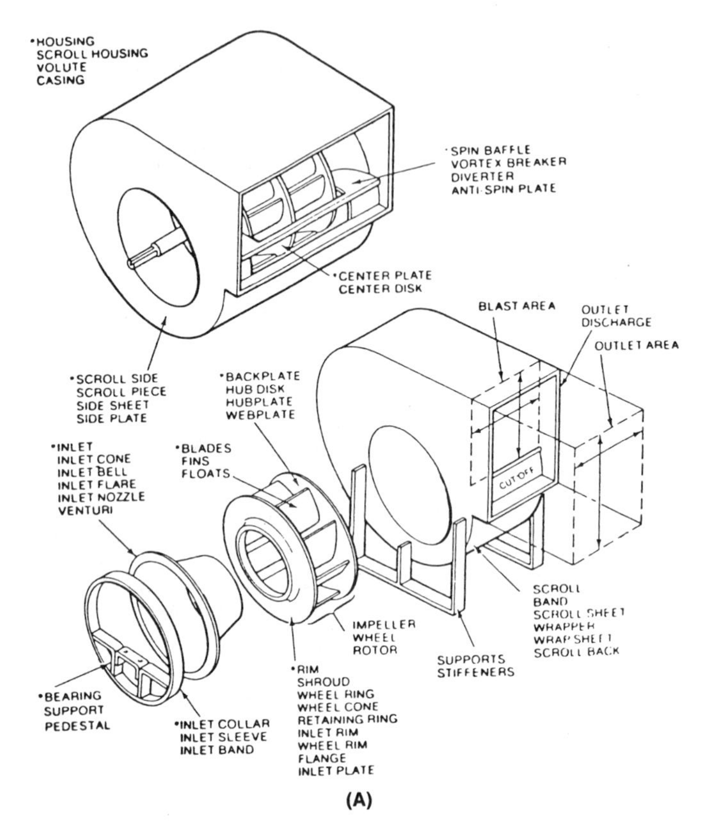

Figure 11. (A) Terminology for centrifugal fan components; (B) terminology for axial and tubular centrifugal fans.

Using Figure 10C for airfoil blades, a value of 0.44 for D_s is obtained and the wheel diameter is:

$$D = D_s \cdot \sqrt{Q}/P_s^{1/4}$$
$$= 0.442(232.8/2.35)$$
$$= 43.6 \text{ inches}$$

Thus far, we have discussed the basic theory of the centrifugal machines and their design. Discussions now turn to fans, blowers, and exhausters. Basic fan components are summarized in Figure 11.

Basically, there are *two types of fans* used in industry, viz, the centrifugal fans (low to medium flows and relatively higher pressures than the axial-flow fans) and axial-flow fans (mostly large volume and low pressure fans).

Centrifugal fans are further subclassified as:

- Radial blade
- Forward-curved blade
- Backward-curved blade
- Airfoil

Axial fans are subdivided into the following classes:

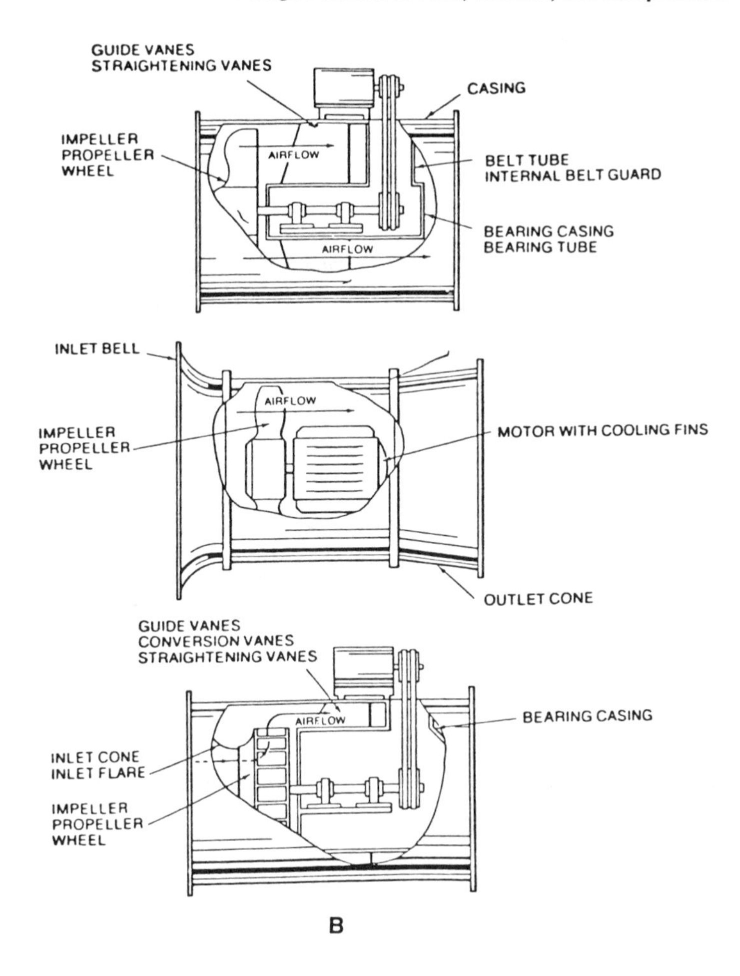

Figure 11. (Continued)

- *Vane-axial Fans*
- *Tube-axial Fans*

The various configurations are shown in Figure 12. Each design is described in detail in the following.

Radial-Flow Fans

These fans may be divided into groups according to their blade shape, drive arrangement, type of inlet, or type of discharge. The blade shapes of radial flow fans may be divided into four general groups (refer to Figure 13):

1. Backward-inclined (a)(b)
2. Backward-inclined radial-tip (c)
3. Radial (d)
4. Forward-inclined (e)(f)

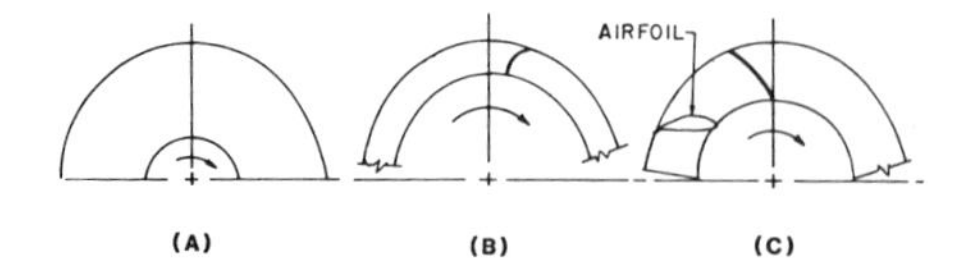

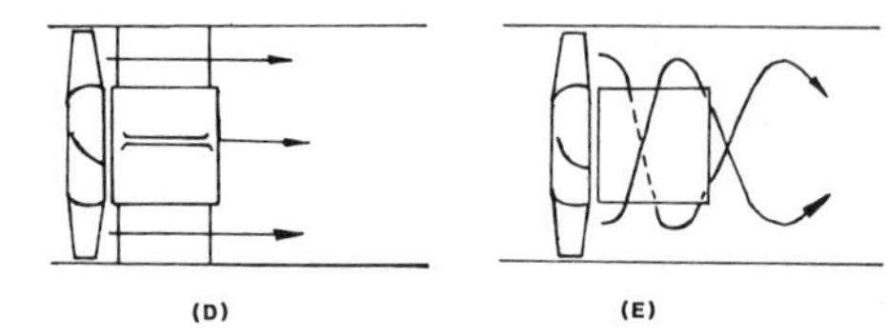

Figure 12. (A) Radial blade design; (B) forward curve blade; (C) backwardly inclined blade and the airfoil design; (D) vane axial fan with discharge guide-vanes; (E) tube axial fan with spiral air flow (no guide-vanes).

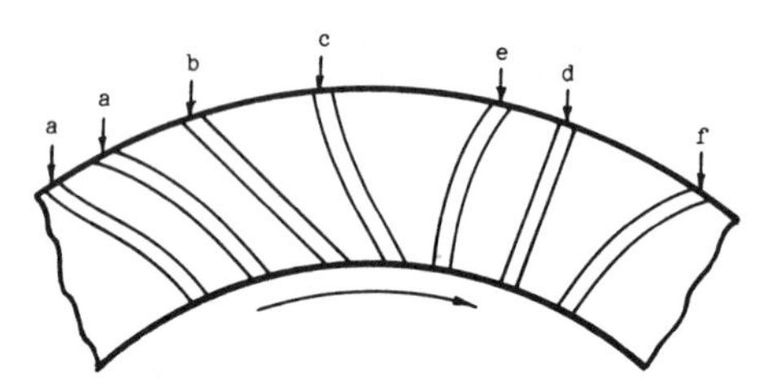

Figure 13. Shows radial fan blade shapes.

For fan wheels of the same size and speed, the alphabetical sequence in Figure 13 parallels a trend toward *increasing* number of blades, inlet diameter, axial width, static pressure, capacity, turbulence, horsepower requirements, and *decreasing* stability and efficiency.

Fan wheels are usually driven by electric motors because of the high efficiency and smooth operating characteristics of motors. Occasionally, however, they are driven by other means, such as gasoline or diesel engines or even steam turbines. The fan wheel may be mounted directly on the driver or may be driven through a rigid, flexible, or fluid coupling or a belt drive. The direct drive is preferred whenever space and speed limitations permit. Eight recommended drive arrangements have been adopted as standard by the National Association of Fan Manufacturers, such as belt drive, wheel overhang, and bearings in pedestal.

The three principal types of inlets are single-width, single-inlet; single-width, double-inlet; and double-width, double-inlet. To obtain double-inlet construction with the same fan, the air is admitted axially from both ends of the shaft, and a material improvement in static pressure capacity and efficiency is obtained. The double-width, double-inlet construction is obtained by placing two single-width, single-inlet fans back to back in a single casing. Inlet design principles are discussed later.

Most radial-flow fans employ a scroll casing whose main functions are first, to collect the individual air streams from the blades and direct them to a single outlet and second, to convert some of the velocity pressure of the gas leaving the blades into static pressure. The conversion is most effective with forward-inclined and radial blades since the gases leaving the impellers of these fans have relatively high velocities. The gases from backward-inclined blades, on the other hand, have relatively low velocities so that scrolls add little to performance.

The demands of different installations may require different directions for the discharge opening relative to the inlet opening. Sixteen different directions have been standardized by the NAFM, such as for example, clockwise, bottom horizontal.

The impeller can be encircled by a diffuser consisting of a series of curved stationary blades that convert some of the velocity pressure into static pressure. However, the cost and space requirements

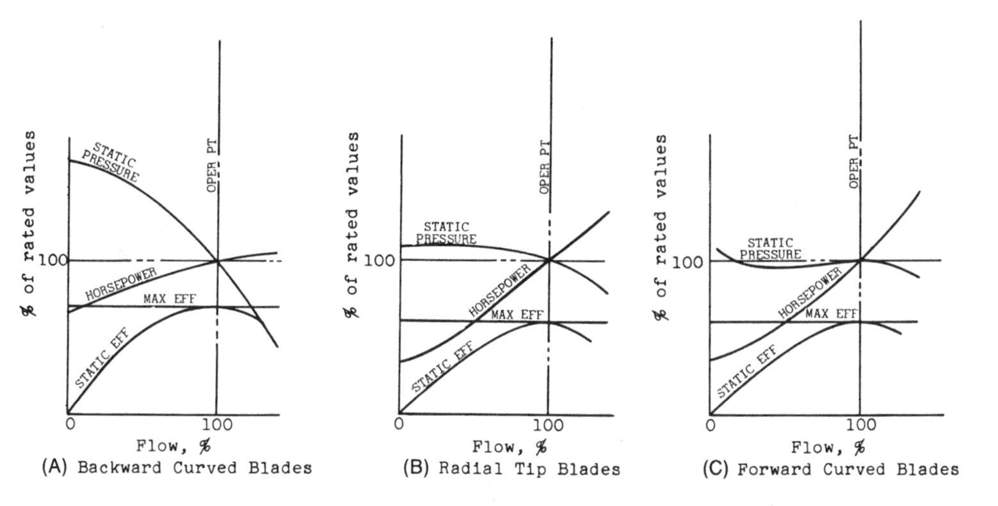

Figure 14. Characteristics of radial-flow fans with scroll. (A) backward curved blades; (B) radial tip blades; (C) forward curved blades.

of such diffusers seldom justify their use. For installations in which there is not room for a scroll or diffuser, or the gas is discharged into a large space, the scroll casing is frequently eliminated. This arrangement is usually called radial or circumferential discharge. Motor and generator cooling applications are usually of this type.

The relative performance of the main types of radial-flow fans are shown in Figure 14. These curves do not show the behavior of any particular fan but rather the average characteristics of fans of these types. The plots report static pressure, static efficiency, and fan horsepower represented as a function of flow at constant speed. These quantities are plotted as percentages of their rated value (except efficiency, which is already dimensionless) in order to show the characteristics of a series of fans, rather than just those of the particular fan tested.

A similar dimensionless method of presenting fan data is in terms of the flow and pressure coefficients. The disadvantage of these methods for design work is the requirement that the quantities of interest must be translated back into dimensional units before being used. A common method of presenting fan data is with the flow in cfm, pressure in inches of water gage, and power in either horsepower or watts.

The pressure curve of Figure 14A shows that, starting with the no-delivery point at left with higher than design pressure, the curve passes through the design point (taken at maximum static efficiency) to the point of free delivery at the far right. Aside from a slight rise that sometimes occurs near the point of no-delivery the curve falls steadily without the dips that frequently occur in the curves forward-inclined blades and axial-flow fans. This pressure characteristic of backward-inclined blades results in a stability of performance that makes these blades (and radial-tip blades with similar characteristics) well suited for fluctuating systems and parallel operations.

The peak value of brake horsepower occurs in the operating range of the fan not far from the point of maximum efficiency. When the driving motor is selected with a power rating equal to this peak value, overloading of the motor is impossible. Thus, the backward-inclined radial fan is considered to be of the nonoverloading type.

Note that the point of design is usually taken at the point of maximum static efficiency. Seventy percent static efficiencies are common in large, well-designed units.

For *forward-inclined blades* the region of decreasing pressure with decreasing flow, apparent in the pressure curve of the forward-inclined blade fan (see Figure 14C) may cause unstable operation when this type of fan is operated in parallel with similar fans or on fluctuating systems. Even in other applications precaution must be taken that the fan does not operate in the unstable range of the pressure curve.

With increasing volume the fan horsepower increases rapidly, giving a maximum horsepower at maximum rate of flow. This maximum requirement is frequently twice the requirement at maximum efficiency. Since radial-flow fans are seldom operated at maximum flow (axial-flow fans operate more efficiently here), this means that power requirements are much larger beyond, than within, the operating range. Thus forward-bladed fans are considered to have overloading characteristics, which is a disadvantage for some applications.

With units having small horsepower requirements a motor is selected capable of supplying the load at free delivery. With large units, however, this procedure is uneconomical and a motor is selected only slightly larger than needed to supply power requirements under operating conditions, provision being made in the installation design to prevent operation at too low static pressures. Thus the operating range of the forward-inclined blade fan, limited at high flows by increasing power requirements and at low flows by unstable operation, is much narrower than that of the backward-blade fan.

The static efficiency curve for the forward-inclined blade fan has its maximum at the point of design. Maximum efficiencies for large well-designed fans with suitable scrolls may be as high as 60 percent as compared with 70 percent for backward-inclined blade fans. Efficiencies of sirocco fans normally range from 35% to 45% when operating at the design point.

The performance characteristics of radial blade fans (refer to Figure 14B) are, in general, intermediate between those of the backward-inclined and forward-inclined blade fans. With a properly designed inlet, the efficiencies can be about as high as for the forward-curved fan. However, the simple radial blade has very low efficiency due to the shock loss at the entrance.

Table 1 summarizes principle operating characteristics of the three subdivisions of radial-flow fans. This table, together with the performance characteristics shown in Figure 14, should provide a clear indication of the fan that is best for a given application.

Airfoil profiles with thick leading edges are generally inefficient for radial-flow fans. The leading and trailing edges of the blades should be sharp for best results. Also, the use of airfoil profiles offers more restriction to flow if the number of blades is large. Thus, in most applications, a blade made from a thin sheet will perform as well as the airfoil blade as long as proper attention is given to air inlet and exit conditions and to design factors, such as the method of fastening blades to the shroud, curvature of blades, and the smoothness of the flow path. The thin-sheet blade costs less and is easier to manufacture than the airfoil profile.

The number of blades is important from two standpoints. First, if the surface friction loss increases, there is too much restriction of the inlet air flow to the blades. If too great a spacing is allowed, the air expands too rapidly for proper diffusion and there is back flow between the blades with resulting inefficiency. There is a fairly standard range for the number of blades used in commercial fans as indicated in the following

Type of Fan	Number of Blades
Backward-inclined blades	4–24
Radial blades	8–28
Forward-inclined blades	16–66

In general, the number of blades should be the minimum number required to obtain proper diffusion. However, for any type fan, there is generally a wide number of blades between too many and too few where the fan operation is about the same.

For a fan with given outer diameter and speed, higher pressures are produced by longer blades. However, as the inner diameter is reduced, the flow becomes restricted. Thus, the inner diameter should not be made so small that more head is lost at the entrance, due to the restriction of flow, than is gained by the increased blade length. Neglecting the head loss due to restriction, it can be inferred that for a given fan with a fixed speed and flow, the head varies as

$$1 - \left(\frac{D_2}{D_1}\right)^2$$

Table 1
Effect of Blade Shape on Radial-Flow Fan Operation

Characteristic	Backward-Inclined	Radial	Forward-Inclined
Number of blades	Smallest	Medium	Largest
Size for same pressure and capacity	Largest	Medium	Smallest
Volume handled for same size and speed	Smallest	Medium	Highest
Range of produced volume of flow	Widest	Medium	Narrowest
Pressure for same size and speed	Lowest	Medium	Highest
Efficiency*	Highest	Medium	Lowest
Usual operating speed	Highest	Medium	Lowest
Overloading characteristics	Nonoverloading	Usually nonoverloading	Will overload under some condition
Reversibility	Not reversible	Reversible (with straight blades)	Not reversible
Suitability for parallel operation	Best	Medium	Worst
Suitability for direct motor drive	Best	Medium	Worst
Self-cleaning characteristics	Nonself-cleaning	Self-cleaning	Nonself-cleaning
Cost for same impeller size	Higher	Lowest (with straight blades)	Higher

* *Radial and forward-inclined have scrolls.*

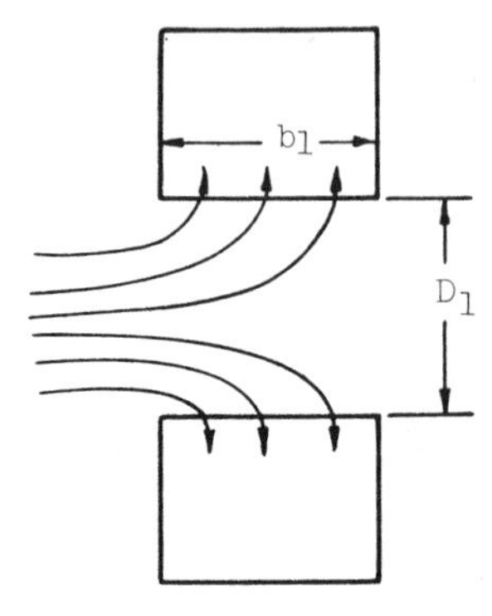

Figure 15. Illustrates that an increase in blade width produces little increase in flow.

As the width of the fan increases, a fan is capable of handling a greater capacity. However, if the blade width is increased beyond the point where the circumferential area at the inner edge of the blade is more than the inlet area of the fan, an increase of blade width will produce little increase in flow, due to the restriction offered by the entrance. This is illustrated in Figure 15. The ratio of entrance area to the area at the inner blade edge is:

$$\frac{A_i}{A_1} = \frac{\pi D_1^2/4}{\pi D_1 b_1} = \frac{D_1}{4b_1} \tag{25}$$

Thus, if the fan width is made appreciably more than a fourth to a half of the inner diameter, poorer performance will result due to throttling at the entrance. For these reasons fans for handling large volumes of flow at small pressures are designed with wide blades with a shallow depth, (large inner diameter). Those for producing high pressures at low flow rates use narrow blades with a greater depth.

Consideration must be given to the orientation of flow into the leading (inner) edge of the blades in order to minimize shock loss at the entrance; that is the angle, β, of the moving fluid relative to the blade at the leading edge should be kept at a minimum. This angle β is shown in Figure 16, where V_{1r} is the radial component of fluid velocity V_1 and u_1 is the relative tangential velocities of the air and blade. The reduced pressure of the air on the suction side of the blade tends to cause separation of the air stream from the blade with resulting turbulence, noise, and inefficiency. This tendency becomes especially important where the blade joins the shroud or backplate which holds the blades in place. Interference between the blade and shroud accentuates the tendency toward separation on the suction side. Ample filleting of the joint between the blade and shroud is therefore of importance in obtaining broader efficiency and pressure characteristics. The leading edge of the blade should be sharp and the angle of attack should be as small as possible.

Noise caused by eddies produced by the fan blades can be decreased only by careful attention to such details as blade shape and entrance and exit conditions, that is, the avoidance of sharp bends and projections into the air stream. Sharp corners should be filleted.

Examples of pressure vs. flow coefficient curves for radial-flow fans are shown in Figure 17. The performance characteristics in these plots are defined as follows:

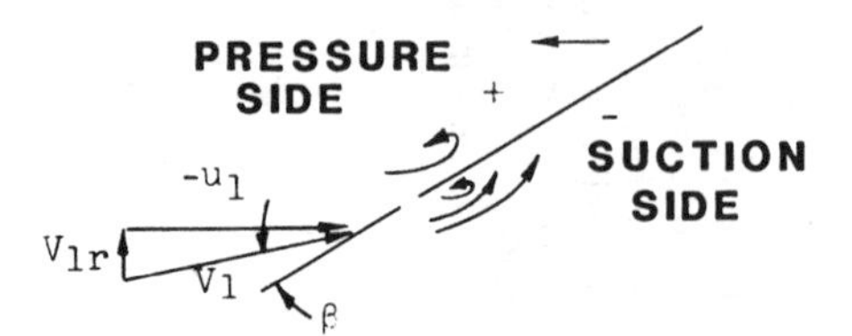

Figure 16. Shows orientation of flow into leading edge of blade.

ψ = pressure coefficient

$$\psi = \frac{\Delta p_s}{\frac{1/2 \rho u_2^2}{g}}$$

ϕ = flow coefficient

$$\phi = \frac{Q}{u_2' A_2}$$

η_s = static efficiency

$$\eta_s = \frac{\Delta p_s \times Q \times 1.57 \times 10^{-4}}{\text{horsepower input to rotor}}$$

d_s = specific diameter

$$d_s = D_c \frac{\Delta p^{*1/4}}{Q^{1/2}} \left[\frac{0.075}{\rho}\right]^{1/4}$$

or

$$d_s = 1.58 \times 10^{-2} \left[\frac{D_c}{A_2^{1/2}}\right] \frac{\psi^{1/4}}{\phi^{1/2}}$$

N_s = specific speed

$$N_s = n \frac{Q^{1/2}}{\Delta p^{*3/4}} \left[\frac{\rho}{0.075}\right]^{3/4}$$

or

$$N_s = 8.05 \times 10^4 \frac{A^{1/2}}{D_2} \frac{\phi^{1/2}}{\psi^{3/4}}$$

Note both specific quantities are defined on the basis of the pressure differential in inches of water. For increased utility in fan selection d_s is defined in terms of D_c a size parameter which characterizes the rotor-diffuser dimensions. Examples of D_c for radial fans are given in Figure 18.

The d_s-N_s curves are identified by numbers referring to specific fans on the ψ vs. ϕ figures. Figure 19 describes the characteristics of the various fans.

Axial-Flow Fans

Axial-flow fans have been divided into three classifications by the National Association of Fan Manufacturers; they are propeller, tube-axial, and vane-axial. The propeller fan is an axial-flow fan which either has no housing, such as, for example, the common office or desk fan, or has a simple ring mounting consisting of a plate with a hole in it. An example of the latter is a kitchen ventilating fan. This fan is especially suited to handling large volumes of flow at low heads, not exceeding one inch of water. The tube-axial fan is an axial-flow fan in a tubular housing without stationary guide vanes. It can be used at heads somewhat higher than those suitable for propeller fans. The vane-axial fan is an axial-flow fan mounted in a tubular housing with stationary guide

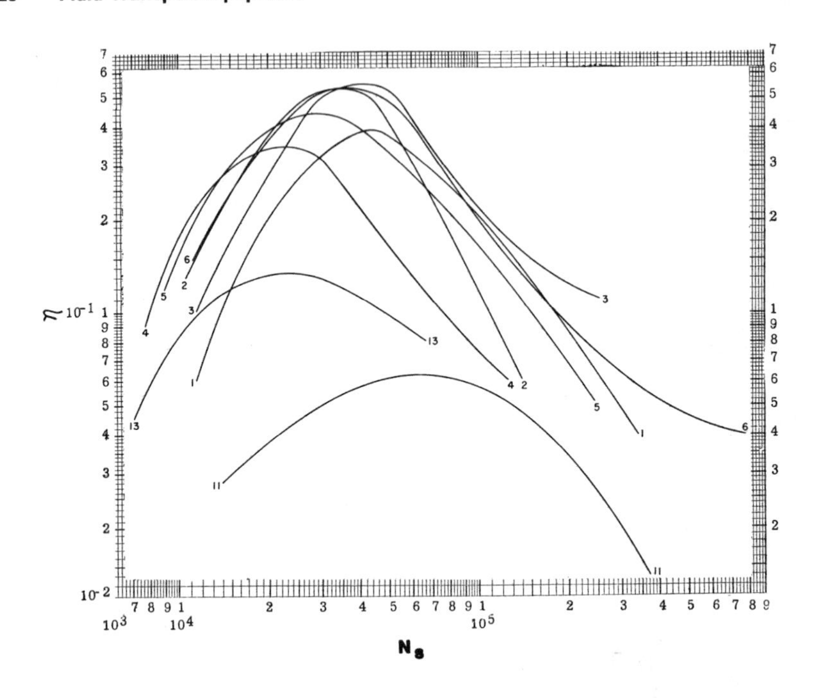

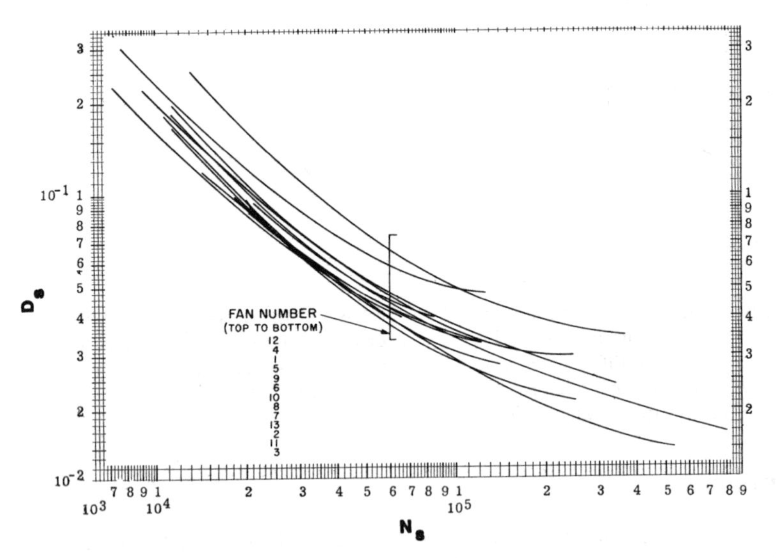

Figure 17. Composite N_s–D_s and efficiency curves for forward blade fans.

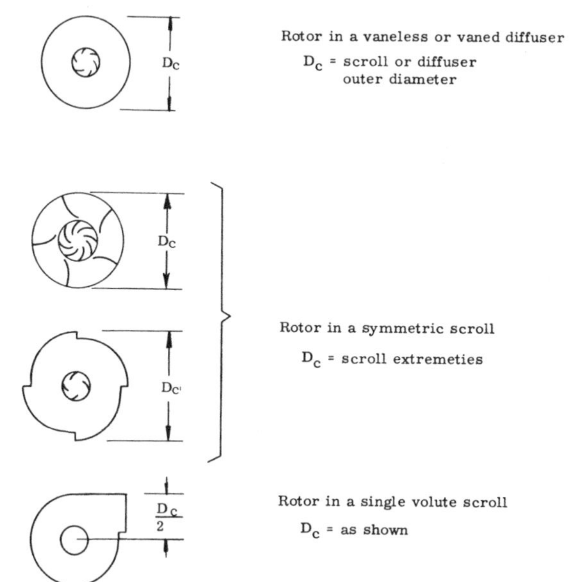

Figure 18. Examples of D_c for radial fans.

vanes, either before or after the fan wheel. The vanes, which may be either fixed or adjustable, direct the air flow to reduce the circulation and thus increase useful head and efficiency. The vane-axial fan is used for higher pressures, while still higher heads can be developed by using several stages.

Performance curves for a typical axial-flow fan are illustrated in Figure 20. These curves show static pressure, static efficiency, and fan horsepower plotted as a function of rate of flow at constant speed.

The pressure characteristics of an axial-flow fan are very similar to those of a forward-inclined radial-flow fan. The no-delivery pressure is high and decreases rapidly with increasing capacity until a point at about 50 to 80 percent of rated capacity is reached. At this position a point of inflection, and on some fans a small rise in pressure, occurs. With further increase in flow the pressure decreases sharply until the free delivery point is reached. The inflection point (also called the breakdown point) represents a transition from the condition of flow at low capacity, in which radial components of velocity are appreciable, to the condition at higher capacity, in which the flow is almost entirely axial. The breakdown at low flows is caused by stalling at the blade entrance. Operation of the fan in this region is likely to be unstable with flow instabilities in the form of pulsations.

Figure 20 shows that the fan horsepower is a maximum at no flow, decreasing with increasing flow while in the unstable region. As the flow becomes stable the horsepower increases to a value less than that for zero flow and then again decreases with larger flows. The slope of the decreasing portions of the curve varies widely among different designs. With some fans, it is practically horizontal, while with others it is very steep. Fans with steep characteristics require large motors.

Axial-flow fans are employed when it is necessary to develop velocity head; hence, the total efficiency rather than the static efficiency is commonly taken as a basis for rating. Total efficiencies of up to 70% may be attained with small enclosed units; however, efficiencies of 50%–60% are more common. Large vane-axial units frequently attain 85% to 90%. Open, household type, fan efficiencies range from 40% for small fans to 50% for large exhaust fans.

The radial flow of gas along the blade of an axial-flow fan lowers the efficiency, pressure, and flow rate of the fan and is one of the prime causes of fan noise. Such radial flow can be eliminated

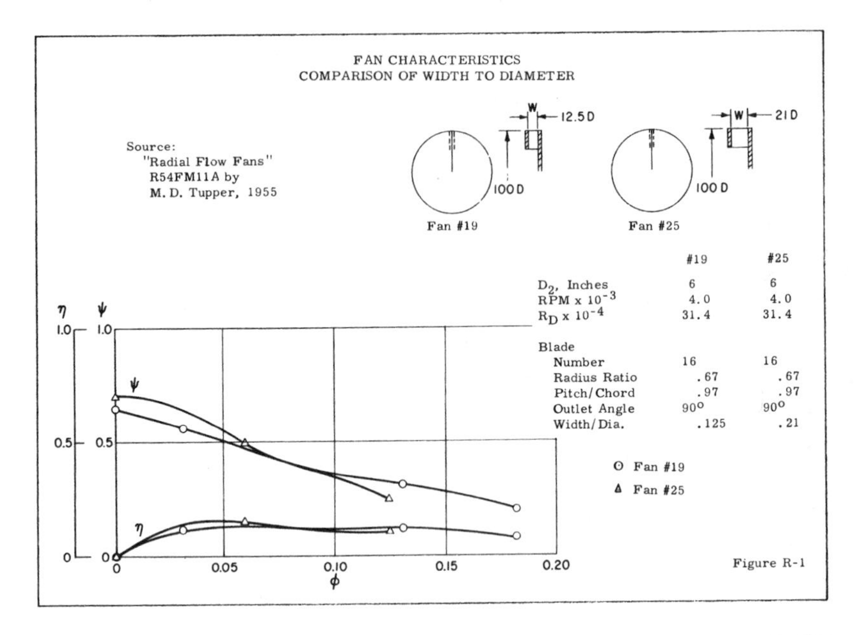

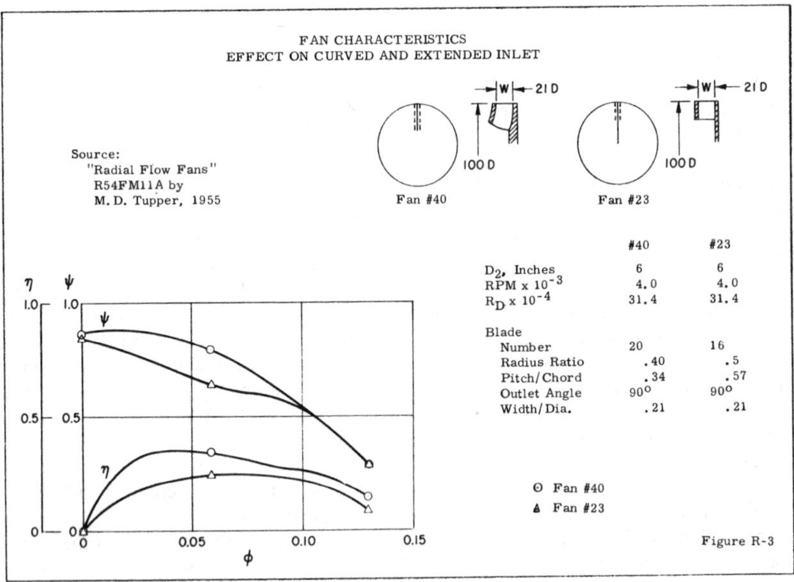

Figure 19. Shows characteristics of radial-flow fans.

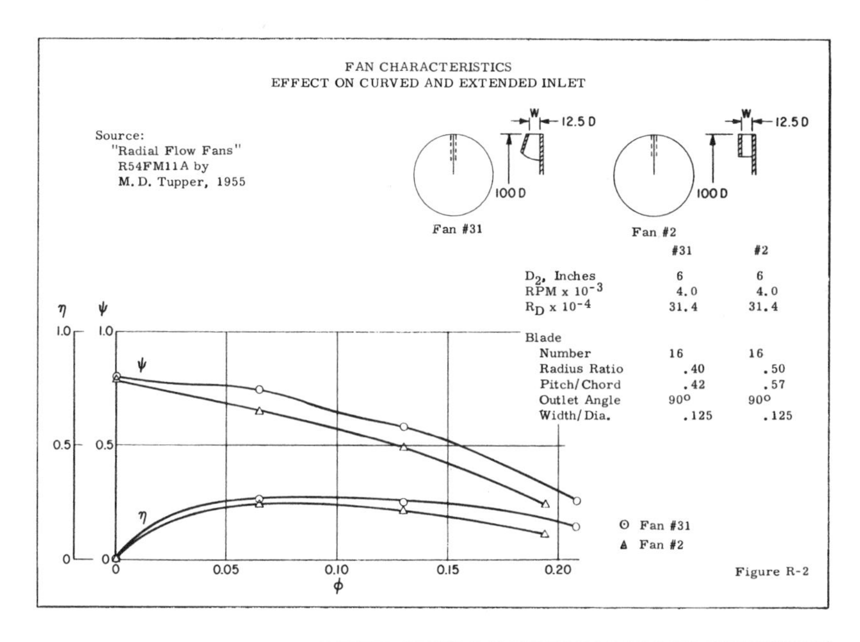

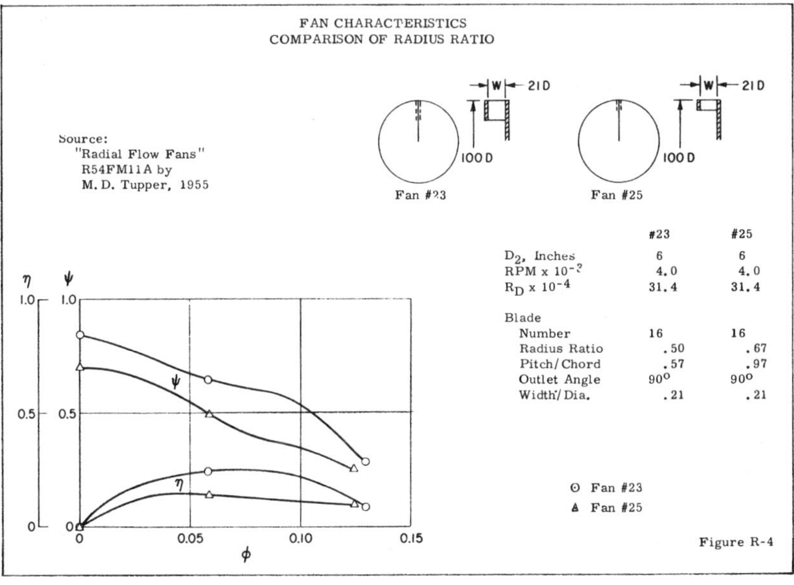

Figure 19. (continued)

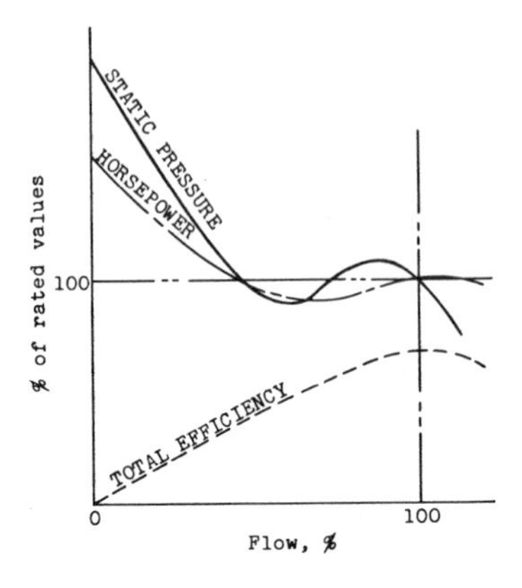

Figure 20. Shows characteristics of a propeller fan.

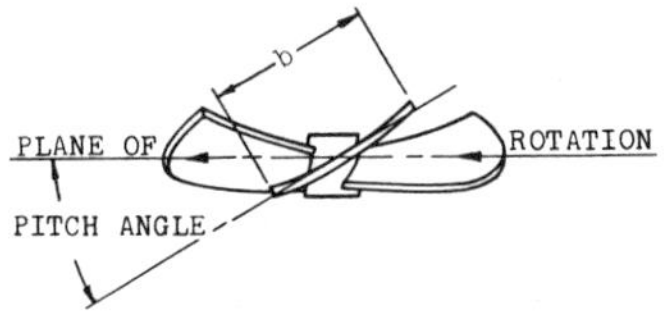

Figure 21. Shows plane of rotation and pitch angle for axial-flow fans.

by designing the blade so that the pressure differences across the blade are the same at all sections. This is described in more detail later.

The blade profile of the higher pressure units, such as vane-axial and tube-axial fans, is usually an airfoil section providing a high lift coefficient. For propeller-type fans, however, the advantage of airfoil blades is small. Many good designs of blades made of thin sheets have been made, as the simplicity of construction outweighs the advantage of an airfoil shape on low pressure fans.

As noted earlier, the pressure developed increases directly with the number of blades. The relation breaks down as the number of blades becomes large, since there is interference between blades and restriction of flow due to the cross-sectional area of the blades themselves. The selection of the correct number of blades is largely a matter of experience. For low pressures, it is usually desirable to have as few blades as possible (two, three, or four being optimum depending on mechanical construction considerations). For higher pressures, using vane-axial or tube-axial construction, a large hub diameter is generally used, as well as a large number of blades (as many as 24 to 26).

The *pitch angle* (also called the "blade angle") is the angle of twist of the blades with respect to the plane of rotation at a given diameter (see Figure 21). Pitch angle varies with the pressure required and with the blade shape. The greater the pitch the greater the pressure developed at a given speed, since the lift coefficient C_L of a section is roughly proportional to the sine of the angle of attack for small angles. As the angle increases, the point of aerodynamic stall is reached, and instability occurs.

The pitch or "helix" angle is varied from hub to tip in order to keep the pressure across the face of the fan constant. In this manner, the lift coefficient times the width times the relative air velocity remains constant. The parts of the blade away from the hub are traveling faster than those near the hub and do not require as great a pitch angle to produce the same pressure.

The pitch angle of a thin plate is often increased from the leading edge to the trailing edge, resulting in a scoop-shaped blade. Under design conditions, the flow through the fan is smooth; however, as the flow decreases, the effective angle of attack of the blades increases. This results in stalling at the blade entrance and consequently noisy operation and low efficiency.

The width of the blades may either increase, decrease, or remain constant with distance from the hub. The width of propeller fan blades usually increases away from the hub. The width of vane-axial blades generally decreases away from the hub. The pressure rise is roughly proportional to the number of blade times the blade width so that as the number of blades decreases the width must be increased for the same pressure rise.

For propeller fans operating at very low heads, the size of the hub diameter can be as small as desired. For tube-axial and vane-axial fans which are used to obtain higher heads, the hub diameter is made large, resulting in relatively short stubby blades.

The reason for increased pitch toward the hub has already been discussed. However, there is a limit to this increase, and it will not be possible to get sufficient pressure rise near the hub if the hub diameter is too small. With a small hub diameter there will be greater pressure at the outer edge than at the hub. This condition can lead to serious back flow if the pressure difference is large. Thus, for tube-axial and vane-axial fans operating with large pressure rises the hub diameter must be made large.

Care must also be given to proper clearances. The clearance between the fan blade and housing should be kept as low as manufacturing practice permits in order to minimize tip losses. For velocity type fans developing little pressures this clearance should be small.

As will be discussed later, there are many applications where varying flow quantities are required. When this is the case, a fan size is selected that provides the maximum desired flow, and some method is furnished for regulating this flow either manually or automatically. Four means are available for regulating the flow of axial-flow fans:

1. Variable operating speed
2. Adjustable outlet dampers
3. Adjustable inlet guide vanes
4. Variable pitch propellers

Variations in pitch may be obtained by constructing the fan with adjustable blades whose pitch can be altered when the fan is stopped. This is done by rotating the blades about a radial line and clamping at the desired position. If variation of pitch while in operation is desired, a more complex system is required.

Of the various methods of control, variable speed is the most efficient, followed by variable pitch and inlet guide vanes, whereas adjustable dampers are wasteful and inefficient. However, variable speed and variable pitch propellers and costly; whereas adjustable inlet guide vanes are moderately expensive and adjustable dampers relatively inexpensive in first cost.

Controllable Pitch Fans and System Design

Variable air volume is a concept in air-conditioning systems where the maximum quantity of cooling air is supplied only to those areas where peak demand conditions exist. For areas not exposed to the maximum demand, a reduced or lesser amount of air is supplied. All this is accomplished under simultaneous operation. Through the utilization of this concept wherein only the needed amount of air is introduced into any occupied space, variable air volume reduces the initial cost as well as the total operating costs of the system. Further improved system acoustics and improved system efficiencies result from this application. Further, associated equipment such as fans, filters, and cooling coils as well as duct runs can be selected smaller than comparable selections for a constant-volume system.

Note that the constant-volume system does provide excellent temperature and humidity control and because of this fact some of its operating disadvantages are often ignored. For constant-volume systems the cooling load on a multiple room, constant-volume reheat system is fixed at a maximum value despite any reduction of requirements of cooling air to the occupied space. Then it follows that because of the reduction in demand for cooling, the total energy required for this space actually increases as the need for cooling·on this space decreases.

The usual design characteristic of constant-volume reheat systems are:

1. Air supply components are sized for the sum of all peak loads on all occupied spaces.
2. Refrigeration systems are sized for the sum of the peak loads.
3. Generally, there exists an energy consumption increase with a decrease in cooling load.

There are certain modifications and certain design variations where some of these disadvantages are offset. This is generally accomplished by a constant volume bypass type duct system where the required air supply temperature variation is obtained by mixing warm room air with a refrigerated supply air system. Figure 22 shows a schematic of a typical supply fan system. The example shown

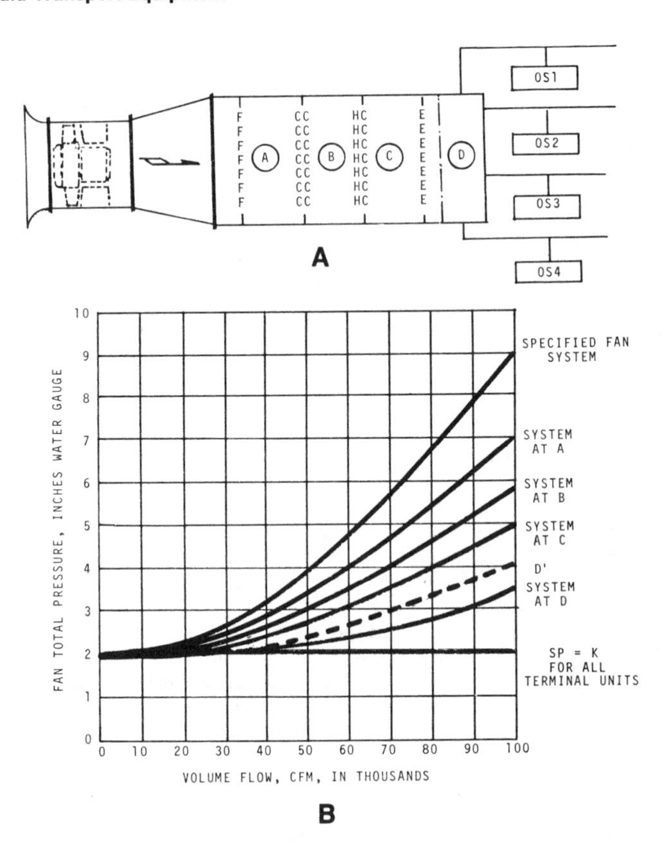

Figure 22. (A) Schematic of supply fan system; (B) plot of system pressure volume characteristics. (Courtesy Joy Manufacturing Co., New Philadelphia, OH.)

is for a single duct system with the usual HVAC apparatuses: the filters, the heating coils, the cooling coils, the eliminators. Also shown on the schematic is a plenum and a number of take-offs to the various occupied spaces. This represents a basic system for any building ventilation system.

The pressure-flow relationships for the system can be analyzed for each of these relations. In this example we consider a design that provides a constant static pressure regardless of flow to all points within the system. Then for the occupied spaces, OS1, OS2, OS3, and OS4, regardless of the total flow, we would like to see each terminal unit receive a constant value of static pressure. This is indicated by the horizontal line on the pressure/volume plot. On this line we next superimpose the various system resistances and the friction losses. The uppermost curve indicates the fan system, i.e., the type of system that the fan will operate against. It is a summation of all the system resistances plus the static pressure that is required for the terminal units.

At major resistance points the system lines are designated as Systems A, B, C, and D. The system line defined by A is characteristic of the system at a point downstream of the filter. The system line defined by B is representative of the system downstream of the cooling coil, and similarly for system line C and system line D.

Note that the pressure differences between F and A, A and B, B and C, and C and D, at any point along these system lines, represents pressure differences we might expect at the specific flow conditions. These represent variable losses and are probably the maximum expected loss. The filter

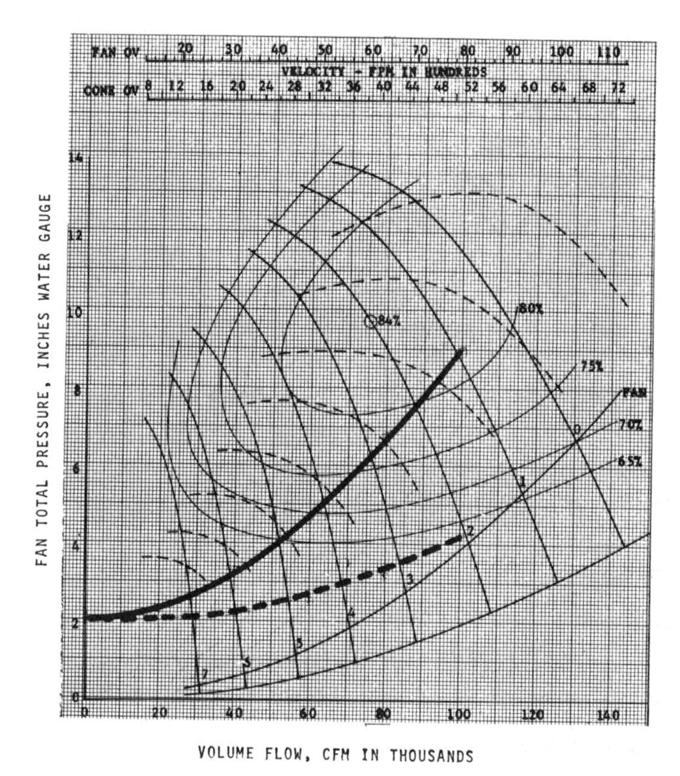

Figure 23. Typical vane axial fan performance curve with system overplots. (Courtesy Joy Manufacturing Co., New Philadelphia, OH.)

resistance is variable. Similarly so are loss variances within the coils as a function of the condition of the coils at the time of flow.

Directing attention to the plenum and the downstream distribution system, losses can be expected to be principally due to duct friction and velocity. These are not expected to change with the condition of the system, or with time.

A duct layout encompasses certain sizes of duct, certain velocity schedules, certain turns and elbows, mechanical restrictions, and some regains. These should not change with time. It can be concluded that the relationship assigned for this plenum and its composite downstream system is a relationship that will hold over the period of time.

The downstream system can be assigned a worst possible condition whence we may deduce that there may exist within a system a condition of load that will demand the maximum operating condition. We denote this worst case system as the system line defined by D′. If then we control on this system, we certainly will satisfy all the possible or known possible demands of the system. This implies that we are going to control or establish control for an idealized system independent of the resistance variations.

Some of the defined system lines are plotted on a typical performance curve of a vane-axial fan. Figure 23 plots the specified fan system and the D′ system, the system we intend to control on. Particularly note that the fan system line will operate over a broad range in volume, say 40%, at a good efficiency level. In fact, for some selected systems, efficiency improvement can result as the system is turned down.

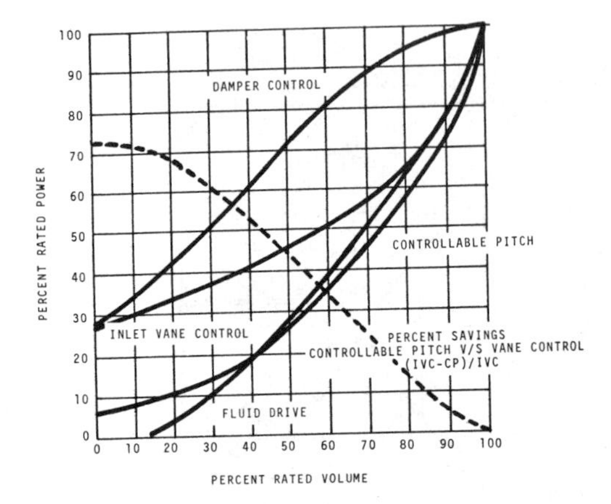

Figure 24. Comparison curves for volume control methods.

As noted earlier, control can be accomplished by any of the following:

1. Damper control
2. Inlet-vane control
3. Variable-speed control
4. Blade-pitch control

The first three items (the damper, inlet vane, and variable speed control) are generally associated with centrifugal and in-line type fans, although they may and have been used on vane-axial fans. The fourth is primarily restricted to vane-axial fans since this fan design lends itself to the blade modulation concept. This means that we must consider part load efficiency of the fan/drive unit when discussing efficiencies.

Figure 24 shows the percentage power consumption levels for reduced volume flows for the noted control devices. The damper and inlet vane control curves are based on expected centrifugal

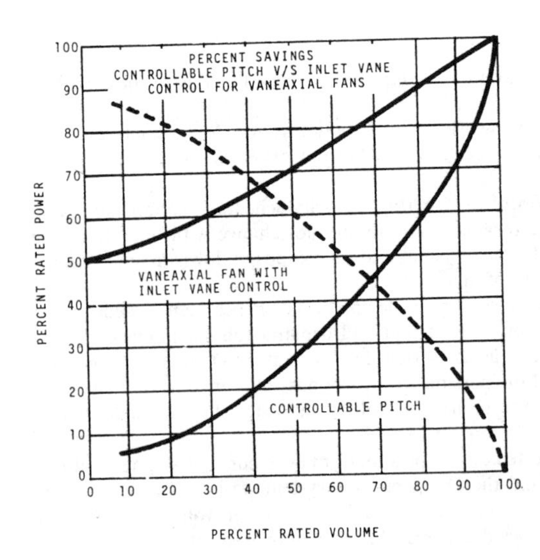

Figure 25. Power-volume comparison curves for vane axial fan with controllable pitch and inlet vane control.

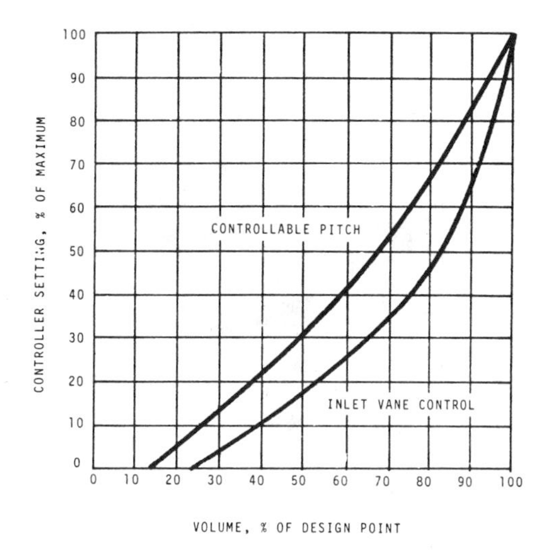

Figure 26. Response characteristics of controllable pitch and centrifugal fans with inlet vane control.

fan performance; the fluid drive can be applied to either centrifugal or axial fan performance; the controllable-pitch fan is strictly applicable to vane-axial fans as stated. It becomes readily apparent that considerable savings in power consumption can be affected through the use of a controllable-pitch vane-axial fan.

Efficient part load operation is a prime advantage of the vane-axial controllable-pitch fans, plus the added feature of reducing generated noise levels at reduced flows; whereas, throttling through inlet vanes increases generated noise levels. When inlet vanes are applied to vane-axial fans, their characteristics are similar to those for the centrifugal fan units. Figure 25 compares the power requirements for controllable-pitch vane-axial fans and vane-axial fans equipped with inlet-vane control.

Note that inlet vane control requires a considerable constriction in air flow before volume reduction takes place. Generally, for a 20% reduction in volume, the inlet vanes must modulate aproximately 55%. For the same volume reduction, the controllable-pitch axivane fan needs to be modulated only 32%. This is indicated generally in Figure 26.

One approach to controlling a system based on a controllable-pitch fan for a variable air volume system is shown in Figure 27. The system makes use of the concept of total pressure.

From Figure 22 a relationship at the plenum and a relation between total flow and total energy were established. This established relationship of total pressure to supply fan flow becomes the control criteria.

To effect control, measurements of the two system variables, total pressure, and fan flow are required. Total pressure is the sum of velocity pressure and static pressure. Flow is measured by velocity pressure equated against the total flow area. Both these measurements are accomplished through the use of a standard laboratory quality pitot tube.

The flow pitot tube output is sensed by the supply fan flow differential transmitter. This transmitter will output a pneumatic signal as a function of the differential. Its output signal is fed to the supply fan flow-pressure controller, and in turn becomes the set point for this controller. In effect, this becomes a floating set point, since it is variable in nature. As the flow increases, the differential increases; hence, this value of set point increases. Similarly as the flow decreases, the differential decreases; and again, the value of the set point will decrease.

The system total pressure is sensed by a pitot tube and the duct pressure transmitter. This output signal is then matched in the flow-pressure controller against the flow set point. The flow-pressure controller is a proportional plus reset controller. In automatic operation, it will position the blade

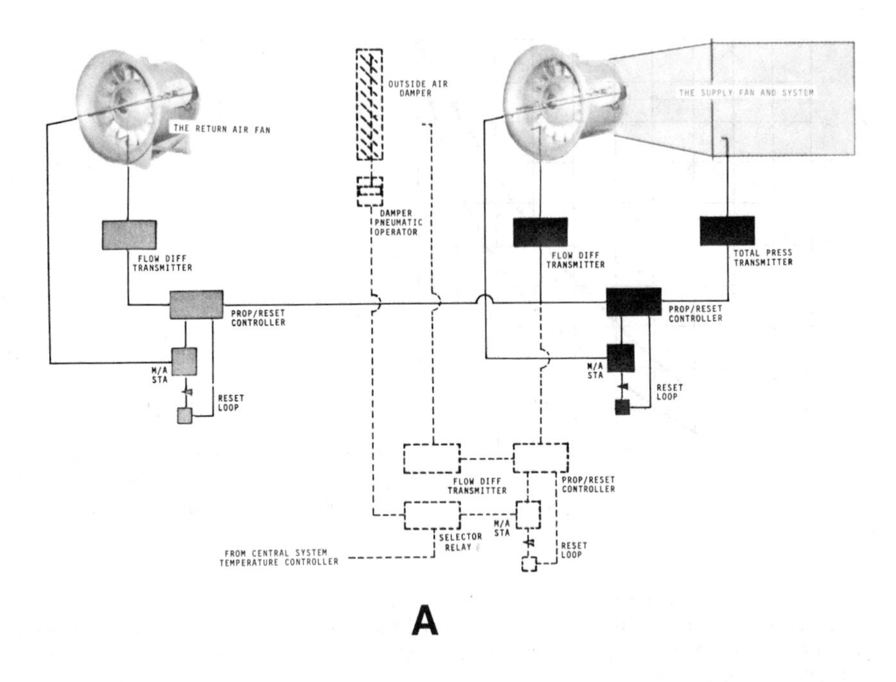

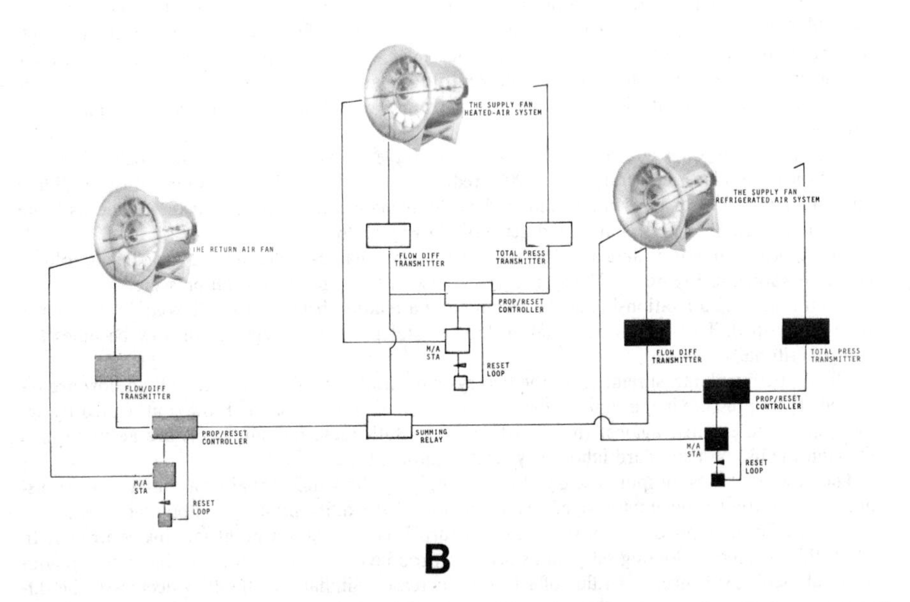

Figure 27. (A) Illustrates control scheme for variable air volume system; (B) Control scheme for controllable pitch with dual duct arrangement. (Courtesy Joy Manufacturing Co., New Philadelphia, OH.)

setting angle on the controllable-pitch fan to the setting necessary to establish the preset flow-total pressure relationship.

Through a few system checks, the balance between supply and return flow can be fairly well established. Further, since return air system pressures and velocities are generally low, it becomes difficult to control this system on the static pressure concept. One approach is to have the return air fan serve as a slave to the supply fan. This is accomplished by metering the return fan flow and comparing it to the supply fan flow. The return air flow is measured via the pitot tube and flow differential transmitter. The output of this transmitter is fed into a proportional plus reset controller whose set point is established by the supply fan flow differential transmitter.

In automatic operation the system will position the return air fan blades to provide the flow necessary to satisfy the supply fan/return fan ratio. We assume that the characteristics of the return air system will not change appreciably with time.

The approach can be extended to the overall system. Consider the control of outdoor air (OA), returned air (RA), and exhaust air (EA). A principal control consideration is that of outdoor air. Generally, this is accomplished through the use of a velocity controller installed in the outdoor air intake duct. This flow can be slaved directly to the supply fan and made to vary directly with and in proportion to the supply flow. When it becomes desirable to draw in large quantities of outdoor air to deviate from this preset flow ratio, control of the OA damper can be transferred directly to the control system temperature controller. Similarly, the returned air and exhaust air can be programmed for control from the basic flow relationship.

The control arrangement for a dual duct variable air volume system is illustrated in Figure 27B. Note we put both the hot deck and the cold deck under flow-total pressure control, each as an independent variable air volume system. Again, each system will operate independently—for maximum system efficiency. To slave the return air fan to both supply fan flows we add the two individual flows, then match the return fan flows to this value.

At first look one might tend to inherently conclude that the two-fan proposal for dual duct systems is probably too costly. In some instances this might be true, but only for the first cost consideration. The flow and pressure system demands are too widely varying to permit a cost generalization. One must equate this proposal against the traditional single-fan two-damper arrangement. Certainly there can be no doubt as to the operating advantages, both from power cost and control response, of the two-fan proposal.

The two-fan system virtually eliminates the problem of VAV unit recirculation due to over pressurization at the minimum or zero demand flow condition. For the single-fan with damper throttling, the plenum total energy is set by the total system flow. Then as the main duct flows vary, the developed duct static pressure varies, increasing as the flow decreases. The two-fan arrangement eliminates this condition. Further discussions are given in References 1 and 2.

We will discuss dual duct systems in more detail as these are often used for providing temperature control for individual zones or spaces. Such systems are often found in office buildings, schools, hospitals, apartment buildings, hotels, and other multi-room structures where highly variable sensible heat loads are prevalent.

Such a system offers several advantages:

1. Individual room or zone temperature control.
2. Being an all-air system, services such as power, water, and drains are required only in the equipment rooms.
3. The system can use all outdoor air when the temperature is low to satisfy cooling loads.
4. All major equipment can be centrally located in a single area for more convenient service and maintenance.
5. Air intakes are located at only one central point.
6. The systems are essentially self-balancing.
7. With proper fan type and control, lower fan power consumption costs can be realized.

A dual-duct system is composed of two air streams—one hot and one cold—connected to a terminal mixing unit, which proportions the hot and cold air in response to a thermostat located in

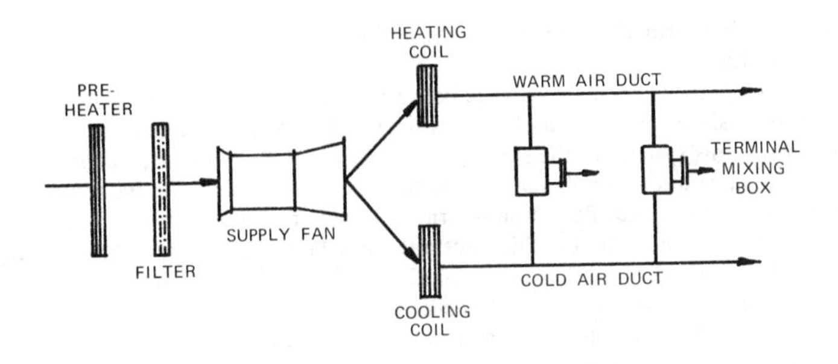

Figure 28. Schematic of single fan, dual-duct supply system.

the space served by that terminal unit. Though a total constant-volume flow of air is always provided for each space or zone, the quantities of warm and cool air making up this total flow vary as space conditions dictate. In effect, then, the warm air system and the cool air system are actually separate variable air volume systems.

Supply air for the dual-duct, constant-volume system can be furnished from either a single source or from two separate sources with independent hot deck and cold deck fans. In the latter case, each system would be treated as a separate variable air volume system. Figure 28 shows a single-fan dual-duct, constant volume supply system. The resistance to flow on the inlet side of the fan will always see a constant-volume flow. Those resistances represented by coils and filters will be variable, to a degree, depending upon the conditions of these items at any particular time of flow. To simplify discussions, assume a constant inlet resistance based upon the worst anticipated conditions.

Looking at the downstream side of the fan, we note that the total flow from the fan separates into two distribution systems—the cold duct system and the warm duct system. The losses through each of these systems, with the exception of the heating coils and cooling coils, are principally due to duct friction and velocity schedules and do not change with the conditions of the system or with time. The resistance through the coils will vary with their condition at any time of flow and, of course, the amount of flow through them.

Although the total volume flow through the discharge system is constant, the flow through its two branches—the warm duct and the cold duct—varies proportionally in response to heating or cooling requirements. Since this is the case, then the resistance in each of these systems *due to flow only* varies as the square of the volume flow.

Since the supply plenum at the discharge side of the fan is common to both duct systems, any pressure buildup in one system is common to both. It follows, then, that the pressure at this point must always be adequate to overcome whatever maximum resistance to flow exists in any leg, plus the minimum static pressure required at the entrance to any terminal unit when maximum flow of either hot or cold air is required at that unit. As earlier, when specifying conditions for fan design, one must determine the total volume flow required and calculate the resistance for the worst possible conditions at the most remote terminal unit. Since the mixing boxes are designed to provide a constant discharge of tempered air with varying duct static pressures, they will, by throttling, induce pressure build-up in the system above that actually required for a particular flow in any particular leg. On a system where there is no capacity control device on the fan, the total pressure differential across the fan will remain constant due to the pressure build-up caused by the terminal units, with resulting constant fan brake horsepower at all conditions of operation.

As an example, consider a dual-duct, constant-volume system requiring a total design flow of 60,000 cfm. In this case, the cold duct is sized at 90% of total flow or 54,000 cfm. At this flow, the total maximum resistance calculates to 8.8 in. wg SP, of which 3-in. wg is on the inlet side of the fan. The hot duct has been sized for 85% of the cold duct area. At 85% of the cold air flow the total maximum resistance if 8.8 in. wg for 46,000 cfm passing through the hot duct system. Figure 29 gives an illustrative performance curve, where the fan brake horsepower for the design condition of 60,000 cfm, 8.8-in. SP (9.7-in. TP) would be 112.

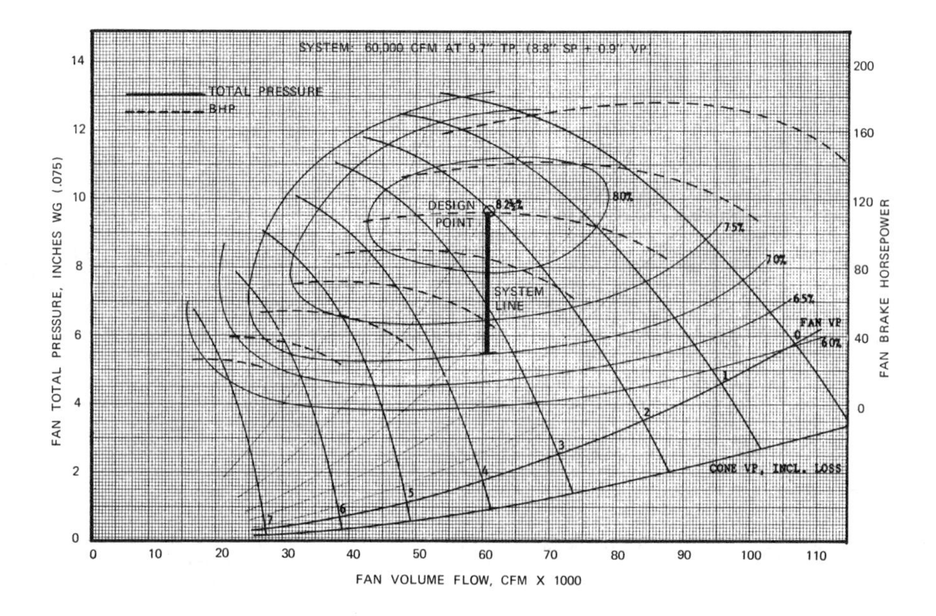

Figure 29. Illustrative fan performance curve.

If the fan remains on a constant blade pitch (a #2 blade setting), its horsepower consumption will always be 112 BHP because the mixing boxes (terminal units) will always maintain a constant total volume flow. As cold air decreases in flow to the units, the warm air increases. The boxes will always maintain a suitable resistance in the system to maintain constant volume.

Since everything in the system on the discharge side of the fan, with the exception of the boxes, if fixed, the resistance to the flow caused by that part of the system varies with the square of the flow. For proper operation it is necessary to maintain a minimum of 1-in. wg constant pressure at each box.

On the inlet side of the fan, the system is somewhat variable due to dirt accumulation on the filters and coils. Assuming, however, a condition of maximum dirt build-up, the resistance through this part of the system stays constant with constant air flow. Hence, the calculated 3-in. inlet resistance can be considered as a constant resistance that does not change during operation of the air-conditioning system.

Although the total volume flow through the discharge system is constant, that total system actually consists of two separate systems. The flow in each of these systems (hot and cold ducts), varies in response to heating or cooling requirements. It then follows that the resistance to flow through each of these systems also varies. Only at times of maximum heating or cooling is the full pressure capability of the fan required.

Assuming a constant resistance at the fan's inlet (3-in. wg) and a constant pressure at the most remote box (1-in. wg) then for the total system design pressure of 8.8-in. wg, 4-in. wg is constant and 4.8-in. wg is variable depending upon the flows in the hot and cold ducts. Under these conditions, a properly controlled controllable-pitch fan would operate as shown in Table 2.

This analysis considers a constant inlet resistance based upon maximum design conditions which includes a pressure drop across the filters with maximum dirt build-up. In practice, the filter resistance will vary with time. The degree of variation depends upon the type of inlet filters, the degree of filtration required, the rate of dirt build-up, and the frequency of filter replacement. Typically, there may be a filter resistance variation of 1-in. wg or greater between dirty and clean filters. If there is less than a 1-in. resistance with clean filters, then the common resistance at that time will

Table 2
Data Tabulation for Constant-Volume, Single-Fan, Dual-Duct System

System Volume Flow		System Resistance		Required Fan TP for Maximum System Resistance (4.9 in.)	Required Fan bhp
Cold Duct	Warm Duct	Cold Duct	Hot Duct		
54,000	6,000	4.8	0.08	9.7	112
50,000	10,000	4.1	0.23	9.0	105
45,000	15,000	3.34	0.51	8.24	98
40,000	20,000	2.65	0.90	7.55	91
35,000	25,000	2	1.42	6.9	85
30,000	30,000	1.48	2.04	6.94	86
25,000	35,000	1.03	2.77	7.67	92
20,000	40,000	0.66	3.6	8.5	100
14,000	46,000	0.32	4.8	9.7	112

only be 2-in. wg at the inlet side of the fan, plus the 1-in. wg minimum SP at the inlet to the most remote box. Under these conditions a controllable-pitch fan would operate as tabulated in Table 3.

These results are graphically plotted on Figure 30. It's interesting to note that full fan capacity is required only at times of maximum heating or cooling loads. Fan brake horsepower is reduced at all other load requirements if the fan is controlled to supply only the energy required by the system.

Note that the fan must work against a constant-volume, variable-pressure system. This system is plotted as a straight vertical line on Figure 29. For minimum fan brake horsepower, the required fan must produce a constant-volume flow against the minimum system resistance at any point of time and at any state of condition of the system. To effect system control, a measure of the volume flow must be made to the preselected level, and the fan's output adjusted to maintain this level. Further, this control must be capable of modulating the fan volume to establish this control with a minimum of disturbance and variation.

Volume flow is measured through the use of a laboratory-type pitot tube and or an annular flow element located in the fan annulus downstream of the rotating stage. Basically either of these units

Table 3
Data Tabulation for Constant-Volume, Single-Fan, Dual-Duct System

System Volume Flow		System Resistance		Required Fan TP for Maximum System Resistance (3.9 in.)	Required Fan bhp
Cold Duct	Warm Duct	Cold Duct	Hot Duct		
54,000	6,000	4.8	0.08	8.7	102
50,000	10,000	4.1	0.23	8.0	96
45,000	15,000	3.34	0.51	7.24	89
40,000	20,000	2.65	0.91	6.55	83
35,000	25,000	2	1.42	5.9	78
30,000	30,000	1.48	2.04	5.95	79
25,000	35,000	1.03	2.77	6.67	84
20,000	40,000	0.66	3.6	7.5	92
14,000	46,000	0.32	4.8	8.7	102

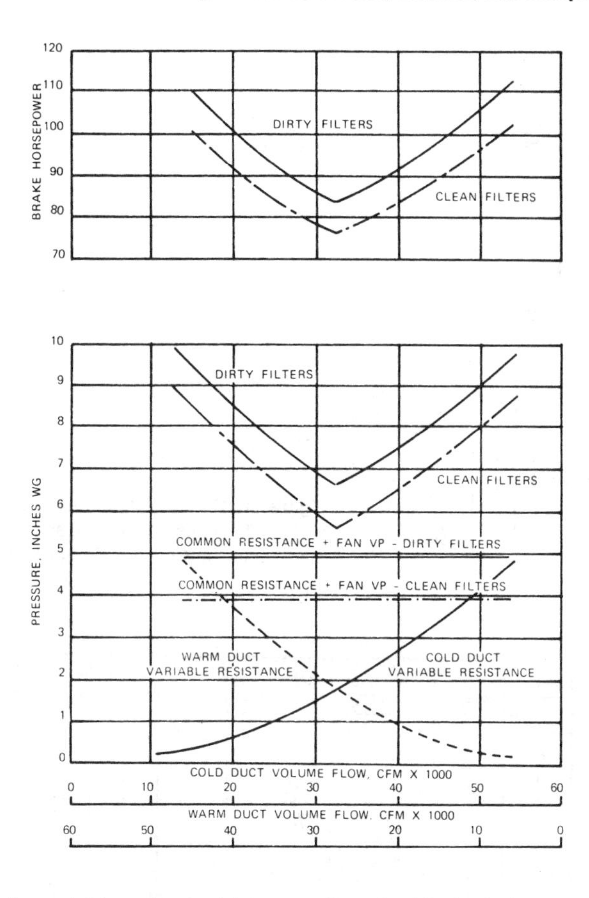

Figure 30. Plots for variable system resistances and required fan brake horsepower, single-fan-dual duct system. (Courtesy Joy Manufacturing Co., New Philadelphia, OH.)

indicate the point SP and TP. Inputting these values into a differential pressure transmitter, provides an output signal that is a function of volume flow. The output signal is then compared to a fixed set point signal in a proportional-plus-reset controller. This controller will then output a signal to the fan-blade positioner-operator to modulate the blade setting angle to a position necessary to satisfy and maintain the desired constant-volume flow.

One control schematic is illustrated in Figure 31. The necessary control elements can be located immediately adjacent to the supply fan, or be integrally attached to the fan casing.

Further discussions can be found in References 3 and 4.

We now direct attention to another frequently designed system, namely, parallel fans for large volume flows. The principal advantage gained from parallel fans is that of increased mechanical reliability, a function of the number of fans being used which is applicable to either constant-volume or variable-air-volume (VAV) systems.

The control of fans for parallel system operation is relatively simple with two basic schemes that are used. The principally used scheme provides control to select the required number of fans to satisfy the system demand and operates each fan at the same volume and pressure, the selected number operating, in effect, in unison. The second scheme, allows the first fan to achieve its maximum blade angle setting, maximum volume output, and then modulates the succeeding fans to meet the system demands. Of the two schemes, the first scheme is the more frequently used.

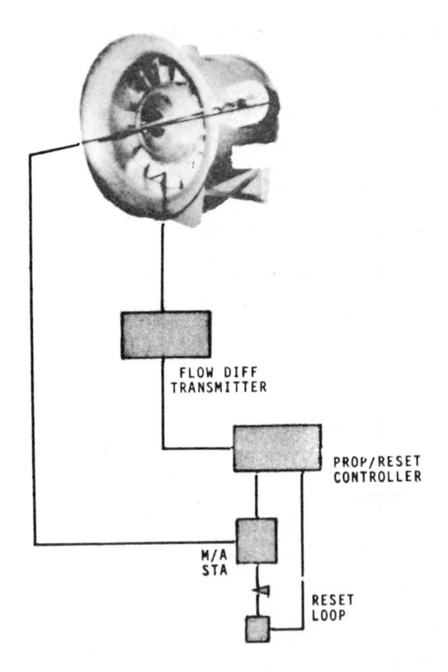

Figure 31. Control schematic for controllable pitch, axivane fan constant volume system. (Courtesy Joy Manufacturing Co., New Philadelphia, OH.)

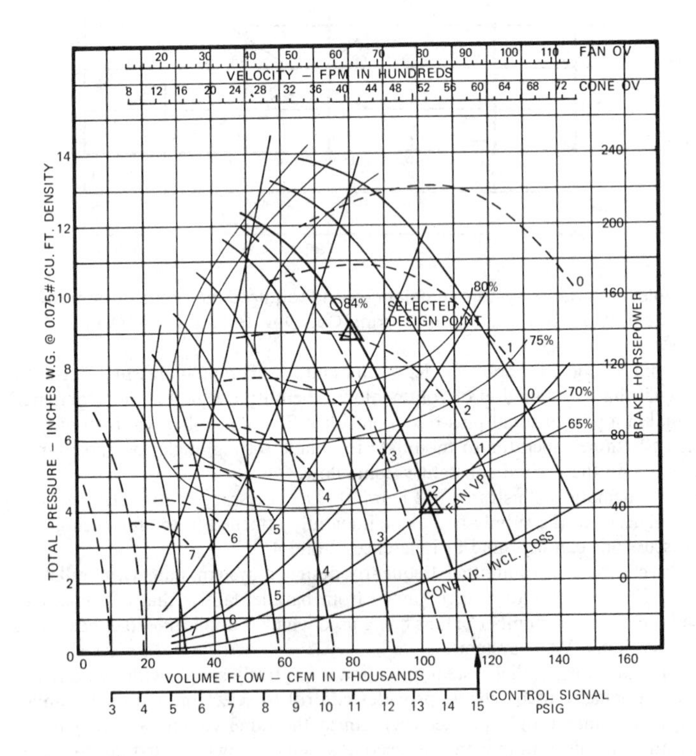

Figure 32. Characteristic plots of control signal vs. blade setting angle. (Courtesy Joy Manufacturing Co., New Philadelphia, OH.)

Control is straightforward with controllable-pitch fans equipped with pneumatic operators and positioners. The control input signal to the positioner establishes a unique position for each signal pressure setting and sets a specific blade angle. Figure 32 provides an illustrative example of how at various control signal inputs unique blade setting angles are established. The resulting volume flow from this set blade angle is a function of the system resistance. At a 15-lb control signal and the #2 blade setting position, with the system resistance at 4 in., the volume flow rate would be 103,000 cfm. However, if the system resistance is 9 in., then the volume flow rate would be 80,000 cfm. Note that each control signal sets a unique blade setting angle and that the resultant performance is a function of the system resistance corresponding to this flow rate.

To illustrate parallel fan operation, consider a system of 240,000 cfm at 9.0-in. TP with a condition of maintaining a 1.0-in. SP at all terminal units regardless of system flow. Then at zero volume flow the system is 1.0-in. TP (1.0-in. SP). This system line deviates from the standard parabolic system in that it does not intersect through the zero-volume/zero-pressure point. Next assume that the fan selection and performance is as per Figure 33. On this system line is plotted the maximum performances for the selected fan. Refer to Figure 33 for 1-fan, 2-fan, and 3-fan operations. The intersection of these fan performance lines with the predefined system line indicates the capability of either single or combined fan performance.

From this curve one can then develop the lines of response for a 3–15 psig control signal range. These plots indicate that a system demand to 108,000 cfm on the system curve will be met with a single operating fan. The 2-fan operation will carry through to 180,000 cfm and the 3-fan operation to 240,000 cfm, the selected design point. Note that for any number of fans in operation, the control

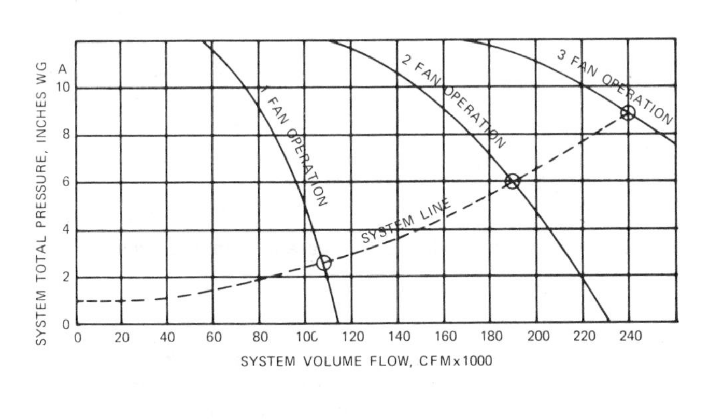

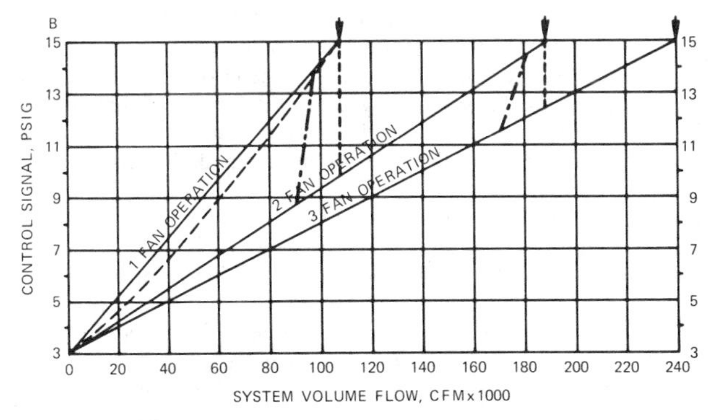

Figure 33. System lines and control signal pressures for fans in parallel operation.

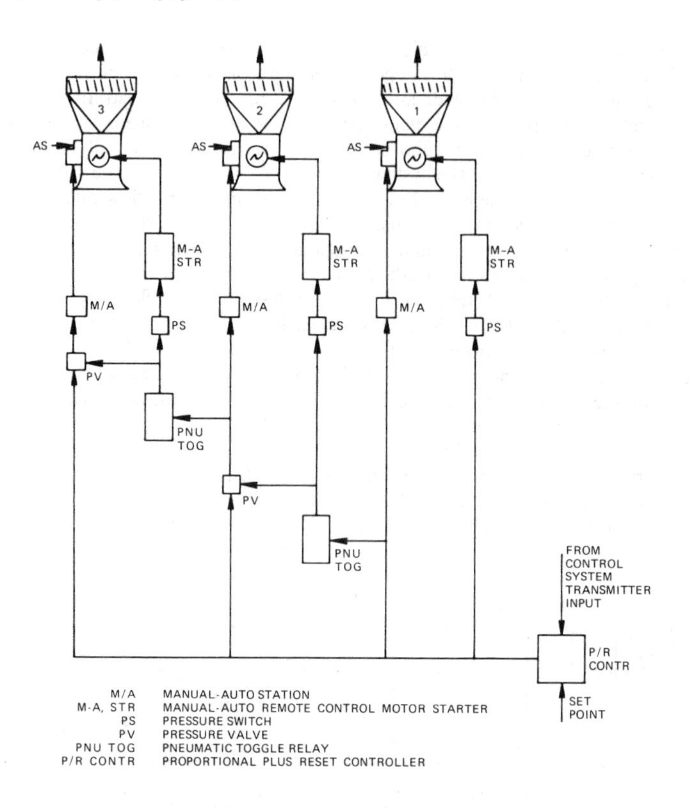

Figure 34. Illustrative control scheme for fans in parallel operation with pneumatic toggling.

signal will always be at a 3–15 psig level. This means that each fan, singly or in any combination, has the capability of modulating its blade angle from minimum to maximum for a 3–15 psig signal range. For each fan 3 psig is the minimum setting and 15 psig is the maximum setting blade angle.

In this example the operation of the multiple-fan arrangement on the system can now be correlated to the 3–15 psig signal range, which is indicated on the plot in Figure 33. Note that the operating lines are indicated as being linear, which is not necessarily the case.

As the control signal reaches the maximum point and the next fan is placed into operation, the control components must be capable of adjusting to meet this decay of control signal. This system operates in either a forward or reverse direction, placing or taking fans out of service and out-putting a volume-pressure level necessary to satisfy the system demand. The required control action is often referred to as pneumatic toggling.

Figure 34 indicates typical theoretical cut-offs for the toggling points, the vertical straight lines for the 1-2 and 2-3 fan operation. To preclude the possibility of a cyclic action, on and off, for the fan being placed into or taken out of service theoretical toggling point cut-offs are not generally used. Instead, overlap is provided for to ensure a stable operation. This is accomplished by reducing the toggling points to some level (shown by the inclined lines). To further ensure operational stability between the 1-2 and 2-3 fan operation, the higher pressure toggle point for each fan is usually selected at different levels, the highest level for the last fan to be placed into service.

An example of the control schematic for a 3-fan system is illustrated in Figure 33. Three fans are installed in parallel, each unit equipped with either a motorized or a gravity backdraft damper. The purpose of the damper is to preclude the possible recirculating of a nonoperating fan. Control can be by either static pressure, variable pressure as a function of volume, or any other specified variable air volume control. For the operation of this parallel system, it is imperative that the output out of the selected transmitters be input into a proportional-plus-reset controller and this input then matched against a setpoint, either fixed or variable, in the P/R. The output of the P/R is then fed through to each of the fans through an individual manual/auto (M/A) station. The M/A stations are specified only to permit either the manual or the automatic mode of operation of each fan.

The output of the P/R is fed directly to the first fan. This line branches into the pneumatic toggle for the second fan. The pneumatic toggle is a relay that will pass through a pneumatic signal, usually the air supply, when the input signal reaches some preselected level. The relay is an adjustable ratio type which employs an input signal feedback. That is, it can be adjusted to maintain this pneumatic output through a range of input signal drop (decay), or to a lower pressure level. It is analogous to a differential pressure switch outputting a signal at some preselected level and holding the signal until a lower level is reached. The output of the pneumatic toggle is then fed to a pressure valve in the signal line to the second fan, which thus modulates the position as a function of the control signal.

This action provides synchronous operation of the first two fans, using the same control signal from the P/R for a specified pressure range. When the P/R signal reaches the maximum level the second pneumatic toggle is energized and opens the signal to the third fan.

The preceding illustrates the response to a variable air-volume system and associated blade positioning. The same control components can be used to automatically start-stop the fan motors in complement. This is also indicated and is effected by taking the output of the pneumatic toggle to a pressure switch which serves as the automatic start-stop feature of the motor line starter.

Variable-Pitch Fans

Variable pitch vane-axial fans enable changing of the pitch, or angle, of the fan blades in response to a flow controller while the fan is rotating. Changing the blade angle essentially provides a new fan characteristic to meet each new system requirement, maintaining high efficiency and conserving energy. These fans are especially suited for use in air-handling systems which require variations in volume and/or pressure to control temperature, humidity, or air quality. They are particularly effective when used in variable air volume air-conditioning systems.

Standard commercial units have foot-mounted motors up-stream of the rotors. This permits flexibility of design with three fan arrangements available. Examples of units are shown in Figure 35.

Figure 36 shows a comparison of several methods of fan capacity control. Note that at 50% flow, the fans require 18% power while inlet vanes require 54% power. These type of fans can control at very low flows and hence, the savings can be even more dramatic for weekend and nighttime set-backs of air-conditioning systems.

In applications where operating air pressure is not available, these fans can be furnished with a mechanical linkage across the vane section. This linkage can be operated manually, or by a positioning-type modulating electric motor. Adjustable pitch fans can be furnished where the fan must be stopped and blades adjusted individually to a new angle. (see Figure 37.)

Fans of this type are available in a flow range of 5,000 to 250,000 cfm and total pressures to 14-in. wg. Fans are usually direct driven at standard motor speeds. In many instances, the most advantageous selection will result in a fan with a smaller diameter or higher velocity than is dictated by the duct design. If so, an additional expansion in area is desirable and an outlet diffuser should be used. A properly designed diffuser offers a further advantage in the conversion of some additional velocity pressure into static pressure, achieving static regain. A diffuser contains a central core of the same diameter as the fan core. This design assures uniform reduction of the air velocity with minimum turbulence as it leaves the fan to achieve optimum static regain. Also, the larger diameter at the diffuser outlet minimizes the loss caused by the abrupt expansion at the end of the central core.

Diffusers are also available in sound attenuating versions. This type of unit serves a dual purpose by providing static regain and sound attenuation in the same device. Figure 38 shows a typical

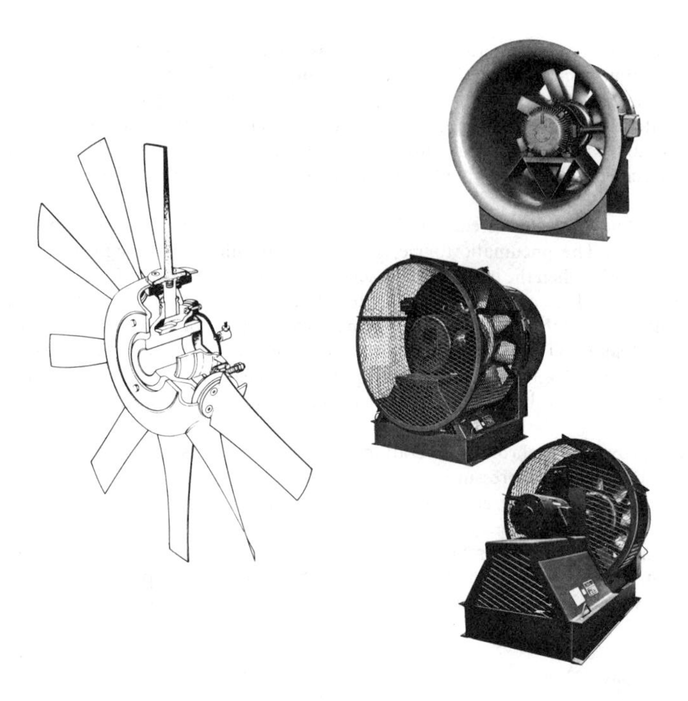

Figure 35. Examples of variable pitch fans. (Courtesy Flakt Products, Fort Lauderdale, FL.)

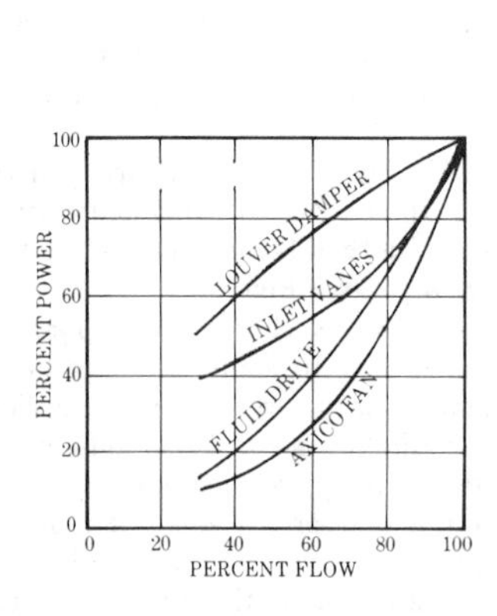

Figure 36. Comparison of different methods of fan control.

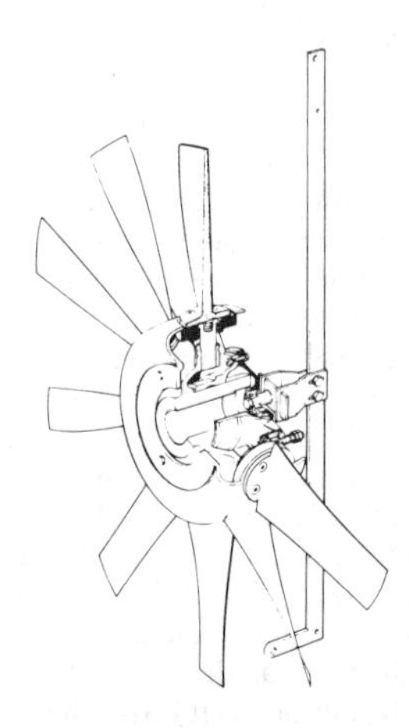

Figure 37. Shows adjustable pitch fan with adjustment mechanism.

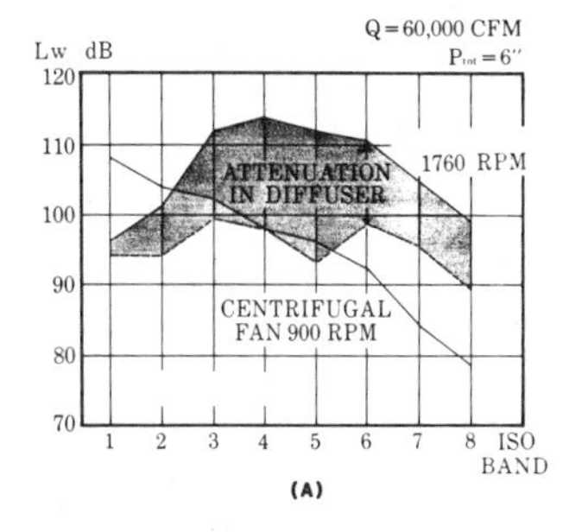

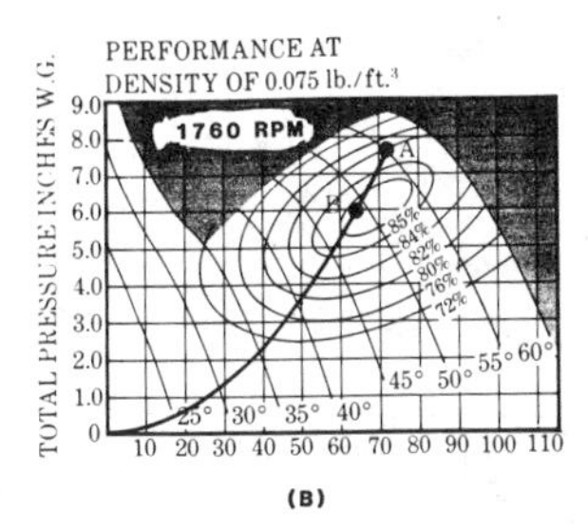

Figure 38. (A) Sound level comparison with diffuser; (B) acoustic performance of diffuser. (Courtesy Flakt Products, Fort Lauderdale, FL.)

example of acoustic diffuser performance. Sound ratings are published as sound power levels in decibels. Figure 38 shows the sound power level of a fan with a capacity of 60,000 cfm at 6-in. wg total pressure contrasted with a backward curved centrifugal fan of equivalent capacity. Note that the centrifugal fan has higher sound power levels in the lower ISO bands, where attenuation is more difficult.

The sound power levels of fans vary as the blade angles change. If a fan choice is made at a maximum selection point, such as "A" in Figure 38B, the sound power level will reduce when the fan is modulated along the system characteristic curve to "B." This is because the fan efficiency increases resulting in smoother air flow. The sound level also decreases as the volume of air flow decreases. This situation does not occur when inlet guide vane dampers are used to reduce fan capacity because of the increase of air turbulence as the inlet dampers are closed. Figure 39 shows typical dimensions and a specification sheet for one manufacturer's unit. Figure 40 shows an exploded view of a typical vane-axial fan.

Air-Conditioning Systems

In this discussion, the air foil bladed centrifugal fan is intended when the term *centrifugal fan* is used. The term *incline fan* will be used to describe a fan utilizing a standard single-width air-foil centrifugal wheel in an in-line housing, as fabricated by most manufacturers. Fan sizing should be based on selection procedures followed by consultants, with the following cost elements to be considered:

1. Initial cost
2. Accessory cost
3. Space cost
4. Structural cost
5. Handling and installation costs
6. Installation flexibility
7. Maintenance cost
8. Operating cost
9. Selection cost
10. Layout cost

In comparing fan types, generalizations can be used, but with caution. The technology of axial fans is sufficiently different from that of centrifugals (in-lines are similar to centrifugals), however, to make carte blanche generalizations can be hazardous.

On the average, the vane-axial fan will cost slightly less than the centrifugal fan and a fair amount less than the in-line. Generalizing on initial cost is particularly difficult since it varies

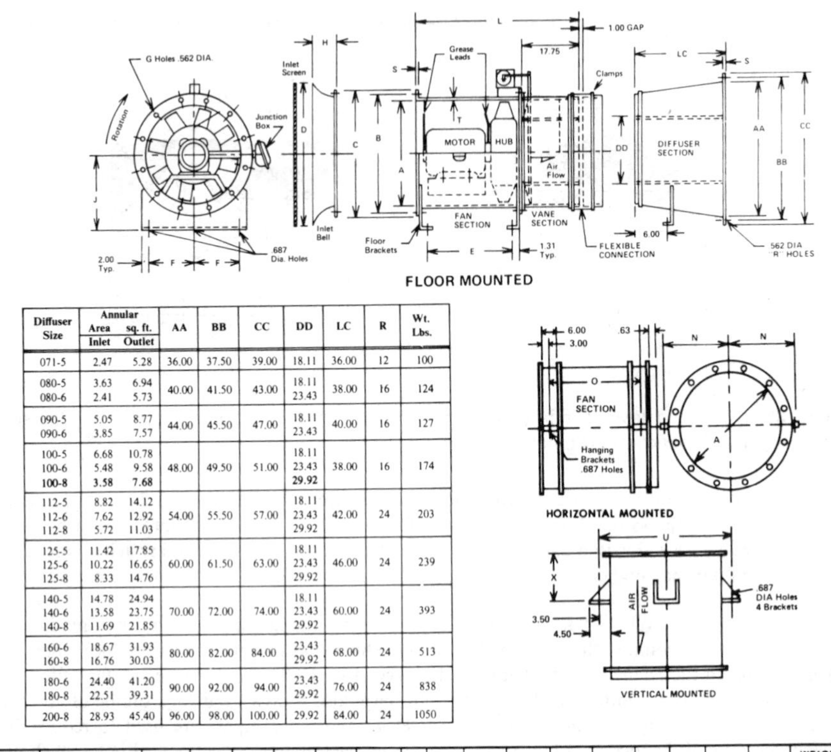

Diffuser Size	Annular Area sq. ft. Inlet	Annular Area sq. ft. Outlet	AA	BB	CC	DD	LC	R	Wt. Lbs.
071-5	2.47	5.28	36.00	37.50	39.00	18.11	36.00	12	100
080-5 080-6	3.63 2.41	6.94 5.73	40.00	41.50	43.00	18.11 23.43	38.00	16	124
090-5 090-6	5.05 3.85	8.77 7.57	44.00	45.50	47.00	18.11 23.43	40.00	16	127
100-5 100-6 **100-8**	6.68 5.48 **3.58**	10.78 9.58 **7.68**	48.00	49.50	51.00	18.11 23.43 **29.92**	38.00	16	174
112-5 112-6 112-8	8.82 7.62 5.72	14.12 12.92 11.03	54.00	55.50	57.00	18.11 23.43 29.92	42.00	24	203
125-5 125-6 125-8	11.42 10.22 8.33	17.85 16.65 14.76	60.00	61.50	63.00	18.11 23.43 29.92	46.00	24	239
140-5 140-6 140-8	14.78 13.58 11.69	24.94 23.75 21.85	70.00	72.00	74.00	18.11 23.43 29.92	60.00	24	393
160-6 160-8	18.67 16.76	31.93 30.03	80.00	82.00	84.00	23.43 29.92	68.00	24	513
180-6 180-8	24.40 22.51	41.20 39.31	90.00	92.00	94.00	23.43 29.92	76.00	24	838
200-8	28.93	45.40	96.00	98.00	100.00	29.92	84.00	24	1050

FAN SIZE	T FRAME MOTOR MIN	T FRAME MOTOR MAX	A	B	C	D	E	F	G	H	J	L	N	O	S	T	U	X	WEIGHT EXCL. MOTOR
071	213	286	27.95	29.85	31.35	37.95	26.37	13.00	12	5.18	20.75	47.00	17.67	22.87	.188	.188	35.34	14.63	320 lbs.
080	213	286	31.50	33.38	34.78	41.50	26.37	14.00	12	5.18	26.00	47.00	19.44	22.87	.188	.188	38.88	14.63	375 lbs.
090	213	405	35.43	37.25	38.70	45.44	34.37	16.00	12	5.18	28.00	55.00	21.41	30.87	.188	.188	42.82	18.63	450 lbs.
100	213	405	39.37	41.25	42.75	49.38	34.37	18.00	12	5.18	30.00	55.00	23.38	30.87	.188	.188	46.76	18.63	572 lbs.
112	213	445S	44.09	46.00	47.47	54.12	38.37	20.00	16	5.18	33.00	59.00	25.74	34.87	.188	.188	51.48	20.63	700 lbs.
125	284	445S	49.21	51.00	52.59	63.22	38.37	23.00	16	7.18	36.00	59.00	28.30	34.87	.188	.188	56.60	20.63	715 lbs.
140	284	445S	55.12	57.62	59.62	69.12	38.37	25.00	24	7.18	39.00	59.00	31.31	34.75	.250	.250	62.62	20.63	805 lbs.
160	284	445S	63.00	65.50	67.50	77.00	38.37	29.00	24	7.18	44.00	59.00	35.25	34.75	.250	.250	70.50	20.63	1054 lbs.
180	284	445S	70.87	73.37	75.37	84.86	38.37	34.00	30	7.18	47.00	59.00	39.18	34.75	.250	.250	78.36	20.63	1246 lbs.
200	284	445S	78.74	81.24	83.24	92.74	38.37	38.00	30	7.18	51.00	59.00	43.12	34.75	.250	.250	86.24	20.63	1560 lbs.

All Dimensions In Inches

Figure 39. Typical specification sheet for a variable-pitch fan. (Courtesy Flakt Products, Fort Lauderdale, FL.)

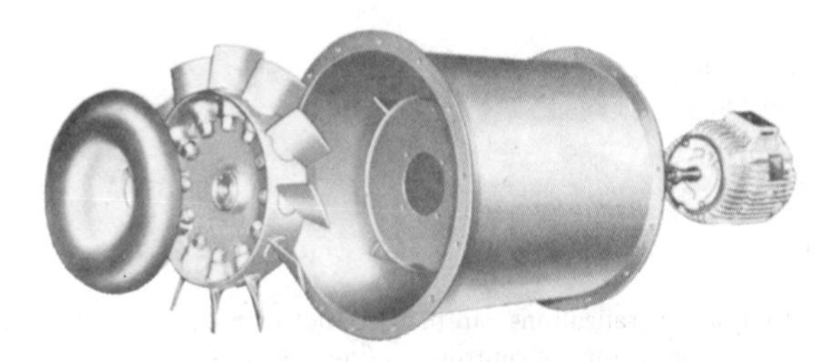

Figure 40. Exploded view of typical vane axial fan. Shows nosepiece, hub, blades, casing and motor. (Courtesy Joy Manufacturing Co., New Philadelphia, OH.)

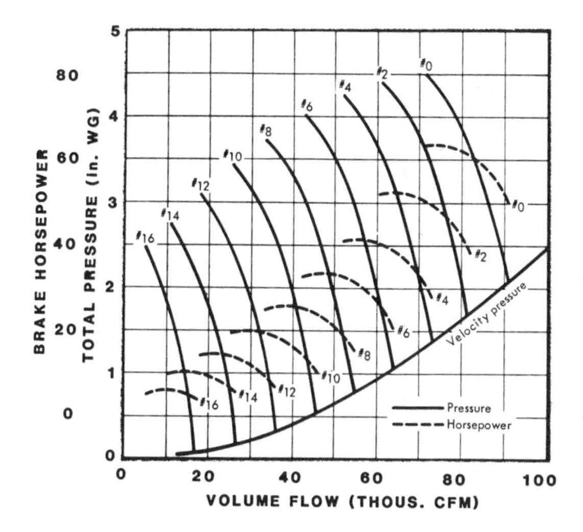

Figure 41. Pressure (solid lines) and horsepower (dashed lines) plotted against volume for different blade settings for a vane axial fan. Blade diameter 54 in.; hub diameter $26\frac{1}{4}$ in.; operating speed 1150 rpm. Fan supplies a 54 in. diameter duct. (Courtesy Joy Manufacturing Co., New Philadelphis, OH.)

considerably as a function of the volume-pressure relationship. Over the broad range of selection, however, vane-axial fans will average approximately 10 percent less than centrifugals and approximately 20% less than in-lines. A high volume, low pressure (e.g., 60,000 cfm at $1\frac{1}{2}$ in. wg) vane-axial could well cost 25% less than a centrifugal and 35% less than an in-line. Conversely, a low volume, high pressure (e.g., 20,000 cfm at 5 in. wg) vane-axial might cost slightly more than a centrifugal and approximately the same as an in-line. These cost comparisons are based on including motor mounting feet and motors for vane-axial fans and motors, V-belt drives, and belt guards for centrifugals and in-lines. Two additional major accessory costs are those for vibration isolation and sound attenuation.

The cost of acoustic treatment will generally be greater for vane-axial fans because of its characteristic of increased noise generation at higher frequencies. In those applications where attenuation is required for centrifugal or in-line fans, the same attenuation will usually satisfy the vane-axial fan since attenuators invariably oversatisfy the demands of high frequency noise. In systems having a relatively high degree of natural attenuation, such as the discharge side of supply fans and the inlet side of return fans, high-frequency noise is naturally attenuated, minimizing or eliminating the need for added attenuation. Installations where line-of-sight contact exists between the fan and an occupied space are those where the vane-axial fan will have an acoustic disadvantage.

Fans most often considered in air-conditioning systems are the vane-axial fan, the air-foil bladed centrifugal fan and the in-line (tubular) centrifugal fan. The specific performance considerations to be reviewed in selection are volume, pressure, and horsepower. Figure 41 presents the performance characteristics of a typical fan. The size designation 54-$26\frac{1}{4}$-1150 indicates a blade (i.e., wheel) diameter of 54 in., a hub diameter of 26-$\frac{1}{4}$ in., and an operating speed of 1,150 rpm. The solid lines represent the useful range of pressure-volume curves. The 0 curve represents the wide open or maximum pitch blade setting of the fan; the 16 curve represents an approximate minimum pitch setting or minimum capability of the fan. Performance for any intermediate blade setting can be interpolated.

The dashed lines indicate the horsepower consumption for each of the blade settings. Again, interpolation between horsepower curves will indicate power requirements for intermediate blade settings.

It should be emphasized that all performance data presented in Figure 41 are at the single full load motor speed of 1,150 rpm. The various performance curves are obtained by varying the pitch or angle of attack of the fan blades. The inherent flexibility is apparent, particularly when compared to the normal limitations of $\pm 10\%$ speed variation with V-belt drives.

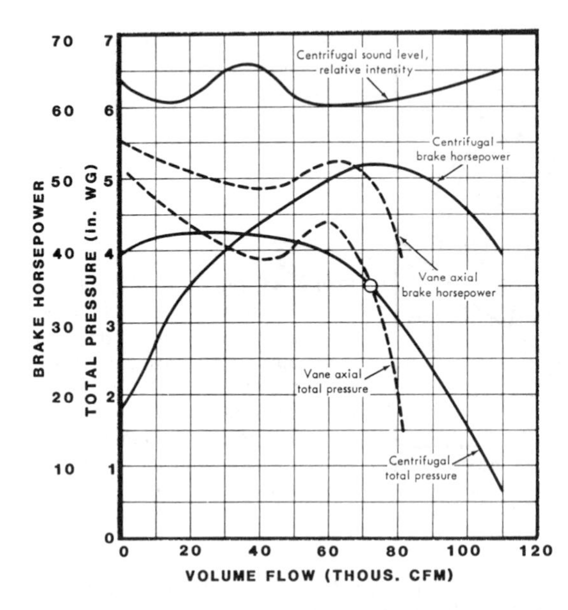

Figure 42. Performance characteristics of different fans.

Both vane-axial fan and axial fans in general, have an inherently high outlet velocity. The axial flow fan develops its static pressure by pushing the air axially through the casing. The outlet diameter of the fan casing closely approximates the blade diameter since the closer this tolerance is held the better the performance characteristics become.

The centrifugal fan relies on a large expanding scroll-type housing to obtain efficient static pressure generation. For equal performance with equivalent wheel diameters, therefore, the vane-axial casing outlet velocity will be substantially higher than that of a comparable centrifugal fan.

If casing outlet velocities must be reduced to meet duct velocities or to reduce discharge shock losses on blow-through systems, outlet cones are usually employed. These cones reduce the velocity to desirable levels with a static region—that is, the conversion of part of the casing velocity presure into useful static pressure at the cone discharge. Since there is a small loss in the cone, total efficiency will drop slightly. Static efficiency will increase as a result of the regain of velocity pressure.

Figure 42 shows one blade setting of a fan extended to the block tight or no-flow condition. The pressure curve reaches a peak, and the fan then enters a stall condition in which air separation from the wheel occurs. This causes the pressure curve to drop before it rises again to the zero flow condition. Operation in this "stall" region is to be avoided since air pulsation and relatively high noise levels will result.

Figure 42 compares a typical centrifugal fan pressure curve. Operation from peak pressure to no flow is stabler. The theoretical pressure curve of the centrifugal rises constantly to block tight, but in practice air separation from the wheel again occurs. The pressure curve collapses, accompanied by higher sound levels. While the centrifugal fan curve is drawn as a smooth line from peak pressure back to block tight, test data do not confirm this operation. The stall can be unstable in centrifugal fans. The sudden increase in noise level in this stall is indicated by the sharp rise in the relative sound intensity curve at the top of Figure 42.

Operation at any point on the backside of the pressure curve is therefore to be avoided, in accordance with the recommendations of most if not all centrifugal fan manufacturers. Initial selection on the backside results in a definite oversizing of the fan. A smaller fan would be less expensive, quieter, stabler, and more efficient.

At times, systems will be installed in which the pressure loss is far in excess of that anticipated in the original design. The result may be operation on the undesirable backside of the pressure

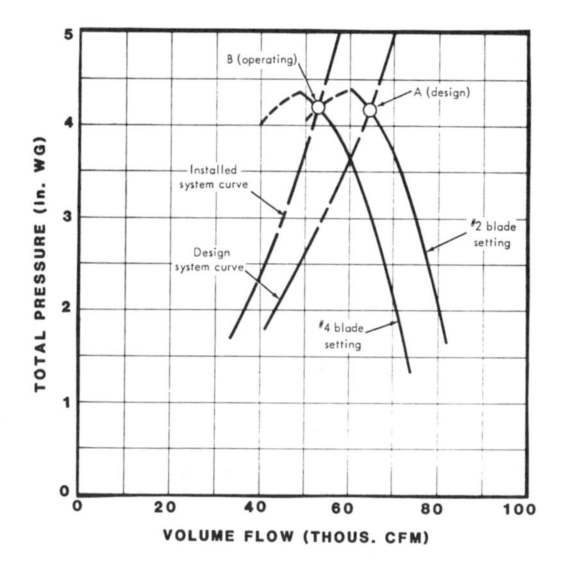

Figure 43. Shows that fan stability is increased with decreasing blade angle.

curve. The higher-than-anticipated system loss will result in a volume output below the design level, a slightly pulsating fan output, and an increased noise level. This usually happens only when an extraordinary system restriction is in evidence, such as a partially closed fire damper, an exceptionally poor outlet condition, a blocked coil, etc. The solution is to eliminate the restriction so as to allow proper operation. Miscalculation of system resistance is not usually of sufficient magnitude to result in operation on the backside of the curve.

Occasionally a system is installed in which the fan operates in the stall area and the resistance cannot be removed. Speedup, while occasionally possible, is generally impractical for any fan type because of excessive power consumption and noise buildup. In many of these cases, however, the reduced volume is satisfactory for system performance, but the noise generation is not. A fan was selected to operate on a one-blade setting for operation at Point A. The installed resistance is such that the fan operates near the stall area at Point B. Assuming that the reduced volume is acceptable, stable performance and an equal- or lower-than-design noise level may be obtained by changing the blade setting (from No. 2 to No. 4 setting curves). Reductions in blade angle will result in an inherently stabler pressure curve. (See Figure 43.)

The centrifugal fan does not have this capability, and the return to stable performance by speed control is not always possible. The system, from the point of view of the fan, can be changed in this case by adding a recirculating duct (see Figure 44). While the system will not receive any more air, the fan will return to stable operation.

Curve steepness is the last evaluating point in discussing the pressure curve. Since even the most carefully calculated system loss is subject to the vagaries of installation, the actual system loss may differ considerably. For most systems, a fan with a vertical pressure curve would be the ideal solution. This would mean that the fan volume could remain constant regardless of pressure loss. Unfortunately, such a fan does not exist. The next best approach is to select a fan with a curve that approaches the ideal curve as closely as possible.

The fan with the steepest curve will be least subject to volume fluctuations with given changes in system loss. This improves system reliability, as well as minimizes the need for the contractor to adjust fan output to meet the desired volume during initial balancing. The design point for selection is 72,200 cfm at 3.5-in. wg total pressure. Alternate system curves are drawn for 5- and 2-in. wg total pressure systems when handling 72,200 cfm. The intersections of these alternate system curves with the fan curves give the actual operating points if the installed system differs from design by these amounts. (See Figure 45.)

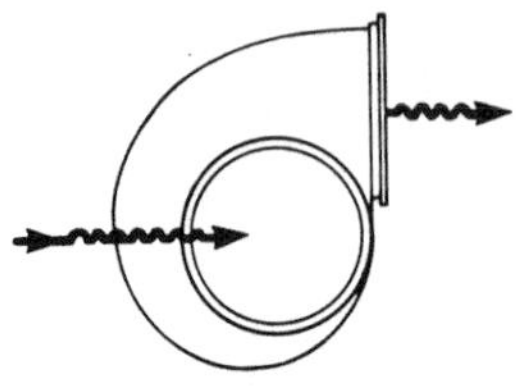

a) Unstable flow condition

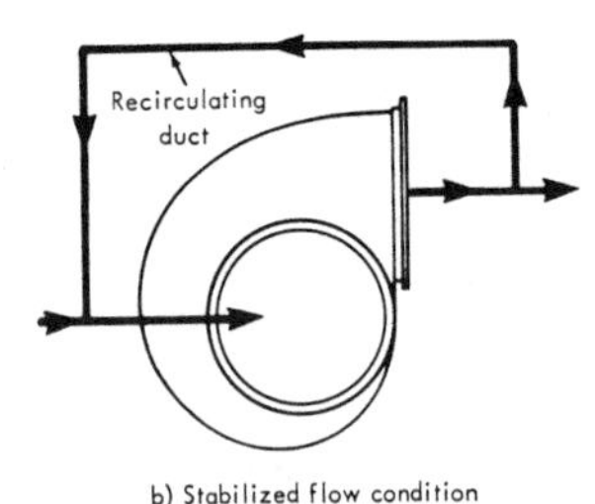

b) Stabilized flow condition

Figure 44. Illustrates that recirculating duct reestablishes stable flow in a centrifugal fan.

Returning to Figure 42, the horsepower curve for a vane-axial curve is shown to decline from the normal selection range to the maximum flow condition much the same as for air foil centrifugals. Throttling back toward block tight from the normal selection range results in a drop in horsepower as the stall portion of the curve is entered, followed by an increase to the no-flow condition. Block tight horsepower varies depending on the blade setting and fan size from slightly less than the maximum horsepower of the normal selection range to 10% above it. This minimal rise at the block tight condition allows normal factors of safety in motor sizing to be applied even if occasional operation at the block tight condition might occur, either planned or not. Block tight condition in actual practice does not mean zero flow. Leakage through a closed fire damper or inlet vane control will still "leak" some minimum amount.

Horsepower consumption at this point will therefore be below the maximum consumption in the normal selection zone. A truly "limiting" horsepower curve is therefore provided. Consumption

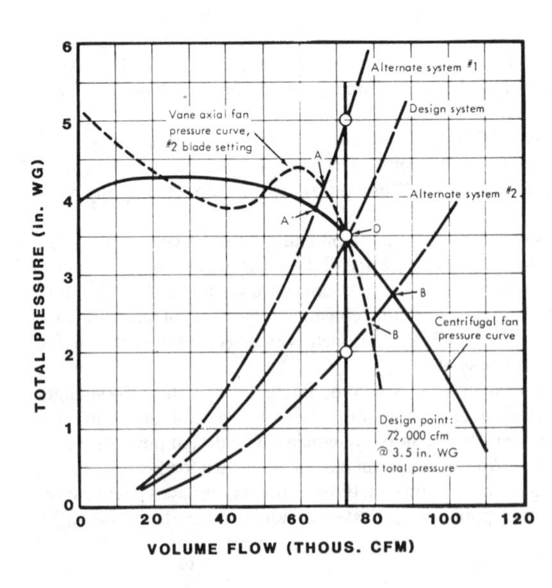

Figure 45. Shows variations in fan performance with system changes.

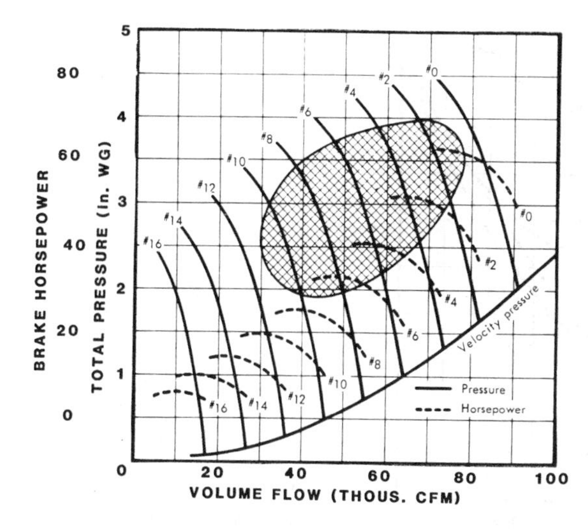

Figure 46. Shows minimum total efficiency of 80% is indicated by cross-hatched area. Obtained by replotting Figure 41.

of the industrial axial fan, on the other hand, usually rises very rapidly at block tight. The centrifugal fan is similar to the vane-axial except that the horsepower curve continues at a low value from the stall area to block tight.

The efficiency curve of the vane-axial is similar in shape to that of the centrifugal except that peak efficiency is carried across a broader range of performance because of the vane-axial's inherent high volumetric capacity.

Ideally, fans should be selected at or near peak efficiency for most comfort air conditioning applications. Selection in this area affords minimum power consumption and optimum sound characteristics. Selection at lower efficiencies may be satisfactory for acoustically noncritical areas.

Figure 46 presents typical pressure curves. Areas of high efficiency are superimposed to present the areas of performance generally recommended. The cross-hatched area represents a minimum of 80% total efficiency and may be considered as "ideal" selection.

Discussion of performance would be incomplete without a review of the devices available to control the output of the fan. This control may be required to maintain a constant volume with a varying system loss, to maintain a constant pressure when volume demand is the variable, or to meet any requirement for combined variation of volume and pressure. Systems requiring such control devices are many. Following are some typical ones in commercial and institutional projects:

1. Variable volume air systems.
2. Dual-duct systems with separate hot and cold deck fans.
3. Garage exhaust systems.
4. Lab hood supply or exhaust systems.
5. Systems programmed for day-night operation.
6. Systems designed to meet present and future conditions.

The first three systems are usually automatically controlled by devices sensing pressure, temperature, or carbon monoxide. Pressure sequencing or step controllers may be used on lab hood supply fans. Day-night operation may be automatic or manual.

This control is commonly provided by the following means:

1. Damper control.
2. Inlet vane control.
3. Variable speed control (e.g., fluid drives, magnetic couplings, and variable speed motors).
4. Fan blade pitch control.

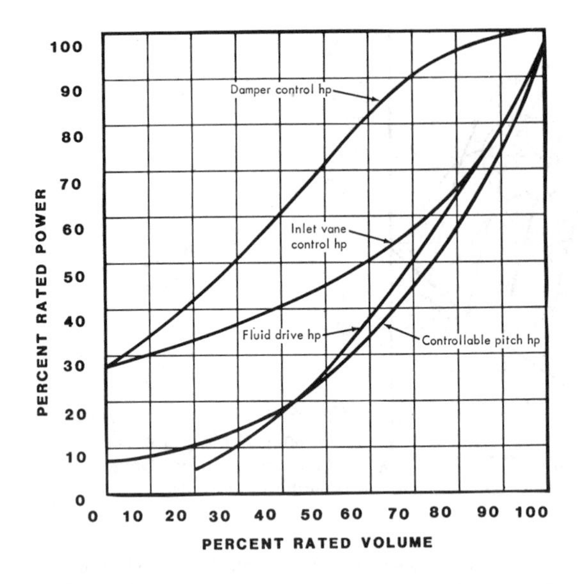

Figure 47. Shows power consumption of various control devices.

The first three are commonly associated with centrifugal and in-line fans.

Dampers may take many forms, including outlet dampers attached directly to the fan outlet, volume dampers in the system for balancing duct runs, and throttling devices in terminal units for final room balancing. They are similar in that they all restrict air flow and thus increase resistance. Its main advantages are simplicity and low cost. High horsepower consumption, relatively nonlinear response, and high noise-generating characteristics are the disadvantages. Damper controls are rarely used today for automatic control of total fan output because of these limitations.

The most commonly used device for automatic control of centrifugal or in-line fan output is the inlet vane control. By prespinning as well as throttling the air prior to its entry into the wheel, reduced output with power savings can be obtained. On the negative side, inlet vane control is relatively inefficient, noisy, and nonlinear in response.

Automatic variation of fan speed through fluid or magnetic couplings or variable speed motors has had only limited application in comfort air conditioning because of high initial cost and increased space demands. Good acoustic characteristics and high efficiency are the prime advantages.

With vane-axial fans each blade can be adjusted individually. With controllable pitch, the fan blades are linked together so that the pitch of all blades may be simultaneously controlled while the fan is in operation. This differs from damper control since changing the blade pitch changes the angle of attack so that more or less air is "bitten off." Efficient and quiet flow with linear response is attainable throughout the range of pitch adjustment.

Efficient part-load operation is one of the prime advantages of controllable pitch. Figure 47 shows the power consumption curves for various control devices. The damper and inlet vane control curves are based on centrifugal fan performance, fluid drive on either centrifugal or axial performance, and controllable pitch on vane-axial performance.

Linear response is also favorable with controllable pitch. Inlet vane control requires a considerable constriction in air flow before effective control takes place. For a 20% reduction in volume, inlet vanes must be closed approximately 55%; for the same reduction, vane-axial blade pitch needs to be changed only 33%.

Closing of inlet vanes exposes sharp vane edges to high velocity air flow, resulting in increased noise levels. These increases have been quantitatively defined [5]. Since there is no restriction of air flow or exposure of sharp edges, the noise levels are lower at reduced load except for a slight increase in low frequency noise at very low loads. This increase is meaningless since the vane-axial fan inherently generates a very small amount of low frequency noise. Inlet vane control on the other hand adds up to 8 dB of sound to the base sound curve in the important third and sixth octave bands at reduced result. The net result is that a vane-axial fan with controllable pitch is acoustically comparable to a centrifugal fan with inlet vane control at reduced load settings in many applications and better in some.

Noise, Vibration, and Mechanical Considerations

Basics of Sound and Analysis

Sound is sensed when pressure waves created by a source impinge on the sensing elements of the ear. Sound pressure is measured in terms of dynes per square centimeter or microbars. The approximate threshold of hearing for a sound at 1,000 Hz is 0.0002 dynes per sq cm; standard atmosphere has a pressure of 1,013,000 dynes per sq cm.

Presentation of sound values is complicated not only by the very small pressures but also by the extremely wide range of audible sound pressure. While the threshold of sound is 0.0002 microbar, the sound pressure at the inlet of a mechanical draft fan might approach 200 microbars—a million times the threshold value. To state this wide range of sound pressures with more usable numbers, the decibel (dB) notation is used to describe sound pressure levels (L_p or SPL).

A decibel value is a number that expresses a ratio of any value to a given reference point. It is presented in logarithmic form to reduce the wide range of sound values to a more usable numbering system. In expressing sound pressure levels by means of decibels, the threshold of sound, 0.0002 microbar, is used as the reference value. Since acoustic energy is proportional to the square of sound pressure, sound pressure levels are defined as $L_p = 10 \log(p/p_r)^2$ or $20 \log p/p_r$, where $p_r = 0.0002$ microbar. The sound pressure level of the threshold of hearing is therefore 0 dB since the log of 0.0002/0.0002 is 0. The sound pressure level of 200 microbars is 120 dB.

Rating fans in terms of sound pressure levels is impractical since sound pressure at any given point near the sound source is affected by the environment. A single fan located in a small, reverberant equipment room will produce higher sound meter measurements than if it were tested in a roof location with no other equipment adjacent to it. The roof location presents a diffuse field for fan sound, and most of the sound generated by the fan continues off into space. The enclosed fan room, however, reflects the sound waves back and forth and raises the meter readings well above the levels measured in the roof test. Yet, the total amount of energy developed by the fan is the same in either case.

Fan ratings are therefore presented in the form of sound power levels, which represent an energy concept rather than a sound pressure concept. Sound power level (L_w or PWL) is a measure of the total energy a fan produces in the form of sound. Sound power is independent of environment while sound pressure is not. Sound power levels are determined by comparing the sound pressure levels of a test fan to those of a known reference source in a given environment and converting back to sound power levels.

The decibel notation is used to express sound power levels as well as sound pressure levels for the preceding reasons. The reference level used is 10^{-12} watt. The sound power level is defined as $L_w = 10 \log W/R_r$, where $W_r = 10^{-12}$ watt.

Doubling of sound power means an increase in the sound power level of 3 dB since $10 \log 2 = 3$. Doubling the sound power level, however, does not mean a doubling of the apparent sound to the human ear. A change of 3 dB from one room to another would hardly be noticed. An increase of approximately 10 dB is received by the human ear as an apparent doubling in sound. This apparent conflict should be understood if one is to gain a feel for the numbers used in expressing sound power levels, particularly when comparing sound power level spectra of different fan types.

Audible sounds are a function of the amplitude, dB, and frequency, Hz, of sound waves. Generally, the audible frequency range of sound may be described as 15 to 20,000 Hz. Expression of a single (overall) sound power level of a noise source such as a fan is rather meaningless. Not only do the amplitudes vary considerably with changes in frequency, but the ear also finds low frequency noise more acceptable than high frequency noise. Sound power levels must therefore be defined by frequency range.

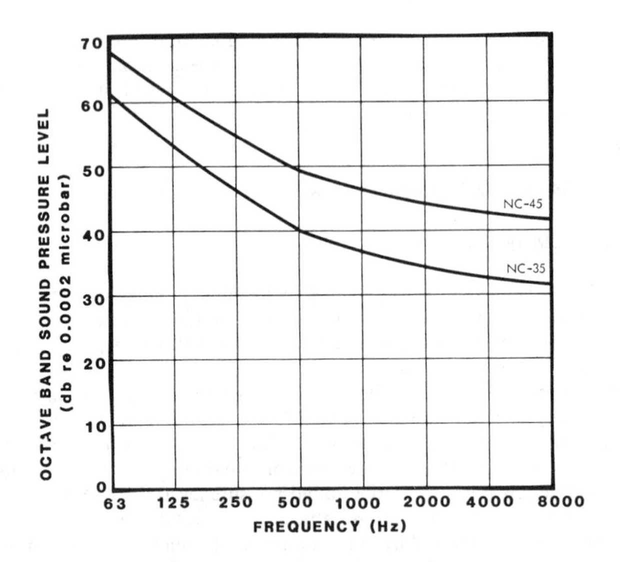

Figure 48. Shows noise criterion curves for limitations of sound pressure levels.

For air-conditioning applications, the range of 45 to 11,200 Hz is used for analysis since this is the range in which the greater portion of objectionable equipment-generated noise occurs. This range is divided into octaves, the first being from 45 to 90 Hz and the remaining seven being approximate multiples of two of the previous bands. The eighth band covers the frequencies of 5,600 to 11,200 Hz. Sound power levels should be presented as dB values referenced to 10^{-12} watts in each of the eight bands. The amplitudes of each frequency within each band are weighted and a single value presented for the band.

Amplitude defines the level of sound while frequency determines the quality of sound. Analysis of each is required to determine acceptable sound levels for the human ear. Since general environmental noise also determines the limits of acceptable equipment sound levels, the environmental conditions must also be examined.

Noise criterion (NC) curves outline the recommended limits of sound pressure levels re 0.0002 microbar for each octave band for various environmental conditions. Figure 48 illustrates the NC 35 and 45 curves. These indicate average allowable sound pressure levels for two general environmental conditions. As an example for air conditioning, the NC 35 (apartment buildings, hotels, executive offices, laboratories, etc.) and NC 45 curves (wash rooms, kitchens, lobbies, banking areas, open offices, etc.) give average limits of sound pressure levels by octave bands. The shape of these curves shows that the human ear finds high frequency sound more objectionable than low frequency sound. For example, the NC 35 curve reveals that the average person in an executive office would find a pressure level of 45 dB acceptable in the third band but objectionable in the sixth band.

In general, the higher the fan outlet velocity and fan speed, the higher the sound generation. While this is true in general when comparing smaller sizes of the same fan type, it does not hold true when comparing different fan types.

Fan noise is a function of the fan design, volume flow, the pressure, and the efficiency of the fan. After selecting a fan for a given application, the proper size selection of that unit must be based on efficiency since the most efficient operating range will also be the quietest. Outlet velocity has no bearing on fan noise and selections made on this basis alone are meaningless. Similarly, noise comparisons of different types of fans on the basis of rotational speed are erroneous. The only valid basis for comparison is the actual sound power levels generated by the different types of fans when these fans are all producing the required volume flow rate and pressure [6].

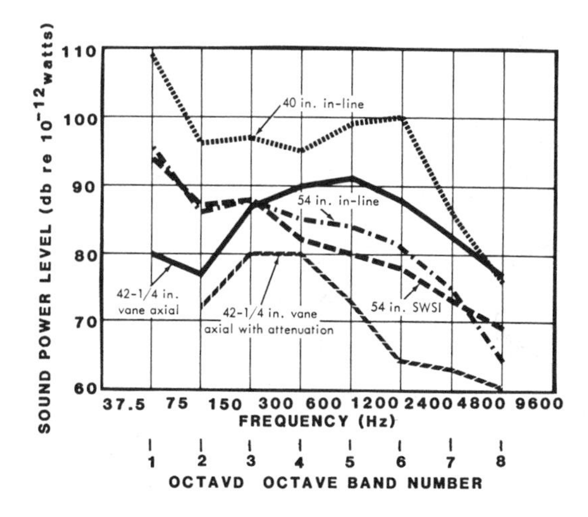

Figure 49. Comparative sound spectra for different fans handling 30,000 cfm of air at $1\frac{1}{2}$ in. W.G. static pressure. (Courtesy Joy Manufacturing Co., New Philadelphia, OH.)

Figure 49 presents comparative spectra for four fan selections for a typical return air application handling 30,000 cfm at $1\frac{1}{2}$ in. wg static pressure. The spectrum of the 40 in. in-line fan is included to show what happens if a fan is severely undersized. Since sound spectra vary with performance and fan size, reference to actual sound power levels of different fan types is suggested for proper evaluation of a given system.

Novick [7] gives the following illustration to compare the difference in spectra between two types of fans (a vane-axial fan and a centrifugal fan). Figures 50 and 51 along with Table 4 present an example taken from Chapter 31 of the 1967 *ASHRAE Guide and Data Book*. Figure 57, page 389 of the Guide, presents two problems based on a given system. The problems are:

1. In the absence of manufacturer's data, calculate the sound power level of the centrifugal supply fan.
2. Based on a given noise criterion for the occupied space, calculate the system attenuation and determine whether the fan is acoustically acceptable.

Table 4 and Figure 51 give the solution in tabular and graphic forms. It should be noted that the system attenuation does not reduce the second and third band sound levels of the centrifugal fan to maximum acceptable levels. The solution was to add a packaged silencer (attenuator) to the system to reduce this low frequency noise to acceptable levels.

The sound curve for a vane-axial fan has been superimposed on the graphic solution for comparison.

Substitution of current manufacturers' data on air foil centrifugals would reduce the centrifugal fan generation in the third band to a point near that of the vane-axial so that the demand for a packaged attenuator would not be as great as the problem indicates. Second band noise, however, would still require the use of a packaged attenuator. Therefore, at least for the parameters of this example, the vane-axial proves to be more desirable acoustically.

Note that natural system attenuation or built-in attenuation reduces high frequency noise more readily than low frequency noise. Therefore, where there is a high degree of natural attenuation, as at the discharge side of most supply fans and the inlet side of many return fans, a sharp reduction in high frequency noise will be effected. For example, a single, lined elbow of 20-in. width can reduce sixth band noise as much as 16 dB while reducing second band noise by only 1 dB [6].

A basic consideration in fan application is for those acoustically critical systems where line-of-sight contact might exist between fan and human ear or for systems that have relatively little natural attenuation. High frequency sound tends to be directional; that is, once directed along a confined path, it does not disperse as readily as low frequency sound. This "focusing" means that high frequency sound will carry over greater distances than low frequency sound.

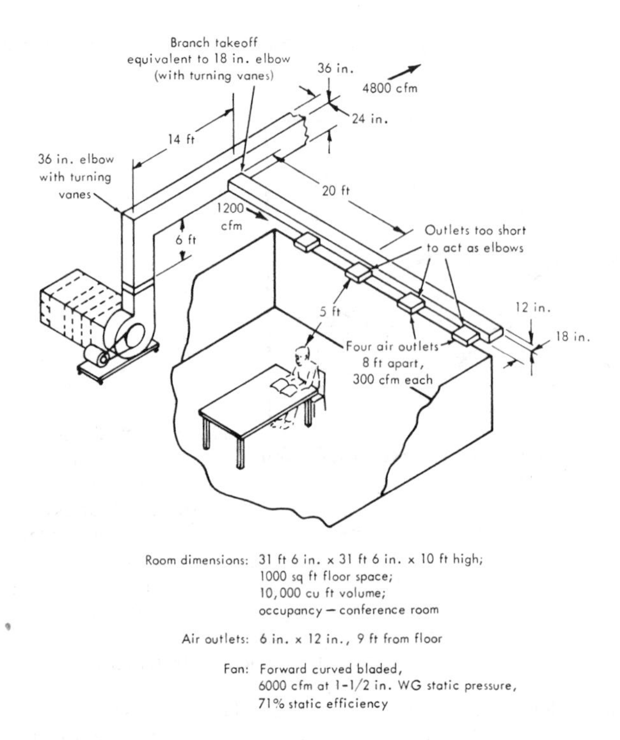

Figure 50. Illustrative system described for *ASHRAE Guide and Data Book* example.

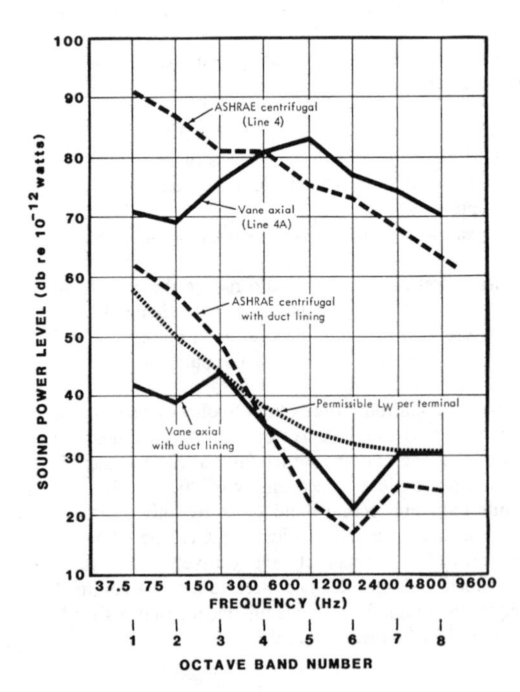

Figure 51. Graphical solution of *ASHRAE Guide and Data Book* example.

Table 4
Tabulated Solutions for *ASHRAE Guide and Data Book* Examples

Line	Item	Reference	Octave bands 1	2	3	4	5	6	7	8
Example 4, Chapter 31, Table 18, p. 390 —										
4	Sound power level of ASHRAE centrifugal fan	Sum of Lines 1, 2, and 3	91	87	81	81	75	73	68	63
(4A)	Sound power level of vane axial fan	Test per AMCA Bulletin 300	71	69	76	81	83	77	74	70
11	Total natural attenuation in branch	Sum of Lines 5 through 10	29	28	25	25	22	22	22	22
12	Sound power level per terminal without sound treatment using ASHRAE centrifugal	Line 4 minus Line 11	62	59	56	56	53	51	46	41
(12A)	Item above using vane axial	Line (4A) minus Line 11	42	41	51	56	61	55	52	48
13	Permissible sound power level per terminal	Line 8 of Table 9 minus 3 db	58	50	44	38	34	32	31	31
14	Minimum attenuation of sound treatment for fan noise using ASHRAE centrifugal	Line 12 minus Line 13	4	9	12	18	19	19	15	10
(14A)	Item above using vane axial	Line (12A) minus Line 13	—	—	7	18	27	23	21	17
Example 6, Chapter 31, Table 20, p. 393 —										
5	Attenuation of 10 ft of 1 in. duct lining	—	0	2	7	21	31	34	21	17
6	Additional attenuation required for ASHRAE centrifugal	Line 14, Example 4, minus Line 5, Example 6	4	7	5	—	—	—	—	—
(6A)	Additional attenuation required for vane axial	Line (14A), Example 4, minus Line 5, Example 6	—	—	—	—	—	—	—	—
Calculation of fan spectra at terminal —										
A	Total system attenuation, natural plus duct lining	Line 11, Example 4, plus Line 5, Example 6	29	30	32	46	53	56	43	39
B	Net spectrum for ASHRAE centrifugal fan	Line 4, Example 4, minus Line A	62	57	49	35	22	17	25	24
C	Net spectrum for vane axial fan	Line (4A), Example 4 minus Line A	42	39	44	35	30	21	31	31

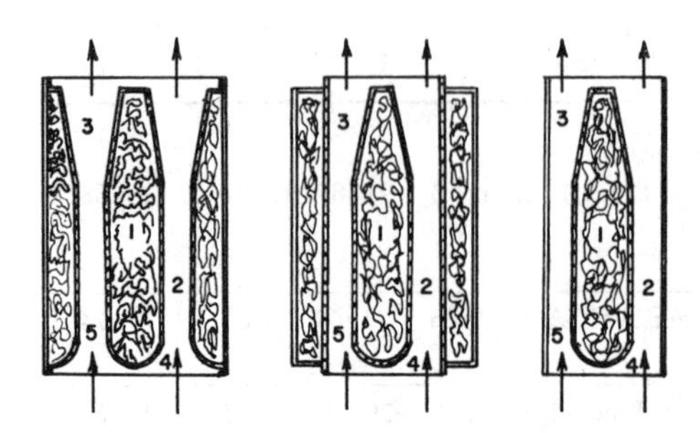

Figure 52. Shows features of packaged silencers.

Three potential problems that may be aggravated by excessive high frequency sound levels are:

1. Transmission outside a building when an exhaust fan is located adjacent to the discharge louver.
2. Transmission outside a building when an open inlet supply fan is located in line with the outside air intake.
3. Transmission into corridors when return louvers are built directly into the equipment room from corridors.

Knowledge of these specific aspects of high frequency sound leads to the proper application of fans. The focusing characteristic can actually be used to advantage since beamed high frequency noise can be deflected by simple baffles or goosenecks to a much greater degree than low frequency noise. Lining of elbows and the use of packaged attenuators offer possible solutions. Packaged attenuators offer a simple, low cost method of reducing fan noise levels, particularly high frequency noise.

Packaged silencers are available from manufacturers as standard fan system components. In general, this is an economical approach to minimizing potential noise problems. For example, a three-foot long silencer can provide as much noise control as 260 feet of lined duct.

Figure 52 shows examples of duct silencers. These systems are equipped with acoustic baffles (1) designed for maximum attenuation at low frequencies, straight-through air passages for maximum air handling at low static pressure drop (2), evase (tapered) exits for static pressure regain (3), a solid, rounded nose for noise reduction (4), and bell-mouth entrances to minimize inlet turbulence (5).

Acoustic performance data on silencers is generally presented in terms of frequency bands one octave wide (identified by the octave band number). Octave bands are also identified by their approximate geometric mid frequency.

Empirical relationships between the noise generated by a fan and its size, static pressure, speed, and capacity have been developed. The change in overall sound power level varies as:

$$70 \ln(\text{size}_2/\text{size}_1) + 50 \ln(\text{speed}_2/\text{speed}_1)$$

$$20 \ln(\text{size}_2/\text{size}_1) + 25 \ln(\text{pressure}_2/\text{pressure}_1)$$

$$10 \ln(\text{capacity}_2/\text{capacity}_1) + 20 \ln(\text{pressure}_2/\text{pressure}_1)$$

These relationships also apply only for a fixed point of rating.

Note that a double-width fan is essentially two fans of the same size, speed, and sound power level; therefore, its sound power level will be 10 ln(2) or 3 dB greater than the smaller one. Caution must be exercised in the use of these guidelines since not all types of fans exactly follow these relations. For example, the radiated sound power of a centrifugal fan is not all concentrated on the blade frequency. The various components in the different frequency bands may vary significantly, while the overall sound level follows these relationships closely. For axial-flow fans, there is a wider range of the vortex noise, hence a slightly greater deviation from these equations will be expected than for centrifugal fans [8].

Fan noise can be separated into two portions: one portion associated with the developed head, and the other associated with the flow. The total radiated sound power is expressed as:

$$W = \frac{C_1 H^3}{(\alpha)} + \frac{C_2 Q^5}{D^{12}(\alpha)^4} \tag{26}$$

where W = overall sound power, watts
H = static head, inches of water
Q = capacity, ft^3/min.
α = aspect ratio (width/diameter)
$C_1 = 0.90 \times 10^{-4}$ (empirical constant)
$C_2 = 5.2 \times 10^{-10}$ (empirical constant)

For minimum noise, the preceding relation indicates a large aspect ratio to be desirable. Also, the noise power decreases with an increase in fan diameter, until the head noise predominates. When the head noise begins to dominate the total noise, the optimum fan diameter is obtained. These observations generally apply to small centrifugal fans and it is therefore risky to extrapolate results to large fans.

We can now discuss the various sources of fan noise in more detail. The first topic concerns aerodynamic noise sources:

Rotational noise.

- *Noise generated by blades.* This is also called blade noise and is caused by the air impulse at a given point on a blade revolving. It is the frequency that determines the fundamental tone of the noise. Methods of noise reduction of this type of noise are:

1. For an axial fan, increase blade width while reducing the intensity of the harmonics relative to the fundamental. The blade thickness has only a slight effect on rotational noise.
2. Figures 53 and 54 illustrate that increasing the number of blades tends to reduce the number of audible harmonics; hence, a decrease in the total rotational noise is achieved. Figure 53 shows a one-blade propeller fan which will receive an impulse each time by passing an element of area in the disk. If another blade is added, as shown in Figure 54, all of the odd harmonics of the one blade fan will be cancelled, and the strength of the even harmonics is doubled. However, the capacity of the fan has also been increased. Thus for the same capacity

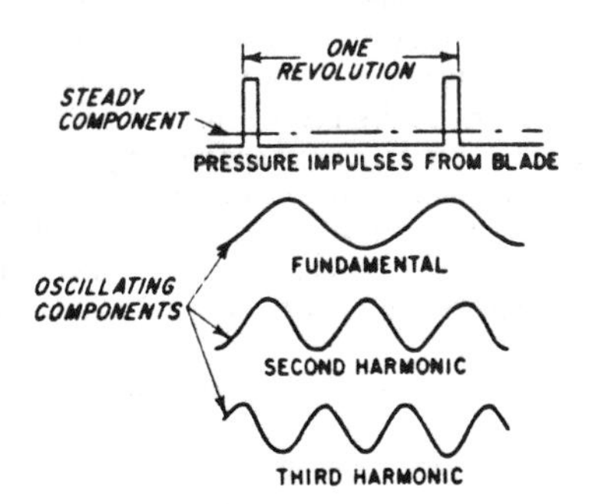

Figure 53. Schematic diagram of the steady and oscillating components of the impulses to air by a single-blade propeller fan.

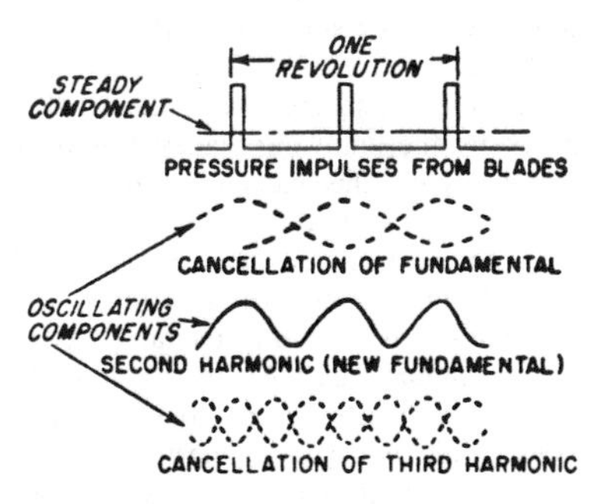

Figure 54. Shows the steady and oscillating components of the impulses to air by a two-blade propeller fan.

and rpm, a smaller fan may be substituted. Doubling the number of blades decreases the total rotational noise by more than 3 dB. The following tabulated values provide more information about noise reduction [9]:

Number of blades on impeller	1	2	4	8	16
Decrease of sound level (dB)	0	3	6	9	12

3. A propeller fan with higher harmonics can be furnished with a properly designed shroud to reduce noise considerably. But if the flow breaks down over part of the shroud, the noise may become worse than that without the shrouded case [8].

- *Axial fan guide vanes effect.* To ensure minimum noise of vane-axial fans, the number of vanes should not equal the number of blades and the vanes should also be spaced as far as practical from the blades. If the same number of guide vanes and blades are called for in the design, the passage of air over the guide vanes will accentuate noise at the blade frequency and its harmonics—especially in the case of close spacings between vanes and blades. An unequal number of vanes and blades will raise the product frequencies, but such tones are comparatively low in intensity [8].
- *Centrifugal fan scroll and cutoff effects.* The scroll of a centrifugal fan may take the form of an involute where the axial clearance increases directly in proportion to the angle traversed. A constant mean velocity of flow around the scroll is required for a uniformly distributed air supply from the blade discharge area. Such a scroll is desirable for minimum noise. A reduction in the exit velocity and an increase in pressure will occur when the scroll clearance is increased. The pressure increase will cause an abrupt pressure change at the cutoff which serves to increase the noise at the blade-passing frequency. The cutoff clearance is also critical. The smaller the clearance, the more rotational noise will be generated at the cutoff. A five-percent wheel diameter clearance is generally considered to be a minimum; and it should be increased somewhat for low-noise applications.

Vortex noise.

- *Noise generated by blades.* The major source of vortex noise is fan blades. A pressure gradient develops across the blade in the direction of its thickness when a blade moves through the air. If the airflow next to the blade is steady, or laminar, this pressure gradient is essentially constant and little noise results. The flow may separate from the convex side of the blade with an incorrectly designed blade profile, thus giving rise to large eddies. However, the eddy formation and the pressure pattern fluctuate rapidly and cause considerable noise. Since vortices will be shed from the trailing edge of the blade and alternately from opposite sides of the blade, a broad-band noise spectrum results due to the random blade size and vortex releasing point. For axial fans, the noise increases with the thickness of the trailing edge due to such vortices; and also, vortex noise increases sharply when the air flow is such that the eddies peeling from one blade are struck by the following blades. For centrifugal fans, the vortex noise happens only if the air completely fills the space between the blades. This effect may be minimized by using the curved trailing edges so that the most of the vortices are shed at the tips of the blades. Proper design of the entrance angle is important for centrifugal fans in order to minimize vortices. In studying vortex formation, hot-wire anemometry is a valuable tool as well as high-speed motion pictures with smoke injection into the fan [8, 10].
- *Air-stream turbulence.* An air stream itself may also generate noise. A common example is when the air stream enters relatively still air in a mixing zone. Vortices will develop because of the velocity gradients existing in the mixing region. A broad-band noise spectrum will develop according to the vortices growing, decaying, and moving in random fashion. An obvious approach to noise reduction is to avoid relatively high speeds in the mixing zones. Also, the presence of any sharp edges or bends will cause an increase in turbulence and noise; and the placement of heat exchangers too close to the fan intake or exhaust will result in increased noise. A more controllable source of turbulent noise is due to obstructions in the air-stream at the inlet or outlet.

Duct effects. An ideal duct system is designed such that all rotors, stators, and the duct have perfect circular symmetry to prevent propagation of rotational noise down the duct axis. Unfortunately, in many duct system, such symmetry is not possible, and an air-borne sound tends to be propagated down the duct. Because of its random nature, no phase cancellation is possible for vortex noise. Therefore, this type of noise is always radiated.

The resonance of fan housings and ducts may be another source of effective radiating noise, since any mechanically resonant system will be excited at its natural frequency of vibration. An inside duct liner is often introduced to reduce air stream noise as well as duct vibration.

As noted earlier, one economical method of noise reduction is to introduce commercial or custom-made silencers. Custom-made silencers can be constructed in maintenance shops to include acoustical labyrinths, parallel baffle silencers, acoustic-lined plenums, acoustic-lined ducts, and acoustic-lined bends.

If duct walls are lined with an absorbent having an absorption coefficient, α, the dB reduction obtainer is given by:

$$\Delta L = 12.6\ P\alpha^{1.4}/A \tag{27}$$

where ΔL = change in sound pressure level
P = perimeter of duct, inches
A = area of duct, in.^{2}
α = coefficient of absorption (this is frequency-dependent so octave band data will be used to determine section loss and length of duct to be treated).

This relation can be applied with an accuracy of a 10% error. Absorption coefficients typically lie between 0.20 to 0.40 for the ducts having cross-sectional dimensions in the ratio of 1:1 to 2:1, and frequencies between 250 to 2,000 cycles per second [11].

Plenum chambers can also be lined with sound absorbing material. Sound level reduction can be found approximately from the following relation:

$$\Delta L = 10 \ln[(\alpha)]A_p/A_b]$$

where α = coefficient of absorption of liner
A_p = area treated on plenum walls
A_b = discharge area of blower

An absorbent-lined bend should add about 5 dB attenuation, with length of treatment at about 5 times the duct width. A large number of commercial silencers are available to suit any fan or duct size to obtain greater attenuation. Since noise travels both upstream and downstream, silencers should be installed on both intake and delivery sides of the fan.

Manufacturers will normally supply data on noise reduction and pressure loss for stated air flows so these units can be purchased readily for specific jobs with a fair degree of assurance that they will perform satisfactorily [11]. A final note is that a flaring-fan outlet or inlet bell may amplify internal fan noise; hence, a properly designed bell is important for a high-velocity blower to reduce aerodynamic noise [8].

Axial-flow compressors. A sequence of multiple-blade propellers (rotors) form an axial-flow compressor. Stators are placed between rotors, possibly at the ends of the array. A circular duct is usually included in the entire assembly. The external noise of a compressor is from the vibration of the case in the immediate vicinity of the rotors. The sound pressure of the rotational noise within the compressor, opposite the rotor tips, can be found from the following equation:

$$P = 213\ \Delta P\left[\frac{(\Delta r) + 1}{(\Delta r) - 1}\ln(\Delta)\right]^{1/2} \tag{28}$$

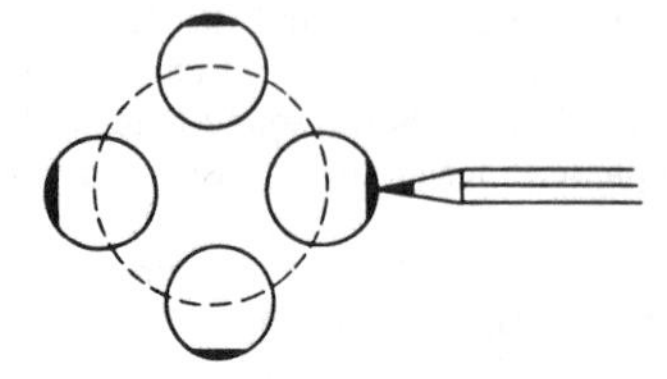

Figure 55. Illustrates the orbit of the shaft centerline.

where P is in units of microbars

and ΔP = pressure rise across the plane of rotation, inches of water
Δr = ratio of inside compressor radius to hub radius

The fundamental frequency is the so-called blade-passing frequency of the rotor:

$$f = \frac{N(rpm)}{60} \text{ cps} \tag{29}$$

where N = number of rotor blades

Then the sound pressure level in this case is given by:

$$L_p = 20 \ln \frac{P}{0.0002} \text{ dB} \tag{30}$$

We now direct attention to the second class of noise sources, namely, nonaerodynamic.

Fan unbalancing. A complete vibration once per revolution may result from fan unbalance. If the fan speed is within the range of 900 to 3,600 rpm, a result of 15 to 60 cps vibrations will form in an inaudible or low frequency range of hearing. But if a fan is mounted rigidly on supports and has a very light housing construction, high-frequency tones may be generated by such vibrations [8].

Rotor balancing can be done in a number of ways; the simplest of which is shown in Figure 55. If the fan is rotating at a low speed and has large deflection, the pencil will mark the heavy spot on the rotor. Weight should then be added to the shaft at the point 180° opposite the pencil mark. This approach illustrates the basic idea of balancing [12].

For large size fans and more accurate requirement, an instrumentation system can be introduced for balancing purposes as shown in Figure 56. This system is relatively simple and yet accurate. The two pairs of vertical and horizontal probes are normally located toward each end of the shaft, usually just inboard and outboard of the bearings. These displacement sensors are positioned 90° apart at the same lateral location, and provide the orbit of the shaft at these particular positions. Normally, the 200-mV peak-to-peak volts per mil of peak-to-peak displacement is used for calibration, with the oscilloscope sensitivity calibrated at 200 mV/cm. Then the actual mils of displacement can be easily read directly on the scope graticule as 1 mil/cm, or equivalently 25.4 micro m/cm. The probe provides both phase reference and speed information required for balancing by observing every single notch or projection on the shaft. The signal of the keyphazer is used not only to synchronize the scope but also to control the scope intensity. A dual beam scope is usually used as a readout device and provides a graphical picture of the balancing procedure [12].

For quiet operation, vibration-isolation mounts should be used as well as flexible connections to duct work. The more perfect the fan balance, the less likely is noise generation from this source.

Bearing noise. This is not a steady noise, since good bearings are not generally a source of objectionable noise. However, well-lubricated sleeve bearings are quieter than ball or roller bearings; where these are damaged or the raceways pitted, a high-frequency noise is usually present. A stetho-

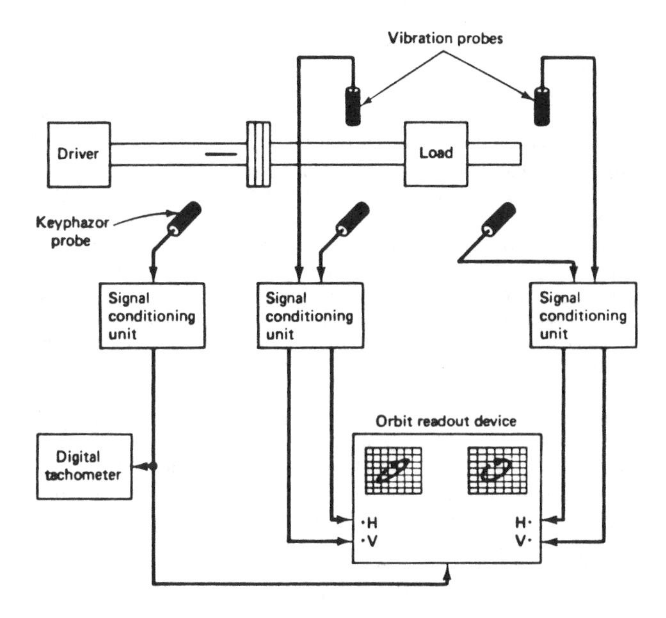

Figure 56. Illustrates a system for balancing.

scope or sensitive vibration indicator may be introduced for detecting purposes. Where it is advisable to use silicone grease in antifriction bearings, some increase in noise must be expected [8].

Structural resonance. A wide range of frequencies is present in most fan noise. If the energy in a given band is high and corresponds to the natural frequency of some part of the fan, the resulting noise may be radiated efficiently.

Added bracing can be used to raise the natural frequency of the part to some higher value, or damping material may be applied to it in order to reduce the noise radiation [8].

Motor noise. If the impeller is mounted directly on the motor shaft, noise of magnetic origin may be radiated by the fan. The usual precautions for isolating the motor feet should be observed in some low speed fans. But for high-speed, high-pressure fans, such a mount is less practical and less important due to the sound intensity emitted being lower than that of the fan noise. Longer blades of the backward-curved-blade type will help in this respect if the direction of the motor is not to be reversed [8].

Couplings. A better alignment usually can eliminate squeaks. Encased, oil-lubricated couplings generally will improve noise levels. Hydraulic couplings usually show a beat frequency if out of balance. The floor on which the coupling is mounted is a good noise radiator from such unbalance.

Belts. If flat and V belts are designed with a low safety factor, some noise may be generated on acceleration. Ordinary chain and straight-spur-gear drives may be quite noisy. Flexible-chain and skewed-gear drives can be encased for oil lubrication to reduce noise level. Pulleys should be well balanced.

Fan Noise Testing

Fan noise measurements can be accomplished by a number of methods. Noise analyses are usually obtained as sound pressure levels (for a given frequency band) at specified locations.

Duct tests. A microphone can be placed within a duct to measure the sound in a moving airstream of appreciable velocity.

- *Windscreen.* To improve the signal-to-noise ratio by keeping the turbulent eddies away from microphone sensing element, a windscreen is usually used over the microphone. Crystal and dynamic microphones often have a built-in windscreen and may not need additional screening for low air velocities. For condenser microphones, screening is required in light air currents.

  It should be noted that microphone wind noise occurs predominantly at lower frequencies, especially with a built-in screen. Therefore, without additional shielding, the overall sound pressure in a moving airstream cannot be measured with an ordinary crystal microphone. Data obtained from the higher octave bands, which are often causing noise complaints, are quite feasible to count to be accurate. Also, measurements can be taken, for a particular series of tests, from the microphone with and without a windscreen for comparisons. Corrections must be made for the sound attenuation offered by the windscreen [8].
- *Correlation technique.* An alternative scheme can be made for an airstream measurement correlation technique. By using two microphones placed side by side in the airstream, electrical outputs can be correlated by suitable electronic equipment. This correlation procedure tends to discriminate against microphone wind noises, which are random and different at each microphone, to provide an improved signal-to-noise ratio. The basic limitation involves the propagation of wind-generated sound from one microphone to the other. The signal-to-noise ratio is improved by (D/2–D/3) dB, if the pressure attenuation for this propagation is D dB. Filters are often used to eliminate the very low frequencies due to the decreasing of microphone cross-talk as the increasing of frequency. Signal-to-noise ratio can be improved about 20 to 30 dB using this method.
- *Acoustic termination.* Sound waves are usually reflected back from a duct's open end, from which longitudinal standing waves will occur within the duct. To prevent the interference with an accurate measurement of sound pressure, an acoustic termination is introduced as indicated in Figure 57. For minimizing reflections at low frequencies, sound-absorptive wedges are required to cover an area equal to that of the duct cross section and identical in shape. A glass-wool blanket type of lining should be installed around all walls, and a flare chosen to provide a minimum of airflow restriction. Since guide vanes will reflect sound, it is recommended not to use them in the flare section. When high-pressure fans are being tested under reduced flow, some noise will be generated at openings. However, the termination will serve as an acoustic plenum chamber which will minimize the noise transmission to the microphone.
- *Vibration transmission.* The noise being measured will increase by the panel resonances caused by vibration. Hence, one must prevent vibrations from being transmitted to the microphone location along the duct. A short section of canvas coupling at about three duct diameters from the fan is recommended to support the remainder of the duct on inflated rubber pads. Duct walls

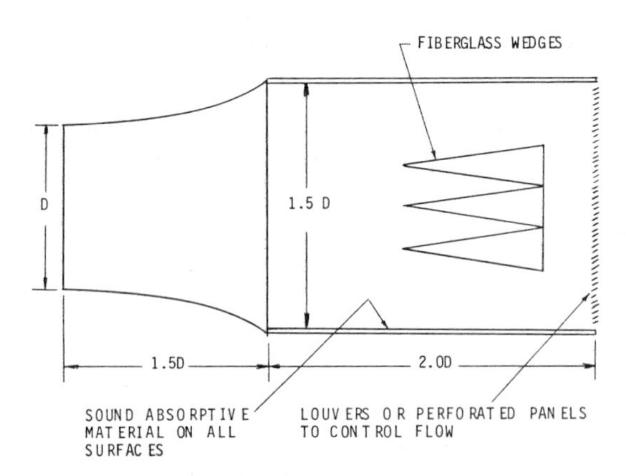

Figure 57. Shows acoustic termination for duct sound tests.

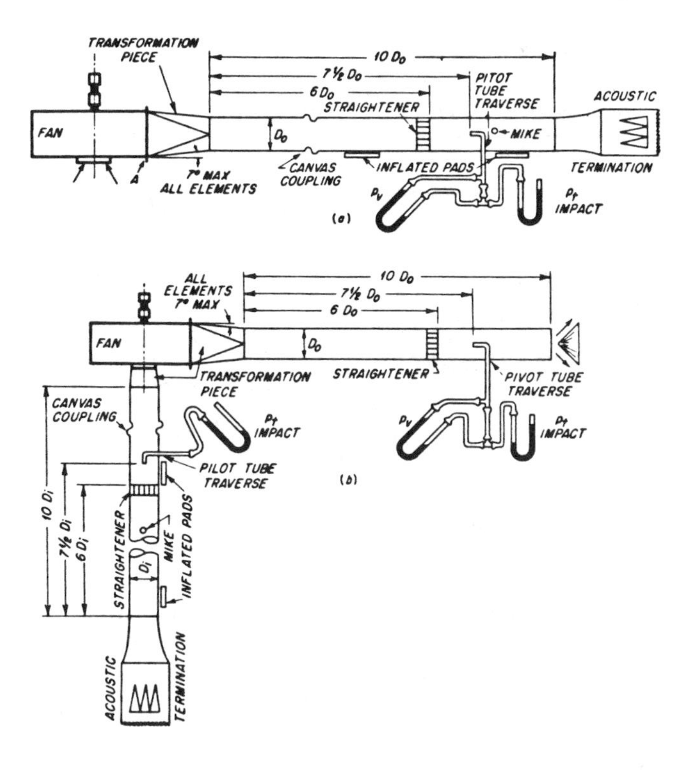

Figure 58. Sound-test specifications for centrifugal fans—single or double inlet with discharge duct and single inlet with inlet duct.

should be smooth, heavy, and as nonresonant as possible. For rectangular ducts, one-inch-thick cement asbestos board is generally very satisfactory.

- *Test specifications.* Figures 58–60 illustrate recommended specifications for measuring the noise of ducted fans, both exhaust and intake. The specifications are based on the National Association of Fan Manufactures, Inc. procedure of airflow tests. Some necessary modifications for sound measurement are noted below:

Exhaust Fans

1. Canvas coupling at $3D_0$.
2. Exhaust end of duct mounted on inflated rubber pads.
3. Reflection-free termination at duct outlet.
4. Microphone at about $8D_0$, along center line of duct.
5. Windscreen over microphone, or twin-microphone and correlator, if required.

Intake Fans

1. Canvas coupling at $8D_0$.
2. Intake end of duct mounted on inflated rubber pads.
3. Reflection-free termination at duct inlet.
4. Microphone at about $3D_0$, along center line of duct.
5. Windscreen over microphone, or twin-microphone and correlator, if required.

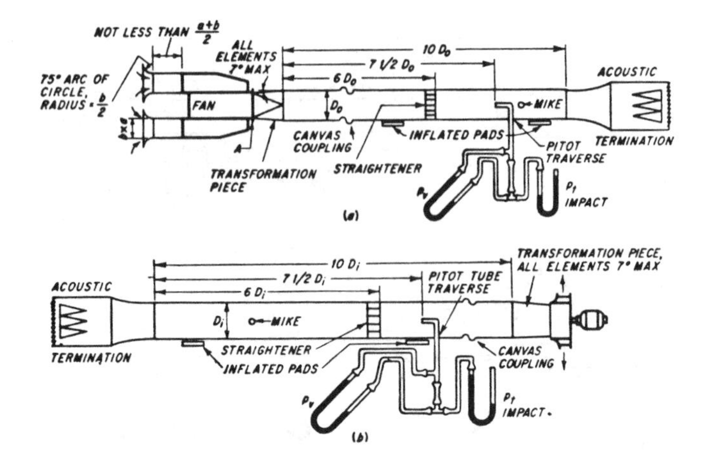

Figure 59. Sound-test specifications for centrifugal fans—single or double inlet with inlet boxes and radial delivery.

Sound pressure level measurements, for any given test condition, should be made for each of the various octave bands. The following equation can be used to estimate the sound power level for each octave band:

$$L_w = L_p + 10 \ln S - 10.6 \text{ dBp} \tag{31}$$

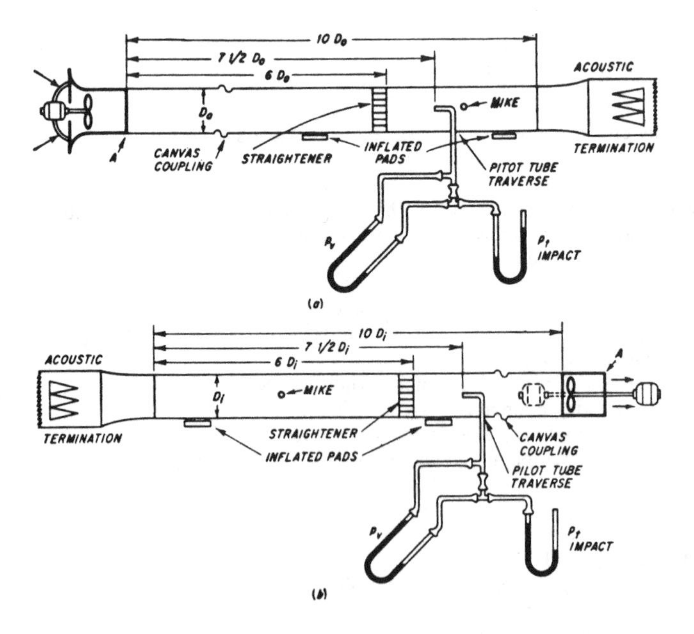

Figure 60. Sound-test specifications for axial fans with discharge and inlet ducts.

where L_w = sound power level, re 1 picowatt (i.e., 1 microwatt)
L_p = sound pressure level within the duct, re 0.0002 microbar
S = cross-sectional area of the duct in square feet

The above expression in metric units is

$$L_w = L_p + 10 \ln S - 0.3 \text{ dBp} \tag{32}$$

The only difference between these two expressions is that S is measured in square meters.

The total sound power, W, in watts can be computed from:

$$W = 10^{-12} \ln^{-1}(L_w/10) \tag{33}$$

For low-frequency measurement, the radiated sound power will be less than that measured within the duct; and for high frequency measurements, the sound power will be radiated from the end of an open duct.

Note that only exhaust- or intake-sound power is measured in the preceding tests. It is determined from the sound power for both directions of airflow which are added, together to obtain the total power.

Figure 61 shows an isolated drive applied to an axial fan. A similar scheme can also be used to centrifugal fans. The purpose of this scheme is to obtain aerodynamic noise only by excluding the motor noise.

At high frequencies, the lack of a perfectly plane sound-wave front across the duct may become a possible source of error. A centrally located microphone usually provides accurate data.

Free-field tests. The measurement of sound at the exit of a duct where it enters a room or free space is referred to as a free-field test. Room fans, unit heaters, and room air conditioners are equipment not intended for use with ducts. Hence, it is more desirable to use such devices for normal conditions of operation.

At high frequencies, sound radiation is usually directive; a number of locations should be chosen for obtaining accurate measurement data.

- *Spherical measurements.* The measuring points should be distributed uniformly over the spherical surface. For the higher octave bands, it is advisable to use a 20-point array.
- *Hemispherical measurements.* One hemisphere of the point-array sets, such as wall-, floor-, or ceiling-mounted equipment, may be used. For the 12-point hemisphere, it is advisable to place the microphone several inches from the room surface in order to minimize pressure doubling effects.
- *Axis of symmetry.* A certain degree of symmetry may be present due to the shape of the noise source, in many spherical or hemispherical tests. A minimization of the number of measurement positions will be the result of such symmetry.

  For a rectangular duct exit, a single point in each of the planes parallel to the base might be chosen to reduce the 12-point array to 4 points.

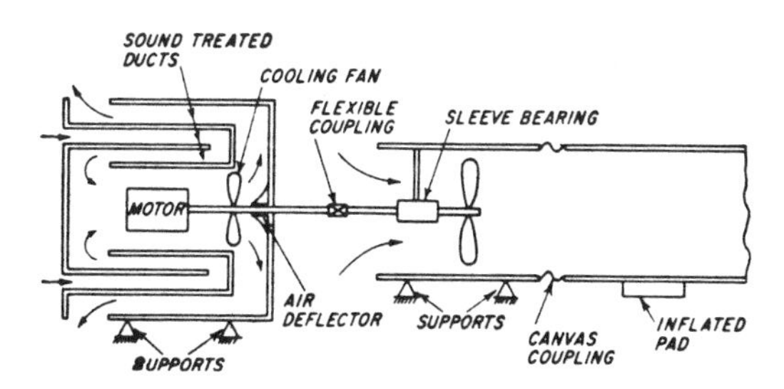

Figure 61. Sound-isolated drive motor for axial-fan duct tests.

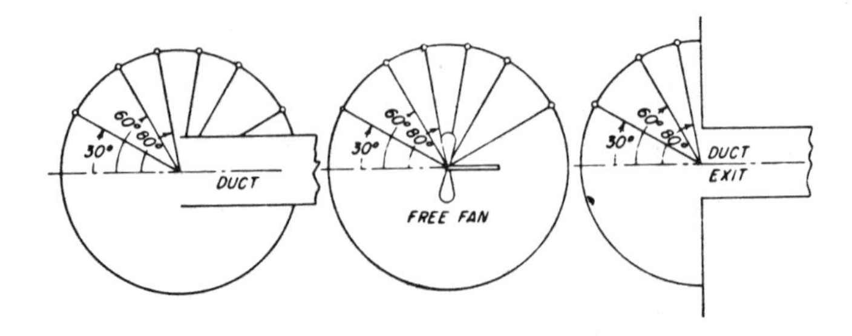

Figure 62. Microphone positions suitable for cases of circular symmetry.

Figure 62 indicates that a number of points may be chosen in a single plane for a circular duct. A microphone is placed in the middle of each zone, which is formed by dividing a hemisphere into three equal sections. For more accurate results, these angles are 33°, 60°, and 80°. A 5-zone hemisphere might be introduced with angles of 25°, 45°, 60°, 73°, and 84° for a highly directive source.

- *Calculation of sound power level.* From the previous schemes, the equal areas of the hypothetical sphere of the hemisphere are obtained due to the microphone positions. Thus, the average sound pressure level over the surface is simply the average of the several readings. A simple arithmetical average is usually sufficient for the total range of less than 6 dB. Corrections should be made in the case of eliminating measurement points by symmetry before averaging. For a rectangular duct exit, an additional 6 dB should be added to the reading obtained in the base plane; and 3 dB to the second plane, etc.

  The following relations show the sound power level by using the average sound pressure level, $\bar{L}_p$:

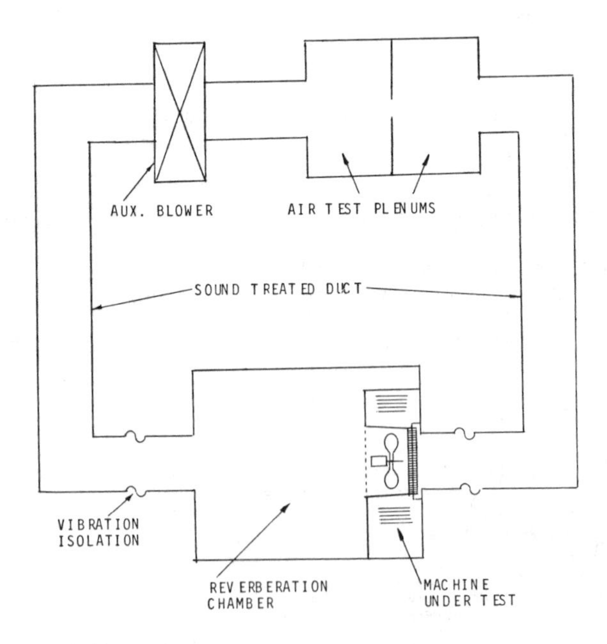

Figure 63. Schematic of a combined reverberation and airflow test.

$$L_w = \bar{L}_p + 20\ \ln(r) + 0.6\ \text{dBp} \qquad \text{(for a sphere)}$$
$$L_w = \bar{L}_p + 20\ \ln(r) - 2.4\ \text{dBp} \qquad \text{(for a hemisphere)} \tag{34}$$

where r = radius of the measuring sphere in feet. For the metric system, the preceding equations become:

$$L_w = \bar{L}_p + 20\ \ln(r) + 10.9\ \text{dBp} \qquad \text{(for a sphere)}$$
$$L_w = \bar{L}_p + 20\ \ln(r) + 7.9\ \text{dBp} \qquad \text{(for a hemisphere)} \tag{35}$$

where r = radius of the measuring sphere in meters [8].

Reverberant-field tests. Reverberant-plenum chambers are used in a test which combines airflow and sound. Sound isolation for flow-control orifices and auxiliary-blower systems are necessary for this kind of test. The sound pressure level due to a given sound source is essentially independent of position in a large reverberant enclosure. Therefore, the sound level is determined solely by the power of the sound source. In practice, it is preferable to use the average sound level for several locations due to the imperfect sound diffusion (see Figure 63).

The sound power level can be determined from the following equation

$$L_w = \bar{L}_p + 10\ \ln(a) - 16.4,\ \text{dBp} \tag{36}$$

where a = total absorption in the room in sabins

or:

$$L_w = \bar{L}_p + 10\ \ln(a) - 6.1,\ \text{dBp} \tag{37}$$

where a = total absorption in the room in square meters

The use of this equation requires that the measurement point, or points, be located in the reverberant field.

Since sound absorption varies with atmospheric conditions, it is important to maintain a constant temperature and humidity.

Sound-control criteria are usually specified in terms of the sound pressure level at the location of interest. The sound pressure level can be computed at any point for a given enclosure and system configuration, if the total radiated sound power is known. The sound power approach has the advantage of independence of installation conditions.

Sound-power spectra can achieve the most accurate sound analyses. It is possible to obtain a reasonable sound-power estimate from the rated horsepower of the fan, if measurement facilities are not available.

The relationship between the overall acoustic power, W_a, and the nameplate power, W_e, is as follow:

$$W_a = \eta W_e \tag{38}$$

where η = efficiency (an approximate value of 1.3×10^{-6} is usually used to give the sound power level within about plus or minus 5 dB for fans in current use)

This yields the following equation for the overall sound power level, re 1 picowatt (i.e., 1 micro-microwatt):

$$L_w = 90 + 10\ \ln(hp),\ \text{dBp} \tag{39}$$

where hp = shaft horsepower

Figure 64 gives typical spectra for different types of fans. The upper curves provide a reasonable factor of safety for acoustic treatment design [8].

For instance, the overall sound power levels of two kitchen exhaust fans:

1. a $\frac{1}{30}$ hp centrifugal fan
2. a $\frac{1}{70}$ hp axial fan

were measured and found to agree with equation $W_a = \eta W_e$ within a few dB. Also, the spectrum shapes agree well with those of Figure 63.

If the fan is coupled to the room by means of ducts, a correction must be made for the effect of various branches in the ductwork (in general, the power divides roughly in proportion to area), for the sound absorption of sound treated duct sections, for the effect of turns, for losses due to sound radiation through duct walls, and for noise generated at outlet grilles [8].

It is difficult to specify an acceptable level for noise generated by fans, since the tolerance varies from individual to individual. Each system must be diagnosed along the lines indicated in order to eliminate unnecessary sources of noise and to achieve a sound design.

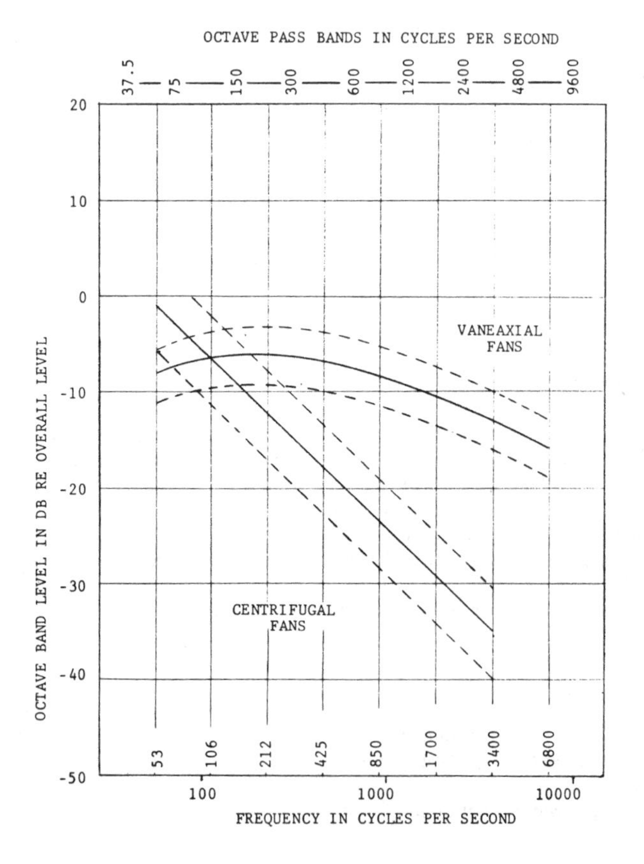

Figure 64. Average sound-power level spectra (in decibels) re overall level, for centrifugal and vane axial fans.

In the design stage, many potential noise sources can be eliminated. As more and more quantitative data become available, especially by applying computer technology to minimize tedious calculation work, a design estimate of noise levels will become a possibility.

It should be noted that it is a mistake to design for more noise reduction than is really needed because the costs increase rapidly as permissible levels are lowered. For existing fan conditions, fan noise reduction methods and devices are recommended to suit various requirements. Costs are still a major consideration. More in-depth coverage of fan noise and system analysis are covered in References 13 through 30.

Fan Dynamics and Vibrations

All rotating equipment impellers have two major forces acting on them: axial force, due to the unbalance of pressures acting on the two sides of the impeller, and centrifugal force. Double inlet fans do not have any unbalance, and therefore do not have much of an axial thrust. For single inlet centrifugal fans, the axial thrust can be calculated as follows (see Figure 65):

$$T = (A_1 - A_s)(P_2 - P_1) \tag{40}$$

where T = axial thrust, lb
A_1 = inlet area, sq in.
A_s = shaft sleeve area, sq in.
P_2 = pressure in the back shroud, psi
P_1 = pressure in the inlet, psi

In the axial fans, the thrust is approximated by the following equation:

$$T = 4.078 \cdot P_t \cdot D_{tip}^2 \tag{41}$$

In *centrifugal fans*, the *thrust* may be approximated as follows:

$$T = 4.078 \cdot P_s \cdot D_i^2 \cdot C$$

where P_t = fan total pressure, in. of water gage
P_s = fan static pressure, in. of water gage
D_{tip} = tip diameter, ft
D_i = inlet diameter, ft
C = constant. If the pressure in the fan housing is positive, then $C = 1.0$, but if the pressure in the fan housing is negative (for example, exhausters), then $C = 2.0$

The amount of the centrifugal force acting on a blade can be determined by:

$$F = mv^2/gr \tag{42}$$

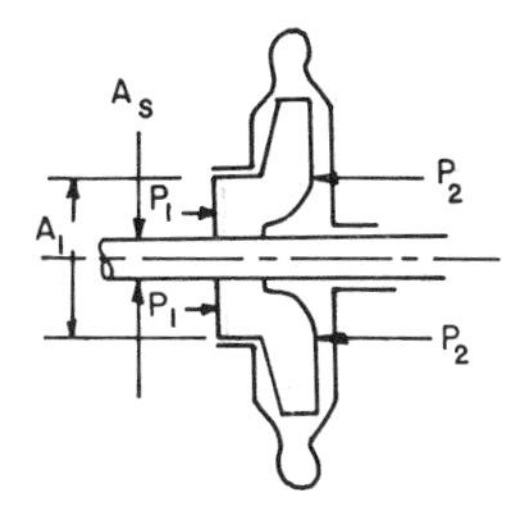

Figure 65. Shows axial thrust on a single inlet, impeller.

where F = centrifugal for, lbf
m = mass of the wheel, lbm
v = tip velocity, ft/sec
r = radius of the wheel, ft
g = gravitational constant, ft-lbm/lbf-sec^2

The preceeding equation can be simplified by replacing 'v' by $2\pi rN/60$:

$$F = mrN^2/2,934$$

The magnitude of the centrifugal force is helpful to the design of the blades and their attachment to the fan hub. There is another force that acts on the housing and that is the net radial force or radial thrust, due to the *volute design.* There are basically two types of volutes, viz. single volute and double volute. In a single volute casing, as shown in Figure 66, the radial force varies along the periphery of the impeller at off-design condition, and the net resultant radial force will act in a direction shown with an arrow. The radial forces are pretty much balanced when the machine is operating at the design point. In a double-volute design (Figure 66), the radial forces are balanced even at off-design conditions. Actually there are two net resultants acting opposite to each other and are almost equal in magnitude. The resultants F_1 and F_2 (Figure 66), tend to balance out even though they are not eliminated. Any unbalanced radial force or load is taken up by the radial bearings which can be either antifriction (ball bearings) or journal sleeve bearings.

If there is any unbalance in the center of gravity and the center of rotation of the wheel, it would cause *vibrations* which are passed on to the bearings and further down to the foundations. For this reason manufacturers recommend that their equipment be mounted on vibration isolating pads. Industry practice is to have *operating speeds* at least 20% above or below the critical speed. The shafts that operate below the critical speed are called the "stiff shaft design" and those that operate

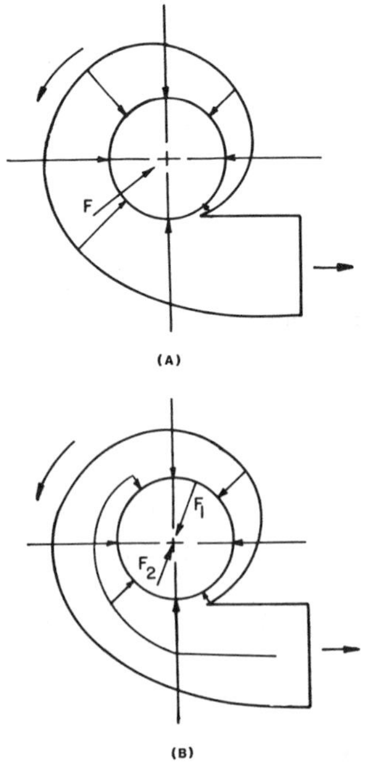

Figure 66. (A) Typical single volute on a scroll showing the radial forces acting at off-design point. F is the net resultant force acting on the casing; (B) shows the double-volute design and radial forces which are balanced. F_1 and F_2 are forces.

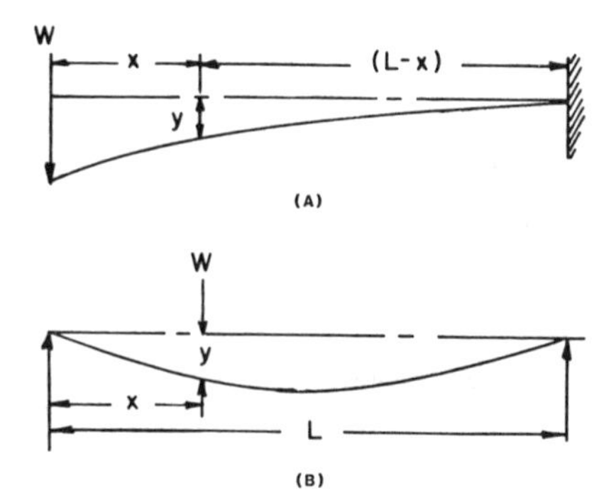

Figure 67. Shows shaft deflection.

above the critical speed are called the "flexible shaft." In flexible shaft design the shaft has to go through the critical speed during the starting and the shutdown of the machinery. The critical speed of a shaft is calculated from

$$N_{cr} = \frac{60}{2\pi}\sqrt{9/y}$$

$$= 187.7/\sqrt{y} \tag{43}$$

where y = static deflection of the shaft, in.

There are other methods for calculating the critical speed of the shaft (e.g., Rayleigh's or Dunkerley's method). The common cases of "overhung shaft" and "simply supported shaft" are discussed in the following. For a single-bearing overhung design, the shaft deflection can be calculated as (refer to Figure 67A):

$$y = -\frac{W}{6E1}(x^3 - 3L^2 + 2L^3) \tag{44}$$

For a two-bearing design, the shaft deflection is given by (see Figure 67B):

$$y = -\frac{W}{48E1}(3L^2 x - 4x^3)$$

$$= -\frac{WL^3}{48E1} \tag{45}$$

This last expression describes the case when the load acts in the center, between the two bearings, i.e., when x = L/2.

Sometimes there are other ways when minor or secondary vibrations are excited. These result from oil-whip, sleeve bearings, couplings, etc. separately or in combination with each other.

Oil whip—Most high speed and high load machines usually use sleeve bearings which are normally oil-lubricated. The oil in these types of bearings sometimes forms what is called oil wedge, which travels around the shaft at about half the speed of the shaft. If this wedge speed happens to coincide with the shaft's first critical speed, the shaft will start to vibrate. Once these vibrations get started, they will not disappear even if the shaft speed is increased to about twice the critical speed. These vibrations are very difficult to eliminate, but the following procedures may help reduce them.

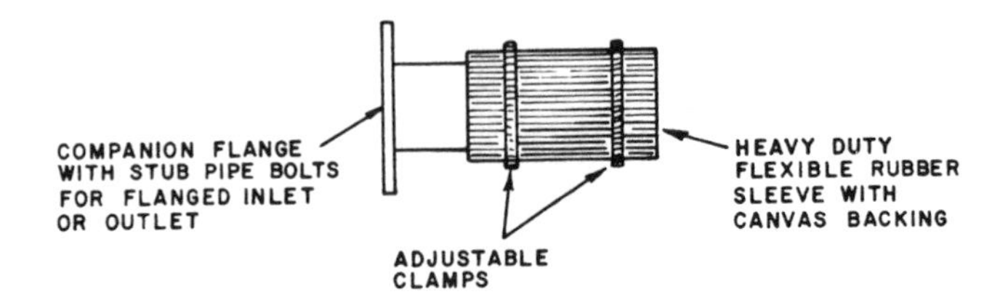

Figure 68. Shows flexible connector for blowers and fans.

1. Switch to anti-friction bearings.
2. Decrease the bearing length to increase the load on the oil wedge and thus reduce the oil-whip-induced vibrations.
3. Do not run the machinery at twice the critical speed.

Sleeve bearings—With sleeve bearings small disturbances are transmitted from the coupling/driver. Other minor fluctuations will easily produce shaft vibrations, since the shaft does not have a continuous contact with the bearings. Sometimes these vibrations disappear if you reduce the clearances between the shaft and the sleeve bearing.

Coupling—These vibrations are usually caused by either misalignment or are transmitted by the driver. Even with the elastic couplings, if the shaft runs above the first critical speed, a slight misalignment will produce periodic disturbances.

Fan housings are usually not as rugged as those of large pumps and compressors, and the manufacturers normally do not allow any loads on the fan housing. Even so, loads imposed by ductwork are sometimes unavoidable. Especially in high temperature cases, the loads due to thermal growth cannot be eliminated. In such cases, the use of flexible connections, or expansion joints at the discharge end or at both the discharge and the inlet connection as well may be used. In the latter case, the fan should be specified with a flanged inlet, so that proper connection with the expansion joint can be made in the field (refer to Figure 68).

It is common practice among consulting engineers to specify spring-supported concrete inertia pads to prevent vibration and noise transmission from mechanical equipment into the building structure. This is a direct result of increased requirements for the control of vibration and structure-borne noise. Another approach is to mount fan units on properly selected vibration isolators. In essence, the problem is to mount the machine (any machine, regardless of the degree of dynamic unbalance, can be considered to be an unbalanced machine) in such a manner that no vibrations will appear in the structure to which it is attached. The classical solution to the theory of vibration isolation appears in many textbooks. However, it is worthwhile to outline in brief the important points of this theory.

Figure 69. Schematic of solid and flexible machine foundations.

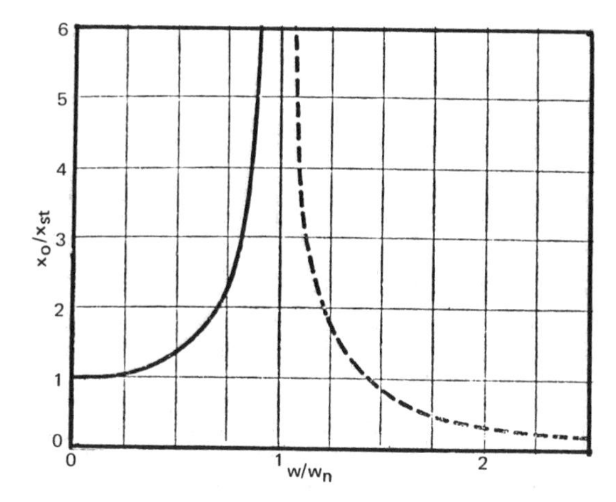

Figure 70. Resonance diagram illustrating system motion under a constant force with variable frequency.

Consider a machine mounted solidly to its supporting structure as depicted in Figure 69A and then consider the same machine attached to the same structure via a flexible system as in Figure 69B. If now the machine displays an unbalanced force, varying sinusoidally, we can compare the resulting structure motion for each case, keeping the force constant, and varying the frequency.

For the solidly connected system, the amplitude of motion will be as shown in Figure 70. The ordinate indicates the ratio of forced amplitude, x_0, to static amplitude, x_{st}, and the abscissa is the rate of forcing frequency, w, to the natural frequency, w_n, of the supporting structure. At $w/w_n = 0$, the deflection of the system will be that of the static deflection only. Where w/w is very large, the deflection approaches zero; the structure simply cannot follow the foricng frequency. At $w/w_n = 1$, the forcing frequency and the natural frequency coincide and deflections become large. This is called "resonance." Since motion and hence stress levels become amplified at or near resonance, this condition must be avoided, of course.

From this plot, it becomes evident that if units are to be solidly attached, the design of the supporting structure must have an extremely low natural frequency. This is not generally possible since this condition of design would predicate other major problems. It is of interest to note that the higher the operating frequency, the lower the transmitted vibration level. It should be noted that in a multiple structure, many different natural frequencies do exist and in all probability cannot be accurately determined; hence, it becomes necessary to eleminate force transmittal to the structure.

For the flexibly connected system, since only the springs are in contact with the structure, the only transmitted force can be the spring force, which is the product of the spring rate and the spring displacement. In Figure 70, if the ordinate is x_q/x_{st}, this item can be equated to kx_0/p_0 or spring force/impressed force. This then will be transmitted force/impressed force or the "transmissibility." We can compute a transmissibility curve for frequency ratios w/w_n (shown in Figure 71).

The ideal transmissibility for a machine is zero. This can be accomplished provided the design of the supporting isolator is such that the natural frequency of the combined system is slow when compared to the disturbing frequency. In other words, the selected isolator must be relatively soft. Figure 71 is a first-order approximation which must be corrected by considering damping. The resulting condition including damping is shown dotted on this curve. Note that damping is advantageous only in the region where w/w_n is less than 1.414. A similar analysis can be made for the vibration resulting from torque reaction. Reactive torque must be transmitted to the foundation. Usually its vibration motion is quite small but, even at this, it can excite structural systems if not properly isolated.

It is therefore possible to achieve a high level of vibration isolation through a flexible mounting. Further, the isolator need only to satisfy the condition of frequency. Its construction can be either

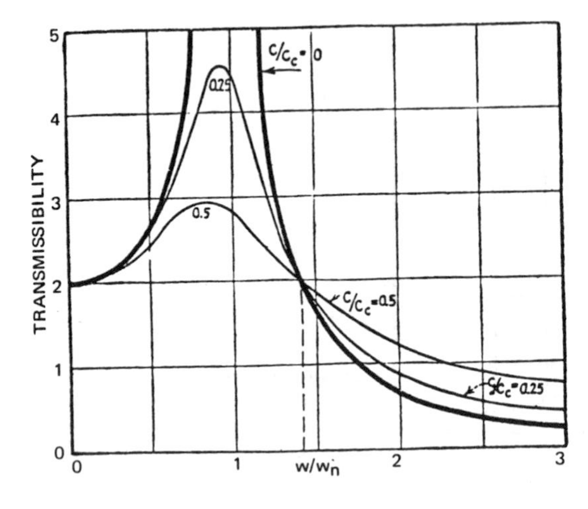

Figure 71. Transmissibility diagram showing possible vibration isolation with flexible supports.

that of a spring or a visco-elastic material; the selection of either to be based on the mechanical properties of the material and its application limit.

Charts are available from which one can graphically determine the isolator design requirement for a desired isolator efficiency. In common use is an isolation efficiency curve prepared by the Lord Manufacturing Company, Erie, Pennsylvania (see Figure 72). With a known forcing frequency and isolator static deflection, one can determine the isolation efficiency. Iterating the variables the optimum and/or desired vibration isolation can be evaluated.

In operation machine systems undergo a series of displacements that result in the following actions:

1. Dynamic unbalance, which causes a motion perpendicular to the axis of rotation.
2. Torque reaction, which causes a rotation about the axis of rotation.
3. Thrust force, which causes a displacement opposite the direction of air flow.

With a flexible supported system, the arrangement of the rotating system must be such so as to be able to handle these displacements without affecting the running alignment. Consider a centrifugal fan unit with two individual axes of rotation. In this case, it is possible to have the motions of the driver and the driven unit out of phase, thus generating severe strains on the drive-train and further creating vibration forces. The units must be solidly mounted on a common base so that the individual axes of rotation remain fixed for all displacements. Isolation is accomplished by flexibly mounting the machine set and the base as a unit (e.g., on an inertia pad).

In contrast, an axial fan carries its own running alignment, regardless of forced excursion or displacement, provided, of course, that the fan design is structurally rigid about all axes and can handle the imposed loads without undue deformation. The axial fan being tubular is inherently strong; however, it must contain a proper balance of structural components, the rigidity of the motor mounting plate, the fit-up, and number of vanes carrying the inner fairing to the outer casing, the thickness of the casing, etc.

Overall Design Considerations

This last subsection summarizes important considerations and guidelines for the preliminary design of fan systems as well as upgrading existing fans. When it is necessary to design and construct a new fan, the following steps are necessary:

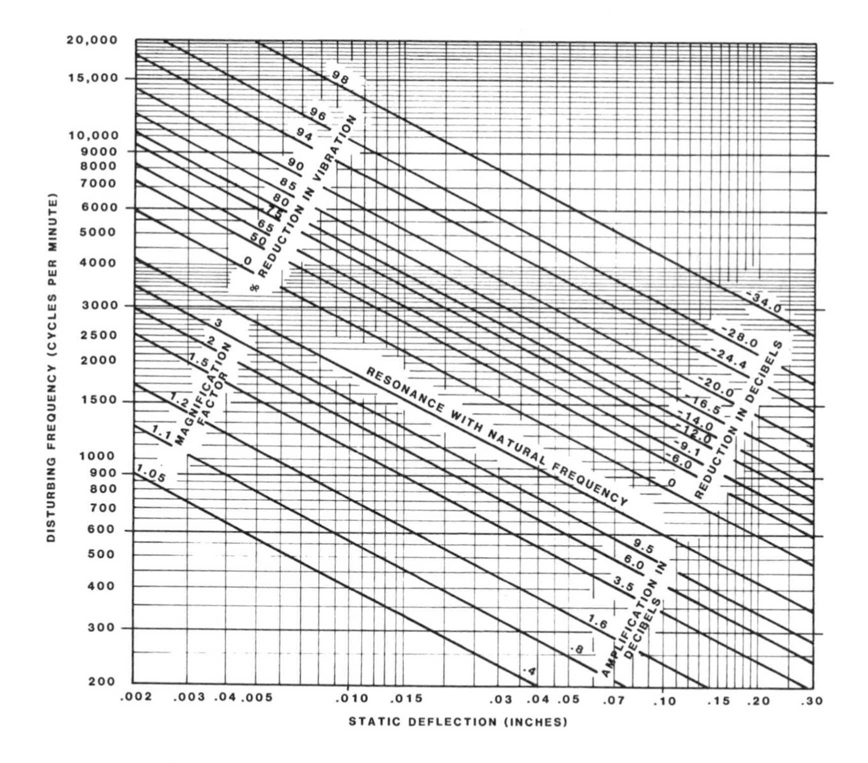

Figure 72. Isolation efficiency curves (Courtesy Lord Manufacturing Co., Erie, PA).

1. Determine the flow required.
2. Determine the pressure the fan must develop.
3. Determine the power requirements of the fan.
4. Select fan type (radial or axial; forward, backward, or radial blades, etc.)
5. Find a particular fan of the type selected in step 4 for which test data are available and which approximate the requirements of steps 1, 2, and 3.
6. Adapt or modify to application.

After the initial choice of a particular fan, it will not fit the job too well; that is, it will be either too fast, too slow, too large, or too small. It is then necessary to modify the fan to fit the application. This modification is made by means of the fan laws which have been established for geometrically similar fans serving similar systems. These laws define the relations between operating parameters. If the performance of the fan thus selected is not adequate, and would be worth much effort to improve it, refinement of the blade profile could be attempted by detailed application of aerodynamic theory.

Although the fan laws were presented earlier, a restatement of them is worthy at this point. Geometrically similar fans are alike in all proportions but different in size. As might be expected, such fans perform in a similar manner. If two such fans were operated in similar systems with a smoke-producing substance injected into the air flow at the fan inlet, photographs of the flow would show that the flow patterns were substantially identical. One important application of the fan laws is the case of predicting the performance of a fan from a geometrically similar fan. The fan laws are also of prime importance in predicting the performance of one fan running at variable speeds.

If two such similar fans, a and b, are taken, the following equations can be written:

$$Q_a = A_a V_a \quad \text{and} \quad Q_b = A_b V_b \tag{46}$$

where $V_a = K_v D_a N_a$
$V_b = K_v D_b n_b$
$A_a = K_A D_a^2$
$A_b = K_A D_b^2$

Combining the preceding expression results in:

$$\frac{Q_a}{Q_b} = \frac{N_a D_a^3}{N_b D_b^3} \tag{47}$$

Further, $p_a = K_p \rho V_a^2$ and $p_b = K_p \rho V_b^2$, and hence:

$$\frac{p_a}{p_b} = \frac{\rho_a N_a^2 D_a^2}{\rho_b N_b^2 D_b^2}$$

$$\frac{\Delta p_{sa}^*}{\Delta p_{sb}^*} = \frac{N_a^2 D_a^2}{N_b^2 D_b^2} \tag{48}$$

or

$$\frac{p_{0a}}{p_{0b}} = \frac{\rho_a N_a^3 D_a^5}{\rho_b N_b^3 D_b^5} \tag{49}$$

The above relationships constitute the so-called fan laws.

It should be emphasized that these fan laws hold only for geometrically similar fans operating in geometrically similar systems.

No simple *general* relationships exist between fans which are not geometrically similar. Many correlations of limited applicability, however, have been established to relate fan performance to variations in particular fan parameters and dimensions. In general these relationships hold only over a very limited range of the variables concerned, but are often quite accurate for small changes.

If performance curves (see Figure 73) for geometrically similar fans are prepared it is found that the shapes of the curves are the same. The fan laws can be used to relate the pressures, flows, speeds, and horsepowers of a point on one curve to another point on the curve for the geometrically similar fan. Such points which are related through the fan laws, are called corresponding points.

At corresponding points of two curves, the pressure coefficients, flow coefficients, specific speeds, and efficiencies of the two fans will be equal. The converse of this statement is not necessarily true. That is, if two geometrically similar fans have the same specific speeds or flow coefficients they are not necessarily operating at corresponding points. Since the efficiency and head curves are not single valued for all fans, two geometrically similar fans might have equal efficiencies or pressure coefficients and be operating at different specific speeds.

We now direct design notes specifically to *radial-flow fans.* These units are best suited for the majority of low peripheral-speed applications. They are inherently high-pressure, low-volume fans; if necessary, however, they may be designed to move large volumes of air at relatively low pressures. Fans equipped with straight radial blades have the advantage of being reversible, a feature which is required in many motor applications, but they also have the disadvantage of low efficiency.

In designing this type of fan, the following conditions must first be determined:

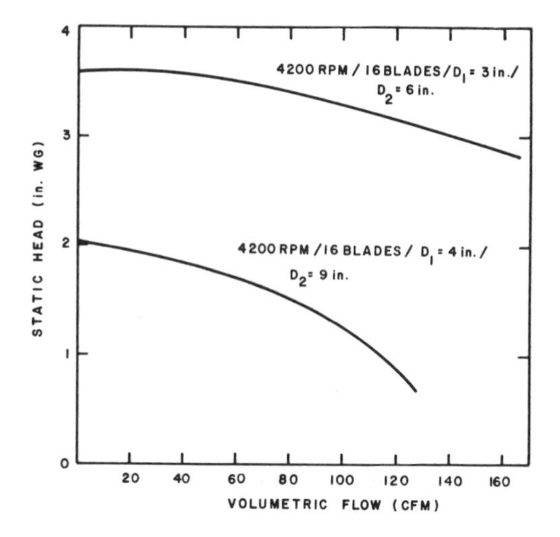

Figure 73. Performance of a radial blade fan.

1. The flow resistance curve of the system.
2. The required air flow for the intended application.

The pressure required to force air through the ventilating system is proportional to the square of the flow. The system resistance is composed of some combination of series and parallel flow paths, as for example, in a motor, the axial flow through air gap and stator slots with radial flow between rotor packages. This resistance, when plotted as resistance head vs. flow, is known as the resistance curve of the system.

The determination of the required flow for cooling is a fundamental design problem based on experience. If a fan design is available which has pressure-vs.-flow curves which give pressure and flow combinations not differing very much from the requirements, the problem is simplified, provided:

1. The fan operates at or near its peak efficiency at the flow point considered.
2. The fan fits into the available space.

Frequently, the system characteristics require a fan for which performance curves are not available. Within limits, modifications of known data can be made to predict th characteristics of the required fan. These modifications are made not only by the use of the fan laws but also by specially determined relationships that are applicable where the fan resistance is a large part of the system resistance.

There are several special relationships for radial-flow fans which are briefly outlined below.

Blade inlet angle, β_1—for zero impact loss at intake, when the inlet air is entering in a radial direction, the tangent of the blade inlet angle β_1 should be equal to the ratio of the radial component of inlet air velocity divided by the blade tip velocity at the inlet periphery.

$$\tan \beta_1 = \frac{V_{1r}}{u_1} = \phi \tag{50}$$

The angle of incidence = 0.

For short-bladed sirocco-type fans, it often is not practicable to meet this requirement. When the angle between the inlet edge of the blade and the inlet stream is large, small variations of this angle have very little effect.

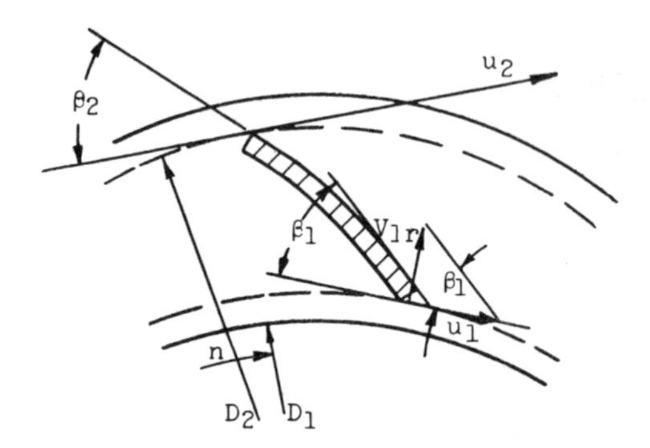

Figure 74. Shows blade exit and β_2.

Blade exit angle, β_2—flow (Q and hence ϕ) is proportional to exit angle, β_2 (see Figure 74), under the following restricted conditions:

1. Blades curve backward with β_2 less than 30°.
2. The fans are identical except for β_2.
3. Back-pressure on the fan near the value for maximum efficiency.

Above $\beta_2 = 30°$ tests indicate that Q is proportional to a fractional power of β_2.

Number of blades, N—for conventional design, the number of blades selected is such that the projected area of the blades projected in a radial direction, covers about 100% of the circumferential area, $\pi D_2 b$. In general, the number of blades increase as the ratio of D_1/D_2 increases, i.e., the shallower the blades, the larger the number.

Where N is reduced below 20, the static pressure, Δp_s^* (and hence ψ) is reduced proportional to $N^{0.7}$, for the following restricted conditions:

1. Exit angle, β_2, between 15° and 90°.
2. Fans identical except for number of blades, N.
3. Range of b/D_2 from 0.10 to 0.11.
4. Range of D_1/D_2 from 0.83 to 0.91.
5. Operation near point of maximum efficiency.
6. Range of blade chord depths from 0.083 to 0.161.

Blade width, b—the blade width, b, affects the flow, Q, but the relationship is not, in general, simple. Blade width and shroud depth are interdependent and there is an optimum combination for each load restriction (system resistance). Similarly, there is an optimum width, b, and inlet area, A, for each shroud depth.

Blade depth—a deeper blade increases the pressure output of a fan, but the relationship with depth is not simple, except at zero flow.

For any fan the pressure developed is commonly expressed as

$$\Delta p_s^* = \frac{\psi (u_2)^2}{2g}, \text{ ft of air} \tag{51}$$

At zero flow, the pressure developed by a fan with no scroll nor diffuser is due solely to the centrifugal force acting on the air, and hence, theoretically, the following relation exists.

$$p_s^* = \left[1 - \left(\frac{D_1}{D_2}\right)^2\right]\left(\frac{u_2'}{4000}\right)^2, \text{ inches of water}$$

where $\left[1 - \left(\frac{D_1}{D_2}\right)^2\right] = \psi_0$

At zero flow, the pressure developed is theoretically dependent only on the tip velocity, u_2, and on the diameter ratio, D_1/D_2, of the blade. The relationship holds regardless of the size of the fan.

Actually slightly higher pressures than those predicted by the preceding equation can be developed at zero flow by the use of a scroll or diffuser, which retains some effect even at zero flow because of leakage through the clearances between the impeller and the adjacent stationary surfaces. The actual pressure coefficient at zero flow ranges from 4% to 14% above predictions.

Shrouds and clearances—clearances between the impeller and the adjacent stationary wall must be sufficiently small, and shrouds must be appropriately designed, or fan performance will be seriously impaired.

Smooth surfaces—roughness, as with cast fans, induces no appreciable deterioration of performance, at least for 6-in. fans. Cast impellers may have thicker blades which reduce free area, however, the leading edge of cast blades is relatively thick, and may therefore give opportunity for performance improvement by appropriate rounding of the edges.

The design of inlets may be as important as the design of the fan itself. This is especially true in the design of a fan by scaling up or down from a similar fan. Inlet conditions should be duplicated, in the chosen scale ratio, as closely as possible. Proper design of inlets leads to not only good performance characteristics, but also minimum noise levels. The inlet design should be chosen both to minimize resistance to flow and to improve the flow lines so that eddies (flow separation from the surface) and entrance shock do not occur.

Inlets and inlet conditions not purposely constructed to provide good flow are so varied in character that very little can be said about them even qualitatively. Each individual case deserves careful consideration.

The type of fan blade and shape of inlet to be used for a given application will depend upon the relative importance of high efficiency and first cost for that application. For instance, optimum performance can be obtained by using a blade shape as shown in Figure 75E; however, cost considerations may dictate the use of an inlet as shown in Figure 75C, D, or even B. An inlet, as shown in Figure 75A should be avoided since poor efficiency and noise will result.

The use of a beveled blade shape, as shown in Figure 75C, compensates for the axial momentum of the air entering the fan which tends to carry the air toward the backplate, P. The increased depth of this blade also increases the developed pressure. Eddies, E, may still exist near the shroud, S; however, unless the inlet surface is faired smoothly to a radial direction, as shown in Figure 75D. Blade shape Figure 75D, is highly desirable because the flow changes gradually from axial to radial.

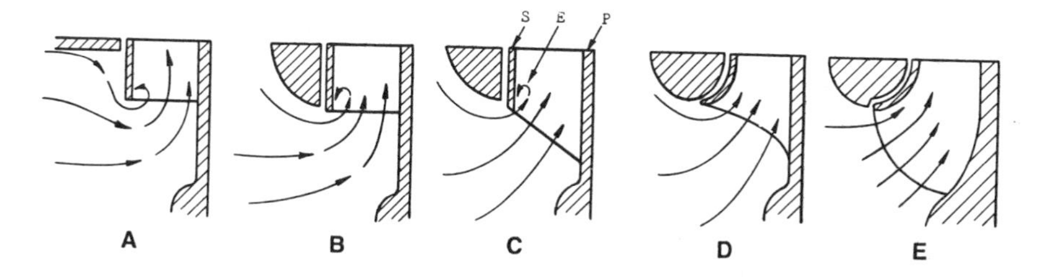

Figure 75. Inlet shapes for radial flow fans: (A) bad; (B) better; (C, D) still better; (E) best.

The area of flow at entrances should not be made so small that appreciable head is lost in this region. Sharp bends and restrictions should be avoided. Design of inlet orifices depends upon the D_0/D_2 ratio of the blades and upon the blade length b_1. In any case, it is highly desirable that $(\pi D_0^2/4)$ be greater than $(\pi D_1 b_1)$. Figure 76 defines the system orientation. This prevents flow separation from shroud, S, and crowding of the flow toward the backplate, P. This requirement results in narrow blades, i.e., small, b_1, but the volume output per inch of blade length is high. For the case where $D_0 = D_1$ the requirement becomes: $D_1 > 4b_1$.

Less conservative ratios are as follows:

	Backward-Turned Blades	Radial Blades	Forward-Turned Blades
D_0/D_2	0.60–0.75 large volume fans 0.2–0.5 pressure fans	0.70–0.80 mechanical draft fans 0.2–0.5 pressure blowers	0.80–0.95
Recommended b_1/D_0	0.40	—	—
Upper limit b_1/D_0	0.55 large volume fans	0.55	0.65

Clearances (refer to Figure 76) between orifices and fans should be as small as manufacturing methods will allow. This is most critical in backward-turned blades.

Twisted radial-turning vanes (stationary) located close to the fan in the inlet stream have been used to give the air an initial rotation in the direction of blade rotation prior to entrance into the blade. Shock of the inlet edges is reduced, resulting in a net increase in efficiency, and general performance is improved. If inlet guide vanes are turned backward (against the rotation) more energy is imparted by the impeller to the fluid and an increase in pressure or capacity or both may result, but at a loss in efficiency. Inlet flow conditions may be predetermined by the use of adjustable inlet guide vanes which can be adjusted to give high efficiency over a wide range of flow. If the entering air has a component of rotation in the direction of fan rotation, it will decrease the relative velocity of the air and fan blades. If it is rotating against the fan, it increases the relative velocity. For this reason, the operating characteristics of a fan rotating at a given speed can be varied widely by giving the entering air a component of rotation.

It often happens that the air entering the fan acquires a component of rotation due to the relative positions of the inlet piping of the fan. This can cause serious deviations from design performance. The use of inlet guide vanes helps to minimize such inlet flow disturbances.

For axial-flow fans operating at relatively high pressures inlet orifice axial location is very important. The axial location is even more important than the shape of the inlet. A sharp-edged orifice properly located will give about the same performance as a well-rounded inlet.

It should be noted that performance curves are based on total pressure (TP) vs. volume flow (cfm). Total pressure (TP) is the sum of the associated velocity pressure (VP) plus (+) the system static pressure (SP). Velocity pressure (VP) from the fan discharge remains constant in a constant

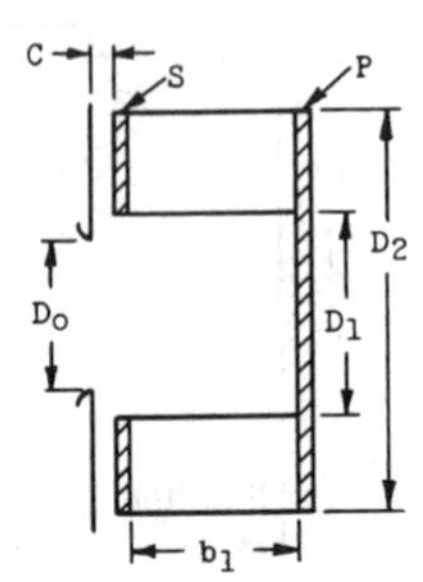

Figure 76. Defines clearances between orifices and fans.

size discharge duct, i.e., fan diameter. However, its value can change as a function of the discharge configuration; duct enlargement or reduction, discharge to plenum, or to atmosphere, either saving or wasting significant levels of power in performing its work. Thus, knowledge of discharge mechanics and accounting for attendant losses and gains in VP and SP are of great importance. Outlet cones can be designed to reduce outlet velocity to a reasonable VP and effect maximum gain of SP and power saving. If it is not possible to incorporate a standard outlet cone, then the duct work (transitions) should be arranged so as to provide reasonable levels of regain. For illustration, consider a fan selected to a duty of 25,000 cfm at 2.5 in. SP and apply it to varying discharge connections. The pertinent velocities associated with the fan system are:

Fan annulus area	3.24 sq ft	7,716 fpm	3.71 in. VP_a	(annulus)
Fan outlet area	7.07 sq ft	3,536 fpm	0.78 in. VP_f	(fan area)
Cone outlet area	10.56 sq ft	2,367 fpm	0.35 in. VP_c	(cone)

Example discharge connections are shown in Figure 77 from whence it can be concluded that cases A and B compare free discharge with and without a cone, and C and D compare delivery to a duct with and without a cone.

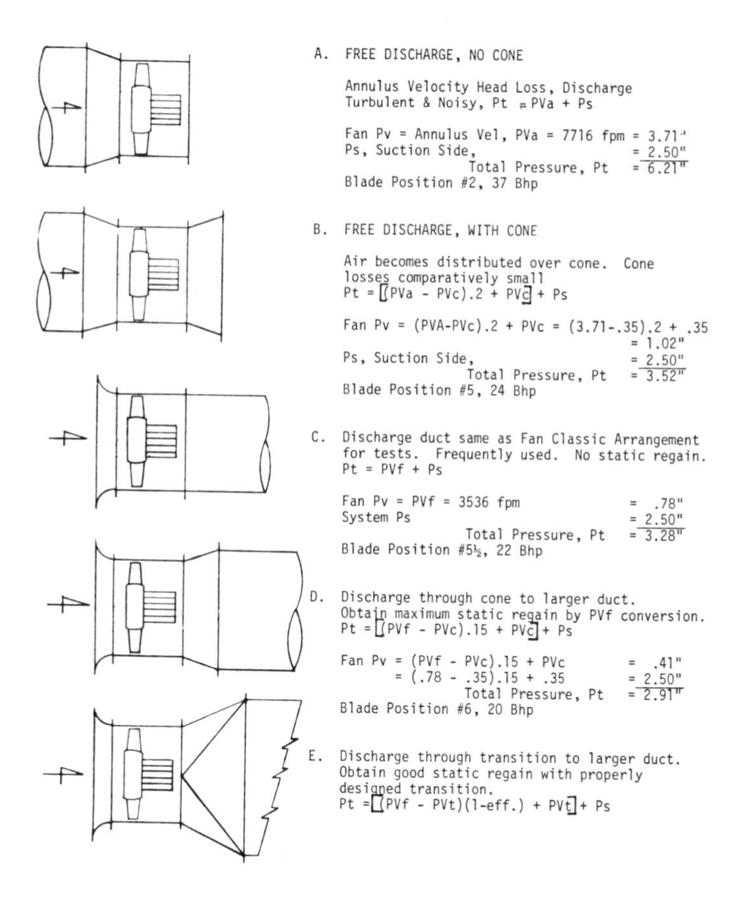

Figure 77. Shows various discharge connection arrangements. (Courtesy Joy Manufacturing Co., New Philadelphia, OH.)

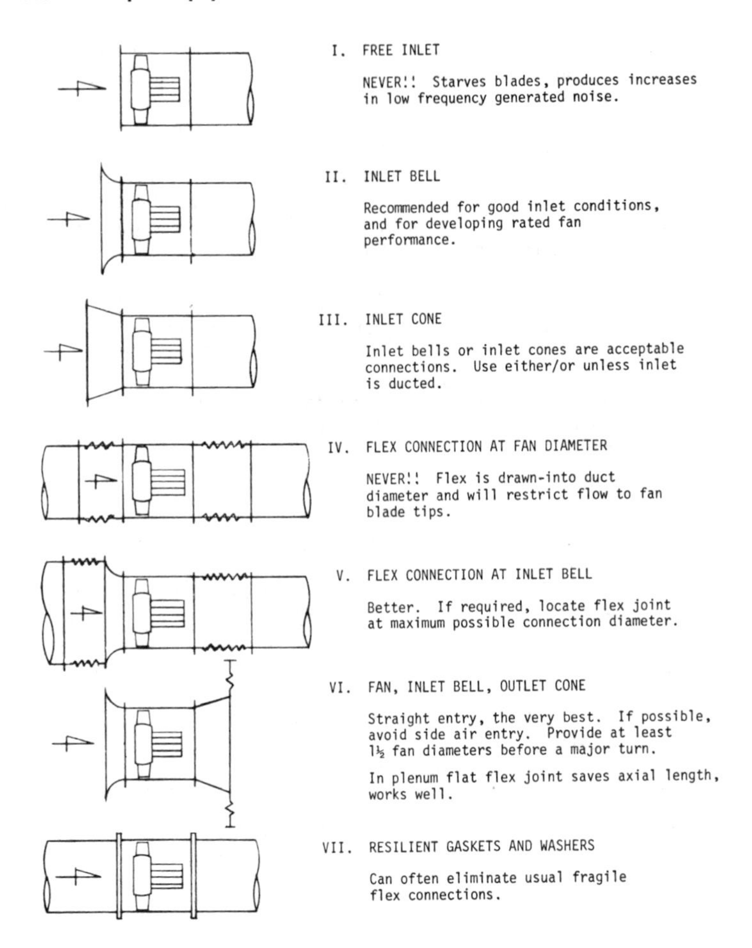

Figure 78. Shows various inlet connections and flexible connections (joints). (Courtesy Joy Manufacturing Co., New Philadelphia. OH.)

Comments relative to inlet connections and flexible connections are shown in Figure 78. Also, Figure 79 shows velocity regain efficiencies for transition sections.

Scrolls, Shrouds, and Clearances

Special scroll designs need to be considered for some applications. For example, it may be necessary to have the air discharge from the fan in two separate directions, in which case, a scroll shape similar to that shown in Figure 80B would be used.

The periphery of a scroll is usually an involute curve generated from a circle (refer to Figure 81). An involute is the curve generated by the end of a string (b-c) which is kept taut while being unwound from a circle called the generating circle (radius a in Figure 81). The equations describing

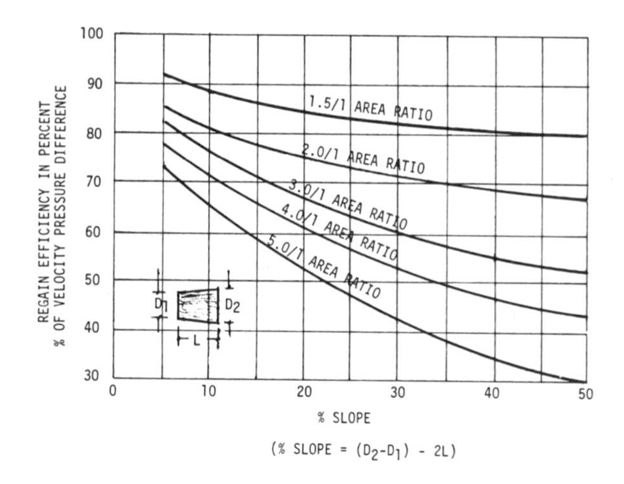

Figure 79. Shows velocity regain efficiences for transition sections. (Courtesy Joy Manufacturing Co., New Philadelphia, OH.)

the curve are:

$$\left.\begin{aligned} x &= a(\cos\theta + \theta\sin\theta) \\ y &= a(\sin\theta - \theta\cos\theta) \end{aligned}\right\} \qquad (52)$$

where θ = the angle, measured from the point of intersection of the involute and the generating circle radius, in radians

a = the radius of the generating circle

These equations enable a point-by-point calculation of the involute when the radius of the generating circle, a, is known. The size and shape of a fan scroll depend upon the radius of the base

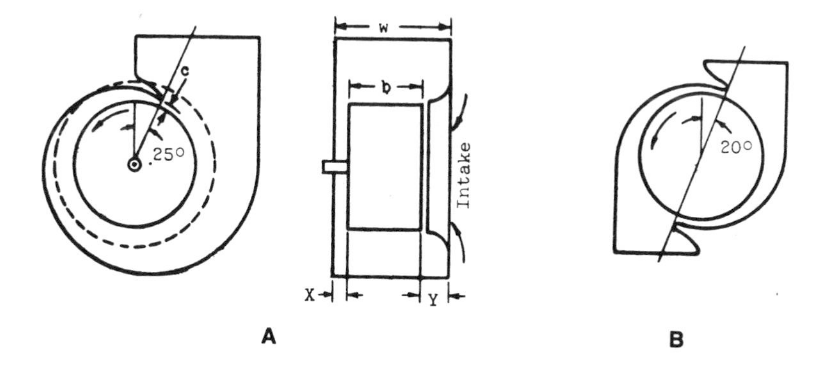

Figure 80. Shows special scroll designs: (A) involute; (B) double discharge scroll.

circle, a, the diameter of the fan impeller plus clearance (fan circle), D_2', the width of the scroll opening, L_s, the expansion angle, α_e, and the cutoff angle, β_{co}.

The relationships between the principal parameters of the scroll are:

$$\left.\begin{aligned} a &= (D_2'/2)\sin\alpha_e \\ L_s &= a(2\pi - \beta_{co} + \text{ctn}\,\alpha_e) = \text{scroll opening} \end{aligned}\right\} \tag{53}$$

Construction of an involute is as follows. The radius of the generating circle, a, is determined by the expansion angle, α_e, and the diameter of the fan plus clearance, D_2'. The involute can then be constructed about the generating circle. The cutoff angle, β_{co}, is measured from $\theta = 0$ to a radial line drawn through the intersection of the involute and the fan circle. A line drawn through the intersection, e (if the fan circle and the radius at θ equal 0 and are tangent to the generating circle, a) intersects the involute at the termination point, c, of the involute. This line determines, L_s, the scroll length at the exit.

The straight portion of the exhaust should extend a length at least equal to the fan diameter, D_2, as shown in Figure 81 for best performance.

The cutoff point of a scroll need not be sharp; it can be a blunt, though rounded design without detriment to the overall performance and with the addition of better quietness of operation. The sharp knife-like edge formerly used often acts like a reed vibrating in a high-velocity gas stream, with consequent noise. Further in the interest of silent operation the cutoff piece can be made of wood, rubber, or some other soft material which acts as a sound dampener.

Fan performance is seriously impaired unless the clearances are kept very small between the fan impeller and the stationary parts so as to seal effectively the inlet air from the exhaust air. If large

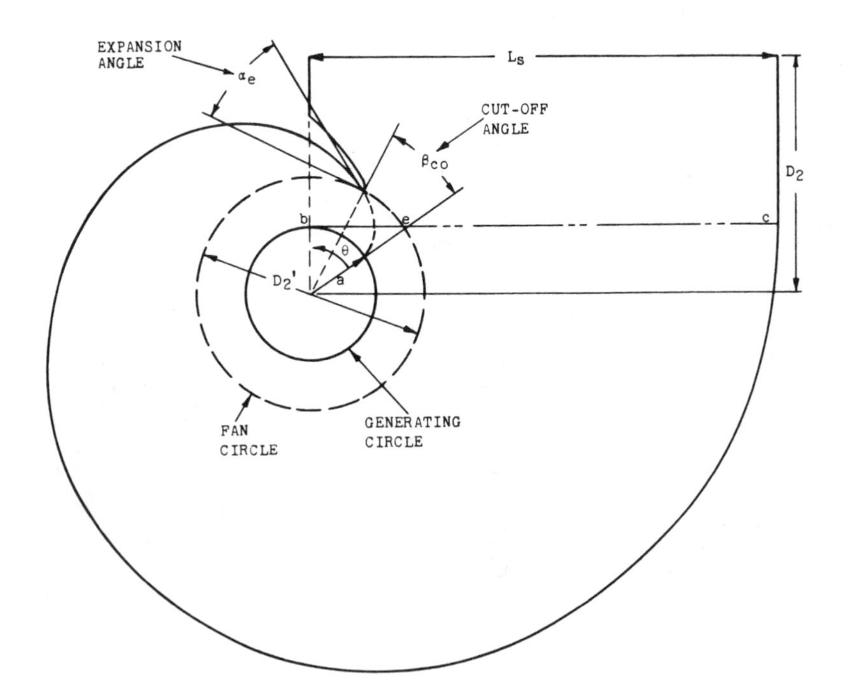

Figure 81. Shows construction of fan scroll.

clearances are used, exhaust air returns to the inlet through the ineffective seal. This recirculation decreases both the pressure and the volume of useful air and increases the power required to drive the fan.

For axial-flow fans, the clearances between the tips of the fan blades and the stationary housing around them is the dimension to be minimized. A clearance of only 0.1% of wheel diameter is a desirable specification to limit satisfactorily the amount of leakage past the blades.

For velocity-type axial-flow fans, however, where the static pressure rise across the blade is not high, the clearance may be greater and often the housing may be cast or rolled to shape without the necessity of a machine finish. Generally, the blade section of the fan housing is machined or sized so that a close running clearance is possible.

For radial-flow fans a retaining ring, either rotating or stationary designed to prevent leakage of exhaust air into the inlet region of a fan is called a shroud. Shroud rings are especially important for low-flow, high-pressure work. Minimum clearance between an open-face fan (i.e., without a rotating shroud) and a stationary inlet or retaining ring may be a satisfactory substitute for a rotating shroud. Rotating retaining rings are usually conical shaped (on the air flow side) for both backward and radial-tip blades for optimum channel dimensions. Refinements such as the shaping of the retaining ring can often be left out with forward curved blades since efficiencies there are generally comparatively low.

The clearance between the inlet and the impeller is important for all types of radial-flow fans, but the quantitative requirements differ for different types of blades. For backward curved blades the inlet clearance is of particular importance and should be kept within the order of 0.4% of the outer diameter. For radial-tip blades, the clearances should be kept to about the same order as for backward-curved blades, but performance is less sensitive to variation of the clearance. With forward-curved blades, a large clearance is not only permissible, but is actually necessary for stability to minimize pulsations.

Blade axial width, b, and shroud radial depth are interdependent and must therefore be considered together. As noted earlier, there is an optimum fan blade depth and shroud depth for each load restriction. There is also an optimum blade width, b, and blade inlet area $A_1 = (\pi b D_1)$ for each shroud depth. If b and A_1 are too small compared to the shroud inlet area for axial flow to the impeller they will offer excessive restriction and use up an excessive amount of the static pressure produced by the fan. If b and A_1 (or more specifically, D_1) are too large compared with diameter at the shroud inlet, the radial component of flow at the blade inlet periphery will be too small and excessive impact loss will result. Figure 82A shows the relative proportions for high-flow, low-pressure work. Figure 82B shows the relative proportions for medium-flow, medium-pressure work. Figure 82C shows the relative proportions for high-pressure, low-flow work.

Where the blades extend inward nearly to the center of the fan (as in Figure 82C) shroud depth rather than blade depth is the important dimension to consider. Stationary shrouds must have close clearances (less than about 0.4% of outer blade diameter) with the blades; pressure outputs and flow, Q, both suffer from too-large clearances.

For multiple-blade diffusers the shroud width, w, should be as nearly the same as the fan width, b, as is possible allowing only for running clearance of the impeller from the housing walls which extend inward past the ends of the fan blades (see Figure 83). It is essential that these walls provide a close-fitting seal. If there is appreciable leakage past the seal the performance will be seriously impaired.

The cutoff clearance, c, (refer to Figure 83) should be as small as possible for best air flow and pressure. However, small clearances will produce increased noise.

The shroud width, w, should be approximately 1.60 times the blade width, b, except for backward leaning blades where a ratio of as high as 2.2 has been found best. If, w, is less than this value the fan capacity drops below its optimum value.

The fan should be located closer to the backplate than to the inlet opening as shown in Figure 84. By this arrangement better performance due to a better distribution of the air over the axial length of the blade is obtained.

The cutoff clearance, c (Figure 84), should be as small as possible; however, a small cutoff clearance is not nearly as important as in the multivane diffuser because the area between the fan and the cutoff portion of the housing is a much smaller portion of the discharge area. Experience indicates

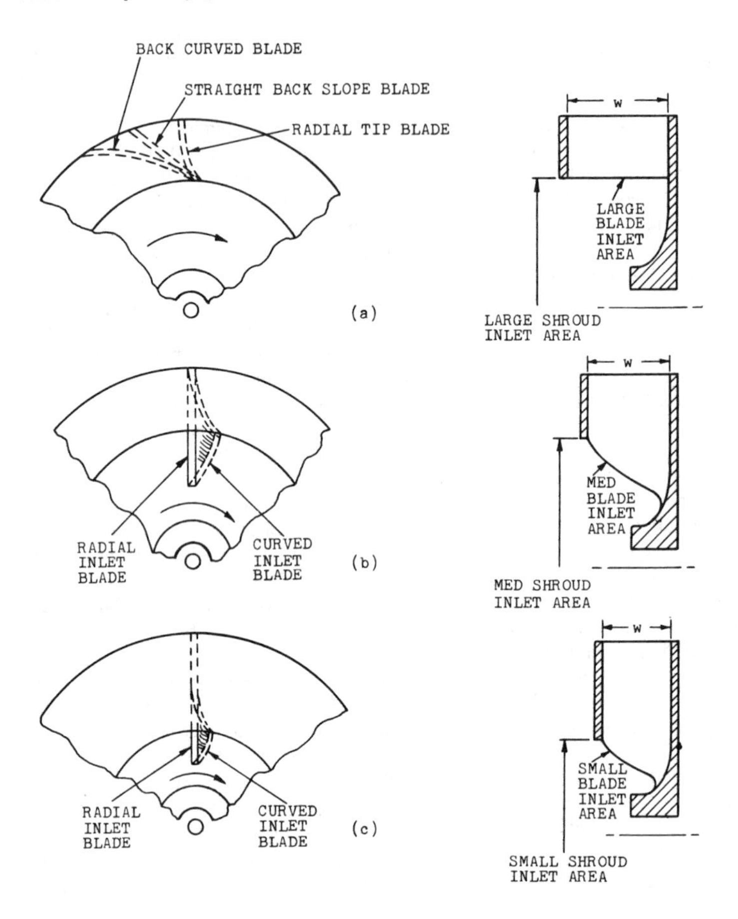

Figure 82. Shows relative proportions for high flow, low pressure work.

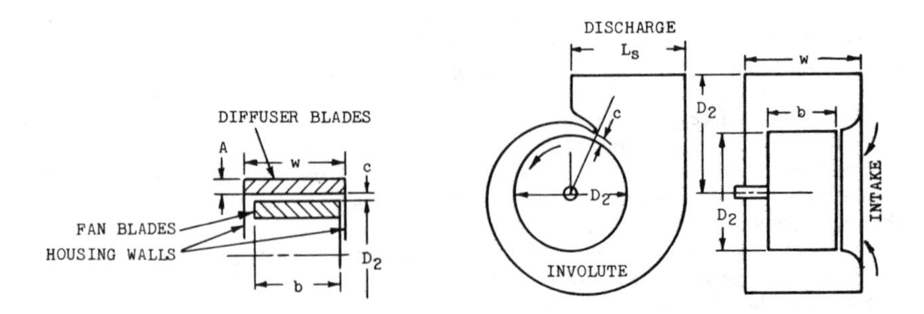

Figure 83. Shows running clearance of impeller from housing.

Figure 84. Shows cutoff clearance in housing.

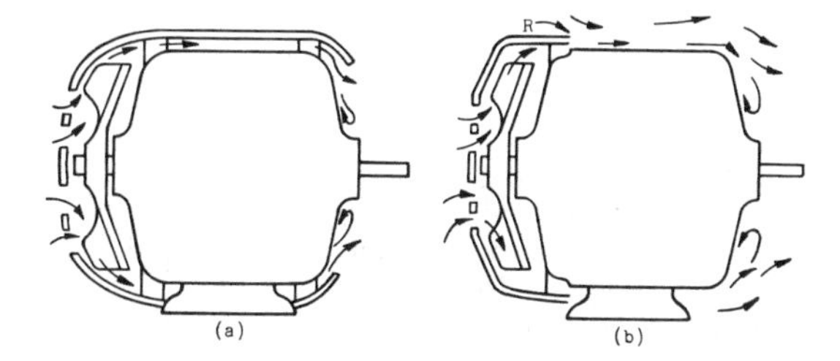

Figure 85. Two types of totally-enclosed fan cooled motors.

that making the clearance about 3% to 7% of the fan impeller outer diameter, D_2, gives satisfactory performance. In special cases, however, it may have to be greater than this because of noise produced by its too great proximity to the blading.

Enclosures for Motors

Totally enclosed motors must rely upon heat removal by convection and by conduction through the mountings. In very small sizes especially for intermittent operation, natural convection (and conduction) may be sufficient. Except for such cases, however, forced convection, and hence fans, must be used. Two types of enclosures for a totally enclosed fan-cooled motor are illustrated in Figure 85; (A) represents good cooling performance but is wasteful of material in the outer housing; (B) represents a closer balance between material utilization and effective cooling.

In "open-type" motors the removal of heat from the windings and laminations may be accomplished by end ventilation where air is circulated by fans at both ends and caused to circulate over the end windings; for relatively large motors, however, it sometimes is preferable to use axial ventilation where the coolant fluid is circulated through axial ducts in the stator and in some cases the rotor. Still another type of ventilation utilizes radial ducts at intervals along the armature. The design of fans for these applications is complicated by the fact that proper direction and baffling of the coolant flow must be achieved for effective cooling.

For open type double-end ventilated small motors the effects of various baffles and other design changes are shown in Figure 86.

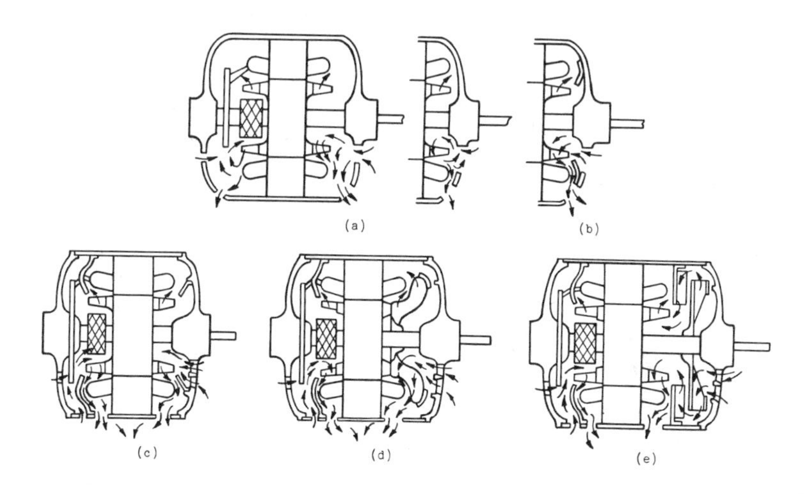

Figure 86. Illustrates double-end ventilated small motors. Systems are shown in order of increasing cooling capacity (i.e., a—least, e—greatest).

COMPRESSION MACHINES

General Information

The ratio of the final pressure, p_2, established by a compression machine to the initial suction gas pressure, p_1, is called the *compression ratio.* Depending on the value of the compression ratio, gas displacement machines usually are classified as *fans* $p_2/p_1 < 1.1$ for displacement of large amounts of gas; *blowers,* $1.1 < p_2/p_1 < 3.0$, for moving gas at relatively high resistances in pipelines; *compressors,* $p_2/p_1 > 3.0$, for establishing high pressures; or *vacuum pumps,* for compressing gas that is below atmospheric pressure so that it can be discharged to the atmosphere.

The types of compression machines most often employed in process plant applications fall into the following classes: centrifugal, axial, reciprocating, and rotary, depending on how the mechanical elements act on the gas. There are two categories of compression machines under which these classes are grouped: dynamic and positive displacement.

Dynamic machines employ rotating vanes or impellers to impart velocity and pressure to the gas and are classified as centrifugal and axial. These machines operate by developing a high gas velocity and converting this pressure in the diffusing flow passage. In general, they tend to have a lower efficiency than positive displacement machines. However, they do operate at relatively high speeds to provide high flow rate in relation to the physical size of the machine. Also, dynamic machines usually have significantly lower maintenance requirements than positive displacement machines.

Positive displacement compressors consist of two types: reciprocating and rotary. Both confine successive volumes of gas within a closed space in which the pressure of gas is increased as the volume of the closed space decreases.

Reciprocating-type compressors have one or more cylinders, each fitted with a piston driven by a crankshaft through a connecting rod. Each cylinder is equipped with suction and delivery valves and means for cooling mechanical parts. Figure 87 illustrates both water-cooled and air-cooled compressor cylinders. Gas is introduced into the cylinder during the suction stroke. At the end of the stroke, the piston's motion is reversed, and gas is compressed and expelled during the delivery stroke. When only one end of the piston acts on the gas, the machine is referred to as a *single-acting compressor.* Machines in which compression is effected at both ends

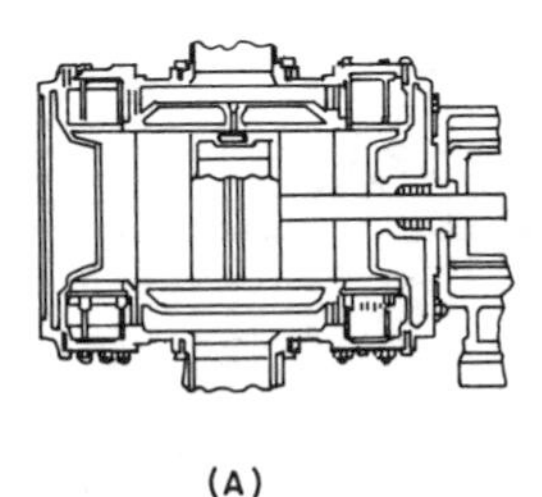

(A)

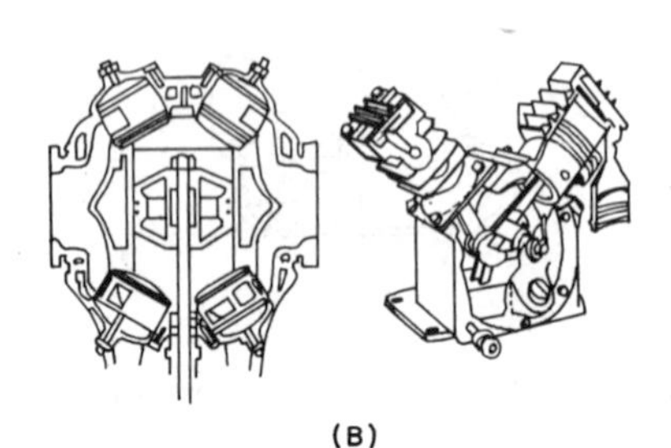

(B)

Figure 87. (A) Water-cooled compressor cylinder; (B) air-cooled compressor cylinder.

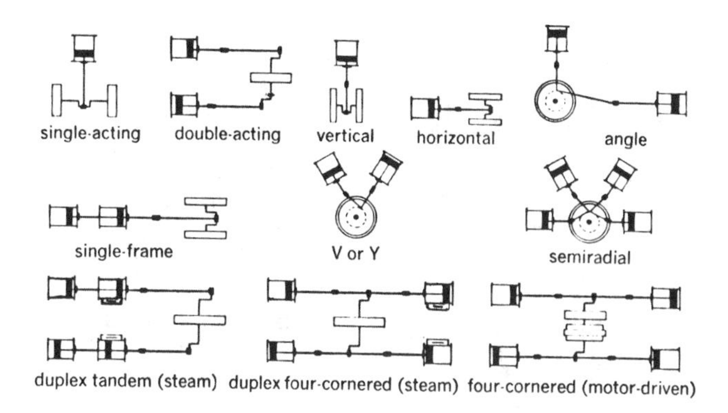

Figure 88. Shows frame arrangements for positive-displacement piston compressors.

of the piston are called *double-acting machines.* They deliver about twice as much gas per cylinder per cycle as the single-acting machines. Figure 88 illustrates various frame arrangements for positive displacement piston compressors. *Single-stage compressors* compress the gas in each cylinder from the initial intake pressure to the final delivery pressure on each working stroke of the piston.

In *two-stage compressors,* the gas is compressed to an intermediate pressure in one cylinder while another cylinder is used to raise the pressure to the final delivery pressure. Machines utilizing two or more stages are called *multistage compressors.*

Vertical and horizontal compressors may be single-cylinder or multicylinder designs. Angle types are multicylinder, with one or more horizontal and vertical compressing elements. Single-frame (straight-line) machines are horizontal or vertical double-acting compressors with one or more cylinders in line with a single-frame having one crank throw and one connecting rod and cross head. The V- or Y-type compressor comprises a two-cylinder, vertical, double-acting machine with cylinders usually at a 45° angle with the vertical. These designs employ a single crank. Semiradial compressors are similar to the V- or Y-type, but have horizontal double-acting cylinders on each side. Duplex compressors have cylinders on two parallel frames attached by a common crankshaft. Duplex-tandem, steam-driven units employ steam cylinders in line with gas cylinders. Duplex four-cornered steam-driven compressors have one or more compressing cylinders on each end of the frame and one or more steam cylinders on the opposite end. Four-cornered motor-driven units have the motor on a shaft between compressor frames.

Reciprocating compressors are capable of gas capacities as high as 100,000 cfm, at pressures in excess of 35,000 psi. Special units with higher capacities or pressures can be custom made. Water generally is used as the coolant for cylinders, intercoolers, and aftercoolers, however, other liquids, including refrigerants, are used.

The *single-stage, horizontal, single-acting compressor* consists simply of a cylinder with a reciprocating piston. Operation is illustrated in Figure 89A. The cylinder is covered from one side with a head in which the suction valve (3) and the delivery valve (4) are located. The piston is connected directly to the rod (5) and crank (6) with a shaft having a flywheel (7). The piston movement from left to right effects gas discharging in the space between the head and piston. Because of the difference in pressure between the suction line and the cylinder, the valve (3) opens and, hence, gas enters the cylinder. On the reverse of the piston stroke, the suction valve closes and the gas in the cylinder compresses to a certain pressure. At that time, the valve (4) opens and the gas is delivered into the discharge piping.

In single-stage, double-acting compressors (Figure 89B), the gas is compressed in a cylinder (1) intermittently from both sides of the piston (2). For one double-stroke piston there are two suctions and two deliveries. The cylinder is fitted with two suction valves (3) and two delivery valves (4). Compressors employing double-acting pistons are more complicated; however, their capacity is

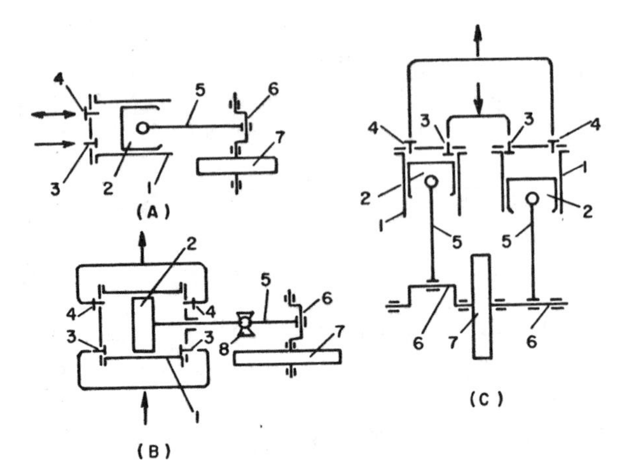

Figure 89. Single-stage piston compressors: (A) one-cylinder single-acting; (B) one-cylinder double-acting; (C) two-cylinder single-acting: 1—cylinder, 2—piston, 3—suction valve, 4—delivery valve, 5—connecting rod, 6—crank, 7—flywheel, 8—crosshead.

two times higher than that of single-acting units of similar size and weight. This increased capacity also may be achieved in *multicylinder* single- and double-acting compressors.

The *two-cylinder single-acting compressor* (Figure 89C) consists of two single-acting compressors driven by one crankshaft, with cranks displaced at 90° or 180° from each other. Cylinders, and sometimes the compressor head, are equipped with waterjackets to remove the heat of compression. Although heat removal is not complete, it does decrease the energy expenses for compression significantly.

Vertical single-stage compressors have a number of advantages over horizontal. They are generally higher-speed units (for horizontal compressors n = 100–240 rot/min; for vertical compressors n = 300–500 rot/min or greater); consequently, their capacity is higher and they require less space. Cylinders and pistons undergo less wear. With horizontal cylinders, especially those of large diameters, local wear on the piston occurs due to the action of gravity force, which results in decreased piston velocity.

Multistage compression is employed when the required pressure ratio, p_2/p_1 is large and it is not practical to perform the entire compression in a single cylinder because of the high temperatures and the adverse effect of clearance volume on the efficiency. Mechanical construction also is difficult because single cylinders would require sufficient strength to withstand the final pressure but be of a large enough size to hold the gas at the initial pressure, p_1.

Staged compression is performed in *multistage* compressors in which the gas passes in series through a number of cylinders of gradually decreasing volume. Hence, the gas is compressed gradually to the final pressure. Multistage machines are provided with intercoolers between stages. Intercoolers are basically heat exchangers that remove the heat of compression from the gas and reduce the temperature to approximately the temperature at the intake. Cooling reduces the volume of gas entering the high-pressure cylinders, reduces the horsepower required for compression and, at high pressures, maintains the temperature within safe operating limits.

Stages typically are arranged in different cylinders, as shown in Figure 90A, B and C, and in a single cylinder with a differential piston Figure 90D. V-type compressor cylinders are arranged at some angle (usually 45°) from the vertical.

Multistage compressors with stages arranged separately in several cylinders may have more than one cylinder in line (Figure 90A). They also may have duplex reciprocating compressors on one end of a frame (Figure 90B). Such compressors are heavy and large and therefore are subject to considerable unbalanced inertia forces. This can limit compressor operation at high speeds.

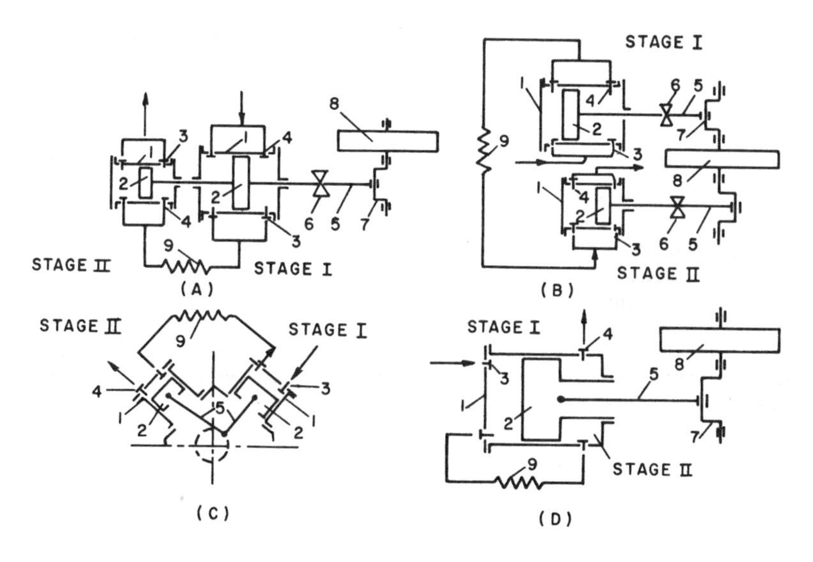

Figure 90. Positive-displacement piston compressors with compression stages in separate cylinders: (A) single; (B) duplex; (C) V-type; (D) differential: 1—cylinder, 2—piston, 3—suction valve, 4—delivery valve, 5—connecting rod, 6—crosshead, 7—crank, 8—flywheel, 9—intercooler.

The disadvantage just noted is eliminated by *horizontally opposed reciprocating compressors* with cylinders on opposite sides of the crankcase. These employ a multithrow type of shaft with one crank throw per cylinder. Such units are dynamically balanced, which increases the crankshaft rotation speed 2–2.5 times. This, in turn, increases the capacity of the compressor. The system's weight, including electric motors, can be 50%–60% less than that with cylinders arranged on one side end.

Compressors with a differential piston may have several compression stages formed by a cylinder surface and a piston of variable (differential) cross section. The ratio between piston cross sections depends on the compression ratio in each stage.

Usually the differential piston is used for two-stage compression in machines of low and medium capacities. In large machines, the differential piston may be wedged because the piston length diameter ratio is small. The compression ratio (defined as the sum of the piston displacement and clearance volume divided by the clearance volume) in each stage is selected to provide the most effective use of cylinder volume, increase the compressor volumetric coefficient, decrease energy consumption, and to establish the gas temperature at the end of compression.

Thermodynamic Principles

The compression of real gases involves changes in pressure, volume, and temperature only. Pressure-volume-temperature interrelations at pressures up to 10^6 N/m^2 (~ 10 atm) can be expressed by the ideal gas law:

$$P\hat{V} = RT$$

where P = pressure
$\hat{V}$ = molar volume
T = absolute temperature
R = gas constant

For pressures exceeding 10^6 N/m^2 ($p > 10$ atm) van der Waals formula more accurately describes the pressure-volume-temperature relation:

$$\left(p + \frac{a}{v^2}\right)(v - b) = RT \tag{54}$$

where p = gas pressure (N/m^2)
v = specific volume of gas (m^3/kg)
$R = 8314/M$ = gas constant (J/kg K)
M = molecular weight (kg/kg-mole)
T = temperature (K)
"a" and "b" = empirical constants for a given gas

In the absence of data to establish "a" and "b," estimates may be made from the critical parameters of the gas, i.e., critical temperature, T_{cr}, and critical pressure, P_{cr}:

$$a = \frac{27R^2T_{Cr}^2}{64P_{cr}} \tag{55a}$$

$$b = \frac{RT_{cr}}{8P_{cr}} \tag{55b}$$

The work expended for compression may be determined from Bernoulli's equation:

$$\hat{\ell} + q = (i_2 - i_1) + \frac{w_2^2 - w_1^2}{2} \tag{56}$$

where $\hat{\ell}$ = work spent for compressing 1 kg gas (in a compressor)

q = heat introduced per 1 kg gas

i_2 and i_1 = enthalpies of gas before and after compression, respectively

Differences in gas velocities before and after compression usually are neglected, i.e., it is assumed that $w_1 = w_2$. The preceeding expression takes the form

$$\hat{\ell} + q = i_2 - i_1 \tag{57}$$

The work expended in the compressor, $\hat{\ell}$, and the heat added, q, produce an increase in the enthalpy of the gas. The compression process may be either adiabatic or isothermal.

In adiabatic compression there is no input or output of heat from the compressor. That is, $q = 0$, the expression simplifies to

$$\hat{\ell} = i_2 - i_1 \tag{58}$$

Work is converted into heat, thus raising the temperature of the gas. The net result is an increase in enthalpy. Hence, for adiabatic compression we can expect the gas temperature to increase to high values.

Under isothermal conditions the gas is compressed at constant temperature. The internal energy of gas remains unchanged, $u_1 = u_2$, and the following relationship applies, $p_1v_1 = p_2v_2$. Consequently, $u_1 + p_1v_1 = u_2 + p_2v_2, i_1 = i_2$, i.e., the gas enthalpies at compression do not change. The following expression describes isothermal compression:

$$\hat{\ell} + q = 0 \quad \text{or} \quad \hat{\ell} = -q \tag{59}$$

The negative sign on q denotes the output of heat. Hence, all the work expended in compression is converted into heat and evacuated from the gas. Temperature, internal energy, and the enthalpy of the gas therefore will not change. This means that the compressor must be cooled at isothermal conditions to evacuate an amount of heat equivalent to the work spent.

Expressing work in units of $kg_f - m$ and enthalpy in kcal/kg, then

$$A\hat{\ell} + q = i_2 - i_1 \tag{60}$$

where $A = \frac{1}{427}$, thermal equivalent of work ($kcal/kg_f \times m$).

Correspondingly, for adiabatic compression

$$\hat{\ell} = \frac{i_2 - i_1}{A} \tag{61}$$

And for isothermal compression:

$$A\hat{\ell} = -q \tag{62}$$

Compression may be illustrated graphically by a temperature-entropy diagram (T-S plot) (Figure 91). Entropy, S, is a thermodynamic function of state of a defined body or substance. It is increased on the addition of heat to the body. The higher the body's temperature, the smaller the degree of entropy increase.

For a reversible process, the increase in entropy due to the addition of heat, q (J/kg), is:

$$\Delta S = \int_{T_1}^{T_2} \frac{\Delta q}{T}, \text{ J/kg-°K} \tag{63}$$

Examining the T-S diagram in Figure 91, note that there are two types of lines, those corresponding to constant pressure (called isobars) and lines corresponding to constant temperatures (isotherms). The line AKB is a boundary curve. The region below the curve corresponds to wet vapor, and the branch AK corresponds to dry saturated vapor. Point K is a critical point. The region to the left of branch AK is the liquid state, and the region to the right of KB is superheated vapor. Because the processes of vaporization and condensation occur at constant pressures and temperatures, the isobars coincide with isotherms in the wet vapor region of the diagram. The condensation of a mixture of wet vapors occurs at variable temperatures, and, hence, isobars do not coincide with isotherms in the wet vapor region.

Also shown in the T-S diagram are lines of constant enthalpies (i = constant). The enthalpies of ideal gases only depend on temperature, and for such gases the lines of constant enthalpy coincide with the isotherms. The enthalpy of a real gas is also a function of pressure, and hence, lines of constant i do not coincide with the isotherms.

Gas compression of a gas can be explained by Figure 91 as follows: For an adiabatic compression $q = 0$ and $\Delta S = 0$, i.e., the process continues without a change in entropy. Consequently, adiabatic compression can be represented by the vertical line 1-2, where point 1 characterizes the state of the gas before compression. Point 1 is located at the intersection of isobar p_1 with isotherm T_1. Point 2 represents the condition of the gas after compression and is located on the isobar corresponding to pressure p_2.

An isothermal compression (temperature is constant) is represented by line 1-2 in Figure 91. Point 2′ characterizes the state of the gas after compression and is located on isobar p_2. The amount of heat evacuated, q, is $T\,\Delta S$, which is denoted by the cross-hatched rectangular area of rectangle a-1-2′-b. The height of this rectangular area is T_1, and the base represents the change in entropy ΔS. We note that entropy decreases in this case, i.e., ΔS is negative. Therefore, the amount of heat also will be negative, that is, the process will be accompanied by an evacuation of heat. The same area a-1-2′-b expresses the work of isothermal compression in thermal units, whereas the area

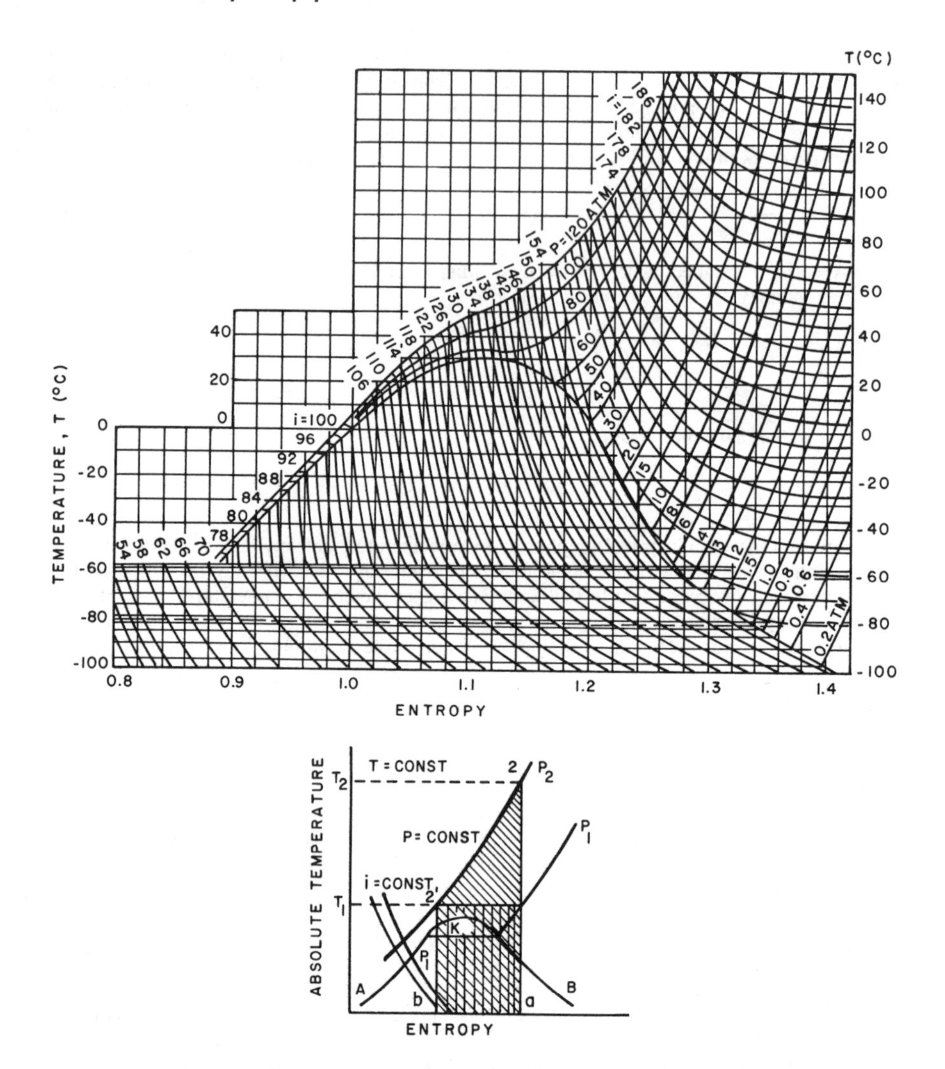

Figure 91. Temperature-entropy diagram for gas compression.

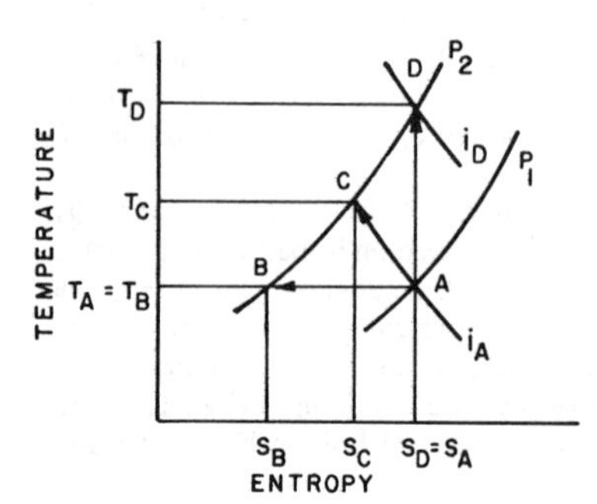

Figure 92. Temperature-entropy diagram illustrating gas compression.

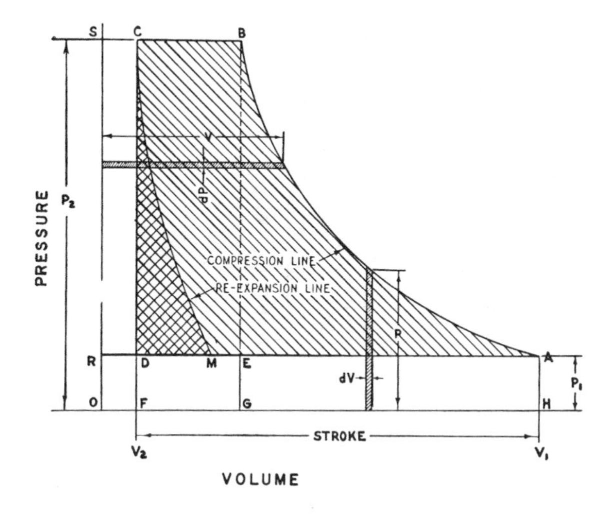

Figure 93. Illustrates single-state compression cycle on a pressure-volume diagram.

a-2-2′-b represents the work of adiabatic compression. Although adiabatic and isothermal compressions are ideal, in practice they may be approached fairly closely.

When real gases are compressed, their volumes, pressures, and temperatures change. In addition, a portion of the heat is dissipated to the atmosphere. This is known as *polytropic* compression. The process of polytropic gas compression from pressure p_1 to p_2 is illustrated by the lines AC on the T-S diagram in Figure 92. The amount of heat generated in the compression of 1 kg of gas is equivalent to the specific work of polytropic compression, $\hat{\ell}_{pol}$, and may be determined approximately from the diagram or analytically through the following relationship:

$$q_{pol} = \hat{\ell}_{pol} \simeq (S_A - S_C)\frac{T_A + T_C}{2} + C_p(T_C - T_A) \tag{64}$$

Let us examine a single-stage compression cycle in detail by considering the pressure-volume diagram shown in Figure 93. Gas enters at point A with pressure, p_1, and volume, v_1, and is compressed along the line AB. The delivery valve opens at point B when pressure p_2 is reached. At point C, the delivery valve closes and the piston begins its return stroke, allowing the pressure to decrease to p_1 along the line CM, opening the suction valve. Suction occurs along the line MA. The total work of compression from a pressure p_1 to a pressure p_2 may be defined as the product of the pressure difference and volume (represented by the shaded area ABCM). To evaluate the work performed along the compression line AB, let $d\hat{\ell} = V\,dP$.

In an actual compression or expansion of a perfect gas, Boyle's law (PV = K, where K is a constant), is applicable, provided the process takes place at constant temperature. If a temperature change occurs, an exponent, n, is included on the volume:

$$p_1 v_1^n = p_2 v_2^n = K = \text{constant} \tag{65}$$

The subscripts refer to different conditions of the same weight of gas. In practice, most compression and expansion curves follow this equation with the exponential term ranging between 1.0 and 1.5. The specific value of n depends on a number of factors, such as the peculiarities of the gas compressed, the specific heats of the gas, degree of cooling, the operating characteristics of the compressor cylinder, and/or the amount of ring leakage. Figure 94 shows different compression curves plotted on logarithmic coordinates, producing straight lines with slopes equal to the tangents of angles A, B, C, etc.

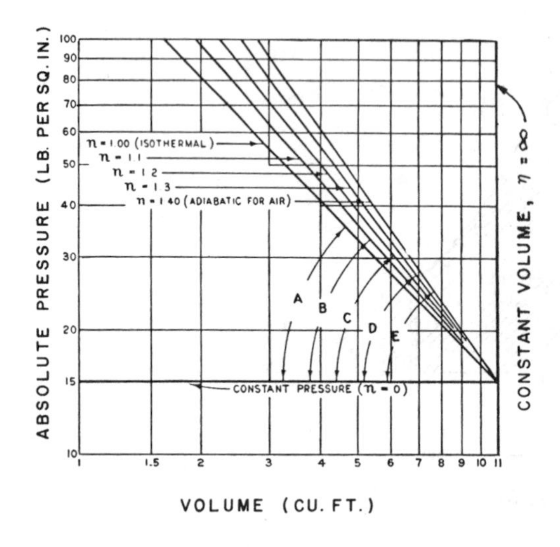

Figure 94. Compression curves as functions of gas volume and pressure.

The isothermal case has a slope of 1.0 and is readily recognized as an equilateral hyperbola. The slopes of the other curves are usually greater than 1.0. When exponent n = 1.0, the pressure-volume change has taken place without any change in temperature. If n is greater than 1, the expansion is polytropic. In an adiabatic expansion or compression, the value of n is the ratio of the specific heats of the gas compressed, C_p/C_v, where C_p is the specific heat at constant pressure, and C_v is the specific heat at constant volume.

Note that $P = KV^n$. By raising both sides of this equation to the 1/n power, we obtain:

$$P_1^{1/n}V_1 = K^{1/n} \tag{66}$$

or

$$V = \left(\frac{K}{P}\right)^{1/n} \tag{67}$$

Substituting

$$\hat{\ell} = K^{1/n} \int_1^2 \frac{dP}{P} \tag{68}$$

Integration produces two solutions, depending on whether the value of n is equal to unity. For isothermal compression (when n = 1.00),

$$\tilde{\ell} = -K \ln\left(\frac{P_2}{P_1}\right) = P_1V_1 \ln\left(\frac{P_2}{P_1}\right) \tag{69}$$

This expression determines the work from zero volume to the line AB, including the area ABSR in Figure 93. Reexpansion takes place on the other side of the cylinder along the line CM, returning to the cycle the work (area CMRS) of opposite sign from the total area ABSR. Hence, the net work is the difference in areas, ABCM.

Isothermal compression is seldom achieved in practice; however, the ideal condition is instructive when comparing the performances of different compressors. For the second case, where $n \neq 1$, i.e.,

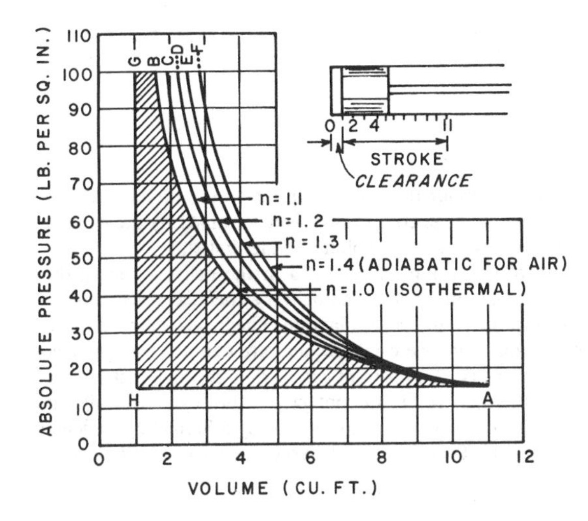

Figure 95. Comparative power requirements at different n-values [31].

polytropic performance,

$$\tilde{\ell} = \frac{n}{n-1} K^{1/n}[P_2^{(n-1)/n} - P_1^{(n-1)/n}]$$
$$= -\frac{n}{n-1} P_1^{1/n} V_1 [P_2^{(n-1)/n} - P_1^{(n-1)/n}] \tag{70}$$

Multiplying the numerator and denominator on the right-hand side by $P_1^{(n-1)}$ we obtain:

$$\tilde{\ell} = \frac{n}{n-1} P_1 V_1 \left[\left(\frac{P_2}{P_1}\right)^{(n-1)/n} - 1\right] \tag{71}$$

Gill [31] compares power requirements of a compressor for different values of n. In the analysis, 10 ft^3 of a gas (air) is compressed from 15 to 100 psia. The predictions of the last expression are shown plotted in Figure 95, where the work of isothermal compression is represented by area ABGHA. The adiabatic value for air occurs at n = 1.4, and, hence, the work area is AFGHA. Comparing these two cases, then, a net savings in work denoted by area AFBA is achieved by isothermal compression over adiabatic compression. In practice, few compressors ever operate adiabatically due to radiation losses, condensation, etc. At the same time, it is impossible to maintain constant temperature, and hence, isothermal compression is also rare.

Figure 95 shows that the power coefficient becomes smaller as the compression cycle approaches the isothermal case. To approximate the isothermal state as closely as possible, compressor cylinders are waterjacketed and intercoolers are provided between stages in multistage compression.

Piston Compressors

The capacity of a piston compressor is defined as the actual amount of gas volume, V, handled by the machine per unit time at standard conditions. *The actual capacity of a gas compressor* is defined as the quantity of gas compressed and delivered, expressed in volume units per unit time at conditions of total temperature, total pressure, and composition prevailing at the compressor inlet.

Capacity always is expressed in terms of the gas intake conditions, rather than in terms of STP.

$$V = \lambda_v Q_T = \lambda_v V_p \tag{72}$$

where λ_v = delivery coefficient

For multicylinder compressors, the capacity computed from this expression should be multiplied by the number of cylinders. The capacity of multistage compressors is determined by the capacity of the first stage. In this case, V_p is the piston displacement volume of the first stage.

The delivery coefficient, λ_v, is the ratio of gas volume discharged in the delivery piping (but reduced to the suction conditions) to the displacement, V_p. The delivery coefficient accounts for all capacity losses, including those not shown in the indicator diagram. The losses detected on the diagram are those related to the gas expansion from the clearance and are denoted by coefficient λ_0. Additional losses are due to the loss of capacity as a result of gas leakage through piston rings, valves, stuffing boxes, and gas expansion, when gas comes in contact with hot cylinder walls and mixes with hot gases in the clearance spaces. These losses are considered by a hermetic coefficient, λ_h, and a thermal coefficient, λ_T, correspondingly. Therefore, the delivery coefficient may be defined as a product of three coefficients:

$$\lambda_v = \lambda_0 \lambda_h \lambda_T \tag{73}$$

For modern compressors, the values of these coefficients are $\lambda_h = 0.95$–0.98 and $\lambda_T = 0.9$–0.98. As noted earlier, the volume of gas taken in by the compressor, V_s, is less than the working vclume of the cylinder, V_p. The ratio of the compressor capacity to its displacement is called the volumetric coefficient:

$$\lambda_0 = \frac{V_s}{V_p} \tag{74}$$

We denote x as the ratio of the difference between total cylinder volume ($V_0 = V_p + \epsilon' V_p$) and the actual gas volume taken in ($V_s = \lambda_0 V_p$) to the piston displacement, V_p:

$$x = \frac{V_0 - V_s}{V_p} = \frac{V_p + \epsilon' V_p - \lambda_0 V_p}{V_p} = 1 + \epsilon' - \lambda_0 \tag{75}$$

where ϵ' is the clearance, defined as the volume remaining in the cylinder at the extreme position of the piston divided by the displacement of the cylinder. Hence, the relation between volumetric coefficient, λ_0, and clearance, ϵ', for a given x is

$$\lambda_0 = 1 + \epsilon' - x \tag{76}$$

Expansion of gas in the clearance space can be assumed to be polytropic, with an exponent, m_p, somewhat less than the exponent of the compression polytrope, m (for two-atomic gases, m is equal to 1.2). Therefore,

$$p_2(\epsilon' V_p)^{m_p} = p_1(V_0 - V_s)^{m_p} = p_1(xV_p)^{m_p} \tag{77}$$

where $xV_p = V_0 - V_s$ is the gas volume after expansion from pressure p_2 to p_1 in the clearance space (Figure 93). From this equation

$$x = \epsilon' \left(\frac{P_2}{P_1}\right)^{1/m_p} \tag{78}$$

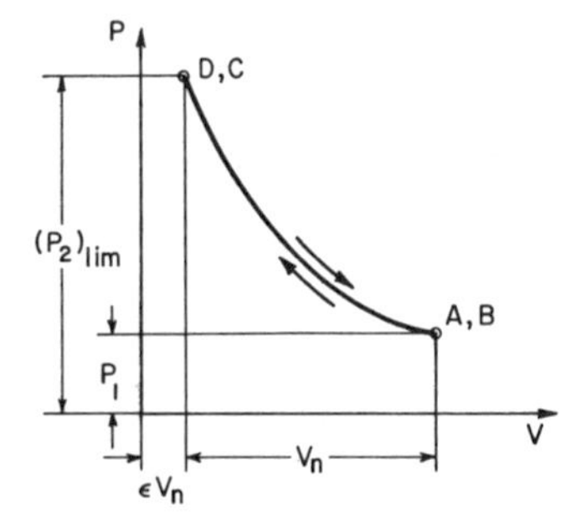

Figure 96. Indicator diagram of compressor operation at the compression limit.

and

$$\lambda_0 = 1 - \epsilon'\left[\left(\frac{P_2}{P_1}\right)^{1/m_p} - 1\right] \tag{79}$$

Thus, the value of volumetric coefficient, λ_0 of the compressor depends on the value of relative clearance, ϵ', the pressure ratio p_2/p_1 and the exponent of polytrope m_p of gas expansion. The capacity of a compressor will be higher at lower pressure ratios and clearances and with a greater exponent of expansion in this space. This last formula shows that the volumetric coefficient, λ_0, decreases with an increased compression ratio and, at a certain value, may approach zero. The compression ratio $p_2/p_1 = 0$ at a volumetric coefficient of zero is referred to as the *compression limit*.

At the limiting compression ratio the expanding as in the clearance space occupies the total volume of the cylinder. The gas suction is terminated and the compressor capacity becomes equal to zero. On the indicator diagram shown in Figure 96, the curves of compression and expansion merge into a single line. Hence, the area of the indicator diagram and, consequently, the indicator horsepower at the compression limit, are zero.

The limit of pressure at the polytropic gas expansion in the clearance is determined from the equation with $\lambda_0 = 0$:

$$1 - \epsilon'\left[\left(\frac{P_2}{P_1}\right)^{1/m_p} - 1\right] = 0$$

or

$$\left(\frac{P_2}{P_1}\right)_{\lambda_0 = 0} = \left(\frac{1}{\epsilon'} + 1\right)^{m_p} \tag{80}$$

The lower compression ratio is assumed to be the limiting case. It may be assumed that compressors with volumetric coefficients less than 0.7 are uneconomical. The corresponding *volumetric limit* of the pressure ratio (p_2/p_1) is obtained from the following equation:

$$1 - \epsilon'\left[\left(\frac{P_2}{P_1}\right)^{1/m_p} - 1\right] = 0.7$$

or

$$\left(\frac{P_2}{P_1}\right)_0 = \left(\frac{0.3}{\epsilon'} + 1\right)^{m_p} \tag{81}$$

The gas temperature after compression in a single-stage compressor should not exceed 150–160°C. At higher temperatures lubrication is difficult due to carbonization of the oil. Furthermore, there is a risk of causing oil mist explosions in the cylinders when gases containing oxygen are being compressed. An expression for the limiting compression ratio due to the gas temperature rise during compression may be obtained by eliminating V between the following equations:

$$\frac{P_1 V_1}{T_1} = \frac{P_2 V_2}{T_2} = K \tag{82}$$

and

$$P_1 V_1^n = P_2 V_2^n = K \tag{83}$$

For adiabatic compression,

$$\left(\frac{P_2}{P_1}\right)_T = \left(\frac{T_2}{T_1}\right)^{n/(n-1)} \tag{84}$$

There are some compressor designs in which the rate of compression is higher than that determined from this expression. In this case, the lubrication is carried out by water injection into the cylinder, which also provides for partial cooling of the gas. In practice, the compression ratio does not exceed a value of 5 (if the cylinder is cooled), with the exception of small compressors, in which the compression ratio may be increased up to 8.

Theoretical horsepower, N_T (wt), is defined as the horsepower required to compress the gas delivered by the compressor through the specified range of pressures. For a multistage compressor (with intercooling between stages), the *theoretical horsepower* assumes equal work in each stage and perfect cooling between stages. Theoretical horsepower may be calculated as a product of compressor capacity, $V\rho$ (kg/sec), and the specific work of compression, $\hat{\ell}$ (J/kg), as estimated from the preceding formulas, and

$$N_T = V\rho\hat{\ell} \tag{85}$$

where V = volumetric capacity of compressor (m^3/sec)

$\rho = \frac{1}{v}$ = gas density (kg/m^3)

If the volumetric capacity of the compressor and the gas density are reduced to conditions at suction, i.e., if $V = V_1$ and $\rho = \rho_1 = 1/v_1$, then

$$N_{T,iso} = P_1 V_1 \ln\left(\frac{P_2}{P_1}\right) \tag{86}$$

$$N_{T,ad} = \frac{\kappa}{\kappa - 1} P_1 V_1 \left[\left(\frac{P_2}{P_1}\right)^{(\kappa-1)/\kappa} - 1\right] \tag{87}$$

$$N_{T,pol} = \frac{m}{m - 1} P_1 V_1 \left[\left(\frac{P_2}{P_1}\right)^{(m-1)/m} - 1\right] \tag{88}$$

Compressor efficiency cannot be evaluated from mechanical efficiency, which is the ratio of the horsepower imparted to the gas and the brake horsepower. Such an evaluation would suggest that the lowest efficiencies require machines with intensive water cooling, as a considerable fraction of the energy of compression is absorbed in the form of heat by cooling water. However, it is well known that the desired increase of gas pressure is simply achieved with a minimum of energy spent in machines in which intensive cooling is provided. Therefore, the efficiencies of compression machines should be based on the *relative thermodynamic efficiency*. In this manner, a comparison can be made between a given machine and the most economical one of the same class.

Machines with water cooling are compared with a conditional machine that compresses gas isothermally. The ratio of the horsepower of an "isothermic" machine, $N_{T, iso}$, to the actual horsepower, N, of a machine equipped with heat removal is called the isothermal efficiency:

$$\eta_{iso} = \frac{N_{T, iso}}{N} \tag{89}$$

Compressors operating without heat removal generate additional heat by friction between moving parts and gas, hydraulic resistances, etc. Gas compression in such machines is effected *polytropically* when heat is transferred from the gas during the compression and the exponent of the polytropic curve is $m > K$. To estimate the thermodynamic efficiencies of such machines, the conditional machine may be assumed to compress the gas adiabatically or *isentropically*.

An "isentropic" machine is the most economical system from this class of machines that work without heat removal. The ratio of compression horsepower of an isentropic machine, $N_{T, ad}$, to the horsepower, N, of a given compressor without heat removal is called the *isentropic* (*adiabatic*) efficiency:

$$\eta_{ad} = \frac{N_{T, ad}}{N} \tag{90}$$

Brake horsepower or shaft horsepower, N_e, is measured horsepower input to the compressor. It is equal to the *theoretical horsepower* defined as horsepower required to compress gas divided by *mechanical efficiency*, which characterizes horsepower losses for mechanical friction in the compressor:

$$N_e = \frac{N}{\eta_{mech}} \tag{91}$$

$$N_e = \frac{\eta_{T, iso}}{\eta_{iso} \times \eta_{mech}} \tag{92}$$

The product of isothermic efficiency and mechanical efficiency is referred to as the total isothermal compressor efficiency:

$$\eta_{T, iso} = \eta_{iso} \times \eta_{mech} \tag{93}$$

The horsepower of an electric motor, N_{em}, is greater than the compressor brake horsepower by the value of the horsepower losses in the transmission and the motor itself. These losses are estimated by the *efficiency of transmission*, η_{tr}, and *efficiency of the electric motor*, η_{em}:

$$N_{em} = \frac{N_e}{\eta_{tr} \times \eta_{em}} \tag{94}$$

The actual horsepower of an electric motor, N_{act}, usually is increased 10%–15%, i.e.,

$$N_{act} = (1.1 \text{ to } 1.15)N_{em}$$

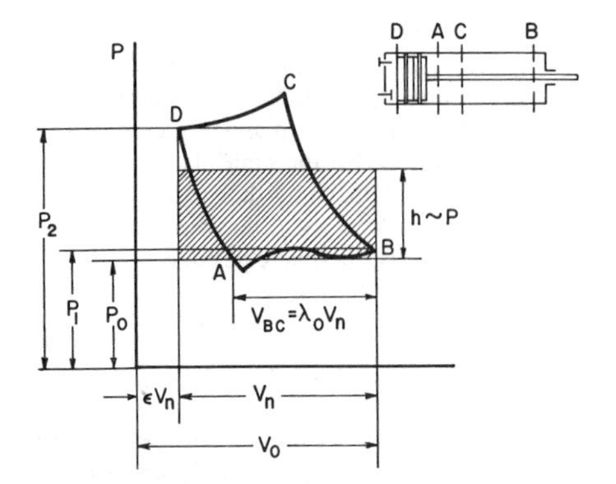

Figure 97. Indicator diagram showing operation of a single-stage piston compressor.

The value of an adiabatic efficiency, η_{ad}, is very close to unity (typically in the range 0.93 to 0.97). Isothermal efficiency is in the range of 0.64 to 0.78, and mechanical efficiency varies from 0.85 to 0.95.

The operational control of piston compressors is illustrated through by the indicator diagram in Figure 97. The diagram summarizes the relationship between pressure and gas volume at intake and delivery for one double piston stroke (i.e., for one rotation of the crankshaft). Point D corresponds to the extreme left piston positon, which, in reality, never approaches the cylinder head.

In tracing the piston's movement through the cylinder, assume that the starting point for the cycle is represented by point D in Figure 97. The gas remaining in the cylinder in the clearance space now begins to expand. This step is represented by curve DA and is accompanied by an increase in volume and a decrease in the gas pressure. This process continues until pressure p_0 in the cylinder becomes somewhat less than pressure p_1 in the suction line. Because of this pressure difference $(p_1 - p_0)$ the suction valve opens and gas enters into the compressor at point A. Gas suction thus continues from point A to B on the diagram. The volume of gas being sucked in, V_s, is proportional to the line AB and may be expressed in terms of fractions of the cylinder working volume, $V_s = \lambda_0 V_p$, where λ_0 is the volumetric efficiency.

When the piston moves to the left suction ceases, and the gas compresses polytropically along curve BC to a pressure somewhat higher than delivery pressure, p_2. On reaching point C, the delivery valve opens. The discharge follows the path of curve CD and is proportional to the delivery volume. When the piston reaches point C, it has compressed the gas to the pressure in the delivery line and must then push the compressed air out through delivery valves into the receiver. Due to the weight of the delivery valves and the tension of the springs holding them in their seats, the cylinder pressure in an actual system rises to a value slightly above that in the receiver just before the valves open. This explains why point C is slightly higher than point D. This pressure gradually drops to the receiver pressure at the end of the stroke.

When the indicator line follows the suction line, the inlet of the compressor is not restricted. Hence, the compressor receives a volume of gas at the suction pressure, represented by the travel of the piston from A to B. This then represents the actual suction capacity of the compressor.

The area enclosed by the indicator diagram is proportional to the useful work performed by the machine in compressing as over a single shaft rotation per unit of piston area. This work corresponds to the *indicated horsepower* of the unit, which may be evaluated from the diagram through the following relation:

$$N_{ind} = \frac{F \times n \times f_{in}}{Sc_{sp}} \tag{95}$$

where F = piston cross-sectional area
n = number of piston double strokes or rotations of the crankshaft
f_{in} = area of indicator diagram
Sc_{sp} = scale of indicator spring

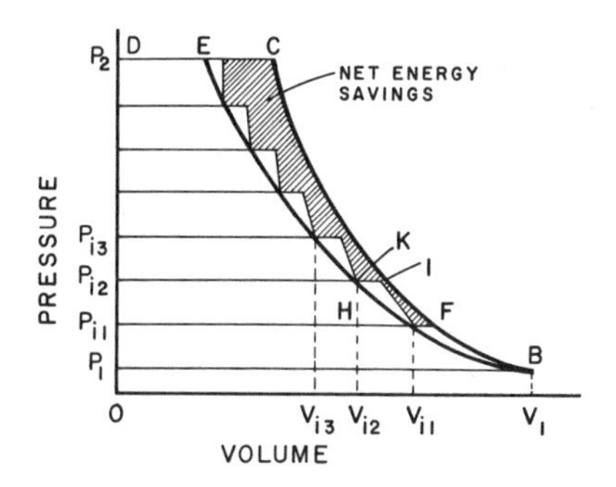

Figure 98. Shows operation of a multistage compression cycle with interstage cooling.

The scale of the indicator spring represents the number of pressure units corresponding to a unit length along the ordinate. Multiplying and dividing this expression by the length of the stroke(s), we obtain

$$N_{ind} = \frac{FSnf_{ind}}{S \times S_{sp}} = V_p p_{ind} \tag{96}$$

where V_p is FSn, i.e., the piston displacement for n rotations, $p_{ind} = f_{in}/(S \times S_{sp})$ is the ratio of the diagram area to the piston stroke, i.e., *the average indicator pressure.*

The average indicator pressure may be computed (based on an appropriate scale) as a height, h, of an equivalent rectangle constructed on the diagram, with the base equal to the diagram length because the volume of piston travel, V_p, for a given compressor at constant F and n is proportional to the length of piston stroke, S. This rectangular area is the shaded region shown in Figure 97. The indicator horsepower for a double-acting compressor is determined for each side of the cylinder separately, and the values obtained then are summed.

Multistage compression is used for preparing gases at high pressures. If the required pressure ratio p_2/p_1 exceeds a value of 4 or 5, it is not practical to carry out the entire compression in a single cylinder (the temperature rise of the compressed gas will become excessive and thus adversely affect efficiency). As noted earlier, to minimize this lower efficiency, it is common practice to divide the compression into several stages, with intercooling between stages. In a multistage compressor, gas passes through a number of cylinders of gradually decreasing volume. This arrangement more closely approaches isothermal conditions, resulting in a considerable savings of power.

An indicator diagram again can be constructed to understand the operation of a multistage compressor (Figure 98). The area ABCD denotes the work performed by compressing isentropically from pressure p_1 to p_2 in a single stage. ABED denotes the work required to perform an isothermal compression. We have, of course, ignored the effect of clearance volume. If a multistage isentropic compression is performed, then the intermediate pressures between stages are p_{i1}, p_{i2}, p_{i3}, etc. To simplify our discussion, we shall assume that the gas can be cooled in an interstage cooler to its first stage intake temperature prior to entering each cylinder.

The compression cycle in Figure 98 is as follows. The piston suction stroke of the first stage is represented by line AB, where gas volume V_1 is admitted at pressure p_1. The gas then is isentropically compressed to pressure p_{i1} (line BF) and then delivered from the first stage at this pressure (line FG). The suction stroke of the second stage is denoted by line GH, where the volume of the gas is reduced in the interstage cooler to V_{i1}. Note that V_{i1} is also the volume of the gas that would have been achieved as a result of an isothermal compression to pressure p_{i1}. In the second stage, the gas again undergoes isentropic compression, but from a pressure p_{i1} to p_{i2} (line HI). Line IJ traces the delivery stroke of the second stage. We then return to the isothermal compression curve (line BE) at point K (i.e., line JK denotes the suction stroke of the third stage).

The overall work performed on the gas is the intermediate area between that for a single-stage isothermal compression and that for an isentropic compression. That is, the net energy savings is denoted by the shaded region in Figure 98.

The total work performed per cycle for an isentropic compression is [32]:

$$W' = P_1V_1\frac{\kappa}{\kappa-1}\left[\left(\frac{P_{i1}}{P_1}\right)^{(\kappa-1)/\kappa} - 1\right] + P_{i1}V_{i1}\frac{\kappa}{\kappa-1}\left[\left(\frac{P_{i2}}{P_{i1}}\right)^{(\kappa-1)/\kappa} - 1\right] + \cdots \tag{97}$$

For perfect interstage cooling, we note from the ideal gas law that:

$$P_1V_1 = P_{i1}V_{i1} = P_{i2}V_{i2} = \cdots$$

Hence, for an n-stage compressor the work performed is:

$$W' = P_1V_1\frac{\kappa}{\kappa-1}\left[\left(\frac{P_{i1}}{P_1}\right)^{(\kappa-1)/\kappa} + \left(\frac{P_{i2}}{P_{i1}}\right)^{(\kappa-1)/\kappa} + \cdots - n\right] \tag{98}$$

The work performed by the compressor will be at a minimum when the partial derivative of W′ with respect to the interstage pressures is zero, i.e., when

$$\frac{\partial W'}{\partial P_{i1}} = \frac{\partial W'}{\partial P_{i2}} = \frac{\partial W'}{\partial P_{i3}} = \cdots = 0$$

Hence, the minimum work for the first stage is:

$$P_1V_1\frac{\kappa}{\kappa-1}\left[\frac{\kappa-1}{\kappa}\left(\frac{P_{i1}}{P_1}\right)^{(\kappa-1)/\kappa} P_{i1}^{-1} + \frac{1-\kappa}{\kappa}\left(\frac{P_{i2}}{P_{i1}}\right)^{(\kappa-1)/\kappa} P_{i1} - 1\right] = 0 \tag{99a}$$

or, on simplification

$$\frac{P_{i1}}{P_1} = \frac{P_{i2}}{P_{i1}} \tag{99b}$$

This expression defines the optimum pressure, p_{i1}, for gas delivery from the first stage. In a similar manner, the optimum value for p_{12} can be obtained. The net result is that the intermediate pressures should be arranged so that the compression ratio is the same in each cylinder and, hence, the work done in each cylinder is the same.

The minimum work of compression in an n-stage compressor becomes

$$P_1V_1\frac{\kappa}{\kappa-1}\left[n\left(\frac{P_2}{P_1}\right)^{(\kappa-1)/\kappa} - n\right] = nP_1V_1\frac{\kappa}{\kappa-1}\left[\left(\frac{P_2}{P_1}\right)^{(\kappa-1)/\kappa} - 1\right] \tag{100}$$

Defining the clearance volumes in the successive cylinders as $C_1, C_2, C_3, \ldots,$ the theoretical volumetric efficiency for the first stage is:

$$V_{cyl} = 1 + C_1 - C_1\left(\frac{P_2}{P_1}\right)^{1/n\kappa} \tag{101}$$

The volume of gas admitted to the first cylinder is:

$$V_{g1} = V_{s1}\left[1 + C_1 - C_1\left(\frac{P_2}{P_1}\right)^{1/n\kappa}\right] \tag{102}$$

The ratio of the volumes of gas admitted to successive cylinders is $(P_1/P_2)^{1/n}$. This assumes that the same mass of gas is sent through each cylinder and that the interstage coolers are perfectly efficient. The volume of gas admitted to the second cylinder is:

$$V_2 = V_{s2}\left[1 + C_2 - C_2\left(\frac{P_2}{P_1}\right)^{1/n\kappa}\right] = V_{s1}\left[1 + C_1 - C_1\left(\frac{P_2}{P_1}\right)^{1/n\kappa}\right]\left(\frac{P_1}{P_2}\right)^{1/n}$$

or

$$\frac{V_{s1}}{V_{s2}} = \frac{1 + C_2 - C_1\left(\frac{P_2}{P_1}\right)^{1/n\kappa}}{1 + C_1 - C_1\left(\frac{P_2}{P_1}\right)^{1/n\kappa}}\left(\frac{P_2}{P_1}\right)^{1/n} \tag{103}$$

The swept volume of gas through each cylinder thus can be calculated from information on V_{s1} and the clearance volume $C_1, C_2, \ldots$.

The preceding equations are based on ideal gas behavior. When dealing with nonideal gases, the change caused by compression can be examined on a temperature-entropy or enthalpy-entropy diagram. Intermediate pressures ($p_{i1}, p_{i2}, p_{i3}, \ldots$) can be chosen on the basis of the same enthalpy change in each cylinder.

For two-stage compression, temperature increase may be determined for each stage separately, taking into account that the temperature depends solely on the compression ratio and the value of exponent n. Because the compression ratio of each stage is the square root of the overall ratio, then

$$T_2 = T_1\sqrt{R^{(\kappa-1)/2\kappa}} = T_1 R_t^{(\kappa-1)/2\kappa} \tag{104}$$

This expression is valid in the case of perfect intercooling, i.e., the gas has to be cooled between stages so that the intake to the high-pressure cylinder is at the same temperature as at the low-pressure intake.

Cylinder sizes may be determined from Boyle's law:

$$\frac{P_2}{P_1} = \frac{V_1}{V_2} = \frac{V_{P_1}\lambda_{V_1}}{V_{P_2}\lambda_{V_2}} \tag{105}$$

where V_P = the displacement of the cylinder
λ_V = the volumetric coefficient

Assuming the same displacement coefficient in both cylinders and taking into account that the displacements are directly proportional to the squares of the cylinder diameters, d, and that the compression ratio in the first stage, p_2/p_1, is equal to the square root of the overall compression ratio, we obtain

$$R_1 = \frac{V_{P_1}}{V_{P_2}} = \frac{d_1^2}{d_2^2} \tag{106}$$

Using the overall compression ratio,

$$\sqrt[4]{R_t} = \frac{d_1}{d_2} \tag{107}$$

This equation assumes that the stroke of both cylinders is the same, hence

$$R_1 = \left(\frac{d_1}{d_2}\right)^2\left(\frac{\lambda_{V_1}}{\lambda_{V_2}}\right) \tag{108}$$

where R_1 is the compression ratio of the low-pressure cylinder.

Substituting the expression for the volumetric coefficient this expression gives

$$R_1 = \left(\frac{d_1}{d_2}\right)^2 \frac{1 - \epsilon_1[R_1^{1/m_P} - 1]}{1 - \epsilon_2[R_2^{1/m_P} - 1]} \quad (109)$$

This equation may be solved by trial and error. See Cheremisinoff et al. [33] for illustrative examples.

Two-stage energy calculations usually are based on polytropic conditions because there would be little reason for two-staging if the isothermal condition were achievable. Two-stage compression may be assumed as two cylinders working separately on different ratios of compression. Hence, the energy required is the sum of the horsepower for the two cylinders added together. For two-stage compression,

$$\hat{\ell} = \frac{n}{n-1} P_1 V_1 [R^{(n-1)/n} - 1] \quad (110)$$

Defining the intake gas in terms of standard cubic feet at 14.73 psia, then

$$\hat{\ell} = 2{,}121 \frac{n}{n-1} V_1 [R^{(n-1)/n} - 1], \text{ ft-lb} \quad (111)$$

In terms of horsepower, V_1 is taken in standard cubic feet of gas handled per minute and divided by 33,000 ft-lb/min (33,000 ft-lb/min is the equivalent of 1 hp).

$$hp = \frac{n}{n-1}\left(\frac{2{,}121}{33{,}000}\right) V_1 [R^{(n-1)/n} - 1]$$

or

$$hp = \frac{0.0643 n V_1}{n-1} [R_1^{(n-1)/n} + R_2^{(n-1)/n} - 2] \quad (112)$$

where V_1 is in cfm.

When the work is divided equally between two states, $R_1 = R_2 = \sqrt{R_t}$,

$$hp = \frac{0.1286 n V_1}{n-1} [R_t^{(n-1)/2n} - 1] \quad (113)$$

The preceding equations apply to double-acting compressors. Displacements of the two cylinders in two-stage compression are evaluated separately. Both capacity and the overall volumetric coefficient are based entirely on the low-pressure cylinder.

If the overall compression ratio exceeds 10, it is often recommended that another stage beyond two-stage compression be introduced to reduce the temperature rise.

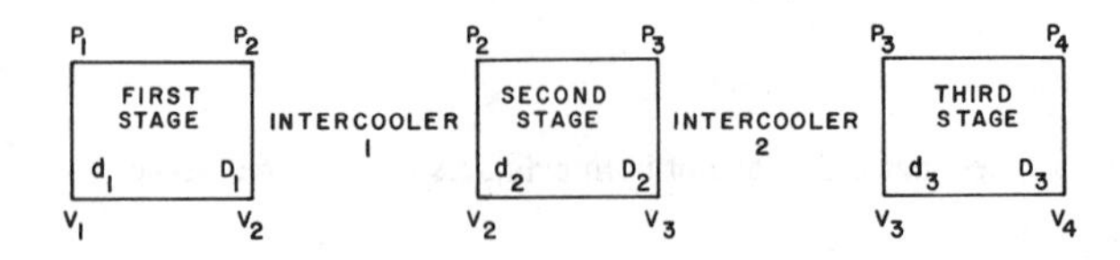

Figure 99. Operating principle for three-stage compression.

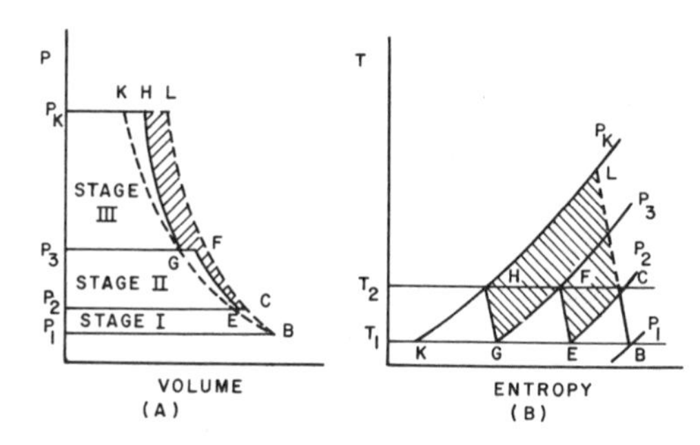

Figure 100. Indicator (A) and entropy (B) diagrams for multistage gas compression.

For equal work in each stage of a three-stage compression,

$$R_1 = R_2 = R_3 = \sqrt[3]{R_t} \tag{114}$$

This equation is valid only for equal clearance in all cylinders and, consequently, equal volumetric coefficients. This is illustrated in Figure 99. Substituting for the pressures corresponding to R_1, R_2, etc., the pressure in the first intercooler becomes

$$P_2 = P_1\sqrt[3]{P_4/P_1}$$

or

$$P_2 = P_1\sqrt[3]{R_t} \tag{115}$$

In a similar manner, the pressure in the second intercooler is

$$P_3 = P_1\sqrt[3]{R_t^2} \tag{116}$$

The temperature increase in a three-stage compression is

$$T_2 = T_1 R_t^{(n-1)/3n} \tag{117}$$

where T_1 and T_2 are the initial and final absolute temperatures in each stage.

The power required in a three-stage compression is

$$hp = \frac{0.1929nV_1}{n-1}\left[R_t^{(n-1)/3n} - 1\right] \tag{118}$$

The indicator diagram for a three-stage compression (with equal compression ratios in each stage) is illustrated in Figure 100. In constructing these diagrams it is assumed that the gas is cooled to its initial temperature in the intercoolers and that the clearances for each stage, along with pressure losses, are zero. Line BC represents compression in the first stage from pressure P_1 to pressure P_2. The gas is then cooled along the line CE (isobar) to initial temperature, T_1. On entering cylinder II, it is compressed to pressure P_3 along the line EF, etc. The process of three-stage compression from pressure P_1 to pressure P_k, with cooling, is expressed by a broken line BCEFGHK.

The line BEGK corresponds to isothermal compression to the pressure P_k in a single-stage compressor, and line BCL represents polytropic compression in the same compressor. From the diagrams, one can see that multistage compression with interstage gas cooling is closer to the isothermal process and, consequently, demands less energy than a single-stage process over the same pressure limits.

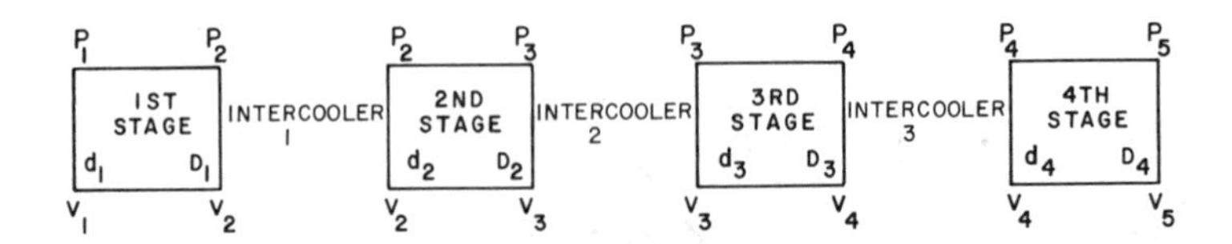

Figure 101. Illustrates four-stage compression.

The area confined inside the indicator diagram and the area under the broken line BCEFGHK on the T-S diagram represent the work of multistage compression (based on an appropriate scale). The shaded area denotes the savings of power over a single-stage. The closer the broken line of multistage compression is to the isotherm, the more stages that are required. However, the economical number of stages generally is limited to five or six. As the number of stages increases, so do capital and operating costs.

For extremely high compression ratios, *four-stage compression* often is used (Figure 101). In the case of equal work at each stage,

$$R_1 = R_2 = R_3 = R_4 = \sqrt[4]{R_t} \tag{119}$$

By substituting in pressures for the compression ratios, an expression for the pressure in each intercooler can be obtained.

For the first intercooler,

$$P_2 = \sqrt[4]{P_1^3 P_5} = P_1 \sqrt[4]{R_t} \tag{120}$$

For pressure in the second intercooler,

$$P_3 = \sqrt{P_1 P_5} = P_1 \sqrt{R_t} \tag{121}$$

And for the pressure in the third intercooler,

$$P_4 = \sqrt[4]{P_1 P_5^3} = P_1 R_t^{3/4} \tag{122}$$

The increase in temperature for four-stage compression is,

$$T_2 = T_1 R_t^{(n-1)/4n} \tag{123}$$

The power required for four-stage compression is

$$hp = \frac{0.2572 n V_1}{n-1}\left[R_t^{(n-1)/4n} - 1\right] \tag{124}$$

where V_1 is in cfm.

The theoretical work of compression becomes minimum when the compression ratios, R, are identical in each stage, i.e.,

$$R = Z\sqrt{P_k/P_1} \tag{125}$$

where Z = the number of stages
P_k = the final pressure
P_1 = the initial pressure

In practice, because of pressure losses between stages (in valves, intercoolers, etc.) the pressure ratio in each stage is somewhat higher than the theoretical prediction. To account for these losses, an empirical correction factor is used

$$R = \psi'\left(\frac{P_k}{P_1}\right)^{1/2} \tag{126}$$

where $\psi' = 1.1\text{–}1.15$ is a coefficient accounting for pressure losses between stages.

The number of compression stages required may be determined from the following equation:

$$Z = \frac{\log P_k - \log P_1}{\log R - \log \psi} \tag{127}$$

To maintain the final gas temperature within allowable limits, a pressure ratio in the range of 2.5 to 3.5 may be assumed. This range generally provides effective use of the cylinder volume, increases the compressor volumetric coefficient, and decreases the energy consumption.

Assuming equal compression ratios in each stage and ideal gas cooling in the intercoolers, the work at each stage will be the same. Then the theoretical work of a multistage compressor in adiabatic compression of 1 kg of gas is

$$\hat{\ell}_{ad} = Z\frac{\kappa}{\kappa - 1}P_1 v_1\left[\left(\frac{P_k}{P_1}\right)^{(\kappa-1)/Z\kappa} - 1\right] \tag{128}$$

The limiting temperature at the end of compression is

$$T_k = T_1\left(\frac{P_k}{P_1}\right)^{(\kappa-1)/Z\kappa} \tag{129}$$

The theoretical volumetric coefficient is

$$\lambda_0 = 1 - \epsilon\left[\left(\frac{P_k}{P_1}\right)^{1/Z\kappa} - 1\right] \tag{130}$$

The theoretical compression work, the temperature at the end of compression, and the volumetric coefficient at polytropic compression may be determined from the preceding equations by changing the exponent from an adiabatic to a polytropic process.

In practice, the work performed by each stage is not always equal because the polytropic exponents and the clearance spaces for each stage differ. In addition, the cooling in intercoolers is neither ideal nor exactly the same.

The major components of typical double-acting compressors are:

1. Frame
2. Cylinder
3. Piston and piston rings
4. Crankshaft
5. Valves
6. Connecting rod
7. Oil reservoir
8. Main bearings
9. Crosshead
10. Piston rod
11. Piston rod packing
12. Cylinder cooking jackets

13. Distance piece
14. Oil scraper rings
15. Oil deflection collar
16. Wrist pin

Some general comments on each follows.

Liners—A cylinder liner is used where changing the diameter of the cylinder is anticipated at some time in the future. The bore diameter may be changed either to meet new capacity requirements or to produce a new refinished cylinder surface after the original surface has been damaged by wear. Liners may also be required when the material used for the bulk of the cylinder will not provide proper wearing properties at the friction interface. This is the case in most steel cylinders.

Cylinder liners are commonly made of cast iron, although special wear resistant or chemically resistant materials may be ordered.

There are two types of liners in general use. The most common is the dry type, which is essentially a shrunk-fit or pressed sleeve within the original bore. These have the disadvantage of reducing the rate of heat rejection to the jacket water. The alternative wet-type liners are designed for jacket water circulation immediately behind the liner. These have a sealed seam between the compression chamber and the jacket which may leak under some extreme circumstances.

Piston rods are subject to repeated compression-tension cycle loading. The surface finish of these rods should be as smooth as possible to avoid fatigue cracking. Heat treating, nitrating, or carburizing is sometimes used to harden the surface of this component.

Valves are the most frequently serviced of all compressor components. Valve designs must strike a compromise between maximizing operating cycles and minimizing valve losses. Maximum operating cycles are attained by valves with minimum lift, and few moving parts. Minimum valve losses (i.e., minimum pressure losses) are attained by valves with high lifts and/or several seating components.

An evaluation of compressor offerings should include a comparison of valve velocities. This parameter is found by dividing the total piston displacement of a cylinder by the total life area of all suction valves. In general, this figure should not exceed 7,500 ft/min. However, very low molecular weight gases, hydrogen in particular, may not induce an adequate valve lift unless velocities are relatively high.

Suction and discharge valves are not interchangeable in function although they may sometimes be interchangeable in port seating dimensions. This is a particular hazard in older units. Care must be exercised by servicemen when replacing inlet and outlet valves, to avoid interchanging the two. These valves may be polarized if desired, by machining or inserting pins to prevent interchanges.

Piston rod packing in heavy duty compressors is almost invariably the full floating mechanical type. The most frequently used ring material is bronze, although micarta, phenolic resins, PTFE, and other materials or combinations of materials may be used for corrosion resistance (refer to Figure 102).

Rod packing will wear out after long running periods even in the cleanest of services. Dirty, wet, or high pressure services will require packing ring replacement much more often.

Distance piece—A distance piece may be installed between the cylinder and frame for one of three reasons.

1. When it is necessary to prevent carryover of frame lubricants into the cylinder, the distance piece assures that no portion of the rod can travel the distance from the frame oil wiper rings to the cylinder packing.
2. In other cases, the distance piece serves as a means of venting the process gas that leaks past the rod packing. Gases may be simply vented to the atmosphere through large ports, or they may be purged from an enclosed distance piece with a stream of inert gas. The latter method is used when hazardous gases are being processed.
3. The distance piece also serves as a service access to the piston rod packing and oil wiper rings.

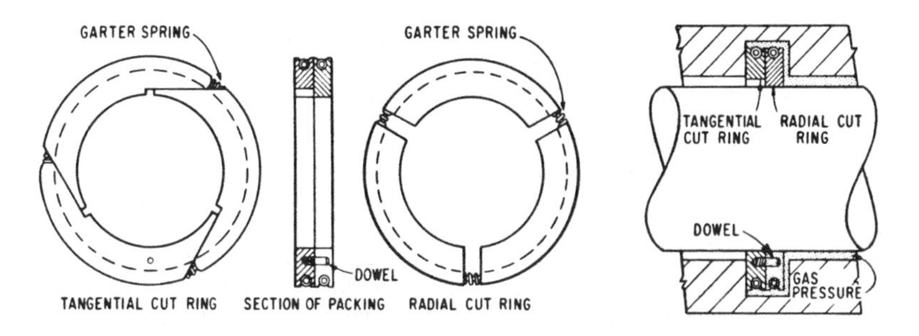

Figure 102. Shows single pair of full floating mechanical packing rings.

Reducing the volumetric capacity of a reciprocating compressor can be accomplished by one of the following means:

- *Start and stop* whereby the pressurized gas is delivered to a storage receiver. Caution must be exercised when sizing a compressor for this type of service to assure that the compressor motor is not subjected to too many starts within a limited time period. Starting is most commonly done while the compressor is unloaded in order to limit the starting torque.
- *Constant speed controls* which include cylinder inlet valve unloading mechanisms, special clearance pockets, and external bypassing.
- *Valve unloaders*, the most commonly used capacity reduction devices, hold the inlet valves open during the compression stroke of the cylinder, thereby preventing compression from occurring. In a double-acting cylinder, unloaders allow capacity reduction steps of 50% and 0% of full acfm. Multi-cylinder units allow still more combinations of loaded and unloaded cylinders (refer to Figure 103).

Care must be exercised in the planning stage when a unit will have valve unloaders in a particular service. A check should be made to guarantee that rod load reversals (compression followed by tension loading) will occur when a cylinder is partially unloaded. In addition, multi-cylinder horizontal opposed units should be unloaded symmetrically to avoid producing excesive unbalanced forces on the crankshaft and frame.

The heat of compression and the heat developed by mechanical friction within the cylinders, can be detrimental to the operation of a compressor. High temperature within cylinder walls can cause lubricant failure and the formation of solid deposits. Excessive and uneven heating can change the dimensions of components resulting in faster wear of valves, rings, and packings. Cylinder castings may even fracture if subjected to uneven heating.

From a thermodynamic viewpoint, heated cylinder surfaces will contribute to a higher horsepower demand and fewer lb/hr of gas delivered. There are two chief cylinder designs used to dissipate this

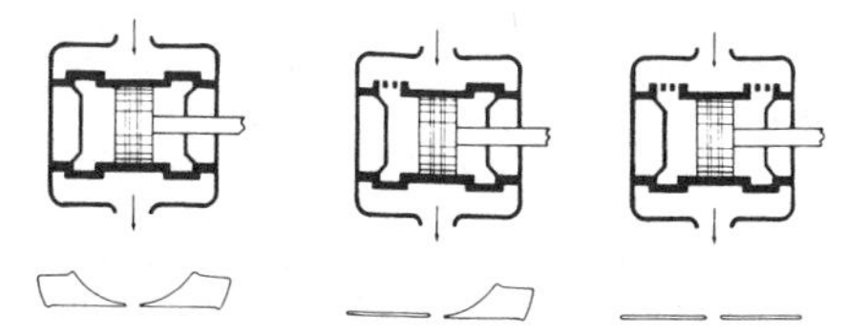

Figure 103. Shows capacity control via valve unloaders. Cylinder on left is loaded on both the head and crank ends (shown by indicator cards). Middle shows head end unloaded for 50% capacity. Cylinder on right has both ends unloaded for 0% capacity.

heat: Air cooled machines have external fins which act to extend the surface area of cylinders in order to transfer heat to the surrounding atmosphere. In most cases, the convention heat-dissipation rate is increased by incorporating an integral fan into the design. Air-cooled cylinder compressors are nearly always small units (less than 100 total hp). They are very common in air package units for limited demand service.

Water-cooled cylinders are the common choice for heavy-duty applications. Cylinder castings include channels for circulating a cooling fluid to maintain a uniform working temperature within the metal walls of the cylinder.

Although the heat dissipation to the cylinder cooling fluid does have a measurable effect on the cylinder discharge temperature as well as the compression efficiency, most general methods of predicting this advantage are not precise. The conservative approach of neglecting the cooling effect of the cylinder on the gas is recommended for all but the most critical process calculations. In those cases, the manufacturer should be consulted for performance data.

In the case of the well-studied air compressor for utility service, the practice has been to assume that 15% to 20% of the accumulated heat that must be dissipated between stages and by aftercoolers, is actually taken by the jacket cooling system.

In order to determine the quantity of water needed for cylinder jackets, an approximate heat rejection rate of 500 Btu/bhp-hr with a 15°F rise in water temperature may be assumed for cast iron cylinders. The heat rejection rate will increase as the cylinder diameter is reduced. The use of dry type cylinder liners, or applications involving gases with low k values (such as natural gas) will reduce the heat rejection rate by some 50%.

Cooling water should never be cold enough to cause condensation within a cylinder. Severe wear or sudden damage to the compressor can result. Incoming water temperature should be 10° to 15°F above the temperature of the incoming gas. An aftercooler or an intercooler may serve as a convenient supply of warm water.

Where only cool water is available, the rate of circulation through the jackets should be controlled to maintain the water outlet temperature at 15° to 20°F above the gas inlet temperature.

The discharge temperature of the cylinder jacket water should be less than 130°F except where condensation within the cylinder may result.

Some compressor applications will require no cooling. These include low-temperature services such as refrigeration, and systems involving low ratios of compression (less than 3) with gases having a low specific heat ratio as in the case of light hydrocarbons. In these cases, the cylinder jackets may be filled with a heat conducting fluid, such as oil or an antifreeze solution, to distribute temperatures evenly throughout the cylinder casing.

A simple thermosyphon system is often used to move the coolant slowly through the jackets. This involves the use of an external coolant reservoir tank and a small amount of piping to allow the coolant to circulate through a closed loop by heat-induced convention.

In many smaller units, splash lubrication is provided by the agitating action of the crankshaft rotating in an oil reservoir. In the single-acting compressors, both the frame bearings and cylinder are lubricated by the same atomized oil mist.

Single-acting splash lubricated compressors require frequent oil changes and occasional oil additions as part of a regular maintenance program.

Larger process compressors have force-feed lubrication systems. The points of lubrication include the frame running gear, crossheads, the cylinders, and packing. In most cases, the running gear and crossheads will be pressure lubricated through a single oil system, while the cylinders are lubricated through an independent oil system. In most cases, the lubricating oil used in the cylinders will not be identical to the oil used in the running gear (Figure 104). This is particularly true in the case of services involving unusual process temperatures, or gases requiring special lubricants.

The frame lubrication system in larger compressors commonly includes a dust-tight crankcase which serves as an oil reservoir, an oil strainer, a small gear-type or centrifugal pump which may be driven from a power take-off on the main drive, and an oil filter. Larger units may use an oil cooler.

Some moderate size units (less than approximately 200 hp) may use a flood type system for lubricating the running gear. Frame oil is carried up by a crankshaft-driven mechanism within the crankcase, and allowed to flow down to lubricate the journals.

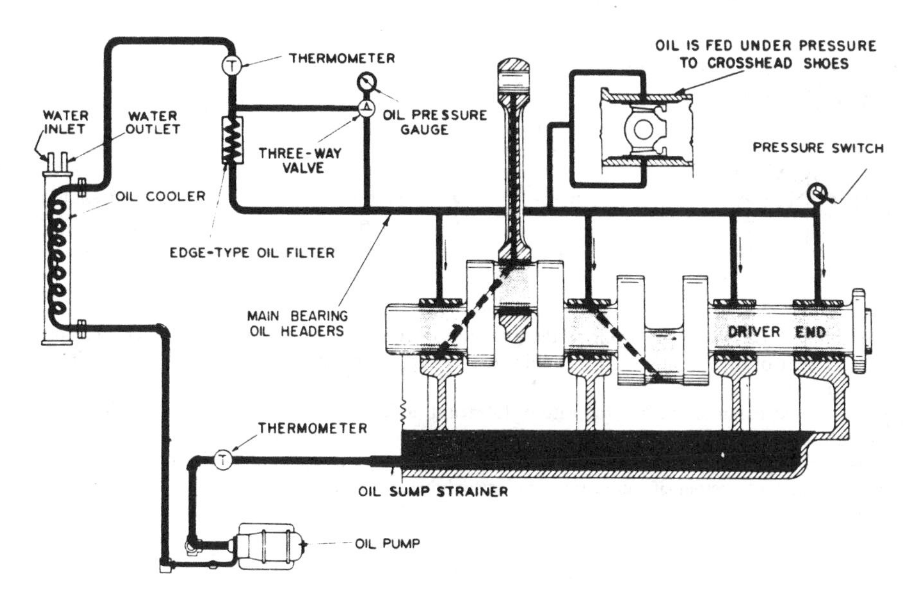

Figure 104. Shows force feel lubrication system. (Courtesy Ingersoll Rand Corp.)

In larger compressors, cylinder lubrication includes a multipoint lubricator which is capable of individual and adjustable flow rates to each point, where lubricants are not recycled.

The points of lubrication within a cylinder include the piston rings, the packing, and on occasion, the valves. Of these, the piston rings are the most critical and are invariably fed through ports in the cylinder bore.

In many cases, the lubricant being fed to the cylinder bore is carried over onto the piston rod and onto the packing in sufficient quantity to eliminate the need for a separate feed to the packing.

Valves usually do not require a separate lubricant feed. Oil is carried to the valve seats with the gas stream. In some installations, oil may be injected into the gas stream before the compressor inlet to guarantee lubrication of the inlet valves.

Care must be exercised in the selection of cylinder lubricating oils. Lubricants are highly individualized from process to process. Halocarbon-compressor oils cannot be interchanged with lubricants for ammonia compressors, utility air compressor oils cannot be interchanged with lubricants for high pressure cylinders, etc.

Thermal decomposition of an incorrectly prescribed lubricating oil in an air compressor can result in accumulations of soot in the discharge piping. These deposits are combustible.

Compressor manufacturers are able to specify the optimum characteristics of a lubricant for a particular machine in a specified service. However, the final recommendation on the exact lubricant to be used should come from a reputable oil supplier who is willing to guarantee the performance of his product.

In general, the break-in period of a new reciprocating compressor will require relatively large amounts of lubricant. Most often, this oil will not be identical to the lubricant that will be used during the process run.

The rate of cylinder feed is not completely predictable. Proper oil feed rates are determined by observation at startup, and every six months or so thereafter.

Minimum cylinder lubrication is a compromise between full lubrication and nonlube design. Full lubrication offers mechanical reliability to minimize downtime, but invariably results in a gas stream contaminated by the lubricant. Nonlubricated cylinders will not contaminate the gas stream to any measurable degree, but are more subject to mechanical wear.

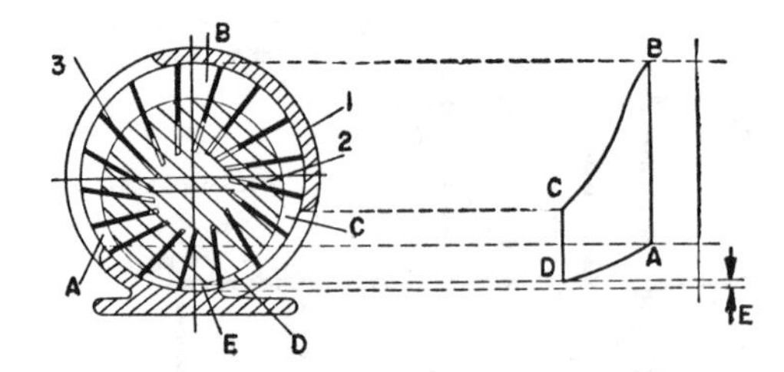

Figure 105. Shows operation of sliding-vane compressor: 1—casing, 2—rotor, 3—sliding blades.

Minimum lube systems incorporate the materials of construction of nonlube cylinders. The mechanical wear of these components is minimized by supplying them with a fine coating of a liquid lubricant.

In the most common design, the rate of lubricant addition is so small that the packing is the only lubricated component. In this case, oil is carried into the cylinder by the rod and slowly migrates to the cylinder walls and valves. The rate of oil feed to the packing itself may be minimized, or eliminated altogether if sufficient crankcase lubricant reaches the packing during normal operation.

Rotary Compressors

Rotary compressors generally are classified as sliding-vane type, two-impeller positive type, screw-type, and liquid-piston type.

A *sliding-vane type compressor* (Figure 105) has a closed casing (1) in which the rotor (2) is located eccentrically relative to the internal surface of the casing. Slits are provided on the rotor through which blades (3) move radially, affected by centrifugal force, which presses them tightly against the internal surface of the casing. Thus, the sickle-shaped working space between the rotor and the casing is divided by the blades into several nonequal chambers. Gas enters from the suction nozzle and fills the chambers. When the rotor turns to position B, suction ceases and the compression cycle now begins. When the chamber rotates to the right, its volume decreases and the gas inside is compressed. The compression cycle terminates when the chamber reaches position C, where it deposits the compressed volume of gas to the discharge piping.

In position D the gas is displaced completely from the working chamber. Clearance is provided between the rotor and the casing. From D to A, gas is expanded in the clearance space. The suction begins at point A, where the cycle begins again.

Rotary sliding-vane compressors are available at operating pressures up to 125 psig and in capacities up to 5,300 cfm. Speeds vary from 3,600 rpm for small-capacity units to 450 rpm in larger systems. Single-stage machines are available at pressures up to 50 psig and vacuums to 29 in. Hg. Two-stage machines are capable of pressures up to 125 psig and vacuums to 29 in. Hg.

The capacity of a rotary sliding-vane compressor, V_1 (m^3/sec), reduced to the suction condition, may be calculated from the following equation:

$$V_1 = 2\ell e n \lambda_v (\pi D - \delta Z) \tag{131}$$

where ℓ = length of a blade, m
e = eccentricity of a rotor, m
n = rotor rotations per unit time, liter/sec
D = inside diameter of the casing, m
δ = blade thickness, m
Z = number of blades (typically 20 to 30)

The relative eccentricity (e/D) typically is 0.06–0.07.

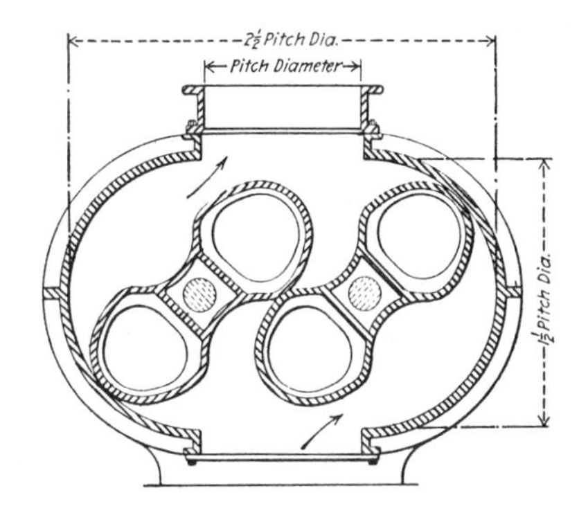

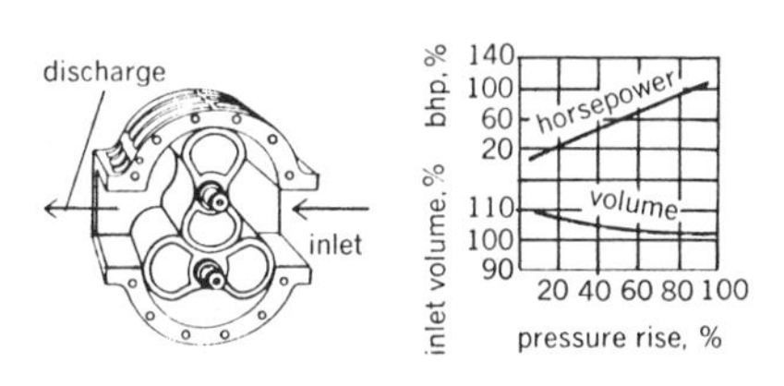

Figure 106. Shows two-impeller compressor along with performance curves for constant speed.

The volumetric coefficient may be computed from

$$\lambda_v = 1 - K\frac{p_2}{p_1} \tag{132}$$

where K = 0.05 for large machines (where capacity exceeds 0.5 m^3/sec)
K = 0.1 for small machines (where capacity is less than 0.5 m^3/sec)

The brake horsepower of a rotary compressor is

$$N_e = \frac{V_1 P_1 \ln(P_2/P_1)}{\eta_{iso}} \tag{133}$$

A *two-impeller, positive-type compressor* (Figure 106) consists of a casing in which two impellers revolve in opposite directions. Each impeller consists of a double-lobe section symmetrical about its shaft. The impellers neither touch each other nor contact the casing. However, clearances are small to minimize leakage.

The rotation of impellers reduces the volume in which the gas is trapped, producing a pressure rise. The gas is drawn through the inlet port and discharged when the impellers pass the outlet port. To maintain the impellers at their proper relative speed, one shaft is driven by the other through a pair of gears.

The pitch diameter of these gears controls the size and capacity of the compressor. The radius of an impeller, or its half length, is designed to be three-quarters of the pitch diameter of the gears. The casing consists of semicylinders separated by a parallel section. The radius of the cylinders is equal to that of the impellers plus clearance. The width of the parallel section is equal to the pitch diameter of the gears plus the clearance.

The horsepower required at the shaft is proportional to the volume and pressure of the gas discharged. It is safe to assume that for each 100 ft^3/min lb pressure of gas discharged, 5 hp is required. The following formulas sometimes are used in calculating the horsepower. To determine the horsepower to drive the two-impeller, positive-type compressor, it is necessary to divide the results obtained by formulas given earlier for the efficiency of the machine. The usual efficiency is between 80% and 90%:

$$hp = \frac{QP_1\left[\left(\frac{P}{P_1}\right)^{1/3} - 1\right]}{11{,}000} \tag{134}$$

This equation is used when it may be assumed that the gas is compressed so quickly that it does not have time to cool to atmospheric temperature:

$$hp = \frac{Q(P - P_1)}{33{,}000} \tag{135}$$

This expression is used for low pressures up to 5 oz.

$$hp = \frac{\text{lb/in.}^2 \times Q}{200} \tag{136}$$

The equation frequently is used for determining the horsepower required to operate the machine. In this last equation, Q represents the volume of gas in cfm displaced by the impellers with no allowance being made for slippage. Note that P_1 is the absolute pressure in the suction line, and P is the discharge pressure, in lb/ft^2. These types of compressors are available for pressure differentials up to about 12 lb/in.^{2} and capacities up to 15,000 ft^3/min.

A *screw-type rotary compressor* (Figure 107), can handle capacities up to about 25,000 ft^3/min at pressure ratios of 4:1 and higher. There is no unbalanced radial load on the rotor shafts. Hence, the magnitude of the pressure rise is not a limiting feature. Relatively small-diameter rotors are available to provide rotative speeds of several thousand rpm. In contrast to the straight-lobe rotary machine, the rotors are male and female whose rotation causes the axial progression of successive sealed cavities. Designs are staged with intercoolers when needed. High-speed operation usually necessitates the use of suction and discharge noise suppressors.

A *liquid-piston compressor* is shown in Figure 108. This design employs any low-viscosity liquid as the compressant to displace the gas handled. A round multiblock rotor forming a series of buckets rotates freely in an elliptical casing containing the liquid, which is carried around by the rotor and is caused to recede from its buckets at the wide point of the ellipse (major axis), permitting the buckets to fill with gas entering from the inlet ports. On reaching the narrow point of the ellipse (minor axis), it surges back into the buckets, compressing the gas and ejecting it through the discharge ports.

Liquid-piston compressors handle up to about 5,000 cfm. Single-stage units can develop pressures to about 75 psi; multistage designs are used for higher pressures.

Another important type of rotary compressor is the liquid ring system. The major components are the rotor, the eccentric or oval-shaped body, and a quantity of liquid formed into a ring within the

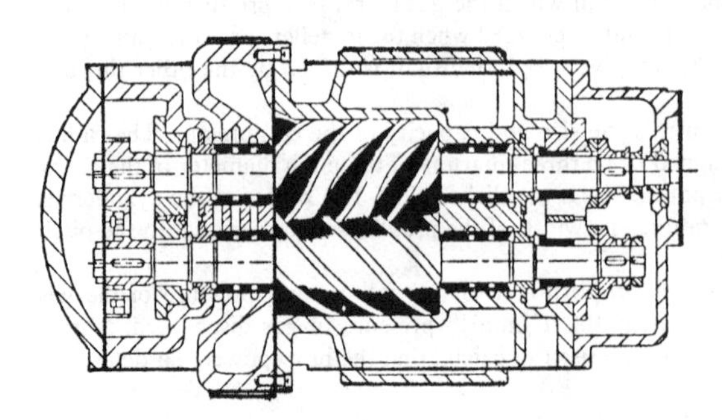

Figure 107. Screw-type rotary compressor.

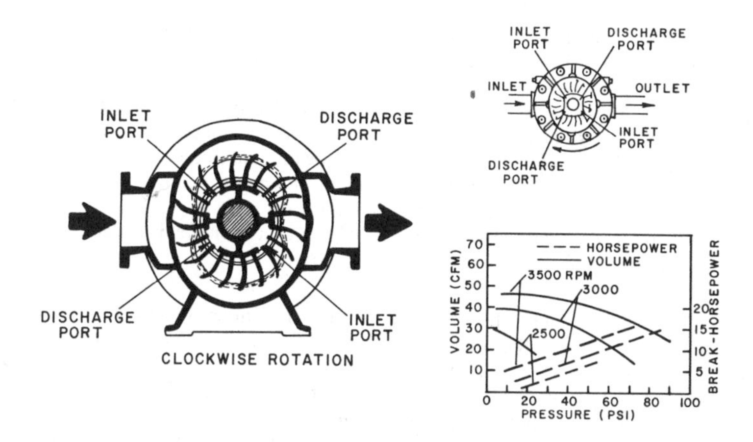

Figure 108. Liquid-piston rotary compressor along with constant-speed performance curves.

body by the turning rotor. The principle of compression is illustrated in Figure 109. As the rotor turns, a series of forward curved rotor vanes induce the liquid within the body to follow the rotor in a cyclical path. The moving liquid forms a ring that aligns with the contours of the body. Each set of blades forms a chamber that changes in volume as the surface of the liquid within the chamber follows the contour of the adjacent body wall. Inlet and outlet ports are located on a central cone surrounding the rotor shaft. This type of compressor should be considered for services involving:

1. *Wet gases.* Slugs of liquid will neither damage nor greatly upset the performance of these machines.
2. *Temperature limitations.* Gas temperature during compression may be closely controlled by maintaining the temperature of the liquid ring medium. Continuous recirculation of the medium through an external heat exchanger is often used.

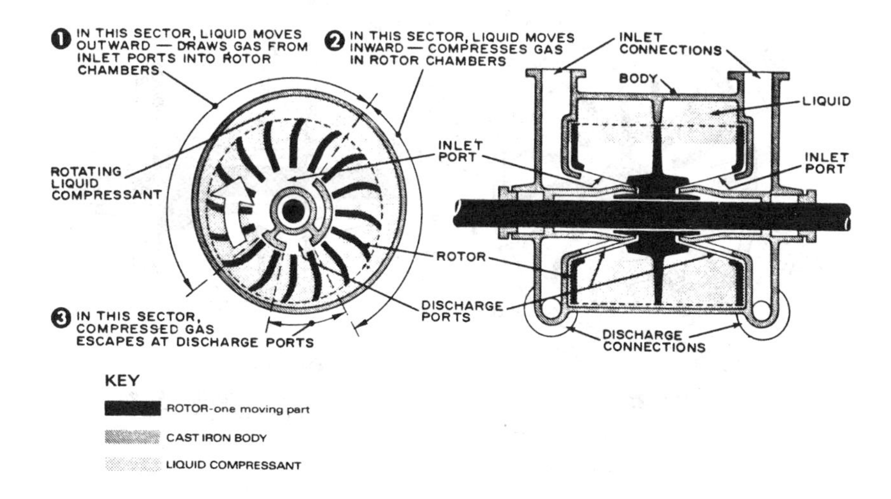

Figure 109. Shows a type CL compressor. (Courtesy Nash Engineering Co.)

3. *Gases with dust.* The principle of operation does not depend on large solid sealing surfaces with careful tolerances. Moreover, the liquid within the compressor acts as a scrubbing medium during compression.
4. *Vapor recovery.* These compressors can serve the function of gas separators by condensing saturated components out of a gas stream while passing the noncondensibles.
5. *Corrosive gases.* The liquid medium coats most of the internal surfaces of the compressor. For example, flooding a compressor built with standard iron construction with an alkaline liquor allows the machine to handle gases with acid vapors.
6. *Limited allowable downtime.* The simplicity of these condensers is reflected in minimal servicing and downtime.

Liquid ring compressors tend to be less power efficient than other designs. Their actual performance characteristics vary with the conditions of service. Standard performance data is published for air service using water as a liquid compressant. The compressor capacity and horsepower for most services can be approximated from the standard tables by using an equivalent cfm with a calculated allowance for the vaporization of the compressant. More precise predictions of performance for gases other than air or a liquid compressant other than water will usually require consideration of the following:

1. Liquid compressant specific gravity, viscosity, specific heat and vapor pressure.
2. Solubility of the gas in the liquid compressant.
3. Effect of condensibles in the gas stream on liquid compressant characteristics. Manufacturers' empirical data describing the effect of these factors on compressor performance is useful when available.

The method of compression and the porting arrangements designed into these compressors permit compression with a near absence of pulsation.

Multiple-stage, single-casing liquid ring compressors are available for increased ratios of compression.

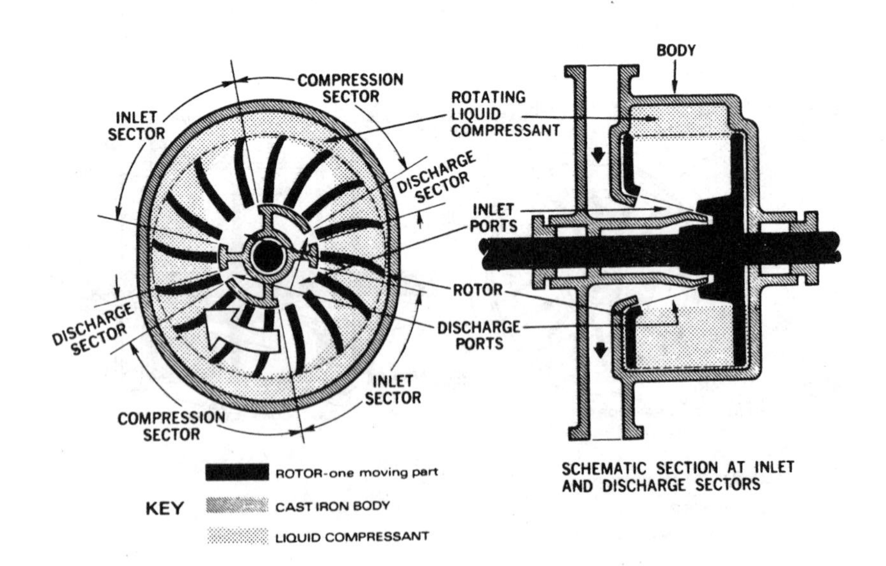

Figure 110. Functional schematic of double lobe pumps.

The *eccentric lobe compressor* provides one compression cycle per chamber per revolution. This design imposes lateral forces on the rotor shaft as compression occurs. It is therefore limited to the lower pressure ranges (approximately 20 psig and under). The eccentric-lobe design tends to be more power efficient than the alternative double-lobe design, and is therefore the preferred style in the higher volume ranges.

The double-lobe design has two cycles of compression per chamber per revolution. Forces on the shaft are balanced, permitting higher ratios of compression than the eccentric lobe design (refer to Figure 110).

Discharge pressure control and capacity control is most often accomplished through an external bypass line. An alternative method of constant speed control used with some units consists of automatically draining part of the liquid "compressant" which allows the units to operate unloaded for a limited period while conserving power. Conventional start-stop control may also be used.

More detailed information and data can be found in References 34–50.

Blowers

The operation of a single-stage centrifugal blower is illustrated in Figure 111. The wheel (2) with blades similar to those in a centrifugal pump, rotates inside the casing (1). The wheel is inserted inside the diffuser (3), where the velocity generated is converted into pressure, partially in the wheel and partially in stationary diffusers following the wheel. The diffusers consist of two circular disks connected by curved blades having a slope opposite to that of the wheel. Gas enters the blower through the nozzle (4) and leaves the blower through the discharge nozzle (5). Figure 112 shows a single-stage blower.

The operating scheme for a multistage blower is illustrated in Figure 113. Contained in the casing (1) are several (usually 3 or 4) wheels (2). The gas passing through the first wheel enters into the guiding diffusers (3) and return channel (4), through which it is guided to the following wheel. The return channel (4) has fixed guiding diaphragms, which impart to the gas velocity and direction. Constant wheel diameters are used in multistage blowers; however, wheel widths decrease in the direction from the first to the last wheel in accordance with the change in gas volume at compression. Thus, gas compression is possible without changing the velocity of rotation or the form of the blades.

In general, the compression ratio in centrifugal blowers does not exceed 3 to 3.5. Consequently, there is no need for cooling. Theoretically, compression in noncooled centrifugal blowers is very close to adiabatic, as shown by line AB on the T-S diagram in Figure 114. The actual energy expended will be greater because of friction losses. The energy spent in overcoming the gas friction is converted almost entirely into heat. Hence, the final gas temperature (T'_2 in Figure 114) is somewhat higher than the temperature, T_2, corresponding to adiabatic compression. The actual compression process from pressure p_1 to p_2 is represented by line AC.

The performance of a blower is characterized by its adiabatic efficiency, η_{ad}, which is the ratio of adiabatic compression work to the actual work expended:

$$\eta_{ad} = \frac{\ell_{ad}}{\ell} = \frac{C_p(T_2 - T_1)}{C_p(T'_2 - T_1)} = \frac{T_2 - T_1}{T'_2 - T_1} \tag{137}$$

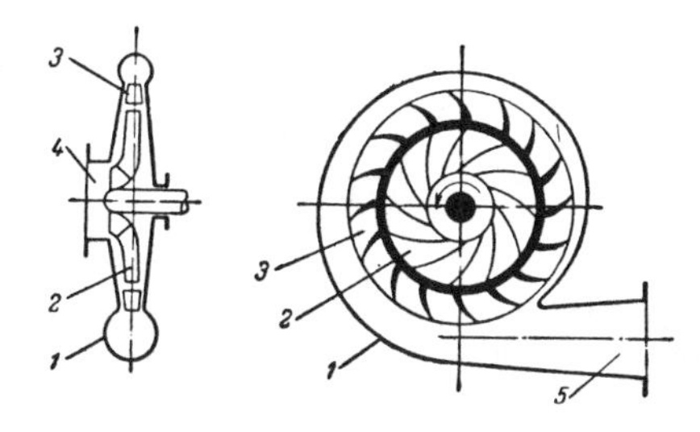

Figure 111. Centrifugal single-stage blower: 1—casing, 2—guiding diffuser, 3—suction nozzle, 4—discharge nozzle.

Figure 112. Single-stage centrifugal blower. (Courtesy North American Manufacturing Co., Cleveland, OH.)

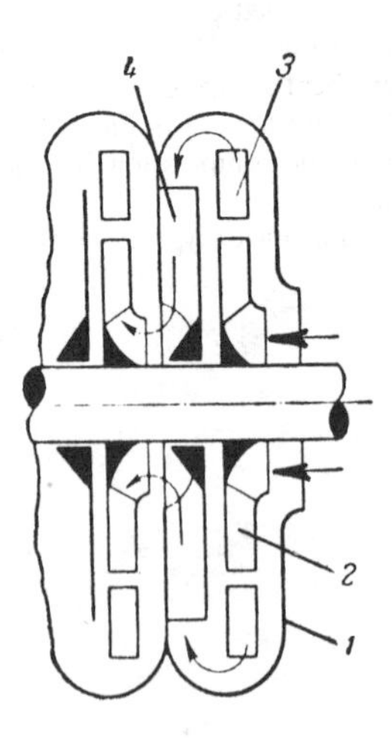

Figure 113. Operation of a multistage blower: 1—casing, 2—wheel, 3—guiding diffusers, 4—return channel.

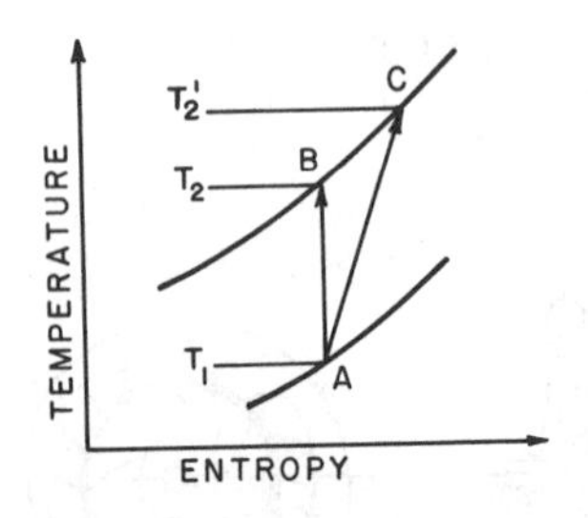

Figure 114. Entropy diagram of gas compression in a centrifugal blower.

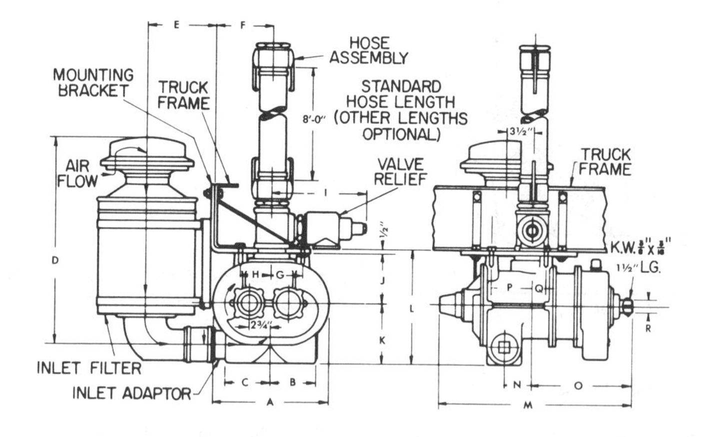

Figure 115. Features of a tractor-mounted rotary blower.

Temperature, T_2 may be calculated from the following equation:

$$T_2 = T_1 \left(\frac{p_2}{p_1} \right)^{(\kappa - 1)/\kappa} \tag{138}$$

and temperature, T'_2, is measured at the exit of the blower.

An additional blower design widely used in pneumatic bulk trailers is the rotary blower. Typical designs consist of helical four-flute gate rotors or two-lobe main rotors. The operating principle of a tractor-mounted unit is illustrated in Figure 115. The screw-type rotor is similar in design to a twin-screw pump. The meshing of the two screw-type rotors is synchronized by timing gears to provide controlled compression of air. Designs are axial flow, positive displacement types. These blowers are adaptable to various types of drives, such as electric motors (constant or variable speed), gasoline or diesel engines, or steam turbines. A V-belt drive, directly connected, or drive-through speed-regulating mechanism may be used. Figure 116 shows a rotary lobe type, axial-flow, positive displacement blower ready for a horizontal installation.

Figure 116. Small rotary blower for horizontal service installation. (Courtesy Gardner-Denver Co., Quincy, IL.)

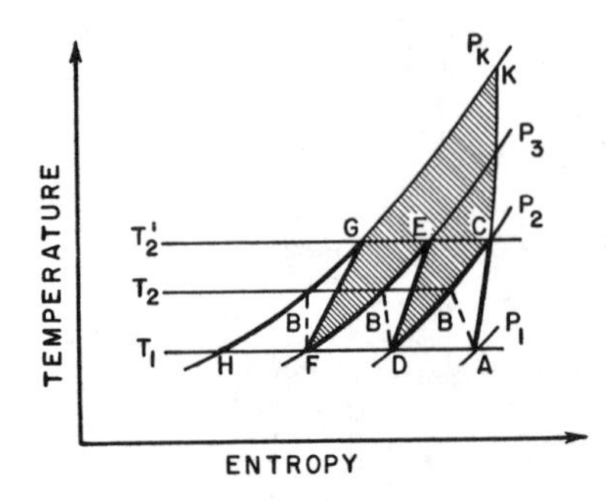

Figure 117. Entropy diagram for gas compression in a multistage centrifugal compressor.

Centrifugal Compressors

Centrifugal compressors are employed when higher compression ratios are needed. Although different from blowers, designs resemble multistage blowers (similar to Figure 113). In contrast to blowers, to achieve greater discharge pressures in compressors, the number of wheels must be increased, wheel sizes (including diameter) must be changed and the velocity of rotation must be raised. The peripheral velocities of the wheels can reach up to 240–270 m/sec or more, with discharge pressures as high as $2.5–3.0 \times 10^6$ N/m^2 (25–30 atm).

Both smaller widths and wheel diameters are needed when adding stages of higher pressure. Centrifugal compressors do resemble blowers, however, in that the diffusers used for converting the gas kinetic energy into potential energy of pressure and the return channels are quite similar.

The working wheels of centrifugal compressors often are sectionalized and, thus, located in two or three casings. As noted in the previous section, high gas compression ratios produce significant increases in the gas temperature and, hence, the need for cooling. Gas cooling is effected either by introducing water in water-cooled diaphragms (located inside the compressor casing) or in external coolers. Cooling in external coolers is more effective and simplifies maintenance of heat transfer surfaces.

Figure 117 provides the entropy diagram of gas compression in a compressor with two external interstage coolers plus final cooling after the last stage. The diagram is interpreted by assuming that the gas is cooled (isobar) to the initial temperature of the gas, with zero pressure losses. The process is illustrated by the broken line ACDEFGH. The shaded area is equivalent to the savings in work obtained over compression without intermediate cooling.

Gas compression in centrifugal compressors is similar to that in blowers. As shown by Figure 117, after compression by the group of noncooled wheels (lines AC, DE and FG), the gas acquires temperature T'_2 which is higher than T_2 at the end of adiabatic compression (point B). The degree at which the process approaches the isothermal case is characterized by the compressor's isothermal efficiency, i.e., the ratio of isothermal work to the work spent:

$$\eta_{iso} = \frac{\hat{\ell}_{iso}}{\hat{\ell}} \tag{139}$$

Isothermal efficiency is in the range of 0.5 to 0.7.

The *axial-flow dynamic compressor* (Figure 118), meets the requirement of large capacity but at relative low compression ratios (3.5 to 4). The characteristics of these machines are quite different from the centrifugal type.

The rotor blades impart velocity and pressure to the gas as the rotor turns, with the velocity being converted to pressure in the stationary blades. Axial-type compressors are driven with steam or gas turbines with velocities of 5,000 to 5,500 rpm. They typically have 10 to 20 stages and operate without gas cooling.

In developing the governing equations for fans, we assumed gas density to be constant. This includes the case of centrifugal acceleration with a change in pressure and purely impulsive action.

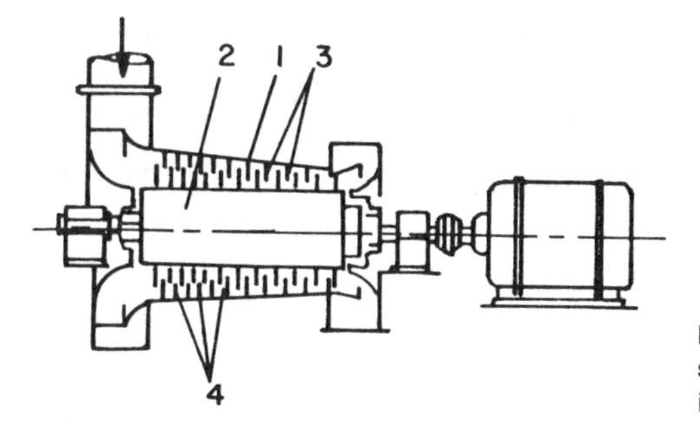

Figure 118. Axial-flow dynamic compressor: 1—casing, 2—rotor, 3—blades, 4—inlet guide blades.

For blowers and compressors, the compression due to centrifugal force within the wheel is significant and therefore must be included in the performance equations.

In developing the equations, we shall assume isothermal conditions because the ratio of compression in each stage is low and intercooling can approximate this state.

The centrifugal force due to a unit area disk of radial thickness, dx, is:

$$dP_x = \frac{w_x u_x^2}{gx} dx = \frac{w_x u^2}{gr^2} x\,dx \tag{140}$$

because

$$u_x = \frac{x}{r} u$$

And

$$\frac{w_x}{w_1} = \frac{P_x}{P_1} \tag{141}$$

where P_x is the absolute pressure in the gas at distance x from the center, and w_1 and P_1 are the weight and pressure, respectively, of the gas at the entrance.

Substituting and dividing by P_x, we obtain

$$\frac{dP_x}{dx} = \frac{w_1 u^2}{P_1 gr^2} x\,dx$$

Then

$$\int_{P_1}^{P_1} \frac{dP_x}{P_x} = \frac{w_1 u^2}{P_1 gr^2} \int_{r_1}^{r} x\,dx$$

Whence

$$\ln\left(\frac{P}{P_1}\right) = \frac{w_1 u^2}{P_1 2gr^2}(r^2 - r_1^2) = \frac{w_1}{P_1}\left(\frac{u^2 - u_1^2}{2g}\right) \tag{142}$$

because

$$r_1 = \frac{r}{u} u_1$$

Regardless of a change in pressure at the outlet the resistance to rotation is due solely to the behavior of the departing jet. This behavior is in a direction opposite to that of the absolute velocity of discharge, $\dot{v}$. However, the component opposed to rotation (that is, in the direction opposite to u) is $w(\dot{v}/g) \cos \theta$, and, as is apparent from the diagrams, $\cos \theta = u + V \cos \beta$. Therefore, the force opposed to rotation is $(w/g)(u + v \cos \beta)$. As the work done by the wheel equals force times distance, then

$$\text{Work} = \frac{w}{g}(u^2 + uV \cos \beta) \tag{143}$$

This is independent of the radial depth, $r - r_1$, of the vanes.

For a gas of uniform density (in low-pressure fans density changes are small), if the machine imparts a head, H, then work is WH,

$$WH = \frac{W}{g}(u^2 + uV \cos \beta)$$

and

$$H = \frac{u^2 + uV \cos \beta}{g}$$

Let R_1 be the final ratio of compression when the gas has been brought to rest after one stage. Then work $= p_a v_a L_n R_1$, where v_a is the volume of gas compressed. Then

$$p_a v_a \ln R_1 = \frac{w_a v_a}{g}(u^2 + uV \cos \beta)$$

and

$$\ln R_1 = \frac{w_a}{p_a}\left(\frac{u^2 + uV \cos \beta}{g}\right) \tag{144}$$

This formula provides the ratio of compression produced by one stage in a centrifugal compressor.

If there are n stages, each contributing an additional ratio $(R_1, R_2, \ldots, R_n)$, then

$$R_n = R_1^n$$

and

$$\ln R_n = n \ln R_1 \tag{145}$$

The tremendous centrifugal force developed in these machines prompts most manufacturers to prefer the outer tips of the propeller blades to be radial ($\beta = 90°$) to avoid cross bending.

In the case in which $\beta = 90°$, Equation 144 simplifies to:

$$\ln R_1 = \frac{w_a u^2}{p_a g} \tag{146}$$

Note that $W = p/53.35t$ and, if t is constant, $w_a/p_a g$ is constant. To adopt the formula to common logarithms, divide the expression by 2.3026.

Obviously, perfect cooling cannot be accomplished. As an example, assume an average temperature of 580°R (120°F).

$$\log R_1 = \left(\frac{1}{2.3026 \times 53.35 \times 580 \times 32.2}\right) u^2 = Ku^2$$

$$\log K = 7.6398$$

Several illustrative examples are given in Reference 33 which expand on the preceding analysis.

NOTATION

- a radius of generating circle of diffuser (ft)
- A area (sq ft)
- A_2, A_F rotor discharge area (sq ft)

  axial flow:

  $$A_2 = \frac{\pi D_2^2}{4}\left(1 - \left(\frac{D_1}{D_2}\right)^2\right)$$

  radial flow: $A_2 = \pi D_2 b$

- A_0 effective area (sq ft)
- b width (ft)
- c clearance (ft)
- C_d discharge coefficient (dimensionless)
- C_D flow resistance coefficient in number of velocity heads loss (dimensionless)
- C_L lift coefficient (dimensionless)
- d particle diameter (inches)
- D_s, d_s specific diameter

  $$\left(\frac{\text{min}^2}{\text{ft}}\right)^{1/4} \times \left(\frac{\text{lb/sq ft}}{\text{in. of water}}\right)^{1/4}$$
  (dimensionless)

  $$d_s = D_c \frac{\Delta p^{*1/4}}{Q^{1/2}}\left(\frac{0.075}{\rho}\right)^{1/4}$$

  $$d_s = 1.58 \times 10^{-2} \frac{D_c}{A_2^{1/2}} \frac{\psi^{1/4}}{\phi^{1/2}}$$

- D diameter (ft)
- D_0 diameter, inlet orifice (ft)
- D_1 diameter rotor inlet or hub (ft)
- D_2 diameter rotor outlet or tip (ft)
- D_c characteristic diameter of fan (ft)
- f Fanning friction factor (dimensionless)

  $$\frac{\Delta p}{\rho V^2/(2g)} = \frac{4L}{D} f$$

- g gravitational acceleration (ft/sec^2)
- h' head (ft of fluid flowing)
- K_A, K_N, K_P, K_V constants (dimensionless)
- L length (ft)
- M Mach number (dimensionless)
- n rotational speed (rpm at test condition)
- N_s specific speed at maximum efficiency

  $$n_s = n \frac{Q^{1/2}}{\Delta p^{*3/4}} \times \left(\frac{\rho}{0.075}\right)^{3/4}$$

  $$n_s = 8.05 \times 10^4 \frac{A^{1/2}}{D_2} \frac{\phi^{1/2}}{\psi^{3/4}}$$

- N number of blades (dimensionless)
- p pressure (lb/sq ft)
- Δp pressure difference (lb/sq ft)
- Δp^* static pressure difference (in. of water)
- P_H power input to air (horsepower)
- P_{kw} power input (kilowatts)
- P_{0a} power output of air (horsepower)

  $$P_{0a} = \frac{\Delta p \, aQ}{33{,}000}$$

- P_w power input (watts)
- Q volumetric flow rate (cfm = cu ft/min)
- r radius (ft)

R system resistance (ft^{-4})
R_D Reynolds number of test data based on diameter (dimensionless)
s specific gravity (dimensionless)
S fan shape factor (dimensionless)
S_1 radius ratio D_1/D_2
S_2 pitch-chord ratio for radial fans

$$S_2 = \frac{2\pi \sin \beta_2}{N \log_e(S_1)}$$

S_3 blade width/rotor diameter

$$S_3 = \frac{b}{D_2}$$

S_4 $\frac{\text{scroll width}}{\text{blade width}} = \frac{b_s}{b}$

S_5 $\frac{\text{scroll diameter}}{\text{blade diameter}} = \frac{D_s}{D_2}$

S_6 $\frac{\text{scroll flow area}}{\text{rotor discharge area}} = \frac{(D_c - D_2)b_s}{2A_2}$

t_C temperature (° Centigrade)
t_F temperature (° Fahrenheit)
T absolute temperature (°K or °R)
u blade velocity at any point (ft/sec)
u' blade velocity at any point (ft/min)
V absolute fluid velocity (ft/sec)
V' absolute fluid velocity (ft/min)
v fluid relative velocity (ft/sec)
v' fluid relative velocity (ft/min)

Greek Symbols

α_e expansion angle (deg.)
β blade angle, measured from tangential (deg., radians)
β_{CO} cutoff angle (deg., radians)
κ, γ ratio of specific heats (dimensionless)
Δ "change of," (dimensionless)
η efficiency (dimensionless)

$$\eta = \frac{\Delta p^* \times Q \times 1.57 \times 10^{-4}}{P_H}$$

θ angle (deg., radians)
μ viscosity, absolute (lb/sec-ft)
ν kinematic viscosity (ft^2/sec)
ρ weight density (lb/cu ft)
ϕ flow coefficient (dimensionless)

$$\phi = \frac{Q}{A_2 u'_2}$$

ψ static pressure coefficient (dimensionless)

$$\psi = \frac{\Delta p_s}{\rho u_2^2/(2g)}$$

ψ_t total pressure coefficient (dimensionless)

$$\psi_t = \frac{\Delta p_t}{\rho u_2^2/(2g)}$$

ψ_0 pressure coefficient at no flow (dimensionless)

Subscripts

1 pertaining to inner blade edge (hub or blade inlet)
2 pertaining to blade outer edge, blade tip or blade outlet
1, 2, 3 also used to distinguish between different fans, shape parameters, resistances, etc.
a air, axial component of
b blade
c characteristic
d diffuser
f fan
i pertaining to fan inlet
m meridional component of
r radial component of
s static, specific, scroll
t tangential component of; total
u blade speed
V velocity
∞ conditions based on mean tangential velocity through blade row

REFERENCES

1. Daryanani, S., et al., "Air Conditioning, Heating and Ventilation," *ACHV Mag.* (March 1966).
2. Bienko, W. J., "Application of Joy Axivane Controllable Pitch Fans and Associated Control Systems for Variable Air Volume Systems," Joy Mfg. Co. Bulletin, New Philadelphia, OH (1980).
3. *ASHRAE Guide and Data Book*, System (1970).
4. *Handbook of Air Conditioning System Design*, Carrier Air Conditioning Co., McGraw-Hill Book Co., 1965.
5. Walters, T. A., and Wendover, W. E., "Predicting Forced Draft Fan Noise," paper presented to ASME Conference, Chicago (April 14, 1964).
6. *ASHRAE Guide and Data Book*, Chapter 4 (1967), p. 58.
7. Novick, H. J., *Heating, Piping and Air Conditioning* (Oct. 1968).
8. Harris, C. M., *Handbook of Noise Control*, McGraw-Hill Book Co., New York, NY, 1957.
9. Eck, B., *Fans, Design and Operation of Centrifugal, Axial-Flow and Cross-Flow Fans*, Pergamon Press Ltd., Elmsford, NY, 1st English Ed., 1973.
10. Doebelin, E. D., *Measurement Systems, Application and Design*, McGraw-Hill Book Co., New York, NY, revised ed., 1975.
11. Hosey, A. D., and Powell, C. H., *Industrial Noise*, U.S. Department of Health, Education and Welfare, Washington, DC 1976.
12. Irwin, J. D., and Graf, E. R., *Industrial Noise & Vibration Control*, Prentice-Hall Inc., Englewood Cliffs, NJ, 1979.
13. *Industrial Noise Control*, American Industrial Hygiene Association, Akron, OH, 3rd Ed., 1975.
14. *Mark's Standard Handbook for Mechanical Engineers*, McGraw-Hill Book Co., New York, NY, 7th Ed., 1967.
15. *General Industry, OSHA Safety and Health Stands (29CFR 1910)*, U.S. Department of Labor, Occupational Safety & Health Administration, OSHA 2206, revised ed., 1981.
16. Grey J., *Noise, Noise, Noise*, The Westminster Press, Philadelphia, PA 1975.
17. Lord, H. W., Gatley, W. S., and Evensen, H. A., *Noise Control for Engineers*, McGraw-Hill Book Co., New York, NY, 1980.
18. Lipscomb, D. M., and Taylor A. C., *Noise Control, Handbook of Principles and Practices*, Van Nostrand Reinhold Co., New York, NY, 1978.
19. Beranek, L. L., *Noise Reduction*, McGraw-Hill Book Co., Inc., New York, NY, 1960.
20. Cheremisinoff P. N., and Young, R. A., *Pollution Engineering Practice Handbook*, Ann Arbor Science Pub., Inc., Ann Arbor, MI, 2nd Printing, 1976.
21. Osborne, W. C., *Fans*, Pergamon Press Ltd., Elmsford, NY, 2nd Ed., 1977.
22. Wallis, R. A., *Axial Flow Fans, Design and Practice*, Academic Press Inc., New York, NY, 1st Published, 1961.
23. Cheremisinoff, P. N., and Cheremisinoff, P. P., *Industrial Noise Control Handbook*, Ann Arbor Science Pub., Inc., Ann Arbor, MI, 1977.
24. Woods, R. I., MIHVE, *Noise Control in Mechanical Services*, Sound Attenuators Ltd. and Sound Research Labs Ltd., Distributed in the U.S.A. by Halsted Press, New York, NY, Reprinted 1976.
25. Cuniff, P. F., *Environmental Noise Pollution*, John Wiley & Sons, Inc., New York, NY, 1977.
26. Miller, R. K., *Handbook of Industrial Noise Management*, The Fairmont Press, Atlanta, GA, 1976.
27. Petrusewicz, S. A., and Longmore, D. K., *Noise and Vibration Control for Industrialists*, American Elsevier Publishing Co., Inc., New York, NY, 1974.
28. L. S. Goodfrend Associates, *Noise from Industrial Plants*, U.S. Environmental Protection Agency, Washington, DC, 1971.
29. Salmon, V. Mills, J. S., and Petersen, A. C., *Industrial Noise Control Manual*, U.S. Department of Health, Education, and Welfare, Public Health Service, Cincinnati, OH 1975.
30. Milne, A., *Noise Pollution, Impact and Countermeasures*, David & Charles, Inc., North Pomfret, VT, 1979.
31. Gill, T. T., *Air and Gas Compressors*, John Wiley & Sons, Inc., NY, 1941.
32. Coulson, J. M., and Richardson, J. F., *Chemical Engineering*, Pergamon Press, Inc., Elmsford, NY, 1962.

33. Azbel, D., and Cheremisinoff, N. P., *Fluid Mechanics and Unit Operations*, Ann Arbor Science Pub., Ann Arbor, MI, 1983.
34. Stepanoff, A. J., *Pumps and Blowers*, John Wiley & Sons, Inc., NY, 1965.
35. Kearton, W. J., *Turboblowers and Compressors*, Pitman Press, NY, 1926.
36. Rollins, J. P. (Ed)., *Compressed Air and Gas Handbook*, 4th ed., Compressed Air and Gas Institute, NY, 1973.
37. Compressed Air and Gas Institute, *Compressed Air Handbook*, McGraw-Hill Book Co., Inc., NY, 1954.
38. *Chemical Engineer's Handbook*, (5th ed.), R. H. Perry and C. H. Chilton (Eds.), McGraw-Hill, 1973, pp. 3–105, 109.
39. *Compressed Air and Gas Data*, C. W. Gibbs, (Ed.) Ingersoll Rand Co., 1969, p. 34–8.
40. Smith, J. M., and Van Ness, H. C., *Introduction to Chemical Engineering Thermodynamics* McGraw-Hill, 1959, pp. 95 to 98.
41. *Chemical Engineer's Handbook*, (5th ed.), R. H. Perry and C. H. Chilton (Eds.) McGraw-Hill, 1973, pp. 3–232 and 3–233.
42. *Engineering Data Book*, Gas Processors Suppliers Association (9th ed.), 1977, pp. 16–11, and 16–14.
43. Benedict, M., Webb, G. B., and Rubin, L. C., *J. Chem. Phys.*, 8:334 (1940).
44. Redlich, O., and Kwong, J. N. S., *Chem. Rev.*, 44:233 (1949).
45. *Compressed Air and Gas Handbook*, 4th ed., J. P. Rollins (Ed.), Compressed Air and Gas Institute, 1973, p. 3–77.
46. *Compressed Air and Gas Data*, C. W. Gibbs (Ed.), Ingersoll Rand Co., 1969, p. 34–117.
47. *Compressed Air and Gas Data*, C. W. Gibbs (Ed.), Ingersoll Rand Co., 1969, p. 4–30.
48. American Petroleum Institute Standard 618, Second Ed., July 1974, Sec. 2.7.
49. Applicable standards may be found in Section VIII of the ASME Boiler and Pressure Vessel Code, in the American Petroleum Institute Standard 618, and in ANSI 31.8.
50. *Gas Processors Suppliers Assn. Engineering Data Book*, (9th ed.), Third rev., 1977, pp. 4–177.

CHAPTER 43

ANALYSIS OF AXIAL FLOW TURBINES

Vijay K. Garg

Department of Mechanical Engineering
Indian Institute of Technology Kanpur
Kanpur, India

CONTENTS

INTRODUCTION, 1335

MERIDIONAL FLOW SOLUTIONS, 1337
Streamline Curvature Method, 1337
The Matrix Method, 1339
Finite Element Methods, 1342

BLADE-TO-BLADE FLOW SOLUTIONS, 1344
Singularity Methods, 1345
Field Methods (Subsonic Blade-to-Blade Calculations), 1345
Transonic Blade-to-Blade Calculations, 1346

THREE-DIMENSIONAL FLOW SOLUTIONS, 1349

CLOSURE: FUTURE TRENDS, 1351

REFERENCES, 1351

INTRODUCTION

Analysis of the flow through a turbomachine is an obvious requirement for design and performance prediction purposes. This flow, however, is extremely complex, being in general three-dimensional, time dependent, compressible, and subjected to viscous shear at all exposed surfaces. The three-dimensional unsteady boundary layers and wakes are affected by tip clearance and other leakage flows, by shock interactions, and by irregularities in the machine boundaries. In addition, the working fluid may be multiphased or a mixture of chemically reacting substances. Clearly, a complete description of such a complicated flow is a formidable task. Most analyses even today follow the classical work of Wu [1] and split the complicated three-dimensional flow into a pair of two-dimensional problems, which are solved numerically. In all such analyses boundary layers are dealt with separately and their effects on the available flow area, fluid deflection, and irreversible losses are incorporated into an otherwise inviscid calculation. Also, calculations are performed for the flow of a single dominant phase or chemical species whose properties are known. Full three-dimensional flow solutions, with or without boundary layers, are still rare; the few that do exist have so far been confined to complex duct flows and turbomachines with fewer blade rows, such as centrifugal compressors and turbines.

Following Wu [1] the three-dimensional flow through a turbomachine is still analysed in terms of two two-dimensional flows on pseudo-orthogonal surfaces. These two flows are:

1. The cascade or blade-to-blade flow, which is usually obtained on surfaces of revolution about the machine axis—the S_1 surfaces of Wu (see Figure 1). Three (or more) S_1 surfaces (hub, mean, and tip) are commonly selected for this flow. The S_1 surfaces are in general twisted in

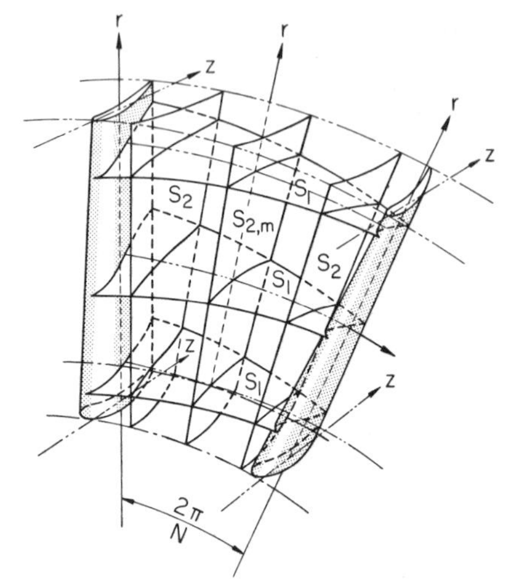

Figure 1. Intersecting S_1 and S_2 surfaces in a blade row [1].

each blade passage making the blade-to-blade solution very difficult. To overcome this problem a blade-to-blade surface of revolution is formed by rotating a streamline from the solution in the representative meridional plane.

2. The meridional- or through-flow on a stream surface spanning the flow from the hub to the casing, and passing between the blades—the S_2 surface of Wu (see Figure 1). In practice, a single representative meridional plane (the mean stream surface S_{2m}) is obtained by pitch averaging the equations of motion across one blade pitch, as described by Horlock and Marsh [2]. The resulting equations describe hypothetical axisymmetric flow, the blade introducing a circumferentially "smeared" blockage due to its finite thickness, a system of body forces necessary for deflecting the flow, and relative total pressure losses due to blade surface boundary layers and the wake.

Various methods have proliferated for flow computations on both the S_1 and S_2 surfaces. These solutions are generally carried out without any iteration between the two surface sets, and, unfortunately, the S_1 surface calculations are frequently made without recognizing stream-tube thickness from the S_2 solution. Recently some calculations involving an iteration between the S_1 and S_2 solutions have been made. There are, however, limits to this approach, and such an iteration can yield only "moderately" three-dimensional flows, that is, flows that demonstrate a single sense of rotation. It is expected [3] that flows that have a strong vortical nature cannot be successfully treated by this approach since it is impossible for the stream surfaces to properly roll up and follow the various vortices. A variety of solutions to the three-dimensional Navier-Stokes equations are reported in the literature; however, application to turbomachinery design and analysis is far from reliable. These techniques deviate widely from the classical Wu approach.

The review article by Japikse [3] gives a good account of the present state of turbomachine flow analysis. Deshpande [4] points out the limitations of the available programs that work well only for a small range of geometries and flow conditions. He also points out that the converged solutions of two different programs may not agree too well. Novak and Hearsey [5] discuss the shortcomings of the approach that takes the meridional solution based on some simplifying assumptions as an input for the blade-to-blade solution without further iteration. Davis and Millar [6] emphasize the sensitivity of the predictions to the cascade loss model used and the need to develope better models.

The foregoing cautionary references notwithstanding, numerical analysis of turbomachinery flow is already accepted as a powerful design tool. Industry probably has developed far more satisfactory programs than are available in the open literature.

In the following sections various methods for obtaining the S_1 and S_2 solutions for axial flow turbines will be reviewed, together with the upcoming three-dimensional flow solutions.

MERIDIONAL FLOW SOLUTIONS

This is the most frequently studied problem in turbomachine analysis. It yields the annular passage definition and airfoil velocity triangle evaluation. Three main approaches have been followed by research workers to obtain the meridional flow solutions. These are the streamline curvature methods, matrix or finite-difference methods, and finite element methods. The last method is the newest, and the experience gained with it is still small. The matrix and streamline curvature methods both resort to finite-differencing of the governing differential equations; however, these equations are used in very different ways in the two methods. At present none of the methods is known to enjoy a clear overall advantage over the others. Each of the preceding methods has also been formulated by more than one group of researchers, and these formulations are not identical. Moreover, most of the analyses are restricted to subsonic flows, though some recent analyses do consider transonic flows.

Streamline Curvature Method

The streamline curvature method (SCM) has been used by Katsanis [7], Novak [8], Frost [9], Davis and Millar [6], Howard and Osborne [10], Kündig [11], Garg [12], and others due to its intrinsic ability to handle odd geometric boundary shapes with ease, and the convenience of quasi-orthogonals (q-Os) in rigorously establishing a general "grid network." Quasi-orthogonals are lines arbitrarily placed across the flow field, roughly normal to the streamlines. The basic theory of the SCM has been described by many authors, particularly by Novak [8]. From the assumption of axial symmetry it is possible to define a series of meridional stream surfaces. These are surfaces of revolution along which fluid particles are assumed to move through the machine. The principle of SCM is to write the equation of motion along lines roughly perpendicular to these stream surfaces (i.e., along q-Os) in terms of the curvature of the surfaces in the meridional plane; see also Smith [13].

A typical q-O and stream surface is illustrated in Figure 2. The acceleration of a fluid particle at any point P in the q-direction may be written as

$$\frac{DV_q}{Dt} = -\frac{V_\theta^2}{r}\sin(\phi + \lambda) + \frac{V_m^2}{r_c}\sin\lambda + V_m\frac{dV_m}{dm}\cos\lambda$$

For inviscid flow the acceleration is related to the pressure gradient by $DV_q/Dt = -(1/\rho)dp/dq$. Using the second law of thermodynamics to express the pressure gradient in terms of total enthalpy and entropy gradients gives, for the gradient of meridional velocity in the q-direction,

$$\frac{d(V_m^2/2)}{dq} = \frac{dH}{dq} - T\frac{dS}{dq} - \frac{1}{2r^2}\frac{d(r^2V_\theta^2)}{dq} + \frac{V_m^2}{r_c}\sin\lambda + V_m\frac{dV_m}{dm}\cos\lambda \tag{1}$$

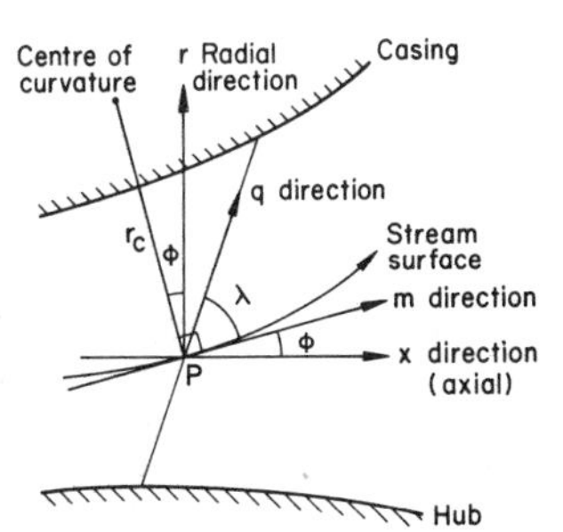

Figure 2. Geometry on a quasi-orthogonal [12].

This equation for the variation of V_m along the q-0 is solved in conjunction with the continuity equation for the flow across the q-O

$$\int_{q-0} 2\pi r\rho V_m \sin\lambda \, dp = \dot{m} \tag{2}$$

The first three terms on the right side of Equation 1 are generally referred to as the radial equilibrium terms and the last two terms as the streamline curvature terms. The limits of integration in Equation 2 take into account the annulus boundary-layer development. The distributions of total enthalpy, entropy, and angular momentum along q-Os can be obtained from the conservation of these quantities along stream surfaces. Within blade rows V_θ can be obtained from V_m, blade rotation and the imposed flow directions corresponding to the mean stream surface geometry. Stagnation enthalpy is then obtained by applying the Euler turbine equation along the stream surfaces, and entropy changes can be obtained from empirical loss correlations. The fluid density, needed for Equation 2, is obtained from the equation of state for the fluid.

The SCM is an iterative means of obtaining a solution for the flow field starting from an initial guess of stream surface shape. This guess gives an estimate of the streamline curvature terms in Equation 1. Starting at the first q-O, Equations 1 and 2 are solved to obtain a new V_m distribution and hence, by interpolation, new points of intersection of the stream surfaces with the q-O. This procedure is repeated at each q-O in turn until a new stream surface pattern is obtained for the whole machine. The streamline curvature terms of Equation 1 can then be updated together with the pitch angles φ, and the next iteration commenced using these updated values. If the new stream surface shapes are used as such to obtain new curvature terms, the iteration is generally unstable. To maintain stability, changes in the curvature terms and in φ must be damped. Over the past decade and a half questions of stability have been raised for many SCMs, problems with running very fine grid sizes have been encountered, and many codes are considered difficult to implement. Wilkinson [14] has examined the stability of the SCM for two-dimensional duct flows and has shown that the optimum damping factor decreases rapidly with the grid aspect ratio. His method can be extended to annular flow with bladed regions, though in practice the large changes in flow conditions both along q-Os and between adjacent q-Os make the method of value only as a guide to the optimum damping factor.

The usual practice [6] is to damp changes of stream surface position. However, Denton [15] and Garg [12] damped changes in the streamline curvature terms, and found that only half as many iterations were required for convergence. It is also usual [6] to recast Equation 1 using the fact that within a blade row V_θ depends on V_m and the blade angles. This introduces additional terms into the equation but allows changes in radial equilibrium terms to be undamped. Denton [15] and Garg [12], however, used Equation 1 directly, and found that the equations and programming were simpler but changes in the radial equilibrium terms had to be damped by a factor $\cos^2\beta$ in order to maintain stability.

An important aspect of the SCM is the use of some curve-fitting technique. The curve-fit procedure ties together the flow field between upstream and downstream locations, thus preserving the ellipticity of the flow field. In fact, Equation 1 is a first-order nonlinear differential equation, so the "ellipticity" depends upon the validity of the curve fit. The SCM involves first-order derivatives along the q-O and along the stream surface. Therefore, quality of the curve fit is extremely important to the method. Wilkinson [16] has shown the advantage of high resolution curve-fitting methods which require heavy damping but rapidly yield a high solution accuracy, and the disadvantages of low resolution spline and least square fits where so much damping is inherent in the estimation of curvature that deceptively little damping is needed in the solution procedure which has an inescapably low accuracy. Various curve fits used so far include parabolas, splines, and quartics.

In the SCM, as in other methods, the net effects due to viscous action are lumped into a total pressure loss and a displacement thickness at the hub and shroud boundaries. The effect of boundary layers on the hub and shroud has been specified either through a blockage factor [6, 12] or by direct calculation [11, 12]. The total pressure loss is calculated by means of a cascade model that generally has two distinct parts for calculation under design and off-design operation of the turbine. Several empirical correlations [17-20] for loss coefficients have been used, but these are based on data which are several decades old and frequently fail to represent the flow field adequately [9]. There is an urgent need to develop better correlations based on recent data that itself is very expensive to obtain. Horlock's review [21] of the loss coefficients used for turbines is now a quarter-century old, and Brown's comparison [22] of the parameters in the loss coefficients developed (mostly for axial flow compressors) in the fifties and sixties is also over a decade old.

The meridional flow solutions found by the SCM have been found [12] to compare reasonably well with the experimental data for a single-stage and a two-stage turbine, but this comparison is by no means exhaustive, and the good agreement may even be fortuitous. More comparisons are also available [9, 15] for axial flow turbines. Before discussing the matrix method, we note that adaptation of the SCM by Frost [9] is unusual in that calculating planes are located not only at axial locations between blade rows, but also within blade rows on the mean stream sheet, as in the matrix method. In principle this should have resulted in improved accuracy of prediction. That it does not appear to have been achieved is probably due to the use of an inadequate loss model as suggested by Davis and Millar [6].

Denton [15] discusses the difficulties encountered in dealing with transonic flow and multiple high pressure ratio stages. In principle the problems of dealing with supersonic flow within blade rows in an axisymmetric solution are very similar to those encountered in duct flow and blade-to-blade calculations described by Novak [23]. Supersonic patches in an unchoked duct flow present no problems, but as choking is approached problems arise due to the extreme sensitivity of velocities to flow rate and due to the occurrence of supersonic and subsonic solutions for the same mass flow. Denton made no attempt to obtain solutions for fully choked flow. He avoided the problem by assuming that the blade throat lies at the trailing edge. This assumption involves little error for highly convergent blade passages but could clearly be in error for impulse-type blading, especially with diverging stream surfaces. Novak's method [23] for obtaining the solution in fully choked flow in a duct can probably be extended to deal with the flow through a single blade row but its use in a multistage machine seems impracticable. In fact, axisymmetric or three-dimensional time marching methods [24, 25] can better handle the problem of supersonic flow within a blade row.

The nagging problem of finding the correct mass flow rate for a specified pressure ratio was given a novel solution by Denton [15]. His method abandons the continuity equation at the blade trailing edges during intermediate iterations, and replaces it with a requirement to obtain a specified or "target" pressure on the midstream surface at each trailing edge. After a few iterations, the solution achieves the specified values of target pressure, but the continuity equation is not satisfied at the trailing edges. The imbalance in mass flow is then used to change the target pressure at the next upstream trailing edge in such a way as to reduce the imbalance; the principle being that the flow rate through any blade row varies more rapidly with the upstream relative total pressure than with the downstream static pressure. This is particularly true for a choked blade row where the downstream static pressure does not affect the mass flow rate. This method of using target pressures enables converged solutions to be obtained even more rapidly than when the correct mass flow is known.

The Matrix Method

The matrix method for meridional flow solutions has been developed by Marsh [26], and Davis and Millar [6]. In this method the flow field is divided into a grid with a set of nodes. At each node, finite difference expressions are written to approximate the differential operators in the governing differential equations of motion and energy. The form of the difference approximation is a strong function of the grid established for the problem. Simple grids lead to very simple finite difference expressions which may achieve high-order accuracy. However, rectangular grids, though simple, provide extreme difficulties at the flow field boundaries in replicating general boundary shapes. Therefore, recent methods have used curvilinear grids conforming to the flow boundaries. In order to achieve an acceptable level of accuracy, a curvilinear grid requires considerably more points (about 10 to 15) in the basic finite difference "stencil" that establishes the value of a finite difference operator at a given point in the flow field. In any case, it is necessary to linearize the system of equations prior to solution by a difference technique. Linearization is accomplished with a variety of techniques ranging from lagging the evaluation of certain terms from one iteration to the next to the utilization of various carefully defined mathematical linearization operations.

The principal equation to be solved [26] is the Poisson differential equation for the stream function ψ

$$\frac{\partial^2\psi}{\partial x^2} + \frac{\partial^2\psi}{\partial y^2} = q(x, y), \tag{3}$$

where the nonlinear q is given by

$$q(x, y) = \frac{\partial\psi}{\partial x}\cdot\frac{\partial}{\partial x}[\ln(r\rho\beta)] + \frac{\partial\psi}{\partial y}\cdot\frac{\partial}{\partial y}[\ln(r\rho\beta)] + \frac{r\rho\beta}{W_x}\left[\frac{\partial\alpha}{\partial y} - T\frac{\partial s}{\partial y} - \frac{W_u}{r}\frac{\partial}{\partial y}(rV_u) - \gamma\right],$$

α, β, and γ have the meanings:

- Duct flow: $\alpha = H$, $\beta = 1$, $\gamma = 0$ (no blades)
- Flow through a stator or rotor row:

$$\alpha = I = H - \omega r V_u, \qquad \beta = B, \qquad \gamma = F_y = F_u \tan\lambda$$

where
- H = the specific stagnation enthalpy
- I = the relative (specific) stagnation enthalpy
- ω = the angular velocity of a blade row
- ρ = the static density
- T = the static temperature
- s = the specific entropy
- r = the radial coordinate
- V_u = the circumferential component of the absolute velocity
- $\bar{W}$ = the velocity vector relative to a blade ($W_u = V_u - \omega r$)

λ and μ ($= \tan^{-1} F_x/F_u$) define the geometry of the mean stream surface within a blade row in the $x - y$ coordinate system, and B is an integrating factor for the continuity equation. It can be physically interpreted as a quantity which is proportional to the local angular thickness of the stream surface, and in the meridional flow analysis, the thickness of the stream surface is assumed to be proportional to the width of the blade passage, that is, B = (circumferential width of the blade passage)/(blade pitch).

Equation 3 holds for the inviscid flow on a stream surface S_2, without any other simplifying assumptions. The shape of the S_2 surface depends upon the solution in the circumferential plane (i.e., S_1 surface) for its direction cosines and thickness distribution, which if determined accurately, lead to solution of Equation 3 without additional simplifications. A brief outline of the steps required for the solution of Equation 3 follows:

1. Place a curvilinear grid between the annulus walls and write finite difference expressions for the derivatives in terms of the function values at the nodes.
2. Estimate the flow properties at each node, evaluate q(x, y) in Equation 3, replace the Laplacian by its finite difference equivalent, and write Equation 3 at each node. This will produce a system of linear simultaneous equations which can be written in matrix form as

$$[A][\psi] = [Q] \tag{4}$$

3. Solve Equation 4 for the vector $[\psi]$.

4. Recompute the vector [Q] by calculating the flow properties at each node from the knowledge of the ψ-field.
5. Repeat Steps 3 and 4 until convergence criterion is satisfied.

Expressing the derivatives in finite-difference form requires that the Taylor series be truncated depending upon the desired accuracy. For third-order accuracy with a curvilinear grid, Davis and Millar [6] required 10 terms, and therefore used 10 nodes to solve for the basic coefficients in the difference equations. In general, this requires that information from four grid lines in both coordinate directions be used, otherwise the 10 equations will not be independent. For some geometries considered by Davis and Millar, it was necessary to use a higher order finite difference stencil (12 nodes) for the first derivative. The finite difference stencils which result from these formulations for irregular grids develop a very complex structure as shown in Figure 3. Such stencils complicate the solution considerably at the boundary nodes. Frequently it is necessary to alter the stencils to replace points which would fall outside the boundary by alternate points inside the flow domain. The multipoint finite-difference operators yield a banded, sparse but extremely large coefficient matrix [A] in Equation 4. To reduce computer storage [A] must be factored into an upper and lower triangular banded matrix before solution. This is necessary since the inverse of [A] is not a banded matrix and core storage requirements would otherwise become excessive. An excellent reference for these considerations is by Davis and Millar [27].

When rotors, stators, or inlet guide vanes occur between calculation planes, the energy level and fluid direction along any streamline is changed. Empirical cascade correlations are used to predict the exit conditions for given inlet conditions, blade geometry, type, and speed. Earlier, we commented upon the inadequacy of such correlations.

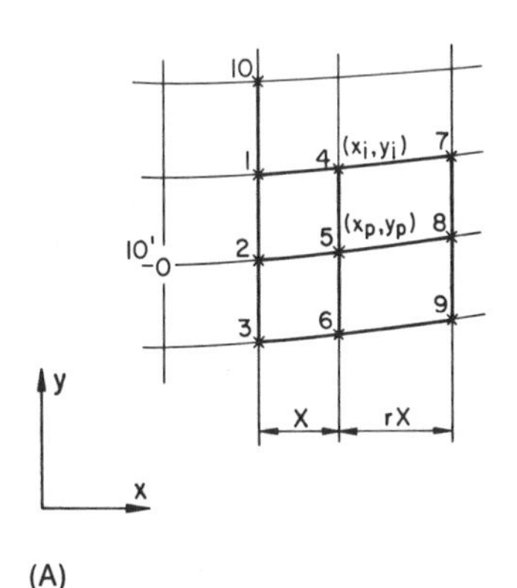

(A)

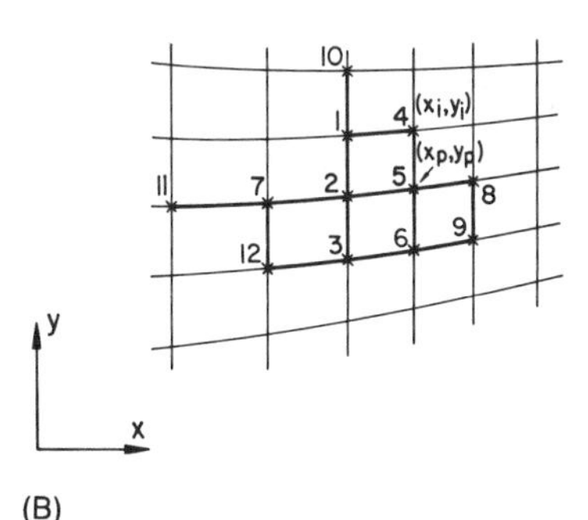

(B)

Figure 3. Finite difference stencils [27]: (A) 10-point stencil on the irregular grid for $\nabla^2 f$; (B) 12-point stencil on the irregular grid for $\partial f/\partial x$.

A very useful comparison between the SCM and matrix methods as applied to three axial-flow compressor applications is described by Davis and Millar [6]. They conclude that while the matrix method requires more care in programming since it involves auxiliary storage and overlay structure in any but a very large computer, it has fewer stability problems, is somewhat more accurate, and provides a better definition (due to a finer grid) than the SCM. However, there is no doubt that the SCM requires fewer nodes than the matrix method to achieve similar accuracy. Though Davis and Millar [6] found that the SCM required a large number of iterations for convergence as compared to the matrix method, Japikse, in a discussion of their work, refers to the possible advantages to be accrued to the SCM using the curve-fitting techniques of Wilkinson [16]. While a drawback of the matrix method is the difficulty experienced in handling slightly supersonic patches in the flow, a clear advantage [27] is the adaptability of one basic program to a variety of configurations. The same basic routines can be used for flow calculations in the meridional and the blade-to-blade surfaces, and should be readily adaptable to axisymmetric intakes and other shapes [6].

A further comparison between the SCM and matrix methods as applied to axial flow turbine calculations is provided by Frost [9]. The calculations were based on Frost's SCM and Marsh's matrix method [26], and yielded very good agreement between the results from the two methods but not with experimental data. Each method required about the same amount of computer time but the SCM used only about half the storage space required for the matrix method.

A contribution to the meridional flow computation by Katsanis [7] combines both the matrix method and the SCM into a single program. The purpose of this approach is to assure a rapid and stable calculation scheme as a base for flows approaching transonic conditions. In cases where transonic solutions are required, Katsanis carries the solution through with an SCM capable of modeling transonic flows. The SCM employed by Katsanis uses spline fits, and it is not obvious whether the precursor matrix method would be required if an alternate method of curve fitting were implemented.

Finite Element Methods

Finite element methods (FEM) for the meridional flow computations have been developed by Adler and Krimerman [28], Oates et al. [29], and Hirsch and Warzee [30, 31]. In this method, the streamline radius of curvature is *not* to be estimated, and arbitrary geometries that present the greatest difficulty for matrix methods are treated with equal ease. Thus, this method does not seem to suffer from the drawbacks of the previous methods.

While Adler and Krimerman [28], and Oates et al. [29] used variational principle formulations, Hirsch and Warzee [30, 31] followed a weighted residual Galerkin approach to develop their set of finite element equations. Oates et al. used a classical actuator disk approach where actuator disks were considered to exist within the flow field and jump conditions were satisfied across the disk. They postulated a variational function which turns out to be merely the integral of the meridional momentum over space. All terms in the variational function were evaluated using the basic governing equations. With the introduction of a finite element grid over the flow field, the variational principle eventually led to a nonlinear system of algebraic equations. Oates et al. [29] solved the system by a Newton-Raphson iteration with the density evaluation lagged by one iteration. Adler and Krimerman [28] used a somewhat similar approach.

In Hirsch and Warzee's method [30, 31], the basic equation is obtained, as in the matrix method, by introducing a stream function, but it is not necessary to reduce it artificially to a Poisson equation since it can be solved in the quasi-harmonic form. The governing equation in the fluid can be written in the form (c.f. Equation 39 of Reference 30),

$$\frac{\partial}{\partial r}\left(C_1 \frac{\partial \psi}{\partial r}\right) + \frac{\partial}{\partial z}\left(C_1 \frac{\partial \psi}{\partial z}\right) + q(r, z) = 0,$$

with boundary conditions

$$C_1 \frac{\partial \psi}{\partial n} + C_2(\psi - \psi_0) = 0,$$

where C_1 and C_2 are coefficients. An approximation of the stream function is to be searched such that the corresponding weighted residual, consisting of the volume and surface residuals, is zero, that is,

$$\int_v W(r, z) \cdot R_v(r, z)\, dv + \int_s W(r, z) \cdot R_s(r, z)\, ds = 0$$

When the solution is exact, the two residuals

$$R_v = \frac{\partial}{\partial r}\left(C_1 \frac{\partial \psi}{\partial r}\right) + \frac{\partial}{\partial z}\left(C_1 \frac{\partial \psi}{\partial z}\right) + q(r, z),$$

$$R_s = C_1 \frac{\partial \psi}{\partial n} + C_2(\psi - \psi_0),$$

are zero at any point in the flow field.

Combining the above equations and integrating by parts leads to the fundamental equation to be solved:

$$\int_\Omega \left[C_1\left(\frac{\partial \psi}{\partial r}\frac{\partial W}{\partial r} + \frac{\partial \psi}{\partial z}\frac{\partial W}{\partial z}\right) - qW\right] d\Omega = 0, \tag{5}$$

where ψ is taken to be the prescribed value ψ_0 on the corresponding part of the boundary, and Ω is the intersection of fluid volume v with the meridional plane. This equation is solved by Hirsch and Warzee [30] with the weight functions W chosen equal to the trial functions introduced in the finite-element process.

The meridional section of the machine is divided into contiguous finite elements connected at given points (nodes). On each element, the unknown function ψ is expressed as

$$\psi = \sum_1^n \psi_i N_i, \tag{6}$$

where n = the number of nodal points for the element (Figure 4: n = 8)
ψ_i = the (unknown) nodal value
N_i = the (known) nodal shape function

Hirsch and Warzee [30, 31] chose the elements to be isoparametric 8-node quadrilaterals (see Figure 4), with corresponding shape functions, N_i, biquadratic in the ξ, η local coordinates. Substituting Equation 6 into 5 with $W(r, z) = N_j(r, z)$, one gets

$$\int_E \left\{C_1\left[\frac{\partial N_j}{\partial r}\sum \psi_i \frac{\partial N_i}{\partial r} + \frac{\partial N_j}{\partial z}\sum \psi_i \frac{\partial N_i}{\partial z}\right] - qN_j\right\} d\Omega = 0,$$

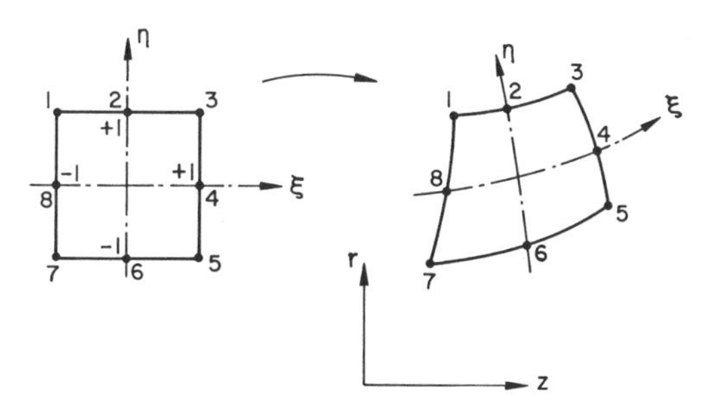

Figure 4. Eight-node isoparametric quadrilateral element [30].

or

$$[K]^e\{\delta\}^e = \{q\}^e \tag{7}$$

with

$$k_{ij}^e = \int_E C_1\left(\frac{\partial N_i}{\partial r}\frac{\partial N_j}{\partial r} + \frac{\partial N_i}{\partial z}\frac{\partial N_j}{\partial z}\right) d\Omega$$

as the "stiffness matrix," and

$$q_i^e = \int_E qN_i\, d\Omega; \qquad \delta_i^e = \psi_i^e$$

Equation 7 is written for each element e. If one assembles all the equations for all elements, the following matrix equation is obtained:

$$[K]\{\delta\} = \{Q\} \tag{8}$$

The solution is effected by initially assuming a stream function distribution, computing all the thermodynamic variables at the nodes on the first inlet station, and then evaluating the same variables throughout the flow field using conventional techniques well known in finite difference procedures (conserving enthalpy or rothalpy and angular momentum in duct regions). As with other finite element techniques, Equation 8 is solved iteratively with an independent density relationship. The authors used NASA correlations for deviations and losses through the rotor passages of axial-flow compressors (single- and multi-staged). Numerical results were found to compare well with the experimental data. The method of solution also allowed for calculation of transonic compressor stages. When compared to the SCM and matrix methods, the FEM is capable of solutions just as accurate and apparently as fast as the other approaches. However, application of the FEM to axial flow turbine calculations is still awaited.

BLADE-TO-BLADE FLOW SOLUTIONS

Classical solutions for the blade-to-blade flow were obtained by conformal transformation [32, 33], but these are restricted to incompressible cascade flow and are not very useful except as test cases for comparing the solutions obtained by numerical methods. "Singularity" methods in which the blades are replaced by vortex and sink-source distributions have been very successful in the study of incompressible flows, but their success is strongly dependent upon an assumption about the trailing edge flow—either that the Kutta condition holds at a sharp trailing edge (i.e., the flow leaves a cusped trailing edge smoothly, with equal velocity and equal pressure from the two surfaces) or that the location of the stagnation point on a rounded trailing edge is known. Linearized compressibility corrections have permitted the extension of these methods to locally sonic flow on the blade. Cascade sweep and lean have also been taken care of. When combined with a detailed boundary-layer solution these methods are computationally economical and have gained almost universal acceptance.

For compressible steady flow in the blade-to-blade plane, the three "field" methods, described earlier for the meridional flow solutions have been used. In the SCM, a differential equation for the gradient of velocity component along the normal, or quasi-normal (q-0), to the streamline is written in terms of the (assumed) radius of curvature of the streamline. This equation is integrated across the blade passage, the constant of integration being determined by the continuity equation. The streamlines are located, the radius of curvature is redetermined, and a solution is obtained by iteration.

In the matrix method, an equation for the stream function is derived and rewritten as a finite difference equation based on a particular form of grid. This equation is solved (simultaneously with an equation for local density) by matrix inversion. A recent method [34] solves the equation for the stream function by the finite element technique but is restricted to subsonic flow through the cascade. While the preceding methods essentially solve the inviscid flow equations for the cascade, with viscous effects accounted for through the use of correlations, a recent field analysis [35] is based on the solution of the full Navier-Stokes equations over the rotor blade-to-blade stream channels. It uses an alternating-direction-implicit (ADI) method to carry out the numerical integration of the elliptic governing equations. However, it considers only the case of laminar flows, and provides flow characteristics within the rotors of a radial inflow turbine and a radial bladed compressor.

Gostelow [36] provides a good review of compressible flow theories for airfoil cascades up to 1972.

Singularity Methods

Singularity methods have been used largely, but not entirely, in the blade-to-blade plane. The essence of these methods is to replace the blade surfaces by singularities (sources, sinks, vortices) to calculate the induced velocity in the field by the (as yet unknown) singularity distribution, and to satisfy the boundary conditions of the inviscid flow (that the velocity normal to the surface is zero).

The earliest solutions were obtained by Schlichting and his colleagues [37] who distributed vortices along the blade camber line and matched the camber surface at three chordwise locations. Separately they placed sources and sinks along the chord line to simulate the blade thickness. The vortex and source-sink distributions were expressed as Fourier series of the chordwise location. The induced velocity, normal to the mean flow, due to these distributions on a single airfoil is known analytically, and Schlichting, et al. calculated numerically the *extra* induced velocity due to adding other airfoils in cascade.

Improvement by Martensen [38] involves the solution of an integral equation for the velocity V(m, n) on the surface of the cascade airfoil (m, n) in the x-y plane. This velocity is found by differentiation of the stream function $\psi(x, y)$, which is of the form

$$\psi(x, y) = \psi_m + \int_0^{\ell} V(m, n) K(x - m, y - n)\, ds, \tag{9}$$

where ψ_m is the mean stream function based on the mean of the upstream and downstream flows, and K is a kernel function. Equation 9 is effectively an integral equation for surface velocity distribution. The method of solution is to satisfy the equation at N points on each side of the known blade surface and to solve by matrix inversion the 2N equations for the velocity.

Recently McFarland [39] has used advanced surface singularity formulations for calculating compressible inviscid flow through a linear cascade of arbitrary blade shapes. It is actually an incompressible flow solution, and compressibility effects are included through the Lieblein-Stockman correction [40]. This simplifies the code but keeps it sufficiently accurate, thus providing a means of rapidly screening blade designs for analysis by more complex codes that have better accuracy. McFarland uses the integral equation formulation of Bristow [41], with the higher order panel formulae as developed by Hess [42], and attributes the discrepancies in the solution for a transonic cascade flow to the use of the compressibility correction.

Field Methods (Subsonic Blade-to-Blade Calculations)

Field methods generally yield more accurate solution for the blade-to-blade flow than the singularity methods but are more complex. Most of the codes currently used model the flow as inviscid, irrotational, and compressible. The flow field is solved using the SCM, matrix method or the FEM. All these methods have been described in detail for the meridional flow solution, and their application to subsonic blade-to-blade flow calculations is straight forward. For transonic blade-to-blade solution, time marching numerical techniques seem to be the best, though recently the FEM has also been used for such calculations. Both the SCM and the matrix method appear to give satisfactory results for isentropic transonic flow, but their validity in a flow with shocks is open to doubt.

As already described, Wilkinson [14, 16] has carried out fundamental evaluations of the SCM, and has applied them to blade-to-blade calculations. His study of convergence and accuracy of the SCM is based on an evaluation of a uniform duct flow field to which a streamline modification is applied along the center streamline and then reduced parabolically across the duct width. The correction required to return the point of maximum displacement to the original uniform conditions is then examined. Based on this model, he derives a relationship for the optimum damping factor, the application of which to the point of maximum curvature in the flow field brings about a solution that converges in the least number of iterations to a prescribed degree of accuracy. Some independent corraboration of Wilkinson's concepts has been developed by Japikse et al. [43] and by Howard and Osborne [10]. For example, very rapid convergence using quartics for complex duct shapes has been achieved. Also, the relationship between number of iterations and minimum curvature has been confirmed for spline solutions.

Matrix methods for the blade-to-blade problem have been developed by Smith and Frost [44] and by Davis and Millar [45]. The latter requires the blade profile geometry, inlet conditions, and flow properties as input and computes velocities and fluid properties at all points of a curvilinear grid placed between the suction and pressure surfaces of the blade passage in the circumferential plane on a surface of revolution. The inviscid, rotational compressible flow field is determined by an iterative solution of the stream function equation in a manner similar to that described earlier.

The FEM for analyzing steady, two-dimensional, inviscid, subsonic flow through an arbitrarily shaped cascade on an arbitrary revolutional surface with a varying channel height has been developed by Shirahata and Daiguji [34]. The flow is, however, assumed to be homentropic and homorothalpic. The Poisson equation for the stream function is solved iteratively by minimization of the functional using linear triangular elements. The method is characterized by improved treatment of the periodic condition in the upstream and downstream regions and the Kutta condition, and by the accurate calculation of velocity by using the least squares method in comparison with the method by Adler and Krimerman [46]. This not only increases the accuracy but also saves computer time.

All the preceding methods are based on an inviscid analysis that is made to interact, at best, with an end wall boundary-layer solution. Although the inviscid analysis is useful for providing a considerable insight into the character of the flow, the neglect of viscous effects is a serious short-coming if detailed quantitative information is desired in order to calculate viscous losses or heat transfer. The approach used to account for viscous flow effects by using the boundary-layer concept in conjunction with an inviscid flow analysis has a number of drawbacks. First, the correct means for matching boundary-layer and inviscid solution do not exist if the inviscid flow is rotational, such as in the case of a curved or a rotating passage. Second, most of the existing interacting boundary-layer techniques are not capable of handling strong interaction mechanisms of the type present in turbomachine rotors. The parabolic flow approximation neglects completely the downstream influences. Consequently important effects such as surface curvature, downstream blockage, and reversed flow regions are totally ignored.

Khalil et al. [35] describe a method for analyzing the viscous flow through turbomachine rotors based on the solution of the full Navier-Stokes equations over the rotor blade-to-blade stream channels. The set of stream surfaces required for the analysis are themselves generated from the solution of the inviscid version of the Navier-Stokes equations as suggested by Wu [1]. An alternating-direction-implicit method is used to carry out the necessary numerical integration of the elliptic governing equations. Results for the flow characteristics within the rotors of a radial inflow turbine and a radial-bladed compressor are provided for the case of laminar flow only. It is claimed that the method is quite general and can deal with a wide range of applications. Possible modifications of the method in order to deal with turbulent flow cases are identified.

Transonic Blade-to-Blade Calculations

Advances in time-marching numerical techniques have made possible reliable transonic cascade calculations. The advantage of using a time-marching technique is that the resulting hyperbolic equations retain their form both in subsonic and supersonic portions of the flow field. Thus mixed subsonic-supersonic flows can be computed with automatic capturing of shock waves. Transonic flow analyses using various compressible flow calculation techniques in the subsonic flow regime and the method of characteristics in the supersonic regime are now obsolete. Besides the time-marching technique, an improved form of the SCM and recently the FEM have been used for transonic flow calculations.

Time-Marching Techniques

Time-dependent solutions of the Euler equations are now widely used for analyzing the flow through turbomachine blade rows. Solutions of the potential flow equation have also recently been extended to compute transonic shocked flow [47]. Although these can be computationally much more efficient than solutions of the Euler equations, the limitation to potential flow rules them out for applications where strong shock waves can occur. Solving the Euler equations is also the most common way of computing fully three dimensional flow in turbomachinery, even for subsonic flow, since it is not generally possible to assume irrotational flow.

The equations are solved in either finite difference or finite volume form. In the former [48, 49] it is usual to transform the computational domain into a uniform rectangular grid and to express the derivatives of the flow variables in terms of values at the nodes of this grid. Specialized numerical techniques (e.g., McCormack or Lax-Wendroff schemes) are required to ensure stability of the integration of the equations through time until a steady state is reached. In the finite volume method [50, 24] equations for the conservation of mass, energy, and momentum are applied to a set of interlocking control volumes formed by a grid in the physical plane. In this manner it is easier to ensure conservation of mass and momentum than in the differential approach but similar numerical schemes are necessary to ensure stability. In comparison to the finite-difference method, the finite volume approach yields a better physical understanding of the flow field owing to the use of a physical grid. This is an important consideration for design engineers who are not usually specialists in numerical analysis. However, preference for either the finite difference or the finite volume scheme is user dependent.

In Denton's opposed-difference method [24] for solving the Euler equations in finite volume form, one takes a very simple and fast first-order scheme and progressively adds on a second or higher order correction as the calculation converges. The method appears to have advantages of speed and simplicity over alternative second-order schemes, and is also extremely "robust". This is exemplified by its wide usage [51–55] for three-dimensional solutions of the Euler equations through blade rows. The same basic algorithm has also been used for flow in the meridional plane by Spurr [56], for unsteady flow by Mitchell [57], and for wet steam flow by Bakhtar and Tochai [58]. The general experience is that satisfactory accuracy can be obtained for most turbine blades, although Singh [59] shows that in some cases the inviscid solution is improved by the iterative addition of a boundary-layer calculation.

Despite the heavy use, and perhaps because of it, Denton's basic scheme [24] has been found to have several drawbacks. They arise mainly from the use of high-order correction factors, which, although ideal for smoothly varying flows, can cause problems at points of discontinuity such as stagnation points and shock waves. The method also becomes unstable when the streamwise component of velocity is negative (i.e., for back flow) and so is unable to deal accurately with the leading-edge flow on blades where the stagnation point lies on the pressure surface. In fact, due to very rapid changes in flow around a leading edge, finite differencing errors are particularly likely to occur there and thus influence the flow on the whole blade surface. It is therefore essential that accurate differencing schemes and sufficient grid points be used around the leading edge. At highly loaded leading edges it turns out that Denton's scheme [24] could produce changes of entropy (stagnation pressure) which have an adverse effect on the blade-surface velocities, particularly on the pressure surface.

In an effort to overcome these difficulties while retaining the advantages of speed and simplicity of the original scheme, Denton has recently revised the method [60]. The new method reduces the problems of shock smearing and lack of entropy conservation. It is based on the original opposed-difference scheme, but this is applied to a new type of grid consisting of quadrilateral elements which do not overlap and have nodes only at their corners. The use of a nonoverlapping grid reduces finite-differencing errors and gives complete freedom to vary the size of the elements. Both these factors help to improve entropy conservation. Considerable savings in computer run time (by a factor of about 3) were obtained by Denton by using a simple multigrid method whereby the solution is advanced simultaneously on a coarse and on a fine grid. The resulting method is claimed to be simpler, faster, and more accurate than its predecessor.

While considerable progress has been made in the calculation of inviscid transonic flows, it is important to consider viscous effects, and in particular the effects of shock and boundary-layer interaction. Singh [59] has shown that considerable enhancement in the quality of flow computations in transonic cascades can be achieved by the inclusion of an integral blade-surface boundary-layer method in an otherwise inviscid time-marching calculation for cascade flows. Recently Singh [55] has applied this boundary layer method to the three-dimensional time-marching technique [51] for the computation of flow in a transonic axial-flow compressor. The boundary-layer calculation forms an integral part of the whole computational scheme, which consists of

1. Use of the time-marching technique to determine the inviscid surface Mach number distributions at specified radial positions from hub to tip for each blade row of the stage.
2. Using these Mach numbers for the blade surface boundary layer displacement thickness calculations.
3. Computing the inviscid Mach numbers with mass flow adjustment (based on the calculated displacement thicknesses) on the blade surfaces.

The boundary-layer computations are done by using integral calculation methods that account for shock and boundary-layer interactions, when they exist. For the transonic axial-flow compressor stage analyzed, good agreement is obtained between the calculated Mach number contours and the experimental results. However, viscous effects are considered only in a limited manner.

Besides the finite-volume approach, other prominent time-marching methods include the two-step Lax-Wendroff scheme [48], the MacCormack scheme [61–63], and the bicharacteristics method [64]. Most of them are applicable to a cascade flow on an arbitrary revolutional flow surface and able to capture the shock waves, but they require long computer time. Meanwhile, the *relaxation methods* that employ the central-difference scheme for subsonic regions and the backward-difference scheme for supersonic regions require less computer time than the preceding time-marching methods, and are applicable to shock capturing, as well. However, their use has been limited to a shockless potential flow through a plane cascade [65].

Streamline Curvature Method

The SCM, traditionally used for through flow or subsonic blade-to-blade flow calculations, has recently been applied to transonic cascade flow computation by Daiguji and Nozaki [66]. In fact, they refer to it as "the method of streamlines" since the basic equations contain no curvature of streamline directly. Their method is based on the SCM developed by Wilkinson [16] with improvements in the velocity calculations, determination of streamlines, flow correction in the upstream and downstream regions, in the stability and convergence rate, and a further shortening of computer time. The method is applicable to flow conditions that deviate considerably from the shockless entry (but no separation bubble), to the transonic flows with locally supersonic regions (but without shock waves), and to the blades that do not have large radius of curvature at the leading edge. Though this method does require less computer time, it cannot match, in its present form the versatility of the time-marching method.

Finite Element Method

While there has been a rapid and widespread development of the FEM for subsonic blade-to-blade calculations based on a Galerkin approximation and direct solution of the system of equations [30, 31], application of the FEM to transonic flow calculations has been delayed due to some problems [67]. With the introduction of artificial viscosity [68] and artificial compressibility [69], however, these difficulties have been overcome. A detailed survey of the techniques devised to overcome these problems is given by Hafez et al. [70]. Finite element solutions for the transonic flow in a cascade are now available [67, 68, 71, 72]. While Ecer and Akay [68, 71, 72] employ a pseudo-time formulation using variational functional-based analyses using the classical Bateman's principle, Deconinck and Hirsch [67] apply an efficient and flexible Galerkin FEM to the full potential equation in artificial compressibility form. In fact, Ecer and Akay [73] have verified that Ritz-type finite elements obtained through the consistent use of Bateman's functional are equivalent to applying a Galerkin formulation to the conservative form of the governing equation.

Ecer and his associate [68, 71, 72] solve the two-dimensional full potential equation for steady transonic flow through cascades, and provide results for the NACA 0012 airfoils at different stagger angles with and without choking. They use a combined shock-fitting and shock-capturing scheme, and discuss the effect of computational grid on the convergence and accuracy of the solution. As expected, they find that a positive artificial viscosity is necessary for convergence of the solution in locally supersonic regions while no artificial viscosity is needed for convergence in locally subsonic regions provided the relaxation factor is positive. For the latter case they also determine bounds for the relaxation factor in order to assure stability of the numerical integration procedure in time, and find that for compressible flows the over-relaxation can be higher as a function of local compressibility compared to incompressible flows.

While Akay and Ecer [72] do provide some improvement of their basic method [68], by far the best FEM for transonic blade-to-blade calculations appears to be that of Deconinck and Hirsch [67]. With the use of artificial compressibility [69], they find that the transonic flow problem can be formulated in a manner that is formally identical to the subsonic formulation. Thus, the need for elaborate integral formulation disappears and a standard Galerkin method can be applied in the whole flow region, independently of the subsonic and supersonic nature of the flow. Their FEM has the following merits: automatic and built-in transformation of the physical body fitted coordinate system to a uniform computational grid; easy and exact introduction of arbitrary mixed-type boundary conditions on curved boundaries without losing accuracy; and the flexibility of changing the order of accuracy in the discretization by changing the type of the element. These features are in addition to the use of multigrid techniques and shock-capturing capacity by means of artificial compressibility. They also maintain the conceptual simplicity and efficiency of the finite difference methods based on the same equation by the use of successive line over-relaxation or approximate factorization for the iterative solution algorithm. Comparison of the classical bilinear second-order accurate elements with third-order accurate biquadratic Lagrange elements [74] shows an improved accuracy for the latter without supplementary cost.

Results for both compressor and turbine cascade flows containing shocks are presented [67] and compared with experimental data. Leading-edge regions with stagnation points are very well simulated without modification of the nose geometry, and excellent shock-capturing properties are observed.

THREE-DIMENSIONAL FLOW SOLUTIONS

In recent years several computational studies have been carried out with an aim to obtain three-dimensional descriptions of the flow pattern in the rotating or stationary blade passages of a turbomachine. Even in the inviscid approximation, however, fully three-dimensional calculations are still rare. Most of the effort has been put in the direction of quasi three-dimensional calculations following the work of Wu [1], where the fully three-dimensional flow is resolved into a succession of two-dimensional calculations.

In Wu's original approach and with the approximation of S_1 surfaces of revolution, one S_2 streamsurface has to be defined along which the throughflow is calculated in interaction with several blade-to-blade S_1 surfaces. The interaction between these two families of surfaces can be quite complicated, since the S_2 surface as well as the thickness of the associated streamsheet can only be determined by the S_1 solutions, which in turn require the knowledge of the through flow in order to provide the boundary conditions (e.g., inlet flow and thermodynamic conditions and outlet pressure) and the thickness of the S_1 streamsheets. The procedure is clearly iterative, and the rapidity of convergence depends upon the degree of interaction and upon its amplitude variation from one iteration to the other. Although the nature of interaction is clearly defined, the definition of S_2 surface presents a basic ambiguity. Wu's method does not provide any answer to this problem, and different choices have been made by various researchers in recent years leading to different computer codes. The differences lie not only in the definition of the S_2 surface but also in the numerical solution technique used.

Senoo and Nakase [75, 76] defined the S_2 surface on the basis of mean camberlines which has been shown [5, 77] not to give reliable results. Bosman and El-Shaarawi [77] generated their S_2 meridional surface from mass-averaged streamlines in each S_1 blade-to-blade surface; these mean streamlines are different from any actual S_1 streamline. Their equations in finite-difference form were solved by a matrix method giving the streamfunction on both S_1 and S_2 surfaces. Novak and Hearsey [5] defined their S_2 surface by the S_1 streamlines separating equal mass flows through the blade passage ($\psi = 0.5$ streamline), and base their solution on a streamline curvature method. Hirsh and Warzee [78, 79] followed a somewhat different approach. They replaced the calculation in the S_2 surface by the computation of the flow in the true meridional plane based on the solution of the exact pitch-averaged equations. These equations contain terms which are the contributions from the nonaxisymmetry of the flow, and which are responsible for the interaction from the S_1 surface. In this way the pitch-averaged flow is exactly solved and provides the boundary conditions for the blade-to-blade calculations. They used a FEM for the solution. Krimerman and Adler [80] considered several hub-shroud surfaces instead of one single "mean" S_2 surface as used in earlier references [5, 77, 78]. This procedure allows the use of nonaxisymmetric blade-to-blade surfaces in contrast to the previously referenced works that are restricted to the assumption of axisymmetric blade-to-blade surfaces. The numerical solution technique used by Krimerman and Adler is based on linear triangular finite elements.

In a recent work, Ucer et al. [81] describe the hub-to-shroud streamsurface with the single equation of Bosman and Marsh [82]. This brings simplicity and uniformity to the hub-to-shroud solutions of different geometries. The blade-to-blade solution is similar to that in References 78–80. However, unlike Hirsch and Warzee's work [78], for two-dimensional axisymmetric hub-to-shroud solution, the geometry of the surface is obtained by taking it parallel to the mean surface of the blades. This allows one to take into account the curvatures of the hub-to-shroud surface in all directions. The through-flow calculations are performed on such surfaces using the loss and deviation correlations for performance prediction. The quasi-three-dimensional solution is started with the blade-to-blade solutions on the hub, mid span, and the tip of the blades. The geometry at the hub-to-shroud surface and its thickness distribution is obtained at the mid streamline of the blade-to-blade solutions. The hub-to-shroud solution is then used to determine the new blade-to-blade surfaces of revolution. The calculation is continued until convergence is reached. Ucer et al. [81] obtain the blade-to-blade surfaces of revolution by rotating the streamlines around the axis, unlike Hirsch and Warzee [78] who locate the surfaces along the finite element boundaries in the meridional plane.

The numerical technique used by Ucer et al. [81] is similar to that used by Hirsch and Warzee [78] but differs considerably from that used by Krimerman and Adler [80] in that isoparametric elements are used. These elements are more efficient than the triangular elements, and are capable of describing the complicated geometries in a better way. Krimerman and Adler [80] also used a different technique in locating the upstream and downstream boundaries in that the boundary is moved away from the blades during the iterations until no further change is observed in the flow field.

All this heavy work on the quasi-three-dimensional calculations notwithstanding, a solution that is based on the coupling of two-dimensional flow solutions is *not* strictly a solution of the three-dimensional flow problem, and the deviation from the true solution is not only difficult to evaluate but could also lead to significant discrepancies between the actual and predicted flow characteristics of a newly designed cascade. In order to alleviate these problems Laskaris [83] has presented a finite element technique that is capable of solving the steady-state hydrodynamic equations of three-dimensional compressible potential flow in rotating or stationary flow regions. The numerical algorithm is based on the Galerkin's method of weighted residuals applied over isoparametric finite elements of the flow region where the velocity potential is approximated by second-order Lagrange polynomials and nodal parameters. The resulting system of nonlinear algebraic equations is solved by the Newton-Raphson iteration scheme. The method is applied to a mixed-flow turbine in subsonic flow of peak Mach numbers near unity. The computed pressure distribution on the blade surface discloses regions of steep pressures recovery where boundary layer separation could occur.

In addition to the preceding strictly inviscid attempts at three-dimensional flow analysis, several viscous flow-field-calculation procedures have also evolved in recent times. However, none of these has yet been applied to the analysis of a turbomachine though some do have the potential. It appears that the complexity of the turbomachine flow, particularly in specifying the numerical grid and boundaries, is presently overwhelming. Thus, only model problems have so far been treated in order to study cascade-type viscous flow fields numerically. Literature is quite abundant these days for such model analyses, and we will refrain from any further discussion on this topic.

CLOSURE: FUTURE TRENDS

In this chapter, we have reviewed a variety of steady and time-dependent, two- and three-dimensional, axial-flow turbine-design-and-analysis numerical techniques. Two-dimensional procedures have shown great progress and today constitute a comprehensive design methodology. Quasi-three-dimensional procedures are also available, though truly three-dimensional methods are still in their infancy.

While analysis of three-dimensional viscous flow through turbomachines is the ultimate goal, there are other problems that need to be worked on (e.g., the extension of some inviscid techniques to truly three-dimensional analysis for both subsonic and (more challenging) transonic flows, and the successful treatment of unsteady flow problems). Moreover, further basic experimental research will be indispensable while developing proper turbulence models.

With the availability of a large variety of computer codes these days, the use of common evaluation cases between different workers is urgently needed since basic choices between the available competitive design codes remain obscure. Besides, the codes are difficult to evaluate by would-be users. Most numerical code developers establish adequate procedures for a specific class of problems; the fundamental accuracy/stability/convergence problems being left for others to study academically, and vice versa. There is thus an urgent need for a conference on the lines of the celebrated 1968 Stanford Conference on Turbulent Flow Calculations.

REFERENCES

1. Wu, C. H., "A General Theory of Three-Dimensional Flow in Subsonic and Supersonic Turbomachines of Axial-, Radial-, and Mixed-Flow Types," NACA TN 2604, Jan. 1952.
2. Horlock, J. H., and Marsh, H., "Flow Models in Turbomachines," *J. Mech. Eng. Sci*, Vol. 13, No. 5 (1971), pp. 358–368.
3. Japikse, D., "Review-Progress in Numerical Turbomachinery Analysis," *J. Fluids Eng. Trans. ASME*, Vol. 98, No. 4 (Dec. 1976) pp. 592–606.
4. Deshpande, R. B., "Certain Properties of Numerical Schemes Used to Solve the Blade-to-Blade Flow Problem," *Int. J. Num. Meth. Eng.* Vol. 12 (1978), pp. 85–92.
5. Novak, R. A., and Hearsey, R. M., "A Nearly Three-Dimensional Intrablade Computing System for Turbomachinery," *J. Fluids Eng. Trans. ASME*, Vol. 99, No. 1 (March 1977), pp. 154–166.
6. Davis, W. R., and Millar, D. A. J., "A Comparison of the Matrix and Streamline Curvature Methods of Axial Flow Turbomachinery Analysis, From a User's Point of View," *J. Eng. Power, Trans. ASME*, Vol. 97, No. 4 (Oct. 1975), pp. 549–560.
7. Katsanis, T., "Use of Arbitrary Quasi-Orthogonals for Calculating Flow Distribution in the Meridional Plane of Turbomachine," NASA TN D-2546, Dec. 1964.
8. Novak, R. A., "Streamline Curvature Computing Procedures for Fluid-Flow Problems," *J. Eng. Power, Trans. ASME*, Vol. 89 (1967), pp. 478–490.
9. Frost, D. H., "A Streamline Curvature Through Flow Computer Program for Analyzing the Flow Through Axial-Flow Turbomachines," ARC R and M 3687, 1972.

10. Howard, J. H. G., and Osborne, C., "A Centrifugal Compressor Flow Analysis Employing a Wake-Jet Passage Flow Model," *J. Fluids Eng. Trans. ASME*, Vol. 99, No. 1 (March 1977), pp. 141–147.
11. Kündig, A. A., "Fast Numerical Procedure to Solve the Meridional Equations of Motion in a Multistage Axial Flow Turbomachine," ASME Paper No. 81-GT-133, 1981.
12. Garg, V. K., "Throughflow Analysis of Axial Flow Turbines," *Comp. Meth. Appl. Mech. Eng.* Vol. 37 (1983), pp. 129–137.
13. Smith, L. H., Jr., "The Radial-Equilibrium Equation of Turbomachinery," *J. Eng. Power, Trans. ASME*, Vol. 88, No. 1, (Jan. 1966), pp. 1–12.
14. Wilkinson, D. H., "Stability, Convergence, and Accuracy of Two-Dimensional Streamline Curvature Methods Using Quasi-Orthogonals," *Proc. Inst. Mech. Engrs.* London, Vol. 184, Part 3G(I) (1969–70), pp. 108–119.
15. Denton, J. D., "Throughflow Calculations for Transonic Axial Flow Turbines," *J. Eng. Power, Trans. ASME*, Vol. 100, No. 2 (April 1978), pp. 212–218.
16. Wilkinson, D. H., "Calculation of Blade-to-Blade Flow in a Turbomachine by Streamline Curvature," ARC R and M 3704, Dec. 1972.
17. Ainley, D. G., and Mathieson, G. C. R., "A Method of Performance Estimation for Axial Flow Turbines, "ARC R and M 2974, 1951.
18. Lenherr, F. K., and Carter, A. F., "Correlations of Turbine Blade Total-Pressure-Loss Coefficients Derived from Achievable State Efficiency Data," ASME Paper No. 68-WA/GT-5, 1968.
19. Dunham, J., and Came, P. M., "Improvements to the Ainley-Mathieson Method of Turbine Performance Prediction," *J. Eng. Power, Trans. ASME*, Vol. 92 (1970), pp. 252–256.
20. Balje, O. E., and Binsley, R. L., "Axial Turbine Performance Evaluation. Part A-Loss-Geometry Relationships," *J. Eng. Power, Trans. ASME*, Vol. 90, No. 4 (Oct. 1968), pp. 341–348.
21. Horlock, J. H., "Review-Losses and Efficiencies in Axial-Flow Turbines," *Int. J. Mech. Sci.* Vol. 2 (1960), pp. 48–75.
22. Brown, L. E., "Axial Flow Compressor and Turbine Loss Coefficients: A Comparison of Several Parameters," *J. Eng. Power, Trans. ASME*, Vol. 94, No. 3 (July 1972), pp. 193–201.
23. Novak, R. A., "Flow Field and Performance Map Computation for Axial flow Compressors and Turbines," AGARD Lecture series No. 83, 1976.
24. Denton, J. D., "A Time Marching Method for Two and Three-Dimensional Blade to Blade Flow," ARC R and M 3775, Oct. 1974.
25. Spurr, A., "Progress in the Development of a Time Marching Method for Through flow calculations," ARC 36907, 1976.
26. Marsh, H., "A Digital Computer Program for the Through-Flow Fluid Mechanics in an Arbitrary Turbomachine Using a Matrix Method," ARC R and M 3509, July 1966.
27. Davis, W. R., and Millar, D. A. J., "A Discussion of the Marsh Matrix Technique Applied to Fluid Flow Problems," *Trans. CASI*, Vol. 5, No. 2 (Sept. 1972), pp. 64–70.
28. Adler, D., and Krimerman, Y., "The Numerical Calculation of the Meridional Flow Field in Turbomachines Using the Finite Element Method," *Israel J. Tech.* Vol. 12 (1974), pp. 268–274.
29. Oates, G. C., Knight, C. J., and Carey, G. F., "A Variational Formulation of the Compressible Throughflow Problem," *J. Eng. Power, Trans, ASME*. Vol. 98, No. 1 (Jan. 1976), pp. 1–8.
30. Hirsch, Ch., and Warzee, G., "A Finite-Element Method for Through Flow Calculations in Turbomachines," *J. Fluids. Eng. Trans. ASME*, Vol. 98, No 3. (Sept. 1976), pp. 403–421.
31. Hirsch, Ch., and Warzee, G., "A Finite Element Method for the Axisymmetric Flow Computation in a Turbomachine," *Int. J. Num. Meth. Eng.* Vol. 10 (1976), pp. 93–113.
32. Merchant, W., and Collar, A. R., "Flow of an Ideal Fluid past a Cascade of Blades," ARC R and M 1893, 1941.
33. Gostelow, J. P., "Potential Flow through Cascades, Extensions to an Exact Theory," ARC CP808, 1964.
34. Shirahata, H., and Daiguji, H., "Subsonic Cascade Flow Analysis by a Finite Element Method," *Bull. JSME*, Vol. 24, No. 187 (Jan. 1981), pp. 29–36.
35. Khalil, I., Tabakoff, W., and Hamed, A., "Viscous Flow Analysis in Mixed Flow Rotors," *J. Eng. Power, Trans. ASME*, Vol. 102, No. 1 (Jan. 1980), pp. 193–201.

36. Gostelow, J. P., "Review of Compressible Flow Theories for Airfoil Cascades," *J. Eng. Power, Trans. ASME*, Vol. 95, No. 4 (Oct. 1973), pp. 281–292.
37. Schlichting, H., and Scholz, N., "Uber die Theoritsche Berechnung der Stromungsverluste eines ebenen Schaufel-gitters," *Ing. Arch.* Bd. XIX, Heft 1, 1951.
38. Martensen, E., "Calculation of Pressure Distribution over Profiles in Cascade in Two-Dimensional Potential Flow by means of a Fredholm Integral Equation," *Arch. for Rat. Mech. and Anal.* Vol. 3, No. 3 (1959), pp. 325.
39. McFarland, E. R., "Solution of Plane Cascade Flow using Improved Surface Singularity Methods," ASME Paper No. 81-GT-169, 1981.
40. Lieblein, S., and Stockman, N. O., "Compressibility Correction for Internal Flow Solutions," *J. Aircraft*, Vol. 9, No. 4 (April 1972), pp. 312–313.
41. Bristow, D. R., "Recent Improvements in Surface Singularity Methods for Flow Field Analysis about Two-Dimensional Airfoils," *Proc. 3rd Computational Fluid Dynamics Conf*, AIAA, 1977, Paper No. 77–641, pp. 95–105.
42. Hess, J. L., "Higher Order Numerical Solution of the Integral Equation for the Two-Dimensional Neumann Problem," *Comp. Meth. Appl. Mech. Eng.*, Vol. 2, Feb. 1973, pp. 1–15.
43. Japikse, D., Wei, S., and Howard, J. H. G., "Calculation of Quasi-Three Dimensional Flow Fields Including Total Pressure Gradients," manuscript.
44. Smith, D. J. L., and Frost, D. H., "Calculation of the Flow Past Turbomachine Blades," *Proc. Inst. Mech. Engrs.* London, Vol. 184, Part 3G(I), 1969–70, paper 27.
45. Davis, W. R., and Millar, D. A. J., "A Matrix Method Applied to the Analysis of the Flow Past Turbomachine Blades," Carleton Univ. Report No. ME/A 72-7, July 1972.
46. Adler, D., and Krimerman, Y., "Calculation of the Blade-to-Blade Compressible Flow Field in Turbo Impellers Using the Finite-Element Method," *J. Mech. Eng. Sci*, Vol. 19, No. 3, (June 1977), pp. 108–112.
47. Farrell, C., and Adamczyk, J., "Full Potential Solution of Transonic Quasi-3D Flow Through a Cascade Using Artificial Compressibility," ASME Paper No. 81-GT-70, 1981.
48. Gopalakrishnan, S., and Bozzola, R., "A Numerical Technique for Calculation of Transonic Flows in Turbomachinery," ASME Paper No. 71-GT-42, 1971.
49. Veuillot, J-P., "Calculation of the Quasi Three-Dimensional Flow in a Turbomachine Blade Row," *J. Eng. Power, Trans ASME*, Vol. 99, No. 1 (Jan. 1977), pp. 53–62.
50. McDonald, P. W., "The Computation of Transonic Flow Through Two-Dimensional Gas Turbine Cascades," ASME Paper No. 71-GT-89, 1971.
51. Denton, J. D., and Singh, U. K., "Time Marching Methods for Turbomachinery Flow Calculation," VonKarman Institute, Brussels, Lecture Series "Application of Numerical Methods to Flow Calculations in Turbomachines," April 1979.
52. Kopper, F., Milano, R., and Vanco, M., "An Experimental Investigation of Endwall Profiling in a Turbine Vane Cascade," *AIAA J.* Vol. 19, No. 8 (Aug. 1981), pp. 1033–1040.
53. Sarathy, K. P., "Computation of Three-Dimensional Flow Fields Through Rotating Blade Rows and Comparison with Experiment," ASME Paper No. 81-GT-121, 1981.
54. Barber, T. J., "Analysis of Shearing Internal Flows," AIAA Paper No. 81-0005, 1981.
55. Singh, U. K., "A Computation and Comparison with Measurements of Transonic Flow in an Axial Compressor Stage with Shock and Boundary Layer Interaction," *J. Eng. Power, Trans. ASME*, Vol. 104, No. 2 (April 1982), pp. 510–515.
56. Spurr, A., "The Prediction of 3-D Transonic Flow in Turbomachinery Using a Combined Throughflow and Blade-to-Blade Time Marching Methods," *Int. J. Heat Fluid Flow*, Vol. 2, No. 4 (Dec. 1980).
57. Mitchell, N., "A Time Marching Method for Unsteady 2-D Flow in a Blade Passage," *Int. J. Heat Fluid Flow*, Vol. 2 No. 4 (Dec. 1980).
58. Bakhtar, F., and Tochai, M., "An Investigation of 2-D Flows of Nucleating and Wet Steam by the Time-Marching Method," *Int. J. Heat Fluid Flow*, Vol. 2, No. 1 (1980).
59. Singh, U. K., "Computation of Transonic Flow in Cascade with Shock and Boundary Layer Interaction," *Proc. 1st Int. Conf. Num. Meth in Laminar and Turbulent Flow*, Swansea, July 1978.

60. Denton, J. D., "An Improved Time Marching Method for Turbomachinery Flow Calculation," *J. Eng. Power, Trans. ASME*, Vol. 105, No. 3 (July 1983), pp. 514–524.
61. Gopalakrishnan, S., and Bozzola, R., "Computation of Shocked Flows in Compressor Cascades," ASME Paper No. 72-GT-31, 1972.
62. Kurzrock, J. W., and Novick, A. S., "Transonic Flow Around Rotar Blade Elements," *J. Fluids Eng. Trans. ASME*, Vol. 97, No. 4 (Dec. 1975), pp. 598–607.
63. Erdos, J. I., Alzner, E., and McNally, W., "Numerical Solution of Periodic Transonic Flow through a Fan Stage," *AIAA J.* Vol. 15, No. 11 (Nov. 1977), pp. 1559–1568.
64. Delaney, R. A., and Kavanagh, P., "Transonic Flow Analysis in Axial-Flow Turbomachinery Cascades by a Time-Dependent Method of Characteristics," *J. Eng. Power, Trans. ASME*, Vol. 98, No. 3 (July 1976), pp. 356–364.
65. Ives, D. C., and Liutermoza, J. F., "Analysis of Transonic Cascade Flow Using Conformal Mapping and Relaxation Techniques," *AIAA J.* Vol. 15, No. 5 (May 1977), pp. 647–652.
66. Daiguji, H., and Nozaki, A., "Analysis of Transonic Cascade Flow using Method of Streamlines," *Bull. JSME*, Vol. 23, No. 182 (Aug. 1980), pp. 1313–1319.
67. Deconinck, H., and Hirsch, Ch., "Finite Element Methods for Transonic Blade-to-Blade Calculation in Turbomachines," *J. Eng. Power, Trans. ASME*, Vol. 103, No. 4 (Oct. 1981), pp. 665–677.
68. Ecer, A., and Akay, H. U., "Investigation of Transonic Flow in a Cascade Using an Adaptive Mesh," AIAA Paper 80–1430, 1980.
69. South, J. C., Keller, J. D., and Hafez, M. M., "Vector Processor Algorithms for Transonic Flow Calculations," *AIAA J.* Vol. 18, No. 7 (July 1980), pp. 786–792.
70. Hafez, M. M., Wellford, L. C., and Murman, E. M., "Finite Elements and Finite Differences for Transonic Flow Calculations," *Finite Elements in Fluids*, Vol. 3, John Wiley, 1978.
71. Akay, H. U., and Ecer, A., "Finite Element Analysis of Transonic Flows in Highly Staggered Cascades," AIAA Paper 81–0210, 1981.
72. Akay, H. U., and Ecer, A., "Transonic Flow Computations in Cascades Using Finite Element Method," ASME Paper 81-GT-4, 1981.
73. Ecer, A., and Akay, H. U., "On the Finite Element Formulation of Mixed Elliptic-Hyperbolic Problems in Fluid Dynamics," *Proc. Int. Conf. on Num. Meth. for Eng.*, Paris, 1978, pp. 315–322.
74. Zienkiewicz, O., *The Finite Element Method in Engineering Science*, London: McGraw-Hill, 1977.
75. Senoo, Y., and Nakase, Y., "A Blade Theory of an Impeller with an Arbitrary Surface of Revolution," *J. Eng. Power, Trans. ASME*, Vol. 93, No. 4 (Oct. 1971), pp. 454–460.
76. Senoo, Y., and Nakase, Y., "An Analysis of Flow Through a Mixed Flow Impeller," *J. Eng. Power, Trans. ASME*, Vol. 94, No. 1 (Jan. 1972), pp. 43–50.
77. Bosman, C., and El-Shaarawi, M. A. I., "Quasi-Three-Dimensional Numerical Solution of Flow in Turbomachines," *J. Fluids Eng. Trans. ASME*, Vol. 99, No. 1 (March 1977), pp. 132–140.
78. Hirsch, Ch., and Warzee, G., "An Integrated Quasi-3D Finite Element Calculation Program for Turbomachinery Flows," *J. Eng. Power, Trans. ASME*, Vol. 101, No. 1 (Jan. 1979), pp. 141–148.
79. Hirsch, Ch., and Warzee, G., "Quasi 3-D Finite Element Computation of Flows in Centrifugal Compressors," In *Performance Prediction of Centrifugal Pumps and Compressors*, S. Gopalakrishnan, et al. (Eds.) ASME, 1980, pp. 69–75.
80. Krimerman, Y., and Adler, D., "The Complete Three-Dimensional Calculation of the Compressible Flow Field in Turbo Impellers," *J. Mech. Eng. Sci.*, Vol. 20, No. 3 (1978), pp. 149–158.
81. Ucer, A. S., Yegen, I., and Durmaz, T., "A Quasi-Three-Dimensional Finite Element Solution for Steady Compressible Flow Through Turbomachines," *J. Eng. Power, Trans. ASME*, Vol. 105, No. 3 (July 1983), pp. 536–542.
82. Bosman, C., and Marsh, H., "An Improved Method for Calculating the Flow in Turbomachines Including a Consistent loss Model," *J. Mech. Eng. Sci.*, Vol. 16, No. 1 (1974), pp. 25–31.
83. Laskaris, T. E., "Finite-Element Analysis of Three-Dimensional Potential Flow in Turbomachines," *AIAA J.* Vol. 16, No. 7 (July 1978), pp. 717–722.

CHAPTER 44

GUIDELINES FOR EFFICIENCY SCALING PROCESS OF HYDRAULIC TURBOMACHINES WITH DIFFERENT TECHNICAL ROUGHNESSES OF FLOW PASSAGES

Jorg Osterwalder and **Lutz Hippe**

Department of Hydraulic Machines and Plants
Technical University
Darmstadt, Federal Republic of Germany

CONTENTS

INTRODUCTION, 1355

EXPERIMENTAL AND THEORETICAL FUNDAMENTALS OF EFFICIENCY SCALING, 1355
- Loss Analyses, 1357
- Loss Distribution Coefficient, 1357
- Resistance Laws, 1359
- Schematic Representation of Scaling Computation, 1359

SIMPLIFIED SCALING PROCESS FOR PRACTICAL APPLICATION, 1359
- Efficiency Scaling $\Delta\eta_{i\wedge}$, 1359
- Determination of Efficiency Loss $\Delta\eta_{i\wedge(ri)}$ Due to Roughness, 1365
- Determination of Technical Roughness $R'_{a(hs)}$ and $R''_{a(hs)}$ Characterizing "Hydraulically Smooth" Surfaces, 1368

COMPUTATION EXAMPLE FOR EFFICIENCY SCALING OF A STORAGE PUMP WITH $n_{q\wedge} = 25$, 1370

NOTATION,

REFERENCES,

INTRODUCTION

Scaling results from the sum of all Re- and roughness-induced changes of individual losses. Computations of this type are, among others, based on the results of loss analyses and the individual resistance laws of all components as well as numerous model tests with pumps, turbines, and pump-turbines. The complex scaling process has already been presented in the mentioned publications. The individual steps of computation and the most important results are presented in the following sections.

EXPERIMENTAL AND THEORETICAL FUNDAMENTALS OF EFFICIENCY SCALING

Changes of $\Delta\eta_i$, $\Delta\psi$, $\Delta\lambda_i$, as shown in Figures 1 and 2, are designated as scaling in general, whereas the efficiency scaling $\Delta\eta_i$ discussed here is to be regarded as the quantity of primary interest. The very complex flow conditions preclude a solely theoretical treatment of the questions under

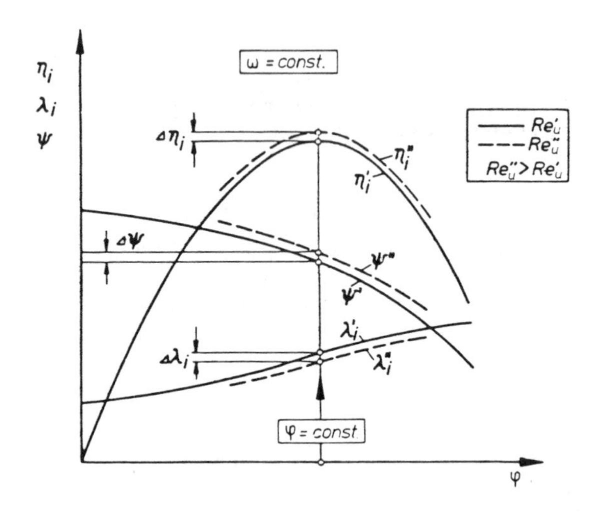

Figure 1. Pump characteristics for illustration of scaling influence on $\Delta\eta_i$, $\Delta\psi$, $\Delta\lambda_i = f(Re_u)$ with φ = constant.

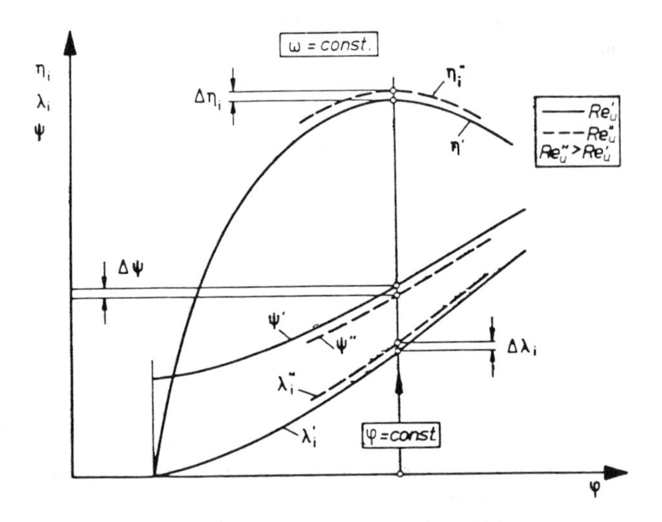

Figure 2. Turbine characteristics for illustration of scaling influence on $\Delta\eta_i$, $\Delta\psi$, $\Delta\lambda_i = f(Re_u)$ with φ = constant.

review. We shall consider the system in a manner such that the loss behavior refers to a three-dimensional, quasi stationary flow with occasionally displaying separation in partly steady, partly rotating flow passages of irregular cross-sections. A point of major interest is at first the determination of the individual loss components and their resistance laws, as reviewed in the following sections.

Loss Analyses

Figures 3 and 4 illustrate the composition of the individual losses. These data refer to power station machines of conventional type. The dimensions of flow passages taken as basis for calculation are average values derived from statistical surveys conducted with the assistance of well-known manufacturers. This also includes data related to the machining method and surface roughness.

Loss Distribution Coefficient

Another task consists in subdividing the individual losses (Figures 3 and 4) into scalable and non-scalable components. Theoretical and experimental investigations provide information on the respective wall friction component $\vartheta_{f(p)}$, which alone is subject to scaling. Scaling tests on complete model machines $[\Delta\eta_i = f(Re)]$ permit determination of the overall scalable component $\vartheta'_f = \sum \vartheta'_{f(p)}$ and facilitate, among other things, the verification of the mentioned individual loss analyses [1–3]. Characteristic data of this process are the loss distribution coefficients

$$v'_p = \frac{\vartheta'_{f(p)}}{\vartheta'_{(p)}} \quad \text{(individual loss)}$$

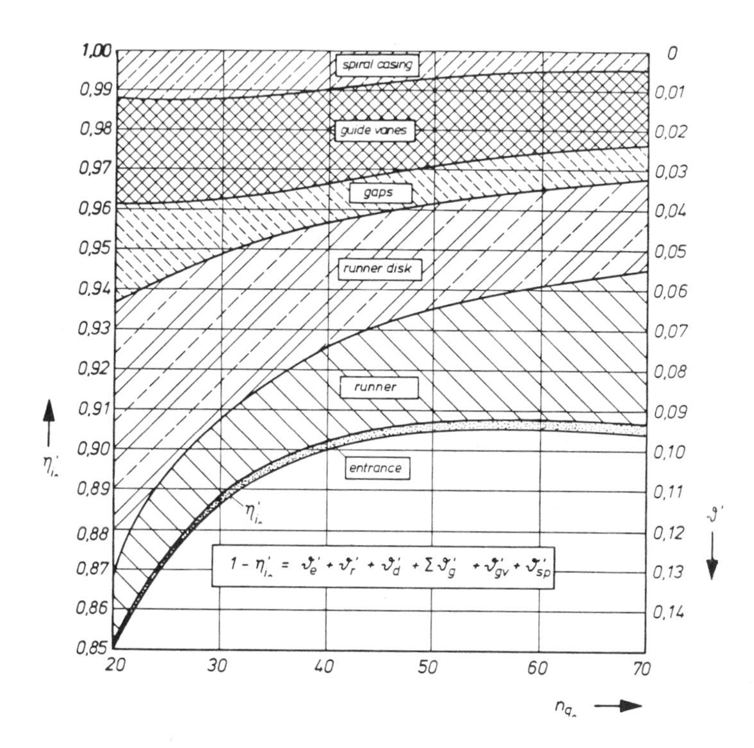

Figure 3. Loss composition of single-stage, single-flow *storage pump model machines* with statistically average representative geometry and "hydraulically smooth" surfaces. The graph refers to the operating range of best efficiency and $Re' = 8 \cdot 0^6$.

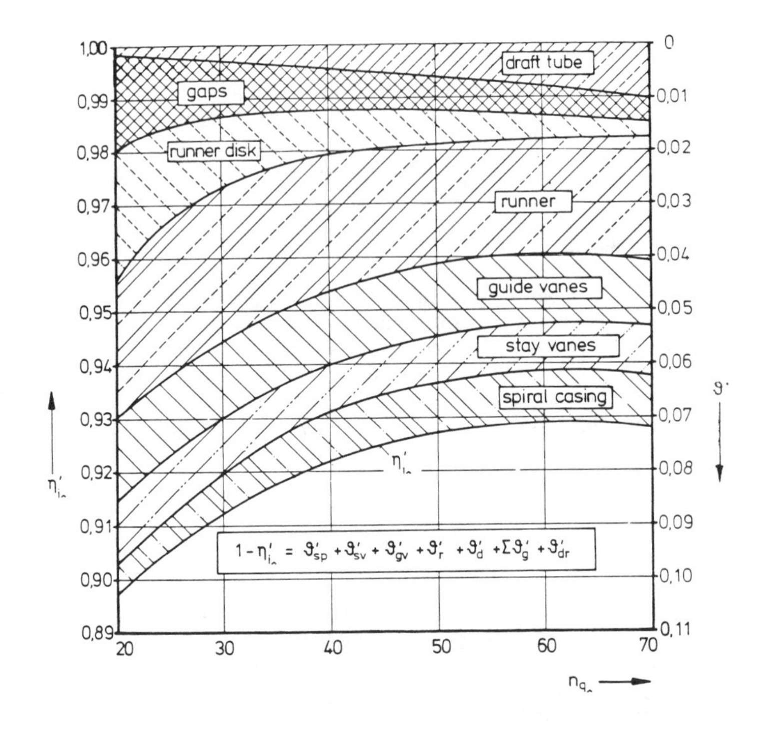

Figure 4. Loss composition of *Francis turbine model machines* with statistically average representative geometry and "hydraulically smooth" surfaces. The graph refers to the operating range of best efficiency and $Re' = 6 \cdot 10^6$.

and

$$V' = \frac{\vartheta'_f}{1 - \eta'_{i\wedge}} \qquad \text{(total loss)}$$

$V'_{\wedge(ref)} = f(n_{q\wedge})$ is shown in Figure 5. It is worth mentioning that $V'_\wedge$ decreases with a growing Reynolds number (cf. p. 1370).

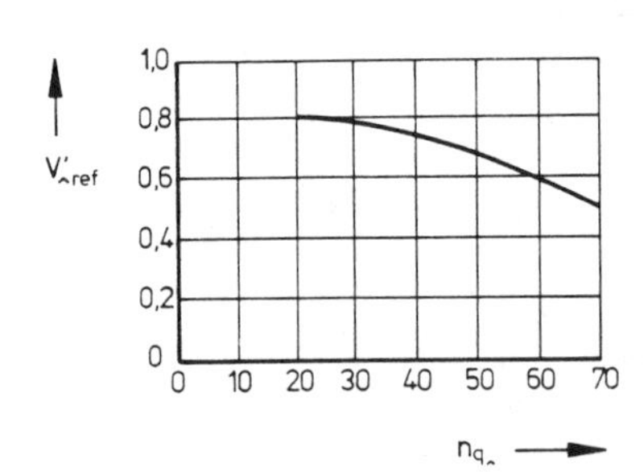

Figure 5. Loss distribution coefficient $V'_{\wedge ref}$ for power station model machines (turbines, pumps, pump turbines) resulting from scaling tests [1] of WG 5. The curve represents average data of numerous laboratory tests.

Resistance Laws

The resistance laws of the individual flow-guiding elements are to be regarded as the basis for the scaling procedure (see Table 1). These laws apply to the range of turbulent flow and to surfaces characterized by the terms "hydraulically smooth" and "technically rough." A prerequisite for their application are data concerning the local Reynolds number and the surface roughness in the flow direction. Information on Re definitions and respective roughness parameters are also given in Table 1.

Schematic Representation of Scaling Computation

The scaling process is reviewed in the following by means of the schematic representation in Figure 6. Proceeding from the model efficiency $\eta'_{i\wedge}$ or total loss $\sum \vartheta_{(p)} = 1 - \eta'_{i\wedge}$ (see Figure 1), the sum of scalable losses is obtained as $\vartheta'_f = [1 - \eta'_{i\wedge}] \cdot V'_\wedge$, with $V'_\wedge$ from Figure 5. The subdivision of ϑ'_f into its individual components $\vartheta'_{f(p)}$ is made in accordance with the section "Loss Distribution Coefficient" and Figure 5. It is worth mentioning that deviations or inaccuracies in this subdivision have no appreciable influence on the main result $\Delta\eta'_{i\wedge}$ [1, 3]. The conversion of scalable losses $\vartheta'_{f(p)}$ from Re′ to Re″ is carried out on the basis of the resistance laws in Table 1 and includes all individual surface roughness parameters. The efficiency scaling sought is:

$$\Delta\eta_{i\wedge} = \sum \vartheta'_{f(p)} - \sum \vartheta''_{f(p)} \quad (1)$$

Computations of this type require specialized knowledge and are very time consuming. In this situation $\Delta\eta_{i\wedge}$ was determined for pumps and turbines of different type ($n_{q\wedge} = 20 - 70$), surface roughness, and Reynolds numbers Re′, (Re″/Re′) by means of electronic data processing. These extensive preparations permit a substantial simplification of scaling computation. The procedure is reviewed in the following section.

SIMPLIFIED SCALING PROCESS FOR PRACTICAL APPLICATION

The following computations are based on the previously mentioned results from electronic data processing which yield numerical values on $\Delta\eta_{i\wedge}$, the efficiency loss $\Delta\eta_{i\wedge(ri)}$ and $\bar{R}_{a(hs)}$ for "hydraulically smooth" surfaces. These parameters are subsequently reviewed in the mentioned sequence.

Efficiency Scaling $\Delta\eta_{i\wedge}$

This is the result of previously mentioned computations (Figure 6), and may be regarded as being well-known. On the other hand $\Delta\eta_{i\wedge}$ can be represented as shown in Equation 2 which is the basis for further considerations as follows:

$$\Delta\eta_{i\wedge} = \sum \vartheta'_{f(p)} - \sum \vartheta''_{f(p)} = V'_\wedge \cdot (1 - \eta'_{i\wedge}) - \left(\frac{c''_w}{c'_w}\right)_{rep} \cdot V'_\wedge \cdot (1 - \eta'_{i\wedge})$$

$$\Delta\eta_{i\wedge} = V'_\wedge \cdot (1 - \eta'_{i\wedge}) \cdot \left[1 - \left(\frac{c''_w}{c'_w}\right)_{rep}\right] \quad (2)$$

Under these circumstances $(c''_w/c'_w)_{rep}$ is the only unknown quantity which refers to mean resistance coefficients representative for all flow passages with any roughness conditions (Table 1).

$$\left(\frac{c''_w}{c'_w}\right)_{rep} = 1 - \frac{\Delta\eta_{i\wedge}}{V'_\wedge \cdot (1 - \eta'_{i\wedge})} \quad (3)$$

Table 1
Resistance Laws and Corresponding Re of Individual Machine Components

Machine Component	Reynolds Number	Resistance Law	Ref.	Characterizing Roughness Parameter	Remarks
Spiral case (turbines)	$Re_{sp} = \frac{\bar{c}_{sp} \cdot D_{sp}}{\nu}$	Technically rough pipes	[5]	$R_{a_{sp}}$	
Spiral case (pumps)	$Re_{sp} = \frac{\bar{c}_{sp} \cdot L_{sp}}{\nu}$	Technically rough flat plates and fully turbulent flow	[2]	$R_{a_{sp}}$	Laminar approach flow lengths are neglected
Stay vanes	$Re_{sv} = \frac{\bar{c}_{sv} \cdot L_{sv}}{\nu}$			$R_{a_{sv}}$	
Guide vanes	$Re_{gv} = \frac{\bar{c}_{gv} \cdot L_{gv}}{\nu}$			$R_{a_{gv}}$	
Runner	$Re_{r} = \frac{\bar{w} \cdot L_{r}}{\nu}$			$R_{a_{r}}$	

Area of runner-side spaces	$Re_u = \frac{u_1 \cdot D_1}{\nu}$ (turbines) $Re_u = \frac{u_2 \cdot D_2}{\nu}$ (pumps)	Experimentally determined resistance laws of rotating runner-like bodies in corresponding housings	[6]	R_{a_d}	Fully turbulent flow and technically rough surfaces
Inlet (pumps)	$Re_e = \frac{c_e \cdot D_e}{\nu}$	Technically rough pipes	[5]	R_{a_e}	Reynolds number and roughness refer to entrance area
Draft tube (turbines)	$Re_{dr} = \frac{c_3 \cdot D_3}{\nu}$			$R_{a_{dr}}$	
Gaps	Re_g results from combination of Re_u and Re_{c_g} according to literature	Rotating gaps respectively technically rough pipes	[13]	R_{a_g}	Assumptions according to [1], [3]: • Model and prototype gap operate in the transition region between "hydraulically smooth" and "fully rough" surfaces with $(R_{a_g}/s)' = (R_{a_g}/s)''$ • In case of labyrinth gaps Q_g is calculated on the basis of the sum of losses of all individual straight gaps.

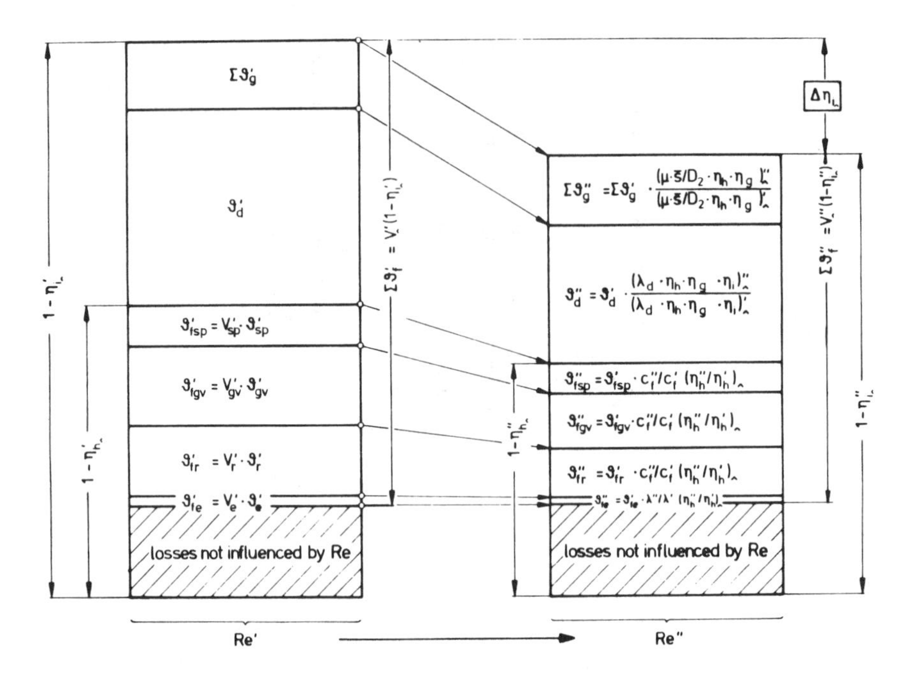

Figure 6. Schematic plotting to demonstrate at the example of a radial pump the method for determination of efficiency scaling $\Delta\eta_{i\wedge}$. The resistance coefficients c_f, λ_d, μ refer to technically rough surfaces and can be taken from References 5, 6, 10, and 13.

The quotient $(c''_w/c'_w)_{rep}$ was calculated for a large number of pumps and turbines of different specific speed $n_{q\wedge}$, loss composition, nominal diameter, Re, and different types of surface roughness. On this basis representative resistance laws can be established which consider the combined effects of all partial losses. Corresponding examples are shown in Figures 7 and 8 for pumps and turbines with $n_{q\wedge} = 30$ and 40. Analogous diagrams were plotted for machines of different specific speed $n_{q\wedge} = 20 - 70$. Deviations due to the difference of machine type ($n_{q\wedge}$) or loss composition are considered by means of K in Figures 9 and 10. With Figures 7 through 10 efficiency scaling can now be determined in a simple manner with the following equation (see computation example at the end of the chapter).

$$\Delta\eta_{i\wedge} = V'_{\wedge} \cdot (1 - \eta'_{i\wedge}) \cdot \left[1 - K \cdot \left(\frac{c''_w}{c'_w}\right)_{rep}\right] \tag{4a}$$

Application of Equation 3 poses some problems when plotting Figures 7 and 8, as $(c''_w/c'_w)_{rep}$ is only known as a quotient, which precludes individual data on $c'_{w\,rep}$ and $c''_{w\,rep}$. This can be solved by assumption of a random numerical value $c'_{w\,rep}$ for Re' on the "hydraulically smooth" curve, and the determination of $c''_{w\,rep}$ is as follows:

$$c''_{w\,rep} = \left(\frac{c''_w}{c'_w}\right)_{rep} \cdot c'_{w\,rep} \tag{4b}$$

Figures 7 and 8 yield numerically correct quotients $(c''_w/c'_w)_{rep}$. The individual data $c''_{w\,rep}$ and $c'_{w\,rep}$ are, however, of a fictitious nature and depend on the mentioned assumption of $c'_{w\,rep}$. This is no disadvantage, as $\Delta\eta_{i\wedge}$ in Equation 4 depends only on the described quotient. It is worth mentioning

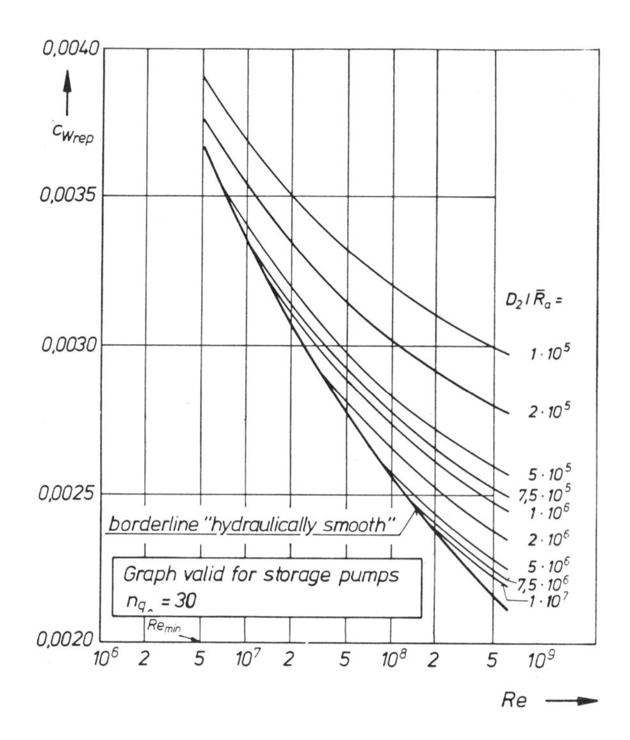

Figure 7. Plotting of resistance coefficients representative for all individual losses. $c_{w\,rep}$ is determined on the basis of Equations 3 and 5 and includes all significant resistance coefficients. The conversion to other $n_{q\wedge}$-values than $n_{q\wedge} = 30$ is carried out with Figure 9.

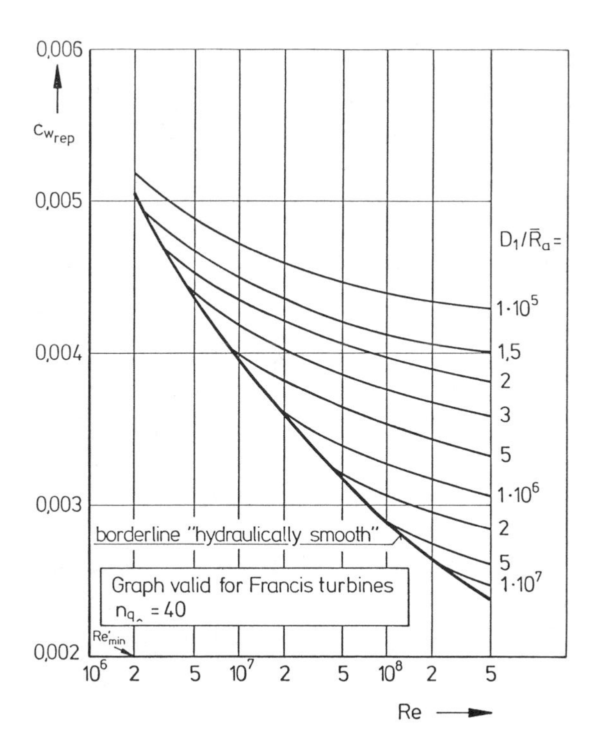

Figure 8. Plotting of resistance coefficients representative for all individual losses. $c_{w\,rep}$ is determined on the basis of Equations 3 and 5 and includes all significant resistance coefficients. The conversion to other $n_{q\wedge}$-values than $n_{q\wedge} = 40$ is carried out with Figure 10.

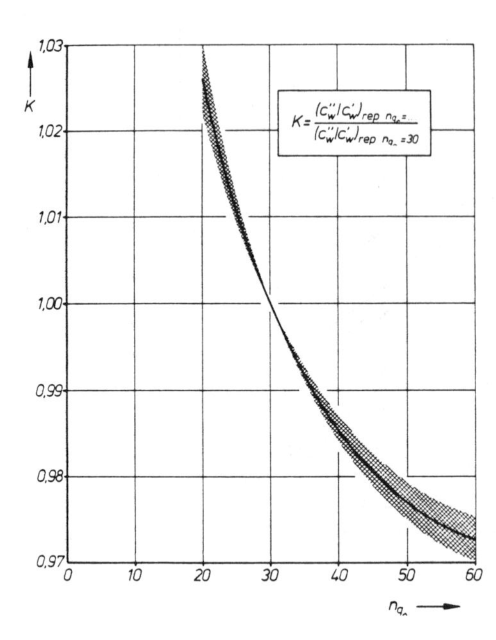

Figure 9. Correction K for conversion of $(c''_w/c'_w)_{rep}$ in Figure 7 to other specific speeds $n_{q\wedge}$ than $n_{q\wedge} = 30$. K is an average value of a scattering range which results from different surface roughnesses $D_2/\bar{R}_a$.

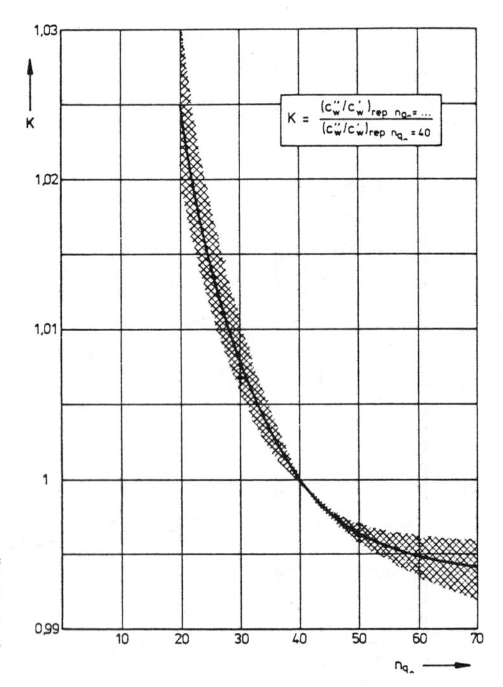

Figure 10. Correction K for conversion of $(c''_w/c'_w)_{rep}$ in Figure 8 to other specific speeds $n_{q\wedge}$ that $n_{q\wedge} = 40$. K is an average value of a scattering range which results from different surface roughnesses $D_1/\bar{R}_a$.

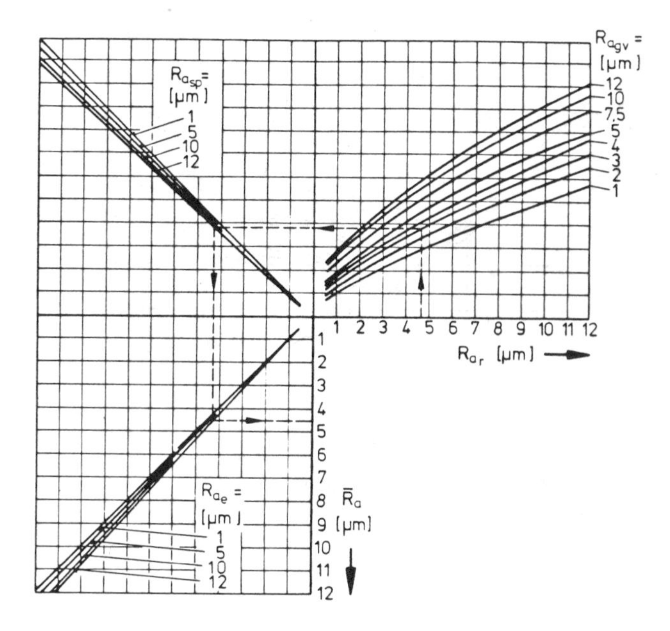

Figure 11. Nomograph for *storage pumps* with $n_{q\wedge} = 20–40$ for easy determination of the average weighted roughness $\bar{R}_a$. The runner-side spaces are assumed to be operated by turning so that roughness influence can be neglected.

that $c'_{w\,rep}$ is in accordance with the resistance laws of flat plates [1, 4], and pipes [5] for Reynolds numbers commonly used in model tests as follows:

$$c'_{w\,rep} = 0.0035 \rightarrow Re' = 8 \cdot 10^6 \text{ (Figure 7)}$$

$$c'_{w\,rep} = 0.0046 \rightarrow Re' = 5 \cdot 10^6 \text{ (Figure 8)}$$

These are weighted average values which satisfy the enhanced significance of the resistance law of "hydraulically smooth" flat plates in comparison with the one of pipes.

The application of scaling law (Equation 4) and particularly the resistance laws, Figures 7 and 8, requires information about the mean representative roughness $D_2/\bar{R}_a$ (pump) or $D_1/\bar{R}_a$ (turbine) of the model and prototype. In accordance to the mentioned determination of $c_{w\,rep}$ the respective results were elaborated by electronic data processing for various surface roughnesses of the individual flow passages. The extensive iteration computations have shown that $\bar{R}_a$ is nearly independent of runner diameter and Reynolds number. The results can be clearly plotted as nomographs (Figures 11 through 14). It should be mentioned that the roughness in the area of the runner-side spaces has practically no influence on $\vartheta_{f(d)}$ in case of turning operation [6] and is disregarded in Figures 11 through 14.

With Equation 4 and Figures 7 through 14 all data required for scaling computation are supplied. The example reviewed later illustrates the process.

Determination of Efficiency Loss $\Delta\eta_{i\wedge(ri)}$ Due to Roughness

According to Figure 15 step-up of efficiency $\Delta\eta_{i\wedge}$ is composed of two steps 1–2 (Re influence → "hydraulically smooth") and 2–x (effect of roughness). For a "hydraulically smooth" model machine

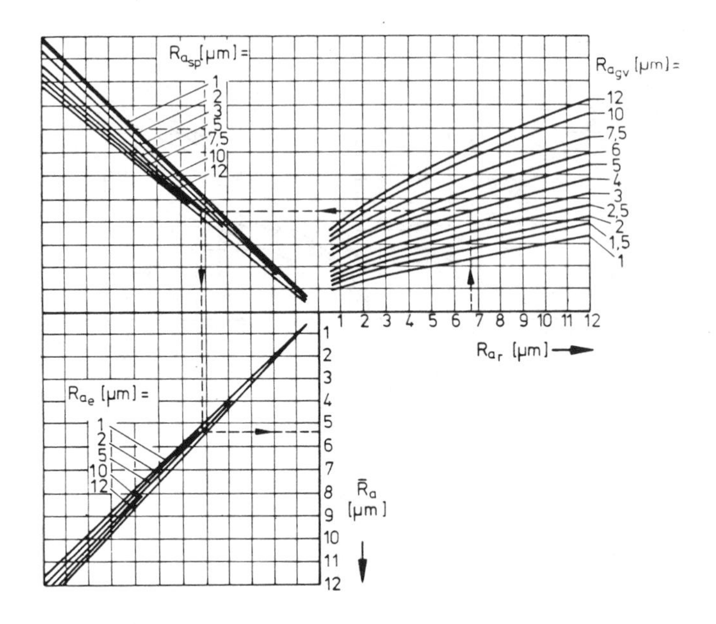

Figure 12. Nomograph for *storage pumps* with $40 < n_{q\wedge} \leq 60$ for easy determination of the average weighted roughness $\bar{R}_a$. The runner-side spaces are assumed to be operated by turning so that roughness influence can be neglected.

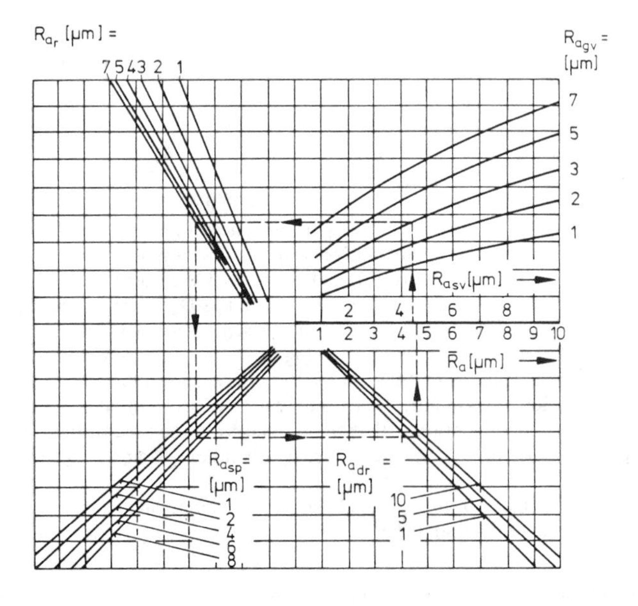

Figure 13. Nomograph for *Francis turbines* with $n_{q\wedge} = 20–40$ for easy determination of the average weighted roughness $\bar{R}_a$. The runner-side spaces are assumed to be operated by turning so that roughness influence can be neglected.

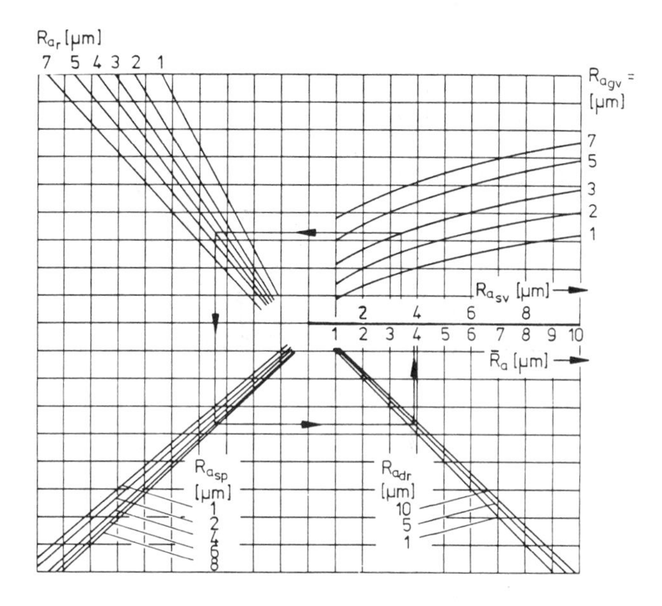

Figure 14. Nomograph for *Francis turbines* with $40 < n_{q\wedge} \leqq 70$ for easy determination of the average weighted roughness $\bar{R}_a$. The runner-side spaces are assumed to be operated by turning so that roughness influence can be neglected.

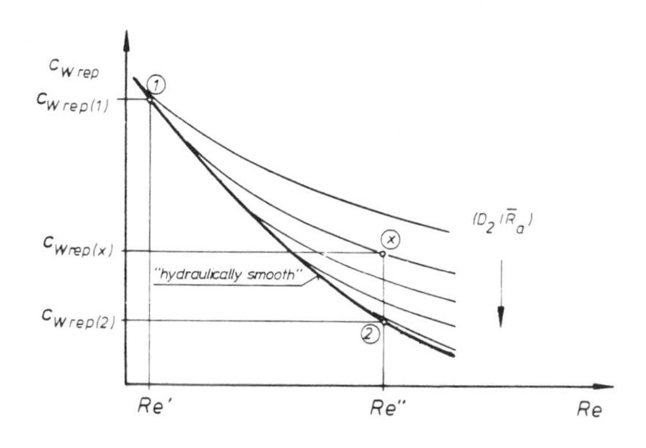

Figure 15. Schematic representation to illustrate the derivation of Equation 5.

1–2: Scaling "hydraulically smooth"

1–x: Scaling from model to prototype with technically rough surfaces

2–x: Representative quantity for efficiency loss $\Delta\eta_{i\wedge(ri)}$ due to roughness influence

roughness influence on the prototype can be determined with Equation 4a as follows:

$$\Delta\eta_{i\wedge(ri)} = \Delta\eta_{i\wedge(hs)} - \Delta\eta_{i\wedge}$$

$$= V'_\wedge \cdot (1-\eta'_{i\wedge}) \cdot \left(1 - K \cdot \frac{c''_{w\,rep(2)}}{c'_{w\,rep(1)}}\right) - \left(V'_\wedge \cdot (1-\eta'_{i\wedge}) \cdot \left(1 - K \cdot \frac{c''_{w\,rep(x)}}{c'_{w\,rep(1)}}\right)\right)$$

$$\Delta\eta_{i\wedge(ri)} = V'_\wedge \cdot (1-\eta'_{i\wedge}) \cdot K \cdot \frac{c''_{w\,rep(2)}}{c'_{w\,rep(1)}} \cdot \left[\frac{c''_{w\,rep(x)}}{c'_{w\,rep(2)}} - 1\right] \quad (5)$$

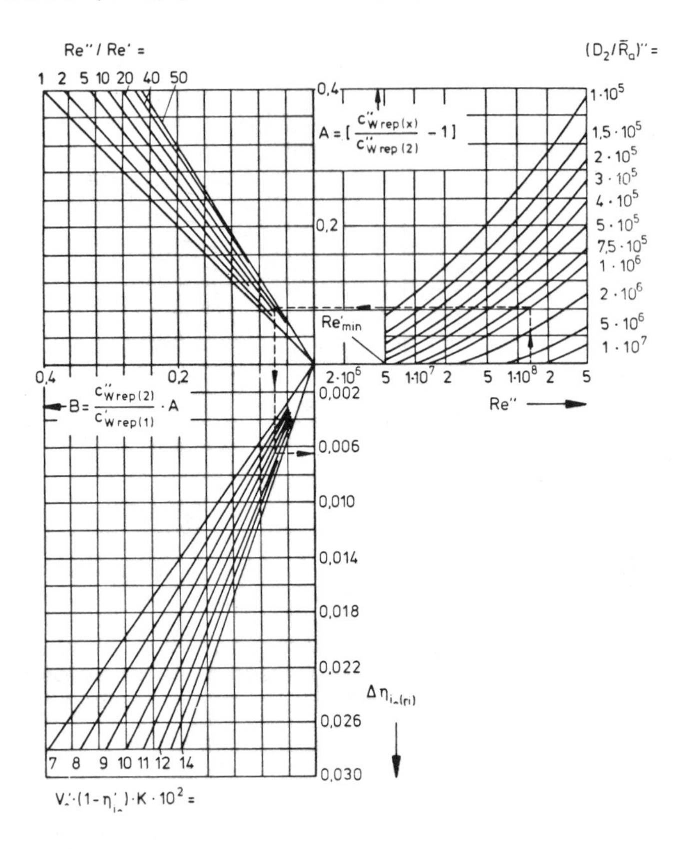

Figure 16. Nomograph for *storage pumps* with $n_{q\wedge} = 20–60$ for determination of efficiency loss $\Delta\eta_{i\wedge(ri)}$ due to roughness influence. This diagram is elaborated on the basis of Equation 5.

This result $\Delta\eta_{i\wedge(ri)}$ can be determined quickly with the aid of Figures 16 and 17. Such data are of interest in conjunction with economy studies in which the costs of surface finishing and energy requirements (pumps) or profit by increased energy production (turbines) are optimized.

Determination of Technical Roughness $R'_{a(hs)}$ and $R''_{a(hs)}$ Characterizing "Hydraulically Smooth" Surfaces

Concern for "hydraulically smooth" surfaces arises because of the importance of surface matching of model machines. The economic viability of surface-finishing quality for model and prototype efficiencies must also be taken into account. The mentioned roughness data can be determined from Figures 16 and 17 for $\Delta\eta_{i\wedge(ri)} = 0$ and are plotted in Figures 18 and 19. They show that $\Delta\eta_{i\wedge(ri)} = 0$ is related to small values $\bar{R}_{a(hs)}$. Thinking about inevitable inaccuracies of scaling computation it seems useful to evaluate roughness data for higher efficiency losses $\Delta\eta_{i\wedge(ri)} = 0.001–0.003$. Corresponding characteristic lines are shown in Figure 19 for the example of a Francis turbine with $n_{q\wedge} = 40$ and $Re''/Re' = 20$. Looking for the influence of the difference $\Delta\eta_{i\wedge(ri)} = 0$ and for instance $\Delta\eta_{i\wedge(ri)} = 0.002$, one recognizes that the resulting differences in $\bar{R}_{a(hs)}$ usually are smaller than 0.3 μm and can be neglected. Consequently it is reasonable to determine $\bar{R}_{a(hs)}$ for $\Delta\eta_{i\wedge(ri)} = 0$ respectively on the basis of the boldly drawn curves on the right side of Figures 18 and 19.

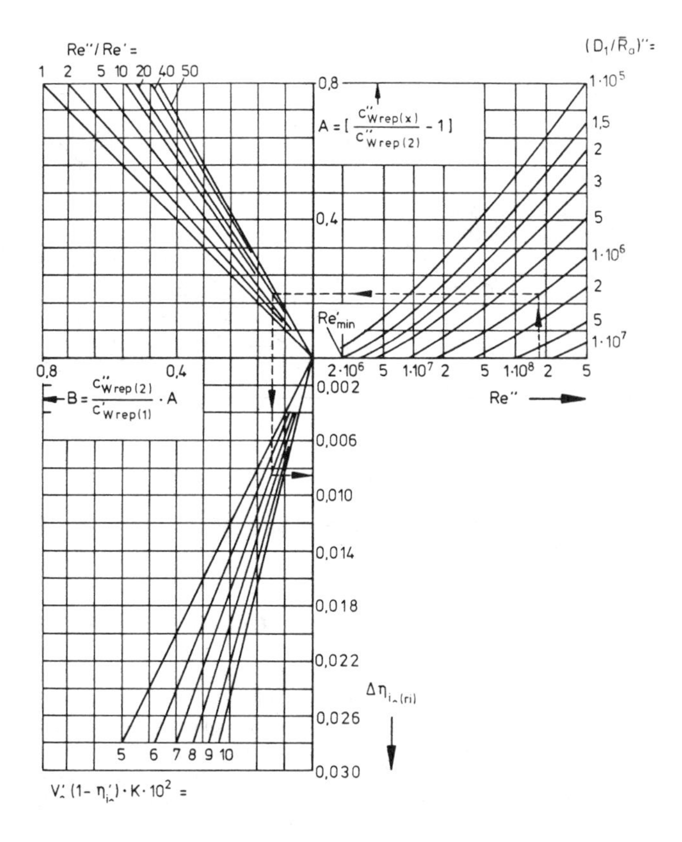

Figure 17. Nomograph for *Francis turbines* with $n_{q\wedge} = 20–70$ for determination of efficiency loss $\Delta\eta_{i\wedge(ri)}$ due to roughness influence. This diagram is elaborated on basis of Equation 5.

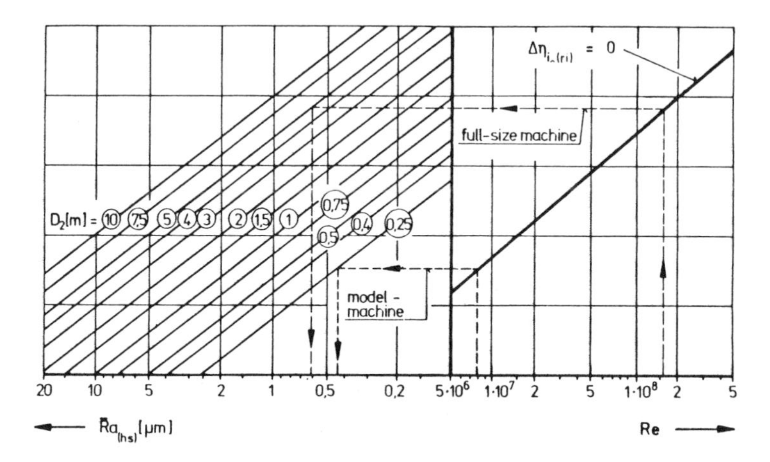

Figure 18. Nomograph for *storage pumps* to determine the representative roughness $\bar{R}_{a(hs)}$ for $\Delta\eta_{i\wedge(ri)} = 0$.

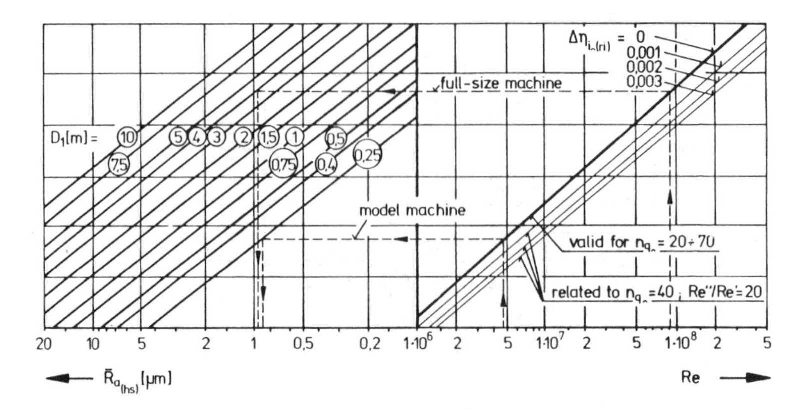

Figure 19. Nomograph for *Francis turbines* to determine the representative roughness $\bar{R}_{a(hs)}$ for a given efficiency loss $\Delta\eta_{i\wedge(ri)} = 0\text{–}0.003$.

COMPUTATION EXAMPLE FOR EFFICIENCY SCALING OF A STORAGE PUMP WITH $n_{q\wedge} = 25$

In agreement with the usual practice $\Delta\eta_{i\wedge}$ is calculated for the point of best efficiency and then taken as a constant value in the whole guarantee range. Consequently the scaling $\Delta\eta_{i\wedge} = \Delta\eta_{i(x)} = \eta''_{i(x)} - \eta'_{i(x)}$ of a random operating point $\eta'_{i(x)}$ with $Re_{(x)}$ has to be computed for the peak efficiency $\eta'_{i\wedge}$ as shown in Figure 20. It is to be considered that $Re' = f(\varphi)$ depends not only on $\psi = f(\varphi)$, but may also be subject to variations of temperature and viscosity. One must also take into account that deviations between Re′ and Re'_{ref} involve a correction of $V'_{\wedge ref}$ (Figure 5) according to Equation 6. This parameter namely applies to the Reynolds number $Re'_{\wedge(ref)}$, which was taken as a basis for the respective scaling tests [9]. Corresponding data are in the case of storage pumps $Re'_{\wedge ref} = 8 \cdot 10^6$ and for turbines $Re'_{\wedge ref} = 7 \cdot 10^6$. For any $Re'_{(x)}$ of model machine the corrected loss distribution

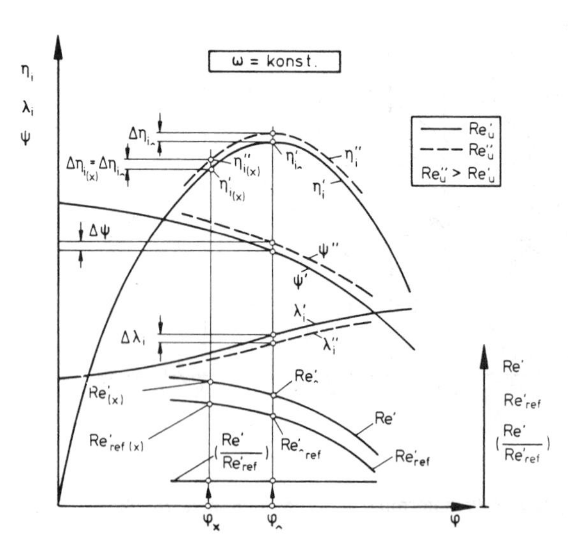

Figure 20. Schematic representation of a pump characteristic to illustrate the efficiency scaling $\Delta\eta_{i(x)}$ for a random operating point.

coefficient $V'_\wedge$ can be determined as follows:

$$V'_\wedge = \frac{1}{1 + \dfrac{1 - V'_{\wedge\,ref}}{V'_{\wedge\,ref}} \cdot \left(\dfrac{Re'_{(x)}}{Re'_{ref(x)}}\right)^{0,16}} \quad (6)$$

In Equation 6 it is assumed that ω, u_2, θ, $v = f(\varphi)$, as usual, are maintained constant during the respective model tests and that $\psi = f(\varphi)$ is not substantially influenced by differences in Reynolds numbers $Re'_\wedge$ and Re'_{ref}. Under these circumstances Equation 7 is valid and the respective quotient may be assumed as being constant (see Figure 20).

$$\left(\frac{Re'_{(x)}}{Re'_{ref(x)}}\right)^{0.16} = \left(\frac{Re'_\wedge}{Re'_{\wedge\,ref}}\right)^{0.16} \quad (7)$$

With the previous explanations the mentioned computation example for scaling of a storage pump can be represented as follows.

Example: Steps of Computation Process Based on the Data in Table 2

1. The mean representative roughness is obtained from Figure 11 as

   $\bar{R}''_a = 4\ \mu m$ and $D''_2/\bar{R}''_a = 1 \cdot 10^6$.

2. From Figure 7 the resistance coefficients result as follows:

   $c'_{w\,rep} = 0.00339$ ($Re' = 9.2 \cdot 10^6$; "hydraulically smooth" curve)

   $c''_{w\,rep} = 0.00270$ ($Re'' = 1.25 \cdot 10^8$; $D''_2/\bar{R}''_a = 1 \cdot 10^6$)

   Hence $(c''_w/c'_w)_{rep} = 0.7965$.

3. Determination of $K = 1.01$ from Figure 9.

4. Figure 5 indicates $V'_{\wedge\,ref} = 0.80$ and from Equation 6 results $V'_\wedge = 0.796$.

5. Using the data from steps 2 through 4 and Equation 4 scaling can be computed as follows:

   $\Delta\eta_{i\wedge} = V'_\wedge \cdot (1 - \eta_{i\wedge})[1 - K \cdot (c''_w/c'_w)_{rep}]$

   $\Delta\eta_{i\wedge} = 0.796 \cdot (1 - 0.873) \cdot [1 - 1.01 \cdot 0.7965] = 0.0197$

   $\eta''_{i\wedge} = \eta'_{i\wedge} + \Delta\eta_i = 0.873 + 0.0197 = 0.8927$

6. With Figures 7 and 16 the efficiency loss due to roughness results in $\Delta\eta_{i\wedge(ri)} = 0.0060$, where:

   $c''_{w\,rep(x)} = c''_{w\,rep} = 0.0027$

   $c''_{w\,rep(2)} = 0.0025$ ($Re'' = 1.25 \cdot 10^8$; "hydraulically smooth" curve)

   $c'_{w\,rep(1)} = c'_{w\,rep} = 0.00339$

7. It is evident from Figure 18 that the conditions for a "hydraulically smooth" prototype $\Delta\eta_{i\wedge(ri)} = 0$ are satisfied with $\bar{R}''_a = 0.52\ \mu m$.

Table 2
Basic Data for Example of Scaling Computation

Object	Assumed Geometrical and Operation Data	Remarks
Storage pump model machine with $n_{q\wedge} = 25$	$D'_2 = 0.52$ m; $D'_0 = 0.25$ m	
	$n' = 24.2\ s^{-1}$; $u'_2 = (D_2 \cdot \pi \cdot n)' = 39.5$ m/s	
	$\nu' = 1.1 \cdot 10^{-6}\ m^2/s$	
	$Y'_\wedge = \psi'_\wedge \cdot (u'^2_2/2) = 819.1\ m^2/s^2\ (\psi'_\wedge = 1.05)$	
	$Re' = (D_0 \cdot \sqrt{2 \cdot Y/\nu})'_\wedge = 9.2 \cdot 10^6$	
	$\eta'_{i\wedge} = 0.873$ (test result)	
	The surface roughness $\bar{R}'_a = 0, 3\ \mu m$ is to be assessed as "hydraulically smooth" according to Figure 19.	
Geometrically similar prototype according to the model scale $M = 7.692$	$D''_2 = 4.0$ m; $D''_0 = 1.923$ m	
	$n'' = 5.5 s^{-1}$; $u''_2 = (D_2 \cdot \pi \cdot n)'' = 69.8$ m/s	
	$\nu'' = \nu'$	
	$Y''_\wedge = \psi''_\wedge \cdot (u''^2_2/2) = 2{,}557.8\ m^2/s^2$	Assumption: $\psi''_\wedge = \psi'_\wedge$
	$Re'' = (D_0 \cdot \sqrt{2 \cdot Y/\nu})''_\wedge = 1.25 \cdot 10^8$	
	Roughness of individual flow passages:	The roughness values are related to the flow direction and can be determined by means of "Rugotestplates" [14] or suitable instrumentation. Remarks with respect to roughness in runner side spaces see Figures 12–15.
	Entrance: $R''_{a_e} = 12\ \mu m$	
	Impeller: $R''_{a_r} = 4.5\ \mu m$	
	Guide vanes: $R''_{a_{gv}} = 3\ \mu m$	
	Spiral casing: $R''_{a_{sp}} = 10\ \mu m$	

The time required for the mentioned computation is of course somewhat longer than that for the usual "hydraulically smooth" scaling procedure [1, 2, 7, 8]. In the respective comparisons it should, however, be considered that additional information on roughness influence is important and justifies an increased effort. It is also worth mentioning that the scaling computation demonstrated for the example of a storage pump can also be carried out for turbines as far as the corresponding diagrams are used. The authors have endeavored to present the process as clearly as possible and hope that the present scaling concept may prove useful in practice.

Acknowledgments

Gratitude is expressed to the members of IAHR—Working Group 5 (the so-called Darmstadt Circle) for their support and valuable suggestions. Thanks is also due the International Association for Hydraulic Research. Rottersdamseweg, Delft, The Netherlands, for permission to reprint from the *IAHR Journal* (No. 2 (1984)).

NOTATION

Dimensionless Coefficients

- c_f coefficient of friction (plates)
- $c_{w\,rep}$ representative coefficient of friction (whole machine)
- K correction constant according to Equation 4
- Re Reynolds number ($Re = D_0 \cdot \sqrt{2Y}/\nu$ (pumps); $Re = D_3 \cdot \sqrt{2Y}/\nu$ (turbines))
- Re_g Reynolds number according to Reference 13
- V loss distribution coefficient
- η efficiency
- ϑ component loss
- λ coefficient of friction (pipes) and also power coefficient
- λ_d coefficient of friction (runner disks)
- μ flow coefficient (gap)
- φ flow coefficient
- ψ pressure coefficient

Operational Data of the Machine

- c absolute velocity (m/s)
- n_q specific speed (min^{-1})
- n rotational speed (s^{-1})
- Q discharge (m^3/s)
- u circumferential velocity related to D (m/s)
- w relative velocity (m/s)
- Y specific energy (Nm/kg, m^2/s^2)

Geometrical Data

- A area (m^2)
- D diameter (m)
- k_{str} roughness parameter according to Reference 10 (m, μm)
- L length (m)
- R_a arithmetical average elevation of roughness (Ra ≡ CLA ≡ AAA), to be converted into R_t (largest elevation of roughness) according to Reference 1 (m, μm)
- $\bar{R}_a$ representative arithmetical average elevation of roughness (whole machine) (m, μm)
- s gap (m)
- ϵ sand roughness according to Reference 12 (m, μm)

Physical Data

- ν kinematic viscosity (m^2/s)
- ρ density (kg/m^3)
- θ temperature (K)

Subscripts

d	runner disk	sv	stay vane
dr	draft tube	^	optimum point
e	entrance or entrance pipe of pump	0	entrance pipe of pumps
ex	exit	1	runner entry
f	friction	2	runner exit
g	gap	3	entrance of draft tube
gv	guide vane	—	average value
h	hydraulic	IAHR	International Association for Hydraulic Research
hs	hydraulically smooth	IEC	International Electrotechnical Commission
i	inner	WG 5	"Darmstadt Circle" or Working Group No. 5 of IAHR, Section on Hydraulic Machinery, Equipment and Cavitation
l	loss		
p	part		
r	runner		
ref	value correlated to $Re = Re_{ref}$		
ri	roughness influence		
sp	spiral casing		

REFERENCES

1. Osterwalder, J., "Efficiency Scale-up for Hydraulic Turbo machines with due Consideration of Surface Roughness," *Journal of Hydraulic Research*, 16 (1978) No. 1, pp. 55–76.
2. Osterwalder, J., and Hippe, L., Studies on Efficiency Scaling Process of Series Pumps, *Journal of Hydraulic Research*, 20 (1982) No. 2, pp. 175–201.
3. Hippe, L., Wirkungsgradaufwertung bei Radialpumpen unter Berücksichtigung des Rauheitseinflusses, dissertation (1984) TH Darmstadt.
4. Schlichting, H., Grenzschichttheorie, Karlsruhe, 1965.
5. Colebrook, C. F., "Turbulent Flow in Pipes, with Particular Reference to the Transition Region between Smooth and Rough Pipe Laws, " *Journal of the Institution of Civil Engineers*, 11 (1938–39), pp. 133–156.
6. Osterwalder, J., and Geis, H., Experimentelle Untersuchungen der Radseitenverluste einer Hochdruck-Francisturbine radialer Bauart, *Maschinenmarkt*, 89 (1983) 31, S. 669–671.
7. Hutton, S. P., Component Losses in Kaplanturbines and the Prediction of Efficiency from Model Tests, *Proc. Inst. Mech. Eng.*, 1954, Vol. 168, pp. 743–762.
8. Pfleiderer, C., Die Kreiselpumpen für Flüssigkeiten und Gase, 4. Auflage, Berlin 1955, S. 158.
9. Osterwalder, J., "Scale-effect," a chapter in Publication No. 193 and 497 of International Elektrotechnical Commission, (1973).
10. Strscheletzky, M., Neuere Erkenntnisse über viskose Unterschichten an rauhen Wänden, Fortschrittsberichte VDI-Zs., Reihe 7, Nr. 43, S. 105 ff.
11. Grein, H., Einige Bemerkungen über die Oberflächenrauhigkeit der benetzten Komponenten hydraulischer Grossmaschinen, Escher-Wyss-News 1 (1975), pp. 1–8.
12. Moody, L. F., Friction factors for pipe flow, Trans. ASME 66, (1946), pp. 671–684.
13. Viano, M., Pertes dans les labyrinthes cylindriques lesses, La Houille Blanche No. 1–1970.
14. Künzli, A., Das Messen und Prüfen von Rauheit an technischen Oberflächen. Schweizer Archiv, Juni 1964.

CHAPTER 45

STABILIZING TURBOMACHINERY WITH PRESSURE DAM BEARINGS

John C. Nicholas

Turbodyne Division
Dresser Industries
Wellsville, New York

CONTENTS

INTRODUCTION, 1375

INERTIA AND TURBULENCE EFFECTS, 1376

SOLUTION TECHNIQUE, 1379

STABILITY, 1381

EXPERIMENTAL RESULTS, 1384

DESIGN OPTIMIZATION, 1388

APPLICATION, 1389

CONCLUSION, 1391

NOTATION, 1391

REFERENCES, 1391

INTRODUCTION

Although tilting pad bearings are currently used in the design of a large percentage of the world's high speed, multistage turbomachinery, there still exists a large class of rotors that are designed or are presently operating with fixed-lobe bearings. Examples include axial compressors, steam turbines, hot gas expanders, and many older model centrifugal compressors. One characteristic of machines that operate successfully with fixed-bore bearings is a low speed-to-weight ratio. This indicates a relatively stiff shaft design and consequently a fundamental mode shape that is nearly rigid. A rigid body mode ensures a significant response at the bearing locations to allow the bearing damping to be effective in suppressing shaft vibrations [1].

A major problem with many plain and axial-groove bearings is that they exhibit a relatively low oil whirl instability threshold speed that produces a reexcitation of the rotor's first fundamental natural frequency [2, 3]. Oil whirl is a high speed and/or light load condition. For example, a plain cylindrical journal bearing with a 6.0 mil radial clearance could not operate above 6,000 rpm without whirling. However, many other fixed-bore bearing designs are effective in suppressing oil whirl and thus can operate above 6,000 rpm without becoming unstable. The most commonly used of these are the pressure dam or step journal bearing [4–11] and the multi-lobe bearing [11–14].

The stability characteristics of plain cylindrical bearings are primarily controlled by the bearing clearance; the tighter the clearance, the higher the instability threshold speed. However, tight clearance bearings present other problems that make them undesirable. For example, oil flow is lower, power loss is higher, and thermal growth of the journal may cause the bearing to wipe. Many bearing-induced instability problems in the field are caused by bearing clearances that have increased due to wear from oil contamination, poor filtration, and/or repeated starts and slow rolling with boundary lubrication.

Thus, it is desirable to analyze and design an effective fixed-bore anti-whirl bearing that is easily manufactured, relatively insensitive to design tolerances, and available for quick retrofits in existing plain bearing inserts. The pressure dam bearing falls into this category. The details of the surface inside the pocket are of secondary importance since the side lands hold the flow and the pressure. The hydrodynamic load created by the pocket provides the increased margin of stability for step bearings compared to plain bearings [6–8]. Finally, the tolerance on the pocket depth is not as critical as lobe-clearance tolerances for multilobe bearings. Tolerances on the lobe radius for multilobe bearings are often on the order of ± 1.0 mils. These tolerances can be relaxed for pocket bearing profiles, which therefore reduces manufacturing costs. Typical pocket depth tolerances are often ± 2.5 mils.

Pressure dam or step journal bearings have long been used to improve the stability of turbomachinery as replacements for plain journal or axial groove bearings. In many cases, these bearings provide a quick and inexpensive fix for machines operating at high speeds near or above the stability threshold. For example, a plain cylindrical axial-groove bearing can easily be removed from a machine displaying subsynchronous vibration. Milling a step in the top pad of the proper size and location may be all that is necessary to eliminate the stability problem. This is much less expensive and faster than installing tilting pad bearings that may require a change in the bearing housing.

A large percentage of the rotors used in the rotating equipment industry operate on vendor installed or retrofitted pressure dam bearings. The most common applications are steam turbines and gear boxes. In high speed gear boxes, the gear loading may vary from several thousand to only a few hundred pounds. This large variance in load is often accompanied by a change in load direction. An example is the gear box between a motor/generator and a start-up steam turbine in a catalytic cracking axial-compressor train. In this particular application, step bearings are used to ensure stable operation when the gear loading is small.

The purpose of this paper is to describe how the pressure dam bearing suppresses oil whirl and to identify the important design parameters necessary to optimize its stability performance. These concepts are extremely important for both rotating equipment vendors and users who design step journal bearings either as original equipment or as retrofits replacing plain bearings.

INERTIA AND TURBULENCE EFFECTS

The majority of oil-lubricated journal bearings operate in the slow viscous flow regime where the viscous forces are much greater than the inertia forces. This is certainly true for a vast majority of the bearings in industrial application where the bearing Reynolds number is usually below 1,500. The Reynolds number is the ratio of inertia forces to viscous forces and, for a journal bearing, is defined as

$$R_e = \frac{\pi R N \rho c}{30\mu} \tag{1}$$

where
- R = journal radius, in.
- N = shaft rotational speed, rpm
- ρ = lubricant density, lb-s^2/in.^{4}
- c = bearing radial clearance, in.
- μ = lubricant viscosity, lb-s/in.^{2}

The analytical solution technique used to solve for the hydrodynamic pressures in a journal bearing neglects inertia forces. Also, turbulence is usually neglected and the flow is assumed laminar. Laminar flow is also a low Reynolds number condition. Thus, these assumptions are valid for most oil-lubricated journal bearings. However, for a pressure dam bearing, the Reynolds number inside the pocket is anywhere from 3 to 20 times the bearing Reynolds number due to the increased pocket

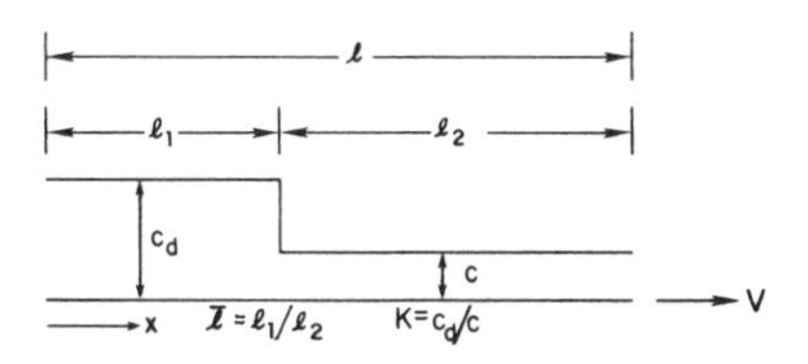

Figure 1. Stepped slider bearing.

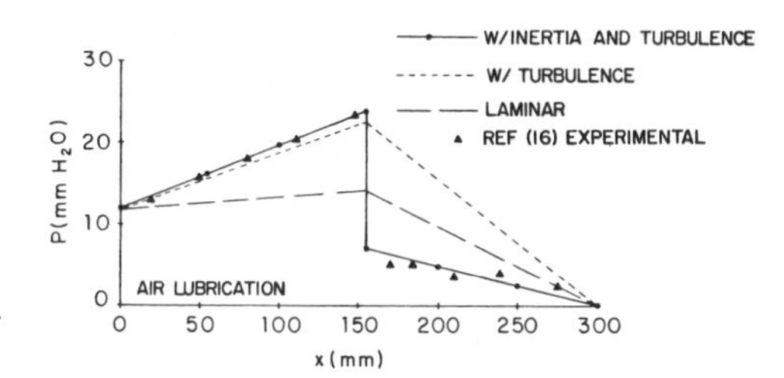

Figure 2. Stepped slider pressure profiles for air lubrication.

clearance. In this region, the flow is turbulent and inertia effects may become significant compared to viscous forces.

To investigate the relative importance of inertia and turbulence, it is instructive to consider a flat-stepped slider bearing as shown in Figure 1. The slider surface is assumed to be infinite in length so that velocities normal into and out of the surface can be ignored. This assumption is not valid for most bearings but it is necessary to include inertia effects in the solution whereas a finite length slider with inertia is an extremely difficult problem. However, the general trends and conclusions based on the infinite solution remain valid for finite length journal bearings. Finite elements are used to solve for the hydrodynamic pressures for the stepped slider [5, 15].

Figure 2 illustrates the hydrodynamic pressure profiles for the infinite flat-stepped slider of Figure 1. Curves are shown for three different analytical techniques. Also, experimental results from Reference 16 are indicated. The analytical solution including inertia and turbulence [5] agrees with the experimental results. Inertialess theory with turbulence over-predicts the load capacity (area under pressure curve) while laminar theory under-predicts the load in the pocket area and over-compensates after the step.

Note that the effect of inertia is to slightly raise the pressure inside the pocket compared to the inertialess case including turbulence. However, after the step the fluid inertia causes a large pressure drop. The overall effect of inertia is to decrease the load-carrying capacity of the stepped slider.

From Figure 2, it appears that inertia must be included to match the experimental results. However, air is the lubricant in Figure 2. Figure 3 shows a comparable plot for oil. Now, the pressure drop after the step is only 10% of the maximum pressure and inertialess theory with turbulence provides a good approximation. Laminar theory again severely underestimates the load.

Load capacity error as a function of film thickness ratio and Reynolds number is plotted in Figure 4 for inertialess theory including turbulence. Oil is the lubricant. For Reynolds numbers below 3,000, the error is less than 10% compared to inertia theory. As the Reynolds number increases, the effect of fluid inertia increases and the error in inertialess theory increases. Note that for film thickness ratios greater than 2.5 the error is negative which means that inertialess theory over-predicts load capacity. Also, negative error implies that, as stated earlier, the effect of fluid inertia is to decrease the load capacity of the slider.

Figure 5 is a series of load curves as a function of pocket clearance ratio, K, and step location, $\bar{\ell}$ (Figure 1). For the stepped slider, these parameters are defined as

$$K = c_d/c \tag{2}$$

$$\bar{\ell} = \ell_1/\ell_2 \tag{3}$$

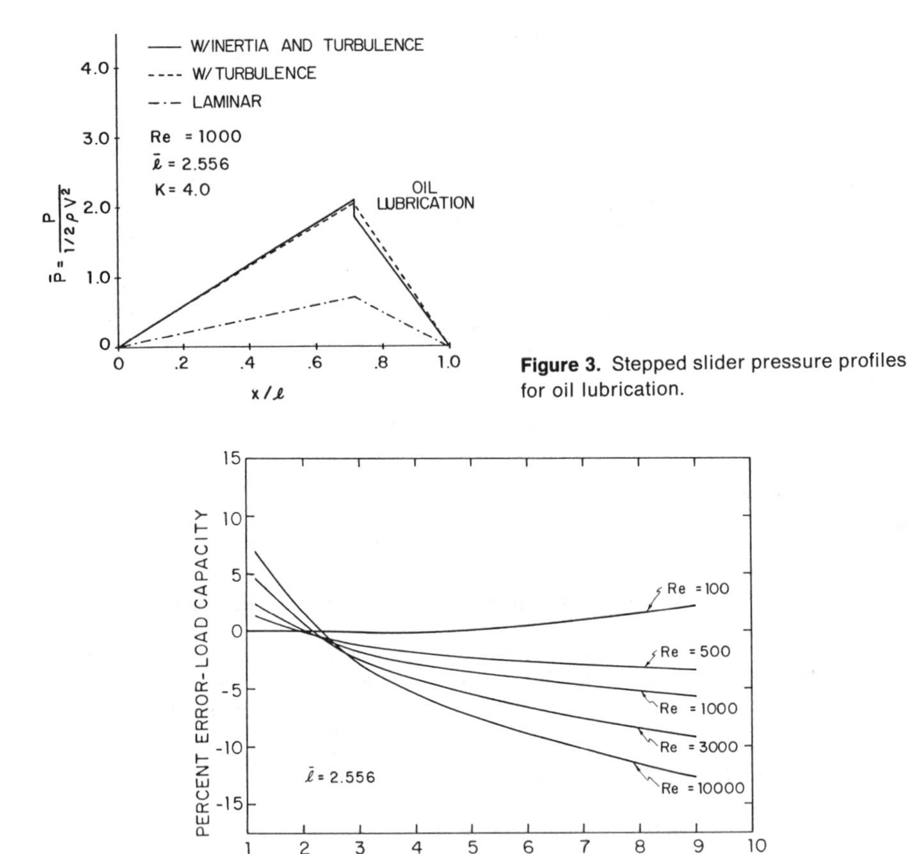

Figure 3. Stepped slider pressure profiles for oil lubrication.

Figure 4. Percent error in load capacity comparing inertia to inertialess theory (both with turbulence effects).

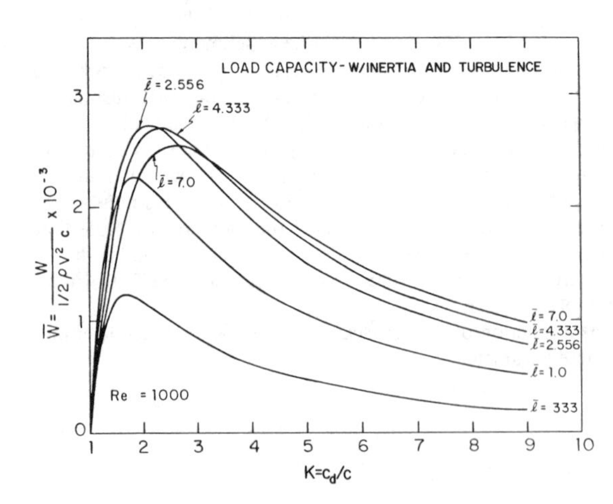

Figure 5. Stepped slider load capacity showing optimum clearance ratio, K, and step location, $\bar{\ell}$.

where c_d = pocket clearance, in.
c = bearing clearance, in.
ℓ_1 = pocket length, in.
$\ell_2 = \ell - \ell_1$, in.
ℓ = total bearing length, in.

Optimum clearance ratios range between 2.0 and 3.0. For K values less than 2.0, the load capacity decreases drastically while for $K > 3.0$, the load capacity decreases much more gradually. This indicates that the tolerance range on the pocket depth should be above optimum to avoid the steep decrease in load for K values less than 2.0.

The optimum step location from Figure 5 is $\bar{\ell} = 2.556$. As $\bar{\ell}$ increases from the optimum value, only a slight decrease in load capacity is evident. However, as $\bar{\ell}$ decreases from the optimum, a large decrease in load capacity results. For $\bar{\ell} = 2.556$, the pocket length, ℓ_1, is 72% of the total length, ℓ (Figure 1). Thus, pocket designs should be 72% of the total length for near optimum load. Pocket lengths greater than 72% are preferable while lengths less than 72% are not recommended.

It is interesting to note that for laminar flow, the optimum length ratio is $\bar{\ell} = 2.556$. However, the optimum clearance ratio is 1.866 for laminar flow. The clearance ratio optimum increases for turbulent flow since, as the pocket depth increases, the Reynolds number inside the pocket increases and the effect of turbulence on load capacity increases. Thus, there is a trade-off between the pure laminar optimum and the increasing effect of turbulence on load capacity as the pocket depth increases.

SOLUTION TECHNIQUE

From the previous section it was determined that for oil lubrication, inertia effects contribute very little (under 10%) to the hydrodynamic load generated by the step for Reynolds numbers less than 3,000. This is well within the operating range and geometry of the majority of the oil-lubricated journal bearings in industrial application. Thus, neglecting inertia but including the effects of turbulence, the finite pressure dam bearing can be analyzed with some degree of confidence and ease.

Most pressure dam bearings have two oil supply grooves located in the horizontal plane as shown in Figure 6. For a downward- (negative y-direction) directed load corresponding to a portion of the rotor weight, a pocket is cut in the upper half of the bearing with the end of the pocket (the step or dam) located in the second quadrant for counter-clockwise shaft rotation. The pocket has side

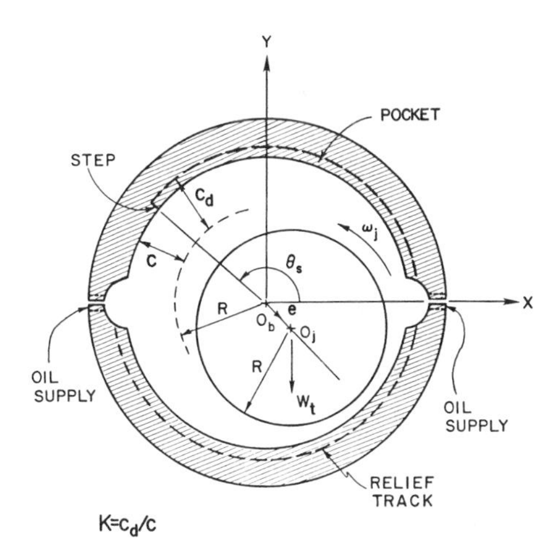

Figure 6. Pressure dam or step journal bearing schematic, side view.

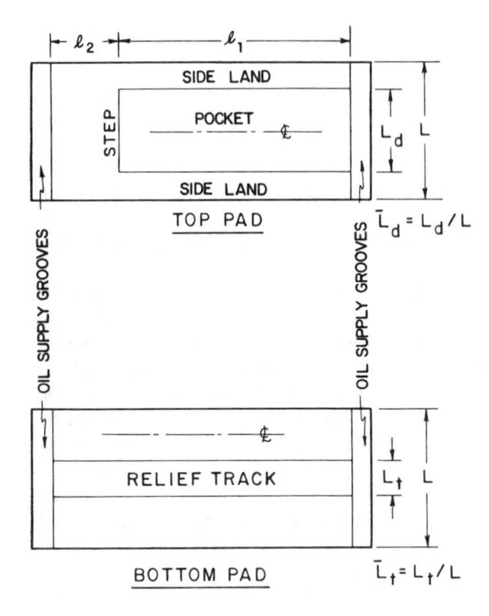

Figure 7. Pressure dam or step journal bearing schematic, top and bottom pads.

lands to hold the pressure and flow as shown in Figure 7. A circumferential relief groove or track is sometimes grooved in the bottom half of the bearing as illustrated in Figures 6 and 7. Both of these effects (dam and relief track) combine to increase the operating eccentricity of the bearing compared to a plain cylindrical bearing.

Investigating the stability characteristics of the pressure dam bearing necessitates the calculation of the bearing's linearized dynamic stiffness and damping coefficients. To this end, a finite element solution technique [15] is again used to solve for the hydrodynamic pressures created by the oil film between the bearing and journal. Each pad of the bearing is divided up into finite elements as indicated in Figure 8 using an automatic mesh generation scheme. The mass rate of flow across the step is conserved without requiring special conditions as with finite differences, an alternate

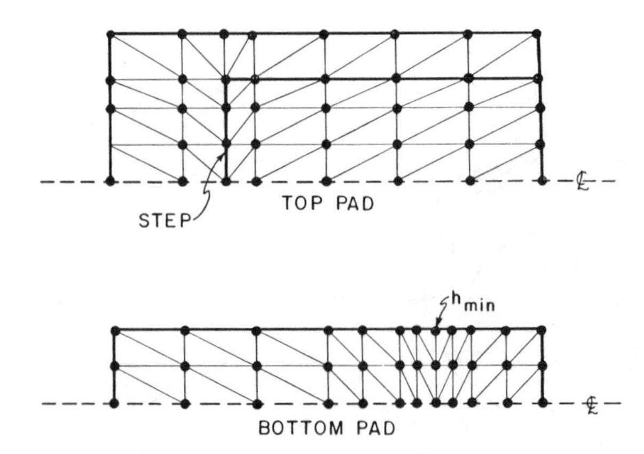

Figure 8. Dividing the pressure dam bearing into finite elements.

solution technique [15]. Nodes in the bottom pad are concentrated around the minimum film thickness where the change in the hydrodynamic pressure is the greatest.

The load capacity of the pressure dam bearing is obtained by integrating the hydrodynamic pressures while linearized dynamic coefficients are determined with small numerical perturbations about the bearing equilibrium position. The equilibrium position is determined by a simple force balance between the resultant hydrodynamic load and the external load. Turbulence effects are included by calculating the local Reynolds number for each element around the bearing and correcting the local viscosity [6].

STABILITY

Prior to in investigating the optimum pressure dam bearing stability configuration, the effect of a stepped pocket on the journal compared to a plain bearing may be easily seen on a bearing eccentricity plot. Figure 9 shows the bearing eccentricity ratio as a function of the Sommerfeld number, a bearing speed parameter. The bearing eccentricity ratio is the amount the journal is offset in the bearing, e (Figure 6) divided by the bearing radial clearance, c. Three curves are plotted on Figure 9. Two different pressure dam bearings with steps located at 125 and 160 degrees are compared to the plain journal bearing curve. All length-to-diameter ratios (L/D) are 1.0.

At high Sommerfeld numbers (light loads and/or high speeds), the journal bearing eccentricity approaches zero and the journal runs centered in the bearing. This condition leads to unstable operation. However, the pressure dam bearing eccentricity either approaches some minimum value or increases as the Sommerfeld number increases.

At high speeds and/or light loads, the step creates a loading that maintains a minimum operating eccentricity. That is, as speed is increased, the bearing eccentricity does not approach zero as it would for plain journal bearings. The eccentricity approaches some minimum value or may even

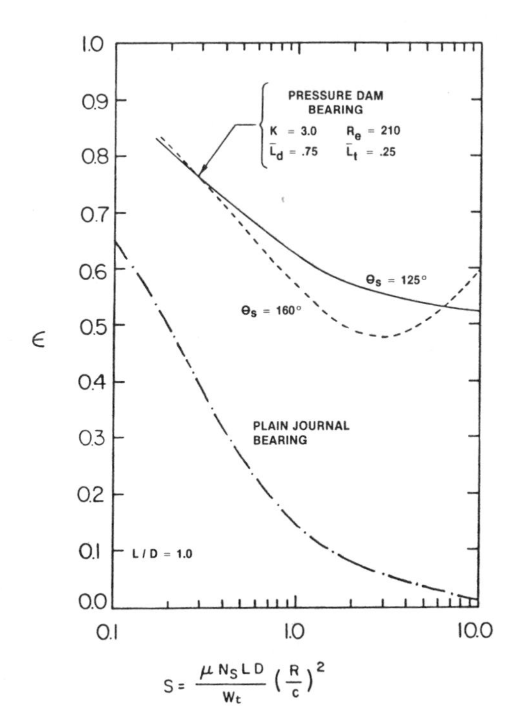

Figure 9. Pressure dam and plain journal bearing eccentricity ratios.

increase with increasing speed due to the step loading. Thus, a properly design step bearing would operate at a moderate eccentricity even at high Sommerfeld numbers. This condition helps to stabilize the pressure dam bearing in the high Sommerfeld number range.

To investigate the stability of a journal in a bearing, the dimensionless stability threshold speed parameter, $\bar{\omega}$ is used [6]. This parameter gives the speed at which the journal becomes unstable in a particular bearing. In this formulation, the shaft is assumed rigid. The effect of a flexible shaft is to lower the instability threshold speed [9]. For horizontal rotors, the threshold speed can be calculated from

$$N_t = 187.6\,\bar{\omega}(c)^{-1/2} \tag{4}$$

where N_t = instability threshold speed, rpm
c = bearing radial clearance, in.
$\bar{\omega}$ = bearing stability parameter, dim

The stability parameter is calculated as a function of the Sommerfeld number from the bearing's stiffness and damping coefficients [9]. Thus, for a given Sommerfeld number corresponding to the bearing's operating conditions, $\bar{\omega}$ may be calculated directly from the dynamic characteristics or determined from a bearing stability plot. The rigid shaft instability threshold speed can then be calculated from Equation 4. For example, for Sommerfeld numbers greater than 0.1, $\bar{\omega} = 2.3$ for a plain journal bearing. For a 6 mil radial clearance and using Equation 4, $N_t = 5{,}570$ rpm. This means that a plain journal bearing operating above S = .1 in a horizontal rotor with a 6 mil radial clearance will become unstable above 5,570 rpm.

Of course, bearings cannot be designed based solely on this parameter. However, insight into step journal bearing design may be gained by examining and comparing $\bar{\omega}$ for different step bearing geometries.

Figure 10 compares the stability characteristics of the plain journal, two-axial groove and grooved lower half bearings to two types of pressure dam bearings. The stability threshold speed, $\bar{\omega}$, is plotted against the Sommerfeld number, S. Also indicated at the top is the bearing eccentricity ratio, ϵ for all five bearings. Bearing numbers 1 and 2 are the plain journal and two-axial groove bearings, respectively. Note that at high Sommerfeld numbers the stability curves for each bearing approach asymptotic values of $\bar{\omega} = 2.3$ (plain journal) and $\bar{\omega} = 2.05$ (2 axial groove).

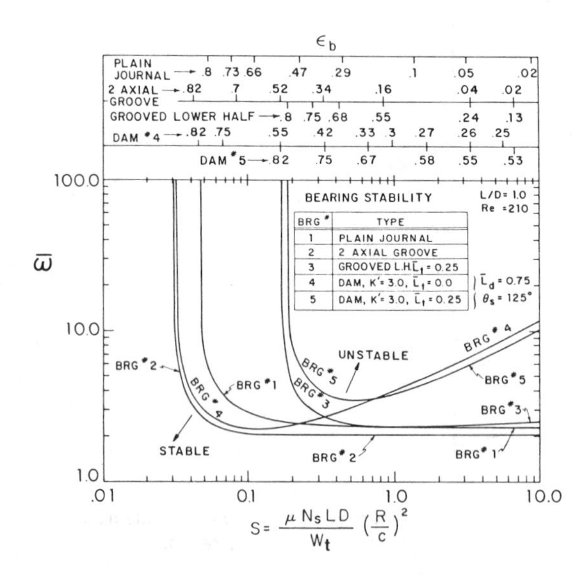

Figure 10. Stability map comparing pressure dam bearing to plain journal, two axial groove and grooved lower half bearings.

Bearing number 3 is the grooved lower half bearing. This bearing is simply a two-axial groove bearing with a circumferential relief track or groove cut in the lower half. In this case, the relief track axial-length ratio (Figures 6 and 7) is $\bar{L}_t = 0.25$ (the relief track is 25% of the bottom pad). A considerable increase in the infinite stability region is evident. That is, the plain journal bearing is theoretically stable at all speeds below a Sommerfeld number of .048 while the grooved lower half bearing increases this range of infinite stability by a factor of 3 to $S \leq 0.17$. The relief track removes part of the bearing load-carrying surface for the bottom pad thereby forcing the bearing to operate at a higher eccentricity ratio. Essentially no increase in stability is seen at high Sommerfeld numbers.

While a bearing designed to operate in the infinite stability region may appear to be advantageous, this design should be avoided since large eccentricities result ($\epsilon \geq 0.8$). A bearing functioning at $\epsilon = 0.8$ at the operating speed may not be able to support the load at idle speed or during slow rolling of the rotor. An exception would be a gear box bearing where the gear load is reduced as speed decreases.

Bearing number 4 is a pressure dam bearing with K = 3.0 (pocket clearance three times as large as the bearing clearance) and $\bar{L}_t = 0.0$ (no relief track). For this case, the stability is increased compared to the journal bearing at high Sommerfeld numbers while the region of infinite stability is less. As discussed previously, at high Sommerfeld numbers, the step forces the journal to operate at a moderate eccentricity. From the top of Figure 10, bearing number 4 operates at an eccentricity ratio of $\epsilon = 0.25$ at S = 5.5. This moderate eccentricity provides the favorable stability characteristics at high Sommerfeld numbers for this step journal bearing.

The effect of varying the clearance ratio, K, on stability is shown in Figure 11. For a pressure dam bearing, the clearance ratio is defined as (see Figure 6):

$$K = c_d/c$$

where c_d = pocket radial clearance, in.
c = bearing radial clearance, in.

For the infinite-stepped slider, the optimum clearance ratio as far as load capacity is concerned is approximately K = 3.0 (Figure 5). Bearing number 3 has a clearance ratio of K = 3.0 and provides

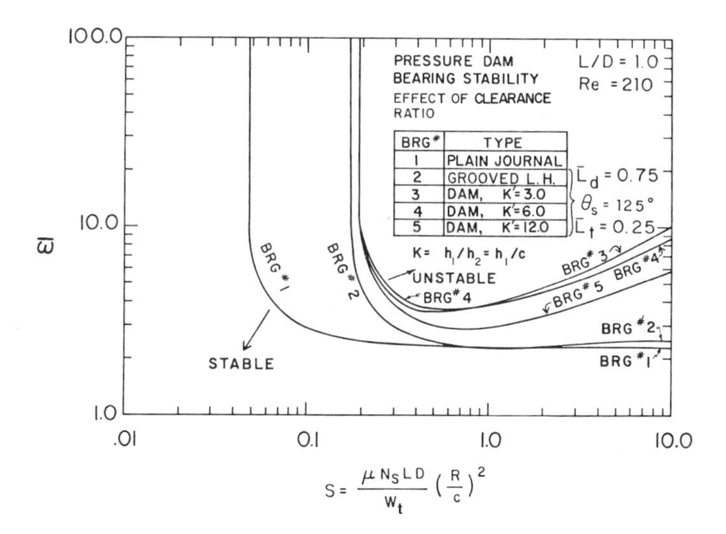

Figure 11. Effect of clearance ratio on pressure dam bearing stability.

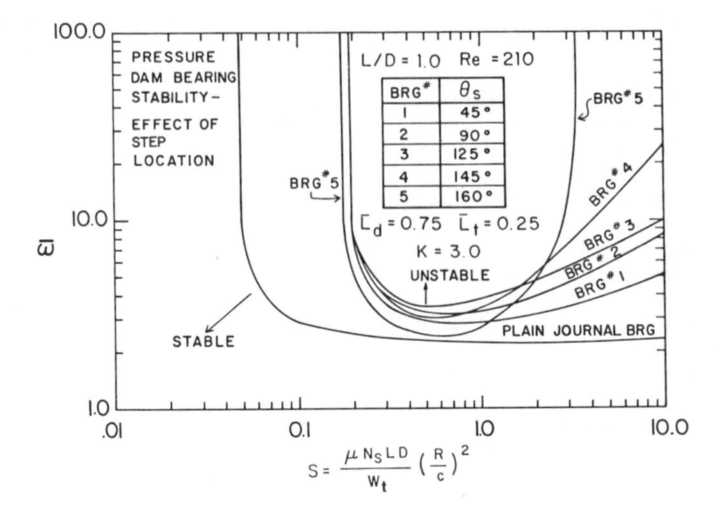

Figure 12. Effect of step location on pressure dam bearing stability.

the best stability characteristics. Bearing number 4 with $K = 6.0$ is only slightly less superior (a 10% decrease at $S = 10.0$). A 40% decrease in stability is evident for step bearing number 5 ($K = 12.0$) when compared to the $K = 3.0$ bearing at $S = 10.0$. The stability curves for bearing number 1 (plain journal) and number 2 (grooved lower half) are also included in Figure 11 for comparison.

Figure 12 illustrates the effect of changing the step location on stability. The step is located by the angle, θ_s (Figure 6) measured with rotation, counter-clockwise from the positive horizontal (+X) axis. From the infinite-stepped slider analysis (Figure 5), the optimum step location as far as load capacity is concerned is $\bar{\ell} = 2.556$. This corresponds to approximately $\theta_s = 125°$. This optimum θ_s does provide the optimum stability for $0.2 \leq S \leq 1.7$. However, for $S > 1.7$, bearing number 5 with $\theta_s = 160°$ is more stable. In fact, this bearing has two regions of infinite stability ($S \leq 0.17$ and $S \geq 3.4$). As θ_s decreases from 160°, stability decreases at high Sommerfeld numbers.

In summary, the optimum Sommerfeld number range for designing a pressure dam bearing to increase stability is $S \geq 2.0$. At these high S values, the important design parameters are the clearance ratio, K, and the step location, θ_s. The optimum clearance ratio in this region is around $K = 3.0$. A slightly larger clearance ratio is recommended to avoid the sudden drop in load capacity for clearance ratios below 3.0. Increasing K from 3.0 to 6.0 reduces stability only slightly in the high Sommerfeld number region (only a 10% decrease in $\bar{\omega}$ at $S = 10.0$).

The optimum step location for stability is between $\theta_s = 125°$ and 160° depending on the Sommerfeld number. For $S \geq 2.0$, $\theta_s = 160°$ is the optimum while for Sommerfeld numbers in the range of $0.2 \leq S \leq 2.0$, 125° is the optimum. A good compromise is $\theta_s = 140°$.

A step journal bearing designed with these recommended K and θ_s values could increase the stability parameter, $\bar{\omega}$ by a factor of 10 or more over a plain journal bearing at high Sommerfeld numbers. Also, the step journal bearing would operate at a moderate eccentricity ratio (between $\epsilon = 0.25$ and 0.5) even though the loading is light and/or speed high.

EXPERIMENTAL RESULTS

This section presents results of an experimental study with a three-mass flexible rotor mounted symmetrically between two bearing supports. Four different step bearing geometries and a two-axial-groove bearing are considered. The pressure dam designs include optimum and off-optimum clearance ratios and step locations. Instability onset speeds are determined experimentally and compared to the analytical predictions.

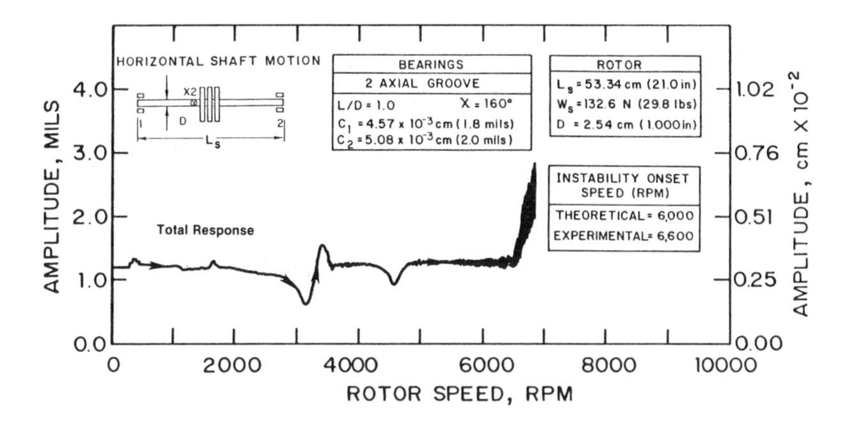

Figure 13. Total response, two axial groove bearings.

The rotor weighs 29.8 lbs, has a bearing span of 21.0 in. and a journal diameter of 1.0 in. A non-contacting probe mounted near the shaft center is used to monitor the horizontal shaft vibration. Additional details of the experimental procedure may be found in References 8, 17, and 18.

Ideally, each bearing was to have a 2.0 mil radial clearance. However, due to difficulties in manufacturing, the radial clearance ranged from 1.8 to 2.5 mils. The clearance was measured cold with a dial micrometer. Several readings were taken and the average value used.

To obtain the experimental instability onset speed, the rotor is accelerated until a large subsynchronous vibration component is observed. Speed-amplitude plots are shown here for the five test cases. In two cases, frequency spectrums are also included.

Figure 13 illustrates the total rotor response with two-axial-groove bearings. Figure 14 is a frequency spectrum for the same case. The subsynchronous component first appears at about N = 6,600 rpm. Note that Figure 14 indicates that the oil whirl instability manifests itself as a reexcitation of the rotor's first fundamental natural frequency of 3,000 cpm.

The theoretically predicted instability onset speed is 6,000 rpm. To obtain this onset speed, the speed-dependent stiffness and damping characteristics are used as input data to a stability computer program that employs a transfer matrix solution. The mass-elastic model of the rotor is also input data to the stability program. A full stability analysis is necessary to accurately predict the threshold speed of a flexible rotor. If Equation 4 is used with c = 1.9 mils and $\bar{\omega} = 2.05$, the resulting threshold

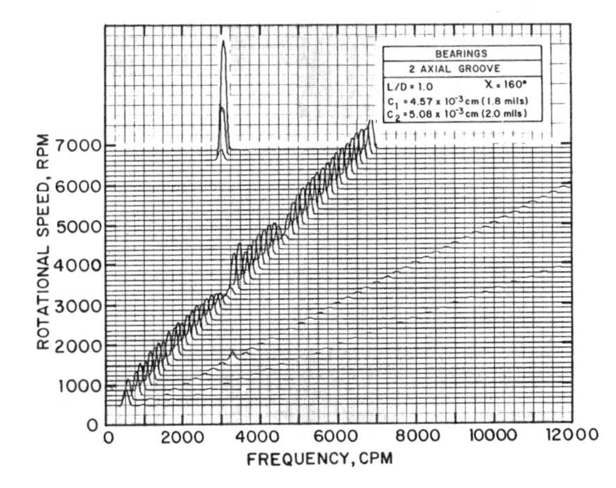

Figure 14. Frequency spectrum, two axial groove bearings.

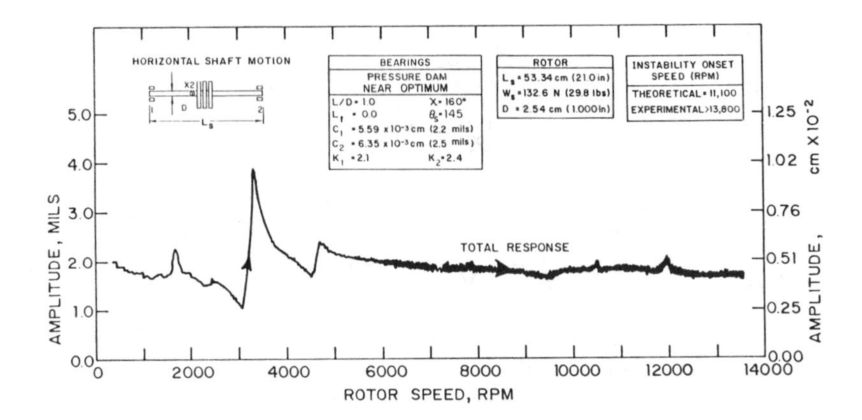

Figure 15. Total response, near optimum pressure dam bearings.

speed is N_t = 8,823 rpm. Thus, the effect of shaft flexibility is to lower the threshold from the rigid rotor prediction of 8,823 rpm down to 6,000 rpm. Caution is necessary when using Equation 4 and the resulting N_t should be viewed as the highest possible instability speed obtainable for the given bearing.

The total response for a near optimum pressure dam bearing design (K = 2.1, and 2.4, and θ_s = 145°) is shown in Figure 15 with a corresponding frequency spectrum in Figure 16. The rotor was run up to maximum speed without a large subsynchronous component appearing. The theoretical prediction is 11,100 rpm.

The effect of increasing the pocket depth from the near optimum case of Figures 15 and 16 is illustrated in Figures 17 and 18. Figure 17 shows the total response with clearance ratios of 6.6 and 8.6. The experimental threshold speed is 8,900 rpm, and the theoretical threshold speed is 8,850 rpm. These speeds are reduced even further for clearance ratios of 11.7 and 8.3 (Figure 18). Both experimental and theoretical threshold speeds are 7,800 rpm for this off-optimum case.

Finally, in Figure 19 near-optimum clearance ratios are used but an off-optimum step location of θ_s = 90° is considered. For this case the experimental threshold speed is 8,600 rpm while the predicted analytical speed is 8,100 rpm.

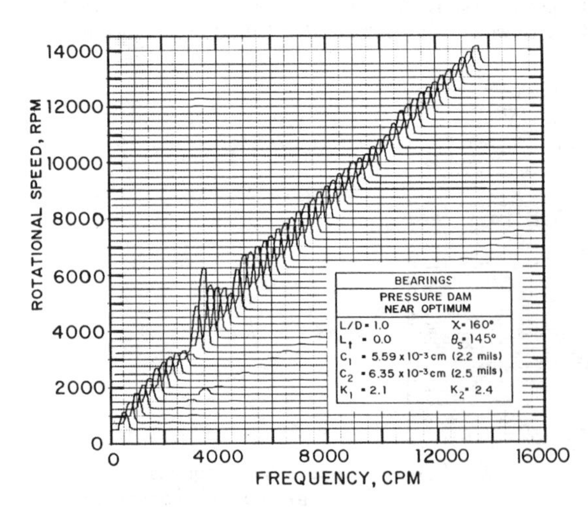

Figure 16. Frequency spectrum, near optimum pressure dam bearings.

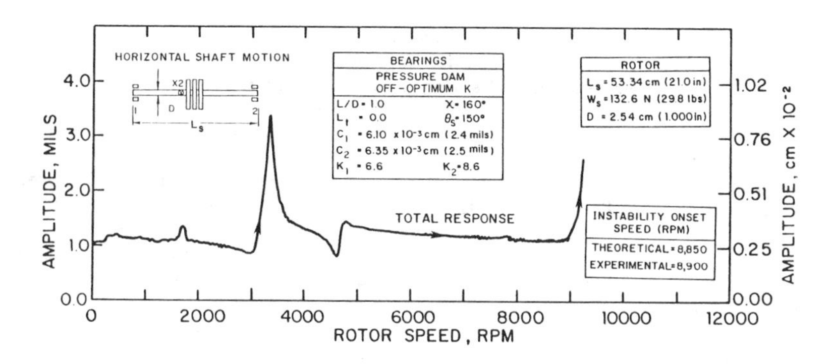

Figure 17. Total response, off-optimum clearance ratio (K = 6.6, 8.6), pressure dam bearings.

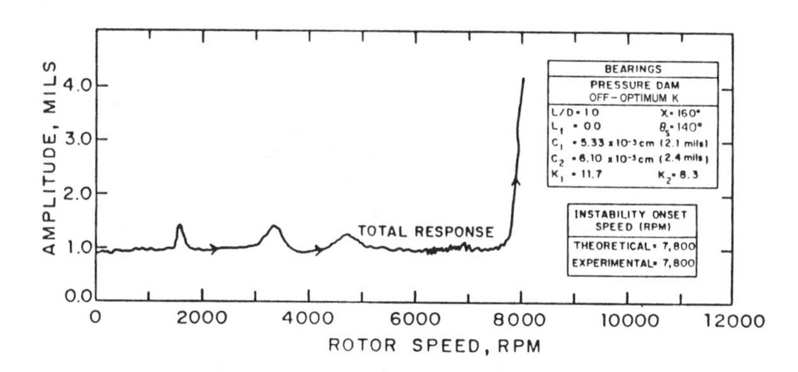

Figure 18. Total response, off-optimum clearance ratio (K = 11.7, 8.3), pressure dam bearings.

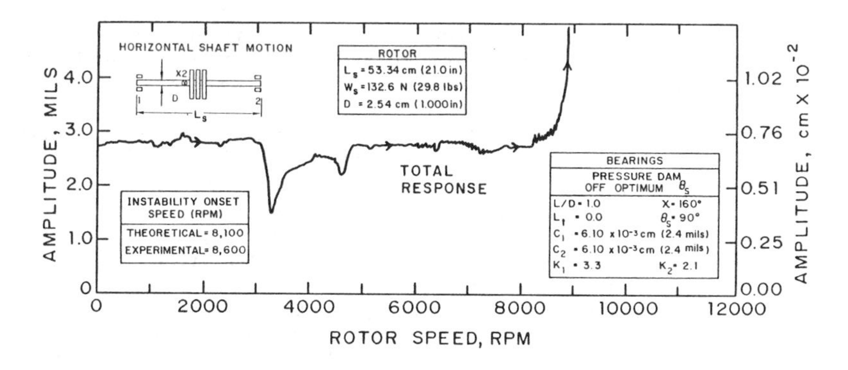

Figure 19. Total response, off-optimum step location, pressure dam bearings.

Table 1
Summary of Experimental-Theoretical Comparison of Pressure Dam Bearing Stability Performance

Bearing Type	K	C (mils)	θ_s (deg)	Instability Threshold Speed: Experimental (rpm)	Instability Threshold Speed: Theoretical (rpm)
Two-axial-groove	—	1.8, 2.0	—	6,600	6,000
Near-optimum pressure dam	2.1, 2.4	2.2, 2.5	145	>13,800	11,100
Off-optimum K pressure dam	6.6, 8.6	2.4, 2.5	150	8,900	8,850
Off-optimum K pressure dam	11.7, 8.3	2.1, 2.4	140	7,800	7,800
Off-optimum θ_s pressure dam	3.3, 2.1	2.4, 2.4	90	8,600	8,100

The results of this study are summarized in Table 1. The theoretical stability analysis predicts the general trends in the experimental data. All step bearing designs increase the instability onset speed over the two-axial-groove case. Comparing all step bearing cases, the near-optimum designs have the highest onset speeds and the off-optimum designs the lowest. These results clearly illustrate the trends discussed in the previous section. As the clearance ratio increases from the near-optimum case, the instability threshold speed decreases. Also, as the location of the step decreases from the near-optimum to $\theta_s = 90°$, a drop in onset speed is evident.

DESIGN OPTIMIZATION

Incorporating all of the data presented in the previous sections, design guidelines can be deduced to optimize the stability performance of pressure dam bearings. These suggested design rules are summarized below:

1. Clearance ratios should range from the optimum of $K = 3.0$ to $K = 6.0$. Values under 3.0 should be avoided due to the sudden loss in load capacity for $K < 2.0$.
2. Steps should be located at about 75% of the total arclength of the pad. For the pressure dam bearing of Figure 6 with two 20° oil feed grooves, 75% of the arclength results in a step located at $\theta_s = 125°$. Larger θ_s values are preferable especially in the higher Sommerfeld number range ($S \geq 2.0$). A reasonable compromise value is $\theta_s = 140°$. Steps should not be located at θ_s values below 125°.
3. The optimum Sommerfeld number range for designing a pressure dam bearing to increase stability is $S \geq 2.0$. In the moderate Sommerfeld number range ($0.2 \leq S \leq 2.0$), the instability onset speed may be increased over a plain bearing by a factor of around 1.5.
4. Designing a bearing with a grooved lower half should be avoided due to the high operating eccentricity ratio. Load problems may develop at idle or slow rolling speeds. Exceptions are for gear box bearings where the gear load decreases with decreasing speed.
5. Since turbulent flow adds energy and consequently load, pockets should be designed to promote turbulence. This may be accomplished by specifying a surface roughness of 125 to 250 micro-inches inside the pocket. This surface roughness often results from the milling process and no additional machining steps are necessary, thereby reducing manufacturing costs. The rough surface should assist in establishing a turbulent flow regime in the pocket area.

6. Since inertia effects decrease the load capacity of the stepped pocket, reducing these effects may be beneficial even though they are of the order of 10% for oil bearings. To accomplish this, the step should not be made sharp. It should remain the radius of the cutting blade used in milling the pocket. Without the sharp step, the effects of fluid inertia may be reduced. Again, this reduces machining costs by eliminating the hand working necessary to sharpen the step.
7. Although not discussed previously, pocket axial lengths should be 65% to 70% of the total axial bearing length.

APPLICATION

A high vibration level in excess of 3.0 mils caused numerous shutdowns of two gas expanders operating under full load. The rotors were supported on plain axial-groove bearings. Each expander drove a separate multistage centrifugal compressor through a gear-type flexible coupling. The two expander-compressor trains are typical of the type of units utilized in the hydrocarbon separation process.

Figure 20 shows a sample frequency spectrum at 5,120 rpm and the large 34 Hz subsynchronous component just prior to trip-out. The 34 Hz vibration grew to around 2.5 mils, and the expander shut down soon after this signature was recorded.

A detailed stability analysis using actual bearing clearances indicated that the rotor-bearing system was unstable as a result of the bearings causing the reexcitation of the expander's first fundamental natural frequency. Utilizing the stepped pocket design guidelines discussed in the previous section, an optimized step bearing was designed to eliminate the oil-whirl-induced instability. A bearing stability plot is shown in Figure 21 showing the original bearing and the optimized pocket design. Note the large increase in instability onset speed for the optimized design compared to the plain axial-groove bearing.

The axial-groove bearings were removed and modified to the optimized pocket design similar to the design analyzed in Figure 21. The resulting frequency spectrum is shown in Figure 22. The 34 Hz component has been suppressed to an acceptable level that is representative of stable units in service. Both units have operated free of high vibration trip-outs since the bearing redesign was incorporated.

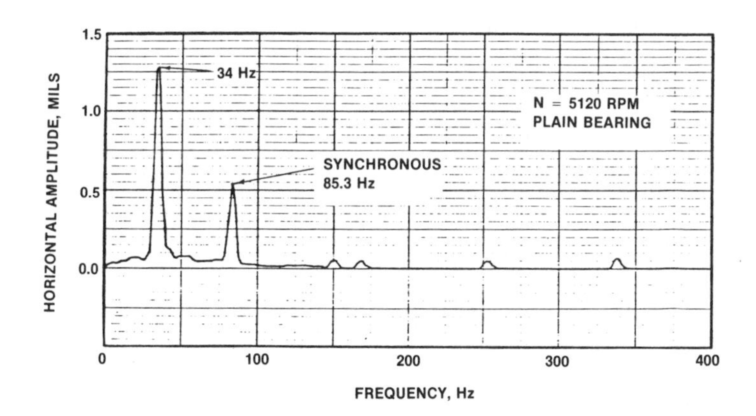

Figure 20. Frequency spectrum showing large 34 Hz component with original axial groove bearings.

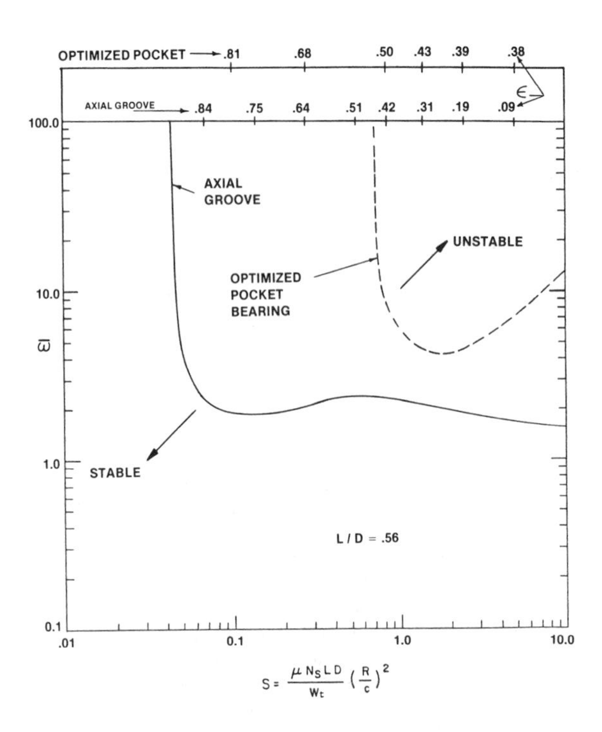

Figure 21. Optimized pocket bearing compared to the axial groove bearing, stability and load capacity.

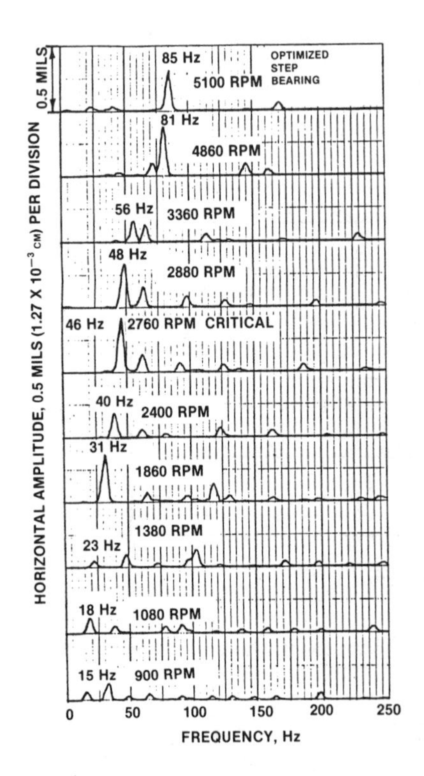

Figure 22. Frequency spectrum showing suppressed 34 Hz component with optimized step bearings.

CONCLUSION

While the pressure dam or step journal bearings will not solve all rotordynamic instability problems, it remains an extremely effective, low cost anti-whirl bearing. If near optimum clearance ratios and step locations are used, many oil whirl instabilities may be eliminated with pocket bearing designs. The design guidelines summarized in this paper should prove extremely useful in designing or retrofitting step bearings in industrial applications.

NOTATION

c	bearing radial clearance, in.
c_d	pocket radial clearance, in.
D	shaft or journal diameter, in.
e	bearing eccentricity, in.
K	c_d/c = clearance ratio, dim
L	bearing axial length, in.
L_s	bearing span, in.
ℓ	$\ell_1 + \ell_2$ = total slider length, in.
ℓ_1	pocket circumferential length (pocket length, slider bearing), in.
ℓ_2	circumferential length minus pocket (bearing length minus pocket, slider bearing), in.
L_d, L_t	pocket axial length, relief track axial length, in.
$\bar{L}_d$, $\bar{L}_t$	pocket axial length ratio, relief track axial length ratio, dim
N, N_s	shaft rotational speed, rpm, rps
N_t	bearing instability threshold speed, rpm
O_b, O_j	bearing, journal center
P	hydrodynamic pressure, slider bearing, lb/in.^{2}
$\bar{P}$	$\dfrac{P}{\frac{1}{2}\rho V^2}$ = nondimensional hydrodynamic pressure, slider bearing
R	shaft or journal radius, in.
R_e	$\dfrac{\pi R N \rho c}{30\mu}$ = bearing Reynolds number, dim
S	$\dfrac{\mu N_s L D}{W_t}\left(\dfrac{R}{c}\right)^2$ = Sommerfeld number
V	slider, velocity, in./s
W_t	bearing external load, lb
W	hydrodynamic load generated by pocket, slider bearing, lb/in.
$\bar{W}$	$\dfrac{W}{\frac{1}{2}\rho V^2 c}$ = nondimensional pocket load, slider bearing
W_s	rotor weight, lb
X, Y	coordinate system for rotating journal in bearing
x	horizontal coordinate, slider bearing, in.

Greek Symbols

ϵ	e/c = bearing eccentricity ratio, dim
θ_s	location of step measured with rotation from positive horizontal (+X) axis, deg
μ	average lubricant viscosity, lb-s/in.^{2}
χ	pad arclength exclusive of oil feed grooves, deg
ρ	average lubricant density, lb-s^2/in.^{4}
$\bar{\omega}$	bearing stability parameter, dim
ω_j	journal rotational speed, s^{-1}

REFERENCES

1. Nicholas, J. C., Gunter, E. J., and Barrett, L. E., "The Influence of Tilting Pad Bearing Characteristics on the Stability of High Speed Rotor-Bearing Systems," *Topics in Fluid Film Bearing and Rotor Bearing System Design and Optimization*, an ASME publication, April 1978.
2. Poritsky, H., "Contribution to the Theory of Oil Whip," *Journal of Applied Mechanics, Trans. ASME* (August 1953) pp. 1153–1161.
3. Hori, Y., "A Theory of Oil Whip," *Journal of Applied Mechanics, Trans. ASME* (June 1959) pp. 189–198.
4. Wilcock, D. F., and Booser, E. R., "Bearing Design and Application," McGraw-Hill, New York, 1957.

5. Allaire, P. E., Nicholas, J. C., and Barrett, L. E., "Analysis of Step Journal Bearings-Infinite Length, Inertia Effects," *ASLE Trans.*, Vol. 22, No. 4 (October 1979) pp. 333–341.
6. Nicholas, J. C., and Allaire, P. E., "Analysis of Step Journal Bearings-Finite Length, Stability," *ASLE Trans.*, Vol. 23, No. 2 (April 1980) pp.197–207.
7. Nicholas, J. C., Allaiire, P. E., and Lewis, D. W., "Stiffness and Damping Coefficients for Finite Length Step Journal Bearings," *ASLE Trans.*, Vol. 23, No. 4 (October 1980) pp. 353–362.
8. Nicholas, J. C., Barrett, L. E., Leader, M. E., "Experimental-Theoretical Comparison of Instability Onset Speeds for a Three Mass Rotor Supported by Step Journal Bearings," *Journal of Mechanical Design, Trans. ASME*, Vol. 102, No. 2 (April 1980) pp. 344–351.
9. Nicholas, J. C., Kirk, R. G., "Selection and Design of Tilting Pad and Fixed Lobe Journal Bearings for Optimum Turborotor Dynamics," *Proceedings of the Eight Turbomachinery Symposium*, Texas A & M University, College Station, Texas, November 1979.
10. Flack, R. D., Leader, M. E., and Allaire, P. E., "Experimental and Theoretical Pressures in Step Journal Bearings," *ASLE Trans.*, Vol. 24, No. 3 (July 1981) pp. 316–322.
11. Allaire, P. E., "Design of Journal Bearings for High Speed Rotating Machinery," *Fundamentals of the Design of Fluid Film Bearings*, an ASME publication. (1979) pp. 45–84.
12. Lund, J. W., and Thomsen, K. K., "A Calculation Method and Data for the Dynamic Coefficients of Oil-Lubricated Journal Bearings," *Topics in Fluid Film Bearing and Rotor Bearing System Design and Optimization*, an ASME publication (April 1978) pp. 1–28.
13. Kirk, R. G., "The Influence of Manufacturing Tolerances on Multi-Lobe Bearing Performance in Turbomachinery," *Topics in Fluid Film Bearing and Rotor Bearing System Design and Optimization*, an ASME publication (April 1978) pp. 108–129.
14. Akkok, M., and Ettles, C. M., "The Effect of Grooving and Bore Shape on the Stability of Journal Bearings," *ASLE Trans.*, Vol. 23, No. 4 (October 1980) pp. 431–441.
15. Allaire, P. E., Nicholas, J. C., and Gunter, E. J., "Systems of Finite Elements for Finite Bearings," *Journal of Lubrication Technology, Trans. ASME*, Vol. 98, No. 2 (April 1977) pp. 187–197.
16. Constantinescu, V. N., and Galetuse, S., "Pressure Drop Due to Inertia Forces in Step Bearings," *Journal of Lubrication Technology, Trans. ASME*, Vol. 98, No. 1 (January 1976) pp. 167–174.
17. Leader, M. E., Flack, R. D., and Allaire, P. E., "Experimental Study of Three Journal Bearings with a Flexible Rotor," *ASLE Trans.*, Vol. 23, No. 4 (October 1980) pp. 363–369.
18. Flack, R. D., Leader, M. E., and Gunter, E. J., "An Experimental Investigation on the Response of a Flexible Rotor Mounted in Pressure Dam Bearings," *Journal of Mechanical Design, Trans. ASME*, Vol. 102, No. 4 (October 1980) pp. 842–850.

CHAPTER 46

FLUID DYNAMICS AND DESIGN OF GAS CENTRIFUGES

A. T. Conlisk

Department of Mechanical Engineering
The Ohio State University
Columbus, Ohio USA

CONTENTS

INTRODUCTION AND OVERVIEW, 1393
The Physical Problem, 1394
Historical Perspective, 1396
Chapter Scope, 1397

THE FLUID MECHANICS OF A GAS CENTRIFUGE, 1398
Governing Equations, 1398
Boundary-Layer Methods, 1401
Limitations of Boundary-Layer Methods, 1405
The Pancake Equation, 1406
Numerical Methods, 1407

SEPARATION ANALYSIS, 1408
Governing Equations, 1408
Boundary Conditions Using Boundary-Layer Methods, 1410

SOLUTIONS TO THE MASS TRANSFER PROBLEM, 1413
Analytical Solution Using Boundary-Layer Methods, 1413
Numerical Solutions to the Mass Transfer Problem, 1422

DESIGN CONSIDERATIONS FOR A GAS CENTRIFUGE, 1426

SUMMARY AND DIRECTION OF FUTURE WORK, 1427

NOTATION, 1428

REFERENCES, 1431

INTRODUCTION AND OVERVIEW

It has been known for many years that gas centrifugation may be used to separate two-component or binary mixtures. Most often this process has been associated with the separation of isotopes, and although other methods are available [1], the centrifugation technique is thought to be the most promising for large-scale production of enriched gas. The motivation for continued research in this area is the physical problem of the separation of uranium isotopes U_{235} and U_{238}; U_{235} is used as fuel for conventional nuclear reactors. Uranium as mined contains only about 0.7% U_{235} and for use in a nuclear reactor, about 3%–5% is said to be required. To enrich the mixture in a gas centrifuge, the uranium is gasified and combined with flourine to form uranium-hexaflouride (UF_6), which is the working fluid. There are many other separation problems to which the gas centrifuge method may be applied; however, since the vast majority of the research in this area has been geared toward separation of uranium isotopes, this review will focus mainly on this problem.

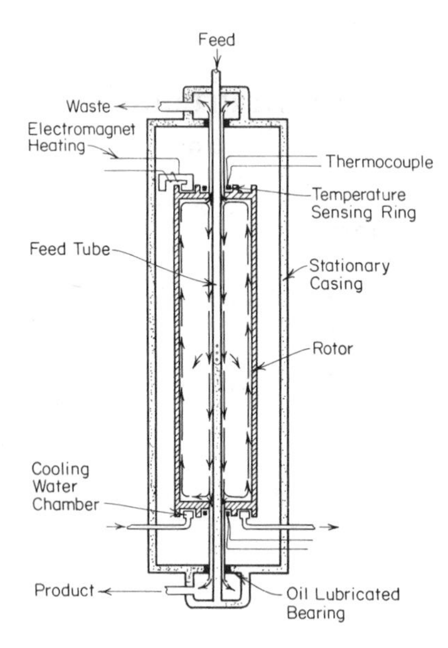

Figure 1. A sketch of the thermally driven Groth-type centrifuge from Olander [6]; the Zippe-type centrifuge is similar but is mechanically driven. The upward (counter-current) flow near the outer wall is a result of the thermal drive and is absent in the presence of the feel flow alone.

At present, uranium is enriched in this country using the gaseous diffusion technology, and this has been the case for over thirty years. However, the gaseous diffusion process requires enormous amounts of power; moreover, capital costs of a gaseous diffusion plant are high. Because the gas centrifuge technology requires only about 5% of the power required for gaseous diffusion and because gas centrifuge plants are less capital intensive, gas centrifugation is expected to provide some of the enriched uranium produced in the United States in the coming years [1].

The gas centrifuge method was investigated extensively during World War II as a means of separating uranium isotopes for use in nuclear weapons. Moreover, with the prospect of nuclear power becoming a significant energy source in this country, research into peaceful uses of the gas centrifuge increased. The classical theory of the gas centrifuge was first presented in the book by Cohen [2] and machines were built by Groth [3], Zippe [4], and Beams [5]; Figure 1 is a sketch of a Groth centrifuge.

It is the purpose of this chapter to describe the basic principles of design and operation of the countercurrent gas centrifuge. In this machine, motion of the gas within the bowl may be driven by a vertical temperature difference, scoops, or baffles for removal of the gas at the end plates, and by the source-sink geometry as well. In particular, we will focus on relatively recent work on the problem dating from approximately 1975 to the present. Detailed historical perspectives of the work prior to this time may be found in References 6, 7, 8 and a recent review of relatively current work may be found in Reference 9. A qualitative description of the gas centrifuge may be found in Reference 10.

The Physical Problem

Two geometries have generally been employed for the countercurrent gas centrifuge. The Zippe or Groth centrifuge consists of an annular region in which fluid is injected into the bowl through a central feed tube. In the centrifuge analyzed by Beams, the geometry is a rotating bowl with no inner feed post; gas is injected into, and withdrawn from the bowl through axisymmetric slots in the end walls. In both these machines gas flow may be induced in the core of the machine which is opposite in direction to that of the flow near the side walls. For this reason the apparatus is called countercurrent. The Zippe and Groth centrifuges will be discussed in detail in the following.

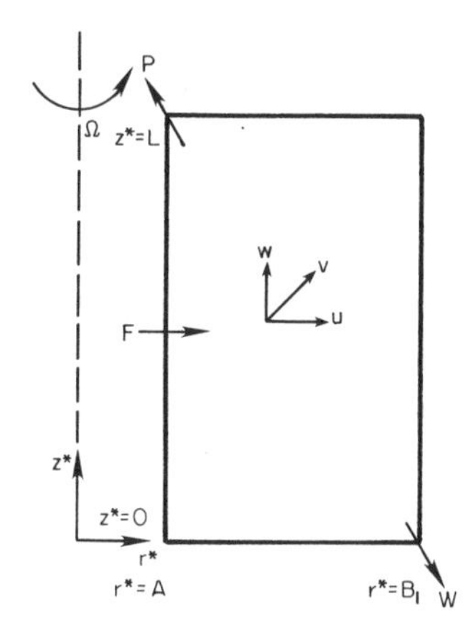

Figure 2. Cross-sectional view of the Groth and Zippe model of the centrifuge. The star quantities denote dimensional quantities and the source-sink geometry shown here corresponds to Figure 1; however, the source-sink geometry depicted here is not to be considered fixed, but is a variable of the problem. Other source-sink configurations will be discussed.

The Zippe or Groth centrifuge may be modeled as a rapidly rotating annulus bounded by radii $r^{*\dagger} = A_1$, B_1 and $z^* = 0, L$ as depicted in Figure 2. To elucidate the physical principle on which the gas centrifuge operates, consider a binary mixture containing the species A and B. Let A denote the lighter (say U_{235}) and B (say U_{238}) the heavier of the two species. The binary mixture is injected through a feed tube at some point on $r^* = A$ (marked F on Figure 1, here shown as $z^* = L/2$). As the fluid is injected into the container, the large radial pressure gradient which is set up as a result of the rotation drives the heavier species to the outer wall of the container. Thus, the region near the inner feed tube is somewhat enriched in the lighter component (P). Waste gas (W) is withdrawn from the container somewhere on $r^* = B_1$. In the absence of a driving mechanism, the fluid rotates in solid body rotation with the container, and it emerges that no vertical separation of the mixture can occur. Thus it is the motion of the gas near the inner feed tube which is the driving mechanism for separation of the mixture. From this fact it is evident that the flow patterns within the centrifuge must play a major role in the separation process.

The single most important aspect of gas centrifugation is the computation of a separation factor. Consider the collection of the lighter species of the binary mixture; for the separation of uranium U_{235}(A) and U_{238}(B) this corresponds to collection of U_{235} at the product port as previously described. The separation factor is defined as

$$\alpha = \frac{\omega_P}{\omega_F} > 1,$$

where ω_p = the mass fraction of species A at the product
ω_F = the mass fraction of species A at the feed

The separation factor must be greater than one for the method to work, and the problem would seem to be to maximize α; however, this is not quite adequate. The magnitude of α is usually close to one and approaches one as the feed flow rate increases. Thus a high value of α is not always desirable if the feed flow rate is too small. The combined effect of throughput and enrichment is embodied in a quantity called the separative power; since the separation factor is very nearly one, it emerges that the separative power of a single machine is very small. What is normally done then is to use many centrifuges in series (and parallel) to effect the desired separation. The problem that emerges is thus to maximize the separative power for a given machine and arrange the array of centrifuges in such a way as to obtain the desired amount of enriched material. In this article, we will focus on the calculation of the individual process separation factor and separative power;

† In this chapter, * denotes a dimensional quantity.

the classical theory of the arrangement of many centrifuges in a cascade is discussed in References 2, 6, and 8.

The problem of describing the details of the gas flow in centrifuges has been primarily an analytical task; this has been the case for two reasons. First, it is extremely difficult to insert diagnostic equipment into a rapidly rotating centrifuge bowl without disturbing the flow field. And second, as will become evident, the most rapid variations in the flow field occur in extremely thin boundary layers which form on the walls of the centrifuge and the flow in these layers cannot normally be resolved to sufficient experimental accuracy. Consequently, this review will be concerned with describing and interpreting the analytical work in gas centrifugation. Although experiments have been conducted on this problem to predict separation factors, published data is extremely rare. Moreover, there are compelling reasons to undertake a theoretical study, not the least of which is the fact that experiments on a test machine are very costly [9]. It is also evident that the results of a theoretical study help to guide the designer in his attempt to improve the performance of a given machine.

Historical Perspective

It is useful at this point to review the development of gas centrifuge research from about World War II on. Essentially, the technical problem involves computing the flow patterns within the centrifuge (fluid mechanics) and then using these flow patterns to solve the mass transfer problem and hence obtain a separation factor. It is the (theoretical) determination of the correct interplay between the fluid mechanics and mass transfer problems that is crucial to the progress of gas centrifuge research. It is this interplay that has puzzled many researchers in this field over the year.

Just after World War II, Cohen [2] published his early theory of the gas centrifuge based on a rather naive understanding of the flow field. He assumed that the radial velocity (u in Figure 2) vanishes and then computed separation factors assuming various forms of the vertical velocity (w in Figure 2). As shown in the next section, these assumptions of the form of the flow patterns are not correct, and it is interesting to note that many other workers in the field have made the same errors [6]. Olander [6] describes the early work of many scientists (from about 1956–1968) in which the effect of either the end walls or the side walls were neglected to produce the so-called "long bowl" and "short bowl" solutions, respectively.

In fairness to the workers in the field at this time, the problem of calculating the flow patterns in a gas centrifuge is extremely difficult and these assumptions appear to have been borne out of the frustration of trying to solve the full governing equations. Moreover, the use of digital computers for solving problems of this type was in its infancy, and computers were not readily available. Nevertheless during this same period, Stewartson [11, 12] was in the process of formulating a theory of incompressible flow in a rapidly rotating container assuming a small but finite departure from solid body rotation. He used what is now termed the method of matched asymptotic expansions in which the governing equations are simplified in regions near the walls (boundary layers) where it emerges that most of the fluid motion takes place. In Stewartson's work none of the *ad hoc* assumptions made by the workers in the gas centrifuge industry were required; moreover, analytic closed-form solutions for the velocity field were obtained. Unfortunately, the workers in the gas centrifuge industry appeared to be unaware of Stewartson's work, and its importance to the gas centrifuge problem was not recognized until the mid 1970s when Japanese workers began to use Stewartson's methods to study the flow in a variety of simplified geometries [13–20]. Then, in the early 1970s, research in the United States and Europe began to accelerate; a number of scientists working on the problem began to meet every couple years or so, and funding by the energy agencies of countries such as France, Japan and the United States began to increase.† In the United States, much of the research in this area has been conducted by the Department of Energy through Union Carbide Corporation and Los Alamos.

---

† The meetings referred to above are the Workshops on Gases in Strong Rotation. Several references to papers presented at these meetings appear in this work.

Just after Stewartson's work, the problem was approached using the "pancake" method [21, 22] which, it emerges, is a model for the flow field within the bowl at extremely high rotational speeds.

In summary then, it appears that the basic physical principles of gas centrifugation were well known by the early 1950s; research centrifuges were built and operated by 1960. However, the problem of calculating the flow patterns in a gas centrifuge, the determination of which are crucial for a detailed understanding of the separation problem, was still not well understood, and it was not until the late 1970s and early 1980's that significant progress in the modeling of a gas centrifuge began to take place [23–29]. It is the relatively recent work on the problem of flow and separation in a gas centrifuge that is the main focus of this article.

Chapter Scope

The primary concern in this article is the discussion of the fluid mechanics and design of the centrifuge bowl; the problem of designing the centrifuge bowl based on a thorough understanding of the underlying fluid mechanics of a gas centrifuge is of primary importance here. Thus such topics as the mechanical design and operation of the bowl and of the cascading arrangements will not be discussed; discussion of these topics appear in earlier reviews ([6, 7, 8] for example).

In the next section, the mass averaged flow patterns which exist in a high speed gas centrifuge are described and the methods used to generate solutions are also discussed. In this section the rapid variation of the flow field which occurs near the walls of the centrifuge is illustrated.

In a later section, the equations to be solved for the detailed mass fraction distribution within the container are derived and attention is given to the boundary conditions of the problem which depend in some sense on the type of method used to solve the problem.

For clarity of presentation, the discussion of the types of solutions which have appeared in the literature comprises a separate section. In that portion of the review the characteristics of the detailed solutions for the mass fraction are described and the behavior of the separation factor as a function of a number of crucial parameters is demonstrated. Particular attention in this section is focused on the discussion of an approximate analytical solution which has recently appeared in the literature [23].

The results presented in the fourth section naturally lead to a discussion of optimizing centrifuge performance and this problem is discussed qualitatively in the fifth section. In the last section, a summary of the review and a discussion of some significant problems which still remain in gas centrifuge technology are presented. Table 1 gives the dimensions and rotation speeds of typical centrifuges.

Table 1
Dimensions and Rotation Speeds of Typical Centrifuges*

Author	Length cm	Outer Radius cm	Peripheral Speed (m/sec)
Zippe [4]	30–38	3.81	300–350
Groth [3]	65–316	9–23	300
GSR Rome	250–500	50	400–600
Soubbaramayer [7]	60	7.35	400
Soubbaramayer [7]	250	25	600
Lopez [42]	62	7.5	400
Wood and Sanders [29]	335.3	9.145	400–700
Merten and Hanel [27]	250	25	600
Conlisk, Foster and Walker Theory [24]	250	25	260

* *Present U.S. machines are believed to be several meters high and operate in the 500–700 m/sec range of peripheral speed [1]. The parameters denoted by GSR are those suggested by participants at the Fluid Workshop on Gases in Strong Rotation held in Rome in 1979.*

THE FLUID MECHANICS OF A GAS CENTRIFUGE

Governing Equations

The basic centrifuge configuration† is indicated in Figures 1 and 2 and consists of an annular region bounded by horizontal end plates at $z^* = 0$ and $z^* = L$. The apparatus is rapidly rotating with angular velocity Ω about the axis of the inner and outer cylinders at $r^* = A_1, B_1$, the motion of the gas within the bowl is that of a compressible gas, and for solid body rotation in the absence of a feed flow, the density distribution in dimensionless form is given by

$$\rho_e = \rho_e^*/\rho_0^* = e^{M^2(r^2 - a^2)/2} \tag{1}$$

where $r = r^*/L$ and M is called the rotational Mach number and is defined by

$$M^2 = \frac{\Omega^2 L^2}{RT_0^*} \tag{2}$$

Here the equilibrium (solid body rotation) state is assumed to be that of a perfect gas at a constant temperature T_0^*; R is the gas constant, and ρ_0^* is the fluid density at $r^* = A_1$.

To obtain the governing equations for the perturbation quantities due to a given driving mechanism, the axisymmetric Navier-Stokes equations and the energy equation are written in reference frame rotating with the bowl; assuming these perturbation quantities are small, the nondimensional equations of motion for the perturbation quantities are given by [23]:

$$\frac{\partial u}{\partial r} + \frac{u}{r} + \frac{\partial w}{\partial z} + rM^2 u = 0 \tag{3}$$

$$\epsilon_f \left\{ u\frac{\partial u}{\partial r} + w\frac{\partial u}{\partial z} - \frac{v^2}{r} + \rho T \right\} - 2v = -rT - \frac{1}{M^2}\frac{\partial p}{\partial r} + \frac{E}{\rho_e(r)}\left\{ \nabla^2 u - \frac{u}{r^2} - \frac{M^3}{3}\frac{\partial}{\partial r}(ru) \right\} \tag{4}$$

$$\epsilon_f \left\{ u\frac{\partial v}{\partial r} + w\frac{\partial v}{\partial z} + \frac{uv}{r} \right\} + 2u = \frac{E}{\rho_e(r)}\left\{ \nabla^2 v - \frac{v}{r^2} \right\} \tag{5}$$

$$\epsilon_f \left\{ u\frac{\partial w}{\partial r} + w\frac{\partial w}{\partial z} \right\} = -\frac{1}{M^2}\frac{\partial p}{\partial z} + \frac{E}{\rho_e(r)}\left\{ \nabla^2 w - \frac{M^2 r}{3}\frac{\partial u}{\partial z} \right\} \tag{6}$$

$$\epsilon_f \, Pr \left\{ u\frac{\partial T}{\partial r} + w\frac{\partial T}{\partial z} \right\} - 4hru = \frac{E}{\rho_e(r)}\nabla^2 T \tag{7}$$

where $h = \dfrac{(\gamma - 1)\, Pr\, M^2}{4\gamma}$

$$\nabla^2 = \frac{1}{r}\frac{\partial}{\partial r}\left(r\frac{\partial}{\partial r} \right) + \frac{\partial^2}{\partial z^2}$$

Here Pr is the Prandtl number and γ is the ratio of specific heats, both of which are assumed to be constant to leading order. In Equations 3 through 7, the representative length scale is L. A representative velocity scale may be defined in terms of the feed flow rate $\dot{m}^*$, and this scale is

$$U_I = \frac{\dot{m}^* E^{-1/2}}{\rho_0^* L^2} \tag{8}$$

† We consider the Zippe-Groth type centrifuge in detail and refer to analysis of the Beams machine where appropriate.

As can be seen there are three major dimensionless parameters which govern the motion of the gas; these are the rotational Mach number defined previously; E, the Ekman number, and ϵ_f, a source-sink Rossby number. The last two parameters are defined by

$$E = \nu/\Omega L^2 \tag{9a}$$

$$\epsilon_f = U_l/\Omega L \tag{9b}$$

Finally, the perfect gas relation for the perturbation quantities is given by

$$p = \rho + T \tag{10}$$

The perturbation quantities p, ρ, T are assumed to be $0(\epsilon_f)$, and are defined by

$$\rho^* = p_e^*(1 + \epsilon_f p)$$

$$\rho^* = \rho_e^*(1 + \epsilon_f \rho)$$

$$T^* = T_0^*(1 + \epsilon_f T)$$

where p_e^*, ρ_e^*, and T_0^* are the distributions for the solid body rotation case.

As is evident from the form of the equations, the general solution to these equations is nearly impossible because of the presence of the nonlinear terms on the left hand side of Equations 4 through 6. However, in centrifuge flows, the feed flow rate $\dot{m}^*$ is very small in order to prevent turbulent mixing which inhibits separation; thus the nonlinear terms may be neglected and on a dimensionless scale the flow rate is assumed to be $0(E^{1/2})$. Moreover, the Ekman number E is extremely small being of $0(10^{-7}–10^{-9})$ in typical cases. This fact may also be exploited to simplify the governing equations as will be seen later in this section. Two points are of interest here. First, the length scale used to nondimensionalize these equations is L, and consequently, the aspect ratio, $1/b = L/B_1$ has been assumed 0(1); however, larger aspect ratios may be considered in which case the radial length scale becomes B_1. In this case the density and Mach number are modified according to

$$\rho_{ep} = \rho_{ep}^*/\rho_0^* = e^{M_p^2(r^2-1)/2}$$

$$Mp^2 = \frac{\Omega^2 B_1^2}{RT_0^*} \tag{11}$$

where ρ_0^* is now the density at $r = 1$ ($r^* = B_1$). This nondimensionalization is necessary when the geometry of interest is a cylindrical can; further, if the Mach number is high enough the gas near the axis of the centrifuge ($r^* \to 0$) is rarefied ($\rho_e \to 0$ there) and using the value of the density there to scale the equations is inappropriate. This situation also occurs when the centrifuge is assumed to be an annulus (Figures 1 and 2) and M (or M_p) increases for fixed inner radius A_1. In Reference 23 and in most previous work [13–20, 23–27], the fluid is assumed to be a continuum and these problems do not arise. The limitations of the continuum theory will be discussed later in this section. Apart from some minor notational differences Equations 3 through 7 have been used as a starting point in the study of centrifuge flows by a vast majority of workers in the field.

Motion may be induced within the bowl of a gas centrifuge by at least three methods. These are:

1. *Differential rotation*, whereby one or more of either the side or the end walls is rotated at a slightly different rate from the rest of the container; this was the type of flow field which Stewartson [11] analyzed and is a model for the scoop drive in the Zippe centrifuge; such a device is called a mechanically driven centrifuge.
2. *An applied thermal gradient*, whereby an externally applied horizontal or vertical temperature gradient occurs. Such a device is called a thermally driven centrifuge (Groth).
3. *An imposed source-sink flow* in which fluid is continuously injected and withdrawn from the container.

Usually one or both of Methods 1 and 2 exist with Method 3, and one consequence of the linearity of the problem is that each effect may be considered separately with the solution for the flow field being the sum of the individual effects.

The boundary conditions associated with the governing equations depend on the driving mechanism of fluid motion; for source-sink flow, the conditions are the no-slip and solid-wall conditions which require

$$u = v = w = 0 \qquad \text{at the walls}$$

In addition, an isothermal temperature distribution is assumed so that in dimensionless form

$$T = 0, \qquad \text{at the walls}$$

For motion driven by a vertical temperature distribution, a thermal Rossby number may be defined according to

$$\epsilon_T = \frac{T_1^* - T_0^*}{T_0^*}$$

where T_0^* and T_1^* are the dimensional temperatures at $z = 0, 1$; the boundary conditions for the perturbation temperature then become

$$T = 0 \qquad \text{at } z = 0$$

$$T = \frac{\epsilon_T}{\epsilon_f} = \lambda_T \qquad \text{at } z = 1$$

Furthermore if the sidewalls at $r = a$ and $r = b$ are maintained at constant temperatures T_a^* and T_b^*, the sidewall thermal conditions may be written

$$T = \lambda_a \lambda_T \qquad \text{at } r = a$$

$$T = \lambda_b \lambda_T \qquad \text{at } r = b$$

where $\lambda_a = (T_a^* - T_0^*)/(T_1^* - T_0^*)$

The definition of λ_b is analogous. Other side-wall thermal conditions have also been investigated.

For motion induced by differentially rotated end plates (a model for the scoops in a mechanically driven machine) a mechanical Rossby number may be defined by

$$\epsilon_m = \frac{\Delta\Omega}{\Omega}$$

where $\Delta\Omega$ is the magnitude of the different rotation. Defining λ_m by

$$\lambda_m = \epsilon_m/\epsilon_f$$

the boundary conditions for v at the end walls may be written

$$v = \lambda_m \lambda_1 r \qquad \text{at } z = 1$$

$$v = \lambda_m \lambda_0 r \qquad \text{at } z = 0$$

Here $\lambda_1 = \pm 1$ and λ_0 is a constant. The conditions for u, w, and T are homogeneous on all walls; the conditions on v are homogeneous on the side walls.

In general, three approaches have been employed to solve the system of governing equations. First, the method of matched asymptotic expansions [23] has been used to exploit the fact that $E \ll 1$ and replace the full equations with simpler equations valid in the boundary layers which form near the walls; it emerges (see below) that most of the fluid motion takes place near the walls so that this simplification is extremely useful. An advantage of this technique is that analytic, closed-form solutions may be obtained for the velocities and temperature in the boundary layers which can then be input into the mass transfer problems as required. The second method used is numerical [24–27] whereby solutions are obtained by digital computer. The disadvantage of numerical techniques is that the cost of computation is often high; also there is some difficulty in observing trends in examining tables of numbers on output. The third approach that has been taken for solution is the pancake method [21, 22, 28, 29]; as employed this method is valid for $M_p^2 = 0(E^{-1/3})$. In this limit all the fluid is contained in a thin boundary layer of width $0(1/M_p^2)$ near the outer wall and the inner core of the centrifuge is rarefied ($\rho_{ep} \to 0$ as $M_p \to \infty$ away from $r^* = B_1$). The method consists of reducing the governing equations to one equation which is valid away from the end walls $z^* = 0, L$.

In what follows we will examine each one of these methods in detail and present results where possible for comparison.

Boundary-Layer Methods

In this method the centrifuge bowl is divided into an inviscid core surrounded by thin boundary layers in which viscous forces are important. The boundary layer structure of a compressible gas in a rotating flow is now well known and is depicted in Figure 3. In general, fluid enters the container through a small axisymmetric slot of width $0(E^{1/2})$ and passes through a vertical boundary layer where $r = a + 0(E^{1/3})$; this inner boundary layer depicted in Figure 3 is present to adjust the vertical velocity to relative rest on the side wall. Fluid which does not pass out of the product port enters a thicker layer of width $0(E^{1/4})$ before being deflected to the end-wall Ekman layers of width $0(E^{1/2})$ [30]. (In honor of their founder the $E^{1/3}$ and $E^{1/4}$ layers are named Stewartson layers.) The purpose of the $E^{1/4}$ layer is to adjust the swirl velocity to relative rest on the side wall. Fluid then passes through the Ekman layer to the outer wall and out of the container. For purely source-sink flow there is no net motion within the core region; only a leading-order swirling motion occurs [23, 31,

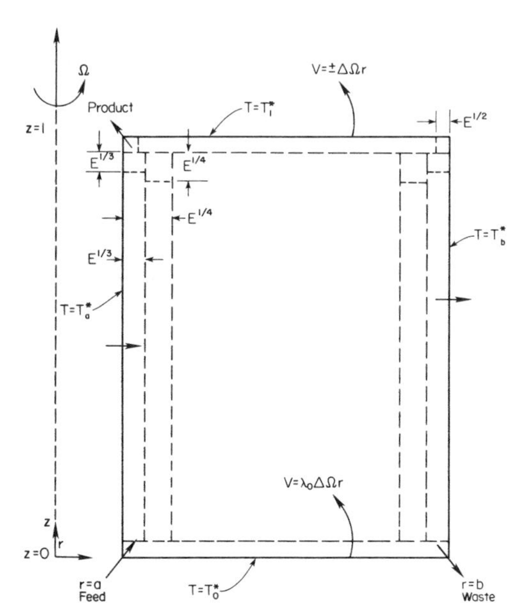

Figure 3. Geometry and coordinate system for a cross section of the model centrifuge. Straight bold arrows indicate possible locations of injection or withdrawal. The container walls are at constant temperature, and the top and bottom plates may be in differential rotation. Shear layer structure is not to scale; note substructure near the upper corners for differential rotation of the top plate. The locations indicated product, feed, and waste correspond to the particular source-sink geometry used in the mass transfer analysis [23].

32]. The primary effect of a small differential rotation of the end plates or a small vertical temperature gradient is to induce a vertical drift within the core region of the centrifuge [11, 13, 23, 33]. It appears that in many situations the presence of either of these latter mechanisms is instrumental in enhancing the enrichment process. Solutions obtained using the analysis described in the following are valid under the limit process $E \to 0$ for M fixed.

To illustrate the types of solutions which arise, we consider the flow in each of the boundary-layer regions and the geostrophic core.

The Geostrophic Flow

In this region the flow variables scale according to

$$\begin{aligned} u &= EU_G(r, z) + \cdots \\ v &= V_G(r, z) + \cdots \\ w &= E^{1/2}W_G(r) + \cdots \end{aligned} \tag{12}$$

and

$$\begin{aligned} T &= T_G(r, z) + \cdots \\ p &= p_G(r) + \cdots \end{aligned} \tag{13}$$

Note from Equation 12 that $u = 0(E)$ in the core and consequently no radial flow may occur to leading order; note however, that a leading order vertical motion may occur as a result of a temperature gradient or differential rotation. For a purely source-sink motion the leading order component of W_G is zero. The core solution depends in general on the type of thermal boundary conditions and many types of solutions have appeared in the literature [14, 16, 18, 23, 24].

The Ekman Layers

The two functions of the Ekman layers are to pass fluid from the inner wall to the outer wall and to bring the velocities to relative rest at the end walls. In this layer, the scaled vertical variable is given by

$$\zeta = \sigma^{1/4}(\rho_e/E)^{1/2}z$$

near $z = 0$ and near $z = 1$,

$$\zeta = \sigma^{1/4}(\rho_e/E)^{1/2}(1 - z)$$

Here the quantity σ is defined by

$$\sigma(r) = 1 + \frac{(\gamma - 1)\,Pr\,M^2r^2}{4\gamma}$$

The solution for the radial and azimuthal velocities are conveniently expressed in terms of a variable $\chi = u + i\sigma^{1/2}v$ and

$$\chi = i\sigma^{1/2}V_G + i\sigma^{1/2}\{V_B - V_G\}e^{-(1+i)\zeta} \tag{14}$$

where $i = \sqrt{-1}$ and V_B is the swirl velocity at the bottom wall. The remarkable feature of the Ekman layer is that it exerts control over any layer of width greater than $0(E^{1/2})$ over it through the Ekman condition which is derived using the continuity equation; the result for the geostrophic core is

$$W_G = \pm\frac{1}{2r\rho_e}\frac{\partial}{\partial r}\{r\sigma^{1/4}\rho_e{}^{1/2}(V_G - V_B)\} \qquad \text{at } z = \frac{1}{2} \mp \frac{1}{2} \tag{15}$$

In particular, forms of Equation 15 must be used when computing solutions to the $E^{1/4}$ layer and $E^{1/3}$ layer.

The $E^{1/4}$-layer

In this layer a scaled boundary-layer variable is defined by

$$\begin{aligned} \xi &= \beta(r - r_0)\mu E^{-1/4} \\ \mu &= \sqrt{2}\sigma^{1/8}(r_0)\rho_e^{1/4}(r_0) \end{aligned} \tag{16}$$

and $r_0 = a$ or b. Substitution of the scaling Equation 16 into the governing equations leads to the solutions

$$\begin{aligned} v_{1/4} &= V_G(r_0)\{1 - e^{-\xi}\} \\ u_{1/4} &= -\mu^2 V_G(r_0)E^{1/2}e^{-\xi}/2\rho_e(r_0) \\ w_{1/4} &= -\beta\mu\sigma^{1/4}(r_0)V_G(r_0)E^{1/4}(z - \tfrac{1}{2})e^{-\xi}/\rho_e^{1/2}(r_0) \\ T_{1/4} &= T_B(r_0, z) - 2hr_0V_G(r_0)(1 - e^{-\xi}) \end{aligned} \tag{17}$$

where $\beta = 1$ at $r_0 = a$
$\beta = -1$ at $r_0 = b$
$T_B(r_0, z)$ = the (fixed) temperature at the side wall

Note that from the first of Equations 17 the swirl velocity has been reduced to zero at $\xi = 0$ ($r = a$).

The solutions Equations 17 are valid for fixed-temperature side walls. Matsuda and Hashimoto [14] consider the case of insulated end plates and conducting side walls, whereas Bark and Hultgren [34] consider the case of insulated side walls and both conducting and insulated end plates. Also, in Equations 17 note that $w_{1/4} \neq 0$ and $u_{1/4} \neq 0$ at $\xi = 0$; consequently an $E^{1/3}$ layer is required to bring these velocities to relative rest. Also $E^{1/4} \times E^{1/4}$ regions are required if the end walls and the side walls are not at the same temperature or if the side walls are differentially rotated; (see References 14 and 23 for the details of the solutions in these corner regions. It should be pointed out that the solutions for the core, the Ekman layer, and the $E^{1/4}$ layer are independent of the method of injection or withdrawal.

The $E^{1/3}$-Layer

The $E^{1/3}$ layer solutions are difficult to obtain because of the method of injection used in the centrifuges. As has already been mentioned fluid enters and leaves the centrifuge through small $0(E^{1/2})$ axisymmetric slots; since $E^{1/3} \gg E^{1/2}$, the $E^{1/3}$ layer sees locations of injection or withdrawal as point sources or sinks, respectively. The match with the $E^{1/4}$ layer indicates that the expansions for the $E^{1/3}$ layer should be

$$\begin{aligned} u_{1/3} &= E^{1/3}\{u_0 + E^{1/12}u_1 + E^{1/6}u_2 + \cdots\} \\ v_{1/3} &= v_0 + E^{1/12}v_1 + E^{1/6}v_2 + \cdots \\ w_{1/3} &= w_0 + E^{1/12}w_1 + E^{1/6}w_2 + \cdots \end{aligned} \tag{18}$$

Similar expansions can be obtained for $T_{1/3}$, $p_{1/3}$. Solutions for u_0, u_0, w_0, u_1, v_1, w_1 are trivial and given in Reference 32 for incompressible flow; the (u_2, v_2, w_2) problem characterizes the layer. The governing equations and boundary conditions for the n = 2 problem are

$$\frac{\partial^3 v_2}{\partial \eta^3} = -2\frac{\partial w_2}{\partial z}, \qquad \frac{\partial^3 w_2}{\partial \eta^3} = 2\frac{\partial v_2}{\partial z}, \qquad \frac{\partial u_2}{\partial \eta} + \frac{\partial w_2}{\partial z} = 0 \tag{19}$$

$$v_2 \sim \eta^2, w_2 \to 0, u_2 \to 1 \qquad \text{as } \eta \to \infty \tag{20}$$

$$u_2 = v_2 = w_2 = 0 \qquad \text{at } \eta = 0 \tag{21}$$

$$w_2 = 0, z = 0, 1 \tag{22}$$

In these equations,

$$\eta = \rho_e^{1/3}(r_0)\sigma^{1/6}(r_0)\beta(r - r_0)E^{-1/3} \tag{23}$$

Here β has the same definition as given previously and r_0 = a or b. Equation 19 subject to Equations 20 through 22 have been solved by Conlisk and Walker [32] for both small $0(E^{1/2})$ slots and finite slots on the side walls or in the corner regions for incompressible flow. As an example of the type of solutions that arise, the solution for a jet emerging from an $0(E^{1/2})$ slot at $z = \phi$ is given by

$$\begin{aligned} u_2 = 1 + 4/\sqrt{3}\,\bar{C} \sum_{n=1}^{\infty} \cos n\pi\phi e^{-\omega_n\eta/2} \sin(\sqrt{3}/2\omega_n\eta + \pi/3) \cos n\pi z \\ - \sum_{\substack{n=-\infty \\ n\neq 0}}^{\infty} \delta(z - \phi - 2n) - 2\bar{C} \sum_{n=-\infty}^{\infty} \delta(z + \phi - 2n) \end{aligned} \tag{24}$$

$$w_2 = 8/\sqrt{3}\,\bar{C} \sum_{n=1}^{\infty} \frac{\cos n\pi\phi}{\omega_n^2} e^{-\omega_n\eta/2} \sin \sqrt{3}/2\omega_n\eta \sin n\pi z \tag{25}$$

In Equation 23 $\bar{C}$ is a source (sink) strength and is proportional to the mass flux passed by the source (sink); since the problem is linear, solutions for any number of sources and sinks may be superposed to yield more complicated flow patterns [32].

In Equations 24 and 25 $\omega_n = (2n\pi)^{1/3}$ and the δ-function singularities are due to the singular behavior of the solution. Ultimately, the vertical flux through the shear layer will be required in the subsequent separation analysis and this is easily obtained from Equation 25 as

$$\begin{aligned} \int_0^{\infty} w_2(\eta, z)\, d\eta &= \bar{C}(1 - z) && \text{for } z > \phi \\ &= -\bar{C}z && \text{for } z < \phi \\ &= \bar{C}(1 - 2z)/2 && \text{for } z = \phi \end{aligned} \tag{26}$$

To illustrate the types of flow patterns which may arise as a result of source-sink motion, on Figure 4 is a cross-sectional view of the centrifuge. On r = a there is a uniform source of fluid and on r = b are two small annular slots at z = 0, 1 of equal strength. Note that fluid emerging from the source is rapidly deflected to the end walls and no net motion occurs in the core. A source-sink geometry more reminiscent of centrifuge application is depicted in Figure 5 where there is a source at z = 0 and a sink of strength 0.5 (cut $\theta = 0.5$) at z = 1. It should be pointed out that the flow patterns depicted here are for incompressible flow and Figure 4 is reproduced from Reference 33. However, the incompressible flow patterns within the $E^{1/3}$ and $E^{1/4}$ layers differ only by a scale factor (compare Equations 19 through 22 with Equations 8 through 12 of Reference 32), and consequently, these figures are qualitatively correct for the compressible case as well. The flow patterns near r = a were first experimentally determined by Hide [31].

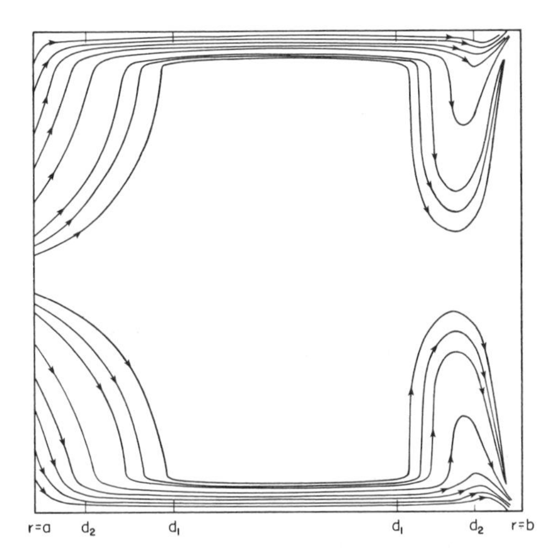

Figure 4. Streamlines for a uniform injection flow at r = a and sinks of equal strength at r = b, z = 0,1. Arrows indicate the direction of flow, and the boundary-layer regions are to scale for $E = 8.0 \times 10^{-4}$; the radial locations labeled d_1 and d_2 indicate the approximate outer boundaries of the $E^{1/4}$ and $E^{1/3}$ layers, respectively.

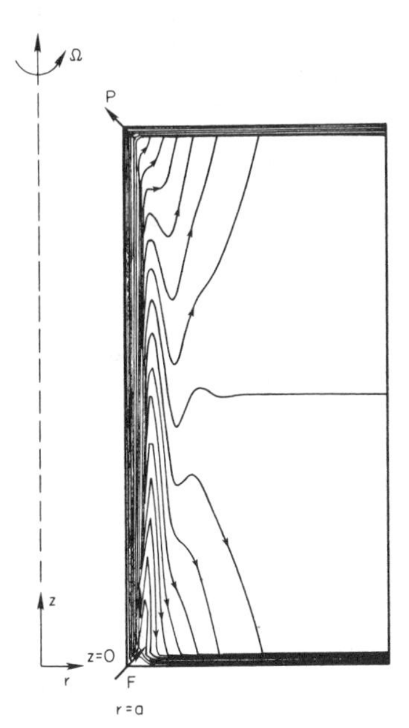

Figure 5. Streamline patterns for feed (F) at z = 0, and product (P) at z = 1; here the cut $\theta = 0.5$. Here the Ekman number $E = 10^{-9}$, and only the shear layer on r = a is shown.

The $E^{1/3}$ layer flow due to differential rotation or vertical temperature gradient will now be considered. Essentially the solution for the velocity field in the $E^{1/3}$ layer for source-sink motion given by Equations 24 and 25 applies for these driving mechanisms with redefinition of the strength of the singularity $\bar{C}$. The details are given in Reference 23.

Limitations of Boundary-Layer Methods

Several limitations on the method of matched asymptotic expansions for the solution of the centrifuge problem emerge; first, the boundary layers are required to be thin, and second, the equations

Table 2
Parameter Restrictions for Validity of the Asymptotic Theory*

A_1/B_1	E/ρ_R	$E^{1/4}$ Layer	Mean Free Path
0.1	3.78×10^{-3}	1.84×10^{-6}	1.03×10^{-1}
0.2	2.75×10^{-3}	5.89×10^{-5}	5.13×10^{-2}
0.3	1.62×10^{-3}	4.47×10^{-4}	3.42×10^{-2}
0.4	7.73×10^{-4}	1.88×10^{-3}	2.57×10^{-2}
0.5	2.98×10^{-5}	5.75×10^{-3}	2.05×10^{-2}
0.6	9.31×10^{-5}	1.43×10^{-2}	1.71×10^{-2}
0.7	2.35×10^{-5}	3.09×10^{-2}	1.47×10^{-2}
0.8	4.81×10^{-6}	6.03×10^{-2}	1.28×10^{-2}
0.9	7.95×10^{-7}	1.09×10^{-2}	1.14×10^{-2}

* *For validity, each entry at a given A/B must be much larger than the corresponding number for E/ρ_R. Here the peripheral speed is 400 m/sec., the diameter is 0.5 m and the aspect ratio is 5.*

must be linear. The first restriction requires $E^{1/4} \ll 1$ since the $E^{1/4}$ layer is the thickest layer in the rotating flow structure. It can be shown [35] that the second restriction requires $\epsilon_f \ll E^{1/4}$.

Two other important restrictions arise as well. In order for a geostrophic core to exist, the outer dimensionless radius, b (1/b is really an aspect ratio) must satisfy the requirement $b = 0(1)$. However, the equations may easily be modified to include the case where $1 \gg b \gg E^{1/4}$ [36, 37]; for $b = 0(E^{1/4})$ and smaller Brouwers [37] has considered the motion which is due to the effects described here.

The last restriction is the requirement that the gas remains a continuum throughout the bowl. This restriction arises when the Mach number increases and we require that the mean free path of the gas near the inner wall be small compared to the Ekman layer thickness there. That requirement translates to

$$E/\rho_{ep} \ll B_1/(A_1 M_p^3)$$

where ρ_{ep} is given by Equation 11. Also the requirement that the $E^{1/4}$ layer remain thin translates, for higher Mach numbers to

$$(E/\rho_{ep})^{1/4} \ll 1$$

where again ρ_{ep} is given by Equation 11. These requirements essentially limit the value of Mach number based on peripheral speed ΩB_1 which can be considered. In Table 2 [36] these restrictions are detailed as a function of the radius ratio A_1/B_1 for a peripheral speed of 400 m/sec. Essentially, this table tells us that for peripheral speed $\Omega B_1 = 400$ m/sec, the validity of this analysis is limited to radius ratio, $A_1/B_1 \gtrsim 0.5$; for $B_1 = 600$ m/sec, it emerges that $A_1/B_1 \geq 0.8$ [36]. These cases correspond to Mach numbers based on peripheral speed of about 4.0 and 6.0. Above these speeds the region near the inner post will usually be rarefied for $A_1/B_1 < 1$.

For fixed aspect ratio as M (or M_p) $\rightarrow \infty$ most of the gas is contained near the outer wall of the centrifuge, and the boundary-layer theory no longer applies near the center post. To consider this situation, the "pancake" theory was developed, and it is this analysis which is considered next.

The Pancake Equation

The pancake equation was first developed by Dr. Lars Onsager in 1961 [22] and formally presented in a report by Carrier and Maslen [38] and in a paper by Carrier [21]. It is named for the fact that most of the fluid is contained in a thin layer or "pancake" near the outer wall of the centrifuge. Wood and Morton [28] and Wood and Sanders [29] have utilized this work and presented flow patterns for differential rotation (or vertical temperature gradient) and source-sink flows, respectively.

The method has been put on a firm theoretical basis by Cole and Schleiniger [39] who show that the pancake equation arises under the limit process

$$M_p^2 \to \infty, \qquad E = \frac{\lambda}{M_p^6} \to 0 \tag{27}$$

with $\lambda = 0(1)$. Here M_p must be based on the outer radius B_1. In this limit, there is a layer of width $0(1/M_p^2) = 0(E^{1/3})$ near the outer wall where the scaled radial variable is given by

$$\eta = M_p^2(r - 1)/2 \tag{28}$$

Here M_p is the peripheral Mach number defined in Equation 11. Also, because the Mach number is so large the equilibrium density distribution varies across the layer as

$$\rho_{ep} = e^{\eta} \tag{29}$$

Substitution of the scaling into the governing Equations 3 through 7 and rearrangement yields the pancake equation according to

$$\{e^{-\eta}(e^{-\eta}\Lambda_{\eta\eta})_{\eta\eta}\}_{\eta\eta} + \lambda^2(G + 4)\Lambda_{zz} = 0 \tag{30}$$

where $\Lambda_\eta = \psi$
ψ = the stream function
$G = M_p^2(\gamma - 1)Pr/\gamma$

The subscripts η denote partial differentiation.

A sketch of the pancake structure for large Mach number is given in Figure 6; notice that in the core region, the gas is rarefied and noncontinuum methods must be used. Also, unlike the Stewartson analysis which is valid for smaller Mach numbers, fluid particles are permitted to pass through the core region from the feed tube on $r = a$ to the outer wall at $r = 1$; recall that in "pancake" theory the radial length scale is taken as the outer radius $r^* = B_1$, and the motion takes place between $r = A_1/B_1$ and $r = 1$. Because of the presence of variable coefficients in Equation 30 solutions are usually computed numerically.

Numerical Methods

An alternative to the asymptotic methods described previously is numerical computation of the flow field; in most of the work in this area, the governing equations are the *unscaled* Navier-Stokes

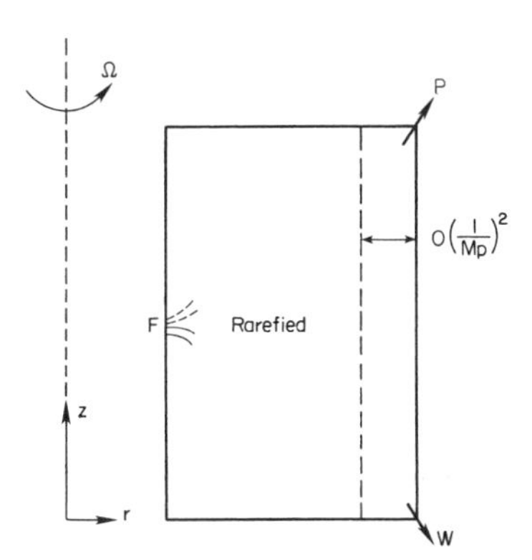

Figure 6. Schematic of the "pancake" structure. Here P, W denotes possible locations of the product and waste ports.

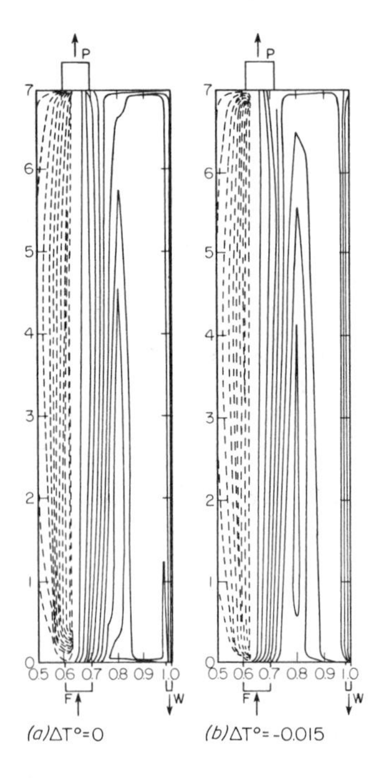

Figure 7. Streamline patterns from Kai [26]; ΔT^* is proportional to the vertical temperature gradient.

equations as discussed in Reference 7. In particular Kai [24, 26], has solved the full nonlinear equations using finite difference techniques, while Lahargue and Soubbaramayer [40] used finite element techniques. The geometry of the centrifuge considered in these two papers is that of a cylindrical can. Previously Bennetts and Jackson [41] reported numerical solutions for incompressible flow in an annulus.

Lopez [42] has considered the flow due to differential rotation of the top disk of a Beams-type centrifuge. He used the linearized equations of motion to solve for the flow field by finite differences; he employed iterative techniques with overrelaxation and his code took 3,000 iterations to converge on a 50-by-60 point mesh. Some of his work is presented in Reference 6. Merten and Hänel [25, 27] use the unsteady equations to calculate the steady solution to the full nonlinear equations. It is interesting to note that they had difficulty in converging to a solution for higher Mach numbers for smaller inner radius indicating that restrictions such as those depicted in Table 2 may apply.

As an example of some of the results which have been reported, on Figure 7 is the Figure 3 of Kai [26] who used a cylindrical can (Beam's model) as a model centrifuge. As depicted in the figure, the gas enters at the feed on the bottom wall, and some of the gas passes out the product port at the top. The rest of the gas passes out the waste port at the bottom corner at the outer wall. Note the presence of a boundary layer on the outer wall. It should be noted, however, that not enough mesh points have been used to properly resolve both Stewartson layers.

SEPARATION ANALYSIS

Governing Equations

Once the flow through the centrifuge bowl has been calculated, the flow patterns may be used to solve the mass transfer problem. The equation which must be solved is well known, and in the notation of References 23 and 43 it is given in dimensionless form by

$$\rho_e(r)E^{-1/2}(\vec{q}\cdot\nabla)\omega_A - \delta\epsilon M^2 H(\omega_A) = \delta\nabla^2\omega_A \tag{31}$$

$$H(\omega_A) = 2\omega_A(1-\omega_A) + r\frac{\partial}{\partial r}\{\omega_A(1-\omega_A)\}$$

where δ = a species flux parameter
ω_A = the mass fraction

$$\delta = E^{1/2}/(\epsilon_f Sc)$$
$$\epsilon = (M_B - M_A)/M_B \tag{32}$$

where M_A and M_B are the molecular weights of species A and B. This equation expresses the fact that the convection of mass and the pressure diffusion of mass balance ordinary diffusion, which is modeled by Fick's law. Traditionally, to solve Equation 31, spurious approximations for the velocity field have been used in the first term of Equation 31 [6, 7, 8].

The boundary conditions associated with Equation 31 are that there is no normal mass flow through solid walls. At radial boundaries the radial mass flux is given by

$$N_{A,r} = -E^{1/2}\delta\left\{\frac{\partial\omega_A}{\partial r} + \epsilon M^2 r_0\omega_A(1-\omega_A)\right\} + \rho_e\omega_A u \tag{33}$$

where r_0 = a or b

At horizontal boundaries, the vertical mass flux is given by

$$N_{A,z} = -E^{1/2}\delta\frac{\partial\omega_A}{\partial z} + \rho_e\omega_A w \tag{34}$$

and for no mass flow across any boundaries

$$N_{A,r} = N_{A,z} = 0 \tag{35}$$

It is useful to examine the solution for the case where the feed flow is so low that the fluid is in solid body rotation ($\delta \to \infty$); furthermore, for separation of uranium isotopes $\epsilon = 0.0085$ and unless $M = 0(1/\sqrt{\epsilon})$, $\epsilon M^2 \ll 1$. In this case the equation 31, subject to Equations 33 and 34 suggests an expansion of the form

$$\omega_A = \omega_F + \epsilon M^2\bar{\omega}_A + \cdots \tag{36}$$

Substitution of Equation 36 into the governing equations yields the following solution for $\bar{\omega}_A$:

$$\bar{\omega}_A = \tfrac{1}{2}(a^2 - r^2) \tag{37}$$

Here ω_A is the light species mass fraction. Note for this case $\vec{q} \to 0$, there is no vertical separation of the mixture and the mixture is not enriched in species A. Thus, the flow patterns which occur within the centrifuge are crucial for enrichment of the mixture, and they cannot be ignored as in much of the early work [6].

From Equation 31 it is apparent that the flow patterns within the centrifuge must be computed before solutions for the species mass fraction can be computed. When numerical solutions to the governing equations are obtained, the boundary conditions are as given by Equation 35; however, when boundary-layer methods are used all of Equations 31, 33, and 34 must be reexamined within each of the layers. These new boundary conditions emerge as compatibility conditions on the specific concentration outside the fluid boundary layers; it will also be shown that to leading order, the mass fraction does not vary across any of the fluid boundary layers in the parameter ranges of interest.

Boundary Conditions Using Boundary-Layer Methods

To develop the correct boundary conditions for the species concentration problem, consider first the geostrophic interior of the centrifuge.It has already been demonstrated that for either a mechanically or thermally driven centrifuge, the radial velocity $u = 0(E)$ and the vertical velocity $w = 0(E^{1/2})$ within the core. Let ω_0 be the leading-order core mass fraction distribution. Then it may easily be shown using Equations 12 and 31 that

$$\rho_e W_G \frac{\partial \omega_0}{\partial z} - \frac{\delta \epsilon M^2}{r} \frac{\partial}{\partial r} \{r^2 \omega_0 (1 - \omega_0)\} = \delta \nabla^2 \omega_0 \tag{38}$$

This equation must be carefully scaled in each of the fluid boundary layers.

Consider first the Ekman layer near $z = 0$ for which the scaled vertical coordinate is

$$\zeta = \sigma^{1/4} (\rho_e / E)^{1/2} z$$

and substitution in Equation 38 yields

$$\frac{\partial^2 \omega_A}{\partial \zeta^2} = (E/\delta^2 \sigma)^{1/2} (\vec{q} \cdot \nabla) \omega_A + 0(E, E/\delta, \epsilon E) \tag{39}$$

where $\vec{q}$ is the dimensionless velocity distribution in the Ekman layer. This equation suggests that the asymptotic expansion of ω_A in the Ekman layer is

$$\omega_A = \bar{\omega}_0(r, \zeta) + (E/\delta^2 \sigma)^{1/2} \bar{\omega}_1(r, \zeta) + \cdots \tag{40}$$

where the series is asymptotic and the error terms in Equation 39 are small only if

$$E^{1/2} \ll \delta \ll E^{-1/2}$$

Substitution of Equation 40 into Equation 39 yields

$$\frac{\partial^2 \bar{\omega}_0}{\partial \zeta^2} = 0, \qquad \frac{\partial^2 \bar{\omega}_1}{\partial \zeta^2} = (\vec{q} \cdot \nabla) \bar{\omega}_0 \tag{41}$$

The boundary conditions on $\bar{\omega}_0$, $\bar{\omega}_1$ are obtained by substitution of Equation 40 into Equation 34 and using the second of Equation 35; this leads to

$$\frac{\partial \bar{\omega}_0}{\partial \zeta} = \frac{\partial \bar{\omega}_1}{\partial \zeta} = 0 \qquad \text{at } \zeta = 0 \tag{42}$$

The solution of the first of Equation 41 which merges smoothly into the geostrophic core solution, is

$$\bar{\omega}_0 = \omega_0(r, 0) \tag{43}$$

and consequently, the species mass fraction is invariant to leading order across the Ekman layer. The solution of the second of Equations 41 satisfying Equation 42 is

$$\bar{\omega}_1 = C_1(r) + \left.\frac{\partial \omega_0}{\partial r}\right|_{z=0} \int_0^{\zeta} (\zeta - t)\bar{u}(r, t)\, dt \tag{44}$$

Here $\bar{u}(r, \zeta)$ is the radial velocity in the Ekman layer and C_1 is an arbitrary function of r. The asymptotic form of Equation 44 is easily found to be

$$\bar{\omega}_1 \sim \left.\frac{\partial \omega_0}{\partial r}\right|_{z=0} \sigma^{1/4}\rho_e^{1/2}\zeta\mathscr{F}_{1/2}(r, 0), \qquad \text{as } \zeta \to \infty \tag{45}$$

where

$$\mathscr{F}_{1/2}(r, 0) = \int_0^{\infty} \bar{u}\, d\zeta \tag{46}$$

is the $0(E^{1/2})$ volumetric flow rate per unit length of circumference in the lower Ekman layer. Expanding ω_0 in a Taylor series about $z = 0$ yields

$$\omega_0(r, z) = \omega_0(r, 0) + z\left.\frac{\partial \omega_0}{\partial z}\right|_{z=0} + 0(z^2) \tag{47}$$

as $z \to 0$ and comparing Equation 45 with Equation 47 using the scaling for ζ given previously, it must be that

$$\delta\frac{\partial \omega_0}{\partial z} = \rho_e\mathscr{F}_{1/2}(r, 0)\frac{\partial \omega_0}{\partial r}, \qquad z = 0 \tag{48}$$

The same analysis near $z = 1$ yields a minus sign on the right-hand side of Equation 48 and the boundary conditions on the two horizontal walls are

$$\mp\rho_e\mathscr{F}_{1/2}(r, z_0)\frac{\partial \omega_0}{\partial r} = \delta\frac{\partial \omega_0}{\partial z} \qquad \text{at } z = z_0 = \frac{1}{2} \pm \frac{1}{2} \tag{49}$$

Similar considerations in the boundary layers on the two vertical surfaces lead to the fact that to leading order the mass fraction does not vary across the layers. That analysis leads to the boundary condition

$$\mp\rho_e(r_0)\mathscr{F}_S(r_0, z)\frac{\partial \omega_0}{\partial z} - \delta\epsilon M^2 r_0\omega_0(1 - \omega_0) = \frac{\partial \omega_0}{\partial r} \tag{50}$$

$$\text{at } r = r_0 = \frac{a + b}{2} \mp \frac{a - b}{2}$$

where $\mathscr{F}_S$ is the sum of the volume fluxes in the Stewartson layers:

$$\mathscr{F}_S(r_0, z) = \mathscr{F}_{1/3}(r_0, z) + \mathscr{F}_{1/4}(r_0, z) \tag{51}$$

The condition on δ which emerges from analysis of the Stewartson layers is stricter than that which arises on analysis of the Ekman layer and is

$$E^{1/4} \ll \delta \ll E^{-1/4} \tag{52}$$

This restriction allows analysis of the feed flow rates of practical interest; for $\delta = 0(E^{-1/4})$, $\epsilon_f = 0(E^{1/4})$ and the fluid equations become nonlinear; for $\delta = 0(E^{1/4})$ the feed flow rate is extremely small.

The boundary conditions for the geostrophic mass fraction distribution are thus given by Equations 49 and 50; however for $\epsilon M^2 \ll 1$ the nonlinear term in Equation 50 is negligible and the boundary conditions contain only first derivatives. In this case the solution to Equation 38 is only unique to within an additive constant, and to pose the problem uniquely requires appending the condition

$$\omega_0 = \omega_F \qquad \text{at } z = z_F,\ r = a \tag{53}$$

where ω_F = the value of the mass fraction at the feed port.

The species mass fraction must satisfy an overall balance equation; this condition is derived by integrating Equation 38 over the domain and using the boundary conditions Equations 49 and 50; this condition is

$$\omega_F = \theta\omega_p + (1 - \theta)\omega_W \tag{54}$$

where ω_p = the mass fraction at the product port
ω_W = the mass fraction at the waste port

For the centrifuge depicted in Figure 1, Conlisk, Foster, and Walker [23] and Conlisk [43] have obtained the velocity field and the fluxes in the boundary layers. The core drift is given by

$$W_G = (q/4r\rho_e)\frac{\partial}{\partial r}\{r^2\rho_e^{1/2}/\sigma^{3/4}\} \tag{55}$$

Here q is a constant which is related to the magnitude of differential rotation and/or differential heating and is defined by

$$q = \lambda_m(\lambda_1 - \lambda_0) - \tfrac{1}{2}\lambda_T$$

where λ_m, λ_1, λ_0, and λ_T were defined earlier. Finally, the flux in the Ekman layer is given by

$$\mathscr{F}_{1/2}(r, z_0) = \frac{(1 - \theta)a}{2r\rho_e} \pm 1/4qr/(\sigma^{3/4}\rho_e^{1/2}), \qquad z_0 = \frac{1}{2} \pm \frac{1}{2} \tag{56}$$

and for injection and withdrawal in the corners (product and feed on r = a; waste on r = b) the fluxes in the shear layers are given by [23, 43]

$$\mathscr{F}_S(a, z) = \frac{(1 + \theta)}{2\rho_e(a)} + qa/(4\sigma^{3/4}) \qquad \text{at } r = a \tag{57}$$

and

$$\mathscr{F}_S(b, z) = (\theta - 1)a/(2b\rho_e(b)) - qb/(4\sigma^{3/4}\rho_e^{1/2}) \qquad \text{at } r = b \tag{58}$$

The last term in each of Equations 57 and 58 is the flux due to differential rotation or vertical temperature gradient.

At this point the problem for the separation of two isotopes has been formulated as a well-posed boundary-value problem. The flow inside the centrifuge has been assumed to be a continuum, which is the case for Mach numbers less than about 6 or 7. Above this range for fixed inner radius, the gas is essentially rarefied near the inner wall and other methods must be used (see Table 2). Several different types of solutions to the governing equations presented here have appeared in the literature. Surprisingly, analytical closed-form solutions for the mass fraction exist for $\delta \ll 1$; numerical solutions based on the asymptotic theory can be computed for $\delta = 0(1)$ [23, 43].

However, most of the solutions that have appeared in the literature are entirely numerical in nature; in this work solutions to the unscaled Navier-Stokes equations are coupled to the mass transfer equation (Equation 31); the boundary conditions employed in this method of solution are Equation 35. These equations are much more difficult to solve than Equation 38 subject to Equations 49 and 50. In the latter case, the required quantities from the fluid mechanics are known once and for all and need not be computed interactively with the mass fraction. Fully numerical solutions have also been obtained using the pancake theory. In the next section we examine these solutions and pay particular attention to the analytical solutions using asymptotic methods for $\delta \ll 1$.

SOLUTIONS TO THE MASS TRANSFER PROBLEM

Analytical Solution Using Boundary-Layer Methods for $\delta \ll 1$

Equation 38 subject to Equations 49 and 50 along with the condition of Equation 53 must be solved to determine the mass fraction distribution within the centrifuge if boundary-layer methods are being employed. If the velocity is obtained using numerical tenchiques, then conditions of Equation 35 should be used replacing Equations 49 and 50. In early work on the problem, analytic solutions were usually found by radially averaging the mass fraction across the centrifuge; this procedure leads to an ordinary differential equation to be solved for the averaged mass fraction [6]. Recently, Conlisk, Foster, and Walker [21] have obtained an approximate analytic solution for the Zippe or Groth centrifuge for $\delta \ll 1$ to Equations 38, 49, 50, 53 without radially averaging, and it is of interest now to describe that work.

The case of small δ (but $\delta \gg E^{1/4}$) corresponds to relatively high flowrates; a value of $\delta = 0.06$ corresponds to a feed flowrate of $\dot{m}^* = 10^{-4}$ kg/sec for a rotational speed of 260 m/sec. In this regime, since $\delta \ll 1$, a classical singular perturbation problem arises. For definiteness, consider the case where the feed lies in the lower corner at $r = a$, $z = 0$; the product is at $r = a$, $z = 1$, and the waste port is at $r = b$, $z = 0$. Conlisk [43] has recently extended the analysis to include the case of the Zippe (or Groth) centrifuge of Figure 1 where the feed port is at $z = \frac{1}{2}$; this source-sink geometry will also be considered here.

The *concentration* boundary-layer structure to be deduced in the analysis is the following. A core distribution is surrounded by concentration boundary layers of thickness $0(\delta^{1/2})$ on $r = a^+$ and $r = b^-$. For the core vertical drift $W_G > 0$ a concentration layer of width $0(\delta)$ exists on $z = 1^-$, while for $W_G < 0$ the horizontal layer occurs at $z = 0+$. Note that by virtue of the fact that $\delta \gg E^{1/4}$ the concentration layers are thicker than any of the velocity boundary layers.

In the core of the centrifuge, the following expansion is assumed for the mass fraction of species A,

$$\omega_A = \omega_F + \delta\epsilon M^2\omega_F(1 - \omega_F)\omega_1(r, z) + \cdots \tag{59}$$

where ω_F is the value of ω_A at the feed port. Substitution in Equation 38 gives the leading order equation

$$\rho_e W_G(r)\frac{\partial \omega_1}{\partial z} = 2 \tag{60}$$

It is convenient to define a constant s by

$$s = \mathrm{sgn}\{W_G(r)\}$$

and substitution of Equation (59) in the boundary condition Equation (49) gives

$$\frac{\partial \omega_1}{\partial r} = 0 \qquad \text{on } z = \frac{1 - s}{2} \tag{61}$$

The solution is

$$\omega_1 = C + 2/(\rho_e W_G)\left\{z - \frac{(1-s)}{2}\right\} \tag{62}$$

where C is a constant to be determined.

The side-wall concentration layers are considered next and define a boundary layer variable $y = \beta(r - r_0)\bar{\delta}^{1/2}$ where $\beta = 1$ for $r_0 = a$ and $\beta = -1$ for $r_0 = b$. Assuming the following expansion,

$$\omega_A = \omega_F + \delta\epsilon M^2\omega_F(1 - \omega_F)\tilde{\omega}_1(y, z) + \cdots \tag{63}$$

it may be shown that,

$$\rho_e(r_0)W_G(r_0)\frac{\partial\tilde{\omega}_1}{\partial z} - \frac{\partial^2\tilde{\omega}_1}{\partial y^2} = 2 \tag{64}$$

The boundary condition at $y = 0$ follows from Equation 50 and is,

$$\frac{\partial\tilde{\omega}_1}{\partial z} = \beta r_0/[\rho_e(r_0)\mathscr{F}_S(r_0)] \qquad \text{at } y = 0 \tag{65}$$

This equation may be integrated along the wall for $\mathscr{F}_S = \text{const.}$, and

$$\bar{\omega}_1 = \beta r_0 z/(\rho_e\mathscr{F}_S) + C' \qquad \text{on } r = a \tag{66}$$

where C′ is a constant. $\mathscr{F}_S$ will be constant on $r = a$ when the feed and product lie in the corner regions of the centrifuge; from Equation 66 it is readily seen that the mass fraction is linearly increasing for $\mathscr{F}_S$ positive. In particular it is desirable to have $\mathscr{F}_S$ small and positive.

Consider now the concentration layer at $r = a+$; since $\omega_A = \omega_F$ at $r = a$, $z = 0$ then $C' = 0$. The solution of Equation 64 which satisfies Equation 66 and matches Equation 62 as $y \to \infty$ is given by

$$\tilde{\omega}_1(y, z) = 2z/(\rho_e W_G) - D_0\left\{z - \frac{(1-s)}{2}\right\}\left\{(2\zeta^2 + 1)\,\text{erfc}\,\zeta - \frac{2}{\sqrt{\pi}}\zeta e^{-\zeta^2}\right\} + (C + D_1)\,\text{erf}\,\zeta + D_2 \tag{67}$$

where the constants D_0, D_1, and D_2 are

$$D_0 = 2/W_G(a) - a/\mathscr{F}_S(a)$$

$$D_1 = \frac{a(s-1)}{2\mathscr{F}_S(a)} \tag{68}$$

$$D_2 = \frac{(s-1)}{2}D_0$$

and the variable ζ is defined by,

$$\zeta = y/2\left\{\frac{\rho_e(a)w_G(a)}{z - (1-s)/2}\right\}^{1/2} \tag{69}$$

This solution describes a boundary layer which originates at the corner $r = a$, $z = (1 - s)/2$ and thickens along $r = a$ toward the other horizontal wall. The value of $\tilde{\omega}_1$ at the top product port is given by

$$\tilde{\omega}_1(a, 1) = a/\mathscr{F}_S(a) \tag{70}$$

and using Equation 66, together with the global species conservation result in Equation 54, the variation for the boundary-layer function on r = b may be shown to be

$$\tilde{\omega}_1(b, z) = \frac{-bz}{\rho_e(b)\mathscr{F}_S(b)} - \frac{\theta a}{(1-\theta)\mathscr{F}_S(a)} \tag{71}$$

This is the boundary condition at y = 0 for the calculation of the concentration layer solution near r = b. This solution is given by replacing a with b in Equations 67 and 69 and by replacing the constants D_0, D_1, D_2 by

$$D'_0 = \frac{1}{\rho_e(b)} 2/W_G(b) + b/\mathscr{F}_S(b)$$

$$D'_1 = \frac{(s-1)b}{2\rho_e(b)/\mathscr{F}_S(b)} + \frac{\theta a}{(1-\theta)\mathscr{F}_S(a)} \tag{72}$$

$$D'_2 = \frac{(s-1)D_0}{2} - \frac{\theta a}{(1-\theta)\mathscr{F}_S(a)}$$

respectively.

A horizontal boundary layer is required at z = (s + 1)/2; defining a boundary-layer variable $\zeta' = -s(z - (1 + s)/2)\delta^{-1}$ and an expansion according to

$$\omega_A = \omega_F + \delta\varepsilon M^2\omega_F(1-\omega_F)\bar{\omega}_1(r, \zeta') + \cdots \tag{73}$$

the leading-order form for Equation 38 is

$$\frac{\partial^2\bar{\omega}_1}{\partial\zeta'^2} = -s\rho_e(r)W_G(r)\frac{\partial\bar{\omega}_1}{\partial\zeta'} \tag{74}$$

The boundary condition at $\zeta' = 0$ is obtained from Equation 49 and

$$\frac{\partial\bar{\omega}_1}{\partial\zeta'} = \rho_e(r)\mathscr{F}_{1/2}\left(r, \frac{s+1}{2}\right)\frac{\partial\bar{\omega}_1}{\partial r} \quad \text{at } \zeta' = 0 \tag{75}$$

while matching to the core solution Equation 62 requires that

$$\bar{\omega}_1 \sim C + \frac{2s}{\rho_e W_G} \quad \text{as } \zeta' \to \infty \tag{76}$$

The solution is given by,

$$\bar{\omega}_1(r, \zeta') = C + \frac{2s}{\rho_e(r)W_G(r)} + \left\{\frac{-2s}{\rho_e(r)W_G(r)} + \frac{D + r^2}{r\rho_e(r)\mathscr{F}_{1/2}(r, (s+1)/2)}\right\}e^{-s\rho_e(r)W_G(r)\zeta'} \tag{77}$$

where D is another constant to be determined.

It may be readily shown by consideration of the extension of the $0(\delta)$ horizontal concentration layer under the $0(\delta^{1/2})$ side-wall layers that for s = 1, the limit of Equation 77 at $\zeta' = 0$, $r \to a$ must be equal to the limit of Equation 67 at $\zeta = 0$, $z \to 1$. This gives a continuous solution in the corner

$r = a+, z = 1-$. If $s = -1$, the side-wall concentration layers originate in the upper corners and grow towards $z = 0$; in this case continuity in the corners (a, 0) and (b, 0) is required. In either case these conditions determine values for the constant C and D in Equation 77 and

$$C\{a\mathscr{F}_{1/2}(a, z_0) - b\rho_e(b)\mathscr{F}_{1/2}(b, z_0)\} = \frac{s+1}{2}\left\{\frac{a^2\mathscr{F}_{1/2}(a,z_0)}{\mathscr{F}_S(a)} + \frac{b^2\mathscr{F}_{1/2}(b, z_0)}{\mathscr{F}_S(b)}\right\} + (b^2 - a^2) + \frac{\theta ab}{1-\theta}\rho_e(b)\frac{\mathscr{F}_{1/2}(b, z_0)}{\mathscr{F}_S(a)} \tag{78}$$

$$D = \frac{(s+1)}{2}\frac{a^2\mathscr{F}_{1/2}(a, z_0)}{\mathscr{F}_S(a)} - a^2 - a\mathscr{F}_{1/2}(a, z_0)C \tag{79}$$

where $z_0 = (s + 1)/2$

A self-consistent solution has been completed here for $E^{1/4} \ll \delta \ll E^{-1/4}$; in terms of the performance of the centrifuge, the principal result is the evaluation of the separation factor from Equation 70 and the result is,

$$\alpha = \frac{\omega_P}{\omega_F} = 1 + \frac{\delta\epsilon M^2 a}{\mathscr{F}_S(a)} \tag{80}$$

where the flux in the side-wall boundary layer is given by Equation 57 which is repeated here according to

$$\mathscr{F}_S(a) = \frac{\theta + 1}{2} + q/(4\sigma(a)^{3/4}) \tag{81}$$

Note that $\mathscr{F}_S(a) = 0$ at $q = q_c$ where

$$q_c = -2(1 + \theta)\sigma(a)^{3/4} \tag{82}$$

and near this critical level of differential rotation and/ or differential heating, Equation 80 is not valid. In this situation the concentration-layer problem on $r = a$ must be reexamined. In particular, consider the case when $\mathscr{F}_S(a) = 0(\delta^{1/2})$; it may be deduced from the boundary condition Equation 50 that in this situation the perturbation in the centrifuge to the feed value ω_F is $0(\delta^{1/2})$ and that all terms in Equation 50 are important to leading order on $r = a$. In addition, the expansions in the core region and boundary layers given by Equations 59 and 63 must be modified to include an $0(\delta^{1/2})$ term.

For the case where the feed and product ports are placed up along the axis, the analysis is similar to that described here. In particular, the result for the separation factor is [43]

$$\alpha = \omega_p/\omega_F = 1 + \frac{\delta\epsilon M^2 a(1 - \omega_F)(z_p - z_F)}{\mathscr{F}_S(a)} \tag{83}$$

where $\mathscr{F}_S$ is the flux in the shear layers for $z_F < z < z_p$; it has the same numerical value as in Equation 57 and consequently, a higher separation factor will always be attained if the feed and product are in the corner regions. Thus the Groth (and Zippe) centrifuge as depicted in Figure 1 is not optimal in this sense. Note also from Equations 80 and 83 that $\alpha = 0(\delta\epsilon M^2)$ and thus increases for decreasing flow rate. This is the familiar tradeoff between high separation factor and low capacity; a small amount of material may be produced at a high-separation factor, but if a higher capacity is desired, many centrifuges are required to multiply the separative effect.

Also, using the expression for M^2 given in Equation 2 and noting that $\epsilon = (M_B - M_A)/M_B$, it is easily shown that $\alpha - 1$ is proportional to $M_B - M_A$ and not $(M_B - M_A)/M_B$ or $(M_B - M_A)/M_B^2$ as

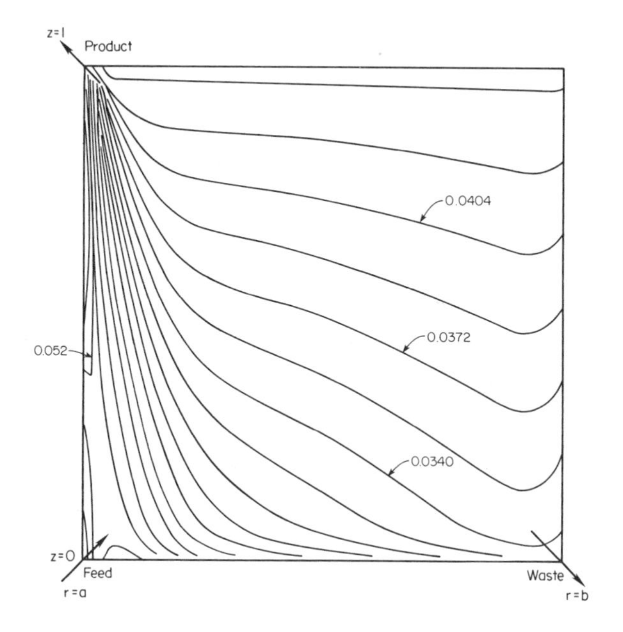

Figure 8. Isomass fraction contours for ω_A for $q = -1$, $M = 1$, $a = 1$, $b = 2$, $\theta = 0.5$ and $\epsilon = 0.01$. Here $\omega_F = 0.05$ and the boundary-layer regions are to scale for $\delta = 0.02$. Contours for ω_A are at intervals of 1.6×10^{-3}. Typical values of the mass fraction are indicated [23].

in some other methods [6]. As a numerical example, for uranium enrichment, $M_B = 352$, $M_A = 349$, and $\epsilon = 0.0085$. For a value of $\delta = 0.06$ and $M = 1$, then $\alpha - 1$ is of the order of 0.0004, which is small indeed. Also from Equation 80 higher separation factors may be obtained by increasing the rotation speed; i.e., increasing M.

As an example of the concentration contours which are described by the analysis, in Figure 8 are isomass fraction contours for the case $\omega_F = 0.05$, $\delta = 0.02$, and $\epsilon = 0.0085$; in addition the cut $\theta = 0.5$ and $a = 1$, $b = 2$. The rotational Mach number $M = 1$, and there is a downward drift in the core region with $q = -1$; note that this value of q is close to, but greater than, the critical value given in Equation 82. To prepare Figure 8, a composite solution consisting of the core and boundary-layer concentration solutions was formed. Generally, the values associated with each contour decrease radially outward in Figure 8. Note the collection of contours near the product port; in addition the presence of the side-wall boundary layers as well as the end-wall concentration layer at $z = 0$ is clearly demonstrated.

The parameter q defined previously is related to the magnitude of differential rotation and/or vertical temperature gradient and is crucial to the separation process. Positive values of q correspond to the physical situation of an upward drift in the core induced by either differential rotation or differential heating, and in this case a net upward motion occurs in the velocity boundary layers at $r = a$. Enrichment always occurs for $q > 0$ and is essentially independent of the inner radius a as well as the value of the outer radius b [23]. As q decreases the enrichment process is enhanced and the separation factor increases to relatively large values, as the critical value of q given by Equation 82 is approached. For $q < 0$ the induced drift in the core is downward, but the net motion in the side-wall velocity boundary layer at $r = a$ is still upward. As $q \rightarrow q_c$ the flux in the boundary layer approaches zero, and separation is enhanced. However, for $q < q_c$, it may be inferred from Equations 81 and 57 that the mixture becomes depleted in species A at the product port; in this regime, the vertical velocity in the boundary layers and the core is negative. The physical reason for this behavior is as follows.

As the binary gas enters at the feed port it rises into the $E^{1/3}$ layer on $r = a$ and a portion moves radially outward toward the $E^{1/4}$ layer. As the gas enters the $E^{1/4}$ layer it is deflected toward the Ekman layers on the end walls whereupon the source-sink part of the flow moves outward toward $r = b$. For $q < 0$, a recirculatory motion in a counter-clockwise direction is set up and superimposed on the source-sink flow. As the gas stream traverses the bottom Ekman layer, depletion of species A occurs and a portion of the gas stream is removed at the waste port; the rest of the stream recirculates up the outer side-wall velocity boundary layer and enters the top Ekman layer. As the gas moves radially inward along the upper Ekman layer, enrichment occurs according to Equation 77. For $q < 0$ but $q > q_c$, the stream in the upper Ekman layer meets a stream of gas which rose directly from the feed port and which is also enriched according to Equation 66 for $\mathscr{F}_S(a) > 0$. Maximum benefit occurs as $\mathscr{F}_S(a) \to 0+$ so that there is only a small upward mass flux in the fluid boundary layers at $r = a$. For $q < q_c$ the effects of differential rotation and/or differential heating exert an increasingly dominant effect on the source-sink flow and a strong and detrimental net recirculation is set up in the container.

The efficiency of a separating unit depends on the combined effects of throughput and enrichment; for example, small throughputs (large δ) in general, yield high separation factors but the total production of enriched gas is very low. The combined effect of throughput and enrichment is embodied in the separative power of the machine, δU; this quantity has been derived in Reference 6, and although the derivation is somewhat obscure, the result is

$$\delta U = \frac{1}{2}\dot{m}^*\theta(1-\theta)\{\alpha_W - 1\}^2 \tag{84}$$

where $\alpha_W = \omega_p(1-\omega_W)/\omega_W(1-\omega_p)$

It is a large separative power which is desirable for a gas centrifuge and not necessarily large separation factors. In the results to follow, the feed mass fraction has been taken as $\omega_F = 0.007$ and $\epsilon = 0.0085$. These values are typical of uranium isotope separation.

As will be apparent in what follows, the aspect ratio 1/b is extremely important for attaining high separative power. The separation factor given by Equation 80 is valid for aspect ratios of 0(1); however the corresponding formula for aspect ratios greater than 1 may easily be shown to be [36]:

$$\alpha = 1 + \frac{\delta\epsilon M_p^2 a}{b\mathscr{F}_S(a)} \tag{85}$$

and has been obtained by taking the radial length scale equal to B_1 and the vertical length scale equal to L. Note that the separation factor increases linearly with aspect ratio, and consequently, the separative power increases quadratically. This behavior is illustrated by the results depicted in Figure 9 where both separation factor and separative power are plotted as a function of aspect

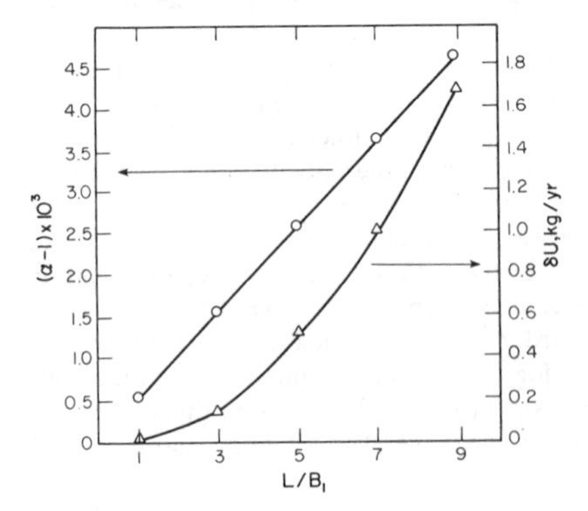

Figure 9. Separation factor and separative power as a function of aspect ratio for feed flow rate 5×10^{-4} kg/sec, $\theta = 0.9$, $A_1 = 0.5B_1$, $B_1 = 0.25$m, $q = -3$, $M = 3$.

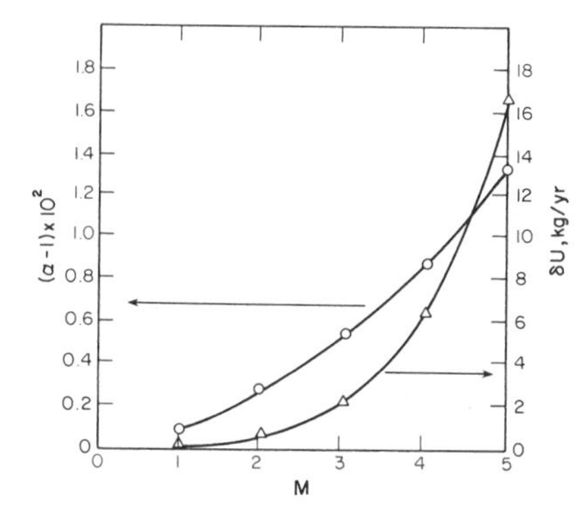

Figure 10. Separation factor and separative power as a function of M for feed flow rate 5×10^{-4} kg/sec, $\theta = 0.9$, $q = -3$, $A_1 = 0.5B_1$, $B_1 = 0.25$m, $L/B_1 = 10$.

ratio for $\dot{m}^* = 5 \times 10^{-4}$ kg/sec ($\delta = 0.012$), $\theta = 0.9$, $A_1 = 0.5B_1$, $B_1 = 0.25$m, $q = -3$, $M = 3$. The properties used are for UF_6 at $T = 320°K$ and a value $M = 3$ corresponds to a peripheral speed of about 260 m/sec. The separative power values are rather small for these values of the parameters.

One way to increase the separative power is to increase the Mach number as is illustrated in Figure 10; here all the parameters are the same as in Figure 9 except the aspect ratio is $1/b = 10$. Hence the largest value of the separative power occurs at $M = 5$ and is 16 kgU/yr. However, it may not be practical to increase the peripheral speed of the bowl; nor may it be practical to increase the aspect ratio. In this case by suitable adjustment of the inner radius $a = A_1/B_1$ (here the radial length scale is taken to be B_1; see Equation 85) and the parameter q, much larger values of the separative power may be obtained.

In Table 3 is also depicted some of the data which have been reported in the literature. The value quoted by Wood and Sanders [29] is obtained using the "pancake theory" for the fluid mechanics and a numerical solution for the mass transfer problem; the results of Soubbaramayer [7] are for a Beams-type centrifuge. It should be pointed out that the characteristics given in Table 3 are not necessarily complete and the specific references should be consulted for details. The numbers quoted by Von Halle [44] are obtained using a radial averaging technique. Also, the number quoted for the Conlisk, Foster, and Walker theory, has been obtained using $q = -6$ and $a = 0.8$; a value of $q = -6$ corresponds to about a 5° temperature difference between the bottom and the top of the centrifuge.

On Figure 11 is depicted the separation factor and separative power as a function of q for the indicated parameters. The dashed line indicates the region in which the present asymptotic theory breaks down; that is, where $\mathscr{F}_S \to 0$; note the rapid increase in separative power near $q = -6$. As discussed previously, increasing the vertical temperature gradient beyond about $q = -11$ sets up a detrimental recirculation within the container and the product port is actually depleted in species A since $\alpha - 1 < 0$ for $q > -11$. Thus there is a value of q such that maximum separation takes place; however, it appears that this value can only be determined by considering the solution as $\mathscr{F}_S \to 0$.

The behavior of separation factor and separative power as a function of cut is depicted in Figure 12; this behavior of the separation factor is typical [6]. The behavior of the separative power, however is not; at these (relatively) high feed flow rates the separative power increases with cut.

A particularly interesting behavior is illustrated in Figure 13 where two variations of separative power which have at one time appeared in the literature are compared with the calculations using the asymptotic theory. The lines ·—· depict the classical separative power variation which has been postulated in the literature; this curve is from the book by Avery and Davies [45] (see also Soubbaramayer [7]) and does not include the effects of flow rate on separation factor. The solid line is that deduced by Von Halle [44] and is based upon radially averaged solutions to the mass transfer problem. The dashed line is the prediction of the present asymptotic theory. The conventional thinking has always been that increasing the feed flow rate will increase separative power; however, this appears to be a function of the type of circulation which is set up within the centrifuge; here

Table 3
Selected Data on Separative Power*

Reference	Length (m)	Radius (m)	Temp (°K)	Cut	Speed (m/sec)	Feed (kg/sec)	δU (kg/yr)
Wood and Sanders [29]	3.35	0.091	300	0.38	Not given	1.32×10^{-5}	38.6
Soubbaramayer [7]	2.5	0.215	310	0.5	600	10^{-4}	13.3–25.6
Conlisk, Foster, Walker Theory	2.5	0.25	320	0.5	260	10^{-4}	47.25
Von Halle [44]	5.0	0.25	320	0.5	600	2.5×10^{-5}	30
Soubbaramayer [7] (optimized)	2.5	0.25	320	0.5	600	10^{-4}	38
Merten and Hänel [27]	2.5	0.25	320	0.5	600	10^{-4}	25.8

* *The Conlisk, Foster, and Walker value is obtained using $a = 0.8$, $q = -6$.*

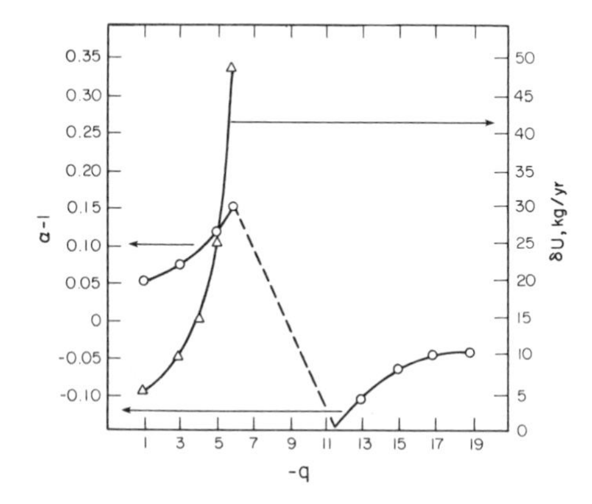

Figure 11. Separation factor and separative power as a function of q for $M = 3$, $A_1 = 0.8B_1$, $B_1 = 0.25m$, $\theta = 0.5$, $1/b = 10$, $m^* = 10^{-4}$ kg/sec. The dashed line indicates the region in which the present theory breaks down, i.e., where $\mathscr{F}_S \rightarrow 0$. Beyond $q = -11$, the product port is depleted and the waste port is enriched in species A; thus beyond $q = -6$ the separative power has no meaning.

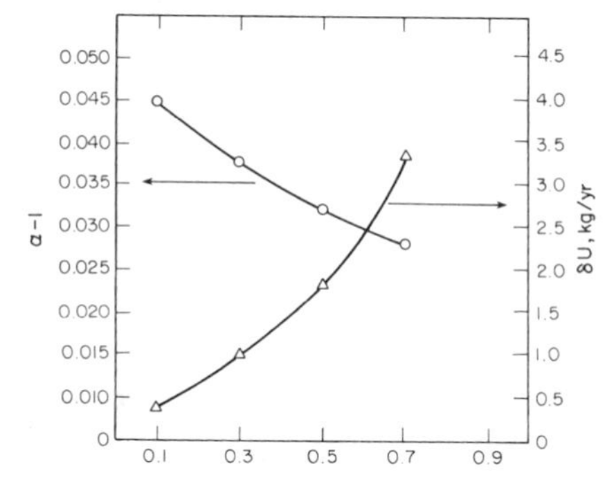

Figure 12. Separation factor and separative power as a function of cut for feed flow rate 10^{-4} kg/sec, $q = -2$, $M = 3$, aspect ratio $= L/B_1 = 10$, $A_1 = 0.5B_1$, $B_1 = 0.25m$.

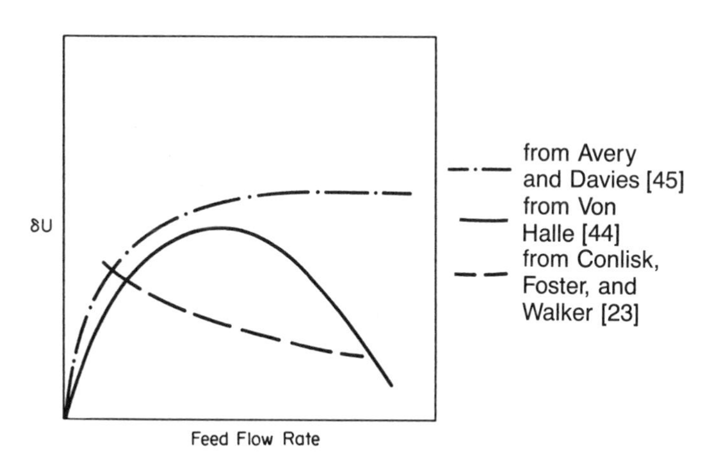

Figure 13. Qualitative representation of three variations of separative power as a function of feed flow rate.

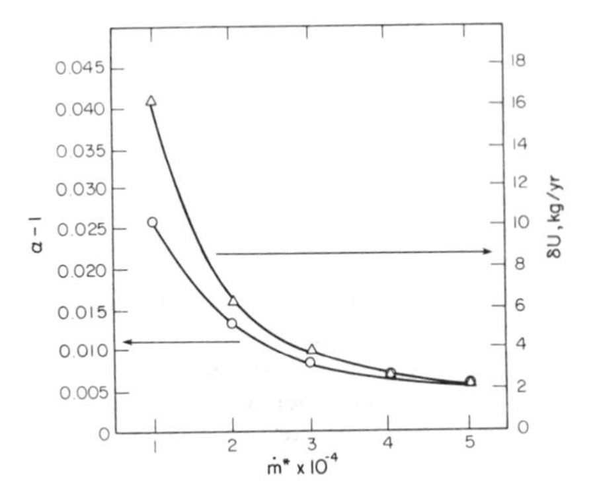

Figure 14. Separation factor and separative power as a function of feed flow rate for $\theta = 0.9$, $M = 3$, $q = -3$, $A_1 = 0.5B_1$, $B_1 = 0.25$m, $L/B_1 = 10$.

α and δU appear to be inversely proportional to feed flow rate. Results of a specific calculation are depicted in Figure 14. This behavior also is obtained at lower flow rates which are considered in the next section. Similar results appear to have been obtained by Kai [46]; moreover, his results depend on the type of circulation set up in the centrifuge.

At this point a solution for the mass fraction of species A has been calculated approximately using the boundary-layer theory, and results for the separation factor and separative power have been obtained for a wide range of parameters. In particular, for fixed rotation speed, the feed flow rate determines the validity of the singular perturbation treatment for $\delta \ll 1$. As δ increases and becomes 0(1), the character of the equations change; moreover, feed flow rates much smaller than those discussed in this section have been considered in the literature. For $\delta = 0(1)$ solutions using the asymptotic theory must be computed numerically, and this leads us naturally to a discussion of the different types of numerical solutions which have appeared in the literature.

Numerical Solutions to the Mass Transfer Problem

As has already been pointed out previously two types of numerical solutions to the mass transfer problem have appeared in the literature; these are, solutions computed interactively with a numerical solution of the flow equations and numerical solutions in which the flow field is previously calculated. Significant contributions in the first type of calculation have been made by Kai [24, 26, 46] and Merten and Hänel [27].

Kai has considered a Beams-type of centrifuge and examined the effect of the magnitude of the vertical temperature gradient; he found that there exists a maximum vertical temperature gradient above which the separation decreases.

Merten and Hänel [27] calculate the solution to the full coupled fluid and mass transfer equations using the unsteady governing equations to obtain the steady solution. They calculated apparently accurate solutions for both the flow field and concentration problem; however, because of the large computer times required in the separative power parameter studies, they used an integral method similar to that used by Soubbaramayer [7] to model the flow field and presented results of those computations. In the paper, the effects of both the thermal and scoop drive mechanisms, feed flow rate, peripheral speed, and aspect ratio or separative power are considered. In comparing the detailed nonlinear solutions with the linear solutions (obtained by neglecting the convective terms in the governing equations), they found that while the flow patterns near the outer wall and the core vertical drift were somewhat different, the separative power values were not greatly affected.

Examples of the second type of calculation have been reported by Nakayama and Torii [47], Torii [48], Soubbaramayer [7], Conlisk, Foster, and Walker [23], and Conlisk [43]. Nakayama and Torii [47] considered a Beams-type centrifuge for the case of no mechanical or thermal driving; the geometry is very similar to the work of Kai [24]. They considered peripheral speeds of 300 m/sec and aspect ratio $1/b = 8$; the feed flow rate is 8×10^{-6} kg/sec. They assumed a velocity and mass fraction distribution which is axially symmetric in order to avoid having to calculate the flow field.

Nevertheless, the authors claim that they obtained good agreement between their work and the experimental results of Beams. Torii [47] extended the analysis to the case of a thermally driven flow field and modeled the feed flow velocity by a fourth-order polynomial. He found results similar to Kai [26] in that beyond a certain value of vertical temperature gradient the separation was degraded. He also determined that the separative power increases as a function of feed flow rate. Soubbaramayer [7] has calculated the separation factor by a finite element technique assuming a quadratic variation of the velocity in the vertical boundary layer; he also found a separative power increase with feed similar to the solid curve in Figure 13.

In the modeling of the flow field in the papers by Nakayama and Torii and Torii and Soubbaramayer, the effect of the flow field is confined to the convective mass transfer term in the governing equations (Equation 31). The flow in the boundary layers however affects the problem also in the boundary conditions Equations 49 and 50 through the mass flux in those layers. This fact has apparently been ignored by the authors just mentioned, although if the Navier-Stokes equations are solved (e.g., Kai, Merten, and Hanel) this criticism does not apply.

Conlisk, Foster, and Walker [23], and Conlisk [43] obtained numerical solutions to Equation 31 subject to Equations 49 and 50 for $\delta = O(1)$. On Figures 15A and B are isomass fraction lines for an aspect ratio $1/b = 5$, $\dot{m}^* = 10^{-5}$ kg/sec; here the outer radius $B_1 = 0.25$m, $A_1/B_1 = 0.5$. Figure 15A corresponds to $q = 0$, and Figure 15B corresponds to $q = 2$; the total mass fraction distribution is plotted, whereas in References 23 and 43 scaled mass fraction lines are plotted. The placement of the feed, product and waste ports are shown; note the similarity in these plots and those of Kai [26] which are depicted on Figure 16.

To illustrate the parametric behavior of the separation factor in this feed flow rate range, in Figure 17 is the separation factor and separative power as a function of cut. Here $M = 3$, $\dot{m}^* = 10^{-5}$ kg/sec, $B_1 = 0.25$, $A_1 = B_1/2$, and the aspect ratio $1/b = 2$. Note the typical decrease of α with cut and the presence of a value of cut around 0.4 which maximizes separative power. The effect of thermal or mechanical driving is depicted in Figures 18 A and B in dimensionless form; these two figures are reproduced from Reference 43 and the source-sink geometry is shown. Note that this is

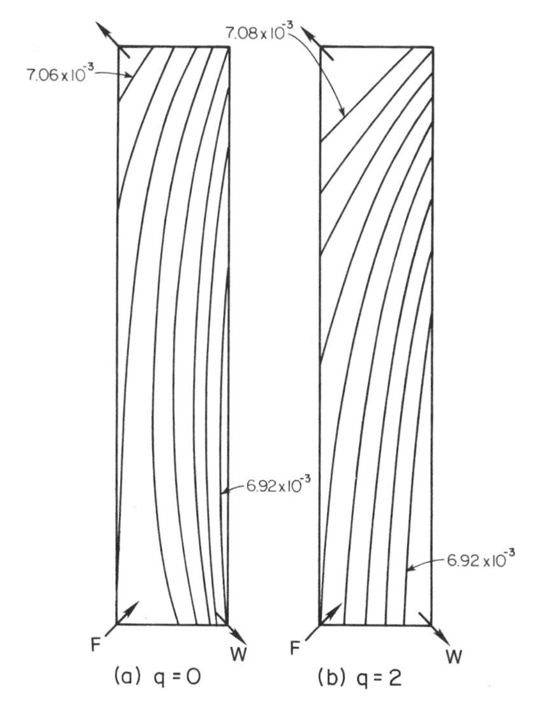

Figure 15. Isomass fraction lines for $M = 3$, $A_1 = 0.5B_1$, $B_1 = 0.25$m, $\theta = 0.5$, $\dot{m}^* = 10^{-5}$ kg/sec, with $B_1/L = 2$. Note the effect of $q \neq 0$ on the isomass fraction lines; for no feed flow and no external drive mechanism, the lines would be vertical. The values of the mass fraction are plotted in equal increments of 2×10^{-5}.

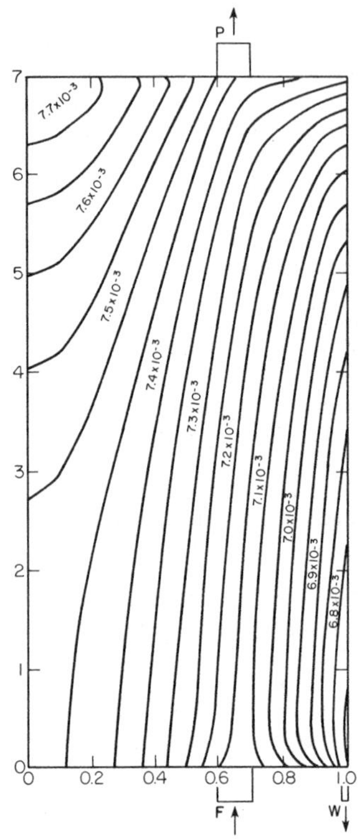

Figure 16. Typical example of isomass fraction lines from Kai [26]. Values of the mass fraction along each line is shown.

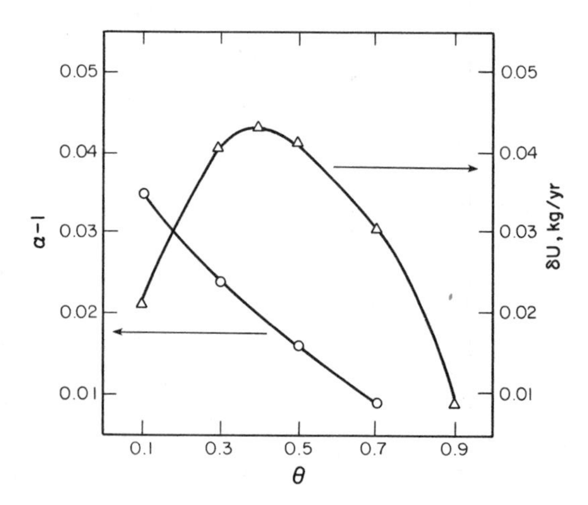

Figure 17. Separation factor and separative power as a function of cut θ for M = 3, $\dot{m}^* = 10^{-5}$ kg/sec, $B_1 = 0.25$m, $A_1 = 0.5B_1$, q = 3, $L/B_1 = 2$.

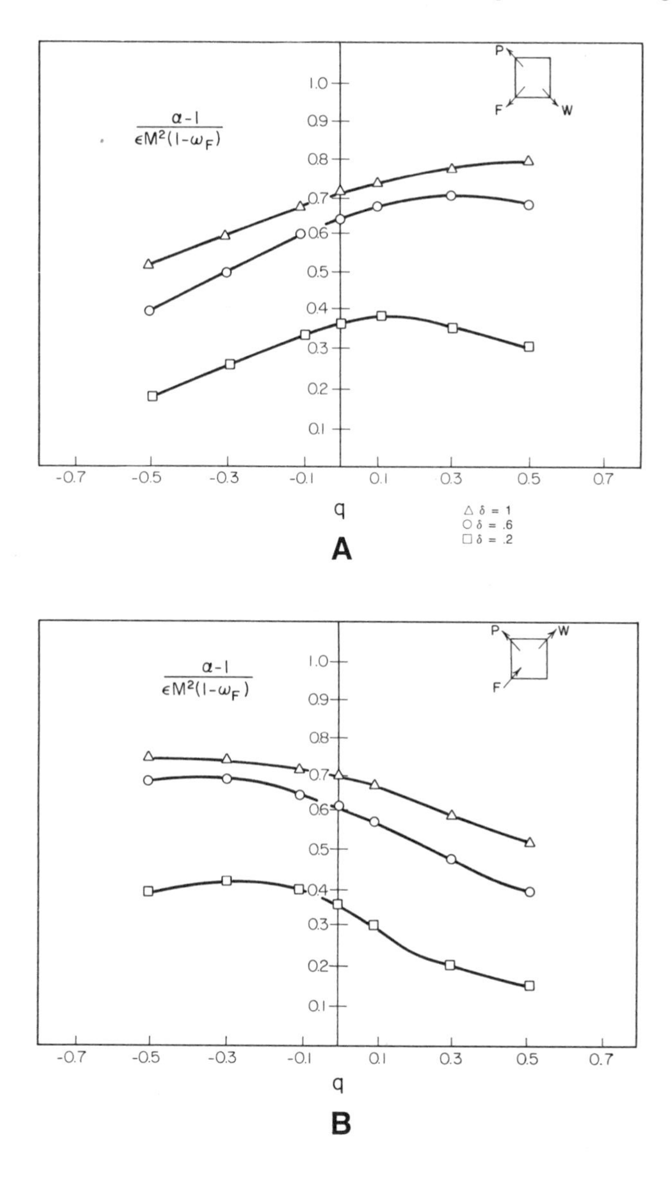

Figure 18. (A) Scaled separation factor as a function q for several values of δ. Here M = 1, a = 1, b = 2, and $\theta = 0.5$; the source-sink geometry is shown. (B) Same as (A) for $z_W = 1$ [43].

for an aspect ratio 1/b = 1/2 and so is for a relatively fat can, although this appears to be typical of larger aspect ratios as well. Note how in this parameter range a desirable value of q (greater or less than zero) is dependent upon the position of waste. Note also the fact that above a certain value of q, the separation factor decreases although for $\delta = 1$ the point does not appear to have been reached. The optimal source-sink geometry in this parameter also appears to be $z_F = 0$, $z_p = 1$, and $z_W = 0$ or 1 as described in Reference 43.

Table 4
Summary of the Parameters of Gas Centrifuge Operation

	Dimensional	Dimensionless
Geometrical	L, A, B, Ω	L/B, A/B
Gas properties	T_0^*, ρ_0^*, p_w^*, D_{AB}, ν, γ	E, M, h, Pr
Flow properties	z_F^*, z_p^*, z_w^*, $\dot{m}^*$, $\dot{m}_p^*$, $\dot{m}_w^*$, ΔT, ΔΩ	z_F, z_p, z_w, ϵ_f, q, θ
Mass transfer	D_{AB}	Sc, δ

DESIGN CONSIDERATIONS FOR A GAS CENTRIFUGE

At this point, having analyzed the flow field and the separative characteristics of a gas centrifuge in some detail for a wide range of parameters, it is possible to elucidate some basic design principles involved in a gas centrifuge. In Table 4 is a summary of the parameters involved, both dimensional and dimensionless, and as can be seen the number of them is staggering; however, progress on design of a typical gas centrifuge may be made on the basis of our previous calculations.

In what follows, the operation of a Zippe- or Groth-type machine with an inner feed post will be considered, but it will be evident that many of the principles elucidated in the following may be carried over to the Beams centrifuge as well. In addition, the mechanical design of the bowl and the associated equipment required for construction of the bowl are not considered here; a lucid discussion of the construction and mechanical design of the early centrifuges is given by London [49].

As has already been stated, it is usually not appropriate to optimize the separation factor alone because the effect of feed flow rate is important in determining the capacity of a given machine. Thus, the separative power δU has been defined to help remedy this situation, and if a choice is to be made between optimizing the separation factor or separative power, separative power should be optimized [6]. Let us assume then, that the temperature and pressure at the outer wall are given; then the gas properties are fixed, namely density ρ_0^*, kinematic viscosity, $\nu = \mu/\rho_0^*$, ratio of specific heats, γ, the Prandtl number, Pr, and the Schmidt number, Sc. These properties are then fixed by the system.

The controllable parameters are given by the rest of the parameters depicted in Table 4. These include, feed flow rate, $\dot{m}^*$; rotation speed Ω; outer radius B_1; inner radius A_1; source-sink geometry z_F^*· z_p^*, z_W^*; product and waste flow rates $\dot{m}_p^*$, $\dot{m}_W^*$; the drive mechanisms ΔT and $\Delta\Omega$, and the cut θ. For a Beams-type centrifuge, of course, the quantity A_1 has no meaning. Based on our previous analysis, it has been determined conclusively, at least for $\delta \ll 1$, that the optimal source-sink geometry for the Zippe or Groth machines is $z_F^* = 0$, $z_p^* = L$, $z_W^* = 0$ (or L really). This leaves us the problem of optimizing the separative power as a function of L, A_1, B_1, Ω, ΔT, $\Delta\Omega$, θ, and $\dot{m}^*$, $\dot{m}_p^*$, $\dot{m}_W^*$. In terms of dimensionless variables we must optimize δU as a function of $L/B = 1/b$, A_1/B_1, M, E, ϵ_f, θ, q, δ. These are eight independent dimensionless parameters corresponding to the eight independent dimensional parameters (two of $\dot{m}^*$, $\dot{m}_p^*$, and $\dot{m}_W^*$, and θ are independent by mass conservation). The rigorous optimization problem is thus seen to be formidable, and it is complicated by the fact that much of the work on this problem is entirely numerical. For an optimization problem it is desirable to have the optimizing function in closed form, and for this reason the work of Reference 23 for $\delta \ll 1$ appears promising. In any case some progress on basic design principles may be made.

Consider first, the aspect ratio L/B_1; the separative power increases as $(L/B_1)^2$, and consequently, the centrifuge should be as long as possible within structural stability constraints [49] (Figure 9). The radius ratio A_1/B_1 is also important, and it appears from the work of Conlisk et al. [36] that A_1/B_1 should be close to 1 within the limits of the analysis. For extremely thin bowls, it appears that the "pancake theory" may apply as $M \to \infty$.

The Mach number varies quadratically as the peripheral speed and thus $\delta U \propto (\Omega B_1)^4$. Consequently, large rotation rates are desirable, again within the constraints of the analysis (Figure 10);

specifically, that the gas remain a continuum. If the Mach number increases above about 6 or 7 depending on the other parameters, then the product port should be moved to the outer wall.

At this point, assume that the dimensional shape of the bowl A_1, B_1, L and the rotation speed Ω have been chosen. Then the parameters L/B_1, A_1/B_1, M, and E are fixed, and the effect of feed flow rate and driving mechanism remain to be considered. Specifying the feed flow rate specifies ϵ_f and hence δ. The separative power decreases as $\dot{m}^*$ increases so that there appears to be a difficulty here. However, it has been indicated that for small values of δ ($\dot{m}^* \sim 10^{-4}$ kg/sec), these larger values of flow rate may be used in conjunction with thermal or mechanical driving to produce rather large values of separative power at moderate values of aspect ratio and Mach number (Figure 11). For smaller flow rates ($\dot{m}^* \sim 10^{-5}$ and smaller) comparable values of the separative power may be obtained only at very high aspect ratios and Mach numbers [29]. The optimal value of thermal or mechanical driving embodied in the value of q depends in a complicated way on other values of the parameters as discussed in References 23, 26, 43.

Lastly the optimal value of the cut is to be considered. It has been reported that for the smaller feed flow rate range, the optimal cut occurs at a value of $\theta \sim 0.4$ (Figure 17 [29, 43]). For larger flow rates the results of Figure 12 indicate that larger cuts are desirable for maximum separative power, although the separation factor decreases.

It should be pointed out that these flow rate design considerations depend on the type of circulation which is set up within the bowl; it appears that the only rigorous optimization procedure which has been described in the literature is that of Soubbaramayer [7]; in that work the effects of the wall thermal conditions are also discussed. It is apparent from these considerations that more work needs to be done in optimizing the operation of a gas centrifuge, although the basic principles of operation appear to be understood, at least in the parameter range in which the gas remains a continuum.

It is to be noted that nothing has been said about the value of ω_F. In most of the work described in this article the value $\omega_F = 0.007$ has been assumed; this is approximately the mass fraction of $U_{235}F_6$ which occurs in nature. However, when many centrifuges are connected in series and/or parallel in order to magnify the individual separation effect, ω_F may be widely varying along the cascade. Typical values may range from $\omega_F \sim 0.007$ to $\omega_F \sim 0.05$. In any case, the analysis described here does not require specification of ω_F for a dimensionless specification of the seperation factor (Figure 18); it does however require a value for ω_F for a value of the actual separative power. The same general comments apply to specification of the individual feed flow rate value for a given centrifuge; because a given centrifuge plant may include thousands of centrifuges, it is important to have solutions for the separation factor over as wide a flow-rate range as possible.

SUMMARY AND DIRECTION OF FUTURE WORK

The present review of the fluid dynamics and design of a gas centrifuge has focused on relatively recent results in the calculation of the mass-averaged flow patterns and concentration distribution within the centrifuge bowl. The work described here has been analytic as opposed to experimental for two reasons. First, it is very difficult to insert diagnostic equipment into a rapidly rotating centrifuge bowl without disturbing the flow field. Second, it is almost impossible to obtain detailed velocity and mass fraction distributions experimentally because most of the variation of the flow quantities takes place in extremely thin layers near the walls of the centrifuge; what experimental data that does exist in published literature is of a global nature is over thirty years old [45, 46].

Despite the experimental obstacles, great progress has been made in the analysis of the problem of separation of isotopes in a gas centrifuge. In this review, it has been demonstrated that the calculation of the detailed mass fraction distribution depends strongly on the ability to calculate the mass-averaged flow field. In particular, three approaches to calculating the flow patterns within a centrifuge have appeared in the literature: analytical techniques used asymptotic expansions, numerical methods, and the "pancake" theory. In all these approaches the major variations in the flow field have been shown to be near the bounding walls of the centrifuge, at least in the case of the Zippe or Groth machines (Figure 5). In the case of the Beams machine, major variations also occur across thin vertical layers which straddle the feed and exit ports (Figure 7). Once the flow patterns have been calculated, the required flow quantities may be substituted into the mass transfer equation, which can then be solved for the pointwise mass fraction distribution in the centrifuge.

The single most important quantity to be calculated from the mass fraction distribution is the separation factor $\alpha = \omega_p/\omega_F$ which measures the degree to which the mixture has been enriched in species A. The variation of separation factor with a number of parameters has been examined, and the results of these parameter studies are a necessary component in the design of the centrifuge bowl. Once the separation factor has been determined, the separative power is easily calculated and it is this quantity which is optimized for gas centrifuge performance.

Assuming that the shape of the centrifuge bowl has been chosen, the critical parameters for design considerations are the Mach number, the type of drive mechanism used to generate the recirculatory motion within the centrifuge, and the feed flow rate. In particular, it has been demonstrated that high Mach numbers are desireable but not necessary to achieve high separative power. In addition, it has been shown that feed flow rates of about 10^{-4} kg/sec, combined with a careful adjustment of the mechanical and thermal drives can yield separative powers comparable to those obtained at extremely high Mach numbers. Moreover, large length-to-diameter ratios have been shown to be desirable but not necessary; in particular, if lower flow rates are used in a particular machine and the magnitude of the thermal and mechanical drives are not chosen carefully, both large Mach number and large aspect ratio may be required. Increasing the Mach number increases the power input required and a large aspect ratio significantly increases material cost.

In the past few years, great progress has been made toward understanding the complex interplay between the flow patterns and the species separation problem in a gas centrifuge. Because prototype experiments are so costly, analytical calculations must play an important role in the overall design process. Yet, a complete understanding of the complicated separation process is elusive, and there remain several areas of possible future research. At the extremely high rotational speeds of present-day centrifuges, the central core is effectively evacuated and is not yet clear from published results what role the central core plays in the separation process. Moreover, widely varying values of the separation factor and separative power have been obtained by many authors for a similar set of parameters. The reason for the variability of these values may be that the relative magnitude of the drive mechanisms in the various analyses are different even though such parameters as feed flow rate are the same. In addition to the question of determining the individual process separation factor, the determination of the proper configuration of individual centrifuges in a cascade is necessary for adequate production of enriched gas. This problem has been considered previously based on an assumed behavior of the separation factor along the cascade. However, since the separation factor depends strongly on feed flow rate (as well as a number of other parameters described here) it is unclear whether this type of analysis can yield a reasonable estimate for the separative capacity of a typical enrichment plant. Finally, it is desirable to compare some of the calculated values of the separation factor with modern experimental data where possible.

Despite these difficulties, analytical gas centrifuge research will continue in the future to provide the necessary design data for the construction of gas centrifuge plants. With the completion of the new gas centrifuge facility being built in Portsmouth, Ohio, in the near future, experimental data may be readily available to guide the analyst in the quest for more accurate design information.

Acknowledgments

The author is grateful to co-workers Professor Michael R. Foster of The Aeronautical and Astronautical Engineering Department at The Ohio State University and Professor J. D. A. Walker of Lehigh University for reading the manuscript and providing helpful criticism.

NOTATION

- a — inner radius of annulus (dimensionless) $= A_1/L$ or A_1/B_1
- A — light species of binary mixture
- A_1 — inner radius of annulus (dimensional)
- b — outer radius of annulus (dimen-

sionless) and aspect ratio ($=B_1/L$ or 1)

- B heavy species of binary mixture
- B_1 outer radius of cylinder (dimensional)
- C constant defined in Equation 62
- $\bar{C}$ strength of fluid source
- C_1 function defined in Equation 44
- C' constant defined in Equation 66
- D constant defined in Equation 77
- D_0 constant defined in Equations 68
- D_1 constant defined in Equations 68
- D_2 constant defined in Equations 68
- D'_0 constant defined in Equations 72
- D'_1 constant defined in Equations 72
- D'_2 constant defined in Equations 72
- D_{AB} diffusion coefficient
- E Ekman number defined in Equation 9(a)
- G constant which appears in Equation 30
- h constant
- H function defined in Equations 31
- L length of centrifuge
- M rotational Mach number defined in Equation 2
- M_p rotational Mach number based on peripheral speed and defined in Equation 11
- $\dot{m}^*$ dimensional absolute feed flow rate
- $\dot{m}_p^*$ dimensional mixture flow rate at the product
- $\dot{m}_w^*$ dimensional mixture flow rate at the waste
- M_A molecular weight of species A
- M_B molecular weight of species B
- $N_{A,r}$ flux of species A in the radial direction defined in Equation 33
- $N_{A,z}$ flux of species A in the vertical direction defined in Equation 34
- p dimensionless pressure
- Pr Prandtl number
- p_G dimensionless pressure in the core of the centrifuge
- $p_{1/3}$ dimensionless pressure in the $E^{1/3}$ layer
- p_e^* dimensional equilibrium pressure distribution
- p_w^* dimensional pressure at the outer wall
- q parameter to indicate magnitude of differential rotation and/or vertical temperature gradient
- q_c critical value of q defined in Equation 82
- r independent variable in the radial direction (dimensionless)
- r^* independent variable in the radial direction (dimensional)
- r_0 parameter used to denote radii a or b
- R gas constant
- s $\mathrm{Sgn}(W_G(r))$
- Sc Schmidt number
- T dimensionless temperature
- T_G dimensionless temperature in the core of the centrifuge
- $T_{1/3}$ dimensionless temperature in the $E^{1/3}$ layer
- $T_{1/4}$ dimensionless temperature in the $E^{1/4}$ layer
- ΔT^* dimensional vertical temperature gradient (K)
- T_0^* dimensional (leading order constant) temperature within the centrifuge (K)
- T_1^* dimensional leading order temperature of the upper wall of the centrifuge
- T_a^* dimensional temperature at $r = a$
- T_b^* dimensional temperature at $r = b$
- u velocity in the radial direction (dimensionless)
- U_I injection velocity based on feed flow rate
- U_G velocity in the radial direction in the core
- $u_{1/3}$ velocity in the radial direction in the $E^{1/3}$ layer
- $u_{1/4}$ velocity in the radial direction in the $E^{1/4}$ layer
- $\bar{u}$ velocity in the radial direction in the Ekman layer
- v velocity in the azimuthal direction
- $v_{1/3}$ velocity in the azimuthal direction in the $E^{1/3}$ layer
- $v_{1/4}$ velocity in the azimuthal direction in the $E^{1/4}$ layer
- V_B azimuthal velocity of the bottom wall
- V_G azimuthal velocity in the geostrophic core
- w velocity in the vertical direction
- $w_{1/3}$ vertical velocity in the $E^{1/3}$ layer
- $w_{1/4}$ vertical velocity in the $E^{1/4}$ layer

W_G vertical velocity in the geostrophic core region
y boundary layer variable defined in Equation 69
z independent variable in the vertical direction (dimensionless)
z^* independent variable in the vertical direction (dimensional)
z_F axial location of feed port
z_p axial location of product port
z_w axial location of waste port

Greek Symbols

α separation factor defined by $\alpha = \omega_p/\omega_F$
α_W parameter defined by Equation 84
β parameter defined by $\beta = 1$ for $r_0 = a$; $\beta = -1$ for $r_0 = b$
γ ratio of specific heats
δ flux parameter defined in Equation 32
δU separative power defined in Equation 84
ϵ $= (M_B - M_A)/M_B$
ϵ_f forced flow Rossby number
ϵ_T thermal Rossby number
ϵ_m mechanical Rossby number
ζ general boundary-layer variable
η $E^{1/3}$ layer variable; also pancake-layer variable
θ cut
λ parameter defined in Equation 27
λ_1 constant proportional to the differential speed of the end wall $z = 1$
λ_0 constant proportional to the differential speed of the end wall $z = 0$
λ_a $=(T_a^* - T_0^*)/(T_1^* - T_0^*)$
λ_b $=(T_b^* - T_0^*)/(T_1^* - T_0^*)$
λ_m $=\epsilon_m/\epsilon_f$
λ_T $=\epsilon_T/\epsilon_f$
Λ pancake equation variable
μ parameter defined in Equation 16; also dynamic viscosity
ξ $E^{1/4}$ layer variable; see Equations 16
ρ dimensionless density
$\rho_e(r)$ equilibrium dimensionless mixture density distribution
ρ_e^* equilibrium dimensional mixture density distribution
ρ_{ep} dimensionless equilibrium mixture density distribution based on outer wall density
ρ_{ep}^* dimensional equilibrium mixture density based on outer wall density
ρ_0^* dimensional mixture density
σ parameter
ϕ axial location of a source on sink on the side-wall of the centrifuge
χ complex Ekman layer variable defined in Equation 14
ψ stream function
ω_0 scaled mass fraction defined in Equation 38
ω_1 perturbation core species mass fraction defined in Equation 59
ω_A absolute species mass fraction
$\bar{\omega}_A$ perturbation species mass fraction defined in Equation 36
ω_F value of mass fraction of species A at the feed port
ω_p value of mass fraction of species A at the product port
ω_W value of mass fraction of species A at waste port
$\bar{\omega}_0$ perturbation species mass fraction defined in Equation 40
$\bar{\omega}_1$ perturbation mass fraction defined in Equation 40; also used as a similar quantity in Equation 73
$\tilde{\omega}_1$ scaled vertical boundary-layer mass fraction defined in Equation 63
Ω rotation speed of the centrifuge
$\Delta\Omega$ difference in rotation speed between the upper and lower wall of the centrifuge

Scripts

$\mathscr{F}_{1/2}(r, z_0)$ volume flux in the Ekman layer at z_0
$\mathscr{F}_{1/3}(r_0, z)$ volume flux in the $E^{1/3}$ layer at r_0
$\mathscr{F}_{1/4}(r_0, z)$ volume flux in the $E^{1/4}$ layer at r_0
$\mathscr{F}_S(r_0, z)$ $= \mathscr{F}_{1/4}(r_0, z) + \mathscr{F}_{1/3}(r_0, z)$

REFERENCES

1. Villani, S., (Ed.) *Uranium Enrichment*, Topics in Applied Physics, 35, Springer-Verlag, New York, 1979.
2. Cohen, K., *The Theory of Isotope Separation*, G.M. Murphy, (Ed.), National Nuclear Energy Series, McGraw-Hill, New York, 1951.
3. Groth, W., *Separation of Isotopes*, London, H., George Newnes, London, 1961 (Chapter 6.)
4. Zippe, G., U.S. Atomic Energy Commission Report, NYO-7348, 1960.
5. Beams, J. W., Snoddy, L. B., and Kuhlthau, A. R., U.N. Conference on Peaceful Uses of Atomic Energy, vol. 4 (1961) p. 428.
6. Olander, D., "Technical Basis of the Gas Centrifuge," in *Adv. Nucl. Sci. Tech., 6* (1972), pp. 105–174.
7. Soubbaramayer, "Centrifugation," in *Uranium Enrichment*, Villani, S., Springer-Verlag, New York (1979), pp. 183–244.
8. Hoglund, R. L., Shacter, J., and Von Halle, E., "Diffusion Separation Methods" in *Encyclopedia* of Chemical Technology, Vol. 7, 3rd edition, Kirk, R. E. and D. F. Othmer (Ed.), Wiley, 1979.
9. Olander, D. R., "The Theory of Uranium Enrichment by the Gas Centrifuge," in *Progress in Nuclear Energy*, Vol. 8, Pergamon Press Ltd., 1981, pp. 1–33.
10. Olander. D. R., "The Gas Centrifuge," *Scientific American*, Vol. 239, 1978, pp.37–44.
11. Stewartson, K., "On Almost Rigid Rotations," *J. Fluid Mech.*,Vol. 3, (1975). pp. 17–26.
12. Stewartson, K., "On Almost Rigid Rotations. Part 2," *J. Fluid Mech.*, Vol. 26 (1966), pp. 131–144.
13. Hashimoto, K., "A Source-Sink Flow of an Incompressible Rotating Fluid," *J. Phys. Soc. Jap.*, Vol. 38 (1975), pp. 1508–1515.
14. Matsuda, T., and Hashimoto, K., "Thermally, Mechanically or Externally Driven Flows in a Gas Centrifuge with Insulated Horizontal End Plates," *J. Fluid Mech.*, Vol. 78 (1976), pp. 337–354.
15. Matsuda, T., and Hashimoto, K., "The Structure of the Stewartson Layers in a Gas Centrifuge Part 1," *J. Fluid Mech.*, Vol. 85 (1978), pp. 433–442.
16. Matsuda, T., Hashimoto, K., and Takeda, H., "Thermally Driven Flow in a Gas Centrifuge with a Insulated Side Wall," *J. Fluid Mech.*, Vol. 73 (1976), pp. 389–399.
17. Matsuda, T., Sakurai, T., and Takeda, H., "Source-Sink Flow in a Gas Centrifuge," *J. Fluid Mech.*, Vol. 67 (1975), pp. 197–208.
18. Matsuda, T., and Takeda, H., "The Structure of the Stewartson Layers in a Gas Centrifuge Part 2: Insulated Side Wall," *J. Fluid Mech.*, Vol. 85 (1978), pp. 443–457.
19. Sakurai, T., and Matsuda, T., "Aerodynamics of a Centrifugal Machine," *J. Fluid Mech.*, Vol. 62 (1974), 727–736.
20. Nakayama, W., and Usui, S., "Flow in Rotating Cylinder of a Gas Centrifuge," *J. Nucl, Sci. Tech.*, Vol. 11 (1974), pp. 242–262.
21. Carrier, G. F., "Phenomena in Rotating Fluids," *Proc. 11th Int. Congress Appl. Mech.*, Munich, Germany, 1964.
22. Onsager, L., "Approximate Solutions of the Linearized Flow Equations," unpublished report, 1965.
23. Conlisk, A. T., Foster, M. R., and Walker, J. D. A., "Fluid Dynamics and Mass Transfer in a Gas Centrifuge," *J. Fluid Mech.*, Vol. 125 (1983), pp. 283–318.
24. Kai, T., "Basic Characteristics of Centrifuges II, Analysis of Fluid Flow in Centrifuges," *J. Nucl. Sci. Technol.*, Vol. 14 (1977), pp. 267–281.
25. Merten, A., and Hänel, D., "Implicit Solution of the Nacier-Stokes Equation for the Flow in a Centrifuge," Proceedings of the Fourth Workshop on Gases in Strong Rotation, Ratz E.(Ed.), August 1981, Oxford, United Kingdom.
26. Kai, T., "Basic Characteristics of Centrifuges III," Analysis of Separation Performance of Centrifuges," *J. Nucl. Sci. Technol.*, Vol. 14 (1977), pp. 506–518.
27. Merten, A., and Hänel, D., "Diffusion and Flow Calculations in Centrifuges Using Full Conservation Equations," Proceedings of the Fifth Workshop on Gases in Strong Rotation, H. G. Wood (Ed.), June 1983, Charlottesville, Va.

28. Wood, H. G., and Morton, J. B., "Onsager's Pancake Approximation for the Fluid Dynamics of a Gas Centrifuge," *J. Fluid Mech.*, Vol. 101 (1980), pp. 1–31.
29. Wood, H. G., and Sanders, G., "Rotating Compressible Flows with Internal Sources and Sinks," *J. Fluid Mech.*, Vol. 127 (1983), pp. 299–313.
30. Ekman, V. W., "On the Influence of the Earth's Rotation on Ocean Currents," *Ark. Mat. Astro. Fys.*, Vol. 2 (1905), pp. 1–52.
31. Hide, R., "On Source-Sink Flows in a Rotating Fluid," *J. Fluid Mech.*, Vol. 32 (1968), pp. 737–764.
32. Conlisk, A. T., and Walker, J. D. A., "Incompressible Source-Sink Flow in a Rapidly Rotating Contained Annulus," *Quart. J. Mech. Appl. Math.*, Vol. 34 (1981), pp. 89–109.
33. Conlisk, A. T., and Walker, J. D. A., "Forced Convection in a Rapidly Rotating Annulus," *J. Fluid Mech.*, Vol. 22 (1982), pp. 91–108.
34. Bark, T., and Hultgren, L., "On the Effects of Thermally Insulating Boundaries on Geostrophic Flows in a Rapidly Rotating Gas," *J. Fluid Mech.*, Vol. 95 (1979), pp. 97–118.
35. Bennetts, D. A. and Hocking, L. M., "On Nonlinear Ekman and Stewartson Layers in a Rotating Fluid," *Proc. R. Soc. Lond. A.*, Vol. 333 (1973), pp. 469–489.
36. Conlisk, A. T., Foster, M. R., and Walker, J. D. A., "Asymptotic Theory of Mass Transfer in a Gas Centrifuge for Small Ekman Number," Proceedings on the Fifth Workshop on Gases in Strong Rotation, H. G. Wood (Ed.), June 1983, Charlottesville, VA.
37. Brouwers, J. J. H., doctoral dissertation, de Technische Hogeschool, Twente Holland, 1976.
38. Carrier, G. F., and Maslen, S. H., "Flow Phenomena in Rapidly Rotating Systems," USAEC Report TID-18065, 1962.
39. Cole, J. D., and Schleiniger, G., "An Asymptotic Study of Pancake Theory," Proceedings of the Fourth Workshop on Gases in Strong Rotation, E. Ratz (Ed.), August, 1981, Oxford, United Kingdom.
40. Lahargue, J. P., and Soubbaramayer, "A Numerical Model for the Investigation of the Flow and Isotope Concentration Field in an Ultracentrifuge," *Compu. Methods Appl. Mech. Eng.*, Vol. 15(1978), pp. 259–273.
41. Bennetts, D. A., and Jackson, D. W., "Source-Sink Flows in a Rapidly Rotating Annulus: A Combined Laboratory and Numerical Study," *J. Fluid Mech.*, Vol. 66 (1974), pp. 684–705.
42. Lopez, S., "On the Effect of Immission of Feed Gas on the Flow Field in a Mechanically Driven Centrifuge," Proceedings of the Third Workshop on Gases in Strong Rotation, G. B. Scuricini (Ed), March 1979, Rome.
43. Conlisk, A. T., "The Effect of Source-Sink Geometry on Enrichment in a Gas Centrifuge," *Physics of Fluids* Vol. 26 (1983), pp. 2946—2957.
44. Von Halle, E., "Procedure for the Calculation of the Separative Performance of a Countercurrent Gas Centrifuge," Proceedings of the Fourth Workshop on Gases in Strong Rotation, E. Ratz (Ed.), August 1981, Oxford, U.K.
45. Avery, D. G., and Davies, E., *Uranium Enrichment by Gas Centrifuge*, Mills and Boon Limited, London, 1973.
46. Kai, T., "Basic Characteristics of Centrifuges," (I) Analysis of Concentration Distribution in Centrifuges," J. At. Energy Soc. Japan, Vol. 17 (1975), pp. 131–140.
47. Nakayama, W. and Torii, T., "Numerical Analysis of Separative Power of Isotope Centrifuges, (I)", *J. Nucl. Sci. Tech.*, Vol. 11 (1974), pp. 495–504.
48. Torii, T. "Numerical Analysis of Thermal Convection on Separative Power of IsotopeCentrifuges, (II), Effect of Thermal Convection on Separative Power," *J. Nucl. Sci. Tech.*, Vol. 14 (1977), pp. 901–910.
49. London, H., *Separation of Isotopes*, George Newnes Limited, London, 1961.

CHAPTER 47

RUPTURE DISC SIZING AND SELECTION

R. A. McIntire

Fike Metal Products, Corporation
Blue Springs, Missouri USA

CONTENTS

INTRODUCTION, 1434

RUPTURE DISC DESIGNS: THEIR APPLICATION AND SELECTION, 1435
The Forward-Acting Disc, 1435
The Reverse-Acting Disc, 1435
Application Considerations, 1435
Guidelines for Selection, 1438

RULES GOVERNING THE USE OF RUPTURE DISCS, 1451
Rupture Discs as Primary Relief Devices, 1451
Rupture Discs as Secondary Relief Devices, 1451
Additional Hazard Considerations, 1451
Rupture Discs in Combination With Pressure Relief Valves, 1451
Burst Pressure Certification, 1452
Manufacturing Design Range, 1452
Stamped Burst Pressure at a Coincident Disc Temperature, 1452

SPECIFYING THE RUPTURE DISC, 1455
Proper Disc Material, 1455
Coincident Burst Temperature, 1455
Operating Pressure Ratio, 1455
Vacuum/Back Pressure, 1455
Cyclic or Pulsing Pressure Duty, 1455
Polymerization Processes, 1455
Fragmentation, 1456
Hydraulic Service, 1457
Burst Pressure/Size/Type of Disc, 1457
Manufacturing Range and Burst Tolerance, 1457

RUPTURE DISC HOLDERS AND HOLDER ACCESSORIES, 1457
The Bolted-Type Assembly, 1457
Union-Type Holders, 1457
Screw-Type Holders, 1460
Holder Accessories, 1460

RUPTURE DISC SIZING, 1462
General Considerations, 1463
Sizing for Liquids, 1464
Sizing for Gases and Superheated Vapors, 1464
Converting Actual Flow Rate to SCFM, 1467
Sizing for Steam Flow, 1467
Sizing for Downstream Piping, 1470
Sample Problems, 1470

INSTALLATION AND MAINTENANCE, 1476,
Installation, 1477
Inspection/Maintenance, 1477

TROUBLE SHOOTING PROBLEM AREAS, 1478
Overtorqued Flange Bolts, 1478
Undertorqued Flange Bolts, 1479
Unseen Damage to Reverse-Acting Rupture Discs, 1480
Exposure to Vacuum Without a Vacuum Support, 1481
"Turtle-Backing", 1481

EXPLOSION VENTING, 1481
The Anatomy of a Dust or Vapor Explosion, 1482
Fundamentals of Explosion Venting, 1482
Types of Vents Available, 1484
Basic Recommendations for Venting, 1486

NOTATION, 1486

INTRODUCTION

Overpressure protection is a fundamental and integral part of all industrial processes. Without it the potential for loss of life and property would be staggering. The proper choice and size of pressure relief devices then become the ultimate goal of the engineer who has the responsibility for the safety of the plant.

A clever approach to the subject of pressure relief is achieved with a device known as *rupture disc.* (See Figure 1.) This rather simple-looking device has found uses in such places as the chemical and petro-chemical industries, steam and nuclear power plants, aerospace hardware, and the food processing industry.

A rupture disc, sometimes referred to as a safety head or burst diaphragm, is a thin membrane of metal, plastic, graphite, or other material designed and tested to hold pressure up to a predictable point. It represents a "weak spot" in a pressure system that will break open to relieve an overpressure condition. Rupture discs usually range in size from about 1/8 to 48 inches in diameter; however, using special techniques and materials, they can be built to 68 inches in diameter or larger.

There are many reasons for using rupture discs in place of other devices. As primary relief, the rupture disc helps eliminate leakage of valuable or toxic gases, and it is useful for protecting valves from highly corrosive products. As a secondary relief device, it provides overpressure protection to process systems subject to unexpected runaway reactions. In general, the rupture disc assembly has a lower first cost than its spring-loaded counterpart, the pressure relief valve, especially in sizes larger than 10-inch nominal diameter. The main disadvantage lies in the fact that it is nonreclosing. Where the pressure relief valve recloses when the pressure is reduced, the rupture disc must be removed from the system and replaced. This becomes an acceptable disadvantage, however, when it is understood that large pressure relief valves can cost as much as 10 or more times the rupture disc device.

Specially designed rupture discs may be used to vent explosions in dust collection systems and hazardous vapor areas. Generally larger and not limited to the traditional round shape, *explosion*

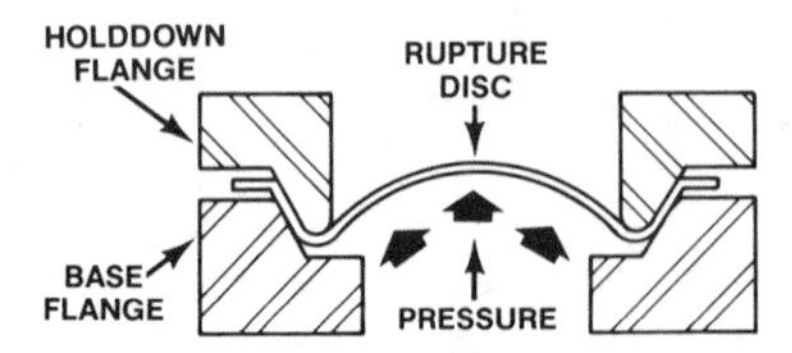

Figure 1. The conventional rupture disc.

vents are designed to burst at very low pressures. Their popularity attests to the fact that they have found good success in limiting damage to expensive plant equipment and preventing injury to personnel.

This chapter will address the rupture disc and explosion vent as pressure relief devices. Emphasis is placed on application, selection, and sizing.

RUPTURE DISC DESIGNS: THEIR APPLICATION AND SELECTION

The rupture disc comes in two basic design configurations; the forward-acting disc and the reverse-acting disc. Each design has unique characteristics that best suit it to its own variety of service conditions. To better understand this form of a pressure relief device, it will be helpful to look at some basics.

The Forward-Acting Disc

The design principle behind the solid metal forward-acting disc, is simply a tension failure of the membrane. As the system pressure approaches the rupture pressure, rapid thinning of the material occurs near the crown of the disc. This leads to a catastrophic material failure and the disc "ruptures." For this type of disc, the rupture pressure is directly proportional to the material thickness and inversely proportional to the diameter.

The solid metal rupture disc will not meet every burst pressure requirement within its upper and lower burst limit. To cover intermediate burst pressures, we must make some adjustments in its characteristics. These adjustments may take the form of additional work done on the material or the addition of an extra component to the disc. In the latter case, a mechanically weakened top section is placed over a seal member that bursts at a lower pressure. The top section is slotted, leaving small tabs to make up the difference between the burst pressure of the seal member and the slotted top section. The burst pressure, then, is a function of both components and can be varied by adjusting the width of the slots. As it turns out, the variations in design that fine tune the burst pressure also provide unique solutions to applications that could not be solved by the solid metal disc alone.

To control fragmentation, the disc may be "scored" or grooved. Scoring is accomplished by forcing a hardened die with a specific pattern, such as a cross or "toilet seat", into the material to some predetermined depth. This places a highly localized stress riser in the material, and the disc ruptures along the path of the score. This type of disc is usually made from thicker material and, as a result, it is less subject to damage.

The Reverse-Acting Disc

The reverse-acting disc is pre-bulged in the same manner as the forward-acting disc. Unlike the tension-loaded disc, however, this disc is installed in the system in the reverse direction. Using the fact that in compression a rupture disc can withstand only about $\frac{1}{3}$ the pressure of the forward-acting disc, it snaps or buckles backward into a stationary knife blade. (See Figure 2A.) This disc has the advantage of opening at lower pressures than the forward-acting disc, and comes eqipped with its own back-pressure protection. The disadvantage is that it is not suitable for liquid systems. Opening of this disc is dependent on the unrestrained expansion of the system media to carry the disc through the knife blade (See Figure 2B.)

Application Considerations

The rupture disc provides the greatest amount of protection to pressure systems if it is applied correctly. Conversely, poor maintenance or no maintenance at all can lull the user into a false sense of security. It is essential, then, that the user know something of how the performance of the rupture disc is influenced by application parameters.

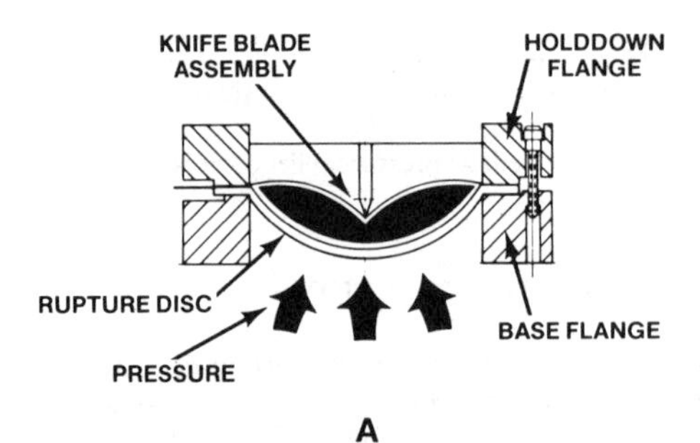

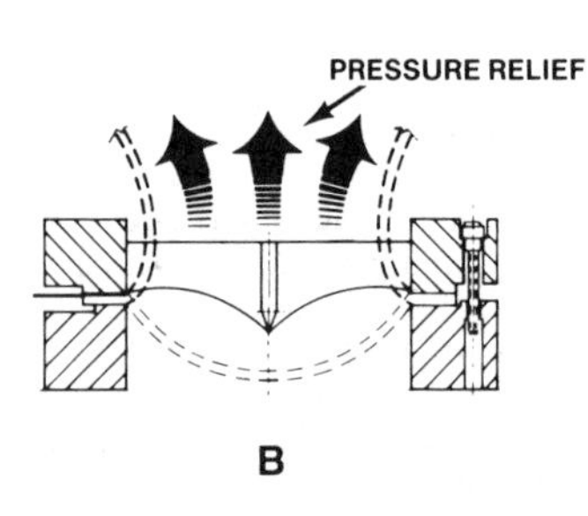

Figure 2. (A) The reverse-acting rupture disc; (B) the reversing action of the disc during burst.

The following information will provide a better understanding of rupture discs and will assist the engineer in providing the best disc for the application. The summaries are general and performance may vary from manufacturer to manufacturer.

Material Selection

Selection of the proper disc material is governed by a combination of factors. These factors include operating temperature, corrosive properties of the process media, size, cyclic service conditions, and burst pressure. Many materials can be used for the same application but some may be better suited to the application than others.

Materials most commonly used are:

300 series stainless steels
Nickel
Silver
Carbon (Graphite)
Copper
Aluminum
Teflon
Monel

Less frequently used materials include:

Platinum
Tantalum
Hastelloy (B and C)
Columbium
Gold
Titanium
Lead
Zirconium

Materials used as coatings and linings include:

Teflon
Polyvinylchloride
Epoxy powder paint
Lead
Gold

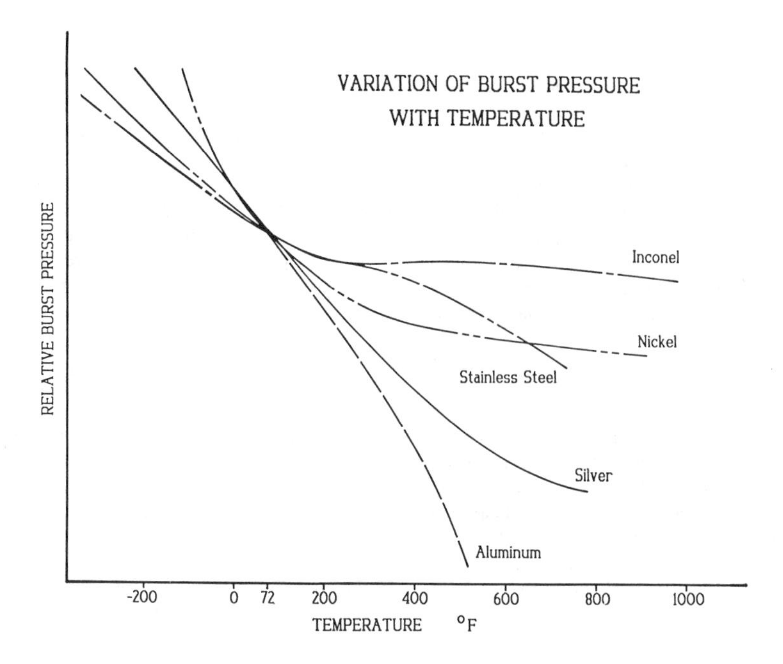

Figure 3. Plot showing the variation of burst pressure with temperature.

The Effects of Temperature

As one might expect, the burst pressure of the forward-acting rupture disc falls as the temperature increases. This holds true for all design variations of this type disc. The same cannot be said of the reverse-acting disc however. Because the crown height of the reverse-acting disc is a function of the burst pressure, the environmental temperature often effects this disc in peculiar ways. In some cases the burst pressure may actually increase as the temperature increases.

Manufacturers rate the rupture disc at ambient temperature (72°F). Any variation from that temperature, either up or down, is represented in terms of percent of the burst pressure at 72°F. A chart showing the variation of temperature with burst pressure rating for forward-acting solid metal discs is shown in Figure 3.

Each of the various materials used in disc fabrication respond differently to changes in temperature and each have different temperature limitations. Maximum temperature ratings have been established over the years and these ratings are shown in Table 1.

Table 1
Recommended Maximum Temperatures for Materials Used in Rupture Discs

Aluminum	250°F
Hasteloy	800°F
Inconel 600	1,000°F
Nickel 200	750°F
Monel	800°F
Stainless steel	600°F
Silver	250°F
Teflon	450°F

Operating Pressure Ratio

This is defined as the ratio of the system operating pressure to the stamped bursting pressure of the rupture disc. Years of experience have shown that the most economical disc life is achieved when the operating pressure ratio is at or near 70%. Ratios in excess of this value increase the likelihood of premature disc failure.

As a standard manufacturing practice, most forward-acting rupture discs are pre-bulged to 70% of the stamped bursting pressure. If the system will operate at a higher ratio than 70%, the manufacturer must know it before the disc is built. Further bugling of the disc after it is installed in the system may alter the burst pressure or contribute to premature failure. The reverse-acting disc is not limited to the 70% operating ratio. In compression, its unique properties permit it to be operated at ratios up to 90% of the burst pressure.

Vacuum Service and Special Environments

When process applications require the rupture disc to be exposed to a vacuum or a back pressure, a mechanical support may be added to the upstream side of the disc to provde the necessary assist. Fitted to the disc at the time of manufacture, most vacuum supports will withstand at least a full vacuum back pressure. The vacuum support may also be used on large, thin, low-pressure discs to prevent damage during shipping and installation.

For special operating conditions and applications, optional equipment such as dust covers may be added to the forward-acting disc to provide protection from damaging downstream environments. This equipment is designed into the disc and has no effect on the burst pressure.

Cyclic Pressure Duty

Quite often the rupture disc will be used in cyclic pressure applications. In such cases, the solid metal pre-bulged disc performs the best; especially when the operating pressure ratio can be kept to 70% or less. The full duty (positive to negative) cycle life of a pre-bulged disc will normally exceed several hundred-thousand pressure cycles, but diminishes significantly as the operating pressure ratio exceeds 70%. The cycle life of scored and composite discs may be somewhat lower depending on the severity of the application. In this case, the scored highly stressed grooves and tabs tend to be more sensitive to fatigue failure at the higher operating pressure ratios.

Effects of Downstream Piping

The rupture disc is normally sized to vent to the atmosphere. Where the installation includes downstream piping, the associated back pressure must be accounted for in the sizing calculations.

Guidelines for Selection

The Conventional "Pre-bulged" Disc

The standard and most commonly used rupture disc is the conventional pre-bulged disc. (See Figure 4.) It is a forward-acting, solid metal disc with a 30° seating surface. Labeled a "pre-bulged" in the early days when it was first developed, this disc has seen service in many applications. It has long been the workhorse of the industry and, no doubt, will be around for a long time to come.

The conventional pre-bulged disc is suitable for virtually all applications where fragmentation is not a problem and a wider manufacturing and slightly larger burst tolerance is acceptable. This is a durable disc and relatively insensitive to cyclic pressure duty when the operating pressure ratio is 70% or less.

Figure 4. Conventional rupture disc before and after rupture (courtesy of Fike Metal Products).

The most common materials of construction include aluminum, nickel, Monel, Inconel, and the 300 series stainless steels. If corrosive environments are anticipated, protective coatings, linings, and coverings may be required to provide the necessary protection. For special applications, the disc may also be made from any of the precious and semi-precious metals.

When the system operates under a vacuum or alternating pressure and vacuum, it may be necessary to add a vacuum support. (See Figure 5) The vacuum support is tightly fitted to the concave side of the disc and will prevent the disc from reversing. If the disc is thin and fragile, or harsh downstream environments threaten the service life, protective rings and covers may be added. These components are spot-welded together to form the sturdy multi-component discs shown in Figures 6 and 7. The protective ring holds the rupture disc tightly to the vacuum support and minimizes wrinkling, while a full cover helps prevent debris from filtering in on top of the disc. All of these devices and coatings are manufactured into the disc and have no effect on the burst pressure. (Tables 2 through 4 give various information on typical conventional rupture discs.)

Figure 5. Vacuum support before and after rupture (courtesy of Fike Metal Products).

Figure 6. Conventional rupture disc with protective ring and vacuum support (courtesy of Fike Metal Products).

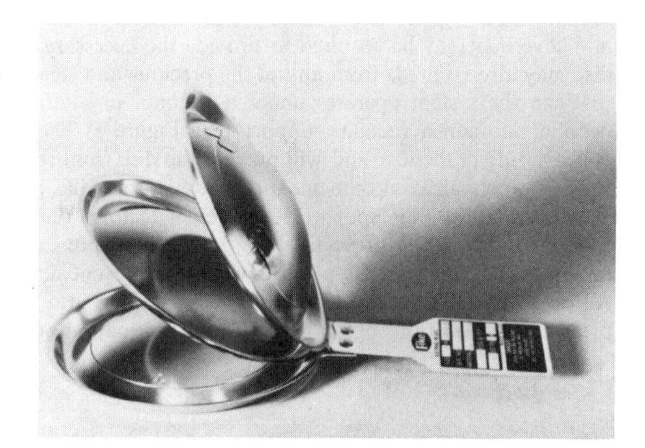

Figure 7. Conventional rupture disc with protective cover and vacuum support (courtesy of Fike Metal Products).

The Composite Rupture Disc

The composite disc consists of two to four elements that form a single unit. From all outward appearances, this disc resembles a conventional disc with a dust cover, but with one marked exception—two components work together to support the burst pressure. (See Figure 8.) The disc is made up of a seal member and a mechanically weakened top section. The seal member holds back the media and transmits part of the pressure load to small tabs on the top section. The burst pressure can be controlled by adjusting the width of the tabs.

The unique feature of this disc is that, in addition to the same materials available for the conventional disc, Teflon can be used for the seal member. This provides somewhat lower burst pressures than the solid metal seals as well as the excellent corrosion resistant properties of Teflon. An extension of this design rearranges the elements and places a second corrosion-resistant element over the composite assembly, effectively isolating the metallic components of the disc. (See Figure 9.)

This disc in all its variations (see Figures 10 and 11) is ideal for applications where high operating pressure-to-vacuum ratio is a problem and where lower burst pressures are required than the conventional disc can provide. This disc also has the unique characteristic of providing a wider range of intermediate burst pressures than the conventional disc. (Tables 5 through 7 provide information on typical composite rupture discs.)

Table 2
Typical Conventional Rupture Disc Burst Pressures at 72°F (22°C)

Rupture Disc (See Maximum Temperature)	Size[2] In.	½	¾	1	1½	2	3	4	6	8	10	12	14	16	18	20	24
	DN	—	20	25	40	50	80	100	150	200	250	300	350	400	450	500	600
Aluminum, Max. Temp. 250° F/121° C	Min.	65	45	34	23	15	10	8	7	5	4	3	3	3	3	3	3
		4.5	**3.1**	**2.3**	**1.6**	**1.0**	**0.7**	**0.6**	**0.5**	**0.3**	**0.3**	**0.2**	**0.2**	**0.2**	**0.2**	**0.2**	**0.2**
	Max.	1500	1000	1000	1000	750	500	400	300	225	175	150	150	150	125	125	100
		103.5	**69.0**	**69.0**	**69.0**	**51.7**	**34.5**	**27.6**	**20.7**	**15.5**	**12.1**	**10.3**	**10.3**	**10.3**	**8.6**	**8.6**	**6.9**
Aluminum, Lead Lined One Side, Max. Temp. 250° F/121° C	Min.	89	61	46	31	18	14	11	9	7	6	5	5	4			
		6.1	**4.2**	**3.2**	**2.1**	**1.2**	**1.0**	**0.8**	**0.6**	**0.5**	**0.4**	**0.3**	**0.3**	**0.3**			
	Max.	1500	1000	700	500	350	250	200	150	100	90	80	70	60			
		103.5	**69.0**	**48.3**	**34.5**	**24.1**	**17.2**	**13.8**	**10.3**	**6.9**	**6.2**	**5.5**	**4.8**	**4.1**			
Aluminum, Vinyl Coat One Side[3]	Min.	65	45	34	23	20	14	11	8	5	5	4	4	4	4	4	4
		4.5	**3.1**	**2.3**	**1.6**	**1.4**	**1.0**	**0.8**	**0.6**	**0.3**	**0.3**	**0.3**	**0.3**	**0.3**	**0.3**	**0.3**	**0.3**
	Max.	1500	1000	1000	1000	750	500	400	300	225	175	150	150	150	125	125	100
		103.5	**69.0**	**69.0**	**69.0**	**51.7**	**34.5**	**27.6**	**20.7**	**15.5**	**12.1**	**10.3**	**10.3**	**10.3**	**8.6**	**8.6**	**6.9**
Aluminum, Vinyl Coat Both Sides[3]	Min.	75	50	34	27	24	17	13	10	6	6	5	5	5	5	5	5
		5.2	**3.4**	**2.3**	**1.9**	**1.7**	**1.2**	**0.9**	**0.7**	**0.4**	**0.4**	**0.3**	**0.3**	**0.3**	**0.3**	**0.3**	**0.3**
	Max.	1500	1000	1000	1000	750	500	400	300	225	175	150	150	150	125	125	100
		103.5	**69.0**	**69.0**	**69.0**	**51.7**	**34.5**	**27.6**	**20.7**	**15.5**	**12.1**	**10.3**	**10.3**	**10.3**	**8.6**	**8.6**	**6.9**
Aluminum, Teflon Coat One Side[3]	Min.	90	60	50	35	28	25	15	10	8	8	7	6	6	6	6	6
		6.2	**4.1**	**3.4**	**2.4**	**1.9**	**1.7**	**1.0**	**0.7**	**0.6**	**0.6**	**0.5**	**0.4**	**0.4**	**0.4**	**0.4**	**0.4**
	Max.	1500	1000	1000	1000	750	500	400	300	225	175	150	150	150	125	125	100
		103.5	**69.0**	**69.0**	**69.0**	**51.7**	**34.5**	**27.6**	**20.7**	**15.5**	**12.1**	**10.3**	**10.3**	**10.3**	**8.6**	**8.6**	**6.9**
Aluminum, Teflon Coat Both Sides[3]	Min.	130	90	75	50	40	30	20	15	10	10	9	8	8	8	8	8
		9.0	**6.2**	**5.2**	**3.4**	**2.8**	**2.1**	**1.4**	**1.0**	**0.7**	**0.7**	**0.6**	**0.6**	**0.6**	**0.6**	**0.6**	**0.6**
	Max.	1500	1000	1000	1000	750	500	400	300	225	175	150	150	150	125	125	100
		103.5	**69.0**	**69.0**	**69.0**	**51.7**	**34.5**	**27.6**	**20.7**	**15.5**	**12.1**	**10.3**	**10.3**	**10.3**	**8.6**	**8.6**	**6.9**
Aluminum, Epon Coat One Side[3]	Min.	75	45	35	25	22	15	12	8	6	6	5	5	5	5	5	5
		5.2	**3.1**	**2.4**	**1.7**	**1.5**	**1.0**	**0.8**	**0.6**	**0.4**	**0.4**	**0.3**	**0.3**	**0.3**	**0.3**	**0.3**	**0.3**
	Max.	1500	1000	1000	1000	750	500	400	300	225	175	150	150	150	125	125	100
		103.5	**69.0**	**69.0**	**69.0**	**51.7**	**34.5**	**27.6**	**20.7**	**15.5**	**12.1**	**10.3**	**10.3**	**10.3**	**8.6**	**8.6**	**6.9**
Aluminum, Epon Coat Both Sides[3]	Min.	90	60	50	35	28	20	15	10	8	8	7	6	6	6	6	6
		6.2	**4.1**	**3.4**	**2.4**	**1.9**	**1.4**	**1.0**	**0.7**	**0.6**	**0.6**	**0.5**	**0.4**	**0.4**	**0.4**	**0.4**	**0.4**
	Max.	1500	1000	1000	1000	750	500	400	300	225	175	150	150	150	125	125	100
		103.5	**69.0**	**69.0**	**69.0**	**51.7**	**34.5**	**27.6**	**20.7**	**15.5**	**12.1**	**10.3**	**10.3**	**10.3**	**8.6**	**8.6**	**6.9**
Copper, Max. Temp. 250° F/121° C	Min.	210	140	110	65	35	28	30	26	37							
		14.5	**9.7**	**7.6**	**4.5**	**2.4**	**1.9**	**2.1**	**1.8**	**2.6**							
	Max.	3000	3000	3000	3000	2500	1500	1000	1000	700							
		206.9	**206.9**	**206.9**	**206.9**	**172.4**	**103.5**	**69.0**	**69.0**	**48.3**							
Copper Lead Lined Max. Temp. 250° F/121° C	Min.	235	155	125	73	40	32	33	28								
		16.2	**10.7**	**8.6**	**5.0**	**2.8**	**2.2**	**2.3**	**1.9**								
	Max.	3000	3000	2000	1500	1000	800	600	500								
		206.9	**206.9**	**137.9**	**103.5**	**69.0**	**55.2**	**41.4**	**34.5**								
Silver, Max. Temp. 250° F/121° C	Min.	245	175	125	85	55	35	25	20	17							
		16.9	**12.1**	**8.6**	**5.9**	**3.8**	**2.4**	**1.7**	**1.4**	**1.2**							
	Max.	3000	3000	3000	3000	2500	2000	1500	1000	500							
		206.9	**206.9**	**206.9**	**206.9**	**172.4**	**137.9**	**103.5**	**69.0**	**34.5**							
Nickel, Max. Temp. 750° F/399° C	Min.	340	260	190	125	75	46	38	33	24	20	20	20	18	18	18	40
		23.5	**17.9**	**13.1**	**8.6**	**5.2**	**3.2**	**2.6**	**2.3**	**1.7**	**1.4**	**1.4**	**1.4**	**1.2**	**1.2**	**1.2**	**2.8**
	Max.	3000	3000	3000	3000	3000	3000	3000	2160	1440	720	720	720	720	720	720	720
		206.9	**206.9**	**206.9**	**206.9**	**206.9**	**206.9**	**206.9**	**149.0**	**99.3**	**49.7**	**49.7**	**49.7**	**49.7**	**49.7**	**49.7**	**49.7**
Monel, Max. Temp. 800° F/427° C	Min.	460	320	250	175	105	80	54	43	37	30	28	26	24	22	20	55
		31.7	**22.1**	**17.2**	**12.1**	**7.2**	**5.5**	**3.7**	**3.0**	**2.6**	**2.1**	**1.9**	**1.8**	**1.7**	**1.5**	**1.4**	**3.8**
	Max.	10000	10000	6000	3000	3000	3000	3000	2160	1440	720	720	720	720	720	720	720
		689.7	**689.7**	**413.8**	**206.9**	**206.9**	**206.9**	**206.9**	**149.0**	**99.3**	**49.7**	**49.7**	**49.7**	**49.7**	**49.7**	**49.7**	**49.7**
Inconel, Max. Temp. 1000° F/538° C	Min.	760	440	410	290	180	130	100	75	40	32	27	23	20	18	16	45
		52.4	**30.3**	**28.3**	**20.0**	**12.4**	**9.0**	**6.9**	**5.2**	**2.8**	**2.2**	**1.9**	**1.6**	**1.4**	**1.2**	**1.1**	**3.1**
	Max.	10000	10000	6000	3000	3000	3000	3000	2160	1440	720	720	720	720	720	720	720
		689.7	**689.7**	**413.8**	**206.9**	**206.9**	**206.9**	**206.9**	**149.0**	**99.3**	**49.7**	**49.7**	**49.7**	**49.7**	**49.7**	**49.7**	**49.7**
316 Stainless Steel Max. Temp. 600° F/316° C	Min.	850	600	485	365	195	135	105	85	65	50	50	48	44	38	30	27
		58.6	**41.4**	**33.5**	**25.2**	**13.4**	**9.3**	**7.2**	**5.9**	**4.5**	**3.4**	**3.4**	**3.3**	**3.0**	**2.6**	**2.1**	**1.9**
	Max.	10000	10000	6000	3000	3000	3000	3000	2160	1440	720	720	720	720	720	720	720
		689.7	**689.7**	**413.8**	**206.9**	**206.9**	**206.9**	**206.9**	**149.0**	**99.3**	**49.7**	**49.7**	**49.7**	**49.7**	**49.7**	**49.7**	**49.7**

Note 1: Burst pressures printed in light are PSIG; bold is BAR G.
Note 2: Consult factory for discs larger than 24" in diameter.
Note 3: All or any one of the members of the three-component disc may be coated with vinyl, maximum temperature 150° F; epon 250° F; or Teflon 450° F. All nickel, monel, inconel and the stainless steel discs when coated with Teflon, vinyl or epon will have slightly higher minimum burst pressures than those listed in Table 5. Maximum temperature application of a disc with a coating depends upon the maximum temperature application of either the metal or the coating, whichever is lower. The maximum temperatures of the metal discs are tabulated above.

Source: Fike Metal Products.

Table 3
Typical Conventional Rupture Disc Manufacturing Range

Specified Rupture Pressure		Mfg. Range % @ 72° F/22° C
PSIG @ 72° F	**BAR G @ 22° C**	
3-6	**.2-.4**	+ 40 to -20
7-10	**.5-.7**	+ 30 to -15
11-15	**.8-1.0**	+ 20 to -10
16-25	**1.1-1.7**	+ 16 to - 8
26-45	**1.8-3.1**	+ 14 to - 7
46-90	**3.2-6.2**	+ 12 to - 6
91-270	**6.3-18.6**	+ 10 to - 5
271-500	**18.7-34.5**	+ 8 to - 4
501-up	**34.6-up**	+ 6 to - 3

Source: Fike Metal Products.

Table 4
Typical Conventional Rupture Disc Normal Burst Tolerance

Marked Rupture Pressure		Rupture Tolerance % @ 72° F/22° C
PSIG @ 72° F	**BAR G @ 22° C**	
3-6	**.2-.4**	+ 22 to -22
7-10	**.5-.7**	+ 15 to -15
11-14	**.8-1.0**	+ 8 to - 8
15-up	**1.0-up**	+ 5 to - 5

Source: Fike Metal Products.

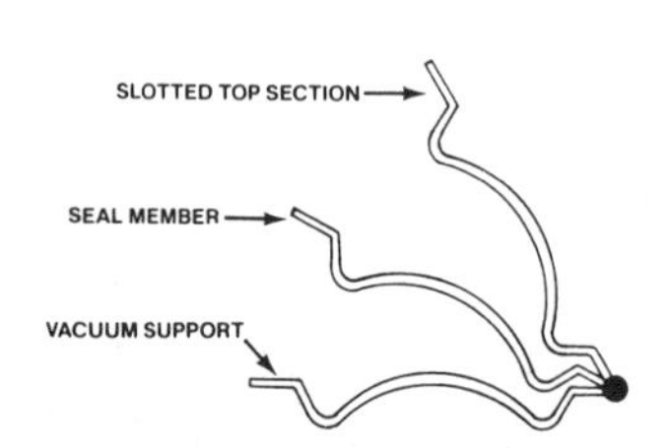

Figure 8. The composite rupture disc.

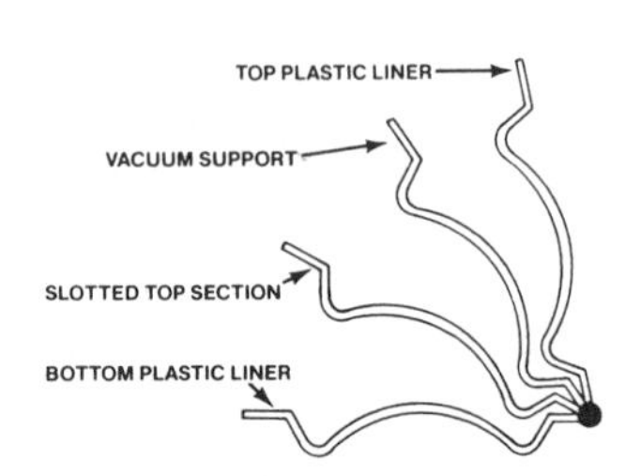

Figure 9. The composite rupture disc with added corrosion protective element.

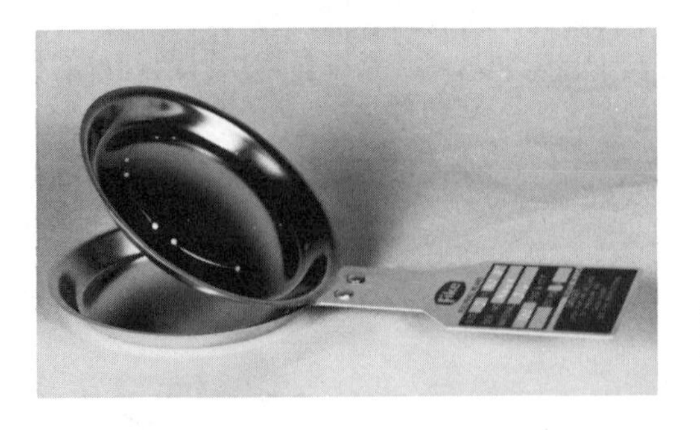

Figure 10. Two-element composite rupture disc with metallic seal (courtesy of Fike Metal Products).

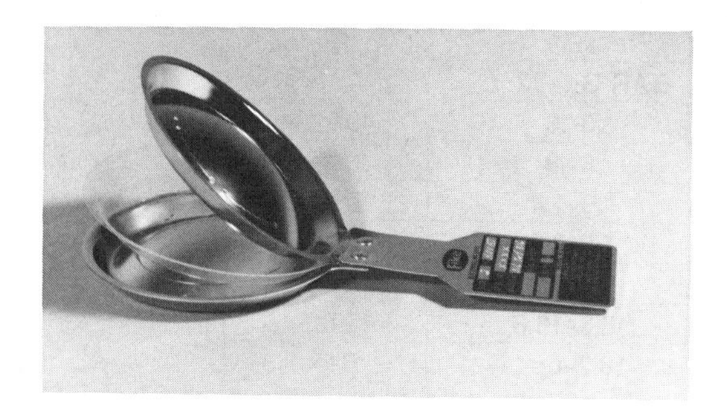

Figure 11. Three-element composite rupture disc with Teflon seal (courtesy of Fike Metal Products).

Table 5
Typical Composite Rupture Disc Burst Pressures at 72°F (22°C)

Seal Member Size[2]	In. **DN.**	1 **25**	1½ **40**	2 **50**	3 **80**	4 **100**	6 **150**	8 **200**	10 **250**	12 **300**	14 **350**	16 **400**	18 **450**	20 **500**	24 **600**
Teflon[3] 450° F/232° C Max. Temp.	Min.	57 **3.9**	38 **2.6**	23 **1.6**	16 **1.1**	15 **1.0**	13 **0.9**	10 **0.7**	8 **0.6**	6 **0.4**	6 **0.4**	5 **0.3**	4 **0.3**	4 **0.3**	4 **0.3**
	Max.	6000 **413.8**	3000 **206.9**	3000 **206.9**	3000 **206.9**	3000 **206.9**	2160 **149.0**	1440 **99.3**	720 **49.7**	720 **49.7**	720 **49.7**	720 **49.7**	720 **49.7**	720 **49.7**	720 **49.7**
Aluminum 600° F/316° C Max. Temp.	Min.	51 **3.5**	35 **2.4**	23 **1.6**	15 **1.0**	12 **0.8**	11 **0.8**	8 **0.6**	6 **0.4**	4 **0.3**	4 **0.3**	4 **0.3**	4 **0.3**	4 **0.3**	4 **0.3**
	Max.	1500 **103.5**	1500 **103.5**	1125 **77.6**	750 **51.7**	600 **41.4**	450 **31.0**	338 **23.3**	263 **18.1**	225 **15.5**	225 **15.5**	225 **15.5**	188 **13.0**	188 **13.0**	150 **10.3**
Aluminum, Lead Lined, One Side	Min.	69 **4.8**	47 **3.2**	27 **1.9**	21 **1.4**	17 **1.2**	14 **1.0**	11 **0.8**	9 **0.6**	8 **0.6**	8 **0.6**	6 **0.4**			
	Max.	700 **48.3**	500 **34.5**	350 **24.1**	250 **17.2**	200 **13.8**	150 **10.3**	100 **6.9**	90 **6.2**	80 **5.5**	70 **4.8**	60 **4.1**			
Aluminum, Teflon Coated, One Side	Min.	75 **5.2**	53 **3.7**	42 **2.9**	38 **2.6**	23 **1.6**	15 **1.0**	12 **0.8**	12 **0.8**	11 **0.8**	9 **0.6**	9 **0.6**	9 **0.6**	9 **0.6**	9 **0.6**
	Max.	1500 **103.5**	1500 **103.5**	1125 **77.6**	750 **51.7**	600 **41.4**	450 **31.0**	338 **23.3**	263 **18.1**	225 **15.5**	225 **15.5**	225 **15.5**	188 **13.0**	188 **13.0**	150 **10.3**
Aluminum, Vinyl Coated, One Side	Min.	51 **3.5**	35 **2.4**	30 **2.1**	21 **1.4**	17 **1.2**	12 **0.8**	8 **0.6**	8 **0.6**	6 **0.4**	6 **0.4**	6 **0.4**	6 **0.4**	6 **0.4**	6 **0.4**
	Max.	1500 **103.5**	1500 **103.5**	1125 **77.6**	750 **51.7**	600 **41.4**	450 **31.0**	338 **23.3**	263 **18.1**	225 **15.5**	225 **15.5**	225 **15.5**	188 **13.0**	188 **13.0**	150 **10.3**
Aluminum, Epon Coated, One Side	Min.	53 **3.7**	38 **2.6**	33 **2.3**	23 **1.6**	18 **1.2**	12 **0.8**	9 **0.6**	9 **0.6**	8 **0.6**	8 **0.6**	8 **0.6**	8 **0.6**	8 **0.6**	8 **0.6**
	Max.	1500 **103.5**	1500 **103.5**	1125 **77.6**	750 **51.7**	600 **41.4**	450 **31.0**	338 **23.3**	263 **18.1**	225 **15.5**	225 **15.5**	225 **15.5**	188 **13.0**	188 **13.0**	150 **10.3**
Silver, 250° F/121° C Max. Temp.	Min.	188 **13.0**	128 **8.8**	83 **5.7**	53 **3.7**	38 **2.6**	30 **2.1**	26 **1.8**							
	Max.	6000 **413.8**	3000 **206.9**	3000 **206.9**	3000 **206.9**	2250 **155.2**	1500 **103.5**	750 **51.7**							
Nickel, 750° F/399° C Max. Temp.	Min.	190 **13.1**	125 **8.6**	75 **5.2**	46 **3.2**	38 **2.6**	33 **2.3**	24 **1.7**	20 **1.4**	20 **1.4**	20 **1.4**	18 **1.2**	18 **1.2**	18 **1.2**	40 **2.8**
	Max.	6000 **413.8**	3000 **206.9**	3000 **206.9**	3000 **206.9**	3000 **206.9**	2160 **149.0**	1440 **99.3**	720 **49.7**	720 **49.7**	720 **49.7**	720 **49.7**	720 **49.7**	720 **49.7**	720 **49.7**
Monel, 800° F/427° C Max. Temp.	Min.	250 **17.2**	175 **12.1**	105 **7.2**	80 **5.5**	54 **3.7**	43 **3.0**	37 **2.6**	30 **2.1**	28 **1.9**	26 **1.8**	24 **1.7**	22 **1.5**	20 **1.4**	55 **3.8**
	Max.	6000 **413.8**	3000 **206.9**	3000 **206.9**	3000 **206.9**	3000 **206.9**	2160 **149.0**	1440 **99.3**	720 **49.7**	720 **49.7**	720 **49.7**	720 **49.7**	720 **49.7**	720 **49.7**	720 **49.7**
Inconel, 1000° F/538° C Max. Temp.	Min.	410 **28.3**	290 **20.0**	180 **12.4**	130 **9.0**	100 **6.9**	75 **5.2**	40 **2.8**	32 **2.2**	27 **1.9**	23 **1.6**	20 **1.4**	18 **1.2**	16 **1.1**	45 **3.1**
	Max.	6000 **413.8**	3000 **206.9**	3000 **206.9**	3000 **206.9**	3000 **206.9**	2160 **149.0**	1440 **99.3**	720 **49.7**	720 **49.7**	720 **49.7**	720 **49.7**	720 **49.7**	720 **49.7**	720 **49.7**
316 Stainless Steel, 600° F/316° C Max. Temp.	Min.	485 **33.5**	365 **25.2**	195 **13.4**	135 **9.3**	105 **7.2**	85 **5.9**	65 **4.5**	50 **3.4**	50 **3.4**	48 **3.3**	44 **3.0**	38 **2.6**	30 **2.1**	27 **1.9**
	Max.	6000 **413.8**	3000 **206.9**	3000 **206.9**	3000 **206.9**	3000 **206.9**	2160 **149.0**	1440 **99.3**	720 **49.7**	720 **49.7**	720 **49.7**	720 **49.7**	720 **49.7**	720 **49.7**	720 **49.7**

Note 1: Burst pressures printed in light are PSIG; bold is BAR G.
Note 2: Consult factory for discs larger than 24" in diameter.
Note 3: Must have a retainer ring to support plastic seal.

Source: Fike Metal Products.

Table 6
Typical Composite Rupture Disc Standard Manufacturing Range

Specified Rupture Pressure		Mfg. Range % @ 72° F/22° C
PSIG @ 72° F	**BAR G @ 22° C**	
4-8	**.3-.6**	+ 40 to -40
9-12	**.6-.8**	+ 30 to -30
13-20	**.9-1.4**	+ 20 to -10
21-45	**1.5-3.1**	+ 16 to - 8
46-90	**3.2-6.2**	+ 12 to - 6
91-270	**6.3-18.6**	+ 10 to - 5
271-up	**18.7-up**	+ 6 to - 3

Source: Fike Metal Products.

Table 7
Typical Composite Rupture Disc Normal Burst Tolerance

Marked Rupture Pressure		Rupture Tolerance % @ 72° F/22° C
PSIG @ 72° F	**BAR G @ 22° C**	
2-3	**.1-.2**	+ 30 to -30
4-6	**.3-.4**	+ 22 to -22
7-10	**.5-.7**	+ 15 to -15
11-14	**.8-1.0**	+ 8 to - 8
15-up	**1.03-up**	+ 5 to - 5

Source: Fike Metal Products.

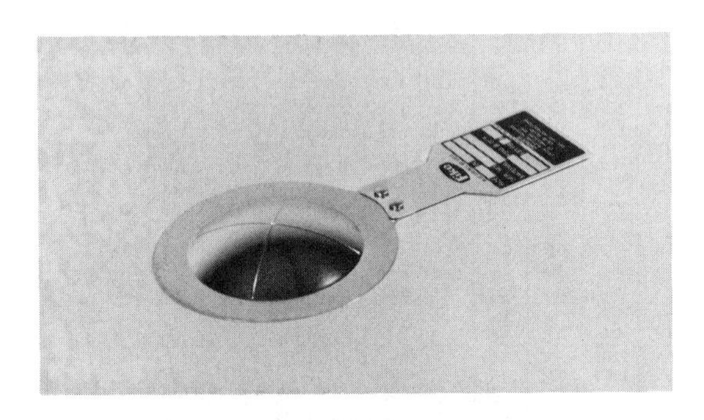

Figure 12. The scored rupture disc (courtesy of Fike Metal Products).

Figure 13. Scored rupture disc after rupture (courtesy of Fike Metal Products).

Table 8
Typical Scored Rupture Disc Burst Pressures at 72°F (22°C)

Size[2]	Inches	½	¾	1	1½	2	3	4	6	8	10	12	14	16	18	20	24
	DN	—	20	25	40	50	80	100	150	200	250	300	350	400	450	500	600
Minimum Burst	Aluminum	45	40	30	25	20	15	15	10	—	—	—	—	—	—	—	—
		—	**2.8**	**2.1**	**1.7**	**1.4**	**1.0**	**1.0**	**0.7**	—	—	—	—	—	—	—	—
	SST	600	400	350	300	200	150	125	100	75	60	50	43	85	75	65	55
		—	**27.6**	**24.2**	**20.7**	**13.8**	**10.4**	**8.6**	**6.9**	**5.2**	**4.1**	**3.4**	**3.0**	**5.9**	**5.2**	**4.5**	**3.8**
	Inconel	350	300	250	200	100	80	65	60	55	44	37	32	65	55	50	45
		—	**20.7**	**17.3**	**13.8**	**6.9**	**5.5**	**4.5**	**4.1**	**3.8**	**3.0**	**2.6**	**2.2**	**4.5**	**3.8**	**3.4**	**3.1**
	Monel	350	300	250	200	100	80	65	60	55	44	37	32	65	55	50	45
		—	**20.7**	**17.3**	**13.8**	**6.9**	**5.5**	**4.5**	**4.1**	**3.8**	**3.0**	**2.6**	**2.2**	**4.5**	**3.8**	**3.4**	**3.1**
	Nickel	250	200	150	100	50	45	40	35	30	24	20	17	30	25	25	20
		—	**13.8**	**10.4**	**6.9**	**3.4**	**3.1**	**2.8**	**2.4**	**2.1**	**1.7**	**1.4**	**1.2**	**2.1**	**1.7**	**1.7**	**1.4**
	Silver	95	90	85	60	30	25	25	20	—	—	—	—	—	—	—	—
		—	**6.2**	**5.9**	**4.1**	**2.1**	**1.7**	**1.7**	**1.4**	—	—	—	—	—	—	—	—
	Tantalum	250	200	140	120	60	45	40	35	45	—	—	—	—	—	—	—
		—	**13.8**	**9.7**	**8.3**	**4.1**	**3.1**	**2.8**	**2.4**	**3.1**	—	—	—	—	—	—	—
Minimum Burst To Withstand Full Vacuum	Aluminum	90	80	60	50	40	40	40	40	—	—	—	—	—	—	—	—
		—	**5.5**	**4.1**	**3.4**	**2.8**	**2.8**	**2.8**	**2.8**	—	—	—	—	—	—	—	—
	SSt	1150	735	565	385	325	255	235	185	115	115	110	95	95	95	95	95
		—	**50.7**	**39.0**	**26.6**	**22.4**	**17.6**	**16.2**	**12.8**	**7.9**	**7.9**	**7.6**	**6.6**	**6.6**	**6.6**	**6.6**	**6.6**
	Inconel	865	550	425	290	245	195	175	135	110	95	90	90	90	90	90	90
		—	**37.9**	**29.3**	**20.0**	**16.9**	**13.4**	**12.1**	**9.3**	**7.6**	**6.6**	**6.2**	**6.2**	**6.2**	**6.2**	**6.2**	**6.2**
	Monel	865	550	425	290	245	195	175	135	110	95	90	90	90	90	90	90
		—	**37.9**	**29.3**	**20.0**	**16.9**	**13.4**	**12.1**	**9.3**	**7.6**	**6.6**	**6.2**	**6.2**	**6.2**	**6.2**	**6.2**	**6.2**
	Nickel	375	250	195	135	80	75	70	70	70	70	70	70	70	70	70	70
		—	**17.2**	**13.4**	**9.3**	**5.5**	**5.2**	**4.8**	**4.8**	**4.8**	**4.8**	**4.89**	**4.8**	**4.8**	**4.8**	**4.8**	**4.8**
	Silver	185	125	95	75	50	50	50	50	—	—	—	—	—	—	—	—
		—	**8.6**	**6.6**	**5.2**	**3.4**	**3.4**	**3.4**	**3.4**	—	—	—	—	—	—	—	—
	Tantalum	360	245	180	130	75	75	75	75	50	—	—	—	—	—	—	—
		—	**16.9**	**12.4**	**9.0**	**5.2**	**5.2**	**5.2**	**5.2**	**3.4**	—	—	—	—	—	—	—

Note 1: Burst pressures printed in light are PSIG, bold is BAR G.
Note 2: Consult Fike for information on discs larger than 24 inches.

Source: Fike Metal Products.

The Scored Rupture Disc

A unique way of effectively controlling fragmentation is through the use of the scored disc. (See Figure 12.) This disc is approximately three times thicker than the conventional disc and is scored in a specific pattern to some predetermined depth. When the disc bursts, opening occurs along the path of the score line as shown in Figure 13.

In general, the minimum burst pressures are nearly the same as those of the conventional rupture disc. Variations are due, in part, to the type of score pattern used, i.e., cross-score, double-hinge, or single-hinge pattern. A useful feature of this disc design, not offered by the conventional disc, is the capacity to withstand a full vacuum without the aid of a vacuum support. The extra metal thickness provides good compressive strength to resist back pressures.

This disc, like the other styles, is available in the same materials as the conventional disc. Some limitations occur, however, when materials, such as Teflon, are not easily scored. For applications involving a corrosive media, protective coatings of spray Teflon, polyvinylchloride, or epoxy may be applied to one or both sides of the disc. (Tables 8 and 9 provide information on scored rupture discs.)

The Reverse-Acting Disc

A major advancement in rupture disc technology came about when the disc was exposed to compression loading. Unlike the forward-acting design, the convex side of the reverse-acting disc is in contact with the process media. In compression, this disc can withstand only about 1/3 the

Table 9
Typical Scored Rupture Disc Manufacturing Ranges

Specified Rupture Pressure		Manufacturing Range %
PSIG @ 72° F	BAR G @ 22° C	
20-45	**1.4-3.1**	+ 14 to -7
46-90	**3.2-6.2**	+ 12 to -6
91-270	**6.3-18.6**	+ 10 to -5
271-500	**18.7-34.5**	+ 8 to -4
501-up	**34.6-up**	+ 6 to -3

Source: Fike Metal Products.

pressure of the forward-acting disc before it collapses. When reversal occurs, the disc snaps backward into a stationary knife blade cutting it open in equal sectors. (See Figure 14.)

It is important to note that not all reverse-acting rupture discs are the same. Unlike the forward-acting conventional disc, the burst and opening characteristics of the reverse-acting disc are sensitive to design. The type of seating arrangement, for instance, effects the minimum burst pressure or the burst tolerance, while the design of the knife blade can have a direct bearing on whether the disc fully opens or not. Each manufacturer applies his own technology, patented design features, and quality controls at the time the disc is built. It becomes the responsibility of the user, then, to match the requirements of his application with the features of the particular disc. (Tables 10 through 12 provide information on typical reverse-acting rupture discs.)

The unique features of this disc include:

- Lower burst pressures than any other type of rupture disc.
- This type of disc will withstand back pressures in excess of its burst pressure.
- The system pressure can generally be operated at up to 90% of the rated burst pressure without premature failure.
- All materials of construction common to other rupture disc styles, including teflon, are available. In addition, protective coatings of teflon, polyvinylchloride, or epoxy resins may be sprayed on the disc for corrosion protection.

Figure 14. The reverse-acting rupture disc before and after rupture (courtesy of Fike Metal Products).

Table 10
Typical Reverse-Acting Rupture Disc Burst Pressures at 72°F (22°C)

Size*	In.	1	1½	2	3	4	6	8	10	12	14	16	18	20	24
	DN	**25**	**40**	**50**	**80**	**100**	**150**	**200**	**250**	**300**	**350**	**400**	**450**	**500**	**600**
Material	Stainless Steel, Inconel, Monel, Nickel														
Minimum Burst Pressure For Full Opening**	PSIG	104	61	54	26	23	15	12	10	7	7	6	5	5	4
	BAR G	**7.2**	**4.2**	**3.7**	**1.8**	**1.6**	**1.0**	**0.8**	**0.7**	**0.5**	**0.5**	**0.4**	**0.3**	**0.3**	**0.3**
Maximum Burst Pressure Available	PSIG	1440	1000	1000	1000	1000	1000	250	200	160	135	120	85	70	45
	BAR G	**99.3**	**69.0**	**69.0**	**69.0**	**69.0**	**69.0**	**17.2**	**13.8**	**11.0**	**9.3**	**8.3**	**5.9**	**4.8**	**3.1**
Minimum Vessel Free Vapor Volume	Cu. Ft.	.4	.4	.435	.955	1.69	3.8	6.75	10.55	15.2	20.67	27	34.2	42.2	60.7
	Cu. M.	**.0113**	**.0113**	**.0123**	**.0270**	**.0479**	**.1076**	**.1911**	**.2987**	**.4304**	**.5853**	**.7646**	**.9684**	**1.195**	**1.719**
Maximum Relief Area	Sq. In.	.78	1.76	3.36	7.4	12.7	29	50	79	113	138	183	234	290	425
	Sq. Cm.	**5.03**	**11.36**	**21.68**	**47.74**	**81.94**	**187.1**	**322.6**	**509.7**	**729.1**	**890.4**	**1181**	**1510**	**1871**	**2742**

Source: Fike Metal Products.

Table 11
Typical Reverse-Acting Rupture Disc Standard Manufacturing Ranges

Specified Rupture Pressure		Mfg. Range %
PSIG @ 72° F	**BAR G @ 22° C**	@ 72° F/22° C
2-3	**.1-.2**	+ 50 to -25
4-6	**.3-.4**	+ 40 to -20
7-10	**.5-.7**	+ 30 to -15
11-15	**.8-1.0**	+ 20 to -10
16-25	**1.1-1.7**	+ 16 to - 8
26-45	**1.8-3.1**	+ 14 to - 7
46-90	**3.2-6.2**	+ 12 to - 6
91-270	**6.3-18.6**	+ 10 to - 5
271-500	**18.7-34.5**	+ 8 to - 4
501-up	**34.6-up**	+ 6 to - 3

Source: Fike Metal Products.

Table 12
Typical Reverse-Acting Rupture Disc Normal Burst Tolerance

Marked Rupture Pressure		Rupture Tolerance
PSIG @ 72° F	**BAR G @ 22° C**	% @ 72° F/22° C
4-5	**.28-.34**	+ 30 to -30
6-8	**.4-.6**	+ 20 to -20
9-12	**.6-.8**	+ 15 to -15
13-17	**.9-1.2**	+ 10 to -10
18-25	**1.2-1.7**	+ 7 to - 7
26-up	**1.8-up**	+ 5 to - 5

Source: Fike Metal Products.

Figure 15. The graphite rupture disc (courtesy of Fike Metal Products).

- This type of disc is ideal for use in isolating pressure relief valves, both from the process media on the inlet and downstream corrosion on the outlet.

Graphite Series Rupture Discs

Graphite rupture discs are forward-acting, nonmetallic rupture discs machined from graphite "logs" then impregnated with resins to make them impervious to most process medias. (See Figure 15.) Generally very fragile, they are suitable for use at standard operating temperatures of −400°F to +300°F and burst pressures from 15 to 300 psig.

The graphite disc is a self-contained device and requires no additional holder. Upon rupture, the membrane portion of the disc fragments completely providing a relief area equivalent to the venting pipe area. For special applications, this disc may be manufactured with vacuum supports, Teflon coating for corrosion protection, or thermal protection for temperatures higher than 300°F. (See Figure 16. Tables 13 through 15 provide information on typical graphite rupture discs.)

Figure 16. Graphite rupture disc with vacuum support (courtesy of Fike Metal Products).

Table 13
Typical Graphite Rupture Disc Dimensions.

	Nominal Burst Diameter	Disc Dimensions A In./Cm.	B In./Cm.	C In./Cm.	C + D In./Cm.
150 ANSI Flange Mounting GD and GDHT Rupture Discs available in the following standard burst pressures[1]: 10-15-20-25-30-40-50-75-100-150 PSIG GDV Rupture Discs available in the following standard burst pressures: 10-15 PSIG	1/25	2.50/6.35	1/2.54	0.88/2.22	2.13/5.40
	1.5/40	3.25/8.26	1.50/3.81	0.88/2.22	2.13/5.40
	2/50	4/10.16	2/5.08	0.88/2.22	2.13/5.40
	3/80	5.25/13.34	3/7.62	0.88/2.22	2.13/5.40
	4/100	6.75/17.15	4/10.16	0.88/2.22	2.13/5.40
	6/150	8.63/21.91	6/15.24	0.88/2.22	2.13/5.40
	8/200	10.88/27.62	8/20.32	1.13/2.86	3.13/7.94
	10/250	13.25/33.66	10/25.40	1.50/3.81	3.75/9.53
	12/300	16/40.64	12/30.48	2/5.08	4.50/11.43
	14/350	17.63/44.77	13.25/33.66	2.25/5.72	5.13/13.02
	16/400	20.13/51.12	15.25/38.74	2.50/6.35	6.38/16.19
300 ANSI Flange Mounting GD Rupture Discs available in the following standard burst pressures: 175-200-225-250-275-300 PSIG	1/25	2.75/6.99	1/2.54	1/2.54	2.25/5.72
	1.5/40	3.63/9.21	1.50/3.81	1/2.54	2.25/5.72
	2/50	4.25/10.80	2/5.08	1/2.54	2.25/5.72
	3/80	5.75/14.61	3/7.62	1.25/3.18	2.50/6.35
	4/100	7/17.78	4/10.16	1.25/3.18	2.50/6.35
	6/150	9.75/24.77	6/15.24	1.75/4.45	3/7.62
	8/200	12/30.48	8/20.32	2.25/5.72	4.25/10.80

Source: Fike Metal Products.

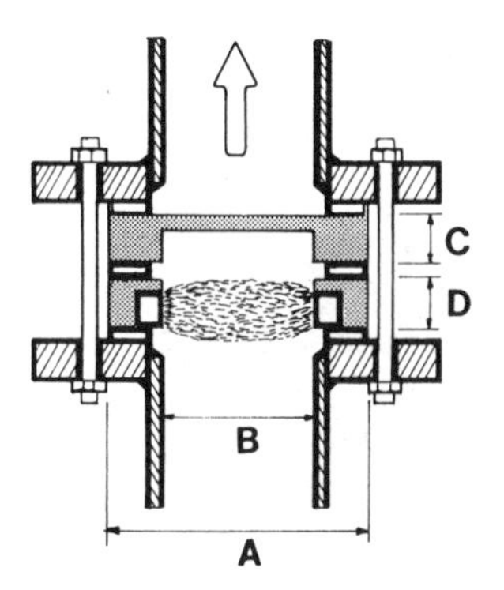

Special Engineered Rupture Disc Applications

Most special rupture disc applications involve some sort of welding process. For these cases electron beam welding and laser welding do a superb job of preventing damage to the disc from the heat of welding. They also provide a means of welding extremely small components. Until the these forms of welding were used for this purpose, the manufacture of small exotic rupture disc

Table 14
Typical Relief Areas for Graphite Rupture Discs

Disc Size	In.	1	1.5	2	3	4	6	8	10	12	14	16
	DN	**25**	**40**	**50**	**80**	**100**	**150**	**200**	**250**	**300**	**350**	**400**
Relief Area GD Discs	$In.^2$	.79	1.77	3.14	7.07	12.6	28.3	50.2	78.5	113.0	137.8	182.6
	$Cm.^2$	**5.10**	**11.42**	**20.26**	**45.62**	**81.3**	**182.6**	**323.9**	**506 5**	**729.1**	**889.1**	**1178.1**
Relief Area GDV Discs	$In.^2$	.79	1.39	2.14	5.57	8.82	17.3	31.8	50.8	69.0	83.3	113
	$Cm.^2$	**5.10**	**8.97**	**13.81**	**35.94**	**56.91**	**111.6**	**205.2**	**327.8**	**445.2**	**537.5**	**729.1**

Source: Fike Metal Products.

Table 15
Typical Minimum Burst Pressures and Tolerances for Graphite Rupture Discs

Nominal Burst Diameter In./**DN**	Minimum Burst Pressure Available PSIG/**BAR G**	Tolerance PSIG/**BAR G**
1/**25**	25/**1.7**	± 1.25/**.09**
1.5/**40**	15/**1.0**	± 1.00/**.07**
2/**50**	10/**.69**	± 1.00/**.07**
3/**80**	7/**.48**	± 0.75/**.05**
4/**100**	5/**.35**	± 0.75/**.05**
6/**150**	3/**.21**	± 0.75/**.05**
8/**200**	1/**.07**	± 0.75/**.05**
10/**250**	1/**.07**	± 0.75/**.05**
12/**300**	1/**.07**	± 0.75/**.05**
14/**350**	1/**.07**	± 0.75/**.05**
16/**400**	1/**.07**	± 0.75/**.05**

Source: Fike Metal Products.

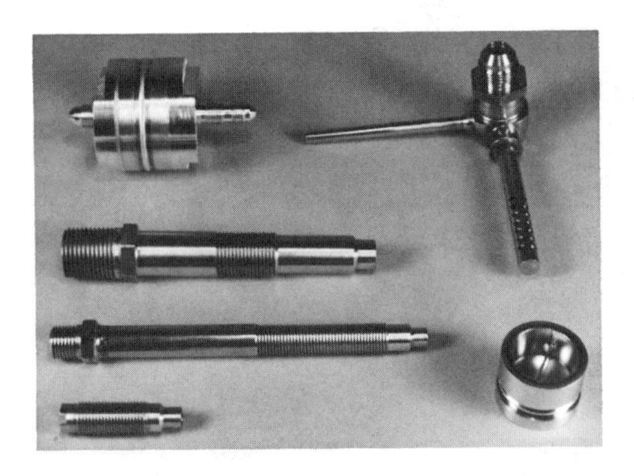

Figure 17. Typical special application rupture disc products (courtesy of Fike Metal Products).

Figure 18. Electron-beam welded rupture disc assembly (courtesy of Fike Metal Products).

designs was simply not possible. Some examples of these products are shown in Figures 17 and 18. Many of these products are used in medical applications, on aircraft and missiles systems, and satellites.

RULES GOVERNING THE USE OF RUPTURE DISCS

The ASME Boiler and Pressure Vessel Code has established rules for the use of rupture discs for overpressure protection. The requirements, spelled out in Section VIII, Div 1 (para UG-125 through UG-134), define the limits of application. The following outline shows how the performance characteristics of rupture disc devices are determined and how they relate to the protection of ASME pressure vessels.

Rupture Discs as Primary Relief Devices

Where a rupture disc is used as the sole pressure relief device, it should be sized to prevent the pressure from rising more than 10% above the maximum allowable working pressure (MAWP) of the vessel. It must be set to operate at a pressure not exceeding the MAWP.

Rupture Discs as Secondary Relief Devices

The ASME Code allows rupture discs to be used in multiples or as a secondary to other pressure relief devices. In this case, when the required capacity is provided in more than one device, they should prevent the pressure from rising more than 16% above the MAWP. In addition, only one device need be set to operate at or below the MAWP. Any additional devices may be set to operate at a pressure no higher than 105% of the MAWP.

Additional Hazard Considerations

When an additional hazard can be created by exposure of a pressure vessel to unexpected sources of external heat such as fire, rupture discs may be used as supplemental pressure relief devices. In this case the set pressure of the device should not exceed 110% of the MAWP of the vessel and the pressure must be prevented from rising more than 21% above the MAWP.

When a rupture disc is primarily intended to protect a pressure vessel that may be subjected to external heat and the vessel has no permanent supply connection, the device should be sized to prevent the pressure from rising more than 20% above the MAWP. In this case the set pressure shall not exceed the MAWP of the vessel.

Rupture Discs in Combination with Pressure Relief Valves

To protect pressure relief valves from corrosion, the ASME Code permits rupture discs to be installed between the pressure relief valve and the pressure vessel. For these applications the space between the rupture disc and the pressure relief valve must have a pressure gauge, try cock, free vent, or suitable telltale indicator to detect a pressure buildup in the event the rupture disc leaks. A pressure buildup between the disc and relief valve may cause the effective relief pressure to rise above the system relief pressure setting.

With this application, the Code requires a combination capacity factor be used as a multiplier to make appropriate changes in the ASME-rated relieving capacity of the pressure relief valve. In lieu of certified lab testing to determine this value, the Code permits a nominal capacity factor of .8 to be used as the multiplier.

A rupture disc may also be installed on the outlet side of a pressure relief valve. This prevents corrosion from the atsmosphere or a common discharge line from reaching the valve internals.

Burst Pressure Certification

To meet the requirements of the ASME Code, the stamped bursting pressure must be the average of two burst tests conducted at the specified disc temperature. The average of the two tests must fall within the manufacturing design range, and the rupture disc must be guaranteed by its manufacturer to burst within ±5% of its stamped bursting pressure and coincident disc temperature.

Manufacturing Design Range

To understand how the stamped burst pressure is determined, it is necessary to understand the concept of "manufacturing design range." Simply defined, it is not practical to build a rupture disc to burst at an exact pressure. Variations in material properties, holder tolerances, and other factors place limitations on this form of metal working. Further, it would be an impossible task to maintain an inventory of rupture disc materials that would service every burst pressure, material, thickness, and diameter application. To cope with these limitations, and still provide rupture discs economically, the manufacturer endeavors to build the disc within a range of burst pressures that is acceptable to the end user. Hence the term "manufacturing design range." The stamped burst pressure must then fall within this manufacturing range. Finally, the manufacturing design range is not constant; low-pressure discs require a larger manufacturing range than smaller, higher pressure discs. This is because at low burst pressures, a manufacturing range of 6% can translate into a fraction of 1 psi.

Stamped Burst Pressure at a Coincident Disc Temperature

The ASME Code permits the stamped burst pressure at a coincident disc temperature to be established by one of the following three methods:

1. At least two sample rupture discs from each lot of rupture discs should be burst at the coincident temperature and one disc burst at room temperature. The stamped burst pressure should be the average of those tests performed at the coincident temperature. The average should not fall outside of the agreed upon manufacturing design range.
2. At least four sample rupture discs, but not less than 5% from each lot, should be burst at four different temperatures, distriubted over the applicable temperature range for which the discs will be used. These data should be used to establish a curve of bursting pressure versus temperature for the lot of discs. The stamped rating at the coincident disc temperature may be interpolated from this curve. (See Figure 19.)
3. For pre-bulged, solid metal discs only, a curve of percentage ratio over a range of temperatures other than ambient, may be established for a particular lot of material. At least two discs, from each lot of discs, made from this lot of material and of the same size as those used in the temperature tests should be burst at ambient temperature to establish the room temperature rating of the lot of material.

The percent change of bursting pressure taken from the curve should be used to establish the stamped rating at the coincident disc temperature for the lot of discs. The stamped burst pressure should not fall outside of the agreed-upon manufacturing design range. (See Figure 20.)

The Code requires the manufacturer to guarantee the rupture disc to burst within ±5% of its stamped bursting pressure at the coincident disc temperature. This means that the burst tests performed in accordance with the preceding three methods must not vary from their average by more than ±5% regardless of the manufacturing design range. This tolerance is referred to as the rupture tolerance. When the Code requires that the set pressure for a primary device shall not exceed the MAWP, it refers to the stamped burst pressure. The ±5% rupture tolerance is permitted to exceed the MAWP if the stamped burst pressure is at the MAWP. The diagrams in Figures 21 and 22 will help to clear up any confusion.

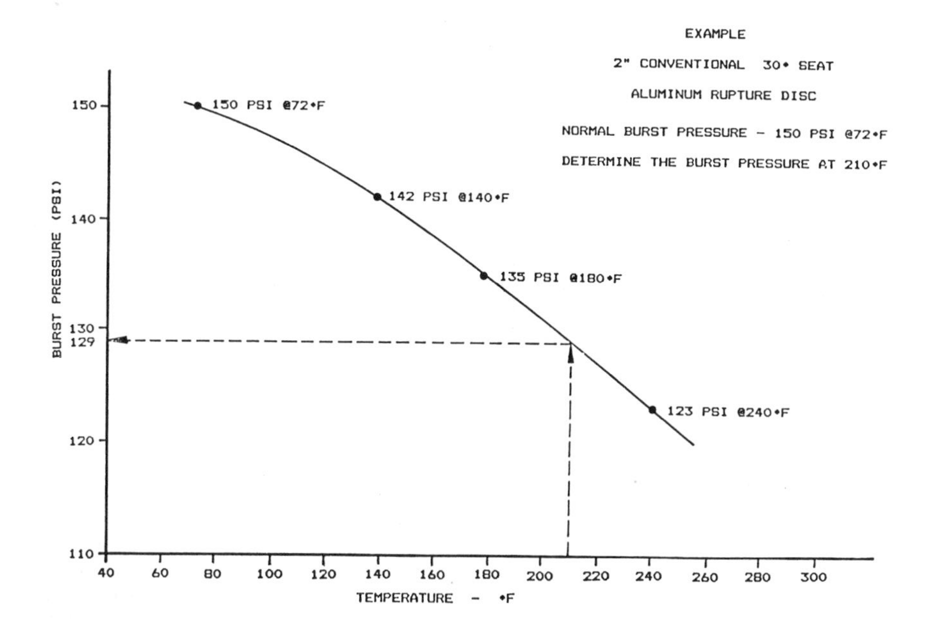

Figure 19. Stamped burst pressure at the coincident disc temperature obtained by Method 2.

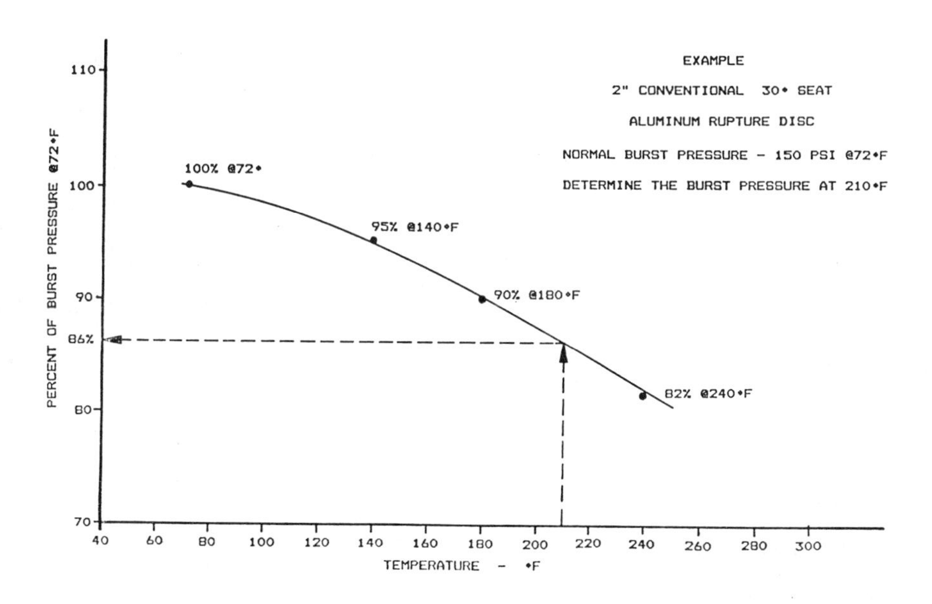

Figure 20. Stamped burst pressure at the coincident disc temperature obtained by Method 3.

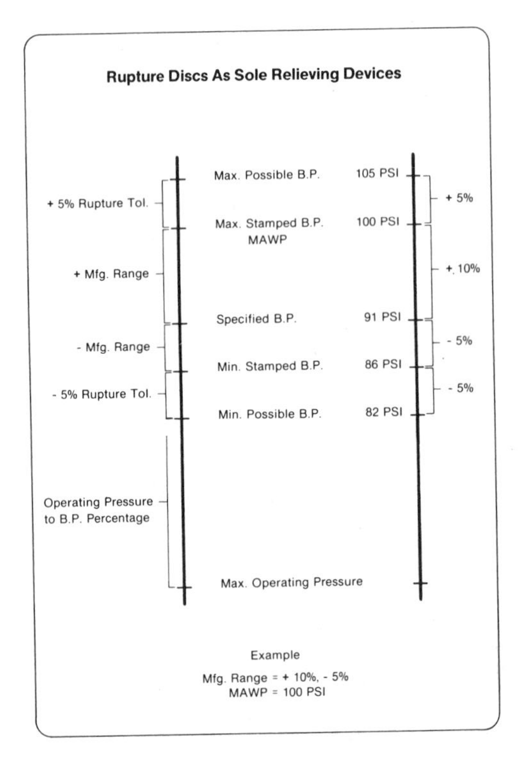

Figure 21. Example showing the relationship between typical burst pressure tolerances and the maximum allowable working pressure (MAWP) for rupture discs when they are used as sole pressure-relieving devices.

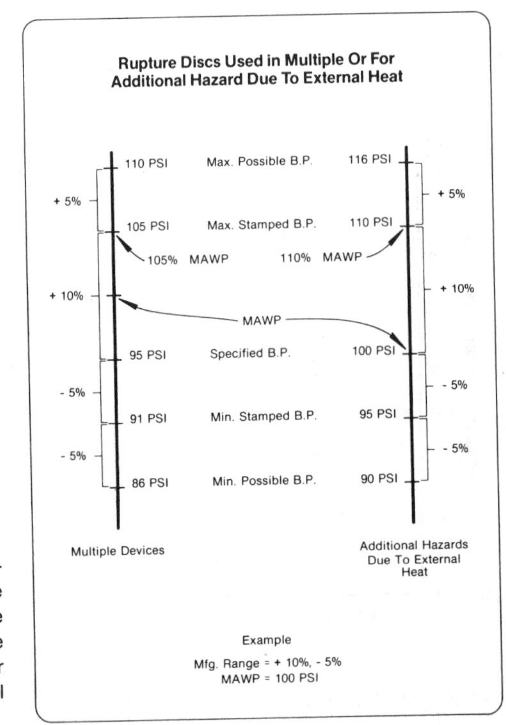

Figure 22. Example showing the relationship between typical burst pressure tolerances and the maximum allowable working pressure (MAWP) for rupture discs when they are used in multiples or for additional hazards due to external heat.

The same line of reasoning applies to devices used in multiples or for additional hazards created by external heat sources. All of these applications relate to the maximum set pressures as allowed by the Code.

SPECIFYING THE RUPTURE DISC

The following information is intended as a guide to the specifying engineer for the selection of rupture discs. The summaries are general but provide the minimum information required by the manufacturer to provide the proper disc for the application.

Proper Disc Material

The engineer should select the proper disc material to withstand the corrosive properties of the process media. In most cases the design engineer will know what materials must be used. If there is doubt about which rupture disc material is right for the process, the manufacturer can assist. Materials with special corrosion-resistant properties such as titanium, tantalum, gold, platinum, and zirconium may also be considered for special process applications.

Coincident Burst Temperature

The engineer must know the system application. It is common practice for engineers to specify a rupture disc that has a bursting pressure and temperature the same as the MAWP of the system. What he may not know, however, is that frequently the disc will "see" temperatures far lower than this. If the disc is placed a significant distance from a pressure vessel, the coincident disc temperature at burst may not be above ambient temperature. So, *it is important that the actual coincident burst temperature be known or accurately predicted.*

Operating Pressure Ratio

The operating pressure ratio must be identified. If the operating pressure will be above 70% of the rated rupture disc burst pressure, this information must be known before the disc is built. The rupture disc manufacturer pre-bulges forward-acting discs to 70% of the rated burst pressure as a standard manufacturing process. Further bulging by the system pressure after installation can result in damage to the disc if it is exposed to a back pressure.

Vacuum/Back Pressure

The engineer should know if the disc will be exposed to a vacuum or back-pressure. When process conditions include a vacuum or back-pressure, a mechanical vacuum support may be required.

Cyclic and Pulsing Pressure Duty

Cyclic pressure duty the disc may see should be identified. As a general rule, cyclic pressures tend to shorten the service life of a rupture disc. In most cases steps can be taken during manufacturing to counter some of these effects.

Polymerization Processes

Polymers tend to coat the pressure side of the disc and may cause a significant rise in burst pressure. The engineer should find out what processes will be used. Manufacturers can work around many problems if they know the characteristics of the process media being relieved.

Fragmentation

In many cases, fragmentation of the rupture disc can plug valves, cause damage and injury and, on occasion, destroy equipment. If the application cannot tolerate the possibility of fragmentation, the manufacturer can supply a disc that predetermines the rupture pattern. In any case, it is good practice to *include the fragmentation requirements* when specifying the rupture disc.

Rupture Disc Specification Worksheet

RUPTURE DISC DEVICE SPECIFICATION SHEET			Data Sheet of		Date		
GENERAL							
1	Tag Number						1
2	Code						2
3	Purpose						3
4	Line Size & Rating						4
5	M.A.W.P. (Vessel Rating)						5
DESIGN CONDITION							
6	Fluid						6
7	Capacity						7
8	MW/SG						8
9	Maximum Operating Pressure						9
10	Coincident Temperature						10
11	Required Relief Area						11
12	Vacuum Service						12
13	Pulsating Service						13
HOLDER SPECIFICATION							
14	Nominal Size						14
15	Rating						15
16	Holder Type—Letter Designation						16
17	Material	Base					17
18		Holddown					18
19		3rd. Comp.					19
20	Accessories						20
21							21
22							22
23							23
24							24
RUPTURE DISC SPECIFICATION							
25	Nominal Size						25
26	Type						26
27	Drawing Number						27
28	Specified						28
29	Burst Pressure						29
30	Manufacturing						30
31	Range						31
32	Rupture Tolerance						32
33	Specified Coincident Temp.						33
34	Material	Top/Cover/Ring					34
35		Seal					35
36		Vacuum Support					36
37	ASME Burst Certifications						37
38							38
39							39
40							40
41							41
42							42

Fike
Metal Products Corp.
704 South 10th Street • P.O. Box 610 • Blue Springs, Missouri 64015 U.S.A
(816) 229-3405 • Telex No. 42263
Printed in U.S.A.

Figure 23. Typical rupture disc specification sheet.

Hydraulic Service

All forward-acting discs are suitable for hydraulic (liquids) service without further considerations. Reverse-acting discs, however, require that a volume of compressible vapor be present in the system to carry the disc through the knife blades. Normally, reverse-acting discs are not recommended for hydraulic service except in special applications.

Burst Pressure/Size/Type of Disc

Always confirm that the size and type disc selected can be provided in the requested burst pressure. It may be necessary to increase the disc size or select an alternate style or material to achieve the required burst pressure.

Correct disc sizing is dependent on using the correct properties of the flowing media. In most cases, the disc manufacturer will have access to the properties of the process media. If the application involves a specialized media, it is necessary that these properties be used to size the disc.

Manufacturing Range and Burst Tolerance

It it important, from the cost and manufacturing standpoint, that as wide a manufacturing range and burst tolerance be specified as possible. In most cases, delivery time and costs can be reduced if it is practical and safe to use a rupture disc with standard manufacturing ranges and tolerances. A typical specification worksheet is shown in Figure 23.

RUPTURE DISC HOLDERS AND HOLDER ACCESSORIES

The typical rupture disc holder is a mechanical two- or three-piece assembly specifically designed to hold the rupture disc. A machined seat between the mating base flange (inlet) and holddown flange (outlet), grips the disc and provides a leak-tight seal to prevent the loss of the process media.

In general, there are three standard types of rupture disc holders used in plant piping systems: bolted type, union type, and screw type.

The Bolted-Type Assembly

Bolted-type rupture disc holders are the most versatile and widely used holders of all three types. They are designed to be used with, or in place of, all ANSI B16.5 flanges, up through 24-inch nominal diameter and larger ANSI and API flanges. The bolted-type holder can be incorporated into a pressure system by welding, threaded connection, or inserted between companion flanges.

The "insert" holder is the most common of all the bolted-type holders. Installed within the bolt circle of the companion flanges, it is smaller and more easily maintained than the other designs. Most "insert" holders are furnished with preassembly screws or side clips to permit assembly on the workbench. Once the disc is in place, the holder may then be installed in the piping system, minimizing the chance of damage to the disc.

Maximum nonshock service pressure ratings generally follow the maximum flange pressure ratings given in the appropriate ANSI and API flange standards. This type of holder is available in most materials including carbon steel, stainless steel, nickel, Monel, Hasteloy, Inconel, brass, and aluminum. (See Table 16 for additional information.)

Union-Type Holders

The union-type rupture disc holder is adaptable to pressure piping systems where space is at a premium and cost is a consideration. (See Figure 24.) They are compact and easier to maintain

Table 16
Typical Dimensions and Configurations for Bolted-Type Holders

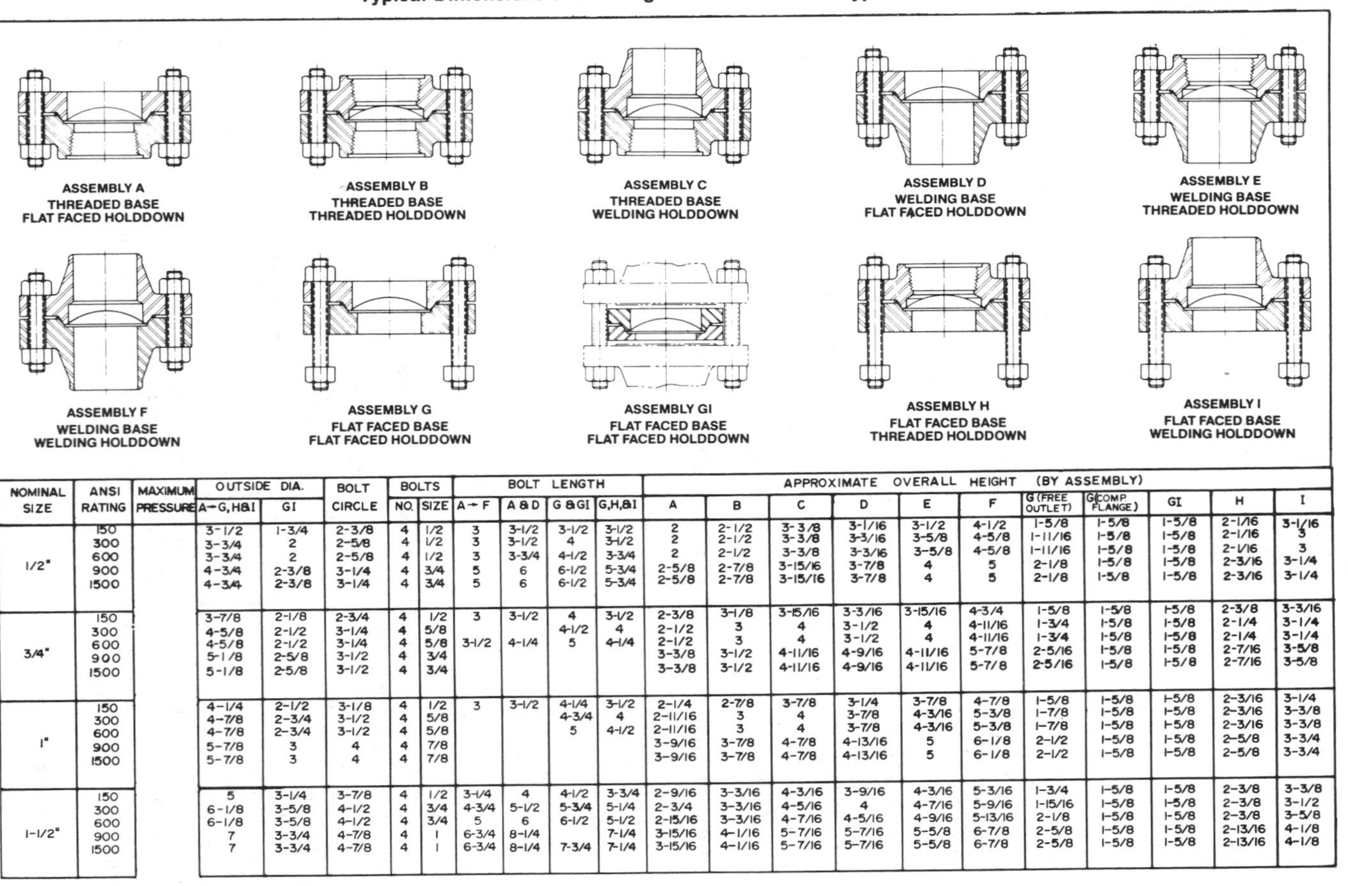

NOMINAL SIZE	ANSI RATING	MAXIMUM PRESSURE	OUTSIDE DIA.		BOLT CIRCLE	BOLTS		BOLT LENGTH				APPROXIMATE OVERALL HEIGHT (BY ASSEMBLY)										
			A–G, H&I	GI		NO.	SIZE	A → F	A & D	G & GI	G, H, & I	A	B	C	D	E	F	G (FREE OUTLET)	G (COMP. FLANGE)	GI	H	I
1/2"	150		3-1/2	1-3/4	2-3/8	4	1/2	3	3-1/2	3-1/2	3-1/2	2	2-1/2	3-3/8	3-1/16	3-1/2	4-1/2	1-5/8	1-5/8	1-5/8	2-1/16	3-1/16
	300		3-3/4	2	2-5/8	4	1/2	3	3-1/2	4	3-1/2	2	2-1/2	3-3/8	3-3/16	3-5/8	4-5/8	1-11/16	1-5/8	1-5/8	2-1/16	3
	600		3-3/4	2	2-5/8	4	1/2	3	3-3/4	4-1/2	3-3/4	2	2-1/2	3-3/8	3-3/16	3-5/8	4-5/8	1-11/16	1-5/8	1-5/8	2-1/16	3
	900		4-3/4	2-3/8	3-1/4	4	3/4	5	6	6-1/2	5-3/4	2-5/8	2-7/8	3-15/16	3-7/8	4	5	2-1/8	1-5/8	1-5/8	2-3/16	3-1/4
	1500		4-3/4	2-3/8	3-1/4	4	3/4	5	6	6-1/2	5-3/4	2-5/8	2-7/8	3-15/16	3-7/8	4	5	2-1/8	1-5/8	1-5/8	2-3/16	3-1/4
3/4"	150		3-7/8	2-1/8	2-3/4	4	1/2	3	3-1/2	4	3-1/2	2-3/8	3-1/8	3-15/16	3-3/16	3-15/16	4-3/4	1-5/8	1-5/8	1-5/8	2-3/8	3-3/16
	300		4-5/8	2-1/2	3-1/4	4	5/8			4-1/2	4	2-1/2	3	4	3-1/2	4	4-11/16	1-3/4	1-5/8	1-5/8	2-1/4	3-1/4
	600		4-5/8	2-1/2	3-1/4	4	5/8	3-1/2	4-1/4	5	4-1/4	2-1/2	3	4	3-1/2	4	4-11/16	1-3/4	1-5/8	1-5/8	2-1/4	3-1/4
	900		5-1/8	2-5/8	3-1/2	4	3/4					3-3/8	3-1/2	4-11/16	4-9/16	4-11/16	5-7/8	2-5/16	1-5/8	1-5/8	2-7/16	3-5/8
	1500		5-1/8	2-5/8	3-1/2	4	3/4					3-3/8	3-1/2	4-11/16	4-9/16	4-11/16	5-7/8	2-5/16	1-5/8	1-5/8	2-7/16	3-5/8
1"	150		4-1/4	2-1/2	3-1/8	4	1/2	3	3-1/2	4-1/4	3-1/2	2-1/4	2-7/8	3-7/8	3-1/4	3-7/8	4-7/8	1-5/8	1-5/8	1-5/8	2-3/16	3-1/4
	300		4-7/8	2-3/4	3-1/2	4	5/8			4-3/4	4	2-11/16	3	4	3-7/8	4-3/16	5-3/8	1-7/8	1-5/8	1-5/8	2-3/16	3-3/8
	600		4-7/8	2-3/4	3-1/2	4	5/8			5	4-1/2	2-11/16	3	4	3-7/8	4-3/16	5-3/8	1-7/8	1-5/8	1-5/8	2-3/16	3-3/8
	900		5-7/8	3	4	4	7/8					3-9/16	3-7/8	4-7/8	4-13/16	5	6-1/8	2-1/2	1-5/8	1-5/8	2-5/8	3-3/4
	1500		5-7/8	3	4	4	7/8					3-9/16	3-7/8	4-7/8	4-13/16	5	6-1/8	2-1/2	1-5/8	1-5/8	2-5/8	3-3/4
1-1/2"	150		5	3-1/4	3-7/8	4	1/2	3-1/4	4	4-1/2	3-3/4	2-9/16	3-3/16	4-3/16	3-9/16	4-3/16	5-3/16	1-3/4	1-5/8	1-5/8	2-3/8	3-3/8
	300		6-1/8	3-5/8	4-1/2	4	3/4	4-3/4	5-1/2	5-3/4	5-1/4	2-3/4	3-3/16	4-5/16	4	4-7/16	5-9/16	1-15/16	1-5/8	1-5/8	2-3/8	3-1/2
	600		6-1/8	3-5/8	4-1/2	4	3/4	5	6	6-1/2	5-1/2	2-15/16	3-3/16	4-7/16	4-5/16	4-9/16	5-13/16	2-1/8	1-5/8	1-5/8	2-3/8	3-5/8
	900		7	3-3/4	4-7/8	4	1	6-3/4	8-1/4		7-1/4	3-15/16	4-1/16	5-7/16	5-7/16	5-5/8	6-7/8	2-5/8	1-5/8	1-5/8	2-13/16	4-1/8
	1500		7	3-3/4	4-7/8	4	1	6-3/4	8-1/4	7-3/4	7-1/4	3-15/16	4-1/16	5-7/16	5-7/16	5-5/8	6-7/8	2-5/8	1-5/8	1-5/8	2-13/16	4-1/8

2"	150	SEE ANSI B16.5—1981 TABLE 2	6	4	4-3/4	4	5/8	4	4-3/4	4-3/4	4	2-9/16	2-3/4	4-1/8	3-15/16	4-1/8	5-7/16	1-7/8	1-5/8	1-5/8	2-1/4	3-3/8
	300		6-1/2	4-1/4	5	8	5/8	4-1/4	5	5	4-1/2	3-1/8	3-1/4	4-9/16	4-7/16	4-5/8	5-7/8	2-1/8	1-5/8	1-5/8	2-1/4	3-9/16
	600		6-1/2	4-1/4	5	8	5/8	4-1/2	5-3/4	5-3/4	5	3-3/8	3-3/8	4-7/8	4-13/16	4-13/16	6-1/4	2-5/16	1-5/8	1-5/8	2-5/16	3-3/4
	900		8-1/2	5-1/2	6-1/2	8	7/8	7	8-3/4	8	7-1/2	4-7/16	4	6-5/16	6-5/8	6-3/16	8-1/2	3	1-5/8	1-5/8	2-9/16	4-15/16
	1500		8-1/2	5-1/2	6-1/2	8	7/8	7	8-3/4	8	7-1/2	4-7/16	4	6-5/16	6-5/8	6-3/16	8-1/2	3	1-5/8	1-5/8	2-9/16	4-15/16
3"	150		7-1/2	5-1/4	6	4	5/8	4-1/4	5-1/4	5-1/4	4-3/4	3-1/8	3-1/4	4-9/16	4-7/16	4-9/16	5-7/8	2-1/8	1-5/8	1-5/8	2-1/4	3-9/16
	300		8-1/4	5-3/4	6-5/8	8	3/4			6-1/2		3-5/8	3-3/4	5-3/16	5-1/16	5-3/16	6-5/8	2-3/8	1-5/8	1-5/8	2-1/2	3-15/16
	600		8-1/4	5-3/4	6-5/8	8	3/4			7-1/4	6-1/2	4-1/16	4-1/8	5-9/16	5-1/2	5-9/16	7	2-5/8	1-5/8	1-5/8	2-11/16	4-1/8
	900		9-1/2	6-1/2	7-1/2	8	7/8			7-3/4					6-5/8		8-1/2	3	1-5/8	1-5/8		4-7/8
	1500		10-1/2	6-3/4	8	8	1-1/8	8-1/4	10-1/2		9				7-3/4		9-3/4	3-3/4	1-7/8	2-1/8		5-3/4
4"	150		9	6-3/4	7-1/2	8	5/8	4-1/4	5-1/4	5-1/4	4-3/4	3-5/16	3-9/16	5	4-11/16	4-13/16	6-3/8	2-1/8	1-5/8	1-5/8	2-7/16	3-13/16
	300		10	7	7-7/8	8	3/4	5-3/4	7	6-1/2	6-1/4	4	4-1/8	5-5/8	5-1/2	5-5/8	7-1/8	2-9/16	1-5/8	1-5/8	2-11/16	4-3/16
	600		10-3/4	7-1/2	8-1/2	8	7/8			8	7-1/4	4-3/4	4-3/4	6-5/8	6-5/8	6-5/8	8-15/16	3	1-5/8	1-5/8	3	4-7/8
	900		11-1/2	8	9-1/4	8	1-1/8			9					7-7/16		9-1/2	3-5/16	1-5/8	1-5/8		5-3/8
	1500		12-1/4	8	9-1/2	8	1-1/4								8-3/8		10-1/4	4-1/8	2-1/8	2-1/8		6
6"	150		11	8-5/8	9-1/2	8	3/4	5-1/4	6-1/4	6-1/4	6	3-13/16	4-1/2	5-15/16	5-1/4	5-15/16	7-3/8	2-7/16	2-1/8	2-1/8	3-1/8	4-9/16
	300		12-1/2	9-3/4	10-5/8	12	3/4			7-1/2		4-7/16	4-1/2	6-5/16	6-1/4	6-5/16	8-1/8	3-1/16	2-1/8	2-1/8	3-1/8	4-15/16
	600		14	10-3/8	11-1/2	12	1	8	10-1/4		9	5-3/4	5-3/4	7-3/4	7-3/4	7-3/4	9-3/4	3-3/4	2-1/8	2-1/8	3-3/4	5-3/4
	900		15	11-1/4	12-1/2	12	1-1/8								9-1/16		11-5/8	4-1/16	2-1/8	3-1/2		6-5/8
8"	150		13-1/2	10-7/8	11-3/4	8	3/4								6-1/16		8-1/2	2-5/8	2-1/8	2-1/8		5-1/8
	300		15	12	13	12	7/8			8					7-1/4		9-1/4	3-3/8	2-1/8	2-1/8		5-7/16
	600		16-1/2	12-1/2	13-3/4	12	1-1/8								9		11-3/16	4-3/16	2-1/8	2-3/8		6-3/8
10"	150		16	13-1/4	14-1/4	12	7/8			7-1/4					6-1/8		8-1/2	2-11/16	2-1/8	2-1/8		5-1/16
	300		17-1/2	14-1/8	15-1/4	16	1	8-1/4	10-1/4	9	8-3/4				7-3/4		9-3/4	3-5/8	2-1/8	2-1/8		5-11/16
12"	150		19	16	17	12	7/8			7-1/4					6-15/16		9-1/2	2-7/8	2-1/8	2-1/8		5-9/16
	300		20-1/2	16-1/2	17-3/4	16	1-1/8	8-3/4	11		9-1/4				8-1/2		10-3/4	3-7/8	2-5/8	2-5/8		6-3/16
14"	150		21	17-5/8	18-3/4	12	1	7	8-1/2		7-3/4				7-1/2		10-1/2	3-1/4	2-7/8	2-7/8		6-5/16
	300		23	19	20-1/4	20	1-1/8			11-1/4					9-1/8		11-3/4	4-1/4	3-7/16	3-7/16		6-15/16
16"	150		23-1/2	20-1/8	21-1/4	16	1								7-9/16		10-1/2	3-3/8	2-7/8	2-7/8		6-5/16.
	300		25-1/2	21-1/8	22-1/2	20	1-1/4								9-1/2		12	4-1/2	3-9/16	3-9/16		7-1/16
18"	150		25	21-1/2	22-3/4	16	1-1/8			9-1/2					8-3/16		11-5/8	3-11/16	2-7/8	2-7/8		6-7/8
	300		28	23-3/8	24-3/4	24	1-1/4								10-1/8		13-1/8	5-1/16	3-3/4	3-11/16		7-5/8
20"	150		27-1/2	23-3/4	25	20	1-1/8			10					8-11/16		12	4	3-5/16	3-1/4		7-1/4
	300		30-1/2	25-1/2	27	24	1-1/4			12-3/4					10-1/2		13-3/8	5-1/16	4-1/8	4-1/16		8
24"	150		32	28-1/8	29-1/2	20	1-1/4			10-3/4					9		12-9/16	4-3/16	3-1/2	3-7/16		7-9/16
	300		36	30-3/8	32	24	1-1/2								11		13-7/8	5-7/16	4-3/8	4-5/16		8-1/4

Source: Fike Metal Products.

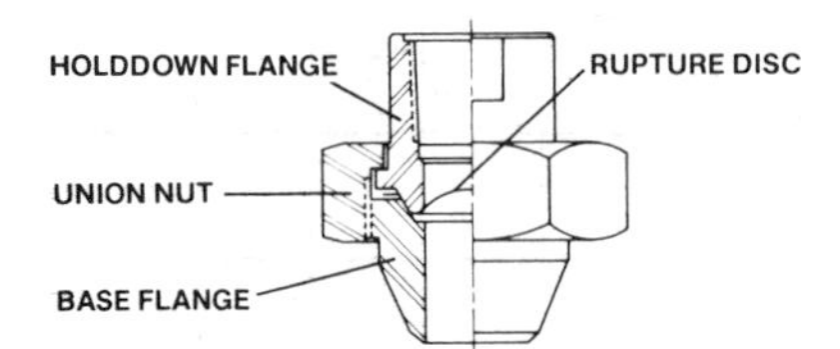

Figure 24. The union-type rupture disc holder (courtesy of Fike Metal Products).

than the bolted-type holder of comparable size. Union-type holders are confined to ½ through 2-inch nominal pipe sizes and the 30° conventional seating arrangement. This holder type is incorporated in the piping system by welding, threaded connection, or any combination of the two.

The union-type holder consists of three components: base (inlet), holddown (outlet), and union nut. Carbon steel is the standard material of construction but other materials are readily available. Typical maximum service pressure ratings, configurations, and dimensions are shown in Table 17.

Screw-Type Holders

Screw-type holders are adaptable to high-pressure equipment such as laboratory pressure equipment, autoclaves, and gas bottles. (See Figure 25.) Standard screw-type holders are generally confined to ½-inch nominal disc size and may use the 30° conventional seat or a "flat seat" arrangement . This holder is incorporated into the piping system exclusively by threaded connection.

The screw-type holder consists of a base (inlet), a holddown ring, and a holddown nut (outlet nut). The holddown ring is necessary to avoid scuffing or wrinkling the rupture disc as the outlet nut is tightened. Screw-type holders are typically provided with an NPT threaded inlet and a variety of outlet styles. They are available in reusable or throw-away configurations. Standard materials of construction include carbon steel, stainless steel, and brass, but other materials are readily available.

Holder Accessories

Many accessories are available for use with the rupture disc and holder. These accessories are often necessary to assist in the installation of the rupture disc, to install additional equipment such

Table 17
Typical Dimensions and Configurations for Union-Type Holders.

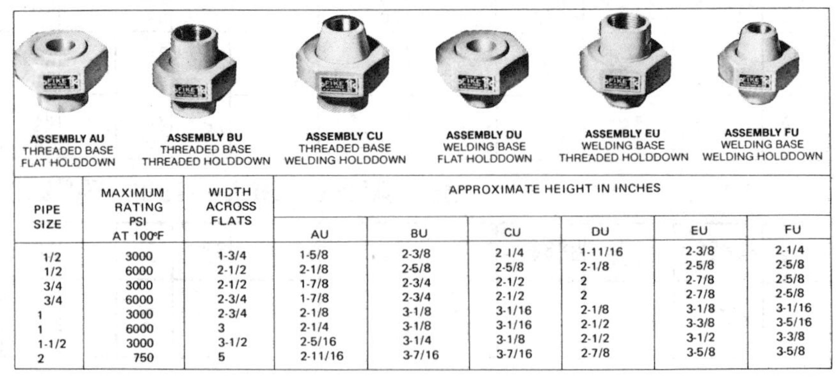

PIPE SIZE	MAXIMUM RATING PSI AT 100°F	WIDTH ACROSS FLATS	APPROXIMATE HEIGHT IN INCHES					
			AU	BU	CU	DU	EU	FU
1/2	3000	1-3/4	1-5/8	2-3/8	2 1/4	1-11/16	2-3/8	2-1/4
1/2	6000	2-1/2	2-1/8	2-5/8	2-5/8	2-1/8	2-5/8	2-5/8
3/4	3000	2-1/2	1-7/8	2-3/4	2-1/2	2	2-7/8	2-5/8
3/4	6000	2-3/4	1-7/8	2-3/4	2-1/2	2	2-7/8	2-5/8
1	3000	2-3/4	2-1/8	3-1/8	3-1/16	2-1/8	3-1/8	3-1/16
1	6000	3	2-1/4	3-1/8	3-1/16	2-1/2	3-3/8	3-5/16
1-1/2	3000	3-1/2	2-5/16	3-1/4	3-1/8	2-1/2	3-1/2	3-3/8
2	750	5	2-11/16	3-7/16	3-7/16	2-7/8	3-5/8	3-5/8

Source: Fike Metal Products.

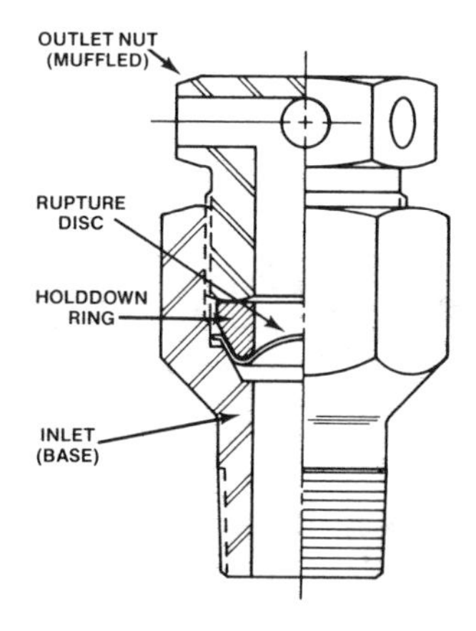

Figure 25. The screw-type rupture disc holder (courtesy of Fike Metal Products).

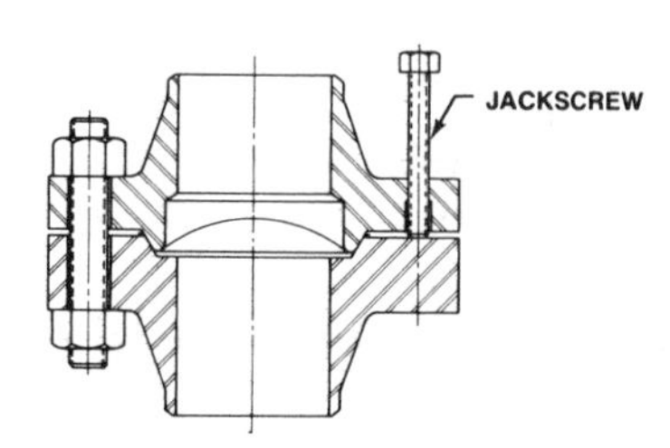

Figure 26. Typical jackscrew installation.

as pressure gages, or to help protect piping equipment. The following discussion is minimal but is included here to inform the reader of the availability of such products.

Jackscrews

Jackscrews provide a means of separating piping flanges to allow for safe and easy installation of fragile rupture discs. They are used only with bolted-type assemblies. (See Figure 26.)

Lifting Lugs

Lifting lugs are primarily used for the handling of large or heavy rupture disc holders. In most cases, the standard nominal $\frac{1}{2}$" lift lug will meet all application requirements, but other sizes may be specified. (See Figure 27.)

Baffle Plate

The baffle plate is used to redirect the discharge flow of the process media from free-vented rupture disc assemblies and to absorb the thrust forces during discharge. (See Figure 28.)

Excess Flow Valve

When rupture discs are used in series or in combination with pressure relief valves, an excess flow valve may be used to vent any slow pressure buildup between the components. When the rupture disc bursts, the overpressure pushes the ball into its seal preventing further flow through the excess flow valve. (See Figure 29.)

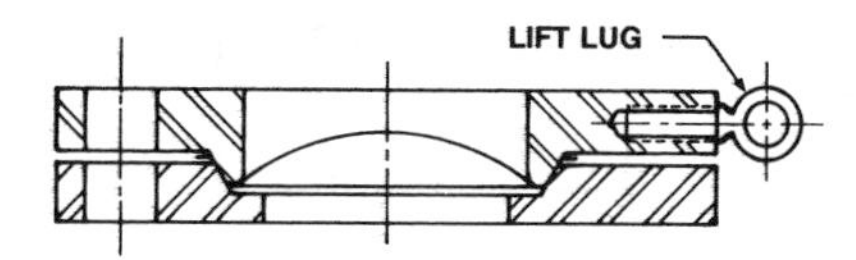

Figure 27. Typical lifting lug installation.

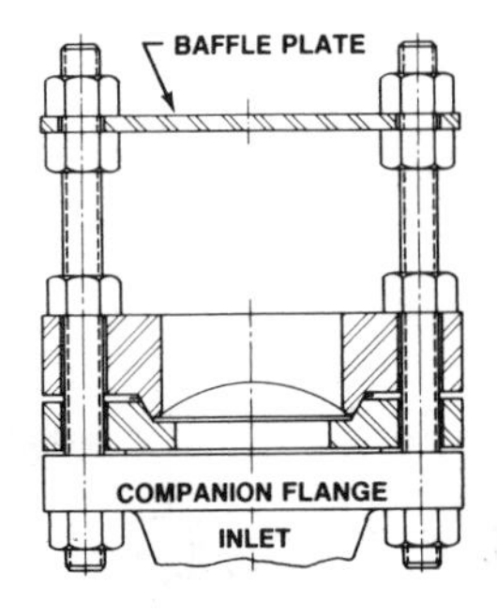

Figure 28. Typical baffle-plate installation.

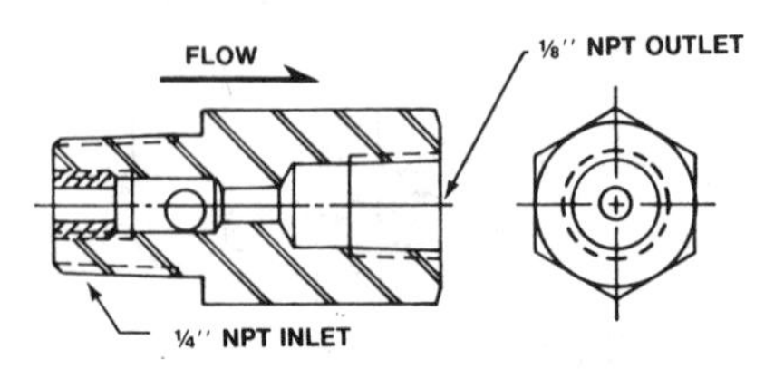

Figure 29. The excess flow valve.

RUPTURE DISC SIZING

Through the years, the rupture disc industry has found that the minimum required relief area can be achieved if the disc is treated as a flat plate orifice. Adequate flow through the disc can then be assured by applying classic momentum and general energy equations.

The following sizing procedures are consistent with Section III and VIII of the ASME Code but should be reviewed for compliance to local codes and laws.

Table 18
Typical Dimensions of Reusable Screw-Type Rupture Disc Holders

Maximum Pressure 3,000 PSIG @ 100° F.

½ 30 SA ½ 30 SB ½ 30 SM

Assembly No.	CONNECTIONS		HEX SIZE ACROSS FLATS		Overall Height
	Inlet	Outlet	A	B	
1/2-30SA	1/4 NPT	Free	1-1/8	1-1/4	2-5/16
1/2-30SA	1/2 NPT	Free	1-1/8	1-1/4	2-5/16
1/2-30SB	1/4 NPT	1/2 NPT	1-1/8	1-1/4	3-1/8
1/2-30SB	1/2 NPT	1/2 NPT	1-1/8	1-1/4	3-1/8
1/2-30SM	1/4 NPT	Muffled	1-1/8	1-1/4	2-1/2
1/2-30SM	1/2 NPT	Muffled	1-1/8	1-1/4	2-1/2

Source: Fike Metal Products.

Table 19
Typical Dimensions of Throw-Away Screw-Type Rupture Disc Holders

Maximum Pressure 1,000 PSIG @ 100° F.

LOC-TITE

S1/210 ASSEMBLIES

ASSEMBLY NUMBER	CONNECTIONS		HEX SIZE ACROSS FLATS	OVERALL HEIGHT
	INLET	OUTLET		
S½10	1/2 NPT	FREE	1-1/8	1-7/16
S½10A	1/4 NPT	FREE	1-1/8	1-7/16
S½10B	3/8 NPT	FREE	1-1/8	1-7/16
S½10C	1/2 NPT	1/2 NPT	1-1/8	2-3/16
S½10D	1/4 NPT	1/2 NPT	1-1/8	2-3/16
S½10E	3/8 NPT	1/2 NPT	1-1/8	2-3/16
S½10F	1/2 NPT	1/4 NPT	1-1/8	2-3/16
S½10G	1/4 NPT	1/4 NPT	1-1/8	2-3/16
S½10H	3/8 NPT	1/4 NPT	1-1/8	2-3/16
S½10K	1/2 NPT	3/8 NPT	1-1/8	2-3/16
S½10L	1/4 NPT	3/8 NPT	1-1/8	2-3/16
S½10M	3/8 NPT	3/8 NPT	1-1/8	2-3/16

Source: Fike Metal Products.

General Considerations

Before discussing the specific aspects of sizing the rupture disc, we must make some assumptions:

- For gases and vapors, the rupture disc assembly is treated as an isentropic nozzle.
- The rupture disc device is assumed to vent to the atmosphere with no downstream piping. If extended lengths of downsteam piping are present, it will be necessary to consider the device as a pipe fitting and size the complete system.
- Upstream flow velocity is negligible. The rupture disc device is assumed to be installed on a large vessel. Thus the velocity of the media through the disc is large compared to the velocity of flow in the vessel.
- The pressure is assumed to increase gradually such as may result from overfilling, thermal expansion, or faulty valving and pressure control equipment. Explosion relief and uncontrolled chemical reactions require special attention.
- Gases and superheated vapors are assumed to have constant specific heat and act as an ideal gas, while saturated and wet vapors are assumed to expand isentropically but are not treated as an ideal gas.
- Assume the coefficient $K = 0.62$ as permitted by the ASME Code. This accounts for entrance and exit losses and losses due to turbulence caused by disc fragments that may protrude into the flow stream.

Sizing for Liquids

The sizing equation for the flow of incompressible fluids through a rupture disc device is derived from Bernoulli's general energy equation and the conservation of momentum. Assuming the fluid enters the device with negligible velocity and with no change in elevation, the velocity at the disc is

$$\bar{V} = \sqrt{2gh_d}, \text{ ft/sec}$$

where $h_d = \frac{Pf}{\gamma}$, ft

From the conservation of momentum equation, the flow rate becomes

$$Q = A\bar{V}, \text{ cu ft/min}$$

$$= KA\sqrt{2g\frac{Pf}{\gamma}}$$

Simplifying and rearranging the equation, the required relief area for any free-flowing liquid through a rupture disc device is given by

$$a = \frac{L\sqrt{SG}}{23.1\sqrt{p}}, \text{ sq in.} \tag{1}$$

The pipe "bore" area at the inlet to the rupture disc device is usually taken as the maximum available relief area. This means that if the calculated relief area is greater than the area of the pipe bore, the pipe is the limiting factor. Otherwise, use the calculated relief area.

Sizing for Gases and Superheated Vapors

From the general energy equation and the conservation of momentum, we have the general equation of flow through an orifice.

$$W = \frac{A\sqrt{2_q(h_0 - h^*)}}{v^*}, \text{ lb/sec} \tag{2}$$

where * represents the conditions at the disc.

Rearranging and combining the ideal gas equation with Equation 2, we have the general rupture disc sizing equations for gases and vapors.

$$W = KP_0a\sqrt{\frac{2g}{RT_0}\left[\frac{k}{k-1}\right]\left[\left(\frac{P_e}{P_0}\right)^{2/k} - \left(\frac{P_e}{P_0}\right)^{k+1/k}\right]}, \text{ lb/sec} \tag{3}$$

$$Q_{act} = \frac{KP_0aRT_0}{2.4\,P_0}\sqrt{\frac{2g}{RT_0}\left[\frac{k}{k-1}\right]\left[\left(\frac{P_e}{P_0}\right)^{2/k} - \left(\frac{P_e}{P_0}\right)^{k+1/k}\right]}, \text{ ACFM} \tag{4}$$

$$Q_s = \frac{KP_0aR520}{2.4(14.7)}\sqrt{\frac{2g}{RT_0}\left[\frac{k}{k-1}\right]\left[\left(\frac{P_e}{P_0}\right)^{2/k} - \left(\frac{P_e}{P_0}\right)^{k+1/k}\right]}, \text{ SCFM} \tag{5}$$

where $\frac{P_e}{P_0} > \left[\frac{2}{k+1}\right]^{k/k-1}$ for subsonic flow

$\frac{P_e}{P_0} \leq \left[\frac{2}{k+1}\right]^{k/k-1}$ for sonic flow

Sizing for Subsonic Flow

When the actual pressure ratio P_e/P_0 is greater than the critical pressure ratio, the flow is *subsonic* and the actual pressure ratio is used in Equations 3 through 5.

The required area of a rupture disc device for any gas or vapor is:

Given the mass flow rate, W, in lb/sec

$$a = \frac{W}{KC_1P_0}\sqrt{\frac{t + 460}{M}}, \text{ sq in.} \tag{6}$$

Given the actual flow rate, Q_{act}, in ACFM

$$a = \frac{Q_{act}}{643.8KC_1}\sqrt{\frac{M}{t + 460}}, \text{ sq in.} \tag{7}$$

Given the flow at standard conditions, Q_s, in SCFM

$$a = \frac{Q_s}{22{,}772KC_1P_0}\sqrt{(t + 460)M}, \text{ sq in.} \tag{8}$$

Given the flow of air at standard conditions, Q_{sa}, in SCFM Air (K = 0.62)

$$a = \frac{Q_{sa}}{115C_1P_0}\sqrt{\frac{t + 460}{520}}\sqrt{\frac{29}{M}}, \text{ sq in.} \tag{9}$$

where $$C_1 = \sqrt{\frac{2g}{1545}\left[\frac{k}{k - 1}\right]\left[\left(\frac{P_e}{P_0}\right)^{2/k} - \left(\frac{P_e}{P_0}\right)^{k+1/k}\right]}$$

The value for C_1 is found in Table 20.

Sizing for Sonic Flow

When the actual pressure ratio P_e/P_0 is less than the critical pressure ratio, the flow is *sonic* and the critical pressure ratio is substituted for P_e/P_0 in Equations 2 through 4. Simplifying and rearranging we have:

Given the mass flow rate, W, in lb/sec

$$a = \frac{W}{KC_2P_0}\sqrt{\frac{t + 460}{M}} \tag{10}$$

Given the actual flow rate, Q_{act}, in ACFM

$$a = \frac{Q_{act}}{643.8KC_2}\sqrt{\frac{M}{t + 460}} \tag{11}$$

Given the flow at standard conditions, Q_s, in SCFM

$$a = \frac{Q_s}{22772KC_2P_0}\sqrt{(t + 460)M} \tag{12}$$

Table 20
Gas Flow Constant C_1 For Subsonic Flow

k ↓ \ P_e/P_o →	.95	.90	.85	.80	.75	.70	.65	.60	.55	.50	.45
1.05	.0440	.0598	.0703	.0777	.0829	.0864	.0884	.0891	.0887	.0871	.0844
1.10	.0441	.0600	.0707	.0783	.0837	.0874	.0897	.0906	.0904	.0891	.0867
1.15	.0441	.0602	.0711	.0788	.0844	.0884	.0908	.0920	.0920	.0909	.0888
1.20	.0442	.0604	.0714	.0793	.0851	.0892	.0919	.0933	.0935	.0927	.0907
1.25	.0442	.0606	.0717	.0798	.0857	.0900	.0929	.0945	.0950	.0943	.0926
1.30	.0443	.0607	.0719	.0802	.0863	.0908	.0938	.0956	.0963	.0958	.0943
1.35	.0443	.0609	.0722	.0805	.0868	.0915	.0947	.0967	.0975	.0973	.0959
1.40	.0444	.0610	.0724	.0809	.0873	.0921	.0955	.0977	.0987	.0986	.0975
1.45	.0444	.0611	.0726	.0812	.0878	.0927	.0963	.0986	.0998	.0999	.0990
1.50	.0445	.0612	.0728	.0816	.0882	.0933	.0970	.0995	.1009	.1011	.1003
1.55	.0445	.0613	.0730	.0819	.0886	.0938	.0977	.1003	.1019	.1023	.1017
1.60	.0445	.0614	.0732	.0821	.0890	.0944	.0983	.1011	.1028	.1034	.1029
1.65	.0446	.0615	.0734	.0824	.0894	.0948	.0989	.1019	.1037	.1040	.1041
1.70	.0446	.0616	.0735	.0826	.0897	.0953	.0995	.1026	.1045	.1054	.1053
2.00	.0448	.0620	.0743	.0839	.0915	.0976	.1024	.1061	.1088	.1104	.1111
2.10	.0448	.0621	.0746	.0842	.0920	.0982	.1032	.1071	.1100	.1119	.1127
2.20	.0448	.0622	.0748	.0845	.0924	.0988	.1040	.1080	.1111	.1131	.1142

Table 21
Gas Flow Constant C_2 For Sonic Flow

k	C_2	k	C_2
1.00	0.0876	1.38	0.0984
1.02	0.0883	1.40	0.0989
1.04	0.0889	1.42	0.0994
1.06	0.0895	1.44	0.0999
1.08	0.0902	1.46	0.1003
1.10	0.0908	1.48	0.1008
1.12	0.0914	1.50	0.1013
1.14	0.0920	1.52	0.1017
1.16	0.0925	1.54	0.1022
1.18	0.0931	1.56	0.1026
1.20	0.0937	1.58	0.1031
1.22	0.0942	1.60	0.1035
1.24	0.0948	1.62	0.1039
1.26	0.0953	1.64	0.1043
1.28	0.0959	1.66	0.1048
1.30	0.0964	1.68	0.1052
1.32	0.0969	1.70	0.1056
1.34	0.0974	2.00	0.1112
1.36	0.0979	2.10	0.1129
		2.20	0.1145

Given the flow of air at standard conditions, Q_{sa}, in SCFM air (K = 0.62)

$$a = \frac{Q_{sa}}{11.4P_0}\sqrt{\frac{t + 460}{520}}\sqrt{\frac{29}{M}}, \text{ sq in.} \tag{13}$$

where $C_2 = \dfrac{520}{3{,}600}\sqrt{k\left(\dfrac{2}{k+1}\right)^{(k+1)/(k-1)}}$

The value for C_2 is found from Table 21.

Converting Actual Flow Rate to SCFM

As defined, a "standard cubic foot" of gas is that quantity of gas which occupies one cubic foot volume at a pressure of one atmosphere (14.7 psi) and a temperature of 60°F. See Table 22 for the specific heat ratio for various gases. Standard flow conditions can be found from actual flow conditions (ACFM) by recalling the perfect gas law.

If a known volume (or volume rate of flow Q_1) of gas at a specified temperature and pressure is allowed to expand to a new volume (or volume rate of flow Q_2) the pressure and temperature relationship is expressed by

$$\frac{P_1Q_1}{T_1} = \frac{P_2Q_2}{T_2}$$

So, to convert from actual flow conditions to standard flow conditions we have

$$Q_s = \frac{520}{14.7} \times \frac{P_{act}}{T_{act}} \times Q_{act} = \text{SCFM} \tag{14}$$

or from actual flow to standard flow of air

$$Q_{sa} = \frac{520}{14.7} \times \sqrt{\frac{M}{29}} \times \frac{P_{act}}{T_{act}} \times Q_{act} = \text{SCFM}_{air} \tag{15}$$

Sizing for Steam Flow

The principles that govern the reversible adiabatic flow of a gas through an orifice also apply to a vapor with limited superheat. However, because the vapor has limited superheat and because its heat capacity is not constant, the critical pressure ratio deviates from the ideal gas relationships. For steam, experience has shown that the following values of the critical pressure ratio, P_e/P_0, are good approximations in their respective regions.

For initially superheated steam,

$$P_e/P_0 = 0.545$$

For initially saturated steam,

$$P_e/P_0 = 0.577$$

Starting again with the general equation of flow, and assuming the steam is expanded isentropically through the orifice, we have the rupture disc sizing equation for steam.

$$A = \frac{144Wv^*}{K223.78\sqrt{h_0 - h^*}}, \text{ sq ft} \tag{16}$$

Table 22
Specific Heat Ratio, k, for Various Gases

GAS OR VAPOR	MOLECULAR WEIGHT	$k = C_P/C_V$
Air	28.97	1.40
Acetic Acid	60	1.15
Acetylene	26.04	1.26
Ammonia	17.03	1.33
Argon	40.00	1.67
Benzene	78.10	1.12
N-Butane	58.12	1.094
ISO-Butane	58.12	1.094
Butene	56.10	1.10
Carbon Monoxide	28.00	1.40
Carbon Disulfide	76.00	1.21
Carbon Dioxide	44.01	1.30
Chlorine	70.90	1.36
Cyclohexane	84.16	1.09
Ethane	30.07	1.22
Ethyl Alcohol	46.07	1.13
Ethyl Chloride	64.50	1.19
Ethyene	28.05	1.26
Helium	4	1.66
Hydrochloric Acid	36.50	1.41
Hydrogen	2.016	1.41
Hydrogen Sulfide	34.07	1.32
Methane	16.04	1.31
Methyl Alcohol	32.04	1.20
Methyl Chloride	50.48	1.20
Natural Gas (Aver.)	19	1.27
Nitric Acid	30	1.40
Nitrogen	28	1.404
Nitrous Oxide	44	1.3
Oxygen	32	1.4
Pentane	72.15	1.07
Propane	44.09	1.13
Sulphur Dioxide	64.06	1.29
Water Vapor	18.02	1.324

where * represents the conditions at the disc.

When the actual pressure ratio is greater than the critical pressure ratio for either region, the flow is *subsonic* and the vlaues of h^* and v^* are taken at the downstream temperature and pressure conditions.

When the actual pressure ratio is less than the critical pressure ratio for either region, the flow is *sonic* and the values for h^* and v^* must be taken at the minimum temperature and pressure in the orifice throat. So that

$$P^* = \text{critical pressure ratio} \times P_0$$

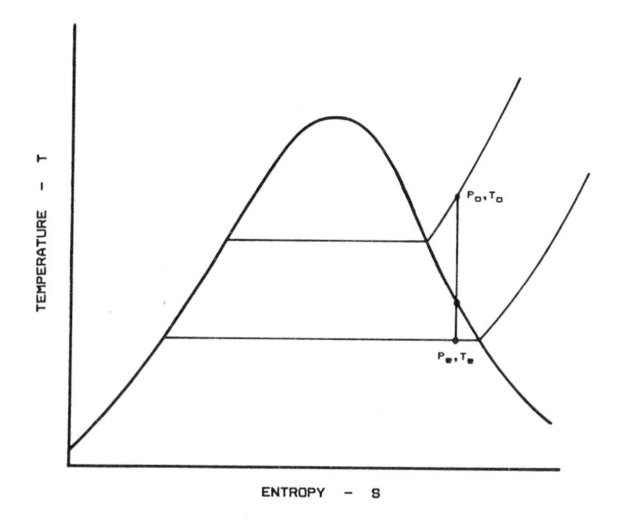

Figure 30. Temperature-entropy diagram showing the isentropic expansion of steam through an orifice.

When the temperature (and pressure) in the orifice throat falls within the saturation region and the steam begins to condense, the values of h* and v* must be corrected for steam quality.
Hence

$$s_0 = s^* = s_f - x_x s_{fg}$$

$$h^* = h_f - x_x h_{fg}$$

$$v^* = v_f - x_x v_{fg}$$

For sonic steam flow through the rupture disc, where the pressure ratio P_e/P_0 is less than the critical pressure ratio, the required flow area may be estimated from the following empirical equations (See Figure 30.)

Initially Superheated Steam

When the actual pressure ratio is less than a critical pressure ratio of 0.545, Napier's empirical equation for superheated steam may be used.

$$a = \frac{70W(1 + 0.00065D)}{0.62P_o} \tag{17}$$

Initially Saturated Steam

Where the actual pressure ratio is less than a critical pressure ratio of 0.577, Napier's empirical equation for initially saturated steam may be used.

$$a = \frac{70W}{KP_0} \tag{18}$$

Initially Wet Steam

Where the actual pressure ratio is less than a critical pressure ratio of 0.577, Napier's empirical equation for initially wet steam may be used.

$$a = \frac{70W(1 - 0.012x)}{KP_0} \qquad (19)$$

Sizing for Downstream Piping

Occasionally it becomes necessary to install the rupture disc in a system with extended lengths of downstream piping. The question then arises as to what L/D ratio represents the rupture disc? Here, the L/D ratio refers to the equivalent-pipe-length-to-diameter ratio.

Because there are so many variables to consider when the rupture disc bursts, the holder and rupture disc are generally assumed to be equivalent to 50–75 pipe diameters. This is roughly equivalent to K = 0.62 as permitted by the ASME code.

Sample Problems

Example 1: Sizing For Liquids

What size rupture disc is required to relieve a liquid under the following conditions?

Specific Gravity, SG = 1.34
Rupture disc burst pressure, p = 374 psig
Flow required, L = 8,300 USgpm

Step 1: Determine the minimum required rupture disc size.

$$a = \frac{L\sqrt{SG}}{23.1\sqrt{p}} = \frac{8300\sqrt{1.34}}{23.1\sqrt{374}}$$

a = 21.5 sq in.

dia = 5.2 in.

Answer: Recommend 6-in. disc.

Example 2: Sizing For Gases (Subsonic Flow)

What size rupture disc is required to relieve a gaseous system under the following conditions?

Rupture disc burst pressure, B.P. = 5 psig
Required flow rate = 12,000 ACFM
Back pressure during venting, P_e = 0 psig (atmospheric)
Flow temperature, t = 385°F
Coefficient, K = 0.62
Media, M = nitrogen (M = 28)

Step 1: Determine the actual pressure ratio.

$$P_e/P_0 = \frac{0 + 14.7}{5 + 14.7} = 0.7462$$

Step 2: Determine the critical pressure ratio.

$$CPR = \left[\frac{2}{k+1}\right]^{k/(k-1)} = 0.528$$

Step 3: Determine if the flow is sonic or subsonic.

$$\frac{P_e}{P_0} > \left[\frac{2}{k+1}\right]^{k/(k-1)}$$

The flow is *subsonic.*

Step 4: Find the flow constant C_1. From Table 20 for

$$\frac{P_e}{P_0} = 0.75$$

and

$$k = 1.4$$

(use the nearest value or interpolate)

$$C_1 = 0.0873$$

Step 5: Determine the required rupture disc size.

$$a = \frac{Q_{act}}{643.8KC_1}\sqrt{\frac{M}{t+460}}$$

$$a = \frac{12{,}000}{643.8(0.62)(0.0873)}\sqrt{\frac{28}{385+460}} = 62.69 \text{ sq in.}$$

dia = 8.93 in.

Answer: Recommended 10-in. disc.

Example 3: Sizing for Gases

What size rupture disc is required to relieve a gaseous system if a back pressure is present? This example is similar to Example 2. The conditions are:

Rupture disc burst pressure, B. P. = 155 psig
Required flow rate = 12000 ACFM
Back pressure during venting, P_e = 140 psig
Flow temperature, t = 385°F
Coefficient, K = 0.62
Media, M = Nitrogen (M = 28)

Step 1: Determine the actual pressure ratio.

$$\frac{P_e}{P_0} = \frac{140 + 14.7}{155 + 14.7} = 0.9116$$

Step 2: Determine the critical pressure ratio.

$$CPR = \left[\frac{2}{k+1}\right]^{k/(k-1)} = 0.528$$

Step 3: Determine if the flow is sonic or subsonic.

$$\frac{P_e}{P_0} > \left[\frac{2}{k+1}\right]^{k/(k-1)}$$

The flow is *subsonic.*

Step 4: Find the flow constant C_1. From Table 20 for

$$\frac{P_e}{P_0} = 0.912$$

and

$$k = 1.4$$

(use the nearest value or interpolate)

$$C_1 = 0.0610$$

Step 5: Determine the required rupture disc size.

$$a = \frac{Q_{act}}{643.8KC_1}\sqrt{\frac{M}{t+460}}$$

$$a = \frac{12000}{643.8(0.62)(0.0610)}\sqrt{\frac{28}{385+460}} = 89.71 \text{ sq in.}$$

dia = 10.69 in.

Answer: Recommend 12-in. disc.

Example 4: Sizing For Gases (Sonic Flow)

What size rupture disc is required to relieve a gaseous system under the following conditions?

Rupture disc burst pressure, B.P. = 150 psig
Required flow rate = 60,000 SCFM
Back pressure during venting, P_e = 0 psig (atmospheric)
Flow temperature, t = 110°F
Coefficient, K = 0.62
Media, M = 1 gaseous helium (M = 4)

Step *1: Determine the actual pressure ratio.*

$$\frac{P_e}{P_0} = \frac{0 + 14.7}{150 + 14.7} = 0.0877$$

Step 2: Determine the critical pressure ratio.

$$CPR = \left[\frac{2}{k+1}\right]^{k/(k-1)} = 0.488$$

Step 3: Determine if the flow is sonic or subsonic.

$$\frac{P_e}{P_0} < \left[\frac{2}{k+1}\right]^{k/(k-1)}$$

The flow is *sonic.*

Step 4: Find the flow constant C_2. From Table 21 for

k = 1.66

(use the nearest value or interpolate)

$C_2 = 0.1048$

Step 5: Determine the required rupture disc size.

$$a = \frac{Q_s}{22{,}772KC_2P_0}\sqrt{(t+460)M}$$

$$a = \frac{60{,}000}{22{,}772(0.62)0.1048(179.7)}\sqrt{(110+460)4} = 10.79 \text{ sq in.}$$

dia. = 3.71 in.

Answer: Recommend 4-in. disc.

Example 5: Sizing For Steam (The General Case)

What size rupture disc is required to relieve a steam system under the following conditions?

Rupture disc burst pressure, B.P. = 85 psig
Coincident burst temperature, t = 600°F
Required flow rate – 20 lbm/sec
Back pressure during venting, P_e = 0 psig (atmospheric)
Coefficient, K = 0.62

Step 1: Determine the actual pressure ratio.

$$\frac{Pe}{P_0} = \frac{0+14.7}{85+14.7} = 0.1474$$

Step 2: Determine if the flow is sonic or subsonic. The flow is *sonic*. By inspection, the pressure ratio is less than either of the critical pressure ratios.

Step 3: Find the properties of steam at the downstream temperature and pressure. From the Steam Tables, we see that the initial temperature is in the superheat range. So the critical pressure ratio

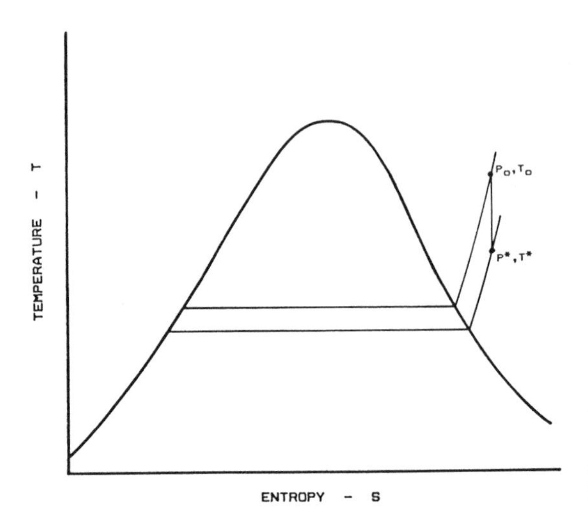

Figure 31. Temperature-entropy diagram illustrating the flow of steam through the rupture disc in Example 5.

is:

$$CPR = \frac{P^*}{P_0} = 0.545$$

The pressure at the throat of the rupture disc is then

$P^* = 54.3$ psia

So that at constant entropy, we have:

$h_0 = 1329.1$ Btu/lbm

$s^* = s_0 = 1.7581$ Btu/lbm R

$v^* = 9.961$ ft^3/lbm

$t^* = 460°$F

$h^* = 1264.0$ Btu/lbm

see Figure 31.
Step 4: Determine the required rupture disc size.

$$a = \frac{144(20)9.961}{0.62(223.78)\sqrt{1{,}329.1 - 1{,}264.0}}$$

$a = 25.63$ sq in.

dia $= 5.71$ in.

Answer: Recommend 6-in. disc.

Example 6: Sizing for Steam

What size rupture disc is required to relieve a steam system under the following conditions?

Rupture disc burst pressure, B.P. = 60 psig
Coincident burst temperature, t = 307°F
Required flow rate = 35 lbm/sec
Back pressure during venting, P_e = 50 psig (atmospheric)
Coefficient, K = 0.62

Step 1: Determine the actual pressure ratio.

$$\frac{P_e}{P_0} = \frac{50 + 14.7}{60 + 14.7} = 0.8661$$

Step 2: Determine if the flow is sonic or subsonic. The flow is *subsonic*. By inspection, the pressure ratio is greater than either of the critical pressure ratios.
Step 3: Find the properties of steam at the downstream temperature and pressure. From the Steam Tables, the steam is saturated. So the critical pressure ratio is:

$$CPR = \frac{P^*}{P_0} = 0.577$$

The pressure at the throat of the rupture disc is the same as the downstream pressure.

$$P^* = 64.7 \text{ psia} \approx 65 \text{ psia}$$

So that at constant entropy, we have:

$$h_0 = 1181.9 \text{ Btu/lbm}$$

$$s^* = s_0 = 1.6259 \text{ Btu/lbm R}$$

$$t^* = 298°F$$

$$s_{liq}(@\ 65 \text{ psia}) = 0.4342 \text{ Btu/lbm R}$$

$$s_{fg}(@\ 65 \text{ psia}) = 1.2032 \text{ Btu/lbm R}$$

$$x_x = \frac{1.6259 - 0.4342}{1.2032} = 0.9904$$

$$v^* = 6.592 \text{ ft}^3/\text{lbm}$$

$$h^* = 1170.4 \text{ Btu/lbm}$$

see Figure 32.
Step 4: Determine the required rupture disc size.

$$a = \frac{144(35)6.592}{(0.62)223.78\sqrt{1{,}181.9 - 1{,}170.4}}$$

$$= 70.61 \text{ sq in.}$$

dia. = 9.48 in.

Answer: Recommend 10-in. disc.

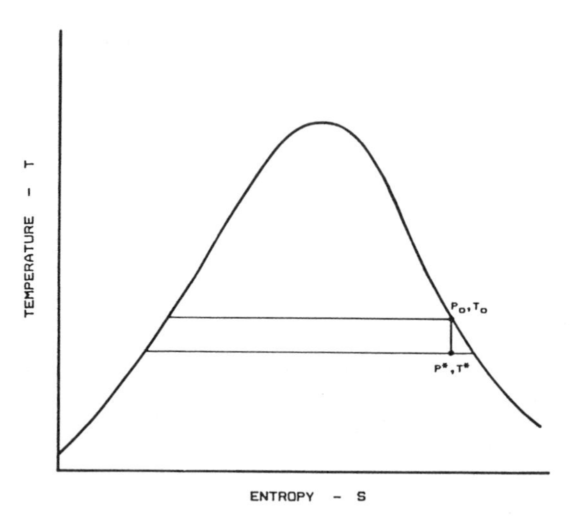

Figure 32. Temperature-entropy diagram illustrating the flow of steam through the rupture disc in Example 6.

Example 7: Sizing for Steam (The Simplified Method)

What size rupture disc is required to relieve a pressure vessel under the following conditions?

Rupture disc burst pressure, B.P. = 110 psig
Required flow rate = 65 lbm/sec
Back pressure during venting, P_e = 0 psig (atmospheric)
Steam quality = 30%
Coefficient, K = 0.62
Media = wet steam

Step 1: Find the flow pressure P_0.

$P_0 = 110 + 14.7 = 124.7$ psia

Step 2: Find the required flow area.

$$a = \frac{70(65)(1 - 0.12(100 - 30))}{0.62 \times 124.7}$$

a = 9.42 sq in.

dia. = 3.46 in.

Answer: Recommend 4-in. disc.

INSTALLATION AND MAINTENANCE

Rupture disc devices provide the most reliable service when they are installed and maintained properly. Rough handling, improper flange bolt torque procedures, and poor maintenance may not only jeopardize costly plant equipment but may also result in serious safety hazards to plant personnel. This section discusses the proper methods of installing and maintaining the rupture disc device.

Installation

1. *Unpack and inspect the components carefully.* Look for nicks, dents, or any damage to the seating surface of the rupture disc holder. Inspect the rupture disc. Carefully clean the disc seating surface if necessary with an approved solvent *Caution:* Handle the disc with extreme care. Finger prints, scratches, or other damage to the disc could result in bursting outside the specified burst tolerance.
2. *Install the rupture disc in the holder.* Place the rupture disc between the inlet (base flange) and outlet (holddown flange) and, if provided, install the preassembly screws or side clips. *Note:* Some rupture discs have flat seats. *Be sure the disc will burst in the direction of the flow arrow on the holder.*
3. *Install the rupture disc device in the pipeline.* Carefully place the rupture disc device between the piping flanges and install the flange bolts. Torque the flange bolts to the values recommended by the rupture disc manufacturer. *Note:* The crown of larger rupture discs may extend beyond the rupture disc holder (holddown flange). Be sure to allow enough clearance between the piping flanges to install the assembly. *Caution: Be sure the flow arrow on the device points in the direction of flow.*
4. *Recommended flange bolt torque procedure.* The flange bolts must be tightened uniformly to ensure even seating of the disc. *Caution:* Flange bolt torque values are usually based on *lightly oiled* screw threads. If a high efficiency thread lubricant is used, the flange bolt torque must be reduced accordingly. Number the bolts sequentially around the flange and lightly hand tighten the nuts until snug. Apply approximately $\frac{1}{4}$ of the recommended torque to Bolt 1 (12 o'clock position). Progress clockwise to the bolts at 6 o'clock, 9 o'clock, and 3 o'clock respectively. Now, starting nearest the 7:30 o'clock position, torque the bolts in the same manner, progressing counter-clockwise applying $\frac{1}{4}$ of the recommended torque to each nut. Then at 3 o'clock, progress clockwise torquing the nuts at each of the four compass points. When all the bolts have been torqued, start the process over applying the second, then third, and finally the fourth increment of the recommended torque value.

   While this procedure seems a bit confusing at first, the installer soon learns that the technique is simply a matter of torquing the opposing bolts while progressing first in one direction then the other. This method is highly effective on stainless steel bolts and nuts where the thread friction factor rapidly increases as the bolt load increases. Continuously working around the flange in the same direction causes the opposing bolt to be torqued to the same values as the other bolts but because of the increased friction factor the resulting bolt load is actually less. The net result is that the rupture disc is not seated evenly.
5. *Final installation details.* For new installations, check to see that the piping is braced to absorb any shock or strong overturning moments when the disc bursts. Install a DANGER sign in a conspicuous location near the "zone of potential danger."

Inspection/Maintenance

Inspection

The rupture disc device should be inspected at least every six months.
Check for:

1. Visible signs of leakage.
2. Corrosion between the two halves of the rupture disc holder.
3. External damage to the rupture disc holder.
4. DANGER sign visibility and condition.

Under normal circumstances, the rupture disc should be replaced once a year. However, operating and environmental conditions will determine the best service life and disc replacement cycle.

Maintenance

Removal of the rupture disc after bursting. Remove the rupture disc device from the piping system. *Note:* Removal of the rupture disc device from the piping system following bursting of the disc may be difficult if the open rupture disc extends into the downstream piping. *Warning:* Disc fragments are "razor" sharp. *Use extreme care when handling the burst disc.* Remove the burst disc from the holder and clean the seating surfaces with an approved solvent. Stubborn dirt or scale may be removed with #4-0 steel wool or #400 grit emery cloth if necessary.

Assembly and installation. Assemble and install the rupture disc as outlined above, using care to prevent damage to the disc and holder. Torque the flange bolts as required.

TROUBLE SHOOTING PROBLEM AREAS

A discussion of rupture discs is not complete without covering the subject of trouble-shooting application and installation problems. This section is devoted to describing the most frequent problems that occur and how to avoid them.

Overtorqued Flange Bolts

Improperly torqued and overtorqued companion flange bolts is a problem that occurs more frequently than any other. It is caused many times by the inexperienced installer, who not having a torque wrench handy, attempts to torque the flange bolts by "feel". Then, just to make sure the system doesn't leak, he tightens the bolts a little more. If the system does leak he learns the hard way that his sense of "feel" was a little out of calibration; he may have doubled or tripled the required bolt load. A similar problem occurs when the installer "goes by the book" to install the device. To make sure the flange nuts turn easy, he adds a little "moly" disulphide, graphite, or other low-friction thread compound to the flange bolts and then he applies the recommended amount of torque. What he may not have realized is that efficient low-friction thread lubes can multiply the required bolt load by as much as 5 times. The end result is that bolt loads of this magnitude will damage delicate rupture discs and deform holder seats.

Holders can be checked for damage by placing a straight edge across the flat gasket face of both halves of the holder. If either half appears "rolled," the holder was overtorqued and should be replaced. (See Figure 33 for an illustration of this procedure.)

When the flange bolts are overtorqued, there is usually enough load on the holder to force the disc farther into the holder seat than it was designed to go. This type of damage is suspected in a conventional holder when the normal $\frac{1}{8}$-in. gap between the base and holddown has been reduced significantly. (See Figure 34.)

Figure 33. A photo illustrating how to check a rupture disc holder for overtorquing (courtesy of Fike Metal Products).

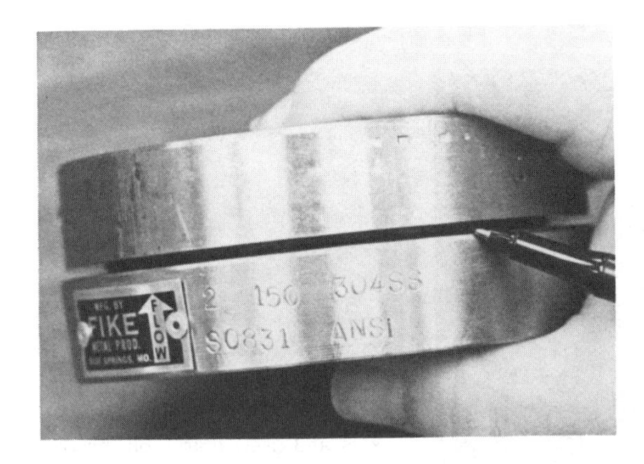

Figure 34. A photo showing the reduced gap between both halves of the rupture disc holder caused by overtorquing of the flange bolts (courtesy of Fike Metal Products).

Figure 35. A photo showing damage to the flat area around the outside diameter of the disc (courtesy of Fike Metal Products).

Most conventional 30° seat rupture discs have a small flat area (approximately $\frac{1}{8}$-in. wide) around the outside diameter of the disc. If there is no evidence of the flat area or the flat area has been "flared" upward parallel to the 30° angle of the seat, the disc has likely been forced too far into the holder—a result of overtorquing. (See Figure 35.)

Undertorqued Flange Bolts

When a rupture disc is installed with too little torque on the flange bolts, the reduced uniform clamping load will permit the disc to slip through the seat and burst prematurely. This problem is evidenced by small "wrinkles" and "draw" marks around the seating surface of the disc as a result of the slippage. The proper balance between flange torque and burst pressure must be achieved or the disc will slip. (See Figure 36.)

A similar problem occurs when the flange bolts are torqued unevenly. In this case the holddown flange and disc are forced into the holder at an angle reducing the clamping load on one side of the disc seating surface. As the pressure increases, the disc may slip at that point and eventually burst prematurely. Notice the wrinkles around the side of the disc opposite the tag and the large wrinkle and flattened tag rivet adjacent to the tag in Figure 37.

Figure 36. A photo showing wrinkles and draw marks on the 30° seating surface of the disc (courtesy of Fike Metal Products).

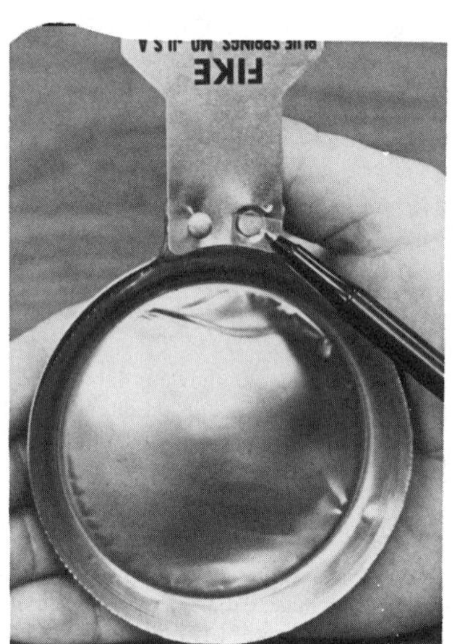

Figure 37. A photo showing wrinkles on one side of the disc (courtesy of Fike Metal Products).

Unseen Damage to Reverse-Acting Rupture Discs

Reverse-acting rupture discs are very sensitive to unseen damage. A common type of unseen damage results when the crown of the disc is "dented." Even though the dent is pushed out and is invisible to the naked eye, the disc will almost always burst at less than the specified burst pressure. It is extremely important then that these discs be handled with care.

Figure 38. A photo showing a rupture disc without a vacuum support that experience a back pressure sufficient to cause reversal (courtesy of Fike Metal Products).

Figure 39. A photo of a rupture disc with the characteristic "turtle-back" damage (courtesy of Fike Metal Products).

Exposure to Vacuum Without a Vacuum Support

When a fragile rupture disc is exposed to a sufficient back pressure (vacuum) it will reverse. Tell-tale signs of disc reversal are evidenced by a wrinkling of the entire disc surface. A vacuum support will prevent this type of damage and thus increase the service life of the disc. (See Figure 38.)

"Turtle-Backing"

The phenomenon of "turtle-backing" results when a rupture disc, pre-bulged to a specified level and fitted with a vacuum support, is exposed to a forward pressure greater than the pre-bulge pressure then subjected to a back pressure. The characteristic wrinkling of the disc resembles the shell of a turtle and will cause the disc to leak or burst prematurely. (See Figure 39.)

EXPLOSION VENTING

Fuel vapors and fine dust particles are combustible in sufficient concentration and turbulence and will "explode" if ignited. In confined spaces, the resulting maximum pressure developed, (P_{max}), will normally grow to seven times the initial pressure of the confined space. Serious safety hazards are then created if the space is not properly vented. The safety of plant personnel and equipment is therefore dependent on the proper choice, sizing, and placement of vents in a closed structure to limit the pressure the structure can withstand. "Explosion venting" is a broad subject that cannot be covered in fine detail in a few pages of text. So, for the purposes of this discussion, we will limit our comments to an overview of the latest technology and a review of the equipment available for effective venting. The exact details of each application must be worked out between the loss-prevention engineer and the equipment manufacturer to provide the most effective protection against damage from overpressure.

The Anatomy of a Dust or Vapor Explosion

When a flammable mixture is ignited, a flame front moves symmetrically outward from the point of ignition. The resulting hot combustion gases expand causing a pressure rise across the flame front and in turn, throughout the confined space. If the expansion process is allowed to continue, it causes a turbulence which stretches and folds the flame, tending to increase the violence of the burning. All these phenomena contribute in accelerating the flame propagation and increasing the pressure build up.

The destructive forces, resulting from a vapor or dust "explosion," increase as the pressure and flame velocity increase. Here we use the term "explosion" for illustration. Later we'll refer to the burning of the fuel-oxidant mixture as a "deflagration." The speed at which the vapor or dust burns is dependent upon the fundamental burning velocity of the fuel consumed and other factors. The rate at which the pressure increases and the general violence of the explosion tend to increase as the strength of the ignition source increases. Locating the ignition source at the geometrical center of confined fuel-oxidant mixture results in development of the highest pressure and rate of pressure rise.

The maximum explosion pressure P_{max} and the maximum rate of pressure rise $(dP/dt)_{max}$ define the violence of the explosion. A typical explosion-pressure-vs-time curve for a vented and unvented fuel/air mixture in a closed vessel is shown in Figure 40.

Fundamentals of Explosion Venting

Definitions

Deflagration and detonation. It is general practice to consider an explosion in one of two modes of combustion: a deflagration or a detonation. A *deflagration* is defined as burning which takes place

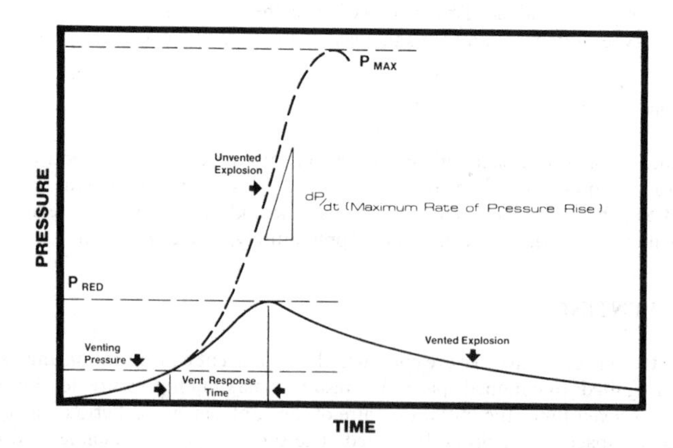

Figure 40. Pressure-vs.-time curve for vented and unvented explosions (courtesy of Fike Metal Products).

at a flame speed below the velocity of sound in the unburned medium. Burning at a flame speed above the velocity of sound is termed a *detonation*. Many detonations initially start as a deflagration but all deflagrations do not progress into detonations. The explosion venting application usually assumes that the overpressure occurs as a deflagration rather than a detonation.

Dust. This is defined as a finely divided solid material usually 420 microns or smaller in diameter.

Explosion. This is the bursting of an enclosure as result of the development of an internal pressure beyond the structural integrity of the enclosure.

Flame speed. This is the speed at which the flame progresses through the unburned mixture. Many factors influence the flame speed, including the chemical composition of the mixture, moisture content, heat of combustion, particle size of the dust or mist, turbulence, and the strength of the ignition source.

Flammability or explosibility limits. This is the upper and lower concentration of gas, dust, or mist in air or other oxidant, below which flame does not propagate when ignited. Limits are usually expressed in terms of percentage by volume of gas in the oxidant. For mists and dusts, the limits are a function of particle size.

Fundamental burning velocity. This is the velocity of the gas normal to the flame front with which the unburned mixture enters the flame. The fundamental burning velocity is a characteristic of the fuel consumed and for most gases, is on the order of 10 in. to 20 in. per second; values for dusts are generally lower. This velocity is determined from laminar flow conditions in carefully controlled laboratory testing.

Rate of pressure rise. This is the amount of pressure increase per unit time usually expressed as dP/dt (psi/sec). The "average rate of pressure rise" is the ratio of the maximum pressure developed to the time interval between the initiation of the deflagration and the maximum pressure. The "maximum rate of pressure rise," $(dP/dt)_{max}$, is taken as the slope of the steepest portion of the pressure-time curve during the development of the deflagration.

Vapor. This is a gas or superheated vapor.

Factors affecting a deflagration or detonation.

Principle factors which affect the development of a deflagration or a detonation include:

Fuel—The fuel may be a gas, mist, or dust or a combination of any of these. In general, gases burn more rapidly than mists or dusts.
Fuel concentration—Most gases have a well defined lower and upper flammability limit and the concentration must be within these limits for a deflagration to occur. For dusts, the upper flammable is not well defined. Some mists can be deflagrated when the temperature of the mist is such that the corresponding vapor pressure will produce a concentration less than the lower flammability limit.
Oxidizer concentration—The oxidizer is normally the oxygen present in the atmosphere. Oxygen concentrations higher than 21% tend to intensify the reaction rate of combustion and increase the probability of detonation.
Fundamental burning velocity and flame speed—The destructive forces, caused by the pressure developed during combustion, tend to increase as the fundamental burning velocity of the fuel mixture increases. The flame speed is most rapid and the highest pressures are developed when the fuel concentration is optimum and uniformly distributed throughout the structure. Flame speeds in a structure may be many times higher than the fundamental burning velocity but never lower. In general, the fundamental burning velocity of dusts is slightly longer than for gases. The longer duration burning time results in a greater total impulse tending to make dust explosions more disastrous than gas explosions.

Ignition source—The rate of pressure development and the maximum rate of pressure rise tend to increase with the strength of the ignition source and turbulence. Ignition can result from a hot surface, flame, or spark and, in many cases, the location of the ignition source cannot be predicted.
Dust particle size—Experimental data shows that the particle size of a dust has little effect on the maximum developed pressure. The rate of pressure rise, on the other hand, increases significantly with a decrease in particle size. In addition, a decrease in particle size lowers the minimum energy required to ignite the dust cloud.

Factors affecting the pressure rate of rise.

Fuel-oxidant mixture—In general, the maximum rate of pressure rise will occur at or near the stoichiometric mixture of the gas or dust and fuel-oxidant.
Volume and shape of the enclosure—The maximum pressure developed during a deflagration results from heating and expansion of the products of combustion and the atmosphere within the enclosure. The maximum pressure therefore is generally unaffected by the size and shape of the vessel. The rate of pressure rise, on the other hand, is proportional to the ratio of the surface area of the vessel to its volume. Extensive tests conducted by Donat (X) and Bartknecht (Y), show that for enclosures with a length-to-diameter ratio of less than 5, the pressure rate of rise for a given fuel-oxidant mixture approximately follows the "cube law." This law states that the product of the pressure rate of rise and the cube root of the volume is constant.

$$(dP/dt)_{max} \times (V^{1/3}) = \text{constant}$$

where $(dP/dt)_{max}$ = maximum rate or pressure rise for dust or gas combustion in a particular vessel

v = volume of the particular vessel.

The constant is referred to as K_{st} for dusts and K_G for gases. For a rate of pressure rise expressed in bar/sec and a vessel volume in m^3, the constant is expressed in $bar \cdot m \cdot sec^{-1}$. On the basis of the cube law, the required venting area for vessels whose length-to-diameter ratio is less than a factor of 5 can be expressed as follows:

$$F_2 = \frac{F_1 V_2 V_1^{1/3}}{V_1 V_2^{1/3}}$$

where F_1 = vent area of the test vessel found necessary to prevent the pressure from exceeding a given value during combustion

F_2 = vent area which will be necessary to prevent the pressure from exceeding the same value on the second vessel

V_1 = volume of the test vessel

V_2 = volume of the second vessel

Types of Vents Available

Through the years many vent designs have been developed for the release of overpressure resulting from deflagrations. As explosion vent technology improved, the various designs fell into general type categories, each with its own strengths and weaknesses. The following descriptions will help the engineer understand the general types of vents available today and how they may be used effectively.

Initially Open Vents

The most effective vent for the release of pressure caused by deflagrations is an initially open vent, an unobstructed opening (open hole). This type of vent, however, provides little protection from the elements, unauthorized entry, or contamination of the product by the outside world. When this is not a problem vents such as large roof openings can protect one-story buildings where lighter-than-air combustible gases can collect near the ceiling.

Open vents are recommended where closed equipment or buildings are not necessary to prevent the loss or contamination of combustible material or where the escape of combustible material will not create a more serious safety hazard.

Unlatched, Obstructed Openings

Vent closures of this type include louvers, hanging doors, and curtains that hang freely in an open vent. Unlike the initially open vent, however, this vent remains closed in normal operation but opens proportionately in response to an overpressure. Because the closure is usually a gravity type, it does not readily permit a positive (or negative) operating pressure. The effectiveness of controlled atmosphere equipment is thus reduced and product or combustible material may escape to the environment. Such closures can be opened to provide nearly unobstructed vents during the operation of equipment, or closed to prevent unauthorized access or when the equipment is not in operation. These vents require detailed engineering analysis to determine the equivalent free vent area and pressure drop during a deflagration.

Sealed Doors and Latchable Closures

To effectively prevent the loss and contamination of combustible products, vent closures may be installed that remain closed until they are "instructed" to open. These vents are kept closed by friction, springs, latches, magnets, and other locking hardward. When an overpressure occurs, the latches release and the closure moves in response to the overpressure.

The effectiveness of this type of vent is a function of the reliability of the latching device, and the mininum presure at which the latch releases. Latchable closures are not generally recommended for low vent pressures or where protection of the structure relies on accurate unlatching pressures to control high rates of pressure rise.

Roof or Wall Panels

In many cases, a portion of a structure can be made to blow out readily if an overpressure occurs. The panels are usually constructed of light-gage sheet metal, corregated plastic paneling, or paper. For large enclosures such as buildings, this type of vent may be the only suitable means of venting the deflagration. These types of vents should weigh less than 1.5 lbs/sq ft.

Rupture Diaphragm Venting Devices

Rupturable explosion venting devices are generally of light-weight low-cost construction, designed to be kept closed at all times and not usually used as doors or inspection ports. This type of vent is typically fitted with a gasket and will open only when the internal pressure exceeds a predetermined point. These vents can be made from metal foil, paper, or plastic film. Vents made from metal foil are usually more durable and can be made to burst at lower pressures than other materials.

The advantages include fast opening, low burst pressure, and relatively small size. This type of vent is generally recommended for dust collectors, tanks, small buildings and other equipment where the combustible material has a high rate of pressure rise and the vent must be opened quickly.

Basic Recommendations for Venting

Venting is a complex subject, its complexity compounded by the lack of essential information and the sometimes unpredictability of a deflagration. It is important, then, that each application be given careful analysis and the proper vent chosen and installed in accordance with current recommended practices.

Years of experience and research have been summarized by the National Fire Protection Association and published in NFPA 68, Guide for Explosion Venting. This guide is written by a committee of technically qualified users and experts in the field of explosion venting, and published by NFPA. Its purpose is to provide the user with the best available criteria for venting of deflagrations. The following summary of the important recommendations and current practices for reducing damage is taken from NFPA 68 (1978 printing), sec. 3–5.

1. Vents are generally required in buildings or enclosures containing operations or processes where dust, gas, or mist may be present in sufficient amounts to create flammable concentrations in air or other oxidizing media.
2. The required venting area is dependent upon the characteristics of the rate of pressure rise, maximum pressure developed, the strength of the enclosure, and design of the vent closure. Emperical methods and nomographs may be used to determine the vent area and are discussed in detail in NFPA 68.
3. Vents should be located as close as possible to potential sources of ignition which may originate the deflagration. Where points of ignition cannot be determined in large enclosures, vents should be evenly distributed.
4. Whenever possible vented products should be directed to a safe location or outside of the enclosure to avoid injury to personnel and minimize damage to property.
5. Vent ducts should be constructed to withstand the maximum pressure of a deflagration. High pressures can develop in long ducts; therefore, the duct length should be minimal and bends should be avoided.
6. Vent designs and placement should take into account maximum wind conditions, operating conditions, such as negative pressures and turbulence, and building safety.
7. Vents should have low weight per unit area. NFPA recommends 2.5 lbs/sq ft maximum.

NOTATION

A	Required flow area in sq ft
ACFM	actual cubic feet per minute.
a	required flow area in sq in.
D	number of degrees of superheat. (burst temperature − saturation temperature)
γ	fluid density in lbm/ft^3
g	gravitational acceleration (32.174 ft/sec^2)
h_d	pressure head in ft
h	enthalpy in Btu/lbm
K	coefficient for sizing (K = 0.62)
k	ratio of specific heats (k = 1.4 for air)
L	liquid flow rate in gal/min
M	molecular weight
MAWP	maximum allowable working pressure
P_0	relieving pressure in psia.
P_e	exit pressure in psia
p	relieving pressure in psig
Q	volumetric flow in ft^3/min
Q_{act}	actual volumetric flow rate in ft^3/min
Q_s	standard volumetric flow rate in ft^3/min
Q_{sa}	standard volumetric flow rate of air in ft^3/min at standard temperature and pressure (60°F and 14.7 psia)
R	gas constant − 1545/M in ft-lbf/lbm °R
s	entropy in Btu/lbm °R
SCFM	standard cubic feet per minute.
SG	specific gravity (water = 1.0, Air = 1.0).
T_0	upstream temperature in °R
T_R	temperature in °R
t	temperature of flowing media at disc relieving pressure in °F
v	specific volume in ft^3/lbm
V	velocity in ft/sec
W	mass flow rate in lbm/sec
X	100 − % steam quality

INDEX

Abrasion factors, 1097, 1104
Absorption damper, 1110, 1112
Absorption pulsation dampers, 1111
Acid egg pump, 990, 995
Acoustic radiation, 1120
Activation energy, 858
Actuators, 1079
Advected eddies, 67
Advection, 55
Aerodynamic effect on jet stability, 500, 501
Aerospace industry, 349, 1153
Agitated liquid-liquid reactor, 863
Agitated vessels, 901, 903
Agitated vessels mixing patterns, 913
Agitation of viscous fluids, 901
Agitator design requirements, 803
Air-jets, 430
Air-sea interaction, 39
Aircraft, 349
Aircraft ejector, 367, 371
Aircraft fuel feed systems, 1153
Airlift pump, 991-994
Alluvial channels, 73
American Society of Civil Engineers, 579
Ammonium hydroxide, 855
Anchor agitated vessels, 902, 903, 907, 913, 916
Anchor impeller, 805, 909
Anchor impellers power correlation, 910
Annual maximum mean daily flows, 282
Annual maximum peaks, 295
Annual maximum peaks—monthly percentage distribution, 302
Antidunes, 77
Apparent Reynolds number, 932, 944, 945
Apparent viscosity, 505
Aqueous systems mixing, 886
Army Corps of Engineers, 274
Atmospheric boundary-layer flow, 47
Atomic fuel recovery, 989
Attentuation
  causes of, 178
  theory, 195
Autocorrelation function, 356
Average bed shear, 74
Average bed shear stress, 79
Axial diffusion, 321
Axial dispersion model, 773, 780, 782
Axial flow, 696, 701, 804
  characteristics of, 805
  impeller, 810
  pattern, 982
  pump, 1134
  turbomachinery, 1155
Axial hydraulic thrust, 1038
Axial jet flow, 383
Axial mixing behavior, 559
Axial pressure gradient, 384
Axial thrust, 1049
Axial turbulence intensity along jet center line, 411
Axially fired furnaces, 369
Axially-symmetric motion, 612
Axisymmetric buoyant jet, 331
Axisymmetric flows, 331, 538
Axisymmetric gas jets, 513
Axisymmetric jet-plate flow, 349
Axisymmetric plume, 324, 326
Axisymmetric jet thermal buoyancy, 328
Axisymmetric shear layer, 396
Axisymmetric vertical jets, 319

Backflow cell model, 776, 780, 781, 796
Backflow rates, 781
Backmixed flow, 773
Backmixing
  degree of, 792
  in horizontal stirred vessel, 795
  in stirred vessels, 772, 782, 788
  with perforated partition plate, 799
Backward shallow water waves, 180
Baffle blocks, 130, 131
  drag coefficient, 129
Baffle configuration, 815
Baffle design, 814
Baffle drag coefficients in supercritical flow, 130
Baffles, 131, 144, 886, 889
Balancing devices, 1051
Balancing drum arrangement, 1051
Bank erosion, 209
Bank stability, 205

Bar-resistance curve, 79, 85
Base factor K, 297
Batch reactor, 711
Batch time, 879
Batchwise jet mixing, 466, 469
Bed load transport, 216
Bed roughness—Manning's coefficient, 207
Bed shear stress, 82, 157
  friction factor, 74
Bedform, 83, 94
  appearance, 72
  evolution, 72
  friction factor, 94
  geometric shape, 79
  geometry, 77, 84, 89
  pressure forces, 80
  steepness, 91
Belanger solution, 127
Bellows pump, 987
Bellows-type metering pumps, 987
Benjamin-Feir type instability, 242
Bernoulli equation, 35, 88, 952, 953, 1190
Bernoulli relation, 6
Bernoulli relation—free surface, 11
Bernoulli theorem, 1004
Bessel function, 29
Bifurcated skew waves, 253
Bifurcations, 245, 251, 254, 256
Bifurcations in ocean-wave evolution, 255
Binomial distribution, 276
Biomechanical applications, 691
Blade width, 918
Blast-wave model, 413
Blending-economical selection, 817
Blowcase pump, 990
Bluff bodies, 78
Bodenstein number, 565
Body of revolution, 427
Boil-off rate, 642, 655
Boiler feed operations, 965
Boiler feed pumps, 960
Bore, 189
Bottom circulation, 586
Boundary layer, 220, 228, 514, 1155, 1177
  development, 156, 1178
  equations, 324, 338, 529
  flow, 324, 344
  investigations, 1170
  separation, 78
  theory, 526
  thickness, 225, 1176
Boundary shear force, 126
Bounded jet, 470
Boussinesq approximation, 11, 18, 20, 331, 441
Boussinesq eddy—viscosity concept, 447
Boussinesq solitary wave, 10
Boussinesq theory, 11
Brake horsepower, 973
Break ratio, 1090
Breaking waves, 242
Breakup
  of jets, 482, 495
  of liquid jet, 482
  pattern of the jet, 485, 498, 504
  phenomena, 483
  time, 488
Breakup length, 483, 484, 488, 491, 494, 501, 506
  of jet, 498, 504
  of laminar jet, 483
  of Newtonian jets, 505
  of water jet, 499
Brewing industry, 471
British Atomic Energy Authority, 471
Brunt-Vaisala frequency, 19
Bubble
  collapse, 1133
  collapse acoustic power, 1133
  collapse velocity, 1124
  dispersion, 773
  formation, 956
  generation, 255
Buoyancy, 48, 58, 342, 683
  -driven motion, 57, 63, 65, 67
  -driven plume, 318
  effects, 67, 326, 334, 701
  effects on the turbulence structure, 67
  -induced flow, 341, 344
  -induced plume, 345
  parameter, 341
  production, 62
Buoyant jet, 317, 318, 325, 326, 336, 345
  flows, 318
  rise, 340
Bypass valves, 1022

Calderbank-Moo Young correlation, 741
Cam drive gear, 1065
Cam pumps, 983
Capacity-correction factor, 1142
Capillary waves, 5
Cargo transfer, 644, 645
Casing geometry, 1003
Catalyst, 688
Cavitating flow, 1155
Cavitating inducer, 1134, 1155
  performance, 1166
Cavitation, 956, 989, 1006, 1072, 1073, 1076, 1119, 1120, 1121, 1126, 1130, 1170
  bubble, 1169
  conditions, 1078

damage prediction, 1128
erosion efficiency, 1133
erosion rate, 1130
failure, 1125
flow regime, 1127
noise, 1127
occurrence, 1134
sigma, 1120
Causes of attentuation, 178
Celerity, 192
Cell formation and stability, 638
Cell models, 755, 776
Central Electricity Generating Board, 1206
Central jet flow velocity, 376
Centrifugal extractor, 495
Centrifugal force, 438, 495, 696, 1006, 1163
Centrifugal impeller, 1153
Centrifugal machines 971
Centrifugal pump, 956, 957, 960, 974, 995, 1001, 1002, 1005, 1121, 1142
characteristics, 1002, 1003
configurations, 1007
performance characteristics, 1140
performance curve shapes, 1010
sizing, 1138
Channel
bed
roughness, 221
shear stress, 222
vegetation, 220
cross section design, 207, 213
-equivalent relative roughness, 103
flow, 221
depth, 225
rough surfaces, 220
geometry, 175
roughness, 106, 107, 225
slope, 106, 264
Chemical conversions, 870
Chemical industry, 349, 960
Chemical process engineering, 1089
Chemical process pump, 961
Chemical pumps, 960, 1153
Chemical reactions, 660, 716, 743, 881
influence of micromixing, 719, 759
Chemical waste products, 317
Chezy coefficient, 176
Chezy formula, 108, 199
Circular buoyant jet, 342
Circular channels, 135
Circular jet, 396, 397, 551
similar treatment, 400
velocity profiles, 392
turbulent, 396
Circulating flow, 467, 937, 939
Circulation
flow rates, 931
flow regime, 550, 555
-interaction model, 867, 869, 870
loops, 852, 871, 873
region, 633
times, 852, 867
velocity, 568
Classical hydraulic jump equation, 150
Closed conduit hydraulic jump, 137
Closed internal eddies, 585
Cnoidal wave, 176
theory, 118
solution, 113
Coalescence, 854, 867
behavior, 853
-dispersion model, 710, 717, 742
events, 880
probability, 710
processes, 877
rates, 857
Coaxial jet, 369, 382, 383
flow, 369
velocities, 383
Coefficient
of discharge, 273
of drag, 81
Combustion, 318, 382, 661
chamber models, 456
chambers, 367, 430
reactions, 765
Complex liquid phase reactions, 759
Composition of LNG mixtures, 641
Compressible displacement thickness, 530
Compressible fluid, 1107
Computer storage memory, 450
Concentration distribution, 714
Concentric cylinder rotational viscometer, 504
Condensation, 767
Conductivity bridge, 53
Conductivity control, 1094
Confined coaxial jets, 367, 384
Confined cross flow, 433
Confined jets, 372, 432
in cross flow, 430
in open ducts, 376
smoke photographs, 432
turbulent, 431
Consecutive competing reactors, 729, 750
Conservation equation, 319
Conservation law, 15, 16
Constant-pressure jets, 372, 374
Constant-speed pump horsepower, 1030
Containment sumps, 1203
Continuity equations, 153, 320, 321, 394
Continuous-flow jet mixing, 472
Continuous-phase flows, 499
Continuous-phase loop, 872
Continuous processing, 818

Continuous reactors residence-time distribution, 709
Continuous stirred-tank reactors, 720
Continuous stirred-reactors mixing earliness, 733
Control-valve pressure differentials, 1142
Convection, 156, 353
  -diffusion equation, 52, 193
  velocity, 353
Convective films, 641
Convective heat and mass transfer processes, 691
Convective heat transfer, 939
Convective mixing, 740, 741
Conventional head, 1038
Conventional inducer, 1169
Cooper-Bosch method, 1163
Copolymers, 558
Core length, 518
Coriolis and centrifugal forces, 1162
Correlation techniques, 360
Corrosion, 661, 1104, 1127
Corrsin microscale, 740
Craya-Curtet number, 375, 378, 379
Credit River, 282
Critical bifurcation-wave steepness, 251
Critical Froude number, 106, 107, 108, 112, 598
Critical Reynolds number, 567
Critical shear stress, 76, 205, 207
Critical velocity, 485, 487, 507
Cross flow distribution, 434
Cross flow Reynolds number, 432
Crystallization, 767
Curl's equation, 714
Curvilinear flow pressure head, 161
Cylindrical solitary waves, 17

Dam-break wave hydraulic resistance, 192
Dam failure problem, 190
Dam reservoir vertical mixing, 631
Damkohler number, 721
Darcy-Weisbach coefficient, 231, 232
Darcy-Weisbach friction factor, 99
Darcy-Weisbach resistance coefficient, 233
Deep fluid theory, 19, 21
Deep water gravity waves—energy spectral density, 44
Deep water surface waves—statistics, 26
Degree day factor, 271
Degree of backmixing, 792
Degree of mixing, 545, 667
Densimetric Froude number, 598
Density distribution, 611
Density distribution function, 577
Density functions, 881
Density ratio, 435
Density stratification, 573, 577
Density stratified fluids—solitary waves, 18
Density stratified reservoirs, 572, 584
Dependence of microscale time constants, 757
Depression wave, 190
Design flood, 267
Design working head, 1019
Destructive interference, 248
Developed head, 1004
Diaphragm metering pump, 987, 1080
Diaphragm pumps, 977, 986, 987, 1065, 1094, 1097
Diaphragm-type process pump, 1087
Diffuser, 1045, 1047
Diffuser vane, 1046
  thickness, 1047
Diffusion, 65, 384, 454, 612, 747, 855
  of jets, 391
  loss factor, 1159
  model, 773
  of plane and circular jets, 400
  stages, 745
Diffusivity coefficient, 210
Digital proportional metering, 1092
Dimensional analysis, 524, 665, 688, 888
Dimensional backflow, 794
Dimensionless boundary-layer equations, 330
Dimensionless eddy diffusivities, 59
Dimensionless groups, 830
Dimensionless growth rate, 478
Dimensionless mixing time, 548, 550, 551, 552, 553, 554, 783, 944
Dimensionless shear stress, 86
Dimensionless spectrum functions, 51
Dimensionless wave
  celerity, 102
  length, 102
  number, 51, 102, 105, 179
Direct-acting steam pumps, 979
Direct-digital proportional metering, 1093
Disc turbine, 888, 891
Disc turbine-angled blade, 893, 896
Discharge
  characteristics, 568
  efficiency, 567
  number, 568
  static head, 1148
  valves, 952
  vessel pressure, 1148
Discontinuous head characteristics, 1011
Disk friction, 1045
Dispersed-phase characteristics, 870
Dispersed-phase properties, 869
Dispersed-phase systems, 880
Dispersion, 18, 19, 189
  in chemical reactions, 854
  geometry, 852, 853

of liquids, 474
model, 781, 867
for waves, 184
Dispersive mixing, 741
Dispersive wave zones, 183
Displacement kinematics, 1070
Displacement pumps, 1071
Dissipation of waves, 26
Dissipation rates, 156
Dissolved gases, 1204
Distribution of drainage areas—Canada, 295
Distribution of wave amplitudes, 41
Disturbance wave, 103
Doppler velocimeter, 53
Double-acting pump, 977, 978, 980
Downcomer, 1193, 1197
Downcomer design, 1200
Downcomer pipework, 1201
Draft tube, 557
Drag
coefficient, 417
effects, 502
force, 132, 440
on one baffle, 132
Drainage basin, 262, 266
Drainage divisions, 302
map of Canada, 294
Drawdown, 1194, 1196, 1199
Drive assembly, 830
Drive gear systems, 1062
Drive gear units, 1079
Drooping head characteristics, 1009
Drooping head curve, 1047
Drop diameter, 500, 858, 861, 863, 865
Drop formation, 483, 494, 496
Drop formation from jet, 491
Drop size, 493, 491, 492, 853, 867, 868
distribution, 492, 852, 856, 880
Drops, 497, 853
Dual jet, 419
mixer, 685
trajectories, 419
Dual tangential jets, 675, 677
Duct configuration, 368
Ducted jet
flow, 368
similarity criteria, 375
Dune surface, 87
Dunes, 71, 75, 76, 87, 89, 93
form resistance, 93
Dynamic-gravity waves, 191
Dynamic-kinematic waves, 186, 187, 193
Dynamic nonlinear wave field, 42
Dynamic pump, 951
Dynamic shock, 189
Dynamic similarity, 823, 830
Dynamic simulation, 863, 873
Dynamic viscosity, 1002
Dynamic wave, 194
Dynamic wave form, 197

Eddy convection velocity, 352
Eddy diffusivities, 53, 60, 64, 396
Eddy diffusivity, 528, 533, 534, 535
of heat, 58
of mass, 58
model, 514
Eddy region under gravity, 88
Eddy transfer coefficients, 533, 535
Eddy viscosity, 374, 385, 530, 573
coefficient, 595, 612
distribution, 455
model, 399
Effect of mass transfer on jet breakup, 494
Effect of viscosity, 484
Effective rainfall, 262
Effective shear stress, 80
Effects of cavitation, 1120
Effects of thermal transport, 344
Efficiency level, 1073
Efficiency of the jump, 123
Elastic modulus, 1125
Elasticities, 1072
Electric actuators, 1090
Enclosed ducts coaxial jets, 381
Enclosed duct configuration, 369
Energy
balance, 156, 644, 857
conservation equation, 329
-containing eddies, 353
dissipation, 134, 1172
equations, 50
exchange, 573
extraction systems, 331
losses, 1120
principle, 190
-spectral density, 27
storage, 331
transfer, 62
Enthalpy, 528
Entrained jet, 470
Entrainment, 124, 157, 353, 375, 378, 420, 440, 814
coefficient, 329, 393, 436, 441
model, 441
velocity, 393, 436
Entrance head, 952
Entry pressure, 1076
Environmental problems, 343
Equations for continuity, 125
Equations of motion, 176, 577, 694
Equipment-friction loss, 1148
Erosion, 745, 1119, 1127, 1132, 1135
curve, 1129

damage effects, 1124
power, 1133
resistance, 1126
Estimating discharge, 273
Estimating floods, 276
Estimating peak, 290
Ethane, 641
Euler equation, 1156, 1157
Euler head rise coefficient, 1173, 1175
Evolution of the jet, 446
Exothermic irreversible reaction 870
Expanded supersonic flows, 520
Exernal gear pump, 981, 982

Far-field noise, 364
Far-field power spectra, 362
Feedback loop, 355
Fermentation, 837, 840, 842, 846
Fermentation specifications, 841
Film-cooling effectiveness, 450
Filtering Remedy and Methodology method, 455
Finite amplitude roll waves, 112
Finite-difference equations, 448, 449
Finite-difference techniques, 374
Finite element analysis of the flow, 90
Finite element formulation of an incompressible rotational inviscid flow, 93
Finite rate reactions, 749
First-order linear system, 873
Fiarst-order reactions, 720
Fissioning, 10
Fixed-mounted mixer, 808
Flashing, 645
Flat-bed friction factor, 84
Flat plate inducers, 1169, 1171, 1175
Floating rings, 1052
Floc particles, 848
Flocculation, 848, 849
Flood
control, 259, 274
frequency analysis, 276, 277, 282
hydrograph, 269
routing, 273
wave, 194
propagation, 175
velocity, 185
Flow
behavior index, 930
configuration, 437
control, 1094
in continuously converging channel, 621
over dams, 273
efficiency, 811
in gradually converging channel, 620
instability, 109, 184
patterns, 396, 585, 597, 804, 814, 816, 831, 903, 928, 1120
in an anchor agitated vessel, 904
in agitated vessels, 902
description, 612
in gas-turbine combustion chambers, 457
in a semi-infinite strip reservoir, 584
records, 292
in a rectangular channel, 138
regime, 73
resistance, 230
separation, 367, 1045
splitter, 960
variation, 1034
visualization, 432, 456, 1167, 1170
Flow-through contraction, 273
Fluid horsepower, 973
Fluid shear rate relationships, 827
Fokker-Planch equation, 43
Forced hydraulic jump, 129
Forced radial hydraulic jump, 143, 144, 147, 160
Forcing frequency, 359
Form drag, 71, 88
Form friction factor, 87, 90, 91
Form resistance, 82
of rigid ripples, 87
Form roughness, 83
Forms of the internal hydraulic jump, 152
Forward shallow water waves, 180
Forward simple wave, 189
Four-bladed inducers, 1176
Four-cylinder diaphragm pump, 1096
Fourier series, 615
coefficients, 581
Fourier transform, 29
Free boundary turbulent flows, 328
Free coaxial jet, 374
Free jet, 351, 360, 375, 382, 385, 434, 470, 538
theory, 162
Free jump velocity distribution, 157
Free radical polymerization, 762, 763
Free stream boundaries, 449
Free surface flows, 182
Free surface hydraulic jumps, 127
Free surface solitary waves, 20
Free vortex motion, 1047
Frequency analysis for Credit River, 283
Frequency distribution, 283
Frequency factors, 281
Friction
effects, 1196
factor, 52, 73, 95, 109, 111
factor effect of the Froude number, 111

factor-Nusselt number correlations, 696
force, 1159
formula (rectangular cross section), 105
head, 952, 1018, 1019
loss, 115, 1149
calculations, 108
in a centrifugal pump, 1008
coefficient, 1160
in rise line, 993
slope, 75
Frictional corrosion, 1072
Froude number, 5, 53, 74, 77, 102, 109, 111, 114, 123, 179, 185, 195, 576, 824
Fuel injection, 412
Fully developed pipe flow, 369, 381
Fully developed plane jet, 398
Fully developed region, 512

Gas bubbles, 1072
Gas jets, 765
Gas-liquid mass transfer, 834, 837, 838
Gas overpower, 837
Gas-phase reactions, 765
Gas-turbine engines, 430
Gaseous jets, 412
Gauging stations, 294, 302
Gaussian distribution, 277
Gaussian error function, 515, 526
Gaussian velocity distribution, 160
Gear pump, 983, 995
Generated gravity-capillary wave, 27
Geometric height, 973
Geometric similarity, 830, 832
Geometrically centered jets, 663
Geometries of impellers, 563
Giant waves, 255
Glass toughening, 349
Grain roughness, 81, 83
Grain shear stress, 83
Gram-Charlier distribution, 38
Grashof number, 324, 326, 640
Gravity-dominated flow, 122
Gravity-wave field, 27
Gravity wavelengths, 194
Gravity waves, 244
Green's law, 9, 15
Growth rate of disturbances, 500
Guide vanes, 135
Gumbel distribution, 277, 278, 281, 285

Hamiltonian, 43,
Hamiltonian system, 42
Harkins-Brown correction factor, 484
Harmonic waves, 186
Hazen-Williams formula, 1018
Head, 811, 1006
Head-capacity curve, 1053, 1054, 1140
Head curve, 973, 1024
flow, 1008
Heat
exchange, 625
flux, 63
transfer, 54, 349, 573, 701, 817, 824, 939
in circular tubes, 701
coefficient, 640, 641, 646, 656, 941
in a combustion chamber, 459
correlations, 936
enhancement, 691
operations, 935
to pseudoplastic liquids, 935
Heated gas jets, 525
Heats of vaporization, 641
Helical coils friction factor, 699
Helical pipes, 696
Helical ribbon agitated vessels, 915, 917, 919–920, 922, 934, 935, 936, 939, 944
flow patterns, 903
velocity distributions, 93
Helical ribbon blenders, 927
Helical ribbon impellers, 901, 908, 910, 911, 912, 913
power correlation, 907
Helium-neon laser, 887
Helmholtz-Kirchoff's hypothesis, 88
Helmoltz's method of free streamline, 89
High-flow-efficiency impellers, 810
High-frequency instability waves, 363
High-speed impinging jet, 359
Holdup, 854
Homogeneous isotropic turbulence, 739, 743
Homogeneous suspension, 886
Horizontal buoyant jet, 341
Horizontal cylindrical vessels flow regimes, 1194
Horizontal density changes, 626
Horizontal plunger pump, 978
Horizontally split pump, 963
Horsepower characteristics, 1011
Horsepower requirements, 1142
Hot-water cavitation, 1123
Hot-wire probe measurement, 1170, 1176
Hubless inducers, 1169
Humber River, 292
Hydraulic Institute Standards, 1142
Hydraulic jump, 122, 123, 124, 129, 132, 160, 164, 166, 167
at an abrupt rise, 148
cavitation potential, 153
circular sections, 132
in closed conduits, 137
with dilution, 150
internal flows, 159
in nonprismatic channels, 139
in nonrectangular channels, 132

in an open channel, 125
pressure fluctuations, 159
solutions, 151
in a trapezoidal channel, 134
turbulence, 153
Hydraulic methods, 272
Hydraulic output, 1071
Hydraulic radius, 79, 176
Hydraulic ram, 990
Hydraulic resistance, 181
of the channel, 214
Hydraulic similitude, 823
Hydraulic structures, 259
Hydraulic turbines, 1134
Hydraulically driven pumps, 988
Hydraulically rough boundary, 81
Hydraulically rough surfaces, 91
Hydrocarbons, 638
Hydrodynamic forces, 71
Hydrodynamic patterns, 749
Hydrodynamic resistance, 482, 492, 500, 502
Hydrodynamic situation, 891
Hydrofoil, 1135
Hydrofoil craft, 1153
Hydrograph methods, 268
Hydrologic simulation models, 268
Hydrology, 259
Hydrostatic law, 100
Hydrostatic pressure, 165

IEM-model, 748
Illinois Urban Drainage Area Simulator, 269
Immiscible liquid-liquid systems, 500
Immiscible non-Newtonian liquid systems, 504
Impeller, 796
arrangement, 1050
correction factor, 794
designs, 806, 1041
discharge pressure, 567
fluid mechanics, 819
function, 811
geometry, 568, 813, 912, 921
mechanical design, 830
pitch, 918
power characteristics, 824, 830
power response characteristics, 803
pumping capacity, 852
spectrum, 812
types, 805, 811
vanes, 1046
Impinging jet, 349, 350
behavior, 364
Impinging region, 351
Impulse response curves, 781
Inclined jets mixing, 552, 555
Incompressible eddy viscosity, 532
Incompressible free turbulent mixing, 539
Incompressible jets, 512, 515
Incompressible Newtonian fluid, 694
Incompressible stratified shear flow, 50
Inducer, 1152, 1153, 1158, 1160
blades, 1177
flow field, 1155
geometry, 1159
performance, 1166
Industrial mixing equipment, 803
Industrial reactors, 767
Infinitely fast single reaction 728
Inflow hydrograph, 275
Influence of nozzle length on jet breakup, 493
Injection flushing, 1103
Inlet flow conditions, 369
In-line piston pumps, 1098, 1104
In-line pumps, 965
Instabilities—three-dimensional, 242
Instability, 237
Instability regions of deep-water waves, 243
Instability waves—noise-generated, 363
Instantaneous reactions, 749
Integral continuity equation, 162
Integral models, 438
Integral momentum equation, 384
Interaction models, 867
Intefacial tension, 853
Intefacial waves, 639
Internal cavity flow, 349
Internal circulation patterns, 751
Internal closed eddies, 587
Internal flow, 122, 153, 167
Internal-flow analytical approaches, 160
Internal flow in the hydraulic jump, 153
Internal flow patterns, 989
Internal gear pump, 982
Internal hydraulic jump, 150
with immiscible fluids, 150
Internal losses, 1008
Internal solitary wave, 21
interactions, 20
Interphase heat and mass transfer, 853
Irreversible reaction, 750
Isothermal expansion, 992

Jacobian, 37
Jet
attachment, 376
breakup, 483
in centrifugal field, 495, 496
column fractures, 418
in a crossflow, 683
development, 517
-edge flow, 349
entrainment, 367
exit geometry, 434
flow, 317, 417

geometries, 677
growth, 444
half-radius, 524
hole flow, 349
impaction, 682
impingement, 349, 354, 432
injection, 660, 677
injector, 673
length, 468, 471, 553, 554
measurements, 515
mixer, 470
design, 661
geometry, 671
mixing, 376, 544, 545, 671
analysis, 532
power requirements, 685
principles, 469
momentum, 520
nozzle, 327
pressure effects, 444
propulsion, 1153
pump, 995
Reynolds number, 467, 519
similarity profiles, 384
stability, 482, 492, 499, 500
analysis, 497
trajectories, 419, 439
velocity, 489, 501, 506
distribution, 368
overall sound pressure level, 360
profiles, 440
Jet-induced circulation flow, 550
Jet-induced flow pattern, 551
Jets
coflowing streams, 400
crossflow, 406, 438, 450
Jet-to-pipe velocity, 667
Jetting conditions, 483
Jetting velocity, 485, 496
Jump length, 128

K-ϵ model calculations, 387
K-ϵ turbulent model, 448
Kadomtsev-Petviashvili equation, 13, 17, 21
Kennedy's velocity formula, 204
Kinematic-dynamic shock, 196
Kinematic fluid velocity, 223
Kinematic pressure, 88
Kinematic shock, 193, 196, 197
Kinematic viscosity, 403
Kinematic wave, 192, 198
amplitude dispersion, 193
equation, 200
theory, 195
velocity, 185
Kinematics of displacement, 1062
Kinetic momentum, 322
Kinetic processes, 855
Kirpich formula, 261, 266
Kolmogorov microscale, 743
Kolmogorov microstructures, 740
Korteweg-de Vries equation, 3, 5
Korteweg-de Vries theory, 3, 6

LNG mixures, 641
LNG tank, 638
Laminar breakup length, 485, 493
Laminar buoyant jets, 318, 327, 328, 345
Laminar flow, 696, 901
Laminar jet regime, 467
Laminar jets, 339, 482, 492
Laminar length, 342
Laminar-to-turbulent jet, 487
Langevin equation, 43
Laplace operator, 577
Large-amplitude solitary waves, 10
Laser doppler velocimeter, 1155
La Spezia incident, 646, 649, 655
La Spezia rollover, 655
Law-of-the-wall, 224, 227, 403
Leakproof pump heads, 1085
Lewis number, 640
Lift pump, 992
Linear density stratification, 573
Linear equations of motion, 39
Linear hydrodynamic stability, 178
Linear surface waves, 39
Linear wave field—steady-state, 28
Linear waves, 31
dispersion, 28
Liquefied natural gas, 637
Liquid crystal method, 912
Liquid inflow, 1204
Liquid jets, 493
in immiscible liquids, 500
penetration, 416
stability theory, 475
Liquid-liquid dispersions, 853
Liquid-liquid jets, 475
Liquid-liquid mass transfer, 834
Liquid-liquid reacting systems, 853
Liquid-liquid reaction vessels, 851
Liquid-liquid reactor, 852, 856
behavior, 854
modeling, 856
simulation, 857
Liquid recirculation, 644
Lockhart-Martinelli parameters, 701
Log-mean driving force, 845
Log-normal distribution, 277, 281
Logarithmic defect law, 229
Long gravity wave, 179, 181, 187, 196
Long-wave theory, 7
Long-wavelength gravity waves, 26

Longitudinal diffusion, 60
Longuet-Higgins stresses, 31
Loop reactor, 557, 559
  flow properties, 567
  mixing behavior, 559
  mixing characteristics, 562
Loss coefficient, 1160
Losses
  in bends, 952
  in suction valves, 952
Low-pass filter, 351
Lower-regime bedforms, 76

Mach disk, 413, 417,
Mach numbers, 417, 516, 517, 1124
Mach reflection, 14
Mach stem, 14
Machinery vibrations, 1126
Macromixing, 766
Macroscopic flow of the jump, 156
Macroscopic momentum equation, 126
Manning equation, 272
Manning formula, 108
Manning-Strickler equation, 75
Manometer, 953
Marangoni number, 495
Marangoni phenomena, 494
Marine propellors, 1134
Marine screw, 563
Markov limit, 43
Mass-conservation equation, 50, 126, 329
Mass conservation law, 574
Mass diffusivity, 320
Mass transfer, 483, 495, 642, 742, 864, 870, 897
  coefficient, 741, 837, 838, 842, 845, 846, 854
  operations, 475
  rates, 639, 837, 853, 859, 860
Mass transport, 48, 65
Material resistance parameters, 1125
Maximum likelihood method, 280
Maximum mean daily discharge, 286
Maximum mean daily flows, 286, 287, 294, 295
Maximum probable flood, 268
Mean baffle forces in a hydraulic jump, 133
Mean circulation time, 545, 555, 757
Mean daily flows, 286, 290
Mean drop diameter, 862
Mean-field closure methods, 387
Mean square surface displacement, 30, 40
Mean square wave amplitude, 36
Mechanical efficiency, 972
Mechanical losses in bearings, packings, 1008
Mechanical seals, 1008, 1052
Mechanically agitated vessel, 852
Mechanically driven diaphragms, 1079
Mechanism for micromixing, 733, 738, 740
Mechanism of circulation, 605
Mechanism of vertical circulation, 618, 633
Median method, 280
Metering errors, 1068
Metering pumps, 987, 1089, 1094
Methane, 638, 641, 651
Method of moments, 279, 280
Method of Monte-Carlo, 715
Michaelis-Menten kinetics, 726
Micro-metering pump, 1069
Micromixing, 707, 708, 716, 719, 720, 743, 745, 755, 760, 762
  characteristics, 758
  intensity, 756, 761
  in a nonuniform stirred tank, 755
  parameters, 720, 730
  simulation, 715
  of single-parameter models, 710
  of successive stages, 738
  theories, 761
  times, 767
Microscale mixing, 822, 834
Microscale shearing effects, 829
Missouri River basin, 81
Mixco column, 791, 793, 794, 795
Mixed flow impellers, 896
Mixed reactor models, 878
Mixer design, 662, 830
Mixer geometry, 661, 675
Mixer power level, 842, 845
Mixer selection, 831
Mixer variables, 831
Mixing, 124, 667, 672
  concentration profiles, 664
  conditions, 682
  criteria, 662
  distance, 663
  duct, 367
  earliness, 737
  effects, 710
  layer, 383, 466
  length hypothesis, 374
  length model, 527
  models, 773, 788
  operations, 817
  patterns, 903, 912
    in an anchor-agitated vessel, 915
  performance, 924
  pressure rise, 369, 370
  process, 778, 832
  pseudoplastic liquids, 944, 945
  quality, 683
  rate, 680
  ratio, 661
  Reynolds number, 568

scale, 785
system geometric parameters, 889
time, 471, 545, 548, 550, 555, 783, 921, 944
of viscous Newtonian liquids, 912
Modeling turbulent jets, 430, 511
Models of variable-density mixing, 527
Moderate amplitude solitary waves, 11
Modified predictor-corrector method, 116
Modified Reynolds number, 891, 932
Molecular diffusion, 742, 854, 912
Molecular mass transfer, 640
Molecular mixing, 773
Molecular weight on influence of mixing, 762
Momentum balance for hydraulic jump, 156
Momentum coefficient, 441
Momentum conservation principles, 189
Momentum equation, 115, 139, 161, 165, 195, 221, 222, 325
Momentum equations for nonlinear terms, 612
Momentum transfer, 367
Monin-Oboukov length scale, 48
Monoclinal flood wave, 195
Monte Carlo Method, 880
Monte Carlo simulation, 879, 881
Monte Carlo simulation models, 879
Motionless mixer, 901
Motive fluid pumps, 989
Movable seals leakage losses, 1073
Multicomponent mixtures, 773
Multicylinder design, 1059
Multicylinder in-line units, 1105
Multicylinder pumps, 1070
Multiple impellers, 792
Multiple piston-type metering pump, 1081
Multiscale perturbation technique, 15
Multistage centrifugal pump, 965, 1038
Multistage horizontally split pump, 964
Multistage pumps, 963
Multi-step reactions, 767

NANA Lewis Research Center, 1167
N-butane, 641
Natural channel Manning roughness coefficient, 272
Natural convection, 334
Natural instability, 359
Navier-Stokes equations, 190
Negatively-buoyant jets, 331
Net positive suction head (NPSH), 1006, 1138
Neutral wave region, 182
Neutral wave zone, 182
Newtonian flow—two-dimensional, 902
Newtonian fluids, 700
Newtonian jet, 503, 505
Nitrogen, 641, 651
Noise
generation of, 360
by a turbulent flow, 360
by instability waves, 363
mechanics, 361
measurements, 1133
radiation, 349, 363
Nonbaffled vessels, 790
Nonbuoyant jet, 318, 321, 325, 332, 336, 338, 665
Noncavitating inducers, 1152
Non-Gaussian statistics, 45
Nonisothermal turbulent jets, 539
Nonlinear waves, 3, 4
equations, 3
field probability distribution, 34
in shallow water, 6
Non-Newtonian behavior, 940
Non-Newtonian effects on the breakup of liquid jet, 497
Non-Newtonian fluids, 487, 703, 1078
Non-Newtonian jets, 506
Non-Newtonian liquids, 505, 928
Non-Newtonian materials, 927
Non-Newtonian systems, 482
Nonprismatic channels, 197
Nonreactive tracers, 748
Nonrigid ripples, form resistance, 93
Normal shock, 412
North Platte River, 275
Nozzle geometry, 369
Nuclear reactors, 1203
heat exchange, 691
Number density continuity equation, 877
Number of stages, 1039
Numerical diffusion, 449
Numerical integration techniques, 125
Numerical models, 460
Numerical simulations, 736
Numerical smearing, 454
Numerical techniques, 339
Nusselt number, 692

Ocean surface waves, 254
Oceanic bubbles, 255
Oçeanography, 28
Olefin, 558
polymerization, 570
One-dimensional flow, 1188, 1193
Onset of choking, 1199
Open-channel flow, 48, 50, 53, 112, 220
unsteady, 175
wave attenuation, 174
Open-channel hydraulic jump, 123
Open-channel sediment transportation, 204
Open-duct flows, 369

Open-duct jet mixing turbulent shear stress distributions, 381
Operating efficiency, 1034
Operating point, 976
Optimal operating efficiency, 1006
Optimum jet-to-pipe diameter, 671
Optimum mixing, 672, 673, 675
  conditions, 671, 674, 685
Optimum pipeline mixing, 680
Orr-Sommerfeld equations, 109
Oscillating discharge pumps, 1072
Oscillating displacement pumps, 1057, 1058, 1060, 1066, 1072
  design, 1078
Oscillating metering pumps, 1079
Oseen's linearized equations of motion, 909
Outflow hydrograph, 275
Outflows from vessels, 1187
Outlet-controlled regime, 1191, 1193
Outlet discharge coefficient, 1191
Overall efficiency, 952
Overall friction factor, 76
Overall sound pressure level, 360
Overland flow, 269

Paddle impellers, 790–793
Paddle stirrer, 892
Paper drying, 349
Parabolic velocity function, 588
Parallel combination model, 779, 782
Parallel piping systems, 1027
Parallel pumping, 1023, 1024
Parseval's theorem, 17
Particle-laden jets, 412
Particle suspension, 897
Peak boil-off, 651
Peak discharge data, 259
Peak flow estimations
  maximumm likelihood method, 285
  median method, 285
  moment method, 285
Peak flows, 258, 260
  in major rivers, 259
Peak index, 295, 304
Peaking horsepower characteristics, 1011
Peaks from urban areas, 268
Peclet-Bodenstein number, 774
Peclet number, 565, 567
Penn State inducers, 1173
Perfect mixed flow, 783, 786, 790
Performance effects, 1120
Period of record, 302
Perturbation analysis, 327
Perturbation frequencies, 244
Phase speed of Stokes waves, 253
Phase transfer catalysis, 855
Phase velocity, 178
Physical-model geometries of combustion chambers, 456
Physiochemical methods, 748
Physiographic characteristics, 262
Pipe deterioration, 1017
Pipe friction, 1017
Pipe-friction loss, 1148
Pipe resonance, 1016
Pipe Reynolds number, 567
Pipeline mixer, 661
Pipeline mixing, 660
Piston glands, 1102
Piston pump, 977
  head, 1085
Piston-type displacement systems, 1079
Piston-type pump heads, 1085
Plane jets, 391, 397, 459
  flow development, 399
  turbulent, 392
  wall, 402
    effect of wall roughness, 404
Plug flow, 779, 782
Plunger pumps, 977, 978
Pneumatic actuators, 1081, 1090
Pneumatic diaphragm pump characteristics, 1098
Poisson equation for pressure, 446
Pollutant conservation equation, 394, 397
Polycondensation, 764, 767
  reactions, 764
Polyesterification, 764
Polyethylene process, 559
Polymer solutions, 487, 504
Polymerization, 767
  reactions, 762
Population balance equation, 878, 881
Population balance techniques, 877
Positive displacement pump, 951
Positive suction pressure, 1138
Potential core region, 512
Potential energy, 1138
Potential flow, 384
Potential theory, 11
Power consumption, 811, 931
  of agitated vessels, 906
  of close-clearance impellers, 906
  of helical ribbon impellers, 906
  of power correlation, 906
  in pseudoplastic liquids, 933
  of shear characteristics, 906
Power correlation, 932, 941
  for Newtonian liquids, 934
Power dissipation, 896
Power law model, 928, 932
Power number, 794, 797, 798, 824, 891, 897
  Reynolds number curve, 820

Power pumps, 979
Power spectrum, 254
Prandtl-Karman resistance, 72
Prandtl's imcompressible turbulent diffusion model, 512
Prandtl's imcompressible turbulent mixing model, 527
Prandtl's mixing length, 162, 394, 399
Prandtl's number, 51, 52, 323, 328, 335, 640
Precipitation, 767
  estimates, 268
  reactions, 761
Predicted velocity distributions, 450
Predicting jets in cross flow, 440
Prediction of jet trajectories, 442
Prediction of normalized total pressure profiles, 453
Pressure recovery, 1193
Pressure distribution on a body of revolution, 426
Pressure drag, 71, 77, 78, 81
Pressure rise coefficient, 1173
Pressure signals of auto-correlations, 357
Pressure variation, finite element analysis, 91
Pressure waves feedback loop, 364
Prismatic channel, 127
  equations of motion, 197
Probability density, 37
  Fokker-Planch equation, 42
  function, 29
  series expression, 34
Probability distribution, 277, 278, 281
Probability of coalescence, 710
Probability wave heights, 32
Process chemicals, 986
Process design requirements, 803
Process optimization, 805
Propagation factor, 178, 181
Propagation in long channels, 200
Propagation of long waves, 5
Propane, 641
Property of traveling waves, 182
Proportional metering, 1090, 1091, 1099
Pseudoplastic liquid, 927, 930
Pseudoplasticity, 940
Pulsation damping, 1110, 1114
Pumps
  affinity laws, 1013
  characteristics, 1002, 1008
  classification, 951
  construction, 1048
  discharge characteristics, 1139
  double casings, 1049
  efficiency, 980, 1013, 1046
  head, 1003, 1020
  horsepower, 1027, 1030
  performance, 974, 1009, 1014, 1033
  performance curves, 1014, 1139
  piping economy, 1150
  selection, 994
  speed, 1024
  system characteristics, 1015
  terminology, 925
  valves, 1086, 1102
  valves leakage losses, 1073
  valves wear particles, 1076
Pumping applications, 995
Pumping equipment, 1030, 1034
  selection, 1001
Pumping network, 953
Pumping power, 701
Pumping systems, 1014
Pycnocline, 18

Quality of distribution, 887
Quality of mixing, 673, 688

Radial flow impellers, 804, 805
Radial flow stilling basins, 143
Radial flow turbine, 821
  velocity patterns, 821
Radial hydraulic jump, 141, 143, 165
Radial split horizontal pumps, 963
Radial stilling basins, 143
Radial stirrers, 892
Radially diverging hydraulic jump, 139
Radio flow-followers, 753
Rainfall duration, 265
Rainfall intensity-frequency-duration maps, 269
Ramjet combustors, 349
Rapid mixing, 551
Rate of erosion, 209
Rate of polymerization, 879
Rate of strain, 351
Rayleigh analysis, 1123
Rayleigh flow, 643, 651
Rayleigh statistics, 45
Reaction kinetics, 857
Reaction time, 847
Reactive tracers, 737
Reactor, 737
  size, 757
Rebounding bubbles, 1124
Reciprocating pumps, 977, 981
Recirculating flow, 368, 469
  dispersion model, 559
Recirculation eddy, 379
Recirculation for a ducted jet, 378
Rectangular channel, 198
  flow, 459

Rectangular hydraulic jump, 127, 142
Rectangular reservoir, 598, 599
  circulation, 582
Recycle model, 550, 555
Reference stream function, 579
Reflection damper, 1110, 1113
Regional frequency analysis, 285
Relative celerity of the shallow water waves, 180
Relative roughness, 91
Remote-head systems, 989
Reservoir flow pattern, 631
Residence time, 734, 773, 867, 869
  distribution, 716, 788, 856
    characteristics, 868
Resistance curve, 82
Resistance effects, 113
Resistance laws by Chezy and Manning, 108
Resonance phenomenon, 354
Resonant frequency, 363
Resonant jet, 363
Resonator, 1113
Retention time, 818
Reverse flow regions, 450
Reversed flow, 367
Reynolds equations
  of motion, 392
  for turbulent flow in open channels, 98
Reynolds stress, 62, 153, 385, 447, 574, 1178
Richardson number, 53, 60, 67, 164
Rigid boundary hydraulics, 72
Ripples, 71, 74, 75, 76, 89
Rise line friction losses, 993
Rising head characteristics, 1009
River navigation, 76
Rocket pump feed systems, 1153
Rocket pump inducers, 1153, 1170
Rocket turbopumps, 1153
Roll wave, 98, 117
  solution, 113, 116
  theory, 112
Rollover, 637, 646, 651, 655
  criteria, 641
  models, 639
  time, 651
Rotary pumps, 981
Rotating disc column, 791, 793
Rough bed channel, 233
Rough channel flow, 223, 225
  velocity profile, 221, 224, 228
Rough channel profiles, 225
Roughness configurations, 73
Roughness Reynolds number, 227
Runge-Kutta method, 52
Runoff degree-day factor, 271
Runoff from snowmelt, 269
S-hydrograph, 265
  method, 266
Salt injection technique, 469
Salt water jets, 340
Sand-bed channels, 95
Saturation temperature, 646
Scalar diffusion, 53, 67
  buoyancy effects, 58
Scale effects, 1128
Scale-up fluid mixer, 829
Scale-up mixing devices, 818
Scale-up relationships, 829, 832
Scaling, 661
Schmidt number, 51, 52, 343, 448
Schrodinger equation, 8
Screw pumps, 982, 983, 984, 995
Seal rings, 977
Second-order rate expression, 880
Second-order Stokes wave, 35
Secondary flows, 458, 696
Sediment charged flows, 216
Sediment-free stable channels, 205
Sediment-stable canal, 203
Sediment-stable channels, 203, 217
Sediment-transport capability, 204
Sediment-transporting channels, 204
Sediment-transporting channels stability, 204
Sediment-transporting stable channels, 208
Sedimentary bedform, 95
Sedimentation problems, 160
Segregated flow, 764
Segregation function, 734, 735, 736
Segregation in single-phase fluids, 738
Selecting a pump, 973
Semi-batch reactor, 730
Semi-infinite strip reservoir, 610
Separation of the boundary layer, 78
Separation point, 89
Separation zones, 78
Series pumping, 1023
Shafts, 830
Shallow water equation, 100, 177
Shallow water theory, 4, 193
Shallow water waves, 182, 185
Sharp-edged nozzle, 529
Shear, 811
  characteristics, 906
  flow, 229
    instability, 26
  layer, 359, 391, 398
    approximations, 371
    curvature, 351, 353
    growth, 353
    problem, 399
    thickness, 351
  mixing, 915
  production, 62

pumping effct, 1162, 1176
rates, 821
stress, 63, 72, 74
due to skin functions, 86
stresses, 1162
thickening, 927
thinning property, 944
velocity, 210, 223
Shield function, 72
Shield parameter, 86
Shock, 186, 989
Shock dampers, 989
Shock formation, 191
Shock shape of transverse gas jets, 415
Shock waves, 1124
Short-circuits, 751
Short-crested waves, 254
Side-entering mixer, 808, 809
Sieve plates, 493
Silt factor, 204
Similarity methods, 324
Similarity parameter, 375
Simple waves, 190
Single-acting piston pump, 977
Single-cylinder pump, 1059, 1071
Single jets, 406
Single oblique jets, 673, 676
Single phase flows, 406, 701
Single phase friction factor correlations, 700
Single screw pumps, 982
Single-stage pumps, 995
Single-station frequency analysis, 282
Sinusoidal pump attentuation, 198
Sizing pumps, 999
Skew waves, 246, 252, 253
of Benjamin-Feir modulations, 246
patterns, 246, 247, 248, 253, 254
interactions, 246
Skin friction coefficients, 1163, 1178
Sliding vane pump, 985, 995
Slope area method, 272
Slope of the energy grade line, 75
Slowly varying solitary waves, 14, 16, 21
Slurries, 1094
Small diameter impeller, 821
Small-scale eddies, 67
Smoke photographs of single jets, 433
Smoke-wire technique, 512
Smooth channel, 109
profiles, 225
Snowmelt equations, 269
Snowmelt flood hydrograph, 271
Snowmelt floods, 269
Snowmelt from snowpack, 269
Soil Conservation Service, 263
Solar energy utilization, 317
Solenoid pump, 1093
Solenoid-type diaphragm metering pump, 1083
Solid agitation, 886
Solid-liquid mass transfer, 897
Solid particles suspensoin, 898
Solids dispersion, 817
Solids suspensions, 815, 817, 833, 897
Solitary interfacial waves, 3
Solitary wave, 3, 4, 9, 12, 14, 15, 20, 118
characteristic length, 7
equations, 5
interactions, 11
solution, 4, 7, 13, 19, 20
two-layer fluids, 19
Solventless polymerizations, 558
Spatial-averaged population balance, 878
Specific energy, 952
Specific speed, 1006, 1010
index, 1014
ranges, 1007
Spectral density, 42
Spectral-equation model, 50, 61
Speed of sound, 362
Spilling breakers, 242
Stability analyses, 480, 498, 505
Stability criterion, 101
Stability of jets, 474, 503
Stability of jets in immiscible liquids, 501
Stability theory, 491
of breakup of jet, 480
Stable channel, 214, 217
geometric properties, 212
shape profile, 206
Stably stratified flow, 54, 58
open-channel, 53
Stages, number of, 1039
Stagnation head, 1162
Stagnation region, 354, 604, 633
Standard baffles, 815
Standard project flood, 268
State of segregation, 748
State variables, 870
Static head, 1019, 1020
available, 1020
Static pressure, 412, 956
Static suction head, 955
Stationary flow devices, 1135
Statistical analysis, 302
Steady flow, 221
of turbulent jets, 444
Steady-state behavior of two-phase reactors, 868
Steady-state nonlinear wave field, 32
Steady-state simulation, 870
Steady-state statistical properties, 27
Steep head characteristics, 1009
Steep head curve, 1011
Steep-sloped prismatic channels, 98

Steep wave train, 238
Stepanoff thermodynamic parameter, 1132
Stilling basin, 131, 153
  decay of turbulence, 158
  design, 143
Stirred reactors, 707, 708
Stirred tank, 751, 868, 869
  internal age distribution, 752
Stirring speed, 757
Stoichiometry, 750
Stokes drift, 44
Stokes expansion, 44
Stokes stream function, 475
Stokes waves, 36, 237, 246, 248
  of Benjamin-Feir modulations, 246, 249
Storage indication method, 273
Storm Water Management Model, 269
Stormer viscosimeter, 843
Straight sediment stable channels, 203
Stratification, 19, 337, 638
Stratified environment, 339
Stratified flows, 47, 618, 619, 632
Stratified holding tanks, 637
Stratified lower-layer flow, 53
Stratified open-channel flows, 52, 55
Stratified outer-layer flows, 48
Stratified reservoir, 577, 605, 627
Stratified shear flow—theoretical work, 48
Stratified short reservoir, 609
Stream function, 90, 322, 577, 579
Streamflow records, 262
Streams in the U.S., 299
Stretching aggregates, 747
Stretching time, 740
Strip integral method, 165, 167
Stripping efficiency, 842
Stroke adjustable metering pumps, 1068
Stroke frequency, 1066, 1075
Stroke reversal, 1097
Strongly-stable flows, 55
Strongly-stable stratification, 57
Strongly-stable stratified flow patterns, 57
Structural requirements, 803
Stuffing box, 1052
Styrene, 763
Subharmonic instabilities, 244
Submerged hydraulic jump, 128, 135, 136, 165
Submerged radial hydraulic jump, 149
Submerged radial hydraulic jump surface profiles, 166
Subsonic jets, 516
Subsonic Mach numbers, 520
Suction casing, 1045
Suction flange pressure, 1076
Suction head, 1206
Suction height, 955
Sulfite oxidation, 846
Sulfuric acid solutions, 855
Supercritical flows, 77
Superficial gas velocity, 838, 844
Supernatent phase, 1194
Supersonic flow, 415, 416
Supersonic injection, 413
Suppression pressure, 1130
Surface circulation, 586, 594, 625
  region, 596, 619
Surface displacement, 36
  non-Gaussian distribution, 38
Surface disturbances, 497, 500
Surface drag, 71, 74
Surface elevation probability density function, 38
Surface forces on immersed bodies, 77
Surface gravity waves, 236
Surface pressure
  distributions, 418
  on a flat plate, 420, 427
  fluctuations, 354
Surface renewal models, 856
Surface roughness, 701
Surface shear stress, 574
Surface tension, 5, 606
Surface vibration, 1120
Surface wave, 236
  field, 30
Surging, 1053, 1054
Suspended sediments, 209, 214
Suspension, 887, 1094
  conditions, 887
  sedimentation, 1103
  valve, 1075
Swirl component, 696
Swirl flow, 691
Swirling jets, 385
Symmetric waves, 252, 253
  patterns, 249
Synthetic unit hydrograph, 262
System head curves, 1019
System resistance curves, 1141

Tangential flow, 913
Tank design, 814
Tanks-in-series model, 776
Tape inserts, 691, 701
  friction factor, 698
Taylor microscale, 754
Tamperature distribution
  in a combustion chamber, 459
  in the jet, 341
Temperature profiles, axisymmetric jet, 340
Tensile strength, 1125
Terminal fall velocity, 210
Terminal roll-wave velocity, 118

Terminal velocity distribution, 597
Theoretical head, 952
Theoretical impeller diameter, 1004
Theory of stratified flows, 573
Thermal boundary layer, 328
Thermal buoyancy, 327
Thermal diffusivity, 321
Thermal plume, 342
Thermal segregation, 765
Thermal shock, 661
Thermal stratification, 622
Thermally buoyant flow, 321
Thermally buoyant jet, 321
Thermally stratified environment, 330
Thermally stratified lower-boundary-layer flows, 47
Thermally stratified open-channel flow, 47
Thermodynamic effects, 1122
Thermodynamic vaporization constant, 1159
Thick-matter piston pumps, 1104
Thoma cavitation parmeter, 1123
Three-bladed inducer, 1176
  turbulent flow field, 1176
Three-cylinder piston pump, 1099
Three-dimensional breaking, 244
Three-dimensional flow, 72
Three-dimensional instability, 245, 251, 254, 256
Three-dimensional mixing, 912
Three-dimensional symmetric waves, 249
Three-dimensional wave breaking, 255
Three-dimensional wave patterns, 245, 246
Three-lobe pumps, 983
Three-parameter log-normal, distribution, 277, 279, 280
Three-rotor-screw pump, 983
Time-averaged fluid density, 517
Time constants, 757
Time-mean velocity distribution, 221
Time of peaks, 295, 304
Top-entering mixers, 807, 809
Top pullout, vertical pump, 965
Total available head, 1020, 1023
Total discharge head, 954
Total efficiency, 1071
Total head, 1148
  calculations, 1144
  of a pump, 976, 1148
Total system head alteration, 1022
Tracer flows, 913
Tracer particles, 902
Trajectories of triple jets, 444
Trajectory analyses, 417, 665
Trajectory of jets, 450
Trajectory prediction, 419
Transfer coefficients, 645
Transient flow, 1188, 1205
Transition region, 368, 512
Transport equations, 385
Transport phenomena, 864
Transport sediment load, 204
Transverse circulation of sediment, 209
Transverse jet, 411, 418
  case, 411
  trajector analysis, 417
  supersonic flows, 412
    liquid jets, 416
Transverse pressure relief, 419
Trapezoidal channels, 132
Trapezoidal hydraulic jump, 135
Trapezoidal stilling basin, 136
Traveling hydraulic jumps, 125, 139
  in closed conduits, 138
  in open channels, 138
Triangular hydrograph, 287, 288, 290, 292
Triple cylinder pumps, 1065
Triplex plunger pump, 979
Triplex pump, 979
Turbine blades, 790
Turbine impellers, 788, 792, 852
Turbomachinery, 1179
  flows, 1155
Turbopump inducer, 1154
Turbulence characteristics, 125
  of the plane jet, 401
  in hydraulic jumps, 157
Turbulence eddy motion enhancement mechanism, 57
Turbulence energy, 54, 62
  budget, 64
Turbulence intensities, 60, 61, 754, 755, 757
Turbulence kinetic-energy, 48, 49
Turbulence macroscale, 754
Turbulence measurements, 381
Turbulence model, 453
Turbulence parameters distribution, 753
Turbulence pressure, 160, 163, 164, 165
Turbulence shear stress, 155
Turbulence structure, 53
Turbulence theory, 738
Turbulence transition, 342
Turbulent compressible jets, 517
Turbulent diffusion, 451
  fluxes, 448
  scalar quantities, 67
Turbulent diffusivity, 454
Turbulent energy, 351
  dissipation, 754
Turbulent flow, 72, 318, 701, 820
  structure, 47
  time-averaged equations, 385
Turbulent-free jets, 511, 512
Turbulent-free mixing, 512
Turbulent heat fluxes, 447

Turbulent intensity, 383
  regime, 470
Turbulent jet(s), 339, 350, 430, 431, 502
  in confined cross flow, 432
  in cross flow, 430, 431
Turbulent kinetic energy, 54, 156, 444, 538
Turbulent length hypothesis, 374
Turbulent mixing, 391, 512, 515, 528
Turbulent shear, 162, 165, 381, 385, 392, 538
  stress tensor, 574, 576
Turbulent stratified flows, 47
Turbulent stresses, 448
Turbulent structure, 453
Twin-screw pump, 983
Twisted pipes, 691
  axial velocity, 697
Twisted-tape inserts, 700
  heat transfer correlations, 702
Two-bladed inducers, 1160, 1170, 1171, 1172
Two-dimensional buoyant jet, 337
Two-dimensional flows, 53, 385, 530, 538
Two-dimensional free shear layer, 359
Two-dimensional jets, 336, 339, 340
Two-dimensional Newtonian flow, 902
Two-dimensional nonbuoyant jets, 333
Two-dimensional plume, 334, 337
Two-dimensional rough channel, 231
Two-dimensional stratified flow,
    kinetic-energy balance equation, 54
Two-dimensional trajectories, 444
Two-dimensional vertical jets, 331
Two-dimensional water waves, 256
Two-dimensional waves, 243, 351
Two-phase flow, 703, 1143
Two-phase liquid-liquid reactor, 857
Two-phase reactors, 854
  steady state, 868
Two-stage overhung pump, 963
Types of impellers, 901
Types of pumps, 1039
Typical centrifugal pump, 975

U-shaped loop reactor, 557
Unconfined free stream, 375
Unconfined internal hydraulic jumps, 152
Uniform flows, 83
Unit hydrograph, 262
  approach, 262
United States Environmental Protection
    Agency, 269
United States Geological Survey, 302
Unstable flows, flow pattern, 58
Unstable turbulent flow, 98
Unstable turbulent channel flow, 98
Unstably stratified flows, 53
Unstably stratified outerlayer, open-channel
    flows, 53
Unsteady-state reaeration, 846
Upper-regime bedforms, 76
Ursell number, 5, 8

Vacuum gauge, 953
Valve bypassing, 1022
Valve closing lag, 1074, 1075
Valve leakage, 1075
Valve throttling control, 1023
Valve wear, 1075
Valves, 1135
Vane configuration, 1045
Vane pumps, 985
Vanishing density stratification, 602, 604
Vapor-liquid equilibrium, 642
Vapor pressure, 1033
Vaporizing film conditions, 649
Variable-density jets, 515, 518, 526, 527
Variable-density turbulent-free jets, 511
Variable speed drive, 1033
  efficiency, 1033
Variable speed pump, 1025, 1033
  horsepower, 1030
Variable stroke frequency, 1090
Vedernikov number, 199
Velocity defect law, 228
Velocity distribution, 928
  effect of roughness, 224
  in an axisymmetric buoyant jet, 329
Velocity fluctuations of spectral distribution,
    352
Velocity head, 952
Velocity measurements in the jet, 411
Velocity profile dimensional analysis, 222
  universal logarithmic, 701
Velocity profiles, 434
  finite element analysis, 91
Velocity reduction function, 227
Velocity scale for plane jets, 401
Venturi bubble collapse, 1125
Venturi cavitation, 1123
Vertical circulation, 572, 573, 578, 579, 584,
    599, 602, 624
  mechanism, 585
  models, 626
Vertical cylindrical vessels flow regimes, 1197
Vertical flow currents, 816
Vertical heat transfer, 626
Vertical jet mixing, 545, 554
Vertically plane motion, 574
Vessel-controlled regime, 1189, 1203, 1204
Vessel designs, 814
Vessel level control, 1206
Vessel outflow maximum discharge, 1205
Vessel outlet-controlled steady-state discharge,
    1205
Vibration analysis, 1052, 1053

Vibration-induced cavitation, 1126
Vibratory horn, 1127
Virtual head, 972
Viscid flow equations, 1152
Viscoelastic liquid, 504
Viscoelastic-to-Newtonian jets breakup length, 504
Viscosity ratios, 688
Viscous dissipation, 48, 49, 319
Viscous effects, 1158
Viscous flow, 1142
  performance characteristics, 1142
Viscous Newtonian liquids, 901
Viscous stress, 392
Viscous sublayer, 228
Volumetric efficiency, 952, 1070, 1107
Volumetric injection flushing, 1102
Volumetric metering pumps, 1078
Voluretric oscillating metering pumps, 1068
Von Karman's constant, 216
Vortex formation, 680

Wake-strength coefficient, 233
Wall jets, 350, 353, 354, 402, 404
Wall shear stress, 1165
Wasate treatment, 848
Water hammer, 956, 1016
Water jet propulsion, 1153
Water purification, 848
Water resource, 267
Water survey, 292
Water-wave equation, 253
Water waves, 27, 41
  non-Gaussian statistics, 26
  steady-state properties, 28
Watershed discharge, 265
Wave
  amplitude, 18, 32, 44, 45
    dimensionless, 114
  attenuation, 182
  basin wave patterns, 240
  breaking, 240, 255
  celerity, 105
  crest, 174
  energy, 255
    flux, 15
  evolution, 242
  field, 246
  front, 194
  generation, frictional effects, 8
  interaction, 40
  motion spectral components, 186
  number, 14, 36, 58, 178, 179, 186, 481, 500, 504
  patterns, 238, 245
  phase difference, 355, 358
    velocity, 185
  phenomenon, 177, 187
  propagation, 4, 15
    lakes, 21
  speed, 5, 12
  superposition, 355
  train, 244
Wave-wave interactions, 40, 43
Waves, 244
Weibull formula, 282
Welding, 349
White-Colebrook formula, 105
White-Colebrook-Thijsse formula, 103
Wide clearance impeller, 915
Wiener-Khintchine relation, 30
Wind action, 573, 574
Wind friction speed, 35
Wind-generated circulation, 574
Wind-generated vertical circulation, 605
Wind-generated waves, 256
Wind influenced vertical circulation, 598
Wind tunnel flows, 55
Wind vertical circulation, 609, 610, 633

Zinc purification, 847